Climate Change 1995

IMPACTS, ADAPTATIONS AND MITIGATION OF CLIMATE CHANGE:
SCIENTIFIC-TECHNICAL ANALYSES

Climate Change 1995

Impacts, Adaptations and Mitigation of Climate Change: Scientific-Technical Analyses

Edited by

Robert T. Watson
Office of Science and Technology Policy, Executive Office of the President

Marufu C. Zinyowera
Zimbabwe Meteorological Services

Richard H. Moss
Battelle Pacific Northwest National Laboratory

Project Administrator
David J. Dokken

Contribution of Working Group II to the Second Assessment Report of the Intergovernmental Panel on Climate Change

Published for the Intergovernmental Panel on Climate Change

Published by the Press Syndicate of the University of Cambridge
The Pitt Building, Trumpington Street, Cambridge CB2 1RP
40 West 20th Street, New York, NY 10011-4211, USA
10 Stanford Road, Oakleigh, Melbourne 3166, Australia

First published 1996

Printed in the United States of America

Library of Congress cataloging-in-publication data available.

A catalog record for this book is available from the British Library.

ISBN 0-521-56431-X Hardback
ISBN 0-521-56437-9 Paperback

Also available from Cambridge University Press:
Climate Change 1995 – The Science of Climate Change
Contribution of Working Group I to the Second Assessment Report of the Intergovernmental Panel on Climate Change. Editors J.J. Houghton, L.G. Meiro Filho, B.A. Callander, N. Harris, A. Kattenberg and K. Maskell. (ISBN 0 521 564336 Hardback; 0 521 564360 Paperback)
Climate Change 1995 – Economic and Social Dimensions of Climate Change
Contribution of Working Group III to the Second Assessment Report of the Intergovernmental Panel on Climate Change. Editors J. Bruce, Hoesung Lee and E. Haites. (ISBN 0 521 560519 Hardback; 0 521 568544 Paperback)

Photo Credits – Photographic imagery embedded in the geodesic globe derived from the following sources: 'The World Bank' CD-ROMs (Aztech New Media Corporation); 'Photo Gallery' CD-ROMs (SoftKey International, Inc.); and the 'Earth, Air, Fire, and Water' CD-ROM (MediaRights, Inc.). The mosquito image was provided by the Iowa State Department of Entomology; hurricane photo courtesy of NOAA. Geodesic globe designer – Mark Sutton.

Contents

Foreword

The Intergovernmental Panel on Climate Change (IPCC) was jointly established by the World Meteorological Organization and the United Nations Environment Programme in 1988, in order to (i) assess available scientific information on climate change, (ii) assess the environmental and socioeconomic impacts of climate change, and (iii) formulate response strategies. The IPCC First Assessment Report was completed in August 1990, and served as the basis for negotiating the UN Framework Convention on Climate Change. The IPCC also completed its 1992 Supplement and "Climate Change 1994: Radiative Forcing of Climate Change and An Evaluation of the IPCC IS92 Emission Scenarios" to assist the Convention process further.

In 1992, the Panel reorganized its Working Groups II and III and committed itself to complete a Second Assessment in 1995, not only updating the information on the same range of topics as in the First Assessment, but also including the new subject area of technical issues related to the economic aspects of climate change. We applaud the IPCC for producing its Second Assessment Report (SAR) as scheduled. We are convinced that the SAR, as the earlier IPCC reports, will become a standard work of reference, widely used by policymakers, scientists, and other experts.

This volume, which forms part of the SAR, has been produced by Working Group II of the IPCC, and focuses on potential impacts of climate change, adaptive responses, and measures that could mitigate future emissions. It consists of 25 chapters covering a wide range of ecological systems and socioeconomic sectors and activities. It also includes brief descriptions of three appendices—two sets of guidelines or methodologies for assessing the potential efficacy of adaptation and mitigation strategies, and an inventory of technology databases and information. The appendices themselves have been or are being published in full as separate stand-alone volumes.

As usual in the IPCC, success in producing this report has depended upon the enthusiasm and cooperation of numerous busy scientists and other experts worldwide. We are exceedingly pleased to note here the very special efforts made by the IPCC in ensuring the participation of scientists and other experts from the developing and transitional economy countries in its activities, in particular in the writing, reviewing, and revising of its reports. The scientists and experts from the developed, developing, and transitional economy countries have given of their time very generously, and governments have supported them in the enormous intellectual and physical effort required, often going substantially beyond reasonable demands of duty. Without such conscientious and professional involvement, the IPCC would be greatly impoverished. We express to all these scientists and experts, and the governments who supported them, our grateful and sincere appreciation for their commitment.

We take this opportunity to express our gratitude to the following individuals for nurturing another IPCC report through to a successful completion:

- Professor Bolin, the Chairman of the IPCC, for his able leadership and skillful guidance of the IPCC
- The Co-Chairs of Working Group II, Dr. R.T. Watson (USA) and Dr. M.C. Zinyowera (Zimbabwe)
- The Vice-Chairs of the Working Group, Dr. M. Beniston (Switzerland), Dr. O. Canziani (Argentina), Dr. J. Friaa (Tunisia), Ing. (Mrs.) M. Perdomo (Venezuela), Dr. M. Petit (France), Dr. S.K. Sharma (India), Mr. H. Tsukamoto (Japan), and Professor P. Vellinga (The Netherlands)
- Dr. R.H. Moss, the Head of the Technical Support Unit of the Working Group, and the talented and dedicated individuals who served as staff, interns, or volunteers during various periods of this assessment: Mr. Shardul Agrawala, Mr. David Jon Dokken, Mr. Steve Greco, Ms. Dottie Hagag, Ms. Sandy MacCracken, Ms. Flo Ormond, Ms. Melissa Taylor, Ms. Anne Tenney, and Ms. Laura Van Wie
- Dr. N. Sundararaman, the Secretary of the IPCC, and his staff including Mr. S. Tewungwa, Mrs. R. Bourgeois, Ms. C. Ettori, and Ms. C. Tanikie.

G.O.P. Obasi

Secretary-General
World Meteorological Organization

Ms. E. Dowdeswell

Executive Director
United Nations Environment Programme

Preface

In June 1993, Working Group II of the Intergovernmental Panel on Climate Change (IPCC) was asked to review the state of knowledge concerning the impacts of climate change on physical and ecological systems, human health, and socio-economic sectors. Working Group II also was charged with reviewing available information on the technical and economic feasibility of a range of potential adaptation and mitigation strategies.

This volume responds to this charge and represents a tremendous achievement—the coordinated contributions of well over a thousand individuals from over 50 developed and developing countries and a dozen international organizations. It includes introductory "primers" on ecological systems and energy production and use; 25 chapters, covering both vulnerability to climate change and options for reducing emissions or enhancing sinks; and three appendices that inventory mitigation technologies and delineate methodologies for assessing impacts/adaptations and mitigation options.

The chapters provide an overview of developments in our scientific understanding since the first IPCC assessments of impacts and response options in 1990, and the supplemental IPCC assessments of 1992. Uncertainties are described, with an eye for identifying both policy significance and research opportunities. In presenting this information, each team of authors has sought to communicate its findings in way that is useful to decisionmakers, research managers, and peers within their field of research; we hope that these audiences, in addition to educators and the general public, will find this volume useful.

Approach of the Assessment

From the earliest stages of the process, participants in the assessment understood the need to confront the fact that confidence in regional projections of temperature, precipitation, soil moisture, and other climate parameters important to impacts models remains low, that uncertainty increases as scale decreases, that patterns of climate change are interwoven with climate variability, and that regional patterns are likely to be affected by both greenhouse gases and anthropogenic aerosols, the latter of which are only now beginning to be incorporated into transient GCM simulations. To provide useful information to decisionmakers, Working Group II needed to find a way to distinguish between uncertainties arising from remaining questions about the responses of systems to a given level or rate of climate change and uncertainties related to the regional-scale climate projections themselves. Consequently, Working Group II decided to focus on *assessing the sensitivity and vulnerability of systems* to a range of climate changes, and only then, having identified response functions and/or potential thresholds, on *evaluating the plausible impacts* that would result from a particular regional climate scenario. In essence, the approach first sought to clarify what was known and unknown about three distinct issues before applying regional climate scenarios to estimate potential impacts. These issues were:

- How *sensitive* is a particular system to climate change—that is, in simplified terms, how will a system respond to given changes in climate? Given the wide range of systems reviewed in this assessment, these relationships are described in a variety of forms, ranging from specification of quantitative functional relationships for some systems (e.g., climate-yield models for agriculture, rainfall-runoff models for hydrological systems, models of energy demand for heating or cooling driven by temperature change) to more qualitative relationships for other systems.
- How *adaptable* is a particular system to climate change—that is, to what degree are adjustments possible in practices, processes, or structures of systems in response to projected or actual changes of climate? This issue is important for both ecological and social systems because it is critical to recognize that both types of systems have capacities that will enable them to resist adverse consequences of new conditions or to capitalize on new opportunities. Adaptation can be spontaneous or planned, and can be carried out in response to or in anticipation of changes.
- Finally, how *vulnerable* is a system to climate change—that is, how susceptible is it to damage or harm? Vulnerability defines the extent to which climate change may damage or harm a system. It depends not only on a system's sensitivity but also on its ability to adapt to new climate conditions. Both the magnitude and rate of climate change are important in determining the sensitivity, adaptability, and vulnerability of a system.

Building on this sensitivity/vulnerability approach, the chapters of the assessment distinguish, to the extent possible, uncertainties relating to remaining questions about the sensitivity, adaptability, or vulnerability of systems to climate change from uncertainties related to the particular regional climate scenarios used in their estimation of potential impacts.

Levels of Confidence

In the course of the assessment, Working Group II also developed a common approach to describe the levels of confidence that author teams were asked to assign to the major findings in the executive summaries of their chapters. Several approaches were considered, and the lead authors finally selected a straight-forward, three-tiered structure:

- *High Confidence*—This category denotes wide agreement, based on multiple findings through multiple lines of investigation. In other words, there was a high degree of consensus among the authors based on the existence of substantial evidence in support of the conclusion.
- *Medium Confidence*—This category indicates that there is a consensus, but not a strong one, in support of the conclusion. This ranking could be applied to a situation in which an hypothesis or conclusion is supported by a fair amount of information, but not a sufficient amount to convince all participating authors, or where other less plausible hypotheses cannot yet be completely ruled out.
- *Low Confidence*—This category is reserved for cases when lead authors were highly uncertain about a particular conclusion. This uncertainty could be a reflection of a lack of consensus or the existence of serious competing hypotheses, each with adherents and evidence to support their positions. Alternatively, this ranking could result from the existence of extremely limited information to support an initial plausible idea or hypothesis.

Readers of the assessment need to keep in mind that while the confidence levels used in the report are an attempt to communicate to decisionmakers a rough sense of the collective judgment by the authors of the degree of certainty or uncertainty that should be associated with a particular finding, they are an imperfect tool. In particular, it should be noted that assigning levels of confidence to research findings is a subjective process; different individuals will assign different levels of confidence to the same findings and the same base of evidence because they demand different standards of proof. Moreover, there are multiple sources of uncertainty, some of which are difficult to identify with precision, leading different individuals to make different judgments. Finally, the amount of evidence that an individual will require to view a finding as "well-established" has been shown to be higher for findings that have high consequence than for findings of lesser consequence or for which less is at stake.

Acknowledgments

We wish to acknowledge the tireless, voluntary efforts of authors, contributors, and reviewers (from universities, private and government laboratories, and industry and environmental organizations). We wish to thank the following talented and dedicated individuals who served as staff, interns, or volunteers at the Working Group II Technical Support Unit during portions of this assessment: Mr. Shardul Agrawala, Mr. David Jon Dokken, Mr. Steve Greco, Ms. Dottie Hagag, Ms. Sandy MacCracken, Ms. Flo Ormond, Ms. Melissa Taylor, Ms. Anne Tenney, and Ms. Laura Van Wie. Without the willingness of all these individuals to give unstintingly of their professional expertise and free time, this assessment would not have been possible. We acknowledge the critical role of many program managers in national and international research programs who supported the work of the authors through grants and release time from other responsibilities. We also note that the volume benefitted greatly from the close working relationship established with the authors and Technical Support Units of Working Groups I and III. Last, but certainly not least, we wish to acknowledge the leadership of the IPCC Chairman, Professor Bert Bolin, and the IPCC Secretary, Dr. N. Sundararaman.

Robert T. Watson
M.C. Zinyowera
Richard H. Moss

Summary for Policymakers:

Scientific-Technical Analyses of Impacts, Adaptations, and Mitigation of Climate Change

A Report of Working Group II of the Intergovernmental Panel on Climate Change

CONTENTS

1. Scope of the Assessment

The charge to Working Group II of the Intergovernmental Panel on Climate Change (IPCC) was to review the state of knowledge concerning the impacts of climate change on physical and ecological systems, human health, and socioeconomic sectors. Working Group II also was charged with reviewing available information on the technical and economic feasibility of a range of potential adaptation and mitigation strategies. This assessment provides scientific, technical, and economic information that can be used, *inter alia*, in evaluating whether the projected range of plausible impacts constitutes "dangerous anthropogenic interference with the climate system," as referred to in Article 2 of the United Nations Framework Convention on Climate Change (UNFCCC), and in evaluating adaptation and mitigation options that could be used in progressing towards the ultimate objective of the UNFCCC (see Box 1).

Box 1. Ultimate Objective of the UNFCCC (Article 2)

"...stabilization of greenhouse gas concentrations in the atmosphere at a level that would prevent dangerous anthropogenic interference with the climate system. Such a level should be achieved within a time frame sufficient to allow ecosystems to adapt naturally to climate change, to ensure that food production is not threatened, and to enable economic development to proceed in a sustainable manner."

2. Nature of the Issue

Human activities are increasing the atmospheric concentrations of greenhouse gases—which tend to warm the atmosphere—and, in some regions, aerosols—which tend to cool the atmosphere. These changes in greenhouse gases and aerosols, taken together, are projected to lead to regional and global changes in climate and climate-related parameters such as temperature, precipitation, soil moisture, and sea level. Based on the range of sensitivities of climate to increases in greenhouse gas concentrations reported by IPCC Working Group I and plausible ranges of emissions (IPCC IS92; see Table 1), climate models, taking into account greenhouse gases and aerosols, project an increase in global mean surface temperature of about 1–3.5°C by 2100, and an associated increase in sea level of about 15–95 cm. The reliability of regional-scale predictions is still low, and the degree to which climate variability may change is uncertain. However, potentially serious changes have been identified, including an increase in some regions in the incidence of extreme high-temperature events, floods, and droughts, with resultant consequences for fires, pest outbreaks, and ecosystem composition, structure, and functioning, including primary productivity.

Table 1: *Summary of assumptions in the six IPCC 1992 alternative scenarios.*

Scenario	Population	Economic Growth	Energy Supplies
IS92a,b	World Bank 1991 11.3 billion by 2100	1990–2025: 2.9% 1990–2100: 2.3%	12,000 EJ conventional oil 13,000 EJ natural gas Solar costs fall to $0.075/kWh 191 EJ of biofuels available at $70/barrel[a]
IS92c	UN Medium-Low Case 6.4 billion by 2100	1990–2025: 2.0% 1990–2100: 1.2%	8,000 EJ conventional oil 7,300 EJ natural gas Nuclear costs decline by 0.4% annually
IS92d	UN Medium-Low Case 6.4 billion by 2100	1990–2025: 2.7% 1990–2100: 2.0%	Oil and gas same as IS92c Solar costs fall to $0.065/kWh 272 EJ of biofuels available at $50/barrel
IS92e	World Bank 1991 11.3 billion by 2100	1990–2025: 3.5% 1990–2100: 3.0%	18,400 EJ conventional oil Gas same as IS92a,b Phase out nuclear by 2075
IS92f	UN Medium-High Case 17.6 billion by 2100	1990–2025: 2.9% 1990–2100: 2.3%	Oil and gas same as IS92e Solar costs fall to $0.083/kWh Nuclear costs increase to $0.09/kWh

[a]Approximate conversion factor: 1 barrel = 6 GJ.
Source: IPCC, 1992: Emissions scenarios for IPCC: an update. In: *Climate Change 1992: The Supplementary Report to the IPCC Scientific Assessment* [J.T. Houghton, B.A. Callander, and S.K. Varney (eds.)]. Section A3, prepared by J. Leggett, W.J. Pepper, and R.J. Swart, and WMO/UNEP. Cambridge University Press, Cambridge, UK, 200 pp.

Human health, terrestrial and aquatic ecological systems, and socioeconomic systems (e.g., agriculture, forestry, fisheries, and water resources) are all vital to human development and well-being and are all sensitive to changes in climate. Whereas many regions are likely to experience the adverse effects of climate change—some of which are potentially irreversible—some effects of climate change are likely to be beneficial. Hence, different segments of society can expect to confront a variety of changes and the need to adapt to them.

Policymakers are faced with responding to the risks posed by anthropogenic emissions of greenhouse gases in the face of significant scientific uncertainties. It is appropriate to consider these uncertainties in the context of information indicating that climate-induced environmental changes cannot be reversed quickly, if at all, due to the long time scales associated with the climate system (see Box 2). Decisions taken during the next few years may limit the range of possible policy options in the future because high near-term emissions would require deeper reductions in the future to meet any given target concentration. Delaying action might reduce the overall costs of mitigation because of potential technological advances but could increase both the rate and the eventual magnitude of climate change, hence the adaptation and damage costs.

Policymakers will have to decide to what degree they want to take precautionary measures by mitigating greenhouse gas emissions and enhancing the resilience of vulnerable systems by means of adaptation. Uncertainty does not mean that a nation or the world community cannot position itself better to cope with the broad range of possible climate changes or protect against potentially costly future outcomes. Delaying such measures may leave a nation or the world poorly prepared to deal with adverse changes and may increase the possibility of irreversible or very costly consequences.

Box 2. Time Scales of Processes Influencing the Climate System

- Turnover of the capital stock responsible for emissions of greenhouse gases: ***Years to decades*** (without premature retirement)
- Stabilization of atmospheric concentrations of long-lived greenhouse gases given a stable level of greenhouse gas emissions: ***Decades to millennia***
- Equilibration of the climate system given a stable level of greenhouse gas concentrations: ***Decades to centuries***
- Equilibration of sea level given a stable climate: ***Centuries***
- Restoration/rehabilitation of damaged or disturbed ecological systems: ***Decades to centuries*** (some changes, such as species extinction, are irreversible, and it may be impossible to reconstruct and reestablish some disturbed ecosystems)

Options for adapting to change or mitigating change that can be justified for other reasons today (e.g., abatement of air and water pollution) and make society more flexible or resilient to anticipated adverse effects of climate change appear particularly desirable.

3. Vulnerability to Climate Change

Article 2 of the UNFCCC explicitly acknowledges the importance of natural ecosystems, food production, and sustainable economic development. This report addresses the potential *sensitivity*, *adaptability*, and *vulnerability* of ecological and socioeconomic systems—including hydrology and water resources management, human infrastructure, and human health—to changes in climate (see Box 3).

Human-induced climate change adds an important new stress. Human-induced climate change represents an important additional stress, particularly to the many ecological and socioeconomic systems already affected by pollution, increasing resource demands, and nonsustainable management practices. The most vulnerable systems are those with the greatest sensitivity to climate changes and the least adaptability.

Most systems are sensitive to climate change. Natural ecological systems, socioeconomic systems, and human health are all sensitive to both the magnitude and the rate of climate change.

Impacts are difficult to quantify, and existing studies are limited in scope. Although our knowledge has increased significantly during the last decade, and qualitative estimates can be developed, quantitative projections of the impacts of climate change on any particular system at any particular location are difficult because regional-scale climate change predictions are uncertain; our current understanding of many critical processes is limited; and systems are subject to multiple climatic and non-climatic stresses, the interactions of which are not always linear or additive. Most impact studies have assessed how systems would respond to climate change resulting from an arbitrary doubling of equivalent atmospheric carbon dioxide (CO_2) concentrations. Furthermore, very few studies have considered dynamic responses to steadily increasing concentrations of greenhouse gases; fewer still have examined the consequences of increases beyond a doubling of equivalent atmospheric CO_2 concentrations or assessed the implications of multiple stress factors.

Successful adaptation depends upon technological advances, institutional arrangements, availability of financing, and information exchange. Technological advances generally have increased adaptation options for managed systems such as agriculture and water supply. However, many regions of the world currently have limited access to these technologies and appropriate information. The efficacy and cost-effective use of adaptation strategies will depend upon the availability of financial resources, technology transfer, and cultural, educational,

Box 3. Sensitivity, Adaptability, and Vulnerability

Sensitivity is the degree to which a system will respond to a change in climatic conditions (e.g., the extent of change in ecosystem composition, structure, and functioning, including primary productivity, resulting from a given change in temperature or precipitation).

Adaptability refers to the degree to which adjustments are possible in practices, processes, or structures of systems to projected or actual changes of climate. Adaptation can be spontaneous or planned, and can be carried out in response to or in anticipation of changes in conditions.

Vulnerability defines the extent to which climate change may damage or harm a system. It depends not only on a system's sensitivity but also on its ability to adapt to new climatic conditions.

Both the magnitude and the rate of climate change are important in determining the sensitivity, adaptability, and vulnerability of a system.

managerial, institutional, legal, and regulatory practices, both domestic and international in scope. Incorporating climate-change concerns into resource-use and development decisions and plans for regularly scheduled investments in infrastructure will facilitate adaptation.

Vulnerability increases as adaptive capacity decreases. The vulnerability of human health and socioeconomic systems—and, to a lesser extent, ecological systems—depends upon economic circumstances and institutional infrastructure. This implies that systems typically are more vulnerable in developing countries where economic and institutional circumstances are less favorable. People who live on arid or semi-arid lands, in low-lying coastal areas, in water-limited or flood-prone areas, or on small islands are particularly vulnerable to climate change. Some regions have become more vulnerable to hazards such as storms, floods, and droughts as a result of increasing population density in sensitive areas such as river basins and coastal plains. Human activities, which fragment many landscapes, have increased the vulnerability of lightly managed and unmanaged ecosystems. Fragmentation limits natural adaptation potential and the potential effectiveness of measures to assist adaptation in these systems, such as the provision of migration corridors. A changing climate's near-term effects on ecological and socioeconomic systems most likely will result from changes in the intensity and seasonal and geographic distribution of common weather hazards such as storms, floods, and droughts. In most of these examples, vulnerability can be reduced by strengthening adaptive capacity.

Detection will be difficult, and unexpected changes cannot be ruled out. Unambiguous detection of climate-induced changes in most ecological and social systems will prove extremely difficult in the coming decades. This is because of the complexity of these systems, their many non-linear feedbacks, and their sensitivity to a large number of climatic and non-climatic factors, all of which are expected to continue to change simultaneously. The development of a baseline projecting future conditions without climate change is crucial, for it is this baseline against which all projected impacts are measured. As future climate extends beyond the boundaries of empirical knowledge (i.e., the documented impacts of climate variation in the past), it becomes more likely that actual outcomes will include surprises and unanticipated rapid changes.

Further research and monitoring are essential. Enhanced support for research and monitoring, including cooperative efforts from national, international, and multi-lateral institutions, is essential in order to improve significantly regional-scale climate projections; understand the responses of human health, ecological, and socioeconomic systems to changes in climate and other stress factors; and improve our understanding of the efficacy and cost-effectiveness of adaptation strategies.

3.1. Terrestrial and Aquatic Ecosystems

Ecosystems contain the Earth's entire reservoir of genetic and species diversity and provide many goods and services critical to individuals and societies. These goods and services include (i) providing food, fiber, medicines, and energy; (ii) processing and storing carbon and other nutrients; (iii) assimilating wastes, purifying water, regulating water runoff, and controlling floods, soil degradation, and beach erosion; and (iv) providing opportunities for recreation and tourism. These systems and the functions they provide are sensitive to the rate and extent of changes in climate. Figure 1 illustrates that mean annual temperature and mean annual precipitation can be correlated with the distribution of the world's major biomes.

The composition and geographic distribution of many ecosystems will shift as individual species respond to changes in climate; there will likely be reductions in biological diversity and in the goods and services that ecosystems provide society. Some ecological systems may not reach a new equilibrium for several centuries after the climate achieves a new balance.

Forests. Models project that a sustained increase of 1°C in global mean temperature is sufficient to cause changes in regional climates that will affect the growth and regeneration capacity of forests in many regions. In several instances this will alter the function and composition of forests significantly. As a consequence of possible changes in temperature and water availability under doubled equivalent-CO_2 equilibrium conditions, a substantial fraction (a global average of one-third, varying by region from one-seventh to two-thirds) of the existing forested area of the world will undergo major changes in broad vegetation types—with the greatest changes occurring in high latitudes and the least in the tropics. Climate change is expected to occur at a rapid rate relative to the speed at which forest species grow,

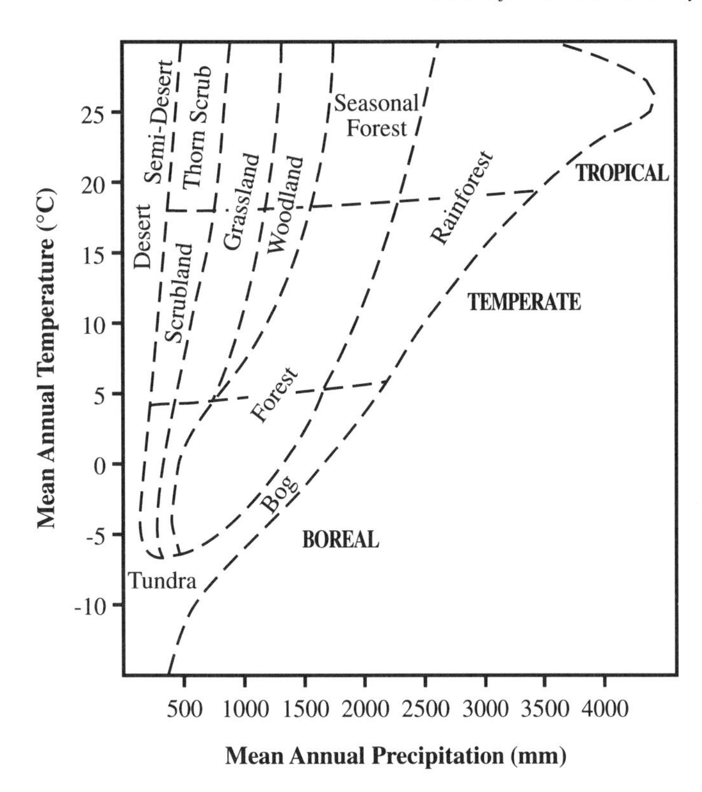

Figure 1: This figure illustrates that mean annual temperature and mean annual precipitation can be correlated with the distribution of the world's major biomes. While the role of these annual means in affecting this distribution is important, it should be noted that the distribution of biomes may also strongly depend on seasonal factors such as the length of the dry season or the lowest absolute minimum temperature, on soil properties such as water-holding capacity, on land-use history such as agriculture or grazing, and on disturbance regimes such as the frequency of fire.

reproduce, and reestablish themselves. For mid-latitude regions, a global average warming of 1–3.5°C over the next 100 years would be equivalent to a poleward shift of the present isotherms by approximately 150–550 km or an altitude shift of about 150–550 m; in low latitudes, temperatures would generally be increased to higher levels than now exist. This compares to past tree species migration rates that are believed to be on the order of 4–200 km per century. Therefore, the species composition of forests is likely to change; entire forest types may disappear, while new assemblages of species, hence new ecosystems, may be established. Figure 2 depicts potential distribution of biomes under current and a doubled equivalent-CO_2 climate. Although net primary productivity could increase, the standing biomass of forests may not because of more frequent outbreaks and extended ranges of pests and pathogens, and increasing frequency and intensity of fires. Large amounts of carbon could be released into the atmosphere during transitions from one forest type to another because the rate at which carbon can be lost during times of high forest mortality is greater than the rate at which it can be gained through growth to maturity.

Rangelands. In tropical rangelands, mean temperature increases should not lead to major alterations in productivity and species composition, but altered rainfall amount and seasonality and increased evapotranspiration will. Increases in atmospheric CO_2 concentration may raise the carbon-to-nitrogen ratio of forage for herbivores, thus reducing its food value. Shifts in temperature and precipitation in temperate rangelands may result in altered growing seasons and boundary shifts between grasslands, forests, and shrublands.

Deserts and Desertification. Deserts are likely to become more extreme—in that, with few exceptions, they are projected to become hotter but not significantly wetter. Temperature increases could be a threat to organisms that exist near their heat-tolerance limits. The impacts on water balance, hydrology, and vegetation are uncertain. Desertification, as defined by the UN Convention to Combat Desertification, is land degradation in arid, semi-arid, and dry sub-humid areas resulting from various factors, including climatic variations and human activities. Desertification is more likely to become irreversible if the environment becomes drier and the soil becomes further degraded through erosion and compaction. Adaptation to drought and desertification may rely on the development of diversified production systems.

Cryosphere. Models project that between one-third and one-half of existing mountain glacier mass could disappear over the next 100 years. The reduced extent of glaciers and depth of snow cover also would affect the seasonal distribution of river flow and water supply for hydroelectric generation and agriculture. Anticipated hydrological changes and reductions in the areal extent and depth of permafrost could lead to large-scale damage to infrastructure, an additional flux of CO_2 into the atmosphere, and changes in processes that contribute to the flux of methane (CH_4) into the atmosphere. Reduced sea-ice extent and thickness would increase the seasonal duration of navigation on rivers and in coastal areas that are presently affected by seasonal ice cover, and may increase navigability in the Arctic Ocean. Little change in the extent of the Greenland and Antarctic ice sheets is expected over the next 50–100 years.

Mountain Regions. The projected decrease in the extent of mountain glaciers, permafrost, and snow cover caused by a warmer climate will affect hydrologic systems, soil stability, and related socioeconomic systems. The altitudinal distribution of vegetation is projected to shift to higher elevation; some species with climatic ranges limited to mountain tops could become extinct because of disappearance of habitat or reduced migration potential. Mountain resources such as food and fuel for indigenous populations may be disrupted in many developing countries. Recreational industries—of increasing economic importance to many regions—also are likely to be disrupted.

Lakes, Streams, and Wetlands. Inland aquatic ecosystems will be influenced by climate change through altered water temperatures, flow regimes, and water levels. In lakes and streams, warming would have the greatest biological effects at high latitudes, where biological productivity would increase, and at the low-latitude boundaries of cold- and cool-water species ranges, where extinctions would be greatest. Warming

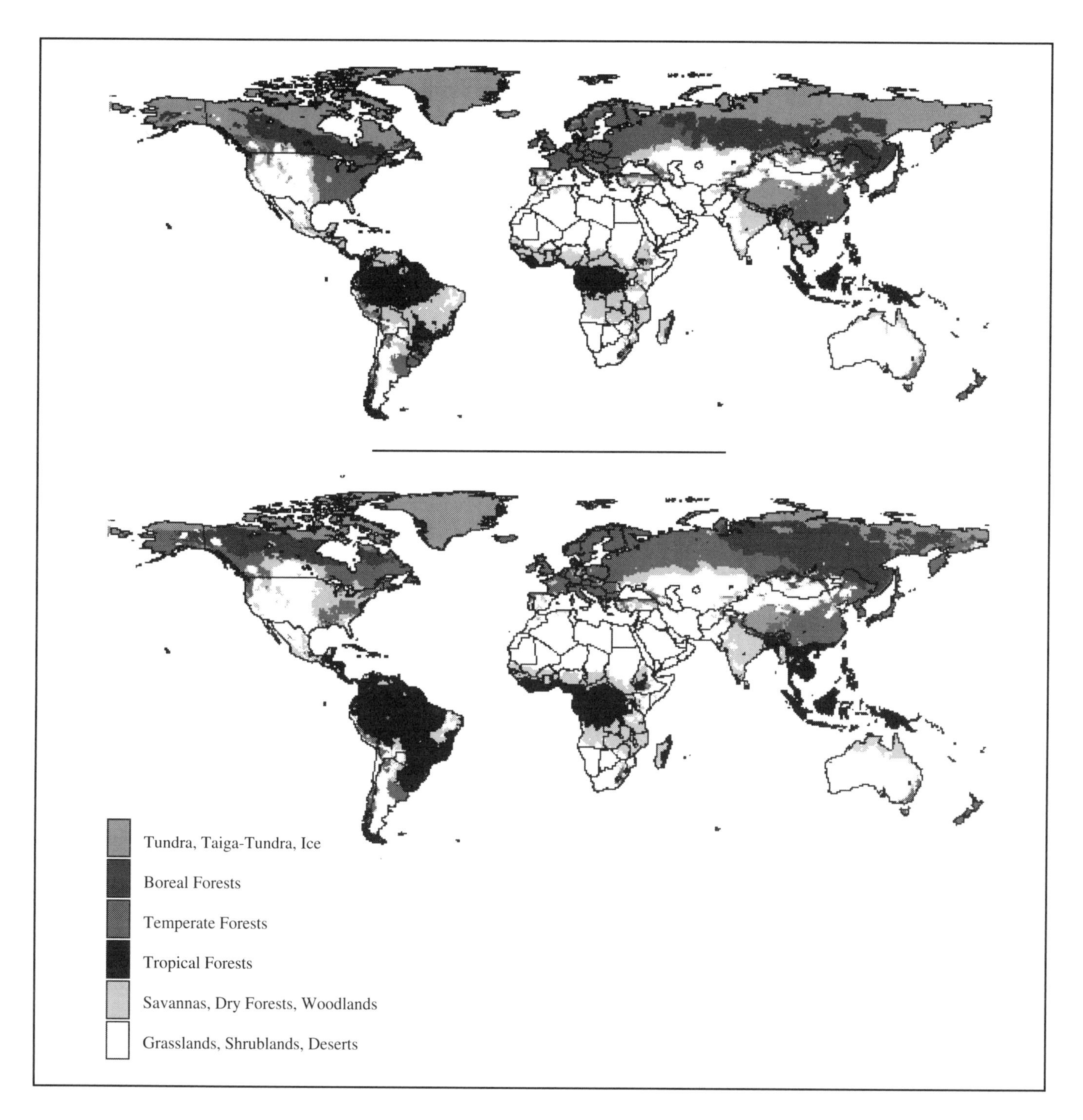

Figure 2: Potential distribution of the major world biomes under current climate conditions, simulated by Mapped Atmosphere-Plant-Soil System (MAPSS) model (top). "Potential distribution" indicates the natural vegetation that can be supported at each site, given monthly inputs of precipitation, temperature, humidity, and windspeed. The lower product illustrates the projected distribution of the major world biomes by simulating the effects of 2 x CO_2-equivalent concentrations (GFDL general circulation model), including the direct physiological effects of CO_2 on vegetation. Both products are adapted from: Neilson, R.P. and D. Marks, 1994: A global perspective of regional vegetation and hydrologic sensitivities from climatic change. *Journal of Vegetation Science*, **5**, 715-730.

of larger and deeper temperate zone lakes would increase their productivity; although in some shallow lakes and in streams, warming could increase the likelihood of anoxic conditions. Increases in flow variability, particularly the frequency and duration of large floods and droughts, would tend to reduce water quality and biological productivity and habitat in streams. Water-level declines will be most severe in lakes and streams in dry evaporative drainages and in basins with small catchments. The geographical distribution of wetlands is likely to shift with changes in temperature and precipitation. There will be an impact of climate change on greenhouse gas release from non-tidal wetlands, but there is uncertainty regarding the exact effects from site to site.

Coastal Systems. Coastal systems are economically and ecologically important and are expected to vary widely in their response to changes in climate and sea level. Climate change and a rise in sea level or changes in storms or storm surges could result in the erosion of shores and associated habitat, increased salinity of estuaries and freshwater aquifers, altered tidal ranges in rivers and bays, changes in sediment and nutrient transport, a change in the pattern of chemical and microbiological contamination in coastal areas, and increased coastal flooding. Some coastal ecosystems are particularly at risk, including saltwater marshes, mangrove ecosystems, coastal wetlands, coral reefs, coral atolls, and river deltas. Changes in these ecosystems would have major negative effects on tourism, freshwater supplies, fisheries, and biodiversity. Such impacts would add to modifications in the functioning of coastal oceans and inland waters that already have resulted from pollution, physical modification, and material inputs due to human activities.

Oceans. Climate change will lead to changes in sea level, increasing it on average, and also could lead to altered ocean circulation, vertical mixing, wave climate, and reductions in sea-ice cover. As a result, nutrient availability, biological productivity, the structure and functions of marine ecosystems, and heat and carbon storage capacity may be affected, with important feedbacks to the climate system. These changes would have implications for coastal regions, fisheries, tourism and recreation, transport, off-shore structures, and communication. Paleoclimatic data and model experiments suggest that abrupt climatic changes can occur if freshwater influx from the movement and melting of sea ice or ice sheets significantly weakens global thermohaline circulation.

3.2. *Hydrology and Water Resources Management*

Climate change will lead to an intensification of the global hydrological cycle and can have major impacts on regional water resources. A change in the volume and distribution of water will affect both ground and surface water supply for domestic and industrial uses, irrigation, hydropower generation, navigation, instream ecosystems, and water-based recreation.

Changes in the total amount of precipitation and in its frequency and intensity directly affect the magnitude and timing of runoff and the intensity of floods and droughts; however, at present, specific regional effects are uncertain. Relatively small changes in temperature and precipitation, together with the non-linear effects on evapotranspiration and soil moisture, can result in relatively large changes in runoff, especially in arid and semi-arid regions. High-latitude regions may experience increased runoff due to increased precipitation, whereas runoff may decrease at lower latitudes due to the combined effects of increased evapotranspiration and decreased precipitation. More intense rainfall would tend to increase runoff and the risk of flooding, although this would depend not only on the change in rainfall but also on catchment physical and biological characteristics. A warmer climate could decrease the proportion of precipitation falling as snow, leading to reductions in spring runoff and increases in winter runoff.

The quantity and quality of water supplies already are serious problems today in many regions, including some low-lying coastal areas, deltas, and small islands, making countries in these regions particularly vulnerable to any additional reduction in indigenous water supplies. Water availability currently falls below 1,000 m^3 per person per year—a common benchmark for water scarcity—in a number of countries (e.g., Kuwait, Jordan, Israel, Rwanda, Somalia, Algeria, Kenya) or is expected to fall below this benchmark in the next 2 to 3 decades (e.g., Libya, Egypt, South Africa, Iran, Ethiopia). In addition, a number of countries in conflict-prone areas are highly dependent on water originating outside their borders (e.g., Cambodia, Syria, Sudan, Egypt, Iraq).

The impacts of climate change will depend on the baseline condition of the water supply system and the ability of water resource managers to respond not only to climate change but also to population growth and changes in demands, technology, and economic, social, and legislative conditions. In some cases—particularly in wealthier countries with integrated water-management systems—improved management may protect water users from climate change at minimal cost; in many others, however, there could be substantial economic, social, and environmental costs, particularly in regions that already are water-limited and where there is a considerable competition among users. Experts disagree over whether water supply systems will evolve substantially enough in the future to compensate for the anticipated negative impacts of climate change on water resources and for potential increases in demand.

Options for dealing with the possible impacts of a changed climate and increased uncertainty about future supply and demand for freshwater include more efficient management of existing supplies and infrastructure; institutional arrangements to limit future demands/promote conservation; improved monitoring and forecasting systems for floods/droughts; rehabilitation of watersheds, especially in the tropics; and construction of new reservoir capacity to capture and store excess flows produced by altered patterns of snowmelt and storms.

3.3. Food and Fiber

Agriculture. Crop yields and changes in productivity due to climate change will vary considerably across regions and among localities, thus changing the patterns of production. Productivity is projected to increase in some areas and decrease in others, especially the tropics and subtropics (Table 2). However, existing studies show that on the whole global agricultural production could be maintained relative to baseline production in the face of climate change modeled by general circulation models (GCMs) at doubled equivalent-CO_2 equilibrium conditions, but that regional effects would vary widely. This conclusion takes into account the beneficial effects of CO_2 fertilization, but does not allow for changes in agricultural pests and the possible effects of changing climatic variability.

Focusing on global agricultural production does not address the potentially serious consequences of large differences at local and regional scales, even at mid-latitudes. There may be increased risk of hunger and famine in some locations; many of the world's poorest people—particularly those living in subtropical and tropical areas, and dependent on isolated agricultural systems in semi-arid and arid regions—are most at risk of increased hunger. Many of these at-risk populations are found in sub-Saharan Africa; south, east, and southeast Asia; and tropical areas of Latin America, as well as some Pacific island nations.

Adaptation—such as changes in crops and crop varieties, improved water-management and irrigation systems, and changes in planting schedules and tillage practices—will be important in limiting negative effects and taking advantage of beneficial changes in climate. The extent of adaptation depends on the affordability of such measures, particularly in developing countries; access to know-how and technology; the rate of climate change; and biophysical constraints such as water availability, soil characteristics, and crop genetics. The incremental costs of adaptation strategies could create a serious burden for developing countries; some adaptation strategies may result in cost savings for some countries. There are significant uncertainties about the capacity of different regions to adapt successfully to projected climate change.

Livestock production may be affected by changes in grain prices and rangeland and pasture productivity. In general, analyses indicate that intensively managed livestock systems have more potential for adaptation than crop systems. This may not be the case in pastoral systems, where the rate of technology adoption is slow and changes in technology are viewed as risky.

Forest Products. Global wood supplies during the next century may become increasingly inadequate to meet projected consumption due to both climatic and non-climatic factors. Boreal forests are likely to undergo irregular and large-scale losses of living trees because of the impacts of projected climate change. Such losses could initially generate additional wood supply from salvage harvests, but could severely reduce standing stocks and wood-product availability over the long term. The exact timing and extent of this pattern is uncertain. Climate and land-use impacts on the production of temperate forest products are expected to be relatively small. In tropical regions, the availability of forest products is projected to decline by about half for non-climatic reasons related to human activities.

Fisheries. Climate-change effects interact with those of pervasive overfishing, diminishing nursery areas, and extensive inshore and coastal pollution. Globally, marine fisheries production is expected to remain about the same; high-latitude freshwater and aquaculture production are likely to increase, assuming that natural climate variability and the structure and strength of ocean currents remain about the same. The principal impacts will be felt at the national and local levels as species mix and centers of production shift. The positive effects of climate change—such as longer growing seasons, lower natural winter mortality, and faster growth rates in higher latitudes—may be offset by negative factors such as changes in established reproductive patterns, migration routes, and ecosystem relationships.

3.4. Human Infrastructure

Climate change and resulting sea-level rise can have a number of negative impacts on energy, industry, and transportation infrastructure; human settlements; the property insurance industry; tourism; and cultural systems and values.

In general, the sensitivity of the energy, industry, and transportation sectors is relatively low compared to that of agricultural or natural ecosystems, and the capacity for adaptation through management and normal replacement of capital is expected to be high. However, infrastructure and activities in these sectors would be susceptible to sudden changes, surprises, and increased frequency or intensity of extreme events. The subsectors and activities most sensitive to climate change include agroindustry, energy demand, production of renewable energy such as hydroelectricity and biomass, construction, some transportation activities, existing flood mitigation structures, and transportation infrastructure located in many areas, including vulnerable coastal zones and permafrost regions.

Climate change clearly will increase the vulnerability of some coastal populations to flooding and erosional land loss. Estimates put about 46 million people per year currently at risk of flooding due to storm surges. This estimate results from multiplying the total number of people currently living in areas potentially affected by ocean flooding by the probability of flooding at these locations in any year, given the present protection levels and population density. In the absence of adaptation measures, a 50-cm sea-level rise would increase this number to about 92 million; a 1-m sea-level rise would raise it to 118 million. If one incorporates anticipated population growth, the estimates increase substantially. Some small island nations and other countries will confront greater vulnerability because their existing sea and

Table 2: *Selected crop study results for 2 x CO_2-equivalent equilibrium GCM scenarios.*

Region	Crop	Yield Impact (%)	Comments
Latin America	Maize	-61 to increase	Data are from Argentina, Brazil, Chile, and Mexico; range is across GCM scenarios, with and without CO_2 effect.
	Wheat	-50 to -5	Data are from Argentina, Uruguay, and Brazil; range is across GCM scenarios, with and without CO_2 effect.
	Soybean	-10 to +40	Data are from Brazil; range is across GCM scenarios, with CO_2 effect.
Former Soviet Union	Wheat Grain	-19 to +41 -14 to +13	Range is across GCM scenarios and region, with CO_2 effect.
Europe	Maize	-30 to increase	Data are from France, Spain, and northern Europe; with adaptation and CO_2 effect; assumes longer season, irrigation efficiency loss, and northward shift.
	Wheat	increase or decrease	Data are from France, UK, and northern Europe; with adaptation and CO_2 effect; assumes longer season, northward shift, increased pest damage, and lower risk of crop failure.
	Vegetables	increase	Data are from UK and northern Europe; assumes pest damage increased and lower risk of crop failure.
North America	Maize Wheat	-55 to +62 -100 to +234	Data are from USA and Canada; range is across GCM scenarios and sites, with/without adaptation and with/without CO_2 effect.
	Soybean	-96 to +58	Data are from USA; less severe or increase with CO_2 and adaptation.
Africa	Maize	-65 to +6	Data are from Egypt, Kenya, South Africa, and Zimbabwe; range is over studies and climate scenarios, with CO_2 effect.
	Millet	-79 to -63	Data are from Senegal; carrying capacity fell 11–38%.
	Biomass	decrease	Data are from South Africa; agrozone shifts.
South Asia	Rice Maize Wheat	-22 to +28 -65 to -10 -61 to +67	Data are from Bangladesh, India, Philippines, Thailand, Indonesia, Malaysia, and Myanmar; range is over GCM scenarios, with CO_2 effect; some studies also consider adaptation.
China	Rice	-78 to +28	Includes rainfed and irrigated rice; range is across sites and GCM scenarios; genetic variation provides scope for adaptation.
Other Asia and Pacific Rim	Rice	-45 to +30	Data are from Japan and South Korea; range is across GCM scenarios; generally positive in north Japan, and negative in south.
	Pasture	-1 to +35	Data are from Australia and New Zealand; regional variation.
	Wheat	-41 to +65	Data are from Australia and Japan; wide variation, depending on cultivar.

Note: For most regions, studies have focused on one or two principal grains. These studies strongly demonstrate the variability in estimated yield impacts among countries, scenarios, methods of analysis, and crops, making it difficult to generalize results across areas or for different climate scenarios.

coastal defense systems are less well-established. Countries with higher population densities would be more vulnerable. For these countries, sea-level rise could force internal or international migration of populations.

A number of studies have evaluated sensitivity to a 1-m sea-level rise. This increase is at the top of the range of IPCC Working Group I estimates for 2100; it should be noted, however, that sea level is actually projected to continue to rise beyond 2100. Studies using this 1-m projection show a particular risk for small islands and deltas. Estimated land losses range from 0.05% for Uruguay, 1% for Egypt, 6% for the Netherlands, and 17.5% for Bangladesh to about 80% for the Majuro Atoll in the Marshall Islands, given the present state of protection systems. Large numbers of people also are affected—for example, about 70 million each in China and Bangladesh. Many nations face lost capital value in excess of 10% of their gross domestic product (GDP). Although annual protection costs for many nations are relatively modest (about 0.1% of GDP), the average annual costs to many small island states total several percent of GDP. For some island nations, the high cost of providing storm-surge protection would make it essentially infeasible, especially given the limited availability of capital for investment.

The most vulnerable human settlements are located in damage-prone areas of the developing world that do not have the resources to cope with impacts. Effective coastal-zone management and land-use regulation can help direct population shifts away from vulnerable locations such as flood plains, steep hillsides, and low-lying coastlines. One of the potentially unique and destructive effects on human settlements is forced internal or international migration of populations. Programs of disaster assistance can offset some of the more serious negative consequences of climate change and reduce the number of ecological refugees.

Property insurance is vulnerable to extreme climate events. A higher risk of extreme events due to climate change could lead to higher insurance premiums or the withdrawal of coverage for property in some vulnerable areas. Changes in climate variability and the risk for extreme events may be difficult to detect or predict, thus making it difficult for insurance companies to adjust premiums appropriately. If such difficulty leads to insolvency, companies may not be able to honor insurance contracts, which could economically weaken other sectors, such as banking. The insurance industry currently is under stress from a series of "billion dollar" storms since 1987, resulting in dramatic increases in losses, reduced availability of insurance, and higher costs. Some in the insurance industry perceive a current trend toward increased frequency and severity of extreme climate events. Examination of the meteorological data fails to support this perception in the context of a long-term change, although a shift within the limits of natural variability may have occurred. Higher losses strongly reflect increases in infrastructure and economic worth in vulnerable areas as well as a possible shift in the intensity and frequency of extreme weather events.

3.5. *Human Health*

Climate change is likely to have wide-ranging and mostly adverse impacts on human health, with significant loss of life. These impacts would arise by both direct and indirect pathways (Figure 3), and it is likely that the indirect impacts would, in the longer term, predominate.

Direct health effects include increases in (predominantly cardiorespiratory) mortality and illness due to an anticipated increase

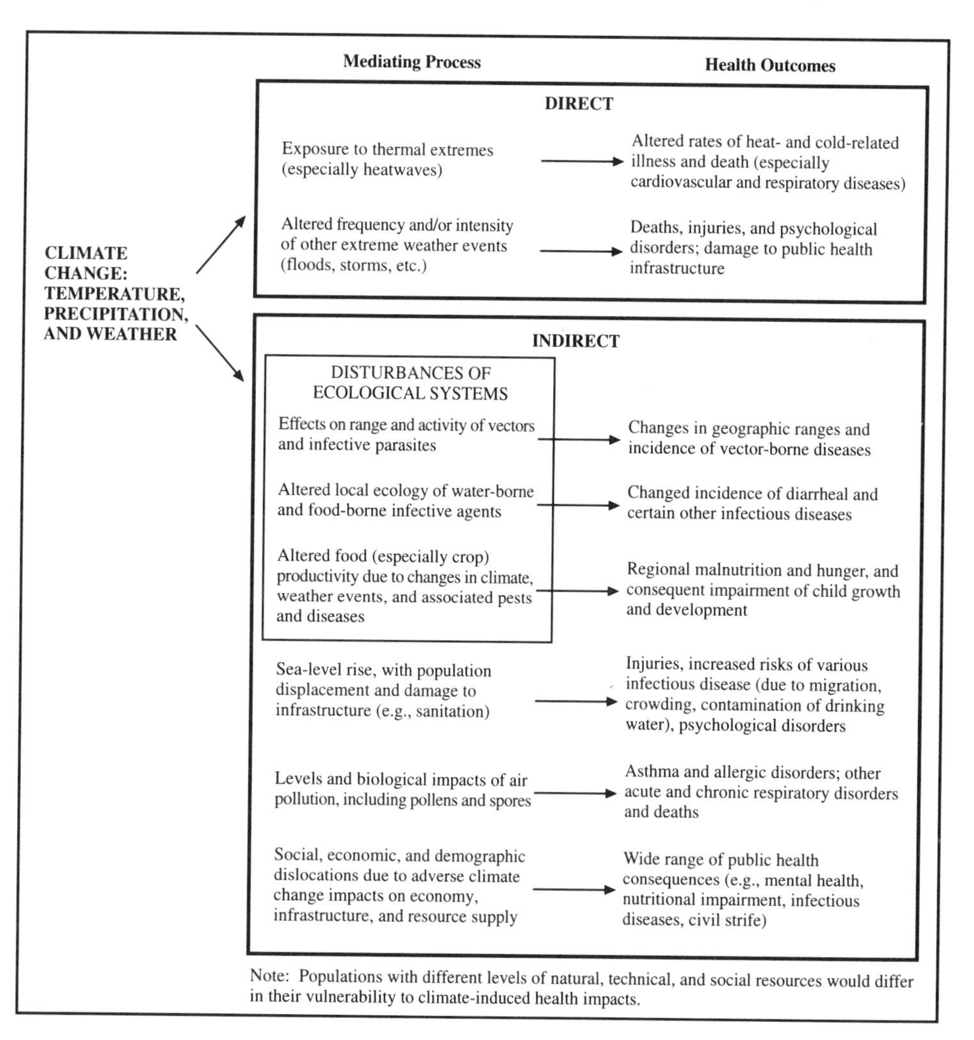

Figure 3: Ways in which climate change can affect human health.

in the intensity and duration of heat waves. Temperature increases in colder regions should result in fewer cold-related deaths. An increase in extreme weather would cause a higher incidence of death, injury, psychological disorders, and exposure to contaminated water supplies.

Indirect effects of climate change include increases in the potential transmission of vector-borne infectious diseases (e.g., malaria, dengue, yellow fever, and some viral encephalitis) resulting from extensions of the geographical range and season for vector organisms. Projections by models (that entail necessary simplifying assumptions) indicate that the geographical zone of potential malaria transmission in response to world temperature increases at the upper part of the IPCC-projected range (3–5°C by 2100) would increase from approximately 45% of the world population to approximately 60% by the latter half of the next century. This could lead to potential increases in malaria incidence (on the order of 50–80 million additional annual cases, relative to an assumed global background total of 500 million cases), primarily in tropical, subtropical, and less well-protected temperate-zone populations. Some increases in non-vector-borne infectious diseases—such as salmonellosis, cholera, and giardiasis—also could occur as a result of elevated temperatures and increased flooding.

Additional indirect effects include respiratory and allergic disorders due to climate-enhanced increases in some air pollutants, pollens, and mold spores. Exposure to air pollution and stressful weather events combine to increase the likelihood of morbidity and mortality. Some regions could experience a decline in nutritional status as a result of adverse impacts on food and fisheries productivity. Limitations on freshwater supplies also will have human health consequences.

Quantifying the projected impacts is difficult because the extent of climate-induced health disorders depends on numerous coexistent and interacting factors that characterize the vulnerability of the particular population, including environmental and socioeconomic circumstances, nutritional and immune status, population density, and access to quality health care services. Adaptive options to reduce health impacts include protective technology (e.g., housing, air conditioning, water purification, and vaccination), disaster preparedness, and appropriate health care.

4. Options to Reduce Emissions and Enhance Sinks of Greenhouse Gases

Human activities are directly increasing the atmospheric concentrations of several greenhouse gases, especially CO_2, CH_4, halocarbons, sulfur hexafluoride (SF_6), and nitrous oxide (N_2O). CO_2 is the most important of these gases, followed by CH_4. Human activities also indirectly affect concentrations of water vapor and ozone. Significant reductions in net greenhouse gas emissions are technically possible and can be economically feasible. These reductions can be achieved by utilizing an extensive array of technologies, and policy measures that accelerate technology development, diffusion, and transfer in all sectors including the energy, industry, transportation, residential/commercial, and agricultural/forestry sectors. By the year 2100, the world's commercial energy system in effect will be replaced at least twice, offering opportunities to change the energy system without premature retirement of capital stock; significant amounts of capital stock in the industrial, commercial, residential, and agricultural/forestry sectors will also be replaced. These cycles of capital replacement provide opportunities to use new, better performing technologies. It should be noted that the analyses of Working Group II do not attempt to quantify potential macroeconomic consequences that may be associated with mitigation measures. Discussion of macroeconomic analyses is found in the IPCC Working Group III contribution to the Second Assessment Report. The degree to which technical potential and cost-effectiveness are realized is dependent on initiatives to counter lack of information and overcome cultural, institutional, legal, financial and economic barriers that can hinder diffusion of technology or behavioral changes. The pursuit of mitigation options can be carried out within the limits of sustainable development criteria. Social and environmental criteria not related to greenhouse gas emissions abatement could, however, restrict the ultimate potential of each of the options.

4.1. Energy, Industrial Process, and Human Settlement Emissions

Global energy demand has grown at an average annual rate of approximately 2% for almost 2 centuries, although energy demand growth varies considerably over time and between different regions. In the published literature, different methods and conventions are used to characterize energy consumption. These conventions differ, for example, according to their definition of sectors and their treatment of energy forms. Based on aggregated national energy balances, 385 EJ of primary energy was consumed in the world in 1990, resulting in the release of 6 Gt C as CO_2. Of this, 279 EJ was delivered to end users, accounting for 3.7 Gt C emissions as CO_2 at the point of consumption. The remaining 106 EJ was used in energy conversion and distribution, accounting for 2.3 Gt C emissions as CO_2. In 1990, the three largest sectors of energy consumption were industry (45% of total CO_2 releases), residential/commercial sector (29%), and transport (21%). Of these, transport sector energy use and related CO_2 emissions have been the most rapidly growing over the past 2 decades. For the detailed sectoral mitigation option assessment in this report, 1990 energy consumption estimates are based on a range of literature sources; a variety of conventions are used to define these sectors and their energy use, which is estimated to amount to a total of 259–282 EJ.

Figure 4 depicts total energy-related emissions by major world region. Organisation for Economic Cooperation and Development (OECD) nations have been and remain major energy users and fossil fuel CO_2 emitters, although their share of global fossil fuel carbon emissions has been declining.

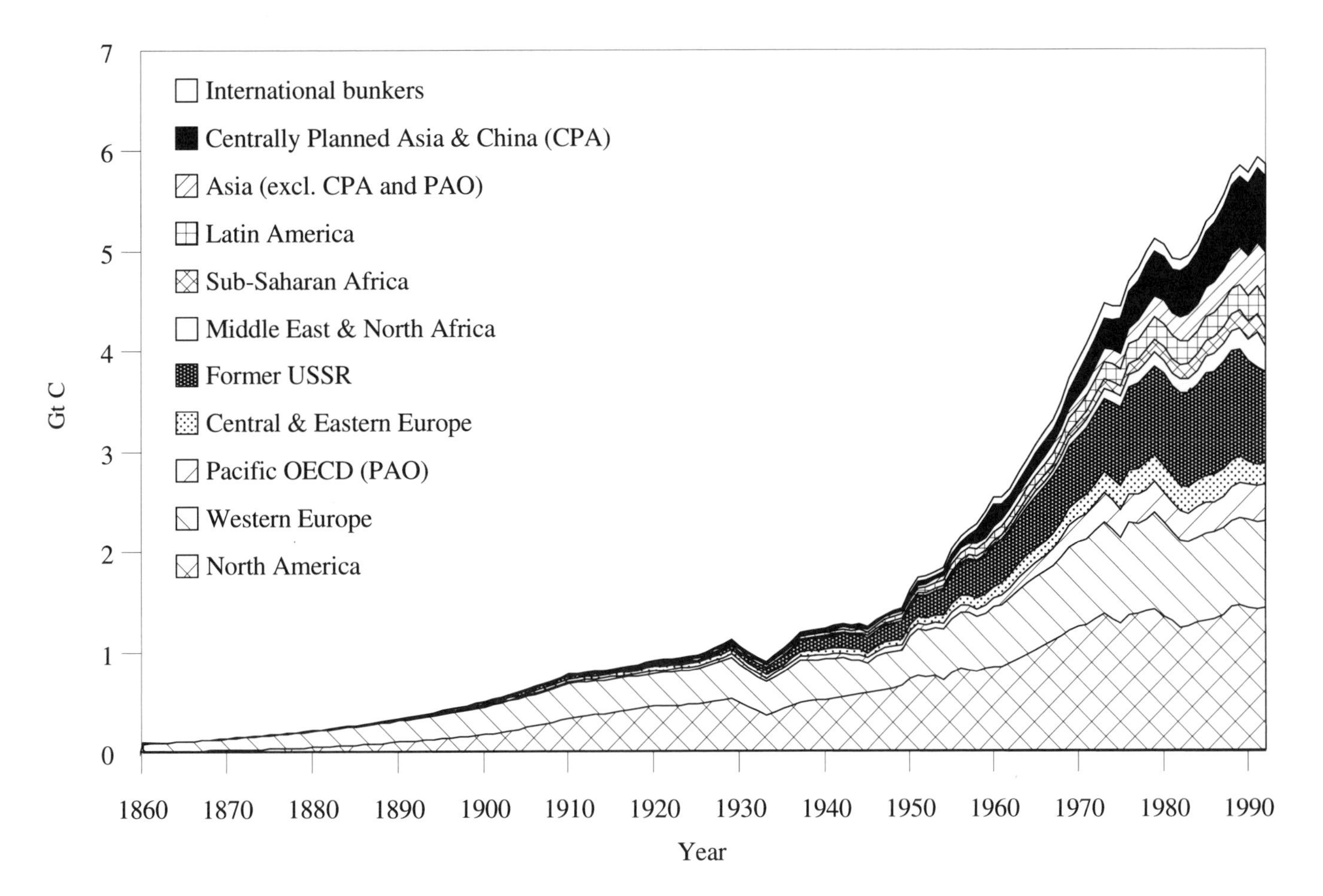

Figure 4: Global energy-related CO_2 emissions by major world region in Gt C/yr. Sources: Keeling, 1994; Marland *et al.*, 1994; Grübler and Nakicenovic, 1992; Etemad and Luciani, 1991; Fujii, 1990; UN, 1952 (see the Energy Primer for reference information).

Developing nations, taken as a group, still account for a smaller portion of total global CO_2 emissions than industrialized nations—OECD and former Soviet Union/Eastern Europe (FSU/EE)—but most projections indicate that with forecast rates of economic and population growth, the future share of developing countries will increase. Future energy demand is anticipated to continue to grow, at least through the first half of the next century. The IPCC (1992, 1994) projects that without policy intervention, there could be significant growth in emissions from the industrial, transportation, and commercial/residential buildings sectors.

4.1.1. *Energy Demand*

Numerous studies have indicated that 10–30% energy-efficiency gains above present levels are feasible at little or no net cost in many parts of the world through technical conservation measures and improved management practices over the next 2 to 3 decades. Using technologies that presently yield the highest output of energy services for a given input of energy, efficiency gains of 50–60% would be technically feasible in many countries over the same time period. Achieving these potentials will depend on future cost reductions, financing, and technology transfer, as well as measures to overcome a variety of non-technical barriers. The potential for greenhouse gas emission reductions exceeds the potential for energy use efficiency because of the possibility of switching fuels and energy sources. Because energy use is growing world-wide, even replacing current technology with more efficient technology could still lead to an absolute increase in CO_2 emissions in the future.

In 1992, the IPCC produced six scenarios (IS92a-f) of future energy use and associated greenhouse gas emissions (IPCC, 1992, 1995). These scenarios provide a wide range of possible future greenhouse gas emission levels, without mitigation measures.

In the Second Assessment Report, future energy use has been reexamined on a more detailed sectoral basis, both with and without new mitigation measures, based on existing studies. Despite different assessment approaches, the resulting ranges of energy consumption increases to 2025 without new measures are broadly consistent with those of IS92. If past trends continue, greenhouse gas emissions will grow more slowly than energy use, except in the transport sector.

The following paragraphs summarize energy-efficiency improvement potentials estimated in the IPCC Second

Assessment Report. Strong policy measures would be required to achieve these potentials. Energy-related greenhouse gas emission reductions depend on the source of the energy, but reductions in energy use will in general lead to reduced greenhouse gas emissions.

Industry. Energy use in 1990 was estimated to be 98–117 EJ, and is projected to grow to 140–242 EJ in 2025 without new measures. Countries differ widely in their current industrial energy use and energy-related greenhouse gas emission trends. Industrial sector energy-related greenhouse gas emissions in most industrialized countries are expected to be stable or decreasing as a result of industrial restructuring and technological innovation, whereas industrial emissions in developing countries are projected to increase mainly as a result of industrial growth. The short-term potential for energy-efficiency improvements in the manufacturing sector of major industrial countries is estimated to be 25%. The potential for greenhouse gas emission reductions is larger. Technologies and measures for reducing energy-related emissions from this sector include improving efficiency (e.g., energy and materials savings, cogeneration, energy cascading, steam recovery, and use of more efficient motors and other electrical devices); recycling materials and switching to those with lower greenhouse gas emissions; and developing processes that use less energy and materials.

Transportation. Energy use in 1990 was estimated to be 61–65 EJ, and is projected to grow to 90–140 EJ in 2025 without new measures. Projected energy use in 2025 could be reduced by about a third to 60–100 EJ through vehicles using very efficient drive-trains, lightweight construction, and low air-resistance design, without compromising comfort and performance. Further energy-use reductions are possible through the use of smaller vehicles; altered land-use patterns, transport systems, mobility patterns, and lifestyles; and shifting to less energy-intensive transport modes. Greenhouse gas emissions per unit of energy used could be reduced through the use of alternative fuels and electricity from renewable sources. These measures, taken together, provide the opportunity for reducing global transport energy-related greenhouse gas emissions by as much as 40% of projected emissions by 2025. Actions to reduce energy-related greenhouse gas emissions from transport can simultaneously address other problems such as local air pollution.

Commercial/Residential Sector. Energy use in 1990 was estimated to be about 100 EJ, and is projected to grow to 165–205 EJ in 2025 without new measures. Projected energy use could be reduced by about a quarter to 126–170 EJ by 2025 without diminishing services through the use of energy efficient technology. The potential for greenhouse gas emission reductions is larger. Technical changes might include reduced heat transfers through building structures and more efficient space-conditioning and water supply systems, lighting, and appliances. Ambient temperatures in urban areas can be reduced through increased vegetation and greater reflectivity of building surfaces, reducing the energy required for space conditioning. Energy-related greenhouse gas emission reductions beyond those obtained through reduced energy use could be achieved through changes in energy sources.

4.1.2. *Mitigating Industrial Process and Human Settlement Emissions*

Process-related greenhouse gases including CO_2, CH_4, N_2O, halocarbons, and SF_6 are released during manufacturing and industrial processes, such as the production of iron, steel, aluminum, ammonia, cement, and other materials. Large reductions are possible in some cases. Measures include modifying production processes, eliminating solvents, replacing feedstocks, materials substitution, increased recycling, and reduced consumption of greenhouse gas-intensive materials. Capturing and utilizing CH_4 from landfills and sewage treatment facilities and lowering the leakage rate of halocarbon refrigerants from mobile and stationary sources also can lead to significant greenhouse gas emission reductions.

4.1.3. *Energy Supply*

This assessment focuses on new technologies for capital investment and not on potential retrofitting of existing capital stock to use less carbon-intensive forms of primary energy. It is technically possible to realize deep emissions reductions in the energy supply sector in step with the normal timing of investments to replace infrastructure and equipment as it wears out or becomes obsolete. Many options for achieving these deep reductions will also decrease the emissions of sulfur dioxide, nitrogen oxides, and volatile organic compounds. Promising approaches, not ordered according to priority, are described below.

4.1.3.1. *Greenhouse gas reductions in the use of fossil fuels*

More Efficient Conversion of Fossil Fuels. New technology offers considerably increased conversion efficiencies. For example, the efficiency of power production can be increased from the present world average of about 30% to more than 60% in the longer term. Also, the use of combined heat and power production replacing separate production of power and heat—whether for process heat or space heating—offers a significant rise in fuel conversion efficiency.

Switching to Low-Carbon Fossil Fuels and Suppressing Emissions. Switching from coal to oil or natural gas, and from oil to natural gas, can reduce emissions. Natural gas has the lowest CO_2 emissions per unit of energy of all fossil fuels at about 14 kg C/GJ, compared to oil with about 20 kg C/GJ and coal with about 25 kg C/GJ. The lower carbon-containing fuels can, in general, be converted with higher efficiency than coal. Large resources of natural gas exist in many areas. New, low capital cost, highly efficient combined-cycle technology has reduced electricity costs considerably in some areas. Natural gas could potentially replace oil in the transportation sector.

Approaches exist to reduce emissions of CH_4 from natural gas pipelines and emissions of CH_4 and/or CO_2 from oil and gas wells and coal mines.

Decarbonization of Flue Gases and Fuels, and CO_2 Storage. The removal and storage of CO_2 from fossil fuel power-station stack gases is feasible, but reduces the conversion efficiency and significantly increases the production cost of electricity. Another approach to decarbonization uses fossil fuel feedstocks to make hydrogen-rich fuels. Both approaches generate a byproduct stream of CO_2 that could be stored, for example, in depleted natural gas fields. The future availability of conversion technologies such as fuel cells that can efficiently use hydrogen would increase the relative attractiveness of the latter approach. For some longer term CO_2 storage options, the costs, environmental effects, and efficacy of such options remain largely unknown.

4.1.3.2. Switching to non-fossil fuel sources of energy

Switching to Nuclear Energy. Nuclear energy could replace baseload fossil fuel electricity generation in many parts of the world if generally acceptable responses can be found to concerns such as reactor safety, radioactive-waste transport and disposal, and nuclear proliferation.

Switching to Renewable Sources of Energy. Solar, biomass, wind, hydro, and geothermal technologies already are widely used. In 1990, renewable sources of energy contributed about 20% of the world's primary energy consumption, most of it fuelwood and hydropower. Technological advances offer new opportunities and declining costs for energy from these sources. In the longer term, renewable sources of energy could meet a major part of the world's demand for energy. Power systems can easily accommodate limited fractions of intermittent generation, and with the addition of fast-responding backup and storage units, also higher fractions. Where biomass is sustainably regrown and used to displace fossil fuels in energy production, net carbon emissions are avoided as the CO_2 released in converting the biomass to energy is again fixed in biomass through photosynthesis. If the development of biomass energy can be carried out in ways that effectively address concerns about other environmental issues and competition with other land uses, biomass could make major contributions in both the electricity and fuels markets, as well as offering prospects of increasing rural employment and income.

4.1.4. Integration of Energy System Mitigation Options

To assess the potential impact of combinations of individual measures at the energy system level, in contrast to the level of individual technologies, variants of a Low CO_2-Emitting Energy Supply System (LESS) are described. The LESS constructions are "thought experiments" exploring possible global energy systems.

The following assumptions were made: World population grows from 5.3 billion in 1990 to 9.5 billion by 2050 and 10.5 billion by 2100. GDP grows 7-fold by 2050 (5-fold and 14-fold in industrialized and developing countries, respectively) and 25-fold by 2100 (13-fold and 70-fold in industrialized and developing countries, respectively), relative to 1990. Because of emphasis on energy efficiency, primary energy consumption rises much more slowly than GDP. The energy supply constructions were made to meet energy demand in (i) projections developed for the IPCC's First Assessment Report (1990) in a low energy demand variant, where global primary commercial energy use approximately doubles, with no net change for industrialized countries but a 4.4-fold increase for developing countries from 1990 to 2100; and (ii) a higher energy demand variant, developed in the IPCC IS92a scenario where energy demand quadruples from 1990 to 2100. The energy demand levels of the LESS constructions are consistent with the energy demand mitigation chapters of this Second Assessment Report.

Figure 5 shows combinations of different energy sources to meet changing levels of demand over the next century. The analysis of these variants leads to the following conclusions:

- Deep reductions of CO_2 emissions from energy supply systems are technically possible within 50 to 100 years, using alternative strategies.
- Many combinations of the options identified in this assessment could reduce global CO_2 emissions from fossil fuels from about 6 Gt C in 1990 to about 4 Gt C/yr by 2050, and to about 2 Gt C/yr by 2100 (see Figure 6). Cumulative CO_2 emissions, from 1990 to 2100, would range from about 450 to about 470 Gt C in the alternative LESS constructions.
- Higher energy efficiency is underscored for achieving deep reductions in CO_2 emissions, for increasing the flexibility of supply side combinations, and for reducing overall energy system costs.
- Interregional trade in energy grows in the LESS constructions compared to today's levels, expanding sustainable development options for Africa, Latin America, and the Middle East during the next century.

Costs for energy services in each LESS variant relative to costs for conventional energy depend on relative future energy prices, which are uncertain within a wide range, and on the performance and cost characteristics assumed for alternative technologies. However, within the wide range of future energy prices, one or more of the variants would plausibly be capable of providing the demanded energy services at estimated costs that are approximately the same as estimated future costs for current conventional energy. It is not possible to identify a least-cost future energy system for the longer term, as the relative costs of options depend on resource constraints and technological opportunities that are imperfectly known, and on actions by governments and the private sector.

The literature provides strong support for the feasibility of achieving the performance and cost characteristics assumed for energy technologies in the LESS constructions, within the next 2 decades, though it is impossible to be certain until the research

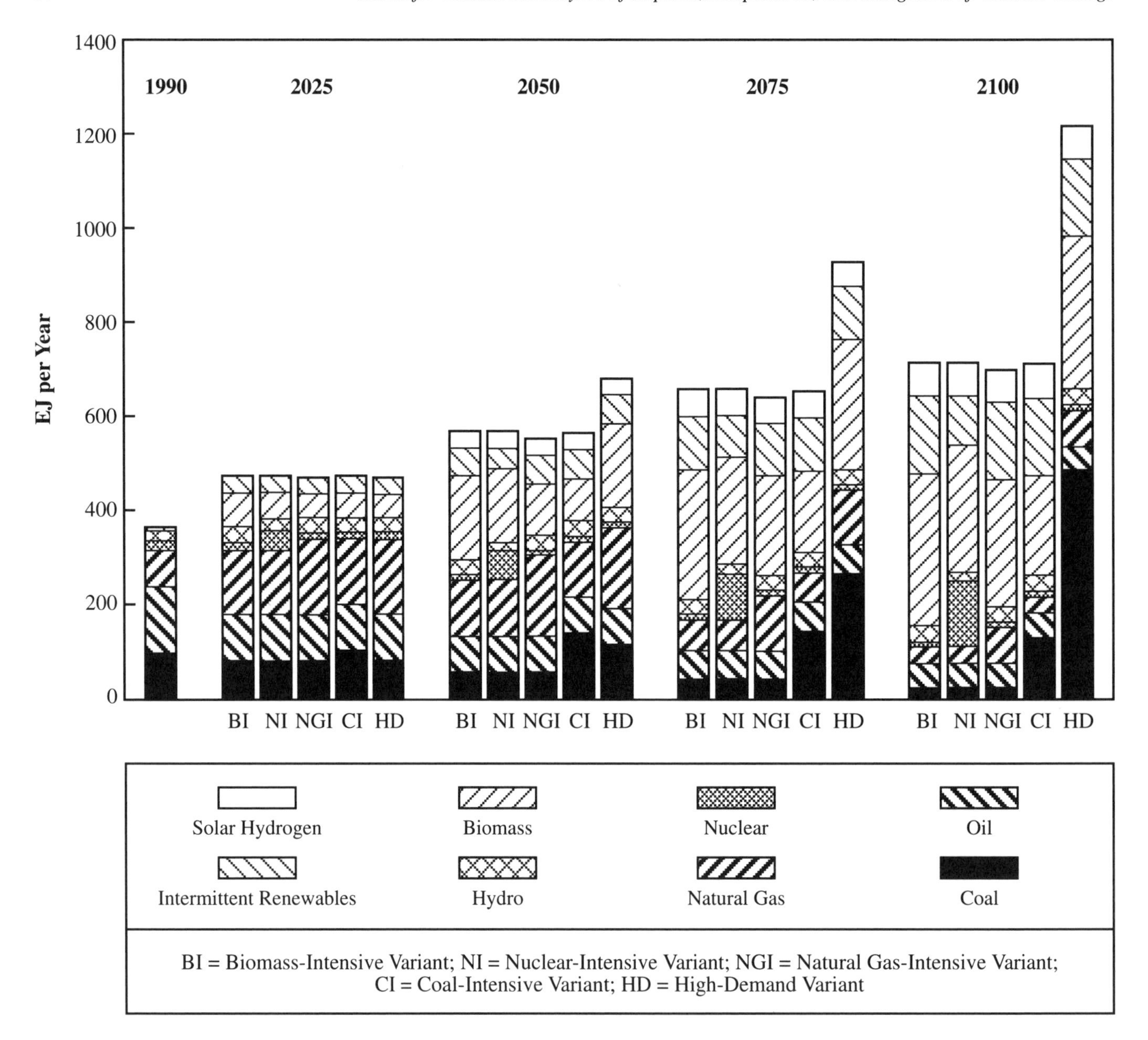

Figure 5: Global primary energy use for alternative Low CO_2-Emitting Energy Supply System (LESS) constructions: Alternatives for meeting different energy demand levels over time, using various fuel mixes.

and development is complete and the technologies have been tested in the market. Moreover, these performance and cost characteristics cannot be achieved without a strong and sustained investment in research, development, and demonstration (RD&D). Many of the technologies being developed would need initial support to enter the market, and to reach sufficient volume to lower costs to become competitive.

Market penetration and continued acceptability of different energy technologies ultimately depends on their relative cost, performance (including environmental performance), institutional arrangements, and regulations and policies. Because costs vary by location and application, the wide variety of circumstances creates initial opportunities for new technologies to enter the market. Deeper understanding of the opportunities for emissions reductions would require more detailed analysis of options, taking into account local conditions.

Because of the large number of options, there is flexibility as to how the energy supply system could evolve, and paths of energy system development could be influenced by considerations other than climate change, including political, environmental (especially indoor and urban air pollution, acidification, and land restoration), and socioeconomic circumstances.

4.2. Agriculture, Rangelands, and Forestry

Beyond the use of biomass fuels to displace fossil fuels, the management of forests, agricultural lands, and rangelands can play an

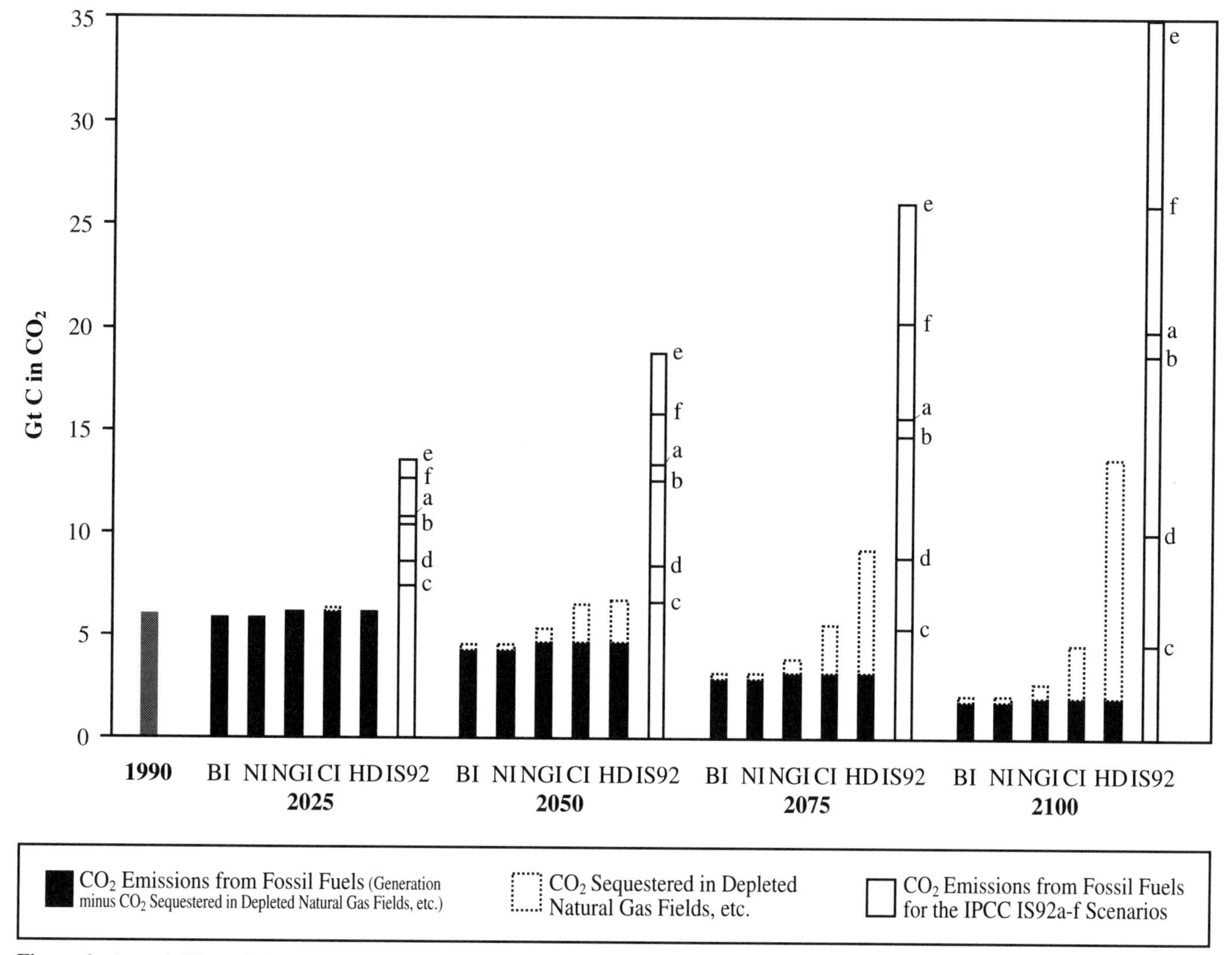

Figure 6: Annual CO_2 emissions from fossil fuels for alternative LESS constructions, with comparison to the IPCC IS92a-f scenarios (see Figure 5 for acronym definitions).

important role in reducing current emissions of CO_2, CH_4, and N_2O and in enhancing carbon sinks. A number of measures could conserve and sequester substantial amounts of carbon (approximately 60–90 Gt C in the forestry sector alone) over the next 50 years. In the forestry sector, costs for conserving and sequestering carbon in biomass and soil are estimated to range widely but can be competitive with other mitigation options. Factors affecting costs include opportunity costs of land; initial costs of planting and establishment; costs of nurseries; the cost of annual maintenance and monitoring; and transaction costs. Direct and indirect benefits will vary with national circumstances and could offset the costs. Other practices in the agriculture sector could reduce emissions of other greenhouse gases such as CH_4 and N_2O. Land-use and management measures include:

- Sustaining existing forest cover
- Slowing deforestation
- Regenerating natural forests
- Establishing tree plantations
- Promoting agroforestry
- Altering management of agricultural soils and rangelands
- Improving efficiency of fertilizer use
- Restoring degraded agricultural lands and rangelands
- Recovering CH_4 from stored manure
- Improving the diet quality of ruminants.

The net amount of carbon per unit area conserved or sequestered in living biomass under a particular forest management practice and present climate is relatively well understood. The most important uncertainties associated with estimating a global value are (i) the amount of land suitable and available for forestation, regeneration, and/or restoration programs; (ii) the rate at which tropical deforestation can actually be reduced; (iii) the long-term use (security) of these lands; and (iv) the continued suitability of some practices for particular locations given the possibility of changes in temperature, water availability, and so forth under climate change.

4.3. Cross-Sectoral Issues

Cross-sectoral assessment of different combinations of mitigation options focuses on the interactions of the full range of technologies and practices that are potentially capable of reducing emissions of greenhouse gases or sequestering carbon. Current analysis suggests the following:

- **Competing Uses of Land, Water, and Other Natural Resources**. A growing population and expanding economy will increase the demand for land and other natural resources needed to provide, *inter alia*, food, fiber, forest products, and recreation services. Climate change will interact with the resulting intensified patterns of resource use. Land and other resources could also be required for mitigation of greenhouse gas emissions. Agricultural productivity improvements throughout the world and especially in developing countries would increase availability of land for production of biomass energy.
- **Geoengineering Options**. Some geoengineering approaches to counterbalance greenhouse gas-induced climate change have been suggested (e.g., putting solar radiation reflectors in space or injecting sulfate aerosols into the atmosphere to mimic the cooling influence of volcanic eruptions). Such approaches generally are likely to be ineffective, expensive to sustain, and/or to have serious environmental and other effects that are in many cases poorly understood.

4.4. Policy Instruments

Mitigation depends on reducing barriers to the diffusion and transfer of technology, mobilizing financial resources, supporting capacity building in developing countries, and other approaches to assist in the implementation of behavioral changes and technological opportunities in all regions of the globe. The optimum mix of policies will vary from country to country, depending upon political structure and societal receptiveness. The leadership of national governments in applying these policies will contribute to responding to adverse consequences of climate change. Governments can choose policies that facilitate the penetration of less greenhouse gas-intensive technologies and modified consumption patterns. Indeed, many countries have extensive experience with a variety of policies that can accelerate the adoption of such technologies. This experience comes from efforts over the past 20 to 30 years to achieve improved energy efficiency, reduce the environmental impacts of agricultural policies, and meet conservation and environmental goals unrelated to climate change. Policies to reduce net greenhouse gas emissions appear more easily implemented when they are designed to address other concerns that impede sustainable development (e.g., air pollution and soil erosion). A number of policies, some of which may need regional or international agreement, can facilitate the penetration of less greenhouse gas-intensive technologies and modified consumption patterns, including:

- Putting in place appropriate institutional and structural frameworks
- Energy pricing strategies (e.g., carbon or energy taxes, and reduced energy subsidies)
- Reducing or removing other subsidies (e.g., agricultural and transport subsidies) that increase greenhouse gas emissions
- Tradable emissions permits
- Voluntary programs and negotiated agreements with industry
- Utility demand-side management programs
- Regulatory programs, including minimum energy efficiency standards (e.g., for appliances and fuel economy)
- Stimulating RD&D to make new technologies available
- Market pull and demonstration programs that stimulate the development and application of advanced technologies
- Renewable energy incentives during market build-up
- Incentives such as provisions for accelerated depreciation and reduced costs for consumers
- Education and training; information and advisory measures
- Options that also support other economic and environmental goals.

Accelerated development of technologies that will reduce greenhouse gas emissions and enhance greenhouse gas sinks—as well as understanding the barriers that inhibit their diffusion into the marketplace—requires intensified research and development by governments and the private sector.

Authors/Contributors

Robert T. Watson, USA; M.C. Zinyowera, Zimbabwe; Richard H. Moss, USA; Roberto Acosta Moreno, Cuba; Sharad Adhikary, Nepal; Michael Adler, USA; Shardul Agrawala, India; Adrian Guillermo Aguilar, Mexico; Saiyed Al-Khouli, Saudi Arabia; Barbara Allen-Diaz, USA; Mitsuru Ando, Japan; Rigoberto Andressen, Venezuela; B.W. Ang, Singapore; Nigel Arnell, UK; Anne Arquit-Niederberger, Switzerland; Walter Baethgen, Uruguay; Bryson Bates, Australia; Martin Beniston, Switzerland; Rosina Bierbaum, USA; Luitzen Bijlsma, The Netherlands; Michel Boko, Republic of Benin; Bert Bolin, Sweden; Suzanne Bolton, USA; Evelyne Bravo, Venezuela; Sandra Brown, USA; Peter Bullock, UK; Melvin Cannell, UK; Osvaldo Canziani, Argentina; Rodolfo Carcavallo, Argentina; Carlos Clemente Cerri, Brazil; William Chandler, USA; Fred Cheghe, Kenya; Chunzhen Liu, China; Vernon Cole, USA; Wolfgang Cramer, Germany; Rex Victor Cruz, Philippines; Ogunlade Davidson, Sierra Leone; Ehrlich Desa, India; Deying Xu, China; Sandra Diaz, Argentina; Andrew Dlugolecki, Scotland; James Edmonds, USA; John Everett, USA; Andreas Fischlin, Switzerland; Blair Fitzharris, New Zealand; Douglas Fox, USA; Jaafar Friaa, Tunisia; Alexander Rauja Gacuhi, Kenya; Wojciech Galinski, Poland; Habiba Gitay, Australia; Peter Groffman, USA; Arnulf Grubler, Austria; Howard Gruenspecht, USA; Steven Hamburg, USA; Timm Hoffman, South Africa; Jarle Inge Holten, Norway; Hisashi Ishitani, Japan; Venugopalan Ittekkot, Germany; Thomas Johansson, Sweden; Zdzislaw Kaczmarek, Poland; Takao Kashiwagi, Japan; Miko Kirschbaum, Australia; Paul Komor, USA; Andrei Krovnin, Russian Federation; Richard Klein, The Netherlands; Shashi Kulshrestha, India; Herbert Lang, Switzerland; Henry Le Houerou, France; Rik Leemans, The Netherlands; Mark Levine, USA; Lin Erda, China; Daniel Lluch-Belda, Mexico; Michael MacCracken, USA; John Magnuson, USA; Gabriel Mailu, Kenya; Joseph Mworia Maitima, Kenya; Gregg Marland, USA; Kathy Maskell, UK; Roger McLean, Australia; Anthony McMichael, Australia/UK; Laurie Michaelis, France; Ed Miles, USA; William Moomaw, USA; Roberto Moreira, Brazil; Patrick Mulholland, USA; Nebojsa Nakicenovic, Austria; Robert Nicholls, UK; Shuzo Nishioka, Japan; Ian Noble, Australia; Leonard Nurse, Barbados; Rispa Odongo, Kenya; Ryousuke Ohashi, Japan; Ezekiel Okemwa, Kenya; Mats Oquist, Sweden; Martin Parry, UK; Martha Perdomo, Venezuela; Michel Petit, France; Warren Piver, USA; P.S. Ramakrishnan, India; N.H. Ravindranath, India; John Reilly, USA; Arthur Riedacker, France; Hans-Holger Rogner, Canada; Jayant Sathaye, USA; Dieter Sauerbeck, Germany; Michael Scott, USA; Subodh Sharma, India; David Shriner, USA; S.K. Sinha, India; Jim Skea, UK; Allen Solomon, USA; Eugene Stakhiv, USA; Oedon Starosolszky, Hungary; Su Jilan, China; Avelino Suarez, Cuba; Bo Svensson, Sweden; Hidekazu Takakura, Japan; Melissa Taylor, USA; Lucien Tessier, France; Dennis Tirpak, USA; Tran Viet Lien, Vietnam; Jean-Paul Troadec, France; Hiroshi Tsukamoto, Japan; Itsuya Tsuzaka, Japan; Pier Vellinga, The Netherlands; Ted Williams, USA; Patrick Young, USA; Youyu Xie, China; Zhou Fengqi, China

Technical Summary: Impacts, Adaptations, and Mitigation Options

Authors/Contributors

Robert T. Watson, USA; M.C. Zinyowera, Zimbabwe; Richard H. Moss, USA; Roberto Acosta Moreno, Cuba; Sharad Adhikary, Nepal; Michael Adler, USA; Shardul Agrawala, India; Adrian Guillermo Aguilar, Mexico; Saiyed Al-Khouli, Saudi Arabia; Barbara Allen-Diaz, USA; Mitsuru Ando, Japan; Rigoberto Andressen, Venezuela; B.W. Ang, Singapore; Nigel Arnell, UK; Anne Arquit-Niederberger, Switzerland; Walter Baethgen, Uruguay; Bryson Bates, Australia; Martin Beniston, Switzerland; Rosina Bierbaum, USA; Luitzen Bijlsma, The Netherlands; Michel Boko, Republic of Benin; Bert Bolin, Sweden; Suzanne Bolton, USA; Evelyne Bravo, Venezuela; Sandra Brown, USA; Peter Bullock, UK; Melvin Cannell, UK; Osvaldo Canziani, Argentina; Rodolfo Carcavallo, Argentina; Carlos Clemente Cerri, Brazil; William Chandler, USA; Fred Cheghe, Kenya; Chunzhen Liu, China; Vernon Cole, USA; Wolfgang Cramer, Germany; Rex Victor Cruz, Philippines; Ogunlade Davidson, Sierra Leone; Ehrlich Desa, India; Deying Xu, China; Sandra Diaz, Argentina; Andrew Dlugolecki, Scotland; James Edmonds, USA; John Everett, USA; Andreas Fischlin, Switzerland; Blair Fitzharris, New Zealand; Douglas Fox, USA; Jaafar Friaa, Tunisia; Alexander Rauja Gacuhi, Kenya; Wojciech Galinski, Poland; Habiba Gitay, Australia; Peter Groffman, USA; Arnulf Grubler, Austria; Howard Gruenspecht, USA; Steven Hamburg, USA; Timm Hoffman, South Africa; Jarle Inge Holten, Norway; Hisashi Ishitani, Japan; Venugopalan Ittekkot, Germany; Thomas Johansson, Sweden; Zdzislaw Kaczmarek, Poland; Takao Kashiwagi, Japan; Miko Kirschbaum, Australia; Paul Komor, USA; Andrei Krovnin, Russian Federation; Richard Klein, The Netherlands; Shashi Kulshrestha, India; Herbert Lang, Switzerland; Henry Le Houerou, France; Rik Leemans, The Netherlands; Mark Levine, USA; Lin Erda, China; Daniel Lluch-Belda, Mexico; Michael MacCracken, USA; John Magnuson, USA; Gabriel Mailu, Kenya; Joseph Mworia Maitima, Kenya; Gregg Marland, USA; Kathy Maskell, UK; Roger McLean, Australia; Anthony McMichael, Australia/UK; Laurie Michaelis, France; Ed Miles, USA; William Moomaw, USA; Roberto Moreira, Brazil; Patrick Mulholland, USA; Nebojsa Nakicenovic, Austria; Robert Nicholls, UK; Shuzo Nishioka, Japan; Ian Noble, Australia; Leonard Nurse, Barbados; Rispa Odongo, Kenya; Ryousuke Ohashi, Japan; Ezekiel Okemwa, Kenya; Mats Oquist, Sweden; Martin Parry, UK; Martha Perdomo, Venezuela; Michel Petit, France; Warren Piver, USA; P.S. Ramakrishnan, India; N.H. Ravindranath, India; John Reilly, USA; Arthur Riedacker, France; Hans-Holger Rogner, Canada; Jayant Sathaye, USA; Dieter Sauerbeck, Germany; Michael Scott, USA; Subodh Sharma, India; David Shriner, USA; S.K. Sinha, India; Jim Skea, UK; Allen Solomon, USA; Eugene Stakhiv, USA; Oedon Starosolszky, Hungary; Su Jilan, China; Avelino Suarez, Cuba; Bo Svensson, Sweden; Hidekazu Takakura, Japan; Melissa Taylor, USA; Lucien Tessier, France; Dennis Tirpak, USA; Tran Viet Lien, Vietnam; Jean-Paul Troadec, France; Hiroshi Tsukamoto, Japan; Itsuya Tsuzaka, Japan; Pier Vellinga, The Netherlands; Ted Williams, USA; Patrick Young, USA; Youyu Xie, China; Zhou Fengqi, China

CONTENTS

1. Scope of the Assessment

The Intergovernmental Panel on Climate Change (IPCC) charged Working Group II with reviewing current knowledge about the impacts of climate change on physical and ecological systems, human health, and socioeconomic sectors. IPCC also asked Working Group II to review available data on the technical and economic feasibility of a range of potential adaptation and mitigation strategies. In producing this report, Working Group II has coordinated its activities with those of Working Groups I and III, and built on the 1990 and 1992 IPCC assessments.

This assessment provides scientific, technical, and economic information that can be used, *inter alia*, in evaluating whether the projected range of plausible impacts constitutes "dangerous anthropogenic interference with the climate system" at the local, regional, or global scales as referred to in Article 2 of the United Nations Framework Convention on Climate Change (UNFCCC), and in evaluating adaptation and mitigation options that could be used in progressing toward the ultimate objective of the UNFCCC (see Box 1). However, the assessment makes no attempt to quantify "dangerous anthropogenic interference with the climate system." Interpreting what is "dangerous" involves political judgment—a role reserved to governments and the Conference of Parties to the UNFCCC.

The UNFCCC's Article 2 explicitly acknowledges the importance of natural ecosystems, food security, and sustainable economic development. This report directly addresses these and other issues important to society, including water resources and human health. It assesses what is known about the impact of climate change on terrestrial and aquatic ecosystems, human health, and socioeconomic systems at varying time and geographic scales, as well as what is known about their vulnerability to climate change. A system's vulnerability to climate change depends on its sensitivity to changes in climate and its ability to adapt. The vulnerability assessment takes into account the different economic and institutional circumstances among developed and developing countries. It also recognizes the strong influence on most of these systems of other human-induced stresses (e.g., population demographics, land-use practices, industrialization, consumption patterns, air and water pollution, and soil degradation). Where possible, this assessment attempts to evaluate the sensitivity of systems and their potential for adaptation to: (i) changes in mean climate; (ii) changes in extreme weather events; (iii) changes in variability; (iv) the rate of climate change; and (v) the effects of elevated CO_2 concentrations on vegetation (via enhanced photosynthesis and water-use efficiency). The report also describes technical guidelines for assessing climate change impacts and adaptations.

This assessment also reviews and analyzes practices and technologies that (i) reduce anthropogenic emissions of greenhouse gases arising from the production and use of energy (industry, transportation, human settlements) and (ii) increase carbon storage (in what are generally called sinks) and reduce greenhouse gas emissions from agricultural, forestry, and rangeland systems. It also describes methodologies for assessing mitigation options and an inventory of technology characterizations.

Box 1. Ultimate Objective of the UNFCCC (Article 2)

"...stabilization of greenhouse gas concentrations in the atmosphere at a level that would prevent dangerous anthropogenic interference with the climate system. Such a level should be achieved within a time frame sufficient to allow ecosystems to adapt naturally to climate change, to ensure that food production is not threatened, and to enable economic development to proceed in a sustainable manner."

2. The Nature of the Issue: Projected Changes in Climate

Earth's climate has remained relatively stable (global temperature changes of less than 1°C over a century) during the last 10,000 years (the present interglacial period). Over this period, modern society has evolved and, in many cases, successfully adapted to the prevailing local climate and its natural variability. Now, however, society faces potentially rapid changes in future climate because of human activities that alter the atmosphere's composition and change the Earth's radiation balance.

Atmospheric concentrations of greenhouse gases (which tend to warm the atmosphere) and aerosols (which in some regions partially offset the greenhouse effect) have increased since the industrial era began around 1750. Carbon dioxide (CO_2) has risen by about 30%, methane (CH_4) by more than 100%, and nitrous oxide (N_2O) by about 15%. These gases now are at greater concentrations than at any time in the past 160,000 years (the period for which scientists can reconstruct historical climates and atmospheric compositions by analyzing ice-core data). The combustion of fossil fuels and, to a lesser extent, changes in land use account for anthropogenic CO_2 emissions. Agriculture is responsible for nearly 50% of human-generated CH_4 emissions and about 70% of anthropogenic N_2O emissions. Although CO_2 emissions far exceed those of CH_4 and N_2O, the global warming potentials of these latter two gases are relatively high. Hence, they represent significant contributors to the anthropogenic greenhouse effect. CO_2 has contributed about 65% of the combined radiative effects of the long-lived gases over the past 100 years; CH_4 and N_2O have contributed about 20 and 5%, respectively.

Most projections suggest that without policies specifically designed to address climate change greenhouse gas concentrations will increase significantly during the next century. Emissions of greenhouse gases and the sulfate aerosol precursor sulfur dioxide (SO_2) are sensitive to growth in population and gross domestic product (GDP), the rate of diffusion of new technologies into the market place, production and consumption patterns, land-use practices, energy intensity, the price and availability of energy, and other policy and institutional developments

Table 1: Summary of assumptions in the six IPCC 1992 alternative scenarios.

Scenario	Population	Economic Growth	Energy Supplies
IS92a,b	World Bank 1991 11.3 billion by 2100	1990–2025: 2.9% 1990–2100: 2.3%	12,000 EJ conventional oil 13,000 EJ natural gas Solar costs fall to $0.075/kWh 191 EJ of biofuels available at $70/barrel[a]
IS92c	UN Medium-Low Case 6.4 billion by 2100	1990–2025: 2.0% 1990–2100: 1.2%	8,000 EJ conventional oil 7,300 EJ natural gas Nuclear costs decline by 0.4% annually
IS92d	UN Medium-Low Case 6.4 billion by 2100	1990–2025: 2.7% 1990–2100: 2.0%	Oil and gas same as IS92c Solar costs fall to $0.065/kWh 272 EJ of biofuels available at $50/barrel
IS92e	World Bank 1991 11.3 billion by 2100	1990–2025: 3.5% 1990–2100: 3.0%	18,400 EJ conventional oil Gas same as IS92a,b Phase out nuclear by 2075
IS92f	UN Medium-High Case 17.6 billion by 2100	1990–2025: 2.9% 1990–2100: 2.3%	Oil and gas same as IS92e Solar costs fall to $0.083/kWh Nuclear costs increase to $0.09/kWh

[a]Approximate conversion factor: 1 barrel = 6 GJ.
Source: IPCC, 1992: Emissions scenarios for IPCC: an update. In: *Climate Change 1992: The Supplementary Report to the IPCC Scientific Assessment* [J.T. Houghton, B.A. Callander, and S.K. Varney (eds.)]. Section A3, prepared by J. Leggett, W.J. Pepper, and R.J. Swart, and WMO/UNEP. Cambridge University Press, Cambridge, UK, 200 pp.

(see Table 1 for a summary of assumptions used in the six IPCC 1992 emissions scenarios). Figure 1 shows plausible ranges of CO_2 emissions, in the absence of emissions abatement policies, projected over the next 100 years in the IPCC IS92 emissions scenarios, which are reevaluated in the Working Group III volume of this 1995 assessment.

Climate models, taking into account greenhouse gases and aerosols, calculate that the global mean surface temperature could rise by about 1 to 3.5°C by 2100. This range of projections is based on the range of sensitivities of climate[1] to increases in greenhouse gas concentrations reported by IPCC Working Group I and plausible ranges of greenhouse gas emissions projected by IPCC in 1992. These projected global-average temperature changes would be greater than recent natural fluctuations and would occur at a rate significantly faster than any since the last ice age more than 10,000 years ago. High latitudes are projected to warm more than the global average. The reliability of regional projections remains low.

Model calculations, however, suggest the following:

- Climate warming will enhance evaporation, and global mean precipitation will increase, as will the frequency of intense rainfall. However, some land regions will not experience an increase in precipitation, and even those that do may experience decreases in soil moisture because of enhanced evaporation. Climate models also project seasonal shifts in precipitation. In general, models project that precipitation will increase at high latitudes in winter, and soil moisture will decrease in some mid-latitude continental regions during the summer.
- Variability associated with the enhanced hydrological cycle translates into prospects for more severe droughts and floods in some places, and less severe droughts and/or floods in other places. As a consequence, the incidence of fires and pest outbreaks may increase in some regions. It remains unclear whether the frequency and intensity of extreme weather events such as tropical storms, cyclones, and tornadoes will change.
- Regional and global climate changes are expected to have wide-ranging and potentially adverse effects on physical and ecological systems, human health, and socioeconomic sectors. These will affect the economy and the quality of life for this and future generations.
- Models project that sea level will increase by about 15 to 95 cm by 2100, allowing for average ice melt, but could be either higher or lower than this range.

[1] Climate sensitivity is the equilibrium change in global annual mean surface temperature due to a doubling of atmospheric concentration of CO_2 (or equivalent doubling of other greenhouse gases).

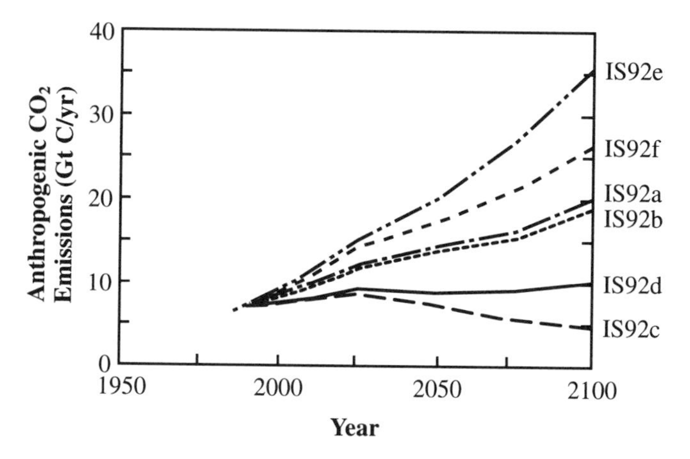

Figure 1: Projected anthropogenic CO_2 emissions from fossil fuel use, deforestation, and cement production for the six IPCC 1992 scenarios (IS92a-f)—from *Climate Change 1994: Radiative Forcing of Climate Change and An Evaluation of the IPCC IS92 Emission Scenarios* (IPCC, 1995).

Policymakers are faced with responding to the risks posed by anthropogenic emissions of greenhouse gases in the face of significant scientific uncertainties. It is appropriate to consider these uncertainties in the context of information indicating that climate-induced environmental changes cannot be reversed quickly, if at all, due to the long time scales associated with the climate system (see Box 2). Decisions taken during the next few years may limit the range of possible policy options in the future, because high near-term emissions would require deeper reductions in the future to meet any given target concentrations. Delaying action might reduce the overall costs of mitigation because of potential technological advances but could increase both the rate and eventual magnitude of climate change, hence the adaptation and damage costs.

Box 2. Time Scales of Processes Influencing the Climate System

- Turnover of the capital stock responsible for emissions of greenhouse gases: ***Years to decades*** (without premature retirement)
- Stabilization of atmospheric concentrations of long-lived greenhouse gases given a stable level of greenhouse gas emissions: ***Decades to millennia***
- Equilibration of the climate system given a stable level of greenhouse gas concentrations: ***Decades to centuries***
- Equilibration of sea level given a stable climate: ***Centuries***
- Restoration/rehabilitation of damaged or disturbed ecological systems: ***Decades to centuries*** (some changes, such as species extinction, are irreversible, and it may be impossible to reconstruct and reestablish some disturbed ecosystems)

Policymakers will have to decide to what degree they want to take precautionary measures by mitigating greenhouse gas emissions and enhancing the resilience of vulnerable systems by means of adaptation. Uncertainty does not mean that a nation or the world community cannot position itself better to cope with the broad range of possible climate changes or protect against potentially costly future outcomes. Delaying such measures may leave a nation or the world poorly prepared to deal with adverse changes and may increase the possibility of irreversible or very costly consequences. Options for adapting to change or mitigating change that can be justified for other reasons today (e.g., abatement of air and water pollution) and make society more flexible or resilient to anticipated adverse effects of climate change appear particularly desirable.

3. Vulnerability to Climate Change: Impacts and Adaptation

This section discusses the sensitivity, adaptability, and vulnerability (see Box 3) of physical and ecological systems, human health, and socioeconomic sectors to changes in climate. It begins with a number of common conclusions, then describes to the extent possible (i) the functions and current status of each system; (ii) the sensitivity of each system to climate change and to other environmental and human-induced factors; and (iii) the vulnerability of each system to climate change, taking into account adaptation options and impediments to adaptation.

Although we have made much progress in understanding the climate system and the consequences of climate change, large uncertainties cloud our view of the 21st century. In particular,

Box 3. Sensitivity, Adaptability, and Vulnerability

Sensitivity is the degree to which a system will respond to a change in climatic conditions (e.g., the extent of change in ecosystem composition, structure, and functioning, including primary productivity, resulting from a given change in temperature or precipitation).

Adaptability refers to the degree to which adjustments are possible in practices, processes, or structures of systems to projected or actual changes of climate. Adaptation can be spontaneous or planned, and can be carried out in response to or in anticipation of changes in conditions.

Vulnerability defines the extent to which climate change may damage or harm a system. It depends not only on a system's sensitivity but also on its ability to adapt to new climatic conditions.

Both the magnitude and the rate of climate change are important in determining the sensitivity, adaptability, and vulnerability of a system.

our understanding of some key ecological processes remains limited. So does our current ability to predict regional climate changes, future conditions in the absence of climate change, or to what degree climate will become more variable. Changes in average conditions, climate variability, and the frequency and intensity of extreme weather events all would have important implications for both ecological and social systems.

3.1. *Common Themes and Conclusions*

Human health, terrestrial and aquatic ecological systems, and socioeconomic systems (e.g., agriculture, forestry, fisheries, and water resources) are all vital to human development and well-being, and are all sensitive to changes in climate. Whereas many regions are likely to experience the adverse effects of climate change—some of which are potentially irreversible—some effects of climate change are likely to be beneficial. Hence, different segments of society can expect to confront a variety of changes and the need to adapt to them. The following conclusions apply to many ecological and socioeconomic systems:

- **Human-induced climate change adds an important new stress**. Human-induced climate change represents an important additional stress, particularly to the many ecological and socioeconomic systems already affected by pollution, increasing resource demands, and nonsustainable management practices. The most vulnerable systems are those with the greatest sensitivity to climate changes and the least adaptability.
- **Most systems are sensitive to climate change**. Natural ecological systems, socioeconomic systems, and human health are all sensitive to both the magnitude and the rate of climate change.
  - *Natural ecosystems*. The composition and geographic distribution of many ecosystems will shift as individual species respond to changes in climate; there will likely be reductions in biological diversity and in the goods and services that ecosystems provide society—for example, sources of food, fiber, medicines, recreation and tourism, and ecological services such as nutrient cycling, waste assimilation, and controlling water runoff and soil erosion. Large amounts of carbon could be released into the atmosphere during periods of high forest mortality in the transition from one forest type to another.
  - *Food security*. Some regions, especially in the tropics and subtropics, may suffer significant adverse consequences for food security, even though the effect of climate change on global food production may prove small to moderate.
  - *Sustainable economic development*. Some countries will face threats to sustainable development from losses of human habitat due to sea-level rise, reductions in water quality and quantity, disruptions from extreme events, and an increase in human diseases (particularly vector-borne diseases such as malaria).
- **Impacts are difficult to quantify, and existing studies are limited in scope**. While our knowledge has increased significantly during the last decade and qualitative estimates can be developed, quantitative projections of the impacts of climate change on any particular system at any particular location are difficult because regional scale climate change projections are uncertain; our current understanding of many critical processes is limited; and systems are subject to multiple climatic and non-climatic stresses, the interactions of which are not always linear or additive. Most impact studies have assessed how systems would respond to climate change resulting from an arbitrary doubling of equivalent atmospheric CO_2 concentrations. Furthermore, very few studies have considered dynamic responses to steadily increasing greenhouse gas concentrations; fewer still have examined the consequences of increases beyond a doubling of equivalent atmospheric CO_2 concentrations or assessed the implications of multiple stress factors.
- **Successful adaptation depends upon technological advances, institutional arrangements, availability of financing, and information exchange**. Technological advances generally have increased adaptation options for managed systems such as agriculture and water supply. However, many regions of the world currently have limited access to these technologies and appropriate information. The efficacy and cost-effective use of adaptation strategies will depend upon the availability of financial resources, technology transfer, and cultural, educational, managerial, institutional, legal, and regulatory practices, both domestic and international in scope. Incorporating climate-change concerns into resource-use and development decisions and plans for regularly scheduled investments in infrastructure will facilitate adaptation.
- **Vulnerability increases as adaptive capacity decreases**. The vulnerability of human health and socioeconomic systems—and to a lesser extent ecological systems—depends upon economic circumstances and institutional infrastructure. This implies that systems typically are more vulnerable in developing countries where economic and institutional circumstances are less favorable. People who live on arid or semi-arid lands, in low-lying coastal areas, in water-limited or flood-prone areas, or on small islands are particularly vulnerable to climate change. Some regions have become more vulnerable to hazards such as storms, floods, and droughts as a result of increasing population density in sensitive areas such as river basins and coastal plains. Human activities, which fragment many landscapes, have increased the vulnerability of lightly managed and unmanaged ecosystems. Fragmentation limits natural adaptation potential and the potential effectiveness of measures to assist adaptation in these systems, such as the provision of migration corridors. A changing climate's near-term effects on ecological and socioeconomic

systems most likely will result from changes in the intensity and seasonal and geographic distribution of common weather hazards such as storms, floods, and droughts. In most of these examples, vulnerability can be reduced by strengthening adaptive capacity.

- **Detection will be difficult and unexpected changes cannot be ruled out**. Unambiguous detection of climate-induced changes in most ecological and social systems will prove extremely difficult in the coming decades. This is because of the complexity of these systems, their many non-linear feedbacks, and their sensitivity to a large number of climatic and non-climatic factors, all of which are expected to continue to change simultaneously. The development of a baseline projecting future conditions without climate change is crucial, for it is this baseline against which all projected impacts are measured. The more that future climate extends beyond the boundaries of empirical knowledge (i.e., the documented impacts of climate variation in the past), the more likely that actual outcomes will include surprises and unanticipated rapid changes.
- **Further research and monitoring are essential**. Enhanced support for research and monitoring, including cooperative efforts from national, international, and multilateral institutions, is essential in order to improve significantly regional-scale climate projections; to understand the responses of human health, ecological, and socioeconomic systems to changes in climate and other stress factors; and to improve our understanding of the efficacy and cost-effectiveness of adaptation strategies.

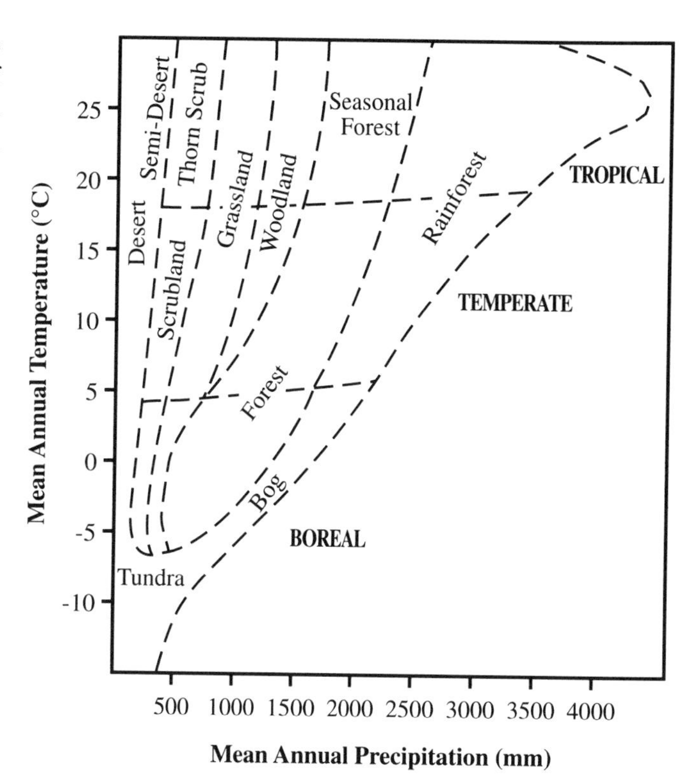

Figure 2: This figure illustrates that mean annual temperature and mean annual precipitation can be correlated with the distribution of the world's major biomes. While the role of these annual means in affecting this distribution is important, it should be noted that the distribution of biomes may also strongly depend on seasonal factors such as the length of the dry season or the lowest absolute minimum temperature, on soil properties such as water-holding capacity, on land-use history such as agriculture or grazing, and on disturbance regimes such as the frequency of fire.

3.2. *Terrestrial Ecosystems*

Ecosystems contain the Earth's entire reservoir of genetic and species diversity, and provide goods and services critical to individuals and societies. These services include (i) providing food, fiber, medicines, and energy; (ii) processing and storing carbon and other nutrients, which affect the atmospheric concentrations of greenhouse gases; (iii) regulating water runoff, thus controlling floods and soil erosion; (iv) assimilating wastes and purifying water; and (v) providing opportunities for recreation and tourism. These systems and the functions they provide are sensitive to the rate and extent of changes in climate. Figure 2 illustrates that mean annual temperature and mean annual precipitation can be correlated with the distribution of the world's major biomes. While the role of these annual means in affecting this distribution appears to be important, it should be noted that the distribution of biomes also strongly depends on seasonal factors and other non-climate conditions, such as soil properties and disturbance regimes.

Changes in climate and associated changes in the frequency of fires and the prevalence of pests could alter various properties of terrestrial ecosystems. These include structure (physical arrangement, density of populations, species composition), function (movement of energy and material within ecosystems), and productivity (rates and magnitudes of carbon fixation and respiration).

Most terrestrial ecosystems are under major pressures from population increases and human decisions about land use, which probably will continue to cause the largest adverse effects for the foreseeable future. However, continuing climate change would induce a disassembly of existing ecosystems and an ongoing assembly of plants and animals into new ecosystems—a process that may not reach a new equilibrium for several centuries after the climate achieves a new balance.

In analyzing the potential impacts of climate change, researchers must consider how species and ecosystems respond to elevated concentrations of atmospheric CO_2. Higher CO_2 concentrations may increase the net primary productivity (NPP) of plants, which would alter species composition by changing the competitive balance among different plants. Although biologists have quantified these effects at the leaf to plant level in controlled settings, such effects are only now being quantified at the ecosystem level. This is because of the complex interactions among plants in their natural environments, which depend

Box 4. Regional Implications of Climate Change for Forests

Tropical forests. It is likely that temperature increases will have a smaller impact on tropical forests than on temperate or boreal forests, because models project that temperatures will increase less in the tropics than at other latitudes. However, tropical forests are very sensitive to the amount and seasonality of rainfall. In general, human activities causing conversion to other land-cover types will likely affect tropical forests more than climate change. If CO_2 fertilization is important, it may lead to a gain in net carbon storage because of the slow rate at which the associated soil respiration increases in this zone.

Temperate forests. In some temperate forests, NPP may increase due to warming and increased atmospheric CO_2. However, in other regions warming-induced water shortages, pest activity, and fires may cause decreased NPP and possible changes in temperate forest distribution. Most temperate forests are located primarily in developed countries that have the resources to reduce the impacts of climate change on their forests through integrated fire, pest, and disease management, and/or encouraging reforestation.

Boreal forests. As warming is expected to be particularly large at high latitudes, and as boreal forests are more strongly affected by temperature than forests in other latitudinal zones, climatic change is likely to have its greatest impact on boreal forests. Increased fire frequency and pest outbreaks are likely to decrease the average age, biomass and carbon store, with greatest impact at the southern boundary, where the boreal coniferous forest is likely to give way to temperate zone pioneer species or grasslands. Northern treelines are likely to advance slowly into regions currently occupied by tundra. The NPP of forests that are not limited by water availability is likely to increase in response to warming, partly mediated by increased nitrogen mineralization. However, there may be a net loss of carbon from the ecosystem because of associated increases in soil organic matter decomposition.

upon changes in factors such as temperature, soil moisture, and the availability of nutrients. To date, the magnitude and persistence of the CO_2 fertilization effect remains unquantified. Despite increases in plant growth anticipated from CO_2 fertilization, the total amount of carbon stored in an ecosystem may still decrease because increased temperature also stimulates the decomposition of dead leaves and soil organic matter.

In addition to the changes within ecosystems, temperature changes and increased CO_2 could result in significant alterations in the overall distribution of the world's biomes. Figure 3 shows the potential distribution of the major world biomes under current climate conditions and a doubled CO_2-equivalent climate-change scenario. The possible effects of climate change on the boundaries of forests and rangelands are discussed further in Boxes 4 and 5. The consequences for some terrestrial ecosystems depend critically on how fast climate zones shift. The rates of these shifts are important in part because different plant species migrate at different rates, depending on their growth and reproductive cycles.

3.2.1. Forests

Forests contain a wide range of species with complex life cycles. These ecosystems contain 80% of all aboveground carbon in vegetation and about 40% of all soil carbon. Forests and forest soils also play a major role in the carbon cycle as sources (e.g., forest degradation and deforestation) and sinks (reforestation, afforestation, and possibly enhanced growth resulting from carbon dioxide fertilization). Forests, particularly in the tropics, harbor as much as two-thirds of the world's biodiversity. They directly affect climate up to the continental scale by influencing ground temperatures, evapotranspiration, surface roughness, albedo, cloud formation, and precipitation.

A variety of biological, chemical, and physical factors affect forest ecosystems. Forest productivity and the number of species generally increase with increasing temperature, precipitation, and nutrient availability. Forests are particularly vulnerable to and may decline rapidly under extreme changes in water availability (either drought or waterlogging). Models project that a sustained increase of 1°C in global mean temperature is sufficient to cause changes in regional climates that will affect the growth and regeneration capacity of forests in many regions. In several instances, this will alter the function and composition of forests significantly. As a consequence of possible changes in temperature and water availability under doubled equivalent-CO_2 equilibrium conditions, a substantial fraction (a global average of one-third, varying by region from one-seventh to two-thirds) of the existing forested area of the world will undergo major changes in broad vegetation types—with the greatest changes occurring in high latitudes.

Climate change is expected to occur rapidly relative to the speed at which forest species grow, reproduce, and reestablish themselves. For mid-latitude regions, an average global warming of 1–3.5°C over the next 100 years would be equivalent to shifting isotherms poleward approximately 150–550 km or an altitude shift of 150–550 m; in low latitudes, temperatures would generally be increased to higher levels than now exist. This compares to past tree species migration rates on the order of 4–200 km per century. Entire forest types may disappear, and new ecosystems may take their places.

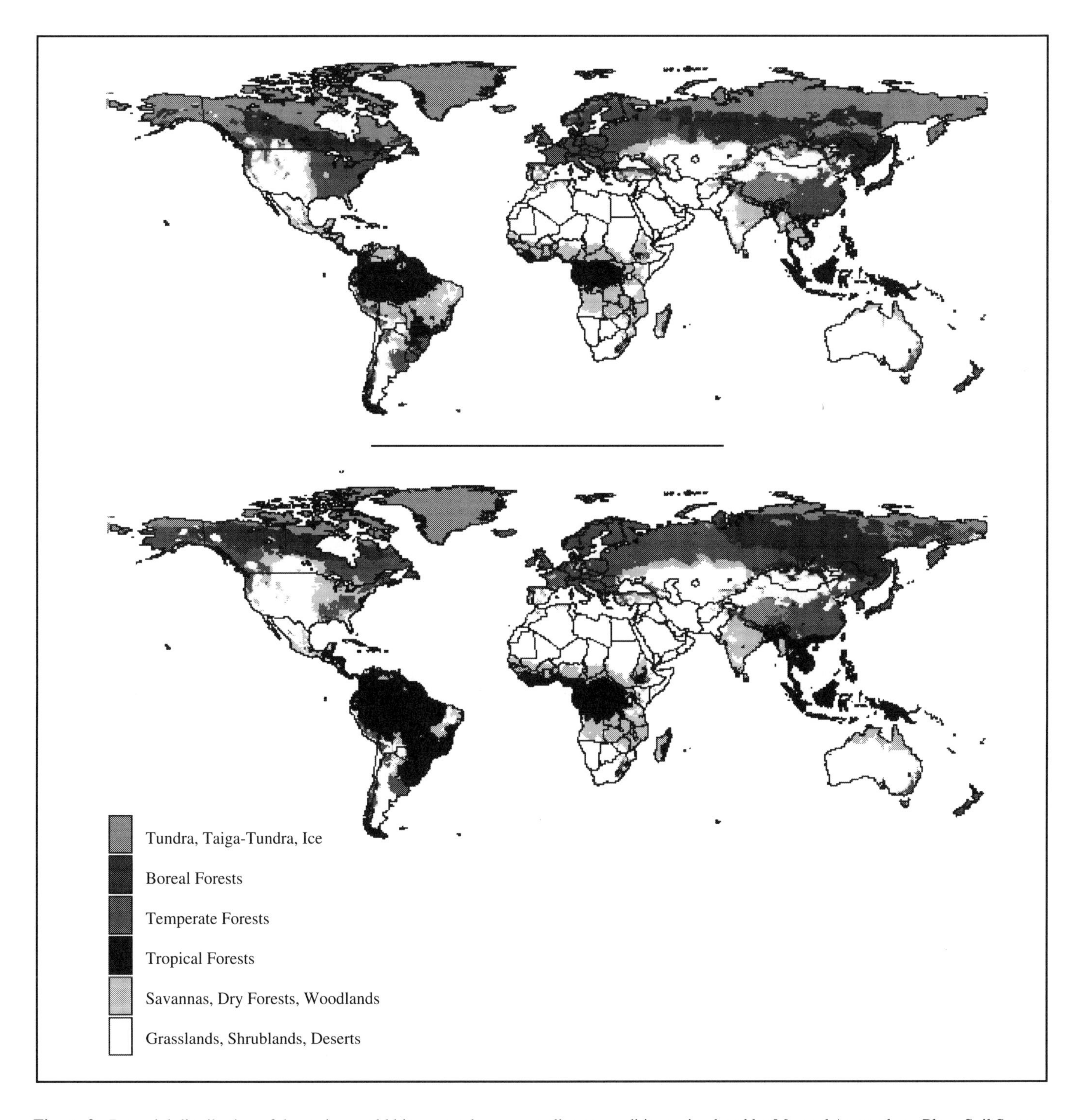

Figure 3: Potential distribution of the major world biomes under current climate conditions, simulated by Mapped Atmosphere-Plant-Soil System (MAPSS) model (top). "Potential distribution" indicates the natural vegetation that can be supported at each site, given monthly inputs of precipitation, temperature, humidity, and windspeed. The lower product illustrates the projected distribution of the major world biomes by simulating the effects of 2 x CO_2-equivalent concentrations (GFDL general circulation model), including the direct physiological effects of CO_2 on vegetation. Both products are adapted from: Neilson, R.P. and D. Marks, 1994: A global perspective of regional vegetation and hydrologic sensitivities from climatic change. *Journal of Vegetation Science*, **5**, 715-730.

Box 5. Regional Implications of Climate Change for Rangelands

Tropical rangelands. Temperature increases *per se* should not lead to major alterations in tropical rangelands, except where infrequent frosts currently limit some species. The most severe consequences could result from altered rainfall (seasonality and amount). Increasing carbon-to-nitrogen ratios could also result in reduced forage quality and palatability. The influence of increasing concentrations of CO_2 on photosynthesis and greater growth would have a major impact on the productivity of these rangelands.

Temperate rangelands. Climate change clearly will alter temperate rangelands and associated savannas and shrublands. Because of the high correlation between rangeland types and climate belts, any shifts in temperature and precipitation will bring corresponding shifts in rangeland boundaries. Continental areas that experience drier conditions during the growing season may see a shift from grasslands to shrublands.

Tundra. Tundra systems should exhibit high sensitivity to climatic warming. Indirect temperature effects, especially those associated with decreases in the amount of frozen soil and with changes in nutrient availability, will result in shifts in species composition. Changes in precipitation and temperature could decrease soil moisture in high-latitude tundra systems, changing some tundra systems from net sinks to net sources of carbon dioxide to the atmosphere while also increasing surface oxidation and decreasing methane flux.

Mature forests are a large terrestrial store of carbon. In general, temperate and tropical forests contain as much carbon aboveground as belowground, but boreal forests contain most of their carbon belowground. It remains unclear whether forests will continue to sequester carbon through growth under less suitable conditions than exist today. Although NPP could increase, the standing biomass of forests may not because of more frequent outbreaks and extended ranges of pests and pathogens, and increasing frequency and intensity of fires. Large amounts of carbon could be released into the atmosphere during transitions from one forest type to another, because the rate at which carbon can be lost during times of high forest mortality is greater than the rate at which it can be gained through growth to maturity.

3.2.2. Rangelands

Rangelands (i.e., unimproved grasslands, shrublands, savannas, deserts, and tundra) occupy 51% of the Earth's land surface. They contain about 36% of its total carbon in living and dead biomass, include a large number of economically important species and ecotypes, and sustain millions of people. Rangelands support 50% of the world's livestock and provide forage for domesticated animals and wildlife.

The amounts and seasonal distribution of precipitation are the primary controls on rangeland carbon cycling and productivity. Water availability and balance play vital roles in controlling the productivity and geographic distribution of rangeland ecosystems; thus, small changes in extreme temperatures and precipitation may have disproportionate effects.

Increases in CO_2 likely will result in reductions of forage quality and palatability because of increasing carbon-to-nitrogen ratios and shrub encroachment. These effects will become more evident in low-latitude rangelands, where the low nutritional value of forage already creates a chronic problem.

Boundaries between rangelands and other ecosystems appear likely to change as a result of direct climate effects on species composition as well as indirect factors such as changes in wildfire frequency and land use. Temperate rangelands will experience these effects the most. Migration rates of rangeland vegetation appear to be faster than for forests.

Precipitation and temperature variations associated with climate change may alter the role of high-latitude tundra systems in the carbon cycle. During the past decade, some tundra systems have shifted from a net sink to a net source of atmospheric CO_2, perhaps due to decreased soil moisture associated with warmer summers (a positive climate feedback). The tundra and taiga (subarctic evergreen forests) also provide 10% of the global atmospheric input of CH_4. Warming-related changes may dry soils and increase surface oxidation, thus decreasing CH_4 releases (a negative climate feedback).

Devising adaptation strategies for rangeland systems may prove difficult in marginal food-producing areas where production is very sensitive to climate change, changing of technology is risky, and the rate of adoption of new techniques and practices is slow. Decreases in rangeland productivity would result in a decline in the overall contribution of the livestock industries to national economies, with serious implications for food production in many developing areas with pastoral economies. Adaptation options in more highly managed pastures include management of forage, animal breeding, pasture renewal, and irrigation.

3.2.3. Deserts and Land Degradation

If the Earth's climate does change as projected in current scenarios, conditions in most deserts are likely to become more extreme—in that, with few exceptions, they are projected to become hotter but not significantly wetter. Temperature increases, in particular, could be a threat to organisms that now exist near their heat-tolerance limits. The impacts of climate change on water balance, hydrology, and vegetation remain uncertain and would probably vary significantly among regions.

Land degradation—which reduces the physical, chemical, or biological quality of land and lowers its productive capacity—already poses a major problem in many countries. Current general circulation models (GCMs) project that in some regions, climate change will increase drought and result in rainfall of higher intensity and more irregular distribution. This could increase the potential for land degradation, including loss of organic matter and nutrients, weakening of soil structure, decline in soil stability, and an increase in soil erosion and salinization. Areas that experience increased rainfall and soil moisture will benefit from the opportunities for more flexible use. However, they may experience stronger leaching, leading to increased acidification and loss of nutrients. In appropriate situations, additions of lime and fertilizers or conservation management policies could correct these deficiencies.

Desertification, as defined by the United Nations Convention to Combat Desertification, is land degradation in arid, semi-arid, and dry sub-humid areas resulting from various factors, including climatic variations and human activities. Droughts may trigger or accelerate desertification by reducing the growth of important plant species. Grazing can strip the land of its cover under these conditions. However, the amount of precipitation needed to sustain growth varies with the temperature, soil moisture capacity, and species. Desertification is more likely to become irreversible if the environment becomes drier and the soil becomes further degraded through erosion and compaction. Adaptation to drought and desertification may rely on the development of diversified production systems, such as agroforestry techniques and ranching of animals better adapted to local conditions. However, adaptation also needs political, social, extension service, and educational inputs.

3.2.4. Cryosphere

Many components of the cryosphere (i.e., snow, ice, and permafrost) are particularly sensitive to changes in atmospheric temperature. The last century has witnessed a massive loss and retreat of mountain glaciers, a reduction in the areal distribution of permafrost, and evidence of later freeze-up and earlier break-up of river and lake ice in many northern countries. These observations are consistent with a 0.5°C increase in the annual global mean temperature during the last century.

Projected changes in climate will substantially reduce the extent and volume of the cryosphere over the next century, causing pronounced reductions in mountain glaciers, permafrost, and seasonal snow cover. By the year 2100, between one-third and one-half of all mountain glaciers could disappear. The reduced extent of glaciers and depth of snow cover would affect the seasonal distribution of river flow, with potential implications for water resources (e.g., hydroelectric generation, agriculture). Data on iceberg calving from ice sheets and expected changes in calving as a result of projected temperature increases are inconclusive, although some scientists suggest that certain ice shelves currently are breaking up systematically. Little change in the extent of the Greenland and Antarctic ice sheets is expected over the next 50–100 years. An increase in temperature would extend the duration of the navigation season on rivers and lakes affected by seasonal ice cover. Projected reductions in the extent and thickness of the sea-ice cover in the Arctic Ocean and its peripheral seas could substantially benefit shipping, perhaps opening the Arctic Ocean as a major trade route. A reduction in the areal extent and depth of permafrost would have serious consequences for a number of human activities. Thawing of permafrost releases CH_4 hydrates.

3.2.5. Mountain Regions

Mountains cover about 20% of continental surfaces and serve as an important water source for most of the world's major river systems. Mountain ecosystems are under considerable stress from humans, and climate change will exacerbate existing conflicts between environmental and socioeconomic concerns.

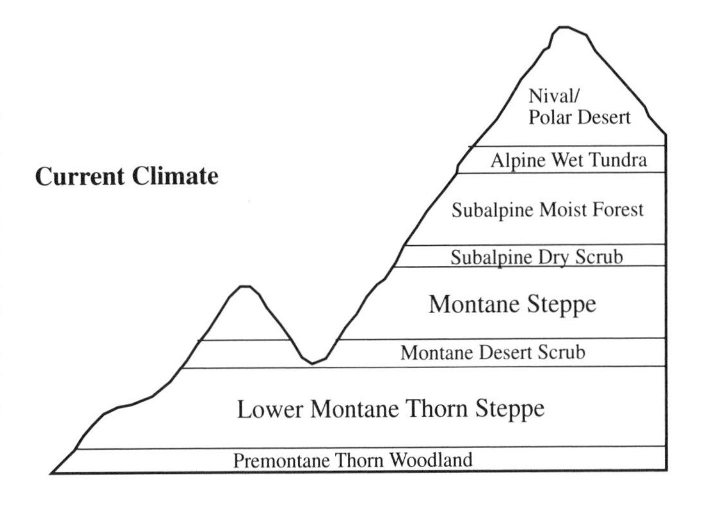

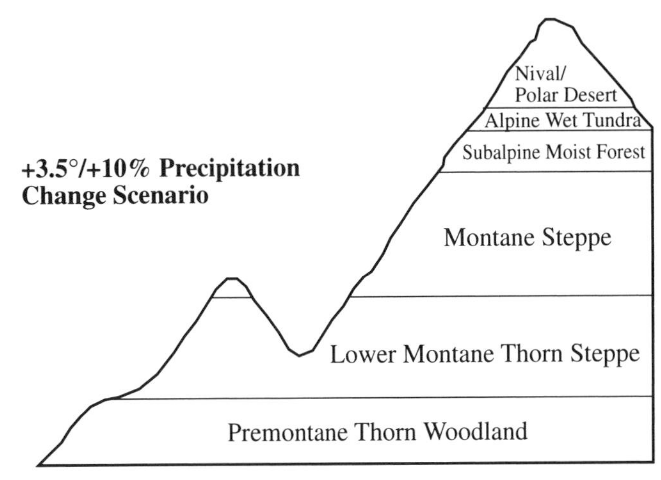

Figure 4: Comparison of current vegetation zones at a hypothetical dry temperate mountain site with simulated vegetation zones under a climate-warming scenario. Source: Beniston, 1994 (see Chapter 5 for complete citation).

Paleologic records indicate that past warming of the climate has caused the distribution of vegetation to shift to higher elevations, resulting in the loss of some species and ecosystems. Simulated scenarios for temperate-climate mountain sites suggest that continued warming could have similar consequences (see Figure 4), thus species and ecosystems with limited climatic ranges could disappear because of disappearance of habitat or reduced migration potential.

In most mountain regions, a warmer climate will reduce the extent and volume of glaciers and the extent of permafrost and seasonal snow cover. Along with possible precipitation changes, this would affect soil stability and a range of socioeconomic activities (e.g., agriculture, tourism, hydropower, and logging). Climate change may disrupt mountain resources for indigenous populations (e.g., fuel and subsistence and cash crops) in many developing countries. Recreational activities, which are increasingly important economically to many regions, also face likely disruptions. People living outside mountain regions who use water originating in them also would experience significant consequences.

Because of their climatic and habitat diversity, mountains provide excellent locations for maintaining biological diversity. In particular, large north-south mountain chains such as the Andes can facilitate migration under changing climate if appropriately managed.

3.3. *Aquatic Ecosystems*

Aquatic ecosystems encompass lakes and streams, non-tidal wetlands, coastal environs, and oceans. These systems are sensitive to changes in temperature, precipitation, and sea level and in turn exert important feedbacks on climate by influencing carbon fluxes. They provide a range of socioeconomic values and benefits—including food, energy, transportation, timber, flood mitigation, erosion control, water supply, and recreation.

3.3.1. *Lakes and Streams*

Climate change will influence lakes and streams through altered water temperatures, flow regimes, and water levels. These will directly affect the survival, reproduction, and growth of organisms; the productivity of ecosystems; the persistence and diversity of species; and the regional distribution of biota. Climatic warming would tend to shift geographic ranges of many species poleward by approximately 150 km for every 1°C increase in air temperature. Changes in the heat balance of lakes would alter their mixing properties, which would have large effects on their primary productivity. Changes in runoff and groundwater flows to lakes and streams would alter the input of nutrients and dissolved organic carbon, which in turn would alter the productivity and clarity of the waters. Changes in hydrologic variability are expected to have greater ecological effects than changes in mean values. For example, increased frequency or duration of flash floods and droughts would reduce the biological diversity and productivity of stream ecosystems.

Although the effects on lake and stream ecosystems will vary with the distribution of climate changes, some general conclusions emerge. Warming would have the greatest biological effects at high latitudes, where biological productivity would increase, and at the low-latitude boundaries of cold- and cool-water species ranges, where appropriate thermal habitat would become more fragmented and extinctions would be greatest. The rate of climatic warming may exceed the rate of shifts in species ranges. Warming of the larger and deeper temperate lakes will increase their productivity and thermal habitat favorable for native fishes, whereas warming of shallow lakes and streams could lead to increased anoxia and reduction in suitable habitat. Deep tropical lakes may become less productive as thermal stratification intensifies and nutrients become trapped below the mixed layer. Lakes and streams in dry, evaporative drainages and in basins with small catchments are most sensitive to changes in precipitation and would experience the most severe water-level declines if precipitation were to decrease. For many lakes and streams, the most severe effects of climate change may be the exacerbation of current stresses resulting from human activities, including increasing demands for consumptive uses, increasing waste effluent loadings and runoff of agricultural and urban pollutants, and altered water balance and chemical input rates from landscape disturbance and atmospheric deposition.

3.3.2. *Non-Tidal Wetlands*

Wetlands are areas of low-lying land where the water table lies at or near the surface for some defined period of time, producing extensive shallow, open water and waterlogged areas. Wetlands exist on every continent (except Antarctica), in both inland and coastal areas, and cover approximately 4–6% of the Earth's land surface. Human activities such as agricultural development, construction of dams and embankments, and peat mining already threaten these ecosystems and have contributed to the disappearance of more than half of the world's wetlands during the last century. Non-tidal (primarily inland) wetlands provide refuge and breeding grounds for many species, including a large number with commercial value; are an important repository of biodiversity; control floods and droughts; and improve water quality. They also serve as a carbon sink. Boreal and subarctic peatlands store an estimated 412 Gt C, or about 20% of the global organic carbon stored in soils.

Climate change will most greatly influence wetlands by altering their hydrologic regimes (increasing or decreasing water availability; changing the depth, duration, frequency, and season of flooding). These changes will affect the biological, biogeochemical, and hydrological functions of wetlands and alter their value to societies for such functions as aquifer recharge, sediment retention, waste processing, and carbon storage. There will be an impact of climate change on greenhouse gas release from non-tidal wetlands, but there is uncertainty regarding the exact effects from site to site; some arctic areas already have shifted from weak carbon sinks to weak CO_2 sources due to drying.

The geographical distribution of wetlands is likely to shift with changes in temperature and precipitation. Warming would severely affect wetlands in arctic and subarctic regions by thawing permafrost, which is key to maintaining high water tables in these ecosystems.

Possibilities for adaptation, conservation, and restoration in response to climate change vary among wetland types and their various functions. For regional and global functions (e.g., trace-gas fluxes and carbon storage), no responses exist that humans can apply at the necessary scale. For wetland functions at the local scale (habitat value, pollution trapping, and, to some degree, flood control), possibilities do exist. However, interest in wetland creation and restoration has outpaced the science and technology needed to successfully create wetlands for many specific purposes.

3.3.3. Coastal Ecosystems

Many commercially important marine species depend on coastal ecosystems for some part of their life cycle, and some coastal ecosystems (e.g., coral reefs, mangrove communities, and beaches) provide an important buffer against storms and surges. Sea-level rise, altered rainfall patterns, and changes in ocean temperature likely will result in additional adverse impacts on coastal ecosystems, particularly where human activities already affect environmental conditions.

Coastal systems are ecologically and economically important, and are expected to vary widely in their response to changes in climate and sea level. Climate change and a rise in sea level or changes in storms or storm surges could result in the displacement of wetlands and lowlands, erosion of shores and associated habitat, increased salinity of estuaries and freshwater aquifers, altered tidal ranges in rivers and bays, changes in sediment and nutrient transport, a change in the pattern of chemical and microbiological contamination in coastal areas, and increased coastal flooding. In many areas, intensive human alteration and use of coastal environments already have reduced the capacity of natural systems to respond dynamically. Other regions, however, may prove far less sensitive and may keep pace with climate and sea-level change. This suggests that future impact analyses must be carried out region by region.

Nonetheless, research findings strongly indicate that some coastal ecosystems face particular risks from changes in climate and sea level. A number of regional studies show the disappearance of saltwater marshes, mangrove ecosystems, and coastal wetlands at a rate of 0.5–1.5% per year over the last few decades. The interaction of climate change, sea-level rise, and human development will further threaten these particularly sensitive ecosystems. Climate-change impacts on coastal ecosystems will vary widely. The sea level may rise faster than sediment accretion rates in many coastal areas; thus, existing wetlands will disappear faster than new ones appear. Ecosystems on coral atolls and in river deltas show special sensitivity to climate and sea-level change. Changes in these ecosystems almost certainly would have major negative effects on tourism, freshwater supplies, fisheries, and biodiversity. Such impacts would further modify the functioning of coastal oceans and inland waters already stressed by pollution, physical modification, and material inputs due to human activities.

Coral reefs are the most biologically diverse marine ecosystems; they also are very sensitive to climate change. Short-term increases in water temperatures on the order of only 1–2°C can cause "bleaching," leading to reef destruction. Sustained increases of 3–4°C above long-term average seasonal maxima over a 6-month period can cause significant coral mortality. Biologists suggest that fully restoring these coral communities could require several centuries. A rising sea level also may harm coral reefs. Although available studies indicate that even slow-growing corals can keep pace with the "central estimate" of sea-level rise (approximately 0.5 cm per year), these studies do not take account of other pressures on coral populations, such as pollution or enhanced sedimentation.

3.3.4. Oceans

Oceans occupy 71% of the Earth's surface. They provide an important component of the climate system, due to their role in controlling the distribution and transfer of heat and CO_2 (internally and with the atmosphere) and in the transfer of water back to the continents as precipitation. They provide vital mineral resources, an environment for living resources ranging from phytoplankton to whales, and support for socioeconomic activities such as transportation, fishing, and recreation.

Oceans are sensitive to changes in temperature, freshwater inputs from continents and atmospheric circulation, and the interaction of these factors with other environmental factors such as ultraviolet-B (UV-B) radiation and pollution (which is especially important in inland seas, bays, and other coastal areas). Climate change may alter sea level, increasing it on average, and could also lead to altered ocean circulation, vertical mixing, wave climate, and reductions in sea-ice cover. As a result, nutrient availability, biological productivity, the structure and functions of marine ecosystems, the ocean's heat- and carbon-storage capacity, and important feedbacks to the climate system will change as well. These changes would have implications for coastal regions, fisheries, tourism and recreation, transport, off-shore structures, and communication. Paleoclimatic data and model experiments suggest that abrupt climatic changes can occur if freshwater influx from the movement and melting of sea ice or ice sheets significantly weakens global thermohaline circulation (the sinking of dense surface waters in polar seas that moves bottom water toward the equator).

3.4. Water

Water availability is essential to national welfare and productivity. The world's agriculture, hydropower production, municipal and industrial water supply, instream ecosystems, water-based recreation, and inland navigation depend on surface and groundwater resources. The quantity and quality of water supplies pose a serious problem today in many regions, including some low-lying coastal areas, deltas, and small islands. Water availability already falls below 1,000 m^3 per person per year—a common benchmark for water scarcity—in a number of countries (e.g., Kuwait, Jordan, Israel, Rwanda, Somalia, Algeria, Kenya). Other nations likely to fall below this benchmark in the next 2 to 3 decades include Libya, Egypt, South Africa, Iran, and Ethiopia. In addition, a number of countries in conflict-prone areas depend on water originating outside their borders (e.g., Cambodia, Syria, Sudan, Egypt, Iraq). This makes them quite vulnerable to any additional reduction in indigenous water supplies. The depletion of aquifers, urbanization, land-cover changes, and contamination exacerbate the problem of water availability.

A changing climate will lead to an intensification of the global hydrological cycle, which determines how precipitation is partitioned between ground and surface water storage (including snow cover), fluxes to the atmosphere, and flows to the oceans. Changes in the total amount of precipitation and in its frequency and intensity directly affect the magnitude and timing of runoff and the intensity of floods and droughts; however, at present, specific regional effects are uncertain. In addition, changes in the hydrological cycle affect land cover and the surface energy balance, thus altering important feedbacks on the climate system.

Relatively small changes in temperature and precipitation, together with the non-linear effects on evapotranspiration and soil moisture, can result in relatively large changes in runoff, especially in arid and semi-arid lands. High-latitude regions may experience increased runoff due to increased precipitation, whereas runoff may decrease at lower latitudes due to the combined effects of increased evapotranspiration and decreased precipitation. Even in areas where models project a precipitation increase, higher evaporation rates may lead to reduced runoff. More intense rainfall would tend to increase runoff and the risk of flooding, although the magnitude of the effect would depend on both an area's rainfall change and its catchment characteristics. This effect could be exacerbated in regions where extensive reductions occur in vegetation (e.g., deforestation, overgrazing, logging). Winter snowfall and spring snowmelt determine the flow rates of rivers and streams in many continental and mountain areas. Should the climate change and the proportion of precipitation falling as snow decrease, widespread reductions in spring runoff and increases in winter runoff seem likely. These changes would have consequences for water storage and delivery systems, irrigation, and hydroelectricity production.

Many factors, including changes in vegetation, population growth, and industrial and agricultural demands complicate an assessment of the potential effects of climate change on water resources. Current understanding suggests that climate change can have major impacts on regional water supplies. At present, general circulation models only provide projections on a large geographic scale. They do not agree on a likely range of changes in average annual precipitation for any given basin or watershed, hence fail to provide sufficient information to assist water managers in making decisions.

A change in the volume and distribution of water would affect all of a region's water uses. The impacts, however, will depend also on the actions of water users and managers, who will respond not only to climate change but also to population growth and changes in demands, technology, and economic, social, and legislative conditions. In some cases—particularly in wealthier countries with integrated water-management systems—these actions may protect water users from climate change at minimal cost. In many others however—particularly those regions that already are water-limited—substantial economic, social, and environmental costs could occur. Water resources in arid and semi-arid zones are particularly sensitive to climate variations because of low-volume total runoff and infiltration and because relatively small changes in temperature and precipitation can have large effects on runoff. Irrigation—the largest use of water in many countries—will be the first activity affected in regions where precipitation decreases. This is because in some regions water used for agriculture costs less than water for domestic and industrial activities. During water shortages, allocations to agriculture will most likely decline before allocations to those other uses.

The increased uncertainty in the future supply and demand of water resources raises a key issue for water management. Countries with high population growth rates are likely to experience significant decreases in per capita water availability even without climate change. The issue is complicated further when projected climate change is taken into account. Based on outputs from transient climate models, hydrological models indicate that per capita water availability would vary widely, from slight increases in percentage of water available per capita to large percentage decreases, for the countries considered. Significant differences in the regional distribution of water deficits and surpluses also may occur within each country.

Options for dealing with the possible impacts of a changed climate and increased uncertainty about future supply and demand for freshwater include:

- More efficient management of existing supplies and infrastructure
- Institutional arrangements (e.g., market and regulatory measures) to limit future demands on scarce water supplies
- Improved hydrological monitoring and forecasting systems and establishment of early warning systems for floods/drought
- Rehabilitation of presently denuded tracts of upland watersheds, especially in the tropics

- Construction of new reservoir capacity to capture and store excess flows produced by altered patterns of snowmelt and storms (although this often would be difficult to plan in the absence of improved watershed climate information).

Water managers, in a continuously adaptive enterprise, respond to changing demographic and economic demands, information, and technologies. Experts disagree about whether water supply systems will evolve substantially enough in the future to compensate for the anticipated negative impacts of climate change and the anticipated increases in demand.

3.5. Food and Fiber

Two broad classes of climate-induced effects influence the quality and quantity of agricultural and forestry yields: (i) direct effects from changes in temperature, water balance, atmospheric composition, and extreme events; and (ii) indirect effects through changes in the distribution, frequency, and severity of pest and disease outbreaks, incidence of fire, weed infestations, or through changes in soil properties. Fisheries respond to direct climate-change effects such as increases in water temperature and sea level and changes in precipitation, freshwater flows, climate variability, and currents. They respond also to indirect effects such as shifts in food supply and the expansion in ranges of red tides and other biotoxins, which could lead to increased contamination of fisheries. The vulnerability of food and fiber production to climate change depends not only on the physiological response of plants and animals but also on the ability of the affected production and distribution systems to cope with fluctuations in yield.

3.5.1. Agriculture

Recent studies support evidence in the 1990 assessment that, on the whole, global agricultural production could be maintained relative to baseline production in the face of climate change modeled by GCMs at doubled-equivalent CO_2 equilibrium conditions. However, more important than global food production—in terms of the potential for hunger, malnutrition, and famine—is the access to and availability of food for specific local and regional populations. Many new crop studies conducted since the 1990 assessment report results that vary widely by crop, climate scenario, study methodology, and site (results are aggregated and summarized in Table 2). Limited ranges for some countries reflect the small number of studies available. At broader regional scales, subtropical and tropical areas—home to many of the world's poorest people—show negative consequences more often than temperate areas. People dependent on isolated agricultural systems in semi-arid and arid regions face the greatest risk of increased hunger due to climate change. Many of these at-risk populations live in sub-Saharan Africa; South, East, and Southeast Asia; and tropical areas of Latin America, as well as some Pacific island nations.

These conclusions emerge from studies that model the effects on agricultural yields induced by climate change and elevated CO_2. These studies presently do not include changes in insects, weeds, and diseases; direct effects of climate change on livestock; changes in soils and soil-management practices; and changes in water supply caused by alterations in river flows and irrigation. Moreover, the studies have considered only a limited set of adaptation measures and are based on yield analyses at a limited number of sites. Failure to integrate many key factors into agronomic and economic models limits their ability to consider transient climate scenarios and to fully address the costs and potential of adaptation. However, increased productivity of crops due to elevated concentrations of CO_2 is an important assumption in most crop modeling studies. Although the mean value response under experimental conditions is a 30% increase in productivity for C_3 crops (e.g., rice and wheat) under doubled-CO_2 conditions, the range is -10 to +80%. The response depends on the availability of plant nutrients, plant species, temperature, precipitation, and other factors, as well as variations in experimental technique.

A wide range of views exists on the potential of agricultural systems to adapt to climate change. Historically, farming systems have adapted to changing economic conditions, technology and resource availability, and population pressures. Uncertainty remains regarding whether the rate of climate change and required adaptation would add significantly to the disruptions resulting from other socioeconomic or environmental changes. Adaptation to climate change via new crops and crop varieties, improved water-management and irrigation systems, and information (e.g., optimal planting times) will prove important in limiting negative effects and taking advantage of beneficial changes in climate. The extent of adaptation depends in part on the cost of the measures used, particularly in developing countries; access to technology and skills; the rate of climate change; and constraints such as water availability, soil characteristics, topography, and the genetic diversity bred into crops. The incremental costs of these adaptation strategies could create a serious burden for some developing countries; some adaptation strategies may result in cost savings for some countries. There are significant uncertainties about the capacity of different regions to adapt successfully to projected climate change. Many current agricultural and resource policies—already a source of land degradation and resource misuse—likely will discourage effective adaptation measures.

Changes in grain prices, in the prevalence and distribution of livestock pests, and in grazing land and pasture productivity all will affect livestock production and quality. In general, analyses indicate that intensively managed livestock systems have more potential for adaptation than crop systems. In contrast, adaptation presents greater problems in pastoral systems where production is very sensitive to climate change, technology changes incur risks, and the rate of technology adoption is slow.

3.5.2. Forestry

Global wood supplies during the next century may become increasingly inadequate to meet projected consumption due to

Table 2: Selected crop study results for 2 x CO_2-equivalent equilibrium GCM scenarios.

Region	Crop	Yield Impact (%)	Comments
Latin America	Maize	-61 to increase	Data are from Argentina, Brazil, Chile, and Mexico; range is across GCM scenarios, with and without CO_2 effect.
	Wheat	-50 to -5	Data are from Argentina, Uruguay, and Brazil; range is across GCM scenarios, with and without CO_2 effect.
	Soybean	-10 to +40	Data are from Brazil; range is across GCM scenarios, with CO_2 effect.
Former Soviet Union	Wheat Grain	-19 to +41 -14 to +13	Range is across GCM scenarios and region, with CO_2 effect.
Europe	Maize	-30 to increase	Data are from France, Spain, and northern Europe; with adaptation and CO_2 effect; assumes longer season, irrigation efficiency loss, and northward shift.
	Wheat	increase or decrease	Data are from France, UK, and northern Europe; with adaptation and CO_2 effect; assumes longer season, northward shift, increased pest damage, and lower risk of crop failure.
	Vegetables	increase	Data are from UK and northern Europe; assumes pest damage increased and lower risk of crop failure.
North America	Maize Wheat	-55 to +62 -100 to +234	Data are from USA and Canada; range is across GCM scenarios and sites, with/without adaptation and with/without CO_2 effect.
	Soybean	-96 to +58	Data are from USA; less severe or increase with CO_2 and adaptation.
Africa	Maize	-65 to +6	Data are from Egypt, Kenya, South Africa, and Zimbabwe; range is over studies and climate scenarios, with CO_2 effect.
	Millet	-79 to -63	Data are from Senegal; carrying capacity fell 11–38%.
	Biomass	decrease	Data are from South Africa; agrozone shifts.
South Asia	Rice Maize Wheat	-22 to +28 -65 to -10 -61 to +67	Data are from Bangladesh, India, Philippines, Thailand, Indonesia, Malaysia, and Myanmar; range is over GCM scenarios, with CO_2 effect; some studies also consider adaptation.
China	Rice	-78 to +28	Includes rainfed and irrigated rice; range is across sites and GCM scenarios; genetic variation provides scope for adaptation.
Other Asia and Pacific Rim	Rice	-45 to +30	Data are from Japan and South Korea; range is across GCM scenarios; generally positive in north Japan, and negative in south.
	Pasture	-1 to +35	Data are from Australia and New Zealand; regional variation.
	Wheat	-41 to +65	Data are from Australia and Japan; wide variation, depending on cultivar.

Note: For most regions, studies have focused on one or two principal grains. These studies strongly demonstrate the variability in estimated yield impacts among countries, scenarios, methods of analysis, and crops, making it difficult to generalize results across areas or for different climate scenarios.

both climatic and non-climatic factors. Assuming constant per capita wood use, analysis of changing human populations suggests that the annual need for timber will exceed the current annual growth increment of 2% by the year 2050. However, growth could increase slightly from warming and enhanced atmospheric CO_2 concentrations, or decrease greatly from declines and mortality of forest ecosystems brought on by climate change.

Tropical forest product availability appears limited more by changes in land use than by climate change, at least through the middle of the next century. By then, projections indicate that growing stock will have declined by about half due to non-climatic reasons related to human activities. This projected decline holds up even after calculating changes in climate and atmospheric composition during this period, which could increase forest productivity and the areas where tropical forests can potentially grow. Communities that depend on tropical forests for fuelwood, nutrition, medicines, and livelihood will most feel the effects of declines in forested area, standing stock, and biodiversity.

Temperate-zone requirements for forest products should be met for at least the next century. This conclusion emerges from projected climate and land-use changes that leave temperate forest covering about as much land in 2050 as today. It further assumes that current harvests increase only slightly, that the annual growth increment remains constant, and that imports from outside increasingly meet the temperate zone's need for forest products—although it is not clear if this assumption can be met given production projections in other zones.

Boreal forests are likely to undergo irregular and large-scale losses of living trees because of the impacts of projected rapid climate change. Such losses could initially generate additional wood supply from salvage harvests, but could severely reduce standing stocks and wood-product availability over the long term. The exact timing and extent of this pattern is uncertain. Current and future needs for boreal forest products are largely determined outside the zone by importers; future requirements for forest products may exceed the availability of boreal industrial roundwood during the 21st century, given the projections for temperate- and tropical-zone forest standing stocks and requirements.

Effective options for adapting to and ameliorating potential global wood supply shortages include the following:

- In the tropics, the greatest progress may result from developing practices and policies that reduce social pressures driving land conversion (e.g., by increasing crop and livestock productivity) and by developing large plantations.
- In temperate areas, application of modern forestry practices to reduce harvest damage to ecosystems, combined with the substitution of nontimber products, could reduce significantly the effect of climate on wood availability.
- In boreal regions, adaptation to potential climate-induced, large-scale disturbances—such as by rapid reforestation with warmth-adapted seeds—appears to be most useful. Increased prices for forestry products seem certain to lead to adaptation measures that will reduce demand, increase harvest intensity, and make tree plantations more economically feasible.

3.5.3. *Fisheries*

Climate-change effects will interact with other stresses on fisheries, including pervasive overfishing, diminishing nursery areas, and extensive inshore and coastal pollution. Overfishing stresses fisheries more than climate change today, but if fisheries management improves and climate changes develop according to IPCC scenarios, climate change may become the dominant factor by the last half of the next century.

Although marine fisheries production will remain about the same globally, projections indicate that higher latitude freshwater and aquaculture production will increase. This assumes, however, that natural climate variability and the structure and strength of ocean currents remains unchanged. Alterations in either would have significant impacts on the distribution of major fish stocks, rather than on global production. Positive effects (e.g., longer growing seasons, lower natural winter mortality, and faster growth rates in higher latitudes) may be offset by negative factors such as changes in established reproductive patterns, migration routes, and ecosystem relationships. Climate change can be expected to have the greatest impact on the following (in decreasing order): (i) freshwater fisheries in small rivers and lakes in regions with larger temperature and precipitation change; (ii) fisheries within exclusive economic zones, particularly where access regulations artificially reduce the mobility of fisher groups and fishing fleets, thus their capacity to adjust to fluctuations in stock distribution and abundance; (iii) fisheries in large rivers and lakes; (iv) fisheries in estuaries, particularly where there are species without migration or spawn dispersal paths, or estuaries affected by sea-level rise or decreased river flow; and (v) high-seas fisheries. Where rapid change occurs due to physical forcing (e.g., changes in currents and natural variability), production will usually favor smaller, low-priced, opportunistic species that discharge large numbers of eggs over long periods. Loss of coastal wetlands could cause a loss of nurseries and have significant adverse effects on fisheries.

If institutional mechanisms do not enable fishers to move across national boundaries, some national fish industries may suffer negative effects. Subsistence and other small-scale fishers, often the most dependent on specific fisheries, will suffer disproportionately from changes. Several adaptation options exist. These include improved management systems, ecological and institutional research, and expansion of aquaculture.

3.6. **Human Infrastructure**

Human infrastructure, a key determinant of a region's productivity and development, includes all varieties of human-made capital and assets—residential and commercial properties, transportation facilities, industries and manufactured goods, coastal embankments, and equipment for energy production and distribution. Human infrastructure and socioeconomic systems—including industry, energy, and transportation—may be affected *directly* through changes in temperature, precipitation, sea level, or increased frequency or intensity of extreme events that can damage exposed infrastructure or affect outputs. Due to the interconnectedness of economic activity, many of the influences of climate change on industry, energy, and transportation are expected to be *indirect* and transmitted by changes in markets sensitive to climate (e.g., energy demand for space heating and cooling) or changes in resources sensitive to climate (e.g., agroindustries and biomass production). Human migration in response to chronic crop failures, regional flooding, or drought may create additional pressure on human infrastructures. These indirect impacts are difficult to quantify or value in monetary terms.

In many sectors, the effects of climate change will amount to less than those resulting from changes in demography, technology, and markets. Yet unexpected changes in climate could occur and much capital is invested in locations that could be affected by such changes. Relocating or defending this infrastructure could require costly actions and a high degree of foresight and coordination.

3.6.1. Industry, Energy, and Transportation

In general, the climate sensitivity of most activities in these sectors is relatively low compared to that of agriculture or natural ecosystems. Certain activities, however, display a greater degree of climate sensitivity than do industry, energy, and transportation as a whole. Subsectors and activities most sensitive to climate change include agroindustry, production of renewable energy such as hydroelectricity and biomass, energy demand, construction, some transportation activities, existing flood mitigation structures, and transportation infrastructure located in many areas, including vulnerable coastal zones and permafrost regions.

In the energy sector, the consequences for hydroelectric power generation will depend upon changes in the balance between the amount and timing of precipitation and evaporation. Increased scarcity of fuelwood in dry and densely populated regions could exacerbate problems caused by deforestation. Peak winter demand for primary energy is projected to decrease due to a reduction in space-heating needs; peak summer demand for electricity may increase with greater cooling requirements in some regions. 'The net effects of these changes in energy demand will be regionally dependent. There is low confidence in estimates of these net effects.

In contrast to research on mitigation options, relatively few studies exist on climate change impacts on and adaptation options for the industry, energy, and transportation sectors. This situation reflects a perception of low vulnerability to climate change for these sectors.

3.6.2. Coastal Zones/Small Islands

Many coastal zones and small islands are particularly vulnerable to direct effects of climate change and sea-level rise. Present estimates of global sea-level rise represent a rate two to five times that experienced during the last 100 years. Recent scientific findings indicate that sea-level rise may deviate from the global average by a factor of two or more due to regional differences in ocean salinity and temperature change.

Sea-level rise can have negative impacts on tourism, freshwater supplies, fisheries, exposed infrastructure, agricultural and dry lands, and wetlands. Impacts may vary across regions; societal costs will greatly depend upon the vulnerability of the coastal system and a country's economic situation.

Climate change clearly will increase the vulnerability of some coastal populations to flooding and erosional land loss. Estimates put about 46 million people per year currently at risk of flooding due to storm surges. This estimate results from multiplying the total number of people currently living in areas potentially affected by ocean flooding by the probability of flooding at these locations in any year, given the present protection levels and population density. In the absence of adaptation measures, a 50-cm sea-level rise would increase this number to about 92 million; a 1-m sea-level rise would raise it to 118 million. If one incorporates anticipated population growth, the estimates increase substantially. Some small island nations and other countries will confront greater vulnerability because their existing sea and coastal defense systems are less well-established. Countries with higher population densities would be more vulnerable. For these countries, sea-level rise could force internal or international migration of populations.

A number of studies have evaluated sensitivity to a 1-m sea-level rise. This increase is at the top of the range of IPCC Working Group I estimates for 2100; it should be noted, however, that sea level is actually projected to continue to rise in future centuries beyond 2100. Studies using this 1-m projection show a particular risk for small islands and deltas. Given the present state of protection systems, estimated land losses range from 0.05% for Uruguay, 1% for Egypt, 6% for the Netherlands, and 17.5% for Bangladesh up to about 80% for the Majuro Atoll in the Marshall Islands. Large numbers of people are also affected—for example, about 70 million each in China and Bangladesh. Many nations face lost capital value in excess of 10% of GDP. Although annual adaptation/protection costs for many nations are relatively modest (about 0.1% of GDP), the average annual costs to many small island states total several percent of GDP. For some island nations, the high cost of providing storm-surge protection would make it essentially infeasible, especially given the limited availability of capital for investment.

3.6.3. Human Settlements

Projected climate change will affect human settlements in the context of changes such as population growth, migration, and industrialization. Settlements where these forces already stress the infrastructure are most vulnerable. Besides coastal communities and those dependent on subsistence, rain-fed agriculture, or commercial fishing, vulnerable settlements include large primary coastal cities and squatter settlements located in flood plains and steep hillsides. Many of the expected impacts in the developing world could occur because climate change may reduce natural resource productivity in rural areas, thus may generally accelerate rural-to-urban migration. Direct impacts on infrastructure would most likely occur as a result of changes in the frequency and intensity of extreme events. These include coastal storm surges, floods and landslides induced by local downpours, windstorms, rapid snowmelt, tropical cyclones and hurricanes, and forest and brush fires made possible in part by more intense or lengthier droughts.

3.6.4. Insurance and Financial Services

Within the financial services sector, property insurance stands most vulnerable to direct climatic influence. A higher risk of extreme events due to climate change could lead to higher insurance premiums or the withdrawal of coverage for property in some vulnerable areas. Changes in climate variability and the risk for extreme events may be difficult to detect or predict, thus making it difficult for insurance companies to adjust premiums appropriately. If such difficulty leads to insolvency, companies may not be able to honor insurance contracts, which could economically weaken other sectors such as banking. The insurance industry currently is under stress from a series of "billion dollar" storms since 1987 (see Table 3), resulting in dramatic increases in losses, reduced availability of insurance, and higher premiums. Some in the insurance industry perceive a current trend toward increased frequency and severity of extreme climate events. Examination of the meteorological data fails to support this perception in the context of a long-term change, although a shift within the limits of natural variability may have occurred. Higher losses strongly reflect increases in infrastructure and economic worth in vulnerable areas, as well as a possible shift in the intensity and frequency of extreme weather events.

Table 3: *"Billion dollar" storms.*

Year	Event	Insured Cost ($B)
1987	"Hurricane" in SE England/NW France	2.5
1988	Hurricane Gilbert in Jamaica/Mexico	0.8
1989	Hurricane Hugo in Puerto Rico/S. Carolina	5.8
1990	European Storms—Four	10.4
1991	Typhoon Mireille in Japan	4.8
1992	Hurricane Andrew in Florida	16.5
1993	"Storm of the Century" in Eastern USA	1.7
1995	Hailstorms in Texas	1.1
1995	Hurricane Opal in Southern USA	2.1

Sources: Munich Re, 1990; Leggett, 1994; PCS, 1995 (see Chapter 17 for complete citations).

Withdrawal of insurance would increase direct financial losses to property owners and businesses unable to obtain insurance, with serious long-term implications for societies and governments. The implications of climate change for financial services outside of property insurance appear less clear; these sectors generally have not acknowledged the potential for such impacts. Adaptation may prove difficult, given the long-term nature of many investors' financial commitments.

3.6.5. Adaptation Options for Human Infrastructure

Because the life cycles of planning and investment for much human infrastructure are shorter than those associated with climate change, adaptation could occur through management and the normal replacement of capital in many sectors, as long as climate change happens gradually. This, however, depends on people and organizations becoming adequately informed about potential impacts and having the financial, technical, and institutional capacity to respond. In the more sensitive sectors, adaptation may need the support of policy measures. Incorporating potential climate and sea-level changes in planning would reduce future risks to human infrastructure. Climate changes, however, could occur suddenly. Uncertainty about the rate and effects of climate change makes some investment decisions difficult. This holds particularly true for planning and investments for infrastructure such as channels, water supply systems, and coastal or river-flooding defenses, which can have lifetimes as long as 100 years.

Many developing countries currently depend on a limited number of crops or on fishing—hence are economically vulnerable to climate change that harms agroindustry. For these countries in particular, diversifying economic activity—coupled with improved management practices such as integrated coastal-zone management and land-use regulation—could constitute an important precautionary response and facilitate successful adaptation to climate change (e.g., by directing populations away from vulnerable locations).

Implementing adaptation measures and integrated management practices requires overcoming constraints that include (but are not limited to) technology and human resource capability; financial limitations; cultural and social acceptability; and political, legal, and other institutional bottlenecks. The literature remains scarce on adaptation; dealing effectively with the issue requires additional research and new methods of risk and probability assessment.

3.7. ***Human Health***

Climate changes and their effects on food security, water supply and quality, and the distribution of ecological systems may have wide-ranging and potentially adverse effects on human health, via both direct and indirect pathways (see Figure 5); it is likely that the indirect impacts would, in the longer term, predominate. Quantifying the potential health impacts of climate change remains difficult. For many effects, forecasting techniques—especially modeling—are just being developed. Furthermore, the extent of climate-induced health disorders depends on numerous coexistent and interacting factors that characterize the vulnerability of the particular population. These include environmental circumstances (such as water purity) and socioeconomic factors (such as nutritional and immune status, population density, and access to health care).

Direct health effects include increases in heat-related (predominantly cardiorespiratory) mortality and illness resulting from an anticipated increase in the intensity and duration of heat waves. Studies in selected cities in North America, North Africa, and East Asia indicate that the annual numbers of heat-related deaths would increase several-fold in response to climate-change projections. Temperature increases in colder regions should result in fewer cold-related deaths.

The incidence of deaths, injuries, psychological disorders, and exposure to chemical pollutants in water supplies would increase if extreme weather events (e.g., droughts and floods) were to become more frequent.

Indirect effects include increases in the potential transmission of vector-borne infectious diseases (e.g., malaria, dengue, Chagas' disease, yellow fever, and some viral encephalitis) caused by extensions of the ranges and seasons of vector organisms. Climate change also would accelerate the maturation of certain infectious parasites (e.g., the malaria organism). Currently, approximately 350 million cases of malaria occur annually, resulting in 2 million deaths. Using first-generation mathematical models, scientists recently forecast the impact of changes in basic climate variables on the global/regional pattern of potential malaria transmission. Approximately 45% of the world's population presently lives in the climate zone where mosquitoes transmit malaria. Projections by models (that entail necessary simplifying assumptions) indicate that the geographical zone of potential malaria transmission in response to world temperature increases at the upper part of the IPCC-projected range (3–5°C by 2100) would increase to approximately 60% by the latter half of the next century. This possible extension in potential transmission area would encroach most on temperate regions. However, actual climate-related increases in malaria incidence (50 to 80 million additional annual cases) would occur primarily in tropical, subtropical, and less well-protected temperate-zone populations currently at the margins of endemically infected areas.

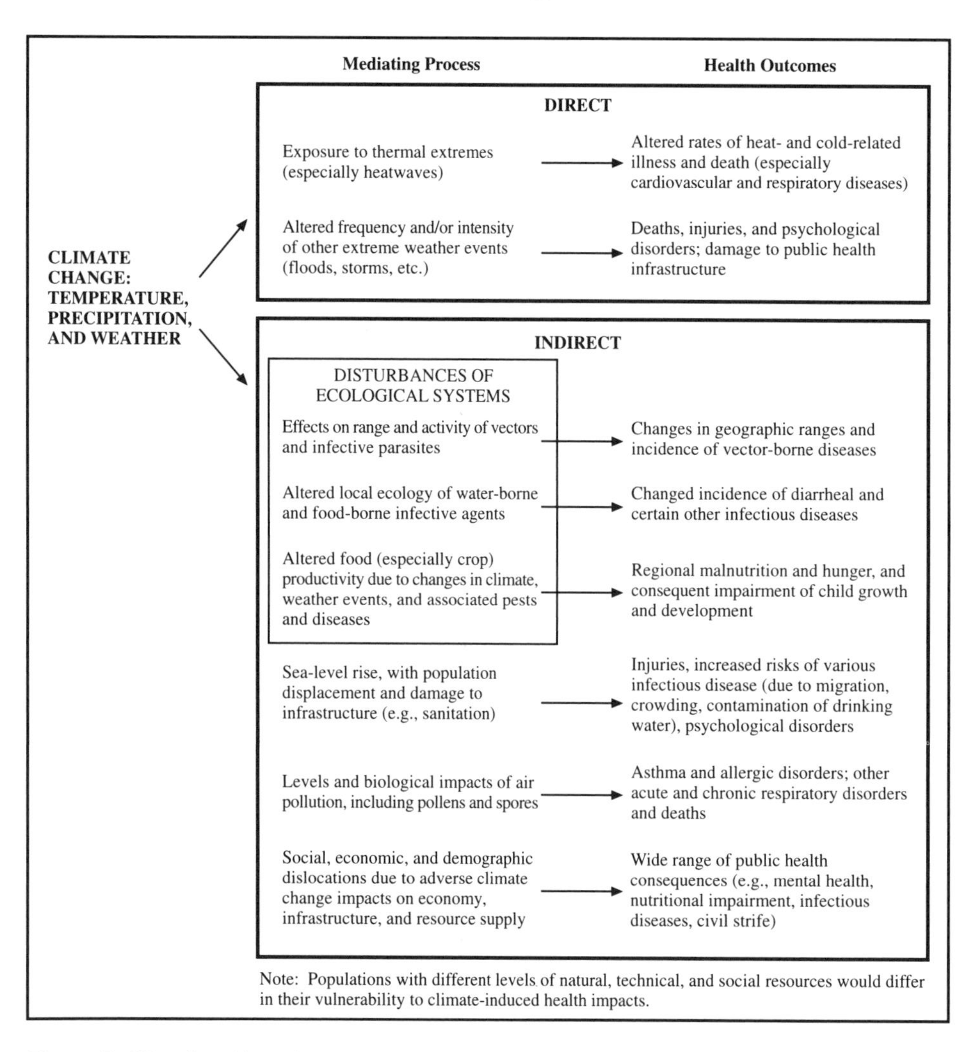

Figure 5: Ways in which climate change can affect human health.

Some increases in non-vector-borne infectious diseases—such as salmonellosis, cholera, and other food- and water-related infections could occur—particularly in tropical and subtropical regions because of climatic impacts on water distribution, temperature, and microorganism proliferation.

Other likely indirect effects include increases in asthma, allergic disorders, cardiorespiratory diseases, and associated deaths. These might result from climate-induced changes in pollens and spores, and from temperature increases that enhance the formation, persistence, and respiratory impact of certain air pollutants. Exposure to air pollution and stressful weather events combine to increase the likelihood of morbidity and mortality.

Though still uncertain, the regional effects of climate change upon agricultural, animal, and fisheries productivity could increase the local prevalence of hunger, malnutrition, and their long-term health impairments, especially in children. Limitations on freshwater supplies also will have human health consequences. A range of adverse public health effects would result from physical and demographic disruptions due to sea-level rise.

Various technological, organizational, and behavioral adaptations would lessen these adverse effects. These include protective technology (e.g., housing, air conditioning, water purification, and vaccination), disaster preparedness, and appropriate health care. However, the tropical and subtropical countries at highest risk from many of these impacts may still lack adequate resources for adaptation.

4. Options to Reduce Emissions and Enhance Sinks of Greenhouse Gases: Mitigation

4.1. Common Themes and Conclusions

Significant reductions in net greenhouse gas emissions are technically possible and can be economically feasible. These reductions can be achieved by employing an extensive array of technologies and policy measures that accelerate technology development, diffusion, and transfer in all sectors, including the energy, industry, transportation, residential/commercial, and agricultural/forestry sectors. By the year 2100, the world's commercial energy system in effect will be replaced at least twice, offering opportunities to change the energy system without premature retirement of capital stock; significant amounts of capital stock in the industrial, commercial, residential, and agricultural/forestry sectors also will be replaced. These cycles of capital replacement provide opportunities to use new, better performing technologies. It should be noted that the analyses of Working Group II do not attempt to quantify potential macroeconomic consequences that may be associated with mitigation measures. Discussion of macroeconomic analyses is found in the contribution of IPCC Working Group III to the Second Assessment Report. The degree to which technical potential and cost-effectiveness are realized is dependent on initiatives to counter lack of information and overcome cultural, institutional, legal, financial, and economic barriers that can hinder diffusion of technology or behavioral changes. The pursuit of mitigation options can be carried out within the limits of sustainable development criteria. Social and environmental criteria not related to greenhouse gas emissions abatement could, however, restrict the ultimate potential of each of the options.

- **Numerous technology options are available for reducing emissions and enhancing sinks**. Promising technologies and measures follow:
  - *Energy demand.* The technical potential for best practice energy technologies to improve energy efficiency relative to present average practice is large; 10–30% efficiency gains at little or no cost in many parts of the world, through conservation measures and improved management practices, and 50–60% is possible, provided that relevant technologies and financing are available.
  - *Industry.* Short-term emissions reductions of about 25% can be obtained in industrialized nations by improving efficiency, recycling materials, and implementing "industrial ecology" practices that use less energy and fewer materials.
  - *Transportation.* Emissions reductions of up to 40% by 2025 could be achieved by changing vehicle engineering to use more efficient drive trains and materials; reducing the size of vehicles; switching to alternative fuels; reducing the level of passenger and freight transport activity by altering land-use patterns, transport systems, mobility patterns, and lifestyles; and shifting to less energy-intensive transportation modes.
  - *Commercial/residential sector.* Cuts of 50% in the projected growth in emissions over the next 35 years—and deeper cuts in the longer term—could be achieved by more efficient lighting, appliances, and space-conditioning systems; reduced heat transfers through walls, ceilings, and windows; and modern control systems such as automatic sensors.
  - *Energy supply.* Options include more efficient conversion of fossil fuels (from the present world average for electric power generation of about 30% to more than 60% in the longer term); switching from high- to low-carbon fossil fuels (coal to oil to gas); decarbonization of flue gases and fuels, coupled with CO_2 storage; increasing the use of nuclear energy; and increased use of modern renewable sources of energy (e.g., biomass for production of electricity and liquid/gaseous fuels, wind, and solar). In the longer term, renewable sources of energy could meet the major part of the world's demand for energy as technological advances offer new opportunities and declining costs.
  - *Land management.* A number of measures could conserve and sequester substantial amounts of carbon (approximately 60–90 Gt C in the forestry sector alone) over the next 50 years, including slowing deforestation, enhancing natural forest generation, establishing tree plantations, and promoting agroforestry. Significant additional amounts could be sequestered by altering management of agricultural soils and rangelands, and by restoring degraded agricultural lands and rangelands. Other practices, such as improving efficiency of fertilizer use or the diet of domesticated ruminants, could reduce emissions of other greenhouse gases such as CH_4 and N_2O.
- **Policies can accelerate reductions in greenhouse gas emissions**. Governments can choose policies that facilitate the penetration of less carbon-intensive technologies and modified consumption patterns. Indeed, many countries have extensive experience with a variety of policies that can accelerate the adoption of such technologies. This experience comes from efforts over the past 20 to 30 years to achieve improved energy

efficiency, reduce the environmental impacts of agricultural policies, and meet conservation and environmental goals unrelated to climate change. Many of these policies appear potentially useful for reducing greenhouse gas emissions. They include energy pricing strategies; changes in agricultural subsidies; provisions for accelerated depreciation and reduced costs for the consumer; tradable emissions permits; negotiated agreements with industry; utility demand-side management programs; regulatory programs, including minimum energy-efficiency standards; market pull and demonstration programs that stimulate the development and application of advanced technologies; and product labeling. The optimum mix of policies will vary from country to country; each nation needs to tailor its policies for local situations and develop them through consultation with those affected. Analysis of the historical experience of different countries with various policy instruments can provide guidance on their strengths and weaknesses.

- **Success is most likely if there are multiple benefits**. Actions to reduce greenhouse gas emissions appear more easily implemented when they are designed to address other concerns that impede sustainable development (e.g., air pollution, traffic congestion, soil erosion).
- **Commitment to further research is essential**. Developing technologies that will reduce greenhouse gas emissions and enhance greenhouse gas sinks—as well as understanding the barriers that inhibit their diffusion into the marketplace—requires a continuing commitment to research.

4.2. *Energy, Industrial Process, and Human Settlement Emissions*

The production, conversion, and end-use of fossil fuel energy results in significant atmospheric releases of greenhouse gases—in particular, CO_2 and CH_4. In the published literature, different methods and conventions are used to characterize energy consumption. These conventions differ, for example, according to their definition of sectors and their treatment of energy forms. Based on aggregated national energy balances, 385 EJ of primary energy was consumed in the world in 1990, resulting in the release of 6 Gt C as CO_2. Of this, 279 EJ was delivered to end users, accounting for 3.7 Gt of carbon emissions as CO_2 at the point of consumption. The remaining 106 EJ was used in energy conversion and distribution, accounting for 2.3 Gt C emissions as CO_2. In 1990, the three largest sectors of energy consumption were industry (45% of total CO_2 releases), residential/commercial (29%), and transport (21%). Of these, transport sector energy use and related CO_2 emissions have been the most rapidly growing over the past 2 decades. The combustion of fossil fuels ranks as the primary human-generated source of SO_2 (the sulfate aerosol precursor), as well as other air pollutants.

Global energy demand has grown at an average annual rate of approximately 2% for almost 2 centuries, although energy demand growth varies considerably over time and between different regions. During that time, the mix of fuels has changed dramatically (see Figure 6). Figure 7 depicts total energy-related emissions by major world region. The Organisation for Economic Cooperation and Development (OECD) nations have been and remain major energy users and fossil fuel CO_2 emitters, although their share of global fossil fuel carbon emissions has been declining. The contribution to global fossil fuel carbon emissions by the former Soviet Union and Eastern Europe (FSU/EE) has grown, although recent economic restructuring has reduced emissions. The diverse developing nations, taken as a group, still account for a smaller fraction of total global CO_2 emissions than industrialized nations (OECD and FSU/EE) due to their lower per capita emission rates (see Figure 8). Most projections indicate that with forecast rates of population and economic growth, the developing world will increase its share of CO_2 emissions as its standard of living increases and as population grows.

Oil and gas reserves worldwide contain approximately 200 Gt C and coal reserves about 600 Gt C. Estimates place remaining, ultimately recoverable, fossil fuel resources at some 4,000 Gt C, roughly three-quarters of it in coal and the rest in conventional and unconventional oil and gas. Atmospheric content totaled about 750 Gt C in 1990. Exhaustion of fossil resources therefore offers no near-term physical-barrier solution to the emissions of energy-related CO_2. The carbon-to-energy intensity ratio of different fossil fuels varies by a factor of almost two: Coal contains 25 Mt C/EJ; oil, 20 Mt C/EJ; and natural gas, 15 Mt C/EJ.Renewable energy sources are sufficiently abundant that they potentially could provide all of the world's energy needs foreseen over the next century. However, economics, technology development, and other practical constraints limit the rate at which the use of renewable energy can expand.

The prices of fuels and electricity influence energy use and fuel choice in all sectors. In many instances, prices do not reflect the full social costs of providing energy; subsidies and other

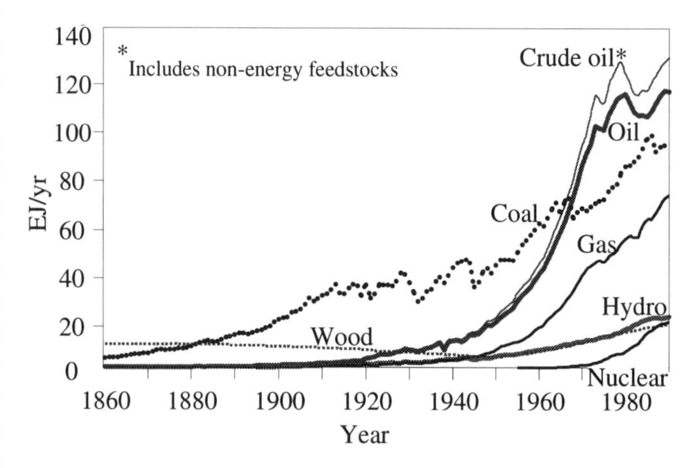

Figure 6: Global primary energy consumption by source, and total in EJ/yr (data for crude oil include non-energy feedstocks). Sources: BP, various volumes; IEA, 1993; Marchetti and Nakicenovic, 1979 (see the Energy Primer for complete citations).

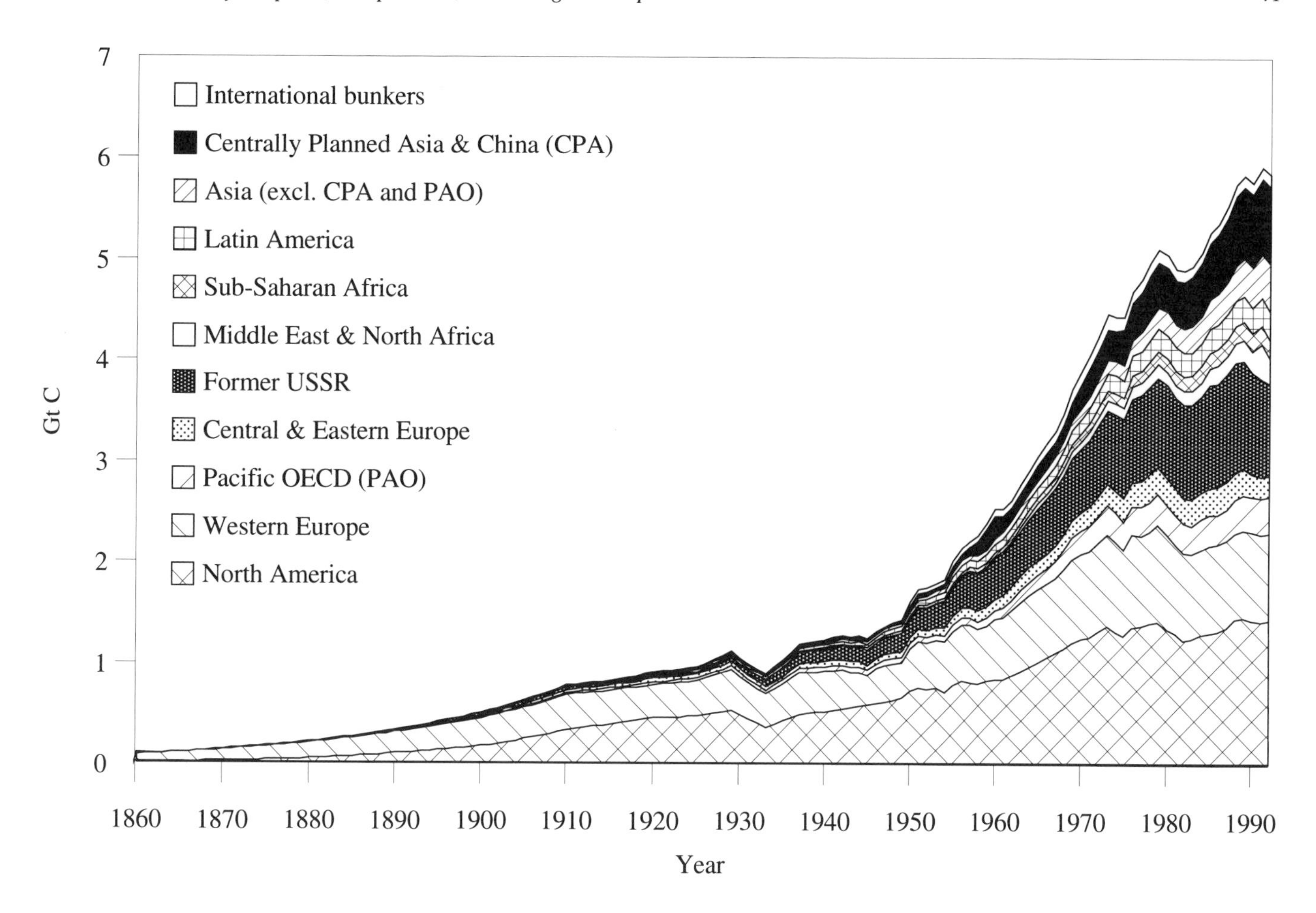

Figure 7: Global energy-related CO_2 emissions by major world region in Gt C/yr. Sources: Keeling, 1994; Marland *et al.*, 1994; Grübler and Nakicenovic, 1992; Etemad and Luciani, 1991; Fujii, 1990; UN, 1952 (see the Energy Primer for complete citations).

interventions restrict the choices made by energy suppliers and users.Price rationalization, voluntary agreements, regulations, and information programs have successfully helped to accelerate the use of more energy-efficient technologies and practices in various areas of end-use, especially in the residential and transport sectors. These measures—when implemented as part of the normal replacement cycles of the world's energy supply infrastructure and energy-use equipment—offer the greatest potential to change the technology and systems now used.

4.2.1. Energy Demand

Numerous studies have indicated that 10–30% energy efficiency gains above present levels are feasible at little or no net cost in many parts of the world through technical conservation measures and improved management practices over the next 2 to 3 decades. Using technologies that presently yield the highest output of energy services for a given input of energy, efficiency gains of 50–60% would be technically feasible in many countries over the same time period. Achieving these potentials will depend on future cost reductions, financing, and technology transfer, as well as measures to overcome a variety of non-technical barriers. The potential for greenhouse gas emission reductions exceeds the potential for energy use efficiency because of the possibility of switching fuels and energy sources. Because energy use is growing world-wide, even replacing current technology with more-efficient technology could still lead to an absolute increase in CO_2 emissions in the future.

In 1992, the IPCC produced six scenarios (IS92a-f) of future energy use and associated greenhouse gas emissions. These scenarios provide a wide range of possible future greenhouse gas emission levels, without mitigation measures. In the Second Assessment Report, future energy use has been reexamined on a more detailed sectoral basis, both with and without new mitigation measures, based on existing studies. Despite different assessment approaches, the resulting ranges of energy consumption increases to 2025 without new measures are broadly consistent with those of IS92. If past trends continue, greenhouse gas emissions will grow more slowly than energy use, except in the transport sector.

The following paragraphs summarize energy efficiency improvement potentials estimated in the IPCC Second Assessment Report. Strong policy measures would be required to achieve these potentials. Energy-related greenhouse gas emission reductions depend on the source of the energy, but

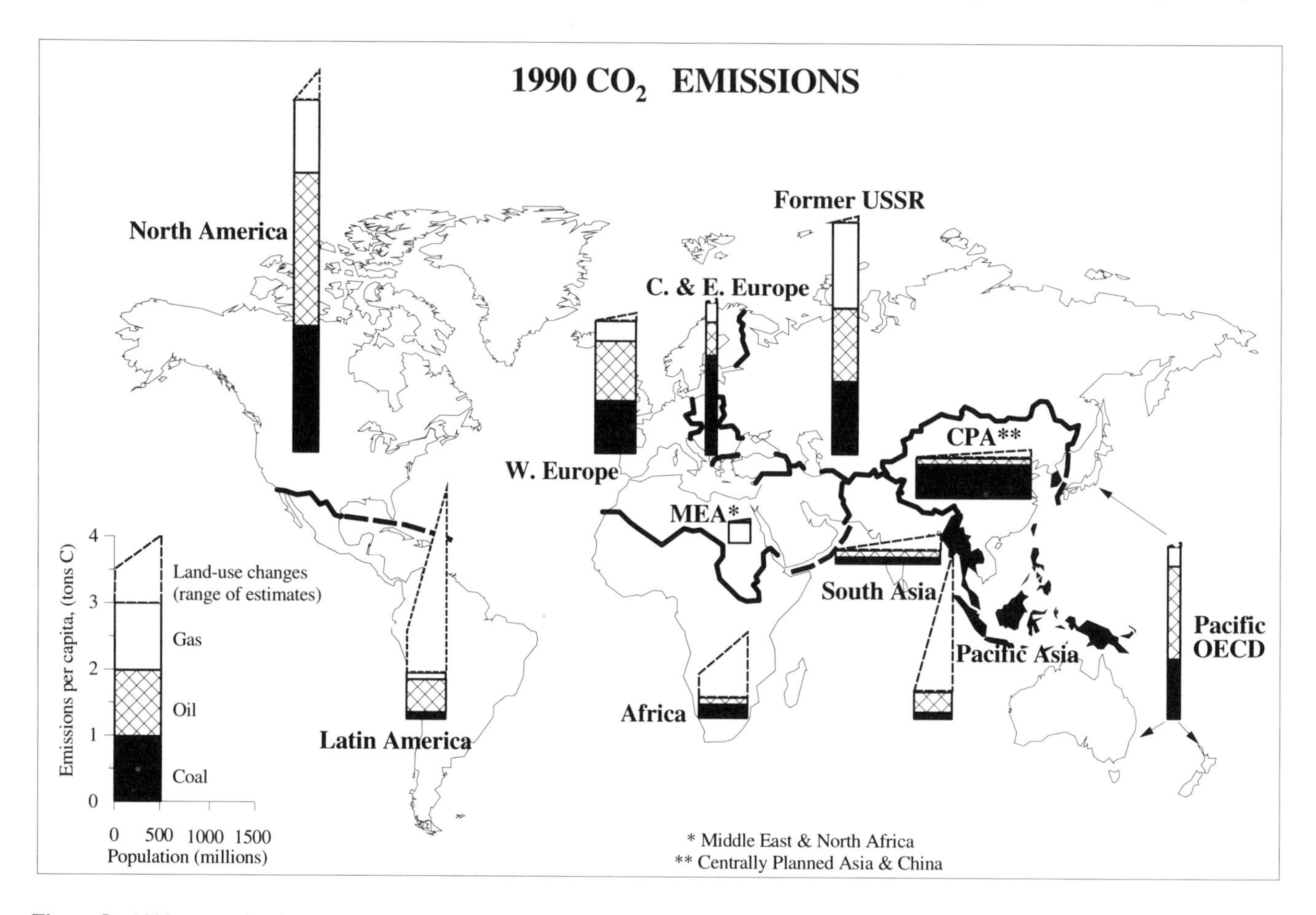

Figure 8: 1990 per capita CO_2 emissions by region and source, fossil fuels, and range for biota sources (includes sustainable use of biomass that does not contribute to atmospheric concentration increase). Sources: IEA, 1993; Marland *et al.*, 1994; Nakicenovic *et al.*, 1993; Subak *et al.*, 1993; IPCC, 1990, 1992; Bos *et al.*, 1992; Houghton *et al.*, 1987 (see the Energy Primer for complete citations).

reductions in energy use will, in general, lead to reduced greenhouse gas emissions.

4.2.1.1. *Emissions mitigation in industry*

Energy use in 1990 is estimated to be 98–117 EJ, and is projected to grow to 140–242 EJ in 2025 without new measures. Developing countries now account for only about one-quarter of global industrial final energy use, but projections put their share of industrial energy demand growth over the next century at more than 90%.

A few basic processes, including the production of iron and steel, chemicals, building materials, and food, account for more than half of all energy use in the industrial sector. Industry has reduced energy intensity impressively over the past 2 decades. In some countries, improvements in energy efficiency have permitted major increases in production with little or no increase in energy use. In contrast to the industrial emissions of developing countries, which continue to increase as their economies grow, developed countries' industrial-sector emissions have stabilized or declined during the past 2 decades. Even the former Soviet Union's industrial emissions have remained stable for a decade. These changes have occurred without any climate policies in place and essentially have arisen from a combination of economic restructuring, shedding of some energy-intensive industries, and gains in the energy efficiency of industrial processes. These developments have combined to decrease overall primary energy intensity (see Figure 9), resulting in stabilized or decreased CO_2 emissions in many industrialized countries (see Figure 10). Because each nation has a unique set of resources, labor, and capital that will influence its development path, this historical perspective does not suggest a particular outcome for developing countries. It does imply that the relationship between CO_2 emissions and GDP is unlikely to retain its present course, given the technological opportunities currently available.

Opportunities now exist to use advanced technologies to reduce emissions significantly in each of the major energy-consuming industries. The short-term potential for energy-efficiency improvements in the manufacturing sector of major industrial nations is around 25%. The potential for greenhouse gas emission reductions is larger. The application of available, highly efficient technologies and practices could significantly reduce industrial energy demand growth.

Efficient use of materials can lower industrial greenhouse gas emissions, and recycling materials can reduce energy

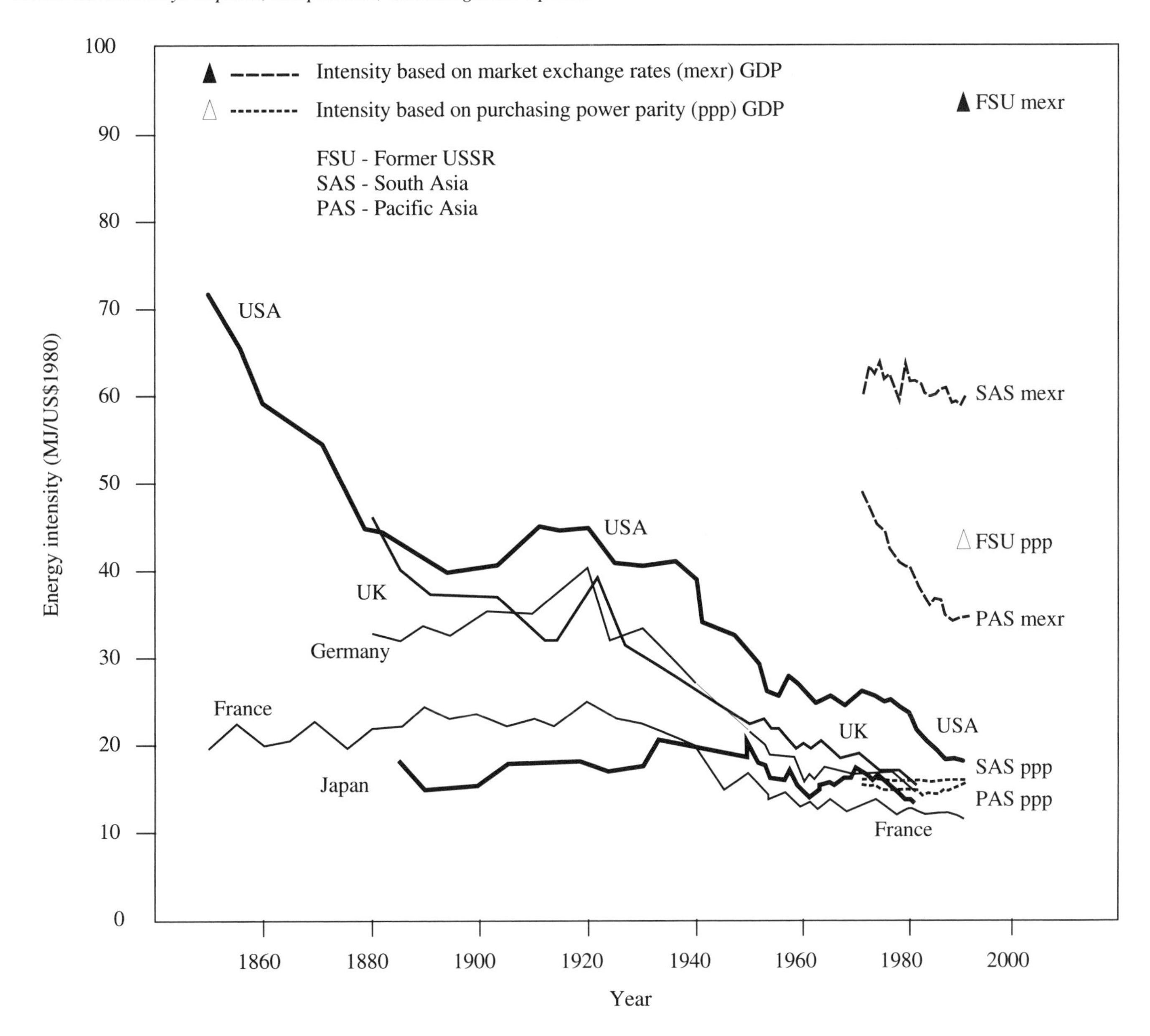

Figure 9: Primary energy intensity (including wood and biomass) of value added in MJ per constant GDP in 1980 dollars [at market exchange rates (mexr) and purchasing power parities (ppp)]. The countries shown account for approximately 80% of energy-related CO_2 emissions. Source: Grübler, 1991 (see the Energy Primer for a complete citation).

requirements and energy-related emissions substantially (see Figure 11). Recycling iron and steel scrap, for example, could cut the energy required per ton of steel produced to half the current level even in Japan, the world's most efficient steel-producing nation. Not all steel scrap is recyclable, however, so recycling provides no panacea. Using less energy-intensive materials offers another option—for example, substituting wood for concrete and other materials as possible.

Some technologies—such as electric motor drives, heating, and evaporation—are used across many industries, thus offering large opportunities for energy savings. Electric motor drives, for example, use more than half of all electricity in the industrial sector in many developing countries, including China and Brazil. Investments in motor-speed controls and more efficient motor components can reduce electricity use in industrial motor applications, with an internal rate of return on investment of 30 to 50%.

Integrating industrial and residential-commercial energy use by establishing energy-management systems that utilize an "energy cascade" presents promising opportunities for increased efficiency—especially in newly industrialized nations, where a systematic approach to efficiency improvements is particularly important. An energy cascade uses successively lower temperature industrial waste heat in a variety of other industrial, residential, and commercial district heating and cooling applications.

Limiting industrial emissions of CO_2 and other greenhouse gases, such as halocarbons, CH_4, and N_2O associated with industrial processes, can also play an important role.

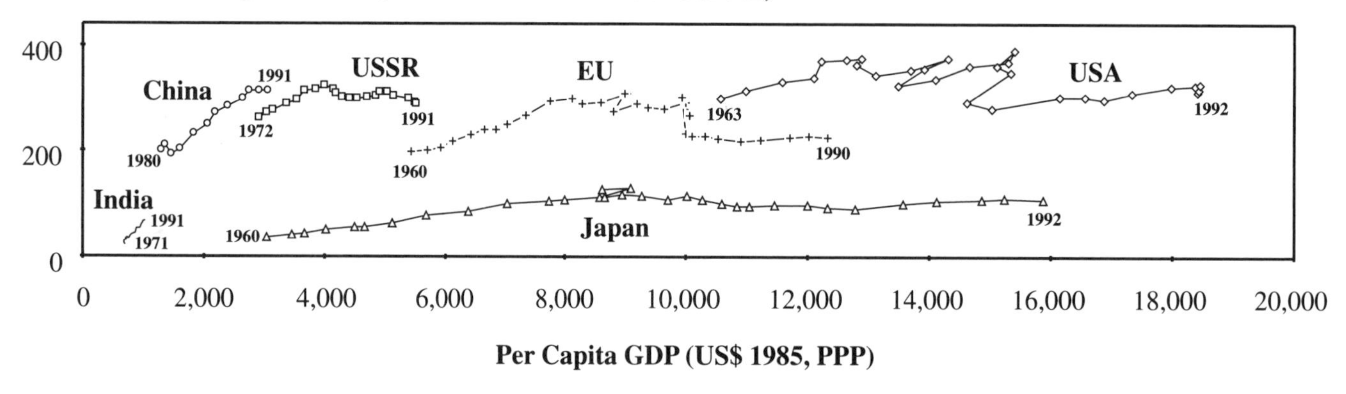

Figure 10: Fossil fuel CO_2 development path for the industrial sectors of the United States, the 15 nations that now comprise the European Union (less the former East Germany), Japan, China, India, and the former Soviet Union. The industrial sector is as defined by OECD, plus CO_2 associated with refineries and the fraction of electricity that is used by industry. The CO_2 values are from OECD (1994), and the ppp values are from Sommers and Heston (1991, 1994); see Chapter 20 for complete citations.

Identifying all the benefits, costs, and potentials for reducing both process and energy greenhouse gases in the industrial sector requires an examination of all aspects of resource use. Developing the capability to carry out full "industrial ecology" analyses represents a major research need.

Implementing more energy-efficient industrial technologies will depend on many factors. Chief among them are financing, energy and materials pricing, market imperfections, research and development, and the extent to which technology transfer occurs from developed countries to developing nations and those with economies in transition.

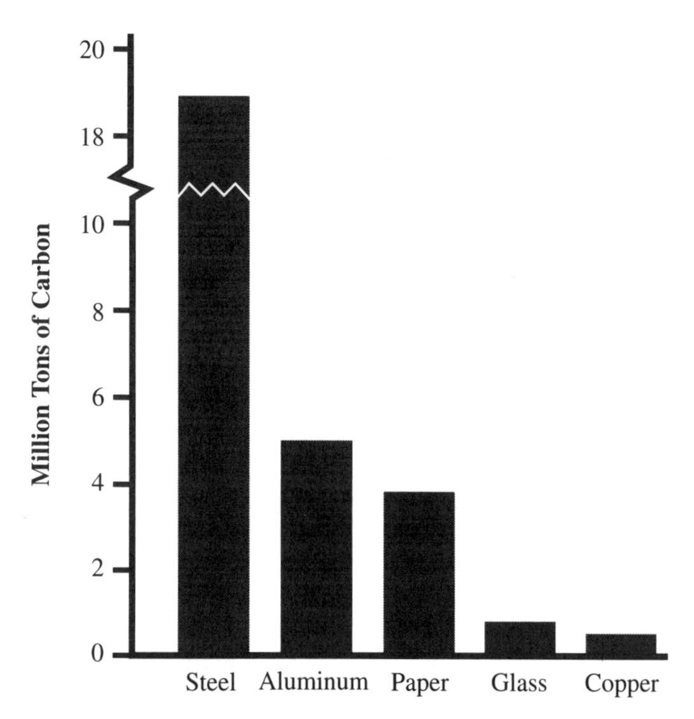

Figure 11: Tons of carbon avoided by OECD countries by increasing recycling by 10%—from *Long-Term Strategies for Mitigating Global Warming* (IIASA, 1993).

4.2.1.2. Emissions mitigation in transportation

Transport energy use in 1990 is estimated to be 61–65 EJ, and is projected to grow to 90–140 EJ in 2025 without new measures (see Figure 12). Energy use by cars, aircraft, and heavy trucks is growing particularly rapidly. Although industrialized countries account for about 75% of current energy use and greenhouse gas emissions from this sector, the greatest growth is expected in developing countries and transition economies. By 2025, these countries could generate the majority of transport-related emissions.

Transportation energy demand historically has been linked closely to GDP growth, although there is also a strong negative correlation with fuel prices (see Figures 13 and 14). However, projected energy use in 2025 could be reduced by about a third to 60–100 EJ through vehicles using very efficient drive-trains, light-weight construction and low-air-resistance design, without compromising comfort and performance. Further energy-use reductions are possible through the use of smaller vehicles; altered land-use patterns, transport systems, mobility patterns, and lifestyles; and shifting to less energy-intensive transport modes. Greenhouse gas emissions per unit of energy used could be reduced through the use of alternative fuels and electricity from renewable sources. These measures, taken together, provide the opportunity for reducing global transport emissions by as much as 40% of projected emissions by 2025. Actions to reduce greenhouse gas emissions from transport can simultaneously address other problems, such as local air pollution.

Realizing these opportunities seems unlikely, however, without new policies and measures in many countries. We have learned much already about the effective use of policies, such as fuel and vehicle taxes and fuel-economy standards, to encourage energy-efficiency improvements. Yet developing cheap, light-weight, recyclable materials and advanced propulsion and vehicle-control systems will require continued research. Policies that affect traffic volume also play an important role. These include road tolls, restriction of car access and parking

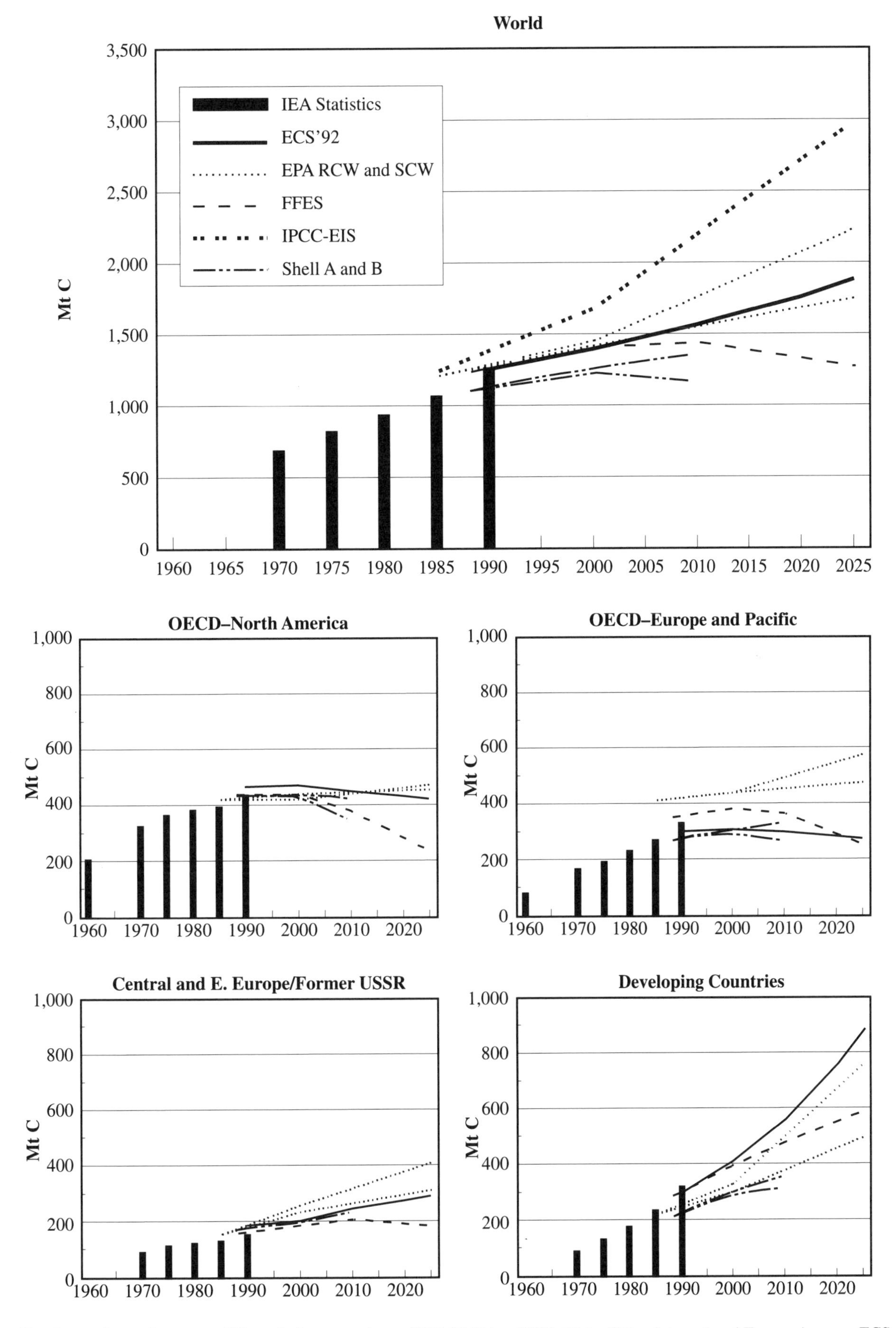

Figure 12: Comparison of transport CO_2 emission scenarios to 2025 (Grübler, 1993). Note: IEA = International Energy Agency; ECS = Environmentally Compatible Energy Strategies; RCW = Rapidly Changing World and SCW = Slowly Changing World; FFES = Fossil-Free Energy System; and EIS = Energy Industry System.

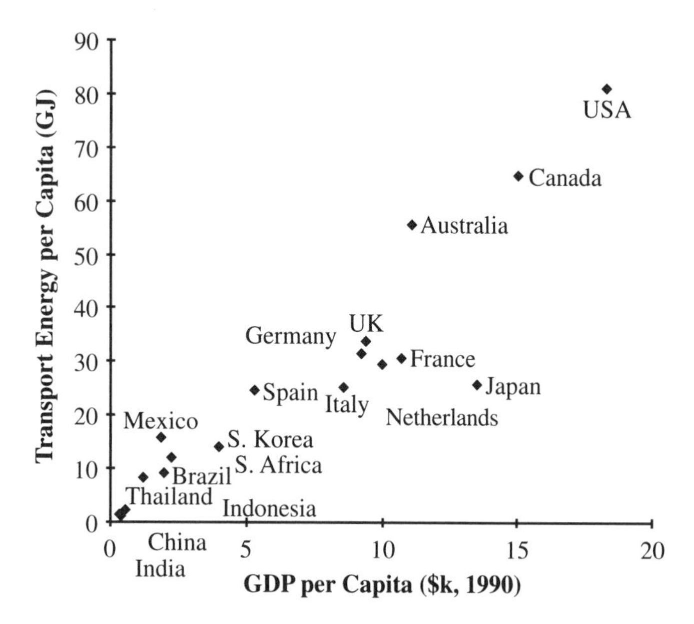

Figure 13: Total transport energy use vs. gross domestic product in 1990, for 18 of the world's largest transport energy users. Excludes Russia, Ukraine, Iran, Saudi Arabia, and Kazakhstan; former West and East Germany data have been combined. Sources: IEA, 1993c, 1993d (see Chapter 21 for complete citations).

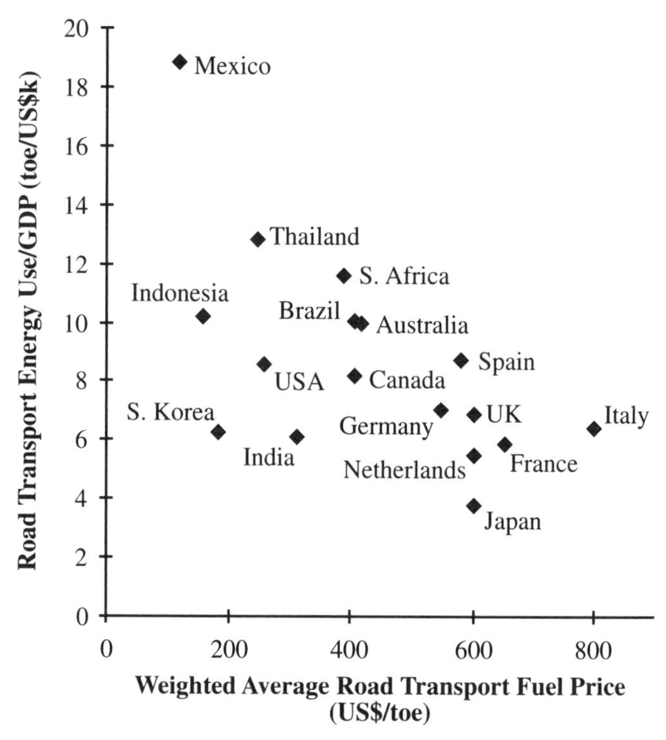

Figure 14: Road transport energy per unit of GDP vs. average fuel price in 1990, for 17 of the world's 20 largest transport energy users. Excludes China, Russia, and Ukraine; former West and East Germany data have been combined. Sources: IEA, 1993c, 1993d, 1994a; ADB, 1994 (see Chapter 21 for complete citations).

in town centers, and the provision of infrastructure for nonmotorized transport in town centers. Several cities in Latin America, Southeast Asia, and Europe have succeeded in stemming growth in car use by employing combined strategies.

Transport plays an increasingly important social and economic role, and measures to reduce emissions may fail if people perceive them as compromising this role. Success in reducing carbon emissions will require integrated approaches and probably will depend on simultaneously addressing other problems such as congestion and air pollution [including emissions of particulates, and of nitrogen oxides (NO_x) and volatile organic compounds (VOCs) that are precursors to tropospheric ozone].

4.2.1.3. *Emissions mitigation in the commercial/residential sector*

Activities in this sector currently account for just over 40% of energy use. Figure 15 shows historical carbon emissions from the developing world, industrialized countries, and the FSU/EE resulting from energy use in the residential and commercial sectors in 1973, 1983, and 1990. Energy use in 1990 is estimated to be 100 EJ, and is projected to grow to 165–205 EJ in 2025 without new measures. Although industrialized countries currently release about 60% of associated emissions, developing countries and the FSU/EE could account for 80% of all growth in building emissions during the next century.

Projected energy use could be reduced by about a quarter to 126–170 EJ by 2025 without diminishing services through the use of energy efficient technology. The potential for greenhouse gas emission reductions is larger. Technical changes might include more efficient space-conditioning and water-supply systems; reducing heat losses through building structures; more efficient lighting; more efficient appliances; and more efficient computers and supporting equipment. In addition, measures to counter trends toward higher ambient temperatures in urban areas (through increased vegetation, greater reflectivity of roofing and siding materials, and better architectural design) can yield significant reductions in the energy required for heating and air conditioning.

This assessment suggests that three types of activities could significantly reduce the growth of building-related emissions:

- Support for energy-efficiency policies, including pricing strategies; individual, meter-based billing for energy use in multiple-family dwellings; regulatory programs including minimum energy-efficiency standards for buildings and appliances; utility demand-side management programs; and market pull and demonstration programs that stimulate the development and application of advanced technologies
- Enhanced research and development in energy efficiency
- Enhanced training and added support for financing efficiency programs in all countries, but especially in developing countries and transition economies.

Many energy-efficiency policies, information/education measures, and research and development programs—carried out primarily in industrialized countries—have achieved significant reductions in energy use. Although the technical and economic potential for further efficiency improvements is high, effective implementation requires well-designed combinations of financial incentives and other government policies. Strategies to reduce emissions likely will prove more effective if they use well-integrated mixes of policies, tailored for local situations and developed through consultation with and participation by those most affected.

4.2.2. *Mitigating Industrial Process and Human Settlement Emissions*

Process-related greenhouse gases—including CO_2, CH_4, N_2O, halocarbons, and SF_6—are released during manufacturing and industrial processes, such as the production of iron, steel, aluminium, ammonia, cement, and other materials. Large reductions are possible in some cases. Measures include modifying production processes, eliminating solvents, replacing feedstocks or materials substitution, increased recycling, and reduced consumption of greenhouse gas-intensive materials. Capturing and utilizing CH_4 from landfills and sewage treatment facilities, and lowering the leakage rate of halocarbon refrigerants from mobile and stationary sources can also lead to significant greenhouse gas emission reductions.

4.2.3. *Energy Supply*

This assessment focuses on new technologies for capital investment and not on potential retrofitting of existing capital stock to use less carbon-intensive forms of primary energy. It is technically possible to realize deep emissions reductions in the energy supply sector in step with the normal timing of investments to replace infrastructure and equipment as it wears out or becomes obsolete. Many options for achieving deep reductions will also decrease the emissions of SO_2, NO_x, and VOCs. Promising approaches, not ordered according to priority, include the following:

- **Greenhouse gas reductions in the use of fossil fuels**
  - *More-efficient conversion of fossil fuels.* New technology offers considerably increased conversion efficiencies. For example, the efficiency of power production can be increased from the present world average of about 30% to more than 60% in the longer term. Also, the use of combined heat and power production replacing separate production of power and heat—whether for process heat or space heating—offers a significant rise in fuel conversion efficiency.
  - *Switching to low-carbon fossil fuels and suppressing emissions.* Switching from coal to oil or natural gas, and from oil to natural gas, can reduce emissions. Natural gas has the lowest CO_2 emissions per unit of energy of all fossil fuels at about 14 kg C/GJ, compared to oil with about 20 kg C/GJ, and coal with about 25 kg C/GJ. The lower carbon-containing fuels can, in general, be converted with higher efficiency than coal. Large resources of natural gas exist in many areas. New, low-capital-cost, highly efficient, combined-cycle

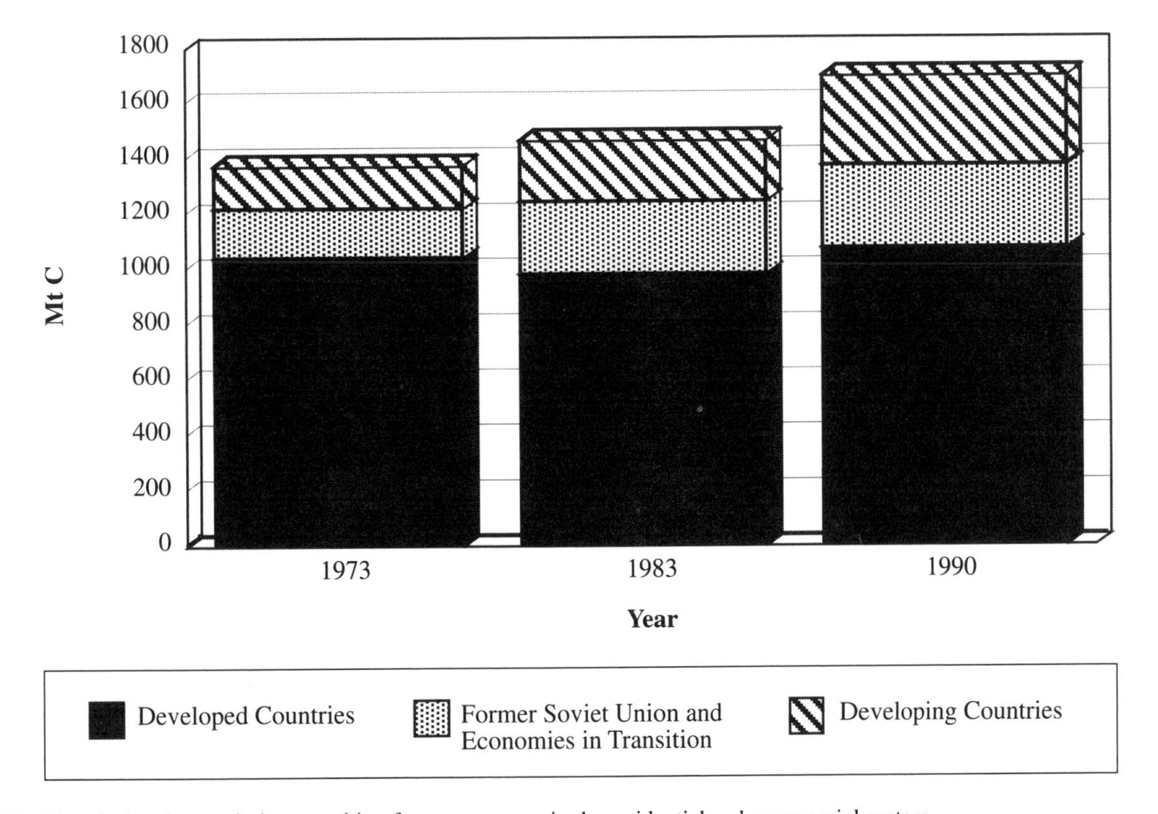

Figure 15: Historical carbon emissions resulting from energy use in the residential and commercial sectors.

technology has reduced electricity costs considerably in some areas. Natural gas could potentially replace oil in the transportation sector. Approaches exist to reduce emissions of CH_4 from natural gas pipelines and emissions of CH_4 and/or CO_2 from oil and gas wells and coal mines.
  - *Decarbonization of flue gases and fuels and CO_2 storage*. The removal and storage of CO_2 from fossil fuel power-station stack gases is feasible, but reduces the conversion efficiency and significantly increases the production cost of electricity. Another approach to decarbonization uses fossil fuel feedstocks to make hydrogen-rich fuels. Both approaches generate a byproduct stream of CO_2 that could be stored, for example, in depleted natural gas fields. The future availability of conversion technologies such as fuel cells that can efficiently use hydrogen would increase the relative attractiveness of the latter approach. For some longer term CO_2 storage options, cost, environmental effects, and efficacy remain largely unknown.
- **Switching to non-fossil fuel sources of energy**
  - *Switching to nuclear energy*. Nuclear energy could replace baseload fossil fuel electricity generation in many parts of the world if generally acceptable responses can be found to concerns such as reactor safety, radioactive-waste transport and disposal, and nuclear proliferation.
  - *Switching to renewable sources of energy*. Solar, biomass, wind, hydro, and geothermal technologies already are widely used. In 1990, renewable sources of energy contributed about 20% of the world's primary energy consumption, most of it fuelwood and hydropower. Technological advances offer new opportunities and declining costs for energy from these sources. In the longer term, renewable sources of energy could meet a major part of the world's demand for energy. Power systems can easily accommodate limited fractions of intermittent generation and, with the addition of fast-responding backup and storage units, also higher fractions. Where biomass is sustainably regrown and used to displace fossil fuels in energy production, net carbon emissions are avoided as the CO_2 released in converting the biomass to energy is again fixed in biomass through photosynthesis. If the development of biomass energy can be carried out in ways that effectively address concerns about other environmental issues and competition with other land uses, biomass could make major contributions in both the electricity and fuels markets, as well as offering prospects of increased rural employment and income.

Future emissions will depend on the nature of the technologies nations choose as they expand and replace existing energy systems.

4.2.4. *Integration of Energy System Mitigation Options*

To assess the potential impact of combinations of individual measures at the energy system level, in contrast to the level of individual technologies, variants of a low CO_2-emitting energy supply system (LESS) are described. The LESS constructions are "thought experiments," exploring possible global energy systems.

The following assumptions were made: World population grows from 5.3 billion in 1990 to 9.5 billion by 2050 and 10.5 billion by 2100. GDP grows 7-fold by 2050 (5-fold and 14-fold in industrialized and developing countries, respectively) and 25-fold by 2100 (13-fold and 70-fold in industrialized and developing countries, respectively), relative to 1990. Because of emphasis on energy efficiency, primary energy consumption rises much more slowly than GDP. The energy supply constructions were made to meet energy demand in (i) projections developed for the IPCC's First Assessment Report in a low energy demand variant, where global primary commercial energy use approximately doubles, with no net change for industrialized countries but a 4.4-fold increase for developing countries from 1990 to 2100; and (ii) a higher energy demand variant, developed in the IPCC IS92a scenario where energy demand quadruples from 1990 to 2100. The energy demand levels of the LESS constructions are consistent with the energy demand mitigation chapters of this Second Assessment Report.

Figure 16 shows combinations of different energy sources to meet changing levels of demand over the next century. The analysis of these variants leads to the following conclusions:

- Deep reductions of CO_2 emissions from energy supply systems are technically possible within 50 to 100 years, using alternative strategies.
- Many combinations of the options identified in this assessment could reduce global CO_2 emissions from fossil fuels from about 6 Gt C in 1990 to about 4 Gt C per year by 2050, and to about 2 Gt C per year by 2100 (see Figure 17). Cumulative CO_2 emissions, from 1990 to 2100, would range from about 450 to about 470 Gt C in the alternative LESS constructions.
- Higher energy efficiency is underscored for achieving deep reductions in CO_2 emissions, for increasing the flexibility of supply-side combinations, and for reducing overall energy system costs.
- Interregional trade in energy grows in the LESS constructions compared to today's levels, expanding sustainable development options for Africa, Latin America, and the Middle East during the next century.

Costs for energy services in each LESS variant relative to costs for conventional energy depend on relative future energy prices, which are uncertain within a wide range, and on the performance and cost characteristics assumed for alternative technologies. However, within the wide range of future energy prices, one or more of the variants would plausibly be capable of providing the demanded energy services at estimated costs

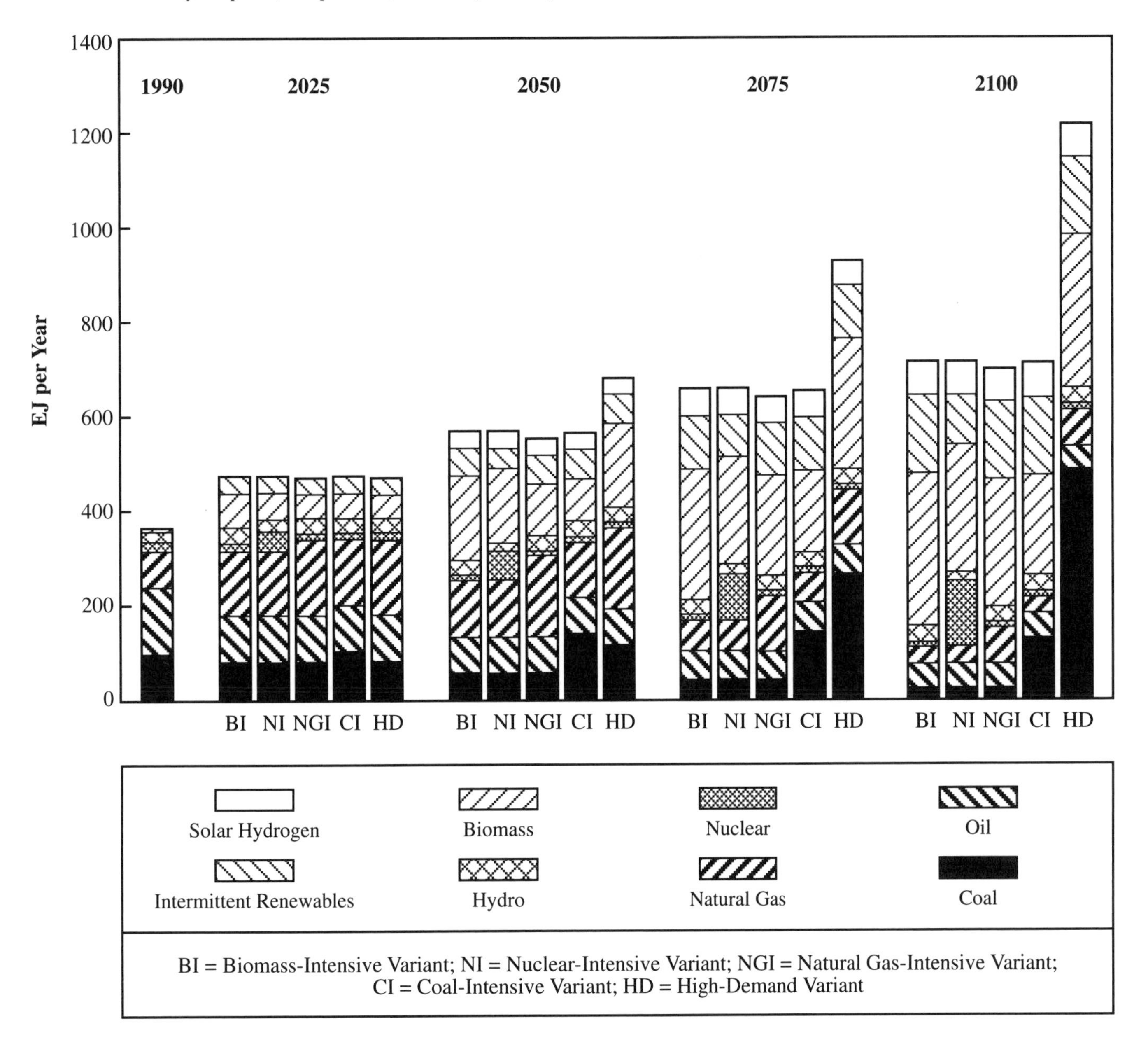

Figure 16: Global primary energy use for alternative Low CO_2-Emitting Energy Supply System (LESS) constructions: Alternatives for meeting different energy demand levels over time, using various fuel mixes.

that are approximately the same as estimated future costs for current conventional energy. It is not possible to identify a least-cost future energy system for the longer term, as the relative costs of options depend on resource constraints and technological opportunities that are imperfectly known, and on actions by governments and the private sector.

The literature provides strong support for the feasibility of achieving the performance and cost characteristics assumed for energy technologies in the LESS constructions, within the next 2 decades, though it is impossible to be certain until the research and development is complete, and the technologies have been tested in the market. Moreover, these performance and cost characteristics cannot be achieved without a strong and sustained investment in research, development, and demonstration. Many of the technologies being developed would need initial support to enter the market, and to reach sufficient volume to lower costs to become competitive.

Market penetration and continued acceptability of different energy technologies ultimately depends on their relative cost, performance (including environmental performance), institutional arrangements, and regulations and policies. Because costs vary by location and application, the wide variety of circumstances creates initial opportunities for new technologies to enter the market. Deeper understanding of the opportunities for emissions reductions would require more detailed analysis of options, taking into account local conditions.

Because of the large number of options, there is flexibility as to how the energy supply system could evolve, and paths of energy system development could be influenced by considerations

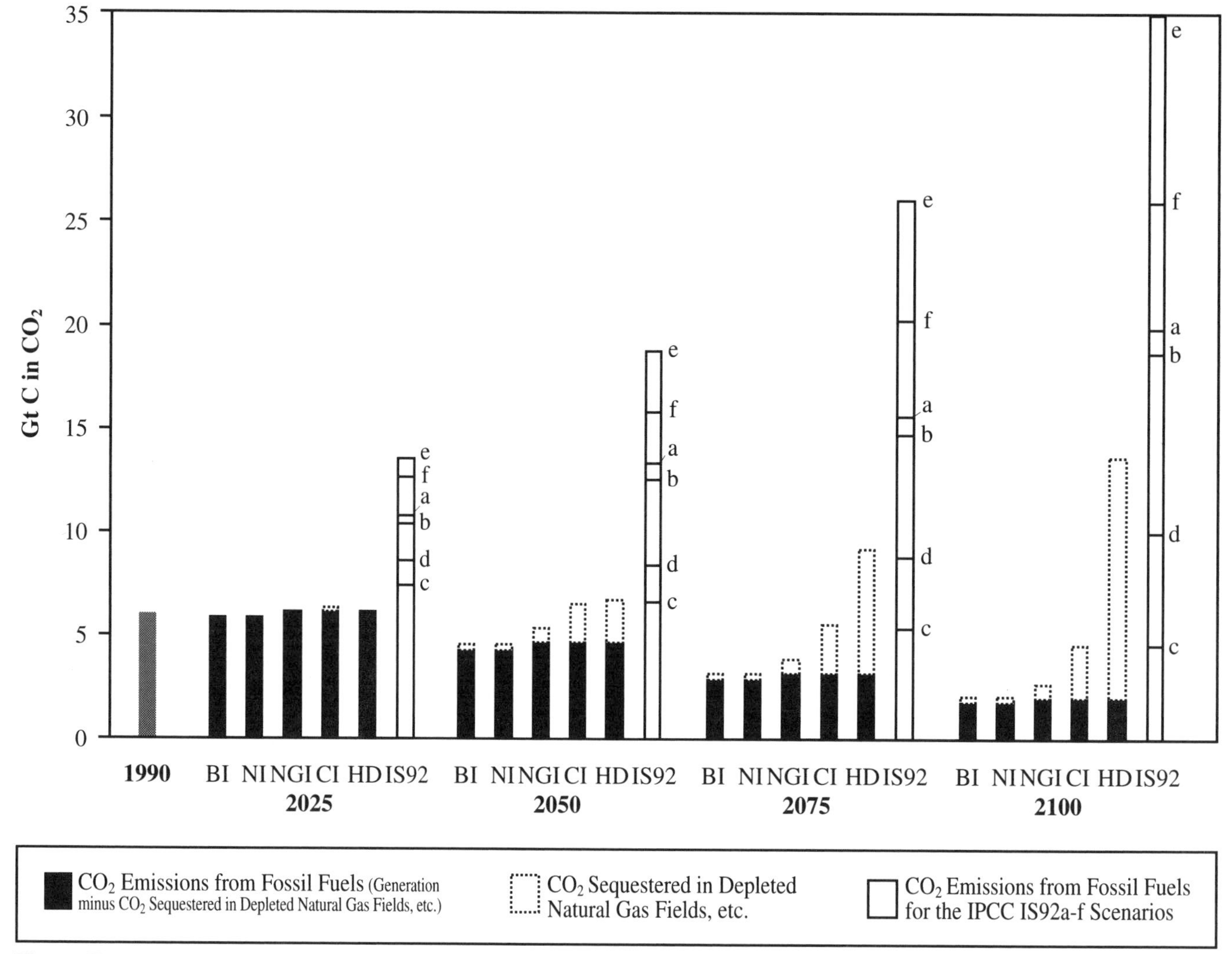

Figure 17: Annual CO_2 emissions from fossil fuels for alternative LESS constructions, with comparison to the IPCC IS92a-f scenarios (see Figure 16 for acronym definitions).

other than climate change, including political, environmental (especially indoor and urban air pollution, acidification, and land restoration) and socioeconomic circumstances.

4.3. Agriculture, Rangelands, and Forestry

Management of agricultural lands, rangelands, and forests can play an important role in reducing current emissions and/or enhancing the sinks of CO_2, CH_4, and N_2O. Measures to reduce emissions and sequester atmospheric carbon include slowing deforestation, enhancing natural forest generation, establishing tree plantations, promoting agroforestry, altering management of agricultural soils and rangelands, restoring degraded agricultural lands and rangelands, and improving the diet of ruminants. Although these are demonstrated, effective measures, a number of important uncertainties linger regarding their global potential to reduce emissions or sequester carbon. The net amount of carbon per unit area conserved or sequestered in living biomass under a particular forest-management practice and present climate is relatively well-understood. The most important uncertainties associated with estimating a global value are (i) the amount of land suitable and available for forestation, regeneration, and/or restoration programs; (ii) the rate at which tropical deforestation can actually be reduced; (iii) the long-term use (security) of these lands; and (iv) the continued suitability of some practices for particular locations given the possibility of changes in temperature, water availability, and so forth under climate change. Management options vary by country, social system, and ecosystem type.

Proposed options for conserving and sequestering carbon and reducing other greenhouse gas emissions in the forestry and agriculture sectors are consistent with other objectives of land management—such as sustainable development, industrial wood and fuelwood production, traditional forest uses, protection of other natural resources (e.g., biodiversity, soil, water), recreation, and increasing agricultural productivity.

4.3.1. Carbon Dioxide

Since the 1990 IPCC assessment, significant new information has emerged concerning the potential of forests to conserve and sequester carbon, as well as the costs of forestry programs to promote this carbon storage. Biomass for the production of electricity or liquid fuels, continuously harvested and regrown in a sustainable manner, would avoid the release of fossil carbon. In the long run, this would offer a more efficient strategy than one based on carbon storage in vegetation and soils, which would saturate with time. The development of this mitigation option will depend critically upon the competitiveness of biomass. However, managing forests to conserve and increase carbon storage offers an effective mitigation option during the transition period of many decades necessary to stabilize atmospheric concentrations of carbon.

In the forestry sector, assuming the present climate and no change in the estimated area of available lands, the cumulative amount of carbon that could feasibly be conserved and sequestered through establishment of plantations and agroforestry, forest regeneration, sustaining existing forest cover, and slowing deforestation over the period 1995–2050 ranges between 60 and 90 Gt C. The literature indicates that the tropics have the potential to conserve and sequester the largest quantity of carbon (80% of the global total in the forestry sector), with more than half due to promoting forest regeneration and slowing deforestation. Tropical America has the largest potential for carbon conservation and sequestration (46% of tropical total), followed by tropical Asia (34%) and tropical Africa (20%). The temperate and boreal zones could sequester about 20% of the global total, mainly in the United States, temperate Asia, the former Soviet Union, China, and New Zealand. Altering the climate and land-use assumptions—for example, by increasing demand for agricultural land in the tropics or accounting for potential consequences of climate change—would reduce these estimates significantly.

Estimates of the full costs of conserving or sequestering carbon in forests should include land values as well as capital, infrastructure, and other costs. However, many of these costs are particularly difficult to generalize given the amenity value of forests, their importance in traditional economies, and their significance in maintaining regional environments and biodiversity. Costs for conserving and sequestering carbon in biomass and soil are estimated to range widely, but can be competitive with other mitigation options. Factors affecting costs include opportunity costs of land, initial costs of planting and establisment, costs of nurseries, the cost of annual maintenance and monitoring, and transaction costs. Direct and indirect benefits will vary with national circumstances and could offset the costs. Additional amounts of carbon could be sequestered in agricultural and rangeland soils over a 50-year period by improved management practices, although less is currently known about the global potential in this sector.

4.3.2. *Methane and Nitrous Oxide*

Nations can achieve significant decreases in CH_4 emissions from agriculture through improved nutrition of ruminant animals and better management of rice paddies (e.g., irrigation, nutrients, new cultivars). Altering the treatment and management of animal wastes and reducing agricultural biomass burning also will decrease CH_4 releases. Combining these practices could reduce CH_4 emissions from agriculture by 25 to 100 Mt/yr. Agricultural sources of N_2O include mineral fertilizers, legume cropping, animal waste, and biomass burning. Using presently available techniques to improve the efficiency of fertilizer and manure could reduce agricultural emissions by 0.3 to 0.9 Mt of nitrogen per year.

4.4. *Cross-Sectoral Issues*

Cross-sectoral assessment of different combinations of mitigation options focuses on the interactions of the full range of technologies and practices that are potentially capable of reducing emissions of greenhouse gases or sequestering carbon.-Current analysis suggests the following:

- *Competing uses of land, water, and other natural resources.* A growing population and expanding economy will increase the demand for land and other natural resources needed to provide, *inter alia*, food, fiber, forest products, and recreation services. Climate change will interact with the resulting intensified patterns of resource use. Land and other resources could also be required for mitigation of greenhouse gas emissions. Agricultural productivity improvements throughout the world and especially in developing countries would increase availability of land for production of biomass energy.
- *Geoengineering options.* Some geoengineering approaches to counterbalance greenhouse-induced climate change have been suggested (e.g., putting solar radiation reflectors in space, or injecting sulfate aerosols into the atmosphere to mimic the cooling influence of volcanic eruptions). Such approaches generally are likely to be ineffective, expensive to sustain, and/or to have serious environmental and other effects which are in many cases poorly understood.

4.5. *Policy Instruments*

Mitigation depends on reducing barriers to the diffusion and transfer of technology, mobilizing financial resources, supporting capacity building in developing countries, and other approaches to assist in the implementation of behavioral changes and technological opportunities in all regions of the globe. The optimum mix of policies will vary from country to country, depending upon political structure and societal receptiveness. The leadership of national governments in applying these policies will contribute to responding to adverse consequences of climate change. Governments can choose policies that facilitate the penetration of less greenhouse gas-intensive technologies and modified consumption patterns. Indeed, many countries have extensive experience with a variety of policies that can accelerate the adoption of such technologies.

This experience comes from efforts over the past 20 to 30 years to achieve improved energy efficiency, to reduce the environmental impacts of agricultural policies, and to meet conservation and environmental goals unrelated to climate change. Policies to reduce net greenhouse gas emissions appear more easily implemented when they are designed to address other concerns that impede sustainable development (e.g., air pollution and soil erosion). A number of policies, some of which may need regional or international agreement, can facilitate the penetration of less greenhouse gas-intensive technologies and modified consumption patterns, including:

- Putting in place appropriate institutional and structural frameworks
- Energy pricing strategies (e.g., carbon or energy taxes, and reduced energy subsidies)
- Reducing or removing other subsidies (e.g., agricultural and transport subsidies that increase greenhouse gas emissions)
- Tradable emissions permits
- Voluntary programs and negotiated agreements with industry
- Utility demand-side management programs
- Regulatory programs including minimum energy-efficiency standards, such as for appliances and fuel economy
- Stimulating research, development, and demonstration to make new technologies available
- Market pull and demonstration programs that stimulate the development and application of advanced technologies
- Renewable energy incentives during market build-up
- Incentives such as provisions for accelerated depreciation and reduced costs for consumers
- Education and training; information and advisory measures
- Options that also support other economic and environmental goals.

Accelerated development of technologies that will reduce greenhouse gas emissions and enhance greenhouse gas sinks—as well as understanding the barriers that inhibit their diffusion into the marketplace—requires intensified research and development by governments and the private sector.

5. Technical Guidelines for Assessing Climate Change Impacts and Adaptations

Working Group II has prepared guidelines to assess the impacts of potential climate change and to evaluate appropriate adaptations. They reflect current knowledge and will be updated as improved methodologies are developed. The guidelines outline a study framework that will allow comparable assessments to be made of impacts and adaptations in different regions/geographical areas, economic sectors, and countries. They are intended to help contracting parties meet, in part, their commitments under Article 4 of the UNFCCC.

Impact and adaptation assessments involve several steps:

- Definition of the problem
- Selection of the methods
- Testing the method
- Selection of scenarios
- Assessment of biophysical and socioeconomic impacts
- Assessment of autonomous adjustments
- Evaluation of adaptation strategies.

Definition of the problem includes identifying the specific goals of the assessment, the ecosystem(s), economic sector(s), and geographical area(s) of interest, the time horizon(s) of the study, the data needs, and the wider context of the work.

The selection of analytical methods(s) depends upon the availability of resources, models, and data. Impact assessment analyses can range from the qualitative and descriptive to the quantitative and prognostic.

Testing the method(s), including model validation and sensitivity studies, before undertaking the full assessment is necessary to ensure credibility.

Development of the scenarios requires, firstly, the projection of conditions expected to exist over the study period in the absence of climate change and, secondly, the projection of conditions associated with possible future changes in climate.

Assessment of potential impacts on the sector(s) or area(s) of interest involves estimating the differences in environmental and socioeconomic conditions projected to occur with and without climate change.

Assessment of autonomous adjustments implies the analysis of responses to climate change that generally occur in an automatic or unconscious manner.

Evaluation of adaptation strategies involves the analysis of different means of reducing damage costs. The methodologies outlined in the guidelines for analyzing adaptation strategies are meant as a tool only to compare alternative adaptation strategies, thereby identifying the most suitable strategies for minimizing the effects of climate change were they to occur.

6. Methods for Assessing Mitigation Options and Inventory of Mitigation Technologies

Working Group II also has prepared guidelines to assess mitigation of options. Recognizing that there are many viable approaches for mitigating greenhouse gas emissions and that countries need to identify those approaches best suited to their needs and conditions, these guidelines provide a range of analytical methods and approaches for assessing mitigation options and developing national mitigation plans and strategies. The types of methods covered include macroeconomic models, decision analysis tools, forecasting methods, costing models, market research methods, and monitoring and evaluation methods.

In addition to descriptions of methods, the guidelines describe several broader aspects of the mitigation options assessment process, including the strategic, analytical, and informational challenges; organizational issues; and important cross-cutting issues such as top-down versus bottom-up modeling, accounting for uncertainty, and incorporating externalities.

The guidelines place a special emphasis on the needs of developing countries and countries with economies in transition. The methods and models addressed are therefore intended to match a range of analytical capabilities and resource constraints.

Research in the preparation of these guidelines has demonstrated that methods to assess mitigation options are available to all countries for use in developing strategies and evaluating programs and projects that (i) support national economic, social, and institutional development goals, and (ii) slow the rate of growth in greenhouse gas emissions. Although involving analytical challenges, these methods have been widely applied in both industrialized and developing countries. Moreover, the Global Environment Facility, other international organizations, and bilateral programs provide resources to assist developing countries and those with economies in transition in obtaining information on, testing, and using these methods. The IPCC recommends that all countries evaluate and use, as appropriate, these draft methods in preparing country studies and their national communications.

In the future, development and application of mitigation assessment methods will result in further improvements in the methods and the capabilities of countries to assess mitigation options. The IPCC, in coordination with other multilateral institutions, could accelerate the dissemination of selected information on assessment methods through seminars, workshops, and educational materials.

Climate Change 1995

Part I

Introductory Materials

Prepared by Working Group II

A

Ecophysiological, Ecological, and Soil Processes in Terrestrial Ecosystems: A Primer on General Concepts and Relationships

MIKO U.F. KIRSCHBAUM, AUSTRALIA

Lead Authors:
P. Bullock, UK; J.R. Evans, Australia; K. Goulding, UK; P.G. Jarvis, UK; I.R. Noble, Australia; M. Rounsevell, UK; T.D. Sharkey, USA

Contributing Authors:
M.P. Austin, Australia; P. Brookes, UK; S. Brown, USA; H.K.M. Bugmann, Germany; W.P. Cramer, Germany; S. Diaz, Argentina; H. Gitay, Australia; S.P. Hamburg, USA; J. Harris, UK; J.I. Holten, Norway; P.E. Kriedemann, Australia; H.N. Le Houerou, France; S. Linder, Sweden; R.J. Luxmoore, USA; R.E. McMurtrie, Australia; L.F. Pitelka, USA; D. Powlson, UK; R.J. Raison, Australia; E.B. Rastetter, USA; R. Roetter, Germany; J. Rogasik, Germany; D.R. Sauerbeck, Germany; W. Sombroek, FAO; S.C. van de Geijn, The Netherlands

CONTENTS

A.1. Introduction

All living organisms in terrestrial ecosystems ultimately depend directly or indirectly upon photosynthesis for their energy requirements. Photosynthesis depends on the absorption of light and the diffusion of CO_2 from the atmosphere to the sites of photosynthesis within leaves. To take up CO_2, plants must open their stomata; this generally results in considerable water loss. Increases in atmospheric CO_2 concentration can reduce water loss and increase photosynthetic carbon gain of most plants. Plant growth is only possible if temperatures are between some lower and upper thresholds; plant growth at most locations is likely to increase with moderate increases in annual mean temperature, mainly due to lengthening of the growing season.

Plants also need inorganic nutrients from the soil, such as nitrate or ammonia and phosphate, and plant response to environmental variables is modified by the availability of these soil nutrients. Furthermore, soil-nutrient availability itself can also be affected by environmental factors.

At any site, there are usually many different plant species that interact with each other and with other organisms in a multitude of different ways. In considering the effect of climate change on any terrestrial ecosystem, it is necessary to consider not only the direct ecophysiological effects in response to climate change but also the ways in which these direct effects are modified by soil feedbacks and biological interactions between different organisms.

Because many of the responses of terrestrial plants in diverse ecosystems are similar, a general description is offered here of some of the major potential impacts of climate change on terrestrial ecosystems (see also Chapter 9, *Terrestrial Biotic Responses to Environmental Change and Feedbacks to Climate*, of the IPCC Working Group I volume). This primer initially provides some overview of effects of the most important climatic driving forces that are likely to change, then discusses effects on soil carbon and nitrogen dynamics, looks at soil fertility as a modifying effect on responses to external driving forces, and includes a brief discussion of soil biological factors. These broad ecosystem responses are modified by ecological interactions between the different organisms in each ecosystem; the primer provides a brief overview of the major ecological factors that must be considered in assessing the impact of climate change on terrestrial ecosystems.

A.2. Climatic Driving Forces

Solar radiation, temperature, precipitation, air humidity, and atmospheric CO_2 concentration are some of the most important external forces that drive ecosystem processes. Precipitation, air humidity, and other meteorological variables—as well as plant and soil variables—together determine water availability for plants. Of these, changes in temperature, water availability, and CO_2 concentration are likely to constitute the most significant changes for terrestrial ecosystems over the coming century. The effects of these changes are discussed in the following sections.

A.2.1. Temperature

The increase in CO_2 and other greenhouse gases is expected to cause an increase in global mean temperature, with larger increases at high latitudes than elsewhere and larger increases during winter than summer (Gates *et al.*, 1992; Greco *et al.*, 1994). Plant growth and health may benefit from increased temperatures because of reduced freezing and chilling damage, but plants may be harmed by increased high-temperature damage. There is some indication that higher mean temperatures in the future will be associated with more variable and more extreme temperatures (Katz and Brown, 1992). However, the increase in global mean temperature during the past half-century has been primarily a result of higher night temperature (Karl *et al.*, 1993).

Net primary productivity (NPP) is generally enhanced by modest increases in temperature, especially in temperate and boreal regions (e.g., Kauppi and Posch, 1985; Cannell *et al.*, 1989; Kokorin *et al.*, 1993; Beuker, 1994). According to a relationship developed by Lieth (1973) based on observed net primary production in a variety of ecosystems, NPP will increase from 1% per °C increase in temperature in ecosystems with a mean annual temperature of 30°C to 10% per °C at 0°C. However, because this relationship is based on observations of NPP under present temperatures, part of the apparent increase could be the result of a correlation between higher solar radiation and higher temperature, so the actual effects of temperature change alone may be smaller.

Effects of increasing temperature on crop yields are more difficult to predict than effects of temperature on NPP because crop yields are not only affected by NPP but also by the phenology of crop development (see also Chapter 13). Increased temperature can speed phenological development, reducing the grain-filling period for crops and lowering yield—as is observed in current conditions (e.g., Monteith, 1981) and in most modeling studies (Warrick *et al.*, 1986; Chapter 13). For example, Wang *et al.* (1992) modeled the growth of wheat with climate change, and found that yield decreased as temperature increased because crop development was hastened. A cultivar from a warmer region, however, responded positively to moderate increases in temperature and gave a larger yield than the cultivar from the colder region under current conditions. This suggests that current cultivars might generally perform more poorly in a warmer climate, but losses could generally be avoided by cultivar substitution.

Extreme temperatures are often more important than average temperatures in determining plant responses (Woodward, 1987). If global warming reduces the frequency of extremely low temperatures, plants may be able to survive at higher latitudes or altitudes, and agricultural plants can be grown for longer periods of the year without the danger of damaging frosts. On the other hand, many plants are adapted to the climate in their current locations, and a general warming could result in premature bud-burst and so, paradoxically, increase frost damage (Hänninen, 1991). Alternatively, plants may experience insufficient chilling exposure so that flowering and fruit and seed production may not proceed at all, or be initiated in a season with inappropriate climate. In another study, increased temperature was found to speed bud-burst, but increasing CO_2 concentrations counteracted the effect, in some locations completely (Murray *et al.*, 1994).

Increased episodes of extremely high temperature can damage plants, especially in conjunction with water shortage. However, if sufficient soil water is available, leaf temperatures can be substantially lower than air temperature as a result of evaporative cooling. Burke *et al.* (1988) report that many plants maintain leaf temperatures within a preferred range, thereby optimizing NPP. While the amount of water consumed for leaf cooling can be substantial and highly variable, typically 200–500 water molecules may be lost for each molecule of CO_2 taken up (Sharkey, 1985). This can lead to insufficient water in plant tissues, resulting in reduced growth. Elevated CO_2 reduces stomatal conductance (Eamus, 1991) and so saves water, but this may further increase the temperature of leaves, over and above increased air temperature.

Plant membranes may be damaged at high temperature (Berry and Björkman, 1980), and membranes in plants are often modified following growth in high temperature. However, this reduces plant performance at low temperature. High temperature can affect other plant processes. Respiration rates are often increased by high temperatures, and plant respiration can account for the loss of a significant fraction of the carbon fixed in photosynthesis. However, plants may acclimate to higher temperature, and the short-term effects of temperature on respiration are often not seen when plants are grown at different temperatures (e.g., Gifford, 1994; Körner, 1995). Also, the response of leaf respiration rate to temperature can be substantially reduced in plants grown in high CO_2 and high temperature (e.g., Wullschleger and Norby, 1992). The interaction between increased temperature and increased CO_2 is substantial, and the effect of warming by itself may be modified or even reversed when CO_2 effects are considered. Other processes—such as the volatilization of hydrocarbons, which could affect carbon balance—show less acclimation to high temperature (Tingey *et al.*, 1991; Lerdau 1991). Hydrocarbon loss from leaves can exceed 10% of the carbon taken up for photosynthesis.

A.2.2. Precipitation and Soil Water Availability

Active physiological processes of plants require an aqueous medium. Higher land plants are able to survive within a generally dry atmosphere because they are covered by a cuticular epidermis that minimizes water loss. For growth, however, plant leaves must take up CO_2 from the atmosphere. This need for a path for diffusion of CO_2 into the leaves also provides a path by which water is lost from leaves. Water loss is regulated by stomata whose aperture is adjusted in response to environmental variables and internal regulators.

If the soil is wet, plants can replace transpired water with water from the soil. If plants continue to extract water from the soil, and if soil water is not replenished by rain or irrigation, then the water content of plants must eventually fall and their physiological function is impaired (Hsiao, 1973; Bradford and Hsiao, 1982). Water-stressed plants restrict further water losses by closing stomata, by adjusting the angle at which leaves are held to minimize light absorption, or by shedding leaves (Passioura, 1982). CO_2 uptake and growth is then reduced or completely prevented (Schulze and Hall, 1982). Water availability at most locations is seasonally variable, with plants experiencing at least temporary drought for some time (Woodward, 1987; Stevenson, 1990). Individual species are adapted to particular water regimes and may perform poorly and possibly die in conditions to which they are poorly adapted (e.g., Hinckley *et al.*, 1981).

Total plant growth under water-limited conditions is essentially given by the product of the amount of water used and water-use efficiency. Short-term water-use efficiency can be expressed as the ratio of the difference of CO_2 concentration between the atmosphere and the sites of photosynthesis to the difference of water-vapor concentration between the sites of evaporation within leaves and the atmosphere. Hence, anything that changes either of these concentration differences can potentially affect water-use efficiency and thus growth under water-limited conditions (Eamus, 1991).

Different plant types (see Box A-1) have different water-use efficiencies. C_3 plants have relatively poor water use efficiency. C_4 plants have higher water-use efficiency because they photosynthesize at lower internal CO_2 concentration and thereby increase the difference in CO_2 concentrations between the atmosphere and the sites of photosynthesis. In CAM plants, CO_2 uptake occurs at night, when leaves are coolest and the leaf to air water vapor concentration difference is smallest. This gives CAM plants the highest water-use efficiency of all plants.

Higher CO_2 concentration generally leads to lower stomatal conductance (Kimball and Idso, 1983; Morison, 1987) and higher leaf photosynthetic rates (Eamus and Jarvis, 1989; Arp, 1991). This improves water-use efficiency, so carbon gain for plants with limited water supply should increase with increasing CO_2 concentration (Rogers *et al.*, 1983; Idso and Brazel, 1984; Tolley and Strain, 1985; Morison, 1987; Eamus and Jarvis, 1989).

Many ecosystems experience shortage of soil water (drought) during some or most of the year, which limits their potential carbon gain (Lieth, 1973; Hinckley *et al.*, 1981; Woodward, 1987). Soil water availability can be related to the ratio of precipitation to potential evapotranspiration, or to other measures of water availability, such as the ratio of actual to potential evapotranspiration. What is important in all these measures is that soil water availability can be affected by changes in either gains (precipitation) or losses (evapotranspiration) of water.

Because warmer air can hold more water, it is likely that increasing temperature will lead to a larger difference between the water-vapor concentration inside leaves and in the air, with

Box A-1. Different Plant Groups

C_3— C_3 is the most basic photosynthetic mechanism. It is called C_3 because the first compound into which CO_2 is incorporated is a compound with three carbon atoms. C_3 plants make up the majority of species globally, especially in cooler or wetter habitats; they include all important tree and most crop species, such as wheat, rice, barley, cassava, and potato.

C_4— C_4 plants have a special CO_2-concentrating mechanism within their leaves by which they can increase the CO_2 concentration to several times above ambient levels. This is done by CO_2 first being incorporated into a 4-carbon compound. This allows these plants to maintain lower intercellular CO_2 concentrations than C_3 plants. C_4 plants tend to grow in warmer, more water-limited regions, and include many tropical grasses and the agriculturally important species maize, sugarcane, and sorghum.

CAM—Crassulacean acid metabolism, CAM, is a variant of C_4 photosynthesis in which CO_2 is not only concentrated but also stored for half a day. CAM plants, such as cacti, often grow in deserts, but they also include more common plants such as pineapple. In these plants, CO_2 uptake occurs at night. The CO_2 is then stored for use in normal photosynthetic reactions during the next day.

little effect on the difference in CO_2 concentrations—thus lowering water-use efficiency. If the absolute humidity increases in line with the saturated humidity at the diurnal minimum temperature, and if both diurnal minimum and maximum temperatures change similarly with global warming, then the leaf-to-air water-vapor concentration differences will increase by about 5–6% per °C warming. However, if nighttime temperatures were to increase more than daytime temperatures—as has been the case over recent decades (Karl *et al.*, 1993)—then the concentration difference will increase by less than 5–6% per °C.

General circulation models (GCMs) suggest that there are likely to be regions where precipitation will increase by more than the global average and where the additional rainfall may be more than sufficient to meet increased evaporative demand, whereas other regions may receive less rainfall than at present (Mitchell *et al.*, 1990; Greco *et al.*, 1994); this is further complicated by feedbacks from the biosphere. Henderson-Sellers *et al.* (1995), for example, show that inclusion of stomatal closure in response to increasing CO_2 concentration led to a reduction in the predicted increase in precipitation from 7.7% for doubled CO_2 to only 5.0%. The timing of water availability within ecosystems may also change. For example, earlier melting of snowpacks may mean that less water is available during summer (Mitchell *et al.*, 1990; Hayes, 1991). There will therefore almost certainly be some regions with improved and others with worse water balances than at present.

Potential evapotranspiration rates can be estimated using a variety of meteorological formulae with varying physical rationales. Some workers have used the Thornthwaite method, which is based on correlations between evapotranspiration and temperature in the current climate (e.g., Thornthwaite, 1948; Le Houerou *et al.*, 1993). Based on the Thornthwaite method, these workers conclude that water may become more limiting with temperature increase in the future (e.g., Gleick, 1987; Rind *et al.*, 1990; Leichenko, 1993). However, the formulae that have the soundest physical bases—the Penman, Priestley and Taylor, and Penman-Monteith equations (Jarvis and McNaughton, 1986; Martin *et al.*, 1989)—predict that potential evapotranspiration would increase with warming by an amount similar to the anticipated increase in precipitation (McKenney and Rosenberg, 1993), provided other factors such as radiation balance and surface resistance do not change. There would then be little change in the global incidence of drought conditions.

Where climate change leads to annual or seasonal changes in water availability, agricultural and forest productivity could change. Significant reductions in soil water availability could lead to forest decline. There could also be indirect problems, such as more floods and greater erosion hazards caused by more intense rainfall. Wind erosion could increase if drought lengthens the time that the ground is bare of vegetation. Changes in the ratio of precipitation to potential evapotranspiration could also affect water discharge into rivers and groundwater reservoirs.

If water runs off the surface instead of infiltrating the soil, the amount available to plants is reduced; this may also lead to erosion. Water infiltration rates are affected by soil texture and structure, slope, vegetation cover, soil surface roughness, surface crusting, and land management. Infiltration is also controlled by soil water content because saturated soils are unable to absorb water and very dry soils can be slow to re-wet. Aggregated soils with good structure facilitate infiltration, and aggregation is strongly affected by the organic matter content of the topsoil. Surface runoff is also strongly dependent on the amount and intensity of rainfall, which in most regions is likely to increase with climate change (e.g., Gordon *et al.*, 1992).

The water-retention capacity of different soil types can influence the intensity of water stress experienced by plants. Soil water retention is significantly affected by soil organic-matter content, particle-size distribution, bulk density, and soil structure. Thus, any decrease in the quantity of soil organic matter as a result of faster decomposition could reduce infiltration rates and soil water retention and accentuate plant water stress.

Waterlogging can develop because of rising groundwater tables or because of the presence of partially permeable layers within the soil profile. Waterlogging can affect plant growth in both agricultural and natural environments by limiting the diffusion of oxygen to plant roots and soil organisms. Excess soil wetness can also be a considerable hindrance in agriculture by impeding soil tillage (Rounsevell and Jones, 1993), and climate change can thus affect crop production by affecting soil workability (Rounsevell *et al.*, 1994).

A.2.3. *Direct Effects of CO_2 Concentration*

Atmospheric CO_2 is a basic substrate for photosynthesis, which underlies plant growth. In the response of plants to CO_2 concentration, it is important to distinguish between plants with the C_3 and C_4 photosynthetic pathways. Increasing CO_2 concentration directly affects photosynthesis in three ways. Firstly, the carboxylating enzyme for carbon reduction in all plants, Rubisco, has a poor affinity for CO_2. The present atmosphere results in CO_2 concentrations in chloroplasts that are well below half-saturation (Farquhar and von Caemmerer, 1982). Consequently, the photosynthetic rate is very responsive to small increases in CO_2 concentration in C_3 plants. Secondly, oxygen competes with CO_2 for the active site on Rubisco, leading to photorespiration (Farquhar and von Caemmerer, 1982). The rise in atmospheric CO_2 concentration will progressively reduce photorespiration and enhance quantum yield of carbon fixation in C_3 plants. This will not occur in C_4 plants, which have a CO_2-concentrating mechanism that already suppresses photorespiration. Thirdly, stomata in many species progressively close as the atmospheric CO_2 concentration increases, thus reducing water loss relative to carbon gain. This gain may be offset to some extent by increased leaf temperature. It has also been argued that CO_2 will alter respiratory activity. Experimental evidence demonstrates both increases and decreases in response to increasing atmospheric CO_2 concentration (e.g., Amthor, 1991; Poorter *et al.*, 1992).

It is well-established that short-term photosynthetic rates in C_3 plants increase by 25–75% for a doubling of CO_2 concentration (Kimball, 1983; Cure and Acock, 1986; Eamus and Jarvis, 1989; Allen, 1990; Bazzaz, 1990; Bowes, 1993; Luxmoore *et al.*, 1993). There are fewer data available for C_4 plants. While some workers have found little response (e.g., Morison and Gifford, 1983; Henderson *et al.*, 1992), others have found increases of 10–25% in photosynthetic rate for a doubling of CO_2 concentration (e.g., Wong, 1979; Pearcy *et al.*, 1982; Polley *et al.*, 1992). The sensitivity of C_3 photosynthesis to CO_2 concentration increases with increasing temperature (Long, 1991; Bowes, 1993; Kirschbaum, 1994); hence, the stimulation of plant growth by increasing CO_2 concentration is likely to be larger at higher temperatures (Idso *et al.*, 1987; Rawson, 1992), with little stimulation and sometimes even inhibition at low temperatures (Kimball, 1983).

The increased photosynthetic rate and decreased water requirement translate into increased growth and crop yield of C_3 plants (Kimball, 1983; Cure and Acock, 1986), increased growth of C_4 plants (Poorter, 1993), and increased tree seedling growth (Luxmoore *et al.*, 1993). However, this connection can be confounded by many factors. Firstly, acclimation of photosynthesis may occur such that photosynthetic capacity is diminished. An average 21% decrease in photosynthetic capacity has been observed in tree species (Gunderson and Wullschleger, 1994), although this may be an artifact of pot size (Arp, 1991; Thomas and Strain, 1991; Sage, 1994). Despite this decline in capacity, photosynthetic rate per unit leaf area was on average 44% higher in elevated than in ambient CO_2 (Gunderson and Wullschleger, 1994). Where reductions in photosynthetic capacity cannot be due to restricted rooting volumes, they are most likely a consequence of nitrogen shortage and reduction in Rubisco activity (e.g., Ceulemans and Mousseau, 1994).

Secondly, it is generally observed that carbon partitioning is altered, resulting in plants in elevated CO_2 having lower leaf area per unit of plant dry weight (e.g., Poorter, 1993), and the proportional increase in aboveground growth is less than the increase in photosynthesis or total growth. Responses may also differ in conjunction with other environmental limitations and between species (Gifford, 1992; Luxmoore *et al.*, 1993).

Because the atmospheric CO_2 concentration has already increased from a preindustrial concentration of about 280 ppmv to about 360 ppmv at present, there should be evidence of increased growth of plants under natural conditions. However, the evidence from tree-ring chronologies is unclear (Innes, 1991). For example, Graumlich (1991) found no growth enhancements at five subalpine sites in the Sierra Nevada (California), but LaMarche *et al.* (1984), Kienast and Luxmoore (1988), Hari and Arovaara (1988), Cook *et al.* (1991), West *et al.* (1993), and Graybill and Idso (1993) observed varying degrees of growth enhancement in recent times compared with preindustrial times. However, part or all of that increase can probably be explained by more favorable temperatures and nitrogen fertilization by moderate levels of industrial pollution, especially as some of the observed increases are far greater than would be expected from CO_2 enrichment alone (Luxmoore *et al.*, 1993).

There is increasing experimental evidence available on the effects of CO_2 enrichment on ecosystem dynamics, although much is still unpublished. The few results available to date give a divergent picture (Körner, 1995). For example, in a warm and nutrient-rich temperate wetland in Maryland, a high and persistent increase in growth has been observed (Curtis *et al.*, 1989; Drake, 1992). Körner and Arnone (1992) observed an 11% increase in biomass over 100 days in an artificial tropical ecosystem; shoot biomass increased by 41% in a California grassland (Jackson *et al.*, 1994), and midseason CO_2 uptake increased substantially in a Swiss alpine grassland (Diemer, 1994). On the other hand, almost no response to CO_2 enrichment has been observed in a cold and nutrient-limited tundra environment (e.g., Tissue and Oechel, 1987). In summarizing ecosystem experiments, Körner (1995) found generally enhanced net CO_2 assimilation but little or no increase in aboveground biomass or leaf area and a likely increase in soil carbon.

Because water-use efficiency can be greatly enhanced by increased CO_2 concentration (Rogers *et al.*, 1983; Tolley and Strain, 1985; Morison, 1987; Eamus and Jarvis, 1989), relative plant responses to increases in CO_2 should be most pronounced under water-limited conditions (e.g., Gifford, 1979; Allen, 1990). This difference between well-watered and water-limited conditions should be most pronounced for C_4 plants (Samarakoon and Gifford, 1995). Growth enhancement by CO_2 is also evident under severe nutrient limitation (e.g., Wong, 1979; Norby *et al.*, 1986; Idso and Idso, 1994; Lutze and Gifford, 1995). In addition to growth responses, it is highly likely that species composition will change as a result of increasing CO_2 concentration (e.g., Bazzaz *et al.*, 1989; Wong and Osmond, 1991).

Results obtained with different species and experimental conditions have shown that plant responses to increasing CO_2 concentration are likely to differ greatly among the ecosystems of the world. An assessment of the role of increasing CO_2 concentration in the global context requires quantification of the various factors that may increase or decrease the response of plants to CO_2 concentration. Most of the feedback effects are still inadequately understood and poorly quantified. However, plant-growth responses to doubled CO_2 concentration do not generally exceed 30% enhancement, even without negative feedback effects. The realized growth enhancements in response to the gradually increasing CO_2 concentration, therefore, are likely to amount to only a small gradual impact on terrestrial ecosystems.

A.3. Soil Processes and Properties

Changes in climate will affect a number of crucial soil processes that will affect the ability of the soil to support particular natural or agricultural communities. The extent of these effects could have far-reaching consequences for the future distribution of fauna and flora, greatly changing distribution patterns and possibly resulting in new combinations of soils and vegetation (Tinker and Ingram, 1994). Soil development is likely to lag behind climate and vegetation change, so that in the medium term of decades to centuries, vegetation classes will probably often occur on soil types on which they are not currently found. It is not clear what consequences this mismatch between vegetation and soils will have for ecosystem function in the longer term.

Soil is formed through the interaction of many variables, the most important being parent material, climate, organisms, relief, and time (Jenny, 1941, 1980; Bridges, 1970; White, 1987). The strength and interactions of these variables differ across the world, producing many different soil types, each forming the basis of different habitats and each with different productive potential. Natural soil-forming processes (pedogenesis) occur slowly, but changes in the physical environment can lead to fundamental changes in soil types. Pedogenesis and the weathering of inorganic soil components in response to climate change have received insufficient scientific attention in the past, although these are very important for the development of new soil types, for nutrient release, and for many of the physical characteristics of soils (Arnold *et al.*, 1990; Brinkman and Sombroek, 1995).

The rates of change of soil processes and properties resulting from climate change are likely to be different for different soil types (Stewart *et al.*, 1990). Scientific assessments of the impacts of climate change on soils have largely been directed at soil processes that will respond most rapidly (over periods of months or years) and are thought to have the greatest effect on ecosystem functioning. These are principally changes in the soil water regime and turnover of organic matter and the related mineralization or immobilization of nitrogen and other nutrients.

Temperature has only marginal effects on reaction rates of most inorganic reactions in the soil, such as ion exchange, adsorption, and desorption, and increasing temperature itself, therefore, is unlikely to be important for the dynamics of inorganic nutrients. A change in soil moisture content, however, could significantly affect rates of diffusion and thus the supply of mineral nutrients such as P and K to plants. This could well alter the species composition of plants in natural systems and may require adjustments to nutrient management and fertilizer use in agriculture.

A.3.1. Carbon Dynamics

The global pool of soil organic matter is estimated to contain about 1500 Gt of carbon (C) (Melillo *et al.*, 1990; Adams *et al.*, 1990; Anderson, 1992; Eswaran *et al.*, 1993). This compares with estimates of 600–700 Gt C in aboveground biomass of vegetation (Melillo *et al.*, 1990; Sombroek, 1990; Anderson, 1992; Schimel *et al.*, 1994), 800 Gt C in the atmosphere, and about 40,000 Gt C in the oceans (Watson *et al.*, 1990; Schimel *et al.*, 1994). Most carbon in soils is associated with organic matter, although carbonate-C can also be significant in calcareous soils, and charcoal may be an important constituent in ecosystems subject to frequent fires. The amount of organic matter in soils is influenced by soil type, land use, and climate affecting the release or sequestration of CO_2 (see Box A-2 and Chapters 23 and 24).

Changes in organic carbon contents of the soil are determined by the balance between carbon inputs and carbon losses by organic-matter decomposition (soil respiration) rates. All terrestrial carbon inputs originate from plant products reaching the soil either as root exudates, dead roots, leaf litter, dead branches, or trees or indirectly as feces or bodies of animals. The annual input of carbon is thereby given by the amount of annual NPP minus the fractions of carbon that are removed from the system (e.g., in agricultural produce), lost during fires, respired by herbivores, or stored in increasing wood volumes on the site.

Box A-2. Terms Defining Carbon Dynamics

The amount of carbon taken up in photosynthesis is defined as the gross primary production (GPP). Some of this carbon is returned to the atmosphere as CO_2 during plant metabolism (autotrophic respiration), giving a net gain of carbon—the net primary production (NPP). Death and shedding of plant parts adds organic carbon to the soil, where it is decomposed by soil animals, fungi, and bacteria (heterotrophic respiration). The difference between NPP and heterotrophic respiration is the net gain or loss of carbon by the ecosystem and is termed the net ecosystem production (NEP).

The great bulk of organic matter reaching the soil is respired by soil organisms within a few years, with the exact time course depending on climatic conditions and litter quality. The remaining organic matter is transformed into different forms with different decomposability. Some of it is highly resistant to decomposition, so it remains in the soil for hundreds to thousands of years even if conditions change greatly. Other material is slightly more labile, and some changes can greatly enhance its decomposition rate, leading to a loss of this fraction over years to decades (see the modeling studies of van Veen and Paul, 1981; Parton *et al.*, 1987; Jenkinson, 1990).

Changes in climatic conditions and land use generally affect both NPP and the rate of organic-matter decomposition. Soil organic matter increases if NPP increases more than decomposition rate, and soil organic matter decreases if decomposition rate increases more than NPP. However, any such changes in soil organic-matter content are very difficult to verify by direct measurements because of the high inherent variability of soil organic-matter content and because likely changes constitute only a very small fractional change of the amounts that are already in the soil.

Comprehensive data for soil organic-matter content in different soils across the Earth have shown that it increases with increasing water availability, and, for a given water status, it increases with decreasing temperature (Post *et al.*, 1982, 1985; Buol *et al.*, 1990). Both NPP and organic-matter decomposition are likely to be enhanced by increasing temperature, as all microbiologically facilitated processes are strongly affected by moisture and temperature. Annual soil respiration rates are likely to increase because of the lengthened season for breakdown of plant material and because increasing temperature strongly stimulates organic-matter decomposition (e.g., Berg *et al.*, 1993; Lloyd and Taylor, 1994; Kirschbaum, 1995), especially in arctic regions subject to permafrost (Reynolds and Leadly, 1992). Organic-matter decomposition is likely to be stimulated more than NPP (Kirschbaum, 1995). Consequently, although global NPP is likely to increase with global warming, soil carbon storage is likely to decrease at the same time, and this could add more CO_2 to the atmosphere (e.g., Schimel *et al.*, 1990; Jenkinson *et al.*, 1991; Thornley *et al.*, 1991; Kirschbaum, 1993, 1995). On the other hand, none of these studies deals with the interactive effect of temperature and moisture limitations on decomposition rates. Should warming generally lead to moisture becoming more limiting for decomposition, then the effect of warming may be less pronounced than is suggested by these studies, which implicitly assume that moisture limitations will remain the same.

Other lines of evidence suggest that the carbon-storage potential of the terrestrial biosphere may not diminish in the future. Terrestrial carbon storage appears to have increased since the last glacial maximum (e.g., Bird *et al.*, 1994), and models of the possible distribution of biomes under future climatic scenarios with their associated observed carbon storage generally suggest increased carbon-storage potential in the future (e.g., Prentice and Fung, 1990; Smith *et al.*, 1992; King and Neilson, 1992).

Experimental work on the effect of increasing CO_2 concentration on soil processes has yielded divergent results (reviewed by van Veen *et al.*, 1991), but modeling studies (e.g., Thornley *et al.*, 1991; Kohlmaier *et al.*,1991; Rastetter *et al.*, 1991; Polglase and Wang, 1992; Gifford, 1992; Kirschbaum, 1993) have consistently led to the result that increasing CO_2 concentration, via increased NPP, would lead to increases in soil carbon storage.

An important feedback is the mineralization of nitrogen and phosphorus (see also Section A.3.2). If increases in NPP lead to enhanced immobilization of nitrogen and phosphorus in soil organic matter, then the higher NPP will subsequently be reduced and the increased carbon input into the soil will not be sustained (Comins and McMurtrie, 1993; Kirschbaum *et al.*, 1994). Soil organic matter will then increase only marginally. On the other hand, if increasing CO_2 concentration can stimulate biological nitrogen fixation and mycorrhizal phosphorus uptake, then large and sustained increases in carbon input and consequently soil organic-matter content are possible. Limited experimental evidence suggests that nitrogen fixation may be enhanced by increasing CO_2 concentration (e.g., Norby, 1987; Arnone and Gordon, 1990; Thomas *et al.*, 1991; Gifford, 1994), so increasing CO_2 concentration may not make nitrogen more limiting as long as nitrogen-fixing plants are present in the ecosystem.

Photosynthesis in C_3 plants is more responsive to CO_2 concentration in warm than in cool conditions. At the same time, equilibrium amounts of soil organic matter change more with temperature at lower than higher temperatures (Jenny, 1980; Post *et al.*, 1982, 1985; Kirschbaum, 1995), so the net balance of effects may lead to soil organic-matter pools increasing in warm regions of the world and decreasing in cool regions (Kirschbaum, 1993).

Land use is significant in determining the balance between soils as a source and as a sink for carbon dioxide. Typically, about half the carbon is lost from soils after conversion of undisturbed forest or grassland to annually plowed cropland (e.g., Schlesinger, 1977, 1986; Buringh, 1984; Allen, 1985)—although this figure varies significantly with region, soil type, and kind of land-use change (e.g., Detwiler, 1986), and there may be greater carbon retention under minimum-tillage practices. A change in land use in the opposite direction can lead to sequestration of carbon in soils (see Chapters 23 and 24), although sequestration of carbon tends to be slower than carbon loss (e.g., King and Neilson, 1992; Smith and Shugart, 1993; Chapter 9, *Terrestrial Biotic Responses to Environmental Change and Feedbacks to Climate*, of the IPCC Working Group I volume). Conversion of forest or grassland soils high in organic matter to arable agriculture will inevitably cause a decline in soil organic-matter content and cause additional CO_2 to be released to the atmosphere. The associated mineralization of nutrients may allow increased plant growth, but large release of inorganic ions, especially if it is not synchronized with crop uptake, can add to the current problem of nitrate leaching into aquifers.

Soils are an important source of methane, contributing more than half the total emissions of 535 ± 75 Mt CH_4 yr^{-1} (Prather *et al.*, 1994; see also Chapter 23). Soil methane comes from four main sources: (1) natural wetland soils (55–150 Mt yr^{-1}); (2) microbial degradation of organic substrates in paddy rice soils (20–100 Mt yr^{-1}); (3) landfills (20–70 Mt yr^{-1}); and (4) termites (10–50 Mt yr^{-1}). Soils also provide sinks for methane of 30 ± 15 Mt CH_4 yr^{-1} (Prather *et al.*, 1994). The capacity of soils to oxidize CH_4 interacts in a complex manner with pH, land use, and the soil nitrogen cycle (Hütsch *et al.*, 1994; King and Schnell, 1994). Future temperature increases in polar regions might lead to the release of methane currently contained in gas hydrates in permafrost regions both on the surface and in submerged regions (Collett *et al.*, 1990; Kvenvolden, 1993).

The balance between the microbial processes of methanogenesis and methane consumption (Knowles, 1993) controls whether soils and paddy systems are a source or sink for methane. These biologically mediated processes are influenced by variables such as organic substrate supply, temperature, hydrologic conditions, pH, redox potential, aeration, and salinity—all of which are affected by climate change. Increasing temperature can alter methane fluxes by changing the rate of methane formation in lakes and wetlands and by altering the ratio of methane synthesis to methane oxidation. In addition, increased NPP provides more substrate for methane production, either from decaying plant matter or root exudations, and will provide more conduits for methane escape from lake and wetland sediments through the emergent plants, especially rice (Schütz *et al.*, 1991).

A.3.2. Soil Nitrogen Dynamics

Nitrogen can be added to the soil as inorganic fertilizer and organic manures and by wet and dry deposition from the atmosphere, or it can also be transferred from the atmosphere by biological nitrogen fixation (e.g., Bradbury and Powlson, 1994). Nitrogen is also released (mineralized) by the microbial decomposition of soil organic matter. Mineral nitrogen may be taken up by plants or re-absorbed by soil microorganisms; it may be leached as nitrate to ground and surface waters or emitted to the atmosphere in gaseous forms after nitrification, denitrification, or volatilization of NH_3 (Bradbury and Powlson, 1994). These processes are strongly influenced by temperature, soil moisture, plant characteristics, and, indirectly, by atmospheric CO_2 concentrations.

Nitrous oxide (N_2O) is formed in soils by denitrification and nitrification. Natural soils emit about 6 Mt N yr^{-1} and cultivated soils 3.5 Mt N yr^{-1}; together they probably contribute more than half of the total N_2O emitted to the atmosphere (Prather *et al.*, 1994; Chapter 23). Much of the recent increase of atmospheric concentration of N_2O is attributed to increased use of legumes and N-fertilizers (Chapter 23). Soils also emit 12 Mt N yr^{-1} as NO_x, which constitutes about 25% of total NO_x emissions (Prather *et al.*, 1994). Numerous environmental variables and agricultural practices influence the biological processes responsible for N-emissions from soils (Armstrong Brown *et al.*, 1995). Environmental variables include soil temperature, moisture content, and aeration status, and agricultural management practices include fertilizer regime, cultivation method, and cropping systems (Chapter 23). Mineralization/immobilization requires moist soil, and denitrification requires water-saturated (anaerobic) soil or, at least, saturated microsites. Denitrification rates consequently increase with increasing soil water content. Soils are also likely to be a small sink for N_2O, although the size of this sink has not yet been satisfactorily quantified.

Nutrient-limited ecosystems, including most natural systems and many subsistence or low-input farming systems, are to some extent buffered against effects of global change: If climatic and atmospheric conditions become more favorable for plant growth, nutrient shortage will impose more serious limitations, whereas if climatic conditions change adversely, nutrients will become less limiting (Shaver *et al.*, 1992). Consequently, NPP and total carbon storage of nutrient-limited systems is likely to be less affected by climate change than that of systems that are not nutrient limited.

Rastetter *et al.* (1992) have argued that NEP can only be positive if the total amount of nutrients in the ecosystem increases, if nutrient distribution changes from components with low ratios of carbon to nutrients to those with higher ratios, or if the carbon-to-nutrient ratio changes within vegetation or the soil.

The total amount of nutrients in an ecosystem may change as a result of altered rates of either inputs or losses. Nutrient losses may be prevented by immobilization in soil organic matter (although this also makes them temporarily unavailable for plants). Nitrogen gains may result from more favorable growth conditions that stimulate biological nitrogen fixation. Other nutrients may be gained through enhanced weathering of soil minerals. Nutrient availability may be increased through more extensive root growth and enhanced activity of mycorrhizal associations.

Levels of nitrogen and sulfur deposition are also increasing in many regions of the world through inputs from industrial pollution and other human activities such as agriculture. This may be having beneficial effects on NPP of many ecosystems, especially chronically nitrogen-deficient forests in high northern latitudes. While low rates of nitrogen input may have beneficial effects, ongoing inputs may reverse the initial gains through development of nutrient imbalances and further acidification of the soil (e.g., Ulrich, 1991, 1994; Linder and Flower-Ellis, 1992; Heath *et al.*, 1993).

A shift of nitrogen from vegetation to the soil has frequently been observed in CO_2-enrichment studies (e.g., Norby *et al.*, 1992; Diaz *et al.*, 1993). This is consistent with the theoretical notion that transiently increased productivity could lead to greater nutrient immobilization in soil organic matter and thereby shift the nutrient capital from the vegetation to the soil (Comins and McMurtrie, 1993; Kirschbaum *et al.*, 1994). Conversely, warming of soils with large organic-matter content may enhance decomposition rates and lead to the mineralization of nutrients (van Cleve *et al.*, 1981, 1990; Melillo *et al.*, 1993). This may stimulate plant growth and redistribute nutrients from soils to plants.

Carbon-to-nutrient ratios within soils are generally fairly stable, although they may differ between different ecosystems. Post *et al.* (1985), for example, document C:N mass ratios ranging from 9 to 30 in a comprehensive analysis of different global systems.

Although climate change may lead to small changes in the nutrient capital of soils or the ratio of carbon to nutrients, large changes are unlikely in the short term. Significant short-term changes in carbon storage are only possible where nutrient distribution changes between the soil and high C:N components, such as wood. Hence, in most systems, nutrient limitations will cause total carbon storage and associated NPP to remain similar to what they are currently. Only in the longer term (centuries) is the nutrient capital of soils likely to change so that climate again provides the essential determinant for NPP and total ecosystem carbon storage.

A.3.3. Soil Biodiversity

Climate change could change the abundance of species within the soil microbial and faunal populations, although the direct effects from changes in soil moisture or temperature will be much smaller than those caused by changes in land use. It is not possible to predict whether there would be a change in biodiversity. However, in mid-northern latitudes (especially Europe), high inputs of industrial nitrogen are associated with major losses of mycorrhizal fungi (e.g., Arnolds and Jansen, 1992). These losses may make forests more vulnerable to drought and disease, which could be further exacerbated by climate change. The increasing concentration of atmospheric CO_2 could also change the composition of organic carbon compounds entering the soil from roots and root exudates, in addition to increasing its quantity. This may alter the species composition of the rhizosphere population—which, in turn, could alter the extent to which plant roots are infected by soil-borne pathogens. Whether such changes would be beneficial or harmful, however, is not known.

The question of whether microflora will change in line with the conditions or lag well behind them has not yet been fully addressed. Previous comparisons of microbial communities across different biomes have been largely confined to fungi and show that species composition is clearly related to biome type, even for the same biome in different continents (Kjoller and Struwe, 1982). Soil organisms with more specific characteristics—such as plant pathogens, symbiotic organisms—and soil fauna, may well be slower in adjusting to new conditions and may generally have slower migration speeds than higher plants. For plant pathogens and symbiotic organisms, there is the additional complication that they are dependent, to varying degrees, upon specific types of vegetation. A vegetation zone shift caused by temperature/precipitation changes will only have reached full equilibrium when both the vegetation and the appropriate microorganisms have established themselves together.

A.4. Ecological Processes

Organisms interact not only with their physical environment but also with other organisms. The complicated sequence of dependencies and interactions among organisms has led to the description of ecology as the study of the "web of life." Impacts on a component of this web may be absorbed by a small part of the web, or they may lead to a cascade of effects throughout the web. This uncertainty makes predictions about the effect of climate change on complex ecological systems very difficult. In the following sections, we discuss some of the processes and interactions that most affect the response of ecological communities to climate change.

A.4.1. Niche

All organisms have preferred places in which to live. These places, called the species' habitat, can be depicted on a map. A species' living requirements can be defined in a more abstract way based on aspects of the environment that define its habitat. These aspects may include a certain range of temperature, precipitation, soil conditions, and so forth. Each of these variables can be thought of as describing an axis in a multidimensional space, and the area or volume that describes the preferred habitat of a species is its niche (e.g., Austin *et al.*, 1990, Begon *et al.*, 1990). A species' "fundamental niche" encompasses all the environmental conditions in which it could potentially grow and reproduce if it were subject to no competition or other effects from other species (e.g., herbivory, disease). The fundamental niche is a consequence of an organism's basic physiological tolerances and ecological traits. In most ecosystems, where organisms compete and interact with others, a species occupies a smaller space, called its "realized niche" (Hutchinson, 1957).

Climate change may cause the distribution of suitable habitats on a map to change. If temperature, moisture, or other climate variables at a particular location change, individuals at that location may find themselves now to be outside their fundamental niche. If a particular location falls outside the fundamental niche of a species, it is certain that individuals of this species will not be able to continue to reproduce and persist at this site. In other cases, the location may still fall within the fundamental niche but outside the realized niche. In this case, it is difficult to predict whether the species will be able to persist at this location because that will depend on which other species are able to persist or invade that site. In other cases, the location may still fall within the realized niche, and it is likely that the species will be able to continue to live at this site—although it may still be affected by changes in the distribution of other species. These assumptions are the basis of a number of models that attempt to predict the redistribution of organisms across the globe as a consequence of climate change (Box, 1981; Emanuel *et al.*, 1985; Prentice *et al.*, 1992, Cramer and Leemans, 1993).

A.4.2. Interactions

A.4.2.1. Competition

Individual organisms compete with each other for essential resources such as light, nutrients, and food, and for specific needs such as nesting space. Climate change may alter the competitive balance between species at a site by differentially affecting their effectiveness in the capture of resources or their efficiency in using them. Climate change may also result in the addition or loss of species from the site, thus changing the outcome of future competitive interactions. Higher CO_2 concentrations are predicted to favor C_3 species over C_4 species. Limited experimental results have shown that, although high CO_2 tends to favor C_3 over C_4 species, there is a wide range of responses and competitive outcomes even among C_3 species (Bazzaz, 1990). The fertilizing effect of widespread nitrogen deposition from industrial areas is a useful analogy for the impacts of climate change on the competitive interactions between species. In areas affected by heavy deposition, rapidly growing species of grasses and forbs are favored over slower growing, nitrogen-efficient shrubs (Chapin, 1980; Field *et al.*, 1992).

A.4.2.2. Herbivory

Several aspects of herbivory might be affected by climate change. Some insect herbivores require cold periods during their life cycle, and these might be affected by warming. Higher rates of carbon fixation (see Section A.2.3) may provide more food for herbivores, but the quality of the food will be affected. It is likely that increased atmospheric CO_2 concentration will lead plants to produce tissue with a higher ratio of carbon to nitrogen (C:N) than under current conditions (Bazzaz, 1990; Field *et al.*, 1992). Many herbivores, especially insects, are limited by not being able to gather enough nitrogen in their diet (Scriber and Slansky, 1981), and a higher C:N ratio will force them to consume more carbon in order to gain a given amount of nitrogen. Some herbivores will increase the amount of tissue they consume to maintain their nitrogen uptake. In this case, damage by herbivores to their plant hosts is likely to increase. In other cases, herbivores will not be able to consume and digest enough plant tissue to maintain their required nitrogen intake. Their nutrition will suffer, and their population numbers will fall; they may even die out (Watt *et al.*, 1995). Some plants may also produce more defensive compounds, such as tannins and phenolics, when grown in elevated CO_2 (e.g., Lavola and Julkunen-Tiitto, 1994). Thus, in those cases, damage by herbivores will decrease in response to increased C:N ratios. Changes in the amount of herbivory will affect the nutrient cycle because nitrogen is cycled more rapidly but is also more likely to be lost to the atmosphere through volatilization when it passes through the guts of herbivores than through litter-decomposing organisms (Chapin and McNaughton, 1989).

Many insect herbivores have a boom–bust population cycle. They build up through several generations, with each generation larger, depleting their plant resources until the population is checked by lack of food, the onset of unsuitable weather (e.g., winter or drought), or natural enemies. If conditions after climate change allow the herbivore population to build up faster, much higher population levels and much greater damage may occur before they are checked. Conversely, some insect herbivores may be disadvantaged by climate change if, for example, the new conditions are more favorable to their enemies or the timing of growth or flowering of their food plants change. There is some evidence that stressful periods such as droughts that lead to reduced photosynthesis also lead to reduced production of compounds that normally help to control insect populations. Thus, some have predicted that under changed climatic conditions, stress periods will be more common and insect outbreaks more frequent (Mattson and Haack, 1987).

A.4.2.3. Other Interactions

There are many other complex interactions within ecological systems. Parasites, disease organisms, and mutualists (i.e., organisms that mutually benefit each other, such as flowering species and their pollinating insects) are subject to the same constraints in relation to niche as other species. Climate change may advantage or disadvantage these species, with significant consequences for their host species. Schemes that help in the assessment of the likely effects of climate change on host–pest interactions are being developed (Landsberg and Stafford Smith, 1992).

A.4.3. Communities and Their Dynamics

A.4.3.1. Community

Any patch of land or ocean contains an assemblage of different species that interact with each other in a variety of different ways. These assemblages are called communities. Just which species are part of a community depends on their niches—that is, whether they can tolerate the physical conditions of the patch and persist in interaction with the other species of the community.

There has long been a debate about the nature of communities and the degree to which they could be described as discrete entities. Communities were once compared with living organisms and described as tightly integrated assemblages of interacting species. Thus, a community was seen as an entity in its own right. If a community was perturbed by the loss or addition of some species, it would tend to recover to its previous composition or change to another discretely recognizable assemblage. The alternative view is that communities are essentially a collection of species, with each species behaving and interacting according to its own physiological and ecological potential. In this view, a community is a more fortuitous collection of species. Species continually invade or are lost from the community as populations fluctuate in response to weather, disturbances, and competitors.

These views about communities represent the end points of a continuum of ideas. Current thinking favors the view that communities are a collection of individual species, but an important ecological research theme is to identify rules that govern community structure (Drake, 1990; Keddy, 1992) and to seek groups of species that are ecological equivalents (sometimes called functional types; Smith *et al.*, 1993).

A.4.3.2. *Succession*

Communities are always changing, both as a result of a changing balance of interactions between the component species and in responses to disturbances. However, there are some patterns in the way communities respond to disturbances and change over time (i.e., succession). Succession was originally described as a sequence of organism-like communities, with one replacing another according to more-or-less strict rules (Clements, 1936). The endpoint of this succession of communities was called the climax community—determined, it was argued, by long-term climatic and soil conditions. Now succession is more often described as the outcome of a series of species losses and invasions as disturbances occur and conditions change (Whittaker, 1975). More emphasis is now placed on the stochastic nature of vegetation change, especially in ecosystems where disturbances such as fire or intense winds are common (Noble and Slatyer, 1980; Shugart, 1984).

The view of communities affects the methods used for predictions of ecological changes in response to global change. If communities change more like "organisms," it would be likely that under climate change one community will be replaced by another already known from elsewhere. This allows simple modeling by matching communities to environmental conditions and assuming that they will redistribute themselves with climate change (see the discussion of Holdridge, etc., in Section A.4.4.2).

However, with a changed climate and disturbance regimes at a given location, conditions for some individuals will remain within their niche space, and they will continue to survive and reproduce at that location. Some species will be lost from the community relatively quickly as existing individuals die under the changed conditions (e.g., extreme temperatures, droughts, waterlogging) or disturbances (e.g., fires). Individuals of other species will be able to survive but not reproduce, and—although some individuals of long-lived species may persist at the site for decades or centuries—the species will eventually be lost from that location.

The loss of individuals from a community creates opportunities for others to invade. However, species best adapted to new conditions will not necessarily be among the early invaders because they may lack the dispersal ability to reach the site. The ability of species to invade new suitable locations varies greatly. Thus, it is unlikely that communities will move en masse. What is more likely is a sequence of invasions and losses leading to completely new communities, as was apparently the case over the past 10,000 years following the last Ice Age (e.g., Davis *et al.*, 1986; Webb and Bartlein, 1992).

A.4.3.3. *Migration and Dispersal*

As the climate changes, zones with suitable habitats for species will move in space. For example, as a result of warmer conditions, a species' niche space will be displaced away from the equator or further up mountains. An important issue in determining the nature of communities in the future is which species will be able to keep up with shifting climate zones.

Most plant species have poor dispersal ability, with the vast majority of seeds falling close to the parent plant. These species migrate very slowly—in most cases, far slower than anticipated shifts in climate zones—and they depend mainly on rare long-distance dispersal events to spread or invade new habitats. The rate at which many species spread after past disturbances, such as the ice ages, can be estimated by mapping the distribution of pollen left behind in bogs, lake sediments, and so forth (e.g., Davis *et al.*, 1986). Most species migration rates, estimated to be 10–30 km per century, are at least an order of magnitude less than the anticipated rates of shift in climate zones of 100 to 600 km over the next century (e.g., Solomon and Cramer, 1993). However, past migration rates are only a poor guide to the future because past rates may have been limited by the rate at which the climate changed. Also, the future potential for species migration is different than it was in the past. On the one hand, landscapes are now much more dissected by human activities than in the past, and this will hinder natural migration rates (di Castri and Hansen, 1992). On the other hand, human activities and technologies, such as motor vehicles and aircraft, provide sources of long-distance dispersal that were not present in the past.

Some species are already adapted to disperse widely and exploit any opportunity for invasion. They are often fast growing and quick to mature. These are often called tramp or ruderal species. They might be expected to do well during any reorganization of communities.

A.4.4. Ecosystems and Biomes

A.4.4.1. Ecosystems

A community and its abiotic environment is called an ecosystem. The abiotic and biotic components of an ecosystem interact, including significant effects of the biotic on the abiotic. For example, plant and animal communities often modify the soils. This process takes many decades or much longer but can eventually result in soils with properties that are closely matched with the communities they support. Where climate change causes significant spatial shifts of the range of species, these relationships may not be retained.

Ecosystem-level feedbacks may either dampen or amplify the growth response to individual climatic variables (Field *et al.*, 1992). The response of the ecosystem may depend on how the complex chain of interactions affects the cycling of nutrients and the efficient use of light and water. For example, if the availability of any one resource, such as carbon, increases, then other resources, such as nutrients, become relatively more limiting. Plants may respond to this by increased root growth, and this allocation shift may reduce the growth response of aboveground biomass to increasing CO_2 concentration.

Furthermore, if growth is determined more by the rate at which nutrients rather than carbon can be obtained, then even total plant production may not be increased. These constraints might eventually restrict the growth response of whole ecosystems to less than might be expected from short-term experiments on single plants or small communities (Körner, 1993). The overall functioning of the ecosystem may sometimes be insensitive to changes in particular species composition. One or more species may be replaced by others with more appropriate temperature optima or responses to CO_2, but the system as a whole may continue to function with broadly the same structure, physiognomy, and physiology.

Conversely, biological nitrogen fixation may be stimulated by more favorable growing conditions and lessen nutrient limitations. In systems that are not strongly nutrient limited, initial growth responses may allow greater leaf-area development, which allows greater light interception and further increases growth responses beyond that predicted from leaf-level responses.

More complex ecosystem responses may also occur. Initially increased growth may lead to reduced palatability of plant tissue for insect herbivores, which may either reduce or increase insect damage (see Section A.4.2.2) and thereby either act as positive or negative feedback on the initial response.

A.4.4.2. Biomes

Most observers readily recognize the similarity of ecosystems in the deserts of the world, no matter where they occur and how far apart they are. There are subtle differences between desert communities, such as the presence of cacti in some deserts and not in others and the diversity of reptiles between different deserts. However, all desert ecosystems operate within similar constraints, and their biota show similar adaptations to deal with these constraints. These similar ecosystems have been described with many different terms (e.g., biomes, life zones, biogeoclimatic zones) and emphases. Many studies relating to the global distribution of biomes and their redistribution under climate change define from 10 to 40 or more classes. Originally, their descriptions were based on the structure and appearance of the vegetation (Walter, 1962), but more recently climatic information has been used (e.g., Box, 1981; Cramer and Leemans, 1993).

A classification of biomes commonly used in global-change research was developed by Holdridge (1947). He used annual biotemperature (a measure related to the sum of temperatures above freezing throughout the year) and total precipitation to classify the vegetation of the world into 14 life zones or biomes. His scheme has been used as the basis for some predictions of the global-scale impacts of changed climates (e.g., Emanuel *et al.*, 1985), although more complex schemes have also been developed based on a wider range of climatic and soil parameters and a wider range of assumptions about the important variables delimiting the vegetation zones of the world (Box, 1981; Prentice *et al.*, 1992; Woodward, 1993; Neilson *et al.*, 1992).

As an example, Figure A-1 shows where some of the world's major biomes are found in relation to temperature and rainfall. In looking at the distribution of any biome in relation to climate, it must be recognized that annual mean values of temperature and precipitation provide only a rough guide to the requirements or tolerances of vegetation classes. In many instances, the distribution of classes also depends on seasonal factors, such as the length of a dry season or the lowest absolute minimum temperature, and on soil factors, such as water-holding capacity.

A.4.4.3. Ecosystem Breakdown

There has been progress in modeling the equilibrium distributions of biomes, which implicitly assumes that communities have time to sort themselves out. But the process of migration and reassembly may well lag behind the rate of climate change. Thus, climate-change studies based on the use of these equilibrium models can provide only a first indication of the direction or magnitude of expected changes.

In addition, there might be rapid breakdown of the existing community structure via the loss of some species and invasion by others. The invading species will tend to be those best

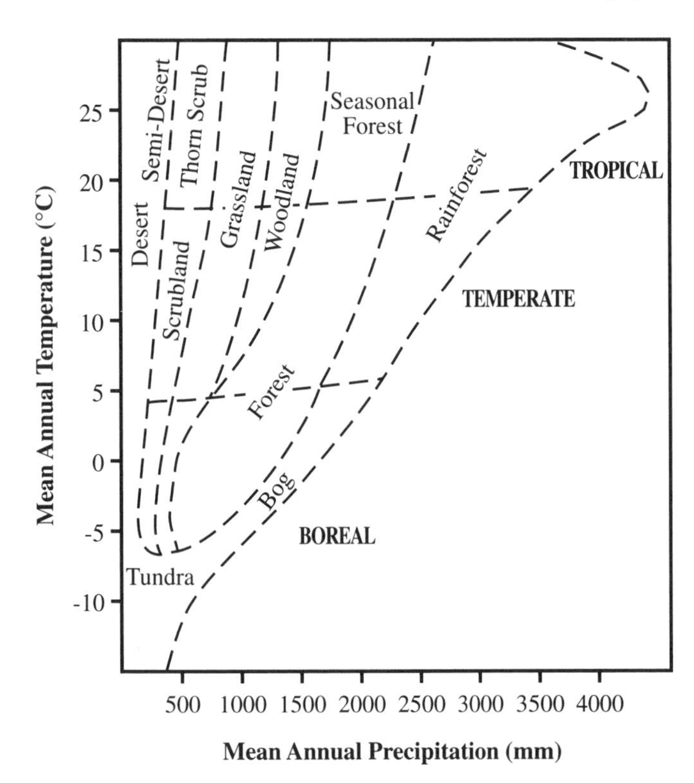

Figure A-1: This figure illustrates that mean annual temperature and mean annual precipitation can be correlated with the distribution of some of the world's major biomes. While the role of these annual means in affecting this distribution appears to be important, it should be noted that the distribution of biomes may also strongly depend on seasonal factors such as the length of the dry or the lowest absolute minimum temperature, on soil properties such as water-holding capacity, on land-use history such as agriculture or grazing, and on disturbance regime such as the frequency of fire (modified from Whittaker, 1975).

adapted to dispersing and invading, but not necessarily those best adapted to the most efficient use of resources in the prevailing conditions. This could lead to transient periods with less efficient use of light and water and less efficient recycling of nutrients. Temporarily lower biomass and transient release of carbon is likely to result from such a breakdown. Considerable scientific uncertainty remains about the length of these transients and the magnitude of the fluxes, but the fluxes could be significant (King and Neilson, 1992; Smith and Shugart, 1993).

A.5. Conclusion

The preceding discussion offers a very brief overview of some of the main factors to be considered in relation to climate change. It describes the response of terrestrial ecosystems to the main aspects of climate change, and how these initial responses are affected by feedbacks from the soil and from interactions between different organisms. The importance of these factors varies greatly among different natural and socioeconomic systems. This will be discussed in greater depth in the following chapters of this report.

References

Adams, J.M., H. Faure, L. Faure-Denard, J.M. McGlade, and F.I. Woodward, 1990: Increases in terrestrial carbon storage from the last glaciation maximum to the present. *Nature*, **348**, 711-714.

Allen, J.C., 1985: Soil responses to forest clearing in the United States and the tropics: geological and biological factors. *Biotropica*, **17**, 15-27.

Allen, L.H., 1990: Plant responses to rising carbon dioxide and potential interactions with air pollutants. *Journal of Environmental Quality*, **19**, 15-34.

Amthor, J.S., 1991: Respiration in a future, higher- CO_2 world. *Plant, Cell and Environment*, **14**, 13-20.

Anderson, J.M., 1992: Responses of soils to climate change. *Advances in Ecological Research*, **22**, 163-210.

Armstrong Brown, S., M.D.A. Rounsevell, V.P. Phillips, and E. Audsley, 1995: Greenhouse gas fluxes and temperate agriculture: a review of environmental and management effects. *Journal of Agricultural Science* (in press).

Arnold, R.W., I. Szabolcs, and V.O. Targulian (eds.), 1990: *Global Soil Change. Report of an IIASA-ISSS-UNEP Task Force on the Role of Soil in Global Change*. International Institute for Applied Systems Analysis, Laxenburg, Austria, 110 pp.

Arnolds, E. and E. Jansen, 1992: New evidence for changes in the macromycete flora of the Netherlands. *Nova Hedwigia*, **55**, 325-351.

Arnone, J.A. and J.C. Gordon, 1990: Effect of nodulation, nitrogen fixation and CO_2 enrichment on the physiology, growth and dry mass allocation of seedlings of Alnus rubra Bong. *New Phytologist*, **116**, 55-66.

Arp, W.J., 1991: Effects of source-sink relations on photosynthetic acclimation to elevated CO_2. *Plant, Cell and Environment*, **14**, 869-875.

Austin, M.P., A.O. Nicholls, and C.R. Margules, 1990: Measurement of the realised qualitative niche: environmental niches in five Eucalyptus species. *Ecological Monographs*, **60**, 161-177.

Bazzaz, F.A., 1990: The response of natural ecosystems to the rising global CO_2 levels. *Annual Review of Ecology and Systematics*, **21**, 167-196.

Bazzaz, F.A., K. Garbutt, E.G. Reekie, and W.E. Williams, 1989: Using growth analysis to interpret competition between a C_3 and a C_4 annual under ambient and elevated CO_2. *Oecologia*, **79**, 223-235.

Begon, M., J.L. Harper, and C.R. Townsend, 1990: *Ecology: Individuals, Populations and Communities*. Blackwell Scientific Publications, Cambridge, UK, 2nd ed., 945 pp.

Berg, B., M.P. Berg, E. Box, P. Bottner, A. Breymeyer, R. Calvo de Anta, M.-M. Couteaux, A. Gallardo, A. Escudero, W. Kratz, M. Madeira, C. McClaugherty, V. Meentemeyer, F. Muñoz, P. Piussi, J. Remacle, A. Virzo de Santo, 1993: Litter mass loss in pine forests of Europe: relationship with climate and litter quality. In: *Geography of Organic Matter Production and Decay* [Breymeyer, A., B. Krawczyk, R. Kulikowski, J. Solon, M. Rosciszewski, and B. Jaworska (eds.)]. Polish Academy of Sciences, Warsaw, Poland, pp. 81-109.

Berry, J.A. and O. Björkman, 1980: Photosynthetic response and adaptation to temperature in higher plants. *Annual Review of Plant Physiology*, **31**, 491-543.

Beuker, E., 1994: Long-term effects of temperature on the wood production of Pinus sylvestris L. and Picea abies (L.) Karst. in old provenance experiments. *Scandinavian Journal of Forest Research*, **9**, 34-45.

Bird, M.I., J. Lloyd, and G.D. Farquhar, 1994: Terrestrial carbon storage at the LGM. *Nature*, **371**, 566.

Bowes, G., 1993: Facing the inevitable: plants and increasing atmospheric CO_2. *Annual Review of Plant Physiology and Plant Molecular Biology*, **44**, 309-332.

Box, E.O., 1981: *Macroclimate and Plant Forms: An Introduction to Predictive Modeling in Phytogeography*. Dr W. Junk Publishers, The Hague, Netherlands, 258 pp.

Bradbury, N.J. and D.S. Powlson, 1994: The potential impact of global environmental change on nitrogen dynamics in arable systems. In: *Soil Responses to Climate Change* [Rounsevell, M.D.A. and P.J. Loveland (eds.)]. NATO ASI Series I: Global Environmental Change, Vol. 23, Springer Verlag, Heidelberg, Germany, pp. 137-154.

Bradford, K.J. and T.C. Hsiao, 1982: Physiological responses to moderate water stress. In: *Physiological Plant Ecology II. Water Relations and Carbon Assimilation* [Lange, O.L., P.S. Nobel, C.B. Osmond, and H. Ziegler (eds.)]. Encyclopedia of Plant Physiology New Series Vol. 12B, Springer-Verlag, Berlin, Germany, pp. 263-324.

Bridges, E.M., 1970: *World Soils*. Cambridge University Press, Cambridge, UK, 89 pp.

Brinkman, R. and W.G. Sombroek, 1995: The effects of global change on soil conditions in relation to plant growth and food production. In: *Global Climate Change and Agricultural Production: Direct and Indirect Effects of Changing Hydrological, Soil and Plant Physiological Processes* [Norse, D. and W.G. Sombroek (eds.)]. John Wiley and Sons, Chichester, UK, pp. 9-24.

Buol, S.W., P.A. Sanchez, J.M. Kimble, and S.B. Weed, 1990: Predicted impacts of climate warming on soil properties and use. In: *Impact of Carbon Dioxide, Trace Gases, and Climate Change on Global Agriculture* [Kimball, B.A., N.J. Rosenberg, and L.H. Allen (eds.)]. ASA Spec. Publ. No. 53, American Society of Agronomy, Madison, WI, pp. 71-82.

Buringh, P., 1984: Organic carbon in world soils. In: *The Role of Terrestrial Vegetation in the Global Carbon Cycle. Measurements in Remote Sensing* [Woodwell, G.M. (ed.)]. SCOPE 23, John Wiley and Sons, New York, NY, pp. 91-109.

Burke, J.J., J.R. Mahan, and J.L. Hatfield, 1988: Crop-specific thermal kinetic windows in relation to wheat and cotton biomass production. *Agronomy Journal,* **80**, 553-556.

Cannell, M.G.R., J. Grace, and A. Booth, 1989: Possible impacts of climatic warming on trees and forests in the United Kingdom: a review. *Forestry,* **62**, 337-364.

Ceulemans, R. and M. Mousseau, 1994: The effects of elevated atmospheric CO_2 on woody plants. *New Phytologist,* **127**, 425-446.

Chapin, F.S. III, 1980: The mineral nutrition of wild plants. *Annual Review of Ecology and Systematics,* **11**, 233-60.

Chapin, F.S. III and S.J. McNaughton, 1989: Lack of compensatory growth under phosphorus deficiency in grazing adapted grasses from the Serengeti Plains. *Oecologia,* **79**, 551-557.

Clements, F.E., 1936: Nature and structure of the climax. *Journal of Ecology,* **24**, 252-284.

Collett, T.S., K.A. Kvenvolden, and L.B. Magoon, 1990: Characterization of hydrocarbon gas within the stratigraphic interval of gas-hydrate stability on the North Slope of Alaska, USA. *Applied Geochemistry,* **5**, 279-287.

Comins, H.N. and R.E. McMurtrie, 1993: Long-term biotic response of nutrient-limited forest ecosystems to CO_2-enrichment; equilibrium behaviour of integrated plant-soil models. *Ecological Applications ,* **3**, 666-681.

Cook, E., T. Bird, M. Peterson, M. Barbetti, B. Buckley, R. D'Arrigo, R. Francey, and P. Tans, 1991: Climatic change in Tasmania inferred from a 1089-year tree-ring chronology of Huon pine. *Science,* **253**, 1266-1268.

Cramer, W.P. and R. Leemans, 1993: Assessing impacts of climate change on vegetation using climate classification systems. In: *Vegetation dynamics and global change* [Solomon, A.M. and H.H. Shugart (eds.)]. Chapman and Hall, New York, NY, pp. 190-217.

Cure, J.D. and B. Acock, 1986: Crop responses to carbon dioxide doubling: a literature survey. *Agricultural and Forest Meteorology,* **38**, 127-145.

Curtis, P.S., B.G. Drake, P.W. Leadley, W.J. Arp, and D.F. Whigham, 1989: Growth and senescence in plant communities exposed to elevated CO_2 concentration on an estuarine marsh. *Oecologia,* **78**, 20-26.

Davis, M.B., K.D. Woods, S.L. Webb, and D. Futyama, 1986: Dispersal versus climate: expansion of Fagus and Tsuga into the Upper Great Lakes Region. *Vegetatio,* **67**, 93-103.

Detwiler, R.P., 1986: Land use change and the global carbon cycle: the role of tropical soils. *Biogeochemistry,* **2,** 67-93.

di Castri, F. and A.J. Hansen, 1992: The environment and development crisis as determinants of landscape dynamics. In: *Landscape Boundaries: Consequences for Biotic Diversity and Ecological Flows* [Hansen, A.J. and F. di Castri (eds.)]. Ecological Studies No. 92, Springer Verlag, New York, NY, pp. 1-18.

Diaz, S., J.P. Grime, J. Harris, and E. Mcpherson, 1993: Evidence of a feedback mechanism limiting plant response to elevated carbon dioxide. *Nature,* **364,** 616-617.

Diemer, M.W., 1994: Mid-season gas exchange of an alpine grassland under elevated CO_2. *Oecologia,* **98,** 429-435.

Drake, B.G., 1992: A field study of the effects of elevated CO_2 on ecosystem processes in a Chesapeake Bay wetland. *Australian Journal of Botany,* **40,** 579-595.

Drake, J.A., 1990: Communities as assembled structures: do rules govern pattern? *Trends in Ecology and Evolution,* **5**, 159-164.

Eamus, D., 1991: The interaction of rising CO_2 and temperature with water use efficiency. *Plant, Cell and Environment,* **14**, 843-852.

Eamus, D. and P.G. Jarvis, 1989: The direct effects of increase in the global atmospheric CO_2 concentration on natural and commercial temperate trees and forests. *Advances in Ecological Research,* **19**, 1-55.

Emanuel, W.R., H.H. Shugart, and M.P. Stevenson, 1985: Climatic change and the broad-scale distribution of terrestrial ecosystem complexes. *Climatic Change,* **7**, 29-43.

Eswaran, H., E. van den Berg, and P. Reich, 1993: Organic carbon in soils of the world. *Soil Science Society of America Journal,* **57**, 192-194.

Farquhar, G.D. and S. von Caemmerer, 1982: Modelling of photosynthetic response to environmental conditions. In: *Physiological Plant Ecology II. Water Relations and Carbon Assimilation* [Lange, O.L., P.S. Nobel, C.B. Osmond, and H. Ziegler (eds.)]. Encyclopedia of Plant Physiology New Series Vol. 12B, Springer-Verlag, Berlin, Germany, pp. 549-588.

Field, C.R., F.S. Chapin III, P.A. Matson, and H.A. Mooney, 1992: Responses to terrestrial ecosystems to the changing atmosphere: a resource-based approach. *Annual Review of Ecology and Systematics,* **23**, 201-235.

Gates, W.L., J.F.B. Mitchell, G.J. Boer, U. Cubasch, and V.P. Meleshko, 1992: Climate modelling, climate prediction, and model validation. In: *Climate Change 1992: The Supplementary Report to the IPCC Scientific Assessment* [Houghton, J.T., B.A. Callander, and S.K. Varney (eds.)]. Cambridge University Press, Cambridge, UK, pp. 96-134.

Gifford, R.M., 1979: Growth and yield of CO_2-enriched wheat under water-limited conditions. *Australian Journal of Plant Physiology,* **6**, 367-378.

Gifford, R.M., 1992: Interaction of carbon dioxide with growth-limiting environmental factors in vegetation productivity: implications for the global carbon cycle. *Advances in Bioclimatology,* **1**, 24-58.

Gifford, R.M., 1994: The global carbon cycle: a viewpoint on the missing sink. *Australian Journal of Plant Physiology,* **21**, 1-15.

Gleick, P., 1987: Regional hydrologic consequences of increases in atmospheric CO_2 and other trace gases. *Climatic Change,* **10**, 137-161.

Gordon, P.H., P.H. Whetton, A.B. Pittock, A.M. Fowler, and M.R. Haylock, 1992: Simulated changes in daily rainfall intensity due to the enhanced greenhouse effect: implications for extreme rainfall events. *Climate Dynamics,* **8**, 83-102.

Graumlich, L., 1991: Subalpine tree growth, climate and increasing CO_2: an assessment of recent growth trends. *Ecology,* **72**, 1-11.

Graybill, D.A. and S.B. Idso, 1993: Detecting the aerial fertilization effect of atmospheric CO_2 enrichment in tree-ring chronologies. *Global Biogeochemical Cycles,* **7**, 81-95.

Greco, S., R.H. Moss, D. Viner, and R. Jenne, 1994: *Climate Scenarios and Socioeconomic Projections for IPCC WGII Assessment*. IPCC, Washington DC, 67 pp.

Gunderson, C.A. and S.D. Wullschleger, 1994: Photosynthetic acclimation in trees to rising atmospheric CO_2: a broader perspective. *Photosynthesis Research,* **39**, 369-388.

Hänninen, H., 1991: Does climatic warming increase the risk of frost damage in northern trees? *Plant, Cell and Environment,* **14**, 449-454.

Hari, P. and H. Arovaara, 1988: Detecting CO_2 induced enhancement in the radial increment of trees. Evidence from northern timber line. *Scandinavian Journal of Forest Research,* **3**, 67-74.

Hayes, J.T., 1991: Global climate change and water resources. In: *Global Climate Change and Life on Earth.* [Wyman, R.L. (ed.)]. Chapman and Hall, New York and London, pp. 18-42.

Heath, L.S., P.E. Kauppi, P. Burschel, H.-D. Gregor, R. Guderian, G.H. Kohlmaier, S. Lorenz, D. Overdieck, F. Scholz, H. Thomasius, and M. Weber, 1993: Contribution of temperate forests to the world's carbon budget. *Water, Air, and Soil Pollution,* **70**, 55-69.

Henderson, S.A., S. von Caemmerer, and G.D. Farquhar, 1992: Short-term measurements of carbon isotope discrimination in several C_4 species. *Australian Journal of Plant Physiology,* **19**, 263-285.

Henderson-Sellers, A., K. McGuffie, and C. Gross, 1995: Sensitivity of global climate model simulations to increased stomatal resistance and CO_2 increase. *Journal of Climatology* (in press).

Hinckley, T.M., R.O. Teskey, F. Duhme, and H. Richter, 1981: Temperate hardwood forests. In: *Water deficits and plant growth. VI. Woody plant communities* [Kozlowski, T.T. (ed.)]. Academic Press, New York, NY, pp. 153-208.

Holdridge, L.R., 1947: Determination of world plant formations from simple climatic data. *Science,* **105**, 367-368.

Hsiao, T., 1973: Plant responses to water stress. *Annual Review of Plant Physiology,* **24**, 519-570.

Hutchinson, G.E., 1957: Concluding remarks. *Cold Spring Harbor Symposium on Quantitative Biology,* **22**, 415-427.

Hütsch, B.W., C.P. Webster, and D.S. Powlson, 1994: Methane oxidation in soil as affected by land use, soil pH and N fertilization. *Soil Biology and Biochemistry,* **26**, 1613-1622.

Idso, S.B. and A.J. Brazel, 1984: Rising atmospheric carbon dioxide concentrations may increase streamflow. *Nature,* **312**, 51-53.

Idso, K.E. and S.B. Idso, 1994: Plant responses to atmospheric CO_2 enrichment in the face of environmental constraints: a review of the past 10 years' research. *Agricultural and Forest Meteorology,* **69**, 153-203.

Idso, S.B., B.A. Kimball, M.G. Anderson, and J.R. Mauney, 1987: Effects of atmospheric CO_2 enrichment on plant growth: the interactive role of air temperature. *Agriculture, Ecosystems and the Environment,* **20**, 1-10.

Innes, J.L., 1991: High-altitude and high-latitude tree growth in relation to past, present and future global climate change. *The Holocene*, **1**, 168-173.

Jackson, R.B., O.E. Sala, C.B. Field, and H.A. Mooney, 1994: CO_2 alters water use, carbon gain, and yield for the dominant species in a natural grassland. *Oecologia,* **98**, 257-262.

Jarvis, P.G. and K.G. McNaughton, 1986: Stomatal control of transpiration: scaling up from leaf to region. *Advances in Ecological Ressearch,* **15**, 1-49.

Jenkinson, D.S., 1990: The turnover of organic carbon and nitrogen in soil. *Philosophical Transactions of the Royal Society London,* **B329**, 361-368.

Jenkinson, D.S., D.E. Adams, and A. Wild, 1991: Model estimates of CO_2 emissions from soil in response to global warming. *Nature,* **351**, 304-306.

Jenny, H., 1941: *Factors of Soil Formation.* McGraw-Hill, New York, NY, 281 pp.

Jenny, H., 1980: *The Soil Resource: Origin and Behavior.* Ecological studies No. 37, Springer Verlag, New York, NY, 377 pp.

Karl, T.R., P.D. Jones, R.W. Knight, G. Kukla, N. Plummer, V. Razuvayev, K.P. Gallo, J. Lindseay, R.J. Charlson, and T.C. Peterson, 1993: A new perspective on recent global warming: asymmetric trends of daily maximum and minimum temperature. *Bulletin of the American Meteorological Society,* **74**, 1007-1023.

Katz, R.W. and B.G. Brown, 1992: Extreme events in a changing climate: variability is more important than averages. *Climatic Change,* **21**, 289-302.

Kauppi, P. and M. Posch, 1985: Sensitivity of boreal forests to possible climatic warming. *Climatic Change,* **7**, 45-54.

Keddy, P.A., 1992: Assembly and response rules: two goals for predictive community ecology. *Journal of Vegetation Science,* **3**, 157-164.

Kienast, F. and R.J. Luxmoore, 1988: Tree ring analysis and conifer growth responses to increased atmospheric CO_2 levels. *Oecologica,* **76**, 487-495.

Kimball, B.A., 1983: Carbon-dioxide and agricultural yield: an assemblage and analysis of 430 prior observations. *Agronomy Journal,* **75**, 779-788.

Kimball, B.A. and S.B. Idso, 1983: Increasing atmospheric CO_2 effects on crop yield, water use and climate. *Agriculture and Water Management,* **7**, 55-72.

King, G.A. and R.P. Neilson, 1992: The transient response of vegetation to climate change: a potential source of CO_2 to the atmosphere. *Water, Air, and Soil Pollution,* **64**, 365-383.

King, G.M. and S. Schnell, 1994. Effect of increasing atmospheric methane concentration on ammonium inhibition of soil methane consumption. *Nature,* **370**, 282-284.

Kirschbaum, M.U.F., 1993: A modelling study of the effects of changes in atmospheric CO_2 concentration, temperature and atmospheric nitrogen input on soil organic carbon storage. *Tellus,* **45B**, 321-334.

Kirschbaum, M.U.F., 1994: The sensitivity of C_3 photosynthesis to increasing CO_2 concentration. A theoretical analysis of its dependence on temperature and background CO_2 concentration. *Plant, Cell and Environment,* **17**, 747-754.

Kirschbaum, M.U.F., 1995: The temperature dependence of soil organic matter decomposition, and the effect of global warming on soil organic carbon storage. *Soil Biology and Biochemistry,* **27**, 753-760.

Kirschbaum, M.U.F., D.A. King, H.N. Comins, R.E. McMurtrie, B.E. Medlyn, S. Pongracic, D. Murty, H. Keith, R.J. Raison, P.K. Khanna, and D.W. Sheriff, 1994: Modelling forest response to increasing CO_2 concentration under nutrient-limited conditions. *Plant, Cell and Environment,* **19**, 1081-1099.

Kjoller, A. and S. Struwe, 1982: Microfungi in ecosystems: fungal occurrence and activity in litter and soil. *Oikos,* **39**, 389-422.

Knowles, R., 1993: Methane: processes of production and consumption. In: *Agricultural Ecosystem Effects on Trace Gases and Global Climate Change.* ASA Special Publication 55, American Society of Agronomy, Madison, WI, pp. 145-156.

Kohlmaier, G.H., M. Lüdeke, A. Janecek, G. Benderoth, J. Kindermann, and A. Klaudius, 1991: Land biota, source or sink of atmospheric carbon dioxide: positive and negative feedbacks within a changing climate and land use development. In: *Scientists on Gaia* [Schneider, S.H. and P.J. Boston (eds.)]. MIT Press, Cambridge, MA, pp. 223-239.

Kokorin, A.O., G.V. Mironova, and N.V. Semenuk, 1993: The influence of climatic variations on the growth of cedar and fir in south Baikal region. In: *Global and Regional Consequences of the Climatic and Environmental Change* [Izrael, Y.A. and Y.A. Anokhin (eds.)] Gidrometeoizdat, St. Petersburg, Russia (in Russian).

Körner, Ch. and J.A. Arnone III, 1992: Responses to elevated carbon dioxide in artificial tropical ecosystems. *Science,* **257**, 1672-1675.

Körner, Ch., 1993: CO_2 fertilization: the great uncertainty in future vegetation development. In: *Vegetation Dynamics and Global Change* [Solomon, A.M. and H.H. Shugart (eds.)]. Chapman and Hall, New York & London, pp. 53-70.

Körner, Ch., 1995: The response of complex multispecies systems to elevated CO_2. In: *Global Change and Terrestrial Ecosystems* [Walker, B.H. and W.L. Steffen (eds.)]. Cambridge University Press, Cambridge, UK (in press).

Kvenvolden, K.A., 1993: Gas hydrates—geological perspective and global change. *Review of Geophysics,* **31**, 173-187.

Landsberg, J. and M. Stafford Smith, 1992: A functional scheme for predicting the outbreak potential of herbivorous insects under global climate change. *Australian Journal of Botany,* **40**, 565-577.

LaMarche, V.C., D.A. Graybill, H.C. Fritts, and M.R. Rose, 1984: Increasing atmospheric carbon dioxide: tree ring evidence for enhancement in natural vegetation. *Science,* **225**, 1019-1021.

Lavola, A. and R. Julkunen-Tiitto, 1994: The effect of elevated carbon dioxide and fertilization on primary and secondary metabolites in birch, Betula pendula (Roth). *Oecologia,* **99**, 315-321.

Le Houerou, H.N., G.F. Popov, and L. See, 1993: *Agro-Bioclimatic Classification of Africa.* FAO Agrometeorological Series Working Paper No. 6, FAO, Rome, Italy, 227 pp.

Leichenko, R.M., 1993: Climate change and water resource availability: an impact assessment for Bombay and Madras, India. *Water International,* **18**, 147-156.

Lerdau, M., 1991: Plant function and biogenic terpene emissions. In: *Trace Gas Emissions from Plants* [Sharkey, T.D., E.A. Holland, and H.A. Mooney (eds.)]. Academic Press, San Diego, CA, pp. 121-134.

Lieth, H., 1973: Primary production: terrestrial ecosystems. *Human Ecology,* **1**, 303-332.

Linder, S. and J. Flower-Ellis, 1992: Environmental and physiological constraints to forest yield. In: *Responses of Forest Ecosystems to Environmental Changes* [Teller, A., P. Mathy, and J.N.R. Jeffers (eds.)]. Elsevier Applied Science, London and New York, pp. 149-164.

Lloyd, J. and J.A. Taylor, 1994: On the temperature dependence of soil respiration. *Functional Ecology,* **8**, 315-323.

Long, S.P., 1991: Modification of the response of photosynthetic productivity to rising temperature by atmospheric CO_2 concentration: has its importance been underestimated? *Plant, Cell and Environment,* **14**, 729-739.

Lutze, J.L. and R.M. Gifford, 1995: Carbon storage and productivity of a carbon dioxide enriched nitrogen limited grass sward after one year's growth. *Journal of Biogeography* (in press).

Luxmoore, R.J., S.D. Wullschleger, and P.J. Hanson, 1993: Forest responses to CO_2 enrichment and climate warming. *Water, Air, and Soil Pollution,* **70**, 309-323.

Martin, P., N.J. Rosenberg, and M.S. McKenney, 1989: Sensitivity of evapotranspiration in a wheat field, a forest, and a grassland to changes in climate and direct effects of carbon dioxide. *Climatic Change,* **14**, 117-151.

Mattson, W.J. and R.A. Haack, 1987: The role of drought in outbreaks of plant-eating insects. *Bioscience,* **37**, 110-118.

McKenney, M.S. and N.J. Rosenberg, 1993: Sensitivity of some potential evapotranspiration estimation methods to climate change. *Agricultural and Forest Meteorology,* **64**, 81-110.

Melillo, J.M., T.V. Callaghan, F.I. Woodward, E. Salati, and S.K. Sinha, 1990: Effects on ecosystems. In: *Climate Change: The IPCC Scientific Assessment* [Houghton, J.T., G.J. Jenkins, and J.J. Ephraums (eds.)]. Cambridge University Press, Cambridge, UK, pp. 283-310.

Melillo, J.M., A.D. McGuire, D.W. Kicklighter, B. Moore III, C.J. Vorosmarty, and A.L. Schloss, 1993: Global climate change and terrestrial net primary production. *Nature,* **363**, 234-240.

Mitchell, J.F.B., S. Manabe, V. Meleshko, and T. Tokioka, 1990: Equilibrium climate change—and its implications for the future. In: *Climate Change: The IPCC Scientific Assessment* [Houghton, J.T., G.J. Jenkins, and J.J. Ephraums (eds.)]. Cambridge University Press, Cambridge, UK, pp. 135-174.

Monteith, J.L., 1981: Climatic variation and the growth of crops. *Quarterly Journal of the Royal Meteorological Society,* **107**, 749-774.

Morison, J.I.L., 1987: Intercellular CO_2 concentration and stomatal responses to CO_2. In: *Stomatal Function* [Zeiger, E., G.D. Farquhar, and I.R. Cowan (eds.)]. University Press, Stanford, CA, pp. 229-251.

Morison, J.I.L. and R.M. Gifford, 1983: Stomatal sensitivity to carbon dioxide and humidity. *Plant Physiology,* **71**, 789-796.

Murray, M.B., R.I. Smith, I.D. Leith, D. Fowler, H.S.J. Lee, A.D. Friend, and P.G. Jarvis, 1994: Effects of elevated CO_2, nutrition and climatic warming on bud phenology in Sitka spruce (Picea sitchensis) and their impact on the risk of frost damage. *Tree Physiology,* **14**, 691-706.

Neilson, R.P, G.A. King, and G. Koerper, 1992: Towards a rule-based biome model. *Landscape Ecology,* **7**, 27-43.

Noble, I.R. and R.O. Slatyer, 1980: The use of vital attributes to predict successional changes in plant communities subject to recurrent disturbance. *Vegetatio,* **43**, 5-21.

Norby, R.J., 1987: Nodulation and nitrogenase activity in nitrogen-fixing woody plants stimulated by CO_2 enrichment of the atmosphere. *Physiologia Plantarum,* **71**, 77-82.

Norby, R.J., C.A. Gunderson, S.D. Wullschleger, E.G. O'Neill, and M.K. McCracken, 1992: Productivity and compensatory responses of yellow-poplar trees in elevated CO_2. *Nature,* **357**, 322-324.

Norby, R.J., E.G. O'Neill, and R.J. Luxmoore, 1986: Effects of atmospheric CO_2 enrichment on growth and mineral nutrition of Quercus alba seedlings in nutrient-poor soil. *Plant Physiology,* **82**, 83-89.

Parton, W.J., D.S. Schimel, C.V. Cole, and D.S. Ojima, 1987: Analysis of factors controlling soil organic matter levels in Great Plains grasslands. *Soil Science Society of America Journal,* **51**, 1173-1179.

Passioura, J.B., 1982: Water in the soil-plant-atmosphere continuum. In: *Physiological Plant Ecology II. Water Relations and Carbon Assimilation* [Lange, O.L., P.S. Nobel, C.B. Osmond, and H. Ziegler (eds.)]. Encyclopedia of Plant Physiology New Series Vol. 12B, Springer-Verlag, Berlin, Germany, pp. 5-33.

Pearcy, R.W., K. Osteryoung, and D. Randall, 1982: Carbon dioxide exchange characteristics of C_4 Hawaiian Euphorbia species native to diverse habitats. *Oecologia,* **55**, 333-341.

Polglase, P.J. and Y.P. Wang, 1992: Potential CO_2-enhanced carbon storage by the terrestrial biosphere. *Australian Journal of Botany,* **40**, 641-656.

Polley, H.W., J.M. Norman, T.J. Arkebauer, E.A. Walter-Shea, D.H. Greegor, Jr., and B. Bramer, 1992: Leaf gas exchange of Andropogon gerardii Vitman, Panicum virgatum L. and Sorghastrum nutans (L.) Nash in a tallgrass prairie. *Journal of Geophysical Research,* **97**, 18,837-844.

Poorter, H., 1993: Interspecific variation in growth response of plants to an elevated ambient CO_2 concentration. *Vegetatio,* **104/105**, 77-97.

Poorter, H., R.M. Gifford, P.E. Kriedemann, and S.C. Wong, 1992: A quantitative analysis of dark respiration and carbon content as factors in the growth response of plants to elevated CO_2. *Australian Journal of Botany,* **40**, 501-513.

Post, W.M., W.R. Emanuel, P.J. Zinke, and A.G. Stangenberger, 1982: Soil carbon pools and world life zones. *Nature,* **298**, 156-159.

Post, W.M., J. Pastor, P.J. Zinke, and A.G. Stangenberger, 1985: Global patterns of soil nitrogen storage. *Nature,* **317**, 613-616.

Prather, M., R. Derwent, D. Ehhalt, P. Fraser, E. Sanhueza, and X. Zhou, 1994: Other trace gases and atmospheric chemistry. In: *Climate Change 1994* [Houghton, J.T., L.G. Meira Filho, J. Bruce, Hoesung Lee, B.A. Callander, E. Haites, N. Harris, and K. Maskell (eds.)]. Cambridge University Press, Cambridge, UK, pp. 72-126.

Prentice, I.C., W. Cramer, S.P. Harrison, R. Leemans, R.A. Monserud, and A.M. Solomon, 1992: A global biome model based on plant physiology and dominance, soil properties, and climate. *Journal of Biogeography,* **19**, 117-134.

Prentice, K.C. and I.Y. Fung, 1990: The sensitivity of terrestrial carbon storage to climate change. *Nature,* **346**, 48-50.

Rastetter, E.B., M.G. Ryan, G.R. Shaver, J.M. Melillo, K.J. Nadelhoffer, J.E. Hobbie, and J.D. Aber, 1991: A general biogeochemical model describing the responses of the C and N cycles in terrestrial ecosystems to changes in CO_2, climate and N deposition. *Tree Physiology,* **9**, 101-126.

Rastetter, E.B., R.B. McKane, G.R. Shaver, and J.M. Melillo, 1992: Changes in C storage by terrestrial ecosystems: how C-N interactions restrict responses to CO_2 and temperature. *Water, Air, and Soil Pollution,* **64**, 327-344.

Rawson, H.M., 1992: Plant responses to temperature under conditions of elevated CO_2. *Australian Journal of Botany,* **40**, 473-490.

Reynolds, J.F. and P.W. Leadley, 1992: Modelling the response of arctic plants to changing climate. In: *Arctic Ecosystems in a Changing Climate, an Ecophysiological Perspective* [Chapin, F.S. III, R.L. Jefferies, J.F. Reynolds, G.R. Shaver, and J. Svoboda (eds.)]. Academic Press, San Diego, USA, pp. 413-438.

Rind, D., R. Goldberg, J. Hansen, C. Rosenzweig, and R. Ruedy, 1990: Potential evapotranspiration and the likelihood of future drought. *Journal of Geophysical Research,* **95**, 9983-10,004.

Rogers, H.H., J.F. Thomas, and G.E. Bingham, 1983: Response of agronomic and forest species to elevated atmospheric carbon dioxide. *Science,* **220**, 428-429.

Rounsevell, M.D.A. and R.J.A. Jones, 1993: A soil and agroclimatic model for estimating machinery work days: the basic model and climatic sensitivity. *Soil and Tillage Research,* **26**, 179-191.

Rounsevell, M.D.A., R.J.A. Jones, and A.P. Brignall, 1994: Climate change effects on autumn soil tillage and crop potential in England and Wales. In: *Proceedings of the 13th ISTRO Conference*, Aalborg, Denmark, pp. 1175-1180.

Sage, R.F., 1994: Acclimation of photosynthesis to increasing atmospheric CO_2: the gas exchange perspective. *Photosynthesis Research,* **39**, 351-368.

Samarakoon, A.B. and R.M. Gifford, 1995: Soil water content under plants at high CO_2 concentration and interactions with the direct CO_2 effects: a species comparison. *Journal of Biogeography* (in press).

Schimel, D.S., W.J. Parton, T.G.F. Kittel, D.S. Ojima, and C.V. Cole, 1990: Grassland biogeochemistry: links to atmospheric processes. *Climatic Change,* **17**, 13-25.

Schimel, D., I. Enting, M. Heimann, T. Wigley, D. Raynaud, D. Alves, and U. Siegenthaler, 1994: CO_2 and the carbon cycle. In: *Climate Change 1994* [Houghton, J.T., L.G. Meira Filho, J. Bruce, Hoesung Lee, B.A. Callander, E. Haites, N. Harris, and K. Maskell (eds.)]. Cambridge University Press, Cambridge, UK, pp. 35-71.

Schlesinger, W.H., 1977: Carbon balance in terrestrial detritus. *Annual Review of Ecology and Systematics,* **8**, 51-81.

Schlesinger, W.H., 1986: Changes in soil carbon storage and associated properties with disturbance and recovery. In: *The Changing Carbon Cycle: A Global Analysis* [Trabalka, J.R. and D.E Reichle (eds.)]. Springer-Verlag, New York, NY, pp. 175-193.

Schulze, E.-D. and A.E. Hall, 1982: Stomatal responses, water loss and CO_2 assimilation rates of plants in contrasting environments. In: *Physiological Plant Ecology II. Water Relations and Carbon Assimilation* (Lange, O.L., P.S. Nobel, C.B. Osmond, and H. Ziegler (eds.)]. Encyclopedia of Plant Physiology New Series Vol. 12B, Springer-Verlag, Berlin, Germany, pp. 181-261.

Schütz, H., P. Schröder, and H. Renneberg, 1991: Role of plants in regulating the methane flux to the atmosphere. In: *Trace Gas Emissions from Plants* [Sharkey, T.D., E.A. Holland, and H.A. Mooney (eds.)]. Academic Press, San Diego, CA, pp. 29-63.

Scriber, J.M. and F. Slansky, 1981: The nutritional ecology of immature insects. *Annual Review of Entomology,* **26**, 183-211.

Sharkey, T.D., 1985: Photosynthesis in intact leaves of C_3 plants: physics, physiology and rate limitations. *Botanical Review,* **51**, 53-105.

Shaver, G.R., W.D. Billings, F.S. Chapin III, A.E. Giblin, K.J. Nadelhoffer, W.C. Oechel, and E.B. Rastetter, 1992: Global change and the carbon balance of arctic ecosystems. *Bioscience,* **42**, 433-441.

Shugart, H.H., 1984: *A Theory of Forest Dynamics*. Springer-Verlag, New York, NY, 278 pp.

Smith, T.M., R. Leemans, and H.H. Shugart, 1992: Sensitivity of terrestrial carbon storage to CO_2-induced climate change: comparison of four scenarios based on general circulation models. *Climatic Change*, **21**, 367-384.

Smith, T.M. and H.H. Shugart, 1993: The transient response of terrestrial carbon storage to a perturbed climate. *Nature*, **361**, 523-526.

Smith, T.M., H.H. Shugart, F.I. Woodward, and P.J. Burton, 1993: Plant functional types. In: *Vegetation Dynamics and Global Change* [Solomon, A.M. and H.H. Shugart (eds.)]. Chapman and Hall, New York, USA, pp. 272-292.

Solomon, A.M. and W.P. Cramer, 1993: Biospheric implications of global environmental change In: *Vegetation Dynamics and Global Change* [Solomon, A.M. and H.H. Shugart (eds.)]. Chapman and Hall, New York, USA, pp. 25-52.

Sombroek, W.G., 1990: Soils on a warmer earth: tropical and subtropical regions. In: *Soils on a Warmer Earth* [Scharpenseel, H.W., M. Schomaker, and A. Ayoub (eds.)]. Elsevier, Amsterdam, Netherlands, pp. 157-174.

Stevenson, N.L., 1990: Climatic control of vegetation distribution: the role of the water balance. *The American Naturalist*, **135**, 649-670.

Stewart, J.W.B., D.W. Anderson, E.T. Elliot, and C.V. Cole, 1990: The use of models of soil pedogenic processes in understanding changing land use and climate conditions. In: *Soils on a Warmer Earth* [Scharpenseel, H.W., M. Schomaker, and A. Ayoub (eds.)]. Elsevier, Amsterdam, Netherlands, pp. 131-132.

Thomas, R.B. and B.R. Strain, 1991: Root restriction as a factor in photosynthetic acclimation of cotton seedlings grown in elevated carbon dioxide. *Plant Physiology*, **96**, 627-634.

Thomas, R.B., D.D. Richter, H. Ye, P.R. Heine, and B.R. Strain, 1991: Nitrogen dynamics and growth of seedlings of an N-fixing tree (Gliricidia sepium (Jacq.) Walp.) exposed to elevated atmospheric carbon dioxide. *Oecologia*, **88**, 415-421.

Thornley, J.H.M., D. Fowler, and M.G.R. Cannell, 1991: Terrestrial carbon storage resulting from CO_2 and nitrogen fertilization in temperate grasslands. *Plant, Cell and Environment*, **14**, 1007-1011.

Thornthwaite, C.W., 1948: An approach towards a rational classification of climate. *Geographical Review*, **38**, 55-94.

Tingey, D.T., D.P. Turner, and J.A. Weber, 1991: Factors controlling the emissions of monoterpenes and volatile organic compounds. In: *Trace Gas Emissions from Plants*. [Sharkey, T.D., E.A. Holland, and H.A. Mooney (eds.)]. Academic Press, San Diego, CA, pp. 93-119.

Tinker, P.B. and J.S.I. Ingram, 1994: Soils and global change: an overview. In: *Soil Response to Climate Change* [Rounsevell, M.D.A. and P.J. Loveland (eds.)]. NATO ASI Series I: Global Environmental Change, Vol. 23. Springer-Verlag, Berlin and New York, pp. 3-12.

Tissue, D.L. and W.C. Oechel, 1987: Physiological response of Eriophorum vaginatum to elevated CO_2 and temperature in the Alaskan tussock tundra. *Ecology*, **68**, 401-410.

Tolley, L.C. and B.R. Strain, 1985: Effects of CO_2 enrichment and water stress on gas exchange of Liquidambar styraciflua and Pinus taeda seedlings grown under different irradiance levels. *Oecologia*, **65**, 166-172.

Ulrich, B., 1991: Introduction to acidic deposition effects: critical deposition rates and emission densities. In: *Acid Depositions in Europe* [Chadwick, M.J. and M. Hutton (eds.)]. Stockholm Environment Institute, Stochkolm, Sweden, pp. 2-16.

Ulrich, B., 1994: Process hierarchy in forest ecosystems—an integrating ecosystem theory. In: *Effects of Acid Rain on Forest Processes* [Godbold, D.L. and A. Hüttermann (eds.)]. Wiley, New York, NY, 419 pp.

van Cleve, K., R. Barney, and R. Schlentner, 1981: Evidence of temperature control of production and nutrient cycling in two interior Alaska black spruce ecosystems. *Canadian Journal of Forest Research*, **11**, 258-273.

van Cleve, K., W.C. Oechel, and J.L. Hom, 1990: Response of black spruce (Picea mariana) ecosystems to soil temperature modification in interior Alaska. *Canadian Journal of Forest Research*, **20**, 1530-1535.

van Veen, J.A., E. Liljeroth, L.J.A. Lekkerkerk, and S.C. van de Gejin, 1991: Carbon fluxes in plant-soil systems at elevated atmospheric CO_2 levels. *Ecological Applications*, **1**, 175-181.

van Veen, J.A. and E.A. Paul, 1981: Organic carbon dynamics in grassland soils. 1. Background information and computer simulations. *Canadian Journal of Soil Science*, **61**, 185-201.

Walter, H., 1962: *Die Vegetation der Erde in ökophysiologischer Betrachung*. Vol. 1, VEB Gustav Fischer Verlag, Jena, Germany, 538 pp.

Wang, Y.P., Jr. Handoko, and G.M. Rimmington, 1992: Sensitivity of wheat growth to increased air temperature for different scenarios of ambient CO_2 concentration and rainfall in Victoria—a simulation study. *Climate Research*, **2**, 131-149.

Warrick, R.A., R.M. Gifford, and M.L. Parry, 1986: CO_2, climatic change and agriculture. In: *The Greenhouse Effect, Climate Change and Ecosystems* [Bolin, B., B.R. Döös, J. Jäger, and R.A. Warrick (eds.)]. SCOPE 29, Wiley, Chichester, UK, pp. 363-392.

Watson, R., H. Rodhe, H. Oeschger, and U. Siegenthaler, 1990: Greenhouse gases and aerosols. In *Climate Change: The IPCC Scientific Assessment* [Houghton, J.T., G.J. Jenkins, and J.J. Ephraums (eds.)]. Cambridge University Press, Cambridge, UK, pp. 1-40.

Watt, A.D., J.B. Whittaker, M. Docherty, G. Brooks, E. Lindsay, and D. Salt, 1995: The impact of elevated CO_2 on insect herbivores. In: *Insects in a Changing Environment: 17th Symposium of the Royal Entomological Society* [Harrington, R. and N.E. Stork (eds.)]. Academic Press, London, UK, pp. 197-217.

Webb, T. III and P.J. Bartlein, 1992: Global changes during the last 3 million years: climatic controls and biotic responses. *Annual Review of Ecology and Systematics*, **23**, 141-173.

West, D.C., T.W. Doyle, M.L. Tharp, J.J. Beauchamp, W.J. Platt, and D.J. Downing, 1993: Recent growth increases in old-growth longleaf pine. *Canadian Journal of Forest Research*, **23**, 846-853.

White, R.E., 1987: *Introduction to the Principles and Practice of Soil Science*. 2nd ed. Blackwell Scientific Publications, Oxford, UK, 244 pp.

Whittaker, R.H., 1975: *Communities and Ecosystems*. 2nd ed. Macmillan Publishing Co., Inc., New York, USA, 387 pp.

Wong, S.C., 1979: Elevated atmospheric partial pressure of CO_2 and plant growth. *Oecologia*, **44**, 68-74.

Wong, S.C. and C.B. Osmond, 1991: Elevated atmospheric partial pressure of CO_2 and plant growth. III. Interactions between Triticum aestivum (C_3) and Echinochloa frumentacea (C_4) during growth in mixed culture under different CO_2, N nutrition and irradiance treatments, with emphasis on below-ground responses estimated using $\partial^{13}C$ value of root biomass. *Australian Journal of Plant Physiology*, **18**, 137-152.

Woodward, F.I., 1987: *Climate and Plant Distribution*. Cambridge University Press, Cambridge, UK, 174 pp.

Woodward, F.I., 1993: Leaf responses to the environment and extrapolation to larger scales. In: *Vegetation Dynamics and Global Changes* [Solomon, A.M. and H.H. Shugart (eds.)]. Chapman and Hall, New York, USA, pp. 71-100.

Wullschleger, S.D. and R.J. Norby, 1992: Respiratory cost of leaf growth and maintenance in white oak saplings exposed to atmospheric CO_2 enrichment. *Canadian Journal of Forest Research*, **22**, 1717-1721.

B

Energy Primer

NEBOJSA NAKICENOVIC, IIASA

Lead Authors:
A. Grübler, IIASA; H. Ishitani, Japan; T. Johansson, Sweden; G. Marland, USA; J.R. Moreira, Brazil; H-H. Rogner, Canada

CONTENTS

B.1. Introduction

This Energy Primer introduces concepts and terms used in the energy-related chapters of this Second Assessment Report that deal with adaptation to and mitigation of climate change. Some of the more important terms also are defined in the Glossary. The Energy Primer also describes some of the more commonly used energy units; most measurements, including energy, are expressed in the International System of Units (SI) and SI-derived units.

Section B.2 describes the global energy system and documents 1990 energy-consumption patterns and carbon dioxide (CO_2) emissions. Thereafter, the concept of energy efficiency is introduced and 1990 global efficiencies are provided along with some estimates of maximum efficiencies that could be achieved under ideal conditions, indicating the theoretical efficiency improvement potential. Section B.3 describes and documents the historical development of energy consumption and associated CO_2 emissions for the world and for major regions. It concludes with a comparison of historical and current energy consumption, estimates of fossil and nuclear energy reserves and resources, and potentials of renewable energy sources. Section B.4 of this Energy Primer gives a brief introduction to the following chapters, which assess energy-related mitigation and adaptation options and measures.

B.2. Energy Systems

B.2.1. The Global Energy System

An *energy system* comprises an energy supply sector and energy end-use. The energy supply sector consists of a sequence of elaborate and complex processes for extracting energy resources, converting these into more desirable and suitable forms of energy, and delivering energy to places where the demand exists. The end-use part of the energy system provides energy services such as cooking, illumination, comfortable indoor climate, refrigerated storage, transportation, and consumer goods. The purpose, therefore, of the energy system is the fulfillment of demand for energy services. Figure B-1 illustrates schematically the architecture of an energy system as a series of

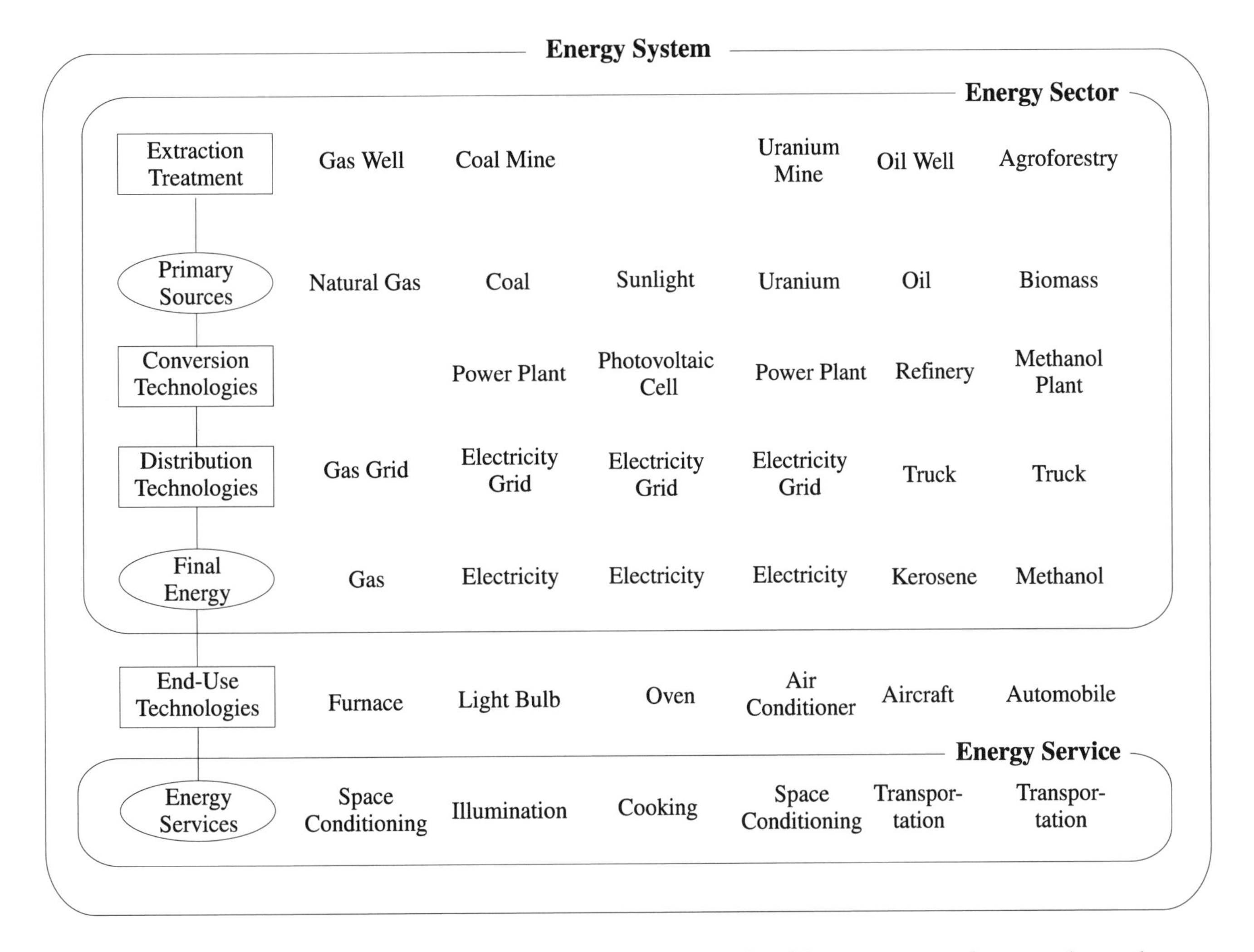

Figure B-1: The energy system: schematic diagram with some illustrative examples of the energy sector and energy end-use and services. The energy sector includes energy extraction, treatment, conversion, and distribution of final energy. The list is not exhaustive, and the links shown between stages are not "fixed" (e.g., natural gas is also used to generate electricity, and coal is not used exclusively for electricity generation). Source: Adapted from Rogner, 1994.

linked stages connecting various energy conversion and transformation processes that ultimately result in the provision of goods and services. A number of examples are given for energy extraction, treatment, conversion, distribution, end-use (final energy), and energy services in the energy system. The technical means by which each stage is realized have evolved over time, providing a mosaic of past evolution and future options.

Primary energy is the energy that is embodied in resources as they exist in nature: the chemical energy embodied in fossil fuels (coal, oil, and natural gas) or biomass, the potential energy of a water reservoir, the electromagnetic energy of solar radiation, and the energy released in nuclear reactions. For the most part, primary energy is not used directly but is first converted and transformed into electricity and fuels such as gasoline, jet fuel, heating oil, or charcoal.

Final energy is the energy transported and distributed to the point of final use. Examples include gasoline at the service station, electricity at the socket, or fuelwood in the barn. The next energy transformation is the conversion of final energy in end-use devices, such as appliances, machines, and vehicles, into *useful energy,* such as work and heat. Useful energy is measured at the crankshaft of an automobile engine or an industrial electric motor, by the heat of a household radiator or an industrial boiler, or by the luminosity of a light bulb. The application of useful energy provides *energy services,* such as a moving vehicle, a warm room, process heat, or light.

Energy services are the result of a combination of various technologies, infrastructures (capital), labor (know-how), materials, and energy carriers. Clearly, all these input factors carry a price tag and, within each category, are in part substitutable for one another. From the consumer's perspective, the important issues are the quality and cost of energy services. It often matters little what the energy carrier or the source of that carrier is. It is fair to say that most consumers are often unaware of the "upstream" activities of the energy system. The energy system is service-driven (i.e., from the bottom up), whereas energy flows are driven by resource availability and conversion processes (from the top down). Energy flows and driving forces interact intimately. Therefore, the energy sector should never be analyzed in isolation: It is not sufficient to consider only how energy is supplied; the analysis also must include how and for what purposes energy is used.

Figure B-2 illustrates schematically the major energy and carbon flows through the global energy system across the main stages of energy transformation, from primary energy to energy services. Energy and carbon estimates represent global averages in 1990. For definitions of energy and carbon-emissions units, see Boxes B-1 and B-2.

In 1990, 385 EJ of primary energy produced 279 EJ of final energy delivered to consumers, resulting in an estimated 112 EJ of useful energy after conversion in end-use devices. The delivery of 112 EJ of useful energy left 273 EJ of rejected energy. Most rejected energy is released into the environment as low-temperature heat, with the exception of some losses and wastes such as the incomplete combustion of fuels.

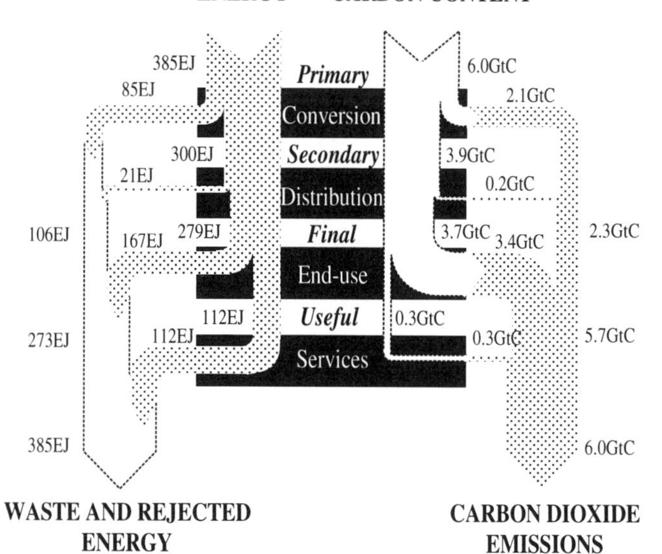

Figure B-2: Major energy and carbon flows through the global energy system in 1990, EJ and Gt C [(billion tons) or Pg C (10^{15} grams) elemental carbon]. Carbon flows do not include biomass. Sources: Marland *et al.*, 1994; IEA, 1993; Marland and Rotty, 1984; Nakicenovic *et al.*, 1993; WEC, 1992a.

More than half of the anthropogenic greenhouse gas (GHG) emissions originate from the energy system (both in terms of mass and in terms of radiative forcing). The predominant gas is CO_2, which represents more than half of the increase in radiative forcing from anthropogenic GHG sources. The majority of this CO_2 arises from the use of fossil fuels, which in turn make up about 75% of the total energy use. The global energy consumption of 385 EJ in 1990 was small compared with the solar radiation of about 5.4 million EJ intercepted annually by Earth. Although small in relation to natural energy flows, the emissions of energy-related GHGs create a danger of anthropogenic interference with Earth's radiative balance (energy budget).

The carbon content of fossil energy in 1990 was about 6 (±0.5) Gt C [(billion tons) or Pg C (10^{15} grams) elemental carbon], of which about 2.3 Gt C were emitted by the energy sector during conversion to fuels and electricity and distribution to final use. The remainder, about 3.7 Gt C, was emitted at the point of end-use. Included are 0.3 Gt C that were extracted from fossil sources without contributing directly to net carbon emissions. This carbon was embodied in durable hydrocarbon-based materials such as plastics, asphalt, lubricants, and pharmaceuticals. The carbon flows in Figure B-2 are simplifications because the carbon in fossil fuels is not completely oxidized to CO_2 during combustion. Eventually, however, most hydrocarbons and other combustion products containing carbon are converted to CO_2. There is some ambiguity concerning the amount of CO_2 emissions from feedstocks embodied in chemical products, as well as the unsustainable use of biomass. Carbon is released by the burning of biomass,

Box B-1. Energy Units and Scales

Energy is defined as the capacity to do work and is measured in joules (J), where 1 joule is the work done when a force of 1 newton (1 N = 1 kg m/s^2) is applied through a distance of 1 meter. Power is the rate at which energy is transferred and is commonly measured in watts (W), where 1 watt is 1 joule per second. Newton, joule, and watt are defined as units in the International System of Units (SI). Other units used to measure energy are toe (ton of oil equivalent; 1 toe = 41.87 x 10^9 J), used by the oil industry; tce (ton of coal equivalent; 1 tce = 29.31 x 10^9 J), used by the coal industry; and kWh (kilowatt-hours; 1 kWh = 3.6 x 10^6 J), used to measure electricity. Figure B-3 shows some of the commonly used units of energy and a few examples of energy consumption levels, along with the Greek names and symbols for factors to power of ten (e.g., exa equals 10^{18} and is abbreviated as E; in 1990, the global primary energy consumption was 385 EJ).

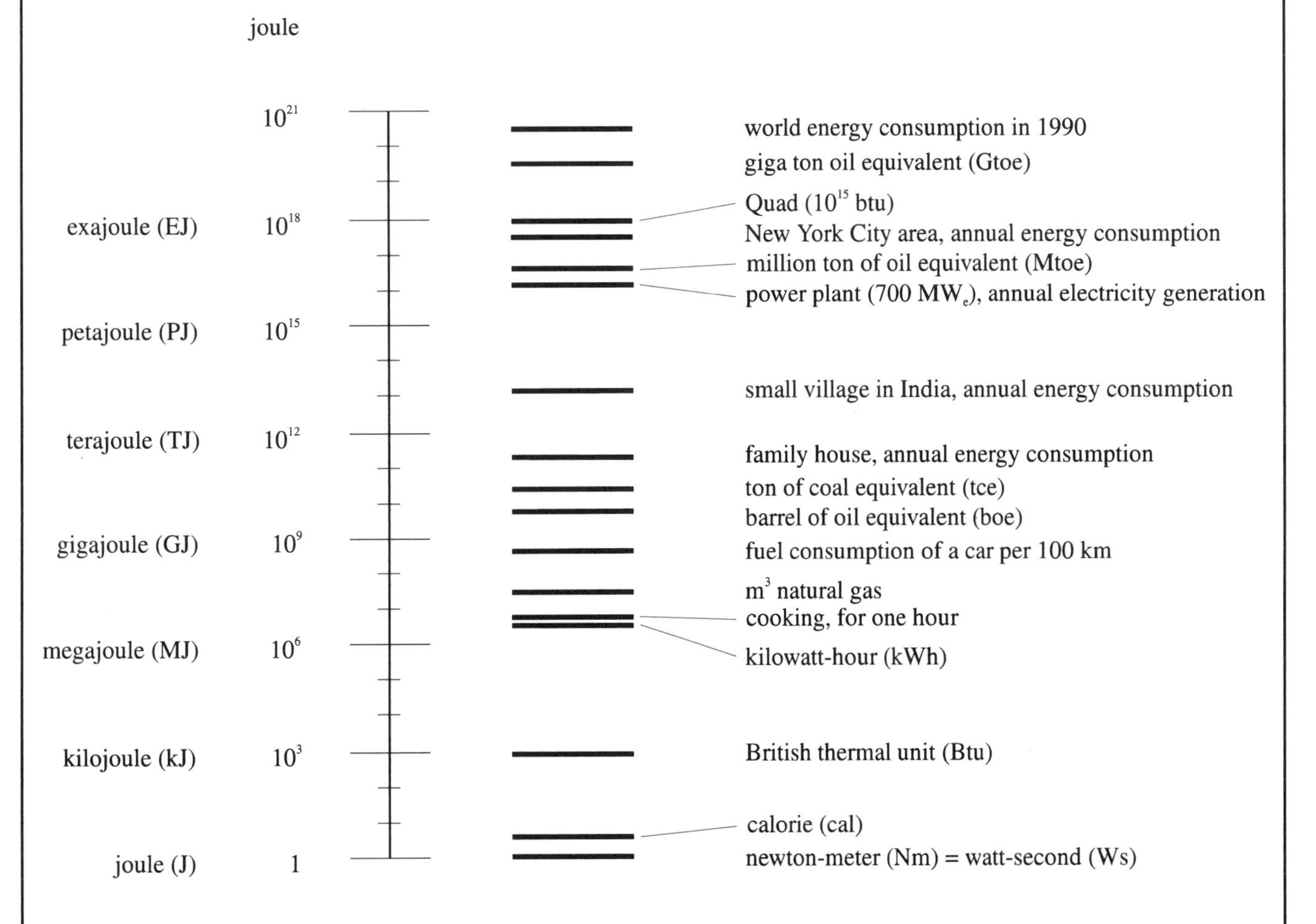

Figure B-3: Examples of some human energy needs and energy conversion devices, in joules, on a logarithmic scale. Sources: Adapted from Starr, 1971; Swedish National Encyclopedia, 1993.

including energy-related uses. Houghton and Skole (1990) estimate the latter to result in gross emissions of 0.7 Gt C per year. The extent of annual net emissions from nonfossil CO_2 is difficult to determine due to forest regrowth and the fact that the majority of biomass use is renewable. For simplification, we assume that feedstocks lead to CO_2 emissions—because even materials like asphalt and durable plastics will eventually be oxidized over a very long time—and that all biomass used as a source of energy is renewable and therefore does not result in net CO_2 emissions.

The 1990 global economic output is estimated at about US\$21 x 10^{12} (World Bank, 1993; UN, 1993); therefore, the average energy intensity was about 18 MJ/US\$, and energy-related carbon intensity was about 250 g C/US\$. In 1990, the average person consumed about 73 GJ of energy and emitted about 1.1 t C (tons carbon or Mg C).

B.2.2. Energy Efficiency

Energy is conserved in every conversion process or device. It can neither be created nor destroyed, but it can be converted from one form into another. This is the first law of thermodynamics. For example, energy in the form of electricity entering

Box B-2. CO_2 Emission Factors

CO_2 emissions are measured in tons (10^6 grams) of elemental carbon. For example, in 1990, global CO_2 emissions were 6 Gt C [(billion tons) or Pg C (10^{15} grams) of elemental carbon]. In the literature, CO_2 emissions also are reported as the mass of the actual molecule (1 kg C corresponds to 3.67 kg of CO_2).

Table B-1: *Carbon-emissions factors for some primary energy sources, in kg C/GJ.*

		OECD/IPCC 1995 (1)	Literature Range	Sources
Wood	HHV	–	26.8–28.4	
	LHV		28.1–29.9	(5),(3)
Peat	HHV	–	30.3	(3)
	LHV	28.9		
Coal (bituminous)	HHV	–	23.9–24.5	(2),(3)
	LHV	25.8	25.1–25.8	(2),(3)
Crude Oil	HHV	–	19.0–20.3	(3),(4)
	LHV	20.0	20.0–21.4	(3),(4)
Natural Gas	HHV	–	13.6–14.0	(2),(5)
	LHV	15.3	15.0–15.4	(2),(5)

Notes: HHV is the higher heating value and LHV the lower; the difference is that HHV includes the energy of condensation of the water vapor contained in the combustion products.
Sources:
(1) OECD/IPCC, 1991 and 1995.
(2) Marland and Pippin, 1990.
(3) Grubb, 1989.
(4) Marland and Rotty, 1984.
(5) Ausubel *et al.*, 1988.

an electric motor results in the desired output—say, kinetic energy of the rotating shaft to do work—and losses in the form of heat as the undesired byproduct caused by electric resistance, magnetic losses, friction, and other imperfections of actual devices. The energy entering a process equals the energy exiting. Energy efficiency is defined as the ratio of the desired (usable) energy output to the energy input. In the electric-motor example, this is the ratio of the shaft power to the energy input electricity. In the case of natural gas for home heating, energy efficiency is the ratio of heat energy supplied to the home to the energy of the natural gas entering the furnace. This definition of energy efficiency is sometimes called *first-law efficiency*.

A more efficient provision of energy services not only reduces the amount of primary energy required but, in general, also reduces adverse environmental impacts. Although efficiency is an important determinant of the performance of the energy system, it is not the only one. In the example of a home furnace, other considerations include investment, operating costs, lifetime, peak power, ease of installation and operation, and other technical and economic factors. For entire energy systems, other considerations include regional resource endowments, conversion technologies, geography, information, time, prices, investment finance, operating costs, age of infrastructures, and know-how.

The overall efficiency of an energy system depends on the individual process efficiencies, the structure of energy supply and conversion, and the energy end-use patterns. It is the result of compounding the efficiencies of the whole chain of energy supply, conversion, distribution, and end-use processes. The weakest link in the analysis of the efficiency of various energy chains is the determination of energy services and their quantification, mostly due to the lack of data about end-use devices and actual patterns of their use.

In 1990, the global efficiency of converting primary energy sources to final energy forms, including electricity, was about 72% (279 EJ over 385 EJ—see Figure B-2). The efficiency of converting final energy forms into useful energy is lower, with an estimated global average of 40% (Nakicenovic *et al.*, 1990;

Gilli *et al.*, 1995). The resulting average global efficiency of converting primary energy to useful energy, then, is the product of the above two efficiencies, or 29%. Because detailed statistics for most energy services do not exist and many rough estimates enter the efficiency calculations, the overall efficiency of primary energy to services reported in the literature spans a wide range, from 15 to 30% (Olivier and Miall, 1983; Ayres, 1989; Wall, 1990; Nakicenovic *et al.*, 1990; Schaeffer and Wirtshafter, 1992; and Wall *et al.*, 1994).

How much energy is needed for a particular energy service? The answer to this question is not so straightforward. It depends on the type and quality of the desired energy service; the type of conversion technology; the fuel, including the way the fuel is supplied; and the surroundings, infrastructures, and organizations that provide the energy service. Initially, energy-efficiency improvements can be achieved in many instances without elaborate analysis through common sense, good housekeeping, and leak-plugging practices. Obviously, energy service efficiencies improve as a result of sealing leaking window frames or installing a more efficient furnace. If the service is transportation—getting to and from work, for example—using a transit bus jointly with other commuters is more energy-efficient than taking individual automobiles. After the easiest improvements have been made, however, the analysis must go far beyond energy accounting.

Here the concept that something may get lost or destroyed in every energy device or transformation process is useful. This "something" is called "availability," which is the capacity of energy to do work. Often the availability concept is called "exergy."[1]

The following example should help clarify the difference between energy and exergy. A well-insulated room contains a small container of kerosene surrounded by air. The kerosene is ignited and burns until the container is empty. The net result is a small temperature increase of the air in the room ("enriched" with the combustion products). Assuming no heat leaks from the room, the total quantity of energy in the room has not changed. What has changed, however, is the quality of energy. The initial fuel-air combination has a greater potential to perform useful tasks than the resulting slightly warmer air mixture. For example, one could use the fuel to generate electricity or operate a motor vehicle. The ability of a slightly warmed room to perform any useful task other than space conditioning is very limited. In fact, the initial potential of the fuel-air combination or the "exergy" has been largely lost.[2] Although energy is conserved, exergy is destroyed in all real-life energy conversion processes. This is what the second law of thermodynamics expresses.

Another, more technical, example should help clarify the difference between first-law (energy) and second-law (exergy) efficiencies. Furnaces used to heat buildings are typically 70 to 80% efficient, with the latest, best-performing condensing furnaces operating at efficiencies greater than 90%. This may suggest that little energy savings should be possible, considering the high first-law efficiencies of furnaces. Such a conclusion is incorrect. The quoted efficiency is based on the specific process being used to operate the furnace—combustion of fossil fuel to produce heat. Because the combustion temperatures in a furnace are significantly higher than those desired for the energy service of space heating, the service is not well-matched to the source, and the result is an inefficient application of the device and fuel. Rather than focusing on the efficiency of a given technique for the provision of the energy service of space heating, one needs to investigate the theoretical limits of the efficiency of supplying heat to a building based on the actual temperature regime between the desired room temperature and the heat supplied by a technology. The ratio of theoretical minimum energy consumption for a particular task to the actual energy consumption for the same task is called exergy or *second-law* (of thermodynamics) *efficiency.*

Consider an example: Providing a temperature of 30°C to a building while the outdoor temperature is 4°C requires a theoretical minimum of one unit of energy input for every 12 units of heat energy delivered to the indoors. To provide 12 units of heat with an 80% efficient furnace, however, requires 12/0.8, or 15, units of heat. The corresponding second-law efficiency is the ratio of ideal to actual energy use (i.e., 1/15 or 7%).

The first-law efficiency of 80% gives the misleading impression that only modest improvements are possible. The second-law efficiency of 7% says that a 15-fold reduction in final heating energy is theoretically possible.[3] In practice, theoretical maxima cannot be achieved. More realistic improvement potentials might be in the range of half of the theoretical limit. In addition, further improvements in the efficiency of supplying *services* are possible by task changes—for instance, reducing the thermal heat losses of the building to be heated via better insulated walls and windows.

What is the implication of the second law for energy efficiencies? First of all, it is not sufficient to account for energy-in

[1] Exergy is defined as the maximum amount of energy that under given (ambient) thermodynamic conditions can be converted into any other form of energy; it is also known as availability or work potential (WEC, 1992b). Therefore, exergy defines the minimum theoretical amount of energy required to perform a given task.

[2] An alternative example: In terms of energy, 1 kWh of electricity and the heat contained in 43 kg of 20°C water are equal (i.e., 3.6 MJ). At ambient conditions, it is obvious that 1 kWh electricity has a much larger potential to do work (e.g., to turn a shaft or to provide light) than the 43 kg of 20°C water. See also Moran, 1989.

[3] For example, instead of combusting a fossil fuel, Goldemberg *et al.* (1988) give the example of a heat pump, which extracts heat from a local environment (outdoor air, indoor exhaust air, groundwater) and delivers it into the building. A heat pump operating on electricity can supply 12 units of heat for 3 to 4 units of electrical energy. The second-law efficiency improves to 25–33% for this particular task— still considerably below the theoretical maximum efficiency. Not accounted for in this example, however, are energy losses during electricity generation. Assuming a modern gas-fired, combined-cycle power plant with 50% efficiency, the overall efficiency gain is still a factor of two compared with a gas furnace heating system.

versus energy-out ratios without due regard for the quality difference (i.e., the exergy destroyed in the process). Minimum exergy destruction means an optimal match between the energy service demanded and the energy source. Although a natural-gas heating furnace may have an energy efficiency of close to 100%, the exergy destruction may be very high, depending on the temperature difference between the desired room temperature and the temperature of the environment. The second-law efficiency, defined as exergy-out over exergy-in, in this natural-gas home heating furnace example is some 7%—that is, 93% of the original potential of doing useful work (exergy) of the natural gas entering the furnace is lost. Here we have a gross mismatch between the natural-gas potential to do useful work and the low-temperature nature of the energy service, namely space conditioning.

There are many difficulties and definitional ambiguities involved in estimating the exergy efficiencies for comprehensive energy source-to-service chains or entire energy systems. There are many examples for the analysis of individual conversion devices; for instance, losses around a thermal power plant are described in Yasni and Carrington (1989). A few attempts have also been made to analyze energy systems efficiencies to useful energy or even to energy services. All indicate that primary-to-service (second-law or exergy) efficiencies are as low as a few percent. AIP (1975) and Olivier and Miall (1983) were among the first to give detailed assessments of end-use exergy efficiencies, including service efficiencies. Ayres (1989) calculates an overall primary exergy to service efficiency of 2.5% for the United States. Wall (1990) estimates a primary-to-useful exergy efficiency in Japan of 21%, and Wall *et al.* (1994) calculate a primary-to-useful exergy efficiency of less than 15% for Italy. Schaeffer and Wirtshafter (1992) estimate a primary-to-useful energy efficiency of 32% and an exergy efficiency of 23% for Brazil. Other estimates include Rosen (1992) for Canada, and Özdogan and Arikol (1995) for Turkey. Estimates of global and regional primary-to-service exergy efficiencies vary from ten to as low as a few percent (Gilli *et al.*, 1990, 1995; Nakicenovic *et al.*, 1990, 1993).

The theoretical potential for efficiency improvements is very large; current energy systems are nowhere close to the maximum levels suggested by the second law of thermodynamics. However, the full realization of this potential is impossible. Friction, resistance, and similar losses never can be totally avoided. In addition, there are numerous barriers and inertias to be overcome, such as social behavior, vintage structures, financing of capital costs, lack of information and know-how, and insufficient policy incentives.

The principal advantage of second-law efficiency is that it relates actual efficiency to the theoretical (ideal) maximum. Although this theoretical maximum can never be reached, low-exergy efficiencies identify those areas with the largest potentials for efficiency improvement. For fossil fuels, this suggests the areas that also have the highest emission-mitigation potentials.

B.3. Energy Use, CO_2 Emissions, and Energy Resources

B.3.1. Past and Present Energy Use

Global primary energy consumption has grown at an average annual rate of about 2% per year for almost 2 centuries—doubling, on average, about every 3 decades. This estimate includes all sources of commercial energy and fuelwood. There is a considerable variation in energy consumption growth rates over time and between different regions. For example, the global fossil energy consumption grew at 5% per year between 1950 and 1970, 3.3% annually between 1970 and 1990, and only 0.3% per year between 1990 and 1994. Emissions and other environmental effects of energy supply and end-use increased at somewhat slower rates than primary energy consumption.

Figure B-4 shows global annual primary energy consumption by source since 1860; Figure B-5 shows the relative shares of each source in total consumption. With the emergence of the coal age and steam power, global primary energy use evolved from a reliance on traditional energy sources, such as fuelwood, to fossil energy. Subsequently, coal was replaced by oil as the dominant primary energy source. Energy conversion also changed fundamentally with internal combustion, electricity generation, steam and gas turbines, and chemical and thermal energy conversion. The dynamics of structural changes in the global energy system, illustrated in Figure B-5, can be characterized by relatively slow rates of change, which are typical for infrastructures. It took about half a century before coal was replaced by crude oil as the dominant global energy source. At the global level, the "time constant" for fundamental energy transitions has been on the order of 50 years. At the regional level and for individual energy technologies and devices, the characteristic time constants are usually shorter, as a result of faster capital turnover, among other factors.

Much of the historical increase in global primary energy consumption has occurred in the more developed countries. About

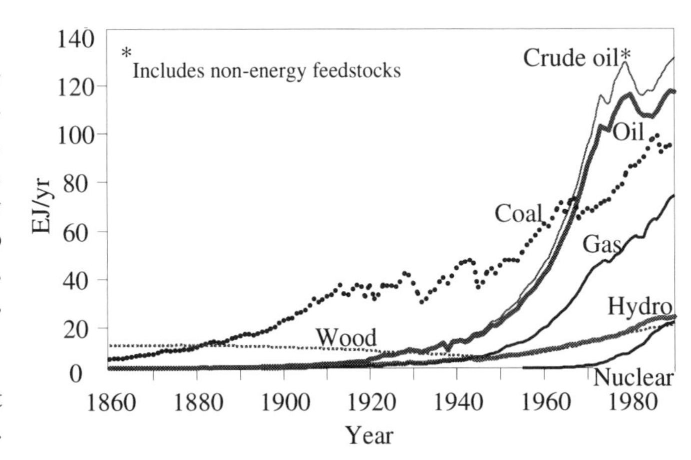

Figure B-4: Global primary energy consumption by source, and total in EJ/yr (data for crude oil include non-energy feedstocks). Sources: BP, various volumes; IEA, 1993; Marchetti and Nakicenovic, 1979.

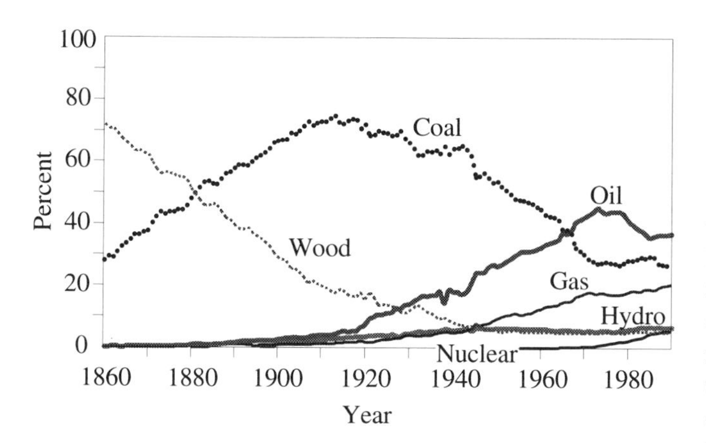

Figure B-5: Shares of energy sources in total global primary energy consumption, in percent of the total.

25% of the world's population consume almost 80% of the global energy. Cumulative consumption is even more unevenly distributed: About 85% of all energy used to date has been consumed by less than 20% of the cumulative global population (measured in cumulative person-years) since 1860. The differences in current per capita commercial energy consumption are more than a factor of 20 between the highest (North America) and the lowest (Africa) energy-consuming regions in the world—but more than a factor of 500 between individual countries. Another important difference is in the structure of energy supply, especially the strong reliance on traditional and noncommercial sources of energy in the developing countries (Hall, 1991).

Table B-2 shows the 1990 energy balance for the world from primary to final energy consumption, by energy carrier and sector. Crude oil is the dominant primary energy source in the world, accounting for 33% of the total, followed by coal and natural gas with 24 and 18% shares, respectively. Hydropower and nuclear energy are regionally important; they contribute more than 19 and 15% to global electricity supply, respectively. About 13% of the final energy is delivered as electricity. The largest final energy share of 38% is taken by oil products, half of them being used in the transport sector and constituting 96% of all the energy needs in this sector. The largest final energy carrier in industry is coal at 30%, accounting for almost 70% of all the direct uses of coal. Two-thirds of primary coal is used for electricity generation. About 30% of natural gas is used for electricity generation; the rest is divided almost equally between industrial uses and those in the household, commercial, and agricultural sectors. Electricity is also almost equally divided between these end uses. Most traditional biomass is used locally, with little or no conversion, and is shown under the "other" sector category in Table B-2. Primary energy consumption is well-documented in both national and international statistics. An exception is the use of traditional noncommercial energy (biomass). Larger uncertainties surround sectoral disaggregations of final energy consumption, due to a lack of detailed statistics in many countries. The numbers on sectoral final energy use given in Table B-2 are estimates. In some cases, alternative estimates are presented in individual chapters that deal with sectoral energy issues.

The historical shifts from traditional energy sources and coal to crude oil and natural gas were accompanied by the development of elaborate conversion systems for the production of more suitable forms of final energy, such as electricity. These structural changes, together with improvements in the performance of individual energy technologies, have resulted in significant efficiency improvements. Efficiency improvements in converting primary sources to final and useful energy forms, along with economic structural change, have contributed to a reduction of specific primary energy needs for generating a unit of economic output, usually measured in terms of gross domestic product (GDP) or gross national product (GNP). This ratio is often called energy intensity. Figure B-6 illustrates the changes in energy intensity for a number of world regions and countries and shows that, on average, 1% less energy per year was required every successive year to generate a unit of economic output. Actual variations of energy intensities and their improvement rates are large—depending, for instance, on the

Table B-2: *Global energy consumption in 1990 by energy source and by sector, in EJ/yr.*

	Coal	Oil	Gas	Nuclear	Hydro[a]	Electricity	Heat	Biomass	Total
Primary	91	128	71	19	21	–	–	55	385
Final	36	106	41	–	–	35	8	53	279
Industry	25	15	22	–	–	17	4	3	86
Transport	1	59	0	–	–	1	0	0	61
Others	10	18	18	–	–	17	4	50	117
Feedstocks[b]	0	14	1	–	–	–	–	0	15

Notes: Primary energy is recovered or gathered directly from natural sources (e.g., mined coal, collected biomass, or harnessed hydroelectricity), then is converted into fuels and electricity (e.g., electricity, gasoline, and charcoal), resulting in final energy after distribution and delivery to the point of consumption.

Sources: IEA, 1993; Hall, 1991, 1993; UN, 1993; WEC, 1983, 1993a, 1993b; Nakicenovic *et al.*, 1993.

[a] Nuclear and hydropower electricity have been converted into primary thermal equivalent, with an average factor of 38.5% (WEC, 1983).

[b] Feedstocks represent non-energy use of hydrocarbons.

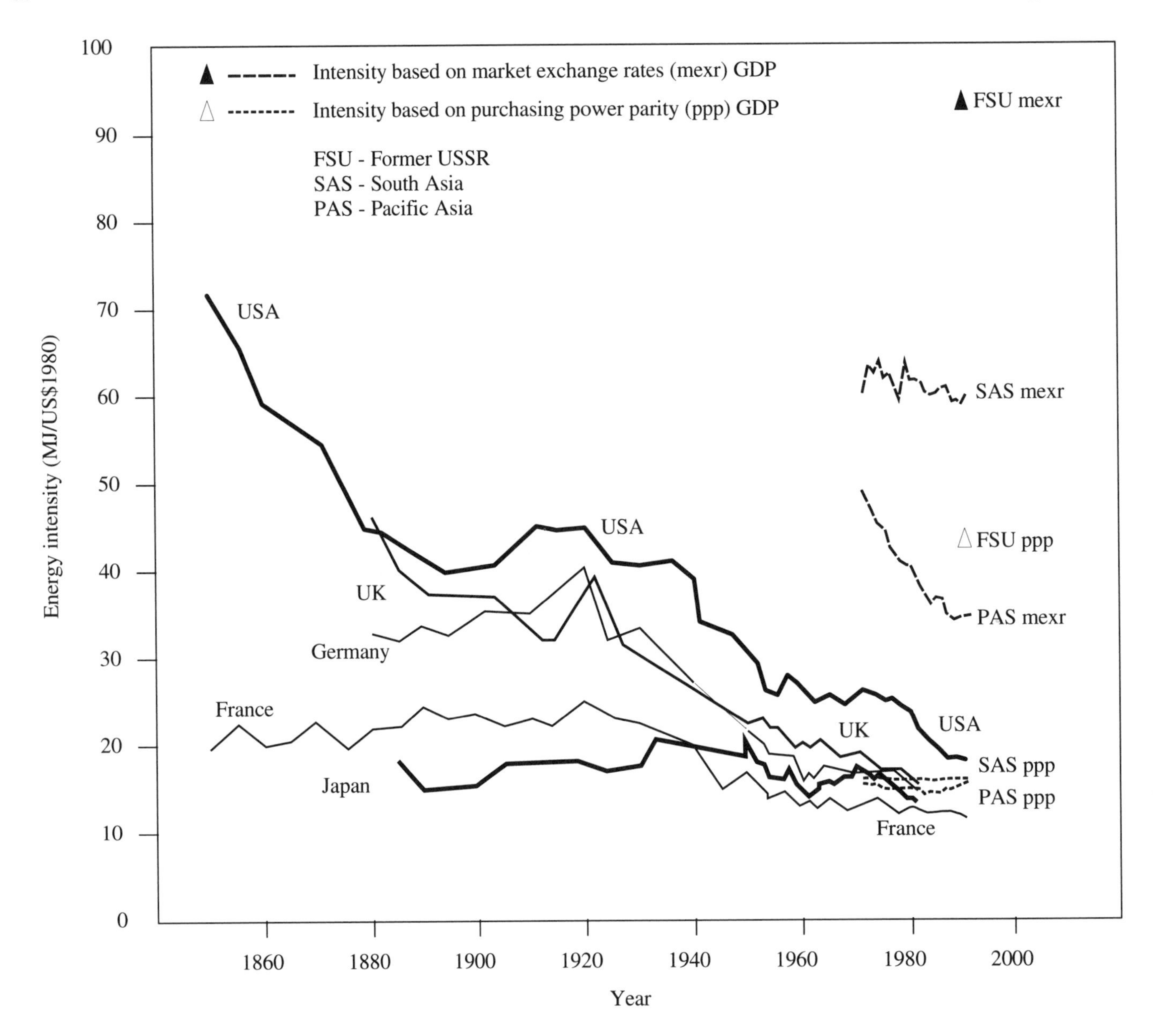

Figure B-6: Primary energy intensity (including wood and biomass) of value added in MJ per constant GDP in 1980 dollars [at market exchange rates (mexr) and purchasing power parities (ppp)]. Source: Grübler, 1991.

measure adopted to compare GDP between countries and regions, geographical factors, energy prices, and policies.

B.3.2. Past and Present CO_2 Emissions

CO_2 emissions from fossil energy consumption in 1990 are estimated at about 6.0 (±0.5) Gt C (Marland *et al.*, 1994; Subak *et al.*, 1993). This represents 70 to 90% of all anthropogenic sources of CO_2 in that year (IPCC, 1992).

Figure B-7 shows fossil energy CO_2 emissions by major world regions (emission factors are given in Box B-2). Developed countries contribute most to present global CO_2 emissions and also are responsible for most of the historical increase in concentrations. Although they are at lower absolute levels, emissions are growing more rapidly in developing countries than in developed regions. The largest single source of energy-related carbon emissions is coal, with about a 43% share, followed by oil with about 39%, and natural gas with 18%. Adding non-energy feedstocks reverses the shares to 40% for coal and 42% for oil. Due to the lack of data, these shares do not include energy-related deforestation or CO_2 emissions from unsustainable use of biomass.

Figure B-8 shows 1990 per capita CO_2 emissions in a number of world regions by source and relates these to the respective population size. Estimates of nonfossil sources of CO_2 are included.[4] The current levels of per capita fossil-fuel carbon

[4] Including CO_2 emissions from land-use changes such as deforestation (1.6±1 Gt C—IPCC, 1992). The extent of annual net emissions from nonfossil CO_2 sources is difficult to determine due to forest regrowth and the fact that the majority of biomass use is renewable.

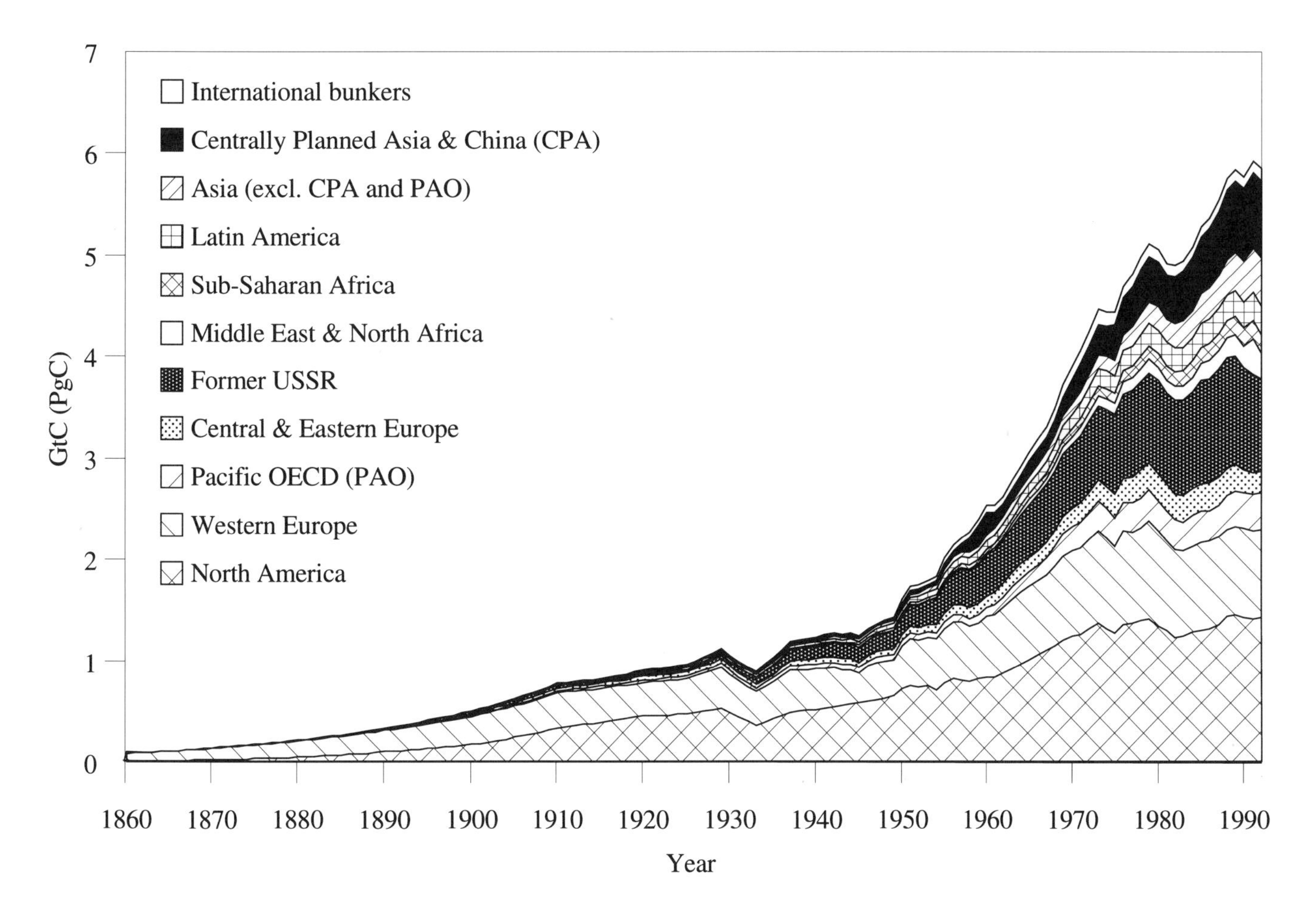

Figure B-7: Global energy-related CO_2 emissions by major world region in Gt C/yr (Pg C/yr). Sources: Keeling, 1994: Marland *et al.*, 1994; Grübler and Nakicenovic, 1992; Etemad and Luciani, 1991; Fujii, 1990; UN, 1952.

emissions in the world regions shown in Figure B-8 differ by a factor of 30. A persistent per capita emission gap remains after including carbon emissions from tropical deforestation, currently estimated to range between 0.6 and 2.6 Gt C/yr throughout the 1980s (IPCC, 1990, 1992; Ferreira and Marcondes, 1991; Houghton *et al.*, 1987).

The CO_2 emission intensity of both energy and economic activities is decreasing. Figure B-9 illustrates the extent of "decarbonization" in terms of the ratio of average carbon emissions per unit of primary energy consumed globally since 1860. The ratio has decreased due to the replacement of fuels with high carbon content, such as coal, by those with lower carbon content, such as natural gas, and by those with zero carbon content, such as nuclear power (see Figure B-5).[5] Energy development paths in different countries and regions have varied enormously and consistently over long periods. The overall tendency has been toward lower carbon intensities, although intensities are currently increasing in some developing countries. At the global level, the reduction in carbon intensity per unit value added has been about 1.3% per year since the mid-1800s—about 1.7% short of that required to offset the growth in global economic output of about 3% per year during that period (hence, global CO_2 emissions have grown at approximately 1.7%/yr).

B.3.3. Energy Reserves, Resources, and Potentials

Energy *occurrences* and their potential recoverability cannot be characterized by a simple measure or single numbers. They comprise quantities along a continuum in at least three, interrelated, dimensions: geological knowledge, economics, and technology. McKelvey (1972) proposed a commonly used diagram with a matrix structure for the classification along two dimensions: decreasing geological certainty of occurrence and decreasing economic recoverability.

Reserves are those occurrences that are identified and measured as economically and technically recoverable with current technologies and prices. *Resources* are those occurrences with

[5] It should be noted that so-called zero-carbon energy sources can result in some CO_2 and other GHG emissions, either because fossil energy is embodied in their construction materials (e.g., concrete in the structures of a nuclear power plant or a hydroelectric dam) or because fossil energy is required for operation and maintenance of energy facilities (e.g., gasoline and diesel vehicles). Some renewable sources also can entail CO_2 and other GHG emissions during operation. Examples include CO_2 emissions from geothermal; CH_4 emissions from anaerobic decay of biomass in flooded hydropower reservoirs; or CH_4, CO, and N_2O emissions from biomass burning.

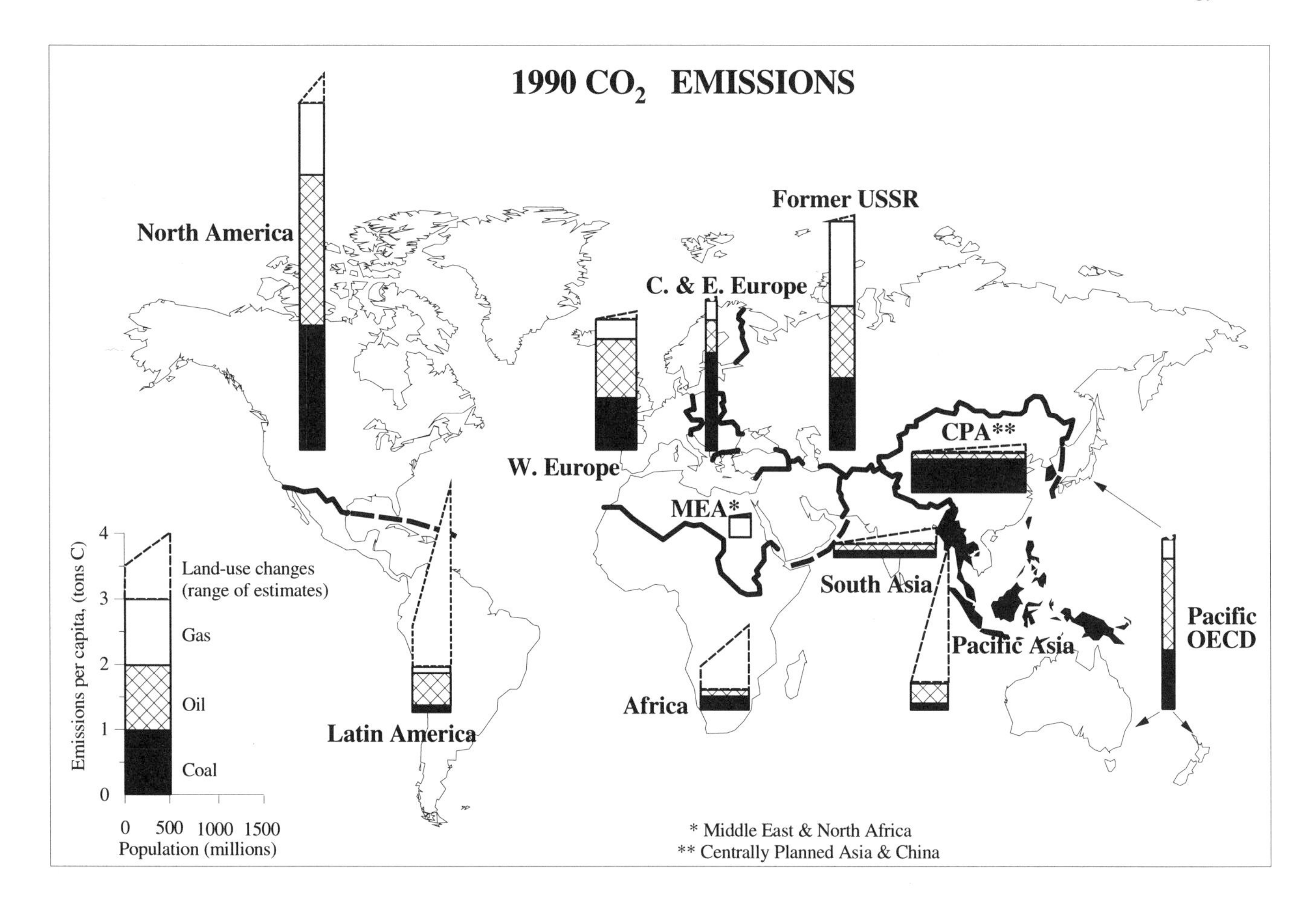

Figure B-8: 1990 per capita CO_2 emissions by region and source, fossil fuels, and range for biota sources (includes sustainable use of biomass that does not contribute to atmospheric concentration increase). Sources: IEA, 1993; Marland *et al.*, 1994; Nakicenovic *et al.*, 1993; Subak *et al.*, 1993; IPCC, 1990, 1992; Bos *et al.*, 1992; Houghton *et al.*, 1987.

less-certain geological and/or economic characteristics, but which are considered potentially recoverable with foreseeable technological and economic developments. The resource base includes both categories.[6] Additional quantities with unknown certainty of occurrence and/or with unknown or no economic significance in the foreseeable future are referred to here simply as "additional occurrences." For example, such additional occurrences include methane hydrates and natural uranium in seawater, both inferred to exist in large quantities but with unknown economic and/or technological means for extraction. Occurrences comprise all of the above three categories: reserves, resources, and additional occurrences.

Improved geological knowledge, both scientific and experimental (e.g., reservoir theories and exploration); improved technology; and changing prices have continuously served to increase the fossil energy resource base and have led to numerous large discoveries. Additions to reserves from resources have historically outpaced consumption. However, transfers from resources to reserves require investments. This adds a financial constraint on the expansion of reserves, so that from an economic point of view it makes little sense to invest in maintaining reserves for more than 20 years of production. For oil, this has indeed been the case. Therefore, there is a lot of exploration that still can be done but that has been deferred to the future for economic reasons. This is an important point when trying to understand energy reserves and why significant discoveries are still being made.

B.3.3.1. Fossil and Nuclear Reserves and Resources

Currently identified global fossil energy reserves are estimated to be about 50,000 EJ. This quantity is theoretically large enough to last 130 years at the 1990 level of global energy consumption of 385 EJ. It is five times larger than cumulative fossil energy consumption since the beginning of the coal era in the mid-19th century. Coal accounts for more than half of all fossil reserves. Table B-3 summarizes past and current consumption levels and estimates of global fossil and nuclear energy reserves, resources, and additional occurrences.

Estimates of resources and additional occurrences of fossil energy are much larger but more uncertain than reserves. Table B-3

[6] The fossil fuel resource-base estimates include potentially recoverable resources of coal, conventional oil, natural gas, unconventional oil (oil shale, tar sands, and heavy crude), and unconventional natural gas (gas in Devonian shales, tight sand formations, geopressured aquifers, and coal seams).

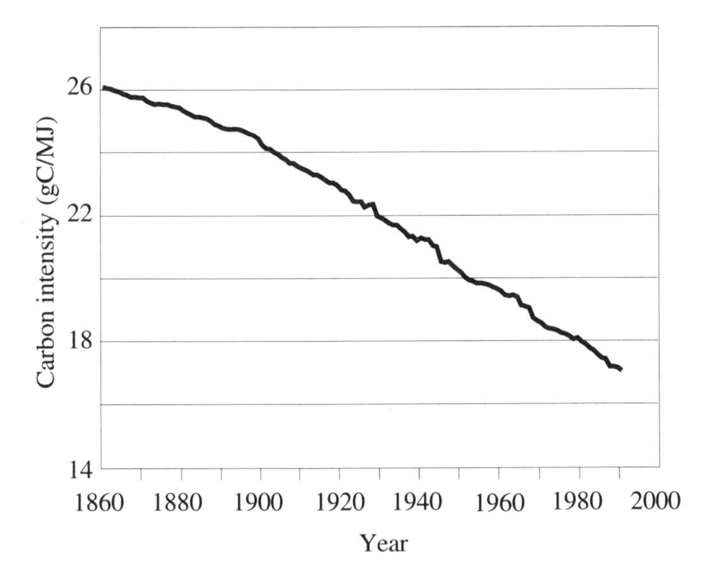

Figure B-9: Global decarbonization of energy since 1860 (including gross carbon emissions from fuelwood), in g C/MJ of primary energy. Source: Nakicenovic *et al.*, 1993.

shows the global fossil resource-base estimate to be about 186,000 EJ, with additional occurrences of almost 1 million EJ. Included in the conventional resources are estimates of ultimately recoverable conventional oil and gas resources remaining to be discovered at 95%, 50%, and 5% probability levels, ranging between from 1,800 and 5,500 EJ for oil and 2,700 and 10,900 EJ for gas (Masters *et al.*, 1991, 1994).

Methane resources are of particular interest because they have the lowest specific carbon emissions per unit energy of all fossil fuels. They are also of interest because methane is the second most important GHG associated with energy use and with anthropogenic activities in general. Because methane has a higher radiative forcing than CO_2 as a GHG, its climate effects are significantly reduced if it is oxidized into CO_2 (i.e., burned) instead of being released into the atmosphere. The reserves of unconventional gas are of the same magnitude as those for oil. The unconventional gas resource base is larger than that of unconventional oil, whereas the conventional resource bases are about the same. Additionally, there are large gas occurrences in the form of hydrates in permafrost areas and offshore continental-shelf sediments—in the range of 800,000 EJ (Kvenvolden, 1993; MacDonald, 1990).

The fossil resource base and additional occurrences are the ultimate global "carbon endowment" available to future generations, a number larger than 25,000 Gt C (Pg C). Fossil energy reserves correspond to 1,000 Gt C—exceeding the current carbon content of Earth's atmosphere (about 770 Gt C, or an

Table B-3: *Global fossil energy reserves, resources, and occurrences, in EJ.*

	Consumption[a]		**Reserves**	**Conventional Resources** Remaining to be Discovered at Probability[b]			**Unconventional Resources**		**Resource**	**Additional**
	1860–1990	1990	**Identified**	95%	50%	5%	Currently Recoverable	Recoverable w/ Techno-logical Progress	**Base**[c]	**Occurrence**
Oil										
Conventional	3343	128	6000	1800	2500	5500			8500	>10000
Unconventional	–	–	7100					9000	16100	>15000
Gas										
Conventional	1703	71	4800	2700	4400	10900			9200	>10000
Unconventional	–	–	6900				2200	17800	26900	>22000
Hydrates[d]	–	–								>800000
Coal	5203	91	25200				13900	86400	125500	>130000
Total	10249	290	50000	>4500	>6900	>16400	>16100	>113200	>186200	>987000
Nuclear[e]	212	19	1800		2300		4100	>6000	>14200	>1000000

Notes: All totals have been rounded; – = negligible amounts; blanks = data not available.
Sources: Nakicenovic *et al.*, 1993; WEC, 1992a; Grübler, 1991; MacDonald, 1990; Masters *et al.*, 1994; Rogner, 1990; BP, various volumes; BGR, 1989; Delahaye and Grenon, 1983.
[a] Grübler and Nakicenovic, 1992.
[b] Masters *et al.*, 1994.
[c] Resource base is the sum of reserves and resources. Conventional resources remaining to be discovered at probability of 50% are included for oil and gas.
[d] MacDonald, 1990.
[e] Natural uranium reserves and resources are effectively 60 times larger if fast breeder reactors are used. Calculated from natural uranium reserves and resources (OECD/NEA and IAEA, 1993) into thermal equivalent for once-through fuel cycle with average factor of 1,700 g per TJ thermal or 4,440 g natural uranium per TJ of electricity (16 Mg natural uranium per TWh_e electricity).

atmospheric concentration of 358 ppm in 1994). The resource base of conventional oil, gas, and coal, with some 3,500 Gt C, is about five times as large as the current atmospheric carbon content.

Uranium reserves, recoverable at costs less than US$130/kg, were evaluated at 3 million tons of natural uranium in January 1993 (OECD/NEA and IAEA, 1993). This corresponds to about 600 EJ of electricity—or about 1,800 EJ thermal equivalent if used in convertor reactors with a once-through fuel cycle (i.e., without reprocessing or final disposal of spent fuel) or to more than 100,000 EJ if used in fast breeder reactors (see Table B-3). In addition, some 4 million tons of natural uranium (corresponding to 2,300 EJ thermal) are known to exist; part of this supply is in countries where recovery costs have not been estimated. Uranium resources recoverable from unconventional ore bodies or as a byproduct amount to some 7 million tons of natural uranium (4,100 EJ thermal). Resources, estimated through geological assessment, amount to some 10 to 11 million tons of natural uranium (5,800 to 6,400 EJ thermal). Additional occurrences that cannot be exploited with current technologies include seawater, with an estimated natural uranium energy content exceeding one million EJ. Thorium reserves and resources are reported only for a few countries, and on that basis are estimated at 4 million tons. Geological information suggests that the resources may be much larger.

B.3.3.2. Renewable Energy Potentials and Natural Flows

In contrast to fossil energy sources, renewable energy forms such as solar, wind, and hydro can be either carbon-free or carbon-neutral. The sustainable use of biomass, for example, is carbon-neutral. Solar photovoltaic electricity generation is carbon-free. One must be careful, of course, to examine the full life cycle of the system when comparing the GHG implications of different energy systems because, for example, all energy systems currently rely on fossil fuels to construct devices, transport material, and dispose of waste.

Figure B-10 provides a schematic illustration of annual global energy flows without anthropogenic interference (Sørensen, 1979), and Table B-4 gives a summary of the annual (global) natural flows of renewable energy worldwide and their technical recovery potentials, as well as estimates for more practical potentials that could be achieved by 2020–2025 with current and near- to medium-term technologies and cost structures. The concept of technical potential can be used in a similar fashion as the concept of energy resources, and potentials by 2020 as the concept of energy reserves. The fundamental difference, of course, is that renewable potentials represent annual flows available, in principle, on a sustainable basis indefinitely, whereas fossil energy reserves and resources, although expanding in time, are fundamentally finite quantities. Life-cycle analyses remain important because although the energy flows are sustainable they still require materials like concrete and copper and the commitment of land and other resources. The renewable energy potentials identified in Table B-4 are theoretically large enough to provide the current primary energy needs for the world, and the technical potentials are large enough to cover most of the conceivable future growth of global energy demand.

Table B-4: *Global renewable energy potentials by 2020–2025, maximum technical potentials, and annual natural flows, in EJ thermal equivalent.*[a]

	Consumption[b]		**Potential by 2020–2025**[c]	**Long-Term Technical Potentials**[d]	**Annual Flows**
	1860-1990	1990			
Hydro	560	21	35–55	>130	>400
Geothermal	–	<1	4	>20	>800
Wind	–	–	7–10	>130	>200000
Ocean	–	–	2	>20	>300
Solar	–	–	16–22	>2600	>3000000
Biomass	1150	55	72–137	>1300	
Total	1710	76	130–230	>4200	>3000000

Sources: Hall *et al.*, 1993; Moreira and Poole, 1993; Grubb and Meyer, 1993; Johansson *et al.*, 1993; Swisher and Wilson, 1993; WEC, 1993b, 1994; Dessus *et al.*, 1992; Grübler and Nakicenovic, 1992; Hall, 1991; IPCC, 1992; Jensen and Sørensen, 1984; Sørensen, 1979.

Notes: All totals have been rounded; – = negligible amounts; blanks = data not available.

[a] All estimates have been converted into thermal equivalent with an average factor of 38.5%.

[b] Grübler and Nakicenovic, 1992.

[c] Range estimated from the literature. Survey includes the following sources: Johansson *et al.*, 1993; WEC, 1993b; Dessus *et al.*, 1992; EPA, 1990. It represents renewable potentials by 2020–2025, in scenarios with assumed policies for enhanced exploitation of renewable potentials.

[d] Long-term technical potentials are based on the Working Group II evaluation of the literature sources given in this table. This evaluation is intended to correspond to the concept of fossil energy resources, conventional and unconventional.

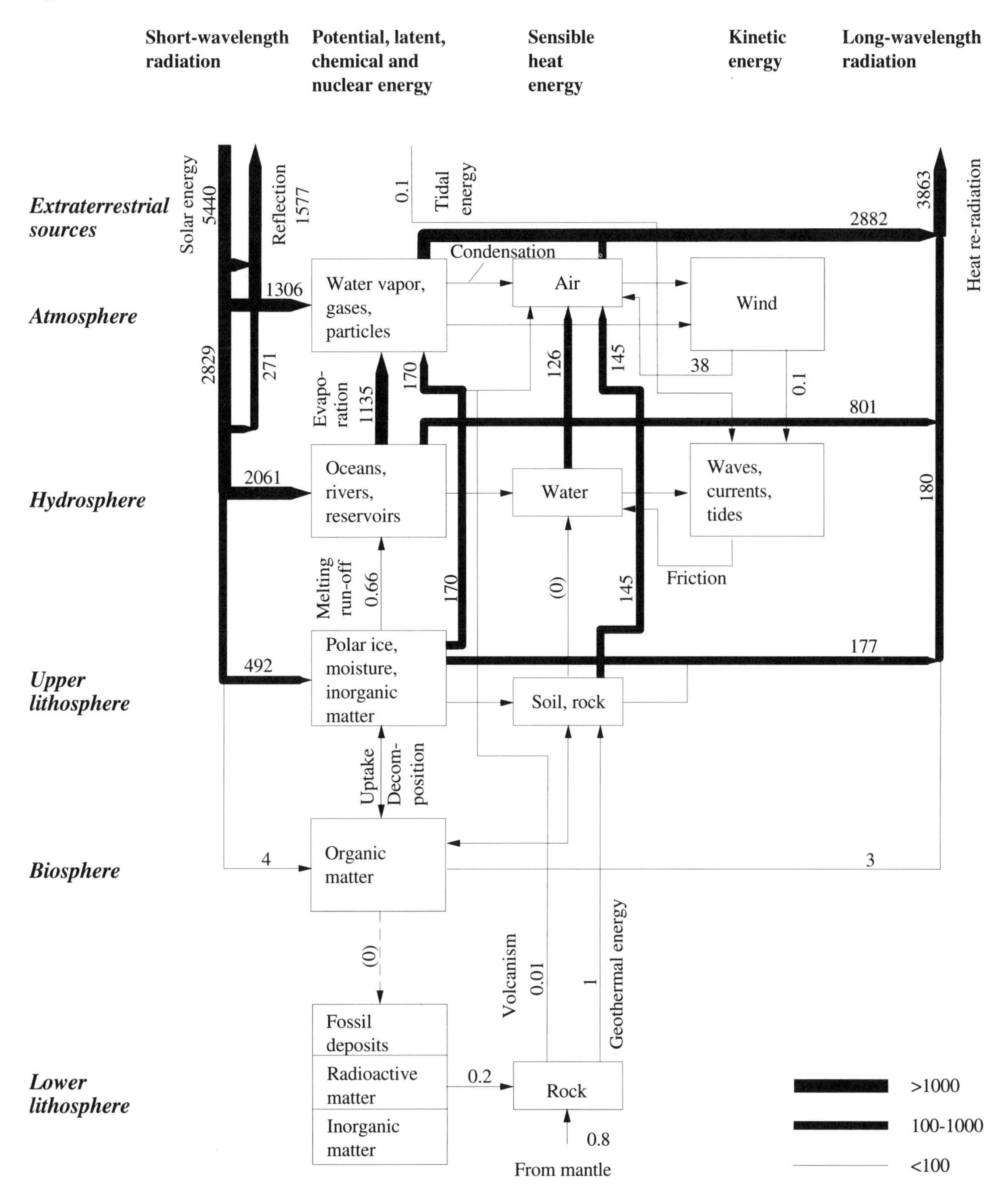

Figure B-10: Global energy balance and flows without anthropogenic interference. The energy flows are in units of 1,000 EJ/yr. Numbers in parentheses are uncertain or rounded. Source: Sørensen, 1979.

Hydropower is currently the most-developed modern renewable energy source worldwide. Table B-4 shows that the maximum technical potential is almost as large as the total final electricity consumption in 1990 as given in Table B-2 (WEC, 1993b; Moreira and Poole, 1993).

The technology to harness geothermal resources is established. Its current total use is about 0.2 EJ of electricity (Arai, 1993; Häfele *et al.,* 1981). There are four types of geothermal occurrences: hydrothermal sources, hot dry rock, magma, and geopressurized sources. The total accessible resource base of geothermal energy to a depth of 5 km is more than 126 million EJ (Palmerini, 1993), but occurrences within easily accessible layers of the crust reduce the technical potential. The annual flow from Earth is estimated at about 800 EJ/yr (Sørensen, 1979). The long-term technical potential could be greater than 20 EJ—especially if deep drilling costs can be reduced, as these are a major limitation to this energy source.

The energy flux of the atmosphere corresponds to about 200,000 EJ/yr of wind energy. The height limitations of wind converters, the distance of offshore sites, and insufficient wind velocities and land use all limit the practical potential. The ultimate potential of wind-generated electricity worldwide could indeed be very large: Some estimates place it at 50 times current global final electricity consumption (Grubb and Meyer, 1993; WEC, 1993b; Cavallo *et al.*, 1993; Gipe, 1991; Häfele *et al.*, 1981). Wind electricity is produced at many sites, and it is often also an economic option for electricity generation. The conversion efficiency is not the real barrier to the successful operation of wind-powered electricity generators. The technological challenge is that wind velocity is not constant in magnitude and direction. To utilize much higher windspeeds offshore, one option is to install floating windmills and to transport the electricity generated directly to the location of consumption or to use it for on-board hydrogen production.

Ocean energy flows include thermal energy, waves, tides, and the sea-freshwater interfaces as rivers flow into oceans. The low temperature gradients and low wave heights lead to an annual flow up to 300 EJ/yr of electricity. The technical potential is about 10 to 100 times smaller (Cavanagh *et al.*, 1993; WEC, 1993b; Baker, 1991; Sørensen, 1979).

All conceivable human energy needs could be provided for by diverting only a small fraction of the solar influx to energy use, assuming that a significantly large area could be devoted to solar energy gathering because of low spatial energy densities. Solar thermal and photovoltaic demonstration power plants are operating in a number of countries. Many gigawatts (GW) of installed electric capacity could be constructed after a few years of development. The main challenge is to reduce capital costs. Other proposals also have been made—for instance, placing solar power satellites in space.

Four general categories of biomass energy resources are used for fuels: fuelwood, wastes, forests, and energy plantations. Biomass wastes originate from farm crops, animals, forestry wastes, wood-processing byproducts, and municipal waste and sewage. The potential of biomass energy crops and plantations depends on the land area available, the harvestable yield, its energy content, and the conversion efficiency. Biomass potentials by 2020–2025 in Table B-4 are based on a literature survey of estimates and scenarios (Johansson *et al.*, 1993; WEC, 1993b; Hall, 1991). The technical potential of biomass energy crops and plantations is especially difficult to estimate. Based on land-use capacity studies, estimates of the land available for tropical plantations range between 580 and 620 million ha (Houghton *et al.*, 1991; Grainger, 1990).

B.4. Energy-Related Chapters

A number of chapters in this report are devoted to the assessment of energy-related impacts of and adaptation to climate change and to energy-related mitigation options. Vulnerability to climate change (including impacts and adaptation) concerning the energy, industry, and transportation sectors is considered in Chapter 11; human settlements are covered in Chapter 12. The general conclusion is that the sensitivity of the energy, industry, and transportation sectors is relatively low, whereas the capacity for autonomous adaptation is expected to be high if climate change is relatively gradual and not too drastic. Infrastructure and activities in these sectors would be susceptible to sudden changes and surprises; however, the subsectors most sensitive to climate change include agroindustry, renewable energy production including hydroelectric generation, construction, and manufacturing heavily dependent on water supplies. The most vulnerable human settlements are located in damage-prone areas of the developing world that do not have resources to cope with impacts.

Energy-related options for controlling the sources and enhancing the sinks of GHGs have an important role, to a varying degree, in all of the mitigation chapters. Chapter 19 assesses energy supply mitigation options, and the following three chapters, 20 through 22, consider individual sectors—industry, transportation, and human settlements, respectively. In addition, mitigation options related to energy supply and use (e.g., biomass) are considered in Chapters 23 and 24 on agriculture and forestry, respectively.

Energy-related emissions account for the largest share of CO_2 sources and have varying importance in the emissions of other GHGs. Global primary energy needs are expected to increase anywhere between 540 and 2,500 EJ by 2100, according to the IS92 IPCC scenarios. A detailed assessment and evaluation of IPCC and other energy and emissions scenarios is provided in *An Evaluation of the IPCC IS92 Emission Scenarios* (Alcamo *et al.*, 1995). With increases in global primary energy, GHG emissions will continue to grow unless they are mitigated. The general conclusion in Chapters 19 to 24 is that the technological potential to achieve significant emission reductions is indeed large, but there are important uncertainties regarding the ease, timing, and cost of implementing mitigation options and measures.

Implementation depends on successful research and development, the existence of the right market and institutional conditions, and timely market penetration, as well as the adoption of new technologies and practices by firms and individuals. Government policies are an important element in the creation of appropriate market conditions and incentive structures.

Chapter 25, the last of the mitigation options chapters, evaluates strategies that emphasize land use and highlights some key cross-cutting themes related to implementation and energy supply and use. It concludes with a discussion of nontraditional mitigation options such as "geoengineering," which might be involved as "last resort options" for the future.

The last two chapters of the report, Chapters 27 and 28, describe methods for assessing mitigation options and offer an inventory of mitigation technologies, respectively. These methods can be used to develop mitigation strategies and evaluate mitigation projects and, together with the inventory, are available to all countries.

References

AIP, 1975: *Efficient Use of Energy.* Conference Proceedings No. 25, American Institute of Physics, New York, NY, 304 pp.

Alcamo, J., A. Bouwman, J. Edmonds, A. Grübler, T. Morita, and A. Sugandhy, 1995: An evaluation of the IPCC IS92 emission scenarios. In: *Climate Change 1994* [Houghton, J.T., L.G. Meira Filho, J. Bruce, Hoesung Lee, B.A. Callander, E. Haites, N. Harris, and K. Maskell (eds.)]. Cambridge University Press, Cambridge, UK, pp. 251-304.

Arai, Y., 1993: Present status and future prospects of geothermal power generation. *Japan 21st*, **38(3)**, 45-46.

Ausubel, J., A. Grübler, N. Nakicenovic, 1988: Carbon dioxide emissions in a methane economy. *Climatic Change,* **12**, 245-263.

Ayres, R.U., 1989: *Energy Inefficiency in the US Economy: A New Case for Conservation.* RR-89-12, IIASA, Laxenburg, Austria, 28 pp.

Baker, C., 1991. Tidal power. *Energy Policy,* **19(8)**, 792-798.

Bos, E., M.T. Vu, A. Levin, and R.A. Bulatao, 1992: *World Population Projections 1992-1993.* Johns Hopkins University Press, Baltimore, MD, 515 pp.

BGR, 1989: *Reserven, Ressourcen und Verfügbarkeit von Energierohstoffen.* Bundesanstalt für Geowissenschaften und Rohstoffe (BGR), Hannover, Germany, 419 pp.

BP, various volumes 1974-1993: *BP Statistical Review of World Energy.* British Petroleum, London, UK.

Cavallo, A., S. Hock, and D. Smith, 1993: Wind energy: technology and economics. In: *Renewable Energy: Sources for Fuels and Electricity* [Johansson, T.B., H. Kelly, A.K.N. Reddy, and R.H. Williams (eds.)]. Island Press, Washington, DC, pp. 121-156.

Cavanagh, J.E., J.H. Clarke, and R. Price, 1993: Ocean energy systems. In: *Renewable Energy: Sources for Fuels and Electricity* [Johansson, T.B., H. Kelly, A.K.N. Reddy, and R.H. Williams (eds.)]. Island Press, Washington, DC, pp. 513-547.

Delahaye, C. and M. Grenon (eds.), 1983: Conventional and unconventional world natural gas resources. CP-83-S4, *Proceedings of the Fifth IIASA Conference on Energy Resources*, IIASA, Laxenburg, Austria, 543 pp.

Dessus, B., B. Devin, and F. Pharabod, 1992: *World Potential of Renewable Energies.* Extraits de la Houille Blanche, Paris, France, 50 pp.

EPA, 1990: *Policy Options for Stabilizing Global Climate Report to Congress: Technical Appendices* [Lashof, D.A. and D.A. Tirpak (eds.)]. USEPA, Washington, DC, pp. A.1-C.52.

Etemad, B., J. Luciani, P. Bairoch, and J.C. Toutain, 1991: *World Energy Production 1800-1985.* Librarie Droz, Geneva, Switzerland, 272 pp.

Ferreira, M.C. and M.E. Marcondes, 1991: Global deforestation and CO_2 emissions, past and present: a comprehensive review. *Energy and Environment,* **2(3)**, 235-282.

Fujii, Y., 1990: *An Assessment of the Responsibility for the Increase in the CO_2 Concentration and Inter-generational Carbon Accounts.* WP-90-55, IIASA, Laxenburg, Austria, 31 pp.

Gilli, P.-V., N. Nakicenovic, and R. Kurz, 1995: *First- and Second-Law Efficiencies of the Global and Regional Energy Systems.* WEC 16th Congress, Tokyo, Japan, 8-13 October, Papers Session 3.1, pp. 229-248.

Gilli, P.-V., N. Nakicenovic, A. Grübler, F.L. Bodda, 1990: *Technischer Fortschritt, Strukturwandel and Effizienz der Energieanwendung—Trends weltweit und in Österreich.* Band 6, Schriftenreihe der Forschungsinitiative des Verbundkonzerns, Vienna, Austria, 331 pp.

Gipe, P., 1991: Wind energy comes of age. *Energy Policy,* **19(8)**, 756-767.

Goldemberg, J., T.B. Johansson, A.K.N. Reddy, and R.H. Williams (eds.), 1988: *Energy for a Sustainable World.* Wiley Eastern Limited, New Delhi, India, 517 pp.

Grainger, A., 1990: Modeling the impact of alternative afforestation strategies to reduce carbon emissions. In: *Proceedings of the Intergovernmental Panel on Climate Change* (IPCC) *Conference on Tropical Forestry Response Options to Global Climate Change, 9-12 January 1990.* Report No. 20P-2003, Office of Policy Analysis, USEPA, Washington, DC, pp. 93-104.

Grubb, M., 1989: On coefficients for determining greenhouse gas emissions from fossil fuel production and consumption. In: *Energy Technologies for Reducing Emissions of Greenhouse Gases, Proceedings of an Experts' Seminar, 1.* OECD, Paris, pp. 537-556.

Grubb, M., and N. Meyer, 1993: Wind energy: resources, systems and regional strategies. In: *Renewable Energy: Sources for Fuels and Electricity* [Johansson, T.B., H. Kelly, A.K.N. Reddy, and R.H. Williams (eds.)]. Island Press, Washington, DC, pp. 157-212.

Grübler, A., 1991: Energy in the 21st century: from resources to environmental and lifestyle constraints. *Entropie,* **164/165**, 29-34.

Grübler, A. and N. Nakicenovic, 1992: *International Burden Sharing in Greenhouse Gas Reduction.* Environment Working Paper No. 55, World Bank Environment Department, Washington, DC, 96 pp.

Hall, D.O., 1993: *Biomass Energy.* Shell/WWF Tree Plantation Review, Study No. 4, Shell International Petroleum Company Limited and World Wide Fund for Nature, London, UK, 47 pp.

Hall, D.O., F. Rosillo-Calle, R.H. Williams, and J. Woods, 1993: Biomass for energy: supply prospects. In: *Renewable Energy: Sources for Fuels and Electricity* [Johansson, T.B., H. Kelly, A.K.N. Reddy, and R.H. Williams (eds.)]. Island Press, Washington, DC, pp. 593-652.

Häfele, W., J. Anderer, A. McDonald, and N. Nakicenovic, 1981: *Energy in a Finite World: Paths to a Sustainable Future (Part 1).* Ballinger, Cambridge, 225 pp.

Häfele, W., 1981: *Energy in a Finite World: A Global Systems Analysis (Part 2).* Ballinger, Cambridge, 837 pp.

Houghton, R.A., R.D. Boone, J.R. Fruci, J.E. Hobbie, J.M. Melillo, C.A. Palm, B.J. Peterson, G.R. Shaver, G.M. Woodwell, B. Moore, D.L. Skole, and N. Meyers, 1987: The flux of carbon from terrestrial ecosystems to the atmosphere in 1980 due to changes in land use: geographical distribution of the global flux. *Tellus,* **39(B)**, 122-139.

Houghton, R.A. and D.L. Skole, 1990: Carbon. In: *The Earth as Transformed by Human Action* [Turner, B.L. II, W.C. Clark, R.W. Kates, J.F. Richards, J.T. Mathews, and W.B. Meyer (eds.)]. Cambridge University Press, Cambridge, UK, pp. 393-408.

Houghton, R.A., J. Unruh, and P.A. Lefèbre, 1991: Current land use in the tropics and its potential for sequestering carbon. *Proceedings of the Technical Workshop to Explore Options for Global Forestry Management*, 24-29 April 1991, Bangkok, Thailand, International Institute for Environment and Development, UK.

IEA, 1993: *Energy Statistics and Balances of Non-OECD Countries 1971-1991.* OECD, Paris, France (three diskettes).

IPCC, 1990: *Climate Change: The IPCC Scientific Assessment.* Cambridge University Press, Cambridge, UK, 365 pp.

IPCC, 1992: *1992 IPCC Supplement,* February, Geneva, Switzerland.

Jensen, J. and B. Sørensen, 1984: *Fundamentals of Energy Storage.* John Wiley and Sons, New York, NY, 345 pp.

Johansson, T.B., H. Kelly, A.K.N. Reddy, and R.H. Williams, 1993: Renewable fuels and electricity for a growing world economy. In: *Renewable Energy: Sources for Fuels and Electricity* [Johansson, T.B., H. Kelly, A.K.N. Reddy, and R.H. Williams (eds.)]. Island Press, Washington, DC, pp. 1-72.

Keeling, C.D., 1994: Global historical CO_2 emissions. In: *Trends '93: A Compendium of Data on Global Change* [Boden, T.A., D. P. Kaiser, R.J. Sepanski, and F.W. Stoss (eds.)]. ORNL/CDIAC-65, Carbon Dioxide Information Analysis Center, Oak Ridge, TN, pp. 501-504.

Kvenvolden, K.A., 1993: A primer on gas hydrates. In: *The Future of Energy Gases* [Howell, D.G. (ed.)]. U.S. Geological Survey Professional Paper 1570, U.S. Government Printing Office, Washington, DC, pp. 279-291.

MacDonald, G.J., 1990: The future of methane as an energy resource. *Annual Review of Energy,* **15**, 53-83.

Marchetti, C. and N. Nakicenovic, 1979: *The Dynamics of Energy Systems and the Logistic Substitution Model.* RR-79-13, IIASA, Laxenburg, Austria, 73 pp.

Marland, G. and R.M. Rotty, 1984: Carbon dioxide emissions from fossil fuels: a procedure for estimation and results for 1950-1982. *Tellus,* **36(B)**, 232-261.

Marland, G. and A. Pippin, 1990: United States emissions of carbon dioxide to the earth's atmosphere by economic activity. *Energy Systems and Policy*, **15**, 319-336.

Marland, G., R.J. Andres, and T.A. Boden, 1994: Global, regional, and national, CO_2 emissions 1950-1991. In: *Trends '93: A Compendium of Data on Global Change* [Boden, T.A., D.P. Kaiser, R.J. Sepanski, and F.W. Stoss (eds.)]. ORNL/CDIAC-65, Carbon Dioxide Information Analysis Center, Oak Ridge National Laboratory, Oak Ridge, TN, pp. 505-581.

Masters, C.D., D.H. Root, and E.D. Attanasi, 1991: World resources of crude oil and natural gas. *Proceedings of the 13th World Petroleum Congress, Buenos Aires.* John Wiley, Chichester, UK, pp. 51-64.

Masters, C.D., E.D. Attanasi, and D.H. Root, 1994: World petroleum assessment and analysis. In: *Proceedings of the 14th World Petroleum Congress Stavanger, Norway.* John Wiley, Chichester, UK, pp. 1-13.

McKelvey, V.E., 1972: Mineral resource estimates and public policy. *American Scientist,* **60**, 32-40.

Moran, M.J., 1989: *Availability Analysis: A Guide to Efficient Energy Use*, 2nd edition. ASMI Press, New York, NY, 260 pp.

Moreira, J.R. and A.D. Poole, 1993: Hydropower and its constraints. In: *Renewable Energy: Sources for Fuels and Electricity* [Johansson, T.B., H. Kelly, A.K.N. Reddy, and R.H. Williams (eds.)]. Island Press, Washington, DC, pp. 73-119.

Nakicenovic, N., L. Bodda, A. Grübler, and P.-V. Gilli, 1990: *Technological Progress, Structural Change and Efficient Energy Use: Trends Worldwide and in Austria.* International part of a study supported by the Österreichische Elektrizitätswirtschaft AG, IIASA, Laxenburg, Austria. 304 pp.

Nakicenovic, N., A. Grübler, A. Inaba, S. Messner, S. Nilsson, Y. Nishimura, H-H. Rogner, A. Schäfer, L. Schrattenholzer, M. Strubegger, J. Swisher, D. Victor, and D. Wilson, 1993: Long-term strategies for mitigating global warming. *Energy*, **18(5)**, 401-609.

Olivier, D. and H. Miall, 1983: *Energy Efficient Futures: Opening the Solar Option.* Earth Resources Research Limited, London, UK, 323 pp.

OECD/IPCC, 1991: *Estimation of Greenhouse Gas Emissions and Sinks.* Final Report from OECD Experts Meeting, 18-21 February 1991. Prepared for Intergovernmental Panel on Climate Change, Revised August 1991, OECD, Paris, France.

OECD/IPCC, 1995: *IPCC Guidelines for National Greenhouse Gas Inventories.* Vol. 3, *Greenhouse Gas Inventory Reference Manual.* OECD and IPCC, UK Meteorological Office, Bracknell, UK.

OECD/NEA, IAEA, 1993: *Uranium 1993 Resources, Production and Demand.* Paris, France.

Özdogan, S. and M. Arikol, 1995: Energy and exergy analysis of selected Turkish industries. *Energy*, **18(1)**, 73-80.

Palmerini, C.G., 1993: Geothermal energy. In: *Renewable Energy: Sources for Fuels and Electricity* [Johansson, T.B., H. Kelly, A.K.N. Reddy, and R.H. Williams (eds.)]. Island Press, Washington, DC, pp. 549-591.

Rogner, H-H., 1990: *Analyse der Förderpotentiale und Langfristige Verfügbarkeit von Kohle, Erdgas und Erdöl.* Studie A.3.1., Enquête-Kommission "Vorsorge zum Schutz der Erdatmosphäre" Deutscher Bundestag, Vol. 4, Fossile Energieträger, Economica-Verlag, Bonn, Germany, pp. 11-86.

Rogner, H-H., 1994: Fuel cells, energy system evolution, and electric utilities. *International Journal of Hydrogen Energy*, **19(10)**, 853-861.

Rosen, M.A., 1992: Evaluation of energy utilization efficiency in Canada. *Energy*, **17**, 339-350.

Schaeffer, R. and R.M. Wirtshafter, 1992: An exergy analysis of the Brazilian economy: from energy products to final energy use. *Energy*, **17**, 841-855.

Sørensen, B. 1979: *Renewable Energy.* Academic Press, London, UK, 683 pp.

Starr, C., 1971: Energy and power. In: *Energy and Power–A Scientific American Book.* W.H. Freeman and Company, San Francisco, CA, pp. 3-15.

Subak, S., P. Raskin, and D. von Hippel, 1993: National greenhouse gas accounts: current anthropogenic sources and sinks. *Climatic Change,* **25**, 15-58.

Swedish National Encyclopedia, 1993: Höganäs Bokfürlaget Bra Böcker, Höganäs, Sweden, pp. 486-491.

Swisher, J. and D. Wilson, 1993: Renewable energy potentials in Nakicenovic, N., *et al.,* Long-Term Strategies for Mitigating Global Warming. *Energy*, **18(5)**, 437-460.

United Nations, 1952: *World Energy Supplies in Selected Years, 1929-1950.* UN, New York, NY.

United Nations, 1993: *1991 Energy Statistics Yearbook.* UN, New York, NY, 448 pp.

Wall, G., 1990: Exergy Conversion in the Japanese Society. *Energy*, **15**, 435-444.

Wall, G., E. Scuibba, and V. Naso, 1994: Exergy Use in the Italian Society. *Energy*, **19**, 1267-1274.

World Bank, 1993: *World Tables.* Johns Hopkins University Press, Baltimore, MD.

WEC, 1983: *Standards Circular No. 1,* 11/83. WEC, London, UK.

WEC, 1992a: *Survey of Energy Resources 16th Edition.* WEC, London, UK.

WEC, 1992b: *Energy Dictionary.* WEC, London, UK.

WEC, 1993a: *Energy for Tomorrow's World.* WEC Commission global report, Kogan Page, London, UK, 320 pp.

WEC, 1993b: *Renewable Energy Resources: Opportunities and Constraints 1990-2020.* WEC, London, UK, pp. 1.1-7.22.

WEC, 1994: *New Renewable Energy Resources: A Guide to the Future.* Kogan Page, London, UK, 391 pp.

Yasni, E. and C.G. Carrington, 1989: The role for exergy auditing in a thermal power station. In: *Proceedings of Annual Meeting of the American Society of Mechanical Engineers*, 13-18 December 1987, Boston, MA [Boehm, R.F. and N. Lior (eds.)]. The American Society of Mechanical Engineers, Heat Transfer Division, Volume 80, New York, NY, pp. 1-7.

Climate Change 1995

Part II

Assessment of Impacts and Adaptation Options

Prepared by Working Group II

1

Climate Change Impacts on Forests

MIKO U.F. KIRSCHBAUM, AUSTRALIA; ANDREAS FISCHLIN, SWITZERLAND

Lead Authors:
M.G.R. Cannell, UK; R.V.O. Cruz, Philippines; W. Galinski, Poland; W.P. Cramer, Germany

Contributing Authors:
A. Alvarez, Cuba; M.P. Austin, Australia; H.K.M. Bugmann, Germany; T.H. Booth, Australia; N.W.S. Chipompha, Malawi; W.M. Ciesla, FAO; D. Eamus, Australia; J.G. Goldammer, Germany; A. Henderson-Sellers, Australia; B. Huntley, UK; J.L. Innes, Switzerland; M.R. Kaufmann, USA; N. Kräuchi, Switzerland; G.A. Kile, Australia; A.O. Kokorin, Russia; Ch. Körner, Switzerland; J. Landsberg, Australia; S. Linder, Sweden; R. Leemans, The Netherlands; R.J. Luxmoore, USA; A. Markham, WWF; R.E. McMurtrie, Australia; R.P. Neilson, USA; R.J. Norby, USA; J.A. Odera, Kenya; I.C. Prentice, Sweden; L.F. Pitelka, USA; E.B. Rastetter, USA; A.M. Solomon, USA; R. Stewart, Canada; J. van Minnen, The Netherlands; M. Weber, Germany; D. Xu, China

CONTENTS

EXECUTIVE SUMMARY

Forests are highly sensitive to climate change. This has been shown by observations from the past, experimental studies, and simulation models based on current ecophysiological and ecological understanding. In particular, the following was concluded:

- Sustained increases of as little as 1°C in mean annual air temperature can be sufficient to cause changes in the growth and regeneration capacity of many tree species. In several regions, this can significantly alter the function and composition of forests; in others, it can cause forest cover to disappear completely (Medium Confidence).
- Suitable habitats for many species or forest types are likely to shift faster with climate change than the maximum natural rate at which many species can migrate and establish. Consequently, slow-growing species, such as late successional species, or those with restricted seed dispersal will be replaced by faster-growing, highly adaptable or more mobile species (High Confidence).
- Forests are particularly vulnerable to extremes of water availability (either drought or waterlogging) and will decline rapidly if conditions move toward one of the extremes (High Confidence).
- Forced by a doubled carbon dioxide (2 x CO_2) climate, global models project that a substantial fraction of the existing forests will experience climatic conditions under which they do not currently exist; eventually, large forested areas will have to change from the current to new major vegetation types (High Confidence). Averaged over all zones, the models predict that 33% of the currently forested area could be affected by such changes; in the boreal zone, one model projects it to be as high as 65% (Medium Confidence). Yet it is currently not possible to predict transient forest responses at a regional to global scale.
- Although net primary productivity may increase, the standing biomass of forests may not increase because of more frequent outbreaks and extended ranges of pests and pathogens and increasing frequency and intensity of fires (Medium Confidence).
- Mature forests are a large store of terrestrial carbon. Because the maximum rate at which carbon can be lost is greater than the rate at which it can be gained, large amounts of carbon may be released transiently into the atmosphere as forests change in response to a changing climate and before new forests replace the former vegetation. The loss of aboveground carbon alone has been estimated to be 0.1–3.4 Gt yr^{-1} or a total of 10–240 Gt (Medium Confidence).

The following regional assessments were primarily based on transient climate-change scenarios for 2050 (Greco *et al.*, 1994).

Tropical Forests

- Tropical forests are likely to be more affected by changes in land use than by climate change as long as deforestation continues at its current high rate (High Confidence).
- Any degradation of tropical forests, whether it is caused by climate or land-use changes, will lead to an irreversible loss in biodiversity (High Confidence).
- CO_2 fertilization may have its greatest effect in the tropics and may lead to a gain in net carbon storage in undisturbed forests, especially in the absence of nutrient limitations (Medium Confidence).
- Tropical forests are likely to be more affected by changes in soil water availability (caused by the combined effects of changes in temperature and rainfall) than by changes in temperature *per se*. Decreases in soil moisture may accelerate forest loss in many areas where water availability is already marginal. In other areas, increasing precipitation may be more than adequate to meet increased evaporative demand and may even lead to erosion (Medium Confidence).

Temperate Forests

- Compared with other latitudinal zones, the potential area for temperate forests is projected to change the least; however, many existing forests will still undergo significant changes in their species composition (High Confidence).
- Water availability will change in many regions, and in some regions where water supply is already marginal, forests may be lost in response to increased summer droughts (Medium Confidence).
- While warming and elevated CO_2 are likely to increase net primary productivity of many forests, net carbon storage may not increase because of the associated stimulation of soil organic matter decomposition by soil warming (Medium Confidence).
- Temperate forests are currently a carbon sink, mainly because of regrowth that started in many regions in the 19th century. However, these forests could become a source if they degrade due to climate change or other causes such as air pollution (Medium Confidence).

- Most temperate forests are located in developed countries with resources to reduce the impacts of climate change on their forests through integrated fire, pest and disease management, and/or encouraging reforestation (Medium Confidence).

Boreal Forests

- Because warming is expected to be particularly large at high latitudes, and boreal forests are more strongly affected by temperature than forests in other latitudinal zones, climate change is likely to have its greatest impact on boreal forests (High Confidence).
- Northern treelines are likely to advance slowly into regions currently occupied by tundra. (High Confidence).
- Increased fire frequency and pest outbreaks are likely to decrease the average age, biomass, and carbon store, with greatest impact at the southern boundary where the boreal coniferous forest is likely to give way to temperate-zone pioneer species or grasslands (Medium Confidence).
- The net primary productivity of forests not limited by water availability is likely to increase in response to warming, partly mediated by increased nitrogen mineralization. However, there may be a net loss of carbon from the ecosystem because of associated increases in soil organic matter decomposition (Medium Confidence).

1.1. Introduction

Forests are among those ecosystems on Earth that remain the least disturbed by human influences. They are of great socioeconomic importance as a source of timber, pulpwood for paper making, fuel, and many non-wood products (see Chapter 15). Furthermore, forests provide the basis for a broad range of other economic and non-economic values, such as resources for tourism, habitat for wildlife, or the protection of water resources. Forests harbor the majority of the world's biodiversity (mainly in the tropics) and, as such, they represent indispensable, self-maintaining repositories of genetic resources. These are essential for improvements in crop and timber selection and in medicine. Finally, aside from special economic interests, forests are also of spiritual importance to many indigenous people.

Globally, forests in 1990 covered about one-fourth of the Earth's land surface (3.4 Gha: FAO, 1993b), although estimates differ due to the exact definition of forests (e.g., 4.1 Gha: Dixon *et al.*, 1994; or, in a very wide definition, 5.3 Gha: Sharma *et al.*, 1992). About 17% of high-latitude, 20% of mid-latitude, and 4% of low-latitude forests can be considered actively managed (see Chapter 24), but only about 100 Mha (~2%) consist of intensively managed plantations (Dixon *et al.*, 1994). The forest regions have been broadly subdivided into latitudinal zones (i.e., the tropical, temperate, and boreal zones).

Forests are ecosystems in which trees interact with each other, with other plants like shrubs and grasses, and with animals or other heterotrophic organisms such as insects or fungi. At a broad scale, forest structure is modified by processes ranging from almost continuous change due to the death of individual trees and the subsequent recruitment of seedlings, to catastrophic events such as fire, insects, wind-fall or logging that kill whole stands of trees simultaneously. At any given moment, a forest represents the outcome of long-lasting past processes, often covering many centuries.

As components of the global climatic system, forests play a major role in the present and projected future carbon budget, since they store about 80% of all aboveground and 40% of all belowground terrestrial organic carbon (e.g., Melillo *et al.*, 1990; Dixon *et al.*, 1994) and can act as sources through deforestation and degradation, as well as sinks through forestation and possibly enhanced growth (see Chapter 23 on the role of forests for mitigation of greenhouse gas emissions). Moreover, forests can directly affect the climate system from local up to continental scales: They influence ground temperatures, evapotranspiration, surface roughness, albedo, cloud formation, and precipitation (e.g., Henderson-Sellers *et al.*, 1988; Gash and Shuttleworth, 1991).

Forest ecosystems respond to environmental changes with time constants ranging from hours to decades and up to millennia (see Box 1-1); they are among the components of the biosphere that respond most slowly to climatic change. The role of forest dynamics in the global climatic system is likely to be long lasting, complex, and difficult to predict. Because of their longevity and because adaptive measures, such as replacement of species, are harder to implement than in agricultural systems, forests may be particularly vulnerable to climatic change.

Sections 1.2.1 and 1.2.2 present assessments of direct forest responses to climatic change, including possible secondary effects on forest pests, fire, and other issues, which may cause forests

Box 1-1. Scales and Equilibrium Assumptions

Climatic changes affect forests on spatial scales ranging from leaves to the canopy and on temporal scales from minutes to centuries; relevant climatic changes occur at all levels, from short-term weather fluctuations creating disturbances (such as frosts) to longer-term changes in average climatic conditions (such as moisture availability or the length of the growing season) or the frequency of extreme events (such as droughts, fires, or intense storms). Current climate models (GCMs) do not fully match these levels, since they are best at simulating average conditions at a relatively coarse spatial resolution and are usually not yet run for longer than a century. Changes in frequencies of extreme events are highly uncertain, as are local-scale climate changes, although both are of high relevance to forests.

Some of the limitations caused by this problem can be overcome in local analyses using downscaling techniques (e.g., Gyalistras *et al.*, 1994), but for global applications the implicit assumption must be made that the probability of extreme events will remain unchanged. Most transient changes in the structure of forests, such as the decline of certain tree species, are driven by a combination of climatic changes and are modified by local, biological interactions acting on temporal scales ranging from months to centuries. It is currently very difficult, therefore, to assess the likely rates of climate-driven, transient change in forests. However, possible future equilibrium conditions can be more adequately predicted.

Equilibrium projections of forest responses implicitly assume the climate to have stabilized at a new steady state, which is not likely to occur soon in reality. However, GCM-derived climate scenarios arbitrarily held constant (e.g., for 2050) allow an assessment of the direction and magnitude of the expected change. Equilibrium projections for future forests, therefore, represent conservative interpretations of minimal changes likely to occur sooner or later and hence include the potential of even greater biospheric changes than the ones currently simulated by the forest models.

Box 1-2. Temperature Thresholds

Trees have widely differing responses to temperature. Some tropical tree species suffer chilling injury at temperatures below +12°C (Lyons, 1973; Lyons *et al.*, 1979), whereas species of colder regions can survive -5°C without ice formation but are sterile at lower temperatures. Classic examples for this phenomenon are *Ilex* and *Hedera* (Iversen, 1944). Broad-leaved evergreen perennials can survive to a limit of about -15°C by supercooling, whereas broad-leaved deciduous trees can supercool to about -40°C (Arris and Eagleson, 1989). Evergreen needle-leaved trees can survive to about -60°C, below which only deciduous species survive. Apart from these killing temperatures, many species require certain minimum numbers of degree days to complete essential life-cycle processes such as bud initiation, pollen formation, flowering, or others (Stern and Roche, 1974). Others require particular sequences of cool temperatures to become frost-hardy at the optimum time and a minimum duration of chilling temperatures to break winter dormancy (Cannell, 1990). Insect pests and other biotic agents that affect forest health may have critical threshold subzero temperatures for winter survival and thermal times to complete a generation. Warming may have positive effects on the growth of many trees and their survival, but by being beneficial to insect pests it also may reduce tree survival or put cold-adapted species at a competitive disadvantage.

either to grow more vigorously or to decline[1] in a changing climate. These assessments are based on the current general understanding of the basic ecophysiological and ecological responses; the latter will be addressed on levels ranging from the stand to the globe. In the three remaining sections, we summarize more specific effects in the tropical, temperate and boreal forest zones.

1.2. Climate and Forests

1.2.1. Sensitivities to Expected Climate Change

Forests are highly dependent on climate in their function (e.g., growth) and structure (e.g., species composition). Forest distribution is generally limited by either water availability or temperature. The ratio of actual evapotranspiration (the amount allowed by available precipitation) to potential evapotranspiration (the amount the atmosphere would take up if soil moisture were not limiting) determines the maximum leaf area index that can be supported (Woodward, 1987). Forests are also usually absent where the mean temperature of the warmest month falls below 10°C (Köppen, 1936) or where the temperature sum above a 5°C threshold is less than 350 degree-days (Prentice *et al.*, 1992).

The survival of many species depends critically on temperature thresholds ranging from +12° to -60°C (e.g., Woodward, 1987; Prentice *et al.*, 1992; Box 1-2). Many species have narrow temperature niches for growth and reproduction. A sustained increase in mean annual temperature of 1°C may cause significant changes in the species composition of stands and hence the distribution of many tree species (Davis and Botkin, 1985; see also Section 1.3). Trees are also sensitive to changes in water availability during the growing season, and leaf area indices, volume growth, and the range boundaries of most tree species are strongly related to water availability (Holdridge, 1967; Hinckley *et al.*, 1981; Gholz, 1982; Austin *et al.*, 1990; Stephenson, 1990).

In addition to thresholds for growth, reproduction, and survival at a given site, there are limitations to the rate at which species can migrate unassisted. Current projected rates of climatic change may exceed these thresholds, as discussed in Section 1.3.5. Climatic warming and associated lower humidities and increase in the frequency and severity of droughts would increase the incidence and severity of wildfires, especially in the boreal region. Changes in fire or storm frequencies are likely to have major impacts on the composition, age-distribution, and biomass of forests (see Sections 1.5.4.5 and 1.6.4.4).

1.2.2. Expected Climatic Changes in Forested Areas

Future forest characteristics are likely to depend on a few specific aspects of the range of climatic changes that could occur. The most relevant are the following:

- Changes in the regional or seasonal pattern of climate, such as the temperature increases that are expected to be greatest at high latitudes, and there, greatest in winter (Greco *et al.*, 1994). Due to this, impacts on forests at high latitudes may be greater than elsewhere.
- Water shortages during the growing season. Decreasing summer precipitation together with increased evaporative demand would lead to decreases in soil water, especially in many mid-latitude regions where water is most critical for growth. It is important to note that water shortages can develop even with unchanged rainfall amounts, due to increasing temperatures causing increased evaporative demand. We expect significant regional variation, with water availability changing only marginally in some regions and improving in others, whereas in many other regions water availability may decrease drastically.

[1] Decline is defined here as "an episodic event characterized by premature, progressive loss of tree and stand vigor and health over a given period without obvious evidence of a single clearly identifiable causal factor such as physical disturbance or attack by an aggressive disease or insect" (Ciesla and Donaubauer, 1994).

- Changes in climate forcing are expected to be one or two orders of magnitude faster than rates of climatic change experienced by forests during most of the past 100,000 to 200,000 years, except, perhaps, during the Younger Dryas Event 10,000 years ago (Dansgaard *et al.*, 1989; Webb III and Bartlein, 1992; Gates, 1993). Such rapid climatic change would have particular impacts on forests. For example, there may be forest decline, interruption of tree life cycles, loss of slowly migrating species, and increasing abundance of more aggressive, early successional species.

1.3. Forests in a Changing Climate

For a detailed assessment of the effects of a changing climate on forest ecosystems, it is necessary to investigate this response to the simultaneous changes in several climatic variables (e.g., temperature, moisture availability, and ambient CO_2 concentrations). Current understanding of ecological relationships and ecophysiological mechanisms allows a comprehensive study of forests at three hierarchical levels of scale, connecting temporal and spatial resolution to the nature of the processes that are being considered (Table 1-1).

At the ecophysiological level, plant organs like stomata and leaves respond almost instantaneously to their environment. This mainly affects forest functions, such as net primary productivity. Some structural aspects, such as leaf area, may respond over a number of years, whereas others, such as species composition, may take centuries to respond to altered conditions. Typically, each species or plant functional type (PFT) is affected differently by climatic change: Some species or PFTs will remain unaffected, some will become more and others less competitive (Smith *et al.*, 1993). Dynamic forest models can be used to simulate this transient behavior of forests (e.g., Shugart, 1984; Prentice *et al.*, 1993). Landscape-scale processes such as the lateral interactions between neighboring patches (e.g., migration or fire) play an important role for possible changes in the local to regional pattern of many forest ecosystems. However, they become less relevant when aggregated to national or continental assessments. At the global level, it is currently possible to investigate only how climatic change might affect the potential geographic equilibrium distribution of biomes—i.e., biogeographic regions (e.g., Emanuel *et al.*, 1985; Prentice *et al.*, 1992)—or to study the major fluxes of trace gases into and out of these biomes (Melillo *et al.*, 1993; Plöchl and Cramer, 1995), with research underway on dynamic modelling (see also Chapter 9, *Terrestrial Biotic Responses to Environmental Change and Feedbacks to Climate*, in the Working Group I volume).

The current models can be used with climatic inputs generated by general circulation models (GCMs) for future climate scenarios, including regionally differentiated, high-resolution scenarios (e.g., Gyalistras *et al.*, 1994). However, current understanding of the physical and biological interactions between environment and organisms is still rather limited, so these techniques enable us only to project future responses as consequences of given assumptions and scenarios and not to make precise forecasts.

1.3.1. Ecophysiological Responses

1.3.1.1. Tree Responses to Temperature and Water Availability

When they are well supplied with water, trees of most temperate and boreal species respond to increased temperature (e.g., from year to year, or when planted at a slightly warmer location) by growing faster (e.g., Cannell *et al.*, 1989; Beuker, 1994), and they reach their largest mass near the warmest boundaries of their geographic ranges (Korzukhin *et al.*, 1989). Generally, there is a positive correlation between net primary productivity and temperature (Box, 1978; Kauppi and Posch, 1985; Kokorin *et al.*, 1993; Morikawa, 1993) or between net primary productivity and actual evapotranspiration (Rosenzweig, 1968; Raich *et al.*, 1991).

However, these potential growth responses to warming may be constrained by other factors. Increasing temperature increases evaporative demand; if rainfall does not increase, more severe water stress will result, which will adversely affect growth and may increase the risk of drought and fire. To what extent this can be compensated by increased water-use efficiency due to increasing CO_2 concentration is not yet known (see Section 1.3.1.2 and Chapter A). Photoperiodic limitations might also

Table 1-1: *Hierarchical levels at which it is currently possible to study and model the impact of climatic change on forest ecosystems. Note that transient responses of trees and forests can currently only be studied at the first two levels.*

Level	Focus	Time		Space	
		Resolution	Scope	Resolution	Scope
Ecophysiological Processes	Plant Metabolism	min..h	1..10 yr	0..100 m	locations
Individuals, Populations, Forest Stands	Ecosystem Dynamics	d..yr	$1..10^3$ yr	~100 m	regions to continents
Plant Functional Types, Vegetation, Vegetation Complexes, Biomes	Biospheric Equilibrium	–	$\geq 10^3$ yr	$\sim 10..10^3$ km	globe

apply in areas where temperature becomes warm enough for tree growth (e.g., Heide, 1974, 1993), and increasing temperature might increase the range of insect pests, which could cause considerable damage.

GCMs suggest a globally averaged increase in precipitation of about 2.5% per degree warming (Mitchell *et al.*, 1990; Greco *et al.*, 1994), which may not be sufficient to meet the increased water requirements of forests. Most GCMs indicate significant declines in net soil moisture over continental areas during the growing season (e.g., Manabe and Weatherald, 1987; Greco *et al.*, 1994). There are likely to be considerable regional differences, however, with some regions likely to receive increased rainfall sufficient to meet increased evaporative demand and other regions receiving even less rainfall than at present. Because the present distribution of species and plant functional types is strongly determined by the total (Holdridge, 1967; Box, 1981; Hinckley *et al.*, 1981; Austin *et al.*, 1990) and seasonal availability of water (Stephenson, 1990; Prentice *et al.*, 1992), any changes in water availability are likely to greatly affect the distribution of species (see Section 1.3.2).

1.3.1.2. Tree Responses to Increased CO_2 Concentration

The importance of CO_2 fertilization continues to be controversial (e.g., Körner, 1993; Idso and Idso, 1994; see also Chapter A). While the response of photosynthesis to CO_2 concentration can be readily observed at the single-leaf or isolated plant level (Kimball, 1983; Cure and Acock, 1986; Mooney *et al.*, 1991; Luxmoore *et al.*, 1993; Wullschleger *et al.*, 1995), it has been argued that this initial benefit may be negated by the various feedbacks in the plant and soil (e.g., Bazzaz and Fajer, 1992; Diaz *et al.*, 1993; Körner, 1993). Quantification of some of these feedbacks has been attempted (e.g., Comins and McMurtrie, 1993; Kirschbaum *et al.*, 1994) but has not yet been possible for the great diversity of natural habitats, in which most of the feedback effects are still inadequately understood and poorly quantified. Experimental approaches have not yet resolved this controversy because of the enormous costs and technical difficulties involved. Only a small number of open-air CO_2 enrichment experiments have been conducted with mature natural populations—and none on forests.

Because atmospheric CO_2 has already increased from a preindustrial concentration of about 280 ppmv to about 360 ppmv at present, increased growth should be observable in the growth of plants under natural conditions. However, the evidence from tree-ring chronologies is unclear, and no generalizations can be made (Innes, 1991). Where growth increases have been observed, part or all of that probably could be explained by more favorable temperatures, water relations, successional age, or nitrogen fertilization by moderate levels of industrial pollution (Innes, 1991; Luxmoore *et al.*, 1993).

When plants have access to limiting amounts of water, growth is limited by the amount of CO_2 that can be obtained in the diffusive gas exchange during photosynthesis. Once the available water has been used up, tissue water potentials fall below threshold levels and growth ceases. Growth under these conditions is determined by the amount of available water multiplied by water use efficiency (WUE). Because WUE can be greatly enhanced by increased CO_2 concentration (Rogers *et al.*, 1983; Tolley and Strain, 1985; Morison, 1987; Eamus and Jarvis, 1989), relative plant responses to increases in ambient CO_2 should be most pronounced under water-limited conditions (e.g., Gifford, 1979; Allen, 1990). While increasing CO_2 concentration may be beneficial for plant growth, some researchers rank it of less importance than changes in temperature and/or precipitation, which can have large impacts when critical thresholds of drought, chilling, or degree-days are reached (e.g., Solomon, 1988).

1.3.1.3. Carbon Storage and Nutrient Availability

Although increasing temperature may lead to higher net primary productivity (NPP), net ecosystem productivity (NEP) may not increase, and may even become negative, because warmer temperatures also greatly stimulate soil organic matter decomposition (e.g., Raich and Schlesinger, 1992; Lloyd and Taylor, 1994; Kirschbaum, 1995; Chapter A). This could release large amounts of CO_2 to the atmosphere (e.g., Jenkinson *et al.*, 1991; Kirschbaum, 1993, 1995; Schimel *et al.*, 1995; Chapter 9, *Terrestrial Biotic Responses to Environmental Change and Feedbacks to Climate*, in the Working Group I volume). However, the direct effect of increasing CO_2 concentrations may partly offset or in some cases even reverse this effect and make NEP positive.

Enhanced decomposition of soil organic matter also should have the effect of mineralizing nutrients—especially nitrogen and phosphorus—that are held in soil organic matter and making them available for plant growth (e.g., Shaver *et al.*, 1992; Melillo *et al.*, 1993). This is likely to be of greatest importance in cool regions of the world, which are mostly nitrogen-limited and often contain large amounts of organic matter (Post *et al.*, 1982). This could sometimes lead to an increase in total carbon storage in systems if nutrients are redistributed from components with low C:N ratios (i.e., soil organic matter) to components with high C:N ratios (i.e., woody stems) (Rastetter *et al.*, 1992; Chapter 9, *Terrestrial Biotic Responses to Environmental Change and Feedbacks to Climate*, in the Working Group I volume). In industrialized regions, nitrogen deposition from the atmosphere may enhance NPP and NEP, leading to increased total carbon storage in forest ecosystems (Kauppi *et al.*, 1992), provided deposition has not reached levels that cause forest decline (Durka *et al.*, 1994).

1.3.2. Species Distributions

Species have responded individually to past environmental changes (Huntley and Birks, 1983; Davis and Zabinski, 1992; Solomon and Bartlein, 1992; Gates, 1993). The set of all possible environmental conditions in which a given species survives

Box 1-3. Ecological Niches and Climatic Change

Any species' survival is influenced by many factors, and the set of environmental conditions in which it can exist and reproduce is called its niche. A distinction must be made between a species' fundamental niche and its realized niche (Hutchinson, 1957; Malanson *et al.*, 1992). The fundamental niche encompasses all environmental conditions in which a species could potentially grow and reproduce with its specific physiological characteristics. The realized niche encompasses those conditions in which a species is actually found. The latter is usually a subset of the fundamental niche due to competition by other species (see Chapter A). Only rarely does the realized niche coincide with the fundamental one (Woodward, 1987; Booth *et al.*, 1988; Austin *et al.*, 1994). Examples of typical environmental factors that determine a species' niche are temperature and precipitation (see Figure 1-1).

Species with narrow niches are potentially very sensitive to climatic changes. A sustained temperature increase of only 1°C could have a major effect on the probability of occurrence of many species (e.g., Arolla pine, Figure 1-1), and in some instances a temperature increase of 2°C can be sufficient to change the environment for some species from very suitable to totally unsuitable (e.g., Whitehead *et al.*, 1993).

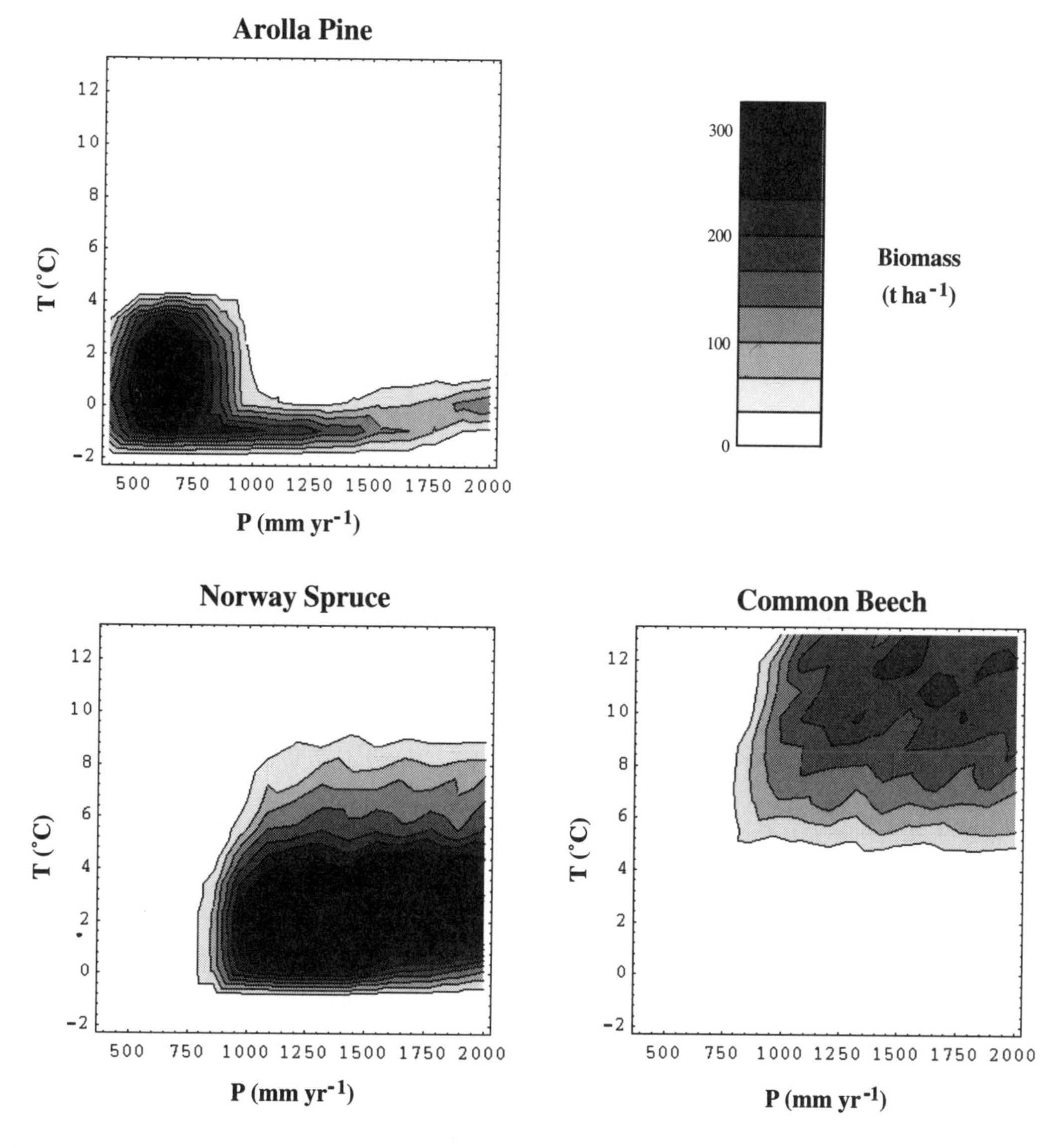

Figure 1-1: Simulated realized niches of three tree species [*Pinus cembra* L. (Arolla Pine), *Picea abies* (L.) KARST. (Norway Spruce), and *Fagus sylvatica* L. (Common Beech)] plotted as biomass versus annual means of temperature (T) and precipitation (P) (after Bugmann, 1994). The realized niches can be rather narrow (e.g., Arolla pine) and are usually smaller than the corresponding fundamental niches. For example, Norway spruce can easily be planted in climatic conditions where it naturally would be outcompeted by beech.

and reproduces (but without evolving—that is, changing its inherited characteristics such as physiological traits), is called its fundamental ecological niche (Hutchinson, 1957; Malanson *et al.*, 1992). In field conditions, species often survive only in a subset of the fundamental niche, the realized niche (Grubb, 1977; Booth *et al.*, 1988; Malanson *et al.*, 1992), owing especially to competitive interactions with other species (see Box 1-3).

With rapid climatic change, conditions may become unsuitable to complete one or more stages of the life cycle, especially if some climate variables were to change significantly more than other variables. For example, pollen and seed development require minimum heat sums and are sensitive to frosts (Stern and Roche, 1974). Seedlings are particularly vulnerable to short-term droughts, saplings to the presence or absence of sunlight, and mature trees to the availability of growing-season soil water. Populations could appear quite healthy while losing the ability to complete their life cycles.

Trees whose seedlings can now survive at a particular site will grow into adults in climates that may be unsuitable in 50–100 years; conversely, adults that could grow in an area in 50–100 years time must grow from seedlings that may be unable to survive current climatic conditions at those sites. A net ecosystem-level impact may be the loss of slow-growing species and the selection of species that complete their life cycles more quickly, such as early successional trees and shrubs. The ability to reach reproductive maturity in a short time favors early successional species that grow in full sun, whereas slower-growing species that begin their life cycles as understory species under closed canopies may be lost. Some model simulations have indicated that this opening of closed forests could result in the loss of three quarters of the trees and aboveground carbon in current temperate-zone forests (e.g., Solomon, 1986).

1.3.3. Transient Responses in Species Compositions

Concerns about the future of forest ecosystems relate not only to the geographic distribution of areas potentially suitable for forests and the performance of trees under different environmental conditions but also to the effect of climate change on the functioning and structure of ecosystems during the transient phase. For instance, it is well known that changing ambient conditions can reduce growth in forest ecosystems (e.g., Solomon and Webb III, 1985; Shugart *et al.*, 1986; Solomon, 1986; Woodwell, 1987; Prentice *et al.*, 1991b; Botkin and Nisbet, 1992; Davis and Zabinski, 1992). Moreover, the magnitude of climate change will subject many species assemblages, within a life cycle of their main species and in most of their distribution area, to climates that now occur outside their current ecological range (e.g., Solomon *et al.*, 1984; Roberts, 1989; Davis and Zabinski, 1992).

The transient response of species to such climatic changes can currently be assessed only with forest succession models (see Box 1-4). Despite some of their deficiencies (e.g., Moore, 1989; Bonan, 1993; Bugmann and Fischlin, 1994; Fischlin *et al.*, 1995), these models can be used to project transient changes in species composition of selected forest types for scenarios of climatic change in the past (Solomon *et al.*, 1981; Lotter and Kienast, 1992) or future (Pastor and Post, 1988; Kienast, 1991; Friend *et al.*, 1993; Bugmann, 1994; Bugmann and Fischlin, 1994; Smith *et al.*, 1994; see Box 1-4). These simulations suggest that climate change could cause widespread tree mortality within a few decades (Solomon *et al.*, 1984; Solomon, 1986; Solomon and West, 1987; Pastor and Post, 1988; Kienast, 1991; Prentice *et al.*, 1991b; Bugmann, 1994; Bugmann and Fischlin, 1994). Solomon and Bartlein (1992) and Pastor and Post (1993), for example, show how lags in population responses to climatic change could result in transient decreases in NPP before better-adapted species eventually replace the original vegetation and result in enhanced growth. It also should be noted that many simulations show species compositions that are not present in existing forests (Bugmann, 1994; Bugmann and Fischlin, 1994; Smith *et al.*, 1994).

Regrowth of better-adapted species or forest types requires many decades to centuries (Dobson *et al.*, 1989; Kienast and Kräuchi, 1991; Bugmann and Fischlin, 1994). Consequently, regions with forests in decline could release large amounts of carbon (e.g., Smith and Shugart, 1993), producing a large transient pulse of CO_2 into the atmosphere (e.g., Neilson, 1993; Chapter 9, *Terrestrial Biotic Responses to Environmental Change and Feedbacks to Climate*, in the Working Group I volume). Whereas some authors have estimated this carbon pulse from aboveground carbon alone to fall within a range of 0.1 to 3.4 Gt yr^{-1} for the annual flux, or 10 to 240 Gt for the accumulated pulse (King and Neilson, 1992), others have estimated a total carbon pulse from above and belowground C as high as 200 to 235 Gt, to be released to the atmosphere during a few decades to a century (see Section 1.3.4; Neilson, 1993; Smith and Shugart, 1993). Such responses, although debatable in their magnitude, are plausible because climatic changes also have been implicated in past episodes of forest and species decline (Cook *et al.*, 1987; Hamburg and Cogbill, 1988; Johnson *et al.*, 1988; Auclair *et al.*, 1992).

1.3.4. Potential Biome Distributions

Patch dynamics models cannot currently be used to simulate the transient behavior of forests in a changing climate on a global scale. This is because they require a large set of species-specific information that is not available from all regions of the world, especially not from the tropical zone. However, on the biospheric level (Table 1-1), several static global vegetation models have recently become available that enable us to make estimates of vegetation-climate equilibria (Box, 1981; Emanuel *et al.*, 1985; Prentice *et al.*, 1992; Smith *et al.*, 1992a; Cramer and Solomon, 1993; Leemans and Solomon, 1993; Monserud *et al.*, 1993b; Tchebakova *et al.*, 1993; Leemans and van den Born, 1994; Neilson *et al.*, 1994) based on earlier related studies (e.g., Köppen, 1936; Holdridge, 1947; Woodward, 1987).

Given any past or present climate, such vegetation models can be used to map the distribution of biomes (e.g., Prentice *et al.*,

Box 1-4. Forest Succession Models

Most forest succession models are based on the gap dynamics hypothesis (Watt, 1947; Bray, 1956; Shugart, 1984). They simulate the establishment, growth, and death of individual trees as a mixture of deterministic and stochastic processes within small—often $^{1}/_{12}$ ha—patches and average the actual forest succession at the ecosystem level from the successional patterns simulated for many individual plots (Shugart, 1984). Earlier work on the potential effects of climatic change on forests had to rely on spatial correlations between forest composition and climatic variables (e.g., Davis, 1986). Similarly, models that do not explicitly account for the differential effects of climatic change on different species may give optimistic projections of the effect of climatic change on ecosystem productivity. The more complex succession models that include the feedbacks between climate and ecosystem processes provide a more pessimistic outlook (see Figure 1–2).

On the other hand, most of these models do not include direct effects of increasing CO_2 concentration, which can ameliorate projected forest decline (e.g., Post *et al.*, 1992). Also, because of the limited availability of data on the fundamental niche of most species, these models are parameterized with information about the realized niche only (Austin, 1992). It has been argued that this might cause succession models to overestimate the extent of forest decline during the transient phase (e.g., Malanson *et al.*, 1992). Despite these caveats, patch models are the best tools currently available to study transient effects during climatic and other changes. Their simulations for current forests in response to past climatic changes provide fairly realistic assessments of likely future conditions (e.g., Solomon and Bartlein, 1992; Bugmann, 1994), although they cannot be interpreted as actual forecasts.

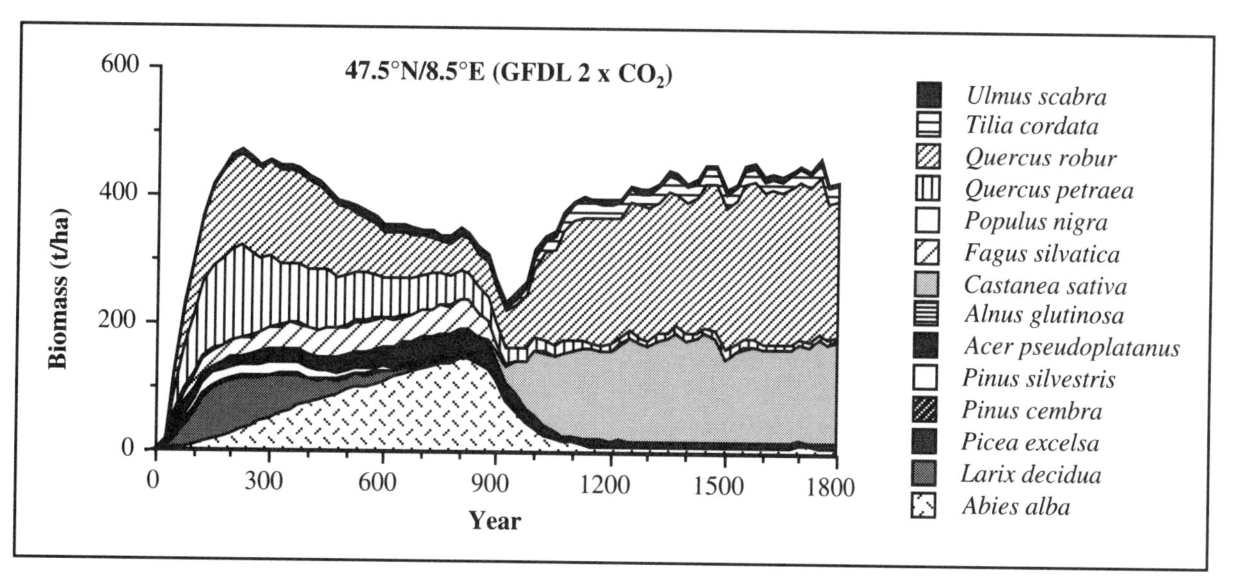

Figure 1-2: Transient response in species compositions simulated by the model FORCLIM-E/P (Bugmann, 1994; Fischlin *et al.*, 1995) at a site in Switzerland (47.5°N/8.5°E) under current climatic conditions (years 0–800) and under a scenario of climatic change derived from output of the GFDL GCM (years 900–1,800). A linear change of climatic parameters was assumed for the years 800–900. The graph shows the average cumulative biomass from 200 simulation runs (after Bugmann and Fischlin, 1996).

1992) or biospheric carbon storage (e.g., Solomon *et al.*, 1993) and to compare the simulated patterns with the present vegetation as provided by the few available global databases (Matthews, 1983; Olson *et al.*, 1983); see Figures 1-3, 1-5 (top), 1-6 (top), and 1-7 (top). Although these models are at an early stage of development (Leemans *et al.*, 1995), there is good statistical agreement between simulated and observed distributions of vegetation classes (e.g., Prentice *et al.*, 1992), except in areas where agriculture dominates. Although global models never can be strictly validated, their recent development toward inclusion of improved bioclimatic driving variables and mechanistic response functions offers increasing confidence in the magnitude of the results they generate. More about the nature and limitations of these models can also be found in Chapter 15.

The more recent models, such as BIOME (Prentice *et al.*, 1992), MAPSS (Neilson *et al.*, 1992), or TVM from IMAGE 2.0 (Leemans and van den Born, 1994), all attempt to simulate vegetation distribution considering ecophysiological traits and their relationship to particular climatic variables. This makes it possible to generate projections of vegetation distributions under past (Prentice, 1992; Prentice *et al.*, 1994; Figure 1-4), present [Prentice *et al.*, 1992; Neilson, 1993; Solomon *et al.*, 1993; Prentice and Sykes, 1995; Figures 1-5 (top), 1-6 (top), and 1-7 (top)], or future climates [Cramer and Solomon, 1993;

Leemans and Solomon, 1993; Neilson, 1993; Solomon *et al.*, 1993; Leemans *et al.*, 1995; Figures 1-5 (bottom), 1-6 (bottom), and 1-7 (middle)]. Apart from being used for climate-change impact assessment studies, they are now also used as a dynamic representation of the land surface for sensitivity studies of GCMs (see Claussen, 1994; Claussen and Esch, 1994; Chapter 9, *Terrestrial Biotic Responses to Environmental Change and Feedbacks to Climate*, in the Working Group I volume).

The models differ in their emphasis on particular processes and relationships (Leemans *et al.*, 1995). For example, BIOME (Prentice *et al.*, 1992) includes physiologically important aspects such as seasonality and moisture balance, and is based on PFTs, which to some extent simulate interspecific competition. MAPSS, on the other hand, couples the rate of transpiration to the conductance of the canopy by simulating maximum leaf-area index and stomatal conductance (Neilson *et al.*, 1994; Neilson and Marks, 1994). The latter makes it possible to incorporate ecophysiological responses such as CO_2 fertilization and WUE, but all species-specific information is lumped. These models use somewhat different vegetation classification schemes [e.g., compare Prentice *et al.* (1992) and Neilson (1993)], which usually makes direct comparisons possible only if some classes are aggregated (compare Figures 1-5 to 1-7 and Table 1-2 with Prentice *et al.*, 1992; Neilson, 1993; or Leemans and van den Born, 1994). Consequently, the vegetation distributions projected for a changed climate may differ substantially (Table 1-2).

One of these models, IMAGE 2.0, also has been used to incorporate the effect of climate change superimposed to the impacts of land-use changes (Alcamo, 1994; Figure 1-7). Its vegetation part (Leemans and van den Born, 1994) makes it possible to assess global forest distributions for the present [Figure 1-7 (top)] as well as the future [e.g., according to the conventional wisdom scenario of Figure 1-7 (middle and bottom)]. However, the potential natural vegetation is essentially modeled in a manner similar to BIOME (Prentice *et al.*, 1992).[2]

Although none of these models deals with transient forest responses, they offer the advantage of providing quantitative estimates for changes in a future climate on the scale of the distinguishable vegetation classes (Table 1-2; Figures 1-3 to 1-7). Due to the limited range of vegetation classes—BIOME and IMAGE, for instance, distinguish 20 (Prentice *et al.*, 1992; Leemans and van den Born, 1994; Prentice *et al.*, 1994); MAPSS distinguishes 33 (Neilson, 1993)—the projected changes (Table 1-2) tend to underestimate actual changes. Changes within a class remain by definition undetected and would have to be modeled by means of more detailed vegetation models.

The total area currently forested is likely to change significantly, if the changes occur according to any of the three models, from its present to a new vegetation class (Table 1-2, D). These estimates range from small changes [e.g., 7.5% (tropical rain forest)] to large ones [e.g., 65.1% (boreal forest)]—both estimated using the BIOME model. Net changes (Table 1-2, D*)—that is, the difference between the total forested area in the future versus the current climate—range from losses of 50.0% (IMAGE/TVM—tropical dry forest) to gains as large as 22.2% (MAPSS—tropical dry forest).

Except for MAPSS in the version shown here (i.e., without considering the partially compensating, increased WUE; Neilson and Marks, 1994), for the tropical zone models project a net

[2] More information about this class of biosphere models can be found in Chapter 9, *Terrestrial Biotic Responses to Environmental Change and Feedbacks to Climate*, in the Working Group I volume, and in Chapters 15 and 24 of this volume.

Table 1-2: *Likely changes in forested areas (Mha) within four biogeographical zones according to three different vegetation models: BIOME (Prentice et al., 1992), MAPSS (Neilson, 1993), and terrestrial vegetation model (TVM) (Leemans and van den Born, 1994) from IMAGE 2.0 (Alcamo, 1994).*

	BIOME 2 x CO_2 (GFDL)			**MAPSS 2 x CO_2 (GFDL)**			**IMAGE-TVM 2050**			**Mean**	
Forest Type	D	s_o	D*	D	s_o	D*	D	s_o	D*	%D	%D*
Tropical Rain	57	706	19	281	1243	-234	129	296	-129	18.8	-14.4
Tropical Dry	153	640	-2	353	528	196	324	294	-309	37.2	-9.3
Temperate	346	1607	544	1007	1039	-155	388	583	-65	35.7	4.5
Boreal	952	511	-379	1231	1117	-529	42	1128	-33	40.4	-17.1

Notes: The first two models are used with a 2 x CO_2 equilibrium climate-change scenario generated by the GFDL GCM (Wetherald and Manabe, 1986). TVM from IMAGE 2.0 generated the changed climate internally, based on comparable assumptions (Conventional Wisdom Scenario) about greenhouse gas concentrations (Alcamo, 1994) and land-use changes as driven by population developments. All numbers relate to a potential forest vegetation in equilibrium with the climate, a steady state that is unlikely to be reached for many centuries. All figures were compiled by using the maximum number of vegetation classes as supported by the models (i.e., more than the 10 classes shown in Figures 1-3 to 1-7). D = Total forested area in transition from current type into a new one; s_o = Total forested area remaining within the same vegetation class; D* = Net change (data compiled by R. Leemans). The last two columns show means over all models as percentages of the currently forested area.

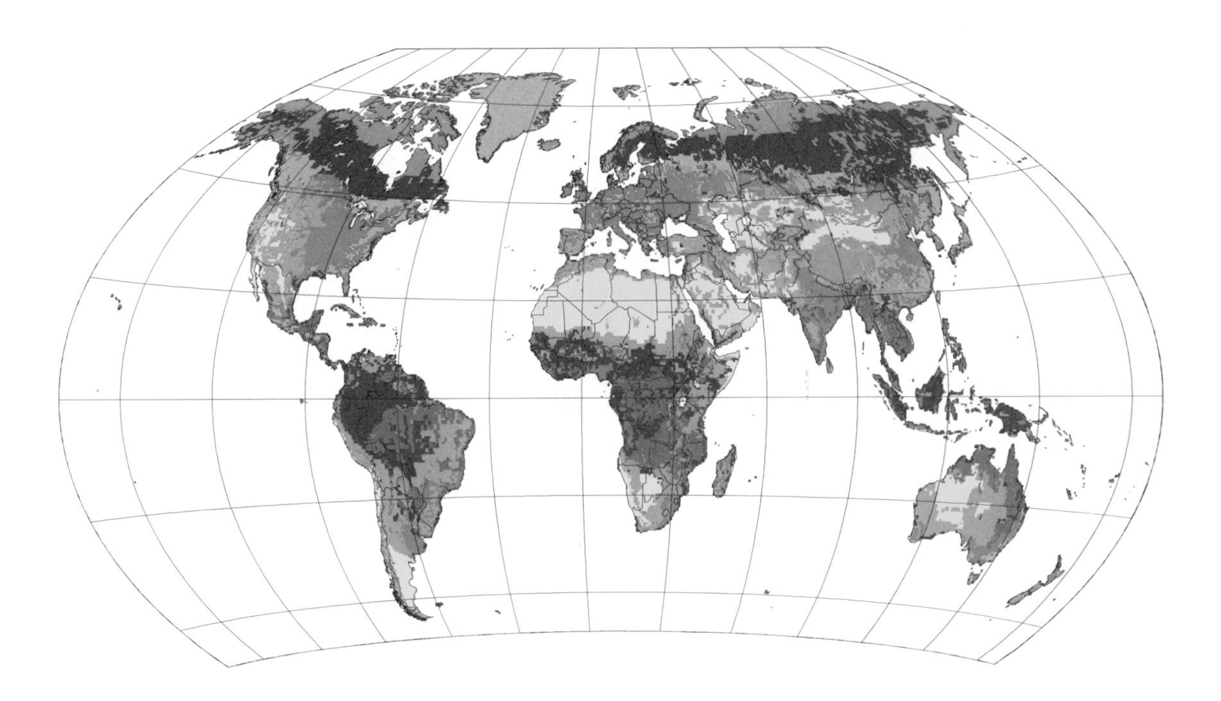

Figure 1-3: Present observed distribution of global vegetation complexes redrawn from the database compiled by Olson *et al.* (1983). For comparison, the vegetation classes have been aggregated to the same classification system as that used for all other model results shown here.

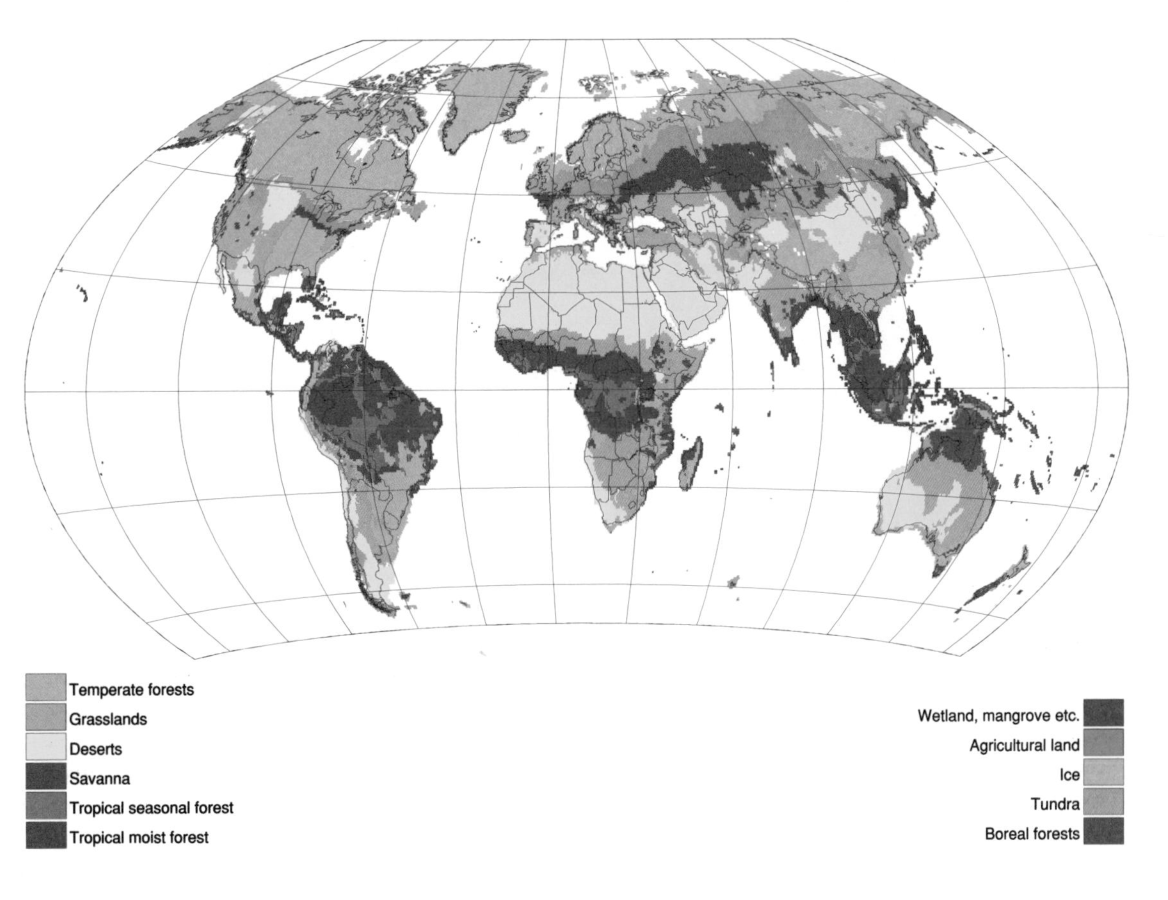

Figure 1-4: Past equilibrium vegetation according to BIOME (Prentice *et al.*, 1992) during the last glacial maximum [i.e., 18,000 years BP (Prentice *et al.*, 1994)]. The climate used for this simulation is derived from a GCM (not from paleoecological data, as this would lead to a circular argument). Note that according to this simulation, boreal forests cover only about 20% (Solomon *et al.*, 1993) of today's potential distribution (Prentice *et al.*, 1994).

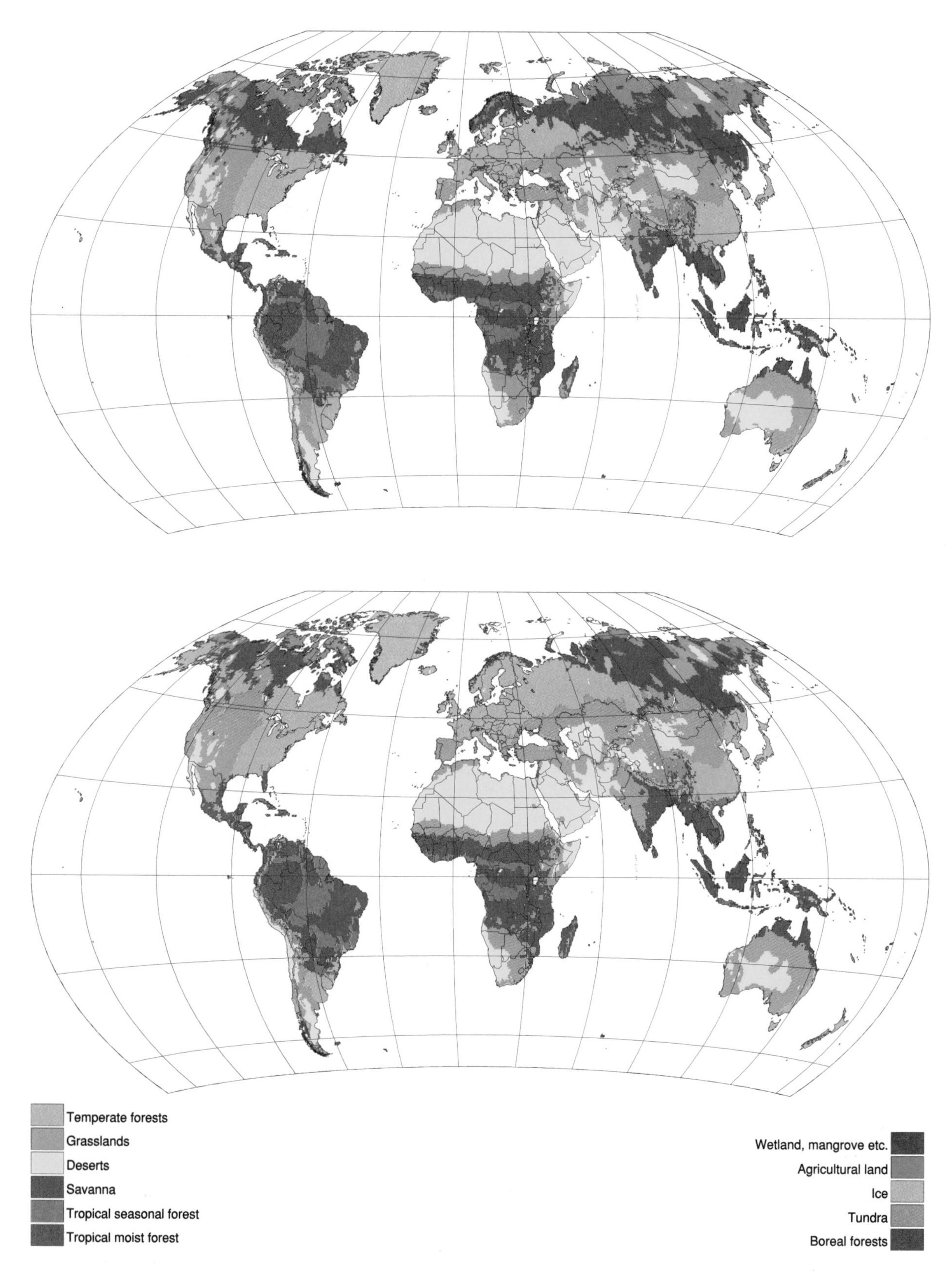

Figure 1-5: Present (top) and future (bottom) potential natural vegetation according to the BIOME model (Prentice *et al.*, 1992; Prentice *et al.*, 1994). The present climate is given by the IIASA climate database (Leemans and Cramer, 1991). The projected shifts in the boundaries of the vegetation classes are due to climatic changes as projected by the difference between a GFDL GCM control run and a 2 x CO_2 scenario (Wetherald and Manabe, 1986). They represent responses of plant functional types to cold tolerance, chilling requirements, and heat and water requirements of global vegetation (Solomon *et al.*, 1993; Prentice and Sykes, 1995).

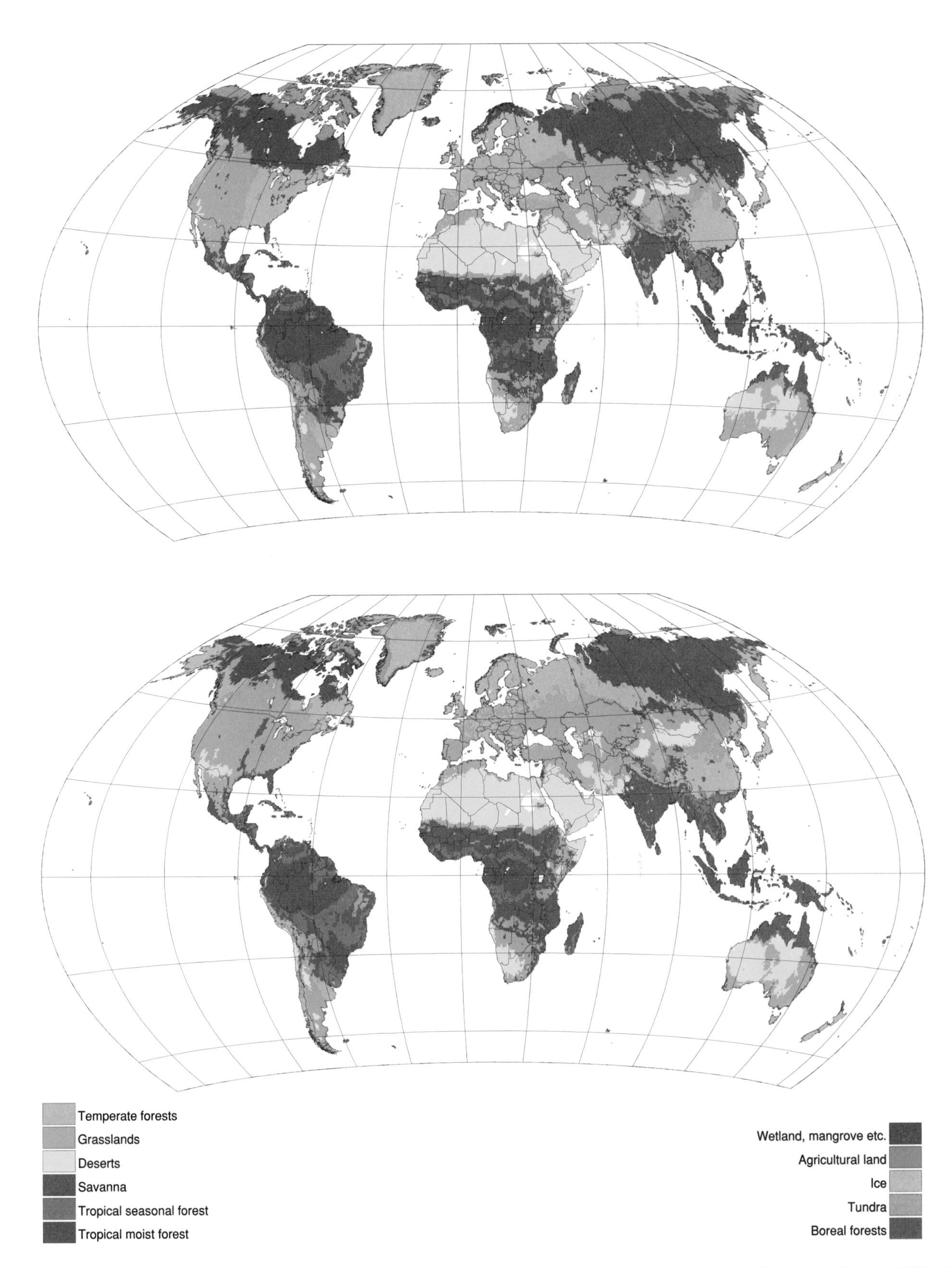

Figure 1-6: Present (top) and future (bottom) equilibrium potential natural vegetation according to the MAPSS model (Neilson, 1993). The present climate is given by the IIASA climate database (Leemans and Cramer, 1991). The projected shifts in the boundaries of the vegetation classes are due to climatic changes as projected by the difference between a GFDL GCM control run and a 2 x CO_2 scenario (Wetherald and Manabe, 1986). The model simulates climate responses due to a simulated steady-state leaf-area index, calculated from a site water and heat balance submodel (Neilson, 1993).

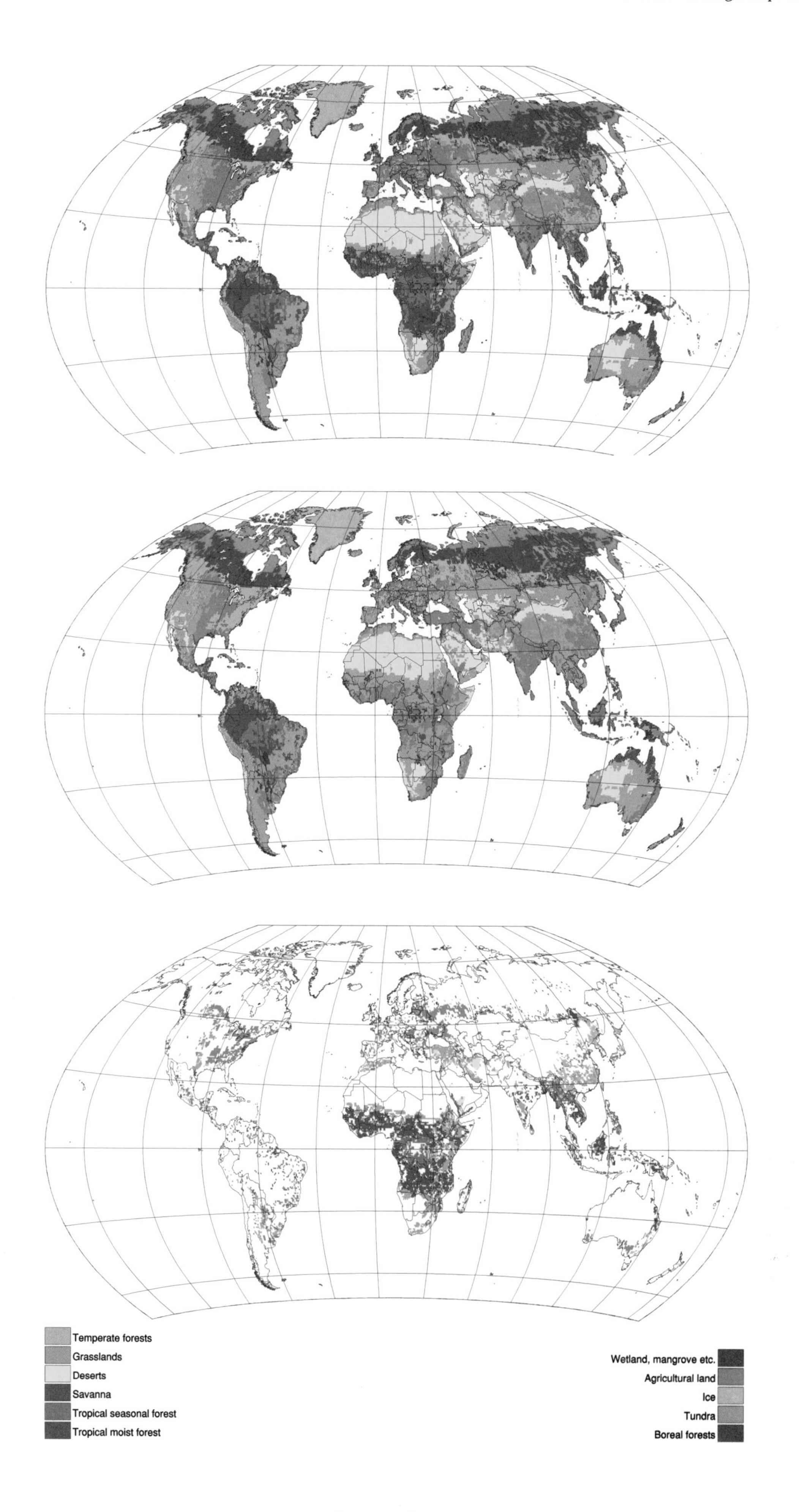
Temperate forests
Grasslands
Deserts
Savanna
Tropical seasonal forest
Tropical moist forest
Wetland, mangrove etc.
Agricultural land
Ice
Tundra
Boreal forests

gain in the potential area of forest distribution. Mainly IMAGE-TVM projects large net losses in the tropical rainforests. These losses are not due to climatic change, however, but are mainly caused by deforestation and other land-use changes (Zuidema *et al.*, 1994).

Although BIOME and MAPSS indicate relatively small net changes (Table 1-2, D*) in tropical dry forests, the actual changes encompass large gross losses (D), which are compensated by similarly large gross gains (not shown in Table 1-2). This is also partly true for the temperate zone, where MAPSS, for example, projects only relatively small net losses (D*); however, the associated gross losses (D) irrespective of gains (not shown) are more than six times larger. Mapped distributions indicate that temperate forests are likely to replace a large area of boreal forest, mainly due to the increase in winter temperatures (Figures 1-5 and 1-6). This allows temperate-zone vegetation to expand poleward into regions from which it is currently excluded by the -40°C threshold for the coldest month (Figures 1-5, 1-6, and 1-7).

The boreal regions, especially, are expected to undergo large vegetation shifts, with both MAPSS and BIOME projecting large changing areas (Table 1-2, D). Both models show large losses in area for current boreal forests, despite their encroachment into current tundra. Shrinkage in total area due to the geographically limited poleward shift leads to a net loss between 379 Mha (25.9%) and 529 Mha (22.5%).[3] The IMAGE model is an exception in this case; it projects much smaller losses of only 33 Mha (2.9%) (Figures 1-5, 1-6, and 1-7). This is mainly due to the structure of the model, in which vegetation change is primarily driven by human land use—which is not an important factor in the boreal zone.

In summary, all models suggest that the world's forests are likely to undergo major changes in the future, affecting more than a third of tropical dry (37.2%), temperate (35.7%), and boreal (40.4%) forests. Except in the temperate zone, the models suggest that there may be a net loss of forest area (Table 1-2, %D*).

Averaged over all zones, this net loss amounts to 9.1% of the currently forested area. This is partly due to the fact that some regions are likely to lose forests for climatic reasons, while climatic gains in other regions might not be realized due to land-use pressures (IMAGE), and some models even predict a net loss (MAPSS). Averaged over all zones and all models, the projections indicate that 33% of the currently forested area is likely to change from the current vegetation class to a new one in response to climate forcing as generated by the GFDL 2 x CO_2 scenario. Compared with other GCMs, the latter represents a medium global warming scenario.

All models tend to underestimate potential changes, since they model only transitions among a limited number of vegetation classes. More importantly, as equilibrium models, none of them reflects the asymmetry between a temporary loss and the subsequent, much slower regrowth of forests in a new climate (see Section 1.3.3).

1.3.5. Past and Future Species Migration Rates

Shifts in the distribution of forest zones take place with significant time lags. The area occupied by different forest types has been quite plastic in the past 100,000 years, responding to changing environments with changes in species associations, structural properties of stands, and areas occupied (Solomon and Webb III, 1985; Webb III and Bartlein, 1992). Unlike the Pleistocene climatic changes, which occurred over thousands of years, future climatic changes of about the same magnitude are expected to take place over 100 years or less.

The change in annual mean temperature that occurs when one moves 100 km poleward may be as high as 0.7°C in mid- and high latitudes, but less at low latitudes. For summer temperatures and toward the interior of the continents, this value may be higher. With altitude, temperature changes of about 0.5–0.7°C per 100 m are also common (see also Chapter 5). With an expected warming between 0.1 and 0.35°C per decade, this means that species would have to migrate 1.5–5.5 km toward the poles per year or increase elevation by 1.5–5.5 m per year in order to remain within similar climatic conditions. Many studies of past changes have estimated natural rates of migrations of trees ranging from 40 to 500 m per year (Davis, 1976, 1981, 1986; Huntley and Birks, 1983; Solomon *et al.*, 1984; Gear and Huntley, 1991; Torrii, 1991). Similarly, Gear and Huntley (1991) calculated from several sites in Britain migration rates for Scots pine of only 40–80 m yr^{-1}. However, for other species, such as white spruce, much faster dispersal rates of up to 1–2 km yr^{-1} also have been reported (e.g., Ritchie and MacDonald, 1986). It is not always clear whether the observed past rates were maximal rates of migration or whether they were limited by the rate at which the climate

< **Figure 1-7:** Present (top), future (middle), and differing (bottom) potential natural vegetation influenced by human land use as generated by the terrestrial vegetation module (TVM; Leemans and van den Born, 1994) from the integrated climate change assessment model IMAGE 2.0 (Alcamo, 1994). The areas shown represent a projection starting with the year 1970 (top) and the internal, dynamic changes calculated by IMAGE 2.0 by the year 2050 (middle). The bottom graphic shows the new class for all areas that are predicted to change from one to another vegetation class. The simulation results generated by IMAGE 2.0 incorporate the effect of land-use changes (e.g., deforestation) as driven by the dynamics of human populations and economic development. Note that IMAGE 2.0, unlike the other models, simulates climatic change independently from a GCM, since it generates its climate internally. The vegetational part (Leemans and van den Born, 1994) of the IMAGE model is similar to BIOME (Prentice *et al.*, 1992) and represents a mixture between a transient and a static response of vegetation to internally as well as externally generated changes.

[3] The apparent lack of coincidence between relative and absolute changes is due to different definitions of the vegetation class "boreal forest." The MAPSS model has a greater area characterized as boreal forest than the BIOME model.

changed (Prentice *et al.*, 1991a; Prentice, 1992). Nevertheless, it is unlikely that future rates of species migration could match those required by the currently expected rates of climatic change in large areas. Pollen, on the other hand, may be dispersed much faster than seeds (e.g., Ennos, 1994), so the movement of genes of different ecotypes within a species might be able to match the speed of climatic change.

Migration of a tree species involves movement of propagules to new locations, establishment of seedlings, growth of individuals to reproductive maturity (which may take from years to decades), and production of new propagules. For long-distance migration, these stages need to be repeated. Several bottlenecks may be encountered: Strong competitors along the route may suppress the completion of full life cycles or specific, mandatory biological symbionts [e.g., the right species of mycorrhizal fungi (Perry *et al.*, 1987, 1990)], or specific pollinators for cross-fertilization may be absent at some point. Seed production, in particular, often depends on a phased sequence of development over at least 2 years, including many steps from floral bud differentiation to seed ripening (Fenner, 1991). Many stages in this sequence could be disrupted by unsuitable climatic conditions (Innes, 1994): With further warming, some species may fail to be chilled sufficiently to release dormancy (Murray *et al.*, 1989; Cannell, 1990), or dormancy may be released too early, which paradoxically could result under some circumstances in greater frost damage (Hänninen, 1991). Conversely, many trees flower profusely following hot, dry conditions (Stern and Roche, 1974). Hence, differential seed production by different species may limit rates of migration and drive substantial changes in species distributions.

Since each stage of a tree's life cycle requires specific environmental conditions, rapid climatic change is less likely to offer sufficiently favorable conditions to complete complex life cycles. This is especially true for some slow-growing species. More flexible species, such as those with wider seed dispersal, or a more invasive growth habit, may be able to move to more favorable habitats, whereas less mobile species are likely to be left behind and would be at a disadvantage as their former habitat becomes climatically unsuitable. Such a decoupling of climate determinants from species and community distributions probably has occurred already many times and at many places during the last millennia (Davis, 1986). The result has been a temporary absence of certain species for hundreds and thousands of years from regions in which they were previously capable of growing. Despite this, species diversity also could be transiently enhanced during a time of change because forests might form a richer mosaic of patches, consisting of some remaining old trees and a variety of invaders that are successful because of locally more favorable conditions.

In the future, human activities may to a certain extent enhance the migration of certain species, especially that of commercially important ones. On the other hand, natural migration of all other forest species will be further hampered by the fragmentation of natural habitats by human infrastructure, farmland, or exotic tree plantations, especially at mid-latitudes.

1.3.6. Biodiversity

To date, only about 1.4–1.7 million of an estimated total of 5–30 million species have been scientifically described (Wilson, 1985, 1992; May, 1990; Ehrlich and Wilson, 1991; Erwin, 1991; Gaston, 1991; Pimm *et al.*, 1995), and fewer than 100,000 are known well (e.g., Wilson, 1988; Pimm *et al.*, 1995). About 20% of the estimated total of 250,000 plant species are woody or tree species (e.g., Groombridge, 1992; Reid, 1992; Pimm *et al.*, 1995). There is poor understanding of the factors responsible for maintenance of high biodiversity (e.g., Grubb, 1977; Stork, 1988; Groombridge, 1992; Norton and Ulanowicz, 1992; Myers, 1995; Pimm *et al.*, 1995), but it is understood that biodiversity can only be managed indirectly through habitat and ecosystem management (e.g., Wilson, 1992). In this respect, forests are of paramount importance because they harbor about two-thirds of all species on Earth (Ehrlich and Wilson, 1991); tropical forests alone harbor at least half of all species (Raven, 1988; Ehrlich and Wilson, 1991; Webb, 1995). Species diversity generally increases strongly as one moves from colder to warmer sites (e.g., Groombridge, 1992), although the reasons for that trend are not well understood. When the size of an ecosystem is reduced to about 10% of its former size, about 50% of the species originally present generally will become extinct. Based on that relationship, it has been estimated, for instance, that a temperature rise of 2°C would cause 10–50% of the animals currently extant in the boreal Great Basin mountain ranges to be lost (Dobson *et al.*, 1989).

Climatic change can affect biodiversity either directly through altering the physiological responses of species or indirectly by altering interspecific relationships (e.g., Woodward, 1987; Tallis, 1990; Peters and Lovejoy, 1992). Whenever these changes lead to a habitat degradation, biodiversity will eventually be adversely affected (e.g., Peters and Lovejoy, 1992; Vitousek, 1994). Furthermore, biodiversity is affected not only by climate change but also by deforestation and other land-use changes that cause further habitat degradation and fragmentation (Janzen, 1988; Dobson *et al.*, 1989; Wilson, 1989; Ehrlich and Wilson, 1991; Myers, 1991, 1993; Postel, 1994; Daily, 1995; Pimm *et al.*, 1995). The combination of climate change with other pressures on ecosystems produces a particularly significant threat to biodiversity due to the following reasons:

- Climate change, together with other causes of habitat degradation, may locally decrease species diversity (Peters and Darling, 1985; Romme, 1991; Myers, 1992; Peters, 1992; Daily, 1995; Pimm *et al.*, 1995; Rind, 1995). Disturbance also may create opportunities for opportunistic or pioneer species, which will become more abundant (Myers, 1993) and, over time, replace many species that are slower growing and require more stable conditions. In some cases, this may lead to a temporary increase in species diversity (e.g., Lugo, 1988; Mooney, 1988; Vitousek, 1988).
- Species may become permanently extinct when local extinction cannot be reversed by reimmigration from

surrounding areas. This problem would be most severe where the climate in a species' reserve changes from being favorable to being completely unsuitable (Peters, 1992), rendering "sanctuaries" into "traps" (Myers, 1993). Thus, an increasing fragmentation of habitats in combination with climate change may cause significant and irreversible species loss, a situation with which many of the current policies on reserves have large difficulties coping adequately (Ehrlich and Wilson, 1991; Botkin and Nisbet, 1992; Franklin *et al.*, 1992; Myers, 1992; Peters, 1992).

Although the effects of climatic change on biodiversity are still poorly understood, many authors anticipate a significant loss of species due to climatic change (e.g., Peters and Lovejoy, 1992; Reid, 1992; Myers, 1993; Vitousek, 1994). Due to its permanency, this must be considered as one of the most important impacts of climate change, not only in economic terms (e.g., Daily, 1995) but also in terms of all other utilitarian and spiritual values that have been attributed to species (e.g., Ehrlich and Wilson, 1991; Sharma *et al.*, 1992; Myers, 1993).

1.3.7. Adaptation

Forests themselves may to some extent acclimate or adapt to new climatic conditions, as evidenced by the ability of some species to thrive outside their natural ranges. Also, elevated CO_2 levels may enable plants to use water and nutrients more efficiently (e.g., Luo *et al.*, 1994). Nevertheless (see Sections 1.3.3 and 1.3.5), the speed and magnitude of climate change are likely to be too great to avoid some forest decline by the time of a CO_2 doubling.

Consideration may therefore be given to human actions that minimize undesirable impacts. Special attention may be given to poor dispersers, specialists, species with small populations, endemic species with a restricted range, peripheral species, those that are genetically impoverished, or those that have important ecosystem functions (Peters and Darling, 1985; Franklin, 1988; Davis and Zabinski, 1992; Franklin *et al.*, 1992). These species may be assisted for a time by providing natural migration corridors (e.g., by erecting reserves of a north–south orientation), but many may eventually require assisted migration to keep up with the speed with which their suitable habitats move with climate change. Some mature forests may be assisted by setting aside reserves at the poleward border of their range, especially if they encompass diverse altitudes, and water and nutrient regimes (Botkin and Nisbet, 1992; Peters, 1992; Myers, 1993), or by lessening pollutant stresses and land-use changes that result in forest degradation (e.g., Vitousek, 1994; Daily, 1995).

1.4. Tropical Forests

1.4.1. Characterization and Key Limitations

Tropical forests cover about 1,900 Mha (FAO, 1993a) and are found between 25°N and 25°S. They are distributed over five continents and are situated mainly in developing countries.

Box 1-5. Tropical Forests and Land-Use Changes

Tropical forests are endangered more by land-use practices than by gradual climatic change. Already at present much of the tropical forest is affected by deforestation due to land conversion and resource use (UNESCO, 1978; FAO, 1982; Brown *et al.*, 1993; Dixon *et al.*, 1994). Such anthropogenic tropical forest disturbances are expected to continue in response to economic development, population growth, and the associated need for agricultural land (DENR-ADB, 1990; Starke, 1994; Zuidema *et al.*, 1994).

Tropical rainforests have been cleared for agriculture since at least 3,000 BP in Africa and 7,000 BP in India and Papua New Guinea (Flenley, 1979). Africa has lost at least 50% of its rainforest, while tropical America and Asia have lost at least 40% (Mabberley, 1992). In addition to deforestation, large areas of previously undisturbed forest are being affected by removal of wood for timber and fuel (Brown *et al.*, 1993; Richards and Flint, 1994). Chapter 24 presents a detailed account of these deforestation trends in tropical forests.

The influence of tropical forests on local and regional climate may be as important as the effects of climate on forests (Lean and Warrilow, 1989). About 20% of the water flux to the atmosphere derives from evapotranspiration from the land, mostly from forested areas with high annual rainfall (Westall and Stumm, 1980; Lean and Warrilow, 1989). Deforestation can significantly reduce global evapotranspiration and increase runoff. This could affect the amount and distribution of precipitation over wide areas (Salati and Jose, 1984). For instance, in the Amazon basin, at least 50% of precipitation originates from evapotranspiration from within the basin (Salati *et al.*, 1979; Salati and Jose, 1984). Deforestation there will reduce evapotranspiration, which could reduce precipitation by about 20%—producing a seasonal dry period and increasing local surface temperatures by 2°C (Gash and Shuttleworth, 1991). This could result in a decline in the area of wet tropical rainforests and their permanent replacement by floristically poorer drought-deciduous or dry tropical forests or woodlands.

The forests consist of rainforests, drought-deciduous forests, and dry forests. In areas with prolonged dry seasons, especially where water limitations are intensified by edaphic conditions (Cole, 1986), there are savannas (Richards, 1966; Odum, 1971; Borota, 1991). Evergreen or partially deciduous forests occur in areas with a mean annual temperature greater than 24°C and with high regular rainfall throughout the year. Most of the closed tropical forests are found in the moist and wet zones, with precipitation to potential evapotranspiration ratios of 1–2 and greater than 2, respectively. The dry zone, with a ratio of precipitation to potential evapotranspiration of less than 1, covers about 40% of the total tropical region. In most of these forests, typical monthly temperatures fall between 24 and 28°C; daily extremes rarely exceed 38°C or fall below 10°C (Longman and Jenik, 1987).

Tropical forests represent about 40% of the world's forested area, containing about 60% of the global forest biomass and one-quarter of total soil carbon (210 Gt C in biomass and 220 Gt C in soils and litter) (FAO, 1982; Longman and Jenik, 1987; Brown *et al.*, 1993; Dixon *et al.*, 1994). Tropical forests cover only 6% of the world's surface but contain about half of all plant and animal species of the world (Bierregaard *et al.*, 1992; Mabberley, 1992; Riede, 1993). Thousands of species in tropical forests are utilized by humans. It is of particular concern that tropical forests are currently being clear-cut, burned, or otherwise degraded by human activity (see Box 1-5).

An important characteristic of tropical forests is that they maintain a tightly closed nutrient cycle and often grow on a mass of very infertile soil, in which only the uppermost few centimeters have substantial amounts of plant-available nutrients. Phosphorus is generally the main limiting nutrient (Jordan, 1985). Therefore, any disturbance that results in loss of nutrients due to leaching, erosion, or timber harvesting can result in decreased growth rates, biomass, and diversity (Whitmore, 1984; Jordan, 1985; Vitousek and Sanford, 1986).

Fire significantly influences the structure, composition, and age diversity of tropical forests. In dry forests, fires tend to be frequent, which excludes fire-sensitive species. In tropical rainforest, fire is usually rare and does not usually spread over wide areas because there is insufficient dry and flammable plant material (Chandler *et al.*, 1983a, 1983b). However, there have been instances in the past when thousands of hectares of tropical rainforest were burned following long dry periods (Goldammer and Seibert, 1990; Goldammer, 1992). Such fires are likely to significantly affect species diversity (Goldammer and Seibert, 1990) and may help prevent forest deterioration by promoting new growth and regrowth (Goldammer and Peñafiel, 1990).

Strong winds associated with tropical cyclones in tropical Asia, Central America, and northern Australia can profoundly influence the structure and floristic composition of forests (Whitmore, 1974; Hartshorn, 1978; Lugo *et al.*, 1983; Longman and Jenik, 1987; DENR-ADB, 1990; Mabberley, 1992; O'Brien *et al.*, 1992). Strong winds frequently damage tree canopies, create gaps in the forest, and modify the forest structure and micrometeorological environment. This may increase litter quality, allow more radiation to reach the forest floor, increase soil temperature, and make more soil water and nutrients available, which could promote the growth of new vegetation (Waring and Schlesinger, 1985).

1.4.2. Projected Climatic Changes

The climatic changes projected for the regions covered with tropical forest over the period 1990–2050 are increases in temperature of around 0.5–1.0°C, with no general change in seasonal amplitude; an increase in rainfall averaged over the region but with highly uncertain regional shifts in rainfall (differing greatly among GCMs), in the range -40% to +60%; and uncertain increases or decreases in soil water content (Greco *et al.*, 1994). These projections imply the potential for an increased occurrence of droughts in some regions and floods in others.

We also must consider the possibility of changes in the frequency of ENSO (El Niño Southern Oscillation; see Glossary) events, which may increase rainfall seasonality in semi-tropical regions and could lead to a longer dry season in some areas and bring more rains in other places. Fire frequencies may increase in some areas in association with drought (Rind *et al.*, 1990). Fires may further reduce local precipitation because fire-emitted aerosols increase the number of cloud condensation nuclei, producing smaller cloud droplets that are less likely to fall as rain (Andreae and Goldammer, 1992).

1.4.3. Impacts of Climate and Land-Use Changes

1.4.3.1. Forest Area, Distribution, and Productivity

All forests are expected to experience more frequent disturbance, with greater and possibly permanent impacts, such as increased soil erosion, and other forms of degradation and nutrient depletion. A study of 54 countries suggested that, between 1990 and 2050, a further 660 Mha are likely to be deforested, reducing the 1990 area by one-third and releasing 41–77 Gt C (Trexler and Haugen, 1995; see also Chapter 24). Rates of deforestation eventually must decrease as less and less of the original forest remains. There are, however, proposals to slow the loss of tropical forests (e.g., Deutscher Bundestag, 1990; UNCED, 1992), and many nations have large-scale plans for the protection or restoration of their forests (e.g., Brazil, India, and China: Winjum *et al.*, 1993).

Global vegetation models do not agree on whether climatic change (in the absence of land-use change) will increase or decrease the total area of tropical forests (depending on the calculation of transpiration and vegetation properties; see Section 1.3.4), but any major shifts in rainfall pattern due to climate and land-use change are certain to change the present distribution of vegetation types within and among biomes (Neilson, 1993). Henderson-Sellers and McGuffie (1994) show that in an enhanced-CO_2 climatic regime, tropical evergreen broadleaf forests could readily re-establish after deforestation. In some areas, decreased rainfall may accelerate the loss of dry forests to savanna, while in others, increased rainfall and increased water-use efficiency with elevated CO_2 may favor the expansion of forests and agroforestry. In both cases, the outcome will be strongly influenced by human activities. Overall, shifts in rainfall patterns in the tropics could increase the rate of conversion of forests to agricultural land by increasing human migration from areas affected by droughts, erosion, or other forms of land degradation to non-degraded and more productive forest land.

The productivities of different areas of tropical forest are likely to increase or decrease in accordance with changes in rainfall, as indicated by simulation studies (Raich *et al.*, 1991). The overall effect of predicted changes in climate (excluding elevated CO_2) on the net primary productivity of tropical evergreen forests may be a decrease due to increased temperature and consequently increased respiration (in the absence of acclimation) and decreased photosynthesis due to increased cloudiness, but elevated CO_2 levels may cause an increase in productivity (in contrast to boreal regions) unless limited by phosphorus supply (Melillo *et al.*, 1993). CO_2 enhancement of photosynthesis and growth, particularly below ground (e.g., Norby *et al.*, 1986, 1992; Luxmoore *et al.*, 1993), is favored by high temperatures in species with C_3 photosynthesis (i.e., in trees) (Long, 1991; Kirschbaum, 1994) and possibly by vegetation disturbance and consequent fast nutrient cycling (Peterson and Melillo, 1985). This is also consistent with studies of the spatial distribution of CO_2 concentrations across the globe, which also suggest that the tropics constitute a biospheric sink that partially offsets the carbon release due to deforestation (Enting and Mansbridge, 1991; Schimel *et al.*, 1995). An alternative view is that elevated CO_2 levels will accelerate carbon and nutrient cycles in tropical ecosystems without increasing growth and biomass (Körner and Arnone, 1992). Researchers are agreed, however, that increased water-use efficiency by plants in response to elevated CO_2 is likely to enhance the productivity of vegetation in the drier tropical regions.

Land-use change is obviously the greatest threat to species diversity of tropical forests, but Cramer and Leemans (1993) speculate that climatic change alone could decrease the diversity of plant types at the boundaries of biomes, particularly in the tropics. Losses are likely to be greatest where the pressures of population and socioeconomic forces are greatest and least controlled (see also Section 1.3.6).

1.4.3.2. Temperature

In general, temperatures in non-montane tropical regions are already high enough for rapid growth year round, and a 2°–3°C increase in temperature alone will have a marginal effect on rates of photosynthesis, growth, decomposition, and nutrient cycling. However, if plants experience temperatures above 35°–40°C over extended periods, especially in combination with water shortage, tissue damage from desiccation and sunscald can occur (Fitter and Hay, 1987). In drier areas, the severity and duration of dry spells could be aggravated by increasing temperature. Also, some species sustain damage from temperatures below 10°–12°C or even below 15°–20°C, and there may be some limited future expansion of those species into regions from which they are currently excluded by cool temperatures (e.g., Smith *et al.*, 1992b).

1.4.3.3. Water

In the semi-arid tropical regions, climate-induced desertification will be a critical issue if precipitation decreases (Mitchell *et al.*, 1990; Greco *et al.*, 1994). Seasonally dry deciduous forests could burn more frequently and be permanently replaced with thorn scrub or savannah vegetation. It should be noted that changes in forest cover can have effects on groundwater supplies, surface runoff, sedimentation, and river flows (Brooks *et al.*, 1991), with potentially serious socioeconomic effects (see also Chapter 5). Also, hydrological changes that include shifts in atmospheric circulation could threaten the survival of cloud forests.

Some evergreen species of the humid forest clearly will be at a disadvantage in those areas that experience more severe and prolonged droughts. Significantly, drought affects the survival of individuals: Those without morphological or physiological adaptations to drought often die. In contrast, an abundance of moisture (in the absence of flooding) more often acts through changed competitive ability. Drought-adapted species are outcompeted by those that lack growth-limiting drought-adaptive traits. Species in moist tropical forests, including economically important hardwoods, are the least drought-adapted in the tropics, and their survival (with the attendant loss in diversity) in some areas must be considered at risk from climatic change.

1.4.3.4. Soil Nutrients

Climatic change (including elevated CO_2 levels) could enhance the supply of nutrients to plants in the tropics by, for instance, increasing root and mycorrhizal growth and thereby increasing access to soil phosphorus and enhancing nitrogen fixation by legumes, which are abundant in the tropics. On the other hand, the more important considerations are likely to be nutrient leaching and soil erosion wherever tree cover is lost (because of droughts or fire) or removed (by logging, clearing, or grazing)—especially where high-rainfall events occur in hilly areas—and immobilization of nutrients in soil organic matter in response to elevated CO_2.

1.4.3.5. Pests and Pathogens

Tropical rainforests contain large numbers of insects and pathogens that can cause serious damage to some plant species and may play a role in regulating species diversity (Coley, 1983). Many factors have been associated with the susceptibility of tropical plants to pests and diseases and with the virility of pests and pathogens. Some of these factors include suboptimal climate, availability of water and nutrients, and the presence of secondary metabolites that act as defensive compounds (Lambert and Turner, 1977; Levitt, 1980a, 1980b; Mattson, 1980; Mattson and Haack, 1987; Jones and Coleman, 1991). Drought stress can sometimes increase host plant suitability due to increases in soluble nitrogen and sucrose (White, 1974), whereas high temperatures and humidities (> 30°C and relative humidity of 50–90%) can decrease the growth rate, survival, and fecundity of some insects (Wilson *et al.*, 1982). Consequently, there is still great uncertainty as to whether the impacts of climate change on the relationship between host plants and pests and pathogens will lead to forest loss or gain.

However, the diversity of species in most tropical forests appears to confer some protection against widespread outbreaks of pests and diseases. The low population density of potential hosts prevents the rapid multiplication and spread of pests and pathogens (Longman and Jenik, 1987). Hence, most cases of debilitating outbreaks of pests and diseases—which can, for instance, affect entire stocks of some mahoganies—occur in plantations, agroforests, or stands that are dominated by one species (UNESCO, 1978).

1.4.3.6. Fire and Wind

It is clear from recent events that wherever there are more droughts, there will be more fires due to the accumulation of more combustible dry organic matter, with major potential economic costs in highly populated areas. In the extreme, recurrent fires will lead to the permanent loss of fire-prone species and the invasion and maintenance of fire-resistant species such as savanna vegetation. Fires, together with the action of wind—especially in regions where storms are frequent—will increase the dominance of pioneer species, including vines and herbs, preventing forests from developing to maturity.

1.5. Temperate Forests

1.5.1. Characterization

Temperate forests occur approximately from 25–50° N and S and are found primarily in developed countries in discontinuous blocks on five continents, sharing the landscape with agricultural land and urban areas. Closed forests originally covered about 1,400 Mha but have been reduced to about 700 Mha, 56% of which is in North America and 24% in Europe (see Chapter 15). Humans have had an impact on almost all of these forests; about 20% are managed for wood production or other uses, and many are being affected by pollutants, with both potentially positive and some obvious negative effects on growth (Innes, 1993).

These forests contain broadleaved and needle-leaved species that may be evergreen or deciduous. The most extensive types of temperate forests are the northern deciduous forests, commonly dominated by members of the *Fagaceae*, with leaf-shedding as an adaptation to winter frosts; the temperate coniferous forests of Europe, western and southeastern North America, and eastern Asia; warm evergreen forests in Australia dominated by *Eucalyptus* species; and other forests in the Southern Hemisphere dominated by *Nothofagus* species.

Temperate forests occur within a range of mean annual temperatures of 6–17°C, where average total precipitation exceeds 500 mm, with broad transition zones to boreal and subtropical forests. Within any region there are often steep gradients in forest types along climate gradients of rainfall and temperature, with change in altitude, and from oceanic to continental areas. At the low-rainfall margin, temperate forests change into savanna-type woodlands or Mediterranean shrublands. Temperate forests exist on a wide range of soils and have diverse ecological properties (Ellenberg, 1971; Bormann and Likens, 1979; Reichle, 1981; Edmonds, 1982).

The standing biomass and carbon content of temperate forests is currently increasing (at 0.2–0.5 Gt C yr^{-1}), largely due to reforestation (about 0.6 Mha yr^{-1}), underharvesting and regrowth after wood removal during the 19th and early 20th centuries (Armentano and Ralston, 1980; Heath *et al.*, 1993; Kurz and Apps, 1993; Sedjo, 1993; Sundquist, 1993; Dixon *et al.*, 1994; Galinski and Küppers, 1994; Kohlmaier *et al.*, 1995), and mainly fertilization effects of nitrogen deposition (Kauppi *et al.*, 1992). The temperate forests therefore are considered to be a carbon sink (Heath *et al.*, 1993). However, if the loss of carbon from decaying forest products and logging debris is taken into account, the net flux of carbon to the atmosphere from northern temperate forests is close to zero (Houghton, 1993) and, in the coming century, increased demand for wood may change the temperate forests to a net carbon source (Heath *et al.*, 1993).

Forest fires occur in most seasonally dry forests, despite control measures, and play an important role in ecosystem dynamics in these forests, especially in parts of North America, Australia, and Mediterranean Europe (Kozlowski and Ahlgren, 1974; Gill *et al.*, 1981; Goldammer and Jenkins, 1990). Fire affects species distribution by favoring fire-resistant species at the expense of those more sensitive to fire. Fire also creates conditions conducive to the establishment of new seedlings.

1.5.2. Key Limitations

The ranges and growth of temperate tree species in wetter, maritime, and high-latitude regions often can be related to the length of the growing season, measured in degree-days, and to absolute minimum temperatures in more continental areas. The temperature niches of individual species are often narrow (Figure 1-1) and are commonly determined by critical thermal requirements of the reproductive cycle (e.g., Pigott and Huntley, 1981).

In drier regions closer to the equator, the existence and growth of temperate forests are controlled largely by water availability. Variation in leaf-area indices and ring widths often is related to available water (Gholz, 1982). A minimum ratio of actual to potential evapotranspiration of approximately 0.65 is considered necessary to support the growth and regeneration of temperate-zone trees (Prentice *et al.*, 1992). Slight shifts to smaller ratios lead to open woodlands, savannahs, and grasslands. Soil water availability also strongly influences forest density, leaf area, growth, and standing biomass.

Seedling establishment is most vulnerable to shifts in precipitation patterns or amounts. Most temperate-zone seedlings can survive and become established only within narrow limits of soil moisture and sunlight. Slight changes in soil moisture can lead to the loss of a season's seedling crop, although losses in tree establishment cohorts may not lead to changes in forest structure for decades.

Like high-latitude populations experiencing low temperature, mid-latitude populations that are often subject to low soil water availability may be very plastic in their response to environmental variation. This plasticity enables temperate-zone trees to survive most weather variations that may be encountered only once every 200 years (Bugmann, 1994). Reproduction at these locations may be possible if conditions in rare years are favorable for the species' reproductive requirements (Pigott and Huntley, 1981).

1.5.3. Projected Climatic Changes

The climatic changes projected for most of the temperate forest region over the period 1990–2050 are increases in both summer and winter temperatures of 1–2°C; regional changes in precipitation in summer and winter, mostly in the ±20% range; and drier soils in summer (mostly with 2–8 mm less water) and, in winter, changes toward drier soils in some regions and wetter soils in others (Greco *et al.*, 1994). These projections imply longer and warmer growing seasons, less extreme subzero temperatures in winter, more frost-free winters in maritime areas, and more summer droughts, particularly in mid-continental regions (Manabe and Weatherald, 1987).

Possibly more important to temperate biotic communities than the magnitude of warming or precipitation change is the speed of climatic change (see Section 1.3.5). The projections imply that it will take less than a century for the summers to be warmer than now throughout the current geographic range of many temperate species.

1.5.4. Impacts of Climatic Change

1.5.4.1. Forest Area, Distribution, and Productivity

As long as the current agricultural surpluses in temperate regions persist, the temperate forest area is likely to be increased by afforestation. In Europe, the potential area for afforestation has been estimated as 44 Mha, and in the United States 100 Mha (Heath *et al.*, 1993).

Climatic change will enable the temperate forest to advance poleward, in many northern areas displacing boreal forest, and also potentially to expand in wet, maritime regions (Kellomäki and Kolström, 1992; Leemans, 1992; Morikawa, 1993). Early successional, pioneer species will be favored, and opportunities will exist for foresters to introduce species and ecotypes from warmer regions (Cannell *et al.*, 1989). However, in drier, continental regions, repeated summer droughts may lead to the loss of temperate forests. The rate of loss in biomass and carbon in these areas could exceed the rate of carbon gain in newly forested areas (Smith *et al.*, 1992a).

The net primary productivity of temperate ecosystems is predicted to increase in response to rising CO_2 concentrations, warming, and increased nitrogen mineralization rates; however, if drier areas, such as those west of the Appalachians, receive just 7% less rainfall, decreased productivity is predicted because of a decrease in the ratio of rainfall to potential evapotranspiration (Running and Nemani, 1991; Melillo *et al.*, 1993; Lüdeke *et al.*, 1995). Similarly, warming of 2°C and slightly decreased precipitation have been predicted to cause forest decline in Missouri (Bowes and Sedjo, 1993). Many experiments have confirmed that elevated CO_2 enhances the growth of young trees and that the effect is sustained over several years (Wullschleger *et al.*, 1995). However, it has proven difficult to detect CO_2-enhanced stem growth of mature trees over the last century (see Section 1.3.1.2 and Chapter A).

Forests that are not in decline may show little change in net carbon storage because, in temperate climates, increases in net primary productivity may be offset by increased soil respiration due to higher temperatures (Thornley *et al.*, 1991; Kirschbaum, 1993). That is, the net ecosystem productivity may not change, and may even decrease. Forests suffering from wildfire, pest outbreaks, or decline events will lose carbon and may become a major source of carbon (King and Neilson, 1992; Smith and Shugart, 1993).

1.5.4.2. Temperature and Water

Species in the temperate zone differ in their temperature optima for growth, their timing and degree of frost hardening and dehardening, their winter chilling requirements, and the number of degree-days needed to complete different stages in their reproductive development (Cannell, 1990). All of these differences are likely to be involved in driving changes in the distribution and productivities of species. Most studies predict substantial change during the next century, but confidence in our predictions is limited by knowledge of adaptive responses of species and the need to consider interactions among responses to temperature, CO_2, changed water relations, and pests and pathogens. Moderate temperature increases alone often can be beneficial, as evidenced by the faster growth and greater seed production of some commercial tree species when transferred to slightly warmer climates. But chilling requirements may not be met, and delayed budburst may limit the amount of light intercepted in the growing season while total respiration increases (LeBlanc and Foster, 1992). It also has been shown that frost hardiness can decrease in response to CO_2 enrichment (Barnes *et al.*, 1996), so that winter damage could even increase.

Species that are growing in regions where growth is limited by water shortages for at least part of the year may be adversely affected by intensification of summer soil water deficits (Greco *et al.*, 1994), as a result of decreases in the ratio of rainfall to potential evapotranspiration (e.g., Addison, 1991).

Warm winters will result in less precipitation falling as snow and reduced regional snow packs, resulting in less carry-over of water from the winter to the growing season (Mitchell *et al.*, 1990). Thus, less water will be available for vegetation in the following growing season, which may lead to drought-induced forest decline (King and Neilson, 1992).

1.5.4.3. Pests and Pathogens

Warming in winter may allow destructive insects and pathogenic fungi to survive at higher latitudes than at present, enabling subtropical or warm-temperate pests and pathogens to invade vegetation from which they are now excluded (Dobson and Carper, 1992). Some insects also will be able to complete more generations per year in warmer climates. Increased incidences of pests and diseases may further limit the growth of stands that are already declining from the effects of climate change or pollution. Summer droughts and other climatic stresses have been associated with outbreaks of bark beetles like the southern pine beetle (*Dendroctonus frontalis*) in southern parts of the United States, bark beetles in western Canada (Kimmins and Lavender, 1992), and bronze birch borers on paper birch in northern Michigan (Jones *et al.*, 1993). Elevated CO_2 can change the palatability of leaves and either promote or discourage insect herbivory (Overdieck *et al.*, 1988; Mueller-Dombois, 1992). In areas where forestry practice has led to the establishment of mono-specific stands, forests are particularly vulnerable to outbreaks of pests and diseases, especially where that combines with poor site quality or exposure to industrial pollutants.

1.5.4.4. Soil Nutrients

Nitrogen supply generally limits the productivity of many temperate forests (Tamm, 1991). Climatic warming will increase rates of turnover of soil nitrogen and carbon. In Europe and the United States, nitrogen fertility also may be improving through inputs from industrial and agricultural pollution. While a low level of nitrogen input may have a beneficial effect (Kauppi *et al.*, 1992), further inputs could reverse the initial gains through the development of nutrient imbalances and further acidification of the soil (Heath *et al.*, 1993).

1.5.4.5. Fire

The projected increase in the incidence of summer droughts in much of the temperate zone will increase the risk of forest fires and extend the hazard to areas that are not now affected, particularly where forests are defoliated or killed by drought, pests, or pathogens. Some species that are not adapted to withstand or regenerate after fire may be lost from fire-affected areas. However, large-scale fires may continue to be rare in the temperate zone and confined to drier parts of North America, Australia, and the Mediterranean region because temperate forests mostly occur in dissected landscapes in countries that can afford fire-control measures.

1.6. Boreal Forests

1.6.1. Characterization

The boreal forest covers approximately 17% of the world's land surface area in a circumpolar complex of forested and partially forested ecosystems in northern Eurasia and North America. It contains about 90 Gt C in living biomass and 470 Gt C in soils and detritus (Dixon *et al.*, 1994); the boreal region is estimated currently to be a sink of 0.4–0.6 Gt C yr^{-1} (Apps *et al.*, 1993; Dixon *et al.*, 1994).

The boreal forest consists primarily of evergreen and deciduous coniferous species and is floristically poor, being dominated by only about 15 tree species in both Eurasia and North America (Nikolov and Helmisaari, 1992). Many species have transcontinental distributions and are adapted to withstand extremes of climate and to regenerate after fire or insect attack. Three forest zones are often recognized (from south to north): closed-crown forest, open-crown forest (or lichen woodland), and forest-tundra. The closed-crown forest borders on steppe/prairie in continental areas, whereas in areas under maritime influence the boreal/temperate forest boundary consists of a relatively species-rich community—often in a mosaic of deciduous species on favorable soils and conifers at less favorable or colder sites (Apps *et al.*, 1993). Much of the boreal forest is embedded in a mosaic of wetlands and peatlands that may act as natural fire breaks. Mosses and lichens also play an important role in boreal ecosystem processes.

The northern limit of the boreal forest is largely determined by temperature (Garfinkel and Brubaker, 1980; Larsen, 1980, 1989; Arno, 1984), while both temperature and water supply determine the forest/steppe and forest/prairie boundaries. In general, the length of the vegetative period (related to the length of the frost-free period or July mean temperature) is of great importance in maritime regions. In more continental regions, drought or extreme subzero temperature may be more important—for instance, the apparent -40°C limit of sap supercooling of hardwood species (Arris and Eagleson, 1989) and the winter or drought tolerance of evergreen conifers. This tolerance may be exceeded in Siberia, which supports large areas of deciduous conifers (*Larix*) (Woodward, 1987). Drought is an important factor in interior Alaska and Siberia, where annual precipitation may be as low as 100–200 mm. Throughout the boreal region, droughts are generally necessary for the onset of fires. Some species, notably *Pinus sylvestris*, are drought-tolerant, whereas others such as *Picea abies* are dominant on wetter soils. A further factor determining the limits of species may be their winter chilling requirements.

The boreal forest consists of a patchwork of small to very large areas that are in various stages of recovery from fire or insect attack (van Cleve *et al.*, 1983a). Stands rarely reach maximum biomass or carbon content (Apps and Kurz, 1994), and many areas are dominated by one or a few species in a narrow age-class range. Also, it may be noted that the boreal forest is less than 12,000 years old and still may be expanding in some areas (Tallis, 1990), and even recovering from cool temperatures in the Little Ice Age (AD 1200–1850) (Campbell and McAndrews, 1993).

1.6.2. Key Limitations

Low air temperatures and small heat sums restrict growth and the production and germination of seed, which is a major factor limiting regeneration in the forest-tundra zone (Henttonen *et al.*, 1986). Unseasonal frosts can damage growth and reproductive cycles. Low soil temperature and permafrosts (in non-oceanic regions) have been demonstrated experimentally to limit growth and nutrient availability (van Cleve *et al.*, 1981, 1983b). Permafrost can restrict root growth and create an impervious layer that impedes soil drainage but can also be responsible for raised areas that permit drainage, producing islands where trees can grow within wetlands.

Low nutrient availability (except after fire) is characteristic of most boreal forests. Low soil temperatures limit the rate of litter decomposition and mineralization (Shaver *et al.*, 1992; Berg *et al.*, 1993; Kobak and Kondrasheva, 1993), and the litter of most coniferous boreal tree species is relatively resistant to decomposition because of its high lignin and low nutrient content. Nitrogen availability in boreal coniferous forests is generally in the range of 5–40 kg N ha^{-1} yr^{-1}, compared to 80–120 kg N ha^{-1} yr^{-1} in northern hardwood stands (Pastor and Mladenoff, 1992).

Natural wildfires are ubiquitous throughout the boreal region because of the buildup of large amounts of litter, much of which is not only resistant to decomposition but is also highly flammable. In the absence of fire-suppression measures, the interval between fires (the fire cycle) ranges from 50 to 200 years from south to north but may be over 1,000 years in wet northern ecosystems (Viereck, 1983; Payette *et al.*, 1989b; Payette, 1992). There are well-established relationships between fire-cycle length, species composition, age-class distribution, and carbon storage (Johnson and Larson, 1991; Kasischke *et al.*, 1995; Kurz *et al.*, 1995).

Boreal forests also are characterized by periodic outbreaks of insect pest populations. For many insect species, outbreaks have been clearly associated with climatic conditions and weather events (e.g., Martinat, 1987; Mattson and Haack, 1987; Volney, 1988), and outbreaks are often most common in the southernmost (warmest) part of the tree-host range (Kurz *et al.*, 1995). On sites where forests have remained unaffected by fire or insect damage for extended periods, they are characterized by multi-aged stands where tree-fall constitutes the most important disturbance. Tree falls create a variety of microenvironments that enhance diversity and affect regeneration patterns (Jonsson and Esseen, 1990; Liu and Hytteborn, 1991; Hofgaard, 1993).

The net primary productivity of boreal forests tends to be low—commonly 3–8 t (dry matter) ha^{-1} yr^{-1}, compared with 7–12 t ha^{-1} yr^{-1} for northern hardwoods (Cannell, 1982; Melillo *et al.*, 1993). Productivity is controlled in a complex way by interactions between various factors as discussed in Section 1.6.4 (Bonan and Shugart, 1989). As a boreal forest stand develops after fire, litter accumulates, more and more nutrients are immobilized in the litter, the depth of thaw is reduced because of the insulating properties of the litter, drainage may then be impaired by permafrost, and moss growth may impair regeneration from seed—some of which may be held within serotinous cones that release their seed only after fire (Bonan, 1992). Fire interrupts this process by burning litter on the forest floor, releasing mineral nutrients, leading to deeper thaw, improving drainage, and often removing the moss layer—all of which improves conditions for seed germination (Landhäusser and Wein, 1993).

1.6.3. Projected Climatic Changes

The climatic changes projected for most of the boreal region over the period 1900–2050 are increases in temperature of around 1–2°C in summer and 2–3°C in winter; regional changes in precipitation in summer and winter, mostly in the ±20% range; and drier soils in summer (averaging about 2–8 mm less water) (Greco *et al.*, 1994). These projections imply longer and warmer growing seasons, appreciably milder winters with the possibility of less extreme minimum temperatures, and less permafrost, which is related to annual mean temperatures.

Most importantly, the changes in temperature, soil water, and vapor pressure deficit may increase the frequency (shorten the return-time) of fires. Flannigan and van Wagner (1991) predict a 40–50% increase in the area burned each year in Canada in a 2 x CO_2 climate scenario, and others have predicted more frequent fires of higher intensity in the forest-tundra (Stocks, 1993; FIRESCAN Science Team, 1995). In Russia, an additional 7–12 million hectares of boreal forest are projected to burn annually within the next 50 years, affecting 30–50% of the land area (Dixon and Krankina, 1993).

1.6.4. Impacts of Climatic Change

There is a general consensus that climatic change will have greater impact on boreal forests than on tropical and perhaps temperate forests, and that more frequent or changed patterns of disturbance by fire and insect pests may be more important agents of change than elevated temperatures and CO_2 levels *per se* (Shugart *et al.*, 1992; Dixon and Krankina, 1993). Overall, the boreal forest is likely to decrease in area, biomass, and carbon stock, with a move toward younger age-classes and considerable disruption at its southern boundary (Neilson *et al.*, 1994; Kurz *et al.*, 1995). CO_2 enrichment itself may have less effect than in warmer climates. In Sections 1.6.4.1 through 1.6.4.4, we consider the likely changes in the distribution and composition of the forest and then elaborate on the factors driving change.

1.6.4.1. Forest Area, Distribution, and Productivity

On its southern border, the boreal forest may give way to northern deciduous forest (or agriculture) in areas with a maritime influence and to grassland or xerophytic steppe vegetation in midcontinental areas, and species shifts may occur in the mid-boreal region (Emanuel *et al.*, 1985; Kellomäki and Kolström, 1992; Rizzo and Wiken, 1992; Dixon and Krankina, 1993; Monserud *et al.*, 1993a; Tchebakova *et al.*, 1994; Prentice and Sykes, 1995). Near tree lines in many areas, there is potential for existing populations of suppressed individuals to grow taller and more vigorously, as has apparently happened in response to past climatic changes (Kullman, 1986; Payette *et al.*, 1989a; Hofgaard *et al.*, 1991).

Over the next century, the potential (or preferred) geographic ranges of species may shift approximately 300–500 km, implying changes in forest-based industries and considerable socioeconomic impacts. In the early Holocene (about 8,000 years ago), when the climate became warmer, fire-adapted hardwood species expanded northward to new sites after fire (Green, 1987). Where northern deciduous species, such as sugar maple, migrate northward, forest productivity may be increased on soils that retain adequate water, whereas productivity may decrease on dry soils where boreal forest may give way to oak-pine savanna (Pastor and Post, 1988).

There is concern that the maximum potential migration rates may be too slow to keep up with the rate of climatic change (see Sections 1.3.3 and 1.3.5)—in which case some researchers consider that there may be areas of transitory forest decline, especially if soils change slowly, are unfavorable for immigrating species, and lack necessary microbes and symbionts (Dixon and Turner, 1991; Davis and Zabinski, 1992; Solomon, 1992; Smith and Shugart, 1993), or where growth by more southern species is limited by photoperiodic constraints. The future of the transitory forest is likely to be determined by increasing occurrence of extended high-intensity wildfires until a new climate-vegetation-fire equilibrium is established (Crutzen and Goldammer, 1993). Other researchers suggest that there may be little forest decline. Intraspecific genetic diversity will buffer change, and species that are no longer in a favorable climate will simply grow and regenerate poorly and be overtaken by invading species either gradually or after disturbance (Malanson *et al.*, 1992; see also Section 1.3.3).

Increasing temperatures are likely to stimulate soil organic matter decomposition and increase nutrient (especially nitrogen) availability, leading to an increase in net primary productivity of non-stressed stands, averaging perhaps 10% in the boreal zone in a 2 x CO_2 climate (Melillo *et al.*, 1993). However, despite increasing productivity, there may be a net carbon loss from the ecosystem because a small temperature rise will greatly enhance decomposition rates (Jenkinson *et al.*, 1991), whereas CO_2 fertilization will be of low effectiveness because of low temperatures (Kirschbaum, 1993). Also, productivity may not increase in dry areas if water limitations were to increase due to increased evaporative demand. Hydrological and landscape changes in the patterns of bogs and forest also may be expected. In the north, melting of permafrost would favor the expansion of wetlands, while drier conditions in the south would lower the water tables (Apps *et al.*, 1993).

In the forest-tundra, rising temperatures are likely to enhance the development and germination of seeds of many species, increasing forest cover and enabling a northward migration to occur, probably after fire (Kullman, 1990; Kellomäki and Kolström, 1992; Landhäusser and Wein, 1993)—but again, hampered to some extent by slow changes in soil conditions (Rizzo and Wiken, 1992). There is clear evidence in the fossil pollen and macrofossil record of expansion and recession of boreal forest in both Eurasia and North America in response to temperature changes over the past 10,000 years (Ritchie, 1987). However, it will take more than 100 years for any new forest areas to mature in the forest-tundra, so the northward expansion of mature boreal forest is likely to be slower than the rate at which it is lost to grassland and temperate deciduous forest at its southern boundary (Rizzo and Wiken, 1992). The tundra itself is likely to become a carbon source in response to warming (Billings *et al.*, 1984; Oechel *et al.*, 1993).

1.6.4.2. Temperature

The productivity of boreal forests—except those at the warmer and drier edge of their species ranges—is likely to respond favorably to increases in temperature, as shown by long-term soil warming experiments and models (van Cleve *et al.*, 1990; Melillo *et al.*, 1993). It has long been known that trees in the boreal forest exhibit positive relationships between annual growth-ring widths and summer temperatures (Mikola, 1962) and between volume growth and number of degree-days (Kauppi and Posch, 1985; Worrell, 1987). Also, most provenance-transfer studies suggest that some southward movement of ecotypes to warmer climates promotes their volume growth (Beuker, 1994; Matyas, 1994). The magnitude of the growth response will depend on the effect of increased temperatures on nutrient availability, evapotranspiration, and the frequency of fires. Subtle interactions may be important. For instance, Bonan *et al.* (1990) and Bonan (1992) show how increased temperatures combined with increased evapotranspiration could result in a faster build-up of litter and a shallower permafrost zone, but an increased probability of fire. The net effect could be to reduce nutrient availability except after fires—whereas on the basis of temperature increases alone, nutrient availability would have been expected to increase (Shaver *et al.*, 1992).

Forest simulation models suggest large shifts in species composition in boreal forests (Shugart *et al.*, 1992). Large increases in temperature would cause annual heat sums to exceed the minimum thresholds for seed production, favoring the northward spread of species, especially in the forest-tundra zone. On the other hand, reduced winter chilling may disrupt both vegetative growth and reproductive processes of species at the southern edges of their ranges (Kimmins and Lavender, 1992).

1.6.4.3. Nutrients

Nitrogen is the nutrient that most limits the productivity of boreal forests (e.g., van Cleve *et al.*, 1983a, 1983b; Tamm, 1991). The critical factor is the rate at which nitrogen is recycled through the litter—which, as noted earlier, is a function of both temperature and litter quality (Berg *et al.*, 1993). Species differ in both their response to nitrogen and their litter quality. *Picea* sites in cold locations have forest floors with a high lignin content, little available nitrogen, and slow decomposition rates, whereas *Betula* or *Populus* sites, especially in warmer locations, have forest floors with less lignin. Thus, whereas small increases in temperature will increase rates of decomposition and nitrogen cycling, large or prolonged temperature increases also will lead to a shift from coniferous to deciduous tree species, and—because of the greater decomposability of deciduous litter—this may have a further positive effect on nitrogen availability, leading to increased productivity. On the other hand, increasing CO_2 concentration could increase the C:N ratio of litter, and the reduced litter quality might slow the rate of nutrient cycling (Taylor *et al.*, 1989; van de Geijn and van Veen, 1993).

1.6.4.4. Fire and Insects

In Canada over recent years, about 1 to 2 Mha have burned each year (Kurz and Apps, 1993), and in Russia, between 1.4 and 10 Mha burned each year between 1971 and 1991 (Dixon and Krankina, 1993). If fires occur very frequently, late successional species become unable to outcompete pioneer species, which will thereby become more common. A threefold increase in wildfire frequency in Canada between a high-fire year (1989) and a reference year (1986) resulted in an 86% reduction in the net ecosystem carbon sink (Kurz *et al.*, 1992). However, fires also have a beneficial effect on subsequent ecosystem function and facilitate regeneration (Landhäusser and Wein, 1993). Fires may have most impact in the forest-tundra zone; indeed, Payette and Gagnon (1985) conclude that the modern forest-tundra boundary of northeastern North America is the result of fires during the last 3,000 years, which were followed by lower temperatures that limited forest regeneration. If future fires were followed by higher temperatures, the fires could be catalysts for a return to coniferous forests that appear to have existed there 7,000–10,000 years ago.

Defoliating insects play an important role in boreal forests, and there are many instances where the proportion of different boreal tree species is related to the intensity of insect pest outbreaks. Pests usually are maintained at a low population equilibrium by a combination of host resistance, natural enemies, and weather conditions. Any one of these factors could be affected by climatic change, particularly tree resistance in response to environmental stress. Once an outbreak exceeds a certain patch size, it can become self-propagating and can spread largely independently of weather conditions. Following climatic warming, it seems very likely that insect outbreaks will expand northward and that new pest and pathogen problems will arise (Kurz and Apps, 1993).

1.7. Research and Monitoring Needs

Our capability to assess the likely fate of the world's forests under altered climatic conditions has been limited because the conceptual modeling framework for such an assessment is still in an early stage of development. It needs to be refined to improve the understanding of climate change impacts at the following three levels: (1) the ecophysiological responses of trees to changing climate and CO_2 concentrations, (2) the relationship between tree growth and transient forest dynamics, and (3) the influence of changing forest characteristics on the global carbon balance and hence their feedback to the greenhouse effect. The predictive power of current modeling approaches decreases from (1) to (3). A consistent research strategy to overcome these limitations needs to be accompanied by a monitoring program that can provide appropriate databases for initialization, calibration, validation, and application of the models. In particular, future work should address the following:

- *Ecophysiology*, specifically the influence of temperature, water availability, ambient CO_2 concentration, photoperiod, and nitrogen availability on the establishment, growth, water use efficiency, stomatal conductance, biomass allocation, and survival of trees under natural conditions. Urgently required also are studies of belowground plant and soil processes such as the decomposition of organic residues under changing environmental conditions. Most previous studies have been limited to short-term responses of young trees and therefore have ignored longer-term processes influenced by more than one factor, as well as the adaptive potential of plants. There is a need for more long-term studies that investigate simultaneously a well-defined set of key factors. The ecophysiological responses should be measured at experimental sites within a range of forest types, and those experiments should follow protocols that support modeling research.
- *Forest dynamics*. Studies should focus on the stability of natural and managed forests under different types of climatic (temperature, water availability) and

chemical (CO_2, nitrogen) regimes, including the influence of changing disturbance conditions (wind, drought, fire, pest frequencies). Because the direction and magnitude of possible climatic change is not known for all factors relevant to ecosystem dynamics, simulations and experimental studies should focus upon a broad range of such conditions. The potential rate of species migration (either assisted or unassisted) needs to be investigated further using appropriate models. A significant part of this activity should be devoted to further development of monitoring strategies that would allow the collection of data on realistic forest responses through all major forest regions.

- *Monitoring*. The development and application of forest models requires improved global databases on the present conditions of the world's forests, as well as their associated site characteristics. It is crucial that these data-gathering activities are intensified, both through the global network of ground observations that can be made in many research institutes worldwide and through the collection of data from spaceborne sensors. For ground-based data collection, standardization and worldwide availability are key issues, as is the continuation of already ongoing observational series. Satellite remote sensing is in a promising stage of development, with respect to both enhanced processing protocols of existing time series of satellite data (such as AVHRR) and the development of sensors with improved capacity to measure ecosystem properties at high spatial and temporal resolution.

To improve the overall modeling framework for the assessment of global forest response to climate change, and to provide an appropriate background for the synthesis of more detailed studies, there is a requirement for process-based terrestrial biosphere models of ecosystem dynamics. These models should build on knowledge from all other levels of model development. Specifically, they should involve a sufficiently high number of plant functional types and forest types to cover the wide range of forests occurring in different climate zones. Various initiatives for the development of such models are currently underway (e.g., Focus 2: Change in Ecosystem Structure, especially activity 2.3. in Steffen *et al.*, 1992; Landsberg *et al.*, 1995). To succeed, they need to be backed up by the research activities listed above.

References

Addison, P.A., 1991: Atmospheric pollution and forest effects. *Revue Forestiere Francaise*, **10**(Hors Serie No. 2), 307-315.

Alcamo, J., 1994: IMAGE 2.0: integrated modeling of global climate change. *Water, Air, & Soil Pollution*, **76(1/2)**, 1-321.

Allen, L.H., 1990: Plant responses to rising carbon dioxide and potential interactions with air pollutants. *Journal of Environmental Quality*, **19**, 15-34.

Andreae, M.O. and J.G. Goldammer, 1992: Tropical wildland fires and other biomass burning: environmental impacts and implications for land-use and fire management. In: *Conservation of West and Central African Rainforests* [Cleaver, K. *et al.* (eds.)]. Abidjan Conference on West Africa's Forest Environment, 5-9 November 1990. The World Bank, Washington, DC, 79-109.

Apps, M.J., W.A. Kurz, R.J. Luxmoore, L.O. Nilsson, R.A. Sedjo, R. Schmidt, L.G. Simpson, and T.S. Vinson, 1993: Boreal forests and tundra. *Water, Air, and Soil Pollution*, **70**, 39-53.

Apps, M.J. and W.A. Kurz, 1994: The role of Canadian forests in the global carbon budget. In: *Carbon Balance of World's Forested Ecosystems: Towards a Global Assessment* [Kanninen, M. (ed.)]. Academy of Finland, Helsinki, Finland, pp. 14-39.

Armentano, T.V. and C.W. Ralston, 1980: The role of temperate zone forests in the global carbon cycle. *Canadian Journal of Forest Research*, **10**, 53-60.

Arno, S.F., 1984: *Timberline: Mountain and Arctic Forest Frontiers*. The Mountaineers, Seattle, WA, 304 pp.

Arris, L.L. and P.A. Eagleson, 1989: Evidence of a physiological basis for the boreal–deciduous ecotone in North America. *Vegetatio*, **82**, 55-58.

Auclair, A.N.D., R.C. Worrest, D. Lachance, and H.C. Martin, 1992: Climatic perturbation as a general mechanism of forest dieback. In: *Forest Decline Concepts* [Manion, P.D. and D. Lachance (eds.)]. APS Press, St. Paul, MN, pp. 38-58.

Austin, M.P., A.O. Nicholls, and C.R. Margules, 1990: Measurement of the realized qualitative niche: environmental niches of five Eucalyptus species. *Ecological Monographs*, **60**, 161-177.

Austin, M.P., 1992: Modelling the environmental niche of plants: implications for plant community response to elevated CO_2 levels. *Australian Journal of Botany*, **40**, 615-630.

Austin, M.P., A.O. Nicholls, M.D. Doherty, and J.A. Meyers, 1994: Determining species response functions to an environmental gradient by means of a beta-function. *Journal of Vegetation Science*, **5**, 215-228.

Barnes, J.D., M. Hull, and A.W. Davison, 1996: Impacts of air pollutants and rising CO_2 in winter time. In: *Plant Growth and Air Pollution* [Yunus, M. and M. Iqbal (eds.)]. Springer Verlag, Berlin, Germany (in press).

Bazzaz, F.A. and E.D. Fajer, 1992: Plant life in a CO_2-rich world. *Scientific American*, **266**, 68-74.

Berg, B., M.P. Berg, E. Box, P. Bottner, A. Breymeyer, R. Calvo de Anta, M.-M. Couteaux, A. Gallardo, A. Escudero, W. Kratz, M. Madeira, C. McClaugherty, V. Meentemeyer, F. Muñoz, P. Piussi, J. Remacle, and A. Virzo de Santo, 1993: Litter mass loss in pine forests of Europe: relationship with climate and litter quality. In: *Geography of Organic Matter Production and Decay* [Breymeyer, A., B. Krawczyk, R. Kulikowski, J. Solon, M. Rosciszewski, and B. Jaworska (eds.)]. Polish Academy of Sciences, Warsaw, Poland, pp. 81-109.

Beuker, E., 1994: Long-term effects of temperature on the wood production of *Pinus sylvestris* L. and *Picea abies* (L.) Karst. in old provenance experiments. *Scandinavian Journal of Forest Research*, **9**, 34-45.

Bierregaard, J.R., T.E. Lovejoy, V. Kapos, A.A. dos Santos, and R.W. Hutchings, 1992: The biological dynamics of tropical rainforest fragments. *Bioscience*, **42**, 859-865.

Billings, W.D., K.M. Peterson, J.O. Luken, and D.A. Mortensen, 1984: Interaction of increasing atmospheric carbon dioxide and soil nitrogen on the carbon balance of tundra microcosms. *Oecologia*, **65**, 26-29.

Bonan, G.B. and H.H. Shugart, 1989: Environmental factors and ecological processes in boreal forests. *Annual Review of Ecology and Systematics*, **20**, 1-28.

Bonan, G.B., H.H. Shugart, and D.L. Urban, 1990: The sensitivity of some high-latitude boreal forests to climatic parameters. *Climatic Change*, **16**, 9-29.

Bonan, G.B., 1992: A simulation analysis of environmental factors and ecological processes in North American boreal forests. In: *A Systems Analysis of the Global Boreal Forest* [Shugart, H.H., R. Leemans, and G.B. Bonan (eds.)]. Cambridge University Press, Cambridge, UK, pp. 404-427.

Bonan, G., 1993: Do biophysics and physiology matter in ecosystem models? *Climatic Change*, **24**, 281–285.

Booth, T.H., H.A. Nix, M.F. Hutchinson, and T. Jovanovic, 1988: Niche analysis and tree species introduction. *Forest Ecology and Management*, **23**, 47-59.

Bormann, F.H. and G.E. Likens, 1979: *Pattern and Process in a Forested Ecosystem: Disturbance, Development and the Steady State: Based on the Hubbard Brook Ecosystem Study*. Springer, New York a.o., 253 pp.

Borota, J., 1991: *Tropical Forests: Some African and Asian Case Studies of Composition and Structure*. Elsevier, New York, NY, 274 pp.

Botkin, D.B. and R.A. Nisbet, 1992: Projecting the effects of climate change on biological diversity in forests. In: *Global Warming and Biological Diversity* [Peters, R.L. and T. Lovejoy (eds.)]. Yale University Press, New Haven, CT, pp. 277-293.

Bowes, M.D. and R.A. Sedjo, 1993: Impacts and responses to climate change in forests of the MINK region. *Climatic Change*, **24**, 63-82.

Box, E., 1978: Geographical dimensions of terrestrial net and gross productivity. *Radiation and Environmental Biophysics*, **15**, 305-322.

Box, E.O., 1981: *Macroclimate and Plant Forms: An Introduction to Predictive Modeling in Phytogeography.* Junk, The Hague a.o., 258 pp.

Bray, J.R., 1956: Gap-phase replacement in a maple-basswood forest. *Ecology*, **37(3)**, 598-600.

Brooks, K.N., P.F. Folliott, H.M. Gregersen, and J.L. Thames, 1991: *Hydrology and Management of Watersheds.* University of Iowa Press, Ames, IA, 2nd. ed., 392 pp.

Brown, S., C.A.S. Hall, W. Knabe, J. Raich, M.C. Trexler, and P. Woomer, 1993: Tropical forests: their past, present and potential future role in the terrestrial carbon budget. *Water, Air, and Soil Pollution*, **70**, 71-94.

Bugmann, H., 1994: *On the Ecology of Mountainous Forests in a Changing Climate: A Simulation Study.* Diss. ETH No. 10638, Swiss Federal Institute of Technology, Zürich, Switzerland, 258 pp.

Bugmann, H. and A. Fischlin, 1994: Comparing the behaviour of mountainous forest succession models in a changing climate. In: *Mountain Environments in Changing Climates* [Beniston, M. (ed.)]. Routledge, London, UK, pp. 237-255.

Bugmann, H. and A. Fischlin, 1996: Simulating forest dynamics in a complex topography using gridded climatic data. *Climatic Change* (in press).

Campbell, I.D. and J.H. McAndrews, 1993: Forest disequilibrium caused by rapid Little Ice Age cooling. *Nature* (London), **366**, 336-338.

Cannell, M.G.R., 1982: *World Forest Biomass and Primary Production Data.* Academic Press, London, UK, 391 pp.

Cannell, M.G.R., J. Grace, and A. Booth, 1989: Possible impacts of climatic warming on trees and forests in the United Kingdom: a review. *Forestry*, **62**, 337-364.

Cannell, M.G.R., 1990: Modelling the phenology of trees. *Silvae Carellica*, **15**, 11-27.

Chandler, C., P. Cheney, P. Thomas, L. Trabaud, and D. Williams, 1983a: *Fire in Forestry.* Vol. I, *Forest Fire Behaviour and Effects.* Wiley and Sons, New York a.o., 450 pp.

Chandler, C., P. Cheney, P. Thomas, L. Trabaud, and D. Williams, 1983b: *Fire in Forestry.* Vol. II, *Forest Fire Management and Organisation.* Wiley and Sons, New York a.o., 298 pp.

Ciesla, W.M. and E. Donaubauer, 1994: *Decline and Dieback of Trees and Forests—A Global Overview.* FAO Forestry Paper No. 120, FAO, Rome, Italy, 90 pp.

Claussen, M., 1994: On coupling global biome models with climate models. *Climate Research*, **4(3)**, 203-221.

Claussen, M. and M. Esch, 1994: Biomes computed from simulated climatologies. *Climate Dynamics*, **9**, 235-243.

Cole, M.M., 1986: *The Savannas: Biogeography and Geobotany.* Academic Press, London, UK, 438 pp.

Coley, P.D., 1983: Herbivory and defensive characteristic of tree species in a lowland tropical forest. *Ecological Monographs*, **53**, 209-233.

Comins, H.N. and R.E. McMurtrie, 1993: Long-term biotic response of nutrient-limited forest ecosystems to CO_2-enrichment; equilibrium behaviour of integrated plant-soil models. *Ecological Applications*, **3**, 666-681.

Cook, E.R., A.H. Johnson, and T.J. Blasing, 1987: Forest decline: modeling the effect of climate in tree rings. *Tree Physiology*, **3**, 27-40.

Cramer, W.P. and A.M. Solomon, 1993: Climatic classification and future global redistribution of agricultural land. *Climate Research*, **3**, 97-110.

Cramer, W.P. and R. Leemans, 1993: Assessing impacts of climate change on vegetation using climate classification systems. In: *Vegetation Dynamics and Global Change* [Solomon, A.M. and H.H. Shugart (eds.)]. Chapman and Hall, New York and London, pp. 190-219.

Crutzen, P.J. and J.G. Goldammer (eds.), 1993: *Fire in the Environment: The Ecological, Atmospheric, and Climatic Importance of Vegetation Fires.* Wiley, Chichester, UK, 400 pp.

Cure, J.D. and B. Acock, 1986: Crop responses to carbon dioxide doubling: a literature survey. *Agricultural and Forest Meteorology*, **38**, 127-145.

Daily, G.C., 1995: Restoring value to the world's degraded lands. *Science*, **269**, 350-354.

Dansgaard, W., J.W.C. White, and S.J. Johnsen, 1989: The abrupt termination of the younger Dryas climate event. *Nature* (London), **339**, 532-533.

Davis, M.B., 1976: Pleistocene biogeography of temperate deciduous forests. *Geoscience and Man*, **13**, 13-26.

Davis, M.B., 1981: Quaternary history and the stability of forest communities. In: *Forest Succession: Concepts and Application* [West, D.C., H.H. Shugart, and D.B. Botkin (eds.)]. Springer Verlag, New York a.o., pp. 132-153.

Davis, M.B. and D.B. Botkin, 1985: Sensitivity of cool-temperate forests and their fossil pollen record to rapid temperature change. *Quaternary Research*, **23**, 327-340.

Davis, M.B., 1986: Climatic instability, time lags and community disequilibrium. In: *Community Ecology* [Diamond, J. and T. Case (eds.)]. Harper & Row Publishers Inc., Cambridge a.o., pp. 269-284.

Davis, M.B. and C. Zabinski, 1992: Changes in geographical range resulting from greenhouse warming: effects on biodiversity in forests. In: *Global Warming and Biological Diversity* [Peters, R.L. and T.E. Lovejoy (eds.)]. Yale University Press, New Haven, CT, pp. 297-308.

DENR-ADB, 1990: *Master plan for forestry development.* Department of Environment and Natural Resources–Asian Development Bank, Manila, Philippines, 523 pp.

Deutscher Bundestag, 1990: *Schutz der tropischen Wälder: Eine internationale Schwerpunktaufgabe (Protecting the Tropical Forests: A High-Priority International Task).* Second report of the Enquete Commission "Vorsorge zum Schutz der Erdatmosphaere," Deutscher Bundestag, Referat Oeffentlichkeitsarbeit, Bonn, Germany, 983 pp.

Diaz, S., J.P. Grime, J. Harris, and E. McPherson, 1993: Evidence of a feedback mechanism limiting plant response to elevated carbon dioxide. *Nature* (London), **364**, 616–617.

Dixon, R.K. and D.P. Turner, 1991: The global carbon cycle and climate change: responses and feedbacks from below-ground systems. *Environmental Pollution*, **73**, 245-262.

Dixon, R.K. and O.N. Krankina, 1993: Forest fires in Russia: carbon dioxide emissions to the atmosphere. *Canadian Journal of Forest Research*, **23**, 700-705.

Dixon, R.K., S. Brown, R.A. Houghton, A.M. Solomon, M.C. Trexler, and J. Wisniewski, 1994: Carbon pools and flux of global forest ecosystems. *Science*, **263**, 185-90.

Dobson, A., A. Jolly, and D. Rubenstein, 1989: The greenhouse effect and biological diversity. *Trends in Ecology and Evolution*, **4(3)**, 64-68.

Dobson, A. and R. Carper, 1992: Global warming and potential changes in host-parasite and disease-vector relationships. In: *Global Warming and Biological Diversity* [Peters, R.L. and T.E. Lovejoy (eds.)]. Yale University Press, New Haven, CT, pp. 201-217.

Durka, W., E.D. Schulze, G. Gebauer, and S. Voerkelins, 1994: Effects of forest decline on uptake and leaching of deposited nitrate determined from ^{15}N and ^{18}O measurements. *Nature* (London), **372**, 765-767.

Eamus, D. and P.G. Jarvis, 1989: The direct effects of increase in the global atmospheric CO_2 concentration on natural and commercial temperate trees and forests. *Advances in Ecological Research*, **19**, 1–55.

Edmonds, R.L., 1982: *Analysis of Coniferous Forest Ecosystems in the Western United States.* Hutchinson Ross Publishing Company, Stroudsburg, PA, 419 pp.

Ehrlich, P.R. and E.O. Wilson, 1991: Biodiversity studies: science and policy. *Science*, **253**, 758-762.

Ellenberg, H. (ed.), 1971: *Integrated Experimental Ecology—Methods and Results of Ecosystem Research in the German Solling Project.* Chapman and Hall/Springer-Verlag, London/Berlin, 214 pp.

Emanuel, W.R., H.H. Shugart, and M.P. Stevenson, 1985: Climatic change and the broad-scale distribution of terrestrial ecosystem complexes. *Climatic Change*, **7**, 29-43.

Ennos, R.A., 1994: Estimating the relative rates of pollen and seed migration among plant populations. *Heredity*, **72**, 250-259.

Enting, I.G. and J.V. Mansbridge, 1991: Latitudinal distribution of sources and sinks of CO_2: results of an inversion study. *Tellus*, **43B**, 156-170.

Erwin, T., 1991: How many species are there?: Revisited. *Conservation Biology*, **5**, 330-333.

FAO, 1982: *Conservation and Development of Tropical Forest Resources.* FAO Forest Paper 37, Food and Agriculture Organization of the United Nations, Rome, Italy, 122 pp.

FAO, 1993a: *Forest Resources Assessment 1990—Tropical Countries.* FAO Forest Paper 112, Food and Agriculture Organization of the United Nations, Rome, Italy, 59 pp.

FAO, 1993b: *Forestry Statistics Today for Tomorrow: 1961-1991...2010.* Report prepared by Statistics and Economic Analysis Staff of the Forestry Department, FAO, Food and Agriculture Organization of the United Nations, Rome, Italy, 46 pp.

Fenner, M., 1991: Irregular seed crops in forest trees. *Quarterly Journal of Forestry*, **85**, 166-172.

FIRESCAN Science Team, 1995: Fire in boreal ecosystems of Eurasia: first results of the Bor forest island fire experiment, fire research campaign Asia-North (FIRESCAN). *Journal of World Resource Review*, **6**, 499-523.

Fischlin, A., H. Bugmann, and D. Gyalistras, 1995: Sensitivity of a forest ecosystem model to climate parametrization schemes. *Environmental Pollution*, **87(3)**, 267-282.

Fitter, A.H. and R.K.M. Hay, 1987: *Environmental Physiology of Plants.* Academic Press, San Diego, CA, 2nd. ed., 421 pp.

Flannigan, M.D. and C.E. van Wagner, 1991: Climatic change and wildfires in Canada. *Canadian Journal of Forest Research*, **21**, 66-72.

Flenley, J.R., 1979: *The Equatorial Rainforest: A Geological History.* Butterworths, London a.o., 162 pp.

Franklin, J.F., 1988: Structural and functional diversity in temperate forests. In: *Biodiversity* [Wilson, E.O. (ed.)]. National Academy Press, Washington, DC, pp. 166-175.

Franklin, J.F., F.J. Swanson, M.E. Harmon, D.A. Perry, T.A. Spies, V.H. Dale, A. McKee, W.K. Ferrell, J.E. Means, S.V. Gregory, J.D. Lattin, T.D. Schowalter, and D. Larsen, 1992: Effects of global climatic change on forests in northwestern North America. In: *Global Warming and Biological Diversity* [Peters, R.L. and T. Lovejoy (eds.)]. Yale University Press, New Haven, CT, and London, UK, pp. 244-257.

Friend, A.D., H.H. Shugart, and S.W. Running, 1993: A physiology-based gap model of forest dynamics. *Ecology*, **74**, 792-797.

Galinski, W. and M. Küppers, 1994: Polish forest ecosystem, the influence of changes in the economic system on the carbon balance. *Climatic Change*, **27**, 103-119.

Garfinkel, H.L. and L.B. Brubaker, 1980: Modern climate-tree-growth relationships and climatic reconstruction in sub-arctic Alaska. *Nature* (London), **286**, 872-874.

Gash, J.H.C. and W.J. Shuttleworth, 1991: Tropical deforestation: albedo and the surface-energy balance. *Climatic Change*, **19**, 123-133.

Gaston, K.J., 1991: The magnitude of global insect species richness. *Conservation Biology*, **5**, 283-296.

Gates, D.M., 1993: *Climate Change and Its Biological Consequences.* Sinauer Associates Inc., Sunderland, MA, 280 pp.

Gear, A.J. and B. Huntley, 1991: Rapid change in the range limits of Scots pine 4,000 years ago. *Science*, **251**, 544-547.

Gholz, H.L., 1982: Environmental limits on aboveground net primary production, leaf area and biomass in vegetation zones of the Pacific Northwest. *Ecology*, **63**, 469-481.

Gifford, R.M., 1979: Growth and yield of CO_2-enriched wheat under water-limited conditions. *Australian Journal of Plant Physiology*, **6**, 367-378.

Gill, A.M., R.H. Groves, and I.R. Noble, 1981: *Fire and the Australian Biota.* Australian Academy of Science, Canberra, Australia, 582 pp.

Goldammer, J.G. and B. Seibert, 1990: The impact of droughts and forest fires on tropical lowland rainforest of eastern Borneo. In: *Fire in the Tropical Biota–Ecosystems Processes and Global Challenges* [Goldammer, J.G. (ed.)]. Springer-Verlag, Berlin-Heidelberg, Germany, pp. 11-31.

Goldammer, J.G. and M.J. Jenkins, 1990: *Fire in Ecosystem Dynamics—Mediterranean and Northern Perspectives.* SPB Academic Publishing, The Hague, The Netherlands, 199 pp.

Goldammer, J.G. and S.R. Peñafiel, 1990: Fire in the pine-grassland biomes of tropical and sub-tropical Asia. In: *Fire in the Tropical Biota–Ecosystems Processes and Global Challenges* [Goldammer, J.G. (ed.)]. Springer-Verlag, Berlin-Heidelberg, Germany, pp. 45-62.

Goldammer, J.G., 1992: *Tropical Forests in Transition: Ecology of Natural and Anthropogenic Disturbance Processes.* Birkhäuser Verlag, Basel a.o., 270 pp.

Greco, S., R.H. Moss, D. Viner, and R. Jenne, 1994: *Climate Scenarios and Socioeconomic Projections for IPCC WG II Assessment.* IPCC - WMO and UNEP, Washington, DC, 67 pp.

Green, D.G., 1987: Pollen evidence for the postglacial origins of Nova Scotia's forests. *Canadian Journal of Botany*, **65**, 1163-1179.

Groombridge, B. (ed.), 1992: *Global Biodiversity: Status of the Earth's Living Resources.* Chapman & Hall, London a.o., 585 pp.

Grubb, P.J., 1977: The maintenance of species-richness in plant communities: the importance of the regeneration niche. *Biological Reviews of the Cambridge Philosophical Society*, **52**, 107-145.

Gyalistras, D., H. von-Storch, A. Fischlin, and M. Beniston, 1994: Linking GCM-simulated climatic changes to ecosystem models: case studies of statistical downscaling in the Alps. *Climate Research*, **4(3)**, 167-189.

Hamburg, S.P. and C.V. Cogbill, 1988: Historical decline of red spruce populations and climatic warming. *Nature* (London), **331**, 428-431.

Hänninen, H., 1991: Does climatic warming increase the risk of frost damage in northern trees? *Plant, Cell and Environment*, **14**, 449-454.

Hartshorn, G.S., 1978: Treefalls and tropical forest dynamics. In: *Tropical Trees as Living Systems* [Tomlinson, P.B. and M.H. Zimmerman (eds.)]. Cambridge University Press, New York, NY, pp. 617-638.

Heath, L.S., P.E. Kauppi, P. Burschel, H.-D. Gregor, R. Guderian, G.H. Kohlmaier, S. Lorenz, D. Overdieck, F. Scholz, H. Thomasius, and M. Weber, 1993: Contribution of temperate forests to the world's carbon budget. *Water, Air, and Soil Pollution*, **70**, 55-69.

Heide, O.M., 1974: Growth and dormancy in Norway spruce ecotypes (Picea abies). I. Interactions of photoperiod and temperature. *Physiologia Plantarum*, **30**, 1-12.

Heide, O.M., 1993: Dormancy release in beech buds (Fagus sylvatica) requires both chilling and long days. *Physiologia Plantarum*, **89**, 187-191.

Henderson-Sellers, A., R.E. Dickinson, and M.F. Wilson, 1988: Tropical deforestation: important processes for climate models. *Climatic Change*, **13**, 43-67.

Henderson-Sellers, A. and K. McGuffie, 1994: Land surface characterisation in greenhouse climate simulations. *International Journal of Climatology*, **14**, 1065-1094.

Henttonen, H., M. Kanninen, M. Nygren, and R. Ojansuu, 1986: The maturation of Scots pine seeds in relation to temperature climate in northern Finland. *Scandinavian Journal of Forest Research*, **1**, 234-249.

Hinckley, T.M., R.O. Teskey, F. Duhme, and H. Richter, 1981: Temperate hardwood forests. In: *Water Deficits and Plant Growth. VI - Woody Plant Communities* [Kozlowski, T.T. (ed.)]. Academic Press, New York, NY, pp. 153-208.

Hofgaard, A., L. Kullman, and H. Alexandersson, 1991: Response of old-growth montane *Picea abies* (L.) Karst. forest to climatic variability in northern Sweden. *New Phytologist*, **119**, 585-594.

Hofgaard, A., 1993: Structure and regeneration patterns in a virgin *Picea abies* forest in northern Sweden. *Journal of Vegetation Science*, **4**, 601-608.

Holdridge, L.R., 1947: Determination of world plant formations from simple climatic data. *Science*, **105**, 367–368.

Holdridge, L.R., 1967: *Life Zone Ecology.* Report, Tropical Science Center, San Jose, Costa Rica, 206 pp.

Houghton, R.A., 1993: Is carbon accumulating in the northern temperate zone? *Global Ecology and Biogeography Letters*, **7**, 611-617.

Huntley, B. and H.J.B. Birks, 1983: *An Atlas of Past and Present Pollen Maps for Europe: 0–13,000 Years Ago.* Cambridge University Press, Cambridge a.o., 667 pp.

Hutchinson, G.E., 1957: Concluding remarks. *Cold Spring Harbor Symposia on Quantitative Biology*, **22**, 415-427.

Idso, K.B. and S.B. Idso, 1994: Plant responses to atmospheric CO_2 enrichment in the face of environmental constraints: a review of the past 10 years' research. *Agricultural and Forest Meteorology*, **69**, 153-203.

Innes, J.L., 1991: High-altitude and high-latitude tree growth in relation to past, present and future global climate change. *The Holocene*, **1**, 168-173.

Innes, J.L., 1993: *Forest Health: Its Assessment and Status.* CAB International, Wallingford, Oxon, UK, 677 pp.

Innes, J.L., 1994: The occurrence of flowering and fruiting in individual trees over 3 years and their effects on subsequent crown condition. *Trees*, **8**, 139-150.

Iversen, J., 1944: *Viscum, Hedera* and *Ilex* as climatic indicators. *Geol Foren Stockholm Forh*, **66**, 463-483.

Janzen, D.H., 1988: Tropical dry forests: the most endangered major tropical ecosystem. In: *Biodiversity* [Wilson, E.O. (ed.)]. National Academy Press, Washington, DC, pp. 130-137.

Jenkinson, D.S., D.E. Adams, and A. Wild, 1991: Model estimates of CO_2 emissions from soil in response to global warming. *Nature* (London), **351**, 304-306.

Johnson, A.H., E.R. Cook, and T.G. Siccama, 1988: Climate and red spruce growth and decline in the northern Appalachians. *Proceedings of the National Academy of Sciences of the United States of America*, **85**, 5369-5373.

Johnson, E.A., and C.P.S. Larson, 1991: Climatically induced change in fire frequency in the southern Canadian Rockies. *Ecology*, **72**, 194-201.

Jones, C.G. and J.S. Coleman, 1991: Plant stress and insect herbivory: towards an integrated perspective. In: *Response of Plants to Multiple Stresses* [Mooney, H.A., W.E. Winner, and E.J. Pell (eds.)]. Academic Press, San Diego, CA, pp. 249-280.

Jones, E.A., D.D. Reed, G.D. Mroz, H.O. Liechty, and P.J. Cattelino, 1993: Climate stress as a precursor to forest decline: paper birch in northern Michigan 1985-1990. *Canadian Journal of Forest Research*, **23**, 229-233.

Jonsson, B.G. and P.-A. Esseen, 1990: Treefall disturbance maintains high bryophyte diversity in a boreal spruce forest. *Journal of Ecology*, **78**, 924-936.

Jordan, C.F., 1985: *Nutrient Cycling in Tropical Forest Ecosystem: Principles and Their Application in Management and Conservation.* Wiley, Chichester a.o., 190 pp.

Kasischke, E.S., N.L. Christensen, Jr., and B.J. Stocks, 1995: Fire, global warming, and the carbon balance of boreal forests. *Ecological Applications*, **5**, 437-451.

Kauppi, P. and M. Posch, 1985: Sensitivity of boreal forests to possible climatic warming. *Climatic Change*, **7**, 45-54.

Kauppi, P.E., K. Mielikäinen, and K. Kuusela, 1992: Biomass and carbon budget of European forests, 1971 to 1990. *Science*, **256**, 70-74.

Kellomäki, S. and M. Kolström, 1992: Simulation of tree species composition and organic matter accumulation in Finnish boreal forests under changing climatic conditions. *Vegetatio*, **102**, 47-68.

Kienast, F., 1991: Simulated effects of increasing atmospheric CO_2 and changing climate on the successional characteristics of Alpine forest ecosystems. *Landscape Ecology*, **5(4)**, 225–238.

Kienast, F. and N. Kräuchi, 1991: Simulated successional characteristics of managed and unmanaged low-elevation forests in Central Europe. *Forest Ecology and Management*, **42**, 49-61.

Kimball, B.A., 1983: Carbon dioxide and agricultural yield: an assemblage and analysis of 430 prior observations. *Agronomy Journal*, **75**, 779-788.

Kimmins, J.P. and D.P. Lavender, 1992: Ecosystem-level changes that may be expected in a changing global climate: a British Columbia perspective. *Environmental Toxicology and Chemistry*, **11**, 1061-1068.

King, G.A. and R.P. Neilson, 1992: The transient response of vegetation to climate change: a potential source of CO_2 to the atmosphere. *Water, Air, and Soil Pollution*, **64**, 365–383.

Kirschbaum, M.U.F., 1993: A modelling study of the effects of changes in atmospheric CO_2 concentration, temperature and atmospheric nitrogen input on soil organic carbon storage. *Tellus*, **45B**, 321-334.

Kirschbaum, M.U.F., 1994: The sensitivity of C_3 photosynthesis to increasing CO_2 concentration—a theoretical analysis of its dependence on temperature and background CO_2 concentration. *Plant, Cell and Environment*, **17**, 747-754.

Kirschbaum, M.U.F., D.A. King, H.N. Comins, R.E. McMurtrie, B.E. Medlyn, S. Pongracic, D. Murty, H. Keith, R.J. Raison, P.K. Khanna, and D.W. Sheriff, 1994: Modelling forest response to increasing CO_2 concentration under nutrient-limited conditions. *Plant, Cell and Environment*, **17**, 1081-1099.

Kirschbaum, M.U.F., 1995: The temperature dependence of soil organic matter decomposition and the effect of global warming on soil organic carbon storage. *Soil Biology & Biochemistry*, **27**, 753-760.

Kobak, K. and N. Kondrasheva, 1993: Residence time of carbon in soils of the boreal zone. In: *Carbon Cycling in Boreal Forest and Sub-Arctic Ecosystems* [Vinson, T.S. and T.P. Kolchugina (eds.)]. EPA, Corvallis, OR, pp. 51-58.

Kohlmaier, G.H., C. Häger, G. Würth, M.K.B. Lüdeke, P. Ramge, F.W. Badeck, J. Kindermann, and T. Lang, 1995: Effects of the age class distribution of the temperate and boreal forests on the global CO_2 source-sink function. *Tellus*, **47B**, 212-231.

Kokorin, A.O., G.V. Mironova, and N.V. Semenuk, 1993: The influence of climatic variations on the growth of cedar and fir in south Baikal region. In: *Global and Regional Consequences of the Climatic and Environmental Change* [Izrael, Y.A. and Y.A. Anokhin (eds.)]. Gidrometeoizdat, St. Petersburg, Russia, pp. 160-175 (in Russian).

Köppen, W., 1936: *Das Geographische System der Klimate.* Gebrüder Bornträger, Berlin, Germany, 46 pp.

Körner, C. and J.A. Arnone III, 1992: Responses to elevated carbon dioxide in artificial tropical ecosystems. *Science*, **257**, 1672–1675.

Körner, C., 1993: CO_2 fertilization: the great uncertainty in future vegetation development. In: *Vegetation Dynamics and Global Change* [Solomon, A.M. and H.H. Shugart (eds.)]. Chapman & Hall, New York, NY, and London, UK, pp. 53-70.

Korzukhin, M.D., A.E. Rubinina, G.B. Bonan, A.M. Solomon, and M.Y. Anotonvsky, 1989: *The Silvics of Some East European and Siberian Boreal Forest Tree Species.* Working Paper WP-89-56, International Institute for Applied Systems Analysis, Laxenburg, Austria, 27 pp.

Kozlowski, T.T. and C.E. Ahlgren, 1974: *Fire and Ecosystems.* Academic Press, New York, NY, 542 pp.

Kullman, L., 1986: Recent tree-limit history of Picea abies in the southern Swedish Scandes. *Canadian Journal of Forest Research*, **16**, 761-771.

Kullman, L., 1990: Dynamics of altitudinal tree-limits in Sweden: a review. *Norsk Geografisk Tidsskrift*, **44**, 103-116.

Kurz, W.A., M.J. Apps, T.M. Webb, and P.J. McNamee, 1992: *The Carbon Budget of the Canadian Forest Sector: Phase I.* Forestry Canada, Northern Forestry Centre, Edmonton, Alberta, 93 pp.

Kurz, W.A. and M.J. Apps, 1993: Contribution of northern forests to the global C cycle: Canada as a case study. *Water, Air, and Soil Pollution*, **70**, 163-176.

Kurz, W.A., M.J. Apps, B.J. Stocks, and W.J. Volney, 1995: Global climate change: disturbance regimes and biospheric feedbacks of temperate and boreal forests. In: *Biospheric Feedbacks in the Global Climate System—Will the Warming Feed the Warming?* [Woodwell, G.M. and F.T. Mackenzie (eds.)]. Oxford University Press, New York, NY, and Oxford, UK, pp. 119-133.

Lambert, M.J. and J. Turner, 1977: Dieback in high site quality *Pinus radiata* stands—the role of sulfur and boron deficiencies. *New Zealand Journal of Forestry*, **7**, 333-348.

Landhäusser, S.M. and R.W. Wein, 1993: Postfire vegetation recovery and tree establishment at the Arctic treeline: climate-change vegetation-response hypotheses. *Journal of Ecology*, **81**, 665-672.

Landsberg, J.J., S. Linder, and R.E. McMurtrie, 1995: *A Strategic Plan for Research on Managed Forest Ecosystems in a Globally Changing Environment.* Global Change and Terrestrial Ecosystems (GCTE) Report No. 4 and IUFRO Occasional Paper 1. GCTE Activity 3.5: Effects of Global Change on Managed Forests—Implementation Plan, GCTE Core Project Office, Canberra, Australia, pp 1-17.

Larsen, J.A., 1980: *The Boreal Ecosystem.* Academic Press, New York a.o., 500 pp.

Larsen, J.A. (ed.), 1989: *The Northern Forest Border in Canada and Alaska.* Springer Verlag, New York a.o., 255 pp.

Lean, J. and D.A. Warrilow, 1989: Simulation of the regional climatic impact of Amazon deforestation. *Nature* (London), **342**, 411-413.

LeBlanc, D.C. and J.K. Foster, 1992: Predicting effects of global warming on growth and mortality of upland oak species in the midwestern United States: a physiologically based dendroecological approach. *Canadian Journal of Forest Research*, **22**, 1739-1752.

Leemans, R. and W.P. Cramer, 1991: *The IIASA Database for Mean Monthly Values of Temperature, Precipitation, and Cloudiness on a Global Terrestrial Grid.* Research Report RR-91-18, IIASA, Laxenburg, Austria, 72 pp.

Leemans, R., 1992: Modelling ecological and agricultural impacts of global change on a global scale. *Journal of Scientific and Industrial Research*, **51**, 709-724.

Leemans, R. and A.M. Solomon, 1993: Modeling the potential change in yield and distribution of the earth's crops under a warmed climate. *Climate Research*, **3**, 79–96.

Leemans, R. and G.J. van den Born, 1994: Determining the potential distribution of vegetation, crops and agricultural productivity. *Water, Air, and Soil Pollution*, **76**, 133-161.

Leemans, R., W.P. Cramer, and J.G. van Minnen, 1995: Prediction of global biome distribution using bioclimatic equilibrium models. In: *Global Change: Effects on Coniferous Forests and Grassland* [Melillo, J.M. and A. Breymeyer (eds.)]. John Wiley and Sons, New York, NY, 57 pp. (submitted).

Levitt, J., 1980a: *Responses of Plants to Environmental Stresses.* Vol. I, *Chilling, Freezing, and High Temperature Stresses.* Academic Press, New York a.o., 2nd. ed., 497 pp.

Levitt, J., 1980b: *Responses of Plants to Environmental Stresses.* Vol. II, *Water, Radiation, Salt, and Other Stresses.* Academic Press, New York a.o., 2nd ed., 607 pp.

Liu, Q. and H. Hytteborn, 1991: Gap structure, disturbance and regeneration in a primeval *Picea abies* forest. *Journal of Vegetation Science*, **2**, 391-402.

Lloyd, J. and J.A. Taylor, 1994: On the temperature dependence of soil respiration. *Functional Ecology*, **8**, 315-323.

Long, S.P., 1991: Modification of the response of photosynthetic productivity to rising temperature by atmospheric CO_2 concentration: has its importance been underestimated? *Plant, Cell and Environment*, **14**, 729-739.

Longman, K.A. and J. Jenik, 1987: *Tropical Forest and Its Environment.* Harlow, Longman Scientific & Technical, Essex, England, 2nd ed., 347 pp.

Lotter, A. and F. Kienast, 1992: Validation of a forest succession model by means of annually laminated sediments. In: *Laminated Sediments* [Saarnisto, M. and A. Kahra (eds.)]. Proceedings of the workshop at Lammi Biological Station, 4–6, June 1990, Geological Survey of Finland, pp. 25–31.

Lüdeke, M.K.B., S. Dönges, R.D. Otto, J. Kindermann, F.W. Badeck, P. Ramge, U. Jäkel, and G.H. Kohlmaier, 1995: Responses in NPP and carbon stores of the northern biomes to a CO_2-induced climatic change, as evaluated by the Frankfurt Biosphere Model (FBM). *Tellus*, **47B**, 191-205.

Lugo, A.E., M. Applefield, D.J. Pool, and R.B. McDonald, 1983: The impact of hurricane David on the forest of Dominica. *Canadian Journal of Forest Research*, **13**, 201-211.

Lugo, A.E., 1988: Estimating reductions in the diversity of tropical forest species. In: *Biodiversity* [Wilson, E.O. (ed.)]. National Academy Press, Washington, DC, pp. 58-70.

Luo, Y., C.B. Field, and H.A. Mooney, 1994: Predicting responses of photosynthesis and root fraction to elevated [CO_2]: interactions among carbon, nitrogen and growth. *Plant, Cell and Environment*, **17**, 1195-1204.

Luxmoore, R.J., S.D. Wullschleger, and P.J. Hanson, 1993: Forest responses to CO_2 enrichment and climate warming. *Water, Air, and Soil Pollution*, **70**, 309-323.

Lyons, J.M., 1973: Chilling injury in plants. *Annual Review of Plant Physiology*, **24**, 445-466.

Lyons, J.M., D. Graham, and J.K. Raison (eds.), 1979: *Low Temperature Stress in Crop Plants: The Role of the Membrane.* Academic Press, New York a.o., 565 pp.

Mabberley, D.J., 1992: *Tropical Rainforest Ecology.* Blackie and Son Ltd., Glasgow and London, UK, 2nd. ed., 300 pp.

Malanson, G.P., W.E. Westman, and Y.-L. Yan, 1992: Realized versus fundamental niche functions in a model of chaparral response to climatic change. *Ecological Modelling*, **64**, 261-277.

Manabe, S. and R.T. Weatherald, 1987: Large-scale changes of soil wetness induced by an increase in atmospheric carbon dioxide. *Journal of Atmospheric Sciences*, **44(8)**, 1211-1235.

Martinat, P.J., 1987: The role of climatic variation and weather in forest insect outbreaks. In: *Insect Outbreaks* [Barbosa, P. and J.C. Schultz (eds.)]. Academic Press, Inc., San Diego, CA, pp. 241-268.

Matthews, E., 1983: Global vegetation and land use: new high-resolution data bases for climate studies. *Journal of Climate and Applied Meteorology*, **22**, 474-487.

Mattson, W.J., 1980: Herbivory in relation to plant nitrogen content. *Annual Review of Ecology and Systematics*, **11**, 119-161.

Mattson, W.J. and R.A. Haack, 1987: The role of drought stress in provoking outbreaks of phytophagous insects. In: *Insect Outbreaks* [Barbosa, P. and J.C. Schultz (eds.)]. Academic Press, Inc., San Diego, CA, pp. 365-407.

Matyas, C., 1994: Modelling climate change effects with provenance test data. *Tree Physiology*, **14**, 797-804.

May, R.M., 1990: How many species? *Philosophical Transactions of the Royal Society of London—Series B*, **330**, 171-182.

Melillo, J.M., T.V. Callaghan, F.I. Woodward, E. Salati, and S.K. Sinha, 1990: Effects on ecosystems. In: *Climate Change—The IPCC Scientific Assessment. Report Prepared for IPCC by Working Group I* [Houghton, J.T., G.J. Jenkins, and J.J. Ephraums (eds.)]. Cambridge University Press, Cambridge a.o., pp. 283-310.

Melillo, J.M., A.D. McGuire, D.W. Kicklighter, B. Moore III, C.J. Vorosmarty, and A.L. Schloss, 1993: Global climate change and terrestrial net primary production. *Nature* (London), **363**, 234-240.

Mikola, P., 1962: Temperature and tree growth near the northern timber line. In: *Tree Growth* [Kozlowski, T.T. (ed.)]. Ronald Press, New York, NY, pp. 265-274.

Mitchell, J.F.B., S. Manabe, V. Meleshko, and T. Tokioka, 1990: Equilibrium climate change—and its implications for the future. In: *Climate Change—The IPCC Scientific Assessment. Report Prepared for IPCC by Working Group I* [Houghton, J.T., G.J. Jenkins, and J.J. Ephraums (eds.)]. Cambridge University Press, Cambridge a.o., pp. 139-173.

Monserud, R.A., O.V. Denissenko, and N.M. Tchebakova, 1993a: Comparison of Siberian palaeovegetation to current and future vegetation under climate change. *Climate Research*, **3**, 143-159.

Monserud, R.A., N.M. Tchebakova, and R. Leemans, 1993b: Global vegetation change predicted by the modified Budyko model. *Climatic Change*, **25**, 59-83.

Mooney, H.A., 1988: Lessons from Mediterranean-climate regions. In: *Biodiversity* [Wilson, E.O. (ed.)]. National Academy Press, Washington, DC, pp. 157-165.

Mooney, H.A., B.G. Drake, R.J. Luxmoore, W.C. Oechel, and L.F. Pitelka, 1991: Predicting ecosystem responses to elevated CO_2 concentrations: what has been learned from laboratory experiments on plant physiology and field observations? *Bioscience*, **41**, 96-104.

Moore, A.D., 1989: On the maximum growth equation used in forest gap simulation models. *Ecological Modelling*, **45**, 63-67.

Morikawa, Y., 1993: Climate changes and forests. In: *The Potential Effects of Climate Change in Japan* [Harasawa, S.N., H. Hashimoto, T. Ookita, K. Masuda, and T. Morita (eds.)]. Center for Global Environmental Research, Tsukuba, Ibaraki 305, Japan, pp. 37-44.

Morison, J.I.L., 1987: Intercellular CO_2 concentration and stomatal responses to CO_2. In: *Stomatal Function* [Zeiger, E., G.D. Farquhar, and I.R. Cowan (eds.)]. University Press, Stanford, CA, pp. 229-251.

Mueller-Dombois, D., 1992: Potential effects of the increase in carbon dioxide and climate: change in the dynamics of vegetation. *Water, Air, and Soil Pollution*, **64**, 61-79.

Murray, M.B., M.G.R. Cannell, and R.I. Smith, 1989: Date of budburst of fifteen tree species in Britain following climatic warming. *Journal of Applied Ecology*, **26**, 693-700.

Myers, N., 1991: Tropical forest; present status and future outlook. *Climatic Change*, **19**, 3-32.

Myers, N., 1992: Synergisms: joint effects of climate change and other forms of habitat destruction. In: *Global Warming and Biological Diversity* [Peters, R.L. and T.E. Lovejoy (eds.)]. Yale University Press, New Haven, CT, and London, UK, pp. 344-354.

Myers, N., 1993: Questions of mass extinction. *Biodiversity and Conservation*, **2**, 2-17.

Myers, N., 1995: Environmental unknowns. *Science*, **269**, 358-360.

Neilson, R.P., G.A. King, and G. Koerper, 1992: Toward a rule-based biome model. *Landscape Ecology*, **7**, 27-43.

Neilson, R.P., 1993: Vegetation redistribution: a possible biosphere source of CO_2 during climate change. *Water, Air, and Soil Pollution*, **70**, 659-673.

Neilson, R.P. and D. Marks, 1994: A global perspective of regional vegetation and hydrologic sensitivities from climatic change. *Journal of Vegetation Science*, **27**, 715-730.

Neilson, R.P., G.A. King, and J. Lenihan, 1994: Modeling forest response to climatic change: the potential for large emissions of carbon from dying forests. In: *Carbon Balance of World's Forested Ecosystems: Towards a Global Assessment* [Kanninen, M. (ed.)]. Academy of Finland, Helsinki, Finland, pp. 150-162.

Nikolov, N. and H. Helmisaari, 1992: Silvics of the circumpolar boreal forest tree species. In: *A Systems Analysis of the Global Boreal Forest* [Shugart, H.H., R. Leemans, and G.B. Bonan (eds.)]. Cambridge University Press, Cambridge a.o., pp. 13-84.

Norby, R.J., E.G. O'Neill, and R.J. Luxmoore, 1986: Effects of atmospheric CO_2 enrichment on the growth and mineral nutrition of *Quercus alba* seedlings in nutrient-poor soil. *Plant Physiology*, **82**, 83-89.

Norby, R.J., C.A. Gunderson, S.D. Wullschleger, E.G. O'Neill, and M.K. McCracken, 1992: Productivity and compensatory responses of yellow-poplar trees in elevated CO_2. *Nature* (London), **357**, 322-324.

Norton, B.G. and R.E. Ulanowicz, 1992: Scale and biodiversity policy: a hierarchical approach. *Ambio*, **21**, 244-249.

O'Brien, S.T., B.P. Hayden, and H.H. Shugart, 1992: Global climatic change, hurricanes, and a tropical forest. *Climatic Change*, **22**, 1750-1790.

Odum, E.P., 1971: *Fundamentals of Ecology.* W.B. Saunders Comp., Philadelphia a.o., 3rd. ed., 574 pp.

Oechel, W.C., S.J. Hastings, G. Vourlitis, M. Jenkins, G. Riechers, and N. Grulke, 1993: Recent change of Arctic tundra ecosystems from a net carbon dioxide sink to a source. *Nature* (London), **361**, 520-523.

Olson, J.S., J.A. Watts, and L.J. Allison, 1983: *Carbon in Live Vegetation of Major World Ecosystems.* Report ORNL-5862, Oak Ridge National Laboratory, Oak Ridge, TN, and Springfield, VA, 152 pp.

Overdieck, D., C. Reid, and B.R. Strain, 1988: The effects of pre-industrial and future CO_2 concentrations on growth, dry matter production and the C/N relationship in plants at low nutrient supply. *Angewandte Botanik*, **62**, 119-134.

Pastor, J. and W.M. Post, 1988: Response of northern forests to CO_2-induced climate change. *Nature* (London), **334(6177)**, 55–58.

Pastor, J. and D.J. Mladenoff, 1992: The southern boreal-northern hardwood forest border. In: *A Systems Analysis of the Global Boreal Forest* [Shugart, H.H., R. Leemans, and G.B. Bonan (eds.)]. Cambridge University Press, Cambridge a.o., pp. 216-240.

Pastor, J. and W.M. Post, 1993: Linear regressions do not predict the transient responses of Eastern North American forests to CO_2-induced climate change. *Climatic Change*, **23**, 111-119.

Payette, S. and R. Gagnon, 1985: Late Holocene deforestation and tree regeneration in the forest-tundra of Québec. *Nature* (London), **313**, 570-572.

Payette, S., L. Filion, A. Delwaide, and C. Bégin, 1989a: Reconstruction of tree-line vegetation response to long-term climate change. *Nature* (London), **341**, 429-432.

Payette, S., C. Morneau, L. Sirois, and M. Desponts, 1989b: Recent fire history of the northern Quebec biomes. *Ecology*, **70**, 656-673.

Payette, S., 1992: Fire as a controlling process in the North American boreal forest. In: *A Systems Analysis of the Global Boreal Forest* [Shugart, H.H., R. Leemans, and G.B. Bonan (eds.)]. Cambridge University Press, Cambridge a.o., pp. 144-169.

Perry, D.A., R. Molina, and M.P. Amaranthus, 1987: *Mycorrhizae*, mycorrhizospheres, and reforestation: current knowledge and research needs. *Canadian Journal of Forest Research*, **17**, 929-940.

Perry, D.A., J.G. Borchers, S.L. Borchers, and M.P. Amaranthus, 1990: Species migrations and ecosystem stability during climate change: the below ground connection. *Conservation Biology*, **4**, 266-274.

Peters, R.L. and J.D.S. Darling, 1985: The greenhouse effect and nature reserves—global warming would diminish biological diversity by causing extinctions among reserve species. *Bioscience*, **35(11)**, 707-717.

Peters, R.L., 1992: Conservation of biological diversity in the face of climate change. In: *Global Warming and Biological Diversity* [Peters, R.L. and T.E. Lovejoy (eds.)]. Yale University Press, New Haven, CT, and London, UK, pp. 59-71.

Peters, R.L. and T. Lovejoy (eds.), 1992: *Global Warming and Biological Diversity*. Yale University Press, New Haven, CT, and London, UK, 386 pp.

Peterson, B.J.J. and J.M. Melillo, 1985: The potential storage of carbon caused by eutrophication of the biosphere. *Tellus*, **37B**, 117-127.

Pigott, C.D. and J.P. Huntley, 1981: Factors controlling the distribution of *Tilia cordata* at the northern limits of its geographical range. III. Nature and causes of seed sterility. *New Phytologist*, **87**, 817-839.

Pimm, S.L., G.J. Russell, J.L. Gittleman, and T.M. Brooks, 1995: The future of biodiversity. *Science*, **269**, 347-350.

Plöchl, M. and W.P. Cramer, 1995: Coupling global models of vegetation structure and ecosystem processes—an example from Arctic and Boreal ecosystems. *Tellus*, **47B(1/2)**, 240-250.

Post, W.M., W.R. Emanuel, P.J. Zinke, and A.G. Strangenberger, 1982: Soil carbon pools and world life zones. *Nature* (London), **298**, 156-159.

Post, W.M., J. Pastor, A.W. King, and W.R. Emanuel, 1992: Aspects of the interaction between vegetation and soil under global change. *Water, Air, and Soil Pollution*, **64(1/2)**, 345-363.

Postel, S., 1994: Carrying capacity: earth's bottom line. In: *State of the World 1994—A Worldwatch Institute Report on Progress toward a Sustainable Society* [Starke, L. (ed.)]. W.W. Norton & Co., New York, NY, and London, UK, pp. 3-21.

Prentice, I.C., P.J. Bartlein, and T. Webb III, 1991a: Vegetation and climate changes in eastern North America since the last glacial maximum. *Ecology*, **72**, 2038-2056.

Prentice, I.C., M.T. Sykes, and W.P. Cramer, 1991b: The possible dynamic response of northern forests to global warming. *Global Ecology and Biogeography Letters*, **1**, 129-135.

Prentice, I.C., 1992: Climate change and long-term vegetation dynamics. In: *Plant Succession: Theory and Prediction* [Glenn-Lewin, D.C., R.K. Peet, and T.T. Veblen (eds.)]. Chapman and Hall, London a.o., pp. 293-339.

Prentice, I.C., W.P. Cramer, S.P. Harrison, R. Leemans, R.A. Monserud, and A.M. Solomon, 1992: A global biome model based on plant physiology and dominance, soil properties and climate. *Journal of Biogeography*, **19**, 117–134.

Prentice, I.C., M.T. Sykes, and W.P. Cramer, 1993: A simulation model for the transient effects of climate change on forest landscapes. *Ecological Modelling*, **65**, 51-70.

Prentice, I.C., M.T. Sykes, M. Lautenschlager, S.P. Harrison, O. Denissenko, and P.J. Bartlein, 1994: Modelling global vegetation patterns and terrestrial carbon storage at the last glacial maximum. *Global Ecology and Biogeography Letters*, **3**, 67-76.

Prentice, I.C. and M.T. Sykes, 1995: Vegetation geography and global carbon storage changes. In: *Biospheric Feedbacks in the Global Climate System—Will the Warming Feed the Warming?* [Woodwell, G.M. and F.T. Mackenzie (eds.)]. Oxford University Press, New York, NY, and Oxford, UK, pp. 304-312.

Raich, J.W., E.B. Rastetter, J.M. Melillo, D.W. Kicklighter, P.A. Steudler, B.J. Peterson, A.L. Grace, B. Moore III, and C.J. Vörösmarty, 1991: Potential net primary production in South America: application of a global model. *Ecological Applications*, **1(4)**, 399–429.

Raich, J.W. and W.H. Schlesinger, 1992: The global carbon dioxide flux in soil respiration and its relationship to vegetation and climate. *Tellus*, **44B**, 81-99.

Rastetter, E.B., R.B. McKane, G.R. Shaver, and J.M. Melillo, 1992: Changes in C storage by terrestrial ecosystems: how C-N interactions restrict responses to CO_2 and temperature. *Water, Air, and Soil Pollution*, **9**, 327–344.

Raven, P.H., 1988: Our diminishing tropical forests. In: *Biodiversity* [Wilson, E.O. (ed.)]. National Academy Press, Washington, DC, pp. 119-122.

Reichle, D.E. (ed.), 1981: *Dynamic properties of forest ecosystems.* Cambridge University Press, Cambridge a.o., 683 pp.

Reid, W.V., 1992: How many species will there be? In: *Tropical Deforestation and Species Extinction* [Whitmore, T.C. and J.A. Sayer (eds.)]. Chapman & Hall, London a.o., pp. 55-73.

Richards, J.F. and E.P. Flint, 1994: A century of land use change in South and Southeast Asia. In: *Effects of Land Use Change on Atmospheric CO_2 Concentrations: Southeast Asia as a Case Study* [Dale, V. (ed.)]. Springer Verlag, New York a.o., pp. 15-66.

Richards, P.W., 1966: *The Tropical Rain Forest: An Ecological Study.* Cambridge University Press, Cambridge, UK, 450 pp.

Riede, K., 1993: Monitoring biodiversity: analysis of Amazonian rainforest sounds. *Ambio*, **22**, 546-548.

Rind, D., R. Goldberg, J. Hansen, C. Rosenzweig, and R. Ruedy, 1990: Potential evapotranspiration and the likelihood of future drought. *Journal of Geophysical Research*, **95(7)**, 9983-10004.

Rind, D., 1995: Drying out. *New Scientist*, **146(1976)**, 36-40.

Ritchie, J.C. and G.M. MacDonald, 1986: The patterns of post-glacial spread of white spruce. *Journal of Biogeography*, **13**, 527-540.

Ritchie, J.C., 1987: *Postglacial Vegetation of Canada.* Cambridge University Press, Cambridge a.o., 178 pp.

Rizzo, B. and E. Wiken, 1992: Assessing the sensitivity of Canada's ecosystems to climatic change. *Climatic Change*, **21**, 37-55.

Roberts, L., 1989: How fast can trees migrate? *Science*, **243**, 735-737.

Rogers, H.H., J.F. Thomas, and G.E. Bingham, 1983: Response of agronomic and forest species to elevated atmospheric carbon dioxide. *Science*, **220**, 428-429.

Romme, W.H., 1991: Implications of global climate change for biogeographic patterns in the greater Yellowstone ecosystem. *Conservation Biology*, **5(3)**, 373-386.

Rosenzweig, M.L., 1968: Net primary productivity of terrestrial communities, prediction from climatological data. *American Naturalist*, **102**, 67-74.

Running, S.W., and R.R. Nemani, 1991: Regional hydrologic and carbon balance responses of forests resulting from potential climatic change. *Climatic Change*, **19**, 342-368.

Salati, E., A. Dall'Olio, E. Matsui, and J.R. Gat, 1979: Recycling of water in the Amazon Basin: an isotopic study. *Water Resources Research*, **15**, 1250-1258.

Salati, E. and P.B. Jose, 1984: Amazon Basin: a system in equilibrium. *Science*, **225**, 129-138.

Schimel, D., I.G. Enting, M. Heimann, T.M.L. Wigley, D. Raynaud, D. Alves, and U. Siegenthaler, 1995: CO_2 and the carbon cycle. In: *Climate Change 1994—Radiative Forcing of Climate Change and an Evaluation of the IPCC 1992 IS92 Emission Scenarios* [Houghton, J.T., L.G. Meira-Filho, J. Bruce, X.Y. Hoesung Lee, B.A. Callander, E. Haites, N. Harris, and K. Maskell (eds.)]. Cambridge University Press, Cambridge, UK, pp. 35-71.

Sedjo, R.A., 1993: The carbon cycle and global forest ecosystem. *Water, Air, & Soil Pollution*, **70**, 295-307.

Sharma, N.P., R. Rowe, K. Openshaw, and M. Jacobson, 1992: World forests in perspective. In: *Managing the World's Forests: Looking for Balance Between Conservation and Development* [Sharma, N.P. (ed.)]. Kendall/Hunt, Dubuque, IA, pp. 17-31.

Shaver, G.R., W.D. Billings, F.S. Chapin III, A.E. Giblin, K.J. Nadelhoffer, W.C. Oechel, and E.B. Rastetter, 1992: Global change and the carbon balance of arctic ecosystems. *Bioscience*, **42(6)**, 433-441.

Shugart, H.H., 1984: *A Theory of Forest Dynamics—The Ecological Implications of Forest Succession Models.* Springer, New York a.o., 278 pp.

Shugart, H.H., M.Y. Antonovsky, P.G. Jarvis, and A.P. Sandford, 1986: CO_2, climatic change, and forest ecosystems. In: *The Greenhouse Effect, Climatic Change and Ecosystems* [Bolin, B., B.R. Döös, J. Jäger, and R.A. Warrick (eds.)]. Wiley, Chichester a.o., pp. 475-522.

Shugart, H.H., R. Leemans, and G.B. Bonan (eds.), 1992: *A Systems Analysis of the Global Boreal Forest.* Cambridge University Press, Cambridge a.o., 565 pp.

Smith, T.M., R. Leemans, and H.H. Shugart, 1992a: Sensitivity of terrestrial carbon storage to CO_2-induced climate change: comparison of four scenarios based on general circulation models. *Climatic Change*, **21**, 367-384.

Smith, T.M., J.B. Smith, and H.H. Shugart, 1992b: Modeling the response of terrestrial vegetation to climate change in the tropics. In: *Tropical Forests in Transition—Ecology of Natural and Anthropogenic Disturbance Processes* [Goldammer, J.G. (ed.)]. Birkhäuser Verlag, Basel, Switzerland, and Boston, MA, pp. 253-268.

Smith, T.M. and H.H. Shugart, 1993: The transient response of terrestrial carbon storage to a perturbed climate. *Nature* (London), **361**, 523–526.

Smith, T.M., H.H. Shugart, F.I. Woodward, and P.J. Burton, 1993: Plant functional types. In: *Vegetation Dynamics and Global Change* [Solomon, A.M. and H. Shugart (eds.)]. Chapman & Hall, New York, NY, and London, UK, pp. 272-292.

Smith, T.M., R. Leemans, and H.H. Shugart, 1994: *The Application of Patch Models of Vegetation Dynamics to Global Change Issues—A Workshop Summary.* Kluwer, Dordrecht, The Netherlands, 22 pp.

Solomon, A.M., D.C. West, and J.A. Solomon, 1981: Simulating the role of climate change and species immigration in forest succession. In: *Forest Succession: Concepts and Application* [West, D.C., H.H. Shugart, and D.B. Botkin (eds.)]. Springer, New York a.o., pp. 154-177.

Solomon, A.M., M.L. Tharp, D.C. West, G.E. Taylot, J.W. Webb, and J.L. Trimble, 1984: *Response of Unmanaged Forests to* CO_2*-induced Climate Change: Available Information, Initial Tests and Data Requirements.* DOE/NBB-0053, National Technical Information Service, U.S. Department of Commerce, Springfield, VA, 93 pp.

Solomon, A.M. and T. Webb III, 1985: Computer-aided reconstruction of late-quaternary landscape dynamics. *Annual Review of Ecology and Systematics*, **16**, 63-84.

Solomon, A.M., 1986: Transient response of forests to CO_2-induced climate change: simulation modeling experiments in eastern North America. *Oecologia*, **68**, 567-579.

Solomon, A.M. and D.C. West, 1987: Simulating forest ecosystem responses to expected climate change in eastern North America: applications to decision making in the forest industry. In: *The Greenhouse Effect, Climate Change, and the U.S. Forests* [Shands, W.E. and J.S. Hoffman (eds.)]. The Conservation Foundation, Washington, DC, pp. 189-217.

Solomon, A.M., 1988: Ecosystem theory required to identify future forest responses to changing CO_2 and climate. In: *Ecodynamics: Contributions to Theoretical Ecology* [Wolff, W., C.-J. Soeder, and F.R. Drepper (eds.)]. Springer, Berlin a.o., pp. 258-274.

Solomon, A.M., 1992: The nature and distribution of past, present and future boreal forests: lessons for a research and modeling agenda. In: *A Systems Analysis of the Global Boreal Forest* [Shugart, H.H., R. Leemans, and G.B. Bonan (eds.)]. Cambridge University Press, Cambridge a.o., pp. 291-307.

Solomon, A.M. and P.J. Bartlein, 1992: Past and future climate change: response by mixed deciduous-coniferous forest ecosystems in northern Michigan. *Canadian Journal of Forest Research*, **22**, 1727-1738.

Solomon, A.M., I.C. Prentice, R. Leemans, and W.P. Cramer, 1993: The interaction of climate and land use in future terrestrial carbon storage and release. *Water, Air, and Soil Pollution*, **70**, 595-614.

Starke, L. (ed.), 1994: *State of the World 1994—A Worldwatch Institute Report on Progress Toward a Sustainable Society.* W.W. Norton and Company, New York, NY, and London, UK, 265 pp.

Steffen, W.L., B.H. Walker, J.S. Ingram, and G.W. Koch, 1992: *Global Change and Terrestrial Ecosystems: The Operational Plan.* IGBP, Stockholm, Sweden, 95 pp.

Stephenson, N.L., 1990: Climatic control of vegetation distribution: the role of the water balance. *American Naturalist*, **135(5)**, 649-670.

Stern, K. and L. Roche, 1974: *Genetics of Forest Ecosystems.* Springer Verlag, Berlin, Germany, 330 pp.

Stocks, B.J., 1993: Global warming and forest fires in Canada. *The Forestry Chronicle*, **69**, 290.

Stork, N.E., 1988: Insect diversity: facts, fiction, and speculation. *Biological Journal of the Linnean Society*, **35**, 321-337.

Sundquist, E.T., 1993: The global carbon dioxide budget. *Science*, **259**, 934-941.

Tallis, J.H., 1990: *Plant Community History—Long Term Changes in Plant Distribution and Diversity.* Chapman and Hall, London a.o., 398 pp.

Tamm, C.O., 1991: *Nitrogen in Terrestrial Ecosystems—Questions of Productivity, Vegetational Changes and Ecosystem Stability.* Springer Verlag, Berlin, Germany, 115 pp.

Taylor, B.R., D. Parkinson, and W.F.J. Parsons, 1989: Nitrogen and lignin content as predictors of litter decay rates: a microcosm test. *Ecology*, **70(1)**, 97-104.

Tchebakova, N.M., R.A. Monserud, R. Leemans, and S. Golovanov, 1993: A global vegetation model based on the climatological approach of Budyko. *Journal of Biogeography*, **20**, 129-144.

Tchebakova, N.M., R.A. Monserud, and D. Nazimova, 1994: A Siberian vegetation model based on climatic parameters. *Canadian Journal of Forest Research*, **24**, 1597-1607.

Thornley, J.H.M., D. Fowler, and M.G.R. Cannell, 1991: Terrestrial carbon storage resulting from CO_2 and nitrogen fertilization in temperate grasslands. *Plant, Cell and Environment*, **14**, 1007-1011.

Tolley, L.C. and B.R. Strain, 1985: Effects of CO_2 enrichment and water stress on gas exchange of *Liquidambar styraciflua* and *Pinus taeda* seedlings grown under different irradiance levels. *Oecologia*, **65**, 166-172.

Torrii, A., 1991: Past Fagus forest's range shifts caused by global climate change in Honshu area, Japan. In: *Transactions of the 102nd Meeting of the Japanese Forestry Society.* Nikkon Ringakkai, Meguro, Tokyo, Japan, pp. 235-237.

Trexler, M.C. and C. Haugen, 1995: *Keeping It Green: Tropical Forestry Opportunities for Mitigating Climate Change.* World Resources Institute, Washington, DC, USA, 52 pp.

UNCED (ed.), 1992: *Statement of Forest Principles.* Final text of agreements negotiated by governments at the United Nations Conference on Environment and Development (UNCED), 3-14 June 1992, Rio de Janeiro, Brazil, United Nations, Department of Public Information, New York, NY, 294 pp.

UNESCO, 1978: *Tropical Forest Ecosystems: A State-of-Knowledge Report.* UNESCO, UNEP & FAO, Paris, France, 683 pp.

van Cleve, K., R. Barney, and R. Schlentner, 1981: Evidence of temperature control of production and nutrient cycling in two interior Alaska black spruce ecosystems. *Canadian Journal of Forest Research*, **11**, 258-273.

van Cleve, K., C.T. Dyrness, L.A. Viereck, J. Fox, F.S. Chapin III, and W.C. Oechel, 1983a: Characteristics of taiga ecosystems in interior Alaska. *Bioscience*, **33**, 39-44.

van Cleve, K., L. Oliver, R. Schlentner, L.A. Viereck, and C.T. Dyrness, 1983b: Productivity and nutrient cycling in taiga forest ecosystems. *Canadian Journal of Forest Research*, **13**, 747-766.

van Cleve, K., W.C. Oechel, and J.L. Hom, 1990: Response of black spruce (*Picea mariana*) ecosystems to soil temperature modification in interior Alaska. *Canadian Journal of Forest Research*, **20**, 1530-1535.

van de Geijn, S.C., and J.A. van Veen, 1993: Implications of increased carbon dioxide levels for carbon input and turnover in soils. *Vegetatio*, **104/105**, 283-292.

Viereck, L.A., 1983: The effects of fire in black spruce ecosystems of Alaska and northern Canada. In: *The Role of Fire in Northern Circumpolar Ecosystems* [Wein, R.W. and D.A. MacLean (eds.)]. Wiley, New York, NY, pp. 201-220.

Vitousek, P.M. and R.L. Sanford, Jr., 1986: Nutrient cycling in moist tropical forest. *Annual Review of Ecology and Systematics*, **17**, 137-167.

Vitousek, P.M., 1988: Diversity and biological invasions of oceanic islands. In: *Biodiversity* [Wilson, E.O. (ed.)]. National Academy Press, Washington, DC, pp. 181-189.

Vitousek, P.M., 1994: Beyond global warming ecology and global change. *Ecology*, **75(7)**, 1861-1876.

Volney, W.J.A., 1988: Analysis of historic jack pine budworm outbreaks in the prairie provinces of Canada. *Canadian Journal of Forest Research*, **18**, 1152-1158.

Waring, R.H. and W.H. Schlesinger, 1985: *Forest Ecosystems—Concepts and Management.* Academic Press, Inc., Orlando, FL, 340 pp.

Watt, A.S., 1947: Pattern and process in the plant community. *Journal of Ecology*, **35**, 1-22.

Webb III, T. and P.J. Bartlein, 1992: Global changes during the last 3 million years: climatic controls and biotic responses. *Annual Review of Ecology and Systematics*, **23**, 141-173.

Webb, S.D., 1995: Biological implications of the middle Miocene Amazon seaway. *Science*, **269**, 361-362.

Westall, J. and W. Stumm, 1980: The hydrosphere. In: *The Handbook of Environmental Chemistry.* Vol.1, *The Natural Environment and the Biogeochemical Cycles* [Hutzinger, O. (ed.)]. Springer Verlag, New York a.o., pp. 17-49.

Wetherald, R.T. and S. Manabe, 1986: An investigation of cloud cover change in response to thermal forcing. *Climatic Change*, **8**, 5-23.

White, T.C.R., 1974: A hypothesis to explain outbreaks of looper caterpillars with special reference to population of *Selidosema suavis* in a plantation of *Pinus radiata* in New Zealand. *Oecologia*, **16**, 279-301.

Whitehead, D., J.R. Leathwick, and J.F.F. Hobbs, 1993: How will New Zealand's forests respond to climate change? Potential changes in response to increasing temperature. *New Zealand Journal of Forestry Science*, **22**, 39-53.

Whitmore, T.C., 1974: *Change with Time and the Role of Cyclones in Tropical Rainforests on Kolomangara, Solomon Islands.* Commonwealth Forestry Institute Paper No. 46, University of Oxford, Oxford, UK, 78 pp.

Whitmore, T.C., 1984: *Tropical Rain Forests of the Far East.* Clarendon Press, Oxford, UK, 2nd. ed., 352 pp.

Wilson, E.O., 1985: The biological diversity crisis: a challenge to science. *Issues in Science and Technology*, **2**, 20-29.

Wilson, E.O. (ed.), 1988: *Biodiversity.* National Academy Press, Washington, DC, 521 pp.

Wilson, E.O., 1989: Threats to biodiversity. *Scientific American*, **261(3)**, 60-66.

Wilson, E.O., 1992: *The Diversity of Life.* Belknap Press of Harvard University Press, Cambridge, MA, 424 pp.

Wilson, K.G., R.E. Stinner, and R.L. Rabb, 1982: Effects of temperature, relative humidity, and host plant on larval survival of the Mexican bean beetle *Epilachna varivetis* Mulsant. *Environmental Entomology*, **11**, 121-126.

Winjum, J.K., R.K. Dixon, and P.E. Schroeder, 1993: Forest management and carbon storage: an analysis of 12 key forest nations. *Water, Air, and Soil Pollution*, **70**, 239-257.

Woodward, F.I., 1987: *Climate and Plant Distribution.* Cambridge University Press, Cambridge a.o., 174 pp.

Woodwell, G.M., 1987: Forests and climate: surprises in store. *Oceanus*, **29**, 71-75.

Worrell, R., 1987: Predicting the productivity of Sitka spruce on upland sites in northern Britain. *Bulletin Forestry Commission* **(No. 72)**, pp. 1-12.

Wullschleger, S.D., W.M. Post, and A.W. King, 1995: On the potential for a CO_2 fertilization effect in forests: estimates of the biotic growth factor, based on 58 controlled-exposure studies. In: *Biospheric Feedbacks in the Global Climate System—Will the Warming Feed the Warming?* [Woodwell, G.M. and F.T. Mackenzie (eds.)]. Oxford University Press, New York, NY, and Oxford, UK, pp. 85-107.

Zuidema, G., G.J. van den Born, J. Alcamo, and G.J.J. Kreileman, 1994: Determining the potential distribution of vegetation, crops and agricultural productivity. *Water, Air, and Soil Pollution*, **76(1/2)**, 163-198.

2

Rangelands in a Changing Climate: Impacts, Adaptations, and Mitigation

BARBARA ALLEN-DIAZ, USA

Principal Lead Authors:
F.S. Chapin, USA; S. Diaz, Argentina; M. Howden, Australia; J. Puigdefábregas, Spain; M. Stafford Smith, Australia

Lead Authors:
T. Benning, USA; F. Bryant, USA; B. Campbell, New Zealand; J. duToit, Zimbabwe; K. Galvin, USA; E. Holland, USA; L. Joyce, USA; A.K. Knapp, USA; P. Matson, USA; R. Miller, USA; D. Ojima, USA; W. Polley, USA; T. Seastedt, USA; A. Suarez, Cuba; T. Svejcar, USA; C. Wessman, USA

Contributing Authors:
W.N. Ekaya, Kenya; J. Ellis, USA; L.D. Incoll, UK; J. Kinyamario, Kenya; C. Magadza, Zimbabwe; T. Oikawa, Japan; O. Sala, Argentina; C. Scoppa, Argentina; N. Maceira, Argentina; R. Rodriguez, Argentina

CONTENTS

EXECUTIVE SUMMARY

Rangelands (i.e., grasslands, shrublands, savannas, hot and cold deserts, and tundra) occupy 51% of the terrestrial land surface, contain about 36% of the world's total carbon in above and belowground biomass (Solomon *et al.*, 1993), include a large number of economically important species and ecotypes, and sustain millions of people. Rangelands support approximately 50% of the world's livestock and provide forage for both domestic and wildlife populations (Briske and Heitschmidt, 1991).

Doubled-CO_2 climate-change scenarios provided for IPCC Working Group II (Greco *et al.*, 1994) were used to analyze the regional impacts of altered climate regimes on rangeland resources. Results indicate:

- Rangeland vegetation is found where mean precipitation, temperature, elevation, and latitude interact to provide sites suitable for herbaceous species, shrubs, and/or open stands of trees. However, carbon cycling, productivity, and species composition in any region is directly related to the highly variable amounts and seasonal distribution of precipitation and is only secondarily controlled by other climate variables. Small changes in the frequency of extreme events may have disproportionate effects on what managers must cope with in rangeland systems (High Confidence).
- CO_2 increases are likely to result in reductions of forage quality and palatability because of increasing carbon to nitrogen ratios. These effects will be more common in lower-latitude rangelands where low nutritional value is already a chronic problem (Medium Confidence).
- Boundaries between rangelands and other biomes are likely to change with changes of climate: directly, through climate-driven changes in species composition, and indirectly, through changes in wildfire regimes, opportunistic cultivation, or agricultural release of the less-arid margins of the rangeland territory. These effects will be more common in the temperate rangelands (High Confidence).
- Climatic warming may cause tundra to become a net source of carbon dioxide. Temperature increases in the tundra will reduce species richness, especially of insect-pollinated forbs (Medium Confidence).
- The extensive use of rangelands suggests that strategies to reduce greenhouse gas emissions must focus on maintaining or increasing carbon sequestration through better soil and vegetation management and reducing methane by altering animal production systems (High Confidence).
- Human adaptive strategies for coping with climate-change effects on rangelands include the preservation of an extensive spatial scale in management units and the development of viable marketing systems capable of absorbing opportunistic variations in number and/or kind and class of animals (Medium Confidence).

2.1. Introduction: Description of the World's Rangeland Resources

Rangelands occupy approximately 51% of the terrestrial surface of the Earth, or 68.5 million km^2 (Lean *et al.*, 1990; Prentice *et al.*, 1992). Rangelands include unimproved grasslands, shrublands, savannas, and hot and cold deserts. For the purposes of this assessment, tundra and improved pastures are included here, while hot deserts are described in Chapter 3.

The primary use of rangelands has been and is for grazing by domestic livestock and wildlife. Rangelands support fifty percent of the world's livestock (WRI, 1992) and provide forage for both domestic and wild animal consumption (Briske and Heitschmidt, 1991). Rangeland management systems vary from nomadic pastoralism to subsistence farming to commercial ranching. Markets are largely externally driven, with extensive social systems in contrast to commercial agriculture. Overgrazing (the result of animal consumption exceeding carry capacity) has been and is common to rangelands throughout the world (WRI, 1992).

Most of the forage consumed by domestic livestock (cattle, sheep, goats, buffalo, and camels) is supplied by rangelands in Africa and Asia (Table 2-1). Other regions of the world support many fewer animals from rangeland forage resources. Numbers and kinds of domestic and wild grazing animals vary by region.

The distribution of the world's rangelands falls on continua from wet to dry, fertile to infertile, and hot to cold (Figure 2-1). Holdridge (1947) arrayed the world's vegetation into life zones based on mean annual growing season temperature (°C), average annual precipitation, an index of precipitation divided by potential evapotranspiration (PET), and latitude (Figure 2-2). Rangelands occupy areas where combinations of temperature, precipitation, PET, and latitude prevent occupation by dense stands of trees. Figure 2.2 is a good display of the types of ecosystems covered in this chapter, although the hot deserts are described in more detail in Chapter 3. Holdridge and others (Walter, 1973; Bailey, 1989; Prentice *et al.*, 1992) recognized that seasonality of precipitation, soil factors, herbivory, and fire are additional important factors in determining local vegetation composition and structure. Interactions between plant-available moisture and available nutrients also are crucial to rangeland structure and composition (Walker, 1993).

Rangeland ecosystems are diverse (Table 2-2), and stores of carbon both above and below ground vary by ecosystem type (Solomon *et al.*, 1993). Carbon (C) stored in the world's rangeland vegetation has been estimated by the BIOME 1.1 model at 749.7 Gt (Solomon *et al.*, 1993), although annual net primary productivity varies widely (Barbour *et al.*, 1987). Carbon in the world's rangeland soils has been estimated at 591.6 Gt, or 44% of the world's total soil carbon (Solomon *et al.*, 1993). Preindustrial levels of soil C (to 20 cm depth) in grasslands alone have been estimated at approximately 96 Gt (Ojima *et al.*, 1993b). Current soil C stocks for these grasslands (based on Ojima *et al.*, 1993b) are estimated at 81–164 Gt, based on land-cover estimates of Bailey (1989) minus estimated land under cultivation in these regions provided by Prentice *et al.* (1992).

In rangelands, the amount and timing of precipitation are the major determinants of community structure and function. Other driving forces that determine plant community composition, distribution, and productivity include temperature, fire, soil type, and herbivory. These driving forces prevent species senescence and allow for periodic rejuvenation by eliminating aboveground biomass and organic debris and thereby liberating nutrients. Most rangelands do not exist in a state of equilibrium, nor do they exhibit linear successional trends; rather, they fluctuate from one state to another depending on rain, fire, grazing, and other anthropogenic factors (Westoby *et al.*, 1989; Stafford Smith and Pickup, 1993; Hobbs *et al.*, 1991; Stafford Smith, 1994). Most rangelands are nonequilibrial within the range of climatic conditions that they currently experience. Underlying trends in climate change may cause some compositional states to become more likely (e.g., the

Table 2-1: *World livestock and their dependence on rangeland resources. Percent of feed from rangeland is an estimate based on figures for the amount of grain fed to livestock; the assumption is that the rest of the feed source supporting the animals is from rangeland [World Resources, 1992–1993 (UNEP/UNDP)].*

	Land Resources (1,000 hectares)		**Animal Units (1,000 animals)**			
Region	Permanent Pasture	Woodland or Other	Cattle	Sheep, Goats	Buffalo, Camel	Feed from Rangeland
Africa	890,889	1,200,565	183,715	372,038	16,877	82%
North, Central America	368,631	779,838	161,050	33,566	9	35%
South America	447,863	237,792	262,254	134,759	1,190	54%
Asia	694,251	1,043,666	389,730	640,938	138,814	84%
Europe	83,177	92,524	124,900	166,023	370	45%
Former Soviet Union	371,500	678,829	118,767	145,588	707	45%
Oceania	436,622	199,523	30,862	223,838	ND	50%

ND = No data.

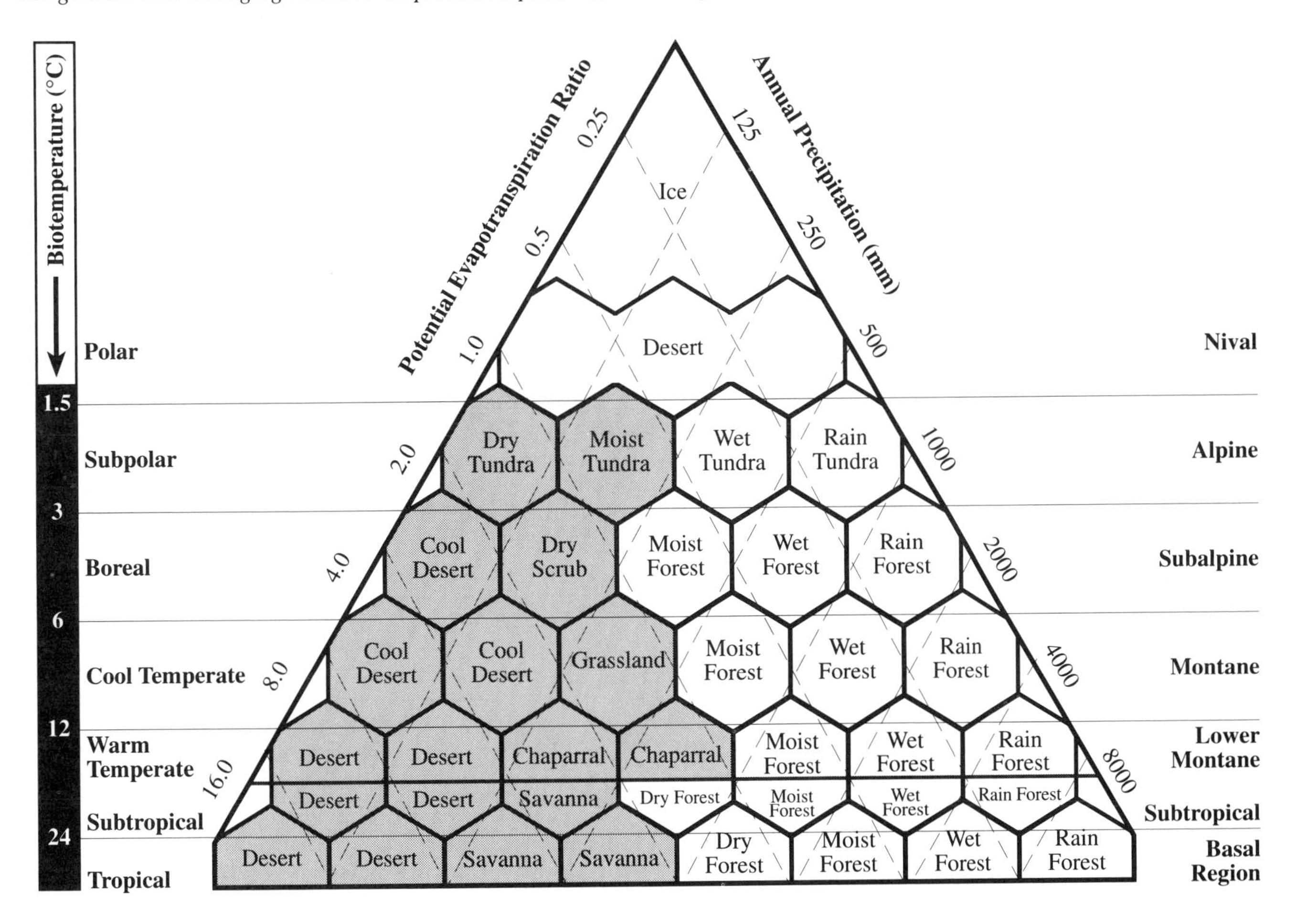

Figure 2-1. Distribution of rangeland types using the Holdridge classification system (Holdridge, 1947) as modified by Cramer and Leemans (1993). Biotemperature is the mean value of all daily mean temperatures above 0°C, divided by 365. The heavy line between Warm Temperate and Subtropical zones represents the boundary for risk of a killing frost. The demand of plants for moisture is represented by mean annual precipitation and potential evapotranspiration ratio. The shaded areas are rangeland types discussed in this chapter.

shrub invasion process is hypothesized to be linked to climate shifts or to grazing in the United States). However, in managed rangeland systems, management is as likely, at present, to maintain the systems in the desired state. In many areas, however, poor management of rangelands enhances the changes likely to be caused by climatic/atmospheric change—for example, by increasing the ratio of woody to herbaceous vegetation.

2.2. Climate Variables

2.2.1. Precipitation and Temperature

Among climatic variables, those related to water availability and water balance appear most influential in controlling the geographic distributions of rangeland vegetation (Stephenson, 1990) and production (Rosenzweig, 1968; Webb *et al.*, 1983; Sala *et al.*, 1988). There is a strong linear relationship between aboveground net primary productivity (ANPP) and annual precipitation in rangeland ecosystems (Rutherford, 1980; Webb *et al.*, 1983; Le Houerou, 1984; Sala *et al.*, 1988; Scholes, 1993), with typically 90% of the variance in primary production accounted for by annual precipitation (Le Houerou and Hoste, 1977; Webb *et al.*, 1978; Walter, 1979; Foran *et al.*, 1982; Sala *et al.*, 1988).

Temperature does not account for a significant fraction of the ANPP variability among years in a particular region (Sala *et al.*, 1988; Lauenroth and Sala, 1992; Sala *et al.*, 1992). However, temperature appears to be the major climatic variable controlling the process of decomposition in grasslands and other rangeland types (Jenkinson, 1977; Meentemeyer, 1984; Burke *et al.*, 1989). It has been shown experimentally that decomposition rates increase rapidly with temperature, provided that there is enough water (Meentemeyer, 1978, 1984; Schimel, 1988; Holland *et al.*, 1992; Ojima *et al.*, 1993a; Parton *et al.*, 1993; Townsend *et al.*, 1992; Burke *et al.*, 1995a).

Temperature also affects C_3/C_4 species composition of grasslands (Sims, 1988) because of differences in optimal growing conditions (see Chapter A). The distribution (Terri and Stowe, 1976; Tieszen *et al.*, 1979) and seasonal activities of C_3 and C_4 grasses (Kemp and Williams, 1980; Hicks *et al.*, 1990) often are highly correlated with temperature. This is not necessarily true for dicotolydenous plants such as legumes, however (Ehleringer and Field, 1993).

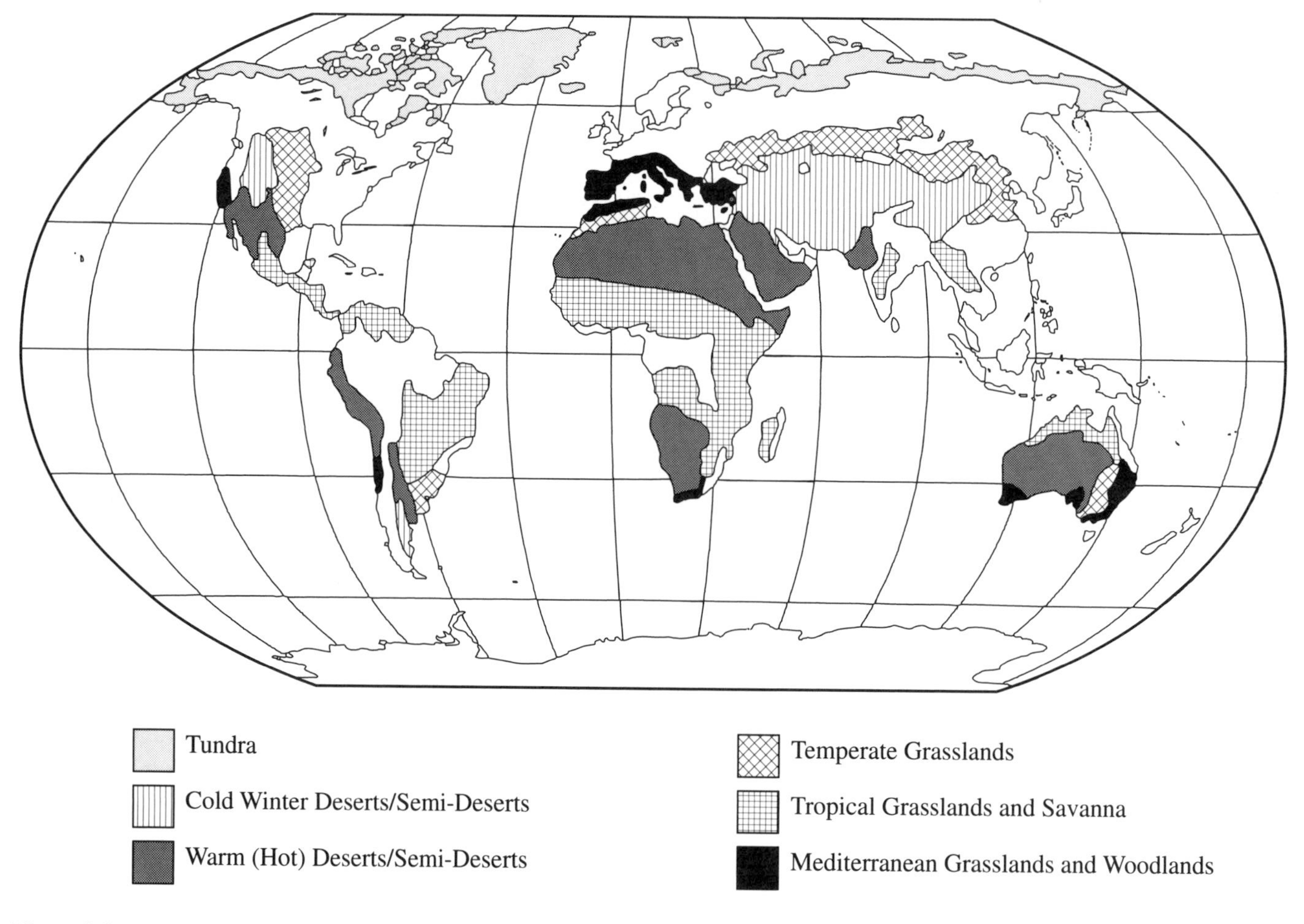

Figure 2-2: Geographic distribution of the world's rangelands (Lean *et al.*, 1990; Bailey, 1989; Prentice *et al.*, 1992).

2.2.2. *Effects of Altered Atmospheric CO_2 Concentrations on Rangelands*

Most research on rangeland plant response to elevated CO_2 has focused on responses of individual plants (Bazzaz and McConnaughay, 1992; Idso and Idso, 1994). A wide range of plant parameters have been measured (see Chapter A), and as a rule, studies show that increasing CO_2 levels increase photosynthesis, water-use efficiency (WUE), above- and belowground productivity, and nitrogen-use efficiency (NUE) and decrease stomatal conductance, transpiration, and whole-plant respiration (Wong *et al.*, 1979; Kimball, 1983; Webb *et al.*, 1983; Strain and Cure, 1985; Morison, 1987; Smith *et al.*, 1987; Idso, 1989; Bazzaz, 1990; Kimball *et al.*, 1990; Bazzaz and McConnaughay, 1992; Kimball *et al.*, 1993; Rozema *et al.*, 1993; Zak *et al.*, 1993; Gifford, 1994; Idso and Idso, 1994; Rogers *et al.*, 1994). For example, controlled-environment studies have repeatedly demonstrated that growth and net photosynthesis of C_3 plants, including woody species such as *Prosopsis glandulosa* (Polley *et al.*, 1994), are highly responsive to CO_2 when it is increased over concentrations representative of the recent and prehistoric past (Gifford, 1977; Baker *et al.*, 1990; Allen *et al.*, 1991; Polley *et al.*, 1993a).

Responses to CO_2 increase are not necessarily consistent across species or under all environmental conditions. Two processes—increased photosynthesis and reduced transpiration—typically contribute to the positive influence of CO_2 on plant WUE. Atmospheric CO_2 directly affects the coupling between climatic water balance and vegetation by altering the efficiency with which plants use water. The effect of CO_2 on water use may be more important in water-limited rangelands than in moist grassland systems or irrigated agricultural land. Elevated CO_2 can result in a slower depletion of soil water in drought-stricken rangeland grasses where there is no change in leaf area due to limitations such as nitrogen availability (Owensby *et al.*, 1993; Campbell *et al.*, 1995a).

Plant responses under variable environmental conditions or in competitive settings do not necessarily match responses under steady-state conditions and/or in the absence of competition. That is, results from studies in pots cannot automatically be assumed relevant to plants in native ecosystems (Nie *et al.*, 1992; Campbell and Grime, 1993; Owensby *et al.*, 1993; Campbell *et al.*, 1995b). Competitive outcome in response to elevated CO_2 can be difficult to predict because of the feedback between litter quality, soil, and plant production (Schimel *et al.*, 1991; Bazzaz and McConnaughay, 1992; Ojima *et al.*, 1993b; Newton *et al.*, 1995). In natural ecosystems, nutrient limitation may influence the response of individual species (Billings *et al.*, 1984; Wedin and Tilman, 1990;

Table 2-2: *Major rangeland vegetation types.*

Type	Area (10^6 km^2)	Major Characteristics	General Climate	Current C Stores (Gt) Above Ground	Below Ground
Cool Semi-Desert	4.91	Sparse shrubs dominate	Cold winters with snow, hard frosts; 50–250 mm annual precipitation; meets <28% of demand	3.0	30.7
Cool Grass Shrub	7.17	C_3 grasses dominate, with or without shrubs	Strong summer/winter peaks in precipitation; averages 400–1600 mm; meets 28–65% of demand	5.8	70.9
Warm Grass Shrub	9.84	C_4 grasses dominate	Strong seasonality in precipitation; dry season of ~5 months; rainfall meets 18–28% of demand	13.0	87.2
Hot Desert	20.70	Shrubs less than 100% cover	Mild winters; 50–250 mm precipitation; rainfall meets <18% of demand	6.3	63.1
Tropical Savanna/ Dry Woodland	17.18	C_4 grasses/sedges dominate, with significant discontinuous cover of trees and shrubs	Strong seasonality in precipitation; rainfall meets 45–80% of demand	78.6	124.8
Xerophytic Woods/ Shrubland	10.63	Widely variable woodlands, sclerophyllous shrubs, and annual grasses	Prolonged summer drought of 2–11 months; rainfall meets 28–45% of demand	45.5	80.9
Tundra	11.61	Dominated by herbaceous species	Temperatures generally <10°C; variable precipitation	5.9	134.0

Conroy, 1992; Comins and McMurtrie, 1993), and complex feedbacks between plant and soil microorganisms may arise (Diaz *et al.*, 1993). In some studies, soil microbial respiration rate is greatly enhanced under lower CO_2 conditions and reduced under higher CO_2 conditions in the laboratory (Koizumi *et al.*, 1991) and in the field (Nakadai *et al.*, 1993). Experimental doubling of CO_2 concentration has relatively little effect on plant growth in Arctic tundra, presumably because of constraints of low nutrient supply (Tissue and Oechel, 1987; Grulke *et al.*, 1990). Increased levels of carbon dioxide may increase individual plant production, but the effect on whole-ecosystem production is less clear, as resource augmentation has been shown to decrease species diversity and differentially affect individual species growth (favoring pioneer species) (Wedin and Tilman, 1990; Bazzaz, 1993).

Plants grown in CO_2-enriched environments tend to have reduced mineral concentrations (except phosphorus) relative to those grown at ambient CO_2 levels (Conroy, 1992; Overdieck, 1993; Owensby *et al.*, 1993). Decreases in nitrogen (N) concentration with elevated CO_2 have been demonstrated with a wide array of species (Conroy, 1992; Coleman *et al.*, 1993; Diaz *et al.*, 1993; Overdieck, 1993). These changes in mineral concentration in plants make them less palatable to herbivores and may even influence herbivore health and production (Fajer *et al.*, 1991). Increased concentrations of carbon-based, nonpalatable secondary compounds could also deter herbivory, although the evidence for this process is weaker (Ayres, 1993; Diaz, 1995). The influence of elevated CO_2 on nutrient cycling must also be considered in light of potential impacts on litter inputs and litter quality. Reduced nitrogen concentrations may reduce the rate of decomposition, thereby slowing nutrient cycling and energy flow (Bazzaz and Fajer, 1992; Schimel *et al.*, 1994). In addition, surplus C may be converted to secondary compounds such as tannins, which decompose slowly.

According to controlled-environment studies, C_3 plants tend to be more responsive to CO_2 than are C_4 ones (see Chapter A). This effect is not always sustained at the community level, however. For example, Owensby *et al.* (1993) report that production of a C_4-dominated tallgrass prairie increased when exposed to elevated CO_2, apparently as a result of higher water-use efficiency, but other studies have detected no effect on production, flowering, or growth rates (Polley *et al.*, 1994; Korner and Miglietta, 1994). Data from grasslands ungrazed for 100 years in Rothamsted, UK, suggest that rising CO_2 does not affect herbage yield (Jenkinson *et al.*, 1994).

Differential species responses to CO_2 could also have implications for species assemblages and plant-community dynamics. Therefore, the outcome of multispecies experiments is not readily predictable from the behavior of individual species (Bazzaz, 1992, 1993; Korner, 1993; Diaz, 1995).

2.2.3. *Other Important Trace Gases*

The atmospheric concentrations of a number of important trace gases apart from CO_2 have increased (Cicerone and Oremland, 1988; IPCC, 1990, 1992), but their effects on rangeland ecosystems are not clear because they do not directly affect the rates of ecosystem processes or community dynamics (Mooney *et al.*, 1987).

A number of factors influence CH_4, NO, and N_2O emissions from rangeland ecosystems: deposition of nitrogen, nitrogen fertilization, changing climate, elevated CO_2, ozone influences on plant production, changing plant community composition, changing animal populations, and changing land use (Schimel *et al.*, 1990, 1991; Ojima *et al.*, 1993a; Parton *et al.*, 1994). Many of these factors are likely to respond non-linearly to climate changes. Changing precipitation and temperature directly affect the production of both NO and N_2O by regulating microbial activity (Firestone and Davidson, 1989). With abundant substrate, increasing temperatures generally increase rates of production, but changing soil moisture could decrease the rates of production. With no change in diet, more animals would result in greater N_2O production (Bouwman *et al.*, 1995). However, all of the other factors listed are likely to change emissions through regulation of the nitrogen cycle. The greatest increase in emissions will occur when nitrogen supply exceeds plant demand, because this directly increases the nitrogenous substrates available for volatilization and leaching.

Many of the factors that influence nitrogen gas emissions also influence net methane emissions, but the influences are not necessarily in the same direction. Conversion of forests to pastures, fertilization, and, probably, nitrogen deposition decrease soil methane consumption (Mosier *et al.*, 1991; Keller *et al.*, 1993; Ojima *et al.*, 1993a). Any factor that enhances primary productivity, such as elevated CO_2, N deposition, or fertilization, is likely to increase methane emissions because methane production is limited by substrate availability (Whiting and Chanton, 1993; Dacey *et al.*, 1994; Valentine *et al.*, 1994; King and Schnell, 1994; Hutsch *et al.*, 1995). The possible extent of this enhancement is not yet clear. On the other hand, factors that decrease primary production (and thus CO_2 uptake) or pH, like ozone and acid rain, could have the opposite effect (Yavitt *et al.*, 1993).

Increases in animal populations capable of emitting CH_4 via enteric fermentation [currently estimated to be 80 Gt/yr (Cicerone and Oremland, 1988)] are likely to increase methane production, but this could be partially offset if diets are improved through management (Lodman *et al.*, 1993; Ward *et al.*, 1993). Methane emissions from rangeland ruminants are largely a function of intake (generally about 80% of the variance), with feed quality accounting for about 5%. Hence, increasing intake is likely to dominate the impacts of increased feed quality. However, in rangelands where feed quality is low, there is often an intake restriction (Poppi *et al.*, 1981); hence, if CO_2 and temperature reduce feed quality, there is likely to be a tradeoff betweeen reduced intake and increased emissions per unit feed. What will happen is that the energy balance in the animal over the maintenance level will decrease, and the productivity of the animal will decrease as a consequence—and thus there will be increased methane emissions per unit product.

Global estimates of NO emissions from soils range from 4–20 Mt of nitrogen annually, representing between 16 and 20% of the global budget (Davidson, 1991; Yienger and Levy, in press). Tropical and subtropical savannas contribute almost 40% of the NO produced in soils globally, with temperate grasslands contributing another 3% (Davidson, 1991). There are no published estimates of NO production from tundras or deserts. Temperate grasslands and tropical savannas each contribute between 10 and 25% of the global N_2O budget (0.5–2.0 Mt N/yr) (Prather *et al.*, 1995).

Tropical areas that have been converted from forests to pastures due to land-use change contribute a variable proportion to both the global NO and N_2O budgets, with elevated emissions for many years following conversion, dropping to below original emissions after 20 years (Keller *et al.*, 1993). Emissions of N_2O from cattle and feedlots contribute an additional 0.2 to 0.5 Mt N annually. Assuming that N_2O emissions from pasture conversion represents as much as 50% of the emission from cultivated soils, grasslands and savannas may contribute as much as 42% of the NO emitted globally (Prather *et al.*, 1995).

Ammonia (NH_3) is a third nitrogen gas species exchanged with the atmosphere. Ammonia plays an important role in aerosol formation, which in turn influences radiative transfer and contributes substantially to N deposition. Approximately 50% of the NH_3 emitted to the atmosphere is produced by livestock, much of which is raised on rangelands or in feedlots, with an additional contribution by emissions from fertilizers (Dentener and Crutzen, 1994).

2.3. Ecosystem Variables

2.3.1. *Herbivory*

Grazing and browsing by domestic and wild animals, including invertebrates, alters the appearance, productivity, and composition of rangeland plant communities. Herbivores mediate species abundance and diversity by differential use of plants commonly susceptible to defoliation (Archer and Smeins, 1991). The role of herbivory and subsequent impacts on rangeland structure and function depends in part on the evolutionary history of grazing in an ecosystem (Payne and Bryant, 1994). Herbivory is a disturbance whether the ecosystem has a short (Milchunas *et al.*, 1988; Milchunas and Lauenroth, 1993) or long (Hobbs and Huenneke, 1992) evolutionary history of grazing.

Plant species that are well-adapted to the prevailing climate, soils, and topography may be competitive dominants of a plant community when herbivore populations are low (Archer, 1994) but may become subordinates or even face extinction as levels of herbivory increase (Archer, 1992). Grazing establishes a new set of boundary conditions for dominance. Species with competitive advantage in a moderately grazed system are generally those that are palatable and have a capacity to sprout rapidly after being eaten. Nonpalatability becomes a competitive advantage only when grazing is excessive. For example, excessive herbivory has been shown to favor unpalatable woody species over graminoids (Archer, 1994). Examples are available from North America (Archer, 1994), Africa (van Vegten, 1983; Skarpe, 1990), northern Australia (Lonsdale and Braithwaite, 1988), and South America (Morello *et al.*, 1971; Schofield and Bucher, 1986). Browsing herbivores play a key role in maintaining grasslands, meadows, and savannas, and they can change closed woodland, thickets, or heathland into open, grass-dominated systems (Sinclair and Norton-Griffiths, 1979; Berdowski, 1987).

The extent of influence that herbivory has on the structure and function of a range system depends strongly on interactions with factors such as previous grazing history, nutrient turnover rate, fire frequency and intensity, soil moisture, and soil compaction. Studies of the interactions of vertebrate grazing with these and other factors on rangelands have illustrated the tremendous influence such interactions have on these systems (Milchunas *et al.*, 1988; Holland and Detling, 1990; Hobbs *et al.*, 1991; Holland *et al.*, 1992; Seastedt *et al.*, 1991; Parton *et al.*, 1994; Dyer *et al.*, 1993; Turner *et al.*, 1993). Invertebrate grazing, a potentially more ubiquitous and chronic factor, has received less attention but is equally important to the evolutionary and ecological organization of these systems (Evans and Seastedt, in press).

Climate change and CO_2 may also interact with grazing effects. Drought favors the concentration of herbivores at sites with more reliable water supply and hence leads to overgrazing in these locations (Heitschmidt and Stuth, 1991; Heady and Childs, 1994). Increasing CO_2 will likely result in lower forage quality because of higher C:N ratios and higher concentration of unpalatable and/or toxic compounds in plants (Fajer *et al.* 1989, 1991; Johnson and Lincoln, 1991). Forage consumption may increase up to a point as a result, because herbivores will need to eat more to meet nutritional demands (Fajer *et al.*, 1991)—although the loss of palatability may reduce grazing pressure on certain species. At present, it is possible to say only that the overall carrying capacity for herbivores may increase, decrease, or remain constant, depending on the balance between eventual increases in productivity and decreases in nutritional value.

2.3.2. Fire

Fire has been a factor in the evolution of grasslands and many other rangelands. Annual net primary productivity is strongly influenced by both fire and climate (Hulbert, 1969; Old, 1969; Towne and Owensby, 1984; Abrams *et al.*, 1986; Sala *et al.*, 1988). Fire alters the structure of grassland vegetation (Knapp, 1984, 1985) and affects nutrient cycling (Raison, 1979; Kucera, 1981; Risser and Parton, 1982; Ojima *et al.*, 1990; Ojima *et al.*, 1994a). Aboveground biomass production in temperate, ungrazed grasslands generally increases following fire during years with normal precipitation (Towne and Owensby, 1984). This increased production has been attributed to release of readily available nitrogen and phosphorus from plant material, increased nitrogen mineralization rates, enhanced nitrogen fixation, and altered microclimatic conditions, including improved light availability (Daubenmire, 1968; Old, 1969; Raison, 1979; Biederbeck *et al.*, 1980; Knapp and Seastedt, 1986; Hulbert, 1988).

By making N and other nutrients more available to soil microorganisms, fire may result in enhanced emissions of N_2O from soil (Anderson *et al.*, 1988). Bouwman (1993, 1994) calculated that from the approximately 12 Mt N/yr remaining on the ground after burning (Crutzen and Andreae, 1990), 20% is volatilized as NH_3, and 1% (0.1 Mt/yr) of the remaining N is emitted to the atmosphere as N_2O (Carras *et al.*, 1994). An estimated 10% of this N is amenable to mitigation. Most of these emissions are short-term effects, which occur directly following fire.

Grasslands subject to annual fires may suffer several long-term effects, such as decreasing soil organic matter, changes in species composition, and long-term loss of nitrogen through volatilization and immobilization (Daubenmire, 1968; Old, 1969; Biederbeck *et al.*, 1980; Kucera, 1981; Risser *et al.*, 1981; Ojima *et al.*, 1990; Ojima, 1994a). Nitrogen limitation is a common characteristic of fire-maintained grassland and rangeland ecosystems (Seastedt *et al.*, 1991).

The influence of climate change on large-scale events, such as wildfire, can only be inferred. However, the predicted increase in climate variability has led several authors to suggest that the frequency and severity of wildfires will increase in grassland/rangeland settings (Graetz *et al.*, 1988; Moore, 1990; Ottichilo *et al.*, 1991; Cherfas, 1992; Torn and Fried, 1992).

2.3.3. Land Use and Management

Land use differs from other factors influencing rangelands because of the inherent involvement of human socioeconomic factors. Human social/economic systems may produce more change in actively managed rangeland ecosystems than any other forces of global change (e.g., Burke *et al.*, 1995b). Model experiments to investigate the feedbacks between land-use change and the climate system suggest strong links (e.g., Dickinson and Henderson-Sellers, 1988).

The most common land-use transformation has been conversion of rangeland to cropland. The pattern of settlement has had important implications for rangeland transformations (Reibsame, 1990). Conventional agriculture has essentially mined many previous rangelands by enhancing decomposition, reducing soil organic matter, and increasing leaching and erosion (e.g., Bauer and Black, 1981; Burke *et al.*, 1989). The net result is that soils must now receive heavy doses of fertilizer and pH buffers to produce acceptable yields in industrialized countries.

Land use can dramatically alter C storage and fluxes, and modification of land-use practices can greatly influence the net flux of soil C, changing rangelands from a source to a sink for atmospheric C. Although conversion to small-grain cropping is not a large deviation from natural grassland ecosystems relative to albedo (except in periods of bare soil), plant stature, growing season, and energy fluxes, conversion of grasslands to cropland can result in a rapid decline in C stores. Up to 50% of soil C can be lost as a result of conversion from native rangeland to crops (Haas *et al.*, 1957; Cole *et al.*, 1989, 1990; Burke *et al.*, 1989).

For example, the potential influences of changing land management and climate-change effects on C storage of grasslands were assessed using the Century model (Ojima *et al.*, 1993b; Parton *et al.*, 1995). Future soil C pools were calculated for a "regressive management" scenario (i.e., removal of 50% of the aboveground biomass during grazed months) and a "sustainable" scenario [i.e., using moderate grazing and burning regimes as specified by Ojima *et al.* (1993b)]. The impact of "regressive" land management resulted in a loss of soil C in all regions after 50 years, with the largest losses in the warm grasslands. The total net loss relative to current condition was projected at 10.8 Gt. When this regressive management is compared to the sustainable management system (i.e., light grazing), the net difference is 37.6 Gt of soil C.

2.4. Issues of Scale

Scale is a particularly important issue in rangelands because of the very large size of management units, which is often caused by low productivity. Thus, rangeland management units incorporate spatial heterogeneity, instead of partitioning it as is usually the case in more intensive agricultural systems (Stafford Smith and Pickup, 1993). Most grasslands and rangelands are in climatic regions that are subject to frequent alterations in essential resource availability. The superimposition of management practices such as fire and grazing adds further variation in resource availability over time. Rangelands are characterized by more complex structure and function than other biomes because controlling factors (i.e., precipitation, energy, nutrients) often fluctuate rapidly over short time periods (Seastedt and Knapp, 1993). This variability in ecosystem behavior will lead to complex responses to global change.

The large areas of rangelands and the numbers of livestock they support play significant roles in biogeochemical and climate feedbacks. Response to climate change hinges on the variable scales at which these systems are used, the spatial heterogeneity incorporated in their management units, and the potentially large number of species that may interact in the primary production system (Stafford Smith, 1993). Limitations on production processes occur diurnally to seasonally through water and nutrient availability and management practices (including grazing pressure and distribution, fire, fertilizer use, and water management). Feedback effects on soils and vegetation composition,

Box 2-1. Impacts of Climate Change on Pastoral Peoples: The Example of Africa

Major shifts in climate—from relatively wet periods to dry periods or droughts—are common in the rangelands of Africa. Long-term rainfall records and other evidence from Africa's rangeland areas show that shifting climate dynamics have prevailed there for at least the last 10,000 years (Nicholson, 1983; Hulme, 1990). Traditional African livestock husbandry practices have been greatly influenced by the highly variable environment (Ellis and Swift, 1988; Behnke *et al.*, 1993; Ellis and Galvin, 1994). Because the economy depends directly and indirectly on livestock products, pastoralists have developed multiple coping mechanisms. Pastoralists cope with climate variability by keeping diverse species of livestock, by temporary emigration, by maintaining economic diversity, and even by allocating seasonal and drought-induced nutritional stress among stronger community members (Coughenour *et al.*, 1985; Galvin, 1992; Galvin *et al.*, 1994). This system requires extensive land area and external markets.

The pastoralists' stabilizing strategies have become constrained in this century primarily as a result of an increasing human population in conjunction with stable or declining livestock populations and decreasing land area. Although much recent literature suggests that livestock populations are on the increase, abundant data for the arid and semi-arid zones show otherwise. East African pastoralists have been unable to expand livestock holdings due to such factors as disease epidemics, recurring droughts, and intertribal raiding (Sandford, 1983; Homewood and Rodgers, 1984; O'Leary, 1984; Arhem, 1985; Sperling, 1987; McCabe, 1990). In addition, dry-season ranges with more reliable water supply have been lost to colonial and African agriculturalists, to game parks, and to game conservation areas (Campbell, 1984; Bekure *et al.*, 1987; Homewood *et al.*, 1987; Little, 1987). Pastoralists also have taken up agriculture in an effort to meet their increasing food demands. Thus, with increasing population pressure on a declining resource base, there is an expanding reliance on the marketing of animals. However, African livestock market infrastructure and stability is still in an early stage of development.

Persistence of the pastoralists' system in the arid and semi-arid ecosystems of Africa will require a clear understanding of the nonequilibrium dynamics of these systems and development of additional coping strategies.

including carbon and nitrogen sequestration, are thus influenced at seasonal, annual, and decadal temporal scales, but are experienced at landscape spatial scales.

2.5. Extreme Events

Increased frequency and magnitude of extreme events is often mentioned as a potential characteristic of future global climate (Easterling, 1990). Because extremes drive rangeland systems (Griffin and Friedel, 1985; Westoby *et al.*, 1989), small changes in the frequency of extreme events may have a disproportionate effect on what management must cope with in these systems. For example, the structure of long-lived perennial communities may change drastically if the frequency of extremes increases significantly, because seedling establishment and mortality of these plants are highly sensitive to extremes (Graetz *et al.*, 1988). Both the stability of forage supply and the balance between temperate and subtropical species are largely controlled by the frequency of extreme climatic events and thus are subject to change in a CO_2-warmed world (Campbell and Grime, 1993). Individual events (e.g., a major rainfall event that recharges deep moisture stores) and extended periods of above- or below-average temperature or rainfall are both likely to be significant (Stafford Smith and Morton, 1990).

Rangelands are systems adapted to a wide range in climate, fire frequencies, and grazing intensities, but anthropogenic fragmentation has already affected rangeland systems such that many systems may be vulnerable to further pressures (Archer, 1994). With the addition of climate change to existing impacts, rangelands may be more susceptible to extreme events such as drought, 100-year floods, and insect outbreaks.

2.6. Boundary Changes

Boundaries among rangeland vegetation types and between rangeland systems and other biomes are likely to change with changes in climate. Shifting boundaries between plant community types (the ecotone), potentially in response to past changes in climate, are well documented in the fossil record (Solomon and Shugart, 1993). Significant changes in the distribution of grasslands and arid lands worldwide in response to climate change are suggested by biogeographical models (Emanuel *et al.*, 1985; Prentice and Fung, 1990; Prentice *et al.*, 1992; Neilson, 1993; Henderson-Sellers and McGuffie, 1995). Specific changes in rangeland ecosystem boundaries will be determined by the nature of climate change, including the frequency and severity of extreme events. As nonequilibrium systems, rangelands may experience structural reorganization, and individual species may change distribution, but there probably will not be any synchronized movement of entire vegetation belts. Land-use change and human interventions will greatly modify the expressions of climate change and may overwhelm them, particularly in tropical and subtropical areas (Parton *et al.*, 1994).

Semi-arid and arid ecosystems may be among the first to show the effects of climate change (OIES, 1991). Their sensitivity to climate change may be due to the current marginality of soil water and nutrient reserves. The droughts of the 1930s and the 1950s in the United States, for example, changed plant production systems (Weaver and Albertson, 1944; Albertson *et al.*, 1957) and led to modification of land surface characteristics. Weaver and Albertson (1944) showed that the distribution of grasslands changed in the United States, with the shortgrass species moving eastward into the midgrass prairie and midgrass species moving eastward into the tallgrass prairie. In addition, these grasslands were invaded by ruderals (weedy species) and other non-native grasses. In Australia, the distribution of *Heteropogon contortus* shifted west in wet decades and eastward in drier periods (Bisset, 1962).

The balance between herbaceous and woody vegetation is sensitive to climate in most grassland/savanna regions (Parton *et al.*, 1994). Changes such as replacement of grasses with woody plants can occur quite quickly (Griffin and Friedel, 1985; McKeon *et al.*, 1989; Westoby *et al.*, 1989)—within a decade or so in response to a mixture of reduced grazing, fire suppression, and climate variability. Fire and grazing regimes, in conjunction with changes in climate characteristics affecting soil moisture status, relative humidity, or drought stress, will have the greatest influence on grassland-woody species boundaries (Neilson, 1986; Hulbert, 1988; Schlesinger *et al.*, 1990; Hobbs *et al.*, 1991; Archer *et al.*, 1995).

Mayeux *et al.* (1991) discuss evidence that open, mostly C_4 grasslands worldwide are becoming increasingly populated by C_3 woody plants—a process abetted, perhaps, by the 25 to 30% increase in atmospheric CO_2 and correlated rise in plant WUE in the last two centuries. There is ample evidence that C_3 woody species have increased in both mesic and arid C_4 grasslands during the historical period of CO_2 increase. The change from shrubland to grassland that would be predicted from climatic diagrams as plant water-use efficiency rose, however, has not been evident. Here, the opposing influence of a possible change to a warmer, drier climate since industrialization—perhaps in combination with grazing by domestic livestock (Hastings and Turner, 1965; Archer *et al.*, 1995)—may have negated the effects of increased CO_2 on vegetation structure.

In some regions, warmer temperatures and increased summer rainfall, with fewer frost days, may facilitate the encroachment of both annual and perennial subtropical C_4 grasses (e.g., *Digitaria sanguinalis, Paspalum dilatatum*) into some temperate C_3 grassland areas. This would likely cause a depression in forage quality and result in a more pronounced warm-season peak in biomass production, with consequent problems for traditional livestock systems. The rate of invasion of these and other weed species is expected to be greatest in more productive grassland and rangeland systems, with high stocking pressures, high utilization, and consequently greater opportunities for seedling establishment and invasion in gaps (Campbell and Hay, 1993; Campbell and Grime, 1993). The invasion rates of C_4 grasses may be offset to some extent by increased atmospheric CO_2 concentration (Campbell and Hay, 1993), but the effects are unclear.

Biogeographical shifts predicted for CO_2 and temperature increases are variable, depending on the vegetation classification used, general circulation model (GCM) assumptions, and assumptions about current vegetation distribution and equilibrium status (Parton *et al.*, 1994). Recent analyses using BIOME 1.1 (Prentice *et al.*, 1992) and IMAGE 2.0 (Alcamo, 1994) were compared to determine potential rangeland boundary shifts under different model assumptions (Table 2-3; see also Chapter 1, Figures 1-3 through 1-7, and Table 1-2).

The results suggest that rangelands are likely to be vulnerable to biogeographic change and that the areal extent of the changes may be significant (Table 2-3). Warm grass/shrub types will likely expand, while tundra ecosystems will most likely contract. The opposing results occur because of differences in the definitions of vegetation types, regional variation, and model assumptions. However, the analysis is instructive.

The direction of carbon-storage changes projected by three GCM models of climate change do agree quite closely under climate-change scenarios (Table 2-4). Analyses indicate that carbon stores are expected to increase in warm grass shrub and tropical savanna/dry woodland types, partly because these vegetation types are expected to expand (Solomon *et al.*, 1993). Conversely, total carbon stores are likely to decline in the other major rangeland types.

Table 2-3: *Potential changes in extent of rangeland cover types from the BIOME 1.1 model (Prentice et al., 1992) and IMAGE 2.0 (Alcamo, 1994). Numbers are in 10^6 km^2 and are rounded, so numbers do not exactly sum.*

	BIOME $2 \times CO_2$ (GFDL)			**IMAGE 2.0 (TVM) 2050**		
Type	B[1]	A[2]	C[3]	B[1]	A[2]	C[3]
Cool Semi-Desert	4.91	2.61	-2.30	2.28	2.28	0.00
Cool Grass Shrub	7.17	3.91	-3.21	5.26	8.12	2.85
Warm Grass Shrub	9.84	17.92	+8.08	17.50	25.98	+8.48
Hot Desert[4]	20.70	20.90	+0.199	16.90	16.46	-0.44
Tropical Savanna/ Dry Woodland	17.18	19.79	+2.61	11.46	3.68	-7.79
Xerophytic Wood/ Shrubland	10.63	11.89	+1.25	8.50	4.69	-3.81
Tundra	11.61	4.40	-7.21	11.38	11.34	-0.04

Note: This table does not display which vegetation types specifically change into another vegetation type. Refer to Figures 1-3 through 1-7 and to Table 1-2 in Chapter 1 for a graphical display of changes in rangeland boundaries between types.
Source: Rik Leemans, RIVM, The Netherlands.
[1] B = Before climate change; numbers vary because of different assumptions used in the classification of vegetation types between the models.
[2] A = After climate change.
[3] C = Change in area.
[4] See Chapter 3.

It is important to compare these impacts to those resulting directly from human activity (Parton *et al.*, 1994). According to Burke *et al.* (1990), the impact of a 2 x CO_2-driven climate-change scenario on grassland soil C levels in the southern Great Plains would be an order of magnitude smaller than the impact of plowing the grassland for crop production. Mosier *et al.* (1991) have shown that plowing and fertilization of grasslands increases N_2O flux by 50% and decreases soil sink strength for CH_4 by 50%. These impacts, too, are much larger than the potential effects of a 2 x CO_2-driven climate change (Burke *et al.*, 1990). While the effects of plowing on soil carbon and trace-gas fluxes may be large on a regional basis, the impact of CO_2 and climate may be similar on a global average. The effects of climate and land-use changes combined could lead to more rapid changes than either would alone.

2.7. Regional Variation

2.7.1. Low-Latitude Regions

Natural tropical grasslands (Box 2-2) are a close second to tropical forests in extent and may equal them in productivity (Hall and Scurlock, 1991). Tropical grasslands and savannas are important as both a source and a sink of C at the regional and world level. Because of their extent, productivity, and capacity for above- and belowground biomass accumulation, the amount of C sequestered is very high (Fisher *et al.*, 1994), (Tables 2-2 and 2-4). On the other hand, they are frequently burned and therefore represent a large efflux of C into the atmosphere. Almost one-fifth of the world's population lives in these ecosystems, many in rural societies that depend on subsistence agriculture (Frost *et al.*, 1986).

Extreme events seem to influence the system more strongly than do average climatic parameters. Highly seasonal water availability and chronically low soil-nutrient availability appear to be the most limiting factors (Medina, 1982; Frost *et al.*, 1986; Solbrig *et al.*, 1992). Natural and anthropogenic fires are a major and apparently very ancient structuring force. In some tropical systems, frequent burning does not lead to major changes or invasion, and the dominants tend to recover fast (Coutinho, 1982; Farinas and San José, 1987; Lewis *et al.*, 1990; Medina and Silva, 1990; Silva *et al.*, 1991). In others, invasion of exotic species is the norm (D'Antonio and Vitousek, 1993). The effect of climate change on tropical rangelands will depend on the balance between increasing aridity favoring C_4 grasses and increasing CO_2 concentration favoring C_3 photosynthesis (Box 2-3). The already low nutritional value of most tropical grassland plants may decrease as a consequence of increased C:N ratio and because of higher

Box 2-2. The Northern South America Savanna

Location: Northern South America (Medina, 1986, 1987; Sarmiento, 1983, 1984; Sarmiento and Monasterio, 1983; Goldstein and Sarmiento, 1987).

Climate: Average daily temperatures ranging from 22°C to more than 35°C; average annual precipitation ranging from 800 to 2000 mm with a strong seasonal distribution, including a dry period of 3–6 months (Medina, 1982; Sarmiento *et al.*, 1985).

Vegetation: Continuous cover of perennial grasses with a significant discontinuous cover of perennial woody plants (Medina, 1982; Medina and Silva, 1990).

Probable biological impacts of climate change: An alteration in the amount and pattern of rainfall, the occurrence of extreme events (e.g., hurricanes, drought), and the El Niño–Southern Oscillation (ENSO) may alter actual functioning of savanna ecosystems. The vegetation structure of the ecosystems located at higher altitudes may become similar to those 300–500 m lower in elevation. During ENSO years, precipitation in northern South America is lower (Aceituno, 1988), increasing the likelihood of drought. Because ENSO events could become more frequent and bring more severe weather under a doubled-CO_2 climate, these systems could fail to function as they do now.

production of secondary metabolites in C_3 plants (Oechel and Strain, 1985; Fajer *et al.*, 1989). Thus, the carrying capacity for herbivores may be reduced in this region.

2.7.2. *Middle-Latitude Regions*

Temperate rangeland systems experience drought stress at some time during the year, but the degree of drought stress and its timing vary among regions. The distribution of different rangeland types is correlated with different climatic patterns, including the seasonal distribution of mean monthly air temperature and precipitation and the ratio of precipitation to potential evapotranspiration (e.g., see Boxes 2-4 and 2-5).

Temperate grasslands and associated savannas and shrublands clearly are vulnerable to climate change. Parton *et al.* (1993) highlight three ways in which changing climate might influence mid-latitude grassland ecosystem function and structure. In some grassland regions, significant reductions in precipitation or increases in temperature may accelerate degradation and lead to the replacement of grassland vegetation by woody species. Second, soil erosion rates may increase as plant cover decreases, leading to a nearly irreversible loss of productive potential. Finally, the above transitions are almost always exacerbated by intense human land use, as occurs in most temperate grassland and savanna regions (OIES, 1991; Archer *et al.*, 1995). Degradation can thus occur through both functional (e.g., reduced primary productivity) and structural (e.g., perennial-to-annual vegetation, shrub encroachment) change (see Boxes 2-6 and 2-7).

Nutrient cycling in temperate grasslands is largely driven by climate. Rates of carbon gain, decomposition, and nutrient cycling are all sensitive to temperature and moisture (Parton *et al.*, 1987; Schimel *et al.*, 1990). Soil organic-matter storage—the primary

Table 2-4: *Rangeland BIOME 1.1 carbon stocks (in Gt) under three 1994 IPCC (Greco et al., 1994) climate scenarios (Max-Planck Institute, Geophysical Fluid Dynamics Laboratory, and UK Meteorological Office). Data are from Solomon et al., 1993; current carbon estimates can be found in Table 2-2.*

Type	ECHAM Climate			GFDL Climate			UKMO Climate		
	AG[1]	BG[2]	C[3]	AG[1]	BG[2]	C[3]	AG[1]	BG[2]	C[3]
Cool Semi-Desert	2.1	21.5	-10.2	1.7	17.6	-14.4	1.5	15.7	-16.5
Cool Grass Shrub	4.1	50.2	-22.3	3.2	39.3	-34.1	2.4	29.5	-44.7
Warm Grass Shrub	17.4	116.7	+33.8	19.7	132.1	+51.6	22.5	150.3	+7.2
Hot Desert	6.2	62.4	-0.8	6.3	63.5	-5.9	6.2	61.7	-1.5
Tropical Savanna/Dry Woodland	75.9	120.4	-7.1	91.5	145.2	+33.3	99.0	157.1	+52.7
Xerophytic Wood/Shrubland	47.1	83.9	+4.6	44.4	79.1	-2.8	46.5	82.7	+2.8
Tundra	4.4	100.8	-34.7	3.3	76.0	-60.6	2.4	55.2	-82.2

[1]AG = Above ground biomass from current climate estimates.
[2]BG = Below ground biomass from current climate estimates.
[3]C = Change in total carbon stocks (Gt) from current climate estimates.

Box 2-3. African Savannas

Location: Nearly continuous distribution in the central African continent (see Figure 2-1).

Climate: Distinct wet (600–1500 mm/yr) and dry (400–800 mm/yr) savannas.

Vegetation: Distinct layers of woody and herbaceous vegetation. In the moist savanna, vegetation is dominated by broad-leafed trees of the genera *Brachystegia* and *Julbernadia* as dominants. Herbivory by large mammals consumes a relatively small fraction of total primary production (5–10%) due to high concentrations of secondary compounds (mainly tannins) and low nitrogen concentrations in the foliage of the woody plants; thus, there is a high fire-fuel load and a short (1–3 year) fire interval. Dry savanna is dominated by fine-leafed trees (mainly *Acacia*) and stoloniferous grasses (Scholes and Walker, 1993). Large mammal herbivory is high (10–50% of total primary production), the fuel load is low, and the inter-fire interval is relatively long (5+ years).

Probable biological impacts of climate change: The major predicted impact is an increase in the frequency and severity of drought (McKeon *et al.*, 1993). An increase in the frequency of drought is likely to lead to episodic die-offs of woody vegetation. This will increase the fire-fuel load and thus potentially increase the frequency and intensity of fires. The greatest impacts are likely to be in the semi-arid savannas, where human populations depend mainly on pastoralism (see Box 2-1). Human responses are likely to include increasing the distribution and security of surface water for livestock. This is likely to further weaken the grass layer, with acceleration of desertification. One strategy to bolster the production of animal protein would be to promote the sustainable use and management of indigenous herbivore species, especially preferential browsers (giraffe, eland, kudu, impala, etc.), many of which are independent of surface water and largely do not compete with domestic grazers.

reserve for nutrients in grasslands—is sensitive to temperature and generally decreases with increasing temperature. In New Zealand, research using both spatial analogue (Tate, 1992) and modeling approaches (Tate *et al.*, 1995) has shown that soil carbon could decline by about 5% per °C increase in regional temperature. If temperatures increase in grasslands worldwide, soil carbon storage is likely to decline (Parton *et al.*, 1994). Declining soil carbon would likely lead to nutrient loss in the longterm, and could lead to degraded hydrological properties and increased erosion. Losses of soil carbon and increases in temperature could also lead to higher trace-gas emissions as the soil N-cycle is accelerated. Thus, carbon storage and nutrient cycling in temperate grasslands could change significantly if climate changes; indeed, interannual variability in climate is clearly reflected in observed and modeled production and decomposition. However, Buol *et al.* (1990) showed that, for soils that experience a wide annual temperature fluctuation, there is no correlation of organic carbon content to mean annual temperature.

Box 2-4. Rio de la Plata Grassland in South America

Location: Argentine Pampas, the Campos of Uruguay, and Southern Brazil.

Climate: Temperate humid to semi-arid. Mean annual temperature 13–18°C, mean annual minimum temperature <13°C. Mean annual rainfall 400–1600 mm, evenly distributed over the year in the east, becoming concentrated in spring and fall in the west (Cane, 1975; Cabrera, 1976; Lemcoff, in Soriano, 1992).

Vegetation: Mostly tall tussock grasses, but perennial grazing lawns are the actual physiognomy. C_3 and C_4 grasses, with the growth of C_3 and C_4 grasses segregated temporally (Cabrera, 1976; Arana *et al.*, 1985; Leon, in Soriano, 1992).

Probable biological impacts of climate change: These grasslands are vulnerable to drought. IPCC climate scenarios (Greco *et al.*, 1994) suggest that mean annual winter (JJA) temperatures may increase 0.5 to 1°C, while mean annual summer (DJF) temperatures may increase as much as 2°C. Generally summer (DJF) precipitation is expected to decline from 0–40% by the year 2050, while winter (JJA) precipitation may increase or decrease ±20% in the region (Greco *et al.*, 1994). If these projections are correct, the crop and animal production for the region could drop drastically. During the 1920s and 1930s, extremely dry conditions caused a shift of agriculture to the east, and the abandoned agricultural land eroded, with the loss of 60,000 km^2 of arable land (Suriano *et al.*, 1992). Unusually dry summers may bring about some long-lasting changes in community structure (Chaneton *et al.*, 1988; Sala, in Soriano, 1992). The projected changes may allow C_4 species to develop earlier in the season, while decreased summer precipitation will likely favor C_3 species.

Box 2-5. Mediterranean-Basin Region

Location: Bordering the Mediterranean Sea, stretching 5000 km eastward from the Atlantic coast between 30 and 45°N.

Climate: Transitional between temperate and dry tropical climates, with a distinct summer drought of variable length. Highly variable annual precipitation.

Vegetation: Evergreen sclerophyllous shrubs that change to heathlands in nutrient-poor soils along an increasing-moisture gradient and into open woodlands and annual grass or dwarf shrub steppes along an increasing-aridity gradient.

Probable biological impacts of climate change: A shift of carbon storage from soil to biomass is likely to occur, with probable negative effects on soil stability and increased erosion (Ojima *et al.*, 1993b). Mediterranean rangelands are expected to undergo a 300–500 km poleward extension of grass and dwarf shrub steppes at the expense of the sclerophyllous shrubland. In Euro-Mediterranean countries, extension of shrubland as a result of agricultural release is expected, with a parallel increase in wildfires and evapotranspiration, and a decrease in livestock grazing with increases of game and other wildlife. In Afro-Asian Mediterranean countries, extension of agriculture and overgrazing in marginal areas, combined with overall smaller P:PET ratios, will probably lead to further degradation of plant cover and soil loss.

Finally, the balance between herbaceous and woody vegetation is sensitive to climate in most grassland/savanna regions, and the productivity and human utility of these systems is very sensitive to this balance (see Boxes 2-5 through 2-8).

2.7.3. *High-Latitude Regions*

Warming is expected to be most pronounced at the poles (Spicer and Chapman, 1990; Long and Hutching, 1991). In the high southern latitudes of South America, annual mean temperature and low and irregularly distributed precipitation (especially as ice and snow) are the main factors constraining grassland productivity (Mann, 1966; Sala *et al.*, 1989; Fernández-A. *et al.*, 1991). Most Southern Hemisphere high-latitude grasslands lie to the south of the 13°C isotherm and are composed of C_3 species. By 2050, virtually all Southern Hemisphere grasslands might lie to the north of that isotherm; thus, the area would be more suitable for C_4 species (R. Rodríguez and G. Magrin, pers. comm., on the basis of Deregibus, 1988). This change in species composition would decrease the nutritional value of the grasslands, further decrease their carrying capacity, and eventually aggravate the overgrazing-erosion problem. On the other hand, dominant C_3 grasses and shrubs may be promoted by elevated CO_2, depending on water availability, and the increased growth may eventually lead to increased production and carrying capacity.

Indirect temperature effects associated with changes in thaw depth, nutrient availability, and vegetation will cause substantial

Box 2-6. Great Basin of North America

Location: Nevada and rain-shadow parts of Oregon, California, Utah, and Arizona in the United States.

Climate: Continental climate with average annual precipitation averaging 50–250 mm. Snow and hard frost are common in winter.

Vegetation: The Great Basin desert is dominated by big sagebrush (*Artemesia tridentata*), although other shrubs dominate on shallow saline soils (*Atriplex* sp.) or in specific soil complexes. Cover is generally low, varying from 5 to 50%.

Some probable biological impacts of climate change: There are a number of possible vegetation changes that could occur as a result of IPCC climate-change scenarios. An increase in precipitation during the growing season may increase the proportion of grasses and forbs over woody plants, which may increase the potential for more frequent fires, further favoring grasses. Cold conditions with increased precipitation during the winter will likely favor drought-tolerant woody species (e.g., *Artemisia* sp.) over herbaceous species. These conditions would also decrease fire frequency, thereby benefiting woody plants over grasses and forbs. Mild, wet winters and dry summers may enhance the spread and dominance of introduced Eurasian annuals such as cheatgrass (*Bromus tectorum*) and Russian thistle (*Salsola kali*). Cheatgrass is a fire-prone species, and previous work has demonstrated that frequent fire-return intervals occur with cheatgrass dominance (Whisenant, 1990). A decrease in total annual precipitation will likely result in shrub steppe communities shifting toward desert shrub.

Box 2-7. European Steppe and Semi-Natural Grassland

Location: Scattered throughout Europe.

Climate: Steppes occupy sites with a continental climate and annual rainfall between 350–500 mm concentrated in the warm season, but with late-summer drought (Walter, 1968). Semi-natural grasslands occupy a wide range of climates, with annual rainfall between 400 mm and 2300 mm (Titlyanova *et al.*, 1990).

Vegetation: Steppes are multi-layered communities (Lavrenko and Karamysheva, 1992), with tall bunch grasses (*Stipa* sp.), short bunch grasses (*Cleistogenes* sp., *Festuca* sp.), and dwarf bunch species (*Carex* sp., dwarf *Stipa* sp.); moss layers are dominant in the northern zone, and ephemeral plants plus lichen layers are dominant in the southern zone. Semi-natural grasslands are secondary plant communities with a multi-layer structure of grasses and forbs (Rychnovska, 1992). Floristic diversity provides semi-natural grasslands better potential adaptation to climate fluctuations; belowground biomass and soil aggregation are the main stabilizing factors.

Probable biological impacts of climate change: Increases of 3 ± 1.5°C within the next century as projected in current scenarios (Greco *et al.*, 1994), could cause a 250–300 km northern displacement of phytoclimatic boundaries (Anderson, 1991) with a substantial reduction of tundra, mires, and permafrost areas and a parallel expansion of grassland and grain cropping into the southern margin of boreal forest zones in Europe. In addition, changes in the concentration of atmospheric gases may alter the competitive relationships in the plant community. For example, indirect competitive effects have been reported in southern Britain (Evans and Ashmore, 1992), where above-normal ozone concentrations caused a decrease in the biomass of upper-layer dominant grasses with a parallel increase in lower-layer forb biomass by reducing competition for light.

changes in Arctic species composition, litter quality, and nutrient availability (Chapin and Shaver, 1985; Oberbauer and Oechel, 1989; Kielland, 1990; Marion *et al.*, 1991; Nadelhoffer *et al.*, 1991; Havstrom *et al.*, 1993; Wookey *et al.*, 1993; Chapin *et al.*, 1995). Increased nutrient availability increases shrub abundance and decreases the abundance of mosses, an important soil insulator (Chapin *et al.*, 1995).

The changes in biotic interactions that will occur with climatic warming are poorly known. Field experiments of ≤10 years duration, simulating a 3°C warming, show an increase in shrub abundance and a decline in species richness (Chapin *et al.*, 1995). Insect-pollinated forbs are expected to decline—suggesting that insect pollinators (Williams and Batzli, 1982) and migrating caribou, which utilize these forbs during lactation, could be adversely affected (White and Trudell, 1980). Warmer summers also increase insect harassment of caribou and cause declines in feeding and summer energy reserves. In contrast, browsing mammals, such as snowshoe hare and moose, may benefit from climatic warming because an increase in shrubs will increase fire frequency and lead to an increase in the proportion of early-successional vegetation.

The presence of trees at high latitudes may be determined more by soil temperature than by air temperature (Murray, 1980), so expected warming of soils and increase in thaw depth could have a strong influence on treeline location. Simulation models suggest that conversion of boreal forest to a treeless landscape would increase winter albedo, reduce energy absorption, and cause a 6°C decrease in regional temperature (Bonan *et al.*, 1992). Conversely, we expect treeline advance to provide a strong positive feedback to regional and global warming.

Regional warming and treeline advance are expected to increase fire frequency, directly releasing CO_2 stored in peat to the atmosphere. Current predictions (Anderson, 1991) indicate that the tundra may change from being a sink to a source of C, with a net flux to the atmosphere of about 100 g C/m^{-2}/yr. This study also projected that about 15% of C storages in soils of peats, tundra, and boreal forests could be released over the next 50 years. Recent work suggests soil drying over the past decade (Maxwell, 1992) has already changed the Alaskan tundra from a net sink to a net source of CO_2 (Anderson, 1991; Oechel *et al.*, 1993; Zimov *et al.*, 1993a, 1993b).

2.8. Modeling Rangeland Ecosystem Response

Several models are currently used to predict rangeland response to climate change and land use. While most global-scale models are correlational—such as those based on the Holdridge Life Zone classification scheme (e.g., Emanuel *et al.*, 1985)—there are many regional-scale models of ecosystem dynamics in rangelands (e.g., Century: Parton *et al.*, 1987, 1988, 1994; GEMS: Hunt *et al.*, 1991; GRASS: Coughenour, 1984; Hurley model: Thornley and Verbene, 1989; Thornley *et al.*, 1991; SPUR: Hanson *et al.*, 1993; STEPPE: Coffin and Lauenroth, 1990; GRASP: McKeon *et al.*, 1990).

These models generally show that when a variety of directional climate change scenarios are imposed on rangeland ecosystems, such as in Century model simulations, effects are detectable but small on a per-unit-area basis relative to changes induced by management activities (Burke *et al.*, 1989)—although climate effects are more substantial globally (Parton

Box 2-8. Australian Tussock Grassland and Steppe

Location: Alluvial plains running southwest from the eastern half of the continent.

Climate: There is an enormous climatic gradient across the continent, from 100–200 mm median annual rainfall in the central part of the region—distributed throughout the year at random, without seasonality—to 800–900 mm along the northern and southern borders, the former concentrated in summer and the latter in winter as a Mediterranean-type climate. In the south, there are a significant number of frosts; winter rains are usually gentle and associated with frontal systems passing around the South Pole. In the north, there is no frost, and rains are often monsoonal and very heavy. Occasionally, northern cyclonic incursions may reach the south, bringing the heaviest rainfall events. Rainfall variability is very high—increasing toward the equator and as annual totals decline—and is greatly affected by ENSO and other climatic systems driven by sea surface temperatures in the Pacific and Indian oceans.

Vegetation: A complex range of communities, including monsoon tallgrass (*Schizachyrium* sp.), subtropical and tropical tallgrass (*Heteropogon contortus*), tussock and midgrass (*Astrebla* sp., *Aristida* sp., *Bothriochloa* sp.), hummock grasslands (*Spinifex* sp.), acacia and eucalyptus shrublands and woodlands, temperate grasslands (*Themeda triandra*, *Danthonia* sp.), annual grasslands (*Echinochloa* sp.), and chenopod shrublands (*Atriplex* sp., *Maireana* sp.).

Probable biological impacts of climate change: Scenarios of future climate (Greco *et al.*, 1994) suggest increased variability and unpredictability in productivity and community composition in the region (McKeon and Howden, 1991; Stafford Smith *et al.*, 1995). A shift toward summer rainfall will have a major effect in southern Australia, where vegetation is currently dominated by winter rainfall (Stafford Smith *et al.*, 1995). Although it is unlikely that mean climatic conditions will move outside their envelope of current possible conditions, extreme drought periods, high intensity rainfalls, and water distribution could become more common (Whetton, 1993; Stafford Smith *et al.*, 1995).

et al., 1995). A recent intercomparison of several of these models showed that the degree to which forests invade grasslands or vice versa depends on the GCM climate scenario used (VEMAP Participants, 1995).

These models, plus recent global-scale mechanistic models, have been used to evaluate the outcome of various management schemes, including fire, grazing, and strip-mine reclamation (e.g., de Ridder *et al.*, 1982; Coughenour, 1984, 1991; Coffin and Lauenroth, 1990; Ellis *et al.*, 1990; Ågren *et al.*, 1991; Hunt *et al.*, 1991; Holland *et al.*, 1992; Hanson *et al.*, 1993; Ojima *et al.*, 1994b). They have also been used to assess potential impacts of various climate scenarios on plant productivity and nutrient cycling (e.g., BIOME-BGC: Running and Hunt, 1993; Century: Parton *et al.*, 1987, 1992, 1994; SPUR: Hanson *et al.*, 1993; TEM: McGuire *et al.*, 1992).

Results from tests of rangeland sensitivity to climate and management using these models support previous discussion of the importance of seasonality and distribution of rainfall, and suggest that rangelands are sensitive to changes in environmental conditions associated with fire and grazing as well as their land-use history.

2.9. Human Adaptation

By the time there is a detectable rise in mean temperature of a degree or two, pastoral societies may already have begun to adapt to the change. Adaptation may be more problematic in some pastoral systems where production is very sensitive to climatic change, technology change is risky (Caceres, 1993), and the rate of adoption of new technology is slow. Decreases in rangeland productivity would result in a decline in overall contribution of the livestock industry to national economies. This would have serious implications for the food policies of many underdeveloped countries and on the lives of thousands of pastoral peoples.

Intervention—in terms of active selection of plant species and controlled animal stocking rates—is the most promising management activity to lessen the negative impact of future climate scenarios on rangelands. Proper rangeland management, including sustainable yield and use of good-condition areas while marginal or poor condition areas are allowed to rest, will become increasingly necessary under climatic conditions projected for the future. A shift in reliance toward more suitable and more intensively managed land areas for food and fiber production could have the dual benefits of greater reliability in food production and lesser detrimental impact of extreme climatic events such as drought on rangeland systems. Improvement and intensification of management of certain areas of rangeland may also have the additional benefit of reducing average methane emissions per head of livestock because of improved feed quality.

One management option for the future is to actively change species composition of selected rangelands. Legume-based grassland systems may become more important in the future because legumes reduce the reliance on fertilizer inputs and improve the nutritive value of forage. The use of leguminous species is important as a means of producing more-sustainable forage systems (Riveros, 1993). Legume species generally show larger yield responses to elevated CO_2 than do grasses at warm temperatures and also show enhanced nitrogen-fixation rates (Crush *et al.*, 1993; Newton *et al.*, 1994; Campbell *et al.*, 1995a). The benefits of leguminous species could be offset to some extent by future increases in damaging UV-B radiation (Caldwell *et al.*, 1989), because legumes may be quite susceptible to UV-B relative to other species (Krupa and Kickert, 1989; Brown, 1994). However, the effects are uncertain at present because there are only limited experimental data available on the effects of future UV-B radiation levels on forage and rangeland legumes.

In some grass-dominated rangeland systems, improved pastures may help people to adapt livestock grazing strategies. In these systems, there is significant intervention—including selection of forage type, selective animal breeding, pasture renewal, irrigation, and other practices (Campbell and Stafford Smith, 1993). This intervention provides opportunities for graziers to adapt systems so they are protected against negative effects of global change and so that any potential benefits are realized. McKeon *et al.* (1993) developed a scheme for adapting rangelands to climate change by linking management decisions to climate variability.

2.10. Mitigation

Opportunities for reducing greenhouse gas (GHG) emissions on rangelands (Table 2-5) include maintaining or increasing carbon sequestration through better soil management (Ojima *et al.*, 1994b), reducing methane production by altering animal-management practices (Cicerone and Oremland, 1988; Howden *et al.*, 1994), and using sustainable agriculture practices on rangelands capable of sustaining agriculture (Mosier *et al.*, 1991).

Productivity and carbon cycling in rangeland ecosystems are directly related to the amounts and seasonal distribution of precipitation and are only secondarily controlled by other climate variables and atmospheric chemistry (Sampson *et al.*, 1993). In fact, rangeland productivity may vary as much as fivefold because of timing and amounts of precipitation (Walker, 1993).

Carbon storage in grasslands, savannas, and deserts is primarily below ground (Table 2-2). Estimates using the BIOME model (Solomon *et al.*, 1993) suggest that in the world's rangelands approximately 595 Gt C is stored in belowground biomass (44% of the world's total), while only about 158 Gt is stored in rangeland aboveground biomass. Thus, good soil management is the key to keeping or increasing C storage and protecting rangeland health (condition). Improving rangeland health, and thus the amount and kind of vegetation, will also reduce methane emission from ruminant animals per unit product by improving the quality of their diet. Total methane emissions from the grazing system will decrease only if there are commensurate reductions in animal numbers in the less-resilient parts of the grazing landscape however (Howden *et al.*, 1994).

Nonsustainable land-use practices such as inappropriate plowing, overgrazing by domestic animals, and excessive fuelwood use are root causes of degradation of rangeland ecosystems (Ojima *et al.*, 1993b; Sampson *et al.*, 1993). Some systems may already be degraded to such an extent that the ability of the ecosystem to recover under better management practices will be greatly hampered. Practices listed in Table 2-5 may prove to be useful in GHG mitigation. None of the practices is relevant for every country, social system, or rangeland type. Unfortunately, because of the high dependence of rangeland function on adequate amounts and timing of rainfall, none of these practices is likely to significantly improve rangeland function and carbon sequestration without adequate rainfall (Heady, 1988; McKeon *et al.*, 1989).

Reduction of animal numbers. Reduction of animal numbers can increase carbon storage through better plant cover (e.g., Ash *et al.*, 1995) and decrease methane emissions. This practice can have a positive ecosystem effect if there is sufficient rainfall, but reduction in animal numbers on rangelands may require alternative sources of food for humans, and thus changes in national and/or regional food-production policies.

Changing the mix of animals. Changing the mix of animals on a given rangeland area can increase carbon storage and decrease methane emissions, but the benefit derived depends on the kind of rangeland and the proposed mix of animals. If a country is considering only cattle and small stock (sheep, goats, etc.), the mix may not be ecologically efficient; rather, it may reflect an economic risk aversion—in bad times, cattle die but goats survive. In this kind of grazing mix, the ecosystem may deteriorate. In contrast, a mix of cattle and wildlife ruminants may be both ecologically and economically efficient.

Changing animal distribution. Changing animal distribution through salt placement, development of water sources, or fencing can increase carbon sequestration through small increases in plant cover overall and improved status of the root system (due to less intense grazing). None of the changes in animal distribution, however, is expected to affect methane production. Animal-management practices will be specific to local and regional production systems; for example, fencing and/or placement of salt may not be useful in herding systems and may potentially interfere with wildlife migration. Costs and practices will vary widely by region.

Agroforestry. In regions where woody species and grasses coexist (Boxes 2-2, 2-3, 2-5, and 2-8), management practices to enhance both woody and herbaceous productivity may increase carbon storage and reduce methane emissions per unit product from domestic and wild ruminants, by improving the quality of the diet.

Watershed-scale projects. Practices involving the development of dams with large-scale water-storage capacity may improve long-term carbon storage and reduce animal methane production per unit product, by improving the quality of the diet through improved animal-management options and improved

Table 2-5: *Practices (mitigation options) to improve rangeland condition or health, and to reduce GHG emissions on rangelands. Possible effects of implementing each practice are listed for two greenhouse gases, and qualitative cost/benefits estimates are provided. Unhealthy, poor-condition rangelands are those lands where soil loss, plant species and cover loss, species invasions, and interrupted and poorly functioning nutrient cycling are the norm. Good condition, healthy rangelands, on the other hand, have nutrient cycling and energy flows intact; soils are not eroding; and plant species composition and productivity is indicative of a functioning ecosystem (NRC, 1994).*

Practice	Healthy	Not Healthy	Carbon	Methane	Biophysical Benefit/Cost	Soc./Cultural Benefit/Cost	Economic Benefit/Cost	General Comments
Reduce animal numbers	No	Yes	Increases carbon sink because of increasing vegetation cover and better root growth	Reduces animal methane production through reduction in total number	Increases plant cover, increases soil organic matter, and improves productivity	Depends on country and value of animals as a social resource	Depends on the value of livestock products to national and/ or local economy	May require changes in national and/or regional food production policies
Change mix of animals	Yes	Yes	Possible increase in carbon sink with change in plant species	Change to native herbivores reduces emissions in some cases	Potential changes in plant species composition	Depends on country and cultural value of specific animal type	Depends on the value of livestock products	Positive effect in general; improves efficiency of utilization
Alter animal distribution by placing salt licks or feed supplements	Yes	Yes	Increases carbon sink because of increasing vegetation cover overall	Feed supplements may decrease methane production	Useless in rangeland areas already high in salt, but very useful in rangelands low in N,P	Appropriate in countries where animals graze extensively, rather than herded	Cost and distribution of salt and supplements	Positive; not applicable for herding systems
Alter animal distribution by placing water resources	Yes	Yes	Potentially increases carbon sink by increasing total vegetation cover	No effect	Developed water resources may not be sustainable	May affect territorial and property boundaries	Motorized water sources often too costly to purchase or maintain	Negative impacts if used to increase numbers of animals
Alter animal distribution by placing fences	No	Yes	Increases carbon sink because of vegetation cover overall	No effect	Benefit is to control domestic animal number and distribution	Depends on country and livestock/ wildlife system	Varies, depending on country and source and kind of materials	Potentially interferes with wildlife migration
Provide livestock protein and phosphorus supplement	Yes	Yes	If supplement increases intake (usual case) and stocking rate is not reduced, soil carbon may decline	Decrease in methane production, especially CH_4 per unit product	May reduce extensive grazing to some degree	Possible where animals are herded	Cost of protein blocks or similar supplement, but often large increases in productivity	Potentially difficult to distribute to local areas

Table 2-5 continued

Practice	Healthy	Not Healthy	Carbon	Methane	Biophysical Benefit/Cost	Soc./Cultural Benefit/Cost	Economic Benefit/Cost	General Comments
Increase native grasses and/or plant-adapted species	No	Yes	Increases carbon sink because of increasing vegetation cover overall	Possible methane reduction if quality of diet increases, but animals eat more	Benefit in retention of native species for gene conservation	Local people rely on native species for medicine and other health-related goods	Depends on the value of livestock and wildlife products, and the value of herbal medicine	Potential unknown benefits from native species; adapted species survive over the long term
Selective application of herbicides	No	Possibly	Potentially increases carbon sink, although not if woody species removed	Potential increase if animal numbers expand	Cost if non-target species affected: water pollution, damage to food chain	Cost if non-target species affected: water pollution, damage to food chain, removal of firewood source	Varies, depending on country and source of herbicide	Costs or benefits depend on meeting of management goals
Mechanical treatment or restoration	No	Possibly	Potentially increases carbon sink	Potential increase if animal numbers expand	Potential for large-scale alteration of soil and vegetation	May not fit pastoral system	Varies with country, depending on availability of equipment	Benefits depend on success of treatment relative to disruption of ecosystem
Plant halophytes (salt-tolerant species)	If appropriate		Increases carbon sink and increases productivity	No known effect	Benefit with increased plant cover and productivity	Benefit with increasing forage production for livestock and wildlife	Cost of planting and maintaining with irrigation	Brings into production otherwise non-productive land
Apply prescribed fire	Yes	Yes	Increases carbon sink and increases productivity in the long term on appropriate rangeland types	Possible methane reduction per unit product by increasing quality of diet	In systems adapted to fire, can increase productivity and maintain nutrient cycling	Use of fire can be part of social system; uses local knowledge	Threat of wildfire and destruction of human resources	Short-term increase in CO_2 to atmosphere; long-term benefits in adapted systems
Implement agroforestry systems	Yes	Yes	Increases carbon sink and increases productivity in the long term on appropriate rangeland types	Possible methane reduction per unit product by increasing quality of diet	Possible benefit with increased plant cover, diversity, and productivity	Potential benefit with change in grass/browse forage mix for livestock and wildlife	Cost of planting and maintaining	Increases carbon storage in trees; benefit in diversity and productivity if adapted species are used

Table 2-5 continued

Practice	Healthy	Not Healthy	Carbon	Methane	Biophysical Benefit/Cost	Soc./Cultural Benefit/Cost	Economic Benefit/Cost	General Comments
Develop large-scale watershed projects	Possibly	Possibly	Increases carbon sink and increases productivity	Methane reduction by increasing quality of diet	Potential for large land disturbance, with benefit to human and animal populations because of regulated and regular water supply	Potential for improved food production, both plant and animal	Cost of dams, etc.; benefit of hydroelectric power	Potential for increased human and animal populations because of increase in water availability

food-production systems. However, such projects are expensive and can result in social and cultural dislocation, local extinction of wildlife, and may result in increases in both human and animal population density.

Although rangelands have been used historically for livestock production for meat, wool, hides, milk, blood, and/or pharmaceuticals, policymakers must realize that an equally important objective for maintaining rangelands may be to maximize the number of domestic animals as a social resource (e.g., in some tribal cultures, wives are bought with livestock). Rangelands also are increasingly affected by human activities for mineral production, construction materials, fuel, and chemicals. Additionally, rangelands provide habitat for wildlife, threatened and endangered species, anthropological sites, ecotourism, and recreational activities. As the human population grows, rangelands also incur increased demand for marginal agriculture production. All of these activities and uses potentially affect rangeland condition—and thus the potential of the ecosystem to sequester carbon.

Using the practices in Table 2-5 requires the recognition that productivity on rangelands is variable and that vegetation and animal response to changes in livestock numbers is not linear. Futhermore, there are system lags, and management activities may result in largely irreversible changes in the rangeland system. Risk-management strategies include long-term low stocking or changing animal numbers annually to track variation in precipitation (McKeon *et al.*, 1993). Either approach requires flexible management response to different events, opportunities, and hazards and whole-system analyses in which the interactions of climate, land management, fire, and plant and animal species and communities will determine the outcome.

2.11. Research Needs

We need to know more about the C and N storage capacity of soils under different conditions of degradation and under different land uses and land management, as well as the potential for increased sequestration of C, especially in whole-system analyses. More information is needed regarding the actual release rates of various greenhouse gases in real environments, across the spectrum of climates, soils, and land uses (rather than the current measurements just from a few points) and land management. Our lack of ability to track plant responses within growing season also affects our ability to accurately estimate rapid changes in greenhouse gas fluxes and remains a major gap in our ability to link management practices with global climate-change issues.

Rangeland research on climate-change effects should emphasize multispecies field experiments. The concept of functional groups might be a way to cope with the overwhelming species diversity of most rangelands, especially grasslands. Models that incorporate nonequilibrium theory in grassland and rangeland ecosystem response would be most useful.

Rangeland research emphasis should be on multidisciplinary projects in which different components (e.g., vegetation, soil, herbivores) and various approaches (e.g., conservation, economic sustainability) are simultaneously considered. Maintenance of long-term monitoring sites is essential for understanding boundary changes (especially C_3/C_4, shrub/grass boundaries), as well as fire, herbivory, and land-use affects on rangeland productivity and diversity. Along with monitoring, there needs to be an ongoing analysis of the results and review of monitoring methods and criteria.

Information on various aspects of climate, community structure and function, and land use needs to be standardized.

At a national level, scientists and policymakers must better understand the implications of land use for global-change issues, because land use and management feeds into global change as well as being influenced by it. The interaction between human population growth and rangeland use must also be better understood. Land use will change in the future as a result of altered climatic extremes (especially those in rainfall and temperature), decreased productivity and soil fertility, and possibly socioeconomic changes spurred by adaptation and mitigation requirements.

Interfacing research on global change effects on agroecosystems with those on rangeland systems would be helpful in that it would promote access to a greater range of databases and models. The International Geosphere Biosphere Programme (IGBP)—Global Change and Terrestrial Ecosystems (GCTE) Project Task 3.1.3 (Pastures and Rangelands) (Campbell and Stafford Smith, 1993; Stafford Smith *et al.*, 1995) is combining the research efforts of grassland scientists in improved pastures and rangeland systems to predict global-change effects on the whole spectrum of different grassland types, taking into account future land-use change.

Finally, detailed analyses of the potential biological effects of climate change in terms of rangeland boundaries, coupled with an economic assessment of potential effects on pastoral peoples and national economies, will improve future assessments of impact, adaptation, and mitigation potentials on rangelands.

References

Abrams, M.E., A.K. Knapp, and L.C. Hulbert, 1986: A ten-year record of aboveground biomass in Kansas tallgrass prairie: effects of fire and topographic position. *American Journal of Botany*, **73**, 1506-1515.

Aceituno, P., 1988: On the functioning of the Southern-Oscillation in the South America sector. Part I: Surface climate. *Mon. Wea. Rev.*, **116**, 505-524.

Ågren, G.I., R.E. McMurtrie, W.J. Parton, J. Pastor, and H.H. Shugart, 1991: State-of-the-art of models of production decomposition linkages in conifer and grassland ecosystems. *Ecological Applications*, **1**, 118-138.

Alcamo, J. (ed.), 1994: *IMAGE 2.0: Integrated Modeling of Global Climate Change*. Kluwer Academic Publishers, Dordrecht, Netherlands, 318 pp.

Allen, Jr., L.H., E.C. Bisbal, K.J. Boote, and P.H. Jones, 1991: Soybean dry matter allocation under subambient and superambient levels of carbon dioxide. *Agronomy Journal*, **83**, 875-883.

Anderson, I.C., J.S. Levine, M.A. Poth, and P.J. Riggan, 1988: Enhanced biogenic emissions of nitric oxide and nitrous oxide following surface biomass burning. *Journal of Geophysical Research*, **93**, 3893-3898.

Anderson, J.M., 1991: The effects of climate change on decomposition processes in grassland and coniferous forests. *Ecological Applications*, **1(3)**, 326-247.

Arana, S.C., N.H. Mailland, S.I. Alonso, N.O. Maceira, and C.A. Verona, 1985: Crecimiento y fenologia de poblaciones vegetales en un pastizal natural. Su significado adaptativo. *RIA (INTA)*, **20**, 105-134.

Archer, S., 1994: Woody plant encroachment into Southwestern grasslands and savannas: rates, patterns and proximate causes. In: *Ecological Implications of Livestock Herbivory in the West* [Vavra, M., W.A. Laycock, and R.D. Pieper (eds.)]. Society for Range Management, Denver, CO, pp. 13-68.

Archer, S., D.S. Schimel, and E.A. Holland, 1995: Mechanisms of shrubland expansion: land use, climate or CO_2? *Climate Change*, **29**, 91-100.

Archer, S. and F.E. Smeins, 1991: Ecosystem-level processes. In: *Grazing Management: An Ecological Perspective* [Heitschmidt, R.K. and J.W. Stuth (eds.)]. Timberline Press, Portland, OR, pp. 109-139.

Arhem, K., 1985: *Pastoral Man in the Garden of Eden*. University of Uppsala, Uppsala, Sweden, 123 pp.

Ash, A.J., S.M. Howden, and J.G. McIvor, 1995: Improved rangeland management and its implications for carbon sequestration. Proc. Vth International Rangelands Congress, Salt Lake City, Utah (in press).

Ayres, M.P., 1993: Plant defense, herbivory, and climate change. In: *Biotic Interactions and Global Change* [Karelva, P.M., J.G. Kingsolver, and R.B. Huey (eds.)]. Sinauer Assoc. Inc., Sunderland, MA, pp. 75-94.

Bailey, R.G., 1986: A world ecoregions map for resource reporting. *Environmental Conservation*, **13(3)**, 195-202.

Bailey, R.G., 1989: Explanatory supplement to ecoregions map of the continents. *Environmental Conservation*, **16(4)**, 307-310.

Baker, J.T., L.H. Allen, Jr., and K.J. Boote, 1990: Growth and yield responses of rice to carbon dioxide concentration. *Journal of Agricultural Science*, **115**, 313-320.

Barbour, J.G., J.H. Burk, and W.D. Pitts (eds.), 1987: *Terrestrial Plant Ecology*. The Benjamin/Cummings Publishing Company, Inc., Menlo Park, CA, 634 pp.

Bauer, A. and A.L. Black, 1981: Soil carbon, nitrogen and bulk density comparisons in two crop land tillage systems after 25 years and in virgin grassland. *Soil Science Society of America Journal*, **45**, 1166-1170.

Bazzaz, F.A. and K.D.M. McConnaughay, 1992: Plant-plant interactions in elevated CO_2 environments. *Australian Journal of Botany*, **40**, 547-563.

Bazzaz, F.A. and E.D. Fajer, 1992: Plant life in a CO_2-rich world. *Scientific American*, **266**, 68-74.

Bazzaz, F.A., 1993: Scaling in biological systems: population and community perspectives. In: *Scaling Physiological Processes: Leaf to Globe* [Ehleringer, J.R. and C.B. Field (eds.)]. Academic Press, Inc., San Diego, pp. 159-166.

Bazzaz, F.A., 1990: The response of natural ecosystems to the rising global CO_2 levels. *Annual Review of Ecological Systems*, **21**, 167-196.

Behnke, R., I. Scoones, and C. Keerven (eds.), 1993: *Range Ecology at Disequilibrium*. Overseas Development Institute, London, UK, 248 pp.

Bekure, S., P.N. de Leeuw, and B.E. Grandin, 1987: *Maasai Herding. An Investigation of Pastoral Production on Group Ranches in Kenya*. International Livestock Center for Africa, Nairobi, Kenya, 158 pp.

Berdowski, J.J.M., 1987: Transition from heathland to grassland initiated by the heather beetle. *Vegetatio*, **72**, 167-173.

Biederbeck, V.O., C.A. Campbell, K.E. Bowren, M. Schnitzer, and R.N. McIver, 1980: Effect of burning on cereal straw, on soil properties and grain yields in Saskatchewan. *Soil Science Society of America Journal*, **44**, 103-111.

Billings, W.D., K.M. Peterson, J.O. Luken, and D.A. Mortensen, 1984: Interaction of increasing atom carbon dioxide and soil nitrogen on the carbon balance tundra microcosms. *Oecologia*, **65**, 26-29.

Bisset, W.J., 1962: The black speargrass (heteropogon contortus) problem of the sheep country in central western Queensland. *Queensland Journal of Agricultural Sciences*, **19**, 189-207.

Bonan, G.B., D. Pollard, and S.L. Thompson, 1992: Effects of boreal forest vegetation on global climate. *Nature*, **359(6397)**, 716-718.

Bouwman, A.F., 1993: Inventory of greenhouse gas emissions in the Netherlands. *Ambio*, **22**, 519-523.

Bouwman, A.F., 1994: Computing land use emissions of greenhouse gases. *Water, Air, and Soil Pollution*, **76**, 231-258.

Bouwman, A.F., K.W. Van der Hoek, and J.G.J. Olivier, 1995: Uncertainties in the global source distribution of nitrous oxide. *JGR-Atmospheres*, **100**, 2785-2800.

Briske, D.D. and R.K. Heitschmidt, 1991: An ecological perspective. In: *Grazing Management: An Ecological Perspective* [Heitschmidt, R.K. and J.W. Stuth (eds.)]. Timber Press Inc., Portland, OR, pp. 11-27.

Brown, J., 1994: Antarctic ozone going fast but some plants can survive. *New Scientist*, **143**, 11.

Buol, S.W., P.A. Sanchez, S.B. Weed, and J.M. Kimble, 1990: Predicted impact of climatic warming on soil properties and use. In: *Impact of Carbon Dioxide, Trace Gases, and Climate Change on Global Agriculture* [Kimball, B.A., N.J. Rosernberg, and L.H. Allen, Jr. (eds.)]. ASA Special Publ. No. 53, American Society of Agronomy, Crop Science Society of America, and Soil Science Society of America, Madison, WI, pp. 71-82.

Burke, I.C., C.M. Yonker, W.J. Parton, C.V. Cole, K. Flach, and D.S. Schimel, 1989: Texture, climate and cultivation effects on soil organic matter content in U.S. grassland soils. *Soil Science Society of America Journal*, **53**, 800-805.

Burke, I.C., T.G.F. Kittel, W.K. Lauenroth, P. Snook, C.M. Yonker, and W.J. Parton, 1990. Regional analysis of the Central Great Plains: sensitivity to climate variability. *Bioscience*, **41**, 685-692.

Burke, I.C., W.K. Lauenroth, and D.P. Coffin, 1995a: Recovery of soil organic matter and N mineralization in semiarid grasslands: implications for the Conservation Reserve Program. *Ecological Applications* (in press).

Burke, I.C., E.T. Elliott, and C.V. Cole, 1995b: Influence of macroclimate and landscape position and management on nutrient conservation and nutrient supply in agroecoystems. *Ecological Applications,* **5**, 124-131.

Cabrera, A.L., 1976: *Regiones Fitogeográficas Argentinas*. ACME, Buenos Aires, Argentina.

Caceres, D.M., 1993: *Peasant Strategies and Models of Technological Change: A Case Study from Central Argentina*. M.Phil. thesis, University of Manchester, Manchester, UK.

Caldwell, M.M., A.H. Termaura, and M. Tevini, 1989: The changing solar ultraviolet climate and the ecological consequences for higher plants. *Trends in Ecology and Evolution*, **4**, 363-367.

Campbell, D.J., 1984: Response to drought among farmers and herders in southern Kajiado District, Kenya. *Human Ecology*, **12(1)**, 35-64.

Campbell, B.D. and J.P. Grime, 1993: Prediction of grassland plant responses to global change. In: *Proceedings of the XVII International Grassland Congress*. pp. 1109-1118.

Campbell, B.D. and R.M. Hay, 1993: Will subtropical grasses continue to spread through New Zealand? In: *Proceedings of the XVII International Grassland Congress*. pp. 1126-1128.

Campbell, B.D. and M. Stafford Smith, 1993: Defining GCTE modelling needs for pastures and rangelands. In: *Proceedings of the XVII International Grassland Congress*. pp. 1249-1253.

Campbell, B.D., G.M. McKeon, R.M. Gifford, H. Clark, M.S. Stafford Smith, P.C.D. Newton, and J.L. Lutze, 1995a: Impacts of atmospheric composition and climate change on temperate and tropical pastoral agriculture. In: *Greenhouse '94* [Pearman, G.I. and M.R. Manning (ed.)]. CSIRO, Melbourne, Australia (in press).

Campbell, B.D., W.A. Laing, D.H. Greer, J.R. Crush, H. Clark, D.Y. Williamson, and M.D.J. Given, 1995b: Prediction of adaptive changes in improved temperate pastures as a result of global change. *Journal of Biogeography* (in press).

Carras, J.N., P.J. Fraser, D.W.T. Griffith, D.F. Hurst, and D.J. Williams, 1994: Trace gas emissions from Australian savannah fires during the 1990 dry season. *Journal Atmospheric Chemistry,* **18(1)**, 33-56.

Chaneton, E.J., J.M. Facelli, and R.J.C. León, 1988: Floristic changes induced by flooding on grazed and ungrazed lowland grasslands in Argentina. *Journal of Range Management*, **41**, 495-499.

Chapin, F.S. III and G.R. Shaver, 1985: Individualistic growth response of tundra plant species to environmental manipulations in the field. *Ecology,* **66**, 564-576.

Chapin, F.S. III, G.R. Shaver, A.E. Giblin, K.G. Nadelhoffer, and J.A. Laundre, 1995: Response of arctic tundra to experimental and observed changes in climate. *Ecology,* **76**, 694-711.

Cherfas, J., 1992: Cloudy issues, burning answers. *The Independent*, **9/7**.

Cicerone, R.J. and R. Oremland, 1988: Biogeochemical aspects of atmospheric methane. *Global Biogeochemical Cycles*, **2**, 299-327.

Coffin, D.P. and W.K. Lauenroth, 1990: A gap dynamics simulation model of succession in a semiarid grassland. *Ecol. Modelling*, **49**, 229-266.

Cole, C.V., I.C. Burke, W.J. Parton, D.S. Schimel, D.S. Ojima, and J.W.B. Stewart, 1990: Analysis of historical changes in soil fertility and organic matter levels of the North American Great Plains. In: *Challenges in Dryland Agriculture. A Global Perspective* [Unger, P.W., T.V. Sneed, W.R. Jordon, and R. Jensen (eds.)]. Proceedings of the International Conference on Dryland Farming, August 1988, Austin, TX, pp 436-438.

Cole, C.V., J.W.B. Stewart, D.S. Ojima, W.J. Parton, and D.S. Schimel, 1989: Modeling land use effects on soil organic matter dynamics in the North American Great Plains. In: *Ecology of Arable Land—Perspectives and Challenges. Developments in Plant and Soil Sciences* [Clarholm, M. and L. Bergstrom (eds.)] Kluwer Academic Publishers, Dordrecht , Netherlands, 39, pp. 89-98.

Coleman, J.S., K.D.M. McConnaughay, and F.A. Bazzaz, 1993: Elevated CO_2 and plant nitrogen use: is reduced tissue nitrogen concentration size-dependent? *Oecologia*, **93**, 195-200.

Comins, H.N. and R.E. McMurtrie, 1993: Long-term response of nutrient limited forests to CO_2 enrichment: equilibrium behavior of plant-soil models. *Ecological Applications*, **3**, 666-681.

Conroy, J.P., 1992: Influence of elevated atmospheric CO_2 concentrations on plant nutrition. *Australian Journal of Botany*, **40**, 445-456.

Coughenour, M.B., 1984: A mechanistic simulation analysis of water use, leaf angles, and grazing in East African graminoids. *Ecological Modelling*, **26**, 203-220.

Coughenour, M.B., J.E. Ellis, D.M. Swift, D.L. Coppock, K. Galvin, J.T. McCable, and T.C. Hart, 1985: Energy extraction and use in a nomadic pastoral ecosystem. *Science*, **230(4726)**, 619-625.

Coutinho, L.M., 1982: Ecological effects of fire in Brazilian cerrado. In: *Ecology of Tropical Savannas* [Huntley, B.J. and B.H. Walker (eds.)]. Springer-Verlag, Berlin, Germany, pp. 273-291.

Crush, J.R., B.D. Campbell, and J.P.M. Evans, 1993: Effect of elevated atmospheric CO_2 levels on nodule relative efficiency in white clover. In: *Proceedings of the XVII International Grassland Congress*. pp. 1131-1133.

Crutzen, P.J. and M.O. Andreae, 1990: Biomass burning in the tropics: impact on atmospheric chemistry and biogeochemical cycles. *Science,* **250**, 1669-1678.

D'Antonio, C.M. and P.M. Vitousek, 1992: Biological invasions by exotic grasses, the grass/fire cycle, and global change. *Annual Review of Ecological Systems*, **23**, 63-87.

Dacey, J.W.H., B.G. Drake, and M.J. Klug, 1994: Stimulation of methane emission by CO_2 enrichment of marsh vegetation. *Nature*, **370**, 47.

Daubenmire, R., 1968: Ecology of fire in grasslands. *Advanced Ecological Research*, **5**, 209-266.

Davidson, E.A., 1991: Fluxes of nitrous oxide and nitric oxide from terrestrial ecosystems. In: *Microbial Production and Consumption of Greenhouse Gases: Methane, Nitrogen Oxides, and Halomethanes* [Rogers, J.A. and W.B. Whitman (eds.)]. American Society for Microbiology, Washington DC.

Dentener, F.J. and P.J. Crutzen, 1994: A three dimensional model of the global ammonia cycle. *Journal of Atmospheric Chemistry*, **19**, 331-369.

Deregibus, V.A., 1988: Importancia de los pastizales naturales en la República Argentina: situación present y future. *Revista Argentina de Producción Animal*, **8**, 67-78.

de Ridder, N., L. Stroosnijder, A.M. Cisse, and H. van Keulen, 1982: PPS course book. Vol. I, Theory. In: *Productivity of Sahelian Rangelands: A Study of the Soils, the Vegetations and the Exploitation of That Natural Resource.* Wageningen Agricultural University, Department of Soil Science and Plant Nutrition, Wageningen, Netherlands, pp. 227-231.

Diaz, S., 1995: Elevated CO_2 responsiveness, interactions at the community level, and plant functional types. *J. Biogeography*, **22** (in press).

Díaz, S., J.P. Grime, J. Harris, and E. McPherson, 1993: Evidence of a feedback mechanism limiting plant response to elevated carbon dioxide. *Nature*, **364**, 616-617.

Dickinson, R.E. and A. Henderson-Sellers, 1988: Modeling tropical deforestation: a study of GCM land-surface parameterizations. *Q. J. R. Meteorol. Soc.*, **114**, 439-462.

Dyer, M.I., C.L. Turner, and T.R. Seastedt, 1993: Herbivory and its consequences. *Ecological Applications*, **3**, 10-16.

Easterling, W.E., 1990: Climate trends and prospects. In: *Natural Resources for the 21st Century* [Sampson, R.N. and D. Hair (eds.)]. Island Press, Washington DC, pp. 32-55.

Ehleringer, J.R. and C.B. Field (eds.), 1993: *Scaling Physiological Processes: Leaf to Globe*. Academic Press, New York, NY, 388 pp.

Ellis, J. and K.A. Galvin, 1994: Climate patterns and land use practices in the dry zones of east and west Africa. *Bioscience*, **44(5)**, 340-349.

Ellis, J.E., K.A. Galvin, J.T. McCabe, and D.M. Swift, 1990: *Pastoralism and Drought in Turkana District, Kenya*. A Report to NORAD, Development Systems Consultants, Inc., Bellvue, CO.

Ellis, J. and D.M. Swift, 1988: Stability of African pastoral ecosystems: alternate paradigms and implications for development. *Journal of Range Management*, **41**, 450-459.

Emanuel, W.R., H.H. Shugart, and M.P. Stevenson, 1985: Climatic change and the broad-scale distribution of terrestrial ecosystems. *Climate Change*, **7**, 29-43.

Evans, P.A. and M.R. Ashmore, 1992: The effects of ambient air on a semi-natural grassland community. *Agric. Ecosyst. Environ.*, **38(1-2)**, 91-97.

Evans, E.W. and T.R. Seastedt, in press: The relations of phytophagous invertebrates and rangeland plants. In: *Rangeland Plant Morphology and Physiology* [Bedunah, D.J. (ed.)]. Society for Range Management, Denver, CO.

Fajer, E.D., M.D. Bowers, and F.A. Bazzaz, 1989: The effects of enriched carbon dioxide atmospheres on plant-insects herbivore interactions. *Science*, **243**, 1198-1200.

Fajer, E.D., M.D. Bowers, and F.A. Bazzaz, 1991: The effects of enriched CO_2 atmospheres on the buckeye butterfly, Junonia coenia. *Ecology*, **72**, 751-754.

Farinas, M. and J.J. San José, 1987. Cambios en el estrato herbáceo de una parcela de sabana protegida del fueo y del pastoreo durante 23 años. Calabozo. Venezuela. *Acta Científica Venezolana*, **36**, 199-200.

Fernández-A, R.J., O.E. Sala, and R.A. Golluscio, 1991: Woody and herbaceous aboveground production of a Patagonian steppe. *Journal of Range Management*, **44**, 434-437.

Firestone, M.K. and E.A. Davidson, 1989: Microbial basis of NO and N_2O production and consumption in soil. In: *Exchange of Trace Gases Between Terrestrial Ecosystems and the Atmosphere* [Andreae, M.O. and D. Schimel (eds.)]. John Wiley and Sons, Chichester, UK.

Fisher, M.J., I.M. Rao, M.A. Ayarza, C.E. Lascano, J.I. Sanz, R.J. Thomas, and R.R. Vera, 1994: Carbon storage by introduced deep-rooted grasses in the South American savannas. *Nature*, **371**, 236-238.

Foran, B.D., G. Bastin, E. Remenga, K.W. Hyde, 1982: The response to season, exclosure and distance from water of three central Australian pasture types grazed by cattle. *Australian Rangeland Journal*, **4**, 5-15.

Frost, P., E. Medina, J.C. Menaut, O. Solbrig, M.J. Swift, and B. Walker (eds.), 1986: Responses of savannas to stress and disturbance: a proposal for a collaborative program of research. In: *Biology International*, Special Issue No. 10. International Union of Biological Sciences, Paris, France, pp. 1-82.

Galvin, K.A., 1992: Nutritional ecology of pastoralists in dry tropical Africa. *American Journal of Human Biology*, **4(2)**, 209-221.

Galvin, K.A., D.L. Coppock, and P.W. Leslie, 1994: Diet, nutrition and the pastoral strategy. In: *African Pastoralist Systems: An Integrated Approach* [Fratkin, E., K.A. Galvin, and E.A. Roth (eds.)]. Lynne Rienner, Boulder, CO, pp. 113-132.

Gifford, R.M., 1977: Growth pattern, carbon dioxide exchange and dry weight distribution in wheat growing under differing photosynthetic environments. *Australian Journal of Plant Physiology*, **4**, 99-110.

Gifford, R.M., 1994: The global carbon cycle: a viewpoint about the missing carbon. *Australian Journal of Plant Physiology*, **21**, 1-15.

Goldstein, G. and G. Sarmiento, 1987: Water relations of trees and grasses and their consequences for the structure of savanna vegetation. In: *Determinants of Tropical Savannas*. IUBS Monograph Series 3, IRL Press Ltd, Oxford, UK, pp. 13-38.

Graetz, R.D., B.H. Walker, and P.A. Walker, 1988: The consequences of climatic change for seventy percent of Australia. In: *Greenhouse. Planning for Climate Change*. CSIRO, Division of Atmospheric Research, Melbourne, Australia, pp. 399-420.

Greco, S., R.H. Moss, D. Viner, and R. Jenne, 1994: *Climate Scenarios and Socioeconomic Projections for IPCC WGII Assessment*. IPCC-WMO and UNEP, Washington DC, 67 pp.

Griffin, G.F. and M.H. Friedel, 1985: Discontinuous change in central Australia: some implications of major ecological events for land management. *Journal of Arid Environments*, **9**, 63-80.

Grulke, N.E., G.H. Riechers, W.C. Oechel, U. Hjelm, and C. Jaeger, 1990: Carbon balance in tussock tundra under ambient and elevated CO_2. *Oecologia*, **83**, 485-494.

Haas, H.J., C.E. Evans, and E.F. Miles, 1957: *Nitrogen and Carbon Changes in Great Plains Soils Influenced by Cropping and Soil Treatments*. Technical Bulletin 1164, USDA, Washington DC, 111 pp.

Hall, D.O. and J. Scurlock, 1991: Tropical grasslands and their role in the global carbon cycle. In: *Facets of Modern Ecology* [Esser, G. and D. Overdick (eds.)]. Elsevier, Amsterdam, Netherlands, pp. 555-573.

Hanson, J.D., B.B. Baker, and R.M. Bourdon, 1993: Comparison of the effects of different climate change scenarios on rangeland livestock production. *Agricultural Systems*, **41(4)**, 487-502.

Hastings, J.R. and R.M. Turner, 1965: *The Changing Mile*. University of Arizona Press, Tucson, AR, 317 pp.

Havstrom, M., T.V. Callaghan, and S. Jonasson, 1993: Differential growth responses of Cassiope tetragona, an arctic dwarf-shrub, to environmental perturbations among three contrasting high- and sub-arctic sites. *Oikos*, **66**, 389-402.

Heady, H.F. (ed.), 1988: *The Vale Rangeland Rehabilitation Program: An Evaluation*. Resource Bulletin PNW-RB-157, USDI Bureau of Land Management, USDA Forest Service, Portland, OR, 151 pp.

Heady, H.F. and R.D. Childs, 1994: *Rangeland Ecology and Management*. Westview Press, Boulder, CO, 519 pp.

Heitschmidt, R.K. and J.W. Stuth, 1991: *Grazing Management. An Ecological Perspective*. Timber Press, Portland, OR, 259 pp.

Henderson-Sellers, A. and K. McGuffie, 1995: Global climate models and 'dynamic' vegetation changes. *Global Change Biology*, **1**, 63-75.

Hicks, R.A., D.D. Briske, C.A. Call, and R.J. Ansley. 1990: Co-existence of a perennial C_3 bunchgrass in a C_4 dominated grassland: an evaluation of gas exchange characteristics. *Photosynthetica*, **24**, 63-74.

Hobbs, R.J. and L.F. Huenneke, 1992: Disturbance, diversity and invasion: implication for conservation. *Conservation Biology*, **6**, 324-337.

Hobbs, R.J., D.S. Schimel, C.E. Owensby, and D.S. Ojima, 1991: Fire and grazing in the tallgrass prairie: contingent effects on nitrogen budgets. *Ecology*, **72**, 1374-1382.

Holdridge, L.R., 1947: Determination of world plant formations from simple climate data. *Science*, **105**, 367-368.

Holland, E.A. and J.K. Detling, 1990: Plant response to herbivory and below ground nitrogen cycling. *Ecology*, **71**, 1040-1049.

Holland, E.A., W.J. Parton, J.K. Detling, and D.L. Coppock, 1992: Physiological responses of plant populations to herbivory and their consequences for ecosystem nutrient flow. *American Naturalist*, **142**, 685-706.

Homewood, K.M. and W.A. Rodgers, 1984: Pastoralism and conservation. *Human Ecology*, **12**, 431-442.

Homewood, K., W.A. Rodgers, and K. Arhem, 1987: Ecology of pastoralism in Ngorongoro Conservation Area, Tanzania. *Journal of Agricultural Science, Cambridge*, **108**, 47-72.

Howden, S.M., 1991: Methane production from livestock. Draft Australian Greenhouse Gas Emissions Inventory 1987-88. *Greenhouse Studies*, **10** (DASET), 15-22.

Howden, S.M., D.H. White, G.M. McKeon, J.C. Scanlan, and J.O. Carter, 1994: Methods for exploring management options to reduce greenhouse gas emissions from tropical pastures. *Climate Change*, **30**, 49-70.

Hulbert, L.C., 1988: Causes of fire effects in tallgrass prairie. *Ecology*, **69**, 46-58.

Hulme, M., 1990: The changing rainfall resources of Sudan. *Transactions, Institute of British Geographers*, **15**, 21-34.

Hunt, H.W., M.J. Trlica, E.F. Redente, J.C. Moore, J.K. Detling, T.G.F. Kittel, D.E. Walter, M.C. Fowler, D.A. Klein, and E.T. Elliott, 1991: Simulation model for the effects of climate change on temperate grassland ecosystems. *Ecological Modelling*, **53**, 205-246.

Hutsch, B.W., C.P. Webster, and D.S. Powlson, 1995: Methane oxidation in soil as affected by land use, soil pH and N fertilization. *Soil Biology and Biochemistry*, **24**, 1613-1622.

Idso, K.E. and S.B. Idso, 1994: Plant responses to atmospheric CO_2 enrichment in the face of environmental constraints: a review of the last 10 years' research. *Aric. For. Meteorol.*, **69**, 153-203.

Idso, S.B. 1989: *Carbon Dioxide and Global Change: Earth in Transition*. IBR Press, Tempe, AR, 292 pp.

IPCC, 1990: *Climate Change: The IPCC Scientific Assessment* [Houghton, J.T., G.J. Jenkins, and J.J. Ephraums (eds.)]. Cambridge University Press, Cambridge, UK, 365 pp.

IPCC, 1992: *Climate Change 1992. The Supplementary Report to the IPCC Scientific Assessment* [Houghton, J.T., B.A. Callander, and S.K. Varney (eds.)]. Cambridge University Press, Cambridge, UK, 200 pp.

Jenkinson, D.S., 1977: Studies on the decomposition of plant material in soil. V. The effects of plant cover and soil type on the loss of carbon from ^{14}C labeled rye grass decomposing under field conditions. *Journal of Soil Science Society of America*, **28**, 424-494.

Jenkinson, D.S., J.M. Potts, J.N. Perry, V. Burnett, K. Coleman, and A.E. Johnston, 1994: Trends in herbage yields over the last century on the Rothamsted Long-Term Continuous Hay experiment. *Journal of Agricultural Science, Cambridge*, **122**, 365-374.

Johnson, R.H. and D.E. Lincoln, 1991: Carbon allocation patterns and grasshopper nutrition—the influence of CO_2 enrichment and soil mineral limitation. *Oecologia*, **87**, 127-134.

Keller, M., E. Veldkamp, A. Weitz, and W. Reiners, 1993: Effect of pasture age on soil trace gas emissions from a deforested area of Costa Rica. *Nature*, **365**, 244-246.

Kemp, P.R. and G.J. Williams III, 1980: A physiological basis for niche separation between Aropyron smithii (C_3) and Bouteloua racilis (C_4). *Ecology*, **61**, 846-858.

Kielland, K., 1990: *Processes Controlling Nitrogen Release and Turnover in Arctic Tundra*. Ph.D. diss., University of Alaska, Fairbanks, AK.

Kimball, B.A., 1983: Carbon dioxide and agricultural yield: an assemblage and analysis of 430 prior observations. *Agronomy Journal*, **75**, 779-788.

Kimball, B.A., J.R. Mauney, F.S. Nakayama, and S.B. Idso, 1993: Effects of elevated CO_2 and climate variables on plants. *Journal Soil Water Conservation*, **48**, 9-14.

Kimball, B.A., N.J. Rosenberg, and L.H. Allen, Jr. (eds.), 1990: *Impact of Carbon Dioxide, Trace Gases, and Climate Change on Global Agriculture*. ASA Spec. Publ. No. 53, American Society of Agronomy, Crop Science Society of America, and Soil Science Society of America, Madison, WI, 133 pp.

King, G.M. and S. Schnell, 1994: Effect of increasing atmospheric methane concentration on ammonium inhibition of soil methane consumption. *Nature*, **370**, 282-284.

Knapp, A.K., 1984: Post-burn differences in solar radiation, leaf temperature and water stress influencing production in lowland tallgrass prairie. *American Journal of Botany*, **71**, 220-227.

Knapp, A.K., 1985: Effect of fire and drought on ecophysiology of Andropogon gerardii and Panicum virgatum in tallgrass prairie. *Ecology*, **66**, 1309-1320.

Knapp, A.K. and T.R. Seastedt, 1986: Detritus accumulation limits the productivity of tallgrass prairie. *Bioscience*, **36**, 662-668.

Koizumi, H., T. Nakadai, Y. Usami, M. Satoh, M. Shiyomi, and T. Oikawa, 1991: Effect of carbon dioxide concentration on microbial respiration in soil. *Ecol. Res.*, **6**, 227-232.

Korner, C., 1993: CO_2 fertilization: the great uncertainty in future vegetation development. In: *Vegetation Dynamics and Global Change* [Solomon, A.M. and H.H. Shugart (eds.)]. Chapman & Hall, New York, NY, pp. 53-70.

Korner, C. and F. Miglietta, 1994: Long term effects of naturally elevated CO_2 on mediterranean grassland and forest trees. *Oecologia*, **99**, 343-351.

Krupa, S.V. and R.N. Kickert, 1989: The greenhouse effect: impacts of ultraviolet-B (UV-B) radiation, carbon dioxide (CO_2) and ozone (O_3) on vegetation. *Environmental Pollution*, **61**, 263-393.

Kucera, C.L., 1981: Grasslands and fire. In: *Fire Regimes and Ecosystem Properties*. U.S. Forest Gen. Tech. Rep. WO-26, USDA, Washington DC, 594 pp.

Lauenroth, W.K. and O.E. Sala, 1992: Long-term forage production of North American shortgrass steppe. *Ecological Applications*, **2**, 397-403.

Lavrenko, E.M. and Z.V. Karamysheva, 1992: Steppes of the former Soviet Union and Mongolia. In: *Natural Grassland*. Elsevier, Amsterdam and New York, pp. 3-59.

Lean, G., D. Hinrichsen, and A. Markham, 1990: *Atlas of the Environment*. Prentice Hall Press, New York, NY, 192 pp.

Le Houerou, H.N., 1984: Rain use efficiency: a unifying concept in arid-land ecology. *Journal of Arid Environments*, **7**, 213-247.

Le Houerou, H.N. and C.H. Hoste, 1977: Rangeland production and annual rainfall relations in the Mediterranean basin and in the African Shelo-Sudanian zone. *Journal of Range Management*, **30**, 181-189.

Lewis, J.P., E.F. Pire, D.E. Prado, S.L. Stofella, E.A. Franceschi, and N.J. Carnevali, 1990: Plant communities and phytogeographical position of a large depression in the Great Chaco, Argentina. *Vegetatio*, **86**, 25-38.

Little, P.D., 1987: Land use conflicts in the agricultural/pastoral borderlands: the case of Kenya. In: *Lands at Risk in the Third World* [Little, P.D., M.M. Horowitz, and A.E. Nyerges (eds.)]. Westview, Boulder, CO, pp. 195-212.

Lodman, D.W., M.E. Branine, B.R. Carmen, P. Zimmerman, G.M. Ward, and D.E. Johnson, 1993: Estimates of methane emissions from manure of United States cattle. *Chemosphere*, **26**, 189-200.

Long, S.P. and P. Hutching, 1991: Primary production in grasslands and coniferous forests with climate change: an overview. *Ecological Applications*, **1**, 139-156.

Londsdale, M. and R. Braithwaite, 1988: The shrub that conquered the bush. *New Scientist*, **15**, 52-55.

Mann, G.F., 1966: *Bases Ecológicas de la Explotación Agropacuaria en América Latina*. OEA, Chile.

Marion, G.M., D.S. Introve, and K. Vaneleve, 1991: The stable isotope geochemistry of $CaCO_3$ on the Tanana River floodplain of interior Alaska, USA—Composition and mechanisms of formation. *Chemical Geology*, **86**, 97-110.

Maxwell, B., 1992: Arctic climate: potential for change under global warming. In: *Arctic Ecosystems in a Changing Climate: An Ecosphysiological Perspective* [Chapin, F.S. III, R.L. Jeffereies, J.F. Reynolds, G.R. Shaver, and J. Svoboda (eds.)]. Academic Press, San Diego, CA, pp. 11-34.

Mayeux, H.S., H.B. Johnson, and H.W. Polley, 1991: Global change and vegetation dynamics. In: *Noxious Range Weeds* [James, L.F., J.O. Evans, M.H. Ralphs, and R.D. Child (eds.)]. Westview Press, Boulder, CO, pp. 62-74.

McCabe, J.T., 1990: Turkana pastoralism. A case against the tragedy of the commons. *Human Ecology*, **18**, 81-104.

McGuire, A.D., J.M. Melillo, L.A. Joyce, D.W. Kicklighter, A.L. Grace, B. Moore III, and C.J. Vorosmarty, 1992: Interactions between carbon and nitrogen dynamics in estimating net primary productivity for potential vegetation in North America. *Global Biogeochemical Cycles*, **6**, 101-124.

McKeon, G.M., S.M. Howden, *et al.*, 1989: The effect of climate change on crop and pastoral production in Queensland. In: *Greenhouse: Planning for Climate Change*. CSIRO, Melbourne, Australia, pp. 546-565.

McKeon, G.M., K.A. Day, S.M. Howden, J.J. Mott, D.M. Orr, W.J. Scattini, and E.J. Weston, 1990: Northern Australian savannas: management for pastoral production. *Journal of Biogeography*, **17**, 355-372.

McKeon, G.M. and S.M. Howden, 1991: Adapting northern Australian grazing systems to climatic change. *Climate Change Newsletter*, **3(1)**, 5-8.

McKeon, G.M., S.M. Howden, N.O.J. Abel, and J.M. King, 1993: Climate change: adapting tropical and sub-tropical grasslands. In: *Proc. of the XVIIth International Grassland Congress, Palmerston North, New Zealand*. pp. 1181-1190.

Medina, E., 1982: Physiological ecology of neotropical savanna plants. In: *Ecology of Tropical Savannas* [Hyntley, B.J. and B.H. Walker (eds.)]. Springer-Verlag, Berlin, Germany, pp. 308-335.

Medina, E., 1986: Forests, savannas and monyane tropical environments. In: *Photosynthesis in Contrasting Environments* [Baker, N.R. and S.P. Long (eds.)]. Elsevier Science Publishers, B.V., Amsterdam and New York, pp. 139-171.

Medina, E., 1987: Nutrients, requirements, conservation and cycles in the herbaceous layer. In: *Determinants of Tropical Savannas*. IVBS Monograph Series No. 3, IRL Press LTD, Oxford, UK, pp. 39-65.

Medina, E. and J. Silva, 1990: Savannas of the northern South America: a steady state regulated by water-fire interactions on a background of low nutrient availability. *Journal of Biogeography*, **17**, 403-413.

Milchunas, D.G., O.E. Sala, and W.K. Lauenroth, 1988: A generalized model of the effects of grazing by large hervibores on grassland community structure. *American Naturalist*, **132**, 87-106.

Mooney, H.A., P.M. Vitousek, and P.A. Matson, 1987: Exchange of materials between the biosphere and atmosphere. *Science*, **238**, 926-932.

Moore, J., 1990: Effects of climate change on rangelands. In: *World Resources*. Oxford University Press, Oxford, UK, pp. 111-116.

Morello, J., N.E. Crudelli, and M. Saraceno, 1971: *La Vegetacion de la Republica Argentina. Los Vinalarer de Formosa (la Colonizodova Leñosa Prosopsis ruscifolia Gris.)*. INTA, Buenos Aires, Argentina.

Morison, J.I.L. 1987. Intercellular CO_2 concentration and stomatal response to CO_2. In: *Stomatal Function* [Zeiger, E., G.D. Farquhar, and I. Cowan (eds.)]. Stanford University Press, Stanford, CA, pp. 229-251.

Mosier, A.R., D. Schimel, D. Valentine, K. Bronson, and W. Parton, 1991: Methane and nitrous oxide fluxes in native, fertilized and cultivated grasslands. *Nature*, **350**, pp. 330-332.

Murray, D.F., 1980: Balsam poplar in arctic Alaska. *Canadian Jounal of Anthropology*, **1**, 29-32.

Nadelhoffer, K.J., A.E. Giblin, G.R. Shaver, and J.A. Laundre, 1991: Effects of temperature and substrate quality on element mineralization in 6 arctic soils. *Ecology*, **72(1)**, 242-253.

Nakadai, T. Koizumi, Y. Usami, M. Satoh, and T. Oikawa, 1993: Examination of the method for measuring soil respiration in cultivated land: effect of carbon dioxide concentration on soil respiration. *Ecol. Res.*, **8**, 65-71.

Neilson, R.P., 1993: Vegetation redistribution: a possible biosphere source of CO_2 during climatic change. *Water, Air, and Soil Pollution*, **70**, 659-673.

Neilson, R.P., 1986: High-resolution climatic analysis and southwest biogeography. *Science*, **232**, 27-34.

Newton, P.C.D., H. Clark, C.G. Bell, E.M. Glasgow, and B.D. Campbell, 1994: Effects of elevated CO_2 and simulated seasonal changes in temperature on the species composition and growth rate of pasture turves. *Annals of Botany*, **73**, 53-59.

Newton, P.D., H. Clark, C.C. Bell, E.M. Glasgow, K.R. Tate, D.J. Ross, G.W. Yeates, and S. Saggar, 1995: Plant growth and soil processes in temperate grassland communities at elevated CO_2. *Global Ecology and Biogeography* (in press).

Nicholson, S.E., 1983: Climate variations in the Sahel and other African regions during the past five centuries. *Journal of Arid Environments*, **1**, 3-24.

Nie, D., H. He, G. Mo, M.B. Kirkham, and E.T. Kanemasu, 1992: Canopy photosynthesis and evapotranspiration of rangeland plants under doubled carbon dioxide in closed-top chambers. *Agriculture Forest Meteorology*, **61**, 205-217.

NRC, 1994: *Rangeland Health*. National Academy Press, Washington DC, 108 pp.

Oberbauer, S.F. and W.C. Oechel, 1989: Maximum CO_2-assimilation rates of vascular plants on an Alaskan arctic tundra slope. *Holarctic Ecology*, **12(3)**, 312-316.

Oechel, W.C., S.J. Hastings, G. Vourlitis, and M. Jenkins, 1993: Recent change of arctic tundra ecosystems from a net carbon dioxide sink to a source. *Nature*, **361(6412)**, 520-523.

Oechel, W.C. and B.R. Strain, 1985: Native species responses to increased carbon dioxide concentration. In: *Direct Effects of Increasing Carbon Dioxide on Vegetation* [Strain, B.R. and J.D. Cure (eds.)]. DOE/ER-0238, U.S. Department of Energy, NTIS, Springfield, VA, pp. 117-154.

OIES, 1991: *Arid Ecosystem Interactions*. Report OIES-6 UCAR, UCAR Office for Interdisciplinary Research, Boulder, CO.

Ojima, D.S., W.J. Parton, D.S. Schimel, and C.E. Owensby, 1990: Simulated impacts of annual burning on prairie ecosystems. In: *Fire in North American Tallgrass Prairies* [Collins, S.L. and L.L. Wallace (eds.)]. University of Oklahoma Press, Norman, OK, pp. 99-118.

Ojima, D.S., D.W. Valentine, A.R. Mosier, W.J. Parton, and D.S. Schimel, 1993a: Effect of land use change and methane oxidation in temperate forest and grassland soils. *Chemosphere*, **2**, 675-685.

Ojima, D.S., W.J. Parton, D.S. Schimel, J.M.O. Scurlock, and T.G.F. Kittel, 1993b: Modelling the effects of climatic and carbon dioxide changes on grassland storage of soil carbon. *Water, Air, Soil Pollution*, **70(1-4)**, 643-657.

Ojima, D.S., D.S. Schimel, W.J. Parton, and C.E. Owensby, 1994a: Long- and short-term effects of fire on nitrogen cycling in tallgrass prairie. *Biogeochemistry*, **23**, 1-18.

Ojima, D.S., K.A. Galvin, and B.L. Turner, 1994b: The global impact of land-use change. *Bioscience*, **44(5)**, 300-304.

Old, S.M., 1969: Microclimate, fire and plant production in an Illinois prairie. *Ecological Monographs*, **39**, 355-384.

O'Leary, M., 1984: Ecological villains or economic victims: the case of the Rendille of northern Kenya. *Desertification Control Bulletin*, **11**, 17-21.

Ottichilo, W.K., J.H. Kinuthia, P.O. Ratego, and G. Nasubo, 1991: *Weathering the Storm—Climate Change and Investment in Kenya*. Environmental Policy Series No. 3, Acts Press, Nairobi, Kenya, 91 pp.

Overdieck, D., 1993. Elevated CO_2 and the mineral content of herbaceous and woody species. *Vegetatio*, **104/105**, 403-411.

Owensby, C.E., P.I. Coyne, J.M. Ham, L.M. Auen, and A.K. Knapp, 1993: Biomass production in a tallgrass prairie ecosystem exposed to ambient and elevated CO_2. *Ecological Applications*, **3**, 644-653.

Parton, W.J., J.M.O. Scurlock, D.S. Ojima, D.S. Schimel, D.O. Hall, and SCOPEGRAM Group Members, 1995: Impact of climate change on grassland production and soil carbon worldwide. *Global Change Biology*, **1**, 13-22.

Parton, W.J., J.M.O. Scurlock, D.S. Ojima, T.G. Gilmanov, R.J. Scholes, D.S. Schimel, T. Kirchner, J.C. Menaut, T. Seastedt, E. Garcia Moya, A. Kamnalrut, and J.L. Kinyamario, 1993: Observations and modeling of biomass and soils organic matter dynamics for the grassland biome worldwide. *Global Biogeochemical Cycles*, **7(4)**, 785-809.

Parton, W.J., B. McKeown, V. Kirchner, and D. Ojima, 1992: *CENTURY Users' Manual*. Natural Resource Ecology Laboratory, Colorado State University, Fort Collins, CO.

Parton, W.J., D.S. Schimel, C.V. Cole, and D.S. Ojima, 1987: Analysis of factors controlling soils organic matter levels in Great Plains Grasslands. *Soil Science Society of America Journal*, **51**, 1173-1179.

Parton, W.J., J.W.B. Stewart, and C.V. Cole, 1988: Dynamics of C, N, P, and S in grassland soils: a model. *Biogeochemistry*, **5**, 109-131.

Parton, W.J., D.S. Schimel, D.S. Ojima, and C.V. Cole, 1994: A general model for soil organic matter dynamics: sensitivity to litter chemistry, texture and management. In: *Quantitative Modeling of Soil Forming Processes* [Bryant, R.B. and R.W. Arnold (eds.)]. SSSA Spec. Publ. 39. ASA, CSSA and SSA, Madison, WI, pp. 137-167.

Payne, N.F. and F.C. Bryant, 1994: *Techniques for Wildlife Habitat Management of Uplands*. McGraw-Hill, New York, 840 pp.

Polley, H.W., H.R. Johnson, B.D. Marino, and H.S. Mayeux, 1993a: Increase in C_3 plant water-use efficiency and biomass over glacial to present CO_2 concentrations. *Nature*, **361**, 61-64.

Polley, H.W., H.B. Johnson, and H.S. Mayeux, 1994: Increasing CO_2: comparative responses of the C_4 grass Schizachyrium and grassland invader Prosopis. *Ecology* (in press).

Poppi, D.P., D.J. Minson, and J.H. Ternouth, 1981: Studies of cattle and sheep-eating leaf and stem fractions of grasses. I. The voluntary intake, digestibility and retention time in reticulo-rumen. *Australian Journal of Agricultural Research*, **32**, 99-108.

Prather, M., R. Derwent, D. Ehhalt, P. Fraser, E. Sanhueza, and X. Zhou, 1995: Other trace gases and atmospheric chemistry. In: *Climate Change 1994: Radiative Forcing of Climate Change and an Evaluation of the IPCC IS92 Emission Scenarios* [Houghton, J.T., L.G. Meira Filho, J. Bruce, H. Lee, B.A. Callander, E. Haites, N. Harris, and K. Maskell (eds.)]. Cambridge University Press, Cambridge, UK, pp. 75-126.

Prentice, K.C., W. Cramer, S.P. Harrison, R. Leemans, R.A. Monserud, and A.M. Solomon, 1992: A global biome model based on plant physiology and dominance, soil properties, and climate. *Journal of Biogeography*, **19**, 117-134.

Prentice, K.C. and I.Y. Fung, 1990: The sensitivity of terrestrial carbon storage to climate change. *Nature*, **346**, 48-51.

Raison, R.J., 1979: Modification of the soil environment by vegetation fires with particular reference to nitrogen transformations: a review. *Plant and Soil*, **51**, 73-108.

Reibsame, W.E., 1990: Anthropogenic climate change and a new paradigm of natural resource planning. *Professional Geographer*, **42(1)**, 1-12.

Risser, P.G. and W.J. Parton, 1982: Ecological analysis of tallgrass prairie: nitrogen cycle. *Ecology*, **63**, 1342-1352.

Risser, P.G., E.C. Birney, H.D. Blocker, S.W. May, W.J. Parton, and J.A. Weins, 1981: *The True Prairie Ecosystem*. Hutchinson Ross, Stroudsburg, PA, 557 pp.

Riveros, F., 1993: Grasslands for our world. In: *Proceedings of the XVII International Grassland Congress*. pp. 15-20.

Rogers, H.H., G.B. Runion, and S.V. Drupa, 1994: Plant responses to atmospheric CO_2 enrichment with emphasis on roots and the rhizosphere. *Environmental Pollution*, **83**, 155-189.

Rosenzweig, M.L., 1968: Net primary productivity of terrestrial communities: prediction from climatological data. *American Naturalist*, **102**, 67-74.

Rozema, J., H. Lambers, S.C. van de Geijn, and M.L. Cambridge, 1993: *CO_2 and Biosphere*. Kluwer Academic Publishers, Boston, MA, 484 pp.

Rutherford, M.C., 1980: Annual plant production-precipitation relations in arid and semiarid regions. *South African Journal of Science*, **76**, 53-56.

Running, S.W. and E.R. Hunt, Jr., 1993: Generalization of a forest ecosystem process model for other biomes, BIOME-BGC, and an application for global-scale models. In: *Scaling Processes Between Leaf and Landscape Levels* [Ehleringer, J.R. and C. Field (eds.)]. Academic Press, Orlando, FL, pp. 141-158.

Rychnovska, M., 1992: Temperate seminatural grasslands of Eurasia. In: *Natural Grassland*. Elsevier, Amsterdam and New York, pp. 125-163.

Sala, O.E., W.K. Laurenroth, and W.J. Parton, 1992: Long-term soil water dynamics in the short-grass steppe. *Ecology*, **73**, 1175-1181.

Sala, O.E., R.A. Golluscio, W.K. Lauenroth, and A. Soriano, 1989: Resource partitioning between shrubs and grasses in the Patagonian steppe. *Oecologia*, **81**, 501-505.

Sala, O.E., W.J. Parton, L.A. Joyce, and W.K. Lauenroth, 1988: Primary production of the central grasslands region of the United States. *Ecology*, **69**, 40-45.

Sampson, R.N., M. Apps, S. Brown, *et al.*, 1993: Workshop summary statement: terrestrial biospheric carbon fluxes—quantification of sinks and sources of CO_2. *Water, Air and Soil Pollution*, **70**, 3-15.

Sandford, S., 1983: *Management of Pastoral Development in the Third World.* John Wiley, New York, NY, 316 pp.

Sarmiento, G., 1983: The savannas of tropical America. In: *Tropical Savannas* [Bourliere, F. (ed.)]. Elsevier, Amsterdam, Netherlands, pp. 245-288.

Sarmiento, G., 1984: *The Ecology of Neotropical Savannas.* Harvard University Press, Cambridge, MA, 235 pp.

Sarmiento, G. and M. Monasterio, 1983: Life forms and phenology. In: *Ecosystems of the World 13. Tropical Savannas* [Bourliere, F. (ed.)]. Elsevier, Amsterdam, Netherlands, pp. 79-108.

Sarmiento, G., G. Goldstein, and F. Meinzer, 1985: Adaptative strategies of woody species in neotropical savannas. *Biological Reviews*, **60**, 315-355.

Schimel, D.S., B.H. Braswell, Jr., E.A. Holland, R. McKeown, D.S. Ojima, T.H. Painter, W.J. Parton, and A.R. Townsend, 1994: Climatic, edaphic, and biotic controls over carbon and turnover of carbon in soils. *Global Biogeochem. Cycles*, **8**, 279-293.

Schimel, D.S., 1988. Calculation of microbial growth efficiency from nitrogen-15 immobilization. *Biogeochemistry*, **6(3)**, 239-243.

Schimel, D.S., T.G.F. Kittel, and W.J. Parton, 1991: Terrestrial biogeochemical cycles—global interactions with the atmosphere and hydrology. Tellus Series A—*Dynamic Meteorology and Oceanography*, **43(4)**, 188-203.

Schimel, D.S., W.J. Parton, T.G.F. Kittel, D.S. Ojima, and C.V. Cole, 1990: Grassland biogeochemistry: links to atmospheric processes. *Climatic Change*, **17**, 13-25.

Schlesinger, W.H., J.F. Reynolds, G.L. Cunningham, L.F. Huenneke, W.M. Jarrell, R.A. Virginia, and W.G. Whitford, 1990: Biological feedbacks in global desertification. *Science*, **247**, 1043-1048.

Schofield, C.J. and E.H. Bucher, 1986: Industrial contributions to desertification in South America. *Trends in Ecology and Evolution*, **1**, 78-80.

Scholes, R.J., 1993: Nutrient cycling in semi-arid grasslands and savannas: its influence on pattern, productivity and stability. In: *Proceedings of the XIV International Grassland Congress.* 2304 pp.

Scholes, R.J. and B.H. Walker, 1993: *An African Savanna—Synthesis of the Nylsvley Study.* Cambridge Studies in Applied Ecology and Resource Management, Cambridge University Press, Cambridge, UK, 306 pp.

Seastedt, T.R. and A.K. Knapp, 1993: Consequences of nonequilibrium resource availability across multiple time scales: the transient maxima hypothesis. *American Naturalist*, **141**, 621-633.

Seastedt, T.R., J.M. Briggs, and D.J. Gibson, 1991: Controls of nitrogen limitation in tallgrass prairie. *Oecologia* (Berlin), **87**, 72-79.

Silva, J.F., J. Raventos, H. Caswell, and M.A. Trevisan, 1991: Population responses to fire in a tropical savanna grass, Andropogon semiberbis: a matrix model approach. *Journal of Ecology*, **79**, 345-356.

Sims, P.L., 1988: Grasslands. In: *North American Terrestrial Vegetation* [Barbour, M.G. and W.D. Billings (eds.)]. Cambridge University Press, New York, NY, 434 pp.

Sinclair, A.R.E. and M. Norton-Griffiths (eds.), 1979: *Serengeti: Dynamics of an Ecosystem.* University of Chicago Press, Chicago, IL, 389 pp.

Skarpe, C., 1990: Shrub layer dynamics under different herbivore densities in an arid savanna, Botswana. *Journal of Applied Ecology*, **27**, 873-885.

Smith, D.D., B.R. Strain and T.D. Sharkey, 1987. Effects of CO_2 enrichment on four Great Basin grasses. *Functional Ecology*, **1**, 139-143.

Solbrig, O., G. Goldstein, E. Medina, G. Sarmiento, and J.F. Silva, 1992: Responses of tropical savannas to stress and disturbance: a research approach. In: *Ecosystem Rehabilitation.* Vol. 2, *Ecosystem Analysis and Synthesis* [Wali, M.K. (ed.)]. SPB Academica Publishing, The Hague, Netherlands, pp. 63-73.

Solomon, A.M., I.C. Prentice, R. Leemans, and W.P. Cramer, 1993: The interaction of climate and land use in future terrestrial carbon storage and release. *Water, Air, and Soil Pollution*, **70**, 595-614.

Solomon, A.M. and H.H. Shugart (eds.), 1993: *Vegetation Dynamics and Global Change.* Chapman and Hall, New York, NY, 337 pp.

Soriano, A., 1992: Rio de la Plata grasslands. In: *Natural Grasslands: Introduction and Western Hemisphere* [Coupland, R.T. (ed.)]. Elsevier, Amsterdam, Netherlands, pp. 367-408.

Sperling, L., 1987: Wage employment among Samburu pastoralists of north central Kenya. *Research in Economic Anthropology*, **9**, 167-190.

Spicer, R.A. and J.L. Chapman, 1990: Climate change and the evolution of high-latitude terrestrial vegetation and floras. *Trends in Ecology and Evolution*, **5(9)**, 279-284.

Stafford Smith, D.M., 1993: *A Regional Framework for Managing the Variability of Production in the Rangelands of Australia. Defining Stocking Rates for Viable and Sustainable Production in the Face of Market and Climatic Variability.* CSIRO/RIRDC Project CSW-28A.

Stafford Smith, D.M., 1994: *Sustainable Production Systems and Natural Resource Management in the Rangelands.* ABARE Outlook Conference.

Stafford Smith, D.M. and G. Pickup, 1993: Out of Africa, looking in: understanding vegetation change. In: *Range Ecology at Disequilibrium* [Behnke, R., I. Scoones, and C. Keerven (eds.)]. Overseas Development Institute, London, UK, pp. 196-244.

Stafford Smith, D.M. and S.R. Morton, 1990: A framework for the ecology of arid Australia. *Journal of Arid Environments*, **18**, 255-278.

Stafford Smith, M., B. Campbell, W. Steffen, S. Archer, and D. Ojima, 1995: *GCTE Task 3.1.3: Global Change Impacts on Pastures and Rangelands. Implementation Plan.* Global Change and Terrestrial Ecosystems Report No. 3, GCTE Core Project Office, Canberra, Australia.

Stephenson, N.L., 1990: Climatic control of vegetation distribution: the role of the water balance. *American Naturalist*, **135**, 649-670.

Strain, B.R. and J.D. Cure (eds.), 1985: *Direct Effects of Increasing Carbon Dioxide on Vegetation.* DOE/ER-0238, Carbon Dioxide Research Division, U.S. Department of Energy, Washington DC, 286 pp.

Suriano, J.M., L.H. Perpozzi, and D.E. Martinez, 1992: El cambio global: tendencias climáticas en la Argentina y el mundo. *Ciencia Hoy*, **3**, 32-39.

Tate, K.R., 1992: Assessment, based on a climosequence of soils in tussock grasslands, of soil carbon storage and release in response to global warming. *Journal of Soil Science*, **43**, 697-707.

Tate, K.R., D.J. Giltrap, A. Parshotam, A.E. Hewitt, and D.J. Ross, 1995: Impacts of climate change on soils and land systems in New Zealand. In: *Greenhouse '94.* CSIRO Australia (in press).

Terri, J.A. and L.G. Stowe, 1976: Climatic patterns and the distribution of C_4 grasses in North America. *Oecologia* (Berlin), **23**, 1-12.

Thornley, J.H.M., D. Fowler, and M.G.R. Cannell, 1991: Terrestrial carbon storage resulting from CO_2 and nitrogen fertilization in temperate grasslands. *Plant, Cell and Environment*, **14**, 1007-1011.

Thornley, J.H.M. and E.L.J. Verberne, 1989: A model of nitrogen flows in grassland. *Plant, Cell and Environment*, **12**, 863-886.

Tieszen, L.L., M.M. Senyimba, S.K. Imbamba, and J.H. Troughton, 1979: The distribution of C_3 and C_4 grasses and carbon isotope discrimination along an altitudinal and moisture gradient in Kenya. *Oecologia* (Berlin), **37**, 337-350.

Tissue, D.L. and W.C. Oechel, 1987: Physiological response to Eriphorum vaginatium to field elevated CO_2 and temperature in the Alaskan tussock tundra. *Ecology*, **68**, 401-410.

Titlyanova, A.A., R.I. Zlotin, and N.R. French, 1990: Changes in structure and function of temperate grasslands under the influence of man. In: *Managed Grasslands (Regional Studies).* Elsevier, Amsterdam and New York, pp. 301-334.

Torn, M.S. and J.S. Fried, 1992: Predicting the impacts of global warming on wildland fire. *Climatic Change*, **21**, 257-274.

Towne, G. and C.E. Owensby, 1984: Long-term effects of annual burning at different dates in ungrazed Kansas tallgrass prairie. *Journal of Range Management*, **37**, 392-397.

Townsend, A.R., P.M. Vitousek, and E.A. Holland, 1992: Tropical soils could dominate the short-term carbon cycle feedbacks to increased global temperatures. *Climate Change*, **22**, 293-303.

Turner, C.L., T.R. Seastedt, and M.I. Dyer, 1993: Maximization of above ground grassland production: the role of defoliation frequency, intensity and history. *Ecological Applications*, **3**, 175-186.

Valentine, D.W., E.A. Holland, and D.S. Schimel, 1994: Ecosystem and physiological controls over methane production in northern wetlands. *JGR-Atmospheres*, **99**, 1563-1571.

Van Vegten, J.A., 1983: Thornbush invasion in a savanna ecosystem in eastern Botswana. *Vegetatio*, **56**, 3-7.

VEMAP Participants, 1995: Vegetation/Ecosystem Modeling and Analysis Project (VEMAP): Assessing biogeography and biogeochemistry models in a regional study of terrestrial ecosystem responses to climate change and CO_2 doubling. *Global Biogeochemical Cycles* (in press).

Walker, B.H., 1993: Rangeland ecology: understanding and managing change. *Ambio*, **22(23)**, 80-87.

Walter, H., 1968: *Die vegetation der Erde in óko-physiologischer Bertrachtung*. Gustav Fischer Verlag, Stuttgart, Germany, 935 pp.

Walter, H., 1973: *Vegetation of the Earth in Relation to Climate and the Eco-Physiological Conditions*. Springer-Verlag, New York, NY, 237 pp.

Walter, H., 1979: *Vegetation of the Earth and Ecological Systems of the Geo-biosphere*. Heidelberg Science Library, Springer-Verlag, New York, NY, 2nd ed., 274 pp.

Ward, G.M., K.G. Doxtader, W.C. Miller, and D.E. Johnson, 1993: Effects of intensification of agricultura practices on emission of greenhouse gases. *Chemosphere*, **26**, 87-94.

Weaver, J.E. and F.W. Albertson, 1944: Nature and degree of recovery of grassland from the great drought of 1933 to 1940. *Ecological Monographs*, **14(4)**, 393-497.

Webb, W.L., W.K. Lauenroth, S.T. Szarek, and R.S. Kinerson, 1983: Primary production and abiotic controls in forests, grasslands, and desert. *Ecology*, **64**, 134-151.

Webb, W., S. Szarek, W.K. Laurenroth, and R. Kinerson, 1978: Primary production and water use in native forest, grassland and desert ecosystems. *Ecology*, **59**, 1239-1247.

Wedin, D.A. and D. Tilman, 1990: Species effects on nitrogen cycling: a test with perennial grasses. *Oecologia*, **84**, 433-441.

Westoby, M., B. Walker, and I. Noy-Meir, 1989: Opportunistic management for rangelands not at equilibrium. *Journal of Range Management*, **42**, 26-274.

Whetton, P., 1993: New climate change scenarios. *Climate Change Newsletter*, **5(2)**, 7-8.

Whisenant, S.G., 1990: Changing fire frequencies on Idaho's Snake River Plains: ecological and management implications. In: *Proceedings Cheatgrass Invasion, Shrub Dieoff, and Other Aspects of Shrub Biology, and Management*. USDA Forest Service Inter-mountain Research Station, Gen. Tech. Report INT-276, pp. 4-10.

White, R.G. and J. Trudell, 1980: Habitat preference and forage consumption by reindeer and caribou near Atkasook, Alaska. *Arctic and Alpine Research*, **12**, 511-529.

Whiting, G.J. and J.P. Chanton, 1993: Primary production control of methane emissions from wetlands. *Nature*, **36**, pp. 794-799.

Williams, J.B. and G.O. Batzli, 1982: Pollination and dispersion of five species of lousewort (Pedicularis) near Atkasook, Alaska. *Arctic and Alpine Research*, **14**, 59-74.

Wong, S.C., I.R. Cowan, and G.D. Farquhar, 1979: Stomatal conductance correlates with photosynthetic capacity. *Nature*, **282**, 424-426.

Wookey, P.A., A.N. Parsons, J.M. Welker, J.A. Potter, T.V. Callaghan, J.A. Lee, and M.C. Press, 1993: Comparative responses of phenology and reproductive development to simulated envrionmental change in sub-arctic and high arctic plants. *Oikos*, **67**, 490-502.

WRI, 1992: *World Resources, 1992-1993: A Guide to the Global Environment, Toward Sustainable Development*. Oxford University Press, New York and Oxford, 400 pp.

Yavitt, J.B., J.A. Simmona, T.J. Fahey, 1993: Methane fluxes in a northern hardwood forest ecosystem in relation to acid precipitation. *Chemosphere*, **26**, 721-730.

Yienger, J.J., and H. Levy II, 1995: Global inventory of soil biogenic NO_x emissions. *JGR-Atmospheres* (in press).

Zak, D.R., K.S. Pregitzer, P.S. Curtis, J.A. Teeri, R. Fogel, and D.L. Randlett, 1993. Elevated atmospheric CO_2 and feedback between carbon and nitrogen cycles. *Plant and Soil*, **151**, 105-117.

Zimov, S.A., I.P. Semiletov, and Y.V. Voropaev, 1993a: Wintertime CO_2 emission from soils of northeastern Siberia. *Arctic*, **46(3)**, 197-204.

Zimov, S.A., G.M. Zimova, S.P. Daviodov, and A.I. Daviodova, 1993b: Winter biotic activity and production of CO_2 in Siberian soils—a factor in the greenhouse effect. *Journal of Geophysical Research—Atmospheres*, **98(3)**, 5017-5023.

3

Deserts in a Changing Climate: Impacts

IAN R. NOBLE, AUSTRALIA; HABIBA GITAY, AUSTRALIA

Contributing Authors:
A.N. Alwelaie, Saudi Arabia; M.T. Hoffman, South Africa; A.R. Saunders, Australia

CONTENTS

EXECUTIVE SUMMARY

Deserts are an environmental extreme. The biota in deserts are at the limit of conditions that allow growth and survival and show very specialized adaptations to aridity and heat. The biomass of plants and animals is low; nevertheless, deserts have a significant species richness and a high endemism. The following statements regarding climate change can be made with certainty:

- Most deserts are likely to become even more extreme if climate changes as projected by current scenarios; most desert regions are expected to become hotter and most will probably not become significantly wetter (Medium Confidence).
- If changes in the frequency or intensity of the rainfall events occur, they are likely to cause changes in the flora and fauna. Higher atmospheric CO_2 concentrations and more intense rainfall events may lead to the opportunity for greater invasion of desert systems, particularly by C_3 plants. More frequent large pulses of rain will allow ephemeral organisms to reproduce more often. Any reduction in the intensity of the rainfall, however, could also be detrimental to this set of organisms due to false starts in their life cycles (Medium Confidence).
- In a few places (e.g., central Australia) conditions may improve, but any net change in rainfall—thus the vegetation—will depend largely on the human management of these regions (Low Confidence).
- Opportunities to mitigate greenhouse gas emissions in desert regions are few (High Confidence).

Human-induced desertification has the potential to counteract any ameliorating effect of climate change on most deserts unless appropriate management actions are taken.

3.1. Introduction

This chapter describes deserts and the main processes that dominate extreme desert environments. It then discusses how these processes might be affected by climate change. Variability in different deserts is described, and the likely impacts on specific areas are explored. The chapter concentrates on the hottest and driest land types and excludes extreme-cold deserts of polar regions.

3.2. Climate and Biology

3.2.1. Definition and Extent

Deserts are characterized by low rainfall that is highly variable both intra- and interannually. Desert air is very dry, and incoming solar and outgoing terrestrial radiation are intense. Large daily temperature fluctuations occur (ca. 30°C), and potential evapotranspiration is high. There is high spatial and temporal variability in desert biota, driven largely by water availability (Noy Meir, 1985).

There are many definitions of deserts, but the common point is that water (or the lack of it) is the dominant factor controlling ecosystem processes. A common definition classifies ecosystems with an annual rainfall of <250 mm as arid and <100 mm as extremely arid (Noy Meir, 1973; Walter, 1985). Both are commonly referred to as deserts, but this chapter concentrates on the extremely arid regions; Chapter 2 deals with most aspects of arid and semi-arid regions. The permanent biota in extremely arid regions have specialized adaptations to cope with the harsh environment—sometimes obtaining their moisture from fog or dew. There also are ephemeral biota that migrate into desert regions or become active from drought-resistant seeds or dormant life stages, to take advantage of the rare rainfall events. Thus, extreme deserts have very distinctive characteristics.

Desert environments arise in several circumstances. Most deserts occur between 20° and 40° latitude, where persistent high-pressure cells bring dry air to the Earth's surface (see Figure 2-2 in Chapter 2). However, the climate depends on the precise balance between this air mass and topography, continentality, and sea currents (Evenari, 1985a). Some deserts are in the rain shadows of major mountain ranges (e.g., the Atacama desert of South America; see also Chapter 5). Others are a great distance from a source of oceanic moisture or are more-or-less continuously swept by winds that have traveled a great distance over land (e.g., the central Asian deserts).

Approximately 30% of the globe's land surface is desert or semi-desert, with almost 5% receiving less than 70 mm annual rainfall, 11% less than 100 mm, and 18% less than 120 mm (Shmida, 1985; Le Houerou, 1992). Some of the extremely arid deserts are shown in Figure 3-1. Human population estimates for the extremely arid regions are not available; however, for regions classified as drylands (i.e., <75 growth days per year)

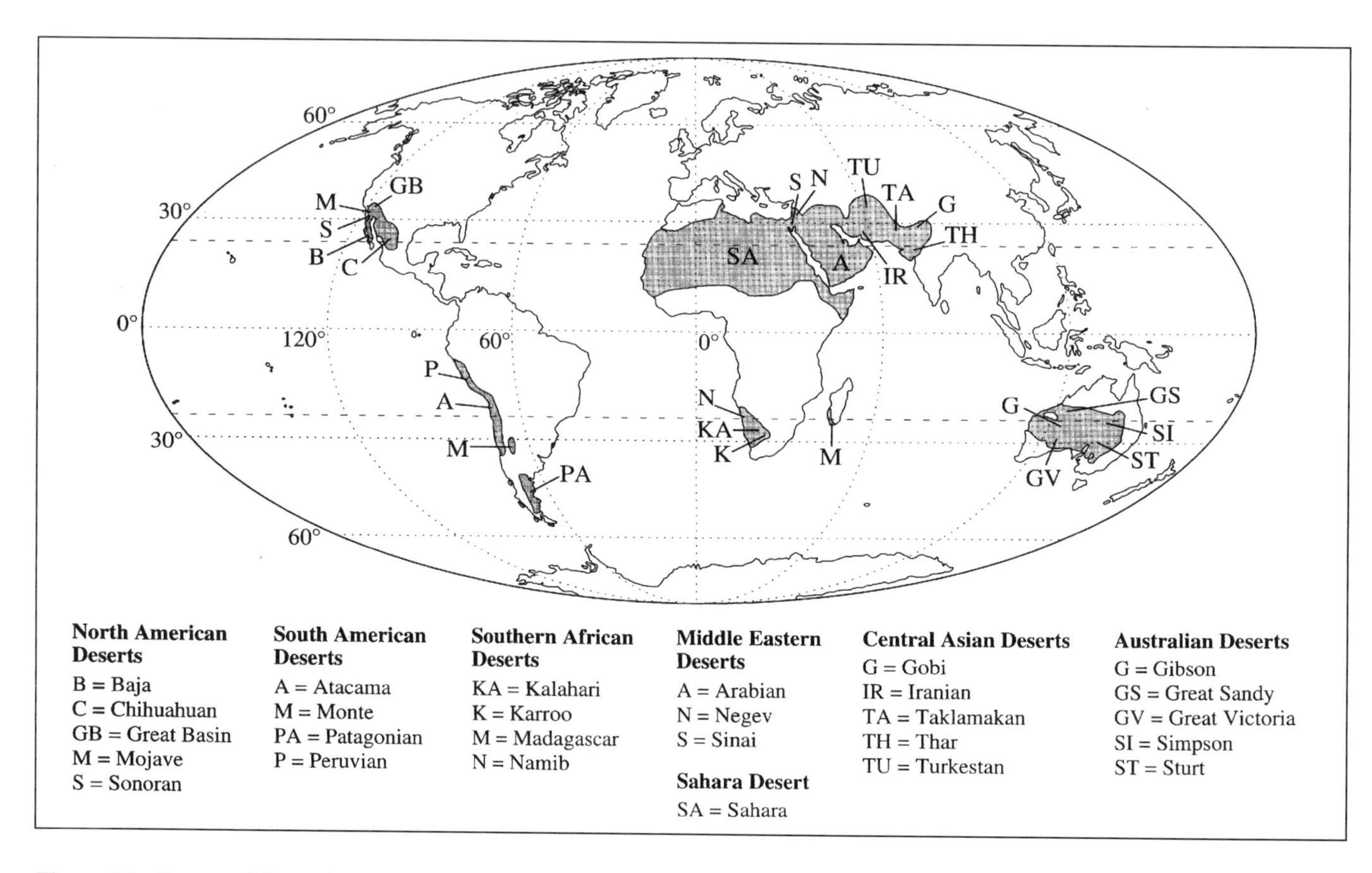

Figure 3-1: Deserts of the world.

it is over 800 million people (World Resources Institute, 1988–89). Many of these people will be affected by phenomena described in this chapter.

3.2.2. *Biological Productivity*

Average aboveground net primary productivity in areas with annual precipitation of <250 mm varies from almost 0 to about 2000 kg/ha/yr (Evenari *et al.*, 1971; Noy Meir, 1973; Hadley and Szarek, 1981; Webb *et al.*, 1983; Le Houerou, 1984). Most productivity estimates are for aboveground vegetation only; there is very little information on belowground plant productivity. Root to shoot ratios differ greatly among desert life forms (Noy Meir, 1973). Ephemeral species often have relatively small root systems (root:shoot < 0.5), whereas some perennials have large belowground storage systems with root:shoot ratios as high as 20. Productivity is directly related to rainfall, often by a simple linear function. Alternatively, it can be expressed as a close relationship with actual evapotranspiration because actual evapotranspiration is closely related to rainfall (Branson *et al.*, 1981). Estimates of water-use efficiency by desert plants vary from 0.3 to 2.0 mg dry matter produced per gram of water transpired, with a threshold of 25 to 170 mm of annual rainfall to sustain any vegetation at all (Noy Meir, 1973; Webb *et al.*, 1978).

3.2.3. *Desert Biodiversity*

Although deserts are not as diverse as some biomes, they have significant diversity—with many groups of organisms and often a high degree of endemism (Le Houerou, 1992). Taxa that are diverse relative to other biomes include predatory arthropods, ants, termites, snakes, lizards, migratory birds, succulents, and annual plants (McGinnies *et al.*, 1968; Huenneke and Noble, 1995). There is substantial variation in the richness of particular taxa among the deserts of different continental areas (Evenari, 1985b; Huenneke and Noble, 1995). Desert biota are unusual in the frequency with which particular morphological or physiological adaptations have evolved independently in different taxonomic groups (Cloudsley-Thompson, 1993). For example, many animals and plants have thick skins or cuticles to reduce water loss; other have hairs or spines to reflect radiation.

Biodiversity—measured as the number of species (i.e., species richness)—is moderately high in semi-arid regions and declines with increasing aridity for most taxa (Shmida, 1985; Pianka and Schall, 1981; Currie, 1991; O'Brien, 1993). For example, Aronson and Shmida (1992) found that plant species richness fell linearly from 116 species at a site with 280 mm mean annual rainfall to 15 species at a site with about 30 mm mean annual rainfall. The number of life forms (e.g., annual grasses, small shrubs) declined with increasing aridity, and only chamaephytes (low-growing perennial plants with buds just above the soil surface; e.g., *Salsola tetrandra*) persisted at the driest sites in all years. In addition, they found very little fluctuation of species richness with variability in annual rainfall at the driest site.

Correlative studies show that the diversity and density of some animal groups are related to vegetation productivity gradients, precipitation, and potential and actual evapotranspiration (Clark, 1980; Maurer, 1985; Owen, 1988; Currie, 1991; Hoffman *et al.*, 1994; Specht and Specht, 1994). However, the links are complex—depending on the taxonomic group, seasonality of precipitation, frequency of droughts, and the scale at which diversity and density are measured. Thus, changes in animal populations in response to changes in climate are very difficult to predict.

3.3. Ecosystem Variables

3.3.1. *Temperature*

The hottest deserts of the world (e.g., central Sahara, Namib) have average monthly temperatures above 30°C during the warmest months, with extremes above 50°C. The diurnal range often is very high, with nights as cold as 10°C. High insolation loads mean that soil surface temperatures often rise to more than 80°C (Cloudsley-Thompson, 1977).

Some high continental deserts have extremely cold winters; e.g., the Chihuahuan Desert in North America has average monthly minimum temperatures below freezing (MacMahon and Wagner, 1985). Cold conditions prevent effective growth of plants in winter; however, snow or moisture often accumulates and can contribute to growth in the warmer months.

3.3.2. *Rain and Soil Moisture*

3.3.2.1. *Factors in Moisture Budget*

Precipitation, soils, and temperature all contribute to the nature of desert ecosystems. Vegetation is largely a function of rainfall modified by soil type (in particular, the coarseness of the soil; Bertiller *et al.*, 1995) because these two factors determine potential soil moisture and water-holding capacity. Higher temperatures usually imply higher vapor-pressure deficits (evaporative demands) and thus higher evaporation and transpiration losses.

The availability of moisture at and near the soil surface is critical for the activity of desert organisms. The water available to plants is dependent on the balance between input from precipitation and losses from direct evaporation from the soil surface, transpiration through the leaf-root systems of plants, and drainage below the root zone. Thus, the effectiveness of a given amount of rain is dependent on how far it infiltrates below the surface; greater penetration reduces evaporative losses. More-intense rainfall events tend to penetrate less and create more runoff than less-intense events. In contrast to most other arid ecosystems, the sandy soils of extreme deserts support higher plant biomass than do clay soils (Walter, 1985) because water penetrates deeper in coarser soils.

To a large degree, water loss due to transpiration is an unavoidable byproduct of the necessity for plants to take up CO_2 from the atmosphere through stomata. Various adaptations have developed to limit this loss (Ehleringer and Monson, 1993). Most C_3 species have a common metabolism but can vary enormously in leaf structure, leaf protection, stomatal density, and control over stomatal activity. For example, many species close their stomata during the hottest part of the day to avoid the highest rates of water loss. Some C_4 species have a special leaf anatomy and metabolism that allows them to concentrate CO_2 within the leaf and thus reduce water loss. Several major groups of plants (mostly succulents) have CAM species, in which most gas exchange is limited to the cooler nighttime hours.

These mechanisms are all of particular importance in desert communities, where water is usually limiting while light and CO_2 are usually in excess. Increased concentrations of atmospheric CO_2 will alter the relative effectiveness of these mechanisms and thus the balance of plant communities. It is expected that this will favor C_3 species over C_4 and CAM species (Skiles and Hanson, 1994), which could make desert communities more subject to invasion and increased dominance by C_3 species.

3.3.2.2. *Pulse System*

Biological activity in most ecosystems is controlled by the availability of light, water, and nutrients. Complex interactions and feedbacks have developed that govern the growth and development of organisms in these systems. In contrast, desert ecosystems are dominated by rainfall events. Rainfall is infrequent, usually highly unpredictable, and often provides moist conditions for only a short period. Deserts have been described as pulse-driven ecosystems (Noy Meir, 1973). This means that the response of desert ecosystems to climate change may be different than that of other systems, where more complex feedbacks and interactions have developed.

Most desert organisms are adapted to one of two survival mechanisms (see Noy Meir, 1973). Some survive in a dormant state between rainfall pulses and take quick advantage of the brief respites. Typical examples are the ephemeral plants, but certain species of crustacea, molluscs, and frogs use similar strategies (Williams and Calaby, 1985). Other organisms—such as perennial plants, reptiles, and some mammals—persist through the dry conditions because they have extremely efficient water and energy conservation strategies (Williams and Calaby, 1985). Some smaller plants and animals are able to persist on the moisture settling as dew or fog (Louw and Seely, 1982).

Climatic changes that increase the frequency of pulses will favor the pulse-adapted (i.e., ephemeral) species if the duration of favorable conditions is not shortened. In the long term, more-frequent pulses of ephemerals could affect the long-lived perennial species either by directly competing for resources or indirectly affecting local fire frequency and grazing intensity (Skarpe, 1992).

If, under climate change, the duration of periods of moist conditions becomes shorter or more variable, the pulse-adapted species may be disadvantaged. They will be triggered into responding more frequently but will often fail to complete their life cycle in the shorter moist periods (Frasier *et al.,* 1985, 1987). For example, ephemeral plants could be triggered into germinating but not have time to set seeds before drying out, thus depleting their seed bank.

3.3.3. *Wind*

Windiness varies from desert to desert, but in general windiness is not significantly different than in nondeserts except that there is little vegetation to moderate the wind near the ground surface. This creates soil instability and is the major factor creating and shaping dune systems and extending desert boundaries (Alwelaie *et al.,* 1993). Some organisms are adapted to feed on the accumulation of organic matter on the slip faces of dunes (Louw and Seely, 1982).

3.3.4. *Fogs*

Fog and dew can each contribute 30 mm or more of moisture per year. Dew can form on more than half the nights of a year in some deserts (Seely, 1979). Thus, fog and dew are an important and relatively reliable source of water for plants and animals alike (e.g., Namib beetles and lizards; Louw and Seely, 1982). In some of the driest coastal deserts (e.g., Atacama or Peruvian deserts), fogs are particularly frequent and provide the majority if not all of the moisture in most years (Rauh, 1985). Alwelaie *et al.* (1993) have suggested that the shrub-dominated vegetation in a desert island in the Red Sea depends on fog for its water supply.

3.3.5. *Nutrient Cycles*

The main sources of nitrogen in deserts are N-fixing organisms on the soil surface such as lichens, cyanobacteria, moss, and fungi. Atmospheric deposition is an important source in some areas. Nitrogen fixation inputs in arid regions of the United States are 25 to 40 kg N/ha/yr (West and Skujins, 1977; Rundel *et al.,* 1982), whereas atmospheric deposition is estimated to be 3 kg N/ha/yr (Peterjohn and Schlesinger, 1990). Assimilation of nitrogen from fog is sometimes considered to be a significant source of N, but Evans and Ehleringer (1994) found that this is not the case in the fog zones of the Atacama Desert.

Rates of decomposition in deserts are slow because warm and moist conditions that favor decomposition are rare and volatilization of N can be high due to low vegetation cover and high evaporation. Thus, many desert systems are N-limited. Some studies have shown that decomposition rates are correlated with rainfall in arid regions (Santos *et al.*, 1984), but others (Whitford *et al.*, 1986) have failed to find a correlation. Steinberger and Shmida (1984) conclude that along a rainfall gradient of 300 mm to 25 mm there is no correlation between decomposition and rainfall and that most decomposition in very dry regions arises from mechanical rather than biotic processes.

Soil nitrogen is unevenly distributed in arid environments and is usually higher in surface soil and beneath larger plants or shrubs than in bare areas (Charley and West, 1977). Most nitrogen fixation occurs near the soil surface, and species that take up most of their water from the surface zone might be at an advantage over deeper-rooted species. However, Evans and Ehleringer (1994) found that this is not the case for deep-rooted *Chrysothamnus nauseosus* individuals compared with neighboring shallow-rooted species. They conclude that water deep in the soil profile also provides an adequate nitrogen source.

3.4. Impacts of Climate Change

3.4.1. *Higher Temperature*

For most desert regions, the temperature increases predicted in the Greco *et al.* (1994) scenarios are typically in the range of 0.5 to 2.0°C, with greater increases in summer. A rise of 2°C without an increase in precipitation will increase potential evapotranspiration by 0.2 to 2 mm per day (Mabbutt, 1989; Chapter 4). Evapotranspiration may increase due to higher leaf temperatures and higher surface and soil evaporation rates, although the direct effects of higher CO_2 concentrations may partly compensate the tendency for increased transpiration.

Nasrallah and Balling (1993) found an average 0.07°C increase per decade in temperature records for the Middle East from 1945–90 with the greatest increases in spring and the smallest in winter. This increase appears to have been slowed by sulfate emissions and increased by human-induced desertification. Mühlenbruch-Tegen (1992) found no evidence of changes in mean daily temperatures during the past 50 years in South Africa but did find changes in daily minimum and maximum temperatures. These could be accounted for by changes in cloudiness. For example, increased cloud cover may decrease maximum temperatures and increase minimum temperatures by reducing insolation during the day and long-wave reradiation at night. Lane *et al.* (1994) found that from 1901–87, mean annual temperatures increased by 0.12°C per decade in the desert southwest region of the United States, but they could not separate an anthropogenic component from natural variability. Changes in land-use practices can affect surface temperatures. Balling (1988) has suggested that an observed 2.5°C temperature difference across the Mexican/U.S. boundary in the Sonoran desert may be due largely to heavier grazing on the Mexican side, which has led to lower vegetation cover and higher albedo.

Many organisms are already near their tolerance limits, and some may not be able to persist under hotter conditions. Higher temperatures in arid regions with cold winters are likely to allow spring growth to begin earlier. In some cases this may result in earlier depletion of water reserves accumulated over the cooler winter, leading to an even longer period of potential drought (Skiles and Hanson, 1994).

3.4.2. *Rainfall*

The Greco *et al.* (1994) scenarios indicate that while some desert regions will receive more rainfall than at present, most will remain extremely arid. In most deserts the predicted additional rainfall is at most only a few centimeters per year, with some models predicting a decrease in many areas. Decreased rainfall is predicted for large parts of the Sahara, northern Arabian, Sonoran, and central and western Asian deserts. Rainfall in the Chihuahuan, southern Arabian, and Atacama deserts is predicted to increase slightly or remain the same, while some of the arid regions of central Australia are predicted to have significant increases in precipitation. The prediction of increased rainfall in Australian desert regions based on the Greco *et al.* (1994) scenarios is contradicted, however, by a general circulation model (GCM) developed by CSIRO in Australia (Whetton *et al.*, 1994). In an alternative approach, Wigley *et al.* (1980) used past warm-weather periods as an indicator of changes in precipitation under global warming. They conclude that precipitation in the arid and desert portions of the United States would decrease under a warmer climate scenario.

The coefficient of annual variation (standard deviation/mean) increases rapidly as mean annual rainfall falls from 300 to 100 mm per year; for sites with 100 mm annual rainfall, the standard deviation can be 0.65 or higher (MacMahon and Wagner, 1985; Fisher, 1994). This means that at sites with 100 mm/yr average rainfall, one year in six will have less than 35 mm of rain. Thus, extreme desert systems already experience wide fluctuations in rainfall and are adapted to coping with sequences of extreme conditions. Initial changes associated with climate change are less likely to create conditions significantly outside the present range of variation.

As CO_2 concentrations increase, plant transpiration is expected to decrease and water-use efficiency to rise (Mooney *et al.*, 1991). Simulation studies for semi-arid and arid sites indicate that water-use efficiency (i.e., carbon fixed per unit water transpired) might increase by up to 50% under a doubled-CO_2 scenario. Even in simulations in which precipitation was assumed to decrease by 10%, standing biomass increased by 15 to 30% (Skiles and Hanson, 1994).

All predictions of changes in precipitation should be treated with caution. None of the current GCMs can reliably predict the distribution of rainfall throughout the year, nor the distribution of rainfall per event. These characteristics are as important as absolute rainfall amounts in the initiation and maintenance of plant growth. Pittock (1988) suggests that in Australian arid areas there may be an increase in the amount of rainfall for a given event. This suggests that pulse species could benefit in these areas because more of the events may result in moisture sufficient for the completion of their life cycles.

3.4.3. *Runoff and Ephemeral Waters*

The impact of changes in temperature and precipitation regimes on runoff and groundwater recharge are complex and depend greatly on the precise timing of events and assumptions about surface conditions. Increases in rainfall intensity, amount per event, and storm clustering will tend to increase runoff in arid and semi-arid ecosystems (Wang and Mayer, 1994; Weltz and Blackburn, 1995). Vegetative cover tends to reduce runoff (Walker, 1991; Weltz and Blackburn, 1995).

A landscape mosaic of runoff and run-on areas is an important component of desert systems. In some deserts with gently sloping surfaces (<0.6°), narrow bands of vegetation ("stripes") develop. These bands trap the runoff from wider bands of almost-bare surface upslope of them (White, 1971). At a larger scale, arid lands often are a mosaic of erosion cells, each consisting of erosional, transfer, and sink areas (Pickup, 1985). The sink areas receive the benefits of most of the precipitation falling on the entire cell and accumulate many of the surface nutrients. Both of these situations will amplify any change in runoff patterns (Noble and Burke, 1995).

Lapenis and Shabalova (1994) have suggested that soil structural change may be an important component of climate change because increased precipitation will lead to more organic matter being accumulated in the soil, improved soil field capacity, enhanced moisture availability, and changed runoff patterns. However, their model shows that this process is slow (ca. 400 yr) in desert regions. Fossil aquifers are important water sources in many deserts but will not be affected in the time scales relevant to humans.

Catchment-scale runoff is difficult to predict. An empirical study that predicted catchment-scale runoff from precipitation and temperature suggests that runoff will decrease in arid watersheds in the western United States (Revelle and Waggoner, 1983). However, Idso and Brazel (1984) show that the effect could be reversed if changes in water-use efficiency under increased CO_2 concentrations were taken into account. Skiles and Hanson (1994) used a more detailed ecosystem model to take into account the seasonal responses of vegetation and found that there was little change, or a decrease, in runoff in simulated semi-arid watersheds. Morassutti (1992) concludes that runoff from three major drainage basins in Australia, including the arid Lake Eyre drainage basin, would increase under scenarios based on the GISS GCM. However, he warns that there are a number of shortcomings in the model and that no trends in precipitation and runoff are discernible from the historical record.

Neilson and Marks (1994) used five GCMs to look at the effect of climate change at the global scale. Their maps suggest that average annual runoff (in the five GCMs) under doubled-CO_2 scenarios for most of the desert areas of the world will remain the same or increase by 0–25 mm at the most. However, this lack of response may be due to their inclusion of Leaf Area Index (LAI) as a major component in the runoff calculations. In all their desert biomes, LAI is <0.1 or not calculable.

Scattered permanent and ephemeral waters are important for migrating and breeding birds. More than 300,000 birds can reproduce on the ephemeral Lake Eyre in central Australia after a major rainfall event that floods the lake— an event that has occurred only four times in the last 100 years. A population of thousands of egrets has been recorded in ephemeral waters following torrential rainfall in the Peruvian coastal deserts (Caviedes, 1984). Ephemeral water bodies can thus be very important to wetland birds (Kingsford and Porter, 1992) in different desert systems. However, very little is known about the effect of ephemeral events on the overall population dynamics of birds.

3.4.4. *Desert Locusts*

The desert locust has a wide distribution in Africa and southwestern Asia, extending over areas with a mean annual precipitation of <50 to 500 mm (Uvarov, 1977). There have been various widely distributed plagues recorded this century (Uvarov, 1977; Pedgley, 1989) that have caused enormous damage. For successful breeding, locusts need warm and moist conditions (Pedgley, 1989); warmer temperatures also hasten their life cycle. Locusts favor ephemeral systems with silty or clayey soil that allows the eggs to survive and provides enough vegetation for the later life stages. However, to reach plague levels they also require meteorological and vegetation conditions that bring previously scattered populations together, allowing them to continue breeding while avoiding natural enemies. Scattered pockets of ephemeral, green vegetation appear to be most suitable.

In the gregarious phase, adult swarms can fly for many days and cover several hundred kilometers, especially in favorable wind conditions (Uvarov, 1977; Pedgley, 1989). Gentle winds allow locusts to take advantage of moist conditions in widely dispersed areas. The most common breeding sites have been recorded during the hot season in the Sahel-Sahara region (Uvarov, 1977). This area has an average annual precipitation of 50–200 mm, and the lush ephemeral vegetation supported by run-on in the wadi systems provides a suitable food source as well as areas suitable for egg-laying.

Other locust species (e.g., the red locust; Walker, 1991) have different behaviors, but in most cases they are best able to take advantage of pulses of moisture availability spread in both space and time. Changes in the frequency of future rainfall events that favor ephemeral or pulse systems could lead to more plagues of locust species. Conversely, in areas that become more continuously wetter or drier, they may be disadvantaged.

3.4.5. *ENSO and Related Issues*

It has been shown that the El Niño Southern Oscillation (ENSO) phenomenon has a wide-ranging effect on regional climates that varies with latitude (Caviedes, 1984). Because major desert areas are driven by anticyclone systems, changes in sea surface temperature have repercussions for the moisture budgets of deserts. For example, El Niño has been associated with droughts in Australia (Nicholls, 1985) and north central Africa (Caviedes, 1984). However, it can also lead to higher precipitation and air temperatures, as in the coastal deserts of Peru and Chile. In these deserts, Caviedes (1984) recorded about a three- to tenfold increase in precipitation and up to a 4°C increase in air temperature during the 1982–83 El Niño. However, the central areas of Peru and the higher elevation areas suffered from drought during the El Niño period. Sea surface temperatures also affect humidity, and some coastal deserts may have more dew and fog associated with ENSO events.

Meehl *et al.* (1993) simulated the effect of ENSO events using a coupled ocean-atmosphere model in combination with a 2 x CO_2 scenario. Their general conclusion is that under doubled CO_2 the ENSO will intensify, and anomalously dry areas will become drier and wet areas will become wetter during ENSO events.

3.5. Biogeographical Shifts

Desert regions are known to have undergone significant changes in the past. For example, the northern Sahara changed from Mediterranean flora at about 10,000 yr BP to its present aridity over several thousand years (Evenari, 1985a). Historical records show little shift in desert climates more recently, although vegetation at the margins of desert regions responds dramatically to variations in seasonal rainfall (Tucker *et al.*, 1991). There has also been massive change in some areas due to human activity (see Chapter 4). Long-term studies of desert communities show that they are highly responsive to small changes in the climate regime in which they grow, and in particular to rare events such as unusually moist or dry periods (Hall *et al.*, 1964; Noble, 1977; Noble and Crisp, 1980; Goldberg and Turner, 1986; Turner, 1990).

Monserud *et al.*'s (1993) projection of global vegetation changes in response to a doubled CO_2-driven climate indicates relatively little change in their desert category (by their definition, about 25% of land surface and thus close to the generally accepted desert and semi-deserts). They found that 82–92% of existing deserts remain so classified under any of the four GCM climates they used. For all vegetation types combined, only 59–66% remain in the same category; deserts were the most stable of the 16 vegetation types they considered.

The Greco *et al.* (1994) scenarios predict moister conditions along the boundaries of only a few deserts. Even if wetter conditions were to prevail, the "greening" of the deserts would often be negated by pressure from the expanding human population and the associated desertification problems (see Chapter 4). Climate amelioration may lead to increased invasion by species able to disperse widely and possibly to increased fire frequency if the occurrence of ephemeral flora becomes more common. Long-range dispersal mechanisms tend to be rare in plants of the extreme deserts (Ellner and Shmida, 1981; Chambers and MacMahon, 1994). Thus, even if deserts with little plant cover (e.g., the central Sahara) were to become more moist, vegetation may take a long time to respond. In some cases, as in Australia, species (especially birds) that are dependent on ephemeral water bodies may show a population increase.

3.6. Mitigation

Despite their extent, extreme deserts store only a small portion of the global C and N pools (much less than 1%); thus, deserts are not a likely sink for excess CO_2. Some small gains may arise through modified land-use and grazing management practices (see Chapter 2).

3.7. Future Needs

There is a lack of information on many aspects of the extreme deserts. The highly pulsed availability of resources and the ephemeral nature of much of the biological activity make them difficult to study. Predictions of the impacts of climate change will depend critically on the precise changes in the seasonality and distribution of rainfall events as much as on the absolute amounts of rainfall.

Acknowledgments

Thanks to Mark Stafford Smith, Raymond Turner, Margo Davies, James White, Richard Moss, and the rest of the IPPC Working Group II Technical Support Unit.

References

Alwelaie, A.N., S.A. Chaudary, and Y. Alwetaid, 1993: Vegetation of some Red Sea Islands of the Kingdom of Saudi Arabia. *Journal of Arid Environments*, **24**, 287-296.

Aronson, J. and A. Shimda, 1992: Plant species diversity along a Mediterranean-desert gradient and its correlation with interannual rainfall fluctuations. *Journal of Arid Environments*, **23**, 235-247.

Balling, R.C., Jr., 1988: The climatic impact of a Sonoran vegetation discontinuity. *Climatic Change*, **13**, 99-109.

Bertiller, M.B., N.O. Elissalde, C.M. Rostagno, and G.E. Defosse, 1995: Environmental patterns and plant distribution along a precipitation gradient in western Patagonia. *Journal of Arid Environments*, **29**, 85-97.

Branson, F.A., G.F. Gifford, K.G. Renard, and R.F. Hadley, 1981: *Rangeland Hydrology* [Reid, E.H. (ed.)]. 2nd ed. Kendall/Hunt Publishing Company, Dubuque, IA, 340 pp.

Caviedes, C.N., 1984: El-Nino 1982-1983. *Geographical Review*, **74**, 268-290.

Chambers, J.C. and J.A. MacMahon, 1994: A day in the life of a seed: movements and fates of seeds and their implications for natural and managed systems. *Annual Review of Ecology and Systematics*, **25**, 263-92.

Charley, J.L. and N.E. West, 1977: Micro-patterns of nitrogen mineralization activity in soils of some shrub-dominated semi-desert ecosystems of Utah. *Soil Biology and Biochemistry*, **9**, 357-366.

Clark, D.B., 1980: Population ecology of *Rattus rattus* across a desert montane forest gradient in the Galapagos Islands. *Ecology*, **61**, 1422-1433.

Cloudsley-Thompson, J.L., 1977: *Man and the Biology of Arid Zones.* Edward Arnold, London, UK, 182 pp.

Cloudsley-Thompson, J.L., 1993: The adaptational diversity of desert biota. *Environmental Conservation*, **20**, 227-231.

Currie, D.J., 1991: Energy and large-scale patterns of animal- and plant-species richness. *The American Naturalist*, **137**, 27-49.

Ehleringer, J.R. and R.K. Monson, 1993: Evolutionary and ecological aspects of photosynthetic pathway variation. *Annual Review of Ecology and Systematics*, **24**, 411-439.

Ellner, S. and A. Shmida, 1981: Why are adaptations for long-range seed dispersal rare in desert plants? *Oecologia*, **51**, 133-144.

Evans, R.D. and J.R. Ehleringer, 1994: Plant $d^{15}N$ values along a fog gradient in the Atacama desert, Chile. *Journal of Arid Environments*, **28**, 189-193.

Evenari, M., 1985a: The desert environment. In: *Ecosystems of the World: Hot Deserts and Arid Shrublands,* vol. 12A [Evenari, M., I. Noy Meir, and D. Goodall (eds.)]. Elsevier, Amsterdam, Netherlands, pp. 1-22.

Evenari, M., 1985b: Adaptations of plants and animals to the desert environment. In: *Ecosystems of the World: Hot Deserts and Arid Shrublands,* vol. 12A [Evenari, M., I. Noy Meir, and D. Goodall (eds.)]. Elsevier, Amsterdam, Netherlands, pp. 79-92.

Evenari, M., L. Shanan, and N.H. Tadmor, 1971: *The Negev: The Challenge of a Desert.* Harvard University Press, Cambridge, MA, 345 pp.

Fisher, M., 1994: Another look at the variability of desert climates, using examples from Oman. *Global Ecology and Biogeography Letters*, **4**, 79-87.

Frasier, G.W., J.R. Cox, and D.A. Woolhiser, 1985: Emergence and survival response of seven grasses for six wet-dry sequences. *Journal of Range Management*, **38**, 372-377.

Frasier, G.W., J.R. Cox, and D.A. Woolhiser, 1987: Wet-dry cycle effects on warm-season grass seedling establishment. *Journal of Range Management*, **40**, 2-6.

Goldberg, D.E. and R.M. Turner, 1986: Vegetation change and plant demography in permanent plots in the Sonoran Desert. *Ecology*, **67**, 695-712.

Greco, S., R.H. Moss, D. Viner, and R. Jenne, 1994: *Climate Scenarios and Socioeconomic Projections for IPCC WGII Assessment.* IPCC-WMO and UNEP, Washington DC, 67 pp.

Hadley, N.F. and S.R. Szarek, 1981: Productivity of desert ecosystems. *BioScience*, **31**, 747-753.

Hall, E.A.A., R.L. Specht, and C.M. Eardley, 1964: Regeneration of vegetation on Koonamore Vegetation Reserve, 1926-1962. *Australian Journal of Botany*, **12**, 205-264.

Hoffman, M.T., G.F. Midgley, and R.M. Cowling, 1994: Plant richness is negatively related to energy availability in semi-arid southern Africa. *Biodiversity Letters*, **2**, 35-38.

Huenneke, L.F. and I.R. Noble, 1995: Biodiversity and ecosystem function: Ecosystem analyses. Arid Lands. In: *Global Biosphere Assessment*, section 7.1.4. UNEP, Cambridge University Press, Cambridge, UK (in press).

Idso, S.B. and A.J. Brazel, 1984: Rising atmospheric carbon dioxide concentrations may increase streamflow. *Nature*, **312**, 51-53.

Kingsford, R.T. and J.L. Porter, 1993: Waterbirds of Lake Eyre, Australia. *Biological Conservation*, **65**, 141-151.

Lane, L.J., M.H. Nichols, and H.B. Osborn, 1994: Time series analysis of global change data. *Environmental Pollution*, **83**, 63-68.

Lapenis, A.G. and M.V. Shabalova, 1994: Global climate changes and moisture conditions in the intracontinental arid zones. *Climatic Change*, **27**, 283-297.

Le Houerou, H., 1984: Rain use efficiency: a unifying concept in arid-land ecology. *Journal of Arid Environments*, **7**, 213-247.

Le Houerou, H.N., 1992: Outline of the biological history of the Sahara. *Journal of Arid Environments*, **22**, 3-30.

Louw, G. and M. Seely, 1982: *Ecology of Desert Organisms.* Longman Group Ltd., London, UK, 194 pp.

Mabbutt, J.A., 1989: Impacts of carbon dioxide warming on climate change in the semi-arid tropics. *Climatic Change*, **15**, 191-221.

MacMahon, J.A. and F.H. Wagner, 1985: The Mojave, Sonoran and Chihuahuan Deserts of North America. *Ecosystems of the World: Hot Deserts and Arid Shrublands,* vol. 12A [Evenari, M., I. Noy Meir, and D. Goodall (eds.)]. Elsevier, Amsterdam, Netherlands, pp. 105-202.

Maurer, B.A., 1985: Avian community dynamics in desert grasslands: observational scale and hierarchical structure. *Ecological Monographs*, **55**, 295-312.

McGinnies, W.G., B.J. Goldman, and P. Paylore (eds.), 1968: *Deserts of the World: An Appraisal of Research into their Physical and Biological Environments.* University of Arizona Press, Tucson, AZ, 788 pp.

Meehl, G.A., G.W. Branstator, and W.M. Washington, 1993: Tropical Pacific interannual variability and CO_2 climate change. *Journal of Climate*, **6**, 42-63.

Monserud, R.A., N.M. Tchebakova, and R. Leemans, 1993: Global vegetation change predicted by the modified Budyko model. *Climatic Change*, **25**, 59-83.

Mooney, H.A., B.G. Drake, R.J. Luxmore, W.C. Oechel, and L.F. Pitelka, 1991: Predicting ecosystem responses to elevated CO_2 concentration. *BioScience*, **41**, 96-104.

Morassutti, M.P., 1992: Australian runoff scenarios from a runoff-model. *International Journal of Climatology*, **12**, 797-813.

Mühlenbruch-Tegen, A., 1992: Long-term surface temperature variations in South Africa. *South African Journal of Science*, **88**, 197-205.

Nasrallah, H.A. and R.C. Balling, Jr., 1993: Spatial and temporal analysis of Middle Eastern temperature changes. *Climatic Change*, **25**, 153-61.

Neilson, R.P. and D. Marks, 1994: A global perspective of regional vegetation and hydrologic sensitivities from climate change. *Journal of Vegetation Science*, **5**, 715-730.

Nicholls, N., 1985: Impact of the Southern Oscillation on Australian crops. *Journal of Climatology*, **5**, 553-560.

Noble, I.R., 1977: Long term biomass dynamics in an arid chenopod shrub community at Koonamore, South Australia. *Australian Journal of Botany*, **25**, 639-653.

Noble, I.R. and I.C. Burke, 1995: Biodiversity and ecosystem function: cross biome comparisons. How does biodiversity influence landscape structure? In: *Global Biosphere Assessment*, section 7.2.5, UNEP. Cambridge University Press, Cambridge, UK (in press).

Noble, I.R. and M.D. Crisp, 1980: Germination and growth models of short-lived grass and forb populations based on long-term photo-point data at Koonamore, South Australia. *Israel Journal of Botany*, **28**, 195-210.

Noy Meir, I., 1973: Desert ecosystems: environment and producers. *Annual Review of Ecology and Systematics*, **4**, 25-51.

Noy Meir, I., 1985: Desert ecosystem structure and function. In: *Ecosystems of the World: Hot Deserts and Arid Shrublands*, vol. 12A [Evenari, M., I. Noy Meir, and D. Goodall (eds.)]. Elsevier, Amsterdam, Netherlands, pp. 93-103.

O'Brien, E.M., 1993: Climatic gradients in woody plant species richness: towards an explanation based on an analysis of South Africa's woody flora. *Journal of Biogeography*, **20**, 181-198.

Owen, J.G., 1988: On productivity as a predictor of rodent and carnivore diversity. *Ecology*, **69**, 1161-1165.

Pedgley, D.E., 1989: Weather and the current desert locust plague. *Weather*, **44**, 168-171.

Peterjohn, W.T. and W.H. Schlesinger, 1990: Nitrogen loss from deserts in the southwestern United States. *Biogeochemistry*, **10**, 67-79.

Pianka, E.R. and J.J. Schall, 1981: Species densities of Australian vertebrates. In: *Ecological Biogeography of Australia* [Keast, A. (ed.)]. W. Junk, The Hague, Netherlands, pp. 1675-1694.

Pickup, G., 1985: The erosion cell—a geomorphic approach to landscape classification in range assessment. *Australian Rangeland Journal*, **7**, 114-121.

Pittock, A.B., 1988: Actual and anticipated changes in Australia's climate.In: *Greenhouse: Planning for Climate Change* [Pearman, G.I. (ed.)]. CSIRO, Melbourne, Australia, pp. 35-51.

Rauh, W., 1985: The Peruvian-Chilean deserts. In: *Ecosystems of the World: Hot Deserts and Arid Shrublands*, vol. 12A [Evenari, M., I. Noy Meir, and D. Goodall (eds.)]. Elsevier, Amsterdam, Netherlands, pp. 239-267.

Revelle, R.R. and P.E. Waggoner, 1983: Effects of a carbon dioxide-induced climate change on water supplies in the western United States. In: *Changing Climate: Report of the Carbon Dioxide Assessment Committee.* National Academy Press, Washington DC, 496 pp.

Rundel, P.W., E.T. Nilson, M.R. Sharifi, R.A. Virginia, W.M. Jarrell, D.H. Kohl, and G.B. Shearer, 1982: Seasonal dynamics of nitrogen cycling for a Prosopis woodland in the Sonoran Desert. *Plant and Soil*, **67**, 343-353.

Santos, P.F., N.Z. Elkins, Y. Steinberger, and W.G. Whitford, 1984: A comparison of surface and buried *Larea tridentata* leaf litter decomposition in North American desert. *Ecology*, **65**, 278-284.

Seely, M.K., 1979: Irregular fog as water source for desert beetles. *Oecologia*, **42**, 213-227.

Shmida, A., 1985: Biogeography of the desert flora. In: *Ecosystems of the World: Hot Deserts and Arid Shrublands*, vol. 12A [Evenari, M., I. Noy Meir, and D. Goodall (eds.)]. Elsevier, Amsterdam, Netherlands, pp. 23-77.

Skarpe, C., 1992: Dynamics of savanna ecosystems. *Journal of Vegetation Science*, **3**, 293-300.

Skiles, J.W. and J.D. Hanson, 1994: Responses of arid and semiarid watersheds to increasing carbon dioxide and climate change as shown by simulation studies. *Climatic Change*, **26**, 377-397.

Specht, A.S. and R.L. Specht, 1994: Biodiversity of overstorey trees in relation to canopy productivity and stand density in the climatic gradient from warm temperate to tropical Australia. *Biodiversity Letters*, **2**, 39-45.

Steinberger, Y. and A. Shmida, 1984: Decomposition along a rainfall gradient in the Judean desert, Israel. *Oecologia*, **82**, 322-324.

Tucker, C.J., H.E. Dregne, and W.W. Newcomb, 1991: Expansion and contraction of the Sahara Desert from 1980 to 1990. *Science*, **253**, 299-301.

Turner, R.M., 1990: Long term vegetation change at a fully protected Sonoran Desert site. *Ecology*, **71**, 464-477.

Uvarov, B.P., 1977: *Grasshoppers and Locusts. A Handbook of General Acridology,* vol. 2. Cambridge University Press, Cambridge, UK, 588 pp.

Walker, B.H., 1991: Ecological consequences of atmospheric and climate change. *Climatic Change*, **18**, 301-316.

Walter, H., 1985: *Vegetation of the Earth and Ecological Systems of the Geo-Biosphere.* 3rd ed. Springer-Verlag, Berlin, Germany, 318 pp.

Wang, D. and L. Mayer, 1994: Effects of storm clustering on water balance estimates and its implications for climate impact assessment. *Climatic Change*, **27**, 321-342.

Webb, W.L., W.K. Lauenroth, S.R. Szarek, and R.S. Kinerson, 1983: Primary production and abiotic controls in forest, grassland and desert ecosystems in the United States. *Ecology*, **64**, 134-151.

Webb, W.L., S.R. Szarek, W.K. Lauenroth, and R.S. Kinerson, 1978: Primary productivity and water use in native forest, grassland and desert ecosystems. *Ecology*, **59**, 1239-1247.

Weltz, M.A. and W.H. Blackburn, 1995: Water budget for south Texas rangelands. *Journal of Range Management*, **48**, 45-52.

West, N.E. and J. Skujins, 1977: The nitrogen cycle in the North American cold-winter semi-desert ecosystems. *Oecologia Planatarum*, **12**, 45-53.

Whetton, P.H., P.J. Raynor, A.B. Pittock, and M.R. Haylock, 1994: An assessment of possible climate change in the Australian region based on an intercomparison of General Circulation Modeling results. *Journal of Climate*, **7**, 441-463.

White, L.P., 1971: Vegetation stripes on sheet wash surfaces. *Journal of Ecology*, **59**, 615-622.

Whitford, W.G., Y. Steinberger, W. Mackay, Z.W. Parker, D. Freckman, J.A. Wallwork, and D. Weems, 1986: Rainfall and decomposition in the Chihuahan Desert. *Oecologia*, **68**, 512-515.

Wigley, T.M.L., P.D. Jones, and P.M. Kelly, 1980: Scenario for a warm, high CO_2 world. *Nature*, **283**, 17-21.

Williams, O.B. and J.H. Calaby, 1985: The hot deserts of Australia. In: *Ecosystems of the World: Hot Deserts and Arid Shrublands*, vol. 12A [Evenari, M., I. Noy Meir, and D. Goodall (eds.)]. Elsevier, Amsterdam, Netherlands, pp. 269-312.

World Resources Institute and International Institute for Environment and Development, 1988-89: *World Resources, 1988-89.* Basic Books, New York, NY, 400 pp.

4

Land Degradation and Desertification

PETER BULLOCK, UK; HENRI LE HOUÉROU, FRANCE

Principal Lead Authors:
M.T. Hoffman, South Africa; M. Rounsevell, UK; J. Sehgal, India; G. Várallyay, Hungary

Contributing Authors:
A. Aïdoud, Algeria; R. Balling, USA; C. Long-Jun, China; K. Goulding, UK; L.N. Harsh, India; N. Kharin, Turkmenistan; J. Labraga, Argentina; R. Lal, USA; S. Milton, South Africa; H. Muturi, Kenya; F. Nachtergaele, FAO; A. Palmer, South Africa; D. Powlson, UK; J. Puidefabregas, Spain; J. Rogasik, Germany; M. Rostagno, Argentina; P. Roux, South Africa; D. Sauerbeck, Germany; W. Sombroek, FAO; C. Valentin, France; W. Lixian, China; M. Yoshino, Japan

CONTENTS

EXECUTIVE SUMMARY

The impact of climate change on soils needs to be considered in parallel with impacts caused by unsustainable land-management practices. In many cases, it is impossible to separate the effects of these impacts; often they interact, leading to a greater cumulative effect on soils than would be predicted from a simple summation of their effects. Findings follow:

- Fundamental soil properties and processes—including organic-matter decomposition, leaching and acidification, salinization, soil water regimes, soil stability, and soil erosion—will be influenced by changes in climate (High Confidence).
- Desertification arises both from human abuse of the land and from adverse climate conditions. Climate-related factors such as increased drought can lead to an increase in the vulnerability of land to desertification and to the escalation of the desertification process (High Confidence).
- Reversing the effects of desertification is not always possible and is more difficult for drier environments with shallower soils (High Confidence).
- Changes in the frequency and intensity of precipitation will have the greatest direct effect on soils via erosion by water. However, future erosion risk is likely to be related more to increases in population density, intensive cultivation of marginal lands, and the use of resource-based and subsistence farming techniques than to changes in precipitation regimes (High Confidence).
- In structurally stable soils, greater precipitation will increase the rate of leaching of basic cations. In the long term, after buffering pools have become exhausted, accelerated leaching could lead to soil acidification (High Confidence).
- Higher temperatures are likely to increase the decomposition rate of organic matter and organic-matter loss (High Confidence). However, too little is known about the influence of increased atmospheric carbon dioxide (CO_2) levels on the amount of organic matter returning to the soil to judge the extent to which soil organic-matter levels will decline.
- Where conditions become more arid, salinization and alkalization are likely to increase because evapotranspiration and capillary rise will be enhanced (High Confidence). Areas with a shallow water table may experience increased salinization if rainfall increases (Medium Confidence).
- Predicted warming may give rise to higher evaporation rates, leading to drier soils and more frequent episodes of severe wind erosion (Medium Confidence).
- Because arid and semi-arid land ecosystems have little ability to buffer the effects of climate variability relative to most other terrestrial ecosystems, they are particularly vulnerable to climate change and may be among the first ecosystems to be affected by global environmental change (Medium Confidence).
- Adaptation to desertification will rely on conventional strategies, such as the use of agroforestry, animals that are better adapted to dry conditions, diverse and multiple production systems, and water- and energy-saving techniques (Medium Confidence).
- Climate change could be beneficial to some semi-arid and sub-humid tropical highlands because of temperature increases and CO_2 fertilization (Medium Confidence).

4.1. Introduction

Land degradation, defined here as "a reduction in the capability of the land to support a particular use" (Blaikie and Brookfield, 1987), is considered to be one of the major problems facing the world (UNEP, 1992). In this chapter, we examine the likely impact of climate change on this global problem. Will degradation of land be exacerbated or reduced? There are many recognized forms of land degradation, including soil erosion, salinization, soil contamination, loss of soil organic matter, decline in nutrient levels, acidification, and loss of soil structure. Desertification currently affects about one-sixth of the world's population and one-quarter of the world's land: 6 to 7 million hectares (Mha) are lost annually due to soil erosion, and up to 20 Mha of irrigated land are affected by salinization (World Resources Institute, 1992). Much of this degradation is undermining economic development and is already irreversible.

With the forecast world population increase, there will be increased pressure on the soil to produce more food; thus, future trends in the potential of land to produce food will be of particular interest. The impact of climate change on the health of the land should be considered in parallel with the effects of existing pressures caused by unsustainable land management—because it is often impossible to separate the effects of these impacts and because their cumulative impact on soils often is greater than a simple summation.

Emphasis in this chapter is given to the effects of climate change on soil erosion, salinization, and desertification, but impacts on other degradation processes capable of being accelerated by climate change also are reviewed. Additional processes of importance include organic-matter loss, nutrient loss, acidification, and soil structural deterioration (Rogasik *et al.*, 1994). In some instances, changes in such processes will lead to the formation of new soil types requiring new types of management and having different use potentials.

4.2. Soil Erosion: Causes, Processes, and Predictions

Soil erosion is the movement and transport of soil by various agents—particularly water, wind, and mass movement—that leads to a loss of soil. Erosion has been recognized by many scientists and some governments as a major problem since the U.S. Dust Bowl of the 1930s (Jacks and Whyte, 1939); the causes and processes involved have been well-researched over the last 60 years, providing voluminous literature.

Clearing soils of their natural vegetation cover and subsequently using them for arable agriculture has been the primary cause of soil erosion onset. Historically, this has resulted in disastrous soil losses, particularly in conjunction with climate variations and extreme weather events (Bork, 1983, 1989). More recently, population pressure in developing countries, as well as changes in land use and management in other parts of the world, have increased erosion susceptibility even further. The annual rates of new soil formation are crudely estimated to be between 2 and 11 t/ha (Wischmeier and Smith, 1978), although experimentally derived rates often are less than this (Lal, 1994a). Globally, rates of soil erosion can exceed these estimated values by 10- to 20-fold, thereby reducing productivity (Crosson and Stout, 1983; Sehgal and Abrol, 1994a, 1994b) and causing sediment and nutrient loading of rivers (Clark *et al.*, 1985).

The main factors influencing soil erosion are rainfall (amount, frequency, duration, and intensity), windspeed (direction, strength, and frequency of high-intensity events), land use and management, topography, and soils and their properties (Morgan, 1986; Hallsworth, 1987). Since the 1950s, significant advances have been made in predicting erosion risk, particularly through model development. Several models that are commonly used include the Wind Erosion Equation (Woodruff and Siddoway, 1965), the Universal Soil Loss Equation (USLE) (Wischmeier and Smith, 1978), EPIC (Williams *et al.*, 1990; Skidmore and Williams, 1991), WEPP (Flanagan *et al.*, 1991), and EUROSEM (Morgan *et al.*, 1992). All of these models have weather/climate components and are capable of being manipulated to predict erosion risk under different climate-change scenarios (e.g., Favis-Mortlock, 1994). Generally, increased rainfall intensity and amount, increased windspeed, and increased frequencies of high-wind events—especially if coupled with increasing droughtiness—are likely to lead to an increase in erosion. However, erosion is site-specific, and different permutations of conditions can increase or decrease it. Research has predicted erosion risk as a function of land use and environmental conditions (Nearing *et al.*, 1994; Skidmore *et al.*, 1994), agronomic quality (Williams *et al.*, 1984), and environmental quality (Knisel, 1980; Lal, 1994a).

Despite scientific advances, the problem of accelerated soil erosion is more serious now than ever before (Lal, 1994c), and erosion hazards are being exacerbated particularly in the tropics and subtropics (Lal, 1994d). Reliable estimates of the current amounts of erosion and future areas at risk are difficult to obtain (Table 4-1; Lal, 1994b), although erosion has been suggested as affecting about 33% of the land currently used for global crop production (Brown and Young, 1990).

4.2.1. Erosion by Water

Climate change is likely to affect soil erosion by water through its effects on rainfall intensity, soil erodibility, vegetative cover, and patterns of land use. Estimates of changes in erosion risk can be made from a knowledge of projected changes in these factors. General circulation models (GCMs) can provide a range of climate scenarios, but these alone are not sufficient to predict future erosion risk, particularly because GCMs are currently poor predictors of changes in rainfall intensity and surface windspeed. In addition to more regionally reliable GCMs, accurate and reliable databases of parameters such as vegetation cover, soil properties, land use, and management

Table 4-1: *Estimates of current global rates of soil erosion and likely future trends.*

	Soil Erosion by Water				Soil Erosion by Wind	
Region	Area[1] (10⁶ ha)	Denudation Rate[2] (mm/yr)	Dissolved Load[3] (10⁶ t/yr)	Future Trends	Area[1] (10⁶ ha)	Future Trends
Africa	227	0.023	201	+	186	+
Asia	441	0.153	1592	+	222	+
South America	123	0.067	603	+	42	-
Central America	46	)	)	+	5	-
		0.055)	758)			
North America	60	)	)	+/-	35	-
Europe	114	0.032	425	+/-	42	+/-
Oceania	83	0.390	293	+	16	+
World	1094	0.079	3872	+	54.8	+

Notes: + = increased risks; - = decreased risks.
[1]Oldeman, 1991-92.
[2]Lal, 1994b.
[3]Walling, 1987.

systems are needed. These databases are important for the assessment of biophysical processes and biomass productivity (Rosenzweig *et al.*, 1993) and are needed at scales appropriate to regional and global degradation processes (Bliss, 1990; Bouwman, 1990; Batjes *et al.*, 1994). Such databases have not yet been developed.

GCMs indicate a marked change in soil moisture regime and, therefore, attendant changes in soil erodibility, land use, and vegetative cover. Regions likely to experience an increase in soil moisture include southern Asia (50 to 100%), South America (10 to 20%), and Oceania (10 to 20%). Regions likely to experience decreases in soil moisture include North America (10 to 50%), sub-Saharan Africa (10 to 70%), and Europe (10 to 60%) (Schneider, 1989). However, with respect to erosion, much will depend on the pattern, intensity, and seasonality of rainfall events.

A potential change in the climate might have positive effects that lead to a decrease in soil erosion risk (Sombroek, 1991–92; Brinkman and Sombroek, 1993) as a result of negative feedback mechanisms. Examples include increased biomass production, increased vegetation cover, and enhanced soil organic-matter content resulting from elevated CO_2 concentrations. However, predicted changes in temperature, rainfall, and soil moisture (IPCC, 1994) suggest that few areas will receive benefits from negative feedback effects. Instead, projected declines in levels of soil organic matter and the weakening of soil structure will make soils increasingly prone to erosion.

Modeled estimates of the effect of climate change on soil erosion depend on assumptions regarding the frequency and intensity of precipitation (Phillips *et al.*, 1993). However, some estimates of future erosion have been made. For example, changes in the U.S. national average for sheet and rill erosion have been modeled under a number of different climate change scenarios. For cropped land, modeled erosion increased by 2 to 16%; in pastureland, predicted changes were -2 to +10%; and in rangeland, modeled erosion changed by -5 to +22% (Phillips *et al.*, 1993). EPIC was used in the UK to show that, although an increase in temperature has little effect on erosion rate, a 15% increase in rainfall leads to a 27% increase in mean annual erosion (Favis-Mortlock *et al.*, 1991; Boardman and Favis-Mortlock, 1993; Favis-Mortlock, 1994).

4.2.2. *Erosion by Wind*

Soil erosion occurs when wind transports soil particles by suspension, surface creep, or saltation over distances ranging from a few centimeters to many kilometers. Wind erosion not only transports soil particles around arid and semi-arid landscapes but provides inputs into ecosystems around the world and may even alter global climatic patterns. Wind erosion is particularly problematic on sandy and organic soils subject to low soil moisture, patchy vegetation, sporadic rainfall, and periodic winds. Even soils resistant to wind erosion can be blown away if the soil is damaged by trampling of animals, loosening by ploughing and tillage, pulverization by traffic (human and animal), and denudation of natural vegetation by the expansion of agriculture, excessive grazing, or fire.

Wind erosion is mainly a feature of arid and semi-arid conditions but may occur in moister zones where soil damage occurs during or just prior to periods of high wind velocity. In some areas, climate change is expected to lead to more droughty soils and less vegetation, both of which make soils more vulnerable to wind erosion (Middleton, 1985;

Hennessy *et al.*, 1986). The predicted warmer climate may give rise to higher evaporation rates, leading to more frequent soil drying whether precipitation increases or not. Consequently, wind erosion may become more frequent. Although the influence of climate and human activities on wind erosion is well-documented, the feedback effect of windblown dust in the atmosphere on the warming of the upper atmospheric layers has not been thoroughly investigated (Nicholson, 1994). Desertification—an advanced result of wind erosion—is considered in Section 4.4.

4.2.3. Mass Movement and Subterranean Erosion

The occurrence of mass movement depends upon the interaction of various factors, including land form, lithology, soil type, rainfall intensity and duration, drainage characteristics, vegetal cover, and human intervention (Rosewell *et al.*, 1991). Where sloping land is subjected to increasing rainfall of high intensity, it will become increasingly subject to mass movement.

There are various forms of subterranean erosion (Dunne, 1990), including piping (Kirkby and Morgan, 1980), intrasoil erosion, and tunnel erosion (Barrow, 1991). As with mass movement, increasing rainfall amounts and intensities will lead to increases in this form of erosion.

4.2.4. Coastal Erosion

The main factors affecting coastal erosion and sediment redistribution are winds, currents, tides, and floods. Rising sea level could increase coastal erosion and the loss of highly productive wetlands—and thus intensify pressure on the remaining land (see Chapter 9).

4.2.5. Adaptive Strategies for Mitigating the Impact of Climate Change on Soil Erosion

More than 60 years of experience are available in the development of conservation techniques and soil protection, following the U.S. Dust Bowl and similar phenomena elsewhere in the world (Kirkby and Morgan, 1980; FAO, 1983; Morgan, 1986; Schwertmann *et al.*, 1989; Lal, 1990). A serious problem, however, is that many of these conservation techniques have not been adopted. Nevertheless, they are available and could be used for or adapted to changing climatic conditions. The most important requirement is that governments recognize the problem of soil erosion and set in motion the appropriate conservation mechanisms.

Future erosion risk is more likely to be influenced by increases in population density, intensive cultivation of marginal lands, and the use of resource-based and subsistence farming techniques than by changes in climate. One can anticipate that erosion, mass movement, and landslides are most likely to increase in and near regions of high population density.

4.3. Salt-Affected Soils

4.3.1. Coastal Salinity

Coastal salinity depends primarily on sea level and its tidal, seasonal, and long-term fluctuations; the temperature, concentration, and chemical composition of seawater; the geology, geomorphology, and relief of the coastal area, including river deltas and estuaries; and climate (temperature, precipitation, rate of evaporation, and spatial and temporal variability in them) (Jelgersma, 1988; Day and Templet, 1989; Pirazzoli, 1989; Szabolcs and Rédly, 1989; Fisher, 1990; Várallyay, 1994). The assessment and prediction of coastal salinity and salt accumulation requires comprehensive information on the hydrological, chemical, and ecological consequences of a potential rise in the eustatic sea level. Consequences include inundation of coastal lowland plains, deltas, and estuaries of big rivers; intrusion of saline seawater or brackish tidal water; rapid erosion of coastlines; a rising water table; and impeded drainage in coastal plains. In general, any rise in the eustatic sea level will result in the territorial extension of coastal salinity under the direct and indirect influences of saline seawater (Titus, 1987; Jelgersma, 1988; Szabolcs and Rédly, 1989; Hekstra, 1989; Pirazzoli, 1989; Scharpenseel *et al.*, 1990; Szabolcs, 1991).

4.3.2. Continental Salt Transport and Salt Accumulation Processes

Impact analysis of climatic scenarios on continental salt transport and salt accumulation processes requires a much more complex approach. There are three principal mechanisms of salinization: salt accumulation, seepage, and wind deposition. Salinization by salt accumulation occurs when leaching is reduced and salt accumulates at the surface or at some depth in the soil profile—where, following erosion, it may become exposed (West *et al.*, 1994). Salinization also can occur when salt is leached into a perched water table and then seeps to a lower point in the landscape (Ballantyne, 1963). Salinization by wind deposition relies on a suitable source of salt deposits. All three of these salinization processes are likely to be affected by climate change. The effects of two possible future climate scenarios—warm and dry versus warm and wet—are considered in Sections 4.3.2.1 and 4.3.2.2. Such climates are likely to show high spatial and temporal variability; various combinations of temperature and precipitation changes are likely in nature (Manabe and Holloway, 1975; Solomon *et al.*, 1987; Szabolcs and Rédly, 1989). Further variations may be caused by the physiographic variability of specific regions (agroecological unit, water catchment area, etc.) (Jelgersma, 1988; Brammer and Brinkman, 1990; Fisher, 1990; Szabolcs, 1990) and by human activities (e.g., agricultural water-management practices and water-conservation practices) (Hekstra, 1989; Szabolcs, 1989; Várallyay, 1990a, 1990b, 1990c).

4.3.2.1. Warm and Dry Climate

Rising temperatures during the summer will lead to higher evapotranspiration and aridity and thus higher concentration of salts in the soil solution. Where the water table is shallow, drying usually is associated with enhanced capillary rise from groundwater to overlying soil horizons. The consequences are increasing salinity in the solid and liquid phases of the soil, salt efflorescence or salt crust on the soil surface, higher salt content in the soil profile, and higher salt concentration in the soil solution. If the lowering of the groundwater table is prevented by horizontal inflow, capillary transport leads to increasing salt accumulation in the overlying soil layers. Seepage from unlined canals and reservoirs and/or filtration losses from over-irrigated or imperfectly irrigated fields will lead to horizontal groundwater inflow and/or saline seep from the surroundings to low-lying areas. These processes result in a rise of the water table due to increased capillary transport and thus to increasing salinity (Várallyay, 1968, 1994).

Higher winter temperatures have much less impact on salinity. For example, changing the ratio of rain to snow and/or having a shorter period of freezing leads to better infiltration concurrent with leaching. If the water table does rise during the temperate-zone winter, there is negligible (or no) capillary transport (because few plants are active), thus no salt accumulation (Várallyay, 1994).

4.3.2.2. Warm and Wet Climate

If the temperature increase is accompanied by a significant increase in rainfall, aridity and the concentration of salts in the soil solution will be reduced. This climatic scenario results in a net downward flow of water in the soil profile, at least for most of the year, associated with leaching, as well as a salinity reduction. However, where there is periodic drying (e.g., in a dry summer), capillary transport of salts can occur from shallow groundwater to overlying soil horizons. Thus, increased climate variability that includes periods of severe drying can lead to temporally sporadic or seasonal salt accumulation.

The accumulation of salts in soils often negatively affects soil properties and processes, including nutrient-holding capacity, nutrient dynamics, bulk density, soil structure, and porosity (see Box 4-1). Salt accumulations create extreme ecological soil environments and reduce the potential of the land for agricultural and many other uses (Kovda and Szabolcs, 1979; Szabolcs, 1990, 1991; Várallyay, 1994).

4.4. Desertification

Desertification is the process of ecological degradation by which economically productive land becomes less productive and, in extreme cases, develops a desert-like landscape incapable of sustaining the communities that once depended on it (Kassas, 1988; Westing, 1994). Desertification has aroused much emotive and scientific interest, and there are many different definitions of the term (Verstraete, 1986). The definition used in this chapter is the one adopted at the United Nations Conference on Environment and Development (UNCED) in June 1992: "...land degradation in arid, semi-arid, and dry sub-humid areas resulting from various factors including climatic variations and human activities" (UNEP, 1992; Hulme and Kelly, 1993). Desertification occurs most frequently in ecosystems that have low rainfall, long dry seasons, recurrent droughts, mobile surface deposits, skeletal soils, and sparse vegetation cover (Le Houérou, 1968; Dregne, 1983; Kassas, 1988).

Box 4-1. The Impact of Salinization on Crop Productivity in India

Soil salinity is an important chemical degradation problem for Indian agriculture, affecting 10 Mha of land—of which 2.5 Mha lie in the Indo-Gangetic Plateau. Saline soils are dominantly observed in coastal areas and in some irrigated areas inland. Sea-level rise would adversely affect the 7,000-km coastal belt of India, comprising 20 Mha of coastal ecosystem. A rise of 70 cm per century (IPCC, 1990) would inundate 25% of the coastal areas of Kerala, parts of Pondicherry, Karaikal, Tamil Nadu, and vast deltaic areas of Sundarbans (West Bengal), increasing coastal salinity and reducing crop productivity. Increased tidal ingression through creeks along the east and west coasts may further aggravate coastal salinity and damage fragile wetland ecosystems. Expected losses of soil productivity will depend on the level of sodicity but would be expected to approximate those listed in Table 4-2.

Table 4-2: *Percent loss in soil productivity.*

Exchangeable Sodium Percentage	Loss of Productivity in Alluvium-Derived Soils (Fluvisols)	Loss of Productivity in Black Soils (Vertisols)
Up to 5	nil	Up to 10
5–15	<10	10–25
15–40	10–25	25–50
>40	25–50	>50

Source: Sehgal and Abrol, 1994b.

Climate-related factors such as increased drought can lead to an increase in the vulnerability of land to desertification and to the escalation of the desertification process. In essence, desertification results from a combination of drought and mismanagement of land—in particular, the disharmony between land use and management on the one hand and the soil and prevailing climate on the other. Desertification may, in turn, affect local and global climate and thus should be viewed in a cause-effect context, as outlined by Hulme and Kelly (1993) and shown in Figure 4-1.

Table 4-3: Precipitation-to-potential evapotranspiration (P:PET) ratios of hyper-arid to sub-humid lands.

P:PET	Regions
< 0.05	Hyper-arid (true climatic desert)
0.05–0.20	Arid (subject to desertification)
0.20–0.45	Semi-arid (subject to desertification)
0.45–0.70	Sub-humid (subject to desertification)

4.4.1. The Climatic Background of Desertification

We use the term "desert" in the sense of "true" or "climatic" desert, which is associated only with the hyper-arid zone. Desertification is a process that occurs mainly in arid to sub-humid climates that have precipitation:potential evapotranspiration (P:PET) ratios shown in Table 4-3.

The area of land occupied by these four zones in which true or induced deserts can occur is 47% of the land mass of the planet (Table 4-4).

The hyper-arid to dry sub-humid zones are subject to strong fluctuations in climate. Data from dendrochronological studies, pollen analyses, lake level surveys, glacier advances and retreats, crop distribution surveys, and grape harvesting dates confirm the occurrence of many fluctuations in the last 2,000 years.

Regional, short-term trends or fluctuations in rainfall are a feature of semi-arid and arid lands. There has been a clear decline in rainfall in parts of Chile, where the influence of the El Niño–Southern Oscillation (ENSO) has been evident since the beginning of the century (Burgos *et al.*, 1991; Santibanez and Uribe, 1994). Rainfall in the African Sahel has failed to reach 1931–1960 mean levels in virtually all of the last 25 years. Although similar dry periods have occurred in the historical and recent geological past, there is evidence that the recent dry period in the Sahel shows more of a tendency to continental-scale dryness (Nicholson, 1994). Increasing rainfall variability

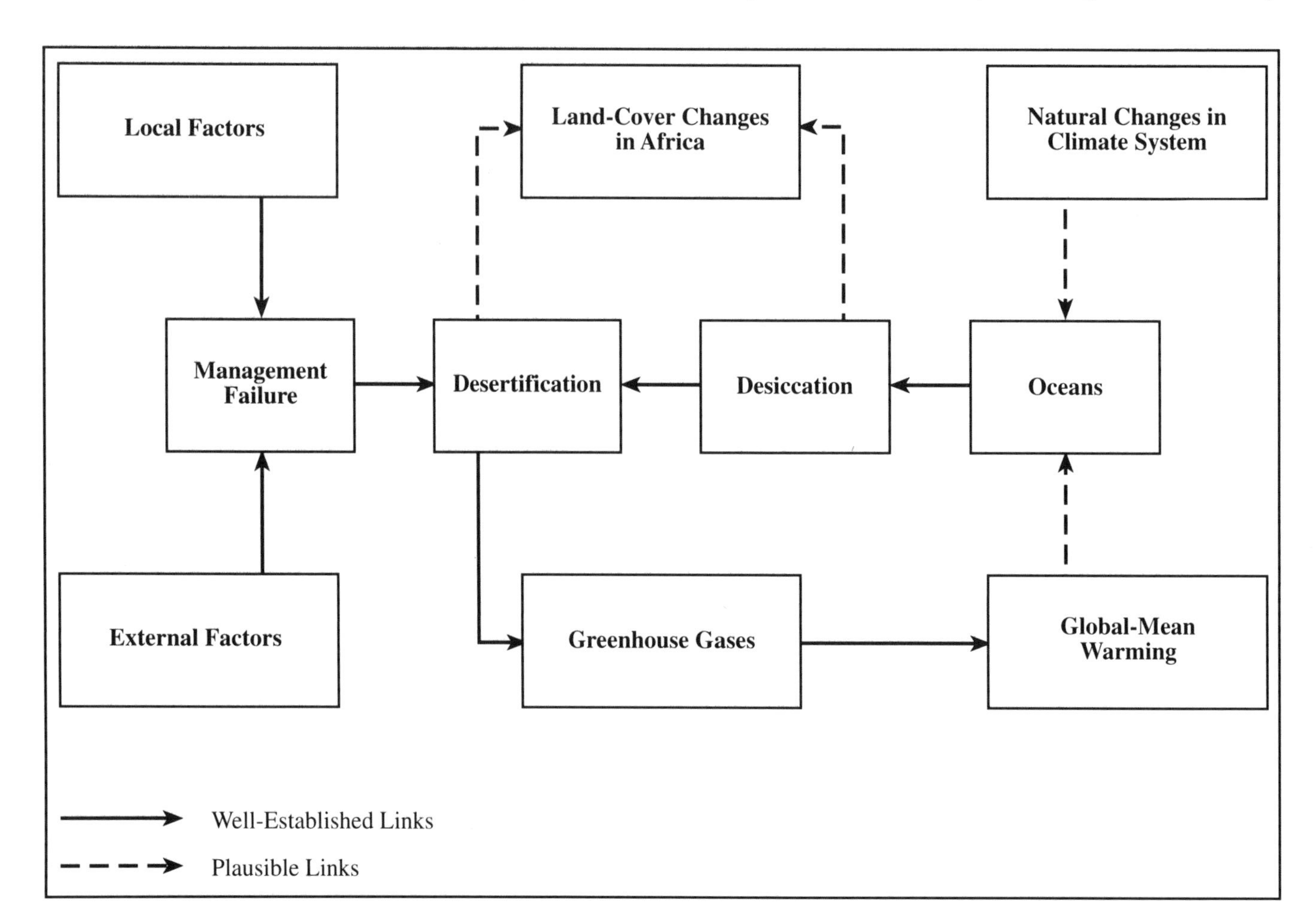

Figure 4-1: A matrix of cause and effect surrounding desertification and the role of climate change (Hulme and Kelly, 1993).

Table 4-4: Regional distribution of world drylands (10^3 km²) (after Oldeman et al., 1990; UNEP, 1992).

Zone	Africa	Asia	Australasia	Europe	North America	South America	Total
Cold	0.0	1082.5	0.0	27.9	616.9	37.7	1765.0
Humid	1007.6	1224.3	218.9	622.9	838.5	1188.1	5100.4
Dry Sub-Humid	268.7	352.7	51.3	183.5	231.5	207.0	1294.7
Semi-Arid	513.8	693.4	309.0	105.2	419.4	264.5	2305.3
Arid	503.5	625.7	303.0	11.0	81.5	44.5	1569.2
Hyper-Arid	672.0	277.3	0.0	0.0	3.1	25.7	978.1
Total	2965.6	4256.0	882.2	950.5	2190.9	1767.5	13012.7

appears to be associated with this; Hulme (1992) found that there were more areas of increased variability than those of reduced variability. Highly variable conditions can trigger desertification. However, other than those areas under the influence of the ENSO and the few areas with longer-term droughts such as the Sahel, there is little evidence for recent changes in drought frequency or intensity in arid and semi-arid lands (see Chapter 3, *Observed Climate Variability and Change*, of the IPCC Working Group I volume).

Analysis of temperature data for semi-arid and arid lands by various groups presents a more confused picture than that for rainfall. For example, in Kenya there was a mean annual temperature rise of 0.4°C over the period 1942–91, perhaps tied to urbanization (Kinuthia *et al.*, 1993). In South Africa, there was no temperature trend over the 1940–90 period (Mühlenbruch-Tegen, 1992). Central Asia did not see a consistent trend over 1900–1980 (Kharin, 1994); there was no trend in southern France, either (Daget, 1992). Over the last 100 years, temperature in North America increased by 0.8°C (Balling, 1994). Chapter 3 of the IPCC Working Group I volume summarizes the global surface temperature trends for the period 1920–93. The future magnitude of increases in temperature in arid and semi-arid lands is likely to be similar to or lower than the global estimate of 0.005°C per annum. The greatest increase in such lands is likely to be experienced above latitudes north and south of 40°.

Whereas most terrestrial ecosystems have some built-in ability to buffer the effects of climate variability, this is not so true of those in arid and semi-arid lands—where even small changes in climate can intensify the already high natural variability and lead to permanent degradation of the productive potential of such lands (OIES, 1991). Arid and semi-arid lands may thus be among the first regions in which ecosystem dynamics become altered by global environmental change (West *et al.*, 1994).

There also is the question of the extent to which desertification itself can exacerbate climate change. It has been suggested that surface air temperature has increased significantly in desertified areas because of changes in land cover, thereby affecting global mean temperature (Balling, 1991). A reduction in surface soil moisture results in more available energy for warming the air because less is used to evaporate water. Hulme and Kelly (1993) consider this effect to be very weak but identify a better-established, though less-direct, link through the effects of changes in carbon sequestration potential and possible changes in soil conditions on emissions of nitrous oxide and methane. Hulme and Kelly (1993) point to the difficulty of quantifying the precise effect of desertification on global warming. Arid-region temperatures and rainfall should be monitored to develop a stronger information base.

4.4.2. The Nature, Causes, and Severity of Desertification

Desertification arises both from human abuse of the land and from adverse climatic conditions such as extended drought (UNEP, 1992)—which may trigger, maintain, or even accelerate the process of dryland degradation. Currently there is disagreement as to whether human impacts or climatic factors are the primary agents responsible for the desertification of the world's arid and semi-arid lands.

Some evidence suggests that human impacts arising from overstocking, overcultivation, and deforestation are primarily responsible for the process. Such conclusions have been reached by scientists working in northwest China (Chao Sung Chiao, 1984a, 1984b); Australia (Perry, 1977; Mabbut, 1978); South America (Soriano, 1983); North America (Dregne, 1983; Schlesinger *et al.*, 1990); Europe (Lopez-Bermudez *et al.*, 1984; Rubio, 1987; Katsoulis and Tsangaris, 1994; Puigdefabregas and Aguilera, 1994; Quine *et al.*, 1994); North and West Africa and the Sahel (Pabot, 1962; Le Houérou, 1968, 1976, 1979; Depierre and Gillet, 1971; Boudet, 1972; Lamprey, 1988; Nickling and Wolfe, 1994; Westing, 1994); East Africa (Lusigi, 1981; Muturi, 1994); and South Africa (Acocks, 1952; Hoffman and Cowling, 1990a; Bond *et al.*, 1994; Dean and McDonald, 1994). Many of these studies suggest that, although desertification is the result of a complex interaction of a number of factors, the direct causes are human actions—which themselves are a function of population density (see Figure 4-2), cultural traditions, land tenure, and other socioeconomic and political factors. Although climate and soil type are important in determining the severity and rate of desertification, it is ignorance or the force of circumstance in

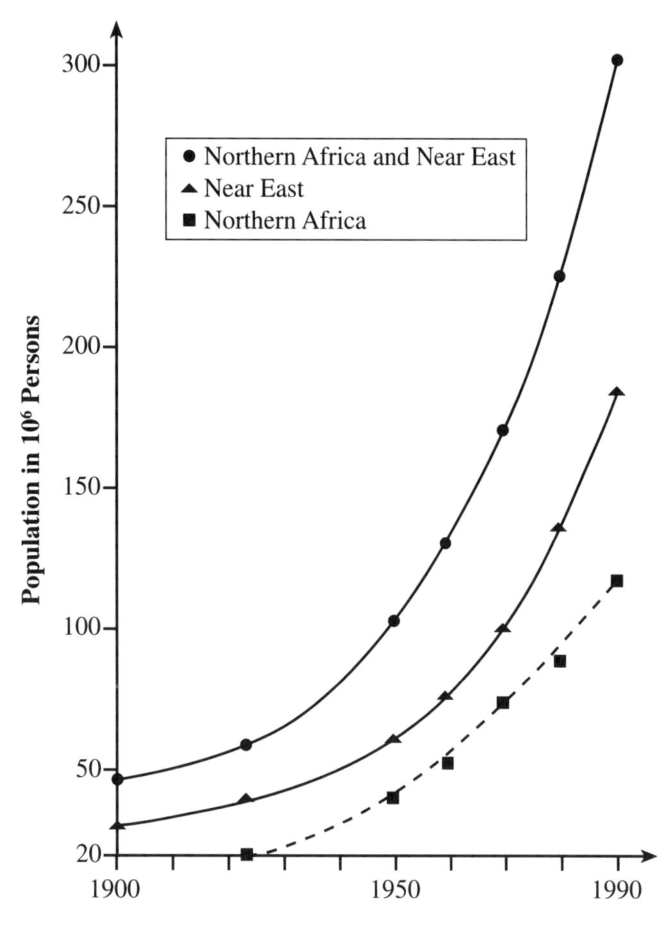

Figure 4-2: Evolution of the human population in northern Africa and the Near East from 1900 to 1990. Sources are various, including FAO yearbooks (Le Houérou, 1992).

failing to match the use and management of the land to the soil and prevailing climate that leads to the removal of soil. Overstocking, deforestation, wood collection, and overcultivation usually are cited as the principal direct causes of the problem; estimates of the percentage of desertified land attributed to each of these factors are available (Table 4-5).

However, a somewhat different view of the causes of desertification also exists. For some Asian environments (e.g., Singh *et al.*, 1994) and particularly in African environments, there is a growing body of literature that emphasizes the impact of extended droughts over the last several decades in the desertification process or suggests that desertification has been overstated due to a lack of adequate information (Hellden, 1988, 1991; De Waal, 1989; Forse, 1989; Mortimore, 1989; Binns, 1990; Hoffman and Cowling, 1990b; Tucker *et al.*, 1991; Grainger, 1992; Thomas, 1993; Dodd, 1994; Pearce, 1994; Thomas and Middleton, 1994). Proponents of this view argue that domestic livestock populations seldom build up to numbers that are damaging for these environments because the periodic droughts that characterize these regions frequently result in high animal mortality (see Behnke *et al.*, 1993; Dodd, 1994; Scoones, 1995).

There also is a great deal of variation in assessments of the nature and severity of the desertification problem, due mainly to the lack of adequate data. Some researchers suggest that almost 20 million km^2 (15% of the land surface of the Earth) are subject to various degrees of desertification; they provide estimates of the extent and severity of the problem for different regions (Dregne, 1983, 1986; Le Houérou, 1992; Table 4-6). In addition, there have been a number of attempts to calculate the rate of expansion of the more arid regions into the more mesic areas. Most of these studies have employed remote sensing, notably with air photos and large-scale satellite images (e.g., SPOT) taken decades apart, combined with detailed maps of ground-control and older vegetation, forest, or range resources. Many of these large-scale studies have suggested that the annual rate of expansion of desertified lands in central Asia, northwest China, northern Africa, and the Sahel ranges from 0.5 to 0.7% of the arid zone (Le Houérou, 1968, 1979, 1989, 1992; Depierre and Gillet, 1971; Boudet, 1972; De Wispelaere, 1980; Gaston, 1981; Haywood, 1981; Floret and Pontainier, 1982; Chao Sung Chiao, 1984b; Peyre de Fabrègues, 1985; Vinogradov and Kulin, 1987; Grouzis, 1988; Rozanov, 1990). Assuming a conservative rate of expansion of 0.5% per annum, this information has been used to suggest an annual increment of desertified land of 80,000 km^2 (see also Hulme and Kelly, 1993). Because 25% of the arid zone already is desertified, this estimate would mean that almost all of the world's arid and semi-arid lands will become desertified within the next century if rates of desert expansion are similar to those that have prevailed over the past 50 years. Complete arid-land desertification could occur earlier if human and livestock populations continue to increase and particularly if the resilience of arid regions is negatively affected by climate change (Hulme and Kelly, 1993).

Table 4-5: *Causes of desertification, in percent of desertified land (after Le Houérou, 1992).*

Regions or Countries	Overcultivation	Overstocking	Fuel and Wood Collection	Salinization	Urbanization	Other
Northwest China	45	16	18	2	3	16
North Africa and Near East	50	26	21	2	1	–
Sahel and East Africa	25	65	10	–	–	–
Middle Asia	10	62	–	9	10	9
United States	22	73	–	5	–	–
Australia	20	75	–	2	1	2

Table 4-6: *Extent and severity of desertification (after Oldeman et al., 1990; Le Houérou, 1992; Le Houérou et al., 1993; UNEP, 1992).*

	Light		**Moderate**		**Strong**		**Severe**	
Region	Area	%	Area	%	Area	%	Area	%
Africa	1180	9	1272	10	707	5.0	35	0.2
Asia	1567	9	1701	10	430	3.0	5	0.1
Australasia	836	13	24	4	11	0.2	4	0.1
North America	134	2	588	8	73	0.1	0	0.0
South America	418	8	311	6	62	1.2	0	0.0
Total	4273	8	4703	9	1301	2.5	75	0.1

Notes: Area desertified in 10^3 km^2; % = area desertified in percent of total drylands; drylands = arid + semi-arid + dry sub-humid.

Some recent research, however, has not supported the expanding-desert hypothesis. First, a number of researchers question the data upon which the severity estimates (see Table 4-6) are based (Hellden, 1988; Thomas, 1993; Dodd, 1994). Second, some argue that the process cannot be accurately described as a degradation front that grows incrementally larger each year. Arid and semi-arid systems are characterized by variable climates subject to large fluctuations in annual rainfall. During favorable or "wet" years there is little evidence to suggest that the desertification front expands (Thomas, 1993). On the contrary, a number of recent studies suggest that vegetation cover at the margins of arid and semi-arid regions both increases and decreases, depending largely on seasonal and annual rainfall totals (Hoffman and Cowling, 1990b; Tucker *et al.*, 1991). The different views regarding the importance of human abuse versus adverse climatic conditions in causing desertification, and the lack of agreement on the scale of the problem, point to the need for a better understanding of the problem and, in particular, better and more comprehensive data on its nature and extent. There is little doubt that desertification is an important environmental problem that needs to be addressed urgently—and that climate change will have an impact on it.

4.4.3. *The Impact of Drought and Desertification on Natural Ecosystems and Rain-Fed Crops*

The impact of drought on natural ecosystems is measured by plant cover and biomass production and by the disruption of food production systems (De Waal, 1989; Mortimore, 1989). Drought reduces the number, phytomass, and ground cover of plants and hence reduces the protection of the soil against erosion (Grainger, 1992). Desertification has much more profound and lasting effects. Desertified soils are subject to extensive water and wind erosion and therefore lose much of their depth and ability to store water and nutrients (Mainguet, 1986; Grainger, 1992). In the worst cases, all perennials are removed, and the soil surface is subjected to large-scale wind and water erosion. Without permanent vegetation protection, the soil surface is eroded by running water, sealed and crusted by raindrop splash, and made increasingly impervious and hence prone to more erosion. Soil surface sealing and encrustation reduces water intake, resulting in a drier environment. Thus, a whole spiral of self-perpetuating edaphic aridity is triggered (Le Houérou, 1969; Floret and Pontainier, 1984). Eventually, all of the soft soil layers are removed, and the situation becomes irreversible.

Schlesinger *et al.* (1990) propose that not only does drought and the loss of vegetation cover lead to desertification, but a mere shift in plant growth form dominance, as a consequence of overgrazing, may drastically reduce productivity. In their model of dryland degradation in the southwestern United States, they suggest that overgrazing results in the redistribution of organic matter and nutrients and is the primary agent responsible for the current conversion of previously productive grasslands to unproductive mesquite (*Prosopis glandulosa*) shrublands. Where resources within semi-arid grasslands previously were homogeneously distributed, overgrazing results in their concentration beneath shrub islands. Not only is the productivity of the land reduced, but positive feedback processes render the changes irreversible (Schlesinger *et al.*, 1990).

Rangelands are likely to be prone to desertification because of their inherent fragility (see Chapter 2). Hanson *et al.* (1993) use three GCMs to assess the effect of climate change on plant and animal production in the Great Plains of North America. Their scenarios predict decreases in plant nitrogen content during summer grazing and decreases in animal production because of increased ambient temperature and decreased forage quality. Using a 1-year time-step livestock-production model, they show for the simulations used that carrying capacities would need to drop from above 6.5/ha to 9/ha to maintain 90% confidence of not overstocking.

Desertification causes soil to lose its ability to support rain-fed crops (El-Karouri, 1986). It inevitably results in emigration as the land cannot sustain the original inhabitants (Westing, 1994). There are indications that as much as 3% of the African population has been permanently displaced, largely as a result of environmental degradation.

4.4.4. *Adaptive Strategies for Mitigating the Impacts of Climate Change on Desertification*

As we have noted, some desertification already is irreversible. Many other examples of desertification are reversible given appropriate legislation, technology, conservation techniques, education, and, most importantly, the will to act. There are various measures to combat the aridity that might lead to desertification. The impact of increased aridity on natural ecosystems and crops would be moderate and in most cases manageable if improved agricultural and management practices—such as the selection of more drought-tolerant species and cultivars, shorter-cycle annual crops, better tillage practices, soil conservation practices, more timely agricultural operations, and more water-conserving crop rotations—were adopted. In natural ecosystems such as rangelands, increased aridity could be mitigated through the use of more appropriate stocking rates and more balanced grazing systems such as deferred grazing, the concurrent utilization of various kinds of stock species and breeds, game ranching, mixed stock and game ranching, commercial hunting, and so forth.

4.4.4.1. *Water Use and Management*

One of the primary factors affecting water availability is the annual distribution of precipitation. There are large differences in water availability in soils within a given land system or landscape, depending on topography, geomorphology, and the nature of the soil and its depth. Many surveys show that water availability in contiguous soils may vary by a factor of one to ten or more (Le Houérou, 1962; Floret and Pontainier, 1982, 1984). The actual distribution of water in soils may be modified considerably by tillage and mulching practices, soil and water conservation techniques, runoff farming, water harvesting, and wadi diversion. Most of these techniques have been known for nearly 3,000 years in the Near and Middle East.

The techniques of runoff farming—which probably date to nearly 3,000 years ago in western China, Iran, Saudi Arabia, Yemen, Jordan, northern Africa, and Spain—have hardly been used outside their own regions but are capable of development elsewhere. Unfortunately, some of these techniques are being abandoned in countries in which they have been in use for centuries, such as southern Tunisia, Libya, northern Egypt, Yemen, and Iran. The techniques of water harvesting—well-developed in the Near East and North Africa during Roman and Byzantine periods (200 BC to 650 AD)—could be developed again within and outside these regions taking care to ensure that the new conditions are suitable for them.

Irrigated farming may not suffer from increased aridity even if water availability becomes more difficult. Improved irrigation practices (e.g., generalized drip irrigation, underground irrigation) can save up to 50% water compared with conventional irrigation systems. Under more-arid future conditions, one could expect a very large increase in the use of irrigation, in addition to the expansion of controlled farming, crop genetic improvement, and the expansion of winter-growing crops that are much less demanding on water.

In most arid and semi-arid countries, wasted water is common, particularly in agriculture (due to inefficient irrigation systems). The amount of waste commonly reaches 50% in many countries, but such waste often is easy to prevent with appropriate techniques. In many countries, water from aquifers (both shallow and deep) is being used at rates that exceed recharge. Increased pumping of water from these aquifers for irrigation will hasten their depletion and threaten the availability of this water for even more vital needs. Future development in many countries will require water savings.

Two ways to save water besides improved irrigation and drainage systems are to utilize water-efficient plants and to develop winter-growing crops. Alfalfa, for instance (a very popular fodder crop), is extremely water-demanding—requiring about 700–1000 kg of water to produce 1 kg of fodder dry matter—whereas some other fodder crops are less water-demanding for the same yield. Winter-growing crops consume much less water than summer crops because PET during winter is only a small fraction of the summer amount.

Drainage water, which has a rather high concentration of salts, can be re-used to grow salt-tolerant crops, including cashcrops such as asparagus or industrial crops such as sugar beet; it also may be used to grow timber and fodder. Some timber and fuelwood species can produce high yields with water having half the salt concentration of seawater, as can a number of fodder crops such as saltbushes, fleshy sainfoin, tall fescue, and strawberry clover (Le Houérou, 1986). There is, however, a potentially detrimental effect of re-using salt-enriched drainage waters—namely, the accelerating salinization of soil, as occurred adjacent to the Nile in Egypt.

There are many techniques that could be used to counteract a moderate increase in climatic aridity, such as those outlined above, but these techniques tend to increase the costs of production and thus alter the conditions of commercial competition of products on the markets.

4.4.4.2. *Land-Use Systems*

Adaptation to drought and desertification has challenged pastoralists, ranchers, and farmers for centuries. Pastoralists and ranchers have drought-evading strategies and farmers have drought-enduring strategies. Drought-enduring strategies include the adoption of a light stocking rate that preserves the dynamics of the ecosystems and their ability to recover after drought, and the utilization of agroforestry techniques whereby fodder shrubs and trees that can store large amounts of feed over long periods of time are planted in strategic locations in order to provide an extra source of feed when drought occurs. Among the species used are saltbush plantation, spineless fodder cacti (*Agave americana*), wattles (*Racosperma* spp), mesquites (*Prosopis* spp), and acacias (*Acacia* spp). These provide and thus encourage a more permanent rather than nomadic existence even when the range is dry and parched.

Another drought-enduring strategy utilized in the United States and South Africa is game ranching of animals that are better adapted to dry conditions than are livestock—including several species of Cervidae, various antelopes, and ostriches. There are at present more than 20 Mha of game ranching in South Africa, Namibia, Botswana, and Zimbabwe (Le Houérou, 1994). The utilization of stock species and breeds that are better adapted to dry conditions is a further possible adaptation. Many East African pastoralists (Somali, Rendille, Gabra, Samburu) have shifted recently from cattle-husbandry to camel-rearing as a response to the deterioration of their environment; others have increased the proportion of goats compared to sheep (e.g., Turkana).

A drought-evading approach to coping with drought and desertification is the application of soil-conservation techniques, including zero or minimum tillage, contour furrowing, pitting, banking, terracing, and benching. Most of these techniques have been known to some ethnic groups and to some regions for a long time, yet they have rarely been extended outside their region of origin.

The overall strategy should be the development of diversified and multiple production systems—including wildlife, combined with livestock and game ranching; commercial hunting; various tourism activities, including so-called green-tourism; runoff farming; poultry production; and nonagricultural activities such as handicraft. Such activities are developed to various levels in a number of countries, such as the drylands of the United States (west of 100° longitude), Australia, South Africa, Namibia, Botswana, Zimbabwe, Kenya, and Tanzania. There is no reason why these practices could not be developed elsewhere.

4.4.4.3. Agroforestry

Agroforestry may play an extremely important role in the development of semi-arid and arid lands and in the struggle against desertification (Le Houérou, 1980; Baumer, 1987; Le Houérou and Pontainier, 1987). Agroforestry techniques have been developed for centuries in some rural civilizations—such as Kejri [*Prosopis cineraria* (L) Druce] in northwest India (Rajasthan); *Faidherbia albida* (Del.) Chev. in various parts of intertropical Africa; saltbushes in various arid zones of the world; *Argania spinosa* (L.) Sk in southwest Morocco; *Quercus ilex* L. and *Q. suber* L. in Spain and Portugal; and Espino (*Acacia caven* Mol.) and Algarrobo (*Prosopis spp*) in Latin America. Such techniques permit high rural population densities in arid zones (e.g., 60–80 people/km^2) with *Faidherbia* in southern Sahel, and similar densities in Rajasthan with *Prosopis cineraria* and in southern Morocco with *Argania* (the latter region receives <150–300 mm/yr rainfall).

Agroforestry therefore should be an integral part of any dryland development strategy for sustainable agriculture. This development should include village woodlots, which—when located in strategic situations on deep soils, and benefiting from some runoff—may produce very high yields as long as the right species of tree are selected and then rationally managed. Such woodlots could and should be an important part of the fight against desertification because fuelwood collection is a major cause of land degradation in many developing countries.

4.4.4.4. Conservation and Biodiversity

The world's semi-arid and arid lands contain a large number of species of plants and animals that are important to humankind as a whole. Due to recent desertification of several regions of the world, many species are endangered or will soon be. The only way to preserve this biological capital for the benefit of humankind is *in situ* conservation projects. However, conservation is expensive, particularly in areas where the demand on the land is acute because of high human population growth.

4.5. Other Forms of Soil Degradation

There are a number of other soil processes that can lead to land degradation under a changing climate. For example, organic-matter levels may decline; leaching may increase, giving rise to loss of nutrients and accelerated acidification; and, in some instances, soil structure may deteriorate. Some of these changes already are taking place as a result of land-use changes, poor farming practices, and industrial emissions. To date, the changes are fairly subtle, and increasing slowly—unlike some forms of erosion and salinization which can occur more quickly.

4.5.1. Declining Organic-Matter Levels

Organic matter probably is the most important component of soils. It influences soil stability, susceptibility to erosion, soil structure, water-holding capacity, oxygen-holding capacity, and nutrient storage and turnover, and it provides a habitat for extremely large numbers of soil fauna and microflora. With the expansion of agriculture and its intensification, soil organic-matter levels have declined. In the United States, for example, many prairie soils have lost more than one-third of their initial organic-matter contents after 100 years or more of cultivation (Papendick, 1994).

A decline of 55–58% in organic matter (equivalent to a loss of 29–32 t/ha) has been noted in the top 10 cm of two alluvial soils in New Zealand after just 2 years of continuous cropping (Shepherd, 1992). Much of this organic matter was lost to the atmosphere as CO_2; this demonstrates the point that some intensively cropped land can potentially release a significant amount of greenhouse gases (Tate, 1992).

Organic matter is susceptible to change with changing climate. Higher temperatures increase rates of decomposition. Whether enhanced atmospheric CO_2 concentrations will give rise to sufficient increases in plant growth to return organic matter to soils and compensate for this loss is still not established (see Chapter A). If CO_2 levels do not lead to compensation, it will be important to implement programs of organic-matter management to maintain adequate soil organic-carbon contents to prevent degradation. This also will benefit carbon sequestration (see Chapters 23 and 24).

4.5.2. *Declining Nutrient Levels*

Loss of nutrients is a common phenomenon in countries with low-input agriculture and also occurs under soils with a high leaching potential and a low buffering capacity.

In soils in which the movement of water is predominantly downward, there will be a tendency for nutrients to be lost. In intensive, high-input agriculture, such losses are compensated by fertilizer additions, but under low-input agriculture, loss of nutrients is a common occurrence—one that is recognized as an increasing problem, particularly in subtropical and tropical countries (Smaling, 1990; Pereira, 1993).

Temperature and precipitation changes could affect soil nutrient levels in several ways. Rising temperatures could act to maintain nutrients within the soil because of increased evaporative forces, and thus reduce leaching. An increase in rainfall generally will give rise to increased nutrient loss. Conversely, a decrease in rainfall may lead to upward movement of nutrients—which, in some cases, can lead to salinization (see Section 4.3). Nutrient loss can be compensated by the use of inorganic and organic fertilizers, but the cost can be prohibitive for subsistence farmers.

4.5.3. *Acidification*

Most countries contain large areas of acidified soils (Wild, 1993). The extent to which acidification will increase with climate change depends on both rainfall and temperature. Significant increases in rainfall would be accompanied by increased leaching and thus acidification (Rounsevell and Loveland, 1992). A decline in rainfall could reduce the extent and intensity of acidification. In the sub-arid and sub-humid zones, soils are subject to seasonal changes—from leaching conditions to evaporative conditions. An altered climate could shift conditions to a dominantly leaching regime involving increased acidification, or to a dominantly evaporative regime involving less acidification but possible salinization. The direction of change will depend on the extent to which rainfall and temperature change.

One of the soil types most affected by acidification is termed acid sulfate soils. These soils suffer from extreme acidity as a result of the oxidation of pyrite when pyrite-rich parent materials are drained (Dent, 1986). Pyrite accumulates in waterlogged soils that are rich in organic matter and contain dissolved sulfates, usually from seawater. When these previously waterlogged soils are drained, oxygen enters the soil system and oxidizes pyrite to sulfuric acid, causing the pH to drop to less than 4.

Several areas of the world contain potential acid sulfate soils, which would be converted to acid sulfate soils if climate change were to lead to the lowering of water tables and increased oxidation (in total over 20 Mha; Beek *et al.*, 1980). The principal areas affected would be coastal areas of West Africa, South America, India, and the Far East.

4.5.4. *Deterioration of Soil Structure*

Soil structure regulates water retention and movement, nutrient transformations and movement, faunal activity and species diversity, and the strength and penetrability of the rooting media (Lal, 1994a). Soil structure affects the ability of the soil to be cultivated, and the quality and durability of seedbeds—and thus plant growth and vigor, grain yield, and grain quality (Shepherd, 1992). The potential for deterioration of soil structure as a result of climate change, in conjunction with poor soil management, has received little attention from the scientific community even though soil structure controls water and air movement and transport processes in the soil profile. Changes to soil structure are notoriously difficult to quantify, partly because of the influence of land use and management; research needs to be directed toward a better understanding of this. However, some potential consequences of climate change can be identified.

A decline in soil organic-matter contents as a result of climate change has important implications for soil quality in general and specifically for soil structural development because there is a strong relationship between organic-matter levels and soil structure quality (see Chapter A). A decline in soil organic-matter levels would cause a decrease in soil aggregate stability, an increase in susceptibility to compaction, lower infiltration rates, increased runoff, and, hence, enhanced likelihood of erosion.

A second important function of soil structure—in relation to clay soils particularly—is their influence on water quality and seepage. Soils with high clay contents, especially those with expanding type clays (smectitic mineralogy), tend to shrink when they are dry and swell as they wet-up again. This behavior results in the formation of large cracks and fissures. Drier climatic conditions would be expected to increase the frequency and size of crack formation in soils, especially those in temperate regions that currently do not reach their full shrinkage potential (Climate Change Impacts Review Group, 1991). One consequence would be more rapid and direct movement of water and dissolved solutes from soil to surface waters or aquifers through so-called "bypass" flow via cracks (Armstrong *et al.*, 1994). The more direct movement of water in this way would decrease the filtering capacity of soil and increase the possibility of nutrient losses and pollution of ground and surface waters. Where animal manures or sewage sludge are applied to land, there would be a greater risk of organic contaminants and microorganisms reaching aquifers.

In the case of acid soils, increased cracking could lead to an increase in the quantity of metals (e.g., aluminum, manganese, iron, heavy metals) entering aquifers. In these situations, the combined effects of changes in soil chemical and biological cycles and increased cracking need to be assessed (Rounsevell and Loveland, 1992). An additional consequence is that the soil profile would not re-wet as before, and the rooting zone could be deprived of valuable water. Such changes also would have important implications for the stability of the foundations of buildings and roads—which would have been designed on the basis that such soils did not reach their shrink-swell capacity (Boden and Driscoll, 1987).

4.6. Monitoring and Research Needs

4.6.1. *Monitoring*

Global estimates of land degradation generally are very crude; as a result, it is difficult to present a meaningful picture of its extent and severity. The most recent attempt by Oldeman *et al.* (1990) helps to indicate the types and severity of land degradation, but because of a lack of adequately established national surveys or a global network of monitoring sites, the results can only be a broad approximation. Compared with air and water, soil has received little attention in terms of monitoring; very few countries have sound data on the quality of their soils, and there often are few or no trend data to indicate whether or how soil quality is changing. There is a need to establish a comprehensive monitoring system for soil quality at a national level that will contribute to the creation of a global picture. In the absence of such a program, there will be general ignorance of changes in soil quality with time, leading to an inability to address strategically any deleterious changes. The limited information that is available suggests that land degradation is a major global problem that urgently needs to be addressed.

4.6.2. *Research*

To date, most research on the impacts of climate change on soils has involved taking knowledge of soil processes and properties gained over some 50 years and predicting what effect a new climatic scenario might have on them. Other research has examined the impacts of climate change on soils in terms of their ability to support current and future crops and natural ecosystems. Both of these approaches, although justifiable on the basis of the need for early predictions of the effects of climate change, often lack rigor.

There are three areas in which research now needs to focus:

- Determining the impact of increased atmospheric CO_2 concentrations, temperature, and changes in other climatic parameters on soil properties and processes, including those involved in land degradation under different land uses in different climatic zones
- Identifying the impact of these climate-induced changes on soil processes and properties and on the ability of the soil to perform its functions (e.g., as a habitat for microflora and fauna, nutrient storage and turnover, water storage and movement)
- Understanding how such changes will affect terrestrial ecosystems, what the response times of such systems to soil changes will be, and what the suitability of changed soils will be for crops and more natural habitats.

A successful outcome to this research requires continuing improvements to terrestrial nutrient-cycle models and GCMs. In particular, much better information is required for rainfall—including amount, frequency, duration, and intensity—and for surface wind, including direction, speed, and frequency of high-wind events. Given improved model output that is more reliable at the regional scale and the research outlined above, governments and land managers will receive increasingly valuable indicators of future conditions.

References

Acocks, J.P.H., 1952 (3rd ed., 1988): *Veld Types of South Africa.* Dept. of Agric. & Water Supply, Pretoria, South Africa, 1 col. map 1/1. 500,000 2 sheets, 146 pp.

Armstrong, A.C., A.M. Matthews, A.M. Portwood, T.M. Addiscott, and P.B. Leeds-Harrison, 1994: Modelling the effects of climate change on the hydrology and water quality of structured soils. In: *Soil Responses to Climate Change* [Rounsevell, M.D.A. and P.J. Loveland (eds.)]. NATO ASI Series I, Global Environmental Change, vol. 23, Springer-Verlag, Heidelberg, Germany, pp. 113-136.

Ballantyne, A.K., 1963: Recent accumulation of salts in the soils of south eastern Saskatchewan. *Canadian Journal of Soil Science*, **43**, 52-58.

Balling, Jr., R.C., 1991: Impact of desertification on regional and global warming. *Bull. American Meteorological Society*, **72**, 232-34.

Balling, R.C., 1994: *Climate Variations in Drylands: An Annotated Bibliography of Selected Professional Journal Articles.* Report to the Intern. Panel of Experts on Desertification, INCHED, Geneva, Switzerland, 33 pp.

Barrow, C.J., 1991: *Land Degradation.* Cambridge University Press, Cambridge, UK, 295 pp.

Batjes, N.H., V.W.P. Van Engelen, J.H. Kauffman, and L.R. Oldeman. 1994: Development of soil databases for global environmental modelling. *Proc. 15th World Congress of Soil Science.* Comm. V, vol. 6a, pp. 40-57.

Baumer, M., 1987: *Agroforesterie et Désertification.* ICRAF, Nairobi, Kenya, and CTA, Wageningen, The Netherlands, 260 pp.

Beek, K.J., W.A. Blokhuis, P.M. Dreissen, N. van Breemen, R Brinkman, and L.J. Pons, 1980: Problem soils: their reclamation and management. Institute for Land Reclamation and Improvement, *Land Reclamation and Water Management. Development, Problems and Challenges,* **27**, 43-72.

Behnke, R.H., I. Scoones, and C. Kerven (eds.), 1993: *Range Ecology at Disequilibrium: New Models of Natural Variability and Pastoral Adaptation in African Savannas.* Overseas Development Institute, London, UK, 248 pp.

Binns, T., 1990: Is desertification a myth? *Geography*, **75**, 106-113.

Blaikie, P. and H. Brookfield, 1987. *Land Degradation and Society.* Methuen, London, UK, 296 pp.

Bliss, N.B., 1990: A hierarchy of soil databases for calibrating models of global climate change. In: *Soils and the Greenhouse Effect* [Bouwman, A.F. (ed.)]. J. Wiley & Sons, Chichester, UK, pp. 523-529.

Boardman, J. and D.T. Favis-Mortlock, 1993. Climate change and soil erosion in Britain. *Geographical Journal*, **159**, 179-183.

Boden, J.B. and R.M.C. Driscoll, 1987: House foundations—a review of the effect of clay soil volume change on design and performance. *Municipal Engineer,* **4**, 181-213.

Bond, W.J., W.D. Stock, and M.T. Hoffman, 1994: Has the Karoo spread? A test for desertification using stable carbon isotopes. *South Afr. J. Sci.*, **90(7)**, 391-397.

Bork, H.R., 1983: Die holozäne Relief-und Bodentwicklung in Lossgebieten-Biespiele aus dem südöstlichen Niedersachsen. In: *Bodenerosion, holozäne und pleistozäne Bodenentwicklung* [Bork, H.R. and W. Ricken (eds.)]. *Catena Supplement,* **3**, 93 pp.

Bork, H.R., 1989: Soil erosion during the past millennium in Central Europe and its significance within the geomorphodynamics of the Holocene. *Catena Supplement*, **15**, 121-131.

Boudet, G., 1972: Désertification de l'Afrique tropicale sèche. *Adansonia ser. 2,* **12(4)**, 505-524.

Bouwman, A.F. (ed.), 1990: *Soils and the Greenhouse Effect.* John Wiley and Sons, Chichester, UK, 302 pp.

Brammer, H. and R. Brinkman, 1990: Changes in soil resources in response to a gradually rising sea-level. In: *Soils on a Warmer Earth* [Scharpenseel, H.W., M. Schomaker, and A. Ayoub (eds.)]. Elsevier, Amsterdam, The Netherlands, pp. 145-156.

Brinkman, R. and W.G Sombroek, 1993: *The Effects of Global Change on Soil Conditions in Relation to Plant Growth and Food Production.* Expert Consultation Paper on "Global climate change and agricultural production: direct and indirect effects of changing hydrological, soil and plant physiological processes." FAO, Rome, Italy, 16 pp.

Brown, L.R. and J.E. Young, 1990: Feeding the world in the nineties. In: *State of the World 1990* [Brown, L.R. *et al.* (eds.)]. Worldwatch Institute Report on Progress Toward Sustainable Society, BLDSC, W.W. Norton and Co. Inc., pp. 59-78.

Burgos, J.J., J. Ponce Fuenzalida, and C.B. Molion, 1991: Climate change predictions for South America. *Climate Change*, **18**, 223-229.

Chao Sung Chiao, 1984a: The sand deserts and the Gobis of China. In: *Deserts and Arid Lands* [Farouk El Baz (ed.)]. Martinus Nijhoff, The Hague, The Netherlands, pp. 95-113.

Chao Sung Chiao, 1984b: Analysis of the desert terrain in China using Landsat imagery. In: *Deserts and Arid Lands* [Farouk El Baz (ed.)]. Martinus Nijhoff, The Hague, The Netherlands, pp. 115-132.

Clark, E.H., J.A. Haverkamp, and W. Chapman, 1985: *Eroding Soils—The Off-Farm Impacts.* The Conservation Foundation, Washington, DC, 252 pp.

Climate Change Impacts Review Group, 1991: *The Potential Effects of Climate Change in the United Kingdom.* Department of the Environment, HMSO, London, UK, 124 pp.

Crosson, P.L. and A.T. Stout, 1983: *Productivity Effects of Cropland Erosion in the United States.* Resources for the Future, Washington, DC, 103 pp.

Daget, P., 1992: *J J Corre: Implications des changement climatiques. Etude de cas: Le Golfe du Lion* [Jeftic, L., J.D. Milliman, and G. Sestini (eds.)]. Edw. Arnold, London, UK, pp. 328-357.

Day, J.W. and P.H. Templet, 1989: Consequences of a sea level rise: implications from the Mississippi delta. *Coastal Management,* **17**, 241-257.

De Waal, A., 1989: *Famine That Kills. Darfur, Sudan, 1984-1985.* Clarendon Press, Oxford, UK, 258 pp.

De Wispelaere, G., 1980: Les photographies aeriennes témoins de la dégradation du couvert ligneux dans un écosystème sahélien sénégalais. Influence de la proximité d'un forage. *Cah.,* **XVIII(3-4)**, *ORSTOM,* 155-166.

Dean, W.R.S. and I.A.W. McDonald, 1994: Historical changes in stocking rates of domestic livestock as a measure of semi-arid and arid rangeland degradation in the Cape Province. *J. of Arid Enviro.* (South Africa), **28(3)**, 281-298.

Dent, D., 1986: *Acid Sulphate Soils: A Baseline For Research and Development.* International Institute for Land Reclamation and Improvement, Publication 39, Wageningen, 204 pp.

Depierre, D. and H. Gillet, 1971: Désertification de la zone sahélienne du Tchad. *Bois et Forêts des Tropiques*, **139**, 2-25.

Dodd, J.L., 1994: Desertification and degradation in sub-Saharan Africa: the role of livestock. *Bioscience*, **44**, 28-35.

Dregne, H.E., 1983: *Desertification of Arid Lands.* Harwood Academ, New York, NY, 242 pp.

Dregne, H.E., 1986: Desertification of arid lands. In: *Physics of Desertification* [El-Baz, F. and M.H.A. Hassan (eds.)]. Martinus Nijhoff, Dordrecht, The Netherlands, pp. 4-34.

Dunne, T., 1990: Hydrology, mechanics and geomorphic implications of erosion by subsurface flow. In: *Groundwater Geomorphology* [Higgins, C.G. and D.R. Coates (eds.)]. Special Paper 252, Geological Society of America, Reston, VA, pp. 1-28.

El-Karouri, M.O.H., 1986: The impact of desertification on land productivity in Sudan. In: *Physics of Desertification* [El-Baz, F. and M.H.A. Hassan (eds.)]. Martinus Nijhoff, Dordrecht, The Netherlands, pp. 52-58.

FAO, 1983: *Keeping the Land Alive: Soil Erosion Its Causes and Cures.* FAO, Rome, Italy, 79 pp.

Favis-Mortlock, D.T., R. Evans, J. Boardman, and T.M. Harris, 1991: Climate change, winter wheat yield and soil erosion on the English South Downs. *Agricultural Systems,* **37**, 415-433.

Favis-Mortlock, D.T., 1994: Modelling soil erosion on agricultural land under a changed climate. In: *Soil Responses to Climate Change* [Rounsevell, M.D.A. and P.J. Loveland (eds.)]. NATO, ASI Series 1, Global Environmental Change, vol. 23, Springer-Verlag, Heidelberg, Germany, pp. 211-215.

Fisher, W.R., 1990: Influence of climatic change on development of problem soils, especially in the alluvial domains. In: *Soils on a Warmer Earth* [Scharpenseel, H.W., M. Schomaker, and A. Ayoub (eds.)]. Elsevier, Amsterdam, The Netherlands, pp. 89-99.

Flanagan, D.C., J.E. Ferris, and L.J. Lane, 1991: WEPP. Hillslope Profile Erosion Model. Version 91.5 User Summary. National Soil Erosion Research Laboratory, USDA, West Lafayette, IN.

Floret, C.H. and R. Pontainier, 1982: *L'aridité en Tunisie pré-saharienne.* Travx & Doc., no. 150, ORSTOM, Paris, France, 544 pp.

Floret, C.H. and R. Pontainier, 1984: Aridité climatique et aridité édaphique. *Bull. Soc. Bot. de France, Act.* ***Bot*** **(2/3/4)**, 265-275.

Forse, B., 1989. The myth of the marching desert. *New Scientist*, **4**, 31-32.

Friend, J.A., 1992. Achieving soil sustainability. *Journal of Soil and Water Conservation,* **47**, 156-157.

Gaston, A., 1981: *La Végétation du Tchad, Évolutions Récentes sous les Influences Climatiques et Humaines.* Thèse Doct.Sc. Uni Paris XII, IEMVT, Maisons-Alfort, 333 pp.

Grainger, A., 1992: Characterization and assessment of desertification processes. In: *Desertified Grasslands: Their Biology and Management* [Chapman, G.P. (ed.)]. Academic Press, London, UK, pp. 17-33.

Grouzis, M., 1988: *Structure, Productivité et Dynamique des Systèmes Écologiques Sahéliens* (Mare d'Oursi, Burkina-Faso). Études et Thèses, ORSTOM, Paris, 336 pp.

Hallsworth, E.G., 1987. *Anatomy, Physiology and Psychology of Erosion.* Wiley, Chichester, UK, 176 pp.

Hanson, J.D., B.B. Baker, and R.M. Bourdon, 1993: Comparison of the effects of different climate change scenarios on rangeland livestock production. *Agricultural Systems*, **41**, 487-502.

Haywood, M., 1981: *Evolution de l'Utilisation des Ferres et de Végétation dans La Zone Soudano-Sahélienne du Projet CIPEA au Mali.* CIPEA/ILCA, Addis Abéba, Ethiopia, 187 pp.

Hekstra, G.P., 1989: Global warming and rising sea levels: the policy implications. *Ecologist*, **19**, 4-15.

Hellden, U., 1988: Desertification monitoring: is the desert encroaching? *Desertification Control Bulletin,* **17**, 8-12.

Hellden, U., 1991: Desertification—time for an assessment? *Ambio*, **20**, 372-383.

Hennessy, J.T., B. Kies, R.P. Gibbons, and J.T. Trombe,1986: Soil sorting by forty five years of wind erosion on a southern New Mexico range. *Soil Science Society of America Journal,* **50**, 391-394.

Hoffman, M.T. and R.M. Cowling, 1990a: Desertification in the Lower Sundays River Valley, South Africa. *J. Arid Environments,* **19(1)**, 105-117.

Hoffman, M.T. and R.M. Cowling, 1990b: Vegetation change in the semi-arid eastern Karoo over the last 200 years: an expanding Karoo, fact or fiction? *South Afr. J. of Science,* **86,** 462-463.

Hulme, M., 1992: Rainfall changes in Africa: (1931-1960 to 1961-1990). *International Journal of Climatology*, **12**, 658-690.

Hulme, M. and M. Kelly, 1993: Exploring the links between desertification and climate change. *Environment*, **35**, 4-19.

IPCC, 1990: *Climate Change: The IPCC Scientific Assessment*. Cambridge University Press, Cambridge, UK, 365 pp.

IPCC, 1994: Climate scenarios and socioeconomic projections for IPCC WG II Assessment [Greco, S. *et al.* (eds.)]. Washington, DC, 67 pp.

Jacks, G.V. and R.O Whyte, 1939: *The Rape of the Earth: A World Survey of Soil Erosion,* Faber and Faber, London, UK, 313 pp.

Jelgersma, S., 1988: A future sea-level rise: its impacts on coastal lowlands. In: *Geology and Urban Development, Atlas of Urban Geology*, vol.1. United Nations Economic and Social Commission for Asia and the Pacific, Bangkok, Thailand, pp. 61-81.

Kassas, M.A.F., 1988: Ecology and management of desertification. *Earth, Changing Geographic Perspectives*, **88**, 198-211.

Katsoulis, B.D. and J.M. Tsangaris, 1994: The state of the Greek environment in recent years. *Ambio*, **23**, 274-279.

Kharin, H.N., G.S. Kalenov, and V.A. Kurochkin, 1993: Map of human induced land degradation in the Aral Sea Basin. *Desertification Control Bulletin*, **23**, 24-28.

Kharin, N.G., 1994: Climate change and desertification in Central Asia (mimeo). Desert Research Institute, Ashgabat, Turkmenistan, 13 pp.

Kinuthia, J.H., P.D. Munah, and C.F.K. Kahoro, 1993: Temperature curves and rainfall trends in Kenya during the last 50 years. Proceed. 1st Internat. Conf. of the African Meteorological Society, Nairobi, Kenya, pp. 164-195.

Kirkby, M.J. and R.P.C. Morgan (eds.), 1980: *Soil Erosion*. John Wiley & Sons, Chichester, UK, 306 pp.

Knisel, W.G. (ed.), 1980: *CREAMS: A Field Scale Model for Chemicals, Runoff and Erosion from Agricultural Management Systems*. USDA Cons. Report 16, Washington, DC, 640 pp.

Kovda, V.A. and I. Szabolcs (eds.), 1979. Modelling of soil salinization and alkalization. *Agrokémia és Talajtan,* **28**, suppl., 1-208.

Lal, R., 1990: *Soil Erosion in the Tropics. Principles and Management*. McGraw-Hill, New York, NY, 580 pp.

Lal, R., 1994a: Sustainable land use systems and soil resilience. In: *Soil Resilience and Sustainable Land Use* [Greenland, D.J. and I. Szabolcs (eds.)]. CAB International, Wallingford, UK, pp. 41-68.

Lal, R., 1994b: Global overview of soil erosion. In: *Soil and Water Science: Keys to Understanding Our Global Environment*. Special Publication 41. Soil Science Society of America, Madison, WI, pp. 39-51.

Lal, R., 1994c: Soil erosion by wind and water: problems and prospects. In: *Soil Erosion Research Methods* [Lal, R. (ed.)]. Soil Water Conservation Society, Ankeny, IA, pp. 11-10.

Lal, R., 1994d: Global soil erosion by water and carbon dynamics. In: *Soils and Global Change* [Lal, R. *et al.* (eds.)]. Lewis Publishers, Chelsea, MI.

Lamprey, H., 1988: Report on the desert encroachment reconnaissance in northern Sudan: 21 October to 10 November 1975. *Desertification Control Bulletin*, **17**, 1-7.

Le Houérou, H.N., 1962: Recherchés sur les conditions hydrides des sols de Sologne et leurs rapports avec la composition du tapis végétal, 5 figs h.t., Archives CEFE/CNRS Montpellier, France, 131 pp.

Le Houérou, H.N., 1968: La désertisation du Sahara septentrional et des steppes limitrophes. *Ann Algér de Géogr.*, **6**, 2-27.

Le Houérou, H.N., 1969: La végétation de la Tunisie steppique avec références aux végétations analogues de l'Algérie de la Libye et du Maroc. *Ann.Instit. Nat. de la Rech. Agron. de Tunisie*, **42(5)**, 1-624, 40 phot, 38 figs, 1 carte coul. 1/500 000 (128 000 km^2) 22, Table. h.t.

Le Houérou, H.N., 1976: The nature and causes of desertization. *Arid Zone Newsletter*, **3**, 1-7, Off. of Arid Land Studies, Tucson, AZ.

Le Houérou, H.N., 1979: La désertisation des régions arides. *La Recherche*, **99(10)**, 336-344.

Le Houérou, H.N. (ed.), 1980: *Browse in Africa, the Current State of Knowledge*. Internat. Livest. Centre for Africa (ILCA), Addis Ababa, Ethiopia, 491 pp.

Le Houérou, H.N., 1986: Salt-tolerant plants of economic value in the Mediterranean Basin. *Reclamation and Revegetation Research*, **5**, 319-341.

Le Houérou, H.N., 1989: *The Grazing Land Ecosystems of the African Sahel*. Ecological Studies no. 75, Springer Verland, Heidelberg, Germany, 287 pp.

Le Houérou, H.N., 1992: An overview of vegetation and land degradation in world arid lands. In: *Degradation and Restoration of Arid Lands* [Dregne, H.E. (ed.)]. Internat. Center for Arid and Semi-Arid Lands Studies (ICASALS), Texas Tech University, Lubbock, Texas, pp. 127-163.

Le Houérou, H.N., 1994: *Drought Tolerant and Water Efficient Fodder Shrubs (DTFS), Their Role as "Drought-Insurance" in the Agricultural Development of the Arid and Semi-Arid Zone in South Africa*. Water Research Commission, Ministry of Agriculture Report KV65/94, Pretoria, South Africa, 139 pp.

Le Houérou, H.N. and R. Pontainier, 1987: *Les Plantations Sylvo-Pastorales dans les Régions Arides de la Tunisie*. Note technique du MAB, no. 18, UNESCO, Paris, France, 81 pp.

Le Houérou, H.N., G.F. Popov, and L. See, 1993: *Agrobioclimatic Classification of Africa*. Agrometeorology Series, Working Paper no. 6, FAO, Rome, Italy, 227 pp.

Lopez-Bermudez, F., A. Romero-Diaz, G. Fisher, C. Francis, and J.B. Thornes, 1984: Erosion y ecologia en la Espana semi-arida (Cuenca de Mula, Murcia). *Cuadernos de Investig. Geofra,* **10(1-2)**, pp. 113-126.

Lusigi, W., 1981: *Combatting Desertification and Rehabilitating Degraded Production Systems in Northern Kenya*. IPAL Techn. Report A-4, MAB Project no. 3, Impact of Human Activities and Land-Use Practices on Grazing Lands. UNESCO, Nairobi, Kenya, 141 pp.

Mabbut, J.A., 1978: *Desertification in Australia*. Water Research Foundation of Australia, Report No. 4, Canberra, Australia, 127 pp.

Mainguet, M., 1986: The wind and desertification processes in the Saharo-Sahelian and Sahelian regions. In: *Physics of Desertification* [El-Baz, F. and M.H.A. Hassan (eds.)]. Martinus Nijhoff, Dordrecht, The Netherlands, pp. 210-240.

Manabe, S. and J.R. Holloway, 1975: The seasonal variation of the hydrologic cycle as simulated by a global model of the atmosphere. *J. Geophys. Res.,* **80**, 1617-1649.

Middleton, N.J., 1985: Effects of drought on dust production in the Sahel. *Nature,* **316**, 431-434.

Morgan, R.P.C., 1986: *Soil Erosion and Conservation*. Longman Press, Harlow, UK, 298 pp.

Morgan, R.P.C., J.N. Quinton, and R.J. Rickson, 1992: *EUROSEM: A User Guide. Version 2*. Silsoe College, Cranfield University, Cranfield, UK, 84 pp.

Mortimore, M., 1989: *Adapting to Drought: Farmers, Famines and Desertification in West Africa*. Cambridge University Press, Cambridge, UK, 299 pp.

Mühlenbruch-Tegen, A., 1992: Long term surface temperature variations in South Africa. *South Afr. J. of Sci*, **88**, 197-205.

Muturi, H.R., 1994: *Temperature and Rainfall Trends in Kenya during the Last 50 Years*. Department Research and Development, Ministry of Research, Technology and Training, Nairobi, Kenya, 15 pp.

Nearing, M.A., L.J. Lane, and V.L. Lopes, 1994: Modelling soil erosion. In: *Soil Erosion Research Methods* [Lal, R. (ed.)]. Soil Water Conservation Society, Ankeny, IA, pp. 127-158.

Nicholson, S.E., 1994: Variability of African rainfall on inter annual and decadal time scales. In*: Natural Climate Variability on Decade-to-Century time Scales* [Matinson, D., K. Bryan, M. Ghil, T. Karl, E. Sarachik, S. Sorooshiam, and L. Talley (eds.)]. National Academy of Sciences, Washington, DC.

Nickling, W.G. and S.A. Wolfe, 1994: The morphology and origin of Nabkhas, region of Mopti, Mali, West Africa. *J. Arid Environments*, **28**, 13-30.

OIES, 1991*: Arid Ecosystems Interactions: Recommendations for Drylands Research in the Global Change Research Program*. Universities Center for Atmospheric Research, Boulder, CO, 81 pp.

Oldeman, L.R., R.T.A. Hakkeling, W.G. Sombroek, 1990: *World Map of Status of Human-Induced Soil Degradation: An Explanatory Note*. Nairobi, Wageningen & UNEP, 3 colour sheets 1/3 000 000, ISRIC, 27 pp.

Oldeman, L.R., 1991-92: Global extent of soil degradation. ISRIC, Bi-annual report, Wageningen, The Netherlands, pp. 19-39.

Pabot, H., 1962: *Comment Briser le Cercle Vicieux de la Désertification dans les Régions Sèches de l'Orient*. FAO, Rome, Italy, 5 pp.

Papendick, R.I., 1994: Maintaining soil physical conditions. In: *Soil Resilience and Sustainable Land Use* [Greenland, D.J. and I. Szabolcs (eds.)]. CAB International, Wallingford, UK, pp. 215-234.

Pearce, F., 1992: Mirage of the shifting sands. *New Scientist*, **136(1951)**, 38-43.

Pearce, F., 1994: Deserting dogma. *Geographical Magazine*, **66(1)**, 25-28.

Pereira, H.C., 1993: Food production and population growth. *Land Use Policy*, **10(3)**, 187-191.

Perry, R.A., 1977: The evaluation and exploitation of semi-arid lands: Australian experience. *Philosophical Transactions of the Royal Society of London,* ser. B, **278**, 493-505.

Peyre de Fabrègues, B., 1985: Quel avenir pour l'élevage au Sahel? *Rev. Elev. & Médec Véter Pays Trop.,* **38(4)**, 500-508.

Phillips, D.L., D. White, and B. Johnson, 1993: Implications of climate change scenarios on soil erosion potential in the USA. *Land Degradation and Rehabilitation*, **4**, 61-72.

Pirazzoli, P.A., 1989: Present and near future global sea-level changes. *Global and Planetary Change*, **1(4)**, 241-258.

Puigdefabregas, J. and C. Aguilera, 1994: The Rambla Honda field site: interactions of soil and vegetation along a catena in semi-arid southeast Spain. In: *Mediterranean Desertification and Land Use (MEDALUS)* [Thornes, J.B. and J. Brandt (eds.)]. John Wiley & Sons, London, UK.

Quine, T.A., A. Navas, D.E. Walling, and J. Machin, 1994: Soil erosion and redistribution on cultivated and uncultivated land near Las Bardenas in the central Ebro River Basin, Spain. *Land Degradation & Rehabilitation*, **5**, 41-55.

Rogasik, J., U. Dämmgen, M. Lüttich, and S. Obenhauf, 1994: Wirkungen physikalischer und chemischer Klimaparameter auf Bodeneigenschaften und Bodenprozesse. *Landbauforschung Volkenrode Teil II Klimaveränderungen und Landbewirtschaftung. Sonderheft,* **148**, 107-139.

Rosenzweig, C., M.L. Parry, G. Fisher, and K. Frohberg, 1993: *Climate Change and World Food Supply*. Research Report No. 3, Environmental Change Unit, University of Oxford, Oxford, UK, 28 pp.

Rosewell, C.J., R.J. Crouch, R.J. Morse, J.F. Leys, R.W. Hicks, and R.J. Stanley, 1991: Forms of erosion. In: *Soils: Their Properties and Management* [Charman, R.E.V. and B.W. Murphy (eds.)]. University Press, Sydney, Australia, pp. 12-35.

Rounsevell, M.D.A. and P.J. Loveland, 1992: An overview of hydrologically controlled soil responses to climate change in temperate regions. *The Journal of the South East England Soils Discussion Group*, **8**, 69-78.

Roux, P.W. and M. Vorster, 1987: Vegetation change in the Karoo. *Proceed. Grassland Soc. of South Afr.*, **18**, 25-29.

Rozanov, B.G., 1990: Assessment of global desertification: status and methodologies. In: *Desertification Revisited* [Odingo, R.S. (ed.)]. UNEP, Nairobi, Kenya, pp. 45-122.

Rubio, J.L., 1987: Desertificacion en la communidad valenciana antecedantes historicos y situacion actual de erosion. *Rev Valenc ed Estud. Autonomicos,* **7**, 231-258.

Santibanez, F. and J. Uribe, 1994: El clima y la desertificacion en Chile. *En Taller. En Nacional del Plan Nacional de Accion para Combatir le Desertificacion*. Univ. de Chile, Santiago, Chile, pp. 17-24.

Scharpenseel, H.W., M. Shoemaker, and A. Ayoub (eds.), 1990*: Soils on a Warmer Earth*. Elsevier, Amsterdam, The Netherlands, 295 pp.

Schlesinger, W.H., 1989: Changes in soil carbon storage and associated properties with disturbance and recovery. In: *The Changing Carbon Cycle Analysis* [Traballen, J.R. and D.E. Reichle (eds.)]. Springer-Verlag, Berlin, Germany, pp. 194-220.

Schlesinger, W.H., J.F. Reynolds, G.L. Cunningham, L.F. Huenneke, W.M. Jarrell, R.A. Virginia, and W.A. Whitford, 1990: Biological feedbacks in global desertification. *Science*, **247**, 1043-1048.

Schwertmann, U., R.J. Rickson, and K. Auerswald (eds.), 1989. Soil erosion protection measures in Europe. *Soil Technology Series 1. Catena.*

Scoones, I. (ed.), 1995. *Living with Uncertainty. New Directions in Pastoral Development in Africa*. International Institute for Environment and Development, London, UK, 210 pp.

Sehgal, J. and I.P. Abrol, 1994a: Soil degradation in India—status and impact. *Transactions. 15th International Society of Soil Science, vol. 7a*. International Society of Soil Sciences, Acapulco, Mexico, 212 pp.

Sehgal, J. and I.P. Abrol, 1994b: *Soil Degradation in India*: *Status and Impact*. Indian Council for Agricultural Research. New Delhi, Oxford, and IBH Publishing Co., PVI, Ltd., 80 pp.

Shepherd, T.G., 1992: Sustainable soil-crop management and its economic implications for grain growers. In*: Proceedings International Conference on Sustainable Land Management* [Henriques, P.R. (ed.)]. Napier, New Zealand, pp. 141-152.

Singh, S., A. Kar, D.C. Joshi, S. Kumar, and K.D. Sharma, 1994: The desertification problem in western Rajasthan. *Annals of the Arid Zone*, **33**, 191-202.

Skidmore, E.C. and J.R. Williams, 1991: *Modified EPIC Wind Erosion Model*. Monograph No. 31, Modelling Plant and Soil Systems, American Society of Agronomy, Madison, WI, pp. 457-469.

Skidmore, E.L., L.J. Hagen, D.V. Armburst, A.A. Durar, D.W Fayrear, K.M. Potter, L.E. Wagner, and T.M. Zobeck, 1994: Methods for investigating basic processes and conditions affecting wind erosion. In: *Soil Erosion Research Methods* [Lal, R. (ed.)]. Soil Water Conservation Society, Ankeny, IA, pp. 245-330.

Smaling, E.M.A., 1990: Two secenarios for the Sub-Sahara. *Ceres,* **126,** 19-24.

Solomon, S.I., M. Beran, and W. Hogg (eds.), 1987: *The Influence of Climate Change and Climatic Variability on the Hydrologic Regime and Water Resources.* IASH Publication No. 168. International Association of Hydrological Sciences, Wallingford, UK.

Sombroek, W.G., 1991-92. The greenhouse effect, plant growth and soils. ISRIC Bi-Annual Report, Wageningen, The Netherlands, pp. 15-18.

Soriano, A., 1983: Deserts and semi-deserts of Patagonia. In: *Temperate Deserts and Semi-Deserts* [West, N.E. (ed.)]. Ecosystems of the World, Vol. 5, Elsevier, Amsterdam, The Netherlands, pp. 423-460.

Szabolcs, I., 1989: *Salt Affected Soils*. CRC Press, Inc., Boca Raton, FL,274 pp.

Szabolcs, I., 1990: Impact of climate change on soil attributes. Influence on salinization and alkalization. In: *Soils on a Warmer Earth* [Scharpenseel, H.W., M. Schomaker, and A. Ayoub (eds.)]. Elsevier, Amsterdam, The Netherlands, pp. 61-69.

Szabolcs, I., 1991: Salinization potential of European soils. In: *Landuse Changes in Europe. Processes of Change, Environmental Transformations and Future Patterns* [Brouwer, F.M., A.J. Thomas, and M.J. Chadwick (eds.)]. Geojournal Library, vol. 18, Kluwer Academic Publishers, Dordrecht, The Netherlands, pp. 293-315.

Szabolcs, I. and M. Rédly, 1989: State and prospects of soil salinity in Europe. *Agrokémia és Talajtan*, **38**, 537-558.

Tate, K.K., 1992: Soils and climate change—a New Zealand perspective. *Global Change: Impacts on Agriculture and Forestry*, Royal Society of New Zealand, **30**, 57-62.

Thomas, D.S.G., 1993: Sandstorm in a teacup? Understanding desertification. *The Geographical Journal*, **159**, 318-331.

Thomas, D.S.G. and N.J. Middleton, 1994: *Desertification: Exploding the Myth*. John Wiley, Chichester, UK, 194 pp.

Titus, J.G., 1987: Causes and effects of a sea level rise. In: *Preparing for Climate Change*. Proc. 1st N-American Conf. on Preparing for Climate Change, 27-29 October 1987, Government Institutes Inc., Rockville, MD, pp. 125-139.

Tucker, C.J., H.E. Dregne, and W.W. Newcomb, 1991. Expansion and contraction of the Sahara desert. *Science*, **253**, 299-301.

UNEP, 1991: *Status of Desertification and Implementation of the United Nations Plan of Action to Combat Desertification*. UNEP, Nairobi, Kenya, 77 pp.

UNEP, 1992: *World Atlas of Desertification*. 69 plates. UNEP, Nairobi & Edward Arnold, London, UK, 69 pp.

United Nations, 1992: Managing fragile ecosystems: combating desertification and drought. Chapter 12 of Agenda 21. United Nations, New York, NY, 22 pp.

Várallyay, G., 1968: Salt accumulation processes in the Hungarian Danube Valley. Trans. 9th ISSS Congress, Adelaide, Australia, vol. I. Angus and Robertson, Ltd., Sydney, Australia, pp. 371-380.

Várallyay, G., 1990a: Influence of climatic change on soil moisture regime, texture, structure and erosion. In: *Soils on a Warmer Earth* [Scharpenseel, H.W., M. Schomaker, and A. Ayoub (eds.)]. Elsevier, Amsterdam, The Netherlands, pp. 39-49.

Várallyay, G., 1990b: Consequence of climate induced changes in soil degradation processes. Trans. 14th ISSS Congress. International Society of Soil Sciences, Kyoto, Japan, pp. 265-270.

Várallyay, G., 1990c: Potential impacts of global climatic changes on soil moisture regime and soil degradation processes. Proc. UNESCO/MAB Int. Seminar "Future Research Trends in MAB," University of Fisheries, Tokyo, Japan, pp. 256-267.

Várallyay, G., 1994: Climate change, soil salinity and alkalinity. In: *Soil Responses to Climate Change* [Rounsevell, M.D.A. and P.J. Loveland (eds.)]. NATO ASI Series I, Global Environmental Change, vol. 23, Springer-Verlag, Heidelberg, Germany, pp. 39-54.

Verstraete, M.M., 1986: Defining desertification: a review. *Climatic Change*, **9**, 5-18.

Vinogradov, B.V. and K.N. Kulin, 1987: Aero-space monitoring of desertification dynamics of Black Lands in Kalmykia, according to repeated surveys. *Problemy Osvoenia Pustyn*, **4**, 45-53.

Walling, D.E., 1987: Rainfall, run-off and erosion of the land: a global view. In: *Energetics of Physical Environment* [Gregory, K.J. (ed.)]. John Wiley, Chichester, UK, pp. 89-117.

West, N.E., J.M. Stark, D.W. Johnson, M.M. Abrams, J.R. Wright, D. Heggem, and S. Peck, 1994: Effects of climate change on the edaphic features of arid and semi-arid lands of western North America. *Arid Soil Research and Rehabilitation*, **8**, 307-351.

Westing, A.H., 1994: Population, desertification and migration. *Environmental Conservation*, **21**, 109-114.

Wild, A., 1993: *Soils and the Environment*. Cambridge University Press, Cambridge, UK, 287 pp.

Williams, J.R., C.A. Jones, and P.T. Dyke, 1984: A modelling approach to determine the relationship between erosion and soil productivity. *Trans ASAE*, **27**, 129-144.

Williams, J.R., P.T. Dyke, W.W. Fuchs, V.W. Benson, O.W. Rice, and E.D. Taylor (eds.), 1990: *EPIC-Erosion/Productivity, Impact Calculator: 2 User Manual.* Technical Bull. 1768, U.S. Department of Agriculture.

Wischmeier, W.H. and D.D. Smith, 1978: *Predicting Rainfall Erosion Losses. A Guide to Conservation Planning*. Agricultural Handbook 537, U.S. Department of Agriculture, New York, NY, 58 pp.

Woodruff, N.P. and F.H. Siddoway, 1965: A wind erosion equation. *Soil Science Society America Proceedings*, **29**, 602-608.

WRI, 1992: *World Resources 1992-93*. Oxford University Press, Oxford, UK, 385 pp.

5

Impacts of Climate Change on Mountain Regions

MARTIN BENISTON, SWITZERLAND; DOUGLAS G. FOX, USA

Principal Lead Authors:
S. Adhikary, Nepal; R. Andressen, Venezuela; A. Guisan, Switzerland; J.I. Holten, Norway; J. Innes, Switzerland; J. Maitima, Kenya; M.F. Price, UK; L. Tessier, France

Contributing Authors:
R. Barry, USA; C. Bonnard, Switzerland; F. David, France; L. Graumlich, USA; P. Halpin, USA; H. Henttonen, Finland; F.-K. Holtmeier, Germany; A. Jaervinen, Finland; S. Jonasson, Denmark; T. Kittel, USA; F. Kloetzli, Switzerland; C. Körner, Switzerland; N. Kräuchi, Switzerland; U. Molau, Sweden; R. Musselman, USA; P. Ottesen, Norway; D. Peterson, USA; N. Saelthun, Norway; Xuemei Shao, China; O. Skre, Norway; O. Solomina, Russian Federation; R. Spichiger, Switzerland; E. Sulzman, USA; M. Thinon, France; R. Williams, Australia

CONTENTS

EXECUTIVE SUMMARY

This chapter focuses on the impacts of climate change on physical, ecological, and socioeconomic systems in mountain regions; this topic was not addressed in the form of an explicit chapter in the 1990 Assessment Report.

Scenarios of climate change in mountain regions are highly uncertain; they are poorly resolved even in the highest-resolution general circulation models (GCMs). Furthermore, mountains perturb atmospheric dynamic and thermal characteristics, thereby establishing locally distinct, topographical microclimates that influence a wide range of environmental factors. With these facts in mind, we can summarize the effects of climate change on mountains as follows.

Mountain Physical Systems

If climate changes as projected in most climate scenarios, we have a high degree of confidence that the length of time that snow packs remain will be reduced, altering the timing and amplitude of runoff from snow, and increasing evaporation. These changes would affect water storage and delivery infrastructure around the world (see also Chapters 10 and 14). Alterations in precipitation regimes will modify these responses. Changes in extreme events (floods and droughts) could affect the frequency of natural hazards such as avalanches and mudslides. Downstream consequences of altered mountain hydrology are likely to be highly significant to economies dependent on this water.

Mountain Ecological Systems

We have a medium degree of confidence as to current understanding related to responses of mountain vegetation to changes in temperature and precipitation, based on paleoenvironmental studies, observations, and experiments. However, predictions of vegetation shifts are complicated by uncertainties in species-specific responses to increased CO_2 as well as uncertainties in projected regional temperature, precipitation, and soil moisture changes. All of these factors influence competitive interactions between species, particularly near ecotones such as timberline where species migrating upslope in response to warming are likely to encounter more slowly responding current inhabitants.

Human land use has contributed to diminished biological diversity in many mountain regions around the world. We have a medium degree of confidence as to the probability that climate change will exacerbate fragmentation and reduce key habitats. There is cause for concern that mountaintop-endemic species may disappear.

Mountain Socioeconomic Systems

In many developing countries, mountains provide food and fuel needed for human survival. We have a high degree of confidence that disruption of mountain resources needed for subsistence would have major consequences, but we have a medium degree of confidence as to the response of mountain populations to such changes.

In areas where people do not depend on mountain environments for subsistence, mountain lands are primarily used for cash crops, mining, timber harvest, and recreational activities. While not critical to human survival, these mountain areas are of increasing local and regional economic significance.

Mountains can be used effectively as conservation reserves in a changing climate because they support a relatively broad distribution of possible climates and a high diversity of habitats within a small physical area. Such reserves are most effective if they include a significant elevation range.

We are confident that competition between alternative mountain land uses is likely to increase under climate-change and population-rise scenarios.

5.1. Mountain Characteristics

5.1.1. *Introduction*

Mountain systems account for roughly 20% of the terrestrial surface area of the globe and are found on all continents. They are usually characterized by sensitive ecosystems and enhanced occurrences of extreme weather events and natural catastrophes; they are also regions of conflicting interests between economic development and environmental conservation. Once regarded as hostile and economically nonviable regions, mountains have attracted major economic investments for industry, agriculture, tourism, hydropower, and communication routes. Most mountain regions except in Antarctica are inhabited; Ives (1992) has pointed out that mountains provide the direct life-support base for about a tenth of humankind and indirectly affect the lives of more than half.

Despite their relatively small surface area, mountains are an integral part of the climate system. As a physical barrier to atmospheric flows, they perturb synoptic patterns and are considered to be one of the trigger mechanisms of cyclogenesis in mid-latitudes. Because of significant altitudinal differences, mountains such as the Himalayas, the Rockies, the Andes, and the Alps exhibit within short horizontal distances climatic regimes similar to those of widely separated latitudinal belts; they also feature increased diversity of species, communities, and habitats over lowland environments. Mountains are also a key element of the hydrological cycle, being the source of many of the world's major river systems. Abrupt changes in existing temperature and precipitation patterns that have led to the present distribution of vegetation, ice, snow and permafrost zones would impact heavily on the unique features of mountain environments. This, in turn, would lead to significant perturbations to the existing socioeconomic structures for populations living within the mountains themselves and indirectly to populations living outside these zones but dependent on them.

Mountains around the world are also significant to local and global economies because of the mineral and timber resources they contain, as well as their hydropower potential. Regional climate change could directly affect the viability and health of commercial timber production and, in an indirect manner, add cost to mining and other mineral extraction and processing activities in mountainous areas.

5.1.2. *Climate Characteristics*

A precise understanding of the climatic characteristics of mountain regions is limited by a lack of observations adequately distributed in time and space and insufficient theoretical attention given to the complex interaction of spatial scales in weather and climate phenomena in mountains (Beniston *et al.*, 1994). Meteorological research has tended to focus on the upstream and downstream influences of barriers to flow and on orographic effects on weather systems (Smith, 1979) rather than on microclimates within the mountain environments themselves. Climatic features relevant to mountain environments include microscale features of the atmosphere that are superimposed on larger scales of motion and the influence of elements of the surface, such as vegetation and geomorphologic features, which can create microclimatic contrasts in surface heating, soil moisture, or snow-cover duration (Geiger, 1965). Isolating macro- and microscale processes in order to determine their relative importance is complicated by inadequate databases for most mountain areas of the world (Barry, 1994).

Four principal factors influence mountain climates—namely, altitude, continentality, latitude, and topography. The role of these factors is summarized schematically in Table 5-1 (from Barry, 1992); the effects refer to responses to an increase in the factor listed. These climatic differences, in turn, influence vegetation type and cover, hydrology, and sometimes geomorphic features. In particular, vegetation distribution in mountain regions is so closely linked to climatic parameters that vegetation belt typology has been widely used to delineate climate influences (Troll, 1968; Lauer, 1979; Monasterio, 1980; Quezel and Schevok, 1981; Klötzli, 1984, 1991, 1994;

Table 5-1: *Climatic effects of the basic controls of mountain climate.*

Factors	Primary Effects	Secondary Effects
Altitude	Reduced air density, vapor pressure; increased solar radiation receipts; lower temperatures	Increased wind velocity and precipitation (mid-latitudes); reduced evaporation; physiological stress
Continentality	Annual/diurnal temperature range increased; cloud and precipitation regimes modified	Snow line altitude increases
Latitude	Daylength and solar radiation totals vary seasonally	Snowfall proportion increases; annual temperatures decrease
Topography	Spatial contrasts in solar radiation and temperature regimes; precipitation as a result of slope and aspect	Diurnal wind regimes; snow cover related to topography

Ozenda, 1985; Quezel and Barbero, 1990; Rameau *et al.*, 1993). Although it would appear simple to discuss how these typologies might be modified as a result of changing global climate, the climate system and its interaction with such landscape components as vegetation, geology, topography, and soil are highly nonlinear (Beniston, 1993, 1994).

5.1.3. *Physical Characteristics*

The various components of the hydrologic cycle are critical links between climate and the life of mountain communities and also influence areas downstream. The timing and volume of rainfall and snowmelt are critical constraints on hydrological systems. Sediment loadings can be exceptionally high in many rivers originating in mountains and reflect geomorphologic processes. Unless slowed down by major lakes within mountains themselves or in adjacent lowlands, or by engineering works designed to control torrent flows, rivers will carry their heavy loads unimpeded into the plains. This can be beneficial for agriculture because these sediments create rich and fertile soils, but excessive loads can have adverse consequences, such as flooding of populated lowland areas.

Influencing the hydrological system are elements of the mountain cryosphere, particularly glaciers and seasonal snow packs. These are of major significance in some of the more elevated mountain chains of the world in terms of water availability in the source region of many of the world's rivers (Steinhauser, 1970). Snow cover in mountains is affected by the following factors, which are all linked to climate:

- Snow cover duration, which has been shown to vary linearly with altitude and is a function of slope orientation (Slatyer *et al.*, 1984)
- Snow depth, which also varies with altitude, orientation, and topography (Witmer *et al.*, 1986; Föhn, 1991)
- Snowmelt runoff, which feeds into the hydrologic system of mountains; this is determined by temperatures and surface energy balance in spring (Collins, 1989; Chen and Ohmura, 1990).

Changes in glacier and permafrost ice are linked to changes in the energy balance at the Earth's surface. Rates of such glacier and permafrost changes can be determined quantitatively over various time intervals, allowing direct comparison with estimated effects of anthropogenic greenhouse forcing (Haeberli, 1994). As a consequence, they are among the clearest signals evident in nature of ongoing warming trends related to the enhanced greenhouse effect (Haeberli, 1990; Wood, 1990; WGMS, 1993). Evidence from borehole profiles in permafrost also helps to determine the rate and magnitude of temperature changes (Mackay, 1990, 1992; Vonder Muehll and Holub, 1995).

A major problem in many mountain regions related to climate change is increased erosion and reduced slope stability. The combination of complex orography with steep slopes, intense rainstorms, and, in some regions, frequent earthquakes, causes a high proportion of mass movement, which eventually finds its way into rivers as heavy sediment load. Any changes to an already fragile environment would impact heavily on geomorphologic processes, resulting in indirect impacts to natural and socioeconomic systems (Innes, 1985). For example, higher precipitation, particularly during extreme events, can augment the risk of erosion. However, the potential for increased erosion depends on a number of other factors linked to topography, geology, soil types, and farming and conservation practices associated with a particular region. Enhanced precipitation can sometimes favor the development of denser vegetation, which in turn prevents soils from being eroded. Precipitation is not only a water source in mountain terrain but also the trigger factor for the occurrence of debris flow, landslide, and slope failure. No general study of the relationship between rainfall and landslides exists at a global scale, but in many specific cases a clear relation has been established between rain duration and induced movements (Noverraz and Bonnard, 1992). Relations between rainfall and induced movements can be direct and immediate or more long-term; in many cases, a long period of rainfall prior to the event has to be taken into account to explain a particular landslide (Bonnard, 1994).

5.1.4. *Ecological Characteristics*

Altitudinal vegetation distribution has traditionally been used to characterize mountain environments. This distribution is

Box 5-1. Climate Variabilty as Reflected in Glacier Regimes in the Former Soviet Union (FSU)

Solomina (1995) presents case studies of glacier fluctuations in the FSU. Major conclusions include:

- The length of many mountain glaciers in the FSU has decreased by up to 4 km during the last two centuries. The maximum degradation has occurred in regions with strong glaciation (the Caucasus, Pamirs, Tien Shan, and Altay).
- Changes in the equilibrium line altitude (ELA: the altitude of a glacier at which snow accumulation is equal to snow ablation) have also been more pronounced in regions with large glaciers, namely the Caucasus, Altay, and the periphery of the Central Asian mountains, and less in other mountain zones of the FSU. The mean ELA is estimated as being 50–80 m lower today than during the Little Ice Age.

strongly influenced by climatic parameters, but not exclusively. Topography, edaphic factors, and, in many areas, human activities and the disturbances they generate strongly interfere with the potential distributions suggested by large-scale climatic parameters alone.

The zonation patterns of vegetation differ according to the location of the mountains along a latitudinal gradient, as shown in Figure 5-1. Tropical mountains typically experience relatively uniform mean temperatures throughout the year, with significant diurnal temperature variation. Mountains in temperate and boreal zones, on the other hand, are characterized by seasonal climates with a well-defined growing season, whereby plants can exploit the short, warm summers. Many plant species at high altitudes are capable of surviving harsh conditions as a result of the onset of a hardening process.

The amount and duration of snow cover influences high mountain vegetation by determining growing season and moisture conditions. In the alpine zones, exposed, windy ridges are generally covered with xeric dwarf shrubs and wind-hardy lichens. Depending on nutrition and soil moisture, more mesic dwarf shrub heaths and mesic low herb meadows are found on lee slopes of intermediate snow cover. The late-exposed snow patches have hygrophilous herb communities. Vascular plants may be completely lacking due to the very short growing season.

Both altitude and latitude contribute to high diversity in species composition in mountain forests: Coniferous trees are common in boreal and temperate mountains, sclerophylous types in Mediterranean zones, deciduous species in oceanic boundaries of continents (e.g., Norway, British Columbia), and ericaceous (East Africa; Hedberg, 1951) or podocarp (Equatorial Andes) species in tropical zones. Furthermore, a unique feature of tropical mountains are cloud forests, which exhibit a high degree of biodiversity.

5.1.5. Human Characteristics

Mountains having a history of relatively dense settlement, such as those in Europe, Japan, and the eastern United States, are facing a decline of traditional agriculture and forestry, which have been economically viable only because of government subsidies. Mountains with a history of less dense settlement, such as the western United States, Canada, South America, Africa, Australia, and New Zealand, have over the past three centuries faced pressure from colonization and immigration, which have introduced scattered but growing settlements. In relatively recent decades, tourism has emerged as a significant income source in mountain regions, as well as a major source of environmental stress (Grötzbach, 1988). While in some regions (e.g., western Europe and Japan) mountains are experiencing depopulation (Yoshino *et al.*, 1980; Price, 1994), in general, land-use pressures are increasing because of competition between refuge use, mineral extraction and processing, recreation development, and market-oriented agriculture, forestry, and livestock grazing (Ives and Messerli, 1989; Messerli, 1989).

Mountain habitats in developing nations, where traditional subsistence agriculture continues, face stress from increasing human population, ameliorated to some extent by short-term or seasonal migration. In many of these regions, pressures exist to develop market-based agricultural systems, changing from a largely local economy to a national and international one. In Africa, most of the mid-elevation ranges, plateaus, and slopes of high mountains are under considerable pressure from commercial and subsistence farming activities (Rongers, 1993). In

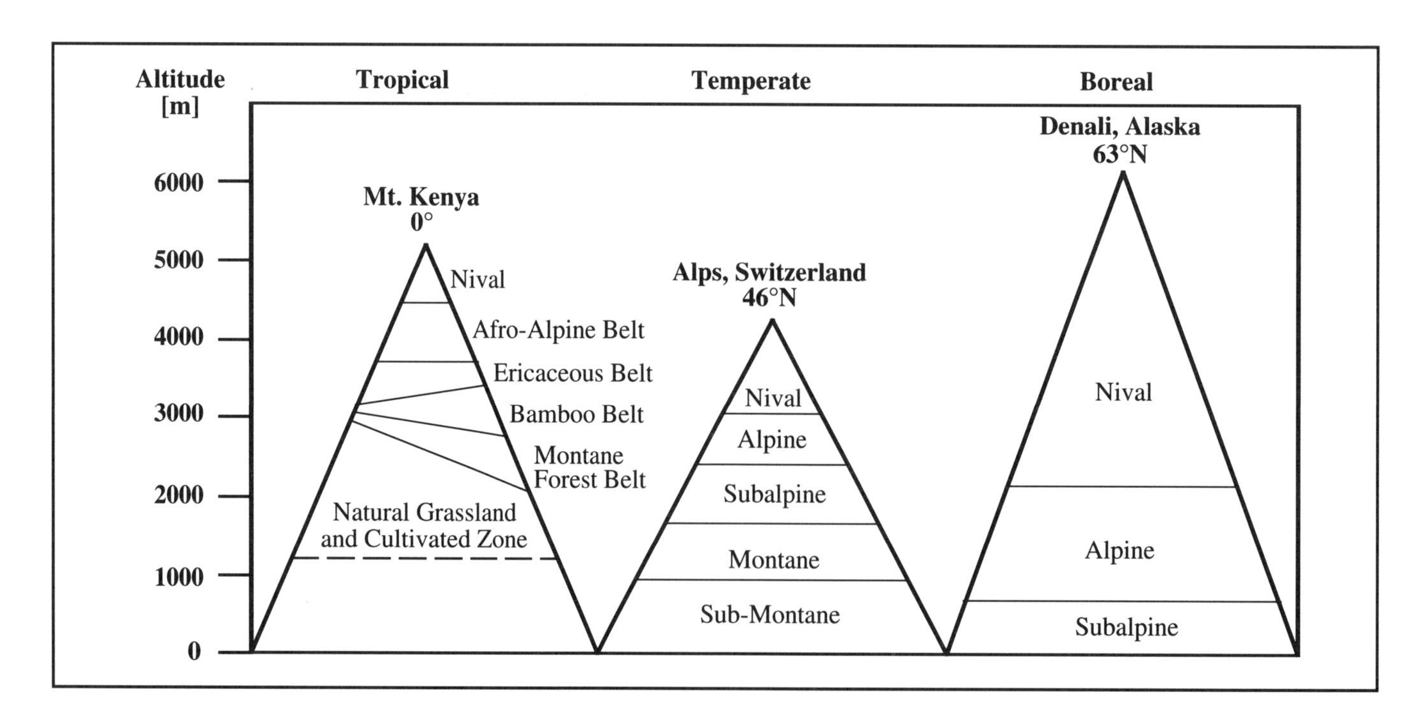

Figure 5-1: Schematic illustration of vegetation zonations in tropical, temperate, and boreal mountains.

Box 5-2. Socioeconomic Characteristics of North American Mountain Regions

The major mountains of North America are the Rockies, the eastern Appalachian chain, and various coastal ranges along the west coast of the continent. Primarily in the United States and in Mexico, the mountains border on semi-arid, highly populated urban areas (Mexico City, Los Angeles, Denver); most North American mountains are sparsely populated, however. Traditional economies have been in the mining and pastoral fields, with limited agriculture where the climate allows it. Forest harvest and replanting has been practiced in many areas. Much of the remaining area, especially in the Western United States and Canada, is set aside for wilderness, where human influence is minimized, and for parks, which are managed (Peine and Martinka, 1992). Because of this rather significant wilderness designation (over 16 million hectares, about 6% of all forested lands in the United States alone), adaptation to the potential impact of global climate change on natural ecosystems and biological diversity is more an administrative management issue than a scientific one (OTA, 1993). One problem is that wilderness areas tend to occupy only the tops of mountains, not the entire mountain, limiting the availability of diverse habitats and hence migration opportunities.

unprotected areas, mountain forests are cleared for cultivation of high-altitude adapted cash crops like tea, pyrethrum, and coffee. Population growth rates in East Africa, among the highest in the world (Goliber, 1985), are concentrated in agriculturally fertile and productive mountain districts (Government of Tanzania, 1979). This is quite evident near Mt. Kenya, Mt. Kilimanjaro, and in the Usambara Mountains (Lundgren, 1980). Tourism is becoming economically important in small, well-defined portions of these regions (Price, 1993).

Particularly in tropical and semi-arid climates, mountain areas are usually wet, cool, and hospitable for human dwelling and commercial cultivation. Human encroachment on mountain regions has reduced vegetation cover, thus increasing soil-moisture evaporation, erosion, and siltation, and thereby adversely affecting water quality and other resources. Impact analyses need to seriously consider not only climate but also the direct anthropogenic influence on mountain regions.

5.2. Impacts of Climate Change

Mountain environments are potentially vulnerable to the impacts of global warming because the combination of enhanced sensitivity to climatic change and limited possibilities for species migration to favorable locations make mountains "islands" in a "sea" of surrounding ecosystems (Busby, 1988). This vulnerability has important ramifications for a wide variety of human uses and natural systems, such as nature conservation, land management, water use, agriculture, and tourism. It also has implications for a wide range of natural systems (hydrology, glaciers and permafrost, ecosystems, and so forth).

Few assessments of the impacts of climate change have been conducted in mountain regions, in contrast to other biomes such as tropical rainforests, coastal zones, and high-latitude or arid areas. There are a number of reasons to explain this, some of which are listed below:

- The dominant feature of mountains (i.e., topography) is so poorly resolved in most general circulation climate models (GCMs) that it is difficult to use GCM-based scenarios to investigate the potential impacts of climate change. There also is a significant lack of comprehensive multidisciplinary data for impact studies, which is one of the prerequisites for case studies of impacts on natural or socioeconomic systems (Kates *et al.*, 1985; Riebsame, 1989; Parry *et al.*, 1992).
- The complexity of mountain systems presents major problems for assessing the potential impacts of climate change. This applies to assessments of changes in both biophysical systems (e.g., Rizzo and Wiken, 1992; Halpin, 1994) and societal systems, the latter in particular because it is difficult to quantify the value of mountain regions in monetary terms (Price, 1990).
- Tourism, which is an increasingly important component of mountain economies, is not an easily defined economic sector and is highly influenced by factors other than climate change.

5.2.1. Scenarios and Methodologies

Mitigation and adaptation strategies to counteract possible consequences of abrupt climate change in mountain regions require climatological information at high spatial and temporal resolution. Unfortunately, present-day simulation techniques for predicting climate change on a regional scale are by no means satisfactory. This is because GCMs generally operate with a rather low spatial resolution over the globe (~300 km) in order to simulate climate trends over a statistically significant period of time. As a result, entire regions of sub-continental scale have been overlooked in terms of their climatological specificities, making it difficult to predict the consequences of climate change on mountain hydrology, glaciers, or ecosystems in a specific mountain region (Giorgi and Mearns, 1991; Beniston, 1994). The situation is currently improving with the advent of high-resolution climate simulations in which the spatial scale of GCMs is on the order of 100 km (Beniston *et al.*, 1995; Marinucci *et al.*, 1995). However, even this resolution is generally insufficient for most impact studies.

A number of solutions exist to help improve the quality of climate data used in impact assessments and economic decisionmaking.

Options include statistical techniques of downscaling from synoptic to local atmospheric scales (Gyalistras *et al.*, 1994); coupling of mesoscale, or limited area, models (LAMs) to GCMs (Giorgi *et al.*, 1990, 1992; Marinucci and Giorgi, 1992; Marinucci *et al.*, 1995); and use of paleoclimatic and geographical analogs (e.g., LaMarche, 1973; Webb *et al.*, 1985; Davis, 1986; Schweingruber, 1988; Graumlich, 1993; Luckman, 1994).

Any meaningful climate projection for mountain regions—and, indeed, for any area of less than continental scale—needs to consider processes acting on a range of scales, from the very local to the global. The necessity of coupling scales makes projections of mountain climates difficult.

5.2.2. *Impacts on Physical Systems*

5.2.2.1. *Impacts on Hydrology*

In spite of limitations in the quality of historical data sets and inconsistencies in projections between GCMs, particularly for precipitation (Houghton *et al.*, 1990), assessments of the potential impacts of climate change on water resources, including snowfall and storage, have been conducted at a variety of spatial scales for most mountain regions (Oerlemans, 1989; Rupke and Boer, 1989; Lins *et al.*, 1990; Slaymaker, 1990; Street and Melnikov, 1990; Nash and Gleick, 1991; Aguado *et al.*, 1992; Bultot *et al.*, 1992; Martin, 1992; Leavesley, 1994). For example, high-resolution calculations for the Alps (Beniston *et al.*, 1995) using a nested modeling approach (GCM at 1° latitude/longitude resolution coupled to a LAM at 20-km resolution) indicate that in a doubled-CO_2 atmosphere wintertime precipitation will increase by about 15% in the Western Alps; this is accompanied by temperature increases of up to 4°C. Summertime precipitation generally decreases over the alpine domain, with July temperatures on average 6°C warmer than under current climatic conditions. Such numerical experiments, while fraught with uncertainty, nevertheless provide estimates of possible future regional climatic conditions in mountains and thereby allow more detailed impact assessments.

Climate change may be characterized by changes in seasonal or annual precipitation, proportions of solid to liquid precipitation, or frequencies of extreme events. Whatever the directions and magnitudes of change, mountain communities and those downstream need to be prepared to implement flexible water management strategies that do not assume that recent patterns will continue. Events in recent history may provide useful guidelines for developing such strategies (Glantz, 1988).

Climate-driven hydrology in mountain regions is determined to a large extent by orography itself; mountain belts produce regional-scale concentration of precipitation on upwind slopes and rain-shadow effects in the lee of mountains and in deep intermontane valley systems, often giving rise to high mountain deserts. Many of the more elevated mountain chains of the world intercept large atmospheric moisture fluxes and produce belts of intense precipitation. Along the southern slopes of the Himalayas, enhancement of monsoonal conditions results in some of the highest annual average precipitation in the world, such as at Cherrapunji, India. The spectrum of variability of hydrological regimes ranges from the predominantly rainforested slopes of Papua-New Guinea to the ice fields of the Patagonian Andes. Climate change will affect the relative importance of these two extreme regimes, as well as the total moisture flux and how it is delivered temporally. Mountains such as the Andes and the Himalayas are source regions for some of the world's largest rivers, such as the Amazon, the Ganges, Irrawady, and Yangtse, and the discharge characteristics of these rivers and their shifts under changed climatic conditions will be largely governed by modified precipitation regimes.

5.2.2.2. *Impacts on Mountain Cryosphere*

The effects of temperature and precipitation changes on glaciers are complex and vary by location. In polar latitudes and at

Box 5-3. Estimates of Cryosphere Response to Climate Change in Asian, Latin American, and African Mountain Chains

Wang (1993) estimates that about 21% of the glacial area will disappear in Northwest China if temperature increases by 1–1.3°C and precipitation decreases by 60–80 mm. Glaciers with areas of less than 0.5 km^2 in the Tianshan and Qilian ranges and permafrost areas in Northeast China and Qing-Zang plateau are expected to disappear if temperature increases by 2°C or more (Zhang, 1992; Wang, 1993).

In the Venezuelan Andes, photographs from 1885 show that the present-day snowline has risen from 4,100 m to more than 4,700 m (Schubert, 1992). These changes in ice extent and in snowline altitude have had important geoecological effects, leading to shifts in vegetation belts and to the fragmentation of previously continuous forest formations. Enhancement of the warming signal in these regions would lead to the disappearance of significant snow and ice surfaces.

On Mount Kenya, the Lewis and Gregory glaciers have shown recession since the late 19th century. Calculations of area and volume loss of the Lewis Glacier for 1978–90 show similar rates to those of the preceding decades. Nevertheless, Hastenrath and Rostom (1990) propose that the retreat in the late 19th century was triggered by a decrease in cloudiness and precipitation, whereas the retreat since the 1920s may be due to an observed warming trend.

very high altitudes of mid-latitudes, atmospheric warming does not directly lead to mass loss through melting/runoff but to ice warming (Robin, 1983). In areas of temperate ice, which predominate at lower latitudes or altitudes, atmospheric warming can directly impact the mass and geometry of glaciers (Haeberli, 1994).

Haeberli (1994) indicates that alpine glacier and permafrost signals of warming trends constitute some of the clearest evidence available concerning past and ongoing changes in the climate system. The glacier fluctuations, in particular, indicate that the secular changes in surface energy balance may well be in accordance with the estimated anthropogenic greenhouse forcing. In fact, the evidence strongly indicates that the situation is now evolving at a high and accelerating rate beyond the range of Holocene (natural) variability. Temperature profiles from permafrost boreholes also confirm the nature of recent warming trends (Vonder Muehll and Holub, 1992; Mackay, 1990, 1992).

In terms of climate-change impacts on snow, Föhn (1991) suggests that one potential effect of global warming in the European Alps might be a delay in the first snowfall and a reduction in the length of snow cover. Analysis of satellite data from the 1980s and early 1990s has shown that lowlands around the Alps experience about 3–4 weeks less snow cover than they historically have had (Baumgartner and Apfl, 1994); this tendency will likely accelerate in a warmer climate. Additionally, snow accumulation and ablation exhibit different temporal patterns than in the past and could be even more irregular in a changed climate. For higher elevations, the total annual snow volume accumulated during the winter has not changed significantly this century, despite the observed global temperature rise.

Investigations concerning the impact of climate change on snow conditions in the French Alps were undertaken with a physically based snow model (CROCUS) coupled to a meteorological analysis system (SAFRAN). The two systems have been validated by comparing measured and simulated snow depth at 37 sites of the area for the period 1981/1991. Sensitivity studies show that lower elevations (i.e., below 1500 m) are extremely sensitive to small changes in temperature, especially in the southern part of the French Alps. Variations of precipitation amount influence the maximum snow depth (or snow water equivalent) much more than snow cover duration (Martin, 1995).

Reduced snow cover will have a number of implications; it will increase early seasonal runoff, leading to drier soil and vegetation in summer and greater fire risk. Further details on impacts on mountain cryosphere are provided in Chapter 7.

5.2.2.3. *Impacts on Extreme Events*

It is uncertain whether a warmer global climate will be accompanied by more numerous and severe episodes of extreme events because current GCM capability to simulate extremes and their altered frequency of occurrence in a changed climate is extremely limited. Enhanced occurrences of intense storms, accompanied by high precipitation and/or winds, would inevitably have significant repercussions on a number of sensitive environmental and socioeconomic systems. If the severity of storms such as "Vivian" (European Alps, 1990, which destroyed vast areas of forested slopes) or the intense precipitation of August 1987 in central Switzerland (which disrupted rail and road traffic in the Gotthard Region, one of the major communication routes across the Alps) were to increase in frequency in a changed climate, populations living in mountains would be faced with significant social and economic hardships.

Other impacts associated with extreme events include fire. Forest fires are likely to increase in places where summers become warmer and drier, as has been projected for the Alps, for example (Beniston *et al.*, 1995). Prolonged periods of summer drought would transform areas already sensitive to fire into regions of sustained fire hazard. The coastal ranges of California, the Blue Mountains of New South Wales (Australia), Mt. Kenya, and mountains on the fringes of the Mediterranean Sea, already subject to frequent fire episodes, would be severely affected. There would be major socioeconomic impacts as well, because many sensitive regions are located close to major population centers (e.g., Los Angeles and the Bay Area in California; the Sydney conurbation in Australia; coastal resorts close to the mountains in Spain, Italy, and southern France).

5.2.2.4. *Impacts on Geomorphologic Processes*

The latitude and altitude of different mountain systems determine the relative amount of snow and ice at high elevations and intense rainfall at lower elevations. Because of the amount of precipitation and relief, and the fact that many of these mountains are located in seismically active regions, the added effect of intense rainfall in low- to middle-altitude regions is to produce some of the highest global rates of slope erosion. Climate change could alter the magnitude and/or frequency of a wide range of geomorphologic processes (Eybergen and Imeson, 1989). The following examples provide an indication of the nature of the changes that might occur with specific changes in climate.

Large rockfalls in high mountainous areas often are caused by groundwater seeping through joints in the rocks. If average and extreme precipitation were to increase, groundwater pressure would rise, providing conditions favorable to increased triggering of rockfalls and landslides. Large landslides are propagated by increasing long-term rainfall, whereas small slides are triggered by high intensity rainfall (Govi, 1990). In a future climate in which both the mean and the extremes of precipitation may increase in certain areas, the number of small and large slides would correspondingly rise. This would contribute to additional transport of sediments in the river systems originating in mountain regions. Other trigger mechanisms for rockfalls are linked to pressure-release joints following deglaciation (Bjerrum and Jfrstad, 1968); such rockfalls may be observed decades after the deglaciation itself, emphasizing the long time-lags involved. Freeze–thaw processes also are very

Box 5-4. Long-Term Records of Climate and Vegetation Dynamics: A Key to Understanding the Future

Paleoenvironmental records represent the only available source of information on the long-term natural variability within the biosphere-atmosphere system. The spatial and temporal resolution of paleoenvironmental data sets has increased in recent decades on annual to millenium scales, and data reflecting vegetation dynamics are now available for many parts of the world (Bradley and Jones, 1992; Wright *et al.*, 1993; Graumlich, 1994). In addition, comparisons of records of past vegetation dynamics to paleoclimatic simulations by GCMs have improved the understanding of the role of climate in governing past vegetation change (Webb *et al.*, 1993). Several major findings of paleoresearch have guided investigations of the effects of future climate change on the Earth's biota. These include changing seasonality, which may result in unexpected vegetation patterns (Prentice *et al.*, 1991); rapid changes in both climate and vegetation with ecosystem-wide implications (Gear and Huntley, 1991); and short-term extreme events that may impact tree population structures (Lloyd and Graumlich, 1995). Thus, the same types of individualistic behavior of species observed in physiological attributes are mirrored in long-term and large-scale vegetation dynamics (Graumlich, 1994).

Future climates are likely to be characterized by combinations of temperature and precipitation that are not replicated on the contemporary landscape. Any predictions of vegetation response to climatic changes must be based on an improved understanding of the relationship between climatic variation and vegetation processes; this is where analyses of high-resolution proxy data can be of use.

important (Rapp, 1960), and several authors have reported possible links between rockfall activity and freeze–thaw mechanisms linked to climate change (Senarclens-Grancy, 1958; Heuberger, 1966).

A further mechanism that would be responsible for decreased slope stability in a warmer climate is the reduced cohesion of the soil through permafrost degradation (Haeberli *et al.*, 1990; see also Chapter 7). With the melting of the present permafrost zones at high mountain elevations, rock and mudslide events can be expected to increase in number and possibly in severity. This will certainly have a number of economic consequences for mountain communities, where the costs of repair to damaged communications infrastructure and buildings will rise in proportion to the number of landslide events. In many mountainous regions, tourist resorts such as those in the Alps and the Rocky Mountains or large urban areas close to mountains (suburbs of South American Andean cities, Hong Kong, or Los Angeles, for example) have spread into high-risk areas, and these will be increasingly endangered by slope instability.

5.2.3. Impacts on Ecological Systems

While the authors acknowledge that climatic change on fauna will be important, priority is given here to vegetation response, particularly because vegetation dynamics will determine future distributions of animal habitats (both directly for herbivores and indirectly for predators).

The potential impacts of future climatic changes on mountain ecosystems and nature reserves have become an increasingly important issue in the study of long-term biodiversity management and protection (Peters and Darling, 1985; McNeely, 1990; Halpin, 1994; OTA, 1993). Projected changes in global temperatures and local precipitation patterns could significantly alter the altitudinal ranges of important species within existing mountain belts and create additional environmental stresses on already fragile mountain ecosystems (Guisan *et al.*, 1995).

On a global scale, plant life at high elevations is primarily constrained by direct and indirect effects of low temperature and perhaps also by reduced partial pressure of CO_2. Other atmospheric influences, such as increased radiation, high wind speeds, or insufficient water supply may come into play, but only on a regional scale (Lüdi, 1938; Fliri, 1975; Lauscher, 1977; Körner and Larcher, 1988; Barry, 1992). Plants respond to these climatological influences through a number of morphological and physiological adjustments, such as stunted growth forms and small leaves, low thermal requirements for basic life functions, and reproductive strategies that avoid the risk associated with early life phases.

5.2.3.1. *Inference from the Past*

Paleoenvironmental information, including tree-ring data and pollen loadings, serves as an indicator of past climate-vegetation relationships. Paleoecology offers insights into the nature of climate-vegetation interactions. The indirect record of the response of organisms to environmental changes of the past is equivalent to results from natural, if unplanned, experiments (Davis, 1989). Because these natural experiments typically include conditions not observed in the 20th century, the paleorecord documents biotic responses to a substantially broader range of environmental variations than can be obtained through observations.

Paleoecological and paleoclimatic studies are of key importance in establishing baselines and are the only means available for determining amplitudes and rates of change of vegetation to natural climate variations. In addition, paleoenvironmental data are essential for model simulations and scenario generation.

5.2.3.2. Ecophysiological Responses

It is known from both common sense and paleoenvironmental research that plant communities respond to a general increase in temperature through a shift toward higher latitudes and altitudes. However, this shift is controlled by ecophysiological processes at the individual plant level, involving direct and indirect effects of temperature increase, photoperiod constraints, and competition processes.

Callaghan and Jonasson (1994) show that direct effects of increased temperature on aboveground processes tend to dominate over indirect effects on belowground processes and that these will be greatest at locations where current climates are extreme (Bugmann and Fischlin, 1994). Reduced duration of snow cover is likely to be important for vegetation in a warmer climate (Körner, 1994).

Heide and coworkers have clearly shown that photoperiod constraints in cold climates may be strong and of overriding importance (Heide, 1985; Solhaug, 1991). For most native species, the precise photoperiodic requirements and their interaction with temperature are not known, thus making it difficult to estimate plant migration based on temperature scenarios alone (Ozenda and Borel, 1991). According to Heide (1989, 1990), some alpine grasses have no photoperiod requirements for initiating growth and are therefore affected by spring frosts. For example, in northern Sweden, Molau (1993) has shown that increased length of the growing season and increased summer temperature are the two components of climate change that will have the greatest impact on reproduction and population dynamics in arctic and alpine plants.

If climatic change leads to warmer and locally drier conditions, mountain vegetation could suffer as a result of increased evapotranspiration. This is most likely for continental and Mediterranean mountains. A change in regional distribution of air humidity could strongly modify the tree species composition of mountain forests; drier conditions, for example, would give continental species an advantage. If changes in water supply were to occur during summer, they would also become effective indirectly by influencing topsoil processes and plant nutrition (Bowman *et al.*, 1993; Baron *et al.*, 1994). When temperatures are low, microbial activity in alpine soils is concentrated in the top few centimeters. This is the part of the soil where nutrient recycling and most of the root growth takes place; when this top layer desiccates, mineral cycling is blocked (Körner, 1989). Regular rainfall is thus a prerequisite for plant nutrition. Because most alpine plants are inherently slow growers, any enhancement of nutrient supply will stimulate the few potentially fast-growing species in a community, eliminate others, and thus cause substantial changes in vegetation structure.

The length and depth of snow cover, often correlated with mean temperature and precipitation, is one of the key climatic factors in alpine ecosystems (Barry and Van Wie, 1974; Aulitzky *et al.*, 1982; Ozenda, 1985; Burrows, 1990; Musselmann, 1994). Snow cover provides frost protection for plants in winter and water supply in spring. Alpine plant communities are characterized by a very short growing season (i.e., the snow-free period) and require water to commence growth. Ozenda and Borel (1991) predict that the most threatened vegetation communities will be those that live in snow beds and in hollows because these groups are subjected to summer drying.

5.2.3.3. Geographical Distribution and Migratory Responses

A general hypothesis presented by Peters and Darling (1985) and others concerns the potential movement of the climatic ranges of species along altitudinal, thermally defined gradients. This conceptual model implies that the boundaries of present species' climatic ranges will respond symetrically to changes in temperature related to the adiabatic lapse rate for a particular mountain site. The general biogeographical rule used to derive this conceptual model is attributed to the "Hopkins bioclimatic law" (MacArthur, 1972; Peters and Darling, 1985), which relates a 3°C change in temperature to a 500 m change in altitude. According to this conceptual model, the expected impacts of climate change in mountainous nature reserves would include the loss of the coolest climatic zones at the peaks of the mountains and the linear shift of all remaining vegetation belts upslope. Because mountaintops are smaller than bases, the present belts at high elevations would occupy smaller and smaller areas, and the corresponding species would have smaller populations and might thus become more vulnerable to genetic and environmental pressure (Peters and Darling, 1985; Hansen-Bristow *et al.*, 1988; Bortenschlager, 1993).

Examples of past extinctions attributed to upward shifts are found in Central and South America, where vegetation zones have moved upward by 1000–1500 m since the last glacial maximum (Flenley, 1979; van der Hammen, 1974). Romme and Turner (1991), in their study on possible implications of climate change for ecosystems in Yellowstone National Park (USA), project species extinctions as a result of fragmentation and shrinking mountaintop habitats.

In the Alps, the main climatic space contraction and fragmentation of plant populations would be in the present alpine and nival belts, where rare and endemic species with low dispersal capacities could become extinct. Because the alpine belt contains a non-negligible part of the endemic alpine flora (15%; Ozenda and Borel, 1991), the potential impact of climate change on floristic diversity in the Alps could be significant. Halpin (1994) states that the process is more complex than suggested above. Even if vegetation belts do not move up as a whole in response to global climate change, the ecological potential of sites will change in relation to shifts in climatic features.

The impact of climatic change on altitudinal distribution of vegetation cannot be analyzed without taking into account interference with latitudinal distribution. Especially at low altitudes, Mediterranean tree species can substitute for sub-montane belt species. In the southern French Alps, if a warming and

Box 5-5. Climate Change Impacts on Vegetation in the Scandinavian Mountains

A simple correlative and qualitative model (Holten and Carey, 1992; Holten, 1995) based on important controlling factors for plant distribution indicates that six continental eutrophic mountain plant species may come under direct threat of extinction in a "Step 2" of climate change impacts on the Fennoscandian range. "Step 2" represents the long-term (several centuries) consequences when climate exceeds the Holocene thermal optimum, resulting in both qualitative and quantitative changes in vegetation. Another 11 species regarded as potentially vulnerable may experience substantial reduction and fragmentation of their populations. The most threatened species are characterized by rarity, low competitive ability, narrow habitat amplitude, and a distributional optimum in the middle alpine zone. Increased drought and overheating will probably have the most negative short-term (50–150 years) effects on mountain plant populations. In the long term (after several centuries), if the boreal forest invades parts of the low-alpine zone, many mountain plant species will be vulnerable to extinction because of their lower competitive ability relative to boreal plants.

a decrease in precipitation were to occur, Ozenda and Borel (1991) predict a northward progression of Mediterranean ecotypes ("steppization" of ecosystems). Kienast *et al.* (1995) predict steppization on no more than 5% of today's forested areas in Switzerland. On the Italian slopes of the Alps, the northward progression of Mediterranean influences would probably be more important. A similar (xeric) change is less likely to take place in the southeastern part of the range (Julian and Carnic Alps), where a much more humid climate exists.

5.2.3.4. Modeling of Communities and Ecosystems Responses

There are a number of ecosystem models currently available that can be used to test the sensitivity of a particular system to processes such as nutrient cycling (e.g., CENTURY, TEM), investigate species composition under changed environmental conditions (e.g., BIOME, DOLY, MAPSS), or assess forest health (e.g., FORET).

A number of modeling studies that were based on forest gap models (Shugart, 1984) have been conducted to assess the impacts of climatic change on forest biomass and species composition in mountainous regions (e.g., Kienast, 1991; Kräuchi and Kienast, 1993; Bugmann, 1994; Bugmann and Fischlin, 1994; Kräuchi, 1994). Although several different models and climate scenarios were used in these studies, they yielded quite similar conclusions regarding the sensitivity of forests in the European Alps. As an example of such a model application, Figure 5-2 illustrates the modeled forest changes in the alpine zone under differing climatic scenarios. Potential regional development of forest vegetation (using the FORSUM model; Kräuchi, 1994) at the Derborence site in the Swiss Alps is illustrated for no change in climate (upper) and for the IPCC IS92A scenario (lower). The forest at this location is currently changing as a result of successional processes. While the overall forest biomass might remain the same, the species composition will be different under the two climates, with a reduction in the competitive dominance of *Picea abies* and an increase in the proportion of broadleaves under the IS92A scenario (Kräuchi, 1994).

Model studies for Australian mountain vegetation show that there is the potential for the expansion of woody vegetation, both trees and shrubs, in response to rising temperature (Grabherr *et al.*, 1994; Williams and Costin, 1994). The treeline may rise by 100 m for each degree increase in mean annual temperature (Galloway, 1989). Elevated summer temperatures may also lead to an expansion of shrub communities (Williams, 1990).

VEMAP (1985), a continental-scale vegetation response study of the United States, considered how three biogeographical models (BIOME2, DOLY, MAPSS; Woodward *et al.*, 1995) respond to doubled-CO_2 climate scenarios. Results on a coarse (0.5° latitude/longitude) grid showed alpine and subalpine regions in the western United States retreating to higher elevation and decreasing in area, while the subalpine montane forest boundary shifted upward. Two models also projected upward shifts in the lower montane boundary, while one produced a lowering of this boundary.

It should be emphasized that there are considerable limitations in present-day simulation techniques for assessing ecosystem response to climate change, in particular the temporal changes of these responses. In general, increases in atmospheric temperature will affect the structure and function of vegetation, as well as species composition where time may not be sufficient to allow species to migrate to suitable habitats (Kienast, 1991; Bugman, 1994; Klötzli, 1994).

5.2.4. Impacts on Human/Socioeconomic Systems

Because of the diversity of mountain economies, from exclusively tourist-based ones to those characterized by centuries-old subsistence agriculture, no single impact study will adequately represent the range of potential socioeconomic responses to climate change. A case-by-case approach is therefore essential to understand how, for example, mountain agriculture may change in Bolivia, tourism may change in Switzerland, or hydropower resources may be affected in New Zealand.

Because humans have influenced mountain ecosystems in many different ways throughout history, anthropogenic impacts generally cannot be dissociated from climate change impacts. Climatic influences are often obscured by the impact of change in land use. An example is the fragmentation of the forest and natural vegetation cover. Because of persistent anthropogenic influences in the past, timberline in mountains such as the Alps has dropped 150–400 m compared to its uppermost position during the postglacial optimum (Holtmeier, 1994). At present, the climatic limit of tree growth in the Alps is situated above the actual forest limit (Thinon, 1992; Tessier *et al.*, 1993). By reducing species diversity and even intraspecies genetic variability of some species, humans have reduced the ability of alpine vegetation to respond to climate change (David, 1993; Peterson, 1994).

5.2.4.1. *Mountain Agriculture*

Mountains contribute a substantial proportion of the world's agricultural production in terms of economic value. Upland regions are characterized by altitudinal climatic gradients that can lead to rapid changes in agricultural potential over comparatively short horizontal distances. Where elevations are high enough, a level eventually will be reached where agricultural production ceases to be profitable or where production losses become unacceptably high. Upland crop production, practiced close to the margins of viable production, can be highly sensitive to variations in climate. The nature of that

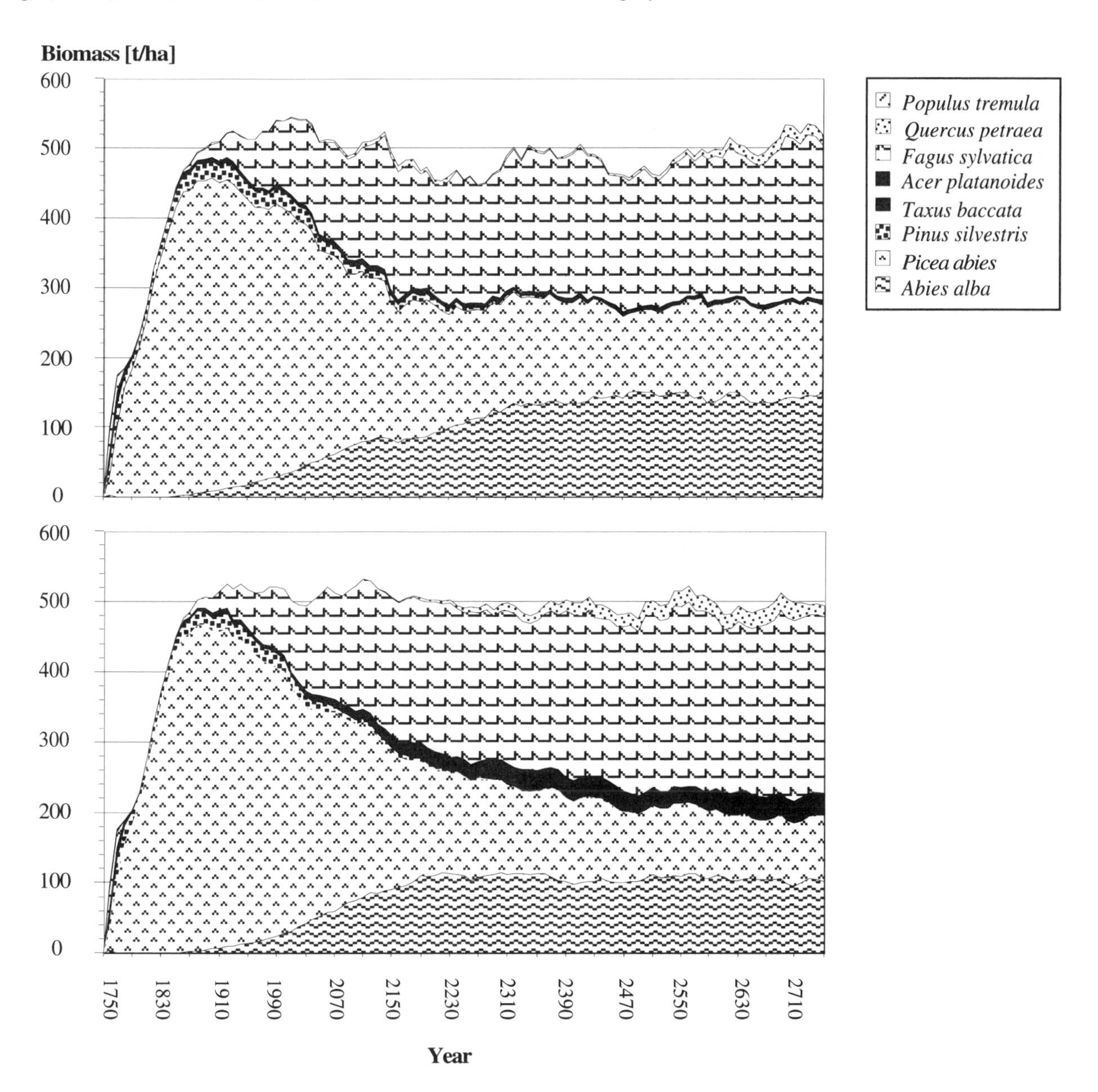

Figure 5-2: Example of simulations of forest response to different climate change scenarios in the alpine zone. Simulated forest succession on a subalpine site (mean annual temperature: 6°C; mean annual precipitation: 1,200 mm; subsoil: limestone and some flysch intersections) assuming today's climatic pattern (upper) and the IPCC IS92a climate scenario (lower; ΔT: +1°C in 2025; ΔT: +3°C in 2100) using FORSUM, a JABOWA/FORET-type gap model (Kräuchi, 1994). The main difference in the transient response in species composition between the baseline climate and the IS92a climate is the reduced importance of *Picea abies* combined with an increasing percentage of deciduous trees. The establishment of *Quercus petraea* and A*cerplatanoides* is particularly stimulated. The site might undergo a transition from *Piceo-Abietion (Adenostylo-Abietetum* or *Equiseto-Abietetum)* to *Abieti-Fagetum.*

Box 5-6. Impacts of Climate Change on Agriculture in Upland Regions of China and Japan

Climatic warming will in general increase agricultural productivity in China (Gao *et al.*, 1993). Under the present climate, crops such as maize and rice suffer from low temperature and frost in northeast and southwest mountainous regions (Ma and Liang, 1990; Li *et al.*, 1993). The yield for crops in these areas is expected to rise with increasing temperatures (Hulme *et al.*, 1992). According to a study by Li *et al.* (1993), a mean warming of 0.4–1.0°C in the mountain regions of southwest China will move the upper boundary for growing grain by 80–170 m, and the growing areas will expand. However, there is uncertainty related to expansion of growing areas because soils are very important limiting factors for grain production. Specific soil properties include physical and chemical conditions that control plant growth, such as permeability, water holding capacity, acidity, soil structure, and density. In addition, the specific varieties that are presently used in mountain agriculture may need to be altered in order to keep up with global warming and thus present an additional economic hardship to the system.

To examine the effect of climatic variability on rice suitability, simple empirical indices have been developed relating rice yield to July/August mean air temperature in upland regions of Tohoku and Hokkaido in northern Japan (Yoshino *et al.*, 1988). In combination with information about regional lapse rates of temperature, estimates have been made of the effects of anomalous weather on yield and on potentially cultivable areas for both cool and warm departures from a current mean climate. The results indicate that a July/August temperature departure of -1.1°C in Hokkaido would reduce yields by some 25% relative to the average, and the potentially cultivable area would contract to elevations below 100 m in a small area of southwestern Hokkaido, where the Sea of Japan acts to ameliorate the climate. A comparable anomaly in Tohoku (-1.0°C) would produce a 13% yield shortfall, and the cultivable area would contract by some 20%, to areas below about 200 m.

Conversely, in warmer-than-average years, higher yields would occur in association with an upward shift in limits of potential cultivation (Carter and Parry, 1994). A temperature rise of more than 3°C would also lead to a yield increase, but while current varieties of rice tend to respond positively to small temperature increments, they are less well adapted to large departures. It is likely that later-maturing varieties would be adopted under much warmer conditions to exploit the greater warmth. Under this scenario, most of the land below 500 m could become viable in Hokkaido, as could land up to 600 m in Tohoku.

sensitivity varies according to the region, crop, and agricultural system of interest. In some cases, the limits to crop cultivation appear to be closely related to levels of economic return. Yield variability often increases at higher elevation, so climate change may mean a greater risk of yield shortfall, rather than a change in mean yield (Carter and Parry, 1994).

Several authors have predicted that currently viable areas of crop production will change as a result of climate change (Alps: Balteanu *et al.*, 1987; Japan: Yoshino *et al.*, 1988; New Zealand: Salinger *et al.*, 1989; and Kenya: Downing, 1992), although other constraints such as soil types may make agriculture unsuitable at these higher elevations. In-depth studies of the effects of climatic change in Ecuador's Central Sierra (Parry, 1978; Bravo *et al.*, 1988) and Papua New Guinea (Allen *et al.*, 1989) have shown that crop growth and yield are controlled by complex interactions among different climatic factors and that specific methods of cultivation may permit crop survival in sites where the microclimates would otherwise be unsuitable. Such specific details cannot be included in GCM-based impact assessments, which have suggested both positive and negative impacts—such as decreasing frost risks in the Mexican highlands (Liverman and O'Brien, 1991) and less productive upland agriculture in Asian mountains, where impacts would depend on various factors, particularly types of cultivars and the availability of irrigation (Parry *et al.*, 1992).

Given the wide range of microclimates already existing in mountain areas that have been exploited through cultivation of diverse crops, direct negative effects of climate change on crop yields may not be too great. While crop yields may rise if moisture is not limiting, increases in the number of extreme events may offset any potential benefits. In addition, increases in both crop and animal yields may be negated by greater populations of pests and disease-causing organisms, many of which have distributions that are climatically controlled.

More important in certain parts of the developing world is the potential for complete disruption of the life pattern of mountain villages that climate change may represent in terms of food production and water management. People in the more remote regions of the Himalayas or Andes have for centuries managed to strike a delicate balance with fragile mountain environments; this balance would likely be disrupted by climate change, and it would take a long time for a new equilibrium to be established. In cases such as this, positive impacts of climate change (e.g., increased agricultural production and/or increased potential of water resources) are unlikely because the combined stressors, including negative effects of tourism, would overwhelm any adaptation capacity of the environment.

Compounded with these effects are those related to augmented duration and/or intensity of precipitation, which would enhance soil degradation (erosion, leaching, and so forth) and lead to loss of agricultural productivity.

5.2.4.2. Hydropower

An important socioeconomic consequence of global warming on the hydrological cycle is linked to potential changes in runoff extremes. Not only the mountain population but also the people in the plains downstream (a large proportion of the world population) presently depend on unregulated river systems and thus are particularly vulnerable to climate-driven hydrological change. Current difficulties in implementing water resource development projects will be compounded by uncertainties related to hydrologic responses to possible climate change. Among these, possible increases in sediment loading would perturb the functioning of power-generating infrastructure.

Sensitivity of mountain hydrology to climate change is a key factor that needs to be considered when planning hydropower infrastructure. In the future, a warmer and perhaps wetter greenhouse climate needs to be considered. The impact of climate on water resources in alpine areas has previously been examined by Gleick (1986, 1987a, 1987b) and Martinec and Rango (1989). Similar studies have related electricity demand to climate (Warren and LeDuc, 1981; Leffler *et al.*, 1985; Maunder, 1986; Downton *et al.*, 1988). However, few have attempted to integrate these impacts of climate change by considering both electricity supply and electricity consumption (Jäger, 1983).

Mountain runoff (electricity supply) and electricity consumption (demand) are both sensitive to changes in precipitation and temperature. Long-term changes in future climate will have a significant impact on the seasonal distribution of snow storage, runoff from hydroelectric catchments, and aggregated electricity consumption. On the basis of a study made in the Southern Alps of New Zealand, Garr and Fitzharris (1994) conclude that according to future climate scenarios (New Zealand Ministry for the Environment, 1990), the seasonal variation of electricity consumption will be less pronounced than at present—with the largest changes in winter, which corresponds to the time of peak heating requirements. There will also be less seasonal variation in runoff and more opportunity to generate power from existing hydroelectric stations. The electricity system will be less vulnerable to climate variability, in that water supply will increase but demand will be reduced. These conclusions suggest that climate change will have important implications for hydroelectricity systems in other mountain areas as well.

5.2.4.3. Commercial Activities: Timber and Mining

Commercial utilization of mountain forests can be affected directly and indirectly by climate change. Direct effects include loss of viability of commercial species, including problems in regeneration and lower seedling survival. Indirect effects relate to disturbances such as fire, insect, and disease losses. These indirect effects depend on the influence of climate on the disturbance agents themselves, as discussed more fully in Chapter 1.

Many of the commercially viable mineral deposits in the world are located in mountain regions. While climate has only a minor direct influence on exploitation of these resources, it may exert a significant indirect influence. Mining causes a surface disruption and requires roads and other infrastructure. Changes in climate that lead to increases in precipitation frequency and/or intensity exacerbate the potential for mass wasting and erosion associated with these developments. Furthermore, the economics of mineral expolitation often requires *in situ* processing of the extracted ore—for example, smeltering and hydrochemical processing. In the latter case, climate, especially precipitation and temperature, is a critical factor in process design.

5.2.4.4. Tourism

Resources required for tourism are climate-dependent—that is, their availability may be affected in the short and long term by variability, extremes, and shifts in climatic means. These resources include the landscapes of natural and anthropogenically influenced ecosystems and climatic conditions that are suitable for specific activities (Price, 1994).

Impacts of climate change on tourism in mountain areas may be divided into two types: direct and indirect. The former would result from changes in the atmospheric resources necessary for specific activities (e.g., clean air, snow). Indirect changes may result from these changes and from wider-scale socioeconomic changes, for example, fuel prices and patterns of demand for specific activities or destinations. Various indirect impacts may also derive from changes in mountain landscapes—the "capital" of tourism (Krippendorf, 1984)—which might lead potential tourists to perceive them as less attractive, and consequently to seek out new locations. There also may be new competition from other tourist locations as climates change, particularly on seasonal timeframes, especially in relation to vacation periods.

5.2.4.4.1. Winter tourism

Scenarios derived from GCMs have been used to examine the possible implications of climate change for skiing in Australia (Galloway, 1988; Hewitt, 1994; Whetton, 1994), Austria (Breiling and Charamza, 1994), eastern Canada (McBoyle and Wall, 1987; Lamothe and Périard, 1988), and Switzerland (Abegg and Froesch, 1994). These studies show that, because the length of the skiing season is sensitive to quite small climatic changes, there could be considerable socioeconomic disruption in communities that have invested heavily in the skiing industry. To some extent, such impacts might be offset by new opportunities in the summer season and also by investment in new technologies, such as snow-making equipment, as long as climatic conditions remain within appropriate bounds. Such investments, following seasons with little snow, have provided some "insurance" in mountain regions of North America. Investment in snow-making equipment has been somewhat less widespread in some European countries such as Switzerland (Broggi and Willi, 1989), despite the fact that seasons with little snow, especially at critical times such as during the Christmas and New Year period, can be economically devastating to mountain communities. Artificial snow-making often raises environmental concern because of the quantities of energy and water required for snow-making, the disturbances generated during the operation of the equipment, and the damage to vegetation observed following the melting of the artificial snow cover.

5.2.4.4.2. *Mountain-protected areas*

Types of tourism vary from "wilderness tourism" in relatively pristine ecosystems to highly developed resorts designed initially for downhill skiing and, increasingly, for year-round use. The question of the management of protected areas in North American mountains under global climate change has been considered by Peine and Martinka (1992), Peterson *et al.* (1990), and Peine and Fox (1995). These studies predict increasing conflicts between different objectives, particularly recreational use and the protection of ecosystems and species.

Strategies advanced for mitigating adverse effects of global climate change include the establishment of refugia (i.e., areas protected by law from human influences). Mountain-protected areas represent a topographically complex environment that is important for the economic and spiritual sustenance of many human cultures. As human populations continue to grow, use of this biologically diverse resource has accelerated, highlighting a strategic role for protected areas that includes both conservation and scientific interests. From a global conservation perspective, protecting representative samples of mountain resources is a high-priority goal. At the same time, scientists are more frequently using protected areas as controls for land-use research and as sites for long-term environmental monitoring. Despite their environmental and scientific value, however, such preserves represent small and fragmented landscapes, making them vulnerable to species migrations and extinctions.

5.3. Future Research and Monitoring Needs

Future research needed to understand and predict effects of climatic change on mountain regions should represent balance and coordination between field studies, including paleoenvironmental data collection; monitoring; experimental studies; and modeling (Guisan *et al.*, 1995). Research requirements include:

- Specific regional field studies (transects, data acquisition, mapping, observations at high elevation, and so on).
- Paleo data to establish baselines, to evaluate responses of ecosystems to natural climate variability, and to provide data for model verification
- Monitoring to establish long-term baseline data, in particular in potentially sensitive regions (e.g., remote areas, high elevations)
- Experimental studies to improve fundamental understanding, to test hypotheses, and to provide empirical information for modeling studies
- Modeling to ameliorate climate scenarios using various downscaling approaches, to improve understanding of how topographic and edaphic variability influence ecosystems and natural resources on the regional scale, and to improve mechanistic modeling of physical, biological, and socioeconomic systems
- Integrated assessment models to address the complex interrelationships of different systems in mountain regions and to provide valuable multidisciplinary information to a range of end-users, including policymakers.

References

Abegg, B. and R. Froesch, 1994: Climate change and winter tourism: impact on transport companies in the Swiss Canton of Graubünden. In: *Mountain Environments in Changing Climates* [Beniston, M. (ed.)]. Routledge Publishing Company, London and New York, pp. 328-340.

Aguado, E. *et al.*, 1992: Changes in the timing of runoff from West Coast streams and their relationships to climatic influences. *Swiss Climate Abstracts*, **special issue**, International Conference on Mountain Environments in Changing Climates, p. 15.

Allen, B., H. Brookfield, and Y. Byron (eds.), 1989: Frost and drought in the highlands of Papua New Guinea. *Mountain Research and Development*, **9**, 199-334.

Andressen, R. and R. Ponte, 1973: *Estudio Integral de las Cuencas de los Rios Chama y Capazón. Sub-proyecto II: Climatologia e Hidrologia.* Universidad de Los Andes, Mérida, Venezuela, 135 pp.

Aulitzky, H., H. Turner, and H. Mayer, 1982: Bioklimatische Grundlagen einer standortgemässen Bewirtschaftung des subalpinen Lärchen–Arvenwaldes. *Eidgenössische Anstalt für forstliches Versuchswesen Mitteilungen*, **58**, 325-580.

Azócar, A. and M. Monasterio, 1980: Caracterización ecológica del clima en el Páramo de Mucubaji. In: *Estudios Ecologicos en los Páramos Andinos* [Monasterio, M. (ed.)]. Ediciones de la Universidad de Los Andes, Mérida, Venezuela, pp. 207-223.

Bach, W., 1987: Appendix—primary data on climatological change based on models. In: *Impact of Climatic Change and CO_2 Enrichment on Exogenic Processes and Biosphere: General, European Workshop on Interrelated Bioclimatic and Land Use Changes, Noordwijkerhout,* vol. C. Kluwer Academic Publishers, Dordrecht, Netherlands, pp. 52-61.

Balteanu, D. *et al.*, 1987: Impact analysis of climatic change in the central European mountain ranges. In: *European Workshop on Interrelated Bioclimatic and Land Use Changes, Noordwijkerhout*, vol. G. Kluwer Academic Publishers, Dordrecht, Netherlands, pp. 73-84.

Baron, J.S., D.S. Ojima, E.A. Holland, and W.J. Parton, 1994: Analysis of nitrogen saturation potential in Rocky Mountain tundra and forests: implications for aquatic systems. *Biogeochemistry*, **27**, 61-82.

Barry, R.G., 1990: Changes in mountain climate and glacio-hydrological responses. *Mountain Research and Development*, **10**, 161-170.

Barry, R.G., 1992a: *Mountain Weather and Climate.* Routledge Publishing Company, London, UK, 2nd ed., 392 pp.

Barry, R.G., 1992b: Climate change in the mountains. In: *The State of the World's Mountains: A Global Report* [Stone, P.B. (ed.)]. Zed Books Ltd., London, UK, pp. 359-380.

Barry, R.G., 1992c: Mountain climatology and past and potential future climatic changes in mountain regions: a review. *Mountain Research and Development*, **12**, 71-86.

Barry, R.G., 1994: Past and potential future changes in mountain environments. In: *Mountain Environments in Changing Climates* [Beniston, M. (ed.)]. Routledge Publishing Company, London and New York, pp. 3-33.

Barry, R.G. and C.C. Van Wie, 1974: Topo- and microclimatology in alpine areas. In: *Arctic and Alpine Environment*. Methuen, London, UK, pp. 73-83.

Baumgartner, M.F. and G. Apfl, 1994: Monitoring snow cover variations in the Alps using the Alpine Snow Cover Analysis System (ASCAS). In: *Mountain Environments in Changing Climates* [Beniston, M. (ed.)]. Routledge Publishing Company, London and New York, pp. 108-120.

Beniston, M., 1993: Prévisions climatiques pour les Alpes: une revue des techniques de régionalisation. *La Météorologie*, **45**, 38-44.

Beniston, M., (ed.), 1994: *Mountain Environments in Changing Climates*. Routledge Publishing Company, London and New York, 492 pp.

Beniston, M., M. Rebetez, F. Giorgi, and M.R. Marinucci, 1994: An analysis of regional climate change in Switzerland. *Theor. and Appl. Clim.*, **49**, 135-159.

Beniston, M., A. Ohmura, M. Rotach, P. Tschuck, M. Wild, and M.R. Marinucci, 1995: *Simulation of Climate Trends over the Alpine Region: Development of a Physically-Based Modeling System for Application to Regional Studies of Current and Future Climate*. Final Scientific Report No. 4031-33250 to the Swiss National Science Foundation, Bern, Switzerland, 200 pp.

Bjerrum, L. and F. Jfrstad, 1968: Stability of rock slopes in Norway. *Norwegian Geotechnical Institute Publication*, **79**, 1-11.

Bonnard, C., 1994: Los deslizamientos de tierra: fenomeni natural o fenomeno inducido por el hombre? In: *First Panamerican Symposium on Landslides*, Guayaquil, Ecuador, vol. 2.

Bortenschlager, S., 1993: Das höchst gelegene Moor der Ostalpen "Moor am Rofenberg" 2760 m. *Festschrift Zoller*, Diss. Bot., **196**, 329-334.

Bowman, W.D., T.A. Theodose, J.C. Schardt, and R.T. Conant, 1993: Constraints of nutrient availability on primary production in two alpine tundra communities. *Ecology*, **74**, 2085-2097.

Bradley, R.S. and P.D. Jones, 1992: *Climate Since AD 1500*. Routledge Publishing Company, London and New York, 692 pp.

Bravo, R.E. *et al.*, 1988: The effects of climatic variations on agriculture in the Central Sierra of Ecuador. In: *The Impact of Climate Variations on Agriculture*. Vol. 2, *Assessments in Semi-Arid Regions* [Parry, M.L., T.R. Carter, and N.T. Konijn (eds.)]. Kluwer Academic Publishers, Dordrecht, Netherlands, pp. 381-493.

Breiling, M. and P. Charamza, 1994: Localizing the threats due to climate change in mountain environments. In: *Mountain Environments in Changing Climates* [Beniston, M. (ed.)]. Routledge Publishing Company, London and New York, pp. 341-365.

Broggi, M.F. and G. Willi, 1989: Beschneiungsanlagen im Widerstreit der Interessen. *CIPRA Publication*, Vaduz, Liechtenstein, 75 pp.

Brown, L.H., 1981: The conservation of forest islands in areas of high human density. *African Journal of Ecology*, **19**, 27-32.

Brun, E., P. David, M. Sudul, and G. Brunot, 1992: A numerical model to simulate snow cover stratigraphy for operational avalanche forecasting. *J. Glaciol.*, **35**, 13-22.

Bucher, A. and J. Dessens, 1991: Secular trend of surface temperature at an elevated observatory in the Pyrenees. *J. Climate*, **4**, 859-868.

Bugmann, H., 1994: *On the Ecology of Mountain Forests in a Changing Climate: A Simulation Study*. Ph.D. Thesis No. 10638, Swiss Federal Institute of Technology (ETH), Zürich, Switzerland, 268 pp.

Bugmann, H. and A. Fischlin, 1994: Comparing the behaviour of mountainous forest succession models in a changing climate. In: *Mountain Environments in Changing Climates* [Beniston, M. (ed.)]. Routledge Publishing Company, London and New York, pp. 204-219.

Bultot, F., 1992: Repercussions of a CO_2 doubling on the water balance—a case study in Switzerland. *Journal of Hydrology*, **137**, 199-208.

Bultot, F., D. Gellens, B. Schädler, and M. Spreafico, 1992: Impact of climatic change, induced by the doubling of the atmospheric CO_2 concentration, on snow cover characteristics—case of the Broye drainage basin in Switzerland. *Swiss Climate Abstracts*, **special issue**, International Conference on Mountain Environments in Changing Climates, p. 19.

Burrows, C.J., 1990: *Processes of Vegetation Change*. Unwin Hyman Publishing, London, UK, 551 pp.

Busby, J.R., 1988: Potential implications of climate change on Australia's flora and fauna. In: *Greenhouse Planning for Climate Change* [Pearman, G.I. (ed.)]. CSIRO, Melbourne, Australia, pp. 387-398.

Callaghan, T.V. and S. Jonasson, 1994: Implications from environmental manipulation experiments for arctic plant biodiversity changes. In: *Arctic and Alpine Biodiversity: Patterns, Causes and Ecosystem Consequences* [Chapin, F.S. and C. Körner (eds.)]. Springer Verlag, Stuttgart, Germany.

Carter, T.R. and M.F. Parry, 1994: Evaluating the effects of climatic change on marginal agriculture in upland areas. In: *Mountain Environments in Changing Climates* [Beniston, M. (ed.)]. Routledge Publishing Company, London and New York, pp. 405-421.

Chalise, S.R., 1994: Mountain environments and climate change in the Hindu Kush-Himalayas. In: *Mountain Environments in Changing Climates* [Beniston, M. (ed.)]. Routledge Publishing Company, London and New York, pp. 382-404.

Chapin, F.S. and Ch. Körner, 1994: Arctic and alpine biodiversity: patterns, causes and ecosystem consequences. *TREE*, **9**, 39-78.

Chapin, F.S. and G.R. Shaver, 1985: Individualistic growth response of tundra plant species to environmental manipulations in the field. *Ecology*, **66**, 564-576.

Chen, J.Y. and A. Ohmura, 1990: On the influence of alpine glaciers on runoff. In: *Hydrology in Mountainous Regions. I. Hydrological Measurements. The Water Cycle* [Lang, H. and A. Musy (eds.)]. IAHS Press, Wallingford, UK, pp. 127-135.

Coetzee, J.A., 1967: Pollen analytical studies in East and Southern Africa. In: *Paleoecology of Africa*, vol. 3. A.A. Balkema, Cape Town, South Africa, 146 pp.

Collins, D.N., 1989: Hydrometeorological conditions, mass balance and runoff from alpine glaciers. In: *Glacier Fluctuations and Climate Change* [Oerlemans, J. (ed.)]. Kluwer Academic Publishers, Dordrecht, Netherlands, pp. 305-323.

Costin, A.B., 1967: Alpine ecosystems of the Australasian region. In: *Arctic and Alpine Environments* [Wright, H.E. and W.H. Osburn (eds.)]. Indiana University Press, Bloomington, IN, pp. 55-87.

David, F., 1993: Altitudinal variation in the response of vegetation to late-glacial climatic events in the northern French Alps. *New Phytol.*, **125**, 203-220.

Davis, M.B., 1981: Quartenary history and the stability of forest communities. In: *Forest Succession* [West, D.C., H.H. Shugart, and D.B. Botkin (eds.)]. Springer-Verlag, New York, NY.

Davis, M.B., 1986: Climatic instability, time lags, and community disequilibrium. In: *Community Ecology* [Diamond, J. and T.J. Case (eds.)]. Harper and Row, New York, NY, pp. 269-284.

Davis, M.B., 1989: Insights from paleoecology on global change. *Bulletin of the Ecological Society of America*, **70**, 222-228.

Decurtins, H., 1992: *Hydrogeographical Investigations in the Mount Kenya Subcatchment of the Ewaso Ng'iro River*. The Institute of Geography, University of Bern, Bern, Switzerland, 23 pp.

Dickinson, R.E., R.M. Errico, F. Giorgi, and G.T. Bates, 1989: A regional climate model for the western United States. *Climatic Change*, **15**, 383-422.

Downing, T.E., 1992: *Climate Change and Vulnerable Places: Global Food Security and Country Studies in Zimbabwe, Kenya, Senegal, and Chile*. Environmental Change Unit, Oxford, UK.

Dix, R.L., and J.D. Richards, 1976: Possible changes in species structure of the subalpine forest induced by increased snowpack. In: *Ecological Effects of Snowpack Augmentation in the San Juan Mountains, Colorado* [Steinhoff, H.W. and J.D. Ives (eds.)]. Colorado State University, Ft. Collins, CO, pp. 311-322.

Downton, M.W., T.R. Stewart, and K.A. Miller, 1988: Estimating historical heating and cooling needs: per capita degree days. *J. Appl. Meteorol.*, **27**, 84-90.

Earle, C.J., L.B. Brubaker, A.V. Lozkin, and P.M. Anderson, 1994: Summer temperature since 1600 for the Upper Koluma Region, Northeastern Russia, reconstructed from tree rings. *Arct. Alp. Res.*, **26**, 60-65.

Emanuel, W.R., H.H. Shugart, and M.P. Stevenson, 1985: Climatic change and the broadscale distribution of terrestrial ecosystem complexes. *Climatic Change*, **7**, 29-43.

Estrada, C. and M. Monasterio, 1988: Ecologia poblacional de una roseta gigante, Espeletia spicata SCH. BIP (Compositae), del paramo desertico. *Ecotropicos*, **1**, 25-39.

Eybergen, J. and F. Imeson, 1989: Geomorphological processes and climate change. *Catena*, **16**, 307–319.

FAO/UNESCO/OMM, 1975: *Estudio Agroclimatologico de la Zona Andina*. FAO Press, Rome, Italy.

Flenley, J.R., 1979: The late quaternary vegetational history of the equatorial mountains. *Progress in Physical Geography,* **3**, 488-509.

Fliri, F., 1975: *Das Klima der Alpen im Raume von Tirol.* Universitätsverlag Wagner, Innsbruck, München, 112 pp.

Föhn, P., 1991: Les hivers de demain seront-ils blancs comme neige ou vert comme les prés? WSL/FNP (éd.), *Argument de la Recherche*, **3**, 3-12.

Franklin, J.F., F.F. Swanson, M.E. Harmon, D.A. Perry, T.A. Spies, V.H. Dale, A. McKee, W.K. Ferrell, J.E. Means, S.V. Gregory, J.D. Lattin, T.D. Schowalter, and D. Larsen, 1991: Effects of global climate change on forests in Northwestern North America. *Northwest Environ. J.*, **7**, 233-257.

Franklin, J.F. and C.T. Dyrness, 1973: *Natural Vegetation of Oregon and Washington.* USDA Forest Service Gen. Tech. Rep. PNW-8, Portland, OR.

Galloway, R.W., 1988: The potential impact of climate changes on Australian ski fields. In: *Greenhouse Planning for Climate Change* [Pearman, G.I. (ed.)]. CSIRO, Aspendale, Australia, pp. 428-437.

Galloway, R.W., 1989: Glacial and periglacial features of the Australian alps. In: *The Scientific Significance of the Australian Alps* [Good, R. (ed.)]. Australian Alps Liaison Committee/Australian Academy of Science, Canberra, Australia, pp. 55-67.

Gao, S., Y. Ding, and Z. Zhao, 1993: The possible greenhouse impact of atmospheric CO_2 content increasing on the agricultural production in future in China. *Scientia Atmospherica Sinica*, **17**, 584-591.

Garr, C.E. and B.B. Fitzharris, 1994: Sensitivity of mountain runoff and hydro-electricity to changing climate. In: *Mountain Environments in Changing Climates* [Beniston, M. (ed.)]. Routledge Publishing Company, London and New York, pp. 366-381.

Gates, D.M., 1990: Climate change and forests. *Tree Physiology*, **7**, 1-23.

Gauslaa, Y., 1984: Heat resistance and energy budget in different Scandinavian plants. *Holarct. Ecol.*, **7**, 1-78.

Gear, A.J. and B. Huntley, 1991: Rapid changes in the range limits of Scots Pine 4,000 years ago. *Science*, **251**, 544-547.

Geiger, R., 1965: *The Climate Near the Ground.* Harvard University Press, Cambridge, MA, 277 pp.

Giorgi, F., 1990: On the simulation of regional climate using a limited area model nested in a general circulation model. *J. Clim.*, **3**, 941-963

Giorgi, F. and L.O. Mearns, 1991: Approaches to the simulation of regional climate change: a review. *Rev. Geophysics*, **29**, 191-216.

Giorgi, F., M.R. Marinucci, and G. Visconti, 1990: Use of a limited area model nested in a general circulation model for regional climate simulations over Europe. *J. Geophys. Res.*, **95**, 18,413-431.

Giorgi, F., M.R. Marinucci, and G. Visconti, 1992: A 2 x CO_2 climate change scenario over Europe generated using a limited area model nested in a general circulation model. II. Climate change. *J. Geophys. Res.*, **97**, 10,011–28.

Glantz, M.H. (ed.), 1988: *Societal Responses to Regional Climatic Change.* Westview Press, Boulder, CO.

Gleick, P.H., 1986: Methods for evaluating the regional hydrologic impacts of global climatic changes. *J. Hydrology*, **88**, 97-116.

Gleick, P.H., 1987a: Regional hydrologic consequences of increases in atmospheric CO_2 and other trace gases. *Climatic Change*, **10**, 137-161.

Gleick, P.H., 1987b: The development and testing of a water balance model for climate impact assessment: modelling the Sacramento Basin. *Water Resources Research*, **23**, 1049-1061.

Goliber, T.J., 1985: Sub-saharan Africa: population pressures and development. *Population Bulletin,* **40**, 1-47.

Golodkovskaya, N.A., 1988: Glacial mudflows on the Central Caucasus: the activity in the last millennium. *Data of Glaciological Studies*, **62**, 71-78

Government of Tanzania, 1979: *Report of Population Census 1978.* Bureau of Statistics, Dar-es-Salaam, Tanzania, 185 pp.

Govi, M., 1990: Conférence spéciale: mouvements de masse récents et anciens dans les Alpes italiennes. *Proc. Fifth Symposium on Landslides*, Lausanne, **3**, 1509–1514.

Grabherr, G. and M. Gottfried, 1992: Global change and terrestrial ecosystems (GCTE): effects of global warming on high alpine vegetation and flora. *Austrian Contributions to the IGBP,* **1**, Austrian Academy of Science, Vienna, 24-26.

Grabherr, G., E. Mahr, and H. Reisigl, 1978: Nettoprimärproduktion in einem Krummseggenrasen (Caricetum curvulae) der Ötztaler Alpen. *Tirol. Oecol. Plant.*, **13**, 227-251.

Grabherr, G., M. Gottfried, and H. Pauli, 1994: Climate effects on mountain plants. *Nature*, **369**, 448.

Graetz, R.D, 1991: The nature and significance of the feedback of changes in terrestrial vegetation on global atmospheric and climatic change. *Climatic Change*, **18**, 147-160.

Graumlich, L.J., 1993: A 1000-year record of temperature and precipitation in the Sierra Nevada. *Quat. Res.,* **39**, 249-255.

Graumlich, L.J., 1994: Long-term vegetation change in mountain environments: paleoecological insights into modern vegetation dynamics. In: *Mountain Environments in Changing Climates* [Beniston, M. (ed.)]. Routledge Publishing Company, London and New York, pp. 167-179.

Graumlich, L.J. and L.B. Brubaker, 1994: Long-term records of growth and distribution of conifers: implications for ecophysiology. In: *Physiological Ecology of North American Forests* [Smith, W. and T. Hinckley (eds.)]. Academic Press, New York, NY.

Grosswald, M.G., and V.N. Orlyankinm, 1979: The Late Pleistocene Pamirs ice sheet. *Data of Glaciological Studies,* **35**, 85-97.

Grötzbach, E.F., 1988: High mountains as human habitat. In: *Human Impact on Mountains* [Allan, N.J.R., G.W. Knapp, and C. Stadel (eds.)]. Rowman and Littlefield, Totowa, NJ, pp. 24-35.

Guhl, E., 1968: Los páramos circundantes de la sabana de Bogotá. Su ecologia y su importancia para el régimen hidrologico de la misma. In: *Geoecology of the Mountainous Regions of the Tropical Americas* [Troll, C. (ed.)]. Ferd. Dümmlers Verlag, Bonn, Germany, pp. 195-212.

Guisan, A., J. Holten, R. Spichiger, and L. Tessier (eds.), 1995: *Potential Ecological Impacts of Climate Change in the Alps and Fennoscandian Mountains.* Publication Series of the Geneva Conservatory and Botanical Gardens, University of Geneva, Geneva, Switzerland, 194 pp.

Gvosdetskii, N.A. and Yu.N. Golubtchikov, 1987: *The Mountains.* USSR Academy of Sciences, Moscow, USSR, 400 pp.

Gyalistras, D., H. von Storch, A. Fischlin, and M. Beniston, 1994: Linking GCM-simulated climatic changes to ecosystem models: case studies of statistical downscaling in the Alps. *Clim. Res.*, **4**, 167-189.

Haeberli, W., 1990: Glacier and permafrost signals of 20th-century warming. *Ann. Glaciol.*, **14**, 99-101.

Haeberli, W., 1994: Accelerated glacier and permafrost changes in the Alps. In: *Mountain Environments in Changing Climates* [Beniston, M. (ed.)]. Routledge Publishing Company, London and New York, pp. 91-107.

Haeberli, W., P. Muller, J. Alean, and H. Bösch, 1990: Glacier changes following the Little Ice Age. A survey of the international data base and its perspectives. In: *Glacier Fluctuations and Climate* [Oerlemans, J. (ed.)]. D. Reidel Publishing Company, Dordrecht, Netherlands, pp. 77-101.

Halpin, P.N., 1994: Latitudinal variation in montane ecosystem response to potential climatic change. In: *Mountain Ecosystems in Changing Climates* [Beniston, M. (ed.)]. Routledge Publishing Company, London and New York, pp. 180-203.

Halpin, P.N., 1995: Global change and natural area protection: management responses and research directions. *Ecol. Appl.* (in press).

Hamilton, A.C., 1993: History of forests and climate. In: *Conservation Atlas of Tropical Forests, Africa.* World Wildlife Fund, Washington, DC.

Hansen-Bristow, K.J., J.D. Ives, and J.P. Wilson, 1988: Climatic variability and tree response within the forest-alpine tundra ecotone. *Annals of the Association of American Geographers*, **78**, 505-519.

Hastenrath, S. and R. Rostom, 1990: Variations of the Lewis and Gregory Glaciers, Mount Kenya, 1978-86-90. *Erdkunde*, **44**, 313-317.

Hedberg, O., 1951: Vegetation belts of East African mountains. *Svensk. Bot. Tidskr.*, **48**, 199-210.

Hedberg, O., 1964a: The phytogeographical position of the afroalpine flora. *Recent Adv. in Bot.*, **117**, 914-919.

Hedberg, O., 1964b: Etudes ecologiques de las flore afroalpine. *Bull. Soc. Royale Botan. de Belgique*, **97**, 5-18.

Heide, O.M., 1985: Physiological aspects of climatic adaptation in plants with special reference to high-latitude environments. In: *Plant Production in the North* [Kaurin, A., O. Junttila, and J. Nilsen (eds.)]. Norweg Univerity Press, Oslo, Norway, pp. 1-22.

Heide, O.M., 1989: Environmental control of flowering and viviparous proliferation in semiferous and viviparous Arctic populations of two Poa species. *Arctic Alpine Res.*, **21**, 305-315.

Heide, O.M., 1990: Dual floral induction requirements in Phleum alpinum. *Ann. Bot.*, **66**, 687-694.

Hengeveld, R., 1989: Theories on species responses to variable climates. In: *Landscape-Ecological Impact of Climatic Change* [Boer, M.M. and R.S. de Groot (eds.)]. IOS Press, Amsterdam, Netherlands, pp. 274-293.

Heuberger, H., 1966: Gletschergeschichtliche Untersuchungen in den Zentralalpen zwischen Sellrain und Oetztal. *Wiss. Alpenvereinhefte*, **20**, 126.

Hewitt, M., 1994: Modeled changes in snow cover in the Australian Alps under enhanced greenhouse gas concentrations. In: *Proc. International Symposium on Snow and Climate*, Geneva, 22-23 September 1994, University of Geneva, pp.19-20.

Holten, J.I., 1995: Effects of climate change on plant diversity and distribution in the Fennoscandian mountain range. In: *Potential Ecological Impacts of Climate Change in the Alps and Fennoscandian Mountains* [Guisan, A., J. Holten, R. Spichiger, and L. Tessier (eds.)]. Publication Series of the Geneva Conservatory and Botanical Gardens, University of Geneva, Geneva, Switzerland, 194 pp.

Holten, J.I., and P.D. Carey, 1992: Responses of climate change on natural terrestrial ecosystems in Norway. *NINA Institute Research Report*, **29**, 1-59.

Holtmeier, F.-K., 1993: Timberline as indicators of climatic changes: problems and research needs. *Paläoklimaforschung/Palaeoclimatic Research*, **9**, 11-22.

Holtmeier, F.-K., 1994: Ecological aspects of climatically caused timberline fluctuations: review and outlook. In: *Mountain Environments in Changing Climates* [Beniston, M. (ed.)]. Routledge Publishing Company, London and New York, pp. 220-233.

Houghton, J.T., G.J. Jenkins, J.J. Ephraums (eds.), 1990: *Climate Change. The IPCC Scientific Assessment.* Cambridge University Press, Cambridge, UK, 365 pp.

Hulme, M., T. Wigley, T. Jiang, Z. Zhao, F. Wang, Y. Ding, R. Leemans, and A. Markham, 1992: *Climate Change Due to the Greenhouse Effect and Its Implications for China.* World Wildlife Fund, CH-1196, Gland, Switzerland.

Huntley, B. 1991: How plants respond to climate change: migration rates, individualism and the consequences for plant communities. *Annals of Botany*, **67**, 15-22.

Hurni, H. and P. Stähli, 1982: Das Klima von Semien, Athiopen Vol. II Kaltzeit bis zur Gegenwart. *Geographica Bernensia*, **13**, University of Bern, Switzerland, pp. 50-82.

Innes, J.L., 1985: Magnitude-frequency relations of debris flows in north-west Europe, *Geografiska Annaler*, **67A**, 23-32.

IPCC, 1992: *Climate Change 1992. The Supplementary Report to the IPCC Scientific Assessment* [Houghton, J.T., B.A. Callander, and S.K. Varney (eds.)]. Cambridge University Press, Cambridge, UK, 200 pp.

IUCN/CDC, 1985: *Agricultural Development and Environmental Conservation in the East Usambara Mountains, Tanzania.* Mission Report of Conservation for Development Centre. IUCN Publications, Gland, Switzerland.

Ives, J.D. and B. Messerli, 1989: *The Himalayan Dilemma.* Routledge Publishing Company, London, UK, 336 pp.

Ives, J.D., 1992: Preface, *The State of the World's Mountains* [Stone, P. (ed.)]. Zed Books, London, UK, pp. xiii-xvi.

Jäger, J., 1983: *Climate and Energy Systems: A Review of their Interactions.* John Wiley and Sons, New York, NY, 231 pp.

Jiang, Y., 1992: The problem of global climate change and the forests in China. *Scientia Silvae Sinicae*, **28**, 431-438.

Kalela, O., 1949: Changes in geographic ranges in the avifauna of Northern and Central Europe in relation to recent changes in climate. *Bird-Banding*, **20**, 77-103.

Kates, R.W., J.H. Ausubel, and M. Berberian (eds.), 1985: *Climate Impact Assessment.* SCOPE 27, John Wiley and Sons, Chichester, UK, 78 pp.

Kasser, P., 1993: Influence of changes in glacierized area on summer run-off in the Porte-du-Scex drainage basin of the Rhône. In: *Symposium on the Hydrology of Glaciers.* Publication No. 95, International Association Scientific Hydrology, Lausanne, pp. 221-225.

Kienast, F., 1989: Simulated effects of increasing atmospheric CO_2 on the successional characteristics of alpine forest ecosystems. In: *Landscape-Ecological Impact of Climatic Change* [Boer, M.M. and R.S. de Groot (eds.)]. IOS Press, Amsterdam, Netherlands, pp. 100-118.

Kienast, F., 1991: Simulated effects of increasing atmospheric CO_2 and changing climate on the successional characteristics of Alpine forest ecosystems. *Landscape Ecology*, **5**, 225-235.

Kienast, F., B. Brzeziecki, O. Wildi, 1995: Long-term adaptation potential of Central European mountain forests to climate change: a GIS-assisted risk assessment. *Forest Ecol. Management* (in press).

Kirchhofer, W. (ed.), 1982: *Climate Atlas of Switzerland.* Swiss Meteorological Institute, Zurich, Switzerland, 125 pp.

Kittel, T.G.F., 1990: Climatic variability in the shortgrass steppe. In: *Climate Variability and Ecosystem Response* [Greenland, D. and L.W. Swift, Jr. (eds.)]. Gen. Tech. Rep. SE-65, USDA Forest Service, Southeastern Region, pp. 67-75.

Klötzli, F., 1984: Neuere Erkenntnisse zur Buchengrenze in Mitteleuropa, Fukarek, Akad. Nauka um jetn. bosne Herc., rad. 72. *Odj. Prir. Mat. Nauka* 21 [P. Festschr (ed.)], Sarajevo, Yugoslavia, pp. 381-395.

Klötzli, F., 1991: Longevity and stress. In: *Modern Ecology: Basic and Applied Aspects* [Esser, G. and D. Overdiek (eds.)]. Elsevier, Amsterdam, Netherlands, pp. 97-110.

Klötzli, F., 1994: Vegetation als Spielball naturgegebener Bauherren. *Phytocoenologia*, **24**, 667-675.

Körner, Ch., 1989: Response of alpine vegetation to global climate change. In: *Landscape-Ecological Impact of Climatic Change* [Boer, M.M. and R.S. de Groot (eds.)]. IOS Press, Amsterdam, Netherlands, pp. 79-87.

Körner, Ch., 1989: The nutritional status of plants from high altitudes: a worldwide comparison. *Oecologia*, **81**, 379-391.

Körner, Ch., 1994: Impact of atmospheric changes on high mountain vegetation. In: *Mountain Environments in Changing Climates* [Beniston, M. (ed.)]. Routledge Publishing Company, London and New York, pp. 155-166.

Körner, Ch. and W. Larcher, 1988: Plant life in cold climates. In: *Plants and Temperature. Symp. Soc. Exp. Biol.*, vol. 42 [Long, S.F. and F.I. Woodward (eds.)]. The Company of Biol. Ltd, Cambridge, UK, pp. 25-57.

Kräuchi, N., 1994: *Modeling Forest Succession as Influenced by a Changing Environment.* Scientific Report Nr. 69 of the Swiss Federal Institute on Forest, Snow, and Landscape, Birmensdorf, Switzerland, 126 pp.

Krauchi, N. and F. Kienast, 1993: Modelling subalpine forest dynamics as influenced by a changing environment. *Water Air Soil Pollut.*, **68**, 185-197.

Krenke, A.N., G.M. Nikolaeva, and A.B. Shmakin, 1991: The effects of natural and anthropogenic changes in heat and water budgets in the central Caucasus, USSR. *Mountain Res. Devel.*, **11**, 173-182.

Krippendorf, J., 1984: The capital of tourism in danger. In: *The Transformation of Swiss Mountain Regions* [Brugger, E.A. *et al.* (eds.)]. Haupt Publishers, Bern, Switzerland, pp. 427-450.

Lai, Z. and B. Ye, 1991: Evaluating the water resource impacts of climatic warming in cold alpine regions by the water balance model: modeling the Urumqi river basin. *Science in China*, **34**, 1362-1371.

Lai, Z., 1992: Analysis of temperature-runoff relationship in the eastern section of Qilian mountain. In: *Memoirs of Lanzhou Institute of Glaciology and Geocryology.* Chinese Academy of Sciences, No. 7, Beijing, China, pp. 84-90.

LaMarche, Jr., V.C., 1973: Holocene climatic variations inferred from treeline fluctuations in the White Mountains, California. *Quat. Res.*, **3**, 632-660.

Lamothe, M. and D. Périard, 1988: Implications of climate change for downhill skiing in Quebec. In: *Climate Change Digest.* Atmospheric Environment Service, Downsview, Canada, pp. 88-103.

Larcher, W., 1980: Klimastress im Gebirge —Adaptationstraining und Selektionsfilter für Pflanzen. *Rheinisch-Westfälische Akad. Wiss. Vorträge*, **N. 291**, 49-88.

Larcher, W., 1982: Typology of freezing phenomena among vascular plants and evolutionary trends in frost acclimation. In: *Plant Cold Hardiness and Freezing Stress* [Li, P.H. and A. Sakai (eds.)]. Academic Press, New York, NY, pp. 417-426.

Lauer, W., 1979: La posicion de los páramos en la estructura del paisaje en los Andes tropicales. In: *El Medio Ambiente Páramo—Actas del Seminario de Mérida, Venezuela* [Salgado-Laboriau, M.L. (ed.)]. Centro de Estudios Avanzados, Caracas, Venezuela, pp. 29-43.

Lauscher, F., 1977: Ergebnisse der Beobachtungen an den nordchilenischen Hochgebirgsstationen Collahuasi und Chuquicamata. In: *Jahresbericht des Sonnblickvereines für die Jahre 1976-1977*, pp. 43-67.

Lauscher, F., 1980: Die Schwankungen der Temperatur auf dem Sonnblick seit 1887 im Vergleich zu globalen Temperaturschwankungen. In: *16 Internat.Tagung Alpine Meteorologie*. Soc. Meteorol. de France, Boulogne-Billancourt, France, pp. 315-319.

Leavesley, G.H., 1994: Modeling the effects of climate change on water resources—a review. In: *Assessing the Impacts of Climate Change on Natural Resource Systems* [Frederick, K.D. and N. Rosenberg (eds.)]. Kluwer Academic Publishers, Dordrecht, Netherlands, pp. 179-208.

LeBlanc, D.C. and J.R. Foster, 1992: Predicting effects of global warming on growth and mortality of upland oak species in the midwestern United States: a physiological based dendroecological approach. *Can. J. For. Res.*, **22**, 1739-1756.

Leemans, R., 1993: The effects of climate change on natural ecosystems in China. In: *Environment and Climate Change: The Challenge for China* [Ding, Y. and A. Markham (eds.)]. Meteorological Press, London, UK, pp. 219-234.

Leffler, R.J., J. Sullivan, and H.E. Warren, 1985: A weather index for international heating oil consumption. In: *World Energy Markets, Stability or Cyclical Change, Proceedings of the 7th Annual North American, International Associations of Energy Economists* [Thompson, W.F. and D.J. De Angelo (eds.)]. Westview Press, Boulder, CO, and London, UK, pp. 630-644.

Li, Y., J. Jiang, G. Long, and Y. Cheng, 1993: The influence of climate warming on rice production in China. In: *The Impact of Climatic Variations on Agriculture and Its Strategic Counter-measure*. Beijing University Press, Beijing, China, pp. 54-130.

Lins, H. *et al.*, 1990: Hydrology and water resources. In: *Climate Change: The IPCC Impacts Assessment* [McG. Tegart, W.J., G.W. Sheldon, and D.C. Griffiths (eds.)]. Australian Government Publishing Service, Canberra, Australia, Chapter 4.

Liu, C. and G. Fu, 1993: Some analyses on climate change and Chinese hydrological regime. In: *Climate Change and Its Impact*. Institute of Geography, CAS, Global Change Study No. l, Series Publication, Meteorological Press, Beijing, China, pp. 205-214.

Liu, D. and D. Zheng, 1990: Geographical studies of the Qing-Zang plateau. In: *Recent Development of Geographical Science in China* [Geographical Society of China (ed.)]. Science Press, Beijing, China, pp. 214-234.

Liverman, D.M. and K.L. O'Brien, 1991: Global warming and climate change in Mexico. *Global Environmental Change: Human and Policy Dimensions*, **1**, 351-364.

Livingstone, D.A. and T. van der Hammen, 1978: Palaeogeography and palaeoclimatology. In: *Tropical Forest Ecosystems: A State of Knowledge Report*. UNESCO/UNEP/FAO, Paris, France, pp. 61-85.

Lloyd, A.H. and L.J. Graumlich, 1995: Dendroclimatic, ecological, and geomorphological evidence for long-term climatic change in the Sierra Nevada, USA. In: *Tree Rings, Environment, and Humanity* [Dean, J.S., D.M. Meko, and T.W. Swetnam (eds.)]. University of Arizona, Tucson, AZ (in press).

Lovett, J.C. and S.K. Wasser (eds.), 1993: *Biogeography and Ecology of the Rain Forest of Eastern Africa*. Cambridge University Press, Cambridge, UK.

Luckman, B.H., 1994: Using multiple high-resolution proxy climate records to reconstruct natural climate variability: an example from the Canadian Rockies. In: *Mountain Environments in Changing Climates* [Beniston, M. (ed.)]. Routledge Publishing Company, London and New York, pp. 42-59.

Lüdi, W., 1938: Mikroklimatische Untersuchungen an einem Vegatationsprofil in den Alpen von Davos III. In: *Bericht. Geobot Forschungsinst. Rübel*, Zürich, pp. 29-49.

Lundgren, L., 1980: Comparison of surface runoff and soil loss from runoff plots in forest and small scale agriculture in the Usambara Mountains, Tanzania. *STOU-Nag.*, **38**, University of Stockholm, Department of Physical Geography, 118 pp.

Lynch, P., G.R. McBoyle, and G. Wall, 1981: A ski season without snow. In: *Canadian Climate In Review—1980* [Phillips, D.W. and G.A. McKay (eds.)]. Atmospheric Environment Service, Toronto, Canada, pp. 42-50.

Ma, S. and H. Liang, 1990: The estimating method of thermal reources and the arrangement of crop variety in the Changbai mountainous region. *Mountain Research*, **8**, 161-166.

MacArthur, R.H., 1972: *Geographical Ecology*. Harper and Row, New York, NY.

MacDonald, G.M., T.W.D. Edwards, K.A. Moser, R. Pienitz, and J.P. Smol, 1993: Rapid response of treeline vegetation and lakes to past climate warming. *Nature*, **361**, 243-257.

Mackay, J.R., 1990: Seasonal growth bands in pingo ice. *Canadian Journal of Earth Science*, **27**, 1115-1125.

Mackay, J.R., 1992: Frequency of ice-wedge cracking 1967-1987 at Garry Island, Western Arctic Coast, Canada. *Canadian Journal of Earth Science*, **29**, 11-17.

Makhova, Y.V. and N.G. Patyk-Kara, 1961: On the history of high mountain vegetation of the Great Caucasus in the Holocene. In: *Paleogeography of the Quaternary Period in the USSR* [Markov, K.K. (ed.)]. USSR Academy of Sciences, Moscow, USSR, pp. 125-130.

Margalitadze, N.A. and K.R. Kimeridze, 1985: Holocene history of the vegetation of the Upper Svanetia. In: *Flora and Vegetation of Svanetia* [Nakhutsrishvili, G.S. (ed.)]. Mezniereba Publishers, Tsilisi, pp. 240-260.

Marinucci, M.R. and F. Giorgi, 1992: A 2 x CO_2 climate change scenario over Europe generated using a limited area model nested in a general circulation model. I: Present day simulation. *J. Geophys. Res.*, **97**, 9,989-10,009.

Marinucci, M.R., F. Giorgi, M. Beniston, M. Wild, P. Tschuck, and A. Bernasconi, 1995: High resolution simulations of January and July climate over the Western Alpine region with a nested regional modeling system. *Theor. and Appl. Clim.* (in press).

Martin, E., 1992: Sensitivity of the French Alps' snow cover to the variation of climatic parameters. *Swiss Climate Abstracts*, **special issue**, International Conference on Mountain Environments in Changing Climates, pp. 23-24.

Martin, E., 1995: *Modélisation de la Climatologie Nivale des Alpes Françaises*. Ph.D. dissertation, Université Paul Sabatier, Toulouse, France, 232 pp.

Martin, E., E. Brun, and Y. Durand, 1994: Sensitivity of the French Alps snow cover to the variation of climatic variables. *Annales Geophysicae* (in press).

Martinec, J. and A. Rango, 1989: Effects of climate change on snow melt runoff patterns. In: *Remote Sensing and Large-Scale Global Processes*. Proceedings of the IAHS Third Int. Assembly, Baltimore, MD, May 1989, IAHS Publ. No. 186, pp. 31-38.

Maunder, W.J., 1986: *The Uncertainty Business*. Methuen and Co. Ltd, London, UK, 420 pp.

McBoyle, G.R. and G. Wall, 1987: The impact of CO_2-induced warming on downhill skiing in the Laurentians. *Cahiers de Géographie de Québec*, **31**, 39-50.

McMurtrie, R.E., H.N. Comins, M.U.F. Kirschbaum, and Y.-P. Wang, 1992: Modifying existing forest growth models to take account of effects of elevated CO_2. *Aust. J. Bot.*, **40**, 657-681.

McNeely, J.A., 1990: Climate change and biological diversity: policy implications. In: *Landscape-Ecological Impact of Climatic Change* [Boer, M.M. and R.S. de Groot (eds.)]. IOS Press, Amsterdam, Netherlands.

Mel'nikova, A.P. and E.K. Bakov, 1989: To the question about degradation of glaciation in the Northern Tien Shan in Holocene. *Data of Glaciological Studies*, **67**, 91-97.

Messerli, P., 1989: *Mensch und Natur im alpinen Lebensraum: Risiken, Chancen, Perspektiven*. Haupt-Verlag, Bern and Stuttgart, 86 pp.

Mikhalenko, V.N. and O.N. Solomina, 1995: Long-term trends of mass balance changes and fluctuations of base glaciers termini in the mountain region of the former USSR (FSU). Part 1, Instrumental records (IAHS). UNESCO Symposium on Glaciers Mass Balances, 14-16 September 1995, Innsbruck, Austria.

Molau, U., 1993: Relationships between flowering phenology and life history strategies in tundra plants. *Arctic and Alpine Research*, **25**, 391-402.

Monasterio, M., 1971: La alta montaoa de America tropical. Estudios Ecologicos en los Ecosistemas de Páramo, Universidad de Los Andes, Facultad de Ciencias, Mérida, Venezuela, 15 pp.

Monasterio, M., 1980a: Los páramos andinos como region natural: caracteristicas biogeográficas generales y afinidades con otras regiones andinas. In: *Estudios Ecologicos en los Páramos Andinos* [Monasterio, M. (ed.)]. Ediciones de la Universidad de Los Andes, Mérida, Venezuela, pp. 15-27.

Monasterio, M., 1980b: Poblamiento humano y uso de la tierra en los altos Andes de Venezuela. In: *Estudios Ecologicos en los Páramos Andinos* [Monasterio, M. (ed.)]. Ediciones de la Universidad de Los Andes, Mérida, Venezuela, pp. 170-198.

Monasterio, M. and J. Celecia, 1991: El norte de los Andes tropicales: sistemas naturales y agrarios en la Cordillera de Mérida. *Coleccion Permanente El hombre y la Biosfera*, UNESCO Publications, Paris.

Monasterio, M. and S. Reyes, 1980: Diversidad ambiental y variacion de la vegetacion en los Páramos de los Andes venezolanos. In: *Estudios Ecologicos en los Páramos Andinos* [Monasterio, M. (ed.)]. Ediciones de la Universidad de Los Andes, Mérida, Venezuela, pp. 47-91.

Monasterio, M. (ed.), 1980: *Estudios Ecologicos en los Paramos Andinos*. Ediciones de la Universidad de los Andes, Mérida, Venezuela, 312 pp.

Musselman, R.C. (ed.), 1994: *The Glacier Lakes Ecosystems Experiment Site*. USDA General Technical Report RM-249, Fort Collins, CO, 94 pp.

Nash, L.L. and P.H. Gleick, 1991: Sensitivity of streamflow in the Colorado Basin to climatic changes. *Hydrology*, **125**, 221-241.

Nekrasov, I.A., E.V. Maksimov, and I.V. Klimovskii, 1973: The last glaciation and the cryolithozone of the Southern Verchojanje. *Knizhnoje Iydatel'stvo*, Jakutsk, 151 pp.

New Zealand Ministry for the Environment, 1990: *Climatic Change, a Review of Impacts on New Zealand*. DSIR Publishing, Wellington, New Zealand.

Nikolov, N.T. and D.G. Fox, 1994: A coupled carbon-water-energy-vegetation model to assess responses of temperate forest ecosystems to changes in climate and atmospheric CO_2. Part I, Model concept. *Environ. Pollut.*, **83**, 251-262.

Noble, I.R., 1993: A model of the responses of ecotones to climate change. *Ecol. Applic.*, **3**, 396-414.

Noverraz, F. and C. Bonnard, 1990: Mapping methodology of landslides and rockfalls in Switzerland. In: *Proc. Sixth International Conference and Field Workshop on Landslides*, Milan, Italy, pp. 43–53.

Noverraz, F. and C. Bonnard, 1992: Le glissement rapide de la Chenanta. In: *Proc. INTERPRAEVENT Symposium*, vol. 2. Swiss Hydrological Service Publication, Bern, Switzerland, pp. 65-76.

Obrebska-Starkel, B., 1990: Recent studies on Carpathian meteorology and climatology. *Int. J. Climatol.*, **10**, 79-88.

Oerlemans, J. (ed.), 1989: *Glacier Fluctuations and Climate Change*. Kluwer Academic Publishers, Dordrecht, Netherlands.

Okishev, P.A., 1982: *Altay Glaciers Dynamics in the last Pleistocene and Holocene*. Tomsk University Press, Tomsk, USSR, 210 pp.

Onipchenko, V.G., 1994: Study area and general description of the investigated communities. In: *Experimental Investigation of Alpine Plant Communities in the Northwestern Caucasus* [Onipchenko, V.G. and M.S. Blinnikov (eds.)]. Veröffentlichungen des Geobotanischen Institutes der ETH, Stiftung Rübel, Zürich, Switzerland, vol. 115, pp. 6-22.

OTA, 1993: *Preparing for an Uncertain Climate*, vols. I and II. OTA-O-567 and 568, U.S. Government Printing Office, Washington, DC.

Overpeck, J.T., P.J. Bartlein, and T. Webb III, 1991: Potential magnitude of future vegetation change in eastern North America: comparisons with the past. *Science*, **254**, 692-699.

Ozenda, P., 1985: *La Végétation de la Chaine Alpine dans l'Espace Montagnard Européen*. Masson, Paris, France, 344 pp.

Ozenda, P. and J.-L. Borel, 1989: The possible responses of vegetation to a global climatic change. Scenarios for Western Europe, with special reference to the Alps. In: *Landscape-Ecological Impact of Climatic Change* [Boer, M.M. and R.S. de Groot (eds.)]. IOS Press, Amsterdam, Netherlands, pp. 221-249.

Ozenda, P. and J.-L. Borel, 1991: *Les Conséquences Ecologiques Possibles des Changements Climatiques dans l'Arc Alpin*. Rapport Futural No. 1, International Centre for Alpine Environment (ICALP), Le Bourget-du-lac, France, 14 pp.

Parry, M.L., 1978: *Climatic Change, Agriculture and Settlement*. Dawson, Folkestone, UK, 214 pp.

Parry, M.L. *et al.*, 1990: Agriculture and forestry. In: *Climate Change: The IPCC Impacts Assessment* [McG. Tegart, W.J., G.W. Sheldon, and D.C. Griffiths (eds.)]. Australian Government Publishing Service, Canberra, Australia, Chapter 2.

Parry, M.L. *et al.*, 1992: *The Potential Socio-Economic Effects of Climate Change in South-east Asia*. United Nations Environment Programme, Nairobi, Kenya.

Peet, R.K., 1978: Latitudinal variation in southern Rocky Mountain forests. *J. Biogeography*, **5**, 275-289.

Peine, J.D. and D.G. Fox, 1995: Wilderness environmental monitoring: assessment and atmospheric effects. In: *The Status of Wilderness Research* [Cordell, K. (ed.)]. General Technical Report, GTR-SE-332, USDA Forest Service, Asheville, NC, 91 pp.

Peine, J.D. and C.J. Martinka, 1992: Effects of climate change on mountain protected areas—implications for management. In: *Proc. IVth World Congress on National Parks and Protected Areas*, Caracas, Venezuela, 10-21 February 1992.

Peine, J.D. and C.J. Martinka, 1994: Impacts of climate change on mountain protected areas: implication for management. In: *The Impact of Climate Change on Ecosystems and Species: Implications for Protected Areas* [Pernetta, J., R. Leemans, D. Elder, and S. Humphrey (eds.)]. IUCN Report, Gland, Switzerland.

Perry, A.H., 1971: Climatic influences on the Scottish ski industry. *Scottish Geographical Magazine*, **87**, 197-201.

Peters, R.L. and J.D.S. Darling, 1985: The greenhouse effect and nature reserves: global warming would diminish biological diversity by causing extinctions among reserve species. *Bioscience*, **35**, 707-717.

Peterson, D.L., 1993: Global environmental change in mountain protected areas: consequences for management. In: *Parks, Peaks and People* [Hamilton, L.S., D.P. Bauer, and H.F. Takeuchi (eds.)]. East-West Center, Honolulu, HI, pp. 29-36.

Peterson, D.L., 1994: Recent changes in the growth and establishment of subalpine conifers in Western North America. In: *Mountain Environments in Changing Climates* [Beniston, M. (ed.)]. Routledge Publishing Company, London and New York, pp. 234-243.

Peterson, D.L., M.J. Arbaugh, L.J. Robinson, and B.R. Derderian, 1990: Growth trends of whitebark pine and lodgepole pine in a subalpine Sierra Nevada forest, California, U.S.A. *Arctic and Alpine Research*, **22**, 233-243.

Peterson, E.B., 1969: Radiosonde data for characterizing a mountain environment in British Columbia. *Ecology*, **50**, 200-205.

Pfister, C., 1985a: *Klimageschichte der Schweiz 1525-1860*, vol. 1. Academia Helvetica, P. Haupt, Bern, Switzerland, 154 pp.

Pfister, C., 1985b: Snow cover, snow lines and glaciers in Central Europe since the 16th century. In: *The Climatic Scene* [Tooley, M.J. and G.M. Sheail (eds.)]. Allen and Unwin, London, UK, pp. 164-74.

Pla Sentis, I., 1988: *Dessarrollo de Indices y Modelos Para el Diagnostico y Prevencion de al Degradacion de Suelos Agricolas en Venezuela*. Publicacion Banco Consolidado, Caracas, Venezuela.

PNUMA, 1990: *Desarrollo y Medio Ambiente en America Latina y el Caribe: Una Vision Evolutiva*. Espania, Madrid, Spain.

Prentice, I.C., P.J. Bartlein, and T. Webb III, 1991: Vegetation and climate change in Eastern North America since the last glacial maximum. *Ecology*, **72**, 2038-2052.

Price, M.F., 1990: Temperate mountain forests: common-pool resources with changing, multiple outputs for changing communities. *Natural Resources Journal*, **30**, 685-707.

Price, M.F., 1992: Patterns of the development of tourism in mountain environments. *GeoJournal*, **27**, 87-96.

Price, M.F., 1993: Patterns of the development of tourism in mountain communities. In: *Mountains at Risk: Current Issues in Environmental Studies* [Allan, N.J.R. (ed.)]. Kluwer Academic Publishers, Dordrecht, Netherlands.

Price, M.F., 1994: Should mountain communities be concerned about climate change? In: *Mountain Environments in Changing Climates* [Beniston, M. (ed.)]. Routledge Publishing Company, London and New York, pp. 431-451.

Quezel, P. and M. Barbero, 1990: Les forêts méditerranéennes: problèmes posés par leur signification historique, écologique et leur conservation. *Acta Botanica Malacitana*, **15**, 145-178.

Quezel, P. and J. Schevok, 1981: Essai de mise en place de la zonation altitudinale des structures forestières. *Ecologia Medit.*, **8**, 389-410.

Rada, F., G. Goldstein, A. Azocar, and F. Torres, 1987: Supercooling along an altitudinal gradient in Espeletia schultzii, a caulescent giant rosette species. *J. Exper. Botany*, **38**, 491-497.

Rameau, J.C., D. Mansion, G. Dumé, A. Lecointe, J. Timbal, P. Dupont, and R. Keller, 1993: *Flore Forestière Française, Guide Ecologique Illustré*. Lavoisier TEC and DOC Diffusion, Paris, France, 2419 pp.

Rapp, A., 1960: Recent developments on mountain slopes in Kärkevagge and surroundings. *Geografiska Annaler*, **42**, 1-158.

Richter, L.K., 1989: *The Politics of Tourism in Asia*. University of Hawaii Press, Honolulu, HI.

Riebsame, W.E., 1989: *Assessing the Social Implications of Climate Fluctuations*. United Nations Environment Programme, Nairobi, Kenya.

Rizzo, B. and E. Wiken, 1992: Assessing the sensitivity of Canada's ecosystems to climatic change. *Climatic Change*, **21**, 37-54.

Roberts, L., 1989: How fast can trees migrate? *Science*, **243**, 735-736.

Robin, G. de Q., 1983: *The Climatic Record in Polar Ice Sheets*. Cambridge University Press, Cambridge, UK, 212 pp.

Robock, A., R. Turco, M. Harwell, T.P. Ackerman, R. Andressen, H.-S. Chang, and M.V.K. Sivakumar, 1993: Use of general circulation model output in the creation of climate change scenarios for impact analysis. *Climatic Change*, **23**, 293-335.

Romme, W.H. and M.G. Turner, 1991: Implications of global climate change for biogeographic patterns in the greater Yellowstone ecosystem. *Conserv. Biol.*, **5**, 373-386.

Rongers, W.A., 1975: Past Wangindo settlements in the Eastern Selous Game Reserve, Tanzania. *Notes and Records*, **77/78**, 21-26.

Rongers, W.A., 1993: The conservation of forest resources of Eastern Africa: past influences, present practices and future needs. In: *Biogeography and Ecology of the Rain Forests of Eastern Africa* [Lovett, J.C. and S.K Wasser (eds.)]. Cambridge University Press, Cambridge, UK.

Rupke, J. and M.M. Boer (eds.), 1989: *Landscape Ecological Impact of Climatic Change on Alpine Regions, with Emphasis on the Alps*. Discussion report prepared for European conference on landscape ecological impact of climatic change, Agricultural University of Wageningen and Universities of Utrecht and Amsterdam, Wageningen, Utrecht and Amsterdam, Netherlands.

Sælthun, N.R., 1992: Modeling hydrological effects of climate change. In: *Climate Change and Evapotranspiration Modelling* [Tallaksen, L. and K.A. Hassel (eds.)]. NHP Report No. 31, The Nordic Coordinating Committee for Hydrology, Oslo, Norway, pp. 73-80.

Salgado-Laboriau, M.L., 1979: Cambios climáticos durante el cuaternario tardio paramero y su correlación con las tierras tropicales calientes. In: *El Medio Ambiente de Páramo* [Salgado-Laboriau, M.L. (ed.)]. Centro de Estudios Avanzados, Caracas, Venezuela, pp. 67-78.

Salgado-Laboriau, M.L., 1984: Late quaternary palynologic studies in the venezuelan Andes. In: *Natural Environment and Man in Tropical Mountain Ecosystems* [Lauer, W. (ed.)]. Erdwissenschattliche Forschung XVIII, Friantz Steiner Verlag, Wiessbaden, Stuttgart, Germany, pp. 279-294.

Salgado-Laboriau, M.L. and C. Schubert, 1976: Palynology of holocene peat bogs from the central venezuelan Andes. *Palaeogeography, Palaeoclimatology, Palaeoecology*, **19**, 147-156.

Salgado-Laboriau, M.L. and C. Schubert, 1977: Pollen analysis of a peat bog from Laguna Victoria (Venezuelan Andes). *Acta Cientifica Venezolana*, **28**, 328-332.

Salgado-Laboriau, M.L., C. Schubert, and S. Valastro, 1977: Palaeocologic analysis of a late quaternary terrace from Mucubaji, Venezuelan Andes. *J. Biogeog.*, **4**, 313-325.

Salinger, M.J., J.M. Williams, and W.M. Williams, 1989: *CO_2 and Climate Change: Impacts on Agriculture*. New Zealand Meteorological Service, Wellington, New Zealand.

Sarmiento, G., 1986a: Ecologically crucial features of climate in high tropical mountains. In: *High Altitude Tropical Biogeography* [Vuilleumier, F. and M. Monasterio (eds.)]. Oxford University Press, Oxford, UK.

Sarmiento, G., 1986b: Los principales gradientes eco-climáticos en los andes tropicales. In: *Ecologia de Tierras Altas (Simposio)* [Forero, E., G. Sarmiento, and C. La Rotta (eds.)]. Anales del IV Congreso Latinoamericano de Botánica, Mededllin, Colombia, pp. 47-64.

Schubert, C., 1974: Late pleistocene Mérida glaciation, Venezuelan Andes. *Boreas*, **3**, 147-152.

Schubert, C., 1992: The glaciers of the Sierra Nevada de Mérida (Venezuela): a photographic comparison of recent deglaciation. *Erdkunde*, **46**, 58-63.

Schweingruber, F.H., 1988: A new dendroclimatic network for Western North America. *Dendrochronologia*, **6**, 171-180.

Senarclens-Grancy, W., 1958: Zur Glacialgeologie der Oetztales und seine Umgebung. *Mittl. Geol. Ges. Wien*, **49**, 257-314.

Serebryanny, L.R., N.A. Golodkovskaya, and A.V. Orlov, 1984: *Glacier Fluctuations and the Processes of Moraine Accumulation in Central Caucasus*. Nauka, Moscow, USSR, 216 pp.

Shi, Y., 1990: Glacier recession and lake shrinkage indicating the climatic warming and drying trend in Central Asia. *Acta Geographica Sinica*, **45**, 1-13.

Shugart, H.H., 1984: *A Theory of Forest Dynamics. The Ecological Implications of Forest Succession Models*. Springer-Verlag, New York, NY, 278 pp.

Skre, O., 1993: Growth of mountain birch (Betula pubescens Ehrh.) in response to changing temperature. In: *Forest Development in Cold Climates* [Alden, J., L. Mastrantonio, and S. Ødum (eds.)]. Plenum Press, New York, NY, pp. 65-78.

Slatyer, R.O., 1989: Alpine and valley bottom treelines. In: *The Scientific Significance of the Australian Alps* [Good, R. (ed.)]. Australian Alps Liaison Committee/Australian Academy of Science, Canberra, Australia, pp. 169-184.

Slatyer, R.O., P.M. Cochrane, and R.W. Galloway, 1984: Duration and extent of snow cover in the Snowy Mountains and a comparison with Switzerland. *Search*, **15**, 327-331.

Slaymaker, O., 1990: Climate change and erosion processes in mountain regions of Western Canada. *Mountain Research and Development*, **10**, 171-182.

Smith, A.P., 1981: Population dynamics of Venezuelan Espeletia. *Smithsonian Contributions in Botany*, **48**, 1-45.

Smith, R.B., 1979: The influence of mountains on the atmosphere. *Advances in Geophysics*, **21**, 87-230.

Smith, R.B., 1986: Current status of ALPEX research in the United States. *Bull. Am. Meteorol. Soc.*, **67**, 310-318.

Snaydon, R.W. and M.S. Davies, 1972: Rapid population differenciation in a mosaic environment. II. Morphological variation in Anthoxantum odoratum. *Evolution*, **26**, 390-405.

Snow, J.W., 1976: The climate of Northern South America. In: *World Survey of Climatology: Climates of Central and South America*, vol. 12 [Schwerdtfeger, W. (ed.)]. Elsevier, Amsterdam, Netherlands, pp. 295-403.

Solhaug, K.A., 1991: Long day stimulation of dry matter production in Poa alpina along a latitudinal gradient in Norway. *Holarct. Ecol.*, **14**, 161-168.

Solomina, O.N., 1995: Equilibrium line altitude shift in the Little Ice Age of the Former Soviet Union mountain glaciers. *Data of Glaciological Studies* (in press).

Solomina, O.N., O.A. Pomortsev, and M.N. Heifets, 1992: Fluctuations of humidity during the last 300 years revealed by dendrochronology. *Water Resources*, **4**, 79-84.

Solomon, A.M., 1986: Transient response of forests to CO_2-induced climate change: simulation modeling experiments in eastern North America. *Oecologia*, **68**, 567-582.

Steinhauser, F., 1970: Die säkularen Änderungen der Schneedeckenverhältnisse in Oesterreich. 66-67 *Jahresbericht des Sonnblick-Vereines*, 1970-1971, Vienna, pp. 1-19.

Street, R.B. and P.I. Melnikov, 1990: Seasonal snow, cover, ice and permafrost. In: *Climate Change: The IPCC Impacts Assessment* [McG. Tegart, W.J., G.W. Sheldon, and D.C. Griffiths (eds.)]. Australian Government Publishing Service, Canberra, Australia, Chapter 7.

Street-Perrott, M., 1994: Paleo-perspectives: changes in terrestrial ecosystems. *Ambio*, **22**, 37-43.

Surova, T.G., L.S. Troitskii, and Ya.-M.K. Punning, 1992: Palaeogeography and absolute chronology of Holocene of polar Urals. *Proceedings of Academy of Sciences of Estonia*, **24**, 79-84.

Tessier, L., J.-L. de Beaulieu, M. Couteaux, J.-L. Edouard, Ph. Ponel, Ch. Rolando, M. Thinon, A. Thomas, and K. Tobolski, 1993: Holocene palaeoenvironments at the timberline in the French Alps—a multidisciplinary approach. *Boreas*, **22**, 244-254.

Thinon, M., 1992: *L'Analyse Pédoanthracologique. Aspects Méthodologiques et Applications*. Ph.D. dissertation, University of Aix-Marseille III, France, 317 pp.

Thornwaite, C.W., 1954: Topoclimatology. In: *Proceedings of the Toronto Meteorological Conference*. Roy. Met. Soc., London, UK, pp. 227-232.

Troll, C., 1968: The cordilleras of the tropical Americas. Aspects of climatic, phytogeographical and agrarian ecology. In: *Geoecology of the Mountainous Regions of the Tropical Americas* [Troll, C. (ed.)]. Ferd. Dümmlers Verlag, Bonn, Germany, pp. 15-56.

UNCTED, 1992: Statement on mountains. In: *Earth Summit*. Kluwer Academic Publishers, Dordrecht, Netherlands, Chapter 13.

Van der Hammen, T., 1974: The pleistocene changes of vegetation and climate in tropical South America. *J. of Biogeography*, **1**, 3-26.

Van der Hammen, T., 1984: Datos eco-climáticos de la transecta Buritaca y alrededores (Sierra Nevada de Santa Marta). In: *La Sierra Nevada de Santa Marta (Colombia), Transecta Buritaca-La Cumbre* [Van der Hammen, T. and P. Ruiz (eds.)]. J. Cramer, Berlin, Germany, pp. 45-66.

Velitshko, A.A. and I.M. Lebedeva, 1974: The attempt of palaeoglaciological reconstruction for the Eastern Pamirs. *Data of Glaciological Studies*, **23**, 109-117.

VEMAP, 1995: Vegetation/Ecosystem Modeling and Analysis Project (VEMAP): Comparing biogeography and biogeochemistry models in a continental-scale study of terrestrial ecosystem responses to climate change and CO_2 doubling. *Global Biogeochemical Cycles* (in press).

Vonder Mühll, D. and P. Holub, 1992: Borehole logging in Alpine permafrost, Upper Engadine, Swiss Alps. *Permafrost and Periglacial Processes*, **3**, 125-132.

Wagner, M.R., 1990: Individual tree physiological responses to global climate scenarios: a conceptual model of effects on forest insect outbreaks. In: *Proceedings of the 1990 Society of American Foresters National Convention*, Washington, DC, pp. 148-153.

Wang, M., G. Zhu, Z. He, and S. Zheng, 1988: The mountain systems in China. *Mountain Research*, **4**, 67-74.

Wang, M., G. Zhu, Z. He, and S. Zheng, 1986: *Mountains in China*. Sichuan Science and Technology Press, Beijing, China.

Wang, Z., 1993; The glacier variation and influence since little ice age and future trends in northwest region, China. *Scientia Geographica Sinica*, **13**, 97-104.

Warren, H.E. and S.K. LeDuc, 1981: Impact of climate on energy sector in economic analysis. *Journal of Applied Meteorology*, **20**, 1431-1439.

Webb III, T., J.E. Kutzbach, and F.A. Street-Perrot, 1985: 20,000 years of global climate change: paleoclimatic research plan. In: *Global Change* [Malone, T.F. and J.G. Roederer, (eds.)]. Cambridge University Press, Cambridge, UK, pp. 182-218.

Webb III, T., P.J. Bartlein, S.P. Harrison, and K.H. Anderson, 1993: Vegetation, lake levels and climate in eastern North America for the last 18,000 years. In: *Global Climates Since the Last Glacial Maximum* [Wright, H.E., J.E. Kutzbach, T. Webb III, W.F. Ruddiman, F.A. Street-Perrott, and P.J. Bartlein (eds.)]. University of Minnesota Press, Minneapolis, MN, pp. 415-467.

Weingarten, B., R.F. Yuretich, R.S. Bradley, and M.L. Salgado-Laboriau, 1990: Characteristics of sediments in an altitudinal sequence of lakes in the Venezuelan Andes: climatic implications. *J. of South American Earth Sciences*, **3**, 113-124.

WGMS, 1993: *Glacier Mass Balance*, Bulletin No. 2 [Haeberli, W., E. Herren, and M. Hoelzle (eds.)]. World Glacier Monitoring Service, ETH Zurich, Switzerland, 74 pp.

Whetton, P.H., 1994: Climate change and snow-cover duration in the Australian Alps. In: *Proc. International Symposium on Snow and Climate*, Geneva, September 22-23, 1994, University of Geneva, pp. 30-31.

White, F., 1983: *The Vegetation of Africa*. UNESCO, Paris, France, 356 pp.

Williams, R.J., 1987: Patterns of air temperature and accumulation of snow in subalpine heathland and grassland communities on the Bogong High Plains, Victoria. *Australian Journal of Ecology*, **12**, 153-163.

Williams, R.J., 1990: Growth of subalpine snowgrass and shrubs following a rare occurrence of frost and drought in south-eastern Australia. *Arctic and Alpine Research*, **22**, 412-422.

Williams, R.J. and A.B. Costin, 1994: Alpine and subalpine vegetation. In: *Australian Vegetation* [Groves, R.H. (ed.)]. Cambridge University Press, Melbourne, Australia, 2nd ed., pp. 467-500.

Williams, M., G. Ingersoll, R. Sommerfeld, and D. Cline, 1992: High resolution measurements of snow covered area and snow water equivalence in the Colorado Front Range. In: *Program and Abstracts: Rocky Mountain Hydrological Research Center 47th Annual Meeting, 1992, August 15, Allenspark, CO*. Environmental Studies in Colorado and Wyoming, Fort Collins, CO, p. 5.

Wimbush, D.J. and A.B. Costin, 1983: Trends in drainage characteristics in the subalpine zone at Kosciusko. *Proceedings of the Ecological Society of Australia*, **12**, 143-154.

Witmer, U., P. Filliger, S. Kunz, and P. Kung, 1986: *Erfassung, Bearbeitung und Kartiering von Schneedaten in der Schweiz, Geographica Bernensia G25*. University of Bern, Bern, Switzerland, 215 pp.

Wood, F.B., 1990: Monitoring global climate change: the case of greenhouse warming. *Bull. Am. Meteorol. Soc.*, **71**, 42-52.

Woodward, F.I., T.M. Smith, and W.R. Emanuel, 1995: A global primary productivity and phytogeography model. *Global Biogeochem. Cycles* (submitted).

Wright, H.E., J.E. Kutzbach, T. Webb III, W.F. Ruddiman, F.A. Street-Perrott, and P.J. Bartlein (eds.), 1993: *Global Climates Since the Last Glacial Maximum*. University of Minnesota Press, Minneapolis, MN.

Yang, Z., 1992: Glacier water resources and effect of glacial water in stream runoff in the Qilian Mountains. In: *Memoirs of Lanzhou Institute of Glaciology and Geocryology*. Chinese Academy of Sciences, No. 7, Beijing, China, pp. 10-20.

Yao, T. and Y. Shi, 1990: Fluctuations and future trends of climate, glaciers and discharge of the Urumqi river in Xinjiang. *Science in China*, **33**, 504-512.

Ye, B. and Z. Lai, 1992: Responses of glaciers to climate change. *Kexue Tongbao*, **37**, 1794-1797.

Yoshino, M., 1975: *Climate in a Small Area*. University of Tokyo Press, Tokyo, Japan, 549 pp.

Yoshino, M., H. Makita, K. Kai, M. Kobayashi, and Y. Ono, 1980: Bibliography on mountain geoecology in Japan, climatological notes. *Tsukuba*, **25**, 1-111.

Yoshino, M., T. Horie, H. Seino, H. Tsujii, T. Uchijima, and Z. Uchijima, 1988: The effects of climatic variations on agriculture in Japan. In: *The Impact of Climatic Variations on Agriculture*. Vol. 1, *Assessments in Cool Temperature and Cold Regions* [Parry, M.L., T.R. Carter, and N.T. Konijn (eds)]. Kluwer Academic Publishers, Dordrecht, Netherlands, pp. 723-868.

Zhang, X., 1992: Climatic series derived from glaciers. In: *Climatic Change and Its Impact in China* [Li, K. (ed.)]. China Ocean Press, Beijing, China, pp. 103-139.

Zhang, Y. and J. Song, 1993: The potential impact of climate change on the vegetation in Northeast China. In: *Climate Change and Its Impact*. Institute of Geography, CAS, Global Change Study No. l, Series Publication, Meteorological Press, Beijing, China, pp. 194-203.

Zhang, Y. and L. Liu, 1993: The potential impact of climate change on vegetation distribution in Northwest China. In: *Climate Change and Its Impact*. Institute of Geography, CAS, Global Change Study No. 1, Series Publication, Meteorological Press, Beijing, China, pp. 178-193.

Zhao, M., 1993: Impact of climate change on physical zones of China. In: *Climate Change and Its Impact*. Institute of Geography, CAS, Global Change Study No. l, Series Publication, Meteorological Press, Beijing, China, pp. 168-177.

Zhao, S., 1983: Physical feature of China's mountain environment and economic problem of its uti lization. *Mountain Research*, **1**, 1-9.

6

Non-Tidal Wetlands

M.G. ÖQUIST, SWEDEN; B.H. SVENSSON, SWEDEN

Principal Lead Authors:
P. Groffman, USA; M. Taylor, USA

Contributing Authors:
K.B. Bartlett, USA; M. Boko, Benin; J. Brouwer, Holland; O.F. Canziani, Argentina; C.B. Craft, USA; J. Laine, Finland; D. Larson, USA; P.J. Martikainen, Finland; E. Matthews, USA; W. Mullié, Holland; S. Page, UK; C.J. Richardson, USA; J. Rieley, UK; N. Roulet, Canada; J. Silvola, Finland; Y. Zhang, China

CONTENTS

EXECUTIVE SUMMARY

Although there are many different systems for defining and classifying wetlands, for this chapter we define wetlands generally as areas of land where the water table is at or near the surface for some defined period of time, leading to unique physiochemical and biological processes and conditions characteristic of waterlogged systems. Wetlands exist in both inland and coastal areas, covering approximately 4–6% of the Earth's land surface. They are found on every continent except Antarctica and in every climate from the tropics to the tundra. This chapter examines the possible impacts of climate change on non-tidal (primarily inland) freshwater wetlands.

Wetlands have many functions that have socioeconomic benefits: They provide refuge and breeding ground for many species, including commercially valuable ones; they are areas of high biodiversity; they control floods and droughts and improve water quality; and they are used for recreation and education. The direct economic value of these benefits varies between regions.

Human activities—such as the conversion of wetlands to agricultural and forest lands, construction of dams and embankments, and peat mining—already pose a serious threat to wetlands worldwide. Mainly as a result of these activities, it is estimated that more than half of the world's wetlands have disappeared during the last century. These anthropogenic effects are most notable in densely populated areas and are expected to increase, especially in developing countries.

We are highly confident that climate change will have its greatest effect on wetlands by altering their hydrologic regimes. Any alterations of these regimes will influence biological, biogeochemical, and hydrological functions in wetland ecosystems, thereby affecting the socioeconomic benefits of wetlands that are valued by humans. Due to the heterogeneity of nontidal wetlands, and because their hydrologic conditions vary greatly within and among different wetland types and sites, the impacts of climate change on these ecosystems will be site-specific. Impacts can be generalized for specific wetland types and, to some degree, wetland regions. However, generalization across wetland types is difficult and cannot be made in terms of locations or wetland categories.

We are highly confident that hydrologic changes or other disturbances that change the vegetation types in wetland areas will affect other wetland functions as well. However, many wetlands have inherently high spatial and temporal variability in plant communities due to climatic variations (e.g., seasonal flooding or drought) and variations in microtopography. We are confident that in some wetlands a changed plant community as a result of climate change will resemble at least some component of the existing community.

We are highly confident that climate change will affect the cycling of carbon in wetlands: Some carbon-sequestering wetlands will change from CO_2 sinks to sources due to a lowering of the water table or increased temperature. Changes in the source/sink relationship of wetlands have already occurred in parts of the arctic region. Climate change leading to an alteration in the degree of saturation and flooding of wetlands would affect both the magnitude and the timing of CH_4 emissions. Drying of northern wetlands could lead to declines in CH_4 emissions.

We are confident that climate change will affect the areal extent and distribution of wetlands, although at present it is not possible to estimate future areal size and distribution of wetlands from climate-change scenarios. Regional studies from east China, the United States, and southern Europe indicate that the area of wetlands will decrease if the climate becomes warmer. Climate warming also would have severe impacts on wetlands in arctic and subarctic regions in this respect because it would result in a melting of permafrost, which is the key factor in maintaining high water tables in these ecosystems.

Adaptation, conservation, and restoration of wetlands in response to climate change varies among wetland types and the specific function being considered. For regional and global functions (e.g., trace-gas fluxes and carbon storage), there are no human responses that can be applied at the scale necessary. For wetland functions that are local in scale (habitat value, pollution trapping, and to some degree flood control), possibilities exist for adaptation, creation, and restoration. However, wetland creation and restoration technologies are just developing, and we do not yet have reliable techniques to create wetlands for many specific purposes.

6.1. Introduction

6.1.1. *Aims and Goals of the Chapter*

This chapter examines the potential impacts of climate change on non-tidal wetland ecosystems and the possible options for responding to these changes. Tidal wetlands are covered in Chapter 9.

This chapter gives particular emphasis to the possible impacts of climate change on the areal extent, distribution, and functions of non-tidal wetlands, in the context of other natural or anthropogenic stressors that are likely to affect these ecosystems simultaneously. In addition to describing the importance of different climate variables and the range of factors that determine the sensitivity of individual wetlands, the chapter uses four case studies to illustrate the effects of climate change on certain defined wetland areas: the Sahel, northern boreal wetlands, Kalimantan (Indonesia), and the Florida Everglades.

This is the first time that IPCC has attempted a detailed assessment of the potential impacts of climate change on the structure and function of wetlands. Previous assessments briefly touched upon wetlands in a qualitative discussion of methane (CH_4) sources and sinks (Melillo *et al.*, 1990) and discussed wetlands in the context of ecosystem responses to increased CO_2 concentrations, illustrated by case studies of the arctic tundra and a salt marsh.

The present assessment is hindered because the literature on wetlands is highly variable in quality and coverage and large gaps in knowledge remain regarding many of their regulating processes. In particular, relatively few studies exist on the impacts of climate change on inland wetlands; most that do exist have been carried out on specific wetland sites and/or have tended to focus on the Northern Hemisphere. These factors are reflected in the examples and conclusions in this chapter and in the emphasis on case studies. Recently, wetlands and wetland-related topics have begun to receive increasingly greater attention, and new information is expected to be published in the near future.

6.1.2. *Definition*

Wetlands exist in both inland and coastal areas, covering approximately 4–6% of the Earth's land surface. A wide variety of wetland definitions are found in the literature. Cowardin *et al.* (1979) argue that there is no single, correct, indisputable, ecologically sound definition for wetlands, primarily because of the diversity of wetlands and because the demarcation between dry and wet environments lies along a continuum. In general, a wetland describes any area of land where the water table is at or near the surface for some defined period of time, leading to unique physiochemical and biological processes and conditions characteristic of shallowly flooded systems (Mitsch and Gosselink, 1993). This chapter will discuss both permanent and temporary wetlands.

6.1.3. *Classification*

Wetlands usually are categorized according to their characteristic vegetation; their location (coastal or inland); the salinity of the water they contain; or other biological, chemical, hydrological, and geographical features. Coastal wetlands are influenced by the ebb and flow of tides and may include tidal salt marshes, tidal freshwater marshes, and mangrove swamps (see Chapter 9).

This chapter covers inland wetlands, or those not subject to tidal influences—including peatlands, swamps, marshes, and floodplains. Peatlands consist of bogs and fens, which may be forested, and are peat-accumulating wetlands in moist climates (peat is partially decomposed plant material). Bogs are acidic, poor in nutrients, and receive water from precipitation only, whereas fens are generally circumneutral, richer in nutrients, and receive water primarily from overland flow and/or groundwater. Swamps or forested wetlands are areas with little or no peat accumulation. Marshes or herbaceous wetlands and floodplains are flooded areas along rivers or lakes (Zoltai and Pollet, 1983).

More than seventy global classification schemes exist internationally. Because the response of wetlands to climate change tends to be site- or region-specific, no existing scheme is useful for this chapter in relating geographic or physical features with climate responses. For this reason, this chapter will focus on describing the climate and other variables that determine the response of individual wetland sites, rather than attempting to correlate responses with particular wetland types.

Many studies have shown that hydrologic parameters are strong controllers of wetland ecosystem structure and function (Gosselink and Turner, 1978; Novitzki, 1989; Kangas, 1990). The source, renewal rate, and timing of the water regime directly control the spatial and temporal heterogeneity of wetland ecosystem structure and function. The hydroperiod—defined as the depth, frequency, duration, and season of flooding—is usually the single most important regulator in wetlands, controlling many of their important characteristics (Lugo *et al.*, 1990a). The hydroperiod is determined by the climate, topography, catchment area, soils, and geology of the region in which the wetland is situated (Armentano, 1990).

For this assessment, we focus on climate-change effects on hydrology as an integrative tool for our analysis. However, these effects are highly site-specific, and there are few general, categorical conclusions that can be drawn. There is extreme hydrological variation between and even within individual wetlands, such as differences in the direction of water flow (vertical, unidirectional, or bidirectional; Lugo *et al.*, 1990b). This variability, coupled with the resolution at which these hydrological differences can be found, reinforces the need to describe wetland responses on a site-by-site basis. It is possible to generalize impacts for specific wetland types and, to some degree, wetland regions, but it is difficult to generalize across different wetland types.

6.1.4. *Global Distribution of Wetlands*

Wetlands are found on every continent except Antarctica and in every climate from the tropics to the tundra (Mitsch and Wu, 1995; Mitsch and Gosselink, 1993). Matthews and Fung (1987) recently conducted extensive surveys to determine the distribution of wetlands on a global scale and estimate that wetlands account for an area of 5.3 x 10^6 km^2, or approximately 4% of the Earth's land surface (Figures 6-1 and 6-2). This estimate is similar to other recent estimates (e.g., Aselmann and Crutzen, 1989) but indicates a possible reduction from previous estimates of around 6% (Bazilivich *et al*, 1971; Maltby and Turner, 1983). However, any estimate of global coverage will depend significantly on the definition of a wetland that is used.

6.1.5. *Current Wetland Stressors*

Wetlands already are threatened by a range of environmental factors, which can be natural or anthropogenic. It is estimated that more than half of the world's wetlands have disappeared since 1900. In the lower 48 states of the United States, approximately 53% of the original wetland area has been lost; 87% of this loss is attributed to agricultural development, 8% to urban development, and 5% to other conversions (Maltby, 1986). The same is valid for most of the developed regions of the world. The status of wetlands in developing countries is currently unknown to a large extent, but population pressures in many regions are steadily increasing the demand for food (Dugan, 1988), which can lead to wetland loss due to agricultural development. Many wetlands, especially in tropical regions, have so far escaped the impacts of human activities owing to their remoteness and unsuitability for agriculture (see Section 6.5.4). However, in recent decades, population pressures and technological advances have extended human influences into previously undisturbed areas (Armentano, 1990). For example, in 1989 it was calculated that only 82% of Indonesia's peat swamp forests remained in their original condition (Silvius, 1989); for some provinces (e.g., South Sumatra), it is predicted that no swamp forest will be left by the year 2000 (PHPA and AWB, 1990). Table 6-1 summarizes the main causes of present-day wetland loss.

6.2. Global Importance of Wetlands

Wetlands have many functions that are considered to have socioeconomic value: They provide refuge and breeding ground for many species, including commercially valuable furbearers, waterfowl, and timber; they often contain a high diversity of species; they control floods and droughts and improve water quality; and they can be used for recreation and education. The socioeconomic value of wetlands will vary from region to region, depending on which wetland functions the local economies regard as valuable. Table 6-2 identifies wetland types with their values.

Some wetlands (usually peatlands) contain potential energy for human consumption. In developing countries with shortages of energy and fuel, peat harvesting can be an attractive financial proposition if extensive peat deposits are available. This can

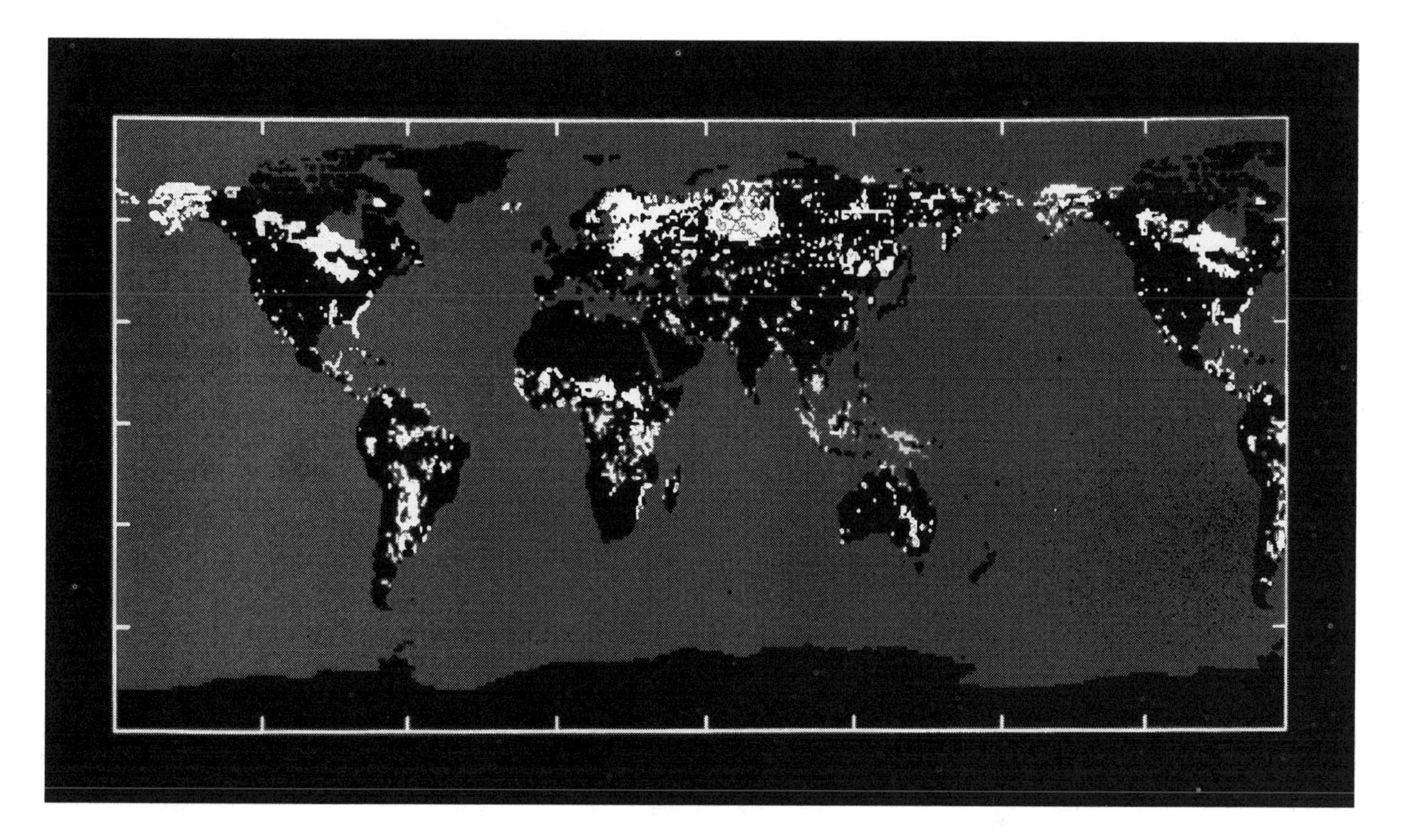

Figure 6-1: Global distribution of wetland ecosystems (modified after Matthews and Fung, 1987). Lighter areas denote wetlands.

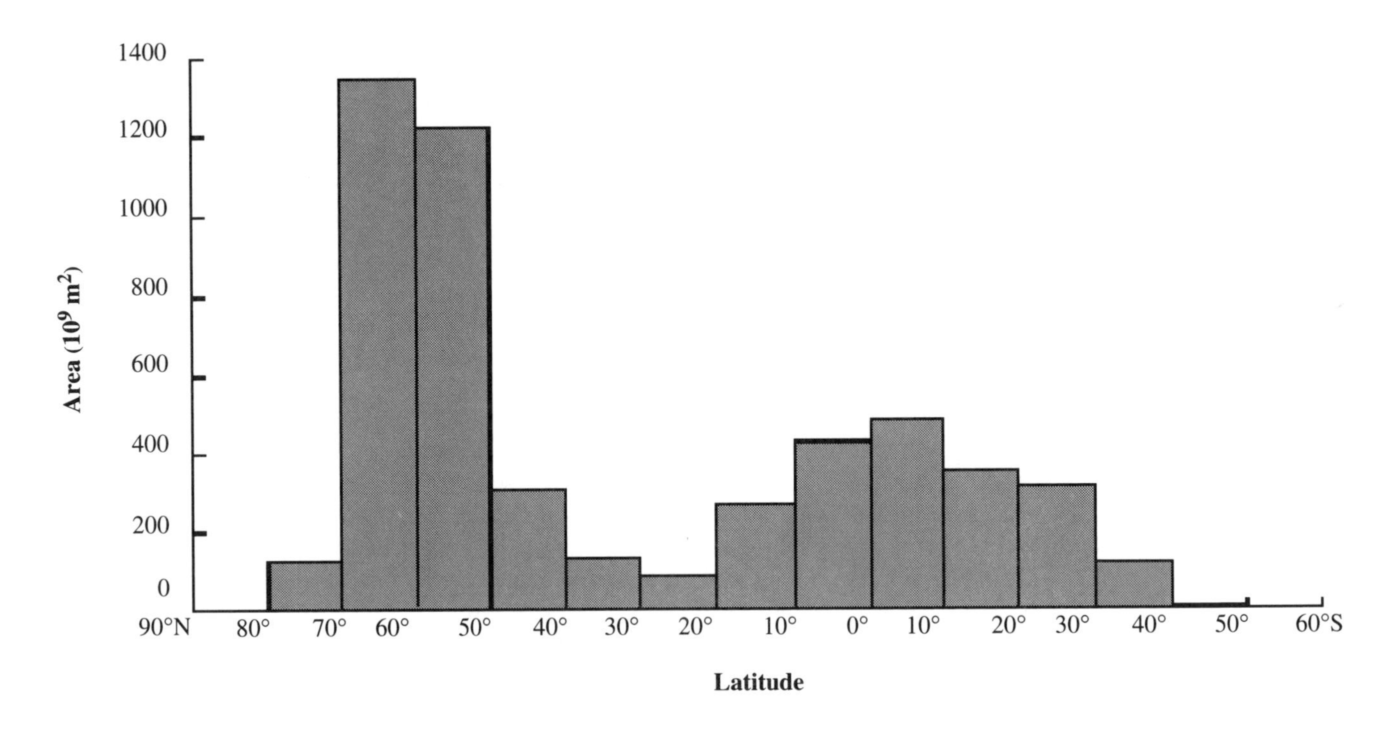

Figure 6-2: Distributions of wetland types along 10° latitudinal belts (modified after Matthews and Fung, 1987).

have the effect of replacing imported energy sources and reducing foreign-exchange requirements (Bord na Móna, 1985). However, large-scale harvesting of peat has led to the destruction of peatland ecosystems, and peat mining has the dual effect of removing a CO_2 sink (but a source of CH_4) and adding to the CO_2 in the atmosphere (see Rodhe and Svensson, 1995).

In the recent past, nonconsumptive benefits of wetlands such as recreation (Mercer, 1990), archaeology (Coles, 1990), education, and science usually were given lower priority in management plans than directly consumptive values because they are highly aesthetic and their values are difficult to quantify (Reimold and Hardinsky, 1979). However, these values have been given much greater attention in recent years, and wetlands worldwide are starting to be considered highly valuable areas to conserve.

6.2.1. *Habitats and Diversity*

The total biodiversity (flora and fauna) of wetlands generally is high in comparison with terrestrial ecosystems. Wetlands provide protective cover and essential feeding, breeding, and maturation areas for a wide range of invertebrates, as well as cold- and warm-blooded vertebrates (Gosselink and Maltby, 1990; Clark, 1979). Some animals are entirely dependent on wetland habitats; others are only partially so. In North America, for example, muskrats and beavers fall into the dependent group, whereas raccoons and various species of deer fall into the nondependent one. In many areas, the remoteness and inaccessibility of wetlands has attracted species that may not be totally wetland dependent but take advantage of the protection and shelter they provide. For example, the Pantanal in Brazil, Paraguay, and Bolivia provides an important habitat for the jaguar (Dugan, 1993).

Probably the best known function of wetlands is as a provider of year-round habitats, breeding areas, and wintering sites for migratory birds, depending on their location (Bellrose and Trudeau, 1988; Dugan, 1993). One example is the prairie pothole region in the United States, where the value of the region as habitat for breeding birds, especially waterfowl, has been thoroughly documented. Annual production is correlated with the number of wet basins for many duck species (Boyd, 1981; Krapu *et al.*, 1983; Reynolds, 1987). More than half of the waterfowl production in North America occurs within this region (Batt *et al.*, 1989).

6.2.2. *Biogeochemical Values*

Wetlands play an important role in the global budgets of carbon (C) and the trace gases CH_4 and nitrous oxide (N_2O). Wetland soils represent a major pool of C that may respond dynamically to climate change (Woodwell *et al.*, 1995). Boreal wetlands, which are extremely susceptible to climate change, are a major contributor to the global CH_4 budget (Matthews and Fung, 1987).

Wetlands are efficient in trapping pollution and processing wastes in human-dominated landscapes. Wetlands have been found to be important "sinks" for pollutants moving from upland areas, preventing their movement into surface water and groundwater (Mitsch and Gosselink, 1993); indeed, artificial

Table 6-1: Causes of wetland losses (modified after Dugan, 1990).

	Floodplains	Marshes	Peatlands	Swamps
Human Actions				
Drainage for agriculture, forestry, and mosquito control	++	++	++	++
Stream channelization for navigation and flood protection		+		
Filling for solid waste disposal, roads, etc.	++	++		
Conversion for aquaculture	+	+		
Construction of dikes, dams, and levees	++	++		
Discharge of toxic compounds and nutrients	++	++		
Mining for peat, coal, gravel phosphate, etc.	+		++	++
Groundwater abstraction	+	++		
Sediment diversion by dams, deep channels, etc.	++	++		
Hydrological alterations by canals, roads, etc.	++	++		
Subsidence due to extraction of groundwater, oil, gas, minerals, etc.	++	++		
Natural Causes				
Subsidence			+	+
Sea-level rise				++
Drought	++	++	+	+
Hurricanes and storms			+	+
Erosion	+		+	
Biotic effects	++	++		

Notes: ++ = common and important cause of wetland loss; + = present but not major cause of wetland loss.

Table 6-2: Wetland values (modified after Dugan, 1990).

	Floodplains	Marshes	Peatlands	Swamps
Function				
Groundwater recharge	++	++	+	+
Groundwater discharge	+	++	+	++
Flood control	++	++	+	++
Erosion control	+	++		
Sediment/toxicant retention	++	++	++	++
Nutrient retention	++	++	++	++
Biomass export	++	+		+
Storm protection	+			
Microclimate stabilization	+	+		+
Water transport	+			
Recreation/tourism	+	+	+	+
Products				
Forest resources	+			++
Wildlife resources	++	++	+	+
Fisheries	++	++		+
Forage resources	++	++		
Agricultural resources	++	+	+	
Water supply	+	+	+	+
Attributes				
Biological diversity	++	+	+	+
Uniqueness to culture	+	+	+	+

Notes: ++ = common and important value of that wetland type; + = less common and important value.

wetlands are being used to treat wastewater. Declines in these functions due to climate change could have important economic and aesthetic implications, particularly in heavily developed areas (Arheimer and Wittgren, 1994).

6.3. Sensitivities and Impacts

6.3.1. Which Wetlands are Most Vulnerable to Climate Change?

It is difficult to determine the vulnerability of specific types of non-tidal wetlands to climate change. One line of reasoning suggests that wetlands in naturally stressed environments appear to tolerate less additional stress than those located in favorable conditions (Lugo and Brown, 1984), meaning that they are more vulnerable to the alterations in hydrological regimes that are expected to result from climate change. By this reasoning, depressional wetlands (found in depressions in the landscape) with small watershed areas and situated in areas where the climate is either dry or wet at present will be most susceptible to these effects (Mitsch and Wu, 1995). In contrast, wetlands along floodplains and lakes should be able to adapt to a changing climate by migrating along river edges up- and downstream as well as up- and downslope to follow water—although the efficiency of such migration will be dependent on a number of factors, including catchment area, topography, and human settlements.

A second line of reasoning suggests that wetland types that have a large degree of inherent exposure to high spatial and temporal variation in environmental conditions may have a greater potential for adaptation to climate change (see Section 6.4).

Arctic and subarctic peatlands will be extremely vulnerable to climate change if warmer temperatures lead to a thawing of the permafrost layer and affect their hydrology through drainage or flooding (Gorham, 1994; Oechel and Vourlitis, 1994; OTA, 1993). These wetlands have a limited capacity to adapt to climate change because it is unlikely that new permafrost areas will form. In addition, non-tidal wetlands located near the coast are vulnerable to changes in climate due to sea-level rise, which would have severe impacts resulting from chemical and hydrological changes caused by intrusion of saline seawater (see Section 6.5.5).

6.3.2. Importance of Different Climate Variables

No single factor determines how climate change will affect individual wetland ecosystems. The variables that are predicted to change include temperature, precipitation, and CO_2 concentrations, resulting cumulatively in changes in water availability. Changes in the frequency and duration of flooding and drought and any alterations in disturbance regimes will be particularly important in determining how the ecological functions of wetlands ultimately are affected.

6.3.2.1. Temperature

Temperature is an important factor controlling many of the ecological and physical functions of wetlands. Primary productivity, microbial activity, and habitat are all controlled to a certain extent by temperature. Temperature also affects evapotranspiration rates, which has implications for the hydrological regime of wetlands by transporting water from the ecosystem to the atmosphere.

6.3.2.2. Precipitation

Precipitation regulates the direct inflow and amount of water to wetland ecosystems. However, the effect of a change in precipitation on a given wetland will depend on the type of wetland and the topographic and geographic characteristics of the region (drainage area, relief, and so forth). For example, very large wetlands, like the Okawango delta in Africa, are supplied with water from a considerable distance. In this case, the spatial variability of climate change could affect the balance between supply and evaporative demand. Further, in wetlands located along floodplains, a change in water availability throughout the drainage area or region will affect flooding and the hydrological regime in complex ways.

Poiani *et al.* (1995) show that the seasonality of precipitation changes is very important to wetland ecosystems. Climate change affecting spring precipitation and runoff may have the greatest impact on wetland hydrology and vegetation. Some modeling studies have indicated that there may be a threshold temperature beyond which changes in precipitation become less important to wetland hydrology. In one study, Poiani and Johnson (1993) conclude that precipitation changes are much less influential on hydrological regimes under a +4°C scenario than under a +2°C scenario.

6.3.2.3. CO_2 Concentration

Current research suggests that elevated CO_2 levels may have a direct fertilization effect on some types of vegetation, leading to higher production rates (Idso and Kimball, 1993). Elevated CO_2 concentrations seem to increase plant tolerance to stress, including photoinhibition, high or low temperature extremes, drought, and waterlogging (Hogan *et al.*, 1991). However, multi-species, intact ecosystems may show complex responses to an increase in CO_2 concentrations (Korner and Arnone, 1992). Elevated CO_2 concentrations in the atmosphere also may affect the rate of evapotranspiration through changes in water usage by plants (Bunce, 1992; Kimball and Idso, 1983). However, the effects of increased CO_2 on transpiration and water-use efficiency appear to decrease as water availability increases or as temperature decreases (Oechel and Strain, 1985). Any changes in transpiration may influence regional water balances and hydrological regimes because vegetation is a critical component of the cycling of water between soils and the atmosphere (Salati and Vose, 1984). Until detailed mechanistic questions regarding the biochemistry and physiology of wetland vegetation can be resolved, the extent to which wetlands will respond to increased CO_2 concentrations remains uncertain (Larson, 1994).

6.3.2.4. *Extreme Events*

Wetlands are sensitive to extreme climate events such as heavy spring flooding and summer drought (Gorham, 1991). Potential changes in precipitation patterns related to the amplitude, periodicity, or frequency of extreme events are critical components in modeling wetland responses to climate change (Poiani and Johnson, 1993). Extreme droughts make wetlands more sensitive to fire (see Sections 6.5.4 and 6.5.5; Gorham, 1994), which could impact ecological functions such as vegetation cover, habitat value, and carbon cycling. At the same time, such biomass burning is likely to result in massive emissions of smoke particles (aerosols) to the atmosphere, which may offset some warming, at least regionally (Penner *et al.*, 1992). Sea-level rise may affect some inland wetlands, causing saltwater intrusion that could result in an encroachment of salt-tolerant wetland communities.

6.3.2.5. *Water Availability and Movement*

Combined changes in temperature, evapotranspiration, precipitation, and CO_2 concentrations ultimately affect the hydrologic regime in a wetland ecosystem. It is critical to consider these combined effects because responses can be nonlinear. For example, it has been found that relatively small differences in precipitation can produce substantial differences in water level, especially for smaller temperature increases (Poiani and Johnson, 1993). Whereas decreased water availability will cause wetland loss due to concomitant drying, climate models suggest that some areas will become wetter—particularly high-latitudes in the winter. Although few studies have examined the potential impacts in these cases, it is conceivable that wetter conditions could be conducive to wetland development in areas not currently occupied by wetlands—as has occurred in areas of the Sahel (see Section 6.5.2).

In some cases, the movement of water in itself can benefit wetlands because it contributes to nutrient transport, aeration, and so forth, and because species can more readily distribute themselves according to the hydrological conditions to which they are adapted. Floods also create currents capable of exporting toxic compounds that otherwise could accumulate in sediments (Lugo *et al.*, 1990a). However, extremely intense hydrological fluxes tend to stress wetland ecosystems.

6.3.3. *Effects on Wetlands Due to Climate Change*

Climate changes affecting the areal extent, distribution, and hydrological regimes of wetlands are expected to have important effects on their ecological functions. These effects must be assessed in light of many uncertainties, particularly concerning the interactions of atmospheric water, surface water, and groundwater components of wetlands (Winter, 1988).

6.3.3.1. *Areal Extent and Distribution*

At present, regional scenarios do not provide adequate information to determine the direction or magnitude of change in the areal extent of wetlands (Gorham, 1991). It seems likely that some wetland regions will become moisture-limited, while other non-wetland areas will develop a climate conducive for wetland development. At this time, however, any estimation of the change in global areal extent would be very uncertain (see Gorham, 1991).

A few studies have attempted to estimate possible regional changes in wetland distribution, although they tend to involve areas that are projected to experience net losses in water availability rather than areas that are expected to experience increases. For example, a study by Brock and van Vierssen (1992) of hydrophyte-dominated wetlands in southern semi-arid regions of Europe concludes that an increase in temperature of 3–4°C would decrease the areal extent of habitat for hydrophytes by 70 to 85% within 5 years—suggesting that wetlands in other semi-arid and arid regions will be very sensitive to climate warming (see Section 6.5.2).

Poiani and Johnson (1993) conducted a study to understand how possible changes in the areal extent of a semipermanent wetland in the prairie pothole region of the United States might affect its ecological functions. For this study, they developed a simulation model for hydrological and vegetation responses. Using output from the Goddard Institute for Space Studies (GISS) general circulation model (GCM) for current and doubled-CO_2 climates (mean monthly air-temperature increase of 3 to 6°C and precipitation ranging from -17% to +29%) in an 11-year simulation, they project a 3% increase in the overall size of this wetland under current climate but a decrease of nearly 12% under the greenhouse scenario. Further, the areas of open water would decrease from 51% at the beginning of the simulation to 0% by the fourth year, allowing emergent plant species to spread over the entire wetland. This change would have serious implications for the wildlife in the region because these areas are extremely important breeding areas for waterfowl (see Section 6.2.1). This study indicates that even if the total decrease in wetland area due to a warmer climate is relatively small (12%), the overall change in wetland characteristics can have severe effects on many of the existing functions and values of wetlands. The model also indicates that wetland size, depth, and vegetative characteristics are more sensitive to changes in temperature than to either increases or decreases in precipitation.

Another study of the prairie pothole region (Larson, in press) examines the relationship between climate variables and the percentage of wet basins, using a multiple linear-regression model. This study focuses on closed-basin wetlands surrounded by either grassland or aspen parkland. The study found that when temperatures increase by 3°C in subsequent model runs

(15 years simulated), the Canadian and U.S. grassland models project declines in the percentage of wet basins of 15% and 28%, respectively. The aspen parkland model, however, projects a decline of 56% in the number of wet basins with increased temperatures. Model response to changes in precipitation were uniform and small across the region. An important consequence of geographical differences in wetland response to temperature is that as waterfowl extend their migrations farther north in drought years, they may face decreased probability of finding suitable wetlands.

A study by Zhang and Song (1993) examines the possible effects of climate change on the areal extent of the wetlands of eastern China, where 75% of the country's wetlands are located. The study examines climate change under six hypothetical climate scenarios (precipitation increasing or decreasing by 10% and temperature increasing 1, 2, and 3°C); the areal extent of herbaceous wetlands declines under all scenarios.

At high northern latitudes, warming could cause a poleward migration of the northern treeline. This shift would decrease the winter albedo because the tree canopy has much lower albedo than exposed snow surfaces (Bonan *et al.*, 1992), greatly affecting regional climate by absorbing more of the sun's incoming energy.

Another implication for northern peatlands is the expected melting of permafrost due to higher temperatures (see Chapter 7). Harriss (1987) concludes that an increase in temperature of 2°C would shift the southern boundary of permafrost in the Northern Hemisphere to the north. The melting of permafrost is likely to have drastic effects on peatland hydrology and landscape patterns, leading to lowered water tables in some areas and flooded thaw lakes in others, as well as to thermokarst erosion (Billings, 1987; Gorham, 1991). It has been suggested (Zoltai and Wein, 1990) that such melting may shift bogs on permafrost back to fens, from which they originated after the warm mid-Holocene period. Vegetetation could shift from black spruce/Sphagnum/lichen (which are typical for bog ecoystems) to grasses, sedges, and reeds. Further, the rate of climate-change impacts on northern peatlands may be such that it causes degradation of southern regions much faster than the northern regions can expand northward (Gorham, 1991). These shifts also would have implications for the carbon cycle, as well as the flux of especially CH_4 from northern peatlands (see Section 6.5.3).

6.3.3.2. Functions

Functions are processes necessary for the self-maintenance of ecosystems, such as primary production, nutrient cycling and decomposition. Wetland functions can be categorized as biological, biogeochemical, or hydrological. These are distinct from, but often translate into, the socioeconomic values perceived by society (Brinson, 1993). None of these categories is exclusive, and each may influence the other; for example, any changes in wetland plant species have far-reaching effects on a wide range of wetland functions due to the unique structural, chemical, and ecological characteristics of different plants.

6.3.3.2.1. Biological functions

Biological functions relate to vegetation, habitats, and species diversity. Climate changes resulting in increased or decreased temperature and water availability will affect the composition and production of vegetation, the quality and areal extent of habitat available for species, and species composition and diversity (Thompson and Hamilton, 1983; Junk, 1983, 1993; Bradbury and Grace, 1983; Reader, 1978; Bernard and Gorham, 1978).

The species composition of plant communities in wetlands is critically affected by water movement and hydroperiod (Lugo *et al.*, 1990a). Often, extended flooding and a longer hydroperiod will result in tree mortality and the replacement of forest by herbaceous vegetation (Lugo *et al.*, 1990a). A long-term lowering of the water table would probably lead to similar changes in the composition and production of the vegetation as found in drainages made for forestry (e.g., in Sweden and Finland).

In forested wetlands subjected to drainage, the lowering of the water table has resulted in an increase in tree-stand volume (Keltikangas *et al.*, 1986; Hånell, 1988). Based on data from several different wetland types given by Hånell (1988), Rodhe and Svensson (1995) calculate that the increase in tree-stand volume could range between 1.2–6.0 kg dry biomass/m^2. Similar results are reported in studies on wetlands in Finland (Ilvessalo and Ilvessalo, 1975).

Changes in wetland plant communities can have important effects on decomposition, nutrient cycling, and plant production functions. The nature and amount of plant litter production strongly influences wetland soil microbial populations (Melillo *et al.*, 1982; McClaugherty *et al.*, 1985; Bowden, 1987). Several studies have found strong links between wetland plant community types and microbial decomposition and nutrient-cycling processes in fens and bogs (Svensson, 1976, 1980; Svensson and Rosswall, 1984; Verhoeven *et al.*, 1990; Verhoeven and Arts, 1992; van Vuuren *et al.*, 1992, 1993; Koerselman *et al.*, 1993). Microbial decomposition of litter and the release or "mineralization" of nutrients contained therein enhances plant productivity and litter quality (nutrient content, degradability; Pastor, 1984). Different plant communities demonstrate different rates of nitrogen availability to plants, carbon storage, and microbial processing of pollutants (Rosswall and Granhall, 1980; Pastor *et al.*, 1984; Morris, 1991; Duncan and Groffman, 1994; Weisner *et al.*, 1994; Schipper *et al.*, 1994).

The physical and chemical characteristics of plants also strongly influence insect species and populations, which provide life-support functions for wetland-dependent birds, fish, and mammals (Mitsch and Gosselink, 1986; Kiviat, 1989). In conservation biology, there is intense interest in the effects of changes in plant species on the habitat value of wetlands (Bratton, 1982; Harty, 1986; Mooney and Drake, 1986; Center *et al.*, 1991; McKnight, 1993).

6.3.3.2.2. *Biogeochemical functions*

Biogeochemical functions include pollution trapping and waste processing, carbon cycling (Svensson, 1986; Armentano and Menges, 1986; Silvola, 1986; Sjörs, 1980; Gorham, 1990, 1991; Miller *et al.*, 1983; Marion and Oechel, 1993), and the flux of greenhouse gases (Svensson *et al.*, 1975; Svensson, 1976; Aselmann and Crutzen, 1989; Matthews and Fung, 1987; Bartlett and Harriss, 1993; Matthews, 1993; Bartlett *et al.*, 1989; Urban *et al.*, 1988; Freeman *et al.*, 1993; Martikainen *et al.*, 1993; Pulliam, 1993).

Wetlands, and especially peatlands, play a significant role in the carbon cycle and presently are net sinks of carbon. A recent compilation of estimates of the amount of carbon held in soil as organic matter (Woodwell *et al.*, 1995) gave a mean of 1,601 Gt, of which about 20% (412 Gt of C) is stored in peatlands. Estimated average accumulation rates for boreal and subarctic peatlands range from 0.05–0.11 Gt C/yr (Armentano and Menges, 1986; Silvola, 1986; Sjörs, 1980; Gorham, 1990, 1991; Miller *et al.*, 1983; Marion and Oechel, 1993). However, some peats may have reached a balance between the degradation and addition rates of organic matter (Malmer, 1992; Warner *et al.*, 1993).

Laboratory experiments and field studies have shown that lowering the water table by 20–30 cm could increase CO_2 fluxes from peat soil 1.5–2.5-fold (Silvola *et al.*, 1985; Moore and Knowles, 1989). However, it seems that changes in the water table are more significant between 0–30 cm than between 30–60 cm (Silvola *et al.*, 1985). Any lowering of the water table would mean, on the average, an extra carbon release of about 100–300 g C/m^2/yr. Further, an increase in temperature of 1–5°C in northern peatlands could decrease carbon accumulation by 10–60% due to enhanced microbial activity (see Section 6.5.3). Very little is known about peatlands in tropical regions, and there is some disagreement about whether tropical peatlands at present function as net carbon sinks or sources (Immirzi and Maltby, 1992; Sorensen, 1993; Sieffermann *et al.*, 1988; see Section 6.5.4). They are, however, large carbon stores, and any significant changes in the degree of storage have implications for carbon cycling.

In the recent geologic past, the tundra was a sink of 0.1–0.3 Gt C/yr (Miller *et al.*, 1983; Marion and Oechel, 1993). However, recent climatic warming in the arctic (see Lachenbruch and Marshall, 1986; Chapman and Walsh, 1993), coupled with the concomitant drying of the active layer and the lowering of the water table, has shifted areas of the arctic from sinks to sources of CO_2 (Oechel *et al.*, 1993). For example, arctic areas that were sinks of 0.1 Gt C/yr now are sources of 0.1–0.6 Gt C/yr (Oechel *et al.*, 1993; Zimov *et al.*, 1993). This illustration of a response of a northern wetland to warming suggests a major change in ecosystem function that may be an early indication of global change in a natural ecosystem.

Although wetland vegetation fixes atmospheric CO_2, biogeochemical processes give rise to other greenhouse gases, such as CH_4. The estimated contribution of wetlands to the annual atmospheric CH_4 burden is 55–150 Mt/yr (Prather *et al.*, 1994). However, the flux measurements on which these estimates are based are biased toward wetlands in northern North America and the Scandinavian countries (Bartlett and Harriss, 1993; Matthews, 1993). Climate change leading to an alteration in the degree of saturation and flooding of wetlands would affect both the magnitude and the timing of CH_4 emissions (see Section 6.5.3.2). Drying of northern wetlands could lead to declines in CH_4 emissions (Roulet *et al.*, 1993; Martikainen *et al.*, 1995). Although emissions of N_2O from wetlands usually are low (Urban *et al.*, 1988; Freeman *et al.*, 1993; Martikainen *et al.*, 1995), a lowering of the water table could increase emissions.

Climate change also can affect the chemical properties of wetlands. Some non-tidal wetlands are saline due to the surplus of evapotranspiration over precipitation. If rates of evapotranspiration increase, there is a risk that more salts will accumulate in these wetlands, which could impair the value of wetlands due to the loss of intolerant species. Wetlands that are presently freshwater could become saline. However, in regions facing a decrease in evapotranspiration, the existing saline wetlands could gradually change as salinity decreases. This will affect the chemical properties of the wetlands—which, in turn, has consequences for biological and ecological characteristics, affecting vegetation, habitat value, and species composition (including the invertebrate community; Swanson *et al.*, 1988).

An increase in temperature also will affect processes such as pollution trapping and waste processing because these processes are regulated by microbial activity and plant uptake, which are important sinks for nutrients (such as nitrogen and phosphorus) and various kinds of pollutants (heavy metals, pesticides and herbicides; Gilliam *et al.*, 1988; van der Valk *et al.*, 1979; Brinson, 1990; Brix and Schierup, 1989; Johnston, 1991; Mitsch and Gosselink, 1993). However, the key features to which these water-quality functions are connected are coupled to the water regimes of wetlands (Kadlec, 1989), so it is impossible to make any general statements about the impact of climate change on these functions.

6.3.3.2.3. *Hydrological functions*

Hydrological functions include flood control (Andriesse, 1988; Novitzki, 1979; Boelter, 1966; Gosselink and Maltby, 1990) and aquifer recharge (Bernaldez *et al.*, 1993). Wetlands temporarily store runoff water, thereby reducing floodwater peaks and protecting downstream areas (Andriesse, 1988; Novitzki, 1979; Boelter, 1966; Gosselink and Maltby, 1990). A reduction of wetland area due to climate change could severely hamper flood-control efforts in some regions. Under certain conditions, climate change could enhance recharge to major aquifers by overlying wetlands—particularly in arid or semi-arid regions, where groundwater is of considerable importance as a source for public water supply and irrigation (Bernaldez *et al.*, 1993; see Section 6.5.2). The contribution of wetlands to groundwater resources depends on the detention of water within the wetland during dry periods; this is likely to be reduced by loss of wetlands (Bernaldez *et al.*, 1993).

At northern latitudes, projected higher spring and winter temperatures would likely affect the amount and timing of runoff from snowmelt and rainfall by changing the patterns of soil freezing and thawing (Mimikou *et al.*, 1991). It also would decrease the ratio of snow to rain, leading to additional changes in spring snowmelt (Cohen, 1986; Gleick, 1987; Croley, 1990).

6.3.4. Interactions Between Climate Change and Other Wetland Stressors

The impacts of climate change on wetlands will interact with other anthropogenic stresses. Due to the importance of hydrological regime and wetland response, we confine ourselves to discussing the cumulative impacts of hydrological changes in wetlands. Apart from climate change, the most common disturbances to the hydrological regimes of wetland ecosystems are alterations in plant communities, storage of surface water, road construction, drainage of surface water and soil water, alteration of ground water recharge and discharge areas, and pumping of ground water. All of the anthropogenic activities and natural causes mentioned in Table 6-1 will, to various extents, impact the hydrology of wetlands. Drainage for agriculture, for example, would cause drastic changes in water level and means near-total destruction of wetland ecosystems, whereas construction of dams for water management could be less severe if efforts to maintain waterflow through the subjected wetlands are made. Due to site-specific responses by wetland ecosystems and the large range of plausible anthropogenic and natural stressors, a quantitative evaluation of them in combination with climate change is difficult. It is conceivable, however, that within the next decades the main threat to wetlands is likely to be due to anthropogenic activities rather than climate change.

6.4. Response Options—Adaptation, Conservation, and Restoration

The prospects for adaptation, conservation, and restoration of wetland ecosystems in response to climate change varies with wetland type and the specific wetland function being considered. For wetland functions that are an aggregated product of regional or global wetland resources (e.g., trace-gas fluxes, carbon storage), there are no human responses that can be applied at the necessary scale. Moreover, changes in these functions as wetlands adapt to climate change are difficult to predict due to the site-specific nature of wetland responses to climate change.

For wetland functions that are more local in scale (e.g., habitat value and pollutant absorption), prospects for adaptation, conservation, and restoration are better than for large-scale functions. However, these prospects will vary strongly with wetland type. Some wetland types have a higher potential for adaptation to climate change due to their inherent exposure to high spatial and temporal variation in environmental conditions. For example, the prairie pothole wetlands discussed in Section 6.3.3.1 respond dynamically and naturally to wide variation in seasonal and annual climate (van der Valk and Davis, 1978; Poiani and Johnson, 1989).

Other wetland types—for example, boreal peatlands—have high spatial variability in plant communities caused by variation in macro- and microtopography (Sjörs, 1950). Hotter, drier, and/or wetter climates will result in a change in the wetland community in these wetlands, but the natural spatial and temporal variation inherent in these systems suggests that the changed community will resemble at least some component of the existing community (Poiani and Johnson, 1989).

Given the site-specific nature of wetland responses to climate change and the importance of inherent variation in fostering potential for adaptation, wetland conservation and restoration efforts should focus on preserving this variation (McNeely, 1990; Leemans and Halpin, 1992; Peters and Lovejoy, 1992). Wetland restoration and creation technologies have a great potential for ameliorating the effects of climate change on wetland functions. However, wetland ecology is complex, and the enthusiasm for wetland creation and restoration has outpaced the scientific understanding and technological development needed to successfully create wetlands for specific purposes (Reed and Brown, 1992; van der Valk and Jolly, 1992). Key areas of concern in wetland creation are how to establish a persistent and resilient assemblage of desired wetland plants, and a lack of understanding of the relationships between different plant assemblages and a range of wetland functions—from microbial pollutant-attenuation mechanisms to sequestration of soil carbon to food-chain support (Zedler and Weller, 1990; Pickett and Parker, 1994).

6.5. Examples and Case Studies

6.5.1. Introduction

This section uses case studies from certain defined wetland areas and regions to demonstrate climate-change impacts on different wetland types and geographic locations. The areas selected for these studies are the Sahel region of Africa, northern wetlands, the Kalimantan of Indonesia, and the Florida Everglades of the United States. These different locations will provide insight into the uses and responses of peatlands, marshes, and floodplains, as well as measures that are presently being taken in some places to restore wetlands affected by human uses and alterations.

The purpose of the case studies is to illustrate the general concepts described in this chapter through specific examples and to identify factors that make wetlands more or less vulnerable to any possible changes. The case studies also illustrate the type of information needed for any given location in order to conduct an assessment of potential risks, as well as the specificity of the information that may be gleaned about possible impacts at this time at individual sites.

Most of the case studies deal with the local functions, uses, and benefits of the wetlands and how these will be affected by climate change: These wetlands are valuable to nearby populations as sources of water, as agricultural land, as habitat for species, and for their other hydrological functions. Some of the case studies deal with the biogeochemical functions of wetlands and the ways in which human alterations and climate-change impacts on wetlands could affect the cycling of CO_2, CH_4, and N_2O to the atmosphere, which would result in global-warming feedback.

6.5.2. *Case Study: The Sahel*

6.5.2.1. *Background*

The Sahelian wetlands are dynamic ecosystems (see Box 6-1). During the past 2 decades, new wetlands of up to 1,800 hectares have formed, while other wetlands have been degraded (Piaton and Puech, 1992; Brouwer and Mullié, 1994a). Wetlands in the Sahel include the floodplains of the large rivers and Lake Chad, as well as thousands of small permanent and temporary wetland ecosystems scattered throughout the region (see Sally *et al.*, 1994; Brouwer and Mullié, 1994a; Windmeijer and Andriesse, 1993). Some of these wetlands, such as the small valley bottoms, contain water only during runoff events.

The small wetlands in the Sahel are very important for agriculture (MHE-DFPP, 1991; MHE-Niger, 1991a; Brouwer and Mullié, 1994b) and are used for dry-season cropping, using moisture left in the soil after the floods have receded, or for small-scale irrigation. In the period 1984–91, between 42,000 and 64,000 hectares of wetlands were used each year in Niger for dry-season cropping, generating an annual income of \$200–\$4,300 per hectare (MAE-Niger, 1993; Raverdeau, 1991; Cherefou Mahatan, 1994). In 1990, $4.1 x 10^6$ hectares of dryland cropping (mostly millet) in Niger generated an income of about \$70 per hectare. This difference in income per hectare is in part a reflection of the quality of the food produced. The small wetlands also have greater production of fish per hectare and greater density of birds than the large wetlands (Brouwer and Mullié, 1994b; Mullié and Brouwer, 1994).

6.5.2.2. *Societal Context*

The human carrying capacity of the Sahel region is already matched or exceeded by population density (van der Graaf and Breman, 1993); dryland agriculture or large-scale migration to other parts of the region are unlikely to be able to relieve the situation. As a result, wetlands will be more sought-after, and pressure for conversion of wetlands to rice fields should increase due to increasing urbanization in West Africa and its effects on the demand for rice.

Droughts also tend to increase pressure on wetlands because they affect the migration patterns of people in the area. During the severe droughts of 1975 to 1988, the number of villages on the Nigerian section of Lake Chad increased from 40 to more than 100 (Hutchinson *et al.*, 1992). Similarly, the use of the Hadejia-Nguru wetlands in Nigeria for agriculture has increased due to droughts. This increase was not foreseen when plans were made for the construction of dams and irrigation projects in the catchment upstream. Wet periods have traditionally meant migration to the normally drier and less-populated northern and western parts of the Département Tahoua, Niger. Because these people often stayed despite less-abundant

Box 6-1. The Wetlands of the Sahel: Effects on Agriculture, Habitat, and Hydrology

Background: The wetlands of the Sahel region of Africa consist of the floodplains of major lakes and rivers, as well as thousands of smaller wetland ecosystems scattered throughout the region. These smaller wetlands are particularly important for agriculture and as a source of income from agriculture; floodplains are important for their hydrological functions. The wetlands in the Sahel are already expected to come under increased pressure for conversion to agriculture and other uses due to urbanization and population growth projected over the next decades.

Possible impacts: With the possible exceptions of eastern Niger and Chad, climate change is expected to decrease the extent of wetlands in the Sahel, due to changes in temperature and precipitation projected by current scenarios. These changes are likely to result in a net loss of water in most of the large rivers in the Sahel over the next 30–60 years, with the exception of the major rivers flowing into Lake Chad. Although few studies exist on the effects on species, the loss of wetlands due to climate change could create a risk of extinction for some local populations of turtles and birds.

Conclusion: Although climate change could have some beneficial effects on the wetland regions of the Sahel, there would be many adverse impacts; some could be potentially irreversible. Even if the predicted decrease in rainfall in the western Sahel were followed by a recovery to present levels or more, part of the damage that is likely to occur in the interim would be very difficult to repair.

rainfall, they increased the pressure on natural resources, including wetlands (DDE-Tahoua, 1993).

6.5.2.3. *Climate Change Expected*

Rainfall in the Sahel region is greater in the north than in the south (see Nicholson, 1978, 1994; Hutchinson *et al.*, 1992). This analysis focuses on the wet season (June–August), when most dryland crops are grown. According to the IPCC Working Group II scenarios (Greco *et al.*, 1994), precipitation in the Sahel zone is projected to decrease by 2020 in the Senegal to Burkina area and increase moderately to considerably in Niger and Chad. Temperature is projected to increase by 0.5–1.5 (0–2.0)°C. For the decade around the year 2050, the three models show less agreement overall, although all indicate that eastern Niger is likely to receive more rain. Temperatures would be similar to those around the year 2020, though possibly somewhat higher in the area of the headwaters of the Senegal and Niger rivers. In the headwaters of the Komadougou Yobé river, rainfall around the year 2020 is expected to be somewhat less than at present. For the year 2050, the scenarios are less conclusive: In the area of the headwaters of the major rivers flowing into Lake Chad, rainfall around 2020 and 2050 is expected to be moderately to considerably greater than now.

6.5.2.4. *Effects on Water Availability and Vegetation*

Overall, these changes suggest that there will be less water in most of the large rivers in the Sahel over the next 30–60 years, with the exception of the major rivers flowing into Lake Chad. This will mean less water available in the floodplains along these rivers, unless there are changes to the management of outflow from dams. Changes to the hydrology of the small wetlands will depend not only on climate change but also on whether they are supplied with surface water or groundwater, on the reaction of the natural vegetation, and on the extent of cropping in their catchment areas.

Recharge to shallow, unconfined groundwater could either increase or decrease as a result of climate change. The groundwater level in southwest Niger increased in the last decade (Leduc *et al.*, in press; Bromley *et al.*, in press a), either as a recovery from the droughts of 1983–84 and/or 1973–74 or as a response to changes in land use. Annual recharge under millet in southwest Niger is on the order of 1–200 mm, a factor ten times greater than under bush and older fallow vegetation (Gaze *et al.*, in press; Bromley *et al.*, in press b). Therefore, an increase in the area sown for millet—as noted by Reenberg (1994) and others—could result in a higher recharge to unconfined groundwater and the wetlands fed by such groundwaters.

The dryland areas should experience less evaporative water loss because less rainfall, at least initially, means less perennial vegetation. However, there will be increased runoff until a new vegetative cover is established. This increase in runoff will likely result in increased erosion (see van Molle and van Ghelue, 1991) and a faster silting up of wetlands. It also could create entirely new wetlands.

Higher temperatures also may adversely affect seedling emergence of millet, the staple cereal over much of the Sahel (see Monteith, 1981)—meaning that less millet would be harvested in the dryland areas and increasing the pressure on wetlands. However, during dry years in eastern Niger, infiltration into the heavier soils in depressions may be so low that cropping becomes unattractive (Reenberg, 1994). During dry years, water is more limiting than nutrients, and cropping on less-fertile upland soils with higher infiltration rates may become more attractive.

Desertification upon drought may result in a loss of wetlands due to moving sands (DDE-Tahoua, 1993; Mahamane Alio and Abdou Halikou, 1993; Framine, 1994). In part, this response can be related to the greater vulnerability of perennial vegetation to desiccation under a monomodal semi-arid rainfall regime (Ellis and Galvin, 1994), when topsoil and seedbanks are washed or blown away during the drought. Droughts may make wetland vegetation more vulnerable to fires intended to improve rangeland vegetation (Hutchinson *et al.*, 1992).

6.5.2.5. *Effects on Biodiversity*

There are few assessments of how these changes could affect local biodiversity. Mullié and Brouwer (1994), following Gibbs (1993), suggest that small wetlands in the Sahel are important for the metapopulation dynamics of certain taxa—meaning that the loss of small wetlands may lead to a significant risk of extinction for local populations of turtles and small birds (Gibbs, 1993). Taxa that are easily transported by wind or birds as adults, eggs, cysts, larvae, and so forth would be subject to less risk (Dumont, 1992; Mullié and Brouwer, 1994; Magadza, 1994).

The importance of wetlands to birds in semi-arid areas can vary greatly from year to year depending on local and regional rainfall (Rose and Scott, 1994). If wetlands in the western Sahel become drier, relatively mobile birds dependent upon wetland habitats will move into wetlands further east (i.e., Niger, northern Nigeria and Cameroon, Chad).

6.5.3. ***Northern Wetlands: Effects on the Carbon Cycle and Trace-Gas Emissions***

6.5.3.1. *Peat Accumulation*

Peatlands are a major store of organic carbon and contain approximately 20% of the total amount of organic carbon stored in soils (see Section 6.3.3.2). A majority of this is in the Northern Hemisphere as carbon stored in the form of peat (see Box 6-2). Peat formation and accumulation in these wetlands is influenced by climate change: A change in climate would lead to changes in the flux of carbon (CO_2 and CH_4) between these ecosystems and the atmosphere, generating a feedback on climate warming.

Box 6-2. Northern Wetlands: Effects on the Carbon Cycle and Trace-Gas Emissions

Background: Peatlands are a major store of organic carbon and contain approximately 20% of the total amount of organic carbon stored in soils. The northern peatlands account for a majority of this as carbon is stored in the form of peat. This case study demonstrates how changes in climate may affect the flux of carbon (CO_2 and CH_4) and nitrous oxide between these ecosystems and the atmosphere, generating a feedback on climate warming.

Possible impacts: Current scenarios suggest that climate change is likely to increase the flux of CO_2 to the atmosphere because temperatures influence whether carbon litter is accumulated into the peat profile or oxidized. In addition, the position of the water table regulates the extent of oxygen penetration into the peat profile. This means that drainage will cause increased decomposition, leading to increased fluxes of CO_2 to the atmosphere, although this effect will decline over time. Further, a decrease in water availability could lead to a decrease in CH_4 emissions from wetlands. Changes in variables such as the areal extent of wetlands and the duration of the active period will determine whether there will be a change in the total CH_4 flux from a wetland. A lowering of the water table would probably not affect nitrous oxide emissions from bogs but could lead to an increase of emissions from fens, although emissions of nitrous oxide from wetlands tend to be low.

Conclusion: Changes in the source/sink relationship have already occurred in wetlands in some parts of the world. Both climate change and human (non-climate) factors are likely to further affect the biogeochemical functions of wetlands.

Wetland vegetation fixes CO_2 from the atmosphere and eventually is added to the top layers of the wetland soil as organic litter. Part of the organic litter is oxidized and emitted as CO_2, and some is accumulated as peat. Several investigations have shown that soil CO_2 efflux from peatlands is strongly related to temperature (Svensson *et al.*, 1975; Svensson, 1980; Glenn *et al.*, 1993; Crill, 1991), although Moore (1986) found a poor correlation between temperature and CO_2 emission rates. Since most of the CO_2 emitted is produced by the upper soil layers (Stewart and Wheatly, 1990), it mainly originates from organic material that has not yet become a part of the peat proper, known as the catotelm. Carbon litter reaching the soil may be either oxidized (emitted as CO_2) or accumulated. Thus, a change in CO_2 emissions will be directly correlated to the portion of organic matter transferred to the catotelm. Because the CH_4 formed will be accompanied by a nearly equal amount of CO_2 (see Gujer and Zehnder, 1978), this relation should hold for most peatland types.

According to one CO_2 efflux temperature-moisture regression model (Svensson, 1980) of the transfer rate of organic matter to the catotelm due to changes in temperature, CO_2 emissions should rise by 12% for each degree Celsius increase in average temperature, given the mean seasonal moisture level. Accordingly, a temperature increase of 1–5°C would result in a 10–60% decrease in the rate at which organic matter is transferred to the catotelm.

Peat accumulation has varied substantially over past millennia (see Malmer, 1992), which is reflected in the quality of peat as a substrate for decomposers. The degradation rate of deep peat is limited by substrate quality rather than by abiotic factors (Hogg *et al.*, 1992). Therefore, the decomposition rate in deep peat will be fairly constant and only marginally affected by changes in temperature. Such constancy would improve the usefulness of the model described above in predicting changes in peat accumulation in response to a temperature change. Changes in hydrology also will influence the accumulation rate of peat because the position of the water table regulates the extent of oxygen penetration into the peat profile. The effect of a lowered water table due to climate change can be compared to the effects noted after drainage of peatlands for forest production. Drainage results in an increased decomposition rate and elevated fluxes of CO_2 to the atmosphere (Silvola *et al.*, 1985; Silvola, 1986; Moore and Knowles, 1989): A 25-cm lowering of the water table gave rise to a twofold increase in CO_2 emissions from peat (Silvola *et al.*, 1985; Moore and Knowles, 1989). Depending on the type of peatland, this elevated flux may reduce carbon accumulation or even reverse the net flux of carbon to make the peatland a net source of atmospheric CO_2. Drained minerotrophic forested peatlands have been reported to respond in the latter way, whereas nutrient-poor peatlands may continue to accumulate carbon at a predrainage level (Laine *et al.*, 1994; see also Tamm, 1951, 1965). Average CO_2 evolution from northern peatlands has been estimated at about 200 gC/m^2/yr (Silvola *et al.*, 1985; Moore 1986, 1989). Following drainage, an elevated CO_2 flow will decline over time (see Armentano and Menges, 1986) owing to substrate depletion as the more easily decomposable fractions of the peat become depleted. However, a drier climate will continue to "drain" the peat successively for a long period; the decline will occur later. To estimate the importance of this, it is assumed that the drainage response reported by Silvola *et al.* (1985) is linear with depth. The increase in CO_2 flows at subsequent drawdowns of 5 cm would then be 40 gC/m^2/yr or another 20% per depth interval.

The scenarios for 2020 and 2050 for the areas of boreal and subarctic peatlands project a temperature increase of 1–2°C and a decrease in soil moisture. Based on this temperature change, it seems reasonable to expect a 25% decrease in the addition of organic matter to the catotelm. It is assumed that this effect would be amplified in response to a decrease in soil moisture. Thus, it is conceivable that the peat accumulation rate will decrease to half of the present rate or even less (i.e., <0.025–0.055 Gt C/yr). Boreal peatlands may even become net sources of atmospheric CO_2. In concluding his discussion of the response of northern wetlands to predicted climate change, Gorham (1991) gives the extreme example of a 1-cm breakdown of the boreal peat layers worldwide. This would result in 2 Gt C/yr, which corresponds to more than a third of the present release of carbon to the atmosphere via fossil fuel combustion. The response in net primary production in relation to climate change is more difficult to predict and may enhance or reduce the effects caused by the estimated changes in the degradation features of peatlands (see Malmer, 1992).

6.5.3.2. *Climatic Controls on Methane Flux*

The net emission of CH_4 from peatlands is dependent on how much CH_4 is formed in the anaerobic parts of the profile and the amount oxidized in the oxic zones. Because the position of the water table and the associated capillary fringe determine the thickness of the zones of production and oxidation of CH_4, the flux of CH_4 is intimately tied to the surface hydrology of the wetland—which in turn is controlled by climate (precipitation and evaporation) and the topographic and geologic setting (surface and subsurface water flow). A decrease in water availability in the peat can lead to a decrease in CH_4 emissions (Whalen *et al.*, 1996; Sundh *et al.*, 1994a, 1994b; Martikainen *et al.*, 1995; Roulet *et al.*, 1993). Deeper penetration of oxygen into the peat also will enhance the capacity of the peat to act as a CH_4-oxidizing filter for CH_4 diffusing from the CH_4-forming sources below. Changes in the direction or magnitude of any or all of the controlling variables discussed above will affect the CH_4 flux. A change in the total CH_4 flux from northern wetlands can be expected if the areal extent of wetlands changes, the duration of the active period changes, and/or the per-unit-area production or oxidation of CH_4 changes.

The relations among moisture content, temperature, and CH_4 flux in individual wetlands have received much attention (Bartlett *et al.*, 1992; Crill *et al.*, 1988; Dise *et al.*, 1992; Moore and Knowles, 1989; Moore and Dalva, 1993; Moore and Roulet, 1993; Svensson, 1976; Svensson and Rosswall, 1984). These relations have been used to estimate qualitatively the year-to-year variation in the flux and the possible direction of change based on changes in temperature and precipitation obtained in 2 x CO_2 scenarios (Table 6-3). Four different approaches have been used to address this issue: (1) correlation of the time series of CH_4 fluxes with the time series of temperature and moisture using interannual data sets; (2) direct observations of changes in CH_4 flux in manipulation experiments that simulate expected changes in wetlands due to climate change; (3) modeling of variability of CH_4 flux using existing climate records and regressions between temperature and CH_4 flux; and (4) modeling of thermal and hydrological regimes of wetlands in 2 x CO_2 climate scenarios and then modeling of change in CH_4 flux using regressions relating CH_4 flux to temperature and moisture in order to predict a change in flux. These studies have shown that the flux of CH_4 is moderately sensitive to changes in temperature and very sensitive to changes in moisture. Using these relative sensitivities as a guide, a qualitative assessment of CH_4 flux from northern wetlands according to six possible climate scenarios is made (Table 6-3). At present, it is not possible to obtain reliable quantitative estimates of the change in flux because the surface hydrology of general circulation models is too coarse to adequately represent the small changes in moisture regime that probably affect the CH_4 flux.

6.5.3.3. *Effects on Nitrous Oxide Emissions*

Emissions of N_2O from northern wetlands are low. *In situ* chamber measurements and laboratory experiments with intact peat cores have revealed emissions below 0.025 g N_2O-N/m^2/yr (Urban *et al.*, 1988; Freeman *et al.*, 1993; Martikainen *et al.*, 1993). A lowering of the water table of bogs will not affect their N_2O emissions, whereas it could strongly increase emissions from fens. Annual emission rates in the range of 0.05–0.14 g N_2O-N/m^2/yr have been reported for drained peat by Martikainen *et al.* (1993) and Freeman *et al.* (1993). The difference between bogs and fens can be explained partly by the fact that drained peat profiles of fens have the capacity to nitrify (Lång *et al.*, 1994). N_2O emissions from drained boreal fens are lower than those from drained agricultural organic soils but 10–100 times higher than the rates from coniferous forest soils (Martikainen *et al.*, 1993).

6.5.4. *Case Study: Kalimantan*

6.5.4.1. *Background*

Kalimantan is one of the largest islands (539,460 km^2) in the Indonesian archipelago (see Box 6-3). The region has a humid tropical climate, with high temperatures and high precipitation. The peatlands of Kalimantan probably play a major role in determining local climate at the present time, although there is no substantial evidence to confirm this.

The largest wetland areas are found in low-lying alluvial plains and basins and flat-bottomed valleys. Most of the freshwater wetlands in the area are forested swamps, specifically either freshwater swamp forests or peat swamp forests (Silvius, 1989). The freshwater swamp forests are rich in epiphytes, rattans, and palms. They provide shelter for a range of rare and endangered species of wildlife, including numerous bird species. The peat swamp forests are a further developmental stage of the freshwater swamp forest. Deep peats are found in the central and western parts of the island (Sieffermann *et al.*, 1988, 1992; Rielly *et al.*, 1992). The peat swamp forests have a relatively high diversity of tree species, but the variety of wildlife tends to be poorer than in freshwater swamp forests (Whitten *et al.*, 1987). Because of the high acidity of the peats and the fact that they are difficult to drain, peat swamp forests are of limited agricultural value (Silvius, 1989). Both swamp types are important watershed areas capable of absorbing and storing excess water and reducing flooding in adjacent areas. They also are an important forestry resource, with many commercially valuable timber species.

The wetlands of Kalimantan currently are deteriorating through the loss of the natural ecosystem, including primary forest cover. Deforestation, drainage, and agriculture all limit the buffering capacity of developed wetlands, causing changes that are long-term and irreversible. Because much of the human settlement at the present time is located in the coastal zone, peat swamp forest on the deeper interior peats has not been subject to large-scale harvesting. These areas are used mainly for timber extraction rather than agriculture; hence, their vegetation cover remains relatively unmodified.

6.5.4.2. *Effects of Temperature Change*

The projected climate change for the region involves an increase in temperature ranging from 0–1.5°C. The effects of

Table 6-3: *Potential changes in CH_4 flux from northern wetlands due to changes in the thermal and moisture regime (adapted with additions from Matthews, 1993).*

Study Description and Location	Change in Thermal and/or Moisture Regime	Observed or Modeled Change in CH_4 Flux	Relative Sensitivity
Field observations of CH_4 flux and temperatures among tundra wetlands of the North Slope (Alaska) with differing moisture levels (1987–1989)[1]	$\Delta T = +4°C$ $\Delta T = +4°C$; wetter $\Delta T = +4°C$; drier	Four-fold increase Four- to five-fold increase Two-fold increase	Large positive sensitivity to temperature increase; small positive sensitivity to moisture change
Field observations of CH_4 flux from permanent tundra wetland sites (Alaska)[2]	4-year variation in temperature and moisture	4 times variation in flux: Flux increased with warmer, wetter conditions	Large positive sensitivity to both temperature and moisture changes
Field observations of CH_4 flux from drained boreal wetlands (Canada)[3]	$\Delta WT = -10$ cm $\Delta WT > -10$ cm	Elimination of CH_4 flux Wetland became small CH_4 sink	Large positive sensitivity to moisture change
Field observations of CH_4 flux from drained boreal wetlands (Finland)[4]	$\Delta WT = -4$ cm $\Delta WT = -20$ cm	Five-fold decrease Elimination of CH_4 flux	Large positive sensitivity to moisture change
Modeling study based on 20th-century historical summer temperature anomalies for five high-latitude wetland regions and a temperature/CH_4 flux regression model[5]	$\Delta T = \pm 2°C$	±15% variance in flux	Moderate sensitivity to temperature
Modeling study simulating change in summer temperature and water table for a northern fen (Canada) in a 2 x CO_2 scenario (+3°C, +1 mm/d P) and temperature/CH_4 flux, and water table/CH_4 flux regression models[6]	$\Delta T \approx +0.8°C$ $\Delta T = +2°C$ $\Delta WT = -14$ cm	+5% increase +15% increase -80% decrease	Moderate sensitivity to temperature; large sensitivity to moisture

Notes: ΔT = change in temperature; ΔWT = change in water table; P = precipitation.
[1]Livingston and Morrissey, 1991.
[2]Whalen and Reeburgh, 1992.
[3]Roulet *et al.*, 1993.
[4]Martikainen *et al.*, 1992.
[5]Harriss and Frolking, 1992.
[6]Roulet *et al.*, 1992.

Box 6-3. The Forested Swamps of Kalimantan: Effects on Habitat, Hydrology, and Carbon Cycling

Background: The wetlands of Kalimantan are found in low-lying alluvial plains and basins and flat-bottomed valleys. Most of the freshwater wetlands in the region are classified as forested swamps, specifically as either freshwater swamp forests or peat swamp forests. The freshwater swamp forests are important in providing shelter for rare and endangered species; both types are important as watersheds and habitats for valuable tree species. They also are carbon sinks but release carbon into the atmosphere when water declines. The Kalimantan wetlands currently are stressed by the loss of the natural ecosystem through deforestation, drainage, and agriculture. These activities limit the buffering capacity of developed wetlands, causing changes that tend to be long-term and irreversible. However, these inland wetlands have escaped some interference because most of the human settlements are located along the coastal zones.

Possible impacts: Increased temperatures are likely to result in a longer period of reduced rainfall because higher temperatures will cause evapotranspiration to exceed precipitation. This is likely to have deleterious effects on the vegetation and hydrology of these wetlands. Climate change also could enhance peat losses in the region that currently result from human interference. On the other hand, increased precipitation in the dry season is likely to be beneficial, as the lower water levels that are typical in this season lead to a net loss of carbon into the atmosphere and an increased risk of fire (one of the greatest threats to their functioning).

Conclusion: It is possible that some measures may be taken in this region in the near future to abate the detrimental impacts caused by human (non-climate) stresses on these wetlands. The extent to which adaptations will be sufficient to counteract changes imposed by a changing climate as well cannot be determined. However, human activities that reduce the resiliency of these wetlands, as well as planned development of previously undisturbed areas of swamp forest, will likely lead to considerably enhanced carbon transfer from the wetlands to the atmosphere, even without climate change.

higher temperature and longer dry periods combine to produce a longer period when evapotranspiration exceeds rainfall and effective rainfall is greatly reduced. This, linked to increasing human activity on peatlands—such as timber extraction, agricultural development, and construction work—could have serious consequences.

Much of the Kalimantan lowland area is subjected to a distinct dry season from July to September or October in which there are high water losses from wetlands as a result of direct evaporation and evapotranspiration. Thus, water levels drop and peat oxidation occurs, with a net loss of carbon to the atmosphere. The spread of fire is one of the greatest threats to the functioning of these wetlands. Peat fires occur frequently in the region, creating palls of smoke sufficiently heavy to close local airports. In 1983–84, fires destroyed 3.5 Mha of both dipterocarp and peat swamp forest in Kalimantan, resulting in a direct economic cost of $2–12 million and an incalculable ecological cost (Maltby, 1986). The risk of fires spreading from cultivated to forest areas increases during the dry season and would increase if the dry season were extended.

6.5.4.3. *Effects of Precipitation Change*

Estimated precipitation changes for the area range from -20% during winter to +40% during the summer. The higher "summer" (i.e., dry-season) values are beneficial to the peatlands because this is the time when they experience the greatest drawdown of the water table. A decrease of precipitation in the wet season could be significant if, as a result, the dry season is extended.

Large-scale removal of the forests and drainage of the underlying peats could prevent further peat formation. The high peats of Kalimantan already appear to be degrading (Sieffermann *et al.*, 1988) and losing carbon directly through oxidation to the atmosphere or indirectly in surface drainage waters, followed by oxidation of carbon compounds at a later stage. Climate change would most surely exacerbate this degradation, leading to peat losses in this region.

6.5.4.4. *Remediation Possibilities*

A forest-management project in Kalimantan currently being sponsored by the British Overseas Development Administration will include suggestions for the sustainable management of the peat swamp forests. The resulting guidelines should help preserve the forest cover on the deeper peats, particularly if extraction methods do not intensify. Indonesian authorities also are introducing stricter regulations and controls on unnatural fires resulting from illegal land clearance and settlement. However, in 1986–89, a feasibility study was carried out to investigate the potential of using deep peat to generate electricity.

The extent to which these potential management changes will be sufficient to counteract the effects of climate change cannot be determined. However, utilization of the forests on these peats and planned development of previously undisturbed areas will likely lead to considerably larger carbon transfer from wetlands to the atmosphere even without climate change.

6.5.5. *Case Study: The Florida Everglades*

6.5.5.1. *Background*

The Everglades is a 500,000-hectare freshwater peatland dominated by vast expanses of sedge and sawgrass, interspersed with shallow-water aquatic communities (sloughs), wet prairies, and tree islands (Loveless, 1959; Gunderson, 1994). Peat accumulation and the subsequent formation of the Everglades began approximately 5,000 years ago as sea-level rise slowed after an initial rapid rise during deglaciation (Gleason and Stone, 1994). As recently as a century ago, the Everglades encompassed more than 1,000,000 hectares, but drainage for agriculture and urban development has resulted in the loss of more than half of the ecosystem (Kushlan, 1989; Davis *et al.*, 1994). The remaining area has been dramatically altered by construction of impoundments, canals, levees, and water-control structures; the system is managed, primarily, as a water source (Light and Dineen, 1994). During the wet season (June–November), excess water from agricultural land and suburban areas is pumped into the Everglades; during the dry season (December–May), the Everglades serves as a water source (DeGrove, 1984). In addition, approximately 50% of the water from the Kissimmee River/Lake Okeechobee complex—the "headwaters" of the Everglades—is diverted by canals to the Atlantic Ocean and the Gulf of Mexico before recharging the wetland (Light and Dineen, 1994). Thus, the present-day Everglades is characterized by a general reduction in the hydroperiod (Fennema *et al.*, 1994; Stephens, 1984; SFWMD, 1992; Walters *et al.*, 1992).

Long-term rates of peat accretion in the Everglades average 0.8–2.0 mm/yr, based on ^{14}C dating of the basal peat (McDowell *et al.*, 1969) and ^{210}Pb dating of peat cores (Craft and Richardson, 1993). However, alterations of the natural hydroperiod and nutrient regimes in the Everglades have resulted in changes in the rate of peat accretion. Areas experiencing reduced hydrology (caused by overdrainage) exhibit lower rates of accretion (1.6–2.0 mm/yr) compared to areas of extended hydroperiod (2.8–3.2 mm/yr) (Craft and Richardson, 1993). Likewise, pollen analysis of peat cores indicates a decrease in the extent of wetland vegetation such as sawgrass and slough and a concurrent increase in terrestrial "weedy" species (ragweed and pigweed) since drainage activities were initiated (Bartow *et al.*, 1994). Thus, future changes in the Everglades ecosystem caused by global warming must be interpreted in the context of recent anthropogenic alterations of hydrology and nutrient regimes (see Box 6-4).

6.5.5.2. *Effects of Sea-Level Rise*

The most immediate effect of climate change will be accelerated sea-level rise, resulting in saltwater intrusion into the lower part of the glades from Florida Bay (Wanless *et al.*, 1994). Increased salinity would result in encroachment of salt-tolerant wetland communities such as mangroves and salt marshes. The areal extent of freshwater communities such as sawgrass, slough, and wet prairie will decrease, and the amount of organic carbon sequestered also will decrease. Another biogeochemical consequence of saltwater intrusion is a shift in anaerobic decomposition away from methanogenesis toward nitrate and sulfate reduction.

6.5.5.3. *Effects of Temperature Change*

Temperature is expected to increase from 0.5 to1.5°C, likely resulting in an increase in evapotranspiration—which may

Box 6-4. The Florida Everglades: Effects on Water Supply and Critical Habitats

Background: The Everglades is a freshwater peatland dominated by sedge and sawgrass, with sloughs, wet prairies, and tree islands. The Everglades is important as a habitat for wildlife, fish, and plant species and as a water source for the neighboring community. It is estimated that drainage for agriculture and urban development in the past century has resulted in a loss of more than half of the ecosystem, and the remaining wetlands have been altered by other construction and so forth.

Possible impacts: Sea-level rise is expected to be perhaps the most important variable that will affect the Everglades, causing saltwater intrusion that is likely to result in an encroachment of salt-tolerant wetland communities. This would decrease the areal extent of the freshwater wetlands, with some effects on anaerobic decomposition. The increased evapotranspiration expected in some seasons would exacerbate this saltwater intrusion. Increased temperature is likely to cause a northward migration of some introduced species but could be conducive for other species. Climate change also is expected to affect the hydrology of the wetlands, causing higher water levels in the winter and lower levels in the summer. This could result in the loss of critical habitats such as sawgrass and wet prairie communities, although these losses are likely to be offset by an increase in woody shrubs and trees.

Conclusion: Overall, the impacts that are projected as a result of climate change would adversely affect the end-users of the ecosystem: waterfowl, fish, and other wildlife; hunters, fishers, and tourists; and surrounding populations that rely upon the Everglades for freshwater resources. However, some action is currently underway to modify the water-control structure in an effort to restore the Everglades.

further exacerbate saltwater intrusion into the Everglades. Although more water will be lost to evapotranspiration, it is likely that oxidation and subsidence of the peat soils will not be dramatically affected because rising sea level will augment the groundwater table, particularly in the southern Everglades. Another consequence of increased temperature may be the northward migration of introduced species such as *Melaleuca quinquenervia*. *Melaleuca,* which has overtaken large areas of the southern Everglades, is seemingly kept in check by frost (Bodle *et al.*, 1994). However, increased temperatures caused by climate change might enable *Melaleuca* to colonize large areas of the northern Everglades.

6.5.5.4. *Effects of Precipitation Change*

Precipitation is projected to decrease during summer (0–20%) and increase during winter (0–20%). The winter months correspond to the dry season (November–May), when approximately 25% (10–15 inches) of the 50 to 60 inches of annual rainfall occurs (MacVicar and Lin, 1984). It is likely that an increase in rainfall during this season will reduce the rate of drawdown that normally occurs during this time. The reduction in summer rainfall in the wet season (May–October) should result in a lowering of the water table compared to current levels. These combined seasonal changes in rainfall should dampen the oscillations between the summer wet season and the winter dry season. As a result, water levels in the Everglades probably will be somewhat higher in the winter and somewhat lower in the summer than they are at present.

A dampening of the annual hydroperiod fluctuation may result in a decrease in the extent of sawgrass, the dominant plant community in the Everglades. Sawgrass communities are partly maintained by fire (Gunderson, 1994); increased rainfall during the dry season may reduce the frequency of fires—in particular, the severe fires that occur in the beginning of the wet season (May) that often burn large areas of the Everglades (Gunderson and Snyder, 1994). It is likely that wet prairie communities also will decrease in extent. These communities, which are important foraging habitat for wading birds (Hoffman *et al.*, 1994), frequently dry down during the spring (Goodrick, 1984). The dampening of the annual hydroperiod fluctuations caused by global warming will likely result in the loss of much of this critical habitat. The decline of sawgrass and wet prairie communities probably will be offset by an increase in woody shrubs and trees. These species are not generally fire-tolerant and often compete more effectively against emergents when water levels are stable.

6.5.5.5. *Socioeconomic Consequences*

It is likely that the greatest impact of climate change will be the loss of freshwater resources that sustain the burgeoning human population of south Florida, as well as the unique Everglades wetland. Competition for this diminishing resource will surely result in a no-win situation for humans and the Everglades under a scenario of global warming and rising sea level.

On a positive note, the U.S. Army Corps of Engineers, which oversees the water resources of south Florida, is evaluating modifications to the system of impoundments, canals, levees, and water-control structures in order to restore the Everglades (and other south Florida ecosystems) while providing for other water-related needs in the region (U.S. ACOE, 1994). This ambitious project, which could cost upward of $2 billion, is designed to increase the spatial extent of wetlands and restore the hydrology and water-quality conditions of the Everglades and other south Florida ecosystems.

6.6. Future Research Needs

Wetlands are highly valued in many areas. The lack of data to fully address their responses to climate change calls for several areas of research in the future:

- Site-specific responses are variable. There is a strong need for a local and regional coupling of climate-change predictions with known responses of specific wetlands, which would allow for modeling of the necessary interactive responses of the hydroperiod, temperature, and water availability at these scales. A network of different wetland sites in different regions of the globe should be established to form the base for such research.
- The feedback on climate by changes in trace-gas flows from wetlands, especially CO_2 and CH_4, upon a climate change calls for a strengthening of ongoing research in this field. This will aid in the judgments necessary for the introduction of adaptation and remediation measures on wetlands.
- The vast "grey" literature existing on different wetland subjects within the frames of local, regional, and country research reports should be examined to further substantiate site-specific responses by different wetlands, including changes in species composition, biogeochemistry, and socioeconomic consequences.

References

Andriesse, J.P., 1988: *Nature and Management of Tropical Peat Soils*. FAO Soils Bulletin 59, FAO, Rome, Italy.

Arheimer, B. and H. Wittgren, 1994: Modelling the effects of wetlands on regional nitrogen transport. *Ambio*, **23**, 378-386.

Armentano, T.V., 1990: Soils and ecology: tropical wetlands. In: *Wetlands: A Threatened Landscape* [Williams, M. (ed.)]. The Alden Press, Ltd., Oxford, UK, pp. 115-144.

Armentano, T.V. and E.S. Menges, 1986: Patterns of change in the carbon balance of organic-soil wetlands of the temperate zone. *J. Ecol.*, **74**, 755-774.

Aselmann, I. and P. Crutzen, 1989: Global distribution of natural freshwater wetlands and rice paddies: their net primary productivity, seasonality and possible methane emissions. *J. Atmos. Chem.*, **8**, 307-358.

Bartlett, D.S., K.B. Bartlett, J.M. Hartman, R.C. Harriss, D.I. Sebacher, R. Pelletier-Travis, D.D. Dow, and D.P. Brannon, 1989: Methane emission from the Florida everglades: patterns of variability in a regional wetland ecosystem. *Global Biogeochemical Cycles*, **3(4)**, 363-374.

Bartlett, K.B., P.M. Crill, R.L. Sass, R.C. Harriss, and N.B. Dise, 1992: Methane emissions from tundra environments in the Yukon-Kuskokwin Delta, Alaska. *J. Geophys. Res.*, **97**, 16,645-660.

Bartlett, K.B. and R.C. Harriss, 1993: Review and assessment of methane emissions from wetlands. *Chemosphere,* **26(1-4)**, 261-320.

Bartow, S., C.B. Craft, and C.J. Richardson, 1994: Historical changes in the Everglades plant community: structure and composition. In: *Effects of Nutrient Loadings and Hydroperiod Alterations in the Water Expansion, Community Structure and Nutrient Retention in the Water Conservation Areas of South Florida.* Annual report to the Everglades Agricultural Area Environmental Protection District, Duke Wetland Center publication no. 94-08, Duke University, Durham, NC, pp. 313-330.

Batt, B.D.J., M.G. Anderson, C.D. Anderson, and F.D. Caswell, 1989: The use of prairie potholes by North American ducks. In: *Northern Prairie Wetlands* [van der Valk, A. (ed.)]. Iowa State Univ. Press, Ames, IA, pp. 204-227.

Bazilivich, N.L., L.Y. Rodin, and N.N. Rozov, 1971: Geophysical aspects of biological productivity. *Soviet Geography*, **15**, 65-88.

Bellrose, F.C. and N.M. Trudeau, 1988: Wetlands and their relationship to migrating and winter populations of waterfowl. In: *The Ecology and Management of Wetlands,* vol. 1 [Hook, D.D., W.H. McKee, Jr., H.K. Smith, J. Gregory, V.G. Burrell, Jr., M.R. DeVoe, R.E. Sojka, S. Gilbert, R. Banks, L.H. Stolzy, C. Brooks, T.D. Matthews, T.H. Shear (eds.)]. Timber Press, Portland, OR, pp. 183-194.

Bernaldez, F.G., J.M. Rey-Benayas, and A. Martinez, 1993: Ecologial impact of groundwater extraction on wetlands (Duro Basin, Spain). *J. Hydrol.*, **141**, 219-238.

Bernard, J.M. and E. Gorham, 1978: Life history aspects of primary production in sedge wetlands. In: *Freshwater Wetlands—Ecological Processes and Management Potential* [Good, R.E., D. Whigham, R.L. Simpson (eds.)]. Academic Press, Inc., New York, NY, pp. 39-52.

Billings, W.D., 1987: Carbon balance of Alaskan tunra and taiga ecosystems: past, present and future. *Quaternary Sci. Rev.*, **6**, 165-177.

Bodle, M.J., A.P. Ferriter, and D.D. Thayer, 1994: The biology, distribution and ecological consequences of Melaleuca quinquenervia in the Everglades. In: *Everglades: The Ecosystem and Its Restoration* [Davis, S.M. and J.C. Ogden (eds.)]. St. Lucie Press, Delray Beach, FL, pp. 341-355.

Boelter, D.H., 1966: Water storage characteristics of several peats *in situ. Soil Science of America Proceedings,* **28**, 433-435.

Bonan, G.B., D. Pollard, and S.L. Thompson, 1992: Effects of boreal forest vegetation on global climate. *Nature,* **359**, 716-718.

Bord na Móna, 1985. *Fuel Peat in Developing Countries.* World Bank Technical Paper Number 14, The World Bank, Washington DC, 146 pp.

Bowden, W.B., 1987: The biogeochemistry of nitrogen in freshwater wetlands. *Biogeochemistry*, **4**, 313-348.

Boyd, H., 1981: Prairie dabbling ducks. *Can. Wildl. Serv. Wildl. Notes*, **9**, 1941-1990.

Bradbury, I.K. and J. Grace, 1983: Primary production in wetlands. In: *Ecosystems of the World.* Vol. 4A, *Mires: Swamp, Bog, Fen, and Moor* [Gore, A.J.P. (ed.)]. Elsevier Sci. Publ., New York, NY, pp. 285-310.

Bratton, S.P., 1982: The effects of exotic plant and animal species on nature preserves. *Natural Areas Journal*, **2**, 3-13.

Brinson, M.M., 1990: Riverine forests. In: *Ecosystems of the World.* Vol. 15, *Forested Wetlands* [Lugo, A.E., M. Brinson, S. Brown (eds.)]. Elsevier Sci. Publ., New York, NY, pp. 87-141.

Brinson, M.M., 1993: *A Hydrogeomorphic Classification for Wetlands.* Wetland Research Program Technical Report WRP-DE-4, U.S. Army Engineers Waterways Experiment Station, Vickenburg, MS, 67 pp.

Brix, H. and H.-H. Schierup,1989: The use of aquatic macrophytes in water-pollution control. *Ambio*, **18**, 100-107.

Brock, T.C.M. and W. van Vierssen, 1992: Climatic change and hydrophyte-dominated communities in inland wetland ecosystems. *Wetlands Ecology and Management*, **2**, 37-49.

Bromley, J., J. Brouwer, and S. Gaze, in press a: The semi-arid groundwater recharge study (SAGRE). In: *Hydrologie et Météorologie de Méso-Échelle dans HAPEX-Sahel: Dispositif de Mesures au Sol et Premiers Résultats.* Cahiers ORSTOM.

Bromley, J., W.M. Edmunds, E. Fellman, J. Brouwer, S.R. Gaze, J. Sudlow, and C. Leduc, in press b: Rainfall inputs and recharge estimate to the deep unsaturated zone of southern Niger. *J. Hydrol.*

Brouwer, J. and W.C. Mullié, 1994a: Potentialités pour l'agriculture, l'élevage, la pêche, la collecte de produits naturels et la chasse dans les zones humides du Niger. In: *Atelier sur les Zones Humides du Niger. Comptes Rendus d'un Atelier à la Tapoa, Parc du 'W,' Niger, du 2 au 5 Novembre 1994* [Kristensen, P. (ed.)]. UICN-Niger, Niamey, Niger, pp. 27-51.

Brouwer, J. and W.C. Mullié, 1994b: The importance of small wetlands in the central Sahel. *IUCN Wetlands Programme Newsletter,* **9**, 12-13.

Bunce, J.A., 1992: Stomal conductance, photosynthesis, and respiration of temperate deciduous tree seedlings grown outdoors at an elevated concentration of carbon dioxide. *Plant Cell Environ.*, **15**, 541-549.

Center, T.D., R.F. Doren, R.L. Hofstetter, R.L. Myers, and L.D. Whiteaker (eds.), 1991: *Proceedings of the Symposium on Exotic Pest Plants.* U.S. Department of the Interior Technical Report NPS/NREVER/NRTR-91/06, Washington DC, 387 pp.

Chapman, W.L. and J.E. Walsh, 1993: Recent variations of sea ice and air temperature in high latitudes. *Bulletin American Meteorological Society*, **74**, 33-47.

Cherefou Mahatan, 1994: *Etude de la Filière des Cultures de Contre-Saison, Zone du PMI.* SNV-Netherlands Organisation for Development Aid, Projet Mares Illela, Niamey, Niger, 80 pp.

Clark, J. 1979. Freshwater wetlands: habitats for aquatic invertebrates, amphibians, reptiles, and fish. In: *Wetlands Functions and Values: The State of Our Understanding* [Greeson, P.E., J.R. Clark, and J.E. Clark (eds.)]. American Water Resources Association Technical Application, Minneapolis, MN, pp. 330-343.

Cohen, S.J., 1986: Impacts of CO_2-induced climatic change on water resources in the Great Lakes basin. *Clim. Change,* **8**, 135-153.

Coles, B., 1990: Wetland archeology: a wealth of evidence. In: *Wetlands: A Threatened Landscape* [Williams, M. (ed.)]. Basil Blackwell, Ltd., Oxford, UK, pp. 145-180.

Cowardin, L.M., V. Carter, F.C. Golet, and E.T. LaRoe, 1979: *Classification of Wetlands and Deepwaer Habitats of the United States.* U.S. Fish and Wildlife Service, FWS/OBS-79/31, U.S. Government Printing Office, Washington DC, 131 pp.

Craft, C.B. and C.J. Richardson, 1993: Peat accretion and N, P and organic C accumulation in nutrient enriched and unenriched Everglades peatlands. *Ecological Applications,* **3**, 446-458.

Crill, P.M., K.B. Bartlett, R.C. Harriss, E. Gorham, E.S. Verry, D.I. Sebacher, L. Madzar, and W.S. Anner, 1988: Methane flux from Minnesota peatlands. *Global Biogeochem. Cycles*, **2**, 371-384.

Crill, P.M., 1991: Seasonal patterns of methane uptake and carbon dioxide release by a temperate woodland soil. *Global Geochem. Cycles*, **5**, 319-334.

Croley, T.E. II, 1990: Laurentian Great Lakes double-CO_2 climate change hydrological impacts. *Clim. Change*, **17**, 27-47.

Davis, S.M., L.H. Gunderson, W.A. Park, J.R. Richardson, J.E. Mattson, 1994: Landscape dimension, composition and function in a changing Everglades ecosystem. In: *Everglades: The Ecosystem and Its Restoration* [Davis, S.M. and J.C. Ogden (eds.)]. St. Lucie Press, Delray Beach, FL, pp. 419-444.

DDE-Tahoua, 1993: *Contribution à l'Élaboration du Plan Quinquennal 1994-1998 du Secteur de l'Environnement.* Direction Départementale de l'Environnement, Département de Tahoua, République du Niger, 12 pp.

DeGrove, J.M., 1984: History of water management in south Florida. In: *Environments of South Florida, Past and Present* [Gleason, P.J. (ed.)]. Miami Geological Society, Miami, FL, pp. 22-27.

Dise, N.B., E. Gorham, and E.S. Verry, 1992: Environmental factors controlling methane emissions from peatlands in northern Minnesota. *J. Geophys. Res.*, **98**, 10,583-594.

Dugan, P., 1988: *The Ecology and Management of Wetlands*, vol. 2 [Hook, D.D., W.H. McKee, Jr., H.K. Smith, J. Gregory, V.G. Burrell, Jr., M.R. DeVoe, R.E. Sojka, S. Gilbert, R. Banks, L.H. Stolzy, C. Brooks, T.D. Matthews, T.H. Shear (eds.)]. Croom Helm, London, UK, pp. 4-5.

Dugan, P.J. (ed.), 1990: *Wetland Conservation: A Review of Current Issues and Required Action.* IUCN, Gland, Switzerland, 96 pp.

Dugan, P., 1993: *Wetlands in Danger.* Reed International Books Limited, Singapore, 187 pp.

Dumont, H.J., 1992: The regulation of plant and animal species and communities in African shallow lakes and wetlands. *Revue d'Hydrobiologie Tropicale*, **25**, 303-346.

Duncan, C.P. and P.M. Groffman, 1994: Comparing microbial parameters in natural and artificial wetlands. *Journal of Environmental Quality*, **23**, 298-305.

Ellis, J. and K.A. Galvin, 1994: Climate patterns and landuse practices in the dry zones of Africa. *BioScience*, **44**, 340-349.

Fennema, R.J., C.J. Neidrauer, R.A. Johnson, T.K. MacVicar, W.A. Perkins, 1994: A computer model to simulate natural Everglades hydrology. In: *Everglades: The Ecosystem and Its Restoration* [Davis, S.M. and J.C. Ogden (eds.)]. St. Lucie Press, Delray Beach, FL, pp. 249-289.

Framine, N., 1994: Pisciculture des zones humides: compatabilité, exploitation et conservation. In: *Atelier sur les zones humides du Niger. Comptes Rendus d'un Atelier à la Tapoa, Parc du 'W,' Niger, du 2 au 5 Novembre 1994* [Kristensen, P. (ed.)]. UICN-Niger, Niamey, Niger, pp. 17-26.

Freeman, C., M.A. Lock, and B. Reynolds, 1993: Fluxes of CO_2, CH_4, and N_2O from a Welsh peatland following simulation of water table draw-down: potential feedback to climate change. *Biogeochemistry*, **19**, 51-60.

Gaze, S.R., J. Brouwer, L.P. Simmonds, and J.Bouma, 1996: Measurement of surface redistribution of rainfall and modelling its effect on water balance calculations for a millet field on sandy soil in Niger. *J. Hydrol.* (in press).

Gibbs, J.P., 1993: Importance of small wetlands for the persistence of local populations of wetland associated animals. *Wetlands*, **13**, 25-31.

Gilliam, J.W., G.M. Chescheir, R.W. Skaggs, and R.G. Broadhead, 1988: Effects of pumped agricultural drainage water on wetland water quality. In: *The Ecology and Management of Wetlands,* vol. 1 [Hook, D.D., W.H. McKee, Jr., H.K. Smith, J. Gregory, V.G. Burrell, Jr., M.R. DeVoe, R.E. Sojka, S. Gilbert, R. Banks, L.H. Stolzy, C. Brooks, T.D. Matthews, T.H. Shear (eds.)]. Croom Helm, London, UK, pp. 183-194.

Gleason, P.J. and P. Stone, 1994: Age, origin and landscape evolution of the Everglades peatlands. In: *Everglades: The Ecosystem and Its Restoration* [Davis, S.M. and J.C. Ogden (eds.)]. St. Lucie Press, Delray Beach, FL, pp. 149-197.

Gleick, P.H., 1987: Regional hydrologic consequences of increases in atmospheric CO_2 and other trace gases. *Clim. Change*, **10**, 137-161.

Glenn, S., A. Heyes, and T. Moore, 1993: Carbon dioxide and methane fluxes from drained peat soils, Southern Quebec. *Global Geochem. Cycles*, **7(2)**, 247-257.

Goodrick, R.L., 1984: The wet prairies of the northern Everglades. In: *Environments of South Florida, Past and Present* [Gleason, P.J. (ed.)]. Miami Geological Society, Miami, FL, pp. 249-289.

Gorham, E., 1990: Biotic impoverishment in northern peatlands. In: *The Earth in Transition: Patterns and Processes of Biotic Impoverishment* [Woodwell, G.M. (ed.)]. CUP N.Y., New York, NY, pp. 65-98.

Gorham, E., 1991: Northern peatlands: role in the carbon cycle and probable responses to climatic warming. *Ecol. Applications*, **1(2)**, 182-195.

Gorham, E., 1994: The future of research in Canadian peatlands: a brief survey with particular reference to global change. *Wetlands*, **14**, 206-215.

Gosselink, J.G. and R.E. Turner, 1978: The role of hydrology in freshwater wetland ecosystems. In: *Freshwater Wetlands—Ecological Processes and Management Potential* [Good, R.E., D. Wigham, R.L. Simpson (eds.)]. Academic Press, Inc., New York, NY, pp. 63-78.

Gosselink, J.G. and E. Maltby, 1990: Wetland losses and gains. In: *Wetlands: A Threatened Landscape* [Williams, M. (ed.)]. Basil Blackwell, Ltd., Oxford, UK, pp. 296-322.

Greco, S., R.H. Moss, D. Viner, and R. Jenne (eds.), 1994: *Climate Scenarios and Socioeconomic Projections for IPCC WG II Assessment.* Intergovernmental Panel on Climate Change Working Group II, 67 pp.

Gujer, W. and A.J.B. Zehnder, 1983: Conversion processes in anaerobic digestion. *Water Sci. Technol.*, **15**, 127-167.

Gunderson, L.H., 1994: Vegetation of the Everglades: determinants of community composition. In: *Everglades: The Ecosystem and Its Restoration* [Davis, S.M. and J.C. Ogden (eds.)]. St. Lucie Press, Delray Beach, FL, pp. 323-340.

Gunderson, L.H. and J.R. Snyder, 1994: Fire patterns in the southern Everglades. In: *Everglades: The Ecosystem and Its Restoration* [Davis, S.M. and J.C. Ogden (eds.)]. St. Lucie Press, Delray Beach, FL, pp. 291-305.

Hånell, B., 1988: Postdrainage forest productivity of peatlands in Sweden. *Can. J. For. Res.*, **18**, 1443-1456.

Harriss, R.C., 1987: Effects of climatic change on northern permafrost. *North. Perspect.*, **15(5)**, 7-9.

Harriss, R.C. and S. Frolking, 1992: The sensitivities of methane emissions from northern freshwater wetlands to global warming. In: *Climate Change and Freshwater Ecocystems* [Firth, P. and S. Fisher (eds.)]. Springer Verlag, New York, NY, pp. 48-67.

Harty, F.M., 1986: Exotics and their ecological ramifications. *Natural Areas Journal*, **6**, 20-26.

Hoffman, W., G.T. Bancroft, and R.J. Sawicki, 1994: Foraging habitat of wading birds in the Water Conservation Area of the Everglades. In: *Everglades: The Ecosystem and Its Restoration* [Davis, S.M. and J.C. Ogden (eds.)]. St. Lucie Press, Delray Beach, FL, pp. 585-614.

Hogan, K.P., A.P. Smith, and L.H. Ziska, 1991: Potential effects of elevated CO_2 and changes in temperature on tropical plants. *Plant, Cell and Environment*, **14**, 763-778.

Hogg, E.H., V.J. Lieffers, and R.W. Ross, 1992: Potential carbon losses from peat profiles: effects of temperature, drought cycles, and fire. *Ecol. Appl.*, **2**, 298-306.

Hutchinson, C.F., P. Warshall, E.J. Arnould, and J. Kindler, 1992: Development in arid lands. lessons from Lake Chad. *Environment*, **34(6)**, 16-20, 40-43.

Idso, S.B. and B.A. Kimball, 1993: Tree growth in carbon dioxide enriched air and its implications for global carbon cycling and maximum levels of atmospheric CO_2. *Global Biogeochemical Cycles*, **7**, 537-555.

Ilvessalo, Y. and M. Ilvessalo, 1975: The forest types of Finland in the light of natural development and yield capacity of forest stands. *Acta Forestalia Fennicia*, **144**, 1-101 (in Finnish with English summary).

Immirzi, P. and E. Maltby, 1992: *The Global Status of Peatlands and Their Role in the Carbon Cycle.* Wetland Ecosystems Research Group, Report No. 11, University of Exeter, UK, 145 pp.

Johnston, C.A., 1991: Sediment and nutrient retention by freshwater wetlands: effects on surface water quality. *Critical Reviews in Environmental Control*, **21**, 491-565.

Junk, W.J., 1983: Ecology of swamps on the middle Amazon. In: *Ecosystems of the World.* Vol. 4B, *Mires: Swamp, Bog, Fen, and Moor* [Gore, A.J.P. (ed.)]. Elsevier Sci. Publ., New York, NY, pp. 269-294.

Junk, W.J., 1993: Wetlands of tropical South America. In: *Wetlands of the World* [Wigham, D.F. *et al.*(eds.)]. Kluwer Academic Publishers, Dordrecht, Netherlands, pp. 679-793.

Kadlec, R.H., 1989: Wetlands for treatment of municipal wastewater. In: *Wetlands Ecology and Conservation: Emphasis in Pennsylvania* [Majumdar, S.K., R.P. Brooks, F.J. Brenner, R.W. Tiner, Jr., (eds.)]. The Pennsylvanian Academy of Sci. Publ., Philipsburg, pp. 300-314.

Kangas, P.C., 1990: Long-term development of forested wetlands. In: *Ecosystems of the World.* Vol. 15, *Forested Wetlands* [Lugo, A.E., M. Brinson, S. Brown (eds.)]. Elsevier Sci. Publ., New York, NY, pp. 25-51.

Keltikangas, M., J. Laine, P. Puttonen, and K. Seppälä, 1986: Peatlands drained for forestry during 1930-1978: results from field surveys of drained areas. *Acta Forestalia Fennica*, **193**, 1-94 (in Finnish with English summary).

Kimball, B.A. and S.B. Idso, 1983: Increasing atmospheric CO_2: effects on crop yield, water use and climate. *Agric. Water Manage.*, **7**, 55-72.

Kiviat, E., 1989: The role of wildlife in estuarine ecosystems. In: *Estuarine Ecology* [Day, J.W. *et al.* (eds.)]. John Wiley & Sons, New York, NY, pp. 437-475

Koerselman, W., M.B. Van Kerkhoven, and J.T.A. Verhoeven, 1993: Release of inorganic N, P and K in peat soils: effect of temperature, water chemistry and water level. *Biogeochemistry*, **20**, 63-81.

Korner, C. and J.A. Arnone III, 1992: Responses to elevated carbon dioxide in artificial tropical ecosystems. *Science*, **257**, 1672- 1675.

Krapu, G.L., A.T. Klett, and D.G. Jorde, 1983: The effect of variable spring water conditions on mallard reproduction. *Auk*, **100**, 689-698.

Kushlan, J.A., 1989: Wetlands and wildlife, the Everglades perspective. In: *Freshwater Wetlands and Wildlife* [Sharitz, R.R. and J.W. Gibbons (eds.)]. CONF 8603101, DOE Symposium Series Number 61, USDOE Office of Scientific and Technical Information, Oak Ridge, TN, pp. 773-790.

Lachenbruch, A.H. and B.V. Marshall, 1986: Changing climate: geothermal evidence from permafrost in the Alaskan Arctic. *Science*, **234**, 689-696.

Laine, J., K. Minkkinen, A. Puhalainen, and S. Jauhiainen, 1994: Effects of forest drainage on the carbon balance of peatland ecosystems. In: *The Finnish Research Programme on Climate Change, Second Progress Report* [Kanninen, M. and P. Heikinheimo (eds.)]. Publication of the Academy of Finland 1/94, Helsinki, Finland, pp. 303-308.

Lång, K., M. Lehtonen, and P. Martikainen, 1993: Nitrification potentials at different pH in peat samples from various layers of northern peatlands. *Geomicrobiology Journal*, **11**, 141-147.

Larson, D.L., 1994: Potential effects of anthropogenic greenhouse gases on avian habitats and populations in the northern great plains. *Am. Midl. Nat.*, **131**, 330-346.

Larson, D.L., in press: Effects of climate on numbers of northern prairie wetlands. *Climate Change*.

Leduc, C., J. Bromley, and P. Schroeter, in press: Groundwater recharge in a semi-arid climate: some results of the HAPEX-Sahel hydrodynamic survey (Niger). *J. Hydrol.*

Leemans, R. and P.N. Halpin, 1992 Biodiversity and global change. In: *Biodiversity Status of the Earth's Living Resources* [McComb, J. (ed.)]. World Conservation Monitoring Center, Cambridge, UK, pp. 254-255.

Light, S.S. and J.W. Dineen, 1994: Water control in the Everglades: a historical perspective. In: *Everglades: The Ecosystem and Its Restoration* [Davis, S.M. and J.C. Ogden (eds.)]. St. Lucie Press, Delray Beach, FL, pp. 47-84.

Livingston, G.P. and L.A. Morrisey, 1991: Methane emissions from Alaskan arctic tundra in response to climate change. In: *International Conferenc on the Role of Polar Regions in Global Change* [Weller, G., C.L. Wilson, and B.A. Severin (eds.)]. Geophysical Institute and Center for Global Change and Arctic Ecosystem Research, University of Alaska, Fairbanks, Alaska, pp. 372-394.

Loveless, C.M., 1959: A study of the vegetation of the Florida Everglades. *Ecology*, **40**, 1-9.

Lugo, A.E. and S. Brown, 1984: The Oklawaha river forested wetlands and their response to chronic flooding. In: *Cypress Swamps* [Ewel, K.C. and H.T. Odum (eds.)]. University Press of Florida, Gainesville, Florida, pp. 365-373.

Lugo, A.E., S. Brown, and M.M. Brinson, 1990a: Concepts in wetland ecology. In: *Ecosystems of the World.* Vol. 15, *Forested Wetlands* [Lugo, A.E., M. Brinson, S. Brown (eds.)]. Elsevier Sci. Publ., New York, NY, pp. 53-85.

Lugo, A.E., S. Brown, and M.M. Brinson, 1990b: Synthesis and search for paradigms in wetland ecology. In: *Ecosystems of the World.* Vol. 15, *Forested Wetlands* [Lugo, A.E., M. Brinson, S. Brown (eds.)]. Elsevier Sci. Publ., New York, NY, pp. 447-460.

MacVicar, T.K. and S.S.T. Lin, 1984: Historical rainfall activity in central and southern Florida. In: *Environments in South Florida, Past and Present* [Gleason, P.J. (ed.)]. Miami Geological Society, Miami, FL, pp. 477-519.

MAE-Niger, 1993: *Annuaire des Statistiques de l'Agriculture et de l'Élevage 1991.* Ministère de l'Agriculture et de l'Elevage, Directions des Etudes et de la Programmation, Service d'Analyse des Politiques et de la Coordination des Statistiques, Niamey, Niger, 111 pp.

Magadza, C.H.D., 1994: Climate change: some likely multiple impacts in southern Africa. *Food-Policy*, **19**, 165-191.

Mahamane Alio and Abdou Halikou, 1993: *Récensement des Mares du Département de Zinder.* Direction Départementale de l'Environnement, Département de Zinder, Zinder, Niger, 12 pp.

Malmer, N., 1992: Peat accumulation and the global carbon cycle. In: *Greenhouse-Impact on Cold-Climate Ecosystems and Landscapes* [Boer, M. and E. Koster (eds.)]. *Catena Suppl.*, **22**, Cremlingen, Catena-Verlag, 97-110.

Maltby, E. and R.E. Turner, 1983: Wetlands of the World. *Geographical Magazine*, **55**, 12-17.

Maltby, E., 1986: Waterlogged wealth. *Earthscan*. Earthscan Publications International Institute.

Marion, G.M. and W.C. Oechel, 1993: Mid- to late-Holocene carbon balance in Arctic Alaska and its implications for future global warming. *The Holocene*, **3**, 193-200.

Martikainen, P.J., H. Nykänen, P.M. Crill, and J. Silvola, 1992: The effect of changing water table on methane flux from two Finnish mire sites. *Suo*, **43**, 237-240.

Martikainen, P.J., H. Nykänen, P.M. Crill, and J. Silvola, 1993: Effect of a lower water table on nitrous oxide fluxes from northern peatlands. *Nature*, **366**, 51-53.

Martikainen, P.J., H. Nykänen, J. Alm, and J. Silvola, 1995: Changes in fluxes of carbon dioxide, methane and nitrous oxide due to forest drainage of mire sites of different trophy. *Plant and Soil*, **168-169**, 571-577.

Matthews, E. and I. Fung, 1987: Methane emissions from natural wetlands: global distribution, area, and environmental characteristics of sources. *Global Biogeochem. Cycles*, **1**, 61-86.

Matthews, E., 1993: Wetlands. In: *Atmospheric Methane: Sources, Sinks, and Role in Global Change* [Khalil, M.A.K. (ed.)]. NATO ASI Series I, Global Environmental Change, vol. 13, Springer Verlag, New York, NY, pp. 314-361.

McClaugherty, C.A., J. Pastor, J.D. Aber, and J.M. Melillo, 1985: Forest litter decomposition in relation to nitrogen dynamics and litter quality. *Ecology*, **66**, 266-275.

McDowell, L.L., J.C. Stephens, and E.H. Stewart, 1969: Radiocarbon chronology of the Florida Everglades peat. *Soil Science Society of America Proceedings*, **33**, 743-745.

McKnight, B.N. (ed.), 1993: *Biological Pollution: The Control and Impact of Invasive Exotic Species.* Indiana Academy of Science, Indianapolis, IN, 261 pp.

McNeely, J.A., 1990: Climate change and biological diversity: policy implications. In: *Landscape Ecological Impacts of Climate Change* [Boer, M.M. and R.S. de Groot (eds.)]. IOS Press, Amsterdam, Netherlands, pp. 406-428.

Melillo, J.M., J.D. Aber, and J.F. Muratore, 1982: Nitrogen and lignin control of hardwood leaf litter decomposition dynamics. *Ecology*, **63**, 621-626.

Melillo, J.M., T.V. Callaghan, F.I. Woodward, E. Salati, and S.K. Sinha, 1990: Effects on ecosystems. In: *Climate Change: The IPCC Scientific Assessment* [Houghton, J.T., G.J. Jenkins, and J.J. Ephraums (eds.)]. Cambridge University Press, Cambridge, UK, pp. 283-310.

Mercer, D.C., 1990: Recreation and wetlands: impacts, conflict and policy issues. In: *Wetlands: A Threatened Landscape* [Williams, M. (ed.)]. Basil Blackwell, Ltd., Oxford, UK, pp. 267-295.

MHE-DFPP, 1991: *Organisation de la Production et de la Commercialisation du Poisson dans le Département de Tahoua.* Ministère de l'Hydraulique et de l'Environnement, Direction de la Faune, de la Pêche et de la Pisciculture, Rapport de Projet, 50 pp. + annexes.

MHE-Niger, 1991: *Les Ressources en Eau du Département de Diffa.* Ministère de l'Hydraulique et de l'Environnement, Direction Départementale de l'Hydraulique de Diffa, Projet PNUD/DCTD NER/86/001, Niamey, Niger, 35 pp. + annexes.

Miller, P.C., R. Kendall, and W.C. Oechel, 1983: Simulating carbon accumulation in northern ecosystems. *Simulation*, **40**, 119-131.

Mimikou, M., Y. Kouvoloulos, G. Cavadias, and N. Vayianos, 1991: Regional hydrologic effects of climate change. *Journal of Hydrology*, **123**, 119-146.

Mitsch, W.J. and J.G. Gosselink, 1986: *Wetlands.* Van Nostrand Reinhold, New York, NY, 539 pp.

Mitsch, W.J. and J.G. Gosselink, 1993: *Wetlands.* Van Nostrand Reinhold, New York, NY, 2nd ed., 722 pp.

Mitsch, W.J. and X. Wu, 1995: Wetlands and global change. In: *Soil Management and Greenhouse Effect* [Lal, R., J. Kimble, E. Levine, and B.A. Stewart (eds.)]. CRC Press, Inc., Boca Raton, Florida, pp. 205-230.

Monteith, J.L., 1981: Climatic variation and the growth of crops. *Quart. J. R. Met. Soc.*, **107**, 749-774.

Mooney, H.A. and J.A. Drake (eds.), 1986: *Ecology of Biological Invasions of North America and Hawaii.* Springer-Verlag, New York, NY, 321 pp.

Moore, T.R., 1986: Carbon dioxide evolution from subarctic peatlands in eastern Canada. *Arctic and Alpine Res.*, **18(2)**, 189-193.

Moore, T.R., 1989: Growth and net production of Sphagnum at five fen sites, subarctic eastern Canada. *Canadian Journal of Botany*, **67**, 1203-1207.

Moore, T.R. and R. Knowles, 1989: The influence of water table levels on methane and carbon dioxide emissions from peatland soils. *Can. J. Soil Sci.*, **67**, 77-81.

Moore, N.T. and M. Dalva, 1993: The influence of temperature and water table on carbon dioxide and methane emissions from laboratory columns of peatland soils. *J. Soil Sci.*, **44**, 651-664.

Moore, T.R. and N.T. Roulet, 1993: Methane flux: water table relations in northern wetlands. *Geophys. Res. Let.*, **20**, 587-590.

Morris, J.T., 1991: Effects of nitrogen loading on wetland ecosystems with particular reference to atmospheric deposition. *Annual Review of Ecology and Systematics,* **22**, 257-279.

Mullié, W.C. and J. Brouwer, 1994: Les zones humides du Niger: écologie, écotoxicologie et importance pour les oiseaux d'eau afrotropicaux et paléarctiques. In: *Atelier sur les Zones Humides du Niger, Comptes Rendus d'un Atelier à la Tapoa, Parc du 'W,' Niger, du 2 au 5 Novembre 1994* [Kristensen, P. (ed.)]. UICN-Niger, Niamey, Niger, pp. 57-74.

Nicholson, S.E., 1978: Climatic variations in the Sahel and other African regions during the past five centuries. *J. Arid Environment*, **1**, 3-24.

Nicholson, S.E., 1994: Recent rainfall fluctuations in Africa and their relationship to past conditions over the continent. *Holocene*, **4**, 121-131.

Novitzki, R.P., 1979: Hydrologic characteristics of Wisconsin's wetlands and their influence on floods, stream flow, and sediment. In: *Wetlands Functions and Values: The State of Our Understanding* [Greeson, P.E., J.R. Clark, and J.E. Clark (eds.)]. American Water Resources Association Technical Application, Minneapolis, MN, pp. 377-388.

Novitzki, R.P., 1989: Wetland hydrology. In: *Wetlands Ecology and Conservation: Emphasis in Pennsylvania* [Majumdar, S.K., R.P. Brooks, F.J. Brenner, R.W. Tiner, Jr. (eds.)]. The Pennsylvanian Academy of Sci. Publ., Philipsburg, pp. 47-64.

Oechel, W.C. and B.R. Strain, 1985: *Native Species Response to Increased Atmospheric Carbon Dioxide Concentrations.* USDOE, NTIS, Springfield, VA, pp. 117-154.

Oechel, W.C., S.J. Hastings, G. Vourlitis, M. Jenkins, G. Riechers, and N. Grulke, 1993: Recent change of Arctic tundra ecosystems from a net carbon dioxide sink to a source. *Nature*, **361**, 520-523.

Oechel, W.C. and G.L. Vourlitis, 1994: The effects of climate change on land-atmosphere feedbacks in arctic tundra regions. *Tree*, **9**, 324-329.

OTA, U.S. Congress, 1993: *Wetlands, Preparing for an Uncertain Climate*, vol. 2. OTA-0-S68, U.S. Government Printing Office, Washington, DC.

Pastor, J., J.B. Aber, C.A. McClaugherty, and J.M. Melillo, 1984: Aboveground production and N and P cycling along a nitrogen mineralization gradient on Blackhawk Island, Wisconsin. *Ecology*, **65**, 256-268.

Penner, J.E., R.E. Dickinson, and C.A. O'Neil, 1992: Effects of aerosol from biomass burning on the global radiation budget. *Science*, **256**, 1432-1434.

Peters, R.L. and T.E. Lovejoy, 1992: *Global Warming and Biological Diversity.* Yale University Press, New Haven, CT, 386 pp.

PHPA and AWB-Indonesia, 1990: Integrating conservation and land-use planning, coastal region of south Sumatra Indonesia. PHPA, Bogor, Indonesia, 25 pp.

Piaton, H. and C. Puech, 1992: *Apport de la Télédétection pour l'Évaluation des Ressources en Eau d'Irrigation pour la Mise en Valeur des Plans d'Eau à Caractère Permanent ou Sémi-Permanent au Niger.* Rapport de synthèse, avec J.Carette, Ecole Polytechnique Fédérale de Lausannem Suisse, Comité Interafricain d'Etudes Hydrauliques, Ouagadougou, Burkina Faso, avec l'aide du Laboratoire Commun de Télédétection CEMAGREF-ENGREF.

Pickett, S.T.A. and V.T. Parker, 1994: Avoiding the old pitfalls: opportunities in a new discipline. *Restoration Ecology*, **2**, 75-79.

Poiani, K.A. and W.C. Johnson, 1989: Effects of hydroperiod on seed-bank composition in semi-permanent prairie wetlands. *Can. J. Bot.*, **67**, 856-864.

Poiani, K.A. and W.C. Johnson, 1993: Potential effects of climate change on a semi-permanent prairie wetland. *Climatic Change*, **24**, 213-232.

Poiani, K.A., W.C. Johnson, T.G.F. Kittel, 1995: Sensitivity of a prairie wetland to increased temperature and seasonal precipitation changes. *Water Resources Bulletin*, **31**, 283-294

Prather, M., R. Derwent, D. Ehhalt, P. Fraser, E. Sanhueza, X. Zhou, 1994: Other trace gases and atmospheric chemistry. In: *Radiative Forcing of Climate Change and an Evaluation of the IPCC IS92 Emission Scenarios* [Houghton, J.T., L.G. Meira Filho, J. Bruce, H. Lee, B.A. Callander, E. Haites, N. Harris, and K. Maskell (eds.)]. Cambridge University Press, Cambridge, UK, pp. 73-126.

Raverdeau, F., 1991: *La Contre Saison au Niger.* Etude des Systèmes de Culture dans le Départements de Tillabery et Dosso, Université de Niamey, Faculté d'Agronomie, Niamey, Niger, 130 pp. + annexes.

Reader, R.J., 1978: Primary production in northern bog marshes. In: *Freshwater Wetlands—Ecological Processes and Management Potential* [Wigham, D., R.E. Good, C.G. Jackson, Jr. (eds.)]. Academic Press, Inc., New York, NY, pp. 53-62.

Reed, S.C. and D.S. Brown, 1992: Constructed wetland design—the first generation. *Water Environment Research*, **64**, 776-781.

Reenberg, A., 1994: Land-use dynamics in the Sahelian zone in eastern Niger—monitoring change in cultivation strategies in drought prone areas. *J. Arid Environments*, **27**, 179-192.

Reynolds, R.E. 1987: Breeding duck population, production and habitat surveys, 1979-1985. *Trans. N. Am. Wildl. Nat. Resour. Conf.*, **52**, 186-205.

Reimold, R.J. and M.A. Hardinsky, 1979: Nonconsumptive use values of wetlands. In: *Wetlands Functions and Values: The State of Our Understanding* [Greeson, P.E., J.R. Clark, and J.E. Clark (eds.)]. American Water Resources Association Technical Application, Minneapolis, MN, pp. 558-564.

Rieley, J.O., R.G. Sieffermann, M. Fournier, and F. Soubies, 1992: The peat swamp forests of Borneo: their origin, development, past and present vegetation and importance in regional and global environmental processes. In: *Proceedings of the 9th International Peat Congress, Uppsala, Sweden,* **1** (special edition of the International Peat Journal), 78-95 .

Rodhe, H. and B.H. Svensson, 1995: Impact on the greenhouse effect of peat mining and combustion. *Ambio*, **24**, 221-225.

Rose, P.M. and D.A. Scott, 1994: *Waterfowl Population Estimates.* IWRB Publ. no. 29, International Waterfowl and Wetland Research Bureau, Slimbridge, UK, 102 pp.

Rosswall, T. and U. Granhall, 1980: Nitrogen cycling in a subarctic ombrotrophic mire. In: *Ecology of a Subarctic Mire* [Sonesson, M. (ed.)]. Ecological Bulletin 30, Stockholm, Sweden, pp. 209-234.

Roulet, N.T., T.R. Moore, J. Bubier, and P. Lafleur, 1992: Northern fens: methane flux and climate change. *Tellus*, **44B**, 100-105.

Roulet, N.T., R. Ash, W. Quinton, and T.R. Moore, 1993: Methane flux from drained northern peatlands: effect of a persistent water table lowering on flux. *Global Biogeochem. Cycles*, **7**, 749-769.

Salati, E. and P.B. Vose, 1984: Amazon basin: a system in equilibrium. *Science*, **225**, 129-138.

Sally, L., M. Kouda, and N. Beaumond, 1994: Zones humides du Burkina Faso. In: *Compte Rendu d'un Séminaire sur les Zones Humides du Burkina Faso* [Sally, L., M. Kouda, and N. Beaumond (eds.)]. IUCN Wetlands Programme, Gland, Switzerland, 290 pp.

Schipper, L.A., C.G. Harfoot, P.N. McFarlane, and A.B. Cooper, 1994: Anaerobic decomposition and denitrification during plant decomposition in an organic soil. *Journal of Environmental Quality*, **23**, 923-928.

SFWMD, 1992: *Surface Water Improvement and Management Plan for the Everglades.* Supporting Information Document, South Florida Water Management District, West Palm Beach, Florida, 472 pp.

Sieffermann, R.G., M. Fournier, S. Truitomo, M.T. Sadelman, and A.M. Semah, 1988: Velocity of tropical peat accumulation in Central Kalimantan Province, Indonesia (Borneo). In: *Proceedings of the 8th International Peat Congress*, vol. 1. International Peat Society, Leningrad, USSR, pp. 90-98.

Sieffermann, R.G., J.O. Rieley, and M. Fournier, 1992: The lowland peat swamps of central Kalimantan (Borneo): a complex and vunerable ecosystem. Proceedings of the International Conference on Geography in the Asian Region, Yogyakarta, Indonesia, 26 pp.

Silvius, M.J., 1989: Indonesia. In: *A Directory of Asian Wetlands* [Scott, D.A. (comp.)]. IUCN, Gland, Switzerland, and Cambridge, UK, pp. 981-1109.

Silvola, J., J. Vaälijoki, and H. Aaltonen, 1985: Effect of draining and fertilization on soil respiration at three ameliorated peatland sites. *Acta Forestalia Fennica*, **191**, 1-32.

Silvola, J., 1986: Carbon dioxide dynamics in mires reclaimed for forestry in eastern Finland. *Ann. Bot. Fenn.*, **23**, 59-67.

Sjörs, H., 1950: Regional studies in north Swedish mire vegetation. *Bot. Notiser*, **1950**, 173-222.

Sjörs, H., 1980: Peat on earth: multiple use or conservation. *Ambio*, **9**, 303-308.

Sorensen, K.W., 1993: Indonesian peat swamp forests and their role as a carbon sink. *Chemosphere*, **27**, 1065-1082.

Stephens, J.C., 1984: Subsidence of organic soils in the Florida Everglades. In: *Environments of South Florida, Past and Present* [Gleason, P.J. (ed.)]. Miami Geological Society, Miami, FL, pp. 375-384.

Stewart, J.M. and R.E. Wheatley, 1990: Estimates of CO_2 production from eroding peat surfaces. *Soil Biol. Biochem*, **22**, 65-68.

Sundh, I., C. Mikkelä, M. Nilsson, and B.H. Svensson, 1994a: Potential aerobic methane oxidation in a Sphagnum dominated peatland—controlling factors and relation to methane emission. *Soil Biol. Biochem.*, **27**, 829-837.

Sundh, I., M. Nilsson, G. Granberg, and B.H. Svensson, 1994b: Depth distribution of microbial production and oxidation of methane in northern boreal peatlands. *Microb. Ecol.*, **27**, 253-265.

Svensson, B.H., A.K. Veum, and S. Kjellvik, 1975: Carbon losses from tundra soil. In: *Ecological Studies Analysis and Synthesis*, vol. 16 [Wielgolaski, F.E. (ed.)]. Fennoscandian Tundra Ecosystems, part 1, Springer-Verlag, Berlin, Germany, pp. 279-286.

Svensson, B.H., 1976: Methane production in tundra peat. In: *Microbial Production and Utilization of Gases (H_2, CH_4, CO)* [Schlegel, H.G., K.G. Gottschal, and N. Pfennig (eds.)]. E Goltze KG, Göttingen, Germany, pp. 135-139.

Svensson, B.H., 1980: Carbon dioxide and methane fluxes from the ombrotrophic parts of a subarctic mire. In: *Ecology of a Subarctic Mire* [Sonesson, M. (ed.)]. Ecological Bulletin 30, Stockholm, Sweden, pp. 235-250.

Svensson, B.H. and T. Rosswall, 1984: *In situ* methane production from acid peat in plant communities with different moisture regimes in a subarctic mire. *Oikos*, **43**, 341-350.

Svensson, B.H., 1986: Methane as a part of the carbon mineralization in a tundra mire. In: *Perspectives in Microbial Ecology* [Megusar, F. and M. Ganthar (eds.)]. Slovene Soc. Microbiology, Ljubljana, Yugoslavia, ISME, pp. 611-616.

Swanson, G.A., T.C. Winter, V.A. Adomaitis, and J.W. LaBaugh, 1988: Chemical characteristics of prairie lake in south-central North Dakota—their potential for influencing use by fish and wildlife. *U.S. Fish. Wildl. Serv. Fish Wildl. Tech. Rep.*, **18**, 1-44.

Tamm, C.O., 1951: Chemical composition of birch leaves from drained mire, both fertilized with wood ash and unfertilized. *Svensk Bot. Tidskr.*, **45**, 309-319.

Tamm, C.O., 1965: Some experiences from forest fertilization trials in Sweden. *Silva Fenn.*, **117(3)**, 1-24.

Thompson, K. and A.C. Hamilton, 1983: Peatlands and swamps of the African continent. In: *Ecosystems of the World.* Vol. 4B, *Mires: Swamp, Bog, Fen, and Moor* [Gore, A.J.P. (ed.)]. Elsevier Sci. Publ., New York, NY, pp. 331-373.

Urban, N.R., S.J. Eisenreich, and S.E. Bayley, 1988: The relative importance of denitrification and nitrate assimilation in mid continental bogs. *Limnology and Oceanography*, **33**, 1611-1617.

U.S. ACOE, 1994: Central and southern Florida project, review study news, December 1994, issue no. 3, U.S. Army Corps of Engineers, Jacksonville District, Jacksonville, FL.

van der Graaf, S. and H. Breman, 1993: *Agricultural Production: Ecological Limits and Possibilities.* Rapports PSS no. 3, Projet Production Soudano-Sahélienne, IER-Mali, CABO-DLO, WAU and IB-DLO, Prepared for the Club du Sahel, Wageningen, Netherlands, 39 pp.

van Molle, M. and P. van Ghelue, 1991: Global change and soil erosion. *Acta Geologica Taiwanica*, **29**, 33-45.

van der Valk, A.G. and C.B. Davis, 1978: Primary production of prairie glacial marshes. In: *Freshwater Wetlands—Ecological Processes and Management Potential* [Good, R.E., D.F. Wigham, R.L. Simpson, and C.G. Jackson, Jr., (eds.)]. Academic Press, Inc., New York, NY, pp. 21-37.

van der Valk, A.G., C.B. Davis, J.L. Baker, and C.E. Beer, 1979: Natural freshwater wetlands as nitrogen and phosphorus traps for land runoff. In: *Wetlands Functions and Values: The State of Our Understanding* [Greeson, P.E., J.R. Clark, and J.E. Clark (eds.)]. American Water Resources Association Technical Application, Minneapolis, MN, pp. 457-467.

van der Valk, A.G. and R.W. Jolly, 1992: Recommendations for research to develop guidelines for the use of wetlands to control rural NPS pollution. *Ecological Engineering*, **1**, 115-134.

van Vuuren, M.M.I., R. Aerts, F. Berendse, and W. De Visser, 1992: Nitrogen mineralization in heathland ecosystems dominated by different plant species. *Biogeochemistry*, **16**, 151-166.

van Vuuren, M.M.I., F. Berendse, and W. De Visser, 1993: Species and site differences in the decomposition of litters and roots from wet heathlands. *Canadian Journal of Botany*, **71**, 167-173.

Verhoeven, J.T.A., E. Maltby, and M.B. Schmitz, 1990: Nitrogen and phosphorus mineralization in fens and bogs. *Journal of Ecology*, **78**, 713-726.

Verhoeven, J.T.A. and H.H.M. Arts, 1992: Carex litter decomposition and nutrient release in mires with different water chemistry. *Aquatic Botany*, **43**, 365-377.

Walters, C.L., L.H. Gunderson, and C.S. Hollings, 1992: Experimental policies for water management in the Everglades. *Ecological Applications*, **2**, 189-202.

Wanless, H.R., R.W. Parkinson, and L.P. Tedesco, 1994: Sea level control on the stability of Everglades wetlands. In: *Everglades: The Ecosystem and Its Restoration* [Davis, S.M. and J.C. Ogden (eds.)]. St. Lucie Press, Delray Beach, FL, pp. 199-223.

Warner, B.G., R.S. Clymo, and K. Tolonen, 1993: Implications of peat accumulation at Point Escuminac, New Brunswick. *Quaternary Research*, **39**, 245-248.

Weisner, S.E.B., P.G. Eriksson, W. Graneli, and L. Leonardson, 1994: Influence of macrophytes on nitrate removal in wetlands. *Ambio*, **23**, 363-366.

Whalen, S.C. and W.S. Reeburgh, 1992: Interannual variations in tundra methane emissions: a 4-year time series at fixed sites. *Global Biogeochem. Cycles*, **6**, 139-159.

Whalen, S.C., W.S. Reeburgh, and C.E. Reimers, 1996: Control of tundra methane emission by microbial oxidation. In: *Landscape Function: Implications for Ecosystem Response to Disturbance, a Case Study in Arctic Tundra* [Reynolds, J.F. and J.D. Tenhunen (eds.)]. Springer Verlag, Berlin, Germany, pp. 257-274.

Whitten, A.J., S.J. Damanik, J. Anwar, and N. Hisam, 1987: *The Ecology of Sumatra.* Gadjah Mada University Press, Yogyakarta, Indonesia, pp. 219-248

Williams, M. 1990: Understanding wetlands. In: *Wetlands: A Threatened Landscape* [Williams, M. (ed.)]. Basil Blackwell, Ltd., Oxford, UK, pp. 1-41.

Windmeijer, P.N. and W. Andriesse 1993: *Inland Valleys in West Africa: An Agro-Ecological Characterization of Rice-Growing Environments.* ILRI Publication no. 52, Institute for Land Reclamation and Improvement, Wageningen, Netherlands, 160 pp.

Winter, T.C., 1988: A conceptual framework for assessing cumulative impacts on hydrology of nontidal wetlands. *Environmental Management,* **12**, 605-620.

Woodwell, G.M., F.T. Mackenzie, R.A. Houghton, M.J. Apps, E. Gorham, and E.A. Davidson, 1995: Will the warming speed the warming? In: *Biotic Feedbacks in the Global Climatic System* [Woodwell, G.M. and F.T. MacKenzie (eds.)]. Oxford University Press, Oxford, UK, pp. 393-411.

Zedler, J.B. and M.W. Weller, 1990: Overview and future directions. In: *Wetland Creation and Restoration: The Status of the Science* [Kusler, J.A. and M.E. Kentula (eds.)]. Island Press, Washington DC, pp. 459-460.

Zhang, Y. and J. Song, 1993: The potential impacts of climate change in the vegetation in northeast China. In: *Climate Change and Its Impact.* Meteorology Press, Beijing, China, pp. 178-193 (in Chinese).

Zimov, S.A., G.M. Zimova, S.P. Daviodov, A.I. Daviodova, Y.V. Voropaev, Z.V. Voropaeva, S.F. Prosiannikov, and O.V. Prosiannikova, 1993: Winter biotic activity and production of CO_2 in Siberian soils: a factor in the greenhouse effect. *Journal of Geophysical Research*, **98**, 5017-5023.

Zoltai, S.C. and F.C. Pollett, 1983: Wetlands in Canada. In: *Ecosystems of the World.* Vol. 4B, *Mires: Swamp, Bog, Fen, and Moor* [Gore, A.J.P. (ed.)]. Elsevier Sci. Publ., New York, NY, pp. 245-268.

Zoltai, S.C. and R.W. Wein, 1990: Development of permafrost in peatlands of northwestern Alberta. In: *Programme and Abstracts, Annual Meeting.* Canadian Association of Geographers, Edmonton, Alberta, Canada, p. 195.

7

The Cryosphere: Changes and Their Impacts

B. BLAIR FITZHARRIS, NEW ZEALAND

Principal Lead Authors:
I. Allison, Australia; R.J. Braithwaite, Denmark; J. Brown, USA; P.M.B. Foehn, Switzerland; W. Haeberli, Switzerland; K. Higuchi, Japan;V.M. Kotlyakov, Russia; T.D. Prowse, Canada; C.A. Rinaldi, Argentina; P. Wadhams, UK; M.-K. Woo, Canada; Xie Youyu, China

Contributing Authors:
O. Anisimov, Russia; A. Aristarain, Argentina; R.A. Assel, USA; R.G. Barry, USA; R.D. Brown, Canada; F. Dramis, Italy; S. Hastenrath, USA; A.G. Lewkowicz, Canada; E.C. Malagnino, Argentina; S. Neale, New Zealand; F.E. Nelson, USA; D.A. Robinson, USA; P. Skvarca, Argentina; A.E. Taylor, Canada; A. Weidick, Denmark

CONTENTS

EXECUTIVE SUMMARY

The cryosphere covers a substantial amount of the Earth's surface and is very sensitive to climate change. This chapter updates the 1990 IPCC Impacts Assessment—which examined snow, glaciers, and permafrost—but also considers snow avalanches and sea, river, and lake ice. Impacts on feedbacks to global climate and sea level caused by changes in the cryosphere are covered in other chapters of the IPCC Second Assessment Report.

This chapter arrives at the following conclusions:

- Many components of the cryosphere are sensitive to changes in atmospheric temperature because of their thermal proximity to melting. The extent of glaciers has often been used as an indicator of past global temperatures (High Confidence).
- Projected warming of the climate will reduce the area and volume of the cryosphere. This reduction will have significant impacts on related ecosystems, associated people, and their livelihoods (High Confidence).
- There will be striking changes in the landscapes of many high mountain ranges and of lands at northern high latitudes (High Confidence). These changes may be exacerbated where they are accompanied by growing numbers of people and increased economic activity (Medium Confidence).

From an examination of past measurements of the cryosphere, the following 20th-century trends are observed:

- Obvious thinning, mass-loss, and retreat of mountain glaciers (High Confidence): The extent of alpine ice in the European Alps probably is more reduced today than at any time during the past 5,000 years.
- Borehole measurements show that permafrost is warming in some areas but not everywhere (Medium Confidence).
- Later freeze-up and earlier break-up dates for river and lake ice in the tundra and boreal lands: These are at least a week different compared to last century (Medium Confidence).
- No convincing evidence of trends in Antarctic or Arctic sea-ice extent (Low Confidence).
- Much variability of seasonal snow from year to year but no definitive trends, except that the areal extent of Northern Hemisphere continental snow cover has decreased since 1987 (High Confidence).
- Little change in the gross features of ice sheets (Medium Confidence).

If projections of climate for the year 2050 are realized (UKMO transient experiment data; Greco *et al.*, 1994), then the following impacts on the cryosphere are likely:

- Pronounced reductions in seasonal snow, permafrost, glacier, and periglacial belts of the world, with a corresponding shift in landscape processes (High Confidence).
- Disappearance of up to a quarter of the presently existing mountain glacier mass (Medium Confidence).
- Increases in the thickness of the active layer of permafrost and the disappearance of extensive areas of discontinuous permafrost in continental and mountain areas (High Confidence).
- Less ice on rivers and lakes. Freeze-up dates will be delayed, and break-up will begin earlier. The river-ice season could be shortened by up to a month. Many rivers within the temperate regions will become ice-free or develop only intermittent or partial ice coverage (Medium Confidence).
- A large change in the extent and thickness of sea ice, not only from warming but also from changes in circulation patterns of both atmosphere and oceans. There is likely to be substantially less sea ice in the Arctic Ocean (Medium Confidence).
- Major changes in the volume and extent of ice sheets and deep, continuous permafrost are unlikely by 2050 because they are very cold and react with longer time lags (High Confidence); however, unforeseen changes in the West Antarctic ice sheet still could occur (Low Confidence).

As a result of these changes in the cryosphere, the following impacts on other systems are expected:

- More water will be released from regions with extensive glaciers (High Confidence). In some semi-arid places near high mountains, such as in central Asia and Argentina, this glacial runoff may increase water resources. In other places, summer water resources may diminish as glaciers disappear (Medium Confidence).
- In temperate mountain regions, reduced snow cover will cause moderation of the seasonal flow regime of rivers, so that winter runoff increases and spring runoff decreases (Medium Confidence). Such changes may benefit the hydroelectricity industry (Medium Confidence).
- Reduced snow cover and glaciers will detract from the scenic appeal of many alpine landscapes

(Medium Confidence). For temperate mountains, less snow will restrict alpine tourism and limit the ski industry to higher alpine areas than at present. Snow seasons will tend to be shorter and less reliable, and there will be detrimental socioeconomic impacts on mountain communities that depend on winter tourism.

- Widespread loss of permafrost over extensive continental and mountain areas will trigger erosion or subsidence of ice-rich landscapes, change hydrologic processes, and release carbon dioxide and methane to the atmosphere (Medium Confidence).
- Cryospheric change will reduce slope stability and increase the incidence of natural hazards for people, structures, and communication links in mountain lands and continental permafrost areas. Buildings, other structures, pipelines, and communication links will be threatened (Medium Confidence).
- Engineering and agricultural practices will need to adjust to changes in snow, ice, and permafrost distributions (High Confidence).
- Thawing of permafrost could lead to disruption of existing petroleum production and distribution systems in the tundra, unless mitigation techniques are adopted. Reduced sea ice may aid new exploration and production of oil in the Arctic basin (Medium Confidence).
- Less sea ice could reduce the renewal of deep waters of the North Atlantic, affect the ocean conveyor system, decrease albedo, and consequently induce climate feedbacks (Low Confidence).
- Improved opportunities for water transport, tourism, and trade at high latitudes are expected from a reduction in sea, river, and lake ice. These will have important implications for the people and economies of the Arctic rim (Medium Confidence).

7.1. Introduction

The cryosphere, which represents all global snow, ice, and permafrost, contains nearly 80% of all freshwater. It includes seasonal snow, mountain glaciers, ice caps, ice sheets, seasonally frozen soils, permafrost, river ice, lake ice, and sea ice (Table 7-1). Permafrost underlies as much as 25% of the global land surface. Seasonal snow has the largest area of any component of the cryosphere; at its maximum in late winter it covers almost 50% of the land surface of the Northern Hemisphere. A huge proportion of the mass of the cryosphere is contained in ice sheets, but at time scales of a century or less they are least sensitive to climate change.

The cryosphere is an important part of Earth's geographical and climate systems, and its components change over diverse time scales. Areas of snow and sea ice expand and contract markedly with the seasons (Table 7-1), and ice sheets underwent vast change as the Earth became warmer or cooler from the last interglacial period. During the previous interglacial—when global temperature was higher than at present—the global cryosphere was smaller, but it had twice its current mass during the last glaciation.

This chapter brings up to date the 1990 IPCC Impacts Assessment that dealt with seasonal snow cover, ice, and permafrost (Street and Melnikov, 1990) and the IPCC 1992 Supplementary Report (Melnikov and Street, 1992). Snow avalanches and sea, river, and lake ice are considered for the first time, and substantial new material has been added on glaciers and permafrost. How each component of the cryosphere is changing and its sensitivity to climate change is assessed, as is the impact of climate change on the cryosphere by about 2050. How these changes in the cryosphere might affect other physical and human systems is also described. Critical information we need to know is identified in the final section.

Although the cryosphere forms an integral part of the climate system, the ways in which it generates important feedbacks into the climate system are not discussed because this information is contained in the Working Group I volume. Similarly, the interaction of ice sheets and sea-level change are considered in detail in the Working Group I volume (see Chapter 7, *Changes in Sea Level*). A summary of ice sheets and climate change is offered here in Box 7-1.

Table 7-1: *Estimate of the size of the cryosphere (modified from U.S. DOE, 1985; Barry, 1985; Street and Melnikov, 1990; Meier, 1993; Gloersen et al., 1992; and Chapter 7, Changes in Sea Level, of the Working Group I volume).*

Source	Area (10^6 km^2)	Ice Volume (10^6 km^3)
Seasonal Snow		
N. Hemisphere winter	46.3	<0.01
N. Hemisphere summer	3.7	
S. Hemisphere winter	0.9	
S. Hemisphere summer	<0.1	
Ice Caps and Glaciers	0.6	0.09
Ice Sheets		
Greenland	1.7	2.95
West Antarctica	2.4	3.40
East Antarctica	9.9	25.92
Antarctic ice shelves	1.6	0.79
Permafrost	25.4	0.16
River and Lake Ice	<1.0	
Sea Ice		
N. Hemisphere winter	16.0	0.05
N. Hemisphere summer	9.0	0.03
S. Hemisphere winter	19.0	0.03
S. Hemisphere summer	3.5	<0.01

7.2. Is the Cryosphere Changing?

7.2.1. *Snow*

Snow tends to be a very transient part of the Earth's surface, often lasting for only a few days or months. Monitoring of global seasonal snow is practical only with satellite remote sensing, so there are no reliable records prior to 1971. Records since then show considerable variability in the Northern Hemisphere continental snow cover from year to year (Figure 7-1), which makes long-term trends difficult to detect. The extent of snow has been less since 1987 (Robinson *et al.*, 1993), with the largest negative snow anomalies occurring in spring. Recent analysis of directly measured snow cover over the North American Great Plains (Brown *et al.*, 1994) reveals

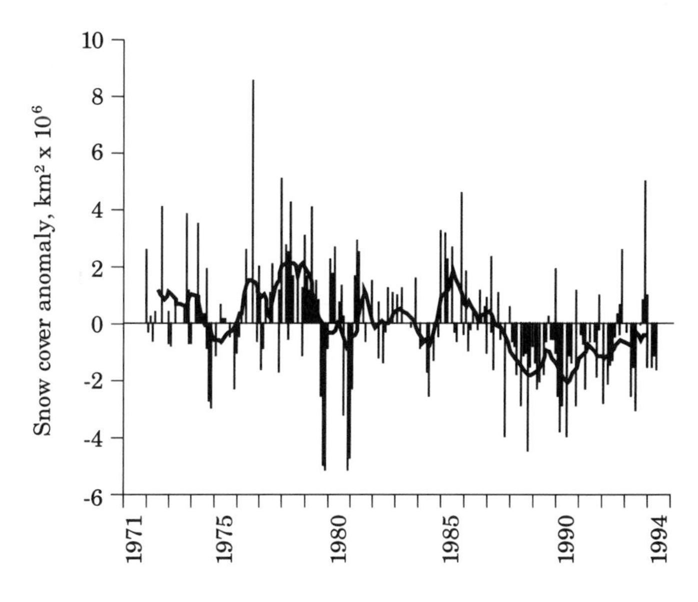

Figure 7-1: Seasonal snow cover anomalies in the Northern Hemisphere for 1971–1994 (after Robinson *et al.*, 1993). Each vertical line represents a season, with the dark line a 12-month running mean.

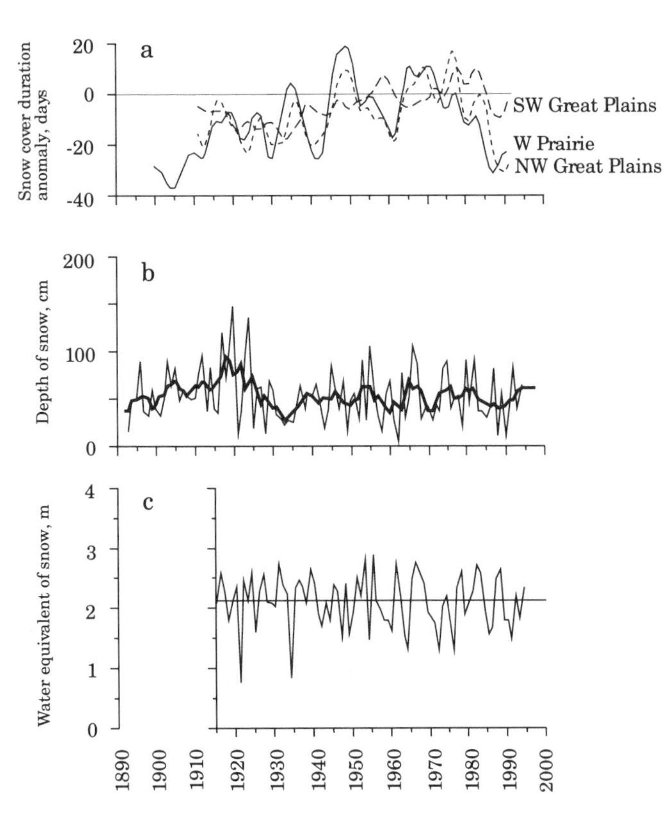

Figure 7-2: Historical variability of snow for three different settings: (a) Snow-cover duration anomalies over the continental interior of North America (after Brown *et al.*, 1994); (b) depth of seasonal snow at an alpine area at Davos, elevation 1560 m, in Switzerland, as measured every 1 January 1893–1994 (updated from Foehn, 1990); and (c) water equivalent of snow near the equilibrium line of a glacier at Claridenfirn, elevation 2700 m, in Switzerland, as measured every spring 1914–1993 (updated from Müller and Kaeppenburger, 1991).

an increase over the past century (Figure 7-2a) but a decline over the Canadian Prairies.

The longest alpine snow-cover time series come from the European Alps. Recently, there has been a trend to lower snow depths in early winter but delayed ablation in spring; however, the century-long record for Davos (Figure 7-2b) shows no obvious trend in snow depth but large interannual variability (Foehn, 1990). This lack of any trend is reinforced by Figure 7-2c, which shows snow water equivalent near the glacier equilibrium line at Claridenfirn (Müller and Kaeppenberger, 1991). Elsewhere, a remarkable decrease in accumulated and maximum snow depth is reported since the 1987–88 winter over a wide area of western Japan. Here, snowfall normally is heavy and is affected by the winter monsoon (Morinaga and Yasunari, 1993).

Few observations assess long-term trends in the structure of snow. In the European Alps, it is summarized by ram (ramsonde) profiles, which are catagorized into six main types. Each type has a certain potential for snow avalanche formation and is linked to a specific winter climate. Over the last 50 years in Switzerland there has been no trend in ram profiles. A few long-term records of avalanche occurrence have been assembled, notably from Iceland (Bjornson, 1980), Europe (Fitzharris and Bakkehoi, 1986), western Canada (Fitzharris and Schaerer, 1980), and the United States (Armstrong, 1978). They demonstrate that although avalanches reach catastrophic proportions in a few winters, there are no clear temporal trends or regular periodicities.

7.2.2. *Ice Caps and Glaciers*

Internationally coordinated, long-term monitoring of glaciers started in 1894 and today involves collection and publication of standardized information on the distribution and variability of glaciers over space (glacier inventories) and time (glacier fluctuations: mass balance, length change). Information on special events (instabilities, catastrophic changes) also is available (UNEP, 1993). The results of long observational series on fluctuations of mountain glaciers represent convincing evidence of past climatic change on a global scale [e.g., IAHS(ICSI)/UNEP/UNESCO, 1988, 1989, 1993, 1994].

Mass loss and retreat of glaciers is common in many mountain areas of the world. Oerlemans (1994) looks at glacial retreat on a global level, using a scaling system to classify different glaciers. He concludes that during the period 1884–1978, mean global glacial retreat corresponded to a calculated warming of 0.66 ± 0.10°C per century. Chapter 7, *Changes in Sea Level*, of the Working Group I volume suggests that glaciers have lost sufficient ice over the last 100 years to raise sea level by 0.2–0.4 mm/yr. Recent analyses by Oerlemans and Fortuin (1992), Meier (1993), and Dyurgerov (1994) all show negative global ice-mass balances during this century. Although the glacial signal appears homogeneous at the global scale, there is great variability at local and regional scales and over shorter time periods of years to decades (Letréguilly and Reynaud, 1989).

Since the end of the Little Ice Age, the glaciers of the European Alps have lost about 30 to 40% of their surface area and about 50% of their ice volume. On average, this rate of icemelt is roughly one order of magnitude higher than the overall mean calculated for the end of the last glaciation and is broadly consistent with anthropogenic greenhouse forcing of 2–3 W/m^2. The recent discovery of a stone-age man from cold ice/permafrost on a high-altitude ridge of the Oetztal Alps confirms the results of earlier moraine investigations: The extent of Alpine ice probably is more reduced today than at any time during the past 5,000 years. Glacier wastage in the European Alps appears to be accelerating (Haeberli, 1994).

In the circum-Arctic, there is a consistent tendency for negative mass balances over the past 30 years. Mass balances of two Canadian ice caps are slightly negative (Koerner and Brugman, 1991), but these are so cold that the main signal is likely to be in the change of firn temperatures. The general picture from West Greenland is that of a strong retreat through this century, with a trim line zone around many glacier lobes. For the North Greenland mountain glaciers and ice caps, the situation is less clear, but the general impression is that changes are small. East

Greenland mountain glaciers seem to have behaved as the West Greenland ones, but surging advances occur frequently in certain regions (Weidick, 1991a-d). The longest glacier record for northern Sweden shows a preponderance of negative mass-balance years since 1946 (Letréguilly and Reynaud, 1989). In Spitsbergen, several glaciers have been losing mass (Hagen and Liestol, 1990), although the extent of glaciers probably was less about 5,000 years ago (Fujii *et al.,* 1990).

In Asia, the area of glaciers in Kazakhstan reduced by 14%, their number diminished by 15%, and the general volume of ice reduced by 11% between 1955 and 1979 (Glazyrin *et al.,* 1986, 1990; Dikikh and Dikikh, 1990; Kotlyakov *et al.,* 1991; Popovnin, 1987). Fluctuations of 224 glaciers in Central Asia from the 1950s to the 1980s can be summarized as retreating (73%), advancing (15%), and stable (12%) (Shi and Ren, 1988, 1990).

The general picture of ice retreat continues for mountains in the tropics. It began around the middle of the 19th century in the Ecuadorian Andes and in New Guinea but only after 1880 in East Africa (Hastenrath, 1994). Schubert (1992) documents fast, ongoing glacier shrinkage in Venezuela, as do Hastenrath and Ames (1995) for the Yanamarey glacier in the Cordillera Blanca of Peru.

The Southern Hemisphere record is not as detailed as that for the Northern Hemisphere. In New Zealand, Ruddell (1990) shows that most glaciers have retreated during the 20th century, some by three kilometers or more. The surface of the Tasman Glacier has thinned by more than 100 m. This widespread recession has reversed for western glaciers since about 1983. In South America, the Upsala glacier has retreated about 60 m/yr over the last 60 years, and this rate seems to be accelerating (Malagnino and Strelin, 1992). The area of the South Patagonian Ice Field has diminished by about 500 km^2 from 13,500 km^2 in 41 years. Surface lowering also has been considerable—more than 100 m at ablation areas of some glaciers (Aniya *et al.,* 1992). The Soler Glacier thinned at a rate of 5.2 m/yr from 1983 to 1985 (Aniya *et al.,* 1992), and the surface of Tyndall Glacier (one of the southernmost outlet glaciers) lowered by 20 m between 1985 and 1990 (a rate of 4.0 m/yr), according to Kadota *et al.* (1992). A few glaciers are advancing: The Pio XI glacier in Patagonia is larger now than it has been at any time in the past 6,000 years (Warren, 1994).

Frontal positions of alpine glaciers of the Dry Valleys in Antarctica have fluctuated, with no apparent trends (Chinn, 1993). On sub-Antarctic Heard Island, in the Indian Ocean, there has been widespread retreat since 1947—with some small glaciers decreasing in area by as much as 65% (Allison and Keate, 1986).

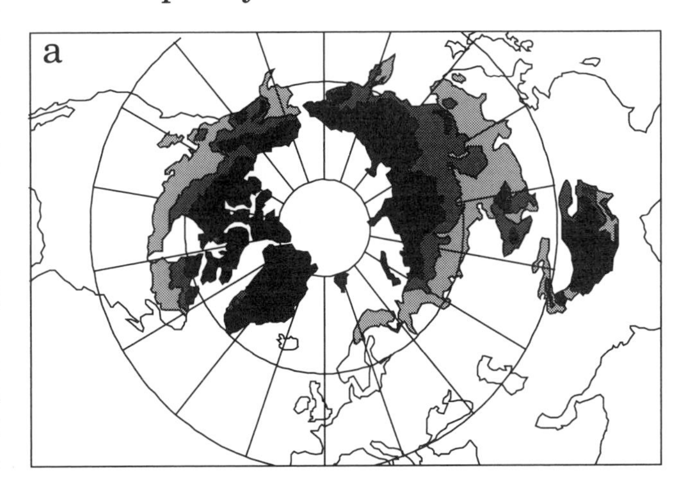

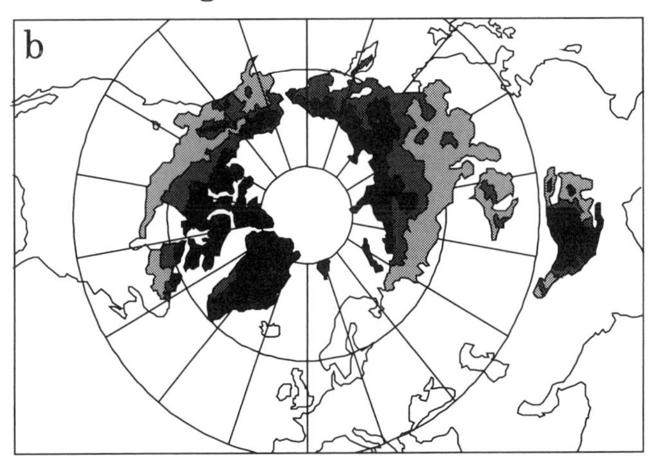

Figure 7-3: Distribution of Northern Hemisphere permafrost for (a) the present and (b) 2050 (based on Nelson and Outcalt, 1987; Anisimov and Nelson, 1995).

7.2.3. *Permafrost*

Permafrost—ground material that remains below freezing—underlies continental areas in the tundra and some boreal lands to considerable depths. It also is present under shallow polar seabeds, in ice-free areas in Antarctica, on some sub-Antarctic islands, and in many mountain ranges and high plateaus of the world (Cheng and Dramis, 1992; Harris and Giardino, 1994; King and Åkermann, 1994; Qiu, 1994). Figure 7-3a shows the contemporary distribution of permafrost in the Northern Hemisphere. Some is relict, having formed during colder glacial periods, but has survived due to the negative heat balance at the ground surface or the very long time it takes for deep permafrost to thaw. There also are many sedimentary structures that show that permafrost was once more extensive than today.

Long-term measurements in deep boreholes in Alaska (Harrison, 1991; Zhang and Osterkamp, 1993), Canada

(Taylor, 1991), and elsewhere (Koster *et al.*, 1994) demonstrate a distinct but spatially heterogeneous warming trend in lowland permafrost. In northern Alaska, Lachenbruch and Marshall (1986) have demonstrated a warming of the permafrost of 2–4°C over the last century. Temperatures along the coast have varied over a range of 4°C (Osterkamp, 1994). During the last two decades, permafrost in Russia and China also has warmed (Pavlov, 1994; Wang and French, 1994). Some discontinuous permafrost in the southern half of Alaska is currently thawing (Osterkamp, 1994).

First attempts are now being made to monitor the long-term evolution of high-mountain permafrost by aerial photogrammetry of permafrost creep, borehole measurements of permafrost deformation and temperature, data archiving from geophysical surface soundings for later repetition, and qualitative analysis of infrared aerial photography (Francou and Reynaud, 1992; Haeberli *et al.*, 1993; Harris, 1990; Vonder Mühll and Schmid, 1993; Wagner, 1992).

7.2.4. *River and Lake Ice*

Chronologies of river and lake ice formation and disappearance provide broad indicators of climate change over extensive lowland areas, in much the same way that glaciers provide indicators for mountains (Palecki and Barry, 1986; Reycraft and Skinner, 1993). River-ice data have been summarized for homogenous hydrologic regions of the former Soviet Union (FSU) over the period 1893–1985 and adjusted to account for any effects of water-resource development (Soldatova, 1992). Although there is appreciable interdecadal variability, there are significant long-term spatial patterns and temporal trends. Freeze-up on rivers such as the Danube, Dnieper, Don, Lower Volga, and rivers of the Black Sea region is now delayed by an average of two to three weeks compared with the early part of the record. Further east, a weaker trend to earlier freeze-up dates is observed for the Yenisey and Lena (Ginzburg *et al.*, 1992).

The same broad-scale pattern is evident for break-up dates (Soldatova, 1993). Break-up on major rivers such as the Upper Volga, Oka, Don, Upper Ob, and Irtysh has advanced by an average of 7–10 days during the last century. In some rivers, such as the Lower Don, the overall result is a reduction in the winter ice season by as much as a month. In Central and Eastern Siberia (e.g., Middle to Lower Yenisey and Upper Lena), some rivers exhibit later break-up dates and, hence, an overall expansion of the ice season. In northern Scandinavia, historical records as far back as 1693 indicate that break-up in the Tornelven river is occurring much earlier during the 20th century than in earlier times (Zachrisson, 1989).

Assel *et al.* (1995) analyze lake-ice freeze-up and break-up in North America using records from 1823–1994 at six sites throughout the Great Lakes. Freeze-up dates gradually become later and ice-loss dates gradually earlier from the beginning of the record to the 1890s but have remained relatively constant during the 20th century. Ice-loss dates at deeper-water environments with mixing of offshore waters were earlier during the 1940s and 1970s but later during the 1960s. Global warming during the 1980s was marked by a trend toward earlier ice-loss dates. Other North American lake-ice chronologies are presented by Hanson *et al.* (1992) and Reycraft and Skinner (1993).

7.2.5. *Sea Ice*

The most effective means of measuring sea-ice extent is through the use of satellite-borne passive microwave radiometers, but these data are available only from 1973 onward. They show no convincing evidence of trends in global sea-ice extent (Gloersen *et al.*, 1992; Parkinson and Cavalieri, 1989; Gloersen and Campbell, 1991). Studies of regional changes in both the Arctic and Antarctic indicate some trends (e.g., Mysak and Manak, 1989; Gloersen and Campbell, 1991; Parkinson, 1992), but longer data sets are needed because they are only of decadal length.

Knowledge of the regional variability of ice thickness comes almost entirely from upward sonar profiling by submarines (Wadhams, 1990a; Wadhams and Comiso, 1992), such as the large-scale maps of mean ice thickness in the Arctic shown in Figure 7-4 (based on Sanderson, 1988; Vinje, 1989; Wadhams, 1992). Variations in ice thickness are in accord with the predictions of numerical models, which take account of ice dynamics and deformation, as well as ice thermodynamics. Measurements from a series of submarine transects near the North Pole show large interannual variability in ice draft over the period 1979–1990 (McLaren *et al.*, 1992). There is evidence of a decline in mean thickness in the late 1980s relative to the late 1970s (Wadhams, 1994). Etkin (1991) found that ice breakup correlated well with melting degree-days in some areas of Hudson Bay and James Bay (Canada), but in other areas of the bays it was more strongly influenced by ice advection, freshwater inflow, and conditions within the air-water-ice boundary layer.

Winter pack ice around Antarctica has major climatic importance. The annual variation in sea-ice area is very large—from 3 x 10^6 km^2 in February to 19 x 10^6 km^2 in September (Gloersen *et al.*, 1992). No change in the thickness of Antarctic sea ice can be detected from the limited information available. The only systematic data have been obtained by repetitive drilling in a region of first-year ice from the eastern Weddell-Enderby Basin. The modal ice thickness is 0.5–0.6 m, and maximum observed keel drafts are about 6 m. In the limited regions of the Antarctic where multiyear ice occurs, there is a preferred ice thickness of about 1.4 m (Wadhams and Crane, 1991; Lange and Eicken, 1991).

7.3. How Sensitive is the Cryosphere to Climate Change?

7.3.1. *Snow*

Snow cover in temperate regions is generally thin (a few meters to a few centimeters) and often close to its melting

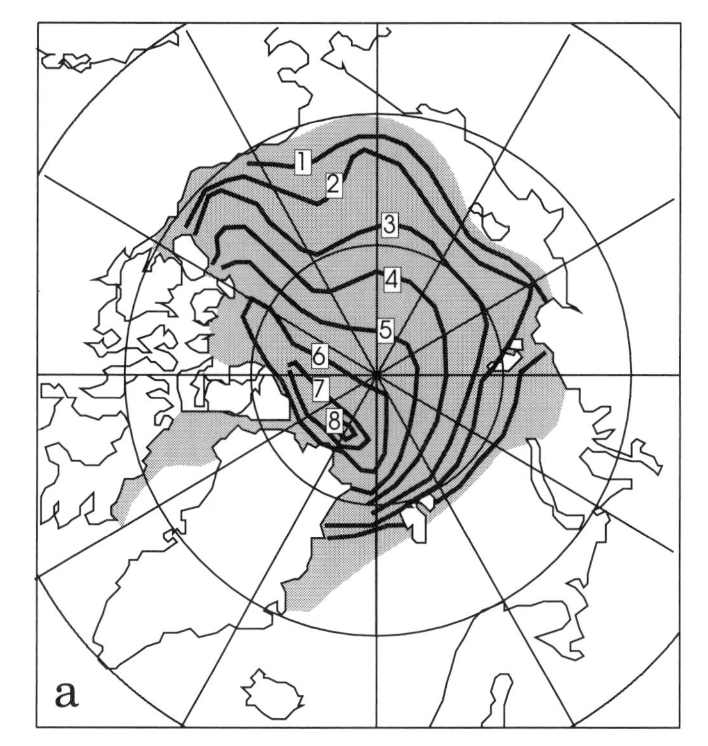

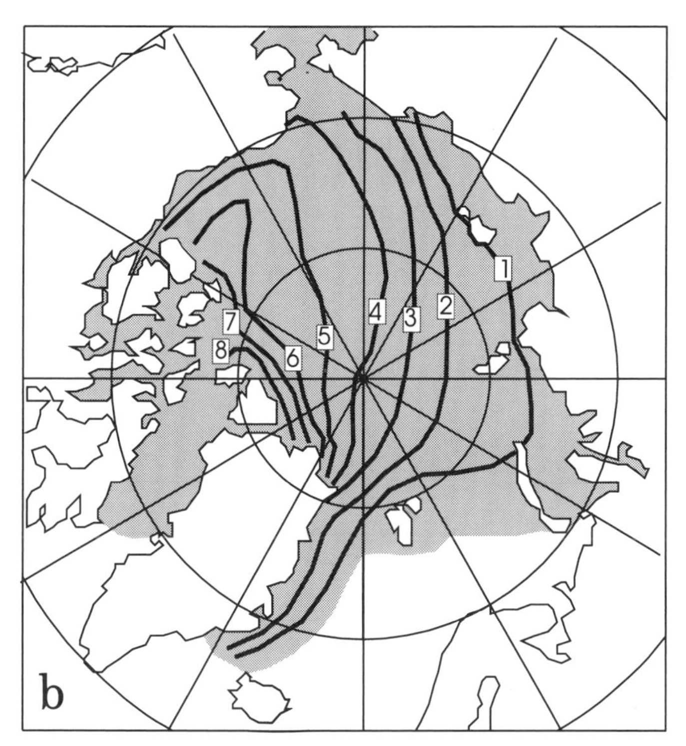

Figure 7-4: Estimated mean thickness of Arctic Basin sea ice in meters for (a) summer and (b) winter (after Sanderson, 1988; Bourke and Garrett, 1987). The data do not include open water, thus overestimating mean ice draft.

point; consequently, both continental and alpine snow covers are very sensitive to climate change. Karl *et al.* (1993) found that a 1°C increase in the annual temperature of the Northern Hemisphere results in a 20% reduction in North American snow cover. Snow accumulation and melt models can estimate water stored as seasonal snow in alpine areas and can be used to examine its sensitivity to climate change. For the Southern Alps of New Zealand, Fitzharris and Garr (1995) found that water stored as seasonal snow declines as temperature increases and precipitation decreases but that the relationship is not linear. Seasonal snow is more sensitive to decreases in temperature than to increases. As precipitation increases, snow accumulation becomes less sensitive to temperature changes. The volume of water stored also depends on catchment hypsometry. In Australia, the duration of snow cover is found to be very sensitive to changes in temperature. Large increases in precipitation (50%) are necessary to offset even a 0.5°C warming (CSIRO, 1994).

Large, catastrophic avalanches are mostly the result of special weather situations lasting for 5 to 10 days. In the Swiss Alps, only nine weather types out of a possible twenty-nine have been responsible for past catastrophic avalanche cycles. These large avalanche episodes are thus sensitive to the frequency of particular weather types. Similar findings also come from Norway (Fitzharris and Bakkehoi, 1986) and Canada (Fitzharris, 1987).

7.3.2. Ice Caps and Glaciers

In the chain of processes linking climate and glacial fluctuations, mass balance is the direct, undelayed reaction; glacier length variation is the indirect, delayed response. Averaged for glaciers with comparable geometry and over time intervals of decades, changes in length provide an integrated and smoothed signal of climate change. At shorter time scales, glaciers will vary markedly with climate change, especially temperature. The nature of the response to warming will vary from glacier to glacier depending on accompanying precipitation change, the mass-balance gradient, and hypsometry.

Most glaciers of the world are more sensitive to changes in temperature than to any other climatic element. In the case of many Asian glaciers, where precipitation occurs mainly during the summer monsoon season, temperature has a double impact. The first impact is an increase in the absorption of solar radiation due to a lowering of the surface albedo as snowfall is converted to rainfall. The second effect is an increase in the energy exchange between the atmosphere and the glacier surface (Ageta and Kadota, 1992). Many maritime glaciers, which have large mass turnover, are more sensitive to changes in precipitation than to temperature. This complex dependence of glacier mass balance on temperature, precipitation, and radiation makes it difficult to define their sensitivity to climate change.

A number of approaches are used to express the sensitivity of glaciers to climate. Where there are long glacier records for calibration, regression equations are used to relate mass balance to summer mean temperature and annual (or winter) precipitation (Laumann and Tvede, 1989; Chen, 1991). Changes in summer ablation rates at the margins of Greenland glaciers are nearly linear with changes in summer temperature, with sensitivities ranging from 0.43–0.57 m/yr/°C (water equivalent; Braithwaite and Olesen, 1990). Although these empirical relationships are valuable for regions in which they have been developed, they are not always applicable to other mountain areas with different climate and terrain. The sensitivity of glacier ablation has been assessed for number of melting degree-days (dd) for a range of glaciers by Braithwaite and Olesen (1989, 1990, 1993). Ablation sensitivity for ice varies from 5.5 to 7.6 mm/dd (Braithwaite and Oleson, 1989). Values for snow range from 1.0 to 5.7 mm/dd (Johannesson *et al.,* 1993). Hydrometeorological models are similar but include precipitation as well as temperature; they also usually estimate the mass balance (e.g., Woo and Fitzharris, 1992) and sometimes the water balance (e.g., Tangborn, 1980). Energy-balance models also have been used to assess the sensitivity of glacier ablation to climate (Oerlemans, 1991; Oerlemans *et al.,* 1993); they are appealing because they are more physically based, but they require input parameters that are difficult to obtain.

Glaciers are sensitive to changes in atmospheric circulation patterns, but it is difficult to quantify the response. The behavior of the Franz Josef glacier in New Zealand this century can be qualitatively explained by changes in the westerlies and shifts in the subtropical high-pressure zone (Fitzharris *et al.,* 1992). A major advance since 1983 appears to be consistent with circulation changes induced by three large El Niño events. In the first half of the 20th century, warming accounted for most of the ice thinning observed on Mount Kenya. However, from 1963 to 1987, greater atmospheric humidity was instrumental in enhancing melt—perhaps a consequence of enhanced evaporation in the Indian Ocean and its subsequent advection over East Africa by the prevailing atmospheric circulation patterns (Hastenrath, 1992, 1994; Hastenrath and Kruss, 1992). In the Canadian High Arctic, mass-balance and ice-core melt studies show that when the circumpolar vortex shifts to the Asian side of the Arctic Ocean and the North American trough is replaced by a blocking ridge, high-arctic glaciers experience high melt (Alt, 1987).

7.3.3. *Permafrost*

The direct influence of climate on permafrost includes the effects of air temperatures and solar radiation. Permafrost also is influenced indirectly by local factors, many of which also have a climate component. These include the thickness and duration of snow cover, the type of vegetation, the properties of the organic layer and soil, and the characteristics of running water (Harris and Corte, 1992; Schmitt, 1993). These factors interact in complex ways, making it difficult to asses the sensitivity of permafrost, and its response, to climate change (Koster, 1991, 1994; Koster and Nieuwenhuijzen, 1992; Koster *et al.,* 1994; Nelson and Anismov, 1993). However, climatic warming usually causes an increase in the thickness of the active layer via melting at the permafrost table (the active layer is the zone of annual freezing and thawing above permafrost). Surface disturbances appear in the first few years, but changes in temperature profiles within the permafrost may be delayed by decades to centuries. The response of permafrost is very dependent on initial ground temperatures and the latent heat (Geo-engineering Ltd, 1995). Displacement of the permafrost base—the final response—takes years to millennia, depending on the depth, thickness, and conductivity of the Earth material. Discontinuous permafrost tends to be most sensitive to climate change because it is usually within one or two degrees of 0°C. Many surface processes that preserve permafrost also are affected by climate change.

Depending on initial conditions, even small changes in climatic regimes cause permafrost to thaw. Simulations performed by Riseborough and Smith (1993) suggest that the rate of thaw and ultimate disappearance of a relatively thin (4.5 m) permafrost profile is highly dependent on the interannual variability of temperature. The broad sensitivity of permafrost to climate change is documented in the former Soviet Union and China, where the permafrost distribution changed substantially during warmer periods of the Quaternary (Kondratjeva *et al.,* 1993; Qiu and Cheng, 1995). The southern limit of lowland permafrost moved at 60 km/°C (Cui and Xie, 1984). In the alpine permafrost of Tibet, the lower elevation changed by 160 m/°C (Cui, 1980). On the northern slope of the Himalayas, the sensitivity is about 80 m/°C (Xie, 1996).

7.3.4. *River and Lake Ice*

Ginzburg *et al.* (1992) and Soldatova (1993) show that ice formation and break-up correlate with air temperature in the preceding autumn and spring months but not with winter temperatures. Spring warming is more important to the timing of break-up than the overall winter severity and peak ice thickness. Assessing ice chronology and seasonal temperature data from Scandinavia, Canada, and the FSU leads to an estimated sensitivity of both freeze-up and break-up of 5 days/°C. The sensitivity for the length of the ice season is about 10 days/°C.

Freeze-up and ice-loss dates for lakes correlate with autumn and winter air temperatures. Assel and Robertson (1995) show that the sensitivity of freeze-up dates for the Great Lakes is approximately 7 days/°C. For sixty-three smaller lakes in Finland, Palecki and Barry (1986) suggest 5.5 days/°C. Anderson *et al.* (1995) show that interannual variation in ice break-up for twenty U.S. lakes can be explained by ENSO.

7.3.5. *Sea Ice*

The thickness of fast ice—which grows in fjords, bays, and inlets in the Arctic; along the open coast in shallow water; and in channels of restricted dimensions—is correlated to the number of degree-days of freezing since the beginning of winter. In pack ice, however, the relationship between ice extent or ice thickness and temperature is less clear. Chapman and Walsh (1993) report that there is a statistically significant decrease in Arctic sea-ice extent in winter correlated with atmospheric warming, but there are no trends in winter Arctic ice extent or in the Antarctic.

The sensitivity of oceanic sea-ice cover to climate change is not well-understood. From the data comparisons made so far by McLaren (1989), Wadhams (1989, 1990a), and McLaren *et al.* (1992), the following tentative conclusions can be drawn.

Ice reaching the Fram Strait via the Trans Polar Drift Stream along routes where it is not heavily influenced by a downstream land boundary shows great consistency in its mean thickness from season to season and from year to year, at latitudes from 84°30'N to 80°N and in the vicinity of 0° longitude. Ice upstream of the land boundary of Greenland shows great changes in mean ice draft due to anomalies in the balance between pressure-ridge formation through convergence and open-water formation through divergence. The overall extent of sea ice is largely determined by ice transport via currents and wind and is not necessarily directly related to *in situ* freezing and melting controlled by temperature (Allison, Brandt, and Warren, 1993; Allison and Worby, 1994). As yet, there is no conclusive evidence of systematic thermodynamic thinning of the sea-ice cover, as might be caused by global warming.

Ocean circulation systems—such as the Atlantic conveyor belt—are thought to be sensitive to changes in sea-ice export from the Arctic (Aagaard and Carmack, 1989). Salt flux from local ice production plays an important role in triggering narrow convective plumes in the central gyre region of the Greenland Sea in winter (Rudels, 1990). Here frazil and pancake ice production due to cold-air outbreaks from Greenland can yield high salt fluxes. There already is evidence that a reduction in ice production has produced an ocean response in the form of reduced volume and depth of convection (Schlosser *et al.*, 1991).

7.4. What Will Be the Impact of Future Climate Change on the Cryosphere?

Outputs from three Transient General Circulation Models (GCMs) in Greco *et al.* (1994) were applied to the major cryospheric regions of the world in order to assess impacts for the decade about 2050. They predict that most regions of the cryosphere will warm by 0.5–2.5°C, but some will be wetter and others drier. The amount of warming is consistent among

Box 7-1. Ice Sheets

The great ice sheets of Antarctica and Greenland have changed little in extent during this century. Dynamic response times of ice sheets to climate change are on the order of thousands of years, so they are not necessarily in equilibrium with current climate. Observational evidence for Antarctic and Greenland ice sheets is insufficient to determine whether they are in balance or have decreased or increased in volume over the last 100 years (see Chapter 7, *Changes in Sea Level*, in the Working Group I volume). Zwally *et al.* (1989) note that the Greenland ice sheet surface elevation is increasing by 0.23 m/yr, but this issue of stability remains contentious (Douglas *et al.*, 1990; Jacobs, 1992). There is evidence for recent increases in snow accumulation in East Antarctica (Morgan *et al.*, 1991). On the other hand, Kameda *et al.* (1990) report that the thickness of this ice sheet has decreased by about 350 m during the last 2,000 years.

If Antarctica were to warm in the future, its mass balance would be positive (see Chapter 7, *Changes in Sea Level*, in the Working Group I volume). The rise in temperature would be insufficent to initiate melt but would increase snowfall. Concern has been expressed that the West Antarctic ice sheet may "surge." Working Group I projects that the probability of this occuring within the next century "may be relatively remote, but not zero." There is considerable doubt regarding the possible dynamic response of ice sheets.

The response of Greenland to warming is likely to be different. Both the melt rates at the margins and the accumulation rates in the interior should increase. The former rate is expected to dominate. Thus, the mass balance would become negative as temperatures rise. Several studies have examined precipitation variations over Greenland (e.g., Bromwich *et al.*, 1993; Kapsner *et al.*, 1995). The likely changes in atmospheric circulation and moisture flux (Calanca and Ohmura, 1994) must be determined before future snowfall can be predicted. Surface topography also plays an important role in determining regional accumulation on the Greenland ice sheet (Ohmura and Reeh, 1991).

Most sheets descend to sea level, where they produce icebergs as pieces break off and float away. These endanger shipping. In Greenland, most iceberg calving comes from a relatively small number of fast-moving tidewater glaciers (Reeh, 1989). Antarctic ice shelves are much larger and grow for periods of 20 to 100 years, regularly calving only relatively small icebergs before enormous pieces break off (Orheim, 1988). For example, giant icebergs have calved from the Filchner, Larsen, Ross, and Shackleton Ice Shelves in recent years. The northerly Wordie Ice Shelf may be systematically breaking up (Doake and Vaughan, 1991), and there have been large changes to the Larsen Ice Shelf culminating in recent large calving rates (Skvarca, 1993, 1994)—including the breakout of a vast iceberg in February 1995. Despite these findings, there is no clear consensus as to whether the frequency of icebergs, and their danger to shipping, will change with global warming.

the three GCMs in most, though not all, cryospheric regions, but predictions for precipitation are discordant among the three GCMs and must be considered less reliable. Assessments of impacts in this chapter generally are made using output from the UKTR model, mainly because of the better resolution of maps in Greco *et al.* (1994) over alpine and polar regions. See Box 7-1 for a discussion of the possible impacts of climate change brought about by the interaction of ice sheets and sea-level change.

7.4.1. Snow

Continental snow cover will be diminished in extent, duration, and depth by the UKTR climate scenarios in Greco *et al.* (1994). Winter snowlines could move further north by 5–10° latitude. The snow season could be shortened by more than a month, depending on snow depth. In North America, climatic warming based on the Boer *et al.* (1992) scenario would cause a 40% decrease in snow-cover duration over the Canadian Prairies and a 70% decrease over the Great Plains (Brown *et al.*, 1994). Using CCM1 model output for the whole Northern Hemisphere, the area of seasonal snow in February may diminish by 6–20%, with a mean decrease of 12×10^6 km^2 (Henderson-Sellers and Hansen, 1995).

Snowfall will begin later and snowmelt will be earlier than at present, so the snow-free season will be extended. The advance in melt time is likely to be less pronounced in the High Arctic, where the snow is so cold that it requires significant warming to produce consistent melt (more than 10°C at many locations). More frequent periods of open water for rivers, lakes, and seas will produce greater snowfall downwind. This will be important near Hudson Bay, the Great Lakes, the Barents Sea, and the Sea of Okhotsk. In the Antarctic, summer temperatures are so low that the present regime of little or no snowmelt will persist.

In alpine areas, the snow line could rise by 100–400 m, depending on precipitation. Higuchi (1991) shows that as warming occurs for many Asian mountains, there is a tendency for rainfall to occur at the expense of snowfall, although the extent of this shift depends on location. Less snow will accumulate at low elevations, but there may be more above the freezing level from any increased precipitation. Martinec *et al.* (1994) and Rango and Martinec (1994) have examined the behavior of a snowmelt-runoff model in various catchments for different climate scenarios. With a rise of 1°C, snow cover would be depleted in winter due to conversion of precipitation to rainfall and increased snowmelt. Five days into the melt season, snow depth would be depleted to the equivalent of 9 days under the present climate. It is estimated that for the New Zealand Alps, total water stored as snow would be reduced by 20% of present for a climate scenario that is 2°C warmer but 10% wetter than present (Garr and Fitzharris, 1994). Impacts of best- and worst-case scenarios on snow cover in the Victorian Alps (Australia) are presented in CSIRO (1994). Simulated average snow cover and the frequency of years with more than 60 days of cover decline at all sites. For the worst case, snow-cover duration at even the highest sites (1,900 m) is halved by 2030 and is near zero by 2070.

7.4.2. Ice Caps and Glaciers

Empirical and energy-balance models both indicate that a large fraction (about one-third to one-half) of presently existing mountain glacier mass could disappear with anticipated warming over the next 100 years (Kuhn, 1993; Oerlemans and Fortuin, 1992). By 2050, up to a quarter of mountain glacier mass could have melted. The scenarios of Greco *et al.* (1994) indicate that some mountain areas will experience an increase in precipitation. Because models demonstrate that increases in temperature usually dominate changes in precipitation, mass balances of glaciers will become negative rather than positive. Glaciers are likely to shrink even where mountains become wetter. An upward shift of the equilibrium line by some 200 to 300 m and annual ice thickness losses of 1 to 2 m are expected for temperate glaciers.

Many mountain chains will lose major parts of their glacier cover within decades. Haeberli and Hoelzle (1995), who have developed algorithms for analyzing glacier inventory data, show that glacier mass in the European Alps could be reduced to a few percent within decades if current warming continues. Nevertheless, the largest alpine glaciers—such as those found around the Gulf of Alaska and in Patagonia, Karakoram, Pamir, Tien Shan, and the Himalayas—should continue to exist into the 22nd century. At high altitudes and high latitudes, glaciers and ice caps may change little in size, but warming of cold firn areas will be pronounced. Their mass balance may be affected through enhanced ablation at low altitudes, while accumulation at higher zones could increase. As an analog, Miller and de Vernal (1992) report that some Arctic glaciers did not shrink during warmer parts of the Holocene but actually grew due to increased precipitation.

There will be pronounced alterations to glacier melt runoff as the climate changes. Glaciers will provide extra runoff as the ice disappears. In most mountain regions, this will happen for a few decades, then cease. For those with very large glaciers, the extra runoff may persist for a century or more and may provide a substantial increase to regional water resources. As the area of ice eventually diminishes, glacial runoff will wane. Tentative estimates have been made for Central Asia (Kotlyakov *et al.*, 1991) based on mass balances from a small number of Tien Shan glaciers for the period 1959–1992. Extrapolation to the whole of Central Asia suggests glacier mass has decreased by 804 km^3 over that time, representing a 15% increase in glacial runoff. Projections to 2100 are presented in Table 7-2; these data assume that glacial runoff varies as a linear function of glacier area.

An alternative approach is based on calculations of ablation and equilibrium-line altitude, according to given climatic scenarios (Kotlyakov *et al.*, 1991; Glazyrin *et al.*, 1990; Dikikh and Dikikh, 1990). Ablation intensifies in Central Asia with

***Table 7-2:** Present and possible future extent of glaciation and glacial runoff in Central Asia (after Kotlyakov et al., 1991).*

Component	Present	2100
Area of glaciers (km²)	115,000	80,000
Decrease of glacier mass (km³/yr)	25	18
Runoff due to precipitation (km³/yr)	73	50
Glacial runoff (km³/yr)	98	68

climate warming because conditions become even more continental. Annual mass balance decreases. These estimates give a high rate of glacier degradation and large alteration of runoff (Figure 7-5). By 2050, the volume of runoff from glaciers in Central Asia is projected to increase threefold.

7.4.3. *Permafrost*

Anisimov and Nelson (1995) have compiled global permafrost maps for the present day and for a 2050 scenario (see Figure 7-3b). The compilation uses a predictive climate-based permafrost model, in which permafrost is classified on the basis of "surface frost index" (Nelson and Anisimov, 1993) and which also considers the influence of snow cover on the soil thermal regime. Calculated contemporary boundaries, based on the Global Ecosystem Database (GED), show good agreement with Figure 7-3a. Table 7-3 provides the areal extent of permafrost distribution in the Northern Hemisphere for two climate scenarios (GED contemporary and 2050). A 16% shrinkage in total permafrost area is projected by 2050. Subsurface conditions may not be in equilirium with the surface area shown. Some may take hundreds of years to respond. The extent of permafrost zones in the Southern Hemisphere, which cover only approximately 0.5 x 10^6 km² in total, also is likely to decrease.

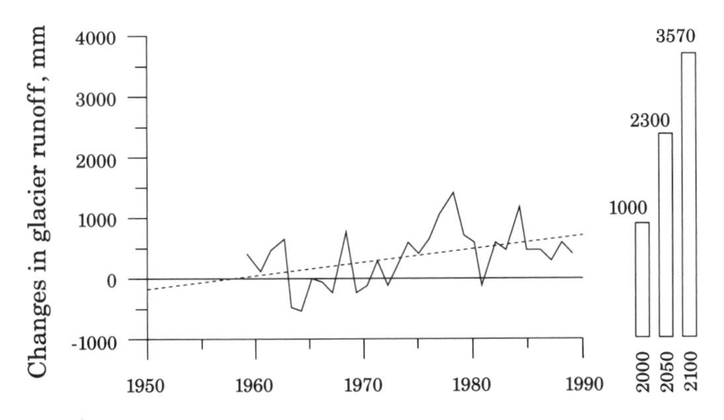

Figure 7-5: Changes in recent and future runoff from Tuyuksu glacier, Zailiyski Alatau, Central Asia (after Kotlyakov *et al.*, 1991). Values represent runoff due to ice loss, expressed as departures from long-term mean.

There will be poleward shift of discontinuous and continuous permafrost zones (Woo *et al.,* 1992; Nelson and Anisimov, 1993). These estimates only treat degradation of near-surface permafrost; more attention must be given to the role of latent heat associated with ablation of ground ice. Simulations conducted by Riseborough (1990) indicate that areas with abundant ground ice, such as western Siberia (Burns *et al.,* 1993), may retain substantial amounts of permafrost that are not in equilibrium with new temperature conditions imposed at the surface. The relict permafrost of the southern part of West Siberia, which may well date from the last glacial epoch, is a present-day example of resistance to changes of temperature in deep-seated, ice-rich permafrost (Kondratjeva *et al.,* 1993).

In areas where permafrost is discontinuous, long-term warming ultimately will lead to its thinning and general disappearance (Wright *et al.,* 1994). Long-term temperature measurements in discontinuous permafrost in Alaska indicate that an increase in surface temperature of 2°C would cause most of the permafrost south of the Yukon River and on the south side of the Seward Peninsula to thaw. In continuous permafrost, a warming of the permafrost, thickening of the active layer, and changes in thaw lake dynamics are likely. Initial degradation will be contemporaneous with the alteration of the climatic signal (Osterkamp, 1994; Kane *et al.,* 1991), but as the thaw plane penetrates more deeply, lag times will increase.

Xie (in press) estimates that the southern boundary of permafrost in northeast China could be at 48°N latitude by the year 2100, and the predominantly continuous permafrost zone may recede to 52°N—similar to that in the Climatic Optimum of the Holocene (Lu Guowei *et al.,* 1993). In Tibet, it is estimated that an air-temperature warming of 3°C would raise the permafrost limit to an elevation of 4,600 m. Continuous frozen ground would disappear, except in the northwestern part of the plateau and areas around the Fenghuo Shan, in the event of a 5°C rise in temperature (Xie, in press).

In the FSU and Europe, reconstructions of the permafrost zone in past warm and cold periods are used as palaeoanalogues (Velichko and Nechaev, 1992; Vanderberghe, 1993). The most significant change in the thermal regime will occur in the high

***Table 7-3:** Calculated contemporary and future areas of permafrost in the Northern Hemisphere (10^6 km²) (based on Nelson and Anisimov, 1993).*

Zone	Contemporary	2050	% Change
Continuous Permafrost	11.7	8.5	-27
Discontinuous Permafrost	5.6	5.0	-11
Sporadic Permafrost	8.1	7.9	-2
Total	**25.4**	**21.4**	**-16**

latitudes within the present-day zones of the tundra and boreal forest. With a rise in temperature of 1°C, the permafrost would be partially preserved to the east of the Pechora river. In the south of Yamal and Gydan, discontinuous permafrost would prevail. Continuous permafrost would be restricted to north of 70°N. It would remain to the east of the Yenisey at the same latitudes. The active layer of fine-grained soils could increase by 20–30 cm. With a temperature rise of 2°C, continuous permafrost would disappear in the north of Europe. From the Lower Ob to the Lower Hatanga, as well as in the Anadyr lowland, only island and discontinuous permafrost would survive. Continuous permafrost would exist only in the lowlands of the Taimyr Peninsula, Lena, lower Kolyma, and lower Indigirka basins. The stratum of seasonal thawing would increase by 40–50 cm. These changes will result in the activation of solifluction, thermokarst and thermoerosion processes, an increase in bogs, and alterations to large tracts of vegetation—which in turn will alter food sources for traditional tundra species within permafrost areas.

Other mathematical models assess permafrost dynamics during warming (MacInnes *et al.,* 1990; Burgess and Riseborough, 1990; Romanovsky, Maximova, and Seregina, 1991; Romanovsky, Garagula, and Seregina, 1991; Osterkamp and Gosink, 1991; Nixon and Taylor, 1994; Nixon *et al.,* 1995). Changes in permafrost temperatures and thaw depths under an increase in annual average air temperatures of 2 and 4°C are simulated by Vyalov *et al.* (1993). Results indicate that warming in the tundra will produce a slight increase in thawing depths, with no radical changes in permafrost conditions. Further south, more marked effects could occur—with a shift to discontinuous permafrost, formation of taliks, and degradation of ground ice. If warming is long-lasting, the permafrost line would recede northward by 500 km along 70°E longitude and by 1,200 km along 100°E longitude. Given an average thawing rate of 100 mm/yr, thawing of the upper 10 m would take a century. Lower limits of permafrost in mountain areas could rise by several hundred meters, although this is dependent on future snow depths (Hoelzle and Haeberli, 1995). Owing to the slow reaction of thermal conditions at depth, pronounced disequilibria are most likely to result over extended time periods and wide areas.

Ice-saturated permafrost forms an impervious layer to deep infiltration of water, maintaining high water tables and poorly aerated soils. Significant increases in active layer depth or loss of permafrost is expected to cause drying of upper soil layers in most regions, as well as enhanced decomposition of soil organic matter. Loss of a sizable portion of more than 50 Gt of carbon in Arctic soils and 450 Gt of carbon in soils of all tundra ecosystems could cause an appreciable positive feedback on the atmospheric rise of carbon dioxide. Marion and Oechel (1993) have examined Holocene rates of soil carbon change along a latitudinal transect across arctic Alaska and conclude that arctic areas will continue to act as a small sink for carbon.

Recent warming and drying apparently have shifted arctic ecosystems from carbon sequestration (as occurred during the Holocene and historical past) to carbon dioxide loss to the atmosphere (Oechel *et al.,* 1993). Estimates of present losses of carbon dioxide from arctic terrestrial ecosystems to the atmosphere range from 0.2 Gt of carbon (Oechel *et al.,* 1993) to much higher values in winter for the Russian Arctic (Zimov *et al.,* 1993; Kolchugina and Vinson, 1993). Malmer (1992) estimates that about 25% or less of the carbon may be released in the form of methane (CH_4). A climatic shift to warmer temperatures in the future would increase the release of CH_4 from deep peat deposits, particularly from tundra soils. It is expected that the release of CO_2 would increase, though not by more than 25% of its present level (Malmer, 1992). Wetter soils could lead to increased methane loss, and a drier tundra might become a sink for atmospheric methane (Christensen, 1991, 1993; Fukuda, 1994).

Large amounts of natural gas—mostly methane—are stored in the form of gas hydrates, although their distribution is not well-known. In the Canadian Arctic Islands and Beaufort-Mackenzie region, analysis of thermal and geophysical logs indicates that 2–4 x 10^3 Gt of methane is stored as hydrates. Decomposition is occurring presently beneath the shelves of the Arctic Ocean in response to the increase of surface temperatures accompanying the recent marine transgression. For the Beaufort Shelf, an estimated 10^5 m^3/km^2 may decompose over the next century (Judge and Majorowicz, 1992; Judge *et al.,* 1994).

There are very few data for the methane content of permafrost itself. Samples obtained near Fairbanks (Kvenvolden and Lorenson, 1993) and the Prudhoe Bay area (Moraes and Khalil, 1993; Rasmussen *et al.,* 1993) suggest substantial variability. Dallimore and Collett (1995) found high methane concentrations in ice-bonded sediments and gas releases suggest that pore-space hydrate may be found at depths as shallow as 119 m. This raises the possibility that gas hydrates could occur at much shallower depth and be more rapidly influenced by climate change than previously thought. Fukuda (1994) estimates methane emissions from melting ground ice in northern Siberia cover the range 2–10 Mt/yr.

7.4.4. River and Lake Ice

Under conditions of overall annual warming, the duration of river-ice cover would be reduced through a delay in the timing of freeze-up and an advancement of break-up (Gray and Prowse, 1993). For freeze-up, higher water and air temperatures in the autumn would combine to delay the time of first ice formation and eventual freeze-up. If there also is a reduction in the rate of autumn cooling, the interval between these two events also will increase. The frequency and magnitude of major frazil-ice growth periods could be reduced. This may alter the types of ice that constitute the freeze-up cover and has implications for hydrotechnical problems associated with particular ice forms.

Many rivers within temperate regions would tend to become ice-free or develop only intermittent or partial ice coverage. Ice growth and thickness would be reduced. In colder regions, the present ice season could be shortened by up to a month by 2050. Warmer winters would cause more mid-winter break-ups as rapid snowmelt, initiated particularly by rain-on-snow events, becomes more common. Warmer spring air temperatures may affect break-up severity, but the results would be highly site-specific because break-up is the result of a complex balancing between downstream resistance (ice strength and thickness) and upstream forces (flood wave). Although thinner ice produced by a warmer winter would tend to promote a thermal break-up, this might be counteracted to some degree by the earlier timing of the event, reducing break-up severity (Prowse *et al.,* 1990).

Changes in the size of the spring flood wave depend on two climate-related factors: the rate of spring warming and the water equivalent of the accumulated winter snowpack. Whereas greater and more-rapid snowmelt runoff favors an increase in break-up severity, the reverse is true for smaller snowpacks and more protracted melt. The final effect on break-up also will depend on the potentially conflicting roles of ice strength and thickness.

For arctic lakes, the duration of ice cover would be shortened (Assel *et al.,* 1995). A longer open-water period, together with warmer summer conditions, would increase evaporative loss from lakes. Some patchy wetlands and shallow lakes owe their existence to a positive water balance and the presence of an impermeable permafrost substrate that inhibits deep percolation. Enhanced evaporation and ground thaw would cause some to disappear. Using a 20-year period of record, Schindler *et al.* (1990) found that climate warming increased the length of the ice-free season by 3 weeks, as well as having numerous indirect effects (see Chapter 10).

If the warming of the 1980s continues unabated over the next 10–20 years, ice cover on the Great Lakes of North America will likely be similar to or less than that during the 1983 ice season, which was one of the mildest winters of the past 200 years. Mid-lake ice cover did not form, and ice duration and thickness were less than normal (Assel *et al.,* 1985). Complete freezing will become increasingly infrequent for larger and deeper embayments in the Great Lakes. The duration of ice cover will decrease as freeze-up dates occur later and ice-loss dates occur earlier. Winters without freeze-up will begin to occur at small inland lakes in the region. Winter lake evaporation also may increase due to the decreased ice cover.

7.4.5. Sea Ice

GCM experiments with simplified treatments of sea-ice processes predict large reductions in sea-ice extent but produce widely varying results and do not portray extent and seasonal changes of sea ice for the current climate very well. Boer *et al.* (1992) estimate that with a doubling of greenhouse gases, sea ice would cover only about 50% of its present area. CCM1 model output presented in Henderson-Sellers and Hansen (1995) projects a 43% reduction for the Southern Hemisphere and a 33% reduction for the Northern Hemisphere. The global area of sea ice is projected to shrink by up to 17 x 10^6 km^2.

Using empirical ice growth-melt models, Wadhams (1990b) predicts that in the Northwest Passage and Northern Sea Route, a century of warming would lead to a decline in winter fast-ice thickness from 1.8–2.5 m at present to 1.4–1.8 m and an increase in the ice-free season of 41–100 days. This effect will be of great importance for the extension of the navigation season in the Russian Northern Sea Route and the Northwest Passage. A possible feedback with snow thickness may alter these relationships. As Arctic warming increases open-water area, precipitation may increase and cause thicker snow cover. The growth rate of land fast ice will decrease, as has been directly observed (Brown and Cote, 1992). But if snow thickness is increased to the point where not all ice is melted in summer, then the protection that it offers the ice surface from summer melt could lead to an increase, rather than decrease, in equilibrium ice thickness.

Predicting future thickness of moving pack ice is a difficult problem, because dynamics (ocean and wind currents), rather than thermodynamics (radiation and heat components), determine its area-averaged mean thickness. Wind stress acting on the ice surface causes the ice cover to open up to form leads. Later under a convergent stress, refrozen leads and thinner ice elements are crushed to form pressure ridges. Exchanges of heat, salt and momentum are all different from those that would occur in a fast ice cover. The effects of variable thickness are very important. The area-averaged growth rate of ice is dominated (especially in autumn and early winter when much lead and ridge creation take place) by the small fraction of the sea surface occupied by ice less than 1 m thick. In fast ice, climatic warming will increase sea-air heat transfer by reducing ice growth rates. However, over open leads a warming will decrease the sea-air heat transfer, so the area-averaged change in this quantity over moving ice (hence its feedback effect on climatic change itself) depends on the change in the rate of creation of new lead area. This is itself a function of a change in the ice dynamics, either driving forces (wind field) or response (ice rheology). Hibler (1989) pointed out a further factor relevant to coastal zones of the Arctic Ocean, such as off the Canadian Arctic Archipelago, where there is net convergence and the mean thickness is very high (7 m or more) due to ridging. Here the mean thickness is determined by mechanical factors, largely the strength of the ice, and is likely to be insensitive to global warming.

Ice dynamics also have other effects. In the Eurasian Basin of the Arctic the average surface ice drift pattern is a current (the Trans Polar Drift Stream) which transports ice across the Basin, out through Fram Strait, and south via the East Greenland Current into the Greenland and Iceland Seas, where it melts. The net result is a heat transfer from the upper ocean in sub-Arctic seas into the atmosphere above the Arctic Basin. A change in area-averaged freezing rate in the Basin would thus cause a change of similar sign to the magnitude of this long-range heat transport.

An identical argument applies to salt flux, which is positive into the upper ocean in ice growth areas and negative in melt areas. Salt is transported northward via the southward (Lemke *et al.*, 1990) ice drift. A relative increase in area-averaged melt would cause increased stabilization of the upper layer of polar surface water, and hence a reduction in heat flux by mixing across the pycnocline. A relative increase in freezing would cause destabilization and possible overturning and convection.

If ice were to retreat entirely from the central gyre region of the Greenland Sea, it may cause deep convection to cease. Already there is evidence from tracer studies (Schlosser *et al.*, 1991) of a marked reduction in the renewal of the deep waters of the Greenland Sea by convection during the last decade. It is not known whether this is part of a natural variation, or a response due to greenhouse warming. If continued, it could have a positive feedback effect on global warming, since the ability of the world ocean to sequestrate carbon dioxide through convection is reduced. Given the complexity of these interactions and feedbacks, it is not at all clear what the overall effect of an air temperature increase on the Arctic ice cover and upper ocean will be. Sensitivity studies using coupled ocean-ice-atmosphere models are required, but results are not yet available.

In the Antarctic, where the sea ice cover is divergent and where land boundaries are less important, it is more reasonable to suppose that the main effect of global warming will be a simple retreat of the ice edge southward. However, Martinson (1990) has demonstrated that even here, a complex set of feedback mechanisms comes into play when the air temperature changes. The balance of lead concentration, upper ocean structure and pycnocline depth adjusts itself to minimize the impact of changes, tending to preserve an ice cover even though it may be thinner and more diffuse.

7.5. What Will Be the Impact of These Cryospheric Changes?

These changes in the cryosphere will affect many natural and managed ecosystems, as well as socioeconomic systems. Only the most notable are considered here. Two regions will be most affected: Temperate mountains and tundra lands (see Boxes 7-2 and 7-3, respectively).

7.5.1. Impact on Hydrology and Water Resources

The impact of climate change on water resources in alpine areas can be large, as shown by Gleick (1987a, 1987b), Martinec and Rango (1989) and Moore (1992), and are discussed at length in Chapter 10. Seasonal changes can be

Box 7-2. Impact on Temperate Mountains

Revegetation of terrain following deglaciation is slow in high-mountain areas. This leaves morainic deposits unprotected against erosion for extended time periods (decades to centuries). There will be increased sediment loads in alpine rivers and accelerated sedimentation in lakes and artificial reservoirs at high altitude. On slopes steeper than about 25–30 degrees, stability problems, such as debris flows, will develop in freshly exposed or thawing non-consolidated sediments.

At places of pronounced glacier retreat, changes in stress distribution and surface temperature conditions in rock walls of deeply cut glacier troughs must also be anticipated, so that massive rock slides will occur in the deglaciated valley (Clague and Evans, 1992, 1994; Evans and Clague, 1993, 1994). Steep hanging glaciers which are partially or entirely frozen to their beds could become less stable. On the other hand, some steep glacier tongues with present-day potential for large ice avalanches will disappear. Lakes dammed by landslides, moraines and glaciers can drain suddenly and produce floods or debris flows orders of magnitude larger than normal stream flow. Processes related to ice retreat such as glacier avalanches, slope instability caused by debuttressing, and glacier floods from moraine or ice-dammed lakes may pose hazards to people, transport routes, and economic infrastrucure in mountain areas. The general tendency in high mountains will be an upslope shifting of hazard zones and widespread reduction in stability of formerly glaciated or perennially frozen slopes (Barsch, 1993; Haeberli, 1992; Gu *et al.*, 1993; Dutto and Mortara, 1992).

Shrinkage of permafrost and snow cover will eliminate snow metamorphism, and avalanche formation in high mountain regions and adjacent lowlands (Keller and Gubler, 1993; Tenthorey, 1992). An increased frequency of catastrophic snow avalanches may occur if sudden cold spells and a mixture of cold and warm air masses are more common in late winter, or if periods of rapid warming are accompanied by heavy rain follow periods of intensive snowfalls. There is insufficient evidence to indicate whether the frequency of such events will change. Avalanches are expected to be less of a hazard than at present.

The empirical basis for assessing alpine hazard probabilities comes from historical documents, statistics of measured time series or traces in nature of past events with long recurrence intervals. A future problem is that these will lose more and more of their significance as climate changes. This is because floods, avalanches, debris flows, and rock falls could have different magnitudes and frequencies than in the present climate, as the condition of the cryosphere in high mountain areas evolves beyond the range of Holocene and historical variability.

Box 7-3. Impact on Tundra Lands

A critical factor influencing the response of tundra to warming depends on the presence of ground ice. Ground ice is generally concentrated in the upper 10 meters of permafrost, the very layers that will thaw first as permafrost degrades. This loss is effectively irreversible, because once the ground ice melts, it cannot be replaced for millenia, even if the climate were to subsequently cool. Response of the permafrost landscape to warming will be profound, but will vary greatly at the local scale, depending on detail of ground ice content. As substantial ice in permafrost is melted, there will be land subsidence. This process of thermokarst erosion will create many ponds and lakes and lead to coastal retreat.

A forerunner of future landscapes can be seen in areas of massive ground ice, such as in Russia, where past climatic warming has altered the landscape by producing extensive flat-bottomed valleys. Ponds within an area of thermokarst topography eventually grow into thaw lakes. These continue to enlarge for decades to centuries, due to wave action and continued thermal erosion of the banks. Liquefaction of the thawed layer will result in mudflows on slopes in terrain that is poorly drained or that contains ice-rich permafrost. On steeper slopes there will also be landslides (Lewkowicz, 1992). Winter discharge of groundwater often leads to ice formation, and this is expected to increase on hillslopes and in the stream channels of the tundra.

Changes in landscape, sea-ice distribution, and river and lake ice could have a major impact on indigenous people who live in Arctic regions and depend upon traditional occupations, food gathering, and hunting (Kassi, 1993; Roots, 1993; Wall, 1993). These include the Inuit of North America and Greenland and the various reindeer herding groups of Eurasia. They depend directly on the living resources of the area and often travel on ice, so their livelihood may be widely affected. Ice roads and crossings are commonly used to link northern settlements. The greatest economic impact is likely to stem from decreases in ice thickness and bearing capacity, which could severely restrict the size and load limit of vehicular traffic (Lonergan *et al.*, 1993). There is likely to be a change in the migration patterns of polar bears and caribou, along with other biological impacts.

There will be considerable impacts of climate change on resource management in the tundra (Wall, 1993). Some infrastructure and mining activities and structures will be threatened by thawing of permafrost. Water resources will change in that the seasonality of river flows will be different. Environmental changes are expected to be greater than for many other places on Earth (Roots, 1993).

marked when the cryosphere is involved in river flow. In their snowmelt runoff simulations, Martinec *et al.* (1994) report that for a 4°C rise in temperature of the Rio Grande basin in the USA, winter runoff increases from 14% of the total annual flow at present to 30%. Summer runoff decreases from 86% of annual flow to 70%. Such changes in snow runoff can affect irrigation water and electricity supply.

Less snow and glacier ice will influence the seasonality of river flow by reducing meltwater production in the warm season. The expected smoothing of the annual runoff amplitude could be both beneficial (e.g., energy production in winter, reduction of summer flood peaks) and adverse (e.g., reduced water supply for summer irrigation in dry areas, more frequent winter floods). As mountain glaciers begin to disappear, then eventually the volume of summer runoff will be reduced due to the loss of ice resources. Consequences for downstream agriculture, which relies on this water for irrigation, are in some places very unfavourable. For example, low and midland parts of Central Asia are likely to gradually change into a more arid, interior desert.

As climatic warming occurs, there will be notable changes in the hydrology of Arctic areas (Woo, 1990). The nival regime runoff pattern will weaken for many rivers in the permafrost region, and the pluvial influence upon runoff will intensify for rivers along the southern margin of the Arctic, regions of Eurasia, and North America. Should climatic change continue, the vegetation will likely be different from today. When the lichens and mosses, that tend to be suppressors of evapotranspiration, are replaced by transpiring plants, evaporative losses will increase. Enhanced evaporation will lower the water table, followed by changes in the peat characteristics as the extensive wetland surfaces become drier (Woo, 1992).

In permafrost regions, increased thawing deepens the active layer, allowing greater infiltration and water storage, especially for rain that falls during the thawed period. Warming of the ground will also lead to the formation of unfrozen zones within the permafrost which provide conduits for groundwater flow. The chemical composition and the amount of groundwater discharge may be changed as sub-permafrost or intra-permafrost water is connected to the surface. In autumn and winter, more groundwater should be available to maintain baseflow, further extending the streamflow season.

Structures such as pipelines, airstrips, community water supply and sewage systems, and building foundations are susceptible to performance problems if existing, frozen foundations or subgrades thaw, even minimally. Special measures would be needed to ensure the structural stability and durability of installations for tourism, mining industry, and telecommunication in permafrost areas affected by climate warming (Anyia *et al.,* 1992; Haeberli, 1992; Vyalov *et al.,* 1993). Transport links could also be affected. For example, the permafrost zone in China contains more than 3,000 km of railway and over 13,000 km of highway. Thawing induced by climate warming will result in serious disruption and increased maintenance costs from ground subsidence, sideslope slumpings, landslides, icings and ice mound growth (Yang, 1985). On the other hand, many northern cities will spend less money on snow and ice clearance. Engineering design criteria will need to be modified to reflect changing snow and frost climates, deepening of the active layer over permafrost, and warming and ultimate disappearance of marginal or discontinuous permafrost. Present permafrost engineering commonly designs for the warmest year in the past 20 years of record (Esch, 1993). Such criteria may need to be reviewed and revised.

Where climate change alters the river-ice regime, substantial effects on the hydrology can be expected that will affect flow, water levels, and storage. For cold continental rivers, many hydrologic extremes, such as low flows and floods, are frequently more a function of ice effects than landscape runoff. At freeze-up, the hydraulic resistance of an accumulating cover can induce sufficiently large hydraulic storage that the river flow falls below that normally expected (Gerard, 1990; Gray and Prowse, 1993). Projected climates will delay the timing of freeze-up and so prolong the autumn low flow period. At break-up, the rapid hydraulic storage and release of water by river-ice jams often forms the most significant hydrologic event of the year. Break-up flood levels typically exceed those that develop during the open-water period, even though actual discharge of water is lower. The impact of climatic warming will be to advance the timing of break, but its effect on break-up flooding is not clear.

Of all river-ice processes, ice-jams are the major source of economic damages, averaging approximately $CDN20–30 million per year in Canada alone (Van Der Vinne *et al.,* 1991). Changes in damages from such events depend on how climate change affects the frequency and severity of river-ice freeze-up and break-up events. Less river ice and a shorter ice season in northward flowing rivers of Canada, Russia and Siberia should enhance north-south river transport. When combined with less sea ice in the Arctic, new opportunites for reorganization of transport networks and trade links will arise. Ultimately those changes could affect trading patterns among Russia, USA, Canada, northern Europe, and Japan.

7.5.2. *Impact on the Hydroelectric Industry*

Altered future climates could have a significant impact on the seasonal distribution of snow storage, runoff into hydroelectric catchments, and aggregated electricity consumption. Garr and Fitzharris (1994) show that in New Zealand the winter (Austral) gap between electricity consumption and generation is reduced for a climate scenario that might be expected about 2050 (Figure 7-6). The electricity system is made less vulnerable to climate variability. Water supply is increased, but demand, which is largely driven by domestic heating, is reduced. There would be less need for new hydro plants and water storage. On the whole, these amount to net benefits. In countries where heating is supplied by natural gas, these changes will be less obvious. Instead the demand may well increase in the summer due to greater use of air conditioners due to a higher frequency of very hot days.

River ice creates a host of hydrotechnical difficulties for the operation of hydropower facilities, ranging from the blockage of trash racks by frazil ice to the curtailment of operations so as to avoid freeze-up flooding. One Canadian hydropower company estimates revenue losses of over $CDN1 million/year associated with the release of sufficient flow to avoid the downstream freeze-off of a major tributary (Foulds, 1988). Major economic savings are likely if the length of the ice season is reduced by climatic warming. Operational problems during the ice covered period will remain, and could even increase because some rivers may experience a higher frequency of winter break-ups.

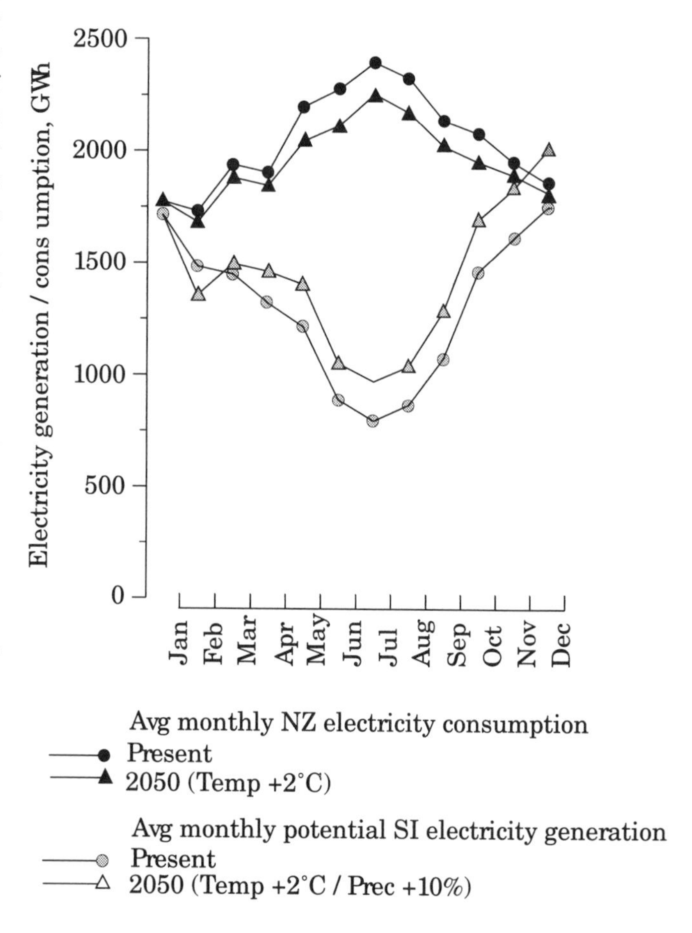

Figure 7-6: Synthesis of hydroelectricity generation and total consumption of electricity for New Zealand: The present compared with 2050 (after Garr and Fitzharris, 1994).

7.5.3. *Impact on Shipping*

In Canada, absence of sea ice south of Labrador would eliminate Canadian Coast Guard ice breaking requirements. This means an annual saving of between $CDN15–20 million. Even larger savings can also be expected in the FSU. The substantial reductions in sea ice in the Arctic Ocean that are expected to occur with climatic warming will increase opportunities for shipping there and open up new trade routes. The effect of annual warming on ice calving, simulated using a simple degree day model (Brown, 1993) shows that for every 1°C of warming there would be a 1° latitude retreat of the iceberg occurrence in the Atlantic Ocean. In the Southern Ocean, any effects of reduced sea ice will be economically less pronounced.

Ice-breakers keep river channels open for ship traffic within Russia on the major northern rivers, and to a limited degree in North America within the Great Lakes system. Considering the high operational costs of ice-breaking, any reduction in the ice season should translate into significant cost-savings. With warming there will be a longer shipping season, allowing increased passage of goods and services and longer time and larger area open for commercial fishing (Reycraft and Skinner, 1993). Changes in fresh colder water in the North Atlantic may change the distribution of fisheries.

7.5.4. *Impact on the Oil Industry*

Projected climatic warming could extend summer open water in the Beaufort Sea an additional 200–800 km offshore (McGillivray *et al.*, 1993). The fetch, and frequency of extreme wave heights, will therefore increase. It is estimated that the frequency of 6m waves would rise from 16% to 39% of the time (McGillivray *et al.*, 1993). Present design requirements for long-lived coastal and offshore structures, such as oil installations, will be inadequate under these conditions.

A possible beneficial effect would be shorter winters disrupting construction, exploration, and drilling programs. A decrease in thickness of first-year ice of 50–70% is projected for the Arctic, which will extend the drilling seasons for floating vessels considerably. Costs of drilling "downtime" on offshore oil and gas drilling explorations could practically be eliminated due to iceberg and sea-ice absence, saving more than $CDN40 million annually (Stokoe, 1988).

7.5.5. *Impact on the Tourist Industry*

Many mountainous regions of the world are used for tourism. A prerequisite for such commercially important activities as skiing is an extended snow cover of sufficient depth. Despite increases in precipitation, warming will decrease the cover, depth, and quality of snow and impair winter tourism in most alpine countries. Subsequent socioeconomic consequences could be detrimental for many mountain communities (Foehn, 1990). In the United States, Cline (1992) estimates ski industry losses from projected warming at $1.7 billion annually. In some countries, ski resorts could be re-established at higher elevations, but this too has associated transportation and emission problems. In others, such as Australia, the ski industry could, under the worst case scenario, be eliminated. Less snow and fewer glaciers on mountains will also diminish the quality of many alpine vistas, where their scenic appeal depends upon their presence in the general landscape. Some countries are still sufficiently high that their mountains will retain snow, so will possess an increasingly scarce, but valuable, scenic and recreational resource.

Reduced sea ice will provide safer approaches for tourist ships and new opportunities for sightseeing around Antarctica and the Arctic.

7.6. What Do We Still Need to Know?

There are many uncertainties in understanding what is currently happening to critical components of the cryosphere, especially ice sheets, sea ice, and permafrost, mainly because existing monitoring systems are inadequate. Critical questions remain about how each component of the cryosphere will react to climate change and a multitude of proposals could be made. The most important are synthesized as follows:

- Climatic scenarios produced by GCMs need to be refined so as to provide more detail for mountain and polar regions. Their output for regions of the globe where snow, ice, and permafrost are dominant needs to be better verified for the present climate. Imprecise estimates of future polar and alpine precipitation, and particularly snow depth, are a major constraint in predicting the behavior of most components of the cryosphere and their impact.
- Research should focus on processes that are driven by interactions between the atmosphere and the cryosphere. Improving understanding of these processes would allow the construction of more sophisticated and realistic climate sensitivity models.
- Monitoring of key components of the cryosphere must continue. The mass balance of the ice sheets of the world is poorly known. Databases need to be further developed and maintained. They provide the benchmark for assessing future change and for model testing. More statistical work is required on existing databases to improve knowledge of cryospheric trends.
- Effects of cryospheric change on other natural systems need to be better understood and quantified, particularly where they affect human communities and economic systems of agriculture, forestry, tourism, transport, and engineered structures. Future response of Arctic sea ice is especially critical because large changes will have profound climatic, economic, trade, and strategic implications.
- New methods are needed to assess probabilities of natural hazard and risk that take into account changing

climate. For hazard mitigation measures in mountain environments of developing countries, transfer of technologies for preparing the necessary assessments would be helpful.

Acknowledgments

The Convening Lead Author was funded by the Climate Management Centre, University of Otago, and by the New Zealand Ministry for Research, Science, and Technology. We wish to thank the many reviewers who made constructive suggestions for improving the chapter.

References

Aagaard, K. and E.C. Carmack, 1989: The role of sea-ice and other freshwater in the Arctic circulation. *Journal of Geophysical Research*, **94**, 485-14-498.

Ageta, Y. and T. Kadota, 1992: Predictions of changes of mass balance in the Nepal Himalaya and Tibetan Plateau: a case study of air temperature increase for three glaciers. *Annals of Glaciology*, **16**, 89-94.

Allison, I.F. and P.C. Keate, 1986: Recent changes in the glaciers of Heard Island. *Polar Record*, **23(144)**, 255-271.

Allison, I.F. and A. Worby, 1994: Seasonal changes of sea-ice characteristics off East Antarctica. *Annals of Glaciology*, **20**, 195-201.

Allison, I., R.E. Brandt, and S.G. Warren 1993: East Antarctic sea ice: albedo, thickness distribution and snow cover. *Journal of Geophysical Research*, **98**, 12,417-429.

Alt, B.T., 1987: Developing synoptic analogs for extreme mass balance conditions on Queen Elizabeth Island ice caps. *Journal of Climatology and Applied Meteorology*, **26**, 1605-1623.

Anderson, W., D.M. Robertson, and J.J. Magnuson, 1995: Evidence of recent warming and ENSO related variation in ice breakup of Wisconsin lakes. *American Society of Limnology and Oceanography* (in press).

Anisimov, O.A. and F.E. Nelson, 1995: Permafrost distribution in the Northern Hemisphere under scenarios of climatic change. *Global and Planetary Change* (submitted).

Aniya, M., R. Naruse, M. Shizukuishi, P. Skvarca, and G. Casassa, 1992: Monitoring recent glacier variations in the southern Patagonia Icefield, utilizing remote sensing data. *International Archives of Photogrammetry and Remote Sensing*, **29(B7)**, 87-94.

Armstrong, B.R., 1978: A history of avalanche hazard in San Juan and Ouray Counties, Colorado, in avalanche control, forecasting and safety. In: *Proceedings of Workshop held in Banff, Alberta*. National Research Council, Canada, p. 199.

Assel, R.A., C.R. Snider, and R. Lawrence, 1985: Comparison of 1983 Great Lakes winter weather and ice conditions with previous years. *Monthly Weather Review*, **113**, 291-303.

Assel, R.A and D.M. Robertson, 1995: Changes in winter air temperatures near Lake Michigan during 1851-1993, as determined from regional lake-ice records. *Limnology and Oceanography*, **40(1)**, 165-176.

Assel, R.A., D.M. Robertson, H. Hoff, and J.H. Segelby, 1995: Climatic change implications from long term (1823-1994) ice records for the Laurentian Great Lakes. (1995). *Annals of Glaciology*, **21** (in press).

Barry, R.G., 1985: The Cryosphere and climate change. In: *Detecting the Effects of Increasing Carbon Dioxide*. DOEIER-0235, U.S. Dept. of Energy, Washington, DC, pp. 109-148.

Barsch, D., 1993: Periglacial geomorphology in the 21st century. *Geomorphology*, **7**, 141-163.

Bjornson, H., 1980: Avalanche activity in Iceland, climatic conditions and terrain features. *Journal of Glaciology*, **26(94)**, 13-23

Boer, G.J., N.A. Mcfarlane, and M. Lazare, 1992: Greenhouse gas-induced climate change simulated with the CCC second-generation general circulation model. *Journal of Climatology*, **5**, 1045-1077.

Braithwaite, R.J. and O.B. Olesen, 1989: Calculation of glacier ablation from air temperatures, West Greenland. In: *Glacier Fluctuations and Climatic Change* [Oerlemans, J. (ed.)]. Kluwer Academic Publishers, Dordrecht, Netherlands, pp. 219-233.

Braithwaite, R.J. and O.B. Olesen, 1990: Increased ablation at the margin of the Greenland ice sheet under a Greenhouse-effect climate. *Annals of Glaciology*, **14**, 20-22.

Braithwaite, R.J. and O.B. Olesen, 1993: Seasonal variation of ice ablation at the margin of the Greenland ice sheet and its sensitivity to climate change, Qamanrssp sermia, West Greenland. *J. Glaciol.*, **39(132)**, 267-274.

Bromwich, D.H., F.M. Robasky, R.A. Keen, and J.F. Bolzan, 1993: Modeled variations of precipitation over the Greenland Ice sheet. *Journal of Climate*, **6(7)**, 1253-1268.

Brown, R.D., M.G. Hughes, and D.A. Robinson, 1994: Characterizing the long term variability of snow cover extent over the interior of North America. *Annals of Glaciology*, **21**, 45-50.

Brown, R.D., 1993: Implications of global climate warming for the Canadian east coast sea ice and iceberg regimes over the next 50-100 years. *Climate Change Digest* 93-03, Canadian Climate Centre, Environment Canada, pp. 1-15.

Brown, R.D. and P. Cote, 1992: Interannual variability of landfast ice thickness in the Canadian High Arctic, 1950-1989. *Arctic*, **45**, 273-284.

Burgess, M.M. and D.W. Riseborough, 1990: Observations on the thermal response of discontinuous permafrost terrain to development and climate change—an 800-km transect along the Norman Wells pipeline. *Proceedings of the Fifth Canadian Permafrost Conference, Collection Nordicana*, **54**, Laval University, 291-297.

Burns, R.A., N.N. Goriainov, J.A. Hunter, A.S. Judge, A.G. Skvortsov, B.J. Todd, and V.M. Timofeev, 1993: Cooperative Russian-Canadian geophysical investigations of permafrost on the Yamal Peninsula, western Siberia. In: *Permafrost: Sixth International Conference, Beijing, China,1993*. Proceedings, vol. 1, South China University of Technology Press, Wushan, Guangzhou, China, pp. 66-71.

Calanca, P. and A. Ohmura, 1994: Atmospheric moisture flux convergence and accumulation on the Greenland Ice Sheet. In: *Snow and Ice Covers: Interactions with the Atmosphere and Ecosystems* (Proceedings of Yokahama Symposia J2 and J5, July 1993). IAHS Publication 223, pp. 77-84.

Chapman, W.L. and J.E. Walsh, 1993: Recent variations of sea ice and air temperature in high latitudes. *Bull. Am. Meteorol. Soc.*, **74**, 33-47.

Chen Ji Yang, 1991: Changes of alpine climate and glacier water resources. *Zuricher Geographische Schriften*, **46**, 196 pp.

Cheng Guodong and F. Dramis, 1992: Distribution of mountain permafrost and climate. *Permafrost and Periglacial Processes*, **3(2)**, 83-91.

Chinn, T.J., 1993: Physical hydrology of the dry valley lakes. In: *Physical and Biochemical Processes in Antarctic Lakes* [Green, W.J. and E.I. Freidmann (eds.)]. Antarctic Research Series, 51, American Geophysical Union, pp. 1-51.

CSIRO, 1994: *Climate Change and Snow Cover Duration in the Victorian Alps*. Report to the Environment Protection Authority, CSIRO Division of Atmospheric Research, Australia, Publication 403, 45 pp.

Cui Zhijiu, 1980: Periglacial phenomena and environmental reconstruction in the Qinghai-Tibet Plateau. In: *Collection of Geological Research Papers for the International Exchange*. Written for the 26th session of the International Geology Conference (Geohydrology, Engineering Geology, Quaternary Geology, Geomorphology), Geological Publishing House, pp. 109-115.

Cui Zhijiu and Xie Youyu, 1984: On the Southern boundary of permafrost and the periglacial environment in Northeast and North China in the late period of the late Pleistocene. *Allagcological Sinica*, **2**, 95-175.

Christensen, T., 1991: Arctic and sub-Arctic soils emissions: possible implications for global climate change. *Polar Record*, **27**, 205-210.

Christensen, T.R., 1993: Methane emission from Arctic tundra. *Biogeochemisry*, **21**, 117-139.

Clague, J.J. and S.G. Evans, 1992: Historic catastrophic retreat of Grand Pacific and Melbern Glaciers, Saint Elias Mountains, Canada: an analogue for the decay of the Cordilleran ice sheet at the end of the Pleistocene? *Journal of Glaciology*, **39(134)**, 619-624.

Clague, J.J. and S.G. Evans, 1994: Formation and failure of natural dams in the Canadian Cordillera. *Geological Survey of Canada Bulletin*, **464**, 35 pp.

Cline, W.R., 1992: Energy efficiency and Greenhouse abatement costs. *Climate Change*, **22(2)**, 95-97.

Dallimore, S.R., 1995: Intrapermafrost gas hydrates from a deep core hole in the Mackenzie Delta, Northwest territories, Canada. *Geology* (in press).

Dikikh, A.N. and L.K. Dikikh, 1990: Water-glacial resources of the Issykkul-Chuiski region. Their modern and future situation. *Water Resources*, **4**, 74-81 (in Russian).

Doake, C.S.M. and D.G. Vaughan, 1991: Rapid disintegration of the Wordie ice shelf in response to atmospheric warming. *Nature*, **350(6316)**, 328-330.

Domack, E.W., A.J. Timothy Jull, and S. Nakao, 1991: Advance of East Antarctic outlet glaciers during the Hypsithermal: implications for the volume state of the Antarctic ice sheet under global warming. *Geology*, **19**, 1059-1062.

Douglas, B.C., R.E. Cheney, L. Miller, and R.W. Agreen 1990: Greenland ice sheet: is it growing or shrinking? *Science*, **248**, 288.

Dutto, F. and G. Mortara, 1992: Rischi connessi con la dinamica glaciale nelle Alpi Italiane. *Geografia Fisica e Dinamica Quaternaria*, **15**, 85-99.

Dyurgerov, M., 1994: *Global Mass Balance Monitoring*. Report to U.S. Department of State, Institute of Geography of Russian Academy of Sciences.

Esch, D.C., 1993: Impacts of Northern Climate Change on Arctic Engineering Practice. In: *Impacts of Climate Change on Resource Management in the North* [Wall, G. (ed.)]. Department of Geography Publications Series, Occasional Paper No. 16, University of Waterloo, Ontario, Canada, pp. 185-192.

Etkin, D.A., 1991: Breakup in Hudson Bay: its sensitivity to air temperatures and implications for climate warming. *Climatological Bulletin*, **24**, 21-34.

Evans, S.G. and J.J. Clague, 1993: Glacier related hazards and climatic change. In: *The World at Risk: Natural Hazards and Climatic Change* [Bras, R.A. (ed.)]. American Institute of Physics Conference Proceedings 277, American Institute of Physics, New York, NY, pp. 48-60.

Evans, S.G. and J.J. Clague, 1994: Recent climate change and catastrophic geomorphic processes in mountain environments. *Geomorphology*, **10**, 107-128.

Fitzharris, B.B., 1987: A climatology of major avalanche winters in Western Canada. *Atmosphere-Ocean*, **25**, 115-136.

Fitzharris, B.B.F. and S. Bakkehoi, 1986: A synoptic climatology of major avalanche winters in Norway. *Journal of Climatology*, **6**, 431-446.

Fitzharris, B.B.F. and C.E. Garr, 1995: Detection of changes in seasonal snow in Southern Alps New Zealand. *Annals of Glaciology* (in press).

Fitzharris, B.B.F., J.E. Hay, and P.D. Jones, 1992: Behaviour of New Zealand glaciers and atmospheric circulation changes over the past 130 years. *The Holocene*, **2(2)**, 97-106.

Fitzharris, B.B.F. and P.A. Schaerer, 1980: Frequency of major avalanche winters. *Journal of Glaciology*, **26(94)**, 43.

Foehn, P.M., 1990: Schnee und Lawinen. In: *Schnee, Eis und Wasser in der Alpen in einer waermeren Atmosphaere, Mitteil.* VAW/ETH Zuerich, Nr. 108, S. 33-48.

Foulds, D.M. (ed.), 1988: *Optimum Operation of Hydro Plants during the Ice Regime of Rivers*. Subcommittee on the Hydraulics of Ice-Covered Rivers, National Research Council of Canada, Ottawa, Canada, 81 pp.

Francou, B. and L. Reynaud, 1992: 10 year surficial velocities on a rock glacier (Laurichard, French Alps). *Permafrost and Periglacial Processes*, **3(3)**, 209-213.

Fukuda, M., 1994: Methane flux from thawing Siberian permafrost (ice complexes)—results from field observations. *EOS*, **75**, 86.

Fujii, Y., K. Kamiyama, T. Kawamura, T. Kameda, K. Izumi, K. Satow, H. Enomoto, T. Nakamura, O.J. Hagen, Y. Gjessing, and O. Watanabe, 1990: 6000-year climate records in an ice core from the Hoghetta ice dome in northern Spitsbergen. *Annals of Glaciology*, **14**, 85-89.

Garr, C.E. and B.B. Fitzharris, 1994: Sensitivity of mountain runoff and hydro-electricity to changing climate. In: *Mountain Environments in Changing Climates* [Beniston, M. (ed.)]. Routledge, London and New York, pp. 366-381.

Geo-Engineering (M.S.T) Ltd., 1995: *Potential Impact of Global Warming in Permafrost in the McKenzie Valley—Results of Geothermal Modelling*. Geological Survey of Canada, Open file 2540, Ottawa, Canada, 40 pp.

Gerard, R., 1990: Hydrology of floating ice. In: *Northern Hydrology, Canadian Perspectives* [Prowse, T.D. and C.S.L. Ommanney (eds.)]. NHRI Science Report No. 1, National Hydrology Research Institute, Environment Canada, Saskatoon, Saskatchewan, Canada, pp. 103-134.

Ginzburg, B.M, K.N. Polyakova, and I.I. Soldatova, 1992: Secular changes in dates of ice formation on rivers and their relationship with climate change. *Soviet Meteorology and Hydrology*, **12**, 57-64.

Glazyrin, G.Y., K.V. Ratsek, and A.S. Tshetinnikov, 1986: Glacial runoff changes of middle Asian rivers due to the possible changes of the climate. *Transaction of SANII*, **117(198)**, 59-70 (in Russian).

Glazyrin G.Y., M.M. Pershukova, and A.V. Yakovlev, 1990: Change of the Kungei-Alatoo chain glaciation within possible changes of the climate. *Transactions of SANIGMI*, **136(217)**, 113-119 (in Russian).

Gleick, P.H., 1987a: Regional hydrologic consequences of increases in atmospheric CO_2 and other trace gases. *Climatic Change*, **10**, 137-161.

Gleick, P.H., 1987b: The development and testing of a water balance model for climate impact assessment: modelling the sacramento basin. *Water Resources Research*, **23(6)**, 1049-1061.

Gloersen, P. and W.J. Campbell, 1991: Recent variations in Arctic and Antarctic sea-ice covers. *Nature*, **352**, 33-36.

Gloersen, P., W.J. Campbell, D.J. Cavalieri, J.C. Comiso, C.L. Parkinson, and H.J. Zwally, 1992: *Arctic and Antarctic Sea Ice, 1978-1987: Satellite Passive-Microwave Observations and Analysis*. SP-511, National Aeronautics and Space Administration (NASA), Washington DC, 290 pp.

Gray, D.M. and T.D. Prowse, 1993: Snow and floating ice. In: *Handbook of Hydrology* [Maidment, D. (ed.)]. McGraw-Hill, New York, NY, pp. 7.1–7.58.

Greco, S., R.H. Moss, D. Viner, and R. Jenne, 1994: *Climate Scenarios and Socioeconomic Projections for IPCC WGII Assessment*. Consortium for International Earth Science Information Network, Washington DC, 67 pp.

Gu Zhongwei, Zhou Youwu, Liang Fenxian, Liang Linheng, and Zhang Qibin, 1993: Permafrost features and their changes in Amur area, Daxinganling Prefecture, North-eastern China. In: *Permafrost: Sixth International Conference, Beijing, China 1993*. Proceedings, vol. 1, South China University of Technology Press, Wushan, Guangzhou, China, pp. 204-209.

Haeberli, W., 1992: Construction, environmental problems and natural hazards in periglacial mountain belts. *Permafrost and Periglacial Processes*, **3(2)**, 111-124.

Haeberli, W., 1994: Accelerated glacier and permafrost changes in the Alps. In: *International Conference on Mountain Environments in Changing Climates, Davos 1992*. Routledge Publishers, London and New York, pp. 1-107.

Haeberli, W., G. Cheng, A.P. Gorbunov, and S.A. Harris, 1993: Mountain permafrost and climatic change. *Permafrost and Periglacial Processes*, **4(2)**, 165-174.

Haeberli, W., M. Hoelzle, F. Keller, W. Schmid, D. Vonder Mühll, and S. Wagner, 1993: Monitoring the long-term evolution of mountain permafrost in the Swiss Alps. In: *Permafrost: Sixth International Conference, Beijing, China 1993*. Proceedings, vol. 1, South China University of Technology Press, Wushan, Guangzhou, China, pp. 214-219.

Haeberli, W. and M. Hoelzle, 1995: Application of inventory data for estimating characteristics of and regional climate change effects on mountain glaciers—a pilot study with the European Alps. *Annals of Glaciology* (in press).

Hanson, H.P., C.S. Hanson, and B.H. Yoo, 1992: Recent Great Lakes ice trends. *Bulletin American Meteorological Society*, **73**, 577-584.

Harris, S.A., 1990: Long-term air and ground temperature records from the Canadian cordillera and the probable effects of moisture changes. *Proceedings Fifth Canadian Permafrost Conference, Collection Nordicana*, pp. 151-157.

Harris, S.A. and A.E. Corte, 1992: Interactions and relations between mountain permafrost, glaciers, snow and water. *Permafrost and Periglacial Processes*, **3(2)**, 103-110.

Harris, S.A. and R. Giardino, 1994: Permafrost in mountain ranges of North America. In: *Permafrost: Sixth International Conference, Beijing, China 1993*. Proceedings, vol. 2, South China University of Technology Press, Wushan, Guangzhou, China.

Harrison, W.D., 1991: Permafrost response to surface temperature change and its implications for the 40,000-year surface history at Prudhoe Bay. *Journal of Geophysical Research*, **96**, 683-695

Hastenrath, S., 1992: Ice flow and mass changes of Lewis Glacier, Mount Kenya, 1986-90; observations and modelling. *Journal of Glaciology*, **38**, 36-42.

Hastenrath, S., 1994: Variations of equatorial glaciers and global change. In: *Proceedings of International Symposium on Seasonal and Long-term Fluctuations of Naval and Glacial Processes in Mountains*. UNESCO-IUGG-IAHS-ICSI, September 1993, Tashkent, Uzbekistan.

Hastenrath, S. and A. Ames, 1995: Recession of Yanamarey Glacier in Cordillera Blanca, Peru, during the 20th century. *Journal of Glaciology*, **41(137)**, 191-196.

Hastenrath, S. and P.D. Kruss, 1992: The dramatic retreat of Mount Kenya's glaciers 1963-87: greenhouse forcing. *Annals of Glaciology*, **16**, 127-133.

Henderson-Sellers, A. and A. Hansen, 1995: *Atlas of Results from Greenhouse Model Simulations*. Model Evaluation Consortium for Climate Assessment, Climate Impacts Centre, Macquarie University, Kluwer Academic Publishers, Dordrecht, Netherlands.

Hibler, W.D. III, 1989: Arctic ice-ocean dynamics. In: *The Arctic Seas. Climatology, Oceanography, Geology, and Biology* [Herman, Y. (ed,)]. Van Nostrand Reinhold, New York, NY, pp. 47-91.

Higuchi, K., 1991: Change in hydrological cycle in cold regions corresponding to "greenhouse" warming. In: *Arctic Hydrology, Present and Future Tasks*. Report 23, Norwegian National Committee for Hydrology, Oslo, Norway, pp. 47-49.

Hoelzle, M. and W. Haeberli, 1995: Simulating effects of mean annual air temperature changes on permafrost distribution and glacier size—an example from the Upper Engadin, Swiss Alps. *Annals of Glaciology* (in press).

IAHS (ICSI)/UNEP/UNESCO, 1988: *Fluctuations of Glaciers, 1900-1985* [Haeberli, W. and P. Muller (eds.)]. Paris, France, 290 pp.

IAHS (ICSI)/UNEP/UNESCO, 1989: *World Glacier Inventory—Status 1989* [Haeberli, W., H. Bösch, K. Scherler, G. Østrem, and C.C. Wallén (eds.)]. Nairobi, Kenya, 390 pp.

IAHS (ICSI)/UNEP/UNESCO, 1993: *Glacier Mass Balance Bulletin No. 2—1990-1991* [Haeberli, W., E. Herren, and M. Hoelzle (eds.)]. WGMS/ETH Zurich, 74 pp.

IAHS (ICSI)/UNEP/UNESCO, 1994: *Glacier Mass Balance Bulletin No. 3—1992-1993* [Haeberli, W., M. Hoelzle, and H. Bösch (eds.)]. WGMS/ETH Zurich, 80 pp.

IPCC, 1990: *Climate Change: The IPCC Impacts Assessment* [McG. Tegart, W.J., G.W. Sheldon, D.C. Griffiths (eds.)]. Australian Government Publishing Service, Canberra, Australia, 268 pp.

IPCC, 1992: *Climate Change 1992. The Supplementary Report to the IPCC Scientific Assessment* [Houghton, J.T., B.A. Callander, and S.K. Varney (ed.)]. Cambridge Univ. Press, Cambridge, UK, 200 pp.

Jacobs, S.S., 1992: Is the Antarctic ice sheet growing? *Nature*, **360**, 29-33.

Johannesson, T., O. Sigurdsson, T. Laumann, and M. Kennett, 1993: Degree-day glacier mass balance modelling with applications to glaciers in Iceland and Norway. *Nordic Hydrological Programme NHP Report*, **33**.

Judge, A.S. and J.A. Majorowicz, 1992: Geothermal conditions for gas hydrate stability in the Beaufort-Mackenzie area: the global change aspect. *Global and Planetary Change*, **6**, 251-263.

Judge, A.S., S.L. Smith, and J.A. Majorowicz, 1994: The current distribution and thermal stability of natural gas hydrates in the polar regions. In: *Proceedings of the Fourth International Offshore and Polar Engineering Conference, April 10-15, 1994, Vol. 1*. International Society of Offshore and Polar Engineers, Osaka, Japan, pp. 307-314.

Kadota, T., R. Naruse, P. Skvarca, and M. Aniya, 1992: Ice flow and surface lowering of Tyndall Glacier, southern Patagonia. *Bulletin of Glacier Research*, **10**, Japanese Society of Snow and Ice, 63-68.

Kameda, T., M. Nakawo, S. Mae, O. Watanabe, and R. Naruse, 1990. Thinning of the ice sheet estimated from total gas content of ice cores in Mizuho Plateau, East Antarctica. *Annals of Glaciology*, **14**, 131-135.

Kapsner, W.R., R.B. Alley, C.A. Shuman, S. Anandakrishnan, and P.M. Grootes, 1995: Dominant influence of Atmospheric Circulation on snow accumulation in Greenland over the past 18,000 years. *Nature*, **373(6509)**, 52-54.

Karl, T.R., P. Ya. Groisman, R.W. Knight, and R.R. Heim, Jr., 1993: Recent variations of snowcover and snowfall in North America and their relation to precipitation and temperature variations. *Journal of Climatology*, **6**, 1327-1344.

Kassi, N., 1993: Native perspective on climate change. In: *Impacts of Climate Change on Resource Management in the North* [Wall, G. (ed.)]. Department of Geography Publications Series, Occasional Paper No. 16, University of Waterloo, Ontario, Canada, pp. 43-49.

Keller, F. and H.U. Gubler, 1993: Interaction between snow cover and high mountain permafrost at Murtel/Corvatsch, Swiss Alps. In: *Permafrost: Sixth International Conference, Beijing, China 1993*. Proceedings, vol. 1, South China University of Technology Press, Wushan, Guangzhou, China, pp. 332-337.

King, L. and H.J. Akermann, 1994: Mountain permafrost in Europe. In: *Permafrost: Sixth International Conference, Beijing, China 1993*. Proceedings, vol. 2, South China University of Technology Press, Wushan, Guangzhou, China, pp. 1022-1027.

Koerner, R.M. and M.M. Brugman, 1991: Mass balance trends in the Canadian High Arctic and Western Cordilleran Glaciers over the past three decades. Paper presented at IAMAP symposium "Climate-dependent dynamics, energy and mass balance of polar glaciers and ice sheets," Vienna, August 1991. Abstract.

Kolchugina, T.P. and T.S Vinson, 1993: Climate warming and the carbon cycle in the permafrost zone of the Former Soviet Union. *Permafrost and Periglacial Processes*, **4**, 149-163.

Kondratjeva, K.A., S.F. Khrutsky, and N.N. Romanovsky, 1993: Changes in the extent of permafrost during the late Quaternary period in the territory of the Former Soviet Union. *Permafrost and Periglacial Processes*, **4**, 113-119.

Koster, E.A., 1991: Assessment of climate change impact in high-latitudinal regions. *Terra*, **103**, 3-13.

Koster, E.A., 1994: Global warming and periglacial landscapes. In: *The Changing Global Environment* [Roberts, N. (ed.)]. Blackwell, Cambridge USA, pp. 127-149.

Koster, E.A. and M.E. Nieuwenhuijzen, 1992: Permafrost response to climatic change. *Catena Supplement*, **22**, 37-58.

Koster, E.A., M.E. Nieuwenhuijzen, and A.S. Judge, 1994: *Permafrost and Climate Change: An Annotated Bibliography*. Glaciological Data Report 27, WDC-A for Glaciology, University of Colorado, Boulder, CO, 94 pp.

Kotlyakov V.M., M.G. Grosswald, M.B. Dyurgerov, and V.L. Mazo, 1991: The reaction of glaciers to impending climate change. *Polar Geography and Ecology*, **15(3)**, 203-217.

Kuhn, M., 1993: Possible future contribution to sea level change from small glaciers. In: *Climate and Sea Level Change: Observations, Projections and Implications* [Warrick, R.A., E.M. Barrow, and T.M.L. Wigley (eds.)]. Cambridge University Press, Cambridge, UK, pp. 134-143.

Kvenvolden, K.A. and T.D. Lorenson, 1993: Methane in permafrost: preliminary results from coring at Fairbanks, Alaska. *Chemosphere*, **26(1-4)**, 609-616.

Lachenbrach, A.H. and B.V. Marshall, 1986: Changing climate—geothermal evidence from permafrost in the Alaskan arctic. *Science*, **234(4777)**, 689-696.

Lange, M.A. and H. Eicken, 1991: The sea ice thickness distribution in the northwestern Weddell Sea. *Journal of Geophysical Research*, **96(C3)**, 4821-4837.

Laumann, T. and A. M. Tvede, 1989: Simulation of the effects of climate changes on a glacier in western Norway. *Norges Vassdrags-og Energiverk. Meddelelse fra Hydrologisk Avdeling*, **72**, 339-352.

Lemke, P., W.B. Owens, and W.D. Hibler III, 1990: A coupled sea ice-mixed layer-pycnocline model for the Weddell Sea. *Journal of Geophysical Research*, **95**, 9527-9538.

Letréguilly, A. and L. Reynaud, 1989: Past and forecast fluctuations of Glacier Blanc (French Alps). *Annals of Glaciology*, **13**, 159-163.

Lewkowicz, A.G., 1992: Factors influencing the distribution and initiation of active-layer detachment slides on Ellesmere Island, Arctic Canada. In: *Periglacial Geomorphology* [Dixon, J.C. and A.D. Abrahams (eds.)]. John Wiley and Sons, New York, NY, pp. 223-250.

Lonergan, S., R. Difrancesco, and M.K. Woo, 1993: Climate-change and transportation in Northern Canada—an intergrated impact assessment. *Climate Change*, **24(4)**, 331-351.

MacInnes, K.L., M.M. Burgess, D.G. Harry, and T.H.W. Baker, 1990: *Permafrost and Terrain Research and Monitoring: Norman Wells Pipeline. Volume II Research and Monitoring Results: 1983-1988*. Environmental Studies Report No. 64, Department of Indian and Northern Affairs Canada, Northern Affairs Program, Ottawa, Canada, 204 pp.

Malagnino, E. and J. Strelin, 1992: Variations of Upsala Glacier in southern Patagonia since the late Holocene to the present. In: *Glaciological Researches in Patagonia (1990)* [Naruse, R. and M Aniya (eds.)]. pp. 61-85.

Malmer, N., 1992: Peat accumulation and the global carbon cycle. *Catena Supplement*, **22**, 97-110.

Marion, G.M. and W.C. Oechel, 1993: Mid- to late-Holocene carbon balance in Arctic Alaska and its implications for future global warming. *The Holocene*, **3**, 193-200.

Martinec, J., A. Rango, and R. Roberts, 1994: Modelling the redistribution of runoff caused by global warming. In: *Effects of Human-Induced Changes on Hydrologigal Systems*. American Water Resources Association, pp. 153-161.

Martinec, J. and A. Rango, 1989: Effects of climate change on snow melt runoff patterns. In: *Remote Sensing and Large-Scale Global Processes*. Proceedings of the IAHS Third International Assembly, Baltimore, MD, May 1989, IAHS Publication No. 186, pp. 31-38.

Martinson, D.G., 1990: Evolution of the Southern Ocean winter mixed layer and sea ice: open ocean deepwater formation and ventilation. *Journal of Geophysical Research*, **95(C7)**, 11,641-654.

McGillivray, D.G, T.A. Agnew, G.A. McKay, G.R. Pilkington, and M.C. Hill, 1993: Impacts of climatic change on the Beaufort sea ice regime: implications for the Arctic Petroleum Industry. *Climate Change Digest* 93-01, Atmospheric Environment Service, Canada, pp. 1-17.

McLaren, A.S., 1989: The under-ice thickness distribution of the Arctic basin as recorded in 1958 and 1970. *Journal of Geophysical Research*, **94**, 4971-4983.

McLaren, A.S., J.E. Walsh, R.H. Bourke, R.L. Weaver, and W. Wittman, 1992: Variability in sea-ice thickness over the North Pole from 1979 to 1990. *Nature*, **358**, 224-226.

Meier, M.F., 1993: Ice, climate and sea level: do we know what is happening? In: *Ice in the Climate System* [Peltier, W.R. (ed.)]. NATO ASI series I, Global Environmental Change, 12, Springer-Verlag, Heidelberg, Germany, pp. 141-160.

Melnikov, P.A. and R.B. Street, 1992: Terrestrial component of the cryosphere. In: *Climate Change 1992. The Supplementary Report to the IPCC Impacts Assessment* [McG. Tegart, W.J. and G.W. Sheldon (eds.)]. Australian Government Publishing Service, Canberra, Australia, pp. 94-98.

Miller, G.H. and A. de Vernal, 1992: Will greenhouse warming lead to Northern Hemisphere ice-sheet growth? *Nature*, **355**, 244-246.

Moraes, F. and M.A.K. Khalil, 1993: Permafrost methane content: 2. Modeling theory and results. *Chemosphere*, **26**, 595-607.

Morgan, V.I., I.D. Goodwin, D.M. Etheridge, and C.W. Wookey, 1991: Evidence from Antarctic ice cores for recent increases in snow accumulation. *Nature*, **354**, 58-60.

Morinaga, Y. and T. Yasunari, 1993: Recent global warming and variation of winter snow cover in Japan. *Science Report, Institute of Geoscience*, University of Tsukuba, Section A, **14**, 49-59.

Müller, H. and G. Kaeppenberger, 1991: Claridenfirnmessungen 1914-1984. *Zürcher Geographische Schriften*, Nr. 40, Geographishe Institut, ETH Zurich (graph updated by H. Lang and U. Steinegger 1995).

Mysak, L.A. and D.K. Manak, 1989: Arctic sea ice extent and anomalies, 1953-84. *Atmosphere and Ocean*, **27(2)**, 346-505.

Nelson, F.E. and O.A. Anisimov, 1993: Permafrost zonation in Russia under anthropogenic climatic change. *Permafrost and Periglacial Processes*, **4(2)**, 137-148.

Nixon, F.M. and A.E. Taylor, 1994: *Active Layer Monitoring in Natural Environments, MacKenzie Valley, Northwest Territories*. Current Research 1994-B, Geological Survey of Canada, Ottawa, Canada, pp. 27-34.

Nixon, F.M., A.E. Taylor., V.S. Allen, and F. Wright, 1995: *Active Layer Monitoring in Natural Environments, Lower McKenzie Valley, Northwest Territories*. Current Research 1995-B, Geological Survey of Canada, Ottowa, Canada, pp. 99-108.

Oechel, W.C., S.J. Hastings, G. Vourlitis, M. Jenkins, G. Riechers, and N. Grulke, 1993: Recent change of Arctic tundra ecosystems from a net carbon dioxide sink to a source. *Nature*, **361(6412)**, 520-523.

Oerlemans, J., 1991: The mass balance of the Greenland ice sheet: sensitivity to climate change as revealed by energy-balance modelling. *The Holocene*, **1(1)**, 40-49.

Oerlemans, J., 1994: Quantifying global warming from the retreat of glaciers. *Science*, **264**, 243-245.

Oerlemans, J. and J.P.F. Fortuin, 1992: Sensitivity of glaciers and small ice caps to greenhouse warming. *Science*, **258**, 115-118.

Oerlemans, J., R.S. van de Wal, and L.A. Conrads, 1993: A model for the surface balance of ice masses: Part II. Application to the Greenland ice sheet. *Z. Gletscherkd. Glazialgeol*, **27/28**, 85-96.

Oerlemans, J. and H.F. Vugts, 1993: A meteorological experiment in the melting zone of the Greenland ice sheet. *Bulletin of the American Meteorological Society*, **74(3)**, 355-365.

Ohmura, A. and N. Reeh, 1991. New precipitation and accumulation maps for Greenland. *Journal of Glaciology*, **37(125)**, 140-148.

Orheim, O., 1988: Antarctic icebergs—production, distribution and disintegration. *Annals of Glaciology*, **11**, 205.

Osterkamp, T.E., 1994: Evidence for warming and thawing of discontinuous permafrost in Alaska. *EOS*, **75(44)**, 85(1994).

Osterkamp, T.E. and J.P. Gosink, 1991: Variations in permafrost thickness in response to changes in climate. *Journal of Geophysical Research*, **96(B3)**,4423-4434.

Osterkamp, T.E., T. Zang, and V.E. Romanovsky, 1994: Evidence for a cyclic variation of permafrost temperatures in Northern Alaska. *Permafrost and Periglacial Processes*, **5**, 137-144.

Palecki, M.A. and R.G. Barry, 1986: Freeze-up and break-up of lakes as an index of temperature changes during the transition seasons: a case study for Finland. *Journal of Climate and Applied Climatology*, **25**, 893-902.

Parkinson, C.L., 1992: Spatial patterns of increases and decreases in the length of the sea ice season in the north polar region, 1979-1986. *Journal of Geophysical Research*, **97(C9)**, 14,377-388.

Parkinson, C.L. and D.J. Cavalieri, 1989: Arctic sea ice 1983-1987: seasonal, regional and interannual variability. *Journal of Geophysical Research*, **94(C10)**, 14,199-523.

Pavlov, A.V., 1994: Climate controls and high-altitude permafrost, Qinghai-Xizang (Tibet) Plateau, China. *Permafrost and Periglacial Processes*, **5**, 101-110.

Popovnin, V.V., 1987: The problem of long term prediction of mountain glaciation evolution and the alternative design for the Dzhankuat glacier. In: *Estimation and Long Term Prediction of Mountain Nature Changing*. Moscow, USSR, pp.128-145 (in Russian).

Prowse, T.D., M.N. Demuth, and H.A.M. Chew, 1990: The deterioration of freshwater ice due to radiation decay. *Journal of Hydraulic Research*, **28(6)**, 685-697.

Qiu Guoqing, 1994: Mountain permafrost in central Asia. In: *Permafrost: Sixth International Conference, Beijing, China 1993*. Proceedings, vol. 2, South China University of Technology Press, Wushan, Guangzhou, China, pp. 1028-1030.

Qiu Guoqing and Cheng Guodong, 1995: Permafrost in China: past and present. *Permafrost and Periglacial Processes*, **6(1)**, 3-14.

Rango, A. and J. Martinec, 1994: Areal extent of seasonal snow cover in a changed climate. *Nordic Hydrology*, **25**, 233-246.

Rasmussen, R.A., M.A.K. Khalil, and F. Moraes, 1993: Permafrost methane content 1: experimental data from sites in northern Alaska. *Chemosphere*, **26**, 591-594.

Reeh, N., 1989: Dynamic and climatic history of the Greenland ice sheet. In: *Quaternary Geology of Canada and Greenland* [Fulton, R.J. (ed.)]. Geological Survey of Canada, Ottawa, Canada, pp. 795-822.

Reycraft, J. and W. Skinner, 1993: Canadian lake ice conditions: an indicator of climate variability. *Climatic Perpectives*, **15**, 9-15.

Riseborough, D.W., 1990: Soil latent heat as a filter of the climate signal in permafrost. In: *Proceedings of the Fifth Canadian Permafrost Conference*. Centre d'etudes nordiques, Universite Laval/National Research Council of Canada, Quebec, Canada, pp. 199-205.

Riseborough, D.W. and M.W. Smith, 1993: Modelling permafrost response to climate change and climate variability. In: *Proceedings of the Fourth International Symposium on Thermal Engineering & Science for Cold Regions* [Lunardini, V.J. and S.L. Bowen (eds.)]. U.S. Army Cold Regions Research and Engineering Laboratory, Special Report 93-22, Hanover, NH, pp. 179-187.

Robinson, D.A., F.T. Keimig, and K.F. Dewey, 1993: Recent variations in Northern Hemisphere snow cover. In: *Proc. Fifteenth Annual Climate Diagnostics Workshop*. NOAA, pp. 219-224.

Romanovsky, V.E., L.S. Garagula, and N.V. Sergina, 1991: Freezing and thawing of soils under the influence of 300- and 90-year periods of temperature fluctuation. In: *Proceedings: International Conference on the Role of the Polar Regions in Global Change,* Fairbanks, AK, June 11-15, 1990 [Weller, G. *et al.* (ed.)]. University of Alaska, Fairbanks, AK, vol. 2, pp. 542-548.

Romanovsky, V.E., L.N. Maximova, and N.V. Sergina, 1991: Paleotemperature reconstruction for freeze-thaw processess during the late Pleistocene through the Holocene. In: *Proceedings: International Conference on the Role of the Polar Regions in Global Change*, Fairbanks, AK, June 11-15, 1990 [Weller G. *et al.* (ed.)]. University of Alaska, Fairbanks, AK, vol. 2, pp. 537-542.

Roots, F., 1993: Climate change—its possible impact on the environment and the people of northern regions. In: *Impacts of Climate Change on Resource Management in the North* [Wall, G. (ed.)]. Department of Geography Publications Series, Occasional Paper No. 16, University of Waterloo, Ontario, Canada, pp. 127-151.

Ruddell, A.R., 1990: The glaciers of New Zealand's Southern Alps: a century of retreat. Paper presented at ANZAAS Congress, Hobart, Australia, February 1990.

Rudels, B., 1990: Haline convection in the Greenland Sea. *Deep-Sea Research*, **37(9)**, 1491-1511.

Sanderson, T.J.O., 1988: *Ice Mechanics Risks to Offshore Structures*. Graham and Trottman, London, UK, 253 pp.

Schindler, D.W., K.G. Beaty, E.J. Fee, D.R. Cruickshank, E.D. de Bruyn, D.L. Findlay, G.A. Linsey, J.A. Shearer, M.P. Stainton, and M.A. Turner, 1990: Effects of climate warming on lakes of the Central Boreal Forest. *Science*, **250**, 967-970.

Schlosser, P., G. Bonisch, M. Rhein, and R. Bayer, 1991: Reduction of deepwater formation in the Greenland Sea during the 1980's: evidence from tracer data. *Science*, **251**, 1054-1056.

Schmitt, E., 1993: Global climatic change and some possible geomorphological and ecological effects in Arctic permafrost environments, Isfjorden and Liefedefjorden, Northern Spitsbergen. In: *Permafrost: Sixth International Conference, Beijing, China 1993*. Proceedings, vol. 1, South China University of Technology Press, Wushan, Guangzhou, China, pp. 544-549.

Schubert, C., 1992: The glaciers of the Sierra Nevada de Merida (Venezuela): a photographic comparison of recent deglaciation. *Erdkunde*, **46**, 59-64.

Shi Yafeng and Ren Binghui, 1988: An introduction to the Glaciers in China. *Science Press*, **75**, 172-174.

Shi Yafeng and Ren Jiawen, 1990: Glacier recession and lake shrinkage indicating a climatic warming and drying trend in Central Asia. *Annals of Glaciology*, **14**, 261-265.

Skvarca, P., 1993: Fast recession of the northern Larsen Ice Shelf monitored by space images. *Annals of Glaciology*, **17**, 317-321.

Skvarca, P., 1994: Changes and surface features of the Larsen Ice Shelf, Antarctica, derived from the Landsat and Kosmos mosaics. *Annals of Glaciology*, **20**, 6-12.

Soldatova, I.I., 1992: Causes of variability of ice appearance dates in the lower reaches of the Volga. *Soviet Meteorology and Hydrology*, **2**, 62-66.

Soldatova, I.I., 1993: Secular variations in river break-up dates and their relationship with climate variation. *Meteorologia i gidrologia*, **9**, 89-96 (in Russian).

Stokoe, P., 1988: Socio-economic assessment of the physical and ecological impacts of climate change on the marine environment of the Atlantic region of Canada. *Climate Change Digest* 88-07, Atmospheric Environment Service, Canada, pp. 1-8.

Street, R.B. and P.I. Melnikov, 1990: Seasonal snow cover, ice, and permafrost. In: *Climate Change: The IPCC Impacts Assesment* [McG. Tegart, W.J., G.W. Sheldon, and D.C. Griffiths (eds.)]. Australian Government Publishing Service, Imprimatur Press, Canberra, Australia, pp. 7-1–7-33.

Sun Jianzhong, 1985: Paleoenvironment in the Northeast China during the last ice age. *China Quaternary Period Research*, **6(1)**.

Tangborn, W., 1980: Two models for estimating climate-glacier relationships in the north Cascades, Washington, U.S.A. *Journal of Glaciology*, **25(91)**, 3-21.

Taylor, A.E., 1991a: Marine transgression, shoreline emergence: evidence in sea-bed and terrestrial ground temperatures of changing relative sea-levels, Arctic Canada. *Journal of Geophysical Research*, **96(B4)**, 6893-6909.

Taylor, A.E., 1991b: Holocene paleoenvironmental reconstruction from deep ground temperatures: a comparison with paleoclimate derived from the 18 O record in an ice core from the Agassiz Ice Cap, Canadian Arctic Archipelago. *Journal of Glaciology*, **37**, 209- 219.

Tenthorey, G., 1992: Perennial ne've's and the hydrology of rock glaciers. *Permafrost and Periglacial Processes*, **3(3)**, 247-252.

UNEP, 1993: *Glaciers and the Environment*. UNEP/GEMS Environment Library No. 9, 23 pp.

U.S. Department of Energy, 1985: *Glaciers, Ice Sheets and Sea Level*. Report of a Workshop held in Seattle, Washington, September 13-15, 1984, DOE/EV/60235-1, 330 pp.

Vandenberghe, J., 1993: Permafrost changes in Europe during the last glacial. *Permafrost and Periglacial Processes*, **4**, 121-135.

Van Der Vinne, G., T.D. Prowse, and D. Andres, 1991: Economic impact of river ice jams in Canada. In: *Northern Hydrology, Selected Perspectives* [Prowse, T.D. and C.S.L. Ommanney (eds.)]. NHRI Symposium No. 6, National Hydrology Research Institute, Environment Canada, Saskatoon, Saskatchewan, Canada, pp. 333-352.

Velichko, A.A. and V.P. Nechaev, 1992: The estimation of permafrost dynamics in Northern Eurasia within the global warming 1992. *Doklady of the Academy of Sciences*, **324(3)**, 667-671 (in Russian).

Vinje, T.E., 1989: An upward looking sonar ice draft series. In: *Proceedings of the 10th International Conference on Port & Ocean Engineering under Arctic Conditions* [Axelsson, K.B.E. and L.A. Fransson (eds.)]. Lulea University, Technology, 1, pp. 178-187.

Vonder Muhll, D.S. and W. Schmid, 1993: Geophysical and photogrammetric investigation of rock glacier Muragl 1, Upper Engadin, Swiss Alps. In: *Permafrost: Sixth International Conference, Beijing, China 1993*. Proceedings, vol. 1, South China University of Technology Press, Wushan, Guangzhou, China, pp. 654-659.

Vyalov, S.S., A.S. Gerasimov, A.J. Zolotar, and S.M. Fotiev, 1993: Ensuring structural stability and durability in permafrost ground at global warming of the Earth's climate. In: *Permafrost: Sixth International Conference, Beijing, China 1993*. Proceedings, vol. 1, South China University of Technology Press, Wushan, Guangzhou, China, pp. 955-960.

Wadhams, P., 1989: Sea-ice thickness in the Trans Polar Drift Stream. *Rapp. P-v Reun Cons. Int. Explor. Mer*, **188**, 59-65.

Wadhams, P., 1990a: Evidence for thinning of the Arctic ice cover north of Greenland. *Nature*, **345**, 795-797.

Wadhams, P., 1990b: Sea ice and economic development in the Arctic Ocean—a glaciologist's experience. In: *Arctic Technology and Economy. Present Situation and Problems, Future Issues*. Bureau Veritas, Paris, France, pp. 1-23.

Wadhams, P., 1992: Sea ice thickness distribution in the Greenland Sea and Eurasian Basin, May 1987. *Journal of Geophysical Research*, **97(C4)**, 5331–5348.

Wadhams, P., 1994: Sea ice thickness changes and their relation to climate. In: *The Polar Oceans and Their Role in Shaping the Global Environment. The Nansen Centennial Volume* [Johannessen, O.M., R.D. Muench, and J.E. Overland (eds.)]. American Geophysical Union., Monograph 85, pp. 337-361.

Wadhams, P. and J.C. Comiso, 1992: The ice thickness distribution inferred using remote sensing techniques. In: *Microwave Remote Sensing of Sea Ice* [Carsey, F. (ed.)]. Geophysical Monograph 68, American Geophysical Union, Washington, pp. 375-383.

Wadhams, P. and D.R. Crane, 1991: SPRI participation in the Winter Weddell Gyre Study 1989. *Polar Record*, **27**, 29-38.

Wagner, S., 1992: Creep of alpine permafrost investigated on the Murtl rock glacier. *Permafrost and Periglacial Processes*, **3(2)**, 157-162.

Wall, G. (ed.), 1993: *Impacts of Climate Change on Resource Management in the North*. Department of Geography Publications Series, Occasional Paper No. 16, University of Waterloo, Ontario, Canada, 245 pp.

Wang Boalai and H.M. French, 1994: Climate controls and high-altitude permafrost, Qinghai Xizang (Tibet) Plateau, China. *Permafrost and Periglacial Processes*, **5**, 87-100.

Warren, C.R., 1994: Against the grain; a report on the Pio XI glacier Patagonia. *Geographical Magazine*, September, **66(9)**, 28-30.

Weidick, A., 1991a: Present-day expansion of the southern part of the Inland Ice. *Rapport Gronlands geoliske Undersogelse*, **152**, 73-79.

Weidick, A., 1991b: Jakobshavn Isbrae during the climatic optimum. *Rapport Gronlands geoliske Undersogelse*, **155**, 67-72.

Weidick, A., 1991c: Neoglacial change of ice cover and the related response to the Earth's crust in West Greenland. *Rapport Gronlands geoliske Undersogelse*, **159**, 121-126.

Weidick, A., 1991d: Present day expansion of the southern part of the Inland Ice. *Rapport Gronlands geoliske Undersogelse*, **152**, 73-79.

Wright, F., M.W. Smith, and A.E. Taylor, 1994: A hybrid model for predicting permafrost occurrence and thickness. *EOS*, **75**, 77.

Woo, M.K., 1990: Consequences of climatic change for hydrology in permafrost zones. *Journal of Cold Regions Engineering*, **4**, 15-20.

Woo, M.K. and B.B. Fitzharris, 1992: Reconstruction of mass balance variations for Franz Josef glacier, New Zealand, 1913 to 1989. *Arctic and Alpine Research*, **24(4)**, 281-290.

Woo, M.K., 1992: Impacts of climate variability and change on Canadian wetlands. *Canadian Water Resources Journal,* **17**, 63-69.

Woo, M.K., A.G. Lewkowicz, and W.R. Rouse, 1992: Response of the Canadian permafrost environment to climatic change. *Physical Geography*, **13**, 287-317.

Xie Youyu, 1985: Preliminary discussion on the climatic conditions for the formation of permafrost in Northeast China and its evolution. *Glacial Cryopedology*, **7(4)**, 323-329.

Xie Youyu, 1996: *Effects of Climatic Change on Permafrost in China*. Institute of Geography, Chinese Academy of Science, Global Change Study No. 2, Series publication (in press).

Yang Hairong, 1985: Measures of preventing and controlling the Roadbed subsidence in the permafrost zone and improving its stability. *Glacial Cryopedology*, **7(1)**.

Zachrisson, G., 1989: Climate variation and ice conditions in the River Tornelven. In: *Conference on Climate and Water*. Publications of the Academy of Finland, vol. 1, Helsinki, Finland, pp. 353-364.

Zhang, T. and T.E. Osterkamp, 1993: Changing climate and permafrost temperatures in the Alaskan Arctic. In: *Permafrost: Sixth International Conference, Beijing, China 1993.* Proceedings, vol. 1, South China University of Technology Press, Wushan, Guangzhou, China, pp. 783-788.

Zimov, S.A., I.P. Semiletov, S.P. Daviodov, Yu.V. Voropaev, S.F. Prosyannikov, C.S. Wong, and Y.-H. Chan 1993: Wintertime CO_2 emission from soils of Northeastern Siberia. *Arctic*, **46**, 197-204.

Zwally, H.J., A.C. Brenner, J.A. Major, R.A. Bindschadler, and J.A. Marsh, 1989: Growth of Greenland ice sheet: measurement and interpretation. *Science*, **246**, 1587-1591.

8

Oceans

VENUGOPALAN ITTEKKOT, GERMANY

Principal Lead Authors:
Su Jilan, China; E. Miles, USA

Lead Authors:
E. Desa, India; B.N. Desai, India; J.T. Everett, USA; J.J. Magnuson, USA; A. Tsyban, Russian Federation; S. Zuta, Peru

Contributing Authors:
E. Aquize, Peru; S. Arnott, USA; P. Ayon Dejo, Peru; D. Binet, France; H.S. Bolton, USA; R. Calienes, Peru; S. Carrasco Barrera, Peru; J.A. Church, Australia; A. Copping, USA; D.L. Fluharty, USA; B.V. Glebov, Russian Federation; K.P. Koltermann, Germany; A.S. Kulikov, Russian Federation; S. Nicol, Australia; P.D. Nunn, Fiji; G.V. Panov, Russian Federation; P.K. Park, USA; A.B. Pittock, Australia; P. Schaefer, Germany; S. Shchuka, Russian Federation; H. Trevino, Peru; D.J. Webb, UK; R. Zahn, Germany

CONTENTS

EXECUTIVE SUMMARY

Global warming as projected by Working Group I of the IPCC will have an effect on sea-surface temperature (SST) and sea level. As a consequence, it is likely that ice cover and oceanic circulation will be affected, and the wave climate will change. The expected changes affect global biogeochemical cycles, as well as ecosystem structure and functions, on a wide variety of time and space scales; however, there is uncertainty as to whether extreme events will change in intensity and frequency. We have a high level of confidence that:

- Redistribution of SST could cause geographical shifts in biota as well as changes in biodiversity, and in polar regions the extinction of some species and proliferation of others. A rise in mean SST in high latitudes should increase the duration of the growing period and the productivity of these regions if light and nutrient conditions remain constant.
- Sea-level changes will occur from thermal expansion and melting of ice, with regional variations due to dynamic effects resulting from wind and atmospheric pressure patterns, regional ocean density differences, and oceanic circulation.
- Changes in the magnitude and temporal pattern of pollutant loading in the coastal ocean will occur as a result of changes in precipitation and runoff.

We can say with a lesser degree of confidence that:

- Changes in circulation and vertical mixing will influence nutrient availability and primary productivity, thereby affecting the efficiency of carbon dioxide uptake by the oceans.
- The oceans' uptake and storage capacity for greenhouse gases will be affected further by changes in nutrient availability in the ocean resulting from other changes in precipitation, runoff, and atmospheric deposition.
- Freshwater influx from the movements and melting of sea ice or ice sheets may lead to a weakening of the global thermohaline circulation, causing unpredictable instabilities in the climate system.

It is presently uncertain whether the frequency and severity of tropical cyclones will increase due to climate change.

The most pervasive effects of global climate change on human uses of the oceans will be due to impacts on biotic resources; transportation and nonliving resource exploitation will be affected to a lesser degree. We can say with a high level of confidence that:

- Increased coral bleaching will occur as a result of a predicted 2°C increase in average global atmospheric temperature by 2050.
- Expanded dredging operations will be necessary to keep major ports open in the Northern Hemisphere, which will increase costs.
- The Northwest Passage and Northern Sea Route of Russia likely will be opened up for routine shipping.
- Growth in the marine instrumentation industry will occur as the need for research and monitoring of climate change increases.

We can say with a lesser degree of confidence that:

- Reduced yields of desirable fish species will occur if average primary productivity decreases.
- If the frequency of tropical storms and hurricanes increases, adverse impacts will be generated for offshore oil and gas activities and for marine transportation in the tropics.
- Marine mineral extraction, except for petroleum hydrocarbons and the marine pharmaceutical and biotechnological industries, is insensitive to global climate change.

Adaptation to the impact of climate change on oceans is limited by the nature of these changes, and the scale at which they are likely to occur:

- No adaptive responses to coral bleaching, even on a regional scale, will be available if average global temperature increases 2°C by 2050. However, reductions in land-based pollution of the marine environment, combined with reductions in habitat degradation/destruction, would produce benefits for fisheries, aquaculture, recreation, and tourism.
- Adaptation options will be available for the offshore oil, gas, and shipping industries if the frequency of tropical storms and hurricanes increases. The options include improved design standards for offshore structures, national and international regulations for shipping, and increased technological capabilities to provide early warning at sea. Governments also can increase attention to institutional design for planning and responding to disasters and acute emergencies.
- Where climate change generates positive effects, market-driven needs will create their own adaptation dynamic. However, adaptation policies will be required to control externalities that are market failures. For instance, opening up both the Northwest Passage and the Russian Northern Sea Route for up to 100 days a year—while a boon to international shipping and consumers in East Asia, North America, and Western Europe—will have to be accompanied by policies designed to limit the total burden of pollutants entering the Arctic environment from ports, ship operations, and accidents.

8.1. Introduction

In this chapter, we attempt to assess the impacts of projected regional and global climate changes on the oceans. Climate change will affect the physical, biological, and biogeochemical characteristics of the oceans at different time and space scales, which should modify their ecological structure and functions and is expected to exert significant feedback controls on the climate system. These changes will be in addition to human activities such as land-use changes, industrialization and urbanization, increased food production, habitat modification, transportation of exotic organisms, and releases of pollutants. The present trends of increasing pollution, development, and overuse are reducing the capacity of coastal oceans and semi-enclosed seas to respond to and compensate for environmental changes that might be brought about by a climate change (GESAMP, 1990).

In providing this assessment, we address questions concerning the functions and characteristics of oceans in relation to their human uses, their responses to large-scale global climate change, the range of available mitigation and adaptation response options, and continuing research and monitoring needs.

8.2. Functions of Oceans

The oceans function as regulators of the Earth's climate and sustain planetary biogeochemical cycles. They also are of significant socioeconomic value as suppliers of resources and products, as sinks for wastes, as areas of recreation and tourism, as a medium for transportation, and as a repository of genetic and biological information. To gain a perspective of the oceans' importance, consider that the global ocean industry (all sectors) realizes about 4% of world gross national product (GNP)—approximately $20 trillion in 1988 (Broadus, 1991).

8.2.1. Climate Regulator

The oceans have significant capacity to store heat and are the largest reservoir of the two most important greenhouse gases—water vapor and carbon dioxide (CO_2). The long timescales of vertical circulation in the oceans result in slow turnover of stored heat and CO_2. The properties of oceans are affected by variability resulting from Sun-Earth orbital changes (e.g., seasonal and Milankovitch cycles); interactions among components of the climate system—for example, the atmosphere-cryosphere—modulate these variations and result in climatic variability on other timescales.

Just less than 60% of the Earth's radiative energy from the Sun is received by the ocean, 80% of which is absorbed in the top 10 m of the ocean. Winds and waves mix down a seasonal surface layer of nearly uniform temperature, salinity, and other properties. This layer extends to tens of meters in the tropics and to several hundred meters in high-latitude seas, where oceanic convections are due to the annual cooling of surface waters and where the winds are stronger and the waves larger. The permanent thermocline lies below the seasonal surface layer, down to about 1000 m depth. Circulation down to the upper thermocline is principally wind-driven and is characterized by basin-scale gyres and intensive western boundary currents such as the Gulf Stream and the Kuroshio. The oceans transport about the same amount of heat poleward as the atmosphere (Trenberth and Solomon, 1994), and these western boundary currents are the most important carriers.

The lower thermocline and the abyssal ocean represent nearly 90% of the volume of the oceans, and most of this water is colder than 5°C. It is dominated by thermohaline circulation driven by differences in density. Globally, the oceanic thermohaline circulation can provide a substantial buffer against greenhouse warming. It incorporates both heat and CO_2 from the atmosphere at higher latitudes and releases them at lower latitudes decades or centuries later. Thus, it plays an important role in controlling the distribution of both heat and greenhouse gases and, as a consequence, modulating global climate.

About 57% of oceanic volume is colder than 2°C. Except for the Arctic Ocean and the nearby part of the Atlantic Ocean, these characteristics can only be derived from Antarctic bottom water (Gordon, 1991). Therefore, a significant part of global thermohaline circulation is driven by the formation and sinking of Antarctic bottom water, principally in the Weddel Sea. As a consequence, the Antarctic Ocean is quite important for the global climate, but its role in contributing variability to climate is not understood. Further, changes in the formation of deep water in the North Atlantic are usually considered to be part of the processes by which the Earth moves from an ice age to an interglacial and back again. This process is explained in Box 8-1.

Studies of gases trapped in polar ice cores have shown that the concentration of atmospheric CO_2 has undergone variations that parallel climatic changes (Barnola *et al.*, 1987). The strong correlation between variations of polar temperature and atmospheric carbon dioxide concentration in comparison with the Sun-Earth orbit-induced glacial-interglacial cycles strongly suggests the existence of feedback mechanisms involving the interplay of biological, chemical, and physical processes (Watson *et al.*, 1990). Because the ocean contains about 60 times more carbon than the atmosphere (Sundquist and Broecker, 1985), variations in atmospheric CO_2 could result from even minor changes in the ocean's physicochemical and biogeochemical characteristics affecting carbon cycling in the sea. Among the proposed explanations for the variability of atmospheric CO_2 contents on these timescales are changes in ocean productivity (Sarnthein *et al.*, 1988) and fluctuations in the quantity of carbon stored in the deep sea—which in turn, are related to changes in ocean circulation (Boyle and Keigwin, 1982; Curry *et al.*, 1988).

The natural fluxes of the exchange of carbon dioxide among atmosphere, ocean, and land biota are much larger than anthropogenic perturbations (IPCC, 1994). Because of complex oceanic carbonate chemistry, the ocean—compared to its reservoir

Box 8-1. Ocean Conveyor Belt

The "conveyor belt" global thermohaline circulation is driven primarily by the formation and sinking of deep water (from 1500 m to the Antarctic bottom water overlying the bottom of the ocean) in the Norwegian Sea (Figure 8-1; Broecker, 1991). This circulation is thought to be responsible for the large flow of upper ocean water from the tropical Pacific to the Indian Ocean through the Indonesian Archipelago—the Indonesian Throughflow (e.g., Broecker, 1991).

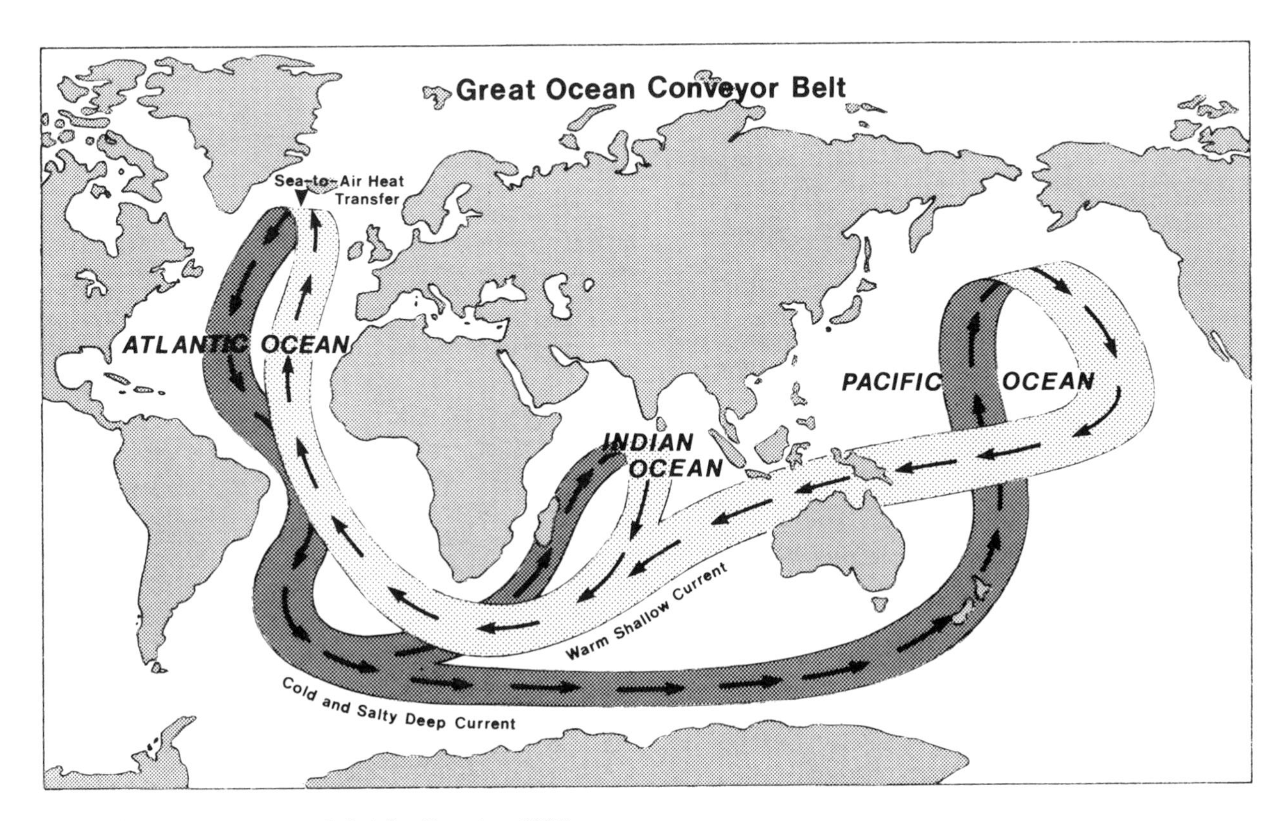

Figure 8-1: Ocean conveyor belt (after Broecker, 1991).

The global conveyor belt thermohaline circulation is controlled by two counteracting forcings operating in the North Atlantic. The *thermal forcing* (high-latitude cooling and low-latitude heating) drives a poleward surface flow, and the *haline forcing* (net high-latitude freshwater gain and low-latitude evaporation) moves in the opposite direction. In today's North Atlantic, the thermal forcing dominates. When the strength of the haline forcing increases due to excess precipitation, runoff, and/or ice melt, the conveyor belt will weaken (i.e., decrease in both poleward warm surface flow and deep-water formation) or even shut down. The variability in the strength of the conveyor belt will lead to climate change in Europe. Climatic variation in other parts of the global ocean also may be influenced by the variability of the conveyor belt (e.g., Gordon *et al.*, 1992; Lehman, 1993). Numerical simulations of global ocean circulation show that small changes in the external forcing at high latitudes can switch off the present conveyor belt and force the ocean to reach another equilibrium state, in which the Antarctic Circumpolar Front represents the sole deep-water source (e.g., Maier-Reimer *et al.*, 1993).

The North Atlantic atmosphere-ocean-cryosphere system appears to have natural cycles of many timescales in switching the conveyor belt. In these cycles, sea ice plays a pivotal role. Periodic movement of excessive sea ice from the Arctic into the Greenland Sea appears to be responsible for the interdecadal variability of the conveyor belt (Mysak *et al.*, 1990), characterized by the freshening of the intermediate water (Dickson *et al.*, 1988). There is no evidence yet that the influx of interdecadal switching extends beyond the North Atlantic Ocean.

Paleorecords show that 10,000 to 80,000 years ago there were millennial oscillations of several degrees of the sea-surface temperature in the North Atlantic (Bond *et al.*, 1993; Figure 8-2), with abrupt temperature shifts in decadal timescales (Taylor *et al.*, 1992). Interaction between the warm ocean surface current and the Scandinavian ice sheets is

Box 8-1 (continued)

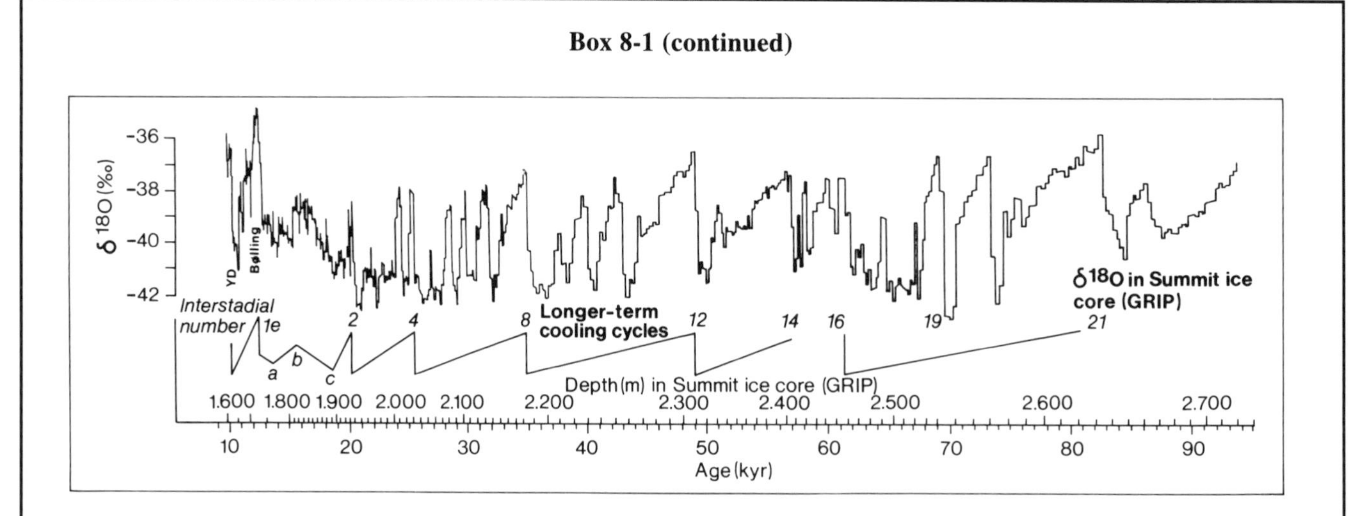

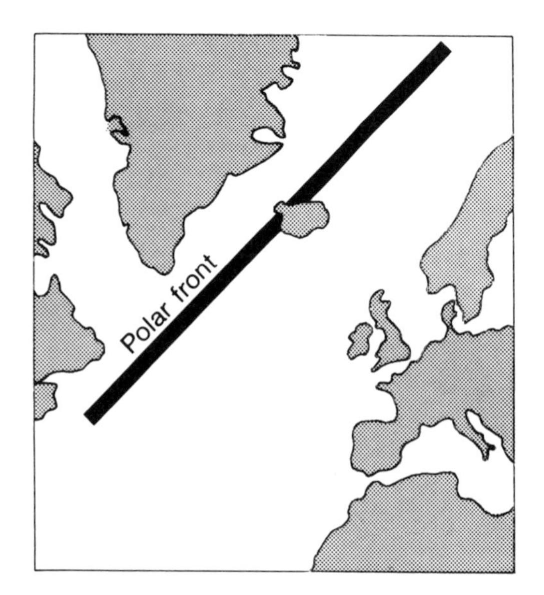

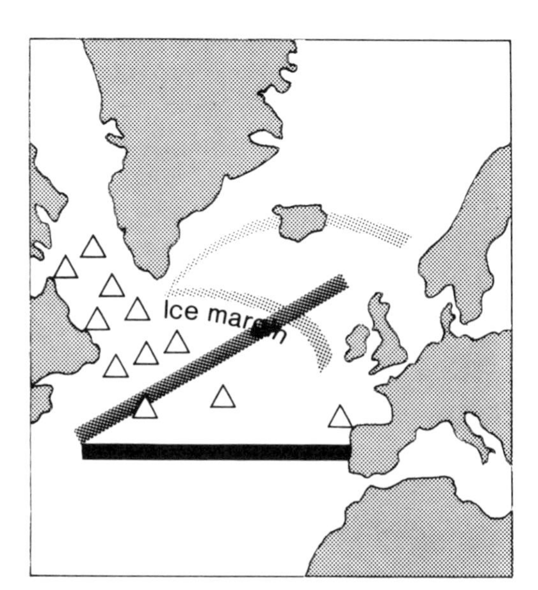

Figure 8-2: Ocean surface temperature in the North Atlantic over the past 80,000 years as inferred from sediment cores (top). These variations may be related to the strength of the "conveyor belt." When the conveyor belt is strong, as in the present, the North Atlantic polar front (i.e., the circulation boundary between cold polar and warm boreal water masses) swings open, allowing warm subtropical water masses to flow on to the north (bottom left). During the last glacial period, the conveyor belt was rapidly weakened many times due to iceberg surges (triangles in bottom right). Then the polar front swung to the east, cutting off the northward flow of warm waters (after Bond *et al.*, 1993; Zahn, 1994).

thought to be the driving mechanism (Bond *et al.*, 1993). These millennial cycles, called Dansgaard-Oeschger cycles, appear to lump into long-term cooling cycles (Dansgaard *et al.*, 1993; Oeschger *et al.*, 1984). Each cooling cycle culminates in an enormous discharge of icebergs into the North Atlantic from massive collapses of Laurentinide ice sheets—called a Heinrich event—followed by an abrupt shift to a warmer climate. The Younger Dryas—an event that happened at the last glacial termination about 11,000 years ago, when there were abrupt temperature decreases in Greenland and Northern Europe—also was a Heinrich event. Similar observations also have been made in the Greenland ice-core record from the last interglacial period (GRIP Project Members, 1993). In as much as the last interglacial seems to have been slightly warmer than the present one, global warming may destabilize the relatively stable climate that the Earth has experienced over the past 8000 years (GRIP Project Members, 1993). There is some doubt, however, about the reliability of the Greenland ice-core record for the last interglacial (e.g., Broecker, 1994). If the last interglacial climate variability did exist, it could have been due to an origin different from fluctuations in the strength of the conveyor belt (e.g., Zahn, 1994).

size—is actually not an efficient sink for increases in atmospheric carbon dioxide concentrations (e.g., Siegenthaler and Sarmiento, 1993). Nevertheless, the ocean is estimated to have taken up about 30% of carbon dioxide emissions arising from fossil fuel use and tropical deforestation between 1980 and 1989 (Siegenthaler and Sarmiento, 1993), thus slowing the rate of greenhouse global warming. The same estimate also leaves about 25% of the emissions unaccounted for. This "missing sink" for carbon dioxide is thought to be located in the terrestrial biosphere (Tans *et al.*, 1990), although Tsunogai *et al.* (1993) suggest that the oceanic intermediate waters are also significant sinks for carbon dioxide.

The global ocean also plays a major role in the global hydrological cycle. The cycling of the oceanic freshwater fraction through advection, evaporation (E), precipitation (P), and—in higher latitudes—the solid-ice phase will be affected by changes in wind systems and oceanic current systems. The freshwater budget of the ocean is still not well enough understood, primarily because of the lack of adequate data, although Schmitt *et al.* (1989) have given some preliminary assessments. Atmosphere-ocean interactions maintaining the overall E-P balance will be affected by changes in circulation, evaporation, precipitation, and the availability of precipitable water in the atmosphere over the ocean. A changing pattern of rainfall over the oceans will cause changes in the rainfall pattern over land, which will in turn have a considerable effect on the salinity of marginal seas.

8.2.2. *Resources and Products*

The oceans provide living and nonliving (minerals and water) resources, opportunities for unconventional energy sources, pharmaceuticals, and marine electronic instrumentation.

8.2.2.1. *Living Resources*

Living resources include fish and shellfish, marine mammals, and seaweeds. In 1990, the world's fish catch was 97 million tons (mt)—14 mt from inland sources and 83 mt from marine sources (FAO, 1992a). However, the annual rate of growth in world marine catch has been declining since 1972 (FAO, 1992b). This decline in the rate of growth, and decline in overall production experienced in 1990, is the combined result of uncontrolled growth in fishing effort and overfishing of important stocks in the Atlantic Ocean. Uncontrolled access to resources and growth in fishing effort supported by large national subsidies are major weaknesses in the world fisheries system as a whole (see Chapter 16). In contrast, the production trend in the world aquaculture industry is the opposite of that in the marine capture fisheries industry. Between 1984 and 1992, world aquaculture production increased 89% and amounted to 23% of the world marine catch with a value of $32.5 billion (in 1992 U.S.$) (Aquaculture Magazine Buyers Guide, 1995). This subject is considered in Chapter 16.

A combination of human activities (e.g., overfishing, pollution of estuaries and the coastal ocean, and the destruction of habitat, especially wetlands and seagrasses) currently exerts a far more powerful effect on world marine fisheries than is expected from climate change. However, fisheries are indeed sensitive to climate change as inferred from paleorecords. The Working Group I report concludes that because marine photosynthesis is limited by nitrogen rather than by carbon, increasing CO_2 contents in the upper ocean is unlikely to have a positive impact on net primary productivity. Rather, primary productivity should respond to changes in nutrient availability resulting from changes in ocean circulation, runoff input, or atmospheric deposition. Because there is substantial uncertainty about whether any of these factors will change with global warming, it is not clear if average primary productivity will be affected. Even if global average primary productivity is negatively affected, there are so many pathways and loops in the chain between primary productivity and upper trophic level species—including a significant microbial loop—and regional conditions vary so widely that it is difficult to say whether reductions in the average yield of commercial fisheries will result.

There are indications that the potential value of the marine pharmaceutical industry is large, although there is no comprehensive study of the market for this industry. The oceans are a potentially enormous source of biological tissue that can be refined and developed into pharmaceuticals, enzyme preparations, gene probes, immunological assays, and new materials. Some of these materials have commercial potential in medicine, agriculture, marine aquaculture, and materials science.

The cleaning and disinfection of contaminated areas using microorganisms (bioremediation) is of potential importance in the oceans, largely on fixed surfaces such as sediments (Bragg *et al.*, 1994; Swannel and Head, 1994). Bioremediation of the marine water column may not be important, largely because most contaminants in the oceans are particle-associated (e.g., sediment particles). Possible exceptions include highly contaminated seawater close to terrestrial hazardous waste disposal sites. To date, efforts to deal with these areas have been few.

The newest class of marine organisms that appears to have considerable commercial potential in the biotechnology industry is the hyperthermophiles (T_{opt} > 85°C) (Kristjansson, 1989). These bacteria live at extremes of pH and temperature that make them attractive to industry because, among other characteristics, they can yield long shelf-lives and tolerate organic solvents and harsh purification processes. Large colonies of hyperthermophiles now have been found at spreading centers in mid-ocean ridges (Jannasch and Wirsen, 1979), in the deep North Sea, and in Alaskan oil reservoirs (Stetter *et al.*, 1993). However, they are difficult to culture in the laboratory. Not surprisingly, the potential value and availability of these hyperthermophiles have stimulated serious efforts to harvest them for industrial applications by biotechnology firms (e.g., Myers and Anderson, 1992).

8.2.2.2. *Nonliving Resources and Unconventional Energy*

The range of products in this category includes petroleum, placer and nearshore deposits (i.e., sand and gravel, sulfur, phosphorite), calcareous oozes including hot brines, manganese nodules, and polymetallic sulfides at spreading centers. The latter two are resources from the deep ocean bed, for which no large-scale commercial recovery has been attempted because it is presently uneconomic.

Petroleum hydrocarbons are by far the most significant contribution of the ocean to the world stock of nonliving (mineral) resources in both quantity and value. Offshore production is estimated to be approximately 26% of terrestrial production for oil, in amount and gross value, and 17% for natural gas (*Oil and Gas Journal*, 1993; *Offshore Journal*, 1993; Oil and Gas Journal Database, 1992).

Broadus (1987) estimates the annual gross value of all other extracted seabed minerals to be more than $600 million—of which tin is the most significant item and the only one that supplies more than 1% of its world market. Sand and gravel, calcium carbonate, sulfur, and other mineral placers are also among the seabed mineral resources that are actively extracted. The harvesting of precious coral, a combined living and mineral resource, supports an industry estimated to be worth more than $50 million a year (Broadus, 1987).

Other resources and products to be found in the ocean include water and unconventional energy resources. In the 1990s, more than 3,500 land-based desalination plants were in operation, producing more than 8 billion liters of water a day (Encyclopedia Britannica, 1993). Thirty percent of the total number of plants were located in the United States, and 20% were in the Middle East. Desalination of sea water is unlikely to be affected by global climate change.

Electricity from tidal power can be produced in only a few areas of the world (e.g., northern France and eastern Canada) where the tidal flux is sufficiently large. The total usable potential is estimated to be on the order of 200 million kW (Charlier and Justus, 1993). The potential for using temperature differentials in the tropical ocean between the surface and the water below the thermocline is quite large. This technique, called ocean thermal energy conversion (OTEC), theoretically can be applied in the latitude belt between 20°N and 20°S, but the area of greatest potential lies between 10°N and 10°S. Total potential electricity production from OTEC ranges between 10^8 and 10^{10} MW (Charlier and Justus, 1993). An upper limit per plant at the 10^7 MW level would amount to 299 quads per year [one quad = 10^{15} Btu (British thermal units)] of electric energy. So far, only two subsidized pilot plants have been in operation.

The oceans are an opaque, three-dimensional medium. The opacity derives from a limited capacity for penetration by either light or electromagnetic waves. The oceans also are characterized by storms, hazards to navigation, and currents of varying strength, so electronic means of sensing, communicating, and managing information are essential (Broadus *et al.*, 1989). The major products here include communication and navigation instruments, sensors, data management instruments, and services. Navies, offshore oil and gas firms, oceanographic research and environmental monitoring entities, commercial shipping and fishing companies, and recreational boaters are the consumers of these products (Broadus *et al.*, 1989). Initial estimates of the size of the world market for marine electronic instrumentation (Hoagland III and Kite-Powell, 1991) are in the range of $3–5 billion for the United States and roughly twice that size for the world (which includes the United States, Western Europe, and Japan).

8.2.3. *Waste Reception and Recycling*

Due to their enormous volume of water, microbial communities, and sediments, the oceans have the capability to receive, dilute, transform, eliminate, store, and recycle massive quantities of wastes from human activities conducted on land and in the atmosphere, as well as on water (Izrael and Tsyban, 1983, 1989; GESAMP, 1986). This capability to accept contaminants, however, is not limitless, and negative consequences can occur—especially in coastal seas and estuaries. The major sources and pathways of pollutants into the oceans are runoff and river inputs of land-based discharges, which contribute an estimated 44% (GESAMP, 1990); atmospheric deposition of land-based pollutants over the oceans, which contributes an estimated 33%; and at sea-activities, including mineral production, maritime transportation, and ocean dumping, which contribute 23% of the total inputs (GESAMP, 1990).

8.2.4. *Recreation and Tourism*

Recreation and tourism is an exceedingly large and rapidly growing market, which exceeded $2 trillion in 1986 (Miller and Auyong, 1991). Tourism accounts for more than 12% of the world's gross product and represents an important socioeconomic activity in the coastal environment (see Chapter 9). The organizational and operational facilities for the tourist industry—which include, for example, hotels, airports, ports, marinas, and the like—affect ecosystems such as estuaries, salt marshes, mangroves, seagrass beds, and other wetlands and coral reefs, all of which are already at risk due to pollution from human activities (Miller and Auyong, 1991). Maintaining a balance between developing tourism and recreation and preserving the aquatic environment is an issue affected by climate change because adverse human impacts on the coastal zone are likely to be amplified by the coastal effects of climate change (see Chapter 9).

8.2.5. *Transportation*

Transportation over water is by far the cheapest of all global transportation media available; consequently, more than 95% of world trade moves by ship. This medium is the least significant contributor of CO_2 per amount of product shipped. The world fleet in 1990 consisted of 424 million gross tons (GT, a volumetric measure) or 667 million deadweight tons (dwt, a measure of carrying capacity). Of that amount, exclusively cargo-carrying ships accounted for 46%; fishing vessels (all types >100 GT) accounted for 19%; and passenger vessels accounted for 18% (Lloyds Register, 1990). These three categories accounted for 83% of the world's fleet.

Marine transportation will be affected by climate change in three respects. First, in regions experiencing increased precipitation and runoff, increased dredging operations will be necessary in ports affected by increased sediment deposition. Second, if the frequency and uncertainty of tropical storms and hurricanes increases, shipping will be adversely affected, loss rates may increase, and freight and insurance rates will rise. Third, if the extent of sea ice is reduced by 20% in northern latitudes—which seems probable given that maximum warming is predicted to occur in high northern latitudes in winter (see the IPCC Working Group I volume, and Chapter 7)—then both the Northwest Passage and the Russian Northern Sea Route will become viable sea routes, thereby facilitating shipping and reducing costs between East Asia and Western Europe.

8.2.6. *Information Function*

In addition to the preceding functions and uses, the oceans represent a natural repository of genetic and biological information. One important effect of climate change is likely to be a net decrease in global biodiversity. In many cases, global warming will act in combination with other human factors—driving species to extinction, narrowing the genetic range within species, and transforming and simplifying ecosystems. Species that are rare, isolated on the edge of their tolerance, genetically impoverished, and in areas undergoing the most abrupt changes are likely to become extinct (Markham *et al.*, 1993).

The biodiversity value of the ocean is particularly important because there are many groups of organisms, even at taxonomic levels above orders, that are exclusively marine. These animals and plants have special status in some ways related to biodiversity because the loss of these species represents a larger impact than the loss of species in widespread taxonomic groups.

8.3. Characteristics of Oceans and Their Responses to Climate Change

In this section we describe the major characteristics of oceans and their likely responses to global climate change scenarios.

8.3.1. *Physical Characteristics*

In the oceans, climate change will be accompanied by changes in temperature, circulation, sea level, ice coverage, wave climate, and extreme events. These changes are expected to have consequences for ecosystem structure and function, as well as for global biogeochemical cycles with feedbacks to the climate system.

8.3.1.1. *Temperature*

Studies of past climate changes show that sea surface temperature (SST) increases along with air temperature, although probably not as much as or as rapidly (IPCC, 1990a). Also, due to greater thermal inertia, changes in oceanic conditions will lag behind changes on the continents [e.g., by 10 years for the North Pacific (Wigley *et al.*, 1985)]. The correlation between SST and land temperatures from a coupled ocean-atmosphere model suggests that SST will increase at a rate generally in line with the increase of mean land-temperature values (Manabe *et al.*, 1991). Exceptions might occur in a belt around Antarctica and in the high-latitude North Atlantic. The modeled SST around Antarctica increases more slowly because of the upwelling deep-water masses. The predicted temperature increase in the high-latitude North Atlantic also is slower, apparently due to a weakening of the conveyor belt thermohaline circulation. The weakened conveyor belt could recover if atmospheric CO_2 concentration leveled off at double the present value but might never recover if it quadrupled (Manabe and Stouffer, 1993), driving the climate system into unknown directions. Exceptions also are expected to occur in arctic and subarctic regions, where the predicted warming may be higher than the global average.

8.3.1.2. *Upwelling and ENSO Events*

Divergence of wind-driven surface ocean currents induces upwelling of water from deeper layers. General circulation models (GCMs) predict a decrease of the meridional (north–south) gradients of sea-surface temperature due to significant warming in polar latitudes, which in turn would lead to a decrease in trade-wind intensity, the strength of the upper ocean currents (Mitchell, 1988), and the area and intensity of upwelling, such as in the equatorial eastern tropical Pacific. According to studies of paleoclimate, the productivity of open oceanic upwelling regions during glacial periods was much higher than during interglacial periods (e.g., Sarnthein *et al.*, 1988; Lapenis *et al.*, 1990). These studies indicate that global warming may be accompanied by a decrease in total productivity of the global ocean (Budyko and Izrael, 1987; Lapenis *et al.*, 1990).

In contrast to model projections, however, observations over large parts of the tropical Atlantic between 1947 and 1986 have shown an increase in the trade winds (Bigg, 1993). Bakun (1990, 1993) suggests that the greenhouse effect will enhance the seasonal warming of continents—leading to a decrease in the pressure over land, an increase in the land–sea pressure difference, and increased alongshore winds. Binet (1988) has observed such effects along the coast of northwest Africa. It appears likely that the strength of both oceanic and coastal upwelling mechanisms could change under conditions of global warming, with profound impacts upon fish species and their production as well as on the climate of the immediate coastal zone (see also Chapter 16).

The El Niño-Southern Oscillation (ENSO) results from fluctuations in the internal interactions within the Earth's ocean-atmosphere system and is the most significant source of interannual variability in weather and climate around the world. The Southern Oscillation (SO) component of ENSO is an atmospheric pattern that extends over most of the global tropics. It principally involves a seesaw in atmospheric mass between regions near Indonesia and the Pacific Ocean region centered near Easter Island. The El Niño component of ENSO is an anomalous warming of the eastern and central tropical Pacific Ocean. In major "warm events," warming extends over much of the tropical Pacific and becomes clearly linked to the atmospheric SO pattern. ENSO events occur every 3 to 10 years, although in recent years the frequency has increased. The influence of ENSO sometimes extends to higher latitudes and has far-reaching climatic and economic consequences around the world (Ropelewski and Halpert, 1987; Zuta, 1986).

Although ENSO is a natural part of the Earth's climate, a major question is whether the intensity or frequency of ENSO events might change as a result of global warming. Historical and paleorecords reveal that ENSO events have changed in frequency and intensity in the past on multidecadal to century timescales (Nicholls, 1992; Anderson *et al.*, 1992; Thompson *et al.*, 1984). It is unclear whether ENSO might change with long-term global warming. Studies of long-term variations in ENSO using models have just begun (Knutson and Manabe, 1994; Kumar *et al.*, 1994), and it is still premature to project the behavior of ENSO events for different climate-change scenarios.

8.3.1.3. Changes in Sea Level

The available estimates of sea-level changes under a changing climate are rather preliminary because of uncertainties both in the projections of climate change itself and in the assessment of the components that influence sea level. The latter include ocean thermal expansion, the effects of changing volumes of polar ice sheets and mountain glaciers, and dynamic effects resulting from ocean circulation, wind and weather patterns, and differences in regional ocean density (Box 8-2).

Revised projections, taking full account of both climate and sea-level components, are not yet available. However, indications are that both global warming and sea-level rise by the year 2100 may be 25–30% less than the IPCC 1990 projections (Wigley and Raper, 1992, 1993). Nonetheless, even the lower estimates of sea-level rise are about two to four times the rate of sea-level rise experienced in the last 100 years.

Rising sea level will have pervasive impacts for coastal zones and their ecosystems. A detailed analysis of the impacts of sea-level rise on the coastal zones may be found in Chapter 9.

8.3.1.4. Ice Cover

Sea ice covers about 11% of the ocean, depending on the season. It affects albedo, salinity, and ocean-atmosphere thermal exchange. The latter determines the intensity of convection in the ocean and, consequently, the mean timescale of deep-ocean processes affecting CO_2 uptake and storage.

Projected changes in climate should produce large reductions in the extent, thickness, and duration of sea ice. Major areas that are now ice-bound throughout the year are likely to have major periods during which waters are open and navigable. Some models even predict an ice-free Arctic (IPCC, 1990b). Melting of snow and glaciers will lead to increased freshwater influx, changing the chemistry of those oceanic areas affected by the runoff. At present, however, there is no convincing evidence of changes in the extent of global sea ice (Gloersen *et al.*, 1992). Studies on regional changes in the Arctic and Antarctic indicate trends of decadal length (e.g., Mysak and Manak, 1989; Parkinson, 1992), often with plausible mechanisms proposed for periodicities of a decade or more. Thus, longer data sets are needed to test whether a genuine long-term trend is developing (see Chapter 7).

8.3.1.5. Wave Climate and Vertical Mixing

Winds and waves are the major forcing factors for vertical mixing; the degree of mixing depends on the vertical density structure. In the past 40 years, there has been an increase in the mean wave height over the whole of the North Atlantic, although it is not certain that global change is the cause of this phenomenon (Bacon and Carter, 1991).

In some permanently stratified anoxic basins, such as the Black Sea, climate change brings with it the possibility of turnover (Mee, 1992). The Black Sea has a very strong density gradient; the only risk of penetration of anoxic bottom waters to the surface would be by elevation of the oxic-anoxic boundary. Although this is controversial, Murray *et al.* (1989) have suggested that the oxic-anoxic boundary in the Black Sea has risen up to 30 m over the last 20 years in association with decreased river flows.

8.3.1.6. Extreme Events

The catastrophic aspects of storms and storm surges are well known, particularly in exacerbating flooding in coastal areas and in erosion and restructuring of coastal formations (see Chapter 9 for details). The causal relationship between sea-surface temperature and the formation of tropical cyclones suggests that the intensity and frequency of tropical cyclones may increase in the future. However, the evidence inferred from models based on such relationships is conflicting (Ryan *et al.*, 1992; Stein and Hense, 1994). A GCM analyzed by Haarsma *et al.* (1993) showed that a doubling of CO_2 concentration increased by about 50% the number of simulated tropical disturbances, as well as

Box 8-2. Changes in Sea Level

Model projections suggest that in the oceans, temperature increase due to climate change is largest in the thermocline (Stouffer *et al.*, 1989) but that the warming signal there is masked by eddies and seasonal variability. However, Bindoff and Church (1992) also found significant warming in the South Pacific below the thermocline, where the signal-to-noise ratio is larger (Figure 8-3). The corresponding sea-level rise caused by the thermal expansion of seawater is 2–3 cm over 22 years—consistent with the estimates of global sea-level rise. Under the predicted global warming, sea-level rise also will be accelerated by both warming of the ocean and possible melting of ice. A 40-cm global mean sea-level rise is projected by 2050 (see the IPCC Working Group I volume). However, global warming also will alter the variations in sea surface caused by dynamic effects resulting from wind and atmospheric pressure patterns, regional ocean density differences, and oceanic circulation (Figure 8-4). Large regional differences in sea-level response can be expected due to the effects of ocean circulation in sea-surface topography (Mikolajewicz *et al.*, 1990). Increases in sea level would be highest in the North Atlantic; in certain regions, such as the Ross Sea, sea level actually would fall (Mikolajewicz *et al.*, 1990). Regional differences in sea-level change are to be expected (see also Church *et al.*, 1991; Cubasch *et al.*, 1994). Therefore, there also will be significant regional differences in sea-level rise with global warming. In addition, it should be emphasized that for coastal environments, it is the relative sea level that matters [i.e., the level of the sea in relation to that of the land (see Chapter 9)].

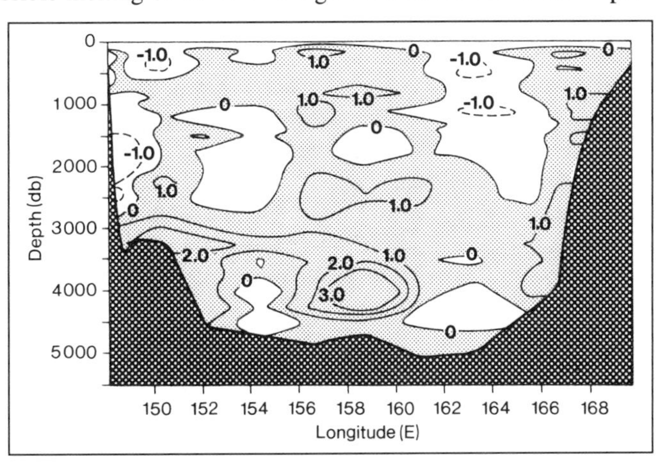

Figure 8-3: The difference of the potential temperature between 1967 and 1990 across 43°S between Australia and New Zealand. The unit of the difference has been normalized by the root-mean-square variability (after Bindoff and Church, 1992).

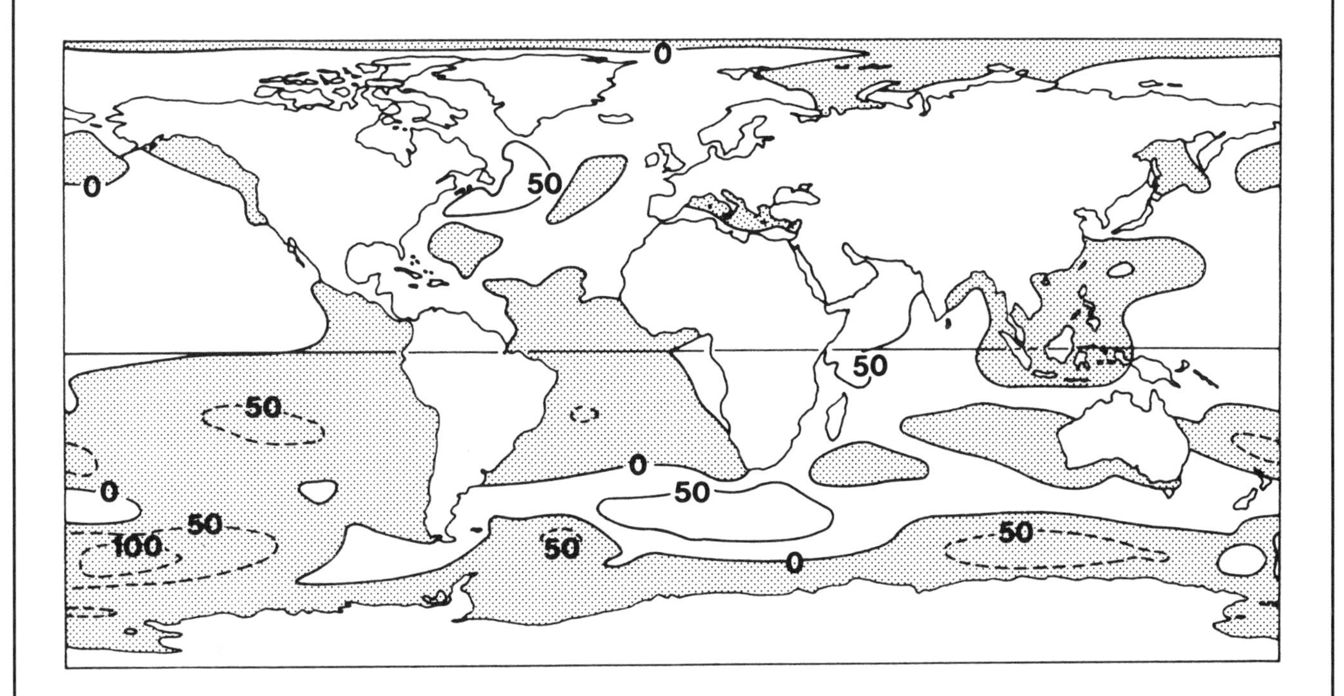

Figure 8-4: Global variations of sea-surface topography (in mm) due to dynamic effects relative to the global mean sea level, as obtained from a transient experiment of increasing carbon dioxide concentrations using a coupled ocean-atmosphere GCM. The results are for years 66–75 of the model run (but cannot be ascribed to a future calendar date). The stippled areas on the map show sea-level rises that are less than the global average (after Gregory, 1993).

the number of intense disturbances, whose maximum windspeed increased by about 20%. The low resolution of the GCM calls for caution in interpreting the simulated changes in the intensity of tropical disturbances (Haarsma *et al.*, 1993). GCM results also indicate a change in cyclone tracks as a response to global warming. Bengtsson *et al.* (1994), on the other hand, report that with a doubling of CO_2 over the next 50 years, the global distribution of storms should be similar to the present geographical position and seasonal variability—although the number of storms is significantly reduced, particularly in the Southern Hemisphere. Tropical cyclone models require an initial disturbance to be present before a full-scale cyclone develops. A GCM, with its coarser horizontal resolution, cannot produce a better resolution than a cyclone model. Correlations also have been found between ENSO (as well as the southeast Asian monsoon) and the regional patterns of tropical cyclone activity (Nicholls, 1984; Evans and Allan, 1992). Unfortunately, it is not yet possible to say how ENSO, and thus tropical cyclones in these regions, will be affected by global warming (see also Section 8.3.1.2).

Although there is uncertainty, the extent of damage caused by great windstorm catastrophes has expanded in recent years. The concentrations of people living in high-risk coastal regions must be considered the main reason for this alarming trend (Berz, 1994). If storms intensify with rising temperatures, there will be adverse consequences for living and nonliving resource exploitation in the ocean and in coastal areas, for marine transportation, and for recreation and tourism. It must be noted, however, that scientific assessment of impacts in this context can differ from economic evaluation of loss because the methodologies used to assess loss are different. From a scientific perspective, damage or injury would be assessed in terms of the components and interrelationships in a particular ecosystem. From an economic perspective, however, loss is defined by value as determined by some market. It is therefore quite possible to get a scientific assessment of low injury to an ecosystem combined with high economic loss value, especially given that the value of waterfront real estate is normally high (see also Chapter 17). The reverse is also true: Major ecological damage might occur with little economic loss.

8.3.2. *Ecological Characteristics and Biodiversity*

Metabolic rates, enzyme kinetics, and other biological characteristics of aquatic plants and animals are highly dependent on external temperatures; for this reason alone, climate change that influences water temperature will have significant impacts on the ecology and biodiversity of aquatic systems (Fry, 1971). The capability of some species to adapt genetically to global warming will depend on existing genetic variation and the rapidity of change (Mathews and Zimmerman, 1990). Species remaining in suboptimal habitats should at least experience reductions in abundance and growth well before conditions become severe enough for extinctions to occur. The resilience of an ecosystem to climate change will be determined to a large extent by the degree to which it already has been impaired by other human activities.

Coastal ecosystems are especially vulnerable in this context. They are being subjected to habitat degradation; excessive nutrient loading, resulting in harmful algal blooms; fallout from aerosol contaminants; and emergent diseases. Human interventions also have led to losses of living marine resources and reductions in biodiversity from biomass removals at increasingly lower trophic levels (Sherman, 1994; Beddington, 1995; Pauly and Christiansen, 1995). The effects on biodiversity are likely to be much less severe in the open ocean than in estuaries and wetlands, where species in shallow, restricted impoundments would be affected long before deep-oceanic species (Bernal, 1991).

The chief biotic effects on individuals of an increase in mean water temperature would be increased growth and development rates (Sibley and Strickland, 1985). If surface temperatures were correlated positively with latitude, and temperature increased, one would expect a poleward shift of oceanic biota. While this may be the general case, there could be important regional variations due to shifts in atmospheric and oceanic circulation. The resulting changes in predator-prey abundance and poleward shifts in species' ranges and migration patterns could, in the case of marine fisheries, lead to increased survival of economically valuable species and increased yield. Such cases have been observed by Wooster and Fluharty (1985) as a result of the large and intense 1983 El Niño.

In high latitudes, higher mean water temperature could lead to an increase in the duration of the growing period and ultimately in increased bioproductivity in these regions. On the other hand, the probability of nutrient loss resulting from reduced deep-water exchange could result in reduced productivity in the long term—again highlighting the importance of changes in temperature on patterns of circulation. Global warming could have especially strong impacts on the regions of oceanic subpolar fronts (Roots, 1989), where the temperature increase in deep water could lead to a substantial redistribution of pelagic and benthic communities, including commercially important fish species.

For tropical latitudes, GCMs predict temperature increases that are half of those predicted for high latitudes, and the impacts in tropical latitudes are less clear. A 1.5°C increase would raise the summertime mean temperature to 30.5°C over much of the tropical and subtropical regions. Most migratory organisms are expected to be able to tolerate such a change, but the fate of sedentary species will be highly dependent on local climate changes. Some corals would be affected (as in the 1983 and 1987 bleaching events), but it is expected that other environmental stresses (e.g., pollution, sedimentation, or nutrient influx) may remain more important factors (e.g., Maul, 1993; Milliman, 1993). Intertidal plants and animals, such as mangroves and barnacles, are adapted to withstand high temperature, and unless the 1.5°C increase affects reproduction, it will have no effect. Similarly, only seagrass beds already located in thermal-stress situations (i.e., in shallow lagoons or near power plant effluents) are expected to be negatively affected by the projected temperature rise. One cannot rule out, however, the possibility of significantly greater tropical warming than 1.5°C. For example, some investigators argue that tropical warming was approximately 5°C from the last glacial maximum to today (Beck *et al.*, 1992). If this value is correct, current GCMs probably underestimate tropical sensitivity.

Some algal species that are growing at the upper limit of their acceptable temperature range may be eliminated by a temperature increase of a few degrees in oceanic surface waters. There is no evidence that such increases will stimulate toxic algal blooms. Other factors, including competition for micro- and macronutrients, predation by zooplankton, and light-limitation, are of much greater significance to the growth rate and survival of marine algae. High levels of nutrients released into the coastal ocean from anthropogenic sources appear to increase the incidence and growth rates of many types of algae, including some toxic species (GESAMP, 1990). For instance, the toxic algae *Heterosigma spp.*, which are responsible for fish kills in net-pen operations, have been linked to nutrient inputs in coastal waters in Japan (Insuka, 1985; Honjo, 1992).

Changes in temperature and salinity are expected to alter the survivorship of exotic organisms introduced through ballast water in ships, especially those species with pelagic larval forms. Introduction of exotic species is a form of biological pollution because, from a human perspective, they can have adverse impacts on ecosystems into which they are introduced and in some cases pose hazards to public health (International Maritime Organizations [IMO], 1991). A classic recent example of the spread of an introduced exotic species is that of the zebra mussel (*Dressena polymorpha*), which was transported to the Great Lakes via transatlantic shipping from the Baltic sea (Mills *et al.*, 1993). Another recent example is that of the ctenophore, *Mnemiopsis leidyi*, which appears to have been introduced to the Black Sea from the east coast of the United States and has experienced explosive development since 1988 (Mee, 1992). Changes in temperature could enhance the potential for the survival and proliferation of exotic species in environments that are presently unfavorable.

Changes also can be expected in the growth rates of biofouling organisms that settle on means of transport, conduits for waste, maritime equipment, navigational aids, and almost any other artificial structure in the aquatic environment. Their species distributions often are limited by thermal and salinity boundaries, which are expected to change with regional changes in temperature and precipitation. Areas that experience warming and reduced precipitation (i.e., salinity increases) likely will have increased problems with biofouling.

Predicted climate change also may have important impacts on marine mammals such as whales, dolphins, and seals, and seabirds such as cormorants, penguins, storm petrels, and albatross. However, it is presently impossible to predict the magnitude and significance of these impacts. The principal effects of climate change on marine mammals and seabirds are expected from areal shifts in centers of food production and changes in underlying primary productivity due to changes in upwelling, loss of ice-edge effects, and ocean temperatures; changes in critical habitats such as sea ice (due to climate warming) and nesting and rearing beaches (due to sea-level rise); and increases in diseases and production of oceanic biotoxins due to warming temperatures and shifts in coastal currents.

Ice plays an important role in the development and sustenance of temperate to polar ecosystems because it creates conditions conducive to ice-edge primary production, which provides the primary food source in polar ecosystems; it supports the activity of organisms that ensure energy transfer from primary producers (algae and phytoplankton) to higher trophic levels (fish, marine birds, and mammals); and, as a consequence, it maintains and supports abundant biological communities (Izrael *et al.*, 1992).

One of the possible beneficial consequences of global warming might be a reduction in the extent and stability of marine ice, which would directly affect the productivity of polar ecosystems. For example, the absence of ice over the continental shelf of the Arctic Ocean would produce a sharp rise in the productivity of this region (Izrael *et al.*, 1992), provided that a sufficient supply of nutrients is maintained. Changes in water temperature and wind regimes as a result of global warming also could affect the distribution and characteristics of polynyas (ice-free areas), which are vital to polar marine ecosystems. In addition, changes in the extent and duration of ice, combined with changes in characteristics of currents—for example, the circumpolar current in southern latitudes—may affect the distribution, abundance, and harvesting of krill. Krill are an important link in the ocean fauna in the Southern Ocean. It is important to understand how, when, and where productivity in the Southern Ocean will change with global warming.

A number of marine organisms depend explicitly on ice cover. For example, the extent of the polar bear's habitat is determined by the maximum seasonal surface area of marine ice in a given year. The disappearance of ice would threaten the very survival of the polar bear, as well as certain marine seals. Similarly, a reduction in ice cover would reduce food supplies for seals and walruses and increase their vulnerability to natural predators and human hunters and poachers. Other animals, such as the otter, could benefit by moving into new territories with reduced ice. Some species of marine mammals will be able to take advantage of increases in prey abundance and spatial/temporal shifts in prey distribution toward or within their primary habitats (e.g., Fraser *et al.*, 1992; Montevecchi and Myers, 1995), whereas some populations of birds and seals will be adversely affected by climatic changes if food sources decline (Polovina *et al.*, 1994) or are displaced away from regions suitable for breeding or rearing of young (Schrieber and Schrieber, 1984).

Animals that migrate great distances, as do most of the great whales and seabirds, are subject to possible disruptions in the distribution and timing of their food sources during migration. For example, it remains unclear how the contraction of ice cover would affect the migration routes of animals (such as whales) that follow the ice front (Izrael *et al.*, 1992). At least some migrating species may respond rapidly to new situations; for example, migrating ducks have altered their routes to take advantage of the recent exploding population of zebra mussels in the Great Lakes (Worthington and Leach, 1992).

While the impacts of these ecological changes are likely to be significant, they cannot be reliably forecast or evaluated. Climate change may have both positive and negative impacts, even on the same species. Positive effects such as extended feeding areas and seasons in higher latitudes, more-productive high latitudes, and lower winter mortality may be offset by negative factors that alter established reproductive patterns, breeding habitat, disease vectors, migration routes, and ecosystem relationships.

8.3.3. *Biogeochemical Cycles*

8.3.3.1. CO_2 Uptake and Biological Productivity

Climate change is expected to have an impact on processes controlling biogeochemical cycling of elements in the oceans, with potential feedbacks on various components of the carbon cycle. These impacts include the uptake of and storage capacity for CO_2 by physicochemical and biological processes. These aspects are considered in Chapter 6, *Climate Models—Projections of Future Climate*, and Chapter 10, *Marine Biotic Responses to Environmental Change and Feedbacks to Climate*, of the Working Group I volume.

Changes in sea-surface temperature can be expected to affect carbon chemistry both directly and indirectly. Changes will occur in the solubility of CO_2 and the remineralization of dissolved organic carbon. However, these changes are not expected to significantly affect atmospheric CO_2 concentrations (IPCC, 1994). Less certain, however, is the role of biological processes in sequestering carbon in the ocean and the potential impact of climate change on these processes.

The major process by which marine biota sequester CO_2 is thought to be controlled by nutrients, not by the concentration of dissolved inorganic carbon. A weakening of vertical and horizontal circulation as a result of warming (e.g., Manabe and Stouffer, 1993) could result in fewer nutrients in surface water, leading to a reduction of biological CO_2 uptake and storage (Volk and Hoffert, 1985). At the same time, less carbon-enriched deep water would be carried to the surface where the CO_2 is released to the atmosphere. The two effects result from the same process but affect the storage of carbon in opposite directions; they dominate models that use constant carbon-to-nutrient ratios to describe the marine biosphere (e.g., Bacastow and Maier-Reimer, 1990). These models project that a reduction of ocean circulation and vertical mixing will increase oceanic carbon storage slightly. A reduction in vertical mixing in the ocean also leads to a reduced downward transport of excess CO_2. It reduces the capacity of the oceans to store excess CO_2 because progressively less water is in contact with the surface over the period of atmospheric perturbation. Current model results suggest that the effects of predicted changes in circulation on the oceanic carbon cycle will not be large (Paillard *et al.*, 1993). However, extrapolation of the long-term impacts of warming on circulation patterns has just begun; hence, the analyses of impacts on the carbon cycle must be viewed as preliminary.

Riebesell *et al.* (1993) claimed recently that phytoplankton growth rates are dependent on the availability of dissolved CO_2 and that this process could have been one of the factors responsible for changes in atmospheric CO_2 contents observed in the ice-core record. Their conclusions have been contested by other researchers who suggest that CO_2-limited photosynthesis cannot be equated with CO_2-limited growth rates (e.g., Turpin, 1993). Further research will be needed to ascertain whether increased CO_2 concentrations are likely to have the effects suggested by Riebesell *et al.* (1993).

Climate change also can affect the role of marine biota in other ways. Shifts may occur in the structure of biological communities in the upper ocean—for example, from coccolithophorids to diatoms—as a result of increased freshwater influx from melting ice and river runoff. This change will alter the ratio of organic carbon to carbonate carbon in the material settling out of the surface layers of the sea (Berger and Keir, 1984; Ittekkot *et al.*, 1991). Biologically mediated carbon storage in the ocean is determined by the rate of transfer of organic matter from the surface layers, which exchange gases with the atmosphere, to the deep ocean, which remains isolated from the atmosphere for up to hundreds of years. The role of this transfer in the context of deep-sea carbon storage has not been adequately addressed in the scientific literature. Recent studies suggest that eolian dust and river-derived mineral matter can enhance carbon fixation at the sea surface by providing essential trace constituents such as iron (e.g., Martin and Fitzwater, 1987; Murray *et al.*, 1994). Furthermore, this material can accelerate the transfer of organic matter to the deep sea by providing "ballast" material for rapidly sinking organic aggregates (Figure 8-5; Ittekkot, 1993). Thus, the efficiency of carbon storage in the deep sea may be affected by a changing climate and associated changes in atmospheric and ocean circulation and nutrient availability, combined with other global changes such as desertification. The magnitude and the direction of these changes remain unclear. Major national and international activities within the framework of JGOFS (Joint Global Ocean Flux Study) and GLOBEC (Global Ocean Ecosystem Dynamics) are expected to provide data to assess the role of biological processes in the removal of CO_2 from the atmosphere into the depths of the oceans.

Most of the organic-carbon burial in modern marine sediments on a global basis occurs in deltas and continental-margin environments (Berner, 1992) that are characterized by inputs from high primary productivity and terrigenous material discharged by the rivers. Both inputs are likely to be influenced by climate change affecting carbon cycling in these regions, with feedbacks to the carbon cycle and hence climate. It is difficult at present to assess the nature and extent of such an influence because other factors such as nutrient loading and river runoff

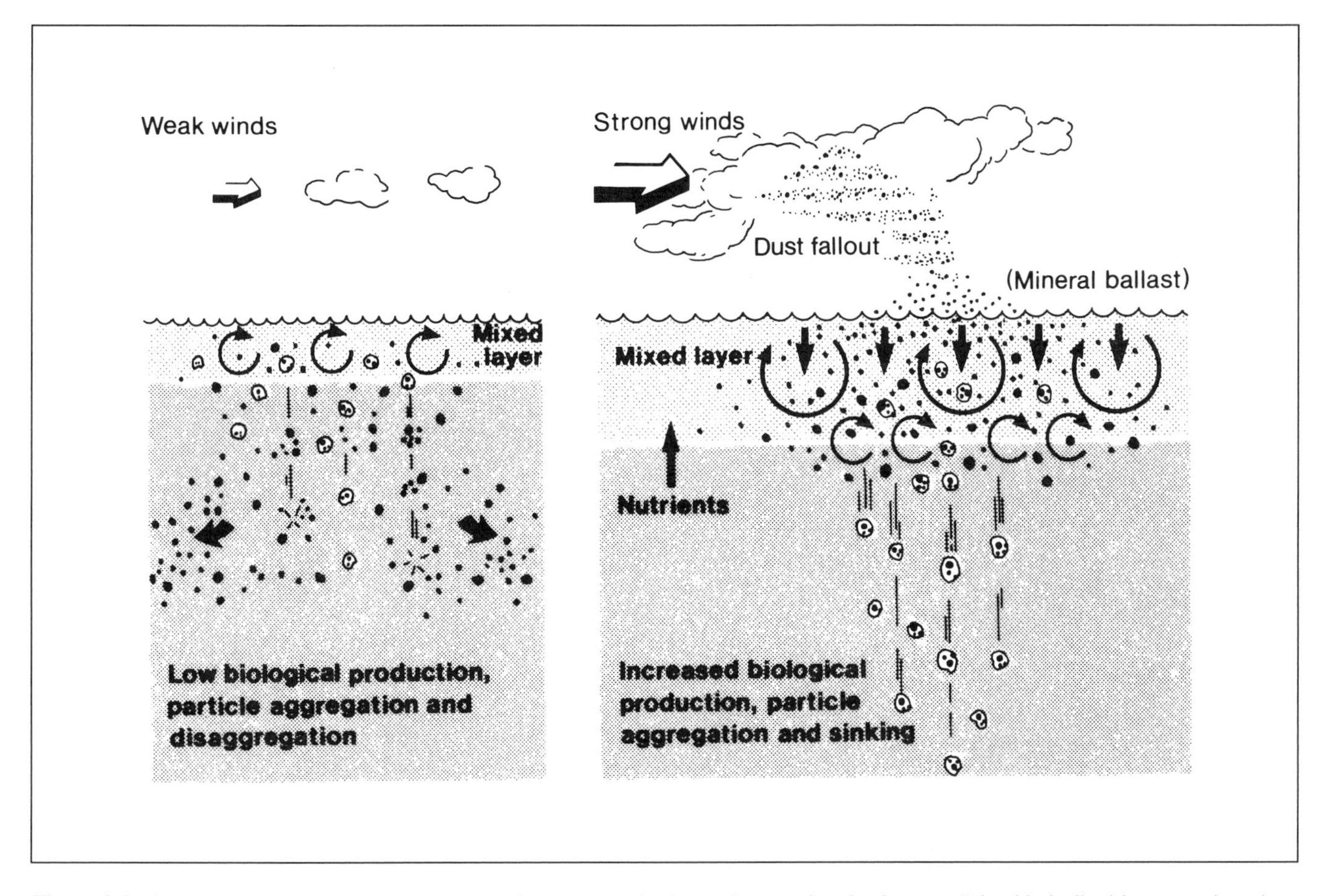

Figure 8-5: Schematic of processes that control organic-matter production and removal to the deep sea ("the abiotically driven organic carbon pump" in the ocean). Dust particles deposited at the sea surface not only introduce essential trace nutrients such as iron (stimulating primary productivity), but also become incorporated into organic aggregates that form high-density particles (the ballast effect) with faster sinking rates—thus accelerating the transfer of newly fixed carbon dioxide to the deep sea. Strong winds enhance the effect as well. The efficiency of this transfer—therefore of carbon storage in the deep sea—is expected to be affected by global warming (after Ittekkot, 1993).

already have a strong impact in these areas and because the role of coastal oceans in the global carbon cycle is poorly understood (Smith and MacKenzie, 1991).

Marine aerosol formation is associated with the production of dimethyl sulfide (DMS) by specific phytoplankton species in the ocean (Charlson *et al.*, 1987). DMS is the most important source of cloud condensation nuclei (CCN). Changes in phytoplankton species due to predicted global warming can have other feedbacks in the climate system, although the direction and quantitative significance of such feedbacks have to be assessed. The role of the oceans in the contribution of other greenhouse gases to the atmosphere—such as, for example, methane and nitrous oxide—is just beginning to be assessed, and the impact of climate change is still uncertain (see Chapter 10, *Marine Biotic Responses to Environmental Change and Feedbacks to Climate*, of the Working Group I volume).

8.3.3.2. *Contaminant Distribution and Degradation*

Contaminant distribution in the ocean will be especially sensitive to climate change because two major pathways of pollutants into the oceans will be affected: river runoff and atmospheric transport. The impacts of global warming on coastal and ocean pollution can be expected to depend on the way global warming will affect the nature and concentrations of pollutants in question, as well as on changes in their persistence in the environment. Changes in the cycling and distribution between phases (dissolved versus sedimentary) could occur under a global warming scenario for sewage, organics, organochlorines, and heavy metals (e.g., Tsyban *et al.*, 1990). In most cases, phase transitions can be expected. A rise in temperature also may result in accelerated biodegradation of global organic pollutants (petroleum and chlorinated hydrocarbons, etc.). This process would promote their removal from the biologically active surface layers of the ocean (Izrael and Tsyban, 1989; Tanabe, 1985). However, the fate of degradation products is less well-known, and there also is the possibility of an increase in the toxicity of at least some of the pollutants under such conditions and with higher temperature. Further, the impact of climate change can be expected to be more severe in coastal seas and landlocked marine basins—such as, for example, the Baltic Sea and the Black Sea, which are already seriously impacted by pollution from human activities.

8.3.4. *Socioeconomic Systems*

The institutional structure of socioeconomic systems is an extremely important determinant of behavior, and it accounts for relative success or failure in system performance (Ostrom, 1990; Haas *et al.*, 1993; Lee, 1993; Putnam, 1993). Institutional structure includes, among other characteristics, patterns of social organization and the distribution of authority.

A survey of the patterns of current social organization on a global basis, however, shows that present patterns and system capabilities are inadequate to respond to climate change. The major drawback is a lack of integrative capacity even at national levels concerning the use of the coastal ocean and even more so with reference to the exclusive economic zone (EEZ) as a whole (Underdal, 1980; Miles, 1989, 1992). This lack of integrative capacity arises because patterns of ocean use have developed largely in isolation from each other and because different technologies have given rise to separate networks, communities, and ways of thinking and doing. These communities have matured into almost fully autonomous sectors, with only weak or no links among them. These sectors are still needed for specialization. What is missing, however, is an integrative overlay in the institutional structure to assess a country's interest in the ocean as a whole and to make judgments about long- and short-term priorities and responses.

8.4. Impacts of Climate Changes on Resources and Products

The potential effects of climate change on resources and products range from significantly negative to significantly positive. These effects and the probability of their occurrences are listed below:

- **Significantly Negative**
  - A predicted 2°C average global atmospheric temperature increase by 2050 will result in increased coral bleaching, which in turn will result in a reduction in coral production. *Rating*: Very probable.
  - Increased precipitation, river runoff, and atmospheric deposition from land-based activities will lead to increased loading of pollutants in coastal waters and an adverse impact on fisheries, coral production, and recreation and tourism. *Rating*: Very probable.
  - If the frequency of tropical storms and hurricanes increases, there will be adverse impacts on offshore oil and gas activities in certain locations and on marine transportation. *Rating*: Uncertain.
- **Mildly Negative**
  - There will be problems in the operation of low-head tidal power plants due to increased sedimentation from increased precipitation and runoff in the Northern Hemisphere. There also will be reduced potential for OTEC in certain locations due to a reduction in differential between surface temperature and temperature below the thermocline. *Rating*: Very probable.
  - There will be increased costs generated by the need to expand dredging operations to keep major ports open in particular locations. *Rating*: Very probable.
  - There will be reduced fisheries yield if average primary productivity decreases. *Rating*: Uncertain. This effect probably will be dwarfed by the combined adverse impacts of overfishing, marine coastal pollution, and habitat destruction. *Rating*: Virtually certain.
- **Neutral**
  - Marine pharmaceutical and biotechnology industries are unlikely to be affected by climate change. *Rating*: Very probable.
  - Nonliving resource exploitation other than petroleum hydrocarbons will not be affected by climate change. *Rating*: Very probable.
- **Mildly Positive**
  - There will be growth in the marine instrumentation industry to facilitate research and monitoring of climate change through expanding capabilities for automatic sensing. *Rating*: Virtually certain.
- **Significantly Positive**
  - There will be increased growth and yield of upper-trophic-level species in commercial fisheries, but this effect will be dwarfed by current overfishing, marine pollution, and habitat destruction. *Rating*: Uncertain.
  - The Northwest Passage and the Northern Sea Route of Russia will be opened for routine shipping, reducing freight rates between East Asia and Western Europe. *Rating*: Very probable.

8.5. Evaluation of the Impacts of Climate Change

8.5.1. *Resources and Products*

In this chapter, the notion of mitigating impacts includes techniques of prevention or reduction in the scope or intensity of predicted effects. Adaptation means coping with or compensating for the rate of damage.

8.5.1.1. *Significantly Negative Impacts*

Reported incidences of coral bleaching appear to be increasing in association with more-frequent ENSO events since 1982–83 (Glynn, 1989; Brown and Ogden, 1993). Coral mortality is positively correlated with the intensity and length of warming episodes (Glynn, 1989; Glynn and Croz, 1990), and recent paleoclimatic investigations of ENSO phenomena show that the coral record can be read as a proxy for increased SSTs, potentially over several thousand years (Cole *et al.*, 1992; Shen, 1993). Corals thrive in a temperature range of 25–28°C, but experimental and observed evidence indicates that mortality can be induced with even a 5°C increase in SST if exposure is prolonged beyond six months (Glynn and Croz, 1990). The time required to induce mortality decreases as SST increases along a gradient of 1–4°C (Glynn, 1989).

Therefore, assuming a 2°C increase in average global temperature by 2050—as predicted by Working Group I—it is doubtful that either mitigation or adaptation would be possible. The magnitude of such an increase, in addition to the rapid rate of change, leaves virtually no margin for corrective action. If maximum temperatures in the tropics exceed an increase of 3°C for extended periods, the impact will be severe. Moreover, corals are already at risk from a wide variety of human activities on land and at sea. This additional risk severely compounds the problem.

Independent of coral bleaching, increased loading of land-based pollutants in the coastal ocean as a result of increased precipitation and atmospheric transport can be mitigated by significantly reducing current loading of land-based pollutants in the marine environment. This very difficult problem has received some scientific and legal attention at global and regional levels since the Stockholm Conference in 1972 and a great deal more since the UN Conference on Environment and Development in Rio de Janeiro in 1992. Indeed, a major international conference on land-based pollution of the marine environment was convened by the United Nations Environment Program (UNEP) in October 1995.

The legal infrastructure at the global level for combating land-based pollution of the marine environment appears to be adequate. The problems to be solved are increasing compliance with existing legal obligations and developing scientifically based policies that effectively respond to the land-based pollution problem (GESAMP, 1990). Effective regulation must be based on an understanding of the fate and transport of contaminants over space and time and the damage caused by these pollutants to the structure and function of aquatic ecosystems. Significant issues include the persistence of contaminants in the aquatic environments, their uptake by commercially important fish and shellfish, the nature of sublethal effects to marine and freshwater organisms, and the risk to humans (Capuzzo, 1990).

In addition, competing legislative and institutional structures at national, regional, and global levels need to be harmonized and rationalized in order to avoid the inconsistencies in policy derived from piecemeal attempts to solve problems. In the formulation of policies, priority must be given to the most-toxic waste streams in order to curtail long-term damage. In this connection, it is important to focus particularly on nonpoint sources of pollution, especially agricultural and urban runoff, and to avoid placing the greatest burdens on estuaries and the coastal ocean. Policy formulation might most effectively emphasize risk assessment and risk management—in support of which the development of long-term monitoring capabilities and standardized databases will be particularly important (GESAMP, 1991).

If the frequency of tropical storms and hurricanes increases, there are no mitigating options available for either the offshore oil and gas industry or the shipping industry. The former will have to adapt by improving the design standards for offshore structures. For marine transportation, the issue of whether the frequency of storms increases is an important question that will affect ship operations, equipment, maintenance, and insurance. One possible adaptation may be in national regulations to avoid areas that are too dangerous or will put biological communities at risk—along with the technological means to provide early warning at sea. If the occurrence of extreme events increases, governments have the option of increasing the attention given to issues of institutional design to respond to disasters and acute emergencies, planning for such responses, and introducing effective training of personnel (Drabek, 1985, 1989).

8.5.1.2. Mildly Negative Impacts

Mitigation and adaptation options are available for the mildly negative impacts. Because the minimum temperature differential required for OTEC plants is 20°C (Charlier and Justus, 1993), the crucial question is to what extent the intermediate water will warm relative to the surface. Should the temperature differential become unfavorable, two corrective measures are at least theoretically possible for floating power plants: relocate the plant, or increase the depth of the cold-water pipe. However, both of these options are seriously constrained by cost, and increasing the depth of the pipe imposes maintenance requirements that are severe below about 450 m (Charlier and Justus, 1993). For marine transportation, if major ports experience significant increases in sediment deposition via river runoff, dredging is the obvious option—with associated increases in the costs of operation.

There is uncertainty with respect to the impact of increased global average temperature of magnitude 2°C on primary productivity, but much less so on the growth and development of species at higher trophic levels. However, the impact of global warming can be swamped by the combined effects of overfishing, increased marine pollution, and habitat degradation/destruction. Again, in this case both mitigation and adaptation options are available.

As detailed in Chapter 16, changes must be made in approaches to management so that fishing effort is controlled and sized to available resources. Control over land-based pollution of the marine environment, combined with control over habitat degradation/destruction, also will go a long way toward creating an environment in which fisheries can prosper.

Moreover, in marine ecosystems where the fish community has been altered through excessive fishing effort—causing cascading effects up the food chain to top predators and down to first and second trophic levels—it is possible to facilitate recovery of depleted species through the introduction of adaptive management techniques (e.g., Sherman, 1994).

8.5.1.3. Neutral Impacts

Except for hydrocarbon extraction in polar areas, marine mineral extraction will be largely insensitive to an increase in average global temperature on the order of 2°C. Hence, mitigation and adaptation options are not required. Similarly, the source material for biotechnological/biomedical uses will not be affected by global warming, because most marine organisms living in the upper layers of the ocean can and do tolerate some degree of seasonal variations in temperature. Those living in the more-constant temperature depths are unlikely to be affected. Global warming will not have an effect on bioremediation applications. The most likely response to higher temperatures will be an increase in growth rate (hence, clean-up rate) of the organisms involved. It must be noted, however, that the changes in biodiversity due to global warming can be expected to have a certain degree of impact on these industries.

8.5.1.4. *Mildly Positive Impacts*

There is no need for mitigation of the mildly positive impacts of global warming; the market-driven need will create its own adaptation, to which the marine instrumentation industry will respond. Growth within the marine instrumentation industry generated by the need to monitor climate change will be reflected in the design of new sensors to facilitate observation, measurement, and monitoring, as well as the adaptation of existing platforms to perform new functions. For instance, offshore oil and gas platforms are now seen as having value for monitoring dimensions of climatic change (MTSJ, 1993). The U.S. National Oceanic and Atmospheric Administration (NOAA) has organized the cooperation of platform operators into a network to provide time-series measurements for the Global Ocean Observation System (GOOS) program in the Gulf of Mexico. Such adaptation appears to be extremely cost-effective.

8.5.1.5. *Significantly Positive Impacts*

Existing human-induced problems will swamp the significantly positive impacts of global warming on fisheries, and the same mitigation options outlined in Sections 8.5.1.1 and 8.5.1.2 relative to negative impacts on fisheries will apply in this case.

On the other hand, opening up the Northwest Passage and the Russian Northern Sea route for routine shipping for as much as 100 days a year (see Chapter 7) will provide significant benefits to many countries in terms of efficiency and speed of service and cost reduction that could translate into lower freight rates. At the same time, because both transportation and hydrocarbon exploitation will be facilitated by the reduction in ice cover in polar seas, the development of these regions for transportation and resource exploitation should be approached with great care. Stringent controls on pollution of these newly accessible regions will be necessary because ice will still retard clean-up possibilities for more than 200 days per year.

8.5.2. **Ecology and Biodiversity**

In some cases, no reliable adaptation is possible for changes in ecology and biodiversity. For example, for large mammals affected by the loss of ice cover, such as polar bears, no adaptation at all will be possible because of the loss of territory and loss of prey. In cases where adaptation options are available, they will vary according to the scale of the changes. If, for example, there is extirpation of a species, then carefully planned reintroductions may work. Survival of the species will then depend on the elimination of the stress factors that caused their extirpation. In the case of global extinctions, there is no adaptation possible. The consequences of an artificial restructuring of an ecosystem or community are at present unpredictable.

In the case of increases in species, eliminating successful exotics is rarely possible. Although the introduction of exotic predators is one possible option, our ability to predict the effects of such introductions is lacking. Depending on management goals, there could be successful introduction of exotics that make the system more useful to humans.

8.5.3. **The Issue of Costs**

Critical to any discussion of adaptation options is the issue of costs. However, at present we do not have sufficiently detailed knowledge of the costs of the most important impacts of climate change on the human uses of the oceans and the available options to mitigate and adapt. Some of the impacts of global climate change are beneficial, some are neutral, and some are adverse. The costs and benefits will be transferred and/or apportioned either by governments or by markets, or by some combination thereof. Such transfers will have price effects, which could affect the competitiveness of ocean industries in relation to each other and their competitors on land. In some cases (e.g., the impacts of sea-level rise), the total burden of cost transfers is likely to be greatest on those least able to bear them (i.e., the poorest countries). It is likely that costs will vary in a nonlinear fashion, sometimes rising much more steeply than the temperature increase with time, especially for flood damages or agricultural impacts.

The point about the competitiveness of ocean industries in relation to their competitors on land requires further elaboration. By and large, it is more costly to work on the ocean than it is on land, and the risk is generally higher. Whether one is dealing with nonliving resources or any other kind of product, the critical issues are relative scarcity, substitutability of materials/products, extension of available resources as a result of technological innovation, or location and exploitation of previously uneconomic resources (Broadus, 1987). Increasing reliance on more costly resources tends to raise the price of extracted materials—prompting substitution, conservation, recycling, and exploration (Broadus, 1987).

Increasing or decreasing scarcity of resources is best reflected in the long-term market price (Broadus, 1987). However, if the practical utilization of a marine resource lies more than two decades into the future, the net present value of that resource is essentially zero because firms will consider that they do not have sufficient control over the relevant factors controlling the cost of production and therefore price. The market's reluctance to act in these circumstances does not mean that nothing will be done. For strategic reasons, some governments may wish to subsidize exploration and even exploitation for varying lengths of time in some of these cases.

8.6. Multistress Factors

We have made the point that the coastal ocean is already under stress as a result of a combination of factors (e.g., increased population pressure in coastal areas, habitat destruction, increased land-based pollution, and increased river inputs of nutrients and other pollutants). In addition, increased UV-B radiation due to stratospheric ozone depletion is expected to impair the resilience of aquatic ecosystems to climate change. Smith *et al.* (1992) demonstrate that increased UV-B radiation has the effect of causing an estimated 6–12% reduction in primary productivity in the surface waters of the Southern Ocean during Austral spring. McMinn *et al.* (1994) conclude, on the other hand, that thick ice cover and the timing of the phytoplankton bloom in the Southern Ocean protect the phytoplankton from the adverse effects of increased UV-B radiation. A more comprehensive treatment of the subject is given by the Ozone Trends Panel (OTP).

Also of importance in the context of the impact of global warming is the debate about whether the phenomenon of coral bleaching is an effect solely of increased water temperature or increased UV-B radiation in the 280–400 nm band. Recent results suggest that bleaching of corals by increased UV-B radiation and by the temperature of surface waters appears to be independent of each other (Gleason and Wellington, 1993). These results are challenged, however, by Dunne (1994), who argues that the experimental controls do not conclusively rule out other factors such as the impact of photosynthetically active radiation (PAR) or synergistic effects between PAR and UV-B radiation on bleaching.

The effects of global climate change therefore will constitute a mixed series of impacts on an already overstressed context, with attendant opportunities for synergistic relationships between the stresses where the climate-change impacts are adverse. Synergy will accelerate the adverse impacts of these stresses. These burdens present a challenge and increase the urgency for the development of integrated coastal management responses at regional and global levels.

These multistress factors also make coastal areas much more vulnerable than the open ocean. Coastal states therefore may wish to consider controlling population density, habitat destruction, and land-based pollution. In this connection, the combination of existing trends of human-induced stress on the coastal environment with potential stresses generated by global climate change requires building the capability for planners and analysts to provide cumulative, integrated impact assessments (Gable *et al.*, 1991). Databases to facilitate such assessments do not yet exist and need to be built. The criteria for designing such databases include the need to "evaluate the exposure, response, risk, and vulnerability of cultural, economic, social, biological, and physiographical implications of global change and its local, basin-wide, regional, and interregional manifestations"; the cumulative effects of environmental variability on various time and space scales; the effects of minor but collectively significant events occurring over long timescales; and the potential for synergistic effects of interacting combinations superimposed on direct effects (Gable *et al.*, 1991).

8.7. Research and Monitoring Needs and Strategies

In this section, we consider research and monitoring needs that will allow better understanding of the characteristics and functions of oceans that are most likely to be affected by a projected climate change; development of methodologies to assess the sensitivity of oceans to climate change; and formulation of pilot studies on potential impacts, thresholds and sensitivities, and mitigation/adaptation strategies:

- *Research activities to better understand processes in the oceans*, in particular the role of the oceans in the natural variability of the climate system at seasonal, interannual, and decadal to century timescales; the role of the Atlantic Ocean in climate variability; the role of the ocean in the hydrological cycle; the role of biological and biogeochemical processes in transient—decades to centuries—carbon storage in the deep sea
- *Long-term monitoring and mapping of*: water-level changes, ice coverage, and thermal expansion of the oceans; sea-surface temperature and surface air temperature; extratropical storms and tropical cyclones; changes in upwelling regimes along the coasts of California, Peru, and West Africa; UV-B radiation, particularly in polar regions, and its impact on aquatic ecosystems; regional effects on distribution of species and their sensitivity to environmental factors; changes in ocean biogeochemical cycles. These activities allow for better understanding of the processes that affect the stability and vulnerability of marine ecosystems and their spatial and temporal variability. They also allow for better assessment of climate change-induced rates and the direction of changes and processes in oceans that already are impacted by other factors.
- *Socioeconomic research activities* to document human responses to global environmental change, such as the establishment of databases on patterns of human responses to global and regional environmental changes; the assessment of the transfer costs and economic effects of global and regional climate change; the development of alternative approaches to mitigation and adaptation, as well as policy design; and assessment of the synergistic effects of sectoral approaches to policy implementation
- *Strategies for implementation* of these monitoring and research needs could include national and international environmental and climate research programs to

consider research and monitoring components specifically designed to investigate the impact of climate change, taking into account the specific regional and national needs based on the pattern of the respective human response to global and environmental changes; and formal coordination between IPCC and other international environmental and climate research and monitoring activities organized by, for example, the International Geosphere-Biosphere Programme (IGBP) and the World Climate Research Programme (WCRP), to pool expertise and efforts to better monitor aspects of climate change and its impacts, as well as to formulate policy options. Because modeling and monitoring depend on actual knowledge and technological capabilities, there also is considerable room for improvement in the level of communication between modelers and monitors in order to improve the state of the art and prediction capabilities.

It must be noted that research programs, by definition, are focused on a topic or problem, have an underlying approach, and promise results in a given time and funding frame. Furthermore, research topics change over longer times. In contrast, monitoring must be protected from such changes because the use of time series depends on the absence of changes in data-collection methods. Institutional research or scientific establishments with a longer breadth, such as government research laboratories, need to be encouraged to get their particular expertise rolled out: long-term maintenance of monitoring and research programs. Especially in view of the need to support long-term observations (observing systems such as GOOS, GCOS, GTOS, etc.) and the complex nature of the issues at hand, it will be necessary to pool the expertise and efforts of individual research institutions and governmental research laboratories actively and intimately.

References

Anderson, R.Y., A. Soutar, and T.C. Johnson, 1992: Long-term changes in El Niño/Southern Oscillation: evidence from marine and lacustrine sediments. In: *El Niño, Historical and Paleoclimatic Aspects of the Southern Oscillation* [Diaz, H.F. and V. Markgraf (eds.)]. Cambridge University Press, Cambridge, UK, pp. 419-433.

Aquaculture Magazine Buyers Guide, 1995: Achille River Corp., Ashville, NC, 4th ed., pp. 8-23.

Bacastow, R. and E. Maier-Reimer, 1990: Ocean-circulation model of the carbon cycle. *Clim. Dynamics*, **4**, 95-125.

Bacon, S. and D.J.T. Carter, 1991: Wave climate changes in the North Atlantic and North Sea. *Int. J. Climatol.*, **11**, 545-558.

Bakun, A., 1990: Global climate change and intensification of coastal ocean upwelling. *Science*, **247**, 198-201.

Bakun, A., 1993: Global greenhouse effects, multidecadal wind trends, and potential impacts on coastal pelagic fish populations. *ICES Mar. Sci. Symp.*, **195**, 316-325.

Barnola, J.M., D. Raynaud, Y.S. Korotkevich, and C. Lorius, 1987: Vostok ice core provides 160,000-year record of atmospheric CO_2. *Nature*, **329**, 408-414.

Beck, J.W., R.L. Edwards, E. Ito, F.W. Taylor, J. Recy, F. Rougerie, P. Joannot, and C. Henin, 1992: Sea-surface temperature from coral skeletal strontium/calcium ratio. *Science*, **257**, 644-647.

Beddington, J., 1995: The primary requirements. *Nature*, **374**, 213-214.

Bengtsson, L., M. Botzet, and M. Esch, 1994: *Will Greenhouse Gas-Induced Warming over the Next 50 Years Lead to Higher Frequency and Greater Intensity of Hurricanes?* Max-Planck Institut für Meteorologie, Report No. 139, MPI, Hamburg, Germany, 23 pp.

Berger, W.H. and R.S. Keir, 1984: Glacial-Holocene changes in atmospheric CO_2 and the deep-sea record. In: *Climate Processes and Climate Sensitivity* [Hansen, J.E. and T. Takahashi (eds.)]. Geophysical Monograph 29, American Geophysical Union, Washington, DC, pp. 337-351.

Bernal, P.H., 1991: Consequences of global change for oceans: a review. *Climatic Change*, **18**, 339-359.

Berner, R.A., 1992: Comments on the role of marine sediment burial as a repository for anthropogenic CO_2. *Global Biogeochem. Cycles*, **6**, 1-2.

Berz, G., 1994: Die Zeichen stehen auf Sturm. *Naturwissenschaften*, **81**, 1-6.

Bigg, G.R., 1993: Comparison of coastal wind and pressure trends over the tropical Atlantic: 1946-1988. *Int. J. Climatol.*, **13**, 411-421.

Bindoff, N.L. and J.A. Church, 1992: Warming of the water column in the southwest Pacific Ocean. *Nature*, **357**, 59-62.

Binet, D., 1988: Role possible d'une intensification des alizes sur le changement de repartition des sardines et sardinelles le long de la cote ouest africaine. *Aquat. Living Resour.*, **1**, 115-132.

Bond, G., W. Broecker, S. Johnsen, J. McManus, L. Labeyrie, J. Jouzel, and G. Bonani, 1993: Correlations between climate records from North Atlantic sediments and Greenland ice. *Nature*, **365**, 143-148.

Boyle, E.A. and L.D. Keigwin, 1982: Deep circulation of the North Atlantic over the last 200,000 years: geochemical evidence. *Science*, **218**, 784-788.

Bragg, J.R., R.C. Prince, E.J. Harner, and R.M. Atlas, 1994: Effectiveness of bioremediation for the Exxon Valdez oil spill. *Nature*, **368**, 413-418.

Broadus, J.M., 1987: Seabed materials. *Science*, **235**, 853-860.

Broadus, J.M., 1991: World trends in ocean technology. In: *Marine Policies Toward the 21st Century: World Trends and Korean Perspectives* [Hong, S.-Y. (ed.)]. Korea Ocean Research and Development Institute, Seoul, Korea, pp. 247-268.

Broadus, J.M., P. Hoagland III, and H.L. Kite-Powell, 1989: Defining a key industry: marine electronics instrumentation. *Sea Technology*, **30**, 59-64.

Broecker, W.S., 1991: The great ocean conveyor. *Oceanography*, **4**, 79-89.

Broecker, W.S., 1994: An unstable superconveyor. *Nature*, **367**, 414-415.

Brown, B.E. and J.C. Ogden, 1993: Coral bleaching. *Scientific American*, **268**, 44-50.

Budyko, M.I. and Yu. A. Izrael (eds.), 1987: *Anthropogenic Climate Change*. Leningrad Hydromteoizdat, 378 pp. (in Russian).

Capuzzo, J.E., 1990: Effects of wastes on the ocean: the coastal example. *Oceanus*, **33**, 39-44.

Charlier, R.H. and J.R. Justus, 1993: *Ocean Energies: Environmental, Economic, and Technological Aspects of Alternative Power Sources.* Elsevier, Amsterdam, The Netherlands, 283 pp.

Charlson, J.L., J.E. Lovelock, M.O. Andreae, and S.G. Warren, 1987: Oceanic phytoplankton, atmospheric sulphur, cloud albedo and climate. *Nature*, **326**, 655-661.

Church, J.A., J.S. Godfrey, D.R. Jackett, and T.L. McDougall, 1991: A model of sea level rise caused by ocean thermal expansion. *J. Climate*, **4**, 438-456.

Cole, J.E., G.T. Shen, R.G. Fairbanks, and M. Moore, 1992: Coral monitors of El Niño/Southern Oscillation dynamics across equatorial Pacific. In: *El Niño: Historical and Paleoclimatic Aspects of the Southern Oscillation* [Diaz, H.F. and V. Markgraf (eds.)]. Cambridge University Press, Cambridge, UK, pp. 349-375.

Cubasch, U., B.D. Santer, A. Hellbach, G. Hegerl, K.H. Höck, E. Maier-Reimer, U. Mikolajewicz, A. Stössel, and R. Voss, 1994: Monte Carlo climate forecasts with a global coupled ocean-atmosphere model. *Clim. Dynamics*, **10**, 1-19.

Curry, W.B., J.C. Duplessy, L.D. Labeyrie, and N.J. Shackleton, 1988: Changes in the distribution of $\partial^{13}C$ of deep water CO_2 between the last glaciation and the Holocene. *Paleoceanography*, **3**, 317-341.

Dansgaard, W., S.J. Johnsen, H.B. Clausen, *et al.*, 1993: Evidence for general instability of past climate from a 250-kyr ice-core record. *Nature*, **364**, 218-220.

Dickson, R.R., J. Meinke, S.A. Malmberg, and A.J. Lee, 1988: The great salinity anomaly in the North Atlantic, 1968-82. *Prog. Oceanogr.*, **20**, 103-151.

Drabek, T.E., 1985: Managing the emergency response. *Public Administration Review*, **45**, 85-92.

Drabek, T.E., 1989: Strategies used by emergency managers to maintain organizational integrity. *Environmental Auditor*, **1**, 139-152.

Dunne, R.P., 1994: Radiation and coral bleaching. *Nature*, **368**, 698.

Encyclopedia Britannica, 1993: Encyclopaedia Britannica Press, Chicago, IL.

Evans, J.L. and R.J. Allan, 1992: El Niño-Southern Oscillation modification to the structure of monsoon and tropical cyclone activity in the Australasia Region. *Int. J. Climatol.*, **12**, 611-623.

FAO, 1992a: *Yearbook of Fishery Statistics: Catches and Landings, 1990*, vol. 70. Department of Fisheries, United Nations, Rome, Italy.

FAO, 1992b: *Review of the Status of World Fish Stocks*. Department of Fisheries, United Nations, Rome, Italy, 114 pp.

Fraser, W.R., W.Z. Trivelpiece, D.G. Ainley, and S.G. Trivelpiece, 1992: Increase in Antarctic penguin populations: reduced competition with whales or a loss of sea-ice due to environmental warming? *Polar Biology*, **11**, 525-531.

Fry, F.E.J., 1971: The effect of environmental factors on the physiology of fish. In: *Fish Physiology,* vol. 6 [Hoar, W.S. and D.J. Randall (eds.)]. Academic Press, New York, NY, pp. 1-98.

Gable, F.J., D.G. Aubrey, and J.H. Gentile, 1991: Global environmental change issues in the western Indian Ocean region. *Geoforum*, **22**, 401-419.

GESAMP, 1986: *Environmental Capacity an Approach to Marine Pollution Prevention*. GESAMP Reports and Studies No. 30, UNEP, Nairobi, Kenya, 50 pp.

GESAMP, 1990: *The State of the Marine Environment*. GESAMP Reports and Studies No. 39, UNEP, Nairobi, Kenya, 111 pp.

GESAMP, 1991: *Global Strategies for Marine Environmental Protection*. GESAMP Reports and Studies No. 45, UNEP, Nairobi, Kenya, 34 pp.

Gleason, D.F. and G.M. Wellington, 1993: Ultraviolet radiation and coral bleaching. *Nature*, **365**, 836-838.

Gloersen, P., W.J. Campbell, D.J. Cavallieri, J.C. Comiso, C.L. Parkinson, and H.J. Zwally, 1992: *Arctic and Antarctic Sea Ice, 1978-1987: Satellite Passive Microwave Observations and Analysis*. SP-511, NASA, Washington, DC, 290 pp.

Glynn, P.W., 1989: Coral mortality and disturbances to coral reefs in the tropical eastern Pacific. In: *Global Ecological Consequences of the 1982-83 El Niño-Southern Oscillation* [Glynn, P.W. (ed.)]. Elsevier, New York, NY, pp. 55-126.

Glynn, P.W. and L.D. Croz, 1990: Experimental evidence for high temperature stress as the cause of El Niño-coincident coral mortality. *Coral Reefs*, **8**, 181-191.

Gordon, A.L., 1991: The Southern Ocean—its involvement in global change. In: *Proc. Conf. on the Role of the Polar Regions in Global Change,* vol. 1 [Weller, G. *et al.* (eds.)]. University of Alaska, Fairbanks, AK, pp. 249-25.

Gordon, A.L., S.E. Zebiak, and K. Bryan, 1992: Climate variability and the Atlantic Ocean. *EOS*, **73**, 161, 164-165.

Gregory, J.M., 1993: Sea level changes under increasing atmospheric CO_2 in a transient coupled ocean-atmosphere GCM experiment. *J. Climate*, **6**, 22-48.

GRIP Project Members, 1993: Climate instability during the last interglacial period recorded in the GRIP ice core. *Nature*, **364**, 203-208.

Haarsma, R.J., J.F.B. Mitchell, and C.A. Senior, 1993: Tropical disturbances in a GCM. *Clim. Dynamics*, **8**, 247-258.

Haas, P.M., R.O. Keohane, and M.A. Levy (eds.), 1993: *Institutions for the Earth*. MIT Press, Cambridge, MA, and London, UK, 426 pp.

Hoagland, P. III and H.L. Kite-Powell, 1991: European advanced maritime electronic instrumentation. *Mar. Policy*, **11**, 431-454.

Honjo, T., 1992: Harmful red tides of Heterosigma akashiwo. In: *Control of Disease in Aquaculture*. Ise, Mie Prefecture, Japan, 29-30 October 1990, NOAA Technical Report NMFS 111, U.S. Department of Commerce, Washington, DC.

IMO, 1991: Marine Environment Protection Committee, MEPC.

Insuka, S., 1985: Results of survey of maximum chlorophyll-a concentrations in coastal waters of Japan. *Bull. Plankton. Soc. Japan*, **32**, 171-178.

IPCC, 1990a: *Climate Change: The IPCC Scientific Assessment* [Houghton, J.T., G.J. Jenkins, J.J. Ephraums (eds.)]. Cambridge University Press, Cambridge, UK, 364 pp.

IPCC, 1990b: *Climate Change: The IPCC Impacts Assessment* [McGTegart, W.J., G.W. Sheldon, D.C. Griffiths (eds.)]. Australian Government Publishing Service, Canberra, Australia, 268 pp.

IPCC, 1994: *Climate Change 1994: Radiative Forcing of Climate* [Houghton, J.T. *et al.* (eds.)]. Cambridge University Press, Cambridge, UK, 339 pp.

Ittekkot, V., 1993: The abiotically driven biological pump in the ocean and short-term fluctuations in atmospheric CO_2 contents. *Global Planet. Change*, **8**, 17-25.

Ittekkot, V., R.R. Nair, S. Honjo, V. Ramaswamy, M. Bartsch, S. Manganini, and B.N. Desai, 1991: Enhanced particle fluxes in Bay of Bengal induced by injection of fresh water. *Nature*, **351**, 385-388.

Izrael, Yu. A. and A.V. Tsyban, 1983: On the assimilative capacity of the World Ocean. Doklady AN SSSR, Vol. 272, N3 (in Russian).

Izrael, Yu. A. and A.V. Tsyban, 1989: *Anthropogenic Ecology of the Ocean*. Gidrometeoizdat, Leningrad, 528 pp. (in Russian).

Izrael, Yu. A., A.V. Tsyban, T.E. Whitledge, C.P. McRoy, and V.V. Shigaev, 1992: Polar marine ecosystems and climate. In: *Results of the Third Joint US-USSR Bering and Chukchi Seas Expedition (BERPAC), Summer 1988* [Nagel, P.A. (ed.)]. U.S. Fish and Wildlife Service, Washington, DC, pp. 7-11.

Jannasch, H.W. and C.O. Wirsen, 1979: Chemosynthetic primary production at East Pacific sea floor spreading centers. *BioScience*, **29**, 592-598.

Knutson, T.R. and S. Manabe, 1994: Impact of increased CO_2 on simulated ENSO-like phenomena. *Geophys. Res. Let.*, **21**, 2295-2298.

Kristjansson, J.K., 1989: Thermophilic organisms as sources of thermostable enzymes. *Trends in Biotechnology*, **7**, 349-353.

Kumar, A., A. Leetmaa, and M. Ji., 1994: Simulations of atmospheric variability induced by sea surface temperatures and implications for global warming. *Science*, **266**, 632-634.

Lapenis, A.G., N.S. Oskina, M.S. Barash, N.S. Blyum, and Ye. V. Vasileva, 1990: The late Quaternary changes in ocean biota productivity. *Okeanologiya/Oceanology*, **30**, 93-101 (in Russian).

Lee, K.V., 1993: Compass and gyroscope: integrating science and politics for the environment. Island Press, Washington, DC, 243 pp.

Lehman, S., 1993: Ice sheets, wayward winds and sea change. *Nature*, **365**, 108-110.

Lloyds Register, 1990: *Statistical Tables*. Lloyd's, London, UK, pp. 3-4, 7-20.

Maier-Reimer, E., U. Mikolajewicz, and K. Hasselmann, 1993: Mean circulation of the Hamburg LSG OGCM and its sensitivity to the thermohaline surface forcing. *J. Phys. Oceanogr.*, **23**, 731-758.

Manabe, S. and R.J. Stouffer, 1993: Century-scale effects of increased atmospheric CO_2 on the ocean-atmosphere system. *Nature*, **364**, 215-218.

Manabe, S., R.J. Stouffer, M.J. Spelman, and K. Bryan, 1991: Transient response of a coupled ocean-atmosphere model for gradual changes of atmospheric CO_2. Part I: Annual response. *J. Clim.*, **4**, 785-818.

Markham, A., N. Dudley, and S. Stolten, 1993: *Some Like It Hot*. WWF International, Gland, Switzerland, 144 pp.

Martin, J.H. and S. Fitzwater, 1987: Iron deficiency limits phytoplankton in the northeast Pacific subarctic. *Nature*, **331**, 341-343.

Mathews, W.J. and E.G. Zimmerman, 1990: Potential effects of global warming on native fishes of the southern great plains and the southwest. *Fisheries*, **15**, 26-32.

Maul, G.A. (ed.), 1993: *Climatic Change in the Intra-Americas Sea*. UNEP/IOC, E. Arnold, London, UK, 384 pp.

McMinn, A., H. Heijnis, and D. Hodgson, 1994: Minimal effects of UV-B radiation on Antarctic diatoms over the past 20 years. *Nature*, **370**, 547-549.

Mee, L.D., 1992: The Black Sea in crisis: a need for concerted international action. *Ambio*, **21**, 278-286.

Mikolajewicz, U., B.D. Santer, and E. Maier-Reimer, 1990: Ocean response to greenhouse warming. *Nature*, **345**, 589-593.

Miles, E.L., 1989: Concepts, approaches, and applications in sea use planning and management. *Ocean Development and International Law*, **20**, 213-238.

Miles, E.L., 1992: Future challenges in ocean management: towards integrated national ocean policy. In: *Ocean Management in Global Change* [Fabbri, P. (ed.)]. Elsevier, London, UK, pp. 595-620.

Miller, M.L. and J. Auyong, 1991: Coastal zone tourism. *Mar. Policy*, **16**, 75-99.

Milliman, J.D., 1993: Coral reefs and their response to global climate change. In: *Climatic Change in the Intra-Americas Sea* [Maul, G.A. (ed.)]. UNEP/IOC, E. Arnold, London, UK, pp. 306-322

Mills, E.L., J.H. Leach, J.T. Carlton, and C.L. Secor, 1993: Exotic species in the Great Lakes: a history of biotic crises and anthropogenic introductions. *J. Great Lakes Res.*, **10**, 1-54.

Mitchell, J.F.B., 1988: Local effects of greenhouse gases. *Nature*, **332**, 399-400.

Montevecchi, W.A. and R.A. Myers, 1995: Prey harvests of seabirds reflect pelagic fish and squid abundance on multiple spatial and temporal scales. *Science*, **117**, 1-9.

MTSJ, 1993: *Marine Technology Society Journal Special Issue*, **27**, 10-77.

Murray, J.W., R.T. Barber, M.R. Roman, and R.A. Feely, 1994: Physical and biological controls on carbon cycling in the equatorial Pacific. *Science*, **266**, 58-65.

Murray, J.W., H.W. Jannasch, S. Honjo, R.F. Anderson, W.S. Reeburgh, Z. Top, G.E. Friedrich, L.A. Codispoti, and E. Izdar, 1989: Unexpected changes in the oxic/anoxic interface in the Black Sea. *Nature*, **338**, 411-413.

Myers, F.S. and A. Anderson, 1992: Microbes from 20,000 feet under the sea. *Science*, **255**, 28-29.

Mysak, L.A. and D.K. Manak, 1989: Arctic sea ice extent and anomalies, 1953-84. *Atmos. Ocean*, **27**, 376-405.

Mysak, L.A., D. Manak, and R. Marden, 1990: Sea ice anomalies observed in Greenland and Labrador seas during 1901-1984 and their relation to an interdecadal Arctic climate cycle. *Clim. Dynamics*, **5**, 111-133.

Nicholls, N., 1984: The Southern Oscillation, sea-surface temperature, and interannual fluctuations in Australian tropical cyclone activity. *J. Climate*, **4**, 661-670.

Nicholls, N., 1992: Historical El Niño/Southern Oscillation variability in the Australasian region. In: *El Niño: Historical and Paleoclimatic Aspects of the Southern Oscillation* [Diaz, H.F. and V. Markgraf (eds.)]. Cambridge University Press, Cambridge, UK, pp. 151-173.

Oeschger, H., J. Beer, U. Siegenthaler, B. Stauffer, W. Dansgaard, and C.C. Langway, 1984: Late glacial climate history from ice cores. In: *Climate Processes and Climate Sensitivity* [Hansen, J.E. and T. Takahashi (eds.)]. Geophysical Monograph 29, American Geophysical Union, Washington, DC, pp. 299-306.

Offshore Journal, 1993: *Statistical Tables*, **53**, 33-34.

Oil and Gas Journal Data Base, 1992: *Statistical Tables*.

Oil and Gas Journal, 1993: *Statistical Tables*, **91**, 37-115.

Ostrom, E., 1990: *Governing the Commons*. Cambridge University Press, Cambridge, UK, 280 pp.

Paillard, D., M. Ghil, and H. le Treut, 1993: Dissolved organic matter and glacial interglacial p CO_2. *Global Biogeochem. Cycles*, **7**, 901-914.

Parkinson, C.L., 1992: Spatial patterns of increases and decreases in the length of the sea ice season in the north polar region, 1979-1986. *J. Geophys. Res.*, **97(C9)**, 14,377-388.

Pauly, S. and V. Christiansen, 1995: Primary production required to sustain global fisheries. *Nature*, **374**, 255-257.

Polovina, J.J., G.T. Mitchum, N.E. Graham, M.P. Craig, E.E. DeMartini, and E.N. Flint, 1994: Physical and biological consequences of a climate event in the central North Pacific. *Fisheries Oceanography*, **3**, 15-21.

Putnam, R.D., 1993: *Making Democracy Work: Civic Traditions in Modern Italy*. Princeton University Press, Princeton, NJ, 258 pp.

Riebesell, U., D.A. Wolf-Gladrow, and V. Smetacek, 1993: Carbon dioxide limitation of marine phytoplankton growth rates. *Nature*, **361**, 249-251.

Roots, E.F., 1989: Climate change: high-latitude regions. *Clim. Change*, **15**, 223-253.

Ropelewski, C.F. and M.S. Halpert, 1987: Global and regional scale precipitation patterns associated with the El Niño/Southern Oscillation. *Mon. Wea. Rev.*, **115**, 1606-1626.

Ryan, B.F., I.G. Watterson, and J.L. Evans, 1992: Tropical cyclone frequencies inferred from Gray's yearly genesis parameter: validation of GCM tropical climates. *Geophys. Res. Let.*, **19**, 1831-1834.

Sarnthein, M., J.C. Duplessy, and M. Fontugne, 1988: Ocean productivity in low and mid latitudes: influences on CO_2 reservoirs of the deep ocean and atmosphere during the last 21,000 years. *Paleoceanography*, **3**, 361-399.

Schmitt, R.W., P.S. Bogden, and C.E. Dorman, 1989: Evaporation minus precipitation and density fluxes in the North Atlantic. *J. Phys. Oceanogr.*, **19**, 1208-1221.

Schrieber, R.W. and E.A. Schrieber, 1984: Central Pacific seabirds and the El Niño Oscillation. *Science*, **225**, 713-715.

Shen, G.T., 1993: Reconstruction of El Niño history from reef corals. *Bull. Inst. fr. etudes andines*, **122**, 125-158.

Sherman, K., 1994: Sustainability, biomass yields, and health of coastal ecosystems: an ecological perspective. *Mar. Ecol. Progress. Ser.*, **112**, 227-301.

Sibley, T.H. and R.M. Strickland, 1985: Fisheries: some relationships to climate change and marine environmental factors. In: *Characterization of Information Requirements for Studies of* CO_2 *Effects* [White, M.R. (ed.)]. DOE/ER-0236, U.S. Department of Energy, Washington, DC.

Siegenthaler, U. and J.L. Sarmiento, 1993: Atmospheric carbon dioxide and the ocean. *Nature*, **365**, 119-125.

Smith, R.C., B.B. Prezelin, K.S. Baker, R.R. Bidigare, N.P. Boucher, T. Coley, D. Karentz, S. MacIntyre, H.A. Matlick, D. Menzies, M. Onderusek, Z. Wan, and K.J. Waters, 1992: Ozone depletion: ultraviolet radiation and phytoplankton biology in Antarctic waters. *Science*, **255**, 952-959.

Smith, S.V. and F.T. Mackenzie, 1991: Comments on the role of oceanic biota as a sink for anthropogenic CO_2 emissions. *Global Biogeochem. Cycles*, **5**, 189-190.

Stein, O. and A. Hense, 1994: A reconstructed time series of the number of extreme low pressure events since 1880. Meteorol. Zeitschrift, N.F. 3. Jg., H. 1, 43-46.

Stetter, K.O., R. Huber, E. Blöchl, M. Kurr, R.D. Eden, M. Fiedler, H. Cash, and I. Vance, 1993: Hyperthermophilic archaea are thriving in deep North Sea and Alaskan oil reservoirs. *Nature*, **365**, 743-745.

Stouffer, R.J., S. Manabe, and K. Bryan, 1989: Interhemispheric asymmetry in climate response to a gradual increase of atmospheric CO_2. *Nature*, **342**, 660-662.

Sundquist, E.T. and W.S. Broecker, 1985: *The Carbon Cycle and Atmospheric* CO_2*: Natural Variations Archean to Present*. Geophysical Monograph 32, American Geophysical Union, Washington, DC, 627 pp.

Swannell, R.P.J. and I.M. Head, 1994: Bioremediation comes of age. *Nature*, **368**, 396-398.

Tanabe, S., 1985: Distribution, behaviour and fate of PCBs in the marine environment. *J. Ocean. Soc. Jap.*, **42**, 358-370.

Tans, P.P., I.Y. Fung, and T. Takahashi, 1990: Observational constraints on the global atmospheric CO_2 budget. *Science*, **247**, 1431-1438.

Taylor, K.C., G.W. Lamorey, G.A. Doyle, R.B. Alley, P.M. Grootes, P.A. Mayewski, J.W.C. White, and L.K. Barlow, 1992: The "flickering switch" of late Pleistocene climate change. *Nature*, **361**, 432-436.

Thompson, L.G., E. Mosley-Thompson, and B.M. Arnao, 1984: El Niño-Southern Oscillation events recorded in the stratigraphy of the tropical Quelccaya ice cap. *Peru. Science*, **226**, 50-53.

Trenberth, K.E. and A. Solomon, 1994: The global heat balance: heat transports in the atmosphere and ocean. *Clim. Dynamics*, **10**, 107-134.

Tsunogai, S., T. Ono, and S. Watanabe, 1993: Increase in total carbonate in the western North Pacific water and a hypothesis on the missing sink of anthropogenic carbon. *J. Oceanogr.*, **49**, 305-315.

Tsyban, A., J. Everett, J. Titus, *et al.*, 1990: World oceans and coastal zones. In: *Climate Change: The IPCC Impacts Assessment,* Report of WG II, Chapter 6 [McGTegart, W.J., G.W. Sheldon, D.C. Griffiths (eds.)]. Australian Government Publishing Service, Canberra, Australia, pp. 6-1–6-27.

Turpin, D.H., 1993: Phytoplankton growth and CO_2. *Nature*, **363**, 678-679.

Underdal, A., 1980: Integrated marine policy: what? why? how? *Mar. Policy*, **7**, 159-169.

Volk, T. and M. Hoffert, 1985: Ocean carbon pumps: analysis of relative strengths and efficiencies in ocean-driven atmospheric changes. In: *The Carbon Cycle and Atmospheric* CO_2*: Natural Variations Archean to Present* [Sundquist, E. and W.S. Broecker (eds.)]. Geophysical Monograph 32, AGU, Washington, DC, pp. 99-110.

Watson, R.T., H. Rodhe, H. Oeschger, and U. Siegenthaler, 1990: Greenhouse gases and aerosols. In: *Climate Change. The IPCC Scientific Assessment* [Houghton, J.T., G.J. Jenkins, and J.J. Ephraums (eds.)]. Cambridge University Press, Cambridge, UK, pp. 1-40.

Wigley, T.M.L. and S.C.B. Raper, 1992: Implications for climate and sea level of revised IPCC emissions scenarios. *Nature*, **357**, 293-300.

Wigley, T.M.L. and S.C.B. Raper, 1993: Future changes in future global mean temperature and sea level. In: *Climate and Sea Level Change: Observations, Projections and Implications* [Warrick, R.A., E.M. Barrow, and T.M.L. Wigley (eds.)]. Cambridge University Press, Cambridge, UK, pp. 111-133.

Wigley, T.M.L., J.K. Angell, and P.D. Jones, 1985: Analysis of the temperature record. In: *Detecting Climatic Effects of Increasing Carbon Dioxide* [MacCracken, M.C. and F.M. Luther (eds.)]. U.S. Department of Energy, Carbon Dioxide Research Division, Washington, DC, pp. 55-90.

Wooster, W.S. and D.L. Fluharty (eds.), 1985: *El Niño North: Niño Effects in the Eastern Subarctic Pacific Ocean*. Washington Sea Grant Program, Seattle, Washington.

Worthington, A. and J.H. Leach, 1992: Concentrations of migrant diving ducks at Point Pelee National Park, Ontario, in response to invasion of Zebra Mussels, Dreissena polymorpha. *Canadian Field-Naturalist*, **106**, 376-380.

Zahn, R., 1994: Core correlations. *Nature*, **371**, 289-290.

Zuta, S., 1986: *The Characteristics of the 1982-83 El Niño off the Pacific Coast of South America*. WMO, Geneva, Switzerland, No. 649.

9

Coastal Zones and Small Islands

LUITZEN BIJLSMA, THE NETHERLANDS

Lead Authors:
C.N. Ehler, USA; R.J.T. Klein, The Netherlands; S.M. Kulshrestha, India; R.F. McLean, Australia; N. Mimura, Japan; R.J. Nicholls, UK; L.A. Nurse, Barbados; H. Pérez Nieto, Venezuela; E.Z. Stakhiv, USA; R.K. Turner, UK; R.A. Warrick, New Zealand

Contributors:
W.N. Adger, UK; Du Bilan, China; B.E. Brown, UK; D.L. Elder, Switzerland; V.M. Gornitz, USA; K. Hofius, Germany; P.M. Holligan, UK; F.M.J. Hoozemans, The Netherlands; D. Hopley, Australia; Y. Hosokawa, Japan; G.A. Maul, USA; K. McInnes, Australia; D. Richardson, UK; S. Subak, UK; M. Sullivan, Australia; L. Vallianos, USA; W.R. White, UK; P.L. Woodworth, UK; Yang Huating, China

CONTENTS

EXECUTIVE SUMMARY

Coastal zones and small islands are characterized by highly diverse ecosystems that are important as a source of food and as habitat for many species. They also support a variety of economic activities—which, in many places, has led to a high rate of population growth and economic development. Many studies indicate that overexploitation of resources, pollution, sediment starvation, and urbanization have:

- Led to a decrease in the resilience of coastal systems to cope with natural climate variability (High Confidence)
- Adversely affected the natural capability of these systems to adapt to changes in climate, sea level, and human activities (High Confidence)
- Led to increased hazard potential for coastal populations, infrastructure, and investment (High Confidence).

As demands on coastal resources continue to increase with a growing population and expanding economic activity, coastal systems continue to face increasing pressures, which often lead to the degradation of these systems. In many parts of the world, for example, coastal wetlands are presently disappearing due to human activities.

Since the IPCC First Assessment Report (1990) and its supplement (1992), the interrelationships between the impacts of climate change and human activities have become better understood. Although the potential impacts of climate change by itself may not always be the largest threat to *natural coastal systems*, in conjunction with other stresses they can become a serious issue for *coastal societies*, particularly in those places where the resilience of natural coastal systems has been reduced. Taking into account the potential impacts of climate change and associated sea-level rise can assist in making future development more sustainable. A proactive approach to enhance resilience and reduce vulnerability would be beneficial to coastal zones and small islands both from an environmental and from an economic perspective. It is also in line with the recommendations of the UN Conference on Environment and Development (UNCED) Agenda 21. Failure to act expeditiously could increase future costs, reduce future options, and lead to irreversible changes.

Important findings by Working Group I that are of relevance to coastal impact assessment include the following:

- Current estimates of global sea-level rise represent a rate that is two to five times higher than what has been experienced over the last 100 years (High Confidence).
- Locally and regionally, the rate, magnitude, and direction of sea-level changes will vary substantially due to changes in ocean conditions and vertical movements of the land (High Confidence).
- It is not possible to say if the intensity, frequency, or locations of cyclone occurrence would change in a warmer world (High Confidence).

Since 1990, there has been a large increase in research effort directed at understanding the biogeophysical effects of climate change and particularly sea-level rise on coastal zones and small islands. Studies have confirmed that low-lying deltaic and barrier coasts and low-elevation reef islands and coral atolls are especially sensitive to a rising sea level, as well as changes in rainfall, storm frequency, and intensity. Impacts could include inundation, flooding, erosion, and saline intrusion. However, it has also been shown that such responses will be highly variable among and within these areas; impacts are likely to be greatest where local environments are already under stress as a result of human activities.

Studies of natural systems have demonstrated, among other things, that:

- The coast is not a passive system but will respond dynamically to sea-level and climate changes (High Confidence).
- A range of coastal responses can be expected, depending on local circumstances and climatic conditions (High Confidence).
- In the past, estuaries and coastal wetlands could often cope with sea-level rise, although usually by migration landward. Human infrastructure, however, has diminished this possibility in many places (High Confidence).
- Survival of salt marshes and mangroves appears likely where the rate of sedimentation will approximate the rate of local sea-level rise (High Confidence).
- Generally, coral reefs have the capacity to keep pace with projected sea-level rise but may suffer from increases in seawater temperature (Medium Confidence).

The assessment of the latest scientific information regarding socioeconomic impacts of climate change on coastal zones and small islands is derived primarily from vulnerability assessments based on the IPCC Common Methodology. Since 1990, many national case studies have been completed, embracing examples of small islands, deltas, and continental shorelines from around the world. These studies mainly utilize a scenario of a 1-m rise in sea level and generally assume the present socioeconomic situation, with little or no consideration of coastal dynamics. There is concern that these studies understate nonmarket values and stress a protection-orientated response perspective. Despite these limitations, these studies provide some important insights into the socioeconomic implications of sea-level rise, including:

- Sea-level rise would have negative impacts on a number of sectors, including tourism, freshwater supply and quality, fisheries and aquaculture, agriculture, human settlements, financial services, and human health (High Confidence).
- Based on first-order estimates of population distribution, storm-surge probabilities, and existing levels of protection, more than 40 million people are estimated to experience flooding due to storm surge in an average year under present climate and sea-level conditions. Most of these people reside in the developing world. Ignoring possible adaptation and likely population growth, these numbers could roughly double or triple due to sea-level rise in the next century (Medium Confidence).
- Protection of many low-lying island states (e.g., the Marshall Islands, the Maldives) and nations with large deltaic areas (e.g., Bangladesh, Nigeria, Egypt, China) is likely to be very costly (High Confidence).
- Adaptation to sea-level rise and climate change will involve important tradeoffs, which could include environmental, economic, social, and cultural values (High Confidence).

Until recently, the assessment of possible response strategies focused mainly on protection. There is a need to identify better the full range of options within the adaptive response strategies: protect, accommodate, and (planned) retreat. Identifying the most appropriate options and their relative costs, and implementing these options while taking into account contemporary conditions as well as future problems such as climate change and sea-level rise, will be a great challenge in both developing and industrialized countries. It is envisaged that the most suitable range of options will vary among and within countries. An appropriate mechanism for coastal planning under these varying conditions is integrated coastal zone management. There is no single recipe for integrated coastal zone management; rather, it constitutes a portfolio of sociocultural dimensions and structural, legal, financial, economic, and institutional measures.

Integrated coastal zone management, which has already started in many coastal countries, is a continuous and evolutionary process that identifies and implements options to attain sustainable development and adaptation to climate change in coastal zones and small islands. Constraints that could hinder its successful implementation include, but are not limited to:

- Technology and human resources capability
- Financial limitations
- Cultural and social acceptability
- Political and legal frameworks.

Continued exchange of information and experience on the inclusion of climate change and sea-level rise within integrated coastal zone management at local, regional, and international levels would help to overcome some of these constraints. In addition, more research is required on the *process* of integrated coastal zone management to improve the understanding and modeling capability of the implications of climate change and sea-level rise on coastal zones and small islands, including biogeophysical effects, the local interaction of sea-level rise with other aspects of climate change, and more complete assessment of socioeconomic and cultural impacts.

9.1. Introduction

Coastal zones and small islands contain some of the world's most diverse and productive resources. They include extensive areas of complex and specialized ecosystems, such as mangroves, coral reefs, and seagrasses, which are highly sensitive to human intervention. These ecosystems are the source of a significant proportion of global food production. Moreover, they support a variety of economic activities, including fisheries and aquaculture, tourism, recreation, and transportation. In recent decades, many coastal areas have been heavily modified and intensively developed, which has significantly increased their vulnerability to natural coastal dynamics and the anticipated impacts of global climate change.

Many attempts have been made to define the "coastal zone" and its land and seaward boundaries. Some definitions are based on physiographic characteristics, such as the extent of tidal influence on the land or the geomorphology of the continental shelf; others simply use a fixed distance from the shoreline. In the case of small islands, the coastal zone could include the entire island. While the boundaries of the coastal zone may or may not coincide with political or administrative boundaries, they rarely coincide with those of areas from which demands on the resources of the coastal area are derived (Ehler, 1993).

Irrespective of how coastal zones are defined, the following characteristics, which are strongly interrelated, often make them distinctive from other areas:

- A high rate of dynamic changes in the natural environment
- A high biological productivity and diversity
- A high rate of human population growth and economic development
- A high rate of degradation of natural resources
- Exposure to natural hazards such as cyclones and severe storms
- The need for management regimes that address both terrestrial and marine issues.

The global importance of coastal zones and small islands in terms of both ecological and socioeconomic values is widely recognized. Many international organizations, including IPCC, have called for action to implement strategies toward better planning and management of coastal areas and resources to prevent them from being degraded and becoming progressively more vulnerable to the potential impacts of climate change and associated sea-level rise. The World Coast Conference, partly held under the auspices of IPCC, has supported the Framework Convention on Climate Change (UNCED, 1992a) and Agenda 21, Chapter 17 (UNCED, 1992b), requiring action on coastal zone management planning, among many other initiatives, by all signatories (WCC'93, 1993).

This chapter presents an assessment of the latest scientific information on the impacts of climate change on coastal zones and small islands and on strategies that countries may wish to apply in response to these impacts. It builds on the previous IPCC assessments carried out in 1990 and 1992 and on the work of the former Coastal Zone Management Subgroup (CZMS) of IPCC. The chapter concentrates on the scientific work completed since 1990, although earlier work is acknowledged where appropriate.

The structure of this chapter is schematically depicted in Figure 9-1. This figure also shows how this chapter relates to other chapters and Working Group reports of this IPCC Second Assessment Report. In Section 9.2, the functions and values of coastal zones and small islands are discussed. Special emphasis is put on the importance of maintaining the proper functioning of coastal systems with regard to sustainable development and their resilience to climate change. Section 9.3 then discusses the likely consequences of climate change on sea-level rise, sea-surface temperatures, and tropical cyclones in the context of coastal zones and small islands. On this basis, Section 9.4 addresses the effects of climate change on biogeophysical systems, and Section 9.5 examines the socioeconomic impacts. Section 9.6 considers response strategies to climate change and the need to integrate them with other coastal management activities. Finally, Section 9.7 identifies needs and opportunities for future research and monitoring that would help nations to respond more appropriately to the likely impacts of climate change on coastal zones and small islands.

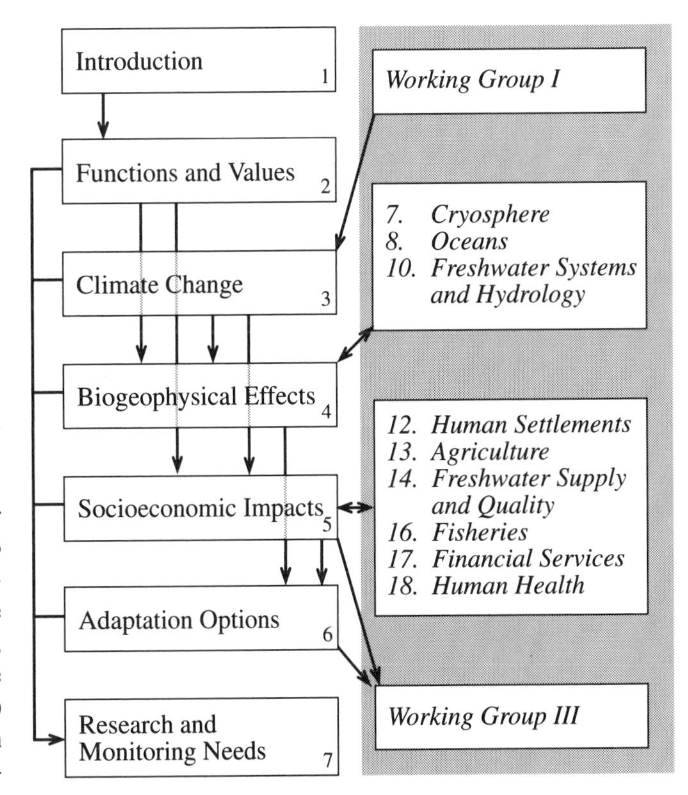

Figure 9-1: Structure of this chapter and interrelationships with other chapters and Working Group volumes of the IPCC Second Assessment Report.

9.2. Functions and Values of Coastal Zones and Small Islands

In large measure, human social and economic well-being depends directly or indirectly on the availability of environmental goods and services provided by marine and coastal systems. Coastal zones and small islands are characterized by highly diverse ecosystems, and a great number of functions are performed over a relatively small area. This concentration of functions, together with their spatial location, makes coastal zones and small islands highly attractive areas for people to live and work in. Coastal human populations in many countries have been growing at double the national rate due to migration to urban coastal centers; it is estimated that 50–70% of the global human population lives in the coastal zone, although there are great variations among countries. The existing rate of socioeconomic development in coastal zones also is unprecedented (WCC'93, 1994).

9.2.1. *Functions and Values*

The total economic value of coastal systems is more than simply the financial value of the coastal resources that they produce. It also includes their role in regulating the environment, their satisfaction of subsistence needs, and their satisfaction of human intellectual and emotional needs. De Groot (1992a) and Vellinga *et al.* (1994) have categorized the environmental functions performed by natural systems as regulation functions, user and production functions, and information functions. *Regulation functions* are crucial in safeguarding environmental quality. They include regulation of erosion and sedimentation patterns, regulation of the chemical composition of the atmosphere and oceans, flood prevention, waste assimilation, maintenance of migration and nursery habitats, and maintenance of biological diversity. Many natural and seminatural coastal systems thus play a fundamental part in the regulation of essential biospheric processes that contribute to the maintenance of a healthy environment and the long-term stability of the biosphere, including the climate system. *User and production functions* are essential in providing many living and nonliving resources that are utilized by human society. These functions include the provision of space and a suitable substrate for human habitation and a variety of socioeconomic activities. Important socioeconomic activities in coastal zones and small islands include tourism and recreation, exploitation of living and nonliving resources (e.g., fisheries and aquaculture; agriculture; extraction of water, oil, and gas), industry and commerce, infrastructure development (e.g., harbors, ports, bridges, roads, sea-defense works), and nature conservation. *Information functions* relate to the part that nature plays in meeting human intellectual and emotional needs. For example, coastal systems can be a source for cultural inspiration, but they also serve as a storehouse for genetic information. Also, scientific understanding of coastal processes and evolution depends on the geological and biological information that coastal systems contain.

Any coastal system can yield values related to the direct, indirect, and future use of the functions described above, as well as non-use, or intrinsic, values (Turner, 1988, 1991; Barbier, 1994). This is illustrated by Figure 9-2. Empirical

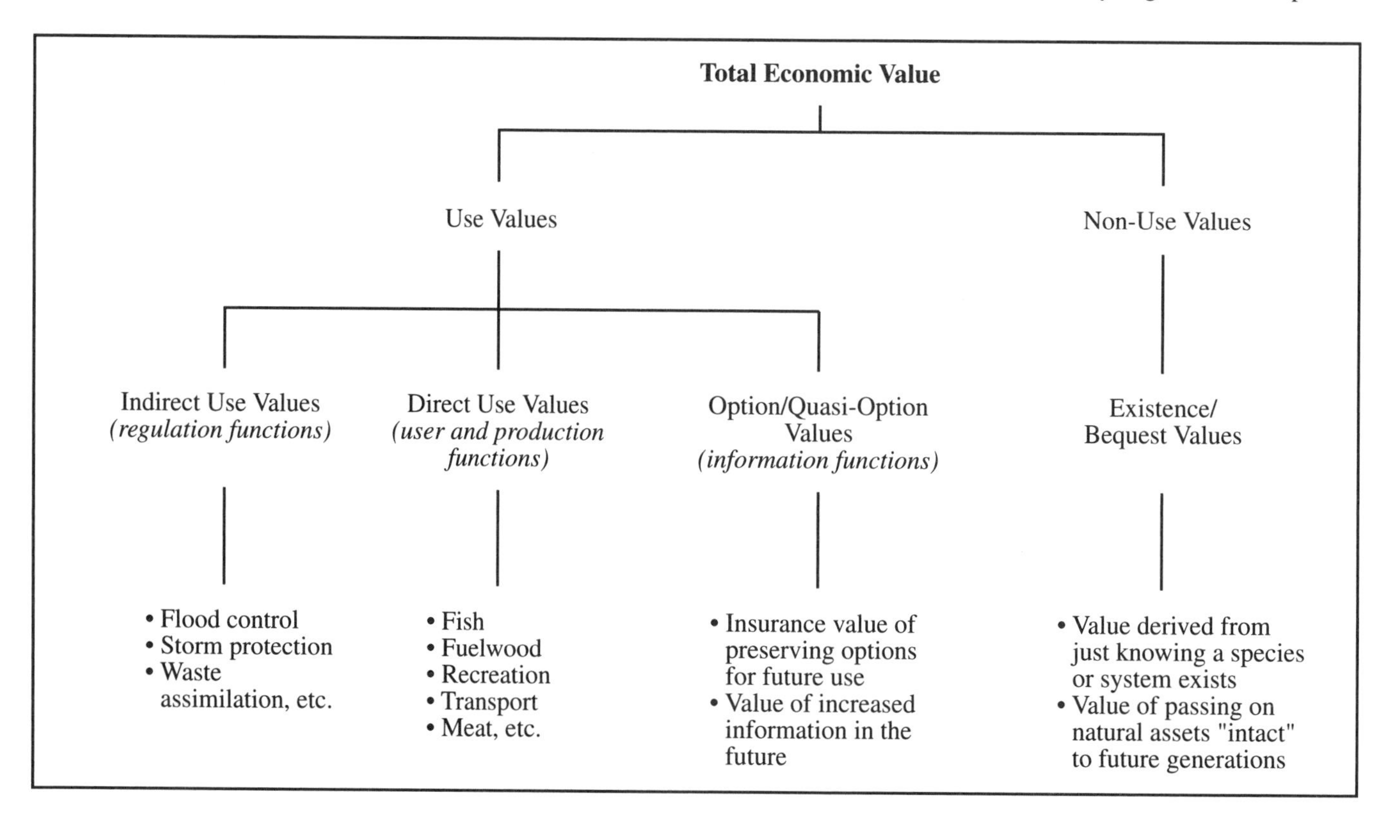

Figure 9-2: Values of coastal systems (adapted from Barbier, 1989).

studies confirm that coastal systems possess significant economic value, in terms of both use and non-use outputs. Mangrove forests, for example, have been shown to sustain more than 70 direct human activities, ranging from fuelwood collection to artisinal fisheries (Dixon, 1989). However, some uses preclude others, so some caution is necessary when the total economic value of such a system is estimated. This caution is also justified by the fact that many functions provide nonmarket goods and services and therefore do not carry appropriate market prices and value. Moreover, individuals and communities may find that nature has a value of its own, independent of human use or perception. They would therefore value nature purely because it exists and would feel a loss if it was damaged or destroyed.

Nonetheless, some studies are available that have produced estimates of the monetary value of coastal functions, using a variety of techniques. For example, recreation and amenity benefits provided by the Broadlands coastal wetland in England are estimated to be around $5 million per year (Bateman *et al.*, 1995). The storm buffering function of the Terrebonne coastal wetlands in Louisiana (United States) has been valued in terms of storm damage avoided. If the wetlands continue to recede at their present rate (3–4 m/yr), the present discounted value of increased expected property damage lies between $2.1–3.1 million (Constanza *et al.*, 1989). The total monetary value of all functions performed in the Ecuadorian Galápagos National Park amounts to $138 million per year (De Groot, 1992a).

9.2.2. *Sustainable Development*

In most coastal nations, a considerable part of gross national product (GNP) is derived from activities that are directly or indirectly connected with coastal zones. Therefore, maintaining the proper functioning of marine and coastal systems is of significant concern to a country's economy. In the shorter term, however, large financial benefits may be available at the expense of longer-term sustainability. Many coastal problems that are currently being encountered worldwide can be attributed to the unsustainable use and unrestricted development of coastal areas and resources. These problems include the accumulation of contaminants in coastal areas, erosion, and the rapid decline of habitats and natural resources. Great care must therefore be taken in planning to avoid overdevelopment and degrading or destroying the very environment that attracted coastal development in the first place.

Nonetheless, increasingly large areas of coasts and small islands have been managed with the aim of maximizing the financial "resource take" provided by the user and production functions performed by coastal systems. However, when one particular function is overexploited, this may not only result in the depletion of the resources that this function provides; it can also reduce the performance of other functions below their full potential. For example, when mangroves are logged and cleared in an unsustainable way, it can be at the expense of functions that enable fish to breed and be caught in the same area or that protect the coast from being eroded. The same applies when the capacity of regulation functions is exceeded—for example, when coastal waters are polluted beyond their waste-assimilative capacity. The maintenance of all functions at a sustainable level would provide higher economic returns over a longer period of time.

One can state that sustainable development in coastal zones and small islands is realized only when it enables the coastal system to self-organize—that is, to perform all its potential functions without adversely affecting other natural or human systems. Climate change could pose an additional threat to the full performance of these functions, compounding the pressures that present-day development activities already place on coastal zone capacities. Overexploitation of resources, pollution, sediment starvation, and urbanization may inhibit or destroy the working of functions that are essential to the provision of goods and services that are difficult to value in monetary terms, or in maintaining the resilience of coastal ecosystems to external stresses, such as climate change. For instance, one important regulation function of natural systems in coastal zones and small islands is the provision of a buffering capacity, protecting the land against the dynamics of the sea. As climate changes and sea level rises, this function will become even more important than it is today, preventing coastal areas from being eroded and inundated as much as they would without this natural protection. Nevertheless, although present-day human activities often result in large financial payoffs in the short term, they may also lead to environmental degradation. Such degradation results in the loss of functions, may increase coastal vulnerability to climate change, and has adverse economic effects on tourism, fishing, and other aspects of the coastal economy. In some cases, the long-term costs of these environmental disruptions may be greater than the long-term benefits of the human activity that caused them.

9.3. Aspects of Climate Change of Concern to Coastal Zones and Small Islands

Sea-level rise and possible changes in the frequency and/or intensity of extreme events, such as temperature and precipitation extremes, cyclones, and storm surges, constitute the components of climate change that are of most concern to coastal zones and small islands. Short-term extreme events are superimposed on long-term changes in CO_2 concentrations, climate, and sea level, and all aspects work in concert to bring about environmental change at regional and local levels. In order to understand fully their interactive effects on the coast, it is paramount to know how the means, variability, and extremes of the range of relevant climatic elements will change at local and regional scales. At such scales—with possibly a few exceptions—the predictive capability of models is currently very low, and knowledge about possible future changes in variability and extremes is meager. This section summarizes the main findings regarding sea-level and climate change that pertain to coastal zones (see the IPCC Working Group I volume for a full discussion).

9.3.1. Sea-Level Rise

9.3.1.1. Global Projections

In 1990, IPCC provided a best estimate and a range of uncertainty of sea-level rise, based on a "business-as-usual" projection of greenhouse-gas emissions for the period 1990–2100. It was estimated that sea level would, on average, rise by about 6 mm/yr, within a range of uncertainty of 3–10 mm/yr (Warrick and Oerlemans, 1990). Subsequent to IPCC90, projections of sea-level rise have been lower, largely as a result of downward revisions in the rate of global warming, which drives sea-level rise (see Section 9.3.2).

Present estimates of sea-level rise have been presented in Chapter 7, *Changes in Sea Level*, of the IPCC Working Group I volume. For a forcing scenario (IS92a) comparable to that of the IPCC 1990 assessment, it is estimated that sea level would, on average, rise by about 5 mm/yr, within a range of uncertainty of 2–9 mm/yr (see Figure 9-3). An important point to bear in mind is that the current best estimates represent a rate of sea-level rise that is about two to five times the rate experienced over the last 100 years (i.e., 1.0–2.5 mm/yr). The current projections of sea-level rise should therefore be of major concern in the context of coastal zones and small islands. Furthermore, model projections show that sea level will continue to rise beyond the year 2100 due to lags in climate response, even with assumed stabilization of global greenhouse-gas emissions (Wigley, 1995).

9.3.1.2. Regional Implications

One cannot assume that changes in sea level at regional and local levels will necessarily be the same as the global-average change, for two broad reasons. First, vertical land movements affect sea level. With respect to the coastal environment, it is relative sea level that is most important—that is, the level of the sea in relation to that of the land. Regionally and locally, vertical land movements can be quite large, even on the decadal time scale. For example, parts of Scandinavia experience uplift (and thus a relative sea-level decline) of about 1 m per century due to the continuing "glacial rebound" following the contraction of the large continental ice sheets at the end of the last Ice Age some 10,000 years ago (Aubrey and Emery, 1993). In contrast, the Mississippi delta is experiencing subsidence (a relative sea-level rise) of about 1 m per century due to consolidation by sediment loading and the diminished supply of additional sediments to the delta required for accretion (Day *et al.*, 1993; Boesch *et al.*, 1994). Locally, tectonic activity, groundwater pumping, and petroleum extraction can cause large and sometimes abrupt changes in relative sea level (Milliman *et al.*, 1989; Han *et al.*, 1995b). Subsidence of urban areas due to groundwater withdrawal has been a significant problem in many locations. In Japan, for example, 2.1 million people live in protected areas below high water due to this cause. Box 9-1 includes four examples of observed sea-level change. Emery and Aubrey (1991) have provided a comprehensive discussion of the

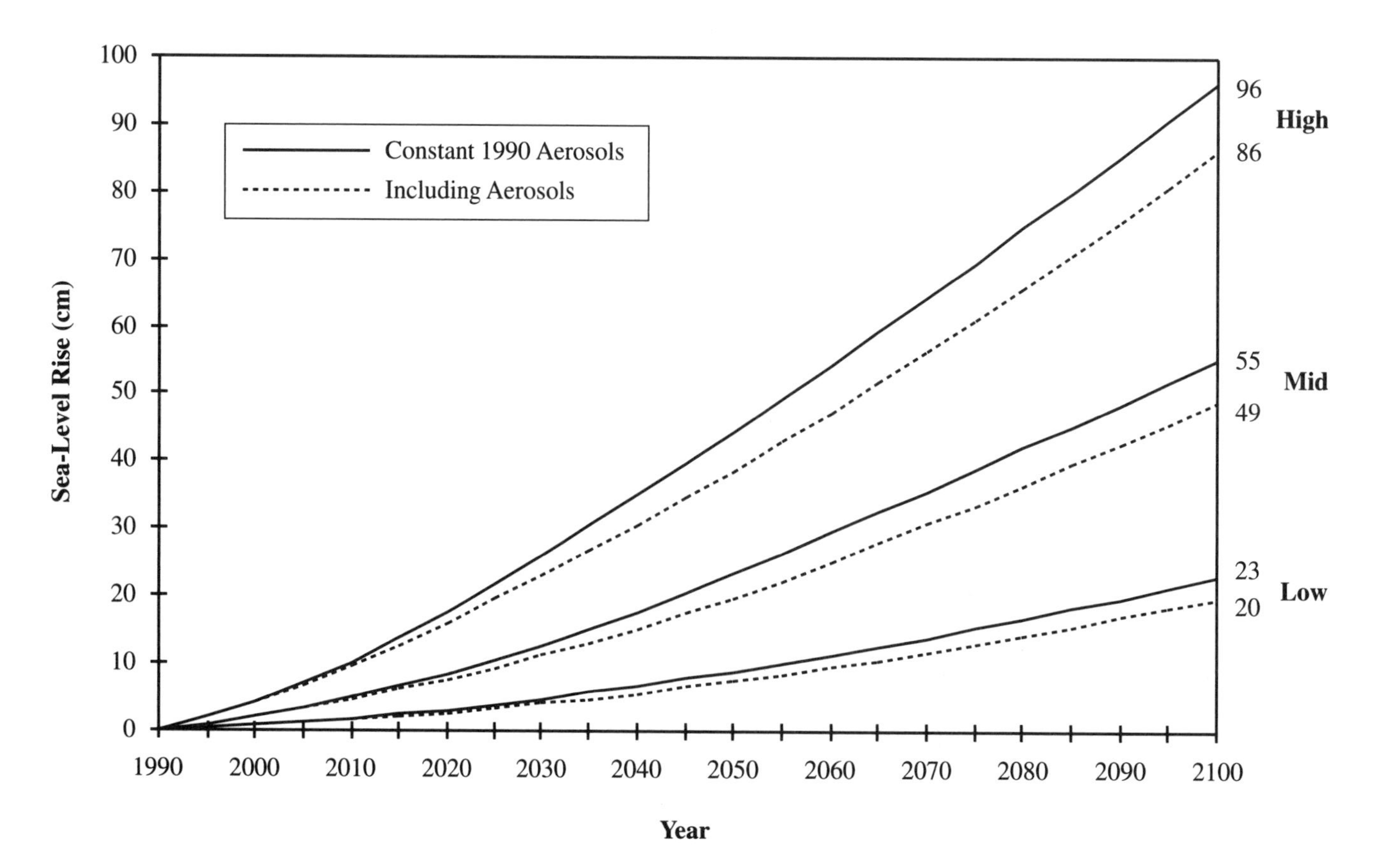

Figure 9-3: 1990–2100 sea-level rise for scenario IS92a (see Chapter 7, *Changes in Sea Level*, of the Working Group I volume).

Box 9-1. Four Examples of Observed Sea-Level Change

Figure 9-4 shows tide-gauge measurements of sea-level changes in Stockholm (Sweden), Fort Phrachula (Bangkok, Thailand), Honolulu (Hawaii, United States), and Nezugaseki (Japan). In a period of rising global sea levels, these four places show different trends in sea-level change due to their being situated in different geological settings. This illustrates the importance of local conditions on relative sea level. In Stockholm, glacial rebound is causing the land to rise out of the sea, resulting in a sea-level fall relative to the land of ~4 mm/yr. In Bangkok, human influence is apparent: Although over the period 1940–1994 relative sea-level rise averages 13.19 ± 0.73 mm/yr, two separate trends can be distinguished; the first, until about 1960, averages ~3 mm/yr, and the second, from 1960 onward, is ~20 mm/yr. The latter trend reflects the increased effort in groundwater pumping since 1960 (Emery and Aubrey, 1991). Nezugaseki shows the dramatic effects of an earthquake, which caused a 15-cm land submergence in 1964, along a coast generally dominated by emergence (Emery and Aubrey, 1991). The Honolulu record is believed to reflect a stable site and it is the only one of the four examples that approximates the global trend.

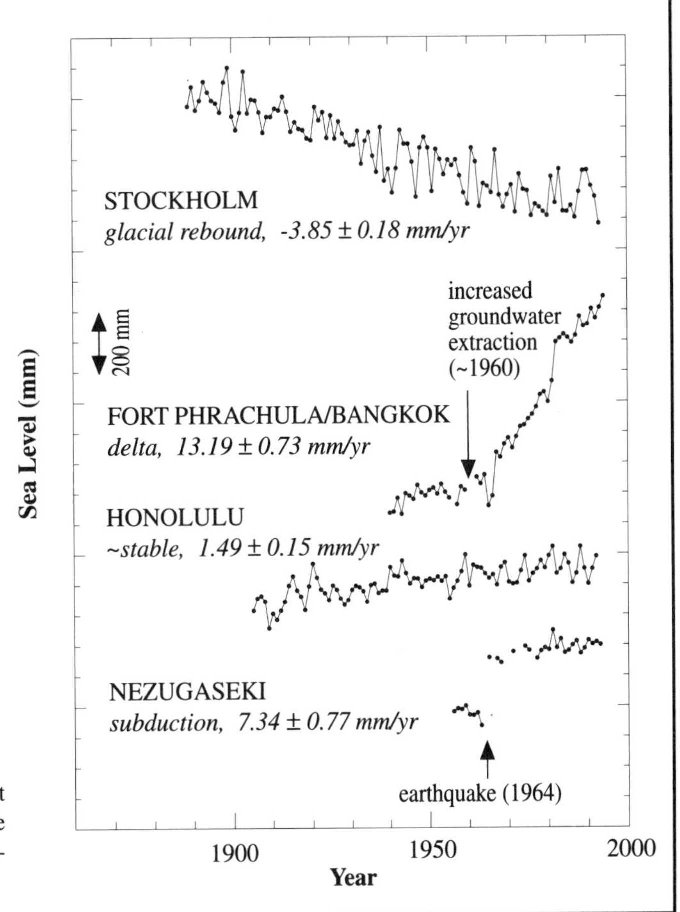

Figure 9-4: Relative sea-level records for Stockholm, Fort Phrachula (Bangkok), Honolulu, and Nezugaseki, indicating the geological settings and observed trends of the four sites (data provided by the Permanent Service for Mean Sea Level, Bidston).

effects of vertical land movements on relative sea level. In assessing coastal impacts, such vertical land movements must be considered in conjunction with climate-related changes in sea level.

Second, there are dynamic effects resulting from oceanic circulation, wind and pressure patterns, and ocean-water density that cause variations in the level of the sea surface with respect to the geoid—that is, the surface that the ocean would describe in the absence of currents, winds, and so forth. These variations have been observed to be on the order of 1 m for major current systems such as the Gulf Stream, Kuroshio, and the Antarctic Circumpolar Current (Pugh, 1987; Levitus, 1982). Climate change may affect ocean circulation, which can result in regional changes of this sea-surface "topography." Recent advances in ocean-atmosphere modeling offer some idea of the possible magnitude of such changes (e.g., Mikolajewicz *et al.*, 1990; Gregory, 1993; Murphy, 1992; Church *et al.*, 1991; Cubasch *et al.*, 1992). Model results show ranges of regional sea-level change that are on the order of two to three times the global-average change (see Box 8-2 in Chapter 8). However, it should be emphasized that confidence in coupled ocean-atmosphere model predictions of regional sea-level changes still must be considered low, as reliable regional scenarios of this effect are not yet available for impact analysis.

9.3.2. *Climate Changes*

One of the certainties regarding global climate change is that the *atmospheric concentration of carbon dioxide* (the major player among the greenhouse gases), now about 360 ppmv, has increased by about 29% over the preindustrial concentration and will continue to increase in the future, even if rather stringent policies on CO_2 emissions were adopted. Although controversial, there is some evidence that increased CO_2 would lead to significant increases in net primary productivity through enhanced photosynthesis (Melillo *et al.*, 1993), which could potentially have far-reaching implications for coastal vegetation and the coastal environment.

Recent projections of *global warming* (Mitchell and Gregory, 1992; Wigley and Raper, 1992; see also Chapter 6, *Climate Models—Projections of Future Climate*, of the IPCC Working Group I volume) are somewhat lower than those presented in IPCC90 (about 0.2°C per decade, as compared to 0.3°C per decade averaged over the next century). Nonetheless, the best-estimate projection still falls within a very large range of uncertainty. Thus, such projections are plausible scenarios of what could occur, not necessarily of what will occur. This caution should be applied even more emphatically to regional details of climate change as they are derived from general circulation models (GCMs). While refinements have been made since

IPCC90, the overall judgment is that confidence in the predicted patterns of climate change at spatial scales relevant to coastal zones and small islands must still be regarded as low (Gates *et al.*, 1992, 1990; Mitchell *et al.*, 1990; McGregor and Walsh, 1993). Nonetheless, there are some broad regional results that are consistently produced by equilibrium and transient GCM experiments that have relevance to the coastal zone. These results include (Mitchell *et al.*, 1990; Gates *et al.*, 1992):

- The mean surface air temperature increases more over land than over oceans (by about a factor of two), and at higher latitudes during winter. The climate of the coastal zone tends to be moderated by ocean temperature and will thus be strongly influenced by sea-surface temperature changes. Northern Hemisphere warming is greater than Southern Hemisphere warming because of the larger extent of land.
- In general, global models show increases in precipitation throughout the year in the high latitudes and during the winter in mid-latitudes. Most models show some increase in Asian monsoon rainfall.
- At scales relevant to coastal zones and small islands, predictions of changes in other climate elements, such as windiness, storminess or radiation, cannot yet be considered reliable.

Although changes in *sea-surface temperatures* (SSTs) are projected to be less than those on land, they are not necessarily less significant. As Edwards (1995) has pointed out, a 2°C change in SST in the tropical and subtropical oceans is considered to be anomalous, but is on the order of temperature changes associated with strong El Niño/Southern Oscillation (ENSO) events. In comparison, the projected change in mean sea-surface temperature for these regions is on the order of 1–2°C by the year 2100. Thus, by 2100, SSTs that are now considered anomalous could well be normal occurrences. Such warming would be unprecedented in the recent geological past.

Tropical cyclones (also known as hurricanes or typhoons, depending on region) affect vast coastal areas in tropical and subtropical countries. Tropical cyclones and associated storm surges can cause enormous loss of life and have devastating impacts on coastal ecosystems and morphology. It is therefore of critical importance to know how the frequency, magnitude, and areal occurrence of such storms will change in a warmer world, if at all. Unfortunately, the evidence from theoretical and numerical models and from observational data is, as yet, inconclusive. While current GCMs provide some indication of possible tropical cyclone formation, identifying "tropical disturbances" (e.g., see Broccoli and Manabe, 1990; Haarsma *et al.*, 1993), they cannot explicitly model such storms at the present grid-scale resolution (Mitchell *et al.*, 1990). However, recent work has been moving in this direction (e.g., Bengtsson *et al.*, 1994). Alternatively, theoretical storm models have been used to examine the maximum storm intensity in relation to SSTs (Emanuel, 1987) but cannot easily be extended to address questions of regional changes in tropical storm intensity or frequency under conditions of climate change (Schlesinger, 1993; Lighthill *et al.*, 1994).

At present, there is no evidence of any systematic shift in *storm tracks*. The tracks are governed by the location of cyclogenesis and prevailing meteorological conditions; thus far, there is no evidence of shift in the preferred locations of cyclogenesis. Empirical studies have found correlations between ENSO and the regional patterns of tropical cyclone activity (as well as the Southeast Asian monsoon, Atlantic hurricanes, Pacific precipitation patterns, and other phenomena) (Nicholls, 1984; Evans and Allan, 1992; see also Chapter 3, *Observed Climate Variability and Change*, of the IPCC Working Group I volume). Progress is being made on model simulations of the present features of ENSO-like events (e.g., Philander *et al.*, 1992), which could lead to a predictive capability for the future. Recent coupled-model simulations under enhanced CO_2 show a tendency toward ENSO-like patterns in the Pacific, with accompanying temperature and precipitation variability (see Chapter 6, *Climate Models—Projections of Future Climate*, of the IPCC Working Group I volume), but realistic simulations of ENSO are not yet possible. The behavior of ENSO is critical in understanding the future coastal effects of both climate change and sea-level rise in the Pacific region and elsewhere (Pittock and Flather, 1993; Pittock, 1993). In short, it is not yet possible to say whether either the intensity or frequency of tropical cyclones (or ENSO) would increase or the areas of occurrence would shift in a warmer world.

Despite the often repeated assertion that *climate variability* could increase in a warmer world, there is little evidence from climate models to support this notion (Gates *et al.*, 1992). However, Working Group I has identified at least one exception that has potentially large implications for coastal areas: Various GCMs consistently predict a higher frequency of convective precipitation in mid- to high-latitude regions of the world. This anticipated change may imply more intense local rainfall, with a decrease in the return period of extreme rainfall events (e.g., Gordon *et al.*, 1992). This could interact with sea-level rise to further increase the likelihood of flooding in low-lying coastal areas (Titus *et al.*, 1987; Nicholls *et al.*, 1995).

Without any change in variability, however, a change in mean value still implies a change in the frequency of *extreme events*. Because such events by definition are at the tails of the probability distribution, the change in return periods can be quite large relative to the change in mean value. In the absence of information about changes in variability, this simple concept has been employed to create scenarios of changes in extremes—for example, for temperature in Britain (Warrick and Barrow, 1991), rainfall in Australia (Pittock *et al.*, 1991), and storm surges in the United States (Stakhiv *et al.*, 1991). As has been pointed out at the beginning of this section, in most cases it is the combination of climate extremes—temperature, precipitation, winds, sea levels—and how they are affected by longer-term changes in climatic means that are especially important for considering the future effects of global warming on coastal zones and small islands. Yet little understanding of the possible interaction of different aspects of climate change in the coastal zone exists.

9.4. Biogeophysical Effects

In the 1990 IPCC Impacts Assessment, Tsyban *et al.* (1990) suggested that the most important aspects of climate change on the coastal zone would be the impact of sea-level rise on coastal residents and marine ecosystems. They argued that a rise in sea level would:

- Inundate and displace wetlands and lowlands
- Erode shorelines
- Exacerbate coastal storm flooding
- Increase the salinity of estuaries, threaten freshwater aquifers, and otherwise impair water quality
- Alter tidal ranges in rivers and bays
- Alter sediment depositional patterns
- Decrease the amount of light reaching water bottoms.

It was recognized then—and highlighted again in the IPCC 1992 supplement (Tsyban *et al.*, 1992) and at the meeting *The Rising Challenge of the Sea* (IPCC CZMS, 1992; O'Callahan, 1994)—that such effects would not be uniform around the world and that certain coastal environments would be especially at risk. These included tidal deltas and low-lying coastal plains, sandy beaches and barrier islands, coastal wetlands, estuaries and lagoons, mangroves, and coral reefs. Small islands became a major focus of concern because some of the more extreme predictions foreshadowed that low atoll and reef islands would completely disappear or become uninhabitable, with the total displacement of populations of several small island nations (Roy and Connell, 1991).

Many of the early studies on the effects of climate change emphasized sea-level rise and were based on a simple inundation model that vertically shifted the land-sea boundary landward by the amount of the projected global rise. However, it has become increasingly clear from recent studies that the geomorphological and ecological responses to a rising sea level will be complex and will also reflect a large number of other factors, including other aspects of climate change. No longer can effects be defined simply in terms of inundation of the sea upon the land, nor by just shifting the land-sea contour by an amount corresponding to the projected vertical increase in global sea level. Biogeophysical effects will vary greatly in different coastal zones around the world because coastal landforms and ecosystems are dynamic and both respond to and modify the variety of external and internal processes that affect them. Effects will depend not only on the local pattern of sea-level rise and climate change (as shown in Section 9.3) but also on the nature of the local coastal environment and on the human, ecological, and physical responsiveness of the particular coastal system being considered (J.R. French *et al.*, 1995).

Since the IPCC 1990 assessment, considerable progress has been made in understanding the effects of sea-level rise and climate change on coastal geomorphological and ecological systems. Studies have shifted from the use of simple, monothematic approaches to more complex yet pragmatic methods (Woodroffe, 1994). Three groups of approaches can be distinguished:

- Retrospective studies concerned with reconstructing past geomorphological and ecological responses to sea-level change in the Holocene, particularly during its rising stage
- Contemporary studies of geomorphological and ecological trends over the past several decades
- Mathematical and simulation modeling of coastal geomorphological and ecological systems using simplified sea-level rise scenarios and process assumptions.

In all three cases, emphasis has been on sea-level rise, with little consideration of other climate-change aspects, although sometimes increased seawater temperatures and storminess have been included. Invariably, global sea-level rise scenarios have been applied, irrespective of their appropriateness at the local or regional level. Further, several authors (e.g., Bird, 1993a; J.R. French *et al.*, 1995) have argued that it is not always appropriate to employ Holocene stratigraphical reconstructions as analogues for the future behavior of coastal systems, primarily because of the modern complications of human impacts that may now have an overriding effect on geomorphological and ecological responses.

In spite of the increased research effort, there is still no generally accepted global typology of coastal types relating to the potential effects of sea-level rise and climate change. There have been some attempts based on the resistance of the coast to environmental forces (e.g., Van der Weide, 1993), some based on both natural and socioeconomic features and processes (Pernetta and Milliman, 1995), and some through the development of a coastal vulnerability index that captures the different characteristics of a coastal region (e.g., Gornitz, 1991). The development of such a typology is clearly an area for substantial international research in the future. Moreover, the emphasis of recent studies on coastal types has been quite uneven. For instance, there has been little research on the potential effects of sea-level rise and climate change on high-latitude coasts, bold coasts, rocky shores, coastal cliffs, coarse clastic coasts, gravel barriers, coastal sand dunes, and seagrass beds. This lack of emphasis, however, does not necessarily imply that anticipated effects on these coastal types are less serious. Studies that show a strong correlation between sea-level rise and erosion of, for example, coastal cliffs, gravel barriers, and sand dunes include Griggs and Trenhaile (1994), Carter and Orford (1993), and Van der Meulen *et al.* (1991), respectively.

There are several comprehensive reviews on the biogeophysical effects of climate and sea-level change on coastal environments (e.g., Bird, 1993b; Oude Essink *et al.*, 1993; Wolff *et al.*, 1993). In addition to these general reviews there is a series of regional summaries covering a large area of the world—including the Mediterranean (Jeftic *et al.*, 1992), European coastal lowlands (Tooley and Jelgersma, 1992), Southeast Asia (Bird, 1993a), the South Pacific (Hay and Kaluwin, 1993), wider Caribbean (Maul, 1993), the Western Hemisphere

(Ehler, 1993), and the Eastern Hemisphere (McLean and Mimura, 1993)—as well as some edited volumes on specific themes, including geomorphic response (J.R. French *et al.*, 1995), coastal wetlands (Parkinson, 1994), and developing countries (Nicholls and Leatherman, 1995a). This section assesses first the biogeophysical effects of climate change on three distinct coastal geomorphic systems, then the effects on two important ecological systems, and finally the effects on coastal biodiversity.

9.4.1. Sedimentary Coasts, Sandy Beaches, Barriers, and Dunes

Open coasts, primarily made up of unconsolidated sands and gravels and exposed to wind and wave action, are common on all inhabited continents and islands of all sizes. About 20% of the world's coast is sandy and backed by beach ridges, dunes, or other sandy deposits. International studies reported by Bird (1985, 1993b) indicate that over the last 100 years about 70% of the world's sandy shorelines have been retreating; about 20–30% have been stable and less than 10% advancing. He has listed at least twenty possible reasons for the prevalence of erosion and has indicated that sea-level rise is only one possibility. Although Stive *et al.* (1990), Leatherman (1991), and others have recognized a causal relationship between erosion and sea-level rise, many attempts to correlate accelerated coastal erosion with global sea-level rise over the last 100 years have not been convincing because of the difficulties in excluding other factors, including human impacts. Analyses of erosional trends on sandy shorelines over the past several decades indicate a predominance of local rather than common explanations—suggesting that, if sea-level rise has been a contributor, its contribution may have been masked by other mechanisms.

Two other approaches have been used to gauge the effect of sea-level rise on sedimentary coasts. First, models have been used to predict beach-profile changes that will result from a rise in water level. Model studies have been reviewed by international expert committees such as the Scientific Committee on Oceanic Research (SCOR, 1991), as well as by individuals (e.g., Healy, 1991; Leatherman, 1991). The best-known model is that of Bruun (1962), who formulated a two-dimensional relationship between rising sea level and the rate of shoreline recession based on the concept of profile equilibrium, which has been the subject of much evaluation (e.g., Dubois, 1992). SCOR (1991) has noted that testing and application of the models for beach response to a long-term rise in sea level have been hampered by significant lag times of beach changes—amounting to months or years—and the importance of other elements of the sediment budget that produce shoreline erosion or accretion irrespective of any sea-level rise. Profile changes assumed by the models have been reasonably well-verified by laboratory and field studies, but the predictive equations are found to yield poor results when the effects of profile lag times and complete sediment budgets are not included in the analysis. One solution to these uncertainties is to determine a range of beach-recession scenarios rather than a single estimate—although SCOR (1991) has concluded that the status of models for the beach response to elevated water levels is far from satisfactory; predictions of the associated shoreline recession rates yield uncertain results; and there is clear need for substantial research efforts (field and laboratory) in this area. A new generation of shoreface-profile evolution models is presently being developed (e.g., Stive and De Vriend, 1995).

Second, morphostratigraphic studies, particularly of sandy barriers, have been undertaken, although frequently these studies predate the recent interest in attempting to predict future coastal response to climate change and sea-level rise. Nevertheless, sandy-barrier responses to rises in sea level in the Holocene can be used as historical analogues. Although transgressive sedimentary sequences, where coastal barriers migrate landward as a result of shoreface erosion and washover, are widespread in North America, Europe, and Australia, other responses to sea-level rise include *in situ* growth (the stationary barrier) and even seaward advance (the regressive sequence). As with other approaches, field-based evolutionary morphostratigraphic models do not yield a consistent response to sea-level rise. Rate of sediment supply and coastal configuration are just two of the other factors that influence how sandy shorelines will respond. In addition to field-based studies, some indication of the complex way that sand barriers have responded to post-glacial sea-level rise has been shown through computer-simulation techniques (Cowell and Thom, 1994; Roy *et al.*, 1994).

Collectively, all of these results suggest that with future sea-level rise there will be tendencies for currently eroding shorelines to erode further, stable shorelines to begin to erode, and accreting coasts to wane or stabilize. Locally, changes in coastal conditions and particularly sediment supply may modify these tendencies (Bird, 1985, 1993b).

9.4.2. Deltaic Coasts, Estuaries, and Lagoons

Deltas form where terrigenous sediment brought down to the coast by rivers accumulates more rapidly than can be removed by waves, tides, and currents. Although there is a wide spectrum of delta types around the world, all are the result of the interaction between fluvial and marine processes. Since ancient times, deltas have been of fundamental importance to civilizations due to the presence of highly productive agricultural lands, fisheries, and human settlement. Many modern delta regions, with their dense populations and intensive economic activities, are now in crisis because of past management practices such as dam, dyke, and canal construction and habitat destruction, which have led to problems such as enhanced subsidence and reduced accretion, salinity intrusion, water quality deterioration, and decreased biological production (Day *et al.*, 1993; Boesch *et al.*, 1994).

Deltaic coasts are particularly susceptible to any acceleration in the rate of sea-level rise (as well as storm frequency or intensity). As Baumann *et al.* (1984) have recognized, delta survival is a battle of sedimentation versus coastal submergence. Most deltas are subsiding under the weight of accumulating sediment, a process that often is enhanced by artificial groundwater withdrawal. Any global sea-level rise will exacerbate existing problems of local submergence. Bird (1993b) has argued that a rising sea level will have two major effects on low-lying deltaic areas: First, it is likely to cause extensive submergence, especially where there is little prospect of compensating sediment accretion. Second, progradation of most deltaic coastlines will be curbed, with erosion becoming more extensive and more rapid.

Similar conclusions have come from a host of case studies around the world, including those reported from Europe and the Mediterranean in Jeftic *et al.* (1992), Tooley and Jelgersma (1992), Poulos *et al.* (1994), and Woodroffe (1994); from the Americas in Day *et al.* (1993, 1994); and from Southeast Asia in McLean and Mimura (1993). While there appears to be general agreement among all of the studies on the implications of reduced sediment discharge, subsidence, and rising sea level, there have been few attempts to determine the relative vulnerability of deltaic regions or to model the effects of sea-level rise on deltas. Exceptions to the former include Ren's (1994) study of the Chinese coast, in which six variables (relief, land subsidence, shoreline displacement, storm surge, tidal range, and coastal defenses) have been used to evaluate risk classes of eight vulnerable areas. Exceptions to the latter include the conceptual model of general deltaic functioning developed by Day *et al.* (1994) and their two-state variable model, which simulates height in sea level and land elevation over time as a function of varying rates of sea-level rise, subsidence, and vertical accretion. Model results of the "date of immersion" (i.e., when sea level equals land elevation) have been produced for several sites in the Mississippi, Camargue, and Ebro deltas. Intradelta variations have been highlighted in the model results. In natural situations, such variations commonly result from variations in subsidence and/or changes in active and passive distributary positions across deltas, as demonstrated for the Nile (Stanley and Warne, 1993) and Rhine-Meuse (Tornqvist, 1993) deltas, respectively.

Studies on the physical response of tidal rivers and estuaries to predicted sea-level rise have covered two main areas—geomorphic changes and saltwater penetration—although the American Society of Civil Engineers (ASCE) Task Committee (1992) has indicated that several hydraulic processes such as tidal range, prism and currents, and sedimentation would also be modified. Bird (1993b) has suggested that estuaries will tend to widen and deepen. This may enhance their role as sediment sinks, causing greater erosion of the neighboring open coast (Stive *et al.*, 1990). However, Pethick (1993) has shown that along the southeast coast of Britain, where relative sea-level rise is already 4–5 mm/yr due to local subsidence, estuarine channels are becoming wider and shallower by local redistribution of sediment as the intertidal profile shifts both upward and shoreward. In some areas, these effects may be offset by increased catchment runoff, greater soil erosion, and increased sediment yield as a result of climate changes. However, as with deltas, critical factors will be relative sea-level change, including local subsidence (e.g., Belperio, 1993), and sediment availability (e.g., Chappell, 1990; Parkinson *et al.*, 1994). In macrotidal estuaries in Northern Australia, channel widening initiated by rising sea level will contribute sediment to the adjacent estuarine plains, which may offset the effect of flooding and lead to steady vertical accretion. One consequence of this would be to endanger backwater swamps and freshwater ecosystems on the estuarine plains (Chappell and Woodroffe, 1994).

The effects of sea-level rise on saltwater penetration in rivers and estuaries have recently been reviewed by Oude Essink *et al.* (1993) and Van Dam (1993), who have suggested that saline water will gradually extend further upstream in the future. More serious is the accelerated effect of saline water intruding into groundwater aquifers in deltaic regions and coastal plains. In these areas, the effect of sea-level rise can be exacerbated by the withdrawal of freshwater, which may result in either subsidence and/or replacement by seawater. Subsidence and landward migration of saltwater are already serious problems in many coastal deltaic areas around the world. Two examples are Myanmar (Aung, 1993) and China (Han *et al.*, 1995b).

9.4.3. *Coral Atolls and Reef Islands*

Coral atolls and reef islands appear especially susceptible to climate change and sea-level rise. Based on the sea-level rise scenarios of the 1980s and the application of simple models, Pernetta (1988) developed an index of island susceptibility for the South Pacific region and concluded that the most susceptible nations included those "composed entirely of atolls and raised coral islands, which will be devastated if projected rises occur," and consequently "such states may cease to contain habitable islands." Three related effects were envisaged: erosion of the coastline, inundation and increased flooding of low-lying areas, and seawater intrusion into the groundwater lens, which would cause reductions in island size, freeboard, and water quality, respectively.

Since that time, a series of vulnerability assessments of atolls and reef islands have been carried out. Studies include the atoll states and territories of Tuvalu, Kiribati, Tokelau, and the Marshall Islands in the Pacific and the Maldives and Cocos (Keeling) Islands in the Indian Ocean (Aalbersberg and Hay, 1993; Woodroffe and McLean, 1992; McLean and D'Aubert, 1993; Holthus *et al.*, 1992; Connell and Maata, 1992; Pernetta, 1992; McLean and Woodroffe, 1993). Generally, these studies have documented the likelihood of more complex and variable responses than initially suggested, recognizing that the balance between reef growth, island accumulation or destruction, and sea-level rise will be locally important. Differences in response can be further expected between islands within and beyond storm belts, between those composed primarily of sand and those of coral rubble, and between those that are or are not anchored to emergent rock platforms. The presence or absence of natural physical shore-protection structures in the form of beachrock or conglomerate outcrops and biotic protection in the form of mangrove or other strand vegetation will also result in different responses between islands.

It is not clear to what extent reef islands will erode or whether sediment from the adjacent reef or lagoon will contribute to the continued growth of islands. McLean and Woodroffe (1993) have envisaged at least three possible responses in the face of sea-level rise: the Bruun response, the equilibrium response, and continued growth, which would result in shoreline erosion, redistribution of sediment, and shoreline accretion, respectively. Each of these processes can be observed on many reef islands today, as well as in the stratigraphic record, suggesting that the factors identified above are significant determinants of island stability. Moreover, as Spencer (1995) has pointed out, coral-island responses to future sea-level rise will vary as a result of constraints on the development of modern reefs and the varying inherited topographies upon which future sea-level will be superimposed.

On small islands, the freshwater lens is an important resource and often is the primary source of potable water on atolls. Recent studies suggest that the first approximation of the response of the freshwater lens to sea-level rise (the Ghyben-Herzberg principle) is not appropriate on small coral islands. The layered-aquifer model—which, among other things, considers geological structure and distinguishes between Pleistocene and Holocene stratigraphic units—is considered more appropriate for assessing freshwater inventories on such islands. If recharge and island width remain constant or expand, freshwater lenses may actually increase in size with a rise in sea level because of the larger volume of freshwater that can be stored in the less-permeable upper (Holocene) aquifer (Buddemeier and Oberdorfer, 1990). On the other hand, if recharge or island width are reduced, a diminution in both freshwater quantity and quality can be expected. In many places, increasing demand and recharge contamination are likely to be more serious issues than freshwater inventory *per se*.

Although recent reviews on coral islands have emphasized their variability and resilience (e.g., Hopley, 1993; McLean and Woodroffe, 1993), such islands remain among the most sensitive environments to long-term climate change and sea-level rise, especially where these effects are superimposed on destructive short-term events such as hurricanes, damaging human activities, and declining environmental quality. In spite of a more optimistic outlook in recent years, Wilkinson and Buddemeier (1994) have maintained that coral-reef islands may be rendered uninhabitable by climate change, especially sea-level rise, and that will necessitate relocation of any remaining human populations (see also Section 9.5).

9.4.4. Coastal Wetlands

Coastal wetlands are frequently associated with deltas, tidal rivers, estuaries, and sheltered bays. Geomorphic and hydrologic changes resulting from sea-level rise will have important effects on these biological communities, as well as on unvegetated tidal flats. The survival of the latter is dependent very much on sediment supply from adjacent river catchments—which, if not provided, will result in substantial loss of such areas. Although Woodroffe (1993) has commented that research on coastal wetlands has concentrated upon reconstructing their development under conditions of sea-level rise during the Holocene, there also have been assessments of contemporary trends and processes and simulation modeling of environmental changes.

Historical studies of temperate salt marshes include those of Allen (1991) and Reed (1990), whereas Pethick (1993) and French (1993) have used current trends and numerical simulation, respectively. Pethick (1993) has shown that salt marshes in southeast England appear to be migrating inland along the estuary but that the natural changes are interrupted by the presence of flood embankments. The result is that loss of the seaward boundaries of these wetlands will continue without compensating landward migration—a process known as coastal squeeze in the United Kingdom. Wolff *et al.* (1993) have concluded that salt marshes have the ability to respond quickly to sea-level rise as long as sedimentation and internal biomass production processes keep pace and as long as the entire marsh can move to higher shore levels or further inland. Provided that it is not constrained by infrastructure, protection works, or other barriers, vertical accretion is likely to neutralize sea-level rise as long as sediment supply is sufficient and horizontal erosion is absent or can be compensated. If not, salt marshes will progressively decline and ultimately disappear. Pethick (1992) has also shown that salt marshes under stable sea level undergo cyclical changes to their seaward boundaries; infrequent high-magnitude storm events erode the edges, while intervening lower-magnitude events allow depositional recovery. An increase in the frequency of storm events as a response to sea-level rise would result in the replacement of such cyclical change by progressive erosion. The sensitivity of certain salt-marsh species to waterlogging and soil-chemical changes also could result in a change in species composition or the migration of vegetation zones (Reed, 1995).

Mangroves grow largely in tidal forests and are characterized by adaptations to unconsolidated, periodically inundated saline coastal habitats. They fringe about 25% of shorelines in the tropics and extend into the subtropics as far north as Bermuda and as far south as North Island, New Zealand. Studies on the effects of sea-level rise on tropical mangrove ecosystems have been primarily of historical nature (reviewed by Woodroffe, 1990; UNEP-UNESCO Task Team, 1993; Edwards, 1995). These studies have shown that extensive mangrove ecosystems became reestablished when sea level stabilized around 6,000 years BP. During the prior rise, mangroves probably survived as narrow coastal fringes, shifting landward with the migrating shoreline. Ellison and Stoddart (1991) and Ellison (1993) have indicated that mangroves in areas of low sediment input in both low-island and high-island settings appear to be unable to accrete vertically as fast as the projected rate of sea-level rise. However, recent evidence from the Florida Keys (Snedaker *et al.*, 1994) has shown that low-island mangroves may be resilient to rates of sea-level rise about twice those suggested as upper limits by Ellison and Stoddart (1991) from their study in Bermuda. It is also apparent that mangrove communities are more likely to survive in macrotidal, sediment-rich environments such as Northern Australia, where strong tidal currents redistribute sediment (Semeniuk, 1994; Woodroffe, 1995), than in microtidal sediment-starved environments such as around the Caribbean (Parkinson *et al.*, 1994).

If the rate of shoreline erosion increases, mangrove stands may tend to become compressed and suffer reductions in species diversity. On the other hand, extensive mangroves in deltaic settings with continuing large inputs of terrigenous sediment are likely to be more resilient to sea-level rise (Edwards, 1995). Thus, different responses can be envisaged in different mangrove settings. Additionally, certain species are likely to be more robust in the face of sea-level rise than others (Ellison and Stoddart, 1991; Aksornkaoe and Paphavasit, 1993).

It is now becoming increasingly clear, as Woodroffe (1993, 1994) and Edwards (1995) have observed, that coastal wetlands (marshes and mangroves) can undergo a number of responses to sea-level rise. Responses may be different in muddy, tide-dominated systems than in more organic systems, in areas of high or low tide range, and in areas of high or low sediment and freshwater input. Thus, the balance between accretion and submergence will be complex, and a range of morphological responses is likely for different coastal types and coastal settings. Although some marshes and mangroves may be under threat from sea-level rise over the next century, human impact has been the major threat up to the present and may be far more important locally than climate change in the long term (Bird, 1993a; WCC'93, 1994). In the case of mangroves, afforestation programs may be one way to compensate for natural or human-induced losses, although experiences in Bangladesh have indicated the difficulties in such a program (Saenger and Siddigi, 1993). There also is some evidence that coastal wetlands may experience loss due to short-term (decadal) acceleration in the rate of sea-level rise (Boesch *et al.*, 1994; Downs *et al.*, 1994).

Whereas mangroves are restricted to the intertidal zone, and salt marsh extends landward into supratidal areas, seagrasses extend subtidally to maximum depths of several tens of meters. Relatively little appears to have been published on the possible effects of climate change and sea-level rise on seagrasses (used here generically to include eelgrasses, turtlegrasses, etc.), although their biology and biogeography have been studied extensively (e.g., Larkum *et al.*, 1989; Mukai, 1993). Edwards (1995) has provided a brief but comprehensive analysis, noting the economic importance of seagrasses, and their ability to trap sediment, accrete vertically, stabilize unconsolidated sediment, slow water movement, and generally serve as natural coastal protection agents. Edwards (1995) has argued that intertidal and shallow seagrass beds (<5 m depth) are most likely to be affected by climate change, particularly by any sustained elevations in sea temperature or increases in freshwater runoff from land. However, the main threat to seagrass habitats is likely to come from increased anthropogenic disturbances, including dredging, overfishing, water pollution, and reclamation. In some parts of the world, seagrass beds are already severely threatened (Fortes, 1988), although elsewhere they have expanded due to eutrophication of estuarine waters.

9.4.5. Coral Reefs

Coral reefs are estimated to cover about 600,000 km^2 of the Earth's surface (Smith, 1978). They are dominated by calcifying organisms that are depositing about 0.6 to 0.9 Gt of calcium carbonate ($CaCO_3$) globally each year (Kinsey and Hopley, 1991). Intuitively, one would think that coral reefs—by precipitating $CaCO_3$ and sequestering carbon—would act as sinks for CO_2, but on the decade to century timescale this is not the case (Smith and Buddemeier, 1992). The calcification process actually generates CO_2 (Ware *et al.*, 1992), and over periods of decades reefs may contribute about 0.02 to 0.08 Gt C/yr as CO_2 to the atmosphere.

The effects of climate change on coral reefs, as well as nonclimatic anthropogenic disturbances, have recently been reviewed by Smith and Buddemeier (1992), Wilkinson and Buddemeier (1994) and Edwards (1995). The global climate-change effects of significance to coral reefs are likely to be increases in seawater temperature and sea-level rise; locally or regionally, changes in storm patterns and coastal currents, as well as changes in rainfall patterns, may have effects on coral communities—for example, through increases in sedimentation.

Coral reefs are particularly sensitive to increases in seawater temperature (Brown, 1987) and increased irradiance (Brown *et al.*, 1994). They respond to the combined effect of irradiance and temperature elevations by paling in color, or bleaching (Brown and Ogden, 1993). Corals do not generally bleach in response to rapid fluctuations in seawater temperature but rather to departures in temperature above their seasonal maximum. If the temperature elevation involves a substantial increase in seawater temperature (3–4°C) for an extended period (>6 months), considerable coral mortality can ensue (Brown and Suharsono, 1990). If, however, the temperature increase is only on the order of 1–2°C and for a limited period, bleached corals may recover but show reduced growth and impaired reproductive capabilities (Brown and Ogden, 1993).

Projected increases in seawater temperatures thus appear to be a major threat to coral reefs. In Indonesia, where severe bleaching took place as a result of seawater warming during an ENSO event in 1983, coral reefs have failed to show continued recovery beyond the initial recovery noted in 1988 (Brown and Suharsono, 1990). Such results have been mirrored in studies in the Galápagos and eastern Panama, where little recovery has been noted since major bleaching in 1982–1983. At sites in the East Pacific, reefs subsequently have shown rapid bioerosion from the destructive grazing activities of sea urchins, and destabilization of reef substrates is anticipated. Full community restoration probably will not occur for several hundred years (Glynn, 1993).

Reef accretion rates—calculated from community calcification rates, growth rates of calcifying organisms, and radiocarbon dating of cores through reefs—range from less than 1 mm/yr to a maximum slightly in excess of 10 mm/yr (Buddemeier and Smith, 1988; Hopley and Kinsey, 1988; Kinsey, 1991). A rate of 10 mm/yr is commonly taken as the consensus value for the maximum sustained vertical reef accretion rate (Buddemeier and Smith, 1988). The present best estimates for global sea-level rise over the next century (see Section 9.3) are well within the range of typical reef accretion rates. Even slowly accreting reef flats should, on average, be able to keep up with this rate of sea-level rise after a lag, provided that other factors such as increased seawater temperatures and damaging anthropogenic influences are not acting simultaneously (Edwards, 1995). Widespread warming of seawater, however, will clearly limit the accretion rates of coral reefs—as exemplified by the Panamanian reef, which before suffering 50% coral mortality as a result of sea-surface warming during an ENSO event was depositing about 10 tonnes $CaCO_3$/ha/yr and now is eroding at a rate of approximately 2.5 tonnes $CaCO_3$/yr, equivalent to vertical erosion of 6 mm/yr (Eakin, 1995).

Smith and Buddemeier (1992) have expected shifts in zonation and community structure associated with the interaction between wave-energy regime and sea level, although far too little is known about the physiological and physical constraints to reef growth (Spencer, 1995). Wilkinson and Buddemeier (1994) have shown that coral reefs have come through episodes of severe climate change in the past and have the necessary resilience to cope with current scenarios of climate change. However, coral reefs near land masses and near large population centers will come under greater human pressure in the future and are likely to be damaged beyond repair.

9.4.6. *Coastal Biodiversity*

Through the biogeophysical effects on coastal geomorphic and ecological systems described in Sections 9.4.1 through 9.4.5, climate change has the potential to significantly affect coastal biological diversity. It could cause changes in the population sizes and distributions of species, alter the species composition and geographical extent of habitats and ecosystems, and increase the rate of species extinction (Reid and Miller, 1989). Coral reefs have the highest biodiversity of any marine ecosystem, with enormous numbers of different species packed into small areas (Norris, 1993). Despite the known and potential value of reef communities and the threats to their health and vigor, the total biodiversity of coral reefs is not known, nor is the fraction of the diversity that is described versus undiscovered. Reaka-Kudla (1995) has estimated that there are about 91,000 described species of coral-reef taxa but argues that undocumented diversity is likely to be much higher. Nonetheless, coral-reef macrobiota represent about 4–5% of the described global biota, although they occupy less than 1% of the Earth's surface. Coral-reef biodiversity is centered around the archipelagos of the Philippines and Indonesia; diversity decreases away from this core. For example, the number of coral species in French Polynesia drops to less than 10% of that in the core area (Wilkinson and Buddemeier, 1994). Similar geographical variations in species richness occur in mangroves and tropical seagrasses (Woodroffe, 1990; Mukai, 1993).

In addition to marine systems such as coral reefs, coastal zones also comprise the adjacent terrestrial environments. Although scientists have long studied the interactions between marine and terrestrial systems and have seen the coastal zone as a discrete entity, it is poorly understood ecologically (Ray, 1991). Among all macroscopic organisms, there are 43 marine phyla and 28 terrestrial phyla; 90% of all known classes are marine (Reaka-Kudla, 1995). Ray (1991) has estimated that 80% of marine phyla occur in the coastal zone, which occupies only 8% of the Earth's surface, and argues that the marine portion of the coastal zone is the most biologically diverse realm on the planet. For example, of the 13,200 species of marine fish, almost 80% are coastal. Tropical coastal zones are particularly rich, with about 182,000 described species (Reaka-Kudla, 1995); they are about twice as rich as temperate coasts.

Coastal zones are sharply subdivided by gradients in geomorphic structure and habitat diversity; this enables them to perform many of the regulation functions and user and production functions outlined in Section 9.2.1. Ray (1991) has identified a nexus between physical processes and ecological pattern and diversity; he suggests that if global warming accelerates during the next few decades, the extent of coastal lagoons, marshes, and so forth will be affected and that these changes will strongly influence the fate of associated biota. In addition, intensive habitat modification on land and deterioration of coastal areas will clearly result in a decline of global biodiversity. Similarly, the capacity of species and ecosystems such as mangroves to shift their ranges and locations in response to climate change will be hindered by human land-use practices that have fragmented existing habitats. The establishment of nature reserves is seen as an option to arrest the decline of coastal biodiversity (Ray and Gregg, 1991; De Groot, 1992b).

9.5. Socioeconomic Impacts

Section 9.4 outlines how the coastal environment can be altered by climate change and sea-level rise; such alterations could have significant effects on functions and values in coastal zones and small islands. This section presents an overview of the related socioeconomic impacts and their evaluation, with an emphasis on the problems of sea-level rise. It highlights the particularly vulnerable situation of low-lying small islands and deltas. The emphasis on sea-level rise reflects a bias of existing studies. A discussion of the human activities in coastal zones that increase vulnerability to climate change also is included.

9.5.1. *Pressures and Management Problems in Coastal Zones and Small Islands*

During the twentieth century, urbanized coastal populations have been increasing because of the many economic opportunities and environmental amenities that coastal zones can provide. The need to protect and enhance the wealth-creation potential of coastal zones has led to widespread coastal construction and modification of natural coastal processes, resulting in losses of coastal habitats, changes in circulation and material flux, and reductions in biological productivity and biodiversity. These pressures are expected to increase substantially in the coming decades (WCC'93, 1994).

Of particular concern is the worldwide destruction and degradation of coral reefs, mangroves, sea grasses, and salt marshes—which, among other things, act as natural barriers against marine erosion processes. The natural response of salt marshes and mangroves to sea-level rise—an upward and landward migration of the intertidal profiles (see Section 9.4)—is inhibited by flood embankments and other human constructions. The result is that erosion of the seaward boundaries of these wetlands will continue without a compensating landward migration, leading to loss of wetland area. Deltaic processes also are being modified. In the United States, for example, the Mississippi River delta was roughly in a state of dynamic balance before the twentieth century, but since then human intervention in the form of large-scale engineering works, levees, dams, canals, and water diversions has effectively starved the wetlands of needed freshwater and sediments and radically altered wetland hydrology. Relative sea level is rising at a rate of up to 1 m per 100 years in this region; up to 100 km^2 of wetlands were lost each year during the 1970s, falling to 50 km^2/yr in the 1980s (Boesch *et al.*, 1994). In Bangladesh, flood defense systems and/or human activities in the Ganges-Bramaputra-Meghna river system may have affected runoff, sediment flow, and deposition rates, with detrimental effects on coastlines, fisheries, and the frequency and severity of inland flooding (Warrick and Rahman, 1992; see also Ives and Messerli, 1989). Small island states (many of which are low-lying) face particularly severe threats, and pollution and mining of coral will further serve to inhibit the capabilities of these countries to respond to sea-level rise.

Human interference in the dynamic processes that affect coastal zones is not restricted to activities within the coastal zone itself. Activities upstream in catchment areas may also play a part. For example, effluent discharging from sewage plants and industrial plants and agricultural runoff can lead to eutrophication, and water-resource schemes (e.g., dams and irrigation systems) can restrict the supply of water, sediment, and nutrients to coastal systems. Sewage and siltation are among the most significant causes of coral-reef and other natural coastal system degradation in the Philippines, Singapore, Malaysia, Indonesia, Sri Lanka, the Pacific islands, Hawaii, the Persian Gulf, the Caribbean, parts of the South American coast, and Cuba (Lundin and Linden, 1993).

The message is clear: Climate-related changes such as accelerated sea-level rise and possibly altered patterns of storm frequency and intensity represent potential *additional* stresses on systems that are already under intense and growing pressure. In addition, there are complex interrelationships and feedbacks between human and environmental driving forces and impacts on the one hand and climate-induced changes and effects on the other. These relationships require considerably more study (Turner *et al.*, 1995b).

9.5.2. *Assessment of Impacts*

Of direct relevance when analyzing socioeconomic impacts is the evaluation of the potential loss of environmental values. As discussed in Section 9.2, a coastal system can yield a number of different values related to the functions and services it provides. A range of methods is available to evaluate these (see Table 9-1).

Table 9-1: *Environmental evaluation methods showing—from left to right—increasing complexity and scale of analysis (adapted from Pearce and Turner, 1992).*

Least Complicated				*Most Complicated*
Financial Analysis	**Economic Cost-Benefit Analysis**	**Extended Cost-Benefit Analysis**	**Environmental Impact Assessment**	**Multi-Criteria Decision Methods**
• Financial profitability criterion • Private costs and revenues • Monetary valuation	• Economic efficiency criterion • Social costs and benefits • Monetary valuation	• Sustainable development principles • Economic efficiency and equity tradeoff environmental standards as constraints • Partial monetary valuation	• Quantification of a diverse set of effects on a common scale, but no evaluation	• Multiple decision criteria • Monetary and nonmonetary evaluation

The more comprehensive the technique, the greater the diversity of information that will be required and yielded to assist the appraisal of policy options from a societal perspective. Therefore, assessing the value of climate-change impacts is not a straightforward issue. More robust techniques are available for deriving use values than for non-use values. Moreover, monetary valuation is not always appropriate when cultural and heritage assets are threatened by climate change and sea-level rise. Small islands often are particularly threatened, including their distinct ways of life and possibly even their distinct cultures. The same applies to heritage and other culturally significant sites on the coast. The core project Land-Ocean Interactions in the Coastal Zone (LOICZ) of the International Geosphere-Biosphere Program (IGBP) (see also Sections 9.6.4 and 9.7) is currently preparing comprehensive guidelines for evaluating coastal functions and services.

Coastal zones and small islands support a range of socioeconomic sectoral activities that can be affected by climate change and sea-level rise. Most of these sectors are covered in separate chapters in this report. Table 9-2 reflects the results of these sectoral assessments as they pertain to direct impacts specifically related to climate change in coastal zones and small islands. The reader is referred to the chapters listed in the table for more detailed discussions of the impacts and vulnerability of each sector.

Tourism also is of great importance to coastal zones and small islands, although it is not covered as a separate sector in this report. Tourism helps to support the economies of many coastal countries (Miller and Auyong, 1991). For many small islands in particular, tourism is the largest contributor to the country's GNP. Coastal tourism can be affected by climate change directly through coastal erosion and changes in weather patterns (e.g., storminess, precipitation, cloud cover). Indirect effects, however, may be just as important. Adverse impacts on freshwater supply and quality, human settlements, and human health will severely affect tourism, as will overdevelopment leading to environmental degradation.

Many efforts have been made in the last few years to assess the implications of climate change and associated sea-level rise on the coastal zone. As part of these efforts, the former Coastal Zone Management Subgroup of IPCC has published a methodology for assessing the vulnerability of coastal areas to sea-level rise (IPCC CZMS, 1991). The framework, called the Common Methodology, has been widely applied as the basis of vulnerability assessment studies. These studies have aimed to identify populations and resources at risk and the costs and feasibility of possible responses to adverse impacts. The vulnerability of many more coastal countries to sea-level rise has been reviewed using other approaches. Examples of assessments of the possible impacts and responses to sea-level rise and climate change can be found in Tobor and Ibe (1990), Parry *et al.* (1992), Bijlsma *et al.* (1993), Warrick *et al.* (1993), Ehler (1993), McLean and Mimura (1993), Qureshi and Hobbie (1994), O'Callahan (1994), WCC'93 (1995), and Nicholls and Leatherman (1995a). A series of studies on Pacific islands has been conducted by the South Pacific Regional Environment Programme, including Holthus *et al.* (1992), Aalbersberg and Hay (1993), McLean and D'Aubert (1993), and Nunn *et al.* (1994a, 1994b).

This section summarizes a number of these vulnerability case studies, with an emphasis on the strengths and weaknesses of those conducted using the Common Methodology or similar approaches.

9.5.2.1. *Vulnerability Assessment and the IPCC Common Methodology*

Vulnerability to impacts is a multidimensional concept, encompassing biogeophysical, socioeconomic, and political factors. The Common Methodology defines vulnerability as "the degree of incapability to cope with the consequences of climate change and accelerated sea-level rise" (IPCC CZMS, 1991). Therefore, analysis of the vulnerability of a coastal area or small island to climate change includes some notion of its *susceptibility* to the biogeophysical effects of climate change and sea-level rise (see Section 9.4), as well as of its natural *resilience*—which is greatly influenced by past, current, and future population and settlement patterns and rates of socioeconomic change. Susceptibility and resilience together determine the natural system's *sensitivity* to anticipated changes. Socioeconomic *vulnerability* is further determined by a country's technical, institutional, economic, and cultural capabilities to cope with or manage the

Table 9-2: *Qualitative synthesis of direct impacts of climate change and sea-level rise on a number of sectors in coastal zones and small islands, based on other chapters in this volume. Chapter numbers are in parentheses.*

	Climate-Related Events				
Impact Categories	Coastal Erosion	Flooding/ Inundation	Saltwater Intrusion	Sedimentation Changes	Storminess
Human Settlements (12)	✓	✓			✓
Agriculture (13)		✓	✓		✓
Freshwater Supply and Quality (14)		✓	✓		
Fisheries (16)	✓	✓	✓	✓	✓
Financial Services (17)	✓	✓			✓
Human Health (18)		✓			✓

anticipated biogeophysical effects and their consequent socioeconomic impacts (Turner *et al.*, 1995a).

The IPCC Common Methodology has aimed to identify "*the types of problems that a country will have to face and, if necessary, the types of assistance that are most needed to overcome these problems.*" Assessments are to "*serve as preparatory studies, identifying priority regions and priority sectors and to provide a first reconnaissance and screening of possible measures*" (IPCC CZMS, 1991). Three boundary conditions and scenarios have been specified in the methodology: the impacts on the natural coastal systems, the impacts on socioeconomic developments, and the implications of possible response strategies for adaptation. The methodology includes consideration of the reference (or present) situation and a rise in sea level of 30 cm to 1 m by the year 2100. These scenarios approximate the low and high estimates of the 1990 IPCC Scientific Assessment. It considers socioeconomic developments by extrapolating 30 years from the present situation. The Common Methodology recommends considering a full range of adaptation options, including at least the extreme options of complete retreat and total protection. To simplify analysis, the method does not consider coastal evolution other than that caused by climate change, nor does it assess the effects of progressive adaptation at the local scale, such as the raising of dikes.

The Common Methodology has helped to focus the attention of many coastal nations on climate change and has contributed to long-term thinking about the coastal zone. On the other hand, a number of problems have been raised concerning the Common Methodology through the experiences of vulnerability assessment case studies (WCC'93, 1994):

- Many case studies have faced a shortage of accurate and complete data necessary for impact analysis. In particular, it often has proven difficult to determine accurately the impact zone in many countries due to the lack of basic data, such as the coastal topography.
- Many studies have found the use of a single global scenario of sea-level rise (1 m by 2100) inappropriate to their respective areas, often due to the lack of more detailed data on coastal elevations; most studies have ignored the spatial distribution of relative sea-level rise and other coastal implications of climate change, largely due to a lack of regional climate scenarios. Future vulnerability assessment would be greatly improved by the availability of regional scenarios for climate change and sea-level rise, including reference, low, and high scenarios.
- Although the Common Methodology has encouraged researchers to take into account the biogeophysical response of the coastal system to sea-level rise, lack of data and models for describing local coastal processes and responses have hindered detailed, quantitative impact assessment. Many case studies have carried out a simple first-order assessment by horizontally shifting the coastline landward by an amount corresponding with the sea-level rise scenario.
- While vulnerability profiles have yielded some useful relative guidance on potential impacts, the Common Methodology has been less effective in assessing the wide range of technical, institutional, economic, and cultural elements present in different localities.
- There has been concern that the methodology stresses a protection-orientated response, rather than considering a full range of adaptation options.
- Market-evaluation assessment frameworks have proved inappropriate in many subsistence economies and traditional land-tenure systems. More attention should be paid to broader socioeconomic evaluation techniques, which include traditional, aesthetic, and cultural values (see Box 9-2).

9.5.2.2. *Vulnerability Assessment Case Studies*

At least 23 country case studies have produced quantitative results that can be interpreted in terms of the IPCC Common

Box 9-2. Cultural Impacts and Alternative Assessments

Conventional impact evaluation techniques and indicators, such as GNP and population at loss, protection costs, and cost-benefit analysis, reflect only one (largely Western) approach for assessing potential damages from climate-related events. This has led to the development of alternative methodologies that seek to assess changes in culture, community, and habitat. In a study of coastal vulnerability and resilience to sea-level rise and climate change, Fiji is considered (Nunn *et al.*, 1994a). The methodology (Yamada *et al.*, 1995) computes a Sustainable Capacity Index based on the sum of ratings of vulnerability and resilience for many categories of cultural, social, agricultural, and industrial impacts at the local, regional, and national levels. Areas with higher concentrations of assets are judged to be more vulnerable, whereas areas with diversity and flexibility in the system—whether natural or managerial—tend to be viewed as more resilient in this analysis. The study has evaluated potential impacts to subsistence economies according to the view that communities in which people feed and clothe themselves with little cash exchange are more vulnerable but that subsistence economies in which staples can be replaced with other crops tend to be more resilient. In addition, cultural sites have been ranked according to the level of national interest in their preservation. The study concludes that subsistence economies and cultural assets are more vulnerable in Fiji and that conventional analyses of relatively high-lying islands such as Fiji would tend to underestimate the potential vulnerability of these areas, given that most people live in the low-lying coastal plain and the majority of cash and subsistence economic activities take place in the low-lying areas.

Methodology. Some results are summarized in Table 9-3, and these show considerable variation in possible impacts from country to country, reflecting that certain settings are more vulnerable than others. This conclusion is widely supported by all of the country studies that are available. Small islands, deltaic settings, and coastal ecosystems appear particularly vulnerable. In addition, developed sandy shores may be vulnerable because of the large investment and significant sand resources required to maintain beaches and protect adjoining infrastructure in the face of sea-level rise (Nicholls and Leatherman, 1995a).

Several caveats are in order so that the following impact estimates can be put into proper perspective (following Section 9.5.2.1). First, the impacts presented assume a 1-m rise in sea level by 2100—which is the high estimate of the IPCC90 business-as-usual sea-level rise scenario—and no other climate change. The latest scientific information, however, suggests a lower global mean sea-level rise (see Section 9.3.1.1). For a number of nations, the impacts of a 50-cm or smaller rise have been examined, including Argentina (Dennis *et al.*, 1995a), parts of north China (Han *et al.*, 1993), Japan (Mimura *et al.*, 1994), Nigeria (G.T. French *et al.*, 1995), Senegal (Dennis *et*

Table 9-3: *Synthesized results of country case studies. Results are for existing development and a 1-m rise in sea level. People affected, capital value at loss, land at loss, and wetland at loss assume no measures (i.e., no human response), whereas adaptation assumes protection except in areas with low population density. All costs have been adjusted to 1990 US$ (adapted from Nicholls, 1995).*

Country/Source	People Affected # people (1000s)	People Affected % Total	Capital Value at Loss Million US$[1]	Capital Value at Loss % GNP	Land at Loss km²	Land at Loss % Total	Wetland at Loss km²	Adaptation/ Protection Costs Million US$[1]	Adaptation/ Protection Costs % GNP
Antigua[2] (Cambers, 1994)	38	50	–	–	5	1.0	3	71	0.32
Argentina (Dennis *et al.*, 1995a)	–	–	>5000[7]	>5	3400	0.1	1100	>1800	>0.02
Bangladesh (Huq *et al.*, 1995; Bangladesh Government, 1993)	71000	60	–	–	25000	17.5	5800	>1000[9]	>0.06
Belize (Pernetta and Elder, 1993)	70	35	–	–	1900	8.4	–	–	–
Benin[3] (Adam, 1995)	1350	25	118	12	230	0.2	85	>400[10]	>0.41
China (Bilan, 1993; Han *et al.*, 1995a)	72000	7	–	–	35000	–	–	–	–
Egypt (Delft Hydraulics *et al.*, 1992)	4700	9	59000	204	5800	1.0	–	13100[11]	0.45
Guyana (Kahn and Sturm, 1993)	600	80	4000	1115	2400	1.1	500	200	0.26
India (Pachauri, 1994)	7100[6]	1	–	–	5800	0.4	–	–	–
Japan (Mimura *et al.*, 1993)	15400	15	849000	72	2300	0.6	–	>156000	>0.12
Kiribati[2] (Woodroffe and McLean, 1992)	9	100	2	8	4	12.5	–	3	0.10
Malaysia (Midun and Lee, 1995)	–	–	–	–	7000	2.1	6000	–	–
Marshall Islands[2] (Holthus *et al.*, 1992)	20	100	160	324	9	80	–	>360	>7.04
Mauritius[4] (Jogoo, 1994)	3	<1	–	–	5	0.3	–	–	–
The Netherlands (Peerbolte *et al.*, 1991)	10000	67	186000	69	2165	5.9	642	12300	0.05
Nigeria (G.T. French *et al.*, 1995)	3200[6]	4	17000[7]	52	18600	2.0	16000	>1400	>0.04
Poland (Pluijm *et al.*, 1992)	240	1	22000	24	1700	0.5	36	1400	0.02
Senegal (Dennis *et al.*, 1995b)	110[6]	>1	>500[7]	>12	6100	3.1	6000	>1000	>0.21
St. Kitts–Nevis[2] (Cambers, 1994)	–	–	–	–	1	1.4	1	50	2.65
Tonga[2] (Fifita *et al.*, 1994)	30	47	–	–	7	2.9	–	–	–
United States (Titus *et al.*, 1991)	–	–	–	–	31600[8]	0.3	17000	>156000	>0.03
Uruguay (Volonté and Nicholls, 1995)[5]	13[6]	<1	1700[7]	26	96	0.1	23	>1000	>0.12
Venezuela (Volonté and Arismendi, 1995)	56[6]	<1	330[7]	1	5700	0.6	5600	>1600	>0.03

[1]Costs have been adjusted to reflect 1990 US$.
[2]Minimum estimates—incomplete national coverage.
[3]Precise year for financial values not given—assumed to be 1992.
[4]Results are linearly interpolated from results for a 2-m sea-level rise scenario.
[5]See also review in Nicholls and Leatherman (1995a).
[6]Minimum estimates—number reflects estimated people displaced.
[7]Minimum estimates—capital value at loss does not include ports.
[8]Best estimate is that 20,000 km² of dry land are lost, but about 5,400 km² are converted to coastal wetlands.
[9]Adaptation only provides protection against a 1-in-20 year event.
[10]Adaptation costs are linearly extrapolated from a 0.5-m sea-level rise scenario.
[11]Adaptation costs include 30-year development scenarios.

al., 1995b), parts of the United Kingdom (Turner *et al.*, 1995a), the United States (Titus *et al.*, 1991), Uruguay (Volonté and Nicholls, 1995), and Venezuela (Volonté and Arismendi, 1995). Second, all of the country studies have assumed that the socioeconomic situation is constant until 2100. This is unrealistic and ignores the rapid coastal development that is occurring with little regard for existing problems, let alone tomorrow's (WCC'93, 1994). However, the mere threat of extensive loss of land and other assets may stimulate macroeconomic effects within national economies. Some assets may be relocated and others may be adapted to reduce the damage implications of climate change. On the other hand, the damage-cost estimates may represent underestimates because they neglect some nonmarket asset values and factors such as the cost of resettlement of coastal populations that cannot be easily protected. Finally, it has been assumed that the rise in sea level will be a slow, gradual process, which may not be the case for all regions. Scientific uncertainties are compounded by the socioeconomic adaptation uncertainties referred to above and by the fact that economic cost estimates are very sensitive to changes in discount rates.

Despite these limitations, these studies have offered some important insights into potential impacts and possible responses to climate change and sea-level rise. Many of the vulnerability assessments emphasize the severe nature of existing coastal problems such as beach erosion, waterlogging, and pollution (e.g., El-Raey *et al.*, 1995; Han *et al.*, 1995b). For many small islands, population pressure and urbanization, coastal pollution, and overexploitation of resources already are critical problems. For deltas and estuaries, changes in sediment supply and distribution are often already causing significant changes in the coastal zone. This reinforces the message that climate change will act on coastal systems that are already under stress.

In addition to accelerated sea-level rise, there is widespread concern about the coastal implications of other aspects of climate change such as changing rainfall and runoff in the catchment area, as well as the effects of changes in storminess and storm surges (e.g., Warrick *et al.*, 1993; McLean and Mimura, 1993). One quantitative vulnerability assessment study exists; it shows that in The Netherlands the costs of avoiding damage related to an adverse 10% change in the direction and intensity of storms may be worse than those of a 60-cm rise in sea level (Peerbolte *et al.*, 1991). This storm-change scenario is arbitrary, but shows that concern is justified and that there is a need for more widespread analysis.

All of the 23 national case studies shown in Table 9-3 project land loss as the sum of dry-land and wetland loss, assuming no protective measures are taken. The estimated losses range from 0.05% of the national land area in Uruguay (Volonté and Nicholls, 1995) to more than 12% of Tarawa, Kiribati (Woodroffe and McLean, 1992); more than 17% of Bangladesh (Huq *et al.*, 1995); and 80% of Majuro atoll, Marshall Islands (Holthus *et al.*, 1992). From fifteen case studies, 63,000 km^2 of wetlands are estimated to be lost. Most of these assessments are based on first-order analyses, and key parameters such as limiting vertical accretion rates and potential for wetland migration often are poorly defined (Nicholls, 1995). The study of the United States has considered wetland migration, estimating that a 50-cm rise in sea level would erode or inundate 38% to 61% of existing coastal wetlands. Assuming that dikes or bulkheads were not built to impede inland migration, new wetland formation on formerly upland areas would reduce the total loss to 17% to 43% (Titus *et al.*, 1991). Therefore, wetland migration is not projected to compensate for losses, even under the most ideal circumstances. Many studies, however, have found that direct human reclamation of wetlands for a range of purposes at present is a much bigger threat than sea-level rise (Nicholls and Leatherman, 1995a).

Fifteen case studies have provided estimates of undiscounted capital value potentially at loss, assuming no protection. Nearly half of the studies have concluded that capital value at loss could exceed 50% of present GNP, illustrating the concentration of infrastructure and economic activity in the coastal zones of many of the countries studied. To counter these impacts, adaptation would be expected (see Section 9.6). Table 9-3 stresses the cost of total protection rather than other possible adaptation options, which may have lower costs. Assuming that costs will accrue uniformly over 100 years, the annual protection costs—as a percentage of present GNP—are highest for the Marshall Islands at 7% (Holthus *et al.*, 1992) and St. Kitts-Nevis at 2.7% (Cambers, 1994). This supports the conclusion that some small islands have a high vulnerability to sea-level rise. However, Kiribati has a similar setting to the Marshall Islands, yet the estimates of protection costs are much smaller, at 0.1% of present GNP (Woodroffe and McLean, 1992). This reflects important differences in assumptions about the meaning of total protection. In Kiribati, local engineers have selected existing low-technology, low-cost gabions to protect the atoll, whereas in the Marshall Islands large and expensive sea walls have been utilized to determine the costs. This comparison shows one of the weaknesses of the Common Methodology and the need to assess a wider range of response options in future vulnerability assessment studies.

In many locations, beaches are likely to require nourishment to protect tourist infrastructure because existing urban and tourist infrastructure could be damaged and destroyed. The amount of sand required to maintain a beach in the face of long-term sea-level rise is uncertain (Stive *et al.*, 1991); in some case studies, the costs of beach nourishment could dominate basic response costs if countries invest in such an adaptation option (Dennis *et al.*, 1995b; Nicholls and Leatherman, 1995a; Volonté and Nicholls, 1995). There is also the question of the availability of sufficient sand resources. The usual source is suitable-grade nearshore deposits, if available. However, the implication of the removal of such deposits must be carefully considered in terms of its effect on the coastal sediment budget and the nearshore wave climate.

In many industrialized countries, the main potential loss from sea-level rise seems to be coastal wetlands, as well as sandy beaches in some countries (e.g., Mimura *et al.*, 1994).

Box 9-3. The Vulnerable Situation of Small Islands

Many small island countries could lose a significant part of their land area with a sea-level rise of 50 cm to 1 m. The Maldives, for example, have average elevations of 1 to 1.5 m above existing sea level (Pernetta, 1992). Although biogeophysical processes may counter land losses (see Section 9.4), the threat of submergence and erosion remains; this could convert many small islands to sandbars and significantly reduce the usable dry land on the larger, more populated islands. Saltwater intrusion and loss of the freshwater lens may be an equally binding constraint on human habitation in some islands, particularly smaller atolls (Leatherman, 1994).

The available case studies have shown that small islands—most particularly, coral atolls such as the Marshall Islands (Holthus *et al.*, 1992)—are heavily oriented toward coastal activities and hence are vulnerable to sea-level rise (e.g., Cambers, 1994; Fifita *et al.*, 1994). At the same time, their relatively small economies may make the costs of adaptation prohibitive. In global terms, the population of small islands is relatively small, but a number of distinct societies and cultures are threatened with drastic changes in lifestyle and possibly forced abandonment from ancestral homelands if sea level rises significantly (Roy and Connell, 1991).

Even the less-vulnerable small islands would suffer significant economic effects from the loss of beach tourism and recreation areas because of sea-level rise and, possibly, more storms leading to increased beach and reef erosion. In 1988, among the Caribbean islands, income from tourism as a percentage of GNP was 69% for Antigua and Barbuda and 53% for the Bahamas; for a dozen other Caribbean islands, tourism revenues make up more than 10% of the GNP (Hameed, 1993). The Indian Ocean islands of the Seychelles and the Maldives also have seen a steady growth in tourism. In 1991, total receipts from tourism generated foreign exchange earnings of $94 million in the Maldives. This represented some 74% of the country's total foreign exchange earnings. Since 1985, tourism has been the single biggest contributor to the GNP of the Maldives. Tourism to developing countries has increased significantly in recent years, and small island developing states have experienced a particularly rapid increase. Tourist numbers to Mauritius, for example, have increased from 1,800 visitors in 1968 to 180,000 in 1988 (UNEP, 1991).

Given accelerated sea-level rise, first-order estimates suggest that substantial investment would be required in some developing countries in order to protect urban areas and maintain related activities such as beach tourism. Nine small island states appear in the list of countries facing the highest coastal protection costs as a percentage of their GNP. The global average percentage required annually for coastal protection is 0.037%; however, for many small islands it is significantly higher—up to 34% for the Maldives (OECD, 1991). To explore the full range of potential responses, more comprehensive assessment of the available adaptation options in these vulnerable settings is urgently required.

However, a change in the frequency, intensity, or distribution of extreme weather events could have implications for urban areas and related capital assets in countries such as Japan, Australia, the United States, and some countries bordering the North Sea.

From the above it is clear that all coastal zones of the world are vulnerable to the range of possible impacts from sea-level rise and other climate-induced impacts, although to different degrees. Studies using other approaches than the Common Methodology support this conclusion. Small island developing states are often judged to be among the most vulnerable countries. In Box 9-3, case material is presented for small islands, centering chiefly on threats to small economies dominated by tourism.

9.5.2.3. Global Vulnerability Assessment

In addition to local and country vulnerability assessments, a Global Vulnerability Assessment (GVA), which provides a worldwide estimate of the socioeconomic and ecological implications of accelerated sea-level rise (Hoozemans *et al.*, 1993) has been conducted, using the same scenarios as the Common Methodology. The GVA has provided estimates of the following impacts: *population at risk*, the average number of people per year subject to flooding by storm surge on a global scale; *wetlands at loss*, the ecologically valuable coastal wetland area under serious threat of loss on a global scale; and *rice production at change*, the changes in coastal rice yields as a result of less-favorable conditions due to sea-level rise in South, Southeast, and East Asia.

Recently, an extension of the GVA has been prepared using a more refined approach to estimate flooding probabilities (Baarse, 1995). Sea-level rise scenarios of both 50 cm and 1 m have been considered. The data sets available for global-scale analysis are limited, and important assumptions are necessary with regard to storm-surge probability and population distribution. Also, increases in wave height and wave run-up have not been taken into account in these analyses, and neither have socioeconomic changes such as population growth. Therefore, the results of both studies must be considered as first-order estimates.

Some conclusions drawn from Hoozemans *et al.* (1993) and Baarse (1995) include:

- Presently, some 200 million people are estimated to live below the "maximum" storm-surge level (the once-per-1000-years storm-surge level). Based on this population estimate, as well as on first-order estimates of storm-surge probabilities and existing levels of protection, 46 million people are estimated to experience flooding due to storm surge in an average year under present conditions. Most of these people live in the developing world.
- The present number of people at risk will double if sea level rises 50 cm (92 million people/yr) and almost triple if it rises 1 m (118 million people/yr).
- The average number of people who will experience coastal flooding more than once per year will increase considerably under both scenarios (80–90% of the respective populations at risk). This estimate underlines that many people will have to adapt to sea-level rise by moving to higher ground, increasing protection efforts or other adaptation options (see Section 9.6).
- Because of regional differences in storm-surge regimes, the increase of flood risk due to sea-level rise is greater than average for the Asian region (especially the Indian Ocean coast), the south Mediterranean coast, the African Atlantic and Indian Ocean coasts, Caribbean coasts, and many of the small islands.
- All over the world, coastal wetlands are presently being lost at an increasingly rapid rate, averaging 0.5–1.5% per year. These losses are closely connected with human activities such as shoreline protection, blocking of sediment sources, and development activities such as land reclamation, aquaculture development, and oil, gas, and water extraction.
- Sea-level rise would increase the rate of net coastal wetland loss. Losses of coastal wetlands of international importance are expected to be greater than average for the coasts of the United States, the Mediterranean Sea, the African Atlantic coast, the coast of East Asia, and the Australian and Papua New Guinean coast.
- Approximately 85% of the world's rice production takes place in South, Southeast, and East Asia. About 10% of this production is located in areas that are considered to be vulnerable to sea-level rise, thereby endangering the food supply of more than 200 million people.
- Less-favorable hydraulic conditions may cause lower rice production yields if no adaptive measures are taken, especially in the large deltas of Vietnam, Bangladesh, and Myanmar.

In summary, the GVA confirms that sea-level rise will have global impacts and reinforces the need for more refined vulnerability assessments at regional and local scales.

9.5.2.4. *Overview of Impact Assessment*

Vulnerability assessment has demonstrated that certain settings are more vulnerable to sea-level rise, including small islands (particularly coral atolls), nations with large deltaic areas, coastal wetlands, and developed sandy shores. However, vulnerability assessment has been less successful in assessing the range of response options to deal with the problems of climate change. Therefore, vulnerability analysis has further utility for countries and areas where none has yet been carried out or where only preliminary studies are available. Even in many areas with a completed vulnerability assessment, an assessment of additional sea-level rise scenarios, scenarios of other impacts of climate change, and a wider range of response options remains necessary. This entails a greater emphasis on local conditions and careful evaluation of progressive adaptation options.

Problems and deficiencies with the Common Methodology have been indicated in several papers in O'Callahan (1994) and McLean and Mimura (1993) [e.g., Kay and Waterman (1993)], and recommendations have been made for integrating vulnerability assessments into the process of coastal zone management (WCC'93, 1994). In order to continue vulnerability assessment studies in a more complete form, approaches should be developed that more readily meet biogeophysical, socioeconomic, and cultural conditions, as well as governmental and jurisdictional arrangements (e.g., Yamada *et al.*, 1995). Common approaches or frameworks that are tailored to the geographic circumstances and needs of each nation should be consistent with the IPCC Technical Guidelines for Assessing Climate Change Impacts and Adaptation (Carter *et al.*, 1994) and need to take into account and correct the weaknesses found with the Common Methodology (McLean and Mimura, 1993; WCC'93, 1994).

9.6. **Response Strategies**

There is no doubt that the threat of climate change and sea-level rise has focused attention on coastal zones and small islands and awakened awareness of the vulnerability of the world's coastal regions in general—and to low-lying coasts, tidal deltas, and small islands in particular. IPCC CZMS (1990, 1992) have distinguished three groups of response strategies: (planned) retreat, accommodate, and protect. The first involves strategic retreat from or the prevention of future major developments in coastal areas that may be impacted. The second includes adaptive responses such as elevation of buildings, modification of drainage systems, and land-use changes. Both strategies are based on the premise that increases in land loss and coastal flooding will be allowed to occur and that some coastal functions and values will change or be lost. On the other hand, these strategies help to maintain the dynamic nature of coastal ecosystems and thus allow them to adapt naturally. The third strategy involves defensive measures and seeks to maintain shorelines at their present position by either building or strengthening protective structures or by artificially nourishing or maintaining beaches and dunes. This strategy could involve the loss of natural functions and values.

As discussed in Section 9.5, vulnerability assessment of various forms provides a range of procedures for a first overview of the consequences of climate change to coastal nations. These procedures also include guidance to a survey of response strategies and a country's capacity to implement those strategies in the context of management and planning of coastal areas. For the majority of coastal nations surveyed, the typical problems posed by sea-level rise (i.e., increased coastal erosion, inundation, flooding, saltwater intrusion) are not uniformly threatening. This does not imply that serious problems will not arise—only that there may be feasible, cost-effective adaptation options.

From Section 9.5, four major areas of concern regarding sea-level rise have emerged: inundation and increased flooding of low-lying islands, inundation and increased flooding of large parts of densely populated deltaic areas, loss of coastal wetlands, and erosion of developed sandy coasts. Each of these areas may require different options to reduce or prevent the prospective adverse impacts associated with biogeophysical changes caused by climate change and sea-level rise. In addition, strategies to reduce the vulnerability of coastal zones and small islands to climate change and sea-level rise should not be seen independently of the resolution of short-term problems arising primarily from human activities. Effective adaptation to climate change and sea-level rise therefore requires a flexible coastal management strategy at all timescales, that incorporates and integrates both short-term and long-term goals.

Although one may argue that there is still a considerable amount of time to implement a response strategy, present-day coastal development often adversely influences the effectiveness of long-term adaptation options (WCC'93, 1994). Moreover, considerable time lags are often involved between the planning and implementation of adaptation options (Vellinga and Leatherman, 1989). The World Coast Conference therefore has concluded that strengthening planning and management capabilities for coastal areas should be delayed no further (WCC'93, 1993, 1994).

Assessing the full social costs of adaptation to climate change is a complicated, often controversial, issue, for which a range of techniques will be required. Apart from estimates of protection costs for a number of countries (see also Table 9-3), this chapter does not offer a comprehensive assessment of these full social costs. This has been done in Chapter 6, *The Social Costs of Climate Change: Greenhouse Damage and the Benefits of Control*, and Chapter 7, *A Generic Assessment of Response Options*, of the IPCC Working Group III volume, which also discuss the techniques available for assessing adaptation costs. A number of cost–benefit studies have been undertaken that attempt to determine optimal response strategies in the face of sea-level rise. Studies include Titus (1991), Nijkamp (1991), Fankhauser (1995), and Turner *et al.* (1995a).

9.6.1. Adaptation Options

There is a wide array of adaptation options that can be employed to retreat, accommodate, and protect. Table 9-4 outlines the options within these three strategies, as listed in the IPCC First Assessment Report (IPCC, 1990).

Box 9-4. Approaches to Coastal Adaptation to Climate Change: Two Examples

West and Central Africa: The coastal nations of West and Central Africa (e.g., Senegal, Gambia, Sierra Leone, Nigeria, Cameroon, Gabon, Angola) have mostly low-lying lagoonal, erosive coasts—and hence are likely to be threatened by sea-level rise, particularly since most of the countries in this area have major, rapidly expanding cities on the coast (Tobor and Ibe, 1990; Adam, 1995; Dennis *et al.*, 1995b; G.T. French *et al.*, 1995). Ibe (1990) has found that large-scale protective engineering measures are impractical in the region because of the high costs to those countries. Instead, low-cost, low-technology, but effective measures—such as permeable nonconcrete floating breakwaters; artificial raising of beach elevations; installation of riprap; timber groins; and so forth—are considered to be more sensible. Ibe (1990) has noted that "fortunately, outside the urbanized centers, the coasts are almost in pristine condition and largely uninhabited. Where coasts are deemed highly vulnerable, total ban of new development is absolutely necessary."

The Netherlands: The Dutch Impacts of Sea-Level Rise on Society (ISOS) study has assessed consequences and the possible responses for water management and flood protection in the Netherlands (Peerbolte *et al.*, 1991). Various scenarios of climate change have been considered, including sea-level rise, changes in river discharges, and changes in storm patterns (i.e., wind direction, storm frequency and intensity). Possible adaptation options, including investments in infrastructure and soft measures such as beach and dune nourishment, have been evaluated and optimized in time for the different sets of scenarios. If sea level were to rise by 60 cm over the next century, $3.5 billion would have to be spent on raising dikes and other safety infrastructure, $500 million on preserving the dune areas, $900 million on adapting flood-prone residential and industrial areas and harbors, and $800 million on adapting water-management facilities. An unfavorable change in storm pattern alone would have the same magnitude of impacts. The occurrence of sea-level rise in combination with the implementation of protective measures will further lead to losses of wetland and intertidal area. These losses cannot be prevented by any realistic additional measures.

Table 9-4: *Response strategies to sea-level rise (IPCC, 1990).*

Present Situation

(Planned) Retreat

Emphasis on abandonment of land and structures in highly vulnerable areas and resettlement of inhabitants

- Preventing development in areas near the coast
- Conditional phased-out development
- Withdrawal of government subsidies

Accommodate

Emphasis on conservation of ecosystems harmonized with the continued occupancy and use of vulnerable areas and adaptive management responses

- Advanced planning to avoid worst impacts
- Modification of land use, building codes
- Protection of threatened ecosystems
- Strict regulation of hazard zones
- Hazard insurance

Protect

Emphasis on defense of vulnerable areas, population centers, economic activities, and natural resources

- Hard structural options
  - Dikes, levees, and floodwalls
  - Sea walls, revetments, and bulkheads
  - Groins
  - Detached breakwaters
  - Floodgates and tidal barriers
  - Saltwater intrusion barriers
- Soft structural options
  - Periodic beach nourishment (beach fill)
  - Dune restoration
  - Wetland creation
  - Littoral drift replenishment
  - Afforestation

Traditionally, the emphasis has been on engineering responses to coastal erosion and protection against flooding, with action often being triggered in response to an extreme event. Now the range of options has expanded to include nonstructural adaptation consisting primarily of zoning, building codes, land-use regulation, and flood-damage insurance, with more emphasis on a precautionary approach. Two different approaches to adaptation to anticipated climate change are discussed in Box 9-4.

IPCC CZMS (1990) has identified the environmental, economic, cultural, legal, and institutional implications of the three response strategies. Vulnerability assessments have since made clear that the extreme options of retreat and full protection highlight the negative effects and overestimate the potential costs and losses from climate change and sea-level rise. Yet adaptation options in low-lying island states (e.g., the Marshall Islands, the Maldives) and for nations with large deltaic areas (e.g., Bangladesh, Nigeria, Egypt, China), which have been identified as especially vulnerable, are problematic because the options have not been fully evaluated but appear limited and potentially very costly. Even without climate change and associated sea-level rise, these nations will continue to experience rapidly increasing vulnerability to natural coastal hazards due to high rates of population growth, increased demands, continued unsustainable exploitation of resources in the coastal zone, and development in upstream catchment areas. Continued natural and possible anthropogenic subsidence (relative sea-level rise) of large river deltas will increase the risk from storm surges. Strategies must be devised to reduce the economic damages and social hazards well before climate change and associated sea-level rise become a significant factor (Han *et al.*, 1995b; Boesch *et al.*, 1994). For these situations, conventional adaptation options will have to be enacted as a first step, while innovative or radical solutions—for example, controlled flooding and sedimentation to harness the natural capability of a delta to respond to sea-level rise—are examined for their effectiveness, environmental impact, social acceptability, and economic efficiency.

Heavily populated areas are primary candidates for structural protection measures such as dikes, sea walls, breakwaters, and beach groins. Because these are expensive options, the use of economic evaluation principles—especially risk assessment and benefit-cost analysis—can provide useful tools in deciding whether to protect or retreat, as well as where such infrastructure investments ought to be placed in order to maximize national or regional social and economic welfare (Carter *et al.*, 1994).

Some of the more detailed recent studies of response strategies to sea-level rise have been accomplished for coastal urban areas. In all instances, the problems of sea-level rise are considered to be serious. Many examples can be found in Frassetto (1991) and Nicholls and Leatherman (1995a), including Venice, Hamburg, London, Osaka, St. Petersburg, Shanghai, Hong Kong, Lagos, Alexandria, Recife, and Tianjin. However, Devine (1992) has argued that the shantytown areas found in many coastal cities may be particularly vulnerable to climate change, and adaptation options are uncertain. Chapter 12 further discusses impacts of and adaptation options to climate change in human settlements, including coastal cities.

Kitajima *et al.* (1993) have undertaken a comprehensive analysis of the range of likely structural measures that would be required to respond to a 1-m sea-level rise at 1,100 Japanese ports, harbors, and neighboring areas, as well as their estimated costs. The cumulative undiscounted costs have been estimated at $92 billion. Of that sum, about $63 billion is for raising port facilities and $29 billion is for adjoining shore protection structures (e.g., breakwaters, jetties, embankments). This cost estimate covers about 25% of Japan's coastline; other residential areas would also have to be protected, if that were the most cost-effective option. Further costs would be incurred to maintain existing standards of protection for populated areas as sea level rises (Mimura *et al.*, 1994). For a rise of 1 m, total costs would exceed $150 million. Further, natural shores could be largely lost from Japan's coast. It should be noted that, on an average annual basis, the costs constitute a small fraction of the GNP.

Adaptation can exploit the fact that coastal infrastructure is not static. There is a turnover of many coastal facilities through major rehabilitation, construction, and technological changes in ports, harbors, and urbanized areas, averaging roughly 25–30 years. Therefore, there will be recurring opportunities to adapt to sea-level rise, especially if the rate is relatively slow and construction and maintenance plans can be taken into account in land-use planning, management, and engineering design criteria (Stakhiv *et al.*, 1991; Yim, 1995). Moreover, experience with allowances for accelerated sea-level rise is limited but growing (e.g., Nicholls and Leatherman, 1995b). The interaction of different aspects of climate change should be considered. For instance, given the likelihood of both sea-level rise and a decrease in the return period of intense rainfall events (see Section 9.3), more consideration of future drainage capacity requirements in low-lying coastal areas may be prudent (Titus *et al.*, 1987; Nicholls *et al.*, 1995).

Some structural examples of proactive adaptation include:

- In the early 1990s, design standards for new seawalls in The Netherlands and eastern England were raised 66 cm and 25 cm, respectively, to allow for accelerated sea-level rise. This has been in response to the IPCC90 best estimate for future sea-level rise; the different magnitudes reflect a 100-year and 50-year planning horizon, respectively.
- The Massachusetts Water Resources Authority has included an additional 46 cm of height in the Deer Island sewage treatment plant. This is a safety factor to maintain gravity-based flows under higher sea levels without the additional costs of pumping.
- In Hong Kong, the West Kowloon reclamation is being built 80 cm above earlier design levels to allow for sea-level rise and/or unanticipated subsidence (Yim, 1995). In this case, costs have increased by less than 1%; future reclamations are expected to be similarly raised, and existing reclamations may be raised as part of the redevelopment cycle.

Enlarged setbacks to allow for expected shoreline recession, most notably in some states in Australia (Caton and Eliot, 1993), would enable planned retreat to accommodate climate change impacts. A variant on fixed setbacks is presumed mobility—whereby coastal residents are allowed to live at the shore but give up their right to protect the shore if it retreats in response to climate change or other causes (Titus, 1991). To counter the coastal squeeze of wetlands and maintain their habitat and flood-buffering functions, managed retreat on estuarine shorelines is increasingly favored in the United Kingdom (Burd, 1995). This involves setting back the line of actively maintained defense to a new line inland of the original and promoting the creation of intertidal habitat on the land between the old and new defenses.

9.6.2. Implementation Considerations

To date, the assessment of possible response strategies has focused mainly on protection. There is a need to better identify the full range of options within the adaptive response strategies: protect, accommodate, and retreat. Identifying the most appropriate options and their relative costs and implementing these options while taking into account contemporary conditions as well as future problems such as climate change and sea-level rise will be a great challenge in both developing and industrialized countries. The range of options will vary among and within countries, and different socioeconomic sectors may prefer competing adaptation options for the same areas. Experience shows that intersectoral conflicts are a major barrier to improved coastal management (WCC'93, 1994). In the present context, they could also be a major barrier to adaptation to climate change. An appropriate mechanism for coastal

Box 9-5. Integrated Coastal Zone Management

ICZM involves comprehensive assessment, setting of objectives, planning, and management of coastal systems and resources, while taking into account traditional, cultural, and historical perspectives and conflicting interests and uses. It is an iterative and evolutionary process for achieving sustainable development by developing and implementing a continuous management capability that can respond to changing conditions, including the effects of climate change. ICZM includes the following:

- Integration of programs and plans for economic development, environmental quality management, and land use
- Integration of programs for sectors such as food production (including agriculture and fishing), energy, transportation, water resources, waste disposal, and tourism
- Integration of all the tasks of coastal management—from planning and analysis through implementation, operation and maintenance, monitoring, and evaluation—performed continuously over time
- Integration of responsibilities for various tasks of management among levels of government—local, state/provincial, regional, national, international—and between the public and private sectors
- Integration of available resources for management (i.e., personnel, funds, materials, equipment)
- Integration among disciplines [e.g., sciences such as ecology, geomorphology, marine biology; economics; engineering (technology); political science (institutions); and law].

planning under these varying conditions is integrated coastal zone management (ICZM) (see Box 9-5).

There is no single recipe for ICZM; rather it constitutes a portfolio of sociocultural dimensions and structural, legal, financial, economic, and institutional measures. There are many approaches as well as diverse institutional arrangements that can be tailored to the particular culture and style of governance. Yet a number of essential prerequisites can be identified (WCC'93, 1994). The first of these is the need for *initial leadership* for the planning process. The initiative may consist of a centrally led "top-down" approach, a community-based "bottom-up" approach, or something in between. The second necessary element of ICZM is the provision of *institutional arrangements*. This may involve creating new institutions but more commonly will involve improving horizontal and vertical linkages between existing ones. Third, *technical capacity* (both technological and human capacities) is necessary for compiling inventories in the planning phase and during the implementation of the program, and for monitoring the changes. The final necessary element of ICZM is *management instruments*. These include tools ranging from directive to incentive-based, all with the aim of encouraging stakeholders to comply with the goals and objectives of the given ICZM program.

At both UNCED and at the World Coast Conference, ICZM has been recognized as the most appropriate process to deal with current and long-term coastal problems, including degradation of coastal water quality, habitat loss, depletion of coastal resources, changes in hydrological cycles, and, in the longer run, adaptation to sea-level rise and other effects of climate change.

The goal of ICZM is not only to address current and future coastal problems but also to enable coastal societies to benefit from a more efficient and effective way of handling coastal development. Most coastal areas are called on to provide multiple products and services. As demands on coastal resources continue to grow with increasing population and economic development, conflicts could become more common and apparent. ICZM should resolve these conflicts and implement decisions on the mix of uses that best serve the needs of society now and in the future. Also, ICZM is important in the context of the increasingly expressed concern for sustainable development. Sustainable use of any natural resource can be achieved only by having in place a set of integrated management tasks that are financed and carried out continuously (WCC'93, 1994).

There is a persuasive case for taking action about climate change now—to institute or expand ICZM and thus comply with the precautionary approach. Although the time lag between planning and investment in integrated (cross-sectoral) management is longer than that for single-sector management, the returns are significantly greater. A proactive approach to ICZM, in order to enhance the resilience of natural coastal systems and reduce vulnerability, would be beneficial from both an environmental and an economic perspective (Jansen *et al.*, 1995). In addition to reducing vulnerability and enhancing the resilience of developed coastal regions, such initiatives also can encompass the large lengths of shorelines that are presently undeveloped but may be subject to significant pressures in the coming decades. By acting now, future development may be designed to be sustainable and to accommodate the potential impacts of climate change and sea-level rise.

9.6.3. Constraints to Implementation

It is important for governments and policymakers to recognize that although a particular response strategy may appear initially to be appropriate, there are constraining factors that can determine how successfully that option can be implemented (SDSIDS, 1994).

The applicability of any option must be evaluated against (among other things) a background of a country's technology and human resources capability, financial resources, cultural and social acceptability, and the political and legal framework. This is not to suggest that these constraints are insurmountable but that decisionmakers must be realistic when considering the range of options available to them.

9.6.3.1. Technology and Human Resources Capability

For many countries, scarcity of (or lack of access to) appropriate technology and trained personnel will impose limits on the adaptation options realistically available. For example, the design, implementation, and maintenance of "state-of-the-art" civil works may be beyond the immediate reach of many developing nations unless there is technical assistance to provide the required technology and human skills. This is highlighted in vulnerability assessments for a number of countries, such as Tonga (Fifita *et al.*, 1994), Bangladesh (Khan *et al.*, 1994), and Belize (Pernetta and Elder, 1993).

Specifically in the case of protection and accommodation, there will be a need for ongoing maintenance and periodic replacement and upgrade. These activities will also require access to the relevant technology and skills to remain effective.

9.6.3.2. Financial Limitations

The implementation of any adaptation option—whether retreat, accommodate, or protect—will necessitate certain financial commitments from governments, although the level of required funding may vary widely from one option to another. In the case of planned retreat, substantial infrastructure would have to be rebuilt and settlements relocated to less-vulnerable areas, at high reinvestment costs. Adjustment strategies might entail acceptance of less-than-ideal circumstances, while simultaneously increasing the costs of reducing flood risks. Protection strategies almost always involve "hard" engineering structures, which are costly both to construct and to maintain. In the Maldives, for example, the present costs of shoreline protection are close to $13,000 per m; in Senegal, Benin, Antigua, Egypt, Guyana, the Marshall Islands, St. Kitts-Nevis, and Uruguay, maintenance of the existing shoreline against a 1-m rise in sea level could require substantial funding compared with the nation's GNP (Nicholls, 1995). However, it should be noted that national responses to climate change will more likely comprise a variable combination of planned retreat, accommodation, and protection; hence, lower-cost responses probably are available in some areas (e.g., Turner *et al.*, 1995a; Volonté and Nicholls, 1995).

Clearly, any combination of response strategies will be largely influenced by monetary considerations, necessitating both short-term investment and a commitment to longer-term maintenance and replacement costs. Many developing countries will find it especially difficult to meet such costs and will increasingly have to turn to donor countries and international agencies for assistance. In Kiribati, for instance, it has been demonstrated that implementation of protection measures especially will almost certainly require external assistance (Abete, 1993). Lack of adequate financial resources will also circumscribe a country's capacity to "purchase" appropriate technology and human skills required for the implementation of various options. Countries should therefore consider designing efficient, least-cost response plans, based on some realistic assessment of what their economies will be able to sustain (WCC'93, 1994).

9.6.3.3. Cultural and Social Acceptability

Although certain options may be technically and financially possible in a given set of circumstances, they may, at the same time, be culturally and socially disruptive. In some societies, resettlement, for example, would lead to dislocation of social and cultural groups and might even involve the loss of cultural norms and values and the assimilation of new ones. Additionally, an option involving planned retreat could mean the loss of access to communally owned resources and land entitlements, which might undermine the entire economic, social, and cultural base of some communities. Other adaptive measures, such as the construction of "hard" engineering structures, could cause the partial or total elimination of access to traditional fishing, hunting, and culturally important sites.

9.6.3.4. Political and Legal Framework

The extent to which a given response strategy can be successfully employed may well be influenced by political and legal considerations (Freestone and Pethick, 1990). Retreat options, for example, might prove infeasible given the policy and legal structures of the "receiving" area. Where international resettlement is indicated, these issues can become even more complex—as demonstrated by the plight of refugees worldwide. Further, some options will be incompatible with existing systems of land tenure and ownership and in some societies would necessitate a fundamental change in arrangements prior to implementation to avoid violating certain rights. Failing this, governments could be called upon to provide substantial compensation to communities for loss of property and resource-use rights.

Strategies that lead to coastal land loss might also have an undesirable impact on a country's Exclusive Economic Zone (e.g., Aparicio-Castro *et al.*, 1990). This could lead to international legal disputes concerning ownership and use of resources. In those circumstances, such options might not only be considered legally unacceptable but politically infeasible as well.

9.6.4. Overcoming the Constraints

As a step toward overcoming these constraints, the World Coast Conference was organized with the objective of bringing together coastal experts and policymakers to identify actions that can be taken to strengthen capabilities for progressive sustainable development and integrated coastal zone management. The conference participants acknowledged that there is an urgent need for coastal states to strengthen their capabilities, in particular with regard to the exchange of information, education and training; the development of concepts and tools; research, monitoring, and evaluation; and funding (WCC'93, 1993). The following are examples of measures that could improve capabilities for developing, implementing, and strengthening national programs for ICZM (WCC'93, 1993):

- Multidisciplinary studies and assessments to determine the potential importance of the coastal zone and its vulnerabilities, particularly those that limit its ability to achieve sustainable development
- An institutional body or mechanism to investigate the need and potential benefits and costs of developing an ICZM program
- A long-term and effective body or mechanism to prepare, recommend, and coordinate the implementation of a permanent ICZM program
- A continuing monitoring and assessment program to collect data, assess results, and identify the need for change or improvement
- An ongoing research program, including an investigation of the potential effects of global climate change, to improve the analytical foundation for the decisionmaking process
- A policy to increase the availability and accessibility of information to all interested parties
- Active support for local initiatives, exchange of practical and indigenous experiences, and enhancement of public participation
- Education, training, and public-awareness efforts to increase the constituency for ICZM
- Coordination of financial support for relevant activities and investigation of innovative sources for additional support.

Effective ICZM can be achieved by coordination among national, regional, and international organizations and institutions. This will help to avoid unnecessary duplication and develop the concepts, tools, and networks needed to facilitate the development and implementation of national programs, which is a complex process that can be accelerated and enhanced through international cooperation. Regional approaches can complement and strengthen activities at the national and international levels.

Various international initiatives have been undertaken to encourage and facilitate coordination and cooperation in both policy and research. Several United Nations organizations and other international governmental and nongovernmental organizations have developed programs aimed at strengthening ICZM capabilities at different levels. An overview of these activities is presented in WCC'93 (1994). In 1993, the IGBP launched its core project LOICZ, which aims to stimulate interdisciplinary scientific coastal research in the context of global change (Pernetta and Milliman, 1995).

Clearly, the wide range of uncertainty in human and natural variables that will affect ICZM emphasizes the need for continued research and monitoring. The results of scientific research and information from monitoring activities need to be integrated into policy development, planning, and decisionmaking throughout the ICZM process.

9.7. Research and Monitoring Needs

Although much has been achieved since the IPCC First Assessment in 1990, this chapter shows that the understanding of the likely consequences of climate change and sea-level rise is still imperfect. This situation can be improved only through a sustained research and monitoring effort, requiring a major commitment of resources at the national, regional, and global levels. The potential problems of small islands, deltas, coastal wetlands, and developed sandy coasts deserve particular attention as part of these efforts. Coastal zones and small islands illustrate the fundamental need for better coupling of research and models from the natural sciences and the social sciences to provide improved analytical capability and information to decisionmakers. An emphasis on understanding the impacts of climate change at the local and regional scales is essential.

There are a number of critical issues and priorities in ongoing research and monitoring, as initiated by IPCC CZMS (1990, 1992), that should be continued over the next few years to enable better decisionmaking concerning the possible impacts of climate change in coastal zones and small islands:

- Development of improved biogeophysical classifications and frameworks of coastal types for climate-change analysis, including the influence of human activities (IGBP-LOICZ has taken an important step in this direction)
- Investigations of geomorphological and biological responses of coastal types and critical ecosystems to climate change and sea-level rise, with specific attention to the response of seagrass to climate change as well as potential changes in sediment budgets
- Improved methodologies for incorporating existing, high-quality historical and geological coastal-change data into response models for climate change
- Improved coastal-processes data (especially in developing countries), based on instrumentation (tide gauges, current meters, wave recorders, etc.), as well as improved capacity to interpret and analyze the data
- Improved databases for vulnerability assessment and adaptation planning on coastal socioeconomic trends, such as population changes and resource utilization and valuation, taking into consideration differences in sociocultural characteristics of countries and ethnic groups

- Extension of new and existing vulnerability assessment studies to include a range of local scenarios of sea-level rise (rather than a single scenario), other possible impacts of climate change such as changing storminess or precipitation, and an assessment of the range of possible adaptation strategies
- Continued education and training relevant to vulnerability assessment and integrated coastal zone management, employing, as far as practicable, standardized methodologies and frameworks.

Some ongoing initiatives have already been undertaken to address these priorities. For example, the Intergovernmental Oceanographic Commission (IOC) coordinates a Global Sea Level Observing System (GLOSS). However, there still are many gaps, and the system requires increased international support and coordination. IGBP-LOICZ aims to stimulate interdisciplinary scientific coastal research in the context of global change. Focus 2 of LOICZ aims to investigate coastal biogeomorphological interactions under different global-change scenarios. It focuses on the interaction between major ecosystem types with the sedimentary environment and aims to assess the implications of ecosystem perturbations on coastal stability with a rise in sea level. Focus 4 is especially relevant for integrated coastal zone management because it addresses the socioeconomic impacts of global change on coastal zones and aims to investigate how improved strategies for the management of coastal resources can be developed.

New initiatives also are required, as has been recognized by participants of the World Coast Conference (WCC'93, 1994). Increased efforts are needed mainly in the social sciences and adaptation to climate change, including the development of:

- Integrated coastal-response models, which seek to combine the interactions of biogeophysical, socioeconomic, and climate-change factors, incorporating the knowledge and technologies of traditional societies and local peoples
- Methods to quantify the benefits of integrated coastal zone management
- A broad framework for the analysis, planning, and management of coastal zones in the context of climate change, recognizing the co-evolution of natural and social systems.

Such a framework would encourage nations in the formulation and implementation of ICZM strategies and programs that are appropriate to climate change and fully take into account the existing environmental, social, cultural, political, governance, and economic contexts. This would help to fulfill the recommendations made by IPCC CZMS (1990).

To facilitate these goals, an international conference to share experience on coastal impacts and adaptation to climate change might be useful, building on the success of earlier meetings of the IPCC Coastal Zone Management Subgroup and the World Coast Conference. Particular targets for such a conference could include:

- An updated assessment of coastal vulnerability to climate change
- Examples of biogeophysical/socioeconomic integration in coastal research and coastal management
- Testing of the framework for ICZM via case studies in a range of countries or regions.

These activities could assist coastal nations in meeting their obligations under Agenda 21 of UNCED.

Acknowledgments

This chapter is only one small step on the road to global sustainable coastal development and could not have been written if many earlier steps had not been taken. For many decades, scientists and policymakers from all over the world have devoted themselves to coastal research and management; the list of references below reflects only a minor part of their efforts. Hence, the authors would like to thank every person and institution who has contributed to the increased understanding of coastal systems and the growing awareness of the importance of their healthy functioning. Special thanks are due to the contributors and reviewers of this chapter and to Mark Naber and Willem Storm (National Institute for Coastal and Marine Management, The Netherlands), who prepared the figures.

References

Aalbersberg, B. and J. Hay, 1993: *Implications of Climate Change and Sea Level Rise for Tuvalu*. SPREP Reports and Studies Series No. 54, South Pacific Regional Environment Programme, Apia, Western Samoa.

Abete, T., 1993: The Kiribati preliminary assessment to accelerated sea-level rise. In: *Vulnerability Assessment to Sea Level Rise and Coastal Zone Management* [McLean, R. and N. Mimura (eds.)]. Proceedings of the IPCC/WCC'93 Eastern Hemisphere workshop, Tsukuba, 3-6 August 1993, Department of Environment, Sport and Territories, Canberra, Australia, pp. 91-98.

Adam, K.S., 1995: Vulnerability assessment and coastal management program in the Benin coastal zone. In: *Preparing to Meet the Coastal Challenges of the 21st Century*, vol. 2. Proceedings of the World Coast Conference, Noordwijk, 1-5 November 1993, CZM-Centre Publication No. 4, Ministry of Transport, Public Works and Water Management, The Hague, The Netherlands, pp. 489-497.

Aksornkaoe, S. and N. Paphavasit, 1993: Effect of sea-level rise on the mangrove community in Thailand. *Malaysian Journal of Tropical Geography*, **24**, 29-34.

Allen, J.R.L., 1991: Salt-marsh accretion and sea-level movement in the inner Severn Estuary: the archaeological and historical contribution. *Journal of the Geological Society*, **148**, 485-494.

Aparicio-Castro, R., J. Castaneda, and M. Perdomo, 1990: Regional implications of relative sea level rise and global climate change along the marine boundaries of Venezuela. In: *Changing Climate and the Coast, Vol. 2* [Titus, J.G. (ed.)]. Proceedings of the first IPCC CZMS workshop, Miami, 27 November –1 December 1989, Environmental Protection Agency, Washington, DC, pp. 385-397.

ASCE Task Committee, 1992: Effects of sea level rise on bays and estuaries. *Journal of Hydraulic Engineering*, **118**, 1-10.

Aubrey, D.G. and K.O. Emery, 1993: Recent global sea levels and land levels. In: *Climate and Sea Level Change: Observations, Projections and Implications* [Warrick, R.A., E.M. Barrow, and T.M.L. Wigley (eds.)]. Cambridge University Press, Cambridge, UK, pp. 45-56.

Aung, N., 1993: Myanmar coastal zone management. In: *Vulnerability Assessment to Sea Level Rise and Coastal Zone Management* [McLean, R. and N. Mimura (eds.)]. Proceedings of the IPCC/WCC'93 Eastern Hemisphere workshop, Tsukuba, 3-6 August 1993, Department of Environment, Sport and Territories, Canberra, Australia, pp. 333-340.

Baarse, G., 1995: *Development of an Operational Tool for Global Vulnerability Assessment (GVA): Update of the Number of People at Risk Due to Sea-Level Rise and Increased Flooding Probabilities*. CZM-Centre Publication No. 3, Ministry of Transport, Public Works and Water Management, The Hague, The Netherlands.

Bangladesh Government, 1993: *Assessment of the Vulnerability of Coastal Areas to Climate Change and Sea Level Rise: A Pilot Study of Bangladesh*. Bangladesh Government, Dhaka, Bangladesh.

Barbier, E.B., 1989: *The Economic Value of Ecosystems: 1—Tropical Wetlands*. LEEC Gatekeeper 89-02, London Environmental Economics Centre, London, UK.

Barbier, E.B., 1994: Valuing environmental functions: tropical wetlands. *Land Economics*, **70**, 155-173.

Bateman, I.J., I.H. Langford, R.K. Turner, K.G. Willis, and G.D. Garrod, 1995: Elicitation and truncation effects in contingent valuation studies. *Ecological Economics*, **12**, 161-179.

Baumann, R.H., J.W. Day, and C. Miller, 1984: Mississippi deltaic wetland survival: sedimentation versus coastal submergence. *Science*, **224**, 1093-1095.

Belperio, A.P., 1993: Land subsidence and sea-level rise in the Port Adelaide Estuary: implications for monitoring the greenhouse effect. *Australian Journal of Earth Sciences*, **40**, 359-368.

Bengtsson, L., M. Botzet, and M. Esch, 1994: *Hurricane-Type Vortices in a General Circulation Model: Part I*. MPI Report No. 123, Max-Planck-Institut für Meteorologie, Hamburg, Germany.

Bijlsma, L., R. Misdorp, L.P.M. de Vrees, M.J.F. Stive, G. Baarse, R. Koudstaal, G. Toms, F.M.J. Hoozemans, C. Hulsbergen, and S. van der Meij, 1993: Changing coastal zones: chances for sustainable development. *Coastline*, **4**, Coastline Special.

Bilan, D., 1993: The preliminary vulnerability assessment of the Chinese coastal zone due to sea level rise. In: *Vulnerability Assessment to Sea Level Rise and Coastal Zone Management* [McLean, R. and N. Mimura (eds.)]. Proceedings of the IPCC/WCC'93 Eastern Hemisphere workshop, Tsukuba, 3-6 August 1993, Department of Environment, Sport and Territories, Canberra, Australia, pp. 177-188.

Bird, E.C.F., 1985: *Coastline Changes: A Global Review*. Wiley-Interscience, Chichester, UK.

Bird, E.C.F., 1993a: Sea level rise impacts in Southeast Asia. *Malaysian Journal of Tropical Geography*, **24**, 1-110.

Bird, E.C.F., 1993b: *Submerging Coasts: The Effects of a Rising Sea Level on Coastal Environments*. John Wiley, Chichester, UK.

Boesch, D.F., M.N. Josselyn, A.J. Mehta, J.T. Morris, W.K. Nuttle, C.A. Simenstad, and D.J.P. Swift, 1994: Scientific assessment of coastal wetland loss, restoration and management in Louisiana. *Journal of Coastal Research*, special issue **20**, 1-103.

Broccoli, A.J. and S. Manabe, 1990: Can existing climate models be used to study anthropogenic changes in tropical cyclone climate? *Geophysical Research Letters*, **17**, 1917-1920.

Brown, B.E., 1987: Worldwide death of corals: natural cyclical events or man-made pollution? *Marine Pollution Bulletin*, **18**, 9-13.

Brown, B.E., R.P. Dunne, T.P. Scoffin, and M.D.A. le Tissier, 1994: Solar damage in intertidal corals. *Marine Ecological Progress Series*, **105**, 219-230.

Brown, B.E. and J.C. Ogden, 1993: Coral bleaching. *Scientific American*, **268**, 64-70.

Brown, B.E. and Suharsono, 1990: Damage and recovery of coral reefs affected by El Niño related seawater warming in the Thousand Islands, Indonesia. *Coral Reefs*, **8**, 163-170.

Bruun, P., 1962: Sea-level rise as a cause of shore erosion. *Journal of the Waterways and Harbors Division*, Proceedings of the American Society of Civil Engineers, **88**, 117-130.

Buddemeier, R.W. and S.V. Smith, 1988: Coral reef growth in an era of rapidly rising sea level: predictions and suggestions for long-term research. *Coral Reefs*, **7**, 51-56.

Buddemeier, R.W. and J.A. Oberdorfer, 1990: Climate change and groundwater reserves. In: *Implications of Expected Climatic Changes in the South Pacific Region: An Overview* [Pernetta, J.C. and P.J. Hughes (eds.)]. UNEP Regional Seas Reports and Studies No. 128, United Nations Environment Programme, Nairobi, Kenya, pp. 56-67.

Burd, F., 1995: *Managed Retreat: A Practical Guide*. Campaign for a Living Coast, English Nature, Peterborough, UK.

Cambers, G., 1994: Assessment of the vulnerability of coastal areas in Antigua and Nevis to sea level rise. In: *Global Climate Change and the Rising Challenge of the Sea* [O'Callahan, J. (ed.)]. Proceedings of the third IPCC CZMS workshop, Margarita Island, 9-13 March 1992, National Oceanic and Atmospheric Administration, Silver Spring, MD, pp. 11-27.

Carter, R.W.G. and J.D. Orford, 1993: The morphodynamics of coarse clastic beaches and barriers: a short- and long-term perspective. *Journal of Coastal Research*, special issue **15**, 158-170.

Carter, T.R., M.C. Parry, S. Nishioka, and H. Harasawa (eds.), 1994: *Technical Guidelines for Assessing Climate Change Impacts and Adaptations*. Report of Working Group II of the Intergovernmental Panel on Climate Change, University College London and Centre for Global Environmental Research, London, UK, and Tsukuba, Japan.

Caton, B. and I. Eliot, 1993: Coastal hazard policy development and the Australian federal system. In: *Vulnerability Assessment to Sea Level Rise and Coastal Zone Management* [McLean, R. and N. Mimura (eds.)]. Proceedings of the IPCC/WCC'93 Eastern Hemisphere workshop, Tsukuba, 3-6 August 1993, Department of Environment, Sport and Territories, Canberra, Australia, pp. 417-427.

Chappell, J., 1990: The effects of sea level rise on tropical riverine lowlands. In: *Implications of Expected Climate Changes in the South Pacific Region: An Overview* [Pernetta, J.C. and P.J. Hughes (eds.)]. UNEP Regional Seas Reports and Studies No. 128, United Nations Environment Programme, Nairobi, Kenya, pp. 28-35.

Chappell, J. and C.D. Woodroffe, 1994: Macrotidal estuaries. In: *Coastal Evolution: Late Quaternary Shoreline Morphodynamics* [Carter, R.W.G. and C.D. Woodroffe (eds.)]. Cambridge University Press, Cambridge, UK, pp. 187-218.

Church, J.A., J.S. Godfrey, D.R. Jackett, and T.J. McDougall, 1991: A model of sea level rise caused by ocean thermal expansion. *Journal of Climate*, **4**, 438-456.

Connell, J. and M. Maata, 1992: *Environmental Planning, Climate Change and Potential Sea Level Rise: Report on a Mission to the Republic of the Marshall Islands*. SPREP Reports and Studies Series No. 55, South Pacific Regional Environment Programme, Apia, Western Samoa.

Constanza, R., S.C. Farber, and J. Maxwell, 1989: Valuation and management of wetland ecosystems. *Ecological Economics*, **1**, 335-361.

Cowell, P.J. and B.G. Thom, 1994: Morphodynamics of coastal evolution. In: *Coastal Evolution: Late Quaternary Shoreline Morphodynamics* [Carter, R.W.G. and C.D. Woodroffe (eds.)]. Cambridge University Press, Cambridge, UK, pp. 33-86.

Cubasch, U., B.D. Santer, A. Hellback, G. Hegerl, K.H. Hock, E. Maier-Reimer, U. Mikolajewicz, A. Stossel, and R. Voss, 1992: *Monte-Carlo Climate Change Forecasts with a Global Coupled Ocean-Atmosphere Model*. MPI Report No. 97, Max-Planck-Institut für Meteorologie, Hamburg, Germany.

Day, J.W., W.H. Conner, R. Constanza, G.P. Kemp, and I.A. Mendelssohn, 1993: Impacts of sea level rise on coastal systems with special emphasis on the Mississippi River deltaic plain. In: *Climate and Sea Level Change: Observations, Projections and Implications* [Warrick, R.A., E.M. Barrow, and T.M.L. Wigley (eds.)]. Cambridge University Press, Cambridge, UK, pp. 276-296.

Day, J.W., D. Pont, C. Ibanez, P.F. Hensel, 1994: Impacts of sea level rise on deltas in the Gulf of Mexico and the Mediterranean: human activities and sustainable management. In: *Consequences for Hydrology and Water Management*. UNESCO International Workshop Seachange '93, Noordwijkerhout, 19-23 April 1993, Ministry of Transport, Public Works and Water Management, The Hague, The Netherlands, pp. 151-181.

De Groot, R.S., 1992a: *Functions of Nature: Evaluation of Nature in Environmental Planning, Management and Decision Making*. Wolters-Noordhoff, Groningen, The Netherlands.

De Groot, R.S., 1992b: Functions and economic values of coastal protected areas. In: *Economic Impact of the Mediterranean Coastal Protected Areas*. Proceedings of MEDPAN, Ajaccio, 26-28 September 1991. Special issue, *MEDPAN Newsletter*, **3**, 67-83.

Delft Hydraulics, Resource Analysis, Ministry of Transport, Public Works and Water Management and Coastal Research Institute, 1992: *Vulnerability Assessment to Accelerated Sea Level Rise, Case Study Egypt*. Delft Hydraulics, Delft, The Netherlands.

Dennis, K.C., E.J. Schnack, F.H. Mouzo, and C.R. Orona, 1995a: Sea-level rise and Argentina: potential impacts and consequences. *Journal of Coastal Research*, special issue **14**, 205-223.

Dennis, K.C., I. Niang-Diop, and R.J. Nicholls, 1995b: Sea-level rise and Senegal: potential impacts and consequences. *Journal of Coastal Research*, special issue **14**, 243-261.

Devine, N.P., 1992: *Urban Vulnerability to Sea-Level Rise in the Third World*. M.A. thesis, State University of New Jersey, New Brunswick, NJ.

Dixon, J.A., 1989: Valuation of mangroves. *Tropical Coastal Area Management*, **4**, 1-6.

Downs, L.L., R.J. Nicholls, S.P. Leatherman, and J. Hautzenroder, 1994: Historic evolution of a marsh island: Bloodsworth Island, Maryland. *Journal of Coastal Research*, **10**, 1031-1044.

Dubois, R.N., 1992: A re-evaluation of Bruun's rule and supporting evidence. *Journal of Coastal Research*, **8**, 616-628.

Eakin, C.M., 1995: Post-El Niño Panamanian reefs: less accretion, more erosion and damselfish protection. In: *Proceedings of the Seventh Coral Reef Symposium*, vol. 1, Guam, 22-27 June 1992. University of Guam Press, Manqilao, Guam, pp. 387-396.

Edwards, A.J., 1995: Impact of climate change on coral reefs, mangroves and tropical seagrass ecosystems. In: *Climate Change: Impact on Coastal Habitation* [Eisma, D. (ed.)]. Lewis Publishers, Boca Raton, FL, pp. 209-234.

Ehler, C.N. (ed.), 1993: *Preparatory Workshop on Integrated Coastal Zone Management and Responses to Climate Change*. Proceedings of the IPCC/WCC'93 Western Hemisphere workshop, New Orleans, 13-16 July 1993, National Oceanic and Atmospheric Administration, Silver Spring, MD.

Ellison, J.C., 1993: Mangrove retreat with rising sea level, Bermuda. *Estuarine, Coastal and Shelf Science*, **37**, 75-87.

Ellison, J.C. and D.R. Stoddart, 1991: Mangrove ecosystem collapse during predicted sea-level rise: Holocene analogues and implications. *Journal of Coastal Research*, **7**, 151-165.

El-Raey, M., S. Nasr, O. Frihy, S. Desouki, and K. Dewidar, 1995: Potential impacts of accelerated sea-level rise on Alexandria Governorate, Egypt. *Journal of Coastal Research*, special issue **14**, 190-204.

Emanuel, K.A., 1987: The dependence of hurricane intensity on climate. *Nature*, **326**, 483-485.

Emery, K.O. and D.G. Aubrey, 1991: *Sea Levels, Land Levels, and Tide Gauges*. Springer-Verlag, New York, NY.

Evans, J.L. and R.J. Allan, 1992: El Niño-Southern Oscillation modification to the structure of the monsoon and tropical cyclone activity in the Australasia region. *International Journal of Climatology*, **12**, 611-623.

Fankhauser, S. 1995: Protection vs. retreat: estimating the costs of sea level rise. *Environment and Planning A*, **27**, 299-319.

Fifita, P.N., N. Mimura, and N. Hori, 1994: Assessment of the vulnerability of the Kingdom of Tonga to sea level rise. In: *Global Climate Change and the Rising Challenge of the Sea* [O'Callahan, J. (ed.)]. Proceedings of the third IPCC CZMS workshop, Margarita Island, 9-13 March 1992, National Oceanic and Atmospheric Administration, Silver Spring, MD, pp. 119-139.

Fortes, M.D., 1988: Mangroves and seagrass beds of East Asia: habitats under stress. *Ambio*, **17**, 207-213.

Frassetto, R. (ed.), 1991: *Impact of Sea Level Rise on Cities and Regions*. Proceedings of the first international meeting "Cities on Water," Venice, 11-13 December 1989, Marsilio Editori, Venice, Italy.

Freestone, D. and J. Pethick, 1990: International legal implications of coastal adjustments under sea-level rise: active or passive policy responses? In: *Changing Climate and the Coast*, vol. 2 [Titus, J.G. (ed.)]. Proceedings of the first IPCC CZMS workshop, Miami, 27 November –1 December 1989, Environmental Protection Agency, Washington, DC, pp. 237-256.

French, G.T., L.F. Awosika, and C.E. Ibe, 1995: Sea-level rise in Nigeria: potential impacts and consequences. *Journal of Coastal Research*, special issue **14**, 224-242.

French, J., 1993: Numerical simulation of vertical marsh growth and adjustment to accelerated sea level rise, North Norfolk, UK. *Earth Surface Processes and Landforms*, **18**, 63-81.

French, J.R., T. Spencer, and D.J. Reed (eds.), 1995: Geomorphic Response to Sea-Level Rise. *Earth Surface Processes and Landforms*, **20**, 1-103.

Gates, W.L., P.R. Rowntree, Q.-C. Zeng, 1990: Validation of climate models. In: *Climate Change: The IPCC Scientific Assessment* [Houghton, J.T., G.J. Jenkins, and J.J. Ephraums (eds.)]. Cambridge University Press, Cambridge, UK, pp. 93-130.

Gates, W.L., J.F.B. Mitchell, G.J. Boer, U. Cubasch, and V.P. Meleshko, 1992: Climate modelling, climate prediction and model validation. In: *Climate Change 1992: The Supplementary Report to the IPCC Scientific Assessment* [Houghton, J.T., B.A. Callander, and S.K. Varney (eds.)]. Cambridge University Press, Cambridge, UK, pp. 97-134.

Glynn, P.W., 1993: Coral reef bleaching: ecological perspectives. *Coral Reefs*, **12**, 1-17.

Gordon, H.B., P.H. Whetton, A.B. Pittock, A.M. Fowler, and M.R. Haylock, 1992: Simulated changes in daily rainfall intensity due to the enhanced greenhouse effect: implications for extreme rainfall events. *Climatic Dynamics*, **8**, 83-102.

Gornitz, V.M., 1991: Global coastal hazards from future sea level rise. *Global and Planetary Change*, **89**, 379-398.

Gregory, J.M., 1993: Sea-level changes under increasing atmospheric CO_2 in a transient coupled ocean-atmosphere GCM experiment. *Journal of Climate*, **6**, 2247.

Griggs, G.B. and A.S. Trenhaile, 1994: Coastal cliffs and platforms. In: *Coastal Evolution: Late Quaternary Shoreline Morphodynamics* [Carter, R.W.G. and C.D. Woodroffe (eds.)]. Cambridge University Press, Cambridge, UK, pp. 425-450.

Haarsma, R.J., F.F.B. Mitchell, and C.A. Senior, 1993: Tropical disturbances in a GCM. *Climatic Dynamics*, **8**, 247.

Hameed, H., 1993: *Sustainable Tourism in the Maldives*. M.Phil. thesis, University of East Anglia, Norwich, UK.

Han, M., N. Mimura, Y. Hosokawa, S. Machida, K. Yamada, L. Wu, and J. Li, 1993: Vulnerability assessment of coastal zone to sea level rise: a case study on the Tianjin coastal plain, North China, by using GIS and Landsat imagery. In: *Vulnerability Assessment to Sea Level Rise and Coastal Zone Management* [McLean, R. and N. Mimura (eds.)]. Proceedings of the IPCC/WCC'93 Eastern Hemisphere workshop, Tsukuba, 3-6 August 1993, Department of Environment, Sport and Territories, Canberra, Australia, pp. 189-195.

Han, M., J. Hou, and L. Wu, 1995a: Potential impacts of sea-level rise on China's coastal environment and cities: a national assessment. *Journal of Coastal Research*, special issue **14**, 79-95.

Han, M., J. Hou, L. Wu, C. Liu, G. Zhao, and Z. Zhang, 1995b: Sea level rise and the North China coastal plain: a preliminary analysis. *Journal of Coastal Research*, special issue **14**, 132-150.

Hay, J.E. and C. Kaluwin (eds.), 1993: *Climate Change and Sea Level Rise in the South Pacific Region*. Proceedings of the second SPREP meeting, Noumea, 6-10 April 1992, South Pacific Regional Environment Programme, Apia, Western Samoa.

Healy, T., 1991: Coastal erosion and sea level rise. *Zeitschrift für Geomorphologie*, **81**, 15-29.

Holthus, P., M. Crawford, C. Makroro, and S. Sullivan, 1992: *Vulnerability Assessment for Accelerated Sea Level Rise Case Study: Majuro Atoll, Republic of the Marshall Islands*. SPREP Reports and Studies Series No. 60, South Pacific Regional Environment Programme, Apia, Western Samoa.

Hoozemans, F.M.J., M. Marchand, and H.A. Pennekamp, 1993: *A Global Vulnerability Analysis: Vulnerability Assessment for Population, Coastal Wetlands and Rice Production on a Global Scale*, 2nd ed. Delft Hydraulics and Ministry of Transport, Public Works and Water Management, Delft and The Hague, The Netherlands.

Hopley, D., 1993: Coral reef islands in a period of global sea level rise. In: *Recent Advances in Marine Science and Technology* [Saxena, N.K. (ed.)]. Proceedings of the Pacific Congress on Marine Science and Technology (PACON), Kona, Hawaii, 1-5 June 1992, PACON International, Honolulu, HI, pp. 453-462.

Hopley, D. and D.W. Kinsey, 1988: The effects of a rapid short-term sea level rise on the Great Barrier Reef. In: *Greenhouse: Planning for Climate Change* [Pearman, G.I. (ed.)]. Brill, Leiden, The Netherlands, pp. 189-201.

Huq, S., S.I. Ali, and A.A. Rahman, 1995: Sea-level rise and Bangladesh: a preliminary analysis. *Journal of Coastal Research*, special issue **14**, 44-53.

Ibe, A.C., 1990: Adjustments to the impact of sea level rise along the West and Central African coasts. In: *Changing Climate and the Coast,* vol. 2 [Titus, J.G. (ed.)]. Proceedings of the first IPCC CZMS workshop, Miami, 27 November –1 December 1989, Environmental Protection Agency, Washington, DC, pp. 3-12.

IPCC, 1990: *Climate Change: The IPCC Response Strategies.* Report of the Response Strategies Working Group of the Intergovernmental Panel on Climate Change, World Meteorological Organization and United Nations Environment Programme, Geneva, Switzerland, and Nairobi, Kenya, 273 pp.

IPCC CZMS, 1990: *Strategies for Adaptation to Sea Level Rise.* Report of the Coastal Zone Management Subgroup, IPCC Response Strategies Working Group, Ministry of Transport, Public Works and Water Management, The Hague, The Netherlands.

IPCC CZMS, 1991: *Common Methodology for Assessing Vulnerability to Sea-Level Rise.* Report of the Coastal Zone Management Subgroup, IPCC Response Strategies Working Group, Ministry of Transport, Public Works and Water Management, The Hague, The Netherlands.

IPCC CZMS, 1992: *Global Climate Change and the Rising Challenge of the Sea.* Report of the Coastal Zone Management Subgroup, IPCC Response Strategies Working Group, Ministry of Transport, Public Works and Water Management, The Hague, The Netherlands.

Ives, J.D. and B. Messerli, 1989: *The Himalayan Dilemma: Reconciling Development and Conservation.* Routledge, London, UK.

Jansen, H.M.A., R.J.T. Klein, R.S.J. Tol, and H. Verbruggen, 1995: Some considerations on the economic importance of pro-active integrated coastal zone management. In: *Preparing to Meet the Coastal Challenges of the 21st Century*, vol. 1. Proceedings of the World Coast Conference, Noordwijk, 1-5 November 1993, CZM-Centre Publication No. 4, Ministry of Transport, Public Works and Water Management, The Hague, The Netherlands, pp. 99-105.

Jeftic, L., J. Milliman, and G. Sestini (eds.), 1992: *Climate Change and the Mediterranean.* Edward Arnold, London, UK.

Jogoo, V.K., 1994: Assessment of the vulnerability of Mauritius to sea-level rise. *Global Climate Change and the Rising Challenge of the Sea* [O'Callahan, J. (ed.)]. Proceedings of the third IPCC CZMS workshop, Margarita Island, 9-13 March 1992, National Oceanic and Atmospheric Administration, Silver Spring, MD, pp. 107-118.

Kahn, M. and M.F. Sturm, 1993: *Case Study Report Guyana: Assessment of the Vulnerability of Coastal Areas to Sea Level Rise.* CZM-Centre Publication No. 1, Ministry of Transport, Public Works and Water Management, The Hague, The Netherlands.

Kay, R. and P. Waterman, 1993: Review of the applicability of the "Common Methodology for Assessment of Vulnerability to Sea-Level Rise" to the Australian coastal zone. In: *Vulnerability Assessment to Sea Level Rise and Coastal Zone Management* [McLean, R. and N. Mimura (eds.)]. Proceedings of the IPCC/WCC'93 Eastern Hemisphere workshop, Tsukuba, 3-6 August 1993, Department of Environment, Sport and Territories, Canberra, Australia, pp. 237-246.

Khan, A.H., S. Huq, A.A. Rahman, M. Shahidullah, A. Haque, S.A. Naqi, M. Rahman, S. Ahmed, S.I. Ali, M.Y. Ali, M. Ahmed, Y. Islam, and F. Mollick, 1994: Assessment of the vulnerability of Bangladesh to sea level rise. In: *Global Climate Change and the Rising Challenge of the Sea* [O'Callahan, J. (ed.)]. Proceedings of the third IPCC CZMS workshop, Margarita Island, 9-13 March 1992, National Oceanic and Atmospheric Administration, Silver Spring, MD, pp. 143-155.

Kinsey, D.W., 1991: The coral reef: an owner-built, high-density, fully-serviced, self-sufficient housing estate in the desert—or is it? *Symbiosis*, **10**, 1-22.

Kinsey, D.W. and D. Hopley, 1991: The significance of coral reefs as global carbon sinks: response to greenhouse. *Palaeogeography, Palaeoclimatology, Palaeoecology, Global and Planetary Change Section*, **89**, 363-377.

Kitajima, S., T. Ito, N. Mimura, Y. Hosokawa, M. Tsutsui, and K. Izumi, 1993: Impacts of sea level rise and cost estimate of countermeasures in Japan. In: *Vulnerability Assessment to Sea Level Rise and Coastal Zone Management* [McLean, R. and N. Mimura (eds.)]. Proceedings of the IPCC/WCC'93 Eastern Hemisphere workshop, Tsukuba, 3-6 August 1993, Department of Environment, Sport and Territories, Canberra, Australia, pp. 115-123.

Larkum, A.W.D., A.J. McComb, and S.A. Shepherd (eds.), 1989: *Biology of Seagrasses.* Elsevier, Amsterdam, The Netherlands.

Leatherman, S.P., 1991: Modelling shore response to sea-level rise on sedimentary coasts. *Progress in Physical Geography*, **14**, 447-464.

Leatherman, S.P., 1994: Rising sea levels and small island states. *EcoDecision*, **11**, 53-54.

Levitus, S., 1982: *Climatological Atlas of the World Ocean.* NOAA Professional Paper 13, U.S. Department of Commerce, Washington, DC.

Lighthill, J., G.J. Holland, W.M. Gray, C. Landsea, G. Craig, J. Evans, Y. Kurihara, and C.P. Guard, 1994: Global climate change and tropical cyclones. *Bulletin of the American Meteorological Society*, **75**, 2147-2157.

Lundin, C.G. and O. Linden, 1993: Coastal ecosystems: attempts to manage a threatened resource. *Ambio*, **22**, 468-473.

Maul, G.A. (ed.), 1993: *Climatic Change in the Intra-Americas Sea.* Edward Arnold, London, UK.

McGregor, J.L. and R.J. Walsh, 1993: Nested simulations of perpetual January climate over the Australian region. *Journal of Geophysical Research*, **98**, 23,283-290.

McLean, R. and A.M. d'Aubert, 1993: *Implications of Climate Change and Sea Level Rise for Tokelau.* SPREP Reports and Studies Series No. 61, South Pacific Regional Environment Programme, Apia, Western Samoa.

McLean, R. and N. Mimura (eds.), 1993: *Vulnerability Assessment to Sea Level Rise and Coastal Zone Management.* Proceedings of the IPCC/WCC'93 Eastern Hemisphere workshop, Tsukuba, 3-6 August 1993, Department of Environment, Sport and Territories, Canberra, Australia.

McLean, R.F. and C.D. Woodroffe, 1993: Vulnerability assessment of coral atolls: the case of Australia's Cocos (Keeling) Islands. In: *Vulnerability Assessment to Sea Level Rise and Coastal Zone Management* [McLean, R. and N. Mimura (eds.)]. Proceedings of the IPCC/WCC'93 Eastern Hemisphere workshop, Tsukuba, 3-6 August 1993, Department of Environment, Sport and Territories, Canberra, Australia, pp. 99-108.

Melillo, J.M., A.D. McGuire, D.W. Kicklighter, B. Moore, C.J. Vorosmarty, and A.L. Schloss, 1993: Global climate change and terrestrial net primary production. *Nature*, **363**, 234-240.

Midun, Z. and S.-C. Lee, 1995: Implications of a greenhouse-induced sea-level rise: a national assessment for Malaysia. *Journal of Coastal Research*, special issue **14**, 96-115.

Mikolajewicz, U., B.D. Santer, and E. Maier-Reimer, 1990: Ocean response to greenhouse warming. *Nature*, **845**, 589-593.

Miller, M.L. and L. Auyong, 1991: Coastal zone tourism. *Marine Policy*, **March**, 75-99.

Milliman, J.D., J.M. Broadus, and F. Gable, 1989: Environmental and economic implications of rising sea level and subsiding deltas: the Nile and Bengal examples. *Ambio*, **18**, 340-345.

Mimura, N., M. Isobe, and Y. Hosokawa, 1993: Coastal zone. In: *The Potential Effects of Climate Change in Japan* [Nishioka, S., H. Harasawa, H. Hashimoto, T. Ookita, K. Masuda, and T. Morita (eds.)]. Center for Global Environmental Research, Environment Agency, Tokyo, Japan, pp. 57-69.

Mimura, N., M. Isobe, and Y. Hosokawa, 1994: Impacts of sea level rise on Japanese coastal zones and response strategies. In: *Global Climate Change and the Rising Challenge of the Sea* [O'Callahan, J. (ed.)]. Proceedings of the third IPCC CZMS workshop, Margarita Island, 9-13 March 1992, National Oceanic and Atmospheric Administration, Silver Spring, MD, pp. 329-349.

Mitchell, J.F.B., S. Manabe, T. Tokioka, and V. Meleshko, 1990: Equilibrium climate change. In: *Climate Change: The IPCC Scientific Assessment* [Houghton, J.T., G.J. Jenkins, and J.J. Ephraums (eds.)]. Cambridge University Press, Cambridge, UK, pp. 131-172.

Mitchell, J.F.B. and J.M. Gregory, 1992: Climatic consequences of emissions and a comparison of IS92a and SA90 (annex). In: *Climate Change 1992: The Supplementary Report to the IPCC Scientific Assessment* [Houghton, J.T., B.A. Callander, and S.K. Varney (eds.)]. Cambridge University Press, Cambridge, UK, pp. 171-176.

Mukai, H., 1993: Biogeography of tropical seagrasses in the Western Pacific. *Australian Journal of Marine and Freshwater Research*, **44**, 1-17.

Murphy, J.M., 1992: *A Prediction of the Transient Response of Climate*. Meteorological Office Climate Research Technical Note 32, UK Meteorological Office, Bracknell, UK.

Nicholls, N., 1984: The Southern Oscillation, sea-surface temperature, and interannual fluctuations in Australian tropical cyclone activity. *Journal of Climate*, **4**, 661-670.

Nicholls, R.J., 1995: Synthesis of vulnerability analysis studies. In: *Preparing to Meet the Coastal Challenges of the 21st Century*, vol. 1. Proceedings of the World Coast Conference, Noordwijk, 1-5 November 1993, CZM-Centre Publication No. 4, Ministry of Transport, Public Works and Water Management, The Hague, The Netherlands, pp. 181-216.

Nicholls, R.J. and S.P. Leatherman (eds.), 1995a: The potential impact of accelerated sea-level rise on developing countries. *Journal of Coastal Research*, special issue **14**, 1-324.

Nicholls, R.J. and S.P. Leatherman, 1995b: Sea-level rise and coastal management. In: *Geomorphology and Land Management in a Changing Environment* [McGregor, D. and D. Thompson (eds.)]. John Wiley, Chichester, UK, pp. 229-244.

Nicholls, R.J., N. Mimura, and J. Topping, 1995: Climate change in South and Southeast Asia: some implications for coastal areas. *Journal of Global Environment Engineering*, **1**, 137-154.

Nijkamp, P., 1991: Climate change, sea-level rise and Dutch defence strategies. *Project Appraisal*, **16**, 143-148.

Norris, E.A. (ed.), 1993: *Global Marine Biological Diversity: A Strategy for Building Conservation into Decision-Making*. Island Press, Washington, DC.

Nunn, P.D., A.D. Ravuvu, W. Aalbersberg, N. Mimura, and K. Yamada, 1994a: *Assessment of Coastal Vulnerability and Resilience to Sea-Level Rise and Climate Change, Case Study: Yasawa Islands, Fiji*. Phase II: Development of methodology. Environment Agency Japan, Overseas Environment Cooperation Centre Japan, South Pacific Regional Environment Programme.

Nunn, P.D., A.D. Ravuvu, E. Balogh, N. Mimura, and K. Yamada, 1994b: *Assessment of Coastal Vulnerability and Resilience to Sea-Level Rise and Climate Change, Case Study: Savai'i Island, Western Samoa*. Phase II: Development of methodology. Environment Agency Japan, Overseas Environment Cooperation Centre Japan, South Pacific Regional Environment Programme.

O'Callahan, J. (ed.), 1994: *Global Climate Change and the Rising Challenge of the Sea*. Proceedings of the third IPCC CZMS workshop, Margarita Island, 9-13 March 1992, National Oceanic and Atmospheric Administration, Silver Spring, MD, pp. 11-27.

OECD, 1991: *Responding to Climate Change: Selected Economic Issues*. Organization for Economic Cooperation and Development, Paris, France.

Oude Essink, G.H.P., R.H. Boekelman, and M.C.J. Bosters, 1993: Physical impacts of sea level change. In: *Sea Level Changes and Their Consequences for Hydrology and Water Management*. UNESCO International Workshop Seachange '93, Noordwijkerhout, 19-23 April 1993, Ministry of Transport, Public Works and Water Management, The Hague, The Netherlands, pp. 81-137.

Pachauri, R.K., 1994: *Climate Change in Asia: India*. Asian Development Bank, Manila, Philippines.

Parkinson, R.W. (ed.), 1994: Sea-level rise and the fate of tidal wetlands. *Journal of Coastal Research*, **10**, 987-1086.

Parkinson, R.W., R.D. de Laune, and J.R. White, 1994: Holocene sea-level rise and the fate of mangrove forests within the Wider Caribbean region. *Journal of Coastal Research*, **10**, 1077-1086.

Parry, M.L., M. Blantran de Rozari, A.L. Chong, and S. Panich (eds.), 1992: *The Potential Socio-Economic Effects of Climate Change in South-East Asia*. United Nations Environment Programme, Nairobi, Kenya.

Pearce, D.W. and R.K. Turner, 1992: *Benefits, Estimates and Environmental Decision-Making*. Organization for Economic Cooperation and Development, Paris, France.

Peerbolte, E.B., J.G. de Ronde, L.P.M. de Vrees, M. Mann, and G. Baarse, 1991: *Impact of Sea Level Rise on Society: A Case Study for The Netherlands*. Delft Hydraulics and Ministry of Transport, Public Works and Water Management, Delft and The Hague, The Netherlands.

Pernetta, J.C., 1988: Projected climate change and sea-level rise: a relative impact rating for the countries of the Pacific basin. In: *Potential Impacts of Greenhouse Gas Generated Climate Change and Projected Sea Level Rise on Pacific Island States of the SPREP Region* [Pernetta, J.C. (ed.)]. ASPEI Task Team, Split, Yugoslavia, pp. 1-10.

Pernetta, J.C., 1992: Impacts of climate change and sea-level rise on small island states: national and international responses. *Global Environmental Change*, **2**, 19-31.

Pernetta, J.C. and D.L. Elder, 1993: Preliminary Assessment of the vulnerability of Belize to accelerated sea-level rise: difficulties in applying the seven step approach and alternative uses of available data. In: *Vulnerability Assessment to Sea Level Rise and Coastal Zone Management*, [McLean, R. and N. Mimura (eds.)]. Proceedings of the IPCC/WCC'93 Eastern Hemisphere workshop, Tsukuba, 3-6 August 1993, Department of Environment, Sport and Territories, Canberra, Australia, pp. 293-308.

Pernetta, J.C. and J.D. Milliman (eds.), 1995: *Land-Ocean Interactions in the Coastal Zone: Implementation Plan*. IGBP Report No. 33, International Geosphere-Biosphere Programme, Stockholm, Sweden.

Pethick, J., 1992: Salt marsh geomorphology. In: *Salt Marshes* [Allen, J.R.L. and K. Pye (eds.)]. Cambridge University Press, Cambridge, UK, pp. 41-62.

Pethick, J., 1993: Shoreline adjustment and coastal management: physical and biological processes under accelerated sea-level rise. *Geographical Journal*, **159**, 162-168.

Philander, S.G.H., R.C. Pacanowski, N.-C. Lau, and M.J. Nath, 1992: Simulation of ENSO with a global atmospheric GCM coupled to a high-resolution tropical Pacific Ocean GCM. *Journal of Climate*, **5**, 308-329.

Pittock, A.B., 1993: Regional climate change scenarios for the South Pacific. In: *Climate Change and Sea Level Rise in the South Pacific Region* [Hay, J.E. and C. Kaluwin (eds.)]. Proceedings of the second SPREP meeting, Noumea, 6-10 April 1992, South Pacific Regional Environment Programme, Apia, Western Samoa, pp. 50-57.

Pittock, A.B. and R.A. Flather, 1993: Severe tropical storms and storm surges. In: *Climate and Sea Level Change: Observations, Projections and Implications* [Warrick, R.A. , E.M. Barrow, and T.M.L. Wigley (eds.)]. Cambridge University Press, Cambridge, UK, pp. 392-394.

Pittock, A.B., A.M. Fowler, and P.H. Whetton, 1991: Probable changes in rainfall regimes due to the enhanced greenhouse effect. In: *Challenges for Sustainable Development*, preprints of papers, vol. 1. International hydrology and water resources symposium, Perth, 2-4 October 1991, The Institution of Engineers, Barton, Australia, pp. 182-186.

Pluijm, M., G. Toms, R.B. Zeidler, A. van Urk, and R. Misdorp, 1992: *Vulnerability Assessment to Accelerated Sea Level Rise: Case Study Poland*. Ministry of Transport, Public Works and Water Management, The Hague, The Netherlands.

Poulos, S., A. Papadopoulos, and M.B. Collins, 1994: Deltaic progradation in Thermackos Bay, Northern Greece and its socio-economic implications. *Ocean & Coastal Management*, **22**, 229-247.

Pugh, D.T., 1987: *Tides, Surges and Mean Sea Level*. John Wiley, Chichester, UK.

Qureshi, A. and D. Hobbie (eds.), 1994: *Climate Change in Asia: Thematic Overview*. Asian Development Bank, Manila, Philippines.

Ray, G.C., 1991: Coastal-zone biodiversity patterns. *Bioscience*, **41**, 490-498.

Ray, J. and W.P. Gregg, 1991: Establishing biosphere reserves for coastal barriers. *Bioscience*, **41**, 301-309.

Reaka-Kudla, M.L., 1995: An estimate of known and unknown biodiversity and potential for extinction on coral reefs. *Reef Encounter*, **17**, 8-12.

Reed, D.J., 1990: The impact of sea level rise on coastal salt marshes. *Progress in Physical Geography*, **14**, 465-481.

Reed, D.J., 1995: The response of coastal marshes to sea-level rise: survival or submergence. *Earth Surface Processes and Landforms*, **20**, 39-48.

Reid, W.V. and K.R. Miller (eds.), 1989: *Keeping Options Alive: The Scientific Basis for Conserving Biodiversity*. World Resources Institute, Washington, DC.

Ren, Mei-e, 1994: Coastal lowland vulnerable to sea level rise in China. In: *Proceedings of the 1993 PACON China Symposium: Estuarine and Coastal Processes* [Hopley, D. and W. Yung (eds.)]. Pacific Congress on Marine Science and Technology, Beijing, 14-18 June 1993, University of North Queensland, Townsville, Australia, pp. 3-12.

Roy, P. and J. Connell, 1991: Climate change and the future of atoll states. *Journal of Coastal Research*, **7**, 1057-1075.

Roy, P.S., P.J. Cowell, M.A. Ferland, and B.G. Thom, 1994: Wave-dominated coasts. In: *Coastal Evolution: Late Quaternary Shoreline Morphodynamics* [Carter, R.W.G. and C.D. Woodroffe (eds.)]. Cambridge University Press, Cambridge, UK, pp. 121-186.

Saenger, P. and N.A. Siddigi, 1993: Land from the sea: the mangrove afforestation program of Bangladesh. *Ocean & Coastal Management*, **20**, 23-39.

Schlesinger, M.E., 1993: Model projections of CO_2-induced equilibrium climate change. In: *Climate and Sea Level Change: Observations, Projections and Implications* [Warrick, R.A., E.M. Barrow, and T.M.L. Wigley (eds.)]. Cambridge University Press, Cambridge, UK, pp. 169-191.

SCOR, 1991: The response of beaches to sea-level changes: a review of predictive models. *Journal of Coastal Research*, **7**, 895-921.

SDSIDS, 1994: *Programme of Action for the Sustainable Development of Small Island Developing States*. Global Conference on the Sustainable Development of Small Island Developing States, Barbados, 25 April–6 May 1994.

Semeniuk, V., 1994: Predicting the effect of sea-level rise on mangroves in Northwestern Australia. *Journal of Coastal Research*, **10**, 1050-1076.

Smith, S.V., 1978: Coral reef area and the contributions of reefs to processes and resources of the world's oceans. *Nature*, **273**, 225-226.

Smith, S.V. and R.W. Buddemeier, 1992: Global change and coral reef ecosystems. *Annual Reviews of Ecology and Systematics*, **23**, 89-118.

Snedaker, S.C., J.F. Meeder, R.S. Ross, and R.G. Ford, 1994: Discussion of Ellison and Stoddart, 1991. *Journal of Coastal Research*, **10**, 497-498.

Spencer, T., 1995: Potentialities, uncertainties and complexities in the response of coral reefs to future sea-level rise. *Earth Surface Processes and Landforms*, **20**, 49-64.

Stakhiv, E.Z., S.J. Ratick, and W. Du, 1991: Risk cost aspects of sea level rise and climate change in the evaluation of shore protection projects. In: *Water Resources Engineering Risk Assessment* [Ganoulis, J. (ed.)]. Springer Verlag, Heidelberg, Germany, pp. 311-335.

Stanley, D.J. and A.G. Warne, 1993: Nile delta: recent geological evolution and human impact. *Science*, **250**, 628-634.

Stive, M.J.F. and H.J. de Vriend, 1995: Modelling shoreface profile evolution. *Marine Geology*, **126**, in press.

Stive, M.J.F., R.J. Nicholls, and H.J. de Vriend, 1991: Sea-level rise and shore nourishment: a discussion. *Coastal Engineering*, **16**, 147-163.

Stive, M.J.F., J.A. Roelvink, and H.J. de Vriend, 1990: Large scale coastal evolution concept. In: *Proceedings of the 22nd Coastal Engineering Conference* [Edge, B.L. (ed.)], Delft, 2-6 July 1990, American Society of Civil Engineers, New York, NY, pp. 1962-1974.

Titus, J.G., 1991: Greenhouse effect and coastal wetland policy: how Americans could abandon an area the size of Massachusetts at minimum cost. *Environmental Management*, **15**, 39-58.

Titus, J.G., C.Y. Kuo, M.J. Gibbs, T.B. la Roche, M.K. Webb, and J.O. Waddell, 1987: Greenhouse effect, sea-level rise and coastal drainage systems. *Journal of Water Resources Planning and Management*, **113**, 216-227.

Titus, J.G., R.A. Park, S.P. Leatherman, J.R. Weggel, M.S. Greene, P.W. Mausel, S. Brown, C. Gaunt, M. Trehan, and G. Yohe, 1991: Greenhouse effect and sea level rise: potential loss of land and the cost of holding back the sea. *Coastal Management*, **19**, 171-204.

Tobor, J.G. and A.C. Ibe (eds.), 1990: *Global Climate Change and Coastal Resources and Installations in Nigeria: Impacts and Response Measures*. Proceedings of a national seminar, Lagos, 20-21 November 1990, Nigerian Institute for Oceanography and Marine Research, Victoria Island, Lagos, Nigeria.

Tooley, M. and S. Jelgersma (eds.), 1992: *Impacts of Sea Level Rise on European Coastal Lowlands*. Blackwell, Oxford, UK.

Tornqvist, T.E., 1993: Holocene alternation of meandering and anastomosing fluvial systems in the Rhine-Meuse delta (Central Netherlands) controlled by sea-level rise and subsoil erodibility. *Journal of Sedimentary Petrology*, **63**, 683-693.

Tsyban, A., J.T. Everett, and J.G. Titus, 1990: World oceans and coastal zones. In: *Climate Change: The IPCC Impacts Assessment* [McG. Tegart, W.J., G.W. Sheldon, and D.C. Griffiths (eds.)]. Australian Government Publishing Service, Canberra, Australia, pp. 6.1-6.28.

Tsyban, A., J.T. Everett, and M. Perdomo, 1992: World oceans and coastal zones: ecological effects. In: *Climate Change 1992: The Supplementary Report to the IPCC Impacts Assessment* [McG. Tegart, W.J. and G.W. Sheldon (eds.)]. Australian Government Publishing Service, Canberra, Australia, pp. 86-93.

Turner, R.K., 1988: Wetland conservation: economics and ethics. In: *Economics, Growth and Sustainable Environments* [Collard, D., D.W. Pearce, and D. Ulph (eds.)]. Macmillan, London, UK, pp. 121-159.

Turner, R.K., 1991: Economics and wetland management. *Ambio*, **20**, 59-63.

Turner, R.K., P. Doktor, and W.N. Adger, 1995a: Assessing the costs of sea level rise. *Environment and Planning A*, **27**, in press.

Turner, R.K., S. Subak, and W.N. Adger, 1995b: Pressures, trends and impacts in the coastal zones: interactions between socio-economic and natural systems. *Environmental Management*, in press.

UNCED, 1992a: *The United Nations Framework Convention on Climate Change*. United Nations Conference on Environment and Development, Rio de Janeiro, 3-14 June 1992.

UNCED, 1992b: *Agenda 21*. United Nations Conference on Environment and Development, Rio de Janeiro, 3-14 June 1992.

UNEP, 1991: *Environmental Data Report*. United Nations Environment Programme, New York, NY.

UNEP-UNESCO Task Team, 1993: *Impact of Expected Climate Change on Mangroves*. UNESCO Reports in Marine Science No. 61, Paris, France.

Van Dam, J.C., 1993: Impact of sea level rise on salt water intrusion in estuaries and aquifers. In: *Sea Level Changes and Their Consequences for Hydrology and Water Management*. UNESCO International Workshop Seachange '93, Noordwijkerhout, 19-23 April 1993, Ministry of Transport, Public Works and Water Management, The Hague, The Netherlands, pp. 49-60.

Van der Meulen, F., J.V. Witter, and W. Ritchie (eds.), 1991: Impact of climatic change on coastal dune landscapes of Europe. *Landscape Ecology*, **6**, 1-113.

Van der Weide, J., 1993: A systems view of integrated coastal management. *Ocean & Coastal Management*, **21**, 149-162.

Vellinga, P., R.S. de Groot, and R.J.T. Klein, 1994: An ecologically sustainable biosphere. In: *The Environment: Towards a Sustainable Future* [Dutch Committee for Long-Term Environmental Policy (ed.)]. Kluwer Academic Publishers, Dordrecht, The Netherlands, pp. 317-346.

Vellinga, P. and S.P. Leatherman, 1989: Sea level rise, consequences and policies. *Climatic Change*, **15**, 175-189.

Volonté, C.R. and J. Arismendi, 1995: Sea-level rise and Venezuela: potential impacts and responses. *Journal of Coastal Research*, special issue **14**, 285-302.

Volonté, C.R. and R.J. Nicholls, 1995: Sea-level rise and Uruguay: potential impacts and responses. *Journal of Coastal Research*, special issue **14**, 262-284.

Ware, J.R., S.V. Smith, and M.L. Reaka-Kudla, 1992: Coral reefs: sources or sinks of atmospheric CO_2? *Coral Reefs*, **11**, 127-130.

Warrick, R.A. and E.M. Barrow, 1991: Climate change scenarios for the U.K. *Transactions of the Institute of British Geographers*, **16**, 387-399.

Warrick, R.A., E.M. Barrow, and T.M.L. Wigley (eds.), 1993: *Climate and Sea Level Change: Observations, Projections and Implications*. Cambridge University Press, Cambridge, UK.

Warrick, R.A. and J. Oerlemans, 1990: Sea level rise. In: *Climate Change: The IPCC Scientific Assessment* [Houghton, J.T., G.J. Jenkins, and J.J. Ephraums (eds.)]. Cambridge University Press, Cambridge, UK, pp. 257-281.

Warrick, R.A. and A.A. Rahman, 1992: Future sea level rise: environmental and socio-political considerations. In: *Confronting Climate Change: Risks, Implications and Responses* [Mintzer, I.M. (ed.)]. Cambridge University Press, Cambridge, UK, pp. 97-112.

WCC'93, 1993: *World Coast 2000: Preparing to Meet the Coastal Challenges of the 21st Century*. Statement of the World Coast Conference, Noordwijk, 1-5 November 1993.

WCC'93, 1994: *Preparing to Meet the Coastal Challenges of the 21st Century*. Report of the World Coast Conference, Noordwijk, 1-5 November 1993, Ministry of Transport, Public Works and Water Management, The Hague, The Netherlands.

WCC'93, 1995: *Preparing to Meet the Coastal Challenges of the 21st Century*. Proceedings of the World Coast Conference, Noordwijk, 1-5 November 1993, CZM-Centre Publication No. 4, Ministry of Transport, Public Works and Water Management, The Hague, The Netherlands.

Wigley, T.M.L., 1995: Global-mean temperature and sea level consequences of greenhouse gas stabilization. *Geophysical Research Letters*, **22**, 45-48.

Wigley, T.M.L. and S.C.B. Raper, 1992: Implications for climate and sea level of revised IPCC emissions scenarios. *Nature*, **357**, 293-300.

Wilkinson, C.R. and R.W. Buddemeier, 1994: *Global Climate Change and Coral Reefs: Implications to People and Reefs*. Report of the UNEP-IOC-ASPEI-IUCN Global Task Team on the Implications of Climate Change on Coral Reefs, IUCN, Gland, Switzerland.

Wolff, W.J., K.S. Dijkema, and B.J. Ens, 1993: Expected ecological effects of sea level rise. In: *Sea Level Changes and their Consequences for Hydrology and Water Management*. UNESCO International Workshop Seachange '93, Noordwijkerhout, 19-23 April 1993, Ministry of Transport, Public Works and Water Management, The Hague, The Netherlands, pp. 139-150.

Woodroffe, C.D., 1990: The impact of sea-level rise on mangrove shorelines. *Progress in Physical Geography*, **14**, 483-520.

Woodroffe, C.D., 1993: Sea level. *Progress in Physical Geography*, **17**, 359-368.

Woodroffe, C.D., 1994: Sea level. *Progress in Physical Geography*, **18**, 434-449.

Woodroffe, C.D., 1995: Response of tide-dominated mangrove shorelines in Northern Australia to anticipated sea-level rise. *Earth Surface Processes and Landforms*, **20**, 65-86.

Woodroffe, C.D. and R.F. McLean, 1992: *Kiribati Vulnerability to Accelerated Sea-Level Rise: A Preliminary Study*. Department of the Arts, Sport, Environment and Territories, Canberra, Australia.

Yamada, K., P.D. Nunn, N. Mimura, S. Machida, and M. Yamamoto, 1995: Methodology for the assessment of vulnerability of South Pacific island countries to sea-level rise and climate change. *Journal of Global Environment Engineering*, **1**, 101-125.

Yim, W.W.-S. 1995: Implications of sea-level rise for Victoria Harbour, Hong Kong. *Journal of Coastal Research*, special issue **14**, 167-189.

10

Hydrology and Freshwater Ecology

NIGEL ARNELL, UK; BRYSON BATES, AUSTRALIA; HERBERT LANG, SWITZERLAND; JOHN J. MAGNUSON, USA; PATRICK MULHOLLAND, USA

Principal Lead Authors:
S. Fisher, USA; C. Liu, China; D. McKnight, USA; O. Starosolszky, Hungary; M. Taylor, USA

Contributing Authors:
E. Aquize, Peru; S. Arnott, Canada; D. Brakke, USA; L. Braun, Germany; S. Chalise, Nepal; C. Chen, USA; C.L. Folt, USA; S. Gafny, Israel; K. Hanaki, Japan; R. Hecky, Canada; G.H. Leavesley, USA; H. Lins, USA; J. Nemec, Switzerland; K.S. Ramasastri, India; L. Somlyódy, Hungary; E. Stakhiv, USA

CONTENTS

EXECUTIVE SUMMARY

The hydrological system is potentially very sensitive to changes in climate. Changes in precipitation affect the magnitude and timing of runoff and the frequency and intensity of floods and droughts; changes in temperature result in changes in evapotranspiration, soil moisture, and infiltration. The resulting changes in surface wetness, reflectivity, and vegetation affect evapotranspiration and the formation of clouds, as well as surface net radiation and precipitation. Meanwhile, the hydrological system is being affected by other, more direct human activities, such as deforestation, urbanization, and water-resource exploitation.

The effects of climate change on hydrological regimes are generally estimated by combining catchment-scale hydrological models with climate change scenarios derived from general circulation model (GCM) output. In addition to the uncertainties associated with GCM simulations, there are three major problems in estimating the hydrological effects of climate change: (1) expressing scenarios at a scale appropriate for hydrological modeling; (2) the considerable errors inherent in climatic and hydrological data used to validate hydrological models; and (3) converting climatic inputs into hydrological responses. Methods have been developed to address most of these problems, but estimates of the hydrological effects of climate change remain very uncertain. This uncertainty is largely due to difficulties in defining credible scenarios for changes in precipitation and assessing vegetation response to a changed atmosphere at appropriate spatial scales.

There are considerable differences in estimated changes in hydrological behavior both between scenarios and between catchments, but it is possible to draw some general conclusions about sensitivities:

- We have a high level of confidence that an increase in air temperature would increase potential evapotranspiration, but the magnitude of increase depends also on changes in net radiation, humidity, windspeed, precipitation and its temporal distribution, and vegetation characteristics. Actual evapotranspiration may increase or decrease according to the availability of soil moisture.
- The effect of a given change in precipitation and evapotranspiration on river runoff and groundwater recharge varies considerably between catchments, depending on climatic regime and (at shorter time scales) catchment physical characteristics. We have a high level of confidence that, in general, the drier the climate, the greater the sensitivity of hydrological regimes to changes in climate.
- Neither climatic nor hydrological change will be equally distributed throughout the year; in some cases it appears that the variability of river flow through the year would increase if climate were to change, and the frequency of both high and low flows would increase. We have a medium level of confidence in this finding, due to variations between catchments and scenarios.
- We have a medium level of confidence that more intense rainfall would tend to increase the occurrence of floods, although the magnitude of this effect would depend not only on the change in rainfall but also on catchment physical and biologic characteristics. These characteristics also may change due to human activity that may or may not be in response to climate change.
- An increase in the duration of dry spells would not necessarily lead to an increase in the occurrence of low river flows and groundwater levels. River flows may be sustained by increased precipitation earlier in the year. The effect of a change in the duration of dry spells depends significantly on catchment physical and biologic characteristics. We have a medium level of confidence in these findings.
- Hydrological regimes in many continental and mountain areas are determined by winter snowfall and spring snowmelt; if climate were to change and the proportion of precipitation falling as snow decreased, we have a high degree of confidence that in such regions there would be a widespread shift from spring to winter runoff. This effect may be modified by changes in the seasonal distribution and amounts of precipitation.
- Very little is known about possible changes in groundwater recharge, which will depend on the balance between changes in opportunities for recharge and the amount of water available for recharge. Therefore, we have a low level of confidence in any conclusions about or projections of changes in groundwater recharge.

Freshwater ecosystems, including lakes and streams (covered in this chapter) and noncoastal wetlands (see Chapter 6), are scattered across the landscape and tightly linked to regional hydrology. The effects of climate change on freshwater ecology will interact strongly with other anthropogenic changes in land use, waste disposal, and water extraction. Changes in lake and stream ecosystems will occur in the context of existing climates and expected changes; effects will differ greatly from place to place. General conclusions follow:

- We have a high level of confidence that climate change will influence freshwater ecosystems directly through changes in water availability by altered flood regimes, water levels, and, in the extreme, the absence of water in streambeds and lake basins.

- We have a high level of confidence that changes in water temperature and thermal structure of freshwaters directly affect the survival, reproduction, and growth of organisms; productivity of ecosystems; persistence and diversity of species; and regional distribution of biota.
- We have a high level of confidence that changes in runoff, groundwater inputs, and direct precipitation to lakes and streams will alter the input of nutrients and dissolved organic carbon, which in turn will alter productivity and clarity. Changes in the duration of lake and stream ice and snow on ice will change mixing patterns, oxygen availability, and the survival and reproductive success of certain organisms.
- We have a medium level of confidence that increases in terrestrial productivity will enhance the productivity of streams by increasing the input of organic detritus and likely increasing the export of CO_2 from lakes to the atmosphere.
- We have a medium level of confidence that climate-induced changes in temperature, ice and snow cover, and biological production will be more intense at higher latitudes—where temperature changes are expected to be greatest—and at the lower-latitude edges of the geographic ranges of many taxa—where extinctions and extirpations will be the greatest, especially where isolation prevents the poleward dispersal of organisms in rivers and streams.
- We have a high level of confidence that water-level declines will be most severe in lakes and streams in dry evaporative drainages and in basins with small catchments—where in many years evapotranspiration losses already exceed precipitation and inputs from groundwater plus overland flow.
- We have a high level of confidence that the effects of an increase in severe flood events may be more damaging in drier climates where soils are more erodible and precipitation-runoff relationships are highly nonlinear. Drought events may be more severe in humid areas that have not experienced frequent droughts in the past.
- We have a high level of confidence that climate changes that increase variability are expected to have greater ecological effects than climate changes associated with a change in average conditions; increased variability in hydrologic conditions with associated flash floods and droughts will reduce the biological diversity and productivity of stream ecosystems.
- We have a high level of confidence that assemblages of organisms will tend to move poleward with warming, with range constrictions and local and global species extinctions occurring at the lower latitudes of their distributions and range extensions and invasions occurring poleward. For smaller, more hydrologically isolated basins, dispersal will lag considerably behind the warming in climate—resulting in near-term reductions in diversity.
- We can say with a lower degree of confidence that warming of larger and deeper temperate lakes may increase their productivity, provided that nutrient inputs are not greatly reduced; in some shallow, stratified lakes, this increase in productivity would lead to a greater likelihood of anoxic conditions in deep water, with subsequent loss of cold-water fauna.
- We have a medium level of confidence that changes in vegetation, in rainfall, and in hydrological regimes affect erosion on hillslopes and in river channels, sedimentation, and channel stability.

Water quality will be affected by changes in water temperature, CO_2 concentration, the processes by which water gets into the stream network and aquifers, and the timing and volume of streamflow. The relative importance of these influences varies between catchments and chemical species, and the influences interact to ameliorate or exaggerate the effects of change. Dissolved oxygen concentrations would not change much if higher water temperatures were associated with increased streamflow, but if streamflow were to decline there would be a large decrease in dissolved oxygen concentrations. In many catchments, stream and groundwater quality is determined by human influences—such as the application of agricultural chemicals, urbanization, and the return of treated sewage. The water quality of streams that already are polluted is likely to be affected by changes in temperature and flow regimes, but changes in land use and the input of pollutants may have a far greater effect than changes in climate.

10.1. Introduction

The hydrological cycle and the climate system are intimately linked. Any change in climate is reflected in changes in key hydroclimatic elements of the hydrological cycle and vice versa: Changes in precipitation affect the magnitude and timing of runoff and groundwater recharge, as well as the frequency and intensity of floods and droughts; changes in temperature result in changes in evapotranspiration, soil moisture, and infiltration conditions, with resulting changes in surface wetness, reflectivity, and vegetation that affect evaporation and the formation of clouds, as well as surface net radiation and—completing the cycle—precipitation.

Freshwater ecosystems include lakes, streams, and wetlands scattered across the landscape and are tightly linked to regional hydrology. These surface freshwater systems constitute a set of scarce resources for humans sensitive to climate change; excluding ice, they constitute only 0.009% of the volume of water on Earth (Goldman and Horne, 1983). Of these surface freshwaters, 99% (by volume) are lakes and 1% are streams. Groundwater is 66 times more abundant than surface freshwaters but still represents only 0.6% of the volume of all surface waters. Freshwater systems greatly influence the distribution of populations and economic growth, add greatly to the biological and ecological diversity of the landscape, and provide a variety of goods and services. The effects of climate change on freshwater ecosystems will interact strongly with anthropogenic changes in land use, waste disposal, and water extraction.

This chapter examines the sensitivities of hydrological and freshwater systems to climate change. The chapter covers four main issues: (1) the potential effects of global warming on components of the hydrological cycle; (2) possible changes in the frequency and magnitude of extreme high and low flows; (3) implications for thermal, chemical, and morphological changes; and (4) consequences for stream and lake ecosystems. The consequences of changes in freshwater systems affecting water resources, ecosystem management, agriculture, power production, and navigation are examined in other chapters of this report.

Although some of the principal linkages between climate and the hydrological system are well understood, predicting the effects of global warming is very uncertain. Current general circulation models (GCMs) work at a spatial resolution that is too coarse for hydrological purposes, producing weather averaged over too large a geographic area and producing average conditions rather than changes in ranges, frequencies, seasonal distributions, and so forth. They do not yet include all of the relevant feedbacks between the land surface and the atmosphere. There is considerable uncertainty in the translation of climate changes into hydrological effects through hydrological models and an inability to maintain consistency between GCM and hydrological model water balances, particularly for evaporation. Different models give different sensitivities to change, and a model calibrated under current conditions may not be appropriate under a changed climate. There still are major uncertainties over the effects of increased CO_2 concentrations on plant water use—and hence transpiration rates—in natural settings at the catchment scale.

Because of the spatial complexity of the Earth's climate-water system, there is no ideal method to delineate hydroclimatic regions that could be used in this chapter. The case studies selected cover all the world's main climatic belts as used by L'vovich (1979)—subpolar, temperate, subtropical, tropical, equatorial, and mountain regions—taking into account the degree of climatic aridity and humidity in the regions covered by the study. The literature available does not cover all regions adequately, and major gaps remain in the regional coverage of published studies and the quality of the data.

10.2. The Hydrological Cycle

10.2.1. The Hydrological Cycle in the Climate System

The hydrological cycle is driven by solar energy and involves water changing its form through the oceans, atmosphere, land and vegetation, and ice and glaciers. After precipitation water reaches the Earth's surface, it either remains on the surface as rain or snow; quickly evaporates; infiltrates into the soil, where it is taken up by vegetation (and eventually transpired through photosynthesis and respiration); moves along the surface or through the soil toward rivers and streams; or percolates to aquifers, recharging groundwater and subsequently draining into rivers and oceans, perhaps after delays of years or centuries. Some water is stored over the winter as snow and over longer time scales as ice.

The partitioning of incoming precipitation between storage, evaporation, infiltration, groundwater recharge, and runoff varies widely between catchments, depending on climatic and catchment characteristics. Vegetation cover has a major effect on the water and energy balance, particularly through interception and evaporation. There are many types of hydrological regimes with different combinations of processes, but it helps to distinguish among three major types: snow-dominated, humid, and semi-arid/arid. In the first regime, a significant proportion of precipitation falls as snow and is stored on the surface until it melts during the spring. In a humid regime, rainfall infiltrates into the soil and reaches the stream network by a variety of routes over different time scales. In semi-arid and arid environments, rainfall is short-lived and generally very intense, and soils tend to be thin. A large proportion of rainfall therefore runs directly off the surface, only to infiltrate into deeper soils downslope or along the river bed.

Hydrological responses vary over time scales from seconds, to days or weeks, through years to decades, and have continuous feedbacks with the atmosphere. In the most general terms, the amount of water present at and beneath the surface affects both the proportion of incoming energy reflected back into the atmosphere and the partitioning of the remaining net radiation between sensible and latent heat. A change in the land surface, and in the hydrological processes operating at the surface, therefore will affect atmospheric processes; the magnitude of the effect depends on the degree and extent of change, as well as local climatic characteristics. It is increasingly clear that hydrological variability can be interpreted in terms of large-scale climatic anomalies—such as those associated with the El Nino/Southern Oscillation (ENSO)—and that there are strong relationships between hydrological anomalies in different parts of the world (e.g., Redmond and Koch, 1991; Aguado *et al.*, 1992; Mechoso and Iribarren, 1992; Simpson *et al.*, 1993; Dracup and Kahya, 1994; Ely *et al.*, 1994).

10.2.2. Anthropogenic Non-Climatic Impacts on the Hydrological Cycle

Human activities are interfering with the hydrological cycle in many regions and catchments. There are many different types of impact; they can be deliberate (perhaps with unanticipated consequences) or inadvertent, and they affect the quantity and quality of water and aquatic biota. Such activities include:

- **River impoundment and regulation**: Dams for flood protection, hydropower purposes, or inland navigation and related control measures generally cause changes in the spatial and temporal distribution of streamflow, which also may have impacts on evaporation and infiltration in areas close to the river bed, as well as on biota. Of particular importance are increases or decreases in river infiltration to groundwater aquifers.
- **Impacts of land use and land-use changes**: Anthropogenic activities that affect the land surface include urbanization; agricultural activities such as irrigation, drainage, land improvement, and the application of agricultural chemicals; deforestation and afforestation; and overgrazing. These activities can cause locally and regionally significant changes in evaporation, water balance, flood and drought frequency, surface and groundwater quantity and quality, and groundwater recharge.
- **Water removal and effluent return**: Water used for municipal, industrial, and agricultural purposes can affect river flows and groundwater levels. This water may subsequently be returned to the hydrological system (although much water used for irrigation is lost by evaporation), but at a lower quality.
- **Large-scale river diversions**: Large-scale diversions of river flow or parts of it, mainly for purposes of irrigation or hydropower generation, can cause serious and manifold changes in the ecosystems of large areas.

10.2.3. Methods for Estimating the Hydrological Effects of Climate Change

There are several possible ways of creating scenarios for climate-change impact studies. Temporal analogs (e.g., Krasovskaia and Gottschalk, 1992) use the past as an analog for the future; spatial analogs substitute one location for another. Many studies have examined the effect of arbitrary changes in climatic inputs on the hydrological system (e.g., Novaki, 1992a, 1992b; Gauzer, 1993; Chiew *et al.*, 1995). Although most studies now use scenarios based on GCM simulations, there is a multitude of uncertainties associated with their use—which are listed here by their relative importance (the first being the most critical):

- Weaknesses of models in coupling the land surface and atmospheric hydrologic cycles and in GCM simulations of regional climate and extremes, particularly with regard to precipitation
- Weaknesses in using GCM simulations to define climate-change scenarios at the spatial and temporal scales required by hydrological models. The spatial resolution of current GCMs is too coarse for their output to be fed directly into hydrological models.
- Weaknesses in simulating changes in hydrological characteristics with given changes in climatic inputs to the hydrological system, which reflect difficulties in developing credible hydrological models with sparse and error-contaminated climatic and hydrological data.

Figure 10-1 attempts to summarize the relative magnitude of the uncertainties. Given that a reasonably realistic hydrological model is used, the greatest uncertainty in hydrological climate-change impact assessment lies in the initial scenarios derived from GCMs—which do not include all of the relevant feedbacks between the land surface and the atmosphere—and the derivation of catchment-scale climate scenarios. Sections 10.2.3.1 through 10.2.3.3 explore further the stages in this cascade of uncertainties.

10.2.3.1. GCMs and Hydrological Requirements

GCMs show progress in simulating the present climate with respect to annual or seasonal averages at large spatial scales ($>10^4$ km^2) but perform poorly at the fine temporal and spatial scales

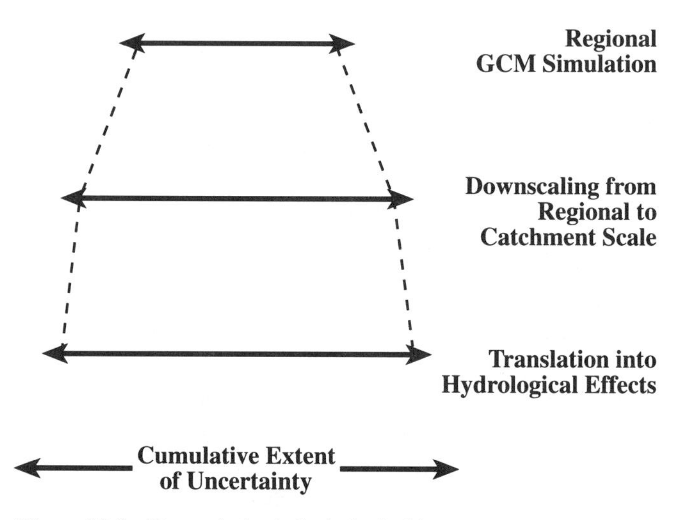

Figure 10-1: Uncertainties in hydrological impact assessment.

relevant to hydrological studies (Askew, 1991; Grotch and MacCracken, 1991). Although GCMs are unanimous in their projections that a doubling of current atmospheric CO_2 concentrations would lead to an increase in global mean temperature and precipitation, they differ in their projections of changes in temperature and precipitation at regional scales that are the same order of magnitude as projected global changes. The direct use of GCM output to drive hydrological models is considered inappropriate due to the coarse resolution of the spatial grids used by current GCMs (relative to the scale of river basins); the simplified GCM representations of topography, land surface and cloud processes, and energy transfer within the oceans; and the simplified coupling of the atmosphere and the oceans. Until recent GCM transient runs, GCM simulations usually were of short duration (less than 30 years) and may not have captured the extreme events of particular interest in flood and drought analyses.

From a hydrological perspective, there is a need to improve the representation of land-surface processes and their interaction with the atmosphere. There now are intensive field studies into the interaction between land-surface hydrological processes and climate at point and regional scales (e.g., HAPEX-MOBILHY: Andre *et al.*, 1986; HAPEX-SAHEL: Goutorbe *et al.*, 1993; EFEDA: Bolle *et al.*, 1993), as well as work to improve models of land-surface processes (e.g., Wood *et al.*, 1992). The World Climate Research Programme (WCRP) sponsored by the World Meteorological Organization (WMO), the Intergovernmental Oceanographic Commission of UNESCO, and the International Council of Scientific Unions has launched a Global Energy and Water Cycle Experiment (GEWEX). The GEWEX Continental-Scale International Project (GCIP) is an initiative to study the water and energy budgets of an extensive geographical area with a large volume of accessible data; the Mississippi River basin has been selected as the primary region of interest. Similar projects have been organized in other areas of the world—such as the Baltic Sea region (BALTEX), the boreal region (BOREAS), and the Himalaya/monsoon region (GAME).

Changes in the temporal distribution of precipitation—even with average precipitation remaining stable—may result in changes in river flow and its extremes, floods and droughts, and water availability for plants and water supply. There is a need to consider changes to the statistical distributions of weather variables rather than just their means because changes in response to climatic warming may be reflected more in extreme events than in average conditions (Katz and Brown, 1992).

10.2.3.2. Creating Scenarios at the Hydrological Scale

GCMs operate at a spatial scale far coarser than hydrological models and do not necessarily simulate regional-scale climate particularly well. It is therefore necessary to downscale GCM results to the catchment scale. Several techniques have been developed to derive scenarios at this spatial scale from GCM output:

- Direct use of GCM-simulated changes in precipitation, temperature, and evaporation, perhaps interpolated to the catchment scale and applied to observed catchment climate data (e.g., Bultot *et al.*, 1992; Arnell and Reynard, 1993; Viner and Hulme, 1994; Kirshen and Fennessey, 1995)
- Stochastic generation of point-scale weather, with stochastic model parameters adjusted according to scenarios derived from GCM output (e.g., Nathan *et al.*, 1988; Cole *et al.*, 1991; Wilks, 1992; Wolock *et al.*, 1993; Bates *et al.*, 1994, 1995; Valdés *et al.*, 1994; Rao and Al-Wagdany, 1995; Tung and Haith, 1995)
- Estimation of catchment-scale weather from large-scale climatic features, such as weather types and mean sea-level pressure fields (e.g., Bardossy and Plate, 1992; Hay *et al.*, 1992; Hughes *et al.*, 1993; von Storch *et al.*, 1993)
- Use of nested regional limited-area models embedded within GCMs to simulate regional climate at a higher spatial resolution (e.g., Leavesley *et al.*, 1992; Hostetler and Giorgi, 1993; McGregor and Walsh, 1994).

10.2.3.3. Hydrological Modeling and Impact Assessment

Hydrological models have been used to a large extent to determine the relative sensitivity of hydrological variables to climatic inputs. These simulations have been useful in the search for amplification effects—that is, whether small changes in one (climatic) variable may cause substantial changes in another variable, thus aggravating water problems particularly in areas that are presently vulnerable. Such a sensitivity analysis allows assessment of the impact of the relative increase or change in temperature and precipitation on changes in hydrological variables of interest (e.g., runoff, evapotranspiration, soil moisture, flood potential). Results of sensitivity analyses are useful in practical "what-if" considerations. Sensitivity analyses also underscore uncertainty in the prediction of future water resources. They may include inconsistencies by ignoring other important climatic variables that may be affected by climate change, such as net radiation, wind, and humidity.

Current approaches to hydrological modeling for climate-change impact assessment include empirical models, water-balance models, conceptual rainfall-runoff models, and physically based, distributed models (Leavesley, 1994). Empirical models consider statistical relations between annual runoff and precipitation, temperature, or potential or actual evapotranspiration for present-day conditions (Revelle and Waggoner, 1983; Arnell, 1992; Duell, 1994). Runoff under changed climate conditions is obtained from regression models by perturbing historical climate series. This approach assumes that the functional form as well as the parameters of a regression model remain valid for a changed climate. Leavesley (1994) has questioned the validity of this approach, and Arnell (1992) has shown that it should be used with extreme caution because the estimated sensitivities to change are very dependent on details of the model. Water-balance and conceptual rainfall-runoff models simulate the movement of water from the time it enters a catchment as precipitation to the time it leaves the catchment as runoff, although rainfall-runoff models consider the flow paths and residence times of water in finer detail (Leavesley, 1994). Klemes (1985) and Becker and Serban (1990) state that calibrated water-balance or conceptual rainfall-runoff models are not directly usable for assessing climate-change impacts because the model parameter estimates are assumed to remain valid for the changed climate. This point was illustrated by Gan and Burges (1990), who found that the calibrated sizes of conceptual moisture storages are not climate-invariant. The size of this problem can be checked, at least approximately, by comparing model prediction errors based on parameters estimated from wet and dry periods within historical records. Physically based models appear to be a superior alternative because their parameters are estimated from field measurements or physically based analyses. These models use physically based process equations to simulate the spatial variability of runoff due to spatial variations in land characteristics such as soil type and topography, as well as rainfall. Although physically based models have the potential to become a universally applicable tool, they come with the price of increased model complexity, data requirements, and computing costs. In practice, there is a paucity of measured field data and a lack of methods for the collection of data at a scale appropriate for such models, limiting their application to studies of small-catchment hydrology (Grayson *et al.*, 1992a, 1992b).

The development of a large-scale model of land-surface hydrology may be started from an even larger-scale continuum of atmospheric processes, which is how GCM researchers were proceeding until recently. In this approach, it is necessary to simplify the spatial distribution of variables of the land surface to an extent that not only distorts their physical significance but provides spatially imprecise inputs to the atmospheric models at the land-atmosphere interface. This problem has been identified by GCM researchers in studies of the regional aspects of changes in climatic variables. Another way to solve the problem is to develop a macroscale model, starting from mesoscale analysis of hydrological processes, as practiced by hydrologists in simulations for water resources management purposes. Although the call for such an approach is being heard more and more frequently, a model of this type is not yet available.

Large-scale hydrological models, which estimate vertical fluxes of the land-surface/atmosphere interface and horizontal fluxes resulting in measurable runoff, using physically based parameters, could be very important in assessing the impacts of climate change and could improve GCM regional simulations through inputs to GCM hydrology. Additional sources of information on these topics include a recent discussion on the type and use of scenarios and hydrological models by Lettenmaier *et al.* (1994), a review of research on climate-change impact on water resources by Chang *et al.* (1992), and the Proceedings of the First National Conference on Climate Change and Water Resources Management (Ballentine and Stakhiv, 1993).

10.3. Effects of Climate Change on the Hydrological Cycle

10.3.1. Introduction

The complex and interacting effects of an increasing concentration of greenhouse gases on the hydrological system are shown in Figure 10-2 (Arnell, 1994). The increase in greenhouse gas concentrations results in an increase in net radiation at the Earth's surface, which results in changes in temperature, rainfall, and evaporation—and hence soil moisture regimes, groundwater recharge, and runoff. Temperature, rainfall, evaporation, and soil moisture affect vegetation growth, as do changes in incoming solar radiation and the atmospheric CO_2 concentration. Higher CO_2 concentrations also may affect plant water use (see Chapter 9, *Terrestrial Biotic Responses to Environmental Change and Feedbacks to Climate*, of the Working Group I volume). The

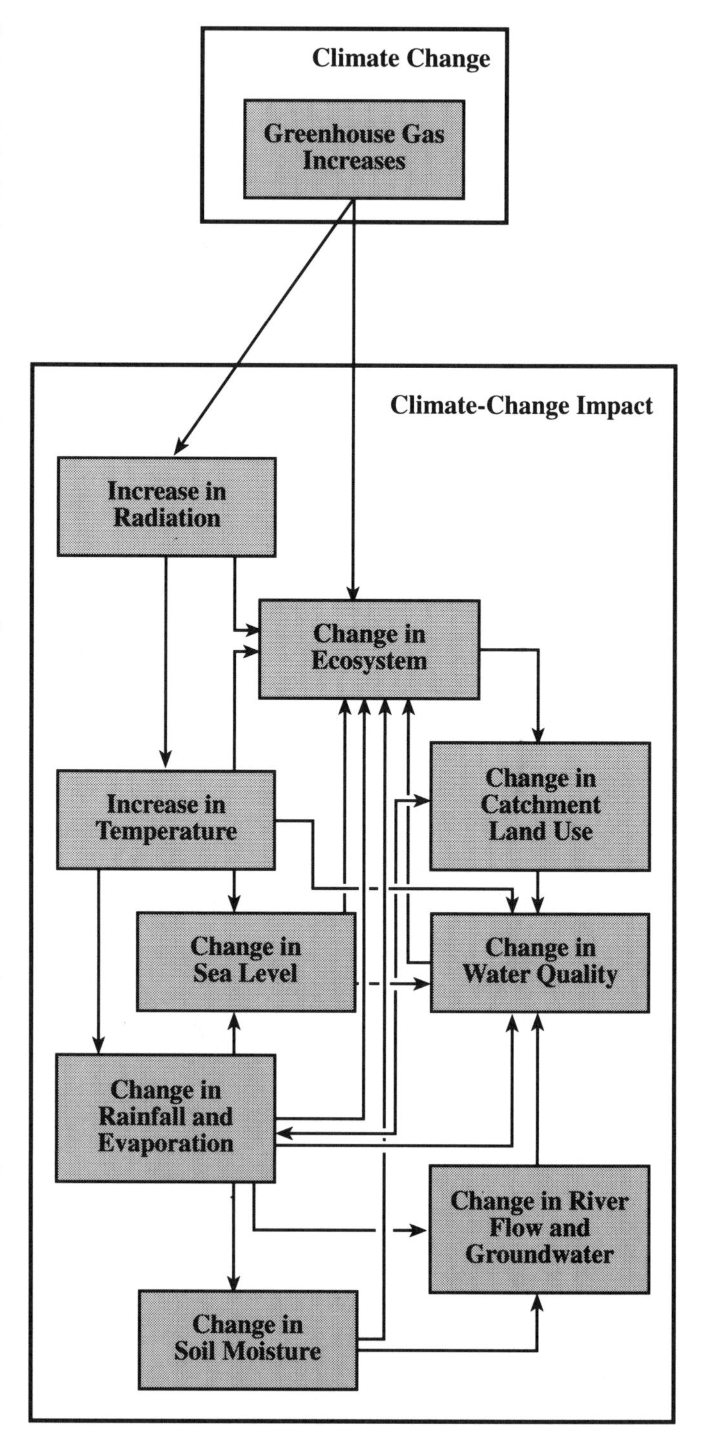

Figure 10-2: Impacts of climate change on the hydrological system (Arnell, 1994).

components of the hydrological system are very clearly linked, and the effects of global warming may be very complex. This complexity is heightened by nonlinear responses and important thresholds in the hydrological system that govern the operation of different types of processes.

The following sections summarize global-warming impacts on various components of the hydrological system. Table 10-1 provides a selection of case studies from the recent literature. Case studies are not available for all regions, but an attempt has been made to list some of the modeling assumptions employed. The results of the studies should be examined in association with the caveats noted in Section 10.2.3 and with the knowledge that hydrological models may not adequately portray the true responses of the modelled processes.

10.3.2. Precipitation

Globally, climate change may result in a wetter world. Present-day climate models project an increase in global mean precipitation of about 3–15% for a temperature increase of 1.5–4.5°C. Although it is likely that precipitation will increase in some areas and decline in others, climate models are not in agreement as to the amount or the regional distribution of precipitation. Projections of regional changes in precipitation are less certain than projections of regional changes in temperature. Higher-latitude regions are expected to experience more precipitation, particularly in the winter. In most cases, this extends to mid-latitudes. Models vary significantly in terms of how the intensity and statistical distribution of rainfall is predicted to shift in the tropics, although little change is expected for the dry subtropics.

10.3.3. Evaporation and Transpiration

Evaporation is a function of the demand for water and the supply of water. It is driven by energy availability, particularly net radiation. An increase in net radiation will increase the demand for evaporation, but this effect will be complicated by changes in the humidity of the air—affecting the ability of the air to accept more water—and by changes in the rate of movement of air across the evaporating surface. In addition, the humidity of air masses is strongly linked to evaporation over land and water bodies, including the oceans. With credible assumptions about increases in radiation and reductions in humidity, a rise of about 2°C could cause an increase in potential evaporation of up to 40% in a humid temperate region (Arnell and Reynard, 1993), but less in a drier environment where changes in humidity are not very important. Higher temperatures will mean that the atmosphere can hold more water—thus enhancing the effects of increased net radiation on evapotranspiration in regions where air humidity currently imposes a constraint on the rate of evaporation (as in humid areas).

The actual evaporation rate often is constrained by moisture availability. A reduction in rainfall or a change in its temporal distribution, and hence in soil moisture, may mean that

Table 10-1: *Summary of recent case studies.*

Region/Country	Investigators	GCMs/Climate Scenarios[1]	Hydrologic Scenarios
Australia			
6 Catchments	Bates *et al.* (1995)	CSIRO 9 (1991)	Increase in magnitude of annual maximum monthly runoff series, with size of increase dependent on hydrological model; increase or decrease in median monthly runoff dependent on location and season.
28 Catchments	Chiew *et al.* (1995)	Hypothetical + BMRC, CCC, CSIRO 9, GFDLH + UKMOH	Annual runoff changes of ±50%; annual soil moisture level changes of -25 to +15%.
China			
Haihe, Huaihe, and Yellow River Basins	Liu *et al.* (1995)	Output from seven GCMs used as input to hydrological models	Decrease in annual runoff of 2–12%.
Finland			
12 Catchments	Vehvilainen and Lohvansuu (1991)	GISS	Increase in mean discharge by 20–50%; considerable increase in mean minumum discharges in winter due to shorter snow-cover period; decrease in mean maximum discharges due to reduced maximum snow water-equivalents; persistent winter snow cover will vanish in southern Finland.
Greece			
Mesochora Catchment	Panagoulia (1992)	GISS	Decreased spring and summer runoff, soil moisture, and mean snow accumulations; increased winter runoff and soil moisture.

Table 10-1 continued.

Region/Country	Investigators	GCMs/Climate Scenarios[1]	Hydrologic Scenarios
Hungary			
Danube Basin	Gauzer (1993)	Hypothetical temperature scenarios	Considerable increase in winter streamflow and decrease in summer applied to September 1991–August 1992 flow; no decrease in spring flood peak, but change in timing.
Drainage Systems	Novaki (1990, 1992a, 1992b)	Hypothetical precipitation and temperature scenarios	1–2°C increase in temperature and 5–20% decrease in annual precipitation may cause 15–30% decrease in drainage discharge at start of spring.
India			
Kolar and Sher Basins	NIH (1994)	Hypothetical scenarios	Sher Basin more sensitive to climate change; impacts large enough in central India to influence storage design and operation.
Japan			
Various	Harasawa (1993)	Not known	Reduced snowfall, hence snowmelt; changed amount and seasonal pattern of water resources; impacts on water quality.
Nepal			
Lantang Kola Catchment	Fukushima *et al.* (1991)	Not known	Increase in river flow of 100% with stable glacier area; +30% if glacier area decreases by 30%.
South Asia			
	Lal (1994)	ECHAM 3–Hamburg	Increase in soil moisture over central India, Bangladesh, and south China in summer; significant decline in soil moisture over central China in winter; increase in runoff over northeastern India, south China, and Indonesia; decline in runoff possible over north China and Thailand.
Switzerland			
Murg Catchment	Bultot *et al.* (1992)	Various sources	Decreased annual deep percolation; increased winter runoff, but little change to annual flow; reduced soil moisture content in summer; shorter spells with snow cover.
Ukraine			
Poles'e River Basin	Nazarov (1992)	Scenarios in Houghton (1991)	Increase in annual runoff of 6%; decrease in evapotranspiration of 2%; decrease in soil moisture storage of 17%.
United Kingdom			
England and Wales Catchments	Arnell (1992)	Hulme and Jones (1989)	Increases in mean annual runoff of 13 to 30% for 10% precipitation increase (15 inches annual rainfall).
United States			
Boston	Kirshen and Fennessey (1993)	GFDL (1988), GISS (1982), OSU (1984–85), UKMO (1986)	Decrease or increase in reservoir-system safe yield, depending on GCM used.
American River, Washington	Lettenmaier (1993)	2–4°C uniform temperature increase	Snow accumulation substantially reduced; high-flow season shifted from spring to winter; peak actual evaporation shifted to spring or early summer due to reduced soil moisture.
Delaware River Basin	McCabe and Wolock (1992)	GFDL, GISS, OSU	Considerably drier conditions than present; future soil moisture decrease partly masked by interannual variability, and amount of decrease varies according to GCM used.
Trinity, Colorado, and Rio Grande Basins	Schmandt and Ward (1993)	Hypothetical scenarios	Difficulty in meeting demand in 2030 under record drought conditions; severe shortages under 2°C increase and 5% decrease in precipitation.
Texas (6 Sites)	Valdés *et al.* (1994)	Various sources	Decrease in mean soil moisture concentration.

[1] Acronyms used in Table 10-1 follow: Bureau of Meteorology Research Center (BMRC), Canadian Climate Center (CCC), Commonwealth Scientific and Industrial Research Organization (CSIRO), Geophysical Fluid Dynamics Laboratory (GFDL), Goddard Institute for Space Studies (GISS), Oregon State University (OSU), and United Kingdom Meteorological Office (UKMO). GFDLH + UKMOH refers to high-resolution GFDL and UKMO models.

actual evaporation would decline even if evaporative demand were to increase.

The evaporative loss of water from vegetation, or transpiration, is influenced by vegetative properties such as albedo, surface roughness, root depth, and stomatal characteristics—and all of these may change with climate. The timing of plant growth and the species of vegetation in a catchment also might change. Plant water use also is related to atmospheric CO_2 concentrations. There are two main effects. First, there is experimental evidence that some groups of plants use water more efficiently when CO_2 concentrations are higher, leading to reduced transpiration (Dooge, 1992; Drake, 1992). Kimball *et al.* (1993) found that spring wheat grown at a CO_2 concentration of 550 ppm had an evaporation rate 11% lower than wheat grown at 370 ppm. The modeling studies of Kuchment and Startseva (1991) show that CO_2 enrichment can lead to a decrease in evapotranspiration despite climate warming and that evapotranspiration responses to changes in climate variables and plant factors can differ greatly from those that consider temperature changes alone. Second, many plants grow more when CO_2 concentrations are higher, and the increased leaf area may compensate for the increased leaf water-use efficiency. These experimental effects may not be reproduced in the field—where plant growth is subject to limitations in nutrients and moisture, as well as competitive interactions—and there is some evidence that plants acclimatize to increased CO_2 concentrations (Wolock and Hornberger, 1991; Dooge, 1992; Hatton *et al.*, 1992; Koerner, 1995; Chapter A; Chapter 9, *Terrestrial Biotic Responses to Environmental Change and Feedbacks to Climate*, of the Working Group I volume).

10.3.4. Soil Moisture

Depending on rainfall or snowmelt intensities, as well as soil capillary and aquifer hydraulic conductivities, water will infiltrate and percolate into the soil. Soil water storage plays a decisive role in evaporation and evapotranspiration of plants. For a given soil depth and soil type, any change in the seasonal distribution of precipitation or its intensities will change soil water storage, runoff processes, and groundwater recharge. Land use, river flow, and groundwater resources are affected. Improved modeling of the hydrological cycle and of these land-surface processes in GCMs is a critical research issue (Chahine, 1992; Dooge, 1992; Rind *et al.*, 1992; Milly and Dunne, 1994).

All climate models show increased soil moisture in high northern latitudes in winter, with some areas of reduction. Most models produce a drier surface in summer in northern mid-latitudes and a reduction in summer precipitation over southern Europe. Reductions over North America are less consistent, and there is greater variation between models over northern Europe and northern Asia. When aerosols are included, the patterns of soil-moisture change in the northern winter are similar but weaker, and summer drying over North America and southern Europe is reduced (see Chapter 6, *Climate Models—Projections of Future Climate*, of the Working Group I volume).

10.3.5. Snow-Cover Accumulation and Melt

Mountainous regions are particularly important in a hydrological context. They are source areas of a significant part of the water resources for many continental areas outside the tropics because of orographically induced high precipitation. Water stored as snow during the winter becomes runoff in the spring. Higher global temperatures in winter would result in more precipitation falling as rain rather than as snow, reducing snow-cover duration and causing the ascent of the snow line. These changes would lead to increased winter runoff and decreased spring and summer runoff, resulting in seasonal changes and possible disruptions in supply in regions without reservoir-based water management.

The melting of snow and glaciers is sensitive to changes in air temperature during the melt season, although temperature is only a rough index of melt energy fluxes—which consist of net radiation, sensible heat, and latent heat. Brun *et al.* (1992) used a physically based snow model to investigate the sensitivity of snow cover in the French Alps to increased incoming energy. They found that at an elevation of 1,500 m above sea level, snow cover can be considered "sensitive" during about half of the days of its existence. This means that even a slight increase in air temperature throughout the winter or a few individual warm spells have important consequences in reducing the snow cover. At 3,000 m above sea level, there is about a 6-month period during which the sensitivity to increased energy input is very weak; the sensitive days constitute about 30% of the total snow-covered period, which essentially represents the melting period. It also was shown that the entire area of the French Alps can be considered to react homogeneously to increases in incoming energy at elevations above 2,400 m above sea level, but that below this elevation the southern part of the Alps reacts more sensitively. This effect is related to latitude effects of air temperature and solar radiation.

Not much is known about evaporation in high mountain areas. Although evaporation seems to be of secondary importance in the water balances for these areas, it may play a considerable role in controlling daily variations in melt rates (Lang, 1981). With increasing altitude, vegetation activity decreases, soil (with its capacity to store water) decreases, and the duration of snow cover increases—together contributing to the general decrease of evaporation with altitude. In a warmer climate, evaporation potential is generally expected to increase, which also is expected for high mountainous terrain.

Mountain orography exerts a great influence on the atmospheric circulation at all spatial scales, which affects the horizontal and vertical distribution of precipitation. This is a point of utmost importance. If the directions of atmospheric moisture advection change (as a consequence of a change in the general route and activity of cyclones), the spatial distribution of precipitation in mountain regions could be strongly modified, compared with the general and large-scale results of present GCM simulations. High mountain areas exposed to strong windward and leeward effects will be affected significantly by such changes.

10.3.6. Groundwater Recharge and Storage

Despite the critical importance of groundwater resources in many parts of the world, there have been very few direct studies of the effect of global warming on groundwater recharge. A change in recharge will depend on the balance between an alteration in the recharge season (or the number of recharge events) and the amount of water available for recharge during the season. In northern Europe, it is uncertain whether increased winter rainfall (as simulated by most GCMs) would make up for a shorter recharge season caused by increased temperatures and evaporation in spring and autumn. In semi-arid and arid areas, where groundwater recharge occurs after flood events, a change in the frequency of rainfall occurrence and the magnitude of those events will alter the number of recharge events.

Using a model of water drainage beyond the rooting zone of plants and assuming rainfall changes of ±20% and unchanged vegetation, Peck and Allison (1988) found that recharge may change by -70% to +230% and -60% to +35% for sites near Khorat in northeastern Thailand (summer-rainfall environment) and near Collie in southwestern Australia (winter-rainfall environment), respectively. A 20% decrease in potential evaporation was found to increase recharge by the same amount or more under low-drainage conditions.

Vaccaro (1992) used a coupled stochastic weather generator and a deep-percolation compartment model to study the sensitivity of groundwater recharge in the Ellensburg basin in Washington state to climatic variability in the historical record and projected climate change. He found that the variability in estimated annual recharge under a climate scenario based on average monthly changes in precipitation and temperature for three different GCMs was less than that estimated from the historical record. The median annual recharge for the scenario was 25% lower than that for the historical simulation.

Wilkinson and Cooper (1993) studied the impact of climate change on aquifer storage and river recharge using a simple model of an idealized aquifer-river system. They found that changes in the seasonal distribution of recharge may have a critical effect on rivers supported by baseflow. The delay in the response of slowly responding aquifers to climate change may be large enough to allow adjustments to water-resources planning over an extended period.

Zhang *et al.* (1994) used a groundwater balance model to study the influence of climatic scenarios on groundwater recharge in the area of Jing-Jin-Tang, China. The results show that a temperature increase of 0.6 to 1.3°C and a precipitation change of -2% to +6% will change groundwater recharge within the range of ±5%.

10.3.7. Streamflow

Even in areas where precipitation increases, higher evaporation rates may lead to reduced streamflow. High-latitude regions may experience increased annual runoff due to increased precipitation, but lower latitudes may experience decreased runoff due to the combined effects of increased evaporation and decreased precipitation. Areas where runoff increases may experience more frequent floods and higher lake and river levels.

Regional changes in streamflow are illustrated by a study examining the impact of global warming on the flow of 33 of the world's major rivers. Miller and Russell (1992) found that all rivers in high northern latitudes showed increased flow of an average of 25% in response to increased precipitation. The largest decreases occurred at low latitudes, with a maximum of -43% for the Indus, -31% for the Niger, and -11% for the Nile due to the combined effects of increased evapotranspiration and decreased precipitation. This study used the NASA/GISS GCM and a 3-year model simulation. It found that the computed runoff for a doubled CO_2 climate depends on the model's treatment of precipitation, evaporation, and soil moisture storage (Miller and Russell, 1992). The authors also discuss the great discrepancies between observed and simulated river flow in some cases, due to oversimplified and unrealistic representations of the hydrological cycle in current GCM models.

Streamflow is highly sensitive to climate change. Shiklomanov (1989) presents a contour map of changes in mean annual runoff of rivers in the former Soviet Union (FSU) with a rise in global air temperature at an early stage of warming (0.5°C). The map shows a 10–20% decrease in the central regions of the European FSU but a 7–10% increase in the northern European FSU and western and middle Siberia. Georgiadi (1991) reports the work of N.A. Speranskaya, which for the same air-temperature rise suggests decreases in mean annual runoff of 6–20% for the Volga and 15% for the Don river systems, an increase of 7–10% for the upper reaches of the Yenisei and Ob river system, and an increase of 5–20% for the Amur river system.

The case studies in this chapter provide a clear indication that dry areas are more sensitive than wet areas to climate variations. Low-flow and dry climates will be more affected than high-flow and humid regions. Warming of the atmosphere alone will decrease streamflow far less than warming accompanied by changes in precipitation. In the case studies presented in this chapter, some areas show increasing annual and monthly flow due to increased precipitation, while other regions show decreased flow due to increased evaporation or decreased precipitation or both. High-latitude regions belong to the former category, lower latitudes more to the latter. Depending on specific scenarios, model simulations may produce an increase or a decrease in annual river flow for the same region, as well as changes in the variability of river flow through the year.

All regions with snowmelt flow regimes point out a shift of seasonal high flow from early summer and spring to the winter season. These regions will very probably have more low-flow problems in the summer. Plain regions with snowmelt floods in the melt season will have fewer meltwater floods in the future scenario—because of decreased snow precipitation—and therefore less snow water-equivalent storage.

Conceptual snow and runoff models have been used to simulate discharge from mountain regions under changed climate conditions (e.g., Kuchment *et al.*, 1987, 1990; Lettenmaier and Gan, 1990; Bultot *et al.*, 1992, 1994; Rango, 1992; Braun *et al.*, 1994). The necessary model parameter values have been evaluated under present-day conditions, usually with the assumption that these parameter values do not change under changed climatic conditions. Most studies show a shift of snowmelt runoff to early spring months at the expense of summer months. In heavily glaciated basins, there is a negative feedback between snow cover and discharge: The smaller the seasonal snow cover, the larger the runoff from glaciers due to increased icemelt caused by reduced albedo as soon as the snow cover disappears (see Roethlisberger and Lang, 1987; Chen and Ohmura, 1990). This aspect generally has not been dealt with in sensitivity studies. Another aspect is the increased loss of glacial areas with increasing summer temperatures. The additional contribution of meltwater to streamflow experienced through decades of strong glacial shrinkage would gradually turn into decreasing streamflow rates. All scenario studies presently available are limited in predictive value because they have distributed the positive temperature change more or less evenly over the year, without taking account of the importance of short-time scale (weather) processes. Detailed discussions of the limitations of GCMs to provide scenarios in alpine regions, and the impact of climate change on these regions, are presented in Beniston (1994) and Chapter 5.

10.3.8. Lakes and Surface Storage

Surface storage waters are important parts of freshwater systems. They often are intensively used for multiple purposes—such as navigation, river flood and low-flow control, recreation, water supply, irrigation, and fishing. To allow an optimal adjustment of water levels, most artificial lakes are regulated, as are natural lakes to some degree. Climate-change impacts on the hydrology of upstream river basins will affect regulated lakes mostly in seasonal and annual regimes and average turnover. Flow and temperature changes also may change the internal stratification of lakes, with consequences to aquatic ecosystems. An increase in air temperature and/or windspeed will cause an increase in lake evaporation.

10.4. Extreme Hydrological Events

10.4.1. Floods

Although the potential impact of climate change on the occurrence of flood disasters has been alluded to frequently in popular accounts of global warming, there have been very few studies addressing the issue explicitly. This is largely because it is very difficult to define credible scenarios for changes in flood-producing climatic events (Weijers and Vellinga, 1995; Beran and Arnell, 1995). Some assessments of the potential effects of global warming on flood occurrence can be made, and it is useful to distinguish among rain-generated floods, snowmelt-generated floods, and hybrids.

Global warming can be expected to produce changes in the frequency of intense rainfall in a catchment for two reasons: (1) There may be a change in the paths and intensities of depressions and storms, and (2) there probably will be an increase in convective activity (Gordon *et al.*, 1992; Whetton *et al.*, 1993). Higher sea-surface temperatures can be expected to increase the intensity of tropical cyclones and to expand the area over which they may develop. The IPCC Working Group I Second Assessment concludes from the analysis of several climate-model experiments that rainfall intensity is likely to increase with increasing greenhouse gas concentrations and that there may be an increasing concentration of rainfall on fewer rain days.

The effects of a change in the frequency of high rainfall will depend not only on the change in rainfall characteristics but also on the characteristics of the catchment. Small catchments with impermeable soils will be very responsive to short-duration intense rainfall; larger catchments, and those with more permeable soils, will be sensitive to longer-duration storm rainfall totals and general catchment wetness.

Changes in the magnitudes of snowmelt and rain-on-snow floods will deserve particular attention in some regions. The magnitudes of snowmelt floods are determined by the volume of snow stored, the rate at which the snow melts, and the amount of rain that falls during the melt period. Many studies have simulated a reduction in snow cover after global warming, but higher precipitation totals still may lead to a greater volume of snow being stored on the catchment. Floods caused by snowmelt may therefore either increase or decrease. A reduction in snow cover also may mean that more rain-generated floods occur during winter, leading to a shift in the seasonal distribution of floods.

A few studies have attempted to quantify changes in flood occurrence. Gellens (1991) used a daily rainfall-runoff model to simulate river flows in three Belgian catchments, assuming an increase in rainfall during winter. He found more frequent floods, with flows remaining above high thresholds for longer periods. The mean annual flood peak increased between 2 and 10% under the scenario used, with the greatest increase in the most-responsive catchment. Bultot *et al.* (1992) used the same model and scenario to estimate possible changes in flood frequency in a small Swiss catchment and found the same results; the mean annual flood increased by 10%.

For the Sacramento-San Joaquin basin in California, Lettenmaier and Gan (1990) found a reduction in snowmelt floods but an increase in the frequency of rain-generated floods, with a consequent increase in the occurrence of a particular discharge being exceeded. Knox (1993) examined a 7,000-year geological record of overbank floods for upper-Mississippi River tributaries in North America. He found that changes in mean annual temperature of about 1–2°C and changes in mean annual precipitation of less than 10–20% can cause large and abrupt adjustments in the magnitudes and frequencies of floods.

Kwadijk and Middelkoop (1994) investigated potential changes in flood risk in the Rhine basin, using both hypothetical change scenarios and scenarios based on GCM simulations, and found large changes in the frequency of given threshold events; an increase in precipitation and a rise in temperature would lead to major increases both in flood frequencies and in the risk of inundation. However, the hydrological simulations were limited to a monthly time-step. At this temporal resolution, the simulation of snow cover is not reliable—especially at low elevations, where the snow cover is built up and disappears again several times, in time scales of days, during one winter.

Liu (1995) simulated increases in the variability of floods over time and in flood frequencies in both northern and southern China. Bates *et al.* (1995) used a stochastic weather generator coupled with two daily rainfall-runoff models to investigate changes in the behavior of annual maximum monthly runoff series for six Australian catchments within a variety of climatic settings. In five cases, the series were noticeably higher for a changed climate than for the present day. The sizes of projected changes were found to depend on the rainfall-runoff model chosen.

There is evidence from climate models and hydrological impact studies that flood frequencies are likely to increase with global warming. The amount of increase is very uncertain and, for a given change in climate, will vary considerably between catchments. In some catchments, floods may become less frequent. There are four main reasons for the large uncertainty: (1) It is very difficult to define credible scenarios, at the catchment scale, for changes in flood-producing precipitation; (2) it is often difficult to model the transformation from rainfall (or snowmelt) to flood in a catchment; (3) available climatic and hydrological records often provide limited information about flood events; and (4) at present, in many cases, it is difficult to differentiate the effects of climatic change from those associated with anthropogenic changes to land use.

10.4.2. Droughts

Drought is a relative term that may refer to a period in which actual moisture supply at a given location cumulatively falls short of climatically average moisture supply; soil moisture is insufficient to meet evapotranspirative demands for the initiation and sustainment of plant growth; below-normal streamflow or above-normal reservoir depletion occurs; or a local or regional economy is adversely affected by dry weather patterns (Rasmusson *et al.*, 1993).

A decrease in average total rainfall, an increase in the frequency of dry spells due to a decline in rain-days, and an increase in potential evapotranspiration due to higher temperatures have the potential to increase drought frequency and severity and to extend their effects into less-vulnerable areas. Increases in the frequency and magnitude of large rainfall events may result in reduced drought potential if they result in higher soil moisture levels and groundwater recharge (Whetton *et al.*, 1993).

The frequency and severity of droughts has received little attention. Four features of most hydrological impact studies make it difficult to assess the effect of climate change on low-flow frequency: (1) Most climate scenarios do not assume any change in interannual climate variability; (2) the rate of occurrence of extreme events in perturbed climate series is influenced strongly by the relative severity of events in the original series; (3) the length of concurrent climatic and hydrological records can be relatively short; and (4) feedback between catchment vegetal cover and atmospheric moisture has been neglected (Savenije and Hall, 1994). Many studies focus on simple descriptive statistics of daily, monthly, seasonal, or annual soil moisture and flow over the simulation period, including minimum annual discharge (Mimikou and Kouvopoulos, 1991); mean minimum discharge (Vehviläinen and Lohvansuu, 1991); annual minimum daily streamflow (Bultot *et al.*, 1992); and flow equalled or exceeded 95% of the time (Arnell and Reynard, 1993; Bultot *et al.*, 1992).

A few studies have attempted to quantify changes in low-flow or drought occurrence and their impacts. Gellens (1991) examined the behavior of low-flow episodes in three Belgian catchments, each with more than 80 years of records. The low-flow behaviors of the catchments were found to respond differently to climate change. Whetton *et al.* (1993) examined changes in drought occurrence in terms of seasonal soil water deficits for nine Australian sites. Results indicate that significant drying may be limited to the south of Australia. These researchers qualify their results by noting that possible changes in ENSO events due to global warming were not considered because no current GCM simulates ENSO effects. The observed interannual variability of rainfall in north and east Australia is strongly influenced by ENSO. Several researchers have investigated the impact of climate change on reservoir reliability using stochastic weather generators or resampling methods to generate long-term sequences of synthetic weather data (e.g., Nathan *et al.*, 1988; Cole *et al.*, 1991; Lettenmaier and Sheer, 1991; Wolock *et al.*, 1993). Descriptions of these and other similar studies appear in Chapter 14.

10.5. Physical and Chemical Changes in Freshwater Ecosystems

This section integrates freshwater ecosystems via hydrology and presents potential effects of climate change on their physical/chemical properties and processes.

10.5.1. Integrated Landscapes of Lakes and Streams via Hydrology

Inland aquatic ecosystems such as freshwater wetlands (Chapter 6) and lakes and streams (covered in this chapter) represent distinct elements of the hydrologic continuum from the atmosphere to the sea. Water flows first downwind as weather systems and then downhill toward the sea. Patterns of atmospheric deposition are closely related to regional- or continental-scale atmospheric "flow paths" (Galloway and Likens, 1979). Freshwater ecosystems occupy landscape depressions where subsurface flows emerge or surface flows and direct precipitation are channeled or retained. Ecological conditions along the freshwater continuum are determined by local factors (soils, vegetation, lake basin shape, stream gradient, temperature) and the legacy provided by spatially and temporally antecedent patches on this flow path. The interplay of climate and catchment controls biogeochemical processes that add particulates and solutes to water during transport in the catchment and within the stream or lake. These materials and physical features regulate biota in streams and lakes, and feedbacks from ecological processes modify water quality and physical features of freshwater ecosystems. Water quality, fisheries, and recreational values of lakes and streams are determined by interactions between the chemical and physical characteristics and the biota of freshwater ecosystems.

Stream ecology has acknowledged the close link between terrestrial catchments and aquatic systems since Hynes (1963, 1975). Progressive longitudinal changes in channel morphology and in dominant sources of water and organic matter, together with downgradient flow, produce predictable ecological patterns along the continuum from small streams to large rivers (Vannote *et al.*, 1980; Minshall *et al.*, 1983). Nutrient and organic-matter input to streams is closely related to terrestrial and wetland processes (Bormann *et al.*, 1969; Likens, 1984; McDowell and Likens, 1988; Wetzel, 1990) and to the flow path of water controlling its dissolved load. The chemistry of water entering streams in Alaska, for example, is dependent not only on the types of terrestrial vegetation through which water flows but also their sequential order (Giblin *et al.*, 1991). Physical and chemical properties of lakes are linked strongly to terrestrial processes via effects on stream discharge and chemistry. Recent studies of lake districts show the importance of landscape influences on internal lake properties (Swanson *et al.*, 1988). In headwater lake districts, biotic diversity and ecological functioning depend upon landscape position; higher drainages receive proportionately more input from direct precipitation and differ markedly from those lower in the landscape that receive larger inputs of water and materials from terrestrial sources (Kratz *et al.*, 1991, 1996).

Changes in climate influence water supply and quality directly through altered precipitation and atmospheric deposition and indirectly by the altered nature of landscape elements through which water and transported substances pass. Many of the most profound effects of climate change on lakes and streams will involve changes in terrestrial ecosystems that alter water and material inputs to freshwaters.

10.5.2. Changes in Streamflow and Lake Water Levels

Streamflow, groundwater flow, and lake levels respond to changes in precipitation and evapotranspiration. Under hotter and drier scenarios, lower runoff and greater evaporation would reduce stream baseflows and could result in periodic drying of some perennial headwater streams. Even in regions that experience increased annual precipitation, reduced summer rains or still-larger increases in evapotranspiration would produce lower baseflows and increased intermittency in summer. Warmer and wetter scenarios—particularly at higher latitudes—would increase runoff, with some intermittent streams becoming perennial and with higher flows altering channel geomorphology. Some lakes could become smaller or dry up entirely as evaporation exceeds water inputs; others could increase in size where precipitation increases more than evaporation.

Long-term, irregular water-level fluctuations occur mainly in semi-arid and dry tropical regions with distinct wet and dry seasons and high interannual variability (John, 1986). Water-level changes are most pronounced in lakes with relatively short retention times, in association with seasonal precipitation patterns. Paleohydrology and paleolimnology are particularly important approaches for obtaining information on paleo lake levels related to climate change. Analyses point out that lake levels and lake hydrology are sensitive to changes in climate (Almquist-Jacobson, 1995; Digerfeldt *et al.*, 1992; Almendinger, 1993; Webb *et al.*, 1944). Lake level response to climatic fluctuations, reconstructed using diatoms for the northern prairie region of the United States, shows such sensitivity (Fritz *et al.*, 1993, 1994; Laird *et al.*, 1996). Moon Lake underwent three major changes in climate: a mid-Holocene

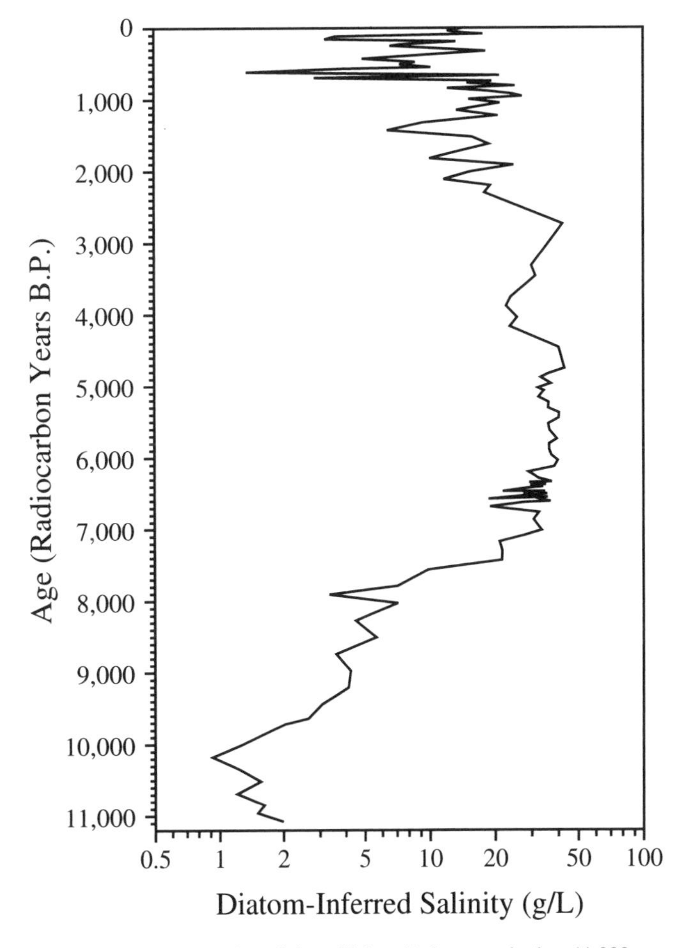

Figure 10-3: Changes in salinity of Moon Lake over the last 11,000 years, inferred from diatom assemblages in sediments (Laird *et al.*, 1996).

period of high salinity indicative of low effective moisture; a transitional period from 4700–2200 before present (BP) with poor diatom preservation; and a late-Holocene period of lower salinity and greater variability, indicating fluctuations in effective moisture (Figure 10-3). Large lakes in the Great Basin of the United States underwent considerable variations in lake levels during the Pleistocene (Benson *et al.*, 1990), and similar variations have been inferred for Mono Lake, California (Stine, 1990).

Lakes have individualistic and often rapid responses to climate changes (Figure 10-4). Lake Titicaca experienced a 6.3-m rise in water level from 1943 to 1986. This increase equals 140 mm per year and exceeds by a factor of 40 the change in mean sea level estimated from global warming. Even Lake Michigan, at temperate latitudes, has oscillated over a 1.7-m range from 1960 to 1993. Fluctuations in lakes with large temporal changes in water level are expected to be exacerbated by climate change.

Larger changes in lake water levels are likely at high latitudes where climate warming scenarios are greatest. In the largest Antarctic desert, the McMurdo Dry Valleys, closed-basin lakes fed by glacier meltwater streams, rose as much as 10 m from 1970 to 1990, or almost 480 mm per year (Chinn, 1993). Marsh and Lesack (1996) project with a hydrologic model and a 2 x CO_2 climate that many lakes in the Mackenzie Delta in the Canadian arctic could disappear in several decades owing to decreased precipitation and flood frequency. Some drainage lakes (no inlets but with an outlet) and seepage lakes (no surface inlet or outlet) in the north central United States (Eilers *et al.*, 1988a) are responsive to changes in precipitation; during recent droughts in the late 1980s, lake levels declined substantially. Some of these lakes operate as drainage lakes during high water and as seepage lakes during dryer periods.

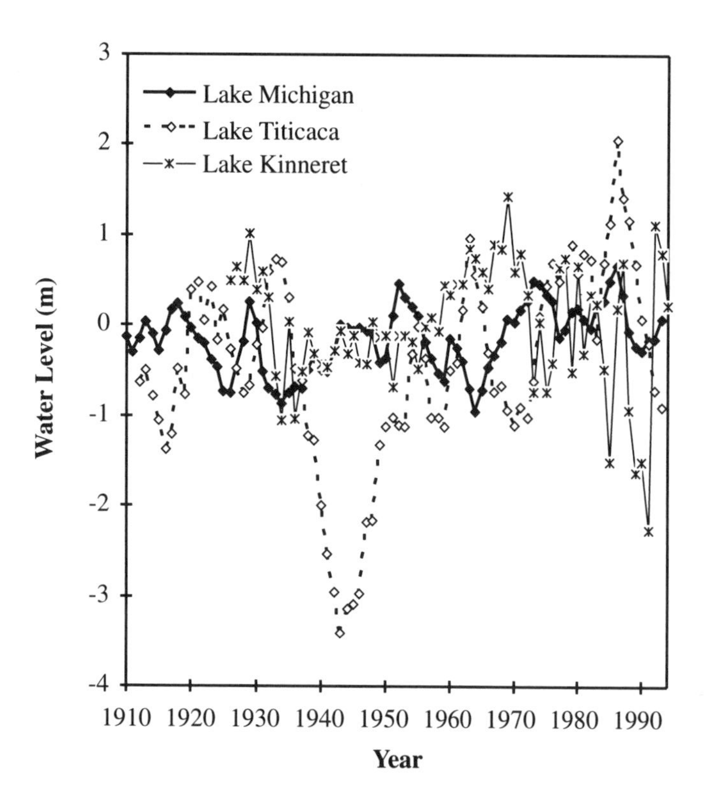

Figure 10-4: Water levels as deviation from the mean for Lakes Kinneret (Middle East), Michigan (North America), and Titicaca (South America) from 1912 to 1993. For Lake Michigan, changes in lake level were calculated as the deviation of the mean annual lake level from the long-term mean lake level (1860–1993) from National Ocean Service, National Oceanic and Atmospheric Administration, USA. For Kinneret and Titicaca, deviation was calculated as the difference between mean annual minimum and maximum water levels and long-term mean [1926–1994 for Kinneret (Serruya, 1978; Gophen and Gal, 1991) and 1912–1993 for Titicaca (Aquize, pers. com.)].

Despite relatively small climatic changes predicted for the tropics, tropical lakes also may be quite sensitive to climate change (see Box 10-1). The level of Lake Victoria (East Africa) rose rapidly in the early 1960s following only a few seasons with above-average rainfall and has remained high since (Sene and Pinston, 1994).

Patterns of water-level change are sensitive to global warming scenarios. For Lake Michigan, 2 x CO_2 scenarios produce a reduced net basin water supply and a lowering of water level of 1 to 2.5 m, depending on the GCM used (Croley, 1990). Assuming that atmospheric CO_2 will double in 33–40 years (see Chapter 6, *Climate Models—Projections of Future Climate*, of the Working Group I volume), this constitutes a decline in water level of 25 to 62 mm/year and exceeds by a factor of about 10 (in the opposite direction) increases in mean sea level estimated from global warming.

10.5.3. Changes in Biogeochemical Fluxes

Physical and biological characteristics of catchments—such as geology, soils and microbes, vegetation, and dominant flow paths of water—control the chemistry of surface water and groundwater. If the water table is close to the surface, flow is routed primarily through organically rich and biologically active upper soil horizons; concentrations of most major ions are lower and concentrations of dissolved organic material higher than if water tables are deep and water moves through lower mineral soil horizons and bedrock before entering lakes or streams. Vegetation and soils, in conjunction with hydrologic flow paths, control the effects of nutrient inputs from atmospheric deposition or fertilizer applications. In agricultural catchments, the effects of fertilizer application on surface-water nutrient concentrations are reduced substantially if water moves through riparian areas where natural vegetation has been preserved (Peterjohn and Correll, 1984). Climate changes that alter hydrologic flow paths or vegetation and soil characteristics will result in changes in groundwater and surface-water chemistry. For streams, these chemical shifts may be episodic or seasonal; for lakes with long residence times, the shifts may be in average chemical characteristics.

Dynamic, process-based water-quality models are needed to estimate effects of climate change because biogeochemical

Box 10-1. African Lakes and Climate Change

African Great Lakes are sensitive to climate variation on time scales of decades to millennia (Kendall, 1969; Livingstone, 1975; Haberyan and Hecky, 1987). Lake Victoria (the world's second-largest freshwater lake by area), Lake Tanganyika (the world's second-deepest lake), and Lake Malawi were closed basins for extended periods in the Pleistocene and Holocene (Owen *et al.*, 1990). Lakes Malawi and Tanganyika were hundreds of meters below their current levels; Victoria dried out completely. Today these lakes are in delicate hydrologic balance and are nearly closed. Only 6% of water input to Tanganyika leaves at its riverine outflow—which was blocked totally when the lake was explored by Europeans (Bootsma and Hecky, 1993).

Higher temperatures would increase evaporative losses, especially if rainfall also declined. Minor declines in mean annual rainfall (10–20%) for extended periods would lead to the closure of these basins even if temperatures were unchanged (Bootsma and Hecky, 1993). Tropical temperatures are increasing; temperatures in the 1980s were 0.5°C warmer than a century earlier and 0.3°C warmer than from 1951–1980. Concurrently, Lake Victoria's epilimnion was warmer by 0.5°C in the early 1990s than in the 1960s (Hecky *et al.*, 1994). Although current climate scenarios project small increases in tropical temperatures, small changes in temperature and water balance can dramatically alter water levels, as well as mixing regimes and productivity.

Recent temperature and rainfall data and GCM simulations indicate increasing aridity in the tropics (Rind, 1995). Increases of 1–2°C in air temperatures could substantially increase the stability of stratification in permanently stratified Tanganyika and Malawi. Their deep waters are continuously warm, but the less than 1°C difference between surface and deep water in warm seasons maintains a density difference that prevents full circulation. Lake Tanganyika's deep water has been characterized as a "relict" hypolimnion that formed under a cooler climate within the past 1,000 years (Hecky *et al.*, 1994). Since then, warming has created a barrier to vertical circulation. Additional warming could strengthen this barrier and reduce the mixing of deep, nutrient-rich hypolimnetic water and nutrient-depleted surface layers; that mixing sustains one of the most productive freshwater fisheries in the world (Hecky *et al.*, 1981).

processes controlling water quality are complex. Such models must accurately represent the hydrologic regime—especially the timing and intensity of infiltration (rainfall and snowmelt), as well as the various flow paths through the catchment and changes in vegetation. Examples of coupled hydrologic-water quality models being developed include the Regional HydroEcological Simulation System (Band *et al.*, 1996), modifications of Topmodel for transport of dissolved organic material (Hornberger *et al.*, 1994), and the Alpine Hydrochemical Model (Wolford and Bales, 1996).

Streams in catchments with low acid-neutralizing capacity periodically may become more acidic. Higher-elevation streams and lakes may receive larger pulses of acidity in the spring if winter snowfall increases or snowmelt occurs over a shorter period—for example, the modeled response with the Alpine Hydrochemical Model for the Emerald Lake catchment in the southern Sierra Nevada region of the United States (Wolford and Bales, 1996). In mountain streams of the southeastern United States, Mulholland *et al.* (1996) project an increase in the frequency or duration of acidic episodes with an expected increase in storm intensity. However, in the northeastern United States, reductions in winter snowpack with climate warming might reduce the intensity and duration of stream acidification in the spring (Moore *et al.*, 1995, 1996). At boreal latitudes, soil drying associated with climate warming can lead to oxidation of reduced forms of sulfur in soils and concurrent acidification of streams (Bayley *et al.*, 1992b; LaZerte, 1993; Schindler *et al.*, 1996; see Box 10-2).

Warmer climates could affect nitrate fluxes through changes in microbial processing rates. Higher temperatures may lead to increased mineralization of organic nitrogen in soil, with increasing amounts available for flushing into lakes and streams. Peak nitrate concentrations in United Kingdom catchments tend to occur following flushing by heavy rainfall after prolonged dry spells; an increase in the duration of dry spells would increase the leachable nitrogen accumulating in soils (Jenkins *et al.*, 1993).

For lakes, position in the local groundwater flow system affects their chemical response to shifting hydrologic conditions. Two mechanisms control drought response in northern Wisconsin (Webster *et al.*, 1996): Longer residence times and evapoconcentration in lower lakes cause conservative solutes to increase in concentration, whereas lesser inflow of solute-rich groundwater in higher lakes decreases solute concentrations. In prairie lakes and wetlands of north-central United States and Canada, differences in local geomorphology, groundwater flow paths, and vegetation produce somewhat different responses in water chemistry to droughts, with salinity increasing in some and decreasing in others (Evans and Prepas, 1996; LaBaugh *et al.*, 1996).

Export of organic matter from terrestrial systems to streams and lakes is expected to change with climate. The amount of wetland drainage strongly influences the concentration of dissolved organic carbon in streams and rivers (Mulholland and Kuenzler, 1979; Eckhardt and Moore, 1990), and changes in wetland distribution and connectivity with streams and lakes

Box 10-2. Climatic Warming Effects: Boreal Lakes and Streams

Boreal soft-water lakes and streams at the Experimental Lakes Area (ELA) in Ontario, Canada, represent numerous ecosystems throughout North America and Eurasia. At ELA, natural variability has been studied for the past 25 years through measurements of climatic, hydrological, physical, chemical, and biological parameters. Analyses of this warmer and drier interval have provided a holistic picture of the potential interactive responses of lakes and streams to climate warming (Schindler *et al.*, 1996, and references therein). The almost 2°C of warming and more than 50% decline in runoff are comparable to GCM predictions for a 2 x CO_2 climate. Biological responses in the lakes were linked to changes in biogeochemical processes in the catchments and physical properties of the lakes. Major changes were in:

- *Hydrochemical fluxes*: Export of many constituents decreased in response to decreased streamflow, weathering rates, and decomposition of organic material; drier soils resulted in shorter periods of elevated water tables and saturated soils. Export of dissolved silica and dissolved organic carbon (DOC) decreased more than 50%. Base cations decreased less, suggesting that dry or burned vegetation was less efficient at retaining base cations than vegetation with adequate moisture. Export of acid anions generally decreased, but acid anions increased relative to base cations—causing lower alkalinity and pH in streams. Soil drying apparently enhanced oxidation and the release of sulfate. Nitrogen and phosphorus flux increased somewhat after drought-enhanced fires and major windstorms, but overall export of these nutrients declined.
- *Lake physics, chemistry, and biology*: The lakes generally became warmer and more transparent, and the increased transparency facilitated the deepening of thermoclines (Figure 10-5). The average ice-free period increased by about two weeks. Even though exports from catchments decreased, concentrations of many inorganic constituents increased because residence time in the lakes and evaporative concentration increased. Unlike streams, lakes became more alkaline because lake processes dominated—with longer residence times specifically from the removal of sulfate by microbial reduction and the dissolution of base cations in lake sediments. Slight increases occurred in phytoplankton biomass and species diversity; primary productivity and chlorophyll concentrations did not change. In one lake, suitable habitat for lake trout was reduced and lake trout disappeared.

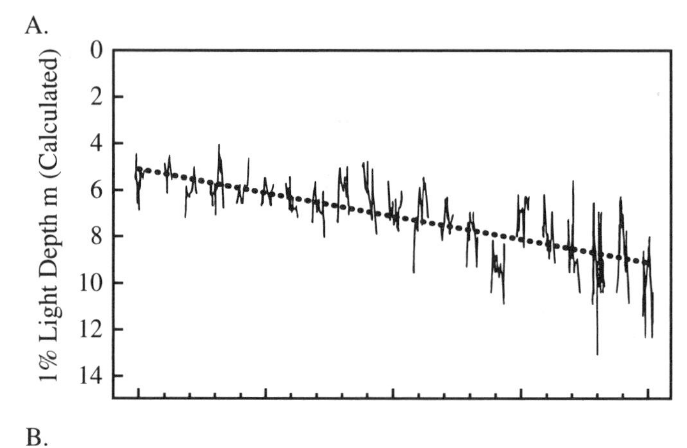

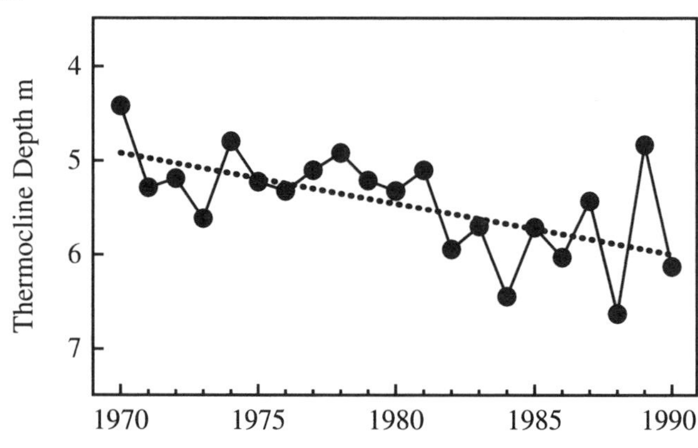

Figure 10-5: (A) Changes in penetration of photosynthetically active radiation (PAR) as the depth of isopleth representing 1% of surface light (equivalent to photic zone depth), from Schindler *et al.* (1996); (B) changes in depth of thermocline.

would alter DOC concentrations. Drought can reduce DOC inputs to streams and lakes on the Canadian Shield (Schindler *et al.*, 1996). Changes in the amount and chemical form of organic-matter inputs can occur because vegetation is altered in the catchment overall or in the riparian zone (Band *et al.*, 1996). Increasing atmospheric CO_2 or climatic warming may increase the rate of organic-matter inputs to streams because riparian plants produce more foliage, and species composition may change (Bazzaz, 1990)—which in turn would increase production of heterotrophic organisms. This might be countered by drier conditions that would reduce organic-matter inputs from riparian zones. Complexities become most apparent when considering interactions between organic matter and nutrient inputs. Higher atmospheric CO_2 concentrations may increase plant foliage but reduce its nitrogen content (Koerner and Miglietta, 1994); rates of organic-matter decomposition and production in detritus-based food webs actually might decline (Meyer and Pulliam, 1992). Under drier climates, reduced runoff from streams that drain wetlands could reduce inputs of dissolved organic carbon but increase inputs of nitrogen from increased organic-matter oxidation. In a simulation of such changes in a Welsh peatland stream, primary production increased (Freeman *et al.*, 1994).

In warmer, drier climates, an increase in wildfires can increase runoff, sediment, and nutrient load (Schindler *et al.*, 1980). Wildfires also cause longer-term (5–50 years) reductions in woody debris and material retention in streams (Minshall *et al.*, 1989).

Observations over a 20-year period of warmer and dryer conditions at the Experimental Lakes Area (see Schindler *et al.*, 1996) illustrate the complex interactions and responses that can be anticipated from climate warming and the value of careful long-term study of aquatic ecosystems to identify critical processes and interactions (Box 10-2).

10.5.4. Changes in Temperature, Dissolved Oxygen and Carbon Dioxide, and Light

Stream water temperature is determined by air temperature, the origin of streamflow (groundwater is often cooler than quick flow in temperate-zone warm seasons), local shading of the stream, and, to a lesser extent, flow volume. Changes in stream temperatures are about 0.9 times as great as changes in air temperature in the UK (Webb, 1992) and the north-central United States (Stefan and Preud'homme, 1993). Simulations from air temperature increases of 1.5° to 4.5°C, using a heat-balance model, produce increases in summer stream temperatures of 2.4° to 4.7°C for Minnesota (Stefan and Sinokrot, 1993). Projected increases in stream temperatures are substantially greater if riparian vegetation is removed, as are stream temperatures where riparian vegetation is experimentally removed (Burton and Likens, 1973). Higher water temperature, along with lower streamflows, could lead to reduced oxygen concentrations in streams receiving organic loading from effluents or in eutrophic streams, where warmer temperatures could stimulate algal blooms.

Thermal regimes in lakes are responsive to climate change because they are controlled by solar radiation, wind velocity, air temperature, humidity, and evaporation, as well as by lake area, depth, and transparency. In north temperate regions, 2 x CO_2 climate scenarios increase summer epilimnion (surface mixed waters) temperatures by 1–7°C, and hypolimnetic (deep cooler waters) temperatures change by -6 to +8°C depending on the size and morphology of the lake (Stefan *et al.*, 1993; DeStasio *et al.*, 1996; Magnuson *et al.*, 1995). Warming increases the duration of warm-season stratification and the sharpness of the metalimnion (the zone of sharp temperature change, >1°C/m, between the epilimnion and hypolimnion). Warming decreases the duration of ice cover or the frequency of winters with winter ice cover. To the south, loss of lake ice cover at some latitudes indicates that lakes that mix twice per year in spring and fall would become monomictic—that is, they would mix through the fall, winter, and spring and stratify longer in the summer. To the north, some lakes that presently are monomictic and mix during summer would stratify in summer and mix twice a year in spring and fall. Some deep lakes that mix twice a year would be less likely to mix completely (McCormick, 1990). Changes in the water level of shallow lakes could change the heat budget or the mixing properties of lakes.

Changes in catchment DOC fluxes would change the mixing regimes of lakes because water clarity and light absorbance are partially functions of DOC concentrations (see Box 10-2). The strong effect of DOC inputs on lake mixing properties appears to be limited primarily to lakes smaller than 500 ha, based on Canadian Shield lakes (Fee *et al.*, 1996).

Climate change could cause changes in the amount of oxygen entering lakes and streams through direct atmospheric exchange or through changes in lake and stream metabolism. With a warmer climate, concentrations of dissolved oxygen are expected to be lower because concentrations at saturation are lower and respiration rates are higher at higher water temperatures. Upland streams in the UK tend to be fast flowing and oligotrophic, with high dissolved oxygen owing to rapid reaeration rates (Jenkins *et al.*, 1993); lowland river systems tend to be slow flowing, with low turbulence and with dissolved oxygen below saturation. Simulations with the QUASAR dynamic water-quality model suggest that higher temperatures and lower streamflow would result in lower oxygen concentrations in UK rivers. Conditions are exacerbated in lowland rivers, which also tend to have agricultural catchments and often are used as disposal routes for sewage and industrial effluents.

Dissolved oxygen levels in lakes, especially in the deep hypolimnetic waters, are responsive to climate warming scenarios. Simulations with 2 x CO_2 climates for Lake Erie in North America (Blumberg and DiToro, 1990) suggest that losses of 1 mg/L in upper layers and 1–2 mg/L in lower layers can be expected, as can an increase in the area of the lake that is anoxic. This is because at warmer lake temperatures, bacterial activity increases in deep waters and surficial sediment, not from a relocation of the thermocline and smaller hypolimnetic volume. Smaller lakes (Figure 10-6) have similar responses to simulations with a 2 x CO_2 climate (Stefan *et al.*, 1993; Stefan and Fang, 1994). In surface waters, dissolved oxygen remains above 7 mg/L and declines by not more than 2 mg/L. In deep hypolimnetic waters, simulated concentrations are as much as 8 mg/L lower in midsummer and never less than 2 mg/L lower. Depletion would occur up to 2 months longer than under baseline conditions owing to longer periods of stratification. Declines in dissolved oxygen are projected to occur in productive (eutrophic) and nonproductive (oligotrophic) lakes but to be more rapid and of longer duration in productive lakes. Simulated declines in dissolved oxygen are more apparent in spring and fall because stratification would be extended into these seasons. Changes in water levels would interact with the mixing properties of shallow lakes and thus their oxygen dynamics.

Dissolved CO_2 concentrations are supersaturated with respect to atmospheric concentrations in most lakes and streams (Kling *et al.*, 1991; Cole *et al.*, 1994)—a result of high CO_2 concentrations in groundwaters and high rates of respiration in many

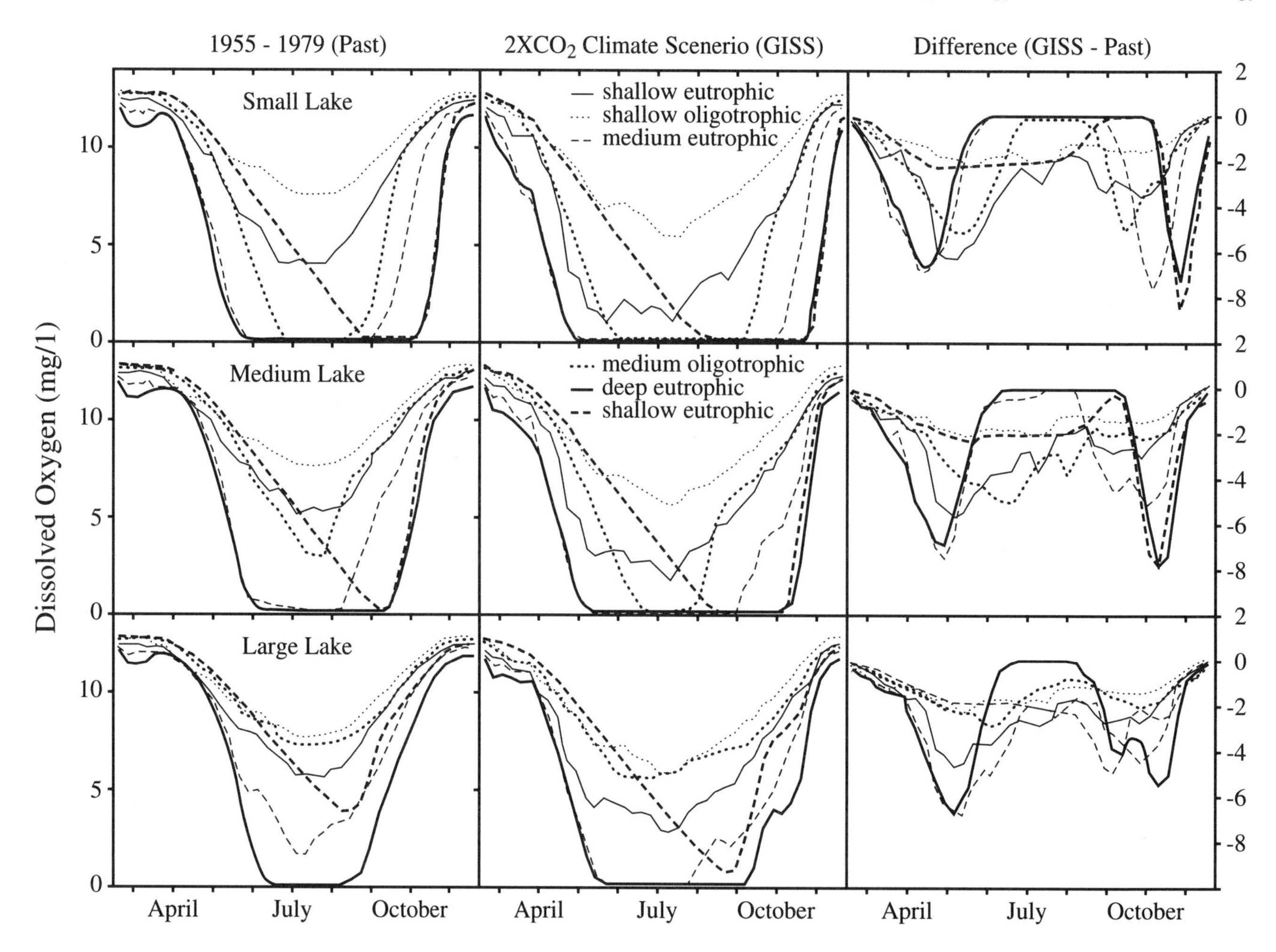

Figure 10-6: Simulated deep-water (hypolimnetic) dissolved oxygen levels in 18 Minnesota lakes (Stefan *et al.*, 1993). Past conditions are on the left, projected 2 x CO_2 climate scenario in the middle, and the difference between future and past climate conditions on the right. Small lakes are on top, medium lakes in the middle, and large lakes on the bottom.

freshwaters. Direct effects of increasing atmospheric CO_2 will be significant only in streams and lakes with low alkalinity and relatively high primary productivity; dissolved CO_2 and HCO_3 concentrations are depleted during daylight, causing increases in pH. In such systems, an increase in atmospheric CO_2 will increase carbon flux from the atmosphere to the water body.

Changes in cloud cover and shading could change light profiles in lakes and streams and affect primary productivity. Light is a major control on photosynthesis in freshwater ecosystems. Reduction in riparian vegetation under drier climates would decrease shading, thus increasing light availability and primary productivity in headwater streams. DOC in freshwaters responds to climate change and alters light availability (see Section 10.5.3 and Box 10-2). High light intensity also can inhibit photosynthesis at the surface, and UV radiation inhibits both algae and invertebrates. Interaction between DOC and UV attenuation is particularly important, given that increased UV radiation is associated with the depletion of ozone. In North American lakes, even small decreases in DOC could significantly enhance the penetration of biologically harmful UV-B radiation into the water column (Scully and Lean, 1994; Williamson *et al.*, 1995).

10.5.5. Erosion, Sedimentation, and River Channel Stability

Changes in vegetation, rainfall, and hydrological regimes affect erosion on hillslopes and in river channels, sediment transport and sedimentation, and river channel stability (Schumm, 1977). Separating the effects of climate fluctuations from changes in land use is difficult because land use also affects hillslope erosion and sediment loads in streams and hence the river channel. Three other general points are apparent from published studies: (1) Much geomorphological activity is episodic, occurring during extreme events; (2) there may be critical thresholds in the geomorphological system, beyond which change is more dramatic; and (3) the effects of climate change work through the geomorphological system over many decades.

Soil and hillslope erosion are sensitive to land-use practices and changes in catchment vegetation. In simulations with no change in land use, increased winter rainfall results in greater erosion from arable lowlands in the UK but lower losses from arable fields in uplands because the warmer climate includes longer growing seasons and hence greater ground cover (Boardman *et al.*, 1990). Hillslope erosion only occurs once rainfall exceeds a critical threshold in an upland UK catchment (Harvey, 1991). If rainfall frequency and/or intensity were to increase significantly and more erosion were to follow, increased river flows would influence channel erosion, with erosion potential increasing for higher flows. Not all sediment generated on hillslopes reaches the river channel, but increased hillslope and channel erosion should increase sediment loads. Sediment loads in the Yellow River (China) increase in simulations of a conceptual hydrologic and sediment model because high intensity rainfalls were more frequent, even though total runoff decreased in the scenarios (Bao, 1994).

Historical analogs provide useful information about possible changes in sediment loads. Sediment loads on the River Garonne (France) during a wet period in the 1840s were four to five times higher than in drier periods (Probst, 1989). River sediment loads in northern regions will be affected not only by flow regime changes but also by temperature rise. If warming raises temperatures above freezing for longer periods, snow and icemelt and permafrost degradation would increase, releasing more sediment into rivers (Woo and McCaan, 1994). Changes in flow regime and sediment loads would change river channel stability; specifics depend not only on changes in inputs but also on the channel's proximity to a change in threshold. A channel at the boundary between braided and meandering would be more sensitive to change than one with a clearly defined, meandering channel. An upland channel might be most sensitive to changes in the supply of sediments from adjacent hillslopes, whereas a lowland channel would be more sensitive to changes in sediment loads from upstream.

Extreme events are most important in triggering channel change (Newson and Lewin, 1991). Given an increase in the frequency of extreme events (Section 10.4), the location of channel instability and sites of erosion and sedimentation should change, although this will depend greatly on local channel characteristics. Changes in channel geomorphology should be greater in unconstrained alluvial channels, owing to their more dynamic geomorphology, than in constrained channels on or near bedrock. However, as noted by Newsom and Lewin (1991), "The interactions between river channel change, sediment transport, climatic and hydrological fluctuations, and human activities are so complex that a simplistic response model relating morphological response to alternative climate-change scenarios is just not practical at present."

10.6. Lake and Stream Biology

This section considers the potential effects of thermal and hydrologic changes on biological rates, life histories, and reproduction, and resulting changes in geographic distribution and biodiversity, as well as possible adaptations to climate change.

Inland waters are perhaps most sensitive to landscape-level effects of climate change and as such serve somewhat as "canaries" for effects on landscapes. Effects are difficult to dissect into their components because lakes and streams aggregate many other influences of human activity. Lakes and streams, as well as wetlands (see Chapter 6), integrate many human and natural events occurring on the landscape. Streams, in a sense, are the arteries of the continents; lakes and wetlands are the integrative sensors of changes in climate, pollution, and land use (Degens *et al.*, 1991). The effects of climate change on terrestrial systems, such as vegetation (see Chapter 1) and land use, are transmitted to and accumulate in freshwater ecosystems. Ecological responses of freshwater to climate change will be exacerbated by other anthropogenic changes.

Climate scenarios used in many of the references cited in this section (and in Section 10.5) differ from the most recent IPCC scenarios. Analyses of climatic warming effects from earlier scenarios may overestimate effects, compared with new analyses with more recent scenarios. Previous IPCC 2 x CO_2 scenarios suggested globally averaged mean annual temperature increases of 1.5° to 4.5°C (IPCC, 1992)—about twice the values from current IPCC scenarios with ramped models and the effect of atmospheric aerosols. Nonetheless, the climate warming effects described in this section are reasonable in direction, if not in magnitude. Current scenarios also suggest that warming will be greater in fall/winter and at night during summer than at other times. Regional differences in future climatic change will be important, and GCM-based global scenarios still do not provide adequate regional scenarios. Regional warming may be considerably less than or greater than globally averaged warming.

10.6.1. Thermal Effects on Biota

Extreme water temperatures can kill organisms. More moderate water temperature variations control physiological rates and behavioral performances, and influence habitat preference (e.g., for fish: Fry, 1947, 1971). Given that species have varying tolerance ranges for temperature, shifts in temperature can produce changes in species composition that can affect the overall productivity of individual freshwater ecosystems and their utility to humans.

Direct effects of climatic warming on the productivity, life history, and reproduction of organisms in streams and lakes will result primarily from increases in average water temperatures and growing season, increases in winter water temperatures and a shorter winter period of cold water or freezing, and increases in maximum summer water temperatures. Indirect effects of warming on productivity and life history include changes in hydrology, especially of extreme events (Section 10.6.2), and in vertical mixing (Section 10.5.4). In streams, ecological effects should be strongest in humid regions where streamflows are less variable and biological interactions control organism abundance (Poff and Ward, 1990). Warming effects should be strong in small, mid-latitude streams where large groundwater discharges currently maintain relatively low maximum water temperatures in summer, as well as in mid-latitude and high latitude streams that experience large increases in annual degree days. Direct thermal effects on lake organisms should be greatest at higher latitudes where the largest changes in temperature are expected.

10.6.1.1. *Freshwater Invertebrates*

Production rates of plankton and benthic invertebrates increase logorithmically with temperature, with rates increasing generally by a factor of 2–4 with each 10°C increase in water temperature, up to about 30°C or more for many organisms (Regier *et al.*, 1990, and citations therein). Macroinvertebrate production rates increased by 3 to 30% per 1°C rise in mean annual water temperature in a comparison of 1,000 stream studies at mid- to high latitudes (Benke, 1993); annual macroinvertebrate production increased and production-to-biomass ratio increased by 3 to 25%.

Regier *et al.* (1990) posit that increases in primary production, zooplankton biomass, and fish yields with temperature indicate that the thermal effects of warming will superficially resemble eutrophication (increasing production from increased nutrients). The effects of these temperature-dependent changes would be least in the tropics, moderate at mid-latitudes, and pronounced in high latitudes. Such increases assume that as temperatures warm, warmer-water assemblages replace the cooler-water ones because in most cases individual species do not continue to increase over the entire temperature range.

Ecosystem production-to-respiration (P:R) ratios decline with warming because respiration increases more rapidly with temperature than photosynthesis does. Even though the photosynthetic rates of many algae increase with temperature—at least up to about 25° to 30°C (Davison, 1991)—ecosystem primary production may increase little because nutrient concentrations and light availability often limit algal production, and these factors also change with climate change. Respiration rates increase with temperature at both the organism and ecosystem levels. At average levels of light and biomass, ecosystem P:R ratio in Arizona declines from 3.0 at 19°C to 1.6 at 30°C (Busch and Fisher, 1981). In Ontario, Canada, the P:R ratio of stream communities on stone surfaces declines from 4 at 4°C to 0.5 at 28°C (Rempel and Carter, 1986). A drop in P:R ratio tends to reduce the average mass of benthic organic matter and/or the export of organic matter downstream (Carpenter *et al.*, 1992). Higher rates of microbial respiration with higher temperatures suggest that food resources for invertebrates feeding on seasonally available detritus from terrestrial vegetation might increase in the short term following its input to streams. However, higher microbial respiration rates will increase organic-matter decomposition rates and may shorten the period over which detritus is available to invertebrates (Rempel and Carter, 1986).

Temperature has strong influences on virtually all physiological and life-history parameters (Table 10-2). That is the reason for increases in secondary production rates with temperature. These direct effects can be used with zooplankton as indicators of biotic vitality in systems undergoing changes in climate. At stressful temperatures, survival and reproduction decline while mortality and development times increase (Roff, 1970; Herzig, 1983; Orcutt and Porter, 1983; Cowgill *et al.*, 1985; Jamieson and Burns, 1988).

Table 10-2: *Response of physiological characteristics and processes of zooplankton and benthic invertebrates to increasing water temperature, based on studies for which temperature remains well below tolerance limits.*

Process	Effect of Increasing Temperature
Life-Stage Development	Faster
Reproduction Rate	Greater
Ingestion Rate	Greater
Growth Rate	Greater
Respiration Rate	Greater
Mortality Rate	Lower
Generation Times	Shorter
Generations per Year	More
Average Body Size	Smaller
Production:Biomass Ratios	Greater

Sources: Lei and Armitage, 1980; Vidal, 1980; Ward and Stanford, 1983; Woodward and White, 1983; Sweeney, 1984; Sweeny *et al.*, 1986; Rempel and Carter, 1987; Short *et al.*, 1987; Jamieson and Burns, 1988; Maier, 1989; McLaren *et al.*, 1989; Abdullahi, 1990; Moore and Folt, 1993; Moore *et al.*, 1995a; Hogg *et al.*, 1995.

Communities in mid-latitude streams dominated by groundwater springs and seepages are particularly susceptible to climatic warming because summer water temperatures are low in these streams and because increases in groundwater temperatures will be approximately equal to increases in average annual air temperature for the region. Thermal optima for many cold-water taxa from the mid- and high latitudes is less than 20°C; summer temperatures may exceed thermal tolerances and reduce production. The growth rate of the stonefly (*Leuctra nigra*) in the UK increases with temperature in experiments up to about 20°C, but survival is reduced by 67% and egg production by 90% at temperatures between 12 and 16°C (Elliott, 1987). In Pennsylvania, experimental increases in stream temperature during autumn—from ambient, near 10°C, to about 16°C—are lethal to 99% of stonefly (*Soyedina carolinensis*) larvae (Sweeney and Vannote, 1986).

Changes in the temporal pattern of warming may have significant and surprising effects because temperature is a cue that stimulates both the production and the release from dormancy of zooplankton over-wintering stages (Korpelainen, 1986; Stirling and McQueen, 1986; Sullivan and McManus, 1986; Marcus, 1987; Hairston *et al.*, 1990; Hairston, 1996; Chen and Folt, 1996). Warming events in autumn could alter the timing or occurrence of resting stages, thus potentially causing the loss of an entire cohort or population and the reduction of eggs in the "seed bank." Resting eggs of the copepod (*Epischura lacustris*) are stimulated to hatch prematurely in autumn by raising temperatures above 15°C (Chen and Folt, 1996).

Invertebrates that can reproduce asexually may be buffered from local extinction at high temperature because they have populations comprising clones with different thermal tolerances. Electrophoretically distinguishable winter and summer clones of *Daphnia magna* from a single pond exhibit large differences in responses to temperatures of 25 and 30°C (Carvalhoe, 1987; LaBerge and Hann, 1990). Winter clones die, whereas summer clones survive and reproduce. This seasonal phenotypic variation in thermal response may increase the ability of these species to adapt. Other species are more likely to go extinct. Glacial relicts (some cold-water species) often lack resting stages and have poor dispersal capabilities. If excessive warming eliminated an entire cohort one year, resting stages would not be available in the sediments to reestablish the population the next year—and local extinction would result. Local extinctions are more likely when warm summer temperatures and anoxia erode the hypolimnetic refuge required by particular species (Dadswell, 1974; Stemberger, 1996).

More-persistent thermal stratification of lakes with warming (Schindler, 1990; DeStasio *et al.*, 1996) could reduce secondary productivity. Greater anoxia in the hypolimnion (Section 10.5.4) may eliminate a refuge from predation or from thermal stress. Warmer epilimnetic temperatures could decrease the nutritional quality of edible phytoplankton (Soeder and Stengel, 1974; Ahlgren *et al.*, 1990; Moore *et al.*, 1995, 1996) or shift the species composition of the phytoplankton community toward less-preferable cyanobacteria and green algae (George and Harris, 1985; Tilman *et al.*, 1986; Moore *et al.*, 1995a).

Temperature increases can reduce the availability of more-nutritious foods for stream invertebrates as well. Water temperatures exceeding 20–25°C reduce diatom taxa (more nutritious) and increase green algae and cyanobacteria (less nutritious) (Patrick, 1969; Lamberti and Resh, 1983).

10.6.1.2. Fish

Body growth and behavioral performances such as swimming ability and foraging success (Bergman, 1987) are controlled by temperature, as are the hatching success of eggs and survival of larvae (Edsall, 1970; Colby and Nepsy, 1981); all are maximum at some intermediate optimum temperature. In North America, freshwater fish have been grouped into three broad thermal groups—called cold-water, cool-water, and warm-water guilds—based on temperature differences in these optima (Hokanson, 1977; Magnuson *et al.*, 1979). As temperatures warm, the performance of each species increases or decreases depending on which side of the optima the temperature began; if dispersal is possible in a heterothermal habitat, each species will have a greater tendency to move into or out of the habitat.

Fish in all three thermal guilds grow faster in 2 x CO_2 climate/lake thermal structure/fish growth simulations for the Laurentian Great Lakes in North America (Hill and Magnuson, 1990), given the assumptions that increased food is available to meet higher metabolic rates and that cooler water refuges are available. For Lake Erie, cold water with sufficient oxygen would not likely be available in the hypolimnion after a warming of this extent (Section 10.5.4), so the assumption of refuges is not always reasonable; for Lakes Michigan and Superior, the assumption is reasonable. (Shallow, unstratified lakes and larger rivers would not be expected to have thermal refuges.) Increased prey appears likely with warming of this extent based on correlation models with inter-lake comparisons of primary production, zooplankton biomass, and fishery yields (Regier *et al.*, 1990).

Population simulations for smallmouth bass (*Micropterus dolomieui*) under a 2 x CO_2 climate in the Laurentian Great Lakes (Shuter in Magnuson, 1989a) included thermal effects on reproduction, hatching success, and growth at all life stages. Simulations were made for warmer (Erie), intermediate (Huron), and colder (Superior) lakes. Warming in the models produced greater young-of-year survival in the intermediate and cold lakes but no change in the warm lake; in all lakes, warming produced an earlier age of maturation, greater young-of-year growth, a longer growing season for adults, greater year-class strengths, and larger fishable populations. Similar results would be expected for sea lamprey (*Petromyzon marinus*) in the Laurentian Great Lakes based on the effects of temperature on hatching success and growth (Holmes, 1990)—except that sea lamprey are a problem there, and considerable funds are spent for lamprey control.

Warmer winter temperatures would increase the winter survival of warmer-water fish and decrease the reproduction of fish that require a cold period for normal gonadal development. Because climate scenarios suggest that warming will be greater in winter than in summer, such influences may be significant for populations at the high- or low-latitudinal edges of their ranges. Winter survival would be enhanced for young-of-year white perch (*Morone americana*) at the northern edge of their range in the Laurentian Great Lakes (Johnson and Evans, 1990). Increases in mean annual temperature from logging activities in a British Columbia (Canada) stream also result in earlier emergence of salmon fry, and a lengthened growing season—and, as a result, increased over-winter survival rates (Holtby, 1988). Warmer winter temperatures are not beneficial for all fish because low winter temperatures for sufficient periods are required for normal gonadal maturation in some species (Jones *et al.*, 1972). For yellow perch—a cool-water fish—the highest percentages of viable eggs produced were 93% after over-wintering at 4°C, 65% at 6°C, and 31% at 8°C.

10.6.1.3. Contaminant Accumulation

Warming of lakes could increase the occurrence of methyl mercury in lakes and the accumulation of mercury in fish. In lake ecosystems, methylation is positively and demethylation negatively related to water temperature; the ratio of methylation to demethylation increases with temperature (Bodaly *et al.*, 1993). In six lakes, 70–80% of the variation in size-adjusted mercury concentrations in fish are associated with temperature: in cisco (*Coregonus artedi*), northern pike (*Esox lucius*), walleye (*Stizostedion vitreum*), and yellow perch, but not for two bottom-feeding fishes, the white sucker (*Catostomus commersoni*), and lake whitefish (*Coregonus clupeaformis*). Mercury concentrations are not associated with other physical or chemical properties of the lakes.

Heavy metals and pesticide accumulation are greater at higher water temperatures (Reinert *et al.*, 1974). In-depth treatment of these processes may be found in Wood and McDonald (1996).

Indirect effects of climatic warming on contaminant accumulation in freshwater plankton are likely to occur as well. Predicted water-chemistry changes, such as a decrease in DOC concentrations, may result in a decrease in chemical binding capacity, as hypothesized by Schindler (1996)—thus causing biotic effects of toxins to increase (Connell and Miller, 1984; Moore *et al.*, 1995, 1996).

10.6.1.4. Ice and Snowmelt Effects

Ice cover in lakes and streams is expected to decrease with climatic warming (see Chapter 7). Reduced durations of ice cover are expected under scenarios with 2 x CO_2 climates; at the lowest latitudes where ice now occurs seasonally, ice is not expected to form at all in many winters (DeStasio *et al.*, 1996). Observed ice durations decreased markedly during a 20-year period of warming in central North America (Schindler *et al.*, 1990). In the Antarctic Dry Valley, ice cover has thinned for some of these permanently ice-covered lakes. Lake Hoare thinned by 20 cm/yr over a 10-year period beginning in 1977; ice cover is now 3.5 m thick (Wharton *et al.*, 1992). Because light attenuation by the ice is a major limiting factor, these climate-related changes are expected to cause shifts in the biota of lakes with substantial periods of ice cover (Doran *et al.*, 1994).

In the Laurentian Great Lakes, loss of winter ice cover results in year-class failure of lake whitefish, because the eggs incubate over the winter and increased turbulence and winter mixing reduce their survival (Brown and Taylor, 1993). On Grand Traverse Bay, Lake Michigan, the number of winters without ice cover has increased in recent years (Assel and Robertson, 1995). If this trend continues—as is expected with greenhouse warming—the lake whitefish are expected to decline in abundance.

In shallow lakes and the backwaters of large rivers, a decrease in ice-cover duration and especially the absence of ice cover would reduce the winter anoxia common at mid- to high latitudes. This could be countered somewhat by lower water levels, which reduce water volumes under the ice and increase the likelihood of winter kill. Winter kill of fish, owing to loss of oxygen under the ice, is a common occurrence in North America and northern Europe; this severe event greatly influences the fish assemblage structure (Tonn, 1990). Small, shallow lakes in midwestern North America that are anoxic in winter and presently have assemblages dominated by the central mud minnow (*Umbra limi*) would be expected to change to ones dominated by northern pike and largemouth bass (Tonn *et al.*, 1990) without ice cover and winter anoxia. In northern Europe, assemblages presently dominated by crucian carp (*Carassius carassius*) would be expected to change to ones dominated by European perch (*Perca fluviatilis*), roach (*Rutilus rutilus*), and other species.

10.6.2. Hydrological Effects on Biota

10.6.2.1. Streams

The largest effects of climate-induced changes in hydrology on productivity in streams and rivers will result from reduction in streamflows predicted for mid-latitudes, changes in the amount and form of winter precipitation and the timing of snowmelt at high elevations, and increases in the magnitude or frequency of extreme events (e.g., floods, droughts).

Reduced streamflows produced by lower precipitation and/or increased evapotranspiration would increase the probability of intermittent flow in smaller streams. Drying of streambeds for extended periods reduces ecosystem productivity because the aquatic habitat is restricted, water quality is reduced (e.g., expanded hypoxia), and intense competition and predation reduce total biomass (Fisher and Grimm, 1991; Stanley and Fisher, 1992). Intermittent streams in Australia and the southwestern United States have invertebrate communities dominated by organisms with resistant life stages and short life cycles (Boulton and Lake, 1992; Gray, 1981). Recovery of benthic invertebrates with the resumption of flow can be slow. More than 4 months were required for recovery of macroinvertebrate biomass following two 12-hour periods of streambed drying in the Colorado River caused by dam operations upstream (Blinn *et al.*, 1995). Effects of drought also can be delayed. Effects of reduced recruitment during a drought in an Australian stream were not observed until the following year (Boulton and Lake, 1992).

For perennial runoff streams (Poff and Ward, 1989), the potential for intermittent flow may be particularly great in relatively humid climates that have low baseflows owing to low groundwater discharges; nearly one-half of such streams in the eastern and southeastern United States may become intermittent with only a 10% decline in annual runoff (Poff, 1992). A 14% decline in average annual precipitation for Alabama is projected to result in declines of 50–60% in minimum 7-day stream flows (Ward *et al.*, 1992)—greatly increasing the possibility that perennial streams will become intermittent, even in this relatively humid region, because groundwater storage is limited.

Reduction in streamflow is likely to reduce the productivity of large flood-plain rivers and low-gradient streams dependent on periodic flooding. Inundation of flood plains provides expanded food-rich habitat and sources of organisms and organic matter for river ecosystems (Welcomme, 1979; Junk *et al.*, 1989; Meyer, 1990). Fish yields are 1.5 to 4 times greater in river-flood plain systems than in equivalent systems without flood-plain inundation (Bayley, 1995). Inputs of organic carbon from flood-plain wetlands account for about 80% of the metabolism in the main channel of a low-gradient stream in Georgia (Meyer and Edwards, 1990). Because organisms in flood-plain rivers and streams are adapted to regular flooding cycles, reduction in flood frequency should have greater effects than increases in flooding. A record flood had little effect on Mississippi River biota, but the absence of a flood during a drought year caused substantial short- and long-term changes in plant and invertebrate communities (Sparks *et al.*, 1990).

Box 10-3. Flood and Drought in Aridland Streams

In aridland streams, biomass and productivity are limited severely by both flood scouring and stream drying (Grimm and Fisher, 1992; Grimm, 1993). Small changes in precipitation may produce large increases in flow variability because runoff response to precipitation is nonlinear (Dahm and Molles, 1992). Longer periods of drought and more-intense storms may produce severe streambank and channel erosion because aridland riparian vegetation is sensitive to the availability of water, and unvegetated soils are highly erodible (Grimm and Fisher, 1992).

Changes in the timing of flood and drought may be more important than changes in the annual averages. In the Sonoran Desert of Arizona, stream communities may be shaped by flash floods and subsequent colonization dynamics; biotic interactions, such as competition for a limiting resource; or morphometric and state changes associated with drought—shrinking ecosystem boundaries and eventual loss of surface water (Figure 10-7). These controlling processes are important because they shape community structure, instantaneous and annual primary and secondary production, and nutrient retention.

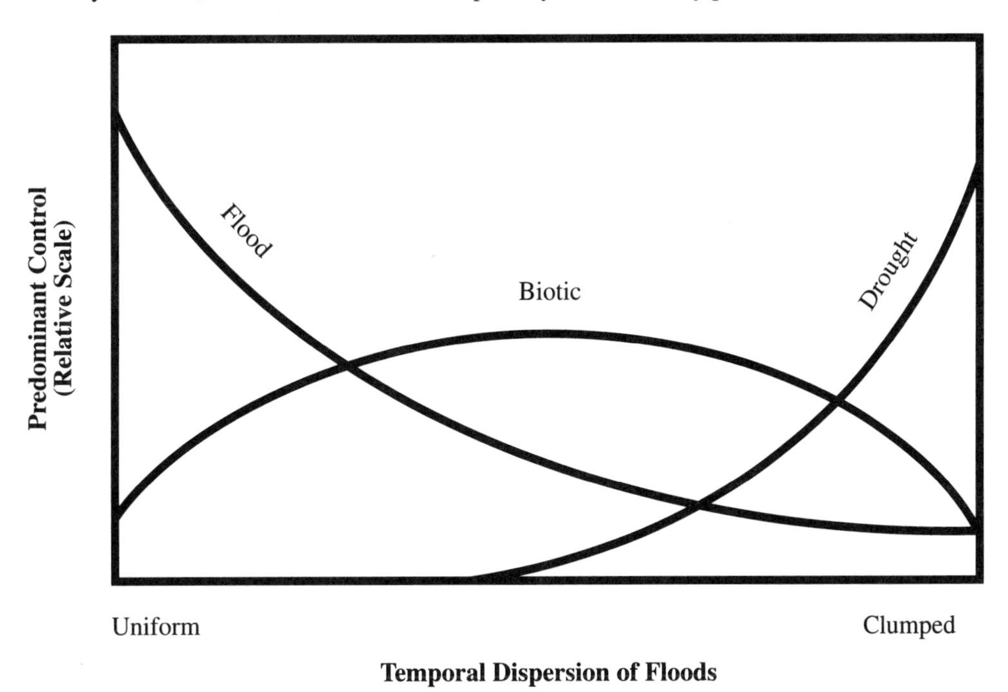

Figure 10-7: Changes in relative importance of major controlling factors in hypothetical years in which annual runoff and storm number are constant but the distribution of floods in time varies.

The contribution of each controlling process to the system's state depends on the temporal pattern of flooding rather than on the total annual discharge. In years with floods evenly distributed in time, drought effects are rare and biotic interactions are moderate. When between-flood periods are long, biotic interactions become more important and drought conditions characterize a substantial proportion of the year (Fisher and Grimm, 1991). In Sycamore Creek, annual runoff in 1970 was equivalent to 1988 but occurred as two flash floods compared to eight in 1988. Consequently, drought dynamics influenced the ecosystem during 76% of 1970 but only 29% of 1988; benthos colonization, succession, and rapid growth and reproduction were prevalent during 55% of 1988 but only 16% of 1970 (Grimm, 1993).

Changes in long-term processes also are important. Catchment-derived fluxes of nitrogen to desert streams are higher in floods that occur after several years of low precipitation (Grimm, 1992), presumably owing to accumulation of nitrate in soils during droughts. Century-scale arroyo-cutting episodes alter drainage patterns, drain wetlands, and shape stream channels (Hastings and Turner, 1965). Small changes in atmospheric circulation patterns exert large effects on the occurrence, timing, and magnitude of convective storms on the Sonoran Desert and thus the flow variability of its streams and rivers.

Increases in flow variability should produce larger effects than changes in mean flow. Large floods scour the beds of small streams and rivers, flushing organisms and detritus downstream and depressing biomass and productivity for some time. New Zealand streams with more-frequent high flows had lower algal biomass and presumably lower primary production than those with more stable hydrographs (Biggs and Close, 1989).

Streams in arid regions such as the Sonoran Desert of Arizona are particularly susceptible to climate-induced changes in flow variability (see Box 10-3).

10.6.2.2. Lakes

Changes in lake water levels (Section 10.5.2; Figure 10-4) have large effects on nearshore biotic assemblages. In Lake Titicaca, South America, submerged macrophytes experience significant mortality when water levels change more rapidly than they can adapt (see Dejoux and Iltis, 1992). In Lake Kinneret in the Middle East, changes in inshore ecosystems occur with rapid water-level fluctuations, even though upper and lower water levels are regulated; the lake is a major reservoir of Israel's freshwater supply. A 4-m water-level decline in Kinneret reduces the stony belt around the lake by 30–94% (Gafny *et al.*, 1992), and the littoral slope changes from steep to slight (Gasith and Gafny, 1990). During low lake levels, wave action affects more areas of soft sediment than at high lake levels, resulting in short-term effects on water quality (Gafny and Gasith, 1989, 1993). Emergent vegetation develops in areas with gentle slope (Gasith and Gafny, 1990). The community of fish breeding in the littoral zone switches from dominance by fish that spawn on stones to those that spawn on sand (Gafny *et al.*, 1992). This may affect year-class strength of the primary planktivorous fish and eventually the entire food web. A warmer and drier climate in the Middle East would exacerbate such changes in inshore aquatic communities.

With declining water levels, lakes might in the short term become more separated from their bordering wetlands. A number of lake fish use these wetlands for spawning and nursery areas (Brazner and Magnuson, 1994). Northern pike, which spawn in flooded sedge meadows in early spring and whose young remain for about 20 days after hatching (Becker, 1983), would be especially damaged by low spring water levels.

Connectivity among lakes would be decreased by the cessation of flow in connecting streams in some lake districts, which could influence community structure and rates of extinction and invasion. For shallow, ice-covered, winter-kill lakes, loss of stream connections can eliminate access of seasonal migrants such as northern pike to adjacent deeper lakes during the winter. Loss of access would eliminate pike from shallow-lake assemblages, and the assemblage would shift toward species more tolerant of low oxygen and intolerant to northern pike predation (Lodge, 1993).

Large water-level changes characteristic of inland waters (Section 10.5.2; Figure 10-4), especially when exacerbated by climate change, have costly effects on urban and agricultural coastlines.

In 1986, agricultural and residential lands surrounding Lake Titicaca became inundated. At the highest water levels, people living in river valleys adjacent to the lake had to move upslope and switch their transport from connecting roads to boats. When the waters receded, the lower elevations began to be used only for agricultural purposes. The costs to adapt to future changes in water levels are $120 million for the first 5 years, based on a plan developed with the help of the European Community (Autoridad Binacional Autonoma de la Cuenca del Sistema TDPS, 1994). For Lake Michigan, bluff erosion during storms at high water results in property loss that includes destruction of residential and other shoreline developments. An extensive study concludes that the cost of controlling water level on all five Laurentian Great Lakes would be prohibitively expensive. Costs associated with adaptation to 2 x CO_2 water-level scenarios also are high. Changnon *et al.* (1989) estimate that increased dredging of harbors for a rather small length of Lake Michigan shoreline, including Chicago, would cost $138 to $312 million if water levels dropped by 1.25 to 2.5 m. Measures such as lowering docks, extending water supply sources and stormwater outfalls added another $132 to $228 million.

These examples demonstrate that water-level fluctuations at present and from climate-change scenarios can be large for lakes. Adaptation to, rather than control of, such large changes appears to be a common historical result in widely different settings and cultures. Such changes challenge the ability of human and natural communities to adapt and, in some cases, may be prohibitively expensive (see Chapter 12).

10.6.3. Species Distributions and Biodiversity

With climate warming, the poleward movement of freshwater communities will be at least as dramatic as the poleward movements of terrestrial vegetation (see Chapter 1). Extinctions and extirpations (local extinctions of species found elsewhere) will occur at the lower latitude boundaries of species distributions; where possible, poleward migrations will occur at the higher latitude boundaries of species distributions. Within geographic ranges, cool- and cold-water assemblages will be reduced in many rivers and shallow, unstratified lakes and ponds; suitable thermal habitat in many deep, stratified lakes will increase for warm-, cool-, and even cold-water organisms (see also Chapter 16).

Biodiversity increases from high to low latitudes (for fish, see Nelson, 1984; for streams, see Allan and Flecker, 1993). One might think that climatic warming, with adequate time, would increase the species diversity of many groups of organisms in mid- and high latitudes. In North America, species densities of fish in quadrates 1° latitude by 1° longitude are better correlated with the climatic factors of the quadrates than with the latitude or longitude of the quadrates (Hocutt and Wiley, 1985). Species density increased with temperature and decreased with aridity, but temperature and aridity together accounted for only 38% of the variation in species density. These general associations for large regions provide little information about the decade to century influences of climate warming on the biodiversity of individual waters. Interaction with more local conditions and local species distributions will be more informative.

10.6.3.1. Edges of Geographic Ranges

Species extinctions and extirpations will occur at the lower latitude boundaries of distributions if summer temperatures increase in streams and shallow, unstratified lakes and ponds and cooler-water refuges are not available. In the southern Great Plains of the United States, summer water temperatures of 38–40°C already approach the lethal limits (less than 40°C) for many native stream fish, most of which are minnows (Matthews and Zimmerman, 1990). Fish aggregate in slightly cooler shaded waters of pools and tributary streams (32–35°C); this crowding induces various stresses, including overexploitation of prey resources. These wide, slow-moving streams have no high-altitude refuges; they flow to the south or east, making escape poleward unlikely. If a 3–4°C warming occurs, many endemic species could become extinct.

Biogeographic distributions of aquatic insects are centered around species' thermal optima (Vannote and Sweeney, 1980). Climatic warming would shift the optimum temperatures poleward and eliminate species near their lower latitude limits. In North America, a 4°C warming is projected to shift stream thermal regimes—and potentially the center of species distributions—about 640 km northward (Sweeney *et al.*, 1992).

At the higher latitude boundaries of species' distributions, organisms should be able to migrate poleward with climate warming, provided the new habitats are accessible through connecting waters. In North America, a 4°C increase in air temperatures is sufficient to move the simulated ranges of smallmouth bass and yellow perch northward across Canada by about 5° latitude, or about 500 km (Shuter and Post, 1990). These simulations include the entire life cycles and seasonality in both thermal and biological models.

10.6.3.2. Within Geographic Ranges

With projected climatic warming, stream fish habitats are predicted to decline across the entire United States by 47% for cold-water, 50% for cool-water, and 14% for warm-water species, independent of influences from other climate-related changes such as reduced stream flow (Eaton and Scheller, 1996). Only a few warm-water fish—bluegill, largemouth bass, channel catfish (*Ictalurus punctatus*), and common carp (*Cyprinus carpio*)—increase markedly in these simulations (Figure 10-8). These simulations assume that waters mix from surface to bottom and local thermal refuges do not exist. Scenarios are derived from air-temperature simulations across

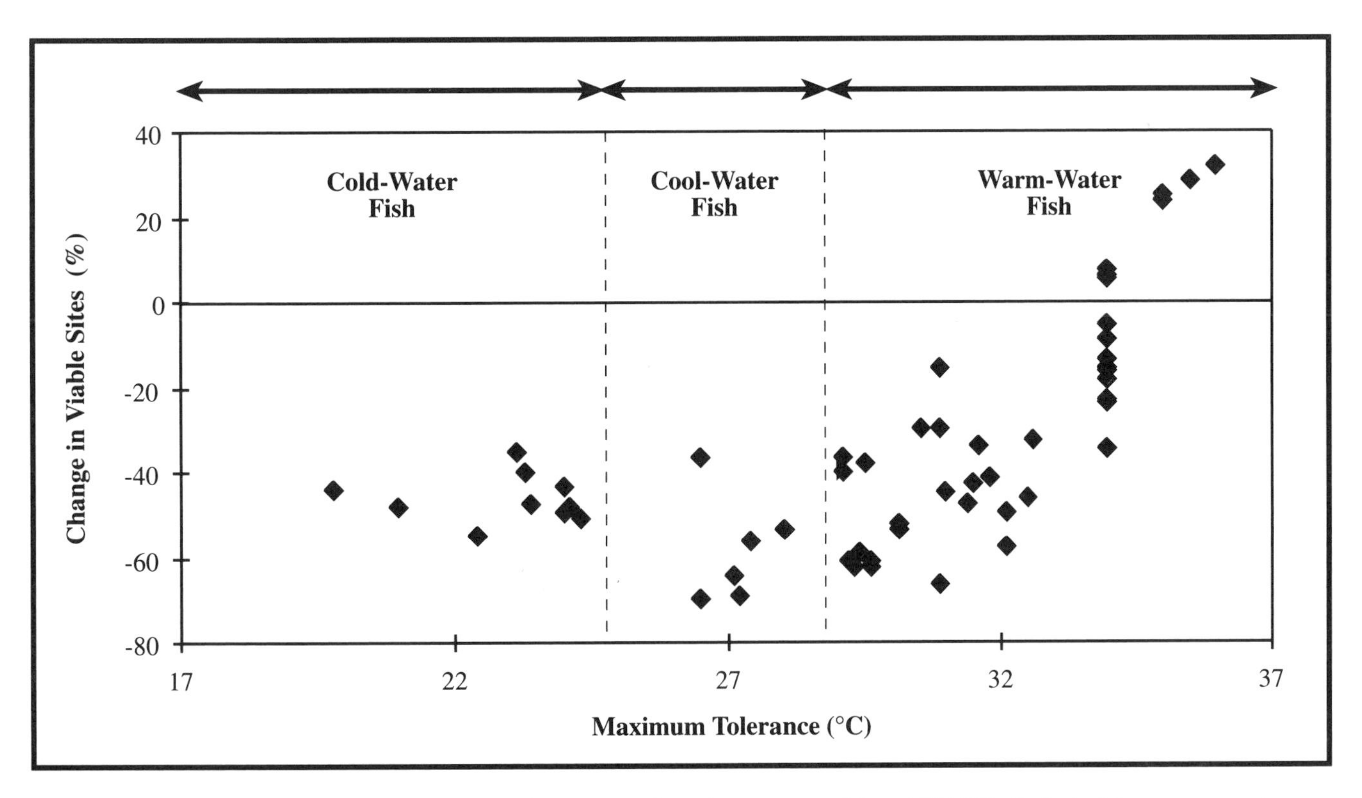

Figure 10-8: Changes in percentage of sites suitable for cold-, cool-, and warm-water fish at 1,700 U.S. Geological Survey stream sites across the United States, under a CCC GCM 2 x CO_2 scenario. Each point is a species plotted against its maximum thermal tolerance. Analysis and definitions are from Eaton and Scheller (1996).

the United States with a 2 x CO_2 climate and a 0.9 factor to convert air to stream temperatures.

Heterogeneous thermal habitats can provide thermal refuges. In laboratory gradients, most freshwater fish spend two-thirds of their time within a 4°C range and all of their time within a 10°C range around their preferred optima (Magnuson *et al.*, 1979; Magnuson and DeStasio, 1996). This ability allows them to seek out survivable or optimum temperatures in lakes and streams that are thermally heterogeneous. In summer, stream fish can move to higher elevations (Rahel *et al.*, 1996), to reaches closer to groundwater sources (Meisner, 1990), or to shaded cooler areas (Matthews and Zimmerman, 1990); in thermally stratified lakes, fish can move downward to deeper, cooler waters (Magnuson and DeStasio, 1996).

In mountain streams of Wyoming, habitat loss is predicted for cold-water fish even with small increases in temperature (Rahel *et al.*, 1996). An increase of 1°C reduces stream habitat for cold-water fishes by 7–16%, 2°C by 15–26%, 3°C by 24–39%, 4°C by 42–54% and 5°C by 64–79%. Remaining enclaves of cold-water fish would exist as smaller fragmented populations, with an increased probability of extinction from ecological disturbances such as fire or drought (Allendorf and Waples, 1987; Mills and Smouse, 1994).

Biotic interactions for stream invertebrates also intensify as flows decline or streams dry; mobile organisms are concentrated into smaller areas, resulting in intense predation and competition and potentially a loss of some taxa and diversity (Carpenter *et al.*, 1992; Grimm, 1993).

In deep, thermally stratified temperate lakes, thermal habitat generally increases with 2 x CO_2 scenarios of global warming, not only for warm- and cool-water fish, but also for cold-water fish (Magnuson *et al.*, 1990; DeStasio *et al.*, 1996; Magnuson and DeStasio, 1996) (Figure 10-9). Similar increases are projected in smaller and larger lakes in Wisconsin for warm- and cool-water fish, but results were equivocal among GCM scenarios for cold-water fish in the smaller lakes. Increases occurred because the length of the growing season increased and because fish could move to deeper, cooler waters when surface waters exceeded preferred temperatures. Deep-water thermal refuges for cold-water fish are maintained in model projections over large latitudinal ranges (McLain *et al.*, 1994), assuming that deep-water oxygen is sufficient.

Changes in deep-water oxygen and other habitat variables may prevent cold-water fish from occupying their thermal niches in a warmer and drier climate (Magnuson and DeStasio, 1996). Increases in water clarity can deepen the thermocline of small lakes, so that deep cold waters are significantly reduced in size (see Sections 10.5.3 and 10.5.4 and Box 10-2), and dissolved oxygen may be reduced in deep waters (see Section 10.5.4 and Figure 10-7). Other relevant analyses exist for northern pike in impoundments (Headrick and Carline, 1993), striped bass in

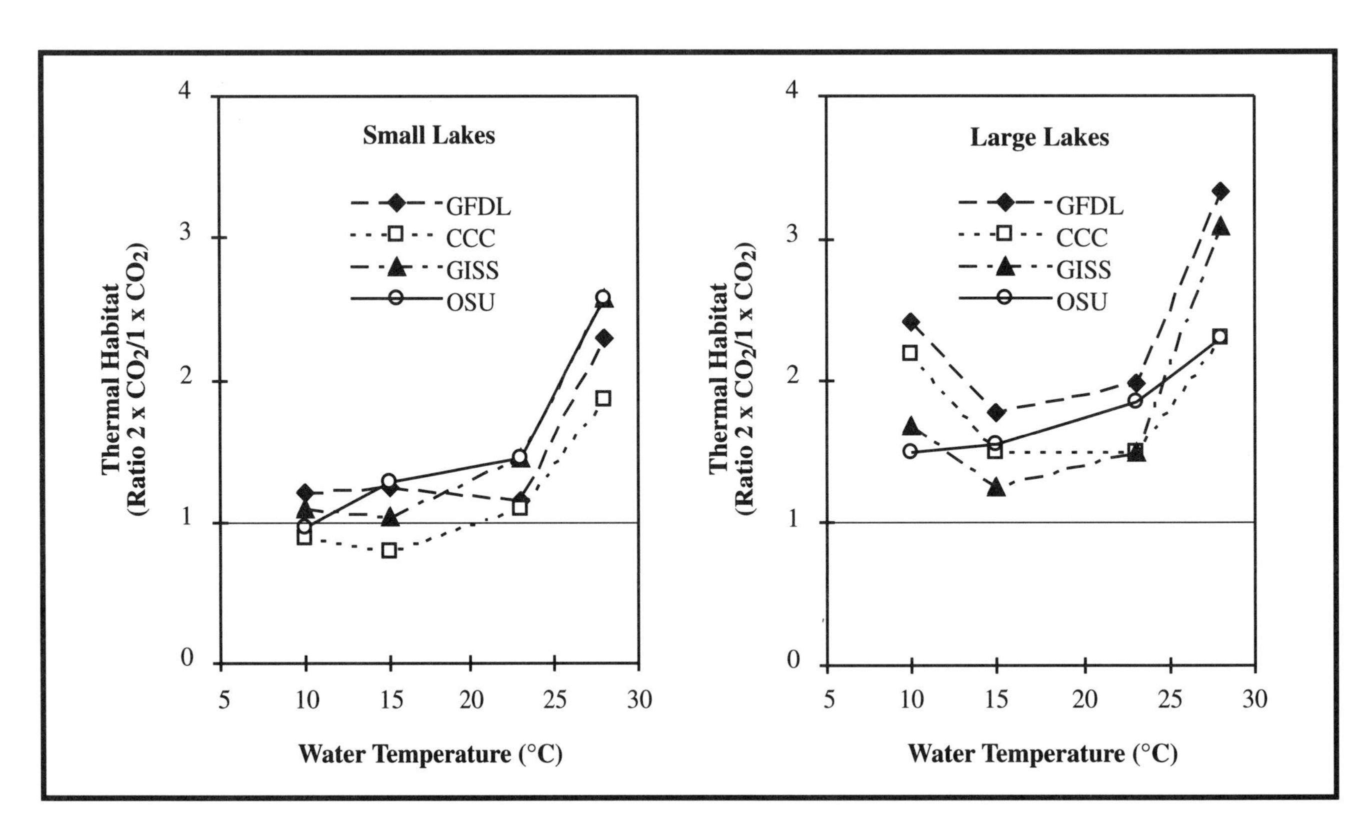

Figure 10-9: Simulated changes in thermal habitat in small (Crystal and Sparkling) and large (Michigan, Mendota, and Trout) stratified temperate lakes for cold- (10 and 15°C), cool- (23°C), and warm-water (28°C) fish under a 2 x CO_2 climate with four GCMs (GFDL, CCC, GISS, and OSU), as modified from Magnuson and DeStasio (1996).

reservoirs (Coutant, 1985, 1987), and anadromous salmon (Crawshaw and O'Connor, 1996).

10.6.3.3. Genetic Adaptation, Dispersal, and Invasions of Exotic Organisms

Rates of climate warming are likely to exceed rates of genetic adaptation for some local populations of aquatic invertebrates, as well as the rate of natural dispersal of warm-tolerant populations that might be expected to replace them (see Box 10-4). Similar projections can be made for fish. Genetic variation in critical and chronic thermal maxima has occurred in separated populations of largemouth bass (Fields *et al.*, 1987), but thermal preference has not changed among populations (Koppelman *et al.*, 1988). Crawshaw and O'Connor (1996) speculate—from mammalian studies and similarities between fish and mammal thermoregulatory anatomy and processes—that genetic shifts in thermal preferences of fish might be fast under certain conditions.

The lack of suitable north–south migration routes (drainages) may limit dispersal, as well as replacement of cold-water fauna with warmer-water fauna. If dispersal rates of arriving warm-water species are low in freshwaters with declining cold-water taxa, the biomass and productivity of the community would be reduced with climate warming, at least initially, because range shifts may lag behind changes in thermal regimes (for stream invertebrates, see Sweeney *et al.*, 1992). This problem would be most restrictive for taxa that are less vagil (e.g., aquatic insects with fragile or short-lived adult stages) and for isolated freshwater ecosystems such as closed-basin lakes and streams. Isolation would be less severe for organisms that can be carried by the wind, such as very small organisms (e.g., algae and microbes) and those with small resting stages (e.g., cladocerans and rotifers), or for organisms with flying adults such as many aquatic insects. Fish dispersal (Magnuson *et al.*, 1989b; Tonn *et al.*, 1990) would be more restricted to water courses, transport by humans, or even rare extreme events (Dennis and Wolff, 1992) such as tornadoes.

Dispersing organisms arriving for the first time in a new ecosystem are exotic (new) to that system. Such invading species often interact with existing species in unexpected ways. Invasions of exotics can disturb the existing community structure and the productivity of species of value to humans. Rates of species invasion would increase poleward as conditions warm (Mandrak, 1989). These invasions can hasten changes associated with climate warming by increasing the rate of decline of existing species (Lodge, 1993) and add to the uncertainty (Magnuson, 1976) of predicting the effects of climate warming.

Box 10-4. Genetic Response to Warming

Genetic variability in mayflies of eastern North America decreases to the north; thus, populations for which global warming may be most pronounced are least-equipped genetically to adapt (Sweeney *et al.*, 1992). Many species of mayflies are weak dispersers and may be unable to move rapidly enough to keep up with poleward migration of isotherms (Sweeney *et al.*, 1992). Increasing temperatures and associated ecological changes may tax their ability to adapt.

Some aquatic invertebrates show high genotypic variability with little differentiation among sites; others vary greatly among sites but exhibit little genetic variability within a site. Hogg *et al.* (1995) tested the hypothesis that populations with little genetic variability would fare less well in a changing environment than those with more genetic variability. The amphipod *Hyalella azteca* disperses poorly but has high genetic variation; a stonefly, *Nemoura trispinosa*, has flying adults but low genetic variation. On one side of a split southern Ontario stream, temperature was raised year-round by 2.0–3.5°C for 3 years. Total invertebrate fauna was reduced by this treatment, but neither *H. azteca* nor *N. trispinosa* declined, although they did show life-history changes. Under warm conditions, the amphipod bred two months earlier and the stonefly emerged two weeks earlier at slightly smaller sizes. The authors postulate that endemic populations with poor dispersal abilities would suffer the long-term consequences of environmental warming.

10.6.3.4. Observed Changes in Biodiversity

Long-term observations as well as observed paleolimnological changes in organisms and in chemical conditions recorded in lake sediments can suggest the kinds of responses of freshwater biodiversity to future changes in climate.

Lake sediments accumulate over time and can contain an interpretable record of the history of the lake's biota from the remains of diatoms, chrysophytes, zooplankton and other aquatic invertebrates, and, rarely, even fish. A variety of organisms have been examined for use in inferring changes in climate; they explicitly indicate changes in lake biota in response to climate change. Diatoms appear to be the most sensitive indicators of past lake conditions; they are being used to examine the history of lakes in closed basins (Juggins *et al.*, 1994) and water levels (Gont *et al.*, 1988). They have been used to examine changes in the duration of ice cover (Smol, 1988). As a consequence of habitat specificity related to temperature (Douglas and Smol, 1995), they may be sensitive indicators of increases in temperature, especially in the high Arctic. Diatom communities respond to changes in salinity (Fritz *et al.*, 1993) that result from changes in precipitation or runoff. Diatoms respond to a number of environmental conditions, and they are sensitive to light and nutrient availability (Kilham *et al.*, 1996). Diatoms and chrysophytes have been especially responsive to past changes in climate and climate-induced changes in water chemistry (Battarbee *et al.*, 1990; Bradbury and Dean, 1993; Cumming *et al.*, 1993; Davis *et al.*, 1994; Smol and Dixit, 1990).

In natural thermal gradients of streams and in experimental streamside channels, the total number of algal species increases with temperature up to 25–30°C, then declines above 30°C as many diatom species are replaced by fewer species of green algae and cyanobacteria (Patrick, 1971; Squires *et al.*, 1979; Lamberti and Resh, 1983). Phytoplankton biodiversity increased slightly over a 20-year period of warmer and drier weather in a Canadian Shield lake (Schindler *et al.*, 1990).

The greatest biodiversity is expected at intermediate disturbance regimes; thus, large increases or reductions in flood frequency or size in streams (Ward and Stanford, 1983; Reice, 1994) or severity of winter kill in lakes (Tonn and Magnuson, 1982) can alter diversity. Analyses of streams in Wisconsin and Minnesota and in New Zealand indicate that fish communities are simpler and comprise more generalist species in streams with more-variable hydrologic regimes (Poff and Allan, 1995; Jowett and Duncan, 1990).

Increases in the frequency or severity of droughts may affect biodiversity in streams more than increases in the size of flood events because droughts result in longer-term habitat loss. Sharp reductions in the diversity of stream invertebrates followed dry years in arid streams of Arizona (Stanley and Fisher, 1992) and Australia (Boulton *et al.*, 1992). Even if streams do not become intermittent, extended periods of low flow would reduce the diversity of organisms intolerant of reduced water quality (Chessman and Robinson, 1987; Boulton and Lake, 1990). Despite short-term reductions in densities and biomass, floods do reduce predation pressure and lessen competition and thus can increase species diversity (Dudgeon, 1993; Grimm, 1993).

10.6.4. Adaptations to Changes in Climate

10.6.4.1. Land Use: Catchment, Riparian/Flood Plain

In smaller streams and rivers, augmentation and protection of riparian vegetation will provide shade and reduce the negative effects of warming. For the north-central United States, Stefan and Sinokrot (1993) estimate that the predicted summer rise in stream temperature will be 6°C higher if streamside vegetation is lost. In the southern United States, the rise in summer stream temperatures is primarily from increased radiation; predicted summer increases in water temperature can be reduced by at least 50% in the more humid regions with more riparian vegetation (Cooter and Cooter, 1990).

10.6.4.2. Hydrological Regime and Water Level

Maintenance of existing flood plains and flood plain-river exchanges or restoration by removing barriers such as levees may help in some cases to detain water during flood events, reducing flood peaks in downstream areas. Flood plains are vital to reducing flow variability and flood peaks in river basins.

Restoration of river and stream channels to more-natural morphologies would mitigate some negative effects of climatic drying and reduced flows. Natural morphologic features such as meanders and variable channel topography with deeper pools provide temporary refuges for organisms intolerant of reduced water depths or streambed drying. Permanent pools are critical refuges for many species when flow ceases in intermittent streams (Williams and Hynes, 1977; Boulton and Lake, 1992).

The already large water-level changes of lakes will be exacerbated and in most regions will be in the direction of lower levels. Shoreline structures and facilities, water intakes, and waste discharge will need to be extended or rebuilt. Human populations in most cases will begin to use the newly exposed lands for agriculture, habitation, and recreation and will have to retreat when water levels rise—with loss of property uses and values (see Chapter 12).

10.6.4.3. Heat-Loading Interactions with Nutrient and Toxic Wastes

Climate changes that produce higher water temperatures and lower flows and water levels will increase the negative effects of wastewater and thermal effluents on freshwater ecosystems. Because warmer temperatures make lakes and streams more productive—in a sense, more eutrophic in their behavior—it may be necessary to reduce alternate causes related to unwanted or nuisance production. Among the options would be to further decrease the loading of nutrients such as phosphorus and nitrogen from sewage treatment plants or to reduce leakage from diffuse agricultural and urban sources. Human adaptations that reduce these effluents—such as recycling or land application of wastewater effluents where appropriate—and increased use of recirculating cooling systems with cooling towers could help. Because the accumulation of many toxic substances in the freshwater food web is greater at higher temperatures, it also may be necessary to further reduce the release of these contaminants or further constrain the ingestion of contaminated fish by humans.

10.6.4.4. Biological Management: Harvest, Removals, Introductions

Zoogeographic responses to the loss of livable habitat at lower latitudes and gains in livable habitat poleward could be assisted by moving organisms poleward where warranted. Although this might save species that are about to be extirpated or go extinct at lower latitudes and expand the ranges at higher latitudes into suitable habitat, the introduction of these species into new habitats will not necessarily be successful. Two problems can be anticipated. First, these species may not establish viable populations in many cases, and the species may be lost before the consequences of introductions become predictable. Second, if they do establish a population they may negatively impact existing species already under stress from climate warming or emigrate into other connected freshwaters and exert unwanted effects there. A common effect of introductions in freshwaters is the extinction of existing species. This also may have negative influences on existing uses of the freshwater ecosystem for fishing and other purposes that rely on water quality.

10.7. Research Needs

Further research is needed to reduce the level of uncertainty in hydrological and freshwater ecosystem impact assessments, focusing mainly on developing credible change scenarios. Specific areas requiring improvements follow:

- The accuracy of and reductions in the discrepancies between GCM simulations at the regional scale
- The understanding and modeling of land–atmosphere exchange processes at a range of spatial and temporal scales
- The understanding of the effects of CO_2 enrichment on plant water use in natural settings at the catchment scale
- Methods for defining credible scenarios for changes in weather patterns leading to flood and drought, including stochastic weather generators and nested regional climate modeling
- The understanding and modeling of hydrological systems under nonstationary climatic conditions, which involve the maintenance and enhancement of monitoring networks; the development of methods for the acquisition of spatially distributed estimates of state variables such as soil moisture, evapotranspiration, and infiltration by remote or inexpensive direct measurements; and the development of credible hydrological models with climate-invariant parameters
- The understanding of the effects of the El Nino/Southern Oscillation (ENSO) and other large-scale atmospheric features on hydrological characteristics, and the changes caused by global warming.

There is a need to reverse the decline in climatic, hydrological, and ecological monitoring networks and services in many countries, particularly coupled precipitation and streamflow monitoring in small catchments that provide whole-system indices of changes in evapotranspiration. Continuing decline will make the detection of climate change impossible in many parts of the world and prevent the production of credible hydrological scenarios.

Research needs in freshwater ecology related to climate change include approaches to a predictive understanding of long-term, slow changes and processes with time lags greater than one year that determine the regional and local behavior of land/water ecosystems. The most important general needs follow:

- Long-term research and monitoring of key physical, chemical, and ecological properties (particularly water temperature and mixing properties; concentrations of nutrients, carbon, and major ions; acid/base status; and populations of key organisms) and processes (e.g., primary production, organic matter decomposition)
- Comparative studies of populations or ecological processes across latitudinal and hydrologic gradients and system types
- Paleo studies using sedimentary records of climate-sensitive parameters aimed at a more complete integration of paleohydrology and paleoecology trends, patterns, and future expectations
- Whole-system experiments altering the thermal, hydrological, or mixing regimes in small lakes and streams, including whole catchments, or in large-scale mesocosms (e.g., lake enclosures, artificial stream channels) to determine the responses of organisms and processes to projected climate changes
- Testing whole-catchment and regional approaches to improve predictive understanding of integrated land/water systems.

References

Abdullahi, B.A., 1990: The effect of temperature on reproduction in three species of cyclopoid copepods. *Hydrobiol*iogy, **196**, 101-109.

Aguado, E., D.R. Cayan, L. Riddle, and M. Roos, 1992: Climate fluctuations and the timing of West Coast streamflow. *Journal of Climate*, **5**, 1468-1483.

Ahlgren, G., L. Lundstedt, M. Brett, and C. Forsberg, 1990: Lipid composition and food quality of some freshwater phytoplankton for cladoceran zooplankters. *Journal of Plankton Research,* **12**, 809-818.

Allan, J.D. and A.S. Flecker, 1993: Biodiversity conservation in running waters. *BioScience*, **43**, 32-43.

Allendorf, F.W. and R.S. Waples, 1996: Conservation genetics of salmonid fishes. In: *Conservation Genetics: Case Histories from Nature* [Avise, J.C. and J.L. Hamrick (eds.)]. Chapman and Hall, New York, NY, pp. 238-280 (in press).

Almendinger, J.E., 1993: A groundwater model to explain past lake levels at Parkers Prairie, Minnesota, USA. *The Holocene*, **3**, 105-109.

Almquist-Jacobson, H., 1995: Lake level fluctuations at Ljustjernen, Central Sweden, and their implications for the Holocene climate of Scandinavia. *Palaeogeography, Palaeoclimatology, and Palaeoecology.*

Andre, J.C., J.P. Goutorbe, and A. Perrier, 1986: HAPEX-MOBILHY: a hydrologic atmospheric experiment for the study of water budget and evaporation flux at the climatic scale. *American Meteorological Society Bulletin*, **67(2)**, 138-144.

Arnell, N.W., 1992: Factors controlling the effects of climate change on river flow regimes in a humid temperate environment. *Journal of Hydrology*, **132**, 321-342.

Arnell, N.W. and N. Reynard, 1993: *Impact of Climate Change on River Flow Regimes in the United Kingdom.* Rep. to Dept. Environ., Inst. Hydrol., Wallingford, UK, 129 pp.

Arnell, N.W., 1994: Hydrological impacts of climate change. In: *The Rivers Handbook*, vol. 2 [Calow, P. and G.E. Petts (eds.)]. Blackwell, Oxford, UK, pp. 173-185.

Askew, A.J., 1991: Climate and water—a call for international action. *Hydrological Sciences Journal,* **36(4)**, 391-404.

Assel, R.A. and D.M. Robertson, 1995: Changes in winter air temperatures near Lake Michigan during 1851-1993, as determined from regional lake-ice records. *Limnology and Oceanography*, **40**, 165-176.

Autoridad Binacional Autonoma de la Cuenca del Sistema TDPS, 1994: *Plan Director Global Binacional.* Autoridad Binacional Autonoma de la Cuenca del Sistema TDPS Publishers, Bolivia, 14 pp.

Ballentine, T.M. and E.Z. Stakhiv (eds.), 1993: *Proc. Climate Change and Water Resources Management.* U.S. Army Institute Water Resources, Fort Belvoir, VA.

Band, L.E., D.S. Mackay, I.F. Creed, and D. Jeffries, 1996: Terrestrial-aquatic coupling: an approach from the terrestrial side. *Limnology and Oceanography*, **41** (in press).

Bao, W.M., 1994: A conceptual flow-sediment coupled simulating model for large basin. *Advanced Water Sciences*, **5(4)**, 287-292 (in Chinese).

Bardossy, A. and E.J. Plate, 1992: Space-time model for daily rainfall using atmospheric circulation patterns. *Water Resources Research*, **28(5)**, 1247-1259.

Baron, J.S., D.M. McKnight, and A.S. Denning, 1991: Sources of dissolved and particulate organic material in Loch Vale watershed, Rocky Mountain National Park, Colorado, USA. *Biogeochemistry*, **15**, 89-110.

Bates, B.C., S.P. Charles, N.R. Sumner, and P.M. Fleming, 1994: Climate change and its hydrological implications for South Australia. *Transactions of the Royal Society of Australia*, **118(1)**, 35-43.

Bates, B.C., A.J. Jakeman, S.P. Charles, N.R. Sumner, and P.M. Fleming, 1995: Impact of climate change on Australia's surface water resources. In: *Proc. Greenhouse 94 Conf.*, 9-14 October 1994, Wellington, New Zealand, CSIRO (in press).

Battarbee, R.W., J. Mason, I. Renberg, and J.F. Talling, 1990: *Paleolimnology and Lake Acidification.* The Royal Society, London, UK, 219 pp.

Bayley, P.B., 1995: Understanding large river-floodplain ecosystems. *BioScience,* **45**, 153-158.

Bayley, S.E., D.W. Schindler, K.G. Beaty, B.R. Parker, and M.P. Stainton, 1992a: Effects of multiple fires on nutrient yields from streams draining boreal forests and fen watersheds: nitrogen and phosphorus. *Canadian Journal of Fisheries and Aquatic Science*, **49**, 584-596.

Bayley, S.E., D.W. Schindler, B.R. Parker, M.P. Stainton, and K.G. Beaty, 1992b: Effect of forest fire and drought on acidity of a base-poor boreal forest stream: similarities between climatic warming and acidic precipitation. *Biogeochemistry*, **18**, 191-204.

Bazzaz, F.A., 1990: The response of natural ecosystems to the rising global CO_2 levels. *Annual Review of Ecological Systems*, **21**, 167-196.

Becker, A. and P. Serban, 1990: *Hydrological Models for Water-Resources Systems Design and Operation.* Operational Hydrol. Rep. No. 34, World Meteorological Organization, Geneva, Switzerland, 80 pp.

Becker, G.C., 1983: *Fishes of Wisconsin.* University of Wisconsin Press, Madison, WI, 1052 pp.

Beniston, M. (ed.), 1994: *Mountain Environments in Changing Climates.* Routledge Publ. Co., London, UK, 461 pp.

Benke, A.C., 1993: Concepts and patterns of invertebrate production in running waters. *Berh. Internat. Verein. Limnol.*, **25**, 15-38.

Benson, L.V., D.R. Currey, R.I. Dorn, K.R. Lajoie, C.G. Oviatt, S.W. Robinson, G.I. Smith, and S. Stine, 1990: Chronology of expansion and contraction of four Great Basin lake systems during the past 35,000 years. *Palaeogeography, Palaeoclimatology, and Palaeoecology*, **78**, 241-286.

Beran, M.A. and N.W. Arnell, 1995: Climate change and hydrological disasters. In: *Hydrology of Disasters* [Singh, V.P. (ed.)]. Kluwer, Dordrecht, The Netherlands (in press).

Bergman, E., 1987: Temperature-dependent differences in foraging abilities of two percids, Perca fluviatilis and Glymnocephalus cernus. *Environmental Biology of Fishes,* **19**, 45-53.

Biggs, B.J.F. and M.E. Close, 1989: Periphyton biomass dynamics in gravel bed rivers: the relative effects of flows and nutrients. *Freshwater Biology*, **22**, 209-231.

Blinn, D.W., J.P. Shannon, L.E. Stevens, and J.P. Carder, 1995: Consequences of fluctuating discharge for lotic communities. *J. North Am. Benthol. Soc.*, **14**, 233-248.

Blumberg, A.F. and D.M. DiToro, 1990: Effects of climate warming on dissolved oxygen concentrations in Lake Erie. *Transactions of the American Fisheries Society,* **119**, 210-223.

Boardman, J., R. Evans, D.T. Favis-Mortlock, and T.M. Harris, 1990: Climate change and soil erosion on agricultural land in England and Wales. *Land Degradation and Rehabilitation,* **2**, 95-105.

Bodaly, R.A., J.W.M. Rudd, R.I.P. Fudge, and C.A. Kelly, 1993: Mercury concentrations in fish related to size of remote Canadian Shield lakes. *Canadian Journal of Fisheries and Aquatic Sciences*, **50(5)**, 980-987.

Bolle, H.-J., J.C. Andre, J.L. Arme, *et al.*, 1993: EFEDA: European field experiment in a desertification-threatened area. *Annales Geophysicae*, **11**, 173-189.

Bootsma, H.A. and R.E. Hecky, 1993: Conservation of the African Great Lakes: a limnological perspective. *Conservation Biology,* **7**, 644-656.

Bormann, F.H., G.E. Likens, and J.S. Eaton, 1969: Biotic regulation of particulate and solution losses from a forest ecosystem. *BioScience,* **19**, 600-610.

Bothwell, M.L., 1988: Growth rate responses of lotic periphytic diatoms to experimental phosphorus enrichment. *Canadian Journal of Fisheries and Aquatic Science*, **45**, 261-270.

Boulton, A.J. and P.S. Lake, 1990: The ecology of two intermittent streams in Victoria, Australia. I. Multivariate analyses of physicochemical features. *Freshwater Biology,* **24**, 123-141.

Boulton, A.J., C.G. Peterson, N.B. Grimm, and S.G. Fisher, 1992: Stability of an aquatic macroinvertebrate community in a multiyear hydrologic disturbance regime. *Ecology*, **73**, 2192-2207.

Boulton, A.J. and P.S. Lake, 1992: The ecology of two intermittent streams in Victoria, Australia. II. Comparisons of faunal composition between habitats, rivers and years. *Freshwater Biology*, **27**, 99-121.

Bradbury, J.P., 1988: A climatic limnologic model of diatom succession for paleolimnological interpretation of verved sediments at Elk Lake, Minnesota. *Journal of Palaeolimnology*, **1**, 115-131.

Bradbury, J.P. and W.E. Dean, 1993: *Elk Lake, Minnesota: Evidence for Rapid Climate Change in the North-Central United States.* Geol. Soc. Amer., Sp. Paper 276, Boulder, CO, 336 pp.

Braun, L.N., W. Grabs, and B. Rana, 1993: Application of a conceptual runoff model in the Langtang Khola Basin, Nepal Himalaya. IAHS Publ. No. 218, pp. 221-237.

Braun, L.N., E. Brun, Y. Durand, E. Martin, and P. Tourasse, 1994: Simulation of discharge using different methods of meteorological data distribution, basin discretization and snow modelling. *Nordic Hydrol*ogy, **25**, 129-144.

Brazner, J.C. and J.J. Magnuson, 1994: Patterns of fish species richness and abundance in coastal marshes and other nearshore habitats in Green Bay, Lake Michigan. *Verh. Internat. Verein. Limnol.*, **25(4)**, 2098-2104.

Brittain, J.E., 1983: The influence of temperature on numphal growth rates in mountain stoneflies (Plecoptera). *Ecology*, **64**, 440-446.

Brown, R.W. and W.W. Taylor, 1993: Factors affecting the recruitment of Lake Whitefish in two areas of northern Lake Michigan. *Journal of Great Lakes Research,* **19**, 418-428.

Brun, E., P. David, M. Sudul, and G. Brunot, 1992: A numerical model to simulate snow-cover stratigraphy for operational avalanche forecasting. *Journal of Glaciology*, **38(128)**, 13-22.

Bultot, F., D. Gellens, M. Spreafico, and B. Schadler, 1992: Repercussions of a CO_2 doubling on the water balance—a case study in Switzerland. *Journal of Hydrology*, **137**, 199-208.

Bultot, F., D. Gellens, B. Schadler, and M. Spreafico, 1994: Effects of climate change on snow accumulation and melting in the Broye catchment (Switzerland). *Climate Change,* **28**, 339-363.

Burton, T.M. and G.E. Likens, 1973: The effect of strip-cutting on stream temperatures in the Hubbard Brook Experimental Forest, New Hampshire. *BioScience*, **23**, 433-435.

Busch, D.E. and S.G. Fisher, 1981: Metabolism of a desert stream. *Freshwater Biology*, **11**, 301-307.

Carpenter, S.R., S.G. Fisher, N.B. Grimm, and J.F. Kitchell, 1992: Global change and freshwater ecosystems. *Annual Review of Ecological Systems*, **23**, 119-139.

Carvalhoe, G.R., 1987: The clonal ecology of Daphnia magna. II. Thermal differentiation among seasonal clones. *Journal of Animal Ecology*, **56**, 469-478.

Chahine, M.T., 1992: The hydrological cycle and its influence on climate. *Nature*, **359**, 373-380.

Chang, L.H., C.T. Hunsaker, and J.D. Draves, 1992: Recent research on effects of climate change on water resources. *Water Resources Bulletin*, **28(2)**, 273-286.

Changnon, Jr., S.A., S. Leffler, and R. Shealy, 1989: Impacts of extremes in Lake Michigan levels along Illinois shoreline: low levels. In: *Appendix H, Infrastructure, The Potential Effects of Global Climate Change on the United States*. U.S. Environmental Protection Agency, Washington, DC.

Chauvet, E. and H. DeCamps, 1989: Lateral interactions in a fluvial landscape: the River Garonne, France. *J. North Amer. Benthol. Soc.*, **8**, 9-17.

Chen, C.Y. and C.L. Folt, 1996: Consequences of fall warming for zooplankton overwintering success. *Limnology and Oceanography*, **41** (in press).

Chen, J. and A. Ohmura, 1990: On the influence of Alpine glaciers on runoff. IAHS Publ. No. 193, pp. 117-125.

Chessman, B.C. and D.P. Robinson, 1987: Some effects of the 1982-1983 drought on water quality and macroinvertebrate fauna in the lower LaTrobe River, Victoria. *Australian Journal of Marine and Freshwater Research*, **38**, 288-299.

Chiew, F.H.S., P.H. Whetton, T.A. McMahon, and A.B. Pittock, 1995: Simulation of the impacts of climate change on runoff and soil moisture in Australian catchments. *Journal of Hydrology*, **167**, 121-147.

Chinn, T.J., 1993: Physical hydrology of the Dry Valley lakes. In: *Physical and Biogeochemical Processes in Antarctic Lakes* [Green, W.J. and E.I. Friedmann (eds.)]. Antarctic Research Series Vol. 59, American Geophysical Union, Washington, DC, pp. 1-53.

Clark, J.S., 1990: Fire occurrence during the last 750 years in northwestern Minnesota. *Ecological Monographs*, **60**, 139-155.

Clark, J.S., 1993: Fire, climate change, and forest processes during the past 2000 years. In: *Elk Lake, Minnesota: Evidence for Rapid Climate Change in the North-Central United States* [Bradbury, J.P. and W.E. Dean (eds.)]. Geol. Soc. Amer., Sp. Paper 276, Boulder, CO, pp. 295-308.

Colby, P.J. and S.J. Nepsy, 1981: Variation among stocks of walleye (Stizostedion vitreum vitreum): management implications. *Canadian Journal of Fisheries and Aquatic Sciences*, **38**, 1814-1831.

Cole, J.A., S. Slade, P.D. Jones, and J.M. Gregory, 1991: Reliable yield of reservoirs and possible effects of climatic change. *Hydrological Sciences Journal,* **36(6)**, 579-598.

Cole, J.J., N.F. Caraco, G.W. Kling, and T.K Kratz, 1994: Carbon dioxide supersaturation in the surface waters of lakes. *Science*, **265**, 1568-1570.

Connell, D.W. and G.J. Miller, 1984: Metals and salts. In: *Chemistry and Ecotoxicology of Pollution*. John Wiley and Sons, Inc., New York, NY, pp. 288-332.

Cooter, E.J. and W.S. Cooter, 1990: Impacts of greenhouse warming on water temperature and water quality in the southern United States. *Climate Research,* **1**, 1-12.

Coutant, C.C., 1985: Striped bass, temperature, and dissolved oxygen: a speculative hypothesis for environmental risk. *Transactions of the American Fisheries Society*, **18**, 161-172.

Coutant, C.C., 1987: Poor reproductive success of striped bass from a reservoir with reduced summer habitat. *Transactions of the American Fisheries Society*, **166**, 154-160.

Covich, A.P., S.L. Fritz, P.J. Lamb, R.G. Marzolf, W.J. Matthews, K.A. Poiani, E.E. Prepas, M. Richman, and T.C. Winter, 1995: Potential effects of climate change on freshwater ecosystems of the western and northern Great Plains of North America: Region 7—Western Great Plains. In: *Regional Assessment of Freshwater Ecosystems and Climate Change in North America* [McKnight, D.M. (ed.)]. USGS-WRD, Boulder, CO, pp. 11-12.

Cowgill, U.M., K.I. Keating, and I.T. Tokahashi, 1985: Fecundity and longevity of Cerodaphnia dubia/affinis in relation to diet at two different temperatures. *Journal of Crustacean Biology,* **5**, 420-429.

Crawshaw, L.I. and C.S. O'Connor, 1996: Behavioral compensation for long-term thermal change. In: *Global Warming—Implications for Freshwater and Marine Fish.* SEB Seminar Series. Cambridge University Press, Cambridge, UK.

Croley, T.E., 1990: Laurentian Great Lakes double-CO_2 climate change hydrological impacts. *Climate Change,* **17**, 27-48.

Cumming, B.F., S.E. Wilson, and J.P. Smol, 1993: Paleolimnological potential of chrysophyte cysts and sponge spicules as indicators of lake salinity. *International Journal of Salt Lake Research*, **2**, 87-92.

Dadswell, M.J., 1974: Distribution, ecology and postglacial dispersal of certain crustaceans and fishes in eastern North America. *Zoology,* **11**.

Dahm, C.N. and M.C. Molles, 1992: Streams in semiarid regions as sensitive indicators of global climate change. In: *Global Climate Change and Freshwater Ecosystems* [Firth, P. and S.G. Fisher (eds.)]. Springer-Verlag, New York, NY, pp. 250-260.

Davis, R.B., D.S. Anderson, S.A. Norton, and M.C. Whiting, 1994: Acidity of twelve northern New England (USA) lakes in recent centuries. *Journal of Palaeolimnology*, **12**, 103-154.

Davison, I.R., 1991: Environmental effects on algal photosynthesis: temperature. *J. Phycol.,* **27**, 2-8.

Degens, E.T., S. Kempe, and J.E. Richey, 1991: *Biogeochemistry of Major World Rivers.* Wiley, New York, NY, 356 pp.

Dennis, J.G. and G. Wolff, 1992: *It's Raining Frogs and Fishes: Four Seasons of Natural Phenomena and Oddities of the Sky*. Harper Perennial, New York, NY, 323 pp.

DeStasio, B.T., D.K. Hill, J.M. Kleinhaus, N.P. Nibbelink, and J.J. Magnuson, 1996: Potential effects of global climate change on small north temperate lakes: physics, fishes, and plankton. *Limnology and Oceanography*, **41** (in press).

Digerfeldt, G., J.E. Almendinger, and S. Bjorck, 1992: Reconstruction of past lake levels and their relation to groundwater hydrology in the Parkers Prairie sandplain, west-central Minnesota. *Palaeogeography, Palaeoclimatology, and Palaeoecology*, **94**, 99-118.

Dooge, J.C.I., 1992: Hydrologic models and climate change. *Journal of Geophysical Research*, **94(D3)**, 2677-2686.

Doran, P.T., R.A. Wharton, Jr., and W.B. Lyons, 1994: Paleolimnology of the McMurdo Dry Valleys, Antarctica. *Journal of Paleolimnology*, **10(2)**, 85-114.

Douglas, M.S.V. and J.P. Smol, 1995: Periphytic diatom assemblages from high Arctic ponds. *J. Phycol.*, **31**, 60-69.

Dracup, J.A. and E. Kahya, 1994: The relationships between U.S. streamflow and La Nina events. *Water Resources Research*, **30(7)**, 2133-2141.

Drake, B.G., 1992: A field study of the effects of elevated CO_2 on ecosystem processes in a Chesapeake Bay wetland. *Aust. J. Bot.*, **40**, 579-595.

Dudgeon, D., 1993: The effects of spate-induced disturbance, predation and environmental complexity on macroinvertebrates in a tropical stream. *Freshwater Biology*, **30**, 189-197.

Duell, Jr., L.F.W., 1994: The sensitivity of northern Sierra Nevada streamflow to climate change. *Water Resources Bulletin*, **30(5)**, 841-859.

Eaton, J.G. and R.M. Scheller, 1996: Effects of climate warming on fish thermal habitat in streams of the United States. *Limnology and Oceanography*, **41** (in press).

Eckhardt, B.W. and T.R. Moore, 1990: Controls on dissolved organic carbon concentrations in streams, southern Quebec. *Canadian Journal of Fisheries and Aquatic Science*, **47**, 1537-1544.

Edsall, T.A., 1970: The effect of temperature on the rate of development and survival of alewife eggs and larvae. *Transactions of the American Fisheries Society,* **99**, 376-380.

Eilers, J.M., D.F. Brakke, and D.H. Landers, 1988a: Chemical and physical characteristics of lakes in the Upper Midwest, United States. *Environmental Science and Technology*, **22**, 164-172.

Eilers, J.M., D.H. Landers, and D.F. Brakke, 1988b: Chemical and physical characteristics of lakes in the Southeastern United States. *Environmental Science and Technology*, **22**, 172-177.

Elliot, J.M., 1987: Temperature-induced changes in the life cycle of Leuctra nigra (Plecoptera: Leuctridae) from a Lake District stream. *Freshwater Biology*, **18**, 177-184.

Ely, L.L., Y. Enzel, and D.R. Cayan, 1994: Anomalous North Pacific atmospheric circulation and large winter floods in the southwestern United States. *Journal of Climate,* **7**, 977-987.

Evans, J.C. and E.E. Prepas, 1996: Potential effects of climate change on ion chemistry and phytoplankton communities in prairie saline lakes. *Limnology and Oceanography*, **41** (in press).

Fee, E.J., R.E. Hecky, S.W. Kasian, and D. Cruikshank, 1996: Potential size-related effects of climate change on mixing depths in Canadian Shield Lakes. *Limnology and Oceanography*, **41** (in press).

Fields, R., S.S. Lowe, C. Kaminski, G.S. Whitt, and D.P. Philipp, 1987: Tical and chronic thermal maxima of northern and Florida largemouth bass and their reciprocal F1 and F2 hybrids. *Transactions of the American Fisheries Society*, **116**, 856-863.

Firth, P. and S.G. Fisher (eds.), 1992: *Global Climate Change and Freshwater Ecosystems*. Springer-Verlag, New York, NY, 321 pp.

Fisher, S.G. and N.B. Grimm, 1991: Streams and disturbance: are cross-ecosystem comparisons useful? In: *Comparative Analysis of Ecosystems: Patterns, Mechanisms and Theories* [Cole, J.C., G.M. Lovett, and S.E.G. Findlay (eds.)]. Springer-Verlag, New York, NY, pp. 196-221.

Fogg, G.E. and B. Thake, 1987: *Algal Cultures and Phytoplankton Ecology*. University of Wisconsin Press, Madison, WI, 269 pp.

Freeman, C., R. Gresswell, H. Guasch, J. Hudson, M.A. Lock, B. Reynolds, F. Sabater, and S. Sabater, 1994: The role of drought in the impact of climatic change on the microbiota of peatland streams. *Freshwater Biology*, **32**, 223-230.

Fritz, S.C., S. Juggins, and R.W. Battarbee, 1993: Diatom assemblages and fonic characterization of lakes of the northern Great Plains, North America: a tool for reconstructing past salinity and climate fluctuations. *Canadian Journal of Fisheries and Aquatic Science*, **50**, 1844-1856.

Fritz, S.C., D.R. Engstrom, and B.J. Haskell, 1994: Little Ice Age aridity in the North American Great Plains: a high-resolution reconstruction of salinity fluctuations from Devil's Lake, North Dakota, USA. *Holocene*, **4**, 69-73.

Fritz, S.C., 1996: Paleolimnological records of climate change in North America. *Limnology and Oceanography*, **41** (in press).

Fry, F.E.J., 1947: Effects of the environment on animal activity. *Publications of the Ontario Fisheries Research Laboratory*, **68**, 1-62.

Fry, F.E.J., 1971: Effects of the environmental factors on the physiology of fish. In: *Fish Physiology: Environmental Factors* [Hoar, W.S. and D.J. Randall (eds.)]. Academic Press, New York, NY, pp. 1-98.

Fukushima, Y., O. Watanabe, and K. Higuchi, 1991: Estimation of streamflow change by global warming in a glacier-covered high mountain area of the Nepal Himalaya. In: *Snow, Hydrology and Forests in High Alpine Areas*. IAHS Publ. No. 205, pp. 181-188.

Gafny, S. and A. Gasith, 1989: Water quality dynamics in the shallow littoral of Lake Kinneret. In: *Environmental Quality and Ecosystem Stability*, vol. IV-B [Spanier, E., Y. Steinberger, and M. Luria (eds.)]. ISEEQS Pub., Jerusalem, pp. 327-336.

Gafny, S., A. Gasith, and M. Goren, 1992: Effect of water level on shore spawning of Mirogrex terraesanctae (Cyprinidae) in lake Kinneret. *Israel. J. Fish. Biol.*, **41**, 863-871.

Gafny, S. and A. Gasith, 1993: Effect of low water level on the water quality of the littoral zone in Lake Kinneret. *Water Science Technology*, **27**, 363-371.

Galloway, J.N. and G.E. Likens, 1979: Atmospheric enhancement of metal deposition in Adirondack lake sediments. *Limnology and Oceanography*, **24(6)**, 1161-1165.

Gan, T.Y. and S.J. Burges, 1990: An assessment of a conceptual rainfall-runoff model's ability to represent the dynamics of small hypothetical catchments. 2. Hydrologic responses for normal and extreme rainfall. *Water Resources Research*, **26(7)**, 1605-1619.

Gasith, A. and S. Gafny, 1990: Effect of water level fluctuation on the structure and function of the littoral zone. In: *Large Lakes: Ecological Structure and Function* [Tilzer, M.M. and C. Serruya (eds.)]. Science-Tech. Pub., Madison, WI, pp. 156-173.

Gauzer, B., 1993: *Effect of Air Temperature Change on the Danube Flow Regime*. VITUKI report, Budapest, Hungary, 11 pp. (in Hungarian).

Gellens, D., 1991: Impact of a CO_2-induced climate change on river flow variability in three rivers in Belgium. *Earth Surface Processes Landforms*, **16**, 619-625.

George, D.G. and G.P. Harris, 1985: The effect of climate change on long-term changes in the crustacean zooplankton biomass of Lake Windermere, UK. *Nature*, **316**, 536-539.

Georgiadi, A.G., 1991: The change of the hydrological cycle under the influence of global warming. In: *Hydrology for the Water Management of Large River Basins* [van de Ven, F.H.M., D. Gutknecht, D.P. Loucks, and K.A. Salewicz (eds.)]. IAHS Publ. No. 201, pp. 119-128.

Giblin, A.E., K.J. Nadelhoffer, G.R. Shaver, J.A. Laundre, and A.J. McKerrow, 1991: Biogeochemical diversity along a riverside toposequence in arctic Alaska. *Ecological Monographs*, **61**, 415-435.

Goldman, C.R. and A.J. Horne, 1983: *Limnology*. McGraw-Hill Book Co., New York, NY, 464 pp.

Goldman, J.C. and E.J. Carpenter, 1974: A kinetic approach to the effect of temperature on algal growth. *Limnology and Oceanography*, **19**, 756-766.

Gont, R.A., L. Lin, and L.E. Ohl, 1988: Holocene lake fluctuations in Pine Lake, Wisconsin. *Wisconsin Academy Science of Arts and Letters*, pp. 107-115.

Gophen, M., 1976: Temperature dependence of food intake, ammonia excetion and respiration in Ceriodaphnia reticulata (jurine), Lake Kinneret, Israel. *Freshwater Biology*, **6**, 451-455.

Gophen, M. and I. Gal, 1992: *Lake Kinneret: Part I, The Lake; Part II, Shores and Sites*. Ministry of Defense Publisher and Kinneret Authority, Water Commission, 335 pp. (in Hebrew).

Gordon, H.B., P.H. Whetton, A.B. Pittock, A.M. Fowler, and M.R. Haylock, 1992: Simulated changes in daily rainfall intensity due to the enhanced greenhouse effect. *Climate Dynamics*, **8**, 83-102.

Goutorbe, J., T. Lebel, and A. Tinga, *et al.*, 1993: HAPEX-Sahel: A large-scale study of land-atmosphere interactions in the semi-arid tropics. *Annales Geophysicae*, **12**, 53-64.

Gray, L.J., 1981: Species composition and life histories of aquatic insects in a lowland Sonoran Desert stream. *Amer. Midl. Nat.*, **106**, 229-242.

Grayson, R.B., I.D. Moore, and T.A. McMahon, 1992a: Physically based hydrologic modeling, 1. Is the concept realistic? *Water Resources Research*, **28**, 2639-2658.

Grayson, R.B., I.D. Moore, and T.A. McMahon, 1992b: Physically based hydrologic modeling, 2. Is the concept realistic? *Water Resources Research*, **28**, 2659-2666.

Gregory, S.V., F.J. Swanson, W.A. McKee, and K.W. Cummins, 1991: An ecosystem perspective of riparian zones. *BioScience*, **41**, 540-551.

Grimm, N.B., 1992: Biogeochemistry of nitrogen in arid-land stream ecosystems. *J. Arizona-Nevada Acad. Sci.*, **26**, 130-146.

Grimm, N.B. and S.G. Fisher, 1992: Responses of arid-land streams to changing climate. In: *Global Climate Change and Freshwater Ecosystems* [Firth, P. and S.G. Fisher (eds.)]. Springer-Verlag, New York, NY, pp. 211-233.

Grimm, N.B., 1993: Implications of climate change for stream communities. In: *Biotic Interactions and Global Change* [Kareiva, P., J. Kingsolver, and R. Huey (eds.)]. Sinauer Associates, Inc., Sunderland, MA, pp. 293-314.

Grimm, N.B., C.N. Dahm, S.W. Hostetler, O.T. Lind, S. Sorooshian, P.L. Starkweather, and W.W. Wurtsbaugh, 1995: Potential responses to climate change of aquatic ecosystems in the Basin and Range, American Southwest and Mexico: Region 8—Basin and Range, and Southwest and Mexico. In: *Regional Assessment of Freshwater Ecosystems and Climate Change in North America* [McKnight, D.M. (ed.)]. USGS-WRD, Boulder, CO, pp. 12-13.

Grotch, S.L. and M.C. MacCracken, 1991: The use of general circulation models to predict regional climatic change. *Journal of Climate*, **4**, 286-303.

Haberyan, K.A. and R.E. Hecky, 1987: The Late Pleistocene and Holocene stratigraphy and paleolimnology of Lakes Kivu and Tanganyika. *Paleogeography, Paleoclimatology and Paleoecology*, **61**, 169-197.

Hairston, Jr., N.G., T.A. Dillon, and B.T. DeStasio, Jr., 1990: A field test for the cues of diapause in a freshwater copepod. *Ecology*, **71**, 2218-2223.

Hairston, N.G., 1996: Zooplankton egg banks as biotic reservoirs in changing environments. *Limnology and Oceanography*, Special Issue (in press).

Hanazato, T. and M. Yasuno, 1989: Effect of temperature in laboratory studies of growth of Chaoborus flavicans (Diptera: Chaoboridae). *Arch. Hydrobiology*, **114**, 497-504.

Hansen, J. *et al.*, 1988: Global climate changes as forecast by the Goddard Institute for Space Studies three-dimensional model. *Journal of Geophysical Research*, **93**, 9341-9364.

Hansen, Jr., N.G. and B.T. DeStasio, Jr., 1988: Rate of evolution slowed by a dormant propagule pool. *Nature*, **336**, 329-342.

Harasawa, H., 1993: Impact of climate change on water resources in Japan. In: *The Potential Effects of Climate Change in Japan* [Nishioka, S. *et al.* (eds.)]. Center Global Environ. Res., Nat. Inst. Environ. Studies, Environment Agency of Japan, Tsukuba, Japan, pp. 1-13.

Harvey, A.M., 1991: The influence of sediment supply on the channel morphology of upland streams: Howgill Fells, North West England. *Earth Surf. Process. Landforms*, **16**, 675-684.

Hastenrath, S. and P.D. Kruss, 1992: Greenhouse indicators in Kenya. *Nature*, **355**, 503.

Hastings, J.R. and R.M. Turner, 1965: *The Changing Mile*. University of Arizona Press, Tucson, AZ, 317 pp.

Hatton, T.J., J. Walker, W.R. Dawes, and F.X. Dunin, 1992: Simulations of hydroecological responses to elevated CO_2 at the catchment scale. *Aust. J. Bot.*, **40**, 679-696.

Hay, L.E., G.J. McCabe, Jr., D.M. Wolock, and M.A. Ayers, 1992: Use of weather types to disaggregate general circulation model predictions. *Journal of Geophysical Research*, **97(D3)**, 2781-2790.

Headrick, M.R. and R.F. Carline, 1993: Restricted summer habitat and growth of northern pike in two southern Ohio impoundments. *Transactions of the American Fisheries Society*, **122**, 228-236.

Hecky, R.E., E.J. Fee, H.J. Kling, and J.M.W. Rudd, 1981: Relationship between primary productivity and fish production in Lake Tanganyika. *Transactions of the American Fisheries Society*, **110**, 336-345.

Hecky, R.E., R.H. Spigel, and G.W. Coulter, 1991: The nutrient regime. In: *Lake Tanganyika and Its Life* [Coulter, G.W. (ed.)]. Oxford University Press, Oxford, UK, pp. 76-89.

Hecky, R.E., F.W.B. Bugenyi, P. Ochumba, J.F. Talling, R. Mugidde, M. Gophen, and L. Kaufman, 1994: Deoxygenation of the deep water of Lake Victoria, East Africa. *Limnology and Oceanography*, **39**, 1476-1481.

Herzig, A., 1983: The ecological significance of the relationship between temperature and duration of embryonic development in planktonic freshwater copepods. *Hydrobiology*, **100**, 65-91.

Hickman, M. and M.A. Reasoner, 1994: Diatom responses to late Quaternary vegetation and climate change, and to deposition of two tephras in an alpine and a sub-alpine lake in Yoho National Park, British Columbia. *Journal of Palaeolimnology*, **11**, 173-188.

Hill, D.K. and J.J. Magnuson, 1990: Potential effects of global climate warming on the growth and prey consumption of Great Lakes fish. *Transactions of the American Fisheries Society*, **119**, 265-275.

Hocutt, C.H. and E.O. Wiley, 1985: *The Zoogeography of North American Freshwater Fishes*. John Wiley and Sons, Inc., New York, NY, 866 pp.

Hogg, I.D., D.D. Williams, J.M. Eadie, and S.A. Butt, 1995: The consequences of global warming for stream invertebrates: a field simulation. *J. Thermal Biology*, **20**, 199-206.

Hokanson, K.E.F., 1977: Temperature requirements of some percids and adaptations to the seasonal temperature cycle. *Journal of the Fisheries Research Board of Canada*, **34**, 1524-1550.

Holmes, J.A., 1990: Sea lamprey as an early responder to climate change in the Great Lakes Basin. *Transactions of the American Fisheries Society*, **119**, 292-300.

Holtby, L.B., 1988: Effects of logging on stream temperatures in Carnation Creek, British Columbia, and associated impacts on the Coho Salmon (Oncorhynchus kisutch). *Canadian Journal of Fisheries and Aquatic Science*, **45**, 502-515.

Hornberger, G.M., K.E. Bencala, and D.M. McKnight, 1994: Hydrological controls on dissolved organic carbon during snowmelt in the Snake River near Montezuma, Colorado. *Biogeochemistry*, **25**, 147-165.

Hostetler, S.W. and F. Giorgi, 1993: Use of output from high-resolution atmospheric models in landscape-scale hydrologic models: an assessment. *Water Resources Research*, **29(6)**, 1685-1695.

Houghton, J.T., 1991: Scientific assessment of climate change: summary of the IPCC Working Group I report. In: *Proc. Second World Climate Conference*. Cambridge Univ. Press, Cambridge, UK, pp. 23-44.

Hughes, J.P., D.P. Lettenmaier, and P. Guttorp, 1993: A stochastic approach for assessing the effect of changes in synoptic circulation patterns on gauge precipitation. *Water Resources Research*, **29(10)**, 3303-3315.

Hulme, M. and P.D. Jones, 1989: *Climate Change Scenarios for the UK Representative to the Institute of Hydrology*. Climatic Research Unit, University of East Anglia.

Hynes, H.B.N., 1963: Imported organic matter and secondary productivity in streams. *Proceedings of the International Congress of Zoology*, **16**, 324-329.

Hynes, H.B.N., 1975: The stream and its valley. *Verh. Internat. Verein. Limnol.*, **19**, 1-5.

IPCC, 1992: *Climate Change 1992: The Supplementary Report to the IPCC Scientific Assessment*. Prepared by IPCC Working Group I [Houghton, J.T., B.A. Callander, and S.K. Varney (eds.)] and WMO/UNEP. Cambridge University Press, Cambridge, UK, 200 pp.

Jacobson, Jr., G.L., 1988: 1. Ancient permanent plots: sampling in paleoecological studies. In: *Vegetation History* [Huntley, B. and T. Webb III (eds.)]. Kluwer Academic Publishers, Dordrecht, The Netherlands, pp. 3-16.

Jamieson, C. and C. Burns, 1988: The effects of temperature and food on copepodite development, growth and reproduction in three species of Boeckella (Copepoda; Calanoida). *Hydrobiology*, **164**, 235-257.

Jenkins, A., M. McCartney, and C. Sefton, 1993: *Impacts of Climate Change on River Water Quality in the United Kingdom*. Rep. to Dept. Environ., Inst. Hydrol., Wallingford, UK, 39 pp.

John, D.M., 1986: The inland waters of tropical West-Africa. An introduction to and botanical review. *Arch. Hydrobiology*, **23**, 1-244.

Johnson, T.B. and D.O. Evans, 1990: Size-dependent winter mortality of young-of-the-year white perch: climate warming and the invasion of the Laurentian Great Lakes. *Transactions of the American Fisheries Society*, **119**, 779-785.

Jones, B.R., K.E.F. Hokanson, and J.H. McCormick, 1972: Winter temperature requirements for maturation and spawning of yellow perch Perca flavenscens (Mitchill). In: *Proceedings of the World Conference towards a Plan of Action for Manking*. Vol. 3, *Biological Balance and Thermal Modifications* [Marois, M. (ed.)]. Pergamon Press, New York, NY, pp. 189-192.

Jowett, I.G. and M.J. Duncan, 1990: Flow variability in New Zealand rivers and its relationship to in-stream habitat and biota. *New Zealand Journal of Marine and Freshwater Research*, **24**, 305-317.

Juggins, S., R.W. Battarbee, S.C. Fritz, and F. Gasse, 1994: The CASPIA project: diatoms, salt lakes, and environmental change. *J. Paleolimn.*, **12**, 191-196.

Junk, W.J., P.B. Bayley, and R.E. Sparks, 1989: The flood pulse concept in river-floodplain systems. *Can. Spec. Publ. Fish. Aquat. Sci.*, **106**, 110-127.

Katz, R.W. and B.G. Brown, 1992: Extreme events in a changing climate: variability is more important than averages. *Climate Change*, **21**, 289-302.

Kendall, R.L., 1969: An ecological history of the Lake Victoria basin. *Ecological Monographs*, **39**, 121-176.

Kilham, S.S., E.C. Theriot, and S.C. Fritz, 1996: Linking planktonic diatoms and climate change using resource theory in the large lakes of the Yellowstone ecosystem. *Limnology and Oceanography*, **41** (in press).

Kimball, B.A., R.L. LaMorte, P.J. Winter, G.W. Wale, and R.L. Garcia, 1993: Effects of free air CO_2 enrichment (FACE) on energy balance and evapotranspiration. In: *Proceedings of Annual Meeting Amer. Soc. Agronomy*. 7 December 1993, Cincinatti, OH.

Kimmel, B.L., O.T. Lind, and L.J. Paulson, 1990: Reservoir primary production. In: *Reservoir Limnology: Ecological Perspectives* [Thornton, K.W., B.L. Kimmel, and F.E. Payne (eds.)]. John Wiley and Sons, Inc., New York, NY, pp. 133-193.

Kirshen, P.H. and N.M. Fennessey, 1995: Possible climate change impacts on water supply of metropolitan Boston. *ASCE J. Water Resour. Plan. Manage.*, **121(1)**, 61-70.

Klemes, V., 1985: *Sensitivity of Water Resource Systems to Climate Variations*. World Climate Appl. Programme, WCP-98, World Meteorological Organisation, Geneva, Switzerland, 142 pp.

Kling, G.W., G.W. Kipphut, and M.C. Miller, 1991: Arctic lakes and streams as gas conduits to the atmosphere: implications for tundra carbon budgets. *Science*, **251**, 298-301.

Knox, J.C., 1993: Large increases in flood magnitude in response to modest changes in climate. *Nature*, **361**, 430-432.

Koerner, C. and F. Miglietta, 1994: Long term effects of naturally elevated CO_2 on mediterranean grassland and forest trees. *Oecologia*, **99**, 343-351.

Koerner, C., 1995: The response of complex multispecies systems to elevated CO_2. In: *Global Change and Terrestrial Ecosystems* [Walter, B.H. and W.L. Steffen (eds.)]. Cambridge University Press, Cambridge, UK.

Koppelman, J.B., G.S. Whitt, and D.P. Philipp. 1988: Thermal preferenda of northern Florida and reciprocal F1 hybrid largemouth bass. *Transactions of the American Fisheries Society*, **117**, 238-244.

Korpelainen, H., 1986: The effects of temperature and photoperiod on life history parameters of Daphnia magna (Crustacea: Cladocera). *Freshwater Biology*, **16**, 615-620.

Krasovskaia, I. and L. Gottschalk, 1992: Stability of river flow regimes. *Nordic Hydrology*, **23**, 137-154.

Kratz, T.K., B.J. Benson, E.R. Blood, G.L. Cunningham, and R.A. Dahlgren, 1991: The influence of landscape position on temporal variability in four North American ecosystems. *Amer. Nat.*, **138**, 355-378.

Kratz, T.K., K.E. Webster, C.J. Bowser, J.J. Magnuson, and B.J. Benson, 1996: The influence of landscape position on lakes in northern Wisconsin. *Freshwater Biology* (in press).

Kuchment, L.S., Y.G. Motovilov, E.L. Muzylev, and N.A. Nazarov, 1987: Estimates of possible variations of snowmelt-runoff characteristics on climate changes, 1987. IAHS Publ. No. 100, pp. 129-138.

Kuchment, L.S., Y.G. Motovilov, and N.A. Nazarov, 1990: Sensitivity of hydrological systems: impact of river basins and climate human activities on hydrological cycle. *M. Nauka*, p. 143.

Kuchment, L.S. and Z.P. Startseva, 1991: Sensitivity of evaporation and soil moisture in wheat fields to changes in climate and direct effects of carbon dioxide. *Hydrological Sciences Journal*, **36(6)**, 631-643.

Kwadijk, J. and H. Middelkoop, 1994: Estimation of impact of climate change on the peak discharge probability of the River Rhine. *Climate Change*, **27**, 199-224.

LaBaugh, J.W., T.C. Winter, G.A. Swanson, D.O. Rosenberry, R.D. Nelson, and N.H. Euliss, Jr., 1996: Changes in atmospheric circulation patterns affect midcontinent wetlands sensitive to climate. *Limnology and Oceanography*, **41** (in press).

LaBerge, S. and B.J. Hann, 1990: Acute temperature and oxygen stress among genotypes of Daphnia pulex and Simocephalus vetulus (Cladocera: Daphniidae) in relation to environmental conditions. *Can. J. Zool.*, **68**, 2257-2263.

Laird, K.R., S.C. Fritz, and E.C. Grimm, 1996: Century-scale reconstruction from Moon Lake, a closed basin lake in the northern Great Plains. *Limnology and Oceanography*, **41** (in press).

Lamberti, G.A. and V.H. Resh, 1983: Geothermal effects on stream benthos: separate influences of thermal and chemical components on periphyton and macroinvertebrates. *Canadian Journal of Fisheries and Aquatic Science*, **40**, 1995-2009.

Lal, M., 1994: Water resources of the South Asian region in a warmer atmosphere. *Advanced Atmospheric Sciences*, **11(2)**, 239-246.

Lang, H., 1981: Is evaporation an important component in high alpine hydrology? *Nordic Hydrology*, **12**, 217-224.

Laybourn-Parry, J., B.A. Abdullahi, and S.V. Tinson, 1987: Temperature-dependent energy partitioning in the benthis copepods Acanthocyclops viridis and Macrocyclops albidus. *Canadian Journal of Fisheries and Aquatic Science*, **40**, 1995-2009.

LaZerte, B.D., 1993: The impact of drought and acidification on the chemical exports from a minerotrophic confer swamp. *Biogeochemistry,* **18**, 153-175.

Leavesley, G.H., M.D. Branson, and L.E. Hay, 1992: Using coupled atmospheric and hydrologica models to investigate the effects of climate change in mountainous regions. In: *Managing Water Resources during Global Change* [Herrmann, R. (ed.)]. Conf. Proc. Amer. Water Resour. Assoc., Bethesda, MD, pp. 691-700.

Leavesley, G.H., 1994: Modeling the effects of climate change on water resources—a review. *Climate Change,* **28**, 159-177.

Lei, C. and K.B. Armitage, 1980: Growth, development and body size of field and laboratory populations of Daphnia ambigua. *Oikos*, **35**, 31-48.

Lettenmaier, D.P. and T.Y. Gan, 1990: Hydrologic sensitivities of the Sacramento-San Joaquin River Basin, California, to global warming. *Water Resources Research*, **26(1)**, 69-86.

Lettenmaier, D.P. and D.P. Sheer, 1991: Climatic sensitivity of California water resources. *ASCE J. Water Resour. Plan. Manage.*, **117(1)**, 108-125.

Lettenmaier, D.P., 1993: Sensitivity of Pacific Northwest water resources to global warming. In: *Proc. Conf. on Climate Change and Water Resources Management* [Ballentine, T.M. and E.Z. Stakhiv (eds.)]. U.S. Army Inst. Water Resources, Fort Belvoir, VA, p. II-41.

Lettenmaier, D.P., G. McCabe, and E.Z. Stakhiv, 1994: Global climate change: effect on hydrologic cycle. In: *Handbook of Water Resources* [Mays, L. (ed.)]. McGraw-Hill, New York, NY.

Likens, G.E., 1984: Beyond the shoreline: a watershed-ecosystem approach. *Internationale Vereinigung fur Theoretische und Angewandte Limnologie*, **22**, 1-22.

Liu, C.Z., 1995: On the sensitivity and vulnerability of water resources to climate change in China. In: *Proceedings of the International Conference on Hydro-Science and Engineering (ICHE-95)*. Volume IIA, pp. 888-895.

Livingstone, D.A., 1975: Late Quaternary climate change in Africa. *Annual Review of Ecology and Systematics,* **6**, 249-280.

Lodge, D.M., 1993: Species invasions and deletions: community effects and responses to climate and habitat change. In: *Biotic Interactions and Global Change* [Kareiva, P.M., J.G. Kingsolver, and R.B. Huey (eds.)]. Sinauer Associates Inc., Sunderland, MA, pp. 367-387.

L'vovich, M.I., 1979: *World Water Resources and Their Future* [English translation by R.L. Nace]. Amer. Geophys. Union, Washington, DC, 415 pp.

Magnuson, J.J., 1976: Managing with exotics—a game of chance. *Transactions of the American Fisheries Society*, **105(1)**, 1-9.

Magnuson, J.J., L.B. Crowder, and P.A. Medvick, 1979: Temperature as an ecological resource. *American Zoology*, **19**, 331-343.

Magnuson, J.J., C.A. Paskowski, F.J. Rahel, and W.M. Tonn, 1989a: Fish ecology in severe environments of small isolated lakes in northern Wisconsin. In: *Symposium Series No. 61* [Sharitz, R.R. and J.W. Gibbons (eds.)]. U.S. DOE Office of Scientific and Technical Information, Oak Ridge, TN, pp. 487-515.

Magnuson, J.J., D.K. Regier, H.A. Holmes, J.D. Meisner, and B.J. Shuter, 1989b: Potential responses of Great Lakes fishes and their habitat to global climate warming. In: *The Potential Effects of Global Climate Change on the United States—Appendix E, Aquatic Resources* [Smith, J.B. and D.A. Tirpak (eds.)]. U.S. EPA, EPA-230-05-89-055, Washington, DC, pp. 2.1-2.42.

Magnuson, J.J., J.D. Meisner, and D.K. Hill, 1990: Potential changes in the thermal habitat of Great Lakes fish after global climate warming. *Transactions of the American Fisheries Society*, **119**, 254-264.

Magnuson, J.J., C.J. Bowser, R.A. Assel, B.T. DeStasio, J.R. Eaton, E.J. Fee, P.J. Dillon, L.D. Mortsch, N.T. Roulet, F.H. Quinn, and D.W. Schindler, 1995: Region 1— Laurentian Great Lakes and precambrian shield. In: *Regional Assessment of Freshwater Ecosystems and Climate Change in North America* [McKnight, D.M. (ed.)]. USGS-WRD, Boulder, CO, pp. 3-4.

Magnuson, J.J. and B.T. DeStasio, 1996: Thermal niche of fishes and global warming. In: *Global Warming—Implications for Freshwater and Marine Fish.* SEB Seminar Series, Cambridge University Press, Cambridge, UK (in press).

Maier, G., 1989: The effect of temperature on the development times of eggs, naupliar and copepodite stages of five species of cyclopid copepods. *Hydrobiology*, **184**, 79-88.

Manabe, S. and R.J. Stouffer, 1980: Sensitivity of a global climate model to an increase of CO_2 concentration in the atmosphere. *Journal of Geophysical Research*, **85**, 5529-5554.

Mandrak, N.E., 1989: Probable invasions of the Great Lakes by new species due to climate warming. *Journal of Great Lakes Research,* **15**, 306-316.

Marcus, N.H., 1987: Differences in the duration of egg diapause of Labidocera aestiva (Copepoda: Calanoida) from the Woods Hole, Massachusetts region. *Biol. Bull.*, **173**, 169-177.

Margalef, R., 1990: Limnology: reconsidering ways and goals. In: *Scientific Perspectives in Theoretical and Applied Limnology* [de Bernardi, R., G. Giussani, and L. Barbanti (eds.)]. Mem. Ist. Ital. Idrobiol. 47, pp. 57-76.

Marsh, P. and L.F.W. Lesack, 1996: Climate change and the hydrologic regime of lakes in the Mackenzie Delta. *Limnology and Oceanography*, **41** (in press).

Matthews, W.J. and E.G. Zimmerman, 1990: Potential effects of global warming on native fishes of the southern Great Plains and the Southwest. *Fisheries*, **15**, 26-32.

McCabe, Jr., G.J. and D.M. Wolock, 1992: Effects of climatic change and climatic variability on the Thornwaite moisture index in the Delaware River basin. *Climate Change,* **20**, 143-153.

McCormick, M.J., 1990: Potential changes in thermal structure and cycle of Lake Michigan due to global warming. *Transactions of the American Fisheries Society*, **119**, 183-195.

McDowell, W.H. and G.E. Likens, 1988: Origin, composition and flux of dissolved organic carbon in the Hubbard Brook Valley. *Ecological Monographs*, **28**, 177-195.

McGregor, J.L. and K. Walsh, 1994: Climate change simulations of Tasmanian precipitation using multiple nesting. *Journal of Geophysical Research*, **99(D3)**, 20,889-905.

McLain, A.S., J.J. Magnuson, and D.K. Hill, 1994: Latitudinal and longitudinal differences in thermal habitat for fishes influenced by climate warming: expectations from simulations. *Verh. Internat. Verein. Limnol.*, **25(4)**, 2080-2085.

McLaren, I.A., J.M. Sevigny, C.J. Corkett, 1989: Temperature-dependent development in Pseudocalanus species. *Canadian Journal of Zoology*, **67**, 559-564.

Mechoso, C.R. and G.P. Iribarren, 1992: Streamflow in southeastern South America and the Southern Oscillation. *Journal of Climate,* **5**, 1535-1539.

Meisner, J.D., 1990: Effect of climatic warming on the southern margins of the native range of brook trout, Salvelinus fontinalis. *Canadian Journal of Fisheries and Aquatic Sciences,* **47**, 1067-1070.

Meyer, J.L., 1990: A blackwater perspective on riverine ecosystems. *BioScience*, **40**, 643-651.

Meyer, J.L. and R.T. Edwards, 1990: Ecosystem metabolism and turnover of organic carbon along a blackwater river continuum. *Ecology*, **60**, 1255-1269.

Meyer, J.L. and W.M. Pulliam, 1992: Modification of terrestrial-aquatic interactions by a changing climate. In: *Global Climate Change and Freshwater Ecosystems* [Firth, P. and S.G. Fisher (eds.)]. Springer-Verlag, New York, NY, pp. 177-191.

Miller, J.R. and G.C. Russell, 1992: The impact of global warming on river runoff. *J. Goephys. Res.*, **97(D3)**, 2757-2764.

Mills, L.S. and P.E. Smouse, 1994: Demographic consequences of inbreeding in remnant populations. *American Naturalist*, **144**, 412-431.

Milly, P.C.D. and K.A. Dunne, 1994: Sensitivity of the global water cycle to the water-holding capacity of land. *Journal of Climate*, **7(4)**, 506-526.

Mimikou, M.A. and Y.S. Kouvopoulos, 1991: Regional climate change impacts. I. Impacts on water resources. *Hydrological Sciences Journal*, **36(3)**, 247-258.

Minshall, G.W., R.C. Petersen, K.W. Cummins, T.L. Bott, and J.R. Sedell, 1983: Interbiome comparison of stream ecosystem dynamics. *Ecological Monographs*, **53**, 1-25.

Minshall, G.W., J.T. Brock, and J.D. Varley, 1989: Wildfires and Yellowstone's stream ecosystems. *BioScience*, **39**, 707-715.

Moore, M.V. and C.L. Folt, 1993: Zooplankton body size and community structure: effects of thermal and toxicant stress. *TREE*, **8**, 178-183.

Moore, M.V., C.L. Folt, and R.S. Stemberger, 1995: Consequences of elevated temperatures for zooplankton assemblages in temperate lakes. *Archiv. Hydrobiology*, **268**, 1-31.

Moore, M.V., M.L. Pace, J.R. Mather, P.S. Murdoch, R.W. Howarth, C.L. Folt, C.Y. Chen, H.F. Hemond, P.A. Flebbe, and C.T. Driscoll, 1996: Potential effects of climate change on freshwater ecosystems of the New England/Mid-Atlantic region. *Hydrological Processes* (in press).

Mortsch, L.D. and F.H. Quinn, 1996: Climate change scenarios for the Great Lakes Basin ecosystem studies. *Limnology and Oceanography*, **41** (in press).

Mulholland, P.J. and E.J. Kuenzler, 1979: Organic carbon export from upland and forested wetland watersheds. *Limnology and Oceanography*, **24**, 960-966.

Mulholland, P.J., G.R. Best, C.C. Coutant, G.M. Hornberger, J.L. Meyer, P.J. Robinson, J.R. Stenberg, R.E. Turner, F. Vera-Herera, and R.G. Wetzel, 1996: Effects of climate change on freshwater ecosystems of the south-eastern United States and Gulf Coast of Mexico. *Hydrological Processes* (in press).

Nathan, R.J., T.A. McMahon, and B.L. Finlayson, 1988: The impact of the greenhouse effect on catchment hydrology and storage-yield relationships in both winter and summer rainfall zones. In: *Greenhouse: Planning for Climate Change* [Perman, G.I. (ed.)]. CSIRO, Melbourne, Australia, pp. 273-295.

Nazarov, N.A., 1992: Estimation of hydrological consequences of possible climatic changes and human activities on the Ukraine Poles'e river basins. In: *Proc. Tenth Symp. Danube Countries, Vol. 2*. Bundesanstalt fuer Gewaesserkunde, Koblenz, Germany, pp. 465-469.

Nelson, J.S., 1984: *Fishes of the World*. John Wiley and Sons, Inc., New York, NY, 2nd ed., 523 pp.

Newson, M.D. and J. Lewin, 1991: Climatic change, river flow extremes and fluvial erosion—scenarios for England and Wales. *Prog. Phys. Geogr.*, **15**, 1-17.

NIH, 1994: Abstract of NIH Report "Response of Indian catchment to expected climate change." National Institute of Hydrology, Roorkee, India.

Novaki, B., 1990: *Effect of Climate Change on Maximum Flows of Winter/Spring*. VITUKI report, Budapest, Hungary, 75 pp. (in Hungarian).

Novaki, B., 1992a: *Climate Variability and Its Hydrological Effects*. VITUKI report, 7613/1/1782, Budapest, Hungary, 12 pp. (in Hungarian).

Novaki, B., 1992b: *Effect of Climate Change on Excess Waters of Winter/Spring and Minimum Flows of Summer/Autumn*. VITUKI report, Budapest, Hungary, 25 pp. (in Hungarian).

Oglesby, R.T., 1977: Phytoplankton summer standing crop and annual productivity as functions of phosphorus loading and various physical factors. *Journal of the Fisheries Research Board of Canada*, **34**, 2255-2270.

Orcutt, Jr., J.D. and K.G. Porter, 1983: Diel vertical migration by zooplankton: constant and fluctuating temperature effects on life history parameters of Daphnia. *Limnology and Oceanography*, **28**, 720-730.

Orcutt, Jr., J.D. and K.G. Porter, 1984: The synergistic effects of temperature and food concentration on life history parameters of Daphnia. *Oecologia*, **63**, 300-306.

Owen, R.B., R. Crossley, T.C. Johnson, D. Tweddle, L. Kornfield, S. Davison, D.H. Eccles, and D.E. Engstrom, 1990: Major low levels of Lake Malawi and their implications for speciation rates in cichlid fishes. *Proceedings of the Royal Society, London*, **240**, 519-553.

Panagoulia, D., 1992: Impacts of GISS-modelled climate changes on catchment hydrology. *Hydrological Sciences Journal*, **37(2)**, 141-163.

Patalas, K., 1975: The crustacean plankton communities of fourteen North American great lakes. *Verhandlungen der Internationale Vereinigung fur Theoretische und Angewandte Limnologie*, **19**, 504-511.

Patrick, R., 1969: Some effects of temperature on freshwater algae. In: *Biological Aspects of Thermal Pollution* [Krendel, P.A. and F.L. Parker (eds.)]. Vanderbilt University Press, Nashville, TN, pp. 161-198.

Patrick, R., 1971: The effects of increasing light and temperature on the structure of diatom communities. *Limnology and Oceanography*, **16**, 405-421.

Peck, A.J. and G.B. Allison, 1988: Groundwater and salinity response to climate change. In: *Greenhouse: Planning for Climate Change* [Perman, G.I. (ed.)]. CSIRO, Melbourne, Australia, pp. 238-251.

Peterjohn, W.T. and D.L. Correll, 1984: Nutrient dynamics in an agricultural watershed: observations on the role of a riparian forest. *Ecology*, **64**, 1249-1265.

Poff, N.L. and J.V. Ward, 1989: Implication of streamflow variability and predictability for lotic community structure: a regional analysis of streamflow patterns. *Canadian Journal of Fisheries and Aquatic Science*, **46**, 1805-1818.

Poff, N.L. and J.V. Ward, 1990: The physical habitat template of lotic systems: recovery in the context of historical pattern of spatio-temporal heterogeneity. *Environmental Management*, **14**, 629-646.

Poff, J.L., 1992: Regional hydrologic response to climate change: an ecological perspective. In: *Global Climate Change and Freshwater Ecosystems* [Firth, P. and S.G. Fisher (eds.)]. Springer-Verlag, New York, NY, pp. 88-115.

Poff, N.L. and J.D. Allan, 1995: Functional organization of stream fish assemblages in relation to hydrologic variability. *Ecology*, **76**, 606-627.

Pollman, C.D. and D.E. Canfield, Jr., 1991: Florida. In: *Acidic Deposition and Aquatic Ecosystems, Regional Case Studies* [Charles, D.F. (ed.)]. Springer-Verlag, New York, NY, pp. 367-416.

Prentice, I.C., 1988: 2. Records of vegetation in time and space: the principles of pollen analysis. In: *Vegetation History* [Huntley, B. and T. Webb III (eds.)]. Kluwer Academic Publishers, Dordrecht, The Netherlands, pp. 17-42.

Probst, J.L., 1989: Hydroclimatic fluctuations of some European rivers since 1800. In: *Historical Change of Large Alluvial Rivers: Western Europe* [Petts, G.E. (ed.)]. Wiley, Chichester, UK, pp. 41-55.

Rahel, F.J., C.J. Keleher, and J.L Anderson, 1996: Habitat loss and population fragmentation for coldwater fishes in the Rocky Mountain region in response to climate warming. *Limnology and Oceanography*, **41** (in press).

Rango, A., 1992: Worldwide testing of the Snowmelt Runoff Model with applications for predicting the effects for climate change. *Nordic Hydrol.*, **23**, 155-172.

Rao, A.R. and A. Al-Wagdany, 1995: Effects of climatic change in Wabash River basin. *ASCE J. Irrig. Drain. Eng.*, **121(2)**, 207-215.

Rasmusson, E.M., R.E. Dickinson, J.E. Kutzbach, and M.K. Cleaveland, 1993: Climatology. In: *Handbook of Hydrology* [Maidment, D.R. (ed.)]. McGraw-Hill, New York, NY, pp. 2.1-2.44.

Redmond, K.T. and R.W. Koch, 1991: Surface climate and streamflow variability in the western United States and their relation to large-scale circulation indices. *Water Resources Research*, **27(9)**, 2381-2399.

Regier, H.A., J.A. Holmes, and D. Panly, 1990: Influence of temperature changes on aquatic ecosystems: an interpretation of empirical data. *Transactions of the American Fisheries Society*, **119**, 374-389.

Reice, S.R., 1994: Nonequilibrium determinants of biological community structure. *American Scientist*, **82**, 424-435.

Reid, S.D., D.G. McDonald, and C.M. Wood, 1996: Interactive effects of temperature and pollutant stress. In: *Global Warming—Implications for Freshwater and Marine Fish*. SEB Seminar Series, Cambridge University Press, Cambridge, UK.

Reinert, R.E., L.J. Stone, and W.A. Willford, 1974: Effect of temperature on accumulaton of methylmecuric chloride and p.p DDT by rainbow trout (Salmo gairdneri). *Journal of the Fisheries Research Board of Canada*, **31**, 1649-1652.

Rempel, R.S. and J.C.H. Carter, 1986: An experimental study on the effect of elevated temperature on the heterotrophic and autotrophic food resources of aquatic insects in a forested stream. *Canadian Journal of Zoology*, **64**, 2457-2466.

Rempel, R.S. and J.C.H. Carter, 1987: Temperature influences on adult size, development, and reproductive potential of aquatic diptera. *Canadian Journal of Fisheries and Aquatic Science*, **44**, 1743-1752.

Resh, V.H., A.V. Brown, A.P. Covich, M.E. Gurtz, H.W. Li, G.W. Minshall, S.R. Reice, A.L. Sheldon, J.B. Wallace, and R. Wissmar, 1988: The role of disturbance in stream ecology. *Journal of the North American Benthological Society,* **7**, 433-455.

Revelle, R.R. and P.E. Waggoner, 1983: Effects of carbon dioxide-induced climatic change on water supplies in the western United States. In: *Changing Climate.* Nat. Acad. Sci., Nat. Acad. Press, Washington, DC, pp. 419-432.

Rind, D., C. Rosenzweig, and R. Goldberg, 1992: Modelling the hydrological cycle in assessments of climate change. *Nature,* **358**, 119-122.

Rind, D., 1995: Drying out the tropics. *New Scientist,* **146**, 36-40.

Roethlisberger, H. and H. Lang, 1987: Glacial hydrology. In: *Glacio-Fluvial Sediment Transfer* [Gurnell, A.M. and J.J. Clark (eds.)]. John Wiley & Sons, Chichester, UK, pp. 207-284.

Roff, J.C., 1970: Aspects of the reproductive biology of the planktonic copepod Limnocalanus macrurus sar, 1863. *Crustaceana,* **22**, 155-160.

Savenije, H.H.G., 1990: New definitions for moisture recycling and the relation with land-use changes in the Sahel. *Journal of Hydrology,* **167**, 57-78, Elsevier, Amsterdam, The Netherlands.

Savenije, H.H.G. and M.J. Hall, 1994: Climate and land use: a feedback mechanism? In: *Proceedings of the Delft Conference on Water and Environment: Key to Africa's Development,* 3-4 June, 1993, Delft, The Netherlands. IHE Report Series, Number 29, pp. 93-108.

Schertzer, W.M. and A.M. Sawchuk, 1990: Thermal structure of the Lower Great Lakes in a warm year: implications for the occurrence of hypolimnion anoxia. *Transactions of the American Fisheries Society,* **119**, 195-209.

Schindler, D.W., R.W. Newbury, K.G. Beaty, J. Prokopowich, T. Ruscznski, and J.A. Dalton, 1980: Effects of a windstorm and forest fire on chemical losses from forested watersheds and on the quality of receiving streams. *Canadian Journal of Fisheries and Aquatic Science,* **37**, 328-334.

Schindler, D.W., 1990: Natural and anthropogenically imposed limitations to biotic richness in freshwaters. In: *The Earth in Transition: Patterns and Processes of Biotic Impoverishment* [Woodwell, G. (eds.)]. Cambridge University Press, Cambridge, UK, pp. 425-462.

Schindler, D.W., K.G. Beaty, E.J. Fee, D.R. Cruikshank, E.R. DeBruyn, D.L. Findlay, G.A. Linsey, J.A. Shearer, M.P. Stainton, and M.A. Turner, 1990: Effects of climatic warming on lakes of the central boreal forest. *Science,* **250**, 967-970.

Schindler, D.W., 1996: Widespread effects of climatic warming on freshwater ecosystems. *Limnology and Oceanography,* **41** (in press).

Schindler, D.W., S.E. Bayley, B.R. Parker, K.G. Beaty, D.R. Cruikshank, E.J. Fee, E.U. Schindler, and M.P. Stainton, 1996: The effects of climatic warming on the properties of boreal lakes and streams at the Experimental Lakes Area, Northwestern Ontario. *Limnology and Oceanography,* **41** (in press).

Schlesinger, M.E. and Z.C. Zhao, 1988: Seasonal climate changes introduced by doubled CO_2 as simulated by the OSU atmospheric GCM/mixed-layer ocean model. *Journal of Climate,* **2**, 429-495.

Schmandt, J. and G.H. Ward, 1993: Climate change and water resources in Texas. In: *Proc. Conf. on Climate Change and Water Resources Management* [Ballentine, T.M. and E.Z. Stakhiv (eds.)]. U.S. Army Inst. Water Resour., Fort Belvoir, VA, pp. II-85 to II-101.

Schumm, S.A., 1977: *The Fluvial System.* John Wiley & Sons, New York, NY, 338 pp.

Scully, N.M. and D.R. Lean, 1994: The attenuation of ultraviolet radiation in temperate lakes. *Arch. Hydrobiology Beih. Ergebn. Limnol.,* **43**, 135-144.

Seaburg, K.G. and B.C. Parker, 1983: Seasonal differences in the temperature ranges of growth of Virginia algae. *J. Phycol.,* **19**, 380-386.

Sellers, W.D. and R.H. Hill, 1974: *Arizona Climate.* University of Arizona Press, Tucson, AZ, 616 pp.

Sene, K.J. and D.T. Pinston, 1994: A review and update of the hydrology of Lake Victoria, East Africa. *Hydrological Sciences Journal,* **39**, 47-63.

Serruya, C., 1978: *Lake Kinneret: Monographiae Biologicae.* W. Junk bv Publishers, The Hague, The Netherlands, 501 pp.

Shiklomanov, I.A., 1989: Climate and water resources. *Hydrological Sciences Journal,* **34(5)**, 495-529.

Short, R.A., E.H. Stanley, J.W. Harrison, and C.R. Epperson, 1987: Production of Corydalus cornutus (Megaloptera) in four streams differing in size, flow, and temperature. *J. North Amer. Benthol. Soc.,* **6**, 105-114.

Shuter, B.J. and J.R. Post, 1990: Climate, population viability and the zoogeography of temperature fishes. *Transactions of the American Fisheries Society,* **119**, 316-336.

Simpson, H.J., M.A. Cane, A.L. Herczeg, S.E. Zebiak, and J.H. Simpson, 1993: Annual river discharge in southeastern Australia related to El Nino-Southern Oscillation forecasts of sea surface temperatures. *Water Resources Research,* **29(11)**, 3671-3680.

Smock, L.A. and E. Gilinsky, 1992: Coastal blackwater streams. In: *Biodiversity of the Southeastern United States* [Hackney, C.T., S.M. Adams, and W.H. Martin (eds.)]. John Wiley and Sons, Inc., New York, NY, pp. 271-314.

Smol, J., 1988: Paleoclimate proxy data from freshwater arctic diatoma. *Verh. Internat. Verein. Limnol.,* **23**, 837-844.

Smol, J.P. and S.S. Dixit, 1990: Patterns of pH change inferred from chrysophue microfossils in Adirondack and northern New England lakes. *Journal of Palaeolimnology,* **4**, 31-41.

Soeder, C. and E. Stengel, 1974: Physico-chemical factors affecting metabolism and growth rate. In: *Algal Physiology Biochemistry* [Stewart, W.P.D. (ed.)]. University of Calif. Press, Berkeley, CA, pp. 714-730.

Sparks, R.E., P.B. Bayley, S.L. Kohler, and L.L. Osborne, 1990: Disturbance and recovery of large floodplain rivers. *Environ. Manage.,* **14**, 699-709.

Squires, L.E., S.R. Rushforth, and J.D. Brotherson, 1979: Algal response to a thermal effluent: study of a power station on the Provo River, Utah, USA. *Hydrobiology,* **63**, 17-32.

Stanley, E.H. and S.G. Fisher, 1992: Intermittency, disturbance, and stability in stream ecosystems. In: *Aquatic Ecosystems in Semiarid Regions: Implications for Resource Management* [Roberts, R.D. and M.L. Bothwell (eds.)]. NHRI Symposium Series 7, Environment Canada, Saskatoon, Saskatchewan, Canada, pp. 271-281.

Stefan, H.G. and B.A. Sinokrot, 1993: Projected global climate change impact on water temperatures in five north-central U.S. streams. *Climatic Change,* **24**, 353-381.

Stefan, H.G. and E.B. Preud'homme, 1993: Stream temperature estimation from air temperature. *Water Resources Bulletin,* **29**, 27-45.

Stefan, H.G., M. Hondzo, and X. Fang, 1993: Lake water quality modeling for projected future climate scenarios. *J. Environ. Qual.,* **22**, 417-431.

Stefan, H.G. and X. Fang, 1994: Model simulations of dissolved oxygen characteristics of Minnesota lakes: past and future. *Environmental Management,* **18(1)**, 73-92.

Steinman, A.D. and C.D. McIntire, 1990: Recovery of lotic periphyton communities after disturbance. *Environ. Manage.,* **14**, 589-604.

Stemberger, R., 1996: Pleistocene refuge areas and postglacial dispersal of copepods of the northeastern United States. *Limnology and Oceanography,* **41** (in press).

Stine, S., 1990: Late Holocene fluctuations of Mono Lake, eastern California. *Palaeogeography, Palaeoclimatology, and Palaeoecology,* **78**, 333-381.

Stirling, G. and D.J. McQueen, 1986: The influence of changing temperature on the life history of Daphniopsis ephemeralis. *Journal of Plankton Research,* **8**, 583-595.

Stoermer, E.F., J.A. Wolin, C.L. Schelske, and D.J. Conley, 1985: An assessment of ecological changes during the recent history of Lake Ontario based on siliceous algal microfossils preserved in the sediments. *J. Phycol.,* **21**, 257-276.

Sullivan, B.K. and L.T. McManus, 1986: Factors controlling seasonal succession of the copepods Acartia hudsonica and A. tonsa in Narragansett Bay, Rhode Island: temperature and resting egg production. *Mar. Ecol. Prog. Ser.,* **28**, 121-128.

Swanson, F.J., T.K. Kratz, N. Caine, and R.G. Woodmansee, 1988: Landform effects on ecosystem patterns and processes. *BioScience,* **38**, 92-98.

Sweeney, B.W., 1984: Factors influencing life history patterns of aquatic insects. In: *Ecology of Aquatic Insects* [Resh, V.H. and D. Rosenberg (eds.)]. Praeger Scientific, New York, NY, pp. 56-100.

Sweeney, B.W. and R.L. Vannote, 1986: Growth and production of a stream stonefly: influences of diet and temperature. *Ecology,* **67**, 1396-1410.

Sweeney, B.W., R.L. Vannote, and P.J. Dodds, 1986: The relative importance of temperature and diet to larval development and adult size of the winter stonefly, Soyedina carolinensis (Plecoptera: Nemoouridae). *Freshwater Biology,* **16**, 39-48.

Sweeney, B.W., J.K. Jackson, J.D. Newbold, and D.H. Funk, 1992: Climate change and the life histories and biogeography of aquatic insects in eastern North America. In: *Global Climate Change and Freshwater Ecosystems* [Firth, P. and S.G. Fisher (eds.)]. Springer-Verlag, New York, NY, pp. 143-176.

Thorp, J.H. and M.D. Delong, 1994: The riverine productivity model: an heuristic view of carbon sources and organic processing in large river ecosystems. *Oikos*, **70**, 305-308.

Tilman, D., R. Kiesling, R. Sterner, S.S. Kilham, and F.A. Johnson, 1986: Green, bluegreen and diatom algae: taxonomic diferences in competitive ability for phosphorous, silicon and nitrogen. *Arch. Hydrobiology*, **106**, 473-485.

Tonn, W.M. and J.J. Magnuson, 1982: Patterns in the species composition and richness of fish assemblages in northern Wisconsin lakes. *Ecology*, **63**, 1149-1166.

Tonn, W.M., 1990: Climate change and fish communities: a conceptual framework. *Transactions of the American Fisheries Society*, **119**, 337-352.

Tonn, W.M., J.J. Magnuson, M. Rask, and J. Toivonen, 1990: Intercontinental comparison of small-lake fish assemblages: the balance between local and regional processes. *American Naturalist*, **136**, 345-375.

Tung, C.-P. and D.A. Haith, 1995: Global-warming effects on New York streamflows. *ASCE J. Water Resour. Plan. Manage.*, **121(2)**, 216-225.

Vaccaro, J.J., 1992: Sensitivity of groundwater recharge estimates to climate variability and change, Columbia Plateau, Washington. *Journal of Geophysical Research*, **97(D3)**, 2821-2833.

Valdes, J.B., R.S. Seoane, and G.R. North, 1994: A methodology for the evaluation of global warming impact on soil moisture and runoff. *Journal of Hydrology*, **161**, 389-413.

Vannote, R.L. and B.W. Sweeney, 1980: Geographic analysis of thermal equilibria: a conceptual model for evaluating the effect of natural and modified thermal regimes on aquatic insect communities. *American Naturalist*, **115**, 667-695.

Vannote, R.L., G.W. Minshall, K.W. Cummins, J.R. Sedell, and C.E. Cushing, 1980: The river continuum concept. *Canadian Journal of Fisheries and Aquatic Science*, **37**, 130-137.

Vehviläinen, B. and J. Lohvansuu, 1991: The effects of climate change on discharges and snow cover in Finland. *Hydrological Sciences Journal*, **36(2)**, 109-121.

Vidal, J., 1980: Physioecology of zooplankton. 1. Effects of phytoplankton concentration, temperature, and body size on the growth rate of Calanus pacificus and Pseudocalamus sp. *Mar. Biol.*, **56**, 111-134.

Vinebrooke, R.D. and P.R. Leavitt, 1996: Effects of UV radiation on periphyton in an alpine lake. *Limnology and Oceanography*, **41** (in press).

Viner, D. and M. Hulme, 1994: *The Climate Impacts of LINK Project: Providing Climate Change Scenarios for Impacts Assessment in the UK.* Climatic Research Unit, University of East Anglia, UK, 24 pp.

Von Storch, H., E. Zorita, and U. Cubasch, 1993: Downscaling of global climate change estimates to regional scales: an application to Iberian rainfall in wintertime. *Journal of Climate*, **6**, 1161-1171.

Wainwright, S.C., C.A. Couch, and J.L. Meyer, 1992: Fluxes of bacteria and organic matter into a blackwater river from river sediments and floodplain soils. *Freshwater Biology*, **28**, 37-48.

Ward, J.V. and J.A. Stanford, 1983: The serial discontinuity concept of lotic ecosystems. In: *Dynamics of Lotic Ecosystems* [Fontaine, T.D. and S.M. Bartel (eds.)]. Ann Arbor Science Publ., Ann Arbor, MI, pp. 29-42.

Ward, A.K., G.M. Ward, J. Harlin, and R. Donahue, 1992: Geologic mediation of stream flow and sediment and solute loading to stream ecosystems due to climate change. In: *Global Climate Change and Freshwater Ecosystems* [Firth, P. and S.G. Fisher (eds.)]. Springer-Verlag, New York, NY, pp. 116-142.

Webb, B.W., 1992: *Climate Change and the Thermal Regime of Rivers.* Report to the UK Department of the Environment. University of Exeter, Exeter, UK, 79 pp.

Webb, T. III, P.J. Barlein, S. Harrison, and K.H. Anderson, 1994: Vegetation, lake levels and climate in the eastern United States since 18,000 yr. B.P. In: *Global Climates Since the Last Glacial Maximum* [Wright, Jr., H.E., J.E. Kutzbach, T. Webb III, W.F. Ruddiman, F.A. Street-Perrott, and P.J. Bartlein (eds.)]. University of Minnesota Press, Minneapolis, MN, pp. 514-535.

Webster, K.E., T.K. Kratz, C.J. Bowser, J.J. Magnuson, and W.J. Rose, 1996: The influence of landscape position on lake chemical responses to drought in northern Wisconsin, USA. *Limnology and Oceanography*, **41** (in press).

Wehr, J.D., 1981: Analysis of seasonal succession of attached algae in a mountain stream, the North Alouette River. *Can. J. Bot.*, **59**, 1465-1474.

Weijers, E.P. and P. Vellinga, 1995: *Climate Change and River Flooding: Changes in Rainfall Processes and Flooding Regimes Due to an Enhanced Greenhouse Effect.* Instituut voor Milieuvraagstukken, Vrije Universiteit, Amsterdam, The Netherlands, R-95/01, 42 pp.

Welcomme, R.L., 1979: *The Fisheries Ecology of Floodplain Rivers.* Longman, London, UK, 317 pp.

Wetzel, R.G., 1990: Land-water interfaces: metabolic and limnological regulators. *Verh. Internat. Verein. Limnol.*, **24**, 6-24.

Wharton, R.A., Jr., C.P. McKay, G.D. Clow, D.T. Andersen, G.M. Simmons Jr., and F.G. Love, 1992: Changes in ice cover thickness and lake level of Lake Hoare, Antarctica: implications for local climate change. *Journal of Geophysical Research*, **97**, 3503-3513.

Whetton, P.H., A.M. Fowler, M.R. Haylock, and A.B. Pittock, 1993: Implications of climate change due to enhanced greenhouse effect on floods and droughts in Australia. *Climate Change*, **25**, 289-317.

Wilkinson, W.B. and D.M. Cooper, 1993: The response of idealized aquifer/river systems to climate change. *Hydrological Sciences Journal*, **38(5)**, 379-390.

Wilks, D.S., 1992: Adapting stochastic weather generation algorithms for climate change studies. *Climate Change*, **22**, 67-84.

Williams, D.D. and H.B.N. Hynes, 1977: The ecology of temporary streams, II: general remarks on temporary streams. *Internat. Revue der Gesamten Hydrobiology*, **62**, 53-61.

Williamson, C.E., 1995: What role does UV-B radiation play in freshwater ecosystems? *Limnology and Oceanography*, **40**, 386-392.

Wolford, R.A. and R.C. Bales, 1996: Hydrochemical modeling of Emerald Lake Watershed, Sierra Nevada, California: sensitivity of stream chemistry to changes in fluxes and model parameters. *Limnology and Oceanography*, **41** (in press).

Wolock, D.M. and G.M. Hornberger, 1991: Hydrological effects of changes in levels of atmospheric carbon dioxide. *J. Forecasting*, **10**, 105-116.

Wolock, D.M., G.J. McCabe, Jr., G.D. Tasker, and M.E. Moss, 1993: Effects of climate change on water resources in the Delaware River Basin. *Water Resources Bulletin*, **29(3)**, 475-486.

Woo, M.K. and S.B. McCann, 1994: Climatic variability, climate change, runoff and suspended sediment regimes in northern Canada. *Phys. Geogr.*, **15**, 201-226.

Wood, E.F., D.P. Lettenmaier, and V.G. Zartarian, 1992: A land surface hydrology parameterization with subgrid variability for general circulation models. *Journal of Geophysical Research*, **97(D3)**, 2717-2728.

Wood, C.M. and D.G. McDonald (eds.), 1996: *Global Warming—Implications for Freshwater and Marine Fish.* SEB Seminar Series, Cambridge University Press, Cambridge, UK.

Woodward, I.O. and R.W.G. White, 1983: Effects of temperature and food on instar development rates of Boeckella symmetrica Sars (Copepoda: Calanoida). *Aust. J. Mar. Freshw. Res.*, **34**, 927-932.

Zhang, S.F. *et al.*, 1994: The impact of climate change on water resources and response strategy in the area of Jing-Jin-Tang. *Advances in Water Sciences*, **7(1)**, 16-27 (in Chinese).

11

Industry, Energy, and Transportation: Impacts and Adaptation

ROBERTO ACOSTA MORENO, CUBA; JIM SKEA, UK

Principal Lead Authors:
A. Gacuhi, Kenya; D.L. Greene, USA; W. Moomaw, USA; T. Okita, Japan;
A. Riedacker, France; Tran Viet Lien, Vietnam

Lead Authors:
R. Ball, USA; W.S. Breed, USA; E. Hillsman, USA

CONTENTS

EXECUTIVE SUMMARY

Introduction

Climate change will have *direct* impacts on economic activity in the industry, energy, and transportation sectors; impacts on *markets for goods and services;* and impacts on the *natural resources* on which economic activity depends. Activities directly sensitive to climate include construction, transportation, offshore oil and gas production, manufacturing dependent on water, tourism and recreation, and industry that is located in coastal zones and permafrost regions. Activities with markets sensitive to climate include electricity and fossil fuel production for space heating and air conditioning, construction activity associated with coastal defenses, and transportation. Activities dependent on climate-sensitive resources include agroindustries (food/drink, forestry-related activity, and textiles), biomass production, and other renewable energy production.

The overall effect of climate change in the industry, energy, and transportation sectors will be the aggregation of a large number of varied individual impacts. The interconnectedness of economic activity means that many of the impacts are indirect and will be transmitted by transactions within and between economic sectors. Energy, water, and agricultural products in particular will transmit climate sensitivity through the economic system.

Findings

There is generally a high level of confidence concerning the sensitivities of specific activities to given changes in given climate variables. Reliable climate scenarios describing climate variables, apart from temperature, do not exist at the regional level, however. There is, therefore, a low level of confidence about the direction as well as the magnitude of some climate impacts when uncertainties about climate change at the regional level are taken into account.

Certain components of the industry/energy/transportation sectors display a greater degree of sensitivity to climate than does the sector as a whole.

- There is a high level of confidence that agroindustries that depend on products such as grain, sugar, and rubber are vulnerable to changes in precipitation patterns and the frequency and intensity of extreme weather events. Agroindustry is of relatively greater significance in many developing countries where, together with agriculture, it constitutes the bulk of economic activity. In areas with low agricultural productivity and high population densities, climate change could have a significant effect on the production of biomass (i.e., living matter that can be burned to meet energy needs).
- There is a high level of confidence that hydroelectric production will be influenced by changes in precipitation and water availability. There is low confidence concerning whether these changes will be beneficial or otherwise because much depends on the relationship between seasonal patterns of precipitation and electricity demand in specific regions.
- It is certain that higher temperatures resulting from climate change will reduce energy needs for space heating and increase those for air conditioning. There is a low level of confidence concerning the balance between these changes. For example, one study of the United States concluded that energy needs could fall by 11% with a 1°C rise in temperature. Another concluded that electricity demand would increase by 4–6% with a rise of 3.7°C.
- There is a high level of confidence that sea-level rise will increase the cost of protecting transportation infrastructure and industrial plants located in coastal regions. Coastal protection will provide a market opportunity for the construction industry.
- There is a high level of confidence that climate change will increase the vulnerability of infrastructures located in permafrost regions.
- There is a high level of confidence that skiing seasons will shorten with consequent impacts on some local economies. There is also a high level of confidence that sea-level rise will affect tourism in beach resort areas.

Although the energy, industry, and transportation sectors are of great economic importance, the climate sensitivity of most activities is low relative to that of agriculture and natural ecosystems, while the capacity for autonomous adaptation is high, as long as climate change takes place gradually. The lifetimes of most assets are short compared to projected time scales for climate change. Consumer goods, motor vehicles, and heating and cooling systems will be replaced several times over the next half century. Even medium-life assets such as industrial plants, oil and gas pipelines, and conventional power stations are likely to be completely replaced, though there will be less opportunity for adaptation. Additional difficulties and costs could arise with long-lived assets, including some renewable energy projects and residential buildings.

The technological capacity to adapt to climate change will be realized only if the necessary information is available and the institutional and financial capacity to manage change exists. Autonomous adaptation cannot be relied upon, and governments may have to set a suitable policy framework, disseminate information about climate change, and act directly in relation to vulnerable infrastructures. Many developing countries are dependent on single crops or on fishing and therefore are economically vulnerable to climate change through impacts on agroindustry. Diversifying economic activity could be an important precautionary response that would facilitate successful adaptation.

Context

In comparison with natural ecosystems and agriculture, little work has been carried out on impacts and adaptation in the industry, energy, and transportation sectors. The literature on adaptation and adaptation policies is particularly sparse. This partly reflects the perception that climate sensitivity is low and that adaptation could take place autonomously. Much of the attention of the policy and research communities has been focused on mitigation actions in the energy and transportation sectors.

Research on climate impacts has focused largely on developed rather than developing countries, even though the latter have less diversified economies that may be more vulnerable to climate change. Within the developed countries, work has focused on a small number of impacts, notably the possible effects of climate change on energy demand. This may reflect the availability of research tools as much as priority-setting.

Further research would be assisted by the generation of climate scenarios covering a fuller range of variables at the regional level. There is a need to broaden the scope of work to cover a wider range of activities (for example, agroindustry) and a wider range of countries, especially developing countries. Studies that begin to draw out the interdependence of economic activities in relation to climate would be particularly helpful.

11.1. Introduction

This chapter reviews literature addressing the potential impacts of climate change on the energy, industry, and transportation sectors and the associated capacity for adaptation. The chapter begins by considering broad issues relating to impacts and adaptation in the three sectors. It moves on to a more precise definition of the sectors, identifying their economic importance and the more important climate sensitivities. There follows a general overview of the different types of studies that have been conducted and the range of methodologies that have been used. The largest part of the chapter is devoted to a detailed review of this literature. For this purpose, activity is classified according to three types of climate sensitivity: activities with markets that are sensitive to climate, activities and processes that may be directly affected by climate change, and activities that are dependent on climate-sensitive resources.

A wide range of individuals and organizations will be affected directly or indirectly by climate change in the energy, industry, and transportation sectors. These include:

- Individuals in their role as consumers or citizens
- Businesses, operating over a wide range of scale and technological sophistication, which may be privately owned or run as state enterprises
- Policymakers concerned with land-use planning and the development of transportation, energy, or industrial infrastructures that may be affected by climate change
- Policymakers making high-level decisions about the adequacy of policies to deal with climate change and the appropriate balance between adaptation and mitigation strategies.

Because global warming and other climatic changes will result from the impact of past economic activity, the first three groups will be affected by climate change independently of current and future mitigation actions. This chapter is aimed at the needs of these groups. The needs of policymakers who are responsible for balancing mitigation and adaptation strategies are addressed, especially in Chapter 6, *The Social Costs of Climate Change: Greenhouse Damage and the Benefits of Control*, and Chapter 7, *A Generic Assessment of Response Options*, in the IPCC Working Group III volume; Nordhaus (1993) also addresses these issues.

11.1.1. Adapting to Climate Change

Adaptation to a changed climate may occur through the actions of individuals and enterprises, or it may be stimulated by policies promoted by those concerned with planning and infrastructure development. "Autonomous adaptation," i.e., action taken by individuals and enterprises on their own initiative, may reduce the need for explicit adaptation policies.

The technological capacity to adapt to climate change depends partly on the rapidity of climate change and the rate of replacement of equipment and infrastructure. On the whole, the lifetimes of assets in the energy, industry, and transportation sectors are short compared to the timescales for change projected using climate models. Table 11-1, based on Skea (1995), shows that, over the 60–70 years during which CO_2 concentrations in the atmosphere might double, many short-lived assets such as consumer goods, motor vehicles, and space heating/cooling systems will be replaced several times, offering considerable opportunities for adaptation. Even medium-life assets such as industrial plants, oil and gas pipelines, and conventional power stations are likely to be completely replaced over such a timescale, though there will be less opportunity for adaptation.

More difficulties could arise with long-lived assets such as certain residential buildings and infrastructure. Some assets (for example, a dam or a tidal barrage) have a design life of more than a century. Long-lived assets may need to function in a climate for which they were not designed. Infrastructure may be modified during its lifetime, however. For example, roads, port facilities, or coastal protection construction may be rebuilt periodically while retaining the same location and basic function. Rebuilding or upgrading offers substantial opportunities to adapt to changing climate conditions. Even medium-life assets such as power stations may be modified substantially during their lifetime—for example, by changing the fuel inputs, improving efficiency, or adding pollution control equipment.

The technological capacity to adapt to climate change will be realized only if the necessary information is available; enterprises and organizations have the institutional and financial capacity to manage change, and there is an appropriate framework within which to operate. In this respect, autonomous adaptation cannot necessarily be relied upon. Governments

Table 11-1. *Examples of asset lifetimes in the energy, industry, and transportation sectors (based on Skea, 1995).*

Short Life (up to 15 years)	
Conventional light bulb	up to 3 years
Cooking stoves in developing countries	2–3 years
Consumer goods	5–10 years
Motor vehicles	10–15 years
Space heating boilers/air conditioning systems	up to 15 years
Fuel supply contracts	up to 15 years
Medium Life (15–50 years)	
Industrial plant	10–30 years
Renewable energy projects (e.g., solar energy)	10–30 years
Commercial/residential buildings	20–30 years
Conventional power plant	30–50 years
Long Life (>50 years)	
Older residential buildings	50 years plus
Infrastructure (roads, railways, port facilities)	50–100 years or longer
Tidal barrage	120 years

may have a role in terms of disseminating information about climate change, setting policy and regulatory frameworks for individual actions, and acting directly to protect vulnerable infrastructure.

Rapid sea-level rise or changes in climate (Horgan, 1993) would limit the scope for autonomous adaptation, put considerable strain on social and economic systems, and increase the need for explicit adaptation policies.

A good indicator of the capacity for climate adaptation is the importance of the climate signal in relation to other pressures for change in the sector concerned. How do the pressures of climate change compare with those of changing demographics, market conditions, technological innovation, or resource depletion? Over periods of half a century or more, many sectors will change beyond recognition, while others may disappear completely. New products, markets, and technologies will also emerge. An industry coping with other, more significant changes may be able to adapt easily to climate change.

Different regions of the world vary greatly in their capacity to adapt. More vulnerable regions include those with less access to new technology, those that rely heavily on single sources of energy, and those dependent on single crops or on fishing. It has been argued (National Academy of Sciences, 1992) that poverty makes people more vulnerable to change and reduces their flexibility to respond. Both individuals and institutions in developing countries are likely to have less capacity to adapt to a changed climate. If climate change takes place gradually and if the potential impacts of climate change are allowed for in the development process, then the prospects for successful adaptation will be enhanced. Diversifying economic activity could be an important precautionary response that would facilitate successful adaptation to climate change.

Individuals and activities can adapt to climate change by migrating towards other climate zones, even if this involves crossing national frontiers. The migration of economic activity could be regarded as a successful adaptation to climate change. Policymakers at the national level, however, will be concerned about the balance of economic activity and autonomous adaptation actions that affect this activity.

11.1.2. Adaptation and Mitigation

Although the balance between adaptation and mitigation strategies is not the focus of this chapter, mitigation is a key concern in the energy sector. Energy plays a pivotal role in the assessment of climate change since fossil fuel combustion is one of the most important sources of greenhouse gas emissions. The energy sector is therefore likely to be one of the main targets for policies and measures aimed at mitigating climate change. Climate change, however, will also modify patterns of energy demand and industrial activity and, hence, greenhouse gas (GHG) emissions. As a result, the amount of mitigation action required to reach any given target for either emissions or concentrations of GHGs will be altered. At the same time, mitigation actions will themselves modify the impacts of climate on the energy sector—for example, by stimulating switches to more or less climate-sensitive energy resources.

A fully consistent treatment of the industry, energy, and transportation sectors would require an "integrated" approach to the assessment of impacts/adaptation and mitigation. Such approaches are now beginning to be developed (for example, Dowlatabadi and Morgan, 1993; Cohan *et al.*, 1994), but few studies take account of feedbacks *within* the energy sector. For the most part this chapter, of necessity, ignores interactions between impacts/adaptation and mitigation actions.

11.2. Characteristics and Sensitivities of the Sectors

Energy, industry, and transportation together cover a wide range of economic activity. Using the international standard industrial classification (ISIC) of all economic activities (United Nations, 1990) as the basis for definition, the sectors cover: manufacturing; mining and quarrying; electricity and gas; construction; transport, storage, and communications; and tourism and recreation.

Among the activities covered in this chapter is agroindustry—the processing of agricultural products into forms that make them suitable for meeting final consumer demands. Food, textiles, and paper are examples of agroindustry. Agriculture, forestry, and fishing, however, are not included. Most of the service sector is not sensitive to climate change, and there are few references in the literature. Notable exceptions are tourism and recreation, which are covered in this chapter, and insurance, which is covered in Chapter 17.

The "energy sector" is spread across a range of ISIC activities and includes: coal, oil, and natural gas production; coke manufacture, refined petroleum products, and nuclear fuel; and electricity generation, electricity and gas transmission, and distribution. Biomass fuels and renewable energy are also covered. Transportation is a major component of the world economy, accounting for 10–20% of gross domestic product (GDP) in most countries. Transportation is growing rapidly; it now accounts for approximately one-fifth of CO_2 emissions from fossil fuel use (Lashof and Tirpak, 1990).

The sensitivity of industry and energy to climate change is widely believed to be low in relation to that of natural ecosystems and agriculture, while adaptability is high (National Academy of Sciences, 1992; Nordhaus, 1993). Different branches of industry, however, vary considerably in their climate sensitivity. Broadly speaking, sectors that are high up the manufacturing chain or are directly dependent on primary resources (agroindustries, renewable energy) display greater sensitivity than do sectors further down the manufacturing chain—for example, engineering. The interactions and interdependence of various activities and systems have been emphasized by some (National Academy of Sciences, 1992; footnote), particularly in relation to adaptive

Table 11-2: Weight of different sectors in industrial production in 1991 (%), based on UNIDO (1993) and FAO (1993).

	World	Developed Countries	Developing Countries	E Europe/ former USSR	W Europe	North America	Central and S America	Africa	Asia	Oceania
Energy and Utilities	19.0	15.1	35.5	12.3	15.7	16.7	21.2	N/A	45.0	20.0
- Coal mining	1.5	1.4	0.7	2.9	1.6	1.6	0.2	N/A	1.2	5.0
- Oil and gas production	8.0	3.9	25.2	1.8	4.2	5.9	8.0	N/A	32.6	2.5
- Petroleum/coal products	2.2	1.3	3.5	3.3	1.2	1.4	6.4	N/A	5.5	0.6
- Electricity, gas, and water	7.3	8.5	6.2	4.3	8.7	7.9	6.7	N/A	5.6	11.9
Agro-Industries	20.2	18.0	22.0	26.9	20.5	17.8	31.8	N/A	16.1	18.1
- Food, beverages, and tobacco	11.1	8.9	13.7	17.0	11.1	7.5	21.8	N/A	8.5	11.0
- Textiles	3.7	2.6	4.5	6.6	3.2	2.2	4.2	N/A	4.9	1.8
- Wood products/furniture	3.0	3.4	1.9	2.5	3.4	4.3	2.4	N/A	1.6	3.5
- Pulp and paper	2.5	3.1	1.9	0.8	2.8	3.8	3.5	N/A	1.1	1.8
Other Industry	60.8	66.9	42.5	60.8	63.8	65.5	46.9	N/A	38.9	61.8
- Mining	2.6	2.3	4.2	1.6	0.9	2.2	3.8	N/A	3.3	24.5
- Chemicals	10.3	10.8	11.7	7.5	11.8	10.7	14.6	N/A	7.5	6.2
- Other	47.9	53.8	26.6	51.7	51.1	52.6	28.5	N/A	28.2	31.1
Total	100.0	100.0	100.0	100.0	100.0	100.0	100.0	100.0	100.0	100.0
% Employed in Agriculture	37.1	5.1	53.6	12.9	6.3	2.2	25.0	61.9	56.5	15.8

Note: N/A = Data not available in original statistical source; figures may not sum exactly due to rounding.

actions. Little work has been carried out that explores these interactions and interdependencies, though some preliminary studies have been carried out (Scheraga *et al.*, 1993).

The magnitude of climate impacts is a function of both the severity of the impact and the importance of the affected sector (National Academy of Sciences, 1992). Table 11-2 shows both the contribution of industry to total economic output in different regions of the world and the contribution of different sectors to industrial output (UNIDO, 1993). Broadly speaking, developing countries are less reliant on industry than are developed countries, but climate-sensitive industrial sectors such as energy and food make a relatively greater contribution to economic output in developing countries. The greater importance is reflected even more strongly in terms of employment (UNIDO, 1993). The number of people working in the food industry in developing market economies rose from approximately 10.5 million in 1970 to 17 million in 1990. Between 1977 and 1989, the rate of growth in developing countries was 5.3% per year as opposed to 2.0% per year in industrialized countries (Table 11-3).

The diverse range of activities covered in this chapter is affected by a correspondingly wide range of climate variables (Table 11-4). Temperature changes over daily and seasonal cycles are of key importance, but other relevant variables include precipitation, humidity, ground moisture, insolation, wind speed, storminess, and sea-level rise. Ideally, impacts and adaptation should be reviewed in the context of climate scenarios covering a wide range of variables at the regional level. The uncertainties in present computer simulations of regional climate change, however, are too large to yield a high level of confidence. This, coupled with the sparse use of climate scenario frameworks in most impacts and adaptation work in the industry, energy, and transportation sectors, means that much of this chapter must be restricted to reviewing climate sensitivities. Recent climate scenarios can be used as benchmarks against which to judge reported sensitivities or the results of impact studies based on different climate assumptions. Many existing impact studies are based on assumptions about benchmark CO_2 doubling that have a range of implications for specific climate

Table 11-3: *Economic importance of production and employment in the food industry, in % (UN Statistical Office, 1993).*

Type of Economy	Weight in Industrial Employment (1990)	Annual Growth Rate of Employment (1977–89)	Annual Growth Rate of Production (1977–89)
World	15.5	1.5	2.6
Developed	9.5	-0.4	2.0
Developing	27.1	2.9	5.3

variables, including temperature and precipitation. The following key points highlight differences between current best estimates of climate change and assumptions that have often been made in the climate impacts literature:

- Globally averaged temperatures, which would, for example, influence energy demand, are projected to rise by 0.15–0.25°C per decade. Many impact studies, however, typically assume a 4.5°C increase by the middle of next century and need to be reviewed with more modest temperature increases in mind. Regional temperature changes may be greater or less than the mean.
- Water availability and agroindustry could be affected by precipitation patterns. Precipitation is likely to increase throughout the year at high northern latitudes. In mid-latitudes, winters could be wetter, while summers could be drier. Precipitation during the Asian monsoon season is likely to increase.
- The frequency of extreme weather events has implications for vulnerable infrastructure in the energy, industry, and transportation sectors. There is little agreement on how storminess might change in a warmer world. There is growing evidence, however, that heavy rain events will become more frequent.
- Sea-level rise could affect coastally located infrastructure. The current estimate of average global sea-level rise is 25–70 cm for the year 2100, which is lower than the range presented by IPCC in 1990 and lower than the range assumed in many impact studies. Sea-level rise at the regional level could be as much as twice or as little as half the global average.

11.3. Overview of the Literature

The sensitivity of the industry, energy, and transportation sectors to climate change has received less attention than has the sensitivity of natural ecosystems or agriculture. This reflects the perception that climate sensitivity is relatively low in the industry and energy sectors (National Academy of Sciences, 1992) and the attention focused on the mitigation of climate change through the reduction of GHG emissions (see Chapters 19–22).

11.3.1. Methods

The studies reviewed in this chapter address:

- The sensitivity of specific activities to weather conditions or sea level
- The potential impacts of climate change on a given sector or activity
- The impacts of climate change on a particular country or region
- The broader economic impacts of climate change, typically by integrating the results of work on several sectors into a single macroeconomic model.

While considerable progress has been made in developing protocols for research on climate impacts (Carter *et al.*, 1994), most studies do not conform to these standards. The methods used in the studies vary considerably in scope and sophistication. The studies do not yet provide a coherent picture of climate impacts in the industry, energy, or transportation sectors.

The studies vary greatly in their use of climate and socioeconomic scenarios. Sensitivity studies make no use of scenarios but simply identify the effect, or the degree of autonomous adaptation, that a given climate stimulus would induce. Some studies have used an historic "analog" of anticipated climate change (Rosenberg and Crosson, 1991), while others (Smith and Tirpak, 1989) have used the results of scenarios generated by climate models. Many studies that use consistent climate scenarios estimate the effects of climate change on *existing* technology and patterns of economic activity. Such exercises have a heuristic value but cannot be taken as a prediction of the actual impacts of climate change because they do not take adequate account of the impacts of technological and economic changes unrelated to climate and the possibility of autonomous adaptation over timescales of decades.

Only the most sophisticated studies use internally consistent scenarios for socioeconomic change as well as climate scenarios. The industry and energy sectors, however, will change significantly over the next half century. The process of economic development will result in changes in the scale, composition, and location of industrial activity. Technological change, changes in resource availability, and environmental constraints likewise will affect activity patterns. The great uncertainties attached to future patterns of activity are reflected in the range of socioeconomic scenarios that have been developed to describe, for example, future energy demand and supply (IPCC, 1992; World Energy Council, 1993a). The failure to take account of socioeconomic change is one of the greatest weaknesses of the impacts/adaptation literature relating to energy, industry, and transportation.

11.3.2. Biases in the Literature

Research is biased toward specific world regions and sectors. Much of the literature is concerned with impacts in developed

Table 11-4: Summary of climate sensitivities in the industry, energy, and transportation sectors.

	Temperature	Precipitation	Windiness	Frequency of Extreme Events	Water Availability	Sea-Level Rise	Other
Agroindustry and Biomass	Impact on inputs from agriculture	Impact on inputs from agriculture		Damage to crops, trees, industrial infrastructure	Impact on inputs (e.g., irrigation)	Impacts on coastally sited fish processing	Main impacts via inputs from agriculture
Renewables	More evaporation from reservoirs	Hydroelectric potential	Wave potential, reservoir evaporation, wind potential	Many renewable systems vulnerable, especially wind turbines, solar systems	Hydroelectric potential	Design of tidal, wave systems	Cloudiness affects solar potential
Energy Extraction	Impact of reduced ice on offshore Arctic operations			Offshore oil and gas		Offshore oil and gas	
Energy Demand	More space heating, less air conditioning		Space heating				More humidity, more air conditioning
Energy Conversion	Slightly less efficient thermal generation				Cooling water availability	Coastal power stations, refineries	
Energy Transport/ Transmission	Pipelines over permafrost vulnerable; lower capacity of power lines	Icing of power lines		Effects on power lines			
Transportation Infrastructure	Vulnerability if permafrost melts; changed freeze-thaw cycles on roads			Effects on roads, railways, bridges		Effects on coastal infrastructure; migration of coastal activity	Changes in movements of agricultural products; settlement patterns
Transportation Operations	Ice and coastal shipping in high latitudes; road maintenance costs; air conditioning in cars	Impact of snow and ice on road and air transport		Safety and reliability of operations (e.g., airports)	Inland navigation		Effects of fog, snow, rain, and ice on operations and safety

Table 11-4 continued

	Temperature	Precipitation	Windiness	Frequency of Extreme Events	Water Availability	Sea-Level Rise	Other
Tourism and Recreation	Shorter skiing seasons	Skiing season		Impact on attraction of mountainous and coastal regions		Impact on beach resorts and marinas	Many climate variables will affect demand for and location of facilities
Construction	Building design	Changed productivity of construction activity (snow, rain)		Impact on construction activity and building design		Larger markets through coastal zone management	
Manufacturing	More demand for air-conditioning equipment; markets for clothing, beverages				Availability of process water for heavy users; industries heavily dependent on hydropower		
Pollution Control	More ozone formation and control needs		Pollution dispersion		Water quality and discharges		

Note: This table identifies impacts and their degree of significance; the direction of impacts and uncertainties are discussed in the text.
Key: Dark grey boxes indicate a significant impact requiring adaptive response at a strategic level; light grey indicates modest impacts requiring adaptive response; and no shading indicates minor impacts.

countries, especially the United States, which was the first country to produce a comprehensive climate effects report (Smith and Tirpak, 1989). Other developed countries have since published comprehensive reviews (UK Climate Change Impacts Review Group, 1991; Nishioka *et al.*, 1993), and some work has been carried out in developing countries (Nguyen *et al.*, 1993). Recent impact studies covering transportation refer only to developed countries such as Canada, New Zealand, and the United States (Nishioka *et al.*, 1992). Such studies generally analyze only the direct impacts of climate on infrastructure and operations. Climate-induced changes in flows of freight and passengers that would affect infrastructure demands have been acknowledged but not quantified.

There are also biases in terms of the topics covered in the literature. The potential impact of climate change on demand for electric power and the consequent impact on investment in new power capacity accounts for a significant proportion of the energy-related literature (e.g., IPCC, 1990b, 1993). The ready availability of information relating electricity demand to weather conditions enables work of some sophistication to be carried out. The topic, however, may not be as important as the volume of literature would suggest.

11.3.3. Framework for the Review

This review classifies economic activity in three ways:

- Economic activity with markets sensitive to climate change
- Economic activity that is directly sensitive to climate
- Economic activity that is dependent on climate-sensitive resources.

Given the wide range of activity covered in this chapter, impacts and adaptation have been covered jointly under each topic.

This framework, which is simpler than that used in the first IPCC Assessment Report (IPCC, 1990b), highlights interactions among different branches of industry and linkages among industry, energy, transportation, and other sectors, such as agriculture and human settlements. These interactions may transmit climate sensitivities through the economic system and are critical to an understanding of the broader impacts of climate change. Table 11-5 splits economic activity into four broad sectors—energy, agroindustry, transport, and other sectors—which are subdivided further into narrower bands of activity. It identifies those classes of climate sensitivity to which each sector is subject and gives a broad indication of the degree of sensitivity. The energy sector and agro-based industries show, in a qualitative sense, the greatest sensitivity. Since energy is an input to most other industrial activity, the impacts of climate change as reflected in energy costs will cascade through the industrial sector. Water (Chapter 14) is another important vector for transmitting climate sensitivity through an economy.

Markets for consumption goods, intermediate goods, capital goods, and transportation services all may be affected by climate change. The task of reinforcing coastal defenses in response to climate change, for example, will fall on the construction industry. Changing flows of freight and passengers will affect investment in new transportation infrastructure. Climate change will influence patterns of energy demand and, consequently, the need for investment in power plant and other supply facilities. Consumer demand for clothing and beverages and, in some parts of the world, air conditioning equipment, will also be altered by climate change.

Economic activity directly sensitive to climate is found in the construction, electricity, oil, gas, transportation, and water sectors. For oil and gas, offshore production is more sensitive than onshore production. In addition, some industrial activity is located in zones that are sensitive to climate change. Oil refining and some types of power generation, for example, tend to be located in coastal zones in order to facilitate the transport of raw materials and products. Transportation is vulnerable through the effects of sea-level rise on coastal infrastructure and the more direct effect of weather on operations. Constructions in the permafrost regions are especially vulnerable to climate change.

Climate-sensitive resources on which industry is dependent include agricultural, forestry, or marine products; water; energy; and various raw materials. Among the sectors and activities that are affected are food, beverages, and tobacco; textiles, leather, and clothing; timber products; pulp and paper; renewable energy; and sectors such as aluminum that are heavily dependent on hydroelectricity.

Table 11-5: *Sectors sensitive to climate.*

	ISIC No.	Markets	Production	Resources
Energy				
Oil and gas production	11	–	**	–
Oil refining	23	–	*	–
Electricity and gas	40	**	*	**
Biomass	N/A	–	–	***
Hydropower, wind	40	–	–	***
Other renewables	40	–	–	*
Agroindustries				
Food, beverages, tobacco	15,16	*	–	***
Textiles	17	*	–	**
Wood and products	20	–	–	*
Pulp and paper	21	–	–	*
Rubber	25	–	–	*
Pharmaceuticals	part of 24	–	–	*
Transport				
Transport, storage, and communications	60–63	*	*	–
Other Industry				
Water industry	41	***	***	***
Construction	45	**	*	–
Tourism and recreation	55 + others	***	–	**

*** = Significant impact requiring adaptive response at a strategic level.
** = Modest impact requiring adaptive response.
* = Minor impact.
N/A= Activity not easily classified within ISIC.

11.4. Economic Activity with Climate-Sensitive Markets

11.4.1. Energy Demand

The two uses of energy most sensitive to climate change are space heating and air conditioning in residential and commercial buildings and agricultural applications such as irrigation pumping and crop drying. Supply companies will adapt by changing the amount of new investment required to meet peak demand on electricity and gas networks and the composition of investment that would most cost-effectively meet changed temporal patterns of demand. Many studies that assess the impact of climate change on building energy demand have been conducted. The majority focus specifically on electricity markets. A smaller number of studies address energy demand in the agricultural sector and impacts on energy supply investments. The impact of climate change on energy demand will be perceptible but modest in relation to the impact of factors such as changes in technology and patterns of economic activity.

This chapter is primarily concerned with the consequences of climate change for the energy supply sector. Chapter 12 focuses on the impacts of changed energy demand in buildings and their occupants. Climate change also could lead to the increased use of air conditioning in motor vehicles; the resulting reduction in fuel efficiency is discussed in Section 11.5.2.

11.4.1.1. Energy Demand for Space Heating and Air Conditioning

11.4.1.1.1. Space heating and air conditioning markets

The mix of energy sources used for space heating varies widely from one region of the world to another but may include electricity, fossil fuels (coal, oil, or gas), or wood. Patterns of energy use in residential and commercial buildings are discussed in Chapter 22 on mitigation options in human settlements. Among the key points are (Hall *et al.*, 1993; Schipper and Meyers, 1992; World Energy Council, 1993a):

- The significant use of biomass fuels in developing countries
- The heavy use of coal in China, Poland, and other countries
- The increase in natural gas use in Europe and North America over the last 2 decades
- An increasing market share for electricity in new homes.

In the commercial sector, air conditioning accounts for a greater proportion of final energy demand than in the residential sector, due partly to internal heat gains from lighting, office equipment, and occupants. The use of air conditioning is still quite low in temperate regions of the world, though its use is growing because the use of electronic equipment is adding to internal heat gains and sealed buildings help to isolate occupants from a noisy or polluted environment (Herring *et al.*, 1988). The use of air conditioning is growing rapidly in a number of developing countries (Schipper and Meyers, 1992). As described in Chapter 22, energy demand in buildings is growing rapidly in developing countries and in countries with economies in transition but is virtually level in the industrialized world.

11.4.1.1.2. Climate sensitivity

The factors determining the use of energy for space heating and air conditioning are discussed in Chapter 12 in relation to human settlements. Energy suppliers, particularly in the electricity and gas sectors, are highly conscious of the link between energy demand and climate because peak capacity needs are determined for extreme weather conditions. In the UK, for example, where natural gas is the dominant heating fuel, annual send-out can vary by as much as ±10% because of variations in annual average temperature. In the UK, the climate-related variability of gas demand since the mid-1980s has been much larger than it was over the period 1950–85 (British Gas plc, 1992).

11.4.1.1.3. The impacts of climate on energy demand

Many national studies assessing the impact of climate change on energy demand have been carried out. The studies cover the United States as a whole (Linder *et al.*, 1989; Niemeyer *et al.*, 1991; Rosenthal, Gruenspecht, and Moran, 1995); Finland (Aittoniemi, 1991); the UK (Parry and Read, 1988; UK Climate Change Impacts Review Group, 1991; Skea, 1992); Japan (Nishinomiya and Kato, 1990; Kurosaka, 1991; Matsui *et al.*, 1993; Nishioka *et al.*, 1993); and New Zealand (Mundy, 1990). Unpublished work on the former Soviet Union also has been carried out. Within the United States, there also have been a number of regional studies covering: Missouri, Iowa, Nebraska, and Kansas—"MINK" (Darmstadter, 1991); the Pacific Northwest (Wade *et al.*, 1989; Scott *et al.*, 1993); the Southeast and New York State (Linder *et al.*, 1989); and the Tennessee Valley (Miller and Brock, 1988). Some work has addressed specific categories of energy demand such as air conditioning (Milbank, 1989; Scott *et al.*, 1994). Chapter 12 also considers some of these studies.

Table 11-6 summarizes the results of regional and national studies that have sought to quantify the relationship between climate change and energy demand. Table 11-6 illustrates the great diversity in climate scenarios, time-frames, the focus in terms of fuels or specific energy end-uses, and the methodologies used. Climate sensitivities have been deduced either from established statistical relationships between weather and energy demand or from projected climate-induced changes in physical parameters such as degree days. The merits of the two approaches are discussed in Chapter 12. Table 11-6 demonstrates the strong emphasis that has been placed on the study of impacts on electricity demand, probably because this issue is "directly analyzable" (National Academy of Sciences,

1992). Although the results of the various studies are very location-dependent, several broad themes emerge:

- **Fossil Fuel Demand**—Climate change will cause the use of fossil fuels for space heating to decline. For example, a temperature increase of 1.3–2.9°C in the UK could reduce the demand for natural gas by 7–20% in 2050 compared with demand without the effects of climate change (UK Climate Change Impacts Review Group, 1991). In the MINK (Missouri, Iowa, Nebraska, Kansas) region of the United States it has been suggested that fossil fuel demand for space heating would decrease by 7–16% as a result of a temperature increase of 0.8°C (Darmstadter, 1991).
- **Electricity Demand**—Whether electricity demand is likely to rise or fall as a result of climate change depends on the relative importance of space heating or air conditioning (Linder *et al.*, 1989). In areas with a high summer load associated with cooling, climate change will result in increased electricity demand. Conversely, where there is a high winter load associated with space heating, demand is likely to fall. In some temperate zones, where air conditioning is growing in importance, it is not clear whether in the long-term the increase in demand for cooling will exceed the reduction in demand for heating (UK Climate Change Impacts Review Group, 1991).
- **Total Energy Demand**—According to a study referring to representative residential buildings conducted by the Japan Architecture Society (1992), temperature rise will lead to a reduction in energy consumption in Sapporo (43°N), whereas in Tokyo (36°N) the reduction of energy for heating in winter is balanced by an increase due to cooling in summer. At Naha (26°N), the overall energy load will be increased by 50 MJ/m^2 due to a 1°C temperature rise. Rosenthal, Gruenspecht, and Moran (1995) concluded that a 1°C global warming would reduce total U.S. energy use associated with space heating and air conditioning by 1 petajoule (PJ), 11% of demand, in the year 2010. Costs would be reduced by $5.5 billion (1991$). A 2.5°C global warming would reduce total costs by $12.2 billion (1993$). This estimate accounts for the latitudinal and seasonal variations in warming but not for daily variations.
- **Peak Demand—**In the UK, peak demand for heating fuels will decline less than total annual demand, leading to a reduced demand load factor. This would be due partly to a shorter heating season (UK Climate Change Impacts Review Group, 1991). The seasonal occurrence of the peak demand for electricity is an important factor. If peak demand occurs in winter, maximum demand is likely to fall, whereas if there is a summer peak, maximum demand will rise. The precise effects are highly dependent on the climate zone (Linder and Inglis, 1989). Climate change may cause some areas to switch from a winter peaking to a summer peaking regime.

11.4.1.1.4. Impacts on investment in electricity supply

Fewer studies have estimated the possible impact of climate change on investment requirements in electricity supply. An exception is the "infrastructure" component (Linder and Inglis, 1989) of the U.S. national climate effects study (Smith and

Table 11-6: *Summary of results of studies relating climate change to energy demand (based on Ball and Breed, 1992).*

Study	Country/ Region	Temperature Change (°C)	Date	Method	Coverage	Change in Annual Demand	Change in Peak Demand
Aittoniemi, 1991	Finland	1.2–4.6			electricity	7–23% down	
Darmstadter, 1991	U.S. MINK	0.81	2030	degree days	agriculture cooling heating	3% up <2% up 7–16% down	
Matsui, 1993	Japan		2050		electricity	5% up	10% up
Rosenthal, Gruenspecht, and Moran, 1995	USA	1	2010	degree days	space heating and cooling	11% down	
Smith and Tirpak, 1989	USA	3.7	2055	weather analog	electricity	4–6% up	13–20% up
UK Climate Change Impacts Review Group, 1991	UK	2.2	2050	degree days	all energy electricity	5–10% down 1–3% down	

Tirpak, 1989). The principal *quantitative* conclusions arising from that study, under one scenario based on large regional temperature rises in the range 3.4–5.0°C by 2055, were that:

- By 2010, peak demand would increase by 29 gigawatts (GW), or 4% of the baseline level for that year. For 2055, peak demand could be increased by 181 GW, or 13% of the baseline level.
- By 2010, additional investment in new capacity attributable to climate change would be 36 GW, or 13% of the new capacity requirements in the baseline scenario. By 2055, the extra capacity needed could be 185 GW, or 16% of new capacity requirements.
- Cumulative generation costs could be 5% higher by 2010 and 7% higher by 2050.

Among the more qualitative conclusions regarding adaptation to climate change were that:

- There could be a need for relatively more peaking, as opposed to baseload, capacity.
- The generation fuel mix would be altered.
- Different regional impacts could lead to more transfers of power from one area to another.
- Construction requirements and greater levels of fuel use may cause further environmental damage.

11.4.1.2. Energy Demand in Agriculture

Changes in energy demand in agriculture will largely be driven by changes in the level of food production on existing land, especially in developing countries. Changes in energy inputs to fertilizers would be an important indirect impact. Climate change, however, would have some direct impacts on energy needs. The MINK study (Darmstadter, 1991) included some approximate estimates of the impacts of a 0.6–0.9°C temperature rise on energy needs for irrigation pumping and crop drying. Energy demand for pumping would rise by around 25%, while demand for crop drying would decline by around 10%. Irrigation pumping demand is likely to be more sensitive to changes in temperature than to changes in precipitation (Wolock *et al.*, 1992). Taking into account the proportion of agricultural energy demand associated with pumping and drying, Darmstadter concluded that energy demand in agriculture in the MINK region would rise by less than 3% as a result of climate change. This conclusion appears not to have been based on any specific assumptions about climate-induced changes in cropping patterns.

11.4.2. Space Heating/Air Conditioning Equipment

Markets for air conditioning equipment are growing independently of climate change. A broader discussion will be found in Kempton (1992) and Andrews (1989). A small number of studies concerning the weather dependency of demand for air conditioners has been conducted. According to new work on patterns of electricity consumption carried out by the Tokyo Electric Company (1994), for example, electricity consumption in the company's supply area rose exponentially during 1990 when air temperatures exceeded 17°C, with most of the increase due to air conditioning. In Japan, sales of air conditioners tend to rise by 40,000 units per day of maximum temperature of over 30°C (Sakai, 1988).

11.4.3. Construction

The construction industry, which is very weather sensitive, embraces architecture, building, and civil engineering. The direct impact of climate change on construction *activity* is addressed in Section 11.5. This section focuses on the impact of climate change on various types of construction and on the level of demand for construction work.

The construction industry carries out work both above and below ground in a wide range of terrains, some sheltered, others exposed. A variety of specific types of construction, including port facilities, are needed in coastal zones. The construction industry also plays an important role in the development of river and other hydrological works. These include dams, water supply systems, and works that safeguard against risks such as flood and urban drainage failure. Other works are required to secure water quality. The industry is also responsible for the civil engineering component, above and below ground, for land-based transportation systems such as roads, railways, oil/gas pipelines, and electric power lines, as well as airports. All of these works are climate sensitive.

11.4.3.1. Climate Sensitivities

The problem of assessing the impacts of likely climate change on construction falls into two distinct parts (UK Climate Change Impacts Review Group, 1991):

- The assessment of the likely impacts of climate change on existing constructions. Modifications needed to counter any unacceptably adverse effects of climate change must be identified. The appropriate timing for modifications must also be decided.
- The assessment of how current design practices might require modification. This assessment should also address design changes that could reduce future emissions of GHGs and other pollutants.

The key climatic risks in relation to constructions are high wind, snow load, driving rain, thermal expansion, excessive rates of weathering, thawing of permafrost, and sea-level rise. In general, risks are created by extreme values and events (for example, very high winds) rather than average conditions. Therefore, the use of climate models to predict changes in the magnitude and frequency of extreme events is a prerequisite for assessment work in this area. There are great uncertainties attached to the future frequency of extreme events at the regional level. Other risks are associated with the increased vulnerability of timber and timber products to insect and fungus attack, the interaction of short wave radiation with cloudiness, and the impact of ambient temperatures on indoor climate (Page, 1990).

11.4.3.2. Adaptation to Sea-Level Rise

The construction industry will be called on to implement adaptation options associated with sea-level rise. The degree to which coastal infrastructure is at risk is described in Chapters 9 and 12. Existing protective constructions, such as breakwaters and sea walls, will also be affected by sea-level rise. A number of examples serve to illustrate the likely consequences.

Titus *et al.* (1991) estimate that about 2,600 km^2 of low-lying land in the United States may need to be protected from sea-level rise. The cost would be $5–13 billion for a 50 cm rise—approximately the current best estimate for the year 2100—and $11–33 billion for a 100 cm rise. Dikes would not be used on barrier islands because of their narrowness and aesthetic considerations. As a result, 420 km^2 of land on Atlantic Coast islands would need to be gradually raised. The cost of elevating buildings would be $15 billion under the 50 cm sea-level rise scenario and $30 billion under the 100 cm scenario. Gradually elevating roads and other infrastructure may be cheaper than building dikes. This has very different implications for the construction industry.

Hata *et al.* (1993) have assessed the impact of sea-level rise on Japanese fishing ports and adjacent coastlines. Quantitative analyses were performed for three different types of fishing ports assuming sea-level rises of 65 cm and 110 cm. It would be necessary to increase the height of the design wave and, consequently, the weight of breakwater structures. Breakwater weight would need to be increased by more than a factor of two assuming a very high sea-level rise scenario of 110 cm. An increase in the structural buoyancy of loading wharves and sea walls would result in a loss of stability. In addition, loading activities and vessel mooring functions would be impaired. In some cases, the surrounding communities themselves would be subject to severe flooding. Kitajima *et al.* (1993) have estimated a cost of $92 billion for protecting Japanese ports, harbors, and adjacent coastal areas against sea-level rise.

An even greater amount of construction activity might be required in Africa, where industry tends to be concentrated in capital cities, many of which are seaports (Tebicke, 1989).

11.4.3.3. Building Design and Climate Change

The technology of building design and construction will continue to evolve over the next half century, permitting adaptation to changed climate conditions. In temperate zones, the use of air conditioning systems has grown (UK Climate Change Impacts Review Group, 1991). In principle, climate change would stimulate the wider introduction of air conditioning systems. In practice, however, this will be influenced by other changes in building design. The use of suitable design techniques could obviate the need for air conditioning systems in many buildings.

A survey by the Japan Architecture Society (1992) identified the following impacts of climate change on urban planning and architecture:

- The effect of wind, solar radiation, and air pollution in urban areas
- The effect of temperature rise and wind upon architecture, indoor climate, and human adaptation
- Effects on the thermal capacity of concrete, asphalt, and other building materials
- Effect on the use of water in buildings.

The greatest negative impacts on constructions are likely to arise from the interaction between sea-level rise and inland water hydrology. The impacts will be in vulnerable areas. The stability of foundations built on shrinkable soils would be affected by increased winter rainfall combined with drier summer soil conditions. Southeastern England, for example, has many properties at risk (Boden and Driscoll, 1987). Building design codes, currently based on historical climate records, may need to be changed in order to anticipate risks assessed in climate impact studies.

11.4.4. ***Demand for Transportation Infrastructure and Services***

Transportation activity and associated energy consumption are growing very rapidly. Recent trends and future projections are discussed in Chapter 21 on mitigation options in the transportation sector. The essential points are:

- Air and highway transport are the fastest growing modes.
- The fastest rates of growth will take place in developing countries.
- Transport-related CO_2 emissions could rise by between 40 and 100% by 2025.

With the exception of electrified rail and pipelines, transportation relies entirely on fossil energy, principally petroleum. Alternative fuels remain a very minor energy source for transportation, although the ethanol-from-biomass experiment in Brazil is notable (Sperling, 1987). Fossil fuels account for a significant share of tons moved in freight transport. In the United States, coal and petroleum constitute 30% of rail carloads, while petroleum, coal, and coke comprise 60% of waterborne ton-miles. Agricultural and food products are second to energy in terms of freight ton-miles. Changes in the location and nature of agricultural activities could have a large impact on the freight transport system.

Climate-induced changes in the distribution of population and economic activity and the consequent effects on the performance of transportation infrastructure and infrastructure needs are of great potential importance. Transportation requirements are closely linked to patterns of human settlement and have been addressed in this context in several infrastructure-based assessments (e.g., Walker *et al.*, 1989). Changes in the nature and location of agricultural production, in the rates of population growth in different regions, in the volumes and types of fossil fuels used, and in tourism and recreational travel can have profound effects on the performance of existing transportation facilities and on requirements to construct new ones. Although it is widely acknowledged that climate change could produce significant redistributions of population and economic activities, particularly agriculture, detailed geographic predictions of these phenomena are lacking (Johnson, 1991; Jansen *et al.*, 1991). Such regional predictions are a prerequisite for quantitative assessment of the impacts of regional redistribution on transportation.

Existing assessments of transportation impacts have recognized the potential significance of changes in geographical patterns of economic activity on the transportation network. Black (1990) notes that even gradual, long-term global warming could cause a major disruption of the movement of goods and people in North America. Irwin and Johnson (1988) suggest that there would probably be a northward spreading of agricultural, forestry, and mining activities, resulting in increased population and intensified settlement patterns in Canada's mid-north and even in Arctic areas. Marine, road, rail, and air links would have to be expanded accordingly. While this would entail substantial extra capital and operating costs, it would also be an economic opportunity.

While increasing temperatures may open up regions to increased development in northern climates, sea-level rise may also force massive migrations from settlements on river deltas and other low-lying areas. Jansen, Kuik, and Spiegel (1991) point out that a possible 26% land loss in Bangladesh would displace 27% of the population and that a 21% land loss in Egypt would displace 19% of the population. The implications of such a population movement for transportation infrastructure requirements have not been addressed.

To date, no quantitative assessments of the impacts of climate on demand for transportation in areas such as tourism and recreation have been made, partly because the necessary regional economic scenarios do not exist (see Section 11.5.9). The current state-of-the-art in intercity passenger and freight transportation modeling is adequate for quantifying major changes in flows and predicting the magnitude of changes in transport networks that would be required to accommodate them (Friesz *et al.*, 1983; Bertuglia *et al.*, 1987). The interactions that would have to be modelled are represented in Figure 11-1. Climate change affects the geography of specific climate-sensitive sectors such as agriculture and tourism, but it also affects population distribution and regional economic growth more generally. Actions taken to mitigate climate change affect other activities, notably the extraction, processing, and distribution of fossil fuels. Changes in the location of these activities

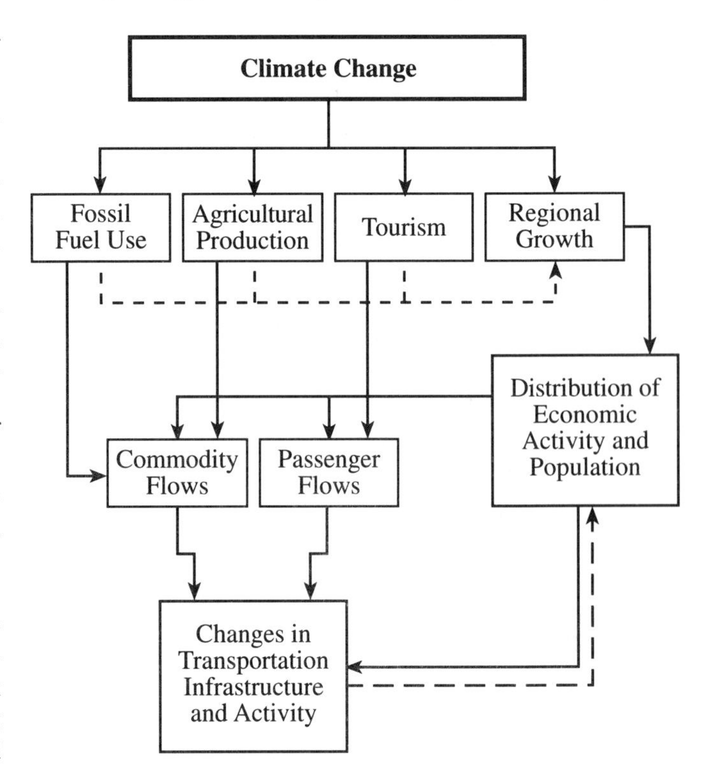

Figure 11-1: Indirect effects of climate change on transportation.

affect the demand for flows of passengers and freight on a regional and global level. These demands then influence the development of transportation infrastructure and transportation movements. In addition, there are feedback effects by which the transportation infrastructure and activity influence patterns of regional growth. The anticipation of climate change would be appropriate in planning investments of a long-lasting infrastructural nature.

11.4.5. Other Market Impacts

The first IPCC assessment report (IPCC, 1990b, Chapter 5, Section 5.2.1.1) reported that climate change could be expected to influence patterns of demand for food, beverages, and clothing. The report suggested that:

- There could be an increased demand for cold drinks such as carbonated beverages, fruit juices, lemonade, and iced tea, and a reduced demand for hot beverages such as hot coffee or tea.
- There could be an increased demand for cotton clothing and certain synthetics and a potential reduction in demand for wools.

At the time, no studies that investigated these possible patterns of change could be cited. This remains the case. While some change in demand for beverages and textile products may be likely as a result of global warming, changes will take place over the long term, and the impacts may be small compared with those induced over a shorter timescale by product innovations and changes in consumer preference.

11.5. Economic Activity Sensitive to Climate

A wide range of economic activity displays some degree of direct sensitivity to climate change. On the whole, the direct impacts of climate change will be minor compared to impacts on resources or markets. The direct effects on activity located on coasts or in permafrost regions could be greater, however. Affected sectors include construction, transportation, energy, and tourism.

11.5.1. Construction

Rain, snow, high winds, and frost all hinder construction. Increased rainfall in winter, which may occur in mid-latitudes, would reduce productivity directly through impacts on working conditions and indirectly through impacts on ground conditions. A warmer winter climate could reduce the negative impacts of frost on construction activity. In areas such as Siberia, however, warming could cause permafrost shrinkage and lead to structural damage. If wind speeds were to increase, there would be a need to pay greater attention to wind protection during construction. Existing work on the variability of the impacts of present climate on the construction process could provide a basis for examining future climate impacts on site production (UK Building Research Establishment, 1990). A study in Japan reported that warm winters and springs have a favorable impact on the ceramic and quarrying industries, while the following have a negative impact: extended periods of rain in early summer, heavy snow, cool and rainy summers, and typhoons (Tokyo District Meteorological Observatory, 1986). Typhoons and other heavy rain, which may increase, can reduce the production of cement tiles and concrete blocks because of their impact on drying processes (Fukuoka District Meteorological Observatory, 1987).

11.5.2. Transportation Operations and Infrastructure

Climate change will have some direct effects on transportation infrastructure and the operation of transportation systems. These may be divided into three categories: the effects of climate on operations, the effects of sea-level rise on coastal facilities, and the effects of climate on infrastructure. These subjects have been included in existing impact studies in which transportation has been treated as one component of human settlements (e.g., Daniels *et al.*, 1992; Walker *et al.*, 1989).

11.5.2.1. Transportation Operations

Global warming will have both negative and positive impacts on the operation and maintenance costs of transportation systems. Studies in temperate and northern climates have generally indicated that higher temperatures will result in lower maintenance costs, especially with fewer freeze-thaw cycles and less snow (Parry and Read, 1988). Black (1990) points out, however, that increased pavement buckling due to longer periods of intense heat also is a possibility. Engineering standards for facility design and maintenance represent a sound, if not completely precise, means of quantifying the relationship between climate and factors such as damage to roads and bridges from freeze-thaw cycles, snow removal costs, salt application, and de-icing. Data from cities with similar climates and engineering handbook standards have been used to quantify the relationship (e.g., Walker *et al.*, 1989).

Inland and coastal water transport is particularly sensitive to droughts, floods, changes in water levels, and icing over of ports and waterways. The MINK study (Frederick, 1991) suggested that lower flows in the Missouri River could shorten the navigation season in the United States Midwest from eight months to five months in six of every ten years. Navigation benefits on the upper Missouri, however, are only $14 million/year, and the MINK study was based on a dry, 1930s-analogue climate rather than a forward-looking climate scenario.

In colder regions, the most significant direct impact of warming is likely to be on inland and coastal water transportation. A longer season for Arctic shipping is likely, with a greater number of frost-free days for northern ports and inland waterways such as the St. Lawrence Seaway (Irwin and Johnson, 1988). Inland waterways, however, may suffer loss of depth for greater periods of seasonal drought, reducing their usefulness for commercial shipping (Black, 1990). A survey of the potential impacts on Canadian shipping suggested net benefits to Arctic and ocean shipping due to deeper drafts in ports and longer navigational seasons, but mixed results for lake and river shipping due to the opposing effects of a longer shipping season but lower drafts (Irwin and Johnson, 1988). In Siberia, many rivers are used as solid roads during winter. Warmer winters would require a shift to water transport or the construction of more all-weather roads. Other climate impacts could arise from changes in snowfall or melting of the permafrost (IBI Group, 1990).

Transportation operations are sensitive to the weather. Fog, rain, snow, and ice slow down transport movements and increase risks of accidents. In addition, maintenance costs and the durability of infrastructure are somewhat dependent on weather events. Changes in the frequency and intensity of catastrophic weather events of short duration (for example, hurricanes, floods, wind shear, and surface movements associated with high rainfall) may have significant impacts on the safety and reliability of transportation. Little evaluation of such factors has been attempted.

Transportation operations are also affected by temperature. Fuel economy is somewhat temperature sensitive, primarily because of the time it takes a vehicle to warm up (Murrell, 1980). The impact on overall fuel economy of a few degrees' increase in average temperature, however, is likely to be a fraction of a percent. Vehicle emissions are also affected by temperature, especially evaporative emissions of hydrocarbons (U.S. EPA, 1989). This is potentially significant because formation of tropospheric ozone, for which hydrocarbons are a precursor, also tends to increase with temperature (see Section 11.5.7).

Increased use of air conditioning in vehicles would have a greater impact on overall fuel economy than would the direct effect of temperature change. U.S. automobiles consume 47 liters of fuel to operate air conditioners for every 10,000 km driven (Titus, 1992). Titus suggests that a CO_2 doubling would increase expenditure on fuel by $1-3 billion annually because of more air conditioning, though many of the specific assumptions used to reach this conclusion are not transparent. CO_2 doubling is assumed to be associated with a 3°C temperature rise. A warmer climate would also increase energy demand associated with refrigerated transport and cold storage (Steiner, 1990).

Interruptions to transportation operations can have significant impacts on industry. During the 1988 drought in the United States, industries that relied on bulk transportation of raw materials and finished products by barge on the Mississippi River found that low water kept more than 800 barges tied up for several months. In 1993, by contrast, floods in the upper Mississippi valley disrupted the barge transportation system. To the extent that industry is moving toward just-in-time production systems, it will become more vulnerable to interruptions for these and other reasons. This issue is also discussed in Chapter 17 in relation to financial services.

11.5.2.2. Sea-Level Rise and Coastal Infrastructure

Transportation infrastructure could be lost to inundation in coastal areas when sea levels rise and shorelines recede, although port infrastructure may be rebuilt sufficiently often to avoid such threats. Many airports are located at low levels. The impacts of sea-level rise have been rigorously examined in a few studies. In a detailed assessment of impacts at six sites on the eastern and Gulf coasts of the United States, Daniels *et al.* (1992) found that impacts varied greatly from city to city. For example, under a low sea-level rise scenario, 40% of the transportation, communications, and utilities infrastructure of Galveston, Texas, would be below mean sea level by 2050, and half by 2100. Under the moderate and high sea-level rise scenarios, approximately half would be below sea level by 2050 and all by 2100 (Daniels *et al.*, 1992, Table 4.7). A similar study of the Daytona Beach, Florida, area concluded that only about 10% of the transportation infrastructure was subject to inundation (Daniels *et al.*, 1992, Table 4.19). The impacts of sea-level rise on transportation infrastructure in low-lying countries such as Bangladesh and Egypt would be massive (U.S. Department of Energy, 1990).

Walker *et al.* (1989) found that considerable damage to roads in the Miami, Florida, area would be caused by a higher water table. If the sea level and water table were to rise by roughly one meter, the subgrade and/or base of many city streets would be subject to a certain amount of saturation given the annual fluctuation in the water table and its proximity to the surface. Structural failure would be caused if a heavy load were to pass over the surface. The cost of raising vulnerable streets in Miami was put at $575/meter for 410 km of road, or $237 million. Walker *et al.* (1989) also found that bridges and causeways that were not inundated would be damaged considerably if not reconstructed. Climate change should be anticipated when constructing or rebuilding transportation infrastructure.

11.5.2.3. Other Infrastructure

Inundation is not the only threat of sea-level rise to transportation infrastructure. Extreme rainfall could have widespread impacts on roads, railways, and other transportation links. As long as rainfall does not become more intense, impacts on urban roads and railways in temperate, tropical, and subtropical zones are likely to be modest. Exceptions could occur in coastal areas where highways and bridges may have to be redesigned for the higher wind pressures of tropical cyclones (Deering, 1994). In mountainous regions, increased intensity of rainfall could increase the risk of mudslides.

11.5.3. Energy Transportation and Transmission

Electricity transmission lines are susceptible to extreme weather events. If the frequency of extreme weather events were to increase, customers would have to accept a less reliable electricity service or pay for the costs of strengthening lines. Storms in Northern Europe in October 1987 led to an average loss of 250 minutes supply to customers of the UK's Central Electricity Generating Board (Electricity Council, 1988). As discussed in Chapter 17, the cost of interruptions to power supply in many service sectors is rising due to the increased use of advanced information and communication technologies. The capacity of electricity transmission lines also drops at higher temperatures. For example, the capacity of the typical line used in the UK falls from 2,720 megavolt-amps (MVA) in winter to 2,190 MVA in summer (Eunson, 1988). Higher temperatures could have minor implications for utilities experiencing a summer peak. Possible thawing of permafrost in Arctic regions may require changes in the design of oil pipelines in order to avoid slumping, breaks, and leaks (Brown, 1989).

11.5.4. Offshore Oil and Gas

Offshore production of oil and gas could be affected to a minor extent by sea-level rise. Royal Dutch Shell has increased the height of North Sea gas platforms above the water level by 1–2 meters to take account of projected sea level rise (National Academy of Sciences, 1992). This will, however, add less than 1% to the total cost of a platform.

Increased wave activity and increased frequency of extreme weather events might have a more significant effect on offshore operations, but little research has been conducted on this topic. Offshore oil and gas production conducted at high latitudes may be assisted by a longer ice-free season (Lonergan, 1989). A recent Canadian study of the Arctic petroleum industry (McGillivray *et al.*, 1993) concluded that *global* temperatures rising by 1–4°C over a period of 50 years and Arctic temperatures rising by two to three times as much would:

- Increase the open water (ice-free) season in the Canadian sector of the Beaufort Sea from 60 to 150 days
- Reduce ice thickness
- Increase the maximum extent of open water in summer from the current 150–200 to 500–800 km offshore
- Increase wave heights (e.g., the proportion of waves in excess of 6 m would rise from 16 to 39%).

As a consequence of reduced ice thickness and more open water, the offshore petroleum industry could experience reduced operating costs. The most critical factor could be ice movement during winter. Increased wave activity, however, would push up design requirements for both offshore structures and associated coastal facilities. In adapting to climate change, the industry would prefer to take a cautious view of these conclusions, discounting the benefits and preparing for increased wave activity (McGillivray *et al.*, 1993).

11.5.5. Thermal Power Generation

The efficiency of electricity generated through both steam and gas cycles would be affected negatively, though in a minor way, by global warming (Ball and Breed, 1992).

Most current thermal electricity generation relies heavily on the availability of cooling water (Smith and Tirpak, 1989; Solley *et al.*, 1988). There is a general trend away from once-through cooling, in which water is returned to rivers, toward evaporative cooling. In the United States, there also is a movement away from river-based power plants toward coastal siting (Smith and Tirpak, 1989). New technologies, such as combined-cycle gas turbines, are reducing the dependence of fossil fuel-fired power generation systems on the availability of cooling water (UK Climate Change Impacts Review Group, 1991). For these reasons, the impact of changes in water availability should be relatively minor.

Power plant output may be restricted because of reduced water availability or thermal pollution of rivers with a reduced flow of water. Events such as these have occurred during droughts in several parts of the world including France and the United States (*Energy Economist*, 1988). Under more extreme temperature conditions, some nuclear plants might shut down to comply with safety regulations (Miller *et al.*, 1992).

11.5.6. Water Availability for Industry

Changing patterns of water activity, particularly at the regional level, are highly uncertain. Industrial manufacturing depends upon water and is affected by its availability in several ways:

- Water is used as an ingredient or processing solvent for pulp and paper, food processing, textiles, and petrochemical refining (see Section 11.6.1).
- Some sectors, such as aluminum and, increasingly, the steel industry, are dependent on electrical energy and in particular on cheap hydroelectricity. Other industries are dependent upon water for process cooling or, indirectly, for cooling water in thermal electric power stations.
- Many industries are dependent upon cheap river and lake barge transportation of bulk raw materials and for shipping finished materials.
- The location of many manufacturers along rivers makes them vulnerable to damage from flooding.

Reductions in rainfall, which are possible during the summer in mid-latitudes, could adversely affect manufacturers in the first three categories by raising the cost of water, energy, and transportation as inputs into their products. Low water levels and higher temperatures could lead to less dissolved oxygen and greater problems with biochemical oxygen demand for industries that utilize water as an ingredient or solvent. This could raise the cost of process water. Efforts to reduce the use of process water through recycling could substantially reduce vulnerability to water shortages.

The global aluminum industry is exceptionally energy intensive and, in 1990, utilized an estimated 280 terawatt-hours (TWh) of electricity (Young, 1992) to convert alumina into 16.3 million metric tons of primary aluminum metal (Plunkert and Sehnke, 1993). Production of aluminum metal is distributed among 41 countries, many of which lack bauxite ore but possess abundant energy resources. The United States is unusual in that only 12.4% of its energy for aluminum production comes from hydroelectricity (Aluminum Association, 1991), whereas most of the other major producers—such as Brazil, Canada, and Ghana—rely almost exclusively on hydroelectricity (Plunkert and Sehnke, 1993). In the six years since 1988, Russian exports have expanded six-fold and now account for 10% of world production (Imse, 1994). Hydroelectricity forms the basis for the development programs and the industrial base of many developing countries. Brazil, for example, now ranks fourth in the world in terms of hydroelectricity production after Canada, the United States, and the former Soviet republics. Developing countries would be particularly hard hit by a reduction in hydroelectricity capacity. Any reduced availability of hydroelectricity could have significant adverse consequences for the industry.

Other industries that might be affected by a decline in regional electricity production are electric arc furnace-based steelmaking, electroplating, and uranium fuel enrichment. Regions suffering a decline in water availability from altered precipitation patterns would see rising water prices. Adaptation would require the development of additional water supply resources, the adoption of less water-intensive processing methods or, alternatively, relocation.

11.5.7. Pollution Control and Climate Change

Some features of climate change could lead to changes in ambient levels of pollution in both air and water. The two possible impacts that have been given brief consideration in the literature are accelerated rates of ozone formation caused by higher temperatures and possible reductions in water quality caused by lower levels of river flow.

Estimates of the impact of temperature change on ozone formation vary. Gery *et al.* (1987) estimate that a temperature rise of 2–5°C would cause ozone concentrations in several major U.S. cities to rise by 1.69 ± 0.37% for every 1°C of warming. Morris *et al.* (1989) estimate that maximum ozone concentrations over a large part of the United States would rise by 1.45 ± 0.50% for every 1°C of warming. One form of adaptation would be to reduce emissions of ozone precursors. Smith and Tirpak (1989) estimate that, in order to maintain air quality, emissions of volatile organic compounds (VOCs) arising from auto exhausts, organic solvents, and gasoline storage would need to be reduced by up to 2% for every 1% increase in ozone concentration. U.S. VOC emissions for the year 2005 have been estimated at approximately 14 million tons (Pechan and Associates, 1990), suggesting that a 2°C temperature rise would add some $1.4–4.2 billion to annual VOC control costs, assuming abatement costs of $1,900–5,500/ton. As discussed in Chapter 12, there are also severe urban air pollution problems in Eastern Europe and in many cities in developing countries. The losses associated with increased air pollution would therefore be geographically widespread (Cline, 1992).

Maintaining water quality under conditions of reduced river flow would similarly require reductions in the discharge of pollutants. Titus (1992) estimated the impacts of an arbitrary CO_2 doubling on water flows in each of the U.S. states and calculated the cost of maintaining water quality. Across the United States, annual water pollution control costs were estimated to rise by $15–52 billion. Titus concluded that maintaining water quality would have a higher cost than problems directly associated with reductions in water quantity. In other parts of the world, however, river flows could increase and water quality would not be reduced.

11.5.8. Coastally Sited Industry

Many industries are located preferentially in coastal zones that are discussed in detail in Chapter 9. In the energy sector, many petroleum refineries and power stations are located in coastal zones because of ready access to supplies of crude oil and fuel or because of the availability of cooling water. In the UK, all oil refineries and more than half of the thermal power stations are located on coastal sites (UK Climate Change Impacts Review Group, 1991). Sea-level rise could result in additional expenditures at existing sites (Smith and Tirpak, 1989; Nishinomiya and Kato, 1990; UK Climate Change Impacts Review Group, 1991). There is a considerable amount of infrastructural investment (transmission lines, transport routes) associated with existing energy facilities, and any major migration of activity is unlikely.

11.5.9. Tourism and Recreation

Tourism and outdoor recreation is one of the most important and rapidly growing service industries throughout the world. In many countries, tourism is a major source of employment. Countries with economies that are highly dependent upon tourism may face great challenges because the resources upon which tourism rests are regionally, nationally, and globally climate-dependent.

Tourism and recreation are sensitive to climate change because part of the industry is closely associated with nature (National Academy of Sciences, 1992). Some parts of the tourism and recreation industry will necessarily migrate as a result of climate change. The Academy panel concluded that the overall effect for a country as large as the United States would probably be negligible, though specific regions could experience adverse or favorable effects. The question of whether the dislocations or relocations caused by climate change will produce costs over and above those that would have been incurred otherwise is important, but difficult to answer.

Two of the most obvious tourism and recreation facilities exhibiting climate sensitivity are skiing and beach resorts. In general, global warming might be expected to reduce the length of the skiing season in many areas and to affect the viability of some ski facilities. On the other hand, the summer recreation season in many areas may be extended. In some coastal areas, the benefits resulting from a longer season may be offset, however, by the loss of economically important beaches and coastal recreational resources, particularly on low-lying and vulnerable tropical islands.

Several studies have projected shorter skiing seasons as a result of climate change. Aoki (1989) found that skiing activity in Japan was highly sensitive to snowfall. In a study of the implications of an effective CO_2 doubling on tourism and recreation in Ontario, Canada, Wall (1988) projected that the downhill ski season in the South Georgian Bay Region could be eliminated, with an annual revenue loss of $36.55 million (Canadian dollars). This outcome assumed a temperature rise of 3.5–5.7°C and a 9% increase in annual precipitation levels. Some of these losses would be offset by an extended summer recreational season. Lamothe and Periard (1988) examined the implications of a 4–5°C temperature rise throughout the downhill skiing season in Quebec. They projected a 50–70% decrease in the number of ski days in Southern Quebec, while ski resorts equipped with snow-making devices would probably experience a 40–50% reduction in the number of ski days.

Because many recreational activities and related facilities are associated with coasts and beaches, sea-level rise may be of special concern to the recreational and tourist industries. Boathouses, residential plots and houses, public buildings, and structures are increasingly under threat due to sea-level rise. The low-gradient recreational beaches characteristic of much of the Atlantic and Gulf coasts of the United States are very vulnerable to erosion. Titus *et al.* (1991) estimate that the cost of sand required to protect major United States recreational beaches from a 50 cm sea-level rise would be $14–21 billion. In addition, elevating infrastructure would cost another $15 billion on the Atlantic coast alone. Several U.S. states have prohibited seawalls on the ocean coast. Texas, Maine, and South Carolina allow public beaches to migrate inland with the shoreline. Property owners must take the risks of an eroding shoreline into account. Sea-level rise could affect fixed waterfront facilities such as marinas and piers. There could be increased erosion of beaches backed by sea walls, leading to a lowering of the beach level and subsequent undermining of the walls. Recreational habitats such as sand dunes, shingle banks, marshlands, soft banks, soft earth cliffs, and coral reefs would also be affected.

A small number of national studies of the potential impacts of climate change on tourism have been conducted. Higher temperatures are likely to stimulate an overall increase in tourism in the UK, with the greatest impact being on holiday activity and some forms of outdoor recreation (UK Climate Change Impacts Review Group, 1991). An increase in sea temperature would increase the pressures of tourism on UK beaches, while coastal erosion may reduce beach area (Baker and Olsson, 1992). A study by the Tokyo District Meteorological Observatory (1986) found that warm springs, falls, and winters, the absence of a rainy season, and very hot, dry summers have a favorable impact on tourism. On the other hand, typhoons, cool and rainy summers, extended periods of rain in the early summer, heavy snow, and cool springs were found to create unfavorable conditions.

11.6. Economic Activity Dependent on Climate-Sensitive Resources

Agroindustry, biomass production, and renewable energy sources depend heavily on climate-sensitive resources and therefore are potentially vulnerable to climate change.

11.6.1. Agroindustry

The impacts of climate change on the primary products of agriculture, forestry, and fishing are carried over to the food and drink sector, industries dependent on forestry products such as pulp and paper, and other sectors, notably textiles. The food and beverage industry is almost completely dependent on agricultural products, with the exception of mineral waters and some soft drinks. The textile and clothing industry is slightly less dependent due to synthetic fibers. In this chapter, these sectors are described collectively as *agroindustry*.

Many assessments of the impacts of climate change on specific crops and on agriculture as a whole have been carried out (Chapter 13). The few conceptual or empirical studies of agroindustry suggest that the sector is indeed vulnerable to climate change. Most of the assessments refer to developed rather than developing countries. Agroindustry, however, is relatively more important in economic terms in developing countries. Moreover, even in the case of industrialized countries, analyses focus on only a few countries, notably the United States. The narrow range of studies carried out is unfortunate because it is virtually impossible to define generic thresholds for the impact of climate variables on agroindustries, even when the same product is derived from the same raw material.

11.6.1.1. Economic Impacts on the Food Industry

11.6.1.1.1. Agriculture-agroindustry linkages

The literature on the impacts of climate change on agroindustry has advanced little since the IPCC Supplement (IPCC, 1993). The U.S. MINK study addresses the linkages between agriculture and agroindustry at a regional level (Rosenberg and Crosson, 1991). Scheraga *et al.* (1993) have examined the consequences of higher agricultural prices resulting from CO_2 doubling using a general equilibrium model. Kane *et al.* (1992) have assessed the economic effects of CO_2 doubling on agricultural commodities at a global level. Some national studies exploring the linkages between primary agricultural products and subsequent processing also have been conducted (UK Climate Change Impacts Review Group, 1991; Antal and Starosolzky, 1990; Comisión de Cambio Climático, 1991). Most of these studies are concerned only with the impact of changes in the mean values of climatic variables on agriculture and agroindustry. Extreme weather events, however, can also carry climate impacts from the primary agriculture sector to agroindustry. The resource base of agroindustry is particularly vulnerable to droughts and tropical cyclones. Such cyclones can also severely damage industrial infrastructures, mainly small and medium-sized factories.

The MINK project (Rosenberg, 1993; Bowes and Crosson, 1993) quantified the links between agricultural production and processing industries. A climate similar to that of the 1930s was superimposed on current and projected patterns of economic activity. The work indicated strong linkages between the regional meat packing industry and feedgrain production and between soybean cultivation and oil production. More than 80% of the outlays by soybean oil mills are accounted for by purchases of oil-bearing crops.

The overall effect of the expected decline in feedgrain production on the regional economy depends on the extent to which the decline affects domestic or export markets. The decline could be as much as 6–10% if drops in production were to affect only domestic markets through the impact on the local meat-packing industry. The authors believe, however, that the economic impact would be closer to 0.6–1.0% because of adaptation on the part of farmers, the possible benefits of higher CO_2 concentrations, and the likelihood that reduced grain output would primarily affect export markets. The higher production costs in MINK under the 1930s climate would weaken the region's competitive position in world grain markets. Over the long term, some animal production might shift to other grain-producing regions less affected by climate change. This could induce a decline in meatpacking in MINK, because these activities tend to be near animal production (Bowes and Crosson, 1993).

Scheraga *et al.* (1993) used a general equilibrium framework to explore the impact of a fairly extreme change in climate on agricultural production costs and the prices of goods and services that use agricultural commodities in the United States. The effects of selected climate impacts on the gross national product (GNP), consumption, investment, and the sectoral composition of output were assessed. A projected decline in crop yields was assumed to raise agricultural prices by more than 20% by 2050. This has an indirect effect on prices in other sectors, as shown in Figure 11-2. The largest price increases occur in agroindustry, notably the food sector, causing food and tobacco output to decline by 10%. Scheraga *et al.* (1993) calculated that the economic burden of climate change would fall disproportionately on lower-income households, which spend a large share of their income on food. This consequence of climate change, applicable to U.S. conditions, could be even more pronounced in developing countries.

Kane *et al.* (1992) examined the impacts of a doubling of atmospheric CO_2 concentrations on world agriculture and the indirect impacts on agroindustry. The study assumed increased precipitation and warming in the high northern latitudes coupled with drying in continental interiors. This work emphasized the importance of trade effects as a result of changes in the relative importance of the agricultural and industrial sectors and the direction and magnitude of the world price effects. Under two scenarios, prices of the main commodities used in food-related agroindustry were projected to rise by between 1 and 37%. The study projected only modest impacts on national economic welfare, which depends on changes in the yield of domestic agriculture, changes in world commodity prices, and the relative strength of the country as a net food importer or exporter. The benefits to producers will be larger than the losses to consumers if the country, like the United States, is a large exporter. This conclusion emphasizes the vulnerability to climate change of agriculture and agroindustries in the majority of developing countries. Rosenzweig and Parry (1994) also note that many cereal-producing countries in the tropical and subtropical zones appear more vulnerable to the potential impact of global warming than are countries in temperate zones.

The link between agriculture and agroindustry has been explored qualitatively in a number of country studies. In the UK, a rise in temperature coupled with increases in rainfall would accelerate the rate of deterioration of food and would change crop production patterns. Reduced rainfall, possible in summer, would reduce production, increasing raw material costs and requiring the import of more raw materials for the food industry (UK Climate Change Impacts Review Group, 1991). In Hungary, an increase in droughts would lead to a decrease in the yield of pastures, which would reduce the economic viability of cattle stocks and the associated meat processing industry (Antal and Starosolzky, 1990).

11.6.1.1.2. Impacts on specific food products

From the literature it is possible to draw more specific examples of the possible indirect impacts of climate change for grain, sugar cane, and fish products.

Grains constitute the world's staple food. In terms of agroindustries they are used in the production of bread, flour, meals, and pasta and form an essential input to the meat processing and dairy industries through animal production. Rosenzweig *et al.* (1993) conclude that climate change will lead to a decrease in cereal production, mainly in developing countries. Warming, coupled with drier climates in the interior of continents in the northern middle latitudes, predicted by most climate models, could have negative impacts on crops and livestock in the United States, Western Europe, and southern Canada, which are the most important grain-producing areas of the world (Kane *et al.*, 1992). Increased precipitation and warming in the high northern latitudes, however, could enhance agricultural production potential.

The northward shift of the North American corn and wheat belt, as a consequence of drier conditions in the Midwest, could bring benefits to Canada, increasing its grain production and export capacity (Smit, 1989). Any shift in the corn belt also would entail large social and environmental costs. Some communities, industries, and the associated physical infrastructure could cease to be viable and might need to be relocated (U.S. Department of State, 1992). Affected agroindustries might include the production of ethanol and sweeteners, both of which are currently expanding.

World sugar production is 100 million tons per year, 60% of which is derived from sugar cane and 40% from sugar beet (UN Statistical Office, 1993). Sugar cane is also used to manufacture distilled alcoholic beverages and ethanol for motor fuel. According to Leemans and Solomon (1993), the distribution of sugar cane in a warming world could be quite similar to that at present, with the possibility of some expansion of the cultivated area. Yields per unit area could be increased by 3.4% and global production by 16.2%. This is based on the assumption that agroindustry associated with sugar cane will be developed intensively for energy production. The expansion of sugar cane for use as an energy source would be aided by the successful development of biomass gasification technology coupled to gas turbines (Larson, 1993). The greater use of sugar cane could form part of climate change mitigation strategies because this would contribute to reductions in greenhouse gas emissions (Acosta and Suarez, 1992; Macedo, 1991).

Sugar cane is produced in a large area of the world extending from 37°N to 31°S. Within that belt there are huge climatic differences (Blume, 1985). Consequently, local conditions must be addressed in assessing climate change impacts. This underlines the impossibility of defining generic thresholds for the impact of climate variables on agroindustries. A study of the productivity of Cuban sugar mills has shown that temperature and, to a greater extent, rainfall patterns are the principal climatic factors

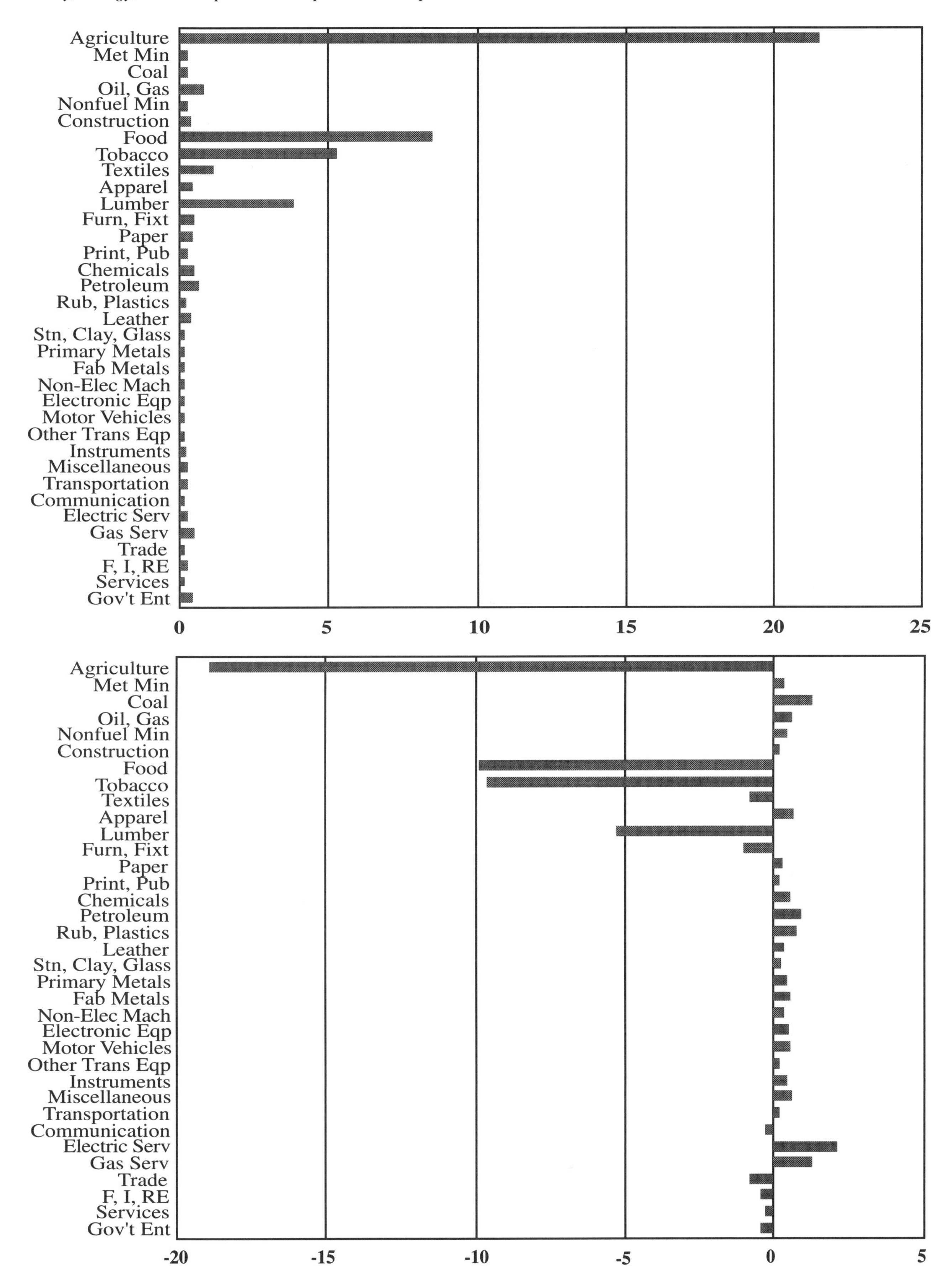

Figure 11-2: Percent change in sector supply prices (top) and domestic output (bottom) for the year 2050, as a consequence of projected decline in crop yields (Scheraga *et al.*, 1993).

influencing sugar production (Comisión de Cambio Climático, 1991). The sugar industry could also be threatened by an increase in the frequency and intensity of hurricanes. Central America, the Caribbean, and Fiji have recently been struck by a number of severe hurricanes that damaged sugar production.

Citrus fruits are the basis for the large-scale production of beverages and other agroindustrial products. The impact of extreme climate events has contributed to important changes in the map of world citrus production. Frosts have a dramatic impact on the production levels of citrus (Buzzandi, 1979). Between 1957 and 1985, the citrus industry in Florida was subject to periodic freezes resulting in widespread production losses (Bulger, 1985; Miller and Glantz, 1988). The freezes in 1980–85 reduced seasonal orange production by about 50%. Those impacts helped to consolidate the expansion of agroindustry based on citrus fruits in Brazil, which was able to increase its exports during a period of rapid industrialization (Castello and Carvalho Filho, 1992).

Climate change would affect fish processing through both the availability of resources and impacts on storage systems. Fish processing comprises the freezing, salting, and drying of fish and the manufacture of fish jelly products, canned products, oil and fat, feed, and fertilizer. Seaweed is also economically significant in some parts of the world. The largest pressures on fish processing in some parts of the world will arise from overfishing and reduction in stocks. Fish jelly products and frozen fish, however, are sensitive to temperature change. Since fish processing industries are usually located along the sea coast, facilities are vulnerable to sea-level rise. For example, Stokoe (1988) notes that fish processing facilities in Canada's Atlantic provinces have a capital value of $1 billion (Canadian dollars), and most are situated at or near the water's edge. They could be vulnerable to inundation or damage as a result of sea-level rise. Reductions in precipitation could reduce potable water needed for fish processing. This is currently a problem due to climate variability, and significant permanent changes could cause additional difficulty.

Higher temperatures and sea-level rise could reduce catches of cod, tuna, and green hata in the Gulf of Tonkin, although the profitability of the shrimp industry could rise due to increased production (Nguyen *et al.*, 1993).

11.6.1.1.3. Key issues for the food industry

The conclusions of the small number of country studies and more general studies can be drawn together:

- Impacts on the food industry depend mainly on changes in the availability and price of agricultural products. Therefore, adaptation measures in agriculture are the key to diminishing any negative impacts on agroindustry.
- Climate change could have a severe impact on regions or countries dependent on single crops. Diversification of economic activity could be an important precautionary response.
- The impacts of climate change for agroindustry will be more pronounced in developing countries than in developed countries because many are located in more vulnerable zones and because their capacity for adaptation is lower.
- The burden of climate change may affect lower-income households disproportionately.
- Climate change could cause a loss of competitiveness in agroindustry in specific countries or regions, especially in developing countries. This could affect the viability of communities, industries, and associated physical infrastructure, resulting in relocation and/or large social and economic costs for the affected regions. On the other hand, some countries or regions might benefit.

11.6.1.2. Industry Dependent on Forestry

Many industrial sectors are dependent on forest products, including wood and timber, pulp and paper, rubber, and pharmaceuticals. Other forest products include nuts and oils. A recent analysis of the potential for products other than timber from a region of the Amazon concluded that their annual market value could be double that of timber itself (Peters *et al.*, 1989). Many forest products originate in developing countries where the commodities play a large role in the local and national economies. The sensitivity of forestry to climate is discussed in Chapter 15.

Forests provide major components for the construction of buildings in most parts of the world. Substantial amounts of furniture and a lesser proportion of other home utensils are manufactured from wood. In Europe, Russia, and North America, indigenous forests support major domestic timber industries. In many developing countries in the tropics, wood is a common property resource used as the predominant building material and for manufacturing agricultural/domestic tools and implements. Significant losses of forests in the North due to climate change would have major economic consequences. In the South, such losses could affect livelihoods, especially in rural areas.

The pulp and paper industry now relies most heavily worldwide on a combination of virgin forests and tree plantations as its source of fiber. Forests with rotation times of 30–50 years or longer are particularly vulnerable to climate change. In North America, the original Canadian boreal forest is still a principal source of newsprint, although short-rotation pine plantations harvested every 20 years in the southeastern United States contribute an increasing share. Europe relies upon boreal plantations in Scandinavia and coniferous plantations in Germany. The vast taiga of Siberia has supplied pulp and fiber for Eastern Europe in the past, but those traditional market relationships have been disrupted. Recent work has raised concern that the very slow regeneration time of the Siberian taiga could lead to major long-term deforestation of the region. Models suggest that changes in albedo accompanying climate change could produce significant *cooling* in the region that would further diminish regrowth.

In industrialized countries, forests are likely to remain the primary source of raw materials for paper manufacture. In developing countries, the per capita demand for paper is still growing rapidly. Some developing countries may need to utilize other sources of raw materials. In 1989, 161 paper mills worldwide utilized bagasse (dry residue) from sugar cane (Silverio *et al.*, 1991). Mexico produced 8.5% of its paper from bagasse in that year (Atchison, 1989). China and other Asian countries use rice stalks in traditional paper making, and there is also the potential for utilizing cellulose fiber from other annual plant sources for pulp. Annual crops such as these are far less vulnerable to climate change.

The main source of natural rubber is the rubber tree, grown principally in Southeast Asia at latitudes below 10°N. More than six million hectares of land were used to produce 4.8 million tons of natural rubber in 1987 (UNIDO, 1993). Extreme temperatures can kill the tree, and relative humidity below 60–65% decreases the production of rubber resin (Tran *et al.*, 1979; Pham, 1973; Le, 1988). Excessive humidity or rainfall would encourage disease and pests (Tran *et al.*, 1979; Tayhieu Station, 1969; Chee, 1977). An increasing frequency of typhoons and tornadoes could cause significant damage to rubber plantations in countries such as Indonesia, the Philippines, Thailand, and Vietnam (Enquête Kommission, 1991; IPCC, 1990a; Tran, 1990). In Malaysia, the main rubber-producing country, yields could decline by 15% under a CO_2 doubling scenario associated with a regional temperature rise of 1–2°C and a 10% increase in rainfall (Asian Development Bank, 1994). Yield loss could rise to 25–40% due to the effects of temperature on the viscosity of resin when the tree is tapped. In Indonesia, the Philippines, coastal South Asia, and Indochina, warming for a CO_2 doubling could be in the range 0.5–3.0°C (CSIRO, 1992a). Adaptation to climate change could involve preparing for new developments outside the 10°N latitude zones. Plantations could be protected from high winds and colder winds by planting dense buffer trees to act as windbreaks. Increased irrigation could be used in Africa and parts of Asia, which might become drier. All of these measures could push up the cost of natural rubber and encourage substitution of synthetic material. This would damage the economies of rubber-growing countries.

Climate change could reduce biodiversity in forested regions, which in turn could reduce the availability of pharmacologically active natural products derived from plants, bacteria, and animals (Wilson, 1992). A quarter of the prescription pharmaceutical drugs dispensed in the United States between 1959 and 1973 had plant-derived active ingredients (Farnsworth and Socjarto, 1985). Plants provide 15–20% of pharmaceuticals in Japan and provided 35–40% in West Germany before unification (Principe, 1989a). The total market value of plant-derived pharmaceutical drugs was an estimated $43 billion in the Organisation for Economic Cooperation and Development (OECD) countries in 1985 (Principe, 1989b).

While it is difficult to make precise links between climate change and species loss, attempts to identify climate-altered conditions that might lead to extinction have been made (Gates, 1993; Wyman, 1991; Rose and Hurst, 1992; Emanuel *et al.*, 1984). Each of these studies predicts major relocation of species and ecosystems. Many species, especially trees, are unlikely to be able to migrate successfully if climate change occurs rapidly.

Surveys suggest that the tropics remain the most potentially productive source of future drugs, especially the Amazon (Peters *et al.*, 1989) and Costa Rica (Tangley, 1990). Most of the focus to date has been on the commercial potential of drugs for markets in industrialized countries, but populations in tropical countries might be disadvantaged if local species upon which they have come to depend for medication were to disappear because of climate change.

11.6.1.3. Textiles

Four fibers form the basis of the textile industry: cotton, wool, cellulosic fibers (rayon), and synthetics. Other fibers—for example, silk and linen—make up a small fraction of the total. Climate has little effect on synthetic fibers, but industries that rely upon natural fibers will be affected in a variety of ways by climate change.

Warming is likely to increase the risk of disease vectors and insect pests for cotton, sheep, and the forest sources of cellulose (Smith and Tirpak, 1989). Altered water availability is most serious for the growing of cotton, which is heavily dependent on irrigation in Central Asia, Egypt, and other parts of the world. Altered precipitation patterns could diminish the usefulness of large reservoirs and water distribution systems. Sheep are most often raised on marginal, drier lands such as in Australia, which accounts for nearly 70% of the world trade in wool for apparel (*Economist*, 1992). The boundaries of suitable grazing regions are likely to shift under any significant alteration of climate. Since many tree species are suitable for producing cellulosic fibers, they are much less vulnerable to altered precipitation patterns.

All fibers require massive amounts of water for processing and dyeing. One ton of wool requires an estimated 200,000 liters of water from shearing to the finished fabric (Watson, 1991). Cellulosic fibers require substantial amounts of water in the pulpmaking process, and cotton and wool require substantial amounts for cleaning. Synthetic fibers require relatively little water after the petroleum refining stage. Approximately 70% of water use is associated with the dyeing of finished cloth (Watson, 1991). Recycling could reduce the degree of dependence on water.

If precipitation patterns were to shift significantly, the production of finished cloth and industrial fibers might shift away from regions that were becoming drier. There might also be a shift toward synthetic and cellulosic fibers, which are less vulnerable. Apparel, carpet, tire cord, and other industries dependent on textile fiber are relocating in response to changing labor markets. This suggests that textile industries located close to supplies of raw materials might respond rapidly to climate-induced changes in availability. While global production might not be greatly altered, the loss of industries might have significant local effects.

11.6.2. Biomass

Biomass is estimated to account for 12–15% of global primary energy consumption (World Energy Council, 1993b; Hall *et al.*, 1993). After oil, coal, and natural gas, biomass is the world's most important source of energy. Trees are the most important source (64%) of biomass fuels in both urban and rural areas. The remainder consists of animal dung and crop residues used in the countryside. A total of 88.5% of wood biomass is used as firewood and the rest as charcoal (Smith *et al.*, 1993). Biomass is the most important source of energy in developing countries, where it accounts for 38% of consumption (Hall *et al.*, 1993). In some of the least industrialized countries, where there is limited access to electricity grids, 90% of energy needs is met by biomass. Financial constraints may also force people to use biomass in cities (see Chapters 12 and 22). Biomass in the domestic sector is used primarily for cooking, although it is also used for space heating in regions with cold winters—for example, Nepal and northern China. Biofuels are transported to cities by draft animals and lorries, sometimes over long distances. In the Sahelian countries of Africa, biofuel trade is a well-organized sector with impressive turnovers (Sow, 1990). Biogas generated by anaerobic fermentation is used mainly in the Asian countryside. According to Dutt and Ravindranath (1993), more than four million family house generators have been installed in China and one million in India. The numbers are growing rapidly.

In industrialized countries, biomass accounts on average for 2.8% of primary energy consumption (Hall *et al.*, 1993). The share is, however, 13% in Scandinavia and 5% in France and the United States (Corté, 1994; and FAO, 1994). In industrialized countries, most biomass is used for household heating in rural areas and small towns. For example, this accounts for 81.5% of biomass consumption in France (Barbier *et al.*, 1994). Usually, fuelwood is burned in stoves in the industrialized countries, though woodchips are becoming more popular for district heating in Austria and Scandinavia. Large amounts of biomass are collected directly by users, bypassing the market system (Riedacker and Robin, 1987).

In both developing and industrialized countries, agroindustries use large amounts of biomass wastes for energy generation. Cogeneration is important in the sugar cane industry in the United States, Cuba, and Mauritius (Turnbull, 1993). These industries do not generally need to buy extra biomass to meet their energy demand. In developing countries, biomass, mainly fuelwood, is also used for drying tea and tobacco, for processing food, for cremation, and for public baths. There is an increasing demand for biofuels in small industries in countries such as India and Brazil (La Rovere, 1994). Large amounts of charcoal based on eucalyptus are used in Brazilian steel mills (Sampaio, 1994).

Biomass is also used for transportation fuels. In Brazil, liquid biofuels accounted for 18% of transportation fuel in 1987 (Goldemberg *et al.*, 1993). There are about fifty facilities manufacturing ethanol fuel from grain, mainly maize, in the United States. In 1987, this accounted for 8% of the U.S. gasoline market (Wyman *et al.*, 1993). Zimbabwe also has started to produce ethanol from sugar cane.

11.6.2.1. Impacts of Climate Change on Domestic Biomass

Fuelwood supply depends very much on forest area and the quantity and pattern of rainfall. As deforestation proceeds around cities in developing countries, mainly to provide more cropland, fuelwood has to be transported greater distances (see also Chapters 12 and 22). Forest yields are likely to be reduced with decreasing rainfall. Thus, in dry and densely populated areas, fuelwood may become more scarce due to a combination of population growth and climate change. The projected deficit of demand over supply could be most critical in tropical Africa (see Chapters 1 and 15).

The growth of annual and perennial plants is likely to increase with CO_2 enrichment and higher temperatures. Forested areas may expand northward, increasing resources in industrialized countries. Biomass resources are expected to increase in Italy (Bindi *et al.*, 1994), for instance, but this depends on the precise nature of any climate change as well as on the crop and the crop system used (see Chapters 1 and 13). Except in regions with reduced rainfall, future biomass supply in industrialized countries is likely to exceed the present demand. As discussed in Chapter 19, more intense biomass production to substitute for fossil fuels may become an important mitigation option in countries with good rainfall regimes. This would also be the case in Latin America and tropical Africa.

Biomass can be converted to methane gas through anaerobic conversion in digestors. The rate of methane production increases at high temperatures and may be interrupted in winter, for example, in North China (Rajabapiah *et al.*, 1993). Global warming may therefore lengthen periods during which high yields can be obtained and extend the areas in which biogas can be produced without heating the digestor, for example, in households.

11.6.2.2. Impacts of Climate Change on Industrial Biomass

The impact of climate change on the availability of agricultural wastes for electricity production and cogeneration will be similar to the impact of climate change on the supply of food and agricultural raw materials. Industries and biofuel production dependent on brought-in supplies of annual biomass crops are more vulnerable. Raw material is more costly to transport than is food.

Large fluctuations in the yield of biomass energy systems could have greater impacts on system reliability and costs than might be estimated by extrapolating experience gained from food crops (Hillman and Petrich, 1994). In 1993, ethanol production from sugar cane in Zimbabwe stopped completely due to an extreme drought. The impact of extreme events on industries dependent on wood biomass are likely to be less dramatic.

11.6.2.3. Adaptation to Gradual Change

In general, decreasing rainfall will lower biomass production, particularly in the low latitude countries. Changes in rainfall intensity and evapo-transpiration will also be relevant. In some tropical and semi-arid countries, adaptation may become necessary before climate change becomes perceptible because of population growth and deforestation around cities. Adaptation could take the form of either switching to new fuels or more efficient production and conversion of biomass.

As forests become more scarce, increasing amounts of fuelwood may be derived from trees planted on private land. Countries that face fuelwood or charcoal shortages and cannot afford or do not want to switch to imported fossil fuels could develop small-scale projects to test the feasibility of new biomass conversion and production techniques. Adaptation will be especially necessary in areas with low land availability or low growth potential—for example, arid and semi-arid regions. Biofuel supply from trees, shrubs, grass, or crop residues can be increased or maintained by the following means:

- Better management of natural resources by giving local populations a stake in sustainably grown forests (Bertrand, 1993)
- Planting more trees on agricultural land and establishing better prices for biofuels derived from these plantations to provide rural incomes and incentives to grow more biomass
- Carrying out research to identify higher yielding species, preferably trees and shrubs, that are easy to propagate and are better adapted to extreme conditions like droughts and acidic or saline soils (Riedacker *et al.*, 1994)
- Increasing the productivity of agricultural land.

More details about these measures are provided in Chapters 1, 13, 15, and 25. There are many technologies that can contribute to the improved conversion of biomass (Antal and Richard, 1991; Mezerette and Girard, 1990).

11.6.2.4. Adaptation to Extreme Events

Energy systems based on a mixture of biofuels and fossil fuels are likely to be less vulnerable to climate change than systems based on a single biofuel. Options that reduce vulnerability include injecting gas derived from biomass into natural gas networks, gasohol (a mixture of gasoline and ethanol), or methanol, which can be derived from biomass, natural gas, or coal. Electricity generation plants that can use either perennial or annual crops also will be less vulnerable (Larson, 1993).

11.6.3. Renewable Energy

As discussed in Chapter 19, the accelerated development of renewable energy systems—for example, solar energy and wind—could help to mitigate emissions of greenhouse gases. Mitigation strategies are likely to have the largest impact on the development of renewable energy, but climate change could modify the potential to some degree:

- By increasing or decreasing the flow of the renewable energy resource being used (e.g., hydroelectric potential)
- By affecting the technology used to collect or convert the resource into a useful form, generally in ways that increase system costs or reduce performance (e.g., if high winds damage photovoltaic installations)
- By affecting willingness to develop renewable energy systems, if climate change places stress on ecosystems or reduces the value of the resource.

Renewable energy systems tend to be developed first at locations that have the best resources and that can be developed to yield useful energy at the lowest cost. Climate change might, at favorable locations, increase the availability of some renewable resources. Others, such as solar radiation, are affected only by changes in cloudiness. Developers of renewable energy systems tend to match equipment carefully to the characteristics of the site, taking account of factors such as the flow of the renewable resource. This both increases the amount of the resource used and reduces the delivered cost of energy. A change in climate can reduce the quality of the match between installed technology and a site's resources, thereby increasing cost, wasting resources, or reducing performance. Hillsman and Petrich (1994) provide a comprehensive review of the possible impacts of climate change on renewable energy systems. Any negative effects of climate change on the operation and cost of renewable energy systems need to be set against positive benefits arising from their role in GHG mitigation strategies.

11.6.3.1. Impacts on Resources

11.6.3.1.1. Hydroelectricity

Hydroelectricity is currently the most exploited renewable energy resource around the world. The principal pathways through which climate could affect hydroelectric resources are through a change in precipitation or the conditions (temperature, insolation, wind, humidity) that affect evaporation from reservoirs. Precipitation can change in quantity, seasonality, and form (for example, snowfall versus rainfall). The percentage of a basin's supply lost to evaporation from reservoirs or lakes depends not only on how much of the basin they cover but also on temperature, wind, and humidity, as well as vegetation cover and the location of a basin's reservoirs in relation to precipitation sources (Cohen, 1987a, 1987b). In an arid region, a change in temperature alone might have less effect on evaporation than it would in a more humid one. All else being equal, reductions in precipitation should also tend to reduce humidity and therefore increase evaporation. An increase in transpiration from vegetation can also reduce the net supply of water to a basin.

An analysis of hydroelectric generating potential in the James Bay region of Quebec under two climate scenarios estimated that an increase in precipitation would outweigh an increase in evaporation, and that generation could increase 6.7–20% (Singh, 1987). Possible responses to decreased production include reliance on other generating sources or the use of demand-side management (*Energy, Economics and Climate Change*, 1992).

Fitzharris and Garr (1995) conclude that changes in precipitation patterns resulting from climate change could result in increased hydroelectric production in the South Island of New Zealand, arising at more useful times of the year. Lettenmaier *et al.* (1992) estimate that a 2–4°C warming in Northern California would reduce snow accumulation and shift peak runoff from spring to winter. This would result in a closer match between runoff and peak power demand in the region, but, on the other hand, water supply would become less reliable. More efficient reservoir management alone could not mitigate the additional risk of floods, and additional reservoir storage for flood flow would be needed (Lettenmaier and Sheer, 1991). Such a response might increase the potential for hydroelectric production during floods but affect the availability of the reservoir system to meet other demands, such as those for irrigation. Where water storage is dominated by snowmelt rather than by reservoirs, river basins are likely to be more sensitive to temperature shifts than to hydrological changes (Lettenmaier and Sheer, 1991).

11.6.3.1.2. Wind

Wind energy depends in part on the temperature gradient from low latitudes to high (Grubb and Meyer, 1993). Estimates of potential climate change based on general circulation models have suggested that the high latitudes should warm relatively more than the tropics (IPCC, 1990a). By itself this would reduce the temperature gradient and global wind resources. The wind resource at some sites, however—such as Altamont Pass in California—is determined largely by a combination of local topography and location, so that the resource there might display less direct sensitivity to changes in global circulation patterns. For existing wind installations, a change in the direction of the prevailing wind also could have a significant effect, depending on the orientation of the array.

Established wind energy systems are potentially vulnerable to changes in the local wind regime because wind energy flux varies with the cube of wind velocity (Cavallo *et al.*, 1993). Windspeeds above average contribute disproportionately to wind generator output. A changed frequency of higher windspeeds would have the greatest impact on output. Baker *et al.* (1990) analyzed long-term wind records at good but undeveloped wind sites and estimated that a 10% change in wind speeds could change wind energy yields by 13–25%, depending on the site and season. In general, if climate change manifests itself as greater variability in wind speeds, then wind energy production will tend to magnify this variability. An increase in wind resources would cause a disproportionately large increase in energy capture, while a decrease would cause a disproportionately large decrease.

Air density decreases with increasing temperature (Cavallo *et al.*, 1993), but an average increase of several degrees at a developed site should reduce output by at most 1–2%. In the high latitudes, where global warming might be more pronounced, air density and hence output could decrease to a greater extent (IPCC, 1990a).

11.6.3.1.3. Other renewable energy sources

Solar thermal and photovoltaic (PV) energy systems are dependent on local conditions (Radesovich and Skinrood, 1989; De Laquil *et al.*, 1993). Systems that concentrate sunlight, using either mirrors to reflect it or lenses to refract it, require direct sunlight. Increases in humidity, haze, or cloudiness reduce the effectiveness of concentrating systems (Kelly, 1993). Global climate change may alter cloud regimes, leading, possibly, to more clouds and less direct solar radiation (Enquête Kommission, 1991). Cloudiness has recently increased over Europe, North America, India, and Australia as a whole (IPCC, 1990a) but has decreased in Southern Australia and in the Sahel (CSIRO, 1992b).

Flat-plate PV systems utilize both direct sunlight and diffuse light scattered by clouds or humidity. Electricity from a flat-plate system in a region with relatively poor conditions might be 40% lower than that from an identical system in a region with excellent conditions (Zweibel and Barnett, 1993). This range is probably much greater than the difference produced by a gradual change in climate. Photovoltaic cells lose about 0.5% of their efficiency per °C above their rated temperature (Kelly, 1993).

Ocean energy systems comprise a variety of technologies, including tidal barrages, wave energy systems, and ocean thermal energy conversion (OTEC). There are few plans to develop ocean energy systems, though their use might grow in the longer term. It would be essential to take account of projected sea-level rise over the very long design life of tidal barrages (UK Climate Change Impacts Review Group, 1991). Wave energy resources at any particular site depend on global wind patterns to generate waves, nearshore winds that affect wave size, and nearshore seabed conditions that affect the refraction and reflection of waves (Cavanagh *et al.*, 1993).

Geothermal resources may be affected by a change in precipitation patterns where the steam resource depends on an aquifer recharged by surface water or precipitation. An assessment of the potential effects of climate change in New Zealand (Mundy, 1990) suggests that increased rainfall could increase the recharge of groundwater for most geothermal fields.

11.6.3.2. Impacts on the Value of Resources

Several renewable energy resources, especially PV systems, often have high value because they are dependable when demand for energy is highest. Some changes in climate could shift demand peaks or change energy yields, thus altering the value of affected systems.

The use of some renewable energy resources must be balanced against other potential uses. For example, in hydroelectric systems, energy production may compete with demands for water to provide wildlife habitat, irrigate crops, maintain navigation, and support recreation. In coastal ocean energy systems, energy production may compete with demands to protect wildlife habitat or support recreation. Climate change could lead the public to place greater value on the nonenergy uses of these resources, leading to a decline in energy production.

11.6.3.3. Impacts on Performance

Climate change could affect renewable energy technologies by changing the performance and the degree of damage incurred during extreme weather events. Adapting to these impacts could alter operation and maintenance costs. The economic impacts, however, are likely to be much lower than those deriving from climate mitigation policies designed to promote renewable energy. The main climate impacts are:

- For hydroelectric systems, precipitation patterns and climatic impacts on vegetation of the watershed could affect rates of siltation and reservoir storage capacity.
- Dust, insects, or ice can reduce wind energy production by about 8% (Lynette, 1989; Lynette and Associates, 1992). Light rainfall may clean the blades and increase energy yields by up to 3%, but heavy rain can reduce energy output by increasing turbulence (Baker *et al.*, 1990).
- Severe weather (tornadoes, hurricanes, snow, and ice) can damage wind machines (Jensen and Van Hulle, 1991), though the use of variable-speed wind turbines shows great promise for reducing this source of vulnerability (Lamarre, 1992).
- Soiling of the reflective surfaces of heliostats through dust can reduce energy capture by solar thermal systems by up to 8% when there is little rainfall (Radesovich and Skinrood, 1989). Efforts to reduce system costs have led to the development of larger heliostats that are more susceptible to wind damage.
- Soiling also affects PV systems (Goossens *et al.*, 1993). Moisture from storms or dew, lightning strikes, overheating, and voltage surges from cloud-induced transients can cause damage to electronic components (Conover, 1989; Kelly, 1993; Boes and Luque, 1993; Smith, 1989). An increase in peak windspeed or in the frequency or severity of storms would require the structures necessary to support PV collectors to be strengthened (Kelly, 1993).
- Ocean energy systems would be susceptible to storm damage (Cavanagh *et al.*, 1993).

11.7. Need for Future Assessments

Many elements of the industry-transportation-energy system are sensitive to extremes of climate rather than average conditions. The impacts of many variables also are specific to geographical regions. If work on the potential impact of climate change on the industry, energy, and transportation sectors is to develop beyond a simple identification of sensitivities, then it will be necessary to generate climate scenarios that cover a wider range of climate variables, including extreme events, and refer to changes at the regional as well as global levels. The capacity to produce such scenarios does not yet exist. Several types of climate impact in the industry, energy, and transportation sectors appear to merit further research.

Developing Country Studies

A number of comprehensive studies of the impacts of climate change on developed countries have been published. Extending this type of work to cover a wider range of developing countries, which are likely to be more sensitive to climate, should have a high priority. Agroindustries merit special attention, given the dependence of many developing countries on agriculture and derivative sectors. Very few studies have been conducted that trace the indirect effects of changes in agricultural activity on associated agroindustries. With a few exceptions, only casual inferences have been drawn concerning the secondary impacts of agricultural productivity on industrial output and employment. This problem can be addressed through economic models that characterize the relationships between agriculture and industrial activity. The linkages between marine ecosystems, fisheries, and fish-processing industries similarly need to be assessed.

Studies at the System Level

Much of the climate sensitivity exhibited by the industries examined in this chapter arises not because of the direct impacts of climate on an activity but because of interdependence *between* activities through either climate-sensitive markets or the use of raw materials and resources that are themselves climate sensitive. Studies that increase understanding of these interdependencies and the indirect consequences of climate change throughout the economic system are an essential complement to studies that focus on specific sectors and activities.

Energy Demand

There is a need for studies that assess the impacts of climate change on energy demand and the impact of GHG mitigation measures within a unified framework. The adoption of mitigation measures such as energy efficiency, renewable energy, or biomass production may influence the vulnerability of the energy system to climate change while, at the same time, the direct impacts of climate change may influence the degree of mitigation action required to attain specific GHG emission objectives. No studies addressing this topic have been identified.

Also, although there have been many studies of the impact of climate on energy demand, there is still considerable uncertainty about the possible links between the adoption of new air-conditioning systems and climate change, particularly in temperate climates where the use of air conditioning is currently marginal.

Transportation

More assessment of impacts on transportation infrastructure and operations is needed. It is especially important to extend studies to developing countries and countries with economies in transition. Issues other than submergence as a result of sea-level rise should be considered. The impacts of temperature, precipitation, and extreme events on both infrastructure and operations should be included.

Transportation issues should be addressed within the wider framework of changes in human settlement patterns. The question of the impacts of climate change on regional transportation systems via the redistribution of population and economic activities has been neglected. There is a need to develop and refine an assessment methodology by adapting existing transportation models. Eventually, such models would be employed to evaluate scenarios of regional growth derived from climate change impact studies of migration, agriculture, industry, and other sectors to determine changes in transportation demand and impacts on transportation networks.

Adaptation

The literature referring to industry, energy, and transportation focuses largely on sensitivities and impacts and is weak on questions of adaptation. There are some areas where research on adaptation could be appropriate. Infrastructure associated with some large-scale renewable energy projects, transportation systems, and buildings has a very long lifetime. It may be necessary to anticipate climate change by assessing how demand for facilities may be affected by climate in the long-term and how climate might influence technical aspects of design. The development of climate guidelines is necessary for the construction and location of coastal structures such as seawalls, harbors, jetties, piers, and causeways that have long lives (Hameed, 1993). Such guidelines would also assist with the design of new buildings, particularly with respect to their energy use characteristics. Improved monitoring of coastal zones, covering sea level, tidal and wave patterns, weather, marine ecosystems, coral reefs, coastal geomorphology, and sedimentology also is required (Crawford, 1993).

References

Acosta, R. and R. Suarez, 1992: *La Caña de Azucar y el Desarrollo Sostenible*. Proceedings of Encuentro de Alta Tecnologia Canaveira, September, Cuba/Saõ Paulo, Araras SP, CETESB SP, 12 pp.

Aittoniemi, P., 1991: Influences of climatic change on the Finnish energy economy. In: *Energy and Environment 1991* [Kainlauri, E., A. Johanson, I. Kurki-Suonio, and M. Geshwiler (eds.)]. American Society of Heating, Refrigerating and Air-Conditioning Engineers Inc., Atlanta, GA.

Aluminium Association Inc., 1991: *Patterns of Energy and Fuel Usage in the US Aluminum Industry: Full Year 1989*. Report to the U.S. Department of Energy, Washington, DC, 11 pp.

Andrews, C., 1989: Anticipating air conditioning's impact on the world's electricity producers. *Energy Journal*, **10(3)**, 107-120.

Antal, E. and O. Starosolzky, 1990: *Role of Climate Change in the Life in Hungary* [Mathe, Gy. (ed.)]. Hungarian Ministry for Environment and Regional Policy, Budapest, Hungary, 24 pp.

Antal, M.J. and J.R. Richard, 1992: A new method for improving the yield of charcoal from biomass. *Proceedings of the 6th European Conference on Biomass*, Athens, Greece, Elsevier, pp. 845-849.

Aoki, Y., 1989: *Climatic Sensitivity of Downhill Ski Demands*. NIES, Tsukuba City, Japan, personal communication.

Asian Development Bank, 1994: *Climate Change in Asia: Eight Country Studies*. Asian Development Bank, Manila, Philippines, 121 pp.

Atchison, J.E., 1989: Producing pulp and paper from sugar cane bagasse: a world wide perspective. *Sugar y Azucar* **84(9)**, 89-100.

Baker, M.M. and L.E. Olsson, 1992: Tourism: A climate-sensitive industry. *Industry and Environment*, **15(3-5)**, 9-16.

Baker, R.W., S.N. Walker, and J.E. Wade, 1990: Annual and seasonal variations in mean wind speed and wind turbine energy production. *Solar Energy* **45(5)**, 285-289.

Ball, R.H. and W.S. Breed, 1992: *Summary of Likely Impacts of Climate Change on the Energy Sector*. U.S. Department of Energy, Washington, DC.

Barbier, C., B. Dessus, S. Lacassagne, and M.C. Zelem, 1994: A. Riahle and P. Radanne le Bois, *Energie en France Evaluation Prospective du Potentiel Mobilisable à l'Horizon 2015 et Ses Conséquences sur l'Environnement*. Les Cahiers du CLIP, Energie et Environnement No. 3, Paris, France, 80 pp.

Bertrand, M. 1993: *La Nouvelle Politique Forestière du Niger et les Marchés Ruraux du Bois Énergie: Innovations Institutionelles, Organisationelles et Techniques*. CIRAD, Montpellier, France, 7 pp.

Bertuglia, C.S., G. Leonardi, S. Occelli, G.A. Rabino, R. Tedai, and A.G. Wilson, 1987: *Urban Systems: Contemporary Approaches to Modeling*. Routledge, Chapman and Hall, New York, NY, 677 pp.

Bindi, M., G. Maracchi, and F. Miglietta, 1994: The effect of climatic change on biomass production in Italy. In: *Biomass for Energy and Industry* [Hall, D.O., G. Grassi, and M. Scheer (eds.)]. Proceedings of the 7th EC Conference, Ponte Press, Bochum, Germany, pp. 484-489.

Black, W.R., 1990: Global warming impacts on the transportation infrastructure. *TR News*, **150**, 2-34. Transportation Research Board, National Research Council, Washington, DC, September-October.

Blume, H., 1985: *Geography of Sugar Cane*. Edition Verlag Dr Albert Bartens, Berlin, Germany. 319 pp.

Boden, J.B. and R.M.C. Driscoll, 1987: House foundations—a review of the effect of clay soil volume on design and performance. *Municipal Engineering*, **4**, 181-213.

Boes, E.C. and A. Luque, 1993: Photovoltaic concentrator technology. In: *Renewable Energy: Source for Fuels and Energy* [Johansson, T.B. *et al.* (eds.)]. Island Press, Washington, DC, pp. 361-401.

Bowes, M.D. and P.R. Crosson, 1993: Consequences of climate change for the MINK economy: impacts and responses. In: *Towards an Integrated Impact Assessment of Climate Change: The MINK Study. Climatic Change,* **24(1-2)**, 131-158.

British Gas plc, 1992: *Financial and Operating Statistics for the Year Ended 31 December 1991*. British Gas, plc, London, UK, 52 pp.

Brown, H.M., 1989: Planning for climate change in the Arctic—the impact on energy resource development. Symposium on the Arctic and Global Change, Climate Institute, 26 October, Ottawa, Ontario, Canada.

Bulger, J.M., 1985: Some cold facts for the Florida citrus grower. *Florida Citrus*. The Citrus Industry, July.

Buzzandi, P.J., 1979: *Coffee Production and Trade in Latin America.* U.S. Department of Agriculture, Foreign Agriculture Service, Washington, DC, May.

Carter, T.R., M.L. Parry, S. Nishioka, and H. Harasawa, 1994: *IPCC Technical Guidelines for Assessing Climate Change Impacts and Adaptations*. CGER-I015-'94, University College London and Center for Environmental Research, National Institute for Environmental Studies, Tsukuba, Japan, 59 pp.

Castello, A.M. and J.J. Carvalho Filho, 1992: Coffee and citrus crops in Brazil and the socioeconomic effects of freezes. In: *Socioeconomic Impacts of Climate Variations and Policy Responses in Brazil* [Magalhaes, A.R. and M.H. Glantz (eds.)]. Esquel Brasil Foundation, Brasília, Brazil, pp. 93-105.

Cavallo, A.J., S.M. Hock, and D.R. Smith, 1993: Wind energy: technology and economics. In: *Renewable Energy: Source for Fuels and Energy* [Johansson, T.B. *et al.* (eds.)]. Island Press, Washington, DC, pp. 121-156.

Cavanagh, J.E., J.H. Clarke, and R. Price, 1993: Ocean energy systems. In: *Renewable Energy: Source for Fuels and Energy* [Johansson, T.B. *et al.* (eds)]. Island Press, Washington, DC, pp. 513-547.

Chee, K.H., 1977: Some new disorders of the Hevea. Proceedings of the RRIM Planters Conference 1977, 12-14 July, Kuala Lumpur, Malaysia, pp. 179-188.

Cline, W.R., 1992: *The Economics of Global Warming*. Institute for International Economics, Washington, DC, 399 pp.

Cohan, D., R.K. Stafford, J.D. Scheraga, and S. Herrod, 1994: The global climate policy evaluation framework. Proceedings of the 1994 A&WMA Global Climate Change Conference, 5-8 April, Phoenix, AR. Air and Waste Management Association, Pittsburgh, PA, 15 pp.

Cohen, S.J., 1987a: Projected increases in municipal water-use in the great lakes due to CO_2-induced climate change. *Water Resources Bulletin,* **23(1)**, 91-101.

Cohen, S.J., 1987b: Sensitivity of water resources in the Great Lakes region to changes in temperature, precipitation, humidity and wind speed. In: *The Influence of Climate Change and Climate Variability on the Hydrologic Regime and Water Resources* [Soloman, S.I., M. Beran, and W. Hogg (eds.)]. IAHS Publication No. 168, International Association of Hydrological Sciences, Wallingsford, UK, 640 pp.

Comisión de Cambio Climático/Academia de Ciencias de Cuba, 1991: *Efectos Potenciales del Cambio Climático Global en Cuba: Una Evalucación Preliminar*. Instituto de Meterologia, Academia de Ciencias de Cuba, Habana, 33 pp.

Conover, K., 1989: *Photovoltaic Operation and Maintenance Evaluation.* Report EPRI GS-6625, Electric Power Research Institute, Palo Alto, CA.

Corté, G., 1994: Développement des filières biomasse énergie dans quelques régions du monde: situation en Europe. In: *Guide Biomasse Energie.* Academia Louvain Laneuve, Louvain, Belgium, pp. 23-44.

Crawford, M.J., 1993: Coastal management in island states: potential uses of satellite imagery, aerial photography and geographic information systems. Proceedings of the IPCC Eastern Hemisphere Workshop, 3-6 August, Tsukuba, Japan, pp. 269-272.

CSIRO, 1992a: *Climate Change Scenarios for South and South-East Asia.* Asian Development Bank Regional Study on Global Environmental Issues, P.B.1 Mordialloc, Australia, 37 pp.

CSIRO, 1992b: Regional impact of the enhanced greenhouse effect on Victoria. In: *Annual Report 1990-91* [Whetton, P.H., A.M. Powler, C.D. Mitchell, and A.H. Pittork (eds.)]. CSIRO, Melbourne, Australia, 50 pp.

Daniels, R.C., V.M. Gornitz, A.J. Mehta, S.C. Lee, and R.M. Cushman, 1992: *Adapting to Sea-Level Rise in the US Southeast: The Influence of Built Infrastructure and Biophysical Factors on the Inundation of Coastal Areas.* ORNL/CDIAC-54, Oak Ridge National Laboratory, Oak Ridge, TN.

Darmstadter, J., 1991: *Energy*. Report V of *Influences of and Responses to Increasing Atmospheric CO_2 and Climate Change: The MINK Project.* Report DOE/RL/01830-H8, U.S. Department of Energy, Washington, DC, 57 pp.

De Laquil, P., D. Kearney, M. Geyer, and R. Diver, 1993: Solar thermal electric technology. In: *Renewable Energy: Source for Fuels and Energy* [Johansson, T.B. *et al.* (eds.)]. Island Press, Washington, DC, pp. 213-296.

Deering, A.M., 1994: Climate change and the insurance industry. *The Earth Observer*, **6(4)**, 41-43.

Dowlatabadi, H. and M.G. Morgan, 1993: A model framework for integrated studies of the climate problem. *Energy Policy,* **21(3)**, 209-221.

Dutt, G.S. and N.H. Ravindranath, 1993: Bioenergy: direct application in cooking. In: *Renewable Energy Sources for Fuels and Electricity* [Johansson, T.B. *et al.* (eds.)]. Island Press, Washington, DC, pp. 787-815.

Economist, 1992: Dumped upon: farmers need fairer, freer trade to prosper. *The Economist: A Survey of Australia*, **323(7753)**, 14.

Electricity Council, 1988: *Annual Report and Accounts 1987/88*. Electricity Council, London, UK, 83 pp.

Emanuel, W.R., H.H. Shugart, and M.P. Stevenson, 1984: Climate change and the broad scale distribution of terrestrial ecosystem complexes. *Climatic Change,* **7**, 29-47.

Energy Economist, 1988: It is not just farmers who suffer from drought. *Financial Times Business Information*, July, pp. 4-7.

Energy, Economics and Climate Change, 1992: New Zealand drought sharpens national energy debate. *Energy, Economics and Climate Change*, **2(7)**, 2-6.

Enquête Kommission German Bundestag, 1991: *Protecting the Earth: A Status Report with Recommendation for a New Energy Policy*. Bonn University, Bonn, Germany, pp. 251-329.

Eunson, E.M., 1988: *Proof of Evidence on System Considerations*. Hinkley Point 'C' Power Station Inquiry, Central Electricity Generating Board, London, UK, September, 60 pp. + app.

FAO, 1993: *Yearbook Production: Vol. 46 1992*. FAO Statistical Series No. 112, Food and Agriculture Organization of the United Nations, Rome, Italy.

FAO, 1994: *Bioenergy for Development: Technical and Environmental Dimensions*, FAO Environment and Energy Paper 13, Food and Agriculture Organization of the United Nations, Rome, Italy, 78 pp. + annexes.

Farnsworth, N.R. and D.D. Socjarto, 1985: Potential consequences of plant extinctions in the United States on the present and future availability of prescription drugs. *Economic Botany,* **39**, 231.

Fitzharris, B. and C. Garr, 1995: Climate, water resources and electricity. In: *Greenhouse 94.* Proceedings of a conference jointly organised by CSIRO, Australia and the National Institute of Water and Atmospheric Research, CSIRO, Melbourne, Australia.

Frederick, K.D. 1991: *Water resources.* Report IV of *Influences of and Responses to Increasing Atmospheric CO_2 and Climate Change: The MINK Project.* U.S. Department of Energy, Washington, DC, 153 pp.

Friesz, T.L., R. Tobin, and P.T. Harker, 1983: Predictive intercity freight network models: the state of the art. *Transportation Research—A*, **17A(6)**, 409-418.

Fukuoka District Meteorological Observatory, 1987: *Survey of Impact of Climate Change upon Local Industries and of Paleoclimatic Data*. Japan Meteorological Society, Tokyo, Japan, 136 pp.

Gates, D.M., 1993: *Climate Change and Its Biological Consequences*. Sinauer and Associates, Sunderland, MA, 280 pp.

Gery, M.W., R.D. Edmond, and G.Z. Whitten, 1987: *Tropospheric Ultraviolet Radiation: Assessment of Existing Data and Effect on Ozone Formation.* U.S. Environmental Protection Agency, Research Triangle Park, NC.

Goldemberg, J., L. Monaco, and I. Macedo, 1993: The Brazilian Fuel Alcohol Program. In: *Renewable Energy: Source for Fuels and Energy* [Johansson, T.B. *et al.* (eds.)]. Island Press, Washington, DC, pp. 841-863.

Goossens, D., Z.Y. Offer, and A. Zangvil, 1993: Wind tunnel experiments and field investigations of eolian dust deposition on photovoltaic solar collectors. *Solar Energy*, **50(1)**, 75-84.

Grubb, M.J. and N.I. Meyer, 1993: Wind energy, resources, systems and regional strategies. In: *Renewable Energy: Source for Fuels and Energy* [Johansson, T.B. *et al.* (eds.)]. Island Press, Washington, DC, pp. 157-212.

Hall, D., F. Rossillo-Calle, R. Williams, and J. Wood, 1993: Biomass for energy: supply prospects. In: *Renewable Energy: Sources for Fuels and Energy* [Johansson, T.B. *et al.* (eds.)]. Island Press, Washington, DC, pp. 593-651.

Hameed, F., 1993: Coastal defence: a Maldivian case study. Proceedings of the IPCC Eastern Hemisphere Workshop, 3-6 August, Tsukuba, Japan, pp. 281-291.

Hata, H., T. Takeuchi, A. Nagano, and H. Kobayashi, 1993: Effect of sea level rise on fishing ports: a case study from Japan. Proceedings of the IPCC Eastern Hemisphere Workshop, 3-6 August, Tsukuba, Japan, pp. 143-150.

Herring, H., R. Hardcastle, and R. Phillipson, 1988: *Energy Use and Energy Efficiency in UK Commercial and Public Buildings up to the Year 2000.* Energy Efficiency Series 6, Energy Efficiency Office, HMSO, London, UK, 173 pp.

Hillsman, E.L. and C.H. Petrich, 1994: *Potential Vulnerability of Renewable Energy Systems to Climate Change.* Report ORNL-6802, Oak Ridge National Laboratory, Oak Ridge, TN, 55 pp.

Horgan, J., 1993: Antarctic meltdown: the frozen continent's ice cap is not as permanent as it looks. *Scientific American*, **266(3)**, 7-12.

IBI Group, 1990: *The Implications of Long-Term Climatic Change on Transportation in Canada.* CCD 90-02, Canadian Climate Change Digest, Atmospheric Environment Service, Environment Canada, Downsview, Ontario, Canada, 8 pp.

Imse, A., 1994: Russia's wild capitalists take aluminium for a ride. *New York Times*, 13 February.

IPCC, 1990a: *Climate Change: The IPCC Scientific Assessment.* Cambridge University Press, Cambridge, UK, 364 pp.

IPCC, 1990b: *Climate Change: The IPCC Impacts Assessment.* Australian Government Publishing Service, Canberra, Australia, 275 pp.

IPCC, 1992: *Climate Change 1992: Supplementary Report to the IPCC Scientific Assessment.* Cambridge University Press, Cambridge, UK, 200 pp.

IPCC, 1993: *Climate Change 1992. Supplementary Report to the IPCC Impacts Assessment.* Australian Government Publishing Service, Canberra, Australia, 102 pp. + app.

Irwin, N. and W.F. Johnson, 1988: *The Implications of Long Term Climatic Changes on Transportation in Canada.* Report prepared for Transport Canada by the IBI Group, Toronto, Canada, 8 pp.

Jansen, H.M.A., O.J. Kuik, and C.K. Spiegel, 1991: Impacts of sea level rise: an economic approach. In: *Climate Change: Evaluating the Socio-Economic Impacts.* OECD, Paris, France, pp. 73-105.

Japan Architecture Society, 1992: *Impact of Architecture upon Global Environment.* Japan Architecture Society, Tokyo, Japan, 92 pp.

Jensen, P.H. and F.J.L. Van Hulle, 1991: Recommendations for a European wind turbine standard load case. In: *Proceedings of the ECWEC'90 Conference* [Petersen, E.L. (ed.)]. Madrid, Spain, pp. 33-38.

Johnson, S.R., 1991: Modeling the economic impacts of global climate change for agriculture and trade. In: *Climate Change: Evaluating the Socio-Economic Impacts.* OECD, Paris, France, pp. 45-71.

Kane, S., J. Reilly, and J. Tobey, 1992: An empirical study of the economic effects of climate-change on world agriculture. *Climatic Change,* **21(1)**, 17-35.

Kelly, H., 1993: Introduction to photovoltaic technology. In: *Renewable Energy: Source for Fuels and Energy* [Johansson, T.B. *et al.* (eds.)]. Island Press, Washington, DC, pp. 297-336.

Kempton, W., 1992: Special issue on air conditioning. *Energy and Buildings,* **18(3-4)**, 171.

Kitajima, S., T. Ito, N. Miura, Y. Hosokawa, M. Tsutsui, and K. Izumi, 1993: Impacts of sea level rise and cost estimate of countermeasures in Japan. Proceedings of the IPCC Eastern Hemisphere Workshop, 3-6 August, Tsukuba, Japan, pp. 115-123.

Kurosaka, H., 1991: Impacts of temperature on electric energy demand for the lighting services. In: *Proceedings of the International Conference on Climate Impacts on the Environment and Society (CIES).* University of Tsukuba, Isiabaki, Japan, January 27-February 1, 1991, WMO/TD No. 435, World Climate Programme, WMO, Geneva, Switzerland, pp. D.7-D.12.

La Rovere, E.L., 1994: Biomass in developing countries: economic competitiveness and social implications. 8th European Conference for Energy, Environment, Agriculture and Industry, 3-5 October 1993, Vienna, Austria, pp. 83-86.

Lamarre, L., 1992: A growth market in wind power. *EPRI Journal*, **17(8)**, 4-15.

Lamothe and Periard, 1988: *Implications of Climate Change for Downhill Skiing in Quebec.* CCD 88-03, Environment Canada, Downsview, Ontario, Canada, 12 pp.

Larson, E.D. 1993: Technology for electricity and fuels from biomass. *Annual Review of Energy and Environment* **18**, 567-630.

Lashof, D.A. and D.A. Tirpak (eds.), 1990: *Policy Options for Stabilizing Global Climate.* Report to Congress, U.S. Environmental Protection Agency, Office of Policy, Planning, and Evaluation, Washington, DC, 810 pp.

Le, Q.H., 1988: *Estimation of the Agrometeorological Condition of Some Main Plants in the Vietnamese Highlands.* IME, Hanoi, Vietnam, 68-78 (in Vietnamese).

Leemans, R. and A.M. Solomon, 1993: Modelling the potential change in yield and distribution of the Earth's crop under a warmed climate. *Climate Research,* **3**, 79-96.

Lettenmaier, D.P. *et al.*, 1992: Sensitivity of Pacific Northwest water resources to global warming. *Northwest Environmental Journal,* **8**, 265-283.

Lettenmaier, D.P. and D.P. Sheer, 1991: Climatic sensitivity of California water resources. *Journal of Water Resources Planning and Management,* **117(1)**, 108-125.

Linder, K.P., M.J. Gibbs, and M.R. Inglis, 1989: *Potential Impacts of Climate Change on Electric Utilities.* Report EN-6249, Electric Power Research Institute, Palo Alto, CA.

Linder, K.P. and M.R. Inglis, 1989: The potential effects of climate change on regional and national demands for electricity. In: *The Potential Effects of Global Climate Change on the United States: Appendix H—Infrastructure* [Smith, J.B. and D.A. Tirpak (eds.)]. U.S. Environmental Protection Agency, May, Washington, DC, 38 pp.

Lonergan, S., 1989: Climate change and transportation in the Canadian Arctic. Climate Institute Symposium on the Arctic and Global Change, 26 October, Ottawa, Ontario, Canada.

Lynette, R., 1989: *Assessment of Wind Power Station Performance and Reliability.* Report EPRI GS-6256, Electric Power Research Institute, Palo Alto, CA.

Lynette, R. and Associates, 1992: *Assessment of Wind Power Station Performance and Reliability.* Report EPRI TR-100705, Electric Power Research Institute, Palo Alto, CA.

Macedo, I.C. 1991: Production and use of sugar cane for energy. ESSET 91, International Symposium on Environmentally Sound Energy Technologies and Their Transfer to Developing Countries and European Economies in Transition, October, Milan, Italy, 18 pp.

Matsui, S. *et al.*, 1993: *Assessment of Impacts of Climate Change on Management of Electric Utilities (1): Development of Impacts Assessment Methods.* Report Y920008, CRIEPI, Tokyo, Japan.

McGillivray, D.G., T.A. Agnew, G.A. McKay, G.R. Pilkington, and M.C. Hill, 1993: *Impacts of Climatic Change on the Beaufort Sea-Ice Regime: Implications for the Arctic Petroleum Industry.* CCD 93-01, Environment Canada, Downsview, Ontario, Canada, 17 pp.

Mezerette, C. and P. Girard, 1990: Environmental aspects of gaseous emissions from wood carbonisation and pyrolysis processes. In: *Biomass Pyrolysis: Liquid Upgrading and Utilisation* [Bridgwater, R.V. and G. Grassi (eds.)]. Elsevier Applied Science, London, UK, pp. 263-288.

Milbank, N., 1989: Building design and use: response to climate change. *Architects Journal,* **96**, 59-63.

Miller, B.A. and W.G. Brock, 1988: *Sensitivity of the TVA Reservoir System to Global Climate Change.* Report WR28-1-680-61, Tennessee Valley Authority Engineering Laboratory, Norris, TN.

Miller, B.A. *et al.*, 1992: *Impact of Incremental Changes in Meteorology on Thermal Compliance and Power System Operations.* Report WR28-1-680-109, Tennessee Valley Authority Engineering Laboratory, Norris, TN, February, 32 pp.

Miller, K.A. and M.H. Glantz, 1988; Climate and economic competitiveness: Florida freezes and the global citrus processing industry. *Climatic Change,* **12(2)**, 135-164.

Morris, R.E., M.W. Gery, M-K. Liu, G.E. Moore, C. Daly, and S.D. Greenfield, 1989: Sensitivity of a regional oxidant model to variations in climate parameters. In: *The Potential Effects of Global Climate Change on the United States* [Smith, J.B. and D.A. Tirpak (eds.)]. EPA-230-05-89, Office of Policy, Planning and Evaluation, U.S. Environmental Protection Agency, Washington, DC.

Mundy, C., 1990: Energy. In: *Climate Change: Impacts on New Zealand*. New Zealand Ministry of the Environment, Wellington, New Zealand.

Murrell, J.D., 1980: *Passenger Car Fuel Economy: EPA and Road*. EPA 460/3-80-010, U.S. Environmental Protection Agency, Motor Vehicle Emissions Laboratory, Ann Arbor, MI, September.

National Academy of Sciences, 1992: *Policy Implications of Greenhouse Warming*. National Academy Press, Washington, DC, 918 pp.

Nguyen, D.N. *et al.*, 1993: *Impacts of Climate Change on Natural Conditions and Socio-Economics in Vietnam*. HMS, Hanoi, Vietnam, July, 10 pp.

Niemeyer, E.V. *et al.*, 1991: *Potential Impacts of Climate Change on Electric Utilities*. Electric Power Research Institute, Palo Alto, CA.

Nishinomiya, S. and H. Kato, 1990: *A Quantitative Assessment of the Potential Impacts of Global Warming on the Electric Power Industry of Japan*. Report T89909, CRIEPI, April.

Nishioka, S., H. Harasawa, H. Hashimoto, T. Okita, K. Masuda, and T. Morita, 1993: *The Potential Effects of Climate Change in Japan*. CGER-I009-'93, Center for Global Environmental Research, National Institute for Environmental Studies, Environment Agency of Japan, Tsukuba, Japan, 93 pp.

Nishioka, S., I. Nazarov, *et al.*, 1992: Energy and industry-related issues, section B.III: Energy; human settlement; transport and industry sectors; human health; air quality; effects of ultraviolet-B radiation. In: *1992 Supplementary Report of Working Group II*. Intergovernmental Panel on Climate Change, Washington, DC, pp. 29-41.

Nordhaus, W.D., 1993: Rolling the dice—an optimal transition path for controlling greenhouse gases. *Resource and Energy Economics*, **15(1)**, 27-50.

Page, J.K. (ed.), 1990: *Indoor Environment: Health Aspects of Air Quality, Thermal Environment, Light and Noise*. World Health Organisation, Geneva, Switzerland.

Parry, M.L. and N.J. Read (eds.), 1988: *The Impact of Climatic Variability on UK Industry*. AIR Report 1, Atmospheric Impacts Research Group, University of Birmingham, Birmingham, UK, 65 pp. + app.

Pechan, E.H. and Associates, 1990: *Ozone Nonattainment Analysis: A Comparison of Bills*, U.S. Environmental Protection Agency, Office of Air and Radiation, Washington, DC.

Peters, C.M., A.H. Gentry, and R.O. Mendelsohn, 1989: Valuation of an Amazonian rainforest. *Nature*, **339(6227)**, 655-656.

Pham, D.B., 1973: *Rubber Growing*. Rural Publishing House, Hanoi, Vietnam, 45 pp. (in Vietnamese).

Plunkert, P.A. and E.D. Sehnke, 1993: *Aluminum, Bauxite, and Alumina: 1991*. Annual Report, U.S. Department of the Interior, Bureau of Mines, Washington, DC, 48 pp.

Principe, P.P., 1989a: The economic significance of plants and their constituents as drugs. *Economic and Medicinal Plant Research*, **3**, 1-17.

Principe, P.P., 1989b: *The Economic Value of Biological Diversity among Medicinal Plants*. OECD Environmental Monograph, Paris, France.

Radesovich, L.G. and A.C. Skinrood, 1989: The power production operation of Solar One, the 10 MWe solar thermal central receiver pilot plant. *Journal of Solar Energy Engineering*, **111(2)**, 145-151.

Rajabapiah, P., S. Jayahumar, and A.K.N. Reddy, 1993: Biogas electricity: the Pura village case study. In: *Renewable Energy Sources for Fuels and Electricity* [Johansson, T.B. *et al.* (eds.)]. Island Press, Washington, DC, pp. 787-815.

Riedacker, A. and S. Robin, 1987: La consommation de bois énergie en France après le second choc pétrolier. *Revue Forestière Française*, **XXXIX(2)**, 81-100.

Riedacker, A., E. Dreye, H. Joly, and C. Pafadnam, 1994: *Physiologie des Arbres et Arbustes en Zones Arides et Semi-Arides*. John Libbey, Montrouge, France, 489 pp.

Rose, C. and P. Hurst, 1992: *Can Nature Survive Global Warming?* World Wildlife Fund, Washington, DC, 59 pp.

Rosenberg, N.J. (ed.), 1993: Towards an integrated assessment of climate change: the MINK study. *Climatic Change*, **24(1-2)**, 1-173, special issue.

Rosenberg, N.J. and P.R. Crosson, 1991: *Processes for Identifying Regional Influences of and Responses to Increasing Atmospheric CO_2 and Climate Change—The MINK Project: An Overview*. DOE/RL/01830T-H5, U.S. Department of Energy, Washington, DC, August, 38 pp.

Rosenthal, D.H., H.K. Gruenspecht, and E.A. Moran, 1995: Effects of global warming on energy use for space heating and cooling in the United States. *Energy Journal*, **16(2)**, 77-96.

Rosenzweig, C. and M.L. Parry, 1994: Potential impact of climate change on world food supply. *Nature*, **367(6450)**, 133-138.

Rosenzweig, C., M.L. Parry, G. Fisher, and K. Frohberg, 1993: *Climate Change and World Food Supply*. Research Report No. 3, Environmental Change Unit, University of Oxford, Oxford, UK, 28 pp.

Sakai, S., 1988: *The Impact of Climate Variation on Secondary and Tertiary Industry in Japan*. Meteorological Research Note 180, Meteorological Society of Japan, Tokyo, Japan, pp. 163-173.

Sampaio, R.S., 1994: The production of steel through the use of renewable energy. 8th European Conference for Energy Environment, Agriculture and Industry, 3-5 October 1993, Vienna, Austria.

Scheraga, J.D. *et al.*, 1993: Macroeconomic modeling and the assessment of climate change impacts. In: *Costs, Impacts and Possible Benefits of CO_2 Mitigation*. IIASA Collaborative Paper Series, CP-93-2, Laxenburg, Austria, June.

Schipper, L. and S. Meyers, 1992: *Energy Efficiency and Human Activity: Past Trends and Future Prospects*. Cambridge University Press, Cambridge, UK, 385 pp.

Schneider, S.H., 1992: Will sea levels rise or fall? *Nature*, **356(6364)**, 11-12.

Scott, M.J., D.L. Hadley, and L.E. Wrench, 1994: Effects of climate change on commercial building energy demand. *Energy Resources*, **16(3)**: 339-354.

Scott, M.J., R.D. Sands, L.W. Vail, J.C. Chatters, D.A. Neitzel, and S.A. Shankle, 1993: *The Effects of Climate Change on Pacific Northwest Water-Related Resources: Summary of Preliminary Findings*. PNL-8987, Pacific Northwest Laboratory, Richland, Washington, 46 pp. + app.

Silverio, N. *et al.*, 1991: *La Diversificacion de la Agroindustria de la Cana de Azucar*. Diversification Project, GEPLACEA, Mexico City, Mexico.

Singh, B., 1987: Impacts of CO_2-induced climate change on hydro-electric generation potential in the James Bay Territory of Quebec. In: *The Influence of Climate Change and Climate Variability on the Hydrologic Regime and Water Resources* [Solomon, S.I., M. Beran, and W. Hogg (eds.)]. IAHS Publication No. 168, International Association of Hydrological Sciences, Wallingford, UK, 640 pp.

Skea, J., 1992: Physical impacts of climate change. *Energy Policy*, **20(3)**, 269-272.

Skea, J., 1995: Energy. In: *Economic Implications of Climate Change in Britain* [Parry, M. and R. Duncan (eds.)]. Earthscan, London, UK, pp. 64-82.

Smit, B. 1989: *Climate Warming and Canada's Comparative Position in Agriculture*. CCD 89-01, Climate Change Digest , Environment Canada, Downsview, Ontaria, Canada, 9 pp.

Smith, J.B. and D.A. Tirpak (eds.), 1989: *The Potential Effects of Global Climate Change on the United States*. EPA-230-05-89, Office of Policy, Planning and Evaluation, U.S. Environmental Protection Agency, Washington, DC, 411 pp. + app.

Smith, K., 1989: *Survey of US Line-Connected Photovoltaic Systems*. Report EPRI GS-6306, Electric Power Research Institute, Palo Alto, CA.

Smith, K.R., M.A. Khalil, R.A. Rasmussen, S.A. Thorneloe, F. Smanegdeg, and M. Apte, 1993: Greenhouse gases from biomass and fossil fuel stoves in developing countries: a Manila Pilot study. *Chemosphere*, **29(1-4)**, 479-505.

Solley, W.B., C.F. Merk, and R.P. Pierce, 1988: *Estimated Use of Water in the United States in 1985*. Circular 1004, U.S. Geological Survey, Washington, DC.

Sow, X. 1990. *Le Bois-énergie au Sahel*. ACCT CTA, Paris, France, 176 pp.

Sperling, D., 1987: Brazil, ethanol and the process of system change. *Transportation Research - A*, **12(1)**, 11-23.

Steiner, T., 1990: Transport. In: *Climatic Change: Impacts on New Zealand*. New Zealand Climate Change Programme, Wellington, New Zealand.

Stokoe, P., 1988: *Socio-Economic Assessment of the Physical and Ecological Impacts of Climate Change on the Marine Environment of the Atlantic Region of Canada—Phase 1*. CCD 88-07, Environment Canada, Downsview, Ontario, Canada, 9 pp.

Tangley, L. 1990: Cataloguing Costa Rica diversity. *BioScience*, **40(9)**, 633-636.

Tayhieu Station on Tropical Plants, 1969: *Influences of Climate on Rubber in the Tayhieu Area*. Rural Publishing House, Hanoi, Vietnam, 38 pp. (in Vietnamese).

Tebicke, H.L., 1989: Climate change, its likely impact on the energy sector in Africa. Remarks presented at IPCC WG2 Section 5 Lead Authors Meeting, 18-21 September, Tsukuba, Japan.

Titus, J.G. 1992: The costs of climate change to the United States. In: *Global Climate Change: Implications, Challenges and Mitigation Measures* [Majumdar, S.K., L.S. Kalkstein, B. Yarnal, E.W. Miller, and L.M. Rosenfeld (eds.)] Pennsylvania Academy of Science, Philadelphia, PA, pp. 385-409.

Titus, J.G., R.A. Park, S. Leatherman, R. Weggel, M.S. Greene, M. Treehan, S. Brown, C. Gaunt, and G Yohe, 1991: Greenhouse effect and sea level rise: the cost of holding back the sea. *Coastal Management*, **19(3)**, 171-204.

Tokyo District Meteorological Observatory, 1986: *Survey of Impact of Climate Change upon Local Industries and of Paleoclimatic Data*. Japan Meteorological Agency, Tokyo, Japan, 159 pp.

Tokyo Electric Company Ltd., 1994: private communication.

Tran, N.K., B.T. Thai, and X.H. Nguyen, 1979: *The Rubber Tree*. Science of Technology Publishing House, Hanoi, Vietnam, 330 pp. (in Vietnamese).

Tran, V.L., 1990: *Climate Variation and Change in Vietnam: Its Impacts on Agriculture*. Centre of Education, Research and Environment Development, Hanoi, Vietnam, 125 pp. (in Vietnamese).

Turnbull, J., 1993: *Strategies for Achieving a Sustainable, Clean and Cost Effective Biomass Resource*. Electric Power Research Institute, Palo Alto, CA, 20 pp.

UK Building Research Establishment, 1990: *Climate and Construction Operations in the Plymouth Area*. BRE/CR 35/90, Building Research Establishment, Watford, UK.

UK Climate Change Impacts Review Group, 1991: *The Potential Effects of Climate Change in the United Kingdom*. Department of the Environment, HMSO, January, London, UK, 117 pp. + annexes.

United Nations, 1990: *International Standard Industrial Classification of All Economic Activities*. Statistical Papers, Series M, No. 4, Rev. 3, New York, NY, 52 pp.

UNIDO, 1993: *Handbook of Industrial Statistics 1993*. United Nations Industrial Development Organisation, Vienna, Austria.

UN Statistical Office, 1993: *Industrial Statistics Yearbook 1991*. Department of International Economic and Social Statistics, United Nations, New York, NY.

U.S. Department of Energy, Office of Policy, Planning and Analysis, 1990: *The Economics of Long Term Global Climate Change: A Preliminary Assessment*. DOE/PE-0096P, Washington, DC, September.

U.S. Department of State, 1992: *National Action Plan for Global Climate Change*, Department of State Publication 10026, Bureau of Oceans and International Environmental and Scientific Affairs, December, 129 pp.

U.S. Environmental Protection Agency, Office of Mobile Sources, 1989: *User's guide to MOBILE4* (Mobile Source Emission Factor Model). EPA-AA-TEB-89-01, Ann Arbor, MI.

Wade, J.E., K. Redmond, and P.C. Klingeman, 1989: *The Effects of Climate Change on Energy Planning and Operations in the Pacific Northwest*. BPA 89-29, Bonneville Power Administration.

Walker, J.C., T.R. Miller, G.T. Kingsley, and W.A. Hyman, 1989: Impact of global climate change on urban infrastructure. In: *The Potential Effects of Global Climate Change on the United States* [Smith, J.B. and D.A. Tirpak (eds.)]. Chapter 2, Appendix H Infrastructure, EPA-230-05-89-058, U.S. Environmental Protection Agency, Washington, DC.

Wall, G., 1988: *Implications of Climate Change for Tourism and Recreation in Ontario*. CCD 88-05, Environment Canada, Downsview, Ontario, Canada, 16 pp.

Watson, J., 1991: *Textiles and the Environment*. The Economist Intelligence Unit, London, UK, 117 pp.

Wilson, E.O., 1992: *The Diversity of Life*. Belknap Press, Harvard University Press, Cambridge, MA, 423 pp.

Wolock, D.M., G.J. McCabe, G.D. Tasker, M.A. Ayers, and L.E. Hay, 1992: Sensitivity of water resources in the Delaware River basin to climate. Proceedings of the Workshop on the Effects of Global Climate Change on Hydrology and Water Resources at the Catchment Scale, 3-6 February, Tsukuba, Japan.

World Energy Council, 1993a: *Energy for Tomorrow's World*. Kogan Page, London, UK, 320 pp.

World Energy Council, 1993b: *Renewable Energy Resources: Opportunities and Constraints 1990-2020*. World Energy Council, London, UK, pp. 1.1-7.22.

Wyman, C., R. Bain, N. Hinman, and D. Stevens, 1993: Ethanol and methanol from cellulosic biomass. In: *Renewable Energy: Source for Fuels and Energy* [Johansson, T.B. *et al.* (eds.)]. Island Press, Washington, DC, pp. 865-923.

Wyman, R. (ed.), 1991: *Global Climate Change and Life on Earth*. Chapman and Hall, New York, NY, 282 pp.

Young, J.E., 1992: Aluminium's real tab. *World Watch,* **5**, 26-33.

Zweibel, K. and Barnett, A.M., 1993: Polycrystalline thin-film photovoltaics. In: *Renewable Energy: Source for Fuels and Energy* [Johansson, T.B. *et al.* (eds.)]. Island Press, Washington, DC, pp. 437-481.

12

Human Settlements in a Changing Climate: Impacts and Adaptation

MICHAEL J. SCOTT, USA

Principal Lead Authors:
A.G. Aguilar, Mexico; I. Douglas, UK; P.R. Epstein, USA; D. Liverman, USA; G.M. Mailu, Kenya; E. Shove, UK

Lead Authors:
A.F. Dlugolecki, UK; K. Hanaki, Japan; Y.J. Huang, USA; C.H.D. Magadza, Zimbabwe; J.G.J. Olivier, The Netherlands; J. Parikh, India; T.H.R. Peries, Sri Lanka; J. Skea, UK; M. Yoshino, Japan

CONTENTS

EXECUTIVE SUMMARY

Climate change will occur against a background of other non-climate environmental factors and socioeconomic factors that could either exacerbate or mitigate the effects of climate change. These other factors may, in many cases, dominate climate change. As compared with the 1990 or 1992 IPCC assessments of impacts, this discussion emphasizes the multiple and interactive pathways by which climate change exacerbates or mitigates the effects of other events that nations may find important. These interactions could occur in unexpected ways, with small changes having disproportionately large outcomes. The major conclusions follow.

Impacts on human settlements from climate change may be indirect, as well as direct. Direct effects of sea-level rise and extreme events are known to be important in coastal zones and island nations. However, many of the impacts on human settlements from climate change are likely to be experienced indirectly through effects on other sectors (for example, changes in water supply, agricultural productivity, and human migration). We have high confidence in the importance of these indirect effects because they depend on well-known mechanisms of social interaction rather than data specific to the climate of the future.

Thresholds beyond which impacts escalate quickly are unique to individual local situations and tend to depend on the degree of adaptive response. For example, the impact of sea-level rise on coastal communities critically depends on the degree to which human lives and assets can be protected, insured, or shifted to new locations that are unthreatened. As discussed in 1990 and 1992, the human settlements most vulnerable to climate change are likely to be in locations already stressed by high rates of population growth, urbanization, and environmental degradation. However, in addition to islands, coastal communities, and communities dependent on marginal rain-fed agriculture or commercial fishing discussed in the previous reports, vulnerable settlements include large primary coastal cities and especially squatter settlements located in flood plains and on steep hillsides. We have medium confidence in this result. The consensus is strong concerning the principle, but the data supporting it apply mainly to analogous circumstances.

Non-climate effects may be more important than climate change. Local environmental and socioeconomic situations are changing rapidly for reasons other than climate change. Worldwide, population growth, industrialization, urbanization, poverty, technological change, and government policy could overwhelm any effects of climate change. We have low confidence in this finding because the result is a matter of some controversy and there is insufficient data on climate-change effects on human settlements relative to those of other disturbances to resolve it.

A significant potential for noncoastal flooding (river basin and local urban flooding) is expected if precipitation intensity increases as a result of climate change. The IPCC Working Group I volume (Chapter 6, *Climate Models–Projections of Future Climate*) states that precipitation intensity may increase and that several models now project increases in higher-rainfall events. If intensity increases, the risk of flooding to settlement infrastructure could be very widely distributed across the planet, not just in coastal zones. One particular problem in providing estimates of damage is specifying the extent of future human intrusion into disaster-prone areas. We have low confidence in this finding because the consensus of climate modelers concerning the effects of climate change on intense rainfall events does not extend to regions and localities and because the degree of adaptation is likely to be important.

Health risks are potentially very large, especially in the informal settlements on the fringes of megacities of the developing world, but the probabilities are difficult to estimate. Current patterns of urbanization, extended into the next century, suggest that the most vulnerable human populations may become even more vulnerable. Economic and environmental refugees may introduce a number of exotic diseases into temperate-zone human settlements. Increased climate variability and associated extreme events can add new breeding sites and new bursts of activity for vector-borne diseases. This finding relies on the best thinking of the community of epidemiologists, but our confidence in the finding varies with the particular disease and location and because necessary supporting environmental data are adequate in some cases and extremely sketchy in others (see also Chapter 18).

Some other important findings are as follows:

- **Migration**—Many of the expected impacts in the developing world will occur because climate change may, by reducing natural resource productivity in rural areas, accelerate rural-to-urban migration, exacerbating already crowded conditions in the cities and further depleting the labor force of the countryside.
- **Water and Biomass**—Global warming can be expected to affect the availability of water resources and biomass, both of which are major energy sources

in many developing countries. Water and biomass resources are already under stress in many of these areas as a result of rising demand. Loss of water and biomass resources may jeopardize energy supply and materials essential for human habitation and energy production.

- **Energy**—Increasing human population and wealth provide a rising energy-demand baseline against which the consequences of changing climate will be played out. Energy demand will be affected by warming, but the direction and strength of the impact will depend on the extent of demands for space heating or cooling and the role of climate-sensitive sources of demand, such as irrigation pumping. Many of the largest increases in baseline demand will occur in developing countries, although not necessarily the largest changes as a result of climate.
- **Adaptation**—Many adaptive mechanisms are available to address each of the potential direct and indirect impacts of climate change on human settlements. The cost and effectiveness of each depend upon local circumstances.

12.1. Introduction

12.1.1. Why Human Settlements

A potentially important way in which climate change could affect human beings is through its effects on human settlements. Settlements, especially cities, have a central role in civilization as the primary generators of human wealth and the engines of social interaction and change. Whether explicitly acknowledged or not, environmentally appropriate economic and social development in settlements is a major goal of national environmental policy throughout the world. According to the United Nations Centre for Human Settlements, a human settlement can be judged on four criteria: (1) the quality of life it offers to its inhabitants; (2) the scale of nonrenewable resource use (including reuse); (3) the scale and nature of the use of renewable resources and the implications of the settlement's demands for sustaining production levels of these resources; and (4) the scale and nature of nonreusable wastes generated by production and consumption activities, the means by which these are disposed of, and the extent to which wastes affect human health, natural systems, and amenity values (Habitat, 1992a). Climate change could affect the sustainability of human settlements either by directly affecting the quality of life in settlements (e.g., by changing the probability of floods or the effects of air-pollution episodes), by modifying the effects of the settlements on their surrounding environments (e.g., by changing the demand for water or changing the assimilation capacity of wetlands), or by changing the economic underpinnings of the settlement (e.g., by changing the productivity of croplands, forests, or fisheries on which the settlement depends).

The 1990 and 1992 IPCC assessments dealt only with the direct effects of climate on human settlements, primarily infrastructure issues. The more subtle effects of human-settlement metabolism may be more important in some instances. Moreover, consideration of the indirect effects of climate change shifts the focus of attention from a narrow and necessarily speculative concern over future impacts on infrastructure to the more general pressing problems of economic and social development. Included in development could be an adaptation policy that includes a variety of adaptive mechanisms.

12.1.2. Guide to the Chapter

This chapter begins with a discussion of the non-climate environmental and socioeconomic factors that are expected to interact with climate change and sea-level rise (see Figure 12-1) in producing effects on human settlements. Some of the non-climate effects (e.g., urbanization or pollution) are likely to be significant enough in many cases to dominate the effects of climate alone but in any case could exacerbate climate effects. The chapter then discusses the possible impacts of climate on various sectors that might be affected (e.g., population, energy demand and supply, infrastructure, and water supplies). Some of these sectors have chapters of their own; however, this chapter emphasizes not only the direct effects of climate on human settlements but also the relationships between human settlements and these other, more directly affected, sectors. Finally, this chapter discusses adaptive responses to climate change—both those that may be taken autonomously by human settlements in response to climate change and other stresses and those that may be undertaken as a matter of deliberate adaptive policy by governments. Some readers will be disappointed that the chapter contains very few quantitative estimates of impact. This is by design. The very variety of human settlements and complexity of their environmental circumstances assure that almost any impact estimates will be local in scope and that both the absolute and relative importance of the various effects will be different in almost every circumstance. Except for sea-level rise, where some international estimates exist, the few estimates in the literature of the impact of climate change on human-settlement infrastructure and inhabitants are local and regional in scope. No study has yet attempted the difficult aggregation of climate and non-climate effects required for a proper analysis. Thus, the quantitative estimates provided in the chapter should be regarded as illustrative and anecdotal in nature.

12.2. Non-Climate Factors

A major difficulty in determining the impact of climate change on human habitat is the fact that many other factors, largely independent of climate change, are also important. In many cases, these other factors are far more important than climate change in terms of the risk they pose for human settlements. These non-climate factors will also increase the vulnerability of some regions to climate change. The most important of these factors include population growth, urbanization and industrialization, technology choices, and government policies. Other social factors, such as cultural clashes and warfare, also play a role. Vulnerability of settlements has to be judged on the basis of the susceptibility of the settlement's resources to damage via climate change, conditioned by the resilience of the resources and the technical, institutional, economic, and cultural capabilities of the settlement to cope with or manage the change

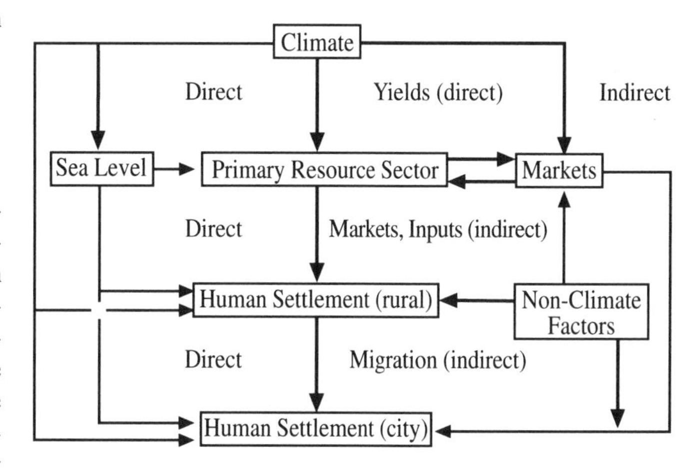

Figure 12-1: Relationship between direct and indirect effects of climate change, sea-level change, and non-climate factors on resource sectors, rural human settlements, and cities.

(IPCC CZMS, 1992; Turner *et al.*, 1994). One can reliably predict that certain developing countries will be extremely vulnerable to climate changes because they are both susceptible to the change and already at the limits of their capacity to cope with climatic events. These include populations in low-lying coastal regions and islands, subsistence farmers, populations in semi-arid grasslands, and the urban poor.

12.2.1. Economic and Social Development

12.2.1.1. Population Growth

The population of the world is estimated at about 5.3 billion. At its peak in the late 1960s and early 1970s, the average annual rate of increase in the world's population was close to 2%. Projections suggest that this rate of increase will fall to less than 1% by the second decade of the 21st century. However, the rate of abatement in the growth rate is expected to vary sharply from region to region. For example, whereas the world's population as a whole is expected to grow at 1.3% per year on average between 2000 and 2025, population is expected to grow at 2.1% in low-income countries but at only 0.2% in the Organization for Economic Cooperation and Development (OECD) countries (World Bank, 1991). Further, the differences in birth rates between rural and urban areas have widened, according to the United Nations World Fertility Survey (1986). Regions already struggling to cope with exploding populations can be expected to be exceptionally vulnerable to climate change.

12.2.1.2. Industrialization and Urban Growth

Since 1950, the number of people living in cities has tripled, increasing by more than 1.25 billion (World Commission on Environment and Development, 1987). By 1980, nearly one in three persons was an urban dweller, and one in ten lived in a city with a million or more inhabitants (a "million city"). By 1990, an estimated 42.65% of the world's people were urban residents. It has been estimated that by 2025 60% of the world's people will be living in towns and cities. In many countries, the fastest-growing areas are in coastal regions and estuaries, many of which are vulnerable to extreme events and sea-level rise (Marco and Cayuela, 1992). Already, two-thirds of the world population lives within 60 km of the coast; this is expected to rise to 75% by the year 2010 (WCC'93, 1994). The developed world has heavily concentrated its people in coastal areas (Friedman, 1984; Handmer, 1989; Marco, 1992; Boissonade and Davy, 1993; Lester, 1993; Skinner *et al.*, 1993; WCC'93, 1994).

Urbanization has been growing much faster and creating higher urban densities in developing countries than in developed countries (Berry, 1990). Although significant differences exist between countries, in developing countries as a group the urban population is growing much faster than in the industrialized countries. In the developing countries, the urban population is growing by an average of 3.6% per year compared to only 0.8% in the industrialized countries, which are believed to have essentially completed their urbanization process and in some cases are de-urbanizing. The uncontrolled expansion of towns and cities in the developing countries has led to overwhelmed transport, communications, water-supply, sanitation, and energy systems. Urban sprawl preempts other land uses near these towns and cities. Between 1980 and 2000, urban areas in developing countries are expected to grow from about 8 million hectares to more than 17 million hectares. The development policy in most developing countries after World War II favored modernization biased toward the urban-industrial sector, especially manufacturing targeted to satisfy domestic demand. In Latin America, for example, "import-substitution industrialization" registered an unprecedented success between 1950 and 1970; the regional gross domestic product (GDP) grew at an annual average rate of 5%. However, rural public investment was neglected, and the rural labor force was economically marginalized. Excess rural labor migrated to urban unemployment and informal activities (PNUMA/MOPU/AECI, 1989, Part III). Some analyses emphasize the attraction of cities ("pull factor") in the rural exodus, while others point out the aggravation of high rural population growth and the expelling process in rural areas ("push factor"). The latter definitely played a fundamental role in migration. In the last decades, rural life conditions have experienced continuous deterioration in relative and absolute terms, and large sectors of peasants have been excluded from the modernization process (Armstrong and McGee, 1985, Chapters 4 and 5), leaving them more vulnerable to climate variability. Growing scales of urban metabolism (urban demands for natural resources and the increase in the inflow and outflow of materials, products, energy, water, people, and wastes) add to pressures on the environment and could exacerbate the effects of climate change. Particularly salient has been the rapid growth of motorized transportation in urban areas.

12.2.1.3. Poverty

In the 1980s, many developing countries not only failed to keep pace with the industrial countries but saw their incomes fall. The living standards of millions in Latin America are now lower than in the early 1970s; in most of sub-Saharan Africa, living standards have fallen to levels last seen in the 1960s. According to some estimates, the percentage of the world's population living in absolute poverty decreased, at least up to the middle of the 1980s, but the absolute number of poor has increased (United Nations, 1989; Gilbert, 1992). It was estimated that in 1985 1.115 billion people lived in poverty in the developing world. That is approximately one-third of the total population of the developing world. Of these, 630 million were extremely poor (World Bank, 1990). Nearly half of the developing world's poor, and nearly half of those in extreme poverty, live in South Asia, followed in order by sub-Saharan Africa, the Middle East and North Africa, Latin America, the Caribbean, and East Asia (World Bank, 1990). What is particularly relevant for environmental management purposes is that poverty has "urbanized": The number of poor in the world increased 11% from 1970 to

1985, but the number of urban poor expanded 73% worldwide and up to 81% in Africa. Poverty compounds the effects of inadequate drinking water, poor sanitation, and housing in crowded buildings constructed in flood-prone areas and other dangerous circumstances (Habitat, 1992b). In addition, it restricts the ability of communities to invest in practices and technologies that may be more environmentally appropriate. In some cases, poverty restricts access to education and information about better practices that are available. Although extremely serious in themselves, all of these factors are likely to compound any negative impacts of climate change.

12.2.2. Technology

12.2.2.1. Infrastructure

The choices that people make concerning which technology is used to provide a given service may have a dramatic impact on the degree to which a given change in climate will affect human settlements. For example, periodic additions of mechanization in North American agriculture during the last century (e.g, tractors, self-propelled harvesters, center-pivot irrigation) have significantly reduced the number of people engaged in agriculture and the accompanying requirements for their economic and social support services. As a consequence, there are now fewer settlements dependent on agriculture or likely to be directly affected by drought, flood, or other weather incidents. Previous choices of technology may also considerably restrict the options available for adaptation. For example, the ordinary operating life of commercial buildings in the United States is about 80 to 90 years (Pierce, 1994), whereas the life of water-supply systems is around 50 years (Internal Revenue Service, 1994). It follows that climate changes occurring in the first half of the 21st century would affect energy and water-supply systems that will mostly look like today's. The viability of entire transportation systems—for example, the bias that exists toward automobiles in many human settlements throughout the world—is influenced strongly by past technological choices and spatial patterns of settlement that grew up around these choices. This, in turn, has significantly influenced the long-term livability and viability of these settlements.

Standard engineering practice calls for safety standards of drains, culverts, bridges, dikes, and dams. Based on historical data, these are built to withstand 25-year, 50-year, 100-year, or 400-year floods. If rainfall intensity increases with climate change (Pittock, 1994), these structures either would have to be built to a higher standard or a greater probability of damage accepted. A structure strong enough to withstand a 25-year flood in the future might have to be strong enough to withstand a flood now considered to be a 100-year event. Today's structures would have a higher risk of failure. Lowering this risk would require investments that compete for very limited discretionary dollars, especially in many developing countries.

There is a huge unsatisfied demand for shelter in developing countries. Indeed, construction of living space is among the more energy-intensive activities in some countries, contributing 17% of India's annual carbon dioxide emissions, for example (Tiwari and Parikh, 1993). In addition, in developed countries, existing buildings contribute a large portion of the primary fuel use and greenhouse gas (GHG) emissions. In the United Kingdom and Western Europe, the percentage may be some 50% (Courtenay, 1992). If mitigation of climate change requires large increases in energy costs, the impacts on this sector and on human settlements could be significant.

12.2.2.2. Environmental Management

The manner in which human settlements dispose of their waste, and even the degree to which they declare certain physical outputs of production and consumption processes to be "waste," reflects a series of social choices. The consequences of those choices on the environment varies widely and may be affected by climate change.

For example, the most common method for disposing of solid waste is land disposal, either sanitary landfill or simply disposing of the waste on unused land. Leakage and runoff from landfills and wastes released to surface waters adversely affect the quality of ground and surface waters. In some cases, an incinerator is used to reduce the volume of the organic component (Habitat, 1992a). To reduce the burden on landfills, resource recovery by one of several strategies can be pursued and may be more likely in less-wealthy settlements (Habitat, 1992a). However, if settlements are poor, less-safe technologies may be used for disposal. If climate changes, the assimilative capacity of airsheds may be reduced because of increased prevalence of stagnant-air episodes, while sea-level rise and potential flooding could restrict the availability of land for safe disposal of solid waste. At the same time, demand for disposal areas continues to rise.

12.2.3. Policy

12.2.3.1. Land and Development Policies

The rights and conditions under which land is held (land-tenure systems) can strongly affect the management of this resource and either promote or discourage adaptation to environmental conditions of all kinds. Settlement history has a great deal to say about this, and land-tenure systems can result in mismatches between population and productive land. Some traditional land-tenure systems have adapted to unique climate or cultural systems and have significant survival value (Oguntoyinbo, 1991). For example, in some West African land-tenure systems, tenure goes with the crop, not the land. This system encouraged the growing of perennial crops, such as trees. Subsequent conversion to a fee simple absolute system (ordinary private ownership) encouraged the planting of commercial row crops, the opposite of the intended result (Rayner and Richards, 1994). Communally managed commercial livestock-grazing schemes appear to have worked better in Africa than fenced-ranching schemes, in that they permit freer adjustment to spatial variability in rainfall (Thompson and Wilson, 1994). In Mexico, *ejido* usufruct land-tenure systems (now undergoing privatization) have been used, in part, to combat the consequences of variability in rainfall (Thompson and Wilson, 1994) but have been more vulnerable to routine drought and generally show chronic low yields in comparison with private lands that are generally in more favorable locations and can attract higher investments in seed, fertilizer, and water supply (Liverman, 1992). This situation has encouraged rural-to-urban migration. In Nepal, forests were historically managed locally by the community. When the state forest-management system was established, both land tenure and management passed to Kathmandu, and deforestation resulted. A mixed system of state ownership and local control seems to have improved forest management (Thompson, Warburton, and Hatley, 1986). In all cases, local knowledge has proved important.

Some land uses can themselves increase flood hazard by lowering the ground level through drainage (Tooley, 1992) and water extraction (Hadfield, 1994). Deforestation exacerbates runoff, and dams prevent the accumulation of silt on river deltas (Clark, 1991; Kreimer and Munasinghe, 1991). Local flooding can be made worse if much of the settlement's surface area is paved (Berry, 1990; Henri, 1991; Marco and Cayuela, 1992), even in years of great drought (Handmer, 1989).

Government actions (such as zoning practices in flood plains to limit settlement) can reduce adverse interactions between development and climate. However, governments may lack the governance capacity or will to adequately regulate land uses and respond to hazards (Beatley, 1994) or may face sometimes-irreconcilable duties to make settlements both affordable and safe. This quandary manifests itself in floodplains in the conflict over making flood insurance affordable and available without, at the same time, encouraging settlement in areas that have to be defended at great cost (Arnell, 1983; Henri, 1991; Murray, 1991; Denlea, 1994). In addition, government infrastructure investments and other actions can themselves create unfavorable environmental outcomes. In Mexico, Australia, and India, for example, government investment in irrigation has sometimes resulted in high soil salinity and difficulties for the farm communities (Liverman, 1992; Leichenko and Westcoat, 1993). Water-management projects established to cope with frequent drought or flooding have also created poorly assimilated refugee populations. For example, the creation of Lake Kariba, between Zambia and Zaire, has marginalized the Tonga people (Magadza, 1991). Numerous other examples are cited by Gleick (1992). Many of the adverse environmental consequences of urban slums occur because residents lack clear legal tenure and because of the lack of land-use controls.

Most fundamental environmental problems of settlements themselves cannot be tackled unless city governments have the capacity (power, resources, and technical expertise) to intervene in the urban land market. While land-use planning and controls can be problematic everywhere, in many developing countries in particular this lack of capacity, lack of land-use controls, and lack of legal alternatives to unauthorized settlements creates a haphazard, sprawling pattern and density of development that will not allow cost-effective infrastructure and service provision. It also promotes illegal housing on dangerous land sites, sprawl over prime agricultural land, and degradation of the natural landscape (Hardoy and Satterthwaite, 1984; Gilbert, 1992). Even when planning norms exist, they may be weak or unconnected with the way in which land is actually appropriated. In some big urban settlements like Mexico City, planning did not have stipulations to regulate the land market, nor was it associated with concrete programs for addressing priority needs and problems, such as housing and services (Aguilar, 1987; Aguilar and Olivera, 1991).

12.2.3.2. Migrant and Refugee Populations

Conflict between groups in society can exacerbate the effects of climate, creating refugee problems of major dimensions. Such conflict has occurred on the Horn of Africa for 20 years, where successive civil wars in Ethiopia, Eritrea, Somalia, and Sudan have repeatedly disrupted both agriculture and food-delivery systems. Human settlements in many rural areas have been abandoned, and thousands have succumbed to famine and disease (Tolba *et al.*, 1992). Adverse climate conditions may compound the plight of refugees (Newland, 1994).

As land and water resources bear increased population and development burdens, disputes over resource use become increasingly acute and may themselves become a source of conflict. For example, in northwest Mexico, the water rights in the Cucurpe area have sometimes become a cause for violence (Liverman, 1992). In the Middle East, Jordan, Israel, and Syria have had great difficulty resolving issues over the distribution of both groundwater and surface waters of the Jordan River Valley (Lowi, 1992; Suhrke, 1993), which contributes to the political instability of that region. Environmental degradation has been a major source of internal social and economic conflict and migration within China (Smil, 1992). Similar (though less extreme) difficulties have attended the distribution of Colorado River water in North America (Glantz, 1988) and Euphrates River water among Turkey, Syria, and Iraq (Gleick, 1992).

In some cases, climate-related environmental disaster has created significant refugee populations, while in other cases the management of resources to prevent disaster has created the problem. These refugee populations are often either socially marginalized or become a source of conflict (Suhrke, 1993). For example, repeated flooding in Bangladesh has resulted in migration of thousands of Bangladeshis to India, where ethnic conflict has resulted (Homer-Dixon *et al.*, 1993; Hazarika, 1993).

12.3. Impacts and Ranges of Sensitivities to Climate Change

12.3.1. Population Migration

If future climates resemble those projected by the general circulation models, wetter coasts, drier mid-continent areas, and sea-level rise may cause the gravest effects of climate change through sudden human migration, as millions are displaced by shoreline erosions, river and coastal flooding, or severe drought (e.g., Xia Guang, 1991). In the developing world, many areas to which people might relocate are likely to have insufficient support services to accommodate the new arrivals. Other consequences can include ethnic tensions and regional overpopulation.

Less dramatically, long-term problems with land fertility arising from the interaction between climate and social and agricultural practices produce rural–rural and rural–urban migration. It is known, for example, that a large number of the rural–rural migrants in Nigeria originate in the very densely populated areas where soils are impoverished and have also been destroyed by erosion (Udo *et al.*, 1990). Other things being equal, more-intense rainfall accompanying climate change could accelerate these soil-erosion and leaching processes.

Where social safety nets do not intervene and where land uses and techniques are inappropriate, mid-continental droughts have contributed to similar migrations in developed countries (Riebsame, 1990).

In much of the developing world, perhaps 90% of the migrants do not have the special skills that would attract a salary high enough to enable them to live in a better-designed and built urban settlement. Consequently, they end up living in informal peri-urban settlements with some serious infrastructure problems, ranging from unhealthy environment and water supply to difficult access to energy, transportation, communications, and shelter. About one-third of the population in many developing-world cities lives in these informal settlements. Some can be enormous. Ciudad Netzahualcoyotl, the large informal settlement of the 1970s (now a more regular, permanent city) in the Valley of Mexico had more than 1.2 million inhabitants in 1990, according to the Population Census. A newer informal settlement that mainly developed during the 1980s is Valle de Chalco, with an approximate population at the beginning of the 1990s of 400,000 inhabitants (Aguilar and Olivera, 1991; Hiernaux, 1991).

12.3.2. Energy

Generally, studies have shown differing location-specific overall aggregate energy impacts, depending on how much energy use is related to residential and office heating and cooling. Climate warming will increase energy consumption for air conditioning and, conversely, lower it for heating. This result was discussed in the 1990 and 1992 assessments for developed countries. This section confirms those earlier findings and adds some insights on the developing and newly industrialized countries. Because agriculture, transportation, and industry all have been allocated their own sections in this report and because Chapter 22 provides considerable background on fuel use in residential and commercial buildings, this section primarily focuses on energy demand in the residential and commercial sectors as influenced by climate change. Change can be expected in climate-sensitive sources of energy such as hydroelectric power, biomass, solar, and wind. Those resources are discussed in much greater detail in Chapter 11.

12.3.2.1. Demand

Space conditioning of buildings is one of the most climate-sensitive uses of energy—especially the use of electricity for residential and commercial air conditioning, and electricity plus other fuels for space heating. Although usually smaller in total magnitude of energy demand, the use of electricity and fuels for irrigation pumping and the use of fuels for drying of agricultural crops also can be significant weather-sensitive demands in some regions (Scott *et al.*, 1993; Darmstadter, 1993). Climate change has negligible direct effect on vehicle performance, but transportation activities are sensitive to immediate impacts from weather and may be indirectly sensitive to gradual shifts in human activities as they respond to climate change and to greenhouse and environmental strategies. For example, if warming should accelerate urbanization tendencies caused by adverse effects on the rural resource base, additional urban congestion could reduce vehicle performance in cities. This performance decline provides incentives for transportation modal shifts (to public transit, for example).

Changes in population, wealth and activities of the population, energy prices, cost and characteristics of technologies used, and daily and seasonal operational patterns for these technologies help drive urban energy demand. For example, discrete zoning of residential and commercial areas may cause higher levels of transportation use (Matsuoka *et al.*, 1992). D.W. Jones (1991) found that urbanization was an important determinant of energy demand, though less important than per capita income and industrialization. Mechanization of food delivery and journey to work are major factors. Because the effects of climate change appear to be small in comparison with potential changes caused by these other factors, adjustments can be made to accommodate climate change at small to moderate cost (Linder and Inglis, 1989). A special consideration in considering the effects of urbanization on energy use is the urban "heat island" effect, in which the replacement of vegetative ground cover in urban settlements with structures, streets, and similar surfaces causes outdoor temperatures within these settlements to be several degrees higher than they otherwise would be (Boodhoo, 1991). Energy consumed for air conditioning may increase heat-island effects (Hanaki, 1993). Heat-island effects and actions to reduce them are further discussed in Chapter 22 in the context of the extra energy that must be expended in buildings to counteract heat-island effects.

The sensitivity of space-heating and air-conditioning demand to climate may be estimated in two ways: (1) by using statistical relationships between demand and climate variables of the type routinely used by energy suppliers in carrying out short-run demand forecasting (an analog approach) and (2) by using underlying physical parameters, such as heating or cooling degree-days, to make estimates based on the physical characteristics of buildings (UK Climate Change Impacts Review Group, 1991). In general, the analogue approach results in a lower estimate of climate sensitivity because it takes account of the fact that building occupants may, through autonomous adaptation, adjust their comfort levels. On the other hand, the analogue approach does not take account of some forms of long-term adaptation, notably the modification of building design and space-heating or air-conditioning systems to take account of changed climate conditions (UK Climate Change Impacts Review Group, 1991). In temperate-zone developed countries, investigators have observed that the net balance of energy demand depends largely on the balance between cooling and heating in the residential and commercial sectors (Linder *et al.*, 1987; Scott *et al.*, 1993; Rosenthal, Gruenspecht, and Mann, 1995). In the commercial sector, air conditioning accounts for a greater proportion of final energy demand, partly because of internal heat gains from lighting, office equipment, and occupants. Generally, in cool areas with limited air conditioning, space-heating fuel use and electricity demand for heating (and, sometimes, total electricity demand) would likely decline (Stokoe *et al.*, 1987; Singh, 1988; Mundy, 1990; Aittoniemi, 1991a, 1991b; UK Climate Change Impacts Review Group, 1991; Scott *et al.*, 1993). Where cooling is already more important than heating or may become more important, the increase in cooling demand will be greater than the decrease in heating demand (Loveland and Brown, 1990).

In developed countries, increases in commercial building space, space conditioning, and office automation drive much of the overall increase in energy demand and an increase in electricity's share (Nishinomiya and Kato, 1989; Herring *et al.*, 1988), while energy-efficiency codes and varieties of green labeling have reduced energy demand. In some cases, climate warming can be expected to further increase both peak and total annual electric power consumption (Linder *et al.*, 1987; Linder and Inglis, 1989; Niemeyer *et al.*, 1991; Nishinomiya and Nishimura, 1993). However, primary energy demand can still either fall (Nishinomiya and Nishimura, 1993) or rise (UK Climate Change Impacts Review Group, 1991), depending on local conditions.

The relationship between cooling energy and temperature appears to be nonlinear because of latent heat of water condensed by active system cooling equipment and lost efficiency of passive and natural ventilation at high temperatures (Millbank, 1989; Hulme, Haves, and Boardman, 1992; Scott, Hadley, and Wrench, 1994). Huang *et al.* (1987) showed better cooling correlations using degree-hours, latent enthalpy hours, latent cooling loads, and adjustments for natural ventilation. In more recent unpublished work, Huang has shown that a uniform 1.7°C increase in temperature causes about a 12% to 20% increase in cooling loads in cities in newly industrialized or developing countries with hot climates. In cities with cooler climates (e.g., Mexico City, Sao Paulo, Beijing), the estimated increases in cooling loads are larger in relative terms but smaller in absolute terms. Cities in warmer climates may experience large relative decreases in their very small heating loads of 50% or more, virtually eliminating the requirement to heat.

Research in developing countries has shown that appliance use in general (OTA, 1991; Figueroa, Ketoff, and Masera, 1992) and air-conditioning saturation in particular depend on urbanization and income levels. Meyers *et al.* (1990) credited the rapid growth in residential energy use in developing and newly industrialized countries to an increase in the number of households and electrification. Market penetration of air conditioners is still low in most countries, but once they become affordable their proliferation can be very rapid, particularly in hot climates. This has happened in the Philippines, Malaysia, Thailand, Taiwan, and Brazil, which also have humid climates (Schipper and Meyers, 1992). In Taiwan, air-conditioner market penetration more than doubled (12% to 29%) between 1979 and 1989, while in Korea, which has a relatively short summer, it grew from zero in 1976 to 9% in 1989. In comparison, the market penetration of air conditioning in Japan is close to 60%. Similar findings are reported by Sathaye and Tyler (1991) and Parker (1991). Urbanization, rising incomes, and warmer climates could combine to increase both air-conditioner market penetration and cost per unit.

12.3.2.2. Supply

Understanding the effects of climate on hydroelectric, biomass, solar, and wind energy is particularly important because renewable energy sources are playing a significant role in the energy planning of many countries. This could become an increasingly important concern in developing countries, many of which are facing serious economic pressures from the need to import conventional energy resources.

Although cooking usually is a more important end use than heating, woodfuels are particularly important in developing countries because of their (typically) low cost and the simplicity and low cost of the end-use technology (Hall *et al.*, 1993). Among the largest potential impacts of climate change on the developing world are the threats in many areas to fuelwood and charcoal, which are principal sources of energy in most sub-Saharan African nations and many other developing countries. For example, cities and towns place large demands on adjacent and distant forests. Bamako, the capital of Mali, is supplied by a zone with a 100-km radius (Simmons, 1989), but most of the 612 tons per day arriving in Delhi, India comes from 700 km away in Madya Pradesh (Hardoy *et al.*, 1992), and charcoal from northern Thailand is exported to Bangladesh (Tolba *et al.*, 1992). Locally, firewood removal can be very important. Wood is known to be scarce around Kinshasa in Zaire, Brazzaville in the Congo, Nairobi in Kenya, Niamey in Niger, Ouagadougo in Burkina Faso, and in much of Nepal (Williams, 1990). More than 90% of the energy in some African countries depends on biomass (fuelwood). Even if climate does not change, the combination of population growth and a declining resource in some cases will bring cost pressure to change the fuels for cooking and heating from wood and charcoal to other fuels (see Chapter 11). Because of uncertainties in water-resource projections derived from current climate models, providing reliable regional projections of future moisture conditions in these countries is very difficult, although researchers are working on the problem (e.g., Magadza, 1993). Analysis of this situation should be a top priority for energy planners. Further details are provided in Chapter 11.

Many developing countries and some developed countries depend significantly on hydroelectric power (World Resources Institute, 1992). Even constant per capita electricity consumption implies that without major additions from technical potential, hydropower could decline significantly as a source of power by the early part of the next century. Management of some hydroelectric power systems could be significantly complicated by reduced snowpacks, which change the seasonality of storage and supply, and greater flood intensities if rainfall increases. Geographically extensive hydroelectric systems may be surprisingly unaffected despite climate sensitivity of some components (Sias and Lettenmaier, 1994). Rising electricity consumption and (perhaps) declining future capability resulting from climate change could force some nations to rely increasingly upon fossil fuels. For additional details concerning the effects of climate change on energy supply, see Chapter 11.

12.3.3. Air Pollution, Waste Management, and Sanitation

Climate change does not directly lead to air, water, or soil-column pollution. However, if changes in climate significantly alter local and regional weather patterns, underlying trends in pollution damage may change nonlinearly. This issue received only brief attention in the 1990 and 1992 assessments. Global monitoring of urban air quality in cities indicates that nearly 900 million people are exposed to unhealthy levels of sulfur dioxide (SO_2) and more than 1 billion are exposed to excessive levels of particulates (Schteingart, 1988). While cities such as Manchester, London, Tokyo, and Frankfurt have seen great improvements in air quality (especially SO_2), other cities in the low latitudes have increased SO_2 levels. Moreover, photochemical smog produced by the reaction of sunlight with ozone and photochemical oxidants, such as peroxyacetal nitrate (PAN) from nitrogen oxides (NO_x) and hydrocarbon emissions, is particularly prevalent in cities in semi-arid regions, such as Los Angeles, Tehran, and Mexico City. In British cities, NO_x is probably of more concern now than sulfur oxides (SO_x). Warming can exacerbate the formation of smog. Global warming appears likely to aggravate tropospheric ozone and other air-quality problems in polluted urban areas, which in turn may increase human respiratory disease. In developing countries, much of the pollution problem comes from small and numerous sources rather than large industrial sources (Parikh, 1992).

Rapid industrialization and high urban growth rates have engendered major air-pollution problems in urban centers (e.g., Brazil, Chile, Mexico, Malaysia, Indonesia, Thailand). In Mexico City, for example, severe pollution stems from 36,000 factories and 3 million motor vehicles emitting 5.5 million tons of contaminants per year (Schteingart, 1988), leading in part to high blood levels of lead in newborns (Schteingart, 1989). Box 12-1 describes the complexity of interacting urban growth and air pollution problems of the Basin of Mexico (Mexico City and surrounding urban area). Similar problems are caused by private diesel buses in Santiago, Chile (Crawford, 1992) and by the use of smoky, high-sulfur coal briquettes in domestic stoves in Chongqing, China (Wang *et al.*, 1994). In all of these cases, if climate change were to add to the number of air-stagnation periods, the effects on human health and economic productivity (from pollution itself or from emergency countermeasures) would be more severe.

Many landfills, abandoned industrial waste dumps, agricultural chemical residues in soils, and polluted river and lake sediments pose long-term potential chemical "time bombs." A chemical time bomb is defined as a time-delayed, nonlinear response of soils, sediments, and groundwaters to stored pollutants under changing climate and land-use conditions (Hejkstra *et al.*, 1993). The metabolism of human settlements is responsible for these potential chemical time bombs—some of which, such as landfills and old industrial sites, lie directly in urban areas. Hydrological modifications, especially if rainfall intensities increase following climate change, could lead to more rapid leaching from old landfills that are not watertight, affecting groundwater tables and thus drinking-water supplies. Land-use modifications driven by changes in water availability or altered economic situations could lead to changes in ground cover that mobilize toxic residues in soils (Hesterberg *et al.*, 1992). In forests, an increase in temperature and a reduction in transpiration by trees leads initially to increased mineralization of organic matter, followed by nitrification and mobilization of hazardous substances (Mayer, 1993).

The rapid urbanization of low-lying areas in many developing countries has resulted in poor sanitary conditions and low water quality. In many cases, this degradation is exacerbated by natural climatic phenomena. For example, in the rapidly growing coastal cities of Lagos, Port Harcourt, and Warri in Nigeria, wastewater and sewage are discharged into open wayside gutters. Flooding during the rainy season produces an unhealthy and noxious environment (Udo *et al.*, 1990). Since these areas are also subject to sea-level rise and perhaps more intense and extended coastal storms under climate change, the triple effects of urbanization, poverty, and climate change could make the environment in cities like these dangerous for human health.

12.3.4. Infrastructure

Except for a reduction in the degree of expected sea-level rise and more emphasis on extreme events, little has changed in our understanding of this issue since the 1992 assessment concerning direct impacts of climate. However, the indirect impacts of climate and socioeconomic change were not discussed in earlier assessments. Although bridges, roads, and buildings can have a physical lifetime of several hundred years when properly maintained, urban infrastructure often has a rather short lifetime because population, urban activity, and other factors often change significantly within shorter time periods than the physical lifetime of the structure. Such socioeconomic change

Box 12-1. The Challenge of Urban Growth and Air Pollution: Mexico City

The metropolitan area of Mexico City (AMCM) consists of more than 2,000 square kilometers, comprising the Federal District (where the government seat is) and 21 adjacent municipalities in the State of Mexico. Growth of 5% per year from 1940 to 1980 and 2% per year for the last 10 years has resulted in 15 million inhabitants in 1990, making the AMCM one of the largest cities of the world. As the country's political, economic, industrial, and social capital, Mexico City contains about 20% of the national population, provides 47% of all jobs in industry, and generates 48% of public investment in social welfare (Aguilar and Sanchez, 1993). It also contains 30% of all the industrial plants in the country (a total of 38,000), which contribute 20% of Mexico City's pollution, and has about 2.5 million motor vehicles that consume about 14 million liters of gasoline and 4 million liters of diesel fuel per day, generating 75% of the pollution in the city (Departmento del Distrito Federal, 1987).

Mexico City's location in a deep valley at approximately 2,300 meters above sea level creates a natural isolation and contributes to frequent thermal inversions (especially in winter), preventing the dispersion of pollutants. Ecological damage has been aggravated by the lack of measures to protect the local environments. GCMs are silent on the question of local winter temperature inversions, but more frequent or persistent inversions would worsen the Mexico City air quality, while less frequent inversions would improve it.

A comprehensive program of action was announced in 1990 to rationalize urban transport, improve the environmental qualities of fuels burned, install pollution-control equipment, and regenerate natural areas (Mendez, 1991). Some of the main actions taken in this new program also would reduce emissions of greenhouse gases:

- Emission of atmospheric pollutants was controlled in service facilities, such as public baths, dry-cleaning shops, and laundries. These places frequently did not comply with technical standards because of their old equipment and a notorious lack of maintenance.
- Industries in the basin were inspected regularly and systematically to verify the correct functioning of their now-mandatory pollution-monitoring and emission-control equipment.
- Checking vehicular exhaust emissions became mandatory in late 1989. From 1991 onward, the regulation was changed to compulsory verification every six months for heavy-use vehicles and yearly for all other vehicles.
- The use of each car was banned one day of the week, according to the numbers on the license plate, to reduce the circulation of vehicles by approximately 500,000 cars, on average, during working days.
- In 1991, unleaded gasoline and catalytic converters became obligatory in all new cars. However, the lower levels of lead increased the amount of unburned oil residues—which, in turn, increased the ozone formed in the air of the city during daylight hours (Bravo *et al.*, 1991).
- The two thermoelectric plants in the basin increased their consumption of natural gas as a substitute for oil.

Despite all these measures, atmospheric pollution has remained dangerous to human health. Public officials acknowledge that employment and productivity in the urban economy is still a priority over environmental preservation (Mumme, 1991). Furthermore, some industrial plants have yet to install antipollution devices. Because of the high cost involved, these plants will depend on state credit programs that are to be implemented in future years to afford these devices.

Additionally, because of the ban on using cars once or twice in a week, some people have bought a second or third car in order to drive on all working days. This has increased the number of the private vehicles by about 20%, as well as the consumption of gasoline. Catalytic converters are optional for used cars, which represent most of the private vehicles in the urban areas. As a result, atmospheric pollution has remained at critical levels in the past five years. Not only have some individual pollutants frequently exceeded what are considered the healthy limits, as is the case with ozone, but the overall air-quality index (abbreviated in Spanish as IMECA) has surpassed 100 points—the air-quality norm recommended by the World Health Organization during most of the year. This fact indicates that at least some pollutants are well above the healthful limit and that long periods of exposure to high pollution levels are notable.

requires renewal or alteration of infrastructure. Adaptation to climatic change becomes an important factor in the design of infrastructures with long lifetimes.

A significant component of the increase in costs associated with risks to infrastructure occurs because the "built environment" has become more valuable over time as investments have occurred. Although some infrastructure has become better designed against natural hazards, risks to infrastructure could increase over time, even without climate change. As with population, human settlement infrastructure has increasingly concentrated in areas vulnerable to flooding, fires, landslides, and other extreme events. Much of the increase in the value of these physical assets has come because of rising per capita wealth (Stavely, 1991; Central Statistical Office, 1994). These assets are often housed in lightly constructed residential property, which in the case of Hurricane Andrew in Florida accounted for 65% of all insured losses (IRC, 1995). Use of modern construction materials (e.g., aluminum sheeting used for roofing) and aesthetic design features without sufficient understanding of their appropriate use has compounded the damages of extreme events. Modern buildings are difficult to repair if structural pillars are damaged (Jakobi, 1993), and exterior and contents damages also can add up to severe losses (FEMA, 1992). Although the amount or proportion of national physical assets exposed to climate hazards is not readily available, it is known that in the United States about $2 trillion in insured property value lies within 30 km of coasts exposed to Atlantic hurricanes (IRC, 1995); in the Netherlands $186 billion lies in the hazard zone. The corresponding total in Japan is $807 billion (WCC, 1993, 1994), and in Australia, $25 billion (uninsured) (Peele, 1988). Elsewhere, the value of exposed assets is not well known, but Hohmeyer and Gartner (1992) updated a 1971 estimate to give $22 trillion. This value may have doubled due to economic growth since the 1970s. Insured values are usually considerably less than total values because of the difficulty and expense of obtaining coverage.

Climate can directly affect infrastructure via atmospheric processes, fire, and flood. Some researchers have found, for example, that return periods for major flood events may increase fourfold by the year 2070 (Gordon *et al.*, 1992; Smith, 1993; Whetton *et al.*, 1993; Bates *et al.*, 1994); that is, the 100-year flood would become the 25-year flood. If so, design and safety standards would have to be revised for culverts, drains, bridges, dikes, and dams. Moreover, existing infrastructure would be more vulnerable to failure (Minnery and Smith, 1994). Lake levels in many regions in the world are known to fluctuate under current climate conditions, in some cases changing more rapidly than settlements can adapt to easily (Dejoux and Iltis, 1992) or influencing water quality (Gafny and Gasith, 1989, 1993). This fluctuation occurs both because of varying precipitation and runoff and because of variations in human withdrawal of water for a variety of purposes. Climate change could add to the list of variables affecting lake levels.

Protection of infrastructure from extreme rainfall events, river flooding, landslides, and coastal flooding could become a more serious problem with climate change, in some cases pointing toward retreat from hazardous areas. As noted in Chapters 9 and 17, many inhabited areas in the world, especially those in coastal zones, are already sensitive to flooding, landslides, and wind damage under existing climate (e.g., see Ojo, 1991). Some coastal areas, most notably river deltas, face relative sea-level rise already due to geological factors; withdrawal of oil, natural gas, and water from geological formations underlying the delta regions; and sediment impoverishment because of water and flood control works. Coastal zone agriculture, coastal mangroves, and coral reefs are important to human settlements in some parts of the world (Morgan, 1993; Yoshino, 1993) and are sensitive to sea-level rise. Based on a population-at-risk concept (populations subject to annual flooding), a 1-meter sea-level rise, and year 2020 populations, Hoozemans *et al.* (1993) note that very large populations are sensitive to sea-level rise in Bangladesh, China, Egypt, India, Mozambique, Pakistan, and Vietnam. Others note particular locations at risk, like Jakarta, Indonesia (Sari, 1994). Excluding Mozambique and Vietnam, projected aggregate annual costs of protection were $290 million for these countries, less than 1% of their gross national product (GNP). Countries with costs of protection above 5% of GNP included Anguilla, Cocos Islands, Gambia, Guinea-Bissau, Guyana, Kiribati, the Maldives, the Marshall Islands, Mozambique, Tokelau, Turks and Caicos, and Tuvalu. Fankhauser (1995) has calculated annualized coastal-protection costs for the world on the order of $1 billion per year for a recent estimate of a 50-cm sea-level rise by the year 2100. Inundation and erosion in OECD countries are about half the world total, accounting for between 0.01 and 0.27% of 1985 GDP in OECD countries (Rijsberman, 1991). More refined analyses have been done of several countries' coastal infrastructure-protection needs, including the Netherlands (UNEP and Government of the Netherlands, 1991) and Japan (Kitajima *et al.*, 1993). Optimal protection strategies have been investigated for several locations, including the Netherlands (UNEP and Government of the Netherlands, 1991). See Chapter 9 for additional details concerning coastal impacts and costs.

Settlements in forested regions in many areas are vulnerable to seasonal wildfires. This includes settlements in temperate-zone regions such as Canada (Forestry Canada, 1991), tropical forested regions such as Borneo (*Economist*, 1994), and mediterranean climates like the state of California in the United States or southern Australia (Cheney, 1979; Foster, 1994). If the climate in these areas should become even drier and warmer, the frequency of fire danger would increase (Street, 1989; White, 1992; Ryan, 1993), although it is possible that fuel buildup under drought conditions would decrease, decreasing fire intensities.

Such information needs to be developed for other climate-related impacts. One potential adaptation to a wetter climate or to a warmer climate in an area vulnerable to river flooding from snowmelt would be to expand or reinforce riverine flood-control systems with dikes, check dams, and expanded storage facilities. Alternatively—or in addition to these measures—land-use planning and regulation can be strengthened for hazardous areas. The costs of these actions have not been estimated in relation to climate change, at least in part because the location of future climate-enhanced flood danger is so poorly understood that any estimate would be largely speculation.

Transportation impacts were discussed in the 1990 and 1992 assessments and are discussed in greater detail in Chapters 11 and 21. The findings are locality-specific and have not changed significantly. Although in the colder regions global warming may encourage the poleward expansion of human settlement and open some winter water travel (Sanderson, 1987; Stokoe *et al.*, 1987), thawing of the permafrost may also disrupt infrastructure and transport (loss of foundations for airports and runways; loss of seasonal ice roads) and adversely affect the stability of existing buildings and conditions for future construction (IBI Group, 1990). If extreme rainfall events become more common, impacts on roads, railways, and other transportation links could become global in reach. Although cost estimates do not exist, they could be very high.

Human muscle power, draft animals, and small watercraft still play a significant role in transportation in many parts of the developing world, especially in the rural areas of those countries located in tropical and subtropical climates. For example, in India, about 10% of tonnage moves by truck, while bullock carts take more than two-thirds (OTA, 1991). Where more frequent or severe heat stress is a possible result of climate change, the work capacity of both humans and draft animals may decline, making transportation less efficient. Where severe weather is more frequent or intense, small craft on rivers, lakes, and other water bodies may be more subject to reduced periods of operation, disrupting farm-to-market distribution systems.

12.3.5. Water Supply

The conclusions for water supply are not dramatically different from the 1990 and 1992 assessments. For example, although some of the details have changed, infiltration of seawater into coastal aquifers is still considered a general potential hazard of sea-level rise, as discussed in Chapter 9.

Changes in water distribution resulting from changed precipitation patterns are important to agriculture, energy, and health. For example, changes in water quality and availability may affect human settlements indirectly through flooding or drought-induced famine and malnutrition. Despite the drilling of many wells, the great Sahelian drought in West Africa during the period 1968 through 1973 (caused by adverse weather compounded by increased human and livestock population pressure on an area of limited long-term biological productivity) resulted in a significant amount of human migration. The 1972–1973 period brought large numbers of rural people from the Niger Republic, Chad, and the far north of Nigeria into the urban areas of Nigeria, including the southern cities of Lagos, Abeokuta, Benin, Warri, Port Harcourt, and Calabar—compounding the urban problems of those cities (Udo *et al.*, 1990). Much of the rapid population growth in the Valley of Mexico has been attributed to a continuous influx of rural residents from drought-ridden agricultural areas (Ezcurra, 1990).

Even under current climate, water availability remains a major and accelerating problem for large human settlements in locations as diverse as California (Vaux, 1991) and Houston, Texas (Scheer, 1986), in the United States; Bombay and Madras in India (Leichenko, 1993); and much of northern urban China (Smil, 1993). Lack of access to clean drinking water is highly correlated with numerous adverse health conditions, including high infant mortality (Parikh, 1992). Water quality can be adversely affected by nonpoint pollution, saltwater intrusion into estuaries, lowered stream flows, and groundwater mining—all problems associated with climate change, human-population growth, economic development, or all of the above (Jacoby, 1989).

Demand for water is affected by price, income, technology, and other influences. Other things being equal, more water is used as temperatures increase. For example, in Japan, urban water demand has been estimated to increase about 3.3% per degree celsius for days when the maximum temperature is over 17°C (Hanaki, 1993). For additional discussion of the effects of climate change on water supplies, see Chapter 14.

12.3.6. Health

The earlier assessments of global climate change in 1990 and 1992 provided little information on the potential effects of climate change on human health. More recent thought reveals several additional causes for concern (see Chapter 18 for details). This section describes the specific role of human settlements. Climate-related changes in human settlements may affect several elements that are critical to human health. Kalkstein (1993) and Kalkstein and Smoyer (1993) suggest that the major potential pathways are heat stress directly causing premature mortality (particularly in nonacclimated, unprotected populations in hotter developing countries); outbreaks of infectious diseases whose ranges change because of changes in human migration patterns and spread by contact or close proximity between infected and noninfected individuals; and effects of infectious diseases whose agents, vectors, hosts, ecological niches, or predators on vectors are affected by climate, or compounded by changes in land use, eutrophication of waters, and other anthropogenic environmental insults, such as acid rain or pesticides. The infectious diseases include several serious tropical diseases. In addition, if water supplies are disrupted as a result of more frequent droughts and floods, the full range of nontropical waterborne diseases such as cholera, typhoid, diarrheal, and helminth diseases (hookworm, roundworm, etc.) spread by unsafe drinking water may come into play more frequently.

Anthropogenically and climatically reduced biodiversity (e.g, from prolonged droughts) is a particular concern because biodiversity provides the prime buffer of ecosystems against stress and provides for the biological control of pests and pathogens. Tropical and waterborne diseases and diseases directly transmitted person-to-person annually account for about a fifth of the total annual world loss of disability-adjusted life years from all causes, including other health conditions, accidents, and war (World Bank, 1993). The disability-adjusted life year is a statistic that combines the loss of years of life to premature death with an index of the lost quality of life produced by varying degrees of persisting disability (e.g., blindness or paralysis). There are adverse implications for worker productivity, trade, transport, and tourism (Epstein, 1994).

Urban settlements, particularly the often (but not always) overcrowded and poorly serviced slums, shanty towns, and squatter settlements of the developing world, provide a potentially excellent breeding ground for disease organisms and vectors, as well as poor sanitation and vulnerable populations in close proximity to each other (WHO Commission on Health and the Environment, 1991). These settlements are often located in the least desirable areas, which may be flood-prone lowlands, hazardous waste sites, downwind from industrial sites, or on steep hillsides; therefore, inhabitants will be more vulnerable to, and weakened by, waterborne disease, pollution, and natural disasters, such as floods and landslips. As some agents of infectious diseases become increasingly drug-resistant (e.g., tubercle bacilli and malarial plasmodia) and vectors resistant to pesticides, the importing of these infectious diseases to highly concentrated, vulnerable populations becomes a significant public-health concern.

Tropical Chagas' disease, a leading cause of heart disease throughout the Americas, is becoming increasingly urban. Reports from Honduras and Chile reflect this rural–urban shift, paralleling the movement of populations. Once prevalent in the blood supply, Chagas can be spread directly by blood transfusions. A greater number of heat waves could increase the risk of excess mortality. Increased heat stress in summer is likely to increase heat-related deaths and illnesses (Kalkstein and Smoyer, 1993). Deaths due to cold exposure and blizzards in colder regions like Canada could decrease, although these gains are unlikely to offset the increases in heat-related deaths (see Chapter 18).

Global warming appears likely to worsen air-pollution conditions, especially in many heavily populated and polluted urban areas. Climate change-induced alterations in photochemical reaction rates among chemical pollutants in the atmosphere may increase oxidant levels, adversely affecting humans and complicating the effects of increased urbanization in locations like Mexico City (Griffith and Ford, 1993) or Tehran (Bonine, 1993). The range of certain vector-borne diseases, such as malaria, yellow fever, and dengue, may extend both in latitude and altitude to new areas and settlements currently at the margins of endemic areas. For example, highland cities such as Harare and Nairobi, which are currently malaria-free, are vulnerable (see Chapter 18 for a more complete discussion).

Still another mechanism through which global warming might affect human health is through harmful algal blooms, which promote diseases like paralytic, diarrheal, and amnesic shellfish poisoning, as well as cholera (Epstein *et al.*, 1993). The increase in blooms worldwide is thought to be a direct consequence of human activities on a local or regional scale that could be reinforced by the existence of environmental conditions such as warmer seawater in coastal areas that promotes growth. Excess nutrients from sewage and fertilizers, over-harvesting of fish and shellfish, and loss of coastal wetlands and coral reefs also contribute to the problem.

12.4. Extreme Events

12.4.1. Types of Events

Climatologically, extreme events are associated with "anomalous" weather, sometimes defined in terms of a "return period" of 10 years, 100 years, or some other value (probabilities equivalent to one event in 10 years, 100 years, etc.). The length of the return period is decided on a case-by-case basis, depending on (1) the climatic phenomenon, such as flood, wind damage, and drought; (2) the climatic region or zone; (3) the season; and (4) the intensity of the impact on human activities, plants, and animals. Extreme amounts/values that cause damage change from year to year as the level of human development and activity in the area changes. Therefore, it is very difficult to define threshold levels of temperature or precipitation that can be used as permanent criteria in any one region or applied everywhere in the world.

One important aspect of extreme events is the apparent randomness and abruptness with which they arrive. Gradual changes in air pollution, acid deposition, desertification, water shortages, salt water intrusion, soil degradation, and deforestation are also serious but tend to arrive slowly enough that local/regional/national authorities can take successful long-term countermeasures.

Extreme events may arrive without any definite cycle or periodicity, so they constitute risks that can be the subject of insurance. For the effects of climate-related extreme events on insurance and the insurance industry, see Chapter 17. Extreme events are often related to disease outbreaks. El Niño/Southern Oscillation events and teleconnected weather patterns across the globe (Glantz *et al.*, 1991) are highly correlated with outbreaks of malaria (Bouma *et al.*, 1994), dengue, and algal blooms (Hallegraeff, 1994), as well as other problems (Glantz *et al.*, 1987). Both droughts and floods are related to outbreaks of vector-borne diseases and agricultural pests. The number of extreme events in terms of economic and insured losses is increasing, at least in part because of the ever-increasing scale of human activity. The amount of insured losses caused by extreme events is increasing at a higher rate than is the amount of economic loss. Insured losses at the beginning of the 1990s are two to three times those of the 1980s (Berz, 1993a, 1993b).

12.4.2. Mechanisms of Effects

To discuss the mechanism of effects of extreme events on human settlements, we refer to the example of typhoon hits in Japan (Fukuma, 1993). Based on the record for the 79 years from 1913 to 1991, 245 typhoons were classified into six grades based on intensity. Changes in the number of deaths are shown in Figure 12-2, with running 20-year averages since 1921. It is clear that the number of deaths is larger for the more severe typhoons and that deaths have been declining in all categories since World War II. This decline is thought to have occurred because increasingly effective countermeasures have been taken during the last 40 years, including more robust infrastructure design standards. These countermeasures appear to be more effective with respect to the mid-class Grade III typhoons than for either the weaker typhoons (where the lower limit of fatalities may have been reached) or the stronger typhoons (where more effective actions still could be taken).

The countermeasures include conservation of mountain slopes, rivers, and coasts; harmonization of institutions and laws related to disaster mitigation; preparation of systems for disaster prevention, including better design; meteorological observation and information systems; promoting people's consciousness of preventing disasters; and development of communication systems for disaster occurrences. Thus, the impact of extreme events on human settlements can be affected by the grade of the extreme event, the level of economic and technological development, and the extent of countermeasures taken. Climate change may affect the intensity or the probability of extreme events. The impacts on human settlements depend in part on the present vulnerability (including coping ability) of the settlements.

12.4.2.1. Floods

Vulnerable regions are some small islands, notably coral atolls; regions hit by tropical cyclones, such as typhoons and monsoons; and the lower reaches of big rivers like the Mississippi, Hwang Ho (Yellow River), Yangtze, and Nile. Delta regions in South and Southeast Asia are particularly vulnerable. Local-scale floods show various effects on rice yield, as shown by Yoshino (1993) for tropical Asia. Flooding is a common extreme event that poses challenges for the insurance industry and the public sector (in planning and protecting infrastructure). Although humans possess considerable adaptive management capability to deal with floods, flooding could become a more common problem with climate change, even if average precipitation decreases (see Section 12.3.4; Chapters 14 and 17; and Chapter 6, *Climate Models–Projections of Future Climates*, of the IPCC Working Group I volume).

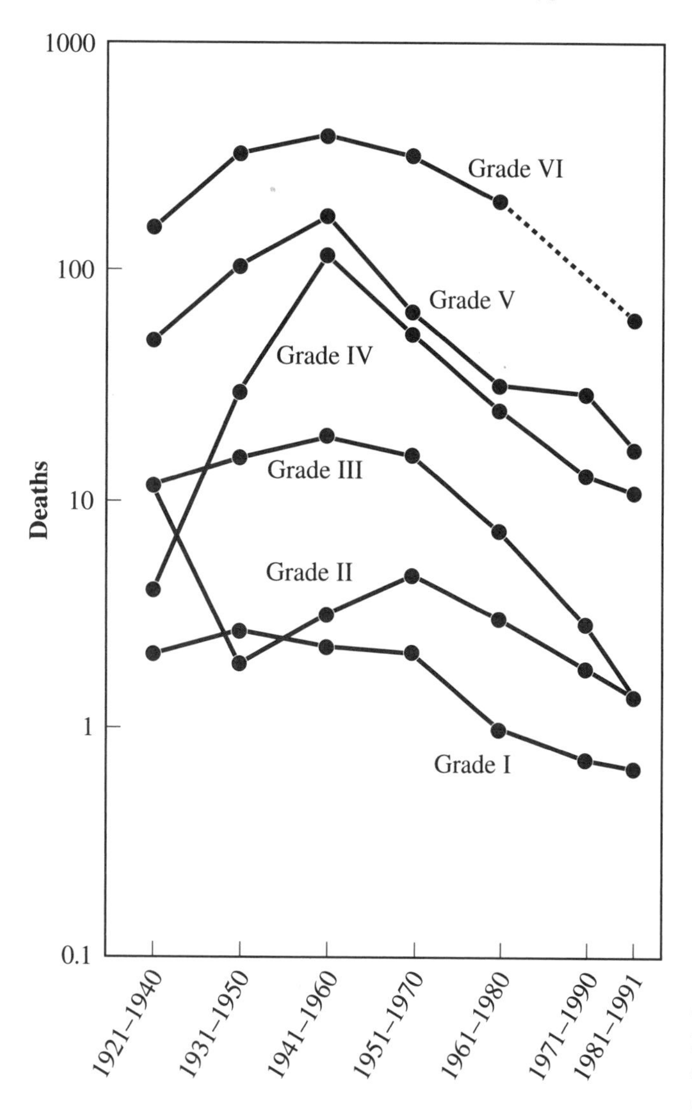

Figure 12-2: Change in numbers of deaths per typhoon hit in Japan during the last 70 years according to Grade I (weak), II (normal), III (strong), IV (stronger), V (violent), and VI (super violent) (Fukuma, 1993).

12.4.2.2. Landslides

Although landslides occur in many regions throughout the world, damaging property and disrupting transportation links, the 1990 and 1992 assessments gave relatively little attention to land stability as a climate-related problem. However, in the poorest hilly cities of the world, many hundreds of thousands of people live in illegal settlements on potentially unstable hillsides especially vulnerable to climate change. Hundreds of people were killed or seriously injured and thousands made homeless by landslides in Rio de Janiero in 1987 and 1992; Medellin, Columbia in 1988; Mexico City in 1990–1992 (Aguilar and Sanchez, 1993); Caracas in 1989 (Hardoy *et al.*, 1992); and Hong Kong in 1925, 1964, 1966, 1972, and 1976 (Styles and Hansen, 1989). A similar degree of susceptibility to landslipping would be found in other areas of deeply weathered rock in tropical and subtropical areas subject to heavy rain.

Many landslides are related to inherently unstable materials that move when subject to extreme rainfalls, such as the Loess Plateau of China (Billard *et al.*, 1992; Derbyshire and Wang, 1994), the quick clays of Quebec, and the deeply weathered crystalline rocks of the tropics and subtropics (Brand, 1989). Others occur in steeplands, such as the Himalayas, the Alps, and the mountains of western Canada and western China, and may be influenced by earthquakes, rapid snowmelt, or extreme precipitation (Brabb, 1991; Haigh, 1994). Relict periglacial landslides occur in much of Northern Europe and North America (Johnson and Vaughan, 1989). Stable under present natural conditions, these landslides are reactivated by urban construction activities and are triggered by heavy rains (Caine, 1980).

Landsliding is regarded as a catastrophic event—one that is large, sudden, and rare on human timescales (Jacobson *et al.*, 1989). Their triggering depends on an earthquake or a rainfall event and is related to antecedent conditions. Alteration in the frequency and magnitude of precipitation related to climate change (caused by changes in tracks of tropical cyclones, for example, or increases in the intensity of rainfall) could alter the probability of landslides and debris flows (Gordon *et al.*, 1992). In some places, currently inactive slides may be reactivated; in others the risks of slipping may be reduced. Greater aridity, on the other hand, could so reduce vegetation or so increase the risk of fire (which reduces vegetation) that mud flows and debris flows could become more frequent. In some temperate-zone regions, for example, landsliding in inland areas is likely to become more pronounced as a result of higher winter rainfall, increased likelihood of summer droughts, and increased summer storm activity (Jones, 1993).

Landslide susceptibility also is being altered by the movement of more people into urban areas in mountain environments and the expansion of cities from plains into the surrounding hills, both for reasons of prestige among the rich and lack of affordable building space for the poor. Both the wealthy and poor thus become potential victims.

12.4.2.3. Wind

Vulnerable regions include areas susceptible to tropical and non-tropical cyclones, including west-facing mountainous areas for polar cyclones and upper Westerlies. Strong gusts can also be associated with the downbursts of tornadoes. The countries most affected by tornadoes are the United States, Canada, and Russia, where local areas can experience serious tornado and hail damage (Phillips, 1990; Paul, 1994), though they occur in many parts of the world (Grazulis, 1991). Tropical cyclones (hurricanes and typhoons) commonly affect South, Southeast, and East Asia and Oceania, as well as Central America, the Caribbean, and parts of Mexico and the United States. Some authors suggest that storm tracks may change, or that severe tropical cyclones may become more severe with climate change (Okamoto, 1991). However, there is little agreement among climate models on either mid-latitude storms or tropical cyclones (see Chapter 6, *Climate Models–Projection of Future Climates*, of the IPCC Working Group I volume; also see Chapter 17 for further details on cyclone damage).

12.4.2.4. Heat Waves

Heat waves may become more common and severe if the climate warms. Climatologically, in the Northern Hemisphere, the southern part of the temperate zone and northern part of the subtropical zone in summer are the most vulnerable. In the Southern Hemisphere, sub-Saharan Africa and Australia are both vulnerable, especially when heat is combined with drought. Aerologically, vulnerable areas are the eastern parts of prolonged, stationary upper troughs, where warm air flow from the tropics can invade the temperate zone. Heat waves appear to increase overall death rates, even in acclimatized populations. They are a particular concern when combined with heat islands and substandard housing in urban areas (see Chapter 18).

12.4.2.5. Cold Waves

Similarly, cold waves may become less common or severe. Climatologically, in the Northern Hemisphere, the northern part of the subtropical zone and the southern part of the temperate zone in winter are the most vulnerable. Episodes of extreme cold and blizzards are major climate concerns for the circumpolar countries like Russia and Canada (e.g., Hage, 1985; Phillips, 1993; RSC/CAE, 1994). In the Southern Hemisphere, Antarctic storms sometimes strongly affect weather on New Zealand's South Island. In both hemispheres, severe snow or ice storms can adversely affect most economic sectors. Aerologically, the western parts of the stationary upper trough, where the cold air can more easily flow from higher latitudes, are the most vulnerable areas. In Asia, for example, cold waves can penetrate to 15° to 20°N latitude (Yoshino, 1989; Yoshino and Kawamura, 1989).

12.4.2.6. Droughts

There are meteorological, hydrological, and agricultural droughts. They occur widely, from the tropics to the high latitudes. In the lower latitudes, they are intensified by prolonged dry seasons caused by anomalous monsoon circulation. In the middle latitudes, anomalies in cyclonic activity are effective. The most vulnerable areas are those under the influence of subtropical anticyclones. Droughts could be more severe at higher temperatures because evapotranspiration would be enhanced, reducing soil moisture. Agricultural settlements in regions such as sub-Saharan Africa, Australia, China, southern Europe, and midcontinental North America are projected to be sensitive to drought conditions. See Chapter 13 for details on agricultural effects. Chapter 14 notes that water distribution systems, power plant cooling, and river navigation can be adversely affected and that settlements in the Middle East and North Africa may be particularly vulnerable to drought.

12.4.2.7. Wildfires

So-called "Mediterranean-Climate" regions are at the most risk because of their dry summer climate. The risk of fires is very high in the summer months (Donguedroit, 1991), when severe storms may also spawn lightning strikes. These may become more common if climate changes (Price and Rind, 1993). The regions surrounding the Mediterranean Sea, California in the United States, and southeastern Australia all have suffered from severe fire damage recently. Fires are also frequent in the taiga regions of Siberia, Canada, and Alaska and in Borneo and Indonesia (during El Niño years). These tendencies are related to recent increases in resident population and tourism, inappropriate urban plans and environmental management, shortages of labor in the timber industry, wasteful logging practices and swidden agriculture, lags in upgrading fire prevention systems, and so forth. They could be exacerbated by global warming if it also results in drying or intensification of El Niño conditions and if fuel is available.

12.4.2.8. High Tides, Storm Surges, and Tsunamis

Global warming results in sea-level rise, so the effects of "normal" extreme events such as high tides, storm surges, and seismic sea waves (tsunamis or "tidal waves") would become more severe. Multiple risks such as earthquake and tsunami or storm surge at high tide are of particular concern.

12.4.2.9. Acute Air-Pollution Episodes

In addition to areas historically famous for air pollution—such as London, Los Angeles, and Mexico City—many developing-world cities are becoming increasingly affected. Air-pollution episodes could become more serious if temperature inversions become more frequent or persistent in these areas.

The effects of extreme events are inherently uncertain and carry with them an element of probability or risk. These risks can and have been calculated for anticipated climate change in some regions, such as coastal Japan (Mimura *et al.*, 1993). More generally, however, many of today's standard engineering and insurance calculations, which depend on historical observation, could be overturned by changes in the intensity or frequency of extreme events that increase the costs, risks, or both costs and risks to lives, infrastructure, and insurance pools.

Extreme events sometimes occur successively or simultaneously. When this happens, as in the case of a series of typhoon hits, the damage is more serious because the area has been made vulnerable by previous damage (e.g., the ground is already waterlogged, or flood-control structures have been weakened). Sometimes a meteorological extreme event occurs immediately following some other form of disaster, more than simply adding to the stress on social and economic conditions. For example, a severe typhoon struck the Philippines just after the eruption of Mt. Pinatubo, causing severe mudflows at the base of the mountain. Such secondary impacts can be relatively serious in comparison with the damage of the original event.

12.5. Adaptation Options

The list of adaptive actions discussed in this section is adapted from United Nations Centre for Human Settlements (Habitat, 1992b). The list includes both actions autonomously taken by humans and deliberate, planned, climate-adjustment actions nurtured by policy. Where data are available, the section also discusses the costs of adaptation. However, not enough focused research has been done on adaptation to permit full comparisons between adaptation and mitigation of climate change or even to fully characterize the costs of adaptation. Although regional and local governments may have limited powers or boundaries that do not correspond to natural physical systems affected by climate change, a holistic approach needs to be taken to planning human settlements. Each settlement must fulfill multiple objectives, of which adaptation to climate will be only one (and in many cases, not even a very important one). Examples of this holistic approach may be found in Douglas (1983) or White (1994). Sector-by-sector adaptations are discussed in Sections 12.5.1 through 12.5.6.

12.5.1. Population Migration

One of the potentially destructive effects from a variety of social and economic perspectives is forced internal or international population movement—both low-key, long-term migrants responding to relative economic opportunity by moving between regions and from rural areas to cities, and "ecological refugees" responding to specific natural disasters. Economic migration can be reduced if economic opportunity and services of civilization can be delivered to the regions of origin and thus prevent the population movement (Calva Tellez, 1992; Rello, 1993). This reduction can be accomplished in part by immigration/emigration policies if regional (especially rural) and national economic development are undertaken at the same time. Controlling the degree of urbanization appears to be doubtful as a solution (Shukla and Parikh, 1992). Decentralizing government administration to secondary cities reduces to some degree the impact of population movements on primary cities because the "pull" effect of government employment is distributed among several locations. Economic dislocation programs, such as disaster assistance, can offset some of the more serious negative consequences of climate change and reduce the number of ecological refugees. Effective land-use regulation can help direct population shifts away from vulnerable locations such as floodplains, steep hillsides, and low-lying coastlines.

12.5.2. Energy

A number of specific actions can be taken to offset the effects of climate change in the energy sector. Increased building-shell efficiency and changes to building design that reduce air-conditioning load show promise (Scott *et al.*, 1994). Though effective, however, some strategies may have other costs (Loveland and Brown, 1990). Air conditioning may offset some of the more deleterious effects of heat waves, although because acclimatization is reduced, this is not necessarily so (McMichael, 1993). Reducing the size of space-heating capacity in response to warmer climate would be a logical adaptive response in more temperate and polar countries and may free up investment funds for other purposes, even within the energy sector. Community design to reduce heat islands (through judicious use of vegetation and light-colored surfaces) (Akbari *et al.*, 1992), reducing motor transportation, and taking advantage of solar resources also should be included in the package of possibilities. Many of these actions, such as urban tree planting and urban mass transit, are typically justified as mitigation options to reduce fossil-fuel use in the urban buildings sector (see Chapter 22) or to reduce the adverse energy-use, congestion, and health consequences of transportation networks that are based on automobiles and motorized two-wheelers (see Chapter 21). Full environmental and health costing of fossil fuels, which can discourage the use of fossil fuels through environmental "adders" (theoretical surcharges on the energy produced with fossil fuels for planning purposes to compare environmental consequences or actual surcharges charged to consumers to discourage their consumption of energy produced with fossil fuels), is sometimes regarded as a mitigative action. However, these same mitigative responses also have adaptive value in that they reduce the impact of warming on urban discomfort caused by heat-island effects.

Some of these actions have unexpected social costs if not planned appropriately. For example, urban trees can be aesthetically attractive while reducing air-pollution and heat-island effects, altering local meteorological conditions, and providing shelter from wind (NCPI, 1992). However, they also can accelerate desiccation of clay soils (Freeman, 1992)—leading to subsidence—and contribute windthrow to property damage during high-wind events (Stavely, 1991; Shearn, 1994). See Chapter 17 for further details.

Until recently, traditional styles of clothing, buildings, and cultural features such as hours of work varied considerably around the world, mirroring the outdoor environment and each region's unique sociocultural and technical response to it (Shove, 1994). Unique cultural response has been overtaken by standardization in hours of work, clothing, and heating and cooling technology, including standardized definitions of human comfort and "ideal" indoor environments (Fanger, 1970; King, 1990; Baker, 1993). Localized culture- and situation-specific responses, such as reintroducing the "siesta" or varying the strategy between the prestige commercial sector (hotels, banks, and offices) and the residential sector, offer the opportunity to have a less energy-intensive response to global warming. Locally specific strategies will reflect the distribution of wealth and status, slowing if not reversing the trend to globalized lifestyle and standardized indoor environment.

12.5.3. Air Pollution, Waste Management, and Sanitation

Reducing industrial pollution is a desirable activity under current climate because it can not only reduce acute health-threatening conditions like heat waves but also can reduce other pollution-related problems, such as the buildup of heavy metals in the environment, loss of aquatic biodiversity, and loss of forested areas. Reductions in the burning of fossil fuels not only have an impact in solving all of the foregoing problems but have mitigative effects on global warming itself. Reducing automotive traffic in sensitive airsheds has a similar effect, as does substitution of more benign fuel types (e.g., natural gas for coal and the use of unleaded gasoline).

Countermeasures for chemical time bombs include a complete inventory and system of monitoring for urban-area landfills and waste sites; tight control of waste flows from source to final resting place for present waste disposal; and dumping of toxic materials only in safe, permitted sites. Such control is difficult to maintain in poor cities that lack adequate urban infrastructure but is extremely important because so many people depend on shallow groundwater wells. In rural areas, control is more difficult, especially for widely used agricultural chemicals. Safe disposal of containers and residual amounts is extremely important but often neglected, particularly where the chemicals are used by laborers for absentee landlords on large estates.

Although increasing human activity and urbanization are likelier to have more important impacts on sewage management than will climate change, the impact of climate change or the necessity of reduction in GHGs may influence sewage-management policy in the direction of water recycling and reducing energy use. Water demand will increase in many urban areas, whereas severe reduction in available potable-water resources because of climate change is predicted in some areas. By necessity, reuse of treated sewage will increase to cope with relative shortage of urban water resources. Limited reuse of sewage has been applied in areas that are subject to dry climate (South Africa, Israel, California, etc.) or in areas with very high density of urban activity (e.g., Japan). These examples show that reuse of treated sewage for flush toilets, industrial use, agricultural use, or even part of the raw water destined for potable water supply is technologically possible. For some uses, waterborne toxins not addressed by conventional treatment can be a concern.

The reuse of treated sewage requires advanced sewage-treatment technology in addition to the more-conventional secondary treatment. There are many options in the advanced treatment category. Depending on the proposed quality of treated water, its cost and consumption of energy and resources vary. Some of the advanced treatments consume much more energy than the conventional sewage treatment processes. Energy-intensive processes often tend to be adopted in the developed countries so that the quality of treated water obtains acceptance from the public. Savings in energy consumption are important and should be taken into account in such cases. On the other hand, inexpensive and less energy-intensive methods are normally chosen in the developing countries. Such a choice is desirable from the view point of the global environment, but high priority should also be given to avoiding hygienic risk either by using a high grade of treatment or by limiting the uses of the treated sewage.

12.5.4. Infrastructure

As described in the 1992 assessment with respect to the coastal zone in particular, available adaptive responses to flooding fall broadly into three categories: retreat, accommodation, and protection. There are various environmental, economic, social, cultural, legal, institutional, and technological implications for each of these options. Retreat could lead to a loss of property, potentially costly resettlement of populations, and, in some notable cases, refugee problems. Accommodation could result in declining property values and in costs for modifying infrastructure. Protection may require significant investment, can have environmental costs to shorelines, and may be ineffective if the frequency or magnitude of extreme events increases. Generally speaking, building codes and other design and construction standards offer one of the most effective ways to limit the effects of extreme events such as tropical cyclones and flooding. As the climate changes, the standards need to be periodically revisited and updated. The supporting analysis for the standards needs to be couched in the language of engineers so that design solutions reflect increased climate uncertainty and risk of natural hazards. Also, developing countries may require assistance in obtaining the necessary knowledge and institutional capacity to develop and implement the standards.

In developed countries, the costs of defending infrastructure from river and coastal flooding have been calculated for numerous sites, but mostly for extreme events under current climate conditions. There have been a growing number of region- and city-specific studies that suggest that climate change could prove costly to major urban areas in coastal zones of developed nations. Examples from the United States (Walker *et al.*, 1989) and Japan (Mimura *et al.*, 1993; see Figure 12-3, which shows typhoon-vulnerable areas of the Tokyo metropolitan area) illustrate the point.

In Japan, protection of coastal infrastructure was estimated to cost $63 billion for raising port facilities (quays, wharfs, jetties, seawalls, etc.), plus $29 billion for coastal facilities (water gates, breakwaters, seawalls, etc.) (Mimura *et al.*, 1993). For an extended discussion of strategies and further international estimates of the vulnerabilities and costs of sea-level rise, see Chapter 9, Hoozemans *et al.* (1993), or IPCC CZMS (1992); also see Chapter 17.

Although no comprehensive international estimates have been made of additional pumping-station, water-supply, or drainage requirements for coastal regions, some analyses have been done of the requirements to maintain water balances in some areas. For example, Arai (1990) analyzed the water balance in Tokyo. A handful of estimates have been done on the costs of adapting to changes in lake levels, in the range of a few hundred million dollars for small areas in Lake Michigan (Changdon *et al.*, 1989) and Lake Titicaca.

Identification and mapping of landslides provides information on what has happened in the past, but such landslide inventory maps at a scale of 1:100,000 or larger probably cover less than 1% of the land and sea areas of the world. Several urban areas with particular landslide problems have set up detailed schemes for landslide mapping and landslide risk assessment. The work of the Geotechnical Control Office in Hong Kong (Brand, 1989), the San Mateo County Gas Project in California (Brabb, 1993), and the Colorado Geologic Survey (Mears, 1977) are good examples of such schemes. However, many of the poorer countries of the tropics have little information. Even in developed countries like Britain, a detailed inventory of landslides was not completed in the 1980s (D.K.C. Jones, 1991). From such inventories, landslide susceptibility maps can be produced.

Landsliding is usually predictable, and a wide range of techniques is available to help urban planners minimize risk through avoidance, control, and improved resistance (Jones, 1993). Landslide susceptibility maps show areas with different potentials for future landslide movement. Real-time warning systems for landslides are rare, but successful operation of a debris-flow warning system for the San Francisco Bay region indicated that, where sufficient telemetric rain gauges are available, adequate warning may be achieved. Many countries, however, rely on structural measures to mitigate debris flows and rock avalanches. In western China, for example, many long training walls restrict the width of debris flows down valley floors. The safety of such measures could require reevaluation. The effectiveness of a good landslide risk inventory is well illustrated in Hong Kong, where urban land-capability assessment and control of building construction are based on assessments of slope stability (Styles and Hansen, 1989).

Building controls do little for structures that existed before the controls were implemented and those that are built illegally. These situations call for a great increase in understanding and awareness, even in wealthy communities. Many of the poor of the world will continue to exist on hazardous hillsides. Forced evacuation and rigid enforcement of zoning regulations may be only temporary measures unless viable, practical, and economic alternatives are provided for the poor.

Increased storminess or rainfall intensity in regions where settlements are expanding onto hillsides featuring relict periglacial landslides will lead to greater frequency of landsliding. Here, more rigid application of geomorphological knowledge in planning decisions is required. Existing structures that are threatened need to be identified. Particular attention in all regions needs to be paid to mining waste, quarry tailings, and other landfills and spoil heaps, which may become unstable and threaten schools, hospitals, dwellings, business premises, and water quality.

Though the extent of adaptive capacity varies among nations and localities, human beings have developed multiple adaptive responses over the centuries to cope with high fire-danger regimes. Major adaptive responses include both fire prevention and fire control and suppression; use of fireproof and fire-retardant materials such as concrete and steel, rather than wood in buildings; development of extensive firefighting networks in settled areas (including personnel and mechanical firefighting equipment such as fire trucks, fire-hydrant systems, and sprinkler systems in buildings); fire-control systems in rural areas, such as controlled burns to limit fuel, fire breaks, aerial fire retardant delivery, and rural fire departments and "smoke jumpers"; and short-term activity controls during high fire-danger weather, such as prohibitions on open burning, commercial activities such as logging, and recreation activities such as hiking, hunting, or use of off-road vehicles. If frequency or intensity of fire danger warrant, another potential adaptive response involves improved spatial planning of communities and some longer-term land-use controls. These provide better isolation of fires and could limit damage to human settlements. If global climate change makes high fire danger a more common or dangerous occurrence in places such as western North America and southeastern Australia, more frequent and intense application of these principles may reduce vulnerabilities to fire. In locations where fire danger is lessened, relaxation of some controls may be possible and desirable.

Integrating transportation and land-use planning can reduce the demand for transportation infrastructure. While largely undertaken for mitigative reasons, integrating transportation and land-use planning could reduce the adverse impacts of air pollution under current climate as well as reduce urban heat islands and assure that future heat waves would be less of a

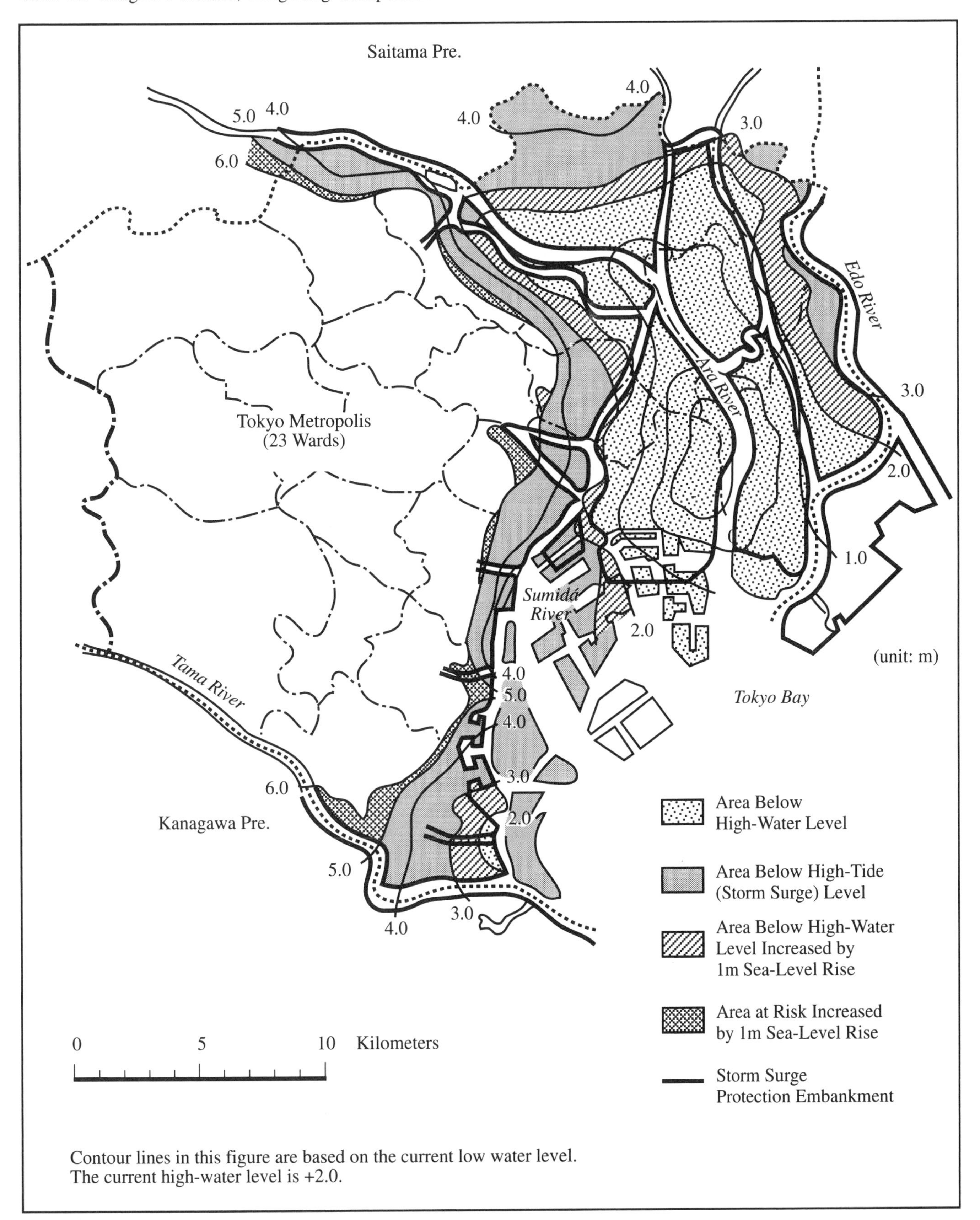

Figure 12-3: Distribution of lowlands and positions of high-tide embankments in Tokyo (Mimura *et al.*, 1993).

health and cooling-energy problem than they otherwise would be. Appropriately timed investments—especially public transit and transportation-management schemes—can improve traffic flow, reduce the need for travel, and emphasize appropriate modal choice (especially foot traffic and bicycles for short trips). These appear particularly promising in reducing local urban air pollution in the developing world (OTA, 1991). In addition, adjusting human settlement patterns may strongly affect transportation in the long run (see Chapter 21).

12.5.5. *Water Supply*

Potential changes in hydrologic balance because of climate change will have an impact on the availability of water resources and eventually on water balance and water supply. However, more significant and acute imbalance between water demand and the water resource is expected in many regions because of the increase both in population and in per capita water demand. Water-shortage problems exist and are becoming increasingly serious even under present circumstances (Laburn, 1993). Severe imbalances are foreseen, especially in developing countries where water demand is increasing rapidly and the institutional system and water supply facilities are still insufficient. Future management must take into account such socioeconomic-driven increases in water demand and the potential decrease in available water resources. Solutions to such imbalances include minimizing net water use and developing new water resources, including limited wastewater reuse. Prevention of water pollution is necessary to maintain a water quality satisfactory enough for potable water supply. See Chapter 14 for additional details on water-management options. The Nordic Freshwater Initiative proposed two key principles for future sustainable development and water use in developing countries (Jønch-Clausen, 1993). The first principle is management at the lowest appropriate levels. Centralized and top-down approaches to water-resource development and management are often insufficient for efficient local management, although the centralized mechanism is good to ensure national economic and social interests. The second principle is to consider water as an economic good. Numerous opportunities exist for improved management of water infrastructure, pricing policies, and demand-side management of supply (Frederick and Gleick, 1989). The cost includes direct cost, opportunity cost, and environmental cost. Charging for water use induces conservation and protection of water resources and develops the consciousness of water management.

12.5.6. *Health*

A range of adaptive mechanisms for offsetting the potential human-health effects of global warming lie in improving certain aspects of health services and other public services that settlements provide in any case (World Bank, 1993). Improved sanitation and water treatment both reduce the spread of waterborne diseases and may provide a measure of safeguard against importing exotic enteric waterborne diseases such as cholera. Effective inoculation programs against tropical diseases are limited to a few arbovirus infections, principally yellow fever and Japanese encephalitis. Vaccination against cholera is about 50% effective for a period of a few months (American Public Health Association,1990). Improved disease-treatment intervention has reduced mortality and morbidity rates from a number of diseases and could prove an effective, if expensive, alternative in some cases. This strategy could take the form of health and relief services' emergency response to specific disease outbreaks, for example. Improved air-pollution control is desirable from a number of perspectives under current climate and could have beneficial effects on reducing the impact of climate warming. Pest suppression/management can in some instances reduce the prevalence of disease vectors or hosts (e.g., the early twentieth century campaign against mosquitos in Panama) and thus offset to some degree the expansion of range of some of these pests.

Finally, disease surveillance could be strengthened and integrated with other environmental monitoring to design early warning systems; develop early, environmentally sound public health interventions; and develop anticipatory societal policies to reduce the risk of outbreaks and subsequent spread of epidemics. For additional details on health considerations, see Chapter 18.

12.6. Needs for Future Research

We are still very much at the infancy of this topic. Information needed to calculate the effects of climate change on human settlements would answer two broad questions, both of which must be answered at the regional and local level: (1) What is the present situation with respect to human population and settlement infrastructure? (2) What are the options for future economic and social development (especially adaptation to climate), and what will determine which option(s) will be adopted?

- **Present Situation**
  - Generally speaking, consistent, useful estimates of infrastructure types and values, especially in flood zones, do not really exist. Databases on regional capital stock are necessary to calculate infrastructure at risk from extreme events. Information is also required on capital turnover rates to determine "natural" replacement rates.
  - Better-founded information is needed to support estimates of the effects of climate on national energy demand, particularly the balance between heating and cooling in temperate-zone countries, which is currently a matter of some controversy. More needs to be done on the inventories and market penetrations of energy-using equipment in the commercial sector in all countries and in the residential sector of developing countries.
- **Future Situation**
  - We need better information on the probability, location, intensity, and consequences of future extreme climate events of all types. Currently, the

information available from climate models on extremes of temperature, precipitation, and windspeed is inadequate to do more than identify this as an issue.

- Cost estimates are needed for nearly every infrastructure and settlement issue. With the exception of the costs of coastal defense and some studies on energy demand, little has been done on the costs associated with the effects of climate change on human settlements. There are only limited databases on the costs of potential urban flooding and landslides, the costs of migration for infrastructure development, and the costs of health intervention programs and air- and water-pollution prevention.
- Data are needed on technology dynamics. Beyond general principles, we know very little about which variables govern the development and adoption of new technologies. Yet this may have a profound impact on the future shape of human settlements.
- We have better data and theories on population fertility, survival, and death, but regional migration is much more difficult to forecast quantitatively.
- Analysis of the future location of economic activity needs a better theory and better data. The general principles underlying locational choice of industry are known. However, it is nearly impossible to extend these principles beyond particular current business decisions to whole industries, especially for many years in the future. Data are needed on the costs of production, potential suppliers and markets, and the costs of transportation, especially in the developing countries.

References

Aguilar, A.G., 1987: Urban planning in the 1980s in Mexico City. Operative process or political facade? *Habitat International,* **11(3)**, 23-38.

Aguilar, A.G. and G. Olivera, 1991: El control de la expansion urbana en la ciudad de Mexico. Conjeturas de un falso planteamiento. *Estudios Demograficos y Urbanos,* **6(1)**, 89-115, El Colgio de Mexico, Mexico (in Spanish).

Aguilar, A.G. and M.L. Sanchez, 1993: Vulnerabilidad y riesgo en la ciudad de Mexico. *Ciudades,* **17**, 31-39, Puebla, Mexico (in Spanish).

Aittoniemi, P., 1991a: Influences of climate change on electricity consumption and production of hydropower in Finland. In: *Proceedings of the International Conference on Climate Impacts on the Environment and Society (CIES),* University of Tsukuba, Ibaraki, Japan, January 27-February 1, 1991. WMO/TD-No. 435, World Meteorological Organization, Geneva, Switzerland, pp. 464-468.

Aittoniemi, P., 1991b: Influences of climatic change on the Finnish energy economy. In: *Energy and Environment 1991* [Kainlauri, E., A. Johansson, I. Kurki-Suonio, and M. Geshwiler (eds.)]. American Society of Heating, Refrigeration, and Air-Conditioning Engineers, Atlanta, GA, 548pp.

Akbari, H., S. Davis, S. Dorsano, J. Huang, and S. Winnett, 1992: *Cooling Our Communities: A Guidebook on Tree Planting and Light-Colored Surfaces.* EPA 22P-2001, LBL-31587, U.S. Government Printing Office, Washington, DC, 217 pp.

American Public Health Association, 1990: *Control of Communicable Diseases.* American Public Health Association, Washington, DC, 15th ed., 532 pp.

Arai, T., 1990: Hydrology in Tokyo. *Geographical Review of Japan,* **63B**, 88-97.

Armstrong, W. and T.G. McGee, 1985: *Theatres of Accumulation: Studies in Asian and Latin American Urbanization.* Methuen, London, UK, 269 pp.

Arnell, N., 1983: *Insurance and Natural Hazards.* Discussion Paper 23, Department of Geography, University of Southampton, Southampton, UK.

Baker, N.V., 1993: Thermal comfort evaluation for passive cooling. In: *Solar Energy in Architecture and Urban Planning* [Foster, N. and H. Scheer (eds.)]. H.S. Stephens & Associates, Felmersham, Bedford, UK, 750 pp.

Bates, B.C., S.P. Charles, N.R. Sumner, and P.M. Fleming, 1994: Climate change and its hydrological implications for South Australia. *Transactions of the Royal Society of South Australia,* **118(1)**, 35-43.

Beatley, T., 1994: Promoting sustainable land use: mitigating natural hazards through land use planning. In: *Natural Disasters: Local and Global Perspectives.* 1993 Annual Forum of National Committee on Property Insurance, Insurance Institute for Property Loss Reduction, Boston, MA, pp. 31-36.

Berry, B.J.L., 1990: Urbanization. In: *The Earth as Transformed by Human Action: Global and Regional Changes in the Biosphere over the Past 300 Years* [Turner, B.L. *et al.* (eds.)]. Cambridge University Press, New York, NY, pp. 103-119.

Berz, G., 1993a: The insurance industry and IDNDR: common interests and tasks. *Newsletter of the United Nations IDNDR,* **15**, 8-10.

Berz, G., 1993b: Global warming and the insurance industry. *Interdisciplinary Science Review,* **18(2)**, 120-125.

Billard, A., T. Muxart, E. Derbyshere, Y. Egels, M. Kasser, and Jingtai-Wang, 1992: Glissements de terrain induits par des pluies dans le loess de la Province de Gansou, China. *Annales de Geographie,* **367**, 520-540.

Boissonade, A. and W. Dong, 1993: Windstorm model with applications to risk management. In: *Natural Disasters: Protecting Vulnerable Communities* [Merriman, P.A. and C.W.A. Browitt (eds.)]. Proceedings of the IDNDR Conference, London, UK, 13-15 October 1993, T. Telford Publishing, New York, NY, pp. 331-343.

Bonine, M.E., 1993: Cities of the Middle East and North Africa. In: *Cities of the World: World Regional Urban Development* [Brunn, S.D. and J.F. Williams (eds.)]. HarperCollins College Publishers, New York, NY, 2nd ed., pp. 305-349.

Boodhoo, Y., 1991: The need to mitigate the impact of climate in the tropics. In: *Proceedings of the International Conference on Climate Impacts on the Environment and Society (CIES),* University of Tsukuba, Ibaraki, Japan, January 27-February 1, 1991. WMO/TD-No. 435, World Meteorological Organization, Geneva, Switzerland, pp. D67-D72.

Bouma, M.J., H.E. Sondorp, and H.J. van der Kaay, 1994: Health and climate change. *The Lancet,* **343**, 302.

Brabb, E.E., 1991: The world landslide problem. *Episodes,* **14(1)**, 52-61.

Brabb, E.E., 1993: The San Mateo County GIS project for predicting the consequences of hazardous geological processes. In: *Geographical Information Systems in Assessing National Hazards* [Carrara, A. and F. Guzzetti (eds.)]. Kluwer Academic Publishers, Boston, MA, pp. 119-121.

Brand, E.W., 1989: Occurrence and significance of landslides in southeast Asia. In: *Landslides—Extent and Economic Significances* [Brabb, E.E. and B.L. Harrod (eds.)]. Balkema, Rotterdam, Netherlands, pp. 303-324.

Bravo, H. *et al.,* 1991: Ozono y lluvia acida en la ciudad de Mexico. *Ciencias,* No. 22, Facultad de Ciencias, UNAM, Mexico, pp. 33-40.

Calva Tellez, J.L., 1992: Efectos de un tratado trilateral de libre comercio en el sector agropecuario Mexicano. In: *La Agricultura Mexicana frente al Tratado de Libre Comercio* [Calva Tellez, J.L. *et al.* (eds.)]. Universidad Autonoma de Chapingo, Mexico, pp. 13-32.

Caine, T.N., 1980: The rainfall intensity duration control of shallow landslides and debris flows. *Geografiska Annaler A,* **62A**, 23-35.

Central Statistical Office, 1994: *General Household Survey (1992).* Central Statistical Office, London, UK.

Changdon, S.A., S. Leffler, and R. Shealy, 1989: Impacts of extremes in Lake Michigan levels along Illinois shorelines: low levels. In: *Potential Effects of Global Climate Change on the United States.* Appendix H, *Infrastructure.* PB90-172313, U.S. Environmental Protection Agency, Washington, DC. National Technical Information Service, Springfield,VA, pp. 3.1-3.48.

Cheney, N.P., 1979: Bushfire disasters in Australia 1945-1975. In: *Natural Hazards in Australia* [Heathiske, R.L. and B.G. Thom (eds.)]. Australian Academy of Science, Canberra, Australia, pp. 72-92.

Clark, J.R., 1991: Coastal zone management. In: *Managing Natural Disasters and the Environment* [Kreimer, A. and B.G. Thom (eds.)]. Colloquium sponsored by the World Bank, 27-28 June 1990, World Bank, Washington, DC.

Crawford, L., 1992: Smog-bound in Santiago. *British Medical Journal*, **305**, 213.

Courtenay, R., 1992: Keynote address on environment. In: *Proceedings of the CIB '92*. World Building Congress, Montreal, Canada, 17-22 May 1992.

Darmstadter, J., 1993: Climate change impacts on the energy sector and possible adjustments in the MINK region. *Climatic Change*, **24(1-2)**, 117-131.

Dejoux, C. and A. Iltis (eds.), 1992. *Lake Titicaca: A Synthesis of Limnological Knowledge*. Kluwer Academic Publishers, Boston, MA, 573 pp.

Denlea, Jr., L.E., 1994: Regulating insurance availability and insurer solvency: are they at cross purposes? In: *Natural Disasters: Local and Global Perspectives*. 1993 Annual Forum of National Committee on Property Insurance, Insurance Institute for Property Loss Reduction, Boston, MA, pp. 10-14.

Departmento del Distrito Federal, 1987: *Programa General de Desarrollo Urbano del Distrito Federal*. Direccion General de Reordenacion Urbana y Proteccion Ecologica, Mexico.

Derbyshire, E. and J. Wang, 1994: China's Yellow River Basin. In: *The Changing Global Environment* [Roberts, N. (ed.)]. Blackwell, Oxford, UK, pp. 440-462.

Douglas, I., 1983: *The Urban Environment*. Edward Arnold, London, UK, 229 pp.

Douguedroit, A., 1991: Influence of a global warming on the risk of forest fires in the French Mediterranean area. In: *The Global Environment* [Takeuchi, K. and M. Yoshino (eds.)]. Springer-Verlag, Berlin and Heidelberg, Germany, 257 pp.

Economist, 1994: Pollution: some vision. *The Economist*, **333(7884)**, 36-39.

Epstein, P.R., T.E. Ford, and R.R. Colwell. 1993. Marine ecosystems. *The Lancet*, **342**, 1216-1219.

Epstein, P.R., 1994: Letter to the *New York Times*, 13 November, p. 14.

Ezcurra, E., 1990. The basin of Mexico. In: *The Earth as Transformed by Human Action: Global and Regional Changes in the Biosphere over the Past 300 Years* [Turner, B.L. *et al.* (eds.)]. Cambridge University Press, New York, NY, pp.577-588.

Fanger, P.O., 1970: *Thermal Comfort*. Danish Technical Press, Copenhagen, Denmark, 244 pp.

Fankhauser, S., 1995: *Valuing Climate Change. The Economics of the Greenhouse Effect*. Earthscan, London, UK, 180 pp.

FEMA, 1992: *Building Performance: Hurricane Andrew in Florida*. Federal Emergency Management Agency, Washington, DC, 470 pp.

Figueroa, M., A. Ketoff, and O. Masera, 1992: *Residential Energy Use and Conservation in Venezuela: Results and Implications of a Household Survey in Caracas*. LBL-30508, Lawrence Berkeley Laboratory, Berkeley, CA, 66 pp.

Forestry Canada, 1991: *Selected Forestry Statistics, Canada, 1991*. Information Report E-X-46, Forestry Canada, Ottawa, Ontario, Canada.

Foster, C., January 7, 1994: Bush fires fan across New South Wales in Australia. *Christian Science Monitor*, **7**, 1.

Frederick, K.D. and P.H. Gleick, 1989: Water resources and climate change [Rosenberg, N.J., W.E. Easterling III, P.R. Crosson, and J. Darmstadter (eds.)]. In: *Greenhouse Warming: Abatement and Adaptation*. Resources for the Future, Washington, DC, pp. 133-143.

Freeman, T.J., 1992: Seasonal foundation movements in clay soil. In: *Proceedings of Fifth DYP Insurance and Reinsurance Group Conference on Changing Weather Patterns*. DYP Group Ltd., London, UK, 14 pp.

Friedman, D.G., 1984: Natural hazard risk assessment for an insurance program. *Geneva Papers on Risk and Insurance*, **9(30)**, 57-128.

Fukuma, Y., 1993: Objective evaluation of preparedness against typhoon, by typhoon classification. *Journal of Meteorological Research*, **45(5)**, 159-196 (in Japanese).

Gafny, S. and A. Gasith, 1989: Water quality dynamics in the shallow littoral of Lake Kinneret. In: *Environmental Quality and Ecosystem Stability*, vol. IV-B [Spanier, E., Y. Steinberger, and M. Luria (eds.)]. ISEEQ Pub., Jerusalem, Israel.

Gafny, S. and A. Gasith, 1993. Effect of low water level on the water quality in the littoral zone of Lake Kinneret. *Water Science and Technology*, **27**, 363-371.

Gilbert, A., 1992: Third world cities: housing, infrastructure, and servicing. *Urban Studies*, **29(3/4)**, 435-460.

Glantz, M.H. (ed.), 1988: *Societal Responses to Regional Climate Change: Forecasting by Analogy*. Westview Press, Boulder, CO, 428 pp.

Glantz, M., R.W. Katz, and M. Krenz (eds.), 1987: *The Societal Impacts Associated with the 1982-83 Worldwide Climate Anomalies*. National Center for Atmospheric Research, Boulder, CO, 105 pp.

Glantz, M.H., R.W. Katz, and N. Nicholls, 1991: *Teleconnections Linking Worldwide Climate Anomalies*. Cambridge University Press, Cambridge, UK, 535 pp.

Gleick, P.H., 1992: Water and conflict. In: *Occasional Paper Series of the Project on Environmental Change and Acute Conflict*, Number 1. University of Toronto and American Academy of Arts and Sciences. Peace and Conflict Studies Program, University of Toronto, Toronto, Ontario, Canada, pp. 3-28.

Gordon, H.B., P.H. Whetton, A.B. Pittock, A.B. Fowler, and M.R. Haylock, 1992: Simulated changes in daily rainfall intensity due to enhanced greenhouse effect: implications for extreme rainfall events. *Climate Dynamics*, **8**, 83-102.

Grazulis, T.P., 1991: *Significant Tornadoes, 1880-1989*. Vol. I, *Discussion and Analysis*. Environmental Films, St. Johnsbury, VT, 1340 pp.

Griffith, E. and L. Ford, 1993: Cities of Latin America. In: *Cities of the World: World Regional Urban Development* [Brunn, S.D. and J.F. Williams (eds.)]. HarperCollins College Publishers, New York, NY, 2nd ed., pp. 225-265.

Habitat, 1992a: People, settlements, environment, and development: improving the living environment for a sustainable future. United Nations Centre for Human Settlements (Habitat), Nairobi, Kenya, 57 pp.

Habitat, 1992b: Improving the living environment for a sustainable future. United Nations Centre for Human Settlements (Habitat), Nairobi, Kenya, 53 pp.

Hadfield, P., 1994: Revenge of the rain gods. *New Scientist*, **20** August, 14-15.

Hage, K.D., 1985. Weather extremes in Alberta: 1880 to 1960. *Climatological Bulletin*, **19(1)**, 3-15.

Haigh, M.J., 1994: Deforestation in the Himalaya. In: *The Changing Global Environment* [Roberts, N. (ed.)]. Blackwell, Oxford, UK, 531 pp.

Hallegraeff, G.M., 1993: A review of harmful algal blooms and their apparent global increase. *Phycologia*, **32(2)**, 79-99.

Hall, D., F. Rossillo-Calle, R. Williams, and J. Wood, 1993: Biomass for energy: supply prospects. In: *Renewable Energy Sources for Fuels and Electricity* [Johansson, T.B. *et al.* (eds.)]. Island Press, Washington, DC, 1160 pp.

Hanaki, K., 1993: Impact on urban infrastructure in Japan. In: *The Potential Effects of Climate Change in Japan* [Nishioka, S. *et al.* (eds.)]. CGER-IOO9-'93, Center for Global Environmental Research, National Institute for Environmental Studies, Tsukuba, Irabaki, Japan, pp. 81-85.

Handmer, J.W., 1989: The flood risk in Australia. In: *Natural Hazards and Reinsurance* [Britton, N.R. and J. Oliver (eds.)]. Proceedings of a Seminar Sponsored by Sterling Offices (Australia), Ltd., Cumberland College of Health Sciences, Lidcombe, NSW, Australia, pp. 45-59.

Hardoy, J.E., D. Mitlin, and D. Satterthwaite, 1992: *Environmental Problems in Third World Cities*. Earthscan, London, UK, 302 pp.

Hardoy, J.E. and D. Satterthwaite, 1984: Third world cities and the environment of poverty. *Geoforum*, **15(3)**, 307-333.

Hazarika, S., 1993: Bangladesh and Assam: land pressures, migration, and ethnic conflict. In: *Occasional Paper Series of the Project on Environmental Change and Acute Conflict*, Number 3. University of Toronto and American Academy of Arts and Sciences. Peace and Conflict Studies Program, University of Toronto, Toronto, Ontario, Canada, pp. 45-65.

Hejkstra, G.P., W.M. Stigliani, and G.R.B. Ter Meulen-Smidt, 1993: Report of the closing session of the SETAC conference, Potsdam, Germany, 24 June 1992; chemical time bombs. *Land Degradation and Rehabilitation*, **4**, 199-206.

Henri, C., 1991: The insurance industry response to flood. In: *Natural Hazards and Reinsurance* [Oliver, J. and N.R. Britton (eds.)]. Proceedings of a Seminar Sponsored by Sterling Offices (Australia), Ltd. Cumberland College of Health Sciences, Lidcombe, NSW, Australia, pp.157-175.

Herring, H., R. Hardcastle, and R. Phillipson, 1988: *Energy Use and Energy Efficiency in UK Commercial and Public Buildings Up to the Year 2000*. Energy Efficiency Series 6, Energy Efficiency Office, HMSO, London, UK, 173 pp.

Hesterberg, D., W.M. Stigliani, and A.C. Imeson (eds.), 1992: *Chemical Time Bombs: Linkages to Scenarios of Socioeconomic Development*. IIASA Executive Report 20 (CTB Basic Document 2), International Institute for Advanced Systems Analysis, Laxenburg, Austria, 28 pp.

Hiernaux, D., 1991: Servicios urbanos, grupos populares y medio ambiente en Chalco, Mexico. In: *Servicios Urbanos, Gestion Local y Medio Ambiente* [Schteingart, M. and L. D'Andrea (eds.)]. Colmex/CERFE, Mexico.

Hohmeyer, O. and M. Gartner, 1992: *The Costs of Climate Change: A Rough Estimate of Orders of Magnitude*. Report to the Commission of the European Communities DGXII. Fraunhofer Institute for Systems and Innovation Research, Karlsruhe, Germany, 60 pp.

Homer-Dixon, T.F., J.H. Boutwell, and G.W. Rathjens, 1993: Environmental change and violent conflict. *Scientific American*, **268(2)**, 38-45.

Hoozemans, F.J.J., M. Marchand, and H.A. Pennekamp, 1993: *A Global Vulnerability Analysis, Vulnerability Assessment for Population, Coastal Wetlands and Rice Production on a Global Scale*. Delft Hydraulics and Ministry of Transport, Public Works and Water Management, Delft and The Hague, Netherlands, 2nd ed., 184 pp.

Huang, J., R.L. Ritschard, J.C. Bull, and L. Chang, 1987: *Climatic Indicators for Residential Heating and Cooling Loads, ASHRAE Transactions*, Vol. 93-1. American Society of Heating, Refrigeration, and Air Conditioning Engineers, Atlanta, GA, 72-111 pp.

Hulme, M., P. Haves, and B. Boardman, 1992: Impacts of climate change. In: *Future Buildings Forum, Innovative Cooling Systems*. Workshop Report, Solihull (UK), 12-14 May 1992.

IBI Group, 1990: *The Implications of Long-Term Climatic Changes on Transportation in Canada*. CCD 90-02, Canadian Climate Digest, Atmospheric Environment Service, Environment Canada, Downsview, Ontario, Canada, pp. 1-8.

IPCC CZMS, 1992: *Global Climate Change and the Rising Challenge of the Sea*. Report of the Intergovernmental Panel on Climate Change, Response Strategies Working Group, Coastal Zone Management Subgroup. Ministry of Transport, Public Works and Water Management, Directorate General Rijkswaterstaat, Tidal Waters Division, The Hague, The Netherlands, 35 pp.

IRC, 1995: *Coastal Exposure and Community Protection—Hurricane Andrew's Legacy*. Insurance Research Council, Wheaton, IL, 48 pp.

IRS, 1994: Table B-2. Table of class lives and recovery periods. In: *Depreciation. For Use in Preparing 1994 Returns*. Publication 534, U.S. Internal Revenue Service, Washington, DC, p. 286.

Jacobson, R.B., A.J. Miller, and J.A. South, 1989: The role of catastrophic geomorphic events in central Appalachian landscape evolution. *Geomorphology*, **2**, 257-284.

Jacoby, H.D., 1989: Likely effects of climate on water quality. In: *Coping with Climate Change* [Topping, J.C. (ed.)]. The Climate Institute, Washington, DC.

Johnson, R.H. and R.B. Vaughan, 1989: The Cows Rocks landslide. *Geological Journal*, **24**, 354-370.

Jakobi, W., 1993: The future of the catastrophic reinsurance market. *Catastrophe Reinsurance Newsletter*, **10**, 4-7; **11**, 18-20; and **12**, 36-39.

Jønch-Clausen, T., 1993: National report of Denmark. In: *Proceedings of the 19th Annual Water Supply Congress, Managing the Global Environment: The Role of the Water Manager.* Budapest, Hungary, 2-8 October 1993.

Jones, D.K.C., 1991: Human occupance and the physical environment. In: *The Changing Geography of the UK* [Johnston, R.J. and V. Gardner (eds.)]. Routledge, London, UK, 2nd ed., pp. 382-428.

Jones, D.K.C., 1993: Slope instability in a warmer Britain. *Geographical Journal*, **159**, 184-195.

Jones, D.W., 1991: How urbanization affects energy-use in developing countries. *Energy Policy*, **19(9)**, 621-630.

Kalkstein, L.S., 1993: Health and climate change: direct impacts in cities. *The Lancet*, **342**, 1397-1399.

Kalkstein, L.S. and K.E. Smoyer, 1993: The impact of climate change on human health: some international implications. *Experientia*, **49**, 969-979.

King, A., 1990: Architecture, capital, and the globalisation of culture. In: *Global Culture* [Featherstone, M. (ed.)]. Sage, Newbury Park, CA, 411 pp.

Kitajima, S., T. Ito, N. Mimura, Y. Hosokawa, M. Tsutsui, and K. Izumi, 1993: Impact of sea level rise and cost estimate of countermeasures in Japan. In: *Proceedings of the IPCC Eastern Hemisphere Workshop on Vulnerability Assessment to Sea Level Rise and Coastal Zone Management*. Tsukuba, Japan, 3-6 August 1994.

Kreimer, A. and M. Munasinghe (eds.), 1991: Managing environmental degradation and natural disasters: an overview. In: *Managing Natural Disasters and the Environment, Colloquium Sponsored by the World Bank, 27-28 June 1990*. The World Bank, Washington, DC, pp. 3-6.

Laburn, R.J., 1993: International report. In: *Managing the Global Environment: The Role of the Water Manager*. Proceedings of the 19th Annual Water Supply Congress, Budapest, Hungary, 2-8 October 1993.

Leichenko, R.M., 1993: Climate change and water resource availability: an impact assessment for Bombay and Madras, India. *Water International*, **18(3)**, 147-156.

Leichenko, R.M. and J.L. Westcoat, 1993: Environmental impacts of climate change and water development in the Indus Delta Region. *Water Resources Development*, **9(3)**, 247-261.

Lester, R.L., 1993: Are there new rules for insurers? Defining roles and obligations for the industry. In: *Catastrophe Insurance for Tomorrow: Planning for Future Adversities* [Britton, N.R. and J. Oliver (eds.)]. Proceedings of a Seminar Sponsored by Sterling Offices (Australia), Ltd. Griffith University, Brisbane, Australia, pp. 11-30.

Linder, K.P., M.J. Gibbs, and M.R. Inglis, 1987: *Potential Impacts of Climate Change on Electric Utilities*. Report 88-2, New York State Energy Research and Development Authority, Albany, NY.

Linder, K.P. and M.R. Inglis, 1989: The potential effects of climate change on regional and national demands for electricity. In: *The Potential Impacts of Global Climate Change on the United States*. Appendix H, *Infrastructure* [Smith, J.B. and D.A. Tirpak (eds.)]. PB90-172313, National Technical Information Service, Springfield, VA, pp. 1.1-1.25.

Liverman, D., 1992: The regional impact of global warming in Mexico: uncertainty, vulnerability, and response. In: *The Regions and Global Warming: Impacts and Response Strategies* [Schmandt, J. and J. Clarkson (eds.)]. Oxford University Press, Oxford, UK, pp. 44-68.

Loveland, J.E. and G.Z. Brown, 1990: *Impacts of Climate Change on the Energy Performance of Buildings in the United States*. OTA/UW/UO, Contract J3-4825.0, Office of Technology Assessment, United States Congress, Washington, DC, 58 pp.

Lowi, M.R., 1992: West Bank water resources and resolution of conflict in the Middle East. In: *Occasional Paper Series of the Project on Environmental Change and Acute Conflict*, Number 1. University of Toronto and American Academy of Arts and Sciences. Peace and Conflict Studies Program, University of Toronto, Toronto, Ontario, Canada, pp. 29-61.

Magadza, C.H.D., 1991: Social impacts of the creation of Lake Kariba. In: *Guidelines of Lake Management*. Vol. 2, *Socio-Economic Aspects of Lake Reservoir Management* [Hashimoto, H. (ed.)]. ILEC/UNDP, Otsu, Japan.

Magadza, C.H.D., 1993: Climate change and water supply security in southern Africa. In: *Proceedings of 4th Science Symposium, Zimbabwe Research Council*. Harare, Zimbabwe.

Marco, J.B., 1992: Flood risk mapping. In: *Coping with Floods, Proceedings of NATO Advanced Studies Institute, October 1992* [Rossi, G., N. Harmancioglu, and V. Yevjevich (eds.)]. E Majorana Centre for Scientific Culture, University of Catania, Italy, 776 pp.

Marco, J.B. and A. Cayuela, 1992: Urban flooding: the flood planned town. In: *Coping With Floods, Proceedings of NATO Advanced Studies Institute, October 1992* [Rossi, G., N. Harmancioglu, and V. Yevjevich (eds.)]. E Majorana Centre for Scientific Culture, University of Catania, Italy, 776 pp.

Matsuoka, Y., T. Morita, and T. Arimura, 1992: Development of efficient and sustainable urban systems to care for the earth. *Environmental Research Quarterly*, **86**, 51-65 (in Japanese).

Mayer, R., 1993: Chemical time bombs related to forestry practice: distribution and behaviour of pollutants in forest soils. *Land Degradation and Rehabilitation*, **4**, 275-279.

McMichael, A.J., 1993: Global environmental change and human population health: a conceptual and scientific challenge for epidemiology. *International Journal of Epidemiology*, **22(1)**, 1-8.

Mears, A.I., 1977: Debris flow hazards analyses and mitigation, an example from Glenwood Springs, Colorado. *Colorado Geologic Survey Information Series*, **8**, Colorado Geologic Survey, Denver, CO, 45 pp.

Mendez, G.F., 1991: Mexico City's program to reduce air pollution. In: *Economical Development and Environmental Protection in Latin America* [Tulchin, J.S. (ed.)]. Lynne Reinner Publishers, Boulder, CO, 143 pp.

Meyers, S., S. Tyler, H. Geller, J. Sathaye, and L. Schipper, 1990: *Energy Efficiency and Household Electric Appliances in Developing and Newly Industrialized Countries*. LBL-29678, Lawrence Berkeley Laboratory, Berkeley, CA, 86 pp.

Millbank, N., 1989: Building design and use: response to climatic change. *Architects Journal*, **96**, 59-63.

Mimura, N., M. Isobe, and Y. Hosokawa, 1993: Coastal zone. In: *The Potential Effects of Climate Change in Japan* [Nishioka, S. *et al.* (eds.)]. CGER-IOO9-'93, Center for Global Environmental Research, National Institute for Environmental Studies, Tsukuba, Irabaki, Japan, pp 57-81.

Minnery, J.R. and D.I. Smith, 1994: Climate change, flooding, and urban infrastructure. Paper presented at Greenhouse 94. Australian–New Zealand Conference on Climate Change, 10-14 October in Wellington, New Zealand. Center for Resource and Environmental Studies, Australian National University, Canberra, Australia, 20 pp.

Morgan, J., 1993: Natural and human hazards. In: *South-east Asia's Environmental Future: The Search for Sustainability* [Brookfield, H. and Y. Byron (eds.)]. United Nations University Press, Tokyo, Japan, and Oxford University Press, Oxford, UK, pp. 286-303.

Mumme, S., 1991: System maintenance and environmental reform in Mexico. Paper presented at the XIII Coloquio de Antropologia e Historia Regional, Zamora, Michoacan, 7-9 August, Mexico.

Mundy, C.J., 1990: Energy sector. In: *New Zealand Climate Change Programme, Climate Change: Impacts on New Zealand*. Ministry for the Environment, Aukland, New Zealand.

Murray, C., 1991: Plans after the flood. *Surveyor*, 7 March, 16-17.

NCPI, 1992: *Natural Disaster Loss Reduction Update. Update 2 (1)*. National Committee on Property Insurance, Boston, MA, 12 pp.

Newland, K., 1994: Refugees: the rising flood. *World Watch*, **7(3)**, 10-20.

Niemeyer, E.V. *et al.*, 1991: *Potential Impacts of Climate Change on Electric Utilities*. Electric Power Research Institute, Palo Alto, CA.

Nishinomiya, S. and H. Kato, 1989: *Potential Effects of Global Warming on the Japanese Electric Industry-Event Tree of Impacts on the Electric Utility Industry Stemming from Climate-Induced Changes in the Natural Environment, Ecosystems, and Human Society*. Central Research Institute of the Electric Power Industry, Otemachi, Tokyo, Japan.

Nishinomiya, S. and Y. Nishimura, 1993: Impact on the energy sector. In: *The Potential Effects of Climate Change in Japan* [Nishioka, S. *et al.* (eds.)]. CGER-IOO9-'93, Center for Global Environmental Research, National Institute for Environmental Studies, Tsukuba, Irabaki, Japan, pp. 71-79.

Oguntoyinbo, J.S., 1991: Climate impact on environment and society in the Sudano-Sahelian Zone of West Africa. In: *Proceedings of the International Conference on Climate Impacts on the Environment and Society (CIES)*, University of Tsukuba, Ibaraki, Japan, January 27-February 1, 1991. WMO/TD-No. 435, World Meteorological Organization, Geneva, Switzerland, pp. D.31-D.35.

Ojo, S.O., 1991: Implications of climate change on environment and man in West and Central Africa. In: *Proceedings of the International Conference on Climate Impacts on the Environment and Society (CIES)*, University of Tsukuba, Ibaraki, Japan, January 27-February 1, 1991. WMO/TD-No. 435, World Meteorological Organization, Geneva, Switzerland, pp. D.25-D.30

Okamoto, K., 1991: Influence of the greenhouse effect on the environment and the society in Japan. In: *Proceedings of the International Conference on Climate Impacts on the Environment and Society (CIES)*, University of Tsukuba, Ibaraki, Japan, January 27-February 1, 1991. WMO/TD-No. 435, World Meteorological Organization, Geneva, Switzerland, pp. D.79-D.84.

Parikh, J., 1992: *Environmental Concerns and Human Development: Small and Numerous Sources of Pollution*. Reprint No. 27-1992, Indira Gandhi Institute of Development Research, Bombay, India, 25 pp.

Parker, D., 1991: *Residential Demand Side Management in Thailand for the IIEC*. Florida State Energy Solar Center, Cape Canaveral, FL.

Paul, A., 1994: Tornados and hail. In: *Proceedings of a Workshop: Improving Responses to Atmospheric Extremes, the Role of Insurance and Compensation*, Toronto, Ontario, Canada, 3-4 October 1994. The Climate Institute, Washington, DC and Environmental Analysis Research Group—Environment Canada, Downsview, Ontario, Canada.

Peele, B.D., 1988: Insurance and the greenhouse effect. In: *Greenhouse: Planning for Climate Change* [Peaman, G.I. (ed.)]. CSIRO Publications, East Melbourne, Victoria, Australia, pp. 588-601.

Phillips, D.W., 1990: *The Climates of Canada*. Ministery of Supply and Services Canada, No. En 56-1/1990E. Hull, Quebec, Canada, 176 pp.

Phillips, D., 1993. *The Day Niagra Falls Ran Dry! Canadian Weather Facts and Trivia*. Key Porter Books, Ltd., Toronto, Canada, 226 pp.

Pierce, B., 1994: *Office of Building Technologies Evaluation and Planning Report*. BNL-52426, Brookhaven National Laboratory, Upton, NY, 127 pp.

Pittock, B., 1994: Modeling of present and future climate: interpolation of regional studies. In: *IPCC Special Workshop on Article 2 of the U.N. Convention on Climate Change*. Fortaleza, Brazil, 17-21 October 1994, IPCC, Geneva, Switzerland, pp. 63-67.

Price, C. and D. Rind, 1993: Lightning fires in a 2 x CO_2 world. In: *,oceedings of the 12th Conference on Fire and Forest Meteorology*, 26-28 October, Jekyll Island, GA. Society of American Foresters, Bethesda, MD, and American Meteorological Society, Boston, MA, 796 pp.

PNUNA/MOPU/AECI, 1989: *Desarrollo y Medio Ambiente en America Latina. Una vision evolutiva*. Programa de Naciones Unidas para el Medio Ambiente, Ministerio de Obras y Urbanismo y Agencia Espanola de Cooperacion Internacional, Madrid, Spain.

Rayner, S. and K. Richards, 1994: I think that I shall never see. . .a lovely forest policy: land use programs for conservation of forests. Paper prepared for IPCC Working Group III Workshop on Policy Instruments and Their Implications, 17 January 1994, Pacific Northwest Laboratory, Washington, DC, 28 pp.

Rello, T., 1993: Ajuste Macroeconomico y Politica Agricola en Mexico. In: *Mexico, Auge, Crisis y Ajuste* [Bazdresch, C. *et al.* (comp.)]. El Trimestre Economico No. 73, Vol. 3, Fondo de Cultura Economica, Mexico, pp. 372-393.

Riebsame, W.E., 1990: The United States Great Plains. In: *The Earth as Transformed by Human Action: Global and Regional Changes in the Biosphere over the Past 300 Years* [Turner, B.L. *et al.* (eds.)]. Cambridge University Press, New York, NY, pp. 561-575.

Rijsberman, F., 1991: Potential costs of adapting to sea level rise in OECD countries. In: *Responding to Climate Change: Selected Economic Issues*. OECD, Paris, France, pp. 11-49.

Rosenthal, D.H., H.K. Gruenspecht, and E. Moran, 1995: Effects of global warming on energy use for space heating and cooling in the United States. *Energy Journal*, **16(2)**, 77-96.

RSC/CAE, 1994. *National Report of Canada*. Prepared for the IDNDR Mid-Term Review and the 1994 World Conference on Natural Disastre Reduction, Yokohama, Japan, 23-27 May 1994.

Ryan, C.J., 1993: Costs and benefits of tropical cyclones, severe thunderstorms, and bushfires in Australia. *Climatic Change*, **25(3-4)**, 353-368.

Sanderson, M., 1987: *Implications of Climatic Change for Navigation and Power Generation in the Great Lakes*. Phase II: *Socioeconomic Assessment of the Implications of Climatic Change for Commercial Navigation and Hydroelectric Power Generation in the Great Lakes-St. Lawrence River System*. The Great Lakes Institute, University of Windsor, Windsor, Ontario, Canada, 21 pp.

Sari, A.P., 1994: *Indonesia Country Report, Climate Change in Asia*. Asian Development Bank, Manila, Philippines.

Sathaye, J. and S. Tyler, 1991: Transitions in household energy use in urban China, India, the Philippines, Thailand, and Hong Kong. *Annual Review of Energy and the Environment*, **16**, 295-335.

Scheer, D.P., 1986: Managing water supplies to increase water availability. In: *U.S. Geological Survey, National Water Summary, 1985-Hydrologic Events and Surface Water Resources, Water Supply Paper 2300*. U.S. Government Printing Office, Washington, DC, pp. 101-112.

Schipper, L. and S. Meyers, 1992: *Energy Efficiency and Human Activity: Past Trends and Future Prospects*. Cambridge University Press, Cambridge, UK, 385 pp.

Schteingart, M., 1988: Mexico City. In: *The Metropolis Era*. Vol. 2, *Megacities* [Dogan, M. and J.D. Kasarda (eds.)]. Sage, Newbury Park, CA, 65 pp.

Schteingart, M., 1989: The environmental problems associated with urban development in Mexico City. *Environment and Urbanization*, **1**, 40-50.

Scott, M.J., D.L. Hadley, and L.E. Wrench, 1994: Effects of climate change on commercial building energy demand. *Energy Sources*, **16(3)**, 339-354.

Scott, M.J., R.D. Sands, L.W. Vail, J.C. Chatters, D.A. Neitzel, and S.A. Shankle, 1993: *The Effects of Climate Change on Pacific Northwest Water-Related Resources: Summary of Preliminary Findings*. PNL-8987, Pacific Northwest Laboratory, Richland, WA, 46 pp.

Shearn, W.G., 1994: Personal lines pricing—insurance or discrimination. In: *Sixteenth UK Insurance Economists Conference, Cripps Hall, University of Nottingham, 20/21 April, 1994*. Department of Insurance, University of Nottingham, Nottingham, UK.

Shove, E., 1994: Threats and defences in the built environment. In: *Perspectives on the Environment 2* [Elworthy, S. (ed.)]. Avebury Press, Aldershot, Hants, UK, 238 pp.

Shukla, V. and J. Parikh, 1992: *The Environmental Consequences of Urban Concentration: Cross-National Perspectives on Economic Development, Air Pollution, and City Size*. Discussion Paper No.72, Indira Gandhi Institute of Development Research, Bombay, India, 36 pp.

Sias, J.C. and D.P. Lettenmaier, 1994: *Potential Effects of Climatic Warming on the Water Resources of the Columbia River Basin*. Water Resources Series Technical Report No. 142, University of Washington Department of Civil Engineering, Seattle, WA, 142 pp.

Simmons, I.G., 1989: *Changing the Face of the Earth: Culture, Environment, History*. Blackwell, Oxford, UK, 487 pp.

Singh, B., 1988: *Prospectives d'un Changement Climatique du a un Doublement de CO_2 Atmospherique pour les Resources Naturelles du Quebec*. Department of Geography, University of Montreal, Montreal, Quebec, Canada.

Skinner, J.L., M.E. Gilliam, and T.M. O'Dempsey, 1993: The new California? Demographic and economic growth in Queensland. In: *Catastrophe Insurance for Tomorrow: Planning for Future Adversities, Proceedings of a Seminar Sponsored by Sterling Offices (Australia), Ltd.* [Britton, N.R. and J. Oliver (eds.)]. Griffith University, Brisbane, Australia, pp. 1-39.

Smil, V., 1992: Environmental change as a source of conflict and economic losses in China. In: *Occasional Paper Series of the Project on Environmental Change and Acute Conflict*, Number 2. University of Toronto and American Academy of Arts and Sciences. Peace and Conflict Studies Program, University of Toronto, Toronto, Ontario, Canada.

Smith, D.I., 1993: The impacts of flooding and storm surge. In: *Climate Impact Assessment Methods for Asia and the Pacific* [Jakeman, A.J. and A.B. Pittock (eds.)]. Australian Government Printing Office, Canberra, Australia, pp. 85-89.

Stavely, J.K., 1991: 1989-All that and more? Australian hazard insurance experience during the period 1989-1991: implications for the industry in a period of economic downturn. In: *Natural and Technological Hazards: Implications for the Insurance Industry* [Britton, N.R. and J. Oliver (eds.)]. Proceedings of a Seminar Sponsored by Sterling Offices (Australia) Ltd. University of New England, Armidale, New South Wales, Australia.

Stokoe, P.K., M. LeBlanc, P. Lane and Associates, Ltd., and Discovery Consultants, Ltd., 1987: *Socio-Economic Assessment of the Physical and Ecological Impacts of Climate Change on the Marine Environment of the Atlantic Region of Canada, Phase I*. School for Resource and Environmental Studies, Dalhousie University, Halifax, Nova Scotia, Canada, 8 pp.

Street, R.B., 1989: Climate change and forest fires in Ontario. In: *Proceedings, 10th Conference on Fire and Forest Meteorology*. Forestry Canada, Ottawa, Ontario, Canada, pp. 177-182.

Styles, K.A. and A. Hansen, 1989: *Territory of Hong Kong, Geotechnical Area Studies Programme GASP Report XII*. Geotechnical Control Office, Civil Engineering Services Department, Hong Kong, 346 pp.

Suhrke, A., 1993: Pressure points: environmental degradation, migration, and conflict. In: *Occasional Paper Series of the Project on Environmental Change and Acute Conflict*, Number 3. University of Toronto and American Academy of Arts and Sciences. Peace and Conflict Studies Program, University of Toronto, Toronto, Ontario, Canada, 67 pp.

Thompson, G.D. and P.N. Wilson, 1994: Common property as an institutional response to environmental variability. *Contemporary Economic Policy*, **12(3)**, 10-21.

Thompson, M., M. Warburton, and T. Hatley, 1986: *Uncertainty on a Himalayan Scale*. Ethnographica, London, UK, 162 pp.

Tiwari, P. and J. Parikh, 1993: *Cost of Carbon Dioxide Reduction in Building Construction*. Discussion Paper 97, Indira Gandhi Institute of Development Research, Bombay, India, 53 pp.

Tolba, M.K., O.A. El-Kholy, E. El-Hinnawi, M.W. Holdgate, D.F. McMichael, and R.E. Munn, 1992: *The World Environment 1972-1992: Two Decades of Challenge*. Chapman and Hall, London, UK, 884 pp.

Tooley, M.J., 1992: Sea level changes and forecasting flood risks. *Geneva Papers on Risk and Insurance*, **17(64)**, 406-414.

Turner, R.K., P. Doktor, and W.N. Adger, 1994: *Assessing the Costs of Sea Level Rise: East Anglian Case Study*. Environment and Planning.

Udo, R.K., O.O. Arcola, J.O. Ayoade, and A.A. Afolayan, 1990: Nigeria. In: *The Earth as Transformed by Human Action: Global and Regional Changes in the Biosphere over the Past 300 Years* [Turner, B.L. *et al.* (eds.)]. Cambridge University Press, New York, NY, pp. 589-603.

UK Climate Change Impacts Review Group, 1991: *The Potential Effects of Climate Change in the United Kingdom*. Department of the Environment, HMSO, London, UK, 124 pp.

UNEP and the Government of the Netherlands, 1991: *Impact of Sea Level Rise on Society. A Case Study for the Netherlands*. Delft Hydraulics Laboratory, Delft, Netherlands, 125 pp.

United Nations, 1989: *1989 Report on the World Social Situation*. United Nations publication E.89.IV.1, United Nations, New York, NY, 126 pp.

United Nations World Fertility Survey, 1986: *Fertility Behavior in the Context of Development*. Population Studies No. 100 (United Nations Publications, Sales no. E.86.XIII.5), United Nations, New York, NY, 383 pp.

U.S. Congress, Office of Technology Assessment (OTA), 1991: *Energy in Developing Countries*. OTA-E-486, U.S. Government Printing Office, Washington, DC, 137 pp.

Vaux, H.J., 1991: Global climate change and California's water resources. In: *Global Climate Change and California: Potential Impacts and Responses* [Knox, J.B. and A.F. Scheuring (eds.)]. University of California Press, Berkeley, CA, pp.69-96.

Walker, J.C., T.R. Miller, G.T. Kingsley, and W.A. Hyman, 1989: The impact of global climate change on urban infrastructure. In: *The Potential Impacts of Global Climate Change on the United States*. Appendix H, *Infrastructure* [Smith, J.B. and D.A. Tirpak (eds.)]. PB90-172313, National Technical Information Service, Springfield, VA, pp.2.1-2.37.

Wang, X., X. Qian, I. Douglas, and H. Mi, 1991: The formation and harmful consequences of acid rain in Chonqing, China. Paper presented at Global Forum '94 Academic Conference, Manchester, UK, June 1994.

WCC'93, 1994: *Preparing to Meet the Coastal Challenges of the 21st Century*. World Coast Conference 1993 Report, Rijkwaterstaat, The Hague, Netherlands, 49 pp.

Whetton, P.H., A.M. Fowler, M.R. Haylock, and A.B. Pittock, 1993: Implications of climate change due to the enhanced greenhouse effect on floods and droughts in Australia. *Climatic Change*, **25**, 289-317.

White, B., 1992: Bushfires in the "Greenhouse." *Trees and Natural Resources*, **34(4)**, 26.

White, R.R., 1994: *Urban Environmental Management: Environmental Change and Urban Design*. John Wiley and Sons, Chichester, UK, 233 pp.

WHO Commission on Health and the Environment, 1992: *Report of the Panel on Urbanization*. World Health Organization, Geneva, Switzerland, 160 pp.

Williams, M., 1990: Forests. In: *The Earth as Transformed by Human Action: Global and Regional Changes in the Biosphere over the Past 300 Years* [Turner, B.L. *et al.* (eds.)]. Cambridge University Press, New York, NY, pp. 179-201.

World Bank, 1990: *World Development Report, 1990*. World Bank and Oxford University Press, New York, NY, 272 pp.

World Bank, 1991: *World Bank Report 1991: The Challenge of Development*. Oxford University Press, New York, NY, 304 pp.

World Bank, 1993: *World Development Report 1993: Investing in Health*. Oxford University Press, New York, NY, 344 pp.

World Commission on Environment and Development, 1987: *Our Common Future*. Oxford University Press, Oxford, UK, 383 pp.

World Resources Institute, 1992: *1992-93 World Resources: Toward Sustainable Development*. Oxford University Press, New York, NY, 385 pp.

Xia Guang, 1991: Chinese population-environment relationship and the climatic impacts on it. In: *Proceedings of the International Conference on Climate Impacts on the Environment and Society (CIES)*, University of Tsukuba, Ibaraki, Japan, January 27-February 1, 1991. WMO/TD-No. 435, World Meteorological Organization, Geneva, Switzerland, pp. D.19-D.24.

Yoshino, M., 1989: Problems in climates and agroclimates for mountain developments in Xishuanbanna, South Yunnan, China. *Geographical Review of Japan*, **62B(2)**, 149-160.

Yoshino, M., 1993: Climatic change and agriculture: problems for the Asian tropics. In: *Southeast Asia's Future: The Search for Sustainability* [Brookfield, H. and Y. Byron (eds.)]. United Nations University Press, Tokyo, Japan, and Oxford University Press, Oxford, UK, 442 pp.

Yoshino, M. and R. Kawamura, 1989: Structure of cold waves over east Asia. In: *Professor Yao Zhensheng Memorial Volume*. Nanjing University, Meteorology Publications, Beijing, China, pp. 340-349.

13

Agriculture in a Changing Climate: Impacts and Adaptation

JOHN REILLY, USA

Lead Authors:
W. Baethgen, Uruguay; F.E. Chege, Kenya; S.C. van de Geijn, The Netherlands; Lin Erda, China; A. Iglesias, Spain; G. Kenny, New Zealand; D. Patterson, USA; J. Rogasik, Germany; R. Rötter, The Netherlands; C. Rosenzweig, USA; W. Sombroek, FAO; J. Westbrook, USA

Contributing Authors:
D. Bachelet, France; M. Brklacich, Canada; U. Dämmgen, Germany; M. Howden, Australia; R.J.V. Joyce, Wales; P.D. Lingren, USA; D. Schimmelpfennig, USA; U. Singh, IRRI, Phillipines; O. Sirotenko, Russia; E. Wheaton, Canada

CONTENTS

EXECUTIVE SUMMARY

Substantial research progress has been made on the impacts of climate change on agriculture since the 1990 IPCC assessment, with new studies for different sites and countries covering many areas of the world. New studies also have been conducted that integrate site, country, and regional impacts to provide information about how global agricultural production and consumption may be affected. On the whole, these studies support the evidence presented in the first IPCC assessment that global agricultural production can be maintained relative to baseline production in the face of climate changes likely to occur over the next century (i.e., in the range of 1 to 4.5°C) but that regional effects will vary widely. Major uncertainties result from the lack of reliable geographic resolution in future climate predictions, difficulties in integrating and scaling-up basic physiologic responses and relationships, and difficulty in estimating farm sector response and adaptation to changing climate as it varies across the world. Thus, while there will be winners and losers stemming from climate impacts on agricultural production, it is not possible to distinguish reliably and precisely those areas that will benefit and those that will lose.

Major conclusions of the assessment cover four areas: (1) The direct and indirect effects of changes in climate and atmospheric constituents on crop yield, soils, agricultural pests, and livestock; (2) estimates of yield and production changes for specific localities, countries, and the world, based on studies that integrate multiple direct and indirect effects; (3) the conditions that determine vulnerability and areas and populations that are relatively more vulnerable to adverse changes; and (4) adaptation potential.

Direct and Indirect Effects

Experimental results, detailed modeling of basic processes, and knowledge of physical and biological processes provide basic understanding of direct and indirect effects of climate on agricultural production:

- The results of a large number of experiments to resolve the effect of elevated CO_2 concentrations on crops have confirmed a beneficial effect. The mean value yield response of C_3 crops (most crops except maize, sugar cane, millet, and sorghum) to doubled CO_2 is +30%. Measured response ranges from -10% to +80%. Only gradually is the basis for these differences being resolved. Factors known to affect the response include the availability of plant nutrients, the crop species, temperature, precipitation, and other environmental factors. Differences in experimental technique also are responsible for variations in the measured response (High Confidence).
- Changes in soils, e.g., loss of soil organic matter, leaching of soil nutrients, and salinization and erosion, are a likely consequence of climate change for some soils in some climatic zones. Cropping practices such as crop rotation, conservation tillage, and improved nutrient management are, technically, quite effective in combating or reversing deleterious effects (High Confidence).
- Livestock production will be affected by changes in grain prices, changes in the prevalence and distribution of livestock pests, and changes in grazing and pasture productivity. Analyses indicate that intensively managed livestock systems have more potential for adaptation than crop systems. In contrast, adaptation may be more problematic in pastoral systems where production is very sensitive to climate change, technology changes introduce new risks, and the rate of technology adoption is slow (see Chapter 2) (Medium Confidence).
- The risk of losses due to weeds, insects, and diseases is likely to increase (Low Confidence).

Regional and Global Production Effects

To evaluate the direct and indirect effects of climate on yield at the farm, regional, or higher levels requires integrated models that consider system interactions. Issues of scale add uncertainty, and higher-order models have not generally taken into account the climatic effects on soils and pests. Despite these limitations, climate change studies show:

- Crop yields and productivity changes will vary considerably across regions. Thus, the pattern of agricultural production is likely to change in a number of regions (High Confidence).
- Global agricultural production can be maintained relative to base production under climate change as expressed by general circulation models (GCMs) under doubled CO_2 equilibrium climate scenarios (Medium Confidence).
- Based on global agricultural studies using doubled CO_2 equilibrium GCM scenarios, lower-latitude and lower-income countries have been shown to be more negatively affected. Again, crop model simulation results vary widely, e.g., ±20% changes in yield, for specific countries and sites across studies and GCM scenarios (Medium Confidence).

Vulnerability

Vulnerability is used here to mean the *potential* for negative consequences given the range of possible climate changes that might reasonably occur and is not a prediction that negative consequences will occur. Vulnerability can be defined at different scales, including yield, farm or farm sector, regional economic, or hunger vulnerability.

- Vulnerability to climate change depends on physical and biological response but also on socioeconomic characteristics. Low-income populations depending on isolated agricultural systems, particularly dryland systems in semi-arid and arid regions, are particularly vulnerable to hunger and severe hardship. Many of these at-risk populations are found in Sub-Saharan Africa, South and Southeast Asia, as well as some Pacific Island countries and tropical Latin America (High Confidence).

Adaptation

Uncertainty remains with regard to the ability of agricultural systems to adapt to climate change. Historically, farming systems have responded to a growing population and have adapted to changing economic conditions, technology, and resource availabilities. It is uncertain whether the rate of change of climate and required adaptation would add significantly to the disruption likely due to future changes in economic conditions, population, technology, and resource availabilities.

- Adaptation to climate change is likely; the extent depends on the affordability of adaptive measures, access to technology, and biophysical constraints such as water resource availability, soil characteristics, genetic diversity for crop breeding, and topography. Many current agricultural and resource policies are likely to discourage effective adaptation and are a source of current land degradation and resource misuse (High Confidence).
- National studies have shown incremental additional costs of agricultural production under climate change which could create a serious burden for some developing countries (Medium Confidence).

13.1. Introduction

Climate change will affect agriculture through effects on crops (Section 13.2); soils (Section 13.3); insects, weeds, and diseases (Section 13.4); and livestock (Section 13.5). Climatic conditions interact with agriculture through numerous and diverse mechanisms. Mechanisms, effects, and responses include, for example, eutrophication and acidification of soils, the survival and distribution of pest populations, the effects of CO_2 concentration on tissue- and organ-specific photosynthate allocation, crop breeding aims, animal shelter requirements, and the location of production (Brunnert and Dämmgen, 1994). Variation of agricultural and climatic conditions across the world leads to different local and regional impacts (Section 13.6). High levels of uncertainty necessitate assessment of vulnerability to the adverse effects of potential climate change (Section 13.7). The effects of climate change on local, national, and regional economies and food supplies depend on future economic and agricultural conditions and on the international transmission of supply shocks through international trade (Section 13.8). Historically, agriculture has proved to be highly adaptive to changing conditions, but uncertainty remains with regard to adaptation to potential climate change (Section 13.9). Emission-control efforts also will likely affect agriculture (Chapter 23), as will competition for land and water resources from other sectors affected by climate change (Chapter 25).

13.2. Climatic Effects on Crop Plants

The main climate variables important for crop plants, as for other plants, are temperature, solar radiation, water, and atmospheric CO_2 concentration. Important differences in temperature sensitivity and response to CO_2 exist among C_3, C_4, and CAM plants (Chapter A). While the basic physiological response of crop plants are no different than other plants of similar types, crop plants have been selected for particular traits and grow under more highly controlled conditions than plants growing in forests (managed and unmanaged) or other ecosystems. Most crop plants are C_3, the principal C_4 exceptions are maize, millet, sorghum, and sugar cane. Crop traits are selected and bred into different varieties to produce high yields for different climate and resource conditions. Most crop plants are grown as annuals (e.g., grains, potatoes, and most vegetable and fiber crops) with the crop species, variety, and planting time chosen by the farmer at the start of each growing season. Tree fruits, coffee, tea, cocoa, bananas, grapes, many forage and pasture crops, and other small fruits are the principal perennials. Nutrients and water which may be limited under natural conditions are more likely to be (or can be) augmented via fertilization, irrigation, and management of crop residue. Competition with other plants is controlled. The primary focus of crop production is the efficient production of harvestable yield, usually of only one component of the plant (e.g., seed, fruit, root, leaf).

13.2.1. Elevated CO_2 and Crops

Experiments concerning crop performance at elevated CO_2 concentrations in general show a positive but variable increase in productivity for annual crops (Kimball, 1983; Strain and Cure, 1985; Cure and Acock, 1986; Allen *et al.*, 1990; Kimball *et al.*, 1990; Lawlor and Mitchell, 1991; Bazzaz and Fajer, 1992; Gifford and Morison, 1993; Koerner, 1993; Rozema *et al.*, 1993; Mooney and Koch, 1994; Rogers *et al.*, 1994). Annual C_3 plants exhibit an increased production averaging about 30% at doubled (700 ppm) CO_2 concentrations; both biomass and seed production show an increase in almost all experiments under controlled conditions (Cure and Acock, 1986; Rogers and Dahlmann, 1993). Although the mean value response (+30% for C_3 crops) has been confirmed, variations in responsiveness between plant species and ecosystems persist (from -10% to +80%), and only gradually is the basis for these differences being resolved. Fewer experiments have been conducted with perennial crops—woody species in particular—but evidence suggests that the growth response would be less than for annual crops; the measured response for C_4 crops is much smaller than that for C_3 crops.

Variations in the growth enhancement among crops and different varieties of the same crop and over years are high, and interactions with nutrient and water availability are complex (e.g., Goudriaan and De Ruiter, 1983; Chaudhuri *et al.*, 1986; Mitchell *et al.*, 1993). The extent and occurrence of physiological adaptations of the photosynthetic apparatus, particularly of perennial plants, to long-term exposure to high CO_2 concentrations—which is more directly relevant to long-term climate change—are still unresolved (Cure and Acock, 1986; Bazzaz and Fajer, 1992; Wolfe and Erickson, 1993). Plants with nitrogen-fixing symbionts (e.g., peas, beans, alfalfa), under favorable environmental conditions for both symbiont and plant, tend to benefit more from enhanced CO_2 supplies than do other plants (Cure *et al.*, 1988). The content of nonstructural carbohydrates generally increases under high CO_2, while the concentrations of mineral nutrients and proteins are reduced (Mooney and Koch, 1994; Rogers *et al.*, 1994; Koerner and Miglietta, in press). Root/shoot ratios often increase under elevated CO_2 levels favoring root crops and also contribute to soil organic matter build-up (Mauney *et al.*, 1992; Mitchell *et al.*, 1993). Crop plants, like other plants, show increased water use efficiency under elevated CO_2 levels, but water consumption on a ground-area basis vs. a leaf-area basis is much less affected. Water use on a ground-area basis can actually increase if leaf area (canopy) increases. The range of water-use efficiencies among varieties for major crops is wide (e.g., Cure and Acock, 1986; Kimball *et al.*, 1993; Sombroek and Gommes, 1993), providing the opportunity to breed or select for improved efficiency. At higher CO_2 levels, plant growth damage done by air pollutants like nitrogen oxide (NO_x), sulfur dioxide (SO_2), and ozone (O_3) is reduced because of partial stomatal closing and other physiological changes (Van de Geijn *et al.*, 1993).

The wide range in estimated responses to elevated CO_2 is due in large part to the different experimental systems employed

(Gifford and Morison, 1993), each with advantages and disadvantages (Krupa and Kickert, 1989). Year-to-year variations in response also occur because of varying weather conditions that more or less favor CO_2 response of crops. The most widely used experimental system is the open-top chamber. Free-air CO_2 enrichment (FACE) experiments are more expensive but attempt to create conditions close to those likely to be experienced in an open field. Initial results from these experiments confirm the basic positive response of crops to elevated CO_2, but studies have been conducted only for a few crops (Mauney *et al.*, 1992). In addition, even the FACE experimental set-up creates a modified area (Kimball *et al.*, 1993; Vugts, 1993) analogous to a single irrigated field within a dry environment.

13.2.2. Temperature, Moisture, and Other Variables

Plant growth and crop yields clearly depend on temperature and temperature extremes, as shown for five major food crops (Table 13-1), and vary for crops with different photosynthetic pathways (Le Houerou *et al.*, 1993). The optimum range for C_3 crops is 15 to 20°C and for C_4 crops 25 to 30°C; for CAM crops a nighttime temperature of 10 to 20°C is optimal. C_4 and CAM crops require a minimum temperature of 10 to 15°C for growth and are relatively sensitive to frost. The minimum temperature for C_3 crops ranges from 5 to 10°C; C_3 crops exhibit variable frost sensitivity. Cumulative temperature above minimum growing temperature is an important determinant of crop phenological development. For annual crops, warmer temperatures speed development, shortening the period of growth and lowering yield if shortened growth period is not fully compensated by more rapid development at the higher temperature (Ellis *et al.*, 1990). The amount of light received during the reproductive stage may determine whether yields fall (Acock and Acock, 1993).

Table 13-1: *Crop physiology: Temperature thresholds (°C) of some major crops of the world (optimum range refers to the entire growing season and narrows under high light and elevated CO_2).*

Crop	Optimum Range	Lower Range	Upper Range	References/ Remarks
Wheat (C_3)	17–23	0	30–35	Burke *et al.*, 1988; Behl *et al.*, 1993; optimal enzyme function and representative seasonal foliage temperature
Rice (C_3)	25–30	7–12	35–38	Le Houerou *et al.*, 1993; Yoshida,1981
Maize (C_4)	25–30	8–13	32–37	Le Houerou, 1993; Decker *et al.*, 1986; Pollak and Corbett, 1993; Ellis *et al.*, 1992; Long *et al.*, 1983
Potato (C_3)	15–20	5–10	25	Haverkort, 1990; Prange *et al.*, 1990
Soybean (C_3)	15–20	0	35	Hofstra and Hesketh, 1969; Jeffers and Shibles, 1969

The variation of temperature requirements and temperature extremes for different cultivars of the same species and among species, however, is quite wide for most crops, as shown for major crops such as wheat, rice, maize, and soybean (Table 13-2). Such variation in requirements provides significant scope for adaptation through switching among existing cultivars and crops or introducing genetic variability through conventional plant breeding.

It has been suggested that higher mean temperatures will be accompanied by higher variability and more frequent occurrence of extremes (Katz and Brown, 1992). Such changes could severely affect plant functioning at low and high mean temperatures (Larcher, 1980; Kuiper, 1993).

Climate variability and resultant interannual variability in yield is a fundamental feature of most cropping systems (Gommes, 1993). In temperate areas, low temperatures and/or frost occurrences are the usual limiting factor for growing season length. In tropical and subtropical areas, where temperature variation is less extreme throughout the year than in temperate areas, season-length and seasonal variation depend on regular patterns of precipitation.

Temperature, solar radiation, relative humidity, wind, and precipitation vary not only over the season and year but on a daily basis. The effects of variability can be both positive and negative depending on the crop and the nature of the variability. Cool-season vegetable crops (e.g., spinach, cauliflower, broccoli) perform best when nights are cool; their quality deteriorates under warmer temperatures. Apples, pears, cherries, and many other tree fruits require winter chilling periods for buds to set but can suffer near-total loss of the fruit crop if late frosts damage the blossoms. Alternate thawing and freezing of the ground can cause the loss of perennial forage crops such as alfalfa. Extreme climatic events (e.g., storms with high winds, flooding, heavy rains, hail, hard late or early frosts) can be responsible for severe or total crop loss.

The anticipated increase in temperatures with global warming can lead to spikelet sterility in rice, loss of pollen viability in maize, reversal of vernalization in wheat, and reduced formation of tubers and tuber bulking in potatoes for areas near critical thresholds (Table 13-3). Yield losses can be severe in these

Table 13-2: *Crop phenology for important crops.*

Crop	Base Temp. (°C)	Max. Devel. (°C)	Emergence to Pre-Anthesis (degree days)	Post-Anthesis to Maturity (degree days)	Reference
Wheat	0	20–25	750–1300	450–1050	Van Keulen and Seligman, 1987; Elings, 1992; Hodges and Ritchie, 1991
Rice	8	25–31	700–1300	450–850	Yoshida, 1981; Penning de Vries, 1993; Penning de Vries *et al.*, 1989
Maize	7	25–35	900–1300	700–1100	Van Heemst, 1988; Pollak and Corbett, 1993; Kiniry and Bonhomme, 1991; Rötter, 1993
Soybean	0	25–35	highly variable	450–750	Wilkerson *et al.*, 1989; Swank *et al.*, 1987

Source: Adapted from Acock and Acock, 1993.
Notes: Base Temp. = minimum temperature for growth; Max. Devel. = optimum temperature for crop development; Emergence to Pre-Anthesis = cumulative degree days needed from crop emergence to pre-anthesis (flowering); Post-Anthesis to Maturity = cumulative degree days required (after flowering) to reach maturity.

cases, if temperatures exceed critical limits for periods as short as 1 hour during anthesis (flowering). Burke *et al.* (1988) found that under ample water supply different crop species manage to maintain foliage temperature within their specific optimum range (thermal kinetic window), thereby maximizing biomass accumulation.

Species-specific optimal temperature ranges are relatively narrow compared to diurnal and seasonal foliage temperature fluctuations. If temperature variability on a daily and seasonal basis were to decrease, crop cultivars in environments with mean temperatures close to cultivar-specific optima could profit from prolonged exposure to optimum temperatures. While variability in some climatic dimensions may increase, the temperature increase since the 1940s has been due mainly to increased nighttime temperature (Kukla and Karl, 1993), thus reducing the diurnal variation of temperature and considerably influencing the nighttime respiration. The response of respiration is theoretically derived and experimentally well-established, but the concept is challenged for situations where plants or crops are grown continuously and have adapted to the higher temperature (Gifford and Morison, 1993).

Table 13-3. *Temperature thresholds: High temperature effects on key development stages of five major arable crops.*

Crop	Effect	Reference
Wheat	Temperature >30°C for more than 8 hours can reverse vernalization	Evans *et al.*, 1975
Rice	Temperature >35°C for more than 1 hour at anthesis causes high percentage spikelet sterility	Yoshida, 1981
Maize	Pollen begins to lose viability at temperatures >36°C	Decker *et al.*, 1986
Potato	Temperatures >20°C depress tuber initiation and bulking	Prange *et al.*, 1990
Soybean	Great ability to recover from temperature stress; critical period in its development unknown	Shibles *et al.*, 1975

Source: Acock and Acock, 1993.

Other environmental changes will interact with changes in climate variables and elevated CO_2 to affect crop yields. Among these are exposure to O_3—tropospheric (surface) concentrations of which have doubled over the past 100 years in the Northern Hemisphere (Feister and Warmbt, 1987; Volz and Kley, 1988; Anfossi *et al.*, 1991) to a level estimated to reduce yields in the range of 1 to 30% (e.g., Ashmore, 1988; Van der Eerden *et al.*, 1988; Bosac *et al.*, 1993; Sellden and Pleijel, 1993)—and exposure to ultraviolet-B (UV-B) radiation, which is expected to increase due to stratospheric ozone depletion. UV-B fluxes depend on cloud cover (Van der Leun and Tevini, 1991) and decrease by a factor of 10 with latitude and increase with altitude. Thus, crops grown at high latitudes and high altitudes are most likely to be affected. Plant sensitivities vary widely by species (Rozema *et al.*, 1991; Tevini, 1993) and across cultivars of species, such as soybeans (Biggs *et al.*, 1981).

Some studies have investigated the combined effect of various stresses including climate, CO_2, O_3, and UV-B on crops (e.g., Goudriaan and De Ruiter, 1983; Chaudhuri *et al.*, 1986; Allen, 1990; Mitchell *et al.*, 1993; Krupa and Kickert, 1993). Results indicate that rice, barley, sorghum, soybean, oats, beans, and peas are relatively more sensitive to UV-B than other major crops and that wheat, maize, potatoes, cotton, oats, beans, and peas are relatively more sensitive to O_3. For other crops, including cassava, sugar cane, sweet potato, grapes, coconut, rye and peanuts, no clear results are available.

Consistent moisture availability throughout the crop growth period is critical. Overall, the hydrological cycle is expected to intensify with higher evaporation, air humidity, and precipitation. Higher temperatures would, at the same time, increase crop water demand. Global studies have found a tendency for increased evaporative demand to exceed precipitation increase

Box 13-1. Modeling Crop Response to Environmental Change

Climatic and other factors strongly interact to affect crop yields. Models have provided an important means for integrating many different factors that affect crop yield over the season (Rötter, 1993). Scaling up results from detailed understanding of leaf and plant response to climate and other environmental stresses to estimate yield changes for whole farms and regions, however, can present many difficulties (e.g., Woodward, 1993).

Higher-level, integrated models typically accommodate only first-order effects and reflect more complicated processes with technical coefficients. Mechanistic crop growth models take into account (mostly) local limitations in resource availability (e.g., water, nutrients) but not other considerations that depend on social and economic response, such as soil preparation and field operations, management of pests, and irrigation.

Models require interpretation and calibration when applied to estimating commercial crop production under current or changed climate conditions (see Easterling *et al.*, 1992; Rosenzweig and Iglesias, 1994); in cases of severe stress, reliability and accuracy in predicting low yields or crop failure may be poor. With regard to the CO_2 response, recent comparisons of wheat models have shown that even though basic responses were correctly represented, the quantitative outcome between models varied greatly. Validation of models has been an important goal (Olesen and Grevsen, 1993; Semenov *et al.*, 1993a, 1993b; Wolf, 1993a, 1993b; Delecolle, 1994; Iglesias and Minguez, 1994; Minguez and Iglesias, 1994).

Further integration of models of crop yield, phenology, and water use with geographic-scale agroclimatic models of crop distribution and economic response has also occurred (e.g., Kaiser *et al.*, 1993; Kenny *et al.*, 1993; Rötter and van Diepen, 1994). Simplified representations of crop response have been used with climate and soil data that are available on a global basis (Leemans and Solomon, 1993). More aggregated statistical models have been used to estimate the combined physical and socioeconomic response of the farm sector (Mendelsohn *et al.*, 1994; Darwin *et al.*, 1995).

Incorporation of the multiple effects of CO_2 in models has generally been incomplete. Some do not include any CO_2 effects and thus may overestimate negative consequences of CO_2-induced changes in climate. Other models consider only a crude yield effect. More detailed models consider CO_2 effects on water use efficiency (e.g., Wang *et al.*, 1992; Leuning *et al.*, 1993). With few exceptions, most models fail to consider CO_2 interactions with temperature and effects on reproductive growth (Wang and Gifford, 1995).

in tropical areas (Rosenzweig and Parry, 1994), but this result varies locally and across climate scenarios. Large changes in the seasonal pattern of rainfall or changes in the consistency of rainfall within the crop growing season would likely matter more for annual crops than changes in annual precipitation and potential evaporation rates. Some studies suggest increased intensity of rainfall, which may result in increased runoff (Whetton *et al.*, 1993). Changes in large-scale atmospheric patterns such as the El Niño/Southern Oscillation (ENSO) and tropical monsoons could cause significant shifts in rainfall patterns, with consequent effects on agricultural production. ENSO events currently have widespread and well-documented impacts on production in Oceania, Latin America, Africa, South and Southeast Asia, and the United States (Folland *et al.*, 1990, 1991; McKeon *et al.*, 1990; Nicholls and Wong, 1990; McKeon and White, 1992; Rimmington and Nicholls, 1993), but climate scientists have not resolved how ENSO events may be affected by long-term climate change.

A variety of models (see Box 13-1) that integrate understanding of crop yield and climate relationships have been used for estimating changes in potential crop yield, zonal production shifts, and regional and global production (see Sections 13.5 and 13.8.2).

13.3. Soil Changes and Agricultural Practices

Climate affects most major soil processes and is a major factor in soil formation; the potential effects of changing climate and higher atmospheric CO_2 on soils are highly interactive and complex (Bouwman, 1990; Buol *et al.*, 1990). Many of the world's soils are potentially vulnerable to soil degradation (e.g., loss of soil organic matter, leaching of soil nutrients, salinization, and erosion) as a likely consequence of climate change (see Chapter 4).

Cropping practices that maintain a more closed ground cover over longer periods—including crop rotation, planting of cover crops, and reduced or minimum tillage—combined with integrated nutrient management are quite effective in combating or reversing current land degradation and would be similarly effective where climate change had the potential to exacerbate land degradation (Brinkman and Sombroek, 1993; Rasmussen and Collins, 1991; Logan, 1991). Brinkman and Sombroek (1993) found that, in most cases, changes in soils by direct human action, whether intentional or unintended, are likely to have a greater impact than climate change. They found that under a transient scenario of climate change, soil physical, chemical, and biological processes will be given time to adapt, thereby counteracting human-induced land degradation.

13.4. Weeds, Insects, and Diseases

Weeds, insects, and pathogen-mediated plant diseases are affected by climate and atmospheric constituents. Resultant changes in the geographic distribution of these crop pests and their vigor in current ranges will likely affect crops. Existing research has investigated climatic determinants of the range of many pests, but the potential changes in crop losses due to climatically driven changes in pests have not been included in most agricultural impact studies (Waggoner, 1983; Stinner *et al.*, 1987; Prestidge and Pottinger, 1990).

13.4.1. Weeds

Weeds will benefit from the "CO_2 fertilization effect" and from improvements in water use efficiency associated with increasing CO_2 concentrations, but the impact on crop production will depend on how enhanced-growth weeds compete with enhanced-growth crops. Of the 86 plant species that contribute 90% of per capita food supplies worldwide, 80 are C_3 plants (Prescott-Allen and Prescott-Allen, 1990), while 14 of the world's 18 worst weeds are C_4 plants (Holm *et al.*, 1977). Experiments generally show C_3 species to benefit from CO_2 enrichment at the expense of C_4 species (Patterson and Flint, 1980, 1990; Carter and Peterson, 1983; Patterson *et al.*, 1984; Wray and Strain, 1987; Bazzaz and Garbutt, 1988).

Regarding temperature, most weeds of warm season crops originate in tropical or warm temperate areas and are responsive to small increases in temperature. For example, the growth of three leguminous weeds increased significantly as day/night temperature increased (Flint *et al.*, 1984). Biomass of C_4 smooth pigweed (*Amaranthus hybridus*) increased by 240% for an approximate 3°C temperature increase; C_4 grasses also showed large increases (Flint and Patterson, 1983; Patterson, 1993).

Accelerated range expansion of weeds into higher latitudes is likely (Rahman and Wardle, 1990; Patterson, 1993) as demonstrated for itchgrass (*Rottboellia cochinchinensis*, Lour.), cogongrass (*Imperata cylindrica*), Texas panicum (*Panicum texanum*), and witchweed (*Striga asiatica*) (Patterson *et al.*, 1979; Patterson, 1995). However, not all exotic weeds will be favored by climatic warming. Patterson *et al.* (1986) found loss of competitiveness under warmer conditions for the southward spread of wild proso millet (*Panicum miliaceum*) in the southwestern United States.

Increasing CO_2 and climate change probably will also affect mechanical, chemical, and natural/biological efforts to control weeds (Patterson, 1993, 1995), which currently cause worldwide crop production losses of about 12% (25% for traditional production systems) (Parker and Fryer, 1975). Environmental factors including temperature, precipitation, wind, soil moisture, and atmospheric humidity influence when herbicides are applied, as well as their uptake and metabolism by crops and target weeds (Hatzios and Penner, 1982; Muzik, 1976). High leaf starch concentrations, which commonly occur in C_3 plants grown under CO_2 enrichment (Wong, 1990), might interfere with herbicide activity. CO_2 enrichment also increases the growth of rhizomes and tubers in C_3 plants (Oechel and Strain, 1985), which could reduce the effectiveness of chemical and mechanical control of deep-rooting, perennial C_3 weeds. On the other hand, increased temperatures and increased metabolic activity tend to increase uptake, translocation, and effectiveness of many herbicides. Natural and biological control of weeds and other pests depends on the synchrony between the growth, development, and reproduction of biocontrol agents and their targets. Such synchrony may be disrupted if climate changes rapidly, particularly if climatic extremes occur more frequently. Global warming could facilitate overwintering of insect populations and favor earlier poleward migrations in the spring, which could increase the effectiveness of biological control of weeds in some cases. Conversely, such enhanced over-winterings would accelerate the spread of viruses by migrating vectors like aphids.

13.4.2. Insect Activity and Distribution

Climate change will affect the distribution and degree of infestation of insect pests through both direct effects on the life cycle of insects and indirectly through climatic effects on hosts, predators, competitors, and insect pathogens. There is some evidence that the risk of crop loss will increase due to poleward expansion of insect ranges. Insect populations faced with modified and unstable habitats created under annual arable agricultural systems generally must diapause (enter a state of dormancy), migrate, otherwise adapt genetically, or die. Insect species characterized by high reproduction rates are generally favored (Southwood and Comins, 1976). Human alteration of conditions that affect host plant survival—irrigation, for example—also affects phytophagous (leaf-eating) insect populations.

Insect life cycle processes affected by climate and weather include lifespan duration, fecundity, diapause, dispersal, mortality, and genetic adaptation. Porter *et al.* (1991) list the following effects of temperature on insects: limiting geographical ranges; over-wintering; population growth rates; number of generations per annum; length of growing season; crop-pest synchronization; interspecific interactions; dispersal and migration; and availability of host plants and refugia. The effects of climate and weather on insect life cycles have been documented for a wide variety of insect pests of agriculture, rangelands, and forests (Dobzhansky, 1965; Fye and McAda, 1972; Tauber *et al.*, 1986; Mattson and Haack, 1987; Kingsolver, 1989; Cammell and Knight, 1991; Harrington and Stork, 1995). Many effects involve changes in the severity of outbreaks following extreme weather events (on the order of hours to weeks). Freezing temperatures are a major factor in mortality, but *Drosophila* sp. insects that survive relatively colder temperatures have been found to be more fecund than cohorts that were not exposed to the low temperatures (Dobzhansky, 1965). Temperatures that exceed critical thresholds frequently have adverse effects on fecundity, as in the cases of bollworm, *Helicoverpa zea* (Boddie); tobacco budworm, *Heliothis virescens* (F.); beet armyworm, *Spodoptera exigua* (Hübner); cabbage looper, *Trichoplusia ni* (Hübner); saltmarsh caterpillar, *Estigmene acrea* (Drury); and pink bollworm, *Pectinophora gossypiella* (Saunders) (Fye and McAda, 1972).

Abnormally cool, wet conditions are associated with high subsequent infestations of cotton by cotton tipworm, *Crocidosema plebejana* (Zeller) (Hamilton and Gage, 1986). November precipitation and April temperature are the best indicators of mean grasshopper densities in Southern Idaho (Fielding and Brusven, 1990). Drought can affect various physiological processes of plants, which may increase the plants' attractiveness and susceptibility to phytophagous insects (Mattson and Haack, 1987). Intense precipitation has been noted as a deterrent to the occurrence and success of oviposition by insects such as the European corn borer, *Ostrinia nubilalis* (Hübner) (Davidson and Lyon, 1987). Abundant precipitation can affect mortality, for example, through drowning of soil-dwelling insects (Watt and Leather, 1986) but is more likely to affect insects indirectly through climatic effects on insect pathogens, predators, and parasites, as has been shown for *H. zea* pupae especially under persistently saturated soil conditions (Raulston *et al.*, 1992).

While the specific ways in which climate change could affect persistent wind patterns such as nocturnal wind jets in the United States and convergence systems such as the Inter-Tropical Convergence Zone (ITCZ) are poorly predicted by GCMs, changes in the strength, timing, and geographical extent of these systems have been hypothesized. Important insect pests of agriculture currently use these systems to disperse widely from decaying habitats to viable habitats (Pedgley, 1982). Corn earworm, *H. zea* moths in the United States (Westbrook *et al.*, 1985; Scott and Achtemeier, 1987; Lingren *et al.*, 1993), grasshoppers or locusts [*Ailopus simulatrix* (Walker)], old world bollworm [*Heliothis armigera* (Hübner)], and whitefly [*Bemisia tabaci* (Gennadius)] disperse via atmospheric transport. In the case of the ITCZ and locusts, the winds also contribute ephemeral precipitation for host plants as reviewed by Joyce (1983) and Rainey (1989). Changes in these systems would affect the speed and range of dispersal of these pests.

Many of the documented effects of climate on insects are based on unusual weather events affecting the severity of insect outbreaks within their normal range or the (unintended) introduction of exotic species into new environments or extension of (host) crops into new environments (e.g., through irrigation). Human-induced local climate change within urban areas provides evidence of how insects can adapt to changes in their environment. Examples of genetic adaptation include: froggatt (*Dacus tryoni*), lucerne flea (*Sminthurus viridis* L.) and sheep blowfly (*Lucilia cuprina*) (Wied.) in Australia after introduction from Europe or with the expansion of crop areas; the European corn borer, *O. nubilalis* (Hübner) and the European spruce sawfly, *Gilpinia hercyniae* (Hartig) in North America after introduction from Europe; and *Drosophila serrata* Mallock across latitudes (Birch, 1965). *Drosophila serrata* Mallock also has responded by natural selection to the changed conditions in large cities (Dobzhansky, 1965). Selection to tighten host-insect interaction [such as codling moth, *Cydia pomonella* (L.), in fruit trees] appears to favor the evolution of isolated host races (Poshly and Bush, 1979). Climatic, agronomic, political, and economic factors jointly led to populations of boll weevil, *Anthonomus grandis grandis* Boheman, which adapted differently across cotton areas in the United States (Terranova *et al.*, 1990). Although insect populations have been redistributed during periods of major climatic change, a continuous intermixing of beetle species' gene pools in the Rocky Mountains of the United States was determined to have prevented speciation (Elias, 1991).

Understanding of insect physiological development and behavior has led to the development of numerical models that estimate insect growth, movement, and mortality in response to potential changes in climate (Goodenough and McKinion, 1992). Minimum and maximum temperatures, required cumulative degree days, and, where important, the effects of drought or wet conditions (which are the basis for such models), have been established for many common agricultural pests (e.g., Fye and McAda, 1972; Davidson and Lyon, 1987). Simulation models predict potential redistribution of insects under simulated climates.

Specific studies of the likely impacts of climate change were reviewed by Cammel and Knight (1991), Porter *et al.* (1991), Sutherst (1990, 1991), and Sutherst *et al.* (1995). They demonstrate that impacts could be severe in many different environments and involve numerous different species of insect pests. Principal concern is with species that can increase their population size by undergoing an extra generation each year in warmer climates or expand their geographical distributions. For example, Porter *et al.* (1991) found that in Europe *O. nubilalis* would shift 1220 km northward if temperature increased by 3° to 6°C by 2025–2070. For a 3°C temperature increase in Japan, Mochida (1991) predicted expanded ranges for tobacco cut worm (*Spodoptera litura*), southern green stink bug (*Nezera viridula*), rice stink bug (*Lagynotomus eleongatus*), lima-bean pod borer (*Etiella zinckenella*), common green stink bug (*Nezera antennata*), soybean stem gall (*Asphondylia sp.*), rice weevil (*Sitophilus oryzae*), and soybean pod borer (*Legumunuvora glycinivorella*), but a decreased range for rice leaf beetle (*Oulema oryzae*) and rice leaf miner (*Agromyza oryzae*). Vegetation subzones were linked with microclimates to extrapolate spruce weevil hazard zones in British Columbia (Spittlehouse and Sieben, 1994). Models that match the presence of particular species with discrete ranges of temperature and precipitation parameters such as CLIMEX may be especially appropriate for projecting the effects of climate change on insect redistribution (Worner, 1988; Sutherst *et al.*, 1995).

Actual insect distributions under climate change will also depend on host distributions (Rainey, 1989), competition with existing species (DeBach, 1965), adaptability to new conditions, and the presence of natural enemies in the area. Because climate effects on insect life cycles frequently depend on extreme events (e.g., freezing, intense precipitation) and climatic features such as the persistent winds of the Inter-Tropical Convergence system, the reliability of predicted redistribution of insects depends, in part, on the reliability of predictions of these features of climate.

13.4.3. Plant Diseases

The occurrence of plant fungal and bacterial pests depends on temperature, rainfall, humidity, radiation, and dew. Climatic conditions affect the survival, growth, and spread of pathogens, as well as the resistance of hosts. Friedrich (1994) summarizes the observed relationship between climatic conditions and important plant diseases. Among these, mild winters have been associated with more rapid and stronger outbreaks of powdery mildew (*Erysiphe grammis*), brown leaf rust of barley (*Puccinia hordei*), and strip rust of cereals (*Puccinia striiformis*) (Meier, 1985). Mild winters combined with very warm weather conditions provide optimal growth conditions for cercosporea leaf spot disease (*Cercosporea beticola*), powdery mildew (*Erysiphe betae*), and rhizomania disease (*Rizomania*) (Treharne, 1989). Warm, humid conditions lead to earlier and stronger outbreaks of late potato blight (*Phytophthora infestans*) (Löpmeier, 1990; Parry *et al.*, 1990). Dry and hot summers generally reduce infestations of most fungal diseases because plant resistance is increased. Summer dryness, particularly in early summer, also decreases rhynchosporium leaf blotch (*Rhynchosporium secalis*) and septoria leaf spot diseases (*Septoria tritici* and *S. nodorum*), but more frequent summer precipitation, particularly heavy storms, would increase incidences of these diseases because rain and rainborne splash water is the means by which disease spores are spread (Royle *et al.*, 1986). Warmer temperatures would likely also shift the occurrence of these diseases into presently cooler regions (Treharne, 1989).

13.5. Animal Agriculture

Climate affects animal agriculture in four ways: through (1) the impact of changes in livestock feedgrain availability and price (e.g., Adams *et al.*, 1990; Bowes and Crosson, 1993; Kane *et al.*, 1993; Rosenzweig and Parry, 1994); (2) impacts on livestock pastures and forage crops (Wilson, 1982; Martin *et al.*, 1991; Easterling *et al.*, 1993; McKeon *et al.*, 1993); (3) the direct effects of weather and extreme events on animal health, growth, and reproduction (Bianca, 1970; Rath *et al.*, 1994); and (4) changes in the distribution of livestock diseases (Stem, 1988; U.S. EPA, 1989).

Generally, the impacts of changes in feedgrain prices or forage production on livestock production and costs are moderated by markets. Impacts of changes in feedgrain supply on the supply of meat, milk, egg, and other livestock products in terms of price increase is substantially less than the initial feedgrain price shock (Reilly *et al.*, 1994). Bowes and Crosson (1993) demonstrated the importance of feed exports or imports into a region in determining downstream impacts on livestock and meatpacking industries. Abel and Levin (1981) found that, for developing country agriculture, livestock are a better hedge against losses than are crops because animals are better able to survive extreme weather events such as drought. The relatively lower sensitivity of livestock to climate change is also documented for the historical case of the U.S. Dust Bowl experience of the 1930s (Waggoner, 1993).

The impact of climate on pastures and unimproved rangelands may include deterioration of pasture quality toward poorer quality, subtropical (C_4) grasses in temperate pastoral zones as a result of warmer temperatures and less frost, or increased invasion of undesirable shrubs—but also potential increases in yield and possible expansion of area if climate change is favorable and/or as a direct result of increasing CO_2 (Martin *et al.*, 1993; McKeon *et al.*, 1993; Salinger and Porteus, 1993; Campbell *et al.*, 1995). (See Chapter 2 for details on possible changes in species composition of rangelands and effects on pastoral agriculture.)

Heat stress has a variety of detrimental effects on livestock (Furquay, 1989), with significant effects on milk production and reproduction in dairy cows (Johnston, 1958; Thatcher, 1974; Khan, 1991; Orr *et al.*, 1993). Swine fertility shows seasonal variation due to seasonal climate variability (Claus and Weiler, 1987). Reproductive capabilities of dairy bulls and boars and conception in cows are affected by heat stress (Egbunike and Elmo, 1978; Cavestany *et al.*, 1985; Berman, 1991). Livestock management and development of breeds better-suited to tropical climates has been a specific consideration (Bonsma, 1949; Du Preez *et al.*, 1990).

Analyses suggest that warming in the tropics and in the subtropics during warm months would likely impact livestock reproduction and production negatively (e.g., reduced animal weight gain, dairy production, and feed conversion efficiency) (Hahn *et al.*, 1990; Baker *et al.*, 1993; Klinedinst *et al.*, 1993; Rath *et al.*, 1994). Results are mixed for impacts in temperate and cooler regions: forage-fed livestock generally do better (due to more forage) but more capital-intensive operations, like dairy, are negatively affected (Parry *et al.*, 1988; Baker *et al.*, 1993; Klinedinst *et al.*, 1993). Warming during the cold periods for temperate areas would likely be beneficial to livestock production due to reduced feed requirements, increased survival of young, and lower energy costs.

Impacts may be minor for relatively intense livestock production systems (e.g., confined beef, dairy, poultry, swine) because such systems control exposure to climate and provide opportunity for further controls (e.g., shading, wetting, increasing air circulation, air conditioning, and alterations of barns and livestock shelters). Livestock production systems that do not depend primarily on grazing are less dependent on local feed sources, and changes in feed quality can be corrected through feed supplements. The fact that livestock production is distributed across diverse climatic conditions from cool temperate to tropical regions provides evidence that these systems are adaptable to different climates.

Many studies of climate and weather impacts on livestock find that the principal impacts are an increased role for management, adoption of new breeds in some cases where climate changes are moderate (for example, Brahman cattle and Brahman crosses are more heat- and insect-resistant than breeds now dominant in Texas and Southern Europe), and introduction of different species in some cases of extreme weather changes (Entwistle, 1974; Hahn, 1988, 1994; Hahn *et al.*, 1990; Baker *et al.*, 1993; Baker, 1994; Klinedinst *et al.*, 1993; Rath *et al.*, 1994).

13.6. Regional Climate Impacts: Studies and Issues

Variation of agriculture systems, climates, resources, and economic characteristics across and within countries may be more important in determining the effects of climate change than differences in climate scenarios themselves. Agricultural policy is an important consideration in most regions. Agricultural policies have had many and changing goals. Climate change is generally not among top policy priorities for agricultural policymakers, but climate change could affect the cost and likelihood of achieving other policy priorities—such as food adequacy and reduction of chronic hunger, improving export competitiveness, assuring regional and national economic and social development, increasing farm income and the viability of rural communities, assuring water availability and quality, reducing or reversing land degradation and soil loss through erosion, and other conservation and environmental objectives.

13.6.1. Africa and the Middle East

While Africa, particularly sub-Saharan Africa, is highly dependent on agriculture, relatively little quantitative work has been done on the impacts of climate change. Little or no effort has been devoted to studying agricultural effects on countries of the Middle East. The available studies for sub-Saharan Africa suggest that critical thresholds are related to precipitation and the length of the growing season, although warmer temperatures and increased radiation may benefit highland areas. National and local assessments providing a detailed understanding of crop-specific responses and regional impacts are still lacking.

Recent studies (Ominde and Juma, 1991; Ottichilo *et al.*, 1991; Downing, 1992; Schulze *et al.*, 1993; Sivakumar, 1993; Magadza, 1994) indicate that most of Africa will be sensitive to climate change, although some regions may benefit from warmer and wetter conditions (Table 13-4). Downing (1992), in analyses for Kenya, Zimbabwe, and Senegal, evaluated the sensitivity to incremental climatic variations suggested by GCM scenarios for the region. Nationally, he estimated that potential food production in Kenya would increase, particularly if rainfall increases, but that the impacts would vary regionally; sub-humid and semi-arid provinces supporting socioeconomically vulnerable groups would be negatively affected even with increases in national food production potential.

Table 13-4: *Selected crop studies for Africa and the Middle East.*

Study	Scenario	Geographic Scope	Crop(s)	Yield Impact (%)	Other Comments
Eid, 1994	GISS, GFDL, UKMO	Egypt	Wheat Maize	-75 to -18 -65 to +6	w/ CO_2 effect; also temperature and precip. sensitivity; adaptation would require heat-resistant variety development
Schulze *et al.*, 1993	+2°C (1)	South Africa	Biomass Maize	decrease increase	Mapped results, not summarized as average change for entire region
Muchena, 1994	GISS, GFDL, UKMO	Zimbabwe	Maize	-40 to -10	w/ CO_2 effect; also temperature and precip. sensitivity; adaptations (fertilizer and irrigation) unable to fully offset yield loss
Downing, 1992	+2/+4°C, ±20% precip.	Zimbabwe Senegal Kenya	Maize Millet Maize	-17 to -5 -70 to -63 decrease	Food availability estimated to decline in Zimbabwe; carrying capacity fell 11 to 38% in Senegal; overall increase for all crops in Kenya with zonal shifts
Akong'a *et al.*, 1988	Historical droughts, sensitivity	Kenya	Maize, livestock	negative effects of drought	Considered broader socioeconomic impacts, small-holder impacts, and policy implications
Sivakumar, 1993	1945–64 vs. 1965–88	Niger, West Africa	Growing season	reduced 5–20 days	Crop variety development, timely climate information seen as important adaptation strategies

Akong'a *et al.* (1994) and Sivakumar (1993) considered the effects of climatic variability (primarily periodic droughts) on agriculture in some areas of the region, finding that such droughts have significant negative effects on production, crop season length, and higher-order social impacts. The persistence of such periodic droughts and the potential for them to change in frequency and severity in the Sahel and in eastern and southern Africa indicate the need for further research to develop adaptive strategies. The effects of greater frequency and severity include growing aridity in the savannas; deforestation and soil erosion in all farming systems but particularly in the humid, sub-humid, and equatorial regions; and salinization of irrigated lands. The economies of countries of North Africa and the Middle East are generally less dependent on agriculture than are those of sub-Saharan Africa. One study for Egypt (Eid, 1994) indicated the potential for severe impacts on national wheat and maize production.

13.6.2. South and Southeast Asia

South and Southeast Asia include the southern portion of Asia, from Pakistan in the west to Vietnam in the east, as well as Indonesia and the Philippines. Seasonal monsoons are a dominant climate feature that affect agriculture. Matthews *et al.* (1994a, 1994b) have estimated the impacts on rice yields for many countries in the region for equilibrium climate scenarios of three major GCMs that predict temperature and precipitation increases for the region. The results show substantial variation in impact across the region and among the GCMs (Table 13-5).

Table 13-5: *Selected crop studies for South and Southeast Asia.*

Study	Scenario	Geographic Scope	Crop(s)	Yield Impact (%)	Other Comments
Rosenzweig and Iglesias (eds.), 1994[1]	GCMs	Pakistan	Wheat	-61 to +67	UKMO, GFDL, GISS, and +2°C, +4°C, and ±20% precip; range is over sites and GCM scenarios with direct CO_2 effect; scenarios w/o CO_2 and w/ adaptation also were considered; CO_2 effect important in offsetting losses of climate-only effects; adaptation unable to mitigate all losses
		India	Wheat	-50 to +30	
		Bangladesh	Rice	-6 to +8	
		Thailand	Rice	-17 to +6	
		Philippines	Rice	-21 to +12	
Qureshi and Hobbie, 1994	average of 5 GCMs	Bangladesh	Rice	+10	GCMs included UKMO, GFDLQ, CSIRO9, CCC, and BMRC; GCM results scaled to represent 2010; includes CO_2 effect
		India	Wheat	decrease	
		Indonesia	Rice	-3	
			Soybean	-20	
			Maize	-40	
		Pakistan	Wheat	-60 to -10	
		Philippines	Rice	decrease	
		Sri Lanka	Rice	-6	
			Soybean	-3 to +1	
			Coarse Grain	decrease	
			Coconut	decrease	
Parry *et al.*, 1992	GISS	Indonesia	Rice	approx. -4	Low estimates consider adaptation; also estimated overall loss of farmer income ranging from $10 to $130 annually
			Soybean	-10 to increase	
			Maize	-65 to -25	
		Malaysia	Rice	-22 to -12	
			Maize	-20 to -10	Maize yield affected by reduced radiation (increased clouds); variation in yield increases; range is across seasons
			Oil Palm	increase	
			Rubber	-15	
		Thailand sites	Rice	-5 to +8	
Matthews *et al.*, 1994a, 1994b	3 GCMs	India	Rice	-3 to +28	Range across GISS, GFDL, and UKMO GCM scenarios and crop models; included direct CO_2 effect; varietal adaptation was shown to be capable of ameliorating the detrimental effects of a temperature increase in currently high-temperature environments
		Bangladesh		-9 to +14	
		Indonesia		+6 to +23	
		Malaysia		+2 to +27	
		Myanmar		-14 to +22	
		Philippines		-14 to +14	
		Thailand		-12 to +9	

[1] Country studies were by Qureshi and Iglesias, 1994; Rao and Sinha, 1994; Karim *et al.*, 1994; Tongyai, 1994; and Escaño and Buendia, 1994, for Pakistan, India, Bangladesh, Thailand, and the Philippines, respectively.

Spikelet sterility emerged as a major factor determining the differential predictions; where current conditions were near critical thresholds, a difference in mean temperature of less than a degree resulted in a positive yield change rapidly becoming a large decline. However, genetic variability among varieties suggests relative ease in adapting varieties to new climate conditions. Temperature effects alone were generally found to reduce yields, but CO_2 fertilization was a significant positive effect.

Brammer *et al.* (1994) conclude that, among other things, the diversity of cropping systems does not allow a conclusion of magnitude or direction of impact to be made for Bangladesh at this time. Parry *et al.* (1992) showed yield impacts that vary across the countries of Thailand, Indonesia, and Malaysia and across growing season. Coastal inundation was also estimated to be a threat to coastal rice and to fish, prawn, and shrimp ponds. These authors estimated that a 1-meter sea-level rise could cause a landward retreat of 2.5 km in Malaysia; such a rise was estimated to threaten 4200 ha of productive agricultural land, an area equal to slightly less than 1% of Malaysia's paddy rice area. Model results showed that under Goddard Institute for Space Studies (GISS) doubled CO_2 climates, erosion rates in three Malaysian river basins increased from 14–40%, and soil fertility declined on average by 2–8%.

13.6.3. East Asia

Several major studies have been conducted for countries in East Asia, including China (mainland and Taiwan), North and South Korea, and Japan. Possible climatic impacts span a wide range depending on the climate scenario, geographic scope, and study (Table 13-6). For China, results show generally negative yield effects but range from less than 10% (Zhang, 1993) to more than 30% (Jin *et al.*, 1994). While finding large changes for all of China, Hulme *et al.* (1992) conclude that to a certain extent, warming would be beneficial, with increasing yield due to diversification of cropping systems. However, they estimated that by 2050, when they expect an average warming for China of 1.2°C, increased evapotranspiration would generally exceed increases in precipitation, thus leading to a greater likelihood of yield loss due to water stress for some rice-growing areas, even as the area suitable for rice increases.

Table 13-6: *Selected crop studies for East Asia.*

Study	Scenario	Geographic Scope	Crops	Yield Impact (%)	Other Comments
Tao, 1992	2 x CO_2 +1°C	China	Wheat Rice Cotton, Fruits, Oil Crops, Potatoes, Corn	-8 -6 -4 to +1	Agricultural productivity loss >5%; included direct CO_2 effect; positive effects in NE and NW; negative in most of the country; no change in SW
Zhang, 1993	+1.5°C	South of China	Rice	-11 to -7	Double-crop; included direct CO_2 effect
Jin *et al.*, 1994	GCMs	South of China	Rainfed Rice Irrigated Rice	-78 to -6 -37 to +15	Range across GISS, GFDL, and UKMO scenarios; no consideration of enrichment effects of CO_2
Sugihara, 1991	2 x CO_2 +3°C	Japan	Rice	+10	
Suyama, 1988	+2°C	Japan	Temperate Grass	-10 to +10	Average +5.6% in productivity for grass; included direct CO_2 effect
Yoshino, 1991	+2°C	Japan	Sugar Cane	-8	Rainfall was reduced 25 to 30%, May to October
Seino, 1994	GISS, GFDL, UKMO	Japan	Rice Maize Wheat	-11 to +12 -31 to +51 -41 to +8	Impacts vary by GCM scenarios and area; included direct CO_2 effect; generally positive in N and negative in S
Horie, 1993	GCMs	Hiroshima and Akita, Japan	Rice	-45 to +30	Estimated 14% increase for all of Japan; range based on different crop models, GCMS, and across sites; included direct CO_2 effect; Akita more favorable than Hiroshima
Matthews *et al.*, 1994a, 1994b	GCMs	South Korea Mainland China Taiwan Japan	Rice	-22 to +14 -18 to -4 +2 to +28 -28 to +10	Range across GISS, GFDL, and UKMO scenarios and crop models; included direct CO_2 effect; varietal adaptation capable of ameliorating the effects of a temperature increase in currently hot environments

For China, warming would likely cause a general northward movement of agroclimatic regions, with certain exceptions in the south where the moisture deficit may increase even more than in the north. The general possibility of increased summer dryness in the continental mid-latitudes suggests the following six areas as most likely to be negatively affected by climate change (Lin Erda, 1994): the area around the Great Wall lying southeast of the transition belt between crop agriculture and animal husbandry; the Huang-Hai Plains where dryland crops like wheat, cotton, corn, and fruit trees are grown; the area north of Huai River including Eastern Shandong that lies along the south edge of the south temperate zone; the central and southern areas of Yunnan Plateau; middle and lower reaches of Yangtze River; and the Loess Plateau. In general, these areas would be at heightened risk of drought and would suffer potential increases in soil erosion. The Yunnan Plateau, with generally abundant rainfall, is subject to alternating droughts and waterlogging; production is sensitive to changes that would increase the variability of climate.

Indices of vulnerability based on physical productivity and socioeconomic capability to adapt show that among China's thirty provinces, Shanxi, Inner Mongolia, Gansu, Hebei, Qinghai, and Ningxia are particularly vulnerable and less able to adapt to climate change. These seven provinces produced 12% of China's total agricultural output value in 1990 (*Statistical Yearbook of Agriculture of China*, 1991). Thus, the areas along the Great Wall and Huang-Hai Plains are areas that are both socioeconomically and agronomically vulnerable to climate change and also are areas where climate projections suggest possible adverse changes in climate.

Climate change will occur against a steadily increasing demand for food in China over the next 55 years (Lu and Liu, 1991a, 1991b). The increased annual cost of government investment only (excluding farmers' additional costs) in agriculture due to climate change through 2050 was estimated at 3.48 billion U.S. dollars (17% of the cost of government investment in agriculture in 1990).

Studies for Japan (Table 13-6) indicate that the positive effects of CO_2 on rice yields would generally more than offset negative climatic effects in the central and northern areas, leading to yield gains; in the southwest, particularly in Kyushu, the rice yield effects were, on balance, estimated to be negative for several climate scenarios (Seino, 1993a, 1993b). Horie (1987) found generally negative effects on rice yield in Hokkaido under the GISS climate scenario when rice variety was not changed but found increased yields if longer maturing varieties were adopted. Horie (1991), under the Oregon State University (OSU) climate scenario, found that rice yields would fall in most areas of the country but that changes in rice variety and other management changes could recover most losses except in the southwest, where the projected increase in temperature of 4.0–4.5°C exceeded the temperature tolerances of japonica rice varieties. Additional considerations for Japanese agriculture are possible changes in flowering and maturation of fruit trees, with potential northward shifts in cultivated areas and changed distribution of insect pests (Seino, 1993a).

13.6.4. *Oceania and Pacific Island Countries*

Oceania includes Australia, New Zealand, Papua New Guinea, and numerous small islands and coral atolls of the Pacific Island Countries (PICs).

Findings for Australia include: (1) poleward shifts in production, (2) varying impacts on wheat including changes in grain quality, (3) likely inadequate chilling for stone fruit and pome fruit and lower fruit quality, (4) increased likelihood of heat stress in livestock, particularly dairy and sheep, (5) increased infestation of tropical and subtropical livestock parasites but possible decreases for other species, (6) livestock benefits due to warmer and shorter winters, (7) increased damage due to floods and soil erosion, (8) increased drought potential with wheat and barley more sensitive than oats, (9) changes in the severity of outbreaks of downy mildew on grapevines and rust in wheat, and (10) beneficial effects of elevated CO_2 levels for many agricultural crops (Hobbs *et al.*, 1988; Nulsen, 1989; Pittock, 1989; Wardlaw *et al.*, 1989; Blumenthal *et al.*, 1991; Wang *et al.*, 1992; CSIRO, 1993; Hennessy and Clayton-Greene, 1995; Wang and Gifford, 1995).

Studies of New Zealand agriculture considered the main effects of climate change on New Zealand's important pastoral agriculture to be: (1) a poleward spread of subtropical pastures, (2) a resultant decrease in the area of temperate pasture, (3) higher yields, (4) altered seasonality of production, (5) spread of growth to higher elevations, and (6) decreased growth in the eastern areas of the North Island due to drier and warmer conditions (Ministry for the Environment, 1990; Salinger and Hicks, 1990; Martin *et al.*, 1991). Models to predict effects on forage production systems are currently being developed and validated. Initial modeling simulations of the effects of doubled-CO_2 climates on pasture yield including the beneficial effects of CO_2 varied from +10 to +77% (Korte *et al.*, 1991; Butler *et al.*, 1991; Martin *et al.*, 1991). The higher figures, however, likely overestimated the beneficial effects of CO_2 (Campbell, 1994). More recent work (Campbell *et al.*, 1995; Newton *et al.*, in press) indicates gains of up to 15%.

Studies have found variable impacts on horticultural crops in New Zealand, with a general poleward shift, including an expansion of the area for subtropical crops but a contraction of area suitable for temperate crops such as apples and kiwi fruit that require winter chilling (Salinger *et al.*, 1990). Studies for maize suggest expansion of the suitable growing area into the Canterbury Plains of the South Island (Kenny *et al.*, 1994; Tate *et al.*, 1994; Warrick and Kenny, 1994).

For small island states, fewer studies have been conducted. Singh *et al.* (1990) conclude that, in general, crop yields would be lower because of reduced solar radiation (from increased cloudiness), higher temperature (leading to shorter growth duration and increased sterility), and water availability (both drought and inundation). Sea water intrusion also could affect some coastal areas. Some of the negative effects, particularly

Table 13-7: Selected studies for Australia and New Zealand.

Study	Scenario	Geographic Scope	Crops	Yield Impact (%)	Other Comments
Campbell, 1994; Campbell *et al.*, 1995; Newton *et al.*, in press	Various	New Zealand	Pasture	increase overall, but decrease in some regions	GCM scenarios and climate sensitivity result in yield increases w/ and w/o CO_2 fertilization; increases of up to 15% w/ CO_2 fertilization; earlier studies found larger increases, but likely overestimated the CO_2 effect; additional findings include altered seasonality of pasture with effects on livestock management
Salinger *et al.*, 1990	*	New Zealand	Temperate Crops and Pasture	increase	Crop shifts of 200m in altitude and 200km poleward; earlier crop maturation; longer frost-free season
McKeon *et al.*, 1988	+2°C, -20% winter precipitation, +30% rest	SW Queensland, Australia	Semi-Arid Perennial Grass, Wheat	+31 (-1, +35) +23 (-6, +35)	Temperature only (precipitation only in parentheses); no CO_2 effect; for grasslands, increased risk from undesirable shrubs and grasses, animal nutrition and health, and soil erosion; new areas may be available
Vickery *et al.*, 1993	+2°C, +10 to +20% precip. Nov. to March	Northern NSW, Australia	Cool Temperate Grazing	not assessed	Expansion of high NPP (3.2 to 4.8 t/ha) area from 59 to 64% of the region; soil and landscape as possible limiting factors not considered
Erskine *et al.*, 1991	GISS, GFDL, UKMO	S. Australia	Wheat	-6 to +13	Included CO_2 to 555 ppm; results w/o CO_2 were -15 to -16%; no varietal adaptation or other management change
Wang *et al.*, 1992	+3°C	Horsham, Victoria, Australia	Wheat	-34 to +65	Included CO_2 to 700 ppm; range across cultivars; losses occurred for early-maturing varieties and stemmed mainly from shortened vegetative growing period
Russel, 1988	+2 to +4°C	Queensland, Australia	Sugar Cane	+9 to +13	Southward expansion possible; also considered similar temperature declines; yields -40 to -17%

*Used climate scenarios as given in Salinger and Hicks (1990), which are mean changes derived from the results of several GCMs.

of C_3 crops, would be offset by the beneficial effects of elevated CO_2. No quantitative estimates of sensitivities or thresholds directly related to agriculture have been reported in major impact studies (e.g., Hughes and McGregor, 1990; Pernetta and Hughes, 1990; Hay and Kaluwin, 1992).

Singh (1994) considers the vulnerability of small island nations to some large changes in climatic conditions that, while not currently predicted by climate models, indicate sensitivities. Increasing aridity (reductions in rainfall and prolonged dry seasons) affecting small islands and the leeward side of bigger islands was estimated to result in general crop failure, migration of human populations, wind erosion, and negative impacts on wildlife. A somewhat less severe drying (prolonging the dry season by 45 days or more) would decrease yields of maize (30 to 50%), sugar cane (10 to 35%), and taro (35 to 75%). Singh (1994) also found that significantly increased rainfall (+50%) during the wet season on the windward side of large islands, while increasing yields of taro (5 to 15%), would reduce yields of rice (10 to 20%) and severely reduce maize yields (30 to 100%). Maize failures from increased precipitation stem from inundation, and rice yield losses stem from increased cloudiness (reduced solar irradiation).

13.6.5. Areas of the Former USSR

Climate impact studies conducted over the past 15 years include those of Zhukovsky and Belechenko (1988), Zhukovsky *et al.* (1992), and Sirotenko *et al.* (1984). These studies did not include the direct effect of increasing atmospheric CO_2. Recent estimates have included coverage of most of the region and have included the CO_2 effect (Menzhulin and Koval, 1994) and other environmental change (Sirotenko and Abashina, 1994; Sirotenko *et al.*, 1991). These new studies also have been the first conducted for this region based on climate scenarios drawn from GCM runs (Table 13-8).

Table 13-8: *Selected crop studies for Russia and the former Soviet republics.*

Study	Scenario	Geographic Scope	Crop(s)	Yield Impact (%)	Other Comments
Menzhulin and Koval, 1994	GISS, GFDL, UKMO	Russia and Former Soviet Republics	Winter and Spring Wheat	-19 to +41	Included CO_2 effect; GISS strongly positive, UKMO negative; temperature sensitivity alone of 2 and 4°C resulted in yield losses of 20 to 30%; impacts also varied widely across the 19 sites studied
Sirotenko and Abashina, 1994; Sirotenko *et al.*, 1991[1]	EMI	Russia	Crop Yield, Grassland Productivity	+10 to+35 -2 to +26	Large positive effect in Southern Volga and Northern Caucasus areas; slightly negative crop yield in S. Krasnoyarsky and Far East; range is w/ and w/o +20% CO_2, +30% tropospheric ozone, and -20% soil humus
	CCC, GFDL		Crop Yield, Grassland Productivity	-14 to +13 -27 to -2	

[1]These studies used a reconstructed Eemian Interglacial (EMI) climate and the Canadian Climate Centre (CCC) and Geophysical Fluid Dynamics Laboratory (GFDL) GCMs. The temperature and precipitation predictions of the GCMs were scaled by the factors 0.51 (CCC) and 0.45 (GFDL) to generate climates that could be observed by the year 2030.

Sirotenko and Abashina (1994) and Sirotenko *et al.* (1991) scaled the equilibrium temperature and precipitation derived from GCMs to generate scenarios applicable to the year 2030 and used the Eemian interglacial period (EMI) climate. In the EMI scenario, warming was greatest in January and precipitation increased substantially in both January and July, whereas the GCM scenarios suggested drier and warmer summer conditions. Changes in potential crop yield and potential productivity of grasses are based on a geoinformation system CLIMATE-SOIL-YIELD and a dynamic-growth crop simulation model (Sirotenko, 1981; Abashina and Sirotenko, 1986; Sirotenko, 1991) (Table 13-8). The results indicate that the climate response of crops and grasses can differ even in sign.

The estimated response of agriculture varied significantly across the region as well as across climate scenarios. Sirotenko and Abashina (1994) and Sirotenko *et al.* (1991) estimate the impacts to be favorable on agriculture of the northern areas of European Russia and Siberia and to cause a general northward shift of crop zones. Actual changes in production would reflect both areal expansion and the yield changes on existing crop areas as reported in Table 13-8. The more arid climate of the Canadian Climate Centre (CCC) and Geophysical Fluid Dynamics Laboratory (GFDL) GCMs was projected to have severe effects on grain production in the steppes of Povolzhye, Northern Caucasus, and the southern portion of Western Siberia, where grain production was estimated to fall by 20–25%. Menzhulin and Koval (1994) simulated yield increases exceeding 50% in the northwestern, central, and eastern regions of Kazakhastan under the GISS scenario, primarily because of increased moisture in these currently arid areas, but these areas did not benefit substantially under the GFDL scenario. Overall, the UK Meteorological Office (UKMO) scenario produced the largest yield declines.

Sirotenko and Abashina (1994) and Sirotenko *et al.* (1991) found that potential increases in ozone and loss of soil organic matter reduced potential yields substantially; when combined with the climate/CO_2 scenarios, grass yields declined by about one-quarter and crop yields by 10% in both CCC and GFDL GCM scenarios. Kovda and Pachepsky (1989) report the potential for significant additional soil loss and degradation resulting from climate change.

Historically, climate variability has been a significant contributor to yield variability in areas of the former USSR. For example, the increase in aridity during the 1930s was estimated to have decreased yields by 25–39% (Menzhulin, 1992). If drier conditions prevail under climate change, similar yield effects may occur.

13.6.6. Latin America

Climate impact studies for Latin America that include the direct effect of CO_2 generally show negative impacts for wheat, barley, and maize but positive impacts for soybeans (Table 13-9). A study of Norte Chico, Chile, suggested decreased yields for wheat and grapes but increases for maize and potatoes (Downing, 1992). The Norte Chico results are difficult to generalize because the climate for Chile exhibits steep temperature gradients from east to west due to the change in altitude, as well as wide variation from north to south.

The largest area with clear vulnerability to climate variability in the region is the Brazilian northeast. Like most agricultural areas of Latin America, this region has a rainy season when crops are grown and a dry season with little or no rain. In the case of the Brazilian northeast, the rainy season is relatively short (3–4 months) and the occurrence of years with no rainy season is frequent. These years are characterized by the occurrence of famine and large-scale migrations to metropolitan areas. Climatic variations that would result in shorter rainy seasons and/or increased

Table 13-9: Selected crop studies for Latin America.

Study	Scenario	Geographic Scope	Crop(s)	Yield Impact (%)	Other Comments
Baethgen, 1992, 1994	GISS, GFDL, UKMO[1]	Uruguay	Barley Wheat	-40 to -30 -30	w/ and w/o CO_2; with adaptation, losses were 15 to 35%; results indicate increased variability
Baethgen and Magrin, 1994	UKMO	Argentina Uruguay	Wheat	-10 to -5	w/ CO_2; high response to CO_2, high response to precipitation
Siquera *et al.*, 1994; Siquera, 1992	GISS, GFDL, UKMO[1]	Brazil	Wheat Maize Soybean	-50 to -15 -25 to -2 -10 to +40	w/ CO_2, w/o adaptation; adaptation scenarios did not fully compensate for yield losses; regional variation in response
Liverman *et al.*, 1991, 1994	GISS, GFDL, UKMO[1]	Mexico	Maize	-61 to -6	w/ CO_2; adaptation only partly mitigated losses
Downing, 1992	+3°C, -25% precip.	Norte Chico, Chile	Wheat Maize Potatoes Grapes	decrease increase increase decrease	The area is especially difficult to assess because of the large range of climates within a small area
Sala and Paruelo, 1992, 1994	GISS, GFDL, UKMO[1]	Argentina	Maize	-36 to -17	w/ and w/o CO_2; better adapted varieties could mitigate most losses

[1]These studies also considered yield sensitivity to +2 and +4°C and -20 and +20% change in precipitation.

frequency of rainless years would have extremely negative consequences for the region.

13.6.7. Western Europe

Simulated yields of grains and other crops have been generally found to increase with warming in the north, particularly when adaptation is considered, but decrease substantially in the Mediterranean area even with adaptation (Table 13-10). Northern yield increases depend on the beneficial effects of CO_2 on crop growth and climate scenarios showing sufficient increases in precipitation to counter higher rates of evapotranspiration. Yield declines in the Mediterranean region are due to increased drought resulting from the combination of increased temperature and precipitation decreases (or insufficient increases to counter higher evapotranspiration).

For many vegetable crops, warmer temperatures will generally be beneficial, with the options and possibilities for vegetable production generally expanding in northern and western areas. For cool-season vegetable crops such as cauliflower, larger temperature increases may reduce the number of plantings possible during cooler portions of the year or decrease production, particularly in southern Europe (Kenny *et al.*, 1993; Olesen and Grevsen, 1993). Decreased yield quality is also possible, with the effect most pronounced in southern Europe.

Warmer winters will reduce winter chilling and probably adversely affect apple production in temperate maritime areas, and could lead to loss of adequate winter chilling for crops such as peaches, nectarines, and kiwi fruit in southern Europe. Significant shifts in areas suitable for different types of grapes also could occur (Kenny and Harrison, 1992).

Among other effects, warming implies reduction in greenhouse costs for horticultural production. However, increased infestations of pests such as the Colorado beetle on potatoes and rhizomania on sugar beet may result from higher temperatures.

Studies have investigated changes in crop potential evapotranspiration (PET) (Le Houerou, 1994; Rowntree, 1990), finding that under higher temperatures the crop growing season would be extended for grain crops, assuming an increase in the number of frost-free days. For southern Europe, the extension of the growing season would likely be insufficient to avoid high summer temperatures by planting earlier. Thus, reduced grain filling period and lower yields are likely. Other studies have explored how the climatically limited range for crops, including maize, wheat, cauliflower, and grapes, would change under various GCMs and other climate scenarios (Carter *et al.*, 1991a, 1991b; Parry *et al.*, 1992; Kenny *et al.*, 1993; Kenny and Harrison, 1993; Wolf, 1993a, 1993b). In general, these studies found a northward shift of crop-growing zones with potential for grain maize to be grown as far north as the UK and central Finland. Increased demand for irrigation and/or increased areas likely to suffer from water deficits, particularly in southern Europe, also were found.

13.6.8. USA and Canada

Studies listed in Tables 13-11 (USA) and 13-12 (Canada) show a wide range of impacts. Much of the wide variation reflects differences among sites. Effects are more likely negative or more severely negative for southern areas and for climate scenarios such as the UKMO GCM scenario in which the temperature increases are large (+5.2°C) or the GFDL scenario in which summer aridity increased.

Rosenzweig (1985) simulated increased areas for wheat production, especially in Canada, under the GISS climate change scenario, while major wheat regions in the United States remain the same (Rosenzweig, 1985). Crosson (1989) found that warmer temperatures may shift much of the wheat-maize-soybean producing capacity northward, reducing U.S. production and increasing production in Canada. Shifting climate zones may result in lower production of corn or wheat and different and more diverse crops because the productivity in the new areas is likely to be limited due to the shallow, infertile soils (CAST, 1992).

Across studies in Table 13-11 for the USA that combined biophysical and economic impacts, market adjustments lessen the impacts of negative yield changes. Different assumptions about changes in U.S. population, income, trade barriers, and institutions were found, in some cases, to determine whether the net economic impact of climate change on the USA was negative or positive. Kaiser *et al.* (1993) found that possible increases in agricultural commodity prices could more than offset farm income loss. Mendelsohn *et al.* (1994) used an econometric approach to directly estimate the impact of climate change on agricultural revenue and asset values. This approach more fully considers potential adaptation to different climates as directly observed across climates that vary as a result of geography. They found that for the United States, warming would generally be beneficial even without the direct effect of CO_2. This approach calculates an equilibrium response after complete adjustment and does not consider price changes.

A number of studies have considered the vulnerability of prairie agriculture to climate change in Canada (Cohen *et al.*, 1992). Factors cited as contributing to vulnerability include the

Table 13-10: *Selected crop studies for Western Europe.*

Study	Scenario	Geographic Scope	Crop(s)	Yield Impact (%)	Other Comments
Oleson *et al.*, 1993	*	Northern Europe	Cauliflower	increase	Quality affected by temperature; longer season
Goudriaan and Unsworth, 1990	+3°C	Northern Europe	Maize (fodder)	increase	Shift to grain production possible
Squire and Unsworth, 1988	+3°C	Northern Europe	Wheat	increase	
Kettunen *et al.*, 1990	GCMs	Finland	Potential Yield	+10 to +20	Range is across GISS and UKMO GCMs
Rötter and van Diepen, 1994	+2°C (winter), +1.5°C (summer)	Rhine Area	Cereals, Sugar Beet, Potato, Grass	+10 to +30	Also +10% winter precipitation; includes direct effect of CO_2; range is across crop, agroclimatic zone, and soil type; decreased evapotranspiration (1 to 12%), except for grass
UK Dept. of Environment, 1992	GCMs +1, +2°C	UK	Grain Horticulture	increase or level increase	Increased pest damage; lower risk of crop failure
Wheeler *et al.*, 1993	*	UK	Lettuce	level	Quality affected; more crops per season possible
Semonov *et al.*, 1993	*	UK France	Wheat	increase or decrease	Yield varies by region; UKMO scenario negative; includes adaptation and CO_2
Delecolle *et al.*, 1994	GCMs** +2, +4°C	France	Wheat Maize	increase or level	Northward shift; w/ adaptation, w/ CO_2; GISS, GFDL, and UKMO GCMs
Iglesias and Minguez, 1993	GCMs**	Spain	Maize	-30 to -8	w/ adaptation, w/ CO_2; irrigation efficiency loss; see also Minguez and Iglesias, 1994
Santer, 1985	+4°C	Italy/Greece	Biomass	-5 to +36	Scenarios also included -10% precipitation
Bindi *et al.*, 1993	+2, +4°C and *	Italy	Winter Wheat	not estimated	Crop growth duration decreases; adaptation (using slower developing varieties) possible

* Climate scenarios included GISS, GFDL, and UKMO and time-dependent scenarios, using GCM methodology, based on emission scenarios proposed by the IPCC in 1990. Composite scenarios for temperature and precipitation were based on seven GCMs and scaled by the global-mean temperature changes associated with the IPCC 1990 emission scenarios for the years 2010, 2030, and 2050 (Barrow, 1993).

** These studies also considered yield sensitivity to +2 and +4°C and -20 and +20% change in precipitation.

prairie's importance as an agricultural producer, located in a marginal climate, constrained by both temperature and precipitation; soil limitations that limit shifting of cropping northward; known sensitivity to climate as evidenced by past drought experiences; and vulnerability to midcontinental drying indicated by GCMs. The effects of the 1988 drought are an indication of the region's sensitivity to climate variability. The effects included dust storms and wind erosion, production declines of 29% (grains) to 94% (hay), falling grain inventories, higher prices, poor pastures for livestock with some movement of cattle to moister areas, higher feed costs for livestock, and farm income reductions of 50% to 78% compared to 1987 figures.

13.7. Regional Summary: Relative Vulnerability

There has been a substantial number of new agricultural impact studies since the 1990 IPCC and the 1992 update, as reviewed in Section 13.6. For countries in sub-Saharan Africa, the Middle East and North Africa, Eastern Europe, and Latin America, however, there are still relatively few studies. For most regions, studies have focused on one or two principal grains. These studies strongly demonstrate the variability in estimated yield impacts among countries, scenarios, methods of analysis, and crops, making it difficult to generalize results across areas or for different climate scenarios. Thus, the ability to extend, interpolate, or extrapolate from the specific climate scenarios used in these studies to "more" or "less" climate change is limited.

Given these uncertainties in both magnitude and direction of impact, a key issue is *vulnerability* to possible climate change. Vulnerability is used here to mean the *potential* for negative consequences that are difficult to ameliorate through adaptive measures given the range of possible climate changes that might reasonably occur. Defining an area or population as vulnerable,

Table 13-11: *Selected U.S. agricultural impact studies.*

Study	Scenario	Geographic Scope	Crops	Yield Impact (%)	Other Comments
Adams *et al.*, 1988, 1990, 1994	GCMs[1]	U.S.	All	increase and decrease	Results vary across GISS, GFDL, and UKMO climates and regions; generally positive for 2°C and negative for 4°C; net economic effects depend on exports and CO_2 effects; increased irrigation; adaptation mitigates losses
Cooter, 1990	GISS	South	Maize	decrease	Potential risk of aquifer contamination
Easterling *et al.*, 1993	1930s analog	Missouri, Iowa, Nebraska, Kansas	Maize Sorghum Wheat Soybean Alfalfa	-23 to -6 -20 to +26 -11 to +17 -26 to +2 -5 to +22	More severe effect w/o CO_2 or adaptation; less severe or increase w/ CO_2 and adaptation
Kaiser *et al.*, 1993, 1994	Mild, severe	Central and southeast states, Minnesota, Nebraska	Maize Wheat Sorghum Soybean	increase and decrease	Climate scenarios included +2.5°C/+10% precipitation and +4.2°C/+20% precipitation; economic adaptation included; northern states less affected; results vary by crop/scenario
Mendelsohn *et al.*, 1994	+2.5°C, +8% precip.	U.S. county level	All	not estimated	Positive effect on crop revenue after long-run adjustment when considering revenue shares as weights that give greater importance to vegetables, fruits, etc.
Mearns *et al.*, 1992a, 1992b	GISS	Kansas	Wheat	increased variability and crop failure	Precipitation more important than temperature in scenarios, except for GISS
Muchow and Sinclair, 1991		Illinois	Grains	increase or slight decrease	Most sensitive to precipitation changes
Rosenzweig *et al.*, 1994	GISS, GFDL, UKMO*	Southeast U.S., Great Lakes U.S. sites	Soybean Maize Wheat	-96 to +58 -55 to +62 -100 to +180	w/ CO_2 effect; UKMO scenario, southern sites more severely affected; average for total USA assessed for wheat (-20 to -2%), maize (-30 to -15%), and soybeans (-40 to +15%); Peart *et al.* (1989), Ritchie *et al.* (1989), and Rosenzweig (1990) are similar studies

[1]These studies also considered yield sensitivity to +2 and +4°C and -20 and +20% change in precipitation.

therefore, is not a prediction of negative consequences of climate change; it is an indication that, across the range of possible climate changes, there are some climatic outcomes that would lead to relatively more serious consequences for the region than for other regions.

Vulnerability depends on the unit of observation and the geographic scale considered. *Yields* are relatively more vulnerable if a small change in climate results in a large change in yield. Evidence suggests that yields of crops grown at the margin of their climatic range or in climates where temperature or precipitation could easily exceed threshold values during critical crop growth periods are more vulnerable (e.g., rice sterility: Matthews *et al.*, 1994a, 1994b).

Farmer or farm sector vulnerability is measured in terms of impact on profitability or viability of the farming system. Farmers with limited financial resources and farming systems with few adaptive technological opportunities available to limit or reverse adverse climate change may suffer significant disruption and financial loss for relatively small changes in crop yields and productivity. For example, semi-arid, cool temperate, and cold agricultural areas may be more vulnerable to climate change and climate variability (Parry *et al.*, 1988).

Regional economic vulnerability reflects the sensitivity of the regional or national economy to farm sector impacts. A regional economy that offers only limited employment alternatives for workers dislocated by the changing profitability of farming is relatively more vulnerable than those that are economically diverse. For example, the Great Plains is one of the U.S. regions most dependent on agriculture, and thus might be the most economically vulnerable to climate change (Rosenberg, 1993).

Hunger vulnerability is an "aggregate measure of the factors that influence exposure to hunger and predisposition to its consequences" involving "interactions of climate change, resource constraints, population growth, and economic development" (Downing, 1992; Bohle *et al.*, 1994). Downing (1992) concluded that the semi-extensive farming zone, on the margin of more intensive land uses, appears to be particularly sensitive to small changes in climate. Socioeconomic groups in such areas, already vulnerable in terms of self-sufficiency and food security, could be further marginalized.

These different concepts of vulnerability include different scales of impact—from crop to individual farmer to food markets to the general economy. Given these various definitions and scales of impact, there are people vulnerable to climate change in most regions. Key characteristics of each of the regions help to suggest those more likely to have vulnerable populations (Table 13-13).

By most measures, many of the populations in sub-Saharan Africa appear most vulnerable. The region is already hot, and large areas are arid or semi-arid; average per capita income is among the lowest in the world and has been declining since 1980; more than 60% of the population depends directly on agriculture; and agriculture is generally more than 30% of gross domestic product (GDP). Relatively little of the cropland

Table 13-12: *Selected Canadian agricultural impact studies.*

Study	Scenario	Geographic Scope	Crop(s)	Yield Impact (%)	Other Comments
Williams *et al.*, 1988[1]	GISS84	Saskatchewan	Spring Wheat	-28 to -18	Large interannual fluctuations underlie mean impacts (e.g., -78% yield impact in extreme year); temperature increase of 3°C offset by +40% precipitation
Mooney, 1990[1]; Mooney *et al.*, 1991[1]	GISS	Manitoba, Alberta, Saskatchewan	Spring Wheat, Multiple Crops	-36 (Manitoba) negative and positive	Similar for other crops; corn and potatoes increased; precipitation derived from analogous region data; greater crop variety, production area increase
van Kooten[1]	CCC	SW Saskatchewan	Spring Wheat	-15 to +2	Positive effects when CCC precipitation used; negative used current norm for precipitation
Arthur and Abizadeh, 1989[1]	GISS, GFDL	Alberta, Manitoba, Saskatchewan	Wheat, Oats, Barley, Flax, Canola, Hay	small decrease to +28	10 out of 12 scenarios resulted in gain in net crop revenue; adapt by planting earlier
Brklacich *et al.*, 1994; Brklacich and Smit, 1992	GISS, GFDL, UKMO[2]	Alberta, Manitoba, Saskatchewan, Ontario	Wheat	-40 to +234	Results varied widely by site and scenario; adaptation and CO_2 were strongly positive effects; Ontario study showed increased net returns, but also variability; N gains and S loses

[1]As reported in Cohen *et al.*, 1992.
[2]These studies also considered yield sensitivity to +2 and +4°C and -20 and +20% change in precipitation.

is irrigated, and much of the agricultural land is used for grazing. Severe famine and starvation have been more prevalent in sub-Saharan Africa than in other regions over recent decades. Political and civil instability have greatly worsened problems. The potential for continued instability is an additional factor that increases vulnerability.

Populations in South Asia are vulnerable because of heavy dependence on agriculture and high population density. Agriculture accounts for more than 30% of GDP in most countries in the region. Each hectare of cropland supports 5.4 people. Tropical storms are an important feature of the climate around which current systems operate. These storms can be destructive but also are the main source of moisture. Changes in their frequency or severity would have significant impacts. The area is already intensely cropped, with 44% of the land area used as cropland. An estimated 31% of cropland is already irrigated, which may reduce vulnerability somewhat, providing water resources remain adequate. An additional factor that may reduce vulnerability in the future is the relative strength of the economy over the past decade. Countries in this area also have been relatively successful in avoiding the more severe effects of food shortages through programs that ensure access to food during potential famine situations. Chronic hunger remains a problem, however, for the poorer segments of the population, particularly in semi-arid and arid parts of the region.

Within East Asia, populations in the more arid areas of China appear most vulnerable to the possibility of mid-latitude continental drying. In general, the region supports a large population

Table 13-13: *Basic regional agricultural indicators.*

	Sub-Saharan Africa	Middle East/ N. Africa	South Asia	SE Asia	East Asia	Oceania	Former USSR	Europe	Latin America	USA, Canada
Agric. Land (%)[1]	41	27	55	36	51	57	27	47	36	27
Cropland (%)[1]	7	7	44	13	11	6	10	29	7	13
Irrigated (%)[1]	5	21	31	21	11	4	9	12	10	8
Land Area (10^6ha)	2390	1167	478	615	993	845	2227	473	2052	1839
Climate	tropical; arid, humid	subtropical, tropical; arid	tropical, subtropical; humid, arid	tropical; humid	subtropical, temp. oceanic, continental; humid	tropical, temp. oceanic, subtropical; arid, humid	polar, continental, temp. oceanic; humid, arid	temp. oceanic, some sub-tropical; humid, arid	tropical, subtropical; mostly humid	continental, subtropical, polar, temp. oceanic; humid, arid.
Pop. (10^6)	566	287	1145	451	1333	27	289	510	447	277
Agric. Pop. (%)	62	32	63	49	59	17	13	8	27	3
Pop/ha Cropland	3.6	3.4	5.4	5.7	12.6	0.5	1.3	3.7	2.9	1.2
Agric. Prod. (10^6t)										
Cereals	57	79	258	130	433	24	180	255	111	388
Roots and Tubers	111	12.5	26	50	159	3	65	79	45	22
Pulses	5.7	4.1	14.4	2.5	6.3	2	6	7	5.8	2.2
S. Cane and Beet	60	39	297	181	103	32	62	144	494	56
Meat	6.7	5.5	5.7	6.4	39.6	4.5	17	42	20.5	33.5
GNP/Cap.[2]	350	1940	320	930	590	13780	2700	15300	2390	22100
Annual Growth[2]	-1.2	-2.4	3.1	3.9	7.1	1.5	N/A	2.2	-0.3	1.7
Ag. (% of GDP)[2]	>30	10–19	>30	20–>30	20–29	<6	10–29	<6	10–19	<6

[1]Agricultural land includes grazing and cropland, reported as a percent of total land area. Cropland is reported as a percent of agricultural land. Irrigated area is reported as a percent of cropland.
[2]GNP is in 1991 U.S.$; annual growth (%/annum) is for the period 1980–1991.

Source: Computed from data from FAO Statistics Division (1992); GNP per capita, GNP growth rates, and agriculture as a share of the economy are from World Bank, *World Development Indicators 1993*, and temperature and climate classes from Rötter *et al.*, 1995. Note: East Asia GNP excludes Japan. Also, regional GNP data generally include only those countries for which data are given in Table 1 in *World Development Indicators*. Countries with more than 4 million population for which GNP data are not available include Vietnam, Democratic Republic of Korea, Afghanistan, Cuba, Iraq, Myanmar, Cambodia, Zaire, Somalia, Libya, and Angola; land areas are in hectares, production is in metric tonnes.

per hectare of cropland (12.6). The rapid economic growth achieved over the past decade, the fact that the region's climate is somewhat cooler, and the diverse sources of food production reduce vulnerability of populations in this region. Japan's GNP provides it with significant capability to limit climatic losses from agriculture compared with other countries in the region.

Southeast Asia combines tropical temperatures with generally ample moisture, but the region is subject to tropical storms. The region supports a large population with a relatively high population density per hectare of cultivated land. For several countries in this region, agriculture contributes more than 30% of GDP. GNP per capita is somewhat higher than that of either South or East Asia, and growth has been substantial over the past decade. However, Table 13-13 excludes several countries in the region due to lack of data. These countries, including Vietnam, Cambodia, and Myanmar, have relatively large populations, and their economic performance has been poorer than others in the region. Populations in these countries may be particularly vulnerable to changes in tropical storms.

The Middle East and North Africa are already very hot and generally arid. The current climate greatly limits the portion of land currently suitable for agriculture. A large share of current cropland is irrigated. Among developing country areas, a relatively smaller share of the population (32%) depends directly on agriculture. Agriculture is quite diverse; fruits, vegetables, and other specialty crops are important. The region is heterogeneous, including relatively wealthy oil-exporting countries, Israel, and several poorer countries—making the regional average economic performance somewhat misleading. Per capita GNP has fallen substantially over the past decade, with declining oil prices and political disruptions partly responsible.

Latin America and countries that now make up the area that was formerly the USSR are similar in terms of per capita GNP. While data are unavailable with regard to economic performance for the area of the former USSR, evidence strongly suggests that the region suffered a decline in per capita GNP over the past decade, as has Latin America. The two regions are also similar in that average population density and population per hectare of cropland are moderate to low. Estimates of potential additional cropland for the world suggest that these two areas could be the source of substantial additional cropland (Crosson, 1995). Thus, expansion of land area or relocation to adapt to climate change is a possible response, partly mitigating vulnerability. The prevalence of childhood malnutrition, infant mortality, and low median age at death are somewhat higher in Latin America than in the area of the former USSR (World Bank, 1993). While both regions are primarily humid, substantial agricultural areas are arid or semi-arid and drought-prone. The notable difference between the two regions is that Latin America is generally already tropical or subtropical, and even though GCMs predict less warming in the tropics, further warming may be deleterious. In contrast, agriculture in large areas of the former USSR is limited by cool temperatures and so may benefit from warming. Arid areas of tropical Latin America, such as northwest Brazil, are particularly vulnerable to changes in ENSO events if they result in less reliable precipitation.

Europe, the USA and Canada, and Oceania have high GNP per capita, the agricultural population is a small share of the total population, and agriculture is in general a small share of the economy. As a result, vulnerability to climate-change-induced hunger or severe economic distress for the overall economy is relatively low. These areas are important for world food production. Mid-continental areas of the U.S. and Canada, the Mediterranean area of Europe, and large areas of Australia are prone to drought, which would be exacerbated if climate change reduced moisture availability or increased the demand for water as occurs in several GCM scenarios. Economic dislocation is likely to be limited to the agricultural sector or to subregions highly dependent on agriculture.

Small island nations, especially where incomes are low, are subject to particular vulnerabilities. Potential loss of coastal land to sea level rise, salt water intrusion into water supplies, damage from tropical storms, and temperature and precipitation change will combine to affect the agriculture of island nations. Sea water inundation and salt water intrusion are not unique to small island nations. However, these problems take on greater importance where coastal area is a high proportion of the total area of the country, alternative sources of fresh water are limited, and the area available for retreat from sea-level rise is limited. Local sources of food are especially important for these countries because transportation costs can be substantial for remote locations with small populations, particularly for highly perishable products. Most attention has been focused on Pacific Island Countries in Oceania, but other island countries such as the Maldives are presented with similar conditions.

13.8. Global Agricultural Issues and Assessments

13.8.1. The Current and Future Agricultural System

Climate change will be only one of many factors that will affect world agriculture. The broader impacts of climate change on world markets, on hunger, and on resource degradation will depend on how agriculture meets the demands of a growing population and threats of further resource degradation. World agriculture has proven in the past to be responsive to the increasing demand for food. Evidence of this is the trend of falling real prices for food commodities (Mitchell and Ingco, 1995, estimated a 78% decline between 1950 and 1992) and the steady growth of worldwide food production over the past 3 decades. Average annual increases were 2.7% per annum during the 1960s, 2.8% during the 1970s, and 2.1% during the 1980s.

Despite global abundance, many countries suffer from disrupted agricultural production and distribution systems, such that famine and chronic hunger are a reality or a distinct threat. While the number of people suffering from chronic hunger has declined (from an estimated 844 million in 1979 to 786 million in 1990; Bongaarts, 1994), the causes of famine are complex, including a lack of the rights and means to obtain food (employment, adequate income, and a public system for responding to famine) (e.g., Sen, 1981, 1993); political systems disrupted by war and unrest; ineffective or misdirected policies; as well as, or in addition to, drought and other extreme climatic events (McGregor, 1994). The nearly 800 million people still estimated to suffer from chronic hunger and malnutrition represent 20% of the population of developing countries, with the percentage as high as 37% in sub-Saharan countries (FAO, 1995). In many situations of chronic hunger, the population is rural and their livelihood depends primarily on agriculture. However, Kates and Chen (1994a, 1994b), though noting the potential risks of climate change beyond those represented in median cases, provide an array of actions that could be undertaken to achieve a food-secure world.

Agricultural and resource policies have important effects on agricultural production, and national governments have intervened in agriculture in many ways and for various reasons (Hayami and Ruttan, 1985). Many developed countries have subsidized agriculture and thereby encouraged production, while intentionally idling land to control surplus stocks of agricultural commodities. Many developing countries have controlled food prices to benefit lower-income food consumers but have thereby discouraged domestic production. Reduction of trade-distorting government interventions in agriculture was part of the recently concluded round of the General Agreement on Tariffs and Trade (GATT). National policies also greatly affect land use and water use, management, and pricing.

Three major studies of the future world food situation suggest that in the absence of climate change, food supply will continue to expand faster than demand over the next 20 to 30 years, with world prices projected to fall (Alexandratos, 1995; Mitchell and Ingco, 1995; Rosegrant and Agcaoili, 1995). Others are less optimistic, citing limits on further land expansion and irrigation, resource degradation, and reduced confidence that the historical rates of increase in yield will continue (Bongaarts, 1994; McCalla, 1994; Norse, 1994).

Factors that will jointly determine whether agricultural supply increases can keep pace with demand include (1) how fast demand will grow, (2) the future availability of land and its quality, (3) the future availability of water, and (4) whether improvements in technology will continue to result in rapid yield growth. Finally, economic growth and development are closely tied to demand growth and, in many developing countries, are also dependent on agricultural development.

13.8.1.1. Demand Growth

Between 1950 and 1990, world population grew at a 2.25% compound annual rate. Through 2025, population is projected to grow at a compound annual rate of between 1.13% and 1.55% (high and low UN variants). The decade of the 1990s is projected to have the largest absolute population addition, with declining additions in subsequent decades (Bongaarts, 1995). Thus, food supply growth could slow by 40–50% from recent decades while maintaining per capita food production levels. Income growth likely will cause demand to grow more rapidly than population and will change the composition of demand, most likely away from food grains and toward meat, fruits, and vegetables. The shift to meat is likely to increase the demand for grain for livestock feed. Increased demand generated by increased income growth would allow more and higher-quality food to be consumed per capita (Mitchell and Ingco, 1995), but this depends on how food consumption is distributed. Beyond 2025, population growth is generally projected to be low as world population is projected to stabilize by around 2075.

13.8.1.2. Land Quantity and Quality

Some estimates suggest that there is much potentially available land (Buringh and Dudal, 1987), but the cost of bringing it under production may be high, with attendant adverse environmental impacts limiting expansion (Crosson, 1995). Intensification of production on existing cropland may worsen land degradation and put additional pressure on water and soil resources. Firm data on the extent and severity of land degradation and its impact on production potential for most of the world are not available (El-Swaify *et al.*, 1982; Dregne, 1988; Nelson, 1988; Lal and Okigbo, 1990), but a recent overview (Oldeman et al., 1990), though still qualitative in economic terms, confirms significant degradation and loss of arable land, especially in Africa (see Chapter 4). Studies disagree on the extent to which intensification affects land degradation (Crosson and Stout, 1983; Brown and Thomas, 1990; Tiffen *et al.*, 1993). Competition for agricultural resources for other uses may also affect the supply and price of land for agriculture. Carbon sequestration, biomass energy production, forest product production, the potential development of new non-food agricultural products, and removal of agricultural land from production for other environmental objectives will affect the amount of land available for food production (see Chapters 23, 24, and 25).

13.8.1.3. Water Supply and Irrigation

Irrigation has contributed significantly to increased production in the past. Currently, 17% of global cropland is irrigated, but this 17% of land accounts for more than one-third of total world food production. An estimated additional 137 million hectares have the potential to be irrigated, compared with the 253 million hectares currently irrigated (World Bank, 1990), but the cost of doing so may be prohibitive. Current water systems in many developing countries achieve low efficiencies of water distribution, and average crop yields are well below potential (Yudelman, 1993; Crosson, 1995; Rosegrant and Agcaoili, 1995). There are environmental and health-related effects of irrigation such as soil salinization and the spread of water-borne diseases that may limit further expansion (Brown and Thomas, 1990; Dregne and Chou, 1990; Jensen *et al.*, 1990; Crosson, 1995). Major factors contributing to these irrigation problems in both developing and developed countries are unpriced and heavily subsidized water resources; inadequate planning, construction, and maintenance of water systems; unassigned water rights or rules that limit the transfer of rights; and conflicts between development and distribution goals (Frederick, 1986; Asian Development Bank, 1991; Moore, 1991; Umali, 1993; Yudelman, 1993; Appendine and Liverman, 1994). Solutions to these problems are available in most cases, and a recent study found that investments in irrigation have been at least as profitable as investments in other agricultural enterprises (World Bank, 1994). Changes in potential irrigation water supply due to climate change have not generally been integrated into agricultural impact studies, with few exceptions (e.g., Rosenberg, 1993). For climatic impacts on water supply, see Chapter 10.

13.8.1.4. Future Yield Growth

Assumed continuation of yield increases due to improving technology and further adoption of existing technologies is uncertain. Gaps between actual and potential yield are cited as evidence of unexploited production potential, but potential yields are rarely, if ever, attained in practice (Tinker, 1985; Plucknett, 1995). Realization of improved varieties depends on continuation of agricultural research and crop breeding systems and the exchange of germplasm (Duvick, 1995).

13.8.1.5. Future Economic Development

The impact of climate change on human populations in terms of famine, chronic hunger, health, and nutrition will depend on how and whether currently poor areas develop over the next 20 to 50 years. The future path of development of currently vulnerable countries remains uncertain. Policy failures, wars, and political and civil unrest are identified causes, but correcting these problems has proved difficult (e.g., van Dijk, 1992; Anand and Ravallion, 1993). Lagging agricultural development has been identified as a consequence of significant policy distortions in many developing countries, conflicting with the industrial sector and limiting the ability of the broader economy to grow (Hayami and Ruttan, 1985; Adelman and Vogel, 1992; Cavallo *et al.*, 1992; FAO, 1995).

13.8.2. Global Climate Impact Studies

Accurate consideration of national and local food supply and economic effects depends on an appraisal of changes in global food supply and prices. International markets can moderate or reinforce local and national changes. In 1988, for example, drought presented a more severe threat because it occurred coincidentally in several of the major grain-growing regions of the world.

While uncertainties continue to exist about the direction of change in global agricultural production resulting from climate change, changes in the aggregate level of production have been found to be small to moderate (Kane *et al.*, 1992; Fischer *et al.*, 1994; Reilly *et al.*, 1994; Rosenzweig and Parry, 1994). Studies show that a disparity in agricultural impact between developed and developing countries can be reinforced by markets (Tables 13-14, 13-15).

Rosenzweig *et al.* (1994) found that in lower-latitude developing countries, cereal grain crop yields and production declined under climate change scenarios ranging from 2.5 to 5.2°C. The study further found that the population at risk of hunger (defined as a measure of food energy availability, which depends on income and food price levels, relative to nutritional requirements) could increase despite adaptation. The study involved agricultural scientists in 18 countries using comparable crop growth models for wheat, rice, maize, and soybean (IBSNAT, 1989) and consistent climate change scenarios (Rosenzweig and Iglesias, 1994). Estimated yield changes were the basis for supply changes in the Basic Linked System (BLS), a world food trade model (Fischer *et al.*, 1988).

Reilly and Hohman (1993) and Reilly *et al.* (1994) used the same national crop yield changes as Rosenzweig *et al.* (1994) in a different trade model and found that agricultural exporters may gain even though their supplies fall as a result of higher world prices. They found that developing countries did worse in economic

Table 13-14: *Change in cereals production under three different GCM equilibrium scenarios (percent from base estimated in 2060).*

Region	GISS	GFDL	UKMO
World Total			
Climate effects only	-10.9	-12.1	-19.6
Plus physiological effect of CO_2	-1.2	-2.8	-7.6
Plus adaptation level 1	0.0	-1.6	-5.2
Plus adaptation level 2	1.1	-0.1	-2.4
Developed Countries			
Climate effects only	-3.9	-10.1	-23.9
Plus physiological effect of CO_2	11.3	5.2	-3.6
Plus adaptation level 1	14.2	7.9	3.8
Plus adaptation level 2	11.0	3.0	1.8
Developing Countries			
Climate effects only	-16.2	-13.7	-16.3
Plus physiological effect of CO_2	-11.0	-9.2	-10.9
Plus adaptation level 1	-11.2	-9.2	-12.5
Plus adaptation level 2	-6.6	-5.6	-5.8

Source: Rosenzweig and Parry, 1994.
Notes: Level 1 adaptation included changes in crop variety but not the crop, the planting date of less than 1 month, and the amount of water applied for areas already irrigated.
Level 2 adaptation additionally included changes in the type of crop grown, changes in fertilizer use, changes in the planting date of more than 1 month, and extension of irrigation to previously unirrigated areas.

Table 13-15: Economic effects of three GCM equilibrium scenarios (billions of 1989 U.S.$).

Region\GCM	With CO_2 and Adaptation GISS	GFDL	UKMO	With CO_2, No Adaptation GISS	GFDL	UKMO	No CO_2, No Adaptation GISS	GFDL	UKMO
Developing									
<$500/Cap.	-0.2	-2.6	-14.6	-2.1	-5.3	-19.8	-56.7	-66.1	-121.1
$500–2000/Cap.	-0.4	-2.9	-10.7	-1.8	-5.1	-15.0	-26.2	-27.9	-48.1
>$2000/Cap.	-0.6	-0.5	-1.0	-0.8	-0.9	-0.3	-6.7	-4.4	-3.9
Eastern Europe	2.4	0.0	-4.9	1.9	-2.0	-11.0	-12.5	-28.9	-57.5
OECD	5.8	0.0	-6.5	2.7	-3.6	-15.1	-13.5	-21.5	-17.6
TOTAL	7.0	-6.1	-37.6	0.0	-17.0	-61.2	-115.5	-148.6	-248.1

Source: Reilly *et al.*, 1994.
Notes: Measured as annual loss or gain in consumer and producer surplus plus change in society's cost of agricultural policies. Columns may not sum to total due to independent rounding. Adaptation is level 1, as in Table 13-14.

terms as a group, but that some developing countries benefitted. The pattern of winners and losers varied among climate scenarios. Moreover, they found that food-importing countries were found, in some cases, to suffer economic loss because of higher world prices even if the country's crop production potential improved.

Table 13-14 also illustrates how trade and adaptation capability can interact. Developing countries' production levels fell more under adaptation level 1 (small changes in planting dates, changes in cultivars, and additional irrigation water for areas already irrigated) than with no adaptation because their estimated capability to adapt was less than in developed countries. Thus, cereal production in developing countries fell further as trade shifted toward developed-country exports at the expense of developing country production. At adaptation level 2 (substantial changes in planting dates, changes in crop, and extension of irrigation systems), this situation was reversed.

Another global modeling effort considered potential yield and distribution of crops based on a crop suitability index (Cramer and Solomon, 1993; Leemans and Solomon, 1993). Although considering only one climate scenario and omitting the effect of CO_2 fertilization, they found that high-latitude regions uniformly benefited from longer growing periods and increased productivity. Other regions either did not benefit significantly or lost productivity.

More recent work considering global agriculture under climate change found far greater potential for global agriculture to adapt to changing climate than earlier studies (Darwin *et al.*, 1995; reported in Reilly, 1995). This study estimated that climate change alone as represented by equivalent doubled-CO_2 climate scenarios of UKMO, GISS, GFDL, and OSU, without consideration of the direct effects of CO_2, would result in global production losses of less than 1% if no additional land area were devoted to agriculture. If new land area could be brought under cultivation, grain production was estimated to increase on the order of 1%.

13.9. Adaptation

Historically, farming systems have adapted to changing economic conditions, technology, and resource availabilities and have kept pace with a growing population (Rosenberg, 1982; CAST, 1992). Evidence exists that agricultural innovation responds to economic incentives such as factor prices and can relocate geographically (Hyami and Ruttan, 1985; CAST, 1992). A number of studies indicate that adaptation and adjustment will be important to limit losses or to take advantage of improving climatic conditions (e.g., National Academy of Sciences, 1991; Rosenberg, 1992; Rosenberg and Crosson, 1991; CAST, 1992; Mendelsohn *et al.*, 1994).

Despite the successful historical record, questions arise with regard to whether the rate of change of climate and required adaptation would add significantly to the disruption likely due to future changes in economic conditions, technology and resource availabilities (Gommes, 1993; Harvey, 1993; Kane and Reilly, 1993; Smit, 1993; Norse, 1994; Pittock, 1994; Reilly, 1994). If climate change is gradual, it may be a small factor that goes unnoticed by most farmers as they adjust to other more profound changes in agriculture stemming from new technology, increasing demand for food, and other environmental concerns such as pesticide use, water quality, and land preservation. However, some researchers see climate change as a significant addition to future stresses, where adapting to yet another stress such as climate change may be beyond the capability of the system. Part of the divergence in views may be due to different interpretations of adaptation, which include the prevention of loss, tolerating loss, or relocating to avoid loss (Smit, 1993). Moreover, while the technological potential to adapt may exist, the socioeconomic capability to adapt likely differs for different types of agricultural systems (Reilly and Hohmann, 1993).

13.9.1. The Technological Potential to Adapt

Nearly all agricultural impact studies conducted over the past 5 years have considered some technological options for adapting to climate change. Among those that offer promise are:

- **Seasonal Changes and Sowing Dates**—For frost-limited growing areas (i.e., temperate and cold areas), warming could extend the season, allowing planting of longer maturity annual varieties that achieve higher yields (e.g., Le Houerou, 1994; Rowntree, 1990). For short-season crops such as wheat, rice, barley, oats, and many vegetable crops, extension of the growing season may allow more crops per year, fall planting, or, where warming leads to regular summer highs above critical thresholds, a split season with a short summer fallow. For subtropical and tropical areas where growing season is limited by precipitation or where the cropping already occurs throughout the year, the ability to extend the growing season may be more limited and depends on how precipitation patterns change. A study for Thailand found yield losses in the warmer season partially offset by gains in the cooler season (Parry *et al.*, 1992).
- **Different Crop Variety or Species**—For most major crops, varieties exist with a wide range of maturities and climatic tolerances. For example, Matthews *et al.* (1994) identified wide genetic variability among rice varieties as a reasonably easy response to spikelet sterility in rice that occurred in simulations for South and Southeast Asia. Studies in Australia showed that responses to climate change are strongly cultivar-dependent (Wang *et al.*, 1992). Longer-season cultivars were shown to provide a steadier yield under more variable conditions (Connor and Wang, 1993). In general, such changes may lead to higher yields or may only partly offset losses in yields or profitability. Crop diversification in Canada (Cohen *et al.*, 1992) and in China (Hulme *et al.*, 1992) has been identified as an adaptive response.
- **New Crop Varieties**—The genetic base is broad for most crops but limited for some (e.g., kiwi fruit). A study by Easterling *et al.* (1993) explored how hypothetical new varieties would respond to climate change (also reported in McKenney *et al.*, 1992). Heat, drought, and pest resistance; salt tolerance; and general improvements in crop yield and quality would be beneficial (Smit, 1993). Genetic engineering and gene mapping offer the potential for introducing a wider range of traits. Difficulty in assuring that traits are efficaciously expressed in the full plant, consumer concerns, profitability, and regulatory hurdles have slowed the introduction of genetically engineered varieties compared with early estimates (Reilly, 1989; Caswell *et al.*, 1994).
- **Water Supply and Irrigation Systems**—Across studies, irrigated agriculture in general is less negatively affected than dryland agriculture, but adding irrigation is costly and subject to the availability of water supplies. Climate change will affect future water supplies (see Chapter 10). There is wide scope for enhancing irrigation efficiency through adoption of drip irrigation systems and other water-conserving technologies (FAO, 1989, 1990), but successful adoption will require substantial changes in how irrigation systems are managed and how water resources are priced. Because inadequate water systems are responsible for current problems of land degradation, and because competition for water is likely to increase, there likely will be a need for changes in the management and pricing of water regardless of whether and how climate changes (Vaux, 1990, 1991; World Bank, 1994). Tillage method and incorporation of crop residues are other means of increasing the useful water supply for cropping.
- **Other Inputs and Management Adjustments**—Added nitrogen and other fertilizers would likely be necessary to take full advantage of the CO_2 effect. Where high levels of nitrogen are applied, nitrogen not used by the crop may be leached into the groundwater, run off into surface water, or be released from the soil as nitrous oxide. Additional nitrogen in ground and surface water has been linked to health effects in humans and affects aquatic ecosystems. Studies also have considered a wider range of adjustments in tillage, grain drying, and other field operations (Kaiser *et al.*, 1993; Smit, 1993).
- **Tillage**—Minimum and reduced tillage technologies, in combination with planting of cover crops and green manure crops, offer substantial possibilities to reverse existing soil organic matter, soil erosion, and nutrient loss, and to combat potential further losses due to climate change (Rasmussen and Collins, 1991; Logan, 1991; Edwards *et al.*, 1992; Langdale *et al.*, 1992; Peterson *et al.*, 1993; Brinkman and Sombroek, 1993; see also Chapter 23). Reduced and minimum tillage techniques have spread widely in some countries but are more limited in other regions. There is considerable current interest in transferring these techniques to other regions (Cameron and Oram, 1994).
- **Improved Short-Term Climate Prediction**—Linking agricultural management to seasonal climate predictions (currently largely based on ENSO), where such predictions can be made with reliability, can allow management to adapt incrementally to climate change. Management/climate predictor links are an important and growing part of agricultural extension in both developed and developing countries (McKeon *et al.*, 1990, 1993; Nichols and Wong, 1990).

13.9.2. The Socioeconomic Capability to Adapt

While identifying many specific technological adaptation options, Smit (1993) concluded that necessary research on their cost and ease of adoption had not yet been conducted.

One measure of the potential for adaptation is to consider the historical record on past speeds of adoption of new technologies (Table 13-16). Adoption of new or different technologies depends on many factors: economic incentives, varying resource and climatic conditions, the existence of other technologies (e.g., transportation systems and markets), the availability of information, and the remaining economic life of equipment and structures (e.g., dams and water supply systems).

Specific technologies only can provide a successful adaptive response if they are adopted in appropriate situations. A variety of issues has been considered, including land-use planning, watershed management, disaster vulnerability assessment, port and rail adequacy, trade policy, and the various programs countries use to encourage or control production, limit food prices, and manage resource inputs to agriculture (CAST, 1992; OTA, 1993; Smit, 1993; Reilly *et al.*, 1994; Singh, 1994). For example, studies suggest that current agricultural institutions and policies in the United States may discourage farm management adaptation strategies, such as altering crop mix, by supporting prices of crops not well-suited to a changing climate, providing disaster payments when crops fail, or prohibiting imports through import quotas (Lewandrowski and Brazee, 1993).

Existing gaps between best yields and the average farm yields remain unexplained, but many are due in part to socioeconomic considerations (Oram and Hojjati, 1995; Bumb, 1995); this adds considerable uncertainty to estimates of the potential for adaptation, particularly in developing countries. For example,

Table 13-16: *Speed of adoption for major adaptation measures.*

Adaptation	Adjustment Time (years)	Reference
Variety Adoption	3–14	Dalrymple, 1986; Griliches, 1957; Plucknett *et al.*, 1987; CIMMYT, 1991; Wang *et al.*, 1992
Dams and Irrigation	50–100	James and Lee, 1971; Howe, 1971
Variety Development	8–15	Plucknett *et al.*, 1987; Knudson, 1988
Tillage Systems	10–12	Hill *et al.*, 1994; Dickey *et al.*, 1987; Schertz, 1988
New Crop Adoption: Soybeans	15–30	FAO, Agrostat (various years)
Opening New Lands	3–10	Medvedev, 1987; Plusquellec, 1990
Irrigation Equipment	20–25	Turner and Anderson, 1980
Transportation System	3–5	World Bank, 1994
Fertilizer Adoption	10	Pieri, 1992; Thompson and Wan, 1992

Baethgen (1994) found that a better selection of wheat variety combined with an improved fertilizer regime could double yields achieved at a site in Uruguay to 6 T/ha under the current climate with current management practices. Under the UKMO climate scenario, yields fell to 5 T/ha—still well above 2.5–3.0 T/ha currently achieved by farmers in the area. On the other hand, Singh (1994) concludes that the normal need to plan for storms and extreme weather events in Pacific island nations creates significant resiliency. Whether technologies meet the self-described needs of peasant farmers is critical in their adaptation (Cáceres, 1993). Other studies document how individuals cope with environmental disasters, identifying how strongly political, economic, and ethnic factors interact to facilitate or prevent coping in cases ranging from the Dust Bowl disaster in the United States to floods in Bangladesh to famines in the Sudan, Ethiopia, and Mozambique (McGregor, 1994). These considerations indicate the need for local capability to develop and evaluate potential adaptations that fit changing conditions (COSEPUP, 1992). Important strategies for improving the ability of agriculture to respond to diverse demands and pressures, drawn from past efforts to transfer technology and provide assistance for agricultural development, include:

- Improved training and general education of populations dependent on agriculture, particularly in countries where education of rural workers is currently limited. Agronomic experts can provide guidance on possible strategies and technologies that may be effective. Farmers must evaluate and compare these options to find those appropriate to their needs and the circumstances of their farms.
- Identification of the present vulnerabilities of agricultural systems, causes of resource degradation, and existing systems that are resilient and sustainable. Strategies that are effective in dealing with current climate variability and resource degradation also are likely to increase resilience and adaptability to future climate change.
- Agricultural research centers and experiment stations can examine the "robustness" of present farming systems (i.e., their resilience to extremes of heat, cold, frost, water shortage, pest damage, and other factors) and test the robustness of new farming strategies as they are developed to meet changes in climate, technology, prices, costs, and other factors.
- Interactive communication that brings research results to farmers—and farmers' problems, perspectives, and successes to researchers—is an essential part of the agricultural research system.
- Agricultural research provides a foundation for adaptation. Genetic variability for most major crops is wide relative to projected climate change. Preservation and effective use of this genetic material would provide the basis for new variety development. Continually changing climate is likely to increase the value of networks of experiment stations that can share genetic material and research results.

- Food programs and other social security programs would provide insurance against local supply changes. International famine and hunger programs need to be considered with respect to their adequacy.
- Transportation, distribution, and market integration provide the infrastructure to supply food during crop shortfalls that might be induced in some regions because of climate variability or worsening of agricultural conditions.
- Existing policies may limit efficient response to climate change. Changes in policies such as crop subsidy schemes, land tenure systems, water pricing and allocation, and international trade barriers could increase the adaptive capability of agriculture.

Many of these strategies will be beneficial regardless of how or whether climate changes. Goals and objectives among countries and farmers vary considerably. Current climate conditions and likely future climates also vary. Building the capability to detect change and evaluate possible responses is fundamental to successful adaptation.

13.10. Research Needs

The continuing uncertainty in projections suggests three critical, high-priority research needs:

- Development and broad application of integrated agricultural modeling efforts and modeling approaches particularly applicable at the regional scale, including increased attention to validation, testing, and comparison of alternative approaches. Climate effects on soils and plant pests, consideration of other environmental changes, and adaptation options and economic responses should be an integrated part of the models rather than treated on an *ad hoc* basis or as a separate modeling exercise. Inclusion of these multiple, joint effects may significantly change our "mean" estimate of impact, and more careful attention to scale and validation should help to reduce the range of estimates for specific regions and countries across different methodologies.
- Development of the capability to readily simulate agricultural impacts of multiple transient climate scenarios. Study of the sensitivities of agriculture to climate change and the impacts of doubled-CO_2 equilibrium scenarios has not led to the development of methods that readily can be applied to transient climate scenarios. To deal credibly with the cost of adjustment, about which there is significant uncertainty, the process of socioeconomic adjustment must be modeled to treat key dynamic issues such as how the expectations of farmers change, whether farmers can easily detect climate change against a background of high natural variability, and how current investments in equipment, education, and training may lead to a system that only slowly adjusts, or adjusts only with high cost and significant disruption. The ability to readily simulate effects under multiple climate scenarios is necessary to quantify the range of uncertainty.
- Evaluation of the effects of variability rather than changes in the "mean" climate, and the implication of changes in variability on crop yields and markets. Extreme events have severe effects on crops, livestock, soil processes, and pests. The more serious human consequences of climate change also are likely to involve extreme events such as drought, flooding, or storms, where agricultural production is severely affected.

State-of-the-art research has begun addressing these areas, and a number of promising approaches have begun to appear in the literature or are expected soon. Most are, as yet, "demonstration" research projects, choosing limited geographic areas where data are more available and considering convenient examples for climate scenarios. Caution in drawing broader policy implications from such studies is warranted because there is little or no basis to make inferences to broader populations, to other locations, or to specific climate scenarios.

Acknowledgments

The Convening Lead Author would like to thank the following individuals for their assistance in chapter preparation: B.D. Campbell, New Zealand; P. Condon, USA; T. Downing, UK; K.J. Hennessy, Australia; D.G. Johnson, USA; W.D. Kemper, USA; J. Lewandrowski, USA; A.B. Pittock, Australia; N. Rosenberg, USA; D. Sauerbeck, Germany; R. Schulze, South Africa; P.B. Tinker, UK; M. Yoshino, Japan.

References

Abashina, E.V. and O.D. Sirotenko, 1986: An applied dynamic model of crop formation for modeling agrometerological support systems. *Trudy ARRI-AM*, **21**, 13-33.

Ackerly, D.D., J.S. Coleman, S.R. Morse, and F.A. Bazzaz, 1992: CO_2 and temperature effects on leaf area production in two annual plant species. *Ecology*, **73**, 1260-1269.

Ackerson, R.C., U.D. Havelka, and M.G. Boyle, 1984: CO_2-enrichment effects on soybean physiology. II. Effects of stage-specific CO_2 exposure. *Crop Science*, **24**, 1150-1154.

Acock, B. and M.C. Acock, 1993: Modeling approaches for predicting crop ecosystem responses to climate change. In: *International Crop Science*, vol. I. Crop Science Society of America, Madison, WI, pp. 299-306.

Adams, R.M., B.A. McCarl, D.J. Dudek, and J.D. Glyer, 1988: Implications of global climate change for western agriculture. *Western Journal of Agricultural Economics*, **13**, 348-356.

Adams, R.M., C. Rosenzweig, R.M. Peart, J.T. Ritchie, B.A. McCarl, J.D. Glyer, R.B. Curry, J.W. Jones, K.J. Boote, and L.H. Allen, Jr., 1990: Global climate change and US agriculture. *Nature*, **345**, 219-224.

Adams, R.M., R.A. Fleming, C.-C. Chang, B.A. McCarl, and C. Rosenzweig, submitted: A reassessment of the economic effects of global climate change on U.S. agriculture. *Climatic Change*.

Adelman, I. and S. J. Vogel, 1992: The relevance of ADLI for sub-saharan Africa. In: *African Development Perspectives Yearbook*, LIT Verlag, Hamburg, pp. 258-279.

Agcaoili, M., and M.W. Rosegrant, 1995: Global and regional food demand, supply and trade prospects to 2010. In: *Population and Food in the Early 21st Century: Meeting Future Food Demand of an Increasing World Population* [Islam, N. (ed.)]. Occasional Paper, International Food Policy Research Institute (IFPRI), Washington, DC, pp. 61-90.

The Agricultural Atlas of the People's Republic of China, 1990. Map Publishing House, Beijing, China.

Akong'a, J., T.E. Downing, N.T. Konijn, D.N. Mungai, H.R. Muturi, and H.L. Potter, 1988: The effects of climatic variations on agriculture in Central and Eastern Kenya. In: *The Impact of Climatic Variations on Agriculture*. Vol. 2, *Assessments in Semi-Arid Regions* [Parry, M.L., T.R. Carter, and N.T. Konijn (eds.)]. Kluwer Academic Press, Dordrecht, Netherlands, pp. 123-270.

Allen, Jr., L.H., 1990: Plant responses to rising carbon dioxide and potential interactions with air pollutants. *Journal of Environmental Quality*, **19**, 15-34.

Allen, S.G., S.B. Idso, B.A. Kimball, J.T. Baker, H.L. Allen, J.R. Mauney, J.W. Radin, and M.G. Anderson, 1990: *Effects of Air Temperature and Atmospheric CO_2 Plant Growth Relationships*. Report TR 048, U.S. Department of Energy/ U.S. Department of Agriculture, Washington, DC, 60 pp.

Alexandratos, N., 1995: The outlook for world food production and agriculture to the year 2010. In: *Population and Food in the Early 21st Century: Meeting Future Food Demand of an Increasing World Population* [Islam, N. (ed.)]. Occasional Paper, International Food Policy Research Institute (IFPRI), Washingon DC, pp. 25-48.

Amthor, J.S., 1989: *Respiration and Crop Productivity*. Springer-Verlag, Berlin, Germany, 215 pp.

Amthor, J.S., 1991: Respiration in a future, higher- CO_2 world. *Plant, Cell and Environment*, **14**, 13-20.

Anand, S. and M. Ravallion, 1993: Human development in poor countries: on the role of private incomes and public services. *Journal of Economic Perspectives*, **7(1)**, 133-150.

Anfossi, D., S. Sandroni, and S. Viarengo, 1991: Tropospheric ozone in the nineteenth century: the montcalieri series. *Journal of Geophysical Research*, **96**, 17,349-52.

Appendini, K. and D. Liverman, 1994. Agricultural policy and food security in Mexico. *Food Policy*, **19(2)**, 149-164.

Arthur, L.M. and F. Abizadeh, 1988: Potential effects of climate change on agriculture in the prairie region of Canada. *Western Journal of Agricultural Economics*, **13**, 216-224.

Ashmore, M.S., 1988: A comparison of indices that describe the relationship between exposure to ozone and reduction in the yield of agricultural crops. *Atmos. Environ.*, **22**, 2060-2061.

Asian Development Bank, 1991: Unpriced resources and the absence of markets. In: *Asian Development Outlook*. Asian Development Bank, Singapore, pp. 239-240.

Baethgen, W.E., 1994: Impact of climate change on barley in Uruguay: yield changes and analysis of nitrogen management systems. In: *Implications of Climate Change for International Agriculture: Crop Modeling Study* [Rosenzweig, C. and A. Iglesias (eds.)]. U.S. Environmental Protection Agency, Uruguay chapter, Washington, DC, pp. 1-13.

Baker, B.B., J.D. Hanson, R.M. Bourdon, and J.B. Eckert, 1993: The potential effects of climate change on ecosystem processes and cattle production on U.S. rangelands. *Climatic Change*, **23**, 97-117.

Baker, J.T. and L.H. Allen, Jr., 1993: Contrasting crop species responses to CO_2 and temperature: rice, soybean and citrus. *Vegetatio*, **104/105**, 239-260.

Baker, J.T., J.H. Allen, and K.J. Boote, 1990: Growth and yield responses of rice to carbon dioxide concentration. *Journal of Agricultural Science*, **115**, 313-320.

Baltensweiler, W. and A. Fischlin, 1986: The larch budmoth in the alps. In: *Dynamics of Forest Insect Populations: Patterns, Causes, Implications* [Berryman, A.A. (ed.)]. Plenum Press, New York, NY, pp. 331-351.

Barbour, D.A., 1986: The pine looper in Britain and Europe. In: *Dynamics of Forest Insect Populations: Patterns, Causes, Implications* [Berryman, A.A. (ed.)]. Plenum Press, New York, NY, pp. 291-308.

Barrow, E.M., 1993: Scenarios of climate change for the European community. *European Journal of Agronomy*, **2**, 247-260.

Bazzaz, F.A. and E.D. Fajer, 1992: Plant life in a CO_2-rich world. *Scientific American*, **26(1)**, 68-74.

Bazzaz, F.A. and K. Garbutt, 1988: The response of annuals in competitive neighborhoods: effects of elevated CO_2. *Ecology*, **69**, 937-946.

Behl, R.K., H.S. Nainawatee, and K.P. Singh, 1993: High temperature tolerance in wheat. In: *International Crop Science*, vol. I. Crop Science Society of America, Madison, WI, pp. 349-355.

Bejer, B., 1986: The nun moth in European spruce forests. In: *Dynamics of Forest Insect Populations: Patterns, Causes, Implications* [Berryman, A.A. (ed.)]. Plenum Press, New York, NY, pp. 211-231.

Bergthorsson, P., H. Bjornsson, O. Dyrmundsson, B. Gudmundsson, A. Helgadottir, and J.V. Jonmundsson, 1988: The effects of climatic variations on agriculture in Iceland. In: *The Impacts of Climatic Variations on Agriculture*. Vol. 1, *Cool Temperate and Cold Regions* [Parry, M.L., T.R. Carter, and N.T. Konijn (eds.)]. Kluwer Academic Press, Dordrecht, Netherlands, pp. 383-512

Berman, A., 1991: Reproductive responses under high temperature conditions. In: *Animal Husbandry in Warm Climates* [Ronchi, B., A. Nardone, and J.G. Boyazoglu (eds.)]. EAAP Publication No. 55, EAAP, pp. 23-30.

Berryman, A.A. and G.T. Ferrell, 1986: The fir engraver beetle in western states. In: *Dynamics of Forest Insect Populations: Patterns, Causes, Implications* [Berryman, A.A. (ed.)]. Plenum Press, New York, NY, pp. 555-577.

Bhattacharya, N.C., D.R. Hileman, P.P. Ghosh, R.L. Musser, S. Bhattacharya, and P.K. Biswas, 1990: Interaction of enriched CO_2 and water stress on the physiology of and biomass production in sweet potato grown in open top chambers. *Plant, Cell and Environment*, **13**, 933-940.

Bianca, W., 1970: Animal response to meterological stress as a function of age. *Biometerology*, **4**, 119-131.

Biggs, R.H., S.V Kossuth, and A.H. Teramura, 1981: Response of 19 cultivars of soybeans to ultraviolet-B irradiance. *Physiol. Plant*, **53**, 19-26.

Bindi, M., M. Castellani, G. Maracchi, and F. Miglieta, 1993: The ontogenesis of wheat under scenarios of increased air temperature in Italy: a simulation study. *European Journal of Agronomy*, **2**, 261-280.

Birch, L.C., 1965: Evolutionary opportunity for insects and mammals in Australia. In: *The Genetics of Colonizing Species* [Baker, H.G. and G.L. Stebbins (eds.)]. Academic Press, New York, NY, pp. 197-211.

Blumenthal, C.S., F. Bekes, I.L. Batey, C.W. Wrigley, H.J. Moss, D.J. Mares, and E.W.R. Barlow, 1991: Interpretation of grain quality results from wheat variety trials with reference to high temperature stress. *Australian Journal of Agricultural Research*, **42**, 325-334.

Bohl, H.G., T.E. Downing, and M.J. Watts, 1994. Climate change and social vulnerability: toward a sociology and geography of food insecurity. *Global Environmental Change*, **4(1)**, 37-48.

Bongaarts, J., 1994: Can the growing human population feed itself? *Scientific American*, **270(3)**, 36-42.

Bongaarts, J., 1995: How reliable are future population projections? In: *Population and Food in the Early 21st Century: Meeting Future Food Demand of an Increasing World Population* [Islam, N. (ed.)]. Occasional Paper, International Food Policy Research Institute (IFPRI), Washington DC, pp. 7-16.

Bonsma, J.C., 1949: Breeding cattle for increased adaptability to tropical and subtropical environments. *Journal of Agricultural Sciences*, **39**, 204 ff.

Borden, J.H., 1986: The striped ambrosia beetle. In: *Dynamics of Forest Insect Populations: Patterns, Causes, Implications* [Berryman, A.A. (ed.)]. Plenum Press, New York, NY, pp. 579-596.

Bornman, J.F., 1993: Impact of increased ultraviolet-B radiation on plant performance. In: *International Crop Science*, vol. I. Crop Science Society of America, Madison, WI, pp. 321-323.

Bosac, C., V.J. Black, C.R. Black, J.A. Roberts, and F. Lockwood, 1993: Impact of O_3 and SO_2 on reproductive development in oilseed rape (Brassica napus L.). I. Pollen germination and pollen tube growth. *New Phytol.*, **124**, 439-446.

Bouwman, A.F., 1990: *Soils and the Greenhouse Effect*. John Wiley and Sons, New York, NY, 575 pp.

Bowes, M.D. and P. Crosson, 1993: Consequences of climate change for the MINK economy: impacts and responses. *Climatic Change*, **24**, 131-158.

Bowman, J., 1988: The "greenhouse effect." In: *The "Greenhouse Effect" and UK Agriculture* [Bennet, R.M. (ed.)]. Center for Agricultural Strategy, University of Reading, Reading, UK, **19**, pp. 17-26.

Brammer, H., M. Asaduzzaman, and P. Sultana, 1994: *Effects of Climate and Sea-Level Changes on the Natural Resources of Bangladesh.* Briefing Document No. 3, Bangladesh Unnayan Parishad (BUP), Dhaka, Bangladesh, 35 pp.

Brinkman, R. and W.G. Sombroek, 1993: The effects of global change on soil conditions in relation to plant growth and food production. *Expert Consultation Paper on Global Climate Change and Agricultural Production: Direct and Indirect Effects of Changing Hydrological, Soil, and Plant Physiological Processes.* FAO, Rome, December 1993, 12 pp.

Brklacich, M., R. Stewart, V. Kirkwood, and R. Muma, 1994: Effects of global climate change on wheat yields in the Canadian prairie. In: *Implications of Climate Change for International Agriculture: Crop Modeling Study* [Rosenzweig, C. and A. Iglesias (eds.)]. U.S. Environmental Protection Agency, Canada chapter, Washington, DC., pp. 1-23.

Brklacich, M. and B. Smit, 1992: Implications of changes in climatic averages and variability on food production opportunities in Ontario, Canada. *Climatic Change*, **20**, 1-21.

Brown, H.C.P. and V.G. Thomas, 1990: Ecological considerations for the future of food security in Africa. In: *Sustainable Agricultural Systems* [Edwards, C.A., R. Lal, P. Madden, R.H. Miller, and G. House (eds.)]. Soil and Water Conservation Society, Ankeny, IA.

Brunnert, H. and U. Dämmgen (eds.), 1994: *Klimaveränderungen und Landbewirtschaftung, Part II, Landbauforschung.* Völkenrode, Spec. vol. **148**, 398 pp.

Bumb, B., 1995: Growth potential of existing technology is insufficiently tapped. In: *Population and Food in the Early 21st Century: Meeting Future Food Demand of an Increasing World Population* [Islam, N. (ed.)]. Occasional Paper, International Food Policy Research Institute (IFPRI), Washingon DC, pp. 191-205.

Buol, S.W., P.A. Sanchez, J.M. Kimble, and S.B. Weed, 1990: Predicted impact of climate warming of soil properties and use. In: *Impact of CO_2, Trace Gases, and Climate Change on Global Agriculture* [Kimball, B.A., N.J. Rosenberg, and L.H. Allen, Jr. (eds.)]. Special Publication 53, American Society of Agronomy, Madison, WI, pp. 71-82.

Buringh, P. and R. Dudal, 1987: Agricultural land use in time and space. In: *Land Transformation in Agriculture* [Wolman, M. and F. Fourneir (eds.)]. John Wiley and Sons, Chichester, UK, pp. 9-24.

Burke, J.J., J.R. Mahan, and J.L. Hatfield, 1988: Crop-specific thermal kinetic windows in relation to wheat and cotton biomass production. *Agronomy Journal*, **80**, 553-556.

Butler, B.M., C. Matthew, R.G. Heerdegen, 1991: The "greenhouse effect"—what consequences for seasonality of pasture production. *Weather and Climate*, **11**, pp.168-170.

Cáceres, D.M., 1993. *Peasant Strategies and Models of Technological Change: A Case Study from Central Argentina.* M. Phil. Thesis, University of Manchester, Manchester, UK, 95 pp.

Cameron, D. and P. Oram, 1994: *Minimum and Reduced Tillage: Its Use in North America and Western Europe and Its Potential Application in Eastern Europe, Russia, and Central Asia.* International Food Policy Research Institute, Washington, DC, 121 pp.

Cammell, M.E. and J.D. Knight, 1991: Effects of climate change on the population dynamics of crop pests. *Advanced Ecological Research*, **22**, 117-162.

Campbell, B.D., 1994: *The Current Status of Climate Change Research in Relation to Pastoral Agriculture in New Zealand.* National Science Strategy Committee for Climate Change, Information Series 7, The Royal Society of New Zealand, Wellington, 30 pp.

Campbell, B.D., G.M. McKeon, R.M. Gifford, H. Clark, D.M. Stafford Smith, P.C.D. Newton, and J.L. Lutze, 1995. Impacts of atmospheric composition and climate change on temperate and tropical pastoral agriculture. In: *Greenhouse 94* [Pearman, G. and M. Manning (eds.)]. CSIRO, Canberra, Australia.

Carter, D.R. and K.M. Peterson, 1983: Effects of a CO_2 enriched atmosphere on the growth and competitive interaction of a C_3 and a C_4 grass. *Oecologia*, **58**, 188-193.

Carter, T.R., M.L. Parry, and J.R. Porter, 1991a: Climate change and future agroclimatic potential. *European International Journal of Climatology*, **11**, 251-269.

Carter, T.R., J.R. Porter, and M.L. Parry, 1991b: Climatic warming and crop potential in Europe: prospects and uncertainties. *Global Environmental Change*, **1**, 291-312.

Carter, T.R., J.H. Porter, and M.L. Parry, 1992: Some implications of climatic change for agriculture in Europe. *Journal of Experimental Botany*, **43**, 1159-1167.

CAST, 1992: *Preparing U.S. Agriculture for Global Climate Change.* Task Force Report No. 119, Council for Agricultural Science and Technology, Ames, IA, 96 pp.

Caswell, M.F., K.O Fuglie, and C.A. Klotz, 1994: *Agricultural Biotechnology: An Economic Perspective.* AER No. 687, U.S. Department of Agriculture, Washington, DC, 52 pp.

Cavallo, D., R. Domenech, and Y. Mundlak, 1992: *The Argentina That Could Have Been: The Costs of Economic Repression.* ICS Press, New York, NY, 192 pp.

Cavestany, D., A.B. El Wishy, and R.H. Foote, 1985: Effect of season and high environmental temperature on fertility of holstein cattle. *Journal of Dairy Science*, **68**, 1471-1478.

Chaudhuri, U.N., E.T. Kanemasu, and M.B. Kirkham, 1989: Effect of elevated levels of CO_2 on winter wheat under two moisture regimes. Report No.50, *Response of Vegetation to Carbon Dioxide*, U.S. Department of Energy, Washington, DC, 49 pp.

Chaudhuri, U.N., R.B. Burnett, M.B. Kirkham, and E.T. Kanemasu, 1986: Effect of carbon dioxide on sorghum yield, root growth, and water use. *Agricultural and Forest Meteorology*, **37**, 109-122.

Christiansen, E. and A. Bakke, 1986: The spruce bark beetle of Eurasia. In: *Dynamics of Forest Insect Populations: Patterns, Causes, Implications* [Berryman, A.A. (ed.)]. Plenum Press, New York, NY, pp. 479-503

Claus, R., and U. Weiler, 1987: Seasonal variations of fertility in the pig and its explanation through hormonal profiles. *Animal Breeding Abstract*, **55**, 963.

CIMMYT, 1991: *CIMMYT 1991 Annual Report: Improving the Productivity of Maize and Wheat in Developing Countries: An Assessment of Impact.* Centro Internacional de Mejoramiento de Maiz y Trigo, Mexico City, Mexico, 22 pp.

Cohen, S., E. Wheaton, and J. Masterton, 1992: *Impacts of Climatic Change Scenarios in the Prairie Provinces: A Case Study from Canada.* SRC Publication No. E-2900-4-D-92, Saskatchewan Research Council, Saskatoon, Canada, 157 pp.

Connor, D.J. and Y.P. Wang, 1993: Climatic change and the Australian wheat crop. In: *Proceedings of the Third Symposium on the Impact of Climatic Change on Agricultural Production in the Pacific Rim* [Geng, S. (ed.)]. Centre Weather Bureau, Ministry of Transport and Communications, Republic of China.

Coop, L.B., B.A. Croft, and R.J. Drapek, 1993: Model of corn earworm (Lepidoptera: Noctuidae) development, damage, and crop loss in sweet corn. *Journal of Economic Entomology*, **86(3)**, 906-916.

Cooter, E.J., 1990: The impact of climate change on continuous corn production in the southern U.S.A. *Climatic Change*, **16**, 53-82.

COSEPUP, 1992: *Policy Implications of Global Warming.* National Academy Press, Washington, DC, 127 pp.

Couvreur, F. and J. Tranchefort, 1993: Effects of climate change on winter wheat, winter barley and forage production in France. In: *The Effect of Climate Change on Agricultural and Horticultural Potential in Europe* [Kenny, G.J., P.A. Harrison, and M.L. Parry (eds.)]. Environmental Change Unit, University of Oxford, Oxford, UK, pp. 137-156.

Cramer, W.P. and A.M. Solomon, 1993: Climatic classification and future global redistribution of agricultural land. *Climate Research*, **3**, 97-110.

Crosson, P., 1989: Climate change and mid-latitudes agriculture: perspectives on consequences and policy responses. *Climatic Change*, **15**, 51-73.

Crosson, P., 1995: Future supplies of land and water for world agriculture. In: *Population and Food in the Early 21st Century: Meeting Future Food Demand of an Increasing World Population* [Islam, N. (ed.)]. Occasional Paper, International Food Policy Research Institute (IFPRI), Washingon DC, pp. 143-160.

Crosson, P. and A. Stout, 1983: *Productivity Effects of Cropland Erosion in the United States.* Resources for the Future, Washington, DC, 176 pp.

CSIRO, 1993: *Agriculture & Greenhouse in South-Eastern Australia.* CSIRO, Department of Conservation and Natural Resources, and Department of Agriculture, Melbourne, Australia, 5 pp.

Cure, J.D. and B. Acock, 1986: Crop responses to carbon dioxide doubling: a literature survey. *Agriculture and Forest Meteorology*, **38**, 127-145.

Cure, J.D., D.W. Israel, and T.W. Rufty, 1988: Nitrogen stress effects on growth and seed yield on nodulated soybean exposed to elevated carbon dioxide. *Crop Science*, **28**, 671-677.

Dahlman, R., 1993: CO_2 and plants, revisited. *Vegetatio*, **104/105**, 339-355.

Dalrymple, D.G., 1986: *Development and Spread of High-Yielding Rice Varieties in Developing Countries*, 7th ed. U.S. Agency for International Development, Washington, DC, 117 pp.

Darwin, R., M. Tsigas, J. Lewandrowski, and A. Raneses, 1995: *World Agriculture and Climate Change: Economic Adaptation*. Report No. AER-709, Economic Research Service, Washington, DC, 86 pp.

Davidson, R.H. and W.F. Lyon, 1987: *Insect Pests of Farm, Garden and Orchard*. John Wiley and Sons, New York, NY, 640 pp.

DeBach, P., 1965: Some biological and ecological phenomena associated with colonizing entomophagous insects. In: *The Genetics of Colonizing Species* [Baker, H.G. and G.L. Stebbins (eds.)]. Academic Press, New York, NY, pp. 287-303.

Decker, W.L., V.K. Jones, and R. Achutuni, 1986: *The Impact of Climate Change from Increased Atmospheric Carbon Dioxide on American Agriculture*. DOE/NBB-0077, U.S. Department of Energy, Carbon Dioxide Research Division, Washington, DC, 44 pp.

Delécolle, R., D. Ripoche, F. Ruget, and G. Gosse, 1994: Possible effects of increasing CO_2 concentration on wheat and maize crops in north and southeast France. In: *Implications of Climate Change for International Agriculture: Crop Modeling Study* [Rosenzweig, C. and A. Iglesias (eds.)]. U.S. Environmental Protection Agency, France chapter, Washington, DC, pp. 1-16.

Dickey, E.C., P.J. Jasa, B.J. Dolesh, L.A. Brown, and S.K. Rockwell, 1987: Conservation tillage: perceived and actual use. *Journal of Soil and Water Conservation*, **42(6)**, 431-434.

Diepen, C.A. van, C. Rappoldt, J. Wolf, and H. van Keulen, 1988: *Crop Growth Simulation Model WOFOST Documentation (Version 4.1)*. SOW-88-01, Centre for World Food Studies, Wageningen, Netherlands, 299 pp.

Dierckx, J., J.R. Gilley, J. Feyn, and C. Belmans, 1987: Simulation of soil water dynamics and corn yields under deficit irrigation. *Irrigation Science*, **30**, 120-125.

Dobzhansky, T., 1965: "Wild" and "domestic" species of Drosophila. In: *The Genetics of Colonizing Species* [Baker, H.G. and G.L. Stebbins (eds.)]. Academic Press, New York, NY, pp. 533-546.

Downing, T.E., 1992: *Climate Change and Vulnerable Places: Global Food Security and Country Studies in Zimbabwe, Kenya, Senegal, and Chile*. Research Report No. 1, Environmental Change Unit, University of Oxford, Oxford, UK, 54 pp.

Dregne, H., 1988: Desertification of drylands. In: *Challenges in Dryland Agriculture: A Global Perspective* [Unger, P., T. Sneed, W. Jordan, and R. Jensen (eds.)]. Texas Agricultural Experiment Station, Amarillo, TX.

Dregne, H. and N.-T. Chou, 1990: Global desertification dimensions and costs. In: *Degradation and Restoration of Arid Lands*. Texas Tech University, Lubbock, TX.

Du Preez, J.H., W.H. Gieseke, and P.J. Hattingh, 1990: Heat stress in dairy cattle and other livestock under southern African conditions: I. Temperature-humidity index mean values during four main seasons. *Onderstepoort Journal of Veterinary Research*, **57**, 77-87.

Duvick, D., 1995. Intensification of known technology and prospects of breakthroughs in technology and future food supply. In: *Population and Food in the Early 21st Century: Meeting Future Food Demand of an Increasing World Population* [Islam, N. (ed.)]. Occasional Paper, International Food Policy Research Institute (IFPRI), Washingon DC, pp. 221-228.

Easterling, W.E. III, P.R. Crosson, N.J. Rosenberg, M. McKenney, L.A. Katz, and K. Lemon, 1993: Agricultural impacts of and responses to climate change in the Missouri-Iowa-Nebraska-Kansas (MINK) region. *Climatic Change*, **24**, 23-61.

Easterling, W.E., N.J. Rosenberg, M.S. McKenney, C.A. Jones, P.T. Dyke, and J.R. Williams, 1992: Preparing the erosion productivity impact calculator (EPIC) model to simulate crop response to climate change and the direct effects of CO_2. *Agricultural and Forest Meteorology*, **59**, 17-34.

Eaton, C., 1988: What is hindering development of agriculture in the South Pacific? The producer's view. *Pacific Economic Bulletin*, **3(2)**, 31-36.

Edwards, J.H., C.W. Wood, D.L. Thurow, and M.E. Ruff, 1992: Tillage and crop rotation effects on fertility status of a hapludult soil. *Soil Science Society of America Journal*, **56**, 1577-1582.

Egbunike, G.N. and A.O. Elemo, 1978: Testicular and epidydimal reserves of crossbred European boars raised and maintained in the humid tropics. *Journal of Reproductive Fertility*, **54**, 245-248.

Eid, H.M., 1994: Impact of climate change on simulated wheat and maize yields in Egypt. In: *Implications of Climate Change for International Agriculture: Crop Modeling Study* [Rosenzweig, C. and A. Iglesias (eds.)]. U.S. Environmental Protection Agency, Egypt chapter, Washington, DC, pp. 1-14.

El-Swaify, S., E. Dangler, and C. Armstrong, 1982: *Soil Erosion by Water in the Tropics*. University of Hawaii, Honolulu, HI.

Elings, A., 1992: The use of crop growth simulation in evaluation of large germplasm collections. PhD thesis, Wageningen Agricultural University, Wageningen, Netherlands, 183 pp.

Elias, S.A., 1991: Insects and climate change: fossil evidence from the Rocky Mountains. *BioSci Am. Inst. Biol. Sci.*, **41(8)**, 552-559.

Ellis, R.H., P. Headley, E.H. Roberts, and R.J. Summerfield, 1990: Quantitative relations between temperature and crop development and growth. In: *Climate Change and Plant Genetic Resources* [Jackson, M. *et al.* (eds.)]. Belhaven, London, UK, pp. 85-115.

Ellis, R.H., R.J. Summerfield, G.O. Edmeades, and E.H. Roberts, 1992: Photoperiod, temperature, and the interval from sowing to tassel initiation in diverse cultivars of maize. *Crop Science*, **32**, 1225-1232.

Entomological Society of America, 1989: *Common Names of Insects and Selected Organisms*. Entomological Society of America, Lanham, MD, 199 pp.

Entwistle, K.W., 1974: Reproduction in sheep and cattle in the Australian arid zone. In: *Studies of the Australian Arid Zone* [Wilson, A.D. (ed.)]. CSIRO, Melbourne, Australia, pp. 85-97.

Escaño, C.R. and L.V. Buendia, 1994: Climate impact assessment for agriculture in the Philippines: simulation of rice yield under climate change scenarios. In: *Implications of Climate Change for International Agriculture: Crop Modeling Study* [Rosenzweig, C. and A. Iglesias (eds.)]. U.S. Environmental Protection Agency, Philippines chapter, Washington, DC, pp. 1-13.

Evans, L.T., I.F. Wardlaw, and R.A. Fischer, 1975: Wheat. In: *Crop Physiology* [Evans, L.T. (ed.)]. Cambridge University Press, London, UK, pp. 101-149.

FAO, 1989: *Guidelines for Designing and Evaluating Surface Irrigation Systems*. Irrigation and Drainage Papers, **45**, FAO, Rome, Italy, 137 pp.

FAO, 1991: *Water Harvesting*. AGL Miscellaneous Papers, **17**, FAO, Rome, Italy, 133 pp.

FAO, 1992: *AGROSTAT Digital Data*. FAO, Statistics Division, Rome, Italy.

FAO, 1995: *World Agriculture: Toward 2010, an FAO Study* [N. Alexandratos (ed.)]. John Wiley and Sons, Chichester, UK, and FAO, Rome, Italy, 488 pp.

Feister, U. and W. Warmbt, 1987: Long-term measurements of surface ozone in the German Democratic Republic. *Journal of Atmospheric Chemistry*, **5**, 1-21.

Fielding, D.J. and M.A. Brusven, 1990: Historical analysis of grasshopper (Orthoptera: Acrididae) population responses to climate in southern Idaho, 1950-1980. *Environmental Entomolology*, **19(6)**, 1786-1791.

Fischer, G., K. Frohberg, M.A. Keyzer, K.S. Parikh, and W. Tims, 1990: *Linked National Models: A Tool for International Food Policy Analysis*. Kluwer Academic Press, Dordrecht, Netherlands, 214 pp.

Fischer, G., K. Frohberg, M.L. Parry, and C. Rosenzweig, 1994: Climate change and world food supply, demand and trade. *Global Environmental Change*, **4(1)**, 7-23.

Fisher, R.A., 1950: Gene frequencies in a cline determined by selection and diffusion. *Biometrics*, **1(4)**, 353-360.

Flint, E.P. and D.T. Patterson, 1983: Interference and temperature effects on growth in soybean (Glycine max) and associated C_3 and C_4 weeds. *Weed Science*, **31**, 193-199.

Flint, E.P., D.T. Patterson, D.A. Mortensen, G.H. Riechers, and J.L. Beyers, 1984: Temperature effects on growth and leaf production in three weed species. *Weed Science*, **32**, 655-663.

Folland, C., J. Owen, M.N. Ward, and A. Coleman, 1991: Prediction of seasonal rainfall in the Sahel region using empirical and dynamical methods. *Journal of Forecasting*, **10**, 21-56.

Folland, C.K., J.R. Karl, and K.Y.A. Vinnikov, 1990: Observed climate variations and change. In: *Climate Change: The IPCC Scientific Assessment* [Houghton, J.T., G.J. Jenkins, and J.J. Ephrams (eds.)]. Cambridge University Press, Cambridge, UK, pp. 195-238.

Frederick, K.D., 1986: *Scarce Water and Institutional Change*. Resources for the Future, Washington, DC, 207 pp.

Friedrich, S., 1994: Wirkung veräderter klimatischer factoren auf pflanzenschaedlinge. In: *Klimaveraenderungen und Landwirtschaft, Part II, Lanbauforsc* [Brunnert, H. and U. Dämmgen (eds.)]. Vlkenrode, Spec. vol. **148**, pp. 17-26.

Fuhrer, J., A. Egger, B. Lehnherr, A. Grandjean, and W. Tschannen, 1989: Effects of ozone on the yield of spring wheat (triticum aestivum L., cv. Albis) grown in open-top chambers. *Environmental Pollution*, **60**, 273-289.

Furquay, J.W., 1989: Heat stress as it affects animal production. *Journal of Animal Science*, **52**, 164-174.

Fye, R.E. and W.C. McAda, 1972: *Laboratory Studies on the Development, Longevity, and Fecundity of Six Lepidopterous Pests of Cotton in Arizona*. Technical Bulletin No. 1454, Agricultural Research Service, U.S. Department of Agriculture, Washington, DC.

Geria, C., 1986: The pine sawfly of central France. In: *Dynamics of Forest Insect Populations: Patterns, Causes, Implications* [Berryman, A.A. (ed.)]. Plenum Press, New York, NY, pp. 377-405.

Gifford, R.M., 1979: Growth and yield of CO_2-enrichment wheat under water-limited conditions. *Australian Journal of Plant Physiology*, **6**, 367-378.

Gifford, R.M. and J.I.L. Morison, 1993: Crop responses to the global increase in atmospheric carbon dioxide concentration. In: *International Crop Science*, vol. I. Crop Science Society of America, Madison, WI, pp. 325-331.

Godden, D. and P. D. Adams, 1992: The enhanced greenhouse effect and Australian agriculture. In: *Economic Issues in Global Climate Change: Agriculture, Forestry, and Natural Resources* [Reilly, J.M. and M. Anderson (eds.)]. Westview Press, Boulder, CO, pp. 311-331.

Gommes, R., 1993: Current climate and population constraints on world agriculture. In: *Agricultural Dimensions of Global Climate Change* [Kaiser, H.M. and T.E. Drennen (eds.)]. St. Lucie Press, Delray Beach, FL, pp. 67-86.

Goodenough, J.L. and J.M. Mckinion (eds.), 1992: *Basics of Insect Modeling*. ASAE Monograph No. 10, American Society Agricultural Engineers, St. Joseph, MI, 221 pp.

Goudriaan, J. and H.E. De Ruiter, 1983: Plant growth in response to CO_2 enrichment at two levels of nitrogen and phosphorus supply: dry matter, leaf area and development. *Netherlands Journal of Agricultural Science*, **31**, 157-169.

Goudriaan, J. and M.H. Unsworth, 1990: Implication of increasing carbon dioxide and climate change for agricultural productivity and water resources. In: *Impact of Carbon Dioxide Trace Gases and Climate Change on Global Agriculture*. ASA Special Publication No. 53, American Society of Agronomy, Madison, WI, pp. 111-130.

Grégoire, J.C., 1986: The greater European spruce beetle. Madison. In: *Dynamics of Forest Insect Populations: Patterns, Causes, Implications* [Berryman, A.A. (ed.)]. Plenum Press, New York, NY, pp. 455-478.

Griliches, Z., 1957: Hybrid corn: an exploration in the economics of technological change. *Econometrica*, **25**, 501-522.

Hackett, C., 1990: Plant ecophysiological information for contingency thinking in the southwest Pacific in face of the greenhouse phenomenon. In: *Implications of Expected Climate Changes in the South Pacific Region: An Overview* [Pernetta, J.C. and P.J. Hughes (eds.)]. UNEP Regional Seas Report and Studies No. 128, UNEP.

Hahn, G.L., 1985: Management and housing of farm animals in hot environments. In: *Stress Physiology on Livestock Ungulates* [Yousef, M.K. (ed.)]. CRC Press, **2**, Boca Raton, FL, pp. 151-174.

Hahn, G.L., P.L. Klinedinst, and D.A. Wilhite, 1990: Climate change impacts on livestock production and management. Paper presented at the American Meteorological Society Annual Meeting, 4-9 February, Anaheim, CA.

Hamilton, J.G. and S.H. Gage, 1986: Outbreaks of the cotton tipworm, Crocidosema plebejana (Lepidoptera: Tortricidae), related to weather in southeast Queensland, Australia. *Environmental Entomolology*, **15**, 1078-1082.

Hardakker, J.B., 1988: Smallholder agriculture in the South Pacific: a few furphies and some home truths. *Pacific Economic Bulletin*, **3(2)**, 18-22.

Harrington, R. and N.E. Stork (eds.), 1995: *Insects in a Changing Environment*. London, Academic Press, 535 pp.

Haverkort, A.J., 1990: Ecology of potato cropping systems in relation to latitude and altitude. *Agricultural Systems*, **32**, 251-272.

Harvey, L.D.D., 1993: Comments on "an empirical study of the economic effects of climate change on world agriculture." *Climatic Change*, **21**, 273-275.

Hatzios, K.K. and D. Penner, 1982: *Metabolism of Herbicides in Higher Plants*. CEPCO iv., Burgess Publ., Edina, MN.

Haukioja, E., S. Neuvonen, S. Hanhimäki, and P. Niemelä, 1986: The autumnal moth in Fennoscandia. In: *Dynamics of Forest Insect Populations: Patterns, Causes, Implications* [Berryman, A.A. (ed.)]. Plenum Press, New York, NY, pp. 163-178.

Havelka,U.D., R.C. Ackerson, M.G. Boyle, and V.A. Wittenbach, 1984a: CO_2-enrichment effects on soybean physiology: I. Effects of long-term CO_2 exposure. *Crop Science*, **24**, 1146-1150.

Havelka,U.D., R.C. Ackerson, M.G. Boyle, and V.A. Wittenbach, 1984b. CO_2-enrichment effects on wheat yield and physiology. *Crop Science*, **24**, 1163-1168.

Haverkort, A.J., 1990: Ecology of potato cropping systems in relation to latitude and altitude. *Agricultural Systems*, **32**, 251-272.

Hay, J.E. and C. Kaluwin (eds.), 1992: *Climate Change and Sea Level Rise in the South Pacific Region*. Proceedings of the Second SPREP Meeting, Apia, Western Samoa, SPREP.

Hayami, Y. and V.W. Ruttan, 1985: *Agricultural Development: An International Perspective*. The Johns Hopkins University Press, Baltimore, MD, 506 pp.

Hennessy, K.J. and K. Clayton-Greene, 1995: Greenhouse warming and vernalisation of high-chill fruit in southern Australia. *Climatic Change*, **30(3)**, 327-348.

Hill, P.R., D.R. Griffith, G.C. Steinhardt, and S.D. Parsons, 1994: *The Evolution and History of No-Till Farming in the Midwest*. T-by-2000 Erosion Reduction/Water Quality Program, Purdue University, West Lafayette, IN, 45 pp.

Hobbs, J., J.R. Anderson, J.L. Dillon, and H. Harris, 1988: The effects of climatic variations on agriculture in the Australian wheat belt. In: *The Impact of Climatic Variations on Agriculture*. Vol. 2, *Assessments in Semi-Arid Regions* [Parry, M.L., T.R. Carter, and N.T. Konijn (eds.)]. Kluwer Academic Press, Dordrecht, Netherlands, pp. 665-753.

Hodges, T. and J.T. Ritchie, 1991: The CERES-Wheat phenology model. In: *Predicting Crop Phenology* [Hodges, T. (ed.)]. CRC Press, Boca Raton, FL, p. 133-141.

Hofstra, G. and J.D. Hesketh, 1969: Effects of temperature on the gas exchange of leaves in the light and dark. *Planta*, **85**, 228-237.

Hogg, D.B. and M. Calderon C., 1981: Developmental times of Heliothis zea and H.virescens (Lepidoptera: Noctuidae) larvae and pupae in cotton. *Environmental Entomology*, **10**, 177-179.

Hogg, D.B. and A.P. Gutierrez, 1982: A model of the flight phenology of the beet armyworm (Lepidoptera: Noctuidae) in central California. *Hilgardia*, **48(4)**, 1-36.

Holm, L.G., D.L. Plucknett, J.V. Pancho, and J.P. Herberger, 1977: *The World's Worst Weeds: Distribution and Biology*. University of Hawaii Press, Honolulu, HI.

Horie, T., 1987: The effect of climatic variation and elevated CO_2 on rice yield in Hokkaido. In: *The Impacts of Climatic Variations on Agriculture*.Vol. 1, *Cool Temperate and Cold Regions* [Parry, M.L., T.R. Carter, and N.T. Konijn (eds.)]. Kluwer Academic Press, Dordrecht, Netherlands, pp. 809-825.

Horie, T., 1991: Growth and yield of rice and climate change. *Agr. Horticul.*, **66**, 109-116.

Horie, T., 1993: Predicting the effects of climatic variation and effect of CO_2 on rice yield in Japan. *Journal of Agricultural Meteorology* (Tokyo), **48**, 567-574.

Howe, C., 1971: *Benefit Cost Analysis for Water System Planning: Water Resources Monograph 2*. American Geophysical Union, Washington, DC, 75 pp.

Hughes, P.J. and G. McGregor (eds.), 1990: *Global Warming-Related Effects on Agriculture and Human Health and Comfort in the South Pacific*. A report to the South Pacific Regional Environment Programme and the United Nations Environment Programme. The Association of South Pacific Environmental Institutions, University of Papua New Guinea, Port Moresby, Papua New Guinea.

Hulme, M., T.Wigley, T. Jiang, Z. Zhao, F. Wang, Y. Ding, R. Leemans, and A. Markham, 1992: *Climate Change Due to the Greenhouse Effect and Its Implications for China*. CRU/WWF/SMA, World Wide Fund for Nature, Gland, Switzerland.

IBSNAT, 1989: *Decision Support System for Agrotechnology Transfer Version 2.1 (DSSAT V2.1)*. Department of Agronomy and Soil Science, College of Tropical Agriculture and Human Resources, University of Hawaii, Honolulu, HI.

Idso, S.B., B.A. Kimball, M.G. Anderson, and J.R. Mauney, 1987: Effects of atmospheric CO_2 enrichment on plant growth: the interactive role of air temperature. *Agricultural Ecosystems and the Environment*, **20**, 1-10.

Idso, S.B., B.A. Kimball, and J.R. Mauney, 1987: Atmospheric carbon dioxide enrichment effects on cotton midday foliage temperature: implications for plant water use and crop yield. *Agronomy Journal*, **79**, 667-672.

Iglesias, A. and M.I. Minguez, in press: Perspectives for maize production in Spain under climate change. In: *Climate Change and Agriculture* [Harper, L., S. Hollinger, J. Jones, and C. Rosenzweig (eds.)]. American Society of Agronomy, Madison, WI.

Isaev, A.S., Y.N. Baranchikov, and V.S. Malutina, 1986: The larch gall midge in seed orchards of south Siberia. In: *Dynamics of Forest Insect Populations: Patterns, Causes, Implications* [Berryman, A.A. (ed.)]. Plenum Press, New York, NY, pp. 29-44.

James, L.D. and R.R. Lee, 1971: *Economics of Water Resources Planning*. McGraw Hill, New York, NY, 272 pp.

Jeffers, D.L. and R.M. Shibles, 1969: Some effects of leaf area, solar radiation, air temperature, and variety on net photosynthesis in field-grown soybeans. *Crop Science*. **9(6)**, 762-764.

Jensen, M., W. Rangeley, and P. Dieleman, 1990: Irrigation trends in world agriculture. In: *Irrigation of Agricultural Crops*. Agronomy Monograph No. 30, ASA-CSSA-SSA, Madison, WI.

Jin, Z., 1993: Impacts of climate change on rice production and strategies for adaptation in the southern China. In: *Climate Change, Natural Disasters and Agricultural Strategies*. China Meteorological Press, Beijing, China, pp. 149-157.

Jin, Z., H.C. Daokou Ge, and J. Fang, 1994: Effects of climate change on rice production and strategies for adaptation in southern China. In: *Implications of Climate Change for International Agriculture: Crop Modeling Study* [Rosenzweig, C. and A. Iglesias (eds.)]. U.S. Environmental Protection Agency, China chapter, Washington, DC, pp. 1-24.

Johnston, J.E., 1958: The effects of high temperature on milk production. *Journal of Heredity*, **49**, 65-68.

Joyce, R.J.V., 1983: Aerial transport of pests and pest outbreaks. *EPPO Bulletin*, **13(2)**, 111-119.

Kahn, H.E., 1991. The effect of summer decline in conception rates on the monthly milk production pattern in Israel. *Animal Production*, **53**, 127-131.

Kaiser, H., S. Riha, D. Wilkes, and R. Sampath, 1993: Adaptation to global climate change at the farm level. In: *Agricultural Dimensions of Global Climate Change* [Kaiser, H.M. and T.E. Drennen (eds.)]. St. Lucie Press, Delray Beach, FL, pp. 136-152.

Kane, S. and J. Reilly, 1993: Reply to comment by L.D. Danny Harvey on "an empirical study of the economic effects of climate change on world agriculture." *Climatic Change*, **21**, 277-279.

Kane, S., J. Reilly, and J. Tobey, 1992: An empirical study of the economic effects of climate change on world agriculture. *Climatic Change*, **21(1)**, 17-35.

Karim, Z., M. Ahmed, S.G. Hussain, and Kh.B. Rashid, 1994: Impact of climate change on production of modern rice in Bangladesh. In: *Implications of Climate Change for International Agriculture: Crop Modeling Study* [Rosenzweig, C. and A. Iglesias (eds.)]. U.S. Environmental Protection Agency, Bangladesh chapter, Washington, DC, pp. 1-11.

Kates, R. and R. Chen, 1994a: Climate change and food security. *Global Environmental Change*, **4(1)**, 37-48.

Kates, R. and R. Chen, 1994b: World food security: prospects and trends. *Food Policy*, **19(2)**, 192-208.

Katz, R.W. and R.G. Brown, 1992: Extreme events in a changing climate: variability is more important than averages. *Climatic Change*, **21**, 289-302.

Kenny, G.J., P.A. Harrison, J.E. Olesen, and M.L. Parry, 1993: The effects of climate change on land suitability of grain maize, winter wheat and cauliflower in Europe. *European Journal of Agronomy*, **2**, 325-338.

Kenny, G.J. and P.A. Harrison, 1993: Analysis of effects of climate change on broadscale patterns of agroclimate in Europe. In: *The Effect of Climate Change on Agricultural and Horticultural Potential in Europe* [Kenny, G.J., P.A. Harrison, and M.L. Parry (eds.)]. Environmental Change Unit, University of Oxford, Oxford, UK, pp. 201-224.

Kenny, G.J., P.A. Harrison, and M.L. Parry (eds.), 1993: *The Effect of Climate Change on Agricultural and Horticultural Potential in Europe*. Environmental Change Unit, University of Oxford, Oxford, UK, 224 pp.

Kenny, G.J. and P.A. Harrison, 1992: Thermal and moisture limits of grain maize in Europe: model testing and sensitivity to climate change. *Climate Research*, **2**, 113-129.

Kenny, G.J. and P.A. Harrison, 1992a: The effects of climate variability and change on grape suitability in Europe. *Journal of Wine Research*, **3**, 163-183.

Kenny, G.J. and J. Shao, 1992b: An assessment of latitude-temperature index for predicting climate suitability of grapes in Europe. *Journal of Horticultural Science*, **67**, 239-246.

Kettunen, L., J. Mukula, V. Pohjonen, O. Rantanen, and U. Varjo, 1988: The effects of climatic variation on agriculture in Finland. In: *The Impact of Climatic Variations on Agriculture*. Vol. 1, *Cool Temperate and Cold Regions* [Parry, M.L., T.R. Carter, and N.T. Konijn (eds.)]. Kluwer Academic Press, Dordrecht, Netherlands, pp. 511-614.

Kimball, B.A., 1983: Carbon dioxide and agricultural yield: an assemblage and analysis of 430 prior observations. *Agronomy Journal*, **75**, 779-788.

Kimball, B.A., J.R. Mauney, F.S. Nakayama, and S.B. Idso, 1993: Effects of elevated CO_2 and climate variables on plants. *Journal of Soil and Water Conservation*, **48**, 9-14.

Kimball, B.A., N.J. Rosenberg, and L.H. Allen, Jr. (eds.), 1990: *Impact of Carbon Dioxide, Trace Gases, and Climate Change on Agriculture*. ASA Special Publication No. 53, American Society of Agronomy, Crop Science Society of America, and Soil Science Society of America, Madison, WI, 133 pp.

Kingsolver, J.G., 1989: Weather and the population dynamics of insects: integrating physiological and population ecology. *Physiol. Ecol.*, **62(2)**, 314-334.

Kiniry, J.R. and R. Bonhomme, 1991: Predicting maize phenology. In: *Predicting Crop Phenology* [Hodges, T. (ed.)]. CRC Press, Boca Raton, FL, pp. 115-131.

Klinedinst, P.L., D.A. Wilhite, G.L. Hahn, and K.G. Hubbard, 1993: The potential effects of climate change on summer season dairy cattle milk production and reproduction. *Climatic Change*, **23(1)**, 21-36.

Knudson, M., 1988: *The Research and Development of Competing Biological Innovations: The Case of Semi- and Hybrid Wheats*. PhD diss., University of Minnesota, St. Paul, MN, 195 pp.

Kobahashi, F., 1986: The Japanese pine sawyer. In: *Dynamics of Forest Insect Populations: Patterns, Causes, Implications* [Berryman, A.A. (ed.)]. Plenum Press, New York, NY, pp. 431-454.

Koerner, C., 1993: CO_2 fertilization: the great uncertainty in future vegetation development. In: *Vegetation Dynamics and Global Change* [Solomon, A.M. and H.H. Shugart (eds.)]. Chapman and Hall, New York, NY, pp. 53-70.

Koerner, C. and F. Miglietta, in press: Long-term effects of naturally elevated CO_2 on Mediterranean grassland and forest trees. *Oecologia*.

Korte, C.J., P.D. Newton, D.G. McCall, 1991: Effects of climate change on pasture growth using a mechanistic simulation model. *Weather and Climate*, **11**, 171-172.

Kovda, V.A. and Ya. A. Pachepsky, 1989: *Soil Resources of the USSR, Their Usage and Recovering: The Report of the VII All-Union Congress of Soil Scientists*. Academy of Sciences of the USSR, Scientific Center of Biological Research, Puschino, USSR, 36 pp.

Kraalingen, D.W.G. van, 1990: Effects of CO_2 enrichment on nutrient-deficient plants. In: *The Greenhouse Effect and Primary Productivity in European Agro-Ecosystems* [Goudriaan, J., H. van Keuten, and H.H. van Laar (eds.)]. Pudoc, Wageningen, Netherlands, pp. 42-54.

Krenzer, E.G. and D.N. Moss, 1975: Carbon dioxide enrichment effects upon yield and yield components in wheat. *Crop Science*, **15**, 71-74.

Kress, L.W., J.E. Miller, H.J. Smith, 1985: Impact of ozone on winter wheat yield. *Environ. Exp. Bot.*, **25**, 211-228.

Krupa, S.V. and R.N. Kickert, 1989: The greenhouse effect: impacts of ultraviolet-B, radiation, carbon dioxide, and ozone on vegetation. *Environmental Pollution*, **61**, 263-293.

Krupa, S.V. and R.N. Kickert, 1993: The greenhouse effect: the impacts of carbon dioxide (CO_2), ultraviolet-B (UV-B) radiation and ozone (O_3) on vegetation (crops). In: *CO_2 and Biosphere* [Rozema, J., H. Lambers, S.C. van de Geijn, and M.L. Cambridge (eds.)]. Kluwer Academic Publishers, Dordrecht, Netherlands, pp. 223-238.

Kuiper, P.J.C., 1993: Diverse influences of small temperature increases on crop performance. In: *International Crop Science*, vol. I. Crop Science Society of America, Madison, WI, pp. 309-313.

Kukla, G. and T.R. Karl, 1993: Nighttime warming and the greenhouse effect. *Envir. Sci. Technol.*, **27(8)**, 1468-1474.

Lal, R. and B. Okigbo, 1990: *Assessment of Soil Degradation in the Southern States of Nigeria*. Environment Department Working Paper No. 39, The World Bank, Washington, DC.

Langdale, G.W., L.T. West, and R.R. Bruce, 1992: Restoration of eroded soil with conservation tillage. *Soil Technology*, **5**, 81-90.

Larcher, W., 1980: *Physiological Plant Ecology*, 2nd ed. Springer-Verlag, Berlin, Germany.

Lawlor, D.W. and A.C. Mitchell, 1991: The effects of increasing CO_2 on crop photosynthesis and productivity: a review of field studies. *Plant, Cell and Environment*, **14**, 807-818.

Lawlor, D.W., A.C. Mitchell, J. Franklin, V.J. Mitchell, S.P. Driscoll, and E. Delgado, 1993: Facility for studying the effects of elevated carbon dioxide concentration and increased temperature on crops. *Plant, Cell and Environment*, **16**, 603-608.

Leemans, R., 1992: Modelling ecological and agricultural impacts of global change on a global scale. *Journal of Scientific & Industrial Research*, **51**, 709-724.

Leemans, R. and A.M. Solomon, 1993: Modeling the potential change in yield and distribution of the earth's crops under a warmed climate. *Climate Research*, **3**, 79-96.

Le Houereou, H.N., 1990: Global change: vegetation, ecosystems and land use in the Mediterranean basin by the twenty-first century. *Israel Journal of Botany*, **39**, 481-508.

Le Houerou, H.N., G.F. Popov, and L. See, 1993: *Agro-Bioclimatic Classification of Africa*. Agrometeorology Series Working Paper, 6, FAO, Research and Technology Development Division, Agrometeorology Group, Rome, Italy, 227 pp.

Lerner, I.M., 1958: *The Genetic Basis of Selection*. John Wiley and Sons, New York, NY, 298 pp.

Leuning, R., Y.P. Yang, D. De Pury, O.T. Denmead, F.X. Dunin, A.G. Condon, S. Nonhebel, and J. Goudriaan, 1993: Growth and water use of wheat under present and future levels of CO_2. *Journal of Agricultural Meteorology*, **48(5)**, 807-810.

Lewandrowski, J.K. and R.J. Brazee, 1993: Farm programs and climate change. *Climatic Change*, **23**, 1-20.

Lin, E., 1994: The sensitivity and vulnerability of China's agriculture to global warming. *Rural Eco-Environment*, **10(1)**, 1-5.

Lincoln, D.E., 1993: The influence of plant carbon dioxide and nutrient supply on susceptibility to insect herbivores. In: *CO_2 and Biosphere* [Rozema, J., H. Lambers, S.C. van de Geijn, and M.L. Cambridge (eds.)]. Kluwer Academic Publishers, Dordrecht, Netherlands, pp. 273-280.

Lingren, P.D., V.M. Bryant, Jr., J.R. Raulston, M. Pendleton, J. Westbrook, and G.D. Jones, 1993: Adult feeding host range and migratory activities of corn earworm, cabbage looper, and celery looper (Lepidoptera: Noctuidae) moths as evidenced by attached pollen. *Journal of Economic Entomology*, **86**, 1429-1439.

Liverman, D., M. Dilley, K. O'Brien, and L. Menchaca, 1994: Possible impacts of climate change on maize yields in Mexico. In: *Implications of Climate Change for International Agriculture: Crop Modeling Study* [Rosenzweig, C. and A. Iglesias (eds.)]. U.S. Environmental Protection Agency, Mexico chapter, Washington, DC, pp. 1-14.

Liverman, D. and K. O'Brien, 1991: Global warming and climate change in Mexico. *Global Environmental Change,* **1(4)**, 351-364.

Logan, T.J., 1991: Tillage systems and soil properties in North America. *Tillage Research* **20**, 241-270.

Long, G.E., 1986: The larch casebearer in the intermountain northwest. In: *Dynamics of Forest Insect Populations: Patterns, Causes, Implications* [Berryman, A.A. (ed.)]. Plenum Press, New York, NY, pp. 233-242.

Long, S.P., T.M. East, and N.R. Baker, 1983: Chilling damage to photosynthesis in young Zea mays. I. Effects of light and temperature variation on photosynthetic CO_2 assimilation. *Journal of Experimental Botany*, **34(139)**, 177-188.

Löpmeier, F.J., 1990: Klimaimpaktforschung aus agrarmeteorologischer. *Sicht. Bayer. Landw. Jarhb.*, **67(1)**, 185-190.

Lu Liangshu and Liu Zhicheng, 1991: *Studies on the Medium and Long-Term Strategy of Food Development in China*. Agricultural Publishing House, Beijing, China, pp. 9-37.

Lu Liangshu and Liu Zhicheng, 1991: *Productive Structure and Developmental Prospects of Planting Industry*. Agricultural Publishing House, Beijing, China, pp. 21-27.

Madden, J.L., 1986: Sirex in Australasia. In: *Dynamics of Forest Insect Populations: Patterns, Causes, Implications* [Berryman, A.A. (ed.)]. Plenum Press, New York, NY, pp. 407-429.

Magadza, C.H.D., 1994. Climate change: some likely multiple impacts in southern Africa. *Food Policy*, **19(2)**, 165-191.

Martin, R.J., C.J. Korte, D.G. McCall, D.B. Baird, P.C.D. Newton, N.D. Barlow, 1991: Impact of potential change in climate and atmospheric concentration of carbon dioxide on pasture and animal production in New Zealand. *Proceedings of the New Zealand Society of Animal Production*, **51**, 25-33.

Martin, R.J., D.B. Baird, M.J. Salinger, P.R. van Gardingen, and D.G. McCall, 1993: Modelling the effects of climate variability and climate change on a pastoral farming system. In: *Proceedings XVII International Grassland Congress*. pp. 1070-1072.

Matthews, R.B., M.J. Kropff, D. Bachelet, 1994a: Climate change and rice production in Asia. *Entwicklung und Ländlicherraum*, **1**, 16-19.

Matthews, R.B., M.J. Kropff, D. Bachelet, and H.H. van Laar, 1994b: The impact of global climate change on rice production in Asia: a simulation study. Report No. ERL-COR-821, U.S. Environmental Protection Agency, Environmental Research Laboratory, Corvallis, OR.

Mattson, W.J. and R.A. Haack, 1987: The role of drought in outbreaks of plant-eating insects. *BioScience*, **37(2)**, 110-118.

Mauney, J.R., K.F. Lewin, G.R. Hendrey, and B.A. Kimball, 1992: Growth and yield of cotton exposed to free-air CO_2 enrichment (FACE). *Critical Review Plant Science*, **11**, 213-222.

McCalla, A.F., 1994: Agriculture and food needs to 2025: Why we should be concerned. Consultative Group on International Agricultural Research, Sir John Crawford Memorial Lecture, 27 October, Washington, DC.

McKenney, M.S., W.E. Easterling, and N.J. Rosenberg, 1992: Simulation of crop productivity and responses to climate change in the year 2030: The role of future technologies, adjustments and adaptations. *Agricultural and Forest Meteorology*, **59**, 103-127.

McKeon, G.M., K.A. Day, S.M. Howden, J.J. Mott, D.M. Orr, W.J. Scattini, and E.J. Weston, 1990: Management for pastoral production in northern Australian savannas. *Journal of Biogeography*, **17**, 355-372.

McKeon, G.M., S.M. Howden, N.O.J. Abel, and J.M. King, 1993: Climate change: adapting tropical and subtropical grasslands. In: *Proceedings XVII International Grassland Congress*, CSIRO, Melbourne, pp. 1181-1190.

McKeon, G.M. and D.H. White, 1992: El Nino and better land management. *Search*, **23**, 197-200.

McGregor, J., 1994. Climate change and involuntary migration. *Food Policy*, **19(2)**, 121-132.

Mearns, L., C. Rosenzweig, and R. Goldberg, 1992a: Effect of changes in interannual climatic variability on CERES-wheat yields: sensitivity and 2 x CO_2 general circulation model studies. *Agricultural and Forest Meteorology*, **62**, 159-189.

Mearns, L.O., C. Rosenzweig, and R. Goldberg, 1992b: *Sensitivity Analysis of the CERES-Wheat Model to Changes in Interannual Variability of Climate*. U.S. Environmental Protection Agency, Washington, DC.

Medvedev, Z.A., 1987: *Soviet Agriculture*. W.W. Norton and Co., New York, NY, 464 pp.

Meier, W., 1985: *Pflanzenschutz im Feldbau*. Tierische Schädlinge und Pflanzenkrankheiten, 8 Auflage, Zurich-Reckenholz, Eidgenössiche Forschungsanstalt fur Landwirtschaftlichen Pflanzenbau.

Mendelsohn, R., W.D. Nordhaus, and D. Shaw, 1994: The impact of global warming on agriculture: a ricardian analysis. *American Economic Review*, **84(4)**, 753-771.

Menzhulin, G.V., 1992: The impact of expected climate changes on crop yields: estimates for Europe, the USSR, and North America based on paleoanalogue scenarios. In: *Economic Issues in Global Climate Change: Agriculture, Forestry, and Natural Resources* [Reilly, J.M. and M. Anderson (eds.)]. Westview Press, Boulder, CO, pp. 353-381.

Menzhulin, G.V. and L.A. Koval, 1994: Potential effects of global warming and carbon dioxide on wheat production in the Former Soviet Union. In: *Implications of Climate Change for International Agriculture: Crop Modeling Study* [Rosenzweig, C. and A. Iglesias (eds.)]. U.S. Environmental Protection Agency, FSU chapter, Washington, DC, pp. 1-35.

Miglieta, F., A. Raschi, R. Resti, and G. Zipoli, 1991: Dry matter production, yield and photosynthesis of soybean grown in natural CO_2 enriched environment. *Agronomy Abstracts*, pp. 22.

Miglieta, F., 1991: Simulation of wheat ontogenesis: I. The appearance of main stem leaves in the field; II. Predicting final main stem leaf number and dates of heading in the field. *Climate Research*, **1**, 145-150, 151-160.

Miglieta, F., 1992: Simulation of wheat ontogenesis: III. Effect of variety, nitrogen fertilization and water stress on leaf appearance and final leaf number in the field. *Climate Research*, **2**, 233-242.

Miglieta, F. and J.R. Porter, 1992: The effects of CO_2 induced climatic change on development of wheat: Analysis and modelling. *Journal of Experimental Botany*, **43**, 1147-1158.

Mínguez, M.I. and A. Iglesias, 1994: Perspectives of future crop water requirements in Spain: the case of maize as a reference crop. In: *Diachronic Climatic Changes: Impacts on Water Resources* [Angelakis, A. (ed.)]. Springer-Verlag, New York, NY.

Ministry for the Environment, New Zealand, 1990: Climatic change: impacts on New Zealand. In: *Implications for the Environment, Economy and Society.* Ministry for the Environment, Wellington, New Zealand.

Mitchell, D.O. and M. Ingco, 1995: Global and regional food demand and supply prospects. In: *Population and Food in the Early 21st Century: Meeting Future Food Demand of an Increasing World Population* [Islam, N. (ed.)]. Occasional Paper, International Food Policy Research Institute (IFPRI), Washingon DC, pp. 49-60.

Mitchell, R.A.C., V.J. Mitchell, S.P. Driscoll, J. Franklin, and D.W. Lawlor, 1993: Effects of increased CO_2 concentration and temperature on growth and yield of winter wheat at two levels of nitrogen application. *Plant, Cell and Environment*, **16**, 521-529.

Mochida, O., 1991: Impact of CO_2-climate change on pests distribution. *Agr. Horticul.*, **66**, 128-136.

Montgomery, M.E. and W.E. Wallner, 1986: The gypsy moth: a westward migrant. In: *Dynamics of Forest Insect Populations: Patterns, Causes, Implications* [Berryman, A.A. (ed.)]. Plenum Press, New York, NY, pp. 353-375.

Mooney, H.A., P.M. Patusek, and P.A. Matson, 1987: Exchange of materials between terrestrial ecosystems and the atmosphere. *Science*, **238**, 926-932.

Mooney, H.A. and W. Koch, 1994: The impact of rising CO_2 concentrations on the terrestrial biosphere. *Ambio*, **23(1)**, 74-76.

Moore, M. R., 1991: The bureau of reclamation's new mandate for irrigation water conservation: purposes and policy alternatives. *Water Resources Research*, **27(2)**, 145-155.

Morison, J.I.L., 1993: Response of plants to CO_2 under water limited conditions. *Vegetatio*, **104/105**, 193-209.

Muchena, P., 1994: Implications of climate change for maize yields in Zimbabwe. In: *Implications of Climate Change for International Agriculture: Crop Modeling Study* [Rosenzweig, C. and A. Iglesias (eds.)]. U.S. Environmental Protection Agency, Zimbabwe chapter, Washington, DC, pp. 1-9.

Muchow, R.C. and T.R. Sinclair, 1991: Water deficit effects on maize yields modeled under current and "greenhouse" climates. *Agronomy Journal*, **83**, 1052-1059.

Muzik, T.J., 1976: Influence of environmental factors on toxicity to plants. In: *Herbicides: Physiology, Biochemistry, Ecology*, 2nd ed., vol. 2 [Audus, L.J. (ed.)]. Academic Press, New York, NY, pp. 204-247.

Nair, K.S.S., 1986: The teak defoliator in Kerala, India. In: *Dynamics of Forest Insect Populations: Patterns, Causes, Implications* [Berryman, A.A. (ed.)]. Plenum Press, New York, NY, pp. 267-289.

Nelson, R., 1988: *Dryland Management: The Land Degradation Problem.* Environment Department Working Paper No. 8, World Bank, Washington, DC, 28 pp.

Nicholls, N. and K.K. Wong, 1990: Dependence of rainfall variability on mean rainfall, latitude and the Southern Oscillation. *Journal of Climatology*, **3**, pp. 163-170.

Norse, D., 1994. Multiple threats to regional food production: environment, economy, population? *Food Policy*, **19(2)**, 133-148.

Nulsen, R.A., 1989: Agriculture in southwestern Australia in a greenhouse climate. In: *Proceedings of the Fifth Agronomy Conference, Australian Society of Agronomy.* Australian Society of Agronomy, Parkville, Australia, pp. 304-311.

Odum, E.P., 1971: *Fundamentals of Ecology*. W.B. Saunders Co., Philadelphia, PA, 574 pp.

Oechel, W.C. and B.R. Strain, 1985: Native species responses to increased atmospheric carbon dioxide concentration. In: *Direct Effects of Increasing Carbon Dioxide on Vegetation* [Strain, B.R. and J.D. Cure (eds.)]. DOE/ER-0238, U.S. Department of Energy, Washington, DC, pp. 117-154.

Okigbo, B.N., 1992: Conservation and use of plant germ plasm in African traditional agriculture and land-use systems. Keynote Address, CTA/IB PG/KARL Seminar on Safeguarding the Genetic Basis of Africa's Traditional Crops, 5-9 Oct, Nairobi, Kenya.

Oldeman, L.R., R.T.A. Hakkelling, and W.G. Sombroek, 1990: *World Map of the Status of Human-Induced Soil Degradation (GLASOD) with Explanatory Note.* International Soil Reference and Information Center, Wageningen, Netherlands, and United Nations Environmental Program, Nairobi, Kenya, 27 pp.

Olesen, J.E., F. Friis, and K. Grevsen, 1993: Simulated effects of climate change on vegetable crop production in Europe. In: *The Effect of Climate Change on Agricultural and Horticultural Potential in Europe* [Kenny, G.J., P.A. Harrison, and M.L. Parry (eds.)]. Environmental Change Unit, University of Oxford, Oxford, UK, pp. 177-200.

Olesen, J.E. and K. Grevsen, 1993: Simulated effects of climate change on summer cauliflower production in Europe. *European Journal of Agronomy*, **2**, 313-323.

Ominde, S.H. and C. Juma, 1991: *A Change in the Weather: African Perspectives on Climatic Change.* ACTS Press, Nairobi, Kenya.

Oram, P.A., and B. Hojjati, 1995: The growth potential of existing agricultural technology. In: *Population and Food in the Early 21st Century: Meeting Future Food Demand of an Increasing World Population* [Islam, N. (ed.)]. Occasional Paper, International Food Policy Research Institute (IFPRI), Washingon DC, pp. 167-189.

Orr, W.N., R.T. Cowan, and T.M. Davison, in press: Factors affecting pregnancy rate in Holstein-Friesian cattle mated during summer in a tropical upland environment. *Australian Veterinary Journal.*

Ottichilo, W.K., J.H. Kinuthia, P.O Ratego, and G. Nasubo, 1991: *Weathering the Storm: Climate Change and Investment in Kenya.* ACTS Press, Nairobi, Kenya.

Paltridge, G. (ed.), 1989: *Climate Impact Response Functions—Report of Workshop Held at Coolfont, WV.* National Climate Program Office, Washington, DC, 52 pp.

Parker, C. and J.D. Fryer, 1975: Weed control problems causing major reductions in world food supplies. *FAO Plant Protection Bulletin*, **23**, 83-95.

Parry, M.L., 1990: *Climate Change and World Agriculture*. Earthscan Publications ltd., London, UK, 157 pp.

Parry, M.L., M. Blantran de Rozari, A.L. Chong, and S. Panich (eds.), 1992: *The Potential Socio-Economic Effects of Climate Change in South-East Asia.* United Nations Environment Programme, Nairobi, Kenya.

Parry, M.L., T.R. Carter, and N.T. Konijn (eds.), 1988: *The Impact of Climate Variations on Agriculture.* Vol. 1, *Assessments in Cool Temperate and Cold Regions.* Kluwer Academic Publishers, Dordrecht, Netherlands, 876 pp.

Parry, M. T.R. Carter, and N.T. Konijn (eds.), 1988: *The Impact of Climatic Variations on Agriculture*. Vol. 2, *Assessment in Semi-Arid Regions.* Kluwer Academic Publishers, Dordrecht, Netherlands, 764 pp.

Parry, M.L., J.H. Porter, and T.R. Carter, 1990: Agriculture: climatic change and its implications. *Trends in Ecology and Evolution*, **5**, 318-322.

Parton, K. and E. Fleming, 1992: Food security: a comment. *Pacific Economic Bulletin*, **7(1)**, 39-40.

Pashley, D.P. and G.L. Bush, 1979: The use of allozymes in studying insect movement with special reference to the codling moth, Laspeyresia pomonella (L.) (Olethreutidae). In: *Movement of Highly Mobile Insects: Concepts and Methodology in Research* [Rabb, R.L. and G.G. Kennedy (eds.)]. North Carolina State University Press, Raleigh, NC, pp. 333-341.

Patterson, D.T., 1993: Implications of global climate change for impact of weeds, insects, and plant diseases. In: *International Crop Science*, vol. I. Crop Science Society of America, Madison, WI, pp. 273-280.

Patterson, D.T., 1995. Weeds in a changing climate. *Review of Weed Science*, **7** (in press).

Patterson, D.T. and E.P. Flint, 1980: Potential effects of global atmospheric CO_2 enrichment on the growth and competitiveness of C_3 and C_4 weed and crop plants. *Weed Science*, **28**, 71-75.

Patterson, D.T. and E.P. Flint, 1990: Implications of increasing carbon dioxide and climate change for plant communities and competition in natural and managed ecosystems. In: *Impact of CO_2, Trace Gases, and Climate Change on Global Agriculture* [Kimball, B.A., N.J. Rosenberg, and L.H. Allen, Jr. (eds.)]. Special Publication 53, American Society of Agronomy, Madison, WI, pp. 83-110.

Patterson, D.T., C.R. Meyer, E.P. Flint, and P.C. Quimby, Jr., 1979: Temperature responses and potential distribution of itchgrass (Rottboellia exaltata) in the United States. *Weed Science*, **27**, 77-82.

Patterson, D.T., E.P. Flint, and J.L. Beyers, 1984: Effects of CO_2 enrichment on competition between a C_4 weed and a C_3 crop. *Weed Science*, **32**, 101-105.

Patterson, D.T., A.E. Russell, D.A. Mortensen, R.D. Coffin, and E.P. Flint, 1986: Effects of temperature and photoperiod on Texas panicum (Panicum texanum) and wild proso millet (Panicum miliaceum). *Weed Science*, **34**, 876-882.

Pedgley, D.E., 1982: *Windborne Pests and Diseases: Meteorology of Airborne Organisms*. Ellis Horwood Ltd., West Sussex, England, UK, 250 pp.

Peart, R., J. Jones, B. Curry, K. Boote, and L.H. Allen, Jr., 1989: Impact of climate change on crop yield in the southeastern USA. In: *The Potential Effects of Global Climate Change on the United States* [Smith, J. and D. Tirpak (eds.)]. Report to Congress, Appendix C-1, U.S. Environmental Protection Agency, Washington, DC, pp. 2.1-2.54.

Penning de Vries, F.W.T., 1993: Rice production and climate change. In: *Systems Approaches for Agricultural Development* [Penning de Vries, F.W.T. *et al.* (eds.)]. Kluwer Academic Publishers, Dordrecht, Netherlands, pp. 175-189.

Penning de Vries, F.W.T., D.M. Jansen, H.F.M. ten Berge, and A. Bakema, 1989: *Simulation of Ecophysiological Processes of Growth in Several Annual Crops*. Pudoc, Simulation Monographs 29, Wageningen, Netherlands, 271 pp.

Pernetta, J.C. and P.J. Hughes (eds.), 1990: *Implications of Expected Climate Changes in the South Pacific Region: An Overview*. UNEP Regional Seas Report and Studies No. 128, United Nations Environmental Program.

Peterson, G.A., D.G. Westfall, and C.V. Cole, 1993: Agroecosystem approach to soil and crop management research. *Soil Science Society of America Journal*, **57**, 1354-1360.

Pieri, C., 1992: *Fertility of Soils: The Future for Farmers in the West African Savannah*. Springer Verlag, Berlin, Germany.

Pittock, A.B., 1989: The greenhouse effect, regional climate change and Australian agriculture. In: *Proceedings of the Fifth Agronomy Conference, Australian Society of Agronomy*. Australian Society of Agronomy, Parkville, Australia, pp. 289-303.

Pittock, A.B., 1993: Regional climate change scenarios for the south Pacific. In: *Climate Change and Sea Level Rise in the South Pacific Region: Proceedings of the Second SPREP Meeting* [Hay, J. and C. Kaluwin (eds.)]. CSIRO, Mordialloc, Australia. pp. 50-57.

Pittock, A.B., 1994: Climate and food supply. *Nature*, **371**, 25.

Pittock, A.B., A.M. Fowler, and P.H. Whetton, 1991: Probable changes in rainfall regimes due to the enhanced greenhouse effect. *International Hydrology and Water Resources Symposium*, pp. 182-186.

Pleijel, H., L. Skarby, G. Wallin, and G. Sellden, 1991: Yield and grain quality of spring wheat (triticum aestivum l., cv. Drabant) exposed to different concentrations of ozone in open-top chambers. *Environmental Pollution*, **69**, 151-168.

Pleijel, H., K. Ojanpera, S. Sutinen, and G. Sellden, 1992: Yield and quality of spring barley (hordeum vulgare l.) exposed to different concentrations of ozone in open-top chambers. *Agricultural Ecosystems and the Environment*, **38**, 21-29.

Plucknett, D.L., 1995: Prospects of meeting future food needs through new technology. In: *Population and Food in the Early 21st Century: Meeting Future Food Demand of an Increasing World Population* [Islam, N. (ed.)]. Occasional Paper, International Food Policy Research Institute (IFPRI), Washingon DC, pp. 207-220.

Plucknett, D.L., N.J.H. Smith, J.T. Williams, and N.M. Anishetty, 1987: *Gene Banks and the World's Food*. Princeton University Press, Princeton, NJ, 483 pp.

Plusquellec, H., 1990: *The Gezira Irrigation Scheme in Sudan: Objectives, Design, and Performance*. World Bank Technical Paper, no. 120, World Bank, Washington, DC.

Pollak, L.M. and J.D. Corbett, 1993: Using GIS datasets to classify maize-growing regions in Mexico and Central America. *Agronomy Journal*, **85**, 1133-1139.

Porter, J.R., 1993: AFRCWHEAT2: a model of the growth and development of wheat incorporating responses to water and nitrogen. *European Journal of Agronomy*, **2**, 69-82.

Porter, J., 1994: *The Vulnerability of Fiji to Current Climate Variability and Future Climate Change*. Climate Impacts Centre, Macquarie University, Australia.

Porter, J.R., P.D. Jameison, and R.D. Wilson, 1993: A comparison of the wheat crop simulation models AFRCWHEAT2, CERES-Wheat and SWHEAT. *Field Crops Research*, **33**, 131-157.

Porter, J.H., M.L. Parry, and T.R. Carter, 1991: The potential effects of climatic change on agricultural insect pests. *Agricultural Forest Meteorology*, **57**, 221-240.

Prange, R.K., K.B. McRae, D.J. Midmore, and R. Deng, 1990: Reduction in potato growth at high temperature: role of photosynthesis and dark respiration. *American Potato Journal*, **67**, 357-369.

Prescott-Allen, R. and C. Prescott-Allen, 1990: How many plants feed the world? *Conservation Biology*, **4**, 365-374.

Prestidge, R.A. and R.P. Pottinger, 1990: *The Impact of Climate Change on Pests, Diseases, Weeds and Beneficial Organisms Present in New Zealand Agricultural and Horticultural Systems*. New Zealand Ministry for the Environment, Wellington, New Zealand.

Qureshi, A. and D. Hobbie, 1994: *Climate Change in Asia: Thematic Overview*. Asian Development Bank, Manila, Philippines, 351 pp.

Qureshi, A. and A. Iglesias, 1994: Implications of global climate change for agriculture in Pakistan: impacts on simulated wheat production. In: *Implications of Climate Change for International Agriculture: Crop Modeling Study* [Rosenzweig, C. and A. Iglesias (eds.)]. U.S. Environmental Protection Agency, Pakistan chapter, Washington, DC, pp. 1-11.

Raffa, K.F., 1986: The mountain pine beetle in western North America. In: *Dynamics of Forest Insect Populations: Patterns, Causes, Implications* [Berryman, A.A. (ed.)]. Plenum Press, New York, NY, pp. 505-530.

Rahman, A. and D.A. Wardle, 1990: Effects of climate change on cropping weeds in New Zealand. In: *The Impact of Climate Change on Pests, Diseases, Weeds and Beneficial Organisms Present in New Zealand Agricultural and Horticultural Systems* [Prestidge, R.A. and R.P. Pottinger (eds.)]. New Zealand Ministry for the Environment, Wellington, New Zealand, pp. 107-112.

Rao, D.G. and S.K. Sinha, 1994: Impact of climate change on simulated wheat production in India. In: *Implications of Climate Change for International Agriculture: Crop Modeling Study* [Rosenzweig, C. and A. Iglesias (eds.)]. U.S. Environmental Protection Agency, India chapter, Washington, DC, pp. 1-10.

Rainey, R.C., 1989: *Migration and Meteorology: Flight Behavior and the Atmospheric Environment of Migrant Pests*. Oxford University Press, New York, NY, 314 pp.

Rasmussen, P.E. and H.P. Collins, 1991: Long-term impacts of tillage, fertiliser and crop residue on soil organic matter in temperate semiarid regions. *Advances in Agronomy*, **45**, 93-134.

Rath, D., D. Gädeken, D. Hesse, and M.C. Schlichting, 1994: Einfluß von klimafaktoren auf die tierproduction. In: *Klimaveränderungen und Landwirtschaft, Part II, Landbauforschung* [Brunnert, H. and U. Dämmgen (eds.)]. Völkenrode, Spec. vol. **148**, pp. 341-375.

Raulston, J.R., S.D. Pair, J. Loera, and H.E. Cabanillas, 1992: Prepupal and pupal parasitism of Helicoverpa zea and Spodoptera frugiperda (Lepidoptera: Noctuidae) by Steinernema sp. in corn fields in the lower Rio Grande valley. *Journal of Economic Entomology*, **85**, 1666-1670.

Reilly, J., 1989: *Consumer Effects of Biotechnology*. AIB No. 581, U.S. Department of Agriculture, Washington, DC, 11 pp.

Reilly, J., 1994: Crops and climate change. *Nature*, **367**, 118-119.

Reilly, J., 1995: Climate change and global agriculture: recent findings and issues. *American Journal of Agricultural Economics*, **77**, 243-250.

Reilly, J. and N. Hohmann, 1993: Climate change and agriculture: the role of international trade. *American Economic Association Papers and Proceedings*, **83**, 306-312.

Reilly, J., N. Hohmann, and S. Kane, 1994: Climate change and agricultural trade: who benefits, who loses? *Global Environmental Change*, **4(1)**, 24-36.

Rimmington, G.M. and N. Nicholls, 1993: Forecasting wheat yield in Australia with the Southern Oscillation index. *Australian Journal of Agricultural Research*, **44**, 625-632.

Ritchie, J.T., B.D. Baer, and T.Y. Chou, 1989: Effect of global climate change on agriculture: Great Lakes region. In: *The Potential Effects of Global Climate Change on the United States* [Smith, J. and D. Tirpak (eds.)]. Report to Congress, Appendix C-1, U.S. Environmental Protection Agency, Washington, DC, pp. 1.1-1.42.

Rogers, H.H., G.E. Bingham, J.D. Cure, J.M. Smith, and K.A. Surano, 1983: Responses of selected plant species to elevated carbon dioxide in the field. *Journal of Environmental Quality*, **12**, 569.

Rogers, H.H., J.D. Cure, and J.M. Smith, 1986: Soybean growth and yield response to elevated carbon dioxide. *Agriculture, Ecosystems and Environment*, **16**, 113-128.

Rogers, H.H. and R.C. Dahlman, 1993: Crop responses to CO_2 enrichment. *Vegetatio*, **104/105**, 117-131.

Rogers, H.H., G.B. Runion, and S.V. Krupa, 1994: Plant responses to atmospheric CO_2 enrichment with emphasis on roots and the rizosphere. *Environmental Pollution*, **83**, 155-189.

Rosenberg, N.J., 1992: Adaptation of agriculture to climate change. *Climatic Change*, **21**, 385-405.

Rosenberg, N.J., 1993: *Towards an Integrated Assessment of Climate Change: The MINK Study*. Kluwer Academic Publishers, Boston, MA, 173 pp.

Rosenberg, N.J. and P.R. Crosson, 1991: *Processes for Identifying Regional Influences of and Responses to Increasing Atmospheric CO_2 and Climate Change: The MINK Project, An Overview*. DOE/RL/01830T-H5, Resources for the Future and U.S. Department of Energy, Washington, DC, 35 pp.

Rosenberg, N.J. and M.J. Scott, 1994: Implications of policies to prevent climate change for future food security. *Global Environmental Change*, **4(1)**, 49-62.

Rosenzweig, C., 1985: Potential CO_2-induced climate effects on North American wheat-producing regions. *Climatic Change*, **7**, 367-389.

Rosenzweig, C., 1990: Crop response to climate change in the southern Great Plains: a simulation study. *Professional Geographer*, **42(1)**, 20-37.

Rosenzweig, C., B. Curry, J.T. Richie, J.W. Jones, T.Y. Chou, R Goldberg, and A. Iglesias, 1994: The effects of potential climate change on simulated grain crops in the United States. In: *Implications of Climate Change for International Agriculture: Crop Modeling Study* [Rosenzweig, C. and A. Iglesias (eds.)]. U.S. Environmental Protection Agency, USA chapter, Washington, DC, pp. 1-24.

Rosenzweig, C. and A. Iglesias (eds.), 1994: *Implications of Climate Change for International Agriculture: Crop Modeling Study*. EPA 230-B-94-003, U.S. Environmental Protection Agency, Washington, DC, 312 pp.

Rosenzweig, C. and M.L. Parry, 1994: Potential impact of climate change on world food supply. *Nature*, **367**, 133-138.

Rosenzweig, C., M.L. Parry, and G. Fischer, 1995: Climate change and world food supply. In: *As Climate Changes: International Impacts and Implications* [Strzepek, K.M. and J.B. Smith (eds.)]. Cambridge University Press, Cambridge, UK (in press).

Rötter, R., 1993: *Simulation of the Biophysical Limitations to Maize Production under Rainfed Conditions in Kenya. Evaluation and Application of the Model WOFOST*. PhD thesis, Universität Trier, Trier, Germany (Materialien zur Ostafrika-Forschung, Heft 12), 297 pp.

Rötter, R. and C.A. van Diepen, 1994: *Rhine Basin Study*. Vol. 2, *Climate Change Impact on Crop Yield Potentials and Water Use*. SC-DLO Report, 85.2, Wageningen and Lelystad, Netherlands, 145 pp.

Rötter, R., W. Stol, S.C. van de Geijn, and H. van Keulen, 1995: *World Agro-Climates. 1. Current Temperature Zones and Drylands*. AB-DLO, Research Institute for Agrobiology and Soil Fertility, Wageningen, Netherlands, 55 pp.

Rowntree, P.R., 1990: Predicted climate changes under "greenhouse-gas" warming. In: *Climatic Change and Plant Genetic Resources* [Jackson, M., B.V. Ford-Lloyd, and M.L. Parry (eds.)]. Belhaven Press, London, UK, pp. 18-33.

Rowntree, P.R., 1990: Estimates of future climatic change over Britain. Part 2: Results. *Weather*, **45**, 79-89.

Royle, D.J., M.W. Shaw, and R.J. Cook, 1986: Patterns of development of Septoria nodorume and S. tritici in some winter wheat crops in western Europe, 1981-1983. *Plant Pathology*, **35**, 466-476.

Rozema, J., H. Lambers, S.C. van de Geijn, and M.L. Cambridge (eds.), 1993: *CO_2 and Biosphere*. Kluwer Academic Publishers, Dordrecht, Netherlands, 484 pp.

Rozema, J., J.V.M. Van de Staaij, V. Costa, J.G.M. Torres Pereira, R.A. Broekman, G.M. Lenssen, and M. Stroetenga, 1991: A comparison of the growth, photosynthesis and transpiration of wheat and maize in response to enhanced ultraviolet-b radiation. In: *Impacts of Global Climatic Changes on Photosynthesis and Plant Productivity* [Abrol, Y.P. *et al.* (eds.)]. Vedams Books International, New Delhi, India, pp. 163-174.

Rummel, D.R., K.C. Neece., M.D. Arnold, and B.A. Lee, 1986: Overwintering survival and spring emergence of Heliothis zea (Boddie) in the Texas southern high plains. *Southwest Entomology*, **11(1)**, 1-9.

Ruttan, V.W., D.E. Bell, and W.C. Clark, 1994. Climate change and food security: agriculture, health and environmental research. *Global Environmental Change*, **4(1)**, 63-77.

Sala, O.E. and J.M. Paruelo, 1994: Impacts of global climate change on maize production in Argentina. In: *Implications of Climate Change for International Agriculture: Crop Modeling Study* [Rosenzweig, C. and A. Iglesias (eds.)]. U.S. Environmental Protection Agency, Argentina chapter, Washington, DC, pp. 1-12.

Salinger, M.J. and Hicks, D.M., 1990: The scenario. In: *Climatic Change: Impacts on New Zealand. Implications for the Environment, Economy and Society*. Ministry for the Environment, Wellington, New Zealand.

Salinger, M.J. and A.S. Porteus, 1993: Climate change and variability: impacts on New Zealand pastures. In: *Proceedings XVII International Grassland Congress*, pp. 1075-1077.

Salinger, M.J., M.W. Williams, J.M. Williams, and R.J. Maritin, 1990: Agricultural resources. In: *Climatic Change: Impacts on New Zealand. Implications for the Environment, Economy and Society*. Ministry for the Environment, Wellington, New Zealand.

Santer, B., 1985: The use of general circulation models in climate impact analysis—a preliminary study of the impacts of a CO_2-induced climatic change on western European agriculture. *Climatic Change*, **7**, 71-93.

Schertz, D.L., 1988: Conservation tillage: an analysis of acreage projections in the United States. *Journal of Soil and Water Conservation*, **43(3)**, 256-258.

Schulze, R.E., G.A. Kiker, and R.P. Kunz, 1993: Global climate change and agricultural productivity in southern Africa. *Global Environmental Change*, **4(1)**, 329-349.

Schwartz, M. and J. Gale, 1984: Growth response to salinity at high levels of carbon dioxide. *Journal of Experimental Botany*, **35**, 193-196.

Scott, R.W. and G.L. Achtemeier, 1987: Estimating pathways of migrating insects carried in atmospheric winds. *Environmental Entomology*, **16(6)**, 1244-1254.

Seino, H., 1993a: Impacts of climatic warming on Japanese agriculture. In: *The Potential Effects of Climate Change in Japan* [Shuzo, N., H. Hideo, H. Hirokazu, O. Toshiichi, and M. Tsuneyuki (eds.)]. Environment Agency of Japan, Tokyo, Japan, pp. 15-35.

Seino, H., 1993b: Implication of climatic warming for Japanese crop production. In: *Climate Change, Natural Disasters and Agricultural Strategies*. Meteorological Press, Beijing, China.

Seino, H., 1994: Implications of climate change for Japanese agriculture: evaluation by simulation of rice, wheat, and maize growth. In: *Implications of Climate Change for International Agriculture: Crop Modeling Study* [Rosenzweig, C. and A. Iglesias (eds.)]. U.S. Environmental Protection Agency, Japan chapter, Washington, DC, pp. 1-18.

Sellden, G. and H. Pleijel, 1993: *Influence of atmospheric ozone on agricultural crops. In: International Crop Science*, vol. I. Crop Science Society of America, Madison, WI, pp. 315-319.

Semenov, M.A., J.R. Porter, and R. Delecolle, 1993a: Climate change and the growth and development of wheat in the UK and France. *European Journal of Agronomy*, **2**, pp. 293-304.

Semenov, M.A., J.R. Porter, and R. Delecolle, 1993b: Simulation of the effects of climate change on growth and development of wheat in the UK and France. In: *The Effect of Climate Change on Agricultural and Horticultural Potential in Europe* [Kenny, G.J., P.A. Harrison, and M.L. Parry (eds.)]. Environmental Change Unit, University of Oxford, Oxford, UK, pp. 121-136.

Sen, A., 1981: *Poverty and Famines: An Essay on Entitlement and Deprivation*. Oxford University Press, London, UK, 257 pp.

Sen, A., 1993: The economics of life and death. *Scientific American*, 40-47.

Shaw, B., 1992: Pacific agriculture: a retrospective of the 1980s and prospects for the 1990s. *Pacific Economic Bulletin*, **7**, 15-20.

Shibles, R.M., I.C. Anderson, and A.H. Gibson, 1975: Soybean. In: *Crop Physiology* [L.T. Evans (ed.)]. Cambridge University Press, London, UK, pp. 151-189.

Singh, U., 1994: Potential climate change impacts on the agricultural systems of the small island nations of the Pacific. Draft, IFDC-IRRI, Los Banos, Philippines, 28 pp.

Singh, U., D.C. Godwin, and R.J. Morrison, 1990: Modelling the impact of climate change on agricultural production in the South Pacific. In: *Global Warming-Related Effects of Agriculture and Human Health and Comfort in the South Pacific* [Hughes, P.J. and G. McGregor (eds.)]. South Pacific Regional Environment Programme and the United Nations Environment Programme, University of Papua New Guinea, Port Moresby, Papua New Guinea, pp. 24-40.

Siqueira, O.E. de, J.R. Boucas Farias, and L.M. Aguiar Sans, 1994: Potential effects of global climate change for Brazilian agriculture: applied simulation studies for wheat, maize, and soybeans. In: *Implications of Climate Change for International Agriculture: Crop Modeling Study* [Rosenzweig, C. and A. Iglesias (eds.)]. U.S. Environmental Protection Agency, Brazil chapter, Washington, DC, pp. 1-28.

Sirotenko, O.D., 1981: *Mathematical Modeling of Hydrothermal Regimes and the Productivity of Agroecosystems*. Gidrometeoizdat, Leningrad, USSR, 167 pp. (in Russian).

Sirotenko, O.D., 1991: Nmntaiinohhar enetema kanmat-ypokak. *Meterologia and Gidrologia*, **4**, 67-73 (in Russian).

Sirotenko, O.D. and E.V. Abashina, 1994: Impact of global warming on the agroclimatic resources and agriculture productivity of Russia (result of the simulation). *Meteorology and Hydrology*, **4**.

Sirotenko, O.D., E.V. Abashina, and V.N. Pavlova, 1984: The estimates of climate variations of crops productivity. *The Proceedings of the USSR Academy of Science: Physics of Atmosphere and Ocean*, **20(11)**, 1104-1110 (in Russian).

Sirotenko, O.D. *et al.*, 1991: Global warming and the agroclimatic resources of the Russian plain. *Soviet Geography*, **32(5)**, 337-384.

Sivakumar, M.V.K., 1993; Global climate change and crop production in the Sundano-Sahelian zone of west Africa. In: *International Crop Science*, vol. I. Crop Science Society of America, Madison, WI.

Smit, B. (ed.), 1993: *Adaptation to Climatic Variability and Change*. Occasional Paper No. 19, University of Guelph, Guelph, Canada, 53 pp.

Solomon, A., 1993: Modeling the potential change in yield and distribution of the earth's crops under a warmed climate. *Climate Research*, **3**, 79-96.

Solomon, A.M. and H.H. Shugart (eds.), 1993: *Vegetation Dynamics and Global Change*. Chapman and Hall, New York, NY.

Sombroek, W. and R. Gommes, 1993: The climate change-agriculture conundrum. Paper presented at the *Expert Consultation on Global Climate Change and Agricultural Production: Direct and Indirect Effects of Changing Hydrological, Soil and Plant Physiological Processes*, FAO, 7-10 December 1993, 300 pp.

Spittlehouse, D.L. and B. Sieben, 1994: Mapping the effect of climate on spruce weevil infestation hazard. In: *Preprints of the 11th Conference on Biometeorology and Aerobiolology*. American Meteorology Society, Boston, MA, pp. 448-450.

Squire, G.R. and M.H. Unsworth, 1988: *Effects of CO_2 and Climatic Change on Agriculture: 1988 Report to the UK Department of the Environment*. University of Nottingham, Nottingham, UK.

The Statistical Yearbook of Agriculture of China, 1991, 1992. Agricultural Publishing House, Beijing, China.

Stem, E., G.A. Mertz, J.D. Stryker, and M. Huppi, 1988: *Changing Animal Disease Patterns Introduced by the Greenhouse Effect: Report of a Preliminary Study to the Environmental Protection Agency*. Tufts University School of Veterinary Medicine, North Grafton, MA, 76 pp.

Stinner, B.R., R.A.J. Taylor, R.B. Hammond, F.F. Purrington, D.A. McCartney, N. Rodenhouse, and G.W. Barrett, 1987: Potential effects of climate change on plant-pest interactions. In: *The Potential Effects of Global Climate Change on the United States* [Smith, J. and D. Tirpak (eds.)]. U.S. Environmental Protection Agency, Washington, DC, pp. 8.1-8.35.

Stockle, C.O., J.R. Williams, N.J. Rosenberg, and C.A. Jones, 1992: A method for estimating the direct and climatic effects of rising carbon dioxide on growth and yield of crops: Part I, Modification of the EPIC model for climate change analysis. *Agricultural Systems*, **38**, 225-238.

Strain, B.R. and J.D. Cure, 1985. *Direct Effects of Increasing Carbon Dioxide on Vegetation*. DOE/ER-0238, Carbon Dioxide Research Division, U.S. Department of Energy, Washington, DC, 286 pp.

Sugihara, S., 1991: A simulation of rice production in Japan by rice-weather production model. *Kikogaku-Kishogaku Kenkyu Hokoku*, **16**, 32-37.

Sutherst, R.W., 1990: Impact of climate change on pests and diseases in Australasia. *Search*, **21**, 230-232.

Sutherst, R.W., 1991: Pest risk analysis and the greenhouse effect. *Review of Agricultural Entomology*, **79**, 1177-1187.

Sutherst, R.W., G.F. Maywald, and D.B. Skarrate, 1995: Predicting insect distributions in a changed climate. In: *Insects in a Changing Environment* [Harrington, R. and N.E. Stork]. Academic Press, London, UK, pp. 59-91.

Suyama, T., 1988: Grassland production, animal husbandry and climate change. *Kisho-Kenkyu Note*, **162**, 123-130.

Swank, J.C., D.B. Egli, and T.W. Pfeiffer, 1987: Seed growth characteristics of soybean genotypes differing in duration or seed fill. *Crop Science*, **27**, 85-89.

Tao, Z., 1993: Influences of global climate change on agriculture of China. In: *Climate Biosphere Interactions*. John Wiley and Sons, New York, NY.

Tate, K.R., D.J. Giltrap, A. Parshotam, A.E. Hewitt, D.J. Ross, G.J. Kenny, R.A. Warrick, 1994: Impacts of climate change on soils and land systems in New Zealand. Presented at Greenhouse 94: An Australian-New Zealand Conference on Climate Change, 9-14 October, Wellington, New Zealand.

Tauber, M.J., C.A. Tauber, and S. Masaki, 1986: *Seasonal Adaptations of Insects*. Oxford University Press, New York, NY, 411 pp.

Tegart, W.J., G.W. McG. Sheldon, D.C. Griffiths (eds.), 1990: *Climate Change: The IPCC Impacts Assessment*. Australian Government Printing Office, Canberra, Australia, 245 pp.

Temple, P.J., R.S. Kupper, R.L. Lennox, and K. Rohr, 1988: Physiological and growth responses of differentially irrigated cotton to ozone. *Environmental Pollution*, **53**, 255-263.

Teramura, A.H., 1983: Effects of ultraviolet-B radiation on the growth and yield of crop plants. *Physiol. Plant.*, **58**, 415-427.

Terranova, A.C., R.G. Jones, and A.C. Bartlett, 1990: The southeastern boll weevil: an allozyme characterization of its population structure. *Southwestern Entomology*, **15(4)**, 481-496.

Tevini, M. (ed.), 1993: *UV-B Radiation and Ozone Depletion. Effects on Humans, Animals, Plants, Microorganisms, and Materials*. Lewis Publishers, Boca Raton, FL, 248 pp.

Thatcher, W.W., 1974: Effects of season, climate and temperature on reproduction and lactation. *Journal of Dairy Science*, **57**, 360-368.

Thompson, T.P. and X. Wan, 1992: *The Socioeconomic Dimensions of Agricultural Production in Albania: A National Survey*. No. PN-ABQ-691, International Fertilizer Development Center, Washington, DC, 87 pp.

Tiffen, M., M. Mortimore, and F. Gichuki, 1993: *More People, Less Erosion: Environmental Recovery in Kenya*. John Wiley and Sons, Chichester, UK.

Tinker, P.B., 1985. Site-specific yield potentials in relation to fertilizer use. In: *Nutrient Balances and Fertilizer Needs in Temperate Agriculture* [Avon, P. (ed.)]. 18th Coll. Int. Potash Inst., Berne, pp. 193-208.

Tobey, J., J. Reilly, and S. Kane, 1992: Economic implications of global climate change for world agriculture. *Journal of Agricultural and Resource Economics*, **17(1)**, 195-204.

Tongyai, C., 1994: Impact of climate change on simulated rice production in Thailand. In: *Implications of Climate Change for International Agriculture: Crop Modeling Study* [Rosenzweig, C. and A. Iglesias (eds.)]. U.S. Environmental Protection Agency, Thailand chapter, Washington, DC, pp. 1-13.

Tonneijck, A.E.G., 1983: Foliar injury responses of 24 bean cultivars (phaesolus vulgaris) to various concentrations of ozone. *Netherlands Journal of Plant Pathology*, **89**, 99-104.

Tonneijck, A.E.G., 1989: Evaluation of ozone effects on vegetation in the Netherlands. In: *Atmospheric Ozone Research and Its Policy Implications* [Schneider, T. *et al.* (eds.)]. Elsevier Science Publishers B.V., Amsterdam, Netherlands, pp. 251-260.

Treharne, K., 1989: The implications of the "greenhouse effect" for fertilizers and agrochemicals. In: *The "Greenhouse Effect" and UK Agriculture* [Bennet, R.M. (ed.)]. No. 19, Center for Agricultural Strategy, University of Reading, Reading, UK, pp. 67-78.

Turner, H.A. and C.L. Anderson, 1980: *Planning for an Irrigation System.* American Association for Vocational Instructional Materials, Athens, GA, 28 pp.

Umali, D.L., 1993: *Irrigation-Induced Salinity: A Growing Problem for Development and the Environment.* Technical Paper No. 215, The World Bank, Washington, DC, 78 pp.

United Kingdom Department of the Environment, 1991: *United Kingdom Climate Change Impacts Review Group: The Potential Effects of Climate Change in the United Kingdom.* Climate Change Impacts Review Group, HMSO, London, UK, 124 pp.

U.S. Congress, Office of Technology Assessment, 1993: *Preparing for an Uncertain Climate*, vol. 1. OTA-O-567, U.S. Government Printing Office, Washington, DC, 359 pp.

U.S. Environmental Protection Agency, 1989: *The Potential Effects of Climate Change on the United States, Report to Congress.* EPA-230-05-89-050, US EPA, Washington, DC.

Valentine, J.W., 1968: Climatic regulation of species diversification and extinction. *Geological Society of America Bulletin*, **79**, 273-276.

Van de Geijn, S.C., J. Goudriaan, J. Van der Eerden, and J. Rozema, 1993: Problems and approaches to integrating the concurrent impacts of elevated carbon dioxide, temperature, ultraviolet-B radiation, and ozone on crop production. In: *International Crop Science*, vol. I. Crop Science Society of America, Madison, WI, pp. 333-338.

Van der Eerden, L.J., A.E.G.Tonneijck, and J.H.M. Weijnands, 1988: Crop loss due to air pollution in the Netherlands. *Environmental Pollution*, **53**, 365-376.

Van der Eerden, L.J., T. Dueck, and M. Perez-Soba, 1993: Influence of air pollution on carbon dioxide effects on plants. In: *Climate Change: Crops and Terrestrial Ecosystems* [Van de Geijn, S.C., J. Goudriaan, and F. Berendse (eds.)]. Agrobiologische Thema's 9, CABO-DLO, Wageningen, Netherlands, pp. 59-70.

Van der Leun, J.C. and M. Tevini, 1991: *Environmental Effects of Ozone Depletion: 1991 Update.* United Nations' Environment Program, Nairobi, Kenya, 52 pp.

Van Dijk, M.P., 1992: What relevance has the path of the NICs for Africa? *African Development Perspectives Yearbook: 1990/91*, LIT Verlag, Hamburg, pp. 43-55.

Van Heemst, H.D.J., 1988: *Plant Data Values Required for Simple and Universal Simulation Models: Review and Bibliography.* Simulation reports CABO-TT, **17**, Wageningen, Netherlands, 100 pp.

Van Keulen, H. and N.G. Seligman (1987): *Simulation of Water Use, Nitrogen Nutrition and Growth of a Spring Wheat Crop.* Pudoc, Simulation Monographs, Wageningen, Netherlands, 310 pp.

Vaux, H.J., 1990: The changing economics of agricultural water use. In: *Visions of the Future: Proceedings of the 3rd National Irrigation Symposium.* American Society of Agricultural Engineers, St. Joseph MI, pp. 8-12.

Vaux, H.J., 1991: Global climate change and California's water resources. In: *Global Climate Change and California* [Knox, J.B. and A.F. Scheuring (eds.)]. University of California Press, Berkeley, CA, pp. 69-96.

Volz, A. and D. Kley, 1988: Evaluation of the Montsouris series of ozone measurements made in the nineteenth century. *Nature*, **332**, 240-242.

Vugts, H.F., 1993: The need for micrometeorological research of the response of the energy balance of vegetated surfaces to CO_2 enrichment. *Vegetatio*, **104/105**, 321-328.

Waggoner, P.E., 1983: Agriculture and a climate changed by more carbon dioxide. In: *Changing Climate: Report of the Carbon Dioxide Assessment Committee.* National Academy Press, Washington, DC, pp. 383-418.

Waggoner, P.E., 1993: Assessing nonlinearities and surprises in the links of farming to climate and weather. In: *Assessing Surprises and Nonlinearities in Greenhouse Warming* [Darmstadter, J. and M.A. Toman (eds.)]. Resources for the Future, Washington, DC, pp. 45-65.

Wang, Y.P. and R.M. Gifford, 1995: A model of wheat grain growth and its application to different temperature and carbon dioxide levels. *Australian Journal of Plant Physiology* (in press).

Wang, Y.P., Jr. Handoko, and G.M. Rimmington, 1992: Sensitivity of wheat growth to increased air temperature for different scenarios of ambient CO_2 concentration and rainfall in Victoria, Australia—a simulation study. *Climatic Research*, **2**, 131-149.

Wardlaw, I.F., I.A. Dawson, P. Munibi, and R. Fewster, 1989: The tolerance of wheat to high temperatures during reproductive growth. I. Survey procedures and general response patterns. *Australian Journal of Agricultural Research*, **40**, 1-13.

Warrick, R.A. and G.J. Kenny, 1994: CLIMPACTS: the conceptual framework and preliminary development. In: *Towards an Integrated Approach to Climate Change Impact Assessment* [Pittock, A.B. and C.D. Mitchell (eds.)]. Report of Workshop, 26 April, Division of Atmospheric Research, CSIRO, Australia.

Watling, D. and S. Chape (eds.), 1992: *Environment Fiji: The National State of the Environment Report.* IUCN, Gland, Switzerland, 154 pp.

Watt, A.D. and S.R. Leather, 1986: The pine beauty in Scottish lodgepole pine plantations. In: *Dynamics of Forest Insect Populations: Patterns, Causes, Implications* [Berryman, A.A. (ed)]. Plenum Press, New York, NY, pp. 243-266.

Webb, A.R., 1991: Solar ultraviolet-B radiation measurement at the earth's surface: techniques and trends. In: *Impact of Global Climatic Changes on Photosynthesis and Plant Productivity* [Abrol, Y.P. *et al.*(eds.)]. Oxford and IBH publ. Co. Pvt. Ltd., New Delhi, India, pp. 23-37.

Westbrook, J.K., W.W. Wolf, S.D. Pair, A.N. Sparks, and J.R. Raulston, 1985: An important atmospheric vehicle for the long range migration of serious agricultural pests of the south central United States. In: *Preprints of the 7th Conf. on Biometeorol. and Aerobiol.* American Meteorological Society, Boston, MA, pp. 281-282.

Wheeler, T.R., J.I.L. Morrison, P. Hadley, and R.H. Ellis, 1993: Whole-season experiments on the effects of carbon dioxide and temperature on vegetable crops. In: *The Effect of Climate Change on Agricultural and Horticultural Potential in Europe* [Kenny, G.J., P.A. Harrison, and M.L. Parry (eds.)]. Environmental Change Unit, University of Oxford, Oxford, UK, pp. 165-176.

Wheeler, T.R., P. Hadley, J.I.L. Morrison, and R.H. Ellis, 1993: Effects of temperature on the growth of lettuce (Latuca sativa L.) and the implications for assessing the impacts of potential climate change. *European Journal of Agronomy*, **2**, 305-311.

Whetton, P.H., A.M. Fowler, M.R. Haylock, and A.B. Pittock, 1993: Implications of climate change due to enhanced greenhouse effect on floods and droughts in Australia and New Zealand. *Climatic Change*, **25**, 289-317.

Wilkerson, G.G., J.W. Jones, K.J. Boote, and G.S. Buol, 1989: Photoperiodically sensitive interval in time to flower of soybean. *Crop Science*, **29**, 721-726.

Wilson, J.R., 1982. Environmental and nutritional factors affecting herbage quality. In: *Nutrional Limits to Animal Production from Pastures* [Hacker, J. (ed.)]. Farmham Royal, CAB International, pp. 111-131.

Wolf, J., 1993a: Effects of climate change on wheat and maize production in the EC. In: *The Effect of Climate Change on Agricultural and Horticultural Potential in Europe* [Kenny, G.J., P.A. Harrison, and M.L. Parry (eds.)]. Environmental Change Unit, University of Oxford, Oxford, UK, pp. 93-119.

Wolf, J., 1993b: Effects of climate change on wheat production potential in the European Community. *European Journal of Agronomy*, **2**, 281-292.

Wolf, J., M.A. Semenov, J.R. Porter, F. Courveur, and J. Tranchfort, 1993: Comparison of results from different models of calculating winter wheat production. In: *The Effect of Climate Change on Agricultural and Horticultural Potential in Europe* [Kenny, G.J., P.A. Harrison, and M.L. Parry (eds.)]. Environmental Change Unit, University of Oxford, Oxford, UK, pp. 157-163.

Wolfe, D.W. and J.D. Erickson, 1993: Carbon dioxide effects on plants: uncertainties and implications for modelling crop response to climate change. In: *Agricultural Dimensions of Global Climate Change* [Kaiser, H.M. and T.E. Drennen (eds.)]. St. Lucie Press, Delray Beach, FL, pp. 153-178.

Wong, S.C., 1990: Elevated atmospheric partial pressure of CO_2 and plant growth. II. Non-structural carbohydrate content in cotton plants and its effect on growth parameters. *Photosynthesis Research*, **23**, 171-180.

Wood, C.W., H.A. Torbert, H.H. Rogers, G.B. Runion, and S.A. Prior, in press: Free-air CO_2 enrichment effects on soil carbon and nitrogen. *Agricultural and Forest Meteorology.*

Woodward, F.I., 1988: Temperature and the distribution of plant species. In: *Plants and Temperature* [Long, S.P. and F.I. Woodward (eds.)]. The Company of Biologists Ltd., University of Cambridge, Cambridge, UK, pp. 59-75.

Woodward, F.I., 1993: Leaf responses to the environment and extrapolation to larger scales. In: *Vegetation Dynamics and Global Change* [Solomon, A.M. and H.H. Shugart (eds.)]. Chapman and Hall, New York, NY, pp. 71-100.

World Bank, 1993: *World Development Report*. Oxford University Press, New York, NY, 329 pp.

World Bank, 1994: *A Review of World Bank Experience in Irrigation*. The World Bank Operations Department, Report 13676, Washington, DC.

World Bank, 1994: Personal communication from Mr. Antte Talvitie to Mr. P. Condon. Central Technical Department, Transportation Division, Washington, DC.

Worner, S.P., 1988: Ecoclimatic assessment of potential establishment of exotic pests. *Journal Economic Entomology*, **81(4)**, 973-983.

Wray, S.M. and B.R. Strain, 1987: Competition in old-field perennials under CO_2 enrichment. *Functional Ecolology*, **1**, 145-149.

Yoshida, S., 1981: *Fundamentals of Rice Crop Science*. The International Rice Research Institute (IRRI), Los Banos, Philippines, 269 pp.

Yoshino, M., 1991: *Global Climate Change in the Coming Half Century and an Estimate of Its Impact on Agriculture, Forestry, Fishery and Human Environment in Japan*. Report on Climatology and Meteorology, University of Tsukuba, **16**, Tsukuba,106 pp.

Yudelman, M., 1993. *Demand and Supply of Foodstuffs up to 2050 with Special Reference to Irrigation*. International Irrigation Institute, Colombo, Sri Lanka, 100 pp.

Zhang, H., 1993: The impact of greenhouse effect on double rice in China. In: *Climate Change and Its Impact*. Meteorology Press, Beijing, China, pp. 131-138.

Zhukovsky, E.E. and G.G. Belchenko, 1988: The stochastic system of crop yields forecasting. *Transactions of the Agrophysical Institute*, **70**, 2-7 (in Russian).

Zhukovsky, E.E., G.G. Belchenko, and T.N. Brunova, in press: The stochastic analysis of the climate change influence on crop productivity potential. *Meteorology and Hydrology Journal*.

14

Water Resources Management

ZDZISLAW KACZMAREK, POLAND

Principal Lead Authors:
N.W. Arnell, UK; E.Z. Stakhiv, USA

Lead Authors:
K. Hanaki, Japan; G.M. Mailu, Kenya; L. Somlyódy, Hungary; K. Strzepek, USA

Contributing Authors:
A.J. Askew, Switzerland; F. Bultot, Belgium; J.Kindler, USA; Z. Kundzewicz, Switzerland; D.P. Lettenmaier, USA; H.J. Liebscher, Germany; H.F. Lins, USA; D.C. Major, USA; A.B. Pittock, Australia; D.G. Rutashobya, Tanzania; H.H.G. Savenije, The Netherlands; C. Somorowski, Poland; K. Szesztay, Hungary

CONTENTS

EXECUTIVE SUMMARY

Water availability is an essential component of welfare and productivity. Much of the world's agriculture, hydroelectric power production, municipal and industrial water needs, water pollution control, and inland navigation is dependent on the natural endowment of surface and groundwater resources. Changes in the natural system availability would result in impacts that generally are greatest in regions that are already under stress, including currently arid and semi-arid areas, as well as areas where there is considerable competition among users. The purpose of water resources management is to ameliorate the effects of extremes in climate variability and provide a reliable source of water for multiple societal purposes. This evaluation of climate-change impacts focuses on the expected range of changes in the hydrological resources base and the sensitivity of the water supply and water demand components of the water management systems to climate change. It also makes an assessment of the viability of adaptive water management measures in responding to these impacts.

There are reasons for water resources managers, especially in developing nations, to be concerned by the results of climate-change scenarios, which show that the freshwater resources in many regions of the world are likely to be significantly affected. In particular, current arid and semi-arid areas of the world could experience large decreases in runoff—hence posing a great challenge to water resources management. Global-change-induced perturbations may follow widespread periodic and chronic shortfalls in those same areas caused by population growth, urbanization, agricultural expansion, and industrial development that are expected to manifest themselves before the year 2020 (High Confidence).

Uncertainties require considerable investment in research in order to improve prediction and adaptive responses. Some uncertainties in assessing the effects of climate change on water resources are:

- Uncertainties in general circulation models (GCMs) and lack of regional specification of locations where consequences will occur
- Insufficient knowledge on future climate variability, which is a basic element of water management
- Uncertainties in estimating changes in basin water budgets due to changes in vegetation and in atmospheric and other conditions likely to exist 50 to 100 years from now
- Uncertainties in future demands by each water sector
- Uncertainties in the socioeconomic and environmental impacts of response measures.

Hence, predicting where water resources problems due to climate change will occur can only be realized on a subcontinental scale at this time. However, water management decisions are made on the localized, watershed scale. Therefore, despite increases in the number of impact assessments and improvements in the new class of transient GCMs, there is little that can be added to the conclusions of the first two IPCC reports on the subject, other than to note the regions and countries most likely to be vulnerable through a combination of increased demands and reductions in available supplies. However, one important addition is the limited but growing analyses of water systems in the developing world. These limited studies seem to suggest that developing countries are highly vulnerable to climate change because many are located in arid and semi-arid regions and most existing water resources systems in these countries are characterized as isolated reservoir systems. Also, there is more evidence that flooding is likely to become a larger problem in many temperate regions, requiring adaptations not only to droughts and chronic water shortages but also to floods and associated damages and raising concerns about dam and levee failures (Medium Confidence).

Water management is a continuously adaptive enterprise, responding to changes in demands, hydrological information, technologies, the structure of the economy, and society's perspectives on the economy and the environment. This adaptation employs four broad interrelated approaches: new investments for capacity expansion; operation of existing systems for optimal use (instream and offstream); maintenance and rehabilitation of systems; and modifications in processes and demands (e.g., conservation, pricing, and institutions). These water management practices, which are intended to serve the present range of climate variability (which in itself is considerable), may also serve to ameliorate the range of perturbations such as droughts that are expected to accompany climate change. However, adaptations come at some social, economic, and environmental costs.

Most of the standard water resources performance criteria—such as reliability, safe yield, probable maximum flood, resilience, and robustness—are applicable in dealing with the impacts of climate change on water resources systems. This is not to suggest that we can become complacent in our response to climate change. The emphasis of water resources management in the next decades will be on responses to increased demands, largely for municipal water supply in rapidly urbanizing areas, energy production, and agricultural water supply. Water management strategies will focus on demand management, regulatory controls, legal and institutional changes, and economic instruments. The principal conclusions are as follows:

- Most of the regional water resources systems in the 21st century, particularly in developing countries, will become increasingly stressed due to higher demand to meet the needs of a growing population and economy, as well as to protect ecosystems (High Confidence).
- Arid and semi-arid watersheds and river basins are inherently the most sensitive to changes in temperature and precipitation (High Confidence).
- Water demand for irrigated agriculture is very sensitive to climate change, especially in arid and semi-arid regions (High Confidence).
- The current generation of transient GCMs, though much improved, does not offer the degree of watershed-specific information or anticipated variability in future climate required to allow robust estimates to be made regarding changes in water availability (High Confidence).
- Water demand management and institutional adaptation are the primary components for increasing the flexibility of water resources systems to meet increasing uncertainties due to climate change (High Confidence).
- Increased streamflow regulation and water management regimes may be necessary to enable water systems to meet their goals (High Confidence).
- Isolated single-reservoir systems are less adaptable to climate change than integrated multiple-reservoir systems (High Confidence).
- Technological innovations and cost-effective technologies have already played a major role in water management; likely future technological changes can serve to mitigate many of the consequences of climate change (Medium Confidence).
- Changes in the mean and variability of water supply will require a systematic reexamination of engineering design criteria, operating rules, contingency plans, and water allocation policies (Medium Confidence).
- Temporal streamflow characteristics appear to be more variable under future climate scenarios, and amplification of extremes appears likely (Low Confidence).

14.1. Introduction

14.1.1. Objectives

Water resources are an important aspect of the world's social, economic, and ecological systems. Agriculture, hydroelectric power production, municipal and industrial water demands, water pollution control, and inland navigation are all dependent on the natural endowment of surface and groundwater resources. Civilizations have flourished and fallen as a consequence of regional climatic changes, and many "hydraulic civilizations" were formed around the need to control river flow. This endowment is not evenly distributed spatially or temporally. With the imbalance of water supply and demand, many nations are in water-scarce situations and face water crises at a local level. From 1940 to 1987, global water withdrawals increased 210%, while the world's population increased by 117% (Gleick, 1993; Shiklomanov, 1993). A global water resources assessment for the year 2025 (Strzepek *et al.*, 1995) suggests that for the United Nations median population forecast of 8.5 billion (a 55% increase over 1990) and a globally balanced economic growth path, global water use may increase by 70%.

The potential for the world to face a water-stressed condition in 2025 under population and economic growth makes assessment of possible water resource impacts associated with climate change an essential component of the IPCC assessment. Changes in hydrological processes are discussed in Chapter 10. This chapter examines the impact on water supply and use and evaluates possible water management response strategies. Water resources management is the interaction of technology, economics, and institutions for the purpose of balancing water supply with water demand and coping with hazards associated with hydrological extremes. The goal of this chapter is to provide an understanding of the sensitivity of the components of water resources systems to potential climate change. Because the water management process occurs at local and regional levels, this chapter cannot provide an assessment of global or continental impacts of climate change; however, it will glean from available literature the sensitivities of various water systems to increases and decreases in river runoff and examine changes in water demands due to changes in regional climate. Because changes in water use will affect many sectors of society and the economy, other chapters in this volume are referenced, such as those addressing wetlands, coastal zones and small islands, energy supply, transportation, human settlements, agriculture and forestry, fisheries, health problems, and financial services.

The main message of this chapter is that climate change will impact the water resources systems of the world but that we will be able to adapt—though at some cost economically, socially, and ecologically. Some analysts (Rogers, 1993a, 1993b; Klemes, 1993; Stakhiv, 1994) feel that current systems will respond well and costs will be minimal; others feel that adaptation will be difficult and in some cases extremely costly. The reason for this disparity is that the sensitivities, impacts, and costs are nonuniformly distributed across the globe. Most analyses have been conducted for regions in the developed world. The limited case studies implemented on river basins in developing countries show that sensitivities, impacts, and costs may be high. Also, much more research is needed on the response of water demands to climate change, especially non-withdrawal uses such as recreation, fisheries, wetlands, ecosystems, and waterfowl protection.

14.1.2. An Overview of Assessment Issues and Concepts

The two previous IPCC assessments outlined the extensive difficulties in conducting meaningful analyses of climate-change impacts on hydrology and water resources (Shiklomanov *et al.*, 1990; Stakhiv *et al.*, 1992). Since then, many studies have been conducted in different basins—almost exclusively in developed countries—but the general conclusions of the earlier IPCC assessments have not changed. The uncertainties of climate-change impact analysis, especially at the catchment scale, remain large.

It is necessary to distinguish between the physical effects of climate change—which are reviewed in Chapter 10—and the impacts, which reflect a societal value placed on a change in some physical quantity. The impact depends largely on the characteristics of the water-use system: In some cases, a large climate-change effect may have a small impact; in others, a small change may have a large impact.

There are many different types of water supply systems in operation in the world. The simplest "system" extracts water from a local stream or village borehole; this is characteristic for most of the developing world and rural areas in many developed countries. Such supply systems, with no storage, are potentially very sensitive to climate change. The next system level consists of a single managed source—which may be a river, reservoir, or aquifer—coupled with a distribution network to provide water to users and possibly also to treat wastes and return effluents to the river. The sensitivity of such a system to climate change will depend on its characteristics—for example, on the storage-to-runoff ratio and on the seasonal distribution of water supply and demand. The most sophisticated systems are integrated networks, comprising several sources and possibly involving the transfer of water over large distances. Such systems usually are found only in developed countries; their sensitivity to climate change will depend on system structure and the degree of utilization.

Most water managers, whether they are with agencies dealing with multiple-reservoir systems or with small utilities dependent on groundwater, are concerned with three issues: new investments for capacity expansion; the operation and maintenance of existing systems; and modifications in water demand (Rogers, 1993b). Most developed countries have completed major capital-intensive developments of water resources infrastructure. Water managers in those countries operate under conditions of stable population and increased pressure for the incorporation of environmental protection objectives into the operation of existing water resources systems. The main issue they face is reallocation of existing water among competing uses. This requires continuous adaptation driven by new hydrologic information, ecological constraints, water-quality standards, and shifts in demands and preferences. Water-supply entities also may wish to explicitly or implicitly reconsider the level of service delivered. Institutional adaptation—consisting of changes in organizations, laws, regulations, and tax codes—may be the most effective means for aligning water demands with available supplies (Frederick, 1993; Rogers and Lydon, 1994; World Bank, 1994). This situation is a reality for managed water systems but less so for unmanaged systems (e.g., wetlands) dependent entirely on river flow, groundwater level, or precipitation. Water managers in developing countries are facing population growth-driven increases in water demands, and these demands are met primarily by increasing the water supply via capital-intensive investments to develop infrastructure. With planning and construction times of 20 to 30 years or more for major water projects, the question asked by many water resources managers in developing countries (Riebsame *et al.*, 1995) is how climate change might impact the design of new water resource infrastructure.

Table 14-1: *Summary of effects of global warming on water supply.*

Effect of Global Warming	Impact on Water Supply Reliability
Change in river runoff	Yield in direct water abstraction Yield in reservoir systems
Change in groundwater recharge	Yield of groundwater supply systems
Change in water quality	Yield of abstraction systems
Rise in sea level	Saline intrusion into coastal aquifers Movement of salt-front up estuaries, affecting freshwater abstraction points
Change in evaporation	Yield of reservoir systems

14.2. Impact of Climate on Water Supply

14.2.1. Introduction

Climate change is likely to have an impact on both the supply of and demands for water. This section focuses on the supply of water, looking at the river catchment scale, the global and regional context, and water quality; Section 14.3 considers impacts on demands. Most climate-change impact studies have taken the form of sensitivity analyses by feeding climate-change scenarios into hydrological models. The outputs of these studies tend to be expressed in terms of changes in the reliable yield of the systems, changes in the volume of water that can be supplied, or changes in the risk of system failure. Virtually all of the studies have simulated what would happen in the absence of adaptation to change. In practice, however, water management authorities will adapt using existing or new management options—as shown to be feasible in the Great Lakes region by Chao *et al.* (1994) and Hobbs *et al.* (1995)—although such adaptation may incur added costs and involve tradeoffs that result in reductions in service for some water users. Only a few studies (Riebsame *et al.*, 1995; Strzepek *et al.*, 1995) have considered factors other than climate change that might affect water resources over the next few decades, such as population growth, economic development, and urbanization.

There are several possible effects of global warming on the amount of water available within a catchment or water supply area; these are summarized in Table 14-1. The relative importance of each characteristic varies considerably among catchments, depending not only on the hydrological change but also on the characteristics of the supply system. For example, a conjunctive-use system involving several reservoirs, river regulation, and groundwater boreholes will be affected differently than a supply system based on direct abstractions from an unregulated river.

Obviously, changes in river runoff will affect the yields of both direct river abstractions and reservoir-based supply systems, and changes in groundwater recharge will affect groundwater yield. Changes in water quality will affect the amount of suitable water available to a supply system. A rise in sea level has two potential effects. First, there is a risk of saline intrusion into coastal aquifers, contaminating the water supply. This is a major potential threat—particularly to small, low-lying islands, whose main source of water frequently is a shallow lens of freshwater lying just a few meters above sea level (see Chapter 9). Second, a rise in sea level would mean that saltwater could penetrate further upstream into an estuary, perhaps threatening low-lying freshwater intake works. The effects of these changes in the amount of water available on water uses—and hence on system risk and reliability—will be influenced by changes in demands.

Section 14.2.2 reviews some published studies on changes in resource availability at the catchment scale (which focus mostly on surface water resources), and Section 14.2.3 broadens the perspective to a regional scale. Section 14.2.4 reviews how changes in water quality affect the availability of water resources. The effects of changes in hydrological regimes and water quality are examined in Chapter 10.

14.2.2. The River Catchment Scale

This section looks at the availability of water supplies under changed climatic conditions. For hundreds of years, people have adapted their habits and economic activities to relatively variable climatic and hydrological conditions—implicitly assuming that the average climatic state and the range of variability are stable.

This assumption may no longer be valid in some regions of the world because of possible alterations in stochastic properties of hydrological time series. Differences in the output of GCMs coupled with the variety of hydrological transfer models make it difficult to offer a reliable region-specific assessment of future water availability. It is doubtful whether the current technique of conducting "worst-case" analyses—wherein the most extreme scenario of a given GCM is used to develop hydrological responses—is useful for a critical appraisal of regional sensitivities to climate change. If anything, this type of analysis skews the evaluation and deflects the search for pragmatic responses. Progress in hydrological sensitivity analyses in developed nations is accompanied by large information gaps for developing countries that are most often affected by aridity and desertification. Although numerous new water resources impact studies have been conducted, few are from Africa, Asia, South America, or developing countries in general.

Studies that have considered possible changes in water supply in specific areas fall into three groups. The first group of studies infers changes in potential supply directly from modeled changes in annual and monthly water balance. Problems in maintaining summer supplies from direct river abstractions may be inferred, for example, if summer river flows are projected to decline (Arnell and Reynard, 1993). The second group of research has considered the sensitivity of hypothetical supply systems—usually single reservoirs—to changes in inputs. The third group of studies has largely been conducted since IPCC's 1992 Supplementary Report and consists of investigations into specific water-supply systems. Some have looked at individual reservoirs or groundwater resource systems; others have examined entire integrated water-supply systems, including real system operating rules. Table 14-2 lists these studies; several are summarized below.

These studies have simulated river flows using conceptual hydrological models but have used a variety of different scenarios. Mimikou *et al.* (1991), Wolock *et al.* (1993), Nash and Gleick (1993), and Kirshen and Fennessey (1995) all examine the effects of arbitrary changes in precipitation and temperature inputs to investigate the sensitivity of their modeled water resources systems to changes in inputs. A 20% reduction in rainfall in the Acheloos basin in Greece, for example, would increase the risk of system failure (inability to provide target supplies) from less than 1% to 38% (Mimikou *et al.*, 1991); similarly, with a 20% reduction in rainfall, the New York City reservoir system in the upper Delaware valley would be in a "state of crisis" between 27% and 42% of the time, depending on temperature increases (Wolock *et al.*, 1993). Nash and Gleick (1993) and Kirshen and Fennessey (1995) additionally use scenarios based on equilibrium GCM experiments, as do Gellens (1995), Kaczmarek and Kindler (1995), Riebsame *et al.* (1995), Salewicz (1995), Strzepek *et al.* (1995), and, in a generalized way, Hewett *et al.* (1993). All of these studies indicate that water resources systems could be very vulnerable to change in climatic inputs and that a small change in inputs could lead to large changes in system performance, but that there is considerable variability between scenarios. Riebsame *et al.* (1995) found isolated single-reservoir systems in arid and semi-arid regions to be extremely sensitive and less able to adapt (greater than 50% decreases in reservoir yields), with economic and ecological crisis conditions developing in some basins under climate change and seasonal flooding problems in others.

The remaining set of studies (Hobbs *et al.*, 1995; Lettenmaier *et al.*, 1995a, 1995b, 1995c, 1995d, 1995e; Steiner *et al.*, 1995; Waterstone and Duckstein, 1995; Shiklomanov *et al.*, 1995), largely undertaken in the United States, have used scenarios based on the three most recent transient GCM simulations (GFDL-tr, UKMO-tr, and MPI-tr) to investigate possible impacts on integrated, multipurpose water resources systems. All of these studies note the difficulties in forecasting meaningful impacts under the wide range of uncertainties inherent in the analysis; in many cases, the GCM simulations did not reproduce current catchment climate very well. However, a general conclusion from the studies is that even with the large variability in future climate represented by the three transient GCM experiments, most of the systems investigated possess the robustness and resilience to withstand those changes, and adequate institutional capacity exists to adapt to changes in growth, demands, and climate. This conclusion is in contrast to that of many other studies—some summarized above—that have found large changes in system reliability under climate change. There are two main reasons for this difference: First,

Table 14-2: *Investigations into effects of global warming on specific water resources systems.*

Location	Reference
Africa: Nile River	Strzepek *et al.* (1995)
Africa: Zambezi	Riebsame *et al.*(1995)
Africa: Zambezi River	Salewicz (1995)
Asia: Ganges and Brahmaputra	Kwadijk *et al.* (1995)
Asia: Indus River	Riebsame *et al.* (1995)
Asia: Mekong River	Riebsame *et al.* (1995)
Belgium	Gellens (1995)
England (southeast)	Hewett *et al.* (1993)
Greece	Mimikou *et al.* (1991)
Poland	Kaczmarek and Kindler (1995)
Russia and Ukraine: Dnipro River	Shiklomanov *et al.* (1995)
Uruguay	Riebsame *et al.* (1995)
USA: Boston area (1)	Kirshen and Fennessey (1995)
USA: Boston area (2)	Lettenmaier *et al.* (1995b)
USA: Colorado River	Nash and Gleick (1993)
USA: Columbia River	Lettenmaier *et al.* (1995c)
USA: Delaware River	Wolock *et al.* (1993)
USA: Great Lakes	Hobbs *et al.* (1995)
USA: Missouri River	Lettenmaier *et al.* (1995d)
USA: Potomac River	Steiner *et al.* (1995)
USA: Rio Grande River	Waterstone and Duckstein (1995)
USA: Savannah River	Lettenmaier *et al.* (1995a)
USA: Tacoma area	Lettenmaier *et al.* (1995c)

the transient scenarios tend to produce smaller changes in climate than the scenarios based on earlier GCMs; second, the transient-GCM studies examine highly integrated systems, which are inherently more robust than the isolated single-reservoir systems investigated in most other studies.

In some countries, water is predominantly taken from rivers, lakes, and reservoirs; in others, it largely comes from aquifers: Only 15% of Norwegian water is taken from aquifers, for example, whereas 94% of Portuguese supplies comes from groundwater. There have been very few studies of changes in groundwater recharge and implications for aquifer yield. Hewett *et al.* (1993) simulate an increase in recharge, and hence an increase in reliable yield, in part of the chalk limestone aquifer in southern England, but different scenarios in the same region suggest a decline in yield. The effect of a sea-level rise on saline intrusion into coastal aquifers has been investigated in a number of small islands and has been found to be potentially significant (see Chapter 9). Studies in Britain, however, have found that although there are many coastal aquifers potentially at risk, a rise in sea level would have little effect on intrusion and yields (Clark *et al.*, 1992). Saline intrusion along estuaries generally has been found to pose limited threats to freshwater intakes because the change in the position of the salt front is small relative to the intertidal range (Wolock *et al.*, 1993; Dearnaley and Waller, 1993).

This section has introduced some of the studies into water resource availability that have been undertaken in the last five years. There are several points to draw in conclusion. First, there are considerable uncertainties in estimating impacts, due partly to uncertainties in climate-change scenarios and partly to difficulties in estimating the effects of adaptations—both autonomous and climate-induced—over the next few decades. Second, there is evidence that isolated, single-source systems are more sensitive to change than integrated, multipurpose systems, which are considerably more robust. Much of the world's water, however, is managed through single-source, single-purpose systems. Third, there is a suggestion that, in countries with well-managed, integrated water resources, the additional pressures introduced by climate change could be met, with some costs, by techniques already in place to cope with changing demands and management objectives. There is little information, however, about the economic and societal costs of this adaptation. Waterstone *et al.* (1995), for example, conclude that institutional adaptation—changes in water laws, organizations, prices, fees, water marketing, and reservoir operating criteria—could serve to ameliorate the combined effects of increasing population and warming in the semi-arid Rio Grande basin in the southwestern United States, and Hobbs *et al.* (1995) believe that conventional management practices for coping with fluctuating lake levels would be capable of mitigating the effects of climate change on the Great Lakes. Fourth, and perhaps most important, some studies (Riebsame *et al.*, 1995; Strzepek *et al.*, 1995) show that water resources in developing countries often are small-scale, isolated, and under considerable stress and may have a difficult time adapting to climate change effectively. Although the latter studies cover a range of hydroclimatic zones, there is still a major gap in our understanding of the impacts of climate change on the less-developed world—a situation that urgently needs to be rectified. The U.S. Country Studies Program currently is cooperatively studying the vulnerability and adaptability of water resources systems in more than 30 developing countries, with results expected in 1996.

14.2.3. The Global and Regional Context

The growing interest in possible consequences of anthropogenic climate change on regional water resources has given rise to a wealth of studies on the sensitivity of water balance to climatic variables. Much less information is available on the economic and societal consequences of projected global warming. The heaviest current pressures on water resources are the increasing population in some parts of the world and increasing concentrations in urban areas. The illusion of abundance of water on the Earth has clouded the reality that in many countries renewable freshwater is an increasingly scarce commodity (Postel, 1992; World Bank, 1992, 1994; Engelman and LeRoy, 1993). Climate change is likely to have the greatest impact in countries with a high ratio of relative use to available renewable supply. Regions with abundant water supplies are unlikely to be significantly affected, except for the possibility of increased flooding. Paradoxically, countries that currently have little water—for example, those that rely on desalination—may be relatively unaffected.

One study (Strzepek *et al.*, 1995) suggests that, although global water conditions may worsen by 2025 due to population pressure, climate change could have a net positive impact on global water resources. This result, presented in Figure 14-1, is based on runoff characteristics obtained for one particular climate scenario (Miller and Russel, 1992) and should be interpreted with caution. Another macroscale study (Kaczmarek *et al.*, 1995)—based on three transient climate scenarios—leads to a similar conclusion for the Asian continent, while suggesting that in Europe changed climatic conditions will be associated with some decrease of per capita water availability.

Following the concept of Falkenmark and Widstrand (1992) of a water stress index based on an approximate minimum level of water required per capita, Engelman and LeRoy (1993) use 1,000 m^3 per person per year as a benchmark for water scarcity around the world. They found that in 1990 about 20 countries, with a total population of 335 million, experienced serious chronic water problems and that by 2025, 31 countries with 900 million inhabitants could fall into this category because of expected population growth. Table 14-3 summarizes the combined impact of population growth and climate change on water availability in selected countries, based on the IPCC (1992a) socioeconomic scenarios and the results of three transient GCM runs. The second column lists per capita water availability for the present (1990); the third column shows water availability for current climatic conditions, reflecting population growth alone to the year 2050. The last column shows the range of

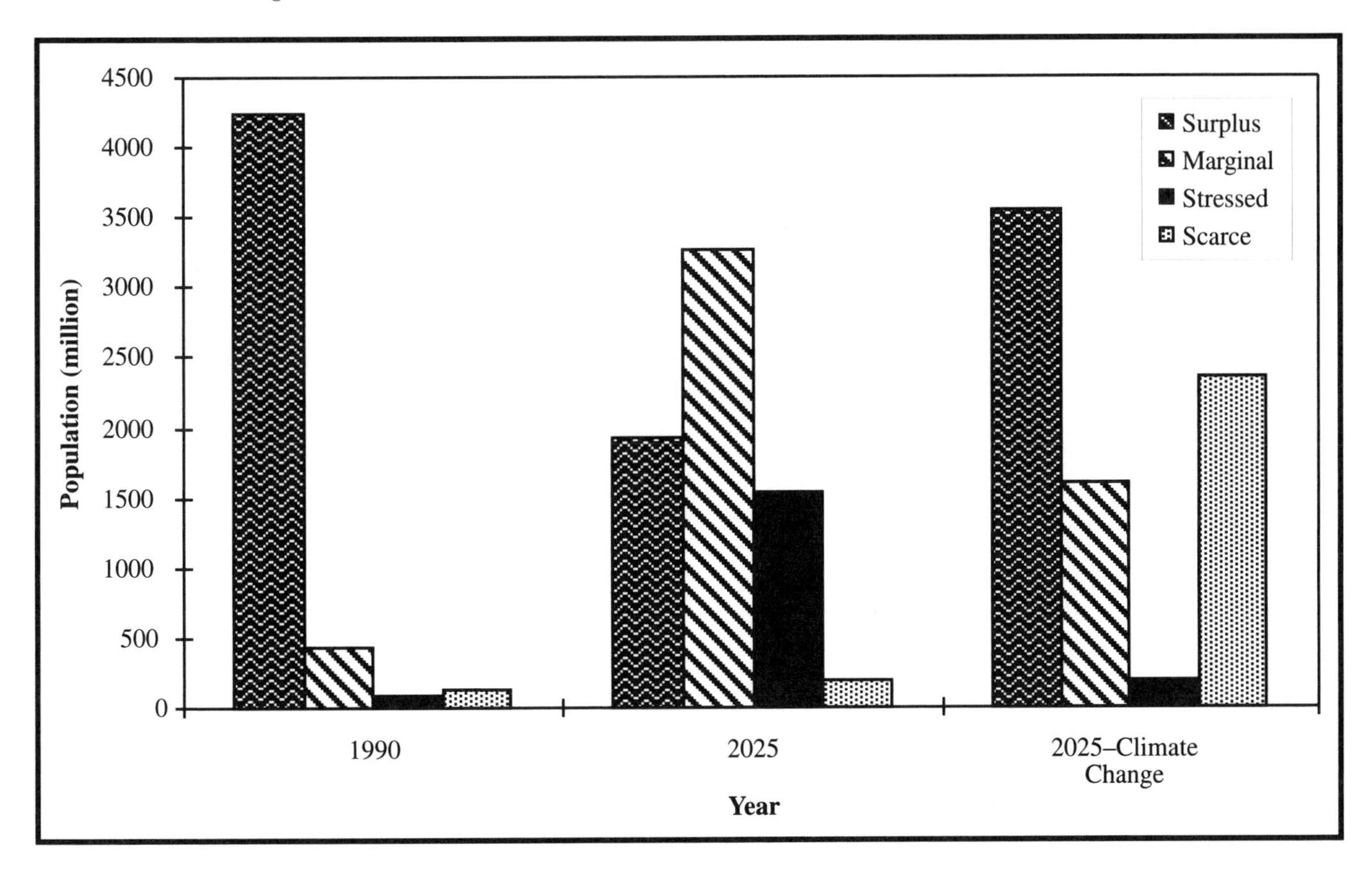

Figure 14-1: Global freshwater vulnerability (Strzepek *et al.*, 1995).

combined effects of population growth and climate change for the three transient scenarios. The sensitivities of national water supplies to changes in temperature and precipitation were estimated by a method proposed by Kaczmarek (1990). It should be added that the future water availability data do not take into account possible changes in water resources systems development (e.g., increased storage, desalination).

The results show that in all countries with high population-growth rates, future per capita water availability will decrease independent of the assumed climate scenario. Large discrepancies may be noted among results obtained for some countries by means of various atmospheric models. This example clearly demonstrates how difficult it would be to initiate water resources adaptation actions based on currently available methods of climate predictions. It can be expected that in many regions of the world, nonclimatic factors will dictate what measures must be undertaken to secure sustainable water supply (Frederick, 1994; Rogers and Lydon, 1994). Predicted climate changes, however, could redistribute water supplies, adding a new, highly uncertain component to the challenge of managing water resources.

Although it is appealing to devise thresholds or benchmarks for water scarcity, such as the one proposed by Falkenmark and Widstrand (1992)—in which economic water scarcity is defined as the condition in which renewable freshwater availability falls below 1,000 m^3 per person per year—it is important to realize that such a threshold is useful only as a rough benchmark for comparison of relative conditions, perhaps to serve as a cautionary flag. Other comparable thresholds have been suggested at 1,700 m^3 per person per year to reflect a condition of water stress, whereas the World Bank (Falkenburg *et al.*, 1990) suggests a 500 m^3 per person per year threshold. On the other hand, Rogers (1992) considers such thresholds not useful in developing water management strategies, noting that both Malta and Israel—with annual per capita water availabilities of 85 m^3 and 460 m^3, respectively—are doing quite well in managing, recycling, and reclaiming their very limited water supplies.

Water availability, food production, population and economic growth, and climate change are linked in a complex way. Conflicts among interests are inherent in regional water management, particularly in regions with scarce water resources. Four objectives important for sustainable water planning may be identified: economic efficiency, environmental quality, equity considerations, and reliability. Transformation in the structure and characteristics of water supply and demand due to climatic and nonclimatic factors may add new aspects to existing social and political problems. Socioeconomic factors greatly influence the ability to solve these problems in the absence of necessary institutions, capital, and technology.

Access to freshwater may be complicated by conflicts arising over rights to water in shared river basins (e.g., Mekong and Nile—Riebsame *et al.*, 1995) and in aquifers that cross international borders (Engelman and LeRoy, 1993). A great number

Table 14-3: Water availability (m^3/yr) in 2050 for the present climatic conditions and for three transient climate scenarios (GFDL, UKMO, MPI).

Country	Present Climate (1990)	Present Climate (2050)	Scenario Range (2050)
China	2,500	1,630	1,550–1,780
Cyprus	1,280	820	620–850
France	4,110	3,620	2,510–2,970
Haiti	1,700	650	280–840
India	1,930	1,050	1,060–1,420
Japan	3,210	3,060	2,940–3,470
Kenya	640	170	210–250
Madagascar	3,330	710	480–730
Mexico	4,270	2,100	1,740–2,010
Peru	1,860	880	690–1,020
Poland	1,470	1,250	980–1,860
Saudi Arabia	310	80	30–140
South Africa	1,320	540	150–500
Spain	3,310	3,090	1,820–2,200
Sri Lanka	2,500	1,520	1,440–4,900
Thailand	3,380	2,220	590–3,070
Togo	3,400	900	550–880
Turkey	3,070	1,240	700–1,910
Ukraine	4,050	3,480	2,830–3,990
United Kingdom	2,650	2,430	2,190–2,520
Vietnam	6,880	2,970	2,680–3,140

of water resources systems are shared by two or more nations. In several cases, there have already been international conflicts. As a result of population pressure, and in cases of negative impacts of climate change, tensions are likely to increase. In order to avoid future conflicts over water use among riparian countries, joint legal agreements should be established. One possibility is to form international water agencies or commissions, with terms of reference including inventory, assessment, monitoring, and apportionment of water resources and due account for possible changes caused by climatic trends. Joint planning is essential for basin-scale water resources development and management in order to cope with negative consequences of climate change. International water agencies should arbitrate on regional water issues and be supported by national legal frameworks to back the regional arbitration accordingly. To be able to fulfill this task, international water agencies in developing countries require well-trained people with knowledge of global-scale processes and, in some cases, external funding for research and development.

14.2.4. Implications of Changes in Water Quality

Water management is concerned not merely with the supply of water but with the supply of water of appropriate quality as well. The definition of "appropriate" varies among uses. Potable water has to be of the highest quality, whereas industry and, to a lesser extent, agriculture can use lower-quality water. Irrigation increasingly employs recycled "dirty" water, raising concern about salinity and public acceptance of wastewater use. Many water management problems around the world today, in fact, relate to the quality of surface water and groundwater. Climate change might exacerbate some of these problems by complicating an already expensive, evolving management process. Water-quality problems usually stem from some form of pollution, ranging from the discharge of untreated sewage into watercourses to the discharge of treated sewage effluent, the leaching of agricultural chemicals, and chemical and thermal pollution from industry. There is a considerable range of experience in dealing with water quality. In some countries, the focus is on preventing poor water quality; in others, effort is directed toward rehabilitation, treatment, sanitation, and public health. In many countries, water quality is hardly addressed at all.

The potential effects of climate change on aquatic ecosystems are reviewed in Chapter 10. Of particular concern for water users are dissolved oxygen content, nitrate and organic pollution concentrations, sediment load, and salinity. The main conclusion is that rivers that presently have poor water quality are likely to be those most affected by changes in temperature, lower flow rates, and increased input of pollutants; climate change is therefore likely to exacerbate water-quality problems in places where such problems already are potentially severe.

Water management agencies in many countries are spending significant sums on maintaining and improving surface and groundwater quality. Although approaches vary, improvement plans tend to include target water-quality and effluent standards, infrastructure for treating effluent returns and polluted water, policies to prevent pollution, and policing. These actions also can be used to maintain and improve water quality under climate change, obviously at some additional cost. For example, it may be necessary to reduce the quantity of treated effluent that can be discharged to a stream with reduced flows, with consequent implications for the discharging organization.

14.3. Impact of Climate on Water Demands

Section 14.2 considered changes in the ability of water resources systems to supply water; this section looks at the demand side. It is useful to distinguish between offstream demands—specifically domestic, industrial, and agricultural demands—and instream demands, such as power generation, navigation, recreation, and ecosystems protection. Increases in water demand are driven by population and economic growth. Demand management has two dimensions: the first long-term, the second in response to short-term shortages. In the long term, one can reduce offstream water demands substantially through technological, economic, legal and administrative, and educational measures. In the short term, demand during temporary supply shortages can be managed through demand reduction measures (such as rationing) and public education.

14.3.1. Agricultural Water Demands

At present, more than 65% of global water withdrawals is for agricultural use; much of this is evaporated and consequently lost to catchment runoff. Irrigation increased significantly until the 1970s, when the rate of expansion fell sharply; since 1980, the annual rate of expansion has been less than 1%—less than the rate of population increase (Postel, 1992). Rapid expansion in the future is unlikely as the cost of developing new schemes increases and investments in irrigation decline. Agricultural irrigation practices are inefficient in many areas of the world. Changes in irrigation technology (such as the use of drop irrigation) often can compensate for anticipated increases in food demands. The effects of climate change on agricultural policy and irrigation requirements are discussed in Chapter 13. It is important to emphasize that the effects of climate change on agricultural demands for water, particularly for irrigation, will depend significantly on changes in agricultural potential, prices of agricultural produce, and water costs.

Both rainfed and irrigated crops will require more water in a warmer world, and this water may not be available through increased precipitation. Allen *et al.* (1991) simulate changes in irrigation demand in the Great Plains region of the United States, showing that the demand for water to irrigate alfalfa would increase, due largely to increases in the length of the crop growing season and crop-water requirements during summer. Another U.S. study—based on a Thornthwaite water-balance model (McCabe and Wolock, 1992)—indicates that, for a broad range of increases in temperature and precipitation, annual irrigation demand increases, even with a 20% increase in precipitation. In a study in Lesotho, Arnell and Piper (1995) simulate an increase in irrigation demands of 7% with a 10% decline in runoff and an increase of more than 20% with a 2°C increase in temperature. They also examine the performance of a hypothetical reservoir supplying irrigation water and find major changes in the reliability of supply. Studies in the UK lead to the conclusion that an increase in temperature of 1.1°C by 2050 may result in an increase in spray irrigation demands of 28%, over and above a projected 75% increase to meet growing demands (Herrington, 1995). Similar calculations in Poland result in a 12% increase in irrigation demands with a 1°C temperature increase and a 1.8% change in water requirements with a 1% precipitation change during the vegetation season. Irrigation demand seems to be more sensitive to changes in temperature than to changes in precipitation. Model results indicate that increased stomatal resistance to transpiration counteracts the effects of temperature increases on irrigation demand.

14.3.2. Municipal Demands

Municipal demands are essentially for domestic and commercial uses. In many developed countries, some components of demand are decreasing due to greater appliance efficiency, but others are increasing as new appliances, such as waste disposal units and automatic washing machines, become more widespread. Herrington (1995) estimates changes in the components of domestic demand in southern England, with and without climatic change and assuming no change in water pricing policies. An increase in per capita demand of 21% is projected between 1991 and 2021 without climate change, with an additional 5% increase due to global warming—largely due to an increase in garden water use. Other studies have found a similar percentage change in domestic and municipal demand due to global warming (Cohen, 1987; Kaczmarek and Kindler, 1989; Kirshen and Fennessey, 1993; Hanaki, 1993; Steiner *et al.*, 1995). Little is known about the impact of climate change on domestic water use in developing countries, but nonclimatic factors—population growth, economic development, water-use efficiency, and water pricing—probably will dominate in shaping trends of future water use in most African and Asian regions.

14.3.3. Industrial Water Use and Thermal and Hydropower Generation

In most developed countries, the demand for water for industrial purposes is declining as traditional major water users such as the steel industry decline in significance and as water is used more efficiently. It should be added, however, that in many cases these water-consuming industrial users are relocating to the developing world, complicating the water resource situation in those regions. Strzepek and Bowling (1995) have found that under a moderate growth assumption, global industrial water use in 2025 may increase by 1.7 to 2.3 times 1990 levels, and most of this growth will occur in the developing world. Climate change is expected to have little direct impact on industrial water use.

A change in water quantity may affect the degree to which demands for cooling water can be satisfied, and a rise in water temperature will reduce the efficiency of cooling systems (Dobrowolski *et al.*, 1995). It also might be more difficult to meet regulatory constraints defining acceptable downstream water temperatures, particularly during extreme warm periods. Several French nuclear power stations were forced to close down or operate well below design capacity during the drought of 1991. A reduction in water availability might lead to an increase in the use of closed-cycle cooling systems as simulated in the Tennessee Valley Authority system (Miller *et al.*, 1993). Changes in water availability and temperature may not cause significant impacts on long-term or annual total production potential—only a 2% decline in annual net system generation was found in the Tennessee Valley—but they could cause short-term operational problems during critical periods.

A change in hydrological regimes has an obvious potential impact on hydropower production but also may affect thermal power generation and general industrial demands for cooling water. Mimikou *et al.* (1991) show a very large change in the risk of being unable to generate design power from hydropower reservoirs in central Greece. In Norway, Saelthun *et al.* (1990) find increased generation potential due largely to a shift in the timing of inflows, mainly as a result of reduced snowfall; the current "waste" of power in the spring is much reduced.

Riebsame *et al.* (1995) examine hydropower generation in four developing-country basins. In the Zambezi basin, the possible reduction of hydropower at Lake Kariba could be replaced by the construction of a new plant at Batoka Gorge, but for a significant cost. Hanemann and McCann (1993) examine the economic cost of changes in hydropower potential in northern California. Under a scenario based on the GFDL equilibrium GCM, annual hydropower production would decrease by 3.8%; production would be 48% higher in January but as much as 20% lower during the peak-load summer months. Water would have to be released during the spring to leave room for flood control. As a result of the reduced hydropower generation, production of power by natural gas would have to increase by 11% to meet the same demands, leading to a $145 million (1993 prices) increase in annual system costs (12.5%). In some cases, the expected shift in the runoff hydrograph, combined with changes in the distribution of irrigation water demands, may lead to reduced energy generation.

14.3.4. Navigation

Changes in flow regimes and lake levels can be expected to affect navigation potential, but there have been very few studies on this topic. The sensitivity of river navigation to extreme conditions was well illustrated on the Mississippi River during the drought of 1988, when river traffic was severely restricted (with consequent effects on the agricultural sector due to difficulties in transporting the grain harvest). High river flows also restrict navigation by increasing energy costs and flooding riverside facilities. Navigation on the river Rhein, for example, is constrained by periods of both high and low flows, and considerable sums are spent on dredging and maintaining channels; a change in sediment load could have major implications for these activities. An increase in temperature, however, would increase the duration of the navigation season on rivers affected by seasonal ice cover. Studies in the Great Lakes (Chao *et al.*, 1994) suggest that a longer shipping season due to a reduction in ice cover would just compensate for lost draft due to lower levels; shipping companies would be able to adjust their operating season in the Great Lakes because many of the raw materials transported are stockpiled.

14.3.5. Recreation and Other Instream Water Uses

There have been very large investments in water-based recreation in many countries, and many large facilities are operated to maximize recreation potential. The effects of global warming on recreation are difficult to determine. Changes in the volume of water stored in a reservoir might affect the use of the reservoir for recreation (Frederick, 1993), and a change in water quality might also affect the recreational use of the water. There are established procedures for estimating the economic benefits of access to recreation, but the sensitivity of recreational use to hydrological characteristics—reservoir storage volume, water quality, and so forth—is not well known; the effects of possible changes therefore are difficult to quantify. One exception is the Great Lakes, where there are clear relationships between beach area, length of recreation season, and recreational benefits: Both beach area and recreation season length would increase under climate warming, resulting in an estimated doubling of recreational benefits, according to Chao *et al.* (1994).

One-tenth of the world's commercial fish yield is obtained from inland waters (Covich, 1993), and recreational fishing also has high economic value. The effects of changes in water temperature, water quality, and river flow regimes on fish populations are outlined in Chapter 10; this section focuses on implications of changes for sport fishing and aquaculture. Hanemann and Dumas (1993) simulate the effects of one global warming scenario on Chinook salmon runs in the Sacramento River, California, and find a reduction in the salmon population largely due to reductions in spawning habitats. Loomis and Ise (1993) estimate the change in the economic value of recreational fishing in the Sacramento River based on an empirical relationship between the number of fish caught and the number of fishing trips made. The reductions in fish population result in an annual loss of recreational benefit of $35 million (1993 values)—a 23% decrease (assuming no change in anglers' willingness to travel to the river).

Stefan *et al.* (1993) conclude that, for the state of Minnesota, the impact of climate change on fishery resources may be significant. Overall fish production may increase, but cold-water fisheries will be replaced in part by warm-water fisheries. This coincides with findings in Poland and Hungary related to the impact of thermal pollution. The comprehensive study of fish yields performed by Minns and Moore (1992) in several hundred watersheds in eastern Canada shows considerable redistribution of fishery capability.

The final instream use to be considered in this section is ecosystem protection. Instream ecosystems demand a certain minimum quantity of water (which may vary throughout the year), and water managers in many countries increasingly are balancing these demands with those of more traditional water users. Several techniques are being developed to estimate instream demands (Stalnaker, 1993); these techniques can be used in principle to estimate the effects on these demands of changes in water availability. However, this has not yet been done, largely because the instream demand models are very uncertain. In principle, it might be possible to maintain certain aquatic ecosystems by managing river flow regimes to minimize changes, but this is perhaps not desirable in an ecological sense because it would create a system that would not be sustainable without human intervention. The general issue of managing the impacts of climate change on natural ecosystems is discussed in Chapter 10 and elsewhere.

14.3.6. Competition Between Demands

Water demand in general increases in all sectors with an increase in temperature; this is a well-accepted consequence of climate change. At the same time, regional and local precipitation changes, which will have important impacts on water demands, are much less clear. Studies of individual sectors and systems show great potential for adapting to water conditions in a changed climate. However, as we move into the 21st century, population and economic pressures may create water-stressed conditions in many parts of the world. Many of the responses being proposed to adapt to climate change require reduction of demands and reallocation of water among water-use sectors.

Present water management is concerned with reconciling competing demands for limited water resources. Currently, these conflicts are solved through legislation, prices, customs, or a system of priority water rights. Change in the amount of water available and water demands is likely to lead in many cases to increased competition for resources. Conflicts may arise between users, regions, and countries, and the resolution of such conflicts will depend on political and institutional arrangements in force. The challenge will be to create integrated demand/supply management systems, as discussed below.

14.4. Management Implications and Adaptation Options

14.4.1. Considerations for Response Strategies

In general, most countries and civilizations have faced water shortages due to natural climate variability, anthropogenic changes and desertification, or overexploitation and pollution of the resource base. Management of water resources inherently entails mitigating the effects of hydrological extremes and providing a greater degree of reliability in the delivery of water-related services. Because different uses have different priorities and risk tolerances, the balance points among them after climate change could be quite different from the present (e.g., hydropower and instream uses may be lost disproportionately compared to water supply). No enterprise is risk-free: Society decides on the level of risk-bearing through the acceptance of certain levels of risk and reliability, as expressed by cost-effective standards and criteria. The marginal cost of reducing each additional increment of risk typically rises rapidly as reliability approaches 100%. Hence, water managers usually deal with 90%, 95%, and 99% levels of reliability as useful performance measures of the available quality and quantity of water.

The same is true in mitigating other natural hazards, especially in traditional approaches to drought mitigation, flood control, and damage mitigation. Given some of the preliminary results of GCM experiments regarding potential changes in rainfall intensity and frequency (Gordon *et al.*, 1992; Whetton *et al.*, 1993), it appears that flood-related consequences of climate change may be as serious and widely distributed as the adverse impacts of droughts. This should raise concerns about dam safety and levee design criteria and spur reconsideration of flood plain management policies. The devastating floods of 1993 in the upper Mississippi River basin have resulted in a U.S. Interagency Floodplain Management Review Committee report (1994) on just such policy issues, exemplifying the need for constant adaptation in the field of water resources management. Comparable adaptations are anticipated in the wake of the most recent European floods of January 1995.

To alleviate human-induced droughts occurring at a regional scale as a consequence of inappropriate land-use practices, wise criteria for land use should be developed to minimize storm-induced runoff and, consequently, minimize erosion and nutrient loss and maximize interception and infiltration. Such strategies would maintain or enhance the recycling of moisture, which at a (sub)continental scale is necessary to sustain rainfall in the region. Most watershed management practices for erosion control and water harvesting do contribute, albeit unintentionally, to the reestablishment of moisture feedback to the atmosphere and, consequently, to the recycling of moisture and rainfall. In view of the vulnerability of regions presently affected by anthropogenic droughts and the dire consequences of resulting desertification, this field of research merits full attention.

The first IPCC reports (1990a, 1990b) contain a discussion of the philosophy of adaptation and a list of adaptation options suited to the range of water management problems that are expected under climate change. Based on a review of the most recent literature, no additional water management actions or strategies unique to climate change have been proposed as additions to the list, other than to note that many nations have pledged to implement action plans for sustainable water resources management as part of their obligation toward Agenda 21. In that respect, the principles laid out in that document would serve as a useful guide for developing a strategy that would enable nations, river basin authorities, and water utilities to prepare for and partially accommodate the uncertain hydrologic effects that might accompany global warming. The World Bank (1993) lays out a framework for water resources management that is expected to serve the needs of developing nations well into the next century and to meet the objectives of Agenda 21.

There are many possibilities for individual adaptation measures or actions. An overview of water supply and demand management options is presented by Frederick (1994) as part of an attempt to develop approaches for dealing with increasing water scarcity. A long-term strategy requires the formulation of a series of plausible development scenarios based on different combinations of population growth assumptions along with economic, social, and environmental objectives (Carter *et al.*, 1994). After these scenarios are established, taking into account the possibility of climate change, a set of alternative long-term strategies for water management must be formulated that consists of different combinations of water management measures, policy instruments, or institutional changes, and is designed to best meet the objectives of a particular growth and development scenario and its consequent CO_2 emissions rate. The range of response strategies must be compared and appraised, each with different levels of service reliability, costs, and environmental and socioeconomic impacts. Some will be better suited to dealing with climate change uncertainty (i.e., more robust and resilient), and others will focus on environmental sustainability. Some are likely to emphasize reliability of supply. The reality is that, after the application of engineering design criteria to various alternatives, the selection of an "optimal" path is a decision based on social preferences and political realities. Engineering design criteria, however, also evolve over time and are updated as new meteorological and hydrological records are extended and the performance of water management systems is tested under varying conditions.

All major institutions that deal with water resources planning and management agree that future water management strategies should include various cost-effective combinations of the following management measures:

- Direct measures to control water use and land use (regulatory, technological)
- Indirect measures that affect behavior (incentives, taxes)
- Institutional changes for improved management of resources
- Improvement in the operation of water management systems
- Direct measures that increase the availability of supply (reservoirs, pipelines)
- Measures that improve technology and the efficiency of water use.

Different strategies apply to different circumstances. Watersheds that have little or no control over natural flows and are largely dependent on precipitation must implement a different set of water management strategies than river basins with a high degree of control in the form of reservoirs, canals, levees, and so forth. Similarly, rapidly urbanizing areas will require different responses than agricultural regions. There is no standard prescribed approach. However, a rational management strategy undertaken to deal with the reasonably foreseeable needs of a region in the absence of climate change, according to the principles espoused in Agenda 21, also will serve to offset many of the possible adverse consequences of climate change.

14.4.2. Implications for Planning and Design

The nature of contemporary water resources management is such that countless numbers of principal factors, economic criteria, and design standards are incorporated simply because of the complexity of integrated water management (e.g., hydropower, ecosystem support, water supply) and objectives (e.g., reliability, costs, safety). Some factors that are routinely assessed inherently represent design thresholds such as the minimum instream flow required for maintaining an aquatic ecosystem or the "probable maximum flood" that is used for most dam-safety risk analyses. The accepted level of water supply reliability of a system is a threshold as well, determined essentially by public preference, economics, and engineering analysis. Planning, by its nature, has inherent risks and develops alternative plans that are packages of complementary actions, project regulations, and management measures that reduce risks in different water-use sectors (with a variety of socioeconomic and environmental impacts and a range of benefits and costs).

Engineering design is largely concerned with performance and reliability. Once a particular plan is selected, engineers ensure that each of the separate components functions as planned and that, collectively, the water management system performs reliably. This was true in Roman times (Frontinus, 98 AD) as well as in contemporary times. Water resources systems are designed to perform reliably over most but not all of the range of anticipated hydrological variability. That reliability criterion is determined through a combination of risk, costs, benefits, environmental impacts, and societal preference. Hence, if climate change alters the frequency, duration, and intensity of droughts and floods, new reliability criteria will evolve over time to adjust for the perceived changes in both availability and use—as will the corresponding types of adaptive behavior.

It also is useful to think of hydrological or watershed response sensitivity to change measured in terms of physical effects. A complementary notion is the susceptibility of various water-use sectors (e.g., hydropower, irrigation, recreation) to incremental changes of outputs (e.g., kilowatt-hours, revenues, visitors). Finally, the vulnerability to failure of a water management system itself—consisting of pipes, pumping stations, reservoirs, and delivery rates—also must be appraised in terms of reliability targets measured as changes in quality, quantity, and probability. To that end, water resources planners, hydrologists, and design engineers have developed a set of practices that explicitly address a range of hydrological, economic, and engineering risk and uncertainty factors and implicitly encompass some notion of sensitivities and thresholds. Fiering (1982) and Hashimoto *et al.* (1982a, 1982b) developed the concepts of robustness (sensitivity of design parameters and economic costs to variability); reliability (a measure of how often a system is likely to fail); resiliency (how quickly a system recovers from failure); vulnerability (the severity of the consequences of failure); and brittleness (the capacity of "optimal" solutions to accommodate an uncertain future). Many if not most of these concepts are analyzed as part of contemporary hydrological and water resources management decisions (Kundzewicz and Somlyody, 1993; Kaczmarek *et al.*, 1995). Riebsame *et al.* (1995) analyze the sensitivities and adaptabilities of five international river basins in the context of the criteria discussed above. The results are presented in Table 14-4.

Table 14-4: *Overall basin sensitivity and adaptability (Riebsame et al., 1995).*

Basin	Hydrological Sensitivity	Structural Robustness	Structural Resiliency	Adaptive Capacity
Uruguay	moderate	high	high	high
Mekong	low	low	high	low
Indus	moderate	high	moderate	high
Zambezi	high	low	low	low
Nile	high	high	low	low

There is a growing tendency to devise management systems that complement supply development with demand management. Nonstructural management measures are increasingly relied on to provide needed robustness without decreasing the overall reliability of a system. Hence, flood-control levees in the United States are now designed to provide varying levels of protection, whereas in the past there was a fixed level of flood protection based on a calculation of the "standard project flood." Today, a levee can be designed to offer protection against a flood with a 2% chance of occurrence (a 50-year return period) in conjunction with a well-organized flood warning and evacuation plan. For example, all 360 U.S. Corps of Engineers reservoirs have both drought contingency plans and flood warning and dam-safety evacuation plans. However, it is important to remember that the combination of numerous design factors, operating rules, reservoir storage allocation decisions, flood forecasting and evacuation planning, and drought contingency planning provides a considerable degree of robustness, resiliency, and flexibility to contend with uncertainty and surprises. Coupled with demand management and institutional and regulatory changes that are needed to cope with anticipated changes in population and demands, water management systems of this scale, if properly managed, offer a well-balanced strategy for dealing with risk and uncertainty, including many of the impacts of climate change. The purpose is to reduce, if not minimize, the adverse social, economic, and environmental consequences of changes in water resources regardless of the agent of change (Stakhiv, 1994), although the costs of adaptation to climate change could be substantial. Most important is the reality that water resources management is an inherently continuous adaptive endeavor at many different levels and spatial scales.

This is not to suggest that we can become complacent in our response to climate change. The challenges and barriers to implementing the water management principles of Agenda 21 are difficult enough to overcome. The water situation in the Middle East and North Africa is precarious and projected to deteriorate as a consequence of population growth and unplanned development. In sub-Saharan Africa (World Bank, 1994), the water situation is expected to worsen due to desert encroachment and drying up of water sources as a result of increased deforestation. Droughts, desertification, and water shortages are a permanent feature of daily life in those countries. The list of nations with water supply problems will expand with the accelerated pace of urbanization. By the year 2030, urban populations will be twice the size of rural populations. By 2000, there will be 21 cities in the world with more than 10 million inhabitants, 17 of which will be in developing countries. It is expected that the population without safe drinking water will increase from 1 billion people in 1990 to 2.4 billion by 2030, assuming a "business-as-usual" scenario. Similarly, population without adequate sanitation will grow from 1.7 billion to 3.2 billion by the year 2030 (World Bank, 1992). Addressing these issues would make it easier to cope with the impacts of climate change when and if they become significant (IPCC, 1992; Goklany, 1992).

On the other hand, the picture may not be as bleak as it appears, in the sense that it is easier and more efficient to organize a water supply, treatment, and delivery system for urban areas with concentrated populations. Also, demands can be more easily managed to promote water-use efficiency. Although municipal and industrial water use will grow, per capita use is likely to decrease and the quality of drinking water will increase with centralized treatment. Many future urban water demands are likely to compete with the irrigated agricultural sector—which uses about 88% of all water withdrawn in Africa, 86% of all water withdrawn in Asia, and 87% of all water withdrawn in the arid Middle East and North Africa (World Bank, 1994). New sources of supply will have to be developed, but existing water use will have to be more effectively managed and efficiently used.

The reality is that increasing water demands will intensify competition for scarce water and further concentrate water use in urban centers. During the next 30 years, these real and complex needs will preoccupy water managers. Water management planning will address these needs, which precede the climate-change signal and thereby serve as a *de facto* adaptation mechanism. Climate change, in its many manifestations, is likely to be a perturbation on what are already difficult and complicated water management problems. Existing drainage systems, water-control structures, and conveyance and distribution systems typically are designed on the basis of design floods or droughts of different return periods and/or annual exceedance probabilities, which are derived from past failures and associated perceived degrees of tolerable risk and economic costs. Because there is a significant turnover in water management infrastructure, with considerable maintenance and major rehabilitation occurring in most countries about every 30 years, it can be expected that the operating capacity of such structures can be made to conform to evolving changes in climate. Analytical tools are available to provide the necessary degree of confidence in the design and operation of such systems in a reliable manner.

14.4.3. Impact on Flood Risk and Management

Although the potential impact of global warming on the occurrence of flood disasters has been alluded to frequently in popular accounts of global warming, there have been very few studies addressing the issue explicitly. This is largely because it is difficult to define credible scenarios for changes in flood-producing climatic events (Beran and Arnell, 1995). Chapter 10 outlines the potential effects of global warming on riverine floods, and Chapter 9 reviews the effects of a rise in sea level and changes in storms on coastal flooding. This section considers the implications of changes in flood occurrence for flood risk and management. There are four major implications: a change in flood loss, a change in flood risk and the standard of service currently provided by flood management and protection schemes, a change in the cost of protecting against floods, and impacts on public and private financial institutions.

The effect of a change in the frequency of floods on flood losses has to be set against many other factors, including population growth, economic development, and expansion onto flood plains. These factors have large impacts, making it difficult to detect climate-related change on flood loss statistics.

It might be easier, however, to detect a change in the standard of service provided by existing flood management schemes. Flood protection works and flood plain land-use plans usually are based on either a risk analysis to determine the most cost-effective level of protection (balancing costs against benefits) or on some legislative or institutional guideline for standard of service. Urban-flood-protection works, for example, typically provide protection against floods with a return period of up to 100 years. A change in flood frequency characteristics can have a very significant effect on flood risk and hence standards of service: A small change in flood magnitudes can have a very large effect on the risk of a particular critical value being exceeded. Beran and Arnell (1995) have shown that, assuming statistical properties similar to those of British rivers, a 10% increase in the mean—with no change in the year-to-year variability of floods—would result in a current 10-year flood occurring on average once every 7 years.

Increased flood frequencies probably would lead to increased expenditures on flood management. Most directly, the costs of providing or improving structural flood defenses would increase; indirect costs to society also may be incurred if larger areas of potentially productive floodplain land are excluded from development by flood plain zoning policies. However, there have been no quantitative studies of the costs of maintaining flood protection in the face of global warming.

Finally, an increase in flood losses may place significant strains on public finances, if damages are covered by public funds, or on the insurance industry. The amount of aid provided varies from country to country, as does the amount of coverage provided by the insurance industry. At the global scale, the international insurance and reinsurance industry may be seriously threatened by the occurrence of a few large, closely spaced storm events.

The Mississippi floods of 1993 and the European floods of 1994/1995 triggered major reviews of flood management policies. It is unlikely that flood management in the 21st century will be similar to flood management in the 20th century; there will probably be an increasing emphasis on adopting nonstructural measures and coping with uncertainty in risk assessments. The future effects of global warming on flood risk and flood management therefore must be seen against this changing institutional and technical background.

14.5. Research Needs

Uncertainties and analytical difficulties confront attempts to quantitatively analyze the direct effects of global warming on water resources demands. Considerable research investment is required in order to improve prediction and adaptive responses in the face of these uncertainties and methodological problems, including:

- Uncertainties in GCMs and lack of regional specification of locations where consequences will occur
- The absence of information on future climate variability
- Uncertainties in estimating changes in basin water budgets due to changes in vegetation and atmospheric and other conditions likely to exist in the future
- Uncertainties in the future demands of each water sector
- Uncertainties in the socioeconomic and environmental impacts of response measures
- Water management criteria under a potential nonstationary climate
- Linkage of water and agricultural sectors through detailed study of the impact of climate change on irrigation
- The impact of land-use and land-cover changes on water management
- The role and impact on groundwater management and conjunctive use.

It is doubtful whether some of these uncertainties can be completely eliminated, and, as with anything in the future, unforeseen elements will arise.

References

Allen, R., F. Gichuki, and C. Rosenzweig, 1991: CO_2-induced climatic changes and irrigation water requirements. *J. Water Res. Plng. and Mgmt.*, **117(2)**, 157-178.

Arnell, N.W. and N.S. Reynard, 1993: *Impact of Climate Change on River Flow Regimes in the United Kingdom.* Institute of Hydrology, Report to UK Department of the Environment Water Directorate, Wallingford, UK, 130 pp.

Arnell, N.W., A. Jenkins, and D.G. George, 1994: *The Implications of Climate Change for the National Rivers Authority.* NRA R and D Report 12, NRA, Bristol, UK, 94 pp.

Arnell, N.W., 1994: *Impact of Climate Change on Water Resources in the United Kingdom: Summary of Project Results.* Institute of Hydrology, Report to UK Department of the Environment, Wallingford, UK, 42 pp.

Arnell, N.W. and B.S. Piper, 1995: *Impact of Climate Change on an Irrigation Scheme in Lesotho.* Working Paper, International Institute for Applied Systems Analysis, Laxenburg, Austria (in press).

Beran, M.A. and N.W. Arnell, 1995: Climate change and hydrological disasters. In: *Hydrology of Disasters* [Singh, V.P. (ed.)]. Kluwer Academic Publishers, Dordrecht, Netherlands (in press).

Carter, T., M. Parry, H. Harasawa, and S. Nishioka, 1994: *IPCC Technical Guidelines for Assessing Climate Change Impacts and Adaptations.* Intergovernmental Panel on Climate Change, Geneva, Switzerland, 59 pp.

Chao, P., B. Hobbs, and E. Stakhiv, 1994: Evaluating climate change impacts on the management of the Great Lakes of North America. In: *Engineering Risk and Reliability in a Changing Environment* [Duckstein, L. and E. Parent (eds.)]. Proc. NATO ASI, Deauville, France, Kluwer Academic Publishers, Dordrecht, Netherlands, pp. 417-433.

Clark, K.J., L. Clark, J.A. Cole, S. Slade, and N. Spoel, 1992: *Effect of Sea Level Rise on Water Resources.* WRc plc. National Rivers Authority R&D Note 74, National Rivers Authority, Bristol, UK.

Cohen, S.J., 1987: Projected increases in municipal water use in the Great Lakes due to CO_2-induced climate change. *Water Resources Bulletin*, **23**, 81-101.

Covich, A.P., 1993: Water and ecosystems. In: *Water in Crisis* [Gleick, P.H. (ed.)]. Oxford University Press, New York, NY, pp. 40-55.

Croley, T.E., 1990: Laurentian Great Lakes double CO_2 climate change: hydrological impacts. *Climate Change*, **17**, 27-47.

Dearnaley, M.P. and M.N.H. Waller, 1993: *Impact of Climate Change on Estuarine Water Quality*. HR Wallingford, Report to UK Department of the Environment, Wallingford, UK, 54 pp.

Dobrowolski, A., D. Jurak, and K.M. Strzepek, 1995: *Climate Change Impact on Thermal Electric Generation of Lake Cooled Power Plants in Central Poland*. International Institute for Applied Systems Analysis, Working Paper (in press).

Engelman, R. and P. LeRoy, 1993: *Sustaining Water: Population and the Future of Renewable Water Supplies*. Population Action International, Washington, DC, 56 pp.

Falkenburg, M., M. Garn, and R. Cesitti, 1990: *Water Resources—A Call for New Ways of Thinking*. The World Bank, INUWS Working Paper, Washington, DC.

Falkenmark, M. and C. Widstrand, 1992: *Population and Water Resources: A Delicate Balance*. Population Reference Bureau, Population Bulletin 47:3, Washington, DC.

Fiering, M., 1982: Estimating resilience by canonical analysis. *Water Resources Research,* **18(1)**, 51-57.

Frederick, K.D., 1994: *Balancing Water Demands with Supplies: The Role of Management in a World of Increasing Scarcity*. Technical Paper No. 189, The World Bank, Washington, DC, 72 pp.

Frederick, K.D., 1993: Climate change impacts on water resources and possible responses in the MINK region. *Climate Change*, **24**, 83-115.

Frontinus, S.E., 98 AD: *The Stratagems and the Aqueducts of Rome* [Bennett, C.E. (trans.)]. Harvard University Press, Cambridge, MA, 483 pp.

Gellens, D., 1995: Sensitivity of the streamflow and of sizing of flood and low flow control reservoirs to the 2 x CO_2 climate change and related hypotheses: study of two catchments in Belgium. In: *Water Resources Management in Face of the Climatic and Hydrologic Uncertainties* [Kaczmarek, Z. *et al.* (eds.)]. Kluwer Academic Publishers, Dordrecht, Netherlands (in press).

Gleick, P.H., 1993: *Water in Crisis*. Oxford University Press, Oxford, UK, 473 pp.

Goklany, I., 1992: *Adaptation and Climate Change*. Presented at the annual meeting of the American Association for the Advancement of Science, Chicago, IL, 6-11 August 1992, 34 pp.

Gordon, H. *et al.*, 1992: Simulated changes in daily rainfall intensity due to the enhanced greenhouse effect: implications for extreme rainfall events. *Climate Dynamics,* **8**, 83-102.

Hanaki, K., 1993: Impact on urban infrastructure in Japan. In: *The Potential Effects of Climate Change in Japan* [Shiuzo Nishioka *et al.* (eds.)]. Center for Global Environmental Research, Tokyo, Japan, pp. 81-85.

Hanemann, W.M. and C.F. Dumas, 1993: Simulating impacts of Sacramento River fall run chinook salmon. In: *Integrated Modelling of Drought and Global Warming: Impacts on Selected California Resources* [Dracup, J. *et al.* (eds.)]. National Institute for Global Environmental Change, University of California, Davis, CA, pp. 69-96.

Hanemann, W.M. and R. McCann, 1993: Economic impacts on the Northern California hydropower system. In: *Integrated Modelling of Drought and Global Warming: Impacts on Selected California Resources* [Dracup, J. *et al.* (eds.)]. National Institute for Global Environmental Change, University of California, Davis, CA, pp. 55-68.

Hashimoto, T., J. Stedinger, and D. Loucks, 1982a: Reliability resiliency and vulnerability criteria for water resource system performance evaluation. *Water Resource Res.,* **18(1)**, 14-20.

Hashimoto, T., D. Loucks, and J. Stedinger, 1982b: Robustness of water resources systems. *Water Resource Res.,* **18(1)**, 21-26.

Herrington, P., 1995: *Climate Change and the Demand for Water*. Report to UK Department of the Environment, University of Leicester, Leicester, UK, 164 pp.

Hewett, B.A.O., C.D. Harries, and C.R. Fenn, 1993: Water resources planning in the uncertainty of climatic change: a water company perspective. In: *Engineering for Climatic Change* [White, R. (ed.)]. T. Telford, London, UK, pp. 38-53.

Hobbs, B.F., P.T. Chao, and J.F. Koonce, 1995: Climate change and management of water levels in the Great Lakes. In: *Proc. of the First National Conference: Climate Change and Water Resources Management*. U.S. Army Corps of Engineers, Fort Belvoir, VA, Chapter 4.

International Joint Commission, 1993: *Methods of Alleviating the Adverse Consequences of Fluctuating Water Levels in the Great Lakes—St. Lawrence River Basin*. A Report to the Governments of Canada and United States, Washington, DC, 53 pp.

IPCC, 1990a: *Climate Change: The IPCC Scientific Assessment* [Houghton, J.T., G.J. Jenkins, and J.J. Ephraums (eds.)]. Cambridge University Press, Cambridge, UK, 365 pp.

IPCC, 1990b: *Climate Change: The IPCC Impacts Assessment* [McG. Tegart, W.J., G.W. Sheldon, and D.C. Griffith (eds.)]. Australian. Gov. Publ. Service, Canberra, Australia, 247 pp.

IPCC, 1992: *The Supplementary Report to the IPCC Impacts Assessment* [McG. Tegart, W.J., G.W. Sheldon, and J.H Hellyer (eds.)]. Australian Gov. Publ. Service, Canberra, Australia, 112 pp.

Kaczmarek, Z., 1990: *On the Sensitivity of Runoff to Climate Change*. WP-90-58, International Institute for Applied Systems Analysis, Laxenburg, Austria, 10 pp.

Kaczmarek, Z. and J. Kindler, 1989: The impacts of climate variability and change on urban and industrial water demand and wastewater disposal. In: *Proc. of the Conference on Climate and Water*, Helsinki, vol. II, Valtion Painatuskeskus, pp. 161-176.

Kaczmarek, Z. and J. Kindler, 1995: National assessment—Poland. In: *Water Resources Management in the Face of Climatic and Hydrologic Uncertainties* [Kaczmarek, Z. *et al.* (eds.)]. Kluwer Academic Publishers, Dordrecht, Netherlands (in press).

Kaczmarek, Z., M. Niestepski, and M. Osuch, 1995: *Climate Change Impact on Water Availability and Use*. WP-95-48, International Institute for Applied Systems Analysis, Laxenburg, Austria, 18 pp.

Kaczmarek, Z., Z. Kundzewicz, and V. Pryazhinskaya, 1995: Climate change and water resources planning. In: *Water Resources Management in the Face of Climatic and Hydrologic Uncertainties* [Kaczmarek, Z. *et al.* (eds.)]. Kluwer Academic Publishers, Dordrecht, Netherlands (in press).

Kirshen, P.H. and N.M. Fennessey, 1995: Possible climate-change impacts on water supply of Metropolitan Boston. *J. Water Resources Planning and Management,* **121**, 61-70.

Klemes, V., 1993: Design implications of climate change. In: *Proc. Conf. on Climate Change and Water Resources Management* [Ballentine, T. and E. Stakhiv (eds.)]. U.S. Army Institute for Water Resources, Fort Belvoir, VA, pp. III.9-III.19.

Kulshreshtha, S., 1993: *World Water Resources and Regional Vulnerability: Impact of Future Changes*. RR-93-10, International Institute for Applied Systems Analysis, Laxenburg, Austria, 124 pp.

Kundzewicz, Z. and L. Somlyódy, 1993: Climatic change impact on water resources a systems view. WP-93-30, International Institute for Applied Systems Analysis, Laxenburg, Austria, 32 pp.

Lettenmaier, D., E. Wood, and J. Wallis, 1994a: Hydro-climatological trends in the continental United States: 1948-88. *J. Climate,* **7**, 586-607.

Lettenmaier, D.P., G. McCabe, and E.Z. Stakhiv, 1994b: Global change: effect on hydrologic cycle. In: *Handbook of Water Resources* [Mays, L. (ed.)]. McGraw-Hill, New York, NY (in press).

Lettenmaier, D.P., A.W. Wood, R.N. Palmer, S.P. Millard, J.P. Hughes, and S. Fisher, 1995a: *Water Management Implications of Global Warming: The Savannah River System*. Report prepared for the U.S. Army Institute for Water Resources, Fort Belvoir, VA.

Lettenmaier, D.P., S. Fisher, R.N. Palmer, S.P. Millard, and J.P. Hughes, 1995b: *Water Management Implications of Global Warming: The Tacoma Water Supply System*. Report prepared for the U.S. Army Institute for Water Resources, Fort Belvoir, VA.

Lettenmaier, D.P., A.E. Kerzur, R.N. Palmer, and S. Fisher, 1995c: *Water Management Implications of Global Warming: The Boston Water Supply System*. Report prepared for the U.S. Army Institute for Water Resources, Fort Belvoir, VA.

Lettenmaier, D.P., D. Ford, and S. Fisher, 1995d: *Water Management Implications of Global Warming: The Columbia River System*. Report prepared for the U.S. Army Institute for Water Resources, Fort Belvoir, VA.

Lettenmaier, D.P., D. Ford, and S. Fisher, 1995e: *Water Management Implications of Global Warming: The Missouri River System*. Report prepared for the U.S. Army Institute for Water Resources, Fort Belvoir, VA.

Lins, H.F. and P.J. Michaels, 1994: *Increasing U.S. Streamflow Linked to Greenhouse Forcing*. EOS, Trans., Amer. Geophys. Union, Washington, DC,, vol. 75, pp. 281/284-285.

Loomis, J. and S. Ise, 1993: Net economic value of recreational fishing on the Sacramento River in 1980. In: *Integrated Modelling of Drought and Global Warming: Impacts on Selected California Resources* [Dracup, J. *et al.* (eds.)]. National Institute for Global Environmental Change, University of California, Davis, CA, pp. 97-106.

McCormick, M.J., 1990: Potential changes in thermal structure and cycle of Lake Michigan due to global warming. *Trans. Amer. Fish. Soc.*, **119**, 183-194.

Matyasovszky, I., I. Bogardi, A. Bardossy, and L. Duckstein, 1993: Space-time precipitation reflecting climate change. *Hydro. Sci. J.*, **38**, 539-558.

McCabe, Jr., G.J. and D.M. Wolock, 1992: Sensitivity of irrigation demand in a humid-temperate region to hypothetical climatic change. *Water Res. Bull.*, **28**, 535-543.

Miller, J.R. and G.L. Russell, 1992: The impact of global warming on river runoff. *J. of Geoph. Res.*, **97**, 2757-2764.

Miller, B.A. *et al.*, 1993: Impacts of changes in air and water temperature on thermal power generation. In: *Managing Water Resources During Global Change: Proc. Int. Symp.* [R. Herman (ed.)]. American Water Resources Association, Bethesda, MD, pp. 439-448.

Mimikou, M., P.S. Hadjisavva, Y.S. Kouvopoulos, and H. Afrateos, 1991: Regional climate change impacts. II: Impacts on water management works. *Hydrol. Sci. J.*, **36**, 259-270.

Minns, C.K. and J.E. Moore, 1992: Predicting the impact of climate change on the spatial pattern of freshwater fish yield capability in eastern Canada. *Climate Change*, **22**, 327-346.

Nash, L.L. and P.H. Gleick, 1993: *The Colorado River Basin and Climatic Change: The Sensitivity of Streamflow and Water Supply to Variations in Temperature and Precipitation.* EPA-230-R-93-009, U.S. Environmental Protection Agency, Washington, DC, 92pp. + app.

Postel, S., 1992: *Last Oasis—Facing Water Scarcity.* W.W. Norton & Co., New York, NY, 239 pp.

Riebsame, W.E. *et al.*, 1995: Complex river basins. In: *As Climate Changes: International Impacts and Implications* [Strzepek, K.M. and J.B. Smith (eds.)]. Cambridge University Press, Cambridge, UK, pp. 57-91.

Rogers, P., 1992: Integrated urban water resources management. In: *Keynote Papers, International Conference on Water and the Environment-Development Issues for the 21st Century*, Dublin.

Rogers, P., 1993a: *America's Water—Federal Roles and Responsibilities.* The MIT Press, Cambridge, MA.

Rogers, P., 1993b: What water managers and planners need to know about climate change and water resources management. In: *Proc. Conference on Climate Change and Water Resources Management* [Ballentine, T. and E. Stakhiv (eds.)]. U.S. Army Institute for Water Resources, Fort Belvoir, VA, pp. I/1-I/14.

Rogers, P. and P. Lydon (eds.), 1994: *Water in the Arab World: Perspectives and Prognoses.* Harvard University Press, Cambridge, MA, 369 pp.

Saelthun, N.R. *et al.* 1990: *Climate Change Impact on Norwegian Water Resources.* Norwegian Water Resources and Energy Administration Publication NR V42, Oslo, Norway, 34 pp. + app.

Salewicz, A., 1995: Impact of climate change on the operation of Lake Kariba hydropower scheme on the Zambezi river. In: *Water Resources Management in the Face of Climatic and Hydrologic Uncertainties* [Kaczmarek, Z. *et al.* (eds.)]. Kluwer Academic Publishers, Dordrecht, Netherlands (in press).

Savenije, H.H.G., 1995: New definitions for moisture recycling and the relation with land-use changes in the Sahel. *Journal of Hydrology*, **167**, 57-78, Elsevier, Amsterdam, Netherlands.

Savenije, H.H.G. and M.J. Hall, 1994: Climate and land use: a feedback mechanism? In: *Proc. of the Delft Conference on Water and Environment: Key to Africa's Development*, Delft, Netherlands.

Shiklomanov, I., H. Lins, and E. Stakhiv, 1990: Hydrology and water resources. In: *The IPCC Impacts Assessment* [McG. Tegart, W.J., G.W. Sheldon, and D.C. Griffiths (eds.)]. Australian Government Publishing Service, Canberra, Australia, pp. 4.1-4.42.

Shiklomanov, I., 1993: World freshwater resources. In: *Water in Crisis, A Guide to the World's Freshwater Resources* [Gleick, P. (ed.)]. Oxford University Press, Oxford, UK, pp. 13-24.

Shiklomanov, I., V. Georgeosky, A. Shereshevsky, *et al.*, 1995: *An Assessment of the Influence of Climate Uncertainty on Water Management in the Dnipro River Basin.* Technical Report, Russian State Hydrological Institute, St. Petersburg, Russia.

Stakhiv, E., H. Lins, and I. Shiklomanov, 1992: Hydrology and water resources. In: *The Supplementary Report to the IPCC Impacts Assessments* [McG. Tegart, W.J. and G.W. Sheldon (eds.)]. Australian Govt. Publ. Service, Canberra, Australia, pp. 71-83.

Stakhiv, E., 1994: Managing water resources for adaptation to climate change. In: *Engineering Risk and Reliability in a Changing Physical Environment* [Duckstein, L. and E. Parent (eds.)]. Proc. NATO ASI, Deauville, Kluwer Academic Publishers, Dordrecht, Netherlands, pp. 379-393.

Stalnaker, C.B., 1993: Evolution of instream flow habitat modelling. In: *The Rivers Handbook*, vol. 2 [Calow, P. and G.E. Petts (eds.)]. Blackwell, Oxford, UK, pp. 276-286.

Stefan, H. *et al.*, 1993: A methodology to estimate global climate-change impacts on lake waters and fisheries in Minnesota. In: *Proc. Conf. on Climate Change and Water Resources Management* [Ballentine, T. and Z. Stakhiv (eds.)]. U.S. Army Institute for Water Resources, Fort Belvoir, VA, pp. II/177-II/193.

Steiner, R., N. Ehrlich, J.J. Boland, S. Choudhury, W. Teitz, and S. McCusker, 1995: *Water Resources Management in the Potomac River Basin Under Climate Uncertainty.* Report prepared for the U.S. Army Institute for Water Resources by the Interstate Commission on the Potomac River Basin, MD.

Strzepek, K., J. Niemann, L. Somlyody, and S. Kulshreshtra, 1995: *A Global Assessment of National Water Resources Vulnerabilities: Sensitivities, Assumptions, and Driving Forces.* IIASA Working Paper (in press).

Strzepek, K. and P. Bowling, 1995: *Global Assessment of the Use of Freshwater Resources for Industrial and Commercial Purposes.* A Study for the United Nations Industrial Development Organization.

World Bank, 1992: *Development and the Environment.* World Development Report 1992, Oxford University Press, Oxford, UK, 308 pp.

World Bank, 1993: *Water Resources Management. A World Bank Policy Paper.* World Bank, Washington, DC, 140 pp.

World Bank, 1994: *A Strategy for Managing Water in the Middle East and North Africa.* MENA Technical Dept., Washington, DC, 96 pp.

U.S. Army Institute for Water Resources, 1994a: *National Study of Water Management During Drought—The Report to the U.S. Congress.* IWR Report, 94-NDS-12, Institute for Warter Resources, Fort Belvoir, VA, 55 pp.

U.S. Army Institute for Water Resources, 1994b: *Lessons Learned from the California Drought (1987-1992)—Executive Summary.* IWR Report, 94-NDS-6, Institute for Warter Resources, Fort Belvoir, VA, 36 pp.

U.S. Army Institute for Water Resources, 1994c: *Water Use Forecasts for the Boston Area using IWR-MAIN.* IWR Report, 94-NDS-11, Institute for Warter Resources, Fort Belvoir, VA, 114 pp. + app.

U.S. Interagency Floodplain Management Review Committee, 1994: *Sharing the Challenge: Floodplain Management into the 21st Century.* Report to the Administration Floodplain Management Task Force, U.S. Government Printing Office, Washington, DC, 191 pp. + app.

Waterstone, M. *et al.*, 1995: *Future Water Resources Management in the Upper Rio Grande Basin.* Report prepared for U.S. Army Institute for Water Resources, Fort Belvoir, VA.

Whetton, P., A. Fowler, M. Haylock, and A. Pittock, 1993: Implications of climate change due to the enhanced greenhouse effects on flood and droughts in Australia. *Climatic Change*, **25**, 289-317.

Wolock, D.M., G.J. McCabe, G.D. Tasker, and M.E. Moss, 1993: Effects of climate change on water resources in the Delawere River basin. *Water Resources Bulletin*, **29**, 475-486.

15

Wood Production under Changing Climate and Land Use

ALLEN M. SOLOMON, USA

Principal Lead Authors:
N.H. Ravindranath, India; R.B. Stewart, Canada; M. Weber, Germany; S. Nilsson, Sweden

Lead Authors:
P.N. Duinker, Canada; P.M. Fearnside, USA; P.J. Hall, Canada; R. Ismail, Malaysia; L.A. Joyce, USA; S. Kojima, Japan; W.R. Makundi, Tanzania; D.F.W. Pollard, Canada; A. Shvidenko, Russia; W. Skinner, Canada; B.J. Stocks, Canada; R. Sukumar, India; Xu Deying, China

CONTENTS

EXECUTIVE SUMMARY

The fundamental question on which this chapter focuses involves forest resource availability and resource consumption: Will the resources from future forests be adequate to meet future needs? Although all forest amenities are of interest, this chapter concentrates primarily upon those for which data are most available, reliable, and measurable: wood and wood products. All of the conclusions depend upon uncertain climate model simulations (resource availability), uncertainty in adaptive changes in forest product use, and uncertain regional human population growth scenarios (resource consumption). Assuming all of these variables change as projected by their respective scenarios, the following forest resource responses are likely (note that none of the conclusions summarized below were considered in the 1990–1992 IPCC Assessments):

- Tropical land use will be much more important than change in climate and atmospheric chemistry in affecting forest product availability there, at least until the middle of the next century (High Confidence). Growing stock in the tropics is projected to decline by about half from anthropogenic deforestation, while consumption is predicted to expand because of a threefold increase of population. The projected standing stock decline appears even after calculation of concomitant changes in climate and atmospheric chemistry, which could increase both productivity and the areas where tropical forests can potentially grow.
- Under these circumstances, tropical forests of densely populated countries in Africa and Asia would be most vulnerable to loss (High Confidence). These areas contain rapidly expanding human populations and concomitant increasing needs for forest products, especially but not entirely for fuelwood. Fuelwood consumption is expected to increase throughout the 55-year study period (High Confidence).
- Temperate-zone wood requirements are likely to continue being met for at least the next century (Medium Confidence). This conclusion is based on assumed climate and land use changes that leave temperate forest covering about as much land by the year 2050 as today. Also, current harvests from plantations and wildland forests are expected to increase only marginally. Additionally, annual growth increments are assumed to remain constant through 2050. Finally, an increasing proportion of requirements for wood is expected to be met by imports from outside the temperate zone.
- Future wood availability in boreal regions is uncertain at best, based on current data, scenarios, and models (High Confidence). Future production is expected to be controlled by increases in climate-induced irregular and large-scale mortality events, which cannot be simulated or predicted with current models. These could generate considerable surplus wood from salvage fellings, but would severely decrease standing stocks over the long term. In addition, current and future harvests in boreal forests are expected to be largely determined by needs outside the zone. The models underlying future projections contain no consideration of these forest economic or trade processes or their implications, a particularly critical omission in boreal regions. Therefore, the available information does not support realistic analysis of future boreal forest product demand.
- Total global wood consumption will exceed the availability expected by the year 2050 (Medium Confidence). Analysis of increasing human populations, with the assumption of constant per capita wood use, suggests that the annual need for timber will exceed the current 2% annual growth increment before the year 2050. Although we assume that the annual increment will remain constant, it could increase slightly from warming and enhanced atmospheric CO_2 concentrations, or it could decrease greatly from growth declines and mortality brought on by climate change.
- Future temperate and possibly tropical zone requirements for boreal forest products may result in serious shortfalls of boreal industrial roundwood during the 21st century based on the analysis of temperate and tropical zone forest standing stocks and requirements (High Confidence).
- The livelihood of indigenous tropical peoples will be adversely affected if climate and land-use change induce forest losses (High Confidence). These peoples depend on a large diversity of forest products that are expected to undergo declines in biodiversity and forest regeneration due to climate change and may be subject to forest degradation apart from land conversion.
- Options for adapting to and ameliorating potential global timber supply shortages in the future include the following actions: In the tropics, the greatest progress may be achieved by reducing social pressures driving land conversion (e.g., by increasing crop and livestock productivity on current agricultural lands) and by development of large tree plantations (High Confidence). In temperate areas, use of modern forestry practices to reduce harvest impacts on ecosystems, combined with substitution of nontimber products for forest products, could reduce climate impacts

significantly (Medium Confidence). In boreal regions, the most useful focus may be on adapting to potential climate-induced large-scale disturbances, such as by rapid reforestation of disturbed areas with warmth-adapted seed provenances (Medium Confidence).

- Research needed to conduct a more accurate assessment of climate change impacts on future forest product availability includes (1) obtaining new data needed to demonstrate and quantify direct effects of atmospheric CO_2 on ecosystem productivity and tree growth (High Confidence), and (2) uniform reporting and assembly of national forest inventory data sets (High Confidence). New mathematical models will be needed especially to (a) simulate transient (lagged) responses of forests to rapid climate change, (b) integrate relationships among climate changes, ecological responses, and economic forces and responses, and (c) simulate regional climate changes and their effects (High Confidence).

15.1. Introduction

This chapter examines the changes induced in human-valued properties and products of forests (primarily wood and wood products) by potential shifts in vigor, productivity, and geographic distribution of tree species populations and of forest ecosystems. These shifts are expected to arise from global changes in climate and atmospheric chemistry. In addition, this chapter examines the roles of other forces—particularly of land use, which may modify or mask effects of changing climate and atmospheric chemistry. The changing commodities and amenities derived from forests are a tertiary response to the initial forcing caused by increased atmospheric concentrations of greenhouse gases. As such, the commodities cannot be estimated reliably until accurate forecasts are made of the intermediate processes and outcomes (climate, atmospheric chemistry, and resulting forest density and growth patterns). On the other hand, we can and do assess the socioeconomic responses to quantitative estimates of changing climate, land use, and forest ecosystem responses that are expected to arise from current practices and processes and from potential future ones.

This chapter focuses on the fundamental question of forest resource availability and resource needs: Given that the availability of growing stocks and related amenities from today's forests are adequate to meet today's needs under today's climate, will the availability from future forests be adequate to meet future needs under predicted future climates? The analysis avoids the more complex issues of economic responses to changes in potential growing stocks (as expected in decreasing prices because of enhanced harvests from use of new technologies, increasing prices from decreasing supplies, and changing prices from product substitution resulting from increasing prices). It concentrates instead on estimating the quantity of growing stock to which economics must be applied and the size of the population needing forest products.

Chapter 1 of this volume evaluates forest geographical, ecological, and productivity responses to changing climate and atmospheric chemistry. Chapter 24 of this volume examines the need for and effects of actions taken to mitigate increasing atmospheric CO_2 concentrations. These include forests subjected to unplanned and nonsustainable forest management practices. A third related chapter in the Scientific Assessment (Chapter 9, *Terrestrial Biotic Responses to Environmental Change and Feedbacks to Climate*, of the IPCC Working Group I volume) defines expected changes in global carbon cycling from shifts in forest geography and productivity under future changes in climate and atmospheric chemistry.

This chapter encompasses wood and wood products and amenities from all forests, whether or not they can be used by humans. Although all forest amenities are of interest, the chapter concentrates primarily upon those for which data are most available and reliable: wood products [roundwood (i.e., that portion of wood and bark removed from forests), industrial roundwood (i.e., that portion of roundwood to be processed further), and fuelwood and charcoal (i.e., that portion of roundwood to be consumed for cooking, heating, and so on)].

The term "availability" of roundwood is used throughout this report, rather than the economically significant term "supply." Roundwood availability (as distinguished from supply or production) depends on the amount of growing stock and on annual growth increments. These are determined by climate, soils, management, and forested area. Only a fraction of growing stocks is normally considered as commercial forest, and only a fraction of commercial forest is available for harvest (supply). Even if harvested wood equalled growing stock—say, by application of advanced technologies—wood products can only be harvested up to the amount of available growing stock. Hence, growing stock is a more conservative estimator of the fundamental limit to harvests than is some smaller quantity, such as sustainable harvest level (see Section 15.5.2).

In the current assessment, we assume that even intensive management will not increase potential harvests beyond the maximum potential natural growing stock. Although the assumption is probably safe, the role of management in increasing future forest biomass (= growing stocks) is examined in detail by Brown *et al.* in Chapter 24. They conclude that during the next 55 years, about 20 Gt (Table 24-4) could be added by forestation to the approximately 1000 Gt (Table 24-1) already present in above- and belowground biomass stocks of forests. This potential 2% contribution by management may not be realized within the realities of limited forest management budgets.

In contrast to variables controlling growing stocks, harvests from forests will depend on the technology available to harvest and market roundwood, and demand for the resulting wood products (and hence, price of the products), as well as the amount of growing stock itself. Demand for wood products ("need" and "consumption" are noneconomic terms used in this report in place of "demand") and related amenities will change with shifts in size and geographic distribution of population, with technological and financial prosperity of end users, with price of the products, with availability of alternative products, and so on. Assessment of these variables is beyond the scope of the current chapter, although they could be profitably examined in the next IPCC assessment report. Here, we assume that per capita consumption remains constant and that population size increases as in the IS92A population projection (IPCC, 1992). In developed countries (e.g., Australia), per capita consumption actually declined during the last decade (FAO, 1992) and could do so in other countries in the future.

15.1.1. 1990 IPCC Forestry Assessment Results

The 1990 IPCC assessment of forestry (IPCC, 1990a) was confined to managed forests—difficult entities to assess because forest areas or volumes are rarely measured as "managed" or "unmanaged," and when they are so divided, the definitions of "managed" vary widely. As a consequence, the 1990 assessment contained few quantitative conclusions. However, it did define uncertainties and issues of importance. In particular, it suggested that forest plantations are probably much more vulnerable to environmental change than are mixed-species and mixed-age stands. The latter contain species and developmental stages that may be less vulnerable to chronic climate change at any given point in time or space, whereas the former are easier for forest pests to invade.

The assessment recommended that impact analysis focus upon transitional climate and forest responses that may be with us for several decades, rather than on a single future long-term "equilibrium" climate. Finally, it recommended that "in the face of great uncertainties about rates and magnitudes of climatic changes, forest responses to potential climate change, and socioeconomic repercussions of climate-induced forest changes, it is prudent to prepare for severe undesirable impacts to ensure that viable and flexible options are implemented to maintain sustainable forest ecosystems." The results of our report are consonant with these two fundamental recommendations. However, unlike earlier efforts, our assessment was developed to go a step further in documenting the level of predictability in future forest production and consumption as constrained by projected changes in climate and land use.

15.1.2. 1995 IPCC Forestry Assessment Strategy

The current chapter approaches the assessment first by quantifying present and future roundwood availability and potential consumption as accurately as possible with available data and models (Section 15.2). Then the validity of these estimates is assessed, first for the globe as a whole (Section 15.2), then within tropical, temperate, and boreal forest regions (Sections 15.3, 15.4, and 15.5). Both validity assessments are based upon less quantifiable but relevant estimators of timber availability and consumption that are not included in the models. These analyses generate definitions of certainties, uncertainties, and unknowns regarding future availability of and needs for timber. Adaptation measures to reduce effects of specific changes are then discussed (Section 15.6). Last, monitoring and modeling products needed to resolve identified uncertainties and unknowns in future IPCC forest product assessments are presented (Section 15.7).

15.2. Present and Future Global Forests

Kirschbaum *et al.* (see Chapter 1) describe the distribution and characteristics of natural forests and discuss the expected responses by forest ecosystems to changing atmospheric chemistry and climate. In Chapter 24, Brown *et al.* examine the potential gain in forest area and density that forest management potentially can provide. In addition, we examine the possibility that forestry and forest management, and several other kinds of land use, may be as important or more important than climate in determining the future spatial and temporal distribution of forests. The interactions of these climatic and management changes will generate impacts that are not necessarily intuitively obvious. Therefore, to be useful, the assessment must at least quantify the changes in wood products should current conditions of change continue unabated (business as usual).

15.2.1. Present Characteristics of Forests

Globally, forests in 1990 covered one-fourth to one-third of the earth's land surface (34 million km^2, FAO, 1993a; 41 million km^2, Dixon *et al.*, 1994; see Chapter 1 for a detailed description). Although the great majority of these forests can be considered as managed to some degree, only about 4% of forests consist of intensively managed plantations (1.35 million km^2, Kanowski *et al.*, 1990; 1.12 million km^2, Dixon, *et al.*, 1994). WRI (1994) cites a doubling of forest plantations during the decade from 1980 to 1990, with three-fourths of all plantations being in Asia. The plantations may produce a disproportionate amount of forest products. About 30% of industrial timber production (industrial roundwood, i.e., non-fuelwood raw logs) in Latin America, for example, may have originated on plantations (McGaughey and Gregerson, 1982). Brown *et al.* (Chapter 24) point out that about 11% of the world's forests undergo active management; this includes 20% of mid-latitude and 17% of high-latitude forests but only about 4% of low-latitude forests.

FAO estimates that globally, forest products were worth about $418 billion in 1991; this amounted to 4% of the gross domestic product (GDP) of developing countries and 1% of the GDP in developed countries (FAO, 1993a). Although most forest products were consumed locally, total annual exports of logs and wood products between 1986 and 1988 were estimated at $68 billion, of which $10 billion was earned by developing countries (Sharma *et al.*, 1992). Many developing (tropical) countries rely heavily on their forest resources for obtaining capital. For example, 1991 forest product exports from Cambodia represented 43% of all trade; from the Congo, 16%; and from Indonesia, 12% (FAO, 1993a).

15.2.2. Projecting Future Forest Characteristics

Numerical projections of forest area and growing stocks require us to combine a spatially explicit mathematical model of forest response to climate and land use with a quantitative scenario of climate and land-use change. Few vegetation models are available from which to choose. There currently are no forest response models that simulate the processes controlling the change of forest productivity or geography over time ("transient" models), and that are globally comprehensive (see Chapter 9, *Terrestrial Biotic Responses to Environmental Change and Feedbacks to Climate*, of the IPCC Working Group I volume).

Static responses of forests—those that would be expected at some time following cessation of directional environmental change (Prentice and Solomon, 1991)—can be projected by several available models. Their fundamental flaw is that dynamic equilibrium of forests cannot be expected for centuries

Box 15-1. Use of Static Models to Estimate Future Forest Production

The application of the BIOME or IMAGE model (Tables 15-1 through 15-4) or other static vegetation models (see Chapter 1 of this volume, and Chapter 9, *Terrestrial Biotic Responses to Environmental Change and Feedbacks to Climate*, of the IPCC Working Group I volume) to estimate potential forest changes under different but stable modeled climate omits potential variables that could be critical to future forest distributions and productivity. The following weaknesses are most important:

- There is no indication that a stable climate will appear in the foreseeable future. Indeed, the global change problem to be assessed involves rapidly changing climate, not stable climate, during the next 50 to 100 years. The static models do not simulate impacts of rapid climate change on slowly responding forests, and current efforts aimed at parameterizations of the process will beg the question.
- Static models do not simulate tree mortality. In a warmer and potentially drier world, the geography of forest ecosystems is expected to shift, as it has many times during climate changes of the past (e.g., Huntley and Webb, 1989). This may be detected initially as negative or positive changes in seedling survival, growth and productivity of established trees, and eventually by forest decline as the environmental variables exceed species tolerances. Such forest decline is likely to be amplified by increases in tree mortality events (dieback), disease and insect outbreaks, fire, and damage associated with extremes in weather, none of which are simulated by the static models.
- Static models assume that trees instantaneously occupy new, climatically suitable areas. However, current projections of climate change appear to be too rapid to allow population geography to adjust: If the Holocene tree species migration rate of 0.2–0.4 km/yr (Davis, 1976, 1983; Huntley and Webb, 1989, suggest 0.1 to 1 km/yr based on examination of extant paleoecological data) records maximum tree migration rates, trees may be unable to attain the 4–6 km/yr rate of climate change expected (Solomon *et al.*, 1984). As a result, large areas of low productivity or degraded forest (forest lacking a full complement of growing stock and tree species, consisting instead of slowly growing and dying trees made ecologically inappropriate by climate change) could be present for centuries.
- Static models assume that new forested areas instantaneously grow to maturity and can be harvested. Thirty to 50 years or more of relatively stable climate are required to reach a harvestable condition of most species once the tree is established. Therefore, most of the new areas capable of supporting forests may consist of young, immature trees. Establishment of these trees under the expected regeneration-unfriendly conditions is potentially an even more serious problem.
- Static models do not account for differences in short- and long-term forest productivity. However, a forest in decline in the short run (50–70 years) will increase the available wood supply. If there are large areas of dead or dying trees, much of this wood probably will be harvested or salvaged. In the long run (>75 years), however, roundwood availability could decrease because of the immaturity of new species populations.
- Static models assume that the soil conditions will be suitable for new tree invasions. However, in boreal regions, permafrost and excess surface water dominate many landscapes, prohibiting uptake of nitrates. Recent glaciation at high latitudes has endowed large areas with thin soils or bare rock. In many tropical areas, prior intensive agriculture has so depleted the soil of phosphorus that tree growth is exceedingly slow. These soil features could form major barriers to immigration and growth even at the slow rates projected by models, making it unlikely that the forest tree populations will be able to migrate at their measured rate of 0.2–0.4 km/yr or reach maturity at their natural rate of 20 to 50 years.

(e.g., Solomon, 1986; Bugmann, 1994), following an environmental equilibrium which itself is not likely in the foreseeable future (see Box 15-1).

Although the static models cannot be used to identify what vegetation units will replace current ecosystems, the models can define which ecosystems will be inappropriate for the new climates and hence are vulnerable to climate change (e.g., Solomon and Leemans, 1990; Chapter 1). This requires only the very simple and robust assumption that climatic limits to forest growth implied by current geography will operate similarly in the future as well. Even this simple assumption may not be entirely correct because of potential changes in climate response patterns of plants induced by increased atmospheric CO_2.

One static model (BIOME 1.1, hereafter BIOME; Prentice *et al.*, 1992,1993) was conceived specifically for assessing geographically distributed responses of vegetation types to expected climate changes. BIOME is unlike other static models in that it is based on specific physiological responses of plants to climate. These responses define a set of plant functional types (PFTs). Each PFT is characterized by a minimal (and different) set of known climate thresholds (minimum winter temperatures, summer warmth, evapotranspiration). BIOME was modified to include climate-dictated potential agricultural land use as a plant functional type (Cramer and Solomon, 1993).

BIOME cannot be used to assess interactions among forestry, land use, and climate change. However, the BIOME vegetation classification has been incorporated into a model of interacting population (including associated natural resource utilization), land use, vegetation, and climate (IMAGE 2.0, hereafter IMAGE; Alcamo, 1994). Land use (agriculture) is modeled as a function of food demand by populations that vary in density over time, and in resource demand from one region to another. Land use is further modeled as constrained by climate and soils.

IMAGE has been used to project the implications of these population-land use-climate interactions on the geography of forests (Zuidema *et al.*, 1994; Leemans and van den Born, 1994). The IMAGE projections generate an annual climate change from greenhouse gas concentrations; by the year 2050, this produces a climate essentially indistinguishable in its forest effects from those produced by the IS1992a future scenario (IPCC, 1992). However, IMAGE has been applied without feedbacks between changed climate and changed wildland vegetation (i.e., BIOME-dictated wildland vegetation does not respond to climate change). Therefore, its effectiveness is reduced where land use is unimportant. Hence, we projected global wood product availability into the future with both BIOME vegetation responses from Solomon *et al.* (1993) and IMAGE land cover change scenarios from Zuidema *et al.* (1994).

15.2.3. Projecting Future Availability of Forest Products

The forest area of ecosystems projected by BIOME for the present and for the year at which climate reflects a doubling of greenhouse gases (GHGs) is shown in Figures 1-5 and 1-8 (see Chapter 1) and is presented in Table 15-1. Note that estimated present forested areas (Dixon *et al.*, 1994) and modeled present forested areas (Solomon *et al.*, 1993) are similar except for a serious discrepancy in the tropics. There, the differences are in both definition (each defines "forest" differently) and in model capability (BIOME is less accurate at distinguishing between tropical forest and tropical savanna; Prentice *et al.*, 1992).

Table 15-1: *Tropical, temperate, boreal, and global forested areas in 10^6 km^2 under different future climate scenarios, at the year of a doubling of atmospheric CO_2, and percentage differences between current and future climates—projected by the BIOME 1.0 model.*

	Present Est.[1]	Area Modeled[2]	GFDL Area	GISS Area	OSU Area	UKMO Area
Tropical Forests	17.6	36.8	40.6	42.9	41.3	41.4
Temperate Forests	10.4	10.6	15.7	15.7	13.3	16.1
Boreal Forests	13.7	16.7	8.4	11.5	13.5	9.1
All Global Forests	41.7	64.1	64.7	70.1	68.1	66.6

[1]From Dixon *et al.*, 1994.
[2]From Solomon *et al.*, 1993.

Table 15-2: *Tropical, temperate, boreal, and total global forested areas in 10^6 km^2 under a single, self-generated future climate and land-use scenario, at different years, projected by the IMAGE 2.0 model (from data of Zuidema et al., 1994).*

	Area of Forest at Year			
	1990	2000	2020	2050
Tropical Forests	27.6	25.0	21.0	14.4
Temperate Forests	5.4	4.1	4.9	5.3
Boreal Forests	13.8	14.3	15.2	15.3
All Global Forests	46.8	43.4	41.1	35.0

The major pattern to emerge from comparison of all four climate scenarios with modeled present area is a moderate increase in area occupied by tropical forests, a strong increase in area occupied by mid-latitude temperate forests, and a similarly severe decline in area occupied by boreal forests. The simulated increase of temperate forest area is almost entirely at the expense of boreal forest area, which is reduced by a lack of very high latitude area to occupy, by loss of area to invading temperate forests at lower latitudes, and by the expansion of agriculture into areas now possessing too short a growing season to allow agriculture.

Table 15-2 presents forested areas in tropical, temperate, and boreal zones projected by IMAGE at 1990 and at 2000, 2020, and 2050. The singular climate scenario is generated within IMAGE, based on annual increments of GHGs generated by modeled energy use, land use fluxes, ocean uptake of CO_2, and so on.

The areas of tropical, temperate, and boreal forests estimated for the present by the two models differ substantially. BIOME initially estimates that 37% more land is in forests than does IMAGE. The differences appear to arise from a much larger agricultural area calculated in IMAGE than in BIOME. It seems likely that BIOME results are less accurate in temperate and tropical areas because IMAGE agricultural area of 1990 is very similar to that measured in FAO statistics (Zuidema *et al.*, 1994), and the current forested area of IMAGE is much closer to that estimated by Dixon *et al.* (1994; see Table 15-1). As one would therefore expect, the greatest difference in 1990 zonal composition between BIOME and IMAGE areal projections is in temperate regions (16.5% vs. 11.6% of forests, respectively) where the greatest density of agricultural land use is found, followed by boreal regions (26.1% vs. 29.6% of forests, respectively) where the lowest density of agricultural land is found. Even if land use were simulated similarly in both models, one would expect the proportions of land in each zone to differ because IMAGE defines the current zones based on national boundaries as well as on the climatic criteria used by BIOME.

The area and geographic distribution of forest changes projected by IMAGE at the year 2050 (Table 15-2) are in sharp contrast to those generated by BIOME at a doubling of GHGs (a point that might be reached by the year 2050; Manabe and Wetherald, 1993). Where BIOME indicates a total global increase in forested area from 1 to 9%, IMAGE calculates a decline of 25%. The large increase in temperate forest area in BIOME simulations becomes a slight decline in IMAGE. Both models project forest gains in temperate regions of North and South America and Europe, but in IMAGE, much land is lost to increased agriculture in Asian temperate regions. A decreasing demand for agricultural land in the developed countries of temperate regions results from slow population growth in Europe and North America and the projection into the future of increases in agricultural efficiency encountered during the past few decades (Zuidema *et al.*, 1994).

The contrast between IMAGE and BIOME projections for the future is greatest in the tropics. Tropical Africa and Asia, where human population growth is greatest, are simulated by IMAGE as losing large amounts of tropical forests and woodlands (Table 15-2). Moderate increases in tropical forests projected by BIOME (Table 15-1) disappear when the need for agricultural land is factored into the calculation, as it is in IMAGE.

Differences in projected boreal forest area are also great. The BIOME losses of 20–50% (Table 15-1) become increases of 10% in boreal forest land (Table 15-2) by IMAGE. BIOME assumes that 50% of the area in which climate permits agriculture will be farmed and that climate changes redistribute mature biomes on the remaining land. IMAGE assumes that declining agricultural demand will allow regrowth of climatically appropriate forests on abandoned farmland but that forests will not be redistributed on undisturbed lands regardless of new climate conditions. This has the effect of eliminating instant forest migration/maturation on lands undisturbed by intensive land use—which are the vast majority of lands in boreal regions, much less in temperate regions, and the minority of lands in many tropical regions. This assumption should be more correct than the assumption of instantaneous migration during the simulated 60 year period (see Box 15-1). Few tree populations in the past 10,000 years have required less than a century to migrate latitudinally across even one of the 55 km pixels simulated by both BIOME and IMAGE (Davis, 1976, 1983; Huntley and Webb, 1989; Huntley, 1990). Fewer still can be expected to form mature, dense populations and forest ecosystems in new locations during that same 55 years.

The land cover and use changes projected by IMAGE appear far more important during this transition period of 60 years than do the effects of concurrent climate change in BIOME. This reflects in part the weak response to climate that IMAGE generates and its assumption that a significant amelioration of climate impacts will result from strong effects of carbon fertilization, water use efficiency, and other feedbacks (Zuidema *et al.*, 1994). This assumption is not included in BIOME. Although there is no definitive evidence that this effect occurs in forest ecosystems, even the greatest benefits claimed for the effect may not be significant (Chapter 1).

Instead, land use must be the most important agent of change in any vegetation replacement scheme based on the relationship between the large magnitude of current global vegetation replacement induced by the current population and the doubling of human population every 35–40 years. We know of no processes (technological or natural) that have been quantified (i.e., that can be included in a simulation) that would ameliorate this enormous impact. Using "back of the envelope" calculations, Sharma *et al.* (1992) conclude that population pressures could reduce forest area 30% by 2025, a value similar to the 24% decline by 2020 estimated by IMAGE (Table 15-2). No other published research of which we are aware takes this land use impact into account in projecting future characteristics of the global terrestrial biosphere.

In sum, without accounting for human demographics and agricultural land use, climate change alone would be responsible for an increase of 1–9% in land suitable for forests, with the largest gains in temperate regions. Projections that simulate future human needs for agricultural land but lack significant impacts of concomitant climate change generate a decline of 25% in forest area by the year 2050, with the greatest losses in the tropics. Because vegetation presence or absence within the envelope of appropriate future climate will be determined by land use at that time, it is the land use projections that must be given the most serious consideration.

The IMAGE projections of forest area have been used to estimate the potential supplies of forest products (from growing stock volume) by assuming that the ratio of forest area to growing stock today will also apply in the future (Table 15-3). We emphasize that the inaccuracies induced by the transient processes discussed in Box 15-1, and those detailed in the following paragraphs, are likely to change this ratio in the future. It is impossible to estimate, however, whether growing stock per unit area will increase or decrease.

Because the volumes in Table 15-3 are related to areas in Table 15-2 (data for the tropical, temperate, and boreal regions in

Table 15-3: *Current and future timber availability (growing stock) in 10^9 m^3, based on 1980 productive forest volume (Sharma, 1992) and percentage changes in forest area in IMAGE (Zuidema et al., 1994).*

	Volume of Forest at Year			
Region	1980	2000	2025	2050
Tropical Forests	146.7	132.8	111.6	76.5
Temperate Forests	41.3	31.4	37.5	40.6
Boreal Forests	90.0	93.2	99.1	99.7
All Global Forests	278.0	257.4	248.2	216.8

Table 15-2 are drawn from individual countries, and in Table 15-3 from the 13 groups of countries), patterns of change are also similar: tropical forest growing stocks decline by almost a half; temperate forests initially decline, then recover by the year 2050; and boreal forests increase slightly over time, with global growing stock volume down by about 22% in the year 2050 (Table 15-3).

The estimates of potential wood availability do not include enhancement of standing stocks from annual forest growth and from tree plantations developed to mitigate atmospheric CO_2 concentrations (see Chapter 24). Measurements of current annual forest growth in tropical regions average 2% of standing stocks (DeBacker and Openshaw, 1972), about 3% of exploitable growing stock in temperate countries, and 1.5% in boreal countries (FAO, 1992). These values would result in an increase in availability of wood products of about 2% (4.24 x 10^9 m^3) above the 216.8 x 10^9 m^3 presented in Table 15-3 by the year 2050.

The annual growth increment assumed in the wood availability projection is unlikely to remain constant. For example, Chapter 24 points out that tropical forest productivity may continue to be degraded by illicit and selective fellings. Also, forest growth may decline and mortality may increase from increasingly inappropriate climate (Solomon and Leemans, 1990; see also Chapter 1). This growth loss also was not included in the calculations that produced Table 15-3. Solomon (1986) used a transient forest response model to calculate an approximate 10% decrease in annual biomass increment at 21 locations in temperate and boreal forests of eastern North America at the time of CO_2 doubling. There were 50% decreases in simulated temperate forests, while boreal sites had increases of severalfold where forests replaced open shrublands.

The opposite effect on annual growth increment is also quite possible (see Chapter 9, *Terrestrial Biotic Responses to Environmental Change and Feedbacks to Climate*, of the IPCC Working Group I volume). Annual growth increment may increase in all zones from the fertilization effect of increasing atmospheric CO_2; increased efficiency of water use; and, particularly at high latitudes, globally warmer temperatures. These enhancements are discussed in detail in Chapter 1, which provides no estimate of the impact on growth increment. However, others (e.g., Melillo *et al.*, 1993) have calculated from carbon flux models that growth increases of 16% may accrue by the time a doubling of atmospheric CO_2 concentrations is reached.

Growing stock volume also might be significantly enhanced through concerted national and international programs to mitigate atmospheric carbon emissions through development of forest plantations for carbon storage and for substitution of fossil fuels with biomass. The maximum increments would vary each year over the life of any proposed program. Chapter 24 uses these estimates, aimed at generating the maximum possible biomass increment (Nilsson and Schopfhauser, 1995), to project potential carbon storage from mitigation measures. Nilsson and Schopfhauser (1995) estimate that about 1.8 x 10^9 m^3 could be incremented annually by the year 2025 (0.7% of supplies shown in Table 15-3), with a peak value of 3.11 x 10^9 m^3 by the year 2055 (1.4% of supplies in year 2050, Table 15-3). This assumes that annual increments will remain constant rather than declining from stress or increasing from CO_2 fertilization.

If the growing stock enhancement for the year 2050 produced by extension of current annual growth increment (4.24 x 10^9 m^3) were combined with that produced by optimum management for the year 2055 (3.11 x 10^9 m^3), forest volumes would decline 19% from 1980 levels by about 2050, rather than declining 22% as calculated in the absence of these factors. Present research results are inadequate to determine whether one should accept or reject either or both of the enhancement factors.

15.2.4. Projecting Future Need for Forest Products

The annual need for forest products must be determined differently from the availability of forest products. One can assume that the need for forest products will be largely independent of climate change and land use effects, at least until available timber supplies have been nearly consumed, forcing prices so high that users cannot afford to purchase them. The simplest assumption is that the need for wood is proportional to population size and associated regional prosperity, although FAO (1993a) depends on gross domestic product (GDP) in more complex and assumption-laden demand models for forest products (Baudin, 1988).

Considering the uncertainties associated with projecting forest product needs to the year 2050, we chose the simpler assumption. Future need for fuelwood/charcoal and industrial roundwood (the two components of roundwood) was calculated (Table 15-4) based on future population estimated for each of the 13 global regions in the IMAGE conventional

Table 15-4: *Current and future need for forest products in 10^9 m^3, based on regional population projections (Alcamo et al., 1994) and regional forest product consumption (FAO, 1993a).*

	Year			
	1990	2000	2025	2050
Industrial Roundwood				
Tropical Regions	0.28	0.35	0.52	0.65
Temperate Regions	0.89	0.95	1.04	1.04
Boreal Regions	0.43	0.45	0.49	0.49
Total Global Volume	1.60	1.75	2.05	2.19
Fuelwood and Charcoal				
Tropical Regions	1.37	1.70	2.63	3.41
Temperate Regions	0.37	0.42	0.49	0.51
Boreal Regions	0.01	0.01	0.01	0.01
Total Global Volume	1.75	2.12	3.13	3.93
Total Volume	3.35	3.88	5.18	6.12

wisdom scenario (Alcamo *et al.*, 1994; Table 3), and the current per capita consumption of forest products in each of these regions (FAO, 1993a).

Current annual global need is approximately 1% of potential global growing stocks (derived from values for product availability in Table 15-3 and for product need in Table 15-4), comparable to values for individual regions calculated by others (e.g., Kauppi *et al.*, 1992; Dixon *et al.*, 1994; Nilsson and Schopfhauser, 1995). Need is projected to reach nearly 3% of global productive standing stocks by the year 2050 (6.12 x 10^9 m^3), as need almost doubles and global growing stock declines by one-fourth or one-fifth. In tropical regions, where as much as 90% of forest production is consumed as fuelwood (FAO, 1993a), need for fuelwood reaches nearly 5% (34.1 x 10^8 m^3) of growing stocks by 2050.

The projected annual need for industrial roundwood and fuelwood/charcoal is well within annual increments projected for growing stocks in boreal and temperate countries. In tropical countries, however, a wood product availability shortfall of some 2% of total standing stocks occurs by 2050; that is, 2% more wood is needed than is being grown each year. If fossil fuels are substituted with fuels from biomass, the shortfall is likely to be more severe. The need for forest products will not necessarily coincide with demand for forested land for agriculture. For example, trees felled for agriculture in Table 15-3 will not all be used to meet the need for forest products in Table 15-4. Hence, the need estimated here is a minimal value and is probably a significant underestimate of growing stock losses to consumption of both roundwood and agricultural land.

Shortages will be more severe in densely populated countries like India, Bangladesh, Pakistan, and China. Shortages may be insignificant in sparsely populated countries such as Brazil, Zaire, and Tanzania. Obviously, continuation of the shortfall would eventually consume the tropical forests in the most impacted regions. Before that occurred, increased value from decreased availability would drive fuelwood prices beyond the reach of all but the richest users, resulting in replacement of fuelwood by less satisfactory products (crop residues, dried livestock dung, etc.) in those areas.

A more sophisticated approach to estimating need—utilizing economic, technological, and political developments to define economic supply and demand, excluding effects of climate-controlled standing stock changes—was generated for the current chapter based on several data sources (Table 15-5).

The economy-based estimates of demand include only industrial roundwood, projected to the year 2020. The approach began with the same current global need (1.60 x 10^9 m^3) as did Table 15-4. By the year 2010, however, economy-based estimators project a much greater need for industrial roundwood than does Table 15-4 (1.75 x10^9 m^3), amounting to 2.15–2.67 x 10^9 m^3 (FAO, 1993a, also projected a consumption of 2.67 x 10^9 m^3 by the year 2010). Need projected by the economy-based method is 2.55–3.16 x 10^9 m^3 by 2020 (compared with the much lower estimate of need of 2.05 x 10^9 m^3 by 2025 in Table 15-4).

Table 15-5: *Economic supply and demand for industrial roundwood to the year 2020 in 10^9 m^3, based on analysis of several authors by Nilsson (1994).*[1]

	Year		
	1991	2010	2020
Global Demand	1.60	2.15–2.67	2.55–3.16
Tropical Supply	0.28	0.27–0.29	0.31–0.33
Temperate Supply	0.81	1.0–1.12	1.14–1.18
Boreal Supply	0.51	0.56–0.73	0.64–0.79
Global Supply	1.60	1.83–2.14	2.09–2.3

[1] Arnold, 1992; Backman, 1994; Backman and Waggener, 1991; FAO, 1991, 1993a, 1993d, 1994; ITTO, 1993; Kallio *et al.*, 1987; Kuusela, 1994; Nilsson *et al.*, 1992a, 1992b, 1994; Perez-Garcia, 1993; Sedjo and Lyon, 1990; Shvidenko and Nilsson, 1994; Thunberg, 1991, 1993; USDA Forest Service, 1990, 1993.

Although the FAO (1993a) consumption projections only extend to the year 2010, they also are based on economic, technological, and political criteria. The FAO generated global fuelwood and charcoal need increases of 31% by 2010 (compared to the 21% increase at 2000, and a 79% increase by 2025 derived from data in Table 15-4). Thus, the economy-based estimates do not disagree, and suggest that population-based estimates of need may be conservatively low.

In summary, the availability of forest products in the year 2050, as limited by climate and land use, appears to be adequate to meet projected needs in temperate and boreal regions but not in tropical regions. There, calculated consumption appears to define a serious shortfall of future availability. This conclusion is based on an obvious increase in demand, especially for fuelwood. Also, an increasing population may convert forests to farms to provide food and may be forced to rely on industrial roundwood for export income. The fuelwood forest resources are primarily of benefit to less-developed tropical countries, and their loss may well be important for populations in the most crowded countries. The developed countries of temperate and boreal regions may not be significantly affected by losses of the roundwood supplies they purchase from the tropics because these can be obtained from regions of low population density (and, hence, of low fuelwood consumption). Additionally, the developed countries are able to substitute other products for the wood products in current use. For example, timber in residential and commercial buildings currently is being replaced by styrofoam blocks filled with reinforced concrete and by steel structures in some parts of the United States and Europe.

15.3. Evaluation of Tropical Wood Availability and Consumption

Analyses of environmental change scenarios presented in Section 15.2 indicate that land use will be very important for tropical forest product availability during the next century. They suggest that the availability of forest products could decline by about one-half from land-use pressures, even though changes in climate and atmospheric chemistry during this period appear likely to increase the area where tropical forests can potentially grow. The analyses in Section 15.2 contrast the potential decline in forest volume with the more than three-fold increase in need for forest products, especially, for fuelwood, to support local subsistence. The scenario indicates that by the year 2050, tropical regions could be subject to serious deficiencies—up to 2% of standing stocks—between annual availability and need. The discussion in Sections 15.3.1 through 15.3.3 evaluates those conclusions.

15.3.1. Critical Considerations

Two primary forces are currently involved in tropical forest change and are likely to continue in importance in the future. First, human population growth in developing countries, which are primarily tropical countries, is 2 to 3% annually, and population is expected to reach 8.7 billion by the middle of the next century (UNFPA, 1992). If tropical regions undergo economic prosperity during the twenty-first century, the need for forest trees and products could increase drastically to generate urban structures and jobs for increasing urban populations. On the other hand, if prosperity falters and large numbers of destitute people seek subsistence unavailable in cities and urban areas, forests and wood supplies are likely to undergo considerable degradation as humans migrate into forests. No matter what future changes in climate and uses of tropical forests occur, effects of the rapidly growing and dense populations in tropical regions will amplify those changes.

Second, the rate of deforestation is very rapid and likely to continue so. It was 15.4 Mha annually (of 1,756 Mha, or 0.8%) for the period 1981–90—higher than in the previous decade (FAO, 1993b). Degradation of remaining forests (such as change from closed to open forests) due to grazing, fires, excessive logging, and fuelwood gathering leads to loss of plant diversity and standing biomass (i.e., forest degradation; Brown *et al.*, 1991; Flint and Richards, 1991). For example, in Africa, much of the change in closed forest cover of 1.5 Mha from 1981 to 1990 was directly attributable to local population growth (FAO 1993b): 34% was from conversion of forests to short fallow agriculture, which reflects the needs of rural populations; 25% was change of closed forests to open forest or forest degradation resulting in loss of tree canopy due to human pressure; 19% was change from forests to fragmented forest interspersed with agrarian land uses, which represents gradual deforestation because these fragments are eventually converted to agriculture; and 16% was change of forest area to other forest cover.

15.3.2. Availability of Tropical Forest Products

Estimates of deforestation rates vary but are uniformly large. FAO (1993a) calculated annual deforestation of 15.4 Mha 1981–90. The IS1992a scenario projected a forest clearing rate of 20 to 23.6 Mha per year by 2025. In that scenario, 73% of all tropical forests were expected to be cleared by 2100 (IPCC, 1992).

In addition, future forest decline may be accelerated by state control over forests and the collapse of traditional community control and management systems over forests (Gadgil and Guha, 1992). This has contributed to forest degradation in the past, and there is no reason to suspect that it will decline in the future. These forces will shape the expected uses of tropical forests for hundreds of years.

The increasing human activity in tropical forests will affect more than just the current standing stocks. Tree felling and wood extraction in the tropics commonly destroy one-tenth to one-third of the advance regeneration and growth in remaining trees (FAO, 1993b). Fragmentation and degradation of forests due to expanding agriculture, need for wood by human population, and livestock grazing pressure will seriously affect forest regeneration, particularly in tropical Asia and Africa. Soil moisture deficits projected for Africa and part of Brazil (IPCC, 1992) will severely hamper forest succession. Increased frequency of forest fire will also reduce the chances of forest regeneration.

Domestic animals are an important factor, reducing forest regeneration by browsing on seedlings and by trampling seedlings and compacting soils. Livestock, including cattle, sheep, and goats, are increasing in Africa and Latin America. In India, for example, the density of domestic livestock is 6 animals/ha of forest (405 million animals in 1982 on 64 Mha). In Africa there is 1 animal/ha of forest (see Figure 15-6 and FAO, 1993b). The total livestock population globally is expected to increase by a factor of three by 2050 (Zuidema *et al.*, 1994). Thus, the total livestock population in Africa would be 1.69 billion by 2050, grazing on declining forests and pasture land.

The IPCC (1992) Response Strategies Working Group has estimated that the need for cropland will increase along with the world's rising population. Such an increase might require about 50% more land to be in crop production by 2025, a figure considerably higher than the estimates generated by the IMAGE model and used in estimating growing stock volume in Section 15.2.3. The cropland in Africa alone is projected to increase from 163 Mha in 1990 to 347 Mha by 2025 (Sauerbeck, 1993).

15.3.3. Need for Tropical Forest Products

Tropical forests provide a range of economic, social, and environmental services to humans. For simplicity, only the major economic products are included below. The product functions could be classified as subsistence (e.g., wood and charcoal used for cooking) and commercial (e.g., industrial wood and sawn wood). Among subsistence products, only in Brazil is charcoal used as a commercial fuel in industry on a large scale. The commercial products, however, are traded within and between countries.

The forest product consumption values estimated in Section 15.2.4 have been expanded in Table 15-6.

Note that estimated need for fuelwood in the tropical countries increases by about 2.5 times to 2050, but in Africa it increases by 3.3 times (Table 15-6). Yet fuelwood availability in Africa, like the rest of the tropics, is expected to decline by about half during the same time period (Table 15-3). Hence, the projected deficit of need over availability of tropical fuelwood may be most critical in tropical Africa. In addition, export of industrial roundwood, which might be used to offset fuelwood deficits, is already very low (Table 15-6), with 89% of African forest harvest being devoted to fuelwood (FAO, 1993a). Examination of tropical forest area changes projected in Table 15-2 and population changes used in the IMAGE projections (Alcamo *et al.*, 1994) indicates that the fuelwood need problem will increase in severity, as the current 0.84 ha of forested land per capita (FAO, 1993a) declines to 0.13 ha per capita by 2050.

The projections made in Table 15-6 do not consider various factors like tropical forest destruction, much of it for increases in livestock, or forest degradation (legal and illicit removal of trees). The projections also do not include afforestation/reforestation, slowing of land needed for agriculture by enhanced crop productivities, or changes in rates of use of different forest products as countries industrialize (such as decline in fuelwood use but increase in use of industrial wood).

Non-timber forest products are particularly important in tropical regions. The tribal and rural communities in tropical countries gather a large range of non-timber forest products for subsistence consumption as well as for commercial purposes (e.g., food, fodder, oil seeds, gum, nuts, rattan, bamboo, and raw material for a range of industrial products, as well as locally used products like baskets, mats, tools, leaf plates). For example, the value of rattan extraction in Southeast Asia was $275 million in 1990; the value of Brazilian nuts was $72 million (FAO, 1993c).

The impacts of climate change itself (forest dieback, retreating forest boundaries, increased insect pest and fire incidence; see Chapter 1) and the adaptation measures to mitigate the effects of climate change such as short-rotation forestry (Chapter 24, this volume), will reduce the availability of non-timber forest products. Declines in biodiversity in tropical forest due to climate change through increased variance in seasonality (Hartshorn, 1992) and increased turnover rates (Phillips and Gentry, 1994) also may reduce the availability of the large diversity of plant and animal products from forestry. This latter process may not occur universally, as shown in certain limited areas of Venezuela (Carey *et al.*, 1994).

The replacement of tropical forests with non-forest land uses, coupled with increased need for forest products, will not necessarily generate losses in tropical forest volumes. For example, a conservation and large afforestation effort has stabilized the area under forest in India at about 64 Mha since 1980 even though the population is growing at more than 2% annually (see Box 15-2).

Plantation forestry for all purposes (mitigation of atmospheric CO_2, wood for pulp, sawnwood, fuelwood, etc.), using degraded lands, could become one of the most important counteragents to deforestation. Estimates in Table 24-4 (Chapter 24) indicate that potential added tropical forest wood stocks could average 300 Mt/year (16.42 Gt over 55 years beginning in 1995).

Section 15.2.3 notes that between 18 and 31 x 10^8 m^3 annually (0.7 to 1.4 % of standing stock at the time of measurement) is the most one can expect from plantations designed to mitigate atmospheric carbon. Yet plantations may be much more significant in alleviating the need for forest products in the future. The area established as plantation in 90 tropical countries by 1990 was 44 Mha (FAO, 1993b), with India accounting for nearly 40% of the total area, followed by Indonesia and Brazil. Tropical countries are implementing large afforestation programs. India averaged 1.44 Mha annually in the decade 1981–1990 (FAO, 1993b) and now is planting about 2 Mha

Table 15-6: *Current and projected forest product needs in the tropics (10^3 m^3), based on per capita consumption estimates from FAO (1993a) projected to 2020 and 2050 with UN (1993) population estimates.*

Product	Region	1991	2010	2020	2050
Industrial Roundwood	All Tropics	237	485	555	791
	Africa	39	55	70	113
	Asia	102	172	193	313
	Latin America	96	258	292	365
Fuelwood and Charcoal	All Tropics	1348	1829	2164	3389
	Africa	446	729	924	1496
	Asia	619	800	900	1496
	Latin America	283	300	340	424
Sawnwood	All Tropics	68	140	159	229
	Africa	5	8	11	18
	Asia	37	59	66	108
	Latin America	26	73	82	103

Box 15-2. Forest Conservation in India

Tropical deforestation for the period 1981–90 is estimated to be 15 Mha annually (FAO, 1993b). In addition, the IMAGE 2 model projects increasing deforestation (Table 15-2) from the simple assumption that larger populations will increase the demand for land, thus increasing deforestation of that land. However, the reality in India presents a contrasting picture.

Since 1980 India has periodically assessed its area under forests through its National Remote Sensing Agency (NRSA). The area under forests was assessed for the periods 1981–83 (FSI, 1988), 1985–87 (FSI, 1990), 1987–89 (FSI, 1992), and the latest for 1989–91 (FSI, 1994).

The area under forest in India since 1982 has been stable at 64 Mha (Table 15-7). The forest conservation achieved in India is significant because the population density in India is high (257 persons/km^2) with a large rural population (627 million in 1991) depending on biomass, and an annual population growth rate of 2.12% during 1981–91.

Notably, the area under dense forest (>40% tree crown cover) increases in each assessment, suggesting an increase in carbon stocks sequestered by Indian forests. Responsible factors include the Forest Conservation Act of 1980, which bans all conversion of tropical forest land to non-forest uses, and the world's largest afforestation program (FAO, 1993b). This exemplifies the adaptation potential of tropical countries to declines in forest areas and demonstrates the limitations of projections of future forest area based on population growth rates, such as those from IMAGE (used to construct Tables 15-2 and 15-3).

Table 15-7: *Area under forest in India, in 10^3 ha, according to assessments of the National Remote Sensing Agency (the year refers to the mid-year of assessment; Ravindranath and Hall, 1994).*

Forest Category	1982	1986	1988	1990
Dense Forest (Crown Cover >40%)	36.1	37.8	38.5	38.6
Open Forest (Crown Cover 10–40%)	27.7	25.7	25.0	25.0
Mangrove Forest	0.4	0.4	0.4	0.5
Total Forested Area	64.2	63.5	63.9	64.1

annually (IPCC, 1992). The dominant species planted in India are largely short-rotation exotics, as they are elsewhere in the tropics (Ravindranath and Hall, 1995).

In India and Brazil, the choice of short-rotation forestry will only meet the need for industrial wood and fuelwood (and charcoal), not for sawn logs for structural uses. Thus, forest plantations (largely softwood) are no substitute for logging in forests for hardwoods. Climate change can worsen the situation by affecting forest regeneration due to warming and decreased availability of soil moisture. There is a possibility that a large area could be planted to hardwoods and long-rotation plantations such as teak.

In summary, the shortfall of wood during the 21st century, projected for tropical areas in Section 15.2.4, appears to underestimate the potential problem. Tropical countries will face serious shortages in forest products required for subsistence (fuelwood, wood and fibers for construction, fruits and nuts for food). However, wise land-use policy may mitigate some of the forest losses.

15.4. Evaluation of Temperate Wood Availability and Consumption

The scenarios discussed in Sections 15.2.3 and 15.2.4 suggest that temperate forests could cover about as much land by 2050 as they do today, and that current consumption (3.1% of standing stock, based on Tables 15-3 and 15-4) may increase only slightly by 2050 (3.8% of standing stock). With a 3–3.5% annual growth increment in temperate zone trees (FAO, 1992), and with an increasing proportion of need being met by imports from outside the temperate zone (FAO, 1993a), the scenario analysis in Section 15.2 indicates that temperate-zone forests are likely to continue meeting the need for forest products on a sustainable basis for at least the next century. The discussion in Sections 15.4.1 through 15.4.3 examines that conclusion in detail.

15.4.1. Critical Considerations

Perhaps the most important characteristic for defining forest product availability in temperate zones, now and in the future, is that temperate zones are most appropriate for mechanized agriculture of monospecific crops; hence, mankind has altered the natural vegetation there both intensively and over a relatively long time. Consequently, at 0.71 billion ha (compiled from FAO, 1993a; Puri *et al.*, 1990; World Bank, 1993; Chadha, 1990; Khattak, 1992; Helvetas and Swiss Development Corp., 1989; Goskomles SSSR, 1990; Kurz *et al.*, 1992; Xu, 1994), temperate forests cover only about half of their potential area, and the remaining forests often differ quite considerably from the original state (Stearns, 1988; Deutscher Bundestag, 1991; Röhrig, 1991).

The prosperity of the developed countries which permits development of mechanized agriculture also gives most countries of the temperate zone greater capability than those in other zones to manage forests to avoid future disturbances (fire, insect attack, mortality from rapid and chronic climate change, and so forth) and to enhance growth (artificial tree migration, plantation establishment, post-establishment thinning and fertilization, and so on).

The capabilities of developed nations which buffer temperate forest product availability from environmental threats also provide a buffering of need for forest products from shortfalls in availability. First, these countries have the wealth to purchase wood products outside the temperate zone. Today, temperate countries consume 80% of the total value of world imports of forest products but supply only 50% of the value of its exports. Second, temperate countries have access to technology that allows product substitution when local availability is inadequate and interzonal trade is too expensive. Pressed fiberboard replaced some weak plywood supplies during the 1970s in the United States (Rose *et al.*, 1987), for example, and currently, increasingly expensive lumber is being replaced in some residential structures with steel, concrete-filled plastic foam, or similar non-forest products.

15.4.2. Availability of Temperate Forest Products

Harvests from temperate-zone forests of all types are used mainly for industrial purposes (Table 15-8), although there are great differences between the two Asian regions in Table 15-8 and the other regions. In Europe, the United States, and the temperate Southern Hemisphere, the industrial wood portion of the total is about 80%, whereas in Asian temperate regions it is only 8.6% (tropical) and 38.0% (temperate) (Table 15-8). Overall, temperate-zone use of forest products focuses on industrial wood.

The scenario analysis above defined no significant climate or land-use related declines in future forest availability. Forest survey measurements are consonant with this analysis. Considering that carbon storage from net ecosystem production in European forests increased about 30% between the early 1970s and the late 1980s and is further increasing, the wood resource is seen as plentiful for the foreseeable future, if no unexpected catastrophic events occur (Kauppi *et al.*, 1992). The remarkable net ecosystem production (NEP) increment in the last two decades may relate less to intrinsic productivity of forests than to increased forest area generated as European agriculture decreased while imports of less-expensive foreign lumber increased (more than one-third of global industrial roundwood imports in 1990; approximately 10% less than Japan imported, and six times the imports by North America; FAO, 1993a). If so, change in land-use policy or in costs of imported roundwood would have a great impact on European NEP and forest availability.

Climate effects may be indirect. In Europe, migration of the Mediterranean tree species to the north may be hampered because most of them are not adapted to the prevailing acidic soil conditions in the northern areas (Ulrich and Puhe, 1993). In the worst case, this may result in losses of some species, with negative effects on biodiversity and wood availability.

Frequent extreme events affecting wide regions—like storms or insect infestations—leading to great amounts of wood that must be harvested, may have a great influence on annual cut and the economic situation of forestry. For example, the price index for timber in the Federal Republic of Germany (FRG) declined from 116% in 1990 to 79% in 1991 as a result of hurricane Wiebke, which caused large windfalls in many parts of

Table 15-8: *Current production and consumption, and projected need for industrial roundwood and fuelwood in 10^6 m^3 (FAO, 1993a).*

Region	Production 1991		Need 1991 and 2010			
	Industrial Roundwood	Fuelwood and Charcoal	Industrial Roundwood 1991	Industrial Roundwood 2010	Fuelwood and Charcoal 1991	Fuelwood and Charcoal 2010
West Europe[1]	138	28	144	207	29	30
East Europe	338	160	323	494	160	102
Temperate/Tropical South Asia	28	299	29	40	299	413
Temperate East Asia	118	193	176	345	193	208
North America[2]	410	86	382	582	86	130
Temperate Southern Hemisphere	42	10	28	44	1	11
Total	1074	776	1082	1712	768	894

[1]Excludes Finland, Norway, and Sweden.
[2]Excludes Canada.

western Europe (Stat. Bundesamt, 1992). Pests can largely dictate where and how many trees must be harvested. If climate-induced changes cause extensive sanitary fellings, this may reduce revenue per unit of wood and hence reduce incentives for intensive management and investment (Pollard, 1993).

One of the most important anthropogenic effects on forest product availability in temperate zones in the past has been intensive land use. Nevertheless, until now in Europe, stable food demand, substantial producer subsidies, and high productivity increases have generated enormous surpluses of agricultural products. As a consequence, the use of arable land decreased substantially in these regions. The situation is similar in the United States and Japan. In the European Community (EC) up to 44 Mha could go out of agriculture without substantial decreases in productivity (Eisenkrämer, 1987; Delorme, 1987). Therefore, in western Europe the common agricultural policy of the EC will be a more important driving force for future changes in land use than climate-induced ecosystem changes (Kitamura and Parry, 1993). This phenomenon is abstracted in IMAGE model projections (Table 15-2) and is responsible for the simulated increase in temperate forest area in Europe (Zuidema *et al.*, 1994).

IMAGE projections suggest a 2% decrease in total area of temperate forests between 1990 and 2050 (Table 15-2). However, if only the temperate deciduous forests are considered, the forested area increases by 60.5 Mha (89%) from 1990 to 2050 (Zuidema, 1994). In contrast, temperate forest area projected by BIOME under four different climate scenarios increases between 25 and 52% by the time the climate resulting from a doubled CO_2 concentration occurs (Table 15-1). If either the BIOME projections for all temperate forests or the IMAGE projections for temperate deciduous forests becomes reality, a wood availability increase of one-tenth to one-half will result.

15.4.3. Need for Temperate Forest Products

Current production and consumption of roundwood in the temperate zone are fairly balanced (Table 15-8). Nevertheless, there are also countries where consumption greatly exceeds production (e.g., Japan, China, Austria, Italy). Projections for 2010 (FAO, 1993a) indicate that the need for industrial roundwood will increase between 38% (temperate South Asia) and 96% (temperate East Asia). It is likely that need will increase everywhere for particular wood types (softwood logs and pulpwood), whereas in some regions (Japan and western Europe) need for all kinds of wood products will increase. As a consequence, future increases in consumption must be met partly by increases of imports. A great portion of these may come from other temperate forested countries, such as Canada, New Zealand, and Chile, but also from plantations in the tropical zone.

Fuelwood consumption in temperate zones is not as important as it is in tropical and subtropical countries. The only exception to this may be China, where 32.5% of the wood production in 1988 was consumed for fuelwood (Xu, 1994). Nevertheless, in temperate South Asia and North America, significant increases (38% and 51% respectively) in fuelwood consumption by 2010 are expected (Table 15-8).

In the developed countries of the temperate zone there is a high and growing need for assets that support a high quality of life. This has produced a change from the wood production point of view to a multiple-use perspective (FAO, 1993c). These needs are not considered in the scenario analysis of Section 15.2 but will generate considerable change in wood availability (much of it being removed from accessibility) and need. The non-wood functions (soil protection, recreation, biodiversity, and cultural uses) and goods and services (fruits, honey, water supply, and so on) are of very high importance. Table 15-9 gives an overview of the importance of main forest functions in some temperate regions.

Table 15-9 shows that in Europe and the United States, other functions than wood production have high importance in relevant areas. For most of the developed temperate-zone countries, it is the explicit policy to emphasize non-wood functions of their forests, especially functions of water protection and recreation (FAO, 1993c). The economic evaluation of these benefits is much more complex than that of wood production because, for most of these values drawn from the forests, supply and demand generally are not regulated by market mechanisms. There are only a few non-timber products that have a direct economic importance for individual countries—for example, cork in Portugal and Spain, chestnuts in Italy, or Christmas trees in Denmark—but they have immediate effects on local people and communities.

Table 15-9: *Importance of forest functions by area as a percentage of total forest land (FAO, 1993c).*

Function	Europe High	Europe Medium	Europe Low	USA High	USA Medium	USA Low	Former USSR High	Former USSR Medium	Former USSR Low
Wood Production	54	24	22	36	41	23	0	18	82
Protection	11	17	72	36	30	34	9	15	76
Water	9	17	74	16	79	5	7	17	76
Hunting	27	55	18	9	45	46	-	-	-
Nature Conservation	4	37	59	8	42	50	0	19	81
Recreation	12	39	49	19	33	48	2	11	87

15.5. Evaluation of Boreal Wood Availability and Consumption

The scenario analysis of Section 15.2 presents a dichotomy in expectations for wood availability in the boreal zone. The climate-only scenarios (Table 15-1) assumed that agriculture occupies 50% of the land permitted by its climatic potential, boreal forest displaces nonforested tundra, and temperate vegetation replaces much of the current lower-latitude range of boreal forests. This scenario suggests that boreal forest areas will shrink between 20 and 50%, depending on the climate change scenario used, by the time a doubling of GHGs occurs sometime in the next century. The scenario driven primarily by land use (Table 15-2) included agricultural land area that was linked to a declining population and demand for food, and allowed little climate-induced change in forest geography, except in the few areas that shifted out of agricultural production. That scenario suggested that boreal forests would increase in area by 10 or 11% by the year 2050.

At the same time, the population-based need estimates indicated a relatively constant need of 0.5% of availability (Tables 15-3 and 15-4)—considerably less than the sustainable harvest to be expected of available supplies, even under the most deleterious climate-only scenario. These are the suppositions to be examined in Sections 15.5.1 through 15.5.3.

15.5.1. Critical Considerations

More than any others, the boreal forests are subject to periodic damage from large-scale disturbances such as fire, long-term drought, windstorm, and insect infestations. Even chronic and uniform climate warming is likely to cause increasing frequencies of irregular, large-scale, and widespread catastrophic disturbance events in which large areas of forest are destroyed in a single growth season. Recovery from these events may take centuries (Payette *et al.*, 1989), and the resulting availability of forest products as sanitation and salvage fellings will thus be very irregular. Therefore, the shortcomings of the static vegetation models described in Box 15-1 are very relevant to boreal forests.

15.5.2. Availability of Boreal Forest Products

Growth increments in natural forests accrue only about 1.5% each year, but active management of the boreal forests yields much higher increment and growing stocks (Kuusela, 1990). In Sweden, active management has raised the average growing stock from 72 m^3/ha in 1925 to 97 m^3/ha in 1985, and the stock is estimated to reach 125 m^3/ha in the year 2050.

The long-term sustainable timber supply from a biological point of view has been estimated with different degrees of sophistication in the boreal countries. The sustainable harvest level is often called the annual allowable cut (AAC). This comprises 910–930 x 10^6 m^3/yr for the next 50–100 years [Runyon (1991), OMNR (1992), and Booth (1993) for Canada; Nilsson *et al.* (1992a) and Hägglund (1994) for Nordic countries; Isaev (1991), Nilsson *et al.* (1992b), and Backman (1994) for the former Soviet Union]. The long-term economic wood supply takes the economic conditions into account and is lower than the AAC (580–630 x 10^6 m^3/yr; *ibid*). Comparison of these values, or of the approximately 100 x 10^9 m^3 of growing stock in boreal forests (90 x 10^9 m^3 tabulated in Table 15-3), to the annual harvests of 750 x 10^6 m^3, indicates that long-term sustainable biological availability substantially exceeds the prospective need for wood in the boreal zone.

Arnold (1992) found a number of broad trends in the long-term future supply of forest products (50-year time horizon) based on analyses of existing studies:

- There will continue to be a major shift from old-growth supply to planted and second-growth supply.
- There will be large resources in boreal forests to support expanded production if need requires.

The following constitutes a broad summary of global forest product availability based on a number of studies (Cardellichio *et al.*, 1988; Kallio *et al.*, 1987; Sedjo and Lyon, 1990; Arnold, 1992; Nilsson *et al.*, 1992a; Thunberg, 1993; Perez-Garcia, 1993; OMNR, 1992; Backman, 1994):

- The availability of coniferous wood (softwood) from North America is expected to decrease due to younger age class structure and tightening environmental regulation.
- Russia will be dealing with a restructuring of its entire society for a long time and will be hard-pressed to sustain timber production at current levels.
- Europe will probably need more than it supplies.

The overall conclusion from these analyses is that the major supply in the near to medium term will come from the traditional wood supply regions, but within 50 years an increase of production through plantation forests may be required (some 150 million ha).

Chapter 1 and Section 15.2 both discuss expected changes in climate and responses by natural forest ecosystems. In a warmer and drier boreal zone, it is likely that geographic boundaries of boreal forest ecosystems will shift (Emanuel *et al.*, 1985a, 1985b; Solomon, 1992; Smith and Shugart, 1993). While the forest ecosystems adapt to climate changes, ecosystems will be in disequilibrium as existing species are replaced by new species more adapted to the environment. Soil conditions that lack the requirements of invading species in many areas will exacerbate this condition. The area of forest affected will greatly depend on the magnitude and rate of the climate change. A likely result of any future climate change is that during the next 50 years the wood availability in the boreal zone will increase. However, it should be pointed out that forest management and protection may have to be intensified substantially during this period, which will be costly (Nilsson and Pitt, 1991; Apps *et al.*, 1993).

As illustrated earlier (Table 15-2), future changes in land use driven by economic factors could increase forest land in the boreal zone in the future (Waggoner, 1994; Nilsson *et al.*, 1992a; Zuidema *et al.*, 1994). Air pollutants are thought to be a problem in the boreal zone today (Nilsson *et al.*, 1992a, 1992b; de Steiguer, 1993); if so, they would affect wood availability as well as the availability of non-wood products in the boreal zone in the mid-term.

In the long run (>50 years), effects of climate change on wood availability could be significant as the availability of wood from existing species runs out and new species are insufficiently developed to provide a new source of supply. Yet in terms of the sustainable use of forests (i.e., of need, discussed in Section 15.5.3), the scenario estimates of availability until 2050 and those that would result from quantification of climate change effects both suggest an excess of availability over need. However, it is also important to emphasize that a significant portion of harvested supply in the future could evolve from irregular, widespread, and catastrophic mortality events that cannot be simulated or predicted by current modeling methods.

15.5.3. Need for Boreal Forest Products

The consumption of forest products by the sparse population of boreal regions can hardly affect boreal forest stocks. Future population estimates suggest little change in population under current immigration policies (e.g., Alcamo *et al.*, 1994). Instead, assessment of needs from boreal regions must focus on trade in forest products and on the resulting economic forces that will continue to define the need for boreal forest products.

Need for forest industry products is determined by changes in population, income, and price and by technological change. Past consumption can only be disaggregated to relevant end uses (preferences) in the more developed countries, and useful predictions can only be made for short periods in the future. Forest products are international commodities, and need and availability in one region depend on developments elsewhere on the globe.

The boreal forests have played an important role as suppliers of raw material for forest industrial products globally and of so-called non-wood benefits. The production of industrial roundwood from the boreal forests in 1991 corresponds to some 33% of world production. The consumption figures presented by FAO for the year 2010 are regarded as strong overestimates among industrial experts; those presented in Table 15-10, calculated as described in Section 15.2.4, may be more realistic.

The most crucial role of boreal forest products is probably the balance of payments. The boreal countries are major net exporters of forest products. More than 20% of the world's forest industry product value stems from the boreal forests. Table 15-11 indicates that for Canada and the Nordic countries (especially Norway and Finland), a very significant portion of their trade involves forest products. It seems likely that wood exports eventually will increase to even greater importance in the former Soviet Union (FSU).

Table 15-10: *Industrial roundwood consumption projected to 2020 in boreal regions, in 10^6 m^3, based on analyses in Table 15-5.*

	Year		
	1991	2010	2020
Coniferous (Softwoods)	452	434–588	496–611
Deciduous (Hardwoods)	60	123–145	141–181
Total Boreal Demand	512	557–733	637–792
Global Need	1602	2145–2674	2551–3156

Arnold (1992) examined long-term trends in need for forest products. Among the most important conclusions relating to boreal wood consumption are the following:

- Global need for industrial wood has been projected to grow by 35 to 75% over the next 50 years.
- Consumption will continue to be concentrated in the developed world, but a substantial part of the increase in consumption will occur in the developing countries.
- Need for roundwood will grow more slowly than need for finished wood products due to technological development.
- Use of nonconiferous (hardwood) species will grow faster than for coniferous (softwood) species.

As in temperate regions, boreal-region forests no longer are viewed as simply a provider of raw materials for industry and a stimulus in the general economy of a country (Westoby, 1962). Today, the nonmarket forest production properties (climate change, biodiversity, global tourism, non-wood products, erosion, water supply) are among the most prominent issues in world forestry (Hyde *et al.*, 1991). The boreal forests promote a large variety of global as well as local and regional non-wood products. However, the extent and value of these services are

Table 15-11: *Economic importance of the forest sector in boreal countries in million U.S.$ (FAO, 1993a).*

	Prod.	**Consum.**	**Import**	**Export**	**% of Trade**	**% of GDP**
Canada	30482	15931	1840	16931	13	5
Nordics[1]	22115	4876	2388	19628	4–36	2–7
FSU[2]	38485	36639	927	2773	4	2
Total	91082	57446	5155	39332		

[1]Norway, Sweden, Finland, and Denmark.
[2]Former Soviet Union.

uncertain because most of them are not connected by any market mechanism. The boreal countries also have stated that future boreal forest policies will place emphasis on non-wood functions.

Economic analyses of the current impact of the non-wood functions on the macro-economy in the boreal countries are limited in scope and uncertain. Hultkrantz (1992) and Solberg and Svendsrud (1992) estimated the value of the non-wood products at 35–75% of the industrial forest sector value (adjusted GDP-calculation) for Sweden and Norway, respectively. Based on the earlier discussion, we conclude that this value will increase in the future in the boreal zone. In the long run, forest management for non-wood products may well become a more valuable activity than logging.

In the future, increased conflicts in the boreal forests are expected between traditional forestry and requirements for wilderness areas, between timber production and biodiversity, and between applied forest technology and landscape development. There will be increased public concern regarding the health and survival of the forests. The forecast reconstruction of the structure of the forest industry will generate regional unemployment and social strain in the boreal countries. There are approximately one million aboriginal people living in the boreal zone. The cultures and economies of these people are intricately adapted to the natural environment, and they depend upon it for self-perpetuation.

The forest industry has restructured extremely quickly during the last 20 years in the boreal countries due to economic development, technological development, market conditions, and internationalization. This kind of structural change is expected to continue in the future. Technological development is also very rapid in the forest sector, especially in the forest industry. Most of the lumber grades known today in the forest industry were not known 20 years ago. This development is also expected to continue in the future.

Many of the changes taking place in the forest sector of the boreal countries are based on the changed end-use preferences. Preference changes, based on shifts in knowledge, values, consumption, and so forth, have happened within a few years (e.g., bleached to unbleached paper, clearcuts to landscape planning), and they influence the whole forest sector. Preference shifts will continue to permeate the future societies of the boreal zone.

What, then, will be the constitution of overall forces on the forest industry? Most obviously, forces other than climate will probably overshadow the impact of ecosystem responses to climate early on (the next 50 years). A major question involves the impact on economic considerations of non-wood functions like biodiversity, landscape conservation, soil carbon capacity, and sustainable watershed development, which could significantly reduce available timber supplies. Unfortunately, we lack benchmarks for measuring the impact of climate change on these functions. In the long run (>50 years), and based on today's knowledge, the situation could be reversed; ecosystem changes induced by shifting climate might overshadow the impact of the other driving forces in the boreal forests.

In summary, the conclusions from scenario analysis about boreal timber availability and need that begin Section 15.5 appear to be very unrealistic. First, timber availability may be more strongly affected by environmental changes in boreal regions than in other zones: increased mortality and slowed forest succession from rapidly warming climate and large-scale disturbances; and enhanced growth from warmer growing season temperatures and increased atmospheric CO_2. These variables are not simulated by the models that produced the scenarios in Section 15.2. Second, boreal forest timber need is controlled by trade in forest products outside the zone. Hence, economic considerations drive need and its subsequent implementation—that is, harvest of timber. The models underlying the scenarios in Section 15.2 contain no consideration of forest economic or trade processes or implications, and therefore are not capable of realistic analysis of boreal forest product need in the future. The economics of wood consumption and the analysis of temperate and tropical zone timber needs, discussed in Sections 15.3 and 15.4, suggest that serious shortfalls of boreal industrial roundwood could occur during the twenty-first century. This would result if future temperate and tropical zone consumption of boreal softwoods expands beyond available wood capacities.

15.6. Adaptation and Coping Options

In the short term, timber supplies from all zones can be readily assured in intensively managed forests. The use of forest plantations to mitigate increasing atmospheric CO_2 and to generate fuels from biomass is a subject of Chapter 24. Unfortunately, past intensive management, especially fire suppression and tree selection at species and intra-specific levels, has created forests that now may be more vulnerable to fire, pests, and pathogens (Schowalter and Filip, 1993), although others dispute that conclusion. Given the current degree of uncertainty over future climates and the subsequent response of forest ecosystems, adaptation strategies (those enacted to minimize forest damage from changing environment) entail greater degrees of risk than do mitigation measures (those enacted to reduce the rate or magnitude of the environmental changes). Emerging principles of risk assessment and management should guide the choice and implementation of "no-regrets," "proactive," and "reactive" options for adaptation (Gucinski and McKelvey, 1992).

15.6.1. Harvest Options

Certain standard harvest options appropriate for ameliorating effects of climate change are suggested:

- **Sanitation Harvests**—Increased timber losses are expected from pests, diseases, and fire because the trees themselves may be stressed by warmer conditions and the greater moisture deficits expected to accompany warming (see Chapter 1), and land clearance for tropical agriculture. Extensive sanitary felling can produce large volumes of timber, reducing revenue per unit of wood and reducing incentives for intensive management, while disrupting future timber availability.

- **Changing Harvesting Method**—Harvesting and even site preparation may be quite dependent on cold winters in northern regions where forests occur on predominantly swamp lands that can only be accessed when frozen (Pollard, 1987). Use of ecosystem protection approaches (e.g., Franklin, 1989; Clark and Stankey, 1991) can be particularly useful in light of multiple-use needs in temperate and boreal regions and the need for mixes of species and ages in all zones to prepare for several different outcomes of environmental change.
- **Shortening Rotations**—Reduction of rotation age is a simple tactic for reducing exposure of maturing timber stocks to deteriorating conditions, as well as increasing opportunities for modifying the genetic makeup of the forest. Gains may be offset by diminished timber quality, a lower mean annual increment, and impacts on other values such as certain game species and aesthetic qualities. Short-rotation plantations in tropical regions are an important option in meeting local fuelwood needs, coincidentally relieving pressure to harvest more pristine forests.
- **Increased Thinning**—The stimulation of tree vigor following stand thinning has special application under increasing moisture deficits. The modification of forest microclimate by thinning offers considerable potential for the management of pests (Amman, 1989; Filip *et al.*, 1992; Paine and Baker, 1993).

15.6.2. Establishment Options

Establishment options that can be quite effective in reducing effects of climate change are selected from common silvicultural practices:

- **Choice of Species in Anticipatory Planting**—Mixed species for current planting should be considered wherever possible, as a means to increase diversity and flexibility in adaptive management. Critical non-timber values such as snow retention, soil stability, water quality, and carbon storage should be taken into account.
- **Vulnerability of Young Stands in Anticipatory Planting**—The planting of species and varieties better adapted to future conditions may well increase vulnerability of the resulting stands during their early establishment. Selection of provenance offers an important tactic to reduce the vulnerability.
- **Assisting Natural Migrations in Protected Areas**—Protected areas are the richest sources of genetic materials and warrant expansion into comprehensive systems. To be effective they must function in landscapes where ecological integrity is sufficient to permit the movement of living organisms; there must be a comprehensive approach to both commodity and reserved lands (Franklin *et al.*, 1990).
- **Assisting Natural Migrations by Transplanting Species**—Large numbers of seedlings raised in nurseries many hundreds of kilometers from planting sites can be planted in the millions per day (Farnum, 1992). Optimism must be tempered by experience, however: Forest managers have been attempting to reforest Iceland for the past 50 years without much apparent success (Loftsson, 1993).
- **Gene Pool Conservation**—Specialist species (indicated by restricted geographic ranges) will be most at risk, and their only chance of surviving may be through conservation in forest reserves, arboreta, and conventional seed banks and cryogenic storage.

15.7. Research Needs

New research products needed to conduct an accurate assessment of socioeconomic impacts of forest responses to climate change have become obvious from the foregoing analysis. The assessments are, and will be, based on quantitative scenarios of changing availability of and need for forest products and amenities. Validity of scenarios is critical if they are to function as descriptions of the implications of current knowledge for future conditions. In all regions, scenarios were inadequate because the effects of increasing CO_2 and forest dieback could not be quantified. In tropical regions, the scenarios were deemed inadequate because they probably underestimated the regional and local importance of land use on timber availability. In temperate regions, the scenarios were incapable of quantifying the buffering effects on timber needs, offered by shifting technology aimed at product substitution and on availability by intense mechanical management of forests. In boreal regions, the scenarios excluded transient responses of timber availability to environmental change and extraregional responses of timber needs by economic processes, which replace local forest product demand there. All regions probably will undergo changes that could not be expected only from the variables considered in this assessment. Thus, research required for an adequate assessment includes both environmental monitoring and data collection, and development of models to project future impacts.

15.7.1. Data Monitoring and Collection

All countries need to produce accurate national forest inventories. Some countries maintain such inventories, but they often use different periods and research designs. For example, in Europe, information from repeated surveys is available only from one-third of the forest area (Kauppi *et al.*, 1992). In addition, existing inventories do not consider regeneration properties, such as frequency of flowering, seed production, or seedling establishment. With regard to climate change, these factors are very important because early life stages are most sensitive to variation in temperature and moisture conditions.

An obvious need involves the collection and analysis of data regarding the beneficial effects of atmospheric CO_2 concentrations. As Chapter 1 points out, information from a single season of monitoring tree seedlings in greenhouse pots is inadequate to document the decadal-scale responses by mixed assemblages of growing and mature trees in ecosystems. Our analysis in Section 15.2 highlights the uncertainties regarding the impacts of these effects, which preclude confidence in estimates we generated of future forest product supplies.

15.7.2. Global Modeling

The analyses generated for this assessment repeatedly demonstrated the absolute necessity for development of models describing transient responses of forests to rapid environmental change. Indeed, as Box 15-1 indicates, the problem to be assessed involves impacts of rapidly changing climate during the next century, not stable climate, and involves variously lagged forest response variables, not responses in lock-step with climate change. We lack models of vegetation response to climate that incorporate the real drivers of future vegetation dynamics: Transient processes involving differential rates of mortality, establishment, and growth; climate control of forest establishment and succession; absence and slow introduction of seeds of trees from provenances, varieties, and species appropriate for the available climate; and so on. For this "transient" assessment, the static model estimates of forest growth over the next 50 to 75 years are woefully inadequate.

A second kind of critically absent model is a forest economics simulation capability that is linked to environmental change. The global trade model developed at the International Institute for Applied Systems Analysis (e.g., Dykstra and Kallio, 1987) is an example of the integrated supply and demand model needed; however, it does not project climate-controlled or land-use-constrained wood availability. Instead, a forest trade model embedded in a global integrated assessment model like IMAGE may be the most appropriate vehicle for calculation of the impacts of climate-ecological-economic relationships.

A third form of model conspicuously absent is a climate model that could provide *regional* estimates of the climatic change implied by increasing GHG concentrations. The existing climate models are not able to generate reliable estimates on regional temperature, precipitation, and hydrology. Yet much of the forest response that is relevant to socioeconomic assessment varies significantly from one locale or region to another. In addition, in order to assess the quantitative implications of ecophysiological processes under chronic climatic change, climate models must describe the regional and seasonal changes of temperature and precipitation.

Acknowledgments

The authors deeply appreciate the critical and constructive reviews of earlier manuscripts by (in alphabetical order) C. Barthod, Martin Beniston, Sandra Brown, Melvin G.R. Cannell, Wolfgang Cramer, Robert H. Gardner, Peter J. Hall, Stephen Hamburg, Mark E. Harmon, Gary Hartshorn, John Innes, Miko Kirschbaum, Christian Körner, G. Landman, S.E. Lee, Jelle van Minnen, P.L. Mitchell, Peter J. Parks, W. Mac Post, Kurt S. Pregitzer, John Riley, Michael T. Ter-Mikaelian, and Jack Winjum. Joseph Alcamo and Rik Leemans provided frequent and valuable discussions on, and data from, the IMAGE 2.0 integrated assessment model.

References

Ahmed, K., 1994: *Renewable Energy Technologies: A Review of Status and Costs of Selected Technologies*. Energy Series Technical Paper No. 240, The World Bank, Washington, DC.

Amman, G.D., 1989: Why partial cutting in lodgepole pine stands reduces losses to maintain mountain pine beetle. Proceedings of a Symposium on Management of Lodgepole Pine to Minimize Losses to Mountain Pine Beetle. Gen. Tech. Rep. INT-262, USDA Forest Service, Washington, DC, pp. 48-59.

Alcamo, J. (ed.), 1994: *IMAGE 2.0: Integrated Modeling of Global Climate Change*. Kluwer Academic Publishers, Dordrecht, Netherlands, 321 pp.

Alcamo, J., G.J. van den Born, A.F. Bouwman, B.J. de Haan, K. Klein Goldewijk, O. Klepper, J. Krabec, R. Leemans, J.G.J. Olivier, A.M.C. Toet, H.J.M. de Vries, and H.J. van der Woerd, 1994: Modeling the global society-biosphere-climate system: Part 2, computed scenarios. *Water, Air and Soil Pollution*, **76**, 37-78.

Apps, M.J., W.A. Kurz, R.J. Luxmoore, L.O. Nilsson, R.A. Sedjo, R. Schmidt, L.G. Simpson, and T.S. Vinson, 1993: Boreal forests and tundra. *Water, Air and Soil Pollution*, **70**, 39-53.

Arnold, M., 1992: *The Long-Term Global Demand for and Supply of Wood*. Paper No. 3, Oxford Forestry Institute, Oxford, UK.

Backman, C., 1994: *The Russian Forest Resource: Physical Accessibility by Economic Region*. Research Report, International Institute for Applied Systems Analysis, Laxenburg, Austria.

Backman, C. and T.R. Waggener, 1991: *Soviet Timber Resources and Utilization: An Interpretation of the 1988 National Inventory*. CINTRAFOR Working Paper 35, University of Washington, Seattle, WA.

Baudin, A., 1988: *Long-Term Economic Development and Demand for Forest Products*. Working Report WP-88-05, International Institute for Applied Systems Analysis, Laxenburg, Austria.

Booth, D.L., 1993: *The Sustainability of Canada's Timber Supply*. Policy and Economics Directorate, Canadian Forestry Service, Ottawa, Canada.

Bourliere, F. (ed.), 1983: *Tropical Savannas, Ecosystems of the World*, Vol. 13. Elsevier Press, Amsterdam, Netherlands.

British Columbia, 1993: *British Columbia Forest Practices Code: Proposed Forestry Practices Rules for British Columbia*. BC Ministry of Forests, Victoria, Canada, 128 pp.

Brown, S., A.J.R. Gillespie, and A.E. Lugo, 1991: Biomass of tropical forests of south and southeast Asia. *Canadian Journal of Forest Resources*, **21**, 111-117.

Bugmann, H.K.M., 1994: *On the Ecology of Mountainous Forests in a Changing Climate: A Simulation Study*. Diss. ETH No. 10638, Swiss Federal Institute of Technology, Zürich, Switzerland, 258 pp.

Cardellichio, P.A., Y.C. Young, C. Binkley, J. Vincent, and D. Adams, 1988: *An Economic Analysis of Short-Run Timber Supply around the Globe*. CINTRAFOR Working Paper No. 18, University of Washington, Seattle, WA.

Carey, E.V., S. Brown, A.J. Gillespie, and A.E. Lugo, 1994: Tree mortality in mature lowland tropical moist and tropical lower montane moist forests of Venezuela. *Biotropica*, **26**, 255-265.

Chadha, S.K., 1990: *Himalaya Environmental Problems*. Ashish Publishing House, New Delhi, India.

Clark, R.N. and G.H. Stankey, 1991: New perspectives or new forestry: the importance of asking the right questions. *Forest Perspectives*, **1**, 9-13.

Cramer, W.P. and A.M. Solomon, 1993: Climatic classification and future distribution of global agricultural land. *Climate Research*, **3**, 97-110.

Davis, M.B., 1976: Pleistocene biogeography of temperate deciduous forests. *Geoscience and Man*, **13**, 12-26.

Davis, M.B., 1983: Quaternary history of deciduous forests of eastern North America and Europe. *Annals of the Missouri Botanical Garden*, **70**, 550-563.

Davis, M.B. and C. Zabinski, 1992: Changes in geographical range resulting from greenhouse warming: effects on biodiversity in forests. In: *Global Warming and Biological Diversity* [Peters, R.L. and T.E. Lovejoy (eds.)]. Yale University Press, New Haven, CT, pp. 297-308.

DeBacker, M. and K. Openshaw, 1972: *Timber Trend Study: Thailand*. FO:DP/THA/69/017, FAO, Rome, Italy.

Delorme, A., 1987: Zur Aufforstung von landwirtschaftlichen Flächen im Hinblick auf den deutschen Holzmarkt und seine zukünftige Entwicklung. *Forstl. Forschungsber. München*, **80**, 20-34.

Deutscher Bundestag (ed.), 1991: *Schutz der Erde*. Bd I + II, Economica Verlag, Bonn, Verlag C.F. Müller, Karlsruhe, Germany.

Dixon, R.K., S. Brown, R.A. Houghton, A.M. Solomon, M.C. Trexler, and J. Wisneiwski, 1994: Carbon pools and flux of global forest ecosystems. *Science*, **263**, 185-190.

Dykstra, D.P. and M. Kallio, 1987b: Introduction to the IIASA forest sector model. In: *The Global Forest Sector: An Analytical Perspective* [Kallio, M., D.P. Dykstra, and C.S. Binkley (eds.)]. John Wiley and Sons, New York, NY, pp. 459-472.

Eisenkrämer, K., 1987: Agrarpolitische Fakten und Handlungszwänge sowie Möglichkeiten und Probleme alternativer Flächennutzungen. *Forstl. Forschungber. München*, **80**, 1-4.

Emanuel, W.R., H.H. Shugart, and M.P. Stevenson, 1985a: Climate change and the broad-scale distribution of terrestrial ecosystem complexes. *Climatic Change*, **7**, 29-43.

Emanuel, W.R., H.H. Shugart, and M.P. Stevenson, 1985b: Climate change and the broad-scale distribution of terrestrial ecosystem complexes: response to comment. *Climatic Change*, **7**, 457-460.

FAO, 1991: *Forest Products: World Outlook Projections*. FAO, Rome, Italy.

FAO, 1992: *The Forest Resources of the Temperate Zone*. Vol. 1, *General Forest Resource Information*. ECE/TIM/62, FAO, Rome, Italy.

FAO, 1993a: *Forestry Statistics Today for Tomorrow: 1961-1991...2010*. FAO, Rome, Italy.

FAO, 1993b: *Forest Resources Assessment 1990: Tropical Countries*. FAO Forestry Paper 112, FAO, Rome, Italy.

FAO, 1993c: *The Forest Resources of the Temperate Zones*. The UN-ECE/FAO 1990 Forest Resource Assessment, Vol. II, FAO, Rome, Italy.

FAO, 1993d: *Forest Products: World Outlook Projections*. FAO, Rome, Italy.

FAO, 1994: *Pulp and Paper Towards 2010*. Paper FO:PAP/94/Inf. 4(a), FAO, Rome, Italy.

Farnum, P., 1992: Forest adaptation to global climate change through silvicultural treatments and genetic improvement. In: *Implications of Climate Change for Pacific Northwest Forest Management* [Wall, G. (ed.)]. Occasional Paper 15: 81-84, Department of Geography, University of Waterloo, Ontario, Canada.

Filip, G.M., B.E. Wickman, R.R. Mason, C.A. Parks, and K.P. Hosman, 1992: Thinning and nitrogen fertilization in a grand fir stand infested with western spruce budworm. II. Tree wound dynamics. *Forest Science*, **38**, 265-274.

Flint, E. and J. Richards, 1991: Historical analysis of changes in land use and carbon stock of vegetation in south and southeast Asia. *Canadian Journal of Forest Resources*, **21**, 91-110.

Franklin, J.F., 1989: Toward a new forestry. *American Forests*, **95(11/12)**, 37-44.

Franklin, J.F., F.J. Swanson, M.E. Harmon, D.A. Perry, T.A. Spies, V.H. Dale, A. McKee, W.K. Ferrell, J.E. Means, S.V. Gregory, J.D. Lattin, T.D. Schowalter, and D. Larsen, 1990: Effects of global climatic change on forests in northwestern North America. In: *Global Warming and Biological Diversity* [Peters, R.L. and T.E Lovejoy (eds.)]. Yale University Press, New Haven, CT, pp. 244-257.

FSI, 1988: *State of Forest Report 1987*. Forest Survey of India, Ministry of Environment and Forests, Dehra Dun, India.

FSI, 1990: *State of Forest Report 1989*. Forest Survey of India, Ministry of Environment and Forests, Dehra Dun, India.

FSI, 1992: *State of Forest Report 1990*. Forest Survey of India, Ministry of Environment and Forests, Dehra Dun, India.

FSI, 1994: *State of Forest Report 1993*. Forest Survey of India, Ministry of Environment and Forests, Dehra Dun, India.

Gadgil, M. and R. Guha, 1992: *This Fissured Land—An Ecological History of India*. Oxford University Press, New Delhi, India.

Goskomles SSSR (State Committee of the USSR Forests), 1990: Forest Fund of the USSR, Moscow, vol. 1, Goskomolesssr, Moscow, Russia, 1005 pp. (in Russian).

Gucinski, H. and R. McKelvey, 1992: Forest management considerations and climatic change in the Pacific Northwest: a framework for devising adaptation/mitigation strategies. In: *Implications of Climate Change for Pacific Northwest Forest Management* [Wall, G. (ed.)]. Occasional Paper 15:135-150, Department of Geography, University of Waterloo, Ontario, Canada.

Hägglund, B., 1994: *The Future of the Swedish Forests*. Royal Swedish Academy of Agriculture and Forestry, Stockholm, Sweden.

Hall, D.O., G. Grassi, and H. Scheer, 1994: *Biomass for Energy and Industry*. Ponte Press, Bochum, Germany.

Harris, L.D. and W.P. Cropper, 1992: Between the devil and the deep blue sea: implications of climate change for Florida's fauna. In: *Global Warming and Biological Diversity* [Peters, R.L. and T.E. Lovejoy (eds.)]. Yale University Press, New Haven, CT, pp. 309-324.

Hartshorn, G., 1992: Possible effects of global warming on the biological diversity in tropical forests. In: *Global Warming and Biological Diversity* [Peters, R.L. and T. Lovejoy, (eds.)]. Yale University Press, New Haven, CT, pp. 137-146.

Heath, L.S., P.E. Kauppi, P. Burschel, H.-D. Gregor, R. Guderian, G.H. Kohlmaier, S. Lorenz, D. Overdieck, F. Scholz, H. Thomasius, and M. Weber, 1993: Contribution of temperate forests to the world's carbon budget. *Soil, Air and Water Pollution*, **70**, 55-69.

Helvetas and Swiss Development Cooperation, 1989: *Country Programme Bhutan*. Swiss Development Cooperation, Bern, Switzerland.

Hultkrantz, L., 1992: National account for timber and forest environmental resources in Sweden. *Environmental and Resource Economics*, **2(3)**, 283-306.

Huntley, B., 1990: European vegetation history: palaeovegetation maps from pollen data—13,000 yr to present. *Journal of Quaternary Science*, **5**, 103-122.

Huntley, B. and T. Webb III, 1989: Migration: species response to climatic variations caused by changes in the earth's orbit. *Journal of Biogeography*, **16**, 5-19.

Hyde, W.F., D.H. Newman, and R.A. Sedjo, 1991: *Forest Economics and Policy Analysis*. World Bank Discussion Paper No. 134, The World Bank, Washington, DC.

IPCC, 1990: *Climate Change: The IPCC Impacts Assessment*. WMO/UNEP, Australian Government Publishing Service, Canberra, Australia.

IPCC, 1992: *Climate Change 1992. The Supplementary Report to the IPCC Scientific Assessment*. WMO/UNEP, Cambridge University Press, Cambridge, UK.

Isaev, A.S. (ed.), 1991: *Forest Management on the Boundary of the XXI Century*. Ecology, Moscow, Russia, 333 pp.

ITTO, 1993: *Analysis of Macroeconomic Trends in the Supply and Demand of Sustainably Produced Tropical Timber from the Asia-Pacific Region*. ITTO, Kyoto, Japan.

Johansson, T.B.J., H. Kelly, A.K.N. Reddy, and R.H. Williams (eds.), 1993: *Renewable Energy: Sources for Fuels and Electricity*. Island Press, Washington, DC.

Kallio, M., D.P. Dykstra, and C.S. Binkley (eds.), 1987: *The Global Forest Sector: An Analytical Perspective*. John Wiley and Sons, New York, NY.

Kanowski, P.J., P.S. Savill, P.G. Adlard, J. Burley, J. Evans, J. Palmer, and P. Wood, 1990: *Plantation Forestry*. World Bank Forest Policy Issues Paper, Oxford Forestry Institute, Oxford University, Oxford, UK.

Kauppi, P.E., K. Mielikäinen, K. Kuusela, 1992: Biomass and carbon budget of European forests 1971-1990. *Science*, **256**, 70-74.

Khattak, A.K., 1992: Development of a model forest management plan for the Panjul forest in western Himalaya (Pakistan). *Forstl. Forschungsber. München*, **117**, 1-175.

Kitamura, T. and M.L. Parry, 1993: Modeling land use/cover change in Europe and northern Asia. Proposal to the IIASA, unpublished.

Kurz, W.A., M.J. Apps, T.M. Webb, P.J. McNamee, 1992: *The Carbon Budget of the Canadian Forest Sector: Phase I*. Information report NOR-X-326, Canadian Forestry Service, Ottawa, Canada.

Kuusela, K., 1990: *The Dynamics of Boreal Coniferous Forests*. Finnish National Fund for Research and Development No. 112, Helsinki, Finland.

Kuusela, K., 1994: *Forest Resources in Europe*. Cambridge University Press, Cambridge, UK.

Leemans, R. and G.J. van den Born, 1994: Determining the potential distribution of vegetation, crops and agricultural productivity. *Water, Air and Soil Pollution*, **76**, 133-161.

Loftsson, J., 1993: Forest development in Iceland. In: *Forest Development in Cold Climates* [Alden, J., J.L. Mastrantonio, and S. Odum (eds.)]. NATO ASI Series A, Life Sciences, Vol. 244, Plenum Press, New York, NY, pp. 453-461.

Manabe, S. and R.T. Wetherald, 1993: Century scale effects of increased atmospheric CO_2 on the ocean-atmosphere system. *Nature*, **364**, 215-218.

McGaughey, S. and H. Gregerson (eds.), 1982: *Forest-Based Development in Latin America*. Inter-American Development Bank, Washington, DC.

Nilsson, S., O. Sallnäs, and P. Duinker, 1992a: *Future Forest Resources of Western and Eastern Europe*. Parthenon Publishing, Lancaster, UK.

Nilsson, S., O. Sallnäs, M. Hugosson, and A. Shvidenko, 1992b: *The Forest Resources of the Former European USSR*. Parthenon Publishing, Lancaster, UK.

Nilsson, S., A. Shvidenko, A. Bondarev, and J. Danilin, 1994: *Siberian Forestry*. Working Paper WP-94-08, International Institute for Applied Systems Analysis, Laxenburg, Austria.

Nilsson, S. and D. Pitt, 1991: *Climate Change in the Forests and Mountains of Europe*. Earthscan Publications Ltd., London, UK.

Nilsson, S. and W. Schopfhauser, 1995: The carbon sequestration potential of a global afforestation program. *Climatic Change* (in press).

OMNR, 1992: *Ontario Forest Products and Timber Resource Analysis*. Ministry of Natural Resources, Sault. Ste. Marie, Ontario, Canada.

Paine, T.D. and F.A. Baker, 1993: Abiotic and biotic predisposition. In: *Beetle-Pathogen Interactions in Conifer Forests* [Schowalter, T.D. and G.M. Filip (eds.)]. Academic Press, London, pp. 61-79.

Payette, S., C. Morneau, L. Sirois, and M. Desponts, 1989: Recent fire history of the northern Quebec biomes. *Ecology*, **70**, 656-673.

Perez-Garcia, J., 1993: *Global Forestry Impacts of Reducing Softwood Supplies from North America*. CINTRAFOR, Working Paper No. 43, University of Washington, Seattle, WA.

Philips, O.L. and A.H. Gentry, 1994: Increasing turnover through time in tropical forests. *Science*, **263**, 954-958.

Pollard, D.F.W., 1987: *Forestry and Climate Change: Facing Uncertainty*. E.B. Eddy Distinguished Lecture Series, November 1987, University of Toronto, Fac. Forestry, Toronto, Canada, pp. 18-39.

Pollard, D.F.W., 1993: personal communication.

Prentice, I.C. and A.M. Solomon, 1990: Vegetation models and global change. In: *Global Changes of the Past* [Bradley, R.S. (ed.)]. OIES, UCAR, Boulder, CO, pp. 365-383.

Prentice, I.C., W. Cramer, S.P. Harrison, R. Leemans, R.A. Monserud, and A.M. Solomon, 1992: A global biome model based on plant physiology and dominance, soil properties and climate. *Journal of Biogeography*, **19**, 117-134.

Prentice, I.C., M.T. Sykes, M. Lautenschlager, S.P. Harrison, O. Denissenko, and P.J. Bartlein, 1993: Modelling global vegetation patterns and terrestrial carbon storage at the last glacial maximum. *Global Ecology and Biogeography Letters*, **3**, 67-76.

Puri, G.S., V.M. Meher-Homji, R.K. Gupta, S. Puri, 1990: *Forest Ecology*. Vol. I. Oxford & IBH Publishung Co. Pvt. Ltd. New Delhi, India.

Ravindranath, N.H. and D.O. Hall, 1994: Indian forest conservation and tropical deforestation. *Ambio*, **23**, 521-523.

Ravindranath, N.H. and D.O. Hall, 1995: *Biomass Energy and Environment: A Developing Country Perspective from India*. Oxford University Press, Oxford, UK (in press).

Röhrig, E., 1991: Temperate deciduous forests. In: *The Ecosystems of the World*, Vol. 7 [Goodall, D.W. (ed.)]. Elsevier, Amsterdam, Netherlands.

Rose, D.W., A.R. Ek, and K.L. Belli, 1987: A conceptual framework for assessing impacts of carbon dioxide change on forest industries. In: *The Greenhouse Effect, Climate Change and U.S. Forests* [Shands, W.E. and J.S. Hoffman (eds.)]. The Conservation Foundation, Washington, DC, pp. 259-276.

Runyon, K.L., 1991: *Canada's Timber Supply: Current Status and Outlook*. Implementation Report EX45, Canadian Forestry Service, Ottawa, Canada.

Sauerbeck, D.R. 1993: CO_2 emissions from agriculture: sources and mitigation potentials. *Water, Air and Soil Pollution*, **70**, 381-388.

Scharpenseel, H.W., 1993: Major carbon reservoirs of the pedosphere. *Water, Air and Soil Pollution*, **70**, 431-442.

Schowalter, T.D. and G.M. Filip, 1993: Bark beetle-pathogen-conifer interactions: an overview. In: *Beetle-Pathogen Interactions in Conifer Forests* [Schowalter, T.D. and G.M. Filip (eds.)]. Academic Press, London, UK, pp. 3-19.

Sedjo, R.A. and K.S. Lyon, 1990: *The Longterm Adequacy of World Timber Supply*. Resources for the Future, Washington, DC.

Sharma, N.P., R. Rowe, K. Openshaw, and M. Jacobson, 1992: World forests in perspective. In: *Managing the World's Forests* [Sharma, N.P. (ed.)]. Kendall/Hunt Publ. Co., Dubuque, IA, pp. 17-31.

Shvidenko, A. and S. Nilsson, 1994: What do we know about Siberian forests? *Ambio*, **23**, 396-404.

Smith, T.M. and H.H. Shugart, Jr., 1993: The transient response of terrestrial carbon storage to a perturbed climate. *Nature*, **361**, 523-526.

Solberg, B. and A. Svendsrud, 1992: Environmental factors and national account of forestry—some findings from Norway. Proceedings IUFRO Centennial Meeting, Berlin/Eberswalde, Germany.

Solomon, A.M., 1986: Transient response of forests to CO_2-induced climate change: simulation experiments in eastern North America. *Oecologia*, **68**, 567-579.

Solomon, A.M., 1992: The nature and distribution of past, present and future boreal forests. In: *A Systems Analysis of the Global Boreal Forest* [Shugart, H.H., G.B. Bonan, and R. Leemans (eds.)]. Cambridge University Press, Cambridge, UK, pp. 291-307.

Solomon, A.M., M.L. Tharp, D.C. West, G.E. Taylor, J.M. Webb, and J.L. Trimble, 1984: *Response of Unmanaged Forests to Carbon Dioxide-Induced Climate Change: Available Information, Initial Tests, and Data Requirements*. TR-009, U.S. Department of Energy, Washington, DC.

Solomon, A.M. and R. Leemans, 1990: Climatic change and landscape ecological response: issues and analysis. In: *Landscape Ecological Impact of Climatic Change* [Boer, M.M. and R.S. de Groot (eds.)]. IOS Press, Amsterdam, Netherlands, pp. 293-316.

Solomon, A.M., I.C. Prentice, R. Leemans, and W.P. Cramer, 1993: The interaction of climate and land use in future terrestrial carbon storage and release. *Water, Air and Soil Pollution*, **70**, 595-614.

Statistisches Bundesamt, 1992: Index der jährlichen erzeugerpreise forstwirtschaftlichen produkte im staatswald der BRD. In: *Zentrale Markt und Preisberichtstelle für Erzeugnisse der Land-, Forst-, und Ernährungwirtschaft*. ZMP-Bilanz 1992 Forst und Holzprodukte Publ., Bonn, Germany.

Stearns, F., 1988: The changing forests of the Lake States. In: *The Lake States Forests* [Shands, W.E. (ed.)]. The Conservation Foundation, Washington, DC, pp. 25-35.

Steiguer, E.J. de, 1993: *Socioeconomic Assessment of Global Change and Air Pollution-Related Forestry Damage*. IUFRO World Series, Vol.4, IUFRO, Geneva, Switzerland, p. 46.

Thunberg, J., 1991: *Framtida Global Rårubalans*. Swedish Pulp and Paper Association, Stockholm, Sweden.

Thunberg, J., 1993: Entering the age of the tree: global forestry trends and their possible impact on the industry. *European Papermaker Magazine*, **1(2)**, 42-45.

Ulrich, B. and J. Puhe, 1993: *Auswirkungen der zukünftigen Klimaveränderungen auf mitteleuropäische Waldökosysteme und deren Rückkopplungen auf den Treibhauseffekt*. Studie für die Enquete-Kommission des Deutschen Bundestages.

UN, 1993: *World Population Prospects: The 1992 Revision*. Department for Economic and Social Information, UN, New York, NY.

UNFPA, 1992: *The State of World Population Fund*. United Nations Population Fund, New York, NY.

USDA Forest Service, 1990: *An Analysis of the Timber Situation in the United States: 1989-2040*. USDA Forest Service, Fort Collins, CO.

USDA Forest Service, 1993: *The 1993 RPA Timber Assessment Update*. USDA Forest Service, Portland, OR.

Waggoner, P.E., 1994: *How Much Land Can Ten Billion People Spare for Nature?* Task Force Report No. 121, Council for Agricultural Science and Technology, Ames, IA.

Westoby, J.L., 1962: *Forest Industries in the Attack on Underdevelopment. The State of Food and Agriculture*. FAO, Rome, Italy.

World Bank, 1993: Staff Appraisal Report, Bhutan, Third Forestry Development Project, Report No. 11666-Bhu, The World Bank, Washington, DC.

WRI, 1994: *World Resources 1994-1995*. Oxford University Press, New York, NY.

Xu, D.Y., 1994: The potential of reducing carbon accumulation in the atmosphere by large-scale afforestation in China and related cost/benefit analysis. *Biomass and Bioenergy* (in press).

Zuidema, G., G.J. van den Born, J. Alcamo, and G.J.J. Kreileman, 1994: Simulating changes in global land cover as affected by economic and climatic factors. *Water, Air and Soil Pollution,* **76**, 163-198.

16

Fisheries

JOHN T. EVERETT, USA

Lead Authors:
A. Krovnin, Russia; D. Lluch-Belda, Mexico; E. Okemwa, Kenya; H.A. Regier, Canada; J.-P. Troadec, France

Contributing Authors:
D. Binet, France; H.S. Bolton, USA; R. Callendar, USA; S. Clark, USA; I. Everson, UK; S. Fiske, USA; G. Flittner, USA; M. Glantz, USA; G.J. Glova, New Zealand; C. Grimes, USA; J. Hare, USA; D. Hinckley, USA; B. McDowall, New Zealand; J. McVey, USA; R. Methot, USA; D. Mountain, USA; S. Nicol, Australia; L. Paul, New Zealand; R. Park, USA; I. Poiner, Australia; J. Richey, USA; G. Sharp, USA; K. Sherman, USA; T. Sibley, USA; R. Thresher, Australia; D. Welch, Canada

CONTENTS

EXECUTIVE SUMMARY

Any effects of climate change on fisheries will occur in a sector that already is characterized on a global scale by full utilization, massive overcapacities of usage, and sharp conflicts between fleets and among competing uses of aquatic ecosystems. Climate-change impacts are likely to exacerbate existing stresses on fish stocks, notably overfishing, diminishing wetlands and nursery areas, pollution, and UV-B radiation. The effectiveness of actions to reduce the decline of fisheries depends on our capacity to distinguish among these stresses and other causes of change. This capacity is insufficient and, although the effects of environmental variability are increasingly recognized, the contribution of climate change to such variability is not yet clear.

While overfishing has a greater effect on fish stocks today than does climate change, progress is being made on the overfishing problem. Overfishing results from an institutional failure to adjust harvesting ability to finite and varying fish yields. Conventional management paradigms, practices, and institutions—inherited from the period when fish stocks were plentiful—are not appropriate for the new situation of generally full exploitation, especially of important fish stocks. Although the Law of the Sea represents an important step in the proper direction, only a few countries have adopted the institutional arrangements needed to regulate the access of fishing fleets to critical areas. The United Nations (UN) Conference on Highly Migrating and Straddling Stocks and the Food and Agriculture Organization (FAO) Code of Conduct for responsible fisheries are likely to accelerate the adoption and effective implementation of regulatory mechanisms. Should climate change develop according to IPCC scenarios, it may become even more important than overfishing over the 50- to 100-year period covered by this 1995 climate assessment.

- Globally, under the IPCC scenarios, saltwater fisheries production is hypothesized to be about the same, or significantly higher if management deficiencies are corrected. Also, globally, freshwater fisheries and aquaculture at mid- to higher latitudes could benefit from climate change. These conclusions are dependent on the assumption that natural climate variability and the structure and strength of wind fields and ocean currents will remain about the same. If either changes, there would be significant impacts on the distribution of major fish stocks, though not on global production (Medium Confidence).
- Even without major change in atmospheric and oceanic circulation, local shifts in centers of production and mixes of species in marine and fresh waters are expected as ecosystems are displaced geographically and changed internally. The relocation of populations will depend on properties being present in the changing environments to shelter all stages of the life cycle of a species (High Confidence).
- While the complex biological relationships among fisheries and other aquatic biota and physiological responses to environmental change are not well understood, positive effects such as longer growing seasons, lower natural winter mortality, and faster growth rates in higher latitudes may be offset by negative factors such as a changing climate that alters established reproductive patterns, migration routes, and ecosystem relationships (High Confidence). Changes in abundance are likely to be more pronounced near major ecosystem boundaries. The rate of climate change may prove a major determinant of the abundance and distribution of new populations. Rapid change due to physical forcing will usually favor production of smaller, low-priced, opportunistic species that discharge large numbers of eggs over long periods (High Confidence). However, there are no compelling data to suggest a confluence of climate-change impacts that would affect global production in either direction, particularly because relevant fish population processes take place at regional or smaller scales for which general circulation models (GCMs) are insufficiently reliable.
- Regionally, freshwater gains or losses will depend on changes in the amount and timing of precipitation, on temperatures, and on species tolerances. For example, increased rainfall during a shorter period in winter still could lead to reduced levels in summer in river flows, lakes, wetlands, and thus in freshwater fisheries (see Chapter 10). Marine stocks that reproduce in freshwater (e.g., salmon) or require reduced estuarine salinities will be similarly affected (High Confidence).
- Where ecosystem dominances are changing, economic values can be expected to fall until long-term stability (i.e., at about present amounts of variability) is reached (Medium Confidence). National fisheries will suffer if institutional mechanisms are not in place that enable fishing interests to move within and across national boundaries (High Confidence). Subsistence and other small-scale fishermen, lacking mobility and alternatives, often are most dependent on specific fisheries and will suffer disproportionately from changes (Medium Confidence).
- Because natural variability is so great relative to global change, and the time horizon on capital replacement

(e.g., ships and plants) is so short, impacts on fisheries can be easily overstated, and there will likely be relatively small economic and food supply consequences so long as no major fish stocks collapse (Medium Confidence).

- An impact ranking can be constructed. The following items will be most sensitive to environmental variables and are listed in descending order of sensitivity (Medium Confidence):
  - Freshwater fisheries in small rivers and lakes, in regions with larger temperature and precipitation change
  - Fisheries within Exclusive Economic Zones (EEZs), particularly where access-regulation mechanisms artificially reduce the mobility of fishing groups and fleets and their capacity to adjust to fluctuations in stock distribution and abundance
  - Fisheries in large rivers and lakes
  - Fisheries in estuaries, particularly where there are species without migration or spawn dispersal paths or in estuaries impacted by sea-level rise or decreased river flow
  - High-seas fisheries.
- Adaptation options providing large benefits irrespective of climate change follow (Medium Confidence):
  - Design and implement national and international fishery-management institutions that recognize shifting species ranges, accessibility, and abundances and that balance species conservation with local needs for economic efficiency and stability
  - Support innovation by research on management systems and aquatic ecosystems
  - Expand aquaculture to increase and stabilize seafood supplies, help stabilize employment, and carefully augment wild stocks
  - In coastal areas, integrate the management of fisheries with other uses of coastal zones
  - Monitor health problems (e.g., red tides, ciguatera, cholera) that could increase under climate change and harm fish stocks and consumers.

16.1. Current Status of Fisheries

16.1.1. World Fisheries Conditions and Trends

Marine fishing generates about 1% of the global economy, but coastal and island regions are far more dependent on fishing. About 200 million people worldwide depend on fishing and related industries for livelihood (Weber, 1994). Marine fish account for 16% of animal protein (5.6% of total) consumption, but developing countries are more dependent on this protein source (Weber, 1993, 1994). Marine catches peaked in 1989 at 85 million tons; freshwater catches were 6.4 million tons, about 7% of the total (FAO, 1993). The potential sustainable yield of marine food fish may be about 100 million tons (Russell and Yonge, 1975); the limit may have been reached. Some 40% of total world production enters international trade (FAO, 1992a), with the top fishing nations being China, Japan, Peru, Chile, Russia, and the United States (FAO, 1993).

World fisheries are characterized by a general state of full or overexploitation of wild stocks and excess harvesting and processing capacities. Some small pelagics, some very-deepwater fish, and some oceanic tunas and Antarctic krill are among several exceptions. Globally, fishing costs are about 20% greater than revenues, with much of the deficit provided by national subsidies (FAO, 1992a). The World Bank (1992) and FAO (1992b) estimate total economic rent and subsidy losses at $79 billion annually, and the situation is likely to worsen.

The quantity of landings is declining (Garcia and Newton, 1994) in 13 of the 15 major marine areas. Only Indian Ocean fisheries continue to increase. Further, there is a deterioration in quality caused by declines in the sizes of highly valued species and a move to bulk landings of lower-value species (Regier and Baskerville, 1986). Thirty percent of global fishery catches are discarded because they are too small, they are prohibited from being landed, or no profitable market exists (Alverson, 1994). This practice of discarding by-catch is more common in industrial fisheries than artisanal ones. Most important stocks are fully or overexploited (Sissenwine and Rosenberg, 1993). In most regions, there is little or no surplus biomass to buffer climate-induced fluctuations in stock abundance relative to current demand. Reduced numbers of year classes also increase stock variability and risk of collapse.

This situation is rooted in the ocean's finite production capacity and in the deficiencies (including concepts and enforcement) of current institutions for adjusting fishing capacities to stock productivity. Present systems were developed when underutilized stocks were available and freely accessible to large-scale fishing operations. Most have not been adjusted to the situation of resource scarcity. Adoption of the UN Convention on Law of the Sea and Exclusive Economic Zones, generally out to 200 miles, has led most countries to control foreign fleets and is a critical step toward improved institutions. However, few countries have taken steps to improve domestic management, even though 90% of production comes from EEZs (Sherman and Gold, 1990; FAO, 1993).

With the advent of EEZs in the mid-1970s, foreign fleets received allocations of stocks that exceeded domestic capacities. Domestic capacity then rapidly increased, often via joint-venture (JV) partnerships with foreign companies. Upon reaching full domestic exploitation through direct replacement and domestic-controlled charters, most JVs were phased out. JVs have not disappeared, however, and foreign fishing continues in some developing regions. Less desirable species and areas or certain species taken by specialized fleets working both international waters and EEZs are most of what remain available to foreign licensed fleets.

Access to high-seas, highly migratory, and straddling (EEZ/high seas) stocks remains mostly open and free, although participants in some fisheries voluntarily limit catches. Without management, fishing cannot be adequately adjusted to average stock abundances, let alone to fluctuations. The 1995 UN Conference on Highly Migrating and Straddling Stocks and the FAO Code of Conduct for responsible fisheries are likely to accelerate the adoption and effective implementation of regulatory mechanisms. With effective management, depleted fisheries could yield another 20 million tons annually (Weber, 1994).

Competition occurs at all levels: countries fishing the same stock, a single country involved in different fisheries in its own or others' waters, and different fishing methods competing within a fishery. In most cases, excess catching power is deployed, and conflicts among competitors often become acute and pervasive.

With few exceptions, governments have not recognized the customary use-rights of traditional fishing communities. Growing demands for fish, water, and space; encroachment by large-scale fishing and aquaculture operations; population concentrations; urban expansion; pollution; and tourism already have harmed small-scale fishing communities in shallow marine waters, lakes, and rivers. These fishers have limited occupational or geographic mobility. With climate change, global and regional problems of disparity between catching power and the abundance of fish stocks will worsen—particularly the interaction between large mobile fleets and localized fishing communities. Aquaculture will develop in new areas, sometimes assisting, sometimes disrupting existing artisanal fishers.

The productivity of freshwaters and sea margins has become stressed by human density and societal actions to benefit non-fisheries sectors. For example, freshwater diversions for agricultural use have resulted in water-level and salinity changes, leading to ecological disasters in the Aral, Azov, and Black seas, and San Francisco Bay (Caddy, 1993; Mee, 1992; Rozengurt, 1992). The artificial opening of the sand barrier at the mouth of the Cote d'Ivoire River to clear floating weeds allowed seawater to enter the lower part of the river and has changed species dominances (Bard *et al.*, 1991; Albaret and Ecoutin, 1991).

On the Nile, the Aswan dam so thoroughly regulates flows that the delta has become degraded ecologically. Local sardine populations that once thrived and provided food for the region have collapsed with the decline in local primary production that depended on the strong surges of flood waters and their pulses of nutrients (Sharp, 1994). Where the Sahelian drought is causing increased salinity in the lower parts of Senegalese rivers, a dam erected near the mouth of the Senegal to stop the rising salinity and ease severe problems to local agriculture prevents fish migration (Binet *et al.*, 1995). Hydroelectric dams in the Dneiper River basin have suppressed spring flows, increased salinities, and left local marshes unflooded at the time of peak fish migration, decreasing fish landings by a factor of five (Mann, 1992).

Problems are markedly more acute in smaller water bodies and fragile coral reefs, especially in areas of high human density. In these areas, habitat degradation often is more important than overfishing. As stresses intensify, impacts that for a long time were limited to freshwaters and littoral areas are now observed in closed and semi-enclosed seas (FAO, 1989b; Caddy, 1993). Some semi-enclosed seas, such as the Mediterranean, Black, Aegean, and Northern Adriatic, are already eutrophic. The diversity of uses in these areas also has introduced "one of the most pervasive and damaging anthropogenic impacts on the world's ecosystem" in the form of nonindigenous species (Mills *et al.*, 1994). Many of these organisms already have extensive invasion histories, are easily transportable, are highly fecund, and tolerate a wide range of environmental conditions. Their damaging effects on indigenous populations may only be enhanced by ecosystem fluctuations accompanying climate change.

Fish habitats are downstream of many impacts, and fish integrate the effects. Fish are symbolic in depicting the health of ecosystems and our ability to manage our resources. Lake Victoria exemplifies this situation and shows how caution must be exercised in introducing nonindigenous species to adapt to, or take advantage of, climate change (see Box 16-1).

Scientists are unable, in most instances, to quantify efficiently each of the man-made and natural stresses on fish stocks. This major constraint on integrated management of water bodies is exemplified by the U.S. fishery for yellowtail flounder on Georges Bank at the southern edge of the species range. The species does not do well when the waters are warm. If there is warming and efforts to rebuild the stock in U.S. waters suffer, it will be difficult to differentiate climate change from fishing as the primary cause (Anthony, 1993).

National and international trends appear to lead to more rational means of management and improved institutions. Among the more advanced concepts is that of Individual Transferable Quotas (ITQs)—in which a science-based catch quota is set and enforced, and fishers can buy and sell percentages of the quota. ITQ systems are being implemented in many areas. Most major fisheries in New Zealand, Canada, Chile, and Iceland; several in Australia; and two in the United States are under ITQ arrangements, with others under discussion (Sissenwine and Rosenberg, 1993).

Box 16-2 contains a regional case study. It is not unique. The steady decline of northeast Atlantic catches since the mid 1970s, the recent banning of cod fishing by Canada, and other situations also could serve as examples.

16.1.2. Aquaculture

Recent growth in total fisheries production is from aquaculture. Aquaculture has grown rapidly during the last few decades and accounts for about 10% of the total world fish production,

Box 16-1. Lake Victoria

Climate change will add to stresses now affecting large lakes and their fisheries. Lake Victoria has reached ecological devastation. Surrounded by urban and agricultural development, it suffers from high nutrient loading, causing eutrophication. Deforestation and erosion are putting sediments into the waters; papyrus swamps are declining with overharvesting; and snails are increasing in abundance. Landing beaches and piers are useless as water continues to recede from the shore. Exposed sand is being harvested, leading to the destruction of spawning grounds and fish refugia. Additionally, introduction of Nile perch contributed to the upset of the productive natural ecosystem and the socioeconomic balance of the region, as summarized below (Barel *et al.*, 1985; Ssetongo and Welcomme, 1985; Acers, 1988; Ogutu-Ohwayo, 1990a, 1990b; Bruton, 1990; Achieng, 1990; Kaufman, 1992):

- The former multispecies fishery has been reduced to three species, due largely to predation by Nile perch and overfishing. Two hundred species have become extinct.
- The decline of perch prey forebodes the collapse of that fishery.
- Subsistence fishers are unable to capitalize to compete in the new fishery. Surviving fishers have adapted by using large-mesh nets to catch perch.
- The local population prefers the native species, as reflected in the low value of perch.
- Perch cannot be sun-dried and must be smoked, requiring extensive use of firewood and leading to deforestation of some islands.
- The most important freshwater fishes in East Africa have vanished from the marketplace.

Box 16-2. The Barents Sea

Luka *et al.* (1991) considered causes of depression of Barents Sea fish stocks and options for recovery. In the late 1970s, total catches reached 4.5 million tons. Then, due to a decreasing abundance of cod, catches dropped, especially sharply in the early 1980s. In the second half of the 1980s, after the fishery on capelin ended, the total catches were about 300,000 tons. The main cause was overfishing, but the unfavorable climatic regime (cooling in 1977–1982) led to poor year classes of all main species. Fishing was very heavy in the 1970s and the first half of the 1980s, despite a declining biomass (Kovtsova *et al.*, 1991; Zilanov *et al.*, 1991). By the second half of the 1980s, herring, capelin, and polar cod—the main food for cod and haddock—were very scarce. This led to slow growth, starvation, and increased mortality. Moreover, predation of cod and haddock on juveniles of deepwater redfish, plaice, and wolffish increased greatly, reducing some strong year classes before they entered the fishery (Kovtsova *et al.*, 1991).

The total Soviet catch was unaffected by these depressed stocks. Fuel was inexpensive, and the fishing fleet moved to other regions [e.g., the Norwegian Sea (blue whiting), Irminger Sea (deepwater redfish), and Antarctica]; thus, there was little impact on the Soviet economy.

By the early 1990s, the Barents Sea spawning biomass was at its historically lowest level. However, there is now a sharp increase in the abundance of cod and capelin, perhaps associated with an observed warming of the northeast Atlantic and adjacent seas.

mostly of higher-valued products. Aquaculture contributes to the resiliency of the fisheries industry, tending to stabilize supply and prices.

Advancements are unevenly distributed across regions, farming systems, and communities. Growth is about 5% annually. The marine component is growing rapidly, but freshwater is still dominant. Aquaculture will not rapidly solve the scarcity of natural fish, and current industry growth will fulfill the demand only for certain commodities, regions, and consumer groups.

Large-scale systems specialize in a few high-value species (e.g., shrimp and salmon) and are developing wherever profitable. Small-scale systems (e.g., for freshwater finfish and shellfish) are restricted to regions where aquaculture and/or settled agriculture have existed for centuries (essentially in eastern and southern Asia and in Europe). This distribution reflects social and institutional factors (notably land and water property regimes) that change very slowly. The need for improved management is great. In Taiwan, for example, inadequate control of expansion of shrimp farming resulted in mass mortalities of stocks and the collapse of production in 1987–89 from 100,000 down to 20,000 tons.

Genetic engineering holds great promise to increase the production and efficiency of fish farming (Fischetti, 1991). However, fishers and resource managers are very concerned about accidental or intentional release of altered and introduced species that might harm natural stocks and gene pools. Around Scandinavia, escapees and nonindigenous reproduction may have reached or exceeded the recruitment of salmon wild stocks (Ackefors *et al.*, 1991). Introduction of pathogenic organisms and antibiotic resistant pathogens is also of concern.

Ranching (in which young fish are released to feed and mature at sea) and fish farming, like their equivalents on land, have self-generated and imposed impediments to success. The activities can compete for coastal space with other uses, and continued expansion can jeopardize the quality and quantity of fish habitat [e.g., through loss of mangroves and wetlands, competition for food with wild stocks, or other factors (NCC, 1989)].

16.2. Climate-Change Impacts

GCM results can be used to make some global inferences, but caution must be used in forecasting fisheries impacts until scientists are able to forecast regional and finer climate and ecosystem processes. Global warming will likely be accompanied by changes in water temperature, precipitation, winds, water and nutrient flows, water level, biogeochemistry, sedimentation, salinity, water mixing, upwelling, ice coverage, and UV-B radiation. Changes in amounts, structures, and timing will affect ecosystem components, including habitats, entire food webs, and species interactions. Chapters 8, 9, and 10 describe these changes and ecosystem impacts.

16.2.1. Freshwater Fisheries

The first indications of reaction by fish species to climate change probably will be among those populations living at the extremes of their temperature tolerance ranges (Holmes, 1990) or residing in streams with high rates of heat transfer from the air. In freshwater, population migration may be limited by watershed constraints (Regier and Holmes, 1991; Shuter and Post, 1990). In large drainage systems, the flexibility for migratory shifts to compatible temperatures will be greater in north–south flowing systems (Meisner, 1990).

It is intuitive to expect poleward range movement for species with climate warming (Shuter and Post, 1990). However,

Box 16-3. Temperature Preferences

Fish cannot regulate their internal temperatures; they try to maintain preferred temperatures through behavior. They seek habitat close to the optimum temperature for growth, foraging success, and protection (Magnuson *et al.*, 1979). Unless a water body has such temperatures for substantial space and time—with other factors also sufficient for well-being—a particular species will not likely thrive (Christie and Regier, 1988). Species have different preferred temperatures, with similarities between species within families. In the figure below (from Lin, 1994), with optimal food and other factors, two salmonid species prefer cold water at or below 25 °C; two percid species prefer cool waters; and two centrarchid species prefer warm waters and tend to "hibernate" in winter when water temperatures fall below 8°C.

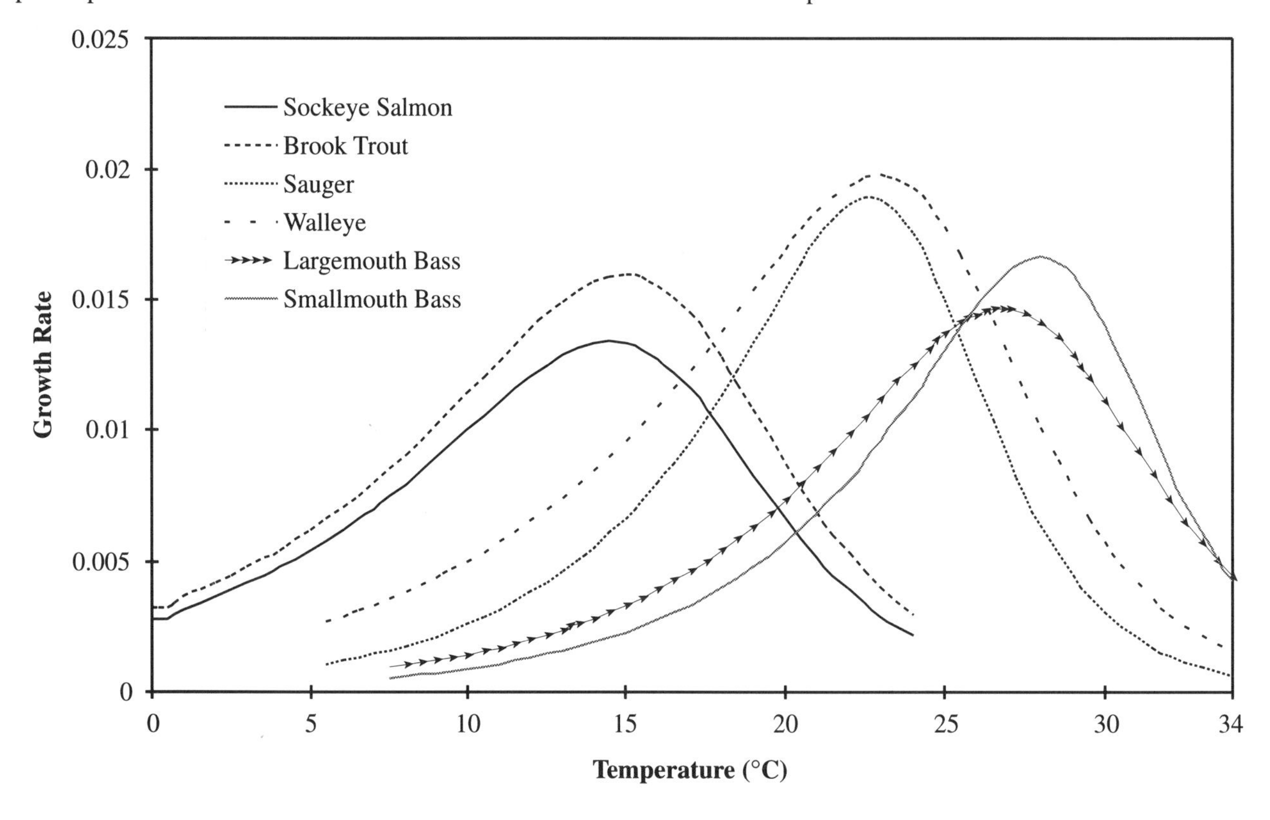

Figure 16-1: Growth rates as a function of temperature.

habitat, food supply, predators, pathogens, and competitors must be within the species' ability to cope. Further, there must be a suitable dispersion route, not blocked by land or some property of the water such as temperature, salinity, structure, currents, or oxygen availability. Movement of animals without a natural dispersal path may require human intervention or hundreds or thousands of years (Kennedy, 1990).

Changes in the amount and timing of precipitation will affect freshwater fisheries production (Welcomme, 1985; Lowe-McConnell, 1987). Thus, expected precipitation increases in high northern latitudes—extending well into mid-latitudes in most cases—in winter (see the Working Group I volume) could provide largely positive benefits overall. However, if accompanied by higher evaporation in summer due to warmer temperatures or loss of habitat due to high flow rates in winter or spring, these benefits may not accrue. Species unable to move or adapt may suffer population declines or become extinct. Management to control critical temperatures of portions of rivers and streams using impoundments, and timing of water releases and runoff may mitigate the impacts of climate change on some species.

Warming should produce a longer growing season for species that have not attained their maximum temperature limits (Shuter and Meisner, 1992), but warming also may lead to reduced productivity in species at or near their maximum limits (Mathews and Zimmerman, 1990). Greater fish production should occur in high-latitude lakes and reservoirs (Schlesinger and McCombie, 1983). Tropical lakes have longer growing seasons and shorter generational times and usually produce higher fish yields than temperate lakes because fish mature at earlier ages. Mean annual air temperature is the most important factor in predicting lake fish production across latitudes (Schlesinger and Regier, 1982), but changing rainfall patterns and flood regimes may have profound effects on river fish (Meisner and Shuter, 1992; McDowall, 1992). In rivers, low oxygen at low flows can severely limit fish production.

Alterations in seasonal climate patterns should change the population distributions in larger lakes. In a study using fish bioenergetics models, certain Great Lakes fisheries were found to have potentially greater yields, whereas others were predicted to collapse (Hill and Magnuson, 1990). White perch—a species of low value—might be among those with increased range and improved recruitment because of a longer growing season and shorter, milder winters (Johnson and Evans, 1990). Yields in southern Lake Michigan and central Lake Erie are projected to remain about the same for lake trout and lake whitefish but increase for walleye. Thermal habitats are expected to increase except for cold-water species in Lake Erie, where the deep water could become anoxic in summer (Magnuson *et al.*, 1990). Large lake fish production could increase about 6% with a 1°C rise in average annual air temperature (Meisner *et al.*, 1987).

Warm-water exotic species may invade large lakes where invasion routes are available (Mandrak, 1989). For example, Coutant (1990) speculates that, with warming, U.S. east coast populations of striped bass, *Morone saxatilis*, could move further north and perhaps enter the Great Lakes—where they might thrive in the lower lakes—and compete with salmonids; in the Gulf of Mexico, Florida coastal waters, and southern U.S. lakes and reservoirs, their existence could be threatened.

Changes in species dominance may occur. Each population has adapted to specific temporal and spatial features. For example, in New Zealand, native species occur throughout the nation, whereas introduced species seem to be temperature-limited in their distribution—suggesting that indigenous species may be more resilient than exotics in a warming climate (New Zealand, 1990). Species invasions and removals occur frequently, even without human causes, but climate and habitat change should accelerate them. Indigenous species will be subject to greater pressures from shifts in ecosystem structures and introductions of opportunistic exotics (Lodge, 1993).

Positive factors associated with greater warming and precipitation at higher latitudes include faster growth and maturation rates, lower winter mortality rates due to cold or anoxia, and expanded habitats with ice retreat. Offsetting negative factors include increased summer anoxia, increased demands for food to support higher metabolism, possible negative changes in lake thermal structures, and reduced thermal habitat for cold-water species. Individual effects are difficult to integrate. However, warm-water lakes generally have higher productivity than cold-water lakes, and existing warm-water lakes will be in areas with the least change in temperature. It is reasonable to expect higher overall productivity from freshwater systems. Most warming should occur during fall, winter, and nighttime during the summer, improving the survival of fish that are less tolerant of cold and having little effect on many species near their upper lethal temperature. Streams may be no more likely to reach high lethal temperatures. Lastly, fishery managers heavily manipulate freshwater fisheries in much of the world. If species mixes continue to be changed to support angler and market preferences and changing habitats, climate-change damages may dampen and benefits may heighten.

16.2.2. Saltwater Fisheries

The linkage between climate and marine fisheries is complex and poorly understood. Physical changes influence recruitment and abundance, the center of distribution and range, and availability by concentrating stocks in fishable locations. There are so many interacting variables (such as currents, upwelling, cloudiness, nutrient flows, salinity, temperature, and solar irradiance) that predictions of changes in primary production are not clear (Bernal, 1991; Mann and Lazier, 1991; see also the Working Group I volume). However, understanding is growing. For instance, in the Pacific, the intensity of the Aleutian low clearly affects primary production (Venrick *et al.*, 1987; Beamish and Bouillon, 1993) and ecosystem production (Polovina *et al.*, 1994).

16.2.2.1. Sea-Level Rise

About 70% of global fish resources depend on near-shore or estuarine habitats at some point in their life cycle (Chambers, 1991; IPCC, 1990). The growing rate of human occupation—with associated pollution—and the high property values of littoral areas, especially in Western countries, will severely constrain the inland displacement of wetlands and other habitats as sea level rises. Fish production will suffer when wetlands and other habitats that serve as nurseries are lost (Costa *et al.*, 1994). Chapter 9 discusses these issues, and Box 16-4 provides an example of the potential effects of sea-level rise on active fisheries.

If sea-level rise is too rapid, natural succession of coastal ecosystems cannot occur, harming many species. With a half-meter rise by 2100—and with protection of all developed areas—more than 10,360 km^2 of wetlands could be lost in the contiguous United States (Park *et al.*, 1989). In the near term, fish production could rise as marshes fragment, flood, die, and decompose, improving fisheries habitat in some cases as the protective land edge increases. More nutrients would become available from leaching of soils and flooded peat. The southeastern United States, where land subsidence compounds sea-level rise, may be in this phase of temporary improvement now (Zimmerman *et al.*, 1991). Eventually, detritus-based food webs would lose the protective habitat and nourishment, as well as wildlife nesting sites and refuge (Kennedy, 1990). In the longer term, by 2050 the overall impact on fisheries will probably be negative. The brown shrimp catch in the U.S. Gulf Coast could fall 25% with a 25-cm rise in sea level (Park, 1991). A 34-cm rise could cause the loss of 40% of the extensive Puget Sound tidal flats—an important habitat for shellfish and waterfowl (Park *et al.*, 1993). In certain areas, already nutrient-rich waters may become eutrophic (Caddy, 1993). Rising sea level may particularly affect coastal ponds as flooding septic systems increase nutrient loading, an important loading mechanism (Valiela and Costa, 1988).

Wetlands, coral reefs, mangroves, and sea grasses require a healthy environment to keep pace with a rising sea, to continue coastal protection benefits, and to serve as fisheries habitat.

Box 16-4. The Caspian Sea

Fluctuation of Caspian Sea water levels provides insight on interacting impacts. An increase in sea level could have a positive economic result for fisheries but a high cost to land-based industry and tourism.

The main species harvested from the Caspian are sturgeon, roach, bream, sander, sazan, herring, and kilkas (Anon., 1989). In the second half of the 1950s, diversion of river waters and unfavorable conditions resulted in a series of low-water years, with reduction of sea area and volume and higher salinities. Roach, bream, sander, and juveniles of all other species declined as their prey decreased in response to the water changes. At the same time, river fish (sheatfish, pike, redeye fish, tench, etc.) increased as suitable spawning habitat expanded.

In the late 1970s, the water level started rising, and bioproductivity increased. Plankton biomass increased threefold and shad—which spawns in the northeast shallow brackish area—recovered (Vodovskaya *et al.*, 1978). By the mid-1980s, high natural reproduction and improvement of the hydrological regime in the northern Caspian resulted in strong year classes of roach, bream, and sander, and they increased to their highest level as favorable conditions stabilized.

As sea level continues to rise and inundate more developed areas, serious environmental implications confront productivity. The Caspian is heavily polluted with oil, phenol, and heavy metals from years of industrial dumping. As land is overwashed, additional oil and pollutants from industrial facilities will further threaten the sea's valuable stocks (Anon., 1994). If the shore is protected extensively, as is planned by Kazakhstan, there also could be significant losses of fish spawning habitat and production.

Sea-level rise with defense of properties, increased freshwater diversion, and human and industrial concentration in coastal areas would be detrimental to fisheries production.

16.2.2.2. Fishery Oceanography

Evidence in support of the view that environmental changes drive many changes in fish stocks has been accumulating in recent years (Mann and Lazier, 1991; Mann, 1992, 1993; Mann and Drinkwater, 1994; Polovina *et al.*, 1995). The question of whether overfishing, environmental change, or a combination of the two is responsible for major declines in fish stocks is still a matter for debate and is situation-specific.

The literature is rich with studies relating historical changes in the abundance and distribution of aquatic organisms to climate changes. The studies show that relatively small changes in climate often produce dramatic changes in the abundance of species—sometimes of many orders of magnitude—because of impacts on water masses and hydrodynamics (e.g., Sharp and Csirke, 1983; Kawasaki *et al.*, 1991; Beukema *et al.*, 1990).

Changes in temperature, winds, currents, salinity, and other physical parameters affect life in the oceans. With warming waters, larger fish tend to move to cooler or more offshore and higher-latitude portions of their habitat, or suffer reproductive stress responses that debilitate individuals and lead to lower viability (Sharp, 1994). Water temperatures will increase least at the equator and more toward the poles. In contrast to tropical and mid-latitude regions, where productivity is mainly nutrient-limited, the basic limiting factors in polar and subpolar regions are light and temperature. Warming in high latitudes should lead to longer growing periods, increased growth rates, and ultimately, perhaps, increases in the general productivity of these regions (Regier *et al.*, 1990). Each species might be affected differently. For example, early research results for northern Pacific salmon indicate that ocean distribution in a warming climate would decline to half its present area by about 2070, and winter habitat could completely disappear for 3 months each year (Welch and Beamish, 1994).

The impacts of warming can be inferred from past displacements of transition zones—for example, the Russell cycle in the western English Channel (Southward, 1980) and the simultaneous discontinuities in the local occurrence of pilchard and herring (Cushing, 1957, 1982). Opposing fluctuations in these fisheries in the Channel and the Bay of Biscay have followed long-term changes in climate for three centuries (Binet, 1988b). Tropical Atlantic albacore recruitment is correlated with temperature anomalies during spawning (Leroy and Binet, 1986).

Warming would affect marine species in shallow, restricted impoundments long before deep oceanic species (Bernal, 1991). While fish can migrate, mollusks usually cannot move to favorable areas, although their progeny may develop and thrive in new locations—or be eliminated if suitable habitats do not develop. One example is the important fishery for soft-shelled clams in the Chesapeake Bay. These clams are essentially absent in warmer estuaries further to the south. They could disappear from the Chesapeake if there is substantial warming (Kennedy, 1990). Conversely, Dickie (1955) showed that Bay of Fundy scallop fluctuation from 1930–1953 followed water temperature but lagged by 6 years (i.e., the time for sea-scallop recruitment). Higher temperatures led to strong year classes (Frank *et al.*, 1990).

Warming and wind changes would affect the distribution and characteristics of polynyas (ice-free areas) and ice edges that are vital to polar ecosystems. Changes in the extent and duration of ice, combined with changes in the polar frontal zone and the circumpolar current in southern latitudes, also may affect the distribution, mass, and harvesting of krill. Krill is the fundamental link in the food web of the southern ocean; much of the fishery it supports is used as feed in aquaculture. Antarctic productivity also is closely related to spring retreat of pack ice. Warming could affect pack-ice movement and extent—in turn, affecting the distribution, abundance, and productivity of krill. However, little change is forecast in Antarctic ice for 100 years (see the Working Group I volume). Changes in wind strength also may affect productivity in the ice edge zone.

In the eastern Pacific, interdecadal natural variation resulted in warm periods of high sardine abundance during the 1940s and 1976–85, separated by cool periods of high anchovy abundance in between and in very recent years (Lluch-Belda *et al.*, 1992). While this suggests a direct correlation of high sardine abundance with warm temperatures, it is likely that there is a more complex relationship. At the time of these eastern Pacific warm periods, the western North Pacific has been generally cool, with very abundant Japanese sardines.

Major shifts in regional and global climate have major effects on the abundance and distribution of fish stocks. The 1976–77 climatic regime shift in the North Pacific Ocean (Beamish, 1995) was apparently the starting point for a high-sardine-abundance regime. Further, this period seems to be giving way to a new period of high anchovy abundance since 1985 (Lluch-Belda *et al.*, 1992). There is a long time lag before scientists recognize these events.

Warmer water and changes in circulation would substantially affect the structure and location of habitats. However, marine organisms as a rule have rather high genetic and behavioral plasticity, allowing them to adapt to constantly changing conditions. This property underlies the relative stability of zoogeographical patterns during natural climate variations (Odum, 1986). However, there may not be high plasticity with respect to temperature-dependent physiological and ecological rates and behavioral thresholds (Lin and Regier, 1995). While the mix of marine species constantly changes with naturally variable climates, the changes forecast by IPCC would happen relatively rapidly and be long-lasting. There will likely be food-chain disruptions that will destabilize certain marine populations (Kawasaki, 1991). Poleward shifts of some, but not all, elements of an ecosystem may result in discontinuous areas of high productivity (Lluch-Belda *et al.*, 1991).

Changes in currents could lead to changes in population location and abundance and the loss of certain populations—or the establishment of new ones—at the periphery of the present species distributions. Current GCMs do not adequately incorporate the ocean, and coupled physical/biological models are in their infancy. To date, only the Semtner-Chervin global ocean circulation model (Semtner and Chervin, 1988) has been generally used to provide information that is directly applicable to local and regional fisheries applications. If only the direction of potential changes in ocean circulation could be known, useful assumptions could be made on the likely impacts on fisheries (Troadec, 1989b).

A general poleward extension of habitats and range of species is likely, but an extension toward the equator may occur in eastern boundary currents. For example, the range of *Sardina pilchardus* prior to World War II was from south Brittany to Morocco. In warm years following the war, sardines were fished up to the North Sea. During the 1970s, new fisheries developed off the Sahara and Mauritania, with small amounts landed as far south as Senegal—perhaps due to upwelling and ecological processes related to tradewind acceleration (Binet, 1988a). Changes in the circulation pattern are likely to induce changes in the larval advection/retention rates in and out of favorable areas and may explain changes in abundance and distribution (Binet, 1988a; Binet and Marchal, 1992). Areal overlaps in closely related species may change in unpredictable directions (Ntiba and Harding, 1993).

Even if the major currents are unchanged (see the Working Group I volume), winds can change local currents and upwelling dramatically. If upwelling-favorable winds intensify, as some authors suggest (Bakun, 1990, 1993a), the mixing of surface and subsurface waters will increase, improving their nutrient contents but perhaps delaying the phytoplankton spring outburst in some areas. The total production and species dominance of the whole food chain will be affected (Aebischer *et al.*, 1990). Conversely, some authors suggest that global warming will lead to a reduced temperature gradient between the poles and the equator—and therefore to decreased oceanic winds and decreased mixing of surface and subsurface waters (Wright *et al.*, 1986).

A physical forcing may have opposite effects in different areas because ecological relationships are not necessarily linear. The 1950–1980 northerly wind increase in western Europe enhanced the Iberian west coast upwelling and is associated with decreased sardine production (Dickson *et al.*, 1988), whereas wind acceleration along the Mauritania coast favored settlement of a new sardine population (Binet, 1988a), possibly illustrating a dome-shaped environment–resource relationship (Cury and Roy, 1989). A gentle breeze enhances the recirculation of nutrients in the euphotic layer and hence the biological production, whereas a strong wind mixes the water and causes phytoplankton to sink below the sunlit layer. Other examples of the influence of changes in upwelling systems and circulation are provided in Box 16-5.

Skud (1982) found that in both marine and freshwater systems, changes in environment can shift species dominance. Dickson and Brander (1993) describe how a change in the west winds resulted in a major influx of West Greenland cod larvae to the 1957 Labrador cod stock. Moreover, winds and warming can interact powerfully—as is demonstrated by the tale of the tilefish. In 1879, the tilefish, responding to a temporary warming of the Atlantic, appeared in large numbers off New England. Three years later, northern gales cooled the water below the

Box 16-5. Upwelling Systems

Changes in circulation and upwelling systems affect the distribution and abundance of pelagic species (Bakun, 1993b), often in interaction with fishing pressure:

- Long-term fluctuations in sardine stocks occur in apparent synchrony off Japan, California, and Chile/Peru (Kawasaki, 1983; Sharp, 1992a, 1992b; Sharp and McLain, 1993). These broad processes link through atmospheric and oceanographic patterns spanning a few decades. Transitions from one pattern to another involve massive declines in populations of coastal upwelling species and the emergence of more oceanic forms to recolonize large expanses of the near-shore environment (*c.f.*, Pauly *et al.*, 1987, 1989; Yañez, 1991).
- Recent changes in the distribution and abundance of west African pelagic stocks correlate with upwelling system modifications (Binet and Marchal, 1992, 1993; Binet and Servain, 1993; Cury and Roy, 1991):
  - A shortened food web resulted in more phytoplankton and fewer zooplankton, benefiting phytoplankton-feeders.
  - New concentrations of *Sardinia pilchardus* appeared off Mauritania and Senegal; pilchard catches declined off northern Morocco.
  - *Sardinella aurita* strongly expanded in the Ivory Coast.
- Similar stock changes are documented from Namibia and South Africa (Crawford *et al.*, 1991; Payne *et al.*, 1987), Argentina and Brazil (Bakun and Parrish, 1990, 1991), California (Parrish and McCall, 1978), Peru/Chile (Pauly and Tsukayama, 1987; Sharp and McLain, 1993), and the subarctic Pacific (Beamish and Bouillon, 1993).

lower thermal limit, causing mass mortality (Freeman and Turner, 1977).

Freshwater flows to the sea lower local salinity, bring nutrients to increase primary production, enhance turbidity, and impact coastal ecosystems. Salinity changes and warming of coastal waters may lead to a latitudinal redistribution of fish stocks, with changes in timing and pathways for migrating species. Increases are expected in evaporation in mid-latitudes and precipitation in low and high latitudes. The 1972–1985 west African drought led to a large-scale ecological change over the west African shelf (Caverivière, 1991). The halving of river flows led to a small salinity increase along the coasts of Senegal and the northern Gulf of Guinea (Mahe, 1993) and greatly reduced sediment loading and turbidity.

In the Casamance—a long, narrow estuary in Senegal—a rainfall shortage that sharply raised salinity caused dramatic changes in fish populations. The total fish biomass remains the same but consists of small, low-priced species. High mortality can occur when the fish cannot escape the salt plug (Albaret, 1987). Shrimp production increased temporarily and then collapsed as salinity increased (Le Reste, 1992).

Higher temperatures and greater water-column stratification can result in less organic material reaching the bottom and thus favor pelagic over demersal fishes (Beamish and Bouillon, 1993). In addition, changes in phytoplankton production, such as shifting downward to smaller species, can lead to longer food webs or benefit smaller fish feeding at lower trophic levels (Frank *et al.*, 1990). Changes in wind, temperature, and precipitation can alter stratification processes and thereby affect the feeding and survival of larval fishes (Owen, 1981, 1989). Some species depend on turbulence at critical stages—for example, to oxygenate eggs or habitats—whereas others require stability to allow surface-water stratification and the concentration of food organisms. Thus, changes in stratification of local water masses could disperse or concentrate food organisms of young fishes (Bernal, 1991). Lasker (1981) hypothesized, and Peterman and Bradford (1987) demonstrated, that the larval mortality of anchovy (*Engraulis mordax)* was negatively correlated to the number of calm periods. Rothschild and Osborn (1988) note that minimal turbulence is useful to enhance the encountering rate of particles and predators, whereas strong turbulence breaks up aggregations of particles on which feeding is possible. This apparent contradiction illustrates the complexity of ecological relationships and the difficulty of prediction.

The nature of climate-change impacts on fish can be inferred, in part, from an analysis of current interannual fluctuations of marine populations. Catch statistics and research surveys can yield information on stock fluctuations. Fish variability has been classified on the basis of taxonomic groups and environments (e.g., Kawasaki, 1980; Sharp, 1986) and by temporal variations (Caddy and Gulland, 1983). These categories describe:

- Highly unstable and unpredictable stocks varying by several orders of magnitude due to sporadic climatic shifts (e.g., capelin, sand eel, and saury)
- Unstable and partly predictable stocks varying by a few to several times due to cyclic climate and fishing variations (e.g., sardines, herrings, and mackerels; Sahelian freshwater fish)
- Stable and predictable stocks varying by tens of percent to a few times due to fishing and climate trends (e.g., bottomfish such as cods and flatfishes and large pelagics such as tunas).

In most cases, temperature increase is not the direct reason for change in aquatic ecosystems and fisheries, although fish may react physiologically or behaviorally. The most crucial problems are associated with the whole complex of ecological changes. Fish have developed many patterns of behavior to deal with climate variability and migrating predators and prey. Some environments call for tightly constrained windows within which reproduction can succeed. Others require constant or prolonged spawning to cope with even finer-grained, less-certain opportunities. Because many fish early-life stages occur in shallow layers and estuaries where various climate components are important (Kawasaki, 1985), there is potential for significant effects on fisheries and dependent communities.

Individual species have adapted strategies to accommodate changes in their environment. Some even have fully diversified local population responses that would seem unlikely. Shad along the eastern United States spawn many batches of smaller eggs in their warmer southern range and, in the less variable north, produce larger eggs that they carry up rivers and spawn in single batches (Leggett and Carscadden, 1978). Thus, changes in patterns of climate variability can be as important as the magnitudes of the changes—and as much a source of uncertainty.

Descriptions of interrelated changes in oceanographic structures and bottomfish populations include: the 1962–1986 cod outburst in the North Sea and the Irish Sea (Hempel, 1978) and the rise and fall, with a maximum in 1945–1950, of Western Greenland cod stocks that were established from recruits of Icelandic stocks during the 1925–1960 North Atlantic warming (Cushing, 1982).

It is difficult to disentangle human and natural climate effects. For example, the environment adversely affected cod off eastern Canada, but the additional burden of heavy fishing probably caused the stock collapse—reducing the biomass, eliminating older year classes, and increasing stock vulnerability (Mann and Drinkwater, 1994).

There may be a synergistic effect between climate change and overfishing. With overfishing, the age composition of a stock becomes increasingly unbalanced: Fewer year classes and mature individuals are available, and fewer reproductive opportunities occur. Overfishing eliminates species characterized by low fecundity, slow growth, and late sexual maturation, while favoring species with opposite strategies, usually of less economic value. It introduces new trophic pathways, possibly produces empty ecological niches, and creates new competitive relationships. For example, there has been a change in fish sizes and species dominance in the Gulf of Biscay since 1800 (Quéro and Emmonet, 1993). Georges Bank also has been the scene of a "biomass flip," with skates and dogfish replacing cod, haddock, and flounder (Rothschild, 1991).

In the case of the Atlantic salmon (Pope *et al.*, 1961), the brown trout (Hardy, 1967), and the North Sea long rough dab (Ntiba, 1990), the larger, older fish lay larger eggs; Blaxter and Hempel (1963) showed that larger Atlantic herring eggs hatch into larger larvae. Larger eggs could influence survival and development and may have important ecological implications (Blaxter, 1988). Recruitment into the fishery is probably balanced on the number of larger and older fish in the population (Bagenal, 1970). Although some species can rebuild quickly (Beamish, 1993), most large species will have difficulty taking advantage of a favorable environmental episode that may come only once in several years. Overfishing also may reduce the genetic resilience of the population, with similar consequences. Climate changes may magnify the effects of overfishing at a time of an inherent instability in world fisheries (Stokes *et al.*, 1993).

Hydrodynamic structures are largely determined by wind fields, currents and fronts, temperature, salinity, and geographic topography. As new hydrographic structures stabilize, new populations may take advantage of the physical features. Colonization is determined by ecological constraints affecting reproduction (Sinclair, 1988). For species dependent on geographical features, there is less flexibility. For example, herring spawning in certain areas disappeared with the collapse of specific herring populations in the 1970s but reappeared in the same locations as the herring recovered (Daan *et al.*, 1990). Dislocations of and changes in density stratifications may affect the timing and location of fish availability for fishing and can be particularly significant in pelagic fisheries (e.g., tuna, salmon, herring, sardines):

- Most Peruvian anchovy concentrations moved into inshore waters after 1972: The percentage of catches made within 10 miles rose from 42% before 1970 to 91% after 1972 (Valdivia, 1978) as the areal extent of the upwelled water dwindled.
- The route by which sockeye salmon return to the Fraser River determines, in part, their availability to Canada and U.S. fisheries. In warm years—and especially in strong El Niño years (see Box 16-6)—80% of returning sockeye may use a cooler, northern route that favors Canadian fishing interests, compared to as little as 10% in some cold years (Miller and Fluharty, 1992).
- High salinity near the coast due to droughts can reduce the area of low-salinity waters, increasing the availability of some species to small-scale fisheries—for example, *Sardinella aurita* in the northern Gulf of Guinea, especially the Ghanaian canoe fishery (Binet, 1982), and *Sardina pilchardus* off northern Morocco (Belvèze and Erzini, 1983).
- Development of the Northeast Atlantic bluefin tuna fishery from 1920 to 1967, with a peak in the 1950s, has been related to the long-term warming of the North Atlantic until the late 1940s and subsequent cooling (Binet and Leroy, 1986).

16.2.2.3. Impacts on Fisheries

Global warming will likely cause collapses of some fisheries and expansions of others. It may be one of the most important factors affecting fisheries now and in the next few hundred years. The level of impact will vary widely and will depend on the complexity of each ecosystem, the attributes and adaptability of

Box 16-6. El Niño: How Short- and Long-Term Climate Events Can Affect Fisheries

Environmental processes, fisheries management, and population dynamics closely interact. The 1972 El Niño, coupled with anchovy overexploitation and resultant recruitment failure, led to a severe crisis in the following years. Although not the direct cause of the collapse (Sharp and McLain, 1993), the El Niño, nested within a much broader set of environmental changes, concentrated the anchovies—enabling the fishery to operate on the entire population, which previously was scattered over a broader area with some portions protected from continuous exploitation (Csirke, 1980). The anchovy collapse was an economic disaster for Peru. The production center for the Humboldt Current shifted southward, from north-central Peru to the Peru/Chile border, with sardines colonizing the coastlines from Ecuador to central Chile.

These short-term effects show how fish species react to some large environmental changes. They may be illustrative of effects from long-term changes. El Niño events, being anomalies riding on the long-term trend, may provide meaningful insights—particularly since some GCM results indicate reduced East–West temperature gradients similar to conditions found in El Niño events (see the Working Group I volume).

There also are less familiar El Niño impacts. Some can last for several years. For example, El Niño Southern Oscillation (ENSO) processes, combined with other natural variability, led to increased prices of salmon from 1980–1984, due to displacement of salmon migratory routes and complex ecological changes from warm-water intrusions along the Pacific coast of North America. Pacific gray whales moved far north of their usual locations, indicating that their habitat had moved northward and shoreward for the first time in decades (Sharp, 1994). The 1982–1983 El Niño led to increased stocks and catches of scallops and oceanic tunas and billfishes in Chile and Peru. The annual migration of whiting along the west coast of the United States and Canada is correlated with water temperature, especially with strong El Niño conditions. In warm years, more stock is available to Canadian fishers (Dorn and Methot, 1990). Temperate Sardinops populations in the Gulf of California are much less available during El Niños, whereas tropical thread herring are much more available (Lluch-Belda *et al.*, 1986).

each species, and the nature of the human communities that depend on them. Defining these contexts is the key to understanding and preparing for climate changes. If climate change occurs on the scale indicated by GCMs, we can expect significant effects, both beneficial and destructive, on the distributions and productivity of valuable regional fisheries and the local industries associated with them (Healey, 1990).

Over the long term, global changes in fish production may maintain some relative balance, but there could be important regional shifts in fishing areas and species compositions, with major socioeconomic impacts. These shifts may be in the relative abundance of particular species between contiguous EEZs or between EEZs and the high seas. Globally, fluctuations in fish abundances and distributions are expected to maintain similar ranges as previously documented due to natural variation (Sharp and Csirke, 1983; Sharp, 1987). In some cases, species may move poleward, but there may be little change evident to someone living in the middle range of a given ecosystem. However, significant differences may occur in habitat suitability and species dominances at the thermal limits of species ranges. Changes in climate near ecosystem borders could mask impacts from harvesting excesses and other anthropogenic changes, generating misguided international disputes. Local communities would be affected by changes in recruitment, distribution, and abundance.

Globally, overfishing and diverse human stresses on the environment will probably continue to outweigh climate-change impacts for several decades. However, fishing impacts are usually reversible—whereas climate-change and habitat-loss impacts are not (at least within the IPCC scenario)—and the overfishing problem may well be solved within the time horizon of this IPCC assessment. For fish production, climate variability is not a new problem. Social policies and controls needed to adapt to global change and those required for managing high-sea stocks are of similar geopolitical magnitude.

In some cases, fisheries on the margin of profitability could prosper or decline. For example, if there is a rapid retreat of sea ice in Antarctica or if the ice has reduced extent, the krill fishery—which is regulated by the three months that the continental shelf is ice-free—could become more attractive to nations not already involved.

In developing countries, fishers can be expected to follow shifting fish distributions—comparable to what happens now within natural climate variability. New fishing sites often have poor or nonexistent provisions for freshwater supply, sanitation, and refrigeration. As a result, seafood handling may be less sanitary than is needed, and cholera or other diseases may occur. Where there are no cold storage facilities, fishers may have to resort to either salting or smoking as alternative means for preservation. Such change can, in turn, lead to local deforestation, increased soil erosion, or loss of mangroves and associated fish nursery habitat. Long-distance migrations of small-scale fishers can be constrained by newly established national frontiers or by confrontation between local fishers and newcomers, as has already happened with west African small-scale fisheries.

In some communities—such as Fortaleza, Brazil—half or more of the fishing fleet may be dependent on wind and sails (Everett, 1994). Reductions in wind strength could cause significant costs to these small-scale fishers operating at the margin of profitability. Conversely, increased wind force in other regions will require more seaworthy vessels or other changes in fishing strategies. There are costs associated with these changes. Such costs are certainly less obvious, and may be less important, when they take place gradually over a longer time period than the life of the fishing vessels and other assets.

The ways in which societies have responded to past changes in the fisheries that support important components of their regional economies are instructive in guiding society toward developing better ways of adapting to the regional impacts of environmental change (Glantz, 1992). As an example, the industry collapses of the Chilean/Peruvian anchovy, the California sardine, and the Alaskan king crab led to varied socioeconomic responses. While the less industrially capitalized Chileans responded to perturbations in the anchovy catch with regulations and an adaptive restructuring of the industry to the more offshore jack mackerel (*Trachurus murphyi*) and secondarily the sardine (*Sardinops sagax*), the Peruvians continued peak extraction until the anchovy catch collapsed and only slowly shifted to alternative fisheries, with serious economic impacts (Caviedes and Fik, 1992). The California sardine bonanza, harvested at increasing levels for half a century, similarly collapsed—with perhaps even greater socioeconomic impacts on the region. When the industry collapsed in the 1960s, most of the fleet converted to less labor-intensive fisheries or was sold for pennies on the dollar; fishing and cannery crews, unneeded in the remaining fisheries, moved north to Alaskan waters, fished only seasonally, or left the industry; and much of the infrastructure was moved to South America, along with the expertise of many fishing captains and cannery operators (Ueber and MacCall, 1992). The 1981 collapse of the Alaskan king crab industry also resulted in economic disaster for the heavily overcapitalized industry, which was faced with foreclosure or diversification. The fleets that survived broadly expanded their fishing grounds, shifted to replacement species (other king crabs and Tanner crabs), or traded their former low-volume/high-value fishery for more high-volume/lower-value fisheries (e.g., pollock) (Wooster, 1992).

With the declaration of extended EEZs, long-range fishing industries have adjusted to the new access conditions through joint-venture agreements, further development of domestic fisheries and aquaculture, development of new processing methods to upgrade lower-value species (e.g., surimi), development of new fisheries within and outside the 200 miles, increased import and reduced consumption of fishery products, and the scrapping of long-distance vessels (Glantz, 1992; Yonesawa, 1981; Kaczynski, 1983a, 1983b). For long-distance fisheries, these adjustments often have entailed considerable costs. The new ocean regime has had more serious impacts on the fisheries sectors of long-distance fishing countries and on those of countries that extended their control over large resources (e.g., Iceland; Glantz, 1992) than would the effects of climate change.

Forecasting by analogy, therefore, can inform about the kind, but not the direction, magnitude, and geographic distribution, of economic and social effects of global change on fisheries. The lessons learned from experience include:

- The magnitude of effects depends on the importance of fisheries in an economy. Although all sectors may benefit from positive effects, the economies of traditional fishing groups and developing countries may be disproportionately impacted by negative effects. With little diversification, their means are limited and they are more vulnerable, particularly in developing countries bounded by major upwelling regions (e.g., Mauritania, Namibia, Peru, Somalia) (Troadec, 1989a).
- Small-scale fisheries, even in industrialized countries, maintain several characteristics of premercantile societies, such as autonomy, self-sufficiency, exploitation of local natural resources, self-limitation of demand, geographic mobility (to offset the law of diminishing returns), solidarity within family, and the original value systems. Industrialized, market societies are characterized by autonomy of individuals, professional specialization, mobility of production factors, development of new activities through technological innovations, and accumulation of profit. Because the economy of these organizations is different, it is difficult to compare their capacity to cope with variability and uncertainty; however, the former's frequent lack of means (poverty) is a major and new handicap.
- When climate change shifts the distribution of a fish stock within an EEZ, or between EEZs, or causes a stock to simply change in abundance within its previous distribution, the resulting mismatch between fish availability and fishing capacity will require greater management flexibility, both nationally and internationally. Because natural variability is so great relative to global change, and the time horizon on capital replacement (ships and plants) is so short, in most situations, there will likely be relatively small economic or food-supply consequences.
- There are three major institutional deficiencies: the allocation of access rights to fisheries and among competing uses of aquatic resources; regional fisheries bodies whose mandate and organization have not adjusted to the new requirements of shared and straddling stocks management; and international institutions whose development is lagging behind the globalization of the economy and uses of the ecosphere (Troadec, 1994).
- When fish stocks decline because of a combination of less favorable environmental conditions and overfishing—and even when there are initially abundant

> stocks—industries often have continued heavy fishing pressure, leading to fishery collapse. With such complex relationships, scientists are often slow to recognize declining resources. Further, under most prevailing institutions, administrators and producers cannot take proper account of the value of use (harvest) rights resulting from resource scarcity when regulating access and planning investments. As a result, industries and governments have been unwilling or unable to accept the advice of scientists and take meaningful action to protect fish stocks and enhance harvest efficiency.

A number of reviews have appeared that attempt to predict the impact of climatic change, based on past events, but of necessity they are generalized and deal more with the direction of ecosystem and population changes than their magnitude (e.g., Costa, 1990; Siegfried *et al.*, 1990; Mann and Lazier, 1991; Carpenter *et al.*, 1992; Shuter and Meisner, 1992; Everett, 1995). They provide a strong foundation for understanding the likely impact of climate changes on future fish production.

The ineffectiveness of many management institutions (Anon., 1994; FAO, 1992b; Neher *et al.*, 1989; Parsons, 1993; Troadec, 1989a) will aggravate the likelihood of collapse. This ineffectiveness stems from insufficient authority, unwillingness to act, or incomplete knowledge about resources and their relationship to the environment (Everett, 1995; Talbot *et al.*, 1994). Because many institutions have not adjusted to new conditions of resource scarcity and have not developed means to value and price the rights to use resources, efficient use of resources will be difficult to achieve.

If future fish abundance becomes less certain, regulatory institutions will have even more difficulty in managing fisheries. Many stocks will risk continued overexploitation. This is already a major cause of variability. Institutional inadequacies may result first in greater difficulties in adapting exploitation and management to climate-induced changes. The current crisis may become worse at first. Then, the worsening of the fisheries crisis could make the need for institutional change more obvious and acceptable.

Gains and losses at all levels of social organization may result from climate-change impacts directly or from human responses to those changes. Some nations, sectors, and groups may have the ability to respond or adapt to climate change, turning it to their advantage (Glantz *et al.*, 1990). Others will not. In general, societies have not coped well with declines in biological productivity, regardless of cause (Glantz *et al.*, 1987; Glantz, 1992). Case studies of societal response provide insight into how people and their governments can better prepare for regional changes in distribution and abundance that might accompany global warming. If, in the face of climate change, leaders desire to maintain fisheries at levels matching society's needs, higher levels of international collaboration will be needed—for which there is no global analog from the past.

16.2.3. Aquaculture

The effects of climate change on aquaculture generally will be positive—through faster growth, lower winter mortality rates, and reduced energy costs in shore-based facilities (Smit, 1993). Nearshore warming could result in increased production in higher-latitude farming operations because of the prolonged availability of nearly optimum air and water temperatures (McCauley and Beitinger, 1992). With expanded regions of warmer water, the economic incentive to develop cultivation of temperate and subtropical species could develop. A decrease in ice cover could substantially relax the geographic limits for commercial operations for species such as salmonids, oysters, and scallops. Major areas with winter ice cover could become ice-free (Frank *et al.*, 1988). Winter temperatures could move above the lower lethal limit for salmonids and allow their aquaculture to expand into northern Nova Scotia, into the Gulf of St. Lawrence, and along Newfoundland's south coast. Oysters and other cultured species would likely increase in productivity with warmer temperatures, through greater spawning success and higher growth rates—if salinity regimes and disease agents do not become limiting (Frank *et al.*, 1990).

Climate warming may prove deleterious to existing aquaculture industries in some areas. A significant increase in summer water temperatures in the Marlborough Sound in New Zealand may have serious implications for the salmon farming industry there because temperatures are already near the maximum for successful cultivation (New Zealand, 1990). Elsewhere, increasing temperatures may augment the growth and spread of disease agents. The limited genetic variability of farm fish may decrease their adaptability and increase their susceptibility to pathogens.

Sea-level rise could damage aquaculture facilities and reduce areas suitable for aquaculture (e.g., for shrimp culture in mangroves in tropical countries or the Bangladesh floodplain, and for shellfish culture in western Europe). However, in some regions, increased areas will become available. For example, in New Zealand, a rising sea is expected to make more areas suitable for oyster farming because many mud flat areas are a little too shallow for intertidal rack cultivation of oysters. This positive benefit may be offset by a warmer sea that would trigger earlier spawning in the oysters, thus reducing their value for the very important Christmas market (New Zealand, 1990). Even in the mangrove situation, all is not clear because land may flood inland of the mangroves, where conditions could be better than the existing mangrove areas that have high-sulfur soils and very acidic conditions.

Culturing fish for release into the wild for augmentation of wild stocks (e.g., cod) or for capture upon return (e.g., salmon) can be affected by a changing climate at the rearing facilities, along the path to the open ocean or large lake, or in the water bodies where growth is to take place. This type of aquaculture may account for half of the global total for aquaculture. Its further growth is hampered by ecological, institutional (e.g., property regimes), and technological factors. Some scientists believe that many bodies of water may be at their biological carrying capacity and cannot support more fish; others believe that there are great opportunities for expansion. In any case, fishery managers and hatchery operators need to consider possible interactive impacts of climate change and augmentation on wild stocks and other users of the ecosystem, such as marine mammals and birds.

New or stronger disease agents might flourish, while fish might have little time to build resistance. Farmed fish also may become increasingly susceptible to antibiotic-resistant strains of pathogens to which they had been accustomed, lowering or completely crippling aquaculture production, especially in developing countries.

In aquaculture facilities, fish often live in shallow structures at all life stages. Eggs and larvae near the surface and near-surface phytoplankton, zooplankton, corals, and wetland plants could be exposed to ultraviolet (UV) levels that could cause genetic abnormalities or death. In aquaculture, shallow ponds and unshielded egg and larval production facilities may be of particular concern; most eggs and larval stages of fish studied by Hunter *et al.* (1982) showed high sensitivity to UV radiation and often died if exposed in shallow containers.

16.2.4. Health Issues

One-fifth of global animal protein intake comes from fish, mostly marine species (Weber, 1994). Although there are large variations by country, fish is generally a much more important part of the diet in developing countries, where animal protein is relatively expensive, but per capita consumption is often highest in wealthier nations (Laureti, 1992). Even in the absence of climate change, fish production growth is not likely, and wealthy nations are importing more fish from developing nations that seek foreign exchange—with nutrition consequences for poor nations. With climate change, food production could change, affecting nutrition and health for societies highly dependent on fisheries.

There are natural toxins in ocean and freshwaters, as on land. The better-known toxins are "red tides"—found in both warm and tropical waters but also colder waters—and ciguatera, most often associated with warm or tropical waters. Red tides refer to high concentrations of toxic dinoflagellates that discolor the sea. Even when not abundant enough to discolor the water, they may toxify shellfish (Sampayo, 1989). Some researchers suggest that one red tide toxin, DSP, can cause liver cancer in humans. As plant toxins are concentrated through bioaccumulation at successive levels of predation, those fish at the top of food webs can contain lethal doses, particularly in fatty organs and tissues. Filter feeders such as mussels, clams, and oysters also concentrate toxins when they consume large quantities of these algae. Similarly, ciguatera poisoning occurs when algal grazers in tropical reefs consume toxin-bearing algae. Fish at the top of the food web, such as barracuda or grouper, can concentrate lethal toxin amounts. Cooking does not destroy these toxins. Ciguatera is usually kept in check by local fishing practices that avoid taking fish from affected areas. These areas usually occur in an erratic distribution. Climate change could alter the local habitats and change the areas where ciguatera toxin is produced, but the changes might be undetected by local fishermen or regulators. Warming would expand the range for ciguatera and other biotoxins and could lead to increased biotoxin poisoning.

A rising sea would lift the water table in low-lying land near the coast, releasing contaminants from dump sites and viruses and bacteria from septic systems into drainage systems and waterways. Such contaminants would enter estuarine and inshore food chains and pose a hazard for human consumers. Governments may close additional areas to the taking of fish and shellfish to protect citizens (IPCC, 1992). Also, some marine species may not have time to build resistance to disease agents that might come with a relatively rapid rise in temperatures. One example may be the susceptibility of North Atlantic green sea urchins to a waterborne disease to which they lack natural resistance at warm temperatures (Frank *et al.*, 1990). It is not just the exposure of organisms to new pathogens that is a concern; there is also the possibility of increasing susceptibility to existing ones as a result of environmental stress.

Changes in precipitation patterns also are expected. Reduced water flows in streams and rivers will concentrate pollutants, rendering some fish unfit for consumption. Areas receiving higher flows may have lower concentrations of pollutants in fish whose use is now restricted (or should be). When increased precipitation comes in the form of pulses and leads to sudden surges of runoff, increased coliform bacteria counts will be reached more often, leading to closures of shellfish beds (New Zealand, 1990). When increased flows of pollutants enter lakes having a high rate of evaporation relative to water discharge, increasing concentrations of pollutants may occur.

Chapter 18 warns that cholera can be more easily transported in ship ballast water when there is a high density of marine algae and that warming could increase the amount and range of algal blooms. Ships often release ballast water in harbors or nearshore areas, thus transporting cholera throughout the world. This is a primary means for the introduction of new cholera strains. Clams, mussels, and oysters can be carriers of cholera, and—since they are often eaten raw—they can transmit cholera to humans. Ballast water seems to be the means of introduction of PSP algae into Tasmanian waters, and reproduction of the algae is helped by warmer, nutrient-rich waters (New Zealand, 1990).

When fishers pursue fish over longer distances and have inadequate infrastructure (e.g., refrigeration, potable water), opportunities for pathogens and spoilage increase. Health issues can cause great economic damage to fisheries because societies will act to keep contaminated food out of the marketplace. Stories about people becoming ill from eating problematic food receive a great deal of publicity, thus depressing markets.

16.2.5. *Infrastructure Issues*

Modern industrial fisheries require complex support, including electronic positioning, weather advisories, transportation, cold storage, ports, processing, markets, electricity, potable water, search and rescue, and provisioning. Any redistribution or increased variability of fishery resources, or of aquaculture production, adds strain on this sector. Natural variability is so great that the provision of these services is generally quite flexible. However, when long-term redistributions occur, economic efficiencies change as the distance between fish and home ports changes. Over time, ports nearer to new fishing areas will gain an advantage. Home ports will change, families will move, and there will be winners and losers. In some cases, the retail market may respond to long-term redistributions, if the predominant species change and the new species require different processing and marketing approaches. Rapid loss of shelf space in markets and absence from restaurant menu listings occurs when species are unavailable. It is difficult to regain these markets, further exacerbating variability in revenues.

A rise in sea level will adversely affect infrastructure that supports fisheries. In less-industrialized areas—where artisanal fishers land catches in numerous remote fishing villages and transport them, often by foot, to commercial centers and tourist motels for sale—a significant rise in sea level will drown much of the existing network of roads and footpaths used by many poor coastal people who financially and nutritionally depend on their daily catch. Sea-level rise also will impact industrialized countries, as implied in the Caspian Sea example (see Box 16-4). Reduction in fresh water supplies due to sea-level rise and increased evaporation—where not offset by increased precipitation—could have significant impacts on fish processing, which uses large amounts of potable water (Smit, 1993).

Some technical adaptations can ease the impacts on industrialized as well as developing countries—such as faster vessels, portable processing plants, alternative transportation infrastructure, and on-board processing. Nevertheless, costs will be incurred to finance the changes, and profitability will be reduced.

16.2.6. *Recreational Fisheries Issues*

There is no significant difference in the impacts of climate change on recreational species as opposed to commercial or subsistence species. Many species are pursued by multiple user groups. However, increased storminess or changes in weather patterns will affect the desirability and effort involved in recreational fishing more than other types of fishing. Changes might include shorter ice-fishing seasons and longer seasons for higher-latitude fisheries. On New Zealand's west coast, expected decreases in westerly winds, higher temperatures, and sea-level rise may lead to a decline in the distribution and abundance of toheroa, a popular shellfish species sought by aboriginal and recreational fishermen (New Zealand, 1990).

Many anglers prefer cold-water species, which are associated with higher expenditures. The ranges of such species can be expected to shift toward higher latitudes. However, if the daytime summer temperatures do not change significantly when compared to changes in nighttime temperatures, the ranges and production of such species may expand.

16.2.7. *Sensitivities, Critical Rates, and Thresholds*

Most fish species are sensitive to the expected climate-change effects. The individuals, or centers of production, can shift rapidly if suitable habitat and paths exist. Fish species are cold-blooded; life processes, such as growth, are faster when warmer (within limits). Many species have narrow ecological niches, but there are many species to fill niches. In a varying environment, species mixes will change. Examples of sensitivities are:

- Fish eggs that rely on a gyre to return them to their habitat on a certain day or week
- Fish eggs in streams or on the sea floor that require a minimum current speed for oxygenation
- Species that require a certain influx of freshwater to induce spawning or to kill predators
- Temperatures above or below the stock's lethal limit.

At the societal level, there are other sensitivities. Species in more stable environments often are more valuable. Fishing interests often can follow fish, but political boundaries or economics can stop pursuit. Even if fishing interests can follow, communities may not. Developing nations dependent on fish as food or export earnings are most sensitive. Examples of societal sensitivities are:

- Immobility of communities or industries dependent on one type of species
- Societies without the money needed to buy replacement foods
- Fishers without the ability to deal with increased storminess.

Information on critical rates, thresholds, and sensitivities exists for only a few important species. Available information includes:

- The relationship between fisheries production and the health or extent of corals, mangroves, and wetlands
- The temperatures governing some fish movements, migrations, reproduction, and mortality (in general, fish are much more sensitive to cooling rates than to warming; Brett, 1980)
- The optimum temperatures needed for certain aquaculture systems
- Freshwater flows required for stock movements, migrations, reproduction, and aquaculture operations
- Nonlinearities: As temperature rises and nutrients increase, eutrophication may suddenly produce critically low values of dissolved oxygen.

Additional responses that have some quantification are:

- The shift in production centers of some oceanic fish in response to temperature changes (e.g., Murawski, 1993)
- The distribution of some fish in relation to northern ice cover (Shuter and Post, 1990)
- Mortality-rate changes of some Antarctic krill and some eastern Pacific pelagic eggs and larvae in response to changes in UV-B radiation (Hunter *et al.*, 1979, 1981; Damkaer *et al.*, 1980, 1981; Damkaer and Dey, 1983; Dey *et al.*, 1988). For some species, the apparent thresholds for development and survival are at current incident UV-B levels (Damkaer *et al.*, 1980). Significant changes in ecosystem function and functioning could occur as species sensitive to UV-B radiation are replaced by more-resistant species (IGBP, 1990).

16.2.8. Storminess

Major storms such as cyclones have major consequences for ecosystems. They cause mixing of ocean waters and internal upwelling over the edges of shelf seas, moving nutrients and organisms into the photic zone—which then helps local food-web production and can be likened to terrestrial plowing (Sharp, 1994). Such mixing, though, can disturb the stratification of water masses and their biological components, on which many species depend as part of their reproductive strategies. Storms also can be disruptive to reefs, mangroves, and shoreline habitats and to egg masses laid on substrate. Storm paths are also important. Path changes can affect the density of plankton forage on which young fish depend, thus modifying the recruitment success of pelagic stocks (Lasker, 1981).

Changes in storm frequency and intensity could have subtle and easily missed impacts in the coastal zone. For example, in regions of small tidal range, such as the Gulf of Mexico, there are wetlands found above the high-tide line that get inundated by storms and weather fronts or distant oceanic events (Swenson and Chuang, 1983). These high super-tidal wetlands are important in supporting marine species. In some interior areas, near and distant storms are the dominant flooding mechanisms and affect the production of shrimp (Childers *et al.*, 1990).

Storms are also of major significance for the fisheries industry. Increased storminess can damage fishery and aquaculture infrastructure and lead to increased costs for the strengthening of facilities. In some cases, increased storminess can lead to longer fishing seasons if ice is broken up or cannot form. Storms are a major threat to the safety of fishers. Increases in intensity or frequency could lead to additional losses in the occupation, which is already among the most dangerous; the fatality rate of commercial fishing is seven times the U.S. industry average (Hart and Perrini, 1984).

16.3. Adaptation Options

Events influencing fisheries are accelerating. Adaptations must be matched with the concept of an already highly stressed ecology and the inherent volatility of fisheries. Further, coincident problems such as pollution, resource degradation, and sea-level rise (or subsidence) also must be addressed.

There is a danger that if climate-change alterations are slow, inexorable, and generally irreversible, impacts resulting from climate change will not be recognized in time. Alternatively, abrupt shifts in climate-influenced factors may be too limiting for species acclimation or management adjustment. Some fisheries involve high capital investment, and others are very important to whole coastal communities. To minimize economic and social disruption, the possibility of permanent long-term changes in the regional availability of traditional fish stocks must be anticipated and new management philosophies developed. Local fishery managers also need to anticipate and take advantage of short-lived benefits (Sharp, 1991). As in any stock management planning, managers must recognize the need for flexible approaches and for greater foresight when establishing policies.

The faster the rate of climate change, the more difficult will be adjustments in human systems. Surprises and divergences from the predicted scenarios are to be expected and must be anticipated by resource and community managers. Preparation for the unexpected is important (New Zealand, 1990). In general, technological innovations are developed faster than are those of a biological or societal nature. Cultural attitudes, which are critical for the adoption of institutional innovations, have the greatest inertia.

Fisheries strategies should concentrate on those that are compatible with sound conservation principles and—like the earlier IPCC response strategies—are beneficial for reasons other than climate change and make sense now in their own right; are economically efficient and cost-effective; serve multiple purposes; are flexible enough to change with new knowledge; are compatible with sustainable economic growth; and are administratively practical (IPCC, 1990).

Table 16-1 provides an overview of key fisheries impacts and whether there might be societal responses or adaptations. An I, P, or T indicates that an important response exists; the right-hand column indicates the type of adaptation or issue involved.

16.3.1. Fisheries

Adaptation to existing climate variability may demonstrate ways to deal with climate change. A Canadian report (Smit, 1993) offers insights. In the short term, the timing of fishing and tasks such as trip preparation and maintenance are adjusted to match the weather. Most full-time fishing interests also have several licenses and gear to allow flexibility in the species and locations fished. Others maintain work options in activities such as construction, tourism, or other sectors, which reduces risk due to climate uncertainty. There also is risk-sharing within family businesses, extended families, and communities, which compensates for temporary setbacks in a particular fishery or enterprise. These may be formal or informal, involving cooperatives or financial institutions, based on a sense of mutual support. The vulnerability of communities to varying stocks, intense fishing pressure, and inappropriate government policies is graphically apparent with the current Canadian moratorium on northern cod, whose stock abundance and distribution may be partly related to water temperature and salinity (Smit, 1993; Mann and Drinkwater, 1994). The policy lesson here is that government fisheries managers should not lock fishing interests into narrow full-time activity within a single fishery. Flexibility in fishing, plus seasonal shifts to other sectors, will be a valuable approach to minimizing climate-change impacts, as long as fishing capacity is consistent with the resource base.

Climate change increases society's need to quickly adjust labor and capital to fish productivity. To have efficient fisheries, the capacity to control human inputs should precede and anticipate the development of technological intensification. This prescription is atypical of societal response. Overcapitalization problems are rooted in the uneven development of biotechnical, administrative, and regulatory systems.

Management systems developed when resources were not under intense fishing or environmental threat have become ill-adapted to resource scarcity. In traditional systems, resource access was regulated through the social structure of rural communities. These systems lose their effectiveness as population pressure and resource uses intensify. They are being supplemented by central legislation, with attendant lag times in addressing problems. These formal regulatory systems have also tended to become ineffective as the need for harvest rationing increases. As resources approach their production limits, administrators and producers have maintained the *status quo* even at the cost of greater economic inefficiency and overfishing. To function most effectively, economic systems need proper institutions (e.g., a market for pricing harvest rights), a property or ownership/stewardship regime for the functioning of that market, and an institutional context to administer the proper functioning of any allocation system based on such rights and markets.

There are several adaptation strategies:

- Develop international agreements for ecosystem research and management and for stock allocation before stocks of fish move due to climate changes. The dramatic decadal fluctuations in marine fisheries yields, considered in light of climate-change concerns,

Table 16-1: *Overview of key fisheries impacts.*

Global Change Impacts	**Certainty**	**Adaptation Example** Institutional/ Policy/Technical	Possible Adaptations
Redistribution of Living Resources	very likely		
– Increase inshore[1]	possible	I/T	– Fish management/industry expansion
– Decrease offshore[1]	possible	I	– Government welfare assistance
– Poleward migration[1]	likely	I/P/T	– Welfare/industry assistance/mobility
– EEZ jurisdiction changes	likely	I/P	– Negotiation/confrontation
Faster Fish Growth Rates	likely		
Extreme Events	possible	I/T	– Forecasts, rescue/hardier equipment
Precipitation Changes	likely	T	– Water desalination for processing
River Temperature Changes	likely	T	– Water release from dams
Aquaculture	possible		
– Sites (freshwater, sea-level rise, temperature)		P/T	– Regulations/species, location
– Growth rates		T	– Species, feed
– UV-B		T	– Shelter
– Diseases		T	– Hygiene, prophylactics
Reduced Ice Cover	likely		
– Freer navigation		T	– New fishing zones, routes, ports
– Less ice strengthening		T	– Ships, facilities
Seafood Health Issues[2]	possible	I/P/T	– Define responsibility, certify safety, establish monitoring system
Flooding from Sea-Level Rise	likely	T	– Build on higher ground, use water-control structures, allow procession
Increased UV-B	likely	T	– Shelter for aquaculture, crew

[1]Generalization difficult.
[2]Increased frequency.

should accelerate the adoption of large marine ecosystems as the appropriate units for fisheries research and management (Sherman and Gold, 1990; Talbot *et al.*, 1994). Some post-United Nations Conference on Environment and Development (UNCED), ecosystem-scale research programs are being implemented (AAAS, 1993). These programs are designed to couple recent advances in ecological monitoring, management, and stress mitigation strategies with needs in developing countries (AAAS, 1990, 1991).

- Modify and strengthen fisheries management policies and institutions and associated fish population and catch-monitoring activities. In some areas, it takes many years to develop and implement changes in fisheries policies, even after the institutions are established.
- In coastal planning and environmental decisionmaking, consider that a healthy environment is a prerequisite for healthy ecosystems.
- Preserve and restore wetlands, estuaries, floodplains, and bottom lands—essential habitats for most fisheries. Cooperate more closely with forestry, water, and other resources managers because of the close interaction between land cover and maintenance of adequate fishery habitat. The adequacy of management practices in all sectors affecting fisheries (e.g., water resources, coastal management) needs to be examined to ensure that proper responses are made as climate changes.
- Monitor and ensure that the habitat needs of marine mammals (as they shift to remain in an optimum environment) are considered from the standpoints of coastal planning and ocean pollution control.
- Promote fisheries conservation and environmental education among fishermen worldwide.
- Be alert to possible increased health problems caused by biotoxins and the release of pollutants from a rising sea level; industry and seafood-safety agencies should take the lead. Take measures to prevent the introduction of undesirable organisms as climate changes.
- Tailor institutional innovations to ecological, technical, economic, and social features. Large-scale and small-scale production systems, in particular, require different regulatory arrangements. There is no standard institutional solution.
- Adapt to fish redistribution with faster vessels, portable processing plants and other onshore infrastructure, and on-board processing.
- Construct and maintain appropriate infrastructure for storm forecasting, signaling systems, and safe refuges for dealing with possible rising sea level and increased storminess.
- Take advantage of reduced need for ice strengthening of vessels and infrastructures in a warmer climate, except perhaps for areas with increased icebergs.
- Foster interdisciplinary research, with scientists meeting periodically to exchange information on observations and research results, and meeting with managers to ensure the proper interpretation of results and the relevance of research.
- In cases of species collapse and obvious ecosystem disequilibrium, restock with ecologically sound species and strains as habitat changes. Great care is needed to avoid long-term ecological damage.
- Consider, cautiously, the use of hatcheries to enhance natural recruitment when climate causes stocks to fall below the ecosystem carrying capacity for a given species. Such enhancement might increase stock productivity (per unit area), reduce recruitment variability (e.g., Héral *et al.*, 1989), and enable the colonization or recolonization of new areas (e.g., Peterman, 1991). Conversely, injudicious use may alter or impoverish the biodiversity and genetic pool of resources and possibly transmit parasites and diseases (Barg, 1992). Research on identifying warm-water species with wide temperature tolerances also could be useful.

16.3.2. Aquaculture

Economic development of coastal communities beyond capture fisheries to tourism and other activities has long been a goal in many areas. Development of aquaculture and tourism will make coastal communities better able to deal with uncertainties of climate change (Smit, 1993). In adapting to climate change, aquaculturists should consider the following:

- Warming will mean generally longer growing seasons and increased rates of biological processes—and often of production.
- Warming will require greater attention to possible oxygen depletion.
- In some areas, the species grown may have to shift to those more tolerant of warmer and perhaps less-oxygenated waters.
- Coastal culture facilities may need to consider the impacts of sea-level rise on facilities and the freeing of contaminants from nearby waste sites.
- Competing wild fisheries production (and indeed, agriculture) may not vary much at the world level, but there could be significant regional changes in quantity and species mix.
- Precipitation, freshwater flows, and lake levels will likely change. Strong regional variations are likely.
- Warming waters could introduce disease organisms or exotic or undesired species before compensating mechanisms or intervention strategies have become established.
- Less ice cover and thinner ice will generally mean less ice damage to facilities and a longer season for production and maintenance.
- Covering culture tanks, or keeping them indoors under controlled light, may be needed more often to protect larvae from solar UV-B.
- Several of the above and other factors, such as competing demand for coastal areas, may argue for technological intensification in ponds and noncoastal facilities.

16.3.3. *Adaptation Constraints*

Technology, trade, and industry change rapidly to take advantage of new situations and adapt to those that are adverse. The success of many entrepreneurs derives from their abilities to adapt rapidly to opportunities and dangers, but those adversely affected by climate change may not be in a position to take advantage of opportunities. Many fish stocks are of supranational scope, but institutions to deal with them are not or lack sufficient authority. Changes in fisheries institutions generally result from crises, yet impacts from a changing climate may be too gradual to trigger needed responses from these institutions—as evidenced by the failure of most governments to respond to the profound impact of the 1976–77 climate change event on Pacific fisheries over the last 20 years (Beamish, 1993). Nations that do not strongly depend on fisheries may react slowly. Also, negative actions such as polluting and overfishing add stress to fish stocks and may reduce society's options.

16.4. Research and Monitoring Needs

Information is most valuable if there are institutions and management mechanisms to use it. Research on improved mechanisms is needed so that fisheries can operate more efficiently with global warming as well as in the naturally varying climate of today. There is relatively little research underway on such mechanisms. While useful inferences can be made about global-change impacts on fisheries, GCMs are inadequate to forecast changes at the regional scale—and even more so at the smaller scales of the spawning and nursery areas that are critical in the relationships between environment and fish populations (Sinclair, 1988).

Knowledge of the reproductive strategies of many species and links between recruitment and environment is poor, but interest has been growing rapidly. Several international, broad-scale research programs are in place, some of which are aimed at providing a scientific basis and linkage to ecosystem sustainability programs (IOC, 1993; Sherman *et al.*, 1992; Wu and Qiu, 1993; Mee, 1992). The growing partnership among funding agencies, marine ecologists, and socioeconomic interests marks an important step toward realization of the UNCED declaration aimed at reversing the declining condition of coastal ecosystems and enhancing the long-term sustainability of marine resources.

The following lists focus on items needed specifically because of climate change. Other types of research, which are prerequisites for dealing with such concerns but which support the day-to-day needs of fisheries managers or relate more to understanding how ecosystems function, are not included.

16.4.1. *Resource Research*

- Determine how fish adapt to natural extreme environmental changes or latitudinal transitions, how fishing affects their ability to survive unfavorable conditions, and how reproduction strategies and environments are linked. Link fishery ecology and regional climate models to enable broader projections of climate-change impacts and improve fishery management strategies. Both of these model types are in their infancy for application to fisheries and require considerable additional specificity, in addition to linking with other models.
- Implement regional and multinational systems to detect and monitor climate change and its impacts—building on and integrating existing research programs. Fish can be indicators of climate change and ecological status and trends (Sharp, 1991). Assemble baseline data now so comparisons can be made later (Sharp, 1989).
- Develop ecological models to assess multiple impacts of human activities. Separate appraisal of individual impacts is no longer sufficient.
- Assess the effect of accrued global changes by compiling and analyzing regional data sources—globally containing information on fisheries distributions and the associated environment—to detect consistent trends over the last century and determine how societies have dealt with these changes.
- Determine the fisheries most likely to be impacted, and develop adaptation strategies.
- Assess the potential leaching of toxic chemicals, viruses, and bacteria due to sea-level rise and how they might affect both fish and the seafood supply.

16.4.2. *Institutional Research*

- Determine institutional changes needed to deal with a changing climate. Such changes are likely the same ones needed for mastering overfishing and coping with the variability and uncertainty of present conditions. Improved institutions would probably reduce stock variability more than climate change would increase it.
- Study the recorded ability of societies—including preindustrial ones, where fisheries are important—to understand how they have adapted their activities when their resources are impacted by climate changes (Glantz and Feingold, 1990).

References

AAAS, 1990: *Large Marine Ecosystems: Patterns, Processes and Yields* [Sherman, K., L.M. Alexander, and B.D. Gold (eds.)]. AAAS Press, Washington, DC, 242 pp.

AAAS, 1991: *Food Chains, Yields, Models, and Management of Large Marine Ecosystems* [Sherman, K., L.M. Alexander, and B.D. Gold (eds.)]. Westview Press, Inc., Boulder, CO, 320 pp.

AAAS, 1993: *Large Marine Ecosystems: Stress, Mitigation, and Sustainability* [Sherman, K., L.M. Alexander, and B.D. Gold (eds.)]. AAAS Press, Washington, DC, 376 pp.

Acers, T.O., 1988: The controversy over Nile perch, Lates niloticus, in Lake Victoria, East Africa. *NAGA*, **11(4)**, 3-5.

Achieng, A.P., 1990: The impact of the Nile perch, Lates niloticus (L.), on the fisheries of Lake Victoria. *Journal of Fish Biology*, **37(A)**, 17-23.

Ackefors, H., N. Johansson, and B. Walhlberg, 1991: The Swedish compensatory programme for salmon in the Baltic: an action plan with biological and economic considerations. In: *Ecology and Management Aspects of Extensive Mariculture: A Symposium Held in Nantes* [Lockwood, S. (ed.)]. ICES Mar. Sci. Symp., 248 pp.

Aebischer, N.J., J.C. Coulson, and J.M. Colebrook, 1990: Parallel long term trend across four marine trophic levels and weather. *Nature*, **347**, 753-755.

Albaret, J.J., 1987: Les peuplements de poisson de la Casamance (Sénégal) en période de sécheresse. *Rev. Hydrobiol. Trop.*, **20**, 291-310.

Albaret J.J. and J.M. Ecoutin, 1991: Communication mer-lagune: impact d'une réouverture de l'embouchure du fleuve Comoé sur l'ichtyofaune de la lagune Ebrié. *J. Ivoir. Océanol. Limnol.*, **1**, 99-109.

Alverson, D.L., M. Freeberg, S. Murawski, and J.G. Pope, 1994: *A Global Assessment of Fisheries Bycatch and Discards.* FAO Fisheries Technical Paper 339, FAO, United Nations, Rome, Italy, 233 pp.

Anon., 1989: *Ichthyofauna and Fishery Resources of the Caspian Sea.* Nauka, Moscow, Russia, 337 pp. (in Russian).

Anon., 1994: Improving the link between fishery science and management: biological, social and economic considerations. In: *ICES Theme Session, St. John, Newfoundland, Canada, September 22 - 27, 1994.*

Anthony, V.C., 1993: The state of groundfish resources off the Northeastern United States. *Fisheries (AFS)*, **18(3)**, 12-17.

Bagenal, T.B., 1970: Fish fecundity and its relationship with stock and recruitment. *ICES/FAO/ICNAF Symposium on Stock and Recruitment*, **No. 18** (mimeo).

Bakun A., 1990: Global climate change and intensification of coastal ocean upwelling. *Science*, **247**, 198-201.

Bakun, A., 1993a: Global greenhouse effects, multi-decadal wind trends, and potential impacts on coastal pelagic fish populations. *ICES Mar. Sci. Symp.*, **195**, 316-325.

Bakun, A. 1993b: The California Current, Benguela Current, and Southwestern Atlantic Shelf ecosystems: a comparative approach to identifying factors regulating biomass yields. In: *Large Marine Ecosystems: Stress, Mitigation, and Sustainability* [Sherman, K., L.M. Alexander, and B.D. Gold (eds.)]. AAAS Press, Washington, DC, pp. 199-221.

Bakun, A. and R.H. Parrish, 1990: Comparative studies of coastal pelagic fish reproductive habitats: the Brazilian sardine (Sardinella aunta). *J. Cons. Explor. Mer.*, **46**, 269-283.

Bakun, A. and R.H. Parrish, 1991: Comparative studies of coastal pelagic fish reproductive habitats: the anchovy (Engraulis anchoita) of the southwestern Atlantic. *ICES J. Mar. Sci.*, **48**, 343-361.

Bard, F.X., D. Guiral, J.-B. Amon Kothias, Ph. K. Koffi, 1991: Syntheses des travaux effectues au C.R.O. sur les vegetations envahissantes flottantes. *J. Ivoir. Oceanol. Limnol.*, **1**, 1-8.

Barg, U.C., 1992: *Guidelines for the Promotion of Environmental Management of Coastal Aquaculture Development*, vol. V. FAO, United Nations, Rome, Italy, 122 pp.

Barel, C.D.N., R. Dorit, P.H. Greenwood, G. Fryer, N. Hughes, P.B.N. Jackson, H. Kawanabe, R.H. Lowe-McConnell, M. Nagoshi, A.J. Ribbink, E. Trewavas, F. Witte, and K. Yamaoka, 1985: Destruction of fisheries in Africa's lakes. *Nature*, **315(6014)**, 19-20.

Beamish, R.J., 1993: Climate change and exceptional fish production off the west coast of North America. *Can. J. Fish. Aquat. Sci.*, **50**, 2270-2291.

Beamish, R.J., 1995: Response of anadromous fish to climate change in the North Pacific. In: *Human Ecology and Climate Change: People and Resources in the Far North* [Peterson, D.L. and D.R. Johnson (eds.)]. Taylor and Francis, Washington, DC, pp. 123-136.

Beamish, R.J. and D.R. Bouillon, 1993: Pacific salmon predictive trends in relation to climate. *Can. J. Fish. Aquat. Sci.*, **50**, 1002-1016.

Belvèze, H. and K. Erzini, 1983: The influence of hydroclimatic factors on the availability of the sardine (Sardina pilchardus) in the Moroccan Atlantic fishery? In: *Proceedings of the Expert Consultation to Examine Changes in Abundance and Species Composition of Neritic Fish Resources* [Sharp, G. and J. Csirke (eds.)]. FAO FIR M/R, **291**, 285-327.

Bernal, P.H., 1991: Consequences of global change for oceans: a review. *Climatic Change*, **18**, 339-359.

Beukema, J.L., W.J. Wolff, and J.J.W.M. Brouns (eds.), 1990: Expected effects of climatic change on marine coastal ecosystems. In: *Developments in Hydrobiology 57*, Kluwer Academic Publishers, Dordrecht, Netherlands, 221 pp.

Binet, D., 1982: Influence des variations climatiques sur la pêcherie des Sardinella aurita ivoiro-ghanéennes: relation sécheresse-surpêche. *Oceanol. Acta*, **5**, 443-452.

Binet, D., 1988a: Role possible d'une intensification des alizés sur le changement de répartition des sardines et sardinelles le long de la Côte ouest africaine. *Aquat. Living Resour.*, **1**, 115-132.

Binet, D., 1988b: French sardine and herring fisheries: a tentative description of their fluctuations since the XVIIIth century. In: *Long Term Changes in Marine Fish Populations* [Wyatt, T. and M.G. Larraneta (eds.)]. Symposium held18-20 November 1986, Bayona Imprenta REAL, Vigo, Spain, pp. 253-272.

Binet, D. and C. Leroy, 1986: La pêcherie du thon rouge (Thynnus thynnus) dans l'Atlantique nord-est était-elle liée au réchauffement séculaire? *ICCAT/SCRS*, **86/52**, 16.

Binet, D. and E. Marchal, 1992: Le developpement d'une nouvelle population de sardinelles devant la Côte d'Ivoire a-t-il été induit par un changement de circulation? *Ann. Inst. océanogr., Paris*, **68(1-2)**, 179-192.

Binet, D. and E. Marchal, 1993: The large marine ecosystems of shelf areas in the Gulf of Guinea: long-term variability induced by climatic changes. In: *Large Marine Ecosystem: Stress, Mitigation, and Sustainability* [Sherman, K., L.W. Alexander, and B. Gould (eds.)]. AAAS Pub. 92-395, American Association for the Advancement of Science, Washington, DC, pp. 104-118.

Binet, D., L. LeReste, P.S. Diouf, 1995: Influence des eaux de ruissellement et des rejets fluviaux sur les ecosystemes et les ressources vivantes des cotes d'Afrique occidentale. In: *FAO Fisheries Technical Paper*, *Effects of Riverine Inputs on Coastal Ecosystems* (in press).

Binet, D. and J. Servain, 1993: Have the recent hydrological changes in the Northern Gulf of Guinea induced the Sardinella aurita outburst? *Oceanologica Acta*, **16(3)**, 247-260.

Blaxter, J.H.S. and C. Hempel, 1963: The influence of egg size on herring larvae (Clupea harengus (L.)). *J. Cons. Perm. Int. Explor. Mer.*, **28**, 211-240.

Blaxter, J.H.S., 1988: Pattern and variety in development. *Fish Physiology*, **X1A**, 1-58.

Brett, J.R., 1980: Thermal requirements of fish—three decades of study, 1940-1970. In: *Biological Problems in Water Pollution* [Tarzwell, C.M. (ed.)]. Transactions 1969 seminar, U.S. Department Health, Education, Welfare, R.A. Taft San. Eng. Center, Cincinnati, Ohio, Techn. Rep. W 60-3, pp. 110-117.

Bruton, M.N., 1990: The conservation of the fishes of Lake Victoria, Africa: an ecological perspective. *Environmental Biology of Fishes*, **27(3)**, 161-175.

Caddy, J.F., 1993: Toward a comparative evaluation of human impacts on fishery ecosystems of enclosed and semi-enclosed seas. *Reviews and Fisheries Science*, **1(1)**, 57-95.M.25.

Caddy, J.F. and J.A. Gulland, 1983: Historical patterns of fish stocks. *Marine Policy*, **7(4)**, 267-278.

Carpenter, S.R., S.G. Fisher, N.B. Grimm, and J.F. Kitchell, 1992: Global change and freshwater ecosystems. *Annu. Rev. Ecol. Syst.*, **23**, 119-139.

Caverivière, A., 1991: L'explosion démographique du baliste (Balistes carolinensis) en Afrique de l'ouest et son évolution en relation avec les tendances climatiques. In: *Pêcheries Ouest Africaines Variabilité, Instabilité et Changement* [Cury, P. and C. Roy (eds.)]. ORSTOM, Paris, France, pp. 354-367.

Caviedes, C.N. and T.J. Fik, 1992: The Peru-Chile eastern Pacific fisheries and climatic oscillations. In: *Climate Variability, Climate Change and Fisheries* [Glantz, M.H. (ed.)]. Cambridge University Press, Cambridge, UK, pp. 355-376.

Chambers, J.R., 1991: Coastal degradation and fish population losses. In: *Stemming the Tides of Coastal Fish Habitat Loss: Proceedings of the Marine Recreational Fisheries Symposium* [Stroud, R.H. (ed.)]. National Coalition for Marine Conservation, Savannah, GA, **14**, pp. 45-51.

Childers, D.L., J.W. Day, and R.A. Muller, 1990: Relating climatological forcing to coastal water levels in Louisiana estuaries and the potential importance of El Nino-southern oscillation events. *Clim. Res.*, **1(1)**, 31-42.

Christie, G.C. and H.A. Regier, 1988: Measures of optimal thermal habitat and their relationship to yields of four commercial fish species. *Can. J. Fish. Aquat. Sci.*, **45**, 301-314.

Costa, M.J., 1990: Expected effects of temperature changes on estuarine fish populations. In: *Expected Effects of Climatic Change on Marine Coastal Ecosystems* [Beukema, J.J. *et al.* (eds.)]. Kluwer Academic Publishers, Dordrecht, Netherlands, pp. 99-103.

Costa, M.J., J.L. Costa, P.R. Almeida, C.A. Assis, 1994: Do eel grass beds and salt marsh borders act as preferential nurseries and spawning grounds for fish? An example of the Mira estuary in Portugal. *Ecological Engineering*, **3**, 187-195.

Coutant, C.C., 1990: Temperature-oxygen habitat for freshwater and coastal striped bass in a changing climate. *Trans. Amer. Fish. Soc.*, **119**, 240-253.

Crawford, R.J.M., L.G. Underhill, L.V. Shannon, W.R. Siegfried, and C.A. Villacastin-Herero, 1989: An empirical investigation of trans-oceanic linkages between areas of high sardine abundance. In: *Long-Term Variability of Pelagic Fish Populations and Their Environment* [Kawasaki, T. and M. Mori (eds.)]. Proceedings of the International Symposium, Sendai, Japan, pp. 14-18.

Csirke, 1980: Recruitment in the Peruvian anchovy and its dependence on the adult population. *Rapp.P.-v. Reun. Cons. int. Explor. Mer.*, **177**, 307-313.

Cury, P. and C. Roy, 1989: Optimal window and pelagic fish recruitment success in upwelling areas. *Can. J. Fish. Aquat. Sci.*, **46**, 670-680.

Cury, P. and C. Roy (eds.), 1991: *Pêcheries Ouest-Africaines: Variabilité, Instabilité et Changement*. ORSTOM, Paris, France, 525 pp.

Cushing, D.H., 1957: The number of pilchards in the Channel. *Fish. Invest. Ser. II*, **21(5)**, 1-27.

Cushing, D.H., 1978: Biological effects of climate change. *Rapp. P.-v Reun. Cons. int. Explor. Mer.*, **173**, 107-116.

Cushing, D.H., 1982: *Climate and Fisheries*. Academic Press, London, UK, 373 pp.

Daan, N., P.J. Bromley, J.R.G. Hislop, and N.A. Nielsen, 1990: Ecology of North Sea fish. *Netherl. Journ. Sea Res.*, **26(2)**, 243-386.

Damkaer, D.M., D.B. Dey, G.A. Heron, and E.F. Prentice, 1980: Effects of UV-B radiation on near-surface zooplankton of Puget Sound. *Oecologia*, **44**, 149-158.

Damkaer, D.M., D.B. Dey, and G.A. Heron, 1981: Dose/dose-rate responses of shrimp larvae to UV-B radiation. *Oecologia*, **48**, 178-182.

Damkaer, D.M and D.B. Dey, 1983: UV damage and photoreactivation potentials of larval shrimp, Pandalus platyceros, and adult euphausids, Thysanoessa raschii. *Oecologia*, **60**, 169-175.

Dey, D.B., D.M. Damkaer, and G.A. Heron, 1988: UV-B dose/dose-rate responses of seasonally abundant copepods of Puget Sound. *Oecologia*, **76**, 321-329.

Dickie, L.M., 1955: Fluctuations in abundance of the giant scallop, Placopecten magellanicus (Gmelin), in the Digby area of the Bay of Fundy. *J. Fish. Res. Bd. Canada*, **12(6)**, 797-857.

Dickson, R.P. and K.M. Brander, 1993: Effects of a changing wind field on cod stocks of the North Atlantic. *Fish. Oceanogr.*, **2(3/4)**, 124-153.

Dickson, R., P.M. Kelly, J.M. Colebrook, W.S. Wooster, and D.H. Cushing, 1988: North winds and production in the eastern North Atlantic. *J. Plankton Research*, **10**, 151-169.

Dorn, M.W. and R.D. Methot, 1990: *Status of the Pacific Whiting Resource in 1989 and Recommendations for Management in 1990*. NOAA Tech. Memo F/NWC-182, U.S. Department of Commerce, Washington, DC, 84 pp.

Everett, J.T., 1994: Personal observation during IPCC workshop in Fortaleza, Brazil, October 1994.

Everett, J.T., 1995: Impacts of climate change on living aquatic resources of the world. In: *The State of the World's Fisheries Resources, Proceedings of the World Fisheries Congress, Plenary Sessions, held in Athens, Greece, 1993*. Oxford & IBH Publishing Co. Pvt. Ltd., New Delhi, India (in press).

FAO, 1989: *Recent Trends in Mediterranean Fisheries, XIXth Session of the GFCM, Livorno (Italy), 27 February-3 March 1989* [Caddy, J.F. (ed.)]. GFCM/RM/7/89/3, FAO, Rome, Italy, 71 pp.

FAO, 1992a: *Fisheries Statistics: Commodities, 1990*, **71**. United Nations, Rome, Italy, 395 pp.

FAO, 1992b: Marine fisheries and the law of the sea: a decade of change. Special chapter of *FAO State of Food and Agriculture, 1992*. FAO Fish. Circ. 853, 69 pp.

FAO, 1993: *Fishery Statistics: Catches and Landings, 1991*, **72**. United Nations, Rome, Italy, 653 pp.

Fischetti, M. 1991: A feast of gene-splicing down on the fish farm. *Science*, **253**, 512-513.

Frank, K.T., R.I. Perry, K.F. Drinkwater, and W.H. Lear, 1988: *Changes in the Fisheries of Atlantic Canada Associated with Global Increases in Atmospheric Carbon Dioxide: A Preliminary Report*. Canadian Technical Report of Fisheries and Aquatic Sciences, 1652, 52 pp.

Frank, K.T., R.I. Perry, and K.F. Drinkwater, 1990: Predicted response of northwest Atlantic invertebrate and fish stocks to CO_2-induced climate change. *Trans. of the Amer. Fish. Soc.*, **119**, 353-365.

Freeman, B.L. and S.C. Turner, 1977: *Biological and Fisheries Data on Tilefish, Lophlatilus chamaeleonticeps Goode and Beans*. Tech. Series Report 5, Sandy Hook Laboratory, Highlands, NJ, 41 pp.

Garcia, S.M. and C.H. Newton, 1994: Responsible fisheries—an overview of FAO policy developments (1945 - 1994). *Mar. Pol. Bul.*, **29(N6-12)**, 528-536.

Glantz, M.H. (ed.), 1992: *Climate Variability, Climate Change and Fisheries*. Cambridge University Press, Cambridge, UK, 450 pp.

Glantz, M.H., R. Katz, and M. Krenz (eds.), 1987: *The Societal Impacts Associated with the 1982-83 Worldwide Climate Anomalies*. National Center for Atmospheric Research, Boulder, CO, 105 pp.

Glantz, M.H., M.F. Price, and M.E. Krenz, 1990: *Report of the Workshop "On Assessing Winners and Losers in the Context of Global Warming."* St. Julians, Malta, 18-21 June 1990. National Center for Atmospheric Research, Boulder, CO, 44 pp.

Glantz, M.H. and L.E. Feingold (eds.), 1990: *Climate Variability, Climate Change and Fisheries*. National Center for Atmospheric Research, Boulder, CO, 139 pp.

Hardy, C.J., 1967: *The Fecundity of the Brown Trout (Salmo trutta Linn.) from the Six Canterbury Streams*. New Zealand Mar. Dept. Fisheries Tech. Rep. 22, Wellington, New Zealand, 14 pp.

Hart, T.E. and F. Perrini, 1984: Analysis of U.S. commercial fishing vessel losses, 1970-1982. Presented at the International Conference on Design, Construction, and Operation of the Commercial Fishing Vessels, Melbourne, FL, 10-12 May 1984.

Healey, M.C., 1990: Implications of climate change for fishery management policy. *Trans. Amer. Fish. Soc.*, **119**, 366-373.

Hempel, G., 1978: North Sea fish stocks: recent changes and their causes. *Rapp. Proc.-v. Reun. Cons. int. Expl. Mer.*, **173**, 145-167.

Héral, M., C. Bacher, and J.-M. Deslous-Paoli, 1989: La capacite biotique des bassins conchylicoles. In: *L'homme et les Ressources Halieutiques. Essai sur l'Usage d'une Ressource Commune Renouvelable* [Troadec, J.-P. (ed.)]. IFREMER, Paris, France, pp. 225-259.

Hill, D.K. and J.J. Magnuson, 1990: Potential effects of global climate warming on the growth and prey consumption of Great Lakes fish. *Trans. Amer. Fish. Soc.*, **119**, 265-275.

Holmes, J.A., 1990: Sea lamprey as an early responder to climate change in the Great Lakes basin. *Trans. Amer. Fish. Soc.*, **119**, 292-300.

Hunter, J.R., J.H. Taylor, and H.G. Moser, 1979: The effect of ultraviolet radiation on eggs and larvae of northern anchovy and Pacific mackerel. *Photochemistry and Photobiology*, **29**, 325-338.

Hunter, J.R., S.E. Kaupp, and J.H. Taylor, 1981: Effects of solar and artificial UV-B radiation on larval northern anchovy, Engraulis mordax. *Photochemistry and Photobiology*, **34**, 477-486.

Hunter, J.R., S.E. Kaupp, and J.H. Taylor, 1982: Assessment of effects of UV radiation on marine fish larvae. In: *The Role of Solar Ultraviolet Radiation in Marine Ecosystems* [Calkins, J. (ed.)]. Plenum Press, New York, NY, 724 pp.

IGBP, 1990: *The IGBP: A Study of Global Change, the Initial Core Projects, 12*. The International Council of Scientific Unions, Graphics Systems AB, Stockholm, Sweden, 217 pp.

IOC, 1993: The case for GOOS. In: *Report of the IOC Blue Ribbon Panel for a Global Ocean Observing System (GOOS)*. IOC/INF-915 Corr., Paris, 23 February, 1993, SC-93/WS3.

IPCC, Working Group II, 1990: *Climate Change, The IPCC Impacts Assessment*. Australian Government Publishing Service, Canberra, Australia, 268 pp.

IPCC, Working Group III, 1990: *Policymakers Summary of the Formulation of Response Strategies*. WMO/UNEP, Geneva, Switzerland.

IPCC, 1992: *1992 IPCC Supplement*. WMO/UNEP, Geneva, Switzerland, 200 pp.

Johnson, T.B. and D.O. Evans, 1990: Size-dependent winter mortality of young-of-the-year white perch: climate warming and invasion of the Laurentian Great Lakes. *Trans. Amer. Fish. Soc.*, **119**, 301-313.

Kaczynski, V., 1983a: Distant water fisheries and the 200 mile economic zone. *Law of the Sea Institute Occasional*, 34, Hawaii University, Honolulu, HI.

Kaczynski, V., 1983b: *Joint Fishery Ventures and Three Ocean Powers*. Report for the Office of Research and Long-Term Assessments, U.S. Department of State, Washington, DC, 85 pp.

Kaufman, L., 1992: Catastrophic change in species-rich freshwater ecosystems: the lessons of Lake Victoria. *Bioscience*, **42(11)**, 846-858.

Kawasaki, T., 1980: Fundamental relations among the selections of life history in the marine teleost. *Bull. Jap. Soc. Scient. Fish.*, **46(3)**, 289-293.

Kawasaki, T., 1983: Why do some pelagic fishes have wide fluctuations in their numbers? Biological basis of fluctuations from the viewpoint of evolutionary ecology. In: *Proc. Expert Consultation on the Changes in Abundance and Species Composition of Neritic Fish Resources, 4 April 18-29, 1983, San Jose, Costa Rica* [Sharp, G.D. and J. Csirke (eds.)]. FAO Fish. Rep. 291, pp. 1065-1080.

Kawasaki, T., 1985: Fisheries. In: *Climate Impact Assessment, Studies of the Interaction of Climate and Society* [Kates, R.W., J.H. Asubel, and M. Berberian (eds.)]. John Wiley & Sons Ltd., New York, NY, pp. 131-153.

Kawasaki, T., 1991: Effects of global climate change on marine ecosystems and fisheries. In: *Climate Change: Science, Impacts, and Policy: Proceedings of the Second World Climate Conference* [Jager, J. and H.L. Ferguson (eds.)]. WMO/UNEP, Cambridge University Press, Cambridge, UK, 578 pp.

Kawasaki, T., S. Tanaka, Y. Toba, and A. Taniguchi (eds.), 1991: *Long-Term Variability of Pelagic Fish Populations and Their Environment*. Pergamon Press, Tokyo, Japan, 402 pp.

Kennedy, V.S., 1990: Anticipated effects of climate change on estuarine and coastal fisheries. *Fisheries*, **15(6)**, 16-24.

Kovtsova, M.V., M.S. Shevelev, and N.A. Yaragina, 1991: Bottom fish stocks state and prospects of their fishery in the Barents Sea. In: *PINRO Complex Fisheries Research in the North Basin: Results and Outlooks* [Sorokin, A.L. (ed.)]. Selected papers, PINRO, Murmansk, pp. 145-165 (in Russian).

Lasker, R., 1981: The role of a stable ocean in larval fish survival and subsequent recruitment. In: *Marine Fish Larvae: Morphology, Ecology and Relation to Fisheries* [Lasker, R. (ed.)]. University of Washington Press, Seattle, WA, pp. 80-89.

Laureti, E. (comp.), 1992: *Fish and Fishery Products. World Apparent Consumption Statistics Based on Food Balance Sheets (1961-1990)*. FAO Fish. Circ. 821(Rev. 2), FAO, Rome, Italy, 477 pp.

Leggett, W.C. and J.E. Carscadden, 1978: Latitudinal variation in reproductive characteristics of American shad (Alosa sapidissima): evidence for population specific life history strategies in fish. *J. Fish. Res. Board Can.*, **35**, 1469-1478.

Le Reste, L., 1992: Pluviométrie et captures de crevettes Penaeus notialis dans l'estuaire de la Casamance (Sénégal) entre 1962 et 1984. *Aquat. Liv. Resour.*, **5**, 233-248.

Leroy, C. and D. Binet, 1986: Anomalies thermiques dans l'Atlantique tropical: conséquences possibles sur le recrutement du germon (Thunnus alalunga). *ICCAT. Rec. Doc. Sci.*, **26(2)**, SCRS/86/46, 248-253.

Lin, P. and H.A. Regier, 1995: Use of Arrhenius models to describe temperature dependence of organismal rates in fish. In: *Proceedings of the Symposium on Climate Change and Northern Fish Populations, Victoria, BC, October 13-16, 1992* [Beamish, R.J. (ed.)]. Special publication of Canadian fisheries and aquatic sciences. National Research Council of Canada, Ottawa, Canada, pp. 211-225.

Lluch-Belda, D., F. Magall, and R.A. Schwartzlose, 1986: *Large Fluctuations in the Sardine Fishery in the Gulf of California: Possible Causes*. CalCOFI Reps., **27**, 136-140.

Lluch-Belda, D., S. Hernendez Velazquez, and R.A. Schwartzlose, 1991: Hypothetical model for the fluctuations of the California sardine populations. In: *Long-Term Variability of Pelagic Fish Populations and Their Environment* [Kawasaki *et al.* (eds.)]. Pergamon Press, New York, NY, pp. 293-300.

Lluch-Belda, D., R.A. Schwartzlose, R. Serra, R. Parrish, T. Kawasaki, D. Hedgecock, and R. Crawford, 1992: Sardine and anchovy fluctuations of abundance in four regions of the world oceans: workshop report. *Fisheries Oceanography*, **1(4)**, 339-347.

Lodge, D.M., 1993: Species invasions and deletions: community effects and responses to climate and habitat change. In: *Biotic Interactions and Global Change* [Kareiva, P.M., J.G. Kingsolver, and R.B. Huey (eds.)]. Sinnauer Assoc., Inc., Sunderland, MA, 559 pp.

Lowe-McConnell, R.A., 1987: *Ecological Studies in Tropical Fish Communities*. Cambridge University Press, Cambridge, MA, 369 pp.

Luka, G.I., G.P. Nizovtsev, M.S. Shevelev, and V.M. Borisov, 1991: Commercial fish stocks of the Barents Sea: causes of depression and prospects of recovery. *Rybnoye khozyaistvo*, **9**, 22-25 (in Russian).

Magnuson, J.J., L.B. Crowder, and P.A. Medvick, 1979: Temperature as an ecological resource. *Amer. Zool.*, **19**, 331-343.

Magnuson, J.J., J.D. Meisner, and D.K. Hill, 1990: Potential changes in the thermal habitat of Great Lakes fish after global climate warming. *Trans. Amer. Fish. Soc.*, **119**, 254-264.

Mahé, G. 1993: *Les Ecoulements Fluviaux sur la Façade Atlantique de l'Afrique. Etude des Eléments du Bilan Hydrique et Variabilité Interannuelle, Anallyse des Situations Hydroclimatiques Moyennes et Extrêmes*. Etudes et Thèses ORSTOM, Paris, France, 438 pp.

Mandrak, N.E., 1989: Potential invasion of the Great Lakes by fish species associated with climate warming. *J. Gr. L. Res.*, **15**, 306-316.

Mann, K.H., 1992: Physical influences on biological processes: how important are they? Benguela trophic functioning [Payne, A.L., K.H. Mann, and R. Kilborn (eds.)]. *S. Afr. J. Mar. Sci.*,**12**, 107-121.

Mann, K.H., 1993: Physical oceanography, food chains and fish stocks: a review. *ICES J. Mar. Sci.*, **50**, 105-119.

Mann, K.H. and K.F. Drinkwater, 1994: Environmental influences on fish and shellfish production in the Northwest Atlantic. *Envir. Rev.*, **2**, 16-32.

Mann, K.H. and J.R.N. Lazier, 1991: *Dynamics of Marine Ecosystems Biological Physical Interactions in the Oceans*. Blackwell Scientific Publications, Boston, MA, 466 pp.

Mathews, W.J. and E.G. Zimmerman, 1990: Potential effects of global warming on native fishes of the southern great plains and the southwest. *Fish.*, **15(6)**, 26-32.

McCauley, W.R. and T. Beitinger, 1992: Predicted effects of climate warming on the commercial culture of the channel catfish, Ictalurus punctatus. *GeoJ.*, **28(1)**, 61-66.

McDowall, R.M., 1992: Global climate change and fish and fisheries: what might happen in a temperate oceanic archipelago like New Zealand. *GeoJ.*, **28(1)**, 29-37.

Mee, L., 1992: The Black Sea in crisis: a need for concerted international action. *Ambio.*, **21(4)**, 1278 -1286.

Meisner, J.D., 1990: Potential loss of thermal habitat for brook trout, due to climatic warming, in two southern Ontario streams. *Trans. Amer. Fish. Soc.*, **119**, 282-291.

Meisner, J.D., J.L. Goodie, H.A. Regier, B.J. Shuter, and W.J. Christie, 1987: An assessment of the effect of climate warming on Great Lakes Basin fish. *J. Gr. L. Res.*, **13**, 340-352.

Meisner, J.D. and B.J. Shuter, 1992: Assessing potential effects of global climate change on tropical freshwater fishes. *GeoJ.*, **28(1)**, 21-27

Miller, K.A. and D.L. Fluharty, 1992: El Niño and variability in the northeastern Pacific Salmon fishery: implications for coping with climate change. In: *Climate Variability, Climate Change and Fisheries* [Glantz, M.H. (ed.)]. Cambridge University Press, Cambridge, UK, pp. 49-88.

Mills, E.L., J.H. Leach, J.T. Carlton, and C.L. Secor, 1994: Exotic species and the integrity of the Great Lakes. *Biosci.*, **44(1)**, 666-676.

Murawski, S.A., 1993: Climate change and marine fish distributions: forecasting from historical analogy. *Trans. Amer. Fish. Soc.*, **122**, 647-658.

NCC, 1989: *Fish Farming and the Safeguard of the Natural Marine Environment of Scotland*. Nature Conservancy Council, Edinburgh, Scotland, 236 pp.

Neher, P.A., R. Arnasson, and N. Mollett (eds.), 1989: Rights based fishing. In: *Proc. NATO Adv. Res. Workshop on Scientific Foundations for Rights Based Fishing, Reykjavik, Iceland, June 27–July, 1988*. Kluwer Academic Publishers, Dordrecht, Netherlands, 541 pp.

New Zealand, 1990: *Climate Change: Impacts on New Zealand*. Ministry for the Environment, Wellington, New Zealand, 244 pp.

Ntiba, M.J., 1990: *The Biology and Ecology of the Long Rough Dab Hippoglossoides platessoides (Fabricius, 1780) in the North Sea*. Ph.D. thesis, University of East Anglia, UK, 156 pp.

Ntiba, M.J. and D. Harding, 1993: The food and the feeding habits of the long rough dab Hippoglossoides platessoides (Fabricius, 1780), in the North Sea. *Neth. J. Sea Res.*, **31**, 189-199.

Odum, G., 1986: *Ecology, 2* [translated by Vilenkin, B.Ya.; Sokolov, V.E. (ed.)]. Moscow, Mir, 376 pp.

Ogutu-Ohwayo, R., 1990a: The decline of the native fishes of lakes Victoria and Kyoga (East Africa) and the impact of introduced species, especially the Nile perch, Lates niloticus, and the Nile tilapia, Oreochromis niloticus. *Environmental Biology of Fishes,* **27(2)**, 81-96.

Ogutu-Ohwayo, R. 1990b: The reduction in fish species diversity in lakes Victoria and Kyoga (East Africa) following human exploitation and the introduction of non-native fishes. *Journal of Fish Biology,* **37(supplement A)**, 207-208.

Owen, R.W., 1981: Patterning of flow and organisms in the larval anchovy environment. In: *Report and Documentation of the Workshop on the Effects of Environmental Variation on the Survival of Larval Pelagic Fishes* [Sharp, G.D. (conv. ed.)]. IOC Workshop Rep. Ser., Unesco, Paris, **28**, 67-200,.

Owen, R.W., 1989: Microscale and fine scale variations of small plankton in coastal and pelagic environments. *J. Mar. Res.*, **47**, 197-240.

Park, R.A., 1991: Testimony before the Subcommittee on Health and Environment, U.S. House of Representatives, *Global Climate Change and Greenhouse Emissions*, SN 102-54, pp. 171-182.

Park, R.A., M.S. Trehan, P.W. Mausel, and R.C. Howe, 1989: The effects of sea level rise on U.S. coastal wetlands. In: *The Potential Effects of Global Climate Change on the United States: Appendix B - Sea Level Rise* [Smith, J.B. and D.A. Tirpak (eds.)]. EPA-230-05-89-052, Washington, DC, U.S. Environmental Protection Agency, pp. 1-1 to 1-55.

Park, R.A., J.K. Lee, and D. Canning, 1993: Potential effects of sea level rise on Puget Sound wetlands. *Geocarto Int.,* **8(4)**, 99-110.

Parrish, R.H. and A.D. MacCall, 1978: *Climatic Variations and Exploitation in the Pacific Mackerel Fishery.* CA Dept. Fish Game Fish. Bull., Sacramento, CA, 167 pp.

Parsons, L.S., 1993: *Management of Marine Fisheries in Canada.* National Research Council of Canada and Department of Fisheries and Oceans, Ottawa, Canada, 702 pp.

Pauly, D. and I. Tsukayama, 1987: *The Peruvian Anchoveta and Its Upwelling Ecosystem: Three Decades of Change.* ICLARM Studies and Reviews 15, Instituto del Mar del Peru (IMARPE), Callao, Peru; Deutsche Gesellschaft fur Technische Zusammenarbeit (GTZ), GmbH, Eschborn, Federal Republic of Germany; and International Center for Living Aquatic Resources Management (ICLARM), Manila, Philippines, 351 pp.

Pauly, D., M.L. Palomares, and F.C. Gayanilo, 1987: VPA estimates of the monthly population length composition, recruitment, mortality, biomass and related statistics of Peruvian anchoveta, 1953 to 1981. In: *The Peruvian Anchoveta and Its Upwelling Ecosystem: Three Decades of Change.* ICLARM Studies and Reviews 15, Instituto del Mar del Peru (IMARPE), Callao, Peru; Deutsche Gesellschaft fur Technische Zusammenarbeit (GTZ), GmbH, Eschborn, Federal Republic of Germany; and International Center for Living Aquatic Resources Management (ICLARM), Manila, Philippines, pp. 142-178.

Pauly, D., D.P. Muck, J. Mendo, and I. Tsukayama, 1989: *The Peruvian Upwelling Ecosystem: Dynamics and Interactions.* ICLARM Conference Proceedings 18, Instituto del Mar del Peru (IMARPE), Callao, Peru; Deutsche Gesellschaft für Technische Zusammenarbeit (GTZ), GmbH, Eschbom, Federal Republic of Germany; and International Center for Living Aquatic Resources Management, Manila, Philippines, 438 pp.

Payne, A.I.L., J.A. Gulland, and K.H. Brink (eds.), 1987: The Benguela and comparable ecosystems. *S. African J. Mar. Sci.*, **5,** 957.

Peterman, R.M., 1991: Density-dependent marine processes in North Pacific salmonids: lessons for experimental designs of large-scale manipulations of fish stocks. In: *Ecology and Management Aspects of Extensive Mariculture: A Symposium Held in Nantes* [Lockwood, S.(ed.)]. ICES Mar. Sci. Symp., pp. 69-77.

Peterman, R.M. and M.J. Bradford, 1987: Wind speed and mortality rate of a marine fish, the northern anchovy (Engraulis mordax). *Science,* **235,** 354-356.

Polovina, J.J., G.T. Mitchum, N.E. Graham, M.G. Craig, E.E. Demartini, and E.N. Flint, 1994: Physical and biological consequences of a climate event in the central North Pacific. *Fish. Oceano.*, **3(1)**, 15-21.

Polovina, J.J., G.T. Mitchum, and G.T. Evans, 1995: Decadal and basin-scale variation in mixed layer depth and the impact on biological production in the Central and North Pacific, 1960-88. *Deep Sea Research* (in press).

Pope, J.A., D.H. Mills, and W.M. Shaerev, 1961: The fecundity of the Atlantic salmon (Salmo salar). *Salm. Fish. Res.*, **26**, 1-12.

Quéro J.-C. and R. Emmonet, 1993: Disparition ou raréfaction d'espéces marines au large d'Arcachon. Actes du IIIé Colloque International "Océanographie du Golfe de Gascogne," *Arcachon-7,8,9 Avril 1992* [Sorbe, J.-C. and J.-M. Jouanneau (eds.)]. Universite de Bordeaux, I, pp. 221-225.

Regier, H.A. and G.L. Baskerville, 1986: Sustainable redevelopment of regional ecosystems degraded by exploitive development. In: *Sustainable Development of the Biosphere* [Clark, W.C. and R.E. Munn (eds.)]. Cambridge University Press, Cambridge, UK, pp. 75-101,.

Regier, H.A., J.A. Holmes, and D. Pauly, 1990: Influence of temperature changes on aquatic ecosystems: an interpretation of empirical data. *Trans. Amer. Fish. Soc.*, **119(2)**, 374-389.

Regier, H.A., J.J. Magnuson, C.C. Coutant, 1990: Introduction to proceedings: symposium on effects of climate change on fish. *Trans. Amer. Fish. Soc.*, **119(2)**, 173-175.

Regier, H.A. and J.A. Holmes, 1991: Impacts of climate change on freshwater fisheries of the Great Plains. In: *Symposium on the Impacts of Climate Change and Variability in the Great Plains* [Wall, G. (ed.)]. Department of Geography Publication Series, Occasional Paper 12, University of Waterloo, Waterloo, Ontario, pp. 301-308.

Rothschild, B.J., 1991: Multispecies interactions on Georges Bank. *ICES Mar. Sci. Symp.*, **193**, 86-92.

Rothschild, B.J. and T.R. Osborn, 1988: Small scale turbulence and plankton contact rates. *J. Plankt. Res.*, **10**, 465-474.

Rozengurt, M.H., 1992: Alteration of freshwater inflows. In: *Stemming the Tide of Coastal Fish Habitat Loss* [Stroud, R.H. (ed.)]. Marine Recreational Fisheries Symposium, 14, National Coalition for Marine Conservation, Savannnah, GA, pp. 73-80.

Russell, S. and M. Yonge, 1975: *The Seas: An Introduction to the Study of Life in the Sea.* Butker & Tanner, London, UK, 283 pp.

Sampayo, M.A., 1989: Red tides off the Portuguese coast. In: *Red Tides: Biology, Environmental Science, and Toxicology* [Okaichi, T., D.M. Anderson, and T. Nemoto (eds.)]. Elsevier Science Publishing Co. Inc., New York, NY, pp. 89-92,

Schlesinger, D.A. and H.A. Regier, 1982: Climatic and morphoedaphic indices of fish yields from natural lakes. *Trans. Amer. Fish. Soc.,* **111,** 141-150.

Schlesinger, D.A. and A.M. McCombie, 1983: *An Evaluation of Climatic Morphoedaphic and Effort Data as Predictors of Yields from Ontario Sport Fisheries.* Ontario Fisheries Technical Report Series, 10, Ministry of Natural Resources, Maple, Ontario, Canada, 14 pp.

Semtner, A.J. and R.M. Chervin, 1988: A simulation of the global ocean circulation with resolved eddies. *J.Geophys. Res.*, **93(C12)**, 15502-15522 and 15767-15775.

Sharp, G.D., 1986: Fish populations and fisheries: their perturbations, natural and man-induced. In: *Ecosystems of the World. 28. Ecosystems of Continental Shelves* [Postma, H. and J.J. Zijlstra (eds.)]. Elsevier, Amsterdam, Netherlands, 421 pp.

Sharp, G.D., 1987: Climate and fisheries: cause and effect or managing the long and the short of it all. In: *The Benguela and Comparable Ecosystems* [Payne, A., J. Gulland, and K. Brink (eds.)]. *S. Afr. J. Mar. Sci.*, **5**, 811-838.

Sharp, G.D., 1989: Climate and fisheries: cause and effect. In: *Long-Term Variability of Pelagic Fish Populations and Their Environment: Symposium Proceedings* [Kawasaki, T., S. Tanaka, Y. Toba, and A. Taniguchi (eds.)]. Pergamon Press, New York, NY, pp. 239-258.

Sharp, G.D., 1991: Climate and fisheries: cause and effect—a system review. In: *Long-Term Variability of Pelagic Fish Populations and Their Environment* [Kawasaki, T., S. Tanaka, Y. Toba, and A. Taniguchi (eds.)]. Pergamon, Oxford, UK, pp. 239-258.

Sharp, G.D., 1992a: *Pieces of the Puzzle: Anecdotes, Time-Series and Insights.* Proceedings Benguela Trophic Functioning Symposium, September 1991, Cape Town. *S. Afr. J. Mar. Sci.*, **12**, 1079-1092.

Sharp, G.D., 1992b: Fishery catch records, ENSO, and longer term climate change as inferred from fish remains from marine sediments. In: *Paleoclimatology of El Niño—Southern Oscillation* [Diaz, H. and V. Markgraf (eds.)]. Cambridge University Press, Cambridge, UK, pp. 379-417.

Sharp, G.D., 1994: Fishery catch records, ENSO, and longer term climate change as inferred from fish remains in marine sediments (manuscript).

Sharp, G.D. and J. Csirke (eds.), 1983: *Proceedings of the Expert Consultation to Examine the Changes in Abundance and Species Composition of Neritic Fish Resources, 18-29 April 1983, San Jose, Costa Rica.* FAO Fish. Rep. Ser. Rome, **291(2-3)**, 1294.

Sharp, G.D. and D.R. McLain, 1993: Fisheries, El Niño-Southern Oscillation and upper ocean temperature records: an eastern Pacific example. *Oceanog.*, **5(3)**, 163-168.

Sherman, K. and B.D. Gold, 1990: Perspectives. In: *Large Marine Ecosystems: Patterns, Processes and Yields* [Sherman, K., L.M. Alexander, and B.D. Gold (eds.)]. AAAS Press, Washington, DC, pp. vii-xi.

Sherman, K., N. Jaworski, and T. Smayda, 1992: *The Northeast Shelf Ecosystem: Stress, Mitigation, and Sustainability, 12-15 August 1991 Symposium Summary.* NOAA Tech. Mem. NMFS-F/NEC-94, U.S. Department of Commerce, Washington, DC, 30 pp.

Shuter, B.J. and J.R. Post, 1990: Climate, population viability, and the zoogeography of temperate fishes. *Trans. Amer. Fish. Soc.*, **119**, 314-336.

Shuter, B.J. and J.D. Meisner, 1992: Tools for assessing the impact of climate change on freshwater fish populations. *GeoJ.*, **28**, 7-20.

Siegfried, W.R., R.J.M. Crawford, L.V. Shannon, D.E. Pollock, A.I.L Payne, R.G. Krohn, 1990: Scenarios for global-warming induced change in the open-ocean environment off the west coast of southern Africa. *S. Afr. J. Sci.*, **86**, 281-285.

Sinclair, M.H., 1988: *Marine Populations. An Essay on Population Regulation and Speciation.* Washington Sea Grant Program, University of Washington Press, Seattle, WA, 252 pp.

Sissenwine, M.P. and A.A. Rosenberg, 1993: Marine fisheries at a critical juncture. *Fisheries (AFS)*, **18(10)**, 6-13.

Skud, B.E., 1982: Dominance in fishes: the relation between environment and abundance. *Sci.*, **216**, 144-149.

Smit, B. (ed.), 1993: *Adaptability to Climatic Variability and Change: Report of the Task Force on Climate Adaptation, Canadian Climate Program.* Occasional Paper 19, Department of Geography, University of Guelph, Guelph, Ontario, 53 pp.

Southward, A.J., 1980: The western English Channel: an inconstant ecosystem? *Nature*, **285**, 361-366.

Ssetongo, G.W. and R.L. Welcomme, 1985: Past history and current trends in the fisheries of Lake Victoria. In: *Report of the Third Session of the Sub-Committee for the Development and Management of the Fisheries of Lake Victoria, Jinja, Uganda, October 1984.* FAO Committee for Inland Fisheries of Africa, FAO Fisheries Report, pp. 123-138.

Stokes, T.K., J.M. McGlade, and R. Law. 1993: *The Exploitation of Evolving Resources.* Proceedings of an International Conference held at Jülich, Germany, 3–5 September, 1991, Springer Verlag, Berlin, Germany, 264 pp.

Swenson, E.M. and W.S. Chuang, 1983: Tidal and subtidal water volume exchange in an estuarine system. *Estuar., Coast., Shelf Science,* **16**, 229-240.

Talbot, Y. *et al.*, 1994: Presentation made to workshop sponsored by the the U.S. Marine Mammal Commission, March 1994, Airlie, VA.

Troadec, J.-P., 1989a: Elements pour une autre strategie. In: *L'homme et les ressources halieutiques, Essai sur l'usage d'une ressource commune renouvelable* [Troadec, J.-P. (sous la dir.)]. IFREMER, Paris, France, pp. 747-795.

Troadec, J.-P., 1989b: *Report of the Climate and Fisheries Workshop.* Organized by M.H. Glantz, National Center for Atmospheric Research, Boulder, CO, 7-8, 16.

Troadec, J.-P., 1994: Scarcity, property allocation, and climate change In: *The Role of Regional Organizations in the Context of Climate Change* [Glantz, M.H. (ed.)]. NATO ASI Series, **114**, 128-135. Springer-Verlag, Berlin, Heidelberg, Germany.

Ueber, E. and A. MacCall, 1992: The rise and fall of the California sardine empire. In: *Climate Variability, Climate Change and Fisheries* [Glantz, M.H. (ed.)]. Cambridge University Press, Cambridge, UK, pp. 31-48.

Valdivia, G.J.E., 1978: The anchoveta and El Nino. *Rapp. P-V. Reun. Cons. Int. Explor. Mer.*, **173**, 196-202.

Valiela, I. and J.E. Costa, 1988: Eutrophication of Buttermilk Bay, a Cape Cod coastal embayment: concentrations of nutrients and watershed nutrient budgets. *Environ. Manage.*, **12**, 539-553.

Venrick, E.L., J.A. McGowan, D.R. Cayan, and T.L. Hayward, 1987: Climate and chlorophyll a: long-term trends in the central north Pacific Ocean. *Science*, **238**, 70-72.

Vodovskaya, V.V., L.M. Shubina, and E.M. Konoplev, 1978: The present state of the Caspian herring stocks and prospects of their exploitation. *Trudy VNIRO*, **131**, 115-123 (in Russian).

Weber, A.P., 1993: *Abandoned Seas: Reversing the Decline of the Oceans*, World Watch Paper 116, World Watch Institute, Washington, DC, 66 pp.

Weber, A.P., 1994: *Net Loss: Fish, Jobs, and the Marine Environment.* World Watch Paper 120, World Watch Institute, Washington, DC, 76 pp.

Welch, D.W. and R.J. Beamish, 1994: *Possible Approaches for Co-operative Pacific Salmon Research.* NPAFC Doc. 95, Dept. of Fisheries and Oceans, Biological Sciences Branch, Pacific Biological Station, Nanaimo, BC, Canada, V9R 5K67, 8 pp.

Welch, D.W., A.I. Chigirinsky, and Y. Ishida, 1995: Upper thermal limits on the late spring distribution of Pacific salmon (Oncorhynchus spp.) in the Northeast Pacific Ocean. *Can. J. Fish. Aquat. Sci.*, **52(5)** (in press).

Welcomme, R.L., 1976: Some general and theoretical considerations on the fish yield of African rivers. *J. Fish. Biol.*, **8**, 351-364.

Welcomme, R.L., 1985: *River Fisheries.* FAO Fish. Tech. Pap. 262, FAO, Rome, Italy, 330 pp.

Wooster, W., 1992: King crab dethroned. *Climate Variability, Climate Change and Fisheries* [Glantz, M.H. (ed.)]. Cambridge University Press, Cambridge, UK, pp. 15-30.

World Bank, 1992: *United Nations Development Programme, Commission of the European Communities and the Food and Agriculture Organization Fisheries Research.* Policy and Research Service 19, The World Bank, Washington, DC, 103 pp.

Wright, D.G., R.M. Hendry, J.W. Loder, and F.W. Dobson, 1986: *Oceanic Changes Associated with Global Increase in Atmospheric Carbon Dioxide: A Preliminary Report for the Atlantic Coast of Canada.* Can. Tech. Rep. Fish. Aquat. Sci. 1426, Dartmouth, Nova Scotia, 78 pp.

Wu, B. and J. Qiu, 1993: Yellow Sea fisheries: from single and multi-species management towards ecosystems management. *J. Oceangr. Huanghai and Bohai Seas*, **11(1)**, 13-17.

Yañez, E., 1991: Relationships between environmental changes and fluctuating major pelagic resources exploited in Chile (1950-1988). In: *Long-Term Variability of Pelagic Fish Populations and Their Environment* [Kawasaki, T., S. Tanaka, Y. Toba, and A. Taniguchi (eds.)]. Pergamon Press, Tokyo, Japan, pp. 301-309.

Yonesawa, K., 1981: Japanese North Pacific fishery at the crossroads. Paper presented to the 21st Annual Meeting of the Law of the Sea Institute, Honolulu, August 1981.

Zilanov, V.K., G.I. Luka, A.A. Glukhov, A.K. Chumakov, Yu.A. Bochkov, and S.S. Drobysheva, 1991: The present state of the Barents Sea ecosystem. In: *Problems of Ecology of Hydrobionts. Collected Papers* [Moiseev, P.A. (ed.)]. VNIIPRKH, Moscow, Russia, pp. 6-14 (in Russian).

Zimmerman, T., R. Minello, E. Klima, and J. Nance, 1991: Effects of accelerated sea-level rise on coastal secondary production. In: *Coastal Wetlands* [Bolton, H.S. and O.T. Magoon (eds.)]. Coastal Zone '91 Conference, Long Beach, CA, Publ. Am. Soc. Civil Engineers, New York, NY, pp. 110-124.

17

Financial Services

ANDREW F. DLUGOLECKI, UK

Lead Authors:
K.M. Clark, USA; F. Knecht, Switzerland; D. McCaulay, Jamaica; J.P. Palutikof, UK; W. Yambi, Tanzania

CONTENTS

EXECUTIVE SUMMARY

Within financial services, the property insurance industry is most likely to be directly affected by climate change, since it is already vulnerable to variability in extreme weather events. The cost of weather-related disasters to insurers has risen rapidly since 1960. For example, the annual insured cost of major windstorms worldwide increased progressively from $0.5 billion in the 1960s to more than $11 billion (constant 1990 dollars) in the early 1990s. This trend has led to restrictions in coverage or steep price increases. Where insurance is unavailable or unaffordable, there are consequences for other economic activities, as well as for consumers and government. New enterprises may not start without insurance. Banks may be exposed to losses where financial transactions are backed by property.

There are several reasons for the escalation in the cost of severe weather. Developed countries have become wealthier. Many more people now live in coastal areas with costly infrastructures. Personal goods and business processes are generally more vulnerable to water damage. The built environment also contributes through inappropriate or incorrect design and construction. The insurance industry has compounded matters by extending the basis of coverage. It is a common perception in the insurance industry that there is a trend toward an increased frequency and severity of extreme climate events. The meteorological literature fails to substantiate this in the context of long-term change, though there may have been a shift within the limits of natural variability.

There is medium confidence that climate change will adversely affect property insurance. More frequent intensive rain events and sea-level rise are material to both storm and flood damage. The climatological literature does not supply sufficient information about future storms in relation to population centers or "combined events," like windstorms associated with heavy rain and storm surges. This could critically alter the pattern of insurance claims. Further, the industry is vulnerable to shifts in the variability as well as the scale of extreme events, and this is still obscure. Meteorological records are too short to calculate return periods of extreme events under current or changing climate conditions. The property insurance sector is very interested in the probability of events two to three years ahead, rather than the timescales associated with climate change. Research on teleconnections where there is a substantial lag between the event and its precursor offers the possibility of improved forward planning for the industry.

Through its experience, the insurance sector can help authorities to improve the response to property damage from extreme events. Control of land use, particularly in floodplains and coastal zones, is essential to prevent values at risk increasing. Information about hazards can be mapped and appropriate physical protections put in place. Necessary improvements to construction design and processes can be identified and incorporated into new building and, if possible, also retrofitted. Many countries now accept that it is not appropriate to account for insurance against extreme events on an annual basis, since funds need to be accumulated over a long period to meet infrequent but severe losses. Where it is not possible to provide private insurance, the industry can still assist with disaster-recovery services. This generally will prove to be a very efficient method compared to other *ad hoc* approaches.

The implications of climate change for financial services outside property insurance cannot be stated with any confidence. Few institutions, if any, were aware of the potential implications when this assessment began. Yet changes in human health may affect the life insurance and pension industries. Banking may be vulnerable to repercussions from property damage. Returns on long-term investments and capital projects may be affected by mitigation measures that alter the economics of whole industries—for example, shifting from carbon fuels to renewable resources. The economies of selected regions, such as coastal zones and islands, may be disadvantaged. Climate change might affect client behavior, or even alter the available portfolio of clients through the changing economics of different industries. Outsiders will be keen to know how the financial sector deploys its funds in the light of climate change. There is a need for increased recognition by the financial sector that climate change is an issue that could affect its future at the national and international level. This could require institutional change.

17.1. Introduction

This chapter commences with a brief overview of financial services. Property insurance is reviewed in more detail as the financial area most vulnerable to direct climatic effects, through multiple claims arising from extreme weather events. The study considers which extreme events are important for property and our limited knowledge of their past and future patterns. Because the future position is unclear, and the previous Assessment Report did not deal extensively with financial services, Section 17.5 discusses the recent upward trend in the cost of extreme events and the criticality of such shocks to the financial system. The techniques property insurers use to adapt to changes in risk also are described, before the chapter looks at how climate change will affect other financial services. Most likely it will be indirectly through the measures taken by other industries or policymakers to adapt to, or mitigate, climate change. Finally, the chapter summarizes cross-cutting implications for policymakers and indicates ways to make future assessments more effective. To the outsider, the discussion may seem biased toward property insurance. This reflects the reality of the direct impacts of climate and the available literature. The indirect impact of climate change outside property insurance is unclear and largely unresearched.

17.2. Financial Services

This section deals with external factors that affect the whole sector. The financial services sector falls into two broad clusters—insurance and banking, which will be described in Sections 17.3 and 17.7, respectively. It has evolved into a highly complex system that recycles "money" among other parts of the economy, including final consumption and capital investment. Governments participate to a varying degree through public debt issues and other investments, and the process operates on timescales from days to decades. The definition used here excludes foreign aid. Clearly, many countries have significant noncash economies; the discussion excludes those activities. Many of the costs of extreme events recorded by the financial sector reflect impacts in other sectors. Although this is in a sense "double-counting," the financial sector transmits these effects to other sectors and may indeed amplify them if it is unable to fulfill other services through lack of capital.

Projecting the impact of climate change strictly requires a view of future society at the relevant horizon. Given the unforeseen changes over the past 100 years, clearly such an attempt for the next 100 years would be highly inaccurate. The alternative adopted here is to identify current trends that may be relevant, without drawing up a coherent "worldview." Technological advances are accelerating, particularly in transportation, telecommunications, and information technology. This is leading to the globalization of society (Kennedy, 1993). Systems are becoming more complex and interconnected, with increases in the scale of operation and associated risk (Giarini and Stahel, 1994). Society is increasingly vulnerable to disruption because consumers are living on credit and businesses are adopting "just in time" techniques. As society becomes more dependent on electricity, any supply disruptions from extreme events become potentially more catastrophic. Economic development has been rapid, bringing with it wealth but also pollution. In many cases, development has occurred in vulnerable locations (see Chapter 12). With the progression from primary to tertiary activities, the impact of climate change on many areas is likely to be indirect and thus more difficult to judge, particularly because social change also has been rapid. Population has burgeoned in many countries, and standards of behavior have altered regarding fundamental issues like the family, crime, and work (Kennedy, 1993). Consumerism and environmentalism have developed strongly as education has spread. Much financial services activity is "driven" by the private automobile, financing its purchase and use. Therefore, mitigating climate change by altering auto use would affect financial services. On the other hand, current trends in auto use cannot continue without putting enormous stress on society (Northcott, 1991). The answer may be to exploit telecommunications as a substitute, but this is not certain.

Trends to deregulate financial services nationally and internationally are putting traditional distribution methods under pressure. Tele-selling will be an increasingly important influence. It is likely that there will be fewer independent operators, often integrated either horizontally or vertically to provide benefits of scale or synergy (Swiss Re, 1992; Muth, 1993). It is possible that clients will key their own information (e.g., through the Internet). Data retrieval will become easier as paper files disappear. Information will become critical for management purposes, not just for recordkeeping. More powerful computers will facilitate the analysis of internal and external databases for sales and underwriting (Dlugolecki, 1995).

Many economies have been plagued by enormous surges in property values, followed by stagnation or even collapse. This makes it difficult to assess assets and liabilities. A second factor that complicates the valuation of property is the generally high rate of inflation between 1960 and 1990. A further difficulty is fluctuating exchange rates in economies with unstable currencies. Such instabilities have led to high interest rates and periods of rapid investment gains. These have diverted attention to "cash flow" strategies (Swiss Re, 1991) and also have created the possibility of insolvency through poor investments (Denlea, 1994). Another area of pressure on financial services is the need to fund environmental claims—seen at its most energetic in the United States, where the Superfund legislation has created huge liabilities for the property-casualty industry (Nutter, 1994). The banking industry also is threatened by "lender liability" as a consequence of repossessing or taking control of polluted land or a polluting operation (Vaughan, 1994).

17.3. The Property Insurance Market

The insurance industry serves to protect other activities from the financial consequences of unexpected events, including natural hazards; therefore, it is directly exposed to climate change (Friedman, 1989; Roberts, 1990). Since 1987, fifteen extreme weather events have cost property insurers and reinsurers $50 billion (Leggett, 1994). In view of these impacts, Sections 17.3 through 17.6 consider property insurance and associated "interruption" coverage. "Property" is defined as buildings and their contents, business and personal goods, and plant and infrastructure; conveyances, livestock, crops, and humans are excluded.

The basic function of insurance is to transfer financial risk from an individual to a group; what is a disaster to the individual can often be managed by larger numbers. Typically, the individual proposes his risk to the underwriter, often with the assistance of an intermediary or broker. In some cases, the risk must be evaluated in great detail, and the insurer may stipulate certain actions (risk management) to reduce the likelihood and scope of potential loss. Finally, the insurer offers specific terms in a policy. The most important terms of the contract involve the price, or premium; any exclusions of events or property not covered by the contract; and limitations on the amount to be paid in a claim. Contracts of insurance are usually renewable. Usually the contract lasts for twelve months and is renegotiated with a view to changes in circumstances, particularly claims experience. Following acceptance of the risk, the insurer must then decide whether to share it with reinsurers. The reinsurer may further spread the risk by "retrocession" to other reinsurers, but this has become less common. The other key activity is settling claims. After a loss, the policyholder reports it to his insurer, who will decide how much investigation is necessary. A loss adjuster is often retained for this purpose. Claims recovery often provides the occasion to implement risk-improvement measures. Policies have become more comprehensive over time, culminating in "all risks" coverage, where the policy stipulates what is excluded. Other significant changes have been to allow for inflation in property values by indexation, and to set aside the effect of depreciation by using replacement values. Finally, policies often allow for economic costs such as alternative accommodation if a claim results in temporary relocation.

17.3.1. *Scale of Activities*

The scale of the industry is measured either by premiums written or capital available. In 1992, the worldwide non-life insurance

Box 17-1. Examples of Property Insurance and Weather Cover

Property insurance is used to varying extents by different nations. Often, consumers are reluctant to purchase sufficient coverage. Lack of capital, culture, and consumer attitude to risk play their parts. In other cases, insurers are unwilling to provide affordable coverage because of an unduly high ratio of risk to return on capital. Lack of flood coverage is the most common absence of insurance. A selection of schemes for natural disasters is discussed briefly below.

Japan
Windstorm—Full coverage became generally available in 1984, with significant deductibles. In 1991, Typhoon Mireille consumed half of the tax-deductible catastrophe reserves insurers had accumulated (Horichi, 1993).

United Kingdom
Flood and subsidence—Coverage for these perils was added to virtually all domestic building policies in the 1960s and 1970s, respectively, at uniform rates. The market is now moving toward risk-based rating (Dlugolecki *et al.*, 1994).

Netherlands
Flood risk—The government accepts responsibility for physical protection. Insurance against flood is not available.

United States
National Flood Insurance Program (NFIP)—Coverage is not privately available. Communities must qualify for the NFIP by undertaking a risk assessment and reduction program (FEMA, 1992a). Problems often arise with determining whether water damage stems from rain or storm surge (Kunreuther *et al.*, 1993).

France
Natural catastrophe—A 1982 French law attached natural catastrophe to all property policies at a centrally determined set of surcharges on the basic premium. Claims are made only if the authorities declare a disaster. In 1990, the scheme was revised to exclude storm, but many practical problems remain, such as the difficulties of separating storm and flood damage and the issue of uninsured property owners (Florin, 1990).

Caribbean
As a result of heavy hurricane damage in the 1980s and 1990s, many insurers and reinsurers withdrew from the area. Remaining reinsurers insisted on separate, increased rates for windstorm, introduction of the "average" clause to eliminate the effects of underinsurance, and high deductibles for catastrophe losses (Murray, 1993; Saunders, 1993).

industry generated premiums of $700 billion. The United States accounts for 44% of the total, the European Union (EU) 29% and Japan 12%, leaving the rest of the world with 15% (Swiss Re, 1994b). These sums meet all claims: liability, accidental damage, fire, and natural hazards, as well as administration costs. Figures on capital are not readily available, but a commercial insurer would generally maintain capital at 40–50% of premiums, giving an equivalent of $280 to $350 billion. This capital provides a reserve to meet all unexpected costs, not just those from extreme weather events, and it is usually subject to a legal minimum requirement (e.g., 17% of premium in the EU). It is reinforced by reinsurers' capacity, but again the capital is required for various classes of claims. The scale of the commercial non-life reinsurance industry is hard to estimate. Based on estimates by Foreman (1989) for the United States, the net premium volume is probably about $100 billion, with capital reserves about the same size. Two-thirds of the capital is required to fund non-property business (Laderman, 1993), leaving approximately $35 billion for natural catastrophes.

17.3.2. Methods of Risk Assessment

Natural hazards are dangerous because of the likelihood of an accumulation of loss through damage to many individual properties (Nierhaus, 1986). Underwriters seek to estimate the maximum probable loss. In fact, any level of loss must have a finite chance of being exceeded; Tiedemann (1991) points out that under certain conditions, a 90% probability of being free from an event for 100 years implies that the event occurs once in 1,000 years. Four methods of a risk assessment are described by Friedman (1984): conventional pricing, two-tier pricing, stage-damage, and simulation.

Conventional pricing is experience-based but with a short "memory," giving a dubious allowance for disasters. Two-tier pricing has a first element based on experience, but also a second element specifically for disasters, set notionally. For stage-damage, the property at hazard is characterized by its key dimensions and vulnerability to different levels of event. Detailed knowledge of the local terrain is then used to model the likely impact on the property of a specific event. In simulation, the portfolio of property is treated at an aggregate level, and the likely damage is assessed by evaluating the damage over a large number of extreme events intended to be typical of the long-run distribution of severity and frequency. This approach overcomes two problems: first, that there are not many catastrophe events in practice, and second, that historic damage reflects very different portfolios (Clark, 1986). The third and fourth methods require considerable data. Topography can affect the damage potential by more than 100% (Georgiou, 1989). Other difficulties are that it may not be practical to keep a schedule of property; losses to production may be caused by events elsewhere; and one "event" may strike several "target" locations (Meek and Tattersall, 1989).

Two other methods are event prediction and parallelisms. As scientific knowledge grows about the conditions that produce hurricanes and other extremes—often months ahead of their occurrence—it is becoming possible to use specific probabilities rather than generalized ones as in simulation (North Atlantic storms: Tinsley, 1988; hurricanes: Clark, 1992; Landsea *et al.*, 1994). In fact, more and more disturbances can be associated with the El Niño Southern Oscillation (ENSO) phenomenon. Alternatively, where data on extremes is lacking, similar conditions may generate a similar distribution of extreme to normal frequency/severity, as for rivers (Meigh *et al.*, 1993).

There now are many applications of Friedman's techniques: storm surge (Stakhiv and Vallianos, 1993), flood (Smith, 1991), and European storms (Schraft *et al.*, 1993). The insurer can study the potential effect of historic events on today's exposure, model hypothetical events, assess the expected cost of natural hazards for a portfolio, and estimate the probability of exceeding some loss level. The financial effects of deductibles or reinsurance can be measured (Boissonade and Dong, 1993). An Applied Insurance Research study of hurricanes gave $53 billion for a class 5 hurricane hitting Miami and $52 billion for a class 4 hitting New York (IRC, 1995). The expected severity of the hazard often is codified by zoning territory to control the amount of risk in any event (Nierhaus, 1986). The CRESTA (Catastrophe Risk Evaluating and Standardized Target Accumulations) system for earthquakes was invented by reinsurers (Foreman, 1989). A multi-peril approach for Australia uses 52 zones for cyclone, thunderstorm, wildfire, and earthquake (Flitcroft, 1989).

17.3.3. Socioeconomic Factors

World population is increasingly concentrated in urban areas, coastal regions, and river valleys (Marco and Cayuela, 1992). The concentration of property in such areas exposes insurers to potentially large losses from extreme events. Already, two-thirds of the world population live within 60 km of the coast; this is expected to rise to 75% by 2010 (IPCC, 1994). The total amount of property assets is not well recorded. Hohmeyer and Gartner (1992) updated a 1971 estimate to give $22 trillion, but, allowing for economic growth, this must be at least $50 trillion by now. In 1988, insured property values were $2 trillion on the Gulf and Atlantic coasts of the United States (IRC, 1995). Most of the insured cost of storms relates to domestic property. This is explained by the sector's greater size and the greater likelihood that industry and government will self-insure their risks from weather. Prosperity has resulted in an increasing stock of personal property, often vulnerable to water, salt, and smoke damage. Parallel changes have occurred in commercial/industrial properties. Changes in organization and practice, and the increased significance of services like tourism, make interruption to business from extreme events more important. See Chapters 11 and 12 for more detail.

Whenever a major flood or storm occurs and householders are uninsured, political pressures build up. Regulators are faced with irreconcilable duties: to make insurance affordable but also available. If premium levels are held down, an influx of claims from high-hazard policies may make the insurer insolvent (Denlea, 1994). A similar dichotomy is evident in the taxation treatment of catastrophe-prone business. The noncatastrophe years are milked for tax revenue, as opposed to allowing the transfer of "profits" to capital reserves (ABI, 1993). In recent years, media attention has highlighted insurance issues (Winters, 1993). Insurers therefore have been forced to settle claims with a minimum of investigation. Independent research shows that many claims are now either exaggerated or totally false (Shearn, 1994).

17.4. Extreme Events and Property Insurance

Severe events are the source of the greatest damage (Olsthoorn and Tol, 1993). For example, floods with a return period greater than 100 years cause more than 50% of the property damage arising from all floods (Smith, 1991). The following climate extremes are of concern to the industry:

River basin and coastal floods: What constitutes a serious flood is in part culture-dependent (Gardiner, 1992). Marco (1992) states that 0.3 m is a "small disturbance," with 1 m causing "a major disaster." The speed of water movement is also important (Smith, 1991).

Drought: The impacts are direct on harvests (which may be insured, and form the basis for futures trading) and indirect through wildfire or building subsidence due to excessive extraction of groundwater (Hadfield, 1994) or drying clay soils (Freeman, 1992).

Windstorm: Maximum gust and storm duration are both important (Schraft *et al.*, 1993), as are gustiness (Muir-Wood, 1993) and the location of the storm track in relation to population centers (Swiss Re, 1994a). A critical determinant of losses is the nonlinear relationship of windspeed to damage (Munich Re, 1993; Schraft *et al.*, 1993). Figure 17-1 shows this for the October 1987 storm that devastated southeast England (Dlugolecki, 1992). Because there is a limit to the damage that can be suffered, the full response curve is sigmoid. Hurricanes produce much higher damage because they are more intense and often accompanied by storm surges, and standards of building control may be poor. "Rainy" storms are more costly than "dry" storms (Bryant, 1991). This may mean an increased storm cost in the future because the atmosphere will generally be more water-laden (see Chapter 6, *Climate Models—Projections of Future Climate*, in the IPCC Working Group I volume). Bryant (1991) also reports the possibility that cyclones can trigger earthquakes.

Convective events and precipitation extremes: Tornadoes, hailstorms, and thunderstorms are expensive despite their localized nature (Munich Re, 1984; Staveley, 1991). Heavy rainfall events can lead to landslides (Ahmad, 1991; Manning *et al.*, 1992).

Temperature extremes: Low temperatures can cause significant property damage due to burst pipes and subsequent escape of water. High- and low-temperature extremes can reduce agricultural and industrial production (Palutikof, 1983) and increase mortality rates.

17.4.1. Present-Day Trends in Extremes

Lengthy base periods and a stable climate are essential to assess return periods (Knox, 1993; Tol, 1993). Reilly (1984) suggests that using 30 years' data could underestimate the true risk by 50%. Marco (1992) notes that current data may not be adequate to assess 100-year flood levels, yet the latter is the design level used for the U.S. Flood Program (with a second, 500-year level). Return periods often are reevaluated in the light of new extremes (Wilkinson and Law, 1990; Tooley, 1992; Anderson and Black, 1993). Attempts to develop long climate time series can be beset by inhomogeneities in the underlying data (von Storch *et al.*, 1993a). The impact of what may be termed "combined events" is important for the industry. Windstorm damage will be greatly increased if there is associated heavy rain or a storm surge (Englefield *et al.*, 1990; Davis and Dolan, 1993). The need for appropriate information has led some insurers to seek solutions through in-house research (Munich Re, 1990, 1993).

There is a large literature on present-day trends in extremes, which is thoroughly reviewed in Chapter 3, *Observed Climate Variability and Change*, of the IPCC Working Group I volume.

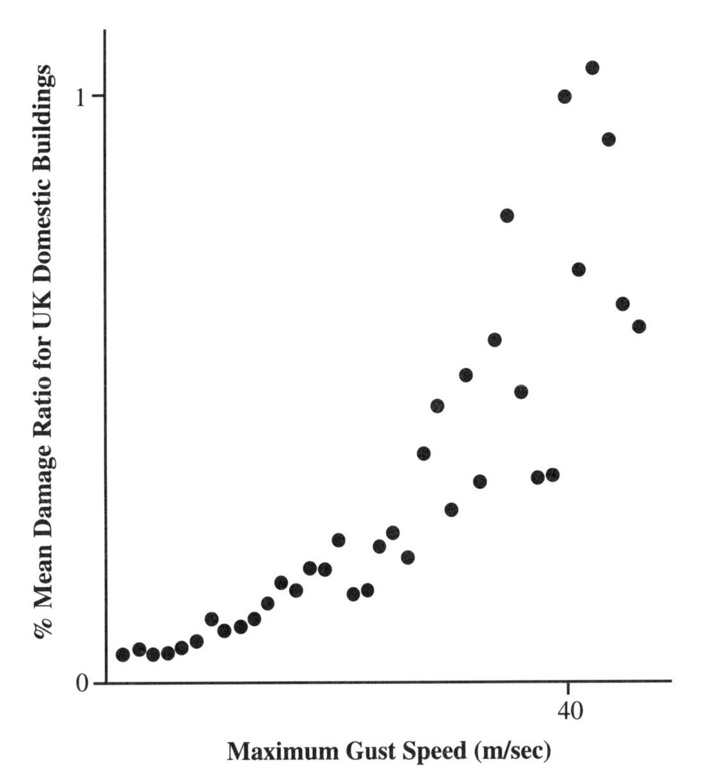

Figure 17-1: Storm cost vs. windspeed (October 1987). Mean damage ratio is based on Association of British Insurers data at postcode-district level (claims cost as a ratio of total value of insured property) for domestic buildings. Windspeed is interpolated from maps in Buller (1988). Each observation represents a postcode sector.

Just as climate varies regionally, we must expect any trends toward changes, or variability, in climate also to vary regionally (Jones and Briffa, 1992). One of the best-studied areas is North Atlantic storms. A study of the literature indicates some of the problems in attempting to determine whether trends are present. Lamb (1991) and Diaz *et al.* (1992) compiled long series of storm and wind records, respectively, but inhomogeneities in data collection are obvious or have been detected (Isemer, 1992). In the absence of reliable long-term records of windspeeds, one approach is to use a proxy variable with a long homogeneous record—for example, trends in the geostrophic wind calculated from mean sea-level pressure data (Palutikof *et al.*, 1992; Schmidt and von Storch, 1993; von Storch *et al.*, 1993b). These authors failed to find any trend that could not be explained by data inhomogeneities. A related insurance study suggests no trend in UK storminess over 60 years (Christofides *et al.*, 1992), but other studies show a correlation between winter storms and temperature (Doberitz, 1991; Dlugolecki, 1992) or a tendency to deeper depressions in the Atlantic in this century (Schinke, 1992). Again, it has been suggested that this is due to inhomogeneities in compiling weather maps (Schmidt and von Storch, 1993).

17.4.2. *Appropriate Information on Future Climate Change and Extreme Events*

The recognized source of information on future climate change is the general circulation model, or GCM (for a review, see Cohen, 1990; Giorgi and Mearns, 1991). Improvements in GCMs in the near future can be expected (e.g., Roeckner *et al.*, 1992), but at present GCMs are of limited use to the financial sector, for the following reasons:

- The length of model run is generally too short for statistical analysis of extreme events. Longer model runs are becoming available (e.g., Manabe and Stouffer, 1993, 1994).
- For analysis of extreme events, model output may be required at the daily timescale. Validation of variables (e.g., daily temperature) generally demonstrates inadequacies (e.g., Palutikof *et al.*, 1995). Statistical downscaling methods offer opportunities to overcome this problem (e.g., Karl *et al.*, 1990; Winkler *et al.*, 1995; Zorita *et al.*, 1995).
- Current models cannot generate sufficient spatial detail. Higher-resolution output from nested Limited Area Models (e.g., Giorgi *et al.*, 1994) needs much more computing.
- There may be a lack of consistency in model results—for example, for tropical and mid-latitude storms (Hewitson and Crane, 1992a, 1992b; von Storch *et al.*, 1993a; von Storch, 1994).

Most literature on climate change at the regional level concentrates on changes in the mean climate and on temperature and precipitation (e.g., Grotch and MacCracken, 1991; Boer *et al.*, 1992; Palutikof *et al.*, 1992). This is of limited use—first, because it is extreme events that have the greater implications for insurance; second, because changes in variables such as windspeed and soil moisture content may be as significant as changes in temperature and precipitation. Even if the mean of a climate variable changes and the variance remains constant, the impact on the frequency of occurrence of extreme events would be nonlinear (Mearns *et al.*, 1984; Parry and Carter, 1985; Wigley, 1985). We cannot assume that the variance will remain the same in a warmer world. Katz and Brown (1992) have shown that the frequency of extreme events is more dependent on changes in the variance than in the mean.

Tropical and mid-latitude storms: GCM results indicate conflicting trends in tropical cyclone frequency and severity. Whereas Haarsma *et al.* (1993) found an increase in the number of tropical storms, Bengtsson *et al.* (1994a, 1994b) found a decrease. Similarly, there is a lack of consensus regarding future trends in mid-latitude storms (see Gates *et al.*, 1992; Hall *et al.*, 1994; Koenig *et al.*, 1993).

Convective events and precipitation extremes: Experiments with GCMs indicate a tendency for increased convective activity in a high-CO_2 world (Hansen *et al.*, 1989; Noda and Tokioka, 1989). This implies an increase in the number of more intense rainstorms and hence in runoff (Mitchell *et al.*, 1990; Gordon *et al.*, 1992; Whetton *et al.*, 1993). An increased frequency of intense rainstorms could lead, in turn, to more landslides.

Droughts: Increases in potential evapotranspiration due to higher temperatures could increase drought potential even in regions where total rainfall increases (Whetton *et al.*, 1993). Several studies project droughtiness in populated areas (Rind *et al.*, 1990: United States; Whetton *et al.*, 1993: South Australia; Palutikof *et al.*, 1994: Mediterranean region).

Temperature extremes: GCM-based regional studies indicate that there will be an increase in high-temperature extremes (Whetton *et al.*, 1993: Australia) and a decrease in low-temperature extremes (Wilson and Mitchell, 1987: Europe).

Coastal flooding: Coastal flooding may be expected to become more common as sea level rises. However, sea level will not rise uniformly around the world (Mikolajewicz *et al.*, 1990; Gregory, 1993; Chapter 7, *Changes in Sea Level*, of the IPCC Working Group I volume). Catastrophic incursions are generally the result of a combined event, such as a high tide associated with onshore winds. In Working Group I's Chapter 7, however, authors indicate that they expect little change in tide and surge as a result of sea-level rise.

17.5. Impact of Extreme Events on Property Insurance

The impact of an extreme event varies depending on the type of insurance system in place and socioeconomic and technological factors. Damage often is uninsured. The most expensive claims have arisen from cyclonic storms. Generally, poorer countries suffer more from the effects of extreme events because their economies are based on primary, weather-vulnerable activities. Less than 0.1% of gross domestic product (GDP) is "burnt up" by catastrophes in United States

(Anderson, 1991), compared to 49% in Tanzania (Nanjira, 1991). Damage to infrastructure may significantly delay recovery (Marco and Cayuela, 1992). Repair costs are often surprisingly high, since the owners may not be in residence; building materials may be in short supply, giving rise to sharp inflationary rises in repair costs; and stricter standards may be enforced (Insurance Services Office, 1994). Increasingly, claims arise indirectly from the disruption caused (business interruption) as well as the direct property damage. Much equipment is vulnerable to water, salt, or smoke damage, and often stocks of goods are minimal, so production quickly ceases (Berz, 1988; Horichi, 1993). Services such as tourism are also prone to interruption from denial of access or from consumer anxiety.

Reinsurers have noted a fourfold increase in disasters since the 1960s (Berz and Conrad, 1993). This is not due merely to better recording, because the major disasters—which account for 90% of the losses and would always be recorded—have increased just as quickly (Swiss Re, 1994a). Much of the rise is due to socioeconomic factors, but many insurers feel that the frequency of extreme events also has increased (Mercantile and General Re, 1992; Berz and Conrad, 1993). At present, the cost of insured catastrophes amounts to about 2.5% of global non-life insurance premiums (Jakobi, 1994), but in Australia it reaches 12%, without allowing for associated administration (Brigstock and Minty, 1993). The UK used to be considered relatively hazard-free. Since 1987, the cost of a single day's storm has twice exceeded £1 billion. Subsidence losses exceeded £500 million in 1990 and 1991, from a base of under £100 million in previous years (Dlugolecki *et al.*, 1995).

17.5.1. International Reinsurers and Storms

Coverage against windstorm is provided in most countries. They cause several types of harm to socioeconomic systems—property damage, lost production, macroeconomic disruption, and personal death, injury, or stress (Anderson, 1991; Smith, 1991). Table 17-1 contains information on the impact of major windstorms worldwide since 1960. To allow a valid comparison for the partial decade of the 1990s, the figures have been expressed as an annual rate in constant values. There is a very clear upward trend in the number and overall cost. In addition, the proportion of insured damage has risen to more than 50% in the 1990s.

Table 17-1: *Major windstorms[a] worldwide: Annual impact 1960–1992.*

	1960s	1970s	1980s	1990s
Number	0.8	1.3	2.9	5.0
Total Damage ($B)	2.0	2.9	3.4	20.2
Insured Cost ($B)	0.5	0.8	1.7	11.3

Notes: Valued at 1990 prices; data from Munich Re & Swiss Re; 1990s include only 1990, 1991, and 1992.
[a] A major windstorm is defined as one costing more than $500M in total damage.

Table 17-2: *"Billion dollar" storms.*

Year	Event	Insured Cost ($B)
1987	"Hurricane" in SE England/NW France	2.5
1988	Hurricane Gilbert in Jamaica/Mexico	0.8
1989	Hurricane Hugo in Puerto Rico/S. Carolina	5.8
1990	European Storms—Four	10.4
1991	Typhoon Mireille in Japan	4.8
1992	Hurricane Andrew in Florida	16.5
1993	"Storm of the Century" in Eastern USA	1.7
1995	Hailstorms in Texas	1.1
1995	Hurricane Opal in Southern USA	2.1

Sources: Munich Re, 1990; Leggett, 1994; PCS, 1995.

Table 17-2 shows insurers had at least one "billion dollar" storm event every year from 1987 to 1993. With such an unexpectedly high frequency, some local insurance companies collapsed, and the international reinsurance market went into shock. These catastrophe losses were one of the reasons for the crisis at Lloyd's, where capacity for all types of insurance fell from £11.4 billion to £8.9 billion between 1991 and 1993 (Tillinghast, 1994). Following the initial reduction in reinsurance availability, there has been recovery as new capital takes advantage of the consequent rise in premium rates, often in offshore tax havens (Hindle, 1993). However, this only has happened by trebling the insurer's retention and increasing the reinsurance charge by a factor of four or five (Laderman, 1993). Reinsurance contracts now have far more restrictive conditions (Powers and Hearn, 1993; Jakobi, 1994). Some hazard-prone regions such as the Caribbean and Pacific now find insurance very expensive or even impossible to obtain (Murray, 1993).

17.5.2. Operational Repercussions

The financial insurance market itself may be disrupted by the effects of storm or flood because of serious failures of power and communications. Companies are tending to focus operations at fewer sites; if any single site is affected, the total impact will be greater. Insurers and loss adjusters may find the increased volume of claims difficult to handle because they probably will be operating on much tighter margins, with fewer staff in reserve (Dlugolecki *et al.*, 1994).

17.5.3. Criticality of Events

Given that it is never possible to guarantee security even with a known loss distribution, the pressure of a worsening loss distribution has serious consequences (Brigstock and Minty, 1993). In such circumstances, the large thrive and the small

Box 17-2. Jamaica and Hurricane Gilbert

On 12 September 1988, Hurricane Gilbert struck Jamaica with winds of 220 kph but no real storm surge. Total losses were about $1 billion, equivalent to about one-third of the annual gross national product (GNP). More than 40% of the housing stock was destroyed, with heavy losses in agriculture, industry, and public services (Siegel and Witham, 1991). Indirect losses were estimated to be up to twice the direct damage, owing to lost production and reduced tourism. The recovery process was hampered by lack of records (Siegel and Witham, 1991), damage to public infrastructure (Alexander, 1993), and scarcity of many goods. Clement (1990) noted that the damage was exacerbated by poor construction practice and breakdown of public order.

The financial consequences were serious. Inflation doubled, and the trade deficit widened from 7% to 39% of GDP (Commonwealth Secretariat, 1991). The currency was devalued (Clement, 1990). Insurance records indicate that 8% of the insured values were destroyed—by no means extreme compared to the 30% caused by Hurricane Iniki in the Pacific (Murray, 1993) but far more extreme than European values. There was a painful readjustment in premium levels, contract terms, and risk-management requirements (Murray, 1993; Saunders, 1993; Nesbeth, 1994), but only when Hurricane Hugo emphasized the loss potential. Jamaica, like the entire Hurricane Belt, had been fortunate for many years; the average recurrence since 1881 for hurricanes had been 5.1 years (Siegel and Witham, 1991).

The Commonwealth Secretariat (1991) noted the benefits of insurance to industry: 90% of plant and 75% of stock was insured. Owing to the heavy use of international reinsurance, only 2% of the insured damage was paid by the local market. This event is typical of the threat to economic development from cyclones (Granger, 1989). Even in Mexico it took twenty months to repair structural damage, with serious consequences for tourism (Clark, 1991b).

survive: The proportion of the U.S. reinsurance market held by the top ten companies rose from 44% in 1985 to 57% in 1992. Clearly, each country will have its own "critical event." At one time it was thought that two hurricane "hits" of $7 billion in one season would be critical to the insurance industry (AIRAC, 1986). However, the industry coped with Hurricane Andrew ($16 billion insured damage, $30 billion in total), although seven companies became insolvent and forty-four sought to reduce their exposure afterward (Insurance Services Office, 1994). Exercises to assess the impact of a major earthquake on the U.S. insurance industry provide some feel for the impact of a major hurricane. Doherty (1993) felt that a $50 billion event would seriously weaken the U.S. insurance industry with its capital base of about $160 billion and leave property owners less well-protected financially. A second study found that the system should be able to cope with an event up to $100 billion because of the immense size of the American economy (GDP in 1990 was $5.5 trillion) (Cochrane, 1993). This study assumed that new capital would enter the industry. It neglected the problems of regional blight and nonavailability of insurance, which would certainly follow. Potentially, as insurers liquidate investments to raise cash, there would be some impact on financial markets, despite reconstruction (Hofmann, 1990).

The global insurance system, of course, will be able to cope with a larger event. However, the global reinsurance industry is much smaller than the direct insurance industry and can only survive by carefully controlling its exposure to any one event or territory. This makes it likely that government would be required to back any scheme for truly catastrophic losses (Gallardo, 1984; Nutter, 1994). However, Cochrane (1993) disagrees, saying that public borrowing is now at its limit and that the emphasis should be on loss prevention. It seems most likely that multiple events would be the cause of a global failure, since the industry has proven fairly robust in response to single catastrophes. Flohn (1981) notes such occurrences in historic times, particularly when climatic change is occurring.

17.6. Adaptation to Climate Change by Property Insurers

Traditionally, insurers have dealt with changes in risk in four ways: restricting coverage so that the balance of risk-sharing shifts toward the insured; transferring risk; physical risk management (before and after the event); or raising premiums. However, in view of the increasing costs of weather claims, insurers now are considering a more fundamental approach—involving better ways of creating the funds required for irregular disasters, education of the affected parties, and cooperation with other functional or regulatory bodies in the property "system" (Horichi, 1993; Paish, 1993; Berz, 1994). Lack of information about extreme events hampers such activity and makes insurers wary of committing their capital.

17.6.1. Restricted Insurance Coverage

Increasing the contribution from the insured encourages better risk management, with consequent administrative savings. This may overcome any "moral hazard" through being insured or defended (Grose, 1992). The drawback could be in fraudulently inflated claims and a potential reduction in the fund from which to meet other claims, should the rating levels be significantly lowered. The obvious method would be to impose significant deductibles (Munich Re, 1993). Alternatives would be the increased use of coinsurance, whereby policyholders share each loss on a predetermined percentage, and reduced upper limits of liability. "Reinstatement as new" coverage needs to be questioned because settlement net of depreciation would more fairly distribute the losses; most consumers must bear depreciation themselves (Shearn, 1994). High-risk items such as satellite dishes could be excluded.

Theoretical studies of risk transfer have identified several conditions that might make insurance nonviable (Berliner and Buhlman, 1986; Peele, 1988; Henri, 1991). These include: inadequate spread of risk, undue influence on the event by the policyholder, insufficient knowledge of the likely loss distribution, inability to define when a loss has occurred and quantify it, and, finally, premiums that are not affordable. The most common areas of nonparticipation are flood damage and damage to growing crops. A change in perception of the risk can lead to withdrawal of participation—for example, following storms in Fiji, Jamaica, and Florida (Clement, 1990; Anderson, 1991; Murray, 1993).

If insurers collectively withdrew coverage entirely for some peril, this would affect financiers, occupiers, the construction industry, and local and national government. New enterprises might never start without being able to protect their capital through insurance. Recovery following a disaster also would be problematic without the influx of insurance funds (IRC, 1995). The authorities might react as happened in Hawaii and Florida in 1993 (Insurance Services Office, 1994) to enforce participation by insurers. Insurers might still participate in the administration of a funding scheme, without bearing the risk (e.g., NFIP in the United States).

Government may decide to become directly involved in important areas such as flood or agricultural insurance where commercial insurers are unwilling (Hueth and Furtan, 1994; Ericksen, 1995). In Bangladesh, farmers require income replacement and recapitalizing, and moneylenders charge high interest rates (Hodgson and Whaites, 1993). Government can also assist with grants and subsidies, and so forth (Foster, 1984). This has several advantages: greater financial strength, less requirement for supervision, mandatory powers to enforce compliance, and recognition of social and international issues such as income distribution and inward investment. There also can be significant disadvantages. Few countries can bear the cost of disaster alone (Mead, 1993). Many losses (e.g., tourism in East Africa; Nanjira, 1991) would not be eligible. Funds may be diverted from more desirable policies. It may encourage hazardous activities, be cumbersome to administer, and undermine the system for standard risks (Porter, 1994). Commercial insurers can help to keep such systems efficient (Dlugolecki, 1993).

Funding can be by disaster relief. This approach is limited in scope, slow to react, undignified, and fraught with abuse. If too generous, it will act as a disincentive to insurance. The goal should be to ensure that insurance coverage is affordable for property owners. Certain propertyowners may decide they have sufficient resources to deal with the hazards themselves (e.g., government and big business) and choose not to insure on a rational basis (Geneva Association, 1992).

17.6.2. Transfer of Risk

Insurers have traditionally shared major exposures to loss with reinsurers. The solvency of the reinsurer is critical (Porro, 1984). Reinsurers now assess whether a given territory will be able to "pay back" its losses within, say, 10 years and carefully control their exposure (Crowley, 1991). Clearly therefore, insurers can only insure great natural perils if they are themselves financially strong, they charge an economic rate to reflect the underlying risk and the partial availability of reinsurance, and they underwrite the risks by acquiring sufficient information and adapting the contract terms to fit the degree of hazard.

A new approach is "catastrophe futures"—contracts issued quarterly at a specific price and redeemable at a price based

Box 17-3. Example of Criticality and Reinsurance

Suppose an insurer covers a portfolio of one million domestic buildings (with contents) valued in total at $100 billion. At an average rate of $0.3%, this gives an annual total premium of $300 million. The rating allows for a catastrophe every 10 years at a claims cost of $100 million, or $10 million per year. The insurer holds free reserves equal to 40% of the premium, equivalent to $120 million. Clearly, the fund can cope with the average catastrophe cost of $10 million per year, but it would be virtually wiped out in the year of the disaster. This is why reinsurance is necessary.

In 1987, the typical excess-of-loss reinsurance protection was equivalent to 30% of the premium at risk (Laderman, 1993). In the above example, we assume that $90 million of the risk was transferred [remember, the risk is actually the value of the property at risk (i.e., $100 billion)]. For this the insurer would pay a reinsurance premium somewhat higher than the annualized catastrophe cost, say $12 million. In the year of a catastrophe, he would now suffer $10 million catastrophe (= $100–90 million), which his resources can accommodate. Clearly, if catastrophes become more frequent, the reinsurer suffers most. However, if catastrophes become larger, the insurer suffers because he faces all the incremental claims above the reinsurer's cutoff point. In fact by 1993, the insurer would have raised his protection to $200 million, but with a deductible of $30 million. At the same time, the reinsurer would have increased his charge to perhaps $75 million (Laderman, 1993)!

upon the actual cost of future catastrophe claims (Bannister, 1993; UNCTAD, 1994). The benefits to the risk bearers are obvious, but so far this approach has not taken off because it lacks an obvious seller to match the buyer (insurer) (Boose and Graham, 1993). Also, insurance regulators need to recognize these contracts as admissible assets (ISO, 1994).

17.6.3. Physical Risk Management

Risk management is "a discipline for living with the possibility that future events may cause harm" (Kloman, 1992). It embraces physical and monetary techniques; we deal with the latter under other headings. Traditionally it has been site-specific, but now there are national and international initiatives [e.g., International Decade for Natural Disaster Reduction (IDNDR)]. Because development has grown very rapidly, there often is no realization of the hazard potential. Also, insurance is not required by the ultimate property owner until later and is often just part of a financial deal. Effective risk management comprises understanding the hazard and eliminating it, if possible; if not, it plans how to modify the impact of an event or the property likely to be at hazard and how to mitigate the loss after the event (Foster, 1984). It involves such techniques as hazard mapping, improved design, simulations, and a full disaster plan including the warning system, emergency response, and recovery (Stone, 1992). Land-use control and information are also essential (Berz, 1994). The issue of coastal defense is considered in detail in Chapter 9.

Within the construction sector, it is essential to identify the standards required and adhere to them. Incorrect construction often is to blame for damage; indeed, one of the great future issues will be how to "retrofit" substandard buildings (Sparks, 1993). This is now done regularly in Fiji, with a scheme that features professional certification of the work and premium discounts (FBSC, 1985; Walker, 1989, 1992). Unless adequate checks and controls are maintained on reinstatement works, the reconstruction may be carried out in a shoddy way—leading to further problems for the future (Norton and Chantry, 1993). It may be necessary to rebuild to higher standards (Peele, 1988).

Storm: Most damage is roof-related or due to poor detailing or wind-blown debris (NCPI, 1992a, 1992b; FEMA, 1992b, 1993). There is always pressure to reduce building costs (Norton and Chantry, 1993). Meek and Tattersall (1989) estimate that higher building standards introduced after Cyclone Tracy in Darwin, Australia would reduce damage by a factor of five. Offshore exposures are an obvious concern. Hurricane Andrew damaged or destroyed 173 energy rigs (Leggett, 1993a).

Flood: This peril is very destructive and strongly localized (Cuny, 1991; Moser, 1992). Yevjevich (1992) lists thirty-three methods to cope with flood—divided into insurance, intensive physical and extensive physical flood-proofing, prediction, and prevention. Alternative accommodation is often required. More stringent planning controls on floodplain development must be adopted (Henri, 1991). Appropriate defense work must be given higher priority in the known problem areas (Anderson, 1991; Shearn, 1994). Upgrading buildings during maintenance (Miller, 1989), reviewing drainage capacity (Robbins, 1993), and landscaping exposed sites (Bush, 1994) could be very effective.

Box 17-4. Risk Management in Tanzania

In recent years, Tanzania has experienced heavy rains and floods. Severe windstorms have been frequent in the lake zone. These have caused financial loss, death, injuries, and the spread of diseases. The government has issued several directives prohibiting construction in the lower lands and has allocated new sites on higher ground. Most of the people cannot afford insurance. The lake zone is the main cotton-growing area. Natural-perils damage to the stored crop and the silos in which they are stored can be covered. The Cotton Act strictly regulates storage construction, and the National Insurance Corporation sends surveyors to advise the cooperative unions on how to minimize loss occurrences. To date, the National Insurance Corporation does not provide growing-crop insurance due to the lack of reliable data.

Risk management involves issues that go beyond the purview of the insurance industry. However, insurers and reinsurers play a leading role in mapping hazards, promoting better design codes, and improving the quality of implementation through trade bodies like the Insurance Institute for Property Loss Reduction (IIPLR) in the United States. At the micro-level, insurers inspect many buildings, making insurance dependent on adopting the inspector's recommendations for risk mitigation. Loss control is generally efficient, with help lines, recommended tradesmen, the use of loss adjusters in emergency panels, instructions through the media, and specialist recovery agencies. However, with the wholesale reduction of staffing in the financial sector, future response may be less effective. It is important that insurers reflect varying degrees of protection in their premium levels, to encourage risk management (Foster, 1984; Natsios, 1991). Infrastructure work such as sea defenses should fall to the public purse. Much of the benefit might accrue to non-property owners, and many property owners do not purchase insurance.

17.6.4. Technical Pricing

The initial response to increased losses might be to increase premiums. However, the decision is not usually simple because the market is increasingly competitive. The buying of market share inevitably leads to adverse results (the "underwriting cycle") (Munkhammar and Themptander, 1984). Individual property rating is based on value at risk, with an assessment of such factors as location, construction, and, for large commercial risks, trade processes. The rate is derived from past losses and modified for irregular events, administration expenses, and profit margin (Brigstock and Minty, 1993). This method is clearly flawed if the size and number of weather related incidents are rising. Reinsurance pricing can modify the direct insurer's strategy significantly (Laderman, 1993). Recent experience will strongly color the consumer's expectations and willingness to pay (Changnon and Changnon, 1990). Owners tend to understate property values, which makes it difficult to ensure that an adequate fund is built up and also makes restitution contentious.

The general principle of insurance is to "share the risk." Henri (1993) points out that using a rate that does not reflect the degree of hazard will result in economically inefficient behavior (adverse selection). Below-average risks will be reluctant to insure, while insurers will shy away from the increasingly heavy pool of risks. A rate varying to reflect the individual exposure to hazard encourages risk management but can result in standard insurance becoming too expensive for some. It is desirable for insurance to be a prerequisite of property purchase/ownership; otherwise damage may not be repaired, leading to a downward spiral of decay. It is easier to achieve tradeoffs with a multi-peril policy (Anderson, 1976), and it has practical benefits in claim handling.

17.6.5. *Funding over Time*

Retrofunding is sometimes used commercially, similar to a banker's line of credit (Jakobi, 1994), but generally it is a government tool, taking the form of relief payments. These payments often do not provide sufficient or efficient assistance (Anderson, 1991; Kunreuther, 1984). The alternative of prefunding (insurance) is therefore coming to the fore. It is important that natural-hazard insurance be treated as a long-term business rather than being accounted and taxed annually. The premiums must provide sufficient surplus to give a reserve for disasters through the troughs of the underwriting cycle. Catastrophe reserves should be accumulated over years, to reduce reliance upon reinsurance arrangements or shareholders' funds and to allow the accumulation of investment gains (Porro, 1984). The build-up of such reserve funds often is exempt from taxation (ABI, 1993; Candel, 1993). Insurers will still require reinsurance to cope with "super-catastrophes" or "multiple hits."

17.6.6. *Improving the Knowledge Base*

The collection and dissemination of information—including education and vocational training—is critical to the success of risk management (Foster, 1984; Ward, 1991). However, there is a serious lack of knowledge everywhere on weather patterns; the effects of climate change on these patterns; the effects of these changes on the potential for property loss and associated consequential costs to society and individuals; the value and specific locations and types of property at risk; and, lastly, a detailed record of the actual losses sustained (insured and uninsured). Insurers will not underwrite climate risks without adequate information on their variability and magnitude. Almost all countries have some form of natural-hazards risk analysis in hand, often prompted by UN initiatives such as IDNDR. Probably the only viable way is to rely on each country to carry out the work for its own hazards but to use common standards—and with central funding for less-developed nations, possibly using the IDNDR as a body to coordinate databases, not just for commercial needs but for the public good (McCulloch and Etkin, 1995). The basic climate data are increasingly costly but are essential both for projecting models of future trends and for claims verification.

17.6.7. *Cooperation*

Instead of dealing with problems piecemeal, financial industries need to influence the decisions of a wide variety of bodies and policymakers at the strategic level:

- The construction industry: regarding methods, standards, and regulation of construction
- Government and its various departments:
  - To resist new development in high risk areas
  - To protect existing developments in high risk areas
  - To plan how insurance can help society cope with extreme events.
- International bodies (e.g., United Nations): Many issues now are transnational (e.g., pollution, competition, communication, climate change). The insurance industry has no international underwriting organization, and this now seems an important gap. Such a body also could coordinate the industry's research work.

17.7. Impacts and Adaptation in Other Financial Services

The bulk of this chapter has dealt with the effects of weather and climate change on the insurance of property. However, other financial services may be affected. Many of the processes involved are similar to property insurance underwriting in that risks are assessed before a financial commitment is made.

17.7.1. *Likely Impacts in Other Non-Life Insurance Activities*

Transportation: This sector is an integral part of modern society. In addition to providing the means for our lifestyles and economies to function, transport goods and services are products in themselves, and the organizations supplying them are major employers. Sensitivity to weather and climate change is considered fully in Chapter 11. Extreme conditions can be very disruptive, causing damage and delay, often compounded by human error in these conditions. There has not been a major published study in this respect. Damage to infrastructure is generally not insured. Damage and particularly injury caused to other users is often insured. Hail damage to vehicles was a particular concern in the Munich hailstorm of 1984 (Munich Re, 1984). However, compensation for delays is not generally covered because of the practical difficulties in proving losses. Travel insurance is an exception, but the sums insured are small.

Liability insurance: Liability insurance has become a rather difficult sector, with the over-use of tort law and increasing emphasis on strict liability combined with environmental degradation claims (Winters, 1993; Nutter, 1994). In dealing with impacts of weather, it is often difficult to prove negligence or to identify a currently pursuable person or organization. There has been some litigation in this field—for example, control of tree-root invasion into neighboring buildings (Institution of Structural Engineers, 1994); control of undergrowth following bush fires in Australia costing over $100 million (Alexander, 1993); limitation of beach erosion, which is often initiated by human action (Bryant, 1991); and even the accuracy of weather forecasts and flood control (Kusler, 1985). At a strategic level, there has even been discussion of national liability for climate change (Stone, 1992; Leggett, 1994), but this does not seem likely to become a commercially insurable risk. An interesting case cited by Changnon and Changnon (1990) involved liability for overselling drought policies, resulting in underwriting losses of $8 million.

Crop insurance: Agriculture is a key sector in many nations, though often it is not a part of the market economy. In most countries, insurers do not offer coverage for growing crops because of the possibility of catastrophic loss and the difficulties of loss control and evaluation. However, schemes do exist. Durham (1994) describes the use of crop insurance in the United States, where a multi-peril-crop insurance policy is sponsored by the federal government, with varying participation by private insurers. Hueth and Furtan (1994) coordinated a detailed investigation of such schemes in the United States, Canada, and other developed economies. They came to the conclusion that crop insurance was not the best way to address farmers' yield uncertainty because of moral hazard (the temptation not to mitigate losses because they will be met by insurance) and adverse selection (the tendency for farmers with worse-than-average risks to insure, knowing that they will make a "profit," while the better risks will not insure). The situation is further complicated by the relatively liberal granting of disaster relief in times of hardship, which discourages the prudent farmer from buying insurance at all. Sugar-crop insurance in Mauritius has been compulsory since 1946 and is government-administered. The scheme was discussed for Tanzania but not adopted (Putty, 1984). Similar schemes exist elsewhere (e.g., Costa Rica), with support from the banking sector (Gallardo, 1984). Climate change is expected to have significant impacts on agriculture (see Chapter 13), so it may be expected that interest in this area will increase.

The various considerations in Section 17.6 apply here. In transportation, weather-related hazards are likely to remain subordinate. There will be more attention to risk management in specific circumstances (e.g., tree owners, coastal property owners) to minimize liability. With regard to crops, private insurers are unlikely to take on large commitments.

17.7.2. Life Insurance and Pensions

By contrast with general insurance, these industries are long-term, highly actuarial in nature, and often linked with savings plans. The contracts are underwritten less rigorously than many general policies, are unusually nonrenewable, and last for many years, until terminated by death of the insured. Investment of the premiums is a major activity, with the emphasis on security over many years. A range of instruments is used, from government bonds to corporate stock, real estate, and fine art. The industry generated $770 billion in premiums in 1992 and is highly influenced by changes in government policy on social insurance and fiscal matters (Swiss Re, 1994b). As with general business, about 85% of the market lies in the United States, EU, and Japan. Although the volume of business is comparable to general insurance, the investible funds are much greater ($3,075 billion in 1989, as against $832 billion for non-life business).

Likely Impacts: The impact of climate change is likely to come in two ways. The first is through changes in morbidity or mortality. Such changes are associated with extremes of heat and cold. Potentially, shifts in mean conditions could lead to the introduction of new pests and diseases or, occasionally, the elimination of old ones (Epstein and Sharp, 1994). There have been many weather-related deaths in developing countries but, in general, without any significant involvement of life insurers. Indirect effects might include food and water shortages (see Chapter 18). The second major impact is in the potential impact on investments. Long-term decisions really need to include climate change as a factor, but this is definitely not happening yet. This will be discussed more fully in Section 17.7.3.

Adaptation: Given that there are many other imponderables such as drugs or crime-related influences on human well-being, it is unlikely that meaningful adjustments to mortality tables can be made. However, investment strategy can be adapted over a shorter period and therefore should be updated periodically to recognize the current views on the impact of climate change (Porro, 1984).

17.7.3. Banking

The other major industry is "banking," which also covers a wide range of activities. Liquid assets (e.g., savings and profits) are invested, either short term to finance debt (e.g., overdrafts) or long-term to finance assets (e.g., mortgages) and provide pensions. This requires banks to provide treasury and investment-management services. Some investments are speculative, with an uncertain return, but often loans are provided in return for fixed interest. This requires similar skills to insurance underwriting, in order to assess the riskiness of the venture or the likelihood of repayment. Loans often are secured on assets such as property (which is itself insured against damage) in order to provide some guarantee of repayment. The majority of loans are short-term (less than five years). Finally, the sector may act to organize the raising of capital for a project or to strengthen the capital base of a company by finding new shareholders. A large and increasing range of financial instruments is used to manage the inherent financial risks (hedging, etc.) (Thompson, 1995).

In relation to this topic, banks fall into three types: international banks (i.e., economically involved all over the world); regional banks; and specialist banks, including building societies, "thrifts," and similar institutions. There has been much talk of product integration between banks and insurers (still forbidden in some countries). However, the cultural differences in treating claims versus bad debts, for example, make this unlikely for the mass consumer market. At corporate level, "risk" and "finance" are almost interchangeable (Seebauer, 1992).

17.7.3.1. Likely Impacts

Banking will be affected indirectly as its customers find their operations, consumption, and/or financial circumstances affected by climate change. Specifically, the banking industry and any investment activity could be affected if property insurers withdraw coverage from property that was the subject for some long-term financial transaction. This already has happened in some small island states, and any coastal site must be challengable (Leggett, 1993a; Dlugolecki *et al.*, 1995). Indeed, the burden of uninsured catastrophe losses could cause loan defaults (Vaughan, 1994). Sectors that may be adversely affected, such as tourism and agriculture in certain regions, may find it difficult to raise finance if uninsured losses are perceived to be a real threat to viability. Moreover, a fall in property values precipitated by a loss of confidence in the local economy could itself trigger a "credit famine," because property is often used as security for loans (Medioli, 1989; Bender, 1991; Thompson, 1995).

If a financial center is unexpectedly exposed to extreme weather events, the added effect associated with being disrupted or even cut off from the nonstop global financial markets could be significant.

It can be argued that for the world banking system there is no critical threshold of climate change because banks would cooperate to reestablish a new basis, as happened after the Third-World debt crisis. The huge scale of global financial flows would outweigh the likely cost of even a series of catastrophes (Dlugolecki and Klein, 1994). At the national level, the banking system could be reestablished with external assistance, but of course this would result in a permanent weakening of the local economy though adjustments in the exchange rate and interest rates. As with insurance, the scale of event will vary from one country to another. For Jamaica, Hurricane Gilbert was extreme enough to cause lasting financial damage (Commonwealth Secretariat, 1991). Cochrane (1993) reports that the U.S. banking system has withstood two shocks of around $300 billion due to rapid changes in oil prices. He concludes that the effect of $100 billion natural catastrophe (in this case, earthquake) could be absorbed. While this might be true at the scale of the U.S. economy, there is no doubt that major local disruption would result, with failures at the micro-level.

Investment activities can be affected indirectly by climate change through the actions of other sectors. In particular, mitigation policies to alter the production and use of energy may have a strong effect on the transport and energy sectors. This may alter the economics of specific technologies and entire nations or regions, with implications for rates of return on investment. Without coherent early guidance, changes in policy may result in suboptimal investment decisions. A more subtle effect on banking may stem from the changing mix of consumption and business activities brought about by climate change (Vaughan, 1994). This will surely affect the requirements for financial services, but no study has been made of this yet.

17.7.3.2. Adaptation

In banking, the key issues will be to anticipate the growing risk exposure of lenders by reduced insurance coverage, to include climate change as one of the factors in assessing the risk in investment and lending opportunities (past and future), to plan for changes in client portfolio, caused by global warming, and to anticipate operational impacts of climate change as outlined in Section 17.5.2.

It will be important to ensure that insurance coverage is available throughout the lifetime of loans. This will mean more stringent lending conditions. Environmentally integrated assessment, with regard to the direct impact on assets and the indirect effect on the borrower's business, will be required. Closer cooperation with insurers, to the extent of partnership or even ownership, is a possibility. Alternatively, it may require government intervention to ensure that alternative protection is available (Britton, 1989; Kunreuther, 1993).

It should be evident that some activities or projects are more likely to be affected by climate change than others. Therefore, this should be taken into account when appraising any investment. In particular, assets with long life spans situated in coastal zones or river basins may be sensitive, and there may be certain stages of construction that are more vulnerable (Anderson, 1991; Bender, 1991). The classic banking solutions to increased risk are similar to those of property insurers: restricted involvement (with consequent difficulties for economic activity) and higher prices, including an environmental risk premium—often not easy to achieve in a competitive environment. More radical approaches are necessary for a fundamental issue like climate change, and enlightened bankers are already exploring such avenues in the light of sustainable development initiatives (Vaughan, 1994). This will initially require a substantial effort to understand the problems and may generate new services to assist those clients most affected, once they have been identified through improved screening. A suitable starting place for such studies will be comprehensive national and international assessments of climate change impact, which allow an integrated appraisal of the many activities supported by financial services (e.g., IPCC, 1990; CCIRG, 1991). Outsiders will be keen to see how the financial sector deploys its funds in the light of climate change. Shareholders and employees will be concerned about "their" capital and how it may be affected by mitigation policies or whether it is being used "responsibly" (ACBE, 1994).

17.7.4. *Brokering*

"Brokering," or arbitrage, provides the link between buyers and sellers of financial instruments, including foreign currency. As financial markets have become more turbulent, the use of "hedging" instruments has increased. This has been accelerated by the trend toward global markets with more deregulation (Swiss Re, 1992).

Likely impacts: A really major catastrophe could affect stock market valuations and the propensity to save. Such impacts currently are largely confined to the impact on insurers' share values and the construction sector.

Adaptation: The role of financial markets is to reallocate risk and to process and disseminate information, and this will be even more necessary with climate change. Research into areas of uncertainty and policymaking will be important to improve the efficiency of resource allocation. Financial rating agencies will devote more attention to this aspect, in particular seeking to quantify company performance on "green" issues (Mueller *et al.*, 1994).

17.8. Implications for Policymakers

One of government's roles is to provide leadership in the management of natural hazards (Towfighi, 1991; Britton and Oliver, 1993). This entails research into occurrence and impacts, followed by implementation of preventative measures. Key hazards are drought, flood, and storm, which often are not insured. Key vulnerable areas are developing countries (Stone, 1992), small island states and coastal zones (IPCC, 1990), cities (Cohen, 1991), and agriculture (IPCC, 1990). The public infrastructure is vital to recovery, but it is often exposed to natural hazards and may require retrofitting for robustness (Kreimer and Munasinghe, 1991; Minnery and Smith, 1995). Much of the cost of climate change will not be insurable [e.g., damage to ecosystems, gradual degradation, loss of economic value of coastal property (Stone, 1992)].

National planning: Short- and long-term objectives are required. Short-term objectives are principally in the area of "good housekeeping" and tightening of existing arrangements to minimize the effects of damage from extreme events. Long-term planning should consider particularly the risk of major inundation (Beatley, 1994). This is an increasing risk as population and investment move to coastal areas or river plains and will be compounded by sea-level rise. The question of how to treat property built on land to be redesignated as hazardous will be contentious (Minnery and Smith, 1995).

Education: All parties in the property market must be fully aware of the potential effects of climate change and fully educated in the means available to combat these impacts so that disaster plans can be prepared at all levels. These parties include property owners and occupiers, architects and builders, insurers, and regulatory authorities.

Integrating financial mechanisms: If the probability of an event is very low, consumers act as if it will never happen, even if the probable maximum loss is very high. It will therefore be necessary to offer incentives (or disincentives) to promote sound risk management by consumers and to introduce mandatory policies for risk-prone locations, backed by appropriate resources for regulation. In many cases, affordability will be an issue.

With better information, it will be possible to assess the risk exposure of individual entities and the nation as a whole. This will serve as a basis for discussion about which risks can be borne by the private sector. Where the risk is too hazardous for private insurance companies, there still are many advantages in involving the private insurance sector in planning for disaster mitigation. Their practical knowledge of marketing a financial product, tariff structures, access to international resources, damage recovery, avoidance of duplicate administration, and fraud control would prove invaluable (Leggett, 1993b; Vellinga and Tol, 1994). Similarly, because of the large-scale financial implications of adaptation or mitigation strategies, it will be important to consider whether and how the financial sector can contribute.

New cooperative institutions are required to draw together the many disciplines needed to tackle the implications of global warming (Kunreuther, 1994; Natsios, 1991; Sykes, 1991). The United Nations' IDNDR and Coastal Zone Management (CZM) provide excellent platforms to begin consultation on information and mitigation strategies (Bender, 1991; Berz, 1994; Clark, 1991a). Even cautious observers support the principle of collaboration for information-gathering in the context of the potential threat of climate change (Ausubel, 1991).

Climate data: The availability and affordability of climate data for research and commercial decisionmaking is critical.

Construction industry: The authorities must work with trade associations to improve building quality and design and consider how to "retrofit" structures if necessary.

Finally, in view of the long-term nature of much financial activity, particularly pension-funding, it would be valuable for governments to give an early indication of how they might modify policy on industries that contribute to, or counter, global warming and then act consistently so that investment can be channeled efficiently (Leggett, 1994). Initially, subsidies may be important for research and development in novel technologies.

17.9. Requirements for Future Assessments

There are several key issues for property insurance and related activities. Better information is required on the past and future location, severity, and frequency of extreme events, especially storms and precipitation. The areas most vulnerable to extreme events need to be identified. Better information is required on the damage-response curve for the various hazards. A standardized database on the cost of extreme events would be extremely useful. Improved techniques are required for applying knowledge of actual events to hypothetical situations to evaluate risks (i.e., simulation).

For other areas of financial services, better information on the likely changes in morbidity or mortality is needed, as is a clearer exposition of how policymakers are likely to seek to adapt to or mitigate climate change.

Finally, it must be said that the current assessment was hampered by the lack of a substantial literature base and lack of awareness on the part of many financial services, organizations, and individuals. Therefore, a program of attention-raising is important.

Acknowledgments

The authors wish to thank all those individuals who have supported this research by providing advice and information—in particular, Professor P. Vellinga, Dr. G. Berz, and all those who attended the workshop on 24/25 March 1994 in Amsterdam. We also wish to thank our governments for their assistance, our own organizations for giving us the opportunity to contribute to this important project, and of course IPCC for providing a sound context. Finally, we thank two people who were indispensable to the smooth completion of the task—Richard Klein in The Netherlands for general administration and Mandy Macdonald for her fast, faultless secretarial work under great time pressure.

References

ABI, 1993: ABI Response to consultative document on equalisation reserves. ABI, London, UK, (unpublished).

ACBE, 1994: What the city should ask. Proceedings of a seminar, November 1994, Advisory Committee on Business and the Environment, Department of Trade and Industry, London, UK, pp. 62.

Ahmad, R., 1991: Landslides triggered by the rainstorm of May 21-22, 1991. *Jamaican Journal of Science and Technology*, **2**, 1-13.

Alexander, D., 1993: *Natural Disasters*. UCL Press, London, UK, 632 pp.

Alexander, H., 1994: *Flooding in Northern Italy—A Report of the November 1994 Floods in Piedmont and an Assessment of Future Risk*. Alexander Howden Group Ltd., London, UK, 8 pp.

AIRAC, 1986: *Catastrophic Losses: How the Insurance System Would Handle Two $7 Billion Hurricanes*. AIRAC, Oak Brook, IL, 73 pp.

Anderson, D.R., 1976: All risks rating within a catastrophe insurance system. *Journal of Risk and Insurance*, **XLIII(4)**, 629-651.

Anderson, M.B., 1991: Which costs more: prevention or recovery? In: *Managing Natural Disasters and the Environment* [Kreimer, A. and M. Munasinghe (eds.)]. Proceedings of a colloquium, 27-28 June 1990, Environment Department, World Bank, Washington, DC, pp. 17-27.

Anderson, J. and A. Black, 1993: Tay flooding: act of God or climate change. *Circulation, Newsletter of the British Hydrological Society*, **38**, 1-4.

Ausubel, J.H., 1991: A second look at the impacts of climate change. *American Scientist*, **79**, 210-221.

Bannister, J. (ed.), 1993: *Insurance Derivatives: A Practical Introduction*. Supplement to Catastrophe Reinsurance Newsletter 9, DYP Group Ltd., London, UK, 8 pp.

Beatley, T., 1994: Promoting sustainable land use: mitigating natural hazards through land use planning. In: *Natural Disasters: Local and Global Perspectives*. 1993 Annual Forum of National Committee on Property Insurance, Insurance Institute for Property Loss Reduction, Boston, MA, pp. 31-36.

Bender, S.O., 1991: Managing natural hazards. In: *Managing Natural Disasters and the Environment* [Kreimer, A. and M. Munasinghe (eds.)]. Proceedings of a colloquium sponsored by the World Bank, 27-28 June 1990, World Bank, Washington, DC, pp. 182-185.

Bengtsson, L., M. Botzet, and M. Esch, 1994a: *Hurricane Type Vortices in a General Circulation Model*. Max Planck Institut fur Meteorologie Report 123 (available from Max Planck Institut fur Meteorologie, Bunestsrasse 55, 20146, Hamburg, Germany), 42 pp.

Bengtsson, L., M. Botzet, and M. Esch, 1994b: *Will Greenhouse Gas-Induced Warming over the Next 50 Years Lead to a Higher Frequency and Greater Intensity of Hurricanes?* Max Planck Institut fur Meteorologie Report 139 (available from Max Planck Institut fur Meteorologie, Bunestsrasse 55, 20146, Hamburg, Germany), 23 pp.

Berliner, B. and N. Buhlmann, 1986: Subjective determination of limits of insurability on the grounds of strategic planning. *Geneva Papers on Risk and Insurance*, **11(39)**, 94-109.

Berz, G., 1988: Climatic change: impact on international reinsurance. In: *Greenhouse: Planning for Climate Change* [Pearman, G.I. (ed.)]. CSIRO Publications, East Melbourne, Victoria, Australia, pp. 579-587.

Berz, G., 1994: Cost of disasters: areas of co-operations with the insurance industry. Presented at UN World Conference on Natural Disaster Reduction, Yokohama, Japan, May, 8 pp.

Berz, G. and K. Conrad, 1993: Winds of change. *The Review*, June 1993, 32-35.

Boer, G.J., N.A. McFarlane, and M. Lazare, 1992: Greenhouse gas-induced climate change simulated with the CCC second-generation general circulation model. *Journal of Climate*, **5**, 1045-1077.

Boissonade, A. and W. Dong, 1993: Windstorm model with applications to risk management. In: *Natural Disasters: Protecting Vulnerable Communities* [Merriman, P.A. and C.W.A. Browitt (eds.)]. Proceedings of the IDNDR Conference, London, 13-15 October, Thomas Telford, London, UK, pp. 331-343.

Boose, M.A. and A.S. Graham, 1993: Predicting trading activity for catastrophe insurance futures. *Journal of Reinsurance*, **1**, 42-55.

Brigstock, C. and D. Minty, 1993: The real cost of catastrophe exposure to insurers. In: *Catastrophe Insurance for Tomorrow: Planning for Future Adversities* [Britton, N.R. and J. Oliver (eds.)]. Proceedings of a seminar sponsored by Sterling Offices (Australia) Ltd., Griffith University, Brisbane, Australia, pp. 51-81.

Britton, N.R., 1989: Community attitudes to natural hazard insurance: what are the salient facts? In: *Natural Hazards and Reinsurance* [Oliver, J. and N.R. Britton (eds.)]. Proceedings of a seminar sponsored by Sterling Offices (Australia) Ltd., Cumberland College of Health Sciences, Lidcombe, NSW, Australia, pp. 107-121.

Britton, N.R. and J. Oliver, 1993: Changing social hazardousness and increasing social vulnerability: whose responsibility? In: *Catastrophe Insurance for Tomorrow: Planning for Future Adversities* [Britton, N.R. and J. Oliver (eds.)]. Proceedings of a seminar sponsored by Sterling Offices (Australia) Ltd., Griffith University, Brisbane, Australia, pp. 31-50.

Bryant, E.A., 1991: *Natural Hazards*. Cambridge University Press, Cambridge, UK, 294 pp.

Buller, P.S.J., 1988: *The October Gale of 1987*. Building Research Establishment, Department of Environment, London, UK, 23 pp.

Bush, D.M., 1994: Coastal hazard mapping and risk assessment. In: *Natural Disasters: Local and Global Perspectives*. 1993 Annual Forum of National Committee on Property Insurance, Insurance Institute for Property Loss Reduction, Boston, MA, pp. 19-26.

Candel, F.M., 1993: Large industrial risks, catastrophes and environmental threats. *Geneva Papers on Risk and Insurance*, **18(69)**, 449-453, Geneva Association, Geneva.

Changnon, S.A. and J.M. Changnon, 1990: Use of climatological data in weather insurance. *Journal of Climate*, **3**, 568-576.

Christofides, S., C. Barlow, N. Michaelides, and C. Miranthis, 1992: *Storm Rating in the Nineties*. General Insurance Study Group, General Insurance Convention of Actuaries, London, UK, 89 pp.

Clark, J.R., 1991a: Coastal zone management. In: *Managing Natural Disasters and the Environment* [Kreimer, A. and M. Munasinghe (eds.)]. Proceedings of a colloquium sponsored by the World Bank, 27-28 June 1990, World Bank, Washington, DC, pp. 115-118.

Clark, J.R., 1991b: Case study: Hurricane Gilbert in Yucatan, Mexico. In: *Managing Natural Disasters and the Environment* [Kreimer, A. and M. Munasinghe (eds.)]. Proceedings of a colloquium sponsored by the World Bank, 27-28 June 1990, World Bank, Washington, DC, pp. 119.

Clark, K., 1986: A formal approach to catastrophe risk assessment and management. *Proceedings of the Casualty Actuarial Society*, **LXXIII(2)140**, 69-92, Boston, MA.

Clark, K., 1992: *Predicting Global Warming's Impact*. Contingencies May/June, Applied Insurance Research, Boston, MA, pp. 1-8.

Clement, D., 1990: *An Analysis of Disaster, Life after Gilbert*. University of the West Indies, Working Paper 37, Kingston, Jamaica, 57 pp.

Climate Change Impact Review Group, 1991: *The Potential Effects of Climate Change in the United Kingdom*. Department of the Environment, HMSO, London, UK, 124 pp..

Cochrane, H., 1993: Banking, oil and earthquake: how severe a blow can the United States economy absorb? In: *Natural Disasters: Perceiving Our Vulnerabilities and Understanding Our Perceptions*. Annual forum of the National Committee on Property Insurance, December 16, Boston, NCPI, Boston, MA, pp. 51-55.

Cohen, M., 1991: Urban growth and natural hazards. In: *Managing Natural Disasters and the Environment* [Kreimer, A. and M. Munasinghe (eds.)]. Proceedings of a colloquium sponsored by the World Bank, 27-28 June 1990, World Bank, Washington, DC, pp. 82-89.

Cohen, S.J., 1990: Bringing the global warming issue closer to home: the challenge of regional impact studies. *Bulletin of the American Meteorological Society*, **71**, 520-526.

Commonwealth Secretariat, 1991: Natural disasters: economic impact and policy options. FMM (91) 10, London, UK (unpublished).

Crowley, K., 1991: Earthquakes—reinsuring the New Zealand risk. *Journal of the Society of Fellows*, **6(1)**, 13-28, Chartered Insurance Institute, London, UK.

Cuny, F.C., 1991: Living with floods: alternatives for riverine flood mitigation. In: *Managing Natural Disasters and the Environment* [Kreimer, A. and M. Munasinghe (eds.)]. Proceedings of a colloquium sponsored by the World Bank, 27-28 June 1990, World Bank, Washington, DC, pp. 62-73.

Davis, J.V., 1992: The past and future of loss financing. *Geneva Papers on Risk and Insurance*, **17(64)**, 355-361.

Davis, R.E. and R. Dolan, 1993: Nor'easters. *American Scientist*, **81**, 428-239.

Denlea, Jr., L.E., 1994: Regulating insurance availability and insurer solvency: are they at cross purposes? In: *Natural Disasters: Local and Global Perspectives*. 1993 Annual Forum of National Committee on Property Insurance, Insurance Institute for Property Loss Reduction, Boston, MA, pp. 10-14.

Diaz, H.F., K. Wolter, and S.D. Woodruff (eds.), 1992: *Proceedings of the International COADS Workshop*, Boulder, CO, 13-15 January 1992. U.S. Department of Commerce, Boulder, CO, 390 pp.

Dlugolecki, A.F., 1992: Insurance implications of climate change. *Geneva Papers on Risk and Insurance*, **17(64)**, July, 393-405, Geneva Association, Geneva, Switzerland.

Dlugolecki, A.F., 1993: The role of commercial insurance in alleviating natural disaster. In: *Natural Disasters: Protecting Vulnerable Communities* [Merriman, P.A. and C.W.A. Browitt (eds.)]. Proceedings of the IDNDR Conference, London, 13-15 October, Thomas Telford, London, UK, pp. 421-431.

Dlugolecki, A.F. and R. Klein (eds.), 1994: *IPCC Workshop on Climate Change and Financial Services*, 24/25 March. Institute for Environmental Studies, Vrije Universiteit, Amsterdam, The Netherlands, 91 pp.

Dlugolecki, A.F., 1995: Insurance demands in a dynamic environment: grappling with the issues. In: *Insurance Viability and Loss Mitigation: Partners in Risk Resolution*. Proceedings of a seminar sponsored by Sterling Offices (Australia) Ltd. Griffith University, Brisbane, Australia (in press).

Dlugolecki, A., C. Elvy, G. Kirby, R. Salthouse, S. Turner, D. Witt, R. Martin, C. Toomer, B. Secrett, J. Palutikof, and D. Clement, 1994: *The Impact of Changing Weather Patterns on Property Insurance*. Chartered Insurance Institute, London, UK, May, 87 pp.

Dlugolecki, A.F., P.J. Harrison, J. Leggett, and J.P. Palutikof, 1995: Implications for insurance and finance. In: *Economic Implications of Climate Change in Britain* [Parry, M.L. and R. Duncan (eds.)]. Earthscan, London, UK (in press).

Doberitz, R., 1992: Vierter Winter Mit Zahlreichen Tiefs Unter 950 hPa Kerndruck in Nordatalantik-keine Zuname der Hurrikanzahlen. *Der Wetterlotse*, **547**, 218-223.

Doherty, N., 1993: Insurance surplus "shock" and its economic implications. In: *Natural Disasters: Perceiving Our Vulnerabilities and Understanding Our Perceptions*. Annual forum of the National Committee on Property Insurance, December 16, Boston, NCPI, Boston, MA, pp. 60-64.

Durham, B., 1994: Hail, drought and disaster. *The Chartered Insurance Institute Journal*, London, **CII(May)**, 22-23.

Englefield, G.J.H., M.J. Tooley, and Y. Zong, 1990: *An Assessment of the Clwyd Coastal Lowlands after the Floods of February 1990*. Environmental Research Centre, University of Durham, UK, 14 pp.

Epstein, P.R. and D. Sharp (eds.), 1994: *Health and Climate Change*. The Lancet, London, UK, 36 pp.

Ericksen, N., 1995: Devolution and co-operation—New Zealand and the resource management act 1991. In: *Co-erce or Co-operate? Rethinking Environmental Management* [May, P. (ed.)]. National Science Foundation, Washington, DC, pp. 83-90.

FBSC, 1985: *Our War Against Cyclones: Guidance to Homeowners on How to Upgrade Existing Homes*. Fiji Building Standards Committee, Commissioner for Insurance, Suva, Fiji.

FEMA, 1992a: *Answers to Questions about the National Flood Insurance Program (FIA-2)*. FEMA, Washington, DC, 50 pp.

FEMA, 1992b: *Building Performance: Hurricane Andrew in Florida*. Federal Insurance Agency, Washington, DC, 93 pp.

FEMA, 1993: *Building Performance: Hurricane Iniki in Hawaii*. Federal Insurance Agency. Washington, DC, 100 pp.

Flitcroft, J.C., 1989: Development and maintenance of a hazard information system. In: *Natural Hazards and Reinsurance* [Oliver, J. and N.R. Britton (eds.)]. Proceedings of a seminar sponsored by Sterling Offices (Australia) Ltd., Cumberland College of Health Sciences, Lidcombe, NSW, Australia, pp. 144-157.

Flohn, H., 1981: Short-term climatic fluctuations and their economic role. In: *Climate and History* [Wigley, T.M.L., M.J. Ingram, and G. Farmer (eds.)]. Cambridge University Press, Cambridge, UK, pp. 310-318.

Florin, P., 1990: L'assurance des tempetes et des catastrophes naturelles en France. *L'Argus*, 14 September, 2340-2348.

Foreman, D.P., 1989: Catastrophe information: its value to reinsurers. In: *Natural Hazards and Reinsurance* [Oliver, J. and N.R. Britton (eds.)]. Proceedings of a seminar sponsored by Sterling Offices (Australia) Ltd., Cumberland College of Health Sciences, Lidcombe, NSW, Australia, pp. 158-167.

Foster, H.D., 1984: Reducing vulnerability to natural hazards. *Geneva Papers on Risk and Insurance*, **9(30)**, 27-56.

Freeman, T.J., 1992: Seasonal foundation movements in clay soil. In: *Proceedings of Fifth DYP Insurance and Reinsurance Group Conference on Changing Weather Patterns*. DYP Group Ltd, London, UK, 14 pp.

Friedman, D.G., 1984: Natural hazard risk assessment for an insurance program. *Geneva Papers on Risk and Insurance*, **9(30)**, 57-128.

Friedman, D.G., 1989: Implications of climate change for the insurance industry. In: *Coping with Climate Change* [Topping, Jr., J.C. (ed.)]. Climate Institute, Washington, DC, pp. 389-400.

Gallardo, G., 1984: The markets for disaster insurance. *Geneva Papers on Risk and Insurance*, **9(31)**, 175-187.

Gardiner, J., 1992: Environmental impact of floods. In: *Coping with Floods* [Rossi, G., N. Harmancioglu, and V. Yevjevich (eds.)]. Proceedings of NATO Advanced Studies Institute, October, E. Majorana Centre for Scientific Culture, University of Catania, Italy, pp. 355-374.

Gates, W.L., J.F.B. Mitchell, G.J. Boer, U. Cubasch, and V.P. Meleshko, 1992: Climate modelling, climate prediction and model validation. In: *Climate Change 1992: The Supplementary Report to the IPCC Scientific Assessment* [Houghton, J.T., B.A. Callander, and S.K. Varney (eds.)]. Cambridge University Press, Cambridge, UK, pp. 97-134.

Geneva Association, 1986: Limits of insurability of risks, from Proceedings of the Second International Conference on Strategic Planning and Insurance, London 1985. *Geneva Papers on Risk and Insurance,* **11(39)**, April, 192 pp.

Georgiou, P.N., 1989: The prediction of wind risk in tropical cyclone regions. In: *Natural Hazards and Reinsurance* [Oliver, J. and N.R. Britton (eds.)]. Proceedings of a seminar sponsored by Sterling Offices (Australia) Ltd., Cumberland College of Health Sciences, Lidcombe, NSW, Australia, pp. 132-143.

Giarini, O. and W. Stahel, 1994: *The Limits to Certainty*. Kluwer, Dordrecht, The Netherlands, 270 pp.

Giorgi, F. and L.O. Mearns, 1991: Approaches to the simulation of regional climate change: a review. *Reviews of Geophysics,* **29**, 191-216.

Giorgi, F., C.S. Brodeur, and G.T. Bates, 1994: Regional climate change scenarios over the United States produced with a nested regional climate model. *Journal of Climate,* **7**, 375-399.

Gordon, H.B., P.H. Whetton, A.M. Fowler, and M.R. Haylock, 1992: Simulated changes in daily rainfall intensity due to the enhanced greenhouse effect: implications for extreme rainfall events. *Climate Dynamics,* **8**, 83-102.

Granger, O.E., 1989: Implications for Caribbean societies of climate change, sea-level rise and shifts in storm patterns. In: *Coping with Climate Change* [Topping, Jr., J.C. (ed.)]. Climate Institute, Washington, DC, pp. 115-127.

Gregory, J.M., 1993: Sea level change under increasing atmospheric CO_2 in a transient coupled ocean-atmosphere GCM experiment. *Journal of Climate,* **6**, 2247-2262.

Grose, V.L., 1992: Risk management from a technological perspective. *Geneva Papers on Risk and Insurance,* **17(64)**, 335-342.

Grotch, S.L. and M.C. MacCracken, 1991: The use of general circulation models to predict regional climate change. *Journal of Climate,* **4**, 286-303.

Haarsma, R.J., J.F.B. Mitchell, and C.A. Senior, 1993: Tropical disturbances in a GCM. *Climate Dynamics,* **8**, 247-257.

Hadfield, P., 1994: The revenge of the rain gods. *New Scientist*, **20**, 14-15.

Hall, N.M.J., B.J. Hoskins, P. Valdes, and C.A. Senior, 1994: Storm tracks in a high resolution GCM with doubled CO_2. *Quarterly Journal of the Royal Meteorological Society*.

Hansen, J., D. Rind, L. Russell, P. Stone, I. Fung, R. Ruedy, and J. Lerner, 1989: Climate sensitivity analysis of feedback mechanisms. In: *Climate Processes and Climate Sensitivity* [Hansen, J. and T. Takahashi (eds.)]. Geophysical Monograph 29, American Geophysical Union, Washington, DC, pp. 130-163.

Henri, C., 1991: The insurance industry response to flood. In: *Natural and Technological Hazards: Implications for the Insurance Industry* [Britton, N.R. and J. Oliver (eds.)]. Proceedings of a seminar sponsored by Sterling Offices (Australia) Ltd., University of New England, Armidale, New South Wales, Australia, pp. 167-173.

Henri, C., 1993: Government philosophies to the uninsured with respect to the impact of major catastrophes. In: *Catastrophe Insurance for Tomorrow: Planning for Future Adversities* [Britton, N.R. and J. Oliver (eds.)]. Proceedings of a seminar sponsored by Sterling Offices (Australia) Ltd., Griffith University, Brisbane, Australia, pp. 157-175.

Hewitson, B.C. and R.G. Crane, 1992a: Regional scale climate prediction from the GISS GCM. *Palaeogeography, Palaeoclimatology, Palaeoecology,* **97**, 249-267.

Hewitson, B.C. and R.G. Crane, 1992b: Regional climate in the GISS Global Circulation Model: synoptic scale circulation. *Journal of Climate,* **5**, 1002-1011.

Hindle, J., 1993: The reinsurer's view: catastrophes are you being served? In: *Climate Change and the Insurance Industry*. Symposium by Greenpeace, 24 May, Greenpeace, London, UK, 4 pp.

Hodgson, R.L.P. and A. Whaites, 1993: The rehabilitation of housing after natural disasters in Bangladesh. In: *Natural Disasters: Protecting Vulnerable Communities* [Merriman, P.A. and C.W.A. Browitt (eds.)]. Proceedings of the IDNDR Conference, London, 13-15 October, Thomas Telford, London, UK, pp. 194-209.

Hofmann, M.R., 1990: Climatic changes and the insurance industry. *Business Insurance,* **43**, 75.

Hohmeyer, O. and M. Gartner, 1992: *The Cost of Climate Change: A Rough Estimate of Orders of Magnitude*. Report to the Commission of the European Communities DGXII, Fraunhofer - Institute for Systems and Innovation Research, Karlsruhe, Germany, 60 pp.

Horichi, S., 1993: Natural disasters and changing environment: can the industry prepare? Address to International Insurance Society, May 1993, available from Tokio Marine and Fire Insurance Co. Ltd, Tokyo, Japan.

Hueth, D.L. and W.H. Furtan (eds.), 1994: *Economics of Agricultural Crop Insurance: Theory and Evidence*. Kluwer Academic Publishers, Dordrecht, The Netherlands, 380 pp.

Institution of Structural Engineers, 1994: *Subsidence of Low Rise Buildings*. Institution of Structural Engineers, London, UK, 105 pp.

Insurance Services Office, 1994: The impact of catastrophes on property insurance. *Catastrophe Reinsurance Newsletter*, **12**, 28-32.

IPCC, 1990: *Impacts Assessment of Climate Change: The Policymakers' Summary of the Report of Working Group II to IPCC*. Commonwealth of Australia, 32 pp.

IPCC, 1994: *Preparing to Meet the Coastal Challenges of the 21st Century.* World Coast Conference 1993 Report, Rijkwaterstaat, The Hague, The Netherlands, 49 pp. and appendices.

IRC, 1995: *Coastal Exposure and Community Protection—Hurricane Andrew's Legacy*. Insurance Research Council, Wheaton, IL (in association with Insurance Institute for Property Loss Reduction), 48 pp.

Isemer, H.J., 1992: Comparison of estimated and measured marine surface wind speeds. In: *Proceedings of the International COADS Workshop*, Boulder, CO, 13-15 January 1992 [Diaz, H.F., K. Wolter, and S.D. Woodruff (eds.)]. U.S. Department of Commerce, Boulder, CO, pp. 143-158.

Jakobi, W., 1994: The future of the catastrophe reinsurance market. *Catastrophe Reinsurance Newsletter,* **10**, 4-7; **11**, 18-20; and **12**, 36-39.

Jones, P.D. and K.R. Briffa, 1992: Global surface air temperature variations during the twentieth century. Part 1: Spatial, temporal and seasonal variations. *The Holocene*, **2**, 165-179.

Karl, T.R. W.-C. Wang, M.E. Schlesinger, and R.W. Knight, 1990: A method of relating general circulation model simulated climate to the observed climate. Part 1: Seasonal statistics. *Journal of Climate,* **3**, 1053-1079.

Katz, R.W. and B.B. Brown, 1992: Extreme events in a changing climate: variability is more important than averages. *Climatic Change,* **21**, 289-302.

Kennedy, P., 1993: *Preparing for the Twenty-First Century*. Fontana Press, Harper Collins Publishers, London, UK, 428 pp.

Kloman, H.F., 1992: Rethinking risk management. *Geneva Papers on Risk and Insurance,* **17(64)**, 299-313.

Knox, J.C., 1993: Large increases in flood magnitude in response to modest changes in climate. *Nature*, **361**, 430-432.

Koenig, W., R. Sausen, F. Sielman, 1993: Objective verification of cyclones in GCM simulations. *Journal of Climate,* **6**, 2217-2231.

Kreimer, A. and M. Munasinghe, 1991: Managing environmental degradation and natural disasters: an overview. In: *Managing Natural Disasters and the Environment* [Kreimer, A. and M. Munasinghe (eds.)]. Proceedings of a colloquium sponsored by the World Bank, 27-28 June 1990, World Bank, Washington, DC, pp. 3-6.

Kunreuther, H., 1984: Causes of underinsurance against natural disasters. *Geneva Papers on Risk and Insurance,* **9(31)**, 206-220.

Kunreuther, H., 1993: Reducing the risks. Is insurance doing all it should? An outsider's view. In: *Catastrophe Insurance for Tomorrow: Planning for Future Adversities* [Britton, N.R. and J. Oliver (eds.)]. Griffith University, Brisbane, Australia, pp. 275-287.

Kunreuther, H.F., N. Ericksen, and J. Handmer, 1993: *Reducing Losses from Natural Disasters through Insurance and Mitigation: A Cross-Cultural Comparison*. Wharton Risk Management and Decision Processess Center Working Paper, University of Pennsylvania, Philadelphia, PA, 27 pp.

Kunreuther, H.F., 1994: The role of insurance and mitigation in reducing disaster losses. In: *Natural Disasters: Local and Global Perspectives*. 1993 Annual Forum of National Committee on Property Insurance, Insurance Institute for Property Loss Reduction, Boston, MA, pp. 27-30.

Kusler, J., 1985: Liability as a dilemma for local managers. *Public Administration Review,* **45**, 118-22.

Laderman, S.K., 1993: The catastrophe reinsurance market: yesterday, today and tomorrow. In: *Catastrophe Insurance for Tomorrow: Planning for Future Adversities* [Britton, N.R. and J. Oliver (eds.)]. Proceedings of a seminar sponsored by Sterling Offices (Australia) Ltd., Griffith University, Brisbane, Australia, pp. 83-98.

Lamb, H., 1991: *Historic Storms of the North Sea, British Isles and Northwest Europe*. Cambridge University Press, Cambridge, UK, 204 pp.

Landsea, C.W., W.M. Gray, P.W. Mielke, Jr., and K.J. Berry, 1994: Seasonal forecasting of Atlantic hurricane activity. *Weather*, **49(8)**, 273-284.

Leggett, J., 1993a: Who will underwrite the hurricane? *New Scientist*, 7 August, 29-33.

Leggett, J., 1993b: Climate change and the future security of the reinsurance market: recent developments and upcoming issues for the industry. *Journal of Reinsurance*, **1**, 73-95.

Leggett, J., 1994: *Climate Change*. Presentation to the 26th Annual Meeting of the Reinsurance Association of America, 29 April, Laguna Niguel, CA, Greenpeace International, London, UK, 53 pp.

Manabe, S. and R.J. Stouffer, 1993: Century-scale effects of increased atmospheric CO_2 on the ocean-atmosphere system. *Nature*, **364**, 215-218.

Manabe, S. and R.J. Stouffer, 1994: Multiple-century response of a coupled ocean-atmosphere model to an increase of atmospheric carbon dioxide. *Journal of Climate*, **7**, 5-23.

Manning, A.S., T. McCain, and R. Ahmad, 1992: Landslides triggered by 1988 Hurricane Gilbert along roads in the Above Rocks area, Jamaica. *Journal of the Geological Society of Jamaica*, **special issue 12**, Natural Hazards in the Caribbean [Ahmad, R. (ed.)], 34-53.

Marco, J.B., 1992: Flood risk mapping. In: *Coping with Floods* [Rossi, G., N. Harmancioglu, and V. Yevjevich (eds.)]. Proceedings of NATO Advanced Studies Institute, October, E. Majorana Centre for Scientific Culture, University of Catania, Italy, pp. 255-276.

Marco, J.B. and A. Cayuela, 1992: Urban flooding: the flood planned town. In: *Coping with Floods* [Rossi, G., N. Harmancioglu, and V. Yevjevich (eds.)]. Proceedings of NATO Advanced Studies Institute, October, E. Majorana Centre for Scientific Culture, University of Catania, Italy, pp. 337-354.

McCulloch, J. and D. Etkin (eds.), 1995: *Improving Responses to Atmospheric Extremes: The Role of Insurance and Compensation*. Proceedings of a workshop, 3-4 October1994, Toronto, Environmental Adaptation Research Group, Environment Canada, Toronto (in association with the Climate Institute, Insurers' Advisory Organisation, Reinsurance Research Council, and other bodies), 223 pp.

Mead, R.D., 1993: The development of a New Zealand local authority natural disasters and emergencies funding facility for infrastructural assets. In: *Catastrophe Insurance for Tomorrow: Planning for Future Adversities* [Britton, N.R. and J. Oliver (eds.)]. Proceedings of a seminar sponsored by Sterling Offices (Australia) Ltd., Griffith University, Brisbane, Australia, pp. 145-156.

Mearns, L.O., R. Katz, and S.H. Schneider, 1984: Extreme high-temperature events: changes in their probabilities with changes in mean temperature. *Journal of Climate and Applied Meteorology*, **23**, 1601-1613.

Medioli, A., 1989: Climate change: the implications for securities underwriting. In: *Coping with Climate Change* [Topping, Jr., J.C. (ed.)]. Climate Institute, Washington, DC, pp. 406-411.

Meek, I.E. and A.R. Tattersall, 1989: Natural hazards: the insurer's viewpoint. In: *Natural Hazards and Reinsurance* [Oliver, J. and N.R. Britton (eds.)]. Proceedings of a seminar sponsored by Sterling Offices (Australia) Ltd., Cumberland College of Health Sciences, Lidcombe, NSW, Australia, pp. 168-175.

Meigh, J.R., J.V. Sutcliffe, and F.A.K. Farquharson, 1993: Prediction of risks in developing countries with sparse river flow data. In: *Natural Disasters: Protecting Vulnerable Communities* [Merriman, P.A. and C.W.A. Browitt (eds.)]. Thomas Telford, London, UK, pp. 315-330.

Mercantile and General Reinsurance (M&G Re), 1992: *Natural Perils in Australia—A Reinsurer's Perspective*. M&G Re, Sydney, Australia, 52 pp.

Mikolajewicz, U., B.D. Santer, and E. Maier-Reimer, 1990: Ocean response to greenhouse warming. *Nature*, **345**, 589-593.

Miller, T.R., 1989: Impact of global climate change on metropolitan infrastructure. In: *Coping with Climate Change* [Topping, J.C. (ed.)]. Climate Institute, Washington, DC, pp. 366-376.

Minnery, J. and D. Smith, 1995: Climate change, flooding and urban infrastructure. In: *Greenhouse '94*. Proceedings of a conference, Wellington, 9-14 October, 1994, CSIRO, Division of Atmospheric Research, Canberra, Australia (in press).

Mitchell, J.F.B., S. Manabe, V. Meleshko, and T. Tokioka, 1990: Equilibrium climate change—and its implications for the future. In: *Climate Change: The IPCC Scientific Assessment* [Houghton, J.T., G.J. Jenkins, and J.J. Ephraums (eds.)]. Cambridge University Press, Cambridge, UK, pp. 131-172.

Moser, D.A., 1992: Assessment of the economic effects of flooding. In: *Coping with Floods* [Rossi, G., N. Harmancioglu, and V. Yevjevich (eds.)]. Proceedings of NATO Advanced Studies Institute, October, E. Majorana Centre for Scientific Culture, University of Catania, Italy, pp. 287-300.

Mueller, K., J. de Frutos, K.-U. Schluessler, and H. Haarbosch, 1994: *Environmental Reporting & Disclosures: The Financial Analyst's View*. European Federation of Financial Analysts' Societies, Basle, Switzerland, 26 pp.

Muir-Wood, R.M., 1993: Windstorm hazard in northwest Europe. In: *Natural Disasters: Protecting Vulnerable Communities* [Merriman, P.A. and C.W.A. Browitt (eds.)]. Proceedings of the IDNDR Conference, London, 13-15 October, Thomas Telford, London, UK, pp. 517-533.

Munich Re, 1984: *Hailstorm*. Munich Re, Munich, Germany, 56 pp.

Munich Re, 1990: *Windstorm—New Loss Dimensions of a Natural Hazard*. Munich Reinsurance Company, Munich, Germany, 116 pp.

Munich Re, 1993: *Winter Storms in Europe—Analysis of 1990 Losses and Future Loss Potential*. Munich Reinsurance Company, Munich, Germany, 55 pp.

Munkhammar, A. and R. Themptander, 1984: Catastrophe insurance. *Geneva Papers on Risk and Insurance*, **9(31)**, 131-134.

Murray, C., 1993: Catastrophe reinsurance crisis in the Caribbean. *Catastrophe Reinsurance Newsletter*, 6 August, DYP Group Ltd, London, UK, pp. 14-18.

Muth, M., 1993: Facing up to the losses. *The McKinsey Quarterly*, **2**, 6-11.

Nanjira, D.D.C.D., 1991: Disaster and development in East Africa. In: *Managing Natural Disasters and the Environment* [Kreimer, A. and M. Munasinghe (eds.)]. Proceedings of a colloquium sponsored by the World Bank, 27-28 June 1990, World Bank, Washington, DC, pp. 82-89.

National Committee on Property Insurance, 1992a: *Natural Disaster Loss Reduction Update*, 1(1). NCPI, Boston, MA, 4 pp.

National Committee on Property Insurance, 1992b: *Natural Disaster Loss Reduction Update*, 2(1). NCPI, Boston, MA, 12 pp.

National Committee on Property Insurance, 1993a: *Natural Disaster Loss Reduction Update*, 2(1). NCPI, Boston, MA, 5 pp.

National Committee on Property Insurance, 1993b: *Natural Disaster Loss Reduction Update*, 2(2). NCPI, Boston, MA, 4 pp.

National Committee on Property Insurance, 1993c: *Natural Disaster Loss Reduction Update*, 2(3). NCPI, Boston, MA, 6 pp.

Natsios, A.S., 1991: Economic incentives and disaster mitigation. In: *Managing Natural Disasters and the Environment* [Kreimer, A. and M. Munasinghe (eds.)]. Proceedings of a colloquium sponsored by the World Bank, 27-28 June 1990, World Bank, Washington, DC, pp. 111-114.

Nesbeth, O., 1994: Address to the Fourteenth Caribbean Insurance Conference, 30 May, General Brokers and Agents Ltd, Nassau, Bahamas, 7 pp. (unpublished).

Nierhaus, F., 1986: A strategic approach to insurability of risks. *Geneva Papers on Risk and Insurance*, **11(39)**, 83-90.

Noda, A. and T. Tokioka, 1989: The effect of doubling the CO_2 concentration on convective and non-convective precipitation in a general circulation model coupled with a simple mixed-layer ocean. *Journal of the Meteorological Society of Japan*, **67**, 1055-1067.

Northcott, J., 1991: *Britain in 2010: The PSI Report*. Policy Studies Institute, London, UK, 364 pp.

Norton, J. and G. Chantry, 1993: Promoting principles for better typhoon resistance in buildings—a case study in Vietnam. In: *Natural Disasters: Protecting Vulnerable Communities* [Merriman, P.A. and C.W.A. Browitt (ed.)]. Thomas Telford, London, UK, pp. 534-546.

Nutter, F.W., 1994: The role of government in the United States in addressing natural catastrophes and environmental exposures. *Geneva Papers on Risk and Insurance*, **19(72)**, 244-256.

Olsthoorn, A.A. and R.S.J. Tol, 1993: *Socio-Economic and Policy Aspects of Changes in the Incidence of Extreme (Weather) Events*. Amsterdam Workshop, 24-25 June 1993, Institute for Environmental Studies, Free University of Amsterdam, The Netherlands, 33 pp.

Paish, A.G.C., 1993: The insurer's response to climate change—an overview. In: *Climate Change and the Insurance Industry*. Symposium by Greenpeace, 24th May, Greenpeace, London, UK, 4 pp.

Palutikof, J.P., 1983: The impact of weather and climate on industrial production in Great Britain. *Journal of Climatology,* **3**, 65-79.

Palutikof, J.P., X. Guo, T.M.L. Wigley, and J.M. Gregory, 1992: *Regional Changes in Climate in the Mediterranean Basin Due to Global Greenhouse Warming*. MAP Technical Report Series 66, Mediterranean Action Plan, United Nations Environment Programme, Athens, Greece, 172 pp.

Palutikof, J.P., C.M. Goodess, and X. Guo, 1994: Climate change, potential evapotranspiration and moisture availability in the Mediterranean Basin. *International Journal of Climatology*, **14**, 853-869.

Palutikof, J.P., J.A. Winkler, C.M. Goodess, and J.A. Andresen, 1995: The simulation of daily time series from GCM output. Part 1: Comparison of model data with observations. Preprint Volume, Sixth Symposium on Global Change Studies, American Meteorological Society, Boston, MA (in press).

Parry, M.L. and T.R. Carter, 1985: The effect of climate variations on agricultural risk. *Climate Change,* **7**, 95-110.

Peele, B.D., 1988: Insurance and the greenhouse effect. In: *Greenhouse: Planning for Climate Change* [Pearman, G.I. (ed.)]. CSIRO Publications, East Melbourne, Victoria, Australia, pp. 588-601.

Porro, B., 1984: Natural disaster insurance, problems of capacity and solvency. *Geneva Papers on Risk and Insurance,* **9(31)**, 167-174.

Porter, R.W. (ed.), 1994: Should governments be insurer of last resort? Alexander & Alexander National Seminar, 1 June, Alexander & Alexander (UK) Ltd, London, UK.

Powers, I.Y. and P. Hearn, 1993: *Coping with the Catastrophe Coverage Crisis*. Emphasis 1993(4), Tillinghast, Towers Perrin, New York, NY, pp. 6-9.

Property Claims Services, 1995: Cat news. In: *Catastrophe Reinsurance Newsletter,* 29 June, Lloyd's of London Press Ltd., London, UK, pp. 1-5.

Putty, M., 1984: *Crop Insurance—The Mauritian Experience*. Seminar on Insurance for Rural Sector, 18-20 July, Insurance Institute of Tanzania, Dar-es-Salaam, Tanzania, 44 pp.

Reilly, F.V., 1984: National flood insurance program flood insurance manual revision rates and rules. In: *Natural Disasters and Insurance* IV Etudes et dossiers No. 77, Geneva Association, Geneva, Switzerland, pp. 1-14.

Rind, D., R. Goldberg, J. Hansen, C. Rosenzweig, and R. Ruedy, 1990: Potential evapotranspiration and the likelihood of future drought. *Journal of Geophysical Research,* **95**, 9983-10,004.

Robbins, J., 1993: Legal and financial liability in disaster compensation at the local level: an Australian case study. In: *Catastrophe Insurance for Tomorrow: Planning for Future Adversities* [Britton, N.R. and J. Oliver (eds.)]. Proceedings of a seminar sponsored by Sterling Offices (Australia) Ltd., Griffith University, Brisbane, Australia, pp. 99-114.

Roberts, T., 1990: Management of risk in a changing environment. In: *Climatic Change—Impacts on New Zealand*. Ministry for the Environment, Wellington, New Zealand, pp. 226-228.

Roeckner, E., K. Arpe, L. Bergtsson, S. Brinkop, L. Dumeril, M. Esch, E. Kirk, F. Lunkeit, M. Ponater, B. Rockel, R. Sausen, V. Schlese, S. Schubert, and M. Windeband, 1992: *Simulation of the Present-Day Climate with the ECHAM Model*. MPI Report 93 (available from Max Planck Institut fur Meteorologie, Bundestrasse 55,20146 Hamburg, Germany), 32 pp.

Saunders, A., 1993: Underwriting guidelines. Presented at the thirteenth Caribbean Insurance Conference, Association of British Insurers, London, UK, 7 pp. (unpublished).

Schinke, 1992: Zum auftreten von zyklonen mit niedrigen kerndrucken in atlantisch—europaischen raum von 1930 bis 1991. Wiss, Zeitschrift der Humboldt Universitat zu Berlin. *R. Mathematik/Naturwiss*, **41**, 17-28.

Schraft, A., E. Durand, P. Hausmann, 1993: *Storms over Europe: Losses and Scenarios*. Swiss Reinsurance Company, Zurich, Switzerland, 28 pp.

Schmidt, H. and H. Von Storch, 1993: German blight storms analysed. *Nature*, **365**, 791.

Seebauer, R., 1992: Bank and insurance companies in the field of financial services. In: *Insurance in a Changing Europe*. Symposium of 18th October, 1991, Gerling-Konzern Globale, Cologne, France, pp. 40-42.

Shearn, W.G., 1994: Personal lines pricing—insurance or discrimination. Sixteenth UK Insurance Economists Conference, Cripps Hall, University of Nottingham, 20-21 April, Department of Insurance, University of Nottingham, UK.

Siegel, S.L. and P. Witham, 1991: Case study: Jamaica. In: *Managing Natural Disasters and the Environment* [Kreimer, A. and M. Munasinghe (eds.)]. Proceedings of a colloquium sponsored by the World Bank, 27-28 June 1990, World Bank, Washington, DC, pp. 166-167.

Smith, D.I., 1991: Extreme floods and dam failure inundation: implications for loss assessment. In: *Natural and Technological Hazards: Implications for the Insurance Industry* [Britton, N.R. and J. Oliver (eds.)]. Proceedings of a seminar sponsored by Sterling Offices (Australia) Ltd., University of New England, Armidale, NSW, Australia, pp. 13-29.

Sparks, P.R., 1993: Hurricane resistance by accident or design. In: *Natural Disasters: Perceiving Our Vulnerabilities and Understanding Our Perceptions*. Annual forum of the National Committee on Property Insurance, December 16, Boston, NCPI, Boston, MA, pp. 24-36.

Stakhiv, E. and L. Vallianos, 1993: *Impacts of Global Warming in Coastal Zones—Accelerated Sea Level Rise and Extreme Events*. Working paper for Working Group II, Intergovernmental Panel on Climate Change (IPCC), Washington, DC, 22 pp.

Staveley, J.K., 1991: 1989—All that and more? Australian hazard insurance experiences during the period 1989-1991: implications for the industry in a period of economic downturn. In: *Natural and Technological Hazards: Implications for the Insurance Industry* [Britton, N.R. and J. Oliver (eds.)]. Proceedings of a seminar sponsored by Sterling Offices (Australia) Ltd., University of New England, Armidale, New South Wales, Australia, pp. 13-29.

Stone, C.D., 1992: Beyond Rio: insuring against global warming. *The American Journal of International Law,* **86**, 445-488.

Swiss Re, 1991: *Results of Non-Life Insurers in 8 Countries*. Sigma (6), Swiss Re, Zurich, Switzerland, 24 pp.

Swiss Re, 1992: *The Insurance Industry in the Context of Finance and Financial Innovation*. Sigma (7), Swiss Re, Zurich, Switzerland, 27 pp.

Swiss Re, 1994a: *Sigma 2*. Swiss Reinsurance Company, Zurich, Switzerland, 48 pp.

Swiss Re, 1994b: *World Insurance in 1992*: Sigma (3), Swiss Re, Zurich, Switzerland, 30 pp.

Sykes, C., 1991: Learning from traditional responses. In: *Managing Natural Disasters and the Environment* [Kreimer, A. and M. Munasinghe (eds.)]. Proceedings of a colloquium sponsored by the World Bank, 27-28 June 1990, World Bank, Washington, DC, pp. 178-179.

Thompson, H., 1995: The financial sector. In: *The Second Assessment Report of the Climate Change Impacts Review Group*. Department of the Environment, HMSO, London, UK (in press).

Tiedemann, H., 1991: Assessing the risk of catastrophes. *Foresight, Journal of the Institute of Risk Management*, August 4-10 (pp. 3-5) and September 5 -10 (pp. 6-8).

Tillinghast, 1994: *Insurance Pocket Book*. NTC Publications, Henley-on-Thames, UK, 213 pp.

Tinsley, B.A., 1988: The solar cycle and the QBO influence on the latitude of storm tracks in the North Atlantic. *Geophysical Research Letters,* **15(5)**, 409-412.

Tol, R.S.J. (ed.), 1993: *Socio-Economic and Policy Aspects of Change in the Incidence and Intensity of Extreme (Weather) Events*. Amsterdam Workshop, 24-25 June 1993, Free University of Amsterdam, The Netherlands, 220 pp.

Tooley, M.J., 1992: Sea level changes and forecasting flood risks. *Geneva Papers on Risk and Insurance,* **17(64)**, 406-414.

Towfighi, P., 1991: Integrated planning for natural and technological disasters. In: *Managing Natural Disasters and the Environment* [Kreimer, A. and M. Munasinghe (eds.)]. Proceedings of a colloquium sponsored by the World Bank, 27-28 June 1990, World Bank, Washington, DC, pp. 106-110.

UNCTAD, 1994: *Review of Developments in the Insurance Market: Alternatives for Insurance of Catastrophes, Environmental Impairments and Large Risks in Developing Countries.* TD/B/CN.4/32, 1 June, Trade and Development Board, Standing Committee on Developing Services Sectors, United Nations Conference on Trade and Development, Geneva, Switzerland, 43 pp.

Vaughan, S., 1994: *Environmental Risk and Commercial Banks.* Discussion paper for UNEP round table on commercial banks and the environment, 26-27 September, United Nations Environment Programme, Geneva, Switzerland, 71 pp.

Vellinga, P. and R.S.J. Tol, 1994: Climate change: extreme events and society's response. *Journal of Reinsurance,* **1**, 59-72.

von Storch, H., J. Guddal, K. Iden, T. Jonsson, J. Perlwitz, M. Reistad, J. de Ronde, H. Schmidt, and E. Zorita, 1993a: *Changing Statistics of Storms in the North Atlantic?* Max-Planck Institut fur Meteorologie Report 116 (available from MPI, Bundestrasse 55, 20146 Hamburg, Germany), 19 pp.

von Storch, H., E. Zorita, and U. Lubasch, 1993b: Downscaling of climate change estimates to regional scales: an application to winter rainfall in the Iberian Peninsula. *Journal of Climate,* **6**, 1161-1171.

von Storch, H., 1994: *Inconsistencies at the Interface of Climate Impact Studies and Global Climate Research.* MPI Report 122 (available from Max Planck Institut fur Meteorologie, Bundestrasse 55, 20146 Hamburg, Germany), 12 pp.

Walker, G.R., 1989: Losses due to the effect of wind and rain on structures during tropical cyclones. In: *Natural Hazards and Reinsurance* [Oliver, J. and N.R. Britton (eds.)]. Proceedings of a seminar sponsored by Sterling Offices (Australia) Ltd., Cumberland College of Health Sciences, Lidcombe, NSW, Australia, pp. 1-17.

Walker, G.R., 1992: Australian achievements in the mitigation of wind damage to housing from tropical cyclones. In: *IDNDR: Australia's Role in the South West Pacific* [Smith, D. and J. Handmer (eds.)]. Resource and Environmental Studies Number 5, Centre for Resource and Environmental Studies, Australian National University, Canberra, Australia, pp. 1-8.

Ward, B., 1991: Training in the Asian-Pacific region. In: *Managing Natural Disasters and the Environment* [Kreimer, A. and M. Munasinghe (eds.)]. Proceedings of a colloquium sponsored by the World Bank, 27-28 June 1990, World Bank, Washington, DC, pp. 135-140.

Whetton, P.H., A.M. Fowler, M.R. Haylock, and A.B. Pittock, 1993: Implications of climate change due to the enhanced greenhouse effect on floods and droughts in Australia. *Climatic Change,* **25**, 289-317.

Wigley, T.M.L., 1985: Impact of extreme events. *Nature*, **316**, 106-107.

Wilkinson, W.B. and F.M. Law, 1990: Engineering design for climate extremes. *Weather*, **45(4)**, 138-145.

Wilson, C.A. and J.F.B. Mitchell, 1987: Simulated climate and CO_2-induced climate change over western Europe. *Climatic Change,* **10**, 11-42.

Winkler, J.A., J.P. Palutikof, J.A. Andresen, and C.M. Goodess, 1995: The simulation of daily time series from GCM output. Part 2: Development of local scenarios for maximum and minimum temperature using statistical transfer functions. Preprint Volume, Sixth Symposium on Global Change Studies, American Meteorological Society, Boston, MA (in press).

Winters, R., 1993: The consumer movements and the impact on insurance: an American point of view. *Geneva Papers on Risk and Insurance,* **18(69)**, 412-425.

Yevjevich, V., 1992: Classification and description of flood mitigation measures. In: *Coping with Floods* [Rossi, G., N. Harmancioglu, and V. Yevjevich (eds.)]. Proceedings of NATO Advanced Studies Institute, October, E. Majorana Centre for Scientific Culture, University of Catania, Italy, pp. 375-392.

Zorita, E., J.P. Hughes, D.R. Lettenmaier, H. von Storch, 1995: Stochastic characterization of regional circulation patterns for climate model diagnostics and estimation of local precipitation. *Journal of Climate* (in press).

18

Human Population Health

ANTHONY J. McMICHAEL, AUSTRALIA/UK

Principal Lead Authors:
M. Ando, Japan; R. Carcavallo, Argentina; P. Epstein, USA; A. Haines, UK; G. Jendritzky, Germany; L. Kalkstein, USA; R. Odongo, Kenya; J. Patz, USA; W. Piver, USA

Contributing Authors:
R. Anderson, UK; S. Curto de Casas, Argentina; I. Galindez Giron, Venezuela; S. Kovats, UK; W.J.M. Martens, The Netherlands; D. Mills, USA; A.R. Moreno, Mexico; W. Reisen, USA; R. Slooff, WHO; D. Waltner-Toews, Canada; A. Woodward, New Zealand

CONTENTS

EXECUTIVE SUMMARY

The sustained health of human populations requires the continued integrity of Earth's natural systems. The disturbance, by climate change, of physical systems (e.g., weather patterns, sea-level, water supplies) and of ecosystems (e.g., agroecosystems, disease-vector habitats) would therefore pose risks to human health. The scale of the anticipated health impacts is that of whole communities or populations (i.e., it is a public health, not a personal health, issue). These health impacts would occur in various ways, via pathways of varying directness and complexity, including disturbance of natural and managed ecosystems. With some exceptions, relatively little research has yet been done that enables quantitative description of these probable health impacts.

It is anticipated that most of the impacts would be adverse. Some would occur via relatively direct pathways (e.g., deaths from heat waves and from extreme weather events); others would occur via indirect pathways (e.g., changes in the range of vector-borne diseases). Some impacts would be deferred in time and would occur on a larger scale than most other environmental health impacts with which we are familiar. If long-term climate change ensues, indirect impacts probably would predominate.

Populations with different levels of natural, technical, and social resources would differ in their vulnerability to climate-induced health impacts. Such vulnerability, due to crowding, food insecurity, local environmental degradation, and perturbed ecosystems, already exists in many communities in developing countries. Hence, because of both the geography of climate change and these variations in population vulnerability, climate change would impinge differently on different populations.

- An increased frequency or severity of heat waves would cause an increase in (predominantly cardiorespiratory) mortality and illness (High Confidence). Studies in selected urban populations in North America, North Africa, and East Asia indicate that the number of heat-related deaths would increase severalfold in response to two general circulation model (GCM)–modeled climate change scenarios for 2050. In very large cities, this would represent several thousand extra deaths annually. Although this heat-related increase in deaths would be partially offset by fewer cold-related deaths, there are insufficient data to quantify this tradeoff; further, this balance would vary by location and according to adaptive responses (Medium Confidence).
- If extreme weather events (droughts, floods, storms, etc.) were to occur more often, increases in rates of death, injury, infectious diseases, and psychological disorders would result (High Confidence).
- Net climate change-related increases in the geographic distribution (altitude and latitude) of the vector organisms of infectious diseases (e.g., malarial mosquitoes, schistosome-spreading snails) and changes in the life-cycle dynamics of both vector and infective parasites would, in aggregate, increase the potential transmission of many vector-borne diseases (High Confidence). Malaria, of which there are currently around 350 million new cases per year (including two million deaths), provides a central example. Simulations with first-generation mathematical models (based on standard climate-change scenarios and incorporating information about the basic dynamics of climatic influences on malaria transmission) predict an increase in malaria incidence in Indonesia by 2070 and—with a highly aggregated model—an increase from around 45% to around 60% in the proportion of the world population living within the *potential* malaria transmission zone by the latter half of the next century. Although this predicted increase in *potential* transmission encroaches mostly into temperate regions, actual climate-related increases in malaria incidence (estimated by one model to be of the order of 50–80 million additional cases annually, relative to an assumed global background total of 500 million by 2100) would occur primarily in tropical, subtropical, and less well protected temperate-zone populations currently at the margins of endemically infected areas. Some localized decreases may also occur (Medium Confidence).
- Increases in non-vector-borne infectious diseases such as cholera, salmonellosis, and other food- and water-related infections also could occur, particularly in tropical and subtropical regions, because of climatic impacts on water distribution, temperature, and microorganism proliferation (Medium confidence).
- The effects of climate change on agricultural, animal, and fisheries productivity, while still uncertain, could increase the prevalence of malnutrition and hunger and their long-term health impairments, especially in children. This would most probably occur regionally, with some regions likely to experience gains, and others losses, in food production (Medium Confidence).
- There would also be many health impacts of the physical, social, and demographic disruptions caused by

rising sea levels and by climate-related shortages in natural resources (especially fresh water) (Medium Confidence).

- Because fossil-fuel combustion produces both carbon dioxide and various primary air pollutants, the climate change process would be associated with increased levels of urban air pollution. Not only is air pollution itself an important health hazard, but hotter temperatures, in urban environments, would enhance both the formation of secondary pollutants (e.g., ozone) and the health impact of certain air pollutants. There would be increases in the frequency of allergic disorders and of cardiorespiratory disorders and deaths caused by various air pollutants (e.g., ozone and particulates) (High Confidence).
- A potentially important category of health impact would result from the deterioration in social and economic circumstances that might arise from adverse impacts of climate change on patterns of employment, wealth distribution, and population mobility and settlement. Conflicts might arise over dwindling environmental resources (Medium Confidence).
- Stratospheric ozone is being depleted concurrently with greenhouse gas accumulation in the troposphere. Although there are some shared and interactive atmospheric processes between disturbances of the stratosphere and troposphere, both they and their health impacts arise via quite distinct pathways. A sustained 10–15% depletion of stratospheric ozone over several decades would cause increased exposure to ultraviolet radiation and an estimated 15–20% increase in the incidence of skin cancer in fair-skinned populations (High Confidence). Lesions of the eye (e.g., cataracts) also may increase in frequency, as might vulnerability to some infectious diseases via adverse effects on immune function (Medium Confidence).

Adaptive options to minimize health impacts include improved and extended medical care services; environmental management; disaster preparedness; protective technology (housing, air conditioning, water purification, vaccination, etc.); public education directed at personal behaviors; and appropriate professional and research training. It also will be important to assess in advance any risks to health from proposed technological adaptations (e.g., exposures that could result from using certain alternative energy sources and replacement chemicals for chlorofluorocarbons; effects of pesticide use on resistance of vector organisms and their predator populations).

There is immediate need for improved and internationalized monitoring of health-risk indicators in relation to climate change. Existing global monitoring activities should encompass health-related environmental and bioindicator-species measurements and, where appropriate, direct measures of human population health status. To assist the evolution of public understanding and social policy, the health sciences must develop improved methods, including integrated predictive models, to better assess how climate change (and other global environmental changes) would influence human health.

In conclusion, the impacts of global climate change, particularly if sustained in the longer term, could include a multitude of serious—but thus far underrecognized—impacts on human health. Human population health is an outcome that integrates many other inputs, and it depends substantially on the stability and productivity of many of Earth's natural systems. Therefore, human health is likely to be predominantly adversely affected by climate change and its effects upon those systems.

18.1. Introduction

18.1.1. Climate Change and Human Population Health: The Nature of the Relationship

Global climate change over the coming decades would have various effects upon the health of human populations (WHO, 1990; Haines and Fuchs, 1991; McMichael, 1993; Last, 1993; *Lancet*, 1994). Because of the nature of the exposures involved, the scale of these climate-related changes would, in general, apply to whole populations or communities, rather than to small groups or individuals. The assessment of health impacts therefore focuses on changes in rates of death or disease in populations.

Many of the health impacts of climate change would occur via processes that are relatively unfamiliar to public-health science. They would not occur via the familiar toxicological mechanisms of localized exposure to environmental contaminants, nor via locally determined influences on the spread of infectious diseases. Instead, many of the impacts would arise via the indirect and often delayed effects of disturbances to natural systems and their associated ecological relationships. For example, changes in background climate may alter the abundance, distribution, and behavior of mosquitoes and the life cycle of the malarial parasite, such that patterns of malaria would change. Climate change also would have varied regional effects on agricultural productivity, so that some vulnerable populations may experience nutritional deprivation. There also would be some rather more readily predictable health impacts, arising, for example, from more frequent or severe heatwaves.

On a wider canvas, several of the world's ecosystems that are important in sustaining human health already have been weakened by damage, habitat loss, and species/genetic depletion. These include agricultural lands and ocean fisheries and the terrestrial ecosystems that influence the transmission of infectious diseases. Climate change may, via various processes, exacerbate those ecosystem disturbances. Because an ecosystem comprises a suite of interacting components, in which member organisms relate to the whole suite rather than to individual parts, the uncoupling of relationships by climate change could initiate a cascade of disturbances that might jeopardize human population health.

The range of potential major types of health impact is shown in Figure 18-1. For simplicity, they have been classified as "direct" and "indirect," according to whether they occur predominantly via the direct impact of a climate variable (temperature, weather variability, etc.) upon the human organism or are mediated by climate-induced changes in complex biological and geochemical systems or by climatic influences on other environmental health hazards.

18.1.2. Forecasting Health Impacts: The Challenge to Health Science

Predictions of future trends in population health are readily made in relation to actual current exposures—for example, future lung cancer rates can be predicted as a function of a population's current cigarette smoking habits (Peto *et al.*, 1994). It is unusual to make predictions on the basis of some anticipated future profile of exposure (e.g., smoking habits in the year 2020), yet this is the nature of the present exercise: Potential health impacts are being assessed in relation to future scenarios of climate change. There are inevitable, multiple uncertainties in such an approach (McMichael and Martens, 1995).

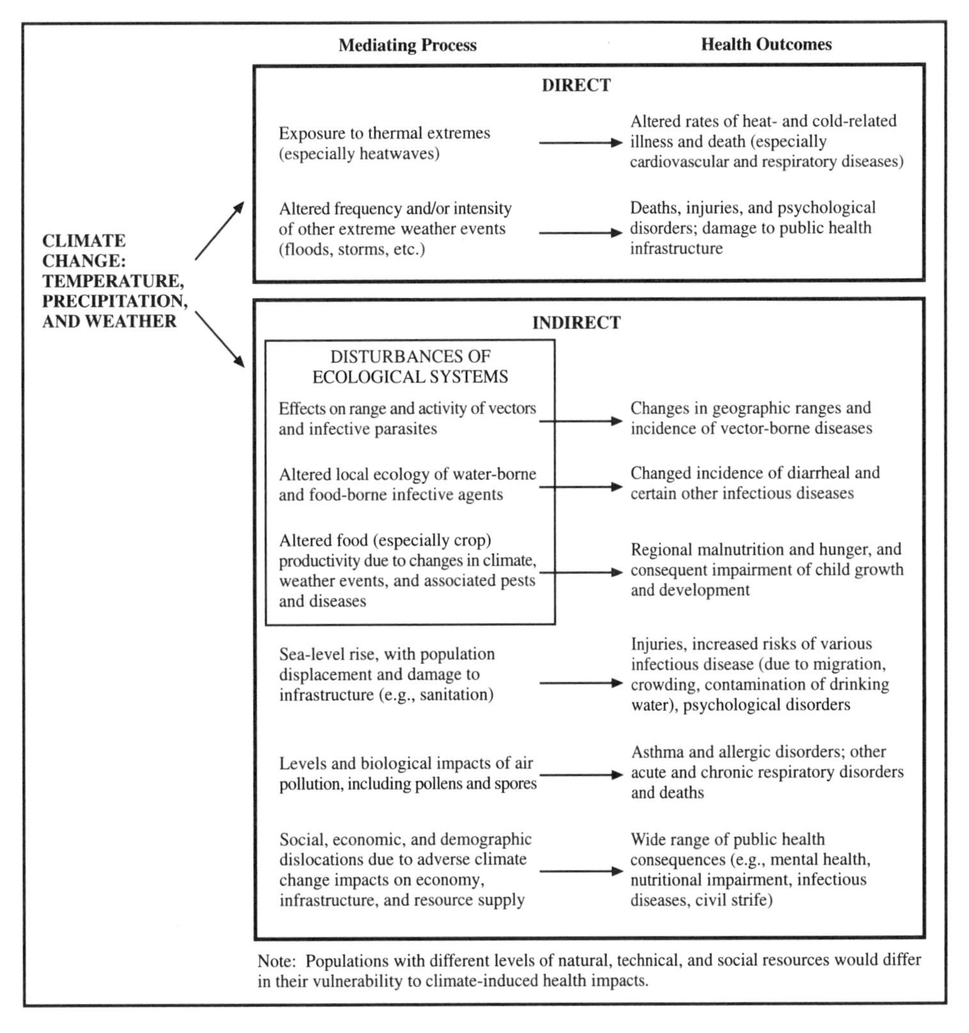

Figure 18-1: Ways in which climate change can affect human health.

Some aspects of climate change and its first-level impacts (on sea level, coastal ecosystems, forests, agriculture, fisheries, etc.), as projected by IPCC Working Group I, would lie outside the range of recorded human health experience. Hence, the forecasting of human health impacts must rely principally upon reference to historical analogy (where available) and reasonable extrapolation, judgment, and the use of integrated mathematical models (Niessen and Rotmans, 1993; McMichael and Martens, 1995). Most of the health impact modeling to date has been at a highly aggregated global or regional level, with no capacity to make finer-grained predictions. Ongoing developments in modeling techniques include the incorporation of complex nonlinear relationships and feedback processes, dealing with uncertainty, and richer use of local detail (Alcamo *et al.*, 1994a, 1994b).

Some health impacts can be forecast by relatively simple extrapolation from empirical epidemiological dose-response data. For example, models to predict the mortality impacts of an increase in heatwaves can be based on existing empirical data. However, predicting the health impact of climate-induced shifts in ecological relationships and habitat boundaries (e.g., malarial mosquitoes, agricultural crops) poses a more complex challenge. Equally complex is the task of predicting the various indirect health impacts of such things as sea-level rise or of civil disruption and enforced migration because of deteriorating environments and dwindling resources.

A further difficulty arises with regional predictions. Not only is the ability to predict regional differences in climate change still limited, but human populations differ greatly in their environmental circumstances, social resources, and preexisting health status. They therefore differ in their vulnerability to climate-induced stresses. Although there is generally insufficient information available to make differentiated assessments of health impacts in different populations, the balance of the assessments published to date is that, because of the geography of the impacts of climate change (particularly in relation to infectious disease transmission and food production) and of population vulnerability, many of the anticipated adverse health impacts would be greater in the world's less-developed regions. Nevertheless, in developed countries, demographic trends including population aging and increasing levels of disability, chronic illness, and coastal retirement may increase the vulnerability of populations.

Although it is tempting to compile a list of discrete health impacts, it is important that the systemic quality of the impact of climate change be understood. There would be many cross-links, including interactions of climate change with other, coexisting environmental changes. Further, the attempts of society to mitigate the health effects of climate change may actually exacerbate some of them. For example, increased use of fertilizers to compensate for a decline in local agricultural production can increase algal blooms and hence the risk of cholera or shellfish poisoning; or the relocation of populations from drought-stricken areas might introduce them to unfamiliar pathogens (or introduce their unfamiliar pathogens). Neither the scope of this chapter nor available scientific knowledge allows comprehensive consideration of these issues.

18.1.3. Sensitivities and Thresholds

Forecasting health impacts requires knowledge of, first, the sensitivity of change in population health status in response to climate change and, second, of any associated thresholds. "Sensitivity" refers to the rate of change in health outcome per unit change in climate (however defined), whereas "threshold" refers to a sudden change in slope or curvature in that dose-response graph (e.g., a certain amount of climate change may be tolerated before an impact on some particular health outcome occurs).

Although there is only limited information on the sensitivities and thresholds of human health response to climate change, some illustrative comments can be made:

- The impact of thermal stress depends on physiological tolerance thresholds being exceeded (see Section 18.2.1). For example, most studies of heat-related mortality show that increased mortality occurs only after a critical temperature has been exceeded for a certain duration. This critical temperature varies geographically, reflecting socioeconomic differences, physiological acclimatization, and cultural-technical adaptation.
- Infectious disease pathogens and insect pests are normally constrained by bioclimatic thresholds. That is, there is a range of climatic conditions with upper and lower thresholds within which a population of organisms is viable (Dobson and Carper, 1993). These thresholds account for seasonal and longer-term fluctuations in the distribution and abundance of most organisms. The uncoupling, by climate change, of previously stable ecological relationships between species would reveal other thresholds as population imbalances pass critical points.
- Changes in the distribution of organisms (e.g., mosquitoes) that spread vector-borne diseases (e.g., malaria) would occur if climate change causes their geographic range to shift. This shift would reflect the critical thresholds of temperature, precipitation, and humidity (i.e., the bioclimatograph) for vector maturation and persistence. For example, the range of *Plasmodium vivax* malaria is limited because the parasite cannot develop inside its mosquito host at temperatures below 14–16°C (Gilles, 1993). Further, blood-feeding arthropods feed and reproduce only above certain temperatures and need less time to complete their life cycle as temperatures increase above that threshold (Curto de Casas and Carcavallo, 1984; Burgos *et al.*, 1994). Threshold effects also apply to the life cycle of the infecting parasite.
- Some animal and plant populations would be unable to migrate or adapt behaviorally to climate change, in part because of constraints imposed by non-climate

variables such as day-length and soil type. Other species with either migratory or dispersive activity (particularly arthropod and weed pests in agriculture and disease vectors that affect human health) would cope with the shifts in climate zones. Humans generally would be less sensitive to changes in background climate because of their capacity to adapt via culture, technology, migration, and behavior.

- The sensitivity of health-outcome response depends on population susceptibility. For example, the impact of a climate-related increase in exposure to infectious agents would depend on prior contact (i.e., herd immunity), on general biological resilience (especially nutritional and immune status), and on population density and patterns of interpersonal contact. Social infrastructure and health-care resources also would condition the impact. In general, the most vulnerable populations or communities would be those living in poverty, with a high prevalence of undernutrition, chronic exposure to infectious disease agents, and inadequate access to social and physical infrastructure.

Table 18-1 lists illustrative examples of the relative sensitivity of selected human health responses to key aspects of climate change. It also summarizes what is known about thresholds in those responses. Further, the "conditionality" of response (i.e., the extent to which it is modulated by other influences) is indicated. This latter criterion is important: Many factors (including the intrinsic vulnerability of the local population) influence the determination of health status, and many of these would condition the health impact of climate change.

Further, many of the impacts of climate change would depend on parameters other than the central changes in mean values. For example, weather variability would be important for extreme events; rates of change would influence the production in ecosystems (e.g., agriculture and fisheries); and the ecology of infectious disease vectors is sensitive to many aspects of climate, including changes in the day-night differential in temperature. These aspects of climate change would vary in their relative importance for different health impacts, as illustrated in Table 18-2. However, information on such details is still rather incomplete.

18.1.4. Major Trends in World Health: Backdrop to Climate Change Impacts

Information about levels and time trends in specific health outcomes facilitates both the prediction and the appraisal of the health impacts of climate change. For example, patterns of malaria and malnutrition are changing around the world in response to various other changes in social, biological, and ecological circumstances. Any predicted impact of climate change upon such health outcomes should be assessed either by differentiating that impact from those of other independent background trends or, if appropriate, by assessing its interactive impact with those other, coexisting influences.

The main contemporary features of world health (World Bank, 1993; Murray and Lopez, 1994) are as follows: (1) near-worldwide increases in life expectancy [with the ex-Eastern Bloc countries standing in sharp recent contrast (Feachem, 1994)], (2) a decline in infant and child mortality in most developing countries, (3) persistent gradients in health status between rich and poor (within and between populations), (4) reductions in certain vaccine-preventable diseases (e.g., polio and measles), (5) increases in the chronic noninfectious diseases of adult life

Table 18-1: *Examples of important aspects of the relationship of selected health outcomes to climate change.*

Health Outcome	**Sensitivity**[a]	**Conditionality**[b]	**Thresholds**
Climate Stress/Mortality	+++	++	Temperature 33°C[c]
Climate Stress/Morbidity	+		
Allergy	++	++	Not applicable
Asthma	+	+	Not applicable
Vector-Borne Diseases (malaria, yellow fever, dengue, onchocerciasis, encephalitis)[d]	++(+)	++++	Temperature isotherms and humidity levels (bioclimatic functions) e.g., 10°C: *Ae.* mosquito, 14–16°C: *P. vivax* parasite
Other Infectious Diseases (e.g., cholera)	+(++)		Temperature thresholds for algal and bacterial proliferation in sea- and freshwater—optimal ranges

++++ = great effect, + = small effect, (+) = possible additional effect

[a] Extent of change in health outcome per unit change in climate (equivalent to "slope" of regression line).

[b] Extent to which sensitivity depends on preceding and coexistent circumstances (i.e., notions of vulnerability/susceptibility and interactive effects).

[c] Based principally on northeastern U.S. data. Critical temperature depends on local climate and population acclimatization.

[d] See also Table 18-3.

Table 18-2: Probable relative impact on health outcomes of the aspects of climate change.

Health Outcome	Change in Mean Temperature	Extreme Events	Rate of Change of Climate Variable	Day-Night Difference
	Aspects of Climate Change			
Heat-Related Deaths and Illness		+++		+
Physical and Psychological Trauma due to Disasters		++++		
Vector-Borne Diseases	+++	++	+	++
Other Infectious Diseases	+	+		
Food Availability and Hunger	++	+	++	
Consequences of Sea-Level Rise	++	++	+	
Respiratory Effects				
– Air Pollutants	+	++		+
– Pollens, Humidity	++			
Demographic Disruption	++	+	+	

Notes: ++++ = great effect, + = small effect; empty cells indicate no known relationship.

(especially heart disease, diabetes, and certain cancers) in urban middle classes in rapidly developing countries, and (6) widespread increases in HIV infection. Rates of disease and death from cigarette smoking are likely to escalate markedly in many countries over the coming decades, as the tobacco industry takes advantage of freer trade and market-based economies (Peto *et al.*, 1994). In many urban populations, drug abuse and violence are increasing.

There appears to be a widespread increase in the tempo of new and resurgent infectious diseases (Levins *et al.*, 1994). This primarily reflects the combination of environmental and demographic changes in the world, plus increases in antibiotic and drug resistance, pesticide resistance, and decreased surveillance (Morse, 1991; CDC, 1994a). The interaction of local climate change with other disruptions of ecosystems may have facilitated various infectious disease outbreaks: the emergence of rodent-borne hantavirus pulmonary syndrome in the United States during 1992–3; various rodent-borne arenaviruses in Africa and South America; the spread of harmful algal blooms—and its association with cholera (which is now affecting more nations worldwide than at any earlier time this century); the rapid resurgence of dengue in the Americas since 1981; and the occurrence of dengue and malaria at higher altitudes than previously recorded (Levins *et al.*, 1994).

Many other important influences on population health are changing over time. For example, new vaccines are being developed, and existing ones are being used more widely; contraception is becoming more widely used, with benefits to maternal and child health; and safe drinking water is becoming available, albeit slowly, to an increasing proportion of householders in poorer countries. Other adverse effects (from cigarettes, drugs, urban traffic, social breakdown, violence, etc.) are increasing widely, and persistent widespread poverty remains a major structural impediment to improved health (WHO, 1995a). Against this complex balance sheet, it is inevitably difficult to estimate the likely net impact on population health status after the additional inclusion of climate change.

18.2. Potential Direct Health Impacts of Climate Change

The direct effects of climate change upon health result from changes in climate characteristics or short-term weather extremes that impinge directly on human biology. The following subsections deal with the health impacts of thermal stress and extreme weather events.

18.2.1. Health Impacts of Altered Patterns of Thermal Stress

Many studies, particularly in temperate countries, have observed a J-shaped relationship (often more generally referred to as a U-shaped relationship) between daily outdoor temperature and daily death rate: Mortality is lowest within an intermediate comfortable temperature range. The graph is not symmetrical; the death rate increases much more steeply with rising temperatures, above this comfort zone, than it does with falling temperatures below that zone (Longstreth, 1989; Kalkstein, 1993; Kunst *et al.*, 1993; Touloumi *et al.*, 1994). Relatedly, death rates in temperate countries appear to be affected across a wider band of decreasing, cold, temperatures than that for increasing, hot, temperatures (Kilbourne, 1992; Kunst *et al.*, 1993). Because death rates in temperate and subtropical zones are higher in winter than in summer (Kilbourne, 1992), it is a reasonable expectation that milder winters in such countries would entail a reduction in cold-related deaths and illnesses. However, since summer-related deaths appear to be more related to temperature extremes than are winter-related deaths, this reduction may not fully offset the heat-related increases. Further, this balance of gains and losses would vary among geographic regions and different populations. The issue of balance is examined further later in this section.

Quantitative interpretation of the impacts of altered daily-temperature distribution also is hampered by the fact that, for both heat-related and cold-related deaths, many of the apparent excess deaths occur in already-vulnerable persons (especially the elderly and the sick). Some analyses indicate that, in the absence of extreme temperatures, many of those persons would have died in the near future. This "mortality displacement" issue also will be considered further.

Global warming is predicted to increase the frequency of very hot days (see Chapter 6, *Climate Models–Projections of Future Climate*, of the IPCC Working Group I volume). The frequency of such days in temperate climates (e.g., USA, UK, Australia) would approximately double for an increase of 2–3°C in the average summer temperature (e.g., CDC, 1989; Climate Change Impacts Review Group UK, 1991). Extensive research has shown that heat waves cause excess deaths (Weihe, 1986; Kilbourne, 1992). Recent analyses of concurrent meteorological and mortality data in cities in the United States, Canada, the Netherlands, China, and the Middle East provide confirmatory evidence that overall death rates rise during heat waves (Kalkstein and Smoyer, 1993; Kunst *et al.*, 1993), particularly when the temperature rises above the local population's threshold value. Therefore, it can be predicted confidently that climate change would, via increased exposure to heat waves, cause additional heat-related deaths and illnesses (Kalkstein, 1993; Haines *et al.*, 1993).

The effect of extreme heat on mortality is exacerbated by low wind, high humidity, and intense solar radiation (Kilbourne, 1992). Indeed, these meteorological elements can be treated synoptically, to evaluate the net impact of weather on human health. For example, recent studies in the United States have described "oppressive" air masses (analogous to the meteorologist's "stagnating" air masses), which represent synoptic meteorological situations that exceed physiological tolerance levels. This approach recognizes that humans principally respond to the umbrella of air that surrounds them rather than to individual meteorological elements (Kalkstein, 1993). It also should be noted that concurrent hot weather and air pollution have interactive impacts on health (Katsouyanni *et al.,* 1993; see also Section 18.3.5).

Healthy persons have efficient heat regulatory mechanisms that cope with increases in temperature up to a particular threshold. The body can increase radiant, convective, and evaporative heat loss by vasodilation (enlargement of blood vessels in the skin) and perspiration (Horowitz and Samueloff, 1987; Diamond, 1991). Further, some acclimatization to persistent oppressive weather conditions can occur within several days (Kilbourne, 1992). Nevertheless, the risk of death increases substantially when thermal stress persists for several consecutive days (Kalkstein and Smoyer, 1993). The elderly and very young are disproportionately affected because of their limited physiological capacity to adapt. Although some individuals die from heat exhaustion or heatstroke, the deaths associated with very hot weather are predominantly associated with preexisting cardiovascular and respiratory disorders, as well as accidents (Larsen, 1990a, 1990b). Although there is less evidence on nonfatal illness episodes, it is a reasonable general assumption that thermal stress also increases rates of morbidity.

Socioeconomic factors may have important modulating effects on thermal stress-related mortality. From studies in the United States, the extent of protection from air conditioning remains unclear (Ellis and Nelson, 1978; Larsen, 1990a; Rogot *et al.*, 1992; Kalkstein, 1993). Of more general importance, people living in poverty, including segments of many urban populations in developing countries, are particularly vulnerable to heat stress. Poor housing, the urban heat island effect, and lack of air conditioning are among the primary causes (Kilbourne, 1989). Complete acclimatization may take up to several years (Babeyev, 1986), rendering immigrants (e.g., rural-to-urban) vulnerable to weather extremes for a considerable time. Ongoing rapid increases in urbanization (see Chapter 12) will increase the number of vulnerable persons.

The question in relation to the abovementioned notion of "mortality displacement" is: Would some of those who die during heat waves have succumbed soon afterward from preexisting frailty or disease (Kalkstein, 1993; U.S. EPA, in press)? Various time-series analyses indicate a "deficit" in daily deaths for up to a month after heat waves (e.g., Figure 18-2), and U.S.-based research suggests that 20–40% of the deaths occurring during heat waves would have occurred within the next few weeks (Kalkstein, 1993). A related uncertainty is whether, as the frequency of heat waves increases, the mortality excess remains constant or whether, as some research suggests, successive heat waves entail a progressive lessening of the associated mortality peak.

Kalkstein (1993) made predictions of heat-related mortality for selected urban populations in North America, North Africa, and East Asia in relation to IPCC-specified climate-change scenarios. This entailed, first, identifying for each population

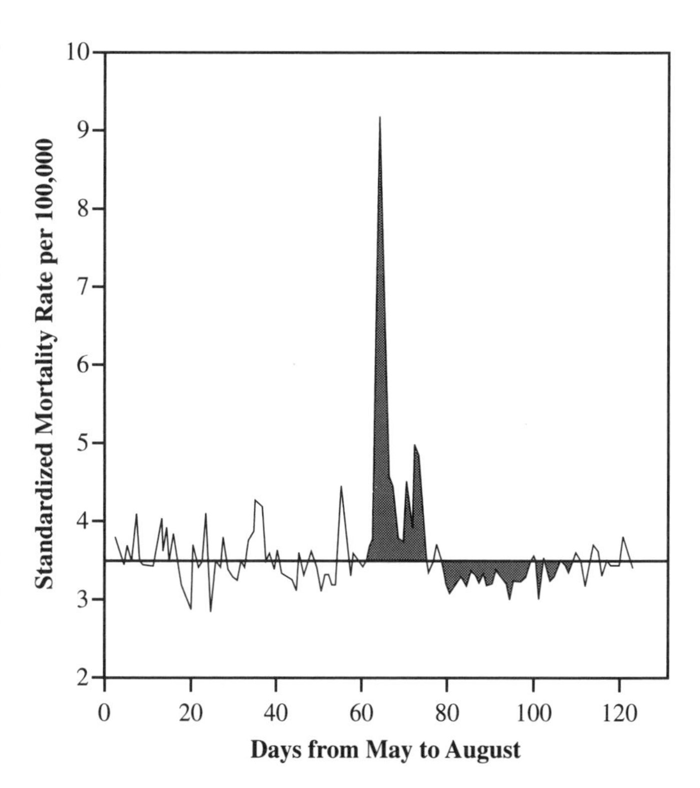

Figure 18-2: Daily summer mortality during a New York heatwave, 1966 (Kalkstein, 1993).

setting the synoptic weather situations that are "oppressive" (i.e., exceed physiological tolerance levels). The annual number of oppressive synoptic situation days was then predicted for the climate-change scenarios, and the annual total of associated excess deaths was estimated. Model-based predictions of the numbers of additional deaths attributable to heat during a typical summer in a future warmer world were thus obtained. Two sets of predicted mortality increases were made: one assuming that the population does not acclimatize and the other assuming partial acclimatization (i.e., complete physiological acclimatization but without improved socioeconomic conditions such as the development of more protective housing).

Consider the population of Atlanta, for example. Presently, Atlanta experiences an average of 78 heat-related deaths each summer. Under the climate projections of the Geophysical Fluid Dynamics Laboratory (GFDL) 1989 (transient) climate-change model, and assuming no change in population size or age profile, this number would increase to 191 in the year 2020 and to 293 in the year 2050. If population acclimatization occurs, the annual total would increase less, to 96 and 147 in those two years. Under the UKTR (transient) model run, the corresponding four projections of heat-related mortality are 20–40% higher than for the GFDL model run: 247 and 436 deaths (unacclimatized, 2020 and 2050) and 124 and 218 deaths (acclimatized, 2020 and 2050). These and other results for selected North American cities, Shanghai, and Cairo indicate that the annual number of heat-related deaths would, very approximately, double by 2020 and would increase several-fold by 2050. Thus, in very large cities with populations displaying this type of sensitivity to heat stress, climate change would cause several thousand extra heat-related deaths annually.

As mentioned earlier, the seasonal death rates in developed countries are highest in winter (Kilbourne, 1992; Tan, 1994). In temperate-zone countries, death rates increase particularly during periods of severe winter weather. However, no single study has yet been published that allows a direct comparison of the anticipated winter gains and summer losses that would accompany global warming. A substantial proportion of winter-related deaths are from cardiovascular disease (Kunst *et al.*, 1993; Langford and Bentham, 1995). It is likely that this increased risk of cardiovascular disease reflects an increased cold-induced tendency for blood to clot (Keatinge *et al.*, 1989), perhaps exacerbated by the fibrinogen-enhancing effect of winter respiratory infections (Woodhouse *et al.,* 1994). The relative importance of respiratory infections and cardiovascular diseases to anticipated reductions in mortality, and how these would vary geographically, remains uncertain. One recent British study has forecast that approximately 9,000 fewer winter-related deaths (estimated to represent a reduction of 2–3%) would occur annually by the year 2050 in England and Wales, under typical climate-change scenarios that entail 2–2.5°C wintertime increases (Langford and Bentham, 1995). Just over half the avoided deaths would be from ischaemic heart disease and stroke, with chronic bronchitis and pneumonia each contributing 5–10%. Other researchers have concluded that a significant portion of the overall winter-related mortality is due to respiratory infections such as influenza (Curwen, 1991). Since these respiratory infections depend upon aerosol transmission—usually in confined, poorly ventilated places—a small rise in winter temperatures should reduce this risk if it encouraged outdoor activities and improved ventilation. However, annual influenza outbreaks do not appear to correlate with mean winter or monthly temperature (CDC, 1994b; Langford and Bentham, 1995).

Overall, the sensitivity of death rates to hotter summers is likely to be greater than to the accompanying increase in average winter temperature. The overall balance is difficult to quantify, and also would depend on the population's capacity for adaptive responses. However, research to date suggests that global warming would, via an increased frequency of heat waves, cause a net increase in mortality and associated morbidity. This conclusion must be qualified by noting that there is an imbalance in the published research—most of which refers to developed, non-tropical, countries.

18.2.2. Health Impacts of Weather Variability and Extreme Events

Global warming may affect ocean currents, air currents, and atmospheric humidity (see Chapter 6, *Climate Models–Projections of Future Climate*, of the IPCC Working Group I volume). Any consequent changes in weather variability may alter the frequency and severity of extreme weather events: bushfires, droughts, floods, storms, and landslides (Gordon *et al.*, 1992; Meehl *et al.*, 1992). However, some of these relationships remain uncertain (e.g., for tropical cyclones; see Lighthill *et al.,* 1994) and have proven difficult to model with GCMs. Such events increase deaths, injuries, infectious diseases, stress-related disorders, and the many adverse health effects associated with social disruption, environmentally enforced migration (Myers, 1993), and settlement in urban slums. Health impacts would be greatest on those communities that are most exposed and have the fewest technical and social resources.

Low-lying, poorly resourced populations would be particularly vulnerable to an increase in frequency of storms and storm surges. In the 1970 Bangladesh cyclone, mortality varied from around 5% inland to almost 50% in coastal communities. Widespread destruction of food supplies may also occur; in 1970, two-thirds of fishing activities along the coasts and plains in Bangladesh were destroyed, along with 125,000 animals (Alexander, 1993). In Andhra Pradhesh, India, in 1970, many victims died when wind and rain caused the collapse of houses (Sommer and Mosely, 1972).

Flash floods and landslides (see Chapter 12) would increase in frequency in regions experiencing increased torrential rainfall. Heavy rains can erode soil, thereby impairing agricultural productivity. Flooding also can affect the incidence of vector-borne diseases. On the one hand, flood waters may wash away mosquito eggs/larvae; on the other hand, residual water may increase mosquito populations and consequent infectious diseases. In southeastern Australia, epidemics of Ross River virus infection follow heavy rains in the Murray-Darling basin (Nicholls, 1993). Flooding also may affect the transmission of diarrheal diseases: In Bolivia, for example, flooding associated with El Niño in 1983 led to a 70% increase in salmonella infections, particularly in children (Telleria, 1986).

Flash flooding is a leading cause of weather-related mortality in the United States (French and Holt, 1989). Damage to homes and displacement of residents may facilitate the spread of infectious diseases because of crowded living conditions. Flooding can contaminate water sources with fecal material or toxic chemicals. Flooding also increases runoff from agricultural lands and urban stormwater systems (Thurman *et al.*, 1991, 1992). Water resources thus are contaminated by toxic chemical wastes, agricultural chemicals, and pathogens (e.g., *Cryptosporidium*). Hazardous exposure may arise from contaminated drinking water and from edible fish that bioaccumulate contaminants.

The many psychosocial effects of natural disasters have been studied in survivors of storms, floods, earthquakes, and fires (e.g., Gregg, 1989). Severe disturbance (including "post-traumatic stress disorder") usually affects only a minority, many of whom recover (de Girolamo and McFarlane, 1994)—although children in affected families may show developmental problems over the ensuing years (Titchener and Frederic, 1976). The level of psychological impact may depend upon the suddenness and unexpectedness of the impact, the intensity of the experience, the degree of personal and community disruption, and long-term exposure to the visual signs of the disaster (Green, 1982; Green *et al.*, 1991). Data from disasters in developing countries are sparse, but a follow-up study after the disastrous 1988 floods in Bangladesh showed increased behavioral disorders in young children (Durkin *et al.*, 1993).

Two final comments on extreme events are appropriate here. First, a major effect of extreme climatic events on human health has been malnutrition and starvation due to severe drought and its consequences (Escudero, 1985). This, in turn, causes increased susceptibility to infection (Tomkins, 1986). Second, there is evidence that environmental degradation, unequal access to resources, and population growth may be potent factors in provoking conflict (Homer-Dixon *et al.*, 1993). By affecting water and food supply, climate change could increase the possibility of violent conflicts in a number of regions.

18.3. Potential Indirect Health Impacts of Climate Change

The indirect effects of climate change upon health are those that do not entail a direct causal connection between a climatic factor (such as heat, humidity, or extreme weather event) and human biology.

18.3.1. Vector-Borne Diseases

The transmission of many infectious diseases is affected by climatic factors. Infective agents and their vector organisms are sensitive to factors such as temperature, surface water, humidity, wind, soil moisture, and changes in forest distribution (Bradley, 1993). This applies particularly to vector-borne diseases (VBDs) like malaria, which require an intermediate organism such as the mosquito to transmit the infective agent. It is therefore projected that climate change and altered weather patterns would affect the range (both latitude and altitude), intensity, and seasonality of many vector-borne and other infectious diseases. In general, increased warmth and moisture would enhance transmission of these diseases. However, any such climate-related redistribution of disease may also entail—perhaps in conjunction with other environmental stresses—some localized reductions in rates of infection.

The sustained (or endemic) transmission of VBDs requires favorable climatic-environmental conditions for the vector, the parasite, and, if applicable, the intermediate host species. For vectors with long life spans (e.g., tsetse flies, bugs, ticks), vector abundance is a major determinant of disease distribution; arthropods that transmit VBDs generally thrive in warmth and moisture. For vectors with short life spans (e.g., mosquitoes, sandflies, blackflies), the temperature-sensitive extrinsic maturation period of the parasite is of critical importance. The geographic distributions of many of the parasites, both unicellular (protozoa) and multicellular (e.g., flukes and worms), correlate closely with temperature (Gillet, 1974; Shope, 1991). The geographic distributions of vector-borne viral infections, such as dengue and yellow fever, are affected by temperature and surface water distribution. For many VBDs, such as plague and hantavirus, rodents act as intermediate infected hosts or as hosts for the arthropod vector (Wenzel, 1994); rodent activity would also tend to increase in a warmer world (Shope, 1991).

In tropical countries, VBDs are a major cause of illness and death. For the major VBDs, estimates of numbers of people at risk and infected, and of VBD sensitivity to climate change, are shown in Table 18-3. While the potential transmission of many such diseases would increase (geographically or from seasonal to year-round) in response to climate change, the capacity to control these diseases also will change. New or improved vaccines can be expected; some vector species can be constrained by use of pesticides. Nevertheless, there are uncertainties and risks here, too: for example, long-term pesticide use breeds resistant strains and kills many predators of pests.

18.3.1.1. Malaria

Malaria remains a huge global public-health problem, currently causing around 350 million new infections annually, predominantly in tropical countries. Malaria is caused by an infective parasite (plasmodium) transmitted between humans by a mosquito vector. Its incidence is affected by temperature, surface water, and humidity (Gill, 1920a, 1920b; Sutherst, 1983). Although anopheline mosquito species that transmit malaria do not usually survive where the mean winter temperature drops below 16–18°C, some higher-latitude species are able to hibernate in sheltered sites. Sporogonic development (i.e., the extrinsic incubation phase of

Table 18-3: *Major tropical vector-borne diseases and the likelihood of change of their distribution with climate change.*

Disease	Vector	Population at Risk (million)[a]	Number of People Currently Infected or New Cases per Year	Present Distribution	Likelihood of Altered Distribution with Climate Change
Malaria	Mosquito	2,400[b]	300–500 million	Tropics/Subtropics	+++
Schistosomiasis	Water Snail	600	200 million	Tropics/Subtropics	++
Lymphatic Filariasis	Mosquito	1,094[c]	117 million	Tropics/Subtropics	+
African Trypanosomiasis (Sleeping Sickness)	Tsetse Fly	55[d]	250,000–300,000 cases per year	Tropical Africa	+
Dracunculiasis (Guinea Worm)	Crustacean (Copepod)	100[e]	100,000 per year	South Asia/ Arabian Peninsula/ Central-West Africa	?
Leishmaniasis	Phlebotomine Sand Fly	350	12 million infected, 500,000 new cases per year[f]	Asia/Southern Europe/Africa/ Americas	+
Onchocerciasis (River Blindness)	Black Fly	123	17.5 million	Africa/Latin America	++
American Trypanosomiasis (Chagas' disease)	Triatomine Bug	100[g]	18 million	Central and South America	+
Dengue	Mosquito	1,800	10–30 million per year	All Tropical Countries	++
Yellow Fever	Mosquito	450	<5,000 cases per year	Tropical South America and Africa	++

+ = likely, ++ = very likely, +++ = highly likely, ? = unknown.
[a]Top three entries are population-prorated projections, based on 1989 estimates.
[b]WHO, 1995b.
[c]Michael and Bundy, 1995.
[d]WHO, 1994a.
[e]Ranque, personal communication.
[f]Annual incidence of visceral leishmaniasis; annual incidence of cutaneous leishmaniasis is 1-1.5 million cases/yr (PAHO, 1994).
[g]WHO, 1995c.

the plasmodium within the mosquito) ceases below around 18°C for *Plasmodium falciparum* and below 14°C for *P. vivax*. Above those temperatures, a small increase in average temperature accelerates the parasite's extrinsic incubation (Miller and Warrell, 1990). Temperatures of 20–30°C and humidity above 60% are optimal for the anopheline mosquito to survive long enough to incubate and transmit the parasite.

Until recent decades, parts of today's developed world were malarious. These included the United States, southern Europe, and northern Australia. In the last century, outbreaks of *P. vivax* malaria occurred in Scandinavia and North America. Although climate change would increase the potential transmission of malaria in some temperate areas, the existing public-health resources in those countries—disease surveillance, surface-water management, and treatment of cases—would make reemergent malaria unlikely. Indeed, malaria is most likely to extend its spread (both latitude and altitude) and undergo changes in seasonality in tropical countries, particularly in populations currently at the fringe of established endemic areas (Martens *et al.*, 1994). Newly affected populations would initially experience high case fatality rates because of their lack of natural acquired immunity.

Box 18-1. Examples of Modeling the Future Impact of Climate Change upon Malaria

Mathematical models have been recently developed for the quantitative prediction of climate-related changes in the potential transmission of malaria (e.g., Sutherst, 1993; Matsuoka and Kai, 1994; Martin and Lefebvre, 1995). Note the use of the important word "potential" here; the models are primarily predicting where malaria could occur as a function of climate and associated environment, irrespective of the influence of local demographic, socioeconomic, and technical circumstances.

A simple model has been used for Indonesia, based on empirical historical data from selected provinces on the relationship of annual average temperature and total rainfall to the incidence of malaria (and dengue and diarrhea) (Asian Development Bank, 1994). The model forecasts that annual malaria incidence (currently 2,705 per 10,000 persons) would, in response to the median climate-change scenario, increase marginally by 2010 and by approximately 25% by 2070. Because of the limited technical information, it is not easy to appraise these particular forecasts.

A more complex integrated global model has been developed by Martens *et al.* (1994). This model takes account of how climate change would affect the mosquito population directly—i.e., mosquito development, feeding frequency, and longevity—and the incubation period of the parasite inside the mosquito (see Figure 18-3). As a highly aggregated model, it does not take account of local environmental-ecological factors, and it therefore cannot be regarded as a source of precise projections. The model's output refers to geographic changes in *potential* transmission (i.e., the range within which both the mosquito and parasite could survive with sufficient abundance for sustained transmission of malaria). This is not, therefore, a projection of actual disease incidence, and it must be interpreted in relation to local control measures, health services, parasite reservoir, and mosquito densities. Further, until such models have been validated against historical data sets, their predictions must be viewed cautiously.

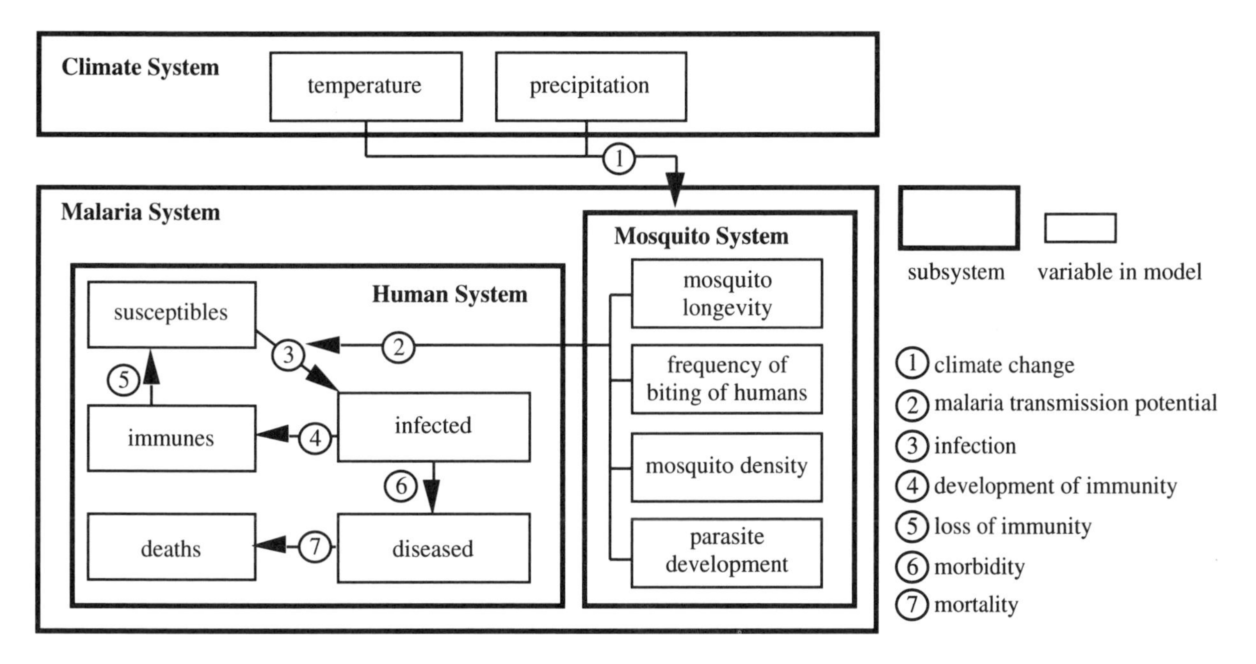

Figure 18-3: Systems diagram of a model designed to assess the impact of climate change on the potential transmission of malaria (adapted from Martens *et al.*, 1994).

This particular model predicts widespread increases in the potential habitat range and the "vectorial capacity" of the mosquito, and therefore potential malaria transmission, in response to climate change. For example, using GCM predictions of 3–5°C increases in global mean temperature by the year 2100, the malaria epidemic potential of the mosquito population is estimated to increase twofold in tropical regions and substantially more than tenfold in temperate climates (Martens *et al.*, 1995a). Overall, there is an increase from around 45% to around 60% of the world population living within the potential malaria transmission zone by the latter half of the next century. However, most of the developed countries have effective control and surveillance measures that should preclude reintroduction of endemic malaria. In the already endemic areas, especially in the subtropics, malaria may increase (although in some hot climates, further temperature increases may shorten the life span of mosquitoes, and local malaria transmission would then decrease). In particular, in adjoining areas of lower endemicity or unstable malaria, the occurrence of infection is far more sensitive

Box 18-1 (continued)

to climate variation, so climate change may have a marked effect on its incidence and stability. Simulations with this model, using several different GCMs, predict a climate-induced increase in the incidence of annual malaria cases of approximately 50–80 million in response to a temperature rise of around 3°C by the year 2100, relative to an assumed approximate base of 500 million annual cases in a 2100 world without climate change (Martens *et al.*, 1995b).

Another recent attempt at aggregated global modeling (Martin and Lefebvre, 1995) has predicted that potential malaria transmission would spread to higher latitudes, while some currently stable endemic areas near the equator would become unstable, leading to reductions in population immunity levels.

Such models, despite their highly aggregated predictions and simplifying assumptions, provide indicative information about the likely impact of climate change on the potential transmission of vector-borne diseases (assuming that other relevant factors remain constant). There is a clear need for validation of these models and for incorporating more extensive detail into them. Meanwhile, this line of research has begun to elucidate the interdependent relationships among climate change, vector population dynamics, and human disease dynamics.

Recent evidence of the responsiveness of malaria incidence to local climate change comes from observations of marked increases in malaria incidence in Rwanda in 1987, when atypically hot and wet weather occurred (Loevinsohn, 1994), and annual fluctuations in falciparum malaria intensity in northeast Pakistan that correlated with annual temperature variations during the 1980s (Bouma *et al.*, 1994). Hence, it is a reasonable prediction that, in eastern Africa, a relatively small increase in winter temperature could extend the mosquito habitat and thus enable falciparum malaria to reach beyond the usual altitude limit of around 2,500 m to the large, malaria-free, urban highland populations, e.g., Nairobi in Kenya and Harare in Zimbabwe. Indeed, the monitoring of such populations around the world, currently just beyond the boundaries of stable endemic malaria, could provide early evidence of climate-related shifts in malaria distribution (Haines *et al.*, 1993).

18.3.1.2. African Trypanosomiasis

African trypanosomiasis, or "sleeping sickness," is transmitted by tsetse flies. The disease is a serious health problem in tropical Africa, being generally fatal if untreated. Research in Kenya and Tanzania shows only a very small difference in mean temperature between areas where the vector, *Glossina morsitans,* does and does not occur. This indicates that a small change in temperature may significantly affect the limits of the vector's distribution (Rogers and Packer, 1993).

18.3.1.3. American Trypanosomiasis (Chagas' Disease)

American trypanosomiasis is transmitted by insects of the subfamily *Triatominae*. It is a major problem in Latin America, with 100 million people at risk and 18 million infected (WHO, 1995c). An estimated 15–20% of infected people develop clinical Chagas' disease. Most of the triatomine vector species need a minimum temperature of 20°C for feeding and reproduction (Curto de Casas and Carcavallo, 1984), but at higher temperatures (28–30°C) they feed more frequently, have a shortened life cycle, and an increased population density (Carcavallo and Martinez, 1972, 1985). At even higher temperatures, the most important vector species, *Triatoma infestans*, doubles its reproductive rate (Hack, 1955).

18.3.1.4. Schistosomiasis

Schistosomiasis is a water-based disease caused by five species of schistosomal flukes. Water snails act as the intermediate host (and, strictly speaking, are not active "vectors"). The infection has increased in worldwide prevalence since mid-century, perhaps largely because of the expansion of irrigation systems in hot climates, where viable snail host populations interact with infected humans (White *et al.,* 1972; Grosse, 1993; Hunter *et al.,* 1993).

Data from both the field and the laboratory indicate that temperature influences snail reproduction and growth, schistosome mortality, infectivity and development in the snail, and human-water contact (Martens *et al.,* 1995b). In Egypt, for example, water snails tend to lose their schistosome infections during winter, but if temperatures increase, snails may mediate schistosomiasis transmission throughout the year (Gillet, 1974; WHO, 1990). Predictive modeling indicates that a change in background temperatures may cause the infection to extend to currently unaffected regions. Fluctuations in temperature may also play an important role in optimizing conditions for the several life-cycle stages of schistosomiasis (Hairston, 1973).

18.3.1.5. Onchocerciasis (River Blindness)

Onchocerciasis, or "river blindness," is a VBD affecting approximately 17.5 million people—some in Latin America, most in West Africa. The vector is a small blackfly of the genus *Simulium,* and the infectious agent is the larva of the *Onchocerca volvulus* parasite. This threadlike worm damages the skin, the lymphatic system, and, in the most extreme cases, the eye. Climate affects the occurrence of onchocerciasis because the vector requires fast-flowing water for successful reproduction (WHO, 1985), and the adult vector can be spread by wind.

A recent simulation study on the potential impact of climate change on blackfly populations in West Africa showed that if temperature and precipitation were to change across parts of the sub-Sahel, as predicted by the Goddard Institute for Space Studies (GISS) GCM (Hansen *et al.,* 1988), blackfly populations may increase by as much as 25% at current breeding sites (Mills, 1995). Since these vectors can travel hundreds of kilometers on wind currents, new habitats in previously unsuitable areas could be quickly colonized by blackflies, introducing onchocerciasis into new areas (Garms *et al.*, 1979; Walsh *et al.*, 1981; WHO, 1985).

18.3.1.6. Trematode Infections

Some trematode infections, such as fascioliasis (a liver fluke that currently affects around 2.4 million persons), would be affected by climate changes because the life cycle and population size of the snail host are very sensitive to temperature. Similarly, the incidence of cercarial dermatitis (skin inflammation) would be increased at higher temperatures. This infection is currently found in Europe (Beer and German, 1994) and the United States (CDC, 1992), where it causes "swimmer's itch"; its incidence has recently increased in Russia (Beer and German, 1993) and in very poor rural communities in developing countries.

18.3.1.7. Vector-Borne Viral Infections

Many vector-borne infective agents are viruses. The human-infecting arboviruses (i.e., arthropod-borne viruses) generally have a mosquito vector. Arboviral infections span a wide clinical spectrum, from those that cause mild feverish illness or subclinical infections to those causing severe and often fatal encephalitis (brain inflammation) or hemorrhagic fever. Under favorable environmental conditions, an arboviral disease can become epidemic (population-wide), from a local endemic base or by its introduction to a previously unaffected area. The distribution and abundance of vectors are influenced by various physical factors (temperature, rainfall, humidity, and wind) and biological factors (vegetation, host species, predators, parasites, and human interventions) (WHO, 1990). Temperature also affects the rapidity of the virus' life cycle—e.g., the extrinsic incubation period for the mosquito-hosted stage of the yellow fever virus varies from several weeks to 8–10 days, depending on temperature.

Increased temperature and rainfall in Australia would influence the range and intensity of various vector-borne viral infections. For example, certain arthropod vectors and natural vertebrate hosts would spread southward and proliferate in response to warming and increased rain, resulting in increased incidence of arboviral infections such as Murray Valley encephalitis (which can cause serious brain damage), Ross River virus (which causes multiple, often long-lasting joint inflammation), and dengue (e.g., Sutherst, 1993; Nicholls, 1993).

Dengue is a severe influenza-like disease, which in some cases may take the form of a hemorrhagic fever, which can cause an average of 15% mortality without proper medical attention. Dengue is transmitted by the *Aedes aegypti* mosquito, as is urban yellow fever. In parts of Asia, dengue also is transmitted by *Ae. albopictus,* which now is colonizing North and South America. Research in Mexico has shown that an increase of 3–4°C in average temperature doubles the rate of transmission of the dengue virus (Koopman *et al.*, 1991). Although there is no clear evidence of regional climatic influence, annual epidemics of dengue have returned to Central America over the past decade (as they did about twenty years ago in Asia), and, in Mexico, dengue has recently spread to previously unaffected higher altitudes (Herrera-Basto *et al.*, 1992). *Ae. aegypti* mosquitoes, once limited to 1,000 meters altitude by temperature in Colombia, have been recently reported above 2,200 meters. The habitat of the mosquito is restricted to areas with a mean midwinter temperature of more than 10°C. Epidemic transmission of dengue is seldom sustained at temperatures below 20°C (Halstead, 1990).

18.3.1.8. Other Vector-Borne Diseases

VBDs are now relatively rare in most developed countries; however, it has been predicted that various VBDs might enter or increase in incidence in the United States because of higher temperatures (Longstreth, 1989; Freier, 1993; Martens *et al.*, 1994); Venezuelan equine encephalitis, dengue, and leishmaniasis could extend into the southern United States; Western equine encephalitis might move further north within the United States (Reeves *et al.,* 1994). The dengue-transmitting *Ae. albopictus* mosquito, which is more cold-hardy than *Ae. aegypti*, is now well established in the United States and may extend toward Canada if temperatures increase.

Climate change would influence the global pattern of VBDs via other disturbances of ecological relationships. For example, it would bring together vertebrate animals of different species and would thereby expose animals to new arthropod vectors. Warming (and rising sea levels) would displace some human populations, perhaps resulting in migration into wilderness areas where zoonotic infectious agents are being transmitted in silent wildlife cycles. Migratory humans would thus be at risk of infection with enzootic (i.e., locally prevalent animal-infecting) agents. Climate-induced changes in ecology also could force the rapid evolution of infectious agents, with newly emergent strains of altered virulence or pathogenicity. Additionally, changes in climate means and variability can disrupt predator/prey ratios, thus loosening natural controls on pests and pathogens.

Two other general points should be noted. First, because vector control methods exist for many of these diseases, developed countries should be able to minimize their impact. Second, however, the quicker "turnover" of the life cycle of parasites at higher temperatures will increase their likelihood of evolving greater resistance to drugs and other control methods. This would pose a particular problem to those tropical countries with high infection rates and limited socioeconomic resources.

18.3.2. Water-Borne and Food-Borne Infectious Diseases

Climatic effects on the distribution and quality of surface water—including increases in flooding and water shortages that concentrate organisms, impede personal hygiene, and impair local sewerage—would influence the risks of diarrheal (including cholera) and dysentery epidemics, particularly in developing countries. Diarrheal diseases can be caused by a large variety of bacteria (e.g., *Salmonella*, *Shigella*, and *Campylobacter*), viruses (e.g., *Rotavirus*), and protozoa (e.g., *Giardia lamblia*, amoebas, and *Cryptosporidium*). Many of these organisms can survive in water for months, especially at warmer temperatures, and increased rainfall therefore could enhance their transport between groups of people. An increased frequency of diarrheal disease is most likely to occur within impoverished communities with poor sanitation. There have been outbreaks of diarrheal disease after flooding in many such settings. If flooding increased, there also would be risks of outbreaks of infection in developed countries within temporary settlements of displaced communities.

The cholera organism, *Vibrio cholerae*, can survive in the environment by sheltering beneath the mucous outer coat of various algae and zooplankton—which are themselves responsive to climatic conditions and to nutrients from wastewater and fertilizers (Epstein, 1992; Smayda, 1990; Anderson, 1992). Increases in coastal algal blooms may therefore amplify *V. cholerae* proliferation and transmission. This might also assist the emergence of new genetic strains of vibrios. Algal blooms also are associated with biotoxin contamination of fish and shellfish (Epstein *et al.*, 1993). With ocean warming, toxins produced by phytoplankton, which are temperature-sensitive, could cause contamination of seafood more often (see also Chapter 16), resulting in increased frequencies of amnesic, diarrheic and paralytic shellfish poisoning and ciguatera poisoning from reef fish. Thus, climate-induced changes in the production of both aquatic pathogens and biotoxins may jeopardize seafood safety for humans, sea mammals, seabirds, and finfish.

Climate change also could create a problem via the warming of aboveground piped-water supplies. In parts of Australia, for example, there has been a seasonal problem of meningoencephalitis caused by the *Naegleria fowleri* amoeba, which proliferates in overland water pipes in summer (NHMRC, 1991). Soil-based pathogens (e.g., the tetanus bacterium and various fungi) would tend to proliferate more rapidly with higher temperature and humidity, depending on the effectiveness of microclimatic homeostatic mechanisms. Higher temperatures would also increase the problem of food poisoning by enhancing the survival and proliferation of bacteria, flies, cockroaches, and so forth in foodstuffs.

18.3.3. Agricultural Productivity and Food Supplies: Effects upon Nutrition and Health

Food, as energy and nutrients, is fundamentally important to health. Malnutrition is a major cause of infant mortality, physical and intellectual stunting in childhood, and immune impairment (thus increasing susceptibility to infections). Currently, around one-tenth of the world's population may be hungry (Parry and Rosenzweig, 1993) and a larger proportion malnourished—although estimates differ according to definition.

Human societies have evolved farming methods to counter various local climatic and environmental constraints on agriculture, especially via irrigation, fertilization, mechanization, and the breeding of better-adapted varieties. Today, as gains in per capita agricultural productivity appear to be diminishing, widespread land degradation accrues, and access to new arable land is declining, the further possibility exists of adverse effects of climate change upon aspects of world food production (Houghton *et al.*, 1990; Kendall and Pimentel, 1994). The impacts of climate change upon crop and livestock yield would be realized within a complex setting that encompasses climate change scenarios, crop yield response, pest population response, demographic trends, patterns of land use and management, and social and economic responses.

18.3.3.1. Modes of Climatic Impact upon Agricultural Productivity

Global warming would alter regional temperature and rainfall. Changes in these two major influences on agriculture, and consequent reductions in soil moisture, could impair the growth of many crops. Increases in the intensity of rainfall in some regions would exacerbate soil erosion. The net global impact of these climate-related changes upon food production is highly uncertain (Reilly, 1994). Although the IPCC assessment is uncertain about the overall impact, it foresees productivity gains and losses in different regions of the world (see Chapter 13). While productivity may increase initially, longer-term adaptations to sustained climate change would be less likely because of the limitations of plant physiology (Woodward, 1987).

Climate change also could affect agriculture by long-term changes in agroecosystems, by an increased frequency and severity of extreme events, and by altered patterns of plant diseases and pest infestations (e.g., Farrow, 1991; Sutherst, 1991). Debate persists over whether enrichment of the atmosphere with carbon dioxide will have a "fertilization effect" (Idso, 1990b; Bazzar and Fajer, 1992; Körner, 1993). Experiments consistently indicate that C_3 plants (e.g., wheat, soya beans, rice, and potatoes) would respond more positively than C_4 plants (e.g., millet, sorghum, and maize), which would be unaffected (see Chapter 13). This effect may be temperature-dependent (Vloedbeld and

Leemans, 1993). Such influences on the climatically optimal mix of crop species would disturb patterns of traditional agriculture in some regions.

18.3.3.2. Impacts upon Food Supplies, Costs, and the Risk of Hunger

Since climate change may threaten food security in poorer countries within the semi-arid and humid tropics (Rosenzweig *et al.*, 1993; see also Chapter 13), poorer countries, already struggling with large and growing populations and marginal climatic conditions, would be particularly vulnerable to food shortages, malnutrition, and demographic disruption. In such countries, there is minimal capacity for adaptive change (Leemans, 1992). Already in Africa, more than 100 million people are "food insecure," many of them in the arid Sahel region. The cost of food on world markets would increase if crop production declined in the world's mid-latitude breadbasket regions. The large minority of the world population that already suffers from malnutrition would then face an increased threat to health from agricultural failure and rising food costs. A recent analysis predicts an extra 40–300 million people at risk of hunger in the year 2060 because of the impact of climate change, on top of a predicted 640 million people at risk of hunger by that date in the absence of climate change (Rosenzweig *et al.*, 1993).

18.3.3.3. Impacts of Climate Change on Non-Cereal Food Production

Climate change may influence the production of noncrop food supplies, including animal productivity. For example, the U.S. Environmental Protection Agency has identified several infectious diseases—such as the horn fly in beef and dairy cattle and insect-borne anaplasmosis infection in sheep and cattle—that could increase in prevalence in response to climate changes (Rosenzweig and Daniel, 1989). An increase in temperature and temperature extremes also could affect the growth and health of farm animals (Furquay, 1989); young animals are much less tolerant of temperature variation than are adult animals (Bianca, 1976).

Changes in ocean temperatures and currents could affect the base of the marine food web and alter the distribution, migration, and productivity of fish species, a major source of protein for many human populations (Glantz, 1992). Increased soil erosion from intensified rainfall raises the turbidity of lakes and rivers, reducing photosynthesis and therefore fish nutrition. As in agriculture, climate change may contribute to the decline of some fisheries and the expansion of others (see Chapter 16).

18.3.4. Health Impacts of Sea-Level Rise

Each of the vast changes in sea level that have occurred during the past million years, before and after ice ages, typically took many thousands of years. The predicted rise of around half a meter over the next century (see Chapter 7, *Changes in Sea Level*, of the IPCC Working Group I volume) would be much faster than anything experienced by human populations since settled agrarian living began. Such a rise would inundate much of the world's lowlands, damage coastal cropland, and displace millions of persons from coastal and small island communities (see Chapter 12).

Much of coastal Bangladesh and Egypt's heavily populated Nile Delta would be flooded. Some low-lying, small island states such as the Maldives and Vanuatu would be at risk of partial immersion, and many other low-lying coastal regions (for example, eastern England, parts of Indonesia, the Florida Everglades, parts of the northeast coast of Latin America) would be vulnerable. The displacement of inundated communities—particularly those with limited economic, technical, and social resources—would greatly increase the risks of various infectious, psychological, and other adverse health consequences.

Sea-level rise could have a number of other effects, of varying directness, upon public health. In some locations, it could disrupt stormwater drainage and sewage disposal. Poverty and the absence of social infrastructure would compound the health consequences of storm damage, disruption of sanitation, and displacement of coastal dwellers. In many places, industrial and agricultural depletion of groundwater already are causing land subsidence, thus decreasing the threshold for impact. Meanwhile, widespread damage to coral reefs is reducing their capacity to buffer shorelines. Rising seas also would cause saltwater to encroach upon freshwater supplies from estuarine and tidal areas. Some changes in the distribution of infectious disease vectors could occur (e.g., *Anopheles sundaicus*, a saltwater vector of malaria).

18.3.5. Climate and Air Pollution: Impacts on Respiratory and Other Health Disorders

The incidence of respiratory disorders—many of which are caused primarily by dusts, noxious gases, allergic reactions, or infections—may be modulated by climate change. Some of these modulatory effects may occur via extreme temperatures or amplification of pollutant levels. Rapid changes in air masses associated with frontal passages may alter the intensity of respiratory illnesses (Ayres, 1990). People with chronic obstructive pulmonary disease (bronchitis and emphysema) often experience exacerbation during winter.

Seasonal allergic disorders would be affected by changes in the production of pollen and other biotic allergens; plant aeroallergens are very sensitive to climate (Emberlin, 1994). Changes in pollen production would principally reflect changes in the natural and agriculturally managed distribution of many plant species—for example, birch trees, grasses, various crops (e.g., oilseed rape, sunflowers), and ragweed species. Hay fever (allergic rhinitis) increases seasonally and may reflect the impact of pollen release. The seasonal distribution and the causation/exacerbation of asthma is more complex. It peaks in the pollen season and increases again later in the year in temperate climates; in the tropics, asthma occurs more frequently in the wet season (LAIA, 1993; *Lancet*, 1985). In many asthmatic individuals, aspects of weather can exacerbate bronchial hyperresponsiveness. For example, the passage of a cold front followed by strong high pressure was found to be associated with unusually high number of asthma admission days in two U.S. cities (Goldstein, 1980). Sandstorms in Kansas (USA) and the Sudan have been accompanied by increases in bronchitis and asthma (Ayres, 1990).

It is well established that exposure to air pollutants, individually and in combinations, has serious public health consequences. For example, exposure to ozone has been shown to exacerbate asthma and impair lung function in children and the elderly (Beckett, 1991; Schwartz, 1994), and both chronic and acute exposures to fine particles are a cause of excess deaths (Dockery *et al.*, 1993; Pope *et al.*, 1995; Schwartz, 1994) even at exposures below prevailing air-quality standards. Since the combustion of fossil fuels is a major source of both carbon dioxide (a major greenhouse gas) and various air pollutants, climate change can be expected to entail more frequent occasions that combine very hot weather with increases in air-pollutant concentrations. In urban environments, the weather conditions that characterize oppressive air masses (see Section 18.2.1) also enhance the concentrations of air pollutants (Seinfeld, 1986); conditions of low wind speed and high humidity occur periodically in which neither heat nor air pollutants are rapidly dispersed. Further, increases in temperature or in ultraviolet irradiation of the lower atmosphere enhance the chemical reactions that produce secondary photochemical oxidant pollutants such as tropospheric ozone (Akimoto *et al.*, 1993; de Leeuw and Leyssius, 1991; Chamiedes *et al.*, 1994).

In many urban settings, studies have shown that daily mortality from cardiovascular and respiratory diseases is a combined function of temperature and air pollutant concentrations. This combination of exposures is also likely to have interactive impacts on health. Indeed, some epidemiological evidence indicates a synergy (a positive interaction) between stressful weather and various air pollutants, especially particulates, upon mortality (e.g., Shumway *et al.*, 1988; Katsouyanni *et al.*, 1993; Shumway and Azari, 1992). The net effect on morbidity/mortality therefore would be greater than anticipated from prior estimates of the separate effects of weather and pollutants.

18.4. Stratospheric Ozone Depletion and Ultraviolet Radiation: Impacts on Health

Stratospheric ozone depletion is a quite distinct process from accumulation of greenhouse gases in the lower atmosphere (troposphere). Depletion of stratospheric ozone has recently occurred in both hemispheres, from polar regions to mid-latitudes (Kerr and McElroy, 1993; see also IPCC Working Group I volume). The major cause of this ongoing depletion is human-made gases, especially the halocarbons (UNEP, 1994).

The problem can be considered alongside climate change for three reasons: (1) several of the greenhouse gases (especially the chlorofluorocarbons) also damage stratospheric ozone; (2) altered temperature in the troposphere may influence stratospheric temperature and chemistry (Rind and Lacis, 1993); and (3) absorption of solar radiation by stratospheric ozone influences the heat budget in the lower atmosphere (see also IPCC Working Group I volume).

Stratospheric ozone absorbs part of the sun's incoming ultraviolet radiation (UVR), including much of the UV-B and all of the highest-energy UV-C. Sustained exposure to UV-B radiation is harmful to humans and many other organisms (UNEP, 1994). It can damage the genetic (DNA) material of living cells and can induce skin cancers in experimental animals. UV-B is implicated in the causation of human skin cancer and lesions of the conjunctiva, cornea, and lens; it may also impair the body's immune system (Jeevan and Kripke, 1993; Armstrong, 1994; UNEP, 1994).

18.4.1. Skin Cancers

Solar radiation has been consistently implicated in the causation of nonmelanocytic and melanocytic skin cancers in fair-skinned humans (IARC, 1992; WHO, 1994b).

Nonmelanocytic skin cancers (NMSCs) comprise basal cell carcinoma (BCC) and squamous cell carcinoma (SCC). The incidence rates, especially of squamous cell carcinoma, correlate with cumulative lifetime exposure to solar radiation (IARC, 1992; Kricker *et al.*, 1995). Studies of the action spectrum (i.e., the relative biological effect of different wavelengths) for skin carcinogenesis in mice indicate that the UV-B band is primarily responsible for NMSC (Tyrrell, 1994). Malignant melanoma arises from the pigment-producing cells (melanocytes) of the skin. Although solar radiation is substantially involved in melanoma causation (IARC, 1992; Armstrong and Kricker, 1993), the relationship is less straightforward than for NMSC; exposure in early life appears to be a major source of increased risk. The marked increases in incidence of melanoma in Western populations over the past two decades (Coleman *et al.*, 1993) probably reflect increases in personal exposure to solar radiation due to changes in patterns of recreation, clothing, and occupation (Armstrong and Kricker, 1994).

The UN Environment Programme predicts that an average 10% loss of ozone (such as occurred at middle-to-high latitudes over the past decade), if sustained globally over several decades, would cause approximately 250,000 additional cases of NMSC worldwide each year (UNEP, 1994). This prediction assumes that a 1% depletion of stratospheric ozone results in a 2.0% (±0.5%) increase in NMSC incidence (80% of which are BCC). Another estimation of this "amplification factor" gives a figure of 2.25% (Slaper *et al.*, 1992; den Elzen, 1994). At higher geographic resolution, Madronich and de Gruijl (1993) predict that persistence of the ozone losses of the 1979–92 period for several decades would cause the incidence of BCC to increase by 1–2% at low latitude (5°), 3–5% at 15–25°, 8–12% at 35–45°, and, at 55–65°, by 13–15% in the northern hemisphere and 20–30% in the south. They estimate that the percentage increases for SCC would be approximately double those for BCC.

18.4.2. *Cataracts and Other Damage to the Eye*

The external epithelial (keratotic) layer of the eye, comprising cornea and conjunctiva, absorbs virtually all UVR of less than 290 nm wavelength. Corneal photokeratitis, pterygium (a growth of the conjunctival epithelium), and climatic droplet keratopathy are thought to be UVR-related (Taylor *et al.*, 1989; Gray *et al.*, 1992; WHO, 1994b). Inside the eye, the lens absorbs much of the residual UVR, and this absorbed radiation may cause cataracts (Taylor *et al.*, 1988; Dahlback *et al.*, 1989; West *et al.*, 1989; WHO, 1994b).

Cataracts (lens opacities) are independent of skin pigmentation (unlike skin cancer). They occur predominantly in old age and cause more than half of the world's estimated 25–35 million cases of blindness (Harding, 1991). In Western countries, 5–10% of people aged over 65 have cataracts (Klein *et al.*, 1992). The prevalence often is much higher among elderly, malnourished persons in poor countries, where micronutrient deficiencies and the metabolic consequences of severe diarrheal episodes may contribute to cataract formation (Harding, 1992). Scientific debate persists over the extent of the influence of UV-B upon cataract formation (Dolin, 1994; WHO, 1994b); some epidemiological studies have found clear-cut positive results, but others have not. The relationship is most evident for cortical and posterior subcapsular cataracts but less so for the more commonly occurring nuclear cataracts.

Ocular photodamage by UVR is enhanced by certain clinical drugs used in photochemical therapy that can cause photosensitizing reactions (Lerman, 1988). Various other photosensitizing medications would render individuals generally more susceptible to adverse health effects from increased exposure to UVR; these medications include psoralens, thiazides, phenothiazines, barbiturates, allopurinol, and retinoic acid compounds (Lerman, 1986).

18.4.3. *Alteration of Immune Function*

Human and animal evidence indicates that UV-B irradiation of skin at quite modest levels causes local and, probably, systemic suppression of immunity (Morison, 1989; Noonan and DeFabo, 1990; Jeevan and Kripke, 1993). Most of the evidence is for local immunosuppression, in which the skin's contact hypersensitivity response is impaired (Giannini, 1986; Yoshikawa *et al.*, 1990; UNEP, 1994). UV-B exposure disturbs the function of the skin's Langerhan cells and stimulates the release of certain cytokines (messenger chemicals) that promote the activity of suppressor T lymphocytes, thus dampening the local immune system (UNEP, 1994).

Evidence for more generalized (i.e., systemic) suppression of immunity comes from studies in humans, which show that sunlight exposure increases the suppressor T cells in blood (Hersey *et al.*, 1983). Although there is evidence in humans of UV-induced changes in the profile of circulating immunologically active lymphocytes for several days to weeks, the extent of systemic immune suppression involved remains uncertain (de Gruijl and van der Leun, 1993). Systemic suppression also occurs in UV-irradiated mice (Kripke, 1981; Jeevan and Kripke, 1990).

Immune suppression would alter susceptibility to infectious diseases (Armstrong, 1994). Exposure to UV-B modifies various immunological reactions in mice that influence the pathogenesis of infectious diseases, such as those due to *Herpes simplex* viruses (Otani and Mori, 1987; Yasumoto *et al.*, 1987), leishmania (Gianinni, 1986; Giannini and DeFabo, 1989), candida (Denkins *et al.*, 1989), and mycobacteria (Jeevan and Kripke, 1989). The relevance of these findings for naturally occurring infectious diseases, and for vaccination efficacy, in humans remains unknown. UNEP (1994) concluded that: "It will be very difficult to assess the role of UV-B radiation on natural infections in human populations. Based on current knowledge, we would predict that an effect of UV-B radiation would manifest as an increase in the severity or duration of disease and not necessarily as an increase in disease incidence."

18.4.4. *Indirect Effects of Ozone Depletion upon Human Health*

An increase in UV-B irradiance is predicted to impair photosynthesis on land and sea (UNEP, 1994). Although the magnitude is uncertain, and may well not be large, there would be at least a marginal reduction in crop yields (Worrest and Grant, 1989) and in the photosynthetic production of biomass by marine phytoplankton, the basis of the aquatic food chain (Smith and Baker, 1989; Smith *et al.*, 1992). Thus, adverse effects of UV-B upon photosynthesis would, to some extent, reduce global food production.

18.5. Options for Adaptation

Various adaptation strategies are possible to reduce the impacts of climate change on human health. Such adaptation could be developed at the population or individual level. The feasibility of adaptation would be constrained for many of the world's populations by a lack of local resources.

At the population level, environmental management of ecosystems (e.g., freshwater resources, wetlands, and agricultural areas sensitive to invasion by vectors), public health surveillance and control programs (especially for infectious diseases), and introduction of protective technologies (e.g., insulated buildings, air conditioning, strengthened sea defences, disaster warning systems) would be important. Improved primary health care for vulnerable populations could play a significant role in reducing a range of health impacts, including some vector-borne and other communicable diseases, and the effects of extreme events. One example is extension of vaccination coverage, although no suitable vaccines exist for some of the diseases most sensitive to climate change (e.g., dengue and schistosomiasis) or for many of the newly emerging infections.

At the individual level, people should be encouraged to refrain from or to limit dangerous exposures (e.g., by use of domestic cooling, protective clothing, mosquito nets). Such behavioral responses could complement any physiological adaptation that might occur spontaneously through acclimatization (to heat stress) or acquired immunity (to infectious diseases).

In view of limitations to the forecasting of health impacts at this stage of our knowledge, an important and practical form of adaptation would be to improve large-scale monitoring and surveillance systems, especially for vulnerable populations and areas. Recently initiated efforts to observe and monitor aspects of the Earth's environment and ecosystems in relation to climate change now should incorporate health-related monitoring (Haines *et al.*, 1993). Advances in climate forecasting and in the regional integration of ecological and health monitoring (including local vulnerability factors) will facilitate development of early-warning systems.

Finally, if health impacts of climate change are probable and serious, then the only effective long-term basis for mitigation lies in primary prevention at the societal level. This would require acceptance of the Precautionary Principle as the foundation of policy response. This, in turn, would suggest some fundamental, and therefore difficult, reorientations of social, economic, and political priorities. Meanwhile, care must be taken that alternative technologies do not introduce new health hazards.

18.6. Research Needs

- Development and validation of integrated mathematical models for the prediction of health impacts. Such models must draw on multiple scientific disciplines and should take maximal account of regional and local influences on the effects being modeled and on their interaction with other environmental stresses.
- Identification and analysis of current or recent settings in which the health impacts of local or regional climate changes (occurring for whatever reason) can be studied. The apparent recent changeable patterns of infectious diseases around the world may afford good opportunities for clarifying and quantifying the influences of climatic factors.
- Incorporation of health-related measurements in global, regional, and local monitoring activities. This would enhance the early detection of shifts in health risks, the evaluation of alternative indices for monitoring health (including the use of sensitive species as bioindicators), and the opportunity to detect and/or examine previously unsuspected or undocumented environment–health relationships.
- Some specific research needs include:
  - Comparison of impacts of heat waves in urban and rural populations, to clarify the relative importance of thermal stress and air pollutants
  - Examination of the interplay between climatic impacts on forests and other terrestrial ecosystems on the range and dynamics of vector-borne disease
  - Study of factors influencing population vulnerability to climate change.

18.7. Concluding Remarks

Forecasting the health impacts of global climate change entails unavoidable uncertainty and complexity. Human populations vary greatly in their vulnerability to climate changes and in their resources for protection and mitigation. Likewise, the responses of infectious disease vectors to changes in climate depend greatly on other concomitant environmental stresses and the adequacy of control measures and health care systems. Meanwhile, population health status continues to be influenced by a rich mix of cultural and socioeconomic factors. Hence, assessing the health impact of climate change requires a systems-based modeling approach that integrates information about climatic factors, other environmental stresses, ecological processes, and social-economic-political inputs and responses.

Alongside the need for improved health impact assessment capability is a precautionary need to develop global, regional, and local monitoring systems for the early detection of climate-induced changes in human health. There have, indeed, been various recent events that, plausibly, might be early signals of such change. The increased heat-related deaths in India in 1995; the changes in geographic range of some vector-borne diseases; the coastal spread of cholera: Could these be early indications of shifts in population health risk in response to aspects of climate change? Of course, it is not possible to attribute particular, isolated events to a change in climate or weather pattern; other plausible explanations exist for each of them, and a number of different factors may combine to produce each event. However, it is important that we begin to assess patterns of change in the various indices of human health that will provide early insight and will assist further the development of predictive modeling.

There is thus a clear need for enhanced research and monitoring activities. This need reflects the assessment that the potential health impacts of climate change, particularly if sustained in the longer term and if generally adverse, could be a serious consequence of the ongoing anthropogenic changes in the composition of Earth's atmosphere.

References

Akimoto, H., H. Nakane, and Y. Matsumoto, 1993: The chemistry of oxidant generation. In: *Chemistry of the Atmosphere: The Impact on Global Change* [Calvert, J.G. (ed.)]. Blackwell Scientific Publications, London, UK, pp. 261-273.

Alcamo, J., G.J.J. Kreilman, M. Krol, and G. Zuidema, 1994a: Modelling the global society-biosphere-climate system, part 1: model description and testing. *Water, Air and Soil Pollution*, **76**, 1-35.

Alcamo, J., G.J. van den Born, A.F. Bouwman, B.J. de Haan, H. Klein-Goldewijk, O. Klepper, R. Leemans, J. Krabec, R. Leemans, J.G.J. Olivier, A.M.C. Toet, H.J.M. de Vries, and H.J. van der Woerd, 1994b: Modelling the global society-biosphere-climate system, part 2: computed scenarios. *Water, Air and Soil Pollution*, **76**, 37-78.

Alexander, D., 1993: *Natural Disasters*. UCL Press, London, UK, 632 pp.

Anderson, D.M., 1992: The Fifth International Conference on Toxic Marine Phytoplankton: a personal perspective. *Harmful Algae News. Supplement to International Marine Science*, **62**, 6-7.

Armstrong, B.K., 1994: Stratospheric ozone and health. *International Journal of Epidemiology*, **23(5)**, 873-885.

Armstrong, B.K. and A. Kricker, 1993: How much melanoma is caused by sun exposure? *Melanoma Research*, **3**, 395-401.

Armstrong, B.K. and A. Kricker, 1994: Cutaneous melanoma. In: *Cancer Surveys*, vol. 19/20 [Doll, R., J.F. Fraumeni, and C.S. Muir (eds.)]. Cold Spring Harbor Laboratory Press, New York, NY, pp. 219-240.

Asian Development Bank, 1994: *Climate Change in Asia: Indonesia Country Report*. Asian Development Bank, Manila, Philippines, pp. 37-42.

Ayres, J.G., 1990: Meteorology and respiratory disease. *Update 1990*, **40**, 596-605.

Babayev, A.B., 1986: Some aspects of man's acclimatization to hot climates. In: *Climate and Human Health, WHO/UNEP/WMO International Symposium*, vol. 2. WMO, Leningrad, pp. 125-126.

Bazzar, F.A. and E.D. Fajer, 1992: Plant life in a CO_2-rich world. *Scientific American*, **266(1)**, 18-24.

Beckett, W.S., 1991: Ozone, air pollution, and respiratory health. *Yale Journal of Biology and Medicine*, **64**, 167-75.

Beer, S.A. and S.M. German, 1993: The ecological prerequisites for a worsening of cerciasis situation in the cities of Russia. *Parazitologia*, **27(6)**, 441-449 [in Russian].

Beer, S.A. and S.M. German, 1994: Ekologicheskie predposylki rasprostraneniia shistosomatidenykh dermatoitov tserkariozov v Moskve i Podomoskov'e. *Medicinschke Parazitologie—Moskau Jan-Mar*, **1**, 16-19.

Bianca, W., 1976: The significance of meteorology in animal production. *International Journal of Biometeorology*, **20**, 139-56.

Bouma, M.J., H.E. Sondorp, and H.J. van der Kaay, 1994: Health and climate change. *The Lancet*, **343**, 302.

Bradley, D.J., 1993: Human tropical diseases in a changing environment. In: *Environmental Change and Human Health* [Lake J., G. Bock, and K. Ackrill (eds.)]. Ciba Foundation Symposium, CIBA Foundation, London, UK, pp. 146-162.

Burgos, J.J., S.I. Curto de Casas, R.U. Carcavallo, and I. Galíndez Girón, 1994: Global climate change influence in the distribution of some pathogenic complexes (malaria and Chagas disease) in Argentina. *Entomologia y Vectores*, **1(2)**, 69-78.

Carcavallo, R.U. and A. Martinez, 1972: Life cycles of some species of Triatoma (Hemipt. Reduviidae). *Canadian Entomology*, **104**, 699-704.

Carcavallo, R.U. and A. Martinez, 1985: Biologica, ecologia y distribucion geografica de los Triatominos Americanos. *Chagas (Special Issue)*, **1**, 149-208.

CDC, 1989: Heat-related deaths—Missouri, 1979–1988. *Morbidity and Mortality Weekly Report*, **38**, 437–439.

CDC, 1992: Cercarial dermatitis outbreak in a state park—Delaware 1991. *Journal of the American Medical Association*, **267(19)**, 2581-2586.

CDC, 1994a: Hantavirus Pulmonary Syndrome—United States, 1993. *Morbidity and Mortality Weekly Report*, **43(3)**, 45-48.

CDC, 1994b: Update: influenza activity—United States and worldwide, 1993-93 season and composition of the 1994-95 influenza vaccine. *Morbidity and Mortality Weekly Report*, **43(10)**, 179-183.

Chameides, W.L., P.S. Kasibhatla, J. Yienger, and H. Levy, 1994: Growth of continental-scale metro-agro-plexes, regional ozone pollution, and world food production. *Science*, **264**, 74-77.

Climate Change Impacts Review Group UK, 1991: *The Potential Effects of Climate Change in the United Kingdom*. HMSO, London, UK, pp. 3-13.

Coleman, M., J. Esteve, P. Damiecki, A. Arslan, and H. Renard, 1993: *Trends in Cancer Incidence and Mortality*. International Agency for Research on Cancer, IARC Scientific Publications No. 121, Lyon, France, 806 pp.

Curto de Casas, S.I. and R.U. Carcavallo, 1984: Limites del triatomismo en la Argentina. I: Patagonia. *Chagas*, **1(4)**, 35-40.

Curwen, M., 1991: Excess winter mortality: a British phenomenon? *Health Trends*, **22**, 169–175.

Dahlback, A., T. Henriksen, S.H.H. Larsen, and K. Stamnes, 1989: Biological UV doses and the effect of ozone layer depletion. *Photochemistry and Photobiology*, **49(5)**, 621-625.

de Girolamo, G. and A. McFarlane, 1994: *Epidemiology of Posttraumatic Stress Disorders: A Comprehensive Review of the Literature*. Unpublished paper.

de Gruijl, F.R. and J.C. van der Leun, 1993: Influence of ozone depletion on incidence of skin cancer: quantitative prediction. In: *Environmental UV Photobiology* [Young, A.R. *et al.* (eds.)]. Plenum Press, New York, NY, pp. 89-112.

de Leeuw, F.A.A. and H.J.v.R. Leyssius, 1991: Sensitivity of oxidant concentrations on changes in UV radiation and temperature. *Atmospheric Environment*, **25A**, 1024-1033.

den Elzen, M., 1994: *Global Environmental Change. An Integrated Modelling Approach*. International Books, Utrecht, Netherlands, 253 pp.

Denkins, Y.D., I.J. Fidler, and M.L. Kripke, 1989: Exposure of mice to UV-B radiation suppresses delayed hypersensitivity to *Candida albicans*. *Photochemistry and Photobiology*, **49**, 615-619.

Diamond, J., 1991: *The Rise and Fall of the Third Chimpanzee*. Radius, London, UK, 360 pp.

Dobson, A. and R. Carper, 1993: Biodiversity. *The Lancet*, **342**, 1096-1099.

Dockery, D.W., C.A. Pope, X. Xu, J.D. Spengler, J.H. Ware, M.E. Fay, B.G. Ferris, and F.E. Speizer, 1993: An association between air pollution and mortality in six U.S. cities. *New England Journal of Medicine*, **329**, 1753-1759.

Dolin, P.J., 1994: Ultraviolet radiation and cataract: a review of the epidemiological evidence. *British Journal of Ophthalmology*, **798**, 478-482.

Durkin, M.S., N. Khan, and L.L. Davidson, 1993: The effects of a natural disaster on child behaviour: evidence for post-traumatic stress. *American Journal of Public Health*, **83**, 1549-1553..

Ellis, F.P. and F. Nelson, 1978: Mortality in the elderly in a heat wave in New York City, August, 1974. *Environmental Research*, **5**, 1-58.

Emberlin, J., 1994: The effects of patterns in climate and pollen abundance on allergy. *Allergy*, **49**, 15-20.

Epstein, P.R., 1992: Cholera and the environment. *The Lancet*, **339**, 1167–1168.

Epstein, P.R., T.E. Ford, and R.R. Colwell, 1993: Marine ecosystems. *The Lancet*, **342**, 1216-1219.

Escudero, J.C., 1985: Health, nutrition and human development. In: *Climatic Impact Assessment* [Kates, R.W., J.H. Ausubel, and M. Berberian (eds.)]. Scope 27, John Wiley and Sons, Chichester, UK, pp. 251–272.

Farrow, R.A., 1991: Implications of potential global warming on agricultural pests in Australia. *EPPO Bulletin*, **21**, 683-696.

Feachem, R., 1994: Health decline in Eastern Europe. *Nature*, **367**, 313-314.

Freier, J.E., 1993: Eastern equine encephalomyelitis. *The Lancet*, **342**, 1281-1282.

French, J.G. and K.W. Holt, 1989: Floods. In: *The Public Health Consequences of Disasters* [Gregg, M.B. (ed.)]. U.S. Department of Health and Human Services, CDC, Atlanta, GA, pp. 69-78.

Furquay, J.W., 1989: Heat stress as it affects animal production. *Journal of Animal Science*, **52**, 164-174.

Garms, R., J.F. Walsh, and J.B. Davis, 1979: Studies on the reinvasion of the Onchocerciasis Control Programme in the Volta River Basin by *Sumulium damnosum s.l.* with emphasis on the Southwestern areas. *Tropical Medicine and Parasitology*, **30**, 345-362.

Giannini, M.S.H., 1986: Suppression of pathogenesis in cutaneous leishmaniasis by UV-irradiation. *Infection and Immunity*, **51**, 838-843.

Giannini, M.S.H. and E.C. DeFabo, 1989: Abrogation of skin lesions in cutaneous leishmaniasis by ultraviolet-B radiation. In: *Leishmaniasis: The Current Status and New Strategies for Control* [Hart, D.T. (ed.)]. NATO ASI Series A: Life Sciences, Plenum Press, London, UK, pp. 677-684.

Gill, C.A., 1920a: The relationship between malaria and rainfall. *Indian Journal of Medical Research,* **37(3)**, 618-632.

Gill, C.A., 1920b: The role of meteorology and malaria. *Indian Journal of Medical Research*, **8(4)**, 633-693.

Gilles, H.M., 1993: Epidemiology of malaria. In: *Bruce-Chwatt's Essential Malariology* [Gilles, H.M. and D.A. Warrell (eds.)]. Edward Arnold, London, UK, pp. 124-163.

Gillett, J.D., 1974: Direct and indirect influences of temperature on the transmission of parasites from insects to man. In: *The Effects of Meteorological Factors Upon Parasites* [Taylor, A.E.R. and R. Muller (eds.)]. Blackwell Scientific Publications, Oxford, UK, pp. 79–95.

Glantz, M.H. (ed.), 1992: *Climate Variability, Climate Change and Fisheries.* Cambridge University Press, Cambridge, UK, 450 pp.

Goldstein, I.F., 1980: Weather patterns and asthma epidemics in New York City and New Orleans, USA. *International Journal of Biometeorology,* **24**, 329-339.

Gordon, H.B., P.H. Whetton, A.B. Pittock, A.M. Fowler, and M.R. Haylock, 1992: Simulated changes in daily rainfall intensity due to the enhanced greenhouse effect: implications for extreme rainfall events. *Climate Dynamics*, **8**, 83-102.

Gray, R.H., G.J. Johnson, and A. Freedman, 1992: Climatic droplet keratopathy. *Surveys of Ophthalmology*, **36**, 241-253.

Green, B.L., 1982: Assessing levels of psychological impairment following disaster: consideration of actual and methodological dimensions. *Journal of Nervous Disease*, **170**, 544-548

Green, B.L., M. Korol, M.C. Grace, M.G. Vary, A.C. Leonard, G.C. Gleser, and S. Smitson-Cohen, 1991: Children and disaster: age, gender, and parental effects of PTSD symptoms. *Journal of the Academy of Child and Adolescent Psychiatry*, **30**, 945-951.

Gregg, M.B. (ed.), 1989: *The Public Health Consequences of Disasters*. U.S. Department of Health and Human Services, CDC, Atlanta, GA, 137 pp.

Grosse, S., 1993: *Schistosomiasis and Water Resources Development: A Reevaluation of an Important Environment-Health Linkage.* The Environment and Natural Resources Policy and Training Project, Working Paper EPAT/MUCIA, Technical Series No. 2., May 1993. University of Michigan Press, Ann Arbor, MI, 32 pp.

Hack, W.H., 1955: Estudio sobre la biologia del Triatoma infestans. *Anales Instituto de Medicina Regional Tucuman,* **4**, 125-147.

Haines, A. and C. Fuchs, 1991: Potential impacts on health of atmospheric change. *Journal of Public Health Medicine*, **13**, 69–80.

Haines, A., P.R. Epstein, and A.J. McMichael, 1993: Global Health Watch: monitoring the impacts of environmental change. *The Lancet,* **342**, 1464-1469.

Hairston, N.G., 1973: The dynamics of transmission. In: *Epidemiology and Control of Schistosomiasis* [Ansari, N. (ed.)]. Karger, Basel, Switzerland, pp. 250-336.

Halstead, S.B., 1990: Dengue. In: *Tropical and Geographical Medicine* [Warren, K. and A.A.F. Mahmoud (eds.)]. McGraw-Hill, New York, NY, 2nd ed., pp. 675-685.

Hansen, J., I. Fung, A. Lacis, S. Lebedoff, D. Rind, R. Ruedy, and G. Russell, 1988: Global climate changes as forecast by the GISS 3-D model. *Journal of Geophysical Research*, **93**, 9341-9364.

Harding, J., 1991: *Cataract: Biochemistry, Epidemiology and Pharmacology.* Chapman Hall, London, UK, 382 pp.

Harding, J., 1992: Physiology, biochemistry, pathogenesis, and epidemiology of cataract. *Current Opinion in Ophthalmology*, **3**, 3-12.

Herrera-Basto, E., D.R. Prevots, M.L. Zarate, J.L. Silva, and J.S. Amor, 1992: First reported outbreak of classical dengue fever at 1,700 meters above sea level in Guerrero State, Mexico, June 1988. *American Journal of Tropical Medicine,* **46(6)**, 649-653.

Hersey, P., G. Haran, E. Hasic, and A. Edwards, 1983: Alteration of T-cell subsets and induction of suppressor T-cell activity in normal subjects after exposure to sunlight. *Journal of Immunology,* **131**, 171-174.

Homer-Dixon, T., J.H. Boutwell, and G.W. Rathjens, 1993: Environmental change and violent conflict. *Scientific American*, **268**, 38-45.

Horowitz, M. and S. Samueloff, 1987: Circulation under extreme heat load. In: *Comparative Physiology of Environmental Adaptations,* vol. 2 [Dejours, P. (ed.)]. Karger, Basel, Switzerand, pp. 94-106.

Houghton, J.T., G.J. Jenkins, and J.J. Ephraums (eds.), 1990: *Climate Change. The IPCC Scientific Assessment.* Cambridge University Press, Cambridge, UK, 365 pp.

Hunter, J.M.L., L. Rey, K.Y. Chu, E.O. Adekolu-John, and K.E. Mott, 1993: *Parasitic Diseases in Water Resources Development: The Need for Intersectoral Negotiation*. WHO, Geneva, Switzerland, 152 pp.

IARC, 1992: *Solar and Ultraviolet Radiation. IARC Monographs on the Evaluation of Carcinogenic Risks to Humans*, vol. 55. IARC, Lyon, France, 316 pp.

Idso, S.B., 1990a: *The Carbon Dioxide/Trace Gas Greenhouse Effect: Greatly Overestimated?* American Society of Agronomy, Madison, WI, pp. 19-26.

Idso, S.B., 1990b: *Interactive Effects of Carbon Dioxide and Climate Variables on Plant Growth.* American Society of Agronomy, Madison, WI, pp. 61-69.

Jeevan, A. and M.L. Kripke, 1989: Effect of a single exposure to UVB radiation on *Mycobacterium bovis* bacillus Calmette-Guerin infection in mice. *Journal of Immunology*, **143**, 2837-2843.

Jeevan, A. and M.L. Kripke, 1990: Alteration of the immune response to Mycobacterium bovis BCG in mice exposed chronically to low doses of UV radiation. *Cellular Immunology*, **130**, 32-41.

Jeevan, A. and M.L. Kripke, 1993: Ozone depletion and the immune system. *The Lancet*, **342**, 1159-1160.

Kalkstein, L.S., 1993: Health and climate change: direct impacts in cities. *The Lancet*, **342**, 1397-1399.

Kalkstein, L.S. and K.E. Smoyer, 1993: The impact of climate change on human health: some international implications. *Experientia*, **49**, 969-979.

Katsouyanni, K., A. Pantazopoulou, G. Touloumi, *et al.*, 1993: Evidence for interaction between air pollution and high temperature in the causation of excess mortality. *Archives of Environmental Health*, **48**, 235-242.

Keatinge, W.R, S.R.K. Coleshaw, and J. Holmes, 1989: Changes in seasonal mortalities with improvement in home heating in England and Wales from 1964 to 1984. *International Journal of Biometeorology*, **33**, 71-76.

Kendall, H.W. and D. Pimentel, 1994: Constraints on the expansion of the global food supply. *Ambio*, **23(3)**, 198-205.

Kerr, R.A. and M.B. McElroy, 1993: Evidence for large upward trends of ultraviolet-B radiation linked to ozone depletion. *Science*, **262**, 1032-1034.

Kilbourne, E.M., 1989: Heatwaves. In: *The Public Health Consequences of Disasters* [Gregg, M.B. (ed.)]. U.S. Department of Health and Human Services, Public Health Service, CDC, Atlanta, GA, pp. 51-61.

Kilbourne, E.M., 1992: Illness due to thermal extremes. In: *Public Health and Preventive Medicine* [Last, J.M. and R.B. Wallace (eds.)]. Maxcy-Rosenau-Last, Norwalk, Appleton Lange, 13th ed., pp. 491–501.

Klein, B.E., R. Klein, and K.L.P. Linton, 1992: Prevalence of age-related lens opacities in a population. The Beaver Dam Eye Study. *Ophthalmology*, **99**, 546-552.

Koopman, J.S., D.R. Prevots, M.A.V. Marin, H.G. Dantes, M.L.Z. Aqino, I.M. Longini, and J.S. Amor, 1991: Determinants and predictors of dengue infection in Mexico. *American Journal of Epidemiology*, **133**, 1168–1178.

Körner, C., 1993: CO_2 fertilization: the great uncertainty in future vegetation development. In: *Vegetation Dynamics and Global Change* [Solomon, A.M. and H.H. Shugart (eds.)]. Chapman and Hall, New York, NY, pp. 9953-9970.

Kricker, A., B.K. Armstrong, D. English, and P.J. Heenan, 1995: A dose response curve for sun exposure and basal cell carcinoma. *International Journal of Cancer*, **60**, 482-488.

Kripke, M.L., 1981: Immunologic mechanisms in UV radiation carcinogenesis. *Advances in Cancer Research*, **34**, 69-81.

Kunst, A.E., C.W.N. Looman, and J.P. Mackenbach, 1993: Outdoor air temperature and mortality in the Netherlands: a time-series analysis. *American Journal of Epidemiology,* **137**, 331-341.

LAIA, 1993*: Seasonal Variations in Asthma.* Lung and Asthma Information Agency Factsheet 93/4, St. George's Medical School, London, UK.

Lancet, 1985: Asthma and the weather. *The Lancet*, **i**, 1079-1080.

Lancet, 1994: *Health and Climate Change*. The Devonshire Press, London, UK.

Langford, I.H. and G. Bentham, 1995: The potential effects of climate change on winter mortality in England and Wales. *International Journal of Biometeorology*, **38**, 141-147.

Larsen, U., 1990a: The effects of monthly temperature fluctuations on mortality in the United States from 1921 to 1985. *International Journal of Biometeorology*, **34**, 136-145.

Larsen, U., 1990b: Short-term fluctuations in death by cause, temperature, and income in the United States 1930 to 1985. *Social Biology,* **37(3-4)**, 172-187.

Last, J.M., 1993: Global change: ozone depletion, greenhouse warming and public health. *Annual Review of Public Health,* **14**, 115-136.

Leemans, R., 1992: Modelling ecological and agricultural impacts of global change on a global scale. *Journal of Scientific and Industrial Research,* **51**, 709-724.

Lerman, S., 1986: Photosensitizing drugs and their possible role in enhancing ocular toxicity. *Ophthalmology*, **93**, 304-318.

Lerman, S., 1988: Ocular phototoxicity. *New England Journal of Medicine,* **319**, 1475-1477.

Levins, R., T. Awerbuch, U. Brinkman, I. Eckhardt, P.R. Epstein, N. Makhoul, C. Albuquerque de Possas, C. Puccia, A. Spielman, and M.E. Wilson, 1994: The emergence of new diseases. *American Scientist,* **82**, 52-60.

Lighthill, J., G. Holland, W. Gray, C. Landsea, J. Evans, Y. Kurihara, and C. Guard, 1994: Global climate change and tropical cyclones. *Bulletin of the American Meteorological Society,* **75(11)**, 2147-2157.

Loevinsohn, M., 1994: Climatic warming and increased malaria incidence in Rwanda. *The Lancet,* **343**, 714-718.

Longstreth, J.A., 1989: Human health. In: *The Potential Effects of Global Climate Change on the United States* [Smith, J.B. and D. Tirpak (eds.)]. EPA-230-05-89-050, U.S. EPA, Washington, DC, pp. 525-556.

Madronich, S. and F.R. de Gruijl, 1993: Skin cancer and UV radiation. *Nature*, **366**, 23.

Martens, W.J.M., J. Rotmans, and L.W. Niessen, 1994: *Climate Change and Malaria Risk: An Integrated Modelling Approach.* GLOBO Report Series No. 3, Report No. 461502003, Global Dynamics & Sustainable Development Programme, Bilthoven, RIVM, 37 pp.

Martens, W.J.M., L.W. Niessen, J. Rotmans, T.H. Jetten, and A.J. McMichael, 1995a: Potential impact of global climate change on malaria risk. *Environmental Health Perspectives,* **103(5)**, 458-464.

Martens, W.J.M., T.H. Jetten, J. Rotmans, and L.W. Niessen, 1995b: Climate change and vector-borne diseases: a global modelling perspective. *Global Environmental Change*, **5(3)**, 195-209.

Martin, P. and M. Lefebvre, 1995: Malaria and climate: sensitivity of malaria potential transmission to climate. *Ambio,* **24(4)**, 200-207.

Matsuoka, Y. and K. Kai, 1994: An estimation of climatic change effects on malaria. *Journal of Global Environment Engineering*, **1**, 1-15.

McMichael, A.J., 1993: Global environmental change and human population health: a conceptual and scientific challenge for epidemiology. *International Journal of Epidemiology*, **22**, 1-8.

McMichael, A.J. and W.J.M. Martens, 1995: The health impacts of global climate change: grappling with scenarios, predictive models, and multiple uncertainties. *Ecosystem Health*, **1(1)**, 23-33.

Meehl, G.A., G.W. Branstator, and W.M. Washington, 1993: Tropical Pacific interannual variability and CO_2 climate change. *Journal of Climate*, **6**, 42-63.

Michael, E. and D.A.P. Bundy, 1995: The global burden of lymphatic filariasis. In: *World Burden of Diseases* [Murray, C.J.L. and A.D. Lopez (eds.)]. WHO, Geneva, Switzerland (in press).

Miller, L.H. and D.A. Warrell, 1990: Malaria. In: *Tropical and Geographical Medicine* [Warren, K.S. and A.A.F. Mahmoud (eds.)]. McGraw-Hill, New York, NY, 2nd ed., pp. 245-264.

Mills, D.M., 1995: A climatic water budget approach to blackfly population dynamics. *Publications in Climatology*, **48**, 1-84.

Morison, W.L., 1989: Effects of ultraviolet radiation on the immune system in humans. *Photochemistry and Photobiology*, **50**, 515-524.

Morse, S.S., 1991: Emerging viruses: defining the rules for viral traffic. *Perspectives in Biology and Medicine*, **34(3)**, 387-409.

Murray, C.J.L. and A.D. Lopez (eds.) 1994: *Global Comparative Assessments in the Health Sector: Disease Burden, Expenditures and Intervention Packages*. WHO, Geneva, Switzerland, 196 pp.

Myers, N., 1993: Environmental refugees in globally warmed world. *BioScience,* **43(11)**, 752-761.

NHMRC, 1991: *Health Implications of Long Term Climatic Change.* Australian Government Printing Service, Canberra, Australia, 83 pp.

Nicholls, N., 1993: El Niño-Southern Oscillation and vector-borne disease. *The Lancet,* **342**, 1284-1285.

Niessen, L.W. and J. Rotmans, 1993: *Sustaining Health: Towards an Integrated Global Health Model.* RIVM Report No. 461502001, Bilthoven, RIVM, 32 pp.

Noonan, F.P. and E.C. DeFabo, 1990: Ultraviolet-B dose-response curves for local and systemic immunosuppression are identical. *Photochemistry and Photobiology,* **52**, 801-810.

Otani, T. and R. Mori, 1987: The effects of ultraviolet irradiation of the skin on herpes simplex virus infection: alteration in immune function mediated by epidermal cells in the course of infection. *Archives of Virology*, **96**, 1-15.

PAHO, 1994: Leishmaniasis in the Americas. *Epidemiological Bulletin*, **15(3)**, 8-13.

Parry, M.L. and C. Rosenzweig, 1993: Food supply and the risk of hunger. *The Lancet*, **342**, 1345-1347.

Peto, R., A.D. Lopez, J. Boreham, M. Thun, and C. Heath, 1994: *Mortality from Smoking in Developed Countries 1950-2000.* Oxford University Press, New York, NY, 553 pp.

Pope, C.A., D.V. Bates, M.E. Raizenne, 1995: Health effects of particulate air pollution: time for reassessment? *Environmental Health Perspectives*, **103**, 472-480.

Reeves,W.C., J.L. Hardy, W.K. Reisen, M.M. Milby, 1994: Potential effect of global warming on mosquito-borne arboviruses. *Journal of Medical Entomology*, **31(3)**, 323-332.

Reilly, J., 1994: Crops and climate change. *Nature*, **367**, 118-119.

Rind, D. and A. Lacis, 1993: The role of the stratosphere in climate change. *Surveys in Geophysics*, **14**, 133-165.

Rogers, D.J. and M.J. Packer, 1993: Vector-borne diseases, models and global change. *The Lancet*, **342**, 1282-1284.

Rogot, E., P.D. Sorlie, and E. Backland, 1992: Air-conditioning and mortality in hot weather. *American Journal of Epidemiology,* **136**, 106-116.

Rosenzweig, C. and M.M. Daniel, 1989: Agriculture. In: *The Potential Effects of Global Climate Change in the United States* [Smith, J.B. and D. Tirpak (eds.)]. EPA-230-05-89-050, U.S. EPA, Washington, DC, pp. 93-122.

Rosenzweig, C., M.L. Parry, G. Fischer, and K. Frohberg, 1993: *Climate Change and World Food Supply.* Research Report No. 3, Environmental Change Unit, Oxford University, Oxford, UK, 28 pp.

Seinfeld, J.H., 1986: *Air Pollution: Physical and Chemical Fundamentals.* MacGraw-Hill, New York, NY, 523 pp.

Schwartz, J., 1994: Air pollution and daily mortality: a review and meta-analysis. *Environmental Research*, **64**, 36-52.

Shope, R., 1991: Global climate change and infectious diseases. *Environmental Health Perspectives*, **96**, 171–174.

Shumway, R.H. and A.S. Azari, 1992: *Structural Modeling of Eepidemiological Time Series.* Final Report ARB A 833-136, Air Resources Board, Sacramento, CA, 67 pp.

Shumway, R.H., A.S. Azari, and Y. Pawitan, 1988: Modelling mortality fluctuations in Los Angeles as functions of pollution and weather effects. *Environmental Research*, **45**, 224-241.

Slaper, H., M.G.H. den Elzen, H.J. van den Woerd, and J. Greef, 1992: *Ozone Depletion and Skin Cancer Incidence, Integrated Modelling Approach.* Report No. 749202001, Bilthoven, RIVM, 45 pp.

Smayda, T.J., 1990: Novel and nuisance phytoplankton blooms in the sea: evidence for a global epidemic. In: *Toxic Marine Phytoplankton* [Graneli, E. *et al.* (eds.)]. Elsevier Science Publishers, New York, NY, pp. 29-40.

Smith, R.C. and K.S. Baker, 1989: Stratospheric ozone, middle ultraviolet radiation and phytoplankton productivity. *Oceanography,* **2**, 4-10.

Smith, R.C., B.B. Prezelin, K.S. Baker, R.R. Bidigare, N.P. Boucher, T. Coley, D. Karentz, S. MacIntyre, H.A. Matlock, D. Menzies, M. Ondrusek, Z. Wan, and K.J. Waters, 1992: Ultraviolet radiation and phytoplankton biology in Antarctic water. *Science*, **255**, 952-959.

Sommer, A. and W.H. Mosely, 1972: East Bengal cyclone of November 1970: epidemiological approach to disaster assessment. *The Lancet*, **I**, 1029-1036.

Sutherst, R.W., 1991: Pest risk analysis and the greenhouse effect. *Review of Agricultural Entomology*, **79(11/12)**, 1177-1187.

Sutherst, R.W., 1993: Arthropods as disease vectors in a changing environment. In: *Environmental Change and Human Health* [Lake, J.V. (ed.)]. Ciba Foundation Symposium, John Wiley and Sons, New York, NY, pp. 124-145.

Tan, G., 1994: The potential impacts of global warming on human mortality in Shanghai and Guangzhou, China. *Acta Scientiae Cicumstantiae*, **14**, 368-373.

Taylor, H.R., S.K. West, F.S. Rosenthal, B. Munoz, H.S. Newland, and E.A. Emmett, 1989: Corneal changes associated with chronic UV irradiation. *Archives of Ophthalmology*, **107**, 1481-1484.

Taylor, H.R., S.K. West, F.S. Rosenthal, M. Beatriz, H.S. Newland, H. Abbey, and E.A. Emmett, 1988: The effect of ultraviolet radiation on cataract formation. *New England Journal of Medicine,* **319**, 1411-1415.

Telleria, A.V., 1986: Health consequences of floods in Bolivia in 1982. *Disasters*, **10**, 297-307.

Thurman, E.M., D.A. Goolsby, M.T. Meyer, D.W. Kolpin, 1991: Herbicides in surface waters of the United States: the effect of spring flush. *Environmental Science and Technology*, **25**, 1794-1796.

Thurman, E.M., D.A. Goolsby, M.T. Meyer, M.S. Mills, M.L. Pomes, D.W. Kolpin, 1992: A reconnaissance study of herbicides and their metabolites in surface water of the mid-western United States using immunoassay and gas chromatography/mass spectrometry. *Environmental Science and Technology*, **26**, 2440-2447.

Titchener, J.L. and T.K. Frederic, 1976: Family and character change at Buffalo Creek. *American Journal of Psychiatry*, **133**, 295-299.

Tomkins, A.M., 1986: Protein-energy malnutrition and risk of infection. *Proceedings of the Nutrition Society*, **45**, 289-304.

Touloumi, G., S.J. Pocock, K. Katsouyanni, D. Trichopoulos, 1994: Short-term effects of air pollution on daily mortality in Athens: a time-series analysis. *International Journal of Epidemiology*, **23(5)**, 957-967.

Tyrell, R.M., 1994: The molecular and cellular pathology of solar ultraviolet radiation. *Molecular Aspects of Medicine*, **15(3)**, 1-77.

UNEP, 1994: *Environmental Effects of Ozone Depletion: 1994 Assessment.* UNEP, Nairobi, Kenya, 110 pp.

U.S. EPA, in press: *Preliminary Assessment of the Benefits to the US of Avoiding, or Adapting to Climate Change.* EPA Climate Change Division, Washington, DC.

Vleodbeld, M. and R. Leemans, 1993: Quantifying feedback processes in the response of the terrestrial carbon cycle to global change—modelling approach of image-2. *Water, Air and Soil Pollution,* **70**, 615-628.

Walsh, J.F., J.B. Davis, and R. Garms, 1981: Further studies on the reinvasion of the Onchocerciasis Control Programme by *Simulium damnosum s.l. Tropical Medicine and Parasitology*, **32(4)**, 269-273.

Weihe, W.H., 1986: *Life Expectancy in Tropical Climates and Urbanization.* Proc. WMO Technical Conference Urban Climate, Mexico City, 26-30 November 1984, WMO-No. 652, WMO, Geneva, Switzerland, 15 pp.

Wenzel, R.P., 1994: A new hantavirus infection in North America. *New England Journal of Medicine*, **330(14)**, 1004-1005.

West, S.K., F.S. Rosenthal, N.M. Bressler, S.B. Bressler, B. Munoz, S.L. Fine, and H.R. Taylor, 1989: Exposure to sunlight and other risk factors for age related macular degeneration. *Archives of Ophthalmology,* **107**, 875-879.

White, G.F., D.J. Bradley, A.U. White, 1972: *Drawers of Water: Domestic Water Use in East Africa.* University of Chicago Press, London, UK, 306 pp.

WHO, 1985: *Ten Years of Onchocerciasis Control in West Africa: Review of Work of the OCP in the Volta River Basin Area from 1974 to 1984.* OCP/GVA/85.1B. WHO, Geneva, Switzerland, 113 pp.

WHO, 1990: *Potential Health Effects of Climate Change: Report of a WHO Task Group.* WHO/PEP/90.10, WHO, Geneva, Switzerland, 58 pp.

WHO, 1994a: *Progress Report Control of Tropical Diseases.* CTD/MIP/94.4, unpublished document.

WHO, 1994b: *The Effects of Solar UV Radiation on the Eye. Report of an Informal Consultation, Geneva, 1993.* WHO, Geneva, Switzerland, 50 pp.

WHO, 1995a: *The World Health Report 1995: Bridging the Gaps.* WHO, Geneva, Switzerland, 118 pp.

WHO, 1995b: *Action Plan for Malaria Control 1995-2000.* Unpublished document.

WHO, 1995c: *Chagas Disease: Important Advances in Elimination of Transmission in Four Countries in Latin America.* WHO Press Office Feature No. 183, WHO, Geneva, Switzerland.

Woodhouse, P.R., K.T. Khaw, M. Plummer, A. Foley, T.W. Meade, 1994: Seasonal variation of plasma fibrinogen and factor VII activity in the elderly: winter infections and death from cardiovascular disease. *The Lancet*, **343**, 435-439.

Woodward, F.I., 1987: Stomatal numbers are sensitive to increases in CO_2 from pre-industrial levels. *Nature*, **327**, 617-618.

World Bank, 1993: *World Development Report 1993.* Oxford University Press, New York, NY, 329 pp.

Worrest, R.C. and L.D. Grant, 1989: Effects of ultraviolet-B radiation on terrestrial plants and marine organisms. In: *Ozone Depletion: Health and Environmental Consequences* [Jones, R. and T. Wigley (eds.)]. John Wiley and Sons, Chichester, UK, pp. 197–206.

Yasumoto, S., Y. Hayashi, and L. Aurelian, 1987: Immunity to herpes simplex virus type 2: suppression of virus-induced immune response in ultraviolet B-irradiated mice. *Journal of Immunology*, **139**, 2788-2793.

Yoshikawa, T., V. Rae, W. Bruins-Slot, J.-W. van den Berg, J.R. Taylor, and J.W. Streilein, 1990: Susceptibility to effects of UVB radiation on induction of contact hypersensitivity as a risk factor for skin cancer in man. *Journal of Investigative Dermatology*, **95**, 530-536.

Climate Change 1995

Part III

Assessment of Mitigation Options

Prepared by Working Group II

19

Energy Supply Mitigation Options

H. ISHITANI, JAPAN; T.B. JOHANSSON, SWEDEN

Lead Authors:

S. Al-Khouli, Saudi Arabia; H. Audus, IEA; E. Bertel, IAEA; E. Bravo, Venezuela; J.A. Edmonds, USA; S. Frandsen, Denmark; D. Hall, UK; K. Heinloth, Germany; M. Jefferson, WEC; P. de Laquil III, USA; J.R. Moreira, Brazil; N. Nakicenovic, IIASA; Y. Ogawa, Japan; R. Pachauri, India; A. Riedacker, France; H.-H. Rogner, Canada; K. Saviharju, Finland; B. Sørensen, Denmark; G. Stevens, OECD/NEA; W.C. Turkenburg, The Netherlands; R.H. Williams, USA; Zhou Fengqi, China

Contributing Authors:

I.B. Friedleifsson, Iceland; A. Inaba, Japan; S. Rayner, USA; J.S. Robertson, UK

CONTENTS

EXECUTIVE SUMMARY

This review focuses on energy supply options that can sharply reduce greenhouse gas (GHG) emissions while providing needed energy services. By the year 2100, the world's commercial energy system will be replaced at least twice, which offers opportunities to change the present energy system in step with the normal timing of the corresponding investments, using emerging technologies in environmentally sound ways. Therefore, this assessment focuses on the performance of these emerging technologies.

Since the preparation of the IPCC 1990 Assessment Report, there have been some significant advances in the understanding of modern technology and technological innovations relating to energy systems that can reduce GHG emissions. Examples of such technologies are found, *inter alia*, in gas turbine technology; coal and biomass gasification technology; production of transportation fuels from biomass; wind energy utilization; electricity generation with photovoltaic and solar thermal electric technologies; approaches to handling intermittent generation of electricity; fuel cells for transportation and power generation; nuclear energy; carbon dioxide (CO_2) sequestering; and hydrogen as a major new energy carrier, produced first from natural gas and later from biomass, coal, and electrolysis.

In the energy supply sector, we conclude with a high degree of confidence that GHG emissions reductions can be achieved through technology options in the following areas (which have been ordered according to type of measure rather than priority):

- **More efficient conversion of fossil fuels**: New technology offers considerably increased conversion efficiencies. For example, the efficiency of power production can be increased from the present world average of about 30% to more than 60% in the longer term. Also, the use of combined heat and power production replacing separate production of power and heat—whether for process heat or space heating—offers a significant rise in fuel conversion efficiency.
- **Switching to low-carbon fossil fuels and suppressing emissions**: Switching from coal to oil or natural gas, and from oil to natural gas, can reduce emissions. Natural gas has the lowest CO_2 emissions per unit of energy of all fossil fuels at ~14 kg/GJ, compared to oil with ~20 kg/GJ and coal with ~25 kg/GJ. The lower carbon-containing fuels can, in general, be converted with higher efficiency than coal. Large resources of natural gas exist in many areas. New, low capital cost, highly efficient, combined-cycle technology has reduced electricity costs considerably in many areas. Natural gas could potentially replace oil in the transportation sector. Approaches exist to reduce emissions of methane from natural gas pipelines and emissions of methane and/or CO_2 from oil and gas wells and coal mines.
- **Decarbonization of flue gases and fuels, and CO_2 storage**: The removal and storage of CO_2 from fossil fuel power-station stack gases is feasible, but reduces the conversion efficiency and significantly increases the production cost of electricity. Another approach to decarbonization uses fossil fuel feedstocks to make hydrogen-rich fuels. Both approaches generate a byproduct stream of CO_2 that could be stored, for example, in depleted natural gas fields. The future availability of conversion technologies such as fuel cells that can efficiently use hydrogen would increase the relative attractiveness of the latter approach. For some longer-term CO_2 storage options, the costs and environmental effects and their efficacy remain largely unknown.
- **Increasing the use of nuclear energy**: Nuclear energy could replace baseload fossil fuel electricity generation in many parts of the world, if generally acceptable responses can be found to concerns such as reactor safety, radioactive-waste transport and disposal, and proliferation.
- **Increasing the use of renewable sources of energy**: Solar, biomass, wind, hydro, and geothermal technologies already are widely used. In 1990, renewable sources of energy contributed about 20% of the world's primary energy consumption, most of it fuelwood and hydropower. Technological advances offer new opportunities and declining costs for energy from these sources. In the longer term, renewable sources of energy could meet a major part of the world's demand for energy. Power systems can easily accommodate limited fractions of intermittent generation, and with the addition of fast-responding backup and storage units, also higher fractions. Where biomass is sustainably regrown and used to displace fossil fuels in energy production, net CO_2 emissions are avoided, as the CO_2 released in energy conversion is again fixed in biomass through photosynthesis. If the development of biomass energy can be carried out in ways that effectively address concerns about other environmental issues and competition with other land uses, biomass could make major contributions in both electricity and fuels markets.

In addition to climate change concerns, there are several concerns of an immediate nature that could be addressed by the energy supply options discussed here (except carbon sequestering); for example, modern renewable sources of energy are often beneficial for local and regional environmental problems (e.g., urban air pollution, indoor air pollution, and acid rain) and for some technologies (especially biomass), rural income and employment generation, and land restoration and preservation. These energy supply options can be pursued to varying degrees within the limits of sustainable development criteria, which may however limit the exploitation of their full technical potential.

A wide range of modular renewable and other emission-reducing technologies are good candidates for cost-cutting through innovation and experience. For such technologies, the cost of the needed research, development, and demonstration (RD&D) and commercialization support is relatively modest (High Confidence).

Some technology options—for example, combined-cycle power generation—would penetrate the current marketplace. To realize other options, governments would have to take integrated action—by improving market efficiency (e.g., by eliminating permanent subsidies for energy), by finding new ways to internalize external costs, by accelerating RD&D on low- and zero-CO_2 emitting technologies, and by providing temporary incentives for early market development for these technologies as they approach commercial readiness. It is concluded with high confidence that the availability, cost, and penetration of technology options will strongly depend on such government action.

To assess the potential impact of combinations of individual measures at the energy system level, in contrast to the level of individual technologies, variants of a low CO_2-emitting energy supply system (LESS) are described. The LESS constructions are "thought experiments" exploring possible global energy systems.

The following assumptions were made: World population grows from 5.3 billion in 1990 to 9.5 billion by 2050 and 10.5 billion by 2100. GDP grows 7-fold by 2050 (5-fold and 14-fold in industrialized and developing countries, respectively) and 24-fold by 2100 (13-fold and 69-fold in industrialized and developing countries, respectively), relative to 1990. Because of emphasis on energy efficiency, primary energy rises much more slowly than GDP. The energy supply constructions were made to meet energy demand in (i) projections developed for the IPCC's First Assessment Report (1990) in a low energy demand variant, where global primary commercial energy use approximately doubles, with no net change for industrialized countries but a 4.4-fold increase for developing countries from 1990 to 2100; and (ii) a higher energy demand variant, based on the IPCC IS92a scenario where energy demand quadruples from 1990 to 2100. The energy demand levels of the LESS constructions are consistent with the energy demand mitigation chapters in this volume.

The analysis of the alternative LESS variants leads to the following conclusions:

- Deep reductions of CO_2 emissions from energy supply systems are technically possible within 50 to 100 years, using alternative strategies.
- Many combinations of the options identified in this assessment could reduce global CO_2 emissions from fossil fuels from about 6 Gt C in 1990 to about 4 Gt C per year by 2050, and to about 2 Gt C per year by 2100. Cumulative CO_2 emissions, from 1990 to 2100, would range from about 450 to about 470 Gt C in the alternative LESS constructions.
- Higher energy efficiency is underscored for achieving deep reductions in CO_2 emissions, for increasing the flexibility of supply-side combinations, and for reducing overall energy system costs.
- Interregional trade in energy grows in the LESS constructions compared to today's levels, expanding sustainable development options for Africa, Latin America, and the Middle East during the next century.

Costs for energy services in each LESS variant relative to costs for conventional energy depend on relative future energy prices, which are uncertain within a wide range, and on the performance and cost characteristics assumed for alternative technologies. However, within the wide range of future energy prices, one or more of the variants would plausibly be capable of providing the demanded energy services at estimated costs that are approximately the same as estimated future costs for current conventional energy. It is not possible to identify a least-cost future energy system for the longer term, as the relative costs of options depend on resource constraints and technological opportunities that are imperfectly known, and on actions by governments and the private sector.

The literature provides strong support for the feasibility of achieving the performance and cost characteristics assumed for energy technologies in the LESS constructions, within the next 1 or 2 decades, although it is impossible to be certain until the research and development is complete and the technologies have been tested in the market. Moreover, these performance and cost characteristics cannot be achieved without a strong and sustained investment in R&D. Many of the technologies being developed would need initial support to enter the market, and to reach sufficient volume to lower costs to become competitive.

Market penetration and continued acceptability of different technologies ultimately depend on their relative cost, performance (including environmental performance), institutional arrangements, and regulations and policies. Because costs vary by location and application, the wide variety of circumstances creates initial opportunities for new technologies to enter the market. Deeper understanding of the opportunities for emissions reductions would require more detailed analysis of promising clusters of options, taking into account local conditions.

Because of the large number of options, there is flexibility as to how the energy supply system could evolve, and paths of energy system development could be influenced by considerations other than climate change, including political, environmental (especially indoor and urban air pollution, acidification, and need for land restoration), and socioeconomic circumstances. Actual strategies for achieving deep reductions might combine elements from alternative LESS constructions. Moreover, there may well be other plausible technological paths that could lead to comparable reductions in emissions. More work is required to provide a comprehensive understanding of the prospects for and implications of alternative global energy supply systems that would lead to deep reductions in CO_2 emissions.

19.1. Introduction

The world energy supply system is huge, and many of the installations have economic lifetimes measured in decades. Annual worldwide investments are on the order of $150 billion. This means that changes will take considerable time to implement. However, within a period of 50–100 years, the entire energy supply system will be replaced at least twice. New investments to replace an old plant or to expand capacity are opportunities to adopt technologies that are more environmentally desirable at low incremental cost. Significant reductions in GHG emissions will not be achieved by a few scattered improvements. New technology must become characteristic for all new investments in order to reduce net carbon emissions significantly. The technologies discussed in this chapter are relevant worldwide and may be introduced more quickly in countries with rapid economic growth than in other countries.

This chapter identifies and assesses energy supply options that can reduce net GHG emissions. However, it does not attempt a comprehensive analysis of all options available for energy system development. There are many ways to reduce emissions. Some are cost-effective now; others will become cost-effective within the next decades; and some options may never be cost-effective, even with massive government support for RD&D and market introduction. Market penetration of different energy sources ultimately depends on the relative costs of fossil and alternative energy forms, as well as other attributes such as abundance, accessibility, institutional arrangements, and regulations and policies that address climate change and other issues. Because costs vary by location and application, the wide variety of circumstances creates initial opportunities for new technologies to enter the market. Deeper understanding of the opportunities for emissions reductions would require more detailed analysis of options, taking into account local conditions. Box 19-1 provides the conventions used throughout this chapter.

19.2. Options to Reduce Greenhouse Gas Emissions

Options to reduce GHG emissions from the energy supply system include improved efficiency in the use of fossil fuels; suppression of GHG emissions; switching to low-carbon fossil fuels; decarbonization of fuels and flue gases, and CO_2 storage and sequestering; and switching to nuclear energy and renewable sources of energy.

19.2.1. More Efficient Conversion of Fossil Fuels to Power and Heat

In this section, technologies for conversion of fossil fuels into heat and electricity—including advanced power-generation technologies, gasification, fuel cells, cogeneration, and the direct use of coal—are discussed.

19.2.1.1. Efficient Power Generation

Large-scale fossil fuel-fired power plants generate power in a steam turbine, a gas turbine, or a combination of the two as a combined cycle. The global average efficiency of fossil-fueled power generation is about 30%, and about 35% in the Organisation for Economic Cooperation and Development (OECD) countries (based on higher heating values, HHV). At a current efficiency of 40%, an increase of 1 percentage point in the efficiency of power generation results in a 2.5% reduction of CO_2 emissions. Figure 19-1 shows the impact of changes in technology and fuels on emissions (Mills *et al.*, 1991). The figure shows that changes in technology can reduce specific emissions by half and changes in fuel by another half, yielding emissions about one-quarter of initial levels.

Environmental and economic concerns have led to major development programs in various countries, in which state-of-the-art, higher-efficiency, combined-cycle, gas-fired stations and coal-fired units operating with supercritical steam cycles have been installed. Such considerations also have led to other major development programs—such as Pressurized Fluidized Bed Combustion (PFBC) and Integrated Gasification Combined Cycles (IGCC) for coal, residual oil, and biomass power generation, which should be commercially available

Box 19-1. Conventions Used

Unless otherwise stated:

Costs are presented in 1990 U.S. dollars and calculated as direct costs on a life-cycle basis, neglecting taxes and assuming a 6% discount rate. Wherever practical, costs are provided for alternative discount rates. Costs for mitigating external effects are included only to the extent required to satisfy existing regulations. There is wide variation in the extent to which external costs have been included in market prices through regulation and/or taxation.

Energy balances are calculated on a life-cycle basis.

Emissions of greenhouse gases [grams of carbon (C) in CO_2 or grams of methane (CH_4)] refer to emissions at the point of use (direct emissions).

The SI system of **units** is used. Weights also are given in metric tons (t).

Heating values. Energy contents of fuels are based on higher heating values (HHV).

before the year 2010 (Jansson, 1991; Menendez, 1992; Wolk *et al.*, 1991). Characteristics of some of these technologies are indicated in Table 19-1. The information in Table 19-1 was compiled for the year 1992 on a consistent basis; however, recent market transactions suggest capital costs that are significantly lower (by 30% or more), especially for natural gas-fired combined-cycle power plants, leading to lower electricity generation costs.

Considering the situation described above, it would be possible to reduce CO_2 emissions from fossil fuel-fired power plants by 25% or more compared to existing generating technology. This

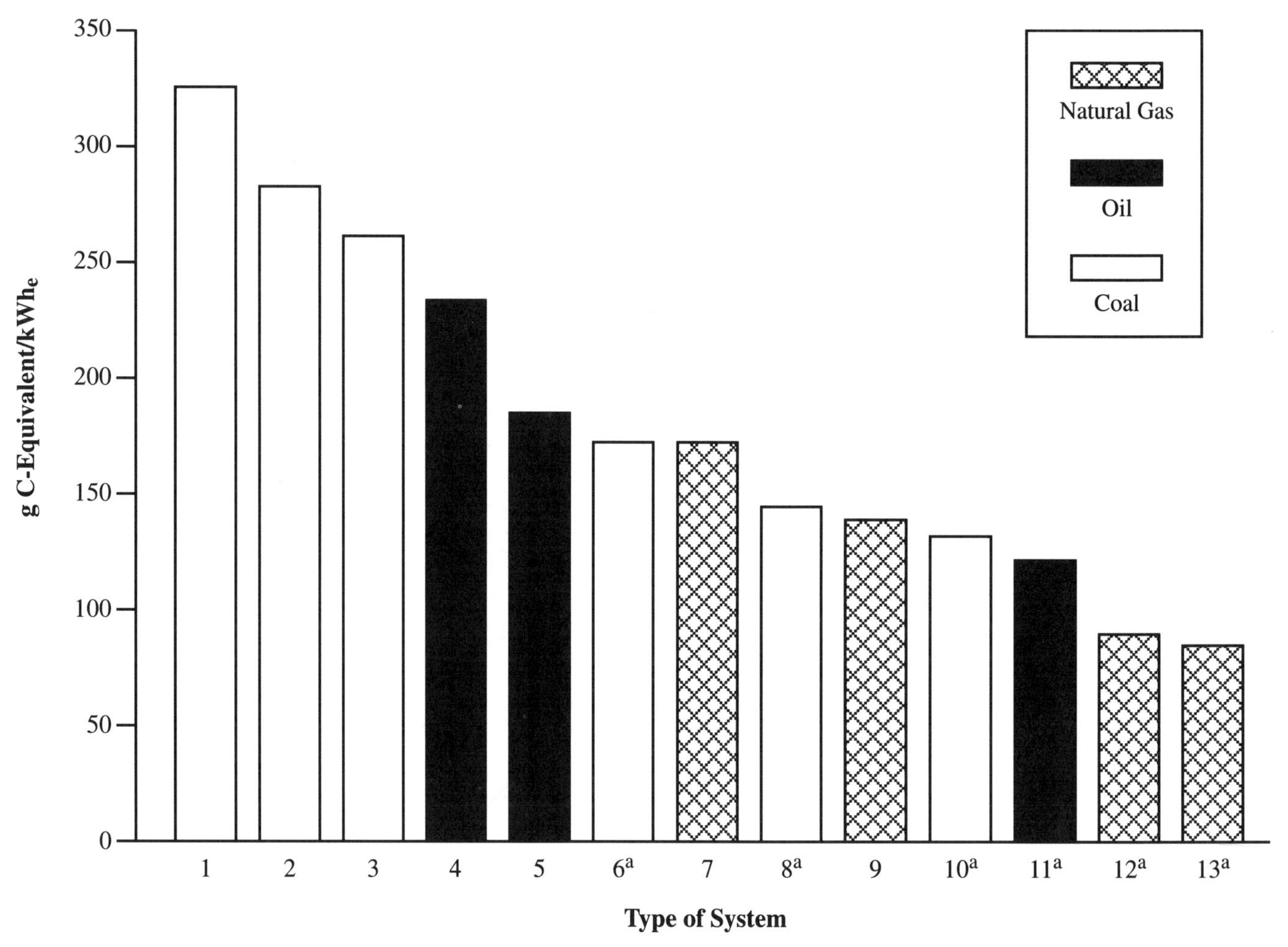

[a]Cogeneration.
[1]Average conventional steam turbine (coal, 34%).
[2]Best available steam turbine (coal, 39%).
[3]Pressurized fluidized bed combustion (coal, 42%).
[4]Average conventional steam turbine (oil, 38%).
[5]Best available combined-cycle gas turbine (oil, 48%).
[6]Cogeneration: Average conventional steam turbine (coal, 78%, 0.50).
[7]Average combined-cycle gas turbine (natural gas, 36%).
[8]Cogeneration: Best available steam turbine (coal, 83%, 0.60).
[9]Best available combined-cycle gas turbine (natural gas, 45%).
[10]Cogeneration: Pressurized fluidized bed combustion (coal, 86%, 0.65).
[11]Cogeneration: Best available steam turbine (oil, 81%, 0.60).
[12]Cogeneration: Steam-injected gas turbine (natural gas, 75%, 0.80).
[13]Cogeneration: Best available combined-cycle gas turbine (natural gas, 77%, 1.0).

Figure 19-1: Greenhouse gas emissions (in grams of C-equivalent per kWh_e) for commercially available alternative fossil fuels and conversion technologies (adapted from Mills *et al.*, 1991). Central-station power plants are compared with cogeneration plants providing both useful heat and power. The energy requirements for electricity production using cogeneration technologies are taken as the total energy supplied minus that which would have been required to produce the heat independently (assuming a boiler efficiency of 90% on an LHV basis). Parenthetical data in the notes above are fuel, efficiency based on higher heating value in percent, and power-to-heat ratio.

reduction could be achieved with state-of-the-art or near-commercialized technology at little technical risk. Major developments in these areas are being actively pursued in a number of countries.

Further opportunities are provided by advanced gas turbines. Higher turbine inlet temperatures have become possible through the development of new materials and improved cooling systems, offering improved efficiency in power generation. Research efforts are concentrated mainly on improved air cooling of the hot gas path of the turbine; new materials (ceramics, composites, and superalloys); and more efficient air cooling systems (Williams and Larson, 1989).

Advanced cycles, in general, couple a higher-temperature thermodynamic cycle to a lower-temperature cycle to increase efficiency. The most important application is the use of the hot exhaust gases from a gas turbine to raise steam for a steam turbine. Major characteristics of the combined cycle are summarized as follows:

- **High efficiency**: The most efficient units now on the market achieve 52% efficiency (net, generator terminals), and 54–55% is expected in the next few years.
- **Low investment costs**: Approximately 30% less than for a conventional steam power plant
- **Good operating flexibility**: Power generation may be adjusted to demand changes relatively easily.
- **Short installation time**: Operation approximately one year after order, making capacity expansion in small steps possible
- **Low environmental impact**: In particular, very low emission levels of NO_x
- **High power-to-heat ratio in a combined heat and power production**: This means that a larger fraction of the energy is produced as more valuable electricity.

Table 19-1: *Characteristics of some systems for fossil fuel-based power generation under European conditions.*

	Coal		**Coal**		**Coal**		**Coal**		**Natural Gas**		**Natural Gas**	
Plant Type	Typical with de-SO_x and de-NO_x		Supercritical with de-SO_x and de-NO_x		IGCC		IGCC with CO_2 Capture		Combined Cycle		Combined Cycle with CO_2 Capture	
Status	Conventional		Established Technology		Demonstration		Available Technology[1]		Established Technology		Available Technology	
Efficiency (%LHV)/ (%HHV)[2,3]	40/38		45(47)/ 43(45)		42(46)/ 40(44)		36/34 [34]		52(55)/ 47(50)		45/41 [44]	
Special Investment Cost ($/kW)[4]	1300		1740		1800		2995		750		1420	
Cost of Electricity (¢/kWh)												
@ Discount Factor (%)	6	10	6	10	6	10	6	10	6	10	6	10
Fuel Price Level A[5]	3.9	4.6	4.1	5.1	4.2	5.2	4.9	6.5	2.4	2.8	4.1	4.9
Fuel Price Level B[6]	4.6	5.3	4.7	5.7	4.8	5.8	5.6	7.2	3.8	4.2	5.8	6.6
Fuel Price Level C[7]	5.0	5.7	5.1	6.1	5.2	6.2	6.1	7.7	4.4	4.8	6.5	7.3
Cost of CO_2 Emission Reduction ($/t C Avoided)[8]	–		–		–		70		–		260	
CO_2 Emission (g C/kWh)	230		200		220		20		110		20	

Sources: Audus and Saroff, 1994; Summerfield *et al.*, 1994; Holland *et al.*, 1994.
[1] All component parts of the technology are available, but have not been demonstrated at scale in this application.
[2] These are typical values for the power production plant and, where relevant, CO_2 capture and disposal. Figures in brackets are LHV efficiencies on a full fuel-cycle basis.
[3] Higher figures in parentheses are for a well-documented, state-of-the-art advanced plant.
[4] For 500-MW net units in mainland northern Europe (overnight build).
[5] Fuel price level A based on gas at 2.2 US$/GJ and coal at 1.2 US$/GJ.
[6] Fuel price level B based on gas at 4.5 US$/GJ and coal at 2.0 US$/GJ.
[7] Fuel price level C based on gas at 5.5 US$/GJ and coal at 2.5 US$/GJ.
[8] On gas-to-gas and coal-to-coal basis (i.e., not including fuel switching gain of coal to gas), using price level B and 10% discount rate.

Because of such prospects, fossil fuel-based thermal power stations could have efficiencies of around 55% (HHV) on average before the middle of the 21st century.

These advanced cycles require clean gaseous fuels, which has stimulated interest in the production of such fuels from coal and oil.

19.2.1.2. Gasification of Fossil Fuels for Power Generation

Gasification of coal and other heavy fossil fuels may be used, for example, to substitute for natural gas, to produce medium calorific gas, or to produce synthetic gas for chemical production.

Fuel flexibility often is cited as a major advantage of gasifiers. Gasifiers can be either air- or oxygen-blown. In the context of power generation, the relative merits of air- and oxygen-blown systems are far from clear. The penalty for producing the required oxygen from air in an oxygen plant is severe because this consumes about 10% of the power produced. On the other hand, the use of oxygen offers the possibility of CO_2 sequestering and suppresses nitrogen oxides (NO_x) (see Section 19.2.3).

Although there are several proprietary gasifiers on the market that are well established in the chemical industry, they are only at the beginning of their development as far as power generation is concerned. Development programs in progress should ensure that, by the year 2000, IGCC systems will be available. Major gasification development programs are underway in Europe, the United States, and Japan.

To provide a sufficiently clean gas, a wide range of gas treatment processes are available from conventional chemical and gas-processing industries. Many of these clean-up processes are capable of delivering a fuel gas with very low levels of undesirable constituents such as sulfur. IGCC systems, therefore, offer great potential to meet strict air pollutant emission limits.

Maintaining the exit gases from the gasifier at their exit temperature—which is commonly in the range 800–1,000°C—during subsequent cleaning of the gas can contribute significantly to the increased efficiency of IGCC systems. This does not hold for systems in which CO_2 is captured because they are based on shift conversion of the gas CO content to CO_2, and this cannot be done at such high temperatures. Much technical effort is being applied in this area. Examples include the development of high-temperature dust-removal systems such as ceramic filters; dry desulfurization systems, such as zinc oxide-based adsorbents; and solid-based ammonia and alkali metal-removal systems. All of these systems have yet to be demonstrated on a significant scale, although recent PFBC work with demonstration-scale hot gas filters has been done in Sweden.

19.2.1.3. Fuel Cells for Power Generation

Fuel cells convert chemical energy into electricity without first burning the fuel to produce heat. Fuel cell power systems are characterized by high thermodynamic efficiency and low levels of pollutant emissions. Almost all fuel cells currently developed for commercialization use pure hydrogen as the fuel. Generally this implies that a primary fuel—for example, coal, natural gas, or biomass—first has to be converted to a hydrogen-rich gas. Fuel-cell systems offer the possibility of small-scale as well as large-scale electricity production at a conversion efficiency from hydrogen ranging from 40–70% (LHV) and more than 80% in cogeneration (de Beer and Nieuwlaar, 1991). Different types of fuel cells are currently being investigated: the alkaline fuel cell (AFC), phosphoric acid fuel cell (PAFC), molten carbonate fuel cell (MCFC), solid oxide fuel cell (SOFC), and solid polymer electrolyte fuel cell (SPEFC) (for an overview, see de Beer and Nieuwlaar, 1991).

Much attention is being given to the development of fuel cells for large-scale power production (more than 200 MW_e), integrating fuel cells with steam turbines or gas turbines. Key uncertainties are the limited fuel-cell lifetime and the relatively high cost of the high-temperature fuel-cell systems. The environmental performance of these systems probably is not much better than the performance of conventional power generation with additional cleaning technologies. Therefore, fuel-cell plants for large-scale power production will have severe competition from advanced IGCC and natural gas, combined-cycle power plants. This situation might change, however, when power production is combined with CO_2 removal; a combined coal-gasifier/fuel-cell system facilitates such removal (Jansen *et al.*, 1994).

Small-scale combined heat and power (CHP) production is the first market segment in which significant market penetration of fuel cell systems is expected. Widespread use would reduce the emission of pollutants significantly.

In the longer term, one can envisage centralized energy systems in which fossil fuels are converted and separated into hydrogen and CO_2 (or carbon black). The latter could be stored; the former piped and utilized for a variety of applications. Low-temperature fuel cells—such as AFC, SPEFC, and PAFC—are especially well-suited for such configurations. Potential markets are CHP in residential and commercial buildings that need low-temperature heat (Ingersoll, 1991; A.D. Little, 1995; Dunnison and Wilson, 1994), as well as transportation applications (see Section 19.2.6.4).

19.2.1.4. Combined Heat and Power Production

CHP production offers a significant rise in fuel efficiency and therefore is of interest in connection with GHG mitigation. CHP has applications in the industrial, residential, and commercial sectors (IEA, 1993a).

Combined production of heat and electricity is possible with all heat machines and fuels (including biomass and solar thermal) from a few kW-rated to large steam-condensing power plants. Heat-plus-power (first-law) efficiencies are typically 80–90%. CHP plants may have an added heat storage that allows the production plant to operate at optimum economy while still covering heat needs. Table 19-2 lists four typical examples of

Table 19-2: *Energy balance and C emissions of CHP plants and separate power and heat production.*

Power and Heat Plant Technology	Energy Balance (GJ) Fuel Input[1]	Power/Heat Output	Fuel Reduction by Introducing CHP	CO_2 Emissions t C	Reduction by Introducing CHP[2]
Large Coal-Fired CHP Plant	100	36/56		2.4	
Large Coal-Fired Power Plant plus Residential Gas Burner	80/59	36/56	28%	2.7	11%
Large Coal-Fired Power Plant plus Residential Coal Burner	80/112	36/56	48%	3.5	31%
Small Biomass-Fueled CHP Plant	100	22/56		0	
Large Coal-Fired Power Plant plus Residential Gas Burner	49/59	22/56	7%	2.0	100%
Small Gas Turbine CHP Plant	100	30/55		1.4	
Large Coal-Fired Power Plant plus Residential Gas Burner	60/58	30/55	15%	2.2	37%
Medium-Sized Gas Engine CHP Plant	100	39/46		1.4	
Large Coal-Fired Power Plant plus Residential Gas Burner	78/48	39/46	21%	2.5	45%

[1]Arbitrarily fixed reference level.
[2]Some of the percentage reductions show the combined effect of CHP and fuel switching.

CHP installations of various sizes, their power and heat production, and reductions in fuel use and CO_2 emissions relative to appropriate alternatives for separate supply of heat and power (Olsen, 1993). Combined-cycle power and heat production provides better thermodynamic performance than single cycles, even if first-law efficiencies are lower, because more electricity is produced. This electricity may in principle be used with heat pumps to generate more heat at a higher temperature.

The employment of CHP is closely linked with the availability or development of district heating (DH) and/or cooling networks and building heat and/or cooling distribution systems, as well as industrial heat loads. DH networks are energy transmission systems suited for the distribution of heat and/or cooling within areas with sufficiently high heat/cooling load densities (Kalkum *et al.*, 1993; Rogner, 1993). Modern water-based heating and/or cooling transmission and distribution systems for cities have low losses, typically 10–15% (WEC, 1991).

A prerequisite for DH is central heating systems in buildings. These are customary in most temperate and some subtropical regions. Central heating is most easily achieved if considered at the planning stage, along with electricity, water and sewage, telephone, and optical fiber lines.

Today, widespread coverage by CHP exists in Denmark and Finland (WEC, 1991; Danish Energy Agency, 1993), and substantial DH exists in Austria, Germany, the Netherlands, Poland, Russia, Sweden, and other countries.

Fossil fuels, industrial wastes, and waste heat from processes are used widely in process industries to generate power and heat. The pulp and paper industry is the largest user of biomass energy through combustion of waste liquors and other wood waste. CHP has been utilized in other industries, but to much lesser extent. Now the situation is changing because of regulatory changes (like the Public Utilities Regulatory Policy Act in the United States) and the good economy of cogeneration. Industrial heat demand often has some special characteristics, such as large and fast load variations, intermittent need for heat and power (mechanical work), and extractions of heat (as steam) at different temperature and pressure levels. Otherwise, CHP technology is very similar to that in DH except for the use of special process and waste-heat boilers, where corrosion may limit the attainable pressure and temperature levels (Saviharju, 1995). Compared with DH plants, the temperature levels of the extracted heat are higher, which results in much lower power yields (lower power-to-heat ratio). These may be enhanced considerably, however, through the use of combined-cycle technology, including IGCC technology. The heat distribution system may serve a single mill or a whole industrial city, ranging to hundreds of MW of heat.

19.2.1.5. Direct Coal Use in Developing Countries and Economies in Transition

Developing countries and economies in transition burn coal directly as fuel for cooking and space heating and fuel for small industries. For example, the direct use of coal accounts for about two-thirds of the coal used in China. Typically, conversion efficiencies are low for these applications, and—because of coal's high content of ash, sulfur, and other pollutants—direct coal use is a major source of local and regional air pollution. A substantial reduction in emissions, including CO_2, can be achieved by using coal in more efficient appliances or by converting coal into a synthetic fuel (or electricity and district heat) before distribution to final users.

Coal conversion to cleaner fuels—for example, town gas—essentially eliminates sulfur and particulate emissions at the point of end-use. Such upstream conversion of coal, however, is capital-intensive, incurs conversion losses, and requires grid distribution systems. Because of capital limitations, the direct use of coal will remain important in many countries for decades (Sun and Li, 1992). Transferring efficient residential and industrial coal appliances to developing countries is as important as transferring coal-cleaning equipment, power generation plants, and other central conversion plants and equipment (Pachauri, 1993; Topper, 1993; Graham-Bryce *et al.*, 1993; Ramakrishnan, 1993; Su and Gu, 1993).

19.2.2. Suppression of GHG Emissions and Fuel Switching

Suppression of CO_2 and methane (CH_4) emissions and switching to low-carbon fuels—for example, from coal to oil and gas, which is a shift to fuels with a lower carbon to hydrogen ratio—offer significant potential for reducing GHG emissions.

19.2.2.1. Suppression of Methane Emissions

Total CH_4 emissions are estimated at 560 ± 90 Mt/yr, of which 70% is from anthropogenic sources and 30% from natural sources (Lelieveld and Crutzen, 1993; IPCC, 1992). About 30% of the anthropogenic sources may be associated with the use of fossil fuels.

CH_4 emissions from coal mining and natural-gas venting, as well as leakage from pipeline and distribution systems, are significant. It has been estimated that the coal industry worldwide contributes 4–6% of global methane emissions (CIAB, 1992). Flaring and venting has been estimated to be about 5% of world natural-gas production (Barn and Edmonds, 1990; U.S. EPA, 1993a). Available technology can reduce emissions from coal mining by 30–90%, from venting and flaring by more than 50%, and from natural-gas distribution systems by up to 80% (U.S. EPA, 1993a). Options for limiting emissions from coal mining; natural-gas production, transmission, and distribution; and landfills (see Chapter 22) may be economically viable in many regions of the world, providing a range of benefits—including the use of CH_4 as an energy source (U.S. EPA, 1990a, 1990b, 1993a, 1993b; Blok and de Jager, 1993).

There are many options for suppressing emissions from the oil and natural gas industries, including capturing and using or recompressing residuals and purged gas, improving gas leakage-detection methods, applying pneumatic devices to control or eliminate venting, repairing or replacing pipelines, and using automatic shutoff valves (U.S. EPA, 1993a). Some of these measures are capital-intensive and could be difficult to carry out in some regions, due to specific geographical as well as economic conditions. Flaring is preferable to venting from a GHG point of view because a molecule of CH_4 leads to a much larger radiative forcing in the atmosphere than a molecule of CO_2.

CH_4 emissions from coal mining depend on several factors, but emissions from underground mines are one order of magnitude higher than from surface mines (Smith and Sloss, 1992). Only emissions from underground mines can be reduced *before* (predegasification), *during* (recovery of ventilation air), and *after* (gob-well recovery, from the highly fractured area of coal and rock that is created by caving in of the mine after mining has been completed) the mining of the coal (U.S. EPA, 1993a). Whereas premining and postmining degasification has been widely applied in many countries, it is not suitable for every coal deposit. The economic feasibility of CH_4 recovery from ventilation air has not yet been demonstrated, although technology is available. In optimal conditions, the combination of these three options—"integrated recovery"—can achieve an emissions reduction of up to 80–90% (Blok and de Jager, 1993; U.S. EPA, 1993a).

19.2.2.2. Flaring of Natural Gas and Alternatives to Flaring

An estimated average of 0.5% or more of natural-gas production is emitted into the atmosphere from upstream oil and gas operations (U.S. EPA, 1990b). In some areas, emissions are assumed to be as high as 15%, of which 6% is emitted from natural-gas use. Such high levels of emissions probably will decrease within 20 to 50 years (Picard *et al*, 1992). Also, many oil production facilities either produce very small amounts of gas in association with oil (after satisfying on-site fuel requirements) or are too far from gas-collecting systems for feasible conservation or reinjection of the gas. Therefore, gas usually is vented or flared if it is uncollected. Options for disposing of or utilizing waste gas at oil production facilities include small-scale on-site power generation, cogeneration, and transport fuel production. These options could reduce emissions by 50–99% (Picard and Sarkar, 1993). The largest reductions of gas venting and flaring from oil and associated gas facilities would result from making these activities economically attractive.

Natural gas is flared to prevent natural gas explosions in the air, to ensure continuous flow, and to keep production smooth in the downstream facilities. Options to reduce the volumes flared in continuous gas flows include using nitrogen as a purge gas and recovery of low-pressure gas. Gas flaring caused by faults in processing could be reduced by improving maintenance. Improving system reliability and storage capacity may reduce emissions of excess gas flared when demand is low. These options vary according to the location and characteristics of the oil industry.

19.2.2.3. Suppression of CO_2 from Natural Gas and Oil Wells

The CO_2 content in natural gas fields or oil wells varies, which significantly affects life-cycle CO_2 emissions. Depending on the ratio of CH_4 to other gases, the extraction and use of natural gas leads to CO_2 emissions of less than 14 kg C/GJ of natural gas (HHV) if the CO_2 content of the natural gas field is less than about 1%. Most fields exploited presently are in this range (Schroder and Schneich, 1986)—for example, the Groningen field in The Netherlands, which contains 0.89% CO_2 (Blok *et al.*, 1989). However, there are also natural gas fields with a much higher CO_2 content. For example, the Krahnberg field in Germany contains 53.4% CO_2 and the Catania field in Italy 48.8% CO_2. Although not yet developed due to high CO_2 content, the Natuna field in Indonesia contains more than 70% CO_2.

Exploitation of such fields requires removal of CO_2 to meet transport and sales specifications, typically less than 2–3%. Ordinarily the CO_2 is emitted to the atmosphere. In the case of the Sleipner Vest field (9.5% CO_2) in Norway, however, the CO_2 removed will be injected into an aquifer at 1,000 m below the main Sleipner platform. Approximately 1 Mt/yr CO_2 (~0.25 Mt/yr of carbon) will be removed from a 100-bar natural gas stream, which is costly but uses well-known technology. The decision to store the CO_2 was stimulated by the introduction of a carbon tax of about $180/t C in Norway (Kaarstad, 1992).

19.2.2.4. Fuel Switching

Switching from coal to oil or natural gas would reduce carbon emissions in proportion to the carbon intensity of the fuel. For example, switching from coal to natural gas would reduce emissions by 40% (see Box B-2 in Chapter B). In addition, the higher energy efficiency achievable with natural gas would reduce emissions further—for example, a shift from coal to natural gas in power generation by 20%.

19.2.3. Decarbonization of Fuels and Flue Gases, CO_2 Storage, and Sequestering

In the longer term, decarbonization would allow continued large-scale use of fossil fuels. Here, decarbonization implies utilization of the energy in carbon with greatly reduced CO_2 emissions. This can be done practically only in large-scale energy conversion facilities. It is logical, therefore, to begin decarbonization efforts in large fossil fuel-burning power stations, which at present account for a quarter of total CO_2 emissions from fossil fuels. Either CO_2 can be captured from flue gases or carbon-containing fuels can be converted to low-carbon, hydrogen-rich fuels before utilizing them (Pearce *et al.*, 1981; Blok *et al.*, 1991, 1992; U.S. DOE, 1993).

If deep CO_2 emission reductions are desired, it would be necessary to extend the effort beyond power generation. This is problematic, however, because most fuels used directly are consumed in small-scale conversion systems in which decarbonization is not practical. This problem might be solved by converting the fossil fuel to a low-carbon, hydrogen-rich fuel or to a carbon-free fuel (essentially hydrogen) and CO_2 in a centralized facility, followed by removal of the CO_2 and distribution of the low-carbon or carbon-free fuel to the consumer (Marchetti, 1989; Blok *et al.*, 1995; Williams, 1996).

In either case—flue gas decarbonization or fuel decarbonization—the captured CO_2 has to be utilized, stored, or isolated from the atmosphere in an environmentally acceptable manner.

19.2.3.1. Decarbonization of Flue Gases

The capture of CO_2 from the flue gases of fossil fuel-fired power plants and its subsequent use or disposal is being actively investigated in a number of countries (Blok *et al.*, 1992; Riemer, 1993; U.S. DOE, 1993; Herzog and Drake, 1993; Aresta *et al.*, 1993; Kondo *et al.*, 1995). Interest in the capture of CO_2 from power plant flue gases started before it was seen as a GHG mitigation option. The driving force was the desire for an inexpensive, readily available source of CO_2 to use in enhanced oil recovery (EOR). Three such plants were built in the United States to provide CO_2 for EOR using natural gas as a primary fuel. Another plant, still in operation, uses coal as a source of CO_2 for the production of soda ash. All of these plants have applied well-established technology to capture the CO_2 from the flue gases.

The use of these technologies incurs a cost and energy penalty. Starting with a conventional coal-fired power plant of 600 MW_e and a coal-to-busbar conversion efficiency of, for example, 41% (LHV), the efficiency might decrease to, for example, 30% if the CO_2 emission is reduced from 230 gC/kWh to about 30 gC/kWh in the modified plant. The costs of electricity production would then increase by about 80%, which is equivalent to $150/t C avoided. Starting with a natural gas-fired, combined-cycle plant of 600 MW_e and a conversion efficiency of 52% (LHV), the efficiency might decrease to 45% if the CO_2 emission is reduced from 110 gC/kWh to about 20 gC/kWh in the modified plant. In this case, the costs of electricity production would increase by about 50%, which is equivalent to $210/t C avoided. R&D efforts are focused on minimizing these penalties (Blok *et al.*, 1992; U.S. DOE,

1993; Herzog and Drake, 1993; Hendriks *et al.*, 1993; Kondo *et al.*, 1995). One option under investigation to reduce costs is the use of oxygen rather than air for combustion to obtain a flue gas that is essentially CO_2. For a coal-fired power station, the removal costs might then be less than \$80/t C avoided (Hendriks, 1994).

19.2.3.2. Decarbonization of Fuels

One frequently suggested scheme for decarbonization of fuels is the development of an IGCC power plant with CO_2 removal. In this scheme, the gasifier off-gas is converted with steam to CO_2 and hydrogen (via $H_2O + CO \longrightarrow CO_2 + H_2$) and then separated to make a fuel gas stream that is essentially hydrogen. Due to its high partial pressure, the CO_2 can be recovered by using a physical solvent for which regeneration requires only a release of pressure. After separation, the hydrogen-rich fuel is burned in a combined cycle to generate electricity. The CO_2 emission factor of the fuel would be smaller than 4 kgC/GJ, compared with 24 kgC/GJ for coal. Starting from an original IGCC plant with a coal-to-electricity efficiency of about 44% (LHV), this efficiency might decrease to about 37%, with the CO_2 emissions reduced from approximately 200 gC/kWh to less than 25 gC/kWh. Due to the recovery, the costs of electricity production might increase by 30–40%. The removal costs would then be less than \$80/t C avoided (Hendriks *et al*, 1993; Hendriks, 1994). These penalties might be reduced substantially if the gasification of coal and the recovery of CO_2 were integrated with the use of fuel cells to generate electricity. Such a configuration could result in a higher conversion efficiency (42–47%, LHV) (Jansen *et al.*, 1992).

The hydrogen-rich fuel also can be used for applications other than power generation, although doing so requires further purification of the fuel (Watson, 1983; Blok, 1991) and the build-up of a corresponding infrastructure (see Section 19.2.6).

Due to the increased costs of the hydrogen fuel compared to the original feedstock, some experts have concluded that decarbonization processes are inherently expensive. For the recovery of CO_2 by steam reforming of natural gas, they calculate costs ranging from \$210 to \$460/t C recovered (Kagoja *et al.*, 1993). It has been suggested that hydrogen has a greater value than the feedstock from which it is produced because it allows the use of more efficient conversion technologies. The costs of decarbonizing fuels therefore must be assessed at the systems level. For example, the production of hydrogen from natural gas or coal and its use in a fuel-cell vehicle typically would lead to lower primary energy consumption than if gasoline derived from crude oil were used in a comparable vehicle (see Section 19.2.6.4). At present, the least costly way to produce hydrogen often involves the use of natural gas as the feedstock; in the process, a stream of pure CO_2 is produced as a byproduct. If this CO_2 could be captured and stored in a nearby exhausted natural-gas field, the costs of avoiding the CO_2 emissions are estimated to be less than \$30/t C (Farla *et al.*, 1992). The net costs could be even less (and possibly close to zero) if possibilities for modest enhanced natural gas recovery as a result of reservoir repressurization by the CO_2 were exploited (Blok *et al.*, 1995).

Carbonaceous materials (fossil fuel, biomass, waste, etc.) also could be converted into chemically stable carbon and hydrogen or hydrogen-rich methanol (Steinberg, 1991). As a result, carbon could be stored instead of CO_2. Although theoretically possible, this approach is very difficult practically.

19.2.3.3. Storage of CO_2

The technology to capture CO_2 is available, but to have an impact on GHG emissions, credible and environmentally acceptable utilization, storage, and/or disposal options are required. Because of the small potential for utilization (Aresta *et al.*, 1993; Herzog and Drake, 1993), several possibilities for underground storage or ocean disposal of CO_2 are being investigated. An overview of storage options is presented in Table 19-3, together with an indication of their potential capacities based on low estimates in the literature. Enhanced oil recovery (EOR) using CO_2 as a miscible flooding agent has a high potential. The CO_2 is partly stored during the recovery of the oil. At present oil price levels, however, the application of CO_2 in EOR is not profitable unless there is access to CO_2 from a cheap source, such as a natural underground source. Examples can be found in the United States, where CO_2 from natural sources is transported hundreds of kilometers by pipeline for EOR use. For a typical case, total pipeline transportation cost of CO_2 is about \$8/t C for a 250-km, 750-mm pipeline for 5.5 Mt C of CO_2 per year (Skovholt, 1993). EOR using CO_2 from power plants might reduce the annual anthropogenic CO_2 emissions by about 1% (Taber, 1993).

Storage in exhausted oil and gas wells is a viable option. The capacity of natural-gas fields to sequester carbon at the original reservoir pressure is generally greater than the carbon content of the original natural gas. As a worldwide average, about twice as much carbon can be stored as CO_2 in depleted reservoirs as was in the original natural gas (Hendriks, 1994). The estimated storage capacity ranges from about 130 Gt C to 500 Gt C, based on different views of recoverable oil and gas. The cost of CO_2 storage in onshore natural-gas fields is estimated to be less than \$11/t C (Hendriks, 1994).

Table 19-3: *Low estimates of CO_2 storage potentials.*

Option	**Potential Global CO_2 Storage Capacity (Gt C)**
Enhanced Oil Recovery	>20
Exhausted Gas Wells	>90
Exhausted Oil Wells	>40
Saline Aquifers	>90
Ocean Disposal	>1200

Another option is storage in saline aquifers (i.e., permeable beds, mostly sandstone), which can be found at different depths all over the world. First estimates of their storage capacity range from about 90 to about 2,500 Gt C, due to different assumptions about the volume of aquifers, the percentage of the reservoir to be filled, the density of CO_2 under reservoir conditions, and the area suitable for CO_2 storage (Riemer, 1993; Hendriks *et al.*, 1993). Depending on the availability of compressed CO_2—for example, from an IGCC plant with CO_2 removal—and other local circumstances, the costs of underground storage onshore might vary from about $7 to $30/t C, transportation costs excluded (Hendriks, 1994). CO_2 storage in aquifers has safety risks and environmental implications. Potential problems include CO_2 escape, dissolution of host rock, sterilization of mineral resources, and effects on groundwater (Riemer, 1993; Hendriks *et al.*, 1993).

The deep ocean is the largest potential repository for CO_2. The oceans contain about 38,000 Gt C and will eventually absorb, after equilibration, perhaps 85% of the CO_2 that is released to the atmosphere from energy conversion processes (Houghton *et al.*, 1990). CO_2 could be transferred directly to the oceans, ideally at a depth of perhaps 3,000 m—from where it would take at least several hundred years before it partly escaped to the atmosphere. Concern over potential environmental impacts, the practical limitations on pipeline depth, and the assurance of adequate retention time suggest that injection at 1,000 m at carefully selected sites could be a realistic option. Costs are probably marginally higher than for subterranean disposal (U.S. DOE, 1993), but—as with most options—the cost of transporting CO_2 to the disposal site often dominates.

Not much is known about the environmental effects of storing CO_2 in the oceans—for example, the impacts on marine life, either directly (as discussed here) or indirectly (via the atmosphere). Preliminary studies indicate that ecological perturbations would be confined to the release area, which would be a small percentage of the whole ocean volume (U.S. DOE, 1993). A maximum deviation of 0.2 pH units for coastal waters has been recommended (U.S. EPA, 1976). This would correspond to a buffer capacity of 1,200 Gt C (Spencer, 1993). These aspects need further research.

19.2.3.4. CO_2 Sequestering

Reforestation with the application of forest management techniques provides a method of offsetting CO_2 emissions (Maclaren *et al.*, 1993; see also Chapter 24). Carbon is captured and stored during the growth time of a forest and would need to be stored for a long time thereafter. Costs are $12–30/t C, assuming a land value of about $800/ha in industrialized countries and about $300/ha in developing countries (Huotari *et al.*, 1993). Other studies indicate costs as low as $3.5/t C (Face Foundation, 1995). At high levels of reforestation, costs can be expected to increase.

Offsetting carbon emissions from a 500 MW_e coal-fired power station (emitting about 0.8 Mt C/yr) by growing a forest on unforested land would require an area of at least 1,700 km^2, on the basis of a productivity of 2–4 tons/ha/yr during 50 years and a storage of 100–200 tons of carbon per hectare in a mature forest.

Growing biomass for energy as a fossil fuel substitute is an alternative to growing biomass for sequestering carbon. The preferred strategy depends on various factors, including the current status of the land and the biomass yields that can be expected. For forests with large standing biomass, the most effective strategy is to protect the existing forest; for land with little standing biomass and low yields, the most effective strategy is to reforest the land for carbon storage. Where high yields can be expected and markets for the biomass are readily accessible, however, often the most effective strategy is to manage the forest as a harvestable energy crop for fossil fuel substitution (Hall *et al.*, 1991; Marland and Marland, 1992).

19.2.4. Switching to Nuclear Energy[1]

In 1992, nuclear power generation totaled 2,030 TWh—about 17% of all electricity or more than 5% of commercial energy consumption worldwide (all statistical data on nuclear power quoted in this section are from IAEA, 1995a; OECD/NEA, 1993). The operational experience of electricity-generating nuclear power plants of all sizes exceeds 6,500 reactor-years. At the beginning of 1993, there were about 425 nuclear power reactors connected to electricity supply networks, with a total installed capacity of about 331 GW_e. More than 30 countries have nuclear power plants in operation or under construction.

If the entire upstream and downstream energy chains for electricity generation are included, nuclear power CO_2 equivalent emissions are 1/10 to 1/100 those of fossil fuel plants. The range depends on assumptions about the competing fossil fuel technology and on the uranium content of ores, the uranium enrichment technology, and the management of radioactive waste (The Netherlands Ministry of Economic Affairs, 1993; Uchiyama and Yamamoto, 1991). Nuclear plants producing electricity reduce CO_2 emissions from the energy sector by about 7% compared to the present world mix of fossil fuel-based power generation (Van de Vate, 1993).

For nuclear energy to play a major role in future GHG mitigation, a considerable expansion must take place. Historically, nuclear power expanded rapidly. Construction starts per year reached a peak in the late 1960s but declined by 90% during the last 6–7 years (see Figure 19-2). Grid connections peaked in the mid-1980s at more than 30 GW/yr and then decreased to a few GW/yr. For nuclear energy to assume an increased role in reducing GHG emissions, growth has to be resumed. It is therefore necessary to analyze the reasons behind the developments summarized in Figure 19-2 and to understand the issues that must be addressed for a revival of the nuclear option.

[1] A supporting document exists for this section (see IAEA/OECD, 1995).

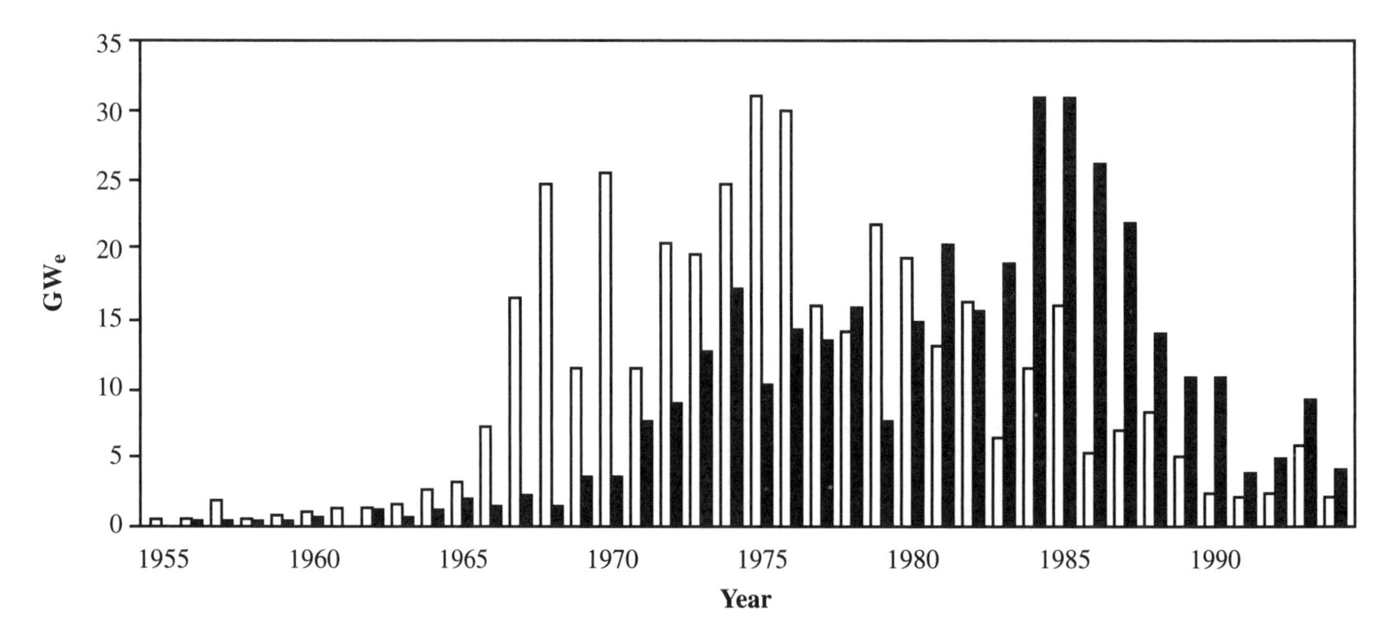

Figure 19-2: Annual nuclear power plant construction starts and connections to the grid, 1955–1994 (IAEA, 1995b). Open bars indicate construction starts; filled bars indicate grid connections.

General acceptance: During the last decades, there has been a decrease in acceptance of nuclear power, especially of building new nuclear power plants. A review of opinion surveys concludes that public concerns about nuclear energy focus on the following issues: doubt about economic necessity, fear of large-scale catastrophes, storage of nuclear waste, and the misuse of fissile material (Renn, 1993; The Netherlands Ministry of Economic Affairs, 1993; Slovic, 1992). Nuclear power expansion has stopped in most countries, but plans for nuclear capacity expansion remain significant in a limited number of countries, such as China, France, Japan, and Korea.

Resource base: The availability of uranium and thorium is unlikely to place a major constraint on the future development of nuclear power (Pool, 1994). Reliance on uranium alone without fast-breeder development could introduce constraints in the longer term if demand for nuclear power were to increase substantially.

Present reactor technology: Several types of reactors have been commercially developed. About 80% of the nuclear units in operation and under construction are light-water reactors (LWR); about 5% are heavy-water reactors (HWR). Gas-cooled reactors (GCR) contribute about 9% of installed capacity, but none are under construction. Liquid metal-cooled reactors (LMR) were conceived and implemented at the start of nuclear power development to use uranium more efficiently, but their deployment has not gained momentum.

Costs: Investment costs for construction of nuclear units are a major component of the total cost of nuclear-generated electricity. They are sensitive to technical parameters; regulation aspects, such as licensing lead times; and, like other capital-intensive options, the interest rate. Nuclear fuel-cycle costs include mining, spent fuel management, and disposal of low-, intermediate-, and high-level radioactive wastes (NEA, 1994). Although final repositories for high-level waste have not yet been implemented, the cost of disposal has been estimated to represent less than 20% of the undiscounted total nuclear fuel cycle cost (4.6% with 5% discount rate, 1% with 10% discount rate) (NEA, 1994). The operation and maintenance of nuclear power plants require industrial, organizational, and regulatory infrastructures, as well as highly qualified manpower—leading to generally higher operating and maintenance costs for nuclear than for fossil fuel-fired power plants.

Direct nuclear generating costs in a number of countries vary from 2.5¢ to 6¢ per kWh_e (IAEA, 1993). The direct cost of electricity from new plants, including waste disposal and decommissioning, is 2.9–5.4¢/kWh_e using a 5% discount rate and 4.0–7.7¢/kWh_e using a 10% discount rate (IEA *et al.*, 1993). Direct costs for new reactors in the Netherlands are an estimated 5.3–6.0¢/kWh_e for a 5% discount rate and 7.0–7.7¢/kWh_e for a 10% discount rate (Beeldman *et al.*, 1993). Direct electricity production costs from a new nuclear power plant are estimated to be lower than or about as high as (IEA *et al.*, 1993) or higher than (Moore and Smith, 1990; Beeldman *et al.*, 1993) those estimated for a natural gas or coal-fired power plant. Uncertainties relating to governmental decisions on such matters as licensing and regulations may affect costs in some countries. Projected levelized costs of baseload electricity generation for plants connected to the grid by the turn of the century indicate that nuclear power will remain an option in several countries where nuclear power plants are in operation or under construction (IEA *et al.*, 1993).

Environmental impact: In routine operation, nuclear power plants and their associated fuel-cycle facilities release small quantities of radioactive materials to the environment. These releases and direct exposure to radiation of workers and the public are monitored and controlled by national authorities, based upon the recommendations of the International Commission for Radiological Protection. The yearly collective dose to the world population from routine nuclear power electricity generation is less than an estimated 1% of the dose due to natural radiation sources (UNSCEAR, 1994).

The safety of nuclear facilities is assessed and regulated by independent authorities reporting to governments, taking into account international codes of practice (INSAG, 1988). The responsibilities and enforcement capabilities of these authorities vary. Before plants are licensed and throughout their operating lives, they are assessed by several techniques, including probabilistic safety assessment (INSAG, 1992). These methods are useful for identifying needs and opportunities for design improvements (U.S. NRC, 1991; INSAG, 1992). There are recognized limitations in applying this method to very-low-probability accident sequences (APS, 1975; Sørensen, 1979b). Safety is considered in design through the principle of defense-in-depth, employing successive barriers to accidental releases of radioactivity. For operational safety, a general appreciation of safety requirements through manpower training and fostering of a safety culture is central (INSAG, 1991). Deficiencies in these matters are believed to be at the root of the Chernobyl accident (INSAG, 1986). The reactor-core meltdown accident at Three Mile Island in the United States demonstrated the benefit of the defense-in-depth approach: Insignificant amounts of radioactivity were released from the plant.

Radioactive waste: All phases of the nuclear fuel cycle, other than the final disposal of spent fuel and high-level radioactive waste, are operating on an industrial scale.

High-level wastes, which contain more than 99% of the radioactivity from nuclear plants, represent after conditioning a few tens of m^3/GW(e)year. If spent fuel is reprocessed, the volume of waste would be one-tenth as large. These high-level wastes are now stored either near the reactors or at reprocessing plants. Geological repositories for high-level waste have been studied in salt and granite, for example (Carlson, 1988). Several countries are studying alternative final repositories of high-level waste, with a view toward implementation early in the next century.

Proliferation: The potential use of nuclear materials and technology for weapons has long been recognized. At the end of 1994, 178 states were parties to the 1970 Treaty on the Non-Proliferation of nuclear weapons (NPT). The treaty places conditions on the transfer of nuclear technology and materials to prevent the development of nuclear weapons. Verification is carried out by independent inspectors from the International Atomic Energy Agency (IAEA), which was established in 1957 (IAEA, 1993). An indication of the impact of the international safeguards regime is the limited number of nuclear weapon states existing at present as compared to fears expressed in the 1960s (Thorne, 1992). However, illicit trafficking of nuclear materials became a concern to governments in 1993–1994 (IAEA, 1994). A 1-GW_e nuclear power plant of the current LWR generation produces about 200 kg of plutonium (Pu) per year; future breeder reactors will produce about 1,500 kg of plutonium per year. Concerns regarding the proliferation risk arise from the accessibility to plutonium because an explosive similar to the Nagasaki bomb could be produced with only about 10 kg of plutonium (Carson Mark, 1993), or with 4 kg of plutonium or less in advanced designs (U.S. DOE, 1994a). Plutonium may be mixed with uranium as a fuel for nuclear power plants and may also be used for nuclear explosives, although the use of power reactor-grade plutonium complicates bomb design. Countries that have produced plutonium-based nuclear explosives have produced weapons-grade plutonium in reactors designed for this purpose (National Academy of Science, 1994). However, the difference in proliferation risk posed by separated weapons-grade plutonium and separated reactor-grade plutonium is small in comparison to the difference between separated plutonium of any grade and unseparated material in spent fuel (National Academy of Sciences, 1994).

New technology: Evolutionary reactor designs are being used in reactors now under construction to provide increased safety, reduced radiation doses to operators, and improved economic performance through reduced construction lead times and reduced operation and maintenance costs (Juhn and Kupitz, 1994). Cost reductions can be obtained by streamlining the reactor's systems and reducing the amount of material and manufactured components required for construction. Reactors with such improved designs are under construction in Canada, France, and Japan.

Designs incorporating more innovative features are being developed in some countries, on the grounds that evolutionary improvements will not suffice to provide the safety that the public and the investors want. In particular, systems with inherently safe characteristics and more passive safety features are being developed to reduce the probabilistic risk of accidents, as well as on-site and off-site impacts in the event of a severe accident. In particular, the concept of Modular High-Temperature Gas-Cooled Reactors—in which graphite is used as the moderator and helium as the coolant—has attractive safety features. A 30-MW_{th} test reactor is under construction in Japan.

Advanced reactors also are being designed and developed to address the challenges of higher reliability and cost reductions while satisfying increasingly demanding safety and waste-management requirements (Kupitz, 1992; IAEA, 1995a). In a recent study for the Netherlands government, researchers have concluded that the majority of these concepts for new power plants are likely to meet current safety criteria (The Netherlands Ministry of Economic Affairs, 1993). The level of safety improvements, however, is unclear (Mårtensson, 1992).

Concerns about the future production of Pu in a world with a large nuclear capacity have stimulated proposals for criteria for diversion-resistant nuclear power (Williams and Feiveson, 1990). Preliminary proposals have been made for significant reductions in plutonium production through the use of fuels such as denaturated thorium—for example, high-energy accelerator devices (Carminati *et al.*, 1993). Interest in LMRs has been revived in view of their potential use in the management and disposal of civil radioactive waste, as well as fissile materials arising from dismantling of nuclear weapons (National Academy of Science, 1994; Chang, 1993). The use of LMRs for transmutation of actinides contained in irradiated fuels from LWRs also has been investigated (NEA, 1995a). Other strategies are being investigated for reducing the period during which radioactive waste repositories need to be controlled. For example, partitioning actinides and transmuting them by neutrons or high-energy protons can substantially reduce the volume of long-lived isotopes (The Netherlands Ministry of Economic Affairs, 1993).

Other concepts are being developed with the objective of enhancing the use of nuclear power for nonelectrical applications. Some nuclear plants supply process and district heat. In the longer term, nuclear energy could be deployed for hydrogen production (Marchetti, 1989) (see Section 19.2.6.4). Market studies for small- and medium-size reactors (SMRs) have shown that nuclear power could in principle be more broadly deployed for district heating, heat supply to industrial complexes (IEA, 1993c), and production of potable water (IAEA, 1992a, 1992b).

In the longer term, it might also be possible to generate electricity from nuclear fusion instead of fission (IFRC, 1990). The primary fuels for fusion reactors—deuterium and lithium—are so abundant that fusion would be a practically inexhaustible source of energy. The conditions required for self-sustaining net power-producing fusion reactions on the Earth have not yet been attained, however. The development and implementation of fusion reactors requires demonstration of scientific, technological, and commercial feasibility. It is not expected that fusion plants could be available on a commercial scale before the second half of the next century (Colombo and Farinelli, 1992).

A major drive will be needed to achieve the early rehabilitation of nuclear energy if it is to contribute to fossil fuel use reductions in the next century. The continuing concern of many members of the general public and many policymakers with regards to safety and proliferation issues, however, may remain a severe constraint on nuclear power generation in many countries (WEC, 1993).

19.2.5. Switching to Renewable Sources of Energy

Renewable sources of energy, including biomass and hydropower, contributed about 20% of the world's primary energy consumption in 1990 (see Table B-2 in Chapter B).

The resource base for renewable energy technologies is very large in comparison to projected world energy needs (see Table B-4 in Chapter B). Impressive technical gains in renewable energy utilization have been made during the past decade (U.S. DOE, 1990; Johansson *et al.*, 1993b; Ahmed, 1994; Kassler, 1994; WEC, 1994).

Renewable sources of energy used sustainably have small emissions of GHGs. The establishment of an infrastructure for large-scale use of renewable sources of energy will generate indirect emissions of GHGs if this infrastructure is put in place using fossil fuels. On the other hand, renewable energy sources could be considered for this, thereby giving rise to very little of such indirect emissions. There are also some GHG emissions associated with the unsustainable use of biomass—for example, from reducing the amount of standing biomass and from decomposition of biomass associated with the establishment of some dams. By and large, the increased use of renewable sources of energy offers substantial reductions of GHG emissions compared to the use of fossil fuels.

19.2.5.1. Hydropower

Hydroelectricity, which depends ultimately on the natural evaporation of water by solar energy, is the only renewable resource used on a large scale for electricity. In 1990, 2,200 TWh were generated, accounting for 18% of the world's total electricity generation, with small or no emissions of GHG.

The energy potential of hydropower is determined by the annual volume of runoff water (47,000 km^3) and by the height it falls before reaching the ocean. Estimates of the theoretical annual potential of world hydroelectricity range from 36,000 to 44,000 TWh, corresponding to 340 to 410 EJ_{th} with conventions from Table B-2 in Chapter B (Raabe, 1985; Boiteaux, 1989; Bloss *et al.*, 1980). This gross theoretical potential is much larger than the technically usable potential, which in turn is substantially larger than the economically exploitable potential. The technically usable potential is an estimated 14,000 TWh/yr (WEC, 1992). The economically exploitable potential—after social, environmental, geological, and other economic constraints are considered—is smaller. The world's long-term economic hydroelectric potential may be on the order of 6,000–9,000 TWh/yr (Moreira and Poole, 1993).

Hydroelectric plants vary in size and share of the world's hydropower potential. Small-scale hydropower, often defined as installations smaller than 10 MW, contributes about 4% of world hydroelectricity (WEC, 1992, 1994). Small plants have a large potential to contribute to socioeconomic development in rural areas. Their potential overall contribution to GHG emissions is limited, however.

Hydropower costs are most commonly measured in terms of the cost per kilowatt installed. With some exceptions, there is surprisingly little systematic data available on the cost of individual plants (Electrobras, 1987; Lazenby and Jones, 1987; Moore and Smith, 1990). Total investment costs for 70 developing countries for the 1990s (Moore and Smith, 1990) suggest that the cost of new hydroelectricity is 7.8¢/kWh.[2]

[2] This is the average value for hydropower plants under construction for this decade. The investment is from Moore and Smith (1990); electricity cost is obtained through average utilization factor (0.42), interest rate (10%), and operational cost (1 ¢/kWh).

An estimate of the value of hydropower must account for its capability to vary the power level over wide ranges, providing a means of handling variations in electricity demand and intermittent power generation from wind and solar sources (Kelly and Weinberg, 1993).

Hydropower is not free of GHG emissions. Bacterial decomposition of biomass in flooded reservoirs produces CH_4 and CO_2. The impact on global warming, in comparison to fossil fuel power generation, depends on the amount and timing of the gases formed. Dams that flood large areas with large quantities of biomass (including underground biomass) generate GHG emissions (Rudd *et al.*, 1993; Pinguelli Rosa and Schaeffer, 1994). The magnitude of this effect can be determined only on a case-by-case basis.

Most hydroplants require a reservoir, which can significantly affect people (through relocation), the terrestrial ecosystem, and the river itself. The social effects of relocation of people are not well known (Cernea, 1988). The area inundated by dams generally has higher agricultural value and a higher population density than the surrounding region as a whole.

The construction phase of a hydroelectric plant has social consequences and direct environmental impacts, such as water diversion, drilling, slope alteration, reservoir preparation, and the creation of an infrastructure for the large workforce (Moreira and Poole, 1993). The social consequences of a large workforce include migration, shanty towns with increased public-health problems, pockets of urban and rural poverty, destruction of the character of local communities, and intense deforestation (Moreira and Poole, 1990). On the positive side, the local transportation network and related industries often grow. The additional infrastructure stimulates regional economic development and has been widely viewed as a regional development tool (Moreira and Poole, 1993). Well-designed installations using modern technology that cascade the water through a number of smaller dams and power plants may reduce the environmental impact of the system (Henry, 1991).

Water quality is a major problem for hydroelectric planners. The inundation of land reduces the production potential for biomass for energy in addition to generating GHGs through biological decomposition (Garzon, 1984). Water management must allow for irrigation (Veltrop, 1991). Sedimentation is a problem in reservoirs and river deltas (Deudney, 1981).

Dams have a significant impact. Dam failure has caused significant losses of life and property (Smets, 1987; Laginha Serafim, 1984). Dams may prevent fish migration and stop water flow, which is a necessity for some fish (Petrere, 1990).

Disturbing aquatic ecosystems in tropical areas can induce indirect environmental effects; for example, increased pathogens and their intermediate hosts may lead to an increase in fatal human diseases such as malaria, schistosomiasis, filariasis, and yellow fever (Waddy, 1973).

The social and environmental constraints on hydropower development must be carefully evaluated in each situation. The evaluation should include the value of avoiding other forms of electricity generation that may bring unwanted social and environmental effects.

19.2.5.2. Biomass

Biomass energy is consumed at an annual rate of 47 EJ (WEC, 1994) to 55 EJ (Hall *et al.*, 1993), mainly for cooking and heating in developing countries. It is used for some small-scale industry, though there is some experience at larger scales.

There is no net atmospheric CO_2 build-up from using biomass grown sustainably because CO_2 released in combustion is compensated for by that withdrawn from the atmosphere during growth. Modern biomass energy also offers the potential for generating income in rural areas (Johansson *et al.*, 1993a). This income could allow developing-country farmers to modernize their farming techniques and reduce the need to expand output by bringing more marginal lands into production (Riedacker and Dessus, 1991). In industrialized countries, biomass production on excess agricultural lands could allow governments eventually to phase out agricultural subsidies (Williams, 1994a).

19.2.5.2.1. Biomass production

Potential biomass energy supplies include municipal solid waste (MSW), industrial and agricultural residues, existing forests, and energy plantations.

MSW: MSW is produced at per capita rates of 0.9–1.9 kg/day in industrialized countries. Energy contents range from 4–13 MJ/kg. Energy can be produced by incineration, biodigestion, or thermochemical gasification to produce electricity, process heat, or fluid fuels. An attraction is the low price to potential users, who may be paid to take the waste because of the high cost of disposal in landfills. If MSW is burned, attention must be given to controlling air pollutants, some of which are carcinogenic compounds (WEC, 1994). Air pollution may be virtually eliminated with some advanced energy-conversion technologies (Chen, 1995).

Industrial and agricultural residues: The energy content of organic waste byproducts of the food, fiber, and forest-product industries is more than one-third of total global commercial energy use (Hall *et al.*, 1993). Some residues should be left at the site, however, to ensure the sustainable production of the main product. If removed and converted to biogas, the nutrients recovered in the digester should be returned to the site. Some recoverable residues would be better used for other purposes, and it will not be practical or cost-effective to recover all residues. Recoverable crop, forest, and dung residues have been estimated to be about 10% of present global commercial energy use (Hall *et al.*, 1993).

Existing forests: The difference between annual increment and harvest for the world's forests has an energy content equal to one-third of world primary energy use (Hall *et al.*, 1993). Large increases in wood recovery for energy, however, would require more intensive forest management and raise concerns about potential loss of biodiversity and natural habitat. Thus, the contribution of existing forests to energy supplies depends on local conditions.

Plantation biomass: Dedicated plantations of woody or herbaceous (annual or perennial) crops offer a large potential for biomass energy. At present there are about 100 million hectares of industrial tree plantations worldwide, of which about 6 million hectares are fast-growing hardwoods, the trees most suitable for energy applications (Hall *et al.*, 1993).

Excess agricultural lands are good candidates for plantations in industrialized countries. The potential for using such lands for biomass energy in Europe has been estimated at 33 Mha by 2020 (Hall, 1994) and 40 Mha for the European Union in the long term (Wright, 1991). Also, it has been estimated that 50–100 Mha of agricultural lands in Europe might become available for other purposes (WRR, 1992). In the United States, idle cropland totaled 33 Mha in 1990 and is projected to grow to 52 Mha by 2030, despite an expected doubling of exports of maize, wheat, and soybeans (SCS, 1989)—largely because of expected increased crop yields.

Concerns about future food supplies (Ehrlich *et al.*, 1993; Kendall and Pimentel, 1994; Brown, 1993) have led some to suggest that land will not be available for biomass production for energy in Africa and other developing regions (Alcamo *et al.*, 1994). The outlook for food production may not be so bleak, however, if agriculture can be modernized in the developing world. With a continuation of historical trends in grain yields, considerable unforested land would be available for biomass energy production in many developing regions despite increased food requirements for a growing population (Larson *et al.*, 1995). Such outcomes depend on the availability of income to modernize agriculture, as well as whether intensified agricultural production can be made environmentally acceptable (see Section 19.3.1.4 and Chapter 25).

Deforested and otherwise degraded lands might be targeted for energy plantations. For tropical regions, Grainger (1988, 1990) has estimated that there are nearly 2,100 Mha of degraded land, of which 30% is theoretically suitable for reforestation. The challenge of restoration is to find a sequence of plantings that can lower ground temperature and restore the organic and nutrient content as well as the moisture level of the soils to attain

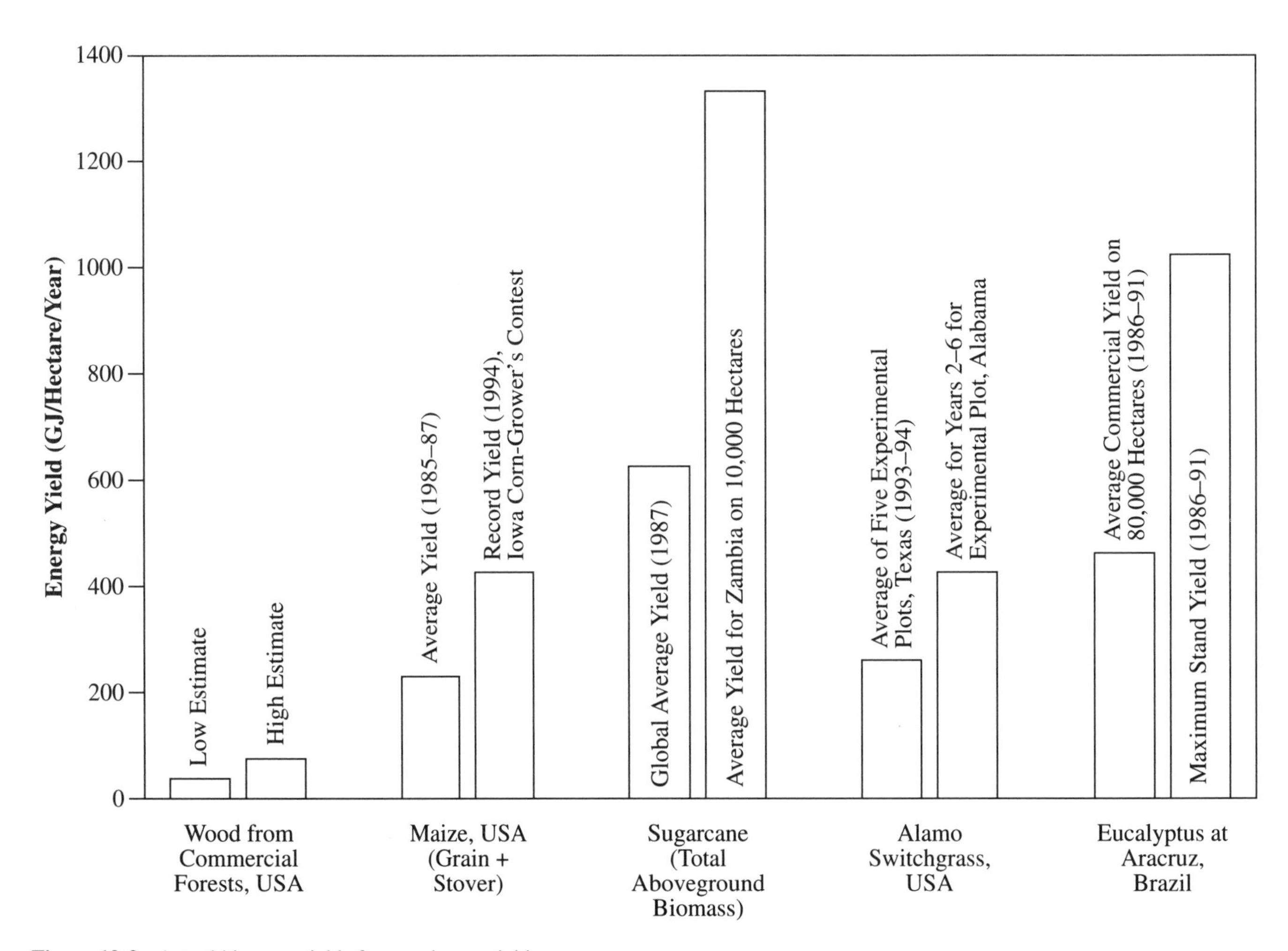

Figure 19-3: Actual biomass yields from various activities.

high and sustainable crop yields (OTA, 1992). Intensive R&D is needed to identify the most promising sites and restoration strategies (Riedacker, 1993; Johansson *et al.*, 1993a).

The objective is for plantations to achieve high sustainable yields at low costs with minimal environmental impact. The characteristics of an ideal energy crop have been described by Goudriaan *et al.* (1991). Biomass energy must be considered in integrated land-use planning (see Chapter 25), with attention given to the economic well-being of the local community and to environmental concerns such as chemical pollution of ground water, soil erosion, and loss of biodiversity and landscape diversity (Beyea *et al.*, 1991; Shell/WWF, 1993; OTA, 1993; Gustafsson, 1994; NBR, 1994; WEC, 1994). Biomass used for energy via thermochemical processes offers much more flexibility than biomass used for food in the choice and mix of species and cultivation practices for addressing environmental concerns because the only biomass attribute of importance then is production cost (Williams, 1994a).

Little biomass has been grown on plantations for energy. Thus, estimates of prospective yields are based on experience with crops grown for food and fiber, limited commercial experience and considerable experimental work with candidate bioenergy crops, and expected improved yields from new technology. The highest biomass yields that have been achieved over large areas are for sugar cane (see Figure 19-3): The 1987 worldwide average, aboveground biomass yield was 36 dry tons per hectare per year (dt/ha/yr); the yield for Zambia (averaged over 10,000 ha) was 77 dt/ha/yr (Hall *et al.*, 1993). At Aracruz in Brazil from 1986 to 1991, yields (stemwood diameter >7 cm) of eucalyptus grown for pulp averaged about 23 dt/ha/yr; the maximum stand yields achieved in the same period were 52 dt/ha/yr (see Figure 19-3). About 11,000 ha of willow plantations have been established for energy in southern Scandinavia. There the yield on the best 100 hectares has averaged 10–12 dt/ha/yr (Ledin *et al.*, 1994) and is projected to be 14–17 dt/ha/yr in 20 years (Christofferson, 1995). In the United States, researchers project that by 2020 yields for hybrid poplar and switchgrass will average 15–20 dt/ha/yr in regions where conditions are favorable for such crops (Walsh and Graham, 1995).

Plantation biomass costs already are favorable in some developing countries. On the basis of commercial experience with eucalyptus in northeast Brazil, an estimated 13 EJ/yr of biomass could be produced on 50 Mha at an average cost for delivered wood chips of $1.7/GJ (Carpentieri *et al.*, 1993). Costs are much higher in industrialized countries. A major U.S. Department of Energy/Department of Agriculture study estimates, however, that a strong and sustained R&D effort could lead by 2020 to 5 EJ/yr of plantation biomass in the United States at a delivered cost of $1.5/GJ or less (Graham *et al.*, 1995). A related study carried out by the U.S. Environmental Protection Agency estimates that land-use competition could increase land rental rates. Thus, the biomass price in 2020 for 5 EJ/yr would be $1.8/GJ under essentially the same technical assumptions about yields and costs (Turnure *et al.*, 1995). For comparison, the U.S. Department of Energy projects that the average price of coal to U.S. utilities will be $1.3/GJ in 2010 (EIA, 1995) under projected demand conditions. Displacement of coal by biomass on a large scale would tend to drive future coal prices to levels that are lower than without this competition.

The energy output–input ratio for plantation biomass depends on crop species, yield, and production technology. Although this ratio can be low for high-quality food crops such as grain, it is generally high for energy crops with good economic prospects. Some estimates are 15:1 for willow (Johansson, 1993); 29:1 for trees grown in short-rotation coppices (Foster and Matthews, 1994); and, in the United States, 11:1 for switchgrass, 12:1 for sorghum, and 16:1 for hybrid poplar (Turhollow and Perlack, 1991).

19.2.5.2.2. Electric power generation

Electricity generation from biomass can take place at scales ranging from a few kilowatts for rural village or agricultural applications, to tens of megawatts for present industrial applications, to hundreds of megawatts for advanced industrial applications.

Producer gas engines: The electrical loads of many poor rural villages of the developing world are in the range of 10–200 kW_e. Producer gas-engine generator sets based on the use of biomass gasifiers coupled to small reciprocating engines are well-matched to these loads (Ravindranath, 1993; Ravindranath and Hall, 1995). These generators often use diesel fuel, but biomass-derived producer gas could replace 75–95% of this diesel fuel.

Modest resources have been committed to developing this technology, mostly since the early 1970s. Until recently, most biomass-producer gas-engine projects failed, largely because of excessive tar formation in the gasifier and maintenance problems posed by tars. These problems have been solved, however; successful field demonstrations have been carried out, and the technology is ready for large-scale commercial applications (Mukunda *et al.*, 1993, 1994). For 100-kW_e units operated on biomass costing $2/GJ, electricity can be produced for 10–15¢/kWh; for low-cost biomass (for example $0.85/GJ for a village plantation in south India), electricity can be produced for less than 10¢/kWh (Larson, 1993), making this technology an attractive option for applications remote from electric grids where biomass is available.

Conventional steam-turbine technology: In many countries, biomass residues are used to generate electricity or CHP with conventional steam turbine technologies. The United States has an installed biomass steam-electric generating capacity of more than 8,000 MW_e (Turnbull, 1993a). Although they tend to be much smaller (typically

20 MW_e) than fossil fuel steam-electric plants, relatively capital-intensive, and energy inefficient (Turnbull, 1993b), commercial biomass steam-electric plants can provide cost-competitive power where biomass prices are low, especially in CHP applications.

Improving steam-turbine cycles: In the near term, steam-electric plant performance can be improved by cofiring large boilers with biomass. This can be achieved by converting existing grate-fired boilers to fluidized bed boilers at modest cost—\$40–100/kW_e, including fuel handling (Saviharju, 1995). Biomass cofiring systems are used extensively in Nordic countries.

Because costly sulfur cleanup is not needed for biomass, optimizing steam-turbine cycle technology can lead to specific capital costs and efficiencies for 100 MW_e plants that are comparable to those for 500 MW_e coal plants requiring sulfur cleanup (RTI *et al.*, 1993). Less capital-intensive, more energy-efficient technologies are needed to make the more abundant and costly biomass sources competitive.

Advanced power cycles involving gasification: Higher efficiency and lower unit capital costs can be realized with advanced conversion technologies. Biomass-integrated gasifier/gas turbine (BIG/GT) cycles are the focus of present development efforts (Williams and Larson, 1993).

Because biomass generally contains little sulfur, gases exiting the gasifier can be cleaned at higher temperatures than is feasible for coal (hot-gas sulfur-removal technology is not commercially proven). Thus, biomass is well-matched to air-blown gasifiers, which tend to be less costly than the oxygen-blown gasifiers used in commercial coal integrated gasifier/gas turbine systems at the scales of interest for biomass power systems. Efficiencies of 40–45% are expected to be achievable (Elliott and Booth, 1993; Consonni and Larson, 1994).

Various projects to demonstrate BIG/GT technology are underway. A 6-MW_e pilot plant is being tested at Värnamo, Sweden. A 30-MW_e commercial demonstration plant is being planned for the northeast of Brazil (with initial operation scheduled for 1997–1998), with support from the Global Environment Facility (Elliott and Booth, 1993). During 1993–1994, two Finnish gasification CHP power demonstration projects (larger than 50 MW_e) were announced, as were three relatively small-scale (7–12 MW_e) demonstration projects (a CHP plant for Denmark and two power-only plants for the UK and Italy) with support from the European Union. Other demonstration projects are being planned in Belgium, the Netherlands, and the United States.

BIG/GT power plants would have low sulfur dioxide (SO_2) particulate, and thermal NO_x emissions, but NO_x emissions could arise from fuel-bound nitrogen (N). NO_x emissions could be kept low by growing biomass with low N content and/or selectively harvesting portions of the biomass having high C/N ratios. For example, trees could be harvested in winter, after the leaves—in which the N is concentrated—have fallen (Ledin and Alriksson, 1992). Stack-gas emission controls might still be needed in areas where there are severe restrictions on NO_x. Gasifier and/or combustor modifications under development offer the promise of suppressing NO_x formation (Leppalahti, 1993). Successful development would obviate the need for stack-gas controls.

The ability of biomass to compete with coal depends on relative fuel prices and relative performances and costs for conversion technologies. A U.S. Environmental Protection Agency assessment exploring prospective advances for both biomass and coal (Turnure *et al.*, 1995) considers future biomass power plants with characteristics that range from a 0.5% lower heat rate and a unit capital cost that is approximately 2% less than for coal plants, to biomass plants characterized by an 8% higher heat rate and a unit capital cost that is 20% higher than for coal plants. Under the first set of conditions, differences in fuel cost alone would drive the economic comparison between biomass and coal; for the second, subsidies in excess of 1¢/kWh would be required to put biomass on a competitive footing with coal, even under relatively optimistic assumptions about biomass production costs.

Other studies suggest that if R&D goals are fully realized for biomass production (Graham *et al.*, 1995) and for the commercialization of advanced BIG/GT technologies designed to exploit the intrinsic characteristics of biomass feedstocks (Williams, 1995a), biomass has good prospects for competing with coal by 2020 in many circumstances, even if the price of biomass is somewhat higher than the price of coal. In the long run, key distinguishing features of biomass and coal IG/GT technologies are likely to arise from the sulfur and moisture contents of the feedstocks.[3] A framework for a quantitative economic comparison highlighting these distinguishing features is offered by fixed-bed integrated gasifier/intercooled steam-injected gas

[3] Even though existing biomass steam-electric plants (fired with forest product and agricultural industry residues or MSW) are characterized by modest scales (25 MW_e a typical plant size), scale is not likely to be a key distinguishing characteristic between biomass and coal plants when advanced BIG/GT technologies using plantation biomass become established in the market, because electric generation costs are not very sensitive to transport distances for the fuel (Faaij *et al.*, 1995; van den Broek *et al.*, 1995). Site-specific analyses of such applications of pressurized BIG/GT systems in the U.S. and Brazil—taking into account the areal distribution of biomass supplies and power plant scale economies—have shown that electricity costs are minimized for capacities in the range 230–320 MW_e (Marrison and Larson, 1995a, 1995b). Moreover, if air-blown gasifiers with hot-gas sulfur cleanup are successfully developed (Corman, 1986), it will be possible to have low-cost coal IG/GT plants at scales much smaller than those optimal for current IG/GT technology based on oxygen-blown gasifiers, for which costs are very scale-sensitive. Even if oxygen-blown gasifiers remain the norm for coal, future coal IG/GT plants may be smaller than the optimal for current technology—because of ongoing progress in reducing costs for air separation generally and in particular for air separation at modest scales, as a result of advances relating to pressure swing adsorption and membrane gas separation systems (Simbeck, 1995).

turbine (IG/ISTIG) systems[4] at the same scale (~110 MW_e) for both coal and biomass (Williams and Larson, 1993). The coal variant of IG/ISTIG, which involves hot-gas sulfur cleanup, has been proposed as a low-cost option for coal (Corman, 1986). The biomass variant does not require sulfur cleanup but does require a biomass dryer. A promising dryer option involves drying biomass in pressurized superheated steam and using the water evaporated from biomass as steam in the BIG/ISTIG system via process integration. For IVOSDIG (involving steam drying of biomass coupled to a steam-injected gas turbine), the first generation of which is being developed in Finland, it has been shown that the plant efficiency (HHV basis) is nearly independent of the fuel moisture content (Hulkkonen *et al.*, 1991). The capital cost penalty for such a steam dryer for BIG/ISTIG is estimated to be $100/kW, about 40% of the estimated incremental cost for hot gas sulfur cleanup for the coal IG/ISTIG. Largely because of the resulting net capital cost advantage to biomass, IG/ISTIG electricity from biomass costing $2/GJ could potentially compete with IG/ISTIG electricity from coal for coal prices in the range $1.4–1.7/GJ (depending on the market value of the sulfuric acid recovered as a byproduct) (Williams, 1995a).

Biomass-integrated gasifier/fuel cell (BIG/FC) systems involving molten carbonate or solid-oxide fuel cells prospectively would be even more energy-efficient than BIG/GT systems and would be especially well-suited for industrial cogeneration applications requiring high-temperature process heat. Although little attention has been given to such systems in development efforts, biomass would, in several respects, be a more attractive feedstock than coal for gasification-based fuel cell applications (Kartha *et al.*, 1994). If successfully developed, BIG/FC systems could extend the range of economically attractive power generation options to smaller scales (~1 MW_e) than would be feasible with BIG/GT systems.

19.2.5.2.3. Biogas production

Simple anaerobic digesters are used in rural areas of some developing countries to produce biogas from manure and crop residues at scales ranging from household to village. They provide fuel for cooking and power (Rajabapaiah *et al.*, 1993), byproducts in the form of fertilizer and feed for pigs and fish farms, and substantial environmental and human health benefits (DEPE, 1992). The future potential is large, particularly if organizational issues are satisfactorily resolved—for example, if local expertise is ensured in operation and maintenance and if users take responsibility for collecting biomass for the biogas plant (Rajabapaiah *et al.*, 1993).

In industrialized countries, the focus is on industrial plants with capacities of several million cubic meters of biogas/yr, compared with 250–300 m^3/yr for household plants in China (Wang *et al.*, 1992) and 10,000 m^3/yr for village-scale plants in India (Rajabapaiah *et al.*, 1993).

In the most common application in industrialized countries, manure and other organic wastes are pretreated and passed through plants for pasteurization, digestion, separation, and gas purification to produce biogas, along with byproducts that include fertilizer and compost. The gas produced in large installations in Denmark is used in gas engines for CHP or in boilers for DH. Experiments with other uses are in progress, including compressed gas for vehicles (Stewart and McLeod, 1980; Danish Energy Agency, 1992).

Economics: Direct costs have declined with experience for Danish plants. Although the government has provided a 25–50% subsidy, the best plant, selling gas at $10 per GJ, would roughly break even in Denmark even if there had been no subsidy (Danish Energy Agency, 1992).

In India, after experiences well short of expectations (Comptroller and Auditor General of India, 1994), cattle-dung biogas plants have now been demonstrated at village scale to be economically viable for providing biogas-generated electricity (~5 kW_e) for lighting and pumping water. Plans are underway to replicate the technology in many villages (Reddy *et al.*, 1994).

Energy balance: For the 10 large Danish biogas plants, the net external energy requirement—considering fuel for transporting manure to the plant and the fertilizer value of the returned residue—is just 0.7% of net production, corresponding to an energy payback of three days (Tafdrup, 1993).

GHG emissions: The average GHG emissions avoided by using the biogas produced by the 10 Danish plants instead of coal for CHP is 17 kgC/m^3 of biomass converted. Moreover, biogas production from manure reduces emissions from spreading manure directly on fields equivalent to about 15 kgC/m^3 of biomass converted, so that total GHG emissions are reduced by 32 kgC/m^3 of biomass converted (Tafdrup, 1993).

19.2.5.2.4. Biofuels for transport

Interest in alternative transport fuels has been driven largely by concerns about oil supply security and urban air quality. These concerns have led Brazil and the United States to adopt policies promoting alternative vehicles and fuels. In light of increasing air pollution in the world's megacities (UNEP/WHO, 1992), such policies are likely to become more widespread.

Traditional biofuels for transport: Efforts to produce biofuels for transport have focused on ethanol from maize, wheat, and sugar cane and on vegetable oils such as rapeseed oil. The most substantial commercial programs are in Brazil and the United States. In 1989, Brazil produced 12 billion liters of fuel ethanol from sugar cane, which was used to power 4.2 million

[4] While IG/ISTIG technology is not currently being developed, its successful development could lead to substantial electricity cost reductions relative to first generation IG/GT technologies for both coal and biomass. Comparing coal and biomass in the context of IG/ISTIG technology represents the competing feedstocks fairly and facilitates a comparison of their key distinguishing characteristics.

cars running on hydrated ethanol and 5 million cars on gasohol, a gasoline-ethanol blend (Goldemberg *et al.*, 1993). In 1993, the United States produced 4 billion liters of ethanol from maize for gasohol applications.

All traditional biomass-derived transport fuels are uneconomic at present. Substantial cost reductions are being made for sugar cane-derived ethanol, however (Goldemberg *et al.*, 1993). Moreover, there are good prospects for making cane-derived ethanol competitive at the present low world oil price if electricity is cogenerated from cane residues using BIG/GT technology along with ethanol from cane juice (Williams and Larson, 1993). In contrast, the prospects are poor for making ethanol economically from grain (Wyman *et al.*, 1993).

Most traditional biofuels are inefficient users of land, with low yields of transport services (vehicle km/ha/yr) compared with what is achievable with advanced technologies using woody biomass feedstocks (Table 19-4). Many also have marginal energy balances. Fossil energy inputs to produce ethanol from wheat, maize, or sugar beets are likely to be comparable to the energy content of the ethanol; for rape methyl ester derived from rapeseed, fossil energy inputs are likely to be about half the energy in the rape methyl ester (Lysen *et al.*, 1992; IEA, 1994b). Greenhouse performance also tends to be marginal. For maize, estimates of net fuel-cycle emissions of CO_2 have ranged from somewhat more (Ho, 1989) to somewhat less (Marland and Turhollow, 1990) than for gasoline. Total life-cycle GHG emissions may even be somewhat higher than for reformulated gasoline, which contains more oxygen to improve combustion and thereby reduce emissions (DeLuchi, 1991). For rape methyl ester, life-cycle CO_2 emissions are less than for fossil fuels (Lysen *et al.*, 1992). There also may be significant emissions of nitrous oxide (N_2O), a powerful GHG, as a result of the conversion of some nitrogen in fertilizer to N_2O. These emissions depend on the type of fertilizer and application (Lysen *et al.*, 1992; Muschalek and Sharmer, 1994), however, and there are large uncertainties (Gosse, 1994; Bernhardt, 1994).

It is now generally believed that alcohol fuels—especially when blended with gasoline and used in flexible-fuel internal-combustion engine vehicles (ICEVs)—offer little or no

Table 19-4: *Energy yield for alternative feedstock/conversion technologies.*

Option	**Feedstock Yield (dry tons/ha/yr)**	**Transport Fuel Yield (GJ/ha/yr)**	**Transport Services Yield[8] (10^3 v-km/ha/yr)**
Rape Methyl Ester (Netherlands)[1]	3.7 of Rapeseed	47	21 (ICEV)
EthOH from Maize (USA)[2]	7.2 of Maize	76	27 (ICEV)
EthOH from Wheat (Netherlands)[3]	6.6 of Wheat	72	26 (ICEV)
EthOH from Sugar Beets (Netherlands)[4]	15.1 of Sugar Beets	132	48 (ICEV)
EthOH from Sugar Cane (Brazil)[5]	38.5 of Cane Stems	111	40 (ICEV)
EthOH, Enzymatic Hydrolysis of Wood (present technology)[6]	15 of Wood	122	44 (ICEV)
EthOH, Enzymatic Hydrolysis of Wood (improved technology)[6]	15 of Wood	179	64 (ICEV)
MeOH, Thermochemical Gasification of Wood[7]	15 of Wood	177	64/133 (ICEV/FCV)
H_2, Thermochemical Gasification of Wood[7]	15 of Wood	213	84/189 (ICEV/FCV)

[1]Per ton of seed: 370 liters of rape methyl ester plus (not listed) 1.4 tons of straw (Lysen *et al.*, 1992).

[2]For wet milling, assuming the U.S. average maize yield, 1989–1992; per ton of grain: 440 liters of ethanol plus (not listed) 0.35 tons of stover (out of 1 ton of total stover, assuming the rest must be left at the site for soil maintenance), 275 kg of corn gluten cattle feed, and 330 kg of CO_2 (Wyman *et al.*, 1993).

[3]Per ton of seed: 455 liters of ethanol plus (not listed) 0.6 tons of straw (Lysen *et al.*, 1992).

[4]Per ton of sugar beet: 364 liters of ethanol (Lysen *et al.*, 1992).

[5]For the average sugar cane yield in Brazil in 1987 (63.3 tons of harvested cane stems, wet weight); per ton of wet cane stems: 73 liters of ethanol (Goldemberg *et al.*, 1993). In addition, (not listed) the dry weight of the attached tops and leaves amounts to 0.092 tons and that for the detached leaves amounts to 0.188 tons per ton of wet stems—altogether some 18 dry tons/ha/yr (Alexander, 1985).

[6]Per ton of feedstock: 338 liters of ethanol plus (not listed) 183 kWh (0.658 GJ) of electricity, present technology; 497 liters of ethanol plus (not listed) 101 kWh (0.365 GJ) of electricity, improved technology (Wyman *et al.*, 1993).

[7]For the indirectly heated Battelle Columbus Laboratory biomass gasifier; per ton of feedstock: 11.8 GJ of methanol or 14.2 GJ of hydrogen; per ton of feedstock, external electricity requirements are 107 kWh (0.38 GJ) for methanol or 309 kWh (1.11 GJ) for hydrogen (Williams *et al.*, 1995a, 1995b).

[8]Fuel consumption rate of the vehicles (in liters of gasoline-equivalent per 100 km) assumed to be 6.30 for rapeseed oil (assumed to be the same as for diesel), 7.97 for ethanol, 7.90 for methanol, and 7.31 for hydrogen used in internal combustion engine vehicles (ICEVs), and 3.81 for methanol and 3.24 for hydrogen used in fuel-cell vehicles (FCVs) (DeLuchi, 1991). Note that 1 liter of gasoline equivalent = 0.0348 GJ, HHV.

air-quality advantages, other than reduction in carbon monoxide (CO) emissions (Calvert *et al.*, 1993). Moreover, reformulated gasolines can meet or surpass the air-pollution reductions of alcohol-gasoline blends (BEST, 1991). With methanol, CO emissions would be reduced, and emissions of volatile organic compounds would be less problematic than for gasoline, but NO_x emissions would probably not be reduced. Ethanol offers lesser air-quality benefits than methanol and may produce more ozone per carbon atom (Calvert *et al.*, 1993). Because of such problems with traditional biofuels, advanced technologies are receiving increased attention.

Advanced biofuels: Advanced biofuels derived from low-cost woody biomass could offer higher energy yields at lower cost and with lower environmental impacts than most traditional biofuels (IEA, 1994b). The advanced biofuel that has received the most attention is ethanol derived from wood via enzymatic hydrolysis (Wyman *et al.*, 1993). For a woody feedstock yield of 15 dry t/ha/yr—which is generally believed to be achievable in large-scale production—the ethanol yield could be more than twice that from grain (Table 19-4). If the U.S. Department of Energy's year-2000 goals for performance and cost are met, energy balances would be favorable; life-cycle emissions of CO_2 for ethanol production and use in ICEVs (in gC/km) would be only about 2% of those from such vehicles operated on reformulated gasoline (Wyman *et al.*, 1993). Furthermore, ethanol will be competitive with gasoline if oil prices are greater than about $25/bbl (Wyman *et al.*, 1993). For comparison, French studies indicate that ethanol from grain will not be competitive until oil prices reach $40–45/bbl (Torck *et al.*, 1988). The major shortcoming of the technology is that it may not lead to air-quality improvements beyond what can be achieved with reformulated gasoline. However, low emissions, along with a two to three-fold gain in fuel economy, might be practically achievable using either fuel with a so-called hybrid car that has an electric drive train, an electric generator driven by a small internal-combustion engine that would provide baseload power, and a small battery or other device for providing peak power (Ross, 1994; Colombo and Farinelli, 1994).

Other advanced fuels include methanol and hydrogen derived from biomass via processes that begin with thermochemical gasification. Used in fuel-cell vehicles (FCVs), they offer good prospects for dealing with the multiple challenges of transportation. Recent advances suggest that the proton-exchange-membrane (PEM) fuel cell is an attractive alternative to the internal combustion engine for cars (AGTD, 1994; Williams, 1993, 1994b), buses, trucks, and trains (see Section 19.2.6.4).

If FCVs were introduced, hydrogen or methanol fuel would be produced initially by steam-reforming natural gas. This is the least-costly route, for which the required technology is commercially available. On the basis of projected prices for natural gas, coal, and biomass, production costs would be comparable for these three feedstocks before 2025 (Williams, 1996).

The prospect that biomass could compete with coal in the longer term reflects the more favorable characteristics of biomass as a feedstock for gasification (Williams *et al.*, 1995a, 1995b). Low sulfur content gives biomass a processing cost advantage compared with coal. In addition, if the biomass gasifier is designed to exploit the much higher reactivity of biomass compared to coal, unit capital costs can be reduced. For both coal and biomass, the initial step involves gasification to produce a synthesis gas (mainly CO and H_2). Coal is gasified in oxygen to produce the desired synthesis gas. The high gasification temperatures needed are achieved by direct heating—that is, by burning some of the coal in place. To take advantage of economies of scale for the oxygen plant, the coal conversion facilities would be large. Because of its much higher reactivity, biomass can be gasified at much lower temperatures—a property that makes it possible to avoid the use of costly oxygen-blown gasifiers and instead to gasify the biomass in steam using indirectly heated gasifiers in smaller facilities.

The yield for methanol or hydrogen produced from woody biomass would be comparable to that for ethanol derived via enzymatic hydrolysis (see Table 19-4). The potential for displacing gasoline used in ICEVs with biomass-derived methanol or hydrogen used in FCVs, however, would be more than twice that for ethanol used in ICEV applications because FCVs would be more than twice as energy-efficient as ICEVs. The FCV thus offers the potential for substantially increasing the role of biomass in transportation, supporting up to seven times as many vehicle-km of transport services per hectare as ethanol derived from grain used in ICEVs (see Table 19-4).

19.2.5.3. Wind Energy

19.2.5.3.1. Stage of development

The technology related to grid-connected wind turbines has become commercially available and mature. The most successful commercial wind turbines have installed capacities of up to 600 kW. A new generation of machines in the 1 MW size range and above is now being investigated. At the end of 1993, the global installed capacity of high-efficiency wind turbines was 3,100 MW, of which approximately 1,700 MW is in the Americas and 1,200 MW in Europe. While many different designs are in use, the "stock average" technology consists of three-bladed, horizontal-axis machines operating at near-fixed rotation speed. In 1993, the global new installation of wind energy was on the order of 550 MW; the annual manufacturing capacity was over 1,000 MW. This capacity could increase rapidly due to the decentralized structure of production in the wind power industry. Technical availability (the capability to operate when the wind is higher than the starting wind speed of the machine) is now typically 95–99%. Small-scale applications of wind energy such as water pumping, battery charging, and stand-alone electricity supply systems also are being developed (Cavallo *et al.*, 1993; WEC, 1994; Sørensen, 1988).

Methods for evaluating wind resources have improved (Troen and Petersen, 1989). In several regions, reviews of resources have been carried out, and siting methods are now being employed regularly to identify the most valuable wind turbine sites (Turkenburg, 1992).

19.2.5.3.2. Technology development

To decrease the cost of electricity, new generations of wind turbines are bigger than current turbines. The reduction of the blade number to two and the introduction of cost-effective power electronic systems to allow for variable-speed operation may further reduce costs.

Intermittent wind power on a large grid can contribute an estimated 15–20% of annual electricity production without special arrangements (see also Section 19.2.6.1). In large utility systems with a small fraction of wind power, savings in fuels and emissions are approximately proportional to the average wind energy input. Employing the so-called loss-of-load-expectation method, the capacity value is typically found to be close to that of a conventional thermal plant with the same energy output at low levels of wind power penetration (supply fraction) and to decline with increasing penetration (van Wijk, 1990; Tande and Hansen, 1991; Grubb and Meyer, 1993). Energy storage facilities probably will be integrated into the utility system, which would allow efficient utilization of wind energy at high penetrations. The wind energy and the storage facility together may supply "firm" power actually substituting for baseload power (Cavallo, 1995; Cavallo and Keck, 1995).

19.2.5.3.3. Economy

The present stock average cost of energy from wind power is approximately 10¢/kWh, although the range is wide. By 2005 to 2010, wind power may be widely competitive with fossil and nuclear power (IAEA, 1991).

A study by Godtfredsen (1993) on new, commercially available wind-power technology gives the distribution of investment cost and the cost of energy. For average new technology, this study shows investment costs of $1,200/kW and electricity production costs of 6¢/kWh. For the best new technology, the figures are $900/kW and 5.5¢/kWh, respectively. The figures are based on an average windspeed of 10 m/s at 10 m above ground level, using single machines or small groups of machines, and do not include long-range transmission costs. Costs could be significantly lower for large wind farms.

The average annual windspeed on the site strongly affects the cost of energy. As a rule of thumb, the wind turbines' production increases with the windspeed to the third power, and the cost of energy decreases accordingly. Wind turbines at very windy sites (for example, coastal regions in northwest Europe) produce electricity at a total cost of 4.0–4.5¢/kWh. Some Danish utility-owned turbines produce electricity at 4.5¢/kWh (Elsamprojekt, 1994).

One set of projections of future costs is presented in Table 19-5. For comparison, Cavallo *et al.* (1993) project a cost of 3.2¢/kWh for 2020, assuming a wind turbine hub height of 50 m, an average wind speed of 5.8 m/s at 10 m, a 6% discount rate, and a useful equipment life of 25 years.

The payback time of the energy invested (energy balance) in the production of the wind turbine is less than one year for average new technology (see Table 19-5).

19.2.5.3.4. Markets

Formulated political goals for the next 20–30 years add up to 150 TWh_e global annual generation. The potential is considerably higher and could be further exploited (see Table B-4 in Chapter B). For example, the World Energy Council presents a "Current

Table 19-5: *Development of wind energy technology: technology level and characteristics, energy payback time for wind turbine, cost of electricity, and barriers to widespread dissemination. Assumptions include discount rates of 6 and 10%, average windspeed at 10 m altitude of 5.5 m/s (roughness class 1), and useful lifetime of 20 years. All costs are in 1990 US$. External costs and CO_2 emissions are not included.*

Technology Level	Technology Characteristics	Energy Balance (months)	Direct Cost (¢/kWh)		Institutional Barriers
			6%	10%	
Stock Average, 1975–90	Three-bladed, induction generator	12	10		kWh-cost
Average New Technology, 1993	Three-bladed, induction generator	9	6	7.4	Cost, public acceptance
Best New Technology, 1993	Three-bladed, induction generator	9	5	6.3	Public acceptance
Near-Term Technology, 2003	Three-bladed, variable speed	6	4.2	5.4	Public acceptance, load management
Long-Term Technology, 2020	Two-bladed, variable speed, flexible structure	6	3.4	4.3	Load management, transmission

Policy Scenario" and an "Ecologically Driven Scenario"—the latter showing an annual production of electricity from wind turbines of close to 1,000 TWh_e in the year 2020 (Van Wijk *et al.*, 1993; WEC, 1994), which is close to 8% of the global annual electricity consumption in 1990.

19.2.5.3.5. Public acceptance

Countries with large numbers of operating wind turbines sometimes experience strong public resistance before installation. Local values, circumstances, and decisionmaking procedures influence the degree of resistance to the noise of turbines, the visual impact on the landscape, the disturbance of wildlife (birds), and the disturbance of telecommunications (Arkesteijn and Havinga, 1992). In areas of Britain and Denmark with wind turbines installed, 70–85% of the citizens are supportive or not concerned (Carver and Page, 1994; DTI, 1993). The little evidence of the impact of turbines on wildlife (Grubb and Meyer, 1993) suggests that it is generally low and species-dependent (Still *et al.*, 1994). Case stories of birds killed by wind turbines have initiated new research that should illuminate the matter. At a 269-unit wind farm near Tarifa, Spain—on the main western bird-migration route between Africa and Europe—dozens of dead birds have been found (*Windpower Monthly*, February 1994). Closure and covering of a nearby waste pit is expected and intended to reduce bird mortality. New blades and gearboxes may generate less noise. Colors, tower type and shape, and number of blades are being studied to improve the appearance of turbines. Still, there are likely to be some sites of exceptional landscape or historic value where wind power plants cannot be accepted.

Experience also indicates how public acceptance can be secured. The most important means are information about wind energy's environmental benefits, a proactive style of communication, and considerate conduct by developers and authorities when wind-farm sites are identified and claimed for wind energy use (Turkenburg, 1992). In Denmark and the Netherlands, private/cooperative ownership of wind turbines also has helped in achieving public acceptance.

19.2.5.4. Solar Electric Technologies

Direct conversion of sunlight to electricity can be achieved by photovoltaic and solar thermal electric technologies.

19.2.5.4.1 Photovoltaic technologies

Photovoltaic (PV) devices made of layers of semiconductor materials convert sunlight directly into electricity. Since they are modular, create no pollution in operation, can be operated unattended, and require little maintenance, PV systems often will be deployed at small scales and close to users. PV technologies can be deployed virtually anywhere—even in areas with frequent cloudiness.

Centralized and distributed grid-connected power generation is most important for GHG emissions reductions. Small-scale applications for rural electrification for lighting, water pumping, refrigeration, and educational purposes are important for development but are less significant for GHG emissions reductions.

For several years, PV has been competitive in stand-alone power sources remote from electric utility grids. It has not been competitive, however, in bulk electric grid-connected applications. System capital costs are $7,000–10,000/kW; the corresponding electricity cost is 23–33¢/kWh, even in areas of high insolation (2,400 $kWh/m^2/yr$). New fossil fuel power plants cost less than 5¢/kWh. PV costs are declining, however. PV module prices in 1992 were one-tenth those in 1976 as cumulative production increased 1,000-fold. There are good prospects for continuing cost reduction, as indicated in Table 19-6 (Kelly, 1993; Ahmed, 1994; INEL *et al.*, 1990; WEC, 1994).

Two basic types of PV devices are flat-plate systems that convert both direct and diffuse radiation; and concentrators that must have direct radiation to work effectively and use mirrors or lenses to concentrate the incident light onto a small area equipped with solar cells. The prospects for major cost reductions are especially good for thin-film PV modules that would be used in flat-plate devices (Carlson and Wagner, 1993; Zweibel and Barnett, 1993) and for sun-tracking, concentrating systems (Boes and Luque, 1993).

Thin-film devices: The active PV materials for thin-film devices are layers 1- to 2-µm thick deposited on appropriate substrate materials (e.g., glass plates) for modules with areas of about 0.5 m^2 or more. Although they are less efficient than the more conventional, thick crystalline silicon devices, the small quantities of active materials required imply a potential for achieving very low costs at acceptable efficiencies (Carlson and Wagner, 1993; Zweibel and Barnett, 1993).

Amorphous silicon (a-Si) is the only thin-film technology established commercially, accounting for 27% of world PV sales of about 63 MW in 1993. The 5% efficiencies of commercial modules are far from the long-term goal of 15% for flat-plate devices (Kelly, 1993). Stabilized efficiencies of 10% have been realized for submodules (Zweibel, 1995). The relative ease of manufacturing a-Si modules (Carlson and Wagner, 1993) may make it possible to design cost-effective a-Si systems that are integrated into building rooftops, windows, and walls. PV electricity generation could become an integral part of the living environment (Hill *et al.*, 1994; Humm and Toggweiler, 1993; Strong and Wills, 1993).

How much it will be feasible to increase a-Si module efficiencies is uncertain. With the current glow-discharge deposition technique, a-Si films must be made ultrathin to compensate for the initial degradation of efficiency following exposure to light. This Staebler-Wronski effect (Carlson and Wagner, 1993) makes efficiencies greater than about 10% difficult. Light-induced efficiency degradation may not be a problem for an alternative hot-wire deposition technique (Vanecek *et*

Table 19-6: *Alternative projections of installed photovoltaic system costs (1990 US$).*

Parameter	U.S. DOE Interlaboratory White Paper[1] Business As Usual	U.S. DOE Interlaboratory White Paper[1] Intensification of RD&D	Williams and Terzian[2] Business As Usual	Williams and Terzian[2] Accelerated Development	Zweibel and Luft[3] Thin-Film Systems
Installed Capital Cost ($/kW)					
2000	3820	2540	4470	3610	3500
2005	–	–	3500	2170	2000
2010	2290	1770	2770	1520	1000
2020	1530	1250	1850	1060	800
2030	1280	1010	–	–	640
Busbar Cost in 2010[4] (¢/kWh)					
@ 2400 kWh/m²/yr	7.6	5.9	9.1	5.0	3.5
@ 1800 kWh/m²/yr	10.1	7.8	12.1	6.7	4.6
@ 1200 kWh/m²/yr	15.1	11.8	18.1	10.0	6.9
Busbar Cost in 2020[4] (¢/kWh)					
@ 2400 kWh/m²/yr	5.1	4.2	6.1	3.5	2.7
@ 1800 kWh/m²/yr	6.8	5.6	8.1	4.7	3.6
@ 1200 kWh/m²/yr	10.2	8.4	12.2	7.1	5.3
Busbar Cost in 2030[4] (¢/kWh)					
@ 2400 kWh/m²/yr	4.2	3.3	–	–	2.2
@ 1800 kWh/m²/yr	5,6	4.5	–	–	3.0
@ 1200 kWh/m²/yr	8.4	6.7	–	–	4.4

[1]Idaho National Engineering Laboratory *et al.*, 1990.
[2]Williams and Terzian, 1993.
[3]Zweibel and Luft, 1993.
[4]Calculated assuming 6% discount rate.

al., 1992; Haage *et al.*, 1994), but it remains to be demonstrated whether high efficiencies can be achieved with this technique (Zweibel, 1995). There are good prospects of achieving the 15% goal with at least one thin-film technology. Besides a-Si, two polycrystalline technologies show great promise: $CuInSe_2$ and CdTe (Zweibel and Barnett, 1993). Although there has not been as much development effort focused on these as for a-Si, progress has been rapid. Between 1977 and 1994, efficiencies of laboratory cells have increased from 6% to about 17% for $CuInSe_2$ and from 8% to about 16% for CdTe (Zweibel, 1995), without serious stability problems. Stable efficiencies for submodules about 0.1 m² in size already are comparable to or greater than what has been realized with a-Si.

Because of its modularity, PV technology is a good candidate for cost-cutting through "learning-by-doing" as well as technological innovation (Cody and Tiedje, 1992; Tsuchiya, 1989). A design exercise exploring what can be achieved by mass-producing near-term $CuInSe_2$ technology (Bradaric *et al.*, 1992) generated estimates that a 50-MW power plant based on 1995 $CuInSe_2$ technology would have an installed cost of $2300/kW, corresponding to a generation cost of about 8¢/kWh in areas with good insolation, and the value of the electricity to the utility would be comparable if the plant were sited where the output is well-correlated with peak electrical demand.

In the transition period to large-scale PV development, the value of PV in grid-connected applications often can be enhanced by locating the PV system close to users—at utility substations and on commercial and residential rooftops—instead of in central stations. For such distributed applications in areas where there is a good correlation between PV output and utility subsystem peak demand, the electricity is worth substantially more to the utility than in central-station configurations (Shugar, 1990; Kelly and Weinberg, 1993; Wenger and Hoff, 1995; Hoff *et al.*, 1995). Moreover, net system costs could be reduced in configurations where PV modules serve dual purposes (for example, being roof tiles as well as providing power). Markets for high-value, grid-connected, distributed PV are likely to be large in both the United States and in many developing countries (Williams and Terzian, 1993).

Although the long-term prospects for thin films are much harder to predict than the prospects for cost-cutting through learning-by-doing, thin-film researchers tend to be optimistic. Thin-film program management at the National Renewable Energy Laboratory in the United States projects that installed system costs will fall to about $1,000/kW by 2010 and ultimately to $600–700/kW (Table 19-6; Zweibel and Luft, 1993).

Concentrating, tracking PV technology: Much higher efficiencies can be achieved with crystalline PV technologies than is possible at present with thin films. Already, efficiencies of 28% for crystalline silicon and 34% for mechanically stacked gallium arsenide/gallium antimony multijunction cells have been achieved, and higher efficiencies are expected. Although such cells are much more costly to make than thin films, the higher costs can be offset by using devices that concentrate the sun's rays by a factor of anywhere from 2 to 1,000. The contribution of the cell cost to the module cost is reduced by this concentrating factor, and the concentrating device is expected to be much less costly per unit area than the cell.

Bradaric *et al.* (1992) have designed a 50-MW_e power plant based on the use of expected 1995-vintage concentrating technology (27.4%-efficient cells used in a two-axis tracking concentrator) brought to technological maturity. Their estimated installed cost is about $3,200/kW for mass-produced modules; the corresponding cost of electricity in a sunny area would be about 9¢/kWh. Whereas high-concentration ratio devices would be most suitably deployed in central-station configurations, low-concentration ratio devices—for example, flat-plate, thin-film devices—would be suitable for many distributed applications, both grid-isolated and grid-connected.

Life-cycle analysis: Life-cycle issues of concern for PV systems are energy payback time, toxic emissions, and the use of scarce materials.

The energy payback time is the ratio of the primary energy required to manufacture 1 m^2 of module plus balance-of-system (support structure, wiring, installation) for the power plant to the rate of primary fossil energy avoided by the electricity production with 1 m^2 of module. Most studies have focused on energy inputs for modules and have found that the energy payback time is strongly correlated with the amount of active PV material used. For silicon devices, the payback time is an estimated 5–10 years for monocrystalline, 3–5 years for polycrystalline, and 0.5–2 years for a-Si modules (von Meier, 1994). For monocrystalline and polycrystalline power plants, the payback is dominated by the energy required to make cells.

For thin-film devices, the difference between current and future technology is expected to be large. The payback for a-Si modules is estimated (for global average insolation of 1,700 $kWh/m^2/yr$) to be 2.5 years for 6%-efficient modules and generally pessimistic conditions versus 0.5 years for 10%-efficient modules and "base case" conditions (van Engelenburg and Alsema, 1994). Assuming future 15%-efficient modules, paybacks are expected to be even lower for polycrystalline modules: 1.6 months for CdTe and 4 months for $CuInSe_2$ (Alsema and van Engelenburg, 1992). For grid-connected systems, the balance-of-system is estimated to contribute an additional 1.8 years with today's technology but only 0.6 years with future systems using 10%-efficient modules (Hagedorn, 1989).

Emissions: Although PV devices emit no pollution in normal operation, some systems involve the use of toxic materials, which can pose risks in manufacture, use, and disposal. Life-cycle risks posed by silicon PV systems are small (Sørensen and Watt, 1993), but the potential hazards posed by the cadmium in CdTe and the selenium in $CuInSe_2$ devices warrant scrutiny—especially air emissions resulting from fires involving rooftop systems and incineration of unrecycled module waste. Such life-cycle emissions per kWh generated would be comparable to those from modern coal-fired power plants (Alsema and van Engelenburg, 1992).

Resources: There are no resource constraints for silicon PV systems. Although thin-film polycrystalline devices use only tiny amounts of active material, some materials are scarce. Of particular concern are the tellurium used in CdTe and especially the indium used in $CuInSe_2$ systems. Recycling might be required in order for these technologies to play major roles in the global energy economy (Alsema and van Engelenburg, 1992).

The societal cost of research, development, and commercialization: PV is the furthest of the renewable energy technologies from being commercially attractive in large-scale applications. How fast costs come down in the future depends on the level of the societal commitment to develop the technology. Williams and Terzian (1993) conclude that it would be worthwhile to accelerate rapidly the pace of development via intensification of the R&D effort and market incentives because the public-sector costs involved are expected to be small compared with the direct societal benefits, measured as the present value of reduced future consumer expenditures on electricity. Williams and Terzian (1993) estimate the public-sector subsidy to develop 400 GW_e of installed PV capacity worldwide by 2020 to be $1 billion/yr for more R&D in the early years, plus a present value of $3.3 billion for future market incentives. These estimates are similar to those made by the World Energy Council in estimating what would be required to commercialize PV technologies: a market stimulation incentive of $2.5–4.0 billion, plus another $5 billion for R&D (WEC, 1994). These costs are very small compared with the costs of commercializing fossil and nuclear technologies, largely because of the small scale of the technology.

19.2.5.4.2. Solar thermal-electric technologies

Solar thermal-electric technologies use mirrors or lenses to concentrate the sun's rays onto a heat exchanger, where a heat-transfer fluid is heated and used to drive a conventional power-conversion system. Applications for solar thermal-electric systems range from central-station power plants to modular, remote power systems. Such systems can be hybridized to run on both solar energy and fossil fuel; some systems contain integral thermal energy storage, allowing dispatchable, solar-only operation. These systems use only direct rays from the sun and are best located in regions of high sunlight intensity. Solar thermal-electric systems have the long-term potential to provide a significant fraction of the world's electricity and energy needs (WEC, 1994; Brower, 1993).

Between 1984 and 1990, nine utility-scale solar thermal power plants—cumulatively over 350 MW_e—were constructed in southern California using solar parabolic-trough technology driving a steam turbine in a hybrid configuration co-fired with natural gas. Developers of this technology in Europe and the United States are pursuing new power plant projects in several developing countries that could result in several hundred megawatts of new installations before the year 2000. In Australia, a tilted, polar-tracking-oriented, parabolic-trough system, with selective surfaces improving absorption and reducing losses and low concentration for higher diffuse radiation acceptance, is being developed that incorporates a storage system to allow high solar capacity factors (Mills and Keepin, 1993).

A solar central receiver—also called a "power tower"—uses a field of sun-tracking mirror assemblies to concentrate sunlight onto a tower-mounted receiver, where a circulating fluid is heated and used to produce power. Power towers typically contain cost-effective thermal energy storage and are intended as central-station power plants for peaking and intermediate load applications in 100–200 MW_e plant sizes (Hillesland and De Laquil, 1988; Hillesland, 1988). The technology is in the demonstration phase and could enter the utility-generation marketplace in the second half of the 1990s. In the United States, the 10-MW_e Solar Two Project, which will demonstrate nitrate salt receiver and storage technology, is scheduled to be operating by 1995 (Von Kleinsmid and De Laquil, 1993). A European industry consortium is planning to develop a 30-MW_e demonstration plant in Jordan using a volumetric air receiver and ceramic storage (De Laquil *et al.*, 1990; Phoebus, 1990; Haeger *et al.*, 1994); this plant could be operational by 1997. In Israel, the Weizmann Institute is developing a pressurized volumetric air receiver that could be used to drive a gas turbine or combined-cycle power plant. Initial receiver testing was encouraging, and scale-up and system integration are now being addressed (Karni, 1994).

Parabolic dishes supplying the external heat for a Stirling heat engine have achieved a maximum net solar-to-electric conversion efficiency of 29.4% and an average daily net efficiency of 22.7% (Washom, 1984). Dish-engine system modules range from 2 to 25 kW, and several companies in the United States and in Europe are developing systems for remote applications as well as grid-connected, distributed-power application (Kubo and Diver, 1993). In the United States, design validation systems are currently undergoing field tests; early commercial systems are expected to be available in 1996. The Australian National University has developed a 50-kW dish concentrator system driving a small, ground-mounted steam power plant (Kaneff, 1994; Stein, 1994).

Table 19-7 lists projections for cost and performance of each solar thermal-electric technology (De Laquil *et al.*, 1993). Parabolic-trough technology has achieved significant cost reductions and current plants have energy costs of 9–13¢/kWh in the hybrid mode. Power towers have significantly lower projected energy costs, 4–6¢/kWh, which can be achieved at production rates of 200 MW/yr without technological breakthroughs. These costs must be proven at commercial scale, however. The thermal-storage feature of the nitrate salt power-tower technology allows electricity to be generated when it is most valuable and, in principle, allows annual solar-only capacity factors of 100%. With high-volume manufacturing, dish-engine systems have a cost potential as low as 5.5¢/kWh in a solar-only mode. They also can be hybridized using most gaseous or liquid fuels to provide continuous operation. In general, all solar thermal-electric technologies require the installation of a few hundred MW_e of systems before they reach their expected cost targets.

Integrated solar, combined-cycle plant configurations, which are being developed for both parabolic-trough and power-tower technology, promise to lower costs significantly for early commercial plants. Capital costs are in the range of $1,000–1,500 per kW, and electricity costs have been calculated to be within 0.5¢/kWh of conventional, natural gas-fired, combined-cycle power plants; the annual solar fraction (the share of demand covered by solar) for these integrated, combined-cycle configurations typically is 20% or lower. Because they reduce both the market entry cost and risk hurdles, such hybrids may lead to earlier introduction of commercial plants. Higher-solar fraction plants can be built as the cost of solar-based electricity approaches the cost of natural gas-based electricity.

Land use, water consumption, compatibility with desert species, and aesthetics are the principal environmental considerations. Because large plants will be best located in desert regions, water consumption is likely to be the most serious environmental issue.

In addition to electricity production, solar thermal systems can provide high-temperature process heat, and central receivers can be used to process advanced fuels and chemicals—for example, hydrogen—which may enhance their long-term potential.

19.2.5.5. Solar Thermal Heating

Solar thermal systems provide heating and hot water for domestic, commercial, or industrial uses (Duffie and Beckman, 1991). Solar cooking and solar cooling cycles are described in Sørensen (1979a). Simple solar thermal systems comprise:

- *Passive solar thermal systems*: Controlled solar collection through building design and the use of proper materials and practices, including smart windows, solar walls, and evaporative cooling (see Chapter 22).
- *Active solar thermal collectors*: Available designs range from unglazed, uninsulated plastic collectors for temperatures just above the ambient temperature to evacuated concentrating collectors for temperatures above 200°C. Flat-plate collectors based on a thermosyphon principle or with forced circulation for 60–80°C dominate at present.
- *Systems for use with solar thermal technologies:* Energy storage in building materials, water (including aquifers), rocks, and soil for individual houses and community systems (Dahlenbäck, 1993), and consideration of fuel backup and proper system management.

Table 19-7: *Levelized energy cost projections.*[1]

	Parabolic Trough			Central Receiver				Dish-Stirling		
	80 MW_e	80 MW_e	200 MW_e	100 MW_e	200 MW_e	200 MW_e	200 MW_e	3 MW_e per year	30 MW_e per year	300 MW_e per year
				first plant	*first plant*	*baseload*	*advanced receiver*	*early remote market*	*early utility market*	*utility market*
Time Frame	Present	1995-2000	2000-2005	1995	2005	2005-2010	2005-2010	1995-2000	2000-2005	2005-2010
Capital Cost Range ($/$kW_e$)	3500–2800	3000–2400	2400–2000	4000–3000	3000–2225	3500–2900	2500–1800	5000–3000	3500–2000	2000–1250
Collector System Typical Cost ($/$m^2$)	250	200	150	175–120	120–75	75	75	500–300	300–200	200–150
Annual Solar-to-Electric Range[2] (%)	13–17	13–17	13–17	8–15	10–16	10–16	12–18	16–24	18–26	20–28
Enhanced Load Matching Method	25% natural gas	25% natural gas	25% natural gas	thermal storage	thermal storage	thermal storage	thermal storage	solar only	solar only	solar only
Solar Capacity Factor Range (%)	22–25	18–26	22–27	25–40	30–40	55–63	32–43	16–22	20–26	22–28
Annual O&M Cost Range (¢/kWh)	2.5–1.8	2.4–1.6	2.0–1.3	1.9–1.3	1.2–0.8	0.8–0.5	1.2–0.8	5.0–2.5	3.0–2.0	2.5–1.5
Solar LEC Range (¢/kWh)	16.7–11.8	17.2–9.8	11.7–7.9	16.1–8.0	10.1–5.8	6.5–4.6	8.2–4.5	32.8–14.6	18.6–8.8	10.6–5.5
Hybrid LEC Range (¢/kWh)	13.0–9.3	13.5–7.9	9.3–6.5	–	–	–	–	–	–	–

[1] From De Laquil *et al.* (1993), which compiled these data from several sources. The levelized energy cost (LEC) calculations are based on a 6% real discount rate; the fixed charge rate for capital is 7.8%.
[2] Typical southwest U.S. site.

Domestic hot water currently is the most important application. For one household, the typical collector area is 3–6 m^2, with a 300-liter water-storage tank. Because the available roof space generally is not a limiting factor, system efficiency is the result of a cost/benefit optimization for the entire system. A typical annual average value is 30–40%. The solar fraction is limited by the seasonal variation in irradiation levels. In equatorial areas it can be 100%, whereas it does not exceed 50% in high-latitude areas without significant storage.

Investment costs for an installed domestic solar hot-water system for one household range from $600 for a 2-$m^2$ thermosyphon system in the Middle East to $6,000 for a 4-$m^2$ pumped system in northwest Europe. Annual system output varies from 300 to 600 kWh_{th}/m^2. At a real rate of return of 6% (10%), heat costs range from $8 ($14) to $80 ($140) per GJ_{th}. The production and use of domestic solar hot-water systems does not have significant environmental effects. The energy payback time is about one year.

Large numbers of solar thermal collectors are installed, notably in Europe, Asia, and North America. An estimated 20–30 million m^2 deliver 40–60 PJ/yr (Nitsch, 1992). The technical potential is much larger. The installed capacity varies greatly, even between countries with similar climate and geography. Thus, attention to implementation issues may help—for example, the adjustment of building codes to accept solar collectors. Several countries offer subsidies in view of the environmental advantages of solar thermal systems (Gregory *et al.*, 1993). Demonstration systems for space heating with up to 7,500 m^2 of collectors have been realized together with seasonal heat storage (Dalenbäck, 1993).

Future prospects include an expected lowering of cost. The main stock of collectors have absorbers of steel with selective coating, glass cover, and aluminum frame and backplate, plus insulating material such as mineral wool. Cheaper constructions use plastic absorbers (polypropylene ribbons), polycarbonate covers, and fiber backplates. Another development is high-efficiency evacuated tubes providing two to three times higher annual collection per unit area. Costs are expected to decline due to higher volume, more efficient manufacturing processes, and reduced installation costs. An important topic for further R&D is the optimization of heat distribution systems with contributions from different heat sources (cogeneration, heat pumps, solar energy, and storage).

19.2.5.6. Geothermal and Ocean Energy

The two remaining renewable sources of energy are geothermal energy and ocean energy.

19.2.5.6.1. Geothermal energy

Geothermal energy resources are made up of both nonrenewable and renewable parts: The nonrenewable is stored from the time of the Earth's formation, and the renewable is due to the isotopic decay of radioactive elements in the Earth's interior.

Electricity is generated from geothermal energy in 21 countries. Commercial production on the scale of hundreds of MW has been undertaken for more than three decades, both for electricity generation and direct heat utilization. The installed capacity is nearly 6,300 MW_e. Four developing countries produce more than 10% of their total electricity from geothermal (El Salvador, 18%; Kenya, 11%; Nicaragua, 28%; the Philippines, 21%). The cost of electric generation is estimated to be around 4¢/kWh_e. Direct use of geothermal water (space heating, horticulture, fish farming, industry, and/or bathing) occurs in about 40 countries; 14 countries have an installed capacity of more than 100 MW_{th}. The overall installed capacity for direct use is about 11,400 MW_{th}, and the annual energy production is about 36 TWh_{th} (Fridleifsson and Frecston, 1994). The production cost for direct use is highly variable but commonly under 2¢/kWh_{th}. In addition, low-temperature heat stored in soil, water, and air can be upgraded by heat pumps and used for space heating and hot water.

Various emissions are associated with geothermal energy, including CO_2, hydrogen sulfide, and mercury. Advanced technologies are almost closed-loop and have very low emissions (WEC, 1994).

In recent decades, installed geothermal energy capacity has grown 10% a year for direct uses and from 8–17% for power generation. The growth is expected to continue but at slightly lower rates (Fridleifsson and Frecston, 1994).

Geothermal energy reserves that might be exploited in the next 2 decades are estimated to be 500 EJ, mostly relatively low-grade heat. Most geothermal systems are hydrothermal, but geopressured, magma, and hot dry rock systems are in the demonstration stage (Palmerini, 1993). Although it can make substantial local contributions in some areas, geothermal energy is unlikely to contribute more than 2% of total global energy requirements, even if hot dry rock geothermal technology is developed at an accelerated rate (Palmerini, 1993).

19.2.5.6.2. Ocean energy

Energy is stored in the tides, waves, and thermal and salinity gradients of the world's oceans. Although the total energy flux of each of these renewable resources is large, only a small fraction of their potential is likely to be exploited in the next 100 years. Ocean energy is spread over a wide area and would require large and expensive plants for collection, and much of the energy is available only in areas far from centers of consumption.

Tidal energy, which uses barrages at sites having a high tidal range, has considerable technical potential (Cavanagh *et al.*, 1993). Many sites with the most potential, however, also have been considered to be of great environmental significance—for example, some are bird habitats (Rodier, 1992; WEC, 1994).

The technology to use wave energy is in its infancy. Near-shore devices are likely to be developed first, but their applicability and potential are limited. Most wave energy is offshore. Because of technological problems and high costs, more powerful large-wave offshore energy plants are unlikely to be deployed for a few decades. Ocean thermal energy conversion (OTEC), which is currently in the prototype stage, is very costly and largely restricted to tropical locations (Cavanagh *et al.*, 1993).

19.2.6. Energy Systems Issues

System design and management issues are important for many new energy technologies that can significantly reduce GHG emissions. The most important are the management of intermittent power generation; the choice of a mix of renewable, fossil, and nuclear electric-generating technologies; the use of electricity transmission technologies; and the use of hydrogen or hydrogen-rich energy carriers.

19.2.6.1. Managing Intermittent Power Generation

Intermittent power-generating supplies can be managed by new load-management techniques; an appropriate mix of dispatchable generating capacity for backing up intermittent sources; interconnecting grid systems; and mechanical, electrochemical, thermal, or other forms of storage (Sørensen, 1984).

If energy users knew the current and expected availability of intermittent energy sources and associated energy prices, they might change their patterns of energy demand. Price signals and related information could be transmitted to users through power lines or other information channels so they could plan their use of electricity accordingly. This kind of demand management can be automated by using intelligent appliances and equipment responding to signals received through the grid according to selections programmed by users.

Energy suppliers might apply transmission planning, including the transfer of electricity over large distances to cope with some of the daily variations of wind and solar energy. The management of interconnected grids could be optimized to allow maximum inputs from variable power sources, as is currently done to make the best use of baseload power plants.

Hydropower plants allow prompt regulation and can back up intermittent energy generators, as can some types of thermal power plants. The ideal thermal complements to intermittent renewable energy plants on grid systems are plants characterized by low unit capital cost (so they can be operated cost-effectively at low capacity factor) and fast response times (so they can adjust to rapid changes in intermittent supply output). Gas turbines and combined cycles satisfy these criteria, but supercritical fossil and nuclear steam-electric plants do not. Thus, nuclear and intermittent renewable power sources are competitive rather than complementary strategies at high grid-penetration levels.

The levels of electricity generation from intermittent sources (for example, from wind turbines or photovoltaic arrays) that can be accepted in a supply system depend strongly on the nature of the system and the measures that manage time variations in load. Typical percentages of wind or solar shares in systems without storage are about 20% for grid systems and up to 50% for large systems with reservoir-based hydro or time-zone variations of loads (Sørensen, 1981, 1987; Grubb and Meyer, 1993; Kelly and Weinberg, 1993).

Short-term storage will ease regulation and possibly improve power quality, but only intermediate and long-term storage will allow large systems to have shares of variable renewable-energy sources that significantly exceed 50% (Jensen and Sørensen, 1984). One important use of storage is to convert a remotely located intermittent electric supply into baseload power to obtain a high level of utilization of electric transmission capacity and thereby reduce the specific cost of long-distance transmission to electric demand centers. The use of compressed-air energy storage (Cavallo, 1995) and even seasonal compressed-air energy storage (Cavallo and Keck, 1995) in conjunction with large wind energy farms has been shown to be an economically promising way to exploit remotely located wind resources.

Round-trip efficiencies for storage range from 65–70% for lead-acid batteries, pumped hydro, and compressed air to 95% for superconducting storage (Jensen and Sørensen, 1984). Some storage technologies pose environmental problems that need to be addressed—including metal contamination from batteries and riverbed erosion problems with water surges from hydropower.

19.2.6.2. Electric Power System Characteristics and Costs

In the decades immediately ahead, many renewable electric systems will be introduced in stand-alone configurations remote from electric utility grids and used in connection with small-scale electrical storage and fossil energy backup systems. In the longer term, however, most renewable electric systems will be connected to electric utility grids. Managing grid-connected renewable electric technologies poses new challenges for utilities. Unlike the thermal power plants that utilities are accustomed to managing, intermittent renewable power plants are not dispatchable. Moreover, many of these technologies produce electricity in plants that are much smaller than those used in today's power systems. Scales range from 1 kW_e for some photovoltaic and fuel-cell systems to 25–300 MW_e for biomass systems because the costs per unit of capacity for such technologies will be relatively insensitive to scale. For these systems, the economies of mass-produced standardized units are more important than the economies achievable with large unit sizes. Many of the smaller-scale technologies will be sited not in central-station plants but at or near customers' premises. Some photovoltaic and fuel-cell systems can be operated unattended and installed even at individual houses. Electricity produced from such "distributed power systems" is worth more to the utility than central-station power whenever the electrical output is highly correlated with the utility peak demand, largely because such siting makes it possible to defer transmission/distribution investments (Shugar, 1990; Hoff *et al.*, 1995).

New analytical tools are being developed to integrate such technologies into electric utility grids and to value and manage power systems that involve combinations of intermittent and dispatchable power sources. One such tool, the SUTIL simulation model (Kelly and Weinberg, 1993), calculates the average cost of electric generation as the result of an hour-by-hour simulation that takes into account demand, the variable output of intermittent renewable equipment, the load-leveling capabilities of hydroelectric facilities, and the dispatching characteristics of alternative thermal-electric plants. Results of a SUTIL simulation of 10 alternative portfolios for a hypothetical utility—based on assumed values of the demand load profile, insolation, fuel costs, and capital costs—are shown in Figure 19-4.[5] Note that the assumed costs for conventional fuels and technologies, particularly for natural gas and for

[5] Since this SUTIL modeling exercise was carried out, present and projected fuel prices and technology costs have changed significantly. For example, compared to the assumed overnight construction costs of $525/kW and $625/kW for conventional and advanced natural gas combined cycle plants, respectively, combined cycle plants are being installed at present in the U.S. at costs as low as $400/kW. As a sensitivity analysis, the SUTIL model was run again with assumed coal and natural gas prices of $1.3/GJ and $3.2/GJ, respectively [equal to national average prices projected for electric utilities in the U.S. in the year 2010 by the U.S. Department of Energy (EIA, 1995)], with an assumed biomass price of $1.5/GJ [an estimated long-run marginal cost for producing 5 EJ/yr of plantation biomass in the U.S. using biomass production technology projected for the year 2020 (Graham *et al.*, 1995)], together with the original SUTIL technology cost assumptions. Although overall generation cost levels are lower with these new fuel price assumptions, the relative costs of alternative investment portfolios differ only modestly from those shown in Figure 19-4. For example, the average generation cost for the updated version of case 9 is 8% higher than for case 4 (3.37¢/kWh vs. 3.12¢/kWh) (Williams, 1995a). Note that differences in marginal costs associated with the introduction of intermittent renewables are significantly greater than is indicated by these average portfolio cost differences.

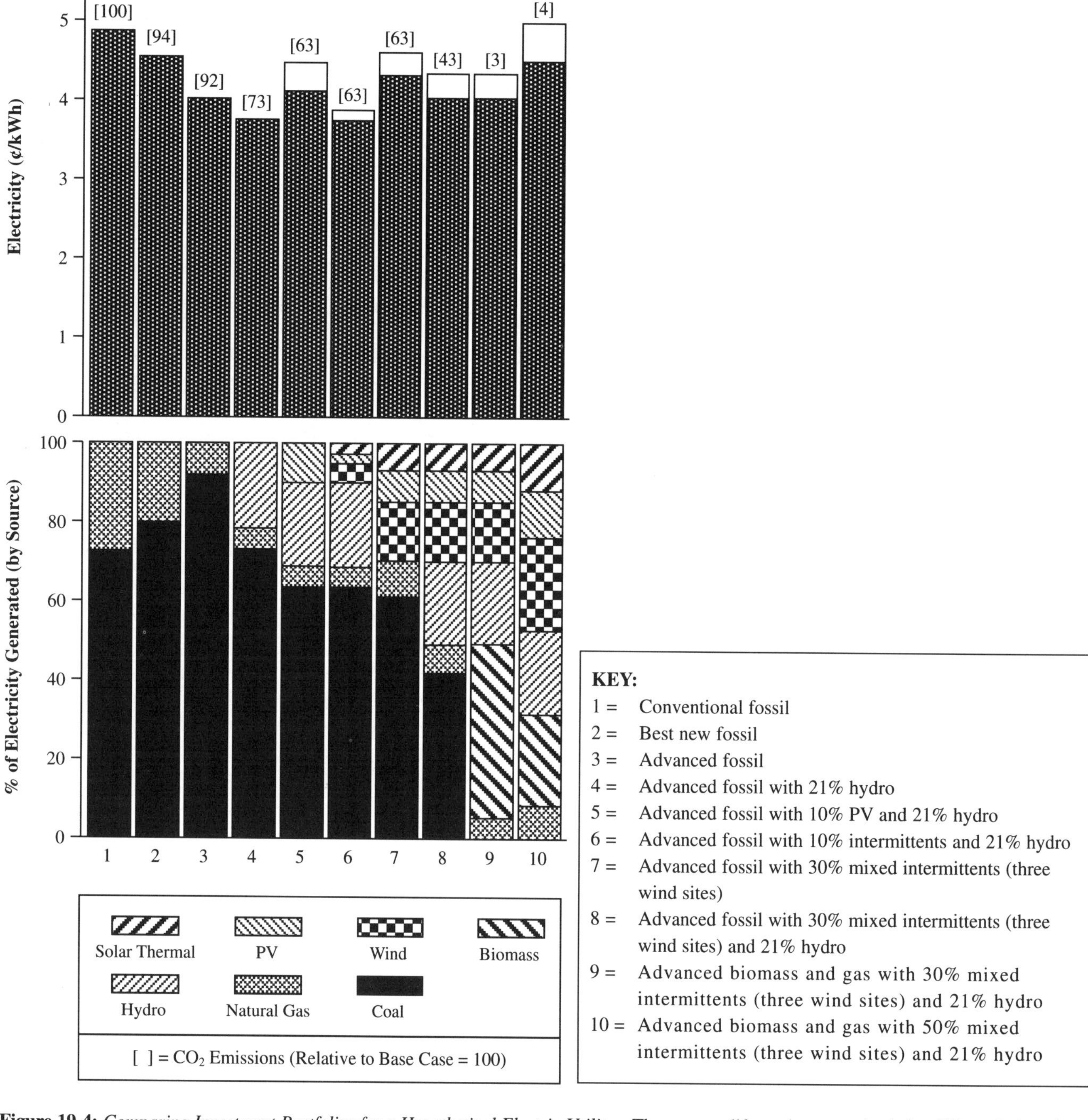

Figure 19-4: *Comparing Investment Portfolios for a Hypothetical Electric Utility*—The average life-cycle cost and relative CO_2 emissions for meeting the annual electricity needs (top) and the fraction of electricity generated by each energy source (bottom) are displayed for alternative configurations of a hypothetical utility. The life-cycle costs were calculated for a 6% discount rate using the SUTIL model (Kelly and Weinberg, 1993). The calculations involved a simulation of the utility that considered the variability of electricity demand and the output of intermittent renewable equipment, the load-leveling capabilities of hydroelectric facilities, and the dispatching of natural gas, coal, and biomass thermal electric plants. The electricity demand profile (that for an actual large utility in northern California) was specified on an hour-by-hour basis throughout the year. The assumed insolation values are for northern California. For the given demand profile, alternative electricity supply portfolios having the same degree of system reliability were constructed. Each portfolio involves specified levels of penetration by intermittent renewable electric and hydroelectric sources, and the model determines the least costly mix of thermal-electric equipment and fuels that would meet the remaining load, assuming alternative levels of technology for the thermal generating equipment and specified 30-year levelized life-cycle fuel prices. For all advanced fossil and renewable energy technologies highlighted in these constructions, it was assumed that R&D is successful in meeting performance and cost goals for the period near 2010. Thirty-year levelized coal, natural gas, and biomass prices were assumed to be \$2.0/GJ, \$4.4/GJ, and \$2.4/GJ, respectively. Assumptions about capital and operations and maintenance costs, and about performance characteristics of alternative technologies, are presented elsewhere (Johansson *et al.*, 1993a). For the six cases displayed on the right, the photovoltaic (pv) systems are sited in distributed configurations; the clear segments at the tops of the bars represent the value of distributed pv power to the utility; the net cost is given by the level at the tops of the shaded bars—the gross cost less the distributed pv benefit.

natural gas-fired combined cycles, are significantly higher than currently prevailing values and projections to 2010 for the U.S. For this set of alternative utility investment portfolios:

- All configurations involving advanced technology (cases 3–10) are less costly than the base case (case 1), which involves conventional fossil fuel technologies.
- There is little variation in cost among the advanced technology options, although there are no less costly options than the advanced fossil fuel options (cases 3 and 4).
- The fraction of electrical energy provided by intermittents can rise up to about 30%, without the use of new electrical storage technology, before costs start rising significantly.
- At high levels of penetration of intermittent renewables, baseload thermal power becomes less important and load-following and peaking power become more important.
- The advanced fossil fuel options offer relatively modest reductions in CO_2 emissions.
- The option offering the greatest reduction in CO_2 emissions [case 9, involving biomass for baseload power and a 30% contribution from intermittent renewables and for which emissions are only 4% of those for the least costly case (case 4)] has an average generation cost that is only 7% higher than for the least costly case.

Although costs for technologies that are not yet commercially available cannot be known precisely, it is plausible—on the basis of what is presently known about the prospective costs of advanced renewable and fossil fuel electric-generating technologies—that in the early decades of the next century, utility planners will be able to assemble alternative electric power systems (involving substantial contributions from renewable energy sources) that generate very low CO_2 emissions. These judgments are based on the assumption that various promising renewable-energy technologies will be targeted for R&D and commercialization programs. At the same time, of course, advances will continue to be made in fossil fuel technologies. In many cases, it will be difficult to provide electricity at lower direct cost than with fossil fuels (for example, because both coal and biomass will be able to exploit the same basic advances in gas turbine and fuel-cell conversion technologies).

19.2.6.3. Electricity Transmission Technology

For transporting electric power over more than about 700 km, DC transmission at high voltage (500 to 1,000 kV) via either overhead lines or underwater cables is less costly than AC transmission. Ohmic losses amount to about 4% of the energy transmitted per 1,000 km, while losses associated with DC/AC conversion amount to about 0.6–0.8% per station. Typically, DC transmission increases the cost of electricity about 7% for a 1,000-km line. The low losses and relatively low costs of long-distance DC transmission make it feasible to transport solar power across multiple time zones and to exploit good hydropower (Moreira and Poole, 1993), wind (Cavallo, 1995), and solar power resources that are remote from demand centers. High transmission capacity utilization can be realized with intermittent renewable resources when used in conjunction with compressed air or other energy-storage schemes (Cavallo, 1995).

19.2.6.4. A Long-Term Electricity/Hydrogen Energy System

Since it was first introduced, electricity has accounted for a growing share of the energy carriers used in the energy economy—a trend that is expected to continue. For the longer term, hydrogen is another high-quality energy carrier that society might adopt for large-scale applications. Both electricity and hydrogen offer good prospects for simultaneously dealing with the challenges facing the energy system in the 21st century—local, regional, and global environmental challenges, as well as security of energy supply (Marchetti, 1989; Rogner and Britton, 1991; Leydon and Glocker, 1992). Both are clean, versatile, and easy to use and can be derived from a wide range of primary energy sources. Because of their very different characteristics, hydrogen and electricity would have complementary roles in the energy economy. Hydrogen can be stored in any quantity, but electricity cannot (though this could change if high-temperature superconductivity could be successfully developed). Electricity can transmit energy without moving material; hydrogen cannot. Hydrogen can be a chemical or material feedstock; electricity cannot. Electricity can process, transmit, and store information; hydrogen cannot.

Hydrogen can be produced from any fossil fuel or from biomass via thermochemical conversion and from any electricity source via electrolysis. Conversion to hydrogen and sequestration of separated CO_2 makes it possible to use fossil fuels with modest life-cycle CO_2 emissions (see Sections 19.2.3.2 and 19.2.3.3). Emissions from the production of hydrogen from biomass grown on a sustainable basis would be especially low and negative if the separated CO_2 were sequestered. For the foreseeable future, thermochemically derived hydrogen will be much less costly than electrolytic hydrogen, even taking into account the costs of sequestering the separated CO_2 (see Figure 19-5).

As is the case for any synthetic fuel, hydrogen generally will cost more to produce than conventional hydrocarbon fuels. Natural gas-derived hydrogen (the least-costly option for at least the next couple of decades) delivered to consumers is likely to be much more costly than gasoline, on a dollar per GJ basis—50% higher for energy prices expected for the United States in 2010 (see Figure 19-5). However, this cost penalty does not imply that hydrogen would not be competitive. The choice of energy carrier, its price, and its source are of little concern to the consumer. The consumer cares about the quality and cost of the energy services provided (Scott, 1993). Just as electricity is competitive as an energy carrier—even though it is much more expensive than coal, oil, or natural gas—high-cost hydrogen could be competitive if the market were to place a high value on hydrogen. It would be difficult for hydrogen to

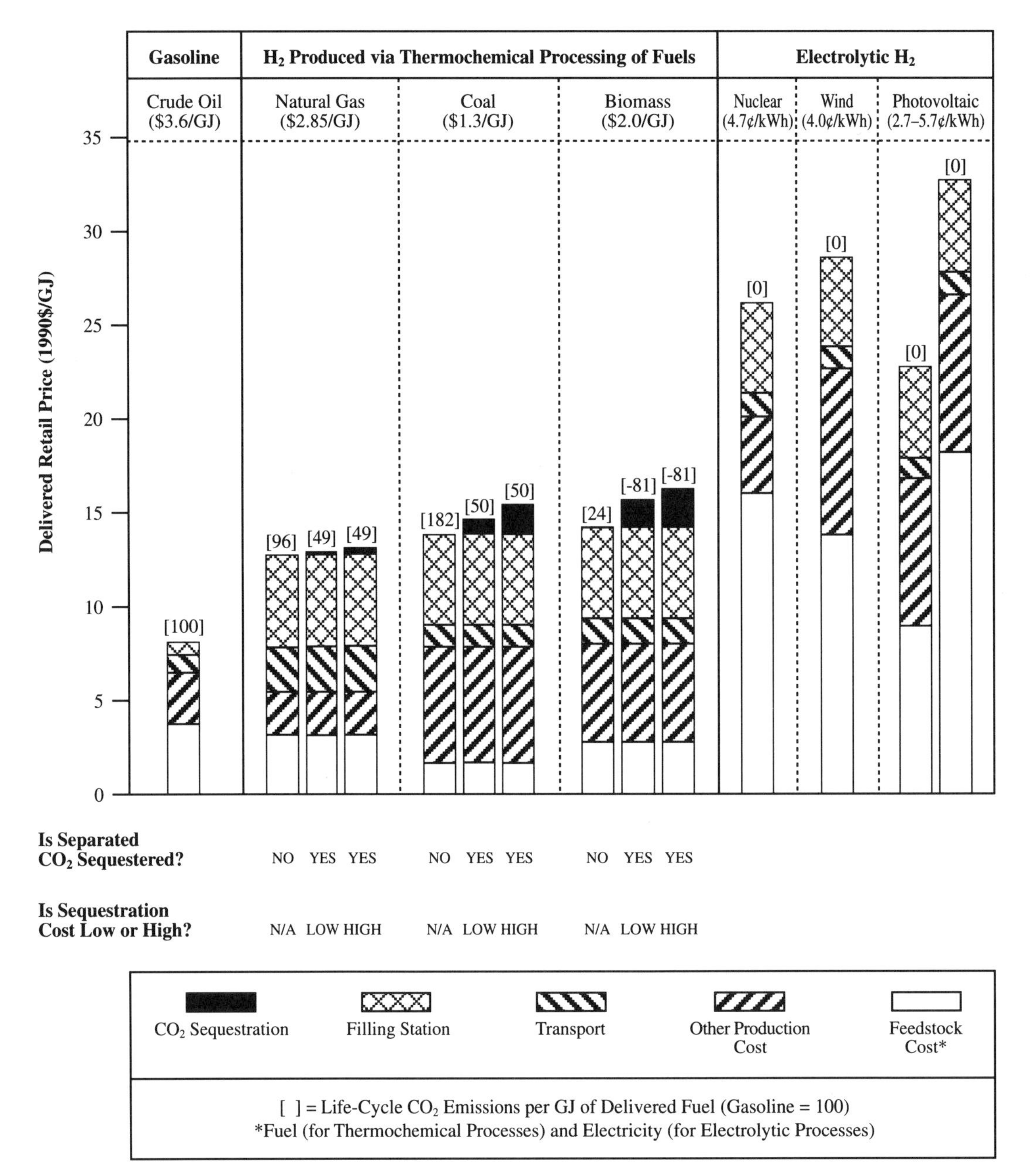

Figure 19-5: *Estimated Life-Cycle Costs to Consumers and CO_2 Emissions per GJ of Hydrogen (H_2) from Alternative Sources for Transport Applications, with a Comparison to Gasoline Derived from Crude Oil*—This figure and Figures 19-6 and 19-7 present alternative measures of cost and life-cycle CO_2 emissions characteristics for systems that provide H_2 derived from natural gas (via steam reforming), coal and biomass (via thermochemical gasification), and electrolytic sources, with comparisons to gasoline (Williams, 1996). Costs were calculated assuming a 10% discount rate, and fuel taxes were excluded. The assumed crude oil, natural gas, and coal prices are, respectively, the world oil price, the U.S. average wellhead price, and the average U.S. electric utility price projected for 2010 by U.S. DOE (EIA, 1995). The consumer gasoline price is for reformulated gasoline, which is estimated to be 16% higher than the price of ordinary gasoline derived from crude at the assumed price of $22/barrel. Plantation biomass prices that are comparable to or lower than the assumed price could be realized in large-scale applications in Brazil today and in the U.S. by 2020 if yield and cost goals for 2020 are realized as a result of successful R&D (see Section 19.2.5.2.1). The assumed electricity costs are for a 600 MW advanced passively safe nuclear plant having an installed cost of $1680/kW (EPRI, 1993); for post-2000 wind technology with an installed cost of $780/kW (Ogden and Nitsch, 1993); and for photovoltaic (pv) systems in areas with high insolation and installed costs of $550–1125 per kW—targets that plausibly could be met using advanced thin-film pv technologies in the long term (Ogden and Nitsch, 1993). Thermochemical production of H_2 generates a byproduct stream of pure CO_2, which can either be released to the atmosphere or compressed and piped to a site where it might be sequestered. The first bar for natural gas, coal, and biomass is the cost when this CO_2 is vented; the second and third bars are for when the separated CO_2 is sequestered, assuming low and high estimates of the sequestering cost (Hendriks, 1994). In the natural gas cases, it is assumed that the H_2 plant is located near a natural gas field and that the separated CO_2 is sequestered in depleted gas wells. For the coal and biomass cases, it is assumed that the separated CO_2 is sequestered in saline aquifers located 250 km from the plant. The numbers at the tops of the bars are the life-cycle CO_2 emissions in kg C per GJ of delivered fuel, relative to life-cycle emissions for gasoline.

compete in markets where energy-conversion equipment designed for hydrocarbon fuels is used. The situation would change markedly if conversion equipment designed to exploit the unique characteristics of hydrogen were used instead.

A class of technologies that could make a hydrogen economy feasible is represented by the fuel cell, which converts fuel directly into electricity without first burning it. The hydrogen fuel cell offers the potential for major contributions in transportation (Williams, 1993; Mark *et al.*, 1994) and in distributed combined heat and power applications (A.D. Little, 1995; Dunnison and Wilson, 1994). The hydrogen fuel cell offers high thermodynamic performance, zero local air-pollution emissions, low maintenance requirements, and quiet operation. In mass production, automobiles powered by the proton exchange membrane (PEM) fuel cell would have much lower costs and much longer ranges between refuelings than battery-powered electric cars (Kircher *et al.*, 1994; Ogden *et al.*, 1994). Powered by either compressed hydrogen or methanol (a hydrogen carrier that would be "reformed" onboard the vehicle into a suitable hydrogen-rich gaseous fuel), PEM fuel-cell cars would require much less fuel and emit much less local pollution than hybrid internal-combustion engine/battery-powered cars, while offering comparable or greater range and comparable or lower life-cycle costs (Kircher *et al.*, 1994; Biedermann *et al.*, 1994).

Because of the much higher fuel economy, the cost of fuel per km of driving a fuel-cell car powered by thermochemically derived hydrogen is likely to be less than 65% of the cost for a gasoline-fired internal combustion engine car of comparable performance (see Figure 19-6)[6], and the total life-cycle cost of owning and operating a fuel-cell car is likely to be slightly less (see Figure 19-7), even though the cost of this hydrogen to the consumer, measured in $/GJ, would be up to about 75% more than for gasoline (see Figure 19-5).[7]

19.3. Low CO_2-Emitting Energy Supply Systems for the World

Each of the options discussed in Section 19.2 has the potential to reduce GHG emissions. To understand better their combined potential contributions to future energy supplies and emissions reductions, assessments are needed at the level of the global energy system. Such assessments should consider internal functional aspects of the energy system (such as the intermittency of solar and wind energy resources) and the linkages between the energy system and other areas for societal concern (such as land-use competition and security issues).

In order to start an assessment at the energy systems level, alternative versions of a Low-Emissions Supply System (LESS) were constructed. These LESS constructions illustrate the potential for reducing emissions by using energy more efficiently and by using various combinations of low CO_2-emitting energy supply technologies—including shifts to low-carbon fossil fuels, shifts to renewable or nuclear energy sources, and decarbonization of fuels, in alternative combinations. Energy supply systems were constructed for this exercise both from "the bottom up" (Section 19.3.1) and "the top down" (Section 19.3.2).

Emphasis in the LESS constructions is on the long term (2025–2100), the prospects for achieving deep reductions in CO_2 emissions from the energy sector with alternative supply mixes, and prospective costs. The need for a long-term focus is stressed in the Working Group I evaluations. Accordingly, emphasis was given to new or improved energy supply technologies that offer the potential for achieving deep reductions in emissions. By the year 2100, the global commercial energy system will have been replaced two to three times—providing many opportunities to change system performance through the use of various new technologies at the time of investment, both for capacity expansion and for replacement.

It is not realistic to probe the deep future taking into account only commercial technologies. The prospective performance and relative cost characteristics for the technologies selected for emphasis in the LESS constructions can be described with a reasonable degree of confidence with present knowledge; all are commercially available, commercially ready, or have good prospects for becoming commercial products within the next 1 or 2 decades, if adequate incentives are provided for the needed R&D and for launching the new industries involved.

To help clarify the options, alternative versions of the LESS were constructed with features that make each option markedly different from the others, and some important features distinguishing the alternatives are highlighted. In the bottom-up constructions, focused attention is given to five LESS variants. Four variants involve a high degree of emphasis on the efficient use of energy. Two of these—which were analyzed in the greatest detail, thus taken to be the reference cases—are a biomass-intensive (BI) variant and a nuclear-intensive (NI) variant. Also, a natural gas-intensive (NGI) variant and a coal-intensive (CI) variant were constructed to explore the extent to which deep reductions in emissions could be achieved using more fossil fuels. A fifth high-demand (HD) variant explores the implication of much greater energy demand growth.

A central finding of the LESS construction exercise is that deep reductions of CO_2 emissions from the energy sector are technically possible within 50 to 100 years, using alternative strategies. Many combinations of the options identified in this assessment could reduce global CO_2 emissions from fossil

[6] As shown in Figure 19-6, the fuel cost per km of thermochemical H_2 would increase only modestly, if the CO_2 separated at the H_2 production plant were isolated from the atmosphere and the extra cost of sequestering this CO_2 were charged to the consumer.

[7] Moreover, because the cost of fuel represents such a small fraction of the total cost of owning and operating a fuel-cell car, the total life-cycle cost for a fuel-cell car operated on electrolytic hydrogen would be less than 7% more than for a gasoline internal combustion engine car, even though the cost of electrolytic hydrogen per km for a PEM fuel-cell car would be up to 50% more costly (see Figure 19-7).

fuels from about 6 Gt C in 1990 to about 4 Gt C per year by 2050, and to about 2 Gt C per year by 2100. Because of the large number of combinations of options, there is flexibility as to how the energy supply system could evolve, and paths to energy system development could be influenced by considerations other than climate change, including political, environmental, and socioeconomic circumstances. However, higher energy demand levels reduce the flexibility for constructing

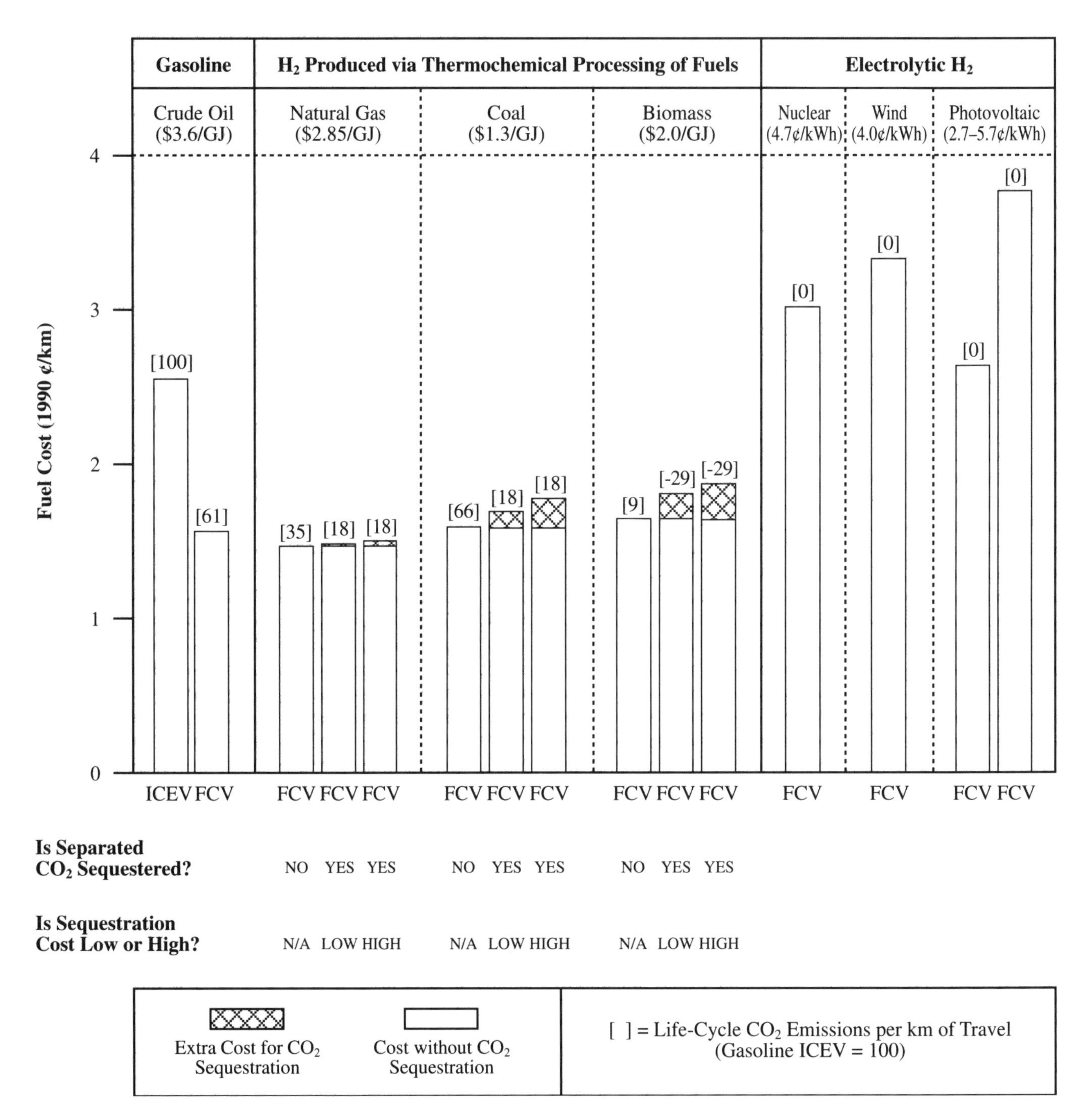

Figure 19-6: *Estimated Life-Cycle Fuel Costs to Consumers and CO_2 Emissions, per km of Driving Fuel Cell Vehicles (FCVs), for H_2 from Alternative Sources for Transport Applications, with a Comparison to Gasoline Derived from Crude Oil and Used in Internal Combustion Engine Vehicles (ICEVs) and FCVs*—The consumer prices and life-cycle CO_2 emissions for H_2 shown in Figure 19-5 are converted here to fuel costs and life-cycle CO_2 emissions per km of driving an FCV, along with a comparison of gasoline costs and lifecycle CO_2 emissions per km of driving, for both ICEV and FCV applications (Williams, 1996). Fuel taxes are not included. The reference gasoline ICEV is a year-2000 version of the Ford Taurus automobile with a fuel economy of 11.0 km/l. The H_2 FCV has performance characteristics that are comparable to those for this ICEV and a gasoline-equivalent fuel economy of 30.4 km/l. Fuel costs are also shown for an FCV operating on gasoline. In this case, the gasoline is converted via partial oxidation onboard the vehicle to a gaseous mixture of H_2 and CO_2, which is a suitable fuel gas for operating the fuel cell; the estimated fuel economy for the gasoline FCV is 18.0 km/l.

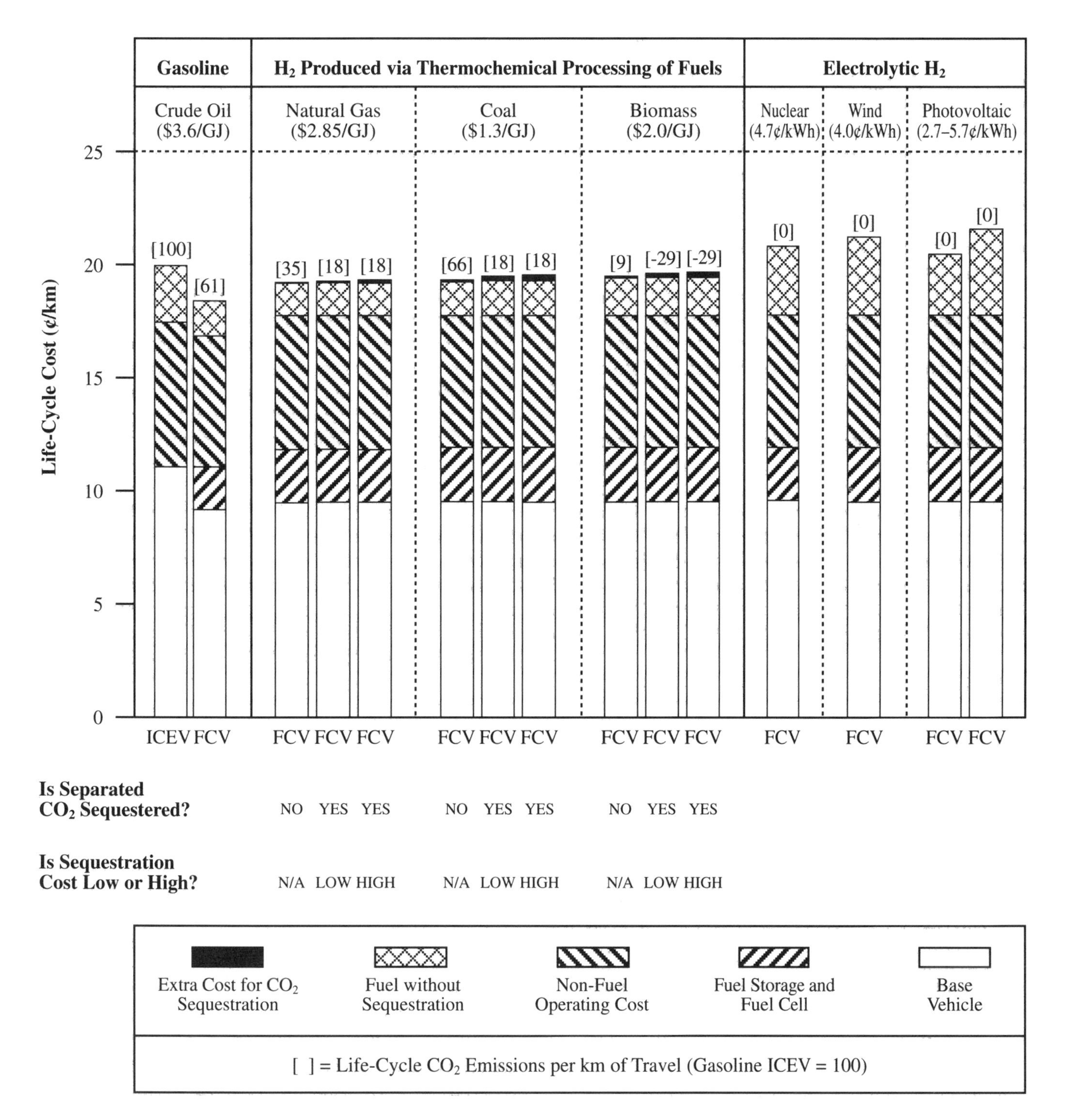

Figure 19-7: *Estimated Life-Cycle Costs to Consumers for Owning and Operating FCVs and CO_2 Emissions, per km of Driving an FCV, for H_2 from Alternative Sources for Transport Applications, with a Comparison to Gasoline Derived from Crude Oil and Used in ICEVs and FCVs*—Vehicle performance and cost characteristics are based on Ogden *et al.* (1994). The estimated fuel-economy characteristics of these vehicles are indicated in Figure 19-6. An operating lifetime of 11 years is assumed for both ICEVs and FCVs; however, FCVs are assumed to be driven 23,000 km per year, compared to 17,800 km per year for ICEVs; this reflects the lower operating costs expected for FCVs operated on thermochemically derived fuels. It is assumed that H_2 is stored onboard vehicles in carbon-fiber-wrapped aluminum tanks at high pressure (550 bar). Because of the bulkiness of gaseous H_2 storage, the H_2 FCV is designed for a range between refuelings of 400 km, compared to 640 km for a gasoline ICEV. The weight of the H_2 FCV is estimated to be 1.3 tons, compared to 1.4 tons for the ICEV. Initial costs are estimated to be $17,800 for an ICEV and $25,100 for an H_2 FCV (in mass production). The initial cost for a gasoline FCV is assumed to be $21,700—the same as the estimated cost for a methanol FCV (Ogden *et al.*, 1994). Retail fuel taxes are included under "other non-fuel operating costs" at the average U.S. rate for gasoline used in ICEVs; to ensure that road tax revenues are the same for all options, it is assumed that retail taxes are 0.75¢ per km for all options (equivalent to 8.2¢ per liter or 31¢ per gallon for gasoline used in ICEVs).

supply-side combinations for deep reductions in emissions and increase overall energy system costs—underscoring the importance of higher energy efficiency.

Costs for energy services in each LESS variant relative to costs for conventional energy depend on relative future energy prices, which are uncertain within a wide range, and on the performance and cost characteristics assumed for alternative technologies. However, within the wide range of future energy prices, one or more of the variants would plausibily be capable of providing the demanded energy services at estimated costs that are approximately the same as estimated future costs for current conventional energy (see, for example, Figures 19-4, 19-6, and 19-7). It is not possible to identify a least-cost future energy system for the longer term, as the relative costs of options depend on resource constraints and technological opportunities that are imperfectly known, and on future actions to be taken by governments and the private sector.

The literature provides strong support for the feasibility of achieving the performance and cost characteristics assumed for energy technologies in the LESS constructions within the next 1 or 2 decades, though it is impossible to be certain until the research and development is complete and the technologies have been tested in the market. Moreover, these performance and cost characteristics cannot be achieved without a strong and sustained investment in R&D. Many of the technologies being developed would need initial support to enter the market, and to reach sufficient volume to lower costs to become competitive.

Market penetration and continued acceptability of different technologies ultimately depends on their relative cost,

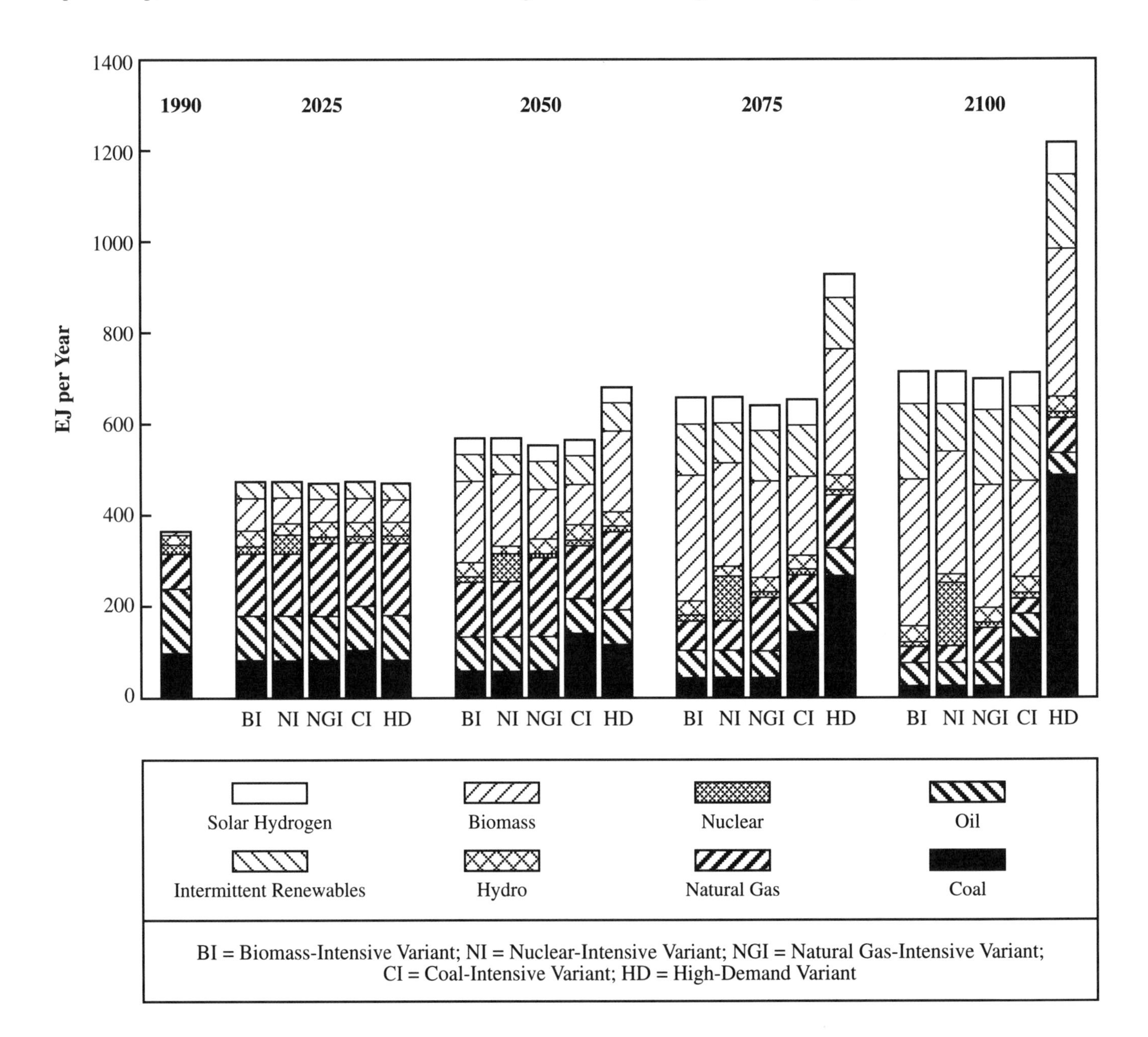

Figure 19-8: Global primary energy use for alternative LESS constructions.

performance (including environmental performance), institutional arrangements, and regulations and policies. Because costs vary by location and application, the wide variety of circumstances creates initial opportunities for new technologies to enter the market. Deeper understanding of the opportunities for emissions reductions would require more detailed analysis of options, taking into account local conditions.

The LESS alternatives are not forecasts; rather, they are self-consistent constructions indicative of what might be accomplished by pursuing particular technological strategies. These alternative paths to the energy future should be regarded as "thought experiments" exploring the possibilities of achieving deep reductions in emissions. Actual strategies for achieving deep reductions might combine elements from alternative LESS constructions. Moreover, there may well be other plausible technological paths that could lead to comparable reductions in emissions. More work is required to provide a comprehensive understanding of the prospects for and implications of alternative global energy supply systems that would lead to deep reductions in CO_2 emissions.

19.3.1. A Bottom-Up Construction for the LESS Reference Cases

The bottom-up LESS reference analysis involved constructing alternative sets of energy supplies for each of 11 regions of the world matched to energy demand levels adopted from a previous IPCC study. The study used a set of demand projections for electricity and for solid, liquid, and gaseous fuels used directly, by world region, for the years 2025, 2050, 2075, and 2100. These projections were developed by the Response Strategies Working Group (RSWG) of IPCC for its 1990 Assessment Report (RSWG, 1990). The RSWG prepared high and low economic growth variants of alternative energy scenarios, including Accelerated Policies (AP) scenarios characterized on the demand side by high rates of improvement in energy efficiency. The energy demand projections for the high

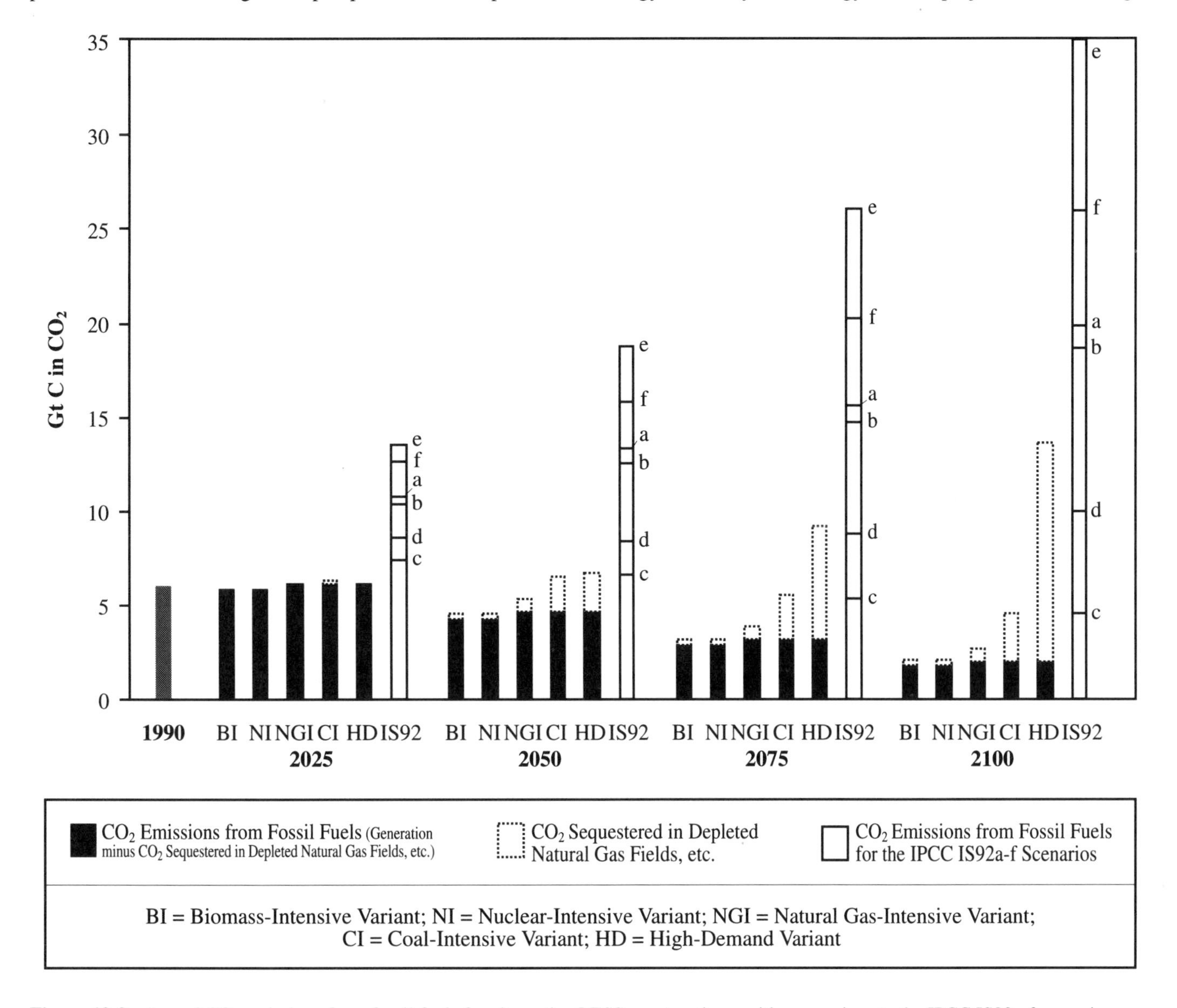

Figure 19-9: Annual CO_2 emissions from fossil fuels for alternative LESS constructions, with comparison to the IPCC IS92a-f scenarios.

economic growth variant of the AP scenarios were adopted for the reference cases of the LESS constructions. These projections are consistent with analysis in the energy demand sections of this report (Chapters 20–22). The demand profiles of the AP scenario were assumed as exogenous inputs to the supply analysis and were not critically reviewed here.

An AP scenario was chosen as the point of departure for the LESS constructions because typically there are large opportunities to reduce CO_2 emissions more cost-effectively via investments in more energy-efficient equipment than via investments in energy supply. The high economic growth variant was chosen to make clear that the potential for emissions reduction illustrated by the LESS constructions is the result of technological choice rather than reduced economic output. The LESS reference constructions are extensions and variants of a methodology developed in a previous study exploring the long-term prospects for renewable energy in the context of the same set of energy demand projections (Johansson *et al.*, 1993a).

Given the AP energy demands for each region, energy supplies were constructed for the LESS variants that are consistent with each region's endowments of conventional and renewable energy sources and general expectations of relative prices.

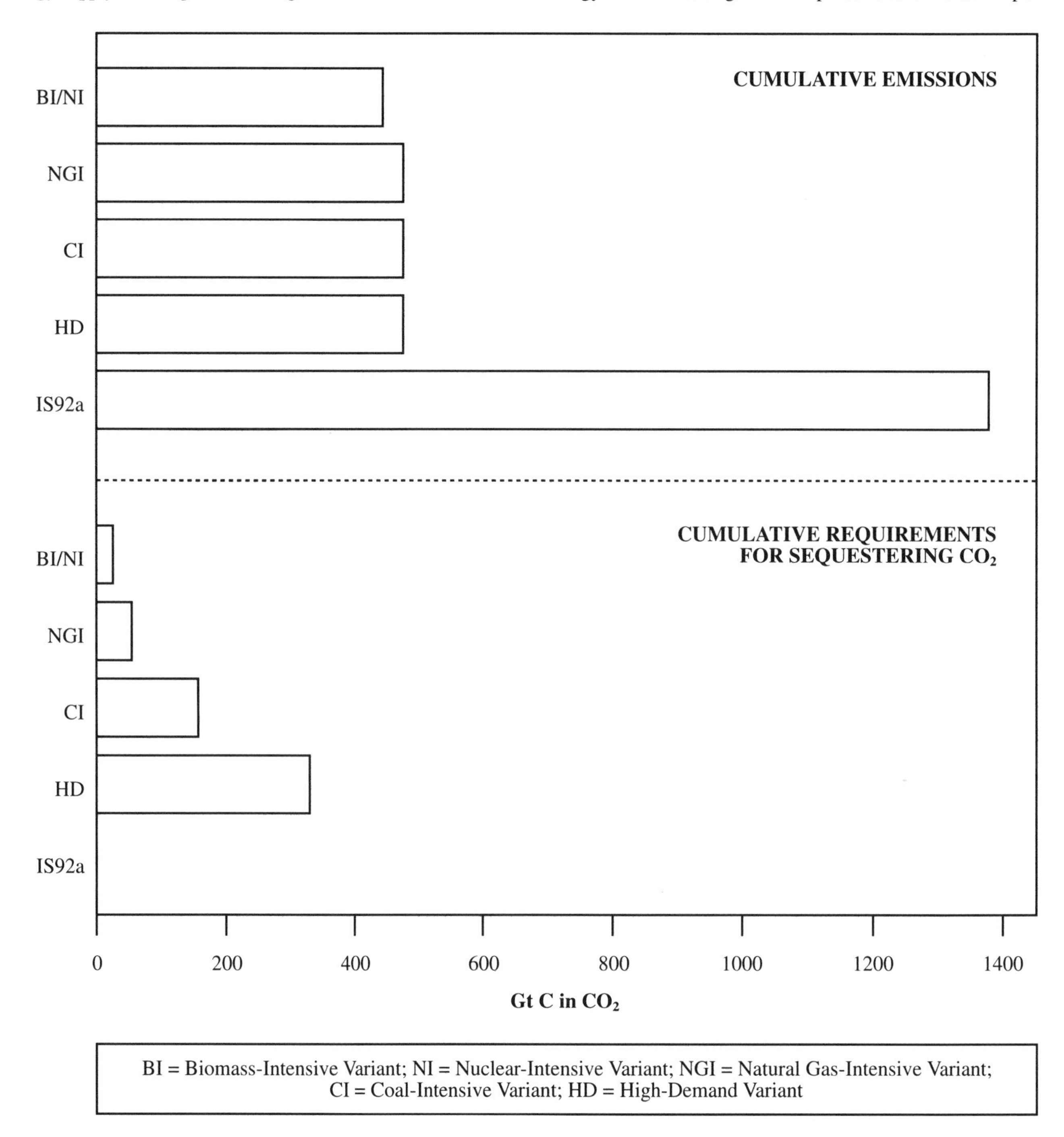

Figure 19-10: Cumulative CO_2 emissions from fossil fuel burning and CO_2 sequestration requirements, 1990–2100, for alternative LESS constructions, with a comparison to the IPCC IS92a scenario (IPCC, 1992).

Highlights of the LESS reference cases are presented in Box 19-2; salient global characteristics of these and other LESS variants are presented in Figures 19-8 to 19-12. Primary energy requirements for one of the reference cases, disaggregated into industrialized and developing regions, are presented in Figure 19-13. Details of the assumed demand projections and supply constructions are in Williams (1995a).

19.3.1.1. *Fossil Fuels in the LESS Reference Cases*

The roles for fossil fuels in the LESS constructions are estimated on the basis of private costs, without consideration of carbon taxes. Particular attention is given to resource constraints on oil and natural gas and local environmental restrictions on the use of coal.

For oil and natural gas, the key assumption is that they will be developed consistent with widely accepted estimates of ultimately recoverable quantities of conventional oil and gas. For undiscovered resources, the mean estimates of the U.S. Geological Survey (Masters *et al.*, 1994) are assumed for regions outside the United States, and U.S. Department of Energy estimates (Energy Information Administration, 1990) are assumed for the United States. Unconventional oil and gas resources are not taken into account. It is assumed that all reserves and estimated recoverable, undiscovered conventional oil and gas resources (some 11,300 EJ of oil and 12,500 EJ of natural gas as of 1993—equivalent to 80 and 160 years of supply at 1990 production rates) will eventually be used for energy.

From a GHG perspective, it matters little how oil and gas consumption evolves over time within this general framework. From an energy policy perspective, it makes sense to shift the mix more to gas in light of comparable amounts of estimated ultimately recoverable conventional resources and a present oil consumption rate that is nearly twice that for gas. To increase the transparency of the analysis, two simplifying assumptions

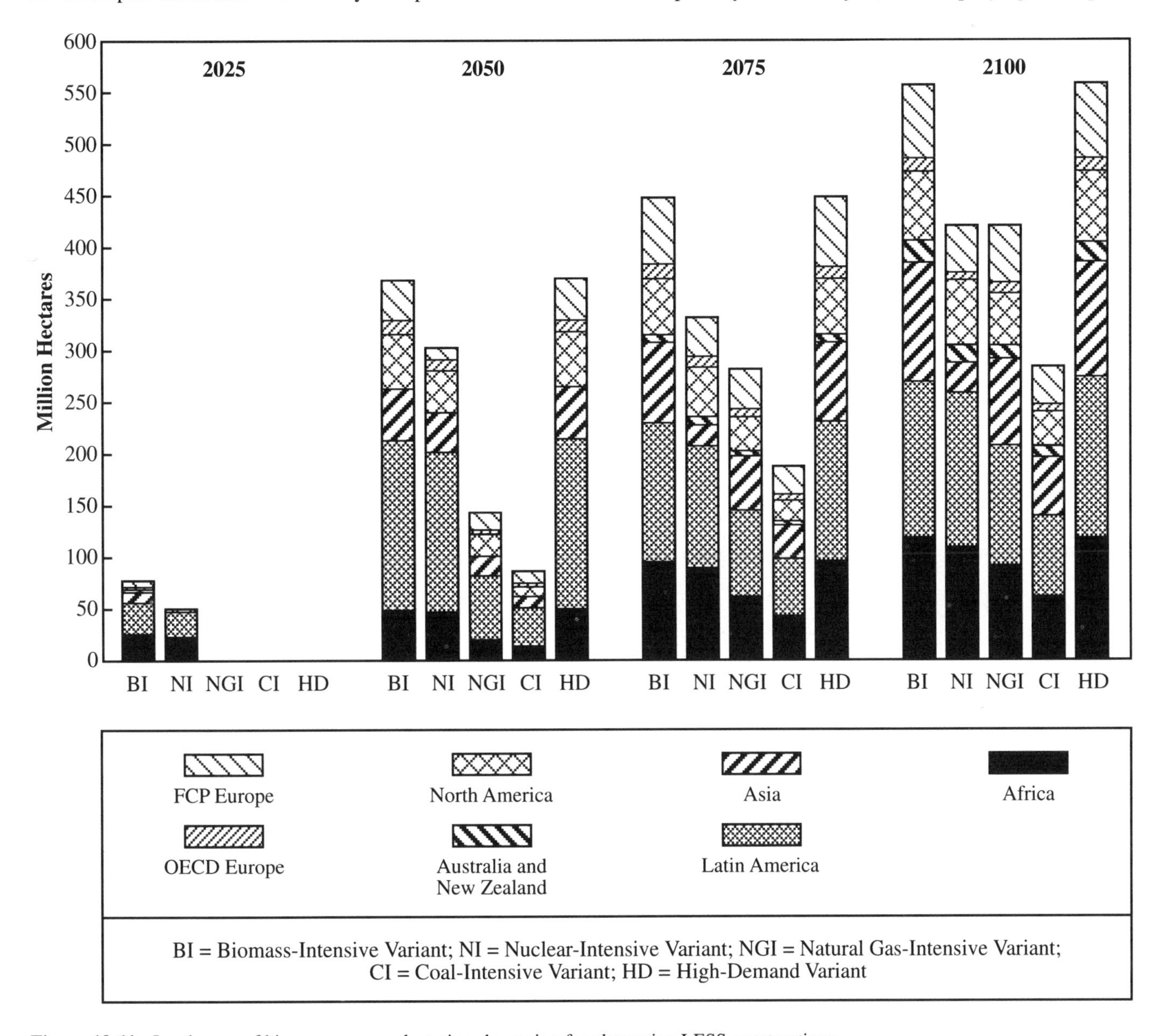

Figure 19-11: Land areas of biomass energy plantations by region for alternative LESS constructions.

were made: For each region, oil production declines at a constant exponential rate such that 80% of the estimated remaining ultimately recoverable resources is used up by 2100; and 80% of the ultimately recoverable natural gas resources also is used up for each region by 2100, but natural gas use first increases to the year 2025 (at the same rates as in the IPCC IS92a scenario) and thereafter declines at a constant exponential rate.

The LESS constructions assume that hydrogen becomes an important energy carrier (for example, for use in fuel-cell vehicles). Hydrogen derived from natural gas often will be one of the least-costly supplies of hydrogen (see Figure 19-5), and sequestration in depleted natural gas fields of CO_2 recovered at hydrogen production plants (containing two-thirds of the carbon in the original natural gas) is used as a decarbonizing strategy (see Figure 19-5). Thus, it is assumed that some natural gas (0% by 2025, 25% by 2050, 50% by 2075, and 75% by 2100) is reformed near the wellhead to hydrogen and that the stream of pure CO_2 generated in conversion is sequestered in depleted natural gas fields, increasing the cost of hydrogen to consumers by 1–3% (see Figure 19-5).

Under these conditions, cumulative CO_2 emissions from 1990 to 2100 from the use of oil and natural gas would be 275 Gt C. If there were no other fossil fuels, it might be feasible to stabilize the atmospheric concentration of CO_2 at or near the present level. IPCC Working Group I has estimated that cumulative CO_2 emissions of 300–430 Gt C in the 21st century would be consistent with stabilization of the atmospheric CO_2 level at a concentration of 350 ppm (IPCC, 1994).

Remaining coal resources are huge, however. Without decarbonization, burning all of the estimated ultimately recoverable coal resources (see Table B-3 in Chapter B) would lead to the

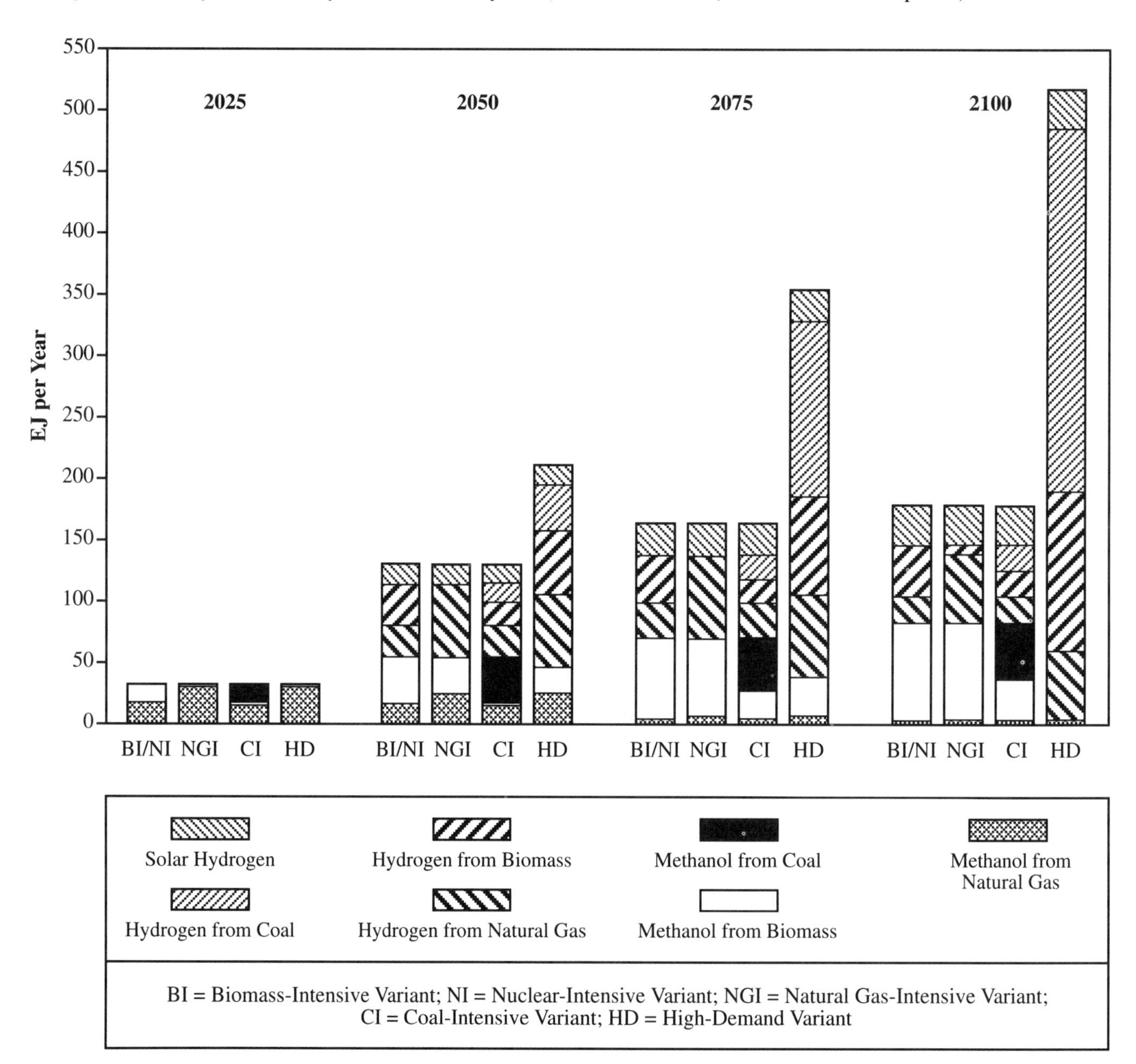

Figure 19-12: Methanol and hydrogen production from alternative sources for alternative LESS constructions.

cumulative release of 3,100 Gt C. Moreover, even if the rate of coal use remained constant at the 1990 level, the burning of coal without decarbonization would add about 270 Gt C to the atmosphere between 1990 and 2100.

The LESS constructions present alternative strategies for reducing emissions from burning coal. In the reference cases, a strategy for reducing coal use is articulated. In the coal-intensive (CI) and high-demand (HD) variants presented later, emissions are reduced instead by pursuing coal decarbonization strategies.

Although coal is abundant and cheap, it is also a dirtier, more difficult-to-use fuel than oil or natural gas. Where there are strict rules to ensure that coal is used in clean ways, coal will face stiff economic competition from many alternative energy sources, both for power generation (see Section 19.3.1.2) and for the production of synthetic fuels that are used directly (see Section 19.3.1.3). Key assumptions underlying the LESS reference cases are that by the time frame of interest (2025–2100), all regions of the world will have adopted environmental standards for using coal equivalent to the most stringent standards now in place in the industrialized world; and, if a nonfossil fuel alternative to coal is available at approximately the same cost for the final product (electricity or synthetic fuel) with these environmental standards under the assumptions about future technology, the alternative is selected. Under these conditions, coal can be plausibly greatly reduced in the LESS reference cases, although in developing countries coal use increases

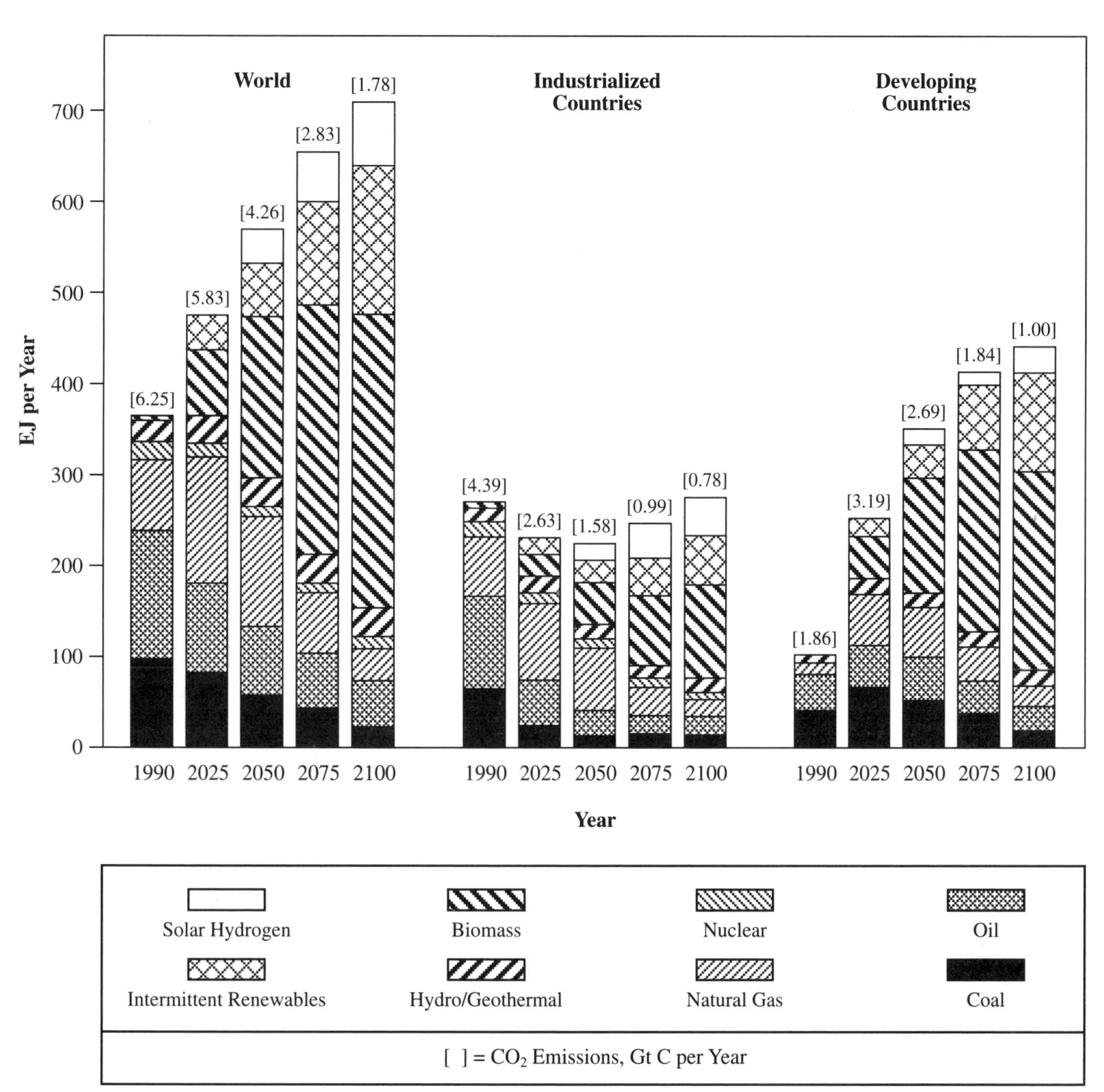

Figure 19-13: Primary commercial energy use by source for the biomass-intensive variant of the LESS constructions, for the world and for industrialized and developing countries.

Box 19-2. Highlights of the LESS Reference Cases (BI and NI Variants)

World population grows from 5.3 billion in 1990 to 9.5 billion by 2050 and 10.5 billion by 2100. Gross domestic product (GDP) grows 6.9-fold by 2050 (5.3-fold and 13.6-fold in industrialized and developing countries, respectively) and 24.6-fold by 2100 (12.8-fold and 68.3-fold in industrialized and developing countries, respectively), relative to 1990. Because of emphasis on energy efficiency, primary energy consumption rises much more slowly than GDP. Global primary commercial energy use roughly doubles, with no net change for industrialized countries but a 4.4-fold increase for developing countries, 1990–2100 (Figure 19-13).

Oil production declines in all regions, at a global average rate of 1.0% per year until 2100, when oil production is produced at 35% of the 1990 rate. Global natural gas production rises 84% by 2025, before beginning a decline at an average rate of 1.8% per year, 2025–2100; by 2100, natural gas is produced at 48% of the 1990 rate. Global coal production declines continually, but in developing countries it first rises 70% by 2025 before beginning to decline (see Figure 19-13). Whereas the decline in oil and gas production is determined by resource constraints, declining coal production is due to competition from nonfossil energy.

Total fossil fuel use stays roughly constant, 1990–2025, but its share in global energy declines from 86% to 67% in this period and to 15% by 2100, as a result of a shift to renewables in the BI variant and to nuclear plus renewables in the NI variant (see Figure 19-8). The NI variant involves increasing nuclear capacity worldwide 10-fold by 2100, so that nuclear accounts for 46% of total electricity in 2100, while hydro, biomass, and intermittent renewables (wind, photovoltaic, and solar thermal-electric power) account for 6%, 10%, and 34%, respectively. In the BI variant, nuclear provides 3% of electricity in 2100; hydro, biomass, and intermittent renewables contribute 10%, 29% and 54%, respectively.

Biomass plays a major role (especially as a feedstock for MeOH and H_2 production and power generation), accounting for 72 EJ or 15% of primary energy in 2025 (47% of biomass is for power generation) and rising to 325 EJ or 46% of primary energy by 2100 (29% for power generation) in the BI variant. In the NI variant, the contribution from biomass is 12% in 2025 (of which 35% is for power generation) and 38% in 2100 (of which 14% is for power generation).

MeOH and H_2 play growing roles as energy carriers in both the BI and NI variants (see Figure 19-12). Their production accounts for 10% of primary energy in 2025, rising to 40% by 2100 (about the same as for electric power generation by then). Natural gas provides nearly 60% of the energy from these energy carriers in 2025, but its share declines to 14% by 2100, while the biomass share increases from about 40% in 2025 to 67% by 2100. Electrolytic H_2 makes no contribution in 2025 but provides 12.5% of the energy from these energy carriers in 2050 and 20% in 2100.

Oil exports from the Middle East decline absolutely but grow as a percentage of global oil consumption—from about 20% in 1990 to more than 25% in 2025 and 33% in 2100. Total energy exports from the Middle East double, 1990–2050, before declining back to the 1990 level by the year 2100 (see Figure 19-14) as a result of growth in exports of natural gas and H_2 derived from both natural gas and solar electricity via electrolysis, which offset the decline in oil exports. Since H_2 is far more valuable than natural gas and oil, the monetary value of Middle East exports increases continually throughout the next century.

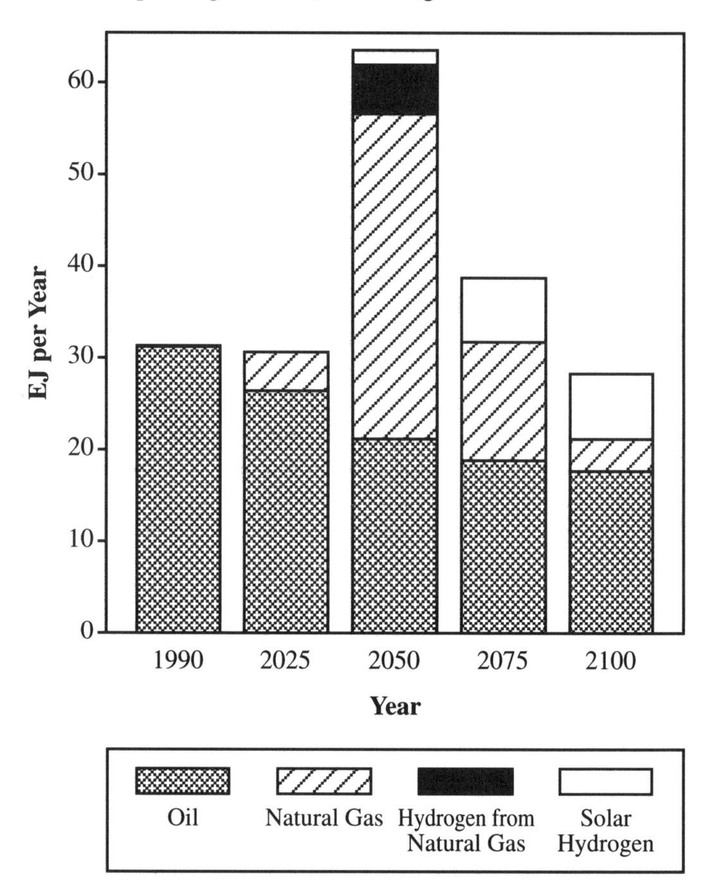

Figure 19-14: Net exports of fuels from the Middle East in the biomass-, nuclear-, and coal-intensive variants of the LESS constructions.

through 2025 before declining (see Figure 19-13 and Box 19-2). If nations elect not to control SO_x or other local pollutant emissions to these levels, prices for coal-derived energy would be lower, and alternatives to coal less attractive in economic terms.

Under these conditions, global CO_2 emissions decline from 6.2 Gt C in 1990 to 5.9 Gt C in 2025 and to 1.8 Gt C in 2100 for both of the LESS reference cases. Cumulative emissions from 1990 to 2100 amount to 448 Gt C, of which 182 Gt C comes from oil, 93 Gt C is from natural gas, and 172 Gt C is from coal.

19.3.1.2. The Electricity Sector in the LESS Reference Cases

Under the demand assumptions of the AP scenario, there is a continuing shift to electricity in the LESS reference cases, with total generation increasing 2.9- and 4.2-fold by 2050 and 2100, respectively. Total primary energy use increases only 1.6- and 1.9-fold, respectively. In the reference cases, the possibilities for achieving deep reductions in emissions both without (BI variant) and with (NI variant) a major expansion of nuclear power are explored.

19.3.1.2.1. The biomass-intensive variant for the electric sector

The BI variant explores the potential for using renewable electric sources in power generation by combining the results of modeling exercises, such as those described in Section 19.2.6.2 using the SUTIL model with considerations of regionally varying resource endowments and constraints. Under these conditions, estimates were made of potential contributions to electric grids from renewables—both intermittent renewables (wind, photovoltaic, and solar thermal-electric technologies) and advanced biomass electric-generating technologies (biomass-integrated gasifier/gas turbine technologies through 2025 and biomass-integrated gasifier/fuel-cell technologies for 2050 and beyond). Advanced gas turbine and fuel-cell technologies fueled with natural gas-derived hydrogen also are stressed. The level of nuclear generation remains constant from 1990 to 2100.

At the global level, 30% of electricity is supplied by intermittent renewables by 2050 in the BI variant, with substantial regional variations (for example, 10% in Africa; 18% in OECD Europe). This appears to be achievable without the use of new storage technologies if these systems are backed up by thermal power systems having characteristics similar to those of natural gas-fired combined cycles and peaking turbines (see, for example, Figure 19-4). Storage technologies make it possible to increase the global average contribution to more than 50% by 2100.

Natural gas is emphasized in the BI variant because of its favorable economics and relatively low CO_2 emissions in combined-cycle and fuel-cell configurations. Also, because natural gas-fired gas turbines and combined cycles have low unit capital costs and can change output levels quickly, they are good complements to intermittent renewable technologies. The share of natural gas in electricity generation increases from 16% in 1990 to 25% by 2050 but falls thereafter to 5% by 2100.

Biomass can provide baseload renewable electricity without ancillary storage. Its contribution to global electricity supply in the BI variant averages one-sixth of the total from 2025 to 2050, increasing to one-quarter from 2075 to 2100. Initial biomass applications involve mainly biomass residues in industrial cogeneration applications, but over time plantation biomass plays an increasing role.

At the global level, coal-based power generation declines nearly 50% by 2050 and 100% by 2100, but it increases 2.7-fold in centrally planned Asia plus South and East Asia from 1990 to 2050 before declining. It is assumed that the rapidly expanding market for coal power there provides a favorable environment for innovation—so energy-efficient, coal-integrated gasifier/gas turbine power cycles become the norm by 2025, and coal-integrated gasifier/fuel-cell technologies become the norm by 2050.

For the power sector as a whole, emissions decline from 1.7 Gt C/yr in 1990 to 1.0 Gt C/yr by 2050 and to 0.06 Gt C/yr by 2100.

The central result of this analysis—that deep reductions in emissions from the power sector are achievable with the BI variant—would not be qualitatively changed by making a significant change in an individual component of the power supply. Consider, for example, the hydropower contribution. By 2050, hydropower generation increases 2.2-fold relative to 1990 and then remains constant. Although this is only 50–80% of the estimated economically exploitable potential (Moreira and Poole, 1993), large hydropower projects are increasingly being challenged on social and environmental grounds. Cutting the hydro increment by half between 1990 and 2050 might be compensated by a 14% increase in the contribution from intermittent renewables by 2050 without fundamentally changing the basic structure of the BI variant and with no increase in global emissions. If the cut in the increment were compensated instead by natural gas power generation, global gas supply requirements in 2050 would go up only 7%, and global emissions would increase only 0.12 Gt C/yr (if the gas were not decarbonized) or 0.03 Gt C/yr (if it were).

Key to the economics of the BI variant are assumptions about relative fuel prices. The results of the SUTIL modeling exercise shown in Figure 19-4 are for electric utility fossil fuel prices projected by the U.S. Department of Energy (U.S. DOE) for the United States in the period near 2010 (Kelly and Weinberg, 1993). Since that modeling exercise was carried out, the coal price projected by U.S. DOE for U.S. electric utilities in 2010 has fallen from \$2.0/GJ to \$1.3/GJ. This reduction in coal price implies a 15% reduction in the busbar cost for the advanced coal power-generation technology to which the advanced biomass power-generation technology was compared in Figure 19-4. Such a sharp drop in coal prices makes it much more difficult for biomass to compete. However, more detailed estimates of future plantation biomass costs have been made since the modeling exercise illustrated in Figure 19-4 was carried out (Graham *et al.*, 1995; Turnure *et al.*, 1995), indicating that biomass prices of \$1.5–1.8/GJ could be expected in the U.S. for plantation production levels up to 5 EJ/yr by 2020 (see Section 19.2.5.2.1). This implies that the electric generation cost reduction for biomass would be about the same as for coal, relative to what was modeled in Figure 19-4 (see footnote 5). But if coal prices turn out to be lower still, as projected in some private forecasts (DRI/McGraw-Hill, 1995), and if biomass conversion technologies turn out to be less efficient and more capital-intensive than coal conversion technologies [e.g., if the more pessimistic assumptions of Turnure *et al.* (1995) relating to biomass turn out to be true], then biomass would require a significant subsidy to compete (see Section 19.2.5.2.2).

The fossil fuel prices for electric utility users assumed in Figure 19-4 are consistent with long-term prices generated independently by the "top-down" global model described in Section 19.3.2. However, other "top-down" models produce different results. In a recent modeling exercise carried out by the Energy Modeling Forum (Gaskins and Weyant, forthcoming), five different models estimated five different average electric-utility coal prices for 2020; these estimates ranged from $1.1 to $2.3/GJ and averaged $1.6/GJ.

The largest uncertainty regarding the prospects for an economically competitive, renewables-intensive electric future probably is the trend in relative fuel prices. However, the BI variant could plausibly provide demanded energy services at costs that are approximately the same as estimated future costs for current conventional technology, unless the lowest estimates for fossil fuels are realized and the costs projected for renewables turn out to be much too low.

19.3.1.2.2. The nuclear-intensive variant for the electric sector

The nuclear-intensive (NI) variant of the reference case involves a revitalization of the nuclear option and extensive deployment of nuclear electric power technology worldwide. Because the power sector is largely decarbonized in the next century under the BI variant (with CO_2 emissions amounting to only 0.06 Gt C/yr in 2100), it is assumed that the energy contributions from fossil fuels are identical for the NI and BI variants. Thus, the NI variant represents an alternative way to achieve the LESS reference case level of CO_2 emissions reduction from the power sector. It involves a major expansion of nuclear power along with a lesser (but still substantial) contribution from renewable electric sources than with the BI variant. The projections for nuclear electric capacity and generation worldwide in the long term would be technically feasible but would require the removal of policy barriers such as moratoria on construction, political decisions to abandon nuclear power, and the absence of support from development banks for nuclear power projects.

The main technical constraints on the growth of nuclear capacity are construction lead times and industrial capabilities for building power plants and fuel-cycle facilities. The availability of sites for nuclear installations—including radioactive waste repositories—was checked by region, taking into account the risk of earthquakes, the need for cooling, and population density.

The nuclear electricity generation projections were derived from penetration curves in each region, based upon the present status and trends of national nuclear programs. The asymptotic share of nuclear power in electricity generation was estimated by region, taking into account the availability of alternative energy sources and the size of the grid-connected electricity network (Semenov *et al.*, 1995).

Under this set of assumptions and constraints, the installed nuclear capacity would grow from the present 330 GW_e to about 3,300 GW_e in 2100 (see Table 19-8). Nuclear power

Table 19-8: *Nuclear power in the nuclear-intensive (NI) variant of the LESS base case.*

	2025			2050			2075			2100		
Region	Nuclear Elect. Gen. (TWh/yr)	Nuclear Share of Total Elect. (%)	Nuclear Capacity (GW_e)	Nuclear Elect. Gen. (TWh/yr)	Nuclear Share of Total Elect. (%)	Nuclear Capacity (GW_e)	Nuclear Elect. Gen. (TWh/yr)	Nuclear Share of Total Elect. (%)	Nuclear Capacity (GW_e)	Nuclear Elect. Gen. (TWh/yr)	Nuclear Share of Total Elect. (%)	Nuclear Capacity (GW_e)
Africa	56	5	8	208	10	31	434	15	66	678	20	103
Asia	1583	21	240	3672	27	556	6479	36	982	9510	45	1442
Australasia and New Zealand	7	5	1	17	10	3	25	15	4	37	20	6
Eastern and Central Europe	1184	30	179	2187	40	331	3349	50	507	4527	60	686
Latin America	204	10	31	849	20	98	1289	30	195	2000	40	303
Middle East	89	10	15	212	12	32	342	15	52	471	17	71
North America	1053	30	159	1525	39	231	2008	48	304	2566	57	389
Western Europe	634	30	96	1090	45	165	1598	60	242	2159	75	327
Total	**4764**	**23**	**721**	**9352**	**30**	**1447**	**15524**	**38**	**2352**	**21948**	**46**	**3327**

would be a major contributor to electricity generation in North and Latin America, Asia, and Europe, with a share ranging from 40–75% by 2100; in Africa, Australia, New Zealand, and the Middle East, the share is less than 20%. Worldwide, nuclear power would provide 46% of electricity by 2100, compared with 17% at present; nuclear's share of electricity would be half as large as that for renewables by 2025, rising to a comparable share by 2100.

With breeder reactors, the availability of natural resources would not place any major constraint on the development of nuclear electricity generation (see Section 19.2.4). It is assumed for the NI variant that the first breeder reactors are commercially deployed in 2025. This would make it possible to support projected nuclear electricity generation over the next century with currently known uranium resources.

At the back end of the fuel cycle, the management of spent fuel would require interim storage capacities and the implementation of final disposal repositories for radioactive waste. Nuclear power plants currently generate about 5 tons of spent fuel per TWh of electricity. With the development of advanced reactors, this quantity will decrease. The NI variant would lead to some 6.3 million tons of accumulated spent fuel by 2100 with current technologies. Efficiency improvements in nuclear fuel utilization—for example, higher burn-up—and deployment of advanced fission reactors are likely to reduce the amount of spent fuel. Because the deployment of breeder reactors is assumed for the NI variant, most spent fuel would be reprocessed, and the volume of high-level waste created would be on the order of 1 $m^3/GW_e/yr$. The accumulated volumes to be disposed of would be some 200,000 m^3 by 2100.

Because the deployment of breeder reactors is assumed for the NI variant, a significant amount of spent fuel would be reprocessed and the fissile materials recycled in reactors. In the long term, the deployment of breeders would reduce plutonium inventories to materials contained in fuel-cycle facilities—that is, reprocessing and mixed-oxide (MOX) fuel fabrication plants or in reactors. The plutonium generation rate in 2100 would be in the range of 0.1–3 million kg/yr, depending on the mix of nuclear technologies. The plutonium inventory would be on the order of 50–100 million kg and would require technical and institutional safeguards to prevent diversions to nuclear weapons purposes (see Section 19.2.4).

The challenges of managing material usable for weapons would be considerably less daunting if the deployment of plutonium breeder reactors could be avoided. This might be possible if uranium could one day be economically recoverable from seawater, if denatured thorium-based fuel cycles could be successfully developed, or if much of the nuclear expansion after 2050 were based on fusion instead of fission. Fusion plants would be less likely to contribute to the acquisition of nuclear-weapons capabilities by subnational groups and would be easier to safeguard against clandestine use for fissile material production by governments (Holdren, 1991).

Nuclear power could be expanded more than is projected for the NI variant if potential nonelectric markets also were taken into account—notably if nuclear energy were used for industrial process heat, for DH, for potable water production, and for electrolytic hydrogen production (see, for example, Figure 19-4).

19.3.1.3. Fuels Used Directly in the LESS Reference Cases

Activities that involve the direct use of fuels account for nearly three-quarters of CO_2 emissions at present. Although the direct use of fuels is projected to grow much more slowly than electricity use, the present high level of emissions and expected rapid growth in transport make the realization of deep reductions in this area a daunting challenge.

The challenge of reducing emissions for fuels used directly is met in the LESS reference cases by shifting to natural gas (with some decarbonization—see Section 19.3.1.1) and biomass and to synthetic fuels derived from them. The options stressed include solid biomass fuel for industry, biogas from dung, and especially methanol and hydrogen derived from natural gas and biomass that would be used mainly in fuel cells.

It is assumed for all the LESS constructions that low-temperature fuel cells come into wide use for transportation and other applications before the end of the first quarter of the next century. Although hydrogen is the preferred fuel for fuel-cell vehicles, it is assumed for the LESS constructions that in the early decades of the next century, the dominant fuel would be methanol (see Figure 19-12), an easy-to-use liquid fuel that poses fewer challenges to the infrastructure than hydrogen. For applications to fuel-cell vehicles, the costs per kilometer of driving to the consumer for methanol derived from natural gas and biomass would be comparable to the costs for hydrogen derived from these fuels (Williams *et al.*, 1995a, 1995b)—and thus to the cost of gasoline derived from crude oil in this period (see, for example, Figure 19-6).

In the first two decades of the next century, natural gas prices are likely to be relatively low. Thus, it is assumed that about 60% of the methanol is derived from natural gas in 2025 (see Figure 19-12). After 2025, it is expected that natural gas prices would rise and natural gas supplies would decline in the BI variant. Thus, there would be a shift in the mix of primary feedstocks from natural gas to biomass and at the same time a shift in the methanol/hydrogen mix toward greater use of hydrogen. The share of hydrogen would grow from 0% in 2025 to more than 50% in 2100 (see Figure 19-12).

Because of the assumption that biomass could be widely competitive in applications where biofuels are used directly, the demand for biomass energy is high in the BI variant. Because nuclear power provides only electric power in the NI variant, the contributions from biomass (see Figure 19-8) and from biomass plantations (see Figure 19-11) are almost as large for this variant as for the BI variant.

Until 2025, most biomass supplies would be residues of the forest-product and agricultural industries and urban refuse. Later, however, the dominant source would be biomass energy farms or plantations. The land area committed to plantations in the BI variant is 83 Mha in 2025 and increases to 572 Mha by 2100. The corresponding land areas for the NI variant are 55 and 433 Mha, respectively. The distribution of land use in plantations by region is indicated in Figure 19-11.

In the LESS reference cases, methanol would become a major commodity traded in international energy markets. Thus, by the middle of the next century, regions such as Latin America and sub-Saharan Africa—where there are large areas potentially available for growing biomass—would become major exporters of biomass-derived methanol.

There are no contributions from electrolytic hydrogen in the LESS reference cases in 2025. Costs for hydrogen derived electrolytically from photovoltaic, wind, or nuclear sources are likely to be much higher than for hydrogen or methanol derived from natural gas or biomass (see Figure 19-5). It is assumed, however, that by 2050 land-use constraints begin to limit the expansion of production of biomass-derived methanol and hydrogen in some areas, so that some electrolytic hydrogen enters the mix (see Figure 19-12 and Box 19-2).

19.3.1.4. Challenges Posed by Biomass Energy in the LESS Reference Cases

Several studies carried out in recent years suggest that "modernized biomass" could play a major role in the world's energy economy in the 21st century (see Box 19-3). Other long-term energy studies have ignored biomass; some that have not expect that biomass energy will evolve more slowly even if biomass energy is encouraged by public policies [for example, WEC's Ecologically Driven Scenario (WEC, 1993, 1994)].

The envisaged development of biomass energy is possible in principle. The assumed average biomass plantation yields—11 dry tons/hectare/year (dt/ha/y) in 2025, rising to 15 dt/ha/y in 2050 and 20 dt/ha/y, 2075-2100—though much higher than the yields for natural forests, are not unreasonable in light of experience with various cultivated crops (see Figure 19-3). Many studies indicate that these yields can be realized at the high net energy output/input ratios assumed for the LESS reference cases (see Section 19.2.5.2). Likewise, although the projected plantation areas are large (see Figure 19-11), even the 572 Mha targeted for the year 2100 in the BI variant is only 12% of the total amount of land in cropland plus permanent pasture (Hall *et al.*, 1993). Yet concerns have been raised about the potential

Box 19-3. The Role of Biomass Energy in Some Recent Global Energy Studies

Five recent studies exploring alternative energy futures provide alternative views of the prospects for biomass energy:

- The World Energy Council projects (WEC, 1994) in its Current Policies Scenario for global energy that traditional (noncommercial) biomass energy use increases from 42 EJ in 1990 to 59 EJ in 2020 and that modern (commercial) biomass energy increases from 5 to 11 EJ. For the WEC Ecologically Driven Scenario (EDS), however, these totals for 2020 are 47 EJ and 25 EJ, respectively [compared to a total biomass energy use rate (all commercial) of 74 EJ/yr in 2025 for the BI variant of the LESS]; the projected 5.5%/yr growth rate for modern biomass energy use, 1990–2020, is regarded as a plausible but ambitious growth schedule, considering all of the new technologies involved and the institutional hurdles that must be overcome.
- Dessus *et al.* (1992) project total world biomass use as 135 EJ in 2020, of which 51% and 17%, respectively, are accounted for by commercial and noncommercial wood recovered from forests, 20% by biomass waste resources, and 12% by biomass crops grown on plantations. Dessus *et al.* (1992) apparently focus their expansion on biomass from forests because this large potential resource might be the most easily exploited potential biomass supply.
- The U.S. Environmental Protection Agency advances a scenario for a greenhouse-constrained world (Rapidly Changing World with Stabilizing Policies) in which the role of commercial biomass energy increases to 136 EJ by 2025 and to 215 EJ by 2050 (Lashof and Tirpak, 1990).
- In their Renewables-Intensive Global Energy Scenario (RIGES), Johansson *et al.* (1993a) project the biomass contribution (all commercial) to be 145 EJ by 2025 and 206 EJ by 2050; plantations account for 55–62% of the total, while forests account for only 5–7%. In the RIGES, the use of modernized biomass was projected to grow rapidly (10%/yr, 1990–2025) because of the multiple benefits it offers.
- In its 1994 long-term global energy scenario exercise, the Shell International Petroleum Company developed a "Sustained Growth" scenario in which it projects that by 2050—as in the BI variant of the LESS—more than half of total world primary energy comes from renewable energy sources. (The projected overall level of renewable energy development is twice the level in the BI variant of the LESS, however, because the overall projected level of energy demand is about twice as high at that time.) In this scenario, the overall contribution from biomass in 2050 is about 15% higher than in the BI variant of the LESS, although about 30% of total biomass in the Shell scenario is for noncommercial uses (Kassler, 1994).

conflict with food production, especially in developing regions (see Section 19.2.5.2 and Chapter 25).

If agricultural production can be modernized and intensified in environmentally acceptable ways, however, large increases in cropland probably will not be needed for food production (e.g., Waggoner, 1994; Smil, 1994). Indeed, low-cost bioenergy could attract industry to rural areas to provide the income growth needed to modernize agriculture (see Section 19.2.5.2). Preliminary analyses indicate good prospects in Brazil for large-scale development of bioenergy plantations without serious competition from food production (Carpentieri *et al.*, 1993). Even in densely populated India, it has been suggested that some 60–70 million hectares of degraded lands might be good for energy plantations. In India, the potential conflict with food production is not expected to be great. Between 1970 and 1990, food production per capita increased while the total land area under cultivation stayed about the same, despite a 60% growth in population. Also, there are good prospects for substantial further gains in food crop yields (Ravindranath and Hall, 1994). Moreover, a preliminary country-by-country analysis exploring the potential for biomass energy production in the context of future food requirements in the developing world projects that there could reasonably be considerable amounts of land committed to biomass energy production in Africa, Asia, and Latin America without posing major conflicts with food production in the period to the year 2025 (Larson *et al.*, 1995). Further investigations of these issues are needed, however.

19.3.1.5. Natural Gas-Intensive Variant of a LESS

The natural gas-intensive (NGI) variant of the LESS constructions (see Figure 19-8) is based on the following assumptions:

- Remaining recoverable natural gas resources are substantially higher than in the reference cases.
- All of the natural gas in excess of that for the references cases is used to make methanol and hydrogen. These displace methanol and hydrogen that would otherwise be produced from plantation biomass in the BI variant to the extent that this is feasible at the higher production levels of natural gas.
- Hydrogen production for this extra natural gas is carried out near depleted natural gas fields, so the stream of CO_2 recovered at the hydrogen production facility can be sequestered there (see Figure 19-5).
- The energy demand levels and structures are the same as for the BI variant.
- Except for the shift from biomass to natural gas for the production of methanol and hydrogen, the energy supply mix is the same as for the BI variant.

In this variant, remaining, ultimately recoverable natural gas resources for the United States are assumed to be the same as in the reference cases. For other regions, the high natural gas resource estimates of Masters *et al.* (1994) are assumed. Thus, total remaining, ultimately recoverable natural gas resources as of 1993 are 17,400 EJ—nearly 40% higher than in the reference cases. As in the reference cases, it is assumed that 80% of these resources are used up by the year 2100. In this case, global natural gas production increases to a level 2.3 times the 1990 level by the year 2050 before global production declines at a constant exponential rate.

In this variant, natural gas would be less costly than in the reference cases, which would make it harder for fuels derived from plantation biomass to compete. In the NGI variant, there would be no plantations in 2025; plantation requirements would be only about 40% of those in the BI variant in 2050 and 75% of those in 2100 (see Figure 19-11).

Annual CO_2 emissions would fall to 2 Gt C/yr by the year 2100 (see Figure 19-9), and cumulative emissions between 1990 and 2100 would be 476 Gt C (see Figure 19-10)—only slightly more than in the reference cases. Cumulative requirements from 1990 to 2100 for sequestering the CO_2 recovered in hydrogen production would be nearly three times as large as in the reference cases (44 Gt C vs. 17 Gt C), but this is small compared with the sequestering capacity of natural gas fields that would be theoretically available by 2100 via natural gas extraction (see Section 19.2.3).

19.3.1.6. Coal-Intensive Variant

Another strategy for achieving deep reductions involves using coal and biomass for methanol and hydrogen production, along with sequestration of the CO_2 separated out at the synthetic fuel production facilities. This strategy is pursued in a way that emphasizes the use of coal relative to biomass, subject to the constraint that the overall level of annual CO_2 emissions is the same as for the NGI variant. It is assumed that for this coal-intensive (CI) variant, all characteristics of the variant other than for methanol and hydrogen production are identical to those for the BI variant (for example, reference-case assumptions about remaining natural gas resources).

The separated CO_2 might be sequestered in saline aquifers or in depleted oil and gas fields (Williams, 1996). Because the cost of the coal or biomass contributes such a small amount to the overall cost of making hydrogen (see Figure 19-5) or methanol (Williams *et al.*, 1995a, 1995b), it would not change the overall economics very much to ship the coal or biomass to a depleted natural gas field site for processing.

This variant offers considerably greater potential for reducing dependence on biomass than is feasible for the NGI variant (see Figure 19-11) because of the specification that CO_2 recovered at hydrogen and methanol conversion facilities be sequestered for biomass as well as coal feedstocks. Costs would be slightly higher because of the added complications of sequestering CO_2 recovered at coal and biomass fuel-processing facilities, but costs for demanded energy services could plausibly still qualify as being approximately the same as estimated future costs for current conventional energy (see Figure 19-6).

Cumulative sequestering requirements for this variant from 1990 to 2100 would be 145 Gt C (see Figure 19-10).

In this variant, global coal use would increase 47% by 2075, followed by a slow decline. Coal use in 2100 would be 31% higher than in 1990.

Two important lessons are highlighted by this variant. First, the production of hydrogen-rich fuels for fuel-cell applications allows coal to have a significant role as an energy source in a GHG-constrained world. Second, in this strategy, coal and biomass play complementary roles. Having some biomass in the energy system allows greater use of coal because for biomass grown on a sustainable basis with sequestration of the CO_2 separated at the conversion facilities, net life-cycle CO_2 emissions are negative (see Figure 19-5).

It is noteworthy that ongoing efforts to commercialize oxygen-blown coal gasifiers for integrated coal gasification/combined cycle power generation help to realize this strategy because the same kind of gasifier is needed for the production of methanol and hydrogen from coal.

19.3.1.7. High-Demand Variant

The LESS reference cases involve a high degree of decoupling of energy demand from economic growth for the entire period to the end of the next century. By 2100, primary energy demand doubles, compared to a quadrupling of energy demand in IPCC's IS92a scenario. The high-demand (HD) variant explores the potential for achieving deep reductions in emissions if energy demand and economic growth follow historical trends in the period beyond 2025. Overall demand for fuels used directly and for electricity are equal to those for the reference cases through the year 2025 but rise thereafter at constant exponential rates to the demand levels for the IS92a scenario by the year 2100. The energy supply base, to which are added the supplies for additional electricity and fuels used directly, is that for the NGI variant.

The major challenge of the HD variant is providing the extra fuels used directly with low emissions. The extra electricity might be provided entirely by some mix of intermittent renewable and nuclear electric sources, so that incremental power generation would pose no CO_2 management problems. To illustrate the possibilities, the HD variant is constructed with all of the incremental electricity provided by intermittent renewables. In principle, all of the incremental power could be provided from these sources. All incremental power requirements would come after the year 2050, by which time storage technology would plausibly be widely available to shape the output of intermittent resources to meet most electrical demand profiles. Also, there do not appear to be physical limits on intermittent resources at these generation levels, although power plant siting constraints in this time frame might limit additional capacity mainly to distributed photovoltaic systems and remotely sited, central-station, renewable electric power plants (Williams, 1995a).

All incremental requirements for fuel used directly are provided by hydrogen derived by thermochemical means from natural gas, biomass, and coal (see Figure 19-12), with sequestering of the CO_2 recovered at the fuel-conversion plants. By 2050 and beyond, the HD variant uses as much primary biomass energy as in the BI variant and as much natural gas as in the NGI variant. The extent of sequestration is dictated by a requirement that annual CO_2 emissions not exceed those for the NGI variant (see Figure 19-9).

Coal use increases 5-fold between 1990 and 2100 (see Figure 19-8); annual sequestration requirements increase from 0 in 2025 to 2.0 Gt C/yr in 2050 and 11.6 Gt C/yr in 2100. Cumulative sequestration requirements from 1990 to 2100 amount to 321 Gt C. In principle, sequestering might be accomplished entirely in natural gas fields until the year 2100 (Williams, 1995a). However, additional secure storage capacity would be needed to continue sequestering CO_2 at high rates and to maintain low emissions levels with continued intensive use of coal well beyond the year 2100.

The achievement of low emissions in this variant appears to be feasible. There would be much less flexibility, however, to choose among alternative options for achieving deep reductions at the high energy-demand levels of the HD variant than under reference-case energy-demand conditions. This exercise underscores the importance of energy efficiency in achieving deep reductions in GHG emissions.

19.3.2. A Top-Down Construction of a LESS

To test the robustness of the bottom-up energy supply analysis indicating that deep reductions in CO_2 emissions could be realized by 2100, a top-down global energy modeling analysis also was carried out for the Working Group (Edmonds *et al.*, 1994), incorporating performance and cost parameters for some of the key energy technologies used in the construction of the BI variant of the reference cases. Six technology cases were modeled using this top-down approach. The trajectories of global energy and the corresponding CO_2 emissions are shown in Figure 19-15.

The Edmonds-Reilly-Barns (ERB) model used here comprises four modules: demand, supply, energy balance, and GHG emissions. The first two modules determine the supply and demand for each of six major primary energy categories in each of nine global regions. The energy balance module ensures model equilibrium in each global fuel market, based on assumptions regarding resources and fuel technologies. The final module calculates energy-related emissions of CO_2. Demand for each fuel is determined by population; labor productivity; energy end-use intensity; energy prices; and energy taxes, subsidies, and tariffs. Energy end-use intensity is a time-dependent index of energy productivity. Demand for energy services in each

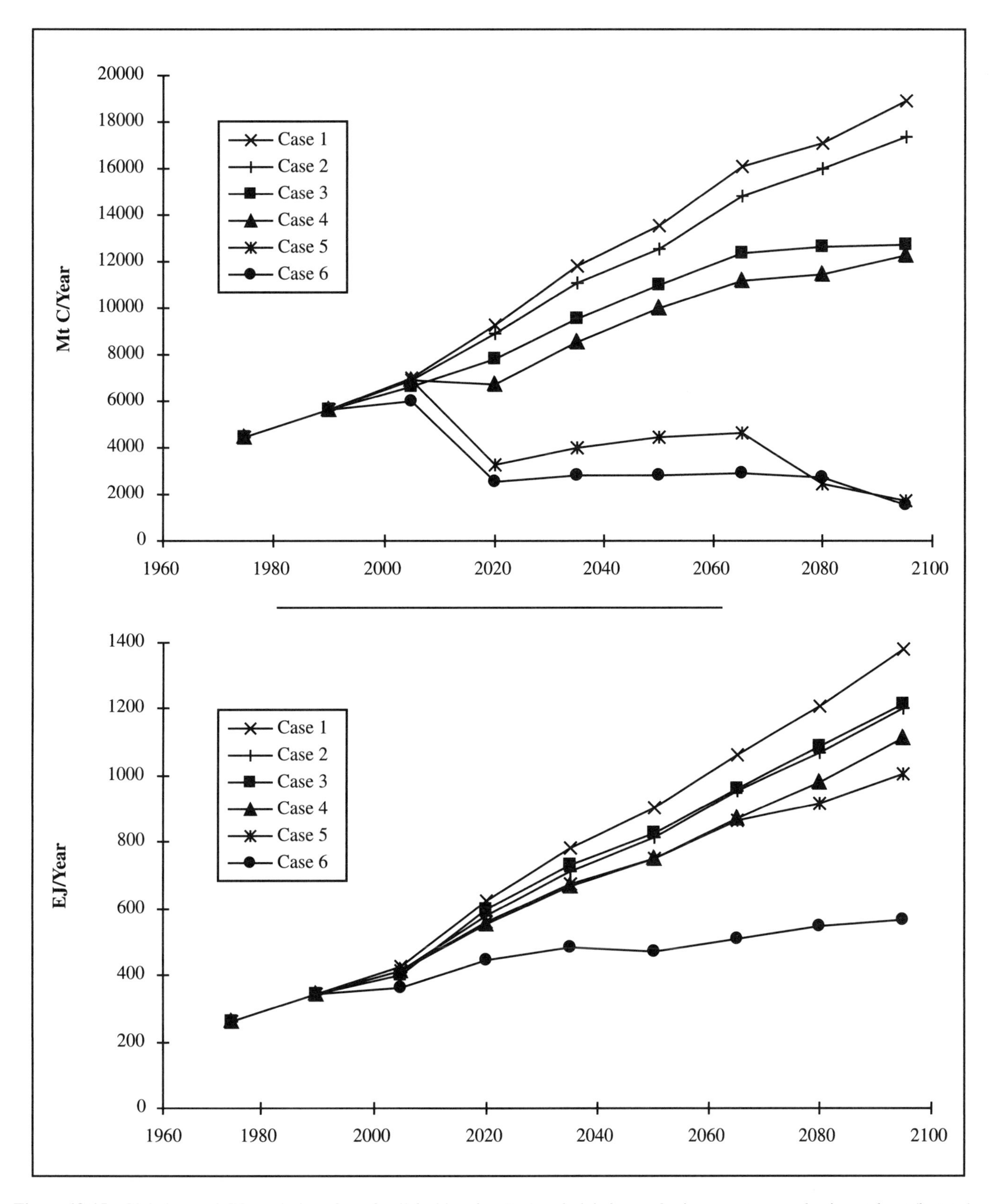

Figure 19-15: Global annual CO_2 emissions from fossil fuel burning (top) and global annual primary energy production and use (bottom) for six alternative cases constructed by Edmonds *et al.* (1994) in a "top-down" modeling exercise aimed at constructing a LESS. Case 1 is the reference scenario, with features very similar to the IPCC IS92a scenario; the exogenous end-use energy intensity improvement rate is 0.5%/yr by 2005, rising to 1%/yr by 2035, and reaching 1.5%/yr by 2065. Case 2 emphasizes energy-efficient power generation from fossil fuels (with efficiencies reaching 66% by 2095), but is otherwise like Case 1. In Case 3, liquefied hydrogen from natural gas, biomass, and electrolytic sources is used in fuel cells for transportation, and solar and wind power become highly competitive. Case 4 is like Case 3 except that compressed hydrogen is used instead of liquefied hydrogen. Case 5 is like Case 4, except that biomass prices are more competitive. Case 6 is like Case 5, except that the exogenous energy end-use intensity improvement rate reaches 2%/yr by 2050.

region's end-use sectors is determined by the cost of providing these services, and by income and population. For OECD regions, energy demand is disaggregated into residential/commercial, industrial, and transportation sectors. Energy supply is disaggregated into renewable and nonrenewable sources. Fossil fuel energy supplies are related to the resource base by grade of the resource, cost of production, and historic production capacity. The rate of technological change on the supply side varies by fuel. The energy balance module chooses energy prices that bring demand and supply into balance, given technological assumptions.

The point of departure for the top-down analysis is a scenario (case 1) that is very similar to IPCC's 1992 IS92a scenario (the same overall levels of population growth, GDP growth, primary energy use, CO_2 emissions, etc.). Five alternative technology cases were then modeled, each incorporating successively more of the characteristics of the technologies assumed for the construction of the BI variant of the reference case.

Two of the options (cases 5 and 6) are characterized by about the same level of CO_2 emissions as the LESS reference case in 2100 [compare Figure 19-15 (top) with Figure 19-9], even though their primary energy-use levels at that time are very different: For case 5 it is 40% higher than for the LESS reference case, and for case 6 it is 20% lower. These energy balances are arrived at with natural gas prices increasing to $8–10/GJ, oil prices increasing to $10–15/GJ, and coal prices increasing to $2–3/GJ by 2100.

All six cases have about the same level of exogenously defined global GDP in 2100 (although global GDP is slightly higher for cases 5 and 6 than for case 1). This shows that if the assumed technological characteristics are realized, deep reductions could be achieved without economic penalty. What is needed is a technology policy that facilitates the development and commercialization of GHG-friendly technologies that offer the potential at maturity of being competitive under market conditions with fossil fuel technologies.

This important result shows that the large differences in outcomes often observed between top-down and bottom-up modeling exercises are not due to irreconcilable differences between these very different approaches, as some have suggested, but rather to differences in assumptions.

19.3.3. Concluding Remarks

Costs for energy services in each LESS variant relative to costs for conventional energy depend on relative future energy prices, which are uncertain within a wide range, and on the performance and cost characteristics assumed for alternative technologies. However, within the wide range of future energy prices, one or more of the variants would plausibly be capable of providing the demanded energy services at estimated costs that are approximately the same as estimated future costs for current conventional energy. In a LESS scenario, substantial reductions in CO_2 emissions would result from the deployment of advanced energy supply technologies, along with more efficient energy-using equipment. Such outcomes appear to be possible given adequate time (several decades) and an economic climate and public- and private-sector policies conducive to the needed innovations. Strong and sustained investments would be needed for research and development, demonstration, and diffusion of energy technologies characterized by low or zero CO_2 emissions; at present, the levels of investment in such activities are low and declining (see Section 19.4).

The LESS constructions show that with modest rather than steep energy demand growth, deep reductions in emissions could plausibly be achieved with a variety of mixes of low-carbon fossil fuels, decarbonized fuels, renewable energy sources, and nuclear energy. Thus, society would not have to pursue all options but would have the flexibility to choose mixes of GHG-friendly technologies based largely on their effects on the local and regional environment, energy security, developmental benefits, and cost. Moreover, society could shift course to a different path if, for unforeseen reasons, an initial choice proved to be problematic or if more attractive options emerged as a result of unexpected technological innovations. Less flexibility would be possible at high levels of energy demand.

Finally, it is hoped that these LESS constructions will encourage others to explore in more detail the advantages and drawbacks of the options described here and to identify and articulate other paths to an energy future characterized by low emissions of CO_2.

19.4. Implementation Issues

The preceding section identified technology options that together could make significant reductions in long-term GHG emissions with projected costs to society of energy services that are approximately the same as estimated future costs for current conventional energy. The ability of any of these technologies to realize its potential depends on the rate and extent of its adoption under widely differing institutional, economic, and natural-resource constraints in various countries and regions. This section briefly addresses some of the associated implementation issues.

A transition to renewable and other low-CO_2-emitting energy technologies is not likely to occur at the pace envisioned under business-as-usual conditions. A variety of problems inhibit their development and deployment. For example, private companies are unlikely to make the investments necessary to develop renewable technologies at a rapid rate because the benefits are distant and not easily captured by individual firms; they also will not invest in large volumes of commercially available technologies to the extent justified by the external benefits (for example, reduced emissions). Also, conventional energy technologies benefit from direct subsidies of more than $300 billion per year worldwide (Koplow, 1993; OECD, 1992; IEA, 1993b; Larsen and Shah, 1994).

Actions conducive to a transition to an energy system based on low CO_2-emitting energy technologies should include policy changes that address seven key areas: energy system planning; financing; technology research and development; technology transfer, adaptation, and deployment; local environmental impacts; capacity building; and institutional arrangements (UNSEGED, 1992; WEC, 1994; Johansson *et al.*, 1993a; UNCNRSEED, 1994, 1995).

Economic instruments such as tradable emissions permits and emissions fees (taxes) could be designed to partly incorporate the true social and environmental cost of existing technologies in the price of the energy that they produce (IEA, 1993b). Others, such as temporary subsidies, could help accelerate the development and implementation of new technologies.

Regulations—defined as legislation or government or private-sector rules (supported by sanctions)—can be designed to influence public and private decisionmakers in their transactions. Regulations might focus on technology performance, influence the use of technology (for example, use of electricity-dispatch rules to prioritize low-emitting technologies), or promote the training of operating personnel through licensing arrangements. Particularly important in the power sector are rules that determine if and under what terms small, independent producers of renewable electricity (including intermittent generation) may gain access to the utility grid. For biomass energy, agricultural, forestry, and land-use policies are important.

For example, in the United States, an effective regulatory instrument—the Public Utilities Regulatory Policy Act (PURPA) in 1978—led to the creation of a competitive, decentralized market. PURPA required electric utilities to buy power from independent producers at the long-term avoided cost. This law is largely responsible for the introduction of 8,000 MW_e from biomass, 1,500 MW_e from wind, 730 MW_e from small-scale hydropower, and 350 MW_e from solar thermal-electric technology (WEC, 1994).

There is a worldwide trend toward increasing competition in the power sector. This will generally be helpful to small, high-efficiency, and more economical cogeneration systems, while discouraging large, less-efficient, and less-economical stand-alone steam turbine-based power plants. On the other hand, this trend will reduce interest in the long term and the measures that need to be taken to bring new technologies to the market.

Research and Development (R&D): High rates of innovation in the energy sector would be needed to realize any of the LESS constructions. The trend in recent years, however, has been declining investment in energy R&D on the part of both the private sector (Williams, 1995b) and the public sector (see Table 19-9). Over the last decade, public-sector support for energy R&D has declined absolutely by one-third and by half as a percentage of GDP. Moreover, government-supported R&D has generally focused on nonrenewable energy technologies. Less than 10% of IEA member governments' support is for renewable energy technologies.

It is important to have a government energy R&D strategy that does not attempt to pick winners. Fortunately, many of the promising technologies for reducing emissions, such as fuel cells and most renewable energy technologies, require relatively modest investments in R&D. This is a reflection largely of the small scale and modularity of these technologies and the fact that they are generally clean and safe (Williams, 1995b). As a result, it should be feasible, even with limited resources for R&D, to support a diversified portfolio of options. It has been estimated that research and development of a range of renewable energy technologies would require on the order of $10 billion (WEC, 1994).

Table 19-9: *Total reported IEA government R&D budgets (columns 1–7; US$ billion at 1994 prices and exchange rates) and GDP (column 8; US$ trillion at 1993 prices).*

Year	(1) Fossil Energy	(2) Nuclear Fission	(3) Nuclear Fusion	(4) Energy Conservation	(5) Renewable Energy	(6) Other	(7) Total	(8) GDP	(9) % of GDP
1983	1.70	6.38	1.43	0.79	1.05	1.08	12.40	10.68	0.12
1984	1.60	6.12	1.44	0.70	1.02	0.99	11.88	11.20	0.11
1985	1.51	6.26	1.42	0.70	0.85	1.04	11.77	11.58	0.10
1986	1.51	5.72	1.31	0.59	0.66	0.94	10.74	11.90	0.09
1987	1.37	4.36	1.23	0.65	0.62	1.04	9.27	12.29	0.08
1988	1.46	3.64	1.13	0.53	0.62	1.19	8.58	12.82	0.07
1989	1.30	4.42	1.07	0.45	0.57	1.33	9.13	13.23	0.07
1990	1.75	4.48	1.09	0.55	0.61	1.15	9.62	13.52	0.07
1991	1.52	4.45	0.99	0.59	0.64	1.39	9.57	13.58	0.07
1992	1.07	3.90	0.96	0.56	0.70	1.28	8.48	13.82	0.06
1993	1.07	3.81	1.05	0.65	0.71	1.38	8.66		
1994	0.98	3.74	1.05	0.94	0.70	1.30	8.72		

Sources: Government energy R&D expenditure data are from IEA (1995); GDP data are from OECD (1994).

Demonstration and Commercialization: R&D programs are necessary but not sufficient to establish new technologies in the marketplace. Commercial demonstration projects and programs to stimulate markets for new technologies also are needed. For a wide range of small-scale, modular technologies—including most renewable energy technologies and fuel cells—energy production costs can be expected to decline with the cumulative volume of production, as a result of "learning by doing" (Williams and Terzian, 1993; WEC, 1994).

The World Energy Council estimates that subsidies on the order of $7–12 billion are needed to support initial deployment of various renewable energy technologies until manufacturing economies of scale are achieved, to compete with conventional options (WEC, 1994). Thus, the World Energy Council estimates that the total investment needed for R&D on and support of initial deployment of renewables to be $15–20 billion. This is 0.1% of the annual global gross national product at the turn of the century and would, of course, be distributed over a couple of decades (WEC, 1994).

Information is a commodity that is especially subject to problems of market failure. One important area for government action is resource evaluation. Unfortunately, sufficiently detailed resource evaluations to facilitate investments are unlikely to be carried out by the private sector. Investigations of a wide range of potentially recoverable renewable resources, especially wind resources and land for energy plantations, would help reduce investment risks considerably (Johansson *et al.*, 1993a; UNCNRSEED, 1994).

In several countries, government encouragement of development and commercialization has been successful in introducing new energy supply technologies in national or regional markets—for example, wind energy in Denmark (Sørensen *et al.*, 1994); grid-connected cogeneration and renewable electric systems in the United States and home-scale photovoltaic systems in Indonesia (WEC, 1994); a sugar cane-based ethanol industry in Brazil (Goldemberg *et al.*, 1993); and a nuclear energy industry in France (Souviron, 1994).

References

AGTD, General Motors Corporation, 1994: *Research and Development of Proton-Exchange Membrane (PEM) Fuel Cell System for Transportation Applications: Initial Conceptual Design Report.* Report prepared for the Chemical Energy Division of Argonne National Laboratory, U.S. Department of Energy.

Ahmed, K., 1994: *Renewable Energy Technologies: A Review of the Status and Costs of Selected Technologies.* Technical Paper Number 240, Energy Series, The World Bank, Washington, DC, 169 pp.

Alcamo, J., C.J. van den Born, A.F. Bouwman, B.J. de Haan, K. Klein Goldeicjk, O. Klepper, J. Krabec, R. Leemans, J.G.J. Olivier, A.M.C. Toet, H.J.M. de Vries, and H.J. van der Woerd, 1994: Modelling the global society-biosphere-climate system. Part 2: Computed scenarios 1994. In: *Image 2.0. Integrated Modelling of Global Climate Change* [Alcamo, J. (ed.)]. Kluwer Academic Publishers, Dordrecht, The Netherlands, pp. 37-78.

Alexander, A.G., 1985: *The Energy Cane Alternative.* Sugar Series, vol. 6, Elsevier, Amsterdam, The Netherlands, 509 pp.

Alsema, E.A. and B.C.W. van Engelenburg, 1992: Environmental risks of CdTe and CIS solar cell modules. Paper presented at the 11th European Photovoltaic Solar Energy Conference, Montreux, Switzerland.

APS, 1975: Report to the American Physical Society by the study group on light-water reactor safety. Published in *Review of Modern Physics*, vol. 47(1).

Aresta, M. *et al.* (eds.), 1993: *Proceedings of the International Conference on Carbon Dioxide Utilization.* Department of Chemistry, University of Bari, Bari, Italy.

Arkesteijn, L. and R. Havinga, 1992: Wind farms and planning: practical experiences in The Netherlands. Conference Proceedings from European Wind Energy Association Special Topic Conference '92: The Potential of Wind Farms, Denmark.

Audus, H. and L. Saroff, 1994: Full fuel cycle evaluation of CO_2 mitigation options for fossil fuel fired power plants. ICCDR-2, Kyoto, Japan, October.

Barn, D.W. and J.A. Edmonds, 1990: An evaluation of the relationship between the production and use of energy and atmospheric methane emissions. U.S. Department of Energy, Washington, DC, pp. 3-7.

Beeldman, M., P. Lako, A.D. Kant, J.N.T. Jehee, and K.F.B. de Paauw, 1993: *Options for Electricity Production.* ECN-C-93-096, Petten, The Netherlands (in Dutch).

Bernhardt, W., 1994: Worldwide experience of Wolkswagen with reference of motor biofuels. European Biofuels Meeting, May, Tours, France.

BEST, National Research Council, 1991: *Rethinking the Ozone Problem in Local and Regional Air Pollution.* National Academy of Sciences, Washington, DC.

Beyea, H.J., J. Cook, D.O. Hall, R.H. Socolow, and R.H. Williams, 1991: *Toward Ecological Guidelines for Large-Scale Biomass Energy Development.* Report of a Workshop for Engineers, Ecologists, and Policymakers convened by the National Audubon Society and Princeton University, National Audubon Society, New York, NY, 28 pp.

Biederman, P., H.G. Duesterwald, B. Hoehlein, U. Stimming, S. Birkle, R. Kircher, C. Noelscher, H. Voigt, and W. Dreenckhahn (Siemens AG, Erlangen, Germany), 1994: Energy conversion chains and legally restricted emissions for road traffic in Germany. In: *Dedicated Conference on the Motor Vehicle and the Environment—Demand of the Nineties and Beyond*, Aachen, Germany, 31 October–4 November.

Blok, K., J. Bijlsma, S. Fockens, and P.A. Okken, 1989: *CO_2 emissiefactoren voor brandstoffen in Nederland.* ECN, Petten, The Netherlands.

Blok, K., 1991: *On the Reduction of Carbon Dioxide Emissions.* Thesis, Utrecht University, Utrecht, The Netherlands.

Blok, K., J. Farla, C.A. Hendriks, W.C. Turkenburg, 1991: Carbon dioxide removal: a review. In: *Proc. ESETT '91*, Milan, Italy, October.

Blok, K. and D. de Jager, 1993: Effectiveness and cost-effectiveness of greenhouse gas emission reduction technologies. In: *Proc. Internat. Symp. on Non-CO_2 Greenhouse Gases, Why and How to Control?* Kluwer Academic Publishers, Maastricht, The Netherlands, 13–15 December, pp. 17–40.

Blok, K., W.C. Turkenburg, C.A. Hendriks, and M. Steinberg (eds.), 1992: *Proceedings of the First International Conference on Carbon Dioxide Removal.* Pergamon Press, Oxford, UK.

Blok, K., R.H. Williams, R.E. Katofsky, and C.A. Hendriks, 1995: Hydrogen production from natural gas, sequestration of recovered CO_2 in depleted gas wells and enhanced natural gas recovery. In: *Proc. 2nd Internat. Symp. on CO_2 Fixation and Efficient Utilization of Energy, Energy International Journal*, 1996 (forthcoming).

Bloss, W.H. *et al.*, 1980: *Survey of Energy Resources.* Prepared for Eleventh World Energy Conference by the Federal Institute of Geosciences and Natural Sciences, Hannover, Germany, WEC, London, UK.

Boes, E.D. and A. Luque, 1993: Photovoltaic concentrator technology. In: *Renewable Energy: Sources for Fuels and Electricity* [Johansson, T.B., H. Kelly, A. Reddy, and R. Williams (eds.)]. Island Press, Washington, DC, pp. 361–401.

Boiteaux, M., 1989: Hydro: an ancient source of power for the future. *Intern. Water Power & Dam Construction*, **41(9)** (September), 10–11.

Bradaric, M., F.E. Davis, W.J. Stolte, and C.R. McGowin, 1992: *The Economics of Photovoltaic Power Generation.* Canadian Electrical Association Conference, Montreal, Canada. See also Stolte, W.J., R.A. Whisnant, and C.R. McGowin, 1993: *Design, Performance, and Cost of Energy from High Concentration and Flat-Plate Utility-Scale Systems.* The latter summarizes the findings in the report Bechtel Group, Inc., 1992: *Engineering and Economic Evaluation of Central-Station Photovoltaic Power Plants.* TR-101255s, EPRI, Palo Alto, CA.

Brower, M., 1993: *Cool Energy: Renewable Solutions to Environmental Problems.* Union of Concerned Scientists, MIT Press, Cambridge, MA, pp. 34–35.

Brown, L.R., 1993: A new era unfolds. In: *State of the World 1993* [Brown, L.R. *et al.* (eds.)]. W.W. Norton, New York, NY, pp. 3–21.

Calvert, J.G., J.B. Heywood, R.F. Sawyer, and J.H. Seinfeld, 1993: Achieving acceptable air quality: some reflections on controlling vehicle emissions. *Science*, **261**, 37-45.

Carlson, D.E. and S. Wagner, 1993: Amorphous silicon photovoltaic systems. In: *Renewable Energy: Sources for Fuels and Electricity* [Johansson, T.B., H. Kelly, A. Reddy, and R. Williams (eds.)]. Island Press, Washington, DC, pp. 403–435.

Carlson, J., 1988: *The Swedish Final Repository for Reactor Waste*. OECD, Paris, France.

Carminati, F., R. Klapish, J.P. Revol, Ch. Roche, J.A. Rubio, and C. Rubbia, 1993: *An Energy Amplifier for Cleaner and Inexhaustible Nuclear Energy Production Driven by a Particle Beam Accelerator.* CERN/AT/93-47(ET), European Organization for Nuclear Research, Geneva, Switzerland.

Carpentieri, E., E. Larson, and J. Woods, 1993: Future biomass-based power generation in Northeast Brazil. *Biomass and Bioenergy*, **4(3)**, 149-173.

Carson Mark, J., 1993: Explosive properties of reactor-grade plutonium. *Science and Global Security*, **4(1)**, 111-128.

Carver, H.A. and D.I. Page, 1994: Public attitudes to the cemmaes wind farm. Wind energy conversion 1994. In: *Proceedings of the 16th BWEA Wind Energy Conference* [Elliot, G. (ed.)]. Mechanical Engineering Publications Ltd., London, UK, pp. 237–240.

Cavallo, A.J., S.M. Hock, and D.R. Smith, 1993: Wind energy: technology and economics. In: *Renewable Energy: Sources for Fuels and Electricity* [Johansson, T.B., H. Kelly, A.K.N. Reddy, and R.H. Williams (eds.)]. Island Press, Washington, DC, pp. 121-156.

Cavallo, A., 1995: High capacity factor wind energy systems. *Journal of Solar Engineering*, **117**, 137-143.

Cavallo, A. and M.B. Keck, 1995: Cost-Effective Seasonal Storage of Wind Energy. In: *Wind Energy*, SED-vol. 16 [Musial, W.D., S.M. Hock, and D.E. Berg (eds.)]. Book No. H00926-1995, American Society of Mechanical Engineers.

Cavanagh, J.E., J.H. Clarke, and R. Price, 1993: Ocean energy systems. In: *Renewable Energy: Sources for Fuels and Electricity* [Johansson, T.B., H. Kelly, A.K.N. Reddy, and R.H. Williams (eds.)]. Island Press, Washington, DC, pp. 513–547.

Cernea, M.M., 1988: *Involuntary Resettlement in Development Projects—Policy Guidelines in World Bank–Financed Projects*. World Bank Technical Paper 80, Washington, DC.

Chang, Y., 1993: *Long-Term and Near-Term Implication of the Integral Fast Reactor Concept*. ORNL, Oak Ridge, TN.

Chen, J., 1995. *The Production of Methanol and Hydrogen from Municipal Solid Waste*. MSE Thesis, Mechanical and Aerospace Engineering Department, CEES Report 289 (March), Princeton University, Princeton, NJ, 247 pp.

Christofferson, L., Swedish Agricultural University, Ultuna, Uppsala, Sweden, 1995: Private communications to T.B. Johansson, February.

CIAB, 1992: *Global Methane Emissions from the Coal Industry*. October, Global Climate Committee, p. 2.

Cody, G.D. and T. Tiedje, 1992: The potential for utility-scale photovoltaic technology in the developed world: 1990-2010. In: *Energy and the Environment* [Abeles, B., A. Jacobson, and Ping Sheng (eds.)]. World Scientific, Teaneck, NJ.

Colombo, U. and U. Farinelli, 1992: Progress in fusion energy. *Annual Review of Energy and the Environment*, **17**, 123-160.

Colombo, U. and U. Farinelli, 1994: The hybrid car as a strategic option in Europe. In: *Dedicated Conference on Supercars (Advanced Ultralight Hybrids)*, 27th ISATA, Aachen, Germany, 31 October–4 November.

Comptroller and Auditor General of India, 1994: Report of the year ended March 1993. Hyderabad, India.

Consonni, S. and E.D. Larson, 1994: Biomass gasifier/aeroderivative gas turbine combined cycles. Paper prepared for the ASME Cogen Turbo Power '94 meeting, Portland, OR.

Corman, J.C., 1986: *System Analysis of Simplified IGCC Plants*. Report prepared for the U.S. Department of Energy by General Electric Company, Corporate Research and Development, Schenectady, NY, ET-14928-13, National Technical Information Service, Springfield, VA.

Dalenbäck, J.-O., 1993: *Solar Heating with Seasonal Storage: Some Aspects of the Design and Evaluation of Systems with Water Storage*. Document D21:1993, Dissertation, Chalmers University of Technology, Göteborg, Sweden.

Danish Energy Agency, 1992: *Update on Centralized Biogas Plants*. DEA, Copenhagen, Denmark, 31 pp.

Danish Energy Agency, 1993: *District Heating in Denmark*. DEA, Copenhagen, Denmark, 57 pp.

de Beer, J. and E. Nieuwlaar, 1991: *De rol van brandstofcellen in de energievoorziening. (The Role of Fuel Cells in Energy Supply).* University of Utrecht, Utrecht, The Netherlands.

De Laquil, P. *et al.*, 1990: *PHOEBUS Project 30 MWe Solar Central Receiver Plant Conceptual Design*. The American Society of Mechanical Engineers, Solar Engineering -1990, New York, NY, pp. 25–30. Presented at the Twelfth Annual ASME International Solar Energy Conference, 1–4 April, Miami, FL.

De Laquil, P., D. Kearney, M. Geyer, and R. Diver, 1993: Solar-Thermal Electric Technology. In: *Renewable Energy: Sources for Fuels and Electricity* [Johansson, T.B., H. Kelly, A.K.N. Reddy, and R.H. Williams (eds.)]. Island Press, Washington, DC, pp. 213–296.

DeLuchi, M., 1991: *Emissions of Greenhouse Gases from the Use of Transportation Fuels and Electricity*. Vol. 1, *Main Text*. ANL/ESD/TM-22, Report prepared for the Argonne National Laboratory, Argonne, IL.

DEPE, Ministry of Agriculture, People's Republic of China, 1992: *Biogas and Sustainable Agriculture: The National Experience. Exchange Meeting on Comprehensive Utilization of Biogas*, Yichang City, Hubei Province, 10-15 October, Bremen Overseas Research and Development Association, Bremen, Germany.

Dessus, B., B. Devin, F. Pharabod, 1992: World potential of renewable energies. *La Hoille Blanche*, **1**, 1-50.

Deudney, D., 1981: *Rivers of Energy: The Hydropower Potential*. Worldwatch Paper 44, Worldwatch Institute, Washington, DC, 55 pp.

DRI/McGraw-Hill, 1995: *World Energy Service: U.S. Outlook*. Lexington, MA, spring/summer.

DTI, 1993: *Solvarmeoversigt*. Danish Technological Institute, Tåstrup, 8 pp.

Duffie, J.A. and W.A. Beckman, 1991: *Solar Engineering of Thermal Processes*, 2nd Edition. John Wiley & Sons, New York, NY, 919 pp.

Dunnison, D.S. and J. Wilson, 1994: PEM fuel cells: a commercial reality. In: *A Collection of Technical Papers: Part 3, 29th Intersociety Energy Conversion Engineering Conference*, Monterey, CA, August 7–11, pp. 1260-1263.

Edmonds, J., M. Wise, and C. MacCracken, 1994: *Advanced Energy Technologies and Climate Change: An Analysis Using the Global Change Assessment Model (GCAM)*. Report prepared for the IPCC Second Assessment Report, Working Group IIa, Energy Supply Mitigation Options, 21 pp.

Ehrlich, P.R., A.H. Ehrlich, and G.C. Daily, 1993: Food security, population, and environment. *Population and Development Review*, **19(1)**, 230 pp.

EIA, 1990: *The Domestic Oil and Gas Recoverable Resource Base: Supporting Analysis for the National Energy Strategy*. SR/NES/90-05, U.S. Department of Energy, Washington, DC.

EIA, 1995: *Annual Energy Outlook 1995, with Projections to 2010*. DOE/EIA-0383(95), U.S. Department of Energy, Washington, DC.

Eletrobras, 1987: *Plano Nacional de Energia Elétrica 1987/2010*. Rel.Geral, December, Rio de Janeiro, Brasil, 269 pp.

Elliott, P. and R. Booth, 1993: *Brazilian Biomass Power Demonstration Project*. Special Project Brief, Shell International Petroleum Company, Shell Centre, London, UK, 12 pp.

Elsamprojekt, 1994: *Wind Energy in Denmark and within the Elsam Utility Area*. Elsamprojekt A/S, Fredricia, Denmark, 6 pp.

EPRI, 1993. *Technical Assessment Guide. Electricity Supply—1993*. EPRI TR-102275-V1R7, Electric Power Research Institute, Palo Alto, CA, June.

Faaij, A., A. Curvers, J. van Doorn, R. van Ree, A. Oudhuis, and L. Waldheim, 1995: Gasification of biomass wastes and residues for electricity production in The Netherlands. In: *Proc. Second Biomass Conf. of the Americas*, Portland, OR, 21–24 August, National Renewable Energy Laboratory, Golden, CO, pp. 594–606.

Face Foundation, 1995: *Face Foundation in Practice*. Arnhem, The Netherlands.

Farla, J., C.A. Hendriks, and K. Blok, 1992: *Carbon Dioxide Recovery from Industrial Processes*. Utrecht University, Utrecht, The Netherlands.

Foster, C. and R. Matthews, 1994: Assessing the energy yield and carbon reduction potential of short-rotation coppice fuelwood. In: *Proceedings of the 8th European Conference on Biomass*, Vienna, Austria, October, Elsevier, pp. 228–239.

Fridleifsson, I.B. and Frecston, 1994: Geothermal energy research and development. *Geothermics*, **L23(2)**, Elsevier Science Ltd.

Garzon, C.E., 1984: *Water Quality in Hydroelectric Projects—Consideration for Planning in Tropical Forest Region*. World Bank Technical Paper 20, Washington, DC.

Gaskins, D. and J. Weyant (eds.), forthcoming: *Reducing Carbon Emissions from the Energy Sector: Cost and Policy Options*. Stanford Press, Stanford, CA.

Godtfredsen, F., 1993: *Sammenligning af danske og udenlandske vindmollers ekonomi (Comparison of the Economy of Danish and Foreign Wind Turbines)*. Risø report Risø-R-662(DA), Risø National Laboratory, DK-4000 Roskilde, Denmark, 39 pp.

Goldemberg, J., L.C. Monaco, and I.C. Macedo, 1993: The Brazilian fuel-alcohol program. In: *Renewable Energy: Sources for Fuels and Electricity* [Johansson, T.B., H. Kelly, A.K.N. Reddy, and R.H. Williams (eds.)]. Island Press, Washington, DC, pp. 841-863.

Gosse, G., 1994: Environmental balance sheets of the motor biofuel channels. European Biofuels Meeting, May, Tours, France, *Blocarburant en Europe Developpement Applications Perspectives 1994-2004*. Actes du ler Forum Europeen sur les Blocarburants, ADEME, Paris, France, pp. 124–131.

Goudriaan, J., M.J. Kropff, and R. Rabbinge, 1991: Mogelijkheden en beperkingen van biomasse als energiebron (Potential and limitations of biomass as an energy source). *Energiespectrum*, June, pp. 171-176.

Graham, R., E. Lichtenberg, V. Roningen, H. Shapouri, and M. Walsh, 1995: The economics of biomass production in the United States. In: *Proceedings of the Second Biomass Conference of the Americas*, Portland, OR, 21–24 August, National Renewable Energy Laboratory, Golden, CO, pp. 1314–1323.

Graham-Bryce, I.J., W.G. Karis, A. Kinoshita, K.M. Sullivan, and G.G. Summers, 1993: IEA Second International Conference on the Clean and Efficient Use of Coal and Lignite, Hong Kong.

Grainger, A., 1990: Modeling the impact of alternative afforestation strategies to reduce carbon emissions. In: *Proceedings of the IPCC Conference on Tropical Forestry Response Options to Global Climate Change*. Report No. 20-P-2003, Office of Policy Analysis, U.S. EPA, Washington, DC.

Grainger, A., 1988: Estimating areas of degraded tropical lands requiring replenishment of forest cover. *International Tree Crops Journal*, **5**, 31-61.

Gregory, J.A., A.S. Bahaj, R.S. Stainton, 1993: *Stimulating Market Success for Solar: A Global Perspective*. Honour paper at ISES World Congress, Budapest, Hungary.

Grubb, M.J. and N.I. Meyer, 1993: Wind energy: resources, systems, and regional strategies. In: *Renewable Energy: Sources for Fuels and Electricity* [Johansson, T.B., H. Kelly, A.K.N. Reddy, and R.H. Williams (eds.)]. Island Press, Washington, DC, pp. 157-212.

Gustafsson, L. (ed.), 1994: Environmental aspects of energy forest cultivation. *Biomass and Bioenergy*, special issue, **6(1/2)**.

Haage, T., S. Bauer, B. Schroeder, and H. Oechsner, 1994: Correlation between improved stability and microstructural properties of a-Si:H and a-Ge:H. Paper presented at the first WCPEC, Waikoloa, HI.

Haeger, M. *et al.*, 1994: PHOEBUS technology program solar air receiver experiment. ASME/JSME/JSES International Solar Energy Conference, San Francisco, CA, Solar Engineering, March.

Hagedorn, G., 1989: Hidden energy in solar cells and photovoltaic power stations. In: *Proceedings of the 9th EC PV Solar Energy Conference*.

Hall, D.O., 1994: Biomass energy options in W. Europe (OECD) to 2050. In: *ECN/IEA/IPCC Workshop on Energy Technologies to Reduce CO_2 Emissions in OECD Europe: Prospects, Competition, Synergy*. Petten, The Netherlands, OECD, Paris, France, pp. 159-193.

Hall, D.O., H.E. Mynick, and R.H. Williams, 1991: Cooling the greenhouse with bioenergy. *Nature*, **353**(September), 11-12.

Hall, D.O., F. Rosillo-Calle, R.H. Williams, and J. Woods, 1993: Biomass for energy: supply prospects. In: *Renewable Energy: Sources for Fuel and Electricity* [Johansson, T.B., H. Kelly, A.K.N. Reddy, and R.H. Williams (eds.)]. Island Press, Washington, DC, pp. 593–652.

Hendriks, C.A., W.C. Turkenburg, and K. Blok, 1993: Promising options to remove carbon dioxide from large power plants. In: *Proceedings of the International Symposium on CO_2 Fixation and Efficient Utilization of Energy*. Tokyo Institute of Technology, Tokyo, Japan, pp. 277–291.

Hendriks, C., 1994: *Carbon Dioxide Removal from Coal-Fired Power Plants*. Kluwer Academic Press, Dordrecht, The Netherlands, 259 pp.

Henry, P., 1991: Improvements for conventional clean energies hydroelectric power. In: *Proc. World Clean Energy Conference*. CMDC, Zürich, Switzerland.

Herzog, H.G. and E.M. Drake, 1993: *Long Term Advanced CO_2 Capture Options*. IEA Greenhouse Gas R&D Programme, Report IEA/93/OE6.

Hill, R., W. Palz, and P. Helm, 1994: *Proceedings of the 12th European Photovoltaic Solar Energy Conference and Exhibition*. Amsterdam, The Netherlands.

Hillesland, Jr., T., and P. De Laquil, 1988: *Results of the U.S. Solar Central Receiver Utility Studies*. VDI Berichte 704.

Hillesland, Jr., T. 1988: *Solar Central Receiver Technology Advancement for Electric Utility Applications*. Phase I Topical Report, Pacific Gas and Electric Company, San Ramon, CA, GM 633022-9 (DOE Contract DE-FC04-86AL38740 and EPRI Contract RP 1478-1), August.

Ho, S.P., 1989: Global impacts of ethanol versus gasoline. Paper presented at the 1989 National Conference on Clean Air Issues and America's Motor Fuel Business, Washington, DC.

Hoff, T.E., H.J. Wenger, and B.F. Farmer, 1995: The value of deferring electric utility capacity investments with distributed generation. *Energy Policy*, November.

Holdren, J.P., 1991. Safety and environmental aspects of fusion energy. *Ann. Rev. Energy Environm.*, **16**, 235-258.

Holland, M. *et al.*, 1994: The full fuel cycle of CO_2 capture and disposal—estimation and valuation of environmental impacts. ICCDR-2, Kyoto, Japan, October.

Houghton, J.T., G.J. Jenkins, and J.J. Ephraums (eds.), 1990: *Climate Change: The IPCC Scientific Assessment*. Cambridge University Press, Cambridge, UK, 365 pp.

Hulkkonen, S., M. Raiko, and M. Aijala, 1991: New power plant concept for moist fuels, IVOSDIG. Paper presented at the ASME International Gas Turbine Aeroengine Congress and Exposition, Orlando, FL, June.

Huotari, J., S. Helyman, and M. Flyktman, 1993: *Indirect Biofixation of CO_2*. CRE/CON 1424 VTT Research Report.

Humm, P. and P. Toggweiler, 1993: *Photovoltaics in Architecture*. Birkhauser Verlag, Basel, Switzerland.

IAEA, 1991: *Keynote Papers*. Senior Expert Symposium on Electricity and the Environment, Helsinki, Finland, 82 pp.

IAEA, 1992a: *Radioactive waste management*. IAEA-STI/PUB/889. Vienna, Austria.

IAEA, 1992b: *Technical and Economic Evaluation of Potable Water Production through Desalination of Sea Water by Using Nuclear Energy and Other Means*. IAEA-TECDOC-666. Vienna, Austria, 152 pp.

IAEA, 1993: *Against the Spread of Nuclear Weapons: IAEA Safeguards in the 1990s*. IAEA, Vienna, Austria, 32 pp.

IAEA, 1994: Resolution by the General Conference of IAEA in 1994. Press release, IAEA, 3 November.

IAEA, 1995a: *Nuclear Power: An Overview in the Context of Alleviating Greenhouse Gas Emissions*. IAEA-TECDOC-793, Vienna, Austria, 41 pp.

IAEA, 1995b: *Nuclear Power Reactors in the World—RDS No. 2*. Vienna, Austria, 78 pp.

IEA, NEA, IAEA, 1993: *Projected Costs of Generating Electricity*. OECD, Paris, France, 192 pp.

IEA, 1993a: *Potential Applications of Combined Heat and Power Generation Systems*. IEA Greenhouse Gas R&D Programme, CRE, Stoke Orchard, UK.

IEA, 1993b: *Taxing Energy: Why and How*. OECD, Paris, France.

IEA, 1993c: *Combined Heat and Power Generation in IEA Member Countries*. IEA/SLT/EC(93)4, Paris, France.

IEA, 1994a: *Greenhouse Gas Emissions from Power Stations*. IEA Greenhouse Gas R&D Programme, CRE, Stoke Orchard, UK, 28 pp.

IEA, 1994b: *Biofuels*. IEA/OECD, Paris, France, 115 pp.

IEA, 1994c: *Energy Policies of IEA Countries: 1993 Review*. OECD, Paris, France.

IFRC, 1990: *Status Report on Controlled Thermonuclear Fusion.* IAEA, Vienna, Austria, 23 pp.

INEL, LANL, ORNL, SNL, and SERI, 1990: *The Potential of Renewable Energy: An Interlaboratory White Paper.* SERI/TP-260-3674, Report prepared for the Office of Policy, Planning, and Analysis, U.S. Department of Energy, Washington, DC.

Ingersoll, J.G., 1991: Energy storage systems. In: *The Energy Sourcebook* [Howes, R. and A. Fainberg (eds.)]. American Institute of Physics, New York, NY, pp. 325–355.

INSAG, 1988: *Basic Safety Principles for Nuclear Power.* IAEA-STI/PUB/802, Vienna, Austria, 72 pp.

INSAG, 1992: *Probabilistic Safety Assessment.* IAEA-STI/PUB/916, Vienna, Austria, 23 pp.

INSAG, 1991: *Safety Culture.* IAEA-STI/PUB/882, Vienna, Austria, 31 pp.

INSAG, 1986: *Summary Report on the Post Accident Review Meeting on the Chernobyl Accident.* IAEA-STI/PUB/740, Vienna, Austria, 106 pp.

IPCC, 1992: *The Supplementary Report to the IPCC Impacts Assessment.* IPCC, Canberra, Australia, 9 pp.

IPCC, 1994: *Radiative Forcing of Climate Change and an Evaluation of the IPCC Emission Scenarios* [Houghton, J.T., L.G. Meira Filho, J. Bruce, Hoesung Lee, B.A. Callander, E. Haites, N. Harris, and K. Maskell (eds.)]. Reports of Working Groups I and III of the IPCC, forming part of the IPCC Special Report to the first session of the Conference of Parties to the UN Framework Convention on Climate Change, Cambridge University Press, Cambridge, UK.

Jansen, D., A.B.J. Oudhuis, and H.M. van Veen, 1992: CO_2 reduction potential of future coal gasification based power generation technologies. *Energy Conversion Management*, **33(5-8)**, 365-372.

Jansen, D., P.C. van der Laag, A.B.J. Oudhuis, and J.S. Ribberink, 1994: *Prospects of Advanced Coal Fuelled Fuel Cell Power Plants.* ECN Petten, The Netherlands.

Jensen, J. and B. Sørensen, 1984: *Fundamentals of Energy Storage.* John Wiley & Sons, New York, NY, 345 pp.

Jansson, S.A., 1991: Status and development potential for PFBC plants. In: *The Environment and Development: Technologies to Reduce Greenhouse Gas Emissions.* Proc. Int. Conf. on Coal, Sydney, Australia, 18-21 November, IEA, Paris, France.

Johansson, T.B., H. Kelly, A.K.N. Reddy, and R.H. Williams, 1993a: Renewable fuels and electricity for a growing world economy: defining and achieving the potential. In: *Renewable Energy: Sources for Fuels and Electricity* [Johansson, T.B., H. Kelly, A.K.N. Reddy, and R.H. Williams (eds.)]. Island Press, Washington, DC, pp. 1-71.

Johansson, T.B., H. Kelly, A.K.N. Reddy, and R.H. Williams (eds.), 1993b: *Renewable Energy: Sources for Fuels and Electricity.* Island Press, Washington, DC, 1160 pp.

Johansson, H., 1993: Energy forestry with effective technology and new breed of salix varieties. In: *Proceedings of the Bioenergy '93 Conference*, Espoo, Finland [Asplund, D. (ed.)]. VTT, Jyuaskyla, Finland, pp. 191–200.

Juhn, P.E. and J. Kupitz, 1994: Role of nuclear power for sustainable development. Proceedings of the International Conference on New Trends of Nuclear System Thermohydraulics, Pisa, Italy, 30 May–2 June, **2**, 615–623.

Kaarstad, O., 1992: Emission-free fossil energy from Norway. In: *Proceedings of the First International Conference on Carbon Dioxide Removal* [Blok, K., W.C. Turkenburg, C.A. Hendriks, and M. Steinberg (eds.)]. Pergamon Press, Oxford, UK, pp. 781–786.

Kagoja *et al.*, 1993: Process evaluation of CO_2 recovery from thermal power plant. JSME-ASME International Conference on Power Engineering, Tokyo, Japan.

Kalkum, B., Z. Korenyl, and J. Caspar, 1993: *Utilization of District Heating for Cooling Purposes. An Absorption Chilling Project in Mannheim, Germany.* Mannheimer Versorgungs- und Verkehrsgesellschaft mbH, Germany, 21 pp.

Kaneff, S., 1994: Configuration, economics and perfomance of distributed dish/central plant solar thermal systmes in the range 2-100 MW_e. Paper presented at the 7th International Symposium on Solar Thermal Concentrating Technologies, Moscow, Russia, 26-30 September (to be published).

Karni, J. *et al.*, 1994: Development of directly irradiated annular pressurized receiver with novel window and absorber. Paper presented at the 7th International Symposium on Solar Thermal Concentrating Technologies, Moscow, Russia, 26-30 September (to be published).

Kartha, S., E.D. Larson, J.M. Ogden, and R.H. Williams, 1994: Biomass-integrated gasifier/fuel cell electric power generation and cogeneration. Prepared for *Fuel Cell Seminar 1994: Demonstrating the Benefits*, San Diego, CA, 28 November–1 December.

Kassler, P., 1994: *Energy for Development.* Shell Selected Paper, Shell International Petroleum Company, London, UK, November, 11 pp.

Kelly, H., 1993: Introduction to photovoltaic technology. In: *Renewable Energy: Sources for Fuels and Electricity* [Johansson, T.B., H. Kelly, A. Reddy, and R. Williams (eds.)]. Island Press, Washington, DC, pp. 297–336.

Kelly, H. and C. Weinberg, 1993: Utility strategies for using renewables. In: *Renewable Energy: Sources for Fuels and Electricity* [Johansson, T.B., H. Kelly, A.K.N. Reddy, and R.H. Williams (eds.)]. Island Press, Washington, DC, pp. 1011-1069.

Kendall, H. and D. Pimentel, 1994: Constraints on the expansion of the global food supply. *Ambio,* **23(3)**, 198-205.

Kircher, R., S. Birkle, C. Noelscher, and H. Voigt, 1994: PEM fuel cells for traction: system technology aspects and potential benefits. In: *Symposium on Fuel Cells for Traction Applications.* Royal Swedish Academy of Engineering Sciences, Stockholm, Sweden, 8 February.

Kondo, J., T. Inui, and K. Wasa (eds.), 1995: *Proceedings of the Second International Conference on Carbon Dioxide Removal.* Pergamon Press, Oxford, UK.

Koplow, D.N., 1993: *Federal Energy Subsidies: Energy, Environment and Fiscal Impact.* The Alliance to Save Energy, Washington, DC.

Kubo, R. and R.B. Diver, 1993: Development of cummins power generation dish-stirling systems for remote power applications. In: *6th International Symposium on Solar Thermal Concentrating Technologies*, Mojocar, Spain. Editorial CIEMAT, Madrid, Spain, pp. 747-762.

Kupitz, J., 1992: *Trends in Advanced Reactor Development.* Paper presented at KAIF meeting, Seoul, Republic of Korea, 21–25 April.

Laginha Serafim, J. (ed.), 1984: *Safety of Dams.* A.A. Balkema, Rotterdam, Boston, 2 vol.

Larsen, B. and A. Shah, 1994: Energy pricing and taxation options for combating the "greenhouse effect." In: *Climate Change: Policy Instruments and Their Implications.* Proceedings of the Tsukuba Workshop on IPCC Working Group III, 17–20 January.

Larson, E., C.I. Marrison, and R.H. Williams, 1995: *CO_2 Mitigation Potential of Biomass Energy Plantations in Developing Regions.* PU/CEES Report, Center for Energy and Environmental Studies, Princeton University, Princeton, NJ.

Larson, E.D., 1993: Technology for electricity and fuels from biomass. *Annual Review of Energy Environm.,* **18**, 567-630.

Lashof, D.A. and D.A. Tirpak, 1990: *Policy Options for Stabilizing Global Climate*, appendices. Report to Congress from the Office of Policy, Planning, and Evaluation, U.S. Environmental Protection Agency, Washington, DC.

Lazenby, J.B.C. and P.M.S. Jones, 1987: Hydroelectricity in West Africa: its future role. *Energy Policy,* **15(5)** (October), 441–455.

Ledin, S. and B. Alriksson, 1992: *Handbook on How to Grow Short Rotation Forests.* IEA Bioenergy Agreement, Task Y, Swedish University of Agricultural Science, Uppsala, Sweden, 184 pp.

Ledin, S., B. Alriksson, H. Rosenquist, and H. Johansson, 1994: Gödsling av Salixodlingar, NUTEK R 1994, 25 pp. (in Swedish).

Lelieveld, J. and P.J. Crutzen, 1993: Methane emissions into the atmosphere—an overview. In: *Proc. IPCC Internat. Workshop on Methane and Nitrous Oxide*, National Institute of Public Health and Environmental Protection, Amersfoot, The Netherlands, 3–5 February, 17 pp.

Leppalahti, J., 1993: Formation and behaviour of nitrogen compounds in an IGCC Process. *Bioresource Technology*, **46**, 65-70.

Leydon, K. and H. Glocker, 1992: *Energy in Europe: A View to the Future* (chapter 2). Analysis and Forecasting Unit of the Directorate General for Energy, European Commission, Brussels, Belgium.

Little, A.D., 1995: *Fuel Cells for Building Cogeneration Applications—Cost/Performance Requirements, and Markets.* Final Report prepared for the Building Equipment Division, Office of Building Technology, U.S. Department of Energy, NTIS, Springfield, VA.

Lysen, E.H., C.D. Ouwens, M.J.G. van Onna, K. Blok, P.A. Okken, and J. Goudriaan, 1992: *The Feasibility of Biomass Production for The Netherlands Energy Economy.* The Netherlands Agency for Energy and the Environment (NOVEM), Apeldoorn, The Netherlands.

Maclaren, J.P., D.Y. Hollinger, P.N. Beets, J. Turland, 1993: Carbon sequestration by New Zealand's plantation forests. *NZ Journal Sci.*, **23(2)**, 194-208.

Marchetti, C., 1989: How to solve the CO_2 problem without tears. *Int. J. Hydrogen Energy,* **14**, 493-506.

Mark, J., J.M. Ohi, and D.V. Hudson, 1994: Fuel savings and emissions reductions from light duty fuel cell vehicles. In: *A Collection of Technical Papers: Part 3, 29th Intersociety Energy Conversion Engineering Conference,* Monterey, CA, 7–11 August, pp. 1425–1429.

Marland, G. and S. Marland, 1992: Should we store carbon in trees? *Water, Air, and Soil Pollution,* **64**, 181-195.

Marland, G. and A. Turhollow, 1990: *CO_2 Emissions from Production and Combustion of Fuel Ethanol from Corn.* ORNL/TN-11180, Environmental Sciences Division, Oak Ridge National Laboratory, Oak Ridge, TN.

Marrison, C.I. and E.D. Larson, 1995a: Cost versus scale for advanced plantation-based biomass energy systems in the U.S. In: *Proc. U.S. EPA Symposium on Greenhouse Gas Emissions and Mitigation Research,* Washington, DC, 27–29 June, National Renewable Energy Laboratory, Golden, CO, pp. 1272–1290.

Marrison, C.I. and E.D. Larson, 1995b: Cost versus scale for advanced plantation-based biomass energy systems in the U.S. and Brazil. In: *Proc. Second Biomass Conf. of the Americas*, 21–24 August.

Mårtensson, A., 1992: Inherently safe reactors. *Energy Policy,* July, **20(7)**, 660–671.

Masters, C.D., E.D. Attanasi, and D.H. Root, 1994: World petroleum assessment and analysis. In: *Proceedings of the 14th World Petroleum Congress, Stavanger Norway*. John Wiley & Sons, New York, NY.

Menendez, J.A.E., 1992: Risk analysis in the development of new technologies of clean coal use. In: *Proc. New Electricity 21, Power Industry Technology and Management Strategies for the 21st Century.* IEA, Tokyo, Japan, May.

Mills, D. and B. Keepin, 1993: Baseload solar power: near-term prospects for load following solar thermal electricity. *Energy Policy,* **21(8)**, 841-857.

Mills, E., D. Wilson, T.B. Johansson, 1991: Getting started: no-regrets strategies for reducing greenhouse gas emissions. *Energy Policy*, July/August, **19(6)**, 256–541.

Moore, E.A. and G. Smith, 1990: *Capital Expenditures for Electric Power in the Developing Countries in the 1990s.* World Bank Industry and Energy Dept. Series, Working Paper No. 21, Washington, DC, 108 pp.

Moreira, J.R. and A.D. Poole, 1990: *Alternativas Energéticas e Amazonia.* Report analysis of the implementation of large energy projects—the case of the electric sector in Brazil, part 4. Ford Foundation, Rio de Janeiro, Brazil, 212 pp.

Moreira, J.R. and A.D. Poole, 1993: Hydropower and its constraints. In: *Renewable Energy: Sources for Fuel and Electricity* [Johansson, T.B., H. Kelly, A.K.N. Reddy, and R.H. Williams (eds.)]. Island Press, Washington, DC, pp. 73–120.

Mukunda, H., S. Dasappa, and U. Srinivasa, 1993: Wood gasification in open-top gasifiers—the technology and the economics. In: *Renewable Energy: Sources for Fuel and Electricity* [Johansson, T.B., H. Kelly, A.K.N. Reddy, and R.H. Williams (eds.)]. Island Press, Washington, DC, pp. 699–728.

Mukunda, H., S. Dasappa, P.J. Paul, N.K.S. Rajan, and U. Srinivasa, 1994: Gasifiers and combustors for biomass. *Energy for Sustainable Development,* **1(3)**, 27-38.

Muschalek, K.I. and K. Scharmer, 1994: The global ecological balance for engine fuel production from vegetable oils. In: *Biomass for Energy and Industry, 7th EC Conference* [Hall, D.O., G. Grassi, and H. Scheer (eds.)]. Ponte Press, Bochum, Germany, pp. 578-583.

National Academy of Sciences, Committee on International Security and Arms Control, 1994: *Management and Disposition of Excess Weapons Plutonium.* National Academy Press, Washington, DC.

NBR, 1994: *Principles and Guidelines for the Development of Biomass Energy Systems.* NREL, Golden, CO.

NEA, 1994: *The Economics of the Nuclear Fuel Cycle.* OECD, Paris, France, 177 pp.

NEA, 1995a: *Proc. 3rd Internat. Information Exchange Meeting on Actinide and Fission Product Partitioning and Transmutation.* OECD, Paris, France, 520 pp.

NEA, 1995b: *Nuclear Energy Data*, OECD, Paris, France, 520 pp.

The Netherlands Ministry of Economic Affairs, 1993: *Nuclear Energy Dossier*. Den Haag, The Netherlands.

Nitsch, J., 1992: Potential, barriers, and market chances for renewable energy sources. *Das Solarzeitalter,* **4/92**, 17–29 (in German).

OECD, 1992: *The Economic Costs of Reducing CO_2 Emissions.* OECD Economic Studies Special Report No. 19, OECD, Paris, France, pp. 141-165.

OECD, 1994: *National Accounts: Main Aggregates.* Vol. I, *1960-1992.* OECD, Paris, France.

OECD/NEA, 1993: *Nuclear Energy Data.* OECD, Paris, France.

Ogden, J.M. and J. Nitsch, 1993: Solar hydrogen. In: *Renewable Energy: Sources for Fuels and Electricity* [Johansson, T.B., H. Kelly, A.K.N. Reddy, and R.H. Williams (eds.)]. Island Press, Washington, DC, pp. 925–1009.

Ogden, J.M., E.D. Larson, and M.A. DeLuchi, 1994: *A Technical and Economic Assessment of Renewable Transportation Fuels and Technologies.* Report to the Office of Technology Assessment, U.S. Congress, Washington, DC, 27 May.

Olsen, F., 1993: *Principles of Combined Heat and Power Generation.* DEA, pp. 18–21.

OTA, 1992: *Technologies to Sustain Tropical Forest Resources and Biological Diversity.* OTA-F-515, U.S. Government Printing Office, Washington, DC.

OTA, 1993: *Potential Environmental Impacts of Bioenergy Crop Production—Background Paper.* OTA-BP-E-118, U.S. Government Printing Office, Washington, DC.

Pachauri, R.K., 1993: *The Economics of Climate Change: A Developing Country Perspective*. The International Conference on the Economics of Climate Change, OECD, IEA, Paris, France.

Palmerini, C.G., 1993: Geothermal energy. In: *Renewable Energy: Sources for Fuels and Electricity* [Johansson, T.B., H. Kelly, A.K.N. Reddy, and R.H. Williams (eds.)]. Island Press, Washington, DC, pp. 549–591.

Pearce, R. and M.V. Twigg, 1981: Coal- and natural gas-based chemistry. In: *Catalysis and Chemical Processes* [Pearce, R. and W.R. Patterson (eds.)]. Blachie & Son Ltd., London, UK, pp. 114–131.

Petrere, M., 1990: Alternativas para o desenvolvimento da Amazonia— a pesca e psicultura. In: *Proceedings of Alternativas para o desenvolvimento da Amazonia.* Brasilia, Brazil, September, pp. 19–20.

Phoebus, 1990: *A 30 MWe Solar Tower Power Plant for Jordan Phase IB - Feasibility Study.* Executive Summary, Phoebus-Consortium, Managing Partner Fichtner Development Engineering.

Picard, D.J., B.D. Ross, and D.W.H. Koon, 1992: *Development of the Inventory: A Detailed Inventory of CH_4 and VOC Emissions from Upstream Oil and Gas Operations in Alberta.* Canadian Petroleum Association, Canada.

Picard, D.J. and S.K. Sarkar, 1993: *A Technical and Cost Evaluation of Options for Reducing Methane and VoC Emissions from Upstream Oil and Gas Operations.* Canadian Association of Petroleum Producers, Canada.

Pinguelli Rosa, L. and R. Schaeffer, 1994: Greenhouse gas emission for hydroelectric reservoirs. *Ambio*, **23**, 164–165.

Pool, T.C., 1994: *Uranium Resources for Long Term, Large Scale Nuclear Power Requirements.* NUEXCO Review, Washington, DC.

Raabe, I.J., 1985: *Hydropower—The Design, Use, and Function of Hydromechanical, Hydraulic, and Electrical Equipment.* VDI Verlag, Dusseldorf, Germany, 684 pp.

Rajabapaiah, P., S. Jayakumar, and A.K.N. Reddy, 1993: Biogas electricity—the Pura Village case study. In: *Renewable Energy: Sources for Fuels and Electricity* [Johansson, T.B., H. Kelly, A.K.N. Reddy, and R.H. Williams (eds.)]. Island Press, Washington, DC, pp. 787–815.

Ramakrishan, K., 1993: Technology options and technology transfers—an Indian experience. IEA Second International Conference on the Clean and Efficient Use of Coal and Lignite, Hong Kong.

Ravindranath, N., 1993: Biomass gasification: environmentally sound technology for decentralized power generation, a case study for India. *Biomass and Bioenergy,* **4**, 49–60.

Ravindranath, N. and D. Hall, 1995: *Biomass Energy and Environment: A Developing Country Perspective from India.* Oxford University Press, Oxford, UK (in press).

Reddy, A.K.N., V. Balu, G.D. Sumithra, A. D'Sa, P. Rajabapaiah, and H.I. Somasekhar, 1994: Replication of rural energy and water supply utilities (REWSUs): an implementation package and proposal. Paper presented at Bioresources '94 Biomass Resources: A Means to Sustainable Development, Bangalore, India, 3–7 October.

Renn, O., 1993: Public acceptance of energy technologies in Europe. Paper presented to the EC-seminar in Venice (DGXII).

Riedacker, A., 1993: La maitrise integree de la gestion des ecosystemes et de l'energie. *Secheresse,* **4(4)**, 265–284.

Riedacker, A. and B. Dessus, 1991: Increasing productivity of agricultural land and forest plantations to slow down the increase of the greenhouse effect. In: *Intensification Agricole et Reboisement dans la Lutte contre le Renforcement de l'Effet de Serre* [Grassi, G., A. Collina, and H. Zibetta (eds.)]. Proceedings of the 6th European Conference on Biomass for Energy, Industry, and Environment, Elsevier Applied Science, London, UK, pp. 228–232.

Riemer, P.W.F. (ed.), 1993: *Proceedings of the IEA Carbon Dioxide Disposal Symposium.* Pergamon Press, Oxford, UK.

Rodier, M., 1992: Electricité de France. In: *Tidal Power: Trends and Developments* [Clare, R. (ed.)]. Inst. of Civil Engineers, London, UK.

Rogner, H.-H. and F.E.K. Britton, 1991: *Energy, Growth & the Environment: Towards a Framework for an Energy Strategy.* Think-piece submitted to the Directorate General for Energy (DG XVII), Commission of the European Communities, Brussels, Belgium.

Rogner, H.-H., 1993: Clean energy services without pain: district energy systems. *Energy Studies Review,* **5(2)**, 114-120.

Ross, M., 1994: *Fuel Economy Analysis for a Hybrid Concept Car Based on Buffered Fuel-Engine Operating at an Optimal Point.* Department of Physics, University of Michigan, Ann Arbor, MI.

RSWG, 1990: *Emissions Scenarios.* Appendix of the Expert Group on Emissions Scenarios (Task A: Under RSWG Steering Committee), U.S. Environmental Protection Agency, Washington, DC.

RTI, EPS, and WEC, 1993: *Whole Tree Energy Design.* Vol. 1, *Engineering and Economic Evaluation.* TR-101564, Electric Power Research Institute, Palo Alto, CA.

Rudd, J.W.M., H. Reed, C.A. Kelly, and R.E. Hecky, 1993: Are hydroelectric reservoirs significant sources of greenhouse gases? *Ambio,* **22**, 246–248.

Saviharju, K., 1995: *Combined Heat and Power Production.* Background paper for IPCC Working Group II. *VTT Energy* (to be published in *VTT Research Notes*), Espoo, Finland, 35 pp.

Schröder, L. and H. Schneich, 1986: *International Map of Natural Gas Fields in Europe.* Niederschasisches Landesamt fur Bodenforschung und Bundesanstalt fur Geowissenschaften und Rohstoffe, Hannover, Germany.

Scott, D.S., 1993: Hydrogen in the evolving energy system. *International Journal of Hydrogen Energy,* **18(3)**, 197-204.

SCS, 1989: *The Second RCA Appraisal: Soil, Water, and Related Resources on Non-Federal Land in the United States—Analysis of Condition and Trends.* U.S. Department of Agriculture, Washington, DC.

Semenov, B.A., L.L. Bennett, E. Bertel, 1995: Nuclear power development in the world. In: *Proceedings of an International Conference on the Nuclear Power Option,* 5–8 September 1994, Vienna, Austria, IAEA, STI/PUB/964, pp. 25–39.

Shell/WWF, 1993: *Plantation Guidelines.* Shell/WWF Tree Plantation Review, Shell International Petroleum Company, Shell Centre, London, UK, and World Wide Fund for Nature, Surrey, UK, 32 pp.

Shugar, D., 1990: Photovoltaics in the utility distribution system: the evaluation of system and distributed benefits. 21st IEEE Photovoltaics Specialty Conference, Las Vegas, NV.

Simbeck, D.R., 1995: Air-blown vs. oxygen-blown gasification—an honest appraisal. In: *Alternate Energy '95.* Council on Alternate Fuels, Vancouver, BC, Canada.

Skovholt, O., 1993: CO_2 transportation system. *Energy Convers. Mgmt.,* **34**, 1095–1103.

Slovic, P., 1992: Perception of risk and the future of nuclear power. In: *Technologies for a Greenhouse Constrained World* [Kuliasha, M.A., A. Zucker, and K.J. Ballew (eds.)]. Lewis Publ., Boca Raton, FL, pp. 349–359.

Smets, H., 1987: Compensation for exceptional environmental damage caused by industrial activities. In: *Insuring and Managing Hazardous Risks.* Springer Verlag.

Smil, V., 1994: How many people can the earth feed? *Population and Development Review,* June, **20(2)**, 255–292.

Smith, I.M. and L.L. Sloss, 1992: *Methane Emissions from Coal.* IEA Coal Research, London, UK, November, 18 pp.

Sørensen, B., 1979a: *Renewable Energy.* Academic Press, London, UK, 683 pp.

Sørensen, B., 1979b: Nuclear power: the answer that became a question. *Ambio,* **8**, 10-17.

Sørensen, B., 1981: A combined wind and hydro power system. *Energy Policy,* March, **9(1)**, 51–55.

Sørensen, B., 1984: Energy storage. *Annual Review of Energy,* **9**, 1–29.

Sørensen, B., 1987: Current status of energy supply technology and future requirements. *Science and Public Policy,* **14**, 252–256.

Sørensen, B., 1988: Optimization of wind/diesel systems. In: *Asian and Pacific Area Wind Energy Conference,* Shanghai, China. Japan Wind Energy Association, Tokyo, Japan, pp. 94–98.

Sørensen, B. and M. Watt, 1993: Lifecycle analysis in the energy field. In: *Energes '93—the 5th International Conference,* Korean Institute of Energy Research, Seoul, Republic of Korea, pp. 66–80.

Sørensen, B., L. Nielsen, S. Pedersen, K. Illum, and P. Morthorst, 1994: *Fremtidens Vedvarende Energisystem (The Future Renewable Energy System.).* The Danish Technology Council, Report 1994/3, Copenhagen, Denmark, 68 pp.

Souviron, J.P., 1994: *Debat National Energie et Environnement Rapport de Synthese aux Ministeres de l'industries, de l'Environnement et de la Recherche & de l'Enseignement superieur,* La Documentation Francaise, Paris, France, 84 pp.

Spencer, D.F., 1993: Use of hydrate for sequestering CO_2 in the deep ocean. National Summer Conference of the American Institute of Chemical Engineers, Seattle, WA.

Stein, W., 1994: Feasibility study for a DSG dish solar power station in Australia. Paper presented at the *7th International Symposium on Solar Thermal Concentrating Technologies,* Moscow, Russia, 26–30 September.

Steinberg, M., 1991: *Biomass and Hydrocarb Technology for Removal of Atmospheric CO_2.* BNL 4410R, Brookhaven National Laboratory, Upton, Long Island, NY.

Stewart, D. and R. McLeod, 1980: *New Zealand Journal of Agriculture,* September, pp. 9–24.

Still, D., B. Little, S.G. Lawrence, and H.A. Carver, 1994: The birds of blyth harbour, wind energy conversion 1994. In: *Proceedings of the 16th BWEA Wind Energy Conference* [Elliot, G. (ed.)]. Mechanical Engineering Publications Ltd., London, UK, June, pp. 241–248.

Strong, S.J. and R.H. Wills, 1993: Building integration of photovoltaics in the United States. In: *Proceedings of the 11th European Photovoltaics Solar Energy Conference and Exhibition,* Montreux, Switzerland, October, pp. 1672–1675.

Su, M. and S. Gu, 1993: Case study: integrating improved coal technologies into energy system of Changshu in China—an IRP approach. IEA Second International Conference on the Clean and Efficient Use of Coal and Lignite, Hong Kong.

Summerfield, I.R. *et al.,* 1994: The full fuel cycle of CO_2 capture and disposal—capture and disposal technologies. ICCDR-2, Kyoto, Japan, October.

Sun, J. and D. Li, 1992: Strategies and policy for electric power development in China. In: *New Electricity 21- An International Conference on Power Industry Technology and Management Strategies for the Twenty-First Century,* Tokyo, Japan.

Taber, J.J., 1993: The supercritical CO_2 extraction of light hydrocarbons in the large-scale miscible displacement process for producing oil from underground reservoirs. In: *JSME-ASME International Conference on Power Engineering,* Tokyo, Japan [Kagoja *et al.* (eds.)], pp. 135–166.

Tafdrup, S., 1993: Environmental impact of biogas production from Danish centralized plants. Paper presented at IEA Bioenergy Environmental Impact Seminar, Elsinore, Denmark.

Tande, J.O.G. and J.C. Hansen, 1991: Determination of wind power capacity value. In: *Proceedings Amsterdam EWEC '91*. Elsevier, Amsterdam, The Netherlands, part I, pp. 158–164.

Thorne, L., 1992: *Nuclear Proliferation and the IAEA Safeguards*. IAEA/PI/A36E/92-02835, Vienna, Austria, pp. 65–72.

Topper, 1993: Improving coal use in developing countries through technology transfer. IEA Second International Conference on the Clean and Efficient Use of Coal Lignite, Hong Kong.

Torck, B. and P. Renault, 1988: Les biotechnologies, un avenir pour une nouvelle chimie et pour l'energie. *Annales des Mines Paris*, October/November, 73–84.

Troen, I. and E.L. Petersen, 1989: *European Wind Atlas*. Risø National Laboratory, Roskilde, Denmark, 656 pp.

Tsuchiya, H., 1989: Photovoltaics cost analysis based on the learning curve. Clean and safe energy forever. In: *Proceedings of the 1989 Congress of the International Solar Energy Society*, Kobe City, Japan.

Turkenburg, W., 1992: On the potential and implementation of wind energy. In: *Proceedings Amsterdam EWEC '91*. Elsevier, Amsterdam, The Netherlands, part II, pp. 171–180.

Turhollow, A.H., and R.D. Perlack, 1991: Emissions of CO_2 from energy crop production. *Biomass and Bioenergy,* **1**, 129–135.

Turnbull, J., 1993a: *Strategies for Achieving a Sustainable, Clean, and Cost-Effective Biomass Resource*. Electric Power Research Institute, Palo Alto, CA, 20 pp.

Turnbull, J., 1993b: Use of biomass in electric power generation: the California experience. *Biomass and Bioenergy,* **4(2)**, 75–84.

Turnure, J.T., S. Winnett, R. Shackleton, and W. Hohenstein, 1995: Biomass electricity: long-run economic prospects and climate policy implications. In: *Proceedings of the Second Biomass Conference of the Americas*, Portland, OR, 21–24 August, National Renewable Energy Laboratory, Golden, CO, pp. 1418–1427.

UNEP/WHO, 1992: *Urban Air Pollution in Megacities of the World*. Blackwell Publishers, Oxford, UK.

UNCNRSEED, 1994: Report of the First Session. UN Economic and Social Council Official Records, Supplement No. 5, Doc. E/C.13/1994/8.

UNCNRSEED, 1995: Report of the Second Session. UN Economic and Social Council Official Records, Supplement No. 5, Doc. E/1995/25.

UNSCEAR, 1994: *Sources and Effect of Ionizing Radiation*. United Nations, New York, NY, 272 pp.

UNSEGED, 1992: *Solar Energy: A Strategy in Support of Environment and Development. A Comprehensive Analytical Study on Renewable Sources of Energy.* Committee on the Development and Utilization of New and Renewable Sources of Energy, A/AC.218/1992/5/Rev.1., 49 pp.

U.S. DOE, 1990: *The Potential of Renewable Energy—An Interlaboratory White Paper*. SERI/TP-260-3674, U.S. Department of Energy, Washington, DC.

U.S. DOE, 1993: *The Capture, Utilization and Disposal of Carbon Dioxide from Fossil Fuel-Fired Power Plants*, vols. I and II. U.S. Department of Energy,Washington, DC.

U.S. DOE, 1994a: *Drawing Back the Curtain of Secrecy: Restricted Data Declassification Policy 1946 to the Present.* RDD-1, U.S. Department of Energy, Washington, DC, 1 June.

U.S. DOE, 1994b. Press Release, Washington, DC, 19 December.

U.S. EPA, 1976: *Quality Criteria for Water*. Report No. EPZ-44019-76-023, Washington, DC.

U.S. EPA, 1990a: *Policy Options for Stabilizing Global Climate*. Report to Congress, Main Report, Washington, DC, December, pp. 3-1 to 5-6, 2-1 to 3-29, 7-11 to 7-18.

U.S. EPA, 1990b: *Methane Emissions and Opportunities for Control.* Workshop Results of IPCC, September (EPA/400/9-90/007).

U.S. EPA, 1993a: *Options for Reducing Methane Emissions Internationally*, Vol. I: Technical Options for Reducing Methane Emissions, Report to Congress, U.S. Environmental Protection Agency, Washington, DC, July, pp. 3-9, 3-20, 3-22, 4-7, and 4-34.

U.S. EPA, 1993b: *Anthropogenic Methane Emissions in the United States.* Report to Congress, U.S. Environmental Protection Agency, Washington, DC, pp. 2-1 to 3-29 and 7-11 to 7-18.

U.S./Japan Working Group on Methane, 1992: Technical Options for Reducing Methane Emissions, Washington, DC, January, pp. 16–71.

U.S. NRC, 1991: *Proceedings from Workshop on PSA Applications and Limitations*. Organized by NEA Committee on Safety of Nuclear Installations, NUREG/CP-0115, Washington, DC.

Uchiyama, Y. and H. Yamamoto, 1991: *Greenhouse Effect Analysis of Power Generation Plants.* Report Y91005, CRIEPI, Tokyo, Japan, 49 pp.

van den Broek, R., A. Faaij, T. Kent, K. Healion, W. Dick, G. Blaney, and M. Bulfin, 1995: Willow firing in retrofitted Irish peat plants. In: *Proc. Second Biomass Conf. of the Americas*, Portland, OR, August.

Van de Vate, J.F., 1993: Electricity generation and alleviating global climate change: the potential role of nuclear power. Paper presented at the UNIPEDE/IEA Conference on Thermal Power Generation and the Environment, Hamburg, Germany, 1–3 September.

van Engelenburg, B.C.W. and E.A. Alsema, 1994: *Environmental Aspects and Risks of Amorphous Silicon Solar Cells*. Report No. 93008, Department of Science, Technology and Society, University of Utrecht, Utrecht, The Netherlands.

van Wijk, A.J.M., J.P. Coelingh, and W.C. Turkenburg, 1993: Wind energy. In: *Renewable Energy Resources: Opportunities and Constraints 1990-2020,* chapter 3. World Energy Council, London, UK, 82 pp.

van Wijk, A.J.M., 1990: *Wind Energy and Electricity Production*. Ph.D Thesis, Utrecht University, Utrecht, The Netherlands.

Vanecek, M., A. Mahan, B. Nelson, and R. Crandall, 1992: Influence of hydrogen and microstructure on increased stability of amorphous silicon. In: *Proceedings of the 11th European Photovoltaics Solar Energy Conference and Exhibition*, Montreux, Switzerland, October.

Veltrop, J.A., 1991: Water, dams and hydropower in the coming decades. *Water Power & Dam Construction*, June, **43(6)**, 37–44.

von Kleinsmid, W. and P. De Laquil, 1993: Solar Two Central Receiver project. In: *6th International Symposium on Solar Thermal Concentrating Technologies*, Mojocar, Spain. Editorial CIEMAT, Madrid, Spain, pp. 641–654.

von Meier, A., 1994: *Manufacturing Energy Requirements and Energy Payback of Crystalline and Amorphous Silicon PV Modules*. Energy and Resources Group, University of California, Berkeley, CA.

Waddy, B.B., 1973: Health problems of man-made lakes: anticipation and realization, Kainji, Nigeria, and Ivory Coast. In: *Man Made Lakes* [Ackermann, W. *et al.* (eds.)] American Geophysical Union, Washington, DC.

Waggoner, P.E., 1994: *How Much Land Can Ten Billion People Spare for Nature?* Council for Agricultural Science and Technology, Ames, IA, 64 pp.

Walsh, M. and R. Graham, 1995: *Biomass Feedstock Supply Analysis: Production Costs, Land Availability and Yields.* Biofuels Feedstock Development Division, Oak Ridge National Laboratory, Oak Ridge, TN, 18 January.

Wang, G., S. Meng, and J. Bai, 1992: Investigations and analysis on comprehensive utilization of family-sized biogas technology in China. In: *Biogas and Sustainable Agriculture: The National Experience*. Exchange Meeting on Comprehensive Utilization of Biogas, Yichang City, Hubei Province, Department of Environmental Protection and Energy, Ministry of Agriculture, People's Republic of China, 10–15 October, pp. 139–151.

Washom, B.J., 1984: *Vanguard I Solar Parabolic Dish Stirling Engine Module.* DOE/AL/16333-2, Advanco Corporation, September.

Watson, A.M., 1983: Use pressure swing adsorption for lowest cost hydrogen. *Hydrocarbon Processing,* March, pp. 91–95.

WEC, 1991: *District Heating/Combined Heat and Power*. WEC, London, UK, 103 pp.

WEC, 1993: *Energy for Tomorrow's World*. St. Martin's Press, New York, NY, 320 pp.

WEC, 1994: *New Renewable Energy Resources—A Guide to the Future.* Kogan Page, London, UK, 391 pp.

Wenger, H.J. and T.E. Hoff, 1995: *A Guide to Evaluate the Cost-Effectiveness of Distributed PV Generation*. Draft final report prepared for Sandia National Laboratories, Albuquerque, NM.

Williams, R.H., 1993: Fuel cells, their fuels, and the U.S. automobile. In: *Proceedings of the First Annual World Car 2001 Conference*, University of California at Riverside, Riverside, CA, 32 pp.

Williams, R.H., 1994a: Roles for biomass energy in sustainable development. In: *Industrial Ecology and Global Change* [Socolow, R.H. *et al.*, (eds.)]. Cambridge University Press, Cambridge, UK, pp. 199–225.

Williams, R.H., 1994b: The clean machine. *Technology Review,* April, pp. 21-30.

Williams, R.H., 1995a: *Variants of a Low CO_2-Emitting Energy Supply System (LESS) for the World.* Report prepared for the IPCC Second Assessment Report, Working Group IIa, Energy Supply Mitigation Options.

Williams, R.H., 1995b: *Making R&D an Effective and Efficient Instrument for Meeting Long-Term Energy Policy Goals.* Directorate General for Energy (DG XVII), European Union, Brussels, Belgium, 19 June, 59 pp.

Williams, R.H., 1996: Fuel decarbonization for fuel cell applications and sequestration of the separated CO_2. In: *Ecorestructuring* [Ayres, R.U. *et al.* (eds.)]. UN University Press, Tokyo, Japan (forthcoming).

Williams, R.H. and E. Larson, 1989: Expanding roles for gas turbines in power generation. In: *Electricity: Efficient End-Use and New Generation Technologies, and Their Planning Implications* [Johansson, T.B., B. Bodlund, and R. Williams (eds.)]. Lund University Press, Lund, Sweden, pp. 503–553.

Williams, R.H. and E. Larson, 1993: Advanced gasification-based biomass power generation. In: *Renewable Energy: Sources for Fuels and Electricity* [Johansson, T.B., H. Kelly, A.K.N. Reddy, and R.H. Williams (eds.)]. Island Press, Washington, DC, pp. 729–785.

Williams, R.H. and E.D. Larson, 1995: Biomass-gasifier gas turbine power generating technology. *Biomass and Bioenergy,* special issue on 1993 EPRI Conference on Strategic Benefits of Biomass and Wastes (in press).

Williams, R.H. and H.A. Feiveson, 1990: Diversion-resistance criteria for future nuclear power. *Energy Policy*, **18(6)**, 543–549.

Williams, R.H. and G. Terzian, 1993: *A Benefit/Cost Analysis of Accelerated Development of Photovoltaic Technology*. PU/CEES Report No. 281, Center for Energy and Environmental Studies, Princeton University, Princeton, NJ, 47 pp.

Williams, R.H., E.D. Larson, R.E. Katofsky, and J. Chen, 1995a: Methanol and hydrogen from biomass for transportation. *Energy for Sustainable Development* , January, **1(5)**, 18–34.

Williams, R.H., E.D. Larson, R.E. Katofsky, and J. Chen, 1995b: *Methanol and Hydrogen Production from Biomass for Transportation, with Comparisons to Methanol and Hydrogen from Natural Gas and Coal.* PU/CEES Report No. 292, Center for Environmental Studies, Princeton University, Princeton, NJ, 46 pp.

Wolk, R.H., G.T. Preston, and D.F. Spencer, 1991: Advanced coal systems for power generation. IEA Conf. on Technology Responses to Global Environmental Challenges, Kyoto, Japan, November.

Wright, D., 1991: *Biomass—A New Future?* Report of the Commission of the European Communities, Brussels, Belgium.

WRR, Wetenschappelijke Raad voor het Regeringsbeleid (The Netherlands Scientific Council for Government Policy), 1992: *Ground for Choices: Four Perspectives for Rural Areas in the European Union.* Report No. 42, Sduuitgeverij, Den Haag, The Netherlands.

Wyman, C., R. Bain, N. Hinman, and D. Stevens, 1993: Ethanol and methanol from cellulosic feedstocks. In: *Renewable Energy: Sources for Fuels and Electricity* [Johansson, T.B., H. Kelly, A.K.N. Reddy, and R.H. Williams (eds.)]. Island Press, Washington, DC, pp. 865–923.

Zweibel, K. and A.M. Barnett, 1993: Polycrystalline thin-film photovoltaics. In: *Renewable Energy: Sources for Fuels and Electricity* [Johansson, T.B., H. Kelly, A. Reddy, and R. Williams (eds.)]. Island Press, Washington, DC.

Zweibel, K. and W. Luft, 1993: *Flat-Plate, Thin-Film Modules/Arrays.* National Renewable Energy Laboratory, Golden, CO.

Zweibel, K., 1995: Thin films: past, present, and future. *Progress in Photovoltaics*, special issue on thin films.

20

Industry

TAKAO KASHIWAGI, JAPAN

Principal Lead Authors:
J. Bruggink, The Netherlands; P.-N. Giraud, France; P. Khanna, India;
W.R. Moomaw, USA

CONTENTS

EXECUTIVE SUMMARY

This chapter provides a summary of options for reducing greenhouse gas emissions from industry, based on a survey of relevant literature. The main purposes are to outline the major sources and trends of emissions from industrial activity and to indicate possible abatement strategies from technological and institutional viewpoints. Emissions from power-generating utilities are not covered; the reader is referred to Chapter 19 for a discussion of this subject. Cogeneration emissions are included in this chapter only to the extent that they take place within an industry. The main findings and conclusions of the chapter follow:

- Industrial sources contribute to greenhouse gas (GHG) emissions in two major ways—fossil fuel combustion for energy and process-related emissions.
- The industrial sector is responsible for more than one-third of global carbon dioxide (CO_2) emissions through energy use. Based on current energy-use patterns, global energy use is expected to rise 75% by the year 2025, with an increasing portion of growth expected to occur in developing countries.
- Several technologies, processes, and new product design concepts exist that could substantially reduce GHG emissions:
  - For industrialized countries, reducing the material content of products, improving energy efficiency, and using fuels with a lower carbon content are essential to lower CO_2 emissions from the industrial sector.
  - For the reindustrializing, transitional countries of eastern and central Europe, as well as the world's developing countries, seizing the opportunity to shift to more advanced technologies will place these countries on a much lower GHG development trajectory.
- Technology transfer from industrial to industrializing countries and the establishment of innovative capacity building in developing countries are expected to generate lower GHG emissions.

Industry Emissions of Greenhouse Gases

Industrial sources of greenhouse gases are related to industrial energy use and to specific industrial production processes. The major industrial trends influencing the growth of future industry-related CO_2 emissions are the aggregate growth rate of industrial output; the structure of industrial growth (in particular the relative share of the energy-intensive, materials-producing subsector); the average energy intensity of specific products; and the fuel mix used in industry. The major challenge for future reductions in the period 1990–2020 is in the area of CO_2 from fossil fuels.

Examples of the historic correlation between growth in per capita gross domestic product (GDP) and fossil fuel CO_2 are examined for several industrialized and developing countries. The industrial-sector contribution of CO_2 varies greatly among countries. The consequences of different technological choices and strategies to date demonstrate that different industrial development paths—with substantially lower greenhouse gas emissions—are possible.

Regarding specific industry emissions, energy-intensive industries such as chemicals, cement, and steel have shown substantial improvements in energy efficiency during the past 20 years, albeit unevenly in different countries. The switch to less carbon-intensive fuels also has continued, so that CO_2 decreases have occurred in some industries during this period. The wood and paper industries in industrialized countries have reduced fossil fuel carbon emissions dramatically by using waste biomass in efficient cogeneration systems. In newly industrializing economies, efficiency gains have been slower because of a lack of economic resources and access to newer technologies. Some industrial processes are beginning to approach thermodynamic energy-efficiency limits; future gains will have to come from materials substitution, process changes, energy supply shifts, and alternative resource and industrial strategies.

Traditional emissions accounting has included all CO_2 emissions from fossil fuels as though they were part of the energy sector. In steel, aluminium, hydrogen, and ammonia production, much of the carbon release actually is process related and could be eliminated by changing production processes. Controlling process-related greenhouse gases, such as halocarbons and hydrohalocarbons, also is important, given their large greenhouse warming potential and long atmospheric lifetimes. Finally, altering chemical manufacturing processes can reduce nitrous oxide (N_2O).

Technical Abatement Options

Different strategies for reducing greenhouse gas emissions deserve attention, particularly fuel substitution—with increased use of lower-carbon fuels, biomass, and renewable energies in industrial processing—and efficiency improvement of energy supply (e.g., cogeneration) and energy use in industrial processes, including less materials-intensive production methods and renewable feedstocks and raw materials. Implementation would be most cost-effective during normal capital stock turnover.

Industries often have high energy-consumption levels, with large amounts of waste heat being released to bodies of water or the atmosphere. Heat cascading can be incorporated into a new or existing factory if careful attention is paid to the temperatures of its various industrial processes. Waste-heat recovery, which usually is aimed at rationalizing specific industrial processes, can be more effective at the level of coordinated manufacturing and energy industrial complexes. Further improvements could involve integrating energy-use issues into urban infrastructure from the initial stages of city planning with the development of support systems such as thermal-management technologies, regulations, and social mechanisms.

The manufacturing sector uses materials and chemicals to generate production and consumption goods, which ultimately are discarded or recycled. Recycling can involve restoring the material to its original use or "cascading" the material by successively downgrading its use into applications requiring lower quality. Materials cascading is well-established in the paper industry, for example, and is effective when recycling products back into the same ones is too energy-intensive or difficult. Generating materials from scrap tends to produce fewer GHGs and is less expensive than the use of primary raw materials, but there may not always be markets for downgraded materials. Emphasis also is needed on technological innovation to upgrade the quality of recycled products.

Analysis of opportunities and priorities entails preventive environmental management based on environmental, material, energy, health, and safety audits. Life-cycle analyses of energy and material flows and costs from an industrial ecosystem perspective also are essential to implement adequate solutions. Industrial systems that efficiently use waste materials and energy could produce major reductions in GHGs. Structural economic changes also are needed, as is a greater emphasis on choosing inherently lower-emitting technologies, designing products for greater reuse and recycling potential, and the replacement of nonrenewable resources with a biologically renewable resource base.

Implementation Aspects

Implementation problems to improve energy and material efficiency in industry may involve information and training aspects, financial and economic conditions, or legal and institutional issues. In the case of industrialized countries, efficiency improvements become more difficult when world energy prices are low. Imposition of ecotaxes to stimulate efficiency in industry is a complex issue because taxes that are not levied on a global scale may provoke industry relocation, which may in turn have adverse effects on emissions efficiency. Systems of internationally traded emission permits and opportunities for joint implementation are other alternatives that require careful consideration, given their economic and political implications.

In developing countries, raising efficiency levels involves complex factors such as the higher capital and foreign-exchange requirements usually associated with more-efficient processes. Moreover, the scale of operations, the scarcity of management resources, and the age composition of equipment creates additional difficulties that are not easily overcome from a purely energy-oriented point of view. Although technology transfer programs could potentially assist in the early adoption of efficient equipment, there is no reason to be more optimistic about their success in the case of energy technology than in the case of other technologies. In all countries, the tendency to look exclusively at short-term financial and economic factors to determine efficiency improvement potential overlooks institutional and organizational issues that are equally important in realizing those potentials. This is particularly true of improvements in materials efficiency, where institutional arrangements for collecting and recycling materials are crucial for a proper functioning of markets.

20.1. Introduction

A range of industrial activities and the use of manufactured products contribute significantly to the buildup of greenhouse gases (GHGs) in the atmosphere. Therefore, any attempt to slow the release of greenhouse gases will require a major restructuring of existing development patterns. The challenge is to determine if existing and future needs (and wants) for goods and services can be met without adversely altering the composition of the atmosphere and the planet's climate system.

The major contribution of greenhouse gases by the industrial sector is from carbon dioxide (CO_2) released by the burning of fossil fuels for energy. In addition, the industrial sector is responsible for significant releases of process-related greenhouse gases, including CO_2, methane (CH_4), nitrous oxide (N_2O), and other industrial gases, of which the most prominent are halocarbons and hydrohalocarbons. Sulfur oxide emissions associated with energy use and from the smelting of sulfide ores appear to be responsible for some cooling as well as for regional environmental damage.

Technological mitigation options for greenhouse gases are increasingly being sought so that societies can continue meeting human needs without jeopardizing the global climate system. This effort must be part of a comprehensive reexamination of industrial activity to address a full range of environmental and economic issues in the context of industrial ecology (i.e., meeting development goals within local and global sustainability capacities). Industrial ecology approaches attempt to emulate natural systems by utilizing waste from one industrial process as feedstock for another; efficiently utilizing "waste heat"; designing industrial processes to minimize energy and materials use; and replacing hazardous substances with more environmentally benign substitutes. Several recent works describe the industrial ecology approach and cite examples of recent attempts to implement it (Socolow *et al.*, 1994; Ayres and Simonis, 1994; Allenby and Richards, 1994).

This chapter attempts to identify existing and possible future technologies and industrial processes that can be adopted by industrial societies—which now produce the majority of greenhouse gases—by transitional economies, and by developing countries where emissions are increasing as efforts are made to improve the standard of living and material well-being of citizens. Technology transfer from industrial countries to industrializing nations and the establishment of innovative capacity-building in developing countries in accordance with the principles of Agenda 21 (UN Conference on Environment and Development, Rio de Janeiro, Brazil, 1992) are essential if the unwanted global consequences of increased atmospheric concentrations of greenhouse gases, sulfate aerosols, and climate change are to be avoided.

20.2. Industrial-Sector Emissions of Greenhouse Gases

20.2.1. Industrial Energy from Fossil Fuels

In 1990, the world industrial sector consumed an estimated 98 EJ of end-use energy (including biomass) and 19 EJ of feedstocks to produce $6.7 x 10^{12} of value added. This consumption resulted in the direct release of an estimated 1,200 Mt C. Electricity and cogenerated process heat added an additional 883 Mt C, for a total industrial-sector contribution of 2,083 Mt C—more than one-third of the total (Grubler and Messner, 1993).

Because countries differ significantly in their methods for recording and maintaining energy and economic data, precise comparisons are not always possible. The industrial use of energy for manufacturing, mining, construction, and feedstocks in Organisation for Economic Cooperation and Development (OECD) countries has fluctuated around 40 EJ since 1970 and typically is only 25–30% of total energy use. Developing countries are estimated to have raised their industrial energy use to 30 EJ by 1988 (35–45% of total energy); the transitional Eastern and Central European economies peaked at around 28 EJ of industrial energy in 1988, but experienced estimated declines of as much as 12% in the next 3 years. The industrial fraction of energy use by the Soviet Union was approximately 40% and that of China 60% in 1988 (Schipper and Meyers, 1992). In part, this variation reflects not only differences in energy intensity but also the more-rapid growth of the industrial sectors of developing countries, the restructuring of OECD economies from manufacturing to service, improved energy efficiency in manufacturing, and the transfer of some energy-intensive industries to developing countries. Within this context, attention should be paid to the potential danger of transferring greenhouse gas-intensive, ecologically damaging processes and production of chlorofluorocarbons (CFCs) to transitional and developing countries.

Worldwide future trends of industrial energy have been projected by inputting data from the Energy Modeling Forum report (EMF, 1994) into the Asian Pacific Integrated Model (AIM) (Morita *et al.,* 1994; Matsuoka *et al.,* 1994). These projections indicate that worldwide industrial energy use could rise from about 107 EJ/yr in 1990 to 140 and 190 EJ/yr in 2025 and 2100, respectively, if emissions are stabilized. These figures would increase to yearly consumption of 242 EJ in 2025 and 500 EJ in 2100 if based on an accelerated-technology scenario (EMF, 1994), where it is assumed that 400 EJ/yr of low-cost biomass becomes available in 2020; 20% of this resource is available at $1.40/EJ and the remaining 80% at $2.40/EJ.

The energy sector as a whole is thought to contribute about 57% of global warming (IPCC, 1990). Energy use in the industrial sector varies vastly among different countries. An examination of data sets that attempt to disaggregate the energy use of particular economies reveals that the structure of economies and the availability of particular energy sources play major roles in the contribution of greenhouse gases by the industrial sector. A summary of energy-related CO_2 releases for the industrial sectors of the 15 largest CO_2-emitting countries is presented in Table 20-1 (di Primio, 1993).

Clearly, improving the energy efficiency of industrial processes can substantially lower the release of fossil-fuel carbon dioxide; subsequent examples illustrate the potential for substantial

Table 20-1: Fossil-fuel carbon dioxide from the 15 largest emitters.[1]

	Total C Emissions (Mt C)	Industrial C Emissions (Mt C)	Industrial % of Total
United States	1394.83	275.4	19.74
USSR[2]	960.76	295.94	30.80
China[3]	519.25	234.29	45.12
Japan	280.40	70.78	25.24
Former West Germany	189.19	31.78	16.80
United Kingdom	154.95	21.36	13.78
India[4]	130.45	37.71	28.90
Poland[2]	126.27	15.60	12.36
Canada	122.00	27.11	22.22
Italy	108.28	20.94	19.34
France	103.72	21.05	20.29
East Germany[2]	88.69	8.87	10.01
South Africa[2]	88.31	18.27	20.69
Mexico[2]	74.17	17.62	23.75
Czechoslovakia[2]	65.63	17.75	27.05

[1]Values are for 1990, unless otherwise noted. Feedstocks are not included in either total carbon data or industrial carbon data; values for electricity inputs into the industrial sector are not included (di Primio, 1993).
[2]1988 (represents latest available data).
[3]1985 (represents latest available data).
[4]1987 (represents latest available data).

reductions. Ultimately, however, thermodynamic free-energy limits preclude further reductions. Any additional lowering of industrial energy emissions can come only by shifting from fossil fuels to non-carbon-based energy sources—such as hydro, nuclear, wind, solar, and geothermal—or by using biomass.

In terms of absolute contribution to global-warming potential, the major sources are CO_2 from fossil-fuel combustion and halocarbons. Cement calcination is the third most important source. Estimates of the quantities of global-warming potential in terms of gigatons of carbon equivalent for these three sources in industrialized, transitional, and developing countries are presented in Table 20-2. It appears that these industrial sources account for close to 30% of total global-warming potential.

20.2.2. *Industrial Processes as Sources of Greenhouse Gases*

In addition to the release of energy-related greenhouse gases, the industrial sector is responsible for the release of a number of process-related greenhouse gases—although estimates of these gases vary in their reliability. They include halocarbons and hydrohalocarbons, CO_2, CH_4, and N_2O. Whereas CO_2 emissions from fossil-fuel combustion represent the largest portion of industry greenhouse gases, the contribution of process-related emissions other than halocarbons is less recognized. Changing the production process has the potential to eliminate all greenhouse gases associated with a particular industrial or manufacturing operation. For the United States, process CO_2 emissions in 1991 are estimated to have amounted to 15.6 Mt C (EIA, 1993), or 1.2% of total fossil-fuel releases. In the comprehensive Oak Ridge data set, non-fuel-related CO_2 is estimated explicitly for cement making and implicitly from coal, natural gas, and petroleum that is oxidized in the manufacturing process (Marland, 1994; IPCC, 1991). Examples of process-related GHG releases are presented in Table 20-3; they are discussed under the appropriate industry group in Section 20.3. In some cases—such as the production of cement, lime, and iron—the release of greenhouse gases is intrinsic to the product, so the only option for GHG reductions is material substitution.

Industrial process-related gases include the following:

- CO_2 from the production of lime and cement (calcination process), steel (coke and pig-iron production), aluminum (oxidation of electrodes), hydrogen (refineries and the chemical industry), and ammonia (fertilizers and chemicals)
- Halocarbons (CFCs), hydrochlorofluorocarbons (HCFCs), and hydrofluorocarbons (HFCs) produced as solvents, aerosol propellants, refrigerants, and foam expanders
- CH_4 from miscellaneous industrial processes (iron and steel, oil refining, ammonia, and hydrogen)
- N_2O from nitric acid and adipic acid (nylon) production
- Carbon tetrafluoride or CFC-14 (CF_4) and hexafluoroethylene or CFC-116 (C_2F_6) from aluminum production (electrolysis).

Estimates of the contribution from other minor sources vary considerably. For CH_4 leakages in industrial processes and industrial-fuel combustion, a range of 1–3 Mt of CH_4 for combustion and 3–20 Mt of CH_4 for processing is reported in the literature, reflecting large uncertainties (Berdowski, 1993). Total global anthropogenic emissions amount to roughly 360 Mt of

Table 20-2: Contribution of industry to global GHG emissions (Gt C-1990).

	Fossil Fuels[1]	Halo-carbons	Cement Production[2]	Total
Industrial Economies	0.72	0.85	0.04	1.61
Transitional Economies	0.48	0.22	0.02	0.72
Developing Economies	0.68	0.22	0.07	0.97
Total	**1.88**	**1.29**	**0.13**	**3.30**

[1]Final demand.
[2]Process emissions.

Table 20-3: Industrial process-related greenhouse gases.

Industrial Process	Halocarbon/ Hydrohalocarbon	CO_2	CH_4	N_2O
Solvents	✔			
Refrigerants	✔			
Foam Blowing	✔			
Cement		✔		
Ammonia		✔	✔	
Hydrogen		✔	✔	
Nitric Acid				✔
Nylon				✔
Aluminum	✔	✔		
Steel		✔	✔	

CH_4. Estimates of industrial N_2O leakages in nitric acid and nylon production (adipic acid) are equally uncertain but now are considered more important than originally thought. A range of 0.5–0.9 Mt of nitrogen—20 to 50% of total anthropogenic N_2O emissions—is mentioned in the literature (Olivier, 1993). Abatement technologies are available that employ conversion to recoverable nitrogen dioxide (NO_2), catalytic dissociation into molecular nitrogen (N_2) and molecular oxygen (O_2), or improved N_2O destruction in specially designed boilers.

Halocarbons not only cause higher radiative forcing, they also deplete stratospheric ozone. Reduction of many halocarbons, including CFCs, already is occurring in industrial and a few developing countries because of the Montreal Protocol and subsequent amendments. The contribution of halocarbons to radiative forcing will begin to decrease after 2000, although concern over illegal imports into industrial countries is growing (*Scientific American*, 1995). Hydrohalocarbon substitutes for CFCs are expected to grow rapidly in the coming decades and will make modest additions to total warming potential.

20.2.3. Past Trends and Future Prospects for Industrial-Sector GHG Emissions

To discuss the historical and future trends in GHG emissions from industrial energy use, it is convenient to distinguish four essential driving forces. The impact of each driving force on GHG emissions can be influenced by specific policies. These four driving forces and their respective policy domains follow:

- Growth in the volume of industrial production (volume effect—macroeconomic policies), which indicates the quantitative impact of industrial growth
- Changes in the sectoral pattern of industrial growth (structural effect—sectoral policies), which indicate the qualitative impact of industrial growth
- Reduction in energy use for specific products (efficiency effect—energy conservation policies), which is a measure of the impact of energy efficiency improvements
- Changes in energy carrier composition (fuel-mix effect—energy supply policies), which indicate the potential to replace coal and oil by natural gas, renewables, or nuclear energy.

To discuss actual trends and policy opportunities with respect to these four driving forces, it is convenient to distinguish between three regional groups of countries: industrial economies (OECD), transitional economies (Eastern Europe and the former Soviet Union), and developing economies (the rest of the world). The discussion centers on manufacturing, which generally accounts for more than 90% of industrial energy demand. Data on other industrial sectors (energy and mining) are not included.

20.2.3.1. Trends for Industrialized (OECD) Countries

Average manufacturing value in industrialized nations grew from 1973 to 1988 at about 2.3% annually. Nevertheless, energy use decreased by about 1.2% annually. Aggregate manufacturing energy intensity was substantially reduced: in Japan by 45%, in the United States by 43%, and in Europe by 34%. Energy intensities for specific industries in the United States tend be higher than those in European countries, whereas those in Japan tend to be lower.

Schipper and Myers (1992) examined the energy/economic value ratio for a variety of industrial sectors for the United States, Japan, and six European countries. All major industrial sectors recorded gains in measured energy efficiency between 1971 and 1988. For this group of eight nations, energy intensity for chemicals declined by 37%, pulp and paper by 27%, building materials by 32%, ferrous metals by 27%, and nonferrous metals by 26%. Surprisingly, less energy-intensive sectors also declined by an average of 37% during the same period. Ferrous metals and pulp and paper require the greatest energy per value added, followed by nonferrous metals, building materials, and chemicals. The major source of these gains was identified as a combination of 1) better housekeeping practices in operation and maintenance, 2) higher cost investments in process equipment, and 3) replacement of older, less efficient technology with new production processes. In some subsectors such as chemicals, there is also some shift to less energy-intensive products. An excellent case study of how multimillion dollar energy efficiency gains have been achieved in the American chemical industry has been given by Nelson (1994).

Several groups have attempted to assess the potential for future savings from energy efficiency gains in the industrial sector of specific economies. The analysis carried out for the Swedish utility Vattenfall identified major industrial energy savings that would lower CO_2 significantly by 2010 (Johansson *et al.*, 1989).

Statistical analysis of the impact of changes in the sectoral composition of industrial production on energy demand show that in this period structural change generally had a continuous, but relatively small, effect in reducing energy demand. Efficiency

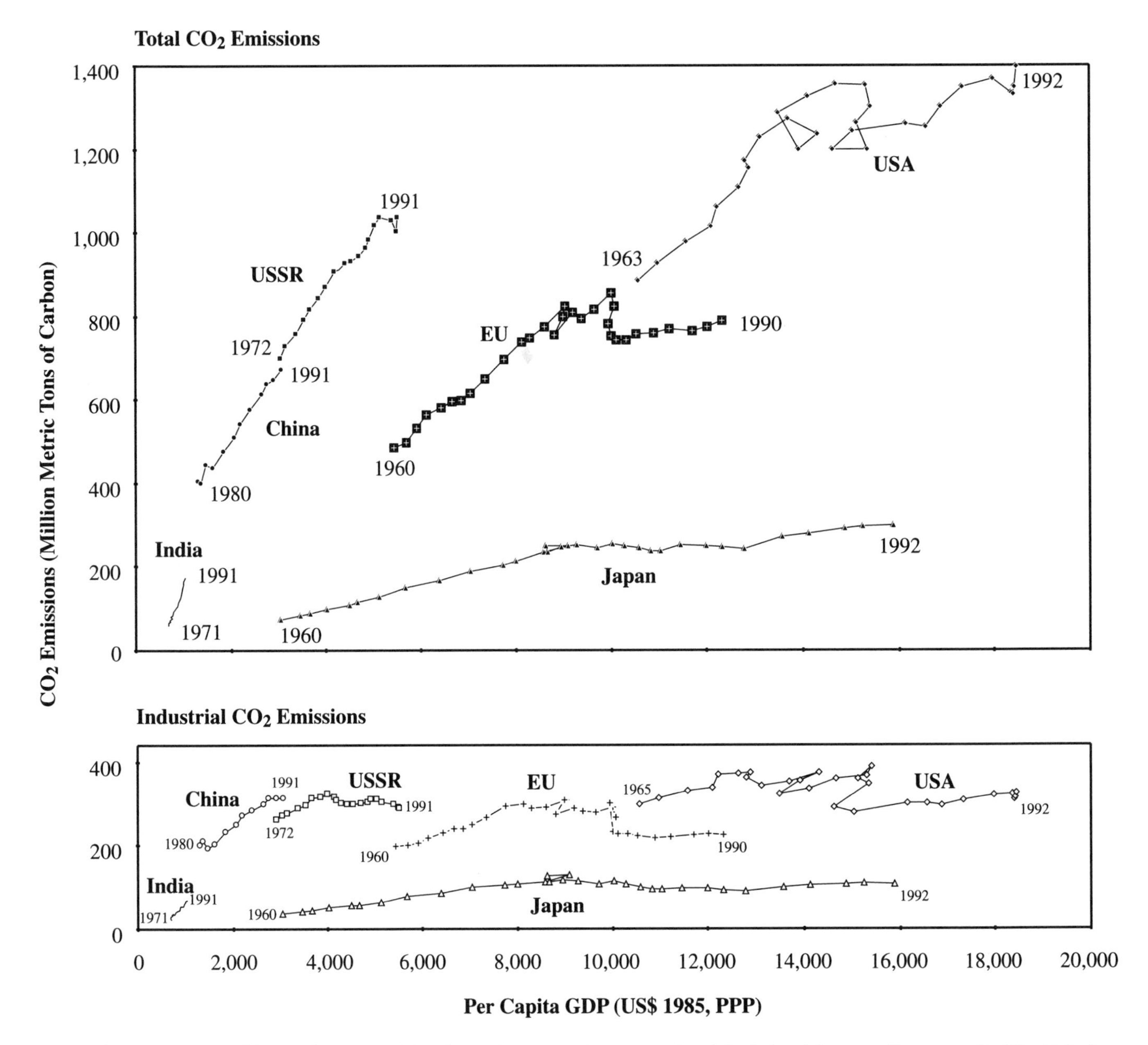

Figure 20-1: Fossil fuel CO_2 development path for the total economy (upper panel) and the industrial sectors (lower panel) of the United States, the 15 nations that now comprise the European Union (less East Germany), Japan, China, India, and the former Soviet Union. The industrial sector is as defined by OECD, plus CO_2 associated with refineries and the fraction of electricity that is used by industry. The CO_2 values are from OECD (1994) and the purchasing power parity (PPP) values are from Summers and Heston (1991, 1994).

improvements have been much more important than structural changes in curtailing energy demand, particularly in the period after 1980 when energy prices were very high (Bending, 1987; Schipper, 1992). Cyclical movements complicate this picture considerably because energy-intensive industries, such as the metallurgical and chemical subsectors, are much more sensitive to business cycles than is aggregate industrial production. Moreover, historical trends are different for fuels and electricity. In general, net improvements in electricity efficiency have been lower than in fuel-use efficiency because of the relatively important role of new electricity-using technologies in manufacturing, which offsets efficiency gains in traditional electrical equipment. Given these energy-demand trends and taking into account the industrial fuel-composition changes toward natural gas, the general trend for GHG emissions from fuel combustion in the manufacturing sector of industrial nations in the past 2 decades has been downward (Torvanger, 1991; Figure 20-1). The overall decrease from 1973 to 1991 was about 15%.

Under business-as-usual conditions, emissions from industrial countries will continue to grow moderately but with a gradual shift away from energy-intensive industries. The materials-producing industries are likely to have slower growth than less energy-intensive industries. This trend is caused partially by saturation effects on the materials demand side (Williams, 1987) and partially by ongoing economic globalization, which may shift energy-intensive industries to countries with major natural resource endowments (e.g., minerals, fossil fuels, and hydropower) and lower environmental priorities (because of lower environmental loads and urgent poverty problems). Nevertheless, the continuing improvement of bulk transportation technology and the growing integration of quantity-oriented production of bulk materials with quality-oriented production of semifinished products may put a substantial brake on the latter trend. Opportunities for further energy conservation are certainly available, especially for newly emerging industries. Fuel-mix effects may tend to increase emissions by a modest amount because the share of coal may increase after the year 2000. The aggregate impact of all effects leads more or less to a stabilization of GHG emissions in OECD industrial countries.

20.2.3.2. Trends for Transitional Economies

Whereas many studies analyze industrial energy demand developments in OECD countries, the historical evidence regarding transitional economies is much scarcer. Data availability and compatibility problems abound. In transitional economies, industry has been very dominant, accounting for approximately half of total energy use. Through the late 1980s, energy demand grew at a rate of about 2% annually, with relatively little progress in energy efficiency or structural change. In the 5 years after it reached a peak in 1988, industrial energy demand dropped sharply from central planning levels, although less so than industrial output. CO_2 is estimated to have dropped by a little more than 25% in the former Eastern bloc between 1988 and 1993 (Brown *et al.*, 1995). Thus, manufacturing energy demand and related GHG emissions have been much more directly coupled to aggregate industrial growth and decline (volume effects) and much less affected by structural and efficiency effects than has been the case in OECD industrialized nations. The trend for transitional economies with respect to GHG emissions had been strongly upward until 1989; at that time, a clear break occurred, and the trend has been sharply downward to the present.

For transitional economies, the rate of economic activity is likely to be more fluctuating than for OECD countries. In the short run, a further decrease in aggregate industrial energy demand can be expected because of the ongoing, dramatic restructuring of the industrial sector—which may affect the energy-inefficient heavy industries in particular. Although some Eastern European countries already are recovering from earlier transition difficulties, this is not yet the case for the former Soviet Union, where further declines are expected (Schipper and Martinot, 1993). At least for the period 1990 to 1995, a sharp drop in GHG emissions is fairly certain. Nevertheless, in the long run, these economies may expand again at a high rate, compensating for their earlier performance lapses. By that time, the relatively strong role of heavy industry will not continue under conditions of a market economy, and the development of the sectoral pattern will be toward reduced GHG emissions. Given excessively high current energy intensities, opportunities for energy conservation are very great. Finally, because of the increasing role of natural gas, the fuel mix is likely to change in a direction favorable for reducing GHG emissions, provided that the high leakage rate of the gas transmission and distribution system can be addressed. The aggregate impact of all effects probably will lead to a small decline in industrial GHG emissions between 1990 and 2020.

20.2.3.3. Trends for Developing Countries

Industrial energy use in developing countries is dominated by China, India, and Brazil. Until 1980, China followed national policies focusing on heavy industry; since 1980, however, the emphasis has shifted toward lighter industries. Industrial growth has been very high—14% annually. At the same time, aggregate manufacturing energy intensity declined considerably—30% between 1980 and 1988 (Huang, 1993). However, structural change accounted for only a small part of this reduction; most of the change was caused by efficiency improvements in all industrial sectors. Industrial growth in India has been slower in the same period but still impressive, at roughly 7% annually. India's energy efficiency increased by about 25%. A larger part of this improvement probably was accounted for by structural changes than was the case in China. The picture for Brazil is altogether different because the growth in export-oriented, energy-intensive industries was very high. From 1973 to 1988, industrial production increased by 65% while energy intensity increased by 25%, leading to a doubling of total energy use. Structural changes between and within industrial sectors were the dominating force, more than compensating for efficiency improvements in individual sectors (Keller, 1991). However, part of this growth has been based on hydroelectric developments and, because of the effects of shifts to biomass alcohol fuel, the additions to GHG emissions are less than energy demand growth.

A study of the Indian industrial sector concluded that energy efficiency gains of 8–10% were possible in the metals industries; 10–15% in chemicals, ceramics, and glass; 20–25% in cement and pulp and paper; and 70–80% in sugar refining. These gains would bring Indian industrial energy intensity more into line with energy-efficient modes of production (TERI, 1991).

With respect to other developing countries, few general conclusions are possible—given the widely diverging industrial developments of Southeast Asian economies versus sub-Saharan economies, for example. The developing world as a whole is experiencing an upward and accelerating trend in industrial GHG emissions.

The largest single additive impact in all respects will come from industrial growth in developing countries. Moreover, because the world's most populous countries—China and India—are dependent to a large extent on domestic coal resources, this impact will be reinforced by an additive fuel-mix impact through increased average CO_2 content of industrial energy use. The impact from sectoral shifts (qualitative structural change) will be additive but relatively small because major expansions in the bulk-materials industries already have taken place. Although the potential for reductions in specific energy use are large, they are likely to meet severe implementation problems, and they will be insufficient to counteract the other effects in a major way. The aggregate effect will lead to a relatively rapid increase in GHG emissions.

20.2.4. Variations Among Regional Groups

It is important to stress that not all economies produce the same levels of greenhouse gases even when they have achieved the same average levels of affluence. An examination of these differences among countries and the particular economic development paths they have followed in getting to their present economic and industrial stage of development can provide useful insights into which policies and strategies may be most effective for producing lower-GHG industrial societies. Figure 20-1 illustrates fossil fuel development paths over the past 30 years for the United States, the 15 nations that now comprise the European Union (less East Germany), Japan, China, India, and the former Soviet Union (Moomaw and Tullis, 1995). These countries accounted for just over 80% of global CO_2 emissions in 1990. Both total and industrial-sector CO_2 emissions are correlated with national GDP, which is expressed as purchasing power parity in 1985 U.S. dollars (Summers and Heston, 1991, 1994). The industrial-sector carbon dioxide data consist of industrial-sector emissions plus refinery emissions and the industrial electricity fraction of each nation's electric power-sector emissions (OECD, 1994). It is clear that different nations have followed very different fossil-fuel trajectories to arrive at their present economic status (Moomaw and Tullis, 1994). The decline in CO_2 in the former Soviet Union is tied to the decrease in industrial output. Whereas total and industrial-sector CO_2 emissions continue to rise for expanding industrializing economies like those of China and India, they have nearly leveled off or decreased for industrial nations. It is important to note that the industrial-sector and total CO_2 emission levels of most OECD countries are below their historic peaks even though their economies and populations have continued to expand.

The expected large impact of volume growth in developing nations does not imply that they will follow the historic industrialization path of industrial nations. Technological developments have changed the optimal scale of production and production-factor requirements dramatically in the past few decades. Moreover, the availability of resources and their relative prices in developing countries are not comparable to historic conditions in industrial countries. Information technology and materials technology have changed industrial processes enormously in the recent past and will continue to do so in the near future. It may well be possible for developing countries to undergo an earlier carbon transition than was the case for present industrial nations.

20.3. Specific Industry Emissions

20.3.1. Basic Metals

The iron and steel, aluminum, copper, and other nonferrous metal industries are extremely energy-intensive (and release process CO_2 and other GHGs, as well). In the United States, they are the leading industrial producers of CO_2, at 63.1 Mt C or 20% of the industrial sector (Marland and Pippin, 1990). Significant efficiency gains have been realized during the past 20 years as new processes have been developed.

20.3.1.1. Iron and Steel

The release of CO_2 from the iron and steel industry in 15 leading industrial nations was estimated to be 151 Mt C in 1990. Carbon inputs are equally divided between energy and chemical feedstocks such as coke. Integrated, primary steel works use approximately 550 kg C per ton of crude steel, whereas electric-arc furnaces using scrap average only 128 kg C per ton of crude steel. The European and Asian steel industries each release approximately 51 Mt C, North America 47 Mt C, and South Africa 2 Mt C. Nationally, the United States leads with 42 Mt C, followed by Japan with 37 Mt C (IISI, 1993).

Iron- and steel-related carbon emissions from the 15 leading producers amount to 2.7% of total global carbon, but their iron and steel production represents only a fraction of the world's total. Several of the largest producers are heavy coal users; they use much-less-energy-efficient technology than that cited above. For example, Japan requires only 19 GJ per ton of steel and the United States 24, but the average for the former Soviet Union is 31, China 38, and India 41 (Chandler, 1985). Additional CO_2 is released in the use of blast-furnace limestone—about 1 Mt C in the United States (Forrest and Szekely, 1991)—and in the removal of CH_4 and other gaseous components from coke. Overall, contributions of the iron and steel industry to carbon emissions could be in the range of 7 to 8% of world totals.

Significant improvements in the efficiency of steelmaking have been made during the past two decades. The Japanese industry has demonstrated a 20% improvement between 1973 and 1990 (IISI, 1993; MITI, 1994). Most of the gains have come about as the result of heat-recovery technologies, although an estimated 40% of process heat is still lost. Improvements have occurred in cokemaking, continuous casting (reducing reheating requirements), and continuous annealing. Top-pressure-recovery turbines also generate electricity from furnace top-gas during iron production. The European coal and steel community is carrying out a research program that suggests that further gains can come from scrap preheating; improved coal and oxygen use; and recycling of coke-oven, blast-furnace, and converter gas (IISI, 1993).

Additional GHG reductions can result from changing iron and steel production processes. Process-related CO_2 for the 15 leading producer countries amounts to 1.3% of global carbon emissions. The most dramatic improvements have come from the adoption of electric-arc furnaces and the efficient evolution of that technology. Using scrap eliminates carbon releases associated with metal reduction, and total carbon use per ton is only about one-fourth that of an integrated facility. Problems from contamination by impurities may limit this method and suggest that the original metallurgical design of steels should consider their recycling potential. Hydrogen rather than carbon reduction of iron ores would produce a dramatic lowering of carbon emissions if inexpensive carbon-free sources such as hydro or solar electrolysis of water could be developed (IISI, 1993). In the future, carbon for steelmaking may come from used tires or organic chemical wastes rather than from coal (Corcoran, 1994; MMT, 1993).

20.3.1.2. Aluminum

The production of bauxite ore is so localized that the top ten national producers are responsible for 92% of the world total (WRI, 1994). Similarly, aluminum production of 18.194 Mt in 1991 is concentrated in the United States (4.12 Mt), the former Soviet Union (2 Mt), Canada (1.83 Mt), Australia (1.24 Mt), Brazil (1.14 Mt), China (0.86 Mt), Norway (0.83 Mt), and Germany (0.74Mt); India (0.44 Mt) and Venezuela (0.6 Mt) also are major producers (Plunkert and Sehnke, 1991). In 1991, 17.2 Mt of aluminum was consumed, 73% of it in 10 top-consuming nations—led by the United States (24%), Japan (14%), Germany (8%), the former Soviet Union (6%), China (5%), and France (4%) (WRI, 1994).

Aluminum refining from bauxite uses vast quantities of electricity and releases CO_2 whenever coal and other fossil fuels are used. An estimated 280 x 10^9 kWh (Young, 1992) was needed to produce 18 Mt of aluminum in 1990 (Plunkert and Sehnke, 1991). The efficiency of electricity use varies from a low of 13,000–15,500 kWh per ton of aluminum in much of Europe, Brazil, Japan, and the United States to a high of 18,000–20,000 kWh/t in Norway, Russia, and Canada. Aluminum from secondary scrap requires only 1,600 kWh/t (Chandler, 1985). In most market economies, hydropower—which is inexpensive and often subsidized—is used as the source of electricity, but in Australia (the leading producer of bauxite), China, India, and the former Soviet Union, coal is a principle source of electricity. Other countries, such as Brazil, Ghana, and Canada, have large hydro projects tied to aluminum production. The United States, which used 859 PJ of energy in aluminum production in 1989, draws 12.4% from hydropower, 3.5% from nuclear, 49.6% from coal, 26.5% from oil, and 8.1% from gas (Aluminum Association, 1991).

Significant improvements in energy efficiency have been realized by the aluminum industry worldwide. In the United States, efficiency gains in 1989 were estimated at 12.1% since 1972, with an additional 9.2% arising from increases in recycled scrap—for a 21.3% reduction in energy use per ton of aluminum (Aluminum Association, 1991).

CO_2 also is released as a process gas from the destructive oxidation of the carbon anode during electrolysis of bauxite ore. Carbon emissions from this source are estimated to be 0.45 kg of carbon per kg of aluminum [some of which can be as carbon monoxide (CO) (Sadoway, 1990)]—which would imply a total release of 8.2 Mt C, or 0.15% of global fossil-fuel carbon in 1991. This carbon apparently is included in existing emission totals as part of the petroleum coke from which the electrodes are made (Marland, 1994). The use of experimental non-carbon electrodes has been demonstrated (Hryn and Sadoway, 1993), and an alternative chloride-based (rather than oxide-based) process with nondissolving graphite electrodes is claimed to release no CO_2 and to be 30% more efficient (Altehnpohl, 1980).

The second class of important aluminum process GHGs are the perfluorocarbons, carbon tetrafluoride and hexafluoroethylene, which are released during electrolysis of bauxite. These gases have an exceptionally strong absorption in the infrared and an estimated lifetime in the atmosphere in excess of 10,000 years. The range of estimates for perfluorocarbons vary from 0.6 to 2.5 kg per ton of aluminum for CF_4 and from 0.06 to 0.25 for C_2F_6 (Abrahamson, 1992; U.S. DOE, 1994; Haupin, 1987). An estimated 28,000 t of CF_4 and 3,200 t of C_2F_6 were released in 1987 (Fabian *et al.*, 1987); CF_4 has been estimated to have contributed 1.7% of the global-warming potential during the 1980s (Lashof and Ahuja, 1990).

Unfortunately, no reliable estimates of total CO_2 release associated with aluminum production exist, but in the United States, the industry accounts for 1.2% of national energy demand and a similar percentage of CO_2 emissions (Aluminum Association, 1991). If this figure prevails worldwide, then aluminum would be responsible for more than 3% of global-warming potential.

20.3.1.3. Copper and Other Nonferrous Metals

In 1992, 9.3 Mt of primary copper was mined worldwide, with nearly 40% coming from Chile and the United States. In addition, 3.4 Mt of lead, 7.1 Mt of zinc, 0.9 Mt of nickel, and lesser amounts of tin and cadmium were produced (WRI, 1994). CO_2 emissions from the U.S. primary and secondary copper industry were estimated at 19 Mt C in 1987, down from 30 Mt C in 1968 (Forrest and Szekely, 1991).

For copper and other nonferrous metals, the major greenhouse gas is CO_2 associated with the energy required for smelting from ore. These metals also are associated with other emissions, such as sulfur dioxide (SO_2). Recycling of metals can greatly reduce the energy required to produce them; substantially lower the CO_2 and SO_2 associated with mining and production; and reduce other air, land, and water pollution. For toxic metals like lead and cadmium, reduced exposure to toxicity is an additional benefit. Nondissipative uses of lead, such as storage batteries, are recycled to a large extent. Large amounts of copper also are recycled, but care must be taken to ensure that impurities in recycled copper do not reduce electrical conductivity where this is important (e.g., in power-transmission lines).

20.3.2. Chemicals

The chemical industry is extremely energy-intensive because of the source of its raw materials—petroleum and natural gas—and because of the energy required to carry out chemical transformations to final products. In the United States, chemicals accounted for 60.3 Mt or 19.6% of industrial-energy CO_2 in 1985, right behind the primary-metals sector. This sector also accounted for most of the 39.4 Mt of nonfossil-fuel carbon released by the industrial sector (Marland and Pippin, 1990). Since the oil shocks of the 1970s, the chemical industry has redesigned processes, significantly reducing energy use and lowering the production of chemical wastes. For example, the Japanese petrochemical industry reports a decrease of 47% in energy per ton of product from 1976 to 1988 though little change during the succeeding 4 years (MITI, 1994). The U.S. chemical industry reduced its energy per unit of output by 21.8% between 1974 and 1992 (Chemical Manufacturers Association, 1994).

20.3.2.1. Halocarbons and Hydrohalocarbons

Halocarbons and hydrohalocarbons currently are the major greenhouse gases associated with industrial activity. These ubiquitous substances are thought to contribute substantially to direct radiative forcing, but this effect may have been largely offset by their depletion of stratospheric ozone (IPCC, 1992). Most CFCs, halons, and solvents (such as carbon tetrachloride and methyl chloroform) are being rapidly phased out to protect the ozone layer. Proposed substitutes like HCFCs and HFCs, while having less ozone-depleting potential because of their shorter atmospheric lifetimes, have considerable global-warming capability. The United States estimates that the HFCs and fluorocarbons (FCs) released in 1990 have a warming potential equivalent to 20 Mt C, but expects this to grow to 45 Mt C by 2000 (U.S. DOE, 1994).

20.3.2.2. Plastics

Plastics and most organic chemicals are derived from petroleum and natural gas. The latest estimate is very much in need of updating, but Marland and Rotty (1984) suggest that 6.7% of liquid fossil fuels are incorporated into plastics, asphalt, and lubricants that are not immediately oxidized to CO_2. Some plastic wastes and lubricants are eventually burned as fuels, but some of this carbon—perhaps one-third—is sequestered in landfills or products. Worldwide, about 1% of natural gas is used for non-ammonia feedstock (Marland and Rotty, 1984). Energy use for plastics production was about 3% of total energy in the United States in 1988 (Franklin Assoc., 1990). An advantage of using more plastics in transportation and in transported containers is their extremely light weight, although petrochemical production is very energy-intensive. Recycling polymers is advantageous from an energy perspective but is difficult because their low density, high bulk, additives, and the mixtures of plastics in the waste stream make them difficult to reprocess. Moreover, innovation in plastics is important and makes standardization difficult. Some standardization on a smaller number of resins or other policy options might help in this matter.

20.3.2.3. Hydrogen and Ammonia

Hydrogen and ammonia are extremely energy-intensive to produce, and they use CH_4 from natural gas as a feedstock. Hydrogen is stripped from CH_4, and the carbon is oxidized and released as CO and CO_2. Marland and Rotty (1984) estimate that about 2% of global natural gas is oxidized in this way. Ogden and Williams (1989) place the energy content of hydrogen production in the United States at about 2% of total energy, which would imply that 5% of natural gas currently is used for hydrogen and ammonia production plus CH_4 leaks.

In the United States, nearly 340 x 10^9 m^3 of hydrogen are used in ammonia production, 900 x 10^9 by refineries, and 25 x 10^9 for methanol production; approximately 5.7 x 10^9 m^3 are produced as merchant hydrogen and by captive users (SRI, 1994). In addition to its main production from CH_4, hydrogen also is produced as a byproduct during oil refining and the electrochemical production of chlorine, sodium hydroxide, and sodium chlorate. Currently, some of this byproduct hydrogen is flared. The current process CO_2 from hydrogen production could be eliminated by using the electrolysis of water as the hydrogen source. At the present time, off-peak hydropower represents the only low-cost electricity supply that is large enough to make a significant contribution toward replacing natural gas as the feedstock. Off-peak power from geothermal, wind, solar (Ogden and Williams, 1989), and ocean thermal energy conversion someday may also be inexpensive enough to produce hydrogen competitively from the electrolysis of water.

20.3.2.4. Other Chemicals

N_2O contributes about 6% to increased radiative forcing, has strong infrared absorption, and has a lifetime of 140 years in the atmosphere. A full understanding of its measured increase in the atmosphere is incomplete. Contributions from fossil-fuel stationary combustion have been shown to be much lower than previously believed (Lyon *et al.*, 1989).

It has been estimated that adipic acid production for nylon accounts for 10% of anthropogenic global N_2O releases (Thiemens and Trogler, 1991), with 3.03 x 10^{-4} kg N_2O released per ton of adipic acid produced (Jacques, 1992). One major manufacturer announced a revised production process that reduces emissions by 98% (*Chemical Week*, 1992). The U.S. production of N_2O associated with adipic acid production has been estimated at 242,000 t, but with the introduction of controls, only 62,000 t is released (U.S. EPA, 1993). Small additional amounts are released in the oxidation of ammonia to nitric acid at a rate of 2 to 20 kg N_2O per ton of ammonia consumed. Controls have been developed and installed in some facilities (*Chemical Week*, 1992). Canada estimates that its annual emissions from this source are only 1 kt and from adipic acid production 30 kt (Jacques, 1992).

A small amount of CO_2 is released in the manufacture and use of lime and soda ash. For the United States, this is estimated to have been 3.26 and 1.14 Mt C, respectively, in 1991 (EIA, 1993).

20.3.3. Pulp and Paper

Because wood and paper are produced from plants that absorb CO_2, they can provide long-term sequestration of carbon in durable products. The major fossil-fuel inputs for paper production are in the harvest and transport of raw materials and in producing the fiber from wood chips. Evaporation of water consumes a very large proportion of the energy in pulp and papermaking. In response to higher energy prices in the 1970s and to comply with clean-air and water requirements in several countries, the paper industry developed efficient cogeneration systems to produce needed steam, hot water, and electricity by burning liquors, bark, and other wood waste. The U.S. paper industry now satisfies 56% of its energy needs by using biofuels in this way (American Forest and Paper Association, 1993). Japan improved the energy efficiency of its pulp and paper industry by 26% between 1980 and 1992 (MITI, 1994). In developing countries, such as Mexico, bagasse from sugar cane provides fiber for papermaking (Gotelli, 1993); China, Japan, and other Asian countries produce paper from rice stalks. Pulp also is used to produce cellulosic textile fibers, such as rayon.

More than 50% of the wood cut in the United States is eventually used for paper and paperboard. Current U.S. paper recycling is 25% (Europe and Japan have achieved rates of more than 50%). A doubling of this rate would reduce the amount of wood cut from U.S. forests by about 17%. Landfills, whose contents are 40% wood and paper products, would benefit similarly in the form of a 13% reduction by weight overall. Thus, paper recycling offers the potential for reducing global warming by increasing forest stands and reducing the amount of atmospheric CH_4 and CO_2 produced from landfills (Zerbe, 1990).

20.3.4. Construction Materials

The production of high-density building materials (such as concrete, glass, brick, and tile) requires high temperatures and hence releases significant quantities of CO_2 during production. Because of their high density, these materials also require large quantities of energy in their transportation.

20.3.4.1. Cement and Concrete

Cement and concrete are key components of commercial and residential construction worldwide. During the past decade, U.S. cement consumption averaged between 75 Mt/yr and 90 Mt/yr (Portland Cement Association, 1990); it is expected to rise to more than 100 Mt/yr by 1997. Worldwide, 1991 cement production reached 1.25 x 10^9 tons (U.S. Bureau of Mines, 1992). Requiring 6.1 GJ per ton of cement (see Table 20-4), cement production is one of the most energy-intensive of all industrial processes.

CO_2 is emitted during cement production in two ways. Approximately 0.75 t of CO_2 is produced per ton of cement from combustion of fossil fuels to operate the rotary kiln. The second source is calcination, in which calcium carbonate ($CaCO_3$) from limestone, chalk, or other calcium-rich materials is heated in kilns to form lime (CaO) by driving off CO_2. This process produces about 0.5 t of CO_2 per ton of cement. Thus, combining these two sources, for every ton of cement produced, 1.25 t of CO_2 is released into the atmosphere—of which 60% comes from energy inputs and 40% from calcination (Griffin, 1987). Worldwide, cement production accounted for approximately 162 Mt of C emissions in 1991, or about 2.6% of total global carbon from oxidation of fossil fuels. The United States annually produces about 9.3 Mt C from cement production, or 6% of global cement-production carbon (CDIAC, 1993).

The carbon intensity of cement is highly dependent on the fuel chosen. Cement production to date has used coal as its primary fuel. As a result, cement production is associated with especially high levels of CO_2, nitrogen oxides (NO_x), and SO_2.

Table 20-4: *Fuel use for cement production (1990).*

Fuel	GJ per Ton of Cement	Percent
Petroleum Products (diesel, gasoline, LPG)	0.06	1.1
Natural Gas	0.5	8.2
Coal and Coke	3.7	60.8
Waste Fuel[1]	0.3	4.9
Electricity[2]	1.5	24.9
Total	**6.1**	**100.0**

Source: Portland Cement Association, 1990.
[1] Waste fuel includes used motor oil, waste solvents, scrap tires, etc.
[2] Electricity figure includes primary energy used to generate electricity.

Producing 80 Mt of cement in the United States in 1992 required about 0.53 EJ, or roughly 0.6% of total U.S. energy use. In terms of its dollar value, however, cement represents only about 0.06% of the U.S. gross national product. Thus, cement production is about 10 times as energy-intensive as the U.S. economy in general. In a number of developing countries, cement production accounts for as much as two-thirds of total energy use.

Process modifications in kiln operation show potential for energy reductions. For example, newer dry-process kilns are more energy-efficient than older wet-process kilns because energy is not required to eliminate moisture. In a modern dry-process kiln, the kiln's exhaust gases are used to preheat the ingredients and drive off moisture. Such a process uses up to 50% less energy than a wet-process kiln (UBC, 1993). Significant reductions in CO_2 emissions from cement production could still be gained by energy-efficiency improvements of cement-kiln operation. Switching to fuels with lower CO_2 content (such as natural gas and agricultural wastes) also can reduce CO_2 emissions; in addition, savings from other techniques (e.g., fly-ash substitution and the use of waste fuels) are possible. Using waste lime from other industries could reduce CO_2 emissions from the calcination process.

20.3.4.2. Brick

Total brick production in the United States has varied recently between about 5 x 10^9 and 8.5 x 10^9 bricks per year. In 1992, production was about 6.25 x 10^9 units. Brick manufacturing constitutes about 96% of total clay-product manufacturing in the United States. Energy estimates for brick production vary widely. In the United States, the most-efficient brick producers, using modern computer-controlled equipment, require between 4.2 and 5.3 MJ/brick. On the other end of the scale, older beehive kilns can consume around 21.1 MJ/brick. Process modifications in brick production could yield large savings in energy use and CO_2 emissions. Technical improvements in burners, air blowers, and "slashing" processes (heating the kiln to excessive temperatures to color bricks for cosmetic reasons) could yield substantial reductions in energy consumption (Brick Institute of America, 1994).

20.3.4.3. Glass

In 1991, the United States produced 20.8 Mt of glass, at 7.4 to 8.4 GJ/t. Of this, 2.2 Mt of flat glass, at 7.4 to 8.4 GJ/t, was manufactured for use in the automotive/transportation and construction industries; 2.0 Mt of fiber glass, at 6.3 to 10.6 GJ/t, was produced for building insulation and a variety of building and product materials. The industry uses mostly natural gas as its energy source, along with some electric furnaces (Ross, 1991).

20.3.4.4. Wood

Wood is a far less energy-intensive building material than steel, concrete, brick, or glass; it provides superior insulating properties; and because of its relatively low density, it has lower transportation-energy requirements. Koch (1992) uses the CORRIM Report (NRC, 1976) to estimate the following embedded energy in residential building materials compared to wood:

- Steel structural building studs (9 x wood)
- Synthetic fiber floor covering (4 x wood)
- Concrete floor (21 x wood)
- Aluminum siding (5 x wood)
- Brick veneer (29 x wood).

An update clearly is needed, to take into account efficiency improvements in each industry.

20.3.5. Food Products and Light Manufacturing

The energy contributions to the food and beverage sector are modest in developing countries but are substantial in industrial nations. Pimintel (1973) has estimated that each U.S. food calorie requires approximately 10 technological calories to grow, transport, refrigerate, and prepare it, partly because the average food travels 2,092 km from the field to the consumer. Significant amounts of energy are expended in refrigeration and food processing, and a large amount of energy is embodied in packaging. For example, in 1991, the United States produced 13.5 Mt of container glass at 7.4 to 8.4 GJ/t, for a total of 114 PJ (Ross, 1994).

In light industry, major GHG reductions are likely to come from improved efficiency in space conditioning and lighting and the replacement of motors and belts with more-efficient ones. In the textile industry, the embodied energy of different fibers suggests that cellulosic fibers may have the lowest values, followed by synthetic petrochemical fibers and conventionally grown natural fibers, which rely heavily on fertilizers and industrial-style agriculture.

20.4. Technical Abatement Options

20.4.1. Fuel Substitution

Fuels with low or zero CO_2-generation potential can be substituted for fuels with high CO_2-generation potential to reduce GHG releases. In general, coal and oil are replaced by natural gas, renewables, and nuclear. For the industrial sector, substitution of coal with natural gas is most relevant. A major study at the International Institute for Applied Systems Analysis (IIASA) on historical trends of energy and fossil-fuel carbon emissions shows that most economies have reduced their energy intensity as industrialization has proceeded and that the carbon content of global energy use has been falling monotonically. The carbon intensity of energy consumed currently is about 0.5 t C/kWyr—down substantially from a value of 0.8 t C/kWyr in 1860 (Nakicenovic *et al.*, 1993).

20.4.1.1. Low-Carbon Fuels

The direct use of low-carbon fuels, such as natural gas, rather than coal in industry can substantially lower carbon dioxide levels. However, the lower price of coal and its much greater abundance have made it the fuel of choice in the absence of incentives or regulations to lower GHGs.

20.4.1.2. Electrification

World economies are becoming more electrified. In the United States, for example, electricity increased from 15.9% of purchased energy in 1974 for six major industrial groups (petroleum and coal products, building materials, paper, chemicals, primary metals, and transportation equipment) to 22.5% in 1988 (GCC, 1993). The implications of this shift for GHG emissions are unclear. Proponents of increased electrification point out that this process often leads to efficiency gains that offset additional CO_2 releases. Additional reductions in GHG emissions are likely to be realized when the direct combustion of fossil fuels is replaced by electrical generating capacity from nuclear, hydro, biomass, or renewable sources. Additional gains can arise when low-temperature industrial process or space heat is supplied by heat pumps rather than by the direct combustion of fossil fuels. Although total energy use in the OECD industrial sector has remained essentially constant during the past 20 years, CO_2 from the industrial sector actually declined by 9.6% between 1970 and 1990 (di Primio, 1993).

On the other hand, electric utilities constitute the largest source of CO_2 in many economies. Di Primio (1993) has identified transformation and distribution losses for electric power production as a principal contribution to CO_2 emissions in both industrial and developing countries, and this percentage has increased with time. The percentage of CO_2 arising from power plants was 22.6% for OECD countries in 1970 but had increased to 29.8% in 1990. For the United States, the increase was even greater: from 23.4% in 1970 to 32.7% in 1990. In the late 1980s, 36.4% of CO_2 came from power plants in India, 19.8% in China, and 32.5% in the former Soviet Union.

In short, the effect of increased electrification on industrial CO_2 releases depends on the particular purpose for which electricity is substituted, the carbon intensity of the source of electric power relative to the direct combustion of fossil fuels, and the structure of a given industrial economy.

20.4.1.3. Biomass

The use of plant material as fuel and feedstock in place of fossil fuels can have a significant effect on the reduction of net CO_2 emissions. Although cellulose releases approximately the same amount of carbon as coal per unit of energy produced, this release of CO_2 is largely offset by the absorption of CO_2 from the atmosphere as biofuel is regrown. Immense potential exists for energy recovery from organic municipal and industrial wastewaters through efficient biomethanation processes (Khanna *et al.*, 1988). Alcohol biomass fuels have played a major role in Brazil (Goldemberg *et al.*, 1993), and plantations could provide significant biofuels in many countries (Hall *et al.*, 1994). A full discussion of liquid fuels and biofuel plantations is included in Chapter 19.

Ironically, although biofuels constitute a major energy source in many developing countries, it is the United States that is the leader in generating electricity from biomass. According to the Electric Power Research Institute (Turnbull, 1993), biomass electric power generation increased from 200 MW in 1979 to 6,000 MW in 1992, plus 1,800 MW of municipal-solid-waste-to-energy facilities. All of the existing generation facilities use conventional steam technology, often at lower temperatures and lower efficiencies than fossil-fuel facilities. Much of this power production from biomass is associated with industries like wood, paper, and food processing that use their own waste. The growth of biomass-fueled facilities in the United States has not been matched in other industrialized countries. This is primarily because of the presence of a regulatory regime in the United States that encourages industrial power production.

Several examples of biomass electric-generating facilities also exist in developing countries. Bagasse is sometimes burned in India, Mexico, and Brazil at sugar refineries for heat and some electricity. A 5-kW diesel generator, modified to burn biomass produced from cow dung and agricultural wastes, produces electricity in Pura, India (Rajabapaiah *et al.*, 1993). The potential for greatly improving the efficiency of biomass power systems lies in the introduction of highly efficient gas turbines coupled to biomass gasifiers, according to Williams and Larson (1993). Their analysis suggests that the 80 cane-growing countries of the developing world have sufficient bagasse and other sugar-cane residues to support such facilities.

20.4.1.4. Renewable Energy Resources

In sunny regions, including many tropical countries, solar drying opportunities exist for grains, specialty fruits and vegetables, sugar cane, and lumber. Solar drying also can be applied to biofuels, whose moisture reduces their effective energy content. For example, approximately half the weight of bagasse from sugar cane is water; when moist, it releases only 9,042 kJ/kg, but when dry, it releases 19,424 kJ/kg (comparable to lignite) (Brown, 1987). In addition, by removing the weight of incorporated water, solar drying can reduce the energy requirements to transport biofuels, lumber, grain, and other products. Existing solar water-heating technology is well-matched to the process-heat levels required by such industries as food processing and textiles and for the curing of plastics, resins, and adhesives.

Few examples can be found of on-site use of solar or wind energy for industrial processes. These intermittent resources would need to be supplemented by grid or conventionally self-generated electricity, used in accord with the availability of wind or solar power, or disengaged from the demand for power through storage devices (Moomaw, 1991). This suggests that electroplating and other electrolytic chemical manufacturing (e.g., the production of chlorine and sodium hydroxide) might also be suitable for self-generated solar or wind energy. Another possible use of wind or solar energy in manufacturing would be for pumping industrial water that could be stored for use on demand (Lonrigg, 1984).

20.4.1.5. *Hydrogen*

Although hydrogen releases no CO_2 when burned or reacted chemically, it is mostly derived from CH_4, which produces CO and CO_2. Several alternatives are possible for producing hydrogen without carbon emissions. It is estimated that it would cost between $9 and $19/GJ to produce hydrogen by off-peak hydroelectric-based electrolysis costing between 1 and 4¢/kWh. Biomass gasification could lower the cost to between $5.9 and $8.5 per GJ (Ogden and Nitsch, 1993). Were the cost to drop sufficiently, considerable reductions in GHGs would occur in the primary-metals industry because of the use of hydrogen to reduce oxide ores.

It has been suggested that photovoltaics are ideally suited as low-voltage DC sources of electricity for the electrolysis of water to produce merchant hydrogen where it is needed (Ogden and Williams, 1989; Ogden and Nitsch, 1993). The auto manufacturer BMW has built a 600-kW photovoltaic test facility with Siemens and the Bavarian government to produce hydrogen automotive fuel.

20.4.2. *Energy Efficiency*

Energy efficiency in industrial plants can be improved through opportunities on the end-use side (such as variable-speed electrical drives and insulation) and opportunities on the conversion side (such as cogeneration applications and advanced boilers).

20.4.2.1. *Efficient Electrical End-Use Devices*

Electrical motors are major consumers of industrial electricity. Several studies in the United States conclude that approximately two-thirds of electricity is consumed by electrical motors, with the industrial sector accounting for between 26 and 30% of the total. These motors are used primarily to drive pumps, compressors, and fans. Major opportunities exist to improve the efficiency of the motors themselves and the systems, connecting shafts, and belts they drive (U.S. DOE, 1980; Baldwin, 1989).

Standard motors operate with an efficiency of 70% for small devices of a few kilowatts, to 92% for large motors of 100 or more kw. High-efficiency motors operate in the range of 83 to more than 95% (Baldwin, 1989). One analysis of an industrial pumping system showed that only 49% of the energy output of the electric motor actually was converted into work to move the liquid (Baldwin, 1989; U.S. DOE, 1980). Optimizing system design rather than simply choosing components can lead to improvements of 60% using existing technology (Baldwin, 1989). A study of Indian industrial efficiency found that replacing traditional power-transmission V belts with modern flat belts could improve efficiency from 85 to 98% (NEERI, 1993). This study, along with Baldwin's, also demonstrates the potential for adjustable-speed electronic drives to better match mechanical load, substantially reducing electricity use.

The U.S. Environmental Protection Agency (EPA) has estimated that 12% of U.S. utility CO_2 emissions—55 Mt C—could be eliminated by replacing existing lighting systems with cost-effective lighting technology (U.S. EPA, 1993). Industry represents one of the largest users of lighting. The voluntary Green Lights program had enrolled more than 1,255 corporations, institutions, and state and local governments by early 1994, all of whom are committed to replacing 90% of their lighting with cost-effective, efficient substitutes within 5 years (U.S. EPA, 1994). Computers are estimated to consume approximately 5% of U.S. electricity; EPA has developed the Energy Star program, which has produced significant declines in the electricity use of new computers and other office equipment now entering the market. The average power demand of different types of computer printers was found to range from as little as 3 to as much as 129 W in one study (Norford *et al.*, 1989). Although these ancillary devices represent a small fraction of total industrial energy use, there is substantial room for significant savings. The surprising success of voluntary programs in the United States suggests that this approach should be more widely utilized in other countries.

20.4.2.2. *Cogeneration and Steam Recovery*

In cogeneration facilities, fuels are burned and the heat produced is used both for generating electricity and for process heat. Cogeneration schemes are of two kinds: topping-cycle and bottoming-cycle. In topping-cycle cogeneration, primary energy (the high-temperature energy, generally about 1,500°C) is used to produce electricity, and the low-temperature heat that emerges from the generator is used for process or space heating (e.g., to dry pulp in a paper mill). In bottoming-cycle cogeneration, primary energy is used to produce heat (e.g., to anneal steel), and the leftover heat is used to generate electricity.

Cogeneration facilities generally use as much as 80% of the heat content of fuels, significantly reducing the amount of CO_2 produced compared to facilities where electricity and heat are produced separately. A variety of technologies now are employed that continue to improve the overall efficiency of cogeneration. Conventional steam turbines that recover a portion of waste heat are being displaced by higher-efficiency and less-costly gas-turbine cogeneration systems and combined-cycle systems that use some of the heat from the gas turbine to run a lower-temperature steam turbine as well as to provide process steam (Williams and Larson, 1989). In the United

States and many European countries, many small-scale cogeneration facilities have been developed in the industrial sector as the result of new technological developments and changes in the laws governing electrical utilities. Industries are now permitted to sell surplus electricity to the utility grid, and utilities are required to purchase it at favorable prices. Since the change in legislation, a majority of the new electric-generating capacity constructed in the United States has been from manufacturing, pulp and paper, and other industrial or independent cogenerators rather than large utilities. The U.S. Office of Technology Assessment has estimated that industrial cogeneration could reduce industrial energy demand by 140 TW/yr in the United States by 2015 (OTA, 1991).

Substantial cogeneration is carried out in Western Europe, especially in Finland, Denmark, the Netherlands, Sweden, and Germany. Because of regulatory constraints and climatic conditions (large air-conditioning demand), Japan was slow to adopt cogeneration relative to other industrial nations. The Japanese electric utility industry law was revised in 1995 and may now facilitate the spread of cogeneration in that country. Because of the large amount of heavy industry and extensive district-heating systems already installed in the former Soviet Union, an enormous potential for major reductions in CO_2 also exists there through the adoption of industrial cogeneration.

20.4.2.3. Waste-Heat Utilization

Even in countries where high levels of industrial energy conservation have already been achieved, a large potential still exists for further industrial heat recovery. Data from a survey conducted in 556 representative factories in Japan indicate that almost half of the surplus heat that currently is discarded has potential for use. From a total of 1.8 EJ surplus energy generated by these factories, about 0.5 EJ already has been used effectively as steam, process gases, or electricity—a 2.7% decrease in primary energy consumption. Unexploited-energy figures for the four main energy-consuming industries in Japan are shown in Table 20-5. The unexploited surplus energy from these industries amounts to almost 1 EJ, of which about 48% has a potential for economic use (Kashiwagi, 1991).

The first step in planning energy utilization is to carefully map the flow of energy in each industry, from primary energy or electricity input to the lowest-temperature waste-heat output. Once these flows are known—including the quantity (e.g., kilowatts) and quality (e.g., temperature levels, voltage, and frequency)—industrial processes can be integrated, initially within each industry, to minimize the input of primary energy and/or electricity.

Cascaded energy use involves fully harnessing the heat produced by fossil-fuel combustion, from its initial 1,800°C down to near-ambient temperatures, with a thermal "down flow" of heat analogous to the downward flow of water in a cascade. If a water flow were to be used to generate electricity, for example, a dam would first be built on a mountain; then a succession of power-generation plants would be set at strategic points down the river to use up heads of water. Once the river water had flowed to sea level, it would have become useless for electricity generation. The heads of water at each generation plant are analogous to temperature gradients in thermal energy; sea level corresponds to ambient temperature. As in the case of river water reaching sea level, heat that has cooled to ambient temperature no longer is usable. Currently, even in the industrial sector, there are extremely few cases of heat being used in multiple stages (cascaded) akin to hydroelectric power generation.

Based on the ideal energy flow among industrial processes classified by their temperature levels, Kashiwagi (1992) suggests a representative example of ideal exergetic utilization of fossil fuel resources in a temperature cascade that utilizes combined heat and power and renewable energy, as shown in Figure 20-2.

In an effort to construct district-heating systems using unexploited energy sources, several countries have been promoting

Table 20-5: *Unexploited energy of the largest energy-consuming industries in Japan (joules).*

			Paper Pulp[2]	Chemical Industry[3]	Cement	Iron and Steel[4]	Total
Energy Flow	**Energy Input**		4.2×10^{17}	1.5×10^{18}	2.6×10^{17}	1.8×10^{18}	4.0×10^{18}
	Unexploited Energy[1]	Usable	1.1×10^{16}	–	2.6×10^{15}	4.9×10^{17}	0.5×10^{18}
		Unusable	1.2×10^{17}	1.9×10^{16}	5.1×10^{16}	3.3×10^{17}	0.5×10^{18}
		Total	1.4×10^{17}	1.9×10^{16}	5.4×10^{16}	8.2×10^{17}	1.0×10^{18}

[1] Amounts of unexploited energy are estimates based on models for the four industrial sectors, thus may differ in precision depending on the model adopted.
[2] Estimates for the paper and pulp industries presuppose increases in effective use of black liquor.
[3] For the chemical industry, even if supply and economic conditions are fulfilled, there are almost no prospects; unexploited energy for the chemical industry refers to the ethylene sector only.
[4] Values of energy input and unexploited energy for iron and steel industries based on examples of combined steel and iron manufacturing.

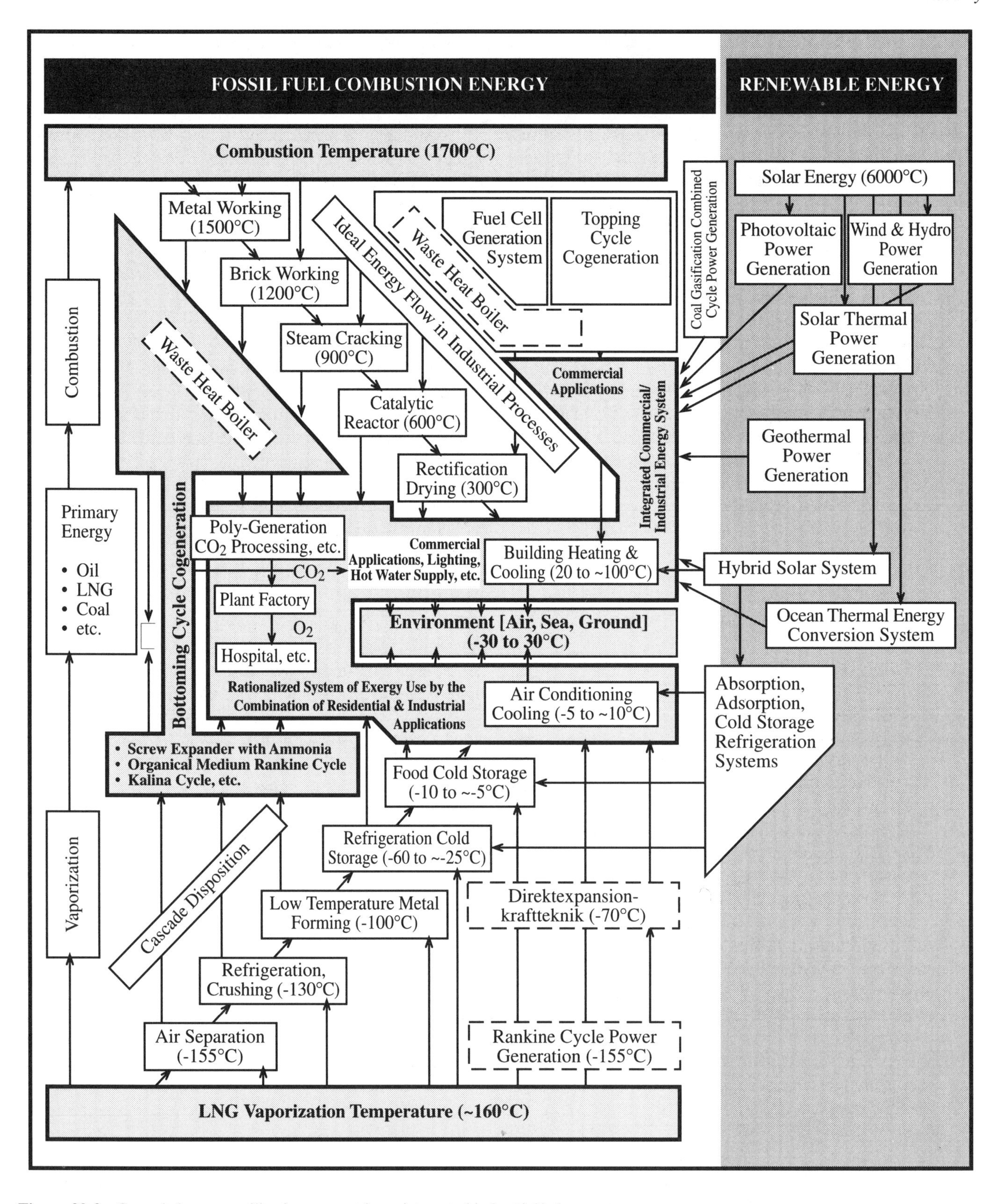

Figure 20-2: Cascaded energy utilization concept in an integrated industrial/urban system.

research and development (R&D) on large-scale, high-efficiency machines; heat-transport technologies; large-scale heat storage; and optimized systems combining these technologies. Expansion of district-heating systems brought on by dramatic increases in the efficiency of unexploited-energy utilization systems is expected to reduce and optimize energy consumption—and, hence, to help shave off the peak of electric power demand.

Thyssen Stahl AG produces 10 Mt of steel each year in Duisburg, Germany, where demand for space heating is high and district-heating systems are well developed. Heat is recovered from its high-temperature furnaces as steam at 480°C and 1.4 MPa for driving electric-power-generating turbines and also as hot water (150°C) for supply to 15,000 of 45,000 nearby households served by the local district-heating system. Because the point-of-use temperature of the hot water must be kept between 90 and 95°C, the operating company of the district-heating system has installed natural-gas boilers to make up for fluctuations in heat demand and/or temporary stoppage of the high-temperature furnaces.

The conception of the Onahama hot-water supply system in Japan dates to the 1960s. A coke factory owned by the Nihon Kasei conglomerate in Onahama, Fukushima Prefecture, had been using seawater to cool high-temperature process gases in its facilities. A feasibility study conducted by Nihon Kasei in 1965 showed that this industrial surplus heat could potentially be used in the nearby Onahama area. Encouraged by these results, the city mayor formed a preparatory committee for the establishment of a hot-water supply company in Onahama, which was run by local associations of commerce, tourism, fishing, and inns. The heat-supply system included a pipeline of more than 10 km, pumps, heat-storage tanks, high-speed filters, and chlorination facilities. Freshwater now is supplied to the factory as a process-gas coolant, with a hot-water output of 10^5 m^3/h at 63°C. After filtering and chlorination, the hot water is pressurized by pumping so that the return pressure always exceeds 10^6 kg/m^2. The point-of-use temperature of this hot water always is maintained at or above 50°C. Monthly water-quality tests carried out by Fukushima Prefecture sanitary

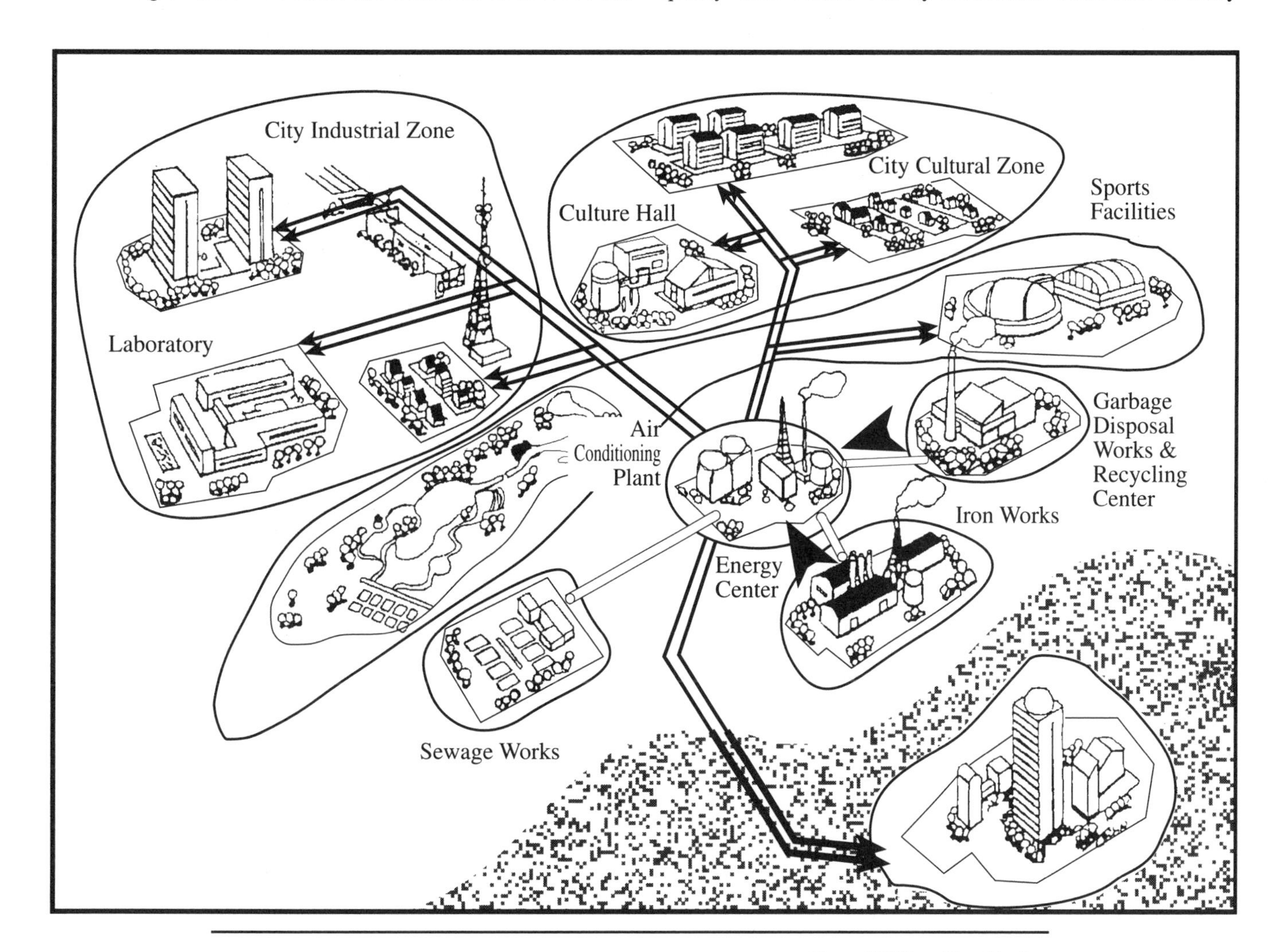

	Half-Million People City	Four-Million Ton Scale Iron Works
Power Consumption	2.0 x 10^8 W	2.5 x 10^8 W
Drainage	1.5 x 10^5 t/day	1.5 x 10^5 t/day
Gas Use (at 41.9 MJ/N m^3)	2.4 x 10^5 Nm^3/day	2.7 x 10^6 N m^3/day
Heat and Hot Water Supply	7.95 x 10^{12} J/day	32.2 x 10^{12} J/day

Figure 20-3: Urban-level energy system with industrial waste-heat utilization.

authorities have consistently shown this hot water to be on par with potable water, which makes it suitable not only for bathing and washing clothes but also for dishwashing.

Large-scale energy conservation will require the establishment of urban-level energy systems supported by industries. Figure 20-3 provides a schematic of this concept, in which a city of 500,000 people would use waste heat from a 4 Mt/yr iron works. An air-conditioning plant and energy center would provide heating and cooling to the city using heat received from the iron works and from a garbage-incineration plant. The figure indicates that the urban demand for energy could be met by using waste heat from the factory. Factories with high energy consumption can use that energy more effectively by conserving energy and using waste heat.

The future technology of broad-area energy utilization systems would involve an advanced cascaded and combined thermal energy-recirculation system based on innovative technology that recovers waste heat from facilities and transports the recovered energy efficiently to remote urban areas. This concept is illustrated in Figure 20-4.

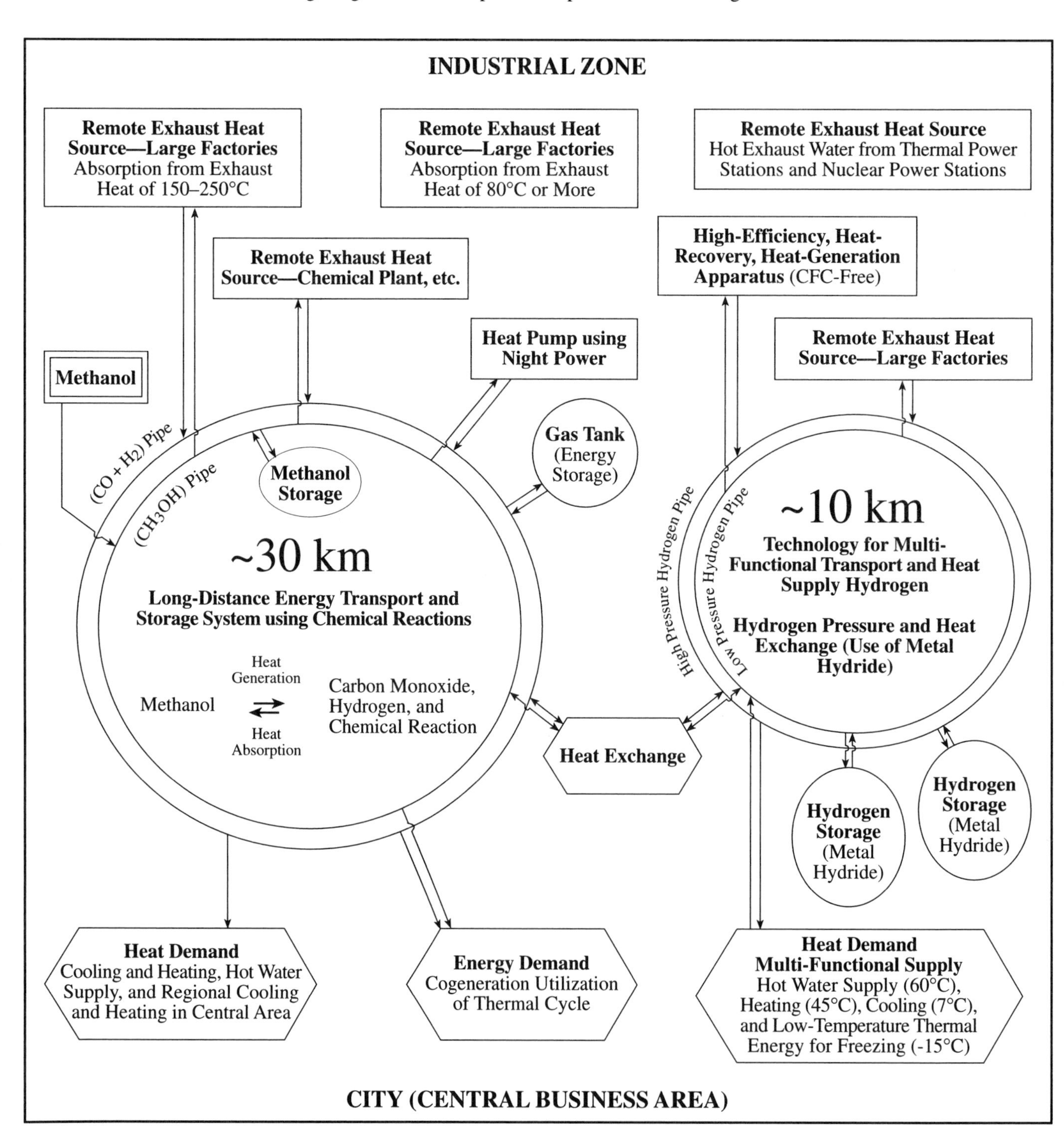

Figure 20-4: Schematic of the area-wide energy utilization network.

To implement this concept, major breakthroughs would be required in two main areas. First, key technologies would have to be developed: heat recovery with a maximum utilization of sensible heat and with a temperature-lifting capability; heat transport and storage technology that employed chemical reactions; and multifunctional heat supply technology. Second, the total system would have to be structured and optimized by matching multiple and wide-area heat sources to the energy demand, which would be supported by industrial energy centers as seen in Figure 20-5.

One estimate for Japan indicates that if unexploited energy in coastal industrial regions near major urban areas were used to the maximum extent, national energy consumption would be reduced by 6% and CO_2 emissions would be reduced by 9% (Kashiwagi, 1992). A system that utilizes high-temperature heat from electric power-producing gas turbines to boost the lower-temperature heat from municipal waste incineration to run steam turbines and provide district heat has been described (Lindemann, 1992; NEDO, 1993). Industrial centers based on sound land-use planning and industrial ecology principles—where waste materials and heat from one production process are utilized by nearby industries—have been attempted only in a few places, such as Kalundborg, Denmark (Frosch, 1995).

20.4.2.4. Materials Recycling and Reuse

When goods are made of materials whose manufacture consumes considerable amounts of energy, the recycling and reuse of those goods can save not only energy but GHGs released to the atmosphere. However, recycling usually involves complex tradeoffs. For example, a car with an aluminum frame is lighter than a car with a steel frame—and thus consumes less energy for the same transportation service—but its own energy content is higher because the energy content of aluminum is higher than that of steel. Using returnable glass bottles instead of plastic bottles for mineral water (i.e., artifacts with a much longer lifetime—the average returnable glass bottle being used 18 times, for example, in Europe) reduces the amount of energy embodied in the materials needed for the same service but increases the energy embodied in collecting, transporting, and washing the bottles. Moreover, lengthening the lifetimes of durable goods may hinder the diffusion of new generations of goods or techniques that emit less CO_2. Assessing the emissions benefits of such actions, therefore, requires careful life-cycle analysis.

Recycling of energy-intensive materials undoubtedly has increased during the past 10 years, mostly in industrialized

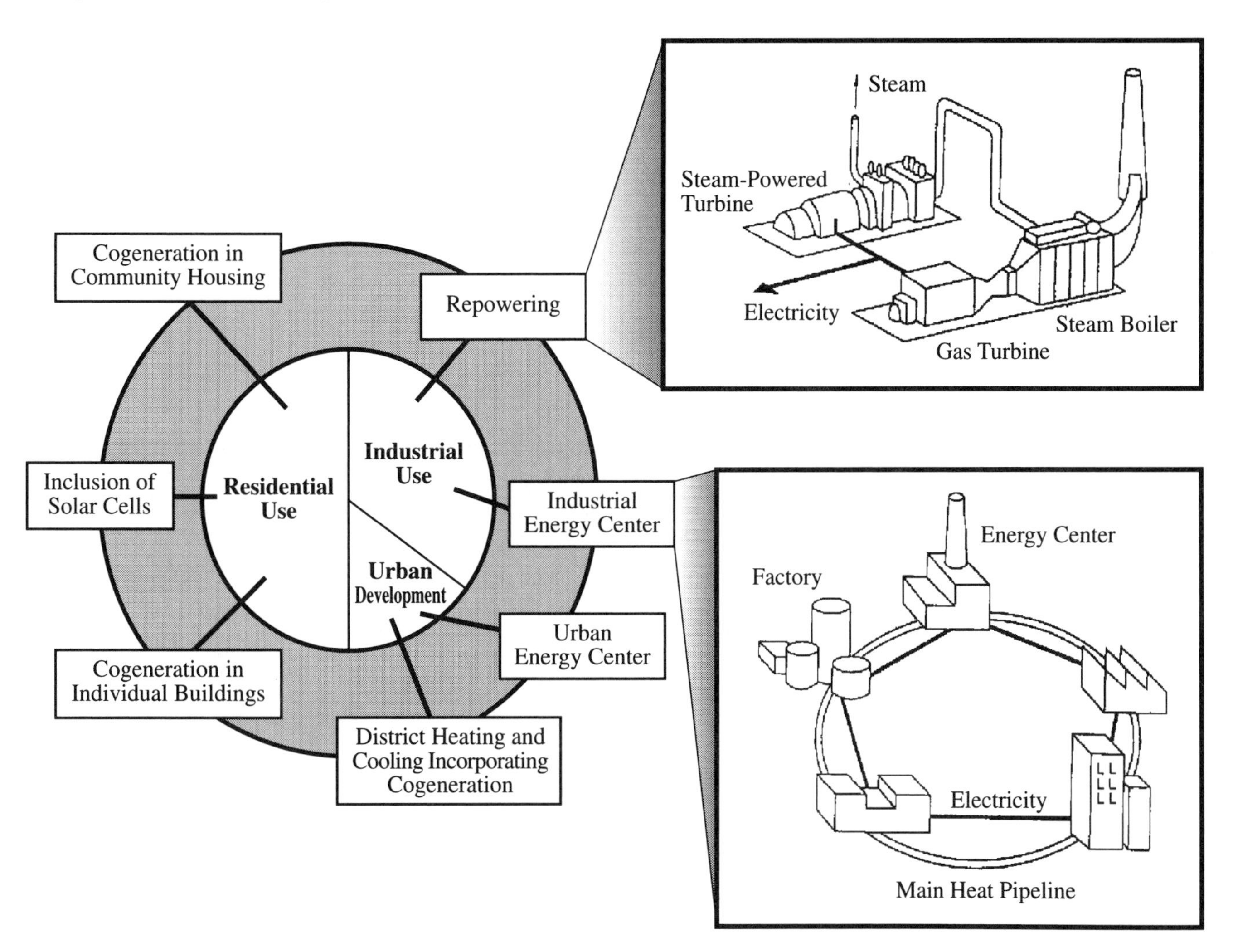

Figure 20-5: The structure of industrial energy centers.

countries. Data on present recycling rates, however, are diverse. The rate of rough steel produced from scrap is about 43% worldwide (IISI, 1993). In OECD countries, the recycling rates of nonferrous metals are 30% for aluminum, 35% for copper, 48% for lead, 25% for zinc, and 20% for tin (Metalgesellshaft, 1988). The recycling rate for paper is estimated to be 30% at the world level, which also is the rate estimated for the present recycling of glass bottles in OECD countries (IIASA, 1993). Reliable rates for plastics are difficult to arrive at, though they seem to be increasing. Recycling data for developing countries are scarce. In some developing countries, such as India, recycling is significant—a great proportion being carried out by the "informal sector" (Konishi, 1995).

Differences in recycling rates among countries yield a first approach to the existing potential. In Japan, for example, rates for paper and glass bottles were expected to reach 55% by 1994 and rates for steel and aluminum cans to reach 60% in 1995. In Europe, voluntary agreements between administrations and business organizations have been signed to reach recycle rates of 60 to 70% for packaging materials and about 75% of the material content of automobiles before the end of the century. All of these figures are well above world averages. Public measures to improve recycling of materials also must strike the right balance among collection, sorting, and reusing activities. This balance is rather unstable at present and has led in extreme cases to a collapse of secondary-materials markets, thus frustrating further developments.

One must recognize that returning the secondary resource into virgin-grade material is not always necessary or desirable. By analogy to heat recovery of lower-temperature heat through cascading processes, materials recovery can be envisioned as a materials-cascading process in which secondary materials may be used most efficiently for different purposes than are virgin materials. To optimize recycling and reuse potential, materials and products must be designed for recovery as well as for their initial intended purpose.

A summary of carbon emissions from primary and secondary materials production for OECD countries is presented in Table 20-6. Primary materials release about four times the CO_2 of secondary materials for all substances except aluminum (where the figures are substantially higher). Carbon savings of 29 Mt for a 10% increase in OECD recycling of each of the materials also is listed.

20.4.4.5. Dematerialization and Materials Substitution

A number of possibilities for reducing industrial greenhouse gases arise from the management of materials rather than of energy. Dematerialization (decreasing the amount of a material or substituting lighter-weight material for it) can play a major role in lowering GHG emissions (Grubler *et al.*, 1993; Herman *et al.*, 1989). Energy-related CO_2 also can be reduced in manufacturing by substituting less energy-intensive materials for more energy-intensive materials. Materials that release significant quantities of process GHGs during production can be replaced with materials that release fewer process GHGs (e.g., replacing aluminum with plastic lowers carbon and fluorocarbon emissions). Developing fundamentally new processes that use intrinsically less energy or are dependent on completely different feedstocks such as dry process for paper making, also can lower CO_2 releases.

The use of plants as a source of chemical feedstock also can reduce CO_2 emissions. Many large wood-products companies already produce chemicals in association with their primary timber or paper production. In India, a major effort to develop a "phytochemical" feedstock base has been underway at the National Environmental Engineering Research Institute. The production of chemicals such as acetone, butandiol, butanol, and isopropanol utilizing water hyacinth, grass, and industrial waste as feedstocks has been demonstrated to be technically feasible. The same institution also has been working to develop processes to improve the production of CH_4, as well as microbial conversion of CH_4 to methanol (Motwani *et al.*, 1993; Juwarkar *et al.*, 1994; Sharma *et al.*, 1994).

20.5. Policy Options

Industrial GHG reductions can be achieved by good housekeeping (operational performance), additional investments in energy-efficient technologies (both conversion and end-use equipment), or redesigning the manufacturing process itself (process innovation and integration). Clearly, the technical and

Table 20-6: *Tons of carbon saved and released in primary and secondary materials production.*

	OECD C Reduction for 10% Increase in Recycling (Mt C/yr)	Primary	Secondary	Primary/ Secondary
Steel	18.9	0.65	0.15	4.33
Aluminum	5.0	3.52	0.09	40.04
Copper	0.44	1.08	0.22	4.96
Glass	0.68	0.20	0.04	4.92
Paper	3.74	0.29	0.07	4.00

Source: IIASA, 1993.

Table 20-7: *Overview of implementation measures and level of impact.*

	Type of Barrier Information and Training	Financial and Economic	Institutional and Legal
Operational Performance	Plant audits Energy management courses	Consultancy grants or subsidies	Energy management centers Engineering consultancy network
Investment in Energy Conversion End-Use Equipment	Product information Specialized courses	Investment subsidies Trade tariff exemptions Market pricing of energy	Performance standards DSM-oriented utilities Demonstration projects
Process Innovation and Integration	Pinch technology Exergy analysis Life-cycle analysis	Internalization of environmental costs	Voluntary agreements R&D programming Land-use planning

financial consequences of these approaches become increasingly important and the role of management becomes correspondingly pervasive. Most industrial facilities and processes have a relatively short lifetime—on the order of a decade or, at the most, 20 years. Hence, there is plenty of opportunity to introduce low-GHG-emitting technology into the manufacturing process as part of normal capital-stock turnover. Unfortunately, under present circumstances, there are no compelling incentives for companies to choose a lower-GHG-emission strategy over a higher one when they are planning new processes or products. Hence, there is a need for additional policy instruments and incentives, some of which are described in this section.

Even when efficiency improvements and greenhouse gas reductions are feasible from a technical perspective, there remain a number of barriers that still may prevent the realization of such improvements. First, a company must be aware of the opportunities that are available. Second, there may be capital constraints such as unfavorable pricing of energy or recycled materials. Third, institutional and legal measures may be necessary where financial and economic incentives remain weak and markets for efficient equipment are just emerging. In some cases, additional research may be needed to develop appropriate energy-efficient technologies, material-efficient applications, or nonpolluting manufacturing processes. Table 20-7 provides a survey of implementation measures according to the type of barrier encountered.

20.5.1. Providing Information and Training

Reduction of greenhouse gases from industrial sources can be promoted by a variety of information-related policy steps on the part of individual firms, industry groups, or governments. Technical guidance can be provided to industrial managers to help identify opportunities to alter techniques of managing energy and materials in the manufacturing process, toward the goal of reducing energy use and GHG emissions.

Consumers often lack information about the environmental consequences of their purchases and seldom consider energy and other life-cycle costs in their purchasing decisions. Energy-efficient appliances, for example, may have higher purchase prices but during their lifetimes cost less for the consumer to operate. The economic disincentive of higher purchase prices might be overcome by better information and labeling about the life-cycle costs and environmental impact of manufactured goods.

20.5.2. Fostering Technology Transfer

Because future global industrial growth will take place largely in the developing and transitional economies, the early transfer of advanced energy-conservation technology may be crucial to curb worldwide GHG emissions. Developing countries might even jump from an early stage of industrial development to an advanced stage in one step. Such leapfrogging makes sense only if the new technology is economically superior to the old technology over its life cycle and if capital is available to purchase it. This may be the case only for isolated, large-scale, turn-key projects that usually are associated with multinational corporations or multinational lending agencies. The applicability of the concept across the spectrum of efficiency options needs careful scrutiny. In many cases, judicious adaptations may be called for, particularly to enable the manufacturing of major components by local industries.

In regard to technology transfer, two issues are still unclear. One is trade barriers in the form of import tariffs and legal barriers for the protection of intellectual-property rights. Such barriers affect the adoption of high-tech components of energy-efficient equipment. A balance must be struck between the interests of developing nations and the interests of industrial firms elsewhere. Ongoing economic globalization and the increasing involvement of foreign investors in many developing countries—including, in the recent past, both China and India—may be symptomatic of changing attitudes on both sides. This development follows in the wake of economic liberalization and shows the importance of institutional aspects for effective technology transfer.

The other issue is the establishment of mechanisms for financial transfers, in particular within the framework of the Global Environmental Facility and Joint Implementation. Energy efficiency is high on the list of priorities in this respect. At present, the future of political modalities for technology transfer still is not clear. Political controversies surrounding joint implementation involve complicated issues regarding responsibilities and credits and are far from resolved.

Nevertheless, joint implementation could potentially be viewed as a mechanism to make private capital available for the transfer of energy-efficient technology. This possibility is particularly true for the industrial sector, where multinational firms are keen to avoid the increasingly steep costs of GHG emissions in some countries if attractive alternatives are available elsewhere. In some cases, utilities from industrial nations are already actively exploring potential opportunities.

There also needs to be a better mechanism for making available new industrial ecology management systems, like that used in Sweden (Graedel *et al.*, 1994), that encourage lower GHG emissions.

20.5.3. Financial and Economic Policies

Conservation opportunities may not be profitable from the perspective of the private investor due to unfavorable pricing of energy or lack of capital. Financial and economic measures are then required.

20.5.3.1. Subsidizing Energy Audits

The first step toward energy improvement requires managerial awareness of the potential benefits of saving energy and an accurate overview of how and where energy is used in the factory and at what costs. Tax credits for plant audits and consulting may help to initiate energy-saving programs by identifying opportunities for good housekeeping and by indicating where energy-saving investments would be technically possible and financially profitable.

20.5.3.2. Providing Fiscal Incentives for the Purchase of Energy-Efficient Equipment

The next step requires a proper market for energy-efficient equipment (cogeneration systems, efficient boilers and furnaces, heat exchangers, electric drives and pumps, insulation materials, and advanced control systems). Fiscal incentives in the form of tax credits may help an emerging market for such equipment to mature. Adequate performance standards and active involvement from utilities will create legal and institutional support at this stage. Finally, in order for decisions about new processes and factories to be guided by energy-cost considerations, attention to plant energy costs will be stimulated by a firm government commitment to long-term energy-demand policies, correcting market distortions from environmental externalities, researching demand-side priorities, and land-use planning (where system-integration aspects are crucial).

Implementation problems for energy-efficiency improvements in developing and transitional economies are more severe than those in industrial countries. In addition to problem areas already indicated, key problems in both developing and transitional economies concern the economy-wide lack of capital, foreign exchange, and industrial energy prices and tariffs that often are below those of the world market. Because energy-saving investments often are characterized by capital and foreign-exchange intensity rather than by labor and resource intensity, their cost-effectiveness tends to be less in those countries than in the industrialized countries. Moreover, developing and transitional economies, in general, are characterized by lower overall productivity, including energy productivity. One should not attempt to solve energy-efficiency problems in isolation from other efficiency problems. Problems related to vintage equipment, scarcity of management skills, small-scale production, or poor technological infrastructure will not be solved by addressing climate-change or energy goals alone.

Market-oriented energy pricing is the first step in raising energy-efficiency priorities in all economies, regardless of their state of development, and already is receiving attention in most developing and transitional economies. The problem of capital and foreign-exchange scarcity has a more permanent and pervasive character. Although taxing environmental externalities is desirable, many industries operate on an international level, so the introduction of such taxes requires international action. Unilateral correction of national market prices can lead to a loss of international competitiveness.

20.5.3.3. Examining Taxation as a Tool

The "material content" of economic growth and its resulting greenhouse effect results from a huge number of interrelated decisions made by economic actors (e.g., business firms and households). Therefore, economic instruments—such as taxes that internalize externalities—seem to be most appropriate policy tools. However, although some assessments have been carried out on the effect of a carbon tax on energy efficiency in different sectors, little work has been done on the effect of such a tax on recycling or material substitutions. We must therefore admit that we do not know the efficiency of economic instruments in this field.

Preliminary results (ADEME, 1993; Gielen, 1993) tend to demonstrate that a carbon tax would mainly influence waste management (increased recycling and/or incineration) and would promote little in the way of material substitutions within products—unless it reaches very high levels and except for products whose material composition has a significant impact on energy consumption in the end-use phase.

Carbon taxes on fuels have been proposed as one way to lower CO_2 emissions from fossil-fuel burning. Interestingly, although the use of carbon taxes on fuels to internalize climate-change costs from that source have been the topic of major discussions, little attention has been given to applying a "greenhouse tax" to process-related releases. In some countries, such as the United States, an "ozone depletion tax" has been applied to CFCs and halons to improve the price competitiveness of substitutes. Such a strategy might be considered for goods whose manufacture produces GHGs.

Internalizing environmental costs in energy prices and tariffs through ecotaxes is particularly problematic for the industrial sector because of the consequences for national competitiveness on international markets. Taxes that are not levied on a global scale may provoke industry relocation, which may adversely affect emissions efficiency as well as international competitiveness. Most countries are hesitant to embark on policy ventures that might endanger their international market position and their attractiveness as industrial locations. Although tax exemptions for specific industries are possible, policymakers are wary about the administrative feasibility and costs of such exemptions. On the other hand, little empirical evidence points to any actual dangers of industrial relocation resulting from ecotax burdens. Because only new investment is affected, the impact of taxation may not be immediately apparent, although the impact may be more pronounced in the future because industries are becoming more international.

Systems of internationally traded emissions permits, carbon taxes, and opportunities for joint implementation are options that require international cooperation. It is difficult for a single nation to impose full environmental cost accounting and remain competitive unless other nations do the same.

20.5.4. Supporting Technical Innovations through Research and Development

During the past two decades, many energy-saving technologies—such as compact fluorescent light bulbs—have been developed with government research support and have successfully penetrated the market. To reach further reductions in energy intensity, more-expensive and less-proven technologies must make a contribution. Energy-conservation programs, therefore, need to devote more attention to research, development, and demonstration than in the past. An appropriate balance between energy-supply R&D budgets and energy-conservation R&D budgets has not yet been achieved. Additional R&D needs to be encouraged in the private sector through appropriate tax incentives.

In several cases, secondary materials similar in quality to primary materials will be available during the coming 10 to 20 years. In these cases, the steady-state rate of recycling will depend mostly on the cost of collecting and sorting. Hence, regulations and policies favoring economic efficiency at the collecting and sorting stage are crucially important. However, for some materials, like plastics, the situation is more problematic. With these materials, obtaining high-quality recycled products or finding uses for recycled low-grade materials seems to remain rather difficult. Hence, public intervention can be essential to promote R&D activities in this field and expand the market for recycled products.

20.5.5. Organizing and Supporting Recycling

Increasing recycling rates is linked to the following measures: limitation of dissipative uses; product design allowing easy-to-treat structures and easy disassembly; standardization of materials to produce more homogeneous scrap; more cost-efficient collecting and sorting; and technical innovation to increase the quality of secondary materials and/or lower their quality requirements in volume markets.

Regulations banning dangerous dissipative uses have been implemented in many OECD countries, particularly for heavy metals. The primary goal obviously was not to limit the greenhouse effect, but the impact on recycling rates was additive. Industrialized countries are putting increasing regulatory pressure on recyclability, mostly driven by solid-waste concerns. This trend is illustrated by recent French and German initiatives concerning packaging wastes. The producers of packaging materials have been made responsible, through voluntary agreements, for setting up organizations (DSD in Germany and Ecoemballage SA in France) to collect, sort, and recycle their materials.

As another example, an ongoing regulatory process within the European Community is aimed at drastically reducing the volumes of industrial and domestic wastes dumped in landfills. The general purpose is to ban the dumping of any type of "unprocessed" waste, thereby limiting landfilling to "ultimate" wastes. Waste "processing" may be recycling and/or incineration with heat recovery and/or power generation—the former being preferred, although it is not clear that clean incineration is always worse for the environment than recycling. For industry, the cost of producing wastes will therefore rise, leading to "material-efficiency" improvements (including internal recycling) within industrial plants.

In many locations, policies dictate that households use different types of receptacles to separate papers, glass, plastics, and organic matter. Such behavior significantly lowers the cost of sorting, improves the homogeneity of scrap, raises the quality of secondary materials and lowers their treatment costs, and ultimately increases the potential rate of recycling. Successes have been reached by several pilot experiences.

Manufacturing materials from scrap tends to be less expensive than from primary raw materials, but secondary materials often cannot be used in all of the applications that primary materials can. Therefore, the maximum rates of recycling depend on the costs of collecting and sorting scrap, as well as the opening of wider markets to secondary materials.

20.5.6. *Creating Voluntary Organizations and Agreements*

Broad voluntary agreements have been fashioned between industry and government to curb energy demand. Such voluntary agreements free governments from the regulatory burden of setting up complicated and inefficient systems of industrial-equipment efficiency standards and operating norms, and enable industries to choose strategies and technologies that are most efficient from the point of view of industry-specific conditions and opportunities. However, the effectiveness of such arrangements clearly depends on the financial conditions facing the industry. When profit positions become less favorable, through internal or external circumstances, the incentive to honor voluntary agreements is likely to suffer.

20.5.7. *Underwriting Demonstration Projects*

Seeing is believing. Often, the amount of uncertainty associated with a new technology being applied to a new situation with untested goals and techniques is too great to convince managers to invest in it—even if theory indicates not only that the investment will be safe but also that it will return dividends throughout its lifetime. However, if those same managers see the technology successfully applied to a test bed under conditions that resemble their company's situation, they might be more convinced that it would work for them. Such demonstrations can be offered by government agencies, trade groups, developers of the technology, utilities, corporations, or other institutions. Cooperative efforts often are most effective.

20.5.8. *Creating Legal and Institutional Instruments*

20.5.8.1. *Developing and Enforcing Standards*

A final approach to reducing GHG emissions from the industrial sector is the development and imposition of standards. This strategy already has been used in such applications as automotive emissions and the energy efficiency of buildings. Such standards can be written by industry groups and incorporated into regulations or legislation. Such standards ensure that feasible goals are set, that the technology to achieve those goals is available or under development, that a level playing field is provided for all participants, that the information needed to comply with the standards is disseminated, and that monitoring and enforcement mechanisms are in place. One of the advantages of such an approach is its flexibility; the standards and the technologies they prescribe can change over time in response to technical advances, economic conditions, and the perception of social and political needs.

20.5.8.2. *Expanding Land-Use Planning*

Opportunities for efficiency improvement through system integration are substantial in densely populated areas of the industrial world, where sources of waste heat and materials and demands for low-temperature heat and waste materials as inputs often are located close together. Such opportunities, however, pose additional implementation barriers because the organizational and regulatory complexities of system-integration projects are particularly demanding. A positive attitude from local authorities and utility planners and an innovative approach in regional planning have proved essential in the few cases in which this approach has been successfully followed.

20.6. Conclusions

The worldwide industrial sector contributes a large percentage of GHGs to the atmosphere. The bulk of this is CO_2 associated with energy use, but significant quantities of process GHGs—including CO_2, CH_4, N_2O, CFCs, and other halo- and hydrohalocarbons—also are released. OECD industrial economies contribute the bulk of fossil fuel CO_2 emissions, but these have remained relatively constant over the past 2 decades. Transitional-economies emissions have declined sharply since 1988 with the contraction of economic output, while developing-economies emissions have grown as their industrial sectors have expanded. A number of opportunities have been identified in the most energy- and GHG-intensive industrial subsectors for more efficient use of energy and materials. Creating industrial ecosystems that make maximal use of recycled secondary materials as feedstocks and take advantage of cascaded lower-temperature heat can lead to substantial reductions in greenhouse gas releases. The use of low-carbon fuels, renewable energy technologies, less energy-intensive materials, and dematerialization can lead to lower GHG emissions in all economies. Shifting industrial practices to reduce GHG emissions (and other environmental pollutants) can occur as part of the regular replacement of capital stock, but information and a variety of economic incentives and other policies are likely to be needed to assure that replacement technologies and processes are chosen that lower GHG releases.

References

Abrahamson, D., 1992: Aluminum and global warming. *Nature*, **356**, 484.

Allenby, B. and D.J. Richards, 1994: *The Greening of Industrial Ecosystems*. National Academy Press, Washington, DC, 259 pp.

Altehnpohl, D., 1980: *Materials in World Perspective*. Springer-Verlag, New York, NY, 60 pp.

Aluminum Association, Inc., Department of Economics and Statistics, 1991: *Patterns of Energy and Fuel Usage in the U.S. Aluminum Industry: Full Year 1989*. Report to the U.S. Department of Energy, 11 pp.

American Forest and Paper Association, 1993: *U.S. Pulp and Paper Industry's Energy Use—Calendar Year 1992*. American Forest and Paper Association, Washington, DC, 19 pp.

Ayres, R. and U. Simonis (eds.), 1994: Industrial metabolism: theory and policy. In: *Industrial Metabolism: Restructuring for Sustainable Development*. United Nations University Press. Tokyo, Japan, pp. 3-20.

Baldwin, S.F., 1989: Energy-efficient electric motor drive systems. In: *Electricity: Efficient End-Use and New Generation Technologies and Their Planning Implications* [Johansson, T.B., B. Bodlund, and R.H. Williams (eds.)]. Lund University Press, Lund, Sweden, pp. 21-58.

Bending, R.C., R.K. Cattell, and R.J. Eden, 1987: Energy and structural change in the United Kingdom and Western Europe. *Annual Review of Energy*, **12**, 185-222.

Berdowski, J.J.M., J.G.J. Olivier, and C. Veldt, 1993: Methane from combustion and industrial processes. In: *Methane and Nitrous Oxide: Methods in National Emissions Inventories and Options for Control.* Proceedings of International IPPC Workshop, RIVM, Bilthoven, Netherlands, pp. 131-142.

Brick Institute of America, 1994: *Brick Production and Shipments Report.* Brick Institute of America, Reston, VA, 5 pp.

Brown, J.G., 1987: *The International Sugar Industry: Development and Prospects*. World Bank, Washington, DC, 69 pp.

Brown, L.R., N. Lenssen, and H. Kane, 1995: *Vital Signs 1995*. Worldwatch Institute, Washington, DC, pp. 66-67.

CDIAC, 1992: *Trends '91: A Compendium of Data on Climate Change*. ORNL/CDIAC-46. Carbon Dioxide Information Analysis Center, Oak Ridge National Laboratory, Oak Ridge, TN, 665 pp.

Chandler, W.U., 1985: *Energy Productivity: Key to Environmental Protection and Economic Progress*. Worldwatch Paper 63, Worldwatch Institute, Washington, DC, 63 pp.

Chemical Manufacturers Association, 1994: Personal communication, 27 January 1994, Washington, DC, 5 pp.

Chemical Week, 1992: Finding the answer in technology, 17 June 1992, p. 79.

Corcoran, E., 1994: A waste not, want not goal: fledgling firm takes a lesson from steelmakers to recycle industrial waste. *The Washington Post*, 22 February, pp. E1, E4.

di Primio, J.C., 1993: *Estimates of Carbon Dioxide Emissions from Fossil Fuels Combustion in the Main Sectors of Selected Countries 1971-1990.* BMFT (Bundesminister fur J Forshung and Technologie), Ikarus Teilproject 9, Julich, Germany, 177 pp.

EIA, 1993: *Emissions of Greenhouse Gases in the United States 1985-1990.* Energy Information Administration, Washington, DC, 99 pp.

EMF, 1994: *Final First Round Study Design for EMF 14: Integrated Assessment of Climate Change*, September 19, 1994. Energy Modeling Forum, Stanford University, Stanford,CA, 30 pp.

Fabian, P., R. Borchers, B.C. Krüger, and S. Lal, 1987: CF_4 and C_2F_6 in the atmosphere. *J. Geophys. Res.*, **92**, 9831-9835.

Forrest, D. and J. Szekely, 1991: Global warming and the primary metals industry. *Journal of Metals*, December, 23-30.

Franklin Associates, 1990: *A Comparison of Energy Consumption by the Plastics Industry to Total Energy Consumption in the US*. Franklin Associates Ltd., November, pp. 4-7.

Frosch, R.A., 1995: The industrial ecology of the 21st century. *Scientific American*, **273(3)**, 178-181.

GCC, 1993: *Energy Efficiency in US Industry: Accomplishments and Outlook.* The Global Climate Coalition, October, Washington, DC, 104 pp.

Gielen, D.J., *et al.*, 1993: *Cost Effective CO_2 Reduction in Coupled Energy and Materials Systems*. Working Paper, ECN Policy Studies, Netherlands.

Giraud, P.N., A. Nadai, ECOBILAN, 1993: *Ecotaxe, Concurrence entre Materiaux et Conception des Objets Techniques*. Research Report for ADEME and the French Ministry of Environment, Paris, France, 150 pp.

Goldemberg, J., L.C. Monaco, and I.C. Macedo, 1993: The Brazilian fuel alcohol program. In: *Renewable Energy: Sources for Fuels and Electricity* [Johansson, T.B., H. Kelly, A.K.N. Reddy, and R. Williams (eds.)]. Island Press, Washington, DC, pp. 841-863.

Gotelli, I.M., 1993: *Using Economic Incentives to Promote Sustainable Industrial Practices: A Case Study of the Sugar Industry in Mexico*. Master's thesis submitted to Professor W.R. Moomaw, 21 December 1993, Fletcher School of Law and Diplomacy, Tufts University, Medford, MA, 122 pp.

Graedel, T.E., I. Horkeby, and V. Norberg-Bohm, 1994: Prioritizing impacts in industrial ecology. In: *Industrial Ecology and Global Change* [Socolow, R., C. Andrews, F. Berkhout, and V. Thomas (eds.)]. Cambridge University Press, Cambridge, UK, pp. 359-370.

Griffin, R.C., 1987: CO_2 release from cement production, 1950-1985. In: *Estimates of CO_2 Emissions from Fossil Fuel Burning and Cement Manufacturing, Based on the United Nations Energy Statistics and the U.S. Bureau of Mines Cement Manufacturing Data* [Marland, G., T.A. Boden, R.C. Griffin, S.F. Huang, P. Karciruk, and T.R. Nelson (eds.)]. Report No. ORNL/CDIAC-25, Carbon Dioxide Information Analysis Center, Oak Ridge National Laboratory, Oak Ridge, TN, pp. 643-680.

Grubler, A., S. Messner, and L. Schrattenholzer, 1993: Emission reduction at the global level. *Energy*, **18**, 539-581.

Hall, D.O., F. Rosillo-Calle, R.H. Williams, and J. Woods, 1994: Biomass for energy: supply prospects. In: *Renewable Energy: Sources for Fuels and Electricity* [Johansson, T.B., H. Kelly, A.K.N. Reddy, and R. Williams (eds.)]. Island Press, Washington, DC, pp. 593-651.

Haupin, W.E., 1987: *Production of Aluminum and Alumina* [Burkin, A.R. (ed.)]. J. Wiley and Sons, Chichester, UK, 241 pp.

Herman, R., S.A. Ardekani, and J.H. Ausubel, 1989: Dematerialization. In: *Technology and Environment* [Ausubel, J.H. and H.E. Sladovich (eds.)]. National Academy Press, Washington, DC, pp. 50-69.

Hryn, J.N. and D.R. Sadoway, 1993: Cell testing of metal anodes for aluminum electrolysis. In: *Light Metals 1993* [Das, S.K. (ed.)]. TMS, Warrendale, PA, pp. 475-483.

Huang, Jin-ping, 1993: Industry energy use and structural change: a case study of the People's Republic of China. *Energy Economics*, **15(2)**, 131-136.

IIASA, 1993: *Long Term Strategies for Mitigating Global Warming*. International Institute for Applied Systems Analysis, Laxenburg, Austria.

IISI, 1993: *Carbon Dioxide and the Steel Industry*. Committee on Environmental Affairs and Committee on Technology. International Iron and Steel Institute, Brussels, Belgium, 68 pp.

IPCC, 1990: *Climate Change: The IPCC Scientific Assessment* [Houghton, J.T., G.J. Jenkins, and J.J. Ephraums (eds.)]. Cambridge University Press, Cambridge, UK, 365 pp.

IPCC, 1992: *Climate Change 1992: The Supplementary Report to the IPCC Scientific Assessment* [Houghton, J.T., B.A. Callendar, and S.K. Varney (eds.)]. Cambridge University Press, Cambridge, UK, 200 pp.

Jacques, A.P., 1992: *Canada's Greenhouse Gas Emissions: Estimates for 1990*. Report EPS 5/AP/4, Environment Protection Series. Environment Canada, Ottawa, Canada, 78 pp.

Johansson, T.B., B. Bodlund, and R.H. Williams (eds.), 1989: *Electricity: Efficient End-Use and New Generation Technologies, and Their Planning Implications*. Lund University Press, Lund, Sweden, 960 pp.

Juwarkar, A., P. Sudhakarbabu, and P. Khanna, 1994: *Production of Biosurfactants—A Case Study*. Chemical Industry Digest, pp. 91-95.

Kashiwagi, T., 1993: *Systems Integration for Rational Use of Industrial Energy*. IPCC Working Group II Subgroup A: Mitigation Options–Industry. Intergovernmental Panel on Climate Change, Geneva, Switzerland.

Kashiwagi, T., 1992: The importance of systems integration in energy technology. In: *International Conference Next Generation Technologies for Efficient Energy Uses and Fuel Switching*. Proceedings of the Conference in Dortmund, 7-9 April 1992. IEA, Dortmund, Germany, pp. 553-564.

Kashiwagi, T., 1991: Present status and prospects for unexploited energy utilization. *Journal of the Fuel Society of Japan*, **70(2)**, 108-116 (in Japanese).

Keller, H.S. and D. Zylberstajn, 1991: Energy-intensity trends in Brazil. *Annual Review of Energy*, **16**, 179-203.

Khanna, P., N. Saraswat, and V.S. Kulkarni, 1988: *Towards Preventive Environmental Policy for Industry in India*. Report, Ministry of Environmental and Forests, Government of India, New Delhi, India.

Koch, P., 1992: Wood vs. non-wood materials in US residential construction. Some energy-related implications. *Forest Products Journal*, **42**, 31.

Konishi, A., 1995: *Waste Paper Trade and Recycling: The Implications for Development and Environment in India*. Ph.D. dissertation, The Fletcher School of Law and Diplomacy, Tufts University, Medford, MA, 270 pp.

Lashof, D.A. and D.R. Ahuja, 1990: *Relative Contributions of Greenhouse Gas Emissions to Global Warming*. Nature, **344**, 529-533.

Lees, E., 1992: Technological developments for energy efficiency and fuel switching in industry. In: *International Conference Next Generation Technologies for Efficient Energy Uses and Fuel Switching*. Proceedings of the Conference in Dortmund, 7-9 April 1992. IEA, Dortmund, Germany, pp. 183-196.

Lindmann, E., 1992: Gasification of industrial wastes. In: *International Conference Next Generation Technologies for Efficient Energy Uses and Fuel Switching*. Proceedings of the Conference in Dortmund, 7-9 April 1992. IEA, Dortmund, Germany, pp. 319-328.

Lonrigg, P., 1984: Use of PV to transport and desalt ground water supplies using brushless DC motors. *Solar Cells*, **85**, 231.

Lyon, R.K., J.C. Kramlich, and J.A. Cole, 1989: Nitrous oxide: sources, sampling and science policy. *Environmental Science and Technology*, **23(4)**, 392-394.

Marland, G., 1994: Personal communication, 25 March 1994.

Marland, G. and A. Pippin, 1990: United States emissions of carbon dioxide to the earth's atmosphere by economic activity. *Energy Systems and Policy*, **14**, 319-336.

Marland, G. and R.M. Rotty, 1984: Carbon dioxide emissions from fossil fuels: a procedure for estimation and results for 1950-1982. *Tellus*, **36B**, 232-261.

Marland, G., R.J. Andres, and T.A. Boden, 1993: Global, regional, and national CO_2 emissions. In: *Trends '93: A Compendium of Data on Global Change* [Boden, T.A., D.P. Kaiser, R.J. Sepanski, and F.W. Stoss (eds.)]. ORNL/CDIAC-65, Oak Ridge National Laboratory, Carbon Dioxide Information Analysis Center, Oak Ridge, TN, pp. 9-85.

Matsuoka, Y., M. Kainuma, and T. Morita, 1994: Scenario analysis of global warming using the Asian-Pacific integrated model (AIM). In: *Integrated Assessment of Mitigation, Impacts and Adaptation to Climate Change* [Nakicenovic, N. *et al.* (eds.)]. IIASA, Laxenburg, Austria, pp. 309-338.

Metalgesellshaft AG, 1988: *Tableaux Statistiques 1977-1987*. Metalgesellshaft AG, Frankfurt, Germany, 471 pp.

MITI, 1990: *Challenges in New Energy Trends*. Agency of Natural Resources and Energy, Ministry of International Trade and Industry, Tokyo, Japan, pp. 59-98 (in Japanese).

MITI, 1993: Comprehensive approach to the new sunshine program. *Sunshine Journal*, **4**, 1-6.

MITI, 1994: *The Basic Statistics for Analysis of CO_2 Mitigation in Japan*. Ministry of International Trade and Industry, Tokyo, Japan.

MMT, 1993: *1993 Annual Report*. Molten Metal Technologies, Waltham, MA, 28 pp.

Moomaw, W.R., 1991: Photovoltaics and materials science: helping to meet the imperatives of clean air and climatic change. *Journal of Crystal Growth*, **109**, 1-11

Moomaw, W.R. and D.M. Tullis, 1994: *Charting Development Paths: A Multicountry Comparison of Carbon Dioxide Emissions in Industrial Ecology and Global Change*. Cambridge University Press, Cambridge, UK, pp. 157-172.

Moomaw, W.R. and D.M. Tullis, 1995: *Industial Sector CO_2 Emissions in Industrial, Transitional and Developing Countries*. Global Development and Environment Institute Report, Tufts University, Medford, MA, pp. 1-34.

Morita, T., Y. Matsuoka, M. Kainuma, *et al.*, 1994: AIM—Asian-Pacific integrated model for evaluating policy options to reduce GHG emissions and global warming impacts. In: *Global Warming Issue in Asia* [Bhattacharya, S.C., *et al.* (eds.)]. Asian Institute of Technology, Bangkok, Thailand, pp. 254-273.

Motwani, M., R. Seth, H.F. Daginawala, and P. Khanna, 1993: Microbial production of 2,3 butanediol from water hyacinth. *Bioresource Technology*, **44**, 187-195.

Nakicenovic, N., A. Grubler, *et al.*, 1993: Long term strategies for mitigating global warming. *Energy*, **18**, 401-419.

NEDO, 1993: *The Environment Friendly Energy Community Project*. NEDO Pamphlet, Tokyo, Japan, pp. 1-6 (in Japanese).

NEERI, 1993: *Environmental Audit of South Indian Viscosc Ltd*. National Environmental Engineering Research Institute, Nagpur, India.

NEERI, 1994: *Value Added Chemicals from Cellulosic Saccharification and Fermentation* [Khanna, P. (ed.)]. National Environmental Engineering Research Institute, Nagpur, India.

Nelson, K., 1994: Finding and implementing projects that reduce waste. In: *Industrial Ecology and Global Change* [Socolow, R., C. Andrews, F. Berkhout, and V. Thomas (eds.)]. Cambridge University Press, Cambridge, UK, pp. 371-382.

Norford, L., A. Rabl, J. Harris, and J. Roturier, 1989: *Electricity: Efficient End-Use and New Generation Technologies, and Their Planning Implications* [Johansson, T.B., B. Bodlund, and R.H. Williams (eds.)]. Lund University Press, Lund, Sweden, pp. 427-460.

Novak, M. and I. Cuhalev, 1993: Indonesian coal for environmental value in Slovenia cogeneration plants. In: *IEA Second International Conference on the Clean and Efficient Use of Coal and Lignite: Its Role in Energy, Environment and Life*. Proceedings of the Conference in Hong Kong, 30 November–3 December 1993. OECD Publications, Paris, France, pp. 697-703.

NRC, 1976: *Renewable Resources for Industrial Materials. A Report of the Committee on Renewable Resources for Industrial Materials Board on Agriculture and Renewable Resources Commission on Natural Resources*. National Research Council, National Academy of Sciences, Washington, DC, 275 pp.

OECD, 1994: *Energy Related CO_2 Emissions, Electricity Production and Consumption 1971-1992*. Organization for Economic Cooperation and Development, Paris, France.

Ogden, J. and J. Nitsch, 1993: Solar hydrogen. In: *Renewable Energy: Sources for Fuels and Electricity* [Johansson, T.B., H. Kelly, A.K.N. Reddy, and R. Williams (eds.)]. Island Press, Washington, DC, pp. 925-1009.

Ogden, J.M. and R.H. Williams, 1989: *Solar Hydrogen: Moving Beyond Fossil Fuels*. World Resources Institute, Washington, DC, 123 pp.

Olivier, J.G.J., 1993: Nitrous oxide emissions from industrial processes. In: *Methane and Nitrous Oxide: Methods in National Emissions Inventories and Options for Control*. Proceedings of International IPPC Workshop, RIVM, Bilthoven, Netherlands, pp. 339-442.

OTA, 1991: *Energy Technology Choices Shaping Our Future*. Office of Technology Assessment. Washington, DC, 148 pp.

Plunkert, P.A. and E.D. Senke, 1993: *Aluminum, Bauxite and Alumina: 1991*. Annual Report, U.S. Department of the Interior, Bureau of Mines, Washington, DC, 48 pp.

Portland Cement Association, 1990: *U.S. Cement Industry Fact Sheet*. Portland Cement Association, Skokie, IL, pp. 14-15.

Pimentel, D., L.E. Hurd, and A.C. Bellotti, 1973: Food production and the energy crisis. *Science*, **182**, 443-449.

Rajabapaiah, P., S. Jayakumar, and A.K.N. Reddy, 1993: Biogas electricity—the Pura Village case study. In: *Renewable Energy: Sources for Fuels and Electricity* [Johansson, T.B., H. Kelly, A.K.N. Reddy, and R.H. Williams (eds.)]. Island Press, Washington, DC, pp. 787-815.

Ramanathan, V., R.J. Cicerone, and H.B. Singh, 1985: Trace gas trends and their potential role in climate change. *J. Geophys. Res.*, **90**, 5547-5566.

Ross, C.P., 1991: *Glass Industry Consulting. Creative Opportunities*. U.S. Glass Industry Report 1991, Laguna Niguel, CA, 4 pp.

Sadoway, D.R., 1990: Metallurgical electrochemistry in nonaqueous media. In: *1990 Elliott Symposium Proceedings*. Massachusetts Institute of Technology, Cambridge, MA, pp. 189-196.

Schaefer, H., 1992: Technology—a key issue to rational and efficient energy use. In: *International Conference Next Generation Technologies for Efficient Energy Uses and Fuel Switching*. Proceedings of the Conference in Dortmund, 7-9 April 1992. IEA, Dortmund, Germany, pp. 629-642.

Schipper, L. and S. Meyers, 1992: *Energy Efficiency and Human Activity: Past, Trends, Future Prospects*. Sponsored by the Stockholm Environmental Institute, Cambridge University Press, Cambridge, UK, pp. 84-110.

Schipper, L. and E. Martinot, 1993: Decline and rebirth: energy demand in the former USSR. *Energy Policy*, **21(9)**, 969-977.

Scientific American, 1995: The treaty that worked—almost. *Scientific American*, **273(3)**, 18-20.

Sharma, S., G. Pandya, T. Chakrabarti, and P. Khanna, 1994: Liquid-liquid equilibria for 2,3 butanediol. *Journal of Chemical Engineering Data*, **39**, 823-826.

Socolow, R., C. Andrews, F. Berkhout, and V. Thomas (eds.), 1994: *Industrial Ecology and Global Change*. Cambridge University Press, Cambridge, UK, 500 pp.

SRI International, 1994: *By-Product Hydrogen Sources and Markets*. Prepared for the National Hydrogen Association 5th Annual U.S. Hydrogen Meeting, March 24, 1994, Barbara Heydorn (consultant), Menlo Park, CA, 15 pp.

Summers, R. and A. Heston, 1991: The Penn world tables (mark V): an expanded set of international comparisons, 1950-1988. *Quarterly Journal of Economics*, **106**, 327-368.

Summers, R. and A. Heston, 1994: *An Expanded Set of International Comparisons, 1950-1992*. World Bank, Washington, DC.

TERI, 1991: *TERI Energy Data Directory and Year Book, 1990-91*. Tata Energy Research Institute, New Delhi, India.

Thiemens, M.H. and W.C. Trogler, 1991: Nylon production: an unknown source of atmospheric nitrous oxide. *Science*, **251**, 932-934.

Torvanger, A., 1991: Manufacturing sector carbon dioxide emissions in nine OECD countries, 1973-1987. *Energy Economics*, **13(2)**, 168-186.

Turnbull, J., 1993: *Strategies for Achieving a Sustainable, Clean and Cost Effective Biomass Resource*. Electric Power Research Institute, Palo Alto, CA, 25 pp.

UBC, 1994: Personal communication, University of British Columbia, Department of Design.

U.S. Bureau of Mines, 1992: Soda ash. In: *Mineral Commodity Summaries,1992*. U.S. Department of the Interior, Bureau of Mines, Washington, DC, pp. 162-163.

U.S. DOE, 1980: *Classification and Evaluation of Electric Motors and Pumps*. DOE/CS-0147, U.S. Department of Energy, Washington, DC, 251 pp.

U.S. DOE, 1994: *The Climate Change Action Plan: Technical Supplement*. DOE/PO-0011, U.S. Department of Energy, Washington, DC, 148 pp.

U.S. EPA, 1993: *Inventory of U.S. Greenhouse Gas Emissions and Sinks, 1990-1993*. EPA230-R-94-014, U.S. Environmental Protection Agency, Climate Change Division, Washington, DC, 170 pp.

Williams, R.H. and E.D. Larson, 1993: Advanced gasification based biomass power generation. In: *Renewable Energy: Sources for Fuels and Electricity* [Johansson, T.B., H. Kelly, A.K.N. Reddy, and R. Williams (eds.)]. Island Press, Washington, DC, pp. 729-786.

Williams, R.H., E.D. Larson, and M.H. Ross, 1987: Materials, affluence and industrial energy use. *Annual Review of Energy*, **12**, 99-144.

World Bureau of Metal Statistics, 1995: *World Metal Statistics Yearbook*. World Bureau of Metal Statistics, Birmingham, UK, 75 pp.

WRI, *et al.*, 1994: *United Nations Environment Programme, United Nations Development Programme. 1994. World Resources 1994-1995*. Oxford University Press, New York, NY, 400 pp.

Young, J.E., 1992: Aluminum's real tab. *World Watch: A Bimonthly Magazine of the Worldwatch Institute*, **5(2)**, 26-33.

Zerbe, J.I., 1990: Biomass energy and global warming. In: *Proceedings of Global Warming—A Call for International Coordination*, 9-12 April 1990, Chicago, IL, pp. 66-82.

21

Mitigation Options in the Transportation Sector

LAURIE MICHAELIS, OECD

Principal Lead Authors:
D. Bleviss, USA; J.-P. Orfeuil, France; R. Pischinger, Austria

Lead Authors:
*J. Crayston, ICAO; O. Davidson, Sierra Leone; T. Kram, The Netherlands;
N. Nakicenovic, IIASA; L. Schipper, USA*

Contributing Authors:
*G. Banjo, Nigeria; D. Banister, UK; H. Dimitriou, Hong Kong; D. Greene, USA;
L. Greening, USA; A. Grübler, IIASA; S. Hausberger, Austria; D. Lister, UK;
J. Philpott, USA; J. Rabinovitch, Brazil; N. Sagawa, Japan; C. Zegras, USA*

CONTENTS

EXECUTIVE SUMMARY

The transport sector—including passenger travel and freight movements by road, rail, air, and water—was responsible for about 25% of 1990 world primary energy use and 22% of CO_2 emissions from fossil-fuel use. It is one of the most rapidly growing sectors. Energy use in 1990 was estimated to be 61–65 EJ. Without new measures, this might grow to 90–140 EJ in 2025. Confidence in this range, and in the outlook for relative contributions of different countries, is low, and future transport energy use will be influenced by government policy. Nevertheless, industrialized countries are expected to continue to contribute the majority of transport-related greenhouse gas emissions until 2025. After 2025, the majority of transport-related emissions may be from countries that are currently developing rapidly or that have economies in transition.

Transport activity increases with rising economic activity, disposable income, access to motorized transport, and falling real vehicle and fuel costs. Societies are becoming increasingly dependent on the car for routine travel, aircraft for long distance travel, and the truck for freight transport. Transport activity, the development of urban and other infrastructure, lifestyles, and patterns of industrial production are all closely interrelated. As well as being linked to economic development, the transport sector imposes burdens on society, including the effects of traffic congestion, accidents, air pollution, and noise. Climate change is currently a relatively minor factor in decisionmaking and policy in the sector, although this could change in the future.

Greenhouse gas emissions could be reduced in all transport modes through changes in vehicle energy intensity or energy source, stimulated by research and development as well as by changes in current patterns of intervention in the markets for vehicles and fuels:

- Improved vehicle energy efficiency might reduce greenhouse gas emissions per unit of transport activity by 20 to 50% in 2025 relative to 1990 without changes in vehicle performance and size.
- If users were prepared to accept changes in vehicle size and performance, transport energy intensity could be reduced by 60 to 80% in 2025.
- With energy-intensity reductions, the use of alternative energy sources could, in theory, almost eliminate greenhouse gas emissions from the transport sector after 2025. A complete transition to zero greenhouse gas emission surface transport is conceivable but would depend on eliminating emissions throughout the vehicle and fuel-supply chain.

In all cases, there is medium confidence in the ranges for these technical potentials, which depend on government policies and decisions made by vehicle and energy suppliers.

The use of motor vehicles might be reduced by changing land-use patterns and lifestyles to reduce the need for goods transport and travel; restrictions on the use of motor vehicles; and fiscal or other measures to discourage motor vehicle use and to encourage the use of nonmotorized transport. No confidence can be placed in estimates for the ultimate potential mitigation through these measures. A few city authorities have used packages of such measures to control traffic and have reduced energy use by 20 to 40%.

Travel-mode switching from car to bus or rail can reduce primary energy use by 30 to 70%, while switching container freight traffic from road to rail can reduce primary energy use by 30%. The percentage of greenhouse gas emission reduction is more than this value if railways are powered with electricity from nonfossil sources. Confidence in the potential for mitigation through mode switching is low because human behavior and choice play a central role, and these factors are poorly understood.

Experts may not agree on the best approach among these options and sometimes give conflicting advice. Nevertheless, there is an emerging consensus that attempts to move traffic to less energy-intensive modes depend on using well-integrated strategies designed specifically for local situations. Measures to encourage nonmotorized transport and public transport tend to work only when combined with measures to reduce the distances people need to travel and discourage energy-intensive vehicle use. Several cities in Latin America, Southeast Asia, and Europe have succeeded in stemming the growth in car use by such combined strategies.

In many circumstances, strategies may not be implemented if they might reduce the benefits provided by transport systems to individuals and firms. Greenhouse gas mitigation strategies will have to address this issue and find ways to meet or change the needs and desires currently met by energy-intensive transport. Preferences in travel behavior are driven by social and cultural factors, as well as by cost-effectiveness in meeting needs.

Policies with a wide variety of social, economic, and environmental objectives—especially those of reducing traffic congestion, accidents, noise, and local air pollution—also can reduce greenhouse gas emissions from transport. Conversely, greenhouse gas mitigation strategies are likely to be more acceptable and successful if they are integrated into wider strategies in transport, urban development, and environment policy.

Appropriate mixes of policies will vary between cities and countries. In small towns with relatively simple infrastructure, provision for nonmotorized transport and alternatives to transport may be particularly important. Cities with rapidly developing infrastructure have an opportunity to manage transport-system development, ensuring the viability of nonmotorized and public transport modes. In countries and cities with highly developed infrastructure and technical capability, changes in the transport system are likely to be slower but may hold more potential to influence vehicle technology and energy sources.

When considering transport-sector mitigation options in engineering, economic, or planning terms, it is important to recognize that there are simplifying assumptions in these three frameworks, so each gives an incomplete picture. None incorporates an accurate understanding of human behavior and choice, although these are essential determinants of greenhouse gas emissions from the sector. Policymakers can address these issues by developing mechanisms and institutions to inform and educate transport stakeholders and involve them in the design and implementation of mitigation strategies.

There are several factors that lead to inertia in the development of transport systems: Technologies and fuels now in the laboratory can require several decades for commercialization; transport infrastructure is developed slowly and has an influence that can last for centuries; stakeholders may be reluctant to change their practices; and transport user behavior and choice also may develop slowly in response to a changing environment. Once transport systems are developed to service the car, truck, and airplane, it is very difficult to reverse the shift away from nonmotorized and public transport. Some immediate actions can be taken to facilitate future mitigation strategies; for example, new infrastructure can be designed to allow for nonmotorized transport. Greenhouse gas mitigation strategies could take some years, even decades, to deliver results and will have to be implemented long before anticipated greenhouse gas reduction needs.

21.1. Introduction

This chapter addresses options for mitigation of greenhouse gas emissions in the transport sector, which includes all types of travel and freight movement by road, rail, air, and water. Fluids—mainly petroleum and natural gas—are moved by pipeline, but greenhouse gas mitigation in pipeline systems is treated in Chapter 19.

Section 21.2 analyzes current emissions of greenhouse gases from transport and their trends, looking at road, rail, water, and air; the contribution of non-CO_2 greenhouse gases; and the patterns in different countries and regions. Section 21.3 reviews the potential for emission reductions through changes in vehicle maintenance and new vehicle design, through changes in vehicle operating practice, and through the introduction of alternative fuels. It also discusses the long-term potential for electrically powered transport and identifies key areas where research and development are needed. Section 21.4 reviews the effects of fiscal, regulatory, planning, and other measures in the transport sector and aims to clarify some of the differences of opinion among experts on the desirability and feasibility of measures.

21.2. Transport and Greenhouse Gas Emissions

21.2.1. Current Emissions

Greenhouse gas emissions from transport result mainly from the use of fossil fuels; the main greenhouse gas produced is CO_2. The transport sector was responsible for about 25% of 1990 world primary energy use and 22% of CO_2 emissions from energy use (including energy use in fuel production; based on IEA, 1993c). These shares are growing in almost all countries. Rising incomes and steady or declining fuel costs have encouraged a 50% growth in world energy use in the sector between 1973 and 1990—an average of 2.4% per year (Orfeuil, 1993). The 65 EJ consumed by transport, including

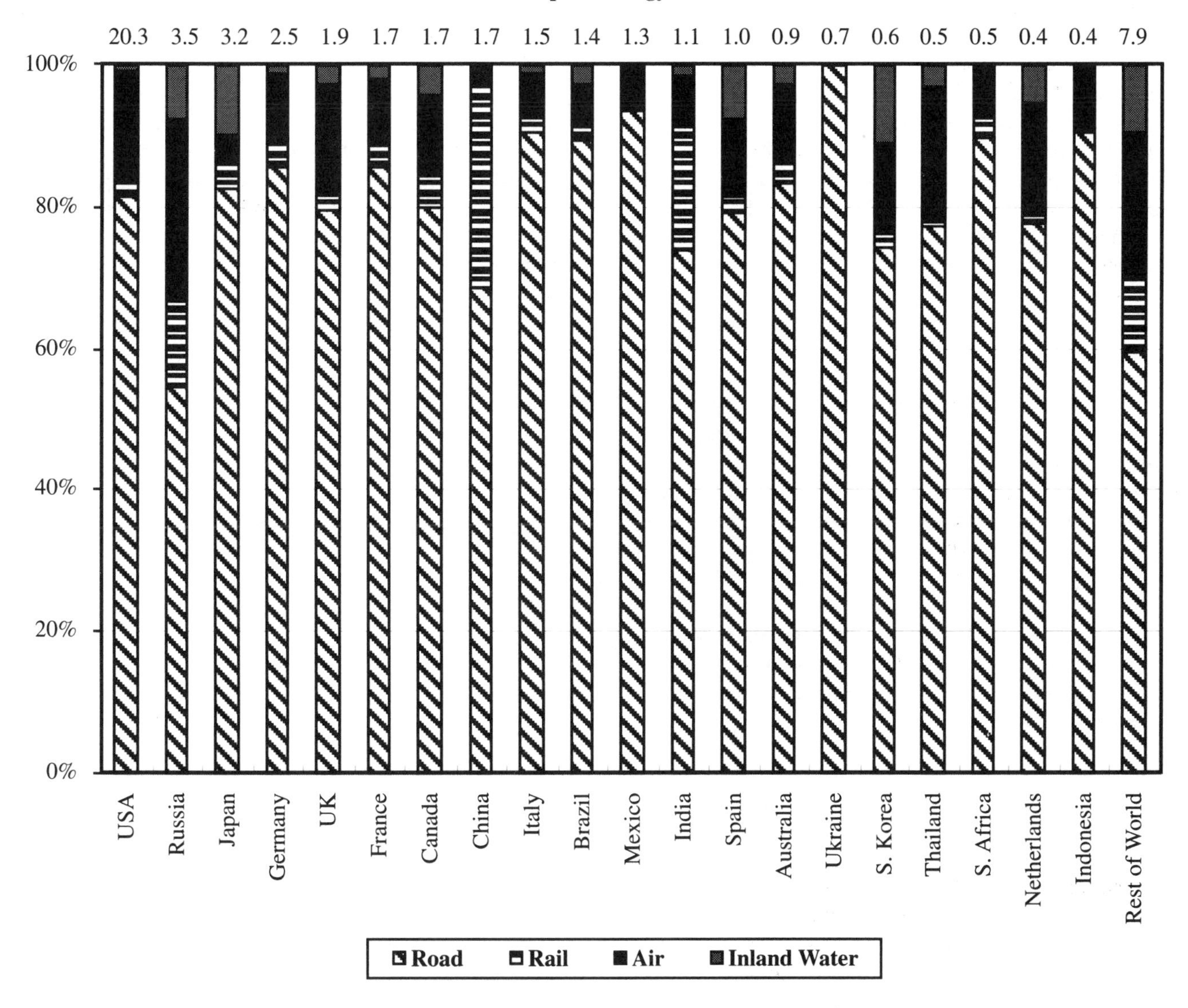

Figure 21-1: Transport-fuel sales, 1990, EJ (IEA, 1993c; Pischinger and Hausberger, 1993).

marine bunker fuel, in 1990 represented 58% of all oil products consumed (IEA, 1993c).

In 1990, about 23 countries' transport sectors are thought to have consumed more than 0.42 EJ in final energy, excluding marine bunkers. Figure 21-1 shows transport-fuel sales, excluding marine bunkers, in the 20 of these countries for which detailed data are available (IEA, 1993c, 1993d). In these countries, road traffic uses roughly 80% of total transport final energy consumption, air traffic 13%, rail 4.4%, and inland water transport 2.6%.

Given the large road-transport share of energy use, greenhouse gas mitigation options discussed in this chapter will focus on this subsector, although some options for air transport also will be addressed.

While CO_2 is the main greenhouse gas emitted by the transport sector, emissions of non-CO_2 greenhouse gases add to the radiative forcing of vehicle emissions. In addition to nitrous oxide (N_2O), methane (CH_4), and chlorofluorocarbons (CFCs), vehicles emit large quantities of carbon monoxide, volatile organic compounds (VOCs), and nitrogen oxides (NO_x) for which global warming potentials (GWPs) are not currently available (see IPCC Working Group I volume, 1995). N_2O contributes about 10% to the radiative forcing of tailpipe emissions of gasoline cars equipped with three-way catalytic converters (based on CEC, 1992; Prigent *et al.*, 1991). Greenhouse gas emissions during car manufacture and disposal add a further 10 to 15% relative to tailpipe CO_2, while releases of CFCs (used mainly in air conditioning) add between 10 and 50%. CO_2 and CH_4 emitted during oil exploration, extraction, processing, and transport contribute 10 to 20% of overall life-cycle forcing caused by vehicles using petroleum products (IEA, 1993a; CEC, 1992; DeLuchi, 1991).

The radiative-forcing effect of non-CO_2 emissions from aircraft engines is very uncertain. In particular, the impact of NO_x emitted at altitudes where subsonic aircraft fly could be more important than equivalent NO_x emissions at the Earth's surface. The effect of aircraft NO_x is uncertain: It could be of similar magnitude or smaller than the effect of CO_2 from aircraft (see Chapter 2.2, *Other Trace Gases and Atmospheric Chemistry,* in the IPCC Working Group I volume). Aircraft also emit carbon monoxide, water vapor, soot and other particles, sulfur gases, and other trace constituents, which have the potential to cause radiative forcing, but the impact of these emissions has not yet been properly assessed.

21.2.2. Projections of Transport Greenhouse Gas Emissions

At least until 2025, CO_2 is likely to remain the main greenhouse gas produced by the transport sector, and emissions will depend mainly on energy use in the sector. Several scenarios of transport energy use to 2025 have been developed; some are shown in Figure 21-2. All of these projections are produced based on some set of assumptions regarding the continuation of historical relationships between transport fuel consumption and variables such as gross domestic product (GDP), fuel prices, and vehicle energy efficiency.

International comparisons reveal a strong correlation between transport energy use and GDP. (The data plotted in Figure 21-3 indicate a GDP elasticity of 0.89, with $R^2 = 0.93$.) Nevertheless, at a given level of GDP, energy use can vary by a factor of two. The growth of transport with income and time is faster for middle-income countries than for very-low-income countries (Button, 1992). Button developed a model for car ownership that has been used by Orfeuil (1993) to indicate that, by 2020—assuming current GDP growth trends continue—developing countries will have a third of the world's car fleet, compared with 14% today.

Many analysts have found a close correlation between road-freight traffic (tonne-km) and GDP (Bennathan *et al.*, 1992), whereas rail freight appears to be almost independent of GDP but is very closely related to country surface area.

Figure 21-4 plots national average road-transport energy use per unit of GDP against fuel prices. It shows that prices may help to explain some, but by no means all, of the variation in energy use per unit of GDP. Variations also may be explained partly by geographic, cultural, and other factors.

Many analysts have criticized the use of historical relationships between the economy and energy use to produce projections. Although car ownership is growing very rapidly in Europe and Asia, Grübler and Nakicenovic (1991) suggest that it may saturate at lower per capita levels than in North America. In this case, global car energy use may not be much higher in 2010 than it was in 1985, assuming that fuel economy continues to improve at historical rates.

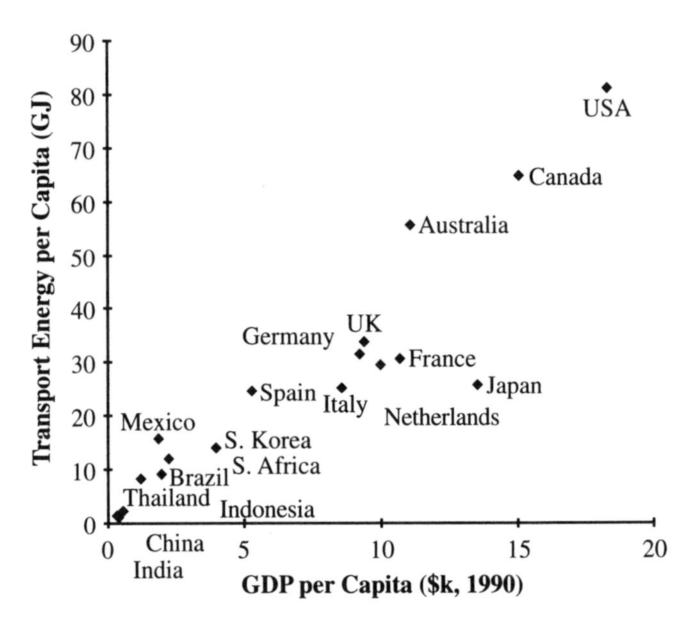

Figure 21-3: Total transport energy use vs. gross domestic product in 1990, for 18 of the world's largest transport energy users. Excludes Russia, Ukraine, Iran, Saudi Arabia, and Kazakhstan; former West and East Germany data have been combined (IEA, 1993c, 1993d).

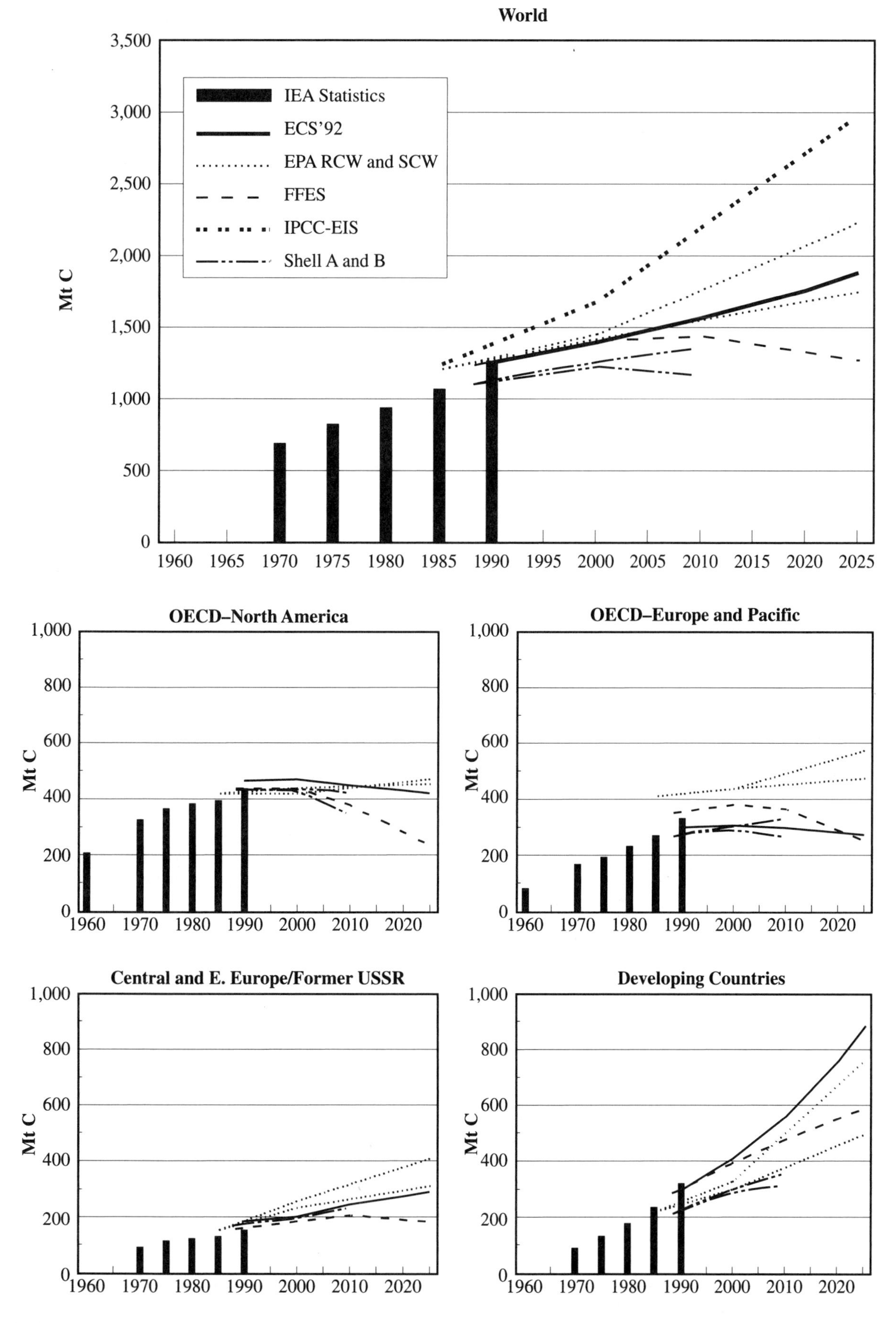

Figure 21-2: Comparison of transport CO_2 emission scenarios to 2025 (Grübler, 1993). Note: IEA = International Energy Agency; ECS = Environmentally Compatible Energy Strategies; RCW = Rapidly Changing World and SCW = Slowly Changing World; FFES = Fossil-Free Energy System; and EIS = Energy Industry System.

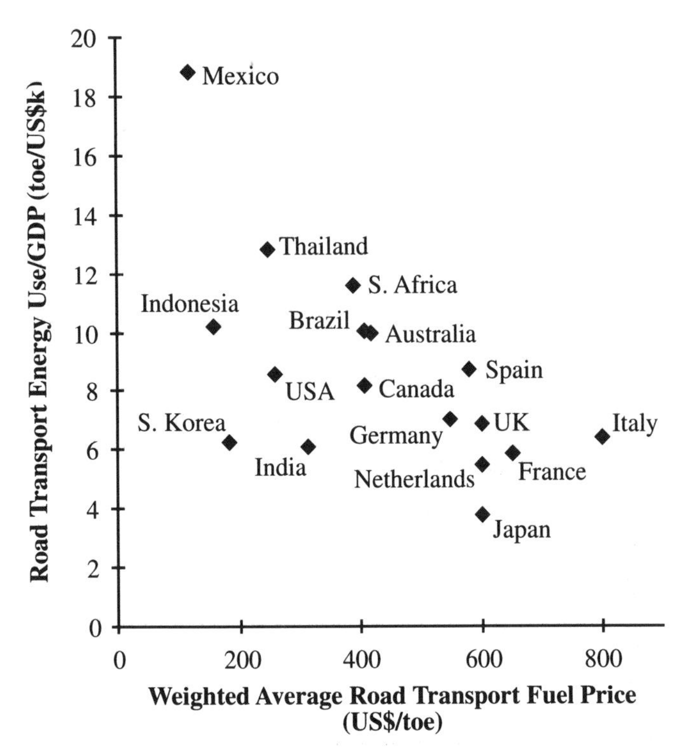

Figure 21-4: Road transport energy per unit of GDP vs. average fuel price in 1990, for 17 of the world's 20 largest transport energy users. Excludes China, Russia, and Ukraine; former West and East Germany data have been combined (IEA, 1993c, 1993d, 1994a; ADB, 1994).

21.2.3. The Role of Transport

Transport is needed for basic survival and social interaction as well as cultural and economic activity and development. Transport systems develop with the socioeconomic situation and the land-use patterns in which they are embedded (Dimitriou, 1992). Climate change is a new concern to be added to the many influences and interests in the transport sector; some of these influences and interests may support the aim of reducing greenhouse gas emissions, while others may be in conflict with it.

People use transport to travel between homes, workplaces, shops, and services; to socialize; and to take vacations. As travel becomes quicker, more flexible, and cheaper, people's options (where to live, work, shop, and socialize) broaden. Some people can use increasing mobility to improve their economic situation and quality of life. For others, especially the poor—particularly those in rural areas—the lack of affordable transport remains an economic and social constraint.

Producers of goods use transport systems to gain access to raw materials and to their markets. As the quality of transport improves and costs fall, the production and distribution of goods becomes cheaper and more efficient. Firms have increasing flexibility in their sources of raw materials and the markets for their products; they have a greater choice of production locations and employees; they also can become more specialized, achieving economies of scale and taking advantage of local conditions suited to their activities.

Transport systems can become an end in themselves; this is perhaps most obvious in the case of the automobile, which has become a symbol of wealth and social status. Tengström (1992) explores the extent to which this technology has shaped modern culture and provides a review of the related literature.

The uses to which transport services are put evolve as incomes rise, costs fall, and priorities change. A number of stages in the use of transport systems can be identified, corresponding roughly to Maslow's hierarchy of needs (Maslow, 1954):

1) Essential survival needs: movement of food, fuel, water, and travel for health care and to escape danger
2) Needs relating to economic security: movement of traded goods—mostly primary commodities—and travel to work and for education
3) Needs relating to social involvement: visiting family and friends, travel to entertainment, and family vacations; goods transport moves beyond essentials to include more manufactured goods
4) Needs relating to self-expression and exploration: tourism and more distant vacations; goods transport moves toward very high value-added goods.

This pattern is reflected in transport statistics and travel surveys, which indicate that, as societies become wealthier, social and leisure travel become more important. The car has been an essential element in this process, allowing increasing flexibility, especially in social relationships. The development of air travel has allowed tourism to flourish.

Transport demand was, until recently, seen in most countries as a social and economic need to be met. At the same time, growing motorized transport use has long been recognized to have negative consequences for society. The most important of these include:

- Traffic congestion
- Accidents
- Noise, vibration, and air pollution
- A wide range of social problems related to the use of land, reduction in nonmotorized transport use, reduction in availability of local services, and so forth, contributing to a declining quality of life for non-car owners.

Measures designed to address these issues provide the main opportunity at present to reduce greenhouse gas emissions from transport. Greenhouse gas mitigation policy is more likely to succeed if it addresses concerns such as these.

21.2.4. Activity and Energy Use in the Transport Sector—Travel Behavior

Table 21-1 shows information from a number of city travel surveys in Asia and Africa. The variation in travel patterns

Table 21-1: *How people travel—percentage of trips by mode in Asia and Africa.*

Country/City	Survey Year	Non-motorized	Private Motorized	Public Transport and Taxi
Asian City Surveys (Midgley, 1994)				
Low Income				
Tianjin	1987	91	–	9
Bombay	1981	26	9	65
Jakarta	1984	40	21	39
Middle Income				
Seoul	1982	12	8	80
Kuala Lumpur	1984	12	46	42
Bangkok	1984	16	24	60
High Income				
Central Tokyo	1988	24	25	51
Greater Tokyo	1988	22	54	24
African City Surveys (Davidson, 1993)				
Abidjan	1988	30	12	51
Dakar	1989	50	17	32
Nairobi	1989	15	25	50
Conakry				
Low Income		55	3	41
Middle Income		27	19	54
High Income		5	57	38

among cities in Asia, even at similar income levels, is quite striking and is much greater than the variation among countries within Europe, for which survey results are provided in Table 21-2. Despite the upward trends in most motorized forms of transport, walking and cycling continue to provide a large proportion of mobility and access needs in many countries.

Several observations can be made based on the European data. First, the variation among countries in distance traveled per day (standard deviation 16% of mean) is about twice as large as that in the number of trips and the time spent traveling (8 and 9%, respectively). It might be concluded from this type of data that (1) people travel to satisfy a certain number of access needs (for work, services, etc.) that do not vary significantly as the transport system changes, and (2) people operate with a time budget and will spend roughly the same amount of time traveling during each day, regardless of the average speed of the transport system. This would imply that measures that make travel faster will tend to increase the distance traveled by people.

Second, in most countries (with the exception of the United Kingdom), public transport is used for longer trips than cars: The public-transport share of trips is smaller than the share of distance. This has been confirmed by other studies (e.g., Birk and Bleviss, 1991). Walking, cycling, and moped trips, not surprisingly, are shorter on average than public-transport and car trips in all countries.

Third, although the car dominates the distance traveled in all countries, the car share of trips is about equal to the share of trips on foot, bicycle, or moped. Thus, although cars appear to dominate European transport according to the distance traveled, they are about equal in importance with travel on foot and on two-wheelers according to the number of trips.

Global road-vehicle fleets grew 140% between 1970 and 1990 (OECD, 1993a). The vehicle population could increase by anything from 60 to 120% by 2025, and 140 to 600% by 2100 (Walsh, 1993b; see Table 21-3). The current annual percentage growth is highest in Southeast Asia, Africa, Latin America, and

Table 21-2: *Daily travel of Europeans according to travel surveys (Salomon et al., 1993).*

				Daily Mean for Respondents			Modal Split (trip/distance-based)			Split by Purpose (trip-based)		
Country	Survey Year	Age Range (yrs)	Period Surveyed	Trips (#)	Distance Traveled (km)	Time Traveling (min)	Walk, Bike, Moped (%)	Public Transport (%)	Car (%)	Work, Education (%)	Shopping, Escorting Children (%)	Social, Leisure (%)
Austria	1983	>6	Weekday	2.9	22	67	40/8	19/34	42/58	40	30–41	18–29
Finland	1986	–	7 day	3.1	–	71	31/6	12/19	57/75	33	34	33
France	1984	>6	7 day	3.1	21	53	41/8	8/17	51/75	38	36	26
Former West Germany	1982	>10	7 day	2.9	30	69	41/8	14/25	45/57	39	32	30
Israel	1984	>8	Weekday	3.0	–	–	37/–	31/–	32/–	43	28	29
Netherlands	1987	>12	7 day	3.4	33	71	47/16	5/12	47/72	29	25	46
Norway	1985	13–74	7 day	3.4	32	71	35/6	11/31	54/63	33	22	45
Sweden	1983	15–84	7 day	3.6	25	–	38/5	12/20	50/70	36	16	48
Switzerland	1984	>10	7 day	3.3	29	70	46/10	12/20	42/70	36	34	30
UK	1986	–	7 day	2.8	23	–	37/9	14/19	49/72	30	40	30

Table 21-3: *Scenarios of world road vehicle population and traffic to 2100 (Walsh, 1993b).*

Year	Vehicle Population (million vehicles) Cars and Light Trucks	Heavy Trucks	Two-Wheelers	Traffic (trillion vehicle-km) Cars and Light Trucks	Heavy Trucks	Two-Wheelers
1990	540	30	110	3.5	0.5	0.3
2030	910–1300	60–90	180–250	5.8–8.2	1.1–1.5	0.5–0.7
2100	1200–3800	110–380	250–800	8.1–24.3	2.0–6.7	0.7–2.2

some central and eastern European countries. Much of the fleet growth in these countries comprises second-hand vehicles imported from the Organisation for Economic Cooperation and Development (OECD) countries, which still have the largest markets for new vehicles. Developing countries account for only 10% of the world's cars (Birk and Bleviss, 1991).

Light-duty vehicle traffic has risen rapidly in recent years with rising incomes and falling real costs of vehicle ownership and use, including fuel costs. The fastest growth has been in Southeast Asia, Africa, Latin America, and most recently in Central and Eastern Europe (Suchorzewski, 1993). Most countries have seen an increase in travel by bus and train from 1965 to 1991, although the growth has been less rapid than that in car traffic (ECMT, 1993; Schipper *et al.*, 1993).

Two-wheelers, especially mopeds with two-stroke engines, are one of the most rapidly growing means of personal transport in parts of South and East Asia and Latin America (Dimitriou, 1992; IEA, 1994c). India has about four times as many two-wheelers as cars (UNESCAP, 1991), and the ratio is expected to increase. Three-wheelers are important means of transport in several Asian cities, where they often are used as taxis.

OECD countries have four cars to every goods vehicle; in non-OECD countries, the ratio is nearer to two cars to every goods vehicle. Goods transport by road rose more during the 1980s than did personal travel. An increasing share of goods transport in industrialized countries has been carried by smaller trucks, as high-value-added goods account for an increasing proportion of production and as the service sector, including retailing, grows faster than the rest of the economy. At the same time, bulk freight is becoming increasingly concentrated in very large trucks (around 40 tonnes gross vehicle weight).

The breakdown by type of vehicle of the 46 EJ of energy used for road transport in 1990 is poorly documented, but Walsh (1993a) estimates that goods transport is currently responsible for about 27% (see Figure 21-5).

Rail freight traffic has increased or remained fairly steady in most countries. Because rail freight is less flexible than road freight, it has tended to become confined to conditions where costs per tonne-km are lower than for road freight; rail terminals exist at, or good road/rail transfer facilities exist near, both the origin and the destination; speed is relatively unimportant; and shipments are regular. Rail freight is most important in large countries where long hauls are more frequent (Bennathan *et al.*, 1992), including the United States, the former Soviet Union, and China. Its market has declined in many small countries. It also has declined as industries have moved toward "lean" manufacturing requiring fast, flexible, "just in time" transport systems. In central and eastern European countries, the process of economic reform also has brought about a large shift of freight from rail to road (GUS, 1993). Meanwhile non-motorized goods transport, which makes a negligible contribution in OECD countries, is important in developing countries, especially in rural areas.

Growth in air travel (passenger-km on scheduled services) has been much greater than economic growth but has been closely linked with it. The growth rate has declined, from 13.4% per year during 1960–1970 to 9.0% per year during 1970–1980 and 5.7% per year during 1980–1990 (ICAO, 1992). An increasing share of air passenger traffic is for leisure purposes. Air freight traffic (tonne-km on scheduled services) has grown

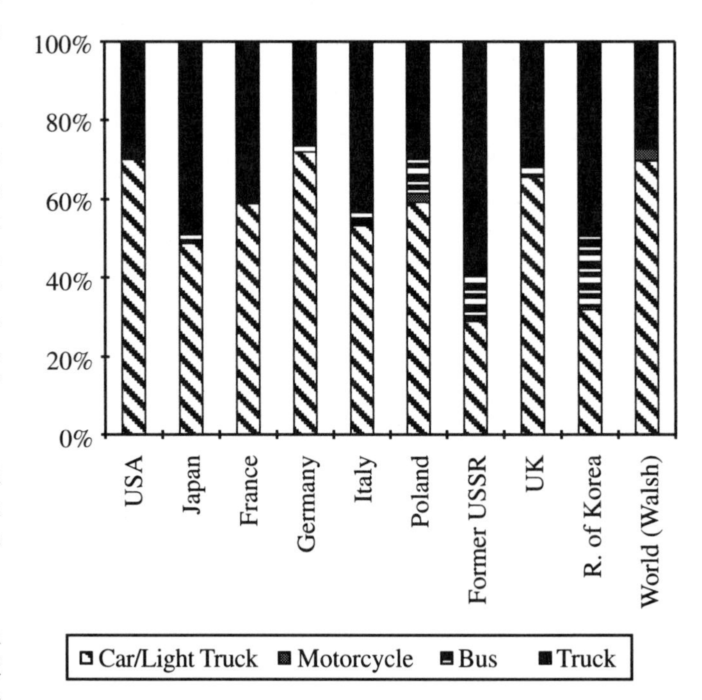

Figure 21-5: Shares of energy use by road vehicles, various years (Schipper *et al.*, 1993; Pischinger and Hausberger, 1993; Walsh, 1993a; Kang, 1989).

more rapidly, at 11.7% per year during 1960–1990 (ICAO, 1992). Half of passenger traffic and about 80% of freight traffic is international (ICAO, 1994a).

Waterborne transport volumes—mostly bulk marine freight—have fluctuated with international trade in bulk commodities. Bulk marine freight grew from 25.3 billion tonne-km in 1981 to 28.8 billion tonne-km in 1991 (OECD, 1993b). Most traffic in 1991 was made up of crude oil and oil products (46%), coal (11%), iron ore (11%), and grain (6%).

21.2.5. Trends in Transport Energy Intensity

So far, this chapter has discussed trends in traffic by different transport modes. Greenhouse gas emissions from transport depend mainly on energy use, which is the product of the energy intensity (energy use per passenger-km or tonne-km) and the level of activity (passenger-km or tonne-km).

Table 21-4 provides estimates of the range of national averages of energy intensity for travel by car, bus, and train in various parts of the world. Differences between countries may be explained by a variety of interlinked factors, including age and design of the vehicle fleet, share of larger or higher-powered cars, quality of maintenance, and level of traffic congestion. All of these variables may be influenced by geographical, social, economic, and other factors.

There are competing influences on energy intensity. High-income countries in North America and Scandinavia have efficient technology but use it partly to provide a higher standard of transport services; vehicles of most types, including buses and trains, tend to be larger, heavier, and more comfortable. Low-income countries, such as China and India, tend to have less-efficient technology—but occupancies are relatively high, leading to low energy intensity. Indeed, average two-wheeler occupancy in Delhi, at 1.6 (Bose and Mackenzie, 1993; UNESCAP, 1991), is higher than car-occupancy levels in many industrialized countries, which can be as low as 1.2. Countries between these two extremes where transport activity is currently increasing rapidly (including southern and eastern European countries, Korea, and Japan) tend to have efficient technology that is used for lower standards of transport service, resulting in quite low energy intensities.

Car-occupancy levels in industrialized countries have declined with rising car ownership. Whereas buses and trucks in developing countries often are overloaded, leading to safety concerns, those in industrialized countries tend to have low occupancies or load factors, so operators and policymakers are concerned to find ways to increase loading.

For cars, there is an estimated 10 to 20% differential between national average on-road fuel economy and the results from official fuel economy tests (IEA, 1993a; Martin and Shock, 1989; Schipper and Tax, 1994). Differences arise because,

Table 21-4: *Passenger transport energy intensity—estimated national averages (Chin and Ang, 1994; Davis and Strang, 1993; Faiz, 1993; Grübler et al., 1993b; CEC, 1992; Schipper et al., 1993; Walsh, 1993a).*

	Light-Duty Passenger Vehicles			**Mopeds**		**Buses**		**Trains**
Country/ Date of Estimate	Fuel Economy[a] (L/100 km)	Load Factor (# people)	Energy Intensity (MJ/pass-km)	Load Factor (# people)	Energy Intensity (MJ/pass-km)	Load Factor (# people)	Energy Intensity (MJ/pass-km)	Energy Intensity[b] (MJ/pass-km)
Sub-Saharan Africa, 1985	20–24	2[c]	3.2–3.8			35–60	0.2–0.33	
China, India, and Thailand, ~1990	11–14	2[c]	1.8–2	1–1.6	0.5–0.8	35[c]	0.35	
Singapore, 1992	9	1.7	1.7	1.2	0.7	n/a	0.6	1.2
Japan and Korea, 1991	10–11	1.4	1.5–1.6	1	0.7–0.8	20[c]	0.65	0.55
United States, 1991	13–14	1.5	2.6			14[c]	0.9	2.75–3.0
Western Europe, 1991	8–11	1.5–1.8	1.2–1.96	1	0.7–0.8	10–25	0.49–1.32	0.75–2.8
Poland, 1991	9	2	1.3	1	0.73	35	0.33	0.32–0.83
Former USSR, 1988	12[c]	2	2			20	0.6	

[a]Estimated gasoline-equivalent fuel consumption per vehicle-km for the national car fleet.
[b]Electricity as primary; efficiency of conversion from primary energy to electricity supplied to locomotives assumed to be 30%.
[c]Very uncertain.

among other things, cars usually are not tested with auxiliary equipment, such as air conditioning, in operation, and some tests do not include cold starts, which can result in excess fuel consumption as high as 50% for short trips in cold weather (Hausberger *et al.*, 1994). In some regions, poor road quality may increase fuel consumption; that increase is an estimated 50% in Russia (Marchenko, 1993).

Car mass and engine size are closely related to energy intensity (Ang *et al.*, 1991). An upward trend in these factors has been observed in several European countries and in Japan (IEA, 1991a; Martin and Shock, 1989). In the United States, a downward trend in engine size and car mass (Difiglio *et al.*, 1990) has been reversed in recent years (NHTSA, 1994).

A more important factor in some countries is the expanding use of "light trucks"; these include small pickup trucks and minibuses. Light trucks accounted for 32% of the personal private vehicle market in the United States in 1990, and their fuel consumption per kilometer is about 36% higher than that of cars (Greene and Duleep, 1993).

The on-road mean fuel consumption per kilometer driven of light-duty passenger vehicles in North America fell by nearly 2% per year between 1970 and 1990, to about 13–14 L/100 km, but remains 30 to 40% higher than that in Europe or Japan (Schipper *et al.*, 1993). In other industrialized countries, changes during the period were quite small (see Figure 21-6), although new-car fuel economy in official tests improved by about 20% during the period (IEA, 1993a).

Most of the cars sold worldwide are either new cars made to designs originating in the OECD or second-hand cars exported

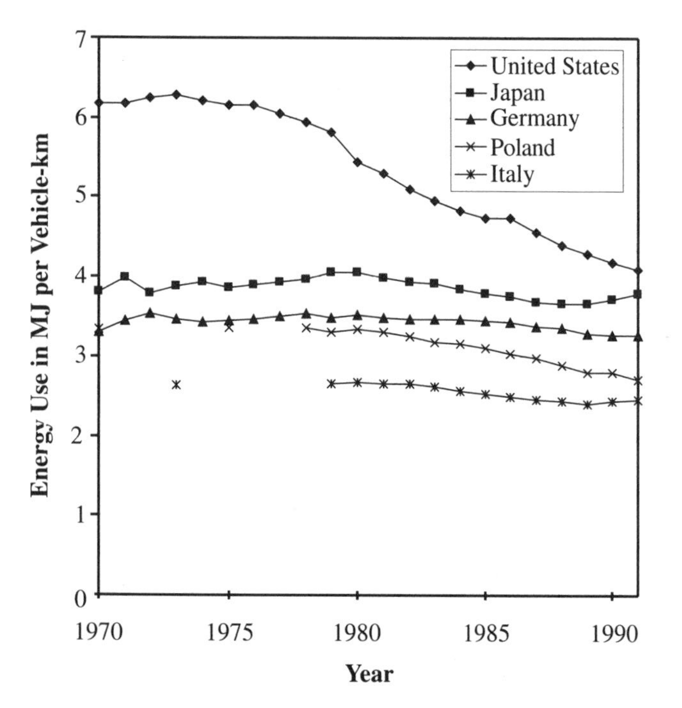

Figure 21-6: Light-duty passenger-vehicle energy intensity (MJ/vehicle-km) (Schipper *et al.*, 1993).

from the OECD. Any changes in vehicle technology, therefore, tend to be driven by the requirements of OECD new-vehicle markets.

While car energy intensity in the OECD has fallen during the past 20 years, the energy intensity of car travel has increased in many countries as a result of declining car occupancy. Meanwhile, the more recent trend is toward higher energy intensity in new cars in countries including the United States, Germany, and Japan (IEA, 1991a, 1993a), and global average fuel economy will not necessarily improve in the near-term future. Nevertheless, projections often incorporate a reduction of 1 to 2% per year (Davidson, 1992; Grübler *et al.*, 1993; IEA, 1991a, 1993a). This can vary considerably within regions: Davidson (1992) gives projections for African countries between 1985 and 2025 ranging from 0.6% per year in Sierra Leone to 1.8% per year in Nigeria.

Few estimates are available of energy intensity in aviation. Flights of 1,000 km or more typically have energy intensity in the range of 1.5 to 2.5 MJ/passenger-km. For short-haul flights, energy intensities can be higher than 5 MJ/passenger-km (ETSU, 1994). World civil air traffic (tonne-kilometers) increased by 150% between 1976 and 1990, while energy consumption rose by only 60%. This situation represents about a 3–3.5% per year reduction in energy intensity (Balashov and Smith, 1992), which is a result of improved aircraft energy efficiency and rising utilization of available seats and cargo capacity.

Walking and cycling only result in greenhouse gas emissions if people eat more food to compensate for the energy they use. This is of the order of 150 kJ/km for walking and 60 kJ/km for cycling (Banister and Banister, 1994; Hughes, 1991). Fossil-fuel use in agriculture, food processing, transport, storage, and cooking is highly variable but on average is of the same order of magnitude as the energy content of the food.

Data on the energy intensity of road freight traffic are hard to obtain and interpret in most countries. Energy use is typically in the range of 3 to 4 MJ/tonne-km on average and 0.7 to 1.4 MJ/tonne-km for heavy trucks (Schipper *et al.*, 1993; CEC, 1992). In countries where services and light industry are growing faster than heavy industry, the share of small trucks or vans in road freight increases; these have high energy intensity compared with large trucks. Along with the increasing power-to-weight ratios of goods vehicles, these trends offset—and in some cases outweigh—the benefits of improving engine and vehicle technology (Delsey, 1991a). The energy intensity tends to be lower in countries with large heavy-industry sectors, where a high proportion of goods traffic is made up by bulk materials or primary commodities.

Average truck energy use per tonne-kilometer of freight moved has shown little sign of reduction during the past 20 years in countries where data are available (Schipper *et al.*, 1993). Although vehicle technology has improved, this improvement has been used partly to increase the power-to-weight ratio of vehicles rather than to reduce energy use. The structure of the road-freight market has shifted in many OECD countries away from medium-sized trucks toward very large trucks for bulk freight and small trucks for retail distribution and high-value goods. Much of the recent growth has been concentrated in small trucks, which can have very high energy use per tonne-kilometer—although perhaps it would be more helpful to measure the goods movements in value rather than weight.

Rail freight often is viewed as a very energy-efficient way of moving goods. Average rail-freight primary energy intensities, at around 0.7–1 MJ/tonne-km, are a quarter of those of average road-freight intensities (CEC, 1992; Schipper *et al.*, 1993; Tanja *et al.*, 1992). However, it is important to compare similar types of goods movements. The energy intensity of rail container freight, at 0.4–0.9 MJ/tonne-km, may not be much smaller than that of road container freight, at 0.7–1.0 MJ/tonne-km (Martin and Shock, 1989; Rigaud, 1989; CEC, 1992).

Marine transport, on average, is the most efficient means of bulk international goods movement, with energy intensities below 0.2 MJ/tonne-km (Bremnes, 1990). Dry bulk goods (e.g., iron ore, cereals, coal, bauxite, and phosphate rock), crude oil, and oil products make up 60 to 70% of goods movements by sea and are moved in the largest, most efficient vessels. There has been a strong trend toward larger vessels (up to 170,000-tonne capacity) to capture economies of scale in both investment and operating costs of ships (Doyle, 1986).

21.3. Reducing Transport Greenhouse Gas Emissions

Greenhouse gas emissions from the transport sector and its energy supply chain can be reduced by policies and measures aimed at:

- Reducing energy intensity through vehicle downsizing, lower power-to-weight ratios, more efficient vehicle technology, changes in vehicle use (load factor, driving style, traffic management, etc.), improvements in infrastructure, or changes in transport mode
- Controlling emissions of carbon monoxide, VOCs, NO_x, N_2O, and methane
- Switching to alternative energy sources with lower full-fuel-cycle greenhouse gas emissions
- Reducing the use of motorized vehicles through switches to nonmotorized transport modes, substitution of transport services with other services (e.g., telecommunications), or reduction in the services consumed.

Section 21.3 provides a review of the potential for technical change to reduce transport greenhouse gas emissions. Section 21.4 addresses the role of policy in changing technology and its use.

21.3.1. The Technical and Economic Potential for Vehicle Changes

Studies of the potential for technical change can give a huge range of results even for a single technology in a single country. For example, in the United States, estimates of potential new-car fuel economy for the late 1990s range from 4.8 to 8.4 L/100 km (DeCicco and Ross, 1993; NRC, 1992; OTA, 1991). Analyses differ partly in the assumptions made but also in the type of potential being considered. It is helpful to distinguish among the *technical potential* (the reduction in greenhouse gas emissions that would result from application of the best available technology), the *economic potential* (the reduction in emissions that can be achieved cost-effectively), and the *"policy" potential* (the reduction in emissions that can reasonably be achieved as a result of policies and measures).

The technical potential usually improves with time as new fuels, prime movers, materials, design techniques, and operating systems become available. The economic and policy potentials do not necessarily improve; they depend on the economic and other priorities of the providers and users of transport services. The extent to which these potentials are achieved depends on a complex interaction among technology, the economy, and choices made by consumers, producers, and policymakers. It is this interaction that most needs to be understood and addressed if policymakers wish to reduce greenhouse gas emissions from the transport sector.

21.3.1.1. Energy-Intensity Potential in Road Vehicles

Reductions in energy use per vehicle-kilometer can be achieved through changes in maintenance practice, vehicle-body design changes, more energy-efficient engine and drive-train designs, and changes in operating practice.

Vehicle maintenance may be inadequate because spare parts and servicing are too expensive or unavailable (Davidson, 1992) or because maintenance is a low priority for drivers. Regular checks on tire pressure, engine oil, and tuning can save energy. Studies on cars have shown a 2 to 10% fuel saving immediately after engine tuning (Davidson, 1992; Martin and Shock, 1989; Pischinger and Hausberger, 1993). Some countries—including Mexico, Korea, and some European countries—and some American states have introduced regular vehicle emission tests, with remedial action when necessary, usually including engine tuning.

Vehicle mass affects energy use for acceleration and in overcoming resistance or friction in the axles, wheels, and tires. In most types of road vehicle, acceleration and rolling resistance each typically account for around a quarter to a third of the useful mechanical energy from the engine, although these shares are larger for city buses and delivery vans. Vehicle mass can be reduced by using advanced materials, improving component design and joining techniques, and reducing vehicle size or engine size. Concept cars have been demonstrated with masses 30 to 40% below those of conventional cars of similar size and performance (Chinaglia, 1991; Delsey, 1991b; Lovins *et al.*, 1993). The technical potential in 2010 for mass reductions without compromising comfort, safety, and performance is probably in the range of 30 to 50% for most vehicle types; this is the target of the United States' Partnership for a New Generation of Vehicles (PNGV). Such a mass decrease would lead to reductions in energy intensity in the region of 15 to 30%, provided engine size is reduced to keep performance constant (ETSU, 1994).

The technical potential in 2010 for reducing *rolling resistance* is probably around 30% for cars but rather less in buses and trucks (ETSU, 1994; DeCicco and Ross, 1993). For most road-vehicle types, a 10% drop in rolling resistance reduces energy intensity by roughly 2 to 3%, provided engine size is reduced (ETSU, 1994).

Air resistance, or drag, accounts for a third to a half of the energy required to move most types of vehicles, although the share is lower for city buses and delivery vans. Changes in vehicle design can reduce it by about 50% for most vehicle types without sacrificing performance—offering a 15 to 30% reduction in energy intensity, provided engine size is reduced (ETSU, 1994). The ultimate limit on drag reduction without reducing vehicle size is probably 70 to 80%, but this would involve radical changes in vehicle shape and performance.

Improved *transmission designs* can reduce energy use by allowing the engine to operate closer to its optimum speed and load conditions. Complete engine load/speed optimization can be achieved with electronically controlled continuously variable transmissions (CVT), although these may be more expensive than conventional gearboxes and currently are available only for small cars. Energy savings in the range of 3 to 10% are possible relative to automatic transmissions (DeCicco and Ross, 1993; Tanja *et al.*, 1992; NRC, 1992).

The three-way *catalytic converter*, combined with electronically controlled fuel injection into the engine air-intake manifold, is rapidly becoming standard for new gasoline cars in medium- and upper-income countries. This technology reduces emissions of carbon monoxide, unburned hydrocarbons, and NO_x from gasoline cars by approximately 60 to 90% relative to a car with a carburetor and no catalytic converter. N_2O emissions are raised approximately fivefold but depend strongly on the age of the catalyst and driving pattern (Ybema and Okken, 1993). The overall effect on radiative forcing of adding a catalytic converter to a gasoline car is small and uncertain; there may be a reduction in some circumstances (CEC, 1992; Pischinger and Hausberger, 1993; Wade *et al.*, 1994).

In *heavy-duty diesel engines*, the main needs for exhaust-gas pollution control relate to particulate and NO_x emissions. Controls such as particulate traps and filters tend to reduce energy efficiency and increase CO_2 emissions, without any compensatory effect from reductions in other greenhouse gases. However, other approaches to reducing engine emissions, such as the use of electronic fuel-injection systems, can lead to reductions in both engine pollution emissions and fuel consumption.

Changes in *gasoline-engine technology* have resulted in gradual improvements in energy efficiency. This progress is likely to continue. Energy-intensity reductions in the range of 15 to 30% are thought to be available with current technology (Bleviss, 1988; CEC, 1992; DeCicco and Ross, 1993; ETSU, 1994; NRC, 1992; Pischinger and Hausberger, 1993). These reductions could arise from a combination of many changes, including improvements in component design and lubrication, improvements in materials, increased use of electronic control systems, and changes in engine design such as the use of three or four valves per cylinder.

Car manufacturers have worked intensively on advanced two-stroke engines with the aim of reaching the noise, emission, and durability standards of existing four-stroke engines. The two-stroke engines would have higher efficiency and power-to-volume and weight ratios than four-stroke engines, allowing for more flexibility in car design and potentially for improvements in fuel economy of around 10%, provided the efficiencies realized were not used to raise power.

Another engine concept is the gasoline direct-injection engine, which might have fuel requirements 10 to 25% lower than those for a conventional engine (Schäpertons *et al.*, 1991); such engines remain far from commercialization however. Lean-burn gasoline engines also may offer 10 to 20% energy savings, but further development will be needed in the engines, or in the development of a lean-burn engine NO_x reduction catalyst, to meet future NO_x emission standards.

Diesel engines in heavy-duty vehicles are already very efficient. The energy efficiency of a large truck engine can approach 40% in use. The potential for further energy savings is probably no more than 10 to 20% in the long term.

Diesel-engine cars, taxis, and vans are widespread in Europe, the Middle East, and Southeast Asia, although they cannot meet some strict U.S. emission standards. Most of these cars have indirect-injection diesel engines and offer 5 to 15% lower fuel consumption (in energy terms) than gasoline cars. Cars with direct-injection diesel engines consume about 10 to 20% less fuel than indirect-injection diesel engines.

Alternative engine designs, including gas turbines and Stirling engines, have received attention from governments and manufacturers over the decades. These engines are cleaner than Otto or diesel engines and are also fuel-flexible, but in the 50 to 300 kW range required for road vehicles, they are currently inefficient and expensive and have poor load-following characteristics. By 2025, improved materials and precision engineering could make these engine types viable for road vehicles. Hybrid engine/electric drivetrains could be used to avoid the need for the engine to match the load on the drivetrain. These are discussed in Section 21.4.

21.3.1.2. Cost of Energy-Intensity Improvements

Several studies (DeCicco and Ross, 1993; Greene and Duleep, 1993; NRC, 1992) use cost models to estimate the potential for reducing the energy intensity of car use in the United States. However, as DeCicco and Ross (1993) observe, it is hard to predict which technical changes will raise manufacturing costs and which will lower them. Changes in materials and manufacturing technique, in particular, frequently result in reduced costs. It is also difficult to carry out cost-effectiveness analysis on changes in vehicle design because many changes have some effect on the appearance if not the performance of a vehicle, and this can influence purchasers' willingness to pay for the changes.

21.3.1.3. Energy-Intensity Potential for Aircraft

Balashov and Smith (1992) estimate that fuel intensity in scheduled air passenger services will improve at an average of 3% per year during the 1990s and 2.5% per year from 2000 to 2010. With traffic expected to grow at 5.5% per year in the 1990s and 5% per year from 2000 to 2010, total fuel consumption could be about 65% higher in 2010 than in 1990. Beyond 2010, fuel-intensity improvements would have to come from the introduction of new engine concepts. An example might be the use of lightweight heat exchangers to provide charge cooling and recuperate exhaust heat from the engine. If such technologies could be applied in aviation, they might, in theory, provide a 20 to 25% energy savings (Grieb and Simon, 1990).

If a new generation of supersonic aircraft is successfully developed and commercialized, it would likely lead to an increase in the average energy intensity of civil air traffic and possibly an increase in traffic.

21.3.1.4. Energy-Intensity Potential for Ships

The energy efficiency of existing ship engines can approach 50%, and only small improvements (5 to 10% savings in fuel used) are anticipated in the future. Hull and propeller design improvements could reduce energy use by a further 10 to 30% (CEC, 1992).

Some existing ships use sails to assist their engines, and one option that has been claimed to save 10 to 20% of ship energy is the use of wind assistance by means of vertical-axis wind turbines (CEC, 1992). These have the added advantage of improving ship stability.

21.3.1.5. Summary of the Potential for Vehicle Energy-Intensity Improvements

By 2010, it may be technically possible to reduce energy intensities for new vehicles of most types by 25 to 50% without

Table 21-5. *Potential for energy efficiency.*[a]

Mode	National Average Load Factors (pass. per seat or tonne load per tonne capacity)[b,d]	National Averages of 1990 Intensity (MJ/pass-km or MJ/tonne-km)[d]	Trend[c,e]	Economic Reduction Potential at Constant Performance[c,e]	Technical Reduction Potential at Constant Performance[c,e]	Technical Reduction Potential: Reduced Speed and Performance[c,e]
			-------- Percentage Change Relative to 1990 Intensity ---------			
Cars	0.25–0.5	1.2–3.1	0 to -30	-20 to -50	-35 to -70	-60 to -80
Buses	0.1–2	0.2–1.3	+10 to -10	0 to -20	-20 to -40	-35 to -60
Trams	0.2–0.8	0.3–1.5	+10 to -10	0 to -20	-20 to -30	-30 to -40
Passenger Trains	0.1–0.8	0.9–2.8	+10 to -10	0 to -20	-25 to -35	-35 to -45
Air Travel	0.5–0.8	1.5–2.5	-10 to -20	-20 to -30	-30 to -50	-40 to -60
Avg. Road Freight	0.2–0.4	1.8–4.5	-10 to -20	-15 to -30	-25 to -50	-40 to -70
Heavy Trucks	0.6–1.1	0.6–1.0	0 to -20	-10 to -20	-20 to -40	-30 to -60
Freight Trains	0.5–0.8	0.4–1.0	0 to -10	-10 to -20	-25 to -35	-30 to -40
Marine Freight	–	0.1–0.4	+10 to -10	+10 to -10	-20 to -30	-30 to -50
Air Freight	n/a	7–15	-10 to -20	-20 to -30	-30 to -50	-40 to -60

[a] Actual potentials depend strongly on vehicle load factors and patterns of use.
[b] Load factors exceeding 1.0 indicate overloading.
[c] 2010 new stock, 2025 fleet average.
[d] Bose and MacKenzie, 1993; CEC, 1992; Davidson, 1992; Hidaka, 1993; Rigaud, 1989; Schipper *et al.*, 1993; UNESCAP, 1989 to 1992, various.
[e] Much of the literature on trends and potentials is focused on cars from 2000 to 2005 (e.g., Difiglio *et al.*, 1990; Greene and Duleep, 1993; DeCicco and Ross, 1993; IEA, 1993a). Values here are the authors' estimates, revised following expert review, based on these and on longer term estimates covering various transport modes in Bleviss (1988), CEC (1992), ETSU (1994), Lovins (1993), Martin and Shock (1989), Pischinger and Hausberger (1993), and Walsh (1993a).

reducing vehicle performance or the quality of transport provided (see Table 21-5). However, the economic potential—the energy-intensity reduction that would be cost-effective—is likely to be smaller than the technical potential. As Table 21-5 indicates, the economic potential for energy savings in cars might be about two-thirds the technical potential. Meanwhile, the trend is for less energy saving than the economic potential; for some key vehicle types, including cars and heavy trucks, it is possible that the fleet average energy intensity might not decrease between 1990 and 2025. Substantial reductions in energy intensity would require new government measures and might entail reductions in vehicle performance.

If the vehicle energy-intensity trends in Table 21-5 are combined with road traffic projections from Walsh (1993b) and air traffic trends from Balashov and Smith (1992) extrapolated to 2025, energy use in the transport sector in 2025 could be 90–140 EJ. This assumes no significant changes in energy use by rail and marine transport. The adoption of energy-efficiency improvements—to achieve the technical potential in energy use at constant performance—would result in energy use in 2025 reduced by around a third to 60–100 EJ.

21.3.1.6. Life-Cycle Greenhouse Gas Emissions

Land-based transport. Several analysts have estimated greenhouse gas emissions from road vehicles on a life-cycle basis that includes vehicle manufacture, vehicle operation, and fuel supply (CEC, 1992; DeLuchi, 1991, 1993b; IEA, 1993a; Pischinger and Hausberger, 1993). CEC (1992) also includes estimates of emissions associated with vehicle disposal and addresses emissions from trains, shipping, and aircraft. All of these sources estimate emissions of CO_2, CO, VOCs, CH_4, NO_x, N_2O, and CFCs, converted to CO_2-equivalents with a variety of GWP estimates. These GWPs all have been superseded by more recent values. Chapter 2.5, *Trace Gas Radiative Forcing Indices* in IPCC Working Group I volume, estimates GWPs as follows for a 100-year time horizon: CH_4, 24.5; N_2O, 320; CFC-12, 8500; HFC-134a, 1300. These values have been used by the authors to recalculate the life-cycle greenhouse gas emissions from various transport modes. GWPs for NO_x, VOCs, and CO are not estimated but are all expected to be greater than zero (Chapter 2.5, *Trace Gas Radiative Forcing Indices* in IPCC Working Group I volume).

The principal greenhouse gas emissions in a vehicle life cycle are CO_2 in the vehicle exhaust, during vehicle manufacture, and in the process of fuel supply; CFCs, primarily as a result of leakage from air conditioning and refrigeration systems; CH_4 emitted during oil extraction and from vehicle disposal where organic wastes are placed in landfills; and N_2O produced during fuel combustion and in the catalytic converters currently used for gasoline engines in many countries.

Vehicles are the main source in some countries of CO, NO_x, and unburned hydrocarbons. Although these are precursors of ozone—and hence are indirect greenhouse gases, GWPs cannot be reliably estimated for them (Chapter 2.5, *Trace Gas Radiative Forcing Indices* in IPCC Working Group I volume). The increasing use of catalytic converters on gasoline-engine cars during the coming 10 to 20 years will reduce emissions of ozone-forming gases and increase emissions of N_2O. The overall greenhouse-forcing impact of this change is likely to be small or neutral.

Air conditioning is likely to be installed in an increasing proportion of cars, although HFC-134a is likely to be used as the main refrigerant instead of CFC-12. At current rates of refrigerant loss, HFC-134a would add about 10% to the life-cycle greenhouse forcing caused by a passenger car. Improved equipment design and maintenance could reduce this loss, perhaps to near zero.

Car air-conditioning systems are powered by electricity generated from the engine alternator at very low efficiency. When operating, they add to vehicle fuel consumption, further increasing greenhouse gas emissions.

For diesel or electric-powered vehicles, emissions of non-CO_2 greenhouse gases contribute a smaller proportion of overall emissions than they do for gasoline cars. Buses, trucks, and trains have much higher utilization rates (distance traveled per year and in their lifetimes), so emissions associated with operation (tailpipe and fuel-supply emissions) dominate life-cycle emissions (CEC, 1992).

Aircraft. Besides CO_2, emissions of NO_x at high altitudes may make a similar contribution to climate change. While CO_2 is emitted in proportion to fuel consumption, factors influencing NO_x emissions from aircraft engines are more complex, and there may be some tradeoff between NO_x emissions and engine efficiency. During the past decade, manufacturers have developed several approaches to reducing aircraft engine NO_x emissions without compromising energy efficiency. New, more complex combustion systems are currently being developed that are anticipated to provide a 30 to 40% reduction of NO_x emissions. These targets appear to have been achieved by one medium-size engine that entered service in 1995, but not by a larger model, owing to major materials and cooling limitations. For the latter, the potential and time scale for achieving the anticipated 30 to 40% reduction are open to question. Research also is underway—primarily aimed at a second generation of supersonic civil aircraft—on systems to achieve greater than 80% reductions. Such systems could start entering service between 2005 and 2010 (Grieb and Simon, 1990).

21.3.2. Operational Influences on Vehicle Greenhouse Gas Emissions

The energy intensity of travel and freight transport is influenced by several factors other than vehicle technology. One of the single strongest influences is the load factor or occupancy of the vehicle, but other factors—including driving style (or speed profile), routing, and traffic conditions—also are important.

Differences in driving style can explain about 20% of variations in energy use by cars, buses, and delivery vans in urban areas (Martin and Shock, 1989; Tanja *et al.*, 1992). The potential for energy saving by "gentler driving" has been estimated to be about 10% in urban areas and 5 to 7% overall (Tanja *et al.*, 1992).

Energy use also can be reduced by traffic-management measures, such as computerized traffic-light control and network and junction design, to reduce congestion and unnecessary stops. Introduction of a traffic-control system in Los Angeles is estimated to have yielded a 12.5% reduction in energy use (Shaldover, 1993). However, energy-use reductions resulting from computerized traffic control may be rapidly reversed because the increase in road network capacity is likely to produce additional traffic.

For commercial vehicles, computerized routing systems can be used to optimize payloads and minimize the time spent and fuel used on the roads. For many haulage firms in industrialized countries, such systems pay for themselves through increased revenue. Some studies indicate that reductions of 25 to 30% in energy use per tonne-kilometer are technically possible (O'Rourke and Lawrence, 1995).

A variety of computerized routing aids are being developed for drivers in general, some providing real-time information on congestion and the availability of alternative modes and routes. Energy-saving potentials in urban road passenger transport could range from a few percent in small towns with little congestion to 30% in large, congested conurbations with effective public-transport alternatives (Shaldover, 1993).

For aircraft, while present flight patterns seek to minimize costs—particularly fuel consumption—there might be scope for operational changes, including alternative routings to avoid greenhouse gas precursor emissions at sensitive altitudes (such as the tropopause), latitudes, and longitudes, or changing seasonal or diurnal patterns that might reduce the impact on atmospheric processes. Civil airlines operate increasingly sophisticated computerized booking systems with multiple tariffs, and the resultant increase in seat occupancy has contributed to falling energy intensity. Further increases in occupancy are likely (Balashov and Smith, 1992).

Speed is an important influence on energy use by all types of vehicles. For road and rail transport, speed limits and vehicle speed limiters are mainly used for safety reasons. Moderate reductions in average road vehicle speed (e.g., from 90 km/h to 85 km/h) can lead to energy savings on the order of 5 to 10% (Tanja *et al.*, 1992). In ships and aircraft, speed is routinely controlled to manage energy costs. Because fuel constitutes 20 to 30% of shipping costs, operators plan voyages using shipping cost models that take account of energy use in choosing the optimum speed. An oil price of $30/barrel can produce optimum speeds 25% lower than a price of $15/barrel; the result is a 33 to 40% energy saving (Doyle, 1986).

21.3.3. Alternative Fuels

Alternative fuels are an important complement to measures that improve fuel economy as a means of reducing greenhouse gas emissions. In the past, alternative-fuel vehicles (AFVs) have been developed as a means of reducing oil consumption; they currently are being promoted as a means of reducing urban air pollution. Some types of AFVs can contribute to meeting both of these goals.

21.3.3.1. Alternative Fuels in Light-Duty Vehicles

For gaseous fuels and alcohols, engine emissions of carbon monoxide are generally lower than with gasoline, and particulate emissions are much lower. Engine NO_x emissions are generally similar to or lower than those from conventional fuels (Gover *et al.*, 1993; IEA, 1993a; OECD, 1993c; Wang, 1995). Emissions of unburned fuel and certain other pollutants, such as formaldehyde, on the other hand, can be higher than for gasoline or diesel fuel. In the short term, some of the main applications for alternative fuels are likely to be in situations where they can substitute for conventional gasoline additives, such as methyl tertiary butyl ether (MTBE), in helping to meet reformulated gasoline requirements (IEA, 1994b).

On a full fuel-cycle basis, alternative fuels from renewable energy sources have the potential to reduce greenhouse gas emissions from vehicle operation (i.e., excluding those from vehicle manufacture) by 80% or more (CEC, 1992; DeLuchi, 1991; IEA, 1993a).

Estimated life-cycle greenhouse gas emission ranges for several fuels, based on a variety of sources, are shown in Table 21-6, which also shows cost estimates for using some of the alternative-fuel vehicles. Some of the fuels listed are already cost-competitive with gasoline in some circumstances. These include diesel, liquefied petroleum gases (LPG), and compressed natural gas (CNG), all of which are in current use. Vehicles operating on these fuels may have full fuel-cycle greenhouse gas emissions about 10 to 30% lower than vehicles operating on gasoline, as shown in Table 21-6. These alternative fuels are used preferentially in high-mileage vehicles, where low fuel costs compensate for high engine and fuel-storage costs. Methanol from natural gas also may be cheaper to use than gasoline in some circumstances and might offer slight greenhouse gas-emission reductions. Methanol from coal also could be cost-competitive with gasoline, but with a 10 to 75% increase in greenhouse gas emissions (DeLuchi, 1991).

Alcohol fuels can be produced in a variety of ways from any of several sources. Life-cycle greenhouse gas emissions depend on the source and conversion technology (DeLuchi, 1993; IEA, 1993a), as Table 21-6 shows. Ethanol is produced from sugar cane for transport use in several countries, including Brazil, Zimbabwe, and Kenya. Full fuel-cycle greenhouse gas emissions are estimated to range from 30 to 50% of those that would be obtained with gasoline where bagasse (crop waste) is used as fuel for conversion facilities to around 80 to 90% where coal is used (estimates based on Goldemberg and Macedo, 1994; IEA, 1994b; Rosenschein and Hall, 1991). N_2O emissions from agriculture form a substantial but uncertain component of these estimates.

Ethanol is produced from corn on a large scale in the United States, where it provides about 0.5% of gasoline demand in energy terms. It is produced from wheat and sugarbeet on a small scale in Europe. Cars using ethanol produced from these food crops would have life-cycle greenhouse gas emissions ranging from 20% of those from gasoline cars when crop waste is used as the conversion fuel (not the current practice anywhere) to nearly 110% where coal is used (Table 21-6; IEA, 1994b).

Hydrogen can be used either in internal combustion engines or in fuel cells. Where hydrogen is produced from renewable sources—either by gasification of biomass or by electrolysis of

Table 21-6: *Life-cycle greenhouse gas emissions and costs for alternative fuel and electric cars (based on DeLuchi, 1992, 1991; IEA, 1993a[a]; CEC, 1992; Pischinger and Hausberger, 1993; Goldemberg and Macedo, 1994; DOE, 1991, 1990, 1989).*

	Greenhouse Gas Emissions in g/km CO_2-equivalent[b]					**Pre-Tax Costs[c]**		
Fuel	Vehicle Manufacture[d]	Fuel Supply[e]	Operation[f]	Total	Vehicle Cost ($)	Fuel Cost ($/L gasoline equiv.)	Fuel Use for Cost Calculation (L/100 km)	Cost in Excess of Gas Vehicle @29US¢/km (US¢/km)
Gasoline	25–27	15–48	182–207	222–282	15168	0.26	7.6	0
Reformulated Gasoline	25–27	17–63	180–193	222–283	15168	0.28–0.30	7.6	0.18–0.32
Diesel	27–29	7–35	139–202	173–266	15168–17443	0.26	6.08	-0.35–3.64
Liquefied Petroleum Gases	26–28	7–20	147–155	180–203	16083–15384	0.19–0.26	7.27	-0.55–1.02
Compressed Natural Gas	29–31	5–68	130–154	164–253	16083–15600	0.18–0.24	7.27	-0.28–0.90
Methanol from Coal	25–27	250	149	424–426	16128–15168	0.25–0.35	7	-0.72–1.45
Methanol from NG	25–27	76	149	250–252	16128–15168	0.25–0.35	7	-0.72–1.45
Methanol from Wood	25–27	25–38	15–16	65–81	16128–15168	0.68–0.82	7	2.30–4.79
Ethanol from Sugar Cane	25–27	30–80	15–16	70–123	16128–15168	0.35–0.38	7	-0.17–1.89
Ethanol from Corn	25–27	50–220	15–16	90–263	16128–15168	0.94–1.03	7	4.61–6.74
Ethanol from Wood	25–27	25–38	15–16	65–81	16128–15168	0.68–0.82	7	2.79–5.27
Liquid Hydrogen ICEV	26–28	0–48	3–12	29–88	19968–18048	0.38–1.44	6.5	4.10–13.97
Liquid Hydrogen FCEV	44–48	0–24	0–5	48–77	20324–30000	0.38–1.44	3.25	6.22–25.64
EV using Electricity Generated from:[g]								
American Average	44–48	135–202	0	179–250				
European Average	44–48	107–160	0	151–208	24768–20928	0.48–0.96	2–3	6.81–14.74
Coal	44–48	180–375	0	224–423				
Oil	44–48	170–255	0	214–303				
Gas (CCGT)	44–48	90–134	0	134–182				
Nuclear	44–48	15	0	59–63				
Hydro/Renewables	44–48[h]	0	0	44–48				

[a] Both the work of the IEA and that of Pischinger are based on a model developed by DeLucchi (1991, 1993).
[b] Average driving cycle based on gasoline car consuming 7 L/100 km; GWPs are for 100-year time horizon (CH_4, 26; N_2O, 270; CO, 3; VOC, 11; NO_x, 0). These differ from the latest IPCC values (IPCC, 1995), but the effect on total life-cycle emissions estimates is negligible.
[c] Based on Renault Clio 1.4 liter, 13800 km/year, 10-year life, 10% d.r.; levelized cost calculations based on IEA, 1993.
[d] Assumes current industrial practices. Ranges reflect differences between regions.
[e] Ranges reflect differences between primary energy sources and conversion technologies.
[f] Ranges reflect differences in vehicle technology, maintenance, and operation.
[g] Emissions based on urban cycle, consuming 200–300 Wh/km from mains (IEA, 1993a; CEC, 1992; DeLucchi, 1993; Eyre and Michaelis, 1991).
[h] DeLucchi (1991) and IEA (1993) ignore emissions associated with electricity-generating plant and electricity-grid construction. In fact, some CO_2 emissions will occur during plant construction, both from energy use and from cement manufacture. Emissions also may result from the flooding of land for hydroelectricity production.

Box 21-1. Ethanol in Brazil

Brazil has used small amounts of ethanol for decades as an octane enhancer in gasoline and as a byproduct market for the sugar industry. However, in the late 1970s, rising oil prices combined with high interest rates and a crash in the world sugar market to create a foreign-debt-servicing crisis. This compelled the Brazilian government to look at ways of reducing petroleum imports.

From 1975 to 1979, the government-sponsored program, Proalcool, increased the ethanol percentage in gasohol to 20%. From 1979 to 1985, Proalcool promoted the use of dedicated ethanol vehicles. In 1986, ethanol vehicles constituted 90% of new car sales, but a further expansion of the program was mothballed because of the drop in the price of oil. In the first phase of the program, the government provided up to 75% subsidies for ethanol-producer investments and assured a 6% return on the investments. At this stage, however, car manufacturers were unwilling to produce ethanol-only vehicles. In the second phase, consumer incentives were introduced for vehicle purchases, and the pump price of ethanol was guaranteed to be no more than 65% that of gasoline. The car industry was encouraged by the government's commitment and started producing ethanol vehicles. Despite initially poor redesign of engines, the consumer take-up was massive.

In late 1980, the world sugar market was improving, and the government began to increase the ethanol price from 40% of the of gasoline price toward the 65% limit. Credit subsidies for distilleries also were suspended. The result was a rapid fall in ethanol-vehicle purchases. The government later regained public confidence by restoring incentives.

Costs of ethanol production in Brazil are around 25 to 28¢/L of gasoline displaced, compared with world market prices of gasoline on the order of 17¢/L (pump costs are usually about 10¢/L higher). The government's current limit on the pump price of ethanol is 80% of the gasoline price.

Sources: Sathaye *et al.* (1989); Goldemberg and Macedo (1994).

water with electricity from renewables—life-cycle greenhouse gas emissions can be very low. The method of hydrogen storage aboard the vehicle can affect life-cycle emissions. If hydrogen is liquefied for storage, the energy input is about one-third of the energy content of the hydrogen; emissions, of course, will depend on how this energy is supplied (DeLuchi, 1993; ETSU, 1994).

Several trials have been carried out, with varying success, of hydrogen cars with Otto-cycle engines. Fuel consumption could, in theory, be 15 to 20% lower than that of gasoline engines on an energy basis (Brandberg *et al.*, 1992; DeLuchi, 1991; ETSU, 1994). Considerable technical development will be required in the various stages of hydrogen production, distribution, storage, and use before it is economical to use in cars and its safety can be fully demonstrated.

21.3.3.2. Alternative Fuels in Heavy-Duty Vehicles

Buses have been operated on LPG, CNG, alcohols, and vegetable oils, and there are several demonstration programs and commercial operations throughout the world. Fewer studies have been carried out into life-cycle greenhouse gas emissions from heavy-duty vehicles than from cars. However, Figure 21-7 shows results from one study (DeLuchi, 1993; IEA, 1993a). This study shows that most of the immediately available alternative fuels for heavy-duty vehicles are unlikely to offer life-cycle greenhouse gas emission reductions. The main reason for this conclusion is that CNG and LPG are used in spark-ignition engines with lower efficiency than existing diesel engines. Alcohols can be used in compression-ignition engines, but expensive fuel additives are required, and spark ignition engines usually are preferred.

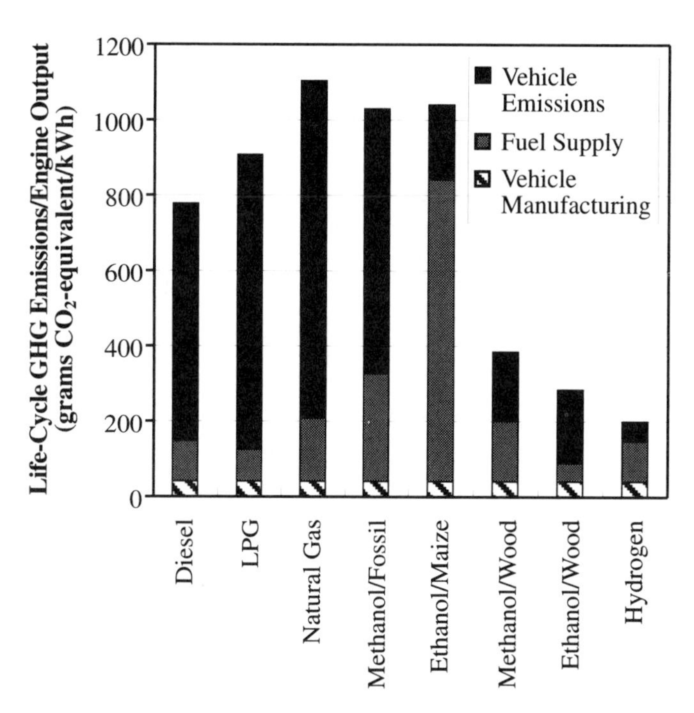

Figure 21-7: Life-cycle greenhouse gas emissions from heavy-duty vehicles with alternative fuels (DeLuchi, 1993; IEA, 1993a).

Box 21-2. CNG in New Zealand

The New Zealand government launched a compressed natural gas (CNG) program in 1979, aiming for 150,000 post-assembly vehicle conversions to CNG by the end of 1985. The main motivations for the program were to enhance New Zealand's energy security and improve its balance of payments. Incentives were provided in the form of grants and loans for vehicle conversion and filling-station development. The cost of car conversion at the time was about NZ$1,500 (US$750 in 1984 prices); the government provided NZ$150. Initial take-up was poor, but it was improved in late 1980 by an increase in the grant to NZ$200 and increased subsidies to fueling stations, along with tax benefits to consumers.

A large increase in government support in 1983 included a low-interest loan for conversions. By 1986, 110,000 vehicles (11% of all cars and light trucks) had been converted, and New Zealand had 400 filling stations; CNG comprised 4.6% of road-transport fuel.

The rise in conversion rate was halted in 1985 when the government reduced support for the program because of perceptions that CNG was sufficiently favored by its low price relative to gasoline, the CNG industry was sufficiently well-established and no longer required government support, and world oil markets were more stable and less emphasis was needed on the development of indigenous energy sources.

The vehicle conversion rate fell from 2,400/month in 1984 to 150/month by 1987. By 1993, CNG sales fell to 42% of their peak level in 1985, contributing only 2% of road-transport fuel. In addition to the withdrawal of government subsidies to CNG, a number of factors may have contributed to this demise. One is the increase in CNG prices from less than half to about two-thirds of the gasoline price, at a time when oil markets have been quite stable. Another possible factor is a negative public perception of CNG resulting from a number of faulty installations in the early 1980s, a cylinder explosion in 1989, the poor performance of dual-fuel vehicles that were not optimized to operate on CNG, and the realization that carrying CNG cylinders substantially reduces luggage space. A third possible factor is a lack of industry enthusiasm for marketing the fuel, bearing in mind that most of the retail outlets are oil-company-owned filling stations.

Sources: Sathaye *et al.* (1989); New Zealand Ministry of Commerce (1994).

In the long term, alternative fuels, such as methanol and ethanol from wood, could be used in heavy-duty vehicles, producing life-cycle emission reductions of more than 50% relative to diesel vehicles. Meanwhile, developments in direct-injection, compression, or spark-ignition engines operating on CNG, LPG, and alcohols could lead to substantial improvements in efficiency with these fuels (CEC, 1992).

Another option not shown on Figure 21-7 is the use of synthetic diesel produced from renewable sources, such as wood. This strategy would allow existing, efficient diesel engine technology to be used with a fuel with very low full-fuel-cycle carbon emissions.

21.3.3.3. Alternative Aviation Fuels

Aircraft carry a large amount of fuel in proportion to their total weight (around half of the takeoff weight for a Boeing 747). In view of this, any alternative fuel has to have a high energy content per unit of mass. LNG and liquid hydrogen fit this requirement, and trials have been carried out with both (Grieb and Simon, 1990; Schäfer *et al.*, 1992). In principle, both LNG and liquid hydrogen aircraft could have lower fuel consumption than conventional aircraft because of the lower weight per unit of energy in the fuel. Energy use depends on the additional weight of the fuel tanks, the design changes required in the aircraft, and the energy required to liquefy the fuel. Overall greenhouse gas emissions from the use of these fuels, including emissions of NO_x and water vapor, cannot yet be predicted. The implications for climate change depend on the GWPs of high-level NO_x and high-level water vapor, which remain uncertain—as do the safety implications of using cryogenic methane and hydrogen in aircraft.

If low-carbon fuels are required for aviation, a safer alternative might be synthetic liquid hydrocarbons from biomass, although even these fuels may have drawbacks, such as poor thermal stability. It will be necessary to demonstrate the safety and reliability of aircraft engines that use synthetic hydrocarbons before they can be adopted for civil aviation purposes.

21.3.4. Electrically Powered Transport Systems

Many analysts believe that the most promising route for the development of transport systems, along with the rest of the energy economy, would be toward the use of either electricity or hydrogen as energy carriers (MacKenzie, 1994; Sperling *et al.*, 1992; Sperling, 1995). This strategy would effectively make final energy use independent of the primary source of energy, allowing renewables such as photovoltaics, wind power, and biomass to be used interchangeably as the electricity or hydrogen source.

Emissions can be eliminated at the point of vehicle operation by the use of some form of electric drive. Power can then be supplied by oxidation of hydrogen in a fuel cell, from an electrical storage device, or directly from the public electricity supply. Electric motors have the advantage of high efficiency at part-load, no fuel consumption when stationary, and the potential for regenerative braking (where the motor is used as a generator to return power to the source or a storage device).

Battery electric vehicles have aroused interest mainly for their potential to reduce urban air pollution. Electric-vehicle technology has been reviewed in detail by several authors (CEC, 1992; DeLuchi, 1991, 1993; ETSU, 1994; IEA, 1993b; MacKenzie, 1994; Sperling, 1995). Life-cycle greenhouse gas emissions depend not only on the primary energy source but also on many aspects of the vehicle technology and the way it is used.

Electric-car technology has been intensively developed in several countries in recent years, mainly as a result of legislation in California. Batteries have so far provided the best combination of energy and power density, along with low cost for on-vehicle electricity storage. Battery-powered cars, buses, and light vans are in use, but they have high costs and poor performance compared with gasoline vehicles, and most types of battery would need replacing several times in the life of a vehicle. The future evolution of this technology and of the market for electric vehicles is very uncertain, depending on factors such as the extent of cost reductions that can be achieved in batteries, electric motors, and control systems; the extent to which battery-recharging rates, energy densities, and power densities can be improved; and market factors, including consumer preference and future legislation to reduce air pollution.

Alternatives to the battery are under development. Whereas batteries store energy in the form of chemicals that react to produce electric power, supercapacitors store energy in the form of electric charge, and flywheel storage devices store energy in the kinetic energy of a rapidly rotating flywheel. These devices may have the potential for much higher power density than batteries, with similar energy-storage density.

Some recent studies in California (Bunch *et al.*, 1993) indicate that, in choosing between conventional and alternative vehicles, car purchasers are likely to be particularly sensitive to the extra cost and shorter range of battery-powered and other alternative-fuel vehicles. Other studies in California (Sperling, 1995) have found that consumer preferences can be modified and that a substantial proportion of drivers might purchase electric vehicles to use for local trips and commuting, keeping their gasoline vehicles for other trips.

The range and durability limitations of batteries are avoided with fuel cells, which produce electric power directly from the oxidation of a fuel carried in a separate storage vessel. The fuel normally is hydrogen, although some fuel-cell types can use methanol or other fuels. Alternatively, methanol can be used to make hydrogen on board the vehicle. Of the various fuel-cell types, the solid polymer fuel cell (SPFC) is probably the strongest near-term candidate for an on-vehicle power source (Appleby and Foulkes, 1989; DeLuchi, 1993; ETSU, 1994). SPFCs have been tested in a van and a bus in Canada, with hydrogen from high-pressure cylinders. They are preferred over other fuel-cell designs because of their robustness, long life, low operating temperature, quick startup, and reasonably high power density.

Currently available SPFCs cost well over $1,000 per kW of output, although some analysts predict future costs as low as $20 per kW (Sperling, 1995). For comparison, gasoline engines cost $20 to 40 per kW. A fuel-cell vehicle probably would use the fuel cell to provide a constant power output, with a battery to store power for peak-load requirements. Costs per kilowatt for such systems would be lower than those for a fuel cell sized for peak loads. Nevertheless, given the technical and cost improvements required, fuel cells are unlikely to be in widespread use for cars before 2025. They will face institutional and other challenges similar to those confronting alternative fuels in general. However, if the various barriers to their use can be overcome, fuel-cell cars could be the predominant mode of motorized travel by 2050.

Many hybrid concepts have been suggested and demonstrated, combining the advantages of electric drives and internal-combustion engines. One approach involves using a low-powered fuel-burning engine to run a generator that provides a constant supply of electric power (see, e.g., Lovins *et al.*, 1993). This is either used directly to power electric motors or fed to a peaking storage device, which might be a battery or flywheel. A major theoretical advantage of such a system is that the engine can operate under its optimum speed and load, and braking energy can, in principle, be recovered and stored. In principle, energy savings in urban driving could amount to 40 to 60%. However, with current technology, energy savings are likely to be very small because efficiency is constrained by energy losses in the electric generator, storage device, and motor; matching the engine and electrical storage device to real on-road power requirements is difficult; and existing batteries are unable to recharge at the rates required to store a useful portion of energy from the engine and from braking. Flywheels may be more appropriate for this purpose, and the concept is under development. Hybrid technology is not likely to achieve substantial market share in the short term without government intervention, in view of the high overall cost of an electric motor, control system, and battery or other storage device. It might be feasible to produce hybrid vehicles on a competitive commercial basis some time between 2005 and 2025.

Direct electric power has long been used in light-rail systems and trolleybuses, as well as in electric trains. Urban bus and light-rail routes involve low speeds and frequent stops. In these circumstances, the features of electric drives are very attractive—in particular the capability for regenerative braking, the avoidance of energy use when stationary, and the absence of emissions at the point of use.

Trolleybuses are in use in numerous cities throughout Europe. They are lighter than battery buses because they do not have to carry an electricity source. They also allow for a more efficient use of grid electricity and a better capability for regenerative braking. Whereas the overall efficiency from grid electricity to motive power for a battery-powered vehicle is about 55%, that of a trolleybus is about 75 to 85%.

Several proposals exist for personal passenger-vehicle systems based on direct electric power. These range from cars supplied with electricity via induction coils placed under the road surface to monorail systems with vehicles powered by linear motors. Some designs involve the use of computer-controlled vehicles for public use. Others provide for the use of private cars off the system using batteries or other power sources. Such systems are potentially an efficient means of using electric power for personal transport.

21.3.5. High-Speed Trains

High-speed trains can have higher energy use per seat-kilometer than traditional trains but much lower energy intensity than cars or aircraft at typical load factors. Their role in greenhouse gas emission reduction is subject to debate and will depend heavily on the situations in which such trains are used: They may attract passengers without reducing the use of other modes, resulting in higher overall energy use and greenhouse gas emissions. Nevertheless, high-speed rail lines between Paris and Lyons in France and between Madrid and Seville in Spain have been remarkably successful in attracting business from the roads and the airlines, and have been profitable.

An alternative to the high-speed train is "maglev": trains that are levitated on a magnetic field rather than having wheels in direct contact with tracks. The only existing maglev systems in use are small-scale local shuttle or transit systems. Maglev trains may have some energy-efficiency advantage compared with wheeled trains at a given speed, but analyses differ on this (Schäfer *et al.*, 1992). Rolling resistance (25% of energy use in a normal high-speed train) is avoided, but energy is required to power the magnets. The main advantage of maglev trains is that they can operate at higher speeds than wheeled trains; the Transrapid line to be built from Hamburg to Berlin will operate at up to 450 km/h. This means that they can compete with air travel over longer distances than existing high-speed trains—potentially as far as 1,500 km. Maglev trains have the disadvantage compared with high-speed trains that, unless provided with auxiliary wheels, they cannot use existing stretches of track to penetrate urban areas and serve destinations without new infrastructure.

21.3.6. National and International Technology Policy

Governments can take various kinds of action to influence technology development and deployment. Some, such as basic research and development, are aimed at bringing technology to the point where industry can further develop it to produce new products and sell them in the market. Others, such as fiscal and regulatory incentives and funding for demonstration programs, are aimed at encouraging industry to carry out its own research or to market different products and encouraging consumers to buy them.

Research and development is usually difficult to plan on a rational, economic basis. Scientific breakthroughs are inherently unpredictable, and even if a technology is successfully developed its market success is also unpredictable. Fundamental research often depends for its existence on some form of government intervention, whether this is direct funding, tax exemptions for firms undertaking research and development, or information brokering. Meanwhile, national investment in research may be justified on cultural grounds as well as on economic grounds. Contributions to international scientific ventures often are motivated by international relations as well as by an interest in cost-effectiveness.

Important areas for fundamental research relating to transport technology include recyclable materials for cheap, robust, lightweight vehicles; materials with special properties for high-efficiency engines, turbines, flywheels, and so forth; advanced conversion technology for liquid fuels from biomass; electrochemistry for batteries and fuel cells; power electronics for electric-propulsion-system management; power-transmission technologies for vehicles powered directly from the mains; and information technology for vehicle and traffic optimization. There is also a need for social and economic research—for example into consumer behavior and choice, including attitudes toward new technology.

Interventions in the market may have the advantage that government does not necessarily have to attempt to "pick winners" among technologies or bear the risk of investment in research that may not bear fruit. However, this risk does have to be borne by manufacturers that wish to remain in the market and have to respond to the government's intervention. One example of a technology-forcing intervention is the California Air Resources Board's strategy for encouraging the development of zero-emission vehicles (see Box 21-3). Meanwhile, a strategy of announcing a timetable for gradually tightening standards or increasing fiscal incentives, aimed at reducing greenhouse gas emissions, will encourage industry both to introduce currently available technology and to develop new technology.

21.4. A Transport-Policy Perspective on Greenhouse Gas Emissions

While technical changes can, in theory, reduce greenhouse gas emissions by 80% or more for a given transport activity, such changes are not occurring under current market conditions. Meanwhile, in most countries, transport-activity growth is faster than the rate of energy-intensity improvement.

Section 21.4 explores different viewpoints that influence policymaking, identifies areas of transport policy that can influence

Box 21-3. CARB Strategy for Encouraging Zero Emissions Vehicles

California's urban air-quality problems have long been recognized as more severe than in other regions of the United States. The California Air Resources Board (CARB) introduced the Clean Fuels and Vehicles Plan in September 1990. The plan imposes progressively lower emission standards on vehicles from 1994 onward. VOC emissions must be cut 80% below 1994 levels by 2000.

Under the legislation, vehicles are given one of four new emission classifications: transitional low-emission vehicle, low-emission vehicle, ultra-low-emission vehicle, or zero-emission vehicle. Manufacturers' sales-weighted VOC emissions calculated from their sales of each vehicle type must not exceed a prescribed level in a given year. A banking and trading system permits manufacturers to earn marketable low-emission credits.

CARB's specifications for reduced summer gasoline volatility, elimination of lead, and the use of detergent additives came into force in 1992. Further tightening of gasoline specifications is expected.

Alternative fuels are being promoted by CARB and through the federal California Pilot Vehicle Program. The fuels are expected to help satisfy low-emission requirements. They include M100, M85, CNG, and LPG. At least 90 Southern California filling stations must supply alternative fuels by 1994, rising to 400 by 1997. From 1994 onward, 200,000 new low-emission vehicles per year (about 10% of the state's new-car fleet) must be sold in California. Alternative-fuel use is also being encouraged through tax exemptions.

Electric vehicles will be required to satisfy the zero-emission-vehicle requirement from 1998 onward. They will have to make up a minimum of 2% of annual new-car sales by companies selling more than 30,000 cars per year in California; the share is scheduled to rise gradually to 10% by 2003. Electric-vehicle use in California is likely to result in greenhouse gas emissions significantly lower than those associated with gasoline-vehicle use because much of the power generation in the state is from non-fossil-fuel sources or natural gas.

greenhouse gas emissions, and examines policy options for emission reduction.

Transport policymakers' central aims are usually to maximize mobility and access to services, enhance economic activity, and improve safety. These aims have to be balanced with energy and environment goals, especially those of reducing urban air pollution and greenhouse gas emissions. At the same time, policymakers have to consider the political acceptability of their measures, as well as their ease of implementation and their effects on equity and social welfare generally.

21.4.1. *Stakeholders and Viewpoints in Policymaking in the Transport Sector*

Policymakers and other stakeholders typically are involved in the transport sector in several different ways:

- Local public-transport authorities or firms make decisions regarding public-transport scheduling and investment, aiming to maximize fare revenue, minimize costs, and in some instances to provide a minimum level of service.
- Local or municipal authorities make many decisions regarding transport policies, urban road layout, parking facilities, urban traffic management, and facilities for walking and cycling. Their objectives include improving access; reducing road congestion, accidents, and noise; and, more recently, reducing air pollution—all within a fixed budget.
- Decisions regarding the development of highways, railways, and other transport infrastructure are usually made by national transport ministries, which often define the parameters within which local transport authorities work and provide funds for infrastructure and services. Transport ministries have traditionally aimed to maximize mobility and access for both freight and motorized passenger transport in a cost-effective manner, sometimes including social costs associated with congestion, accidents, noise, and air pollution.
- Vehicle emission standards are often negotiated between government departments of environment, transport, and industry. Emission standards for road vehicles and aircraft (ICAO, 1993) are, in many instances, agreed between governments—in consultation with industry and other interest groups—with the mediation of intergovernmental organizations.
- Vehicle and fuel taxes are of interest to transport, industry, and energy departments, but they often are determined by finance ministries.
- Transport research and development is undertaken by a range of interests, including various government departments, universities, and other research institutes, as well as vehicle manufacturers and transport-system operators.

- Transport-system users make day-to-day decisions regarding vehicle purchase and utilization but normally have limited influence on transport policy.

Differences in the priorities of decisionmakers have much to do with differences in training, approach, and worldview associated with the professions most involved in each area of policy. Dimitriou (1992) provides a characterization of fields of expertise:

- Engineers often emphasize the operational efficiency of transport systems and usually offer solutions based on technology or the design of infrastructure.
- Economists usually emphasize economic efficiency, aiming to maximize the benefits minus the costs of the transport system. Their solutions tend to focus on improving the functioning of the market.
- Other social scientists, such as sociologists, political scientists, anthropologists, and development planners, are less likely than engineers or economists to use formalized concepts of optimization. Their solutions are likely to focus on institutions, community involvement, and consultation.
- Physical planners (architects, city planners, and transport planners) tend to view transport issues as part of a wider picture of urban development. Their solutions tend to relate to the spatial organization and design of infrastructure.

Politicians, environmentalists, and other interest groups (including transport-user associations) may draw on a variety of these viewpoints and solutions in identifying their preferred transport strategies. To bring transport policy to bear on the aim of reducing greenhouse gas emissions, it will be important to bring the different professions and viewpoints together to develop robust strategies that satisfy all of their criteria and make use of all of their insights and techniques.

21.4.2. Planning, Regulation, and Information Measures

Traditional approaches to transport planning have concentrated on providing sufficient roads to carry projected volumes of traffic and on smoothing the traffic flow. Many planners now recognize that it is impossible to provide sufficient roads to carry unrestrained projections of traffic in urban and perhaps some interurban areas, and that extra road provision can stimulate traffic growth, leading to congestion elsewhere (Goodwin *et al.*, 1991). Increasingly, the objectives of transport are seen as moving people and providing access to services, jobs, and homes rather than moving vehicles. The planning emphasis is on alternatives such as improving the traffic flow, encouraging the use of public transport and nonmotorized transport, and discouraging car use.

Improving the traffic flow in cities to reduce stop-start driving can reduce the energy use per vehicle-kilometer as well as emissions of local pollutants (Ang *et al.*, 1991; Hausberger *et al.*, 1994; Joumard *et al.*, 1990; Tanja *et al.*, 1992). Many measures to improve traffic flow, such as computerized traffic-signal control, actually increase the capacity of the road network so that traffic volumes may increase unless traffic levels are restrained in some way. Provided that traffic is restrained—for example, by limited parking provision, tolls, or fuel taxes or through constraints on vehicle access—the introduction of computerized signal control may lead to energy savings of 10 to 15% (Ang, 1992).

Encouraging public-transport use and nonmotorized modes. Travel by bus, train, foot, or bicycle usually involves lower greenhouse gas emissions per km than travel by car (see Section 21.2.5) and also can reduce wider social and environmental damage. People's willingness to use low-energy-intensity transport modes depends partly on the quality of travel by those modes. Planners can influence this by providing infrastructure, such as pedestrian and cycle paths, bus lanes, and railways. Public-transport use can be increased by providing better information about services, reducing fares, and improving the quality of the service. Flexible services, such as shared taxis and "dial-a-ride" minibuses, can play a role. Improvements in public transport rarely have proved to be an effective means of stemming the rise in private-transport use unless combined with measures to limit the attraction of car travel, such as access and parking constraints. Experience in Hannover, Zurich, and other cities indicates that such combined approaches can be successful (Brög, 1993; Goodwin, 1985; Ott, 1993). Lessons learned in these cities may have particular relevance for cities in Central and Eastern Europe (Suchorzewski, 1993).

A UK study in the 1970s found that the degree of hilliness and perceived safety were the main determinants of the share of trips by bicycle in a number of towns (Rowell and Fergusson, 1991, quoting Waldman, 1977). While it is hard to modify hills, safety and security are important keys to maintaining and promoting levels of cycling and walking. Safety for cyclists can be improved by providing physically separated cycle lanes on and off roads and by giving cyclists priority over motorized traffic at junctions. Security against cycle theft can be improved by providing appropriate parking facilities. Safety for pedestrians can be improved by providing and maintaining footpaths, providing lighting at night, and ensuring safe means of crossing roads. For rickshaws and carts, similar facilities are effective—in particular separated lanes on roads so they do not have to compete for and disrupt traffic.

Discouraging car use. Cities that have succeeded in promoting nonmotorized and public-transport use have done so partly by making car access more difficult. This has been accomplished in two main ways: pedestrianization of town centers, and "traffic calming"—most frequently in residential streets—with a combination of engineering, design, psychological, and architectural methods to slow vehicles down, alter the "balance of power" away from vehicles and toward pedestrians, and restrict car use.

Approaches include the conversion of existing road space to cycle and bus lanes or "high occupancy" lanes for shared cars;

restrictions on parking; limits on vehicle access to certain city areas or on movement between parts of a city; and motor vehicle-free zones. Restrictions on vehicles can result in evasive actions by drivers that lead to higher levels of traffic, energy use, and pollution in other parts of the city. However, if measures are planned well, residents and local businesses frequently are pleased with the results (Ott, 1993). The best schemes have resulted in increased safety, a more pleasant street environment, and improved retail trade (Hass-Klau, 1990, 1993). Poorly designed schemes can make emissions worse (for example, speed bumps that are too far apart encourage acceleration and deceleration), but in general these policies are seen as reinforcing the attractiveness of public transport and deterring traffic growth. The effectiveness of a strategy may hinge on consultation with the local community on its preferred approaches. Flexibility also can be an essential component: Counterproductive effects cannot always be anticipated, and it is important to monitor the results of measures to identify necessary changes.

Urban structure affects the distance people travel to meet their needs. Travel patterns may be influenced by many factors, including the size of settlements, proximity to other settlements, location of workplaces, provision of local facilities, and car ownership. Changes to settlement planning and regulation have been proposed as means of reducing car use and encouraging public transport and nonmotorized travel (Newman and Kenworthy, 1989). However, the link between settlement patterns and energy use for transport is controversial, and analysts have produced conflicting recommendations (Banister, 1992). While there does appear to be an inverse correlation between urban density and transport energy use (Armstrong, 1993), experts disagree on the desirability and effectiveness of densifying existing dispersed populations to reduce energy use. The trend toward lower urban density may result from people purchasing homes in locations with lower land prices where they have an improved quality of life. Attempts to reverse this trend may depend on addressing the welfare implications directly through measures to improve the physical and social environment in high-density areas. Policies that directly discourage the use of energy-intensive transport modes are likely to be more effective in reducing energy use.

In the short term, the greatest potential for urban planning to affect transport energy use is in rapidly developing cities where the car is still a minority transport mode. In such cities, infrastructure development focused on provision for cars and trucks could accelerate the growth in use of these modes and the decline in use of less energy-intensive modes. Provision for nonmotorized modes in new infrastructure is an important enabling factor for policies to encourage the use of these modes and discourage car use.

In the long term, around 2025 to 2050 and beyond, changes in travel culture and lifestyle combined with changes in urban layout might lead to substantial reductions in motorized travel in North American and Australian cities. The potential reduction in European cities is smaller because a higher proportion of access needs already are met by walking and cycling.

21.4.3. National Transport Policy

National transport policymakers influence the decisions taken by local transport and urban authorities through regulations, grants, planning guidelines, provision of information, and other means. National governments can contribute to local initiatives by providing coordination, information, and encouragement to local authorities considering measures that affect energy use and greenhouse gas emissions.

National governments may have a responsibility for a variety of regulatory measures that can influence vehicle fuel economy and greenhouse gas emissions. These measures might include speed limits, vehicle emission standards, and maintenance checks, as well as more direct measures, such as vehicle fuel-economy standards (see Box 21-4).

Drivers are more likely to change their vehicle choices and driving behavior in ways that reduce environmental impact if they understand that impact and are motivated to reduce it (Jones and Haigh, 1994). National measures can include advertizing and information campaigns, as well as the incorporation of energy- and environmental-awareness elements in the school curriculum and driver training. Standards in advertizing also are normally set at a national level. Information-based policies are likely to be most effective if they are maintained over long periods and are linked to changes in fiscal and regulatory policy.

National governments may have a direct involvement in the provision of rail and bus services through ownership, subsidies, franchises, and regulation. Many of the means of encouraging passengers to use these services are similar to those that apply to urban public transport; the most effective measures are likely to be those that discourage car use.

National transport policymakers can influence freight-transport patterns through measures applied to road-freight vehicles and their use. Vehicle licensing fees are often a significant element in road-haulage costs and can be used as a measure either to discourage road freight or to encourage the use of vehicles with lower power-to-weight ratios. Vehicle standards, including safety requirements, weight limits, and emission standards, also can influence vehicle choice and mode choice.

Where goods can be shifted from road to rail, the energy saving per tonne-kilometer can be zero to 50%. A variety of approaches have been developed to allow freight to be transferred easily between road and rail. The greatest energy savings are likely to be obtained with multimodal containers that can be moved between truck trailers and flat rail wagons. Cost savings can be significant for very long trains (100 wagons or more) traveling long distances because driver costs are greatly reduced.

21.4.4. Alternatives to Travel

Many of the needs that are met by the transport system could, in principle, be met by other means. Telecommunications, for

Box 21-4. Fuel-Economy Standards

In theory, the most economically efficient way of reducing car fuel consumption is through fiscal measures (which are discussed in Section 21.4.5). Higher fuel taxes tend to lead drivers to economize through driving more carefully, reducing the distance they drive, and choosing more energy-efficient vehicles. However, fuel taxes are difficult to impose in some countries. Consumers often have inadequate information about the cars they are buying and may not respond optimally to the price change. More complex price-based measures, such as taxes (or subsidies) linked to vehicle fuel economy, may be more effective in the long term, because they give consumers a direct incentive at the time of purchase to choose an energy-efficient vehicle.

An alternative or complement to taxation is to use regulations to require car manufacturers to sell more energy-efficient cars. There are advantages and disadvantages to this approach. Standards avoid the large transfer payments that might be required with a tax approach, given the low price elasticity of transport-fuel demand. On the other hand, improving fuel economy can lead to reduced driving costs and increased travel (the "rebound effect"), so that, while energy use and greenhouse gas emissions may be reduced overall, other environmental and social impacts of transport may be increased. Vehicle standards are sometimes criticized because they provide industry with insufficient flexibility to find the best technical solutions to complex problems. In the U.S. Corporate Average Fuel Efficiency (CAFE) program, this problem is partly addressed by allowing manufacturers freedom to determine the fuel economy of individual models, provided that the average fuel economy of their total sales of cars meets the standard. Companies that do not meet the required average fuel economy have to pay fines. Meanwhile, very-high-consuming cars are subject to a "gas guzzler tax," which is effective in limiting the fuel economy of cars sold (DeCicco and Gordon, 1995). A "sipper rebate" for very-low-consuming cars has been discussed, though never enacted. Manufacturers of alternative-fuel vehicles can claim credits against their corporate-fuel-economy average, allowing them to produce cars with higher fuel intensity.

While CAFE has generated considerable debate in the United States, Greene (1990) finds that the policy was a real constraint on vehicle manufacturers from 1978 to 1989 and that it had twice as much effect on car fuel economy as fuel price. However, the rebound effect was such that somewhere between 5 and 30% of the fuel-economy gain was lost through increased mileage (Berkowitz *et al.*, 1990; Greene, 1992; Jones, 1993).

example, might allow many trips to be avoided. The growth in service industries and associated office-based employment means that many types of work could be carried out in the home or in small regional offices connected to central offices via computer systems, faxes, and telephones. Similarly, an increasing number of transactions can be carried out through telecommunication systems. Much of the existing transport infrastructure in Europe and the United States was developed before such possibilities existed. Thus, culture has developed around travel to work, to shop, and for other purposes. Conversely, in many developing countries, telecommunication infrastructure is likely to precede efficient transport infrastructure. This offers opportunities for cheap access to employment and services that would not otherwise be available.

Telecommunications are likely to substitute for travel only to a limited extent. Indeed, improved communications may encourage more travel, as people develop closer working and social relationships over longer distances. At the upper end of the income scale, mobile telephones and electronic mail allow executives to maintain communication with their offices wherever they are, giving them freedom to travel while keeping in touch. However, at the other end of the scale, telecommunication systems may provide a means to economic growth without the traditional associated costs in infrastructure and the social and environmental burdens of rising levels of transport activity. In the long term, virtual-reality-based communication might replace business and other travel, although the extent of this possible substitution cannot now be predicted.

21.4.5. Economic Influences and Fiscal Measures

Roads in most countries are built and maintained by governments and are available, subject to regulations mainly related to safety, for anyone who wishes to use them. While the costs of road provision are recovered through fuel or vehicle taxes in some countries, in others taxes are insufficient to cover the costs. MacKenzie *et al.* (1992) estimate that road users in the United States pay only 60% of infrastructure costs through taxes and fees. Apart from tolls for special kinds of infrastructure (bridges, tunnels, and motorways), direct payment for road use, or "road pricing," occurs only in a few, mostly experimental, situations.

In Europe, vehicle purchases constitute about 90% of total investment in road transport, with infrastructure investment only 10% (ECMT, 1992). Infrastructure in rail transport accounts for 65 to 80% of total investment, with rolling stock only 20 to 35%. There is no straightforward way to allocate infrastructure costs between freight and passenger transport, but indicative costs for roads in Europe would be US$0.015 to 0.025 per passenger-km and US$0.1 to 0.2 per tonne-km (based on ECMT, 1992). These represent about 5 to 10% of passenger transport costs and a third of freight transport costs, respectively. In Western Europe, direct government expenditures on infrastructure related to car use are usually fully recovered from users through fuel or other taxes, although these taxes may not be explicitly intended to provide for a "road fund" as in the United States. Infrastructure costs for road-freight transport in some countries may not be fully recovered through taxes and fees. Charges reflecting these costs might add 10% to 30% to road-freight costs (Blok, 1991).

Most studies find that, when social and environmental damage are included, the costs not paid in money outlays by car and truck users have about the same order of magnitude as the costs paid (Button, 1990; DeLuchi *et al.*, 1994). Table 21-7 shows results based on several studies in OECD countries. The work by Button (1990) is based on earlier work by Quinet.

From the point of view of transport and environment policy, it is important to understand which of the costs of transport are born by users and which are not. Where users do not bear the full cost of transport, they are likely to make more use of transport services than is economically efficient (Button, 1994). DeLuchi *et al.* (1994) find that, in the United States, motor-vehicle users bear about 70% of the overall cost of vehicle use (see Table 21-8). Most of the remainder relates to free parking provision and accident damage not paid for by insurance.

Environmental external costs amount to \$0.02 to \$0.1 per km driven by cars (Bleijenberg, 1994; DeLuchi *et al.*, 1994; IEA, 1993a; Kågeson, 1994; MacKenzie *et al.*, 1992). Kågeson (1994) gives total external costs for Germany in 1993 of 0.012 European Currency Unit (ECU) (US\$0.015) per tonne-km in trucks. External costs for trains are about 0.005 ECU (US\$0.006) per passenger-km and 0.004 ECU (US\$0.005) per tonne-km.

Table 21-7: *Estimates of total social cost (as percentage of GDP) of transport in OECD countries.*

Cost Item	Road[a]	Other Modes[a]	All Transport[b]
Noise	0.10	0.01	0.3
Local Pollution	0.40		0.4
Total Pollution			1–10[c]
Accidents	2.00		1.5–2
Total Travel Time	6.80	0.07	8.5 (of which 2–3 is due to road congestion)
Use Expenditure	9.00	0.30	–
Total	**18.30**	**4.71**	–

[a] Button, 1990.
[b] Quinet, 1994.
[c] This depends on the assumed damage costs associated with global climate change.

Table 21-8: *Costs of motor vehicle use in the United States (DeLuchi et al., 1994).*

Cost Item	Percentage of Driving Costs
Money Costs of Driving	**40–45**
Paid by user	30
Public roads and services	5
(not recovered through taxes and fees)	(2)
Private sector costs not paid by user (mostly unpriced parking)	5–10
Non-Money Costs of Driving	**55–69**
Borne by user (includes travel time)	40
Not borne by user (mostly accidents)	15–20
(environmental effects on health)	(2–10)
Subtotal—due to road dust	1–7
Subtotal—due to fuel/exhaust emissions	1–3

The incorporation of nonmoney costs of driving as user fees or taxes is one of the most commonly discussed measures to reduce traffic congestion and pollution. The response of transport energy demand to costs, especially fuel prices, has been extensively studied since the early 1970s. Researchers find that:

- A 10% increase in fuel prices results in a 1 to 6% short-term reduction in demand, according to most studies (Dahl and Sterner, 1991; Fowkes *et al.*, 1993a; Tanja *et al.*, 1992), although the effect of price increases is not symmetrical with the effect of price decreases (Dargay, 1993). In the long term, fuel-price increases encourage the use of more energy-efficient vehicles, and a 10% increase in gasoline prices can lead to a 5 to 16% reduction in demand. The response to fuel price falls as incomes rise, all else being equal (Goodwin, 1992; Greening *et al.*, 1994). Where several fuels compete and can be used in the same vehicles, small relative price changes can have a large effect on fuel choice (Greene, 1989; IEA, 1993a).
- A 10% increase in car price leads to a 1 to 5% reduction in total fuel consumption (Tanja *et al.*, 1992), but there is a complex interaction between the price of fuel, the prices of new and second-hand cars of different sizes, car ownership, and car use (Mogridge, 1983). Differential vehicle taxes related to energy efficiency can be one of the most economically efficient means of encouraging the use of energy-efficient vehicles (DeCicco and Gordon, 1995).
- Car and truck use is not affected noticeably when the costs of other transport modes fall, although rising public-transport and rail costs may encourage users to switch modes. Rail and bus travel do depend strongly on the cost of car travel. Studies of the effects of transport costs have been carried out for different transport modes in many situations (Oum *et al.*, 1990;

Fowkes *et al.*, 1993a, 1993b). In the United Kingdom, these studies indicate that a 10% increase in bus or urban metro fares reduces patronage by 3 to 4% in the short term and around 7% in the long term.
- Various factors in addition to transport costs and income affect travel activity, including household size, the occupation of the head of the household, household makeup, and location (Hensher *et al.*, 1990; Jansson, 1989; Walls *et al.*, 1993). People in higher-skilled occupations, requiring higher levels of education, are more price- and income-responsive in their transport energy demand than people in lower-skilled occupations (Greening and Jeng, 1994; Greening *et al.*, 1994). Families are more price- and income-responsive in the early years of childrearing than in the later stages.

Road-use charges can, in theory, help allocate available road space in the most economically efficient way. Road-use fees can be used to reduce both the explicit subsidies for road building and maintenance and the implicit subsidies associated with the externalities of driving. By increasing the variable cost of driving, such changes can encourage travelers to share vehicles, travel shorter distances, or use alternative modes. They are expected to be most effective when charged to the driver at the time and point of use. Polak *et al.* (1994) have analyzed the long-term impact of the Area Licensing Scheme in Singapore and found that a 10% increase in the cost of peak-period travel resulted in a 7% reduction in peak traffic in the short run and a 12% reduction in the long run. Toll rings in the Norwegian cities Oslo and Trondheim have reduced traffic by 4 and 8%, respectively (Polak and Meland, 1994; Ramjerdi, 1992).

Parking fees can be an important component of any transport strategy. In many cities, employees are provided with free car parking at work. Studies in the United States (DOE, 1994) indicate that offering people in large cities the cash value of their parking place as an alternative would raise the effective cost of their trip to work by 116%. Of those offered the "cash-out," 23% would accept it in the long term and choose alternative transport modes to commute to work.

Many other market-based strategies have been proposed, ranging from "pay at the pump" insurance charges collected through filling stations to fuel-consumption permits that would be tradable between manufacturers.

Revenues from charges imposed on cars and trucks can be used to subsidize transport modes with lower social costs and greenhouse gas emissions. Nevertheless, while keeping fares low may maintain existing ridership, it is unlikely to attract many additional users from private transport without other measures.

21.4.5.1. Vehicle Taxes

Vehicle taxes are traditional sources of government revenue and controls on imports. Cars usually are taxed at a higher rate than buses or trucks—a policy that promotes commercial activity as opposed to private car ownership. Many countries base purchase taxes or license fees on engine size or vehicle weight; this can be a powerful instrument to encourage the purchase of small, energy-efficient cars. In some African countries, taxes are used to discourage imports of second-hand vehicles, while, in others, vehicles older than five years are banned (Davidson, 1993).

Recently, several governments have considered the possibility of using "feebates": taxes for vehicles with high fuel consumption along with rebates for vehicles with low fuel consumption. Such a scheme is in operation in Ontario, Canada, with a tax ranging from Can$75 to 4,400 (about US$55 to 3,300) on new cars with fuel consumption over 6 L/100 km sold in the province; purchasers of cars with fuel consumption under 6 L/100 km receive a Can$100 (about US$75) rebate (Canada, 1994).

21.4.5.2. Fuel Taxes

Road-transport fuel taxes mostly are used as a means of revenue raising, often to cover the costs of infrastructure and other transport services. Differential taxes are used in many countries to encourage the use of cleaner fuels, such as unleaded gasoline and low-sulfur diesel, or alternative fuels, such as LPG, CNG, and alcohols. Taxes are usually higher on gasoline than on diesel, a cross-subsidy from car users to truck operators that has a side effect of encouraging the use of diesel cars. The implications for CO_2 emissions are not clear: Diesel cars on average have lower CO_2 emissions per km than gasoline cars, but since the fuel is cheap, their owners will tend to drive their cars farther than do gasoline car owners. On the other hand, if gasoline taxes are higher than they otherwise would be, gasoline-car owners will tend to drive less.

In the case of air travel, contracting states of the International Civil Aviation Organization have agreed on a policy that fuel used for international operations should be exempt from all taxes (ICAO, 1994b). The world average price for aviation fuel for international scheduled services in 1991 was US$0.2/L (ICAO, 1994c). A 10% increase in fuel prices would lead to roughly a 1.5% increase in total passenger-travel costs and might be expected to lead to roughly a 1% short-term decrease in travel (based on ICAO, 1992). The longer-term effect of fuel prices on aircraft energy efficiency is expected to be quite large because the aviation industry pays close attention to life-cycle cost-effectiveness in considering aircraft design.

21.4.6. Combining Measures

Success in reducing greenhouse gas emissions will depend on using combinations of different measures. This applies in particular to measures that aim to restrict travel or vehicle use. Constraints on car use can have unintended effects; parking controls implemented alone could lead to increased travel (NOVEM, 1992). People making short trips are more likely to be discouraged by parking difficulties than those making long trips. Displacing short-trip traffic creates more parking space for long-trip traffic. Other measures such as fuel taxation, energy-efficiency standards, or more general road pricing could be needed to avoid an increase in energy use for long trips. Similarly, if air travel is constrained by airport congestion, the promotion of high-speed rail as an alternative to short-haul air travel may reduce the number of short-haul flights, opening airport takeoff and landing slots for long-haul flights and encouraging additional energy use.

Combinations of measures, including information and education, are needed to be effective in bringing about modal shifts. Drivers tend to underestimate the costs of car travel (partly because they only notice fuel costs) and overestimate the cost and inconvenience of travel by public transport (Brög, 1993).

Many cities have attempted to implement integrated transport policies, using a wide range of measures to reduce traffic and encourage the use of more energy-efficient, low-emission vehicles (ECMT/OECD, 1995). Usually, these attempts are hard to assess because it is difficult to judge what would have occurred if the policies had not been implemented. One analysis of the effect of the transport policy in Singapore is summarized in Box 21-5.

Box 21-5. Singapore: Effects of Integrated Transport Policy

Singapore is a small island state with 2.8 million people in an area of 633 km^2 (44/ha). Since the early 1970s, it has adopted a variety of measures to control the traffic problems associated with high population density and rapid economic growth.

Computerized traffic-signal systems have been widely implemented in the central business district (CBD).

The *Area Licensing Scheme* (ALS), a road-pricing scheme introduced in 1975, was aimed at reducing morning peak traffic in the CBD. Drivers were required to purchase windscreen stickers, which were checked on entering the ALS zone. The program immediately reduced the number of vehicles entering the zone during the morning peak and shifted many people's morning commute habits. The success of the scheme led to its extension to include evening peak hours in 1989 and then to the whole day in 1994. In 1996, an electronic road-pricing system will replace the ALS.

The *Weekend Car Scheme* was introduced in 1991. Owners of cars registered under the scheme can normally drive only on weekends and receive a rebate on vehicle registration fees and import duty. They can purchase day licenses to operate their cars during the week in peak or off-peak hours.

Fiscal measures, including high import duties, vehicle registration fees, and annual road taxes, have been implemented to discourage car ownership. In 1994, import duties and registration fees amounted to 195% of car import values.

The *Vehicle Quota System* was introduced in 1990, limiting new registrations of cars and other vehicles. New-vehicle buyers have to bid for quota allocations in a monthly public auction.

Road tax increases with engine capacity, encouraging the purchase of small, energy-efficient cars.

Fuel tax is approximately 40 US¢/litre.

Public transport is of high quality, with buses providing a 20 km/h average service. There is a 67-km mass rapid-transit system, with more than half of Singapore's homes and work locations within 1 km of the route.

The *road network* has been upgraded and the capacity expanded constantly to provide more efficient transport links and to maximize the effectiveness of the road system.

Settlement planning is systematic, with colocation of homes, shops, schools, recreational facilities, factories, and offices in each of seventeen new towns or housing estates.

The result of this combination of measures on traffic and energy use has been estimated by Ang (1992).

Table 21-9: *Estimated 1990 fuel consumption in Singapore without car constraint policies (in millions of L).*

	Gasoline	**Diesel**
Actual Consumption	741	465
Impact of not having policy		
– Passenger traffic increase	+153	
– Modal shift	+218	-84
– Shift to larger cars	+52	
– Traffic congestion	+122	+77
Consumption without Policy	1286	458
Estimated Impact of Policy on Consumption	-42%	+2%

Source: Ang, 1992, 1993.

As the Singapore example illustrates, measures of many different types can be coordinated to achieve one set of policy objectives. Many other examples can be found, including Curitiba, Zurich, Hannover, Oxford, and Portland, Oregon, where coordinated policies have been used to slow or even reverse the growth in car use (Rabinovitch, 1993; Brög, 1993; Ott, 1993; ECMT/OECD, 1995).

Appropriate mixes of policies will vary between cities and countries. Birk and Zegras (1993) provide case studies of four Asian cities to illustrate different approaches to an integrated strategy for managing the environmental effects of transport. They summarize the potential for policy coordination in a matrix, on which Table 21-10 is based.

21.4.7. Implementation

Greenhouse gas emission reduction in the transport sector, more than in any other sector, depends on obtaining cooperation among the various stakeholders or interest groups who are able to take action or who might be affected by policies. Almost all members of society and all organizations have some involvement in the transport system. There are numerous overlaps between transport policy and other areas of government policy. Inevitably, there will be some clashes of interest, but there also will be areas of agreement. Where government departments and other institutions, organizations, and interest groups can agree on economic, environmental, and social aims, there will be more scope for a coordinated approach to transport policy. Coordination among regions is also important.

The following steps are important in any attempt to implement transport policies.

Understanding the current system and its evolution. Ideally this step would include:

- Carrying out surveys of vehicle flow rates and occupancies, analyzing the records of commercial transport operators, and requesting them or requiring them to keep records
- Employing economists and social scientists to evaluate the importance of different aspects of the transport system to its users
- Establishing processes for consultation with transport users and their representatives
- Examining past trends and evaluating a range of possible futures (a multidisciplinary team is more likely to produce a realistic view of the future than economists, planners, or engineers alone.)
- Considering uncertainty by developing several scenarios against which strategies can be assessed; a robust strategy is one that produces an acceptable outcome in all scenarios.

These activities represent a significant and ongoing commitment of resources to data collection and monitoring that may not always be possible. Information-collection efforts can be approached in stages and have to be designed in the context of the resources available.

Considering a wide range of measures. These might include taxes and fees, alternative-fuel promotion, standards for fuels and vehicles, land-use and infrastructure changes, and promotion of communication technologies.

Evaluation of options would ideally involve a cost-benefit analysis, taking account of all of the social benefits and costs of each measure. While social accounting might become a viable and valuable tool at some time in the future, such techniques are not currently available, and some analysts doubt whether they ever will be—or should be—developed. In the absence of a single quantitative evaluation technique, policymakers can make use of multidisciplinary teams to identify costs and benefits of the various measures and to provide a qualitative evaluation.

Consulting stakeholders. In addition to the importance of stakeholder consultation to obtain information about their needs and activities, measures in the transport sector are more likely to be effective if the people who will be influenced by the outcome are involved in the decision-making process. It is important for stakeholders to be given access to information in the development of transport policy and to be allowed to comment on and possibly correct it as the understanding of the system is developed. In many countries, some form of public consultation is a statutory requirement in the planning process. Consultation can be one of the most valuable steps in decisionmaking more generally, as it may generate new ideas, can help to select the most satisfactory outcomes, and can help to give the eventual users of the system a sense of ownership.

Monitoring and adjustment. Even the best-planned measures are likely to have unexpected outcomes. These occurrences are opportunities to learn and can be dealt with, provided sufficient flexibility is allowed for in the plans. A key to success, therefore, is to plan responsiveness into the decisionmaking process by continuing data collection and analysis to monitor the effects of measures and to allow decisionmakers to make follow-up decisions.

References

Ang, B.W., 1992: Restraining automobile ownership and usage and the transportation energy demand: the case of Singapore. *The Journal of Energy and Development*, **XVII(2)**, 263-290.

Ang, B.W., 1993: An energy and environmentally sound urban transport system: the case of Singapore. *International Journal of Vehicle Design*, **14(4)**.

Ang, B.W. and S.T. Oh, 1988: Transport and traffic management schemes and energy saving in Singapore. *Energy—The International Journal*, **13(2)**, 141-148.

Ang, B.W., T.F. Fwa, and C.K. Poh, 1991: A statistical study of automobile fuel consumption. *Energy—The International Journal*, **16(8)**, 1067-1077.

Appleby, A.J. and F.R. Foulkes, 1989: *Fuel Cell Handbook*. Van Nostrand Reinhold, New York, NY.

Table 21-10: *Matrix of transport system improvement options (derived from Birk and Zegras, 1993).*

	Situation		
Options	*High level of existing infrastructure and strong influence on technology*	*Rapidly developing infra-structure and relatively little influence on technology*	*Small towns, relatively simple infrastructure, little influence on technology*
Reducing Vehicle Emissions	Impose vehicle standards or emission-related taxes/rebates Set up inspection and maintenance programs Driver and police training for efficient driving and traffic control Public-awareness campaigns Research, development, and demonstration (RD&D)	Work with other cities and regions to encourage use of improved vehicles Set up inspection and maintenance programs Driver and police training for efficient driving and traffic control Public-awareness campaigns	Work with other cities and regions to encourage use of improved vehicles Set up inspection and maintenance programs Driver and police training for efficient driving and traffic control Public-awareness campaigns
Shifting to Cleaner or Alternative Fuels	Fuel-quality standards, mandates, and taxes Evaluate options for alternative fuels, especially in large fleets including public transport RD&D	Work with other cities and regions to encourage use of cleaner fuels Evaluate options for alternative fuels, especially in large fleets including public transport	Work with other cities and regions to encourage use of cleaner fuels Evaluate options for alternative fuels, especially in large fleets including public transport
Shifting to Modes with Lower Emissions	Provide facilities for nonmotorized modes Consider nonmotorized streets and zones Exclusive bus lanes and bus/tram priority Consider subsidies/fare controls for public transport Consider investment in public transport	Exclusive bus lanes and bus priority Segregate nonmotorized modes Consider possibility of rail conversion to exclusive bus lanes Consider subsidies/fare controls for public transport Consider investment in public transport	Begin to give priority to mass transit in urban transport policy Ensure access to public transport Exclusive bus lanes and bus priority in new development Segregate nonmotorized modes
Transport Demand Management	Regulate or charge for parking Consider vehicle-free zones or restrictions Consider road-user fees, and vehicle and fuel taxes Investigate telecommuting, etc., as alternatives to transport Carpool and high-occupancy vehicle incentives	Restrict on-street parking Consider vehicle-free zones or restrictions Consider road-user fees, and vehicle and fuel taxes Investigate improved communication as alternative to transport Carpool and high-occupancy vehicle incentives	Formal parking controls or fees Consider vehicle-free zones or restrictions Investigate long-term role of road-user fees to manage traffic Investigate improved communication as alternative to transport Car-pool and high-occupancy vehicle incentives
Transport Infrastructure Development	Computerized traffic control Maintain and provide infra-structure for nonmotorized and public transport Segregated nonmotorized and public transport corridors	Improved traffic control Maintain and provide infra-structure for nonmotorized and public transport Segregated nonmotorized and public transport corridors Consider bypasses	Improved traffic control Give priority to public transport and nonmotorized transport in new infrastructure
Land-Use Planning	Consider deemphasizing central zone as major focus of activity Encourage mixed use, non-motorized/public transport-oriented suburban development Promote mixed use and non-motorized/public transport access in urban redevelopment Discourage parking provision in development and redevelopment	Desegregate land use; allow mixed use in central business district Promote mixed use and non-motorized/public transport access in new development Encourage mixed use, non-motorized/public transport-oriented suburban development	Ensure transport development that enhances rather than hampers economic health of central business district Take transport trends into account and plan transport services in coordination with other development

Armstrong, D.M., 1993: Transport infrastructure, urban form and mode usage: an econometric analysis based on aggregate comparative data. In: *Regional Science Association Thirty Third European Congress, Moscow, 24-27 August 1993*. Northern Ireland Economic Research Centre, 48 University Road, Belfast, N. Ireland, BT7 1NJ, 19 pp.

ADB, 1994: *Energy Indicators of Developing Member Countries of ADB*. Asian Development Bank, Manila, Philippines.

Balashov, B. and A. Smith, 1992: ICAO analyses trends in fuel consumption by world's airlines. *ICAO Journal*, **August**, pp. 18-21.

Banister, D., 1992: Energy use, transport and settlement patterns. In: *Sustainable Development and Urban Form* [Breheny, M.J. (ed.)]. Psion, London, UK, pp. 160-181.

Bennathan, E., J. Fraser, and L.S. Thompson, 1992: *What Determines Demand for Freight Transport?* Policy Research Working Paper WPS 998, The World Bank, Washington, DC, 29 pp.

Berkowitz, M.K., N.T. Gallini, E.J. Miller, and R.A. Wolfe, 1990: Disaggregate analysis of the demand for gasoline. *Canadian Journal of Economics*, **XXIII(2)**, 253-275.

Birk, M.L. and D.L. Bleviss, 1991: *Driving New Directions: Transportation Experiences and Options in Developing Countries*. International Institute for Energy Conservation, Washington, DC, 106 pp.

Birk, M.L. and P.C. Zegras, 1993: *Moving Toward Integrated Transport Planning*. International Institute for Energy Conservation, Washington, DC, 126 pp.

Bleijenberg, A., 1994: The art of internalising. In: *Internalising the Social Costs of Transport* [European Conference of Ministers of Transport]. OECD, Paris, France, pp. 95-112.

Bleviss, D.L., 1988: *The New Oil Crisis and Fuel Economy Technologies*. Quorum Books, Newport, CT, 268 pp.

Blok, P.M., 1991: Prospects for a shift in modal split. In: *Freight Transport and the Environment* [European Conference of Ministers of Transport]. OECD, Paris, France, pp. 129-140.

Bose, R.K. and G.A. Mackenzie, 1993: Transport in Delhi: energy and environmental consequences. In: *Industry and Environment* [United Nations Environment Programme, Paris, France], **16(1-2)**, 21-24.

Brandberg, Å., M. Ekelund, A. Johansson, and A. Roth, 1992: *The Life of Fuels: Motor Fuels from Source to End Use*. Ecotraffic, Stockholm, Sweden, 179 pp.

Bremnes, P.K., 1990: *Exhaust Gas Emission from International Marine Transport*. Marintek, Trondheim, Norway, 27 pp.

Brög, W., 1993: Behaviour begins in the mind: possibilities and limits of marketing activities in urban public transport. In: *European Conference of Ministers of Transport Round Table 92 on Marketing and Service Quality in Public Transport* [European Conference of Ministers of Transport]. OECD, Paris, France, pp. 5-78.

Bunch, D.S., M.A. Bradley, T.F. Golob, R. Kitamura, and G.P. Occhiuzzo, 1993: Demand for clean-fuel vehicles in California: a discrete choice stated preference approach. *Transportation Research A*, **27A(3)**, 237-253.

Button, K., 1990: Environmental externalities and transport policy. *Oxford Review of Economic Policy*, **6(2)**, 61-75.

Button, K., 1994: Overview of internalising the social costs of transport. In: *Internalising the Social Costs of Transport* [European Conference of Ministers of Transport]. OECD, Paris, France, pp. 7-30.

Canada, 1994: *Canada's National Report on Climate Change*. Environment Canada, Ottawa, Canada, 144 pp.

CEC, 1992: *Research and Technology Strategy to Help Overcome the Environmental Problems in Relation to Transport. (SAST Project No 3.) Global Pollution Study*. Report EUR-14713-EN [Commission of the European Communities Directorate-General for Science, Research and Development], CEC, Brussels and Luxembourg, 193 pp.

Chin Then Heng, A. and B.W. Ang, 1994: Energy, environment and transport policy in Singapore. In: *Sectoral Meeting on Energy, Environment and Urban Transport*. UNESCAP, Hong Kong, 78 pp.

Chinaglia, L., 1991: Fuel economy improvement of passenger cars. In: *Low Consumption/Low Emission Automobile. Proceedings of an Expert Panel*. OECD, Paris, France, pp. 121-126.

Dahl, C. and T. Sterner, 1991: A survey of econometric gasoline demand elasticities. *International Journal of Energy Systems*, **11(2)**, 53-76.

Dargay, J.M., 1993: Demand elasticities: a comment. *Journal of Transport Economics and Policy*, **27**, 87-90.

Dargay, J.M. and P.B. Goodwin, 1994: Estimation of consumer surplus with dynamic demand changes. In: *22nd European Transport Forum, London, Volume 1*. The PTRC International Association, Hadleigh, Essex, UK, pp. 1-10.

Davidson, O.R., 1993: Opportunities for energy efficiency in the transport sector. In: *Energy Options for Africa: Environmentally Sustainable Alternatives* [Karakezi, S. and G.A. Mackenzie (eds.)]. Zed Books Ltd., London, UK, pp. 106-127.

Davidson, O.R., 1992: *Transport Energy in Sub-Saharan Africa: Options for a Low-Emissions Future*. Report No. 267, Princeton University Center for Energy and Environmental Studies, Princeton, NJ, 122 pp.

Davis, S.C. and S.G. Strang, 1993: *Transportation Energy Data Book: Edition 13*. Oak Ridge National Laboratory, Oak Ridge, TN, 215 pp.

DeCicco, J. and D. Gordon, 1995: Steering with prices: fuel and vehicle taxation as market incentives for higher fuel economy. In: *Transportation and Energy: Strategies for a Sustainable Transportation System* [Sperling, D. and S. Shaheen (eds.)]. American Council for an Energy-Efficient Economy, Washington, DC and Berkeley, CA, pp. 177-216.

DeCicco, J. and M. Ross, 1993: *An Updated Assessment of the Near-Term Potential for Improving Automotive Fuel Economy*. American Council for an Energy-Efficient Economy, Washington, DC and Berkeley, CA, 99 pp.

Delsey, J., 1991a: Growth of vehicle power and environmental consequences. In: *Freight Transport and the Environment* [European Conference of Ministers of Transport]. OECD, Paris, France, pp. 77-88.

Delsey, J., 1991b: How to reduce fuel consumption of road vehicles. In: *Low Consumption/Low Emission Automobile*. Proceedings of an Expert Panel, OECD, Paris, France, pp 95-104.

DeLuchi, M.A., 1991: *Emissions of Greenhouse Gases from the Use of Transportation Fuels and Electricity*. Vol. 1, *Main Text*. ANL/ESD/TM-22, Vol. 1, Center for Transportation Research, Argonne National Laboratory, Argonne, IL, 142 pp.

DeLuchi, M.A., 1992: *Hydrogen Fuel-Cell Vehicles*. Research Report UCD-ITS-RR-92-14, Institute of Transportation Studies, University of California, Davis, CA, 158 pp.

DeLuchi, M.A., 1993: *Emissions of Greenhouse Gases from the Use of Transportation Fuels and Electricity*. Vol. 2, *Appendixes A-S*. ANL/ESD/TM-22, Vol. 2, Center for Transportation Research, Argonne National Laboratory, Argonne, IL.

DeLuchi, M.A., D. McCubbin, J. Kim, S.-L. Hsu, and J. Murphy, 1994: *The Annualized Social Cost of Motor-Vehicle Use, Based on 1990-1991 Data*. Institute of Transportation Studies, University of California, Davis, CA.

Difiglio, C., K.G. Duleep, and D.L. Greene, 1990: Cost effectiveness of future fuel economy improvements. *The Energy Journal*, **11(1)**, 65-85.

Dimitriou, H.T., 1992: *Urban Transport Planning: A Developmental Approach*. Routledge, London, UK, pp. 184-217.

DOE, 1989: *Assessment of Costs and Benefits of Flexible and Alternative Fuel Use in the U.S. Transportation Sector*. Technical Report Three, *Methanol Production and Transportation Costs*. U.S. Department of Energy, Washington, DC, 46 pp.

DOE, 1990: *Assessment of Costs and Benefits of Flexible and Alternative Fuel Use in the U.S. Transportation Sector*. Technical Report Five, *Costs of Methanol Production from Biomass*. U.S. Department of Energy, Washington, DC, 19 pp.

DOE, 1991: *Second Interim Report of the Interagency Commission on Alternative Motor Fuels*. U.S. Department of Energy, Washington, DC, 66 pp.

DOE, 1994: *The Climate Change Action Plan: Technical Supplement*. DOE/PO-0022, U.S. Department of Energy, Washington, DC, 148 pp.

Doyle, G., 1986: *Marine Transport Costs and Coal Trade*. IEA Coal Research, London, UK.

ECMT, 1992: *Investment in Transport Infrastructure in the 1980s*. OECD, Paris, France, 350 pp.

ECMT/OECD,1995: *Urban Travel and Sustainable Development*. OECD, Paris, France, 238 pp.

ETSU, 1994: *Appraisal of UK Energy Research, Development, Demonstration and Dissemination*. Vol. 7, *Transport*. HMSO, London, UK, pp. 481-638.

Eyre, N.J. and L.A. Michaelis, 1991: *The Impact of UK Electricity, Gas and Oil Use on Global Warming*. Report AEA-EE-0211, Energy Technology Support Unit, Harwell, Oxfordshire, UK, 50 pp.

Faiz, A., 1993: Automotive emissions in developing countries—relative implications for global warming, acidification and urban air quality. *Transportation research A*, **27A(3)**, 167-186. Pergamon, Oxford, UK.

Foley, G. and G. Barnard, 1982: *Biomass Gasification in Developing Countries.* Earthscan, London, UK, 175 pp.

Fowkes, A.S., C.A. Nash, J.P. Toner, and G. Tweddle, 1993a: *Disaggregated Approaches to Freight Analysis: A Feasibility Study.* Working Paper 399, Institute of Transportation Studies, University of Leeds, UK, 133 pp.

Fowkes, A.S., N. Sherwood, and C.A. Nash, 1993b: *Segmentation of the Travel Market in London: Estimates of Elasticities and Values of Travel Time.* Working Paper 345, Institute of Transportation Studies, University of Leeds, UK, 17 pp.

Goldemberg, J. and I.C. Macedo, 1994: Brazilian alcohol program: an overview. *Energy for Sustainable Development,* **1(1)**, 17-22. International Energy Intitiative, Bangalore, India.

Goodwin, P.B., 1985: *Changes in Transport Users' Motivations for Modal Choice: Passenger Transport. European Conference of Ministers of Transport Round Table 68.* OECD, Paris, France, pp. 61-90.

Goodwin, P., S. Hallett, F. Kenny, and G. Stokes, 1991: *Transport: the New Realism.* Rees Jeffreys Road Fund Report 624, Transport Studies Unit, University of Oxford, Oxford, UK.

Goodwin, P.B., 1992: A review of new demand elasticities with special reference to short and long run effects of price changes. *Journal of Transport Economics and Policy,* **26(2)**, 155-170

Gover, M., G. Hitchcock, D. Martin, and G. Wilkins, 1993: Alternative fuels: a life cycle analysis. In: *26th International Symposium on Automotive Technology and Automation. Aachen 1993.* ISBN 0 947719 61 X, Automotive Automation Ltd., Croydon, England, UK, pp. 283-290.

Greene, D.L., 1989: Motor fuel choice: an econometric analysis. *Transportation Research A,* **23A(3)**, 243-253. Pergamon, Oxford, UK.

Greene, D.L., 1990: CAFE or price?: an analysis of the effects of federal fuel economy regulations and gasoline price on new car MPG, 1978-89. *The Energy Journal,* **11(3)**, 37-57.

Greene, D.L., 1992: Vehicle use and fuel economy: how big is the "rebound" effect? *The Energy Journal,* **13(1)**, 117-143.

Greene, D.L. and K.G. Duleep, 1993: Costs and benefits of automotive fuel economy improvement: a partial analysis. *Transportation Research A,* **27A(3)**, 217-235. Pergamon, Oxford, UK.

Greening, L.A. and H.T. Jeng, 1994: Life-cycle analysis of gasoline expenditure patterns. *Energy Economics,* **16(3)**, 217-228.

Greening, L.A., H.T. Jeng, J. Formby, and D.C. Cheng, 1994: Use of region, life-cycle and role variables in the short-run estimation of the demand for gasoline and miles travelled. *Applied Economics,* **27(7)**, 643-655.

Grieb, H. and B. Simon, 1990: Pollutant emissions of existing and future engines for commercial aircraft. In: *Air Traffic and the Environment—Background, Tendencies and Potential Global Atmospheric Effects* [Schumann, U. (ed.)]. Springer Verlag, Berlin, Germany.

Grübler, A. and N. Nakicenovic, 1991: *Evolution of Transport Systems: Past and Future.* RR-91-8, International Institute for Applied Systems Analysis, Laxenburg, Austria, 97 pp.

Grübler, A., S. Messner, L. Schrattenholzer, and A. Schäfer, 1993: Emission reduction at the global level. *Energy,* **18(5)**, 539-581.

Grübler, A, 1993: Transportation sector: growing demand and emissions. *Pacific and Asian Journal of Energy,* **3(2)**, 179-199.

GUS, 1993: *Rocznik Statystyczny 1993* (Annual Statistics 1993). GUS, Warsaw, Poland.

Hass-Klau, C., 1990: *The Pedestrian and City Traffic.* Belhaven Press, London, UK, 277 pp.

Hass-Klau, C., 1993: Impact of pedestrianisation and traffic calming on retailing. A review of the evidence. *Transport Policy,* **1(1)**, 21-31

Hausberger, S. *et al.,* 1994: *KEMIS—A Computer Program for the Simulation of On-Road Emissions Based on the Characteristic Driving Behaviour.* Institute for Internal Combustion Engines and Thermodynamics, Technical University, Graz, Austria.

Hensher, D.A., F.W. Milthorpe, and N.C. Smith, 1990: The demand for vehicle use in the urban household sector: theory and empirical evidence. *Journal of Transport Economics and Policy,* **24(2)**, 119-137.

Hidaka, S., 1993: Modal shifts and energy efficiency. *Energy in Japan,* **122** (July), 8-21. The Institute of Energy Economics, Tokyo, Japan.

ICAO, 1992: *Outlook for Air Transport to the Year 2001.* Circular 237-AT/96, ICAO, Montreal, Canada, 47 pp.

ICAO, 1993: *Annex 16 to the Convention on International Civil Aviation.* Vol. II, *Aircraft Engine Emissions.* Second Edition - July 1993, ICAO, Montreal, Canada, 55 pp.

ICAO, 1994a: *The World of Civil Aviation, 1993-1996.* Circular 250-AT/102, ICAO, Montreal, Canada, 118 pp.

ICAO, 1994b: *ICAO's Policies on Taxation in the Field of International Air Transport.* Doc 8632-C/968, ICAO, Montreal, Canada, 13 pp.

ICAO, 1994c: *Regional Differences in Fares, Rates and Costs for International Air Transport, 1991.* Circular 248-AT/101, ICAO, Montreal, Canada, 39 pp.

IEA, 1991: *Fuel Efficiency of Passenger Cars.* OECD, Paris, France, 91 pp.

IEA, 1993a: *Cars and Climate Change.* OECD, Paris, France, 236 pp.

IEA, 1993b: *Electric Vehicles: Technology, Performance and Potential.* OECD, Paris, France, 201 pp.

IEA, 1993c: *Energy Statistics and Balances of Non-OECD Countries, 1990-1991.* OECD, Paris, France, 659 pp.

IEA, 1993d: *Energy Balances of OECD Countries, 1990-1991.* OECD, Paris, France, 218 pp.

IEA, 1994a: *Energy Prices and Taxes. First Quarter, 1994.* OECD, Paris, France, 397 pp.

IEA, 1994b: *Biofuels.* OECD, Paris, France, 115 pp.

IEA, 1994c: *Energy in Developing Countries: A Sectoral Analysis.* OECD, Paris, France, 130 pp.

IPCC, 1995: *Climate Change 1994.* Cambridge University Press, Cambridge, UK, 339 pp.

Jansson, J.O., 1989: Car demand modelling and forecasting. *Journal of Transport Economics and Policy,* **23(2)**, 125-140.

Jones, C.T., 1993: Another look at U.S. passenger vehicle use and the 'rebound' effect from improved fuel efficiency. *The Energy Journal,* **14(4)**, 99-143.

Jones, P.M. and D. Haigh, 1994: Reducing traffic growth by changing attitudes and behaviour. In: *22nd European Transport Forum, London.* The PTRC International Association, Hadleigh, Essex, UK.

Joumard, R., L. Paturel, R. Vidon, J.-P. Guitton, A.-I. Saber, and E. Combet, 1990: *Emissions Unitaires de Polluants des Véhicules Légers.* INRETS Report No. 116, Institut National de Recherche sur les Transports et leur Sécurité, Bron, France, 120 pp.

Kågeson, P., 1994: Effects of internalisation on transport demand and modal split. In: *Internalising the Social Costs of Transport* [European Conference of Ministers of Transport]. OECD, Paris, France, pp. 77-94.

Kang, S.-J., 1989: Chapter 5, Energy demand in the transportation sector. In: *Sectoral Energy Demand in the Republic of Korea.* Economic and Social Commission for Asia and the Pacific, Regional Energy Development Programme Report (RAS/86/136), United Nations Development Programme, New York, NY.

Lazarus, M., L. Greber, J. Hall, C. Bartels, S. Bernow, E. Hansen, P. Raskin, and D. Von Hippel, 1993: *Towards a Fossil Free Energy Future.* Stockholm Environment Institute, Boston, MA, and Greenpeace International, Amsterdam, Netherlands, 239 pp.

Lovins, A.B., J.W. Barnett, and L.H. Lovins, 1993: *Supercars, The Coming Light-Vehicle Revolution.* Rocky Mountain Institute, Snowmass, CO, 32 pp.

MacKenzie, J.J., 1994: *The Keys to the Car.* World Resources Institute, Washington, DC, 128 pp.

MacKenzie, J.J., R.C. Dower, and D.D.T. Chen, 1992: *The Going Rate: What It Really Costs to Drive.* World Resources Institute, Washington, DC, 32 pp.

Marchenko, D., 1993: Bad roads result in 150 billion losses. *Financial News,* 7 May, Moscow, Russia.

Martin, D.J. and R.A.W. Shock, 1989: *Energy Use and Energy Efficiency in UK Transport up to the Year 2010.* Energy Efficiency Series No. 10, Energy Efficiency Office, Department of Energy, HMSO, London, UK, 410 pp.

Maslow, A., 1954: *Motivation and Personality.* Harper and Row, New York, NY.

Metz, N., 1993: Emission characteristics of different combustion engines in the city, on rural roads and on highways. *The Science of the Total Environment,* **134 (1993)**, 225-235.

Midgley, P., 1994: *Urban Transport in Asia: An Operational Agenda for the 1990s.* World Bank Technical Paper Number 224, The World Bank, Washington, DC, 98 pp.

Mogridge, M.J.M., 1983: *The Car Market.* Psion, London, UK, 208 pp.

NHTSA, 1994: *Data on Diskette.* U.S. Department of Transportation, Washington, DC.

NRC, 1992: *Automotive Fuel Economy: How Far Should We Go?* National Academy Press, Washington, DC, 259 pp.

New Zealand Ministry of Commerce, 1994: *Role of CNG in Contributing to the Government's Energy Policies.* Internal Report, Ministry of Commerce, Wellington, New Zealand.

Newman, P. and M. Kenworthy, 1990: *Cities and Automobile Dependence.* Gower, Aldershot, UK, 388 pp.

NOVEM, 1992: *Transport Policy, Traffic Management, Energy and Environment.* Consultants Report to the IEA, Paris, by Netherlands Agency for Energy and the Environment, Utrecht, Netherlands, 35 pp.

OECD, 1993a: *OECD Environmental Data: Compendium 1993.* OECD, Paris, France, 324 pp.

OECD, 1993b: *Maritime Transport 1992.* OECD, Paris, France, 184 pp.

OECD, 1993c: *Choosing an Alternative Transportation Fuel.* OECD, Paris, France, 169 pp.

OTA, 1991: *Improving New Car Fuel Economy: New Standards, New Approaches.* OTA-E-504, OTA, Washington, DC, 115 pp.

Orfeuil, J.-P. 1993. *Eléments pour une Prospective Transport, Énergie, Environnnement.* Institut National de Recherche sur les Transports et leur Sécurité, Arceuil, France.

O'Rourke, L. and M.F. Lawrence, 1995: Strategies for goods movement in a sustainable transport system. In: *Transportation and Energy: Strategies for a Sustainable Transportation System* [Sperling, D. and S. Shaheen (eds.)]. American Council for an Energy-Efficient Economy, Washington, DC and Berkeley, CA, pp. 59-76.

Ott, R. 1993. Traffic in Zürich. In: *International Conference on Travel in the City: Making It Sustainable.* OECD/ECMT, Paris, France, pp. 135-144.

Oum, T.H., W.G. Waters II, and Jong Say Yong, 1990: *A Survey of Recent Estimates of Price Elasticities of Demand for Transport.* Policy, Planning and Research Working Paper WPS 359, The World Bank, Washington, DC, 56 pp.

PADECO Co. Ltd., 1993: *Non-Motorized Vehicles in Asian Cities.* Part I, *Inventory of Needs and Opportunities (Main Report).* Draft Final Report prepared for The World Bank, June 1993, The World Bank, Washington, DC.

Pischinger, R. and S. Hausberger, 1993: *Measures to Reduce Greenhouse Gas Emissions in the Transport Sector.* Institute for Internal Combustion Engines and Thermodynamics, Technical University, Graz, Austria, 107 pp.

Polak, J.W. and S. Meland, 1994: An assessment of the effects of the Trondheim Toll Ring on travel behaviour and the environment. In: *Towards an Intelligent Transport System: Proceedings of the First World Congress on Applications of Transport Telematics and Intelligent Vehicle-Highway Systems, Volume 2.* ERTICO, Artech House, London, UK, pp. 994-1008.

Polak, J.W., P. Olszewski, and Y.-D. Wong, 1994: Long term effects on travel behaviour of the Singapore area licensing scheme. In: *Proceedings, 7th International Conference on Travel Behaviour.* Imperial College, London, UK, 11 pp.

Prigent, M., G. de Soete, and D. Dozière, 1991: The effect of aging on nitrous oxide N_2O formation by automotive three-way catalysts. *Catalysis and Automotive Pollution Control,* **II**, 425-436.

Quinet, E., 1994: The social costs of transport: evaluation and links with internalisation policies. In: *Internalising the Social Costs of Transport* [European Conference of Ministers of Transport]. OECD, Paris, France, pp. 31-76.

Rabinovitch, J., 1993. Urban public transport management in Curitiba, Brazil. *Industry and Environment* [United Nations Environment Programme, Paris, France], **16(1-2)**, 18-20.

Ramjerdi, F., 1992: *Impacts of the Cordon Toll in Oslo-Akershus, Based on a Panel Study of 1989-1990.* TRU-0257-92. Transport Economics Institute, Oslo, Norway, 87 pp.

Rigaud, G., 1989: Influence du prix de l'energie petroliere sur le marche des transports. In: *14th Congress of the World Energy Council, Sept 1989, Montreal.* World Energy Council, Montreal, Canada, Session 3.5, Paper 2, 23 pp.

Rosenschein, A.D. and D.O. Hall, 1991: Energy analysis of ethanol production from sugarcane in Zimbabwe. *Biomass and Bioenergy,* **1(4)**, 241-246.

Salomon, I.P., P.H.L. Bovy, and J.-P. Orfeuil (eds.), 1993: *A Billion Trips a Day. Tradition and Transition in European Mobility Patterns.* Kluwer Academic Publishers, Dordrecht, Netherlands, p. 40.

Sathaye, J. *et al.,* 1989: Promoting alternative transportation fuels: the role of government in New Zealand, Brazil and Canada. *Energy,* **14(10)**, 575-584.

Schäfer, A., L. Schrattenholzer, and S. Messner, 1992: *Inventory of Greenhouse-Gas Mitigation Measures: Examples from the IIASA Technology Data Bank.* Working Paper WP-92-85, International Institute for Applied Systems Analysis, Laxenburg, Austria, 66 pp.

Schäpertons, H. *et al.,* 1991: *VW's Gasoline Direct Injection (GDI) Research Engine.* SAE Technical Paper Series 910054, Society of Automotive Engineers, Warrendale, PA, 7 pp.

Schipper, L., M.J. Figueroa, L. Price, and M. Epsey, 1993: Mind the gap: the vicious circle of measuring automobile fuel use. *Energy Policy,* **December**, 1173-1190.

Schipper, L.J. and W. Tax, 1994: Yet another gap? *Transport Policy,* **1(1)**, 1-20.

Shaldover, S.E., 1993: Potential contributions of intelligent vehicle/highway systems (IVHS) to reducing transportation's greenhouse gas production. *Transportation Research,* **27A(3)**, 207-216.

Sperling, D., 1995: *Future Drive: Electric Vehicles and Sustainable Transportation.* Island Press, Washington, DC, 175 pp.

Sperling, D.Z., L.J. Schipper, and M.A. DeLuchi, 1992: Is there an electric vehicle future? In: *The Urban Electric Vehicle.* OECD, Paris, France, pp. 373-380.

Suchorzewski, W., 1993: The effects of Warsaw's rising car travel. *The Urban Age,* **2(1)**.

Tanja, P.T., W.C.G. Clerx, J. van Ham, T.J. de Ligt, A.A.W.G. Mulders, R.C. Rijkeboer, and P. van Sloten, 1992: *EC Policy Measures Aiming at Reducing CO_2 Emissions in the Transport Sector.* TNO Policy Research, Delft, Netherlands, 255 pp.

Tengström, E., 1992: *The Use of the Automobile: Its Implications for Man, Society and the Environment.* Swedish Transport Research Board, Stockholm, Sweden, 142 pp.

UNESCAP, 1989 to 1992: *National Reports for China, India, Indonesia, The Republic of Korea, Lao Republic, Malaysia, Maldives, Myanmar, Nepal, The Philippines, Sri Lanka, Thailand and Vietnam.* United Nations Development Programme, New York, NY.

UNESCAP, 1991: *Sectoral Energy Demand in India.* RAS/86/136, United Nations Development Programme, New York, NY.

Wade, J., C. Holman, and M. Fergusson, 1994: Passenger car global warming potential: current and projected levels in the UK. *Energy Policy,* **22(6)**, 509-522.

Walls, M.A., A.J. Krupnick, and H.C. Hood, 1993: *Estimating the Demand for Vehicle-Miles-Travelled Using Household Survey Data. Results from the 1990 Nationwide Personal Transportation Survey.* Working Paper, ENR 93-25, Resources for the Future, Washington, DC.

Walsh, M.P., 1993a: Highway vehicle activity trends and their implications for global warming: the United States in an international context. In: *Transportation and Global Climate Change* [Greene, D.L. and D.J. Santini (eds.)]. American Council for an Energy-Efficient Economy, Washington, DC and Berkeley, CA, 357 pp.

Walsh, M.P., 1993b: Global transport scenarios. In: *Towards a Fossil Free Energy Future: The Next Energy Transition.* Greenpeace International, Amsterdam, Netherlands, Technical Annex, 93 pp.

Wang, M.Q., 1995: Emission reductions of alternative fuel vehicles: implications for vehicle and fuel price susbidies. In: *Transportation and Energy: Strategies for a Sustainable Transportation System* [Sperling, D. and S. Shaheen (eds.)]. American Council for an Energy-Efficient Economy, Washington, DC and Berkeley, CA, pp. 117-138.

Ybema, J.R. and P.A. Okken,1993: *Full Fuel Chains and the Basket of Greenhouse Gases. Integrated Analysis of Options to Reduce Greenhouse Gas Emissions Related to Energy Use in the Netherlands.* ECN-C—93-050, ECN, Petten, Netherlands, 79 pp.

22

Mitigation Options for Human Settlements

MARK D. LEVINE, USA

Principal Lead Authors:
H. Akbari, USA; J. Busch, USA; G. Dutt, Argentina; K. Hogan, USA; P. Komor, USA; S. Meyers, USA; H. Tsuchiya, Japan

Lead Authors:
G. Henderson, UK; L. Price, USA; K.R. Smith, USA; Lang Siwei, China

CONTENTS

EXECUTIVE SUMMARY

This chapter provides a summary of current knowledge of options for reducing the emissions of greenhouse gases (GHGs) in human settlements. The largest portion of GHG emissions in human settlements is in the form of carbon dioxide (CO_2) from energy use in buildings (including emissions from power plants that produce electricity for buildings), amounting to about 1.7 billion tons of carbon. The other three major sources of GHG emissions are methane from urban solid waste (equivalent to 135–275 million tons of carbon in the form of CO_2), methane from domestic and industrial wastewater (200–275 million tons of carbon equivalent), and a variety of GHGs produced through the combustion of biomass in cookstoves throughout the developing world (estimated to be 100 million tons of carbon equivalent). This chapter does not cover CO_2 emissions from the combustion of biomass; this complex topic is treated in Chapters 15 and 24.

Regarding CO_2, we note the following:

- Residential buildings contributed 19% of total global emissions of CO_2 in 1990.
- Commercial buildings contributed an additional 10% of total global emissions of CO_2 in 1990.
- Industrial (primarily OECD) countries produced 63% of global CO_2 emissions in 1990; 19% came from the former Soviet Union and Eastern Europe and 18% from the developing world.
- Overall growth in emissions of CO_2 from buildings was slightly over 1% per year from 1973 to 1990. Almost all of this growth took place in the developing world and in the former Soviet Union and Eastern Europe.

Key Factors Affecting the Growth of Greenhouse Emissions

The potential for the greatest growth in CO_2 emissions—in both percentage and absolute terms—is in the developing world, where per capita energy consumption in human settlements is very low. Even in the industrialized countries, however, if policies to minimize such emissions are not enacted and rigorously carried out, the already high levels of CO_2 emissions can increase. The most significant factors influencing the growth of GHG emissions in human settlements are likely to be efficiency of energy use, carbon intensity of fuels used directly in human settlements or to produce electricity, population growth, the nature of development in the developing world, the nature and rate of global economic growth, and implementation of policies that are directed toward fulfilling national commitments to reducing GHG emissions.

Technical and Economic Potential for Reducing GHG Emissions

Many cost-effective technologies are available to reduce energy consumption and hence CO_2 emissions. Some examples include more efficient space-conditioning systems; improved insulation and reduced air leakage in windows, walls, and roofs—leading to reduced heat losses; and more efficient lighting and appliances (refrigerators, water heaters, cook stoves, etc.). In addition, measures to counter trends toward higher ambient temperatures in urban areas through increased vegetation and greater reflectivity of roofing and siding materials can yield significant reductions in space-cooling energy requirements in warm climates. Finally, technologies for capturing methane gases and converting them to useful purposes exist and are cost-effective in many applications. Other technologies that can reduce or prevent the formation of methane (e.g., in landfills) are also increasingly available.

Policy Options

Policy options for reducing the growth of carbon emissions from human settlements include energy pricing strategies, regulatory programs, utility demand-side management programs, demonstration and commercialization programs, and research and development. Each type of program has been carried out, primarily in industrialized countries, and many have achieved significant energy savings.

Because there has been considerable experience with these policies in many industrialized countries and because the technical and economic potential for energy savings is still high even after many years of improved energy efficiency, improvements in energy efficiency will be possible for many years. The developing world has even greater opportunities to improve energy efficiency. However, resources need to be made available (especially collaborations in training activities and institution building) so that these countries can develop the expertise to bring about higher efficiency.

Energy Scenarios: Potential Impacts of Energy Efficiency

Many "business-as-usual" energy scenarios postulate a 2% annual growth in buildings' energy use—much like that observed during the past several decades. Under the assumption that developing-country economies continue their growth at current rates and that industrialized economies continue their growth at a slower rate than in the past, this 2% annual growth in energy use in buildings assumes significant continued improvement in energy efficiency.

Aggressive energy-efficiency scenarios for buildings show reductions in overall energy demand growth of 0.5 to 1.0% per year. Over 35 years (1990 to 2025), a 2% annual growth rate leads to a doubling of energy use; 1.5% per year leads to a 68% increase; and 1% per year leads to a 42% increase. Thus, energy efficiency alone could contribute about half of the reductions needed to maintain 1990 levels of CO_2 emissions. Such energy efficiency scenarios, however, will require strong and significant policy measures, well beyond what has been adopted to date. In addition to energy and economic policies, the growth of CO_2 emissions will also depend critically on worldwide population growth.

During the same period (through 2025), technologies for control of methane emissions from landfills and wastewater can achieve relative emissions reductions comparable to those of energy-efficiency measures.

Scenarios for the longer term that are aimed at increasing the efficiency of energy use while providing needed energy services suggest that radical transformations in the ways energy is used are possible. A plausible case can be made for a society that meets human needs and aspirations that is not nearly as energy and resource intensive as today's society.

22.1. Introduction

We define human settlements as cities, towns, villages, and even sparsely settled rural dwellings and collections of dwellings. We are concerned with GHG emissions attributable to CO_2 emissions resulting from the production and distribution of fossil fuel and electricity needed for all energy-using activities that take place within residential buildings, CO_2 emissions from energy use in commercial buildings, and the release of methane gas to the atmosphere by processes currently used for waste disposal. This chapter excludes all transport and industrial sources of GHG emissions because these are treated in other chapters.

First we describe the nature of GHG emissions from human settlements to better understand what actions could stabilize or reduce these emissions. Next we review estimates of GHG emissions from human settlements; describe the key factors that affect the growth of these emissions; assess the technical and economic potential for reducing these emissions; and describe policy measures at the local, national, and international levels that can help reduce emissions. We have grouped countries into three categories: industrialized countries (OECD and several newly industrialized countries), the former Soviet Union and Eastern Europe (FSU/EE), and developing countries.

22.2. Historic Trends in GHG Emissions from Human Settlements

Carbon dioxide from energy use in buildings amounts to about 1.7 billion tons of carbon. The other three major sources of GHG emissions are methane from urban solid waste (equivalent to 135–275 million tons of carbon in the form of CO_2), methane from domestic and industrial wastewater (200–275 million tons of carbon equivalent), and a variety of GHGs produced through the combustion of biomass in cookstoves throughout the developing world (estimated to be 100 million tons of carbon equivalent).

22.2.1. CO_2 Emissions

Figure 22-1 shows trends in CO_2 emissions associated with residential and commercial consumption of fossil fuels for the three groups of countries. The data do not include emissions of CO_2 from biomass fuels; although their use in the residential sector of many developing countries is considerable, past and current consumption levels are very uncertain.[1] Further, it is difficult to determine what fraction of biofuels consumption represents net emissions. The IPCC Draft Guidelines for National Greenhouse Gas Inventories require that net CO_2 emissions from burning biomass fuels be treated as zero because these releases are considered within the category Land-Use Change and Forestry.

The majority (63%) of (non-biomass) global CO_2 emissions from the residential/commercial sector in 1990 came from the industrialized countries. The developing countries and FSU/EE each accounted for about 19% of the total, with those shares rising since the 1970s. Whereas CO_2 emissions from the industrialized countries in 1990 were at approximately the same level as in 1973, emissions from the developing countries and FSU/EE have grown because of population growth and an increase in per capita levels of energy services. The growth in FSU/EE emissions reflects a change in heating technology from stoves to district heat.

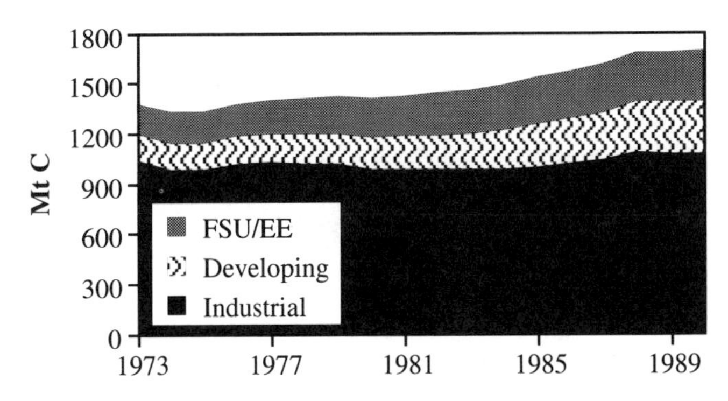

Figure 22-1: Global CO_2 emissions from residential/commercial energy consumption (Scheinbaum and Schipper, 1993; Cooper, 1993; Meyers *et al.*, 1993b).

Figure 22-2 shows that residential buildings account for two-thirds of total CO_2 emissions from buildings in nine industrialized countries. Whereas emissions from energy use in residential buildings in 1990 were slightly below their 1973 level, emissions from commercial buildings have risen. This increase reflects both the growth in floor space and the important role played by electricity.

Estimates for 12 major developing countries for 1985 show the residential sector accounting for 70% of total residential/commercial CO_2 emissions (Sathaye and Ketoff, 1991). Because of uncertain definitions of residential and commercial building space in developing countries, these figures are highly uncertain. The residential sector is estimated to have used 75% of final residential/commercial energy consumed in the former Soviet Union in 1988 (Cooper and Schipper, 1991).

Figure 22-2 also shows emissions for residential energy disaggregated by end-use (Scheinbaum and Schipper, 1993). Space heating accounts for more CO_2 emissions than any other end-use, but its share of the total has declined. Energy use for heating per square meter of living area has decreased considerably in North America and Western Europe since 1973, mainly because of improvements in existing homes and the entry of new, more energy-efficient homes and heating equipment into the stock (Schipper and Meyers, 1992). The share of electric appliances in energy consumption has grown considerably

[1] Hall (1991) derives national estimates of biomass consumption for nearly all developing countries; the total for 1988 for all sectors amounts to 36 EJ, not far below the 47 EJ of commercial energy consumption. Human settlements account for an estimated 80–90% of total biomass use.

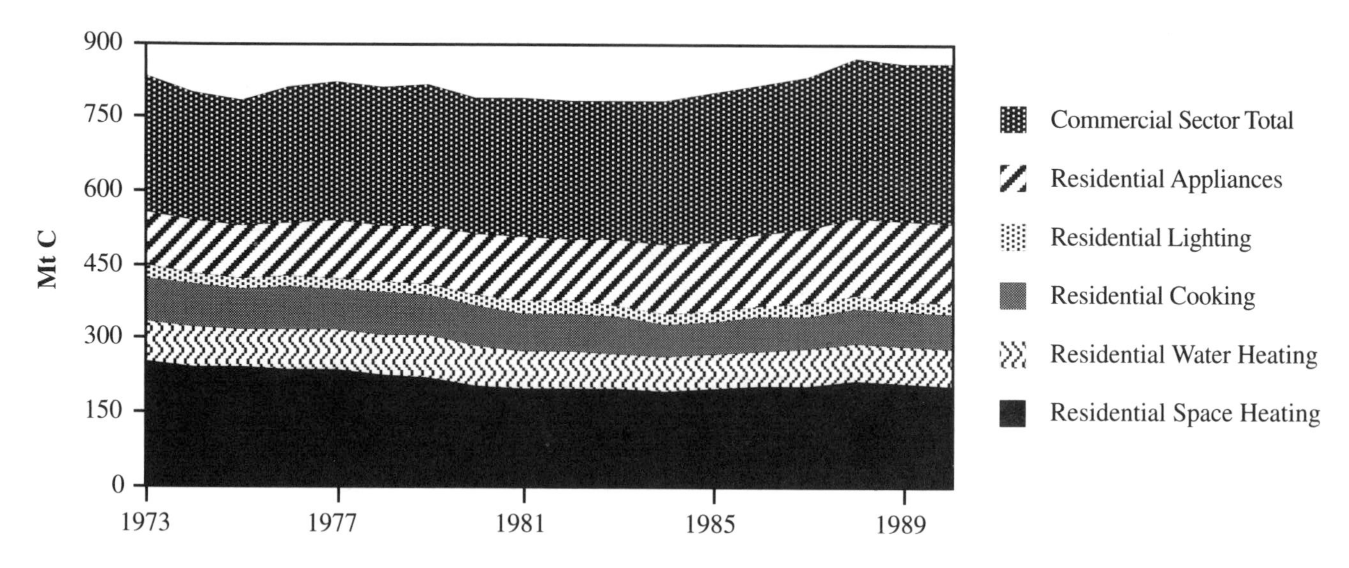

Figure 22-2: CO_2 emissions from residential and commercial energy consumption in the United States, Japan, former West Germany, France, United Kingdom, Italy, Sweden, Norway, and Denmark (Scheinbaum and Schipper, 1993; Cooper, 1993; Meyers *et al.*, 1993b; Meyers, 1994).

since the early 1970s, as rising levels of ownership and increases in the size and features of some appliances have had more impact on energy consumption than have gains in appliance efficiency (Schipper and Hawk, 1991).

The evolution of emissions (or energy use) by end-use for the commercial sector of the industrial countries and for both the residential and commercial sectors of the FSU/EE and the developing countries is less well known. Space heating dominates energy use in both the residential and commercial sectors of the FSU/EE, with an estimated share of around 75% (Cooper and Schipper, 1991). In the developing countries, space heating accounts for only about 20% of total CO_2 emissions, with virtually all of that originating in China (Liu, 1993). Emissions associated with energy use by residential appliances and commercial-sector space conditioning have increased in the past decade.

Using the data and estimates described above, we estimate that the residential and commercial sectors accounted for 19% and 10%, respectively, of global CO_2 emissions from the use of fossil fuels in 1990 (Figure 22-3).

22.2.2. *Non-CO_2 Emissions*

Regardless of the method of harvest, burning of biofuels (and biomass in general) results in net emissions of methane (CH_4), carbon monoxide (CO), nitrous oxide (N_2O), and nitrogen oxides (NO_x) in addition to CO_2. Of these, methane (CH_4) is the most significant because of its larger global warming potential (GWP)—which is 24.5 times that of CO_2—and the large volume of emission compared to the other gases.[2] The data of Hall (1991) on global biofuels combustion indicate that methane emissions from this source in 1988 amounted to approximately 14 million tons. The GWP is about 100 million tons of carbon-equivalent, which compares to global CO_2 emissions from human settlements of approximately 1.7 billion tons carbon (C) (for a 100-year time horizon).

Methane from landfills and from the disposal of domestic and industrial wastewater constitutes the other main source of methane emissions associated with human settlements. Total global methane release from these sources is estimated to be 50 to 80 million tons, as discussed in Section 22.4.4. This amount corresponds to the equivalent of 335 to 535 million tons of carbon-equivalent (using a GWP based on a 100-year time horizon).

22.3. Factors Affecting Future Growth of GHG Emissions from Human Settlements

Future levels of GHG emissions from human settlements will be shaped by four basic factors: population, per capita level of energy-using services, energy intensity of the technologies used to provide those services, and energy sources used by those technologies.[3]

22.3.1. *Population*

Population growth rates depend on a complex mix of cultural, economic, and technological factors, as well as government policies. Declining birth rates are linked to increased economic security, improved health care, and improved opportunities for women, as

[2] We use a GWP of 24.5 for methane based on the IPCC report *Radiative Forcing of Climate Change*, 1995. The GWP corresponds to a GWP of 1 for CO_2 over a 100-year time horizon. In order to convert this into tons of carbon equivalent, the GWP is multiplied by the ratio of the atomic weight of carbon to the molecular weight of CO_2. Thus, to convert tons of methane to tons of atmospheric carbon equivalent, we multiply tons of methane by $24.5 \times 12/44 = 6.7$.

[3] If the energy source is electricity, the level of CO_2 emissions also depends on the mix of fuels used for electricity generation and the efficiency of generation and delivery.

well as access to birth-control techniques. The pace of urbanization, which depends in part on the extent of economic development in rural areas, also will affect birth rates. The willingness of countries to adopt and implement policies that limit population growth will play a major role in affecting future population levels.

Projections of population growth vary with the assumptions made about fertility rates and other factors. Recent projections from the World Bank (1993) show an average annual growth between 2000 and 2025 of 0.3% and 0.4% in the OECD and FSU/EE, respectively, but 1.4% in the developing countries. Among the developing countries, projected growth is much slower in China (0.8% per year)—which has and is expected to continue strict population-control policies—than in middle- and low-income countries (1.5% and 1.7%, respectively).

22.3.2. *Activity Levels in the Residential Sector*

The level of energy-using activity in the residential sector depends greatly on income growth and distribution, household size, and the cost of housing and home appliances. The real cost of most appliances has declined over time, which means that households outside the industrialized countries can acquire them at lower income levels than was the case for industrial-country households in the past. Levels of energy-using activity do not expand linearly with income because saturation of major energy-using equipment begins to appear at higher income levels.

22.3.2.1. Industrialized Countries

The number of households is increasing faster than population because of a decline in household size. Decline in household size will have the largest impact in Japan, which has much larger households today than the other industrialized countries.

Growth in per capita ownership of major appliances, which pushed household electricity use up considerably in the 1970s, will have a much smaller impact in the future because ownership of refrigerator-freezers, freezers, color televisions, and clothes washers is approaching saturation.

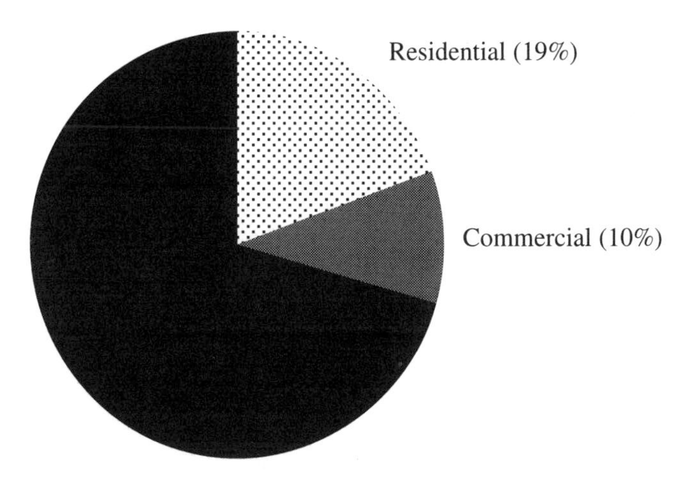

Figure 22-3: Shares of residential and commercial sectors in global CO_2 emissions from energy consumption, 1990. Global emissions in 1990 are estimated as 5.75 billion tons of carbon, using the ratio of 1990 to 1985 global commercial energy consumption, and 1985 global CO_2 emissions given by the IPCC (1991). See text for sources of residential/commercial CO_2 emissions.

22.3.2.2. Developing Countries

The combination of rising population, urbanization, increases in per capita income, and further spread of electrification in rural areas will lead to tremendous growth in demand for residential energy services in the developing countries. Household size will fall with urbanization and decline in fertility rates, as has already occurred in newly industrialized economies in Asia. As in other parts of the world, this decline in household size will increase per capita energy use.

Changes in equipment and fuels will have a major impact on cooking, currently the most important residential end-use in the developing countries. The transition from biomass fuels to kerosene and liquefied petroleum gas (or electricity in some cases) already has occurred to a large extent in urban Latin American households and is proceeding rapidly in urban Asian households (Sathaye and Tyler, 1991). In much of sub-Saharan Africa, on the other hand, the transition to non-biomass fuels has been slowed because of a decline in incomes and fuel distribution problems. The transition will be much slower among rural households that have greater access to biomass resources as well as lower incomes.

Where electricity is not available, kerosene is generally used for lighting, providing much lower lighting levels and consuming far more energy than does electric lighting (Fitzgerald *et al.*, 1991). Thus, electrification will reduce energy use for lighting; at the same time, it will provide an enormous potential for growth in appliance ownership, especially in Asia and Africa. Market penetration of TV sets already is relatively high among electrified households, but penetration of refrigerators is still low (20 to 25%) even in middle-income countries like Thailand and the Philippines—and is lower still in populous countries such as India, China, and Indonesia (Meyers *et al.*, 1990). Refrigerator ownership will grow rapidly as the economies of these countries expand. The other major appliance whose penetration is likely to grow considerably is the automatic clothes washer.

A key uncertainty is the extent to which air conditioning will grow. Its market penetration currently is very low: 2% of homes in the Philippines, 1% in Thailand, and 4% in Brazil. However, the experience of Taiwan, where the proportion of dwellings with air conditioning rose from 12% to 29% of households in the 1980s, suggests that the use of air conditioning could rise in warm climates as households reach upper middle-income levels.

22.3.2.3. Former Soviet Union/Eastern Europe

Overall, increased demand for energy services in the residential sector in the FSU/EE will depend critically on how soon the region can reestablish its various economies. Following fundamental restructuring of the economies and economic recovery, the demand for increased energy services will be largely driven by two factors: increasing housing area per capita (which could rise very rapidly under favorable economic conditions) and increasing demand for appliances. Population is growing slowly, but household size (currently between 3 and 3.5 persons) will fall as more housing is built. House area may grow, particularly if private initiatives lead to increased construction of low-rise and detached (or semi-detached) housing. Central heating penetration will increase, particularly outside of large cities with district heat. Ownership of various electric appliances also will grow somewhat, as will their size and features.

22.3.3. Activity Levels in Commercial Buildings

22.3.3.1. Industrialized Countries

Between 1973 and 1988, the service sector grew 1.3 times as fast as total gross domestic product (GDP) in the United States, 1.4 times as fast in Japan, and 1.8 times as fast in the former West Germany. The service sector is expected to continue to grow more rapidly than GDP in the industrialized countries for the foreseeable future.

Expansion of service-sector floor space is likely to be especially strong in health-care (because of aging populations) and leisure-related buildings (because of increased leisure time among workers and growth in the numbers of retired persons). Both of these sectors are relatively energy-intensive.

22.3.3.2. Developing Countries and Former Soviet Union/Eastern Europe

Large increases in service-sector floor area will occur in the developing countries. Many types of services that have fed the growth of the commercial sector in industrialized countries are just beginning to expand in many developing countries. Hotels and other facilities for tourists also will grow substantially. In the public sector, growth in population will require a substantial increase in education and health-care buildings.

The FSU/EE region currently has less than 5 m^2 of service-sector area per capita—about half the level of Italy (Cooper and Schipper, 1991). Office and retail space will increase considerably to meet the demands of the emerging private sector. Growth also will occur in lodging and restaurants.

22.4. Potential for Reducing GHG Emissions

22.4.1. Residential Buildings

The potential for cost-effective improvement in energy efficiency in the residential sector is high in all regions and for all end-uses. Although a large majority of the net global CO_2 emissions from the residential sector is from industrialized countries, their share of the technical potential for improvement is somewhat smaller for two reasons. First, considerable improvement in the energy efficiency of existing homes, new homes, and equipment in the industrial countries has already taken place over the past two decades, prompted both by higher energy prices and by policies and programs. Thus, the potential for further cost-effective gains, while still considerable, is less than it was in the past.

Second, growth in both the number of households and in equipment stocks per household is increasing much faster in the developing countries than in the industrialized countries, and the average efficiency of new equipment is lower because of the need to keep initial costs low. Much of the new equipment uses electricity, and a good share of the growth in ownership is occurring in countries whose power supply is likely to be dominated by coal. Thus, there is much potential to affect future CO_2 emissions by improving the efficiency of residential buildings and appliances in these countries.

22.4.1.1. Space Conditioning

Space conditioning (heating and cooling) accounts for about half of residential CO_2 emissions from North America and Europe; the largest portion of this is heating rather than cooling. In contrast, relatively little energy is consumed for space conditioning in developing countries, with the important exception of China. This is due, in part, to the much warmer climate in the majority of developing countries and, in part, to the lower level of economic development. As these countries develop, air conditioning will consume rapidly increasing amounts of energy.

The energy consumed for space conditioning can be reduced through increases in the thermal integrity of buildings (i.e., improvements in walls, roofs, and windows), increases in the efficiency of space-conditioning equipment, and improved controls for such equipment. Other ways to alter buildings to reduce energy use include reductions in leaks in ducts carrying hot or cold air to the conditioned space (of particular importance in North America), design changes (e.g., the use of passive solar design), and environmental changes (e.g., the adoption of increased shading and wind breaks from trees and other vegetation). Finally, behavioral factors, such as the timing and level of indoor-temperature settings, have a great deal of influence on the energy use of residences.

Switching from high-CO_2 to low- or no-CO_2 energy sources also can reduce CO_2 emissions from space heating. In the industrialized countries, a significant shift to natural gas and nuclear-generated electricity has already occurred during the past 20 years; further movement toward gas is likely in Western Europe. Eastern Europe and China have considerable potential for switching away from inefficient coal stoves to gas or district heating.

22.4.1.1.1. Building thermal integrity

Opportunities to enhance the energy efficiency of a building shell occur throughout a building's lifetime. Prior to construction, designs with proper orientation, adequate insulation levels, overhangs, and high-quality windows will reduce energy use. In construction, proper sealing and adequate and well-distributed insulation will reduce losses through building shells. Lastly, retrofits—such as the addition of insulation, and storm doors and windows, as well as reduction of thermal bypasses and air leakage—often can save considerable energy. Reducing infiltration is particularly important in many climates; techniques now exist to do this while minimizing the impact on indoor-air quality. So-called "low-energy" homes have demonstrated the technical feasibility of reducing heating requirements to very low levels through the use of high levels of insulation, passive solar design techniques, and other measures, but these have achieved only limited market penetration to date.

Low-emissivity (or low-e) coatings (clear coatings added to glass surfaces) allow the transmission of solar radiation into the interior but reduce radiative heat losses (Rosenfeld and Price, 1992). The addition of a low-e coating can reduce the heat losses of a double-pane window by about one-third. Low-e windows cost 10 to 20% more than regular windows and therefore are generally cost-effective. A double-pane window with gas-filled spaces and two suspended reflective films inside can reduce heat loses by 75%. Improvements in window frames—e.g., greater use of thermal breaks to limit conduction losses through the frame—offer another opportunity for energy savings.

A study for the former West Germany that evaluated homes of different vintages in five building types found that, on average, investments that save 40% of baseline heating energy would be cost-effective even when future energy prices are low (Ebel *et al.*, 1990). A 50% reduction in energy expenditure, however, costs considerably more and would be cost-effective only when energy prices are high.

In the United States, recent studies estimate that energy savings of 30 to 35% could be attained between 1990 and 2010 through retrofits in dwellings built before 1975, but only about half of these retrofits would be cost-effective (EIA, 1990; Koomey *et al.*, 1991). The energy-savings potential for dwellings built between 1975 and 1987 is somewhat less than that for those built before 1975. In the United States, an estimated 25% of residential heating and cooling energy use is associated with losses through windows (Bevington and Rosenfeld, 1990). Analysis for the U.S. housing stock suggests that most cost-effective, currently available energy-saving window systems could reduce energy losses through windows by two-thirds (Frost *et al.*, 1993).

Among industrialized countries, Sweden has gone the furthest in institutionalizing high levels of thermal integrity in new homes (Schipper *et al.*, 1985). Adoption of Swedish-type practices, which rely heavily on factory-built components, in the rest of Western Europe and North America would probably bring a reduction of at least 25% in the space-heating requirements of new dwellings relative to those built in the late 1980s (Schipper and Meyers, 1992).

Building practices in Eastern Europe have resulted in energy efficiency that is well below Western levels (Cooper and Schipper, 1991; Meyers *et al.*, 1993b). Although there are no reliable estimates for the magnitude of energy savings that could be achieved by improving the thermal characteristics of building shells, the potential is likely to be much greater than for Western Europe. Particularly large improvements are possible through reducing air infiltration, increasing roof insulation, and improving the performance of windows. Adequate metering and occupant control systems are needed.

In developing countries where the use of air conditioning is rising, improvements in building thermal integrity can reduce cooling requirements. A study for Thailand found that installing 7.5 cm of insulation in the attic of a typical single-family house would reduce air-conditioning requirements 30% (Parker, 1991). For new homes, building design, reflective materials, and landscaping strategies that minimize solar gains in the summer and enhance natural ventilation can eliminate the need for mechanical cooling even in warm climates and reduce cooling loads in hot, humid climates.

China has a large number of households in regions requiring heating, a large potential for growth in heating demand as restrictions on heating are eased, and heavy reliance on and inefficient use of coal. Considerable potential exists to improve building thermal integrity (especially in new buildings). Since indoor temperatures in most homes are lower than desired, some or even most of the savings from efficiency will go toward greater indoor comfort rather than reduced energy use. Even so, increases in wall and ceiling insulation and in the thermal characteristics of windows can reduce energy use by 40% relative to mid-1980s practice, while allowing a considerable increase in indoor temperatures (Huang, 1989). The government has introduced a standard that calls for new buildings in cities to be designed to use 30% less heating energy relative to 1980 practice, with implementation in the early stages (Siwei and Huang, 1992).

22.4.1.1.2. Space-heating equipment and air distribution systems

In 1992, the United States set a minimum efficiency of 78% for new gas-fired, warm-air furnaces. However, units using "condensing" technology, in which the latent heat of water in the flue gas is recovered, are far more efficient—in the range of 90 to 97%. At present, sales of condensing furnaces are 20 to 25% of total sales, even though their price is considerably higher (typically about $600 more than noncondensing units). The cost-effectiveness of these furnaces depends on the climate and the cost of fuel; measured energy savings from condensing-furnace installations in colder climates in the United States yield simple paybacks of 4 to 7 years (Cohen *et al.*, 1991).

Electric air-source heat pumps are about twice as efficient as electric resistance heaters, and technological improvements could increase the coefficient of performance (COP; the heat delivered divided by the energy consumed by the pump) to as high as 5 in moderate climates (Morgan, 1992). Ground-source heat pumps are even more efficient than air-source heat pumps (U.S. EPA, 1993f). Heat pumps are most effective where both heating in winter and cooling in summer are needed.

Recent research has indicated that, for houses in the United States in which hot or cold air from the furnace or air conditioner is carried by ducts in contact with the outside, leaks and thermal losses from the ducts are typically 30 to 40% of the energy carried by the ducts (Modera, 1993). New techniques are under development to treat ductwork in existing and new houses to reduce or eliminate these leaks (Jump and Modera, 1994; Modera *et al.*, 1992; Treidler and Modera, 1994; and Proctor *et al.*, 1993).

In the FSU/EE, the majority of residential complexes are heated through district heating systems, with or without cogeneration. In addition to significant technical efficiency improvements in these often-antiquated heating systems and better insulation of pipes that carry heat to and among buildings and apartment units, energy savings can be realized through improvement of the operation and control of heating systems. Lack of metering and controls for district heat discourages household conservation efforts. Options include repairing inoperative radiator valves and installing thermostats and individual apartment meters. The latter cannot be easily installed in all buildings, however, because heating systems piped in series (where the output of one radiator is the input to another) require the installation of a bypass pipe. Building heat-distribution systems can be improved through controls that adjust hot-water temperature in response to outdoor temperature, shut off the hot water when no space heating is necessary, or set back temperatures at night (U.S. Congress, OTA, 1993b).

In China, 75% of heated residential space is heated by coal-burning stoves, and 25% is supplied through small boilers or district-heating systems (Liu, 1993). The efficiency of central-heating systems in China often is quite low, and there are many opportunities for cost-effective efficiency improvements (Liu, 1993). Conversion from coal-burning stoves to more modern heating systems appears to increase energy use; however, it does greatly reduce indoor air pollution, provide better control of temperature, and greatly reduce personal effort and inconvenience. Thus, trends to modernize heating in urban areas of China could actually increase carbon emissions as compared with direct coal-burning stoves, but that increase can be moderated by investments in efficient equipment.

22.4.1.1.3. Air-conditioning equipment

Efficiency improvements for air conditioners include better internal insulation in the equipment, larger heat exchangers, higher evaporator temperatures, dual-speed or variable-speed compressor motors to reduce on-off cycling, more efficient rotors and compressors, advanced refrigerants, and more sophisticated electronic sensors and controls (Morgan, 1992). A typical central air conditioner in the United States purchased in 1990 was 36% more efficient than a 1976 model (Levine *et al.*, 1992). The most efficient models in the market are 40% more efficient than the average new model; 20% efficiency gains over the average new model are cost-effective in many regions of the United States (Levine *et al.*, 1992; and Koomey *et al.*, 1991).

In the developing countries in particular, considerable improvement in the efficiency of air conditioners is possible through more widespread use of design options that are common in the industrialized countries (Meyers *et al.*, 1990). In Thailand, for example, the average home air conditioner draws 1.6 kW, whereas the best units require less than 1 kW to provide the same cooling capacity (Parker, 1991).

22.4.1.2. Water Heating

As with space heating, water heating is a major end-use, mainly in North America and Europe. Increased insulation of water heaters, electronic ignition of gas water heaters, and higher efficiency gas burners all promise significant savings over conventional technology. Air-source heat-pump water heaters can provide very high efficiency in warm climates; exhaust-air heat pumps (in which the heat from the exhaust air in ventilation systems is pumped into the stored hot water) are another promising option. Ground-source heat pumps (for space conditioning and heating water) can further increase efficiency and allow application into colder climates. In Eastern Europe, separate provision of domestic hot water from space heating—to reduce the enormous summertime losses in large-scale heat systems that provide only domestic hot water—can contribute energy savings.

Although water heating is not a major end-use in developing countries, it is becoming more common. Options similar to those common in the industrialized countries are available, but increased insulation for storage water heaters yields smaller benefits in warm climates. Solar water heaters are cost-effective in many areas, provided demand for hot water is sufficient, but their high initial cost is a barrier.

22.4.1.3. Lighting

In most developing countries, lighting is the most important electric end-use. Expanded use of compact fluorescent lamps

(CFLs), which require 20 to 25% of the electricity of standard incandescent lamps to produce the same light output, could have a major impact there. Studies of the potential for CFLs in India and Brazil have shown that their use would be highly cost-effective from national and utility perspectives (Gadgil and Jannuzzi, 1991). The cost of avoided peak installed electric capacity from the use of CFLs rather than incandescent lamps is as low as 10% and 15% of the cost of new installed capacity for Brazil and India, respectively, but subsidized electricity prices and the high initial cost of CFLs limit their attractiveness to households. The extent to which these lamps spread will depend on programs to overcome the first-cost barrier.

Replacing a kerosene wick lamp with a 16-watt compact fluorescent lamp increases light output 22-fold, while reducing the fuel-use rate by a factor of 8, even taking into account energy losses in generating electricity (van der Plas, 1988). The potential for cost-effective energy savings in electric lamps is especially large for households that do not yet have access to electricity. According to one estimate, some 2.1 billion people (about 35% of the world total) do not have electricity (Efforsat and Farcot, 1994); they consume an estimated 0.29 EJ of kerosene (Dutt and Mills, 1994; Dutt, 1994).

22.4.1.4. Cooking

Cooking is a relatively minor end-use in the industrialized countries and Eastern Europe, but it is the largest home energy use in most developing countries. Its predominant role is caused not only by the low saturation of other equipment and the lack of space-heating demand but also by the low conversion efficiency of the biomass stoves upon which most households rely. Traditional stoves are only 12 to 18% efficient.

The impact of the transition from biomass fuels to kerosene and liquid petroleum gas (LPG) on global warming depends on the source of the biofuels (whether harvested sustainably or not), the magnitude of products of incomplete combustion (PIC) from biomass stoves, and the global warming potential (GWP) of PIC, which can be considerable. PIC from biomass stoves include CO, methane, and a range of volatile organics. As a mixture, the total PIC GWP from a typical biofuel stove is usually greater than the GWP of the same amount of carbon as CO_2. Exactly how much greater depends on whether indirect as well as direct warming effects are included in the GWP calculations, the time horizon used, and the particular PIC mixture. A kerosene stove produces far less PIC; in addition, much less carbon is involved because a kerosene stove is more efficient and the fuel has more energy per carbon atom than does biomass. In a measured experiment based on cooking the same meal on a kerosene stove and a wood stove, the combined GWP of CO_2 (assuming nonsustainable harvest) and PIC (assuming a 20-year time horizon and including indirect warming effects) was about five times greater for a wood stove than for a kerosene stove (Smith *et al.*, 1993). Using a 100-year time horizon would reduce the effects of combusting biomass in a wood stove to about 2.5 times that of a kerosene stove. These factors are highly uncertain because the GWP of PICs is not known accurately (IPCC, 1994).

Fuel used for cooking a standard meal can be reduced by 30 to 40% through improved wood-burning stoves (Leach and Gowan, 1987; Smith, pers. comm.). An additional 50% fuel savings could be realized by switching to a kerosene stove (Dutt and Ravindranath, 1993). Although producing a significant savings, the switch from wood to kerosene reduces fuel use less than expected because some of the fuel savings is taken back through improved cooking (Fitzgerald *et al.*, 1991). Where fuelwood is gathered and traditional cookstoves are homemade, there are no direct economic benefits to the households of switching to kerosene for cooking. In urban areas, or wherever fuelwood is sold, improved fuelwood stoves as well as kerosene stoves are likely to have far lower life-cycle cost (Dutt and Ravindranath, 1993). The diffusion of improved fuelwood stoves often has failed, however, because they did not fulfill cooking requirements or because of cultural factors, especially in rural areas (Piacquadio Losada, 1994).

Biogas, derived from the anaerobic decomposition of crop wastes and dung, is an alternative cooking fuel, either at the household level or at a community level. The production and use of biogas can result in no net CO_2 emissions and overall reduction in methane emissions compared to spontaneous decay. However, production of biogas as a cooking fuel is not economical compared to improved fuelwood or kerosene stoves (Dutt and Ravindranath, 1993).

Ethanol produced from sugar-cane fermentation is an alternative cooking fuel in areas with surplus sugar-cane production. At the lower end of the range of ethanol production costs, around $7/GJ (Ahmed, 1994), ethanol could be cost-competitive with other cooking alternatives (Dutt and Ravindranath, 1993). There are no net emissions of CO_2 from the use of biogas or ethanol.

Solar ovens are used to replace biomass fuels in many developing countries. Use of these ovens reduces the harvesting of carbon-sequestering trees and eliminates GHG emissions from burning wood or substitute fuels, such as kerosene and LPG. Currently, an estimated 100,000 solar ovens are in use in China, and more than 300,000 in India. Other countries with large numbers of solar ovens include Kenya, Costa Rica, Jamaica, Guatemala, Belize, El Salvador, and Honduras.

In the industrialized countries, there is relatively limited potential (10 to 20%) to improve the energy efficiency of primary cooking devices. In these countries, cooking energy use per household is likely to decline somewhat because of the proliferation of small kitchen appliances and changes in cooking habits (e.g., increased use of microwave in place of infrared heating systems).

22.4.1.5. Electric Appliances

Because most home appliances have lifetimes of 10 to 20 years, changes in new appliances shape the intensity of the stock rather quickly. Such changes can be significant in developing countries, where stocks are growing rapidly.

A series of analyses conducted to support setting energy-efficiency standards in the United States established considerable potential for cost-effective efficiency improvement for most major electric appliances (U.S. DOE, 1989a, 1989b, 1990, 1993). For new refrigerators in the United States, the average efficiency (measured in terms of refrigerated volume per unit of electricity consumption) has increased by almost 200% between 1972 and 1994 (Association of Home Appliance Manufacturers, 1995). Energy use has not declined in proportion to the increase in efficiency because of new features such as icemakers and larger model sizes. Standards applied in 1993 reduced electricity use by 28% relative to the average model produced in 1989 (Turiel *et al.*, 1991). Advanced compressors, evacuated panel insulation, and other features have the potential to produce commercial and cost-effective refrigerators that consume half as much electricity per unit volume as those that meet the 1993 standard. Such a refrigerator would consume 20% as much electricity as a typical 1972 U.S. model. Recent studies found similar results for European refrigerators and freezers (Lebot *et al.*, 1991; GEA, 1993).

For clothes washers and dishwashers, analysis for the United States found that design options that reduce energy use by 30% (including energy to heat water) are cost-effective (EIA, 1990). For clothes washers, a change from vertical-axis to horizontal-axis technology would reduce energy use by about two-thirds relative to the baseline (mainly because of much lower use of hot water). Increasing the spin speed during the spin-dry cycle of a clothes washer can reduce drying energy use by 30 to 50% (because much less energy is expended in mechanical water removal than in thermal water removal). Many European washers are horizontal axis and use higher spin speeds than are standard in the United States. For clothes dryers, the cost-effective reduction in energy use is only 15%, but much greater savings (about 70% relative to the baseline) are possible through the use of a heat-pump dryer. The latter has a significantly higher cost, but a prototype has been developed and successfully tested.

In Eastern Europe, inefficient components are currently being used in most domestic appliances. Refrigeration appliances produced in Poland consume substantially more energy than similar models made in Western Europe (Meyers *et al.*, 1993a). Production facilities are in need of substantial modernization to produce higher quality and more-efficient appliances.

In the developing countries, the most important appliance to target for efficiency improvement in the near term is the refrigerator, whose saturation is growing rapidly. In the longer term, air conditioners will be of great importance. The majority of refrigerators and air conditioners sold in the developing world are well below the state of the art, although the lack of test data makes detailed knowledge of efficiency and energy use impossible (Meyers *et al.*, 1990). A study that considered a variety of appliances in Indonesia estimated that the use of best-available cost-effective technology would result in a reduction in energy use of 20 to 30% relative to the 1988 stock average (Schipper and Meyers, 1991).

22.4.1.6. The Overall Potential for Residential Energy Savings

Several studies at the national and regional levels in the industrial countries have developed estimates of future residential energy use if greater use is made of energy-efficient technologies and practices. Estimating cost-effective savings is a difficult task. Technology is just one of many factors affecting energy use, and the effects of technological change may be masked by population increases, demographic shifts, and other factors. Technology is not stagnant; costs, performance, and efficiencies change as technology is improved and refined. The diversity of the building stock, climatic variations, and uncertainty over future energy costs all make estimating the economic potential for energy savings an uncertain exercise. Furthermore, there are several measures of cost-effectiveness, and one can consider different perspectives—such as the consumer, the utility, and society as a whole. Finally, one can vary the values of inputs in economic calculations, notably the discount rate.

Bearing in mind the inevitable uncertainties and effects of the different definitions of "cost-effective," it is nevertheless useful to review estimates of the energy savings that could result from greater use of cost-effective technologies.[4]

One study estimates that, relative to the predicted business-as-usual 2010 consumption, 13% of the energy used in U.S. homes could be saved with cost-effective technologies; 26% could be saved with technically feasible, but not necessarily cost-effective, technologies (EIA, 1990). These estimates are at the low end of the range of cost-effective savings found in other studies, probably partly due to conservative assumptions concerning building shell retrofits.

One assessment of the residential efficiency potential for electrical end-uses in the United States looked in detail at measures for all end-uses and different building types (Koomey *et al.*, 1991). The study estimates 40% cost-effective savings for the year 2010, where savings were calculated from a baseline with energy efficiency fixed at today's levels, current energy prices were used, and no new technology was assumed (see Figure 22-4).[5]

[4] This refers to technologies that save energy at a cost lower than the average cost of energy or the cost of new supply, depending on the measure of cost-effectiveness used. The cost may be calculated using a discount rate of 6–10% real, reflecting the customers' cost of capital in an industrialized country, or 3–4% real if a social discount rate is used.

[5] As noted, the baseline assumes that buildings and appliances existing in 1990 remain at 1990 efficiency levels (no retrofits) and that all new homes and appliances that enter the stock remain at the efficiency of new systems in 1990. Thus, stock turnover reduces average energy intensities in the baseline case even though new and existing devices are frozen at their 1990 efficiencies.

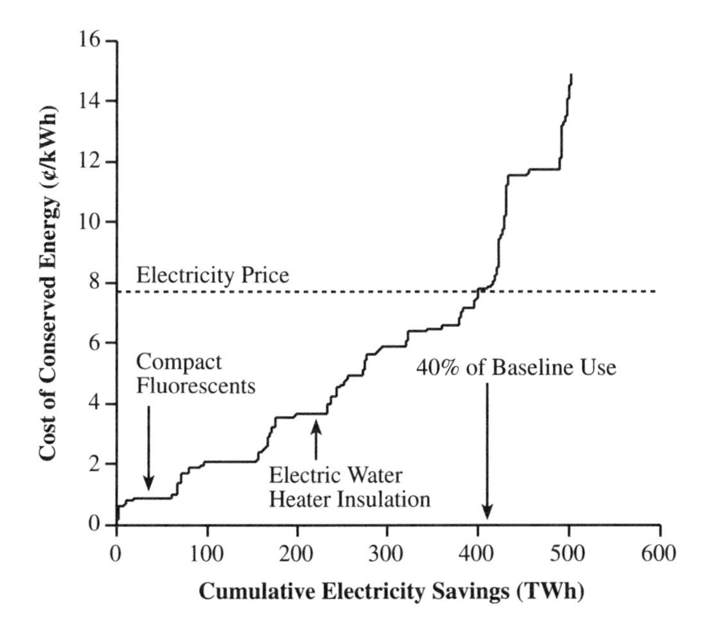

Figure 22-4: Conservation supply curve for electricity in the U.S. residential sector (each "step" represents a specific electricity conservation measure).

Studies conducted in Japan have assessed cost-effective energy savings potential in the year 2010 and thereafter (Japanese Environment Agency, 1992; Tsuchiya, 1990). The greatest energy savings for residential buildings would come from reduction of heating demand through greater insulation in the building envelope and double-glazed windows; provision of space heating and cooling, along with hot water, through multifunction heat pumps; and introduction of solar water heaters. Without energy-efficiency improvements, CO_2 emissions from residential energy use in Japan are projected to increase to 150% of 1990 levels by 2010; with a strong effort to implement cost-effective efficiency measures, CO_2 emissions are projected to be 107 to 123% of 1990 levels. In the longer term (2020 and beyond)—with more time for turnover of building and equipment stock—energy-efficiency measures are projected to be able to bring CO_2 emissions to 97% of 1990 levels.

The energy savings and costs associated with a variety of measures that could be applied for existing and new housing in former West Germany, France, Italy, the United Kingdom, and the Netherlands are evaluated by Krause *et al.* (1994). Cost-effective savings of 40 to 45% in heating requirements are considered possible for gas- or oil-heated residential buildings existing in 1985 that survive to 2020. Cost-effective savings of 28 to 35% are estimated for existing buildings with district heating systems. For residential buildings built between 1985 and 2020, heating energy requirements could be reduced in a cost-effective manner by as much as 70%.

Although the modern sector of developing countries shares many of these opportunities, there is an important difference: The energy consumption in developing countries is much lower but generally has higher rates of growth, because of population increase as well as the increase in energy services per capita. Thus, opportunities are more significant in new buildings and equipment than in retrofits. There also are opportunities unique to developing countries. The overall potential for CO_2 emissions reduction through energy-efficiency improvement and fuel switching is illustrated in Table 22-1 for India. Emissions reductions of 40 Mt (35%) from 1987–88 emission levels are projected. By the year 2010, with a 1.46-fold increase in population and a doubling of the per capita demand for energy services, CO_2 emissions would scale up to 336 Mt with no mitigation but only to 217 Mt with successful mitigation.

Table 22-1: *Potential for CO_2 emissions reductions in India.*

	CO_2 Emissions (Mt) (1987–1988)	**Reduction Potential (%)**	**CO_2 Emissions (Mt) with Mitigation (1987–1988)**
Biomass Combustion	47.6	35	30.9
Kerosene Lamps	8.0	87.5	1.0
Incandescent Lamps	8.4	50	4.2
Fluorescent Lamps	5.3	25	4.0
Fans	7.9	20	6.3
Refrigerators	5.3	50	2.7
Commercial Light	11.8	25	8.9
All Other	20.6	20	16.5
Total	**114.9**	–	**74.5**

Notes: Estimates based on background analysis conducted for Govinda Rao, Dutt, and Philips (1991). Some 16% of all biomass fuels used were logs; this is assumed to be nonrenewable harvesting. Kerosene lamps are assumed to be replaced by fluorescent lamps; the CO_2 emissions of fluorescent lamps are shown in the kerosene-lamp row. Far larger emissions reductions are technically possible with existing technology (e.g., refrigerator energy use can fall by 80% or more, commercial lighting use could be reduced by over 50%, etc.).

22.4.2. *Commercial Buildings*

The commercial sector is especially diverse, even within a single country, with a wide range of building sizes, growth rates, fuels used, operating hours, functions served within buildings, amenity levels, and climates. Larger commercial buildings differ from residential buildings because they tend to have more complex space-conditioning systems (often with mechanical ventilation systems to maintain indoor air quality) and are usually internal-load dominated—meaning that space-conditioning demands arise largely from activities within the building (e.g., people, lighting, and equipment) rather than from exterior conditions. In industrialized countries, commercial buildings tend to be relatively large (e.g., in the United States, 80% of the commercial floor area is in buildings larger than 1,000 m^2) and designed for high amenity levels in terms of thermal and lighting conditions. In developing countries, buildings are smaller, with fewer amenities—though the trend is toward the styles of industrialized countries. Commercial building types include offices, hotels, retail stores, schools, health-care facilities, food-service and sales buildings, warehouses, theaters, museums, and religious buildings. The commercial sector includes buildings in both the public and private sectors.

Electricity is the major form of energy consumed in commercial buildings worldwide, dominating all end-uses except space and water heating and cooking. In the U.S. and OECD commercial sector, electricity meets about 70% of demand in terms of resource energy (EIA, 1994). Within the OECD, the share of electricity (on a resource energy basis) in the commercial-sector fuel-mix ranges from a low of 40% in hospitals in Japan to a high of 80% in hotels and restaurants in Norway.

The major commercial sector end-uses include lighting, space heating, space cooling, ventilation, service water heating, office equipment and other plug loads, refrigeration, and cooking. Data on national average end-use shares of energy in commercial buildings are rare. For the United States, Figure 22-5 shows the end-use breakdown for resource energy (Belzer *et al.*, 1993). In developing countries, the end-use shares for energy differ considerably from those in industrialized countries. For example, lighting is estimated to account for 50% of total electricity in Indian commercial buildings, with space conditioning and refrigeration accounting for 40% and other end-uses the remaining 10% (Nadel *et al.*, 1991).

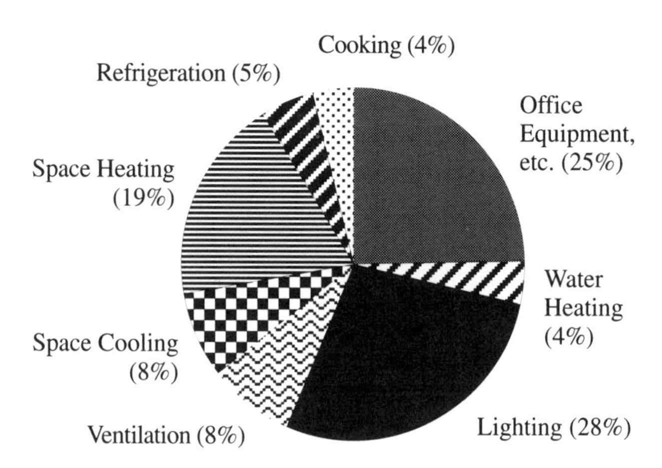

Figure 22-5: Shares of U.S. commercial resource energy by end-use, 1989.

22.4.2.1. *Space Conditioning*

22.4.2.1.1. *Building thermal integrity*

Although larger commercial buildings tend to be internal load-dominated, important savings opportunities nonetheless exist in the design of the building envelope. Windows often represent the largest pathway for heat gain and loss through both conductive and radiative heat transfer. Strategies for reducing energy use associated with windows include: reducing window area, installing external and internal window shading devices, using multiple-pane windows, injecting rare-earth gases between window panes, and applying low-emittance coatings to window glazing surfaces (Sullivan *et al.*, 1992). Low-emittance windows can reduce thermal heat transfer while still transmitting visible light into the interior space, making this technology particularly effective in daylighting applications.

Heat loss and gain through opaque walls and roofs can be reduced by adding insulation and using high solar-reflectance materials for exterior surfaces, though the effectiveness of these measures diminishes for larger buildings with small surface-to-volume ratios.

Shading low-rise commercial buildings by planting trees for sun and wind protection can reduce cooling and heating loads, respectively (Akbari *et al.*, 1992). Through careful window sizing and placement, interior design, and building siting, natural ventilation can be used in some circumstances to augment or supplant the need for mechanical cooling (e.g., Bauman *et al.*, 1992; Boon-Long *et al.*, 1992).

22.4.2.1.2. *Space-heating equipment and air distribution systems*

Systems for heating, ventilating, and air conditioning (HVAC) account for 40 to 50% of electricity use in commercial buildings in some industrialized countries, such as the United States and Germany (IEA, 1989). In other industrialized countries, such as Japan, and in many developing countries, HVAC represents less than 25% of electricity use in commercial buildings.

For commercial buildings, there have been moderate improvements in the efficiency of larger air-conditioning systems during the past decade or so. Levine *et al.* (1992) estimate the potential for cost-effective efficiency improvements in chillers in the United States to be 25 to 38% with currently available technology. Advanced technology—more efficient condensers and improved heat exchangers, wider applicability of large water-cooled chillers, development and application of evaporative cooling techniques, and application of gas-driven chillers—has the potential to increase energy-savings opportunities over the next decades.

For space heating, the typical electric system is either electric-resistance or heat-pump technology, the latter being far more efficient though less common. The typical gas or oil system is either an atmospherically vented furnace or boiler. Improvements in heat exchange to extract the heat of condensation from flue gases, power-venting, and pulsed-combustion technology increase gas-furnace efficiency from a typical 50 to 60% to upward of 90% (Krauss, 1992).

District heating involves the production of heat in a central plant, which is then distributed in the form of steam or hot water to many buildings via underground pipes. Such systems are common in Europe. In Denmark, for instance, district heating is used to meet almost half of the space-heating needs in buildings (U.S. Congress, OTA, 1993b). The efficiency of district heating depends on the system used to produce and distribute the heat. When collected as waste heat from some industrial process or in a cogeneration application, overall efficiency can be very high relative to more traditional on-site space-heating methods.

Energy for air and water transport within buildings can be reduced by employing efficient motors and impeller designs, by good duct and pipe design to reduce static pressure, and by allowing fans and pumps to operate at speeds that closely match thermal loads. With this last approach, a variable-air-volume (VAV) HVAC system with variable-speed drives on the fans is a significant efficiency improvement over constant-volume systems. The savings from VAV range from 30 to 80% (Usibelli *et al.*, 1985). The efficiency advantages of the VAV system have been so well recognized in the United States that more than three-quarters of central systems installed in new buildings are now VAV (Pietsch, 1992). VAV systems are still uncommon in most regions of the developing world.

Greatly increased use of economizers, heat exchangers, and control systems can yield very significant energy savings. Economizers switch the system to the use of outside air when the outdoor temperature is low enough to cool the building. Heat exchangers reclaim heat from exhaust air from space heating, from waste heat in hot-water circulation, or from the ground. Control systems vary from simple zone temperature control with thermostats to comprehensive energy-management control systems that control a variety of systems in the building in addition to HVAC.

22.4.2.2. Water Heating

For many old water-heating systems, combined hot water and space-heating systems are typically highly inefficient. Replacement by stand-alone systems can save 65% (Nadel *et al.*, 1993). The same technologies for improving gas-fired boilers for space heating are applicable for water heating. Electric heat-pump water heaters also are an option for commercial water heating applications, where COPs as high as 5 are available (Abrams, 1992). The heat rejected from food-storage or air-conditioning systems can, under the proper circumstances, be cost-effectively reclaimed for water heating.

22.4.2.3. Lighting

Commercial lighting is provided by three types of systems: incandescent, fluorescent, and high-intensity discharge (HID). As with residential applications, replacing incandescent lamps with CFLs in commercial buildings is a viable savings option and, because of higher usage, can be even more economically attractive in commercial applications.

Fluorescent lighting systems are the most common type of lighting in commercial buildings. Data from OECD countries show fluorescent system market shares for lighting in commercial buildings ranging from 59% in Italy to 90% in Norway (IEA, 1991). In the U.S. commercial sector, fluorescent lamps are in use in 76% of the floor area; incandescent and HID lamps are in use in 19% and 6% of the floor area, respectively (EIA, 1992a). In India, the share of fluorescent lighting is estimated to be around 80% (Nadel *et al.*, 1991).

Five technical options are available to improve fluorescent lamp efficiency: higher surface-area-to-volume ratio, reduced wattage, increased surface area, better phosphors, and reflector lamps (Mills and Piette, 1993).

A lamp ballast is needed to provide a suitable starting voltage, thereafter limiting current flow during operation of fluorescent (and mercury-vapor) lamps. Ordinary magnetic ballasts dissipate about 20% of the total power entering a fixture (Geller and Miller, 1988). In some developing countries, poor-quality ballasts may dissipate as much as 30% (Turiel *et al.*, 1990). More efficient electromagnetic ballasts (also known as core/coil ballasts) make use of better materials to reduce ballast losses to about 10%. Solid-state electronic ballasts cut ballast losses even further and also increase lamp efficacy (lumens of light output per watt of power input) because of high-frequency operation. Such ballasts increase the efficiency of the ballast/lamp system by approximately 20 to 25% relative to that of a system with an ordinary ballast (Verderber, 1988).

Lighting-fixture efficiencies vary from less than 40% to about 92% (Mills and Piette, 1993). Specular reflective surfaces inside a fluorescent lamp fixture can increase the amount of light emitted from a fixture, thereby permitting the removal of lamps (delamping) in retrofit applications or permitting fixtures with fewer lamps to be used in new applications.

HID lamps are used primarily for industrial and outdoor lighting, where very high lighting intensity is required. There are five main types of HID lamps (with efficacy provided in lumens per watt): self-ballasted mercury vapor (20), externally ballasted mercury vapor (38.5), metal halide (54.2), high-pressure sodium (82.5), and low-pressure sodium (136.4) (Levine *et al.*, 1992). Generally, the most common upgrade is to replace mercury-vapor lamps with high-pressure sodium lamps, for which the lighting quality is little changed.

A number of energy-saving lighting controls are now on the market, including multilevel switches, timers, photocell controls, occupancy sensors, and daylight dimming systems. In addition, "task lights" (small lights that illuminate only the work surface) can reduce general lighting needs. These measures typically result in savings of 10 to 15% for photocell controls, 15 to 30% for occupancy sensors, and up to 50% in perimeter zones for daylighting systems (Mills and Piette, 1993; Eley Associates, 1990; and Rubenstein and Verderber, 1990).

Studies of cost-effective energy savings for lighting in commercial buildings in different countries have produced a range of savings estimates: 35% for the United States (Atkinson *et al.*, 1992); between 36 and 86% for five countries in Western Europe (Nilsson and Aronsson, 1993); 70% in Thailand (Busch *et al.*, 1993); 22% in Brazil (Jannuzzi *et al.*, 1991); and 35% in India (Nadel *et al.*, 1991). All of these studies show substantial savings opportunities; results differ in many cases because of differing assumptions.

22.4.2.4. *Office Equipment*

Office equipment accounts for one of the fastest-growing end-uses for energy in commercial buildings. This equipment includes computers, monitors, printers, photocopiers, facsimile machines, typewriters, telecommunications equipment, automatic teller machines, cash registers, medical electronics, and other miscellaneous plug loads. Much of this equipment is left on when not in use during the day, overnight, and on weekends, often consuming energy at close to full load. Laptop computers, which rely on battery power, use special low-power microprocessors and have technology built into the microprocessor that automatically switches it to a low-power mode when not in use, returning quickly to full capability when needed. Recognizing this, the U.S. Environmental Protection Agency (EPA) instituted the "Energy Star" program to promote efficiency improvements first in computers but eventually to include other office electronics. For instance, personal computers with the capability of switching to a low-power mode of 30 watts or less (about 75% less than current models) qualify for the EPA logo that identifies high-efficiency equipment (U.S. Congress, OTA, 1993a). A recent analysis of the impact of changes in the energy efficiency of office equipment in New York state found that today's Energy Star equipment could reduce office equipment energy use by 30% by the year 2000 and by 43% by 2010 (Piette *et al.*, 1995).

22.4.2.5. *The Overall Potential for Commercial Energy Savings*

A number of efforts have been undertaken to combine detailed information about individual technologies for energy saving, as described above, with economic data and information about the existing building stock and planned additions to estimate overall energy-savings potential in commercial buildings in various countries. The analyses have many complexities and uncertainties, including:

- The wide variety of building types among commercial buildings and the difficulty of generalizing to the whole sector
- Differences in opportunities for energy efficiency between new and existing buildings
- Interactions among multiple energy-conservation measures
- Comprehensiveness of the study and criteria used to select measures
- Economic criteria for cost-effectiveness.

As Figure 22-6 shows, a commercial-sector electricity conservation study conducted for the state of New York found economic electricity-savings potential in the commercial sector to be 47% of present consumption from the societal perspective (Miller *et al.*, 1989). At the national level, the Electric Power Research Institute (EPRI) projected from 22 to 49% maximum technical potential for commercial electricity savings (Faruqui *et al.*, 1990). Also at the national level, Rosenfeld *et al.* (1993) found an economic potential of 55% of present electricity consumption in the commercial sector. These savings are for a future year (typically 2020), and they represent savings from current levels of energy efficiency. They are, however, based on current technology and costs, so additional potential savings will be possible over time if energy prices rise in real terms and new energy-efficiency technologies become available.

Such energy-conservation supply-curve studies are much more rare for natural-gas use in commercial buildings; a compilation of assessments by U.S. gas utilities found between 8 and 24% economic potential in their service territories (Goldman *et al.*, 1993).

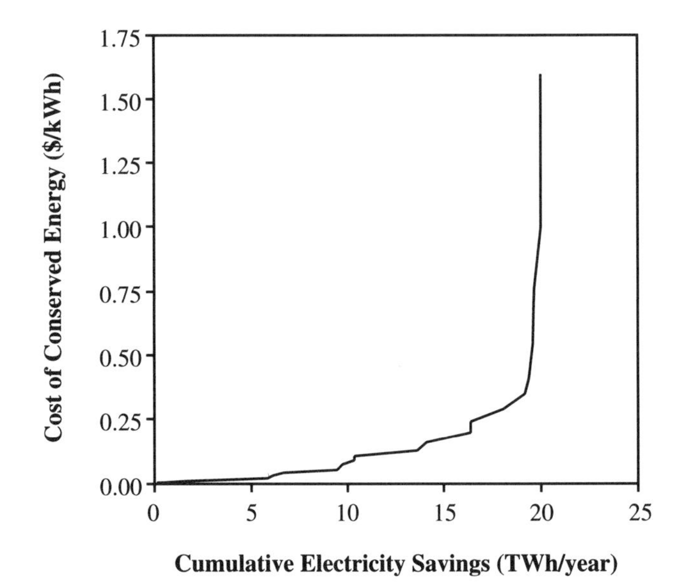

Figure 22-6: Conservation supply curve for electricity in New York state's commercial sector.

In Canada, the economic savings potential for all energy used in offices and retail outlets was estimated at 40 and 28%, respectively (Peat, Marwick, Stevenson, & Kellogg, 1991).

An assessment of energy savings potential for commercial buildings in Japan concluded that energy-efficiency measures, primarily for space conditioning and lighting, could reduce emissions of CO_2 in 2010 from 125% of 1990 levels (without efficiency gains) to 112% of these levels (Japanese Environment Agency, 1992; Tsuchiya, 1990). The longer term potential was about 70% of 1990 levels.

A recent analysis of the Swedish commercial sector found that total electricity use could remain constant through 2010, despite 45% growth in energy services, through implementation of cost-effective retrofit and new construction measures combined with various government and utility incentive programs (Swisher *et al.*, 1994). A Danish project found a large number of electricity-saving measures in commercial buildings (Gjelstrup *et al.*, 1989). One study, which considered lighting, ventilation, pumping, cooling, refrigeration, and other technologies, found that 24% of public-sector (schools, hospitals, etc.) electricity could be saved by behavior and management changes, and an additional 18% reduction was possible with available technologies (Johansson and Pedersen, 1988). Most or all of these savings were thought to be cost-effective. A study of the private service sector found a somewhat lower conservation potential: 10 to 18% through behavior and management changes and an additional 15% reduction through retrofit (Nielsen, 1987).

Among developing countries, a study of economic electricity-savings potential in Brazil estimated 25% cost-effective savings potential in the commercial sector over the next 20 to 30 years (Geller, 1991). In Thailand's large building sector, a comprehensive analysis of efficiency measures in offices, hotels, and retail buildings calculated savings from 45 to 56% (Busch, 1990). Across Southeast Asia, more than 200 audits of existing commercial buildings have indicated average savings from proposed energy-conservation measures between 10 and 20% (Loewen, 1992).

22.4.3. Community-Level Measures: Reducing Urban Heat Islands

The urbanization of the natural landscape—roads, parking lots, bridges, dams, houses, and high-rises—has dramatically altered waters, soils, and vegetation. Replacing vegetation and soil with concrete and asphalt reduces the landscape's ability to lower daytime temperatures through evapotranspiration and eliminates the benefits of shade. The use of dark-colored materials on roads, buildings, and other surfaces creates entire cities that absorb, rather than reflect, incoming solar energy. The combination of reduced reflectivity (called "albedo") and reduced vegetation has resulted in a temperature difference between urban and rural areas that is greatest in late afternoon and early evening, when roads, sidewalks, and walls begin to release the heat they have stored throughout the day.

Throughout the past century, increasing rates of urbanization and industrialization have exacerbated the heat-island effect. Summer temperatures in urban areas are now typically 1 to 3°C higher than in their rural surroundings. Peak temperatures in Los Angeles, for instance, have risen by 3°C in the past 50 years, and mean summertime temperatures in Washington, D.C., have climbed 2°C during the past 80 years. Today, tropical cities are often markedly warmer than their surroundings. Winter nighttime heat islands of 9°C in Mexico City and 6°C in Bombay and Poona, India have been measured (Jauregui, 1984). A comprehensive summary of the heat-island effect in U.S., European, and East Asian cities is provided in Yoshimo (1975, 1991) and Oke (1978).

Heat islands can have either beneficial or detrimental impacts on energy use, depending on geography, climate, and other factors. In warm to hot climates, however, heat islands exacerbate cooling energy use in the summer. For U.S. cities with populations larger than 100,000, electricity use for cooling will increase 10 to 15% for every 1°C increase in temperature (Akbari *et al.*, 1992).[6]

Because urban temperatures during summer afternoons have increased by 1 to 2°C in the past 4 decades, 10 to 20% of the current urban electricity for the cooling of buildings is used to compensate for the heat-island effect alone. Expanded urbanization throughout the world will lead to increasing effects of urban heat islands in the coming decades.

In addition to increasing cooling energy use, heat islands and long-term urban warming affect the concentration and distribution of urban pollution because heat accelerates the chemical reactions that lead to high ozone concentrations [see Akbari *et al.* (1990) for an assessment of effects on ozone levels in Los Angeles].

22.4.3.1. Heat-Island Mitigation

Two factors contributing to urban heat islands can be altered: the amount of vegetation and the solar reflectance of surfaces.

22.4.3.1.1. Vegetation

Trees affect urban climates and building energy use through shading, lowering of windspeeds, and evapotranspiration.

The cooling energy use of two similar sites in Sacramento, California—one with and one without shade trees—was monitored; the shaded site consumed 30% less energy (Akbari *et al.*, 1993). Actual savings at any particular site will depend on

[6] In a typical city in the United States, cooling accounts for 40% of urban electricity use. Thus, total use is increased (reduced) by 4.5 to 6% for every 1°C increase (reduction) in temperature. Peak demand is typically increased (reduced) by about 60% of this amount.

the nature of the shade trees and the climate. In addition, a tree can transpire up to 380 liters of water a day. In a hot, dry location, this produces a cooling effect similar to that of five average air conditioners running for 20 hours (Akbari *et al.*, 1986). In a hot, humid location, however, evapotranspiration is not an effective cooling process.

Considerably greater effects on cooling can be achieved through large-scale tree-planting programs that exert an indirect effect on the community through increased evapotranspiration. Evapotranspiration and shading effects together can reduce air temperatures by as much as 8°C. In Nanjing, China, for instance, after 34 million trees were planted in the late 1940s, average summer temperatures dropped almost 3°C (Garbesi *et al.*, 1989).

22.4.3.1.2. *Solar reflectance of surfaces*

The practice of using high solar-reflectance surfaces to keep buildings and outdoor urban areas cool is not new. In many tropical and Mediterranean countries, particularly those with large amounts of sunshine, the traditional architecture has numerous examples of light-colored walls, roofs, and streets. However, in the many industrialized countries—and increasingly throughout the world—architects and urban planners have overlooked this energy-conscious design principle, relying instead on mechanical air conditioning to maintain comfort during the summer months.

Changing the albedo of a building can reduce direct cooling requirements for houses, single-story industrial buildings, and small commercial buildings significantly. For example, 20% reductions in cooling energy requirements have been estimated for a single-story house (Taha *et al.*, 1988). Measurements made on a sunny day in November in Berkeley, California, showed that the noontime temperature of a dull black surface was 63°C, whereas that of a white acrylic paint surface was only 25°C (Rosenfeld *et al.*, 1994). Similar results would be expected in any location on a hot, sunny day.

Indirect effects (those resulting from the generally lowered temperatures produced by wide-scale increases in albedo) appear to have a larger effect than direct ones. A major program for modifying albedo could reduce urban temperatures by 2 to 3°C, resulting in an estimated reduction of total urban electricity requirements of 8 to 17% and total urban cooling requirements of 21 to 42% (Taha *et al.*, 1988).

22.4.3.2. *Costs and Benefits of Urban Heat-Island Mitigation*

Through direct shading and evapotranspiration, trees reduce summer cooling energy use in buildings at only about 1% of the capital cost of avoided power plants plus air-conditioning equipment (Akbari *et al.*, 1988). Water consumption and tree-maintenance costs are typically a very small fraction of the costs of avoided electricity, except in hot and very dry climates where water is scarce (Akbari *et al.*, 1992). In the latter climates, most of the savings can be achieved through increased albedo combined with drought-resistant vegetation. Increasing the albedo of a city may be very inexpensive if the change is made at the time of routine maintenance. Buildings are typically repainted every 10 years, and high solar-reflectance paint may be used for no additional cost. Similarly, many roof types, including single-ply membrane and built-up roofing, are available in high solar-reflectance materials for no additional cost.

It has been estimated that a nationwide program of planting urban shade trees and lightening surfaces in the United States could, by the year 2015, reduce cooling energy use by 20%. This annual savings of 108 TWh, worth $7 billion (in 1990 dollars), would prevent the emission of 27 Mt of carbon.[7] (For comparison, the recent Climate Change Action Plan for the United States calls for a carbon reduction of 108 Mt by the year 2000.) Applying similar measures to urban areas in wealthy countries in the rest of the world could approximately double these savings to 50 Mt of carbon by 2015 or 2020. As in the United States, such savings would require very aggressive programs for 20 years or longer.

22.4.4. ***Methane Emissions from Waste Disposal and Wastewater***

22.4.4.1. *Magnitude of Emissions*

The anaerobic decomposition of wastes disposed of in landfills or open dumps is a major global source of methane. Recent estimates suggest that 20 to 40 million tons of methane, or about 10% of global methane emissions from human-related activities, are emitted from this source annually (U.S. EPA, 1993a).[8] The large uncertainties in these estimates derive from a lack of information about the amount of organic material actually disposed of in landfills or open dumps by different countries, waste-management practices employed, the portion of organic wastes that decompose anaerobically, and the extent to which these wastes will ultimately decompose (Bingemer and Crutzen, 1987; Orlich, 1990; RIVM, 1993; OECD, 1991; U.S. EPA, 1993a).

Methane emissions result largely from urban areas throughout the world. The industrialized countries of the world contribute about two-thirds of these methane emissions, EE/FSU about 15%, and developing countries about 20% (U.S. EPA, 1993a). Ten countries represent about 60 to 70% of global methane emissions from solid-waste disposal—with the United States

[7] The analysis assumes that 40% of buildings are eligible for increased vegetation and reflectivity. The potential savings from direct effects is about 11%. The indirect effect is based on a 2.2°C citywide cooling and is estimated to be about 12%. The combined direct and indirect effects are then about 20% (Rosenfeld, 1994).

[8] Methane emissions of 20 to 40 million tons per year are equivalent to about 135 to 275 million tons of carbon per year using a GWP of 24.5 for methane, as discussed in Section 22.2.2.

representing about 33%, or around 10 million tons (U.S. EPA, 1993b). With continuing trends in population growth and urbanization, developing countries could account for 30 to 40% of methane emissions from this source by 2000 (U.S. EPA, 1993c).

Methane emissions also result from domestic and industrial wastewater disposal. Recent estimates suggest that 30 to 40 million tons of methane (200 to 275 million tons of carbon in CO_2) are emitted annually from wastewater disposal, primarily from the disposal of industrial wastes (RIVM, 1993; U.S. EPA, 1993a). Regional contributions have not been estimated because of the lack of data.

22.4.4.2. Reducing Methane Emissions

With available technologies and practices, methane emissions may be reduced either by recovering and using the gas or by reducing the source of methane (U.S. EPA, 1993d). The former approach is most suitable for existing solid-waste disposal sites, which often emit methane (often called landfill gas) for 10 to 30 years or more (RIVM, 1993). Frequently, more than 50% of the generated methane can be recovered (Bhide *et al.*, 1990). In many cases, the recovered methane can be used for heating or electricity generation, which is already common in many countries. Although more expensive, landfill gas also can be purified and injected into a natural-gas pipeline or distribution system; there are several such projects in the United States. In Brazil, purified landfill gas has been used as a fuel to power a fleet of garbage trucks and taxicabs. High-rate anaerobic processes for the treatment of liquid effluents with high organic content (sewage, food-processing wastes, etc.) can help reduce uncontrolled methane emissions and are particularly suited to the warmer climates of most developing countries (Lettinga and van Haandel, 1993).

Options for reducing the source of methane emissions include recycling paper products, composting, and incineration (U.S. EPA, 1993d). Paper products make up a significant part of solid waste in industrialized countries (e.g., 40% in the United States) and a growing proportion in some urban centers of developing countries (typically 5 to 20%) (U.S. AID, 1988; Volger, 1984). A variety of recycling processes, differing in technical complexity, can turn this waste into material indistinguishable from virgin products. Composting—an aerobic process for treating moist organic wastes that generates little or no methane—is most applicable in developing countries, where this type of waste makes up a larger fraction of the total; the residue has fertilizer value. As land-disposal costs rise, incineration (often combined with combustion-energy recovery) has become more cost-effective, especially in lower moisture landfills found in industrialized countries.

Existing methane-recovery and source-reduction projects represent a small portion of the potential. Recent studies suggest that global methane emissions from solid-waste disposal can be reduced by about 30% through the widespread use of existing technologies and practices, making both economic and environmental sense (Richards, 1989; U.S. EPA, 1993c, 1993e). A large part of these reductions could take place in the United States, where a pending landfill rule is expected to reduce emissions nationwide by about 6 million tons, or 60%, by the year 2000.[9] Methane recovery is more economical at large waste-disposal sites located close to large urban areas. Fortunately, such sites represent the majority of methane emissions in many countries.

22.5. Policy Options

22.5.1. Buildings

Greater use of available, cost-effective technologies to increase energy efficiency in buildings could lead to sharp reductions in emissions of CO_2 and other gases contributing to climate change. However, policies to promote these technologies often are needed.[10] Consumers and other decision-makers often do not invest in efficiency, even though it appears to offer life-cycle cost savings, for several reasons, including:

- Many buildings are rented, causing a separation between those purchasing energy-using equipment and those paying to operate the equipment.
- Individuals pursue several goals when making energy-related decisions (such as minimizing capital outlays), but very few pursue the goal of minimizing life-cycle or societal costs (see Komor and Wiggins, 1988).
- In new construction, where the greatest and most cost-effective savings are available, the builder does not have to face paying energy bills and often works to minimize initial cost.
- Energy costs are relatively low (e.g., about 1% of salary costs in a typical office in industrialized countries), so those concerned with cost reduction often focus their attention elsewhere.

Analysis of U.S. energy-efficiency choices in the residential sector suggests that markets often behave as if consumers had a discount rate of 50% or more (Ruderman *et al.*, 1987; U.S. Congress, OTA, 1992b). Discount rates implicit in purchasing decisions involving tradeoffs between capital and operating expenses can be 150% or higher among the poorest Indian households.

[9] The landfill rule is designed to reduce emissions of nonmethane organic compounds and air toxics to reduce their adverse air-quality and human-health impacts. It will reduce methane as a side benefit. Greater reductions in methane are expected from this rule than would result if only projects for which reductions in methane releases to the environment were economically viable were undertaken.

[10] For a recent extensive discussion of market failures and energy-efficiency policy issues for the United States, see Levine *et al.* (1994). For a discussion of similar issues related to energy markets in developing countries, see Reddy (1991).

The existence of numerous untapped opportunities for energy savings throughout the world suggests that current market conditions alone will not ensure full implementation of these opportunities. Moreover, energy production and use has significant environmental impacts and other externalities—effects not captured in price.

Yet enthusiasm for policy change must be tempered with a recognition that attempts to increase energy efficiency through regulation or other governmental action may have unanticipated administrative or other costs. Also, current levels of energy efficiency reflect consumer preferences given existing economic incentives, levels of information, and conditions in the market. Finally, there is little consensus on the best policies to promote energy efficiency.

In recent years, many countries have experimented with policies to promote energy efficiency, and evaluation of these efforts has yielded a rich history that can guide future policy efforts. This section describes and summarizes experiences with a range of policy options, including financial incentives, such as energy taxes and rebates; regulations, such as codes and standards; changes in utility regulation; research and development; and improved consumer information.

22.5.1.1. Financial Incentives

A basic policy strategy to motivate greater energy efficiency is to decrease the expense and/or increase the financial benefits of saving energy. Specific options include energy price reform and energy taxes, grants and low-interest loans for building retrofits, rebates for the purchase of energy-efficient appliances, tax credits for energy-efficiency investments, and inclusion of energy efficiency in innovative loans and mortgages.

Energy, especially electricity, is often underpriced; its true costs to society are higher than the price paid by consumers. For example, a World Bank review of energy-pricing policies in the developing world found the average market price of electricity was 4.3¢/kWh, whereas the average cost of new supply was 8.0¢/kWh (Mashayeki, 1990).

Reasons for this often glaring discrepancy between market (consumer) prices and true costs include the following:

- Government agencies set prices in response to multiple goals (provision of basic energy services regardless of ability to pay, promotion of specific policies like industrial development, etc.).
- The marginal cost of electricity varies with system load, and marginal cost pricing is technically difficult.
- There is little agreement on the environmental costs of energy production, distribution, and consumption (EPRI, 1991).

Despite these factors, some countries are attempting to move energy prices to include social costs. In the United States, for example, some utilities are incorporating environmental damage costs, including those caused by CO_2, into planning decisions (CECA, 1993). Many Central and Eastern European countries, in transition from centrally planned to market-driven economies, are instituting large energy price increases in an attempt to close the gap between societal and market prices. Several European countries are considering or implementing carbon taxes on fossil fuels (CECA, 1993).

There also are circumstances in which electricity prices are far above private marginal cost. For example, advanced turbines using natural gas can produce electricity at costs considerably lower than current costs in most utilities in the United States. This has led to a push for major restructuring of the industry in the United States.

Several key points about energy taxes are worthy of note:

- Energy taxes can raise a tremendous amount of revenue, often at relatively low administrative costs. For example, a 0.1 cent (1 mill) per kWh tax on electricity used in buildings in the United States would raise about $1.7 billion annually (EIA, 1993).
- There is only moderate agreement on the effects of price changes on energy consumption or behavior. Estimates of the price elasticity of energy demand (defined as the percent change in energy consumption resulting from a 1% change in price) range from -0.3 to -1.0 for the buildings sector (Energy Modeling Forum, 1981).
- Taxes may have undesired distributional effects. For example, in the United States, low-income households spend a larger fraction of their income on energy services than do higher income households. Therefore, an energy tax would be regressive (USDOC, 1993). Energy taxes also can be an instrument of income redistribution. Many countries tax gasoline and use a part of the revenues for social expenditures, such as universal health care and education. For example, a small tax on urban electricity consumption in Argentina is used to finance the expansion of rural electrification.

Financial incentives to increase energy efficiency—including grants and low-interest loans for building retrofits, rebates for the purchase of energy-efficient equipment, and tax credits for efficiency investments—account for the majority of public-sector expenditures on energy efficiency in many countries. For example, 84% of the U.S. Department of Energy (DOE) budget devoted exclusively to buildings energy conservation (including research and development) is in the form of grants for retrofits to existing buildings; these grants totaled $230 million in 1991 (U.S. Congress, OTA, 1992a). Similarly, electric and gas utilities are spending an increasing fraction of their demand-side budgets on rebates and direct-payment programs.

Several countries have used innovative technology-procurement programs to promote the development and adoption of energy-efficient technologies. This approach is known as market transformation because its intent is to transform the market from one using standard-efficiency technologies to one using highly efficient ones. For example, the Swedish government offered a prize for the development of a window with energy losses half that of a conventional triple-glazed window. To ensure marketability, the contest also required maximum visible light transmission, low weight, and low noise transmission. The window that was developed has a typical payback of about 3 to 4 years compared to a double-pane window and 6 to 8 years compared to a triple-glazed window. The winning company received a cash prize, and additional financial incentives of guaranteed first orders for the new model were offered to promote market penetration of the winning design. The United States used a similar approach to promote the development and commercialization of a highly energy-efficient refrigerator.

Financial incentive programs have achieved mixed results. Some programs, such as direct payments to manufacturers for the production of highly efficient products, appear to be quite successful (Lee and Bennett, 1992). Others, such as tax credits, appear to have limited success (U.S. Congress, OTA, 1992a). Program success appears to be tied to the visibility and timeliness of the incentive. For example, a direct cash payment for the desired behavior (e.g., purchase of a highly efficient appliance) often is more effective than a credit for future energy or tax liability, which can be seen as uncertain and less tangible.

22.5.1.2. Building Codes

An increasing number of countries are requiring minimum levels of energy efficiency in new construction. As of 1992, at least 30 countries had some type of building energy code (Janda and Busch, 1992). These codes range from voluntary goals for certain types of buildings to comprehensive requirements covering all aspects of building energy use.

A recent field study found that about half of the new commercial buildings in the United States have significant energy-code violations, such as incorrectly sized HVAC systems (Baylon, 1992). This results largely from code complexity and the resulting poor understanding of code requirements by designers, contractors, and regulators. New code requirements must be accompanied by training and education, or compliance will suffer. There is some evidence that simpler building codes do receive greater compliance [e.g., the New Zealand code (Isaacs and Trethowen, 1985)].

The actual savings achieved as a result of energy codes depends on the reliability and effectiveness of technologies employed, the level of code compliance, and the manner in which a building is operated after it is occupied. Building commissioning (inspecting energy-using systems and measuring energy use after construction) is one way to help ensure that actual building energy use meets design goals.

22.5.1.3. Appliance Standards

A small but growing number of countries require minimum energy-efficiency levels for energy-using appliances. For example, the United States in 1987 set requirements for residential appliances and in 1992 extended standards to many commercial building appliances (U.S. Congress, OTA, 1992a). The European Community (EC) is considering standards for residential appliances (Lebot *et al.*, 1992; GEA, 1993).

Appliance standards eliminate the least-efficient new appliances by setting minimum energy-efficiency levels for new units. In the United States, appliance efficiency standards already enacted are projected to reduce electricity needs by 22 GW from 1990 to 2015 (Levine *et al.*, 1994).

22.5.1.4. Utility Regulation

Energy production, conversion (such as coal into electricity), and transportation for use in human settlements is dominated by government-operated or government-regulated entities because of the large capital requirements of energy supply systems, the importance of these systems to national security, and the natural monopoly status of many energy-distribution systems. For energy used in buildings (predominantly electricity and natural gas), service typically is provided by government-owned or regulated utilities.

In the past, these utilities viewed their role as providing dependable electric and gas supplies at a reasonable cost; they were not involved in how the energy was used. In recent years, however, uncertainty over future demand, plant siting constraints, environmental regulations, and other concerns have put increasing pressure on utilities to better plan their future capacity needs. One result of these forces is the emergence of a concept of utility planning called integrated resource planning (IRP). IRP has been an especially important aspect of utility planning in the United States during the past decade.

A basic tenet of IRP is that consumers do not require energy *per se* but rather energy services (e.g., lighting, heating, and cooling) and are therefore best served if these services are provided at the lowest overall cost. For example, it may be less expensive for a utility to install energy-efficient lights in offices than to build a new power plant to meet the demand of less-efficient lights. The service provided is the same, but the overall cost to provide it may be lower. IRP is the process of evaluating demand and supply options together to determine how to meet energy-service needs at the lowest cost.

Beginning in the late 1980s, changes in utility regulation in the United States to promote or require utilities to perform IRP led many utilities to invest in energy efficiency rather than in new energy supplies (Krause and Eto, 1988). These regulatory changes included limiting disincentives for efficiency—for example, by decoupling revenues from sales—and providing positive incentives for efficiency. In response to these changes, in 1992 U.S. utilities invested $2.36 billion in demand-side management (DSM) programs—which include energy efficiency as well as peak shifting, peak reduction, and other activities intended to affect the timing or amount of customer energy use (Hirst, 1994). This investment was directed at a wide range of programs, including rebates to utility customers for the purchase of energy-efficiency measures, information programs providing audits and technical assistance, research and development, and funds for energy-service companies that bid against one another to provide energy-efficiency measures to customers (who typically pay a portion of the costs of the demand-side measures).

By one estimate, utility-run DSM programs in the United States led to national reductions in electricity peak demand of 3.7 to 4.2% from 1988 to 1990—about 20 GW of summer on-peak demand (EPRI, 1990). The cost-effectiveness of these investments is somewhat uncertain. A recent analysis of the costs and measured energy savings of 20 utility DSM programs for lighting, representing a total investment of $250 million, estimates the cost of the avoided electricity at 4.7 ± 1.9¢/kWh, compared with 6.9¢/kWh for average current U.S. electricity prices (Eto *et al.*, 1994). In many cases, DSM is less expensive than traditional supply-side options, even in markets such as the United States in which considerable effort has been undertaken by means other than utility programs to increase energy efficiency.

More recently, forces that are leading to deregulation of the utility industry in the United States have reduced U.S. utilities' interest in promoting DSM. The future of large-scale utility DSM programs in the United States is much less certain as a result of possible regulatory changes, although various approaches have been proposed to continue such activities even in a much more deregulated environment. Because the United States has been a leader in this area, many other countries are closely watching these developments, and their own efforts are likely to be influenced by the evolution of utility DSM in the United States.

22.5.1.5. Research and Development

Research and development (R&D) is the process that generates and refines new energy-efficient technologies. In general, only large industries and governments have the resources and interest to conduct R&D. The building industry, in contrast, is highly fragmented. For example, single-family residential construction firms in the United States alone number more than 90,000 (U.S. Congress, OTA, 1992a). This fragmentation makes it difficult for the industry to pool its resources to conduct R&D.

Government-supported R&D has played a key role in developing and commercializing energy-efficient technologies. Low-e windows, electronic ballasts, and high-efficiency refrigerator compressors are examples of widely used technologies whose origins can be traced to public-sector funding of research. Maintaining public support for energy-efficiency R&D (with effective mechanisms to weed out unsuccessful projects and avoid duplication of private investment) will help ensure the availability of the next generation of energy-efficient technologies.

Much of the R&D in industrialized countries is directly applicable to developing countries. Yet there are problems specific to developing countries that do not exist or are not central to industrialized countries. Examples include house design and construction in hot, humid climates for poor people; energy embodied in house construction, which often is greater than the energy used during house operation; and improved fuelwood and kerosene stoves. While researchers in industrialized countries can contribute to these tasks, also needed is an R&D infrastructure based in developing countries, as well as collaboration among different developing countries that share similar problems. Examples of such collaborative efforts include the Foundation for Woodstove Dissemination (based in Nairobi, with regional focal points in various countries) and the International Energy Initiative (with principal offices in Bangalore and São Paulo).

22.5.1.6. Information Programs

Energy information can be imparted in many forms—including labels and rating systems, demonstration programs, energy audits, and workshops. For example, in the United States and Australia, all major energy-using appliances carry energy labels showing estimated annual energy consumption (IEA, 1993).

Evidence is growing that information programs by themselves have only limited effects on behavior or on energy use. One review concludes that "informational programs are not sufficient to induce individuals to engage in resource conserving behaviors" (Katzev and Johnson, 1987). Information programs are built on the premise that people will generally do what is cost-effective if they know what specific opportunities exist. However, consumers and other decisionmakers often define "cost-effective" differently than do analysts, and consumers often lack the incentive or motivation to use energy-efficient technologies. In such cases, information alone will have little effect. Information programs generally are more effective if they are targeted at specific people and specific behaviors; combined with other programs, such as incentives; and evaluated regularly, with the results of these evaluations used to improve the program.

Programs that establish the performance of energy-using equipment often are prerequisites for other policy approaches. Other programs, such as appliance standards, rebates, and building codes, depend on a credible energy rating.

A new type of information program combines site-specific technical information, public commitment, and positive feedback to encourage cost-effective lighting retrofits. The Green Lights program, run by the U.S. EPA, has had considerable success in encouraging large corporations to undertake comprehensive lighting retrofits. This voluntary program offers participating companies positive publicity and free technical support. In exchange, the participants must agree to implement all cost-effective lighting retrofits. By one estimate, this program had reduced electricity demand by 70 MW as of December 1993 (U.S. EPA, 1994). A similar program exists for personal computers: Those meeting the EPA's energy efficiency standard are allowed to be labeled as "Energy Stars" (see Section 22.4.2.4). This is seen as a marketing advantage, and many manufacturers have increased the efficiency of their computers to meet the voluntary standard.

22.5.1.7. Special Policy Options for Developing Countries

Besides policy options shared with industrialized countries, developing countries require training, institution-building, capital, and promotion of rural development to increase energy efficiency in buildings (Munasinghe, 1991).

Training efforts need to address all major elements of energy efficiency and related technologies for reducing GHG emissions, including in-depth studies in engineering, economics, public policy, and management. The training needs to be both theoretical and practical. Much of the training will probably best be done within the developing countries, but substantial assistance from industrialized countries in creating and carrying out training courses will be necessary.

In some cases, new institutions (such as the energy-efficiency centers that have been created in Russia, Eastern Europe, and China) will be needed to coordinate a broad range of activities. Such institutions are needed to make certain that energy efficiency and alternative energy investments are treated equally with traditional supply investments and make it possible to "bundle" a large number of demand-side investments into a large investment so that they can attract capital (Gadgil and Sastry, 1992).

Capital needs to be made available for energy-efficiency projects. This availability of capital depends in large measure on the success of training and institution-building—which will enable developing countries to identify viable projects for investment—as well as on the financial climate of the developing nation. The multinational development banks and the Global Environment Facility—a joint program of the United Nations Development Programme, The World Bank, and the United Nations Environment Programme—can assist greatly in providing large-scale "demonstrations" of energy-efficiency projects in developing countries.

Effective rural development will reduce both migration to urban areas and population growth. Examples of important thrusts for rural development include provision of electricity to all households; improved fuelwood stoves, as well as stoves and fuel supply for kerosene and alcohol; adequate water supply and sewage; adequately matched end-use devices and energy sources (e.g., kerosene stoves but not lamps; biogas for electricity generation but not gas lamps; fluorescent, not incandescent, lamps) (Dutt, 1992); identification of leapfrogging opportunities in appliances (e.g., refrigerators) and other energy-using equipment; and use of high-rate anaerobic processes for wastewater treatment.

There are efforts underway through international organizations, overseas development programs, and developing countries using their own resources (China, Thailand, and Brazil are notable examples) that represent an important start in creating the infrastructure for much more efficient use of energy in developing nations.

22.5.1.8. Summary/Conclusions

Table 22-2 summarizes key characteristics of the options discussed in Section 22.5.1. In addition, several lessons have

Table 22-2: *Key characteristics of selected policy options.*

Policy	New Building Construction	**Can Affect** Existing Building Retrofit	Appliance Selection	**Energy Savings Potential**[1]	**Direct Cost to Government**
Energy Taxes	Yes	Yes	Yes	High	Negative
$ Incentives	Yes	Yes	Yes	High	High[2]
Building Codes	Yes	(3)	(4)	Medium	Low
Appliance Standards	No	No	Yes	High	Low
IRP	Yes	Yes	Yes	High	Low
R&D	(5)	(5)	No	Variable	Medium
Information Programs	Yes	Yes	Yes	Low	Low

[1]Energy savings will, of course, vary; this table shows *potential*, and assumes aggressive implementation (e.g., high energy taxes).
[2]If an incentive is offered by a utility, the direct cost is borne by the utility (which is, in many cases, run by the government).
[3]Some cities, such as San Francisco, require existing buildings to meet energy codes as a condition of change in ownership.
[4]Most code requirements apply to the building shell only; however, some codes apply to appliances as well.
[5]R&D effects are long-term.

been learned from past experience that apply to all options, as follows:

- All policies should be evaluated frequently to determine their effectiveness.
- Many policies work in combination with others. Mutually reinforcing regulatory, information, incentive, and other programs offer the best hope for achieving significant portions of the cost-effective energy-efficiency potential.
- Finally, policy experimentation should be encouraged on a small scale. The process of experimenting, evaluating, improving, and expanding can result in policies that will successfully implement the wealth of technologies now available for reducing the emissions of gases contributing to climate change.

22.5.2. *Heat Islands*

Afforestation efforts or purchases of forested lands to protect them from deforestation have been undertaken to sequester carbon. However, no systematic and comprehensive effort has been made to plant trees or increase the reflectivity of urban areas to mitigate the impact of heat islands. For such an effort to be undertaken, the following steps need to be carried out:

- Create test procedures, ratings, and labels for cool materials.
- Assemble a database on all measures to reduce urban heat islands.
- Incorporate cool roofs and shade trees into building codes in climates where their impact is beneficial.
- Offer utility rebates or other incentives to beat the standards.
- Establish demonstration centers in which the concept of community-wide planting of trees and increased reflectivity is demonstrated to reduce cooling requirements.

22.5.3. *Methane Reduction*

An awareness of the economics of alternatives for methane recovery and use is often lacking among government officials and potential developers in various countries. Experience with techniques for methane recovery and use for power generation also is lacking. Appropriate demonstration projects can help overcome this problem.

Often, many groups are responsible for different aspects of waste management. Any one weakness in the chain of responsibility can cause failures in the overall waste-management system. Furthermore, different groups generally are responsible for energy generation, fertilizer supply, and waste management. In many places, laws and regulations about waste disposal are either unclear, not enforced, or not supportive of measures to collect methane and use it productively.

These problems have been overcome in some communities by organizing joint management groups for waste management, fertilizer supply, and energy generation (U.S. EPA, 1993b). This single entity, which can be private or public, needs to have the capability and authority to deal with these matters simultaneously.

In many developing countries, projects with higher initial costs that are otherwise profitable are not undertaken because of lack of capital for investment (often resulting from uncertain financial institutions that make capital difficult to attract). Also, as noted, many developing countries have highly subsidized energy prices that make methane recovery not cost-effective. Financing of methane recovery (such as that carried out in China under the auspices of the Global Environment Facility) can help solve these problems, as can the deregulation of energy prices.

22.6. Scenarios

The scenarios discussed below focus on GHG emissions from energy consumption in buildings because buildings represent 75% of the emissions treated in this chapter. It is likely, however, that two of the other three sources of GHG emissions from human settlements—methane from landfills and wastewater—could be controlled as much as CO_2 emissions resulting from energy use in buildings. The third source, PICs from burning of biomass in cookstoves, is likely to be largely eliminated over the long term as biomass is converted into or replaced by modern fuel forms.

We focus on two time periods: the intermediate term (2025) and the long term (2075 to 2100). There has been sufficient work on the development of scenarios to make observations regarding the intermediate time period. Comments about the long term are necessarily more speculative.

22.6.1. *Historical Perspective*

Table 22-3 shows total energy use and CO_2 emissions attributed to residential and commercial buildings. This information is presented for the years 1973, 1983, and 1990 for three major regions of the world (OECD, FSU/EE, and developing countries) and for the world as a whole. Table 22-4 is derived from the data in Table 22-3 and shows the annual growth rates of energy consumption and CO_2 emissions for three time periods: 1973 to 1983, 1983 to 1990, and 1973 to 1990. The key points follow:

- Average energy growth in buildings between 1973 and 1990 worldwide was 2.5% per year, compared with an average increase in energy use in all sectors of 2.7% per year during the same period.
- Average annual increase in CO_2 emissions was 1.2% per year. Thus, the carbon emissions increased an average of 1.3% less per year than buildings energy use between 1973 and 1990, indicating an improvement in carbon intensity (carbon emissions per unit of buildings energy use).

Table 22-3: Primary energy use and CO_2 emissions from energy use in residential and commercial buildings (1973–1990).[1]

	Residential		Commercial		Total	
	Energy (EJ)	Emissions (Mt C)	Energy (EJ)	Emissions (Mt C)	Energy (EJ)	Emissions (Mt C)
OECD						
1973	31	694	17	347	48	1041
1983	33	626	20	355	53	981
1990	36	666	25	407	61	1073
FSU/EE						
1973	9	133	3	44	12	177
1983	14	197	6	66	20	263
1990	15	225	8	75	23	300
Developing Countries						
1973	6	109	2	47	7	156
1983	11	154	3	66	13	220
1990	15	225	4	97	19	322
World						
1973	45	936	22	438	67	1374
1983	57	977	29	487	86	1464
1990	66	1116	37	579	103	1695

[1] Primary energy data are from WEC (1995) and supporting documentation for that report. Residential emissions data for industrialized countries are based on extrapolation of data for nine countries (Schienbaum and Schipper, 1993); commercial emissions data for industrialized countries are based on unpublished energy use data for the same nine countries collated from national sources by Lawrence Berkeley Laboratory. Emissions for developing countries are based on extrapolation of energy use data for 16 developing countries based on unpublished energy use data collated from national sources by Lawrence Berkeley Laboratory. Emissions for FSU/EE are based on data for the former Soviet Union (Cooper, 1993) and Poland (Meyers *et al.*, 1993b).

- In the OECD, there was almost no increase in carbon emissions from buildings energy use between 1973 and 1990 (0.2 % per year); however, the annual growth of carbon emissions from energy consumption went from -0.6% (1973 to 1983) to +1.3% (1983 to 1990).
- In the FSU/EE, energy use in buildings increased relatively steadily from 1973 to 1990; however, the rate of growth in carbon emissions from energy used in buildings declined from 4.0% per year (1973 to 1983) to 1.9% per year (1983 to 1990). After 1990, energy use in the region declined because of the economic disintegration experienced as an aftermath of the breakup of the Soviet Union.
- The average increase in energy use in buildings in developing countries from 1973 to 1990 was 5.8%; the annual increase in carbon emissions was 4.4%.
- Overall, residential buildings use about twice as much energy and are responsible for about twice the carbon emissions as commercial buildings. However, annual energy use in commercial buildings has grown over 35% faster between 1973 and 1990, and their annual carbon emissions have grown 60% faster.

22.6.2. *The Intermediate Term (2020–2025)*

None of the well-known global energy scenarios has been devoted only to residential and commercial buildings for the period 2020 to 2025.[11] However, many scenarios have treated all sectors in terms of both energy and CO_2 emissions (IPCC, 1995). We review the following cases: (1) IPCC's 1992 ISb scenario (Pepper *et al.,* 1992); (2) the most recent World Energy Council (WEC) cases (WEC, 1993); and (3) a study for the Global Energy Efficiency Initiative (GEEI) (Levine *et al.,* 1991). We present total energy because these cases do not provide sufficient information to obtain energy used by buildings. We also present an estimate of energy use in buildings and associated CO_2 emissions that are roughly consistent with the three sets of scenarios.[12]

22.6.2.1. Reference Cases

A comparison of the reference cases developed by these three groups reveals surprising similarities among them. The energy

[11] A recent World Energy Council report, not yet widely distributed, provides detailed scenarios for residential and commercial buildings to 2020 (WEC, 1995).

[12] In all of the cases considered, energy in buildings approximately tracks total energy use, as best as can be determined from the information provided. Energy demand growth in residential and commercial buildings is somewhat above that of all end-use sectors for the developing countries and somewhat below that for the industrialized countries. Therefore, we multiply the results for all end-use sectors by 0.30 to obtain estimates of energy use and emissions associated with residential and commercial buildings.

Table 22-4: *Annual average growth rates (%) of primary energy use and CO_2 emissions from energy use in residential and commercial buildings (1973–1990).*

	Energy	Emissions
OECD		
1973–1983	1.0	-0.6
1983–1990	2.0	1.3
1973–1990	*1.4*	*0.2*
FSU/EE		
1973–1983	5.3	4.0
1983–1990	2.2	1.9
1973–1990	*4.0*	*3.2*
LDC		
1973–1983	6.1	3.5
1983–1990	5.4	5.6
1973–1990	*5.8*	*4.4*
World		
1973–1983	2.5	0.6
1983–1990	2.6	2.1
1973–1990	*2.5*	*1.2*

demand in 2025 is between 1.65 and 2.05 times that in 1990, and the average annual growth rate is between 1.5 and 2.0% per year. All three groups produce a 2025 reference case with energy use very nearly double 1990 levels, corresponding to an annual growth rate of 1.9 to 2.0%.

On the one hand, this close agreement among the different groups on an expected reference case for energy in the year 2025 is surprising because of the tremendous number of unknowns about future energy demand. As noted in Section 24.3, a large number of factors influence energy demand growth. The average annual growth in energy demand since 1973 has been about 2.4% per year (with energy in buildings growing at 2.1% per year during the period). Thus, the similarity among the scenarios reveals that the best guess about the future is that it will be much like the past. In this case, it appears that analysts implicitly assume that the next three decades will be much like the two decades since the oil embargo.

The WEC case shows almost the entire energy growth during the period occurring in the developing world. The IPCC and GEEI cases have about 70% of energy growth happening in the developing world. In spite of this concentration of growth, energy per capita in the developing world in all of the reference cases remains a small fraction of that in the industrialized countries. For example, in IPCC reference cases IS92a and IS92b, energy per capita in developing countries grows from 10% of that of industrialized countries in 1990 to 17.5% in 2025.

22.6.2.2. *Efficiency Cases*

WEC provides two energy-efficiency policy cases. One assumes that very significant efforts are made throughout the world to achieve energy efficiency. This case results in a reduction in energy growth by 0.6% per year from the reference case. The other is an extremely high-efficiency case with a "rate of reduction in energy intensity far in excess of anything achieved historically" and a "very low increase in energy demand in the developing countries." In this second case, energy growth diminishes an additional 0.6% per year, resulting in an increase in energy demand of 0.9% per year. In this case, the energy consumption in the OECD declines 25%, that in the former Soviet Union and Eastern Europe remains about the same as in 1990, and that in the developing world doubles.

The GEEI energy-efficiency case reduces annual energy demand growth by 0.73%, yielding an annual increase of 1.1% per year. This case assumes a very significant push for energy efficiency in both the industrialized and the developing countries. In the GEEI efficiency case, energy demand in the OECD is essentially unchanged between the present and 2025, whereas energy demand in the developing world almost doubles.

IPCC does not provide any policy cases. However, the U.S. Environmental Protection Agency produced policy scenarios that are useful in a general way in indicating global energy-efficiency opportunities for 2025 and beyond (U.S. EPA, 1990). The Rapidly Changing World scenarios most nearly approximate the WEC and GEEI cases considered above. Policies promoting energy efficiency reduce the annual growth rate of energy by 0.5% per year, from 1.95 to 1.43% annually, from the present to 2025. In the other major pair of reference and policy cases analyzed by EPA, the Slowly Changing World scenarios, energy demand was reduced by 0.41% per year, but from a much lower growth rate of 1.07% per year.

Although there is tremendous uncertainty about the future demand for energy (IPCC, 1995), the reference case of key analyses hovers around 2% per year energy growth for this time period. Under these assumptions (business much as it has been during the past 2 decades), with considerable emphasis on energy efficiency in these reference cases and most of the energy growth in the developing world, aggressive policies to increase energy efficiency could reduce annual demand growth by 0.5 to 1.0%.

What would be required to maintain carbon emissions from energy at current levels between 1990 and 2025? These studies suggest that, in a reference case with energy demand growing at an average of 2% per year during the period, strong policies to promote energy efficiency could be responsible for about 30% to 60% of the effort needed to achieve zero growth in carbon emissions from energy.[13] The remaining 40 to 70% reductions in carbon emissions would need to come from fuel switching—from fuels of higher to those of lower or zero carbon intensity (i.e., greater use of natural gas, renewable energy sources, or nuclear energy)—to achieve 1990 carbon emissions

[13] A 2% annual energy growth for 35 years will double energy demand; a 1.5% annual growth over this period results in a 68% increase in energy demand; a 1% annual growth increases energy demand by 41%.

in 2020 under the reference-case assumptions. Another way of saying this is that aggressive policy to promote energy efficiency over the next three to four decades could, under typical reference-case assumptions, contribute about half of the effort needed to maintain carbon emissions from energy use at current levels.

22.6.3. The Long Term (2025–2100)

Scenarios for energy demand to the year 2100 vary dramatically (IPCC, 1995). Carbon emissions vary much more than energy consumption because energy of widely varying carbon contents can be used. In reviewing a large number of global scenarios for the year 2100, the IPCC (1995) notes that energy intensity varies by less than a factor of 3. More specifically, the different scenarios calculate (or assume) an energy intensity index varying from about 0.2 to 0.55 for 2100, whereas the index is 1.0 today (the index is a measure of energy use per unit gross world product).

Thus, in this very aggregate view, if gross world product grows to about 4.5 times 2025 levels by 2100 (as assumed in IPCC IS92a and b), then global energy use by 2100 might be expected to grow by a factor of about two to three times that of 2025. Higher and lower growth rates of gross world product per capita and of population would increase or decrease energy growth in this period accordingly. The most extreme of the six IPCC cases has gross world product per capita growing somewhat more than twice and less than half the size of the base case in 2100; population projections also show significant variations by 2100.

This review of long-term scenarios suggests that one might expect energy use in 2100 to be about two to three times that of 2025 if the world economy grows at a modest rate (e.g., 2%) during the period and population growth declines to about 0.4% per year. Lower economic and population growth could significantly lower the energy projection, and higher growth rates would increase it. We believe that this review of energy scenarios applies well to energy in buildings as well. Most long-term scenarios for energy show buildings using 30 to 35% of total energy, both at present and in the long term.

Goldemberg *et al.* (1987a, 1987b, 1988) describe end-use oriented strategies for industrialized countries that suggest ways in which per capita energy use can decline by substantial amounts. They also provide energy strategies for developing countries that increase per capita energy use to meet needed energy services and avoid inefficient energy uses. While the authors carry their analysis only to 2020, their ideas apply well to a longer term view in which major transformations in all energy-using stock will take place.

A low-energy future for buildings in 2100 could include the following:

- Virtual elimination of space heating in all climates by means of building shells with very high resistance to heat loss or gain, involving high-insulation walls, ceilings, and floors and triple-pane windows with transparent, heat-reflecting films; wide use of passive solar designs; and mass-produced components (walls, ceilings, etc.) with very low infiltration rates
- Reduced need for space cooling and dehumidification because of improved building design and use of passive cooling; provision of cooling and dehumidification by very high-efficiency systems; and use of low-resistance ducts and high-efficiency variable-speed motors for pumps
- Water heating that uses active solar systems combined with the use of waste heat from refrigerators
- Advanced lamp technology combined with lighting controls, task lighting, and greater use of natural light, reducing lighting energy requirements to 10 to 15% of today's levels
- A dramatic decline in refrigerator/freezers' energy use through the use of high insulating walls and doors, efficient compressors, advanced motors, and so forth. Energy use declines to 250 kWh/year by 2025. After 2025, entirely new ways of storing and cooling food could further reduce energy requirements.
- Residential and low-rise commercial buildings that become net producers of energy as roofs incorporate photovoltaic panels and small biofuel power plants provide backup electricity.

Such a future could result in long-term declines in residential energy use in the industrialized world and relatively modest added requirements in the developing world after 2025. Energy use in commercial buildings, on the other hand, will likely continue to increase much longer than in residential buildings.

One scenario for low energy use in residential and commercial buildings in 2100 has been developed by Lazarus *et al.* (1993). In this scenario, delivered energy use for residences in 2100 is very close to today's level.[14] Delivered energy for commercial buildings is triple today's level. All of this increase occurs in the developing world; delivered energy use in commercial buildings in the OECD declines from today's levels.

References

Abrams, D.W. and Assoc., 1992: *Commercial Water Heating Applications Handbook*. Report No. EPRI TR-100212, Electric Power Research Institute (EPRI), Palo Alto, CA.

Ahmed, K., 1994: *Renewable Energy Technologies: A Review of the Status and Costs of Selected Technologies*. Technical Paper No. 240, The World Bank, Washington, DC.

Akbari, H., S. Bretz, J. Hanford, D. Kurn, B. Fishman, and H. Taha, 1993: *Monitoring Peak Power and Cooling Energy Savings of Shade Trees and White Surfaces in the Sacramento Municipal Utility District (SMUD) Service Area: Data Analysis, Simulations, and Results*. Lawrence Berkeley Laboratory Report No. LBL-34411, Berkeley, CA, 96 pp.

[14] Primary energy is higher than today because of increasing electrification, brought about in part by the elimination of the use of fossil fuels in the scenario.

Akbari, H., S. Davis, S. Dorsano, J. Huang, and S. Winnett (eds.), 1992: *Cooling Our Communities—A Guidebook on Tree Planting and Light-Colored Surfacing*. Lawrence Berkeley Laboratory Report No. LBL-31587, Berkeley, CA, 217 pp.

Akbari, H., A. Rosenfeld, and H. Taha, 1990: *Summer Heat Islands, Urban Trees, and White Surfaces*. Lawrence Berkeley Laboratory Report No. LBL- 28308, Berkeley, CA, 8 pp.

Akbari, H., J. Huang, P. Martien, L. Rainer, A. Rosenfeld, and H. Taha, 1988: The impact of summer heat islands on cooling energy consumption and global CO_2 concentration. Proceedings of ACEEE 1988 Summer Study on Energy Efficiency in Buildings, Asilomar, CA, **5**, 11-23.

Akbari, H., H. Taha, J. Huang, and A. Rosenfeld, 1986: Undoing summer heat island can save giga watts of power. Proceedings of ACEEE 1986 Summer Study on Energy Efficiency in Buildings, Santa Cruz, CA, **2**, 7-22; also Lawrence Berkeley Laboratory Report No. LBL-21893, Berkeley, CA, 16 pp.

Anon., 1995: *US-Sino Refrigerator Project* (brochure). U.S. Environmental Protection Agency, National Environmental Protection Agency (China).

Association of Home Appliance Manufacturers, 1995: Refrigerators: energy efficiency and consumption trends. July 27 memo.

Atkinson, B., J.E. McMahon, E. Mills, P. Chan, T.W. Chan, J.H. Eto, J.D. Jennings, J.G. Koomey, K.W. Lo, M. Lecar, L. Price, F. Rubenstein, O. Sezgen, and T. Wenzel, 1992: *Analysis of Federal Policy Options for Improving United States Lighting Energy Efficiency: Commercial and Residential Buildings*. Lawrence Berkeley Laboratory Report No. LBL-31469, Berkeley, CA, 171 pp.

Bauman, F.S., D. Ernest, and E.A. Arens, 1992: The effects of surrounding buildings on wind pressure distributions and natural ventilation in long building rows. In: *ASEAN-USAID Buildings Energy Conservation Project Final Report*. Vol. II, *Technology* [Levine, M.D. and J.F. Busch (eds.)]. Lawrence Berkeley Laboratory Report No. LBL-32380, Vol. II, Berkeley, CA, pp. 3.1-3.34.

Baylon, D., 1992: Commercial building energy code compliance in Washington and Oregon. Proceedings of the ACEEE 1992 Summer Study on Energy Efficiency in Buildings, American Council for an Energy-Efficient Economy, Washington, DC, p. 6-1.

Belzer, D.B., L.E. Wrench, and T.L. Marsh, 1993: *End-Use Energy Consumption Estimates for United States Commercial Buildings, 1989*. Pacific Northwest Laboratory Report No. PNL-8946, Richland, WA, 278 pp.

Bevington, R. and A.H. Rosenfeld, 1990: Energy for buildings and homes. *Scientific American*, **263(3)**, 77-86.

Bhide, A.D., S.A. Gaikwad, and B.Z. Alone, 1990: Methane from land disposal sites in India. *International Workshop on Methane Emissions from Natural Gas Systems, Coal Mining and Waste Management Systems*. Funded by Environment Agency of Japan, U.S. Agency for International Development, and U.S. Environmental Protection Agency, Washington, DC.

Bingemer, H.G. and P.J. Crutzen, 1987: The production of methane from solid wastes. *Journal of Geophysical Resources*, **92**.

Boon-Long, P., T. Sucharitakul, C. Tantaitti, T. Sirathanapanta, P. Ingsuwan, S. Pukdee, and A. Promwangkwa, 1992: Simulation of natural ventilation in three types of public buildings in Thailand. In: *ASEAN-USAID Buildings Energy Conservation Project Final Report*. Vol. II, *Technology* [Levine, M.D. and J.F. Busch (eds.)]. Lawrence Berkeley Laboratory Report No. LBL-32380, Vol. II. Berkeley, CA, pp. 4.1-4.12.

Busch, J.F., P. du Pont, and S. Chirarattananon, 1993: Energy-efficient lighting in Thai commercial buildings. *Energy, The International Journal*, **18(2)**, 197-210.

Busch, J.F., 1990: *From Comfort to Kilowatts: An Integrated Assessment of Electricity Conservation in Thailand's Commercial Sector*. Doctoral diss., Energy and Resources Group, University of California, Berkeley, CA, and Lawrence Berkeley Laboratory Report No. 29478, Berkeley, CA, 121 pp.

CECA, 1993: *Incorporating Environmental Externalities into Utility Planning: Seeking a Cost-Effective Means of Assuring Environmental Quality*. Consumer Energy Council of America, Washington, DC, 271 pp.

Cohen, S., C. Goldman, and J. Harris, 1991: *Measured Energy Savings and Economics of Retrofitting Existing Single-Family Homes: An Update of the BECA-B Database*. Lawrence Berkeley Laboratory Report No. LBL-28147, Berkeley, CA, 82 pp.

Cooper, R.C., 1993: *Fuel Use and Emissions in the USSR: Factors Influencing CO_2 and SO_2*. Ph.D. diss., University of California at Berkeley, Berkeley, CA, 365 pp.

Cooper, R.C. and L.J. Schipper, 1991: The efficiency of energy use in the USSR, an international perspective. *Energy*, **17(1)**, 1-24.

Dutt, G.S., 1994: Illumination and sustainable development. Part I: Technology and economics. *Energy for Sustainable Development*, **1(1)**, 23-35, May.

Dutt, G.S. and E. Mills, 1994: Illumination and sustainable development. Part II: Implementing lighting efficiency programs. *Energy for Sustainable Development*, **1(2)**, 17- 27, July.

Dutt, G.S. and N.H. Ravindranath, 1993: Bioenergy: direct applications in cooking. In: *Renewable Energy: Sources for Fuels and Electricity* [Johansson, T.B. *et al.* (eds.)]. Island Press, Washington, DC, pp. 653-697.

Dutt, G.S., 1992: Comparing alternative biogas technologies. Presented at the Workshop on Materials Science and Physics of Non-Conventional Energy Sources, Buenos Aires, Argentina, September-October 1992.

Ebel, W. *et al.*, 1990: *Energiesparpotentiale im Gebauedebestand*. Institut für Wohnen und Umwelt, Darmstadt, Germany.

Efforsat, J. and A. Farcot, 1994: Les lampes portables solaires. *Systemes Solaires*, No. 100, March-April, 15-21.

EIA, 1994: *Annual Energy Outlook 1994*. U.S. Department of Energy, Washington, DC, 185 pp.

EIA, 1993: *Annual Energy Review 1992*. DOE/EIA-0384(92), U.S. Department of Energy, Washington, DC, 223 pp.

EIA, 1992a: *Electric Power Annual 1992*. DOE/EIA-0348(92), U.S. Department of Energy, Washington, DC, 183 pp.

EIA, 1992b: *Lighting in Commercial Buildings*. DOE/EIA-0555(92)/1, U.S. Department of Energy, Washington, DC, 96 pp.

EIA, 1990: *Energy Consumption and Conservation Potential: Supporting Analysis for the National Energy Strategy*. SR/NES/90-02, U.S. Department of Energy, Washington, DC, 261 pp.

Eley Associates, J. Benya, and R. Verderber, 1990: *Advanced Lighting Guidelines*. California Energy Commission Report No. P400-90-014, Sacramento, CA, 175 pp.

Energy Modeling Forum, 1981: *Aggregate Elasticity of Energy Demand*. EMF-4, Stanford, CA.

EPRI, 1991: *Environmental Externalities: An Overview of Theory and Practice*. Electric Power Research Institute, CU/EN-7294, Palo Alto, CA.

EPRI, 1990: *Impact of Demand-Side Management on Future Customer Demand: An Update*. Electric Power Research Institute, CU-6953, Palo Alto, CA, 80 pp.

Eto, J., E. Vine, L. Shown, R. Sonnenblick, and C. Payne, 1994: *The Cost and Performance of Utility Commercial Lighting Programs: A Report from the Database on Energy Efficiency Programs (DEEP) Project*. Lawrence Berkeley Laboratory Report No. LBL-34967, Berkeley, CA, 81 pp.

Faruqui, A., M. Mauldin, S. Schick, K. Seiden, G. Wikler, and C.W. Gellings, 1990: *Efficient Electricity Use: Estimates of Maximum Energy Savings*. Electric Power Research Institute Report No. EPRI CU-6746, Palo Alto, CA, 88 pp.

Fitzgerald, K.B., D. Barnes, and G. McGranahan, 1991: Interfuel substitution and changes in the way households use energy: the case of cooking and lighting behavior in urban Java. *Pacific and Asian Journal of Energy*, new series **1(1)**, 21-49.

Frost, K., D. Arasteh, and J. Eto, 1993: Energy losses and carbon emissions due to windows in the residential sector. *World Resources Review*, **5(3)**, 351-361.

Gadgil, A. and G. Jannuzzi, 1991: Conservation potential of compact fluorescent lamps in India and Brazil. *Energy Policy*, **19(5)**, 449-463.

Gadgil, A.J. and M.A. Sastry, 1992: Stalled on the road to the market: analysis of field experience with a project to promote lighting efficiency in India. Proceedings of the ACEEE 1992 Summer Study on Energy Efficiency in Buildings, American Council for an Energy-Efficient Economy, Vol. 2, Washington, DC, pp 57-70.

Garbesi, K., H. Akbari, and P. Martien (eds.), 1989: *Controlling Summer Heat Islands*. Proceedings of the Workshop on Saving Energy and Reducing Atmospheric Pollution by Controlling Summer Heat Islands, Berkeley, CA; also Lawrence Berkeley Laboratory Report No. LBL-27872, Berkeley, CA, 343 pp.

GEA, 1993: *Study on Energy Efficiency Standards for Domestic Refrigeration Appliances, Final Report*. Commission of the European Communities.

Geller, H., 1991: *Efficient Energy Use: A Development Strategy for Brazil.* American Council for an Energy-Efficient Economy, Washington, DC and Berkeley, CA, 164 pp.

Geller, H. and P. Miller, 1988: *1988 Lighting Ballast Efficiency Standards: Analysis of Electricity and Economic Savings*. American Council for an Energy-Efficient Economy Report No. A883, Washington, DC and Berkeley, CA, 13 pp.

Gjelstrup, G., A. Larsen, L. Nielsen, K. Oksbjerg, and M. Togeby, 1989: *Elbesparelser i Danmark*. AKF Forlaget, Copenhagen, Denmark.

Goldemberg, J., T.B. Johansson, A.K.N. Reddy, and R.H Williams, 1988: *Energy for a Sustainable World.* Wiley Eastern Limited, New Delhi, India, 375 pp.

Goldemberg, J., T.B. Johansson, A.K.N. Reddy, and R.H Williams, 1987a: *Energy for a Sustainable World*. World Resources Institute, Washington, DC, 119 pp.

Goldemberg, J., T.B. Johansson, A.K.N. Reddy, and R.H Williams, 1987b: *Energy for Development*. World Resources Institute, Washington, DC, 73 pp.

Goldman, C., G.A. Comnes, J. Busch, and S. Wiel, 1993: *Primer on Gas Integrated Resource Planning*. National Association of Regulatory Utility Commissioners (NARUC), Washington, DC, 280 pp.

Govinda Rao, R., G. Dutt, and M. Philips, 1991: *The Least Cost Energy Path for India: Energy Efficient Investments for the Multilateral Development Banks*. International Institute for Energy Conservation, Washington, DC.

Hall, D.O., 1991: Biomass energy. *Energy Policy*, October, **19(8)**, 711-737.

Hirst, E., 1994: *Costs and Effects of Electric-Utility DSM Programs: 1989 through 1997*. Oak Ridge National Laboratory Report No. ORNL/CON-392, Oak Ridge, TN, 33 pp.

Huang, Y.J., 1989: *Potentials for and Barriers to Building Conservation in China.* Lawrence Berkeley Laboratory Report LBL-27644, Berkeley, CA, 20 pp.

IEA, 1993: *Energy Balances of OECD Countries, 1990-1991*. OECD/IEA, Paris, France, 218 pp.

IEA, 1991: *Energy Efficiency and the Environment*. OECD/IEA, Paris, France.

IEA, 1989: *World Energy Statistics and Balances, 1971-1987*. OECD/IEA, Paris, France.

IPCC, 1995: *Radiative Forcing of Climate Change and an Evaluation of the IPCC IS92 Emission Scenarios*. Cambridge University Press, Cambridge, UK, 339 pp

IPCC, 1994: *Radiative Forcing of Climate Change, Summary for Policymakers*. Report to IPCC from the Scientific Assessment Working Group (WGI), Washington, DC, 28 pp.

IPCC, 1991: *Energy and Industry Subgroup Report*. U.S. EPA Office of Policy, Planning and Evaluation, Climate Change Division, Washington, DC.

Isaacs, N. and H.A. Trethowen, 1985: *But We Have to Insulate—A Survey of Insulations Levels in New Zealand*. ASHRAE Transactions, vol. 91.

Janda, K. and J. Busch, 1992: Worldwide status of energy standards for buildings. Proceedings of the ACEEE 1992 Summer Study on Energy Efficiency in Buildings, Vol. 6. American Council for an Energy-Efficient Economy, Washington, DC, pp. 103-105.

Jannuzzi, G.D.M., A. Gadgil, H. Geller, and M.A. Sastry, 1991: Energy efficient lighting in Brazil and India: potential and issues of technology diffusion. Proceedings of the 1st European Conference on Energy-Efficient Lighting, Stockholm, Sweden, pp. 339-355.

Japanese Environment Agency, 1992: *Handbook on the Technology to Arrest the Global Warming*, vol. 1-5 [Global Environment Division (ed.)]. Daiichi Gooki Co., Japan.

Jauregui, E., 1984: Tropical urban climates: review and assessment. Proceedings of the technical conference on Urban Climatology and Its Applications with Special Regards to Tropical Areas, Mexico City, 26-30 November 1984, pp. 26-45.

Johansson, M. and T. Pedersen, 1988: *Tekniske elbesparelser*. AKF Forlag, Copenhagen, Denmark.

Jump, D. and M. Modera, 1994: *Impacts of Attic Duct Retrofits in Sacramento Houses*. Lawrence Berkeley Laboratory Report No. LBL-35375, Berkeley, CA, 14 pp.

Katzev, R. and T. Johnson, 1987: *Promoting Energy Conservation*. Westview Press, Boulder, CO, 218 pp.

Koomey, J., C. Atkinson, A. Meier, J.E. McMahon, S. Boghosian, B. Atkinson, I. Turiel, M.D. Levine, B. Nordman, and P. Chan, 1991: *The Potential for Electricity Efficiency Improvements in the United States Residential Sector*. Lawrence Berkeley Laboratory Report No. LBL-30477, Berkeley, CA, 48 pp.

Komor, P. and L. Wiggins, 1988: Predicting conservation choice: beyond the cost-minimization assumption. *Energy, The International Journal,* **13(8)**, 633-645.

Krause, F. and J. Koomey, with D. Oliver, G. Onufrio, and P. Radanne, 1994: *Energy Policy in the Greenhouse.*Vol. 2, part 5, *Space Heating Efficiency in Buildings*. Prepared for the Dutch Ministry of Housing, Physical Planning and Environment, International Project for Sustainable Energy Paths, El Cerrito, CA, 150 pp.

Krause, F. and J. Eto, 1988: *Least-Cost Utility Planning Handbook for Public Utility Commissioners*. Vol. 2, *The Demand Side: Conceptual and Methodological Issues*. National Association of Regulatory Utility Commissioners (NARUC), Washington, DC, 86 pp.

Krauss, W., M. Had, and M.S. Lobenstein, 1992: *Commercial Gas Space Heating Equipment: Opportunities to Increase Energy Efficiency*. Center for Energy and the Environment Report No. CEUE/TR91-3-CM., Minneapolis, MN.

Lazarus, M., L. Greber, J. Hall, C. Bartels, S. Bernow, E. Hansen, P. Raskin, and D. Von Hippel, 1993: *Towards a Fossil Free Energy Future: The Next Energy Transition.* Stockholm Environment Institute—Boston Center, Boston, MA.

Leach, G. and M. Gowen, 1987: *Household Energy Handbook*. The World Bank, Washington, DC.

Lebot, B., A. Szabo, and H. Despretz, 1991: *Gisement des Economies d'Energie du Parc Européen des Appareils Electroménagers obtenues par une Reglementation des Performances Energétiques*. Report to the European Economic Community, Director General for Energy. Sofi Antopolis: Agence Français pour la Matrîse de L'Energie, Paris, France.

Lebot, B., A. Szabo, and P. Michel, 1992: Preparing minimum energy efficiency standards for European appliances. Proceedings of the ACEEE 1992 Summer Study on Energy Efficiency in Buildings, American Council for an Energy-Efficient Economy, Washington, DC, pp. 2-135.

Lee, L. and R. Bennett, 1992: Achieving energy efficiency in manufactured housing through direct resource acquisition. Proceedings of the ACEEE 1992 Summer Study on Energy Efficiency in Buildings, Vol. 6. American Council for an Energy-Efficient Economy, Washington, DC, pp. 147-154.

Lettinga, G. and A.C. van Haandel, 1993: *Anaerobic Digestion for Energy Production and Environmental Protection: Renewable Energy:Sources for Fuels and Electricity* [Johansson, T.B., H. Kelly, A.K.N. Reddy, and R.H. Williams (eds.)]. Island Press, Washington, DC, 1160 pp.

Levine, M.D., E. Hirst, J. Koomey, J.E. McMahon, and A. Sanstad, 1994: *Energy Efficiency, Market Failures, and Government Policy*. Lawrence Berkeley Laboratory Report No. LBL-35376, Berkeley, CA, 39 pp.

Levine, M.D., H. Geller, J. Koomey, S. Nadel, and L. Price, 1992: *Electricity End-Use Efficiency: Experience with Technologies, Markets, and Policies throughout the World.* Lawrence Berkeley Laboratory Report No. LBL-31885, Berkeley, CA, 94 pp.

Levine, M.D., A. Gadgil, S. Meyers, J. Sathaye, J. Stafurik, and T. Wilbanks, 1991: *Energy Efficiency, Developing Nations, and Eastern Europe: A Report to the U.S. Working Group on Global Energy Efficiency.* International Institute for Energy Conservation, Washington, DC, 47 pp.

Liu, F., 1993: *Energy Use and Conservation in China's Residential and Commercial Sectors: Patterns, Problems, and Prospects*. Lawrence Berkeley Laboratory Report No. LBL-33867, Berkeley, CA, 90 pp.

Loewen, J.M. (ed.), 1992: *ASEAN-USAID Buildings Energy Conservation Project Final Report*. Vol. III, *Audits*. Lawrence Berkeley Laboratory Report No. LBL-32380, Vol. III, Berkeley, CA, 24 pp.

Mashayeki, A., 1990: *Review of Electricity Tariffs in Developing Countries during the 1980s*. Energy Series Paper No. 32, World Bank, Industry and Energy Department, Washington, DC, 39 pp.

Meyers, S., L. Schipper, and B. Lebot, 1993a: *Domestic Refrigeration Appliances in Poland: Potential for Improving Energy Efficiency*. Lawrence Berkeley Laboratory Report No. LBL-34101, Berkeley, CA, 25 pp.

Meyers, S., L. Schipper, and J. Salay, 1993b: *Energy Use in Poland, 1970-1991: Sectoral Analysis and International Comparison*. Lawrence Berkeley Laboratory Report No. LBL-33994, Berkeley, CA, 50 pp.

Meyers, S., S. Tyler, H.S. Geller, J. Sathaye, and L. Schipper, 1990: *Energy Efficiency and Household Electric Appliances in Developing and Newly Industrialized Countries.* Lawrence Berkeley Laboratory Report LBL-29678, Berkeley, CA, 50 pp.

Miller, P., J.H. Eto, and H.S. Geller, 1989: *The Potential for Electricity Conservation in New York State.* New York State Energy Research and Development Authority Report No. 89-12, Albany, NY, 304 pp.

Mills, E., and M.A. Piette, 1993: Advanced energy-efficient lighting systems: progress and potential. *Energy, The International Journal*, **18(2)**, 75-97.

Modera, M., 1993: Characterizing the performance of residential air distribution systems. *Energy and Buildings*, **20**, 65-75.

Modera, M., J. Andrews, and E. Kweller, 1992: *A Comprehensive Yardstick for Residential Thermal Distribution Efficiency.* Lawrence Berkeley Laboratory Report No. LBL-31579, Berkeley, CA, 18 pp.

Morgan, D., 1992: *Energy Efficiency in Buildings: Technical Options.* Congressional Research Service (CRS) Report No. 92-121 SPR, The Library of Congress, Washington, DC.

Munasinghe, M., 1991: Energy-environmental issues and policy options for developing countries. In: *Global Warming: Mitigation Strategies and Perspectives from Asia and Brazil* [Pachauri and Behl (eds.)]. Tata-McGraw Hill Publishing Co., New Delhi, India.

Nadel, S., D. Bourne, M. Shepard, L. Rainer, and L. Smith, 1993: *Emerging Technologies to Improve Energy Efficiency in the Residential and Commercial Sectors.* California Energy Commission Report No. P400-93-003, Sacramento, CA.

Nadel, S., V. Kothari, and S. Gopinath, 1991: *Opportunities for Improving End-Use Electricity Efficiency in India.* American Council for an Energy-Efficient Economy Report No. I911, Washington, DC.

Nielsen, B., 1987: *Vurdering af elbesparlsespotentiale: Handel og service.* Dansk Elvaerkers Forenings Udredningsinstitute (DEFU), Teknisk rapport 257, Lynbgy, Denmark.

Nilsson, P.-E. and S. Aronsson, 1993: Energy-efficient lighting in existing non-residential buildings: a comparison of nine buildings in five countries. *Energy, The International Journal*, **18(2)**, 115-122.

Oke, T.R., 1978: *Boundary Layer Climates.* Methuen, London, UK, 435 pp.

OECD, 1991: *Estimation of Greenhouse Gas Emissions and Sinks.* Final Report from the OECD Experts Meeting, prepared for the Intergovernmental Panel on Climate Change, Paris, France.

Orlich, J., 1990: Methane from land disposal sites in India. *International Workshop on Methane Emissions from Natural Gas Systems, Coal Mining and Waste Management Systems*, funded by Environment Agency of Japan, U.S. Agency for International Development, and U.S. Environmental Protection Agency, Washington, DC.

Parker, D., 1991: *Residential Demand Side Management for Thailand.* International Institute for Energy Conservation, Bangkok, Thailand, 107 pp.

Peat, Marwick, Stevenson & Kellogg, and Marbeck Resource Consultants, 1991: *The Economically Attractive Potential for Energy Efficiency Gains in Canada.* Case Study #3—Commercial, Prepared for Energy, Mines, and Resources Canada, Ottawa, Ontario, Canada.

Pepper, W., J. Leggett, R. Swart, J. Wasson, J. Edmonds, and I. Mintzer, 1992: Emission scenarios for the IPCC, an update. Prepared for the Intergovernmental Panel on Climate Change (IPCC), Working Group I.

Piacquadio Losada, M.N., 1994: Primal structures. *Lokayan Bulletin*, special issue: Tribal Identity, **10(5/6)**, 21-39.

Pietsch, J.A., 1992: *End-Use Technical Assessment Guide.* Vol. 2, *Electricity End Use.* Part 2, Commercial Electricity Use. Electric Power Research Institute Report No. EPRI CU-7222, Palo Alto, CA, 350 pp.

Piette, M.A., M. Cramer, J. Eto, and J. Koomey, 1995: *Office Technology Energy Use and Savings Potential in New York.* Prepared for the New York State Energy Research and Development Authority and Consolidated Edison Company of New York, Inc., NYSERDA Report 95-10 (also LBL-36534).

Proctor, J., M. Blasnik, T. Downey, M. Modera, G. Nelson, and J. Tooley, 1993: Leak detectors: experts explain the techniques. *Home Energy*, **10(5)**, 26-31.

Reddy, A.K.N., 1991: Barriers to improvements in energy efficiency. *Energy Policy*, special issue [J. Sathaye (ed.)], **19(10)**, 953-961.

Richards, K.M., 1989: Landfill gas: working with Gaia. *Biodeterioration Extracts* no. 4, Energy Technology Support Unit, Harwell Laboratory, Oxfordshire, UK.

RIVM, 1993: *Proceedings of International IPCC Workshop on Methane and Nitrous Oxide: Methods in National Emissions Inventories and Options for Control* [van Amstel, A.R. (ed.)]. Amersfoort, The Netherlands.

Rosenfeld, A.H., H. Akbari, S. Bretz, B.L. Fishman, D.M. Kurn, D. Sailor, and H. Taha, 1994: Mitigation of urban heat islands: materials, utility programs, updates. *Energy and Buildings*, **22(1995)**, 255-265.

Rosenfeld, A., C. Atkinson, J. Koomey, A. Meier, R.J. Mowris, and L. Price, 1993: Conserved energy supply curves for United States buildings. *Contemporary Policy Issues*, **11(1)**, **45-68.**

Rosenfeld, A. and L. Price, 1992: Options for reducing carbon dioxide emissions. In: *Global Warming: Physics and Facts* (American Institute of Physics Conference Proceedings 247). AIP, Washington, DC, pp. 261-291.

Rubenstein, F. and R. Verderber, 1990: *Automatic Lighting Controls Demonstration.* Pacific Gas and Electric Company, Customer Systems Report No. 008.1-89-24, San Francisco, CA.

Ruderman, H., M.D. Levine, and J.E. MacMahon, 1987: The behavior of the market for energy efficiency in residential appliances including heating and cooling equipment. *The Energy Journal*, **8(1)**, 101-124.

Sathaye, J. and A. Ketoff, 1991: CO_2 emissions from major developing countries: better understanding the role of energy in the long-term. *The Energy Journal*, **12(1)**, 161-171.

Sathaye, J. and S. Tyler, 1991: Transitions in urban household energy use in Hong Kong, India, China, Thailand and the Philippines. *Annual Review of Energy*, **16**, 295-335.

Scheinbaum, C. and L.J. Schipper, 1993: Residential sector carbon dioxide emissions in OECD countries, 1973-1989: a comparative analysis. In: *Proceedings of the 1993 ECEEE Summer Study "The Energy Efficiency Challenge for Europe"* [Ling, R. and H. Wilhite (eds.)]. Rungstedgård, Denmark, pp. 255-268.

Schipper, L.J., and S. Meyers, with R. Howarth and R. Steiner, 1992: *Energy Efficiency and Human Activity, Past Trends, Future Prospects.* Cambridge University Press, Cambridge, UK, 352 pp.

Schipper, L.J., and D. Hawk. 1991: More efficient household electricity use: an international perspective. *Energy Policy*, **19**, 244-265.

Schipper, L.J., and S. Meyers, 1991: Improving appliance efficiency in indonesia. *Energy Policy*, **19(6)**, 578-588.

Schipper, L.J., S. Meyers, and H. Kelly, 1985: *Coming In from the Cold: Energy-Wise Housing in Sweden.* Seven Locks Press, Cabin John, MD, 125 pp.

Siwei, L. and J. Huang, 1992: Energy conservation standard for space heating in Chinese urban residential buildings. *Energy, the International Journal*, **18(8)**, 871-892.

Smith, K.R., M.A.K. Khalil, R.A. Rasmussen, *et al.*, 1993: Greenhouse gases from biomass and fossil fuel stoves in developing countries: a Manila pilot study. *Chemosphere*, **26(1-4)**, 479-505.

Sullivan, R., D. Arasteh, R. Johnson, and S. Selkowitz, 1992: The influence of glazing selection on commercial building energy performance in hot and humid climates. In: *ASEAN-USAID Buildings Energy Conservation Project Final Report.* Vol. II, *Technology* [Levine, M.D. and J.F. Busch (eds.)]. Lawrence Berkeley Laboratory Report No. LBL-32380, Vol. II, Berkeley, CA, pp. 7.1-7.15

Swisher, J., L. Christiansson, and C. Hedenström, 1994: Dynamics of energy efficient lighting. *Energy Policy*, **22(7)**, 581-594.

Taha, H., H. Akbari, A. Rosenfeld, J. Huang, 1988: Residential cooling loads and the urban heat island: the effect of albedo. *Energy and Environment*, **23(4)**, 271-283; also Lawrence Berkeley Laboratory Report No. LBL-24008, Berkeley, CA, 27 pp.

Treidler, B. and M. Modera, 1994: *New Technologies for Residential HVAC Ducts.* Lawrence Berkeley Laboratory Report No. LBL-35445, Berkeley, CA, 29 pp.

Tsuchiya, H. 1990: CO_2 reduction scenario for Japan. *Nikkei Science*, **20(11)**, 6-15.

Turiel, I., D. Berman, P. Chan, T. Chan, J. Koomey, B. Lebot, M.D. Levine, J.E. McMahon, G. Rosenquist, and S. Stoft, 1991: United States residential appliance energy efficiency, present status and future policy directions. In: *State of the Art of Energy Efficiency: Future Directions* [Vine, E. and D. Crawley (eds.)]. American Council for an Energy-Efficient Economy, Washington, DC, pp. 199-227.

Turiel, I., B. Lebot, S. Nadel, J. Pietsch, and L. Wethje, 1990: *Electricity End Use Demand Study for Egypt.* Lawrence Berkeley Laboratory Report No. LBL-29595. Berkeley, CA, 39 pp.

U.S. AID, 1988: *Prospects in Developing Countries for Energy from Urban Solid Wastes.* Bioenergy Systems Report, Office of Energy.

U.S. Congress, Office of Technology Assessment (OTA), 1993a: *Energy Efficiency: Challenges and Opportunities for Electric Utilities*. OTA-E-561, U.S. Government Printing Office, Washington, DC, 193 pp.

U.S. Congress, Office of Technology Assessment (OTA), 1993b: *Energy Efficiency Technologies for Central and Eastern Europe*. OTA-E-562, U.S. Government Printing Office, Washington, DC, 129 pp.

U.S. Congress, Office of Technology Assessment (OTA), 1992a: *Building Energy Efficiency*. OTA-E-518, U.S. Government Printing Office, Washington, DC, 159 pp.

U.S. Congress, Office of Technology Assessment (OTA), 1992b: *Fueling Development*. OTA-E-516, U.S. Government Printing Office, Washington, DC, 286 pp.

U.S. DOC, 1993: *Statistical Abstract of the United States 1992.* Bureau of the Census, Washington, DC, p. 567.

U.S. DOE, 1993: *Technical Support Document: Energy Conservation Standards for Consumer Products,* volumes I-III. Assistant Secretary, Energy Efficiency and Renewable Energy, Office of Codes and Standards, DOE/EE-0009, Washington, DC, 278 pp.

U.S. DOE, 1990: *Technical Support Document: Energy Conservation Standards for Consumer Products: Dishwashers, Clothes Washers, and Clothes Dryers*. Assistant Secretary, Conservation and Renewable Energy, Office of Codes and Standards, DOE/CE-0299P, Washington, DC, 213 pp.

U.S. DOE, 1989a: *Technical Support Document: Energy Conservation Standards for Consumer Products: Refrigerators and Furnaces*. Assistant Secretary, Conservation and Renewable Energy, Building Equipment Division, DOE/CE-0277, Washington, DC, 216 pp.

U.S. DOE, 1989b: *Technical Support Document: Energy Conservation Standards for Consumer Products: Room Air Conditioners, Water Heaters, Direct Heating Equipment, Mobile Home Furnaces, Kitchen Ranges and Ovens, Pool Heaters, Fluorescent Lamp Ballasts & Television Sets,* vol. I. Assistant Secretary, Conservation and Renewable Energy, Office of Codes and Standards, DOE/EE-0009, Washington, DC.

U.S. EPA, 1994: *Green Lights Third Annual Report*. EPA 430-R-94-005, Washington, DC, 25 pp.

U.S. EPA, 1993a: *International Anthropogenic Methane Emissions: Estimates for 1990*. Report to Congress [Adler, M.J. (ed.)]. Office of Policy Planning and Evaluation, Washington, DC, 292 pp.

U.S. EPA, 1993b: *Anthropogenic Methane Emissions in the United States: Estimates for 1990.* Report to Congress [Hogan, K.B. (ed.)]. EPA 430-R-93-003, Office of Air and Radiation, Washington, DC, 260 pp.

U.S. EPA, 1993c: *Options for Reducing Methane Emissions Internationally.* Vol. II, *International Opportunities for Reducing Methane Emissions*. Report to Congress [Hogan, K.B. (ed.)]. EPA 430-R-93-006, Office of Air and Radiation, Washington, DC, 217 pp.

U.S. EPA, 1993d: *Options for Reducing Methane Emissions Internationally.* Vol. I, *Technological Options for Reducing Methane Emissions*. Report to Congress [Hogan, K.B. (ed.)]. EPA 430-R-93-006 A, Office of Air and Radiation, Washington, DC, 261 pp.

U.S. EPA, 1993e: *Opportunities to Reduce Anthropogenic Methane Emissions in the United States*. Report to Congress [Hogan, K.B. (ed.)]. EPA 430-R-93-012, Washington, DC, 383 pp.

U.S. EPA, 1993f: *Space Conditioning: The Next Frontier: The Potential of Advanced Residential Space Conditioning Technologies for Reducing Pollution and Saving Consumers Money*. EPA 430-R-93-004, Office of Air and Radiation, Washington, DC, 245 pp.

U.S. EPA, 1990: *Policy Options for Stabilizing Global Climate.* Report to Congress [Lashoff, D.A. and D.A. Tripak (eds.)]. USA. EPA 21P-2003.1, Washington, DC, 250 pp.

Usibelli, A., S. Greenberg, M. Meal, A. Mitchell, R. Johnson, G. Sweitzer, F. Rubinstein, and D. Arasteh, 1985: *Commercial Sector Conservation Technologies*. Lawrence Berkeley Laboratory Report No. LBL-18543, Berkeley, CA, 306 pp.

van der Plas, R., 1988: *Domestic Lighting*. Industry and Energy Department Working Paper, Policy, Planning, and Research Series No. 68, World Bank, Washington, DC.

Verderber, R., 1988: *Status and Application of New Lighting Technologies*. Lawrence Berkeley Laboratory Report No. LBL-25043, Berkeley, CA.

Volger, J.A., 1984: Waste recycling in developing countries. In: *Managing Solid Wastes in Developing Countries* [Holmes, J. (ed.)]. John Wiley and Sons, Ltd., New York, NY.

World Bank, 1993: *World Development Report 1993*. The World Bank, Washington, DC.

World Energy Council, 1995: *Energy Efficiency Improvement Utilizing High Technology: An Assessment of Energy Use in Industry and Buildings*. Prepared by M.D. Levine, N. Martin, L. Price, and E. Worrell. WEC, London, UK.

World Energy Council, 1993: *Energy for Tomorrow's World—The Realities, the Real Options and the Agenda for Achievement*. St. Martin's Press, New York, NY, 320 pp.

Yoshino, M., 1991: Developments of urban climatology and problems today. *Energy and Buildings*, **15(1-2)**, 1-10.

Yoshino, M., 1975: *Climate in a Small Area*. University of Tokyo Press, Tokyo, Japan.

23

Agricultural Options for Mitigation of Greenhouse Gas Emissions

VERNON COLE, USA

Principal Lead Authors:
C. Cerri, Brazil; K. Minami, Japan; A. Mosier, USA; N. Rosenberg, USA; D. Sauerbeck, Germany

Lead Authors:
J. Dumanski, Canada; J. Duxbury, USA; J. Freney, Australia; R. Gupta, India; O. Heinemeyer, Germany; T. Kolchugina, Russia; J. Lee, USA; K. Paustian, USA; D. Powlson, UK; N. Sampson, USA; H. Tiessen, Canada; M. van Noordwijk, Indonesia; Q. Zhao, China

Contributing Authors:
I.P. Abrol, India; T. Barnwell, USA; C.A. Campbell, Canada; R.L. Desjardin, Canada; C. Feller, France; P. Garin, France; M.J. Glendining, UK; E.G. Gregorich, Canada; D. Johnson, USA; J. Kimble, USA; R. Lal, USA; C. Monreal, Canada; D. Ojima, USA; M. Padgett, USA; W. Post, USA; W. Sombroek, Netherlands; C. Tarnocai, Canada; T. Vinson, USA; S. Vogel, USA; G. Ward, USA

CONTENTS

EXECUTIVE SUMMARY

Agriculture accounts for one-fifth of the annual increase in anthropogenic greenhouse warming. Most of this is due to methane (CH_4) and nitrous oxide (N_2O); agriculture produces about 50 and 70%, respectively, of their anthropogenic emissions. This chapter on mitigation options confirms earlier IPCC estimates and quantifies potential contributions by agriculture to reduce greenhouse gas emissions.

Between 400 and 800 Mt C/yr could be sequestered worldwide in agricultural soils by implementation of appropriate management practices to increase productivity—including increased input of crop residues, reduced tillage, and restoration of wasteland soils. However, soil carbon (C) sequestration has a finite capacity over a period of 50–100 years, as new equilibrium levels of soil organic matter are established.

Biofuel production on 10–15% of the land area currently in agricultural use or in agricultural set-asides could substitute for 300–1,300 Mt of fossil fuel C per year. Recovery and conversion of crop residues could substitute for an additional 100–200 Mt fossil fuel C per year. However, the possible offsets by increased N_2O emissions need to be considered.

Energy use by agriculture relative to farm production has decreased greatly since the 1970s. Fossil fuel use by agriculture in industrialized countries, although constituting only 3–4% of overall consumption, can be further reduced by, for example, minimum tillage, irrigation scheduling, solar drying of crops, and improved fertilizer management.

Significant decreases in CH_4 emissions from agriculture can be achieved through improved nutrition of ruminant animals and better management of paddy rice fields. Additional CH_4 decreases are possible by altered treatment and management of animal wastes and by reduction of biomass burning. These combined practices could reduce CH_4 emissions from agriculture by 15–56%.

The primary sources of N_2O are mineral fertilizers, legume cropping, and animal waste. Some N_2O also is emitted from biomass burning. N_2O emissions from agriculture could be reduced by 9–26% by improving agricultural management with available techniques.

Ranges in estimates of potential mitigation reflect uncertainty in the effectiveness of recommended management options and in the degree of future implementation globally. Comprehensive model-based analyses and improved global databases on soils, land use, and trace gas fluxes are needed to reduce the uncertainty in these estimates.

Farmers will not voluntarily adopt greenhouse gas (GHG) mitigation techniques unless they improve profitability. Some techniques, such as no-till agriculture or strategic fertilizer placement and timing, already are being adopted for reasons other than concern for climate change. Proposed options for reducing GHG emissions are consistent with maintaining or increasing agricultural productivity.

23.1. Introduction

Agriculture contributes significantly to anthropogenic emissions of carbon dioxide (CO_2), methane (CH_4), and nitrous oxide (N_2O). Based on IPCC data (IPCC, 1994) and additional estimates described below, agriculture accounts for about one-fifth of the annual increase in radiative forcing (Figure 23-1). Land-use changes related to agriculture especially in the tropics, including biomass burning and soil degradation, are also major contributors.

The large influence of agriculture needs to be traced back to its individual sources so that effective mitigations can be sought. Options such as those proposed by the International Workshop on Greenhouse Gas Emissions from Agricultural Systems (U.S. EPA, 1990a), and others that have been proposed more recently, are shown in Table 23-1.

The IPCC First Assessment Report (IPCC, 1990a, 1990b) concluded that land-use changes are the most important source of anthropogenic CO_2 after fossil fuel combustion. Improving the productivity of existing farmland was proposed to mitigate these emissions. Regarding the other two trace gases, it was suggested that CH_4 from rice fields could be reduced by 20–40% and CH_4 from ruminant animals by 25–75%. Improving Nitrogen (N) fertilizer efficiency and minimizing N surpluses were cited as important approaches for reducing agricultural N_2O emissions, but no quantitative assessment was made at that time.

Subsequent IPCC assessments (IPCC, 1991, 1992) corroborated most of the earlier suggestions, but estimates of CH_4 emissions from rice were considerably reduced, and animal manure was identified as an additional CH_4 source. The importance of N_2O emissions from heavily fertilized soils was further emphasized, but the quantities involved still remained highly uncertain.

Since then, the availability of new regional and global databases (FAO, 1990a, 1990b, 1990c; Eswaran *et al.*, 1993; Sombroek *et al.*, 1993), additional site-specific measurements, and the utilization of more thoroughly tested simulation models (Raich *et al.*, 1991; Ojima *et al.*, 1993c; Parton *et al.*, 1993; Potter *et al.*, 1993) have improved the possibilities for quantification. However, information concerning greenhouse gas (GHG) fluxes from agricultural systems is still limited.

Global food production needs and farmer/society acceptability considerations suggest that mitigation technologies should meet the following general guidelines: (1) agricultural production levels will be maintained or enhanced in parts of the world where food production and population demand are in delicate balance; (2) additional benefits will accrue to the farmer (e.g., reduced labor, reduced or more efficient use of inputs); (3) agricultural products will be accepted by local consumers.

Table 23-1 summarizes greenhouse gas mitigation options related to agriculture. These options are grouped into different activities, and the list includes a qualitative assessment of their relevance for the three trace gases CO_2, CH_4, and N_2O. Emphasis in this chapter will be given to quantifying these mitigation potentials.

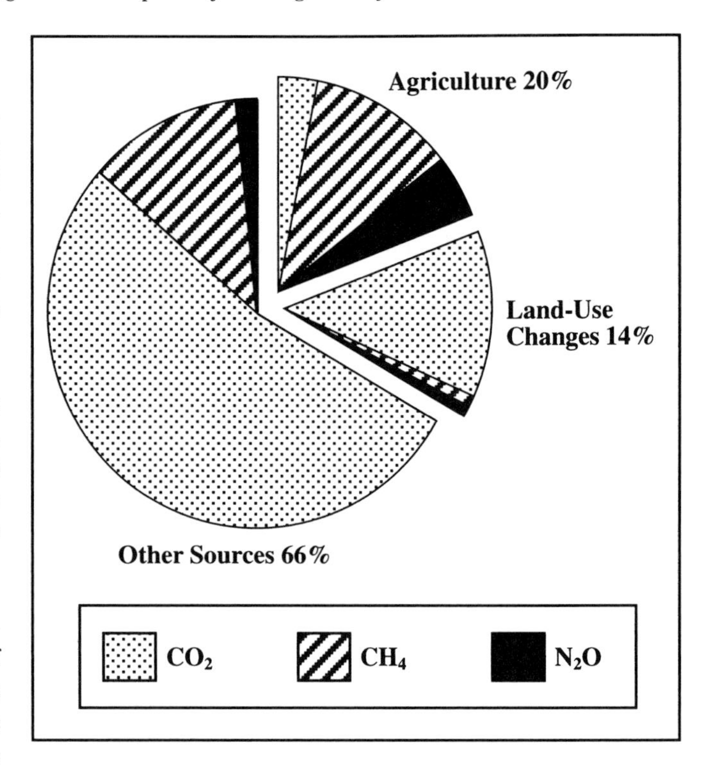

Figure 23-1: Proportions of the annual increase in global radiative forcing attributable to agriculture and agriculture-related land-use changes.

23.2. Carbon Dioxide

Options to mitigate CO_2 emissions from agriculture and land-use changes include the reduction of emissions from present sources and the creation and strengthening of carbon sinks. Increasing the role of agricultural land as a sink for CO_2 includes C storage in managed soils and C sequestration after reversion of surplus farm lands to natural ecosystems (Follett, 1993). Biofuel production on agricultural lands has a considerable potential for mitigation of CO_2 emissions by providing C offsets to use of fossil fuel.

23.2.1. Land-Use Changes

23.2.1.1. Land-Use Changes in the Tropics

Biomass burning and loss of soil C associated with the conversion of native ecosystems to agricultural use in the tropics is believed to be the largest non-fossil fuel source of CO_2 input to the atmosphere. The net release of CO_2 from land-use conversion is thought to be in the range of 1.6 ± 1.0 Gt C/yr (IPCC, 1994). Of the C losses attributed to land use, soil C loss has been estimated to account for 20–40% (Detwiler, 1986;

Houghton and Skole, 1990). Recent data, however, suggest that soil C losses following deforestation may have been overestimated, particularly for forest conversions to pasture, where soil C can recover to levels equal to or higher than native forest within a few years (Lugo and Brown, 1993; Cerri *et al.*, 1994). However, a reduction in land conversion rates remains an important CO_2 mitigation option (see Chapter 19).

Population increase, unresolved land tenure issues, desire for higher living standards, and other sociopolitical factors drive the demand for new cropland in the tropics. However, the suitability and management of this new cropland is often poor, resulting in soil degradation that exerts additional pressure to convert new lands to agriculture. Currently only half of the conversion of tropical forests to agriculture contributes to an increase in productive agricultural area; the other half is used to replace previously cultivated land that has been degraded and abandoned from production (Houghton, 1994). The only way to break out of this cycle is through more sustainable use and improved productivity of existing farmland (Sanchez *et al.*, 1990) and better protection of native ecosystems.

The mitigation options associated with land-use changes are strongly related to major climatic zones. The most significant opportunities appear to be in the humid tropics and in tropical wetlands. In sub-humid zones, much of the land area has already been converted to permanent agriculture. However, improved management and productivity on these lands could help to reduce agricultural expansion (and hence deforestation) in humid zones, especially in Latin America and Africa. In the

Table 23-1: *Options for direct and indirect mitigation of greenhouse gas emissions from agriculture.*

	CO_2	CH_4	N_2O
1. Land Conversion and Management			
– Reduced deforestation rate	H	M	M
– Pasture immediately after deforestation	M		L
– Conversion of marginal agricultural land to grassland, forest, or wetland	M		L
2. Agricultural Land Utilization and Management			
– Restoring productivity of degraded soils	H	L	
– More intensive use of existing farmland	M	L	L
– Restrict use of organic soils	H		L
– Conservation tillage	M		L
– Reduction of dryland fallowing	M		
– Diversified rotations with forage crops	M		L
3. Biofuels			
– Energy crops for fossil fuel substitution	H		
– Agroforestry	L		
– Windbreaks and shelterbelts	L		
– Agroindustrial wastes for fossil fuel substitution	L		
4. Recycling of Livestock and Other Wastes			
– Recycling of municipal organic wastes	L	L	M
– Biogas use from liquid manures		M	
5. Animal Husbandry			
– Supplementing low-quality feed		M	
– Increasing feed digestibility		L	
– Production-enhancing agents		L	
6. Rice Cropping Systems			
– Irrigation management		M	L
– Nutrient management		M	L
– New cultivars and other		M	L
7. Plant Nutrient Management			
– Improved fertilizer use efficiency	L		H
– Nitrification inhibitors			M
– Legume cropping to bolster system productivity			M
– Integrating crop and animal farming			L
8. Minimizing Overall N Inputs			
– Reduced protein inputs in animal feed			M
– Reduced protein consumption by society		M	M

Note: Mitigation Potential: L = Low, M = Medium, H = High.

semi-arid zone, soil and biomass C stocks are smaller and differ less as a function of land use; therefore, the scope for CO_2 mitigation through changing land-use patterns is more limited.

23.2.1.2. Reversion of Agricultural Land in Temperate Zones

In temperate regions there is little development of new agricultural lands, and in regions with food surpluses (e.g., United States, Canada, Western Europe) the agricultural land base is being reduced. A similar situation may occur in the longer term for countries in Eastern Europe and the former Soviet Union (FSU) as per area productivity increases. Thus, the reversion of marginal agricultural land to forest (including shelterbelts and plantations), grassland, and wetlands represents a potential for C sequestration.

Rates of C accumulation in reverted agricultural soils vary greatly depending on climate and soil conditions, the vegetation type established, and the degree of management (Table 23-2).

Upon release from agricultural production, C sequestration would continue only until soils reached a new equilibrium value, most of which would be realized over a 50–100 year period. An exception is the case of reversion to wetlands, where the buildup of organic soils provides a more sustained C sink. About 8 Mha of organic soil-wetlands are currently under cultivation in the temperate zone. As of 1980, these soils were estimated to provide a net C source of ~5 Mt/yr, but represent a potential sink of ~55 Mt C/yr in their native state, with sustained rates of C accumulation of 20–225 g C/m²/yr (Amentano and Menges, 1986).

Currently about 25 Mha in the United States, Canada, and the European Union (EU) (~15% of total cropland) has been taken out of production by government set-aside programs. The U.S. Conservation Reserve Program, started in 1985, is scheduled to begin expiring in 1996, and 75% of the land is expected to return to agriculture by 2005 unless the program is extended. The EU agricultural set-aside programs are for 1–5 years and can include annual cropping for non-food production (e.g., oilseed for fuel), although separate programs exist for tree planting. Because C stocks are likely to be depleted again when lands are returned to cultivation, short duration set-asides have little or no effect on C sequestration. If soils are left uncultivated and allowed to revert to native vegetation, C contents in upper soil horizons could eventually reach levels comparable to their precultivation condition. Considering the ~640 Mha of cropland with real or potential surpluses (United States, Canada, FSU, Europe, Australia, and Argentina) and assuming recovery of the soil C originally lost due to cultivation (25–30% per Davidson and Ackerman, 1993), a permanent set-aside of 15% of the land area might sequester 1.5–3 Gt C.

A large-scale reversion or afforestation of agricultural land is possible only if adequate supplies of food, fiber, and energy can be obtained from the remaining area. This is currently possible in the EU and United States through intensive farming systems. However, if farming intensity changes because of environmental concerns or changes in policy (Enquete Commission, 1995; Sauerbeck, 1994a), this mitigation option may no longer be available.

23.2.2. Carbon Sequestration in Agroecosystems

Losses of soil C as a consequence of cultivation are ubiquitous and well-documented (Haas *et al.*, 1957; Greenland and Nye, 1959; Davidson and Ackerman, 1993). Historical losses of C observed in many soils were due, in part, to low production levels, erosion, inadequate fertilization, removal of crop residues, and intensive tillage. Improved management is capable of

Table 23-2: *Examples of C sequestration rates with conversion of agricultural land to forest and grasslands in some temperate locations.*

Region	Land-Use Conversion	Time Period (yr)	Soil Depth Considered (cm)	Mean Soil C Accumulation (g/m²/yr)
UK	Abandonment to deciduous forest[a,b]	~100	23	50
UK	Planted grassland[c]	15	15	75
Canada	Abandonment to native shortgrass prairie[d]	24	15	13
USA	Grassland seeded for set-aside program[e]	5	100	15–120
USA	Abandonment to native shortgrass prairie[f]	50	10	3
New Zealand	Planted grassland[g]	18	20	100

[a]Jenkinson, 1971.
[b]Jenkinson *et al.*, 1992.
[c]Tyson *et al.*, 1990.
[d]Dormaar and Smoliak, 1985.
[e]Gebhart *et al.*, in press.
[f]Burke *et al.*, submitted.
[g]Haynes *et al.*, 1991.

Table 23-3: *Soil C contents (1 m) and historic soil C loss from cultivated soils worldwide.*

Soil Group[a]	Cultivated Area (Mha)	C Mass—Virgin[b] (Gt C)	C Mass—Cultivated (Gt C)	Soil C Loss[c] (Gt C)
Forest Soils	822	85	63	22
Grassland Soils	438	58	43	15
Wetland Soils	128	54	43	11
Volcanic Soils	31	7	5	2
Other Soils	308	18	13	5
Total	**1727**	**222**	**167**	**55**

[a] Soil groups are aggregated according to the global distribution of cultivated soils in major FAO soil groups by Bouwman (1990). Aggregations included the following soil types: Forest (Acrisols, Podzoluvisols, Ferralsols, Luvisol, Cambisol, Nitosol, Podzol, Chromic Luvisol, Planosol), grassland (Chernozem, Phaeozem, Greyzem, Kastanozem, Vertisol), wetland (Gleysol, Histosol), volcanic (Andosol), and other (Lithosol, Arenosol, Regosol, Solonchak, Solonetz, Xerosol). Some soil types (e.g., Acrisols) that contain subgroups occurring in different vegetation types (e.g., forest vs. grassland) were aggregated under a single soil group (e.g., forest) designation.

[b] C contents, to 1m depth, were calculated separately for each FAO soil type (excluding Histosols) using data from Sombroek *et al.* (1993), then aggregated. C contents for Histosols (112 kg/m^2) were from Amentano and Menges (1986).

[c] Carbon loss due to cultivation of mineral soils was assumed to be 26% of C in the top 30 cm, based on the average reported by Davidson and Ackerman (1993). Losses were calculated separately for the surface layer of soil groups given by Sombroek *et al.* (1993), then aggregated. Estimates of C lost from cultivation of Histosols to 1990 (8 Gt C) were from Amentano and Menges (1986). C mass in cultivated soils was calculated as the difference between uncultivated C contents and C loss estimates.

increasing C levels on present agricultural soils and reducing losses from newly cultivated soils. In general, high residue production, perennial forage crops, elimination of bare fallow, and reduced tillage will promote C sequestration.

An estimate based on the distribution of cultivated soils across major soil groups, their associated C contents, and average loss rates due to cultivation yields a current global stock in cultivated lands of 167 Gt C and a historical loss from these soils of 55 Gt C (Table 23-3). Comparable estimates have been obtained by Houghton and coworkers (Houghton *et al.*, 1983; Houghton and Skole, 1990) from a more detailed analysis of historical land-use data. They estimated global losses of soil C since 1860 to be 30 Gt C (15% of the total 170 Gt C lost from vegetation and soils), or 41 Gt C since the beginning of settled agriculture (Houghton and Skole, 1990). These land use-based estimates do not include C loss from drainage of wetland soils, which could account for the higher value calculated in Table 23-3.

These global estimates of C loss from cultivated soils provide a reference level for the C sequestration that might be achieved through improved management. Increasing C levels in artificially drained wetland soils (e.g., Histosols) is unlikely except in cases where they are reverted to wetlands. Therefore, the potential to increase C levels in cultivated systems is largely restricted to upland soils. Assuming a recovery of one-half to two-thirds of historic C losses as a reasonable upper limit, the global potential for C sequestration in cultivated soils over the next 50–100 years would be on the order of 20–30 Gt. This would require increased production and major improvements in management on much of the world's cultivated areas, particularly in less-developed regions. Under current conditions, it is estimated that there is little or no net C sequestration in temperate soils as a whole, and cultivated soils in the tropics are probably a net source of C (Cole *et al.*, 1993; Sauerbeck, 1993). To identify the most important management factors and constraints on C sequestration, more detailed regional information is provided in Sections 23.2.2.1 through 23.2.2.3.

23.2.2.1. Tropical Agroecosystems

The decrease in soil C with permanent cultivation is of the same order as that for temperate regions, often 20–50% or more, although C losses typically are more rapid in the tropics. It is estimated that arable land and pasture use in the tropics currently contributes a net C flux of 90–230 Mt C annually (Lal and Logan, in press). Decreases in soil C are a result of lower organic matter inputs relative to native systems, enhanced decomposition rates, and/or erosion. Management to increase soil C levels is a high priority for improving the productivity and sustainability of tropical agricultural systems (Swift and Sanchez, 1984). Opportunities to increase soil C levels include improved management of presently cultivated land and the restoration of degraded lands. The latter will include conversion to nonagricultural uses such as forest or biofuel production, but some areas may be suitable for agricultural production as well.

Management practices to increase soil C stocks include reduced tillage, increased production and crop residue return, perennial crops (including agroforestry), and reduced bare fallow frequency. Introduction of deep-rooting cultivars of tropical grasses also shows promise (Fisher *et al.*, 1994). However, there are economic, educational, and sociological constraints to improved soil management in much of the tropics. Many tropical farmers cannot afford or have limited access to purchased inputs such as fertilizer and herbicides. Crop residues often are needed for livestock feed, fuel, or other household uses, which reduces C inputs to soil (Elwell, 1993). On the other hand, development efforts that seek to increase the sustainability and productivity of tropical agriculture will be largely compatible with CO_2 mitigation needs. To the extent that improved management is based on significantly increased fossil fuel consumption, however, benefits for CO_2 mitigation will be decreased.

Increasing soil C levels depends on increasing crop productivity and input of organic matter to the soil (through irrigation, soil fertility improvement, multiple-cropping, controlled grazing) and/or decreasing decomposition rates (through reducing tillage, residue quality, pH, temperature, moisture). The relative importance of these factors will vary according to climate and soil type.

In much of the semi-arid tropics, the predominant land use is pastoral, and grazing control is an important option for maintenance of soil C. Data from northeastern Brazil suggest that improved pastures can maintain C levels comparable to those of native systems (1–1.5% C), which are roughly twice those in shifting cultivation systems (Tiessen, pers. comm.). Water availability is the main determinant of productivity, and the use of reduced tillage and mulching to increase available water and to reduce surface erosion can promote increased soil C (Lal, 1986). In areas where cropping and livestock are closely integrated, such as parts of the west African Sahel and the Miombo zone in southern Africa, better efficiency of manure utilization and supplemental commercial fertilizer can increase productivity and soil C (Feller and Garin, pers. comm.). Overall, however, C stocks are low even under native ecosystems, and therefore CO_2 mitigation options are limited.

In sub-humid and humid zones, CO_2 mitigation potentials are higher. With greater water availability, the physical condition and fertility of soils become more important as production constraints. The inherent physical conditions of these soils are generally good but can be degraded with intensive tillage. Reduced tillage, mulch farming, alley cropping, and other agroforestry practices can contribute to maintenance of soil structure and reduction of erosion and can provide increased organic matter inputs and reduced decomposition rates (by reducing soil temperatures and soil disturbance). In wetland soils, especially paddy fields, low aeration slows decomposition and leads to high C contents, especially in continuously wet soils. When soils dry up during part of the cropping cycle, C contents will decrease unless green manures or large amounts of crop residues (e.g., wheat) are produced in the dry season. The widespread burning of rice straw reduces organic C inputs to the soil. Higher C stocks in wetland soils generally increase CH_4 emissions, although better water management can help control CH_4 emissions.

Restoration of degraded lands is a significant mitigation option. Worldwide there are about 1,200 Mha of moderately to severely degraded lands, which constitute a large potential C sink (Oldeman *et al.*, 1990). In the tropical and subtropical regions of China, about 48 Mha is classified as "wasteland," of which 80% is considered suitable for forestry and 8% and 6% could be restored as cropland and pasture, respectively (Zhao, pers. comm.). Based on field experiments, net C increases in soils reclaimed for agriculture are on the order of 2.4 kg C/m^2 over a 10-year period, which represents a net C sink of 92 Mt C. Higher increases could be obtained for the larger area suitable for reforestation. In India, more than 100 Mha are classified as degraded and greatly depleted in soil C. Experiments have shown that salt- and alkali-affected soils have a relatively high potential for C sequestration if suitable tree and grass species and water management measures are used (Gupta and Rao, 1994). These authors estimated that a restoration of 35 Mha of wasteland area in India could sequester up to 2 Gt C.

23.2.2.2. Temperate Agroecosystems

Most temperate agricultural soils have been cultivated for decades to centuries, and most probably have attained a C content close to equilibrium value. Very little land in the temperate zone is currently being converted to agriculture. For industrialized agriculture (i.e., United States, EU), soil C probably is increasing slightly due to increased productivity and hence greater crop residue inputs, improved residue management, and reduced tillage. In less industrialized agriculture (e.g., FSU) and in areas that have been brought into cultivation more recently, soil C probably is decreasing slightly. As a first approximation, it is reasonable to assume that temperate agricultural soils are not a large source or sink of C under current practices (Cole *et al.*, 1993; Sauerbeck, 1993). However, soil organic C in permanently cropped fields can be increased through a number of management practices, including greater returns of organic materials to soil, decreased periods of fallow, use of perennial and winter cover crops, recycling of organic wastes, reduced tillage, erosion control, and agroforestry.

Soil C levels are closely tied to the rate of C return from crop residues and other sources. Numerous long-term field experiments demonstrate that for many soils, organic C levels are directly proportional to the annual rate of C input (Rasmussen and Collins, 1991; Paustian *et al.*, 1995). Increasing crop production through better nutrient management, reduced fallow periods, and improved cultivars can increase C inputs to soil if crop residues are retained.

Summer fallow is used extensively in semi-arid areas of Canada, the United States, Australia, and the FSU to offset rainfall variability and increase soil water storage. Eliminating or reducing summer fallow through better water management could significantly increase C in semi-arid croplands and decrease soil erosion (Janzen, 1987; Campbell *et al.*, 1990). During fallow periods, mineralization of soil organic C generally is faster than under a crop, and there is no input of crop residues (e.g., in a spring wheat-fallow rotation, there may be only 4 months of crop cover per 24 months).

Greater use of perennial forage crops can significantly increase soil C levels, due to high root C production, lack of tillage disturbance, and protection from erosion. Increases of up to 100 g C/m^2/yr have been documented for cultivated land planted to grassland (see Table 23-2). Where climate permits, winter cover crops decrease erosion and provide additional inputs of C, thereby increasing soil organic carbon (SOC). Simulation of the U.S. cornbelt with the EPIC model suggested an additional C sequestration of 400 g C/m^2 in 100 years from the use of winter cover crops (Lee *et al.*, 1993).

Large applications of manure can increase soil C as much as can reversion to natural vegetation. For example, in the Broadbalk Wheat Experiment at Rothamsted, UK, the application of farmyard manure (FYM) at the very high rate of 35 t/ha annually since 1843 has increased %C in topsoil (0–23 cm) from 0.92% (measured in 1881) to 2.8%. However, such an application rate is not possible on a large scale, and the off-site impacts of large manure application need to be considered.

Reduced- or no-till systems often (but not always) increase soil C. Reduced tillage generally causes organic matter to be concentrated near the surface, but this does not necessarily represent an increase within the total profile (Powlson *et al.*, 1987). However, several studies have shown genuine increases in SOC as a result of reduced tillage (e.g., Dick *et al.*, 1986; Saffigna *et al.*, 1989; Balesdent *et al.*, 1990; Ismail *et al.*, 1994). For the United States, Kern and Johnson (1993) estimate that an increase in the use of reduced tillage from current levels of 27% to 76% of cultivated area by the year 2020 would result in a net C sequestration of 0.2–0.3 Gt C versus a net loss of 0.2 Gt C with current practices. Gaston *et al.* (1993) estimate that no-till management of 181 Mha of climatically suitable land in the FSU could sequester as much as 3.3 Gt C.

In the FSU, significant effort has been made to prevent losses of soil organic matter by using straw mulch and shelterbelts, and terracing steep slopes. Planting of 18 Mha of shelterbelts and afforestation of 53 Mha of degraded landscapes and heavily polluted lands would allow 16 Gt C to be sequestered and conserved over 100 years (160 Mt C/yr)(Kolchugina and Vinson, pers. comm.).

23.2.2.3. Regional Analyses

A rigorous assessment of soil C changes and the potential impacts of various mitigation strategies requires the integration of information on land-use and management practices, soils, and climate at regional scales. The growing availability of national and, to a lesser extent, global databases, together with the use of simulation modeling, provide the framework for such analyses. While no such global-scale analyses of C in agricultural systems have yet been done, three regional analyses for areas in North America illustrate the utility of such an approach, as well as problems and information constraints.

Impacts of different management on soil C content for the U.S. cornbelt region over a 100-year period were assessed by Lee *et al.* (1993) using the EPIC model (Sharpley and Williams, 1990). This study suggested continued losses of C under the current mix of tillage practices. However, widespread use of no-till plus winter cover crops resulted in a net accumulation of 0.10 Mt C/yr. Lee *et al.* (1993) projected that the net effect of a shift to widespread use of no-till plus winter cover crops could conserve 3.3 Mt C/yr over the next 100 years.

The potential C sequestration in Canadian agricultural soils over the next 50 years was assessed by Dumanski *et al.* (pers. comm.) using long-term field data and the Century model (Parton *et al.*, 1987). They assumed that cropland area would remain within 5% of the current area; that there would be a major reduction in summer/fallow in the Chernozemic soil zone; and that cropping practices would be intensified, with increased fertilizer use, improved residue management with reduced tillage, and better erosion control. Over a 50-year period they projected an increased storage of 22 Mt C due to reduction of summer fallow and 69 Mt C from increased hay crops in rotations. Proper fertilization and erosion control through zero-tillage and other measures were projected to further increase C. Summing these management treatments, C equivalent to 3.4% of Canada's present CO_2 emissions could be sequestered annually in improved croplands.

Potential impacts of alternative management practices on soil C in the major agricultural regions of the central United States were analyzed by Donigian *et al.* (1994). They conducted a detailed analysis combining geographic databases on climate, soils, and land use with Century model (Parton *et al.*, 1987) simulations. Management scenarios included increased use of cover crops, increased adoption of reduced tillage, and impacts of land set-aside, as well as potential increases in crop production. This analysis projected an increase in net soil C, primarily due to increases in productivity and crop residue inputs. Thus, a continuation of current trends in agricultural practices would increase C storage by 25 to 50 Mt C per year in a region comprising 70% of U.S. cropland (87 Mha).

A global-level analysis using approaches similar to those outlined above is needed in order to better evaluate potential C changes, including an analysis of management effects and interactions with potential climate change. However, databases on land use and management, soils, and climate for many developing countries are not adequate to provide a reliable framework for analysis. Uncertainties about the distribution of land-use types and management regimes in relation to soils and climate and the behavior of soil organic matter below the surface horizon are important topics that need to be addressed. The current efforts of Food and Agriculture Organization (FAO) to improve such databases at both national and global levels should be expanded.

23.2.3. Fossil Energy Use by Agriculture

Fossil energy use by agriculture is about 3–4.5% of the total consumption in the developed countries of the world (CAST, 1992; Haas *et al.*, 1995a, 1995b; Enquete Commission, 1995). Thus, for the primary farm production sector, the mitigation potentials through reduced fuel consumption are relatively small. The ratio of energy use to farm production has decreased markedly since the "oil crisis" of 1973 sharply increased the costs of energy to all sectors including agriculture (Darmstadter, 1993) and continues to decrease at this time (U.S. Government Executive Office, 1995). Further reductions may be expected by expanded use of minimum tillage, irrigation scheduling, solar drying of crops, and improved fertilizer management.

Tillage and harvest operations account for the greatest proportion of fuel consumption in intensive cropping systems. Considerable energy savings are possible because the fuel requirements using no-till or reduced tillage are 55% and 78%, respectively, of that for conventional moldboard plow tillage (Frye, 1984).

The fixation of atmospheric N into synthetic fertilizer requires about 1.2 kg of fossil C equivalents for each kg of fixed N. Therefore, the present global consumption of about 80 Mt fertilizer N corresponds to the consumption of 100 Mt fossil C per year. Although the N fertilizer use in developed countries may not increase much further, it is predicted to double in developing countries by the year 2025 (IPCC, 1992; Sauerbeck, 1994a). Accordingly, the energy required for manufacturing N fertilizers will increase to about 150 Mt fossil C annually. In Southeast Asia, for example, a 30% increase in the use of fossil fuel is considered necessary by the year 2000, if food production is to follow population and economic growth. Thus, optimizing N use efficiency and minimizing N surpluses provide mitigation of CO_2 emissions in addition to reducing N_2O emissions.

High-intensity animal production has become the biggest consumer of fossil energy in modern agriculture (van Heerwarden *et al.*, 1992; Enquete Commission, 1995). This needs to be considered more critically, as do recent emissions comparisons between conventional and alternative agricultural systems (Enquete Commission, 1995). As an example, Haas *et al.* (1995a, 1995b) calculated that organic farm systems in Germany emit only 39% of the overall fossil C required by conventional farms. This is due mainly to the replacement of mineral N fertilizers by legume cropping, balanced animal stocking rates, and a much lower consumption of feed concentrates (Koepf *et al.*, 1988; Enquete Commission, 1995). Even energy inputs per ton of harvested crop were lower by 20–60% (Haas *et al.*, 1995a, 1995b), although this depends on yield levels and cannot be generalized (Nguyen and Haynes, 1995).

Fuel requirements by the food sector as a whole (including processing, preservation, storage, and distribution) account for 10–20% of total fossil energy consumption (Pimentel *et al.*, 1990; CAST, 1992; Haas *et al.*, 1995a, 1995b). Thus, it may be worthwhile to reconsider the extent of food transport and the overall nutritional habits of the societies in the developed world. For instance, reducing animal protein consumption in Europe and the United States by only one-half of its present excess would decrease N fertilizer requirements by about one-half (Sauerbeck, 1994b; Isermann and Isermann, 1995).

23.2.4. Biofuel Production

Both non-food crops and many conventional agricultural crops produce biomass that is valuable as a feedstock for either energy supply or industrial products. Most of these crops require soil and management conditions similar to food crops and compete for limited resources. The extent to which their production will be expanded in the future depends on the development of new technologies, their economic competitiveness with traditional food and fiber crops as well as with conventional petrochemical feedstocks, and social and political pressures for more renewable and biodegradable materials.

The greatest agricultural potential for mitigating CO_2 lies in increasing the amount and variety of plant biomass used directly for energy production (herein defined as biofuels; see also Chapter 19) as a substitute for fossil energy. This increase could be realized by substituting biofuel crops for other agricultural crops (particularly those in surplus supply), by growing them on lands held in agricultural set-aside programs, or by intermixing biofuel plants with food or forage plants in an agroforestry system.

There are also significant opportunities to utilize crop residues and byproducts for the production of energy to replace fossil fuels. These vary widely, however, in terms of the feasibility of collecting and transporting residues, as well as the extent to which crop residues can be removed from fields without adversely affecting soil C levels and site productivity. In general, it is estimated that only 50% of the residues can be removed without affecting future soil productivity, and only 25% should be considered as recoverable for energy purposes (Sampson *et al.*, 1993). Hall *et al.* (1993) estimate that for the world's major crops (wheat, rice, corn, barley, and sugarcane), a 25% residue recovery rate would amount to an annual C amount of 357 Mt. Much of this would be used for heating and cooking, without substituting for coal or petroleum use. Based on assumptions for energy conversion and degree of substitution for fossil C, crop residues could offset 100–200 Mt fossil C.

Other opportunities include converting marginal or surplus crop and pasture land to forest, increasing the use of forest biomass, noncommercial thinning of industrial timber and paper production wastes, and using recycled wood and paper products for biofuels. Many of the agricultural biofuels considered could be advantageously combined with forest biofuels to stagger harvest dates, reduce storage facility needs, create a more uniform year-round feedstock supply, and reduce the collection radius needed for a feasible utilization facility such as a utility-scale electrical power station.

The CO_2 mitigation potential of a large-scale global agricultural biofuel program can be significant (Johansson *et al.*, 1993). Assuming that 10–15% of the world's cropland area could be made available, fossil fuel substitutions in the range of 0.3 to 1.3 Gt C/yr have been estimated. This estimate does not include the indirect CO_2 offsets of biofuel production through increasing C storage in standing woody biomass, and possibly by increased soil C sequestration.

23.2.4.1. Dedicated Biofuel Crops

Dedicated energy plants, including short-rotation woody crops (SRWC), perennial herbaceous energy crops (HEC), and annuals such as whole-plant cereal crops, could be sustainably grown on 8–11% of the marginal to good cropland in the temperate zone (Table 23-4; Sampson *et al.*, 1993). For example, in the EU it has been estimated that 15–20 Mha of good agricultural land will be surplus to food production needs by the year 2010 (Scurlock *et al.*, 1993; Flaig and Mohr, 1994). This would be equivalent to 20–30% of the current cropland area.

Due to increasing agricultural demand in the tropics, a lower percentage of land is likely to be dedicated to energy crops, so a reasonable estimate may be 5–7% (Sampson *et al.*, 1993). In total, however, there could be a significant amount of land available for biofuel production (FAO, 1994), especially from marginal land and land in need of rehabilitation.

In the United States and Europe, dedicated energy crop yields of about 5 t C/ha/yr currently are achievable from good cropland, and yields of around 9 t C/ha/yr are believed possible by the year 2030 (Wright and Hughes, 1993). In Brazil, SRWC yields in the range of 20 to 30 t C/ha/yr have been reported from trial plots (Betters *et al.*, 1992), representing a doubling of average yields from 30 years ago. Generally, however, average tropical crop yields are lower, ranging from 6 t C/ha/yr currently to perhaps 12 t C/ha/yr in the future (Table 23-4; Sampson *et al.*, 1993).

Given the assumptions in Table 23-4, the range of estimated C emission reductions from energy crops in the tropics is 160 to 513 Mt C/yr. In the temperate regions, C emission reduction could potentially range from 85 to 493 Mt C/yr (Sampson *et al.*, 1993). In addition, agroforestry systems, where trees are grown in intensively managed combinations with food or feed crops, have potential emission reductions of 10 to 55 Mt C/yr in temperate and 46 to 205 Mt C/yr in tropical regions.

23.2.4.2. Biodiesel and Bioethanol

Recently, the use of vegetable oil crops for the production of biodiesel has attracted considerable attention in the United States and the EU (Scurlock *et al.*, 1993). Biodiesel can be burned directly in modified diesel engines or can be used in conventional diesel engines after conversion into methyl or ethyl esters (Vellguth, 1983; Schwab *et al.*, 1987; Sims, 1990). However, it currently costs considerably more to produce than petroleum diesel, so it is not likely to see expanded usage unless technological breakthroughs, ecological considerations, or government subsidies alter the economic situation

Table 23-4: *Potential reduction of C emissions when replacing fossil energy carriers by biofuels of agricultural origin under 2 x CO_2 conditions. Ranges indicate low and high estimates (modified from Sampson et al., 1993).*

Agric. Biofuel Option	Land Area (Mha)[a]	Net C Yield (t C/ha/yr)[b]	Net C Amount (Mt/yr)	Energy Use (%)[c]	Energy Substitution Factor[d]	C Emissions Reduction (Mt/yr)
Dedicated Energy Crops						
Temperate[e]	26–73	5–9	130–657	100	0.65–0.75	85–493
Tropical[f]	41–57	6–12	246–684	100	0.65–0.75	160–513
Temperate Shelterbelts[g]	13–26	2–4	26–104	75	0.50–0.70	10–55
Tropical Agroforestry[h]	41–65	3–6	123–390	75	0.50–0.70	46–205
Total			**525–1835**			**301–1266**

[a]Assuming about 10–15% of world cropland to be available for biofuels. The 10% estimate agrees with Hall *et al.* (1993).
[b]Based on information in Flaig and Mohr (1994), Graham *et al.* (1993), Hall *et al.* (1993), Sampson *et al.* (1993), and others.
[c]Assumed percentage for energy utilization.
[d]Assumed substitution factors for fossil fuel according to Sampson *et al.* (1993).
[e-h]Assumed percentages of total cropland.
[e]8–11% temperate.
[f]5–7% tropical.
[g]2–4% temperate.
[h]5–8% tropical.

(Kleinhanss, 1993; Sampson *et al.*, 1993; Scurlock *et al.*, 1993; Flaig and Mohr, 1994).

Generally, crops from which only the oil, starch, or sugar are used are of limited value in reducing CO_2 emissions, due to the low net energy produced and the relatively high fossil fuel inputs required (Marland and Turhollow, 1991; Flaig and Mohr, 1994). Providing one energy equivalent as rapeseed oil requires about 0.5–0.6 equivalents as fossil fuel, and if bioethanol is produced from grains or root crops, the net energy gain may be as low as 13–20% (Leible and Wintzer, 1993; Graef *et al.*, 1994). However, when using sugarcane, the cane waste provides most of the production energy, resulting in a ratio of ethanol energy to the input of fossil fuel of about 5.2 (Goldemberg *et al.*, 1993). Similarly, palm oil obtained in mills driven by residue fuel at yields of 4–5 t/ha can compete with diesel costs at an energy output versus fossil energy input ratio of 9:5 (Wood and Conley, 1991). The burning of whole-plant biomass as an alternative to fossil fuel results in the most significant CO_2 mitigation (Leible and Wintzer, 1993; Reinhardt, 1993; Kaltschmitt and Becher, 1994), although the actual net effect depends on the plant yield and composition and on the intensity of the cropping system.

23.2.4.3. Overall Fossil Fuel Offsets

Estimating biomass energy production potentials requires assumptions not only about available land, productivity, plant species, and percent of the crop to be used but also about collection and transport, conversion efficiencies, and fuel substitution factors. Table 23-4 was developed as an estimate of the primary energy that could be substituted over the next few decades as a result of agricultural biomass production. Assumptions were made about the relative conversion efficiency of individual fuels and regarding which fossil fuel was to be substituted for by the biomass (Sampson *et al.*, 1993). Overall, agricultural biofuels (energy crops, agroforestry, and crop residues) have the potential to substitute for 0.40 to 1.50 Gt fossil fuel C per year.

23.2.5. Summary of CO_2 Mitigation in Agriculture

Potential C mitigation options in agriculture are significant in relation to anthropogenic emission rates (Table 23-5). The agricultural sector can reduce CO_2 increases in the atmosphere by reducing agriculturally related emissions, sequestering C in

Table 23-5: *Summary of CO_2 mitigation potential for agriculture, expressed as decreases in net C emission rates or as net C storage rates, calculated on an annual basis or accumulated for a 50-year period.*

Category	Annual (Gt C)	Cumulative (Gt C)
Reducing C Emissions		
– Reduction in fossil energy use by agriculture in industrialized countries[a] (assuming 10–50% reduction in current use)	0.01–0.05	0.5–2.5
Increasing C Sinks		
– Increasing soil C through better management of existing agricultural soils (globally)[b]	0.4–0.6	22–29
– Increasing soil C through permanent set-aside of surplus agricultural land in temperate regions		
1) Upland soils[c]	0.015–0.03	0.75–1.5
2) Wetland restoration[d]	0.006–0.012	0.3–0.6
– Restoration of soil C on degraded lands[e] (assuming restoration of 10–50% of global total)	0.024–0.24	1.2–12
Fossil C Offsets		
– Biofuel production from dedicated crops[f]	0.3–1.3	15–65
– Biofuel production from crop residues[g]		
Total Potential CO_2 Mitigation	**0.86–2.44**	**45–122**

[a] Based on current use of 3–4.5% of the total fossil C emission (2.8 Gt C/yr; OECD, 1991) by industrialized countries and an arbitrary reduction range of 10–50%.

[b] Assuming a recovery of one-half to two-thirds of the estimated historic loss (44 Gt) of C from currently cultivated soils (excluding wetland soils) over a 50-year period.

[c] Based on an estimated C sequestration of 1.5–3 Gt over a 100-year period, from a 15% set-aside of cultivated soils (~640 Mha), in industrialized countries with current or potential production surpluses; annual and cumulative rates given as 1 and 50% of that total, respectively.

[d] Based on restoration of 10–20% of former wetland area (8 Mha) now under cultivation in temperate regions.

[e] Assuming potential C sequestration of 1–2 kg C/m^2 over a 50-year period, on an arbitrary 10–50% of moderately to highly degraded land (1.2 x 10^9 ha globally; Oldeman *et al.*, 1990).

[f] Values from Table 23-4.

[g] Based on 25% recovery of crop residues and assumptions on energy conversion and substitution.

soils, and producing biofuel to replace fossil fuels. However, most of the options dealing with land use and soil C sequestration are limited in duration, in that vegetation and soils (under a given set of environmental and management conditions) have a finite capacity to sequester C. Calculations of C sink increases are based on estimates of the difference between current C stocks and those possible under improved management, considering that most of the increase in C would occur within a 50- to 100-year time frame. An exception is C accumulation in wetlands, where C increases can be sustained for much longer periods. Reductions in fossil C consumption by agriculture and the production of biofuels are mitigation options that can, in principle, be maintained indefinitely.

There is a high degree of uncertainty in our estimates concerning both flux rates and C storage capacity, as well as in the level at which various mitigation options could be implemented. Since the latter is largely a function of policy, several of the mitigation potentials have been expressed in terms of arbitrary ranges (10–50%) in implementation, representing possible lower and upper limits for developing policy scenarios.

Large amounts of C could be sequestered in soils (23–44 Gt over a 50-year period) through improved management of agricultural land, permanent set-asides, and restoration of degraded lands. Increasing soil C levels has additional benefits in terms of improving the productivity and sustainability of agricultural production systems. There are potential costs associated with promoting C storage, including fossil fuel requirements (e.g., fertilizers), lost production (e.g., set-aside programs), and additional labor and financial requirements (e.g., land restoration), which may constrain the potential for increasing C storage. Direct fossil fuel use by agriculture is a relatively minor portion of society's total consumption; therefore, even high reductions in use within agriculture have a modest mitigation potential.

The effects of potential climate change and CO_2 enrichment on mitigation have not been explicitly dealt with in the estimates of mitigation potentials. We recognize the importance of these factors, but the present level of knowledge is insufficient to provide quantitative estimates that would incorporate the complex interactions among CO_2, climate change, and land use and management. Analysis of the effects of these factors on C balance and trace-gas emissions requires an integrated model and data analysis at the global scale.

23.3. Methane and Nitrous Oxide

Successful development and implementation of mitigation strategies for agricultural sources of CH_4 and N_2O require an understanding of the effects of land-use change and agricultural practices on fluxes of these gases and on controlling mechanisms. Current knowledge (Batjes and Bridges, 1992; Beese, 1994; Granli and Bockman, 1994; Kroeze, 1994; Mosier *et al.*, in press; Smith *et al.*, 1994) falls short of these criteria but is sufficient to identify key systems/practices and geographic areas to target, as well as likely mitigation technologies. Proposed mitigation options need to be evaluated within the context of farm production systems in order to ensure that interactions and/or feedbacks are accounted for. There is a potential for tradeoffs between CH_4 and N_2O that are only partly assessed in this document.

Sections 23.3.1.1 and 23.3.1.2 provide a new assessment of agricultural sources and sinks of CH_4 and sources of N_2O. Appraisal of mitigation potential in agricultural systems is based on these estimates and the management technology available. The sources of CH_4 considered important include ruminant animals, rice production, animal waste, and biomass burning, while agricultural land use impacts the aerobic soil sink for atmospheric CH_4. The direct and indirect production of N_2O in fertilized systems are considered to be the major agricultural source of N_2O. Other sources—such as biomass burning, enhanced production of N_2O after burning, and increased N_2O production after forest clearing for agriculture—are smaller and are given less rigorous attention. Animal feeding operations are considered minor N_2O sources that are not yet adequately characterized.

23.3.1. *Agricultural Sources and Sinks of CH_4 and N_2O*

23.3.1.1. *Methane*

Biological generation of CH_4 in anaerobic environments, including enteric fermentation in ruminants (Johnson *et al.*, 1993), flooded rice fields, and anaerobic animal waste processing, is the principal source of CH_4 from agriculture. Biomass burning associated with agriculture also contributes to the global CH_4 budget (Delmas, 1993). The primary sink for CH_4 is oxidation with hydroxyl radicals in the troposphere (Crutzen, 1981); in addition, an aerobic soil sink of 10–20% of CH_4 emissions is now evident (Reeburgh *et al.*, 1993).

23.3.1.1.1. *Ruminant animals*

Methane emissions from domestic ruminants are estimated to be about 80 Mt/yr, with a range of 65–100 Mt/yr (IPCC, 1992; Hogan, 1993). Cattle and buffalo account for about 80% of the global annual CH_4 emissions from domestic livestock (Hogan, 1993). Non-ruminant livestock make a relatively small contribution (Crutzen *et al.*, 1986; Gibbs and Leng, 1993; Johnson *et al.*, 1993).

Methane emissions associated with enteric fermentation in ruminants range from 3–8% of gross feed energy intake (Gaedeken *et al.*, 1990; Leng, 1991; U.S. EPA, 1992; Gibbs and Leng, 1993; Johnson *et al.*, 1993), producing 25–37 l of CH_4 per kg dry matter intake (Shibata, 1994). However, for the vast majority of the world's domestic ruminants consuming a wide range of diets under common production circumstances, CH_4 emissions fall near 6% of diet gross energy (range of 5.5 to 6.5%) (Johnson *et al.*, 1993). Restricting the amount of high-quality diet to one-half or less of voluntary consumption, as is frequently done experimentally, can double these percentage losses; however, this seldom occurs in practice. The only production group found markedly different from 6% is the approximately 25 million head of cattle fed very high concentrate diets for about 120 d prior to slaughter (primarily U.S. feedlots), where the emissions average about 3.5% of diet gross energy.

The relatively constant CH_4 emissions as percent of diet for most of the world's livestock has important implications for mitigation strategy. Strategies targeting increased available energy per unit of feedstuff will decrease intake required per unit of product (meat, milk, etc.). Improved feeding and animal management is the only strategy, to date, that has consistently reduced methane emissions. Almost all improved livestock management practices reduce methane per unit of product, and while increased diet energy availability is usually central to the improved practice, a whole array of interrelated inputs are required (protein, minerals, vitamins, improved genetics, reproductive efficiency, animal health, predator control, etc.). The trend worldwide is toward these improved technologies, limited mostly by the market price of animal products.

Examples of methane mitigations that have been analyzed include:

- A variety of feed additives (e.g., antibiotics, ionophore antibiotics and steroids) increase growth rate and increase the feed efficiency of beef cattle, resulting in 5–15% less methane per unit of product (U.S. EPA, 1992).
- Somatotropin (bST) to increase milk production, recently approved in the United States and in several other countries, is projected to reduce methane by 9% if adopted for all dairy cows in the United States (Johnson *et al.*, 1991).
- Treatment of cereal straws with ammonia is a much-researched method and increases digestibility and intake, but increases in methane per kg of straw have been found (Birkelo *et al.*, 1986).

Estimated potentials for adoption of these practices and associated CH_4 reductions are shown in Table 23-6. The greatest opportunity for reducing CH_4 emissions from ruminants is through feed supplementation of cattle and buffalo in Africa, Asia, and Latin America (Leng, 1991; Lin *et al.*, 1994). Supplementation of the diets of native cattle/buffalo in India has been shown to decrease CH_4 emissions per liter of milk produced by a factor of three and per ton of live weight gain by a factor of six (Leng, 1991).

Opportunities for reducing CH_4 emissions from intensively managed cattle are somewhat limited because the CH_4 production per unit of cattle feed is small with a high-quality diet. For dairy cattle, estimated reductions are 10% by genetic improvement, 10% through use of bovine growth hormone (if permitted), and 4% through improved feed formulation (CAST, 1992; U.S. EPA, 1992). For beef cattle, pharmaceuticals being developed that promote protein gain at the expense of fat could reduce CH_4 emissions by as much as 20% (CAST, 1992). Longer-term opportunities include the production of twins to reduce the need for breeding animals and biotechnological approaches to modify rumen fermentation (CAST, 1992; U.S. EPA, 1992). Overall, the combination of these mitigation options could decrease CH_4 emissions from ruminant animals by approximately 30% (Table 23-6). However, increasing food quality may result in increased N_2O emissions and increased energy use, thereby decreasing the effect of the estimated decreases in CH_4 emissions (Ward *et al.*, 1993).

23.3.1.1.2. Animal waste

Significant emissions of CH_4 also occur from animal waste, varying with waste type and management practice (Safley *et al.*, 1992; U.S. EPA, 1993). In general, manures from animals having a high-quality diet have higher potential to generate CH_4 than manures from animals having a low-quality diet. Actual CH_4 emission values depend on the amount of manure produced, its potential to generate CH_4, manure handling practices, and climate.

Global CH_4 emissions from livestock manure were an estimated 20–30 Mt/yr (Safley *et al.*, 1992). More recently, the U.S. Environmental Protection Agency estimated this source to contribute 10–18 Mt/yr (Gibbs and Woodbury, 1993); a value of 14 Mt CH_4/yr is used here. Manure management systems that store manure under anaerobic conditions contribute about 60% of CH_4 of this source because CH_4 is produced during anaerobic decomposition of organic materials in manure (U.S. EPA, 1994). Methane lost from anaerobic digestion constitutes a wasted energy source that can be recovered by using manure management and treatment practices adapted for CH_4 collection (Hogan, 1993). With current technology, CH_4 emissions can be reduced by 25 to 80%. Hogan (1993) identified CH_4 control options that include (Table 23-6):

- **Covered Lagoons**—This option is associated with large-scale, intensive farm operations that are common in North America, Europe, and regions of Asia and Australia. Covered animal-waste lagoons have the potential to recover completely the CH_4 produced from anaerobic waste fermentation. A reasonable estimate is that 40% of the CH_4 from these regions could be mitigated using this approach.
- **Small-Scale Digesters**—These digesters are designed to utilize anaerobic decomposition of organic materials for CH_4 recovery, typically in small-scale operations. Wang *et al.* (1994) noted that about 10 million such biogas digesters are in use in China. These digesters also seem applicable to parts of Africa and South America. Here we assume an efficiency of 70%.
- **Large-Scale Digesters**—Larger, more technically advanced digesters can be integrated with management practices at large livestock operations.

Table 23-6: *Estimated effect of management practices on CH_4 emissions from ruminant livestock, livestock manures, and flooded rice.*

Mitigation Practice		Estimated Decrease due to Practice (Mt CH_4/yr)
Ruminant Livestock		
– Improving diet quality and nutrient balance		25 (10–35)[a]
– Increasing feed digestibility		2 (1–3)
– Production-enhancing agents		2 (1–6)
– Improved animal genetics		–
– Improved reproduction efficiency		–
	Total	**29 (12–44)**
Livestock Manures		
– Covered lagoons		3.4 (2–6.8)
– Small digesters		1.7 (0.6–1.9)
– Large digesters		–
	Total	**5.1 (2.6–8.7)**
Flooded Rice		
– Irrigation management[e]		5[b] (3.3–9.9)
– Nutrient management		10[c] (2.5–15)
– New cultivars and other cultural practices		5[d] (2.5–10)
	Total	**20 (8–35)**

[a] Range of estimates.
[b] About 50% of total rice area and 60% of total rice grain produced is under irrigation (Neue, 1992). Methane production is higher in continually flooded systems and with greater biomass production (Sass *et al.*, 1991). Sass *et al.* (1992) showed that draining the field at specific times decreased CH_4 production by 88% without decreasing rice yield. The effect will probably not be better than 50% when used in the field. The ability to control flooding and drainage will be available to no more than 30% of the total production.
[c] Lindau *et al.* (1993) showed that CH_4 emissions can be decreased by adding sodium sulfate (by 28–35%) or coated calcium carbide (by 36%) with urea compared to urea alone, and by using ammonium sulfate (20%) in place of urea. With all these approaches, CH_4 production could be decreased by about 20%. We assume that this potential can be utilized for all rice production.
[d] Further mitigation potentials are likely to exist in optimizing rice cultivars and other management practices (Neue, 1992; Sass *et al.*, 1992; Lin *et al.*, 1994; Sass, 1994). In controlled experiments, a decrease of about 20% was demonstrated for some of these approaches. An estimated 10% reduction can be achieved from global rice production. Composting rice straw before field application also decreases CH_4 from rice fields (Yagi and Minami, 1990).
[e] Drying the soil surface during the cropping season may increase N_2O emissions. The net effect of both gases needs to be evaluated before the practice can be recommended.

23.3.1.1.3. Rice production

The major pathways of CH_4 production in flooded soil are the reduction of CO_2 with H_2, fatty acids, or alcohols as hydrogen donor and the transmethylation of acetic acid or methyl alcohol by methane-producing bacteria (Conrad, 1989). In flooded rice fields, the kinetics of microbial reduction processes are strongly affected by the composition and texture of soil and its content of inorganic electron acceptors (Neue, 1992). The period between flooding and the onset of methanogenesis can be different for various soils (Sass *et al.*, 1992). As methane production occurs only under highly reduced conditions, on the order of -200 mv redox potential, intermittent flooding or mid-season drainage decreases CH_4 emissions (Sass *et al.*, 1992; Yagi and Minami, 1993). The amount of CH_4 produced in a system is highly dependent upon the quantity of organic carbon available, either from added rice straw (Schutz *et al.*, 1989; Yagi and Minami, 1990; Sass *et al.*, 1991; Neue *et al.*, 1994; Nouchi *et al.*, 1994b) or green manures (Lauren and Duxbury, 1993; Lauren *et al.*, 1994). Rice plants influence CH_4 emissions by providing substrate for root exudation and decay (Schutz *et al.*, 1989; Sass *et al.*, 1991).

There are three processes of CH_4 release into the atmosphere from rice paddies. Methane loss as bubbles is generally a mechanism during the early stages of plant growth and during weeding operations. Diffusion loss of CH_4 across the water surface is another but is a relatively slow process. The third process, transport through rice plant aerenchyma and release to the atmosphere through the shoot nodes, which are not subject to stomatal control, is generally the most important emission mechanism (Cicerone *et al.*, 1983; Seiler *et al.*, 1984; Nouchi *et al.*, 1990, 1994a, 1994b). During the course of the rice-growing season, a large portion of the CH_4 that is produced in the flooded soil is oxidized before it escapes to the atmosphere (Schutz *et al.*, 1989; Sass *et al.*, 1992).

Revised emission estimates: Minami (1993) proposed the methodology for estimating global CH_4 emission from rice fields and presented a CH_4 emission range of 10 to 113 Mt CH_4/yr from world rice paddy fields. A review of CH_4 studies in China, India, Japan, Thailand, the Philippines, and the United States (Sass, 1994) tightened the range of projected CH_4 emissions from rice fields. In his calculations, Sass combined the data for total area of rice paddies with the flux estimates published in various chapters of Minami *et al.* (1994) to produce Table 23-7. For China and India, the annual CH_4 flux estimates were specified by Wang *et al.* (1994) and Parashar *et al.* (1994), respectively. In the other cases, the figures are based on the minimum and maximum reported emission averages. The rice areas in the countries shown in Sass' estimate represent 63% of the total world rice paddy area and result in a total annual CH_4 emission of 16 to 34 Mt. Extrapolating these data to the world, Sass estimates total CH_4 emissions from rice fields to range between 25.4 and 54 Mt/yr, with 50 Mt/yr as a best-guess global emission value. This value is near the IPCC (1992) best estimate of 60 Mt CH_4/yr, but the range indicates that the actual rate may be lower. Wang *et al.* (1994) estimated the annual CH_4 emissions from rice in China to range between 13 and 17 Mt/yr. Lin *et al.* (1994) estimated a slightly lower value from rice in China of <12 Mt CH_4/yr.

Table 23-7: *Estimates of methane emissions from rice fields.*

Country	Total Area of Rice Paddies (Mha)	Total Rice Grain Yield (Mt/yr)	CH_4 Emission (Mt/yr)
China	32.2	174.7	13–17[a]
India	42.2	92.4	2.4–6[b]
Japan	2.3	13.4	0.02–1.04[c]
Thailand	9.6	19.2	0.5–8.8[d]
Philippines	3.5	8.9	0.3–0.7[e]
USA	1.0	6.4	0.04–0.5[f]
Other	54.6	158.5	9.2–20
World Total	**147.5**	**473.5**	**25.4–54**[g]

[a] Wang *et al.* (1994). CH_4 values based on an area-weighted summation for specific regions of the country.
[b] Parashar *et al.* (1994). CH_4 calculated as in (a).
[c] Yagi *et al.* (1994). CH_4 values based on measured minimum and maximum emission rates from the country.
[d] Yagi *et al.* (1994). CH_4 calculations as in (c).
[e] Neue *et al.* (1994). CH_4 calculations as in (c).
[f] Sass and Fischer (1994). CH_4 calculations as in (c).
[g] World total emission rate obtained by an area scaling of the total of the emission rates measured (Sass, 1994).

Estimates of CH_4 emission reduction: Applying the following major management options to global rice production could decrease CH_4 production in rice: (1) water management, (2) nutrient management, (3) cultural practices, and (4) new rice cultivars (Table 23-6).

23.3.1.1.4. Biomass burning

The burning of biomass results in the emission of CH_4 because of incomplete combustion. Emission factors for CH_4 (i.e., the fraction relative to emitted CO_2) vary greatly (0.1% to 2.5%) depending on whether the fire is hot, flaming, or smoldering combustion (Levine *et al.*, 1993). In response to declining agricultural yields and increasing population pressures, farmers in many regions convert forests to cropping land, and many of their techniques involve burning. For example, shifting cultivation requires that forests be cut, and logging debris and unwanted vegetation burned; the land is then farmed for several years, then left fallow to rejuvenate. Savanna and rangeland biomass is often burned to improve livestock forage. Agricultural residues are also burned in the field to return nutrients to the soil or reduce shrubs on rotational fallow lands. Such agriculture-related burning may account for 50% of the biomass burned annually. In addition, about 50% of the worldwide crop residues are burned in small-scale cooking and heating stoves (Hao *et al.*, 1988).

Estimates indicate that 8,700 Mt dry matter/yr (Andreae, 1991) of biomass and 1 to 5% of the world's land (U.S. EPA, 1990b) is burned. Of these estimated CH_4 emissions, those from tropical forest clearing for agriculture, savanna burning, and agricultural crop residue burning are portions of total biomass burning that can be attributed to agricultural practices. Those sources total about 22 Mt CH_4/yr. Burning of crop lands, grasslands, and forests may be reduced through sustained land management programs and the promotion of different land-use practices, including the following:

- Increasing the productivity of existing agricultural lands
- Lengthening the rotation times and improving the productivity of shifting agriculture
- Increasing grassland management
- Incorporating crop residues into soil
- Increasing the use of crop residues as household fuel (remembering that a balance between fuel use and maintenance of soil fertility must be maintained)
- Replacing annual or seasonal crops with trees.

Using various combinations of these practices, an estimated 50% of the CH_4 emitted annually from burning of agricultural wastes (~2.7 Mt/yr) and 20% of other agricultural burning (~3.3 Mt/yr) could be eliminated, providing a total potential mitigation of ~6 Mt CH_4/yr.

23.3.1.1.5. Methane oxidation in soil

Land-use changes and other human-induced alterations of C and N cycles during the past centuries appear to have decreased

CH_4 oxidation in aerobic soils (Ojima *et al.*, 1993), increased N deposition on temperate forest soils (Steudler *et al.*, 1989), and decreased CH_4 uptake by 30 to 60%. Methane oxidation was decreased by about 50% by tilling a semi-arid grassland even when no N fertilizer was ever applied (Mosier *et al.*, 1991). The decrease in CH_4 oxidation in soils when forests or grasslands are converted to agricultural use has been observed in tropical (Keller *et al.*, 1990, 1993) and temperate (Steudler *et al.*, 1989; Mosier *et al.*, 1991; Dorr *et al.*, 1993; Dobbie and Smith, 1994) environments. The decrease seems to be greater as the intensity of the agricultural practices increases. Ojima *et al.* (1993) estimated that land-use changes during the past 200 years have decreased the global temperate soil sink for CH_4 by 20–30%.

A portion of this effect can be attributed to inhibition of CH_4 uptake by inorganic N (Steudler *et al.*, 1989; Keller *et al.*, 1990; Mosier *et al.*, 1991; Nesbit and Breitenbeck, 1992; Hütsch *et al.*, 1993). Additions of inorganic N have been shown to reduce CH_4 uptake in many, but not all, cases (Steudler *et al.*, 1989; Mosier *et al.*, 1991; Adamsen and King, 1993; Cochran *et al.*, 1995).

Reeburgh *et al.* (1993) estimated the global aerobic soil sink to be about 40 Mt CH_4/yr. From a review of available CH_4 uptake data, Minami *et al.* (1993) constrain total terrestrial CH_4 consumption between 7 and 78 Mt/yr. Insufficient information is currently available to recommend agricultural practices to increase the oxidation of CH_4 in cultivated soils.

23.3.1.2. Nitrous Oxide

Nitrous oxide is produced primarily by microbial processes in the soil (Bouwman, 1990). Anthropogenic emission of N_2O occurs as a result of land conversion to agriculture and is likely to be most intensive in agricultural systems that have high N input. Because soil production is the major agricultural source of N_2O, this topic is emphasized (Granli and Bockman, 1994). Agricultural N_2O emissions are thought to arise from fertilization of soils with mineral N and animal manures (this N is partly recycled mineral N and relocated soil mineral N), N derived from biological N fixation (legume crops and free-living N-fixing microbes), and from enhanced soil N mineralization (Duxbury and Mosier, 1993). Information is available only to assess the first three sources (Mosier and Bouwman, 1993; Isermann, 1994a). Nitrous oxide also is directly evolved during biomass burning, and produced in soil after burning, and enhanced emissions arise during conversion of tropical forest to agriculture (Batjes and Bridges, 1992).

A variety of factors control rates of the two microbial processes (nitrification and denitrification) that produce N_2O and N_2O yield. Important variables are soil water content, temperature, nitrate or ammonium concentrations, available organic carbon for denitrification, and pH. Because interactions among the physical, chemical, and biological variables are complex, N_2O fluxes from agricultural systems are highly variable in both time and space (Duxbury and McConnaughey, 1986; Smith, 1990; Beese, 1994; Clayton *et al.*, 1994; Kroeze, 1994; McTaggart *et al.*, 1994). Consequently, prediction of N_2O emissions associated with a unit of N applied to a specific field or fixed by legumes (Mosier, 1993) is not yet reliable. Such predictive capabilities are needed because N_2O emissions derived from agriculture are >75% of the anthropogenic sources (Isermann, 1994a).

23.3.1.2.1. Revised N_2O emission estimates from agricultural soils

Estimates of N inputs to agricultural soils and associated N_2O production are presented in Tables 23-8 and 23-9, respectively. The data are grouped into seven regions of the world. These estimates consider both N_2O directly emitted from agricultural fields and the indirect emissions that occur during parts of the year other than the cropping season, after the N leaves the field. The N fixation contribution does not include N_2O produced in legume pastures. Australia and New Zealand, for

Table 23-8: *Estimated nitrogen applied annually to agricultural lands as synthetic fertilizers and animal wastes, and land area cropped with pulses and soybeans.*

	Synthetic N Consumed	Manure N Produced	Manure N Used as Fertilizer		Harvested Area of Pulses + Soybeans
Region	Mt	Mt	% of Total Manure N	Mt	ha x 10^6
Africa	2.1	20.9	50	10.5	12.8
North and Central America	13.1	7.8	70	5.5	28.1
South America	1.7	21.9	50	11.0	23.5
Asia	37.3	37.4	70	26.2	48.5
Europe	13.6	12.3	90	11.1	4.0
Oceania	0.9	0.5	30	1.5	1.4
Former Soviet Union	8.7	10.1	90	9.1	6.6
Total	**77.4**	**115.3**		**74.9**	**124.9**

***Table 23-9:** Estimates of direct and indirect emissions of N_2O from application of fertilizer N (synthetic or animal waste) to agricultural soils and from soils growing biological N-fixing crops (Mt N_2O-N/yr).*

	Estimated N_2O from				
Region	Mineral N	Animal Waste	N-Fixation	**Total**	**Range**
Africa	0.04	0.21	0.05	0.30	0.15–0.45
North and Cental America	0.26	0.11	0.11	0.48	0.24–0.72
South America	0.03	0.22	0.09	0.34	0.17–0.51
Asia	0.75	0.52	0.19	1.46	0.73–2.19
Europe	0.27	0.22	0.02	0.51	0.26–0.77
Oceania	0.01	0.03	0.01	0.05	0.03–0.08
FSU	0.17	0.18	0.03	0.30	0.19–0.57
Total	**1.53**	**1.49**	**0.50**	**3.50**	**1.8–5.3**

example, contain large areas of pasture land that include legumes as part of the pastoral system. Little data are available for other parts of the globe.

The estimate of synthetic fertilizer N inputs to agricultural soils and the land area of harvested pulses and soybeans are based on country data published by FAO (1990a, 1990b). These values are subject to uncertainty, as are the highly uncertain animal manure N values adapted from Safley (1992) and Bouwman *et al.* (in press). The quantities used as fertilizer for each region are shown in Table 23-8. These numbers are based on estimates of animal distribution and management systems for each region and are not very reliable because of the lack of information.

In addition to including multiple N input sources and N_2O derived from N-fixation, a revised method for calculating the contribution of N_2O from agriculture is used for the estimates in Table 23-9. Earlier estimates generally were based upon assessments derived from reviews of published N_2O emissions data (Bouwman, 1990; Eichner, 1990). More recently, Bouwman *et al.* (in press) reviewed the literature again and presented another assessment of N_2O emissions. They noted that loss of N_2O from agricultural soils may be presented in three ways: (1) the total loss during the period covered by the measurements; (2) the difference between fertilized and control plot, which is referred to as "fertilizer-induced N_2O loss"; and (3) the total loss calculated as a percentage of fertilizer N applied.

Bouwman *et al.* (in press) estimated the total emissions of N_2O from a regression equation: total annual direct field N_2O-N loss = 1 + 0.0125 * N-application (kg N/ha). The value of 1 kg N_2O-N/ha represents the background N_2O-N evolved; the 0.0125 factor accounts for the contribution from fertilization. This estimate includes N sources from a variety of mineral and organic N fertilizers and was based on long-term data sets. The total flux represents N_2O from all sources: native soil N, N from recent atmospheric deposition, past years' fertilization, N from crop residues, N_2O from subsurface aquifers, and current N fertilization.

As explained in Mosier (1993), soil management and cropping systems and unpredictable rainfall inputs affect N_2O emissions more than mineral N sources. As a result, for the purpose of estimating N_2O production from fertilization, we do not use different multiplication factors for different fertilizer types. Limited data also indicate that organic N sources such as animal manures and sewage sludge induce larger N_2O emissions per unit of N added to the soil than does mineral N (Bouwman, 1990, 1994a; Benckiser and Simarmata, 1994). Because of the lack of adequate parallel experiments that cover the range of possibilities of mineral and organic N applications, a single conversion coefficient is used for all sources.

Although these emission estimates are variable, the range is lower than suggested in the OECD/OCDE (1991) calculation methodology. Experience in conducting field flux experiments suggests that much narrower constraints can be placed on the N_2O flux predictions. Bouwman (1994a) estimated that 1.25 ± 1.0% of the applied N was directly emitted, as N_2O encompasses approximately 90% of the direct contributions of fertilization to N_2O emissions.

The indirect contribution of fertilizer N to N_2O emissions, apart from the fields where fertilizer is applied, also must be considered. Based upon the discussions of Duxbury *et al.* (1993), Mosier (1993), and Isermann (1994a), and the large amounts of N_2O frequently found in subsurface aquifers (Bowden and Bormann, 1986; Minami and Ohsawa, 1990), an estimated additional 0.75% of N applications will eventually be evolved to the atmosphere as N_2O resulting from N leaching, runoff, and nitrogen oxides (NO_x) and ammonia (NH_3) volatilization.

The direct and indirect N_2O-N emissions from application of mineral or organic N total approximately 2 ± 1% annually (Mosier *et al.*, in press). This estimate is expected to encompass more than 90% of field situations. Nitrous oxide from biological N-fixation is calculated by multiplying the area of land used for growing pulses plus soybeans in each region by 4 kg N/ha (Duxbury *et al.*, 1982; Galbally *et al.*, 1991).

23.3.1.2.2. Mitigation of N_2O from agricultural soils

A significant fraction of the N_2O evolved from agricultural systems could be avoided if some combination of agricultural management practices listed in Table 23-10 were adopted worldwide. These practices are recommended mainly to improve synthetic fertilizer and manure N use efficiency. The underlying concept in limiting N_2O emissions is that if fertilizer N (all N applied to improve crop growth) is utilized better by the crop, the amount of N needed to meet the growing demand for food will be less; therefore, less N_2O will be produced and less N will leak from the system (Isermann, 1994b; Sauerbeck, 1994a, 1994b). Some of these practices, such as use of nitrification inhibitors, have been shown to have a direct effect on decreasing N_2O emissions in field studies (Aulakh *et al.*, 1984; Minami *et al.*, 1990; Bronson *et al.*, 1992). Ryden (1981) and McTaggart *et al.* (1994) have shown that timing of application of different types of synthetic fertilizer with seasonal water distribution can limit N_2O production.

From Table 23-10 the amount of N_2O-N that is amenable to management is estimated. In these calculations, it is assumed that two-thirds of the N_2O from N applied as manure or synthetic fertilizer is directly emitted from agricultural systems. The remaining one-third is emitted indirectly as a result of runoff, nitrate leaching, and transfer of N to other sections of the ecosystem through NH_3 and NO_x emissions. Only the direct emissions are readily amenable to control by on-farm management, but management options that decrease the amount of external N needed to produce a crop also will decrease indirect N_2O production. It was also assumed that N_2O produced directly from biological N fixation cannot be managed. These estimates were based upon estimated fertilizer use and animal N production in FAO (1990b); thus, they represent estimates for that year and are not future projections.

By better matching N supply to crop demand and more closely integrating animal waste and crop-residue management with crop production, N_2O emissions could be decreased by about 0.38 Mt N_2O-N. Further improvements in farm technology, such as use of controlled-release fertilizers, nitrification inhibitors, timing, and water management, should lead to improvements in N use efficiency and further limit N_2O production by an estimated 0.30 Mt N_2O-N. A total potential reduction of global N_2O emissions from agricultural soils is thus 0.7 (0.34 to 1.0) Mt N_2O-N/yr.

23.3.1.2.3. Release due to biomass burning

Bouwman (1993, 1994b) reviews the status of N_2O formation during biomass burning and estimates a global emission of 0.1 to 0.3 Mt N/yr. This calculation is based on 0.7 ± 0.3% of the N content of the material burned being lost as N_2O (Lobert *et al.*, 1990; Hurst *et al.*, 1994). The estimate includes only those emissions related to savanna burning and deforestation (Crutzen and Andreae, 1990; Hao *et al.*, 1990). A mean value of 0.2 Mt N/yr to represent the N_2O emitted directly from biomass burning related to agriculture is used here. The mitigation potential is difficult to assess because if crop residue is returned to the soil, part of the N mineralized will be converted to N_2O. A significant

Table 23-10: *List of practices to improve efficiency of use of synthetic fertilizer and manure N in agriculture and expected reduction of N_2O emissions assuming global application of mitigation practices (Mt N/yr).*

Practice Followed	Estimated Decrease in N_2O Emissions
Match N Supply with Crop Demand	0.24[a]
– Use soil/plant testing to determine fertilizer N needs	
– Minimize fallow periods to limit mineral N accumulation	
– Optimize split application schemes	
– Match N application to reduced production goals in regions of crop overproduction	
Tighten N Flow Cycles	0.14[b]
– Integrate animal and crop production systems in terms of manure reuse in plant production	
– Maintain plant residue N on the production site	
Use Advanced Fertilization Techniques	0.15[c]
– Controlled-release fertilizers	
– Place fertilizers below the soil surface	
– Foliar application of fertilizers	
– Use nitrification inhibitors	
– Match fertilizer type to seasonal precipitation	
Optimize Tillage, Irrigation, and Drainage	0.15[d]
Total	**0.68**

[a] Assumed that fertilizer N use efficiency can be increased to save 20% of N applied in North America, Europe, and FSU (Doerge *et al.*, 1991; CAST, 1992; Isermann, 1994a; Peoples *et al.*, 1995).

[b] Tightening N cycles may decrease the need for 20% of the N that is used currently in North America, Europe, and FSU, thus saving 20% of fertilizer and reducing N_2O from manure by the same amount where applicable (Buresh *et al.*, 1993; Isermann, 1994a).

[c] Controlled-release fertilizers (Minami, 1994), nitrification inhibitors (Bronson *et al.*, 1992; Keerthisinghe *et al.*, 1993; McTaggart *et al.*, 1994; Minami, 1994), and matching fertilizer type with seasonal precipitation can decrease N_2O emissions 40–90%. We assume that 10% of all fertilizer-derived N_2O production can be decreased by 50%.

[d] There is little published data to confirm this assumption (Granli and Bockman, 1994). A conservative assumption of a 5% decrease that can be achieved globally is used.

decrease in the amount of agricultural biomass that is burned could be achieved by composting the material before it is returned to the field. It is not known, however, how much N_2O is released during composting. Burning also may make N and other nutrients more available to soil microorganisms and result in enhanced emissions of N_2O from soil (Anderson *et al.*, 1988; Anderson and Poth, 1989). Bouwman (1993) and Bouwman *et al.* (in press) calculate that from about 12 Mt N/yr remaining on the ground after burning (Crutzen and Andreae, 1990), 20% is volatilized as NH_3 and that 1%, 0.1 Mt/yr, of the remaining N is emitted to the atmosphere as N_2O. Because of this uncertainty, we expect a mitigation potential of about 10% of the N_2O-N associated with burning of biomass from agriculture.

23.3.1.2.4. Conversion of tropical forest

Conversion of tropical forests to pastures and arable land may contribute an important amount of N_2O to the atmosphere (Keller *et al.*, 1993; Bouwman *et al.*, in press). While fluxes may increase by a factor of 5–8 in the first few years after forest clearing (Luizao *et al.*, 1989; Keller *et al.*, 1993), fluxes gradually decrease during the following 10–20 years (Garcia-Mendez *et al.*, 1991; Keller *et al.*, 1993). Bouwman (1994b) estimates that about 0.4 Mt of additional N_2O-N is emitted annually. We estimate that a 20% reduction from existing levels can be achieved.

23.3.2. Summary of Methane and Nitrous Oxide Emissions and Potential Decreases

Mitigation options are available that could result in significant decreases in CH_4 and N_2O emissions from agricultural systems. If implemented, they are likely to increase rather than decrease crop and animal productivity. Implementation has the potential to decrease CH_4 emissions from rice, ruminants, and animal waste by 30–40% (Table 23-11). The key to decreasing N_2O emissions is improving the efficiency of plant utilization of fertilizer N. This could decrease N_2O emissions from agriculture by almost 20%. Using animal waste to produce CH_4 for energy and digested manure for fertilizer may at some time be cost effective. Economic analyses of options proposed should show positive economic as well as environmental benefits.

23.4. Economic Feasibility of Mitigation Options

As shown in the preceding sections of this chapter, there are policies and tools that, if put into effect, can reduce net carbon emissions from agriculture and increase sequestration. Carbon emissions and capture in agriculture can be brought into balance, but agriculture always will be a net source of N_2O and CH_4. Practices are described in this chapter that would reduce these emissions but not eliminate them. In the case of N_2O, these practices include changes in the timing and placement of fertilizer and use of nitrification inhibitors and fertilizer forms that slow the release of N. In the case of CH_4, these practices include shortening the time during which rice paddies are inundated, altering feeding and husbandry practices to diminish emissions per unit of animal product produced, and utilizing animal wastes for biogas production.

Implementation of these practices will require decisions at many different levels. For example, development of a biofuel

Table 23-11: *Estimated potential impact of mitigation options on CH_4 and N_2O emissions from agriculture.*

Source	Estimated Amount Emitted (Mt/yr)	Potential Decrease (Mt/yr)	(%)
CH_4			
– Ruminant animals	80 (65–100)[a]	29 (12–45)	36
– Animal waste	14 (10–18)	3 (2–7)	21
– Rice paddies	50 (20–60)	20 (8–35)	40
– Biomass burning	22 (11–33)	6 (1.5–4.5)	27
Total	**166 (106–211)**	**58 (24–92)**	**35**
N_2O-N			
– Mineral fertilizer	1.5 (0.5–2.5)	0.3 (0.15–0.45)	20
– Animal wastes	1.5 (0.5–2.5)	0.3 (0.15–0.45)	20
– N-fixation	0.5 (0.25–0.75)	–	–
– Biomass burning	0.2 (0.1–0.3)	0.02 (0.01–0.03)	10
– Soils after burning	0.1 (0.05–0.2)	0.01 (0.005–0.015)	10
– Forest conversion	0.4 (0.1–1)	0.08 (0.04–0.12)	20
Total	**4.2 (1.5–7.25)**	**0.71 (0.36–1.1)**	**17**

[a]Range of estimates.

industry will require changes in infrastructure, institutions, and regional and national policies. At a different scale and level of decisionmaking, the rancher can help reduce CH_4 emissions by providing his/her animals with feed additives. Even the nomadic herder can do this by providing urea-containing blocks for grazing animals to lick. What are the incentives and disincentives for them to do so? Here we examine the feasibility for agricultural producers and the likelihood that they will adopt GHG reduction methods described in this chapter.

It seems reasonable that the world's farmers, ranchers, and pastoralists will not volunteer to implement practices proposed to mitigate greenhouse-forced climate change. This will happen only if the producer is convinced that profitability will improve if these practices are implemented. Incentives such as subsidies can be created to encourage their adoption, and penalties for nonadoption can be imposed.

Examples of mitigation efforts include the following:

- No-till agriculture that increases C storage in the soil is one example of a GHG-mitigating practice that meets the criteria for successful, unforced adaption. No-till practices are used increasingly in the United States in the production of corn, soybeans, and wheat. No-till accounted for about 10–14% of the total acreage of these crops in 1992, double what it had been 5 years earlier.
- Nutrient management practices that result in lower N application rates should reduce emission of N_2O. Soil testing, fertilizer placement, timing, class of fertilizer, and inhibitors are practices that supply nutrients in better accordance with plant physiologic demands. These practices are more likely to be economically feasible on crops that have high N demands (e.g., corn, cotton, and wheat). Because these practices already are being adopted to some extent, only minimal institutional programs may be needed to significantly increase their level of use.
- Reducing the number of days that rice fields are flooded will require large land and water resource investments to provide supplemental storage to change time of flooding. Economic feasibility will be dependent on many unique characteristics of the project site. Such cost estimates will need to address financing and repayment of any structural developments and how such developments change hydrology of water systems and impact other users, in addition to adoption costs incurred by individual producers. As there are no obvious on-farm benefits, and there are adoption costs for this practice, institutional intervention to provide economic incentives or a mandate requiring the practices will likely be needed.
- Increasing quality in livestock feeds is recommended as a means of reducing CH_4 emission per unit of animal product. In developed economies, where there is a high consumption of red meats and dairy products, most livestock animals already are fed higher quality roughage and concentrates. Producers have the knowledge and technical expertise to improve the quality of feed if it can be shown to increase their profits. In developing or underdeveloped countries, land needed for higher-quality roughage production often competes with the food needs of humans. Livestock rely on crop residue and other coarse roughage produced on land not suited for more intensive uses. To add concentrates to feed and to supplement pasture and range with improved crop species adds to production costs. Also, land resources may not be able to support improved species. Whether the benefits generated exceed these costs can best be determined through cost/benefit case studies involving local conditions. Local production relationships, input prices, and local markets will affect long-term economic feasibility.

In general, practices that recover investment cost and generate a profit in the short term are preferred over practices that require a long term to recover investment costs (Rahm and Huffman, 1984). Practices that have a high probability associated with expected profits are desired over practices that have less certainty about their returns. When human resource constraints or knowledge of the practice prevent adoption, public education programs can improve the knowledge and skills of the work force and managers to help advance adoption. Crop insurance or other programs to share the risk of failure due to natural disaster can aid the adoption of practices that increase productivity or expected returns.

23.5. Uncertainties and Future Research Needs

Uncertainties in our present assessment stem from two main sources. Both of these sources point to future research needed for improving our assessment of mitigation options. One source is the inherent unpredictability of future conditions that are controlled primarily by social, economic, and political forces. These conditions include such things as future trends in fossil fuel usage and the degree to which various mitigation strategies will be implemented—decisions that hinge on factors such as economic conditions, environmental awareness, and political will. The most effective way of dealing with these uncertainties is by developing analytical tools that can incorporate socioeconomic factors as potential scenarios in model gaming exercises. Currently, such tools for agricultural systems on a global basis do not exist.

The second source of uncertainty involves deficiencies in our scientific understanding of GHG processes, as well as inadequacies in the information base needed to apply the knowledge that we do have. Gaps in basic understanding can only be addressed through basic research; however, our current ability to assess GHG mitigation in agricultural systems is probably more constrained by a lack of baseline data, in an organized and usable format, than it is by insufficient scientific understanding. Thus, the research needs outlined below focus on the need for compiling and analyzing baseline information:

- **Carbon Sequestration in Agricultural Systems**—The assessment of potentials to increase C stocks in agricultural systems could be improved by using a structured, model-based analysis with global coverage. Suitable models for such a task currently exist and have been used for regional-level analyses (see Section 23.2.2.3). The elements that are lacking are: (1) spatial databases linking climate, soils, and land use and management, which are needed as model inputs; and (2) reliable experimental data to calibrate and/or verify model predictions. A compilation of agricultural land-use information to develop a classification and mapping of agroecological/management zones for the world is sorely needed. Existing long-term agricultural experiments can provide information to evaluate model predictions for different management systems, and soil and climate conditions. Efforts are underway to establish networks of long-term experiment sites and data (Paul *et al.*, 1995; Powlson *et al.*, 1995), but they are still at an early stage.
- **Bioenergy Production from Agricultural Lands**—Research needed to improve assessments of CO_2 mitigation potential through increased use of biofuels includes better technical knowledge of biofuel production and energy conversion efficiencies, as well as information on socioeconomic factors affecting the utilization of biofuels. These needs include: (1) improved information on the actual C feedstock value of forest, agroforestry, and agricultural management systems; (2) better data on energy inputs for the production of wooden goods and tree-derived chemicals and their substitutes; (3) better data on land availability, including cultural, social, and political factors that may preclude some lands from use for C offset projects; and (4) better data, including economic analyses, for the use and efficiency of biofuels, particularly where that usage is conducted outside markets.
- **CH_4 and N_2O Emissions from Agricultural Systems**—Research needed to improve assessments of CH_4 and N_2O mitigation options includes improved synthesis and coordination of existing information and additional field measurements:
  - Available field emission/consumption data need to be carefully assimilated so that comparisons of data sets can be made on a uniform basis. Those data sets that were collected over an insufficient period of time, used inadequate methodology, or are from nonrepresentative systems should be omitted.
  - Existing data need to be applied to validate and calibrate process-based models. Model estimates of gas fluxes should incorporate soil, cropping system, climate, and fertilizer management influences.
  - Field data on gas fluxes are still woefully inadequate. Research needs include year-round field flux measurements in a variety of soils, climates, and cropping systems to compare the impact of management on gas fluxes and to determine the tradeoffs between CH_4 and N_2O flux when management options are exercised. Assessment of entire cropping sequences (e.g., rice-wheat-rice) are needed. The combined use of different flux measurement techniques is needed to evaluate systems over time and space.

Efforts to improve national, regional, and global estimates of gas fluxes are best accomplished through combined efforts. Unfortunately, other than for organizational exercises, little national or international funding has become available to conduct the needed research.

References

Adamsen, A.P.S. and G.M. King, 1993: Methane consumption in temperate and subarctic forest soils: rates, vertical zonation, and responses to water and nitrogen. *Appl. Environ. Microbiol.*, **59**, 485-490.

Amentano, T.V. and E.S. Menges, 1986: Patterns of change in the carbon balance of organic soil-wetlands of the temperate zone. *Ecology*, **74**, 755-774.

Andreae, M.O., 1991: Biomass burning: its history, use, and distribution and its impact on environmental quality and global climate. In: *Global Biomass Burning: Atmospheric, Climate, and Biospheric Implications* [Levine, J.S. (ed.)]. MIT Press, Cambridge, MA, pp. 3-21.

Aulakh, M.S., D.A. Rennie, and E.A. Paul, 1984: Gaseous nitrogen losses from soils under zero-till as compared with conventional-till management systems. *J. Environ. Qual.*, **13**, 130-136.

Balesdent, J., A. Mariotti, and D. Boisgontier, 1990: Effect of tillage on soil organic carbon mineralization estimated from ^{13}C abundance in maize fields. *J. Soil Sci.*, **41**, 587-591.

Batjes, N.H. and S.M. Bridges (eds.), 1992: World inventory of soil emission potentials. Proc. Int. Workshop WAG, 24-27 August 1992, WISE-Report 2, Internat. Soil Reference and Information Centre, Wageningen, 122 pp.

Beese, F., 1994: Gasfoermige Stickstoffverbindungen. In: Studienprogramm Vol. 1 Landwirtschaft, Project D [Enquete Kommission "Schutz der Erdatmosphaere" des Deutschen Bundestages (ed.)]. Economica Verlag GmbH, Bonn, Germany, 88 pp.

Benckiser, G. and T. Simarmata, 1994: Environmental impact of fertilizing soils by using sewage and animal wastes. *Fertilizer Research*, **37**, 1-22.

Betters, D., L.L. Wright, and L. Couto, 1992: Short rotation woody crop plantations in Brazil and the United States. *Biomass and Bioenergy*, **1**, 305-316.

Bockman, O.C., O. Kaarstad, O.H. Lee, and I. Richards, 1991: Energieverbrauch in der Landwirtschaft. In: *Planzenernaehrung und Landwirtschaft*. Nask Hydro, Oslo, Norway, pp. 177-183.

Bouwman, A.F., 1990: Global distribution of the major soils and land cover types. In: *Soils and the Greenhouse Effect* [Bouwman, A.F. (ed.)]. John Wiley and Sons, Chichester, UK, pp 47-59.

Bouwman, A.F., 1993: The global source distribution of nitrous oxide. In: *Methane and Nitrous Oxide* [van Amstel, A.R. (ed.)]. RIVM Report No. 481507003, National Institute of Public Health and Environmental Protection, Bilthoven, Netherlands, pp. 261-272.

Bouwman, A.F., 1994a: *Method to Estimate Direct Nitrous Oxide Emissions from Agricultural Soils*. Report 773004004, National Institute of Public Health and Environmental Protection, Bilthoven, Netherlands, 28 pp.

Bouwman, A.F., 1994b: Estimated global source distribution of nitrous oxide. In: *CH_4 and N_2O: Global Emissions and Controls from Rice Fields and Other Agricultural and Industrial Sources* [Minami, K., A. Mosier, and R. Sass (eds.)]. NIAES Series 2, Yokendo Publishers, Tokyo, Japan, pp. 147-159.

Bouwman, A.F., K.W. van Der Hoek, and J.G.J. Oliver: Uncertainty in the global source distribution of nitrous oxide. *J. Geophs. Res.* (in press).

Bowden, W.B. and F.H. Bormann, 1986: Transport and loss of nitrous oxide in soil water after forest clear-cutting. *Science*, **233**, 867-869.

Bronson, K.F., A.R. Mosier, and S.R. Bishnoi, 1992: Nitrous oxide emissions in irrigated corn as affected by nitrification inhibitors. *Soil Sci. Soc. Am. J.*, **56**, 161-165.

Buresh, R.J., T.T. Chua, E.G. Castillo, S.P. Liboon, and D.P. Garrity, 1993: Fallow and sesbania effects on soil nitrogen dynamics in lowland rice-based cropping systems. *Agron. J.*, **85**, 316-321.

Burke, I.C., W.K. Lauenroth, and D.P. Coffin, 1995: Recovery of soil organic matter and N mineralization in previously cultivated shortgrass steppe. *Ecological Applications* (submitted).

Campbell, C.A., R.P. Zentner, H.H. Janzen, and K.E. Bowren, 1990: *Crop Rotation Studies on the Canadian Prairies*. Research Branch Agriculture Canada, Publication 1841/E, Ottawa, Canada, 133 pp.

CAST, 1992: *Preparing U.S. Agriculture for Global Climate Change*. Task Force Report. No. 119. P.E. Waggoner, Chair. Council for Agricultural Science and Technology, Ames, IA. 96 pp.

Cerri, C.C., M. Bernoux, and G.J. Blair, 1994: Carbon pools and fluxes in Brazilian natural and agricultural systems and the implications for the global CO_2 balance. In: *Proceedings of Int. Soil Sci. Soc., Vol. 5a*. Commission IV, Acapulco, Mexico, July 10-16, 1994, pp. 399-406.

Cicerone, R.J., J.D. Shetter, and C.C. Delwiche, 1983: Seasonal variation of methane flux from a California rice paddy. *J. Geophys. Res.*, **88**, 7203-7209.

Clayton, H., J.R.M. Arah, and K.A. Smith, 1994: Measurement of nitrous oxide emissions from fertilized grassland using closed chambers. *J. Geophys. Res.*, **99**, 16,599-607.

Cochran, V.L., S.F. Schlentner, and A.R. Mosier, 1995: CH_4 and N_2O flux in subarctic agricultural soils. In: *Advances in Soil Sciences: Soil Management and Greenhouse Effect* [Lal, R., J. Kimble, E. Levine, and B. Stuart (eds.)]. CRC Press Inc., Boca Raton, FL, pp. 179-186.

Cole, C.V., K. Flach, J. Lee, D. Sauerbeck, and B. Stewart, 1993: Agricultural sources and sinks of carbon. *Water, Air, Soil Pollut.*, **70**, 111-122.

Conrad, R., 1989: Control of methane production in terrestrial ecosystems. In: *Exchange of Trace Gases Between Terrestrial Ecosystems and the Atmosphere* [Andreae, M.O. and D.S. Schimel (eds.)]. John Wiley and Sons, Chichester, UK, pp. 39-58.

Crutzen, P.J., 1981: Atmospheric chemical processes of the oxides of nitrogen including nitrous oxide. In: *Denitrification, Nitrification and Atmospheric Nitrous Oxide* [Delwiche, C.C. (ed.)]. Wiley, New York, NY, pp. 17-44.

Crutzen, P.J., I. Aselmann, and W. Seiler, 1986: Methane production by domestic animals, wild ruminants, other herbivorous fauna, and humans. *Tellus*, **38B**, 271-284.

Crutzen, P.J. and M.O. Andreae, 1990: Biomass burning in the tropics: impact on atmospheric chemistry and biogeochemical cycles. *Science*, **250**, 1669-1678.

Darmstadter, J., 1993: Climate change impacts on the energy sector and possible adjustments in the MINK region. In: *Towards an Integrated Impact Assessment of Climate Change: The MINK Study* [Rosenberg, N.J. (ed.)]. *Paper 5, Climatic Change*, **24**, 117-129.

Davidson, E.A. and I.L. Ackerman, 1993: Changes in soil carbon inventories following cultivation of previously untilled soils. *Biogeochemistry*, **20**, 161-164.

Delmas, R., 1993: An overview of present knowledge of methane emission from biomass burning. In: *Methane and Nitrous Oxide Methods in National Emission Inventories and Options for Control* [van Amstel, A.R. (ed.)]. RIVM Report No. 481507003, National Institute of Public Health and Environmental Protection, Bilthoven, Netherlands, pp. 171-185.

Detwiler, R.P., 1986: Land use change and the global carbon cycle: the role of tropical soils. *Biogeochemistry*, **2**, 67-93.

Dick, W.A., D.M. Van Doren, Jr., G.B. Triplett, Jr., and J.E. Henry, 1986: *Influence of Long-Term Tillage and Rotation Combinations on Crop Yields and Selected Soil Parameters. II. Results Obtained for a Typic Fragiudalf Soil*. Research Bulletin 1181, The Ohio State University, Ohio Agricultural Research and Development Center, Wooster, OH, 34 pp.

Dobbie, K.E. and K.A. Smith, 1994: Effect of land use on the rate of uptake of methane by surface soils in Northern Europe. *Annales Geophysicae,* **Supplement II, Part II**, 12:C388.

Doerge, T.A., R.L. Roth, and B.R. Gardner, 1991: *Nitrogen Fertilizer Management in Arizona*. College of Agriculture, University of Arizona, Tucson, AZ, 87 pp.

Donigian, Jr., A.S., T.O. Barnwell, Jr., R.B. Jackson IV, A.S. Patwardhan, K.B. Weinrich, A.L. Rowell, R.V. Chinnaswamy, and C.V. Cole, 1994. *Assessment of Alternative Management Practices and Policies Affecting Soil Carbon in Agroecosystems of the Central United States*. Report EPA/600/4-94/067, U.S. EPA, Athens, GA, 194 pp.

Dormaar, J.F. and S. Smoliak, 1985: Recovery of vegetative cover and soil organic matter during revegetation of abandoned farmland in a semi-arid climate. *Journal of Range Management*, **38**, 487-491.

Dorr, H., L. Katruff, and I. Levin, 1993: Soil texture parameterization of the methane uptake in aerated soils. *Chemosphere*, **26**, 697-713.

Duxbury, J.M., D.R. Bouldin, R.E. Terry, and R.L. Tate III, 1982: Emissions of nitrous oxide from soils. *Nature*, **298**, 462-464.

Duxbury, J.M. and P.K. McConnaughey, 1986: Effect of fertilizer source on denitrification and nitrous oxide emissions in a maize field. *Soil Sci. Soc. Am. J.*, **50**, 644-648.

Duxbury, J.M. and A.R. Mosier, 1993: Status and issues concerning agricultural emissions of greenhouse gases. In: *Agricultural Dimensions of Global Climate Change* [Kaiser, H.M. and T.E. Drennen (eds.)]. St. Louis Press. Delray Beach, FL, pp. 229-258.

Duxbury, J.M., L.A. Harper, and A.R. Mosier, 1993: Contributions of agroecosystems to global climate change. In: *Agricultural Ecosystem Effects on Trace Gases and Global Climate Change* [Harper, L.A., A.R. Mosier, J.M. Duxbury, and D.E. Rolston (eds.)]. ASA Special Publication Number 55, Madison, WI, pp. 1-18.

Eichner, M.J., 1990: Nitrous oxide emissions from fertilized soils: summary of available data. *J. Environ. Qual.*, **19**, 272-280.

Elwell, H.A., 1993: Development and adoption of conservation tillage practices in Zimbabwe. In: *Soil Tillage in Africa: Needs and Challenges*. Chapter 10, FAO Soils Bulletin 69, Rome, Italy, pp. 129-164.

Enquete Commision, 1995: *Protecting Our Green Earth. How to Manage Global Warming through Environmentally Sound Farming and Preservation of the World's Forests*. Economica Verlag, Bonn, Germany, 683 pp.

Eswaran, H., E. van den Berg, and P. Reich, 1993: Organic carbon in soils of the world. *Soil Sci. Soc. Am. J.*, **57**, 192-194.

FAO, 1990a: *Fertilizer Yearbook*. Volume 39. FAO Statistics Series No. 95. FAO, Rome, Italy, 127 pp.

FAO, 1990b: *Production Yearbook*. Volume 43. FAO Statistics Series No. 94. FAO, Rome, Italy, 346 pp.

FAO, 1990c: *The State of Food and Agriculture 1990*. David Lubin Memorial Library Cataloging in Publication Data, FAO, Rome, Italy.

Fisher, M.J., I.M. Rao, M.A. Ayarza, C.E. Lascano, J.I. Sanz, R.J. Thomas, and R.R. Vera, 1994: Carbon storage by introduced deep rooted grasses in the South American savannas. *Nature*, **371**, 236-238.

Flaig, H. and H. Mohr (eds.), 1994: *Energie aus Biomasse - eine Chance fuer die Landwirtschaft*. Springer, Berlin-Heidelberg-New York, 376 pp.

Follett, R.F., 1993: Global climate change, U.S. agriculture, and carbon dioxide. *J. Prod. Agric.*, **6**, 181-190.

Frye, W.W., 1984: Energy requirement in no-tillage. In: *No Tillage Agricultural Principles and Practices* [Phillips, R.E. and S.H. Phillips (eds.)]. Van Nostrand Reinhold, New York, NY, pp. 127-151.

Gaedeken, D., D. Rath, and D. Sauerbeck, 1990: Methane production in ruminants. In: *USEPA-IPCC Proc. Workshop on Greenhouse Gas Emissions from Agricultural Systems, Vol. 2*. USEPA-OPA (PM221), Washington, DC, 9 pp.

Galbally, I.E., P.J. Fraser, C.P. Meyer, and D.W.T. Griffith, 1992: Biosphere-atmosphere exchange of trace gases over Australia. In: *Australia's Renewable Resources: Sustainability and Global Change* [Gifford, R.M. and M.M. Barson (eds.)]. Bureau of Rural Resources, Proceedings No. 14, P.J. Grills, Commonwealth Printer, Canberra, Australia, pp. 117-149.

Gaston, G.G., T. Kolchugina, T.S. Vinson, 1993: Potential effect of no-till management on carbon in the agricultural soils of the former Soviet Union. *Agric., Ecosyst., Environ.*, **45**, 295-309.

Gebhart, D.L., H.B. Johnson, H.S. Mayeux, and H.W. Polley: The CRP increases soil organic carbon. *J. Soil and Water Conserv.* (in press).

Gibbs, M.J. and R.A. Leng, 1993: Methane emissions from livestock. In: *Methane and Nitrous Oxide* [van Amstel, A.R. (ed.)]. RIVM Report No. 481507003, National Institute of Public Health and Environmental Protection, Bilthoven, Netherlands, pp. 73-79.

Gibbs, M.J. and J.W. Woodbury, 1993: Methane emissions from livestock manure. In: *Methane and Nitrous Oxide* [van Amstel, A.R. (ed.)]. RIVM Report No. 481507003, National Institute of Public Health and Environmental Protection, Bilthoven, The Netherlands, pp. 81-91.

Goldemberg, J., L.C. Monaco, and I.C. Macedo, 1993: The Brazilian fuel-alcohol program. In: *Renewable Energy: Sources for Fuels and Electricity* [Burnham, L. (exe. ed.)]. Island Press, Washington, DC, pp. 841-863.

Graef, M., G. Vellguth, J. Krahl, and A. Munack, 1994: Fuel from sugar beet and rape seed oil—mass and energy balances for evaluation. Proceedings of the 8th European Conference on Biomass for Energy, Environment, Agriculture and Industry, 3-5 October 1994, Vienna, Austria, pp. 1-4.

Graham, R.L., L.L. Wright, and A. Turhollow, 1993: The potential for short rotation woody crops to reduce CO_2 emissions. *Climatic Change*, **22**, 223-238.

Granli, T. and O.C. Bockman, 1994: Nitrous oxide from agriculture. *Norwegian Journal of Agricultural Sciences,* **Supplement No. 12**, 128.

Greenland, D.J. and P.H. Nye, 1959: Increases in carbon and nitrogen contents of tropical soils under natural fallows. *J. Soil Sci.*, **10**, 284-298.

Gupta, R.K. and D.L.N. Rao, 1994: Potential of wastelands for sequestering carbon by reforestation. *Current Science*, **66**, 378-380.

Haas, H.J., C.E. Evans, and E.F. Miles, 1957: *Nitrogen and Carbon Changes in Great Plains Soils as Influenced by Cropping and Soil Treatments.* Technical Bulletin No. 1164 USDA, State Agricultural Experiment Stations, Washington, DC, 111 pp.

Haas, G., U. Geier, D.G. Schulz, and U. Koepke, 1995a: Klimarelevanz des Agrarsektors der Bundesrepublik Deutschland: Reduzierung der Emission von Kohlendioxid. Berichte ueber Landwirtschaft 73, Landwirtschaftsverlag GmbH, Muenster-Hiltrup, Germany, pp. 387-400.

Haas, G., U. Geier, D.G. Schulz, and U. Koepke, 1995b: Vergleich Konventioneller und Organischer Landbau—Teil I: Klimarelevante Kohlendioxid-Emission durch den Verbrauch fossiler Energie. Berichte ueber Landwirtschaft 73, Landwirtschaftsverlag GmbH, Muenster-Hiltrup, Germany, pp. 401-415.

Hall, D.O., F. Rosillo-Calle, R.H. Williams, and J. Woods, 1993: Biomass for energy supply prospects. In: *Renewables for Fuels and Electricity* [Johansson, B.J., H. Kelly, A.K.N. Reddy, and R.H. Williams (eds.)]. Island Press, Washington, DC, pp. 583-652.

Hao, W.M., D. Scharffe, and P.J. Crutzen, 1988: Production of N_2O, CH_4, and CO_2 from soils in the tropical savanna during the dry season. *J.Atmos Chem.*, **7**, 93-105.

Hao, W.M., M.H. Liu, and P.J. Crutzen, 1990: Estimates of annual and regional releases of CO_2 and other trace gases to the atmosphere from fires in the tropics, based on the FAO statistics for the period 1975-1980. In: *Fire in the Tropical Biota* [Goldhammer, J.G. (ed.)]. Springer Verlag, Berlin, Germany. *Ecological Studies*, **84**, 440-462.

Haynes, R.J., R.S. Swift, and R.C. Stephen, 1991: Influence of mixed cropping rotations (pasture-arable) on organic matter content, water stable aggregation and clod porosity in a group of soils. *Soil & Tillage Research*, **19**, 77-81.

Hogan, K.B., 1993: Methane reductions are a cost-effective approach for reducing emissions of greenhouse gases. In: *Methane and Nitrous Oxide: Methods in National Emissions Inventories and Options for Control* [van Amstel, A.R. (ed.)]. RIVM Report No. 481507003, Bilthoven, Netherlands, pp. 187-201.

Houghton, R.A., J.E. Hobbie, J.M. Melillo, B. Moore, B.J. Peterson, G.R. Shaver, and G.M. Woodwell, 1983: Changes in the carbon content of terrestrial biota and soils between 1860 and 1980: a net release of CO_2 to the atmosphere. *Ecol. Monogr.*, **53**, 235-262.

Houghton, R.A. and D.L. Skole, 1990: Carbon. In: *The Earth as Transformed by Human Action* [Turner, B.L., W.C. Clark, R.W. Kates, J.F. Richards, J.T. Mathews, and W.B. Meyer (eds.)]. Cambridge University Press, New York, NY, pp. 393-408.

Houghton, R.A., 1994: The worldwide extent of land-use change. *BioScience*, **44**, 305-313.

Hütsch, B.W., C.P. Webster, and D.S. Powlson, 1993: Long-term effects of nitrogen fertilization on methane oxidation in soil of the broadbalk wheat experiment. *Soil Biol. Biochem.*, **25**, 1307-1315.

Hurst, D.F., D.W.T. Griffith, J.N. Carras, D.S. Williams, and P.J. Fraser, 1994: Measurement of trace gases emitted by Australian savanna fires during the 1990 dry season. *J. Atmos. Chem.,* **18**, 33-56.

IPCC, 1990a: *Climate Change: The IPCC Scientific Assessment.* IPCC Working Group I [Houghton, J.T., G.J. Jenkins, and J.J. Ephraums (eds.)] and WMO/UNEP. Cambridge University Press, Cambridge, UK, 365 pp.

IPCC, 1990b: *Climate Change: The IPCC Impacts Assessment.* IPCC Working Group II [McG. Tegart, W.J., G.W. Sheldon, and D.C. Griffiths (eds.)] and WMO/UNEP. Australian Government Publishing Service, Canberra, Australia, 275 pp.

IPCC, 1991: *Climate Change: The IPCC Response Strategies.* IPCC Working Group III and WMO/UNEP. Island Press, Covelo, CA, 273 pp.

IPCC, 1992: *Climate Change 1992: The IPCC Supplementary Report to the IPCC Scientific Assessment.* Supporting Material, Working Group III, Response Strategies, Subgroup AFOS, WMO/UNEP, Geneva, pp. 14-22.

IPCC, 1994: *Radiative Forcing of Climate Change.* The 1994 Report of the Scientific Assessment Working Group of IPCC, Summary for Policymakers. WMO/UNEP, Geneva, Switzerland, 28 pp.

Isermann, K., 1993: Territorial, continental and global aspects of C, N, P and S emissions from agricultural ecosystems. In: *NATO Advanced Research Workshop (ARW) on Interactions of C,N, P and S Biochemical Cycles.* Springer-Verlag. Heidelberg, Germany, ASI Series, pp. 79-121.

Isermann, K., 1994a: Agriculture's share in the emission of trace gases affecting the climate and some cause-oriented proposals for sufficiently reducing this share. *Environ. Pollut.*, **83**, 95-111.

Isermann, K., 1994b: *Ammoniak.* Enquete-Kommission "Schutz der Erdatmosphaere" (ed.) Landwirtschaft, Studienprogramm, Economica Verlag, Bonn, Germany.

Isermann, K. and R. Isermann, 1995: Present situation, demands, possible solutions and outlooks for a sustainable agriculture, human nutrition, waste and wastewater management before the background of nutrient balances within the EU [Barrage, A. and X. Edelmann (eds.)]. R'95 Congress (Recovery/Recycling/Re-Integration) Proc. IV, Eidgen, Materialpruefungs - u. Forschungsanstalt (EMPA), Duebendorf, Switzerland, pp. 151-156.

Ismail, I., R.L. Blevins, and W.W. Frye, 1994: Long-term no-tillage effects on soil properties and continuous corn yields. *Soil Sci. Soc. Am. J.*, **58**, 193-196.

Janzen, H.H., 1987: Soil organic matter characteristics after long-term cropping to various spring wheat rotations. *Can. J. Soil Sci.*, **67**, 845.

Jenkinson, D.S., 1971: The accumulation of organic matter in soil left uncultivated. Rothamsted Exp. Sta. Ann. Rep. for 1970, part 2, pp. 113-137.

Jenkinson, D.S., D.D. Harkness, E.D. Vance, D.E. Adams, and A.F. Harrison, 1992: Calculating net primary production and annual input of organic matter to soil from the amount and radiocarbon content of soil organic matter. *Soil Biology and Biochemistry,* **24**, 295-308.

Johansson, B.J., H. Kelly, A.K.N. Reddy, and R.H. Williams (eds.), 1993: *Renewables for Fuels and Electricity.* Island Press, Washington, DC, 1160 pp.

Johnson, D.E., G.M. Ward, and J. Torrent, 1991: The environmental impact of the use of bST in dairy cattle. *J. Environ. Qual.*, **21**, 157-162.

Johnson, D.E., T.M. Hill, G.M. Ward, K.A. Johnson, M.E. Branine, B.R. Carmean, and D.W. Lodman, 1993: Ruminants and other animals. In: *Atmospheric Methane: Sources, Sinks, and Role in Global Change* [Khalil, M.A.K. (ed.)]. Springer-Verlag, New York, NY, pp. 199-229.

Kaltschmitt, M. and S. Becher, 1994: Biomassenutzung in Deutschland—Stand und Perspektiven. In: *Thermische Nutzung von Biomasses* [BMELF (ed.)]. Schriftenreihe "Nachwachsende Rohstoffe," Vol. 2, Landwirtschaftsverlag GmbH, Muenster-Hiltrup, Germany, pp. 9-25.

Keerthisinghe, D.G., J.R. Freney, and A.R. Mosier, 1993: Effect of wax-coated calcium carbide and nitrapyrin on nitrogen loss and methane emission from dry-seeded flooded rice. *Biol. Fertil. Soils*, **16**, 71-75.

Keller, M., M.E. Mitre, and R.F. Stallard, 1990: Consumption of atmospheric methane in soils of central Panama: effects of agricultural development. *Global Biogeochem. Cycles*, **4**, 21-27.

Keller, M., M.E. Veldkamp, A.M. Weltz, and W.A. Reiners, 1993: Effect of pasture age on soil trace-gas emissions from a deforested area of Costa Rica. *Nature*, **365**, 244-246.

Kern, J.S. and M.G. Johnson, 1993: Conservation tillage impacts on national soil and atmospheric carbon levels. *Soil Sci. Soc. Am. J.*, **57**, 200-210.

Kleinhanss, W., 1993: Pflanzenoele als Treibstoff—Erzeugung, Nutzung, Perspektiven. In: *Energie aus Biomasse—eine Chance fuer die Landwirtschaft* [Flaig, H. and H. Mohr (eds.)]. Springer, Berlin-Heidelberg-New York, pp. 67-83.

Koepf, H., S. Kaffka, and F. Sattler, 1988: *Naehrstoffbilanz und Energiebedarf im landwirtschaftlichen Betriebsorganismus.* Verl. Freies Geistesleben, Stuttgart, Germany, 62 pp.

Kroeze, C., 1994: Anthropogenic emissions of nitrous oxide (N_2O) from Europe. *Sci. Total Environ.*, **152**, 189-205.

Lal, R., 1986: Soil surface management in the tropics for intensive land use and high and sustained production. *Adv. Soil Sci.*, Vol. 5, Springer-Verlag, Inc., New York, NY, 109 pp.

Lal, R. and T.J. Logan: Agricultural activities and greenhouse gas emissions from soils of the tropics. In: *Soil Management and Greenhouse Effect* [Lal, R. *et al.* (eds.)]. Lewis Publishers, Chelsea, MI (in press).

Lauren, J.G. and J.M. Duxbury, 1993: Methane emissions from flooded rice amended with green manure. In: *Agroecosystem Effects on Radiatively Important Trace Gases and Global Climate Change* [Harper, L.A., A.R. Mosier, J.M. Duxbury, and D.E. Rolston (eds.)]. ASA Special Publication No. 55, Madison, WI, pp. 183-192.

Lauren, J.G., G.S. Pettygrove, and J.M. Duxbury, 1994: Methane emissions associated with a green manure amendment to flooded rice in California. *Biogeochem.*, **24**, 53-65.

Lee, J.J., D.L. Phillips, and R. Liu, 1993: The effect of trends in tillage practices on erosion and carbon content of soils in the US corn belt. *Water, Air, Soil Pollut.*, **70**, 389-401.

Leible, L. and D. Wintzer, 1993: Energiebilanzen bei nach-wachsenden Energietraegern. In: *Energie aus Biomasse - eine Chance fuer die Landwirtschaft* [Flaig, H. and H. Mohr (eds.)]. Springer, Berlin-Heidelberg-New York, pp. 67-83.

Leng, R.A., 1991: *Improving Ruminant Production and Reducing Methane Emissions from Ruminants by Strategic Supplementation.* USEPA report 400/1-91/004, Office of Air and Radiation, Washington, DC.

Levine, J.S., W.R. Cofer, and J.P. Pinto, 1993: Biomass burning. In: *Atmospheric Methane: Sources, Sinks and Role in Global Change* [Khalil, M.A.K. (ed.)]. Springer-Verlag, Berlin, Heidelberg, Germany, pp. 230-253.

Lin, E., H. Dong, and Y. Li, 1994: Methane emissions of China: agricultural sources and mitigation options. In: *Non-CO_2 Greenhouse Gases* [van Ham, J. *et al.* (eds.)]. Kluwer Academic Publishers, Dordrecht, Netherlands, pp. 405-410.

Lindau, C.W., P.K. Bollich, R.D. DeLaune, A.R. Mosier, and K.F. Bronson, 1993: Methane mitigation in flooded Louisiana rice fields. *Biol. Fertil. Soils*, **15**, 174-178.

Lobert, J.M., D.H. Scharffe, W.M. Hao, and P.J. Crutzen, 1990: Importance of biomass burning in the atmospheric budgets of nitrogen - containing gases. *Nature*, **346**, 552-554.

Lugo, A.E. and S. Brown, 1993: Management of tropical soils as sinks or sources of atmospheric carbon. *Plant and Soil*, **149**, 27-41.

Mann, L.K., 1986: Changes in soil carbon storage after cultivation. *Soil Sci.*, **142**, 279-288.

Marland, G. and A.F. Turhollow, 1991: CO_2 emissions from the production and combustion of fuel ethanol from corn. *Energy*, **16**, 1307-1316.

McTaggart, I., H. Clayton, and K. Smith, 1994: Nitrous oxide flux from fertilized grassland: strategies for reducing emissions. In: *Non-CO_2 Greenhouse Gases* [van Ham, J. *et al.* (eds.)]. Kluwer Academic Publishers, Dordrecht, Netherlands, pp. 421-426.

Minami, K. and A. Ohsawa, 1990: Emission of nitrous oxide dissolved in drainage water from agricultural land. In: *Soils and the Greenhouse Effect* [Bouwman, A.F. (ed.)]. John Wiley and Sons, New York, NY, pp. 503-509.

Minami, K., T. Shibuya, Y. Ogawa, and S. Fukushi, 1990: Effect of nitrification inhibitors on emission of nitrous oxide from soils. *Trans 14th Int. Congr. Soil Sci.*, **2**, 267-272.

Minami, K., J. Goudriaan, E.A. Lantinga, and T. Kimura, 1993: The significance of grasslands in emission and absorption of greenhouse gases. Proc. 17th Int. Grassland Congr., pp. 1231-1238.

Minami, K., 1993: Methane from rice production. In: *Methane and Nitrous Oxide* [van Amstel, A.R. (ed.)]. Proceedings IPCC Workshop, Bilthoven, Netherlands, pp. 143-162.

Minami, K., 1994: Effect of nitrification inhibitors and slow-release fertilizer on emission of nitrous oxide from fertilized soils, In: *CH_4 and N_2O: Global Emissions and Controls from Rice Fields and Other Agricultural and Industrial Sources* [Minami, K., A. Mosier, and R. Sass (eds.)]. NIAES Series 2. Yokendo Publishers, Tokyo, Japan, pp. 187-196.

Minami, K., A. Mosier, and R. Sass (eds.), 1994: *CH_4 and N_2O: Global Emissions and Controls from Rice Fields and Other Agricultural and Industrial Sources.* NIAES Series 2, Yokendo Publishers, Tokyo, Japan, 234 pp.

Mosier, A.R., D.S. Schimel, D. Valentine, K.F. Bronson, and W.J. Parton, 1991: Methane and nitrous oxide fluxes in native, fertilized, and cultivated grasslands. *Nature*, **350**, 330-332.

Mosier, A.R., 1993: Nitrous oxide emissions from agricultural soils. In: *Methane and Nitrous Oxide: Methods in National Emission Inventories and Options for Control Proceedings* [van Amstel, A.R. (ed.)]. National Institute of Public Health and Environmental Protection, Bilthoven, Netherlands, pp. 273-285.

Mosier, A.R. and A.F. Bouwman, 1993: Working group report: nitrous oxide emissions from agricultural soils. In: *Methane and Nitrous Oxide: Methods in National Emission Inventories and Options for Control Proceedings* [van Amstel, A.R. (ed.)]. National Institute of Public Health and Environmental Protection, Bilthoven, Netherlands, pp. 343-346.

Mosier, A.R., J.M. Duxbury, J.R. Freney, O. Heinemeyer, and K. Minami, 1995: Nitrous oxide emission from agricultural fields: assessment, measurements and mitigation. *Plant and Soil* (in press).

Nesbit, S.P. and G.A. Breitenbeck, 1992: A laboratory study of factors influencing methane uptake by soils. *Agric. Ecosyst. Environ.*, **41**, 39-54.

Neue, H.U., 1992: Agronomic practices affecting methane fluxes from rice cultivation. In: *Trace Gas Exchange in a Global Perspective* [Ojima, D.S. and B.H. Svensson (eds.)]. *Ecol. Bull., Copenhagen*, **42**, 174-182.

Neue, H.U., R.S. Lantin, R. Wassmann, J.B. Aduna, M.C.R. Alberto, and M.J.F. Andales, 1994: Methane emission from rice soils of the Philippines. In: *CH_4 and N_2O: Global Emissions and Controls from Rice Fields and Other Agricultural and Industrial Sources* [Minami, K., A. Mosier, and R. Sass (eds.). NIAES, Yokendo Publishers, Tokyo, Japan, pp. 55-63.

Nguyen, M. L. and R.J. Haynes, 1995: Energy and labour efficiency for three pairs of conventional and alternative mixed cropping (pasture-arable) farms in Canterbury, New Zealand. *Agricultue, Ecosystems and Environment*, **52**, 163-172.

Nouchi, I., S. Mariko, and K. Aoki, 1990: Mechanisms of methane transport from the rhizosphere to the atmosphere through rice plant. *Plant Physiol.*, **94**, 59-66.

Nouchi, I., 1994a: Mechanisms of methane transport through rice plants. In: *CH_4 and N_2O: Global Emissions and Controls from Rice Fields and Other Agricultural and Industrial Sources* [Minami, K., A. Mosier, and R. Sass (eds.)]. NIAES, Yokendo Publishers, Tokyo, Japan, pp. 87-104.

Nouchi, I., T. Hosono, K. Aoki, and K. Minami, 1994b: Seasonal variation in methane flux from rice paddies associated with methane concentration in soil water, rice biomass and temperature, and its modeling. *Plant and Soil*, **161**, 195-208.

OECD/OCDE, 1991: *Estimation of Greenhouse Gas Emissions and Sinks.* Final report from the OECD Experts Meeting, 18-21 February 1991, Prepared for Intergovennmental Panel on Climate Change, Washington, DC, Revised August, 1991.

Ojima, D.S., D.W. Valentine, A.R. Mosier, W.J. Parton, and D.S. Schimel, 1993a: Effect of land use change on methane oxidation in temperate forest and grassland soils. *Chemosphere*, **26**, 675-685.

Ojima, D.S., B.O.M. Dirks, E.P. Glenn, C.E. Owensby, and J.M.O. Scurlock, 1993b: Assessment of C budget for grasslands and drylands of the world. *Water, Air, Soil Pollut.*, **70**, 95-109.

Ojima, D.S., W.J. Parton, D.S. Schimel, J.M.O. Scurlock, and T.G.F. Kittel, 1993c: Modeling the effects of climatic and CO_2 changes on grassland storage of soil C. *Water, Air, Soil Pollut.*, **70**, 643-657.

Oldeman, L.R., V.W.P. van Engelen, and J.H.M. Pulles, 1990: The extent of human-induced soil degradation. In: *World Map of the Status of Human-Induced Soil Degradation: An Expanatory Note* [Oldeman, L.R., R.T.A. Hakkeling, and W.G. Sombroek (eds.)]. International Soil Reference and Information Centre, Wageningen, Netherlands.

Parashar, D.C., A.P. Mitra, S.K. Sinha, P.K. Gupta, J. Rai, R.C. Sharma, N. Singh, S. Kaul, G. Lai, A. Chaudhary, H.S. Ray, S.N. Das, K.M. Parida, S.B. Rao, S.P. Kanungo, T. Ramasami, B.U. Nair, M. Swamy, G. Singh, S.K. Gupta, A.R. Singh, B.K. Saikia, A.K.S. Barua, M.G. Pathak, C.P.S. Iyer, M. Gopalakrishnan, P.V. Sane, S.N. Singh, R. Banerjee, N. Sethunathan, T.K. Adhya, V.R. Rao, P. Palit, A.K. Saha, N.N. Purkait, G.S. Chaturvedi, S.P. Sen, M. Sen, B. Arkrkar, A. Banik, B.H. Subbarary, S. Lal, and S. Venkatramani, 1994: Methane budget from Indian paddy fields. In: *CH_4 and N_2O: Global Emissions and Controls from Rice Fields and Other Agricultural and Industrial Sources* [Minami, K., A. Mosier, and R. Sass (eds.)]. NIAES. Yokendo Publishers, Tokyo, Japan, pp. 27-39.

Parton, W.J., D.S. Schimel, C.V. Cole, and D.S. Ojima, 1987: Analysis of factors controlling soil organic matter levels in Great Plains grasslands. *Soil Sci. Soc. Am. J.*, **51**, 1173-1179.

Parton, W.J., J.M.O. Scurlock, D.S. Ojima, T.B. Gilmanov, R.J. Scholes, D.S. Schimel, T.B. Kirchner, J.-C. Meanut, T. Seastedt, E. Garcia Moya, A. Kamnalrut, and J.I. Kinyamario, 1993: Observations and modeling of biomass and soil organic matter dynamics for the grassland biome worldwide. *Global Biogeochemical Cycles*, **7**, 785-809.

Paul, E.A., K. Paustian, E.T. Elliott, and C.V. Cole (eds.), 1995: *Soil Organic Matter in Temperate Agroecosystems. Long-Term Experiments in North America*. Lewis Publishers, Chelsea, MI (in press).

Paustian, K., H.P. Collins, and E.A. Paul, 1995: Management controls on soil carbon. In: *Soil Organic Matter in Temperate Agroecosystems. Long-Term Experiments of North America* [Paul, E.A., K. Paustian, E.T. Elliott, and C.V. Cole (eds.)]. CRC/Lewis Publishers, Chelsea, MI (in press).

Peoples, M.B., A.R. Mosier, and J.R. Freney, 1995: Minimizing gaseous loss of nitrogen. In: *Nitrogen Fertilization in the Environment* [Bacon, P.E. (ed.)]. Marcel Dekker Inc., New York, NY, pp. 565-602.

Pimentel, D., W. Dazhong, and M. Giampietro, 1990: Technological changes in energy use in U.S. agricultural production. In: *Agroecology: Researching the Ecological Basis for Sustainable Agriculture* [Gliessman, S.R. (ed.)]. Ecological Studies: Analysis and Synthesis (USA), **78**, 305-321, Springer, New York, NY.

Potter, C.S., J.T. Randerson, C.B. Field, P.A. Matson, P.M. Viousek, H.A. Mooney, and S.A. Klooster, 1993: Terrestrial ecosystem production: a process model based on global satellite and surface data. *Global Biogeochemical Cycles*, **7**, 811-842.

Powlson, D.S., P.C. Brookes, and B.T. Christensen, 1987: Measurement of soil microbial biomass provides an early indication of changes in total soil organic matter due to straw incorporation. *Soil Biol. Biochem.*, **19**, 159-164.

Powlson, D.S., P. Smith, and J.U. Smith (eds.), 1995: *Evaluation of Soil Organic Matter Models Using Existing Long-Term Data Sets*. NATO ASI Series, Springer Verlag (in press).

Rahm, M.R. and W.E. Huffman, 1984: The adoption of reduced tillage: the role of human capital and other variables. *Am. J. Agric. Econ.*, **66**, 405-413.

Raich, J.W., E.B. Rastetter, J.M. Melillo, D.W. Kicklighter, P.A. Steudler, B.J. Peterson, A.L. Grace, B. Moore III, and C.J. Vorosmarty, 1991: Potential net primary productivity in South America: application of a global model. *Ecol. Applic.*, **1**, 399-429.

Rasmussen, P.E. and H.P. Collins, 1991: Long-term impacts of tillage, fertilizer, and crop residue on soil organic matter in temperate semi-arid regions. *Adv. Agron.*, **45**, 93.

Reeburgh, W.S., S.C. Whalen, and M.J. Alpern, 1993: The role of methylotrophy in the global methane budget. In: *Microbial Growth on C_1 Compounds*. pp. 1-14.

Reinhardt, G.A., 1993: Energie- und CO_2-Bilanzierung nachwachsender Rohstoffe. Theoretische Grundlagen und Fallstudie Raps. Vieweg, Wiesbaden, Germany.

Ryden, J.C., 1981: N_2O exchange between a grassland soil and the atmosphere. *Nature*, **292**, 235-237.

Saffigna, P.G., D.S. Powlson, P.C. Brookes, and G.A. Thomas, 1989: Influence of sorghum residues and tillage on soil organic matter and soil microbial biomass in an Australian Vertisoil. *Soil Biol. Biochem.*, **21**, 759-765.

Safley, L.M., M.E. Casada, J.W. Woodbury, and K.F. Roos, 1992: *Global Methane Emissions from Livestock and Poultry Manure*. USEPA Report 400/1-91/048, Office of Air and Radiation, Washington, DC, 68 pp.

Sampson, R.N., L.L. Wright, J.K. Winjum, J.D. Kinsman, J. Benneman, E. Kursten, and J.M.O. Scurlock, 1993: Biomass management and energy. In: *Terrestrial Biospheric Carbon Fluxes: Quantification of Sinks and Sources of CO_2* [Wisniewski, J. and R.N. Sampson (eds.)]. Kluwer Academic Publishers, Dordrecht, Netherlands, pp. 139-162.

Sanchez, P.A., C.A. Palm, and T.J. Smyth, 1990: Approaches to mitigate tropical deforestation by sustainable soil management practices In: *Soils on a Warmer Earth* [Scharpenseel, H.W., M. Schomaker, and A. Ayoub (eds.)]. Elsevier, Amsterdam, Netherlands, pp. 211-220.

Sass, R.L., 1994: Short summary chapter for methane. In: *CH_4 and N_2O: Global Emissions and Controls from Rice Fields and Other Agricultural and Industrial Sources* [Minami, K., A. Mosier, and R. Sass (eds.)]. NIAES, Yokendo Publishers, Tokyo, Japan, pp. 1-7.

Sass, R.L. and F.M. Fisher, 1994: CH_4 emission from paddy fields in the United States gulf coast area. In: *CH_4 and N_2O: Global Emissions and Controls from Rice Fields and Other Agricultural and Industrial Sources* [Minami, K., A. Mosier, and R. Sass (eds.)]. NIAES, Yokendo Publishers, Tokyo, Japan, pp. 65-77.

Sass, R.L., F.M. Fisher, F.T. Turner, and M.F. Jund, 1991: Methane emissions from rice fields as influenced by solar radiation, temperature, and straw incorporation. *Global Biogeochem. Cycles*, **5**, 335-350.

Sass, R.L., F.M. Fisher, Y.B. Wang, F.T. Turner, and M.F. Jud, 1992: Methane emission from rice fields: the effect of floodwater management. *Global Biogeochem. Cycles*, **6**, 249-262.

Sauerbeck, D., 1993: CO_2 emissions from agriculture: sources and mitigation potentials. *Water, Air, Soil Pollut.*, **70**, 381-388.

Sauerbeck, D., 1994a: Die Landwirtschaft als Verursacherin und Betroffene moeglicher Klimaveraenderungen. Klimaforschung in Bayern. In: *Rundgespraeche der Kommission fuer Oekologie* [Bayrische Akademie der Wissenschaften (ed.)]. Bd. 8, Verlg. F. Pfeil, Muenchen, pp. 151-168.

Sauerbeck, D., 1994b: Nitrogen fertilization and nitrogen balance—environmental consequences and limitations. In: *The Terrestrial Nitrogen Cycle as Influenced by Man* [Mohr, H.U. and K. Muentz (ed.)]. *Nova Acta Leopoldina (Halle)*, **70**, 441-447.

Schutz, H., A. Holzapfel-Pschorn, R. Conrad, H. Rennenberg, and W. Seiler, 1989: A 3-year continuous record on the influence of daytime, season and fertilizer treatment on methane emission rates from an Italian rice paddy. *J. Geophys. Res.*, **94**, 16,405-16.

Schwab, A.W., M.O. Bagby, and B. Freeman, 1987: Preparation and properties of diesel fuels from vegetable oils. *Fuels*, **66(10)**, 1372-1378.

Scurlock, J.M.O., D.O. Hall, J.I. House, and R. Howes, 1993: Utilizing biomass crops as an energy source: a European perspective. In: *Terrestrial Biospheric Carbon Fluxes: Quantification of Sinks and Sources of CO_2* [Wisniewski, J. and R.N. Sampson (eds.)]. Kluwer Academic Publishers, Dordrecht, Netherlands, pp. 499-518.

Seiler, W., R. Conrad, and D. Scharffe, 1984: Field studies of methane emission from termite nests into the atmosphere and measurements of methane uptake by tropical soils. *J. Atmos. Chem.*, **1**, 171-186.

Sharpley, A.N. and J.R. Williams, 1990: *EPIC—Erosion/Productivity Impact Calculator: 1. Model Documentation*. U.S. Department of Agriculture Technical Bulletin No. 1768, Washington, DC, 235 pp.

Shibata, M., 1994: Methane production in ruminants. In: *CH_4 and N_2O: Global Emissions and Controls from Rice Fields and Other Agricultural and Industrial Sources* [Minami, K., A. Mosier, and R. Sass (eds.)]. NIAES, Yokendo Publishers, Tokyo, Japan, pp. 105-115.

Sims, R.E.H., 1990: Tallow esters and vegetable oil as alternative diesel fuels—a review of the New Zealand programme. *Solar and Wind Technology*, **7(1)**, 31-36.

Smith, K.A., 1990: Greenhouse gas fluxes between land surfaces and the atmosphere. *Progress in Phys. Geogr.*, **14**, 349-372.

Smith, K.A., G.P. Robertson, and J.M. Melillo, 1994: Exchange of trace gases between the terrestrial biosphere and the atmosphere in the mid latitudes. In: *Global Atmospheric–Biospheric Chemistry* [Prinn, R.B. (ed.)]. Plenum Press, New York, NY, pp. 179-203.

Sombroek, W.G., F.O. Nachtergaele, and A. Hebel, 1993: Amounts, dynamics and sequestering of carbon in tropical and subtropical soils. *Ambio*, **22**, 417-426.

Steudler, P.A., R.D. Bowden, J.M. Melillo, and J.D. Aber, 1989: Influence of nitrogen fertilization on methane uptake in temperate forest soils. *Nature*, **341**, 314-316.

Swift, M.J. and P.A. Sanchez, 1984: Biological management of tropical soil fertility for sustained productivity. *Nature Res.*, **20**, 2-10.

Tice, T.F. and F.M. Epplin, 1984: Cost-sharing to promote use of conservation tillage. *J. Soil Water Conserv.*, **39**, 395-397.

Tyson, K.C., D.H. Roberts, C.R. Clement, and E.A. Garwood, 1990: Comparison of crop yields and soil conditions during 30 years under annual tillage or grazed pasture. *J. Agric. Sci.*, **115**, 29-40.

U.S. EPA, 1990a: *Greenhouse Gas Emissions from Agricultural Systems*. Proc. IPCC-RSWG-AFOS Workshop, Vol. 1 and 2, USEPA-OPA, Washington, DC, IPCC/WMO, Geneva, Switzerland.

U.S. EPA, 1990b: *Policy Options for Stabilizing Global Climate. Draft Report to Congress* [Tirpak, D. and D. Lashof (eds.)]. U.S. Environmental Protection Agency, Washington, DC, 810 pp.

U.S. EPA, 1992: *Anthropogenic Methane Emissions in the US: Estimates for 1990.* Report No. 430-R-93-003, U.S. Environmental Protection Agency, Office of Air and Radiation, Washington, DC.

U.S. EPA, 1993: Methane emissions from livestock manure. In: *Global Methane Emissions Report to Congress.* Prepared by Climate Change Division, Office of Policy, Planning and Evaluation, Environmental Protection Agency, Washington, DC, Review Draft.

U.S. EPA, 1994: *International Anthropogenic Methane Emissions: Estimates for 1990.* EPA 20-R-93-010, U.S. Environmental Protection Agency, Office if Policy, Planning and Evaluation, Washington, DC, 2122 pp.

U.S. Government Executive Office, 1995: *Economic Report of the President.* U.S. Government Printing Office, Washington, DC, 407 pp.

van Heerwarden, K., 1992: Agriculture and the greenhouse effect. *Change,* **II**, 17-20.

Vellguth, G., 1983: Performance of vegetable oils and their monoesters as fuels for diesel engines. Proceedings of the International Off-Highway Meeting and Exposition, Milwaukee, WI, September 12-15, Society of Automotive Engineers, pp. 1-10.

Wang, Mingxing, Dai Aiguo, Shangguan Xingjian, Ren Lixin, Shen Renxing, H. Schutz, W. Seiler, R.A. Rasmussen, and M.A.K. Khalil, 1994: Sources of methane in China. In: *CH_4 and N_2O: Global Emissions and Controls from Rice Fields and Other Agricultural and Industrial Sources* [Minami, K., A. Mosier, and R. Sass (eds.)]. NIAES, Yokendo Publishers, Tokyo, Japan, pp. 9-26.

Watson, R.T., H. Rodhe, H. Oeschger, and U. Siegenthaler, 1990: Greenhouse gases and aerosols. In: *Climate Change: The IPCC Scientific Assessment* [Houghton, J.T., G.J. Jenkins, and J.J. Ephraums (eds.)]. Cambridge University Press, Cambridge, UK, pp. 1-40.

Watson, R.T., L.G. Meira Filho, E. Sanhueza, and T. Janetos, 1992: Greenhouse gases: sources and sinks. In: *Climate Change 1992: The Supplementary Reports to the IPCC Scientific Assessment* [Houghton, J.T., B.A. Callander, and S.K. Varney (eds.)]. Cambridge University Press, Cambridge, UK, pp. 25-46.

Wood, B.J. and R.H.V. Conley, 1991: The energy balance of oil palm cultivation. Proceedings 1991 PORIM Int. Palm Oil Conference, pp. 130-143.

Wright, L.L. and E. Hughes, 1993: U.S. carbon offset potential using biomass energy systems. In: *Terrestrial Biospheric Carbon Fluxes: Quantification of Sinks and Sources of CO_2*, [Wisniewski, J. and R.N. Sampson (eds.)]. Kluwer Academic Publishers, Dordrecht, Netherlands, pp. 483-498.

Yagi, K. and K. Minami, 1990: Effect of organic matter application on methane emission from some Japanese paddy fields. *Soil Sci. Plant Nutr.*, **36**, 599-610.

Yagi, K. and K. Minami, 1993: Spatial and temporal variations of methane flux from a rice paddy field. *Biogeochemistry of Global Change* [Oremland, R.S. (ed.)]. Chapman and Hall, New York, NY, pp. 353-368.

Yagi, K., H. Tsuruta, K. Mianmi, P. Chairoj, and W. Cholitkul, 1994: Methane emission from Japanese and Thai paddy fields. In: *CH_4 and N_2O: Global Emissions and Controls from Rice Fields and Other Agricultural and Industrial Sources* [Minami, K., A. Mosier, and R. Sass (eds.)]. NIAES, Yokendo Publishers, Tokyo, Japan, pp. 41-53.

24

Management of Forests for Mitigation of Greenhouse Gas Emissions

SANDRA BROWN, USA

Principal Lead Authors:
J. Sathaye, USA; Melvin Cannell, UK; P. Kauppi, Finland

Contributing Authors:
P. Burschel, Germany; A. Grainger, UK; J. Heuveldop, Germany; R. Leemans, The Netherlands; P. Moura Costa, Brazil; M. Pinard, USA; S. Nilsson, Sweden; W. Schopfhauser, Austria; R. Sedjo, USA; N. Singh, India; M. Trexler, USA; J. van Minnen, The Netherlands; S. Weyers, Germany

CONTENTS

EXECUTIVE SUMMARY

Three categories of promising forestry practices that promote sustainable management of forests and at the same time conserve and sequester carbon (C) are considered in this chapter: (1) management for conservation of existing C pools in forests by slowing deforestation, changing harvesting regimes, and protecting forests from other anthropogenic disturbances; (2) management for expanding C storage by increasing the area and/or C density in native forests, plantations, and agroforestry and/or in wood products; and (3) management for substitution by increasing the transfer of forest biomass C into products such as biofuels and long-lived wood products that can be used instead of fossil-fuel based products. Since the 1992 assessment, significant new information has been developed that improves estimates of the quantities of C that can be conserved or sequestered—and the associated implementation costs of forest sector mitigation strategies—and better identifies limits to the amount of lands available for such mitigation strategies.

- The most effective long-term (>50 years) ways in which to use forests to mitigate the increase in atmospheric CO_2 are to substitute fuelwood for fossil fuels and for energy-expensive materials. However, over the next 50 years or so, substantial opportunities exist to conserve and increase the C store in living trees and wood products (High Confidence).
- Under baseline conditions (today's climate and no change in the estimated available lands over the period of interest), the cumulative amount of C that could potentially be conserved and sequestered over the period 1995–2050 by slowing deforestation (138 Mha) and promoting natural forest regeneration (217 Mha) in the tropics, combined with the implementation of a global forestation program (345 Mha of plantations and agroforestry), would be about 60 to 87 Gt—equivalent to 12–15% of the projected (IPCC 1992a scenario) cumulative fossil fuel C emissions over the same period (Medium Confidence).
- The annual C gain from the above program would reach about 2.2 Gt/yr by 2050, or about four times the value of 0.5 Gt C/yr estimated in the 1992 assessment. The gradual increase over time occurs because of the time it takes for programs to be implemented and the relatively slow rate at which C accumulates in forest systems (Medium Confidence).
- Uncertainty associated with the C conservation and sequestration estimates is caused mainly by high uncertainty in estimating land availability for forestation and regeneration programs and the rate at which tropical deforestation can actually be reduced; estimates of the net amount of C per unit area conserved or sequestered under a particular management scheme are more certain (High Confidence).
- The tropics have the potential to conserve and sequester the largest quantity of C—45–72 Gt—more than half of which would be due to promoting natural forest regeneration and slowing deforestation. Tropical America has the largest potential for C conservation and sequestration (46% of the tropical total), followed by tropical Asia (34%) and tropical Africa (20%). The temperate and boreal zones could sequester about 13 Gt and 2.4 Gt, respectively—mainly in the United States, temperate Asia, the former Soviet Union (FSU), China, and New Zealand (Medium Confidence).
- The cumulative cost—excluding land costs and other transaction costs—to conserve and sequester the above amounts of C range from US$247 billion to $302 billion at a unit cost of about $2–8/t C. These unit costs are considerably lower than those in the 1992 assessment, which ranged from $8/t C in tropical latitudes to $28 in non-U.S. Organisation for Economic Cooperation and Development (OECD) countries. Transaction costs may significantly increase these estimated costs (Low Confidence).
- Costs per unit of C sequestered or conserved generally increase from low- to high-latitude nations (High Confidence) and from slowing deforestation and promoting regeneration to establishing plantations (Low Confidence). The latter trend may not hold if transaction costs of slowing deforestation are excessive.
- These cost estimates, although benefitting from improved data and methodology since the first assessment, generally represent only the cost of direct forest practices. These costs could be several times higher if land and opportunity costs and/or the costs of establishing infrastructure, protective fencing, training programs, and tree nurseries were included; on the other hand, costs could be offset by revenues from timber and non-timber products. No complete cost estimates are available (Medium Confidence).
- Under conditions of climate and land-use change as projected by the IMAGE 2.0 model ("conventional wisdom" scenario, akin to the IPCC 1992a scenario), the carbon conservation and sequestration potential may be somewhat less than estimated under baseline conditions because less land may be available in the tropics, unless land was secured for mitigation measures or sustainable agriculture/agroforestry systems were widely adopted, and sequestration by new forests in temperate and boreal regions may be offset by transient decline and the loss of carbon from existing forests in response to climate change (Medium Confidence).

24.1. Introduction

Forest ecosystems merit consideration in biogenic mitigation strategies because they can be both sources and sinks of CO_2, the most abundant greenhouse gas (GHG). Currently the world's forests are estimated to be a net C source, primarily because of deforestation and forest degradation in the tropics. Temperate and boreal forests are a C sink because many are recovering from past natural and human disturbances, and they are actively managed (Dixon *et al.*, 1994a). However, there is the potential to lessen projected C emissions by protecting and conserving the C pools in existing forests; to create C sinks by expanding C storage capacities, by increasing the area and/or C density of native forests, plantations, and agroforests, and by increasing the total pool of wood products; and to substitute fossil fuels with fuelwood from sustainably managed forests, short-lived wood products with long-lived wood products, and energy-expensive materials with wood (Grainger, 1988; Dixon *et al.*, 1991; IPCC, 1992; Winjum *et al.*, 1992a, 1992b; Nilsson and Schopfhauser, 1995; Trexler and Haugen, 1995). This chapter reviews the potential magnitude of forest-based CO_2 stabilization options based on various assessments in recent years; the costs to implement such programs; and the effects of a changed climate, atmospheric composition, and human demographics on the potential amount of C conserved and sequestered. The chapter also suggests how to devise improved assessments to formulate practical strategies.

The first supplemental report to IPCC (Houghton *et al.*, 1992) suggested that aggressive forest-sector mitigation strategies involving planting trees on 1 Gha (Gha = 10^9 ha) of land, combined with phasing out net deforestation by 2025, could create a net C sink of about 0.5 Gt/yr (10^{15} g = 1 Pg) by 2050—a value that could then be maintained through the rest of the 21st century. Since 1992, new research results and information have improved the accuracy of estimates of the quantities of C that potentially can be conserved (maintain C on the land) and sequestered (increase C on the land) through the implementation of forest-sector mitigation strategies, provided new estimates of the costs associated with the mitigation options, and made more-accurate estimates of the land available for such strategies. A review and analysis of new research results and information are the main purposes of this chapter.

Scope of the Chapter

This chapter reviews the potential to manage present and potential forest lands capable of supporting tree cover to conserve and sequester C. Urban forests are not included because they contain and accumulate a very small amount of C compared to other forest lands, although they can contribute to reduced energy consumption (Rowntree and Nowak, 1991). Forest management here includes an array of practices in native forests and on nonforested lands—such as protection, forestation (afforestation and reforestation), intermediate silvicultural treatments (e.g., thinning, fertilization), harvesting, and agroforestry—that promote sustained production of goods and services. These promising forestry practices are considered in arriving at national, regional, and global estimates of forest-sector potential for mitigating the accumulation of primarily CO_2, and to some degree other GHGs, in the atmosphere.

Forest lands are divided into three latitudinal belts: high or boreal (approximately 50–75° N and S latitude), mid or temperate (approximately 25–50° N and S latitude), and low or tropical (approximately 0–25°N and S latitude). Nations or regions are grouped into these belts on the basis of the approximate geographic location of their forests.

The key parameters in any assessment of mitigation strategies are the amount of C per unit area of land that can be conserved or sequestered in vegetation and soil under given site conditions and a given management option; the time period over which this C can be conserved or sequestered; the amount of suitable and available land; the mitigation costs; and the different lifetimes of the end wood products. Literature data and discussions on these factors are presented in this chapter, although large uncertainties exist in some data (Houghton *et al.*, 1993; Iverson *et al.*, 1993; Trexler and Haugen, 1995). Estimates of the mitigation potential and costs of the various options are generated from literature sources. Because no studies to date have addressed the mitigation potential of forests under a changed climate and atmosphere, the mitigation potential could be assessed only under a baseline condition, with no effects of climate change or increased atmospheric CO_2. More briefly, the effects of a changed climate, atmospheric composition, demographics, and land use on the mitigation potential are considered based on interpretation of trends from a global integrated assessment model (Alcamo, 1994). A discussion of new research directions to improve assessments concludes the chapter.

24.2. Role of Forests in the Global Carbon Cycle

24.2.1. Status and Change in Forest Area

Forests cover about 4.1 Gha of the Earth (Dixon *et al.*, 1994a, with revisions based on Kolchugina and Vinson, 1995). Most of the forests are in the low latitudes (43%), followed by the high latitudes (32%) and mid-latitudes (25%). Forest plantations are currently estimated to occupy about 0.1 Gha of land. Although the technology for managing forests is well developed, today only about 11% of the world's forests are managed for goods and services (World Resources Institute, 1990; Winjum *et al.*, 1992a). The extent of management, however, varies by region: About 20% of the mid-latitude forests, 17% of the high-latitude forests, and less than 4% of the low-latitude forests are managed. In addition, large areas of land technically suitable for forests are degraded or are otherwise underproducing because of human misuse (Winjum *et al.*, 1992a). Even though some degraded lands are unsuitable for forestry, there is considerable potential to mitigate CO_2 by better management of forest lands for C conservation, storage, and substitution. However, a balance between objectives for mitigation and other uses of forests must be achieved (see Section 24.3).

The status and areas of forests change, even in the absence of human interference. However, humans influence the pace and extent of change as forests are subjected to controlled and uncontrolled uses (overharvesting and degradation); large-scale occurrence of wildfire; fire control; pest and disease outbreaks; and conversion to non-forest use, particularly agriculture and pastures. At the same time, some areas of harvested and degraded forests or agricultural and pasture lands are abandoned and revert naturally or are converted to forests or plantations. In high latitudes, the area of forests is undergoing little change (Kolchugina and Vinson, 1995). In mid-latitudes, there is a net gain of about 0.7 Mha/yr of forests, mostly in Europe and China (Dixon *et al.*, 1994a). Furthermore, many of the forests in high and mid-latitudes have been harvested (clear cut or selective cut) in the far to near past and are now generally in a stage of regeneration and regrowth (Apps and Kurz, 1991; Birdsey, 1992; Heath *et al.*, 1993; Kauppi *et al.*, 1992, 1995; Kolchugina and Vinson, 1993, 1995).

Low-latitude forests are experiencing high rates of loss—currently estimated to be about 15.4 Mha/yr during 1980–90, but with large uncertainties (FAO, 1993). Much of the deforested area is converted to new agricultural or pasture lands, which often replace degraded agricultural lands that may or may not be capable of supporting tree cover (Brown, 1993; Dale *et al.*, 1993). However, in a few tropical countries, deforestation has decreased during the last decade [e.g., India (Ravindranath and Hall, 1994; see also Section 24.3.1.1), Brazil (Skole and Tucker, 1993), and Thailand (Dixon *et al.*, 1994a)]. In addition to deforestation, large areas of forests are harvested and degraded. For example, about 5.9 Mha/yr of low-latitude forests were logged during 1986–90, and most logging occurred in mature forests (83%) rather than secondary forests (FAO, 1993). These harvested forests can regenerate and accumulate C if they are protected or are relatively inaccessible to human populations, but many of them become degraded (e.g., Lanly, 1982; Brown *et al.*, 1993b, 1994). Forest degradation, resulting in a loss of biomass C, occurs through damage to residual trees and soil from poor logging practices, log poaching, fuelwood collection, overgrazing, and anthropogenic fire (Goldammer, 1990, 1993; Brown *et al.*, 1991, 1993b; FAO, 1993; Flint and Richards, 1994). Similar anthropogenic disturbances most likely have occurred and are still occurring in other forest regions of the world. This means that few forested areas in the world are presently undisturbed by humans; this fact has implications for their present role in the global C cycle and future C sequestration potentials (Lugo and Brown, 1986, 1992; Brown *et al.*, 1992; Wood, 1993).

24.2.2. Forest Carbon Pools and Flux

The world's natural forests contain vast quantities of organic C, with an estimated 330 Gt C in vegetation (live and dead, above and belowground), 660 Gt C in soil (mineral soil plus organic horizon) (Table 24-1), and another 10 Gt C in plantations. Estimates of all C pool components, using published factors (see Dixon *et al.*, 1994a), were made in arriving at these C pool estimates. However, some components are poorly known, such as the C pool in woody detritus and slash and dead roots—which undoubtedly adds to the uncertainty in the estimated total C pool. Most of the C pool in vegetation is located in the low-latitude forests (64%), whereas most of the soil C pool is located in high-latitude forests (52%). Country-level analyses demonstrate that forests already play an important role in the C budget of some countries by offsetting significant amounts of fossil fuel emissions (Box 24-1).

Table 24-1: *Estimated C pools and flux in forest vegetation (above and belowground living and dead mass, including woody debris) and soils (O horizon and mineral soil to 1-m depth) in forests of the world. Dates of estimate vary by country and region, but cover the decade of the 1980s. Estimates are based on complete C budgets in all latitudes, using data from original source or from adjustments for completeness.*

Latitudinal Belt	**C Pools (Gt)**		**C Flux (Gt/yr)**
	Vegetation	Soils	
High			
FSU[1]	46	123	+0.3 to +0.5
Canada[2]	12	211	+0.08
Alaska[3]	2	11	*
Subtotal	**60**	**345**	**+0.48 ± 0.2**
Mid			
USA[3]	15	21	+0.1 to +0.25
Europe[4]	9	25	+0.09 to +0.12
China[5]	17	16	-0.02
Australia[6]	18	33	trace
Subtotal	**59**	**95**	**+0.26 ± 0.1**
Low			
Asia[7]	41–54	43	-0.50 to -0.90
Africa[8]	52	63	-0.25 to -0.45
America[8]	119	110	-0.50 to -0.70
Subtotal	**212**	**216**	**-1.65 ± 0.40**
Total	**331**	**656**	**-0.9 ± 0.5**

* Included with USA.
[1] FSU = Former Soviet Union; Kolchugina and Vinson, 1993, 1995.
[2] Apps and Kurz, 1991; Kurz and Apps, 1993; Kurz *et al.*, 1992.
[3] Birdsey, 1992; Birdsey *et al.*, 1993; Dixon *et al.*, 1994a; Turner *et al.*, 1995a.
[4] Dixon *et al.*, 1994a; Kauppi *et al.*, 1992; includes Nordic countries.
[5] Xu, 1992.
[6] Gifford *et al.*, 1992.
[7] Brown *et al.*, 1993b; Dixon *et al.*, 1994a; FAO, 1993; Houghton, 1995.
[8] Dixon *et al.*, 1994a; FAO, 1993; Houghton, 1995.

Box 24-1. Estimates of Current C Sequestration Rates by Forests in Relation to the Size of their C Pool and Annual Emissions from Fossil Fuel Burning

Some countries that signed the Framework Convention on Climate Change have made detailed calculations of the amounts of C currently being sequestered by their forests, taking into account the historic rates of forest planting and harvesting, and the dynamics of forest growth and the flow of C to litter and forest product pools (see Table 24-2 for a sampling). In Britain, New Zealand, and India, C is being sequestered as a result of recent forestation programs—whereas in Finland, C is being sequestered by natural regeneration and regrowth of forests because the annual growth is greater than the annual harvest. In Canada, carbon sequestration is occurring as a result of recovery from previous fires. In 17 West European countries, forestry offsets from 1–2% (The Netherlands, Britain, Germany) to about 90% (Sweden) of fossil fuel emissions (Burschel *et al.*, 1993; Kauppi and Tomppo, 1993).

Table 24-2: *Carbon storage and C sequestration rates in forests, and national C emissions.*

Country	Year	Carbon Stored in Trees and Litter (Mt)	Fossil Fuel–C Emissions (Mt/yr)	Rate of C Removed by Forests (Mt/yr)[1]	Source
Britain	1990	60[2]	164	2.5 (1.5)	Cannell and Dewar, 1995
New Zealand	1990	113	8	3.5 (44)	Maclaren and Wakelin, 1991
Finland	1992	978	18	5.0 (28)	Karjalainen and Kellomäki, 1993
Germany	1990	1500–2000	268	5.4 (2)	Federal Ministry for Env., 1994
Canada	1986	12000	136	51.0 (37.5)	Kurz *et al.*, 1992
India	1986	10000	137	5.0 (3.6)	Makundi *et al.*, 1996
Poland	1990	1113	131	8.0 (6)	Galinski and Kuppers, 1994
USA	1990	18585	1300	80 (6)	Turner *et al.*, 1995a

[1]Value in parentheses is the percentage of fossil fuel emissions removed by forests.
[2]Plantation forests only; all forests and woodlands contain ~87 Mt C in the trees alone (Milne, pers. comm.).

Mid- and high-latitude forests are currently estimated to be a net C sink of about 0.7 ± 0.2 Gt/yr because forests at these latitudes are, on average, composed of relatively young classes with higher rates of net production as they recover from past disturbances such as abandonment of agricultural land, harvesting, and wildfires; a larger proportion of these forests are actively managed (i.e., established, tended, and protected); and some areas may be responding to increased levels of atmospheric CO_2 and nitrogen (N) (fertilization effect) (Apps and Kurz, 1991; Birdsey, 1992; Kauppi *et al.*, 1992, 1995; Xu, 1992; Heath *et al.*, 1993; Kolchugina and Vinson, 1993, 1995; Kurz and Apps, 1993; Turner *et al.*, 1995a). Because secondary forests in the mid-and high latitudes are rebuilding C pools, there is a finite potential over which this C sequestration can occur. For example, the current C sink in European forests may disappear within 50 to 100 years (Kauppi *et al.*, 1992), although others suggest that it may take forests as long as several centuries to millennia to reach a C steady state in all components (Lugo and Brown, 1986).

Low-latitude forests are estimated to be a relatively large net C source of 1.6 ± 0.4 Gt/yr in 1990 (Table 24-1), caused by deforestation, harvesting, and gradual degradation of the growing stock. Although this is the best estimate available in the literature, we use this value with recognition that there are many reasons to believe that the uncertainty is larger than shown (Lugo and Brown, 1992). Unlike the high- and mid-latitude forests—where estimated C fluxes are based, for the most part, on data from periodic national inventories (i.e., field measurements)—the estimated C flux for low-latitude forests is based on a model that tracks only forests that are cleared or harvested with regrowth. In the model, C accumulates in regrowing forests for up to 50–100 years. Furthermore, the model assumes that all other forests not reportedly affected by humans during the period of model simulation (about 1850–1990) are in C steady state (Houghton *et al.*, 1987). Recent work questions this steady-state assumption and implies that the net tropical C flux could be higher or lower than that reported here, depending upon the relative contribution of forest lands that are still gaining C through recovery from past human disturbances or are losing C through continued human use (Brown *et al.*, 1992; Lugo and Brown, 1992).

The error terms associated with the C flux estimates in Table 24-1 are derived from the range of values resulting from the use of different assumptions in the C budgets for a given country or region. They do not represent errors derived from statistical procedures. Error enters the flux estimation procedure through uncertainties and biases in the primary data, and these compound as the data are combined to draw inferences (Robinson, 1989). Many estimates for components of the forest-sector C budget are probably known no better than ±30% of their mean, and others may be known no better than ± 50% or more of their means (Robinson, 1989). These errors are compounded in making global estimates of C flux—perhaps to large proportions—but to what extent is presently unknown. Clearly, there is a need to apply error-estimation techniques to calculations of the forest-sector C budgets to provide more precise estimates of the C flux.

The global average net C flux and uncertainty term from the world's forests reported here of -0.9 ± 0.5 Gt/yr (a net C source; Table 24-1) is less than that reported in Schimel *et al.* (1995) of -1.1 ± 1.1 Gt/yr. The main reason for this difference is in the interpretation and use of results from the literature. A major difference exists between the respective estimates used for the C flux of boreal forests, particularly for the FSU. In this chapter (Table 24-1), the studies of Kolchugina and Vinson (1993, 1995) were used, which include a complete forest-sector C budget and use more current data; Schimel *et al.* (1995) relied more on the work by Melillo *et al.* (1988) and Krankina and Dixon (1994). These two latter studies were not used in arriving at the estimate in Table 24-1 for the mean C flux for boreal forests because the Melillo *et al.* (1988) results were for an earlier period (prior to 1980) and the Krankina and Dixon (1994) study did not consider all components of the C budget. The main source of difference in the uncertainty term is the error assigned to the tropical net flux: We report an uncertainty term around the tropical flux of ± 0.4 Gt/yr based on Houghton (1995), whereas Schimel *et al.* (1995) report a term of ± 1.0 Gt/yr based on an assumption that the uncertainty was greater than 50% for reasons given above. However, as discussed above, none of these uncertainty terms may reflect the true precision of the estimates.

Substitution of the net C flux for forests reported here (Table 24-1) into the global C budget (Schimel *et al.*, 1995) results in an imbalance of 1.2 ± 1.0 Gt/yr. Because the primary data for C budgets for temperate and boreal countries originate from national forest inventories, any increased growth of forests due to CO_2 and N fertilization and climatic effects is already included in the net flux estimates. In other words, the reported C sink for mid- to high-latitude forests (Table 24-1) includes all these factors already because the data, for the most part, come from repeated forest inventories. In contrast, the tropical forest C flux is based on a model and not on repeated forest inventories. Furthermore, the model does not include effects of CO_2 and N fertilization and climate. This leads to the conclusion that a large part of the imbalance in the global C budget must be due to a C sink in tropical latitudes, which also has been suggested by others (Lugo and Brown, 1992; Taylor and Lloyd, 1992; Schimel *et al.*, 1995). This could be due to a combination of stimulated regrowth from CO_2 and N fertilization and climate, as well as more extensive forest regrowth. It is clear that to resolve this issue, repeated national forest inventories, with permanent plots, are needed in tropical latitudes.

Most biomass burning in tropical forests is intentional and is associated with land-clearing practices. However, wildfires also occur in tropical moist and dry forests; these also are largely of anthropogenic origin (Goldammer, 1990). Factors contributing to increased wildfires in tropical moist forests are the drying of organic materials (fuel) on the forest floor of degraded forests (increased exposure to solar radiation) and changes in the microclimate of forest remnants surrounded by deforested areas (Fearnside, 1990; Kaufman and Uhl, 1990). Significant areas of temperate and boreal forest also are burned by wildfires of natural or anthropogenic origin or prescribed (controlled) fires (Levine, 1991; Auclair and Carter, 1993; Dixon and Krankina, 1993).

The destruction of forest biomass by burning releases, in addition to CO_2, GHGs that are byproducts of incomplete combustion—namely, methane (CH_4), carbon monoxide (CO), nitrous oxide (N_2O), and nitrogen oxide (NO_x), among others. Whereas complex accounting models and forest inventories are needed to estimate the losses and gains of C over different timescales, the emissions of these other gases from biomass burning are instantaneous, absolute transfers from the biosphere to the atmosphere (Crutzen *et al.*, 1979; Crutzen and Andreae, 1990). Globally, biomass burning contributes about 10% of total annual CH_4 emission, 10–20% of total annual N_2O emission, and about half of the CO emission—and so has a significant effect on atmospheric chemistry, especially on tropospheric ozone levels (Houghton *et al.*, 1992). Biomass burning also transfers a fraction (up to 10%) of the C to an inert form (charcoal) with a turnover time that is practically infinite.

Many boreal forests grow on peat or organic soils that contain very large amounts of C. Undisturbed anaerobic, northern peatlands are sinks for CO_2 and sources of CH_4 (Matthews and Fung, 1987; see Chapter 6). Drainage of these soils to improve forest productivity virtually stops CH_4 emissions but initiates rapid CO_2 loss by aerobic decomposition. Draining peat soils for forest establishment can produce a C loss from these soils that exceeds C stored in the forest if 20–30 cm of peat decompose as a result of the drainage (Cannell *et al.*, 1993). There also are vast areas of forested peatlands in the tropics; how they will be affected by drainage is largely unknown (see Chapter 6).

In addition to managing forest vegetation to conserve or sequester C, there also is an opportunity to manage forest soils for the same purposes (Johnson, 1992; Lugo and Brown, 1993; Dixon *et al.*, 1994a). Regional and national programs to conserve soil, including organic matter, have been implemented worldwide (Dixon *et al.*, 1994a). Management practices to maintain, restore, and enlarge forest soil C pools include (after Johnson, 1992) enhancement of soil fertility; concentration of agriculture and reduction of slash-and-burn practices; preservation of wetlands; minimization of site disturbance during harvest operations to retain organic matter; forestation of degraded and nondegraded sites; and any practice that reduces soil aeration, heating, and drying. Several long-term experiments demonstrate that C can accrete in the soil at rates of 0.5 to 2.0 t/ha/yr (Dixon *et al.*, 1994a).

24.3. Carbon Mitigation Options

It should be emphasized at the outset that the objective of the practices presented here—that is, to foster C conservation and sequestration in forests—is but one of a variety of objectives for forest management that needs to be balanced with other objectives. However, most forest-sector actions that promote C conservation and sequestration make good social, economic, and ecological sense even in the absence of climate-change considerations. Other objectives for managing forests include sustainable development, industrial wood and fuel production, traditional forest uses, protection of natural resources (e.g., biodiversity, water, and soil), recreation, rehabilitation of damaged lands, and the like; C conservation and sequestration resulting from managing for these objectives will be an added benefit. For example, although the primary reasons for the establishment of plantations on non-forested land have been economic development, provision of new wood resources (e.g., Portugal, Swaziland), replacement of diminishing or less-productive natural forests (e.g., Australia, Brazil, Malaysia), import substitution (e.g., United Kingdom, Zimbabwe), generation of export income (e.g., Chile, New Zealand), or rehabilitation projects (Evans, 1990; Kanowski and Savill, 1992; Kanowski *et al.*, 1992), they also are considered an important means for sequestering C.

There are basically three categories of forest management practices that can be employed to curb the rate of increase in CO_2 in the atmosphere. These categories are:

1) Management for conservation (prevent emissions)
2) Management for storage (short-term measures over the next 50 years or so)
3) Management for substitution (long-term measures).

The goal of conservation management is mainly to conserve existing C pools in forests as much as possible through options such as controlling deforestation, protecting forests in reserves, changing harvesting regimes, and controlling other anthropogenic disturbances such as fire and pest outbreaks. The goal of storage management is to expand the storage of C in forest ecosystems by increasing the area and/or C density of natural and plantation forests and increasing storage in durable wood products. Substitution management aims at increasing the transfer of forest biomass C into products (e.g., construction materials and biofuels) that can replace fossil-fuel-based energy and products, cement-based products, and other building materials.

24.3.1. Conservation Management

24.3.1.1. Controlling Deforestation

Slowing the rate of loss and degradation of existing forests could reduce CO_2 emissions substantially. The most significant C conservation clearly would occur in the tropics, where each Mha of deforestation produces about a 0.1 Gt C net flux (flux from Table 24-1, divided by area of deforestation of 15.4 Mha/yr). Because the burning of biomass usually accompanies deforestation, slowing deforestation also would reduce emissions of other GHGs.

Reducing tropical deforestation and forest degradation rates would require action to reduce the pressures for land and commodities while increasing the protection of remaining forests for the purposes of conservation and timber production. Most deforestation and degradation is caused by the expansion and degradation of arable and grazing lands and subsistence and commodity demand for wood products—which in turn are a response to the underlying pressures of population growth, socioeconomic development, and political forces. Thus, programs to reduce deforestation must be accompanied by measures that increase agricultural productivity and sustainability, as well as initiatives to slow the rate of population growth (Grainger, 1990; Waggoner, 1994; see also Chapters 13, 15, and 23) and deal with the socioeconomic and political issues.

Although reducing deforestation in the tropics may appear to be a difficult task, there are countries where this is happening (e.g., Brazil, India, Thailand). An example of note is India, where net deforestation has been reduced significantly. Despite its high population density and growth rate, India has succeeded in stabilizing the area under forest during the last decade at about 64 Mha, about 19% of its land area (Ravindranath and Hall, 1994). This stabilization does not mean deforestation of native forests has ceased but rather that loss of native forests is balanced by establishment of plantations (see Chapter 15). This has been achieved by strong forest conservation legislation, a large forestation program, and community awareness.

Past international efforts to curb deforestation, such as the Tropical Forestry Action Plan, have met with limited success (Trexler and Haugen, 1995). Major factors are the absence of comprehensive agricultural policies that meet the needs of resource-poor farmers and the growing global demand for food, fiber, and fuel for the increasing human population (Brown, 1993; Grainger, 1993). Deforestation has been viewed as largely a forestry and conservation problem, and tackling the symptoms rather than the root causes is an incomplete strategy. Global action to mitigate emissions of C by conserving C pools may lead to more interest and success in controlling deforestation and making agriculture more sustainable.

24.3.1.2. Protection and Conservation of Forests

In recent years, there has been significant expansion of "protected areas" into areas of both mature and secondary forests for conservation of biodiversity and sustainable timber production (presently about 10% of the forest land; World Conservation Monitoring Centre, 1992). Carbon pools should remain the same or increase in size in these areas, depending on their present age-class distribution. New protected areas should include those that contain large C pools, such as forests growing on peat soils at high and low latitudes, and high-biomass old-growth forests.

Two biologically significant forest regions that have received attention as C reservoirs in recent years are the Amazon Basin of Brazil and the forests of the FSU (Fearnside, 1992; Kolchugina and Vinson, 1993, 1995; Krankina and Dixon, 1992). The largest contiguous area of C-dense forests in the world is found in the FSU. These forests are subject to future environmental degradation and harvesting, accelerating the loss of C (Brown *et al.*, 1996). Protection of these forests from harvest without regeneration, uncontrolled fires, and pollutants is a priority of the FSU, but infrastructure and resources there are underdeveloped.

The management of tropical forests for sustainable timber production is likely to increase over the coming decades due to collective action under the auspices of the second International Tropical Timber Agreement (agreed to in January 1994). It also is likely that a trend toward management for sustainable timber production in all of the world's forests will occur in the future. Using forests for sustainable timber production—including extending rotation cycles, reducing waste, implementing soil conservation practices, and using wood in a more C-efficient way—ensures that a large fraction of forest C is conserved. Paper recycling is another strategy with the potential to reduce harvest levels and promote greater C conservation (Turner *et al.*, 1995b).

24.3.2. Storage Management

Storage management means increasing the amount of C stored in vegetation (living, above and belowground biomass), soil (litter, dead wood, mineral soil, and peat where important), and durable wood products. Increasing the C pool in vegetation and soil can be accomplished by protecting secondary forests and other degraded forests whose biomass and soil C densities are less than their maximum value and allowing them to sequester C by natural or artificial regeneration and soil enrichment. Other approaches are to establish plantations on non-forested lands; promote natural or assisted regeneration in secondary forests, followed by protection; or increase tree cover on agricultural or pasture lands (agroforestry) for environmental protection and local needs. The C pool in durable wood products can be increased by expanding demand for wood products at a faster rate than the decay of wood and by extending the lifetime of wood products. These measures include timber treatment and the production of long-lasting particle boards (Elliott, 1985).

Sequestering C by storage management produces only a finite C sequestration potential in vegetation and soils, beyond which little additional C can be accumulated. The process may take place over a time period on the order of decades to centuries, depending upon the present age-class of forest, the attainable maximum C density, forest type, species selection, and latitudinal zone. In the long run, this is less helpful than substitution options (see Section 24.3.3), given the expected continuous need to offset future C emissions.

Expansion of C pools through the establishment of plantations is becoming less socially and politically desirable, especially with the global concern for biodiversity and other social, cultural, land-tenure, and economic factors (Nilsson and Schopfhauser, 1995). However, in many situations plantations are the only option, and they can increase local biodiversity through the reestablishment of native species in the understory when they are established on highly degraded lands and are subject to no further management (Lugo *et al*,. 1993; Parrotta, 1993; Allen *et al.*, 1995). These forests can then contribute to the development goals of national forest sectors. Furthermore, if more native forests are to be protected and/or harvesting levels reduced, plantation establishment may become more necessary to offset wood reductions (see Chapter 15).

24.3.3. Substitution Management

Substitution management, which has the greatest mitigation potential in the long term (>50 years) (Marland and Marland, 1992; Swisher, 1995), views forests as renewable resources. It focuses on the rate of C sequestration or the transfer of biomass C into products that substitute for or lessen the use of fossil fuels, rather than on increasing the C pool itself (Grainger, 1990; Mixon *et al.*, 1994). This approach involves extending the use of forests for wood products and fuels obtained either by establishing new forests or plantations or by increasing the growth of existing forests through silvicultural treatments (Table 24-3). However, a consideration of growth rates and initial standing stocks of biomass C is critical in determining which existing forests should be used for this purpose (Marland and Marland, 1992). For example, it is better not to convert forests with a large initial standing biomass C and slow growth rates (e.g., old-growth forests) to managed stands because it may take a very long time (up to centuries) until the net C sequestered returns to its initial value (Harmon *et al.*, 1990; Marland and Marland, 1992)—or never, if they are harvested on a rotational basis (Dewar, 1991; Vitousek, 1991; Dewar and Cannell, 1992; Cannell, 1995). In contrast, forests with high growth rates and low-to-medium initial biomass C standing stocks are amenable for conversion to managed forests, with considerable quantities of C sequestered if the harvested wood is directly used (Marland and Marland, 1992). When presented with cleared or disturbed forest lands, any forest management practice that increases the C pools and cycles the C by harvesting wood for substitution of energy-intensive products or fossil fuels will remove CO_2 from the atmosphere on a continuing basis.

In the case of forests established on non-forested lands for energy products such as fuelwood, not only is there an increase in the amount of C stored on the land (Grainger, 1990; Schroeder, 1992) but, if the wood burned as fuel displaces fossil fuel usage, it creates an effective rate of C sequestration in unburned fossil fuels (Hall *et al.*, 1991; Sampson *et al.*, 1993; see also Chapter 19). There must, however, be a net energy return in the total system (Hall *et al.*, 1986; Herendeen and Brown, 1987). The extent to which fuelwood plantations are able to displace fossil fuel use in developed countries will depend on the continued development of highly efficient technologies (e.g., Williams and Larson, 1993) for converting

Table 24-3: Sequestration potential of different forest types (from Nabuurs and Mohren, 1993).

Forest Type	Long-Term Average Quantity of C in All Living Biomass and Forest Products (t/ha)	Long-Term Average Quantity of C in Litter, Dead Wood, and Soil to 100 cm (t/ha)	Average Net Annual Rate of C Accumulation[1] (t/ha/yr)
Tropical Forests			
Heavily logged evergreen rainforest	144	92	2.4
Selectively logged evergreen rainforest	207	102	2.9
Logged rainforest hampered by vines	125	92	0.8
Heavily logged semi-evergreen rainforest	76	76	1.1
Selectively logged semi-evergreen rainforest	151	98	2.0
Pinus caribaea in Brazil and Venezuela	89	90	5.1
P. elliotii in Brazil	111	80	3.9
Temperate Forests			
Picea in central Europe	137	117	2.0
Pseudotsuga in northwest USA	196	143	3.4
P. radiata in New Zealand and Australia	126	97	4.5
P. taeda in southeast USA	59	81	3.2
Mixed deciduous in central Europe	110	105	1.4
Broadleaf forests on old agricultural land	62–111	75–84	2.2–3.4
Boreal Forests			
Picea in Russia	53	139	1.0

[1] C sink over first rotation through net primary production minus decomposition of soil organic matter, litter, dead wood, logging slash, and products.

wood into clean forms of energy like electricity. The extent to which fuelwood plantations will be established on degraded lands in developing countries for the generation of electricity will depend more on other incentives such as rural employment and income generation.

When forests are used to produce sawtimber, plywood, or other industrial wood products, C can be sequestered for long periods. The length of time depends on how the timber is treated and used. The production of wood products often requires much less energy than does production of alternative products like steel, aluminum, and concrete, and there can be a large energy return on investment in wood products. For example, the substitution of composite solid-wood products with load-bearing capacities for steel and concrete can save large amounts of fossil fuels. However, an analysis of the full life cycles of wood products is required to appreciate the impact on net C storage and net C emissions. Over long time periods, the displacement of fossil fuels either directly or through production of low-energy-intensive wood products is likely to be more effective in reducing C emissions than physical storage of C in forests or forest products. For example, substitution of wood grown in plantations for coal in the generation of electricity can avoid C emissions by an amount up to four times the amount of C sequestered in the plantation, depending upon the time period over which coal resources are expected to last (Ravindranath and Hall, 1995). Furthermore, if the wood resource is derived from sustainably managed plantations, other benefits, such as rural jobs and land rehabilitation (if the project is on degraded lands), will accrue (Ravindranath and Hall, 1995).

24.4. Assessment of C Mitigation Options

The potential land area available for the implementation of forest management options for C conservation and sequestration is a function of the technical suitability of the land to grow trees and the actual availability as constrained by socioeconomic factors. Determining the technical suitability of land is conceptually less difficult, being based generally on the region's climatic and edaphic characteristics. However, moving from what is suitable to what is actually available is more difficult because of institutional, economic, demographic, and cultural factors—all of which influence present and future land-use decisions (Trexler and Haugen, 1995).

This section is divided into two parts: steady-state potential, based on estimates of the amount of C that could be conserved and sequestered when all actually available lands (in some cases, only technically suitable lands are considered) are under management and contain their maximum C density for a given practice; and transient potential, based on estimates of the amount of C that could be conserved and sequestered on all available land subject to reasonable rates of establishment of forest management practices over time.

24.4.1. Steady-State Potential

Most previous estimates of the C sequestration potential are simply the product of the suitable or available land area and the maximum or time-averaged C density (e.g., Grainger, 1988; Postel and Heise, 1988; Sedjo and Solomon, 1989; Winjum *et al.*, 1992a). Using this approach and approximate estimates of areas of land available for reforestation and natural and assisted regeneration, Winjum *et al.* (1992a) estimate that 2.2–5.6 Gt C could be sequestered in the entire high-latitude zone over a 50-year period. In Russia alone, Krankina and Dixon (1994) estimate a potential sequestration of 4.5 Gt C over a 50-year period by replacing hardwood stands with conifers, increasing the productivity of existing forests, and establishing plantations on technically suitable lands. Furthermore, they estimate that exercising fire management and reducing harvest could account for an additional potential C savings of 7.2 Gt over a 50-year period.

On estimates of available land in mid-latitudes, 13.5–27.0 Gt C could be sequestered by forestation and natural regeneration over a 50-year period (Winjum *et al.*, 1992a). In the United States alone, an aggressive program of tree planting on marginal agricultural lands (crops and pastures), increasing windbreaks and shelter belts, and forest management could sequester up to 15 Gt C over about a 40-year period (Sampson, 1992).

At low latitudes, a combination of forestation, agroforestry, and natural or assisted forest regeneration on an estimated 300 to 600 Mha of land considered to be available by Winjum *et al.* (1992a) could conserve and sequester about 36–71 Gt C over 50 years. If all technically suitable lands are considered, more than double this amount could be conserved and sequestered by plantation establishment, converting arable/degraded lands to agroforestry, and protecting existing forests (Houghton *et al.*, 1993; Iverson *et al.*, 1993; Unruh *et al.*, 1993).

Recent country-specific estimates for China, India, Mexico, and Thailand indicate that the available land for establishment of plantations and agroforestry is 172 Mha, 175 Mha, 44 Mha, and 7 Mha, respectively (Masera *et al.*, 1995; Ravindranath and Somashekar, 1995; Wangwacharakul and Bowonwiwat, 1995; Xu, 1995). These estimates of land availability are based on national forest-cover targets for China and India and on technical availability for the other two countries. The C sequestration potential on these lands could be up to about 5 Mt/yr.

24.4.2. Transient Potential

Although the studies cited in Section 24.4.1 set a useful upper bound for the potential for C conservation and sequestration, only a few studies consider the issues of land availability, socioeconomic barriers, the timescale over which C may be sequestered, or realistic or feasible rates of forest establishment or regeneration. At the present rate of successful plantation establishment in the tropics of only 1.8 Mha/yr (FAO, 1993), it would take many decades to achieve the levels of sequestration suggested by Winjum *et al.* (1992a). Similar time frames would be needed for the establishment of agroforestry, as well as large areas of mid- and high-latitude plantations. Once established, reaching maximum C storage would then take up to many decades.

A study by Grainger (1990) focuses on the tropics only. He considers a range of scenarios for conserving and sequestering C, including a combination of forestation and deforestation control, and assumes that all wood harvested from the new plantations would be converted into long-term storage as industrial wood. The study does not consider woody detritus or belowground C, is not regionally explicit, and gives only a partial treatment to land availability. The analysis suggests that sequestering about 3 Gt C/yr (the rate at which it was increasing in the atmosphere in 1980) would require 600 Mha of new forest plantations—roughly the area of degraded lands regarded by Grainger (1988) as physically suitable for plantation establishment.

24.4.2.1. Analytical Approach

Nilsson and Schopfhauser (1995) and Trexler and Haugen (1995) are the only studies suitable for global analysis of the mitigation potential of forests. These studies have been chosen because they are global in nature, include an extensive literature review of the land availability issue, and include feasible rates of establishment of management options. These two studies have been combined to arrive at a global estimate of the potential amount of C that could be conserved and sequestered by different regions of the Earth on an annual and cumulative basis between 1995 and 2050. Nilsson and Schopfhauser (1995) estimate the potential for C sequestration through a feasible global forestation program (Table 24-4). Trexler and Haugen (1995) focus on the tropics only and include the options of slowing deforestation and natural or assisted regeneration of land (followed by protection). Both studies assume aggressive, but unspecified, policy and financial interventions in the forestry sectors, with no future change in climate that might interfere with the proposed strategies. The combined analysis uses data for forestation from Nilsson and Schopfhauser (1995) and data for slowing deforestation and regeneration from Trexler and Haugen (1995).

Nilsson and Schopfhauser (1995) estimate for different countries/regions the amount of land likely to be available, feasible annual planting rates, likely growth rates, and rotation lengths (Table 24-4). They use a growth model, without intensive management, to estimate the quantity of C fixed in aboveground and belowground biomass, litter, and soil organic matter for the period 1995 to 2100. Because the focus of the study was to estimate how much C could be sequestered by a global forestation

Table 24-4: Regional estimates of land availability, average mean annual increment (MAI), rotation length, and planting rate for a global forestation program, including establishment of plantations and agroforestry, to sequester C (data from Nilsson and Schopfhauser, 1995).

Region/ Country	Land Available[1] (Mha)	MAI (m^3/ha/yr)	Rotation Length (yr)	Planting Rate[2] (Mha/yr)
High Latitudes				
Canada[3]	28.3	2.5–8.0	60	1.14
Nordic	0.35	5	60	0.014
FSU	66.5	3	80	1.66
Mid-Latitudes				
USA	21.0	6–15	15–40	0.70
Europe	7.74	6–10	20–60	0.31
China	62.5	2.3	80	2.5
Asia	12.5	12	40	0.50
South Africa	1.9	16	30	0.075
South America	4.6	15	25	0.18
Australia	4.3	6–23	30	0.123
New Zealand	5.0	25	25	0.1
Low Latitudes				
Tr. America	40.8	8–25	20	0.74
Tr. Africa	31.6	8–16	30	0.58
Tr. Asia	57.7	8–16	20	1.05

[1]Full details of sources are provided in Nilsson and Schopfhauser (1995); many of the sources originate from individual countries.
[2]Includes rate of establishment of both plantations and agroforestry systems.
[3]Canada includes not satisfactorily restocked (NSR) forest areas in addition to marginal agricultural lands (Van Kooten, 1991); the low end of the MAI was used for NSR forests.

program only, they make no assumptions about the life expectancy of the wood produced. Further, the amount of C sequestered by this program will be realized only if the forests are harvested at their designated rotation lengths. Considerably higher yields than those reported in Table 24-4, though often possible in plantations, require more intensive management than is likely to be achieved for a large-scale program such as that proposed by Nilsson and Schopfhauser (1995). Similar arguments can be made for shortening the rotation times.

For most high- and mid-latitude countries, the area of actually available land is equated with that technically suitable, or about 215 Mha (Table 24-4). The rather large amount of land available in Canada is from a combination of large areas of not satisfactorily restocked forest lands (19.7 Mha) and marginal agricultural lands (8.6 Mha; Van Kooten, 1991). In low-latitude countries, Nilsson and Schopfhauser estimate actual availability (130 Mha) to be only 6% or so of those lands deemed suitable (2,228 Mha) because of additional cultural, social, and economic constraints (Trexler and Haugen, 1995). The global estimate of 345 Mha of actually available lands for plantations and agroforestry is similar to the low end of the estimated range of 375 to 750 Mha offered by Winjum *et al.* (1992a) for the same forestry practices.

The assumed establishment rates of 8.3 Mha/yr for plantations and 1.4 Mha/yr for agroforestry [estimated from land available, plantation period, and rotation length; see Nilsson and Schopfhauser (1995) for more details] are not unrealistic based on present establishment rates. For example, the assumed annual rate of plantation establishment in the tropics used in Nilsson and Schopfhauser is about 65% of the actual rate for 1980–90 (FAO, 1993). For China, the assumed rate of 2.5 Mha/yr also is lower than the reported rate of 3.9 Mha/yr (Xu, 1992).

Trexler and Haugen (1995) use country-level estimates for each decade from 1990 to 2050 for 52 tropical countries accounting for virtually all of the tropical forests. For each country and decade (based on detailed country-by-country analysis), they estimate current and projected future deforestation rates, the potential reduction in deforestation based on feasible implementation of alternative land uses, and the area presently available for natural or assisted forest regeneration (native forests) followed by protection, as well as likely rates of implementation. Based on these estimates, Trexler and Haugen (1995) project that by the year 2050, deforestation could be reduced by only 20% of the business-as-usual scenario—equivalent to about 138 Mha—and a further 217 Mha of land will be available for natural or assisted regeneration. They also estimate the change in aboveground biomass C associated with each land-use change. We have added estimates of belowground biomass, soil, and litter C (see footnote 2 to Table 24-5) to be consistent with the study of Nilsson and Schopfhauser (1995).

24.4.2.2. Quantities of C Conserved and Sequestered

Together, the studies suggest that 700 Mha of land might be available globally for C conservation and sequestration programs (345 Mha for plantations and agroforestry, 138 Mha for slowed tropical deforestation, and 217 Mha for natural and assisted regeneration of tropical forests). This amount of land could conserve and sequester 60 to 87 Gt C by 2050 (Table 24-5). Globally, forestation and agroforestry account for 50% of the total (38 Gt C), with about 20% of this accumulating in soils, litter, and belowground biomass (Nilsson and Schopfhauser, 1995). The amount of C that could be conserved and sequestered by forest-sector practices by 2050 under baseline conditions is equivalent to about 12 to 15% of the total fossil fuel emissions over the same time period (IPCC 1992a scenario).

The tropics have the potential to conserve and sequester by far the largest quantity of C (80%), followed by the temperate zone (17%), and the boreal zone (3% only). More than half of the tropical sink would be caused by natural and assisted regeneration followed by forest protection and slowed deforestation. Other analyses have shown that forest conservation and natural regeneration is potentially easier, cheaper (see Section 24.5), and more acceptable to the local population than plantation-based forestation (Deutscher Bundestag, 1990; Grainger, 1990). Forestation and agroforestry would contribute less than half of the tropical total, but without them, regeneration and slowed deforestation would be highly unlikely (Trexler and Haugen, 1995).

Table 24-5: *Global estimates of potential amount of C that could be sequestered and conserved by forest management practices between 1995 and 2050 (from Nilsson and Schopfhauser, 1995; Trexler and Haugen, 1995).*

Latitudinal Belt	**Country/ Region**	**Practice**	**C Sequestered and Conserved (Gt)**
High	Canada	Forestation	0.68
	Nordic Europe		0.03
	FSU		1.76
	Subtotal		**2.4**
Mid	Canada	Forestation	0.43
	USA		3.07
	Europe		0.96
	China		1.70
	Asia		2.19
	South Africa		0.44
	South America		1.02
	Australia		0.31
	New Zealand[1]		1.7
	Subtotal		**11.8**
	USA	Agroforestry	0.29
	Australia		0.36
	Subtotal		**0.7**
Low	Tr. America	Forestation	8.02
	Tr. Africa		0.90
	Tr. Asia		7.50
	Subtotal		**16.4**
	Tr. America	Agroforestry	1.66
	Tr. Africa		2.63
	Tr. Asia		2.03
	Subtotal		**6.3**
	Tr. America	Regeneration[2]	4.8–14.3
	Tr. Africa		3.0–6.7
	Tr. Asia		3.8–7.7
	Subtotal		**11.5–28.7**
	Tr. America	Slow Deforestation[2]	5.0–10.7
	Tr. Africa		2.5–4.4
	Tr. Asia		3.3–5.8
	Subtotal		**10.8–20.8**
	Total		**60–87**

[1] This estimate differs from the one made by Maclaren (1996), because it includes multiple rotations and different data for the C budget components.

[2] Includes an additional 25% of aboveground C to account for belowground C in roots, litter, and soil (based on data in Nilsson and Schopfhauser, 1995; Brown *et al.*, 1993b); range in values is based on the use of low and high estimates of biomass C density resulting from the uncertainty in these estimates.

The total quantity and annual rates of C conservation and sequestration would vary among countries or regions (Table 24-5; Figure 24-1). The FSU accounts for more than 70% of the C sequestration potential in the boreal zone (Table 24-5). At mid-latitudes, the greatest potential for sequestering C would be in the United States (about 3.4 Gt), followed by temperate Asia (about 2.2 Gt), and China and New Zealand (1.7 Gt each) (Figure 24-1a). In low latitudes, tropical America would have the greatest potential for C conservation and sequestration (27 Gt), followed by tropical Asia (20 Gt) and tropical Africa (12 Gt) (Table 24-5; Figure 24-1b). Forestation and regeneration would likely have the greatest potential in tropical America and Asia, whereas agroforestry would be the most important activity in tropical Africa (Nilsson and Schopfhauser, 1995).

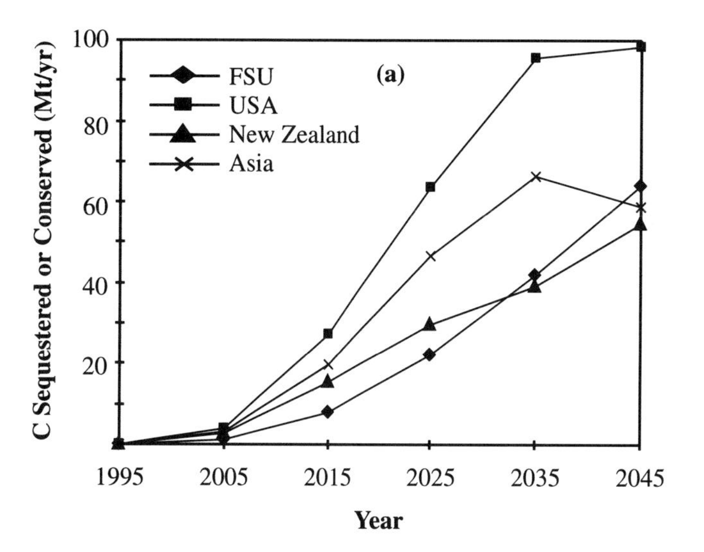

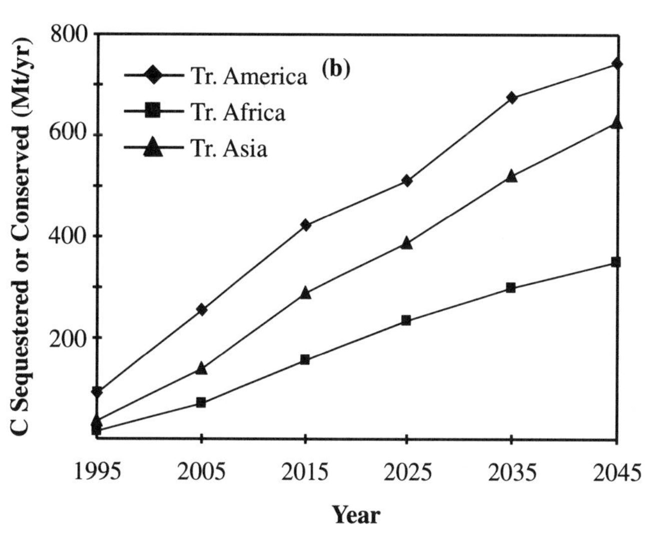

Figure 24-1: Average annual rates of C conservation and sequestration per decade through implementation of forest management options listed in Table 24-5 (a) by four countries or regions of the high- and mid-latitudes with the highest total sequestration rates and (b) for the three tropical (Tr.) regions.

Annual rates of C conservation and sequestration from the practices listed in Table 24-5 would increase over time and reach 2.2 Gt/yr by 2050 (Figure 24-2a), with C accretion in the tropics dominating the flux. Carbon savings from slowed deforestation and regeneration would be the highest initially, but from 2025 onwards—when plantations would reach their maximum C accretion—they would sequester practically identical amounts as forestation (Figure 24-2b). During this period, tropical deforestation would continue, and the tropics would remain a net C source, albeit gradually diminishing. By about 2030, the tropics would become a C sink (Trexler and Haugen, 1995). On a global scale, forests could turn from a source to a sink by about 2010 due to C conserved in other zones.

In summary, the tropics appear to have the greatest long-term potential for C conservation and sequestration by—in decreasing order of importance—protecting lands for natural and assisted regeneration, slowing deforestation, forestation, and agroforestry. The mid-latitudes also could make a significant contribution, but the potential for the high latitudes appears to be limited. The uncertainty in the estimated mitigation potential of forests has not, at present, been estimated. Like the uncertainty associated with the estimated C flux from the world's forests (see Section 24.2.2), the uncertainty in the estimated mitigation potential is likely to be high. The factors causing the highest uncertainty are the estimated land availability for forestation projects and regeneration programs and the rate at which tropical deforestation can be actually reduced. The next most uncertain term is the amount of C that can be conserved or sequestered in tropical forests; there is considerable debate about how much biomass C tropical forests contain (Brown and Lugo, 1992). The net amount of C per unit area that can be sequestered in a global forestation program is more certain. However, as discussed in Section 24.2.2, all errors are certain to be compounded when making global estimates.

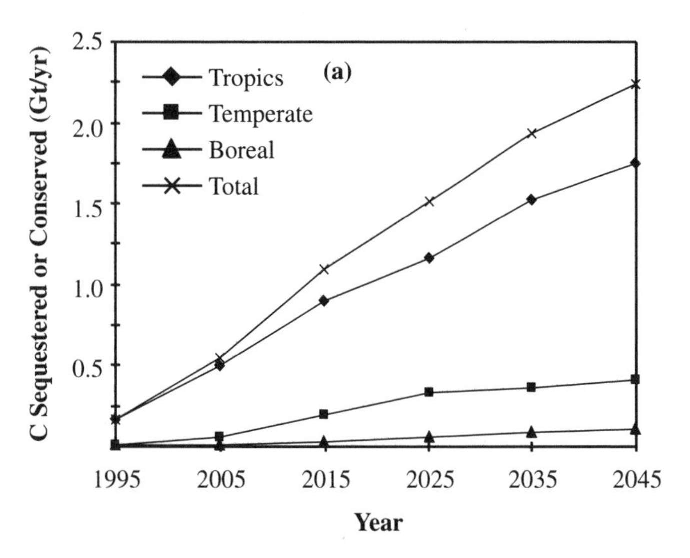

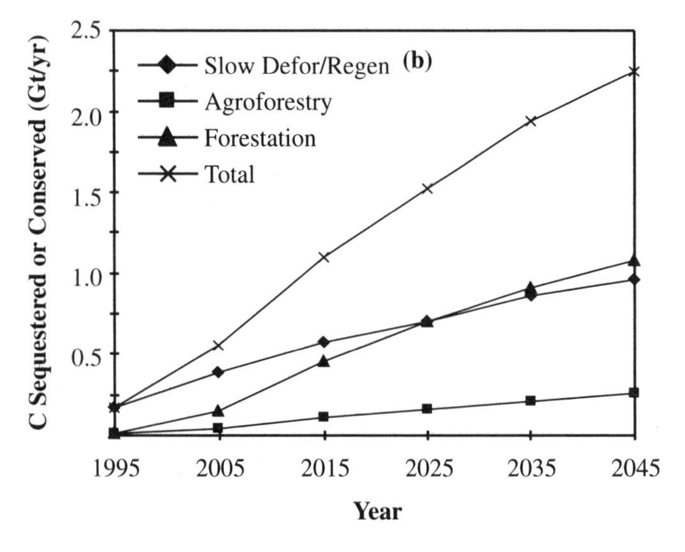

Figure 24-2: Average annual rate of C sequestration and conservation per decade through implementation of the forest management options listed in Table 24-5 by (a) latitudinal region and (b) forest management practice. Note that Defor = deforestation and Regen = natural and assisted regeneration.

The contribution of forestry to mitigation of CO_2 emissions would be considerably higher if the wood produced were assumed to be used as a substitute for fossil fuels (Section 24.3.3; Hall *et al.*, 1991; Sampson *et al.*, 1993). For example, for the forestation program described here (Table 24-5), the quantity of biomass that could potentially be produced over the 55-year period was 147 billion m^3, which is equivalent to about 39 billion tons of coal (W. Schopfhauser, pers. comm.). If the wood were substituted for coal over the same time period, the C emissions avoided would be about 29 Gt, or about 77% of the C sequestered in the forestation program (37.6 Gt) (Table 24-5).

A number of projects to conserve and sequester C along the lines described above are now being jointly implemented between developed and developing countries. They vary from C conservation by protecting forests or developing sustainable forest management practices to increasing C pools through forestation and agroforestry (Box 24-2).

24.5. Project Costs and Benefits of C Conservation and Sequestration

The previous IPCC assessment (IPCC, 1992) reported regional average annual costs of about $8/t C for tropical forestation and reduction of deforestation, increasing to about $28/t C for forestation in non-U.S. OECD countries. Costs for establishing a forest plantation, excluding the opportunity cost of land, were estimated to range between $230 and $1000/ha (Sedjo and Solomon, 1989), with an average cost of $400/ha.

Unit cost estimates have been improved since the earlier IPCC report (see also Chapter 8, *Estimating the Costs of Mitigating Greenhouse Gases*, and Chapter 9, *A Review of Mitigation Cost Studies*, of the IPCC Working Group III volume): (1) They have been estimated for individual countries rather than by regions or for the globe as a whole; (2) they have been developed for several types of mitigation options; (3) other cost

Box 24-2. Jointly Implemented Forestry Offset Projects to Sequester C Emissions

There are several examples of C-offset projects around the world; the main characteristics of six of these projects are highlighted in Table 24-6 (adapted from Dixon *et al.*, 1993a; Kinsman and Trexler, 1993).

Table 24-6: *Jointly implemented forestry offset projects.*

Location	Main Sponsor[1]	Forestry Option	Total Cost (10^6 US$)	Carbon Sequestered (Mt)	Duration (yr)
Guatemala	AES Thames	Sustainable Agroforestry	14	15.5–58	10
Malaysia (1)	ICSB/NEP	Harvest Modification	0.45	0.3–0.6	2–3
(2)	ICSB/FACE	Forest Rehabilitation	14	5–7.5	25
Paraguay	AES Barbers Point	Protection, Sustainable Agroforestry	2–5	13	30
Russia (Saratov)	USEPA/OSU	Forestation	0.2	0.04	50
USA (Oregon)	PacifiCorp	Sustainable Forestry	0.1/yr	0.06/yr	65

[1]AES Thames is a subsidiary of Applied Energy Services (AES) Corporation (Virginia, USA); NEP = New England Power (Massachusetts, USA); FACE = Forest Absorbing Carbon Dioxide Emissions, a project of the Dutch Electricity Generation Board; ICSB = Innoprise Corporation Sdn Bhd (Yayasan Sabah, Malaysia); USEPA/OSU = U.S. Environmental Protection Agency in cooperation with Oregon State University (T. Vinson, pers. comm.); PacifiCorp is an electric utility company in Oregon, USA.

Brief details of two of these offset projects follow:

- The cooperative venture between Innoprise Corporation Sdn Bhd (ICSB), the largest logging concession holder in the state of Sabah, Malaysia, and the Forest Absorbing Carbon Dioxide Emissions (FACE) Foundation promotes the planting of trees to absorb CO_2 from the atmosphere and offset emissions from power stations. The objective is to rehabilitate 25,000 ha of degraded logged forests by enrichment planting with dipterocarps—long-lived local tree species valued for their timber—and by reclaiming degraded areas using indigenous fast-growing pioneer trees. Forest fruit trees also are being interplanted to improve the forest's value for wildlife. Over a 60-year rotation, the rehabilitated forest is expected to sequester at least 200 t C/ha more than degraded logged forest; thus, a total of 5 Mt C will be sequestered. Results achieved so far are very encouraging, with survival rates around 87% and growth rates of 1.2 cm dbh per year (P. Moura Costa, pers. comm.). The long-term nature of the project (25 years) will allow the maintenance and silvicultural treatments required to sustain the growth rates to be achieved.
- For the Reduced-Impact Logging Project, New England Power (NEP) provided funds to ICSB to train personnel and implement a set of harvesting guidelines in 1,400 ha of ICSB concessions. In return, NEP can claim the C retained due to these efforts as a C offset. The project aims to reduce by half the damage to residual trees and soil during timber harvesting. Thus, less woody debris will be produced, decompose, and release CO_2, meaning more C will be retained on-site in living trees (Putz and Pinard, 1993). Conventional selective logging in the concession provides the base for comparison. Controlling logging damage is expected to result in the retention of 25–45 t C/ha after 2 years, and a total of 35–63 kt C are expected to be sequestered, at a cost of $8–14/t C.

components, including maintenance, land rental (opportunity costs), and monitoring and evaluation, are now being addressed; (4) incremental costs have been estimated by some researchers; and (5) analytical methods have been improved to provide better insights and techniques for the evaluation of costs and benefits. Despite these improvements, most estimates do not discount the C flows. Thus, net C storage occurring at any time has the same economic value.

The IPCC (1992) report does not attempt to quantify the benefits of mitigation options to reduce or store C. The valuation of C and other benefits is essential to the successful implementation of mitigation options—particularly for developing countries, where C storage may not be a sufficient inducement for local dwellers to maintain forestation projects. There is no global consensus at present on the monetary value of reducing a unit of atmospheric C.

Incremental costs of comparable projects or programs are defined as the difference between the total costs of an alternative and those of a baseline project that satisfies the same service, such as the demand for food, fuel, fiber, watershed protection, and habitat (Ahuja, 1993; K. King, 1993). Under the UN Framework Convention on Climate Change, a developing country may seek the incremental costs of a project to reduce C emissions or sequester C from developed countries. For the forest sector, however, few estimates are available for the incremental costs of GHG emissions reduction.

24.5.1. Reducing Deforestation and Protecting Forests

The unit cost of reducing C emissions by decreasing deforestation and protecting forests tends to be low (Table 24-7) as the C density of forested areas tends to be relatively high. Cost estimates, often obtained from government budgets, include the small direct cost of protection but generally not the comparatively larger opportunity cost of land. None of the estimates includes the potentially substantial, though as yet unknown, cost of providing an alternative livelihood and a viable lifestyle to those responsible for deforestation. Furthermore, present-day costs of forest protection will likely increase in the future as the amount of land put under protection increases (Adams *et al.*, 1993; Moulton and Richards, 1990).

Most of the costs reported in Table 24-7 are under \$3/t C, but range from tens of cents/t C to about \$15/t C. These costs exclude the value of local benefits that may be derived from protected forests, such as watershed protection, maintenance of biodiversity, education, tourism, and recreation; their inclusion would further offset some of the costs and in many instances result in net benefits (Dixon *et al.*, 1993b). For example, in Thailand, local benefits are estimated at \$2/ha—which would almost offset the \$2.1/ha direct cost of protection (Wangwacharakul and Bowonwiwat, 1995). Because direct benefits offset direct costs, the opportunity cost of land (present value \$44–89/ha) represents the cost of avoiding deforestation.

Table 24-7: *Cost of forest protection or reducing deforestation.*

Country	Cost (\$/t C)
Brazil	2.3[1]–4[2]
Cote d'Ivoire	8[1]
Indonesia	15[1]
Thailand	0.4–0.8[3]
Mexico	1–6[4]
India	0.5[5]
Central America	1–3[6]
Russia	1–3[7]

[1]Darmstadter and Plantinga, 1991.
[2]Cline, 1992.
[3]Based on Wangwacharakul and Bowonwiwat (1994), which includes government budget for protection and opportunity cost of land for agriculture production.
[4]Based on data in Masera *et al.* (1995); lower bound based on government budget for protection, and higher bound on cost of protection of tropical evergreen forests in Tabasco.
[5]Based on \$5/ha cost for a tiger sanctuary and 50 t C/ha of biomass density.
[6]Swisher (1991) estimate based on cost of protected areas reported in the Tropical Forest Action Plan (TFAP) proposals for Costa Rica, Honduras, and Panama.
[7]Krankina and Dixon, 1994.

24.5.2. Expansion of C Pools and C Transfer to Products

Subsequent to the IPCC report (1992), several studies have been done that better determine the cost of expanding C pools and C transfer to products (Table 24-8). The studies have focused on deriving cost estimates of mitigation options for individual countries. The costs generally include initial or establishment costs per ton of C but do not include opportunity costs or maintenance, monitoring, and evaluation costs. The methods for obtaining cost figures include literature surveys, government sources, personal questionnaires, and some field data. At the same time, researchers have refined the methodology for estimating the amount of C sequestered by a particular mitigation option and better identified the individual components of the total cost (Sathaye *et al.*, 1995). This identification has helped improve the transparency of reported cost estimates.

Earlier studies had reported point estimates for cost of C sequestration. Moulton and Richards (1990), however, developed a cost function to reflect the rise in costs associated with large-scale tree planting rather than a simple point estimate, and refined the tree plantation establishment cost estimates by differentiating costs associated with location and site considerations. Based on their data, Cline (1992) estimates costs as rising from \$12/t C for the first 100 Mt of C sequestered to \$41/t C for sequestering between 700 and 800 Mt/yr. Adams *et al.* (1993) report somewhat higher costs than Moulton and Richards (1990) (Table 24-8). The Moulton and Richards (1990) analysis has been expanded by adding the potential effects of forestation on altered timber supply, stumpage prices, harvest levels, and ultimately C sequestration (Turner *et al.*, 1995b). The latter approach couples a forest economics model, a forest inventory model, and a forest C model to make a 50-year projection of the forest sector of the U.S. economy. Coupling of models permits analysis of more-complex mitigation options such as increased paper recycling, alternative harvesting schemes, and tree planting on agricultural land. One drawback of these models is that they ignore new uses of wood—such as for power or heat generation—that would minimize the impact on traditional timber markets.

Several studies conducted for developing countries (Table 24-8) evaluate the cost of sequestering C using options such as agroforestry, long- and short-rotation plantations, natural regeneration, forest management, and silvicultural practices. Based on an estimate of the technically available land area in a given country, the tropical studies have developed cost curves that show increasing marginal costs (\$/t C) with higher sequestration. The curves for India, China, and Thailand, for example, indicate that the unit cost for sequestering C on 80% of the technically available area would be less than \$10/t C.

Table 24-8: *Initial cost of expanding carbon sinks by different regions and practices.*

Region/Country	Practice	Cost[1] (US$/t C)	Source
Boreal	Natural Regeneration[2]	5 (4–11)	Dixon *et al.*, 1994b
	Reforestation	8 (3–27)	
Temperate	Natural Regeneration[2]	1	Dixon *et al.*, 1994b
	Afforestation	2 (1–5)	
	Reforestation	6 (3–29)	
Tropical	Natural Regeneration[2]	1 (1–2)	Dixon *et al.*, 1994b
	Agroforestry	5 (2–11)	
	Reforestation	7 (3–26)	
Central America	Regeneration	4	Swisher, 1991
	Agroforestry	4	
	Plantations	13	
Argentina	Reforestation	31	Winjum *et. al.*, 1993
	Afforestation	18	
Australia	Reforestation	5	Winjum *et. al.*, 1993
Brazil	Reforestation	10	Winjum *et. al.*, 1993
	FLORAM	3–8[3]	Andrasko *et. al.*, 1991
Canada	Reforestation	11	Winjum *et al.*, 1993
	Regeneration	6	
China	Reforestation	10	Winjum *et. al.*, 1993
	Forest Management	3–4	Xu, 1995
	Eucalypt Plantations	8	
	Agroforestry	6–21	
Germany	Reforestation	29	Winjum *et. al.*, 1993
India	Reforestation	15	Winjum *et. al.*, 1993
	Regeneration	2	Ravindranath and Somashekhar, 1995
	Teak Plantations	3	
	Agroforestry	9	
Malaysia	Reforestation	5	Winjum *et. al.*, 1993
Mexico	Reforestation	4	Winjum *et. al.*, 1993
	Plantations	5–11	Masera *et al.*, 1995
	Forest Management	0.3–3	
South Africa	Reforestation	9	Winjum *et. al.*, 1993
Thailand	Teak Plantation	13–26	Wangwacharakul and Bowonwiwat, 1995
	Eucalypt Plantation	5–8	
	Agroforestry	8–12	
USA	Reforestation	5	Winjum *et. al.*, 1993
	Afforestation	2	
	Various Options	5–43[4]	Moulton and Richards, 1990
	Various Options	19–95[5]	Adams *et al.*, 1993
FSU	Reforestation	6	Winjum *et al.*, 1993
	Regeneration	5	
Russia	Plantations	1–8	Krankina and Dixon, 1994

[1] Forest components for sequestering C vary by source: Dixon *et al.* (1994b), Krankina and Dixon (1994), and Winjum *et al.* (1993) include only C in vegetation; Xu (1995), Ravindranath and Somashekhar (1995), Wongwacharakul and Bowonwiwat (1995), and Masera *et al.* (1995) include vegetation and soil C; Swisher (1991), Moulton and Richards (1990), and Adams *et al.* (1993) account for C in vegetation, soil, and litter.

[2] Values in parentheses are interquartile ranges.

[3] Figures vary depending on land rental costs per ha from $400 to 1,000; FLORAM = Florestales Amazonia.

[4] Marginal costs include planting and land rental costs.

[5] Includes land rental costs.

24.5.2.1. Carbon Components

All the above cost estimates account for above and belowground biomass C, but not all account for C in soil, fine and woody detritus, understory, and wood products. Estimates of the distribution of C between aboveground vegetation and soils vary significantly by ecosystem and by bioclimatic and edaphic conditions (Brown *et al.*, 1993a; Sampson *et al.*, 1993). By excluding many of the other C components, the studies reported in Table 24-8 overestimate the unit costs.

The studies summarized in Table 24-8 ignore another potentially large C benefit—that is, the possibility of substituting wood from sustainable plantations for fossil fuels (see Section 24.3.3; Hall *et al.*, 1991; Sampson *et al.*, 1993; see also Chapter 19). The extent to which wood products can displace fossil fuel depends on the products and applications for which wood is substituted. Fossil fuel savings may be considerable when wood is substituted for energy-intensive aluminum, steel, or concrete in construction (Burschel *et al.*, 1993). When harvesting wood to directly displace fossil fuels, the critical elements become the rate of tree productivity and the efficiency with which wood can be harvested and substituted for fossil fuels (Marland and Marland, 1992; see also Section 24.3.3).

24.5.2.2. Cost Components

The unit cost estimates reported in Table 24-8 include the cost to initiate a forest sector project—such as planting stock costs, planting labor, and supervision—but do not include the opportunity cost of land and growing stock, annual maintenance costs, and monitoring and evaluation costs. Of these, the latter two are generally a small fraction of the initial cost. The opportunity cost of land and growing stock, however, could significantly increase the unit cost estimates. Land rental costs are estimated at $148/ha by Moulton and Richards (1990) for the United States; a land purchase price is estimated between $400 and $1,000/ha by Sedjo and Solomon (1989). Land prices are likely to be lower in developing countries. For example, in Thailand, the present value of the opportunity cost of land is estimated to be between $44 and $89/ha (Wangwacharakul and Bowonwiwat, 1995). For degraded lands that are suitable for reforestation, which are widespread across many countries, the land price may be close to zero.

Winjum and Lewis (1993) demonstrate the significance of including the opportunity cost of the forest stock. Using the value of growing stock and revenues (i.e., negative costs), they show that without including growing stock costs, the median values (negative costs) of storing C are -$48/t C for temperate and -$32/t C for tropical plantations. With the growing stock costs included, the median values of storing C increase to -$22 and -$24/t C, respectively. Inclusion of the opportunity cost of land and of forest stock therefore may increase the cost in developed countries two to threefold, but the increase may be smaller in developing countries where degraded lands are available.

24.5.2.3. Monetary Benefits

Implementation of mitigation options will result in direct benefits derived from timber and non-timber forest products. The commercial value of these products will vary with the site and location of a project—particularly in tropical countries, where non-timber forest products provide sustenance to the local dwellers. Additional indirect benefits (nature conservation, recreation, etc.) will add to a project's value. Appropriate distribution of the total benefits among the project beneficiaries, however, is necessary to ensure the project's survival.

Trading in C credits can serve to offset the cost of an international project. Several forest sector C offset projects (Box 24-2) have been established and maintained at costs well below, or at net benefits to, the modeled estimates reported by Dixon *et al.* (1993a).

Monetary timber benefits alone often can more than offset the project costs (Winjum and Lewis, 1993). The cost offset varies depending on inclusion of indirect benefits, such as forest stock value. Other estimates of the benefit of C sequestration for developing countries (Ravindranath and Somashekar, 1995; Roslan and Woon, 1995; Wangwacharakul and Bowonwiwat, 1995; Xu, 1995) confirm that product revenue from many plantation and agroforestry projects could completely offset costs. Schroeder (1992) points out that agroforestry projects could be implemented for C sequestration across many ecological zones and practices at a positive present value of net benefit ranging from $54/ha to $6,000/ha, assuming that product prices remain unchanged in the future. The above estimates suggest that it would be possible to achieve C sequestration at a net benefit or negative cost.

The distribution of benefits plays an important role in assuring the success of either large forestation programs or individual projects. Adams *et al.* (1993) illustrate that a U.S. forestation program to sequester C on agricultural land will bestow higher benefits to agricultural producers and landowners from higher commodity prices, but the decline in timber prices will reduce timber-sale profits for private forest owners. If the distribution of benefits is to be equitable, the government will have to compensate private commercial tree planting to prevent farmers from displacing present tree plantations. For example, Saxena (1989) has shown that the losers in a rural forestry development program in India will be the local forest officers, whereas both the rural rich and the poor will gain to varying degrees from non-timber benefits and timber sales.

It is important that participants in a program aimed at sequestering C gain appropriate and adequate benefits to ensure sustained C storage. Winnett *et al.* (1993) suggest that there is a risk that the behavior of landowners will change if increased wood supply lowers stumpage prices. The result is less incentive to invest in intensive forest management. These studies, however, do not consider the case where wood is sold in energy markets rather than in traditional markets for timber.

Not harvesting timber from a forestation program is another way to avoid the impact on timber prices. For example, Barker *et al.* (1995) estimate that converting 4 Mha of marginal cropland to forests and allowing 2 Mha of bottomland cropland to revert to hardwood wetlands would sequester 850 Mt C in the United States over a 50-year period, at the same time extending wildlife and wetland habitats (non-timber benefits). This example is estimated to cost about $45/t C. There are similar opportunities in Europe, where about 15% of cropland currently is set aside to reduce agricultural surpluses.

24.5.3. Carbon Sequestration through Silviculture

Silvicultural practices (thinning, fertilization, improved harvesting, genetic tree improvement, etc.) are directed to increasing both the growth and the quality of the forest resource. Increased growth *per se* does not increase mean C storage and, indeed, may decrease C storage over a given area if the growth increase is associated with a shift to a younger age-class distribution (Turner *et al.*, 1995b). Many silvicultural practices are designed to improve some aspect of the wood and/or tree form and may do little to increase C sequestration, but the longer-term, indirect effects may have significant C implications. For example, management techniques that allow greater portions of the forest to be harvested and converted to long-lived wood products may sequester a greater mass of C in the long term in both the forest itself and in the forest product stock.

Some silvicultural practices—such as thinning, extending rotation ages, and retaining high levels of coarse woody debris after harvesting—increase or at least stabilize soil C pools and tend to maintain more C on forest lands. These practices have received considerable attention in relation to maintaining biodiversity and soil productivity (Swanson and Franklin, 1992), but their C storage benefits have not been well characterized. Reduction of logging damage to residual trees can also reduce the associated releases of C (see Box 24-2).

A few studies have examined the economics of C sequestration by silvicultural practices. Hoen and Solberg (1994) assessed the efficiency of various strategies to sequester C in biomass of Norwegian forests. Carbon sequestration attributable to thinning and fertilization of stands was $71/incremental ton of C captured, even where current logging levels were maintained. Modeling efforts by Marland and Marland (1992) demonstrate that the relative C benefit of different silvicultural strategies can change dramatically depending on site-dependent characteristics such as forest growth rate, site occupancy at the time management is implemented, and the manner and efficiency with which forest products will be used.

24.5.4. Global Costs

The IPCC report (1992) estimates that the forestation costs (undiscounted) for offsetting between 5 and 26% of the 5.5 Gt C being released annually from fossil fuel combustion would range from $2.4 billion/year to $12 billion/year if accomplished over a 10-year planting period, with declining costs for longer planting periods. Dixon *et al.* (1991) report that the marginal initial cost to sequester global C increases gradually to $10/t C for a storage level of 70 Gt C, which was about 90% of the identified storage potential. Beyond 70 Gt C, the marginal cost increases rapidly. The corresponding total cost to store 70 Gt C amounts to $230 billion. Country-specific marginal cost estimates for Brazil, Russia, and the United States (Dixon *et al.*, 1991) and for India (Ravindranath and Somashekhar, 1995), China (Xu, 1995), Central America (Swisher, 1991), and Thailand (Wangwacharakul and Bowonwiwat, 1995) confirm that between 50 and 90% of the C storage potential can be tapped at an initial cost of less than $10/t C.

Using the mean unit costs for individual options by latitudinal region (from data in Tables 24-7 and 24-8), the cumulative cost (undiscounted) of conserving and sequestering the quantity of C given in Table 24-5 (for individual options, corrected for vegetation only) ranges from $247 billion to $302 billion, at an average unit cost ranging from $4.6 to $3.7/t C, respectively (Table 24-9). Average unit cost decreases as more C is conserved by slowing deforestation and forest regeneration because these are the lowest-cost options. For the forestation program alone, the unit cost would be $6.4/t C, and the total cost would decrease to $253 billion. The estimated unit cost in Table 24-9 for the forestation program is higher than that reported by Nilsson (1995), but it is consistent with new findings by others (e.g.,

Table 24-9: *Global costs of conserving and sequestering C based on the estimates in Table 24-5.*

	Zone/Forestry Options[1]						
	Low			**Mid**		**High**	
	D/R	AF	PL	AF	PL	PL	**Total**
Cost ($/t C)	2	5	7	5	6	8	3.7–4.6[2]
Total Costs (10^9 $)							
– Low estimate	44[3]	27	97	3	60	17	247
– High estimate	99[3]	27	97	3	60	17	302

[1] D/R = slowing deforestation and regeneration; AF = agroforestry; PL = plantations.
[2] Weighted average cost per unit (total costs/total C).
[3] Total costs based on low and high C conservation estimates given in Table 24-5.

Table 24-8). However, Nilsson (1995) argues that these costs may be underestimated by severalfold if additional costs such as establishing infrastructure, protective fencing, education and training, and tree nurseries were included. On the other hand, if timber and other products generate revenue, then the capital investment for the second and subsequent rotations will be derived from the revenue from the previous rotation, and the incremental capital investment will be a fraction of the cost. Discounting future costs also significantly reduces the total and average costs estimated above. Assuming an annual discount rate of 3% reduces the range of total costs to $77 to $99 billion and the average unit cost to $1.4 to $1.2/t C.

Land costs still tend to be excluded from cost analyses; their inclusion could increase the cost severalfold. In addition, the cost and benefit estimates presented here do not cover indirect or nonquantifiable items, such as changes in biodiversity, water resources and soil erosion, and the livelihood of forest dwellers. Their inclusion would provide a more realistic picture of the dislocation caused by strategies to protect forests or reforest suitable lands. However, even if present cost estimates for C mitigation by forestry were doubled or tripled, they would still be considerably lower than a proposed U.S. fossil-fuel tax of $100–350/t C (Rubin *et al.*, 1992).

24.6. Impacts of Future Climate, Atmospheric CO_2, Land Use, and Human Population on C Conservation and Sequestration

The direct (increased concentration of CO_2) and indirect (changes in temperature, moisture regime, growing season length, etc.) effects of a changing atmosphere and climate on forest ecosystems are discussed in more detail in Chapters 1 and A. This section discusses how these changes and future changes in human demographics are likely to affect amounts and rates of C conservation and sequestration. Because projections concerning the effects of climate change and human activities are very uncertain, only their likely range of impacts is presented here.

Environmental factors such as future climate change, increases in atmospheric CO_2, increased mobilization of other elements such as nitrogen and sulfur, and other pollutants such as NO_x and tropospheric ozone are likely to have the greatest impacts on mid- and high-latitude forests (Apps *et al.*, 1993). In the low latitudes, human factors such as changing demographics, increased demand for agricultural land, economic growth, technology, and resource management policies—all of which have the potential to lead to high rates of land-use change—are expected to be the dominant forces on forests and could overwhelm any changes caused by future environmental conditions (Brown *et al.*, 1993a; see also Chapters 1 and 15).

Each of the promising forest management options for mitigation of C emissions is likely to be affected differently under a changed climate and human population density. For regeneration and slowed deforestation in the tropics, demand by an increasing human population for more land for agriculture and wood products (e.g., for industrial and energy use) at the expense of forest cover is likely to have a major effect on land availability; direct and indirect effects of climate change on land-use potentials may be less important in comparison (Brown *et al.*, 1993a). In the mid and high latitudes, where changes in land use are relatively stable at present, the direct and indirect effects of climate change are likely to be more important (see Chapters 1 and 15). For forestation, the key factors are how a changed climate and atmosphere will affect the suitability and availability of lands for plantation and agroforestry establishment, as well as the effects on species selection, rates of tree growth, and other pathways for sequestering C (in soil, litter, roots, etc.). The impacts of the various factors are discussed briefly in Sections 24.6.1 and 24.6.2.

24.6.1. *Impact on Land Suitability and Availability*

The effects of future climate change on the redistribution of global forest biomes are uncertain; consequently, there are large differences in projected changes in forest area for a 2 x CO_2 climate (see Chapters 1 and 15). Furthermore, most models project instantaneous change in potential vegetation distributions only. The development of the IMAGE 2.0 model—which is one of the first efforts to link submodels of the terrestrial system, atmosphere-ocean system, and energy-industry system with geographical specificity—attempts to overcome some of the problems in projecting changes in distribution of vegetation (Alcamo, 1994). Output from the "conventional wisdom" scenario of this model (akin to the IPCC 1992a scenario with respect to projected fossil fuel use; for other aspects of the model, see Alcamo, 1994, and Chapter 25) has been used to address the possible impacts of changed land-cover patterns on C conservation and sequestration potentials. Model output has been used to provide an estimate of likely changes in land suitability (i.e., potential vegetation based on the ability to grow trees, determined by climatic and edaphic factors) and land availability (a reduction in land suitability based on the need for land to grow crops, provide fuels, and so forth for the increasing human population) (see also Chapter 15). The area of land suitable for forest-sector practices for mitigation could increase globally, with the larger gains occurring in high and low latitudes mainly because of warmer temperatures and an extended growing season (Table 24-10; see also Chapter 1). However, the potential increase in suitable lands is based on climatic factors alone and does not consider whether soils will be suitable for forest establishment. This is potentially problematic in the high latitudes, where the boreal forests are projected to migrate into the tundra; tundra soils have characteristics that can retard forest establishment.

Considering the possibility of climate change, forest-sector programs aimed at C conservation and sequestration should be targeted toward those areas, tree species, and practices that are most likely to succeed, even if climate change occurred. The best targets from this perspective of risk may be the humid tropical zone and, to a lesser extent, the boreal zone, where conditions are most likely to become favorable for tree growth in new

***Table 24-10:** Likely change in areas of land suitability (technically suitable to grow trees based on edaphic and climatic factors) and availability (technically suitable lands constrained by social, cultural, economic, and political factors) by 2050, brought about by changes in climate and human demographics (data are from the "Conventional Wisdom" scenario, a simulation of the IMAGE 2.0 model[1]).*

Latitudinal Belt	Change in Land[2] Suitability	Availability
High (>50°)	1.21	1.08
Mid (25–50°)	1.15	1.05
Low (0–25°)	1.31	0.64

[1] Alcamo, 1994; model outputs grouped as in table provided by R. Leemans, RIVM, The Netherlands.
[2] Change computed as area of lands in 2050 divided by area in 1990.

plantations (but see discussion above regarding soils) (G.A. King, 1993; Table 24-10). However, in the tropical zone, a consideration of the human element and its impact on forests could change this risk perspective. Almost all of the increase in suitable lands in the high and mid-latitudes has the potential to become available (Table 24-10). In contrast, gains in land suitability in low latitudes might be unrealized because of the need for their use for agriculture, and so forth; potentially, half of the present available lands could be lost (Table 24-10). Based on the change in land availability alone, the amount of C that can potentially be sequestered in mid and high latitudes would likely change little, with possible slight gains. The potentially large decrease in land availability in low latitudes could significantly reduce the amount of C conserved or sequestered in this region.

It has been proposed that during the transient response—in the first 100 years or so following climate warming—the processes of forest dieback and other disturbances could have major impacts on the C balance of forests (Smith and Shugart, 1993; see also Chapter 1). Gains in C sequestration from forestation programs at mid and high latitudes could be offset in the transient period by the release of C from unmanaged forests, resulting from the processes of migration and regeneration being slower than dieback and other disturbances. However, dieback problems in plantation forests are likely to be relatively less important because these will be managed forests with relatively short rotations, and species substitution could occur.

24.6.2. Impact on Rates of C Conservation and Sequestration

Results from the IMAGE 2.0 model—which incorporates the effects of CO_2 fertilization, temperature, and moisture on plant growth and soil processes—are used as examples of how these factors may affect C conservation and sequestration rates (Klein Goldewijk *et al.*, 1994; also see Chapter 25 for more details on IMAGE 2.0). A discussion of the effects of these factors on ecosystems also is presented in Chapters 1 and A. The values in Table 24-11 are derived from the IMAGE 2.0 model to indicate how much the net C flux from forest ecosystems (or net ecosystem production) could change (percent change from baseline conditions) by 2050; it is assumed that plantations would respond in the same manner. The percent changes in this table do not include regions where land-cover change occurred; they include only those that affect the cycling of C in regions that were forested in 1970 and still forested in 2050 (referred to as stable forested regions). Differences between simulations are governed by interactions among land cover, the biosphere, climate, and CO_2 concentrations.

Forests growing in mid- and high-latitudinal belts could potentially increase their net C uptake in response to a changed climate and atmosphere (Table 24-11, compare entries 1 and 8). In contrast, the net C uptake in low-latitude forests under the

***Table 24-11:** Estimates of magnitudes of direct and indirect effects of global change on net C flux from stable forested regions (i.e., areas that were forested in 1970, the baseline, and still forested in 2050). Data were generated by the IMAGE 2.0 model (van Minnen et al., 1995).*

Effect	Latitudinal Belts (% Change)[1] Low	Mid	High
1. Baseline—all feedbacks to 1970 values	202	-2	-21
2. CO_2 fertilization and water-use efficiency	215	10	-11
3. Temperature and moisture feedback on plant growth	129	143	212
4. Temperature feedback on soil respiration	193	-46	-87
5. 2 and 4	210	-32	-75
6. 3 and 4	154	117	190
7. 2 and 3	136	153	220
8. 2, 3, and 4	174	129	202

[1] All net C fluxes are from the atmosphere to the biota; change in net flux was calculated as the difference between net C flux in 2050 and 1970, divided by the net flux in 1970 x 100.

combined effect of all factors (entry 8) could be lower than under baseline conditions. In other words, the region of the world that has the greatest potential for C conservation and sequestration (Table 24-5) under present climate conditions has the potential to lose some of this benefit under a changed climate and atmosphere. Although tropical forests have the largest potential to respond to CO_2 fertilization and water-use efficiency (Table 24-11, entry 2), this could be potentially offset by the response of plant growth and soil respiration to changed temperature and soil moisture (Table 24-11, entry 6).

In mid and high latitudes, a changed climate could have a large effect on the net C uptake rates of forests and thus rates of C sequestration due mostly to greater stimulation by increased temperatures and moisture on plant growth (Table 24-11, entry 3) than of soil respiration (Table 24-11, entry 4). Under no change in climate, the contribution by forests in these latitudes to the future global amount of C sequestered and conserved would be about 20% (Table 24-5). The results in Table 24-11 suggest that a potential exists to increase this share substantially, particularly given that the amount of land available is likely to increase or at least remain constant (Table 24-10).

In conclusion, under the conditions simulated by the IMAGE 2.0 model, which considers the entire suite of global change factors, there is a potential that C conservation and sequestration by forest management in the tropical region would become less important in the future, mainly due to loss of available land—unless policies are instituted to secure the required land areas for mitigation measures. In contrast, sequestration of C in temperate and boreal forest lands could play a larger role because available lands are likely to remain constant or increase and net rates of C uptake could increase.

24.7. Research and Data Needs

To improve our ability to estimate the mitigation potential of forestry practices, increased efforts are needed in the following areas:

- Realistic land-use modeling at national scales to determine trends in and constraints to forest cover, agricultural land needs, and land availability potentials. Variables such as present and projected population growth, agricultural productivity, forest growth rates, and energy demand should be linked with national trends in land use in these models so that assessments can be made of lands technically and actually available for C conservation and sequestration projects; possible interventions and their direct and indirect effects on future land uses; and key barriers that might be encountered in attempting to implement forestry options for mitigation.
- Improvements in the economic methodology for valuing all costs, including land, and especially benefits associated with forest management options for C conservation, storage, and substitution.
- Improved information about how different silvicultural and other management practices for major forest types and plantation species, growing under different climate/edaphic regimes, affect the dynamics, distribution, and retention of C in forests.
- Better understanding of the efficiency with which wood is converted into wood products, the life expectancy of wood products, and the energy balance of wood products versus that for alternative products that wood displaces.

References

Adams R., D. Adams, J. Callaway, C. Chang, and B. McCarl, 1993: Sequestering carbon on agricultural land: social cost and impacts on timber markets. *Contemporary Policy Issues,* **11**, 76-87.

Ahuja, D., 1993: *The Incremental Cost of Climate Change Mitigation Projects*. GEF Working Paper 9, GEF, Washington, DC, USA, 24 pp.

Alcamo, J. (ed.), 1994: *IMAGE 2.0: Integrated Model of Global Change*. Kluwer Academic Publishers, Dordrecht, The Netherlands, 318 pp.

Allen, R.B., K.H. Platt, and S.K. Wiser, 1995: Biodiversity in New Zealand plantations. *New Zealand Forestry*, **39(4)**, 26-29.

Andrasko, K., K. Heaton, and S. Winnett, 1991: Evaluating the costs and efficiency of options to manage global forests: a cost curve approach. In: *Technical Workshop to Explore Options for Global Forestry Management.* International Institute for Environment and Development, Conference April 24-30, 1991, Bangkok, Thailand, pp. 216-233.

Apps, M.J. and W.A. Kurz, 1991: Addressing the role of Canadian forests and the forest sector activities in the global carbon balance. *World Resources Review,* **3**, 333-344.

Apps, M.J., W.A. Kurz, R.J. Luxmoore, L.O. Nilsson, R.A. Sedjo, R. Schmidt, L.G. Simpson, and T.S. Vinson, 1993: Boreal forests and tundra. *Water, Air, and Soil Pollution,* **70**, 39-54.

Auclair, A.N.D. and T.B Carter. 1993: Forest wildfires as a recent source of CO_2 at northern latitudes. *Canadian Journal of Forest Research,* **23**, 1538-1536.

Barker, J.R., G.A. Baumgardner, D.P. Turner, and J.J. Lee, 1995: Carbon dynamics of the Conservation and Wetland Reserve Programs. *Journal Soil and Water Conservation* (in press).

Birdsey, R.A., 1992: *Carbon Storage and Accumulation in US Forest Ecosystems.* General Technical Report WO-59, USDA Forest Service, Washington, DC, USA.

Birdsey, R.A., A.J. Plantinga, and L.S. Heath, 1993: Past and prospective carbon storage in the United States forests. *Forest Ecology and Management,* **58**, 33-40.

Brown, S., 1993: Tropical forests and the global carbon cycle: the need for sustainable land-use patterns. *Agriculture, Ecosystems, and Environment*, **46**, 31-44.

Brown, S., A.J.R. Gillespie, and A.E. Lugo, 1991: Biomass of tropical forests of south and southeast Asia. *Canadian Journal of Forest Research*, **21**, 111-117.

Brown, S., C.A.S. Hall, W. Knabe, J. Raich, M.C. Trexler, and P. Woomer, 1993a: Tropical forests: their past, present, and potential future role in the terrestrial carbon budget. *Water, Air, and Soil Pollution,* **70**, 71-94.

Brown, S., L.R. Iverson, A. Prasad, and D. Liu, 1993b: Geographical distribution of carbon in biomass and soils of tropical Asian forests. *Geocarto International,* **8**, 45-60.

Brown, S., L.R. Iverson, and A.E. Lugo, 1994: Land use and biomass changes of forests of Peninsular Malaysia during 1972-82: use of GIS analysis. In: *Effects of Land Use Change on Atmospheric CO_2 Concentrations: Southeast Asia as a Case Study* [Dale, V.H. (ed.)]. Springer-Verlag, New York, NY, USA, pp. 117-144.

Brown, S. and A.E. Lugo, 1992: Aboveground biomass estimates for tropical moist forests of the Brazilian Amazon. *Interciencia,* **17**, 8-18.

Brown, S., A.E. Lugo, and J. Wisniewski, 1992: Missing carbon dioxide. *Science,* **257**, 11.

Brown, S., A.Z. Shvidenko, W. Galinski, R.A. Houghton, E.S. Kasischke, P. Kauppi, W.A. Kerz, I.A. Nalder, and V.A. Rojkov, 1996: Forests and the global carbon cycle: past, present, and future role. In: *The Role of Forest Ecosystems and Forest Management in the Global Carbon Cycle* [Apps, M. and D. Price (eds.)]. NATO ARW Series, Springer-Verlag, New York, NY, USA (in press).

Burschel, P., E. Kursten, B.C. Larson, and M. Weber, 1993: Present role of German forests and forestry in the national carbon budget and options to increase. *Water, Air, and Soil Pollution,* **70**, 325-340.

Cannell, M.G.R., 1995: *Forests and the Global Carbon Cycle in the Past, Present and Future.* Research Report No. 3, European Forest Institute, Joensuu, Finland, 66 pp.

Cannell, M.G.R. and R.C. Dewar, 1995: The carbon sink provided by plantation forests and their products in Britain. *Forestry,* **68**, 35-48.

Cannell, M.G.R, R.C. Dewar, and P.G. Pyatt, 1993: Conifer plantations on drained peatlands in Britain: a net gain or loss of carbon? *Forestry*, **66**, 353-368.

Cline, W., 1992: *The Economics of Global Warming.* Institute for International Economics, Washington, DC, USA.

Crutzen, P.J. and M.O. Andreae, 1990: Biomass burning in the tropics: impacts on atmospheric chemistry and biogeochemical cycles. *Science,* **250**, 1669-1678.

Crutzen, P.J., L.E. Heidt, J.P. Krasnec, W.H. Pollock, and W. Seiler, 1979: Biomass burning as a source of atmospheric gases CO, H_2, NO_2, NO, CH_3Cl, and CO_2. *Nature,* **282**, 253-256.

Dale, V.H., R.A. Houghton, A. Grainger, A.E. Lugo, and S. Brown, 1993: Emissions of greenhouse gases from tropical deforestation and subsequent use of the land. In: *Sustainable Agriculture and the Environment in the Humid Tropics.* National Research Council, National Academy Press, Washington, DC, USA, pp. 215-262.

Darmstadter, J. and A. Plantinga, 1991: *The Economic Cost of CO_2 Mitigation: A Review of Estimates for Selected World Regions.* Discussion Paper ENR91-06, Resources for the Future, Washington, DC, USA.

Deutscher Bundestag (ed.), 1990: *Protecting the Tropical Forests a High-Priority International Task.* 2nd Report of the Enquete-Commission, Bonn, Germany.

Dewar, R.C. 1991: Analytical model of carbon storage in the trees, soils and wood products of managed forests. *Tree Physiology,* **8**, 239-258.

Dewar, R. and M. Cannell, 1992: Carbon sequestration in the trees, products and soils of forest plantations: an analysis using UK examples. *Tree Physiology,* **11**, 49-71.

Dixon, R.K. and O.N. Krankina, 1993: Forest fires in Russia: carbon dioxide emissions to the atmosphere. *Canadian Journal of Forest Research,* **23**, 700-705.

Dixon, R.K., P.E. Schroeder, and J.K. Winjum, 1991: *Assessment of Promising Forest Management Practices and Technologies for Enhancing the Conservation and Sequestration of Atmospheric Carbon and Their Costs at the Site Level.* EPA/600/3-91/067, U.S. Environmental Protection Agency, Office of Research and Development, Washington, DC, USA.

Dixon, R.K., K.J. Andrasko, F.G. Sussman, M.A. Lavinson, M.C. Trexler, and T.S. Vinson, 1993a: Forest sector carbon offset projects: near-term opportunities to mitigate greenhouse gas emissions. *Water, Air, and Soil Pollution,* **70**, 561-577.

Dixon, R.K., J.K. Winjum, and P.E. Schroeder, 1993b: Conservation and sequestration of carbon: the potential of forest and agroforest management practices. *Global Environmental Change*, **2**, 159-173.

Dixon, R.K., S. Brown, R.A. Houghton, A.M. Solomon, M.C. Trexler, and J. Wisniewski, 1994a: Carbon pools and flux of global forest ecosystems. *Science*, **263**, 185-190.

Dixon, R.K., J.K. Winjum, K.J. Andrasko, J.J. Lee, and P.E. Shroeder, 1994b: Integrated systems: assessment of promising agroforest and alternative land-use practices to enhance carbon conservation and sequestration. *Climatic Change,* **30**, 1-23.

Elliott, G.K., 1985: Wood properties, and future requirements for wood products. In: *Attributes of Trees as Crop Plants* [Cannell, M.G.R. and J.E. Jackson (ed.)]. Institute of Terrestrial Ecology, Edinburgh, UK, pp. 545-552.

Evans, J., 1990: *Plantation Forestry in the Tropics.* Clarendon Press, Oxford, UK.

FAO, 1993: *Forest Resources Assessment 1990. Tropical Countries.* FAO Forestry Paper 112, Rome, Italy.

Fearnside, P., 1990: Fires in the tropical rain forests of the Amazon Basin. In: *Fire in the Tropical Biota: Ecosystem Processes and Global Challenges* [Goldammer, J.G. (ed.)]. Ecological Studies 84, Springer Verlag, Berlin, Germany, pp. 106-116.

Fearnside, P., 1992: *Carbon Emissions and Sequestration in Forests: Case Studies from Seven Developing Countries.* Vol. 2, *Greenhouse Gas Emissions from Deforestation in the Brazilian Amazon.* LBL-32758 UC-402, Energy and Environment Division, Lawrence Berkeley Laboratory, University of California, CA, USA.

Federal Ministry for the Environment, 1994: *Climate Protection in Germany.* First report of the Government of the Federal Republic of Germany pursuant to the United Nations Framework Convention on Climate Change. Bundesministerium fuer Umwelt, Naturschutz, and Reaktorsicherheit, Bonn, Germany.

Flint, E.P. and J.F. Richards, 1994: Trends in carbon content of vegetation in South and Southeast Asia associated with changes in land use. In: *Effects of Land Use Change on Atmospheric CO_2 Concentrations: Southeast Asia as a Case Study* [Dale, V.H. (ed.)]. Springer-Verlag, New York, NY, USA, pp. 201-300.

Galinski, W. and M. Kuppers, 1994: Polish forest ecosystems: the influence of changes in the economic system on the carbon balance. *Climatic Change,* **27**, 103-119.

Gifford, R.M., N.P. Cheney, J.C. Noble, J.S. Rusell, A.B. Wellington, and C. Zammit, 1992: Australian land use, primary production of vegetation, and carbon pools in relation to atmospheric carbon dioxide concentration. In: *Australia's Renewable Resources: Sustainability and Global Change* [Gifford, R.M. and M.M. Barson (eds.)]. CSIRO, Division of Plant Industry, Canberra, Australia, pp. 151-187.

Goldammer, J.G. (ed.), 1990: *Fire in the Tropical Biota: Ecosystem Processes and Global Challenges.* Ecological Studies 84, Springer Verlag, Berlin, Germany.

Goldammer, J.G., 1993: Historical biogeography of fire: tropical and subtropical. In: *Fire in the Environment: The Ecological, Atmospheric, and Climatic Importance of Vegetation Fires* [Crutzen, P.J. and J.G. Goldammer (eds.)]. John Wiley and Sons, Chichester, UK, pp. 297-314.

Grainger, A., 1988: Estimating areas of degraded tropical lands requiring replenishment of forest cover. *International Tree Crops Journal,* **5(1/2)**, 31-61.

Grainger, A., 1990: Modelling the impact of alternative afforestation strategies to reduce carbon emissions. In: *Proceedings of the Intergovernmental Panel on Climate Change (IPCC) Conference on Tropical Forestry Response Options to Global Climate Change*, Sao Paulo, Brazil, 9-12 January 1990. Report No. 20P-2003, Office of Policy Analysis, U.S. Environmental Protection Agency, Washington, DC, USA, pp. 93-104.

Grainger, A., 1993: *Controlling Tropical Deforestation.* Earthscan Publications Ltd., London, UK.

Hall, C.A.S., C.J. Cleveland, and R. Kaufmann, 1986. *Energy and Resource Quality: The Ecology of the Economic Process.* Wiley Interscience, John Wiley and Sons, New York, NY, USA.

Hall D., H. Mynick, and R. Williams, 1991: Cooling the greenhouse with bioenergy. *Nature*, **353**, 11-12.

Harmon, M.E., W.K. Ferrell, and J.F. Franklin, 1990: Effects on carbon storage conversion of old-growth forests to young forests. *Science,* **247**, 699-702.

Heath, L.S., P.E. Kauppi, P. Burschel, H.D. Gregor, R. Guderian, G.H. Kohlmaier, S. Lorenz, D. Overdieck, F. Scholz, H. Thoasius, and M. Weber, 1993: Contribution of temperate forests to the world's carbon budget. *Water, Air, and Soil Pollution,* **70**, 55-70.

Herendeen, R. and S. Brown, 1987: A comparative analysis of net energy from woody biomass. *Energy*, **12**, 75-84.

Hoen, H.F. and B. Solberg, 1994: Potential and economic efficiency of carbon sequestration in forest biomass through silviculture management. *Forest Science*, **40**, 429-451.

Houghton, J.T., B.A. Callander, and S.K. Varney (eds.), 1992: *Climate Change 1992. The Supplementary Report to the IPCC Scientific Assessment.* Cambridge University Press, Cambridge, UK, 200 pp.

Houghton, R.A., 1995: Effects of land-use change, surface temperature, and CO_2 concentration on terrestrial stores of carbon. In: *Biotic Feedbacks in the Global Climate System* [Woodwell, G.M. and F.T. MacKenzie (eds.)]. Oxford University Press, New York, NY, USA, pp. 333-350.

Houghton, R.A., R.D. Boone, J.R. Fruci, J.E. Hobbie, J.M. Melillo, C.A. Palm, B.J. Peterson, G.R. Shaver, G.M. Woodwell, B. Moore, D.L. Skole, and N. Myers, 1987: The flux of carbon from terrestrial ecosystems to the atmosphere in 1980 due to changes in land use: geographic distribution of the global flux. *Tellus,* **39B**, 122-139.

Houghton, R.A. and D.L. Skole, 1990: Carbon. In: *The Earth as Transformed by Human Action* [Turner, B.L., W.C. Clark, R.W. Kates, J.F. Richards, J.T. Matthews, and W.B. Meyer (eds.)]. Cambridge University Press, New York, NY, USA, pp. 393-408.

Houghton, R.A., J.D. Unruh, and P.A. Lefebvre, 1993: Current land cover in the tropics and its potential for sequestering carbon. *Global Biogeochemical Cycles,* **7**, 305-320.

IPCC, 1992: Climate Change: *The IPCC Response Strategies*. WMO/UNEP, Island Press, Covelo, CA, USA, 273 pp.

Iverson, L.R., S. Brown, A. Grainger, A. Prasad, and D. Liu, 1993: Carbon sequestration in tropical Asia: an assessment of technically suitable forest lands using geographic information systems analysis. *Climate Research*, **3**, 23-38.

Johnson, D.W., 1992: Effects of forest management on soil carbon storage. *Water, Air, and Soil Pollution,* **64**, 83-120.

Kanowski, P.J. and P.S. Savill, 1992: Forest plantations: towards sustainable practice. In: *Plantation Politics: Forest Plantations in Development* [Sargent, C. and S. Bass (eds.)]. Earthscan, London, UK, pp. 121-151.

Kanowski, P.J., P.S. Savill, P.G. Adlard, J. Burley, J. Evans, J.R. Palmer, and P.J. Wood, 1992: Plantation forestry. In: *Managing the World's Forests* [Sharma, N.P. (ed.)]. Kendall/Hunt, Dubuque, IA, USA, pp. 375-402.

Karjalainen, T. and S. Kellomäki, 1993: Carbon storage in forest ecosystems in Finland. In: *Carbon Balance of World's Forested Ecosystems: Towards a Global Estimate* [Kanninen, M. (ed.)]. Proceedings of the IPCC AFOS Workshop, May 11-15, 1992, Joensuu, Finland, Publication of the Academy of Finland 3/93, Helsinki, Finland.

Kaufman, J.B. and C. Uhl, 1990. Interaction of anthropogenic activities, fire, and rain forests in the Amazon basin. In: *Fire in the Tropical Biota: Ecosystem Processes and Global Challenges* [Goldammer, J.G. (ed.)]. Ecological Studies 84, Springer Verlag, Berlin, Germany, pp. 117-134.

Kauppi, P.E., K. Mielikainen, and K. Kuusela, 1992: Biomass and carbon budget of European Forests, 1971 to 1990. *Science,* **256**, 70-74.

Kauppi, P.E., E. Tomppo, and A. Ferm, 1995: C and N storage in living trees within Finland since 1950s. *Plant and Soil,* **168-169**, 633-638.

Kauppi, P.E. and E. Tomppo, 1993: Impact of forests on net national emissions of carbon dioxide in West Europe. *Water, Air, and Soil Pollution,* **70**, 187-196.

King, G.A., 1993: Conceptual approaches for incorporating climatic change into the development of forest management options for sequestering carbon. *Climate Research*, **3**, 61-78.

King, K., 1993: *The Incremental Cost of Global Environmental Benefits*. GEF Working Paper 5, GEF, Washington, DC, USA, 20 pp.

Kinsman, J.D. and M.C. Trexler, 1993: Terrestrial carbon management and electric utilities. *Water, Air, and Soil Pollution,* **70**, 545-560.

Klein Goldewijk, K., J.G. van Minnen, G.J.J. Kreileman, M. Vloedbeld, and R. Leemans, 1994: Simulating the carbon flux between the terrestrial environment and the atmosphere. *Water, Air, and Soil Pollution,* **76**, 199-230.

Kolchugina, T.P. and T.S. Vinson, 1995: Role of Russian forests in the global carbon cycle. *Ambio,* **24**, 258-264.

Kolchugina, T.P. and T.S. Vinson, 1993: Carbon sources and sinks in forest biomes of the former Soviet Union. *Global Biogeochemical Cycles,* **7**, 291-304.

Krankina, O.N. and R.K. Dixon, 1992: Forest management in Russia: challenge and opportunities in the era of peristroika. *Journal of Forestry,* **90**, 29-34.

Krankina, O.N. and R.K. Dixon, 1994: Forest management options to conserve and sequester terrestrial carbon in the Russian Federation. *World Resources Review,* **6**, 88-101.

Kurz, W.A. and M.J. Apps, 1993: Contribution of northern forests to the global C cycle: Canada as a case study. *Water, Air, and Soil Pollution,* **70**, 163-176.

Kurz, W.A., M.J. Apps, T. Webb, and P. MacNamee, 1992: *The Carbon Budget of the Canadian Forest Sector: Phase 1*. ENFOR Information Report NOR-X-326, Forestry Canada Northwest Region, Edmonton, Canada.

Lanly, J.P., 1982: *Tropical Forest Resources*. FAO Forestry Paper 30, FAO, Rome, Italy.

Levine, J. (ed.), 1991: *Global Biomass Burning: Atmospheric, Climatic, and Biospheric Implications.* MIT Press, Cambridge, MA, USA.

Lugo, A.E. and S. Brown, 1986: Steady-state terrestrial ecosystems and the global carbon cycle. *Vegetatio,* **68**, 83-90.

Lugo, A.E. and S. Brown, 1992: Tropical forests as sinks of atmospheric carbon. *Forest Ecology and Management,* **48**, 69-88.

Lugo, A.E. and S. Brown, 1993: Management of tropical soils as sinks or sources of atmospheric carbon. *Plant and Soil,* **149**, 27-41.

Lugo, A.E., J.A. Parrotta, and S. Brown, 1993: Loss in species caused by tropical deforestation and their recovery through management. *Ambio,* **22**, 106-109.

Maclaren, J.P., 1996: Plantation forestry—its role as a carbon sink; conclusions from calculations based on New Zealand's plantation forest estate. In: *The Role of Forest Ecosystems and Forest Management in the Global Carbon Cycle* [Apps, M. and D. Price (eds.)]. NATO ARW Series, Springer-Verlag, New York, NY, USA (in press).

Maclaren, J.P. and S.J. Wakelin, 1991: *Forestry and Forest Products as a Carbon Sink in New Zealand.* Forestry Research Bulletin No. 162, Forestry Research Institute, Rotorua, New Zealand.

Makundi, W., J. Sathaye, and O. Masera, 1996: Carbon emissions and sequestration in forests: summary of case studies from seven developing countries. *Climatic Change* (in press).

Marland, G. and S. Marland, 1992: Should we store C in trees? *Water, Air and Soil Pollution,* **64**, 181-195.

Masera, O., M.R. Bellon, and G. Segura, 1995: Response options for sequestering carbon in Mexico's forests. *Biomass & Bioenergy,* **8(4/5)** (in press).

Matthews, E. and I. Fung, 1987: Methane emission from natural wetlands: global distribution, area, and environmental characteristics of sources. *Global Biogeochemical Cycles,* **1**, 61-86.

Melillo, J.M., J.R. Fruci, R.A. Houghton, B. Moore III, and D.L. Skole, 1994: Land-use change in the Soviet Union between 1850-1980: causes of a net release of CO_2 to the atmosphere. *Tellus,* **40B**, 166-178.

Mixon, J.W., M.F. Patrono, and N.D. Uir, 1994: On the economics of tree farming and carbon storage. *The Science of the Total Environment,* **152**, 207-212.

Moulton, R.J. and K.R. Richards, 1990: *Costs of Sequestering Carbon through Tree Planting and Forest Management in the United States.* General Technical Report WO-58, USDA Forest Service, Washington, DC, USA, 48 pp.

Nabuurs, G.-J. and G.M.J. Mohren, 1993: *Carbon Fixation through Forestation Activities.* IBN Research Report 93/4, Institute for Forestry and Nature Research (IBN-DLO), Wageningen, The Netherlands.

Nilsson, S., 1995: Valuation of global afforestation programs for carbon mitigation: an Editorial essay. *Climatic Change*, **30**, 249-257.

Nilsson, S. and W. Schopfhauser, 1995: The carbon-sequestration potential of a global afforestation program. *Climatic Change*, **30**, 267-293.

Parrotta, J.A., 1993: Secondary forest regeneration on degraded tropical lands: the role of plantations as "foster ecosystems." In: *Restoration of Tropical Forest Ecosystems* [Lieth, H. and M. Lohmann (eds.)]. Kluwer Academic Publishers, Dordrecht, The Netherlands, pp. 63-73.

Postel, S. and L. Heise, 1988: *Reforesting the Earth.* Worldwatch Paper 83, Worldwatch Institute, Washington, DC, USA.

Putz, F.E. and M.A. Pinard, 1993: Reduced-impact logging as a carbon-offset method. *Conservation Biology,* **7**, 755-757.

Ravindranath, N.H. and D.O. Hall, 1994: Indian forest conservation and tropical deforestation. *Ambio,* **23**, 521-523.

Ravindranath, N.H and D.O. Hall, 1995: *Biomass, Energy, and Environment: A Developing Country Perspective from India.* Oxford University Press, Oxford, UK.

Ravindranath, N.H. and B.S. Somashekhar, 1995: Potential and economics of forestry options for carbon sequestration in India. *Biomass & Bioenergy,* **8(4/5)** (in press).

Richards K., R. Moulton, and R. Birdsey, 1993: Costs of creating carbon sinks in the US. *Energy Conversion and Management,* **34(9-11)**, 905-912.

Robinson, J.M., 1989: On uncertainty in the computation of global emissions from biomass burning. *Climatic Change,* **14**, 243-262.

Roslan, B.I. and W.C. Woon, 1995: Economic response to carbon emission and carbon sequestration for the forestry sector in Malaysia. *Biomass & Bioenergy,* **8(4/5)** (in press).

Rowntree, R.A. and D.J. Nowak, 1991: Quantifying the role of urban forests in removing atmospheric carbon dioxide. *Journal of Arboriculture,* **17**, 269-275.

Rubin, E.S., R.N. Cooper, R.A. Frosch, T.M. Lee, G. Marland, A.H. Rosenfeld, and D.D. Stine, 1992: Realistic mitigation options for global warming. *Science*, **257**, 148-149, 261-266.

Sampson, R.N., 1992. Forestry opportunities in the United States to mitigate the effects of global warming. *Water, Air, and Soil Pollution*, **64**, 157-180.

Sampson, R.N., L.L. Wright, J.K. Winjum, J.D. Kinsman, J. Benneman, E. Kursten, and J.M.O. Scurlock, 1993: Biomass management and energy. *Water, Air, and Soil Pollution*, **70**, 139-159.

Sathaye J., W. Makundi, and K. Andrasko, 1995: A comprehensive mitigation assessment process for the evaluation of forestry mitigation options. *Biomass and Bioenergy,* **8(4/5)** (in press).

Saxena, N.C., 1989: Forestry and rural development. *South Asia Journal,* **3(1&2)**. Sage Publications, New Delhi, India.

Schimel, D., I.G. Enting, M. Heimann, T.M.L. Wigley, D. Rayneud, D. Alves, and U. Seigenthaler, 1995: CO_2 and the carbon cycle. In: *Climate Change 1994: Radiative Forcing of Climate Change and an Evaluation of the IPCC IS92 Emission Scenarios* [Houghton, J.T., L.G. Meira Filho, J. Bruce, H. Lee, B.A. Callander, E. Haites, N. Harris, and K. Maskell (eds.)]. Cambridge University Press, Cambridge, UK, pp. 35-71.

Schroeder, P.E., 1992: Carbon storage potential of short rotation tropical tree plantations. *Forest Ecology and Management,* **50**, 31-41.

Sedjo, R., 1983: *The Comparative Economics of Plantation Forestry: A Global Assessment.* Resources for the Future/Johns Hopkins University Press, Baltimore, MD, USA.

Sedjo, R. and A.M. Solomon, 1989: Climate and forests. In: *Greenhouse Warming: Abatement and Adaptation* [Rosenberg, N.J., W.E. Easterling, P.R. Crosson, and J. Darmstadter (eds.)]. Resources for the Future, Washington, DC, USA, pp. 105-120.

Skole, D. and C.J. Tucker, 1993: Tropical deforestation and habitat fragmentation in the Amazon: satellite data from 1978 to 1988. *Science,* **260**, 1905-1910.

Smith, T.M., and H.H. Shugart, 1993: The transient response of terrestrial carbon storage to perturbed climate. *Nature,* **361**, 523-526.

Swanson, F.J. and J.F. Franklin, 1992: New forestry principles from ecosystem analysis of Pacific Northwest forests. *Ecological Applications,* **2**, 262-274.

Swisher, J.N., 1991: Cost and performance of CO_2 storage in forestry projects. *Biomass & Bioenergy*, **1**, 317- 328.

Swisher, J.N., 1995: Forestry and biomass energy projects: bottom-up comparisons of CO_2 storage and costs. *Biomass & Bioenergy*, **8(4/5)** (in press).

Taylor, J.A. and J. Lloyd, 1992: Sources and sinks of atmospheric CO_2. *Australian Journal of Botany*, **40**, 407-418.

Trexler, M.C. and C. Haugen, 1994: *Keeping It Green: Evaluating Tropical Forestry Strategies to Mitigate Global Warming.* World Resources Institute, Washington, DC, USA.

Turner, D.P., G. Koerper, and J.J. Lee, 1995a: A carbon budget for forests of the United States. *Ecological Applications*, **5**, 421-436.

Turner, D.P., G.J. Koerper, M. Harmon, and J.J. Lee, 1995b: Carbon sequestration by forests of the United States: current status and projections to the Year 2040. *Tellus,* **41B**, 232-239.

Unruh, J.D., R.A. Houghton, and P.A. Lefebvre, 1993: Carbon storage in agroforestry: an estimate for sub-Saharan Africa. *Climate Research,* **3**, 39-52.

Van Kooten, G.C., 1991: *Economic Issues Relating to Climate Change Effects on Canada's Forests.* Working Paper 151, Department of Agricultural Economics, University of British Columbia, Vancouver, Canada.

van Minnen, J.G., K. Klein Goldewijk, and R. Leemans, 1995: Modelling the response of terrestrial ecosystems to global environmental change. *Journal of Biogeography*, **22** (in press).

Vitousek, P., 1991: An analysis of forests as a means for counteracting the buildup of carbon dioxide in the atmosphere. *Journal of Environmental Quality,* **20**, 348-357.

Waggoner, P.E., 1994: *How Much Land Can Ten Billion People Spare for Nature*? Task Force Report No. 121, Council for Agricultural Science and Technology, Ames, IA, USA, 64 pp.

Wangwacharakul, V. and R. Bowonwiwat, 1995: Economic evaluation of CO_2 response options in forestry sector: Thailand. *Biomass & Bioenergy,* **8(4/5)** (in press).

Williams, R.H. and E.D. Larson, 1993: Advanced gasification-based biomass power generation. In: *Renewable Energy Sources for Fuels and Electricity* [Johansson, T.B., H. Kelly, A.K.N. Reddy, and R.H. Williams (eds.)]. Island Press, Washington, DC, USA, pp. 729-786.

Winjum, J.K., R.K. Dixon, and P.E. Schroeder, 1992a: Estimating the global potential of forest and agroforest management practices to sequester carbon. *Water, Air, and Soil Pollution,* **64**, 213-227.

Winjum, J.K., R.A. Meganck, and R.K. Dixon, 1992b: Expanding global forest management: an easy first approach. *Journal of Forestry,* **91**, 38-42.

Winjum, J.K., R.K. Dixon, and P.E. Schroeder, 1993: Forest management and carbon storage: an analysis of 12 key forest nations. *Water, Air, and Soil Pollution,* **70**, 239-257.

Winjum, J.K. and D.K. Lewis, 1993: Forest management and the economics of carbon storage: the nonfinancial component. *Climate Research,* **3**, 111-119.

Winnett, S.M., R.W. Haynes, and W.G. Hohenstein, 1993: Economic impacts of individual climate change mitigation options in the U.S. forest sector. *Climate Research,* **3**, 121-128.

Wood, D., 1993: Forests to fields, restoring tropical lands to agriculture. *Land Use Policy,* **April**, 91-107.

World Conservation Monitoring Centre, 1992: *Global Biodiversity, Status of the Earth's Living Resource.* Chapman Hall, London, UK.

World Resources Institute, 1990. *World Resources 1990-91.* Oxford University Press, Oxford, UK, 383 pp.

Xu, D., 1992: *Carbon Emissions and Sequestration in Forests: Case Studies from Seven Developing Countries.* Vol. 3, *India and China.* LBL-32759 UC-402, Lawrence Berkeley Laboratory, University of California, CA, USA.

Xu, D., 1995: The potentiality of reducing carbon accumulation in the atmosphere by large-scale afforestation in China and related cost/benefit analysis. *Biomass & Bioenergy,* **8(4/5)** (in press).

25

Mitigation: Cross-Sectoral and Other Issues

RIK LEEMANS, THE NETHERLANDS

Lead Authors:

S. Agrawala, India; J.A. Edmonds, USA; M.C. MacCracken, USA; R. Moss, USA; P.S. Ramakrishnan, India

CONTENTS

EXECUTIVE SUMMARY

Comprehensive evaluation of possible opportunities for mitigation of greenhouse gas (GHG) emissions requires consideration of a complex set of interlocking issues that together determine feasibility and implementability. Chapters 19 through 24 describe opportunities and challenges for GHG mitigation in six sectors of natural and socioeconomic systems, providing an encompassing summary of possible options. An important additional perspective comes from consideration of issues and opportunities that cut across the options presented for the individual sectors. These cross-cutting issues include whether energy end-use mitigation options can realistically reduce demands to levels that can be met by alternate supply systems at low emissions, whether the many competing pressures for the use of land can be reconciled, and whether nontraditional "geoengineering" offers any plausible options that may be employed to intentionally counterbalance anthropogenic climatic forcings and change. This chapter summarizes the limited information that is available for addressing these questions.

Energy Implications for Low-Emissions GHG Scenarios

The energy supply and end-use sectors are collectively responsible for more than half of the world's total anthropogenic GHG emissions. For most situations, there are more candidate mitigation technologies and options than can be realistically adopted and implemented in a given national or regional setting. In addition to considering purely technological aspects of implementation, however, decisionmakers need to consider corollary benefits and potential side effects of individual mitigation measures. Additional factors to consider include how these options may complement or conflict with each other and with key national and subnational objectives, and which of the candidate options can be most readily implemented by the country's institutions and social and economic structure.

Many emissions-reduction estimates of individual mitigation measures draw from diverse studies based on limited data from specific countries or regions; they often use widely varying social and economic growth assumptions. It is not possible to predict future energy demand reductions in the aggregate with any certainty—as is indicated by the very large range of energy demands described in the end-use demand mitigation chapters in this assessment. However, studies indicate that it is feasible to reduce energy demand in individual end-uses and even in the aggregate—hence to reduce substantially fossil fuel-related emissions against baseline trends. The extent of actual reductions will depend on numerous factors, including energy prices, government policies, continued research and development of energy-efficient devices, and societal and behavioral trends, as well as other environmental concerns. Reductions in energy demand could be coupled with low-emissions energy supply systems to further reduce emissions. Current options in the energy supply arena that would be able realistically to meet energy demand at low emissions will be limited if energy demands are not constrained. Thus, a more flexible energy future is likely to be one where energy supply and end-use mitigation measures are used in conjunction to achieve low emissions.

Issues Related to Land Use and Land Cover

A wide range of future mitigation options and strategies involving various uses of land areas has been suggested. The primary issue in evaluating these options is whether the world can continue to support an increasing population with its growing needs for food and fiber and, at the same time, expand the amount of land used for production of biomass for energy. Analyses of land-use trends and patterns make clear that substantial land can be made available for biomass energy resources only if high rates of improvement in agricultural productivity continue throughout the world. Alternatively, moderate increases in agricultural productivity, reduced population growth rates, and reduced emphasis on meat consumption in the future could provide an adequate global food supply with reduced requirements for agricultural land. Even making optimistic assumptions, however, there will be regional dislocations and imbalances as demand for food, limitations in agricultural productivity, and increasing populations come into conflict. Important stresses on land availability for agriculture, biomass plantations, and other land uses will emerge if population growth continues unabated, high-meat dietary preferences remain stable or increase, the rate of increase in agricultural productivity slows or stops, land degradation continues, or the development and penetration of new technologies cannot be extended to developing countries.

A significant problem in more thoroughly documenting, understanding, and projecting the potential for demands on the world's land resources is that changes in the extent and patterns of land cover and land use are not well-documented. Different global databases indicate large differences for comparable classes of land use and/or land cover. To carry through a thorough assessment of global land resources and their (potential) uses and productivity, significant research is needed—beginning with the development of state-of-the-art land use and land-cover databases that span the world. With such information, assessments of possible mitigation options and strategies related to land use and cover could be carried out using a geo-referenced framework that would enable determination of competing land-use activities and efficiently account for the effects of land degradation and the consequences for other environmental concerns, especially biodiversity.

Concepts for Counterbalancing Climatic Change

A review of conceptual approaches for counterbalancing anthropogenic climate change through geoengineering indicates that many options entail important adverse environmental consequences. Thus, these approaches do not provide an alternative that would readily permit the continued and expanded use of carbon-based fuels. Proposed concepts for geoengineering that have been examined include, for example, the deployment of solar radiation reflectors in space and the injection of sulfate aerosols into the atmosphere to mimic the cooling influence of volcanic eruptions. Most of these approaches are likely to be expensive to sustain and/or to have serious, but poorly understood, side effects, making them unattractive as possible mitigation options. Projections of the chemical and climatic effects of implementing such approaches are, for most options, at least as uncertain as those for future development of technologies and agricultural productivity. Geoengineering options also generally impose added costs on society and become essentially permanent commitments. Although it is perhaps appropriate to keep geoengineering approaches in reserve in case of unexpectedly rapid climatic change, investments in nontraditional approaches would seem much more effective if directed toward developing fossil fuel-free energy sources rather than at efforts to counterbalance the consequences of fossil fuel use.

Integrated Assessment of Mitigation Potential

Comprehensive assessment of different combinations of mitigation options can only be carried out in an integrated manner, considering both the direct and coupled implications of various choices—including, for example, the resulting changes in the fluxes of all greenhouse gases at local, regional, and global scales. The development and testing of capabilities for quantitative analysis, including integrated assessment modeling, are still in early stages of development. As a consequence, there is, at present, no single "right approach" to integrating an analysis across the various issues and disciplines. Models and related analyses, for example, must make tradeoffs between the level of sectoral detail they can include and the level of complexity and data requirements that can be realistically handled. Further, Integrated Assessment Models (IAMs) are only as good as the underlying socioeconomic assumptions and the necessary information on sectoral impacts, adaptation, and mitigation strategies. Across these issues information is lacking in many regions, particularly in developing countries. Over the next few years, the development of adequate databases and the improvement of analysis and assessment capabilities are essential to providing the information needed for the difficult choices that are being faced by decisionmakers.

25.1. Introduction

Mitigation options are measures, methodologies, and technologies to reduce emissions and enhance sinks of greenhouse gases or to otherwise limit climatic change and consequences resulting from human activities. This broad definition encompasses a large diversity of mitigation options. Chapters 19–24 of this volume present and discuss the most important options for a number of individual sectors, including energy supply, industry, transportation, human settlements, agriculture, and forestry. Such a compilation of information on specific measures to abate GHG emissions is an important first step in the analysis of mitigation options. In itself, however, it does not address the important need to analyze the overall, cross-sectoral potential for reduction of emissions within a country or region, the ways in which specific mitigation options interact with other options and with national or regional goals, and the implications of different combinations of mitigation options for the national or regional economy and resource base.

To address these broad cross-sectoral issues, integrated analysis of mitigation options is required. This chapter presents some of the cross-cutting themes central to such an analysis. Chapter 27 and its related appendices provide detailed information and methodologies that will be of use to national decisionmakers in analyzing mitigation options and developing national strategies for GHG emissions reduction. Within each country these options need to be integrated with other key national objectives—including, for example, promotion of rural development, increasing economic growth, generation of new employment opportunities, or improvement of environmental conditions.

There are several conceptual issues related to a cross-sectoral analysis of mitigation options. This chapter addresses some of these issues by focusing on four major questions:

- *What important cross-sectoral issues arise in efforts to reduce GHG emissions from the energy sector?*

  The energy sector is likely to be the major focus of GHG emissions mitigation in most countries. While an evaluation of options in individual sectors indicates that there are many opportunities to reduce fossil-fuel carbon emissions substantially below those in 1990 for particular activities or processes, it is not possible to determine how much total emissions can be reduced without combining these individual options into a comprehensive strategy or scenario. In the aggregate, how much can energy demand be reduced? What supply-side options can be relied upon to meet these energy demands? Which combinations of demand management options and alternative energy supply systems have the greatest potential for reducing emissions and meeting other national/regional goals? Section 25.2 describes the major challenges and approaches to analyzing mitigation options in the energy sector.

- *What specific constraints on land use and availability may limit mitigation options focusing on land management?*

  A wide range of future mitigation options and strategies have been suggested that involve various uses of land. The main issue is whether the world can continue to support an increasing population, with its growing needs for food and fiber, and, at the same time, expand the amount of land used for production of biomass for energy—while preserving environmental resources for other purposes, such as the maintenance of ecosystems, conservation of biodiversity, regeneration and storage of freshwater, assimilation of wastes, and so forth. Section 25.3 summarizes the issues and problems confronting efforts to assess land-use options and scenarios, including the lack of reliable global information on current and projected patterns of land cover and land use.

- *What ideas for large-scale "cross-sectoral" options exist for attempting to deliberately counterbalance potential human-induced climate change?*

  Suggested approaches to counterbalance anthropogenic climate change through geoengineering include, for example, the deployment of solar radiation reflectors in space and the injection of sulfate aerosols into the atmosphere to mimic the cooling influence of volcanic eruptions. Section 25.4 reviews what is known about these concepts, pointing out that based on current information, they appear to be very costly, very uncertain, and/or have significant side effects relative to measures to reduce GHG emissions.

- *Can we assess the overall effectiveness of mitigation options and strategies?*

  Comparison and assessment of different sets of mitigation options is best carried out in an integrated manner, considering fluxes of all GHGs at local, regional, and global scales. Such an assessment also should include the effects of economic activities and trade and potential climate-change feedbacks. Section 25.5 provides an overview of the issues that need to be considered as part of such a comprehensive assessment and includes a brief discussion of integrated models that are currently being developed.

25.2. Energy Implications for Low-Emissions Greenhouse Gas Scenarios

Energy supply and end-use sectors are collectively responsible for more than half of the world's total anthropogenic greenhouse gas emissions. Therefore, mitigation policies for these sectors are critical to achieving any reasonable future stabilization of GHG concentrations in the atmosphere. In 1990, the carbon (C) content of fossil energy was about 6 Gt C. Energy supply and end-use demand sectors both contributed significantly: 2.4 Gt C were emitted during energy conversion and distribution, while about 3.6 Gt C were emitted at the point of end-use (see Chapter B).

Options for reducing energy-related emissions, therefore, fall under two broad categories: those that reduce GHG emissions while providing energy services at projected levels of demand and those aimed at reducing energy demand itself in key end-use sectors. Previous chapters of this report investigate many aspects of these measures in great detail: Chapters 20 through 22 examine options to reduce energy use and process-related GHG emissions in the industrial, transportation, and residential/commercial sectors, respectively, and Chapter 19 discusses technologies to enhance fossil fuel conversion efficiencies and describes Low Emissions Supply Systems (LESS) to meet future energy demands. Chapter 28 and its companion appendix contain a detailed inventory of more than one hundred GHG mitigation technologies for the energy supply and end-use sectors, providing information on key parameters such as performance and environmental characteristics, capital and operating costs, and infrastructure requirements. Also, while specific mitigation policy measures and implementation challenges are discussed in the individual sectoral mitigation chapters, a more generic assessment of decisionmaking under uncertainty, mitigation costs, and macroeconomic measures can be found in Chapters 2, *Decisionmaking Framework to Address Climate Change,* Chapter 9, *A Review of Mitigation Cost Studies,* and Chapter 11, *An Assessment of Economic Policy Instruments to Combat the Enhanced Greenhouse Effect,* of the Working Group III volume.

25.2.1. Cross-Sectoral Implications

While the energy-related mitigation discussion in this assessment is both exhaustive and reflective of the current "state-of-the-science," such a compartmentalized framework of analysis does not address three very important cross-cutting questions:

- How can national decisionmakers rank and choose among mitigation options?
- Can sets of end-use mitigation options taken together realistically reduce future energy demands to levels that are assumed in the construction of energy-supply scenarios that lead to low emissions?
- What are the implications of high-biomass energy futures on other land-use requirements, including the provision of food for the world's increasing population?

Sections 25.2.2 through 25.2.5 provide a roadmap of the kinds of cross-cutting issues and tradeoffs that need to be considered in attempting to answer these questions.

25.2.2. Selecting Mitigation Options

Mitigation agendas are typically laid out by national, regional, and project-level decisionmakers, often with limited resources. A detailed description of mitigation options is therefore of limited use without a proper set of guidelines on how to select the promising options in a particular setting.

The first step in such an assessment is to establish the scope and overall goal(s) of the mitigation strategy. Considerations include whether the strategy is to meet particular emission reduction targets; assess specific technologies or policies; or identify measures that best integrate with key national and subnational objectives, such as increasing economic growth and self-reliance, reducing unemployment and social inequalities, or promoting rural development. Decisionmakers also must determine if they wish to target carbon dioxide (CO_2) only or several GHGs; which of the candidate options can be feasibly implemented by the country's institutional structure; and whether the options being considered will be available and cost-effective in their country within the timeframe under consideration.

25.2.3. Constructing Energy Demand Mitigation Scenarios

In the context of energy demand, the screening criteria described above can be used to identify, rank, and combine promising demand-side options into one or more scenarios. These mitigation scenarios can then be evaluated against the backdrop of a "no-policy," baseline scenario. Such disaggregated descriptive frameworks have been used for energy mitigation scenario analysis at national and regional levels (Lazarus *et al.*, 1995). The question thus emerges whether the range of technology options discussed in the energy end-use chapters of the present report can be used to construct energy demand scenarios in different sectors and regions that can then be aggregated to global levels. Such an analysis has not been attempted here, for the following reasons:

- **Consistency of assumptions**: The end-use chapters in this report review a diverse set of studies on the mitigation potentials of various technologies and policies. These studies often are based on limited data from specific countries or regions and use widely varying social and economic growth assumptions [including gross domestic product (GDP), population growth, prices, productivity, exchange rates, technology diffusion, and market regimes]. Aggregating them or extrapolating their projections to other regions could produce misleading results.
- **Accounting for corollary benefits, potential side effects, and offsetting trends**: Sectoral mitigation strategies operate in a complex societal fabric where they may influence, or be influenced by, other social and environmental constraints and, in many cases, even other mitigation policies. Cleaner and more efficient vehicles, for example, will reduce local pollution and have beneficial impacts on human health; many demand-side management measures may also lower consumer expenditures on energy services. Such corollary benefits would make these options more attractive than from a purely GHG mitigation

standpoint. On the other hand, mitigation options also may have potential side effects. For example, heat cascading systems that result in better waste-energy utilization might require establishing industries and human settlements in close proximity and therefore might stress local water resources and possibly aggravate concerns about air, noise, and water pollution. In addition, the emissions reduction potential of many mitigation options can be offset by societal and behavioral trends. Improved household and car efficiency measures, for example, may have some of their mitigation potential offset by decreasing household size and car occupancy. This is because while less energy may be consumed on a per house or per car basis, more houses and cars will be required for a given number of people. A technology-based aggregation of mitigation potentials is unlikely to incorporate many of these interactions, which may play important roles in determining the feasibility, competitiveness, and achievable mitigation potential of various options in a given setting.

25.2.4. Consistency of Energy Supply Mitigation Scenarios with Energy Demand Projections

Many projections of future energy supply—such as the Low Emissions Supply Systems (LESS) described in Chapter 19—suggest that GHG emissions can be reduced substantially relative to 1990 levels by using a number of fuel-mix choices in the long term. However, these emissions futures hinge critically on the validity of the assumed set of future energy demands.

Future demands depend not only on a number of demographic and socioeconomic factors but also on the extent to which successful reductions can be achieved from the implementation of energy conservation and other demand-side management schemes. Many energy supply projections already assume substantial reductions in demand from baseline trends as a result of these end-use mitigation measures. The demand projections for the LESS constructions, for example, are based on a high economic growth, accelerated policy variant (i.e., including GHG mitigation policies) of the SA90 scenarios developed by the IPCC in 1990 (Bernthal, 1990). The long-term energy demands in this scenario are significantly less than those projected by IS92a, the median in the "nonintervention" (i.e., assuming no explicit GHG mitigation policies) scenario-set developed by the IPCC in 1992 (Leggett *et al.*, 1992). The SA90 scenarios did not consider the range of technological and societal options considered in the end-use chapters of the present report. The question thus emerges: Are the SA90 accelerated policy demands assumed in the construction of the LESS scenarios consistent with reductions achievable in various end-use sectors? If so, can these demand reductions be achieved through mitigation measures that have a net benefit to society, even in the absence of climate change, or are they likely to be realized only at significant cost to society?

To answer these questions adequately would require the development of energy demand scenarios based on the technological options discussed in the end-use mitigation chapters of this report. As explained earlier, such an analysis has not been attempted here. However, short of the development of such scenarios, some conclusions can still be drawn.

First, the end-use demand mitigation chapters indicate that the range of future sectoral demands is very wide indeed. In the transportation sector, for example, if small, energy-efficient cars become fashionable and desirable and if urban traffic congestion and other concerns drive policy leading to greater use of mass transit, then it might be possible to achieve energy demands for transport consistent with SA90 through mitigation options that benefit society even in the absence of climate change. The same, however, cannot be said in a world in which incomes rise, personal vehicles proliferate, car occupancies decline, fossil fuels remain inexpensive, and additional transportation networks are built to meet the growing numbers of vehicles. Similarly, end-use demands for residential and commercial buildings consistent with the SA90 accelerated policy case are conceivable if three conditions are met: (1) energy prices rise over time sufficiently to influence consumer behavior; (2) governments enact strong policies to promote energy efficiency; and (3) research and development of energy-efficient devices is supported and continues to yield viable new products. A number of reference energy scenarios with such variations have been constructed (Alcamo *et al.*, 1995).

Second, unconstrained energy demands are likely to limit the suite of energy supply options that realistically would be available to meet these demands at low emissions. This would lower the flexibility and may increase the vulnerability of the energy supply system. For example, the LESS projections include a "high-demand" variant based on demands similar to those in the median "nonintervention" IS92a scenario (Edmonds *et al.*, 1994). Conceivably, low-emissions supply systems still could be developed that meet IS92a demands for the year 2100 at annual emissions approximately half of the 1990 levels and one-fifth of the IS92a emissions projections for 2100. However, such low-emissions supply systems would need to rely on a substantial increase in coal production (e.g., coal production for LESS "high-demand" constructions for the year 2100 in Chapter 19 is almost five times the 1990 levels) and require extensive CO_2 sequestering in aquifers and natural gas fields—a situation that may not be sustainable as gas fields become filled to capacity. Besides, other energy sources, including biomass, are likely to be pushed close to the limits of their realizable potential to meet high demands. A more flexible and possibly lower-cost energy future is likely to be one where energy supply and end-use mitigation options are used in conjunction to achieve low emissions.

25.2.5. Low Emissions Energy Supply and Land Use

Many alternative energy supply futures require that a substantial portion of the world's future energy needs be met by "modern" biomass. The biomass-intensive variant of LESS, in fact, targets almost 550 Mha of land for energy cultivation by the year 2100. This raises a number of concerns, including whether biomass plantations would conflict with land needed to meet regional food requirements for increasing populations, whether biomass cultivation on degraded lands would really be achievable at productivity levels assumed in the LESS constructions, and whether intensive cultivation practices would be needed that in turn might stress local water resources. As Section 25.3 shows, these concerns further complicate the gamut of issues concerning planning for current and future uses of land and land cover.

25.3. Issues Related to Land Use and Land Cover

Many mitigation options depend upon changes in land management, increasing carbon sequestration potential, and the use of renewable energy such as biomass (e.g., Chapters 19, 22, and 24). Many mitigation policies in the transportation sector also require changes in land-use patterns to reduce the need for goods, transport and travel. All of these mitigation options involve land, land use, and land cover. In the present context, "land" refers to the Earth's surface, including the soil and its geology and hydrology; the biosphere; and the lowest layer of the atmosphere. "Land cover" includes only the upper-soil layers and vegetation, while "land use" refers to the explicit purpose for which humans exploit land and its vegetative cover and consists of a series of activities that alter ecological and physical properties of land (Turner *et al.*, 1993). A clear distinction between land use and cover is important for assessing land-related mitigation options because the availability of land, water, and related commodities is limited, and different land uses compete with each other.

This section discusses issues related to competition for land. Although land required for human settlement, infrastructure, and other uses is important locally and regionally, the cumulative effect of these uses is of lesser importance globally. The following discussion focuses on land required for agriculture—the most important land use worldwide.

Land use and its changes are heterogeneous, both spatially and temporally. The global significance of land use arises as a result of the cumulative effects of local and regional changes in land cover. These changes are not necessarily land-cover conversions. An altered management could change the properties of land cover (e.g., enhance plant growth through fertilization) without changing the overall structure or vegetation category (Turner *et al.*, 1990). The consequences of such modifications seldom appear in regional and global environmental assessments but are important in altering emissions from land use (Leemans and Zuidema, 1995). Coarse land-cover patterns have been used most often to analyze historic and current land-cover change. These assessments are strongly limited by the quality of the available land-cover data. Nonetheless, some insights have emerged.

25.3.1. Patterns of Land Cover

Potential usage of land is determined by the local climate, soil, and land cover. Climate is the major constraint and controls the global patterns of vegetation structure and (potential) species composition (Walter, 1985; Woodward, 1987). Temperature and precipitation define the major latitudinal zones (e.g., boreal, temperate, and tropical), whereas seasonality of these factors largely defines the biomes (e.g., deserts, rangelands, and forests) within these zones. Variation between years, deviations

Box 25-1. Land-Cover Classifications

Research into many aspects of global change requires reliable, geo-referenced information on global land cover for a well-defined time period (Townshend, 1992). Several classification schemes have been developed to describe land-cover patterns. Schemes have been based on structural (e.g., trees vs. shrubs), physiognomic (e.g., deciduous vs. evergreen), floristic (e.g., oak-hickory forest), or bioclimatic (e.g., boreal forest) classes, or mixtures thereof. Imprecise terminology, such as "woodlands," is often used to label classes. Many classification schemes also include indices that are not directly related to land cover but to other environmental variables, such as climate (e.g., tropical rainforest). Such deficiencies are not problematic when a unique classification is applied throughout an analysis, but many global data sets are mixtures from different sources, using similar terminology but largely different criteria (Leemans *et al.*, 1995). One of the few globally comprehensive classifications is the hierarchical UNESCO classification (1973). This hybrid classification (physiognomic and structural characteristics at higher levels, species composition at lower levels) was specially developed for the description of natural vegetation at a climax stage; this is probably one of the reasons that it has never resulted in a global assessment of land cover. Despite their problems, UNESCO vegetation maps have been produced for several regions (e.g., White, 1983). Since the development of the UNESCO classification, no such generally accepted scheme for land cover has been developed, although several approaches are emerging, mainly based on technologies involving remote sensing (Running *et al.*, 1994). These approaches have led already to more comprehensive regional estimates of deforestation patterns (e.g., Skole and Tucker, 1993) and land-cover classifications (Defries and Townshend, 1994). Several international organizations, including UNEP and FAO, have recently formulated requirements for adequate land-use and land-cover classifications (Mücher *et al.*, 1993; UNEP/GEMS, 1993, 1994) and concluded that the implementation of such schemes should be prioritized in order to assure consistent land-use evaluations (Fresco *et al.*, 1994).

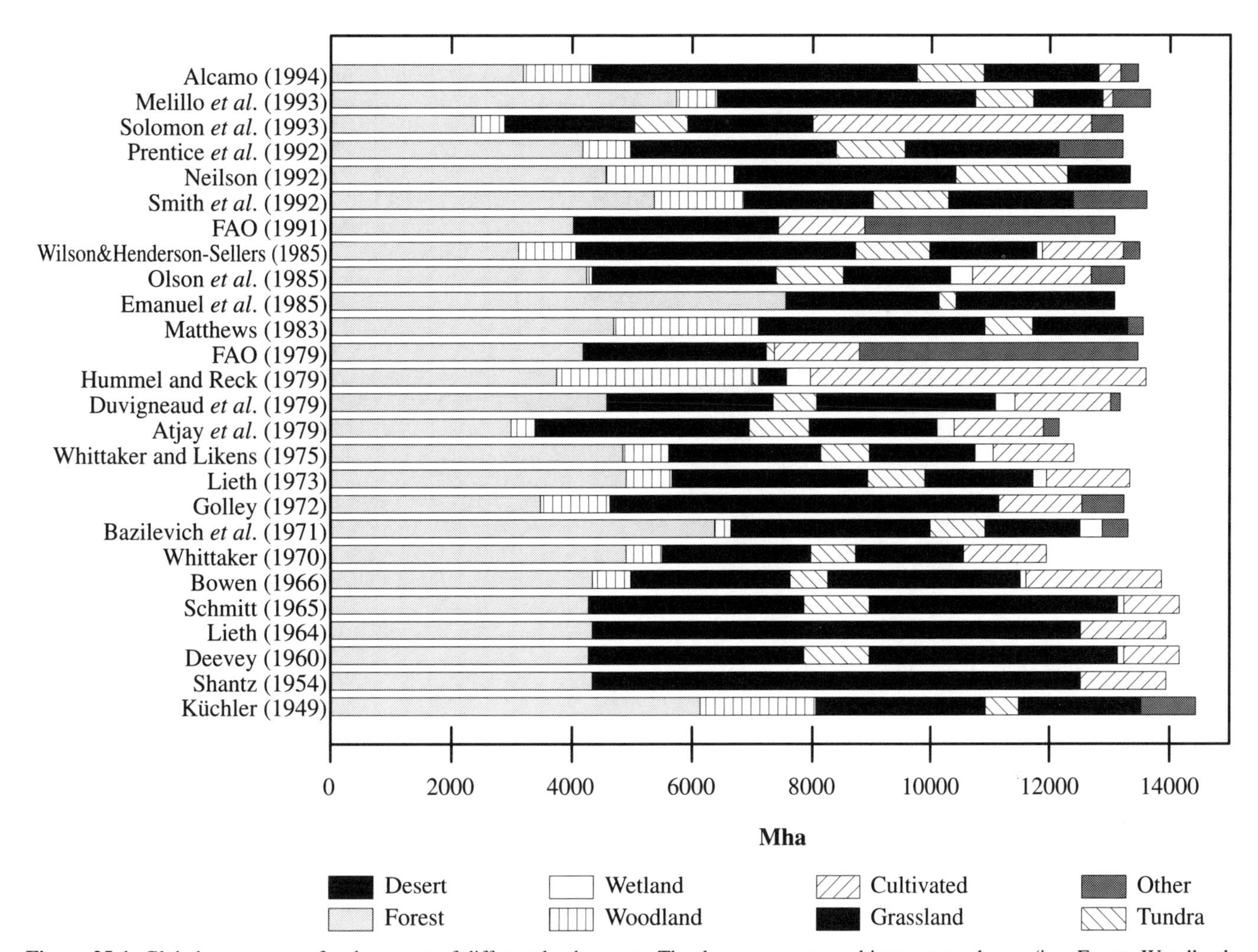

Figure 25-1: Global assessments for the extent of different land covers. The data are aggregated into coarse classes (i.e., Forest, Woodland, Grassland, Tundra, Desert, Wetland, Cultivated, and Other) from the original publications or data sets.

from climatic means, and extreme events (e.g., severe temperatures, storms, and droughts) influence plant growth and survival and lead to successional changes and further differentiate land cover locally. The history of natural events and human-induced changes determines the actual land cover in any region.

Several data sets of global land cover have been compiled. Most of these data banks are based on simple land-cover classification schemes, generally having between five and fifty classes worldwide (see Box 25-1). Despite their relative simplicity, significant differences exist among the different global land-cover databases. Figure 25-1 presents an aggregation of the different land-cover classes in each database into broad, but comparable, classes. These databases are derived from statistical sources, national maps, and/or other sources, such as climate atlases. This has led to the large variation in total extent of each class (Figure 25-1). The most apparent difference between these databases is for cultivated lands, for which estimates range from 0 to 55 x 10^6 km^2. Databases without cultivated land have focused primarily on natural (or potential) vegetation prior to human activities (e.g., Prentice *et al.*, 1992; Neilson *et al.*, 1992), while the largest values for this class (e.g., Solomon *et al.*, 1993) stem from evaluations of land that could potentially be used for cultivation. The most authoritative estimate for cultivated land at present is likely that of FAO (1991), which gives 1450 Mha. Unfortunately, however, all of these estimates are only gross approximations. There are no accurate figures for many countries; much of the information is based on averages or ill-defined categories; and the global compilations are generated from a multitude of sources (Buringh and Dudal, 1987; Defries and Townshend, 1994; Leemans *et al.*, 1995). One of the major limitations is that spatial patterns are often not well-depicted. Linkages between land-cover and other geo-referenced databases—such as climate, topography, and soils—could assist in removing some of the inconsistencies of the existing land-cover databases (Loveland *et al.*, 1991; Running *et al.*, 1994). Improvements in these linkages are a critical constraint in the use of land-cover data in mitigation assessments.

In spite of these shortcomings, many of these estimates are cited and used in different climate-change assessments. The differences in the estimates become important because the use of a different database can lead to significantly different results, and hence conclusions. Thus, global assessments always must be concerned with inconsistencies, limitations, and errors in these data sets. When used with care, these data sets can be used to determine possible impacts of climate change (e.g., Emanuel *et al.*, 1985), simulate global carbon cycling (e.g., Melillo *et al.*,

1993; Prentice *et al.*, 1994), and parameterize land-surface properties in climate models (e.g., Hummel and Reck, 1979; Wilson and Henderson-Sellers, 1985). However, these data sets remain inadequate for describing local and regional properties of ecosystems fully, and consequently for assessing and evaluating mitigation potential accurately.

25.3.2. Societal Uses of Land

Land and land cover provide many resources—including food, fodder, fiber, and biomass—and functions, including biodiversity conservation, water regeneration, waste accumulation, and flood buffering. Human activities have led to significant changes in land cover, and such changes are expected to continue to change the land surface (Houghton, 1994; Table 25-1). On continental, regional, and local scales there are large differences in the rate and nature of changes in land cover (Turner *et al.*, 1990; Meyer and Turner, 1994).

Approximately 11% of the land surface has been converted to cropland, and roughly 25% is occupied by pasture (FAO, 1991); 6% is legally protected in conservation areas and reserves (Morris, 1995). About half of the remaining forests and woodlands are managed secondary forests or plantations. In addition to the readily apparent changes in land cover, much of the land has been degraded or is at risk of degradation, including topsoil loss, nutrient depletion, acidification, and compaction. Recent estimates are 750 Mha for light degradation, 910 Mha for moderate degradation, and 310 Mha for severe degradation (World Resources Institute, 1992; Oldeman, 1993). Damaged and degraded lands can generally be rehabilitated to a productive state, but not always restored to their desired use (Barrow, 1991; Brown and Lugo, 1994). All of these transformations have had a large impact on natural processes, biodiversity, and the resilience of ecosystems, and directly affect the potential for carbon storage and other mitigation options involving land cover.

How much land and water is needed to provide adequate food, fodder, and fiber? What are the competing demands on those resource bases? At what rate is land degradation reducing their availability? Such questions have to be considered when land-focused mitigation options compete with existing land uses.

Table 25-1: *Changes in land cover and population from 1700 to 1980 (from Houghton et al., 1983).*

Year	Forests and Woodlands (Mha)	Grassland and Pastures (Mha)	Croplands (Mha)	Population (millions)
1700	6220	6860	270	680
1850	5970	6840	540	960
1920	5680	6750	910	1650
1950	5390	6780	1170	2500
1980	5050	6790	1480	4500

In general, mitigation options in the agriculture and forestry sectors focus on moving toward a more sustainable use of available resources by enhancing the sequestration potential of the ecosystems involved. Many of these options could well have other positive side effects, including pollution reduction, slowing the rate of land degradation, and biodiversity conservation. The proposed increase in biomass utilization to offset fossil fuel use illustrates the potential for competing demands for land. The suggested quantities of biomass can theoretically be provided by ecosystems globally—for example, through biomass plantations that can be established on many types of lands (Hall and Overend, 1987). However, minimizing costs and energy used in transportation would favor growing biomass close to urban areas; it is in such places where the greatest demands on land will occur, creating a possible conflict between land for biomass plantations and for other uses.

25.3.2.1. Global Food Production

Food security has long been a major concern, and several analyses have been conducted to evaluate whether the future food supply will be adequate to feed the growing population (e.g., FAO, 1993; Bongaarts, 1994; Dyson, 1994; Kendall and Pimentel, 1994; Smil, 1994; Waggoner, 1994). Many of these studies suggest that the land could provide food for the increasing population if current trends in agricultural productivity driven by improved management, use of high-yielding varieties, and changed cropping systems can be sustained. Other studies, however, suggest that increasing land degradation, limited water supply, and dependence on energy-intensive fertilizers, combined with a growing population, have already started to lead to a decline in per capita food production and land availability (e.g., Kendall and Pimentel, 1994). Continued downward trends would render the global food supply much less secure than appears from many analyses (Kendall and Pimentel, 1994).

The results of the above analyses are particularly sensitive to assumptions about dietary preferences. Different consumption patterns can lead to very different gross requirements. The amount of food consumed per capita differs significantly by country and region. Diets consisting mainly of grains and vegetables (e.g., China; Table 25-2) require different quantities and types of agricultural land compared to diets containing more meat and dairy products (e.g., United States, Table 25-2). A meat-based diet for the entire world population would clearly exceed current agricultural capabilities (Waggoner, 1994). A conceptually simple means to provide food for the increasing population would be to alter dietary habits. Although a slight shift from red meat toward poultry is occurring in the developed world (thereby reducing total feed grain requirements), red meat consumption is increasing in many other parts of the world (World Resources Institute, 1992). Bringing about dietary changes may be difficult to implement.

Land suitability for agriculture is dependent upon environmental factors, including climate, terrain, soil, water resources, and nutrient availability. These suitability patterns will shift under

Table 25-2: *Regional and global consumption of foods and grain (from Kendall and Pimental, 1994).*

Food/Feed[a]	**USA**	**China**	**World**
Food Grains	77	239	201
Vegetables	129	163	130
Fruits	46	17	53
Meat and Fish	88	36	47
Dairy Products	258	4	77
Eggs	14	7	6
Fats and Oils	29	6	13
Sugars and Sweeteners	70	7	25
Food Total	711	479	552
Feed Grains	663	126	144
Grand Total	1374	605	696
Calories[b]	3600	2662	2667

[a]In kg capita^{-1} yr^{-1}.
[b]In capita^{-1} day^{-1}.

climatic change. Although the consequences of climatic variability are not always adequately addressed (if at all), the potential global distribution of crops is well-understood and can be modeled for current and changed climate and atmospheric composition (e.g., Brinkman, 1987; Leemans and Solomon, 1993; Rosenzweig and Parry, 1994). Such studies indicate that 65% of the land area is suitable for agriculture (Leemans and Solomon, 1993; Solomon *et al.*, 1993) but that the amount of land actually available is considerably lower because of unsuitable soils and terrain. As a result, the land potentially available for agricultural production is about 3000 Mha (approximately 20% of total land area, of which about 50% is already cultivated; see FAO, 1991).

Potential biomass productivity and, consequently, yield (which is only a fraction of total biomass production) can be estimated from growing-season characteristics and plant type (e.g., C_3 or C_4 plants). These parameters help determine photosynthetic, respiratory, and direct CO_2 responses for a given land area. When suboptimal water and nutrient availability (mainly nitrogen and phosphorus) are accounted for, the attainable yield of biomass ranges from 60–90% of the theoretical potential yield (Figure 25-2). The attainable yield is further reduced by competition for resources (including light); weeds; diseases; pests; and local air, water, and soil pollution. Proper land management, however, can help alleviate some of these reductions. In most cases, less-efficient agricultural practices can explain the large differences in yield and potential increases therein for different countries (see Plucknett, 1994).

Considering that agricultural productivity for many regions is still far below attainable yield levels, there is a large potential to enhance food production (e.g., FAO, 1993), but land degradation strongly affects attainable and actual productivities. The complex interactions among different environmental and management factors make it immensely difficult to project long-term sustainable agricultural productivities. Unfortunately, many future food and biomass assessments strongly focus on potential productivity (e.g., Hall and Overend, 1987; Waggoner, 1994), leading to overly optimistic conclusions.

25.3.2.2. Global Biomass Production for Energy

Biomass production competes with land required for food production and for other uses. Food production requires the most productive lands, whereas biomass crops can probably be grown on less-productive lands as well (Hall and Overend, 1987). However, growing these crops on less-productive land would lead (at least initially) to lower yields. It is often argued that biomass plantations can assist in the rehabilitation or improvement of degraded agricultural land and that these lands could be used in a profitable and sustainable manner (e.g., Bongaarts, 1994; Hall *et al.*, 1994). Although this is probably true for regions with lightly and moderately degraded soils (about 16.5 x 10^6 km^2), plantations are unlikely to be useful in regions with severely degraded soils (about 3.1 x 10^6 km^2). Rehabilitation in such regions is a long-term and difficult process, especially if the original functions of the system (i.e., biological productivity) are to be restored (Barrow, 1991; Brown and Lugo, 1994).

Biomass crops have been demonstrated to be economically feasible in many different regions (e.g., Hall, 1991; Carpentieri *et al.*, 1993). However, large increases in biomass requirements would require a significant quantity of additional land (Alcamo *et al.*, 1994). In many regions this would pose no problem—due to, for example, the use of existing biomass sources, such as municipal waste, and crop and forestry residues. However, in regions where food supply already is stressed and will continue to be stressed in the near future (e.g., sub-Sahelian zone: FAO, 1993), biomass probably is not a viable option.

Whereas most emphasis in biomass energy has been directed toward terrestrial systems, proposed alternative sources include micro-algae, marine algae, and halophytes, which grow rapidly in saline lands. Micro-algae are single-celled, fast-growing plants that grow well over a wide range of environmental conditions. In regions with inexpensive flatlands, shallow ponds can be constructed where these algae can reproduce (Wyman, 1994), but this technique is currently quite expensive. Because their productivity is usually high, the potential use of marine algae for the conversion of solar energy is large (Orr and Sarmiento, 1992), but their natural extent is limited to regions of high nutrients (i.e., areas of upwelling). Currently, cultivation of marine algae is commercially viable only for specific purposes (pharmaceutical, chemical, and food products), because it is too expensive for large-scale biomass production (Bird, 1987). Saline lands in coastal zones and arid regions could produce biomass using halophytes (Glenn *et al.*, 1993). Although halophytes could assist in slowing or rehabilitating degraded

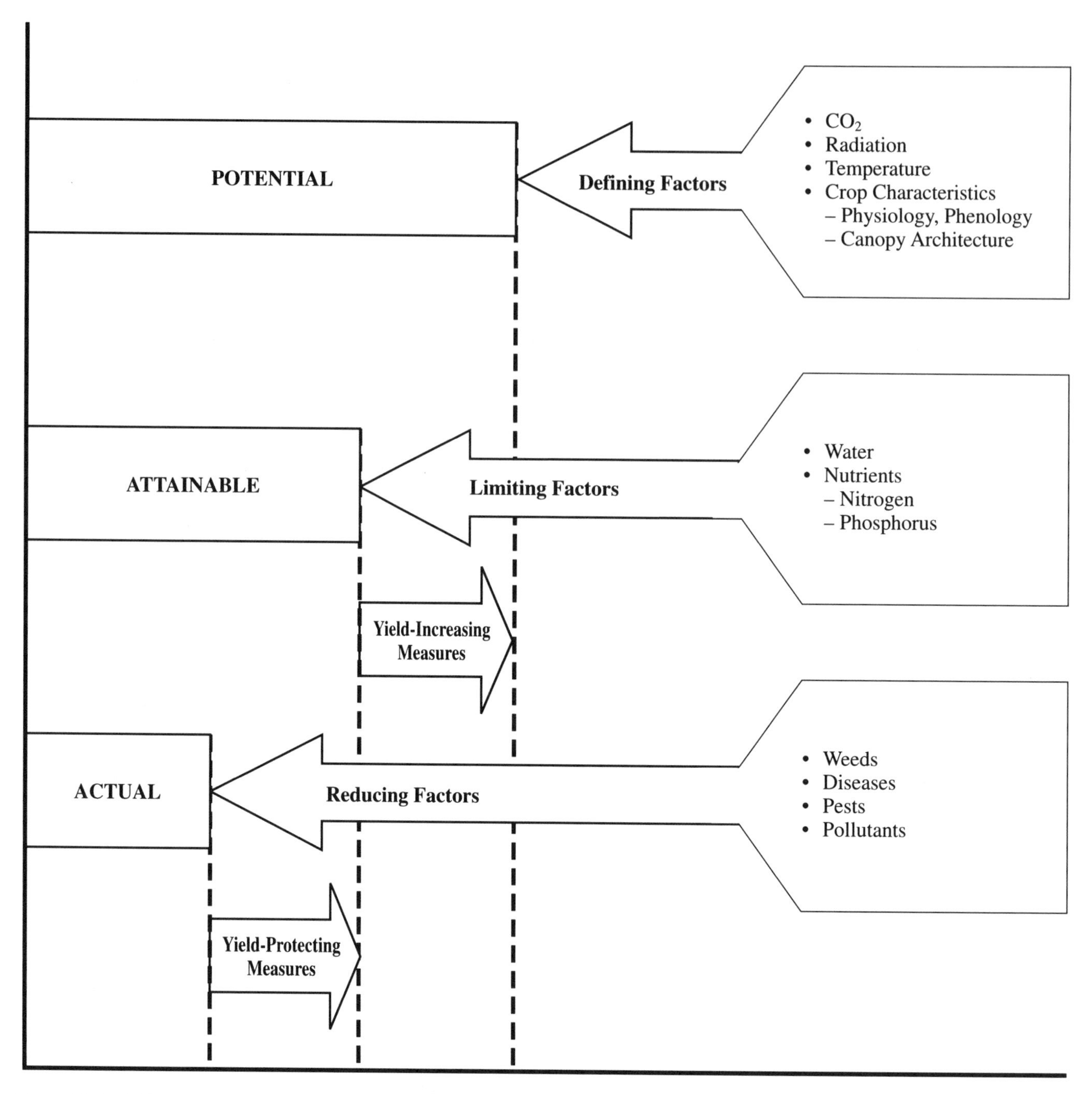

Figure 25-2: Different levels of agricultural productivity (after Rabbinge and van Ittersum, 1994).

arid lands, their productivity is too low for them to be a significant source of biomass in arid regions.

25.3.3. *Land Use and Mitigation: An Assessment*

Article 2 of the UN Framework Convention on Climate Change (FCCC) does not directly address the issue of mitigation. Mitigation options, however, are primary means for meeting the first part of the objective of the Article (i.e., stabilization of GHG atmospheric concentrations). Mitigation options also can help address the second part of the objective (i.e., ecosystem adaptive capabilities, food security, and sustainable economic development). The land-use and land-cover consequences of mitigation options also are linked strongly with UNCED's Agenda 21 and international treaties on desertification, forests, and biodiversity.

The low quality of available land-use and land-cover data and the lack of understanding of the complex human driving forces behind changes in these properties limit the validity of analyses of current land uses and make future projections speculative (Riebsame *et al.*, 1994). Discussions on land availability for biomass production are therefore not yet compelling but are promising enough to support additional, more-comprehensive research. A preliminary conclusion is that if current trends in agricultural production can be continued, competition for land is not likely to be a significant problem globally but could generate serious problems locally and/or regionally. The biggest challenge is to increase agricultural productivity in the least-productive regions and simultaneously to provide room for additional uses and functions for that land.

Much of the scientific effort required to improve these assessments needs to be directed toward integrating important social, economic, and cultural driving forces of land use with physical and ecological (including agricultural) options for land use and land cover (Turner *et al.*, 1993; Meyer and Turner, 1994). For example, large-scale development of biomass plantations could limit degradation of dry lands (e.g., by desertification) and enhance C sequestration, but also could have negative impacts on biodiversity in other regions as a result of increased deforestation or restrictions on the types of afforestation that are feasible (Ramakrishnan *et al.*, 1994). Unfortunately, such an integrated approach is only in the early stages of development (e.g., Alcamo, 1994; Edmonds *et al.*, 1994; Morita *et al.*, 1994; Rotmans *et al.*, 1994).

25.4. Concepts for Counterbalancing Climatic Change

Climate change induced by GHG emissions is an accidental result of human activities. The question, therefore, arises as to whether there are practical, large-scale, deliberate actions that might be taken to counterbalance these inadvertent changes. Such actions, sometimes referred to as "geoengineering" the climate, differ from traditional mitigative actions—which attempt to reduce the causes of the perturbing influences—in that the geoengineering approaches allow the emissions but seek to negate or reverse their influence or effects. While some geoengineering options have been suggested, in general they would require significant efforts relative to implementing alternative energy technologies or to moderating and avoiding the actions causing the greenhouse emissions.

There are four fundamental approaches to geoengineering that have been suggested to limit the perturbing effects and impacts of GHG emissions: (1) accelerating the removal of greenhouse gases from the atmosphere; (2) altering the Earth's radiation balance to compensate for the effects of the increase in concentrations of GHGs; (3) moderating the climatic response to increasing GHGs by counteracting the positive feedback processes in the atmosphere-ocean-land surface system; and (4) counteracting the harmful effects of the changes that do occur.

In reviewing these options, it should be recognized that analysis of these concepts is only schematic. For several of these ideas, effectiveness cannot yet be evaluated, leaving much for further exploration. Coupled to the rather significant costs and potential environmental side effects that are likely to be involved, starting down such a path is most appropriately considered a "last resort" option.

25.4.1. Accelerating the Removal of Greenhouse Gases from the Atmosphere

Direct removal of GHGs from the atmosphere is generally impractical—the concentrations being so dilute that it would be comparatively more practical, although still quite costly, to remove GHGs from the emissions stream at their sources. As an example of the impracticality of such direct efforts, Viggiano *et al.* (1995) analyzed a proposal to remove chlorofluorocarbons (CFCs) already released to the atmosphere using negative-ion chemistry techniques and found that the energy requirements brought the cost of the proposal well above the costs of approaches such as limiting the uses and emissions of CFCs.

Proposed approaches for accelerating the natural removal processes for CO_2 include enhancing the storage of carbon by the terrestrial biosphere and enhancing the pumping of carbon to the deep ocean by the oceanic biosphere. In these schemes, solar energy and photosynthesis would be used to power a removal process facilitated by human intervention. The advantages and disadvantages of possible approaches involving the enhancement of terrestrial carbon storage (e.g., in forests or soils), and the potential for deriving biomass fuels through these efforts, are covered in Chapters 23 and 24; approaches focusing on enhancing ocean removal are addressed here.

The oceans cover more than twice as much area as the land, are uninhabited by humans, and contain much more carbon than the terrestrial biosphere (~40,000 vs. ~2,200 Gt C, respectively)—making the potential for enhancing oceanic storage seemingly significant. In the oceans, the biosphere acts to slowly pump carbon into the deep ocean, while the net effect of the oceanic circulation is to bring carbon back to the surface to be released into the atmosphere. In preindustrial (but postglacial) times, these biological and circulation fluxes were apparently in quite close balance, keeping the atmospheric concentration nearly constant. With human activities releasing about 7 Gt C per year to the atmosphere, the net flux to the deep ocean is estimated to have increased by about 2 Gt C per year (primarily by increased downward transport of carbon). Enhancing this circulation-based removal process to accommodate a greater fraction of fossil-fuel carbon emissions would reduce the atmospheric burden, but such changes would be very difficult to achieve (although the natural downward flux will rise slowly as the CO_2 concentration increases).

Because of the chemical buffering of the atmospheric CO_2 concentration by the ocean, reduction of the atmospheric concentration would require extensive fertilization of the ocean to promote biological carbon uptake. Studies indicate that iron fertilization of the Southern Ocean would lead to only modest additional oceanic uptake of carbon. A recent iron-fertilization experiment on an 8 x 8 km oceanic plot sought to enhance biological carbon uptake but concluded that, although growth was initially enhanced, the net effect was negligible due to losses in the food chain (Martin *et al.*, 1994). However, preliminary results from a more recent iron-fertilization study found a 30- to 40-fold increase in chlorophyll in the eastern equatorial Pacific (Monastersky, 1995a, 1995b); the extent of the potential long-term effects and the consequences for nutrient cycles are not yet known. Even a major fertilization effort might well be equivalent to only the increase in the CO_2 concentration occurring over about a 10-year period (Joos *et al.*, 1991; Peng and Broecker, 1991). Such fertilization might also induce major side effects, further making this approach problematic.

Because carbon dissolved in the ocean also is present in inorganic form, a second conceivable means of increasing oceanic uptake would be by increasing oceanic alkalinity. However, this could be a practical possibility only if the weathering of rocks and river runoff were increased substantially (e.g., by a factor of 30: Flannery *et al.*, 1995)—a solution that is unlikely to be feasible.

25.4.2. *Altering the Earth's Radiation Balance*

Several approaches have been suggested to counterbalance the additional trapping of infrared radiation of GHGs by reducing the available solar energy. A number of analyses (e.g., Manabe and Wetherald, 1980; Hansen and Lacis, 1990) have suggested an approximate equivalence in the influence of changes in radiative forcing by infrared and solar radiation; however, more-recent simulations that include the effects of the oceans show that the equivalence is less than perfect in terms of the latitudinal and seasonal patterns of the climatic response (Hansen *et al.*, 1994). Nonetheless, at least some of the solar-reduction schemes could be latitudinally and perhaps seasonally tailored to achieve essential equivalence, so that the resulting radiative change would be nearly equal and opposite. Comparisons described below generally assume the need for a reduction in solar radiation of 1%, which would roughly counterbalance a 50% increase in CO_2 concentration. Note, however, that to be effective over time, the intensity of the counterbalancing effort would have to increase continuously to match the increasing GHG effect, requiring a continually increasing, and major, societal commitment.

Reducing solar radiation reaching the top of the atmosphere is conceptually possible by putting mirrors either in Sun-synchronous or near-Earth orbits. The first approach would involve placing a 2,000 km-diameter solar radiation deflector at the first Sun-Earth Lagrange point (1.5 million km from Earth), as suggested by Early (1989). Although this would require significant initial efforts—possibly including a construction base on the Moon—it would be relatively easy to sustain, would have few inadvertent consequences, and could be incrementally controlled or removed in the event of unexpected side effects. The National Academy of Sciences (1992) explored a near-Earth option involving orbiting mirrors. They estimated that counterbalancing the effects of a 50% increase in the CO_2 concentration would require placing about 55,000 mirrors, each measuring 10 x 10 km, into orbit; such objects—in addition to being difficult to control—would eclipse the Sun, the Moon, and the stars roughly 1% of the time from the view of a person looking upwards. Placing reflecting or absorbing aerosols in orbit also would be possible, but the amounts would need to be continuously replenished to make up for relatively rapid removal by the solar wind and atmospheric reentry.

In addition to other disadvantages, these extraterrestrial geo-engineering options would require significant up-front funding that could be used alternatively to develop various renewable energy sources. For example, the funding could be used to develop extraterrestrial technologies such as solar-power satellites (NAS, 1981) or to locate solar collectors on the Moon that would beam energy to Earth to provide a substitute for fossil energy; the cost likely would be comparable to lofting satellites that would simply diminish incoming solar energy.

Within the atmosphere, several concepts have been suggested for reducing solar radiation by the requisite amount; each approach has its advantages and disadvantages (National Academy of Sciences, 1992; Flannery et al., 1995). About a trillion reflective balloons—each several meters in diameter and floating in the upper stratosphere—would neither create partial eclipses nor significantly affect stratospheric chemistry but likely would be hard to design and hard to keep near the equator. The multitude of injected aluminum particles from rocket exhaust and the burn-up of reentering spacecraft may already be having a very minor influence of this type; see Brady *et al.* (1994) and TRW (1994) for an evaluation of their effects on the ozone layer. Continuous injection of sulfate aerosols into the stratosphere could be carried out to create the equivalent of a very large volcanic eruption. While this could be achieved at low cost using large artillery pieces for injection (see National Academy of Sciences, 1992), volcanic aerosols have tended to deplete stratospheric ozone. The aerosols would also whiten the skies because they scatter radiation forward more effectively than they reflect it. Injection of sufficient amounts of sooty aerosols into the stratosphere would cause warming in the stratosphere while cooling the troposphere—much like a suggested "nuclear winter" (Turco *et al.*, 1983; Pittock *et al.*, 1989). Due to the infrared effects resulting from stratospheric warming, the amount of soot injected would need to be quite significant. In this case, the sky would dim rather than whiten. Even though direct effects on ozone chemistry might be avoided because pure soot particles can be made unreactive to stratospheric ozone, the increased temperatures in the stratosphere might tend to decrease ozone concentrations. Other inadvertent environmental consequences of such measures remain poorly researched.

As appears to be happening inadvertently as a result of sulfur dioxide emissions from fossil fuel combustion (Charlson *et al.*, 1991; Kiehl and Briegleb, 1993; Taylor and Penner, 1994), injection of sulfate aerosols into the troposphere can lead to a counterbalancing effect to GHGs (National Academy of Sciences, 1992; IPCC, 1995; Flannery *et al.*, 1995). Such aerosols act in the clear sky by reflecting and scattering radiation (creating the white haze so evident over industrial regions). They also may have an effect in cloudy regions by brightening the clouds or enhancing cloud extent or lifetime.

Due to the short lifetimes of tropospheric aerosols, this approach would require injection of sulfate aerosols in amounts much greater than those currently emitted by fossil fuel power plants—leading to acid deposition, visibility impairment, ecosystem damage, and structural damage.

Increases in the albedo of the surface could be used to increase reflection of solar radiation back to space. While this may be practical in cooling residences and urban areas (see Chapter 22), countering the effect of a 50% increase in CO_2 would require covering roughly 10% of the Earth's land area or 5% of the ocean with a substance as reflective as new snow; such a change would be virtually impossible and highly disruptive of the surface climate.

25.4.3. *Altering Climatic Feedback Mechanisms*

Much of the predicted climatic response to GHGs is due to the amplifying effects of positive climatic feedback mechanisms, particularly the increase in the atmospheric water-vapor concentration. If these feedback mechanisms could be countered, the extent of global warming could be greatly reduced.

Possible approaches have not been carefully considered, but might include reducing the rate of evaporation of water (e.g., by coating or covering water surfaces) to reduce the intensity of water vapor feedback; increasing the extent and reflectivity of clouds (e.g., adding sulfates to decrease cloud-droplet size); enhancing the intensity of the oceanic thermohaline circulation (which would cool low latitudes and might promote increased heat loss in high latitudes); or altering atmospheric chemistry (e.g., by reducing tropospheric ozone to reduce its positive greenhouse effect). These proposed ideas have not been studied in even a preliminary way; however, each of these concepts likely has important side effects (e.g., altering precipitation patterns) that are as significant as the inadvertent perturbation to be avoided. Moreover, the positive feedbacks that they are intended to counteract have been poorly quantified; therefore, it is not yet possible to assess the effectiveness of these approaches.

25.4.4. *Countering Harmful Effects*

Means to counter at least some of the harmful consequences of greenhouse warming also have been suggested. For example, the predicted sea-level rise could be reduced by coating the polar ice caps to reduce melting or by pumping sea-water up onto East Antarctica as part of a giant snow-making operation; however, such projects would create enormous energy demands. Alternatively, it might be possible (e.g., by manipulating sea ice cover) to redirect Southern Ocean storms to increase snowfall onto East Antarctica; in fact, such a snow build-up may happen naturally (see Chapter 7, *Changes in Sea Level*, in the Working Group I volume).

If warmer oceans increase typhoon intensity or frequency, it may be possible to alter their tracks or reduce their strength by cloud seeding or by limiting evaporation via releasing oil slicks on the ocean. Locally, increasing the reflectivity of the surface (urban whitening) can reduce warming influences. Overall, however, acting after nature has amplified the original direct forcing of the GHGs is a relatively difficult and very inefficient option, and one that could well lead to unintended side effects.

25.4.5. *Assessment of Climate Adjustment Options*

There are at least as many uncertainties and complications involved in pursuing geoengineering options as in projecting the inadvertent climatic change of GHGs. In addition to scientific uncertainties, the UN Convention on the Prohibition of Military or Any Other Hostile Use of Environmental Modification Techniques, which entered into force 5 October 1978, may introduce complications if some countries may be negatively impacted while others benefit. Geoengineering, therefore, should be considered a viable response option only if imminent and especially rapid or threatening inadvertent change be found as a result of projected greenhouse warming (e.g., if collapse of the polar ice caps and concomitant sea-level rise are imminent). For this reason, exploring the advantages and shortcomings of the range of approaches and alternatives is a useful precaution but is not justified at the expense of the development of more practical and economical approaches.

If undertaken as a complement to reducing or slowing an unexpectedly rapid onset of inadvertent influences, geoengineering activities might be required only for a few decades. However, if undertaken as an alternative to actions to limit increasing emissions of greenhouse gases, geoengineering options would have to continue for many centuries—becoming a new, formidable, and quite possibly very costly societal responsibility. Assumption of this responsibility would need to be considered very carefully because suddenly halting such actions, even unintentionally, would, at least for most approaches, cause a rapid climatic readjustment toward the higher temperature state projected as a consequence of the inadvertent activities (i.e., the Earth would face the "climatic shock" of sudden greenhouse warming). Such a situation might well lead to more detrimental impacts than the slower warming projected if geoengineering options were not undertaken.

25.5. Integrated Assessment of Mitigation Potential

A number of evaluations of the effectiveness of various mitigation options and strategies have been published (e.g., Nakícenovíc *et al.*, 1994). Many of these mitigation assessments are based on models with limited scope, often with a strong emphasis on the energy sector (e.g., Dowlatabadi, 1994). Although some of these models include rudimentary simulations of global biogeochemical and physical processes and are suited for the evaluation of specific sectoral mitigation options, they remain of little use for evaluating combinations of mitigation options and the relations with major cross-sectoral issues, such as land use and water availability.

Many of these assessments also focus only on CO_2, neglecting other GHGs. Such partial assessments can be misleading because—through well-established linkages with other biological, chemical, and physical processes—reduction of CO_2 emissions could lead to enhanced emissions of other trace gases such as CH_4 (e.g., Dacey *et al.*, 1994). Furthermore, competing land uses often are neglected or poorly evaluated. Determination of the effectiveness of mitigation options and strategies should be based on an approach involving the simultaneous evaluation of several trace gases, in which the diverse aspects of land use are guaranteed, and in which important physical and biogeochemical processes, linkages, and feedbacks are included (Leemans, 1995). This requirement should result in the combined analysis of the Earth system (Ojima, 1993), including technological and socioeconomic models (Meyer and Turner, 1994).

Such integration can be successful only if the relevant properties and dimensions of each domain (e.g., geosphere, biosphere, anthroposphere) are addressed. For example, early carbon-cycle models aggregated all land covers into a few classes characterized by globally averaged parameters (e.g., Goudriaan and Ketner, 1984). Although such an approach is straightforward to implement, it does not allow assessment of local and regional consequences of global change (see Chapter 24) and is therefore unsuitable for evaluation of the efficacy of mitigation options. In current assessment models, continental-scale regions are assumed to be homogeneous. Thus, these models are not capable of representing the wealth of socioeconomic components and market forces that cause and react to change at the community, state, and national level. Similar arguments can be developed for all other domains. To address the heterogeneous response of biogeochemical processes, a georeferenced approach has to be adopted. More-recent models have taken such an approach (e.g., Melillo *et al.*, 1993; Klein Goldewijk *et al.*, 1994). The integrated models required to evaluate mitigation options and strategies should be robust and convey confidence in their simulation of local to regional processes, while simultaneously considering regional to global characteristics.

25.5.1. *Components of Integrated Assessment Models*

Integrated Assessment Models (IAMs) can in principle serve three purposes: (1) They can help assess potential responses to climate change, either by comparing the costs of response options to the benefits of avoided impacts or by comparing the relative effectiveness and costs of alternate response options; (2) they provide an overview of the cross-sectoral linkages and tradeoffs that can facilitate a more systematic evaluation of policy options; and (3) they can help evaluate the importance of climate change relative to other socioeconomic concerns.

As discussed in Chapter 10, *Integrated Assessment of Climate Change: An Overview and Comparison of Approaches and Results*, of the Working Group III volume, a number of IAMs have been developed that are integrated over different dimensions and to different degrees. "Full-scale" IAMs seek to address the linkages and feedbacks among human activities, managed and unmanaged ecosystems, emissions and atmospheric composition, and climate change and sea level. The various models differ in the level of detail and complexity considered both within and between the modules, the underlying socioeconomic assumptions used, and the manner in which physical and socioeconomic uncertainties are addressed.

25.5.2. *Illustrative Examples and Results*

IAMs can be used to evaluate the implications of extensive biomass plantations on land use and availability. This section discusses simulation results using two very different IAMs as examples: IMAGE 2.0 (Alcamo *et al.*, 1994) and MiniCAM 2.0 (Edmonds *et al.*, 1995). The basic structures of the models are described in Boxes 25-2 and 25-3, respectively. IMAGE 2.0 is a targets-based IAM with considerable regional and sectoral detail of potential physical impacts. These impacts, however, are not given economic values. Also, the model does not include explicit representations of uncertainty. It does, however, account for land use and changes in land cover through physical and biogeochemical feedbacks, such as changes in albedo, terrestrial carbon storage, and enhanced plant growth. MiniCAM 2.0, on the other hand, is designed to balance the costs and benefits of climate-change policies. In this model, constraints on human activities are explicitly represented and costed out. However, the model has a more aggregated representation of climate-change impacts.

The IMAGE 2.0 baseline scenario uses the population and economic growth assumptions from the IPCC IS92a scenario (Leggett *et al.*, 1992); other scenario input variables come from a variety of sources (Alcamo *et al.*, 1994). As part of this baseline, it is assumed that biomass is used for the generation of 208 EJ energy worldwide in 2100 (and 74 EJ in 2050), thereby reducing the dependence on fossil fuels. The basic assumption for this baseline is that most of this biomass would be taken from readily available sources, such as agricultural and forestry residues, municipal waste, and so forth. A second scenario assumes that only 60% of this demand is readily available, while the remaining 40% has to come from specific biomass crops. This "biomass crop scenario" results globally in an increased demand for agricultural land. A third scenario assumes that no modern biomass is used and that an equivalent amount of the energy demand would be satisfied by oil instead. This "no biomass scenario" does not alter land-cover patterns. The three scenarios differ in their regional and global patterns of fluxes, sources, and sinks of greenhouse gases (Table 25-3).

The three scenarios illustrate different possibilities for increases in the emissions of individual GHGs to the atmosphere as a consequence of interactions of land-use and energy options (Table 25-3). The biomass crop scenario would lead to a large increase in agricultural land cover, although there would be large regional differences in area and timing

Box 25-2. The IMAGE 2.0 Model

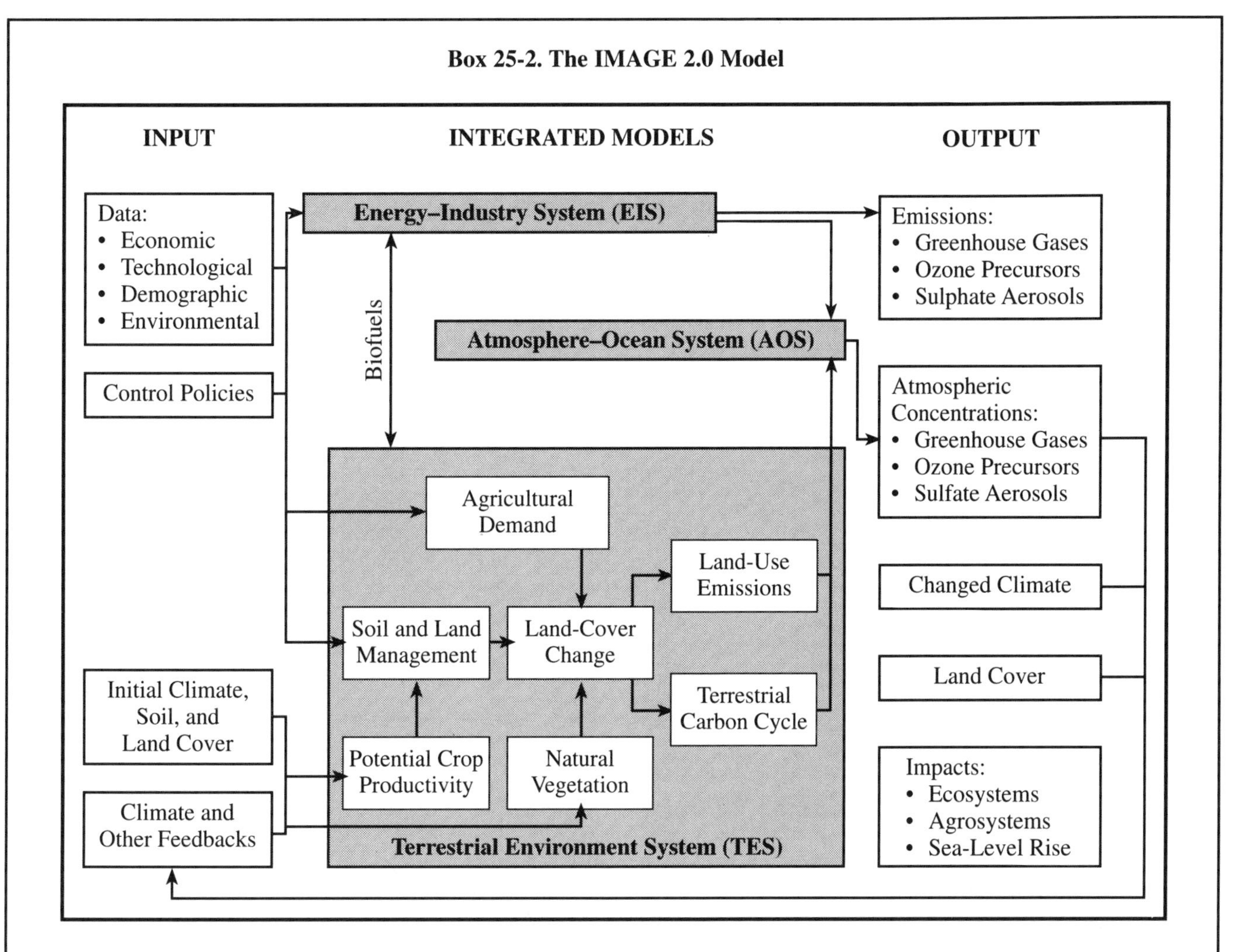

IMAGE 2.0 (Integrated Model to Assess the Greenhouse Effect) is a multidisciplinary, integrated model of climate change. The model is designed to provide support for comprehensive evaluations of national and international policies concerning the build-up of GHGs. The model consists of three fully linked components: Energy Industry System (EIS), Terrestrial Environment System (TES), and Atmosphere Ocean System (AOS). EIS divides the world into 13 regions and computes emissions in each as a function of energy consumption and industrial production. The EIS-models are designed to investigate the effectiveness of different fuel mixes and improved energy efficiencies. TES defines the role of the terrestrial biosphere using vegetation-climate-soil interactions and determining the impacts of land-use change. The dynamic simulation of land-use and land-cover change is an important component of TES. These changes are simulated on a 0.5-degree global grid. Each grid cell is characterized by its climate, topography, soil, and land cover. Changes in natural vegetation are simulated with the BIOME model (Prentice *et al.*, 1992). Potential productivity for eight major crops is computed using the "agro-ecological zone" approach developed by UN-FAO (Brinkman, 1987; Leemans and Solomon, 1993). This approach provides the potential distribution and productivity of crops and natural vegetation over the grid. For each region, the demand for land-based commodities (crops, livestock, biomass, lumber, and fuelwood) is calculated. The per capita consumption is derived from an assumed elasticity between consumption and per capita income. Population, dietary preferences, and socioeconomic factors control changes in demand. Agricultural data for the period 1970–1990 from FAO were used to parameterize the demand functions and to calibrate the geo-referenced area of cultivated land from Olson *et al.* (1985) with observed regional productivities. TES then reconciles the demand for land with its potential through a set of heuristic rules that reflect key driving factors, including proximity to infrastructure, population, and productivity of land. This relatively simple model results in rich patterns of land use and cover change for each region. These changed patterns are used to calculate the GHG emissions and local responses of ecosystems with respect to the terrestrial carbon cycle. TES allows evaluation of the impact of environmental changes on ecosystems and agriculture and the effectiveness of mitigation policies. Such evaluation can be achieved comprehensively because of the systemic linkages with other components of IMAGE 2. TES is linked with EIS through the demand for fuelwood and biomass. TES and AOS are indirectly linked through changes in soil moisture and albedo, mediated by land-cover change. All emissions are combined in AOS to determine atmospheric concentrations of GHGs, while accounting for atmospheric chemistry and oceanic carbon uptake.

Box 25-3. The MiniCAM 2.0 Model

MiniCAM 2.0 is an integrated assessment model with four major components: Human Activities, Atmospheric Composition, Climate and Sea Level, and Ecological Systems. The model considers energy and land-use change both explicitly and interactively. The energy system model is the Edmonds-Reilly-Barns (ERB) model (Edmonds and Reilly, 1985), while the agriculture land-use model (ALM) was developed explicitly for MiniCAM 2.0. ALM partitions land into managed and less-managed systems. The managed lands are used intensively for human settlement and infrastructure and extensively for growing crops, raising livestock, managed forests, or biomass cultivation for energy use. Less-managed lands are partitioned into ecological categories. The allocation of lands to human settlement and infrastructure is determined by population and income, and this use of land takes precedence in the model over all other uses of managed land. Extensive land uses are determined by expected profitability—which in turn depends on plant productivities, product prices, technology, fertilization, atmospheric CO_2 concentrations, climate, population, income, taxes, tariffs, and subsidies. Less-managed lands include those that are "parked" (i.e., excluded from use for managed activities) and those that are potentially available for managed uses.

The boundary between the managed and less-managed systems is determined by the expected profitability of managed lands in general. Within ALM, global markets are established for each of the major traded commodities—crops, livestock, and forest products—and a world price is established that clears international markets. Biomass for energy use is determined interactively with the ERB. Because biomass is used as an energy resource, its demand and price are determined in the ERB while its supply is determined in the ALM. Changes in land allocations determine net trace-gas fluxes from the terrestrial biosphere, while the ERB determines energy-related emissions. Other emissions, such as those from cement manufacture and from CFCs and their substitutes, are handled exogenously.

(Alcamo *et al.*, 1994). This increase would take up less carbon than the land cover the biomass crops replace. This would limit carbon sequestration in ecosystems and enhance emissions of CH_4 and other GHGs. The realized climate change does not necessarily differ significantly for these three scenarios.

The MiniCAM 2.0 model also has been used to address issues of conflicting land use in cases where low fossil fuel emissions objectives are pursued (Edmonds *et al.*, 1995). The model was used to examine the implications for land-use competition resulting from the technologies described in the LESS constructions in Chapter 19. The principal finding of the analysis was that with biomass productivities as high as that considered in LESS, negative impacts from the competition for land use are minimal. That is, increases in emissions from land-use change are likely to be less than 10%, and per capita consumption of crops and livestock also are likely to be within 10% of their reference values. The results, however, are sensitive to several assumptions, such as the productivity of biomass energy plants, the rate of technological progress in agriculture, and the rates of population and income growth.

Table 25-3: *Globally averaged summary of IMAGE 2.0 results for the baseline, biomass crops, and no-biomass scenarios. The scenarios differ in the assumptions regarding the source of 74 or 208 EJ of energy in 2050 and 2100, respectively (after Alcamo et al., 1994).*

Year and Scenario	Change in Atmospheric Concentration (%)		Change in Agricultural Productivity (% ha^{-1})	Change in Agricultural Area (%)	Change in Forest Area (%)	Change in Average Surface Temperature (°C)	
	CO_2	CH_4				N. Hemisphere	S. Hemisphere
1990	358 ppm	1.7 ppm	100	2670 Mha	4720 Mha	14.2	13.0
2050							
Baseline	+46	+47	+72	+9	-26	+1.4	+1.0
Biomass Crops	+49	+53	+72	+30	-32	+1.5	+1.0
No Biomass	+57	+41	+72	+9	-26	+1.4	+1.0
2100							
Baseline	+117	+35	+108	+14	-27	+2.4	+1.8
Biomass Crops	+129	+41	+108	+65	-31	+2.7	+2.0
No Biomass	+139	0	+108	+15	-27	+2.4	+1.9

25.6. Concluding Remarks

A number of cross-sectoral linkages become evident only when an integrated analysis of climate and socioeconomic systems is used to complement sectoral analyses. The field of integrated assessment modeling is still under development, with various models being used to investigate different plausible ways to integrate across various disciplines. At present there is no single "right" approach to integrated assessment modeling, and models often have to make tradeoffs between the level of sectoral detail they can include and the level of complexity and data requirements that can be handled realistically. Further, Integrated Assessment Models are only as good as the underlying socioeconomic assumptions and information on sectoral impacts, adaptation, and mitigation strategies. As discussed throughout this report, there are key gaps in the understanding of the climate system, how climate change would impact the various sectors, and the feasibility and possible ancillary impacts of various adaptation and mitigation options. Across these issues, information from developing countries is particularly lacking.

The discussion in this chapter is, at best, a roadmap of the types of cross-sectoral issues and tradeoffs decisionmakers will need to consider in evaluating the various mitigation options. While the analysis has focused on energy and land use—which are two key arteries cutting across the various sectors—issues relating to society's demands for water and how water resources would be affected by different mitigation strategies have not yet been addressed in the same level of detail. Further, many cross-cutting impacts of mitigation policies also would be transmitted through trade and market forces. An analysis of these, however, falls beyond the mandate of this Working Group.

Acknowledgments

We appreciate suggestions and comments on earlier drafts by the lead authors of the other mitigation chapters, and intensive efforts on this chapter by S. Rayner, D. Shriner, and M. Rounsevell. Without their guidance, the amalgamation of distinct and diverse topics in this chapter would not have been possible. We appreciate the assistance in preparing and editing different versions of the manuscript provided by Ms. V. Malée and Ms. E. Sulzman. The development of this chapter was facilitated by funds provided to the Convening Lead Author by the Dutch Ministry on Housing, Physical Planning and Environmental Protection.

References

Alcamo, J. (ed.), 1994: *IMAGE 2.0: Integrated Modeling of Global Climate Change*. Kluwer Academic Publishers, Dordrecht, The Netherlands, 318 pp.

Alcamo, J., G.J. van den Born, A.F. Bouwman, B. de Haan, K. Klein Goldewijk, O. Klepper, R. Leemans, J.A. Olivier, B. de Vries, H. van der Woerd, and R. van den Wijngaard, 1994: Modeling the global society-biosphere-climate system, part 2: Computed scenarios. *Water Air Soil Pollut.*, **76**, 37-78.

Alcamo, J., A.F. Bouwman, J. Edmonds, A. Grübler, T. Morita, and A. Sugandhy, 1995: An evaluation of the IPCC IS92 emission scenarios. In: *Climate Change 1994: Radiative Forcing of Climate Change and an Evaluation of the IPCC IS92 Emission Scenarios* [Houghton, J.T., L.G. Meira Filho, J. Bruce, H. Lee, B.A. Callander, N. Harris, and K. Maskel (eds.)]. Cambridge University Press, Cambridge, UK, pp. 247-304.

Atjay, G.L., P. Ketner, and P. Duvigneaud, 1979: Terrestrial primary production and phytomass. In: *The Global Carbon Cycle* [Bolin, B., E.T. Degens, S. Kempe, and P. Ketner (eds.)]. Wiley and Sons, New York, NY, pp. 129-187.

Barrow, C.J., 1991: *Land Degradation*. Cambridge University Press, Cambridge, UK, 295 pp.

Bazilevich, N.I., L.Y. Rodin, and N.N. Rozov, 1971. Geographical aspects of biological productivity. *Soc. Geogr. Rev. Transl.*, **12**, 293-317.

Bernthal, F.M. (ed.), 1990: *Climate Change: The IPCC Response Strategies*. World Meteorological Organization/United Nations Environment Program, Geneva, Switzerland, 270 pp.

Bird, K.T., 1987: Cost analysis of energy from marine biomass. In: *Seaweed Cultivation for Renewable Resources* [Bird, K.T. and P.H. Benson (eds.)]. Elsevier, Amsterdam, The Netherlands, pp. 327-350.

Bongaarts, J., 1994: Can the growing human population feed itself? *Sci. Am.*, **1994**, 18-24.

Bowen, H.J.M., 1966: *Trace Elements in Biochemistry*. Academic Press, London, UK, 268 pp.

Brady, B.B., E.W. Fournier, L.R. Martin, and R.B. Cohen, 1994: *Stratospheric Ozone Reactive Chemicals Generated by Space Launches Worldwide*. Aerospace Report No. TS-94(4231)-6, The Aerospace Corporation, El Segundo, CA, 25 pp.

Brinkman, R., 1987: Agro-ecological characterization, classification and mapping: Different approaches by the international agricultural research centres. In: *Agricultural Environments: Characterization, Classification and Mapping* [Bunting, A.H. (ed.)]. C.A.B. International, Wallingford, UK, pp. 31-42.

Brown, S. and A.E. Lugo, 1994: Rehabilitation of topical land: A key to sustaining development. *Restorat. Ecol.*, **2**, 97-111.

Buringh, P. and R. Dudal, 1987: Agricultural land use in space and time. In: *Land Transformation in Progress* [Wolman, M.G. and F.G.A. Fournier (eds.)]. John Wiley and Sons, New York, NY, pp. 9-43.

Carpentieri, A.E., E.D. Larson, and J. Woods, 1993: Future biomass-based electricity supply in Northeast Brazil. *Biomass and Bioenergy*, **4**, 149-173.

Charlson, R.J., J. Langer, H. Rodhe, C.B. Leovy, and S.G. Warren, 1991: Perturbation of the northern hemisphere radiative balance by backscattering of anthropogenic sulfate aerosols. *Tellus AB*, **43**, 152-163.

Dacey, J.W.H., B.G. Drake, and M.J. Klug, 1994: Stimulation of methane emission by carbon dioxide enrichment of marsh vegetation. *Nature*, **370**, 47-49.

Deevey, Jr., E.S., 1960: The human population. *Sci. Am.*, **203**, 195-204.

Defries, R.S. and J.R.G. Townshend, 1994: NDVI-Derived land cover classifications at a global scale. *Int. J. Remote Sens.*, **15**, 3567-3586.

Dowlatabadi, H., 1994: Integrated assessment of climate change: An incomplete overview. In: *Integrative Assessment of Mitigation, Impact, and Adaptation to Climate Change* [Nakícenovíc, N., W.D. Nordhaus, R. Richels, and F.L. Toth (eds.)]. International Institute for Applied System Analysis, Laxenburg, Austria, pp. 105-117.

Duvigneaud, P., B. Bolin, E.T. Degens, and S. Kempe, 1979: The global biogeochemical carbon cycle. In: *The Global Carbon Cycle* [Bolin, B., E.T. Degens, S. Kempe, and P. Ketner (eds.)]. Wiley and Sons, New York, NY, pp. 1-56.

Dyson, T., 1994: Population growth and food production: Recent global and regional trends. *Populat. Developm. Rev.*, **20**, 397-411.

Early, J.T., 1989: Space-based solar screen to offset the greenhouse effect. *J. Brit. Interplanet. Soc.*, **42**, 567-569.

Edmonds, J. and J. Reilly, 1985: *Global Energy: Assessing the Future*. Oxford University Press, New York, NY, 317 pp.

Edmonds, J., M. Wise, and C. MacCracken, 1994: *Advanced Energy Technologies and Climate Change: An Analysis Using the Global Change Assessment Model (GCAM)*. PNL-report PNL-9798, UC-402, Pacific Northwest Laboratory, Richland, WA, 41 pp.

Edmonds, J., M. Wise, R. Sands, C. MacCracken, and H. Pitcher, 1995: *Agriculture, Land-Use, and Energy: An Integrated Analysis of the Potential Role of Biomass Energy for Reducing Potential Future Greenhouse-Related Emissions*. Pacific Northwest Laboratory, Washington, DC, (in press).

Emanuel, W.R., H.H. Shugart, and M.P. Stevenson, 1985: Climatic change and the broad-scale distribution of terrestrial ecosystems complexes. *Climatic Change*, **7**, 29-43.

FAO, 1979: *FAO Production Year Book*. FAO Statistics Series 22, vol. 32, Food and Agricultural Organization of the United Nations, Rome, Italy, pp. 45-56.

FAO, 1991: *Agrostat PC. Land Use*. Computerized Information Series 1/3, Food and Agriculture Organization of the United Nations, Rome, Italy, October, 18 pp. (+Diskette: Version 2.0, April 1995).

FAO, 1993: *Agriculture: Towards 2010*. Conference C93/94, Food and Agriculture Organization of the United Nations, Rome, Italy, November, 362 pp.

Flannery, B.P., G. Marland, W. Broecker, H. Ishatani, H. Keshgi, H. Komiyama, M. MacCracken, N. Rosenberg, M. Steinberg, and T.M.L. Wigley, 1995: Geoengineering climate. In: *The Engineering Response to Climate Change* [Watts, R. (ed.)]. Lewis Publishers, Chelsea, MI (in press).

Fresco, L.O., L. Stroosnijder, J. Bouma, and H. van Keulen (eds.), 1994: *The Future of the Land: Mobilizing and Integrating Knowledge for Land Use Options*. John Wiley and Sons, Chichester, UK, 409 pp.

Glenn, E., V. Squires, M. Olsen, and R. Frye, 1993: Potential for carbon sequestration in the drylands. *Water Air Soil Pollut.*, **70**, 341-355.

Golley, F.B., 1972: Energy flux in ecosystems. In: *Ecosystem Structure and Function* [Wiens, J.A. (ed.)]. Oregon University Press, Corvallis, OR, pp. 69-90.

Goudriaan, J. and P. Ketner, 1984: A simulation study for the global carbon cycle, including man's impact on the biosphere. *Climatic Change*, **6**, 167-192.

Hall, D.O. and R.P. Overend (eds.), 1987: *Biomass: Regenerable Energy*. John Wiley and Sons, Chichester, UK, 594 pp.

Hall, D.O., 1991: Biomass energy. *Ener. Pol.*, **19**, 711-737.

Hall, D.O., F. Rosillo-Calle, and J. Woods, 1994: Biomass utilization in households and industry: Energy use and development. *Chemosphere*, **29**, 1099-1119.

Hansen, J.E. and A.A. Lacis, 1990: Sun and dust versus greenhouse gases: An assessment of their relative roles in global climate change. *Nature*, **346**, 713-719.

Hansen, J., M. Sato, A. Lacis, and R. Reudy, 1994: Climate impacts of ozone change. In *Proceedings of IPCC Hamburg Meeting*, May 1993.

Houghton, R.A., J.E. Hobbie, J.M. Melillo, B. Moore, B.J. Peterson, G.R. Shaver, and G.M. Woodwell, 1983: Changes in the carbon content of terrestrial biota and soils between 1860 and 1980: A net release of CO_2 to the atmosphere. *Ecol. Monogr.*, **53**, 235-262.

Houghton, R.A., 1994: The worldwide extent of land-use change. *Bioscience*, **44**, 305-313.

Hummel, J.R. and R.A. Reck, 1979: A global surface albedo model. *J. Appl. Meteorol.*, **18**, 239-253.

Joos, F., J.L. Sarmiento, and U. Siegenthaler, 1991: Estimates of the effect of southern ocean iron fertilization on atmospheric CO_2 concentrations. *Nature*, **349**, 772-775.

Kendall, H.W. and D. Pimentel, 1994: Constraints on the expansion of the global food supply. *Ambio*, **23**, 198-205.

Kiehl, J.T. and B.P. Briegleb, 1993: The relative roles of sulfate aerosols and greenhouse gases in climate forcing. *Science*, **260**, 311-314.

Klein Goldewijk, K., J.G. van Minnen, G.J.J. Kreileman, M. Vloedbeld, and R. Leemans, 1994: Simulating the carbon flux between the terrestrial environment and the atmosphere. *Water Air Soil Pollut.*, **76**, 199-230.

Küchler, A.W., 1949: A physiognomic classification of vegetation. *Ann. Ass. Amer. Geog.*, **39**, 201-210.

Lazarus, M., C. Heaps, and P. Raskin, 1995: *LEAP—Long-Range Energy Alternatives Planning System: User Guide for Version 95.0*. Stockholm Environment Institute, Boston, MA, 220 pp.

Leemans, R., 1995: Incorporating land-use change in Earth system models. In: *Proceedings of the First IGBP-GCTE Science Conference* [Walker, B. (ed.)]. Cambridge University Press, Cambridge, UK (in press).

Leemans, R. and A.M. Solomon, 1993: The potential response and redistribution of crops under a doubled CO_2 climate. *Clim. Res.*, **3**, 79-96.

Leemans, R. and G. Zuidema, 1995: The importance of changing land use for global environmental change. *Trends Ecol. Evol.*, **10**, 76-81.

Leemans, R., W. Cramer, and J.G. van Minnen, 1995: Prediction of global biome distribution using bioclimatic equilibrium models. In: *Effects of Global Change on Coniferous Forests and Grassland* [Melillo, J.M. and A. Breymeyer (eds.)]. J. Wiley and Sons, New York, NY (in press).

Leggett, J., W.J. Pepper, and R.J. Swart, 1992: Emissions scenarios for the IPCC: An update. In: *Climate Change 1992. The Supplementary Report to the IPCC Scientific Assessment* [Houghton, J.T., B.A. Callander, and S.K. Varney (eds.)]. Cambridge University Press, Cambridge, UK, pp. 71-95.

Lieth, H., 1964: Versuch einer kartographischer darstellung der produktivität der pflanzandecke auf der erde. In *Geogr. Taschenbuch. Franz Steiner Verlag*. Wiesbaden, pp. 72-80.

Lieth, H., 1973: Primary production: Terrestrial ecosystems. *Human Ecol.*, **1**, 303-332.

Loveland, T.R., J.W. Merchant, D.O. Ohlen, and J.F. Brown, 1991: Development of a land-cover characteristics database for the conterminous U.S. *Photgram. Engin. Rem. Sens.*, **57**, 1453-1463.

Manabe, S. and R.T. Wetherald, 1980: On the distribution of climatic change resulting from an increase in CO_2 content of the atmosphere. *J. Atmos. Sci.*, **37**, 99-118.

Martin, J.H., K.H. Coale, K.S. Johnson, S.E. Fitzwater, R.M. Gordon, S.J. Tanner, C.N. Hunter, V.A. Elrod, J.L. Nowicki, T.L. Coley, R.T. Barber, S. Lindley, A.J. Watson, K. Vanscoy, C.S. Law, M.I. Liddicoat, R. Ling, T. Stanton, J. Stockel, C. Collins, A. Anderson, R. Bidigare, M. Ondrusek, M. Latasa, F.J. Millero, K. Lee, W. Yao, J.Z. Zhang, G. Friederich, C. Sakamoto, F. Chavez, K. Buck, Z. Kolber, R. Greene, P. Falkowski, S.W. Chisholm, F. Hoge, R. Swift, J. Yungel, S. Turner, P. Nightingale, A. Hatton, P. Liss, and N.W. Tindale, 1994: Testing the iron hypothesis in ecosystems of the equatorial Pacific Ocean. *Nature*, **371**, 123-129.

Matthews, E., 1983: Global vegetation and land use: New high-resolution data bases for climate studies. *J. Clim. App. Meteorol.*, **22**, 474-487.

Melillo, J.M., A.D. McGuire, D.W. Kicklighter, B. Moore III, C.J. Vorosmarty, and A.L. Schloss, 1993: Global climate change and terrestrial net primary production. *Nature*, **363**, 234-239.

Meyer, W.B. and B.L. Turner II (eds.), 1994: *Changes in Land Use and Land Cover: A Global Perspective*. Cambridge University Press, Cambridge, UK, 537 pp.

Monastersky, R., 1995a: Iron surprise: Algae absorb carbon dioxide. *Science News*, **148**, 53.

Monastersky, R., 1995b: Iron versus the greenhouse. *Science News*, **148**, 220-222.

Morita, T., Y. Matsuoka, M. Kainuma, K. Kai, H. Harasawa, and L. Dong-Kun, 1994: *Asian-Pacific Integrated Model to Assess Policy Options for Stabilizing Global Climate*. AIM report 1.0, National Institute for Environmental Studies, Tsukuba, Japan, July, 23 pp.

Morris, D.W., 1995: Earth's peeling veneer of life. *Nature*, **373**, 25.

Mücher, C.A., T.J. Stomph, and L.O. Fresco, 1993: *Proposal for a Global Land Use Classification*. Final Report LUIS, FAO, ITC and WAU, Rome, Enschede, and Wageningen, February, 37 pp.

Nakícenovíc, N., W.D. Nordhaus, R. Richels, and F.L. Toth (eds.), 1994: *Integrative Assessment of Mitigation, Impact, and Aadaptation to Climate Change*. International Institute for Applied System Analysis, Laxenburg, Austria, 669 pp.

National Academy of Sciences, 1981: *Electric Power from Orbit: A Critique of a Satellite Power System*. National Academy Press, Washington, D.C., 332 pp.

National Academy of Sciences, National Academy of Engineering and Institute of Medicine, 1992: *Policy Implications of Greenhouse Warming: Mitigation, Adaptation, and the Science Base*. National Academy Press, Washington, DC, 918 pp.

Neilson, R.P., G.A. King, and G. Koerper, 1992: Toward a rule-based biome model. *Landscape Ecol.*, **7**, 27-43.

Ojima, D. (ed.), 1993: *Modeling the Earth System*. UCAR/Office for Interdisciplinary Earth Studies, Boulder, CO, 472 pp.

Oldeman, L.R., 1993: Global extent of soil degradation. In: *Soil Resilience and Sustainable Land Use* [Greenland, D.J. and I. Szabolcs (eds.)]. C.A.B. International, Wallingford, UK, pp. 99-118.

Olson, J., J.A. Watts, and L.J. Allison, 1985: *Major World Ecosystem Complexes Ranked by Carbon in Live Vegetation: A Database*. Report NDP-017, Oak Ridge National Laboratory, Oak Ridge, TN, September, 164 pp.

Orr, J.C. and J.L. Sarmiento, 1992: Potential of marine macroalgae as a sink for CO_2: Constraints from a 3-D general circulation model of the global ocean. *Water Air Soil Pollut.*, **64**, 405-422.

Peng, T.-H. and W.S. Broecker, 1991: Dynamical limitations on the Antarctic iron fertilization strategy. *Nature*, **349**, 227-229.

Pittock, A.B., T.P. Ackerman, P.J. Crutzen, M.C. MacCracken, C.S. Shapiro, and R.P. Turco, 1989: *Environmental Consequences of Nuclear War: Physical and Atmospheric Effects*. John Wiley and Sons, Chichester, UK, 350 pp.

Plucknett, D.L., 1994: *Science and Agricultural Transformation*. IFPRI Lecture Series, International Food Policy Research Institute, Washington, DC, September 1993, 34 pp.

Prentice, I.C., W. Cramer, S.P. Harrison, R. Leemans, R.A. Monserud, and A.M. Solomon, 1992: A global biome model based on plant physiology and dominance, soil properties and climate. *J. Biogeogr.*, **19**, 117-134.

Prentice, I.C., M.T. Sykes, M. Lautenschlager, S.P. Harrison, O. Denissenko, and P.J. Bartlein, 1994: Modelling global vegetation patterns and terrestrial carbon storage at the last glacial maximum. *Glob. Ecol. Biogeogr. Let.*, **3**, 67-76.

Rabbinge, R. and M.K. van Ittersum, 1994: Tension between aggregation levels. In: *The Future of the Land: Mobilizing and Integrating Knowledge for Land Use Options* [Fresco, L.O., L. Stroosnijder, J. Bouma, and H. van Keulen (eds.)]. John Wiley and Sons, Chichester, UK, pp. 31-40.

Ramakrishnan, P.S., J. Campbell, L. Demierre, A. Gyi, K.C. Malhotra, S. Mehndiratta, S.N. Rai, and E.M. Sashidharan, 1994: Ecosystem rehabilitation of the rural landscape in south and central Asia: An analysis of issues. In: *Special Publication UNESCO Regional Office of Science and Technology for South and Central Asia (ROSTCA)* [Hadley, M. (ed.)]. New Delhi, India, 29 pp.

Riebsame, W.E., W.B. Meyer, and B.L. Turner II, 1994: Modeling land use and cover as part of global environmental change. *Climatic Change*, **28**, 45-64.

Rosenzweig, C. and M.L. Parry, 1994: Potential impact of climate change on world food supply. *Nature*, **367**, 133-138.

Rotmans, J., M.B.A. van Asselt, A.J. de Bruin, M.G.J. den Elzen, J. de Greef, H. Hilderink, A.Y. Hoekstra, M.A. Janssen, H.W. Köster, W.J.M. Martens, L.W. Niessen, and H.J.M. de Vries, 1994: *Global Change and Sustainable Development: A Modelling Perspective for the Next Decade*. RIVM report no. 461502004, National Institute of Public Health and Environmental Protection, Bilthoven, June, 80 pp.

Running, S.W., T.R. Loveland, and L.L. Pierce, 1994: A vegetation classification logic based on remote sensing for use in global biogeochemical models. *Ambio*, **23**, 77-81.

Schmitt, W.R., 1965: The planetary food potential. *Ann. NY Acad. Sci.*, **118**, 645-718.

Shantz, H.L., 1954: The place of grasslands in the Earth's cover of vegetation. *Ecology*, **35**, 143-145.

Skole, D. and C. Tucker, 1993: Tropical deforestation and habitat fragmentation in the Amazon: Satellite data from 1978 to 1988. *Science*, **260**, 1905-1910.

Smil, V., 1994: How many people can the Earth feed. *Populat. Developm. Rev.*, **20**, 255-292.

Smith, T.M., R. Leemans, and H.H. Shugart, 1992: Sensitivity of terrestrial carbon storage to CO_2 induced climate change: Comparison of four scenarios based on general circulation models. *Climatic Change*, **21**, 367-384.

Solomon, A.M., I.C. Prentice, R. Leemans, and W.P. Cramer, 1993: The interaction of climate and land use in future terrestrial carbon storage and release. *Water Air Soil Pollut.*, **70**, 595-614.

Taylor, K.E. and J.E. Penner, 1994: Response of the climate system to atmospheric aerosols and greenhouse gases. *Nature*, **369**, 734-737.

Townshend, J.R.G., 1992: *Improved Global Data for Land Application: A Proposal for a New High Resolution Data Set.* IGBP-Report No.20, International Geosphere-Biosphere Programme, Stockholm, Sweden, 87 pp.

TRW, 1994: *The Impact of Deorbiting Space Debris on Stratospheric Ozone*. TRW Space and Electronics Group, P.D. Lohn and E.Y. Wong, El Segundo, CA, 83 pp.

Turco, R.P., O.B. Toon, T.P. Ackerman, J.B. Pollack, and C. Sagan, 1983: Nuclear winter: Global consequences of multiple nuclear explosions. *Science*, **222**, 1283-1293.

Turner, B.L. II, W.C. Clark, R.W. Kates, J.F. Richards, J.T. Mathews, and W.B. Meyer (eds.), 1990: *The Earth as Transformed by Human Action: Global and Regional Changes in the Biosphere over the Past 300 Years*. Cambridge University Press, Cambridge, UK, 713 pp.

Turner, B.L. II, R.E. Kasperson, W.B. Meyer, K.M. Dow, D. Golding, J.X. Kasperson, R.C. Mitchell, and S.J. Ratick, 1990: Two types of global environmental change: Definitional and spatial-scale issues in their human dimensions. *Global Environ. Change*, **1**, 14-22.

Turner, B.L., R.H. Moss, and D.L. Skole, 1993: *Relating Land Use and Global Change: A Proposal for an IGBP-HDP Core Project*. IGBP Report No. 24 and HDP Report No. 5, International Geosphere-Biosphere Programme and the Human Dimensions of Global Environmental Change Programme, Stockholm, Sweden, February, 65 pp.

UNEP/GEMS, 1993: *Vegetation Classification*. Report of the UNEP-HEM/WCMC/GCTE preparatory meeting, Charlottesville, Virginia, GEMS Report Series No. 19, United Nations Environment Programme, Nairobi, Kenya, 21 pp.

UNEP/GEMS, 1994: *Report of the UNEP/FAO Expert Meeting on Harmonizing Land Cover and Land Use Classifications*. GEMS Report Series No. 25, United Nations Environment Programme, Nairobi, Kenya, March, 43 pp.

UNESCO, 1973: *International Classification and Mapping of Vegetation*. Report, United Nations Educational, Scientific and Cultural Organization, Paris, France, 35 pp.

Viggiano, A.A., R.A. Morris, K. Gollinger, and F. Arnold, 1995: Ozone destruction by chlorine: The impracticality of mitigation through ion chemistry. *Science*, **267**, 82-84.

Waggoner, P.E., 1994: *How Much Land Can Ten Billion People Spare for Nature?* Task Force Report no. 121, Council for Agricultural Science and Technology, Ames, IA, February, 64 pp.

Walter, H., 1985: *Vegetation of the Earth and Ecological Systems of the Geo-Biosphere*. Springer-Verlag, Berlin, Germany, 318 pp.

White, F., 1983: *Vegetation Map of Africa*. UNESCO, Paris, France, 350 pp.

Whittaker, R.H., 1970: *Communities and Ecosystems*. MacMillian, New York, NY, 270 pp.

Whittaker, R.H. and G.E. Likens, 1975: The biosphere and man. In: *Primary Productivity of the Biosphere* [Lieth, H. and R.H. Whittaker (eds.)]. Springer-Verlag, Berlin, Germany, pp. 305-328.

Wilson, M.F. and A. Henderson-Sellers, 1985: A global archive of land cover and soils data for use in general circulation climate models. *J. Climatol.*, **5**, 119-143.

Woodward, F.I., 1987: *Climate and Plant Distribution*. Cambridge University Press, Cambridge, UK, 174 pp.

World Resources Institute, 1992: *World Resources 1992-93: A Guide to the Global Environment*. Oxford University Press, New York, NY, 385 pp.

Wyman, C.E., 1994: Alternative fuels from biomass and their impact on carbon dioxide accumulation. *Appl. Biochem. Biotechnol.*, **45-46**, 897-915.

Climate Change 1995

Part IV

Technical Appendices

Prepared by Working Group II

26

Technical Guidelines for Assessing Climate Change Impacts and Adaptations

TIMOTHY CARTER, FINLAND; MARTIN PARRY, UK;
SHUZO NISHIOKA, JAPAN; HIDEO HARASAWA, JAPAN

Contributing Authors:
R. Christ, UNEP; P. Epstein, USA; N.S. Jodha, Nepal; E. Stakhiv, USA;
J. Scheraga, USA

CONTENTS

EXECUTIVE SUMMARY

Working Group II of IPCC has prepared Guidelines to assess the impacts of potential climate change and to evaluate appropriate adaptations. They reflect current knowledge and will be updated as improved methodologies are developed. The Guidelines outline a study framework which will allow comparable assessments to be made of impacts and adaptations in different regions/geographical areas, economic sectors, and countries. The Guidelines are intended to help contracting parties meet, in part, their commitments under Article 4 of the UN Framework Convention on Climate Change.

Impact and adaptation assessments involve several steps:

- Definition of the problem
- Selection of the method
- Testing the method
- Selection of scenarios
- Assessment of biophysical and socioeconomic impacts
- Assessment of autonomous adjustments
- Evaluation of adaptation strategies.

Definition of the problem includes identifying the specific goals of the assessment; the ecosystem(s), economic sector(s), and geographical area(s) of interest; the time horizon(s) of the study; the data needs; and the wider context of the work.

The selection of analytical method(s) depends upon the availability of resources, models, and data. Impact assessment analyses can range from the qualitative and descriptive to the quantitative and prognostic.

Testing the method(s), including model validation and sensitivity studies, before undertaking the full assessment is necessary to ensure credibility.

Development of the scenarios requires, firstly, the projection of conditions expected to exist over the study period in the absence of climate change and, secondly, the projection of conditions associated with possible future changes in climate.

Assessment of potential impacts on the sector(s) or area(s) of interest involves estimating the differences in environmental and socioeconomic conditions projected to occur with and without climate change.

Assessment of autonomous adjustments implies the analysis of responses to climate change that generally occur in an automatic or unconscious manner.

Evaluation of adaptation strategies involves the analysis of different means of reducing damage costs. The methodologies outlined in the Guidelines for analyzing adaptation strategies are meant as a tool only to compare alternative adaptation strategies and thereby identify the most suitable strategies for minimizing the effects of climate change were they to occur.

26.1. Objectives

These Guidelines, which are a further development of those previously published (Carter *et al.*, 1992), provide a means for assessing the impacts of potential climate change and of evaluating appropriate adaptations. They reflect current knowledge and will be updated as improved methodologies are developed. They do not aim to prescribe a single preferred method, but provide an analytical outline composed of a number of steps. A range of methods is identified at each step. Where possible, the merits and drawbacks of different methods are briefly discussed, with some suggestions on their selection and use.

The ultimate purpose of the Guidelines is to enable estimations of impacts and adaptations, which will allow comparable assessments to be made for different regions/geographical areas, sectors, and countries. The Guidelines are intended to help contracting parties meet, in part, commitments under Article 4 of the UN Framework on Climate Change.

26.2. Approaches

A general framework for conducting a climate impacts and adaptations assessment contains seven steps:

- Definition of the problem
- Selection of the method
- Testing the method
- Selection of scenarios
- Assessment of biophysical and socioeconomic impacts
- Assessment of autonomous adjustments
- Evaluation of adaptation strategies.

At each step, a range of study methods is available. These are described and evaluated in the following sections. For reasons of brevity, however, only the essence of each method is introduced, along with references to sources of further information.

26.3. Step One—Definition of the Problem

This involves identifying the goals of the assessment, the exposure unit of interest, the spatial and temporal scope of the study, the data needs, and the wider context of the work.

26.3.1. Goals of the Assessment

It is important to be precise about the specific objectives of a study, as these will affect the conduct of the investigation. For example, an assessment of the hydrological impacts of future climatic change in a river catchment would have quite different requirements for data and expertise if the goal is to estimate the capacity for power generation than if it is to predict changes in agricultural income as a result of changes in the availability of water for irrigation.

26.3.2. Exposure Unit to be Studied

The exposure unit (i.e., the impacted object) to be assessed determines, to a large degree, the type of researchers who will conduct the assessment, the methods to be employed, and the data required. Studies can focus on a single sector or activity (e.g., agriculture, forestry, energy production, or water resources), several sectors in parallel but separately, or several sectors interactively.

26.3.3. Study Area

The selection of a study area is guided by the goals of the study and by the constraints on available data. Some options are reasonably well-defined, including governmental units, geographical units, ecological zones, and climatic zones. Other options requiring more subjective selection criteria include sensitive regions and representative units.

26.3.4. Time Frame

The selection of a time horizon for study is also influenced by the goals of the assessment. For example, in studies of industrial impacts the planning horizons may be 5–10 years, while investigations of tree growth may require a 100-year perspective. However, as the time horizon increases, the ability to accurately project future trends declines rapidly. Most climate projections and scenarios rely on general circulation models (GCMs), which are subject to uncertainties. Projections of socioeconomic factors such as population, economic development, and technological change need to be made for periods exceeding 15–20 years.

26.3.5. Data Needs

The availability of data is probably the major limitation in most impact and adaptation assessment studies. The collection of new data is an important element of some studies, particularly for monitoring purposes regarding expected climate changes, but most rely on existing sources. Thus, before embarking on a detailed assessment, it is important to identify the main features of the data requirements—namely, the variables for which data are needed, the time period, spatial coverage and resolution of the required data, the sources and format of the data and their quantity and quality, and the data availability, cost, and delivery time.

26.3.6. Wider Context of the Work

In order to assist policymakers in evaluating the wider significance of an assessment, it is important to place it in the context of similar studies and of the political, economic, and social system of the region.

26.4. Step Two—Selection of the Method

A variety of analytical methods can be adopted ranging from qualitative descriptive studies, through more diagnostic and semi-quantitative assessments, to quantitative and prognostic analyses. Any single impact assessment may contain elements of one or more of these types. Four general methods can be identified: experimentation, impact projections, empirical analog studies, and expert judgment.

26.4.1. Experimentation

In the physical sciences, a standard method of testing hypotheses or of evaluating processes of cause and effect is through direct experimentation. In the context of climate impact and adaptation assessment, however, experimentation has only a limited application. Clearly it is not possible physically to simulate large-scale systems such as the global climate. Only where the scale of impact is manageable, the exposure unit measurable, and the environment controllable can experiments be usefully conducted (e.g., gas enrichment experiments with plants).

26.4.2. Impact Projections

One of the major goals of climate impact assessment, especially concerning aspects of future climatic change, is the prediction of future impacts. A main focus of much recent work has been on impact projections, using an array of mathematical models to extrapolate into the future. First-order effects of climate are usually assessed using biophysical models, and second- and higher order effects using a range of biophysical, economic, and qualitative models. Finally, attempts also have been made at comprehensive assessments using integrated systems models.

26.4.2.1. Biophysical Models

Biophysical models may be used to evaluate the physical interactions between climate and an exposure unit. There are two main types: Empirical-statistical models and process-based models. Empirical-statistical models are based on the statistical relationships between climate and the exposure unit. Process-based models make use of established physical laws and theories to express the dynamics of the interactions between climate and an exposure unit.

26.4.2.2. Economic Models

Economic models of several types can be employed to evaluate the implications of first-order impacts for local and regional economies. The main types of models are firm-level (which depict a single firm or enterprise), sectoral (which simulate behavior within a specific economic sector), and macroeconomic (which simulate entire economies).

26.4.2.3. Integrated Systems Models

Integrated systems models represent an attempt to combine elements of the modeling approaches described above into a comprehensive model of a given regionally or sectorally bounded system. Two main approaches to integration can be identified: the aggregate cost-benefit approach, which is more economically oriented, and the regionalized process-based approach, which focuses more on biophysical effects.

26.4.3. Empirical Analog Studies

Observations of the interactions of climate and society in a region can be of value in anticipating future impacts. The most common method employed involves the transfer of information from a different time or place to an area of interest to serve as an analogy. Four types of analogy can be identified: Historical event analogies, historical trend analogies, regional analogies of present climate, and regional analogies of future climate.

26.4.4. Expert Judgment

A useful method of obtaining a rapid assessment of the state-of-knowledge concerning the effects of climate on given exposure units is to solicit the judgment and opinions of experts in the field. Literature is reviewed, comparable studies identified, and experience and judgment used in applying all available information to the current problem.

26.5. Step Three—Testing the Method

Following the selection of the assessment methods, it is important that these are tested in preparation for the main evaluation tasks. Three types of activity may be useful in evaluating the methods: feasibility studies, data acquisition and compilation, and model testing.

26.5.1. Feasibility Studies

These usually focus on a subset of the study region or sector to be assessed. Such case studies can provide information on the effectiveness of alternative approaches, of models, of data acquisition and monitoring, and of research collaboration.

26.5.2. Data Acquisition and Compilation

Data must be acquired both to describe the temporal and spatial patterns of climate change and their impacts and to develop, test, and calibrate predictive models. Data collection may rely on existing information obtained and compiled from different sources, or require the acquisition of primary data through survey methods, direct measurement, or monitoring.

26.5.3. Model Testing

The testing of predictive models is, arguably, the most critical stage of an impact assessment. Most studies rely almost exclusively on the use of models to estimate future impacts. Thus, it is crucial for the credibility of the research that model performance is tested rigorously. Standard procedures should be used to evaluate models, but these may need to be modified to accommodate climate change. Two main procedures are recommended: validation and sensitivity analysis. Validation involves the comparison of model predictions with real world observations to test model performance. Sensitivity analysis evaluates the effects on model performance of altering its structure or parameter values, or the values of its input variables.

26.6. Step Four—Selection of the Scenarios

Impacts are estimated as the differences between two states: environmental and socioeconomic conditions expected to exist over the period of analysis in the absence of climate change, and those expected to exist with climate change.

26.6.1. Establishing the Present Situation

In order to provide reference points with which to compare future projections, three types of "baseline" conditions need to be specified: the climatological, environmental, and socioeconomic baselines.

26.6.1.1. Climatological Baseline

The climatological baseline is usually selected according to the following criteria:

- Representativeness of the present-day or recent average climate in the study region
- Of sufficient duration to encompass a range of climatic variations
- Covering a period for which data on all climatological variables are abundant, adequately distributed, and readily available
- Including data of sufficient quality for use in evaluating impacts.

It is recommended that the current standard World Meteorological Organization (WMO) normal period (1961–90) be adopted in assessments where appropriate.

26.6.1.2. Environmental Baseline

The environmental baseline refers to the present state of other nonclimatic environmental factors that affect the exposure unit. Examples include groundwater levels, soil pH, extent of wetlands, and so on.

26.6.1.3. Socioeconomic Baseline

The socioeconomic baseline describes the present state of all the nonenvironmental factors that influence the exposure unit. The factors may be geographical (e.g., land use), technological (e.g., pollution control), managerial (e.g., forest rotation), legislative (e.g., air quality standards), economic (e.g., commodity prices), social (e.g., population), or political (e.g., land tenure). All of these are liable to change in the future, so it is important that baseline conditions of the most relevant factors are noted.

26.6.2. Time Frame of Projections

A critical consideration for conducting impact experiments is the time horizon over which estimates are to be made. Three elements influence the time horizon selected: the limits of predictability, the compatibility of projections, and whether the assessment is continuous or considers discrete points in time.

26.6.2.1. Limits of Predictability

The time horizon selected depends primarily on the goals of the assessment. However, there are obvious limits on the ability to project into the future. Climate projections, since they are a key element of climate impact studies, define one possible outer limit on impact projections. GCM estimates seldom extend beyond about 100 years, due to the uncertainties attached to such long-term projections and to constraints on computational resources. This fixes an outer horizon at about 2100. In many economic assessments, on the other hand, projections may not be reliable for more than a few years ahead.

26.6.2.2. Compatibility of Projections

It is important to ensure that future climate, environment, and socioeconomic projections are mutually consistent over space and time. It is important to be clear about (i) the relative timing of increases in greenhouse gas concentrations and climate change, and (ii) the relative timing of a 2 x CO_2 compared to a 2 x CO_2 "equivalent" atmosphere.[1] With regard to the former, there is a lag time of several decades in the response of the climate system to increases in greenhouse gas concentrations. With regard to the latter, a 2 x CO_2 "equivalent" atmosphere occurs earlier than a 2 x CO_2 atmosphere because gases such as CH_4, N_2O, and tropospheric O_3 also contribute to radiative forcing.

[1] A 2 x CO_2 "equivalent" atmosphere is one where the radiative forcing due to changes in all greenhouse gases (CO_2, CH_4, N_2O, O_3, halocarbons) is the same as that of an atmosphere where the concentration of CO_2 has doubled with the concentration of other greenhouse gases remaining unchanged.

26.6.2.3. Point in Time or Continuous Assessment

A distinction can be drawn between considering impacts at discrete points in time in the future and examining continuous or time-dependent impacts. The former are characteristic of many climate impact assessments based on doubled-CO_2 equivalent scenarios. In contrast, transient climatic scenarios allow time-dependent phenomena and dynamic feedback mechanisms to be examined and socioeconomic adjustments to be considered.

26.6.3. Projecting Environmental Trends in the Absence of Climate Change

The development of a baseline describing conditions without climate change is crucial, for it is this baseline against which all projected impacts are measured. It is highly probable that future changes in other environmental factors will occur even in the absence of climate change, which may be of importance for an exposure unit. Examples, as appropriate, include changes in land use, changes in groundwater level, and changes in air, water, and soil pollution. Most factors are related to, and projections should be consistent with, trends in socioeconomic factors. Greenhouse gas concentrations may also change, but these would usually be linked to climate (which is assumed unchanged here).

26.6.4. Projecting Socioeconomic Trends in the Absence of Climate Change

Global climate change is projected to occur over time periods that are relatively long in socioeconomic terms. Over that period it is certain that the economy and society will change, even in the absence of climate change. Official projections exist for some of these changes, as they are required for planning purposes. These vary in their time horizon from several years (e.g., economic growth, unemployment), through decades (e.g., urbanization, industrial development, agricultural production), to a century or longer (e.g., population).

26.6.5. Projecting Future Climate

In order to conduct experiments to assess the impacts of climate change, it is first necessary to obtain a quantitative representation of the changes in climate themselves. No method yet exists of providing confident predictions of future climate. Instead, it is customary to specify a number of plausible future climates. These are referred to as "climatic scenarios," and they are selected to provide climatic data that are spatially compatible, mutually consistent, freely available or easily derivable, and suitable as inputs to impact models.

There are three basic types of scenario of future climate: synthetic scenarios, analog scenarios, and scenarios from general circulation models.

26.6.5.1. Synthetic Scenarios

A simple method of specifying a future climate is to adjust the baseline climate in a systematic, though essentially arbitrary, manner. Adjustments might include, for example, changes in mean annual temperature of ±1, 2, 3°C, etc., or changes in annual precipitation of ±5, 10, 15%, etc., relative to the baseline climate. Adjustments can be made independently or in combination. In this way information can be obtained on the following:

- *Thresholds or discontinuities* of response that might occur under a given magnitude or rate of change. These may represent levels of change above which the nature of the response alters (e.g., warming may promote plant growth, but very high temperatures cause heat stress).
- *Tolerable climate change*, which refers to the magnitude or rate of climate change that a modeled system can tolerate without major disruptive effects (sometimes termed the "critical load"). This type of measure is potentially of value for policy, as it can assist in defining specific goals or targets for limiting future climate change.

One of the main drawbacks of the approach is that adjustments to combinations of variables may not to be physically plausible or internally consistent.

26.6.5.2. Analog Scenarios

Analog scenarios are constructed by identifying recorded climatic regimes that may serve as analogs for the future climate of a given region. These records can be obtained either from the past (temporal analogs) or from another region at the present (spatial analogs).

Temporal analogs are of two types: those based on past instrumental observations (usually within the last century) and those based on proxy data, using paleoclimatic indicators such as plant or animal remains and sedimentary deposits (from the more distant past geological records). The main problem with this technique concerns the physical mechanism and boundary conditions that would almost certainly be different between a warmer climate in the past and a future greenhouse gas-induced warming.

Spatial analogs require the identification of regions today having a climate analogous to the study region in the future. This approach is severely restricted, however, by frequent lack of correspondence between other nonclimatic features of two regions that may be important for a given impact sector (e.g., daylength, terrain, soils, or economic development).

26.6.5.3. Scenarios from General Circulation Models

Three-dimensional numerical models of the global climate system (including atmosphere, oceans, biosphere, and cryosphere) are the only credible tool currently available for simulating the physical processes that determine global climate. Although simpler models also have been used to simulate the radiative effects of increasing greenhouse gas concentrations, only GCMs, possibly in conjunction with nested regional models, offer the possibility to provide estimates of regional climate change, which are required in impact analysis.

GCMs produce estimates of climatic variables for a regular network of grid points across the globe. Results from about 20 GCMs have been reported to date (e.g., see IPCC, 1990, 1992). However, these estimates are highly uncertain because of some important weaknesses of GCMs. These include poor model representation of cloud processes; a coarse spatial resolution (at best employing grid cells of some 250-km horizontal dimension); generalized topography, disregarding some locally important features; and a simplified representation of land-atmosphere and ocean-atmosphere interactions. As a result, GCMs are currently unable to reproduce accurately even the seasonal pattern of present-day climate at a regional scale. Thus, GCM outputs represent, at best, broad-scale sets of possible future climatic conditions and should not be regarded as predictions.

GCMs have been used to conduct two types of experiments for estimating future climate: equilibrium-response and transient-forcing experiments. The majority of experiments have been conducted to evaluate the equilibrium response of the global climate to an abrupt increase (commonly, a doubling) of atmospheric concentrations of CO_2. A measure that is widely used in the intercomparison of various GCMs is the climate-sensitivity parameter. This is defined as the global mean equilibrium surface air temperature change that occurs in response to an increase in radiative forcing due to a doubling of atmospheric CO_2 concentration (or equivalent increases in other greenhouse gases). Values of the parameter obtained from climate model simulations generally fall in the range 1.5–4.5°C (IPCC, 1992). Knowledge of the climate sensitivity can be useful in constructing climate change scenarios from GCMs.

Recent work has focused on fashioning more realistic experiments with GCMs—specifically, simulations of the response of climate to a transient forcing. These simulations offer several advantages over equilibrium-response experiments. First, the specifications of the atmospheric perturbation are more realistic, involving a continuous (transient) change over time in greenhouse gas concentrations. Second, the representation of the oceans is more realistic, the most recent simulations coupling atmospheric models to dynamic ocean models. Finally, transient simulations provide information on the rate as well as the magnitude of climate change, which is of considerable value for impact studies.

The following types of information are currently available from GCMs for constructing scenarios:

- Outputs from a "control" simulation, which assumes fixed greenhouse gas concentrations, and an "experiment," which assumes future concentrations. In the case of equilibrium-response experiments, these are values from multiple-year model simulations for the control and 2 x CO_2 (or equivalent increases in other greenhouse gases) equilibrium conditions. Transient-response experiments provide values for the control equilibrium conditions and for each year of the transient model run (e.g., 1990 to 2100).
- Values of surface or near-surface climatic variables for model grid boxes characteristically spaced at intervals of several 100 km around the globe.
- Values of air temperature, precipitation (mean daily rate), and cloud cover, which are commonly supplied for use in impact studies. Data on radiation, windspeed, vapor pressure, and other variables are also available from some models.
- Data averaged over a monthly time period. However, daily or hourly values of certain climatic variables, from which the monthly statistics are derived, may also be stored for a number of years within the full simulation periods.

26.6.6. Projecting Environmental Trends with Climate Change

Changes in environmental conditions not due to climatic factors already should have been incorporated in the development of the environmental trends in the absence of climate changes; the only changes in these trends to be incorporated here are those due solely to climate change. The two factors most commonly required in assessments are greenhouse gas concentrations and sea-level rise. Future changes in these are still under discussion, but the estimates reported by the IPCC may serve as a useful basis for constructing scenarios (IPCC, 1990). Other factors that are directly affected by climate (such as river flows, runoff, erosion) would probably require full impact assessments of their own, although some might be incorporated as "automatic adjustments" in projections.

26.6.7. Projecting Socioeconomic Trends with Climate Change

The changes in environmental conditions that are attributable solely to climate change serve as inputs to economic models that project the changes in socioeconomic conditions due to climate change both within the study area and, where relevant and appropriate, outside it over the study period. All other changes in socioeconomic conditions over the period of analysis are attributable to nonclimatic factors and should have been included in the estimation of socioeconomic changes in the absence of climate change.

26.7. Step Five—Assessment of Impacts

Impacts are estimated as the differences over the study period between the environmental and socioeconomic conditions projected to exist without climate change and those that are projected with climate change. Assessments may include the elements described in the following subsections.

26.7.1. Qualitative Description

The success of this method rests on the experience and interpretive skills of the analyst, especially the analyst's ability to consider all factors of importance and their interrelationships. Formal methods of organizing qualitative information also exist (e.g., cross-impact analysis).

26.7.2. Indicators of Change

These are particular regions, activities, or organisms that are intrinsically sensitive to climate, and that can provide an early or accurate indication of effects due to climate change.

26.7.3. Compliance to Standards

This may provide a reference or an objective against which to measure the impacts of climate change. For example, the effect on water quality could be gauged by reference to current water quality standards.

26.7.4. Costs and Benefits

These should be estimated quantitatively to the extent possible and expressed in economic terms. This approach makes explicit the expectation that a change in resources and resource allocation due to climate change is likely to yield benefits as well as costs. It can also examine the costs or benefits of doing nothing to mitigate potential climate change.

26.7.5. Geographical Analysis

Impacts vary over space, and this pattern of variation is of concern to policymakers operating at regional, national, or international scales because these spatial differences may have consequent policy and planning implications. The geographical depiction of the effects of climate change using geographical information systems is one method of describing impacts.

26.7.6. Dealing with Uncertainty

Uncertainties pervade all levels of a climate impact assessment, including the projection of future greenhouse gas emissions, atmospheric greenhouse gas concentrations, changes in climate, their potential impacts, and the evaluation of adjustments. There are two methods that attempt to account for these uncertainties: uncertainty analysis and risk analysis.

26.7.6.1. Uncertainty Analysis

Uncertainty analysis comprises a set of techniques for anticipating and preparing for the impacts of uncertain future events. It is used here to describe an analysis of the range of uncertainties encountered in an assessment study.

26.7.6.2. Risk Analysis

Risk analysis deals with uncertainty in terms of the risk of impact. Risk is defined as the product of the probability of an event and its effect on an exposure unit. Since extreme events produce the most significant impacts, there is value in focusing on the changing probability of climatic extremes and of their impacts. Another form of risk analysis—decision analysis—is used to evaluate response strategies to climate change. It can be used to assign likelihoods to different climatic scenarios, identifying those response strategies that would provide the flexibility, at least cost (minimizing expected annual damages), that best ameliorates the anticipated range of impact.

26.8. Steps Six and Seven—Assessment of Autonomous Adjustments and Evaluation of Adaptation Strategies

Impact experiments are usually conducted to evaluate the effects of climate change on an exposure unit in the absence of any responses that might modify these effects and that are not already automatic or built in to future projections. Two broad types of response can be identified: mitigation and adaptation.

26.8.1. Mitigation and Adaptation

Mitigation or "limitation" attempts to deal with the causes of climate change. It achieves this through actions that prevent or retard the increase of atmospheric greenhouse gas concentrations by limiting current and future emission from sources of greenhouse gases and enhancing potential sinks. The evaluation of mitigation policies is outside the scope of these Guidelines.

Adaptation is concerned with responses to both the adverse and positive effects of climate change. It refers to any adjustment—whether passive, reactive, or anticipatory—that can respond to anticipated or actual consequences associated with climate change. It thus implicitly recognizes that future climate changes will occur and must be accommodated in policy.

26.8.2. Steps in Evaluation of an Adaptation Strategy

A broad framework for the evaluation of adaptation strategies to cope with climate change can be identified:

1) Define the objectives
2) Specify the climate impacts of importance
3) Identify the adaptation options
4) Examine the constraints
5) Quantify measures and formulate alternative strategies
6) Weigh objectives and evaluate tradeoffs
7) Recommend adaptation measures.

26.8.2.1. *Defining the Objectives*

Any analysis of adaptation must be guided by some agreed overall goals and evaluation principles. Two examples of general *goals* commonly propounded are (i) the promotion of sustainable development and (ii) the reduction of vulnerability. These are open to various interpretations however, so specific objectives need to be defined that complement the goals. *Objectives* are usually derived either from public involvement, from stated public preferences, by legislation, through an interpretation of goals such as those stated above, or any combination of these.

26.8.2.2. *Specifying the Climatic Impacts of Importance*

This step involves an assessment, following the methods outlined elsewhere above, of the possible impacts of climate variability or change on the exposure unit. Where climatic events are expected that will cause damage, these need to be specified in detail so that the most appropriate adaptation options can be identified.

26.8.2.3. *Identifying the Adaptation Options*

The main task of assessment involves the compilation of a detailed list of possible adaptive responses that might be employed to cope with the effects of climate. The list can be compiled by field survey and interviews with relevant experts, and should consider all practices currently or previously used, as well as possible alternative strategies that have not been used, and newly created or invented strategies.

Six types of strategy for adapting to the effects of climate have been identified:

- *Prevention of loss*, involving anticipatory actions to reduce the susceptibility of an exposure unit to the impacts of climate
- *Tolerating loss*, where adverse impacts are accepted in the short term because they can be absorbed by the exposure unit without long-term damage
- *Spreading or sharing loss*, where actions distribute the burden of impact over a larger region or population beyond those directly affected by the climatic event
- *Changing use or activity*, involving a switch of activity or resource use to adjust to the adverse as well as positive consequences of climate change
- *Changing location*, where preservation of an activity is considered more important than its location, and migration occurs to areas that are more suitable under the changed climate
- *Restoration*, which aims to restore a system to its original condition following damage or modification due to climate.

Numerous options exist for classifying adaptive measures; but generally, regardless of the resources of interest (e.g., forestry, wetlands, agriculture, water), the prospective list may include among other management measures the following:

- Legal
- Financial
- Economic
- Technological
- Public education
- Research and training.

26.8.2.4. *Examining the Constraints*

Many of the adaptation options identified in the previous step are likely to be subject to legislation or to be influenced by prevailing social norms, which may encourage, restrict, or totally prohibit their use. Thus, it is important to examine closely, possibly in a separate study, what these constraints are and how they might affect the range of feasible choices available.

26.8.2.5. *Quantifying the Measures and Formulating Alternative Strategies*

The next step is to assess the performance of each adaptation measure with respect to the stated objectives. It may be possible, if appropriate data and analytical tools exist, to use simulation models to test the effectiveness of different measures under different climatic scenarios. Historical and documentary evidence, survey material, or expert judgment are some other alternative sources of this information. Uncertainty analysis and risk assessment are also considered at this stage. This step is a prelude to developing strategies that maximize the level of achievement of some objectives while maintaining baseline levels of progress toward the remaining objectives.

26.8.2.6. *Weighing Objectives and Evaluating Tradeoffs*

This is the key evaluation step, where objectives must be weighted according to assigned preferences, then comparisons made between the effectiveness of different strategies in meeting these objectives. Standard impact accounting systems can be used in the evaluation. For example, a four-category system might consider (i) national economic development, (ii) environmental quality, (iii) regional economic development, and (iv) other social effects. Selection of preferred strategies then requires the determination of tradeoffs among the categories.

26.8.2.7. *Recommending Adaptation Measures*

The results of the evaluation process should be compiled in a form that provides policy advisors and decisionmakers with information on the best available adaptation strategies. This should include some indication of the assumptions and uncertainties involved in the evaluation procedure, and the rationale used (e.g., decision rules, key evaluation principles, national and international support, institutional feasibility, technical feasibility) to narrow the choices.

26.9. Obtaining a Copy of the Guidelines

Contact information to obtain the complete *IPCC Technical Guidelines for Assessing Climate Change Impacts and Adaptations* follows:

National Institute for Environmental Studies
Center for Global Environmental Research
16-2 Onogawa, Tsukuba
Ibaraki 305
Japan
+81.298.58.2645 (fax)
snishiok@nies.go.jp (e-mail)

University College London
Department of Geography
26 Bedford Way
London WC1H OAP
UK
+44.171.916.0379 (fax)

References

Carter, T.R., M.L. Parry, S. Nishioka, and H. Harawasa, 1992: *Preliminary Guidelines for Assessing Impacts of Climate Change.* Environmental Change Unit, Oxford, UK, and Centre for Global Environmental Research, Tsukuba, Japan, 27 pp.

IPCC, 1990: *Climate Change: The IPCC Scientific Assessment.* Prepared by IPCC Working Group I [Houghton, J.T., G.J. Jenkins, and J.J. Ephraums (eds.)] and WMO/UNEP. Cambridge University Press, Cambridge, UK, 365 pp.

IPCC, 1992: *Climate Change 1992: The Supplementary Report to The IPCC Scientific Assessment.* Prepared by IPCC Working Group I [Houghton, J.T., B.A. Callander, and S.K. Varney (eds.)] and WMO/UNEP. Cambridge University Press, Cambridge, UK, 200 pp.

IPCC, 1994: *IPCC Technical Guidelines for Assessing Climate Change Impacts and Adaptations.* Prepared by IPCC Working Group II [Carter, T.R., M.L. Parry, H. Harasawa, and S. Nishioka (eds.)] and WMO/UNEP. CGER-IO15-'94. University College–London, UK, and Center for Global Environmental Research, Tsukuba, Japan, 59 pp.

27

Methods for Assessment of Mitigation Options

DENNIS A. TIRPAK, USA

Lead Authors:
M. Adler, USA; D. Bleviss, USA; J. Christensen, Denmark; O. Davidson, Sierra Leone; D. Phantumvanit, Thailand; J. Rabinovitch, Argentina; J. Sathaye, USA; C. Smyser, USA

CONTENTS

27.1. Introduction

This chapter is substantially different in focus from most of the preceding chapters on greenhouse gas (GHG) mitigation options. The preceding chapters focus on characterizing, from a global perspective, the full array of available mitigation options. This chapter instead addresses the analytical methods and processes for selecting and analyzing those mitigation options that best suit the specific needs, conditions, and national goals of individual countries. Its purpose is to help policy analysts and decisionmakers, especially in developing countries and countries with economies in transition, to obtain the objective information they need on mitigation options and to assist them in developing coherent national plans and strategies.[1]

This chapter is a summary of a broader set of mitigation guidelines entitled *Methods for Assessment of Mitigation Options*, which is being published as a separate appendix to this report. The chapter summarizes several key points from these guidelines:

- The broad challenges facing decisionmakers and analysts in conducting an effective mitigation options assessment
- The mitigation assessment process, including organizational issues and analytical steps
- Some of the key cross-cutting issues involved in the mitigation assessment process
- The range of analytical methods available to meet most countries' needs and capabilities.

The full guidelines in *Methods for Assessment of Mitigation Options* contain the following documents:

- **Technical Report**—A detailed examination of the methods and issues involved in an assessment of mitigation options and the development of national mitigation plans and strategies
- **Appendix I: Technical Methods**—A catalog of analytical methods, describing in detail their purpose, appropriate applications, potential drawbacks, and references for further information
- **Appendix II: Resources Guide—**A detailed reference guide to other climate-related studies and programs, including a guide to databases and analytical models commonly used for mitigation options assessments
- **Appendix III: Case Studies**—A set of case studies illustrating mitigation assessment processes and key analytical methods and approaches employed by different developing and transition countries
- **Appendix IV: Mitigation Assessment Handbook**—Detailed descriptions of a limited set of basic models that most countries could use in assessing their mitigation options.

As mentioned, the chapter focuses on the analytical needs of developing countries and countries with economies in transition. Because these countries have many other pressing national issues that take precedence over global-environment concerns, they often lack basic information and analytical capabilities for assessing mitigation options or organizing them into coherent national plans and strategies. For that reason, this chapter and its accompanying guidelines emphasize simple, readily available analytical methods and procedures (while not ignoring more sophisticated methods).

It should be pointed out that this chapter and the accompanying guidelines do not prescribe particular analytical methods or approaches to assessing mitigation options. Rather, the material presents a range of methods and approaches from which countries can select to meet their own needs and conditions. Developing countries and countries with economies in transition that need more practical assistance and support selecting and using these methods may be able to obtain such support through one of several multilateral and bilateral climate country-study programs being conducted.[2]

Before proceeding, it is worth defining a few key terms as they are used in this chapter:

- **Mitigation option**: A technology, practice, or policy that reduces or limits emissions of GHGs or increases their sequestration (This chapter does not consider measures to adapt to climate change.)
- **Mitigation options assessment**: The analytical process of identifying, selecting, and organizing mitigation options into a coherent national plan
- **Mitigation methods**: Analytical tools used to assess the impact and performance, costs and benefits, and social/political/institutional desirability of a mitigation option.

27.2. Challenges in a Mitigation Options Assessment

Before discussing specific methods or analytical steps, some of the broad challenges that analysts and policymakers face in organizing an assessment of mitigation options are worth noting. We have organized these into three broad categories:

- **Strategic challenges**: Strategic challenges affect the overall objectives of a mitigation options assessment and the process of selecting and implementing preferred options. Four strategic challenges stand out:
  - Integrating climate-change mitigation with other key national objectives which can require a clear set of national priorities, along with an analytical

[1] More specific information on methods for analyzing options that involve land-use changes—especially in the forestry and agriculture sectors, including biomass energy—is provided in Chapter 25.

[2] Climate country-study programs support developing countries and countries with economies in transition in developing national climate responses. A few examples include the UNEP/RISØ Climate Country Studies effort, the Asian Development Bank ALGAS project, and bilateral activities including those of the United States, Germany, Japan, and the Nordic countries.

process that consistently assesses options in light of these priorities
- Recognizing institutional constraints and deciding whether to fit a mitigation assessment process within these constraints or to seek additional institutional resources
- Relying on regional cooperation to address transboundary issues and to pool resources to lead to more effective assessments.
- Planning for future financing of mitigation options, particularly for developing and transition countries, and the role that the Global Environment Facility and others will play in providing these resources.

- **Analytical challenges**: Analytical challenges are those directly related to assessing mitigation options. Much of this chapter is about specific analytical issues, but analysts may face at least three broad challenges:
  - Employing the appropriate analytical methods effectively to address the specific needs, conditions, and capabilities of a country
  - Accommodating particularly dynamic economies and economies in transition in order to apply specific methods, especially forecasting, effectively
  - Accounting for ancillary costs and benefits and efficiency and equity issues to enhance the acceptability of options.
- **Informational challenges**: Several informational challenges often must be addressed in conducting comprehensive mitigation options assessments:
  - Gaining access to sources of information about technologies, costs, and country specific performance factors
  - Extracting useful data or information from the sources that are available
  - Accurately converting "imported" data or other information to properly reflect the operating conditions of the country in which the information is used.

27.3. Analytical Framework and Levels of Decisionmaking

We have chosen to organize the discussion of methods by national, cross-sectoral, sectoral, program, project, and technology decisionmaking levels, as illustrated in Figure 27-1. This framework is intended to reflect the links between different levels of an economy, the types of decisions made at those different levels, the information required to make those decisions, and, finally, the methods needed to obtain that information. Different organizational approaches are possible, but this framework highlights the importance of understanding the types of decisions to be made when selecting methods.

The four shaded levels in Figure 27-1 (i.e., the cross-sectoral, sectoral, program, and project levels) are the primary focus of this chapter. The national goals level—which addresses a country's national priorities and policies—is outside the purview of this report, although clear priorities and policies contribute to the effectiveness of a mitigation options assessment. Additional information on methods and issues associated with setting national priorities and policies can be found in the IPCC Working Group III volume of the Second Assessment Report. The technology assessment level is covered by the other mitigation chapters in this report and the Technology Characterization Inventory appendix. The decisionmaking levels of direct concern are described below

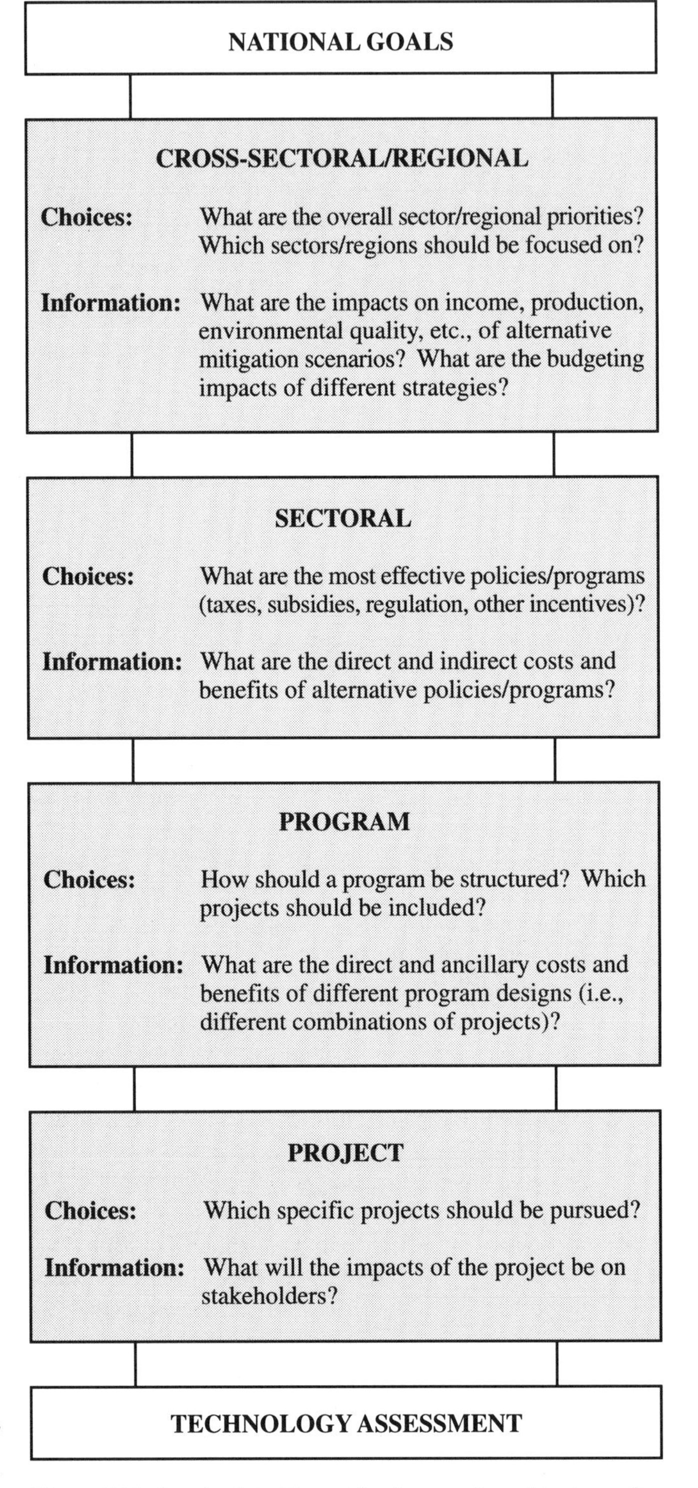

Figure 27-1: Levels of decisionmaking for assessing mitigation options.

(more specific discussion of the methods themselves is contained in Section 27.6):

- **Cross-sectoral level**: At the cross-sectoral level, alternative sectoral and cross-sectoral (e.g., recycling) mitigation options are compared to determine their impact on an economy and other national interests or goals. Typically this involves compiling information on sectoral activities that produce GHGs, such as energy production, or sequester those emissions, such as forestry activities. This information is integrated with national goals and priorities, and resources and responsibilities are allocated to produce a comprehensive, workable plan. The types of methods used to assess and compare options at this level include macroeconomic models, decision-analysis tools, costing methods, and forecasting tools.
- **Sectoral level**: The sectoral level involves analysis to compare the impacts of individual programs or large-scale projects within a specific sector principally for the purpose of prioritizing options and determining sectoral resource requirements. For example, the emission reduction potential and cost per ton of carbon dioxide (CO_2) reduced might be compared for options such as energy-efficiency standards, building codes, and rebates on high-efficiency appliances. As in the cross-sectoral level, costing and forecasting tools are important, although they focus on more detailed program and project information. In addition, integrated analytical tools are becoming available for certain sectors such as energy and forestry that can compare relatively disparate options on a more common basis.
- **Program level**: Program-level analysis compares the costs and impacts of individual projects or bundles of projects for the purpose of developing mitigation programs. A program might consist of a series of projects to introduce improved forestry-management practices into a region. As in cross-sectoral and sectoral analysis, project costing and forecasting tools are important. Technology costing and performance, market research, and monitoring and evaluation tools also are important.
- **Project level**: At the project level, the purpose of analysis is typically to identify and characterize specific project options in terms of their financial costs, technical performance, environmental characteristics, and so forth. Identifying options may involve literature reviews, case studies, and expert judgment. Characterizing options often relies extensively on engineering estimates, performance assessments, and project evaluations, for example.

27.4. Organizing a Mitigation Options Assessment

27.4.1. Organizational Steps

A systematic approach to organizing an assessment process can contribute significantly to the quality and effective communication of results and effective implementation of recommended options. The approach can be developed by addressing four key questions:

- **How should assessment objectives be prioritized and who should be involved?** Mitigation options should be measured not only on the basis of their effectiveness in reducing GHGs but also on the basis of their contribution to other national objectives, such as sustainable development, rural development, or local environmental protection. Multiple stakeholders will be affected by these choices, so involving them in the process can be important.
- **What are the questions being answered by the analysis?** Effective decisionmaking requires the proper information; thus, mitigation analysis must focus on the proper questions. These questions will be defined broadly by national objectives but will vary depending on the decisionmaking level.
- **Which organizations will have institutional responsibility for the analysis?** Institutional capabilities are needed to gather data, select options to analyze, apply the necessary analytical methods, and communicate results. Key tasks involved in assigning the appropriate institutions to the assessment include identifying the available institutional resources, matching institutions to the research questions to be asked, coordinating among institutions, linking the analysis to the decisionmaking process, and designing feedback on the process to possible stakeholders.
- **Which organizations will have responsibility for implementing results?** Although the focus of this chapter is on methods for analysis of mitigation options, implementation issues cannot be completely separated in the discussion. Four factors to consider in the selection of options include the adequacy of implementation capabilities, institutional support for implementation, coordination of implementation, and evaluation and accountability.

27.4.2. Analytical Steps

Typically, a mitigation assessment process will follow a series of steps, each of which produces information for decisionmakers at different levels. The manner in which these steps are performed will reflect each country's resources, objectives, and decisionmaking process. Generally, these steps can be characterized as follows:

- **Baseline development**: Baselines, or "business-as-usual" scenarios, describe the growth in emissions expected as a result of economic growth, population growth, and other factors, assuming that no explicit mitigation policies are adopted.
- **Identification and characterization of alternative technologies and policies**: This may include ranking options with respect to costs, energy consumption, carbon sequestered, and ancillary impacts, among others.

- **Development of alternative scenarios**: Alternative scenarios include future projections of emissions [usually based on the global warming potential (GWP) of all GHGs] and costs, assuming that mitigation actions are taken.
- **Estimation of incremental costs and benefits of options**: Incremental costs and benefits are the difference between the baseline and alternative scenarios.
- **Assessment of the impacts of alternative GHG mitigation scenarios**: Impacts to consider could include macroeconomic, social, and environmental costs and benefits, and equity.
- **Sensitivity analysis**: The sensitivity of results to uncertainties in data or alternative assumptions is often assessed to ensure that the results are reliable and credible.

27.5. Key Methodological Issues

The choice and use of most analytical methods involve a number of methodological issues that decisionmakers and analysts should be aware of when considering a mitigation options assessment. Some of these issues are unique to certain methods, while others cut across many methods. This section outlines nine key methodological issues that span a number of methods:

1) **Top-down versus bottom-up analysis**: Significant controversy has existed between top-down and bottom-up approaches to estimating the costs of mitigation options. Each approach estimates mitigation costs differently, often with significantly different results. The top-down approach generally looks at macroeconomic impacts of alternative mitigation scenarios on income, consumption, or gross domestic product (GDP). A key assumption in the top-down approach is that economies are in equilibrium and that mitigating GHGs, therefore, generally entails some net cost to an economy. This assumption is often questioned by bottom-up analysts. Top-down models also have had problems accounting for different rates of technological change. The bottom-up approach looks at the microeconomic costs of and changes in demand due to individual mitigation options. It generally assumes that there are inefficiencies in the marketplace that allow for cost-effective mitigation options. Bottom-up models have had problems accounting for consumer behavior and administrative costs. In general, the top-down approach yields higher cost estimates for GHG mitigation than the bottom-up approach. Both approaches can serve important purposes, however, and improvements in both approaches are leading to more balanced results (Borero *et al.*, 1991; Krause *et al.*, 1993; UNEP, 1992). See IPCC Working Group III volume, Chapter 8, *Estimating Costs of Mitigating Greenhouse Gases,* Section 8.3.3., for a more thorough description of this issue.
2) **Matching method to objective**: Choosing the appropriate method for the analytical task depends on matching methods to objectives. For example, one objective for an expanding economy may be minimizing the impacts of GHG mitigation on the industrial sector. Sectoral impacts may be more appropriately measured using a macroeconomic analysis of options rather than an engineering cost approach. Alternatively, if an objective is to identify all technologies or policies that are cost-effective, a combination of engineering and cost-effectiveness methods may provide the best approach.
3) **Specifying the baseline and alternative scenarios**: Scenarios are used to portray future GHG emissions and economic activity without mitigation options in place (baseline or business-as-usual scenarios) or with policies in place (mitigation scenarios). Several issues in developing scenarios include the degree to which all sources and sinks are covered; the reasonableness of assumptions on economic growth, technology change, and market imperfections; the consistency of assumptions across analytical steps; and the capacity to incorporate feedbacks. Clearly defined scenarios are important for providing decisionmakers with useful guidance on the allocation of national resources. Scenarios also can play a significant role in the allocation of international funding for climate mitigation projects—for example, through the Global Environment Facility and through joint implementation projects.
4) **Accounting for uncertainty**: Most analyses are affected by uncertainty due to poor data quality, dynamic economies, economies with little historical market data, and so forth. Taking this uncertainty into account in estimating costs and impacts may be more practical than trying to achieve complete accuracy. This can be done by using "range" rather than "point" estimates, using expert judgment where good quantitative data are lacking, use of sensitivity analyses, and so on. Using these approaches will not eliminate uncertainty but will allow decisionmakers to attach greater confidence to those options and policies that yield favorable results (Stokey and Zeckhauser, 1978).
5) **Defining a common measure for comparing options**: A comprehensive and internally consistent assessment requires a common measure or "metric" for option comparison where different types of costs are involved or relevant impacts extend beyond the direct GHG reductions.
6) **Accounting for time in the analysis of costs and benefits**: The flows of costs and benefits from different mitigation options over time are often discounted to their present values so that they can be compared in consistent, present-value terms. The choice of discount rates has important implications but can be difficult to make in practice. Selection of a discount rate may depend on how displaced private uses are estimated, how imperfections in capital markets are

accounted for, and, perhaps most importantly, to what extent a country is willing to forgo current consumption and investment in favor of future environmental protection. Higher discount rates generally favor current consumption, while lower discount rates generally favor future environmental protection. Sensitivity analyses with different discount rates often are employed to estimate their different impacts (Stokey and Zeckhauser, 1978).

7) **Accounting for externalities and ancillary impacts**: Virtually every mitigation option will produce some positive or negative externality and/or ancillary impact. Whether quantifiable or not, these impacts represent real costs or benefits and should be factored into the assessment process.
8) **Data requirements**: A comprehensive mitigation assessment can require detailed information on literally hundreds of options. Few industrialized countries have access to all the data that would be required, but data availability is an even more serious constraint on developing countries. The cost of data collection is an important factor to weigh in developing a mitigation assessment.
9) **Understanding the limits of models**: Quantitative models, from simple spreadsheet costing models to sophisticated macroeconomic models, can be very useful for managing complex analytical tasks. However, the accuracy of the results from these models depends on the quality of data, the choice of assumptions, the appropriateness of the model to the analytical task, and so forth. All of these require good judgment and skill on the part of the analyst. The simple conclusion is that models are not a substitute for good analysis.

27.6. Analytical Methods

Table 27–1 illustrates the relationship between analytical methods and the different decisionmaking levels at which they are used. As shown, the same types of methods often are used at various decisionmaking levels (although the way methods are applied may be different at different levels). This section provides an overview of key analytical methods, including a brief description of the types of information the methods provide:

- **Macroeconomic analysis**: Macroeconomic analysis often is used to describe the current structure of an economy, to predict future economy-wide conditions and their impact on GHG emissions, and to analyze changes in these conditions that could arise from taking actions to mitigate GHG emissions (Borero *et al.*, 1991; Cline, 1992).
- **Decision analysis**: Decision-analysis methods provide a structure for integrating sectoral and cross-sectoral GHG mitigation objectives with other national priorities. Basic decision analysis involves identifying potential options, measuring the potential impacts of those options (i.e., valuing the outcomes of various policy scenarios), and, based on this information, selecting the best options.
- **Costing analysis**: Mitigation costing methods estimate and compare costs and benefits of mitigation options to identify the set of actions that maximizes economic, social, and environmental benefits or minimizes the analogous costs of reducing GHGs. Four basic costing methods for ranking mitigation options are cost-benefit, cost-effectiveness, financial, and cost or supply curve analysis. More advanced methods include simulation and optimization models (Beaver and Huntington, 1991; Borero *et al.*, 1991; Cline, 1992; NAS, 1991; Stokey and Zeckhauser, 1978; UNEP, 1992, 1993). See the IPCC Working Group III volume for a detailed discussion of costing issues.
- **Forecasting**: The acceptability of a mitigation strategy, program, or project is determined by calculating the difference between "what would happen anyway" and "what would happen if mitigation measures were undertaken." The definition and development of these scenarios play a significant role in mitigation assessment. Forecasting is used here as a general term applied to methods used in developing time-dependent scenarios, such as econometric, time-series analysis, and end-use forecasting models.
- **Integrated planning**: Integrated planning approaches provide a structure for complex regional or sectoral assessments of multiple mitigation options. These methods typically are process or decision analytic-based and

Table 27-1: *Matching methods to decisionmaking levels.*

Method/Level	Cross-Sectoral/ Regional	Sectoral Level	Program Level	Project Level
Macroeconomic	✖			
Decision Analysis	✖	✖	✖	✖
Forecasting	✖	✖	✖	
Costing Analysis	✖	✖	✖	✖
Integrated Planning		✖		
Market Research			✖	
Monitoring and Evaluation			✖	✖
Options Identification				✖
Options Characterization				✖

can be designed to produce both quantitative and qualitative results. A variety of quantitative methods can be used to support integrated planning. Integrated planning methods include integrated resource planning (IRP), used increasingly in the power sector; integrated transportation planning (ITP); and integrated forestry and agriculture planning (IIEC, 1994; Sathaye *et al.*, 1994). (For more information on integrated planning in forestry and agriculture, see Chapter 25.)

- **Market research**: Market research is an important analytical method in program design. Market research gathers information from prospective consumers of a particular product or service through focus groups, survey techniques, product testing, and so forth.
- **Monitoring and evaluation**: Mitigation programs and projects should be monitored and evaluated for their actual costs and impacts. The methods used often are the same used to assess a program or project in the first place, such as cost-effectiveness analysis and surveys. The key difference is the need to establish a monitoring and evaluation system during program or project design. This information can help determine whether initial objectives or targets are being achieved, how the program or project can be redesigned to improve results, and how future projects should be designed.
- **Options identification**: Many different mitigation options are available, as this report demonstrates. The analyst needs to identify those options that are most likely to suit the needs and conditions of his or her country. Typically, options are screened against a series of criteria such as technological maturity, commercial availability, and technical performance. Information on potentially suitable options can come from case studies, literature reviews, general opportunity studies, and the judgment of specialized experts.
- **Options characterization**: Once data have been obtained on specific mitigation options, a system is needed for characterizing options to ensure that the data obtained can provide usable information. A variety of methods are available to analysts, including cost curve analysis, estimates of tons of GHG avoided, and engineering assessments.

27.7. Conclusions

The information in this chapter and the accompanying technical documents provide a cross-section of the large body of knowledge and experience available to analyze mitigation options. The information demonstrates that:

- Methods are available to analysts and decisionmakers at different levels of government to assess technology and policy options to mitigate GHGs from all significant sources, as well as their sequestration by sinks. These methods address the energy, industry, transportation, agriculture, and forestry sectors.
- Analytical, technical, and institutional resources are available from many multilateral and bilateral organizations to assist developing countries and countries with economies in transition in assessing mitigation options and strategies. Other economic development assistance programs that support investment decisions in energy production and end-use efficiency, forestry management, transportation, and agriculture also may demonstrate ancillary GHG emission mitigation benefits.
- Mitigation analyses have been and are being applied in developed, developing, and transitional-economy countries. These experiences demonstrate that appropriate mitigation assessment methods can be applied to meet the individual needs and conditions of all countries.

As a result of the available information and the many international and national efforts to assess projects, programs, and national mitigation strategies—as described in more detail below—it is possible to conclude that:

- The availability of mitigation methods and assistance enables all countries to develop strategies and evaluate programs and projects that support national economic, social, and institutional development goals and can slow the rate of growth in GHG emissions. The process of conducting a mitigation options assessment presents challenges to all countries. Projects identified in assessments may be difficult to implement because of a variety of economic, social, and institutional barriers in all countries.
- Development and application of mitigation assessment methods in all sectors and at all levels is an ongoing effort that will result in further improvements in the tools and the capabilities of countries to assess mitigation options. The IPCC, in coordination with other multilateral institutions, could accelerate the dissemination of selected information on assessment methods through seminars, workshops, and educational materials.

Methods are available for assessing mitigation options: The total body of information and other resources available internationally on analytical methods is indeed extensive:

- The technical report accompanying this chapter identifies more than 40 often-used methods for analyzing mitigation options. Many more exist that can address specific issues or adapt to different levels of analytical capability. The technical report also provides countries with approaches for organizing a mitigation options assessment, from identifying national goals to estimating the costs of specific technology options. As with specific analytical methods, assessment processes can take many different forms, depending on the country's needs and conditions.

- From the accompanying appendices, countries also can obtain:
  - More detailed information on the actual application of many of the key methods, including information on appropriate uses for the different methods, potential pitfalls, and where to obtain further assistance
  - See how mitigation options analyses have been conducted in four case studies, ranging from actual experiences with mitigation costing techniques to the mitigation planning process of an individual developing country
  - Obtain step-by-step instructions in developing a mitigation plan using a well-defined set of analytical tools and planning approaches
  - Learn about the many mitigation assessment activities being conducted by other countries and international institutions.

Analytical and technical resources are available to support mitigation analysis. Experience demonstrates that methods are available and appropriate: The fact that mitigation analyses are being conducted in many types of countries and that resources are available to assist countries with their needs is demonstrated by the following:

- The United Nations Environment Programme (UNEP) and the United Nations Development Programme (UNDP) provide support to about twenty-nine developing countries to coordinate climate country studies that include mitigation options assessments. Additional countries are expected to be added. The assessments developed through these studies will provide models of different mitigation strategies, including least-cost mitigation strategies, that other countries will be able to follow.
- The UK, Germany, the Netherlands, and the United States together will have supported more than seventy-five climate country studies in developing countries and countries with economies in transition. (The total number of countries involved is somewhat fewer due to overlap.) Many of these studies are developing mitigation strategies as one component. It is also noteworthy that a number of the studies place significant emphasis on developing local institutional and analytical capabilities to assess and implement mitigation options.
- The pilot phase of the Global Environment Facility allocated about $250 million for more than thirty global-warming mitigation projects. These funds were leveraged with additional World Bank and bilateral assistance funds. These projects will soon begin to yield a wealth of information and practical experience with different approaches to mitigating GHGs.

27.8. Obtaining a Copy of the Guidelines

Means to obtain the full guidelines appendix follow:

U.S. Environmental Protection Agency
Climate Policy and Programs Division
401 M Street, NW
Suite 2122
Washington, DC 20460
Attention: Michael Adler
202.260.9013 (voice) • 202.260.6405 (fax)
adler.michael@epamail.epa.gov (e-mail)

References

Beaver, R. and H. Huntington, 1991: A comparison of aggregate energy demand models being used for global warming policy analysis. Paper presented to the American Statistical Association, Energy Modeling Forum, Stanford University, Stanford, CA.

Borero, G., R. Clarke, and L.A. Winters, 1991: *The Macroeconomic Consequences of Controlling Greenhouse Gases: A Survey—Summary*. Department of Economy, University of Birmingham, Birmingham, UK.

Cline, W.R., 1992: *The Economics of Global Warming*. Institute for International Economics, Washington, DC.

IIEC, 1994: *A Methodology for Greenhouse Gas Mitigation in the Transport Sector*. IIEC, Washington, DC.

Krause, F., E. Haites, R. Howarth, J. Koomey, 1993: Cutting carbon emissions: burden or benefit? The economics of energy-tax and non-price policies. In: *Energy Policy in the Greenhouse*, vol. II, part 1. International Project for Sustainable Energy Paths, El Cerito, CA.

NAS, 1991: *Policy Implications of Greenhouse Warming: Report of the Mitigation Panel*. Panel on Policy Implications of Greenhouse Gas Warming, National Academy Press, Washington, DC.

Sathaye, J., W. Makundi, and K. Andrasko, 1994: An approach to the evaluation of the costs and benefits of forestry mitigation options. *Biomass and Bioenergy*.

Stokey, E. and R. Zeckhauser, 1978: *A Primer for Policy Analysis*. W.W. Norton & Company, Inc., New York, NY.

UNEP, 1992: *UNEP Greenhouse Gas Abatement Costing Studies, Analysis of Abatement Costing Issues and Preparation of a Methodology to Undertake National Greenhouse Gas Abatement Costing Studies, Phase One Report*. UNEP Collaborating Centre on Energy and Environment, RISO National Laboratory, Denmark.

UNEP, 1993: *UNEP Greenhouse Gas Abatement Costing Studies, Analysis of Abatement Costing Issues and Preparation of a Methodology to Undertake National Greenhouse Gas Abatement Costing Studies, Phase Two Report*. UNEP Collaborating Centre on Energy and Environment, RISO National Laboratory, Denmark.

28

Inventory of Technologies, Methods, and Practices

DAVID G. STREETS, USA

Principal Lead Authors:
W.B. Ashton, USA; K. Hogan, USA; P. Wibulswas, Thailand; T. Williams, USA

CONTENTS

28.1. Introduction

The inventory is a database containing information on technologies, methods, and practices that can help to limit emissions of greenhouse gases. The inventory contains information on performance characteristics and applications, capital and operating costs, environmental characteristics (including emissions of greenhouse gases), and infrastructure requirements. The information is compiled in a two-page, standardized format for each technology. The inventory contains information on 105 technologies, including energy supply technologies (44), energy end-use technologies (47), agricultural and forestry practices (12), and other techniques (2). Table 28-1 lists all the technologies in the inventory. The inventory is published as an appendix volume to the Second Assessment Report.

The rationale for the inventory is straightforward. As the global climate change debate shifts toward studying the need for action, it is increasingly important for energy and environmental planners to have access to technical and economic information on the options that are available to reduce emissions and mitigate climate change. This information should be current, consistent, reliable, and widely available. In some countries, very good technology data and sophisticated analytical tools already exist. Often, however, information available from national or regional studies cannot be directly compared or adapted to new conditions because it has been collected in different formats or is based on different assumptions. Some countries, particularly developing and newly industrializing countries, lack sufficient resources to explore the benefits of these technologies on their own. In these cases, planners may rely on data of uncertain provenance, which can lead to unbalanced assessments of technology options, questionable data quality, and, ultimately, poorly informed investments. The need for wide access to a standardized data set drove the assembly of this inventory.

The inventory is designed for use by planners and analysts needing to evaluate options for reducing greenhouse gas emissions that are potentially attractive for near- and medium-term applications (say, the next 5–15 years). Some applications for which the inventory would be well suited are:

- Selection of technological options for use in different parts of the world
- National or regional economic development planning
- National or regional studies of greenhouse-gas emission reduction policies
- Macroeconomic analyses of large-scale climate change mitigation policies.

Although the inventory can be of value in prescreening technological options for specific applications, it is *not* intended to support the following types of studies:

- Engineering design or installation studies
- Site-specific applications
- Detailed technology transfer evaluations.

The inventory does not make any recommendations concerning specific types of technological systems or vendors.

Information on experience with advanced technologies in industrialized countries has been supplemented with information on applications of these technologies in developing countries. As the inventory develops, the aim is to increasingly make it a global resource and make its ownership world-wide. In the interim, users may replace inappropriate data elements with country-specific information that meets their local needs. In the longer term, the inventory will be made available to users in a computerized version. The software will provide guidance and data to permit conversion of data elements, especially costs, to local conditions. The inventory will be improved and extended in the future as a result of applications studies, such as the U.S. Country Studies Program, and government reviews of the material presently included. These additional items will include greater regional discrimination of data ranges, efficiencies achieved in recent practice, and comparative indications of development status.

The following section includes three examples illustrating the type of information contained in the inventory. The format of information for each technology in the inventory follows the outline provided in Table 28-2.

28.2. Emission Reduction Technologies

Carbon dioxide is emitted in large quantities by human activities, mostly as a result of combustion of fossil fuels and vegetation. Many techniques have been demonstrated to limit emissions of carbon dioxide from fuel-burning activities by improving the efficiency with which stored energy is converted to useful energy. The inventory characterizes many such approaches that have higher energy supply efficiencies than conventional systems. These approaches cover direct energy production in the industrial sector, as well as electricity generation, transmission, and distribution. Some of the technologies described are particularly suitable for the kinds of fuel resources, energy demands, and infrastructure found in developing countries. Appropriate technologies for developing countries, such as efficient cook stoves, are included in the inventory. Additionally, energy supply technologies that can replace fuel-burning technologies and essentially eliminate the direct production of carbon dioxide are described. Emphasis is placed on renewable energy supplies (biomass, solar, hydro, and wind energy). However, because some industrializing countries are considering nuclear energy as an option to meet electricity demand without adding to carbon dioxide emissions, several advanced nuclear energy options are included.

A range of technologies can improve the efficiency with which energy is consumed in various applications. Energy end-use technologies for buildings, transportation, and industry are included. Many of these technologies can provide substantial improvements in energy efficiency over conventional approaches, and the capital investments often can be recovered

Table 28-1: *List of technologies contained in the inventory.*

A. Energy Supply—Fossil Fuel Technologies
A.1 Atmospheric Fluidized-Bed Combustion (AFBC)
A.2 Coal Beneficiation
A.3 Coal-Water Mixtures
A.4 Combined Cycle
A.5 Combustion Turbines
A.6 Diesel Cogeneration
A.7 Fuel Cells
A.8 Industrial Cogeneration
A.9 Gas/Oil-Fired Steam Units
A.10 Integrated Gasification Combined Cycle (IGCC)
A.11 Inter-cooled Steam Injected Gas Turbine (ISTIG/STIG)
A.12 Pressurized Fluidized-Bed Combustion (PFBC)
A.13 Pulverized Coal-Fired Power Plant
A.14 Slagging Combustors
A.15 Oil and Natural Gas: Reduced Venting and Flaring of Gas during Production
A.16 Oil and Natural Gas: Improved Compressor Operations
A.17 Oil and Natural Gas: Improved Leak Detection and Pipeline Repair
A.18 Oil and Natural Gas: Low-Emission Technologies and Practices
A.19 Coal Mining: Enhanced Gob Well Recovery
A.20 Coal Mining: Pre-Mining Degasification
A.21 Coal Mining: Integrated Recovery

B. Energy Supply—Renewable Energy Technologies
B.1 Biomass-Fired Power Generation
B.2 Flat-Plate Photovoltaics
B.3 Geothermal Electric
B.4 Municipal Solid Waste (MSW) Mass Burn
B.5 Pelletized Biomass Combustion
B.6 Biogas by Anaerobic Digestion
B.7 Small-Scale Hydro
B.8 Solar Ponds
B.9 Solar Thermal Electric-Parabolic Trough
B.10 Solar Thermal Electric-Central Receiver
B.11 Solar Thermal Electric-Parabolic Dish/Stirling Engine
B.12 Wind Energy Conversion Systems
B.13 Tidal Energy
B.14 Ocean Wave Energy
B.15 Landfills: Gas Recovery and Utilization Techniques

C. Energy Supply—Nuclear Technologies
C.1 Light Water Reactors
C.2 Heavy Water Reactors
C.3 Liquid Metal Fast Reactors
C.4 Gas-Cooled Reactors

D. Energy Supply—Energy Transfer Technologies
D.1 Efficient Electrical Transformers
D.2 Electric Transmission and Distribution Systems
D.3 Thermal Energy Storage Systems
D.4 High-Voltage Direct-Current Transmission

E. Energy End-Use—Transportation Technologies
E.1 Advanced Signalization
E.2 Battery Electric Vehicles (EVs)
E.3 Compressed Natural Gas (CNG) Vehicles
E.4 Continuously Variable Transmission (CVT)
E.5 Direct-Injection Diesel Engines
E.6 Efficient Jet Aircraft
E.7 Efficient Tires
E.8 Ethanol Vehicles
E.9 Fuel-Cell Electric Vehicles (FCVs)
E.10 Two-Stroke Spark-Ignition Engines
E.11 Urban Transit Systems

F. Energy End-Use—Buildings Technologies
F.1 Adjustable Speed Electric Motor Drives
F.2 Advanced Insulation: Gas Filled Panels
F.3 Compact Fluorescent Lights
F.4 Electronic Ballasts
F.5 Efficient Electric Motors
F.6 Efficient Refrigerators
F.7 Energy Management Systems (EMS)
F.8 Glazing: Daylight Control
F.9 Glazing: Insulating
F.10 Glazing: Solar Control
F.11 Glazing: Switchable
F.12 Heat Pump Water Heaters
F.13 High-Albedo Materials
F.14 Landscaping
F.15 Lighting Controls
F.16 Room Air Conditioners (Window-Type)
F.17 Solar Domestic Water Heaters
F.18 Efficient Cooking Stoves

G. Energy End-Use—Industrial Technologies
G.1 Anaerobic Biological Treatment of Waste
G.2 Biofiltration of Gases
G.3 Cement Particle High-Efficiency Air Classifiers
G.4 Ceramic Recuperators
G.5 Continuous Pulp Digesters
G.6 Continuous Steel Casters
G.7 Distillation Control Systems
G.8 Electric Motors Variable Speed Drives—Industrial
G.9 Gas Membrane Separators
G.10 Heat Exchanger Enhancement Techniques
G.11 High-Efficiency Welding Power Supply
G.12 Mechanical Dewatering
G.13 Metal Parts Cleaning
G.14 Pinch Technology
G.15 Pipe Cross Reactors
G.16 Plating Waste Concentrators
G.17 Pulse Combustion Boilers
G.18 Textile Dyeing Vacuum System

H. Agricultural and Forestry Practices
H.1 Reduction in Use of Nitrogen Fertilizer and Animal Manure
H.2 Reduction in Tillage of Agricultural Soils
H.3 Reforestation/Afforestation and Prevention of Deforestation
H.4 Increasing Efficiency/Intensity of Forest Management
H.5 Substitution of Fossil Fuels with Sustainably-Grown Fuelwood
H.6 Increasing Agroforestry Endeavors
H.7 Livestock: Improved Nutrition/Mechanical and Chemical Feed Processing
H.8 Livestock: Improved Nutrition/Strategic Supplementation
H.9 Livestock: Production-Enhancing Agents
H.10 Livestock Manure: Covered Lagoons
H.11 Livestock Manure: Small-Scale Digesters
H.12 Livestock Manure: Large-Scale Digesters

I. Other Techniques
I.1 Landfills: Reducing Landfilling of Waste
I.2 Wastewater Treatment

Table 28-2: *Data elements contained in the inventory.*

General Characteristics
Sector
Applications
Typical Size
Design Fuels
Performance Measure
Design Lifetime
Construction Time
Development Status
Cost Information
Source and Year
Capital and Installation Cost
Nonfuel Operating Cost
Fuel Cost
Environmental Characteristics
Waste Streams
Air Emissions
Greenhouse Gas Emissions
Site-Specific Issues
Retrofit Potential
Infrastructure Requirements
Operating Personnel
Maintenance Personnel
Infrastructure Needs
References

rapidly through energy cost savings and lower operating and maintenance costs. Improved products or better energy services often are ancillary benefits. In many cases, the technologies are easily transferred to developing countries and simple to install and use. The reduction in greenhouse gas emissions that can be gained through the use of higher efficiency end-use technologies depends on the type of fuel used to produce the displaced energy and the efficiency of the primary energy conversion process.

Although research is underway on a variety of techniques for capturing carbon dioxide once it has been produced and then utilizing it or disposing of it, such systems are not included in this inventory because they are not likely to be of significant commercial potential in the near- to medium-term time frame of this inventory.

Methane is emitted from a diverse set of human-related activities that currently represent about 70% of global emissions. Importantly, methane emissions from these activities represent the waste of a valuable fuel; the methane often can be recovered under conditions where the saved fuel justifies the investment. Sixteen currently available technologies and practices are identified and characterized in the inventory. These technologies and practices have been demonstrated to use otherwise wasted methane profitably, have been implemented to some extent already, and could be implemented on a much more widespread basis. They are economically viable under a range of conditions, represent different levels of technical complexity and capital needs, and should be adaptable to a wide variety of country conditions. Furthermore, these technologies and practices are generally attractive options because of the many other benefits that they provide—benefits that are consistent with the development goals of many countries.

The inventory contains characterizations for a range of major methane sources that spans energy and agricultural activities covering about 70% of anthropogenic emissions: natural gas systems, coal mining, waste disposal, wastewater treatment, and domesticated livestock. Efforts still are required to develop and demonstrate options for reducing methane emissions from rice cultivation and biomass burning, and therefore these methane sources are not reflected in the inventory at present. The latest research efforts in these areas have been documented by the U.S./Japan Working Group on Methane (1992).

28.2.1. *Example 1: Energy Supply—Fossil Fuel Technologies*

Pressurized Fluidized-Bed Combustion

Pressurized fluidized-bed combustion (PFBC) is an advanced technology for burning coal that offers high thermal efficiency and low environmental emissions. The system operates at pressures of 6–16 atmospheres and typically utilizes smaller combustion chambers than a conventional furnace because of its more efficient steam production. Due to its high thermal efficiency, PFBC can reduce the quantity of carbon dioxide emitted per unit of energy produced. Therefore, it is a potentially useful technology for mitigating emissions generated in the production of electricity. The inventory identifies the anticipated size of commercial PFBC units (200–400 MW_e), their application (electric power generation), and the fuels that can be used. Performance measures of the technology are provided in terms of the operating thermal efficiency of current demonstration plants (33–42%) and of projected second-generation plants (45–50%).

The capital cost of current demonstration plants ($1900–$3200/kW) is compared with the expected costs of a mature, commercial unit (about $1500/kW) and a second-generation plant (about $1000/kW). Nonfuel operating and maintenance costs range between 2.0 and 2.5¢/kWh. Fuel costs can range anywhere between 2.2 and 34¢/kWh, depending on the type of fuel chosen. The environmental performance of PFBC is very good; it generates relatively small quantities of solid waste, which can be safely landfilled or sold as a byproduct, and it produces low emissions of sulfur dioxide and nitrogen oxides. Most importantly for global climate change, emissions of carbon dioxide are estimated to be in the range of 930–1060 g CO_2/kWh, lower than conventional pulverized coal plants. Labor requirements to operate a PFBC plant (25–89, depending on size) and to maintain it (40–60) are presented, together with relevant information on infrastructure requirements, such as the need for connection to an electricity transmission grid.

28.2.2. Example 2: Energy End-Use Technologies

Compact Fluorescent Lights

Compact fluorescent lights are three to four times more energy efficient than incandescent bulbs and last ten times as long. They can therefore reduce the amount of electricity required to deliver a given amount of lighting. The reduction in carbon dioxide emissions will depend on the fuel used to generate the electricity. Compact fluorescent units are available with screw-in adaptors for use in incandescent lamp fixtures or as dedicated fixtures. Compact fluorescents also are available as integrated lamp/ballast/adapter units; however, since the ballasts typically outlive the lamps, integration of ballast and lamp results in the ballast being discarded before the end of its useful life. Lower wattage twin tubes cost about $5 per lamp. The more luminous quad tubes cost about $9 per lamp. Integrated units cost about $15–25 per lamp. Utility rebates in the range of $4–15 per lamp typically are available in the U.S.

Compact fluorescent lights are slightly bigger and heavier than incandescent bulbs. Thus, they may not be a suitable replacement in every application. Performance suffers in tight fixtures, and under-lit spaces may result if thermal effects are not considered in system planning. Preheat lamps have a 1- to 2-second delay before lighting, although rapid-start systems are becoming available.

28.2.3. Example 3: Energy Supply—Renewable Energy Technologies

Landfills: Gas Recovery and Utilization Techniques

Recovering gas from landfills reduces methane emissions and allows the energy value of the recovered methane to be used profitably. The recovery technologies are relatively straightforward: Gas wells are drilled into an area of disposed waste, and collection systems are installed. These systems typically recover 50–85% of the gas generated. The recovered gas is about 50% methane, a medium-Btu fuel. It typically requires some processing to remove particulate matter, water, and corrosive compounds. The costs of the technology depend on the costs for recovery systems, labor, and utilization systems. Landfill gas recovery provides additional benefits in the form of reduced explosion hazard and protection of local air and water quality. The best opportunities are for large waste-disposal sites that have a nearby demand for medium-quality fuel.

28.3. Other Technology Inventories

Several other initiatives are under way aimed at gathering and disseminating information on technologies for limiting greenhouse gas emissions or increasing greenhouse gas uptake. Efforts will be made to coordinate this inventory with these other systems.

The Environmentally Compatible Energy Strategies (ECS) Project at the International Institute for Applied Systems Analysis in Laxenburg, Austria, has developed a comprehensive inventory of technological options for mitigating and adapting to possible global warming (Schafer *et al.*, 1992). The supporting data base, CO2DB, contains detailed descriptions of the technical, economic, and environmental performance of approximately four hundred technologies. Efficiency improvements, conservation, enhanced use of low-carbon fuels, carbon-free sources of energy, afforestation, and enhancement of carbon sinks are included.

At the Forschungszentrum Julich GmbH in Germany, the IKARUS Project has been involved in the development of computer models and a database for comparing strategies for reducing the energy-related emissions of greenhouse gases, particularly carbon dioxide (Katscher, 1993). The technology database in the IKARUS system contains basic technical, economic, and environmental data on individual technologies that can be used to construct technology chains and feed to sector-specific simulation models. The goal is to be able to assess mitigation possibilities for the German energy system in 1989, 2005, and 2020.

The International Energy Agency (IEA) has two related activities. The IEA established the Centre for the Analysis and Dissemination of Demonstrated Energy Technologies (CADDET), an Implementing Agreement that offers an international network to exchange information on demonstrations of energy-saving end-use technologies for all energy consumers (IEA, 1991). CADDET maintains *The CADDET Register,* a computerized database providing information on international energy demonstration projects. Also, the IEA has implemented the Energy Technology Systems Analysis Programme (ETSAP), an Implementing Agreement that also distributes technology information to IEA member countries (IEA, 1992).

An interagency joint project on databases and methodologies for comparative assessment of different energy sources for electricity generation (DECADES) has been established by the International Atomic Energy Agency (IAEA), in collaboration with several other international organizations, with the objective of enhancing capabilities for incorporating health and environmental issues in the comparative assessment of different electricity generation chains and strategies (IAEA, 1995). The DECADES project includes a technology inventory that characterizes energy chains for electricity generation from fuel extraction to waste management, as well as an information system that provides user-oriented access to electronic databases.

28.4. Obtaining a Copy of the Inventory

Copies of the inventory may be obtained by contacting the Convening Lead Author or through the U.S. Country Studies

Program. The Convening Lead Author may be contacted at the following address:

Argonne National Laboratory
Decision and Information Sciences Division, DIS/900
9700 South Cass Avenue
Argonne, Illinois 60439-4832
USA
708.252.3448 (voice) • 708.252.3206 (fax)
streetsd@smtplink.dis.anl.gov (e-mail).

References

International Atomic Energy Agency, 1995: *Computer Tools for Comparative Assessment of Electricity Generation Options and Strategies.* International Atomic Energy Agency, Vienna, Austria, 80 pp.

International Energy Agency, 1991: *CADDET Brochure.* Centre for the Analysis and Dissemination of Demonstrated Energy Technologies, Organization for Economic Cooperation and Development, Paris, France.

International Energy Agency, 1992: *IEA/ETSAP News.* Energy Technology Systems Analysis Programme, Annex IV: Greenhouse Gases and National Energy Options, The Netherlands Energy Research Foundation, Petten, The Netherlands.

Katscher, W., 1993: *IKARUS: Instruments for Greenhouse Gas Reduction Strategies.* Interim Summary Report for Project Phase 3, Forschungszentrum Julich GmbH, Germany, 76 pp.

Schafer, A., L. Schrattenholzer, and S. Messner, 1992: *Inventory of Greenhouse-Gas Mitigation Measures: Examples from the IIASA Technology Data Bank.* Working Paper WP-92-85, International Institute for Applied Systems Analysis, Laxenburg, Austria, 66 pp.

U.S./Japan Working Group on Methane, 1992: *Technological Options for Reducing Methane Emissions.* Background document for the Response Strategies Working Group, Intergovernmental Panel on Climate Change, 198 pp.

Appendix A

EXPERT REVIEWERS OF THE IPCC WORKING GROUP II VOLUME

Argentina

Osvaldo F. Canziani	IPCC WG2 Subgroup C Co-Chair
Rodolfo Carcavallo	Centro de Investigaciones Biometeorologicas
Susan Curto de Casas	Centro de Investigaciones Biometeorologicas
Sandra Myrna Diaz	Universidad Nacional de Córdoba
Nestor Maceira	INTA
Rafael Oscar Rodriguez	INTA
César Mario Rostagno	CONICET/CENPAT
Carlos Scoppa	Centro de Investigacion de Recursos Naturales Instituto
Pedro Skvarca	Instituto Antàrctico Argentino

Australia

M.P. Austin	CSIRO
John A. Church	CSIRO
Peter Curson	Macquarie University
Derek Eamus	Northern Territory University
Roger Farrow	CSIRO
Margaret Friedel	CSIRO
Habiba Gitay	Australian National University–Canberra
Ann Henderson-Sellers	Climate Impacts Center
David Hopley	James Cook University of North Queensland
Mark Howden	Department of Primary Industries and Energy
Robert Kay	Department of Planning and Urban Development
Paul Kesby	Department of the Environment, Sport, and Territories
G.A. Kile	CSIRO
Miko Kirschbaum	CSIRO
Paul E. Kriedemann	CSIRO
J. Landsberg	CSIRO
Gregory M. McKeon	Queensland Department of Primary Industries
Roger McLean	Australian Defence Force Academy
Tom McMahon	University of Melbourne
R.E. McMurtrie	University of New South Wales
Stephen Nicol	Australian Antarctic Division
Ian Noble	Australian National University–Canberra
J.M. Oades	University of Adelaide
A. Barrie Pittock	CSIRO
Ian Poiner	CSIRO
R.J. Raison	CSIRO
Alison Saunders	Australian National University
Marjorie Sullivan	Huonbrook Environment and Heritage
Hugh Taylor	University of Melbourne
Ron Thresher	CSIRO
George R. Walker	Alexander Howden Reinsurance Brokers, Ltd
Richard Williams	CSIRO

Austria

Bo R. Döös	Global Environmental Management
Arnulf Grübler	IIASA
Stefan Hausberger	University of Technology–Graz
Nebojsa Nakicenovic	IIASA
Yuri Sinyak	IIASA
Johan F. van de Vate	International Atomic Energy Agency
Jelle G. van Minnen	IIASA
Wei Zhihong	IIASA

Bahamas

D. Klean	Bahamas General Insurance

Barbados

Leonard Nurse	Coastal Conservation Unit

Belgium

Dirk Boeye	University of Antwerp
Franz Bultot	RMIB
Gaston R. Demaree	RMIB
Daniel Gellens	RMIB
Yilma Seleshi	RMIB
Oscar L.J. Vanderborght	IGBP

Botswana

Peter P. Zhou	EECG Consultants

Brazil

Antonio R. Magalhaes	Esquel Brazil Foundation
Roberto Moreira	Biomass User Network

Canada

Heather Auld	Environment Canada
Richard J. Beamish	Pacific Biological Station
M. Brklacich	Carleton University
James P. Bruce	IPCC WG3 Co-Chair
C.A. Campbell	Agriculture Canada
Stewart J. Cohen	AES

R.L. Desjardin	Agriculture Canada
David Etkin	AES
Gregory M. Flato	University of Victoria
Edward G. Gregorich	Agriculture Canada
R.E. Hecky	Freshwater Institute
Deborah Herbert	Environment Canada
R. César Izaurralde	University of Alberta
S.N. Kulshreshtha	University of Saskatchewan
John M. Last	University of Ottowa
Beth Lavender	Environmental Adaptation Research Group
Antoni G. Lewkowicz	University of Ottawa
K.H. Mann	Bedford Institute of Oceanography
C. Monreal	Agriculture Canada
Robert D. Moore	Simon Fraser University
R.E. (Ted) Munn	University of Toronto
Keith Puckett	Environment Canada
H-H. Rogner	University of Victoria
Nigel Roulet	McGill University
Wayne Rouse	McMaster University
David W. Schindler	University of Alberta
Brian J. Shuter	Ontario Ministry of Natural Resources
Michael Sinclair	Bedford Institute of Oceanography
Robert B. Stewart	Canadian Forestry Service
C. Tarnocai	Agriculture Canada
Alan E. Taylor	Geological Survey of Canada
Michael T. Ter-Mikaelian	Ontario Ministry of Natural Resources
George Walker	Sterling Offices
David Waltner-Toews	University of Guelph
David Welch	Pacific Biological Station
E. Wheaton	Saskatchewan Research Council
Rodney R. White	University of Toronto
Gordon J. Young	Wilfrid Laurier University

China

Cheng Goudong	Chinese Academy of Sciences
Harry Dimitriou	The University of Hong Kong
Du Bilan	China Institute for Marine Development Strategy
Guanri Tan	Zhongshan University
Li Peiji	Chinese Academy of Glaciology and Geocryology
Lin Erda	Agrometeorology Institute
Liu Chunzhen	Hydrological Forecasting and Water Control Center
Shi Yafeng	Chinese Academy of Sciences
Su Jilan	Second Institute of Oceanography
Wang Lixian	Forestry University of Beijing
Xu Bamin	China Automotive Technology and Research Center
Xu Deying	Chinese Academy of Forestry
Xuemei Shao	Chinese Academy of Science
Yang Huating	China Institute for Marine Development Strategy
Yi Zhang	Chinese Academy of Sciences
Youyu Xie	Chinese Academy of Sciences
Zhou Fengqi	Energy Research Institute

Colombia

Myles J. Fisher	Centro International de Agriculture Tropical

Cuba

Roberto Acosta Moreno	COMARNA
A. Alvarez	Forestry Research Institute
Avelino G. Suarez	Institute of Ecology and Systematics

Denmark

Sven Jonasson	University of Copenhagen
Charles Nielson	ELSAM
John R. Porter	The Royal Veterinary and Agricultural University
Bent Sorensen	Roskilde University
Joel N. Swisher	UNEP
Anker Weidick	Geological Survey of Greenland

Finland

Heikki Henttonen	Finnish Forest Research Institute
Antero Jarvinen	University of Finland
Seppo Kärkkäinen	Technical Research Center of Finland
J. Laine	University of Helsinki
Risto Lemmela	National Board of Waters and the Environment
P.J. Martikainen	National Public Health Institute
Ilkka Savolainen	Technical Research Center of Finland
J. Silvola	University of Joensuu

France

Dominique Bachelet	Center d'Etude Spatiale du Rayonnement
C. Barthod	Forest Health Department
Denis Binet	IFREMER
Fernand David	St. Jérôme Institut Méditerranéen
G. Dedieu	LERTS
Bernard Devin	IPSN/CEA
Christian Feller	ORSTOM
Alberte Fischer	LERTS
Michel Fournier	University of Perpignan
Patrice Garin	CIRAD-SAR
Joel Guiot	Institut Mediterraneen d'Ecologie et de Paleoecologie
G. Landmann	Forest Health Department
E. Le Floc'h	CEFE
Henry Le Houerou	CEFE
Jean Maley	Université des Sciences et Techniques du Languedoc
Michel Petit	IPCC WG2 Subgroup D Co-Chair
Arthur Riedacker	INRA
J. Sircoulon	ORSTOM
Michel Thinon	St. Jérôme Institut Méditerranéen

Germany

Alfred Becker	Potsdam Institute of Climate Impact Research
Joachim Benz	University of Kassel
Gerhard Berz	Munich Reinsurance Company
L.N. Braun	Bavarian Academy of Sciences
Harald Bugmann	Potsdam Institute of Climate Impact Research
Wolfgang P. Cramer	Potsdam Institute for Climate Impact Research
Ulrich Dämmgen	Agricultural Institute for Climate Research
J.G. Goldammer	University of Freiburg
Hartmut Grassl	WMO
Karl Hofius	Federal Institute of Hydrology
Friedrich-Karl Holtmeier	University of Muenster
Venugopalan Ittekkot	University of Hamburg
Hans-Jürgen Jäger	Justus-Liebig Universität Giessen
Eberhard Jochem	Fraunhofer-Institute of Systems and Innovation Research
Michael Knorrenschild	Centre for Environment and Health
Klaus Peter Koltermann	Federal Maritime and Hydrographic Agency
Hans-Jurgen Liebscher	Federal Institute of Hydrology
Dieter Overdieck	Institut für Ökologie der Holzpflanzen
Rudolf Petersen	Wuppertal Institut für Klima
Jutta Rogasik	Agricultural Institute for Climate Research
Dieter R. Sauerbeck	Federal Ministry of Environment
Petra Schaefer	University of Hamburg
H.-W. Scharpenseel	University of Hamburg
Florian Scholz	Holzwirtschaft Institut für Forstgenetik
G. Spatz	University of Kassel
Horst Sterr	University of Oldenburg
Hans von Storch	Max-Planck Institut für Meteorologie
Michael Weber	Ludwig-Maximilians Universität München
Gerd-Rainer Weber	Gesamtverband des Deutschen Steinkohlenberghaus
H. Erich Wichmann	Institüt für Epidemiologie
Rainer Zahn	GEOMAR

Ghana

Nii Boi Ayibotele	Water Resources Research Institute

Hong Kong

Wyss W.-S. Yim	The University of Hong Kong

Hungary

Odön Starosolszky	Water Resource Research Center
Karoly Szesztay	Consultant

Iceland

Ingvar Friedleifsson	United Nations University

India

I.P. Abrol	Indian Council of Agriculture Research
Shardul Agrawala	IPCC WG2 TSU
Elgar Desa	National Institute of Oceanography
L.N. Harsh	Central Arid Zone Research Institute
Shashi M. Kulshrestha	Independent
Dileep Kumar	National Institute of Oceanography
P.S. Ramakrishnan	School of Environmental Sciences
K.S. Ramasastri	National Institute of Hydrology
N.H. Ravindranath	Indian Institute of Sciences
S. M. Seth	National Institute of Hydrology
Subodh K. Sharma	IPCC WG2 Subgroup A Co-Chair
S.K. Sinha	Ministry of Environment and Forests

Indonesia

Sri Soewasti Soesanto	Ministry of Health

Israel

Uri Shamir	Israel Institute of Technology

Italy

Tim Aldington	FAO
Andrew Bakun	FAO
William M. Ciesla	FAO
Mario Contaldi	Agency for New Technology, Energy, and the Environment
Francesco Dramis	Dipartimento di Scienze della Terra
Ugo Farinelli	Agency for New Technology, Energy, and the Environment
Serge Garcia	FAO
Francesco Miglietta	IATA-CNR
Freddy Nachtergaele	FAO
Sergio Scalcino	ENEL Spa
W. G. Sombroek	FAO
GianCarlo Tosato	Agency for New Technology, Energy, and the Environment
Robin Welcomme	FAO

Japan

Elisa Boelman	Tokyo University of Agriculture and Technology
Keisuke Hanaki	University of Tokyo
Keiji Higuchi	Nagoya University
Yasushi Hosokawa	Ministry of Transport
Atsushi Inaba	National Institute for Resources and Environment
Hisashi Ishitani	University of Tokyo
Takao Kashiwagi	Tokyo University of Agriculture and Technology
Hideki Kawai	Federation of Electric Power Companies

Toshi Kitazawa	Tokyo Marine and Fire Insurance Co., Ltd.
Ryuji Mathuhashi	University of Tokyo
Nobuo Mimura	Ibaraki University
Hisayoshi Morisugi	Gifu University
Shuzo Nishioka	Environment Agency
Shunji Ohta	Waseda University
Takehisa Oikawa	University of Tsukuba
Toshiichi Okita	Obirin University
Naoto Sagawa	IEEJ
Hidekazu Takakura	Ministry of International Trade and Industry
J. Takemura	Energy Conservation Center
Hiroshi Tsukamoto	IPCC WG2 Subgroup A Co-Chair
Itsuya Tsuzaka	Global Industrial and Social Progress Research Institute
Mitsutsune Yamaguchi	Tokyo Marine and Fire Insurance Co., Ltd.
Masatoshi Yoshino	Aichi University
Uchijima Zembei	Ochanomizu University

Kenya

Michael Baumer	UNEP
W. Franklin G. Cardy	UNEP
Fred C. Cheghe	Ministry of Research, Technical Training, and Technology
Alexander Ranja Gacuhi	Ministry of Research, Technical Training, and Technology
J.K. Kinyamario	University of Nairobi
Gabriel M. Mailu	Ministry of Research, Technical Training, and Technology
Harun R. Muturi	Ministry of Research, Technical Training, and Technology
Micheni J. Ntiba	University of Nairobi
J.A. Odera	KEFRI
Rispa Achieng Odongo	Commission for Higher Education
Ezekiel Okemwa	Kenya Marine and Fisheries Research Institute
Ruth K. Oniang'o	Jomo Kenyatta University
G. Schneider	UNEP

Korea

Ji-Chil Ryu	Korea Energy Economics Institute

Malawi

N.W.S. Chipompha	Department of Forestry

Malaysia

K.S. Kannan	Universiti Teknologi Malaysia

Maldives

Humdun Hameed	Ministry of Planning and Environment

Mexico

A. G. Aguilar	Instituto de Geografia
Victor J. Jaramillo	Universidad Nacional Autonoma de Mexico
Daniel Lluch-Belda	Centro de Investigaciones Biologicas del Noreste, S.C.
Carlos Montaña	Instituto de Ecologia A.C.
Ana Rosa Moreno	Pan American Health Organization
Gerardo Segura	Universidad Nacional Autónoma de Mèxico

Nepal

Sharad P. Adhikary	Water and Energy Commission Secretariat
S.R. Chalise	ICIMOD

New Zealand

Bryan W. Christmas	Department of Health
Willem de Lange	University of Waikato
B. Blair Fitzharris	University of Otago
G.J. Glova	NIWA
Vincent Gray	Coal Research Association
Terry Healy	University of Waikato
Wayne Hennessey	Coal Research Association
Gavin J. Kenny	University of Waikato
Robert M. Kirk	University of Canterbury
Robert McDowall	National Institute of Water and Atmospheric Research
Kathleen McInnes	University of Waikato
Ian Owens	University of Canterbury
Larry Paul	MAF Fisheries
Peter L. Read	Massey University
Kevin Russell Tate	Manaaki Whenua Landcare Research
Richard Warrick	University of Waikato
Rob Whitney	Coal Research Association
Alastair Woodward	Wellington Medical School

Niger

Christian Valentin	ORSTOM

Nigeria

Larry F. Awosika	Nigerian Institute for Oceanography and Marine Research
George A. Banjo	University of Lagos
Peter Nwilo	University of Lagos

Norway

Annika Hofgaard	NINR
Paul Hofseth	Ministry of Environment
Jarle Inge Holten	NINR
Gorill Kristiansen	The Research Council of Norway
Nils R. Saelthun	Norwegian Water Resources and Energy Administration
Christina Skarpe	NINR
Oddvar Skre	Norwegian Forest Research Institute

Peru

S. Carrasco Barrera	Instituto del Mar del Peru
Patricia Ayon Dejo	Instituto del Mar del Peru

Philippines

Rex Victor Cruz	University of Phillipines–Los Banos
Lewis H. Ziska	International Rice Research Institute

Poland

Wojciech Galinski	Research Institute of Forestry
Zdzislaw Kaczmarek	Polish Academy of Science
Czeslaw Somorowski	Academy of Agriculture
Wojciech Suchorzewski	Warsaw University of Technology
Ryszard B. Zeidler	Polish Academy of Sciences

Portugal

Maria Jose Costa	Dep Zoologia e Antropologia

Republic of Benin

Michel Boko	National University of Benin

Russia

Oleg Anisimov	State Hydrological Institute
B.V. Glebov	RAS
Nikolai Grave	RAS
Martin G. Khublaryan	RAS
A.O. Kokorin	Institute of Global Climate and Ecology
V.M. Kotlyakov	RAS
Andrei S. Krovnin	Russian Federal Research Institute of Fisheries and Oceanography
A.S. Kulikov	RAS
G. Menzulin	State Hydrological Institute
Elena V. Milanova	Moscow State University
G.V. Panov	RAS
S.A. Shchuka	RAS
Mikhail Semenov	Long Ashton Research Station
Oleg Sirotenko	All-Russian Research Institute of Agricultural Meteorology
Olga N. Solomina	RAS
Argenta A. Titlyanova	Institute of Soil Sciences and Agrochemistry

Saudi Arabia

Saiyed Al-Khouli	King Abdulaziz University
Abdullah N. Alwelaie	Imam Muhammed Ibn Saudi Islamic University

Senegal

Christian Floret	ORSTOM

Sierra Leone

Ogunlade R. Davidson	University of Sierra Leone

Singapore

Beng-Wa Ang	National University of Singapore

South Africa

Richard M. Cowling	Institute for Plant Conservation, University of Cape Town
Robert J.M. Crawford	Sea Fisheries Research Institute
M.T. (Timm) Hoffman	National Botanical Institute
S.J. Dean Milton	University of Cape Town
Anthony Palmer	Roodeplaat Grasland Institute
P.W. Roux	Department of Agriculture
Mike Rutherford	National Botanical Institute
Roland Schafe	University of Natal
Robert J. Scholes	CSIR
Roland E. Schulze	University of Natal

Spain

Juan Puigdefábregas	Estacion Expérimental de Zonal Aridas

Sri Lanka

Sarath W. Kotagama	The Open University of Sri Lanka

Sweden

Sten Bergström	Swedish Meteorological and Hydrological Institute
Bert Bolin	IPCC Chairman
Malin Falkenmark	Natural Science Research Council
Bjorn Holmgren	Abisko Research Station
Sven A. Jansson	ABB Carbon AB
Thomas B. Johansson	UNDP
Sune Linder	University of Agricultural Sciences
Ulf Molau	Department of Systematic Botany
Mats Oquist	Swedish University of Agricultural Sciences
Bo Svensson	Swedish University of Agricultural Sciences

Switzerland

Bernard Aebischer	ETH–Zürich
Irene Aegerter	Swiss Association of Electric Power Producers and Distributors
A. Arquit-Niederberger	ProClim
Arthur J. Askew	WMO
Martin Beniston	IPCC WG2 Subgroup C Co-Chair
Heinz Blatter	ETH–Zürich
Christian Bonnard	ETH–Zürich
Eric Durand	Swiss Reinsurance Company
Danny L. Elder	World Conservation Union
Andreas Fischlin	ETH–Zürich
Fritz Gassmann	Paul Scherrer Institute
Patricia Geissler	Conservatoire et jardin botaniques de la Ville de Geneve
Heini Glauser	Swiss Energy Foundation
Gregory Goldstein	WHO
Stefan Husi	Swiss Energy Forum
John L. Innes	Swiss Federal Institute for Forest, Snow and Landscape Research
Frank Kloetzli	Geobotanik
Christian Körner	University of Basel
Norbert Kräuchi	Swiss Federal Institute for Forest, Snow and Landscape Research

Zbigniew W. Kundzewicz	WMO
Herbert Lang	ETH–Zürich
Peter Marti	Metron AG
Peter Meyer	Swiss Reinsurance Company
Kenneth Mott	WHO
Jaromir Nemec	ETH–Zürich
Atsumu Ohmura	ETH–Zürich
A.G.C. Paish	World Fire Statistics Centre
Daniel Perruchoud	Institute of Terrestrial Ecology
John C. Rodda	WMO
Rudolf Slooff	WHO
Rodolphe Spichiger	University of Geneva
Jürg Trüb	Swiss Reinsurance Company
Matthias Weber	Swiss Reinsurance Company

Tanzania

Datius G. Rutashobya	Ministry of Water, Energy, and Minerals

Thailand

Kansri Boonpragob	Ramkhamhaeng University
Dhira Phantumvanit	Thailand Environment Instutite

The Netherlands

Luitzen Bijlsma	National Institute for Coastal and Marine Management
A.F. Bouwman	National Institute of Public Health and Environmental Protection
Niels Daan	Netherlands Institute for Fisheries Investigations
Rudolf S. de Groot	Wageningen Center for Environment and Climate Studies
Leo P.M. deVrees	National Institute for Coastal and Marine Management
J. Dronkers	National Institute for Coastal and Marine Management
L.O. Fresco	Wageningen Agricultural University
Jan Goudriaan	Wageningen Agricultural University
Bill Hare	Greenpeace International
C.H.R. Heip	Netherlands Institute of Ecology
Frank M.J. Hoozemans	Delft Hydraulics
Michael Johann	Greenpeace International
E.H. Kampelmacher	WHO
Richard J.T. Klein	National Institute for Coastal and Marine Management
Eduard Koster	University of Utrecht
Martin C. Kroon	Ministry of Housing, Physical Planning and Environment
Rik Leemans	National Institute of Public Health and Environmental Protection
Pim Martens	University of Limburg
Reimund Roetter	Department of Land Evaluation Methods
Marcel J.F. Stive	Waterloopkundig Laboratorium/Delft Hydraulics
Richard S.J. Tol	Vrije Universiteit
J. van der Weide	Delft Hydraulics
Pier Vellinga	IPCC WG2 Subgroup B Co-Chair

Tunisia

Jaafar Friaa	IPCC WG2 Subgroup D Co-Chair

Turkmenistan

Nikolas Kharin	Desert Research Institute

United Kingdom

W. Neil Adger	University of East Anglia
R. Anderson	St. George's Medical School
Leonie J. Archer	Oxford Institute of Energy Studies
Nigel Arnell	University of Southampton
Edward B. Barbier	University of York
Max Beran	Institute of Hydrology
Keith J. Beven	Lancaster University
Peter M. Blackman	British Bankers Association
Roger Hignett Booth	Shell International Petroleum Co.
Christophe Bourillon	World Coal Institute
Phil Brookes	Institute of Arable Crops Research
Barbara Brown	University of Newcastle upon Tyne
Peter Bullock	Cranfield University
Melvin G.R. Cannell	Institute of Terrestrial Ecology
S.M. Cayless	Department of the Environment
David B. Clement	Toplis and Harding
Andrew F. Dlugolecki	General Accident, Fire and Life Assurance Corp.
Thomas Downing	University of Oxford
Hans DuMoulin	Energy, Technology, and the Environment
Alasdair Edwards	Centre for Tropical Coastal Management Studies
Inigio Everson	British Antarctic Survey
Malcolm John Fergusson	Institute for European Environmental Policy
Bernard Fisher	The University of Greenwich
Kevin Gilman	Institute of Hydrology
M.J. Glendining	Institute of Arable Crops Research
Steve Goldthorpe	CRE Group Ltd.
Phil B. Goodwin	University of Oxford
Andrew S. Goudie	University of Oxford
Keith Goulding	Institute of Arable Crops Research
Ken Gregory	Centre for Business and the Environment
James A. Harris	University of East London
Patrick Holligan	Plymouth Marine Laboratory
Ian Hughes	CRE Group Ltd.
Mike Hulme	University of East Anglia
B. Huntley	University of Durham
Michael Jefferson	WEC
David S. Jenkinson	Institute of Arable Crops Research
W.R. Keatinge	Queen Mary and Westfield College
Susan E. Lee	University of Sheffield
Jeremy Leggett	Greenpeace International
David Lister	DRA
Kathy Maskell	IPCC WG1 TSU

Anthony J. McMichael	London School of Hygiene and Tropical Medicine
Eddy Mesritz	Energy, Technology, and the Environment
Laurie Michaelis	OECD
Peter L. Mitchell	University of Sheffield
Chris Nash	University of Leeds
Robert J. Nicholls	Middlesex University
Timothy O'Riordan	University of East Anglia
Susan Owens	University of Cambridge
S. Page	University of Leicester
A.G.C. Paish	City University Business School
Martin Lewis Parry	University College–London
Peter J.G. Pearson	University of Surrey
John Pethick	Institute of Estuarine and Coastal Studies
John W. Polak	University of Oxford
David Powlson	Institute of Arable Crops Research
David Richardson	Ministry of Agriculture, Fisheries, and Food
J. Rieley	University of Nottingham
J.S. Robertson	ETSU
David Satterthwaite	International Institute for Environment and Development
Catherine A. Senior	Hadley Centre for Climate Prediction and Research
W. Graham Shearn	Sun Alliance Insurance
R.A.W. Shock	Greentie Liaison
Jim F. Skea	University of Sussex
Keith A. Smith	SAC Edinburgh
Susan Subak	University of East Anglia
Clive J. Swinnerton	National Rivers Authority
P. Bernard Tinker	University of Oxford
R. Kerry Turner	University of East Anglia
David Webb	Institute of Oceanographic Sciences
W. Rodney White	HR Wallingford
Brian W. Wilkinson	Institute of Hydrology
F. Ian Woodward	University of Sheffield
Philip L. Woodworth	Bidston Observatory

United States of America

Dean Abrahamson	University of Minnesota
Michael Adler	USEPA
Barbara Allen-Diaz	University of California–Berkeley
Richard B. Alley	Pennsylvania State University
Clinton J. Andrews	Princeton University
Richard W. Arnold	USDA
Shelley Arnott	University of Wisconsin
Raymond A. Assel	NOAA
Robert C. Balling	Arizona State University
Thomas O. Bamwell, Jr.	USEPA
David Banks	American Petroleum Institute
Thomas O. Barnwell, Jr.	USEPA
Jill Baron	Colorado State University
Roger Barry	University of Colorado
K.B. Bartlett	University of New Hampshire
Peter Beedlow	USEPA
Charles Bentley	University of Wisconsin
Rosina Bierbaum	OSTP
Suzanne Bolton	NOAA
Dermont Bouchard	USEPA
Nyle C. Brady	UNDP and The World Bank
Anthony J. Brazel	Arizona State University
Wallace Broecker	Lamont-Doherty Geological Observatory
Sandra Brown	USEPA
Russell Callender	NOAA
William Chandler	Battelle Pacific Northwest Laboratory
F. Stuart Chapin III	University of California
Art Chappelka	Auburn University
Robert Chen	CIESIN
Lars Christersson	University of Washington
Stephen Clark	NOAA
Vernon Cole	Colorado State University
C.B. Craft	Duke University
Christina Cundari	Bionetics Corporation
Jon Cusler	Association of Wetland Managers
Nadine Cutler	University of Washington
Roger Dahlman	DOE
Ann M. Deering	Environmental Technology and Telecommunications, Ltd.
Naomi Detenbeck	USEPA
Robert K. Dixon	USEPA
David Jon Dokken	IPCC WG2 TSU
Bert Drake	Smithsonian Environmental Research Cener
Thomas Drennen	Cornell University
John G. Eaton	USEPA
James A. Edmonds	Battelle Pacific Northwest Laboratory
Charles N. Ehler	NOAA
James Ellis	Colorado State University
John Everett	NOAA
Charles Feinstein	ENVGC
Christopher B. Field	Carnegie Institution of Washington
Shirley Fiske	NOAA
Klaus W. Flach	Soil Conservation Service
Glenn A. Flittner	NOAA
D.L. Fluharty	University of Washington
Jon Foley	University of Wisconsin
Ronald F. Follett	USDA
James Fouts	NIEHS
Douglas G. Fox	U.S. Forest Service
Ladeen Freimuth	Tufts University
Lew Fulton	DOE
Mary Gant	USEPA
Bronson Gardner	Global Climate Coalition
Robert H. Gardner	Appalachian Environmental Laboratory
Henry L. Gholz	University of Florida
Michael H. Glantz	NCAR
Bill Glassley	U.S. Army Corps of Engineers

Name	Affiliation
B. D. Goldstein	Environmental and Occupational Health Science Institute
M.A. Gonzalez-Meler	Smithsonian Environmental Research Center
Vivien Gornitz	Columbia University
Lester D. Grant	USEPA
Steve Greco	IPCC WG2 TSU
David L. Greene	ORNL
Peter M. Groffman	Institute of Ecosystem Studies
Howard K. Gruenspecht	DOE
Pat Halpin	University of Virginia
Steven P. Hamburg	Brown University
Mark E. Harmon	Oregon State University
Gary Spencer Hartshorn	WWF
Jonathan Haskett	USDA
Stefan Hastenrath	University of Wisconsin
Dexter Hinckley	USEPA
Richard (Skea) Houghton	Woods Hole Research Center
Dale Jamieson	University of Colorado–Boulder
Norris Jeffrey	Alaska Fisheries Science Center
D. Gale Johnson	University of Chicago
Dale W. Johnson	Desert Research Institute
Carol A. Johnston	Natural Resources Research Institute
Russell O. Jones	American Petroleum Institute
Chris Justice	NASA
M.R. Kaufmann	U.S. Forest Service
W. Doral Kemper	USDA
Bruce Kimball	USDA
John Kimble	USDA
Janusz Kindler	The World Bank
Timothy Kittel	UCAR
Robert W. Knecht	University of Delaware
Paul Komor	Office of Technology Assessment
Lee Ann Kozak	Southern Company Services
Bob Lackey	USEPA
Rattan Lal	Ohio State Universtiy
D. Larson	Northern Prairie Science Center
Daniel Lashof	Natural Resources Defense Council
Neil Leary	USEPA
Stephen Leatherman	University of Maryland
George H. Leavesley	USGS
Henry Lee	USEPA
Dennis Lettenmaier	University of Washington
Mark D. Levine	Lawrence Berkeley Labortory
Richard Levins	Harvard School of Public Health
Gene E. Likens	Center for Ecosystem Studies
P.D. Lingren	USDA
Harry F Lins	USGS
Orie Loucks	Miami University
Robert Luxmoore	ORNL
Michael C. MacCracken	USGCRP
John J. Magnuson	University of Wisconsin
David C. Major	Social Science Research Council
Adam C. Markham	WWF
Danny Marks	USEPA
Gregg Marland	ORNL
E. Matthews	NASA
George A. Maul	Florida Institute of Technology
Elaine McCormick-Watt	Tufts University
Sam McNaughton	Syracuse University
James P. McVey	NOAA
Assefa Mehretu	Michigan State University
R.D. Methot	NOAA
Steve Meyers	Lawrence Berkeley Laboratory
William K. Michener	Joseph W. Jones Ecological Research Center
R. Ben Mieremet	NOAA
Edward L. Miles	University of Washington
Alan S. Miller	University of Maryland
John D. Milliman	The College of William and Mary
Deborah Mills	University of Delaware
William Moomaw	Tufts University
Richard H. Moss	IPCC WG2 TSU
David Mountain	NOAA
Lee Mulkey	USEPA
Robert Musselman	U.S. Forest Service
Ron Neilson	USDA
Frederick E. Nelson	State University of New York–Albany
Robert J. Noiman	University of Washington
Richard J. Norby	ORNL
Frank W. Nutter	Reinsurance Association of America
Walter Oechel	San Diego State University
Dennis Ojima	Colorado State University
Eduardo P. Olaguer	Global Climate Coalition
David M. Olszyk	USEPA
Gerald T. Orlob	University of California–Davis
Flo Ormond	IPCC WG2 TSU
Tom E. Osterkamp	University of Alaska
Ludmila B. Pachepsky	Duke University
Paul Kilho Park	NOAA
Richard Park	ABT Associates, Inc.
Peter J. Parks	Rutgers University
Anand Patwardhan	Carnegie Mellon University
Eldor Alvin Paul	Michigan State University
David L. Peterson	University of Washington
Donald L. Phillips	USEPA
Julia Philpott	International Institute for Energy Conservation
Rick Piltz	USGCRP
Louis F. Pitelka	EPRI
Warren T. Piver	NIEHS
Wilfred Mac Post	ORNL
Kurt S. Pregitzer	Michigan Technical University
Jonas Rabinovitch	UNDP
Edward B. Rastetter	Marine Biological Laboratory
Greg Rau	U.S. Army Corps of Engineers
Charles Raymond	University of Washington
Steve Rayner	Battelle Pacific Northwest Laboratory
Daniel Reifsnyder	U.S. Department of State
John M. Reilly	USDA
William K. Reisen	University of California

Michael A. Replogle	Environmental Defense Fund
Richard Reppert	U.S. Army Institute for Water Resources
Curtis J. Richardson	Duke University
Jeffrey E. Richey	University of Washington
Martyn J. Riddle	International Finance Corporation
Jens Rosebrock	ENVGC
Norman J. Rosenberg	Batelle Pacific Northwest Laboratory
Cynthia Rosensweig	Goddard Institute for Space Studies
Jayant Sathaye	Lawrence Berkeley Laboratory
Joel Scheraga	USEPA
David Schimmelpfennig	USDA
Richard Schmalensee	Massachusetts Institute of Technology
Michael J. Scott	Battelle Pacific Northwest Laboratory
Gary D. Sharp	Cooperative Institute for Research in the Integrated Ocean Sciences
Kenneth Sherman	NOAA
Robert Shope	Yale University
David Shriner	OSTP
Tom Sibley	University of Washington
Anne Sigleo	USEPA
Upendra Singh	International Fertilizer Development Center
Sarah Siwek	Los Angeles County Metropolitan Transportation Authority
Robert Socolow	Princeton University
Allen M. Solomon	USEPA
Andrew R. Solow	Woods Hole Oceanographic Institution
Eugene Z. Stakhiv	U.S. Army Institute for Water Resources
John Steele	U.S. Army Corps of Engineers
Heinz Stefan	Massachusetts Institute of Technology
Meyer Steinberg	Brookheaven National Laboratory
B.A. Stewart	West Texas A&M University
Kenneth Strzepek	University of Colorado
Melissa Taylor	IPCC WG2 TSU
Anne Tenney	IPCC WG2 TSU
Ronald M. Thom	Battelle Marine Sciences Labatory
Valerie Thomas	Princeton University
David T. Tingey	USEPA
Dennis Tirpak	USEPA
D. Mark Tullis	Tufts University
Bill L. Turner II	George Perkins Marsh Institute
Robert Twilley	University of Southern Louisiana
Limberios Vallianos	U.S. Army Corps of Engineers
Nico van Breemen	Institute of Ecological Studies
Laura Van Wie	IPCC WG2 TSU
Monish Verma	Tufts University
T. Vinson	Oregon State University
Hal Walker	USEPA
Gerald Ward	Colorado State University
Richard H. Waring	Oregon State University
Robert T. Watson	IPCC WG2 Co-Chair
Sheila K. West	Johns Hopkins University
Walt Whitford	USEPA
Diane Wickland	NASA
Mark Wigmosta	Pacific Northwest Laboratories
Edward R. Williams	DOE
Michael D. Williams	Los Alamos National Laboratory
Robert Williams	Princeton University
S. Jeffress Williams	USGS
Jack Winjum	USEPA
Julian Wolpert	Princeton University
George Woodwell	Woods Hole Research Center
P. Christopher Zegras	International Institute for Energy Conservation

Uruguay

Walter Baethgen	IFDC

Venezuela

Rigoberto Andressen	University of Merida
Itamar Galindez Giron	University of Los Andes
Luis Jose Mata	Instituto de Mecanica de Fluidos FI-UCV
Martha Perdomo	IPCC WG2 Subgroup B Co-Chair
Hernán Pérez Nieto	Comision Nacional de Oceanologia
Ignacio Salcedo	DEN/UFPE

Vietnam

TranViet Lien	Institute of Meteorology and Hydrology

Western Samoa

Chalapna Kaluwin	South Pacific Regional Environment Programme

Zimbabwe

Chris H.D. Magadza	University of Zimbabwe
M.C. Zinyowera	IPCC WG2 Co-Chair

Editorial Support

David F. Brakke	University of Wisconsin–Eau Clair
Oma Perrin Bowling	University of Colorado
Rachel Duncan	Green College–Oxford
David A. Gooding	Ecological Society of America
Elisabeth Heseltine	Communication in Science
Robert Lande	Johns Hopkins University
Frederick M. O'Hara	Consultant in Technical Communication
Elizabeth Sulzman	University Corporation for Atmospheric Research
Stephen Threlkeld	The University of Mississippi
Patrick Young	Consultant

Appendix B

GLOSSARY OF TERMS

acclimatization
physiological adaptation to climatic variations

active layer
the top layer of soil in permafrost that is subjected to seasonal freezing and thawing

adaptability
the degree to which adjustments are possible in practices, processes, or structures of systems to projected or actual changes of climate; adaptation can be spontaneous or planned, and can be carried out in response to or in anticipation of changes in conditions

afforestation
forest stands established artificially on lands that previously have not supported forests for more than 50 years

agroclimatic
climatic conditions as they relate to agricultural production; discrete set of zones each of which identifies areas capable of like types and levels of agricultural production

airshed
the mass of air associated with a usually enclosed or otherwise bounded area, like a cove or valley

albedo
the surface reflectivity of the globe

algal blooms
a reproductive explosion of algae in a lake, river, or ocean

alpine
the biogeographic zone made up of slopes above timberline and characterized by the presence of rosette-forming herbaceous plants and low shrubby slow-growing woody plants

anaerobic
living, active, or occurring in the absence of free oxygen

anaerobic digestion
fermentation processes conducted in the absence of oxygen

anthropogenic
caused or produced by humans

anticyclone system
a system of winds that rotates about a center of high atmospheric pressure (clockwise in the Northern Hemisphere and counterclockwise in the Southern Hemisphere)

aquifer
a stratum of permeable rock that bears water

arbovirus
any of various viruses transmitted by arthropods and including the causative agents of dengue fever, yellow fever, and some encephalitis

arid lands
ecosystems with <250 mm precipitation per year

autonomous adaptation
adaptation that occurs without specific human intervention

bagasse
the dry, fibrous residue remaining after the extraction of juice from the crushed stalks of sugar cane; used as a source of cellulose for some paper products and as a source of energy

ballast
a device that takes electricity from the line and transforms that electricity to the proper current and voltage for starting and operating a fluorescent lamp

baseline scenario
the set of predicted levels of economic growth, energy production and consumption, and greenhouse gas emissions assumed as the starting point for an analysis of mitigation options

baseload
that part of total energy demand that does not vary over a given period

biodiversity
the number of different species or functional groups of flora and fauna found in an area or ecosystem

biofuels
fuels obtained as a product of biomass conversion (e.g., alcohol or gasohol)

biogas
a gas composed principally of a mixture of methane and carbon dioxide produced by anaerobic digestion of biomass

biogeography
the study of the geographical distribution of living organisms

biomass
the total quantity of living matter in a particular habitat; plant and organic waste materials used as fuel and feedstock in place of fossil fuels

biome
a grouping of similar plant and animal communities into broad landscape units that occur under similar environmental conditions

bottom-up modeling
a modeling approach that arrives at economic conclusions from an analysis of the effect of changes in specific parameters on narrow parts of the total system

breeder reactor
a reactor that produces from a non-fissile material a fissile material (e.g., plutonium) identical to the one it consumes and in greater quantity (i.e., the ratio of fissile material by material consumed is greater than unity)

building stock
the residential and/or commercial structures extant in a society or a geographic area

busbar
conductor for collecting and distributing electric current; frequently used to denote electricity generation output, capacity, or generating costs of power plants at output or capacity rates of electricity delivered to the busbar

C_3 plants
plants that produce a three-carbon compound during photosynthesis, including most trees and agricultural crops such as rice, wheat, soybeans, potatoes, and vegetables

C_4 plants
plants that produce a four-carbon compound during photosynthesis; mainly of tropical origin, including grasses and the agriculturally important crops maize, sugar cane, millet, and sorghum

calcining
the heating of calcium carbonate to high temperatures to thermally decompose it into carbon dioxide and calcium oxide

calving
the breaking away of a mass of ice from a floating glacier, ice front, or iceberg

CAM
variant of the C_4 photosynthetic pathway in which most gas exchange occurs at night; occurs primarily in succulents (e.g., cacti)

carbon intensity
CO_2 emissions per unit of energy or economic output

carbon sequestration
the biochemical process through which carbon in the atmosphere is absorbed by biomass such as trees, soils, and crops

carbon sinks
chemical processes that absorb carbon dioxide

carbon stocks
the amount of carbon that is stored in carbon sinks

carbon tax
a levy exacted by a government on the use of carbon-containing fuels for the purpose of influencing human behavior (specifically economic behavior) to use less fossil fuels

carrying capacity
the number of individuals in a population that the resources of a habitat can support

catchment
area having a common outlet for its surface runoff

catotelm
the layers of peat generally not subjected to oxic conditions; also called peat proper

Chagas' disease
a parasitic disease caused by the *Trypanosoma cruzi* and transmitted by triatomine bugs in the Americas, with two clinical periods: Acute (fever, swelling of the spleen, edemas) and chronic (heart disorder that may produce high fatality, or digestive syndrome)

CO_2 fertilization
the enhancement of plant growth as a result of elevated atmospheric CO_2 concentration

coefficient of performance
the ratio of useful thermal energy output per unit of energy input; a measure of the thermal efficiency of an energy-conversion device

cogeneration
production of electricity and heat, both as main output, by a combined process (combined heat and power) from energy sources (e.g., from coal, oil, gas, or nuclear power) in an installation such as a thermal plant or block heating station

combined cycle
electricity generation using one or more gas-turbine generator units whose exhaust gases are fed to a waste-heat boiler, which may or may not have a supplementary burner; the steam raised by the boiler is used to drive one or more steam-turbine generator units

communicable disease
infectious disease caused by transmission of an infective biological agent (virus, bacterium, protozoan, or multicellular macroparasite)

consumer surplus
a measure of the value of consumption beyond the price paid for the goods

converter reactor
a device in which a self-sustaining nuclear fission chain reaction can be maintained and controlled (fission reactor)

cryosphere
all global snow, ice, and permafrost

decarbonization
reduction of carbon intensity of an energy process or a system

demand-side management
any activity designed to alter the customer's timing or use of electricity, natural gas, or other energy form; such actions are designed to control the demand upon that utility by manipulating the consumers' use of electricity to limit electricity usage during peak periods, shift usage from demand peaks to demand valleys, build load, conserve overall energy usage, and/or otherwise change the demand placed upon the utility

dematerialization
decreasing the amount of a material employed in a product by employing a design that exercises economy in material usage or by substituting a lighter material for the original

Dengue fever
an infectious viral disease spread by mosquitoes, the first infection of which is often called breakbone fever and is characterized by severe pain in joints and back, fever, and rash; a subsequent infection is usually characterized by fever, bleeding from bodily orifices, and sometimes death

desert
an ecosystem with <100 mm precipitation per year

diapause
period of suspended growth or development and reduced metabolism in the life cycle of many insects, when organism is more resistant to unfavorable environmental conditions than in other periods

digestor
a tank designed for the anaerobic fermentation of biomass; a vessel in which substances are softened or decomposed, usually for further processing

disposable income
the amount of income available for consumers to spend on discretionary items

dissolved load
the amount of particles in a stream or other water source that arises as a result of erosion

district heat
thermal energy transmitted through pipelines in the form of heated water or steam to point of consumption

diurnal climate
a climate with uniform amplitudes of temperature throughout the year

econometric
an approach to studying a problem through use of mathematical and statistical methods in the field of economics to develop and verify theories

economies in transition
national economies that are moving from a period of heavy government control toward lessened intervention, increased privatization, and greater use of competition

ecotax
a levy exacted by a government for the purpose of influencing human behavior (specifically economic behavior) to follow an ecologically benign path

ecotone
transition area between adjacent ecological communities (e.g., between forests and grasslands), usually involving competition between organisms common to both

ecotopic
tendency or involving adjustment to specific habitat conditions

edaphic
of or relating to the soil; factors inherent in the soil

El Niño
an irregular variation of ocean current that, from January to February, flows off the west coast of South America, carrying warm, low-salinity, nutrient-poor water to the south; does not usually extend farther than a few degrees south of the Equator, but occasionally it does penetrate beyond 12°S, displacing the relatively cold Peruvian current; usually short-lived effects, but sometimes last more than a year, raising sea-surface temperatures along the coast of Peru and in the equatorial eastern Pacific Ocean, having disastrous effects on marine life and fishing

embodied energy
the energy (natural gas, oil, coal, etc.) that is required to produce a manufactured good and that is thereby included in the finished product or service

emission factor
a coefficient that relates actual emissions to activity data as a standard rate of emission per unit of activity

endemic infection
a sustained, relatively stable, pattern of infection within a specified population

energy conversion
energy transformation with a change in the form of energy

energy efficiency
ratio of energy output of a conversion process or of a system to its energy input; also known as first-law efficiency

energy intensity
ratio between the consumption of energy to a given quantity; usually refers to the amount of primary or final energy consumed per unit of gross domestic or national product

energy service
the application of useful energy to tasks desired by the consumer such as transportation, a warm room, or light

energy transformation
production of energy involving no change in the physical state of the form of energy

engineering cost
the direct materials and labor costs associated with projects; these estimates are used in assessments of the economic feasibility of mitigation projects

enhanced oil recovery
advanced methods for recovering oil from reservoirs in addition to that recoverable by conventional primary and secondary recovery methods; enables larger proportion of oil *in situ* to be exploited from a reservoir

enthalpy
energy content per unit mass

epidemic
appearance of an abnormally high number of cases of infection in a given population; can also refer to noninfectious diseases (e.g., heart disease) or to acute events such as chemical toxicity

eutrophication
the process by which a body of water (often shallow) becomes (either naturally or by pollution) rich in dissolved nutrients with a seasonal deficiency in dissolved oxygen

evapotranspiration
loss of water from the soil both by evaporation from the surface and transpiration from the plants growing thereon

exergy
the maximum amount of energy that under given (ambient) thermodynamic conditions can be converted into any other form or energy; also known as availability or work potential

exergy efficiency
the ratio of (theoretical) minimum exergy input to actual input of a process or a system; also known as second-law efficiency

fallow
land left unseeded after plowing; uncultivated

fast ice
sea or lake ice that remains tied to the coast (usually less than 2 m above sea level)

feedback
when one variable in a system triggers changes in a second variable that in turn ultimately affects the original; a positive feedback intensifies the effect, and a negative reduces the effect

feedstock
raw material (e.g., oil products or natural gas) used as input into industrial processes for manufacturing of materials or consumer goods (e.g., plastics, fertilizer, etc.); energy feedstocks are often included in total energy consumption even though they represent so-called non-energy consumption

final energy
the energy supplied to the consumer to be converted to useful energy (e.g., electricity at the socket, gasoline at the service station, or fuelwood in the barn); sometimes also called available energy, but the term should be avoided due to possible confusion with availability or exergy

first-law efficiency
see energy efficiency

fluidized bed
a bed of fuel and non-combustible particles set in vigorous, turbulent motion by the combustion air blowing upward throughout the bed; the non-combustible particles are generally coal ash or a sulfur-absorbent (acceptor) such as limestone

fluidized bed combustion
a method of burning a fuel with non-combustible particles in a state of suspension by the upward flow of the combustion air through the fluidized bed

fly-ash
the mineral content of coal released as particulate matter upon its combustion

food calorie
1000 (technical) calories

forest
an ecosystem in which the dominant plants are trees; woodlands are distinguished from forests by their lower density of trees

forestation
generic term for establishing forest stands by reforestation and afforestation

forest decline
premature, progressive loss of tree and stand vigor and health

frazil ice
fine spicules or plates of ice in suspension in water

fuel cells
devices for the conversion of chemical energy to electrical energy

general equilibrium analysis
an approach that considers simultaneously all the markets in an economy, allowing for feedback effects between individual markets

geomorphic
of or related to the form of the Earth or its surfaces

germ plasm
the reproductive cells of an organism, in particular the portion of the cells involved in heredity

greenhouse gas
any gas that absorbs infrared radiation in the atmosphere

gross primary production
the amount of carbon fixed in photosynthesis by plants

ground ice
ice present within rock, sediments, or soil

groundwater recharge
process by which external water is added to the zone of saturation of an aquifer, either directly into a formation or indirectly by way of another formation

halocarbons
chemicals containing carbon and members of the halogen family

heat cascading
process integration aimed at effectively utilizing thermal energy at all temperature levels by successively using waste heat from higher temperature processes as heat sources for lower temperature ones; also applicable to cold sources at below-ambient temperatures, if reversed

heat content
the amount of heat per unit mass released upon complete combustion

heath
any of the various low-growing shrubby plants of open wastelands, usually growing on acidic, poorly drained soils

heat island
an area within an urban area characterized by ambient temperatures higher than those of the surrounding area because of the absorption of solar energy by materials like asphalt

heat pump
a device capable of extracting heat from a lower temperature source (e.g., the external environment) and releasing thermal energy to a higher temperature heat sink (e.g., an installation requiring heating); this temperature lift is achieved by using an internal fluid loop that undergoes changes in phase (evaporation and condensation) and operating pressure (expansion and compression)

herbaceous
flowering, non-woody plants

herbivore
an animal that feeds on plants

higher heating value
quantity of heat liberated by the complete combustion of a unit volume or weight of a fuel assuming that the produced water vapor is completely condensed and the heat is recovered; also known as gross calorific value

hydroperiod
the depth, frequency, duration, and season of wetland flooding

ice cap
a dome-shaped glacier covering a highland area (considerably smaller in extent than ice sheets)

ice jam
an accumulation of broken river or sea ice caught in a narrow channel

ice sheet
a mass of snow and ice of considerable thickness and large area greater than 50,000 km^2

ice shelf
a floating ice sheet of considerable thickness attached to a coast (usually of great horizontal extent with a level or gently undulating surface); often a seaward extension of ice sheets

icing
a sheet-like mass of layered ice formed by the freezing of water as it emerges from the ground or through fractures in river or lake ice

immunosuppression
reduced functioning of an individual's immune system

incidence
the number of cases of a disease commencing, or of persons falling ill, during a given period of time within a specified population

incinerator
a furnace or container for burning waste materials

income elasticity
the expected percentage change in the quantity demand for a good given a 1% change in income

industrial ecology
the set of relationships of a particular industry with its environment; often refers to the conscious planning of industrial processes so as to minimize their negative interference with the surrounding environment (e.g., by heat and materials cascading)

industrialization
the conversion of a society from one based on manual labor to one based on the application of mechanical devices

infiltration
flow of water through the soil surface into a porous medium

infrastructure
the basic installations and facilities upon which the operation and growth of a community depend, such as roads; schools; electric, gas, and water utilities; transportation and communications systems; and so on

inoculation
the introduction of a pathogen or antigen into a living organism to stimulate the production of antibodies

integrated resource planning
a system in which a utility considers all means of meeting the energy-service needs of its customers (on both the supply and demand sides) and to select the mix of actions that meets those needs at the least cost to the consumers, possibly including external costs like environmental degradation

isohyet
a line on a map or chart indicating equal rainfall

joule
unit of energy; 1 joule (J) is the work done when the point of application of a force of 1 newton (1 N = 1 kg m/s^2) is displaced through a distance of 1 m in the direction of the force

land use
the purpose an area of the Earth is put to (e.g., agriculture, forestry, urban dwellings, or transportation corridors) or its character (e.g., swamp, grassland, or desert)

lapse rate
the rate of temperature decrease with increase in altitude

leaching
the removal of soil elements or applied chemicals through percolation

legume
plants that through a symbiotic relationship with soil bacteria are able to fix nitrogen from the air (e.g., peas, beans, alfalfa, clovers)

lichen
symbiotic organisms consisting of an alga and fungus important to the weathering and breakdown of rocks

life-cycle cost
the cost of a good or service over its entire lifetime

lignite
brown coal; a low-grade coal having a heat content slightly higher than peat

low emissivity
a property of materials that hinders or blocks the transmission of a particular band of radiation (e.g., that in the infrared)

lower heating value
quantity of heat liberated by the complete combustion of a unit volume or weight of a fuel assuming that the produced water remains as a vapor and the heat of the vapor is not recovered; also known as net calorific value

macroeconomic
pertaining to a study of economics in terms of whole systems, especially with reference to general levels of output and income and to the interrelations among sectors of the economy

maglev
a mode of transport that uses magnetic force to suspend a vehicle above its supporting structure, thereby limiting or eliminating friction between the vehicle and the ground

malaria
endemic or epidemic parasitic disease caused by species of the genus *Plasmodium* (protozoa) and transmitted by mosquitoes of the genus *Auopheles*; produces high fever attacks and systemic disorders, and kills ~2 million people every year

market equilibrium
the point at which demand for goods and services equals the supply; often described in terms of the level of prices, determined in a competitive market, that "clears" the market

market penetration
the percentage of all its potential purchasers to which a good or service is sold per unit time

materials cascading
successive downgrading of a material's use into applications requiring lower quality

methanogenesis
the creation of methane from its constituent molecules

mitigation
an anthropogenic intervention to reduce the emissions or enhance the sinks of greenhouse gases

monsoon
the season of the southwest wind in India and adjacent areas that is characterized by very heavy rainfall

montane
the biogeographic zone made up of relatively moist, cool upland slopes below timberline and characterized by the presence of large evergreen trees as a dominant life form

moraine
an accumulation of Earth and stones carried and finally deposited by a glacier

morbidity
the rate of occurrence of disease or other health disorder within a population, taking account of the age-specific morbidity rates; health outcomes include, for example, chronic disease incidence or prevalence, rates of hospitalization, primary care consultations, disability-days (e.g., of lost work), and prevalence of symptoms

mortality
the rate of occurrence of death within a population within a specified time period; calculation of mortality takes account of age-specific death rates, and can thus yield measures of life expectancy and the extent of premature death

net ecosystem production
the net gain or loss of carbon from an ecosystem or region

net primary production
the increase in plant biomass or carbon of a unit of a landscape; gross primary production (all carbon fixed through photosynthesis) minus plant respiration equals net primary production

newton
the unit of force required to accelerate 1 kg of mass 1 m/s^2

nitrification
the oxidation of ammonium salts to nitrites and the further oxidation of nitrites to nitrates

NO_x
any of several oxides of nitrogen

non-tidal wetlands
areas of land not subject to tidal influences where the water table is at or near the surface for some defined period of time, leading to unique physiochemical and biological processes and conditions characteristic of water-logged systems

northern wetlands
wetlands in the boreal, subarctic, and arctic regions of the northern hemisphere

orography
the branch of physical geography that deals with mountains and mountain systems

pack ice
any area of sea, river, or lake ice other than fast ice

paleoecology
the branch of ecology concerned with identifying and interpreting the relationships of ancient plants and animals to their environment

pancake ice
new ice about 0.3 to 3-m in diameter, with raised rims about the circumference from striking other pieces

peat
unconsolidated soil material consisting largely of partially decomposed organic matter accumulated under conditions of excess moisture or other conditions that decrease decomposition rates

perfluorocarbons
organic chemicals containing only carbon and fluorine (e.g., carbon tetrafluoride and hexafluoroethylene)

permafrost
perennially frozen ground that occurs wherever the temperature remains below 0°C for several years

phenology
the study of natural phenomena that recur periodically (e.g., blooming, migrating) and their relation to climate and seasonal changes

photochemical smog
a mix of photochemical oxidant air pollutants produced by the reaction of sunlight with primary air pollutants, especially hydrocarbons

photoperiodic response
response to the lengths of alternating periods of light and dark as they affect the timing of development

photosynthate
the product of photosynthesis

photovoltaic
capable of producing a voltage when exposed to radiant energy, especially light

physiographic
of, relating to, or employing a description of nature or natural phenomena

phytophagous insects
insects that feed on plants

potential evapotranspiration
maximum quantity of water capable of being evaporated in a given climate from a continuous stretch of vegetation (i.e., includes evaporation from the soil and transpiration from the vegetation of a specified region in a given time interval, expressed as depth)

potential production
estimated production of a crop under conditions when nutrients and water are available at optimum levels for plant growth and development; other conditions such as daylength, temperature, soil characteristics, etc., determined by site characteristics

prevalence
the proportion of persons within a population who are currently affected by a particular disease

price elasticity
the responsiveness of demand to the cost for a good or service; specifically, the percentage change in the quantity consumed of a good or service for a 1% change in the price for that good or service

primary energy
the energy that is embodied in resources as they exist in nature (e.g., coal, crude oil, natural gas, uranium, or sunlight); the energy that has not undergone any sort of conversion

producer surplus
returns beyond the cost of production that provide compensation for owners of skills or assets that are scarce (e.g., agriculturally productive land)

radiative forcing
a change in average net radiation at the top of the troposphere resulting from a change in either solar or infrared radiation due to a change in atmospheric greenhouse gases concentrations; perturbance in the balance between incoming solar radiation and outgoing infrared radiation

rangeland
unimproved grasslands, shrublands, savannas, and tundra

redox potential
the relative capacity of an atom or compound to donate or accept electrons, expressed in volts; higher numbers denote more powerful oxidizers

reference scenario
the set of predicted levels of economic growth, energy production and consumption, and greenhouse gas emissions (and underlying assumptions) with which other scenarios examining various policy options are compared

reforestation
forest stands established artificially on lands that have supported forests within the last 50 years

regenerative braking
a process wherein the motion of a vehicle is slowed or stopped, and the energy of motion is captured and stored for future reuse

reserves
those occurrences of energy sources or minerals that are identified and measured as economically and technically recoverable with current technologies and prices

resources
those occurrences of energy sources or minerals with less certain geological and/or economic/technical recoverability characteristics, but that are considered to become potentially recoverable with foreseeable technological and economic development

respiration
the metabolic process by which organisms meet their internal energy needs and release CO_2

rotary kiln
a rotating cylinder heated to dry or chemically transform its contents

runoff
water (from precipitation or irrigation) that does not evaporate or seep into the soil but flows into rivers, streams, or lakes, and may carry sediment

ruderal
pertaining to or inhabiting highly disturbed sites; weedy

salinization
the accumulation of salts in soils

saltation
the transportation of particles by currents of water or wind in such a manner that they move along in a series of short intermittent leaps

seasonal climate
a climate characterized by both warm and cold periods through the year

second-law efficiency
see exergy efficiency

semi-arid lands
ecosystems that have >250 mm precipitation per year, but are not highly productive; usually classified as rangelands

sensitivity
the degree to which a system will respond to a change in climatic conditions (e.g., the extent of change in ecosystem composition, structure and functioning, including net primary productivity, resulting from a given change in temperature or precipitation)

sequestration
to separate, isolate or withdraw; usually refers to removal of CO_2 from atmosphere by plants or by technological measures

set-aside program
a generic term covering a variety of government programs—primarily in the U.S., Canada, and Europe that require farmers to remove a portion of their acreage from production for purposes of controlling yield, soil conservation, etc.

shelterbelt
a natural or artificial forest maintained for protection against wind or snow

silt
unconsolidated or loose sedimentary material whose constituent rock particles are finer than grains of sand and larger than clay particles

slip faces
the lee side of a dune where the slope approximates the angle of rest of loose sand (usually ~33°)

snowpacks
a seasonal accumulation of slow-melting snow

soil erosion
the process of removal and transport of the soil by water and/or wind

southern oscillation
a large-scale atmospheric and hydrospheric fluctuation centered in the equatorial Pacific Ocean; exhibits a nearly annual pressure anomaly, alternatively high over the Indian Ocean and high over the South Pacific; its period is slightly variable, averaging 2.33 years; the variation in pressure is accompanied by variations in wind strengths, ocean currents, sea- surface temperatures, and precipitation in the surrounding areas

sphagnum moss
a genus of moss that covers large areas of wetlands in the northern hemisphere; sphagnum debris is usually a major constituent of the peat in these areas

stakeholders
the entities that will be affected by a particular action or policy

stomata
the minute openings in the epidermis of leaves through which gases interchange between the atmosphere and the intercellular spaces within leaves

succession
transition in the composition of plant communities following disturbance

superconduction
the flow of electric current without resistance in certain metals, alloys, and ceramics at temperatures near absolute zero degrees Kelvin, and in some cases at temperatures hundreds of degrees above absolute zero (high-temperature superconduction)

susceptibility
probability for an individual or population of being affected by an external factor

sustainable
a term used to characterize human action that can be undertaken in such a manner as to not adversely affect environmental conditions (e.g., soil, water quality, climate) that are necessary to support those same activities in the future

symbionts
organisms that live together to mutual benefit [e.g., nitrogen-fixing bacteria that live with a plant (legume)]

synoptic
relating to or displaying atmospheric and weather conditions as they exist simultaneously over a broad area

taiga
coniferous forests of northern North America and Eurasia

talik
a layer of unfrozen ground occurring between permafrost and the active layer

technological calorie
the amount of heat needed to raise the temperature of 1 g of water 1°C at 15°C

thermodynamic free-energy limit
the minimum possible amount of energy required for producing a substance from its components, corresponding to the energy required to break down and/or create chemical bonds between molecules; the stronger the bond, the more energy is required to synthesize or break down a substance, which limits the possibility of lowering energy consumption by industrial process energy-efficiency improvements

thermohaline circulation
circulation driven by density gradients, which are controlled by temperature and salinity

thermokarst
irregular, hummocky topography in frozen ground caused by melting of ice

timberline
the upper limit of tree growth in mountains or high latitudes

transpiration
the emission of water vapor from the surfaces of leaves or other plant parts

tsunami
a large tidal wave produced by a submarine earthquake, landslide, or volcanic eruption

unexploited energy
energy (usually thermal, at near-environmental temperature) that is discarded in spite of its potential for use; often existing in large amounts, but with a low heat value/work potential

upwelling
transport of deeper water to the surface, usually caused by horizontal movements of surface water

urbanization
the conversion of land from a natural state or managed natural state (such as agriculture) to cities

useful energy
the energy drawn by consumers such as heat from their own appliances after conversion of final energy (e.g., mechanical energy at the crankshaft of an automobile engine or an industrial electric motor; the heat of a household radiator or an industrial boiler, or the luminosity of a light bulb)

usufruct
the legal right of using and enjoying the fruits or profits of something belonging to another

vector
an organism, such as an insect, that transmits a pathogen from one host to another

vulnerability
the extent to which climate change may damage or harm a system; it depends not only on a system's sensitivity, but also on its ability to adapt to new climatic conditions

wadi
a water course that is dry except during the rainy season; the stream or flush that runs through it

waste heat
excess heat from industrial or other processes that is either discarded or used in other processes requiring lower temperature heat sources

water use efficiency
carbon gain in photosynthesis per unit water lost in evapotranspiration; can be expressed on a short-term basis as the ratio of photosynthetic carbon gain per unit transpirational water loss, or on a seasonal basis as the ratio of net primary production or agricultural yield to the amount of available water

winter dormancy
period without biochemical activity in plant tissues

Appendix C

ACRONYMS AND ABBREVIATIONS

AAAS	American Association for the Advancement of Science
AAC	Annual Allowable Cut
ABI	Association of British Insurers
AC	Alternating Current
ACBE	Advisory Committee on Business and the Environment
ACOE	Army Corps of Engineers
ADB	Asian Development Bank
AECI	Agencia Española de Cooperacion Internacional
AES	Applied Energy Services
AES	Atmospheric Environment Service
AFBC	Atmospheric Fluidized-Bed Combustion
AFC	Alkaline Fuel Cell
AFV	Alternative Fuel Vehicle
AGTD	Allison Gas Turbine Division, General Motors Corporation
AID	Agency for International Development
AIM	Asian Pacific Integrated Model
AIP	American Institute of Physics
AIRAC	All Industry Research Advisory Council
ALM	Agriculture Land-Use Model
ALS	Area Licensing Scheme
AMCM	Metropolitan Area of Mexico City
ANPP	Aboveground Net Primary Productivity
AOS	Atmosphere Ocean System
AP	Accelerated Policies
APS	American Physical Society
ASCE	American Society of Chemical Engineers
AVHRR	Advanced Very High-Resolution Radiometer
BCC	Basal Cell Carcinoma
BEST	Board on Environmental Studies and Toxicology
BI	Biomass-Intensive
BIG/FC	Biomass-Integrated Gasifier/Fuel Cell
BIG/GT	Biomass-Integrated Gasifier/Gas Turbine
BLS	Basic Linked System
BMRC	Bureau of Meteorology Research Center
BNL	Brookhaven National Laboratory
BP	Before Present
BWEA	British Wind Energy Association
CADDET	Centre for the Analysis and Dissemination of Demonstrated Energy Technologies
CAFE	Corporate Average Fuel Efficiency
CAM	Crassulacean Acid Metabolism
CARB	California Air Resources Board
CAST	Council for Agricultural Science and Technology
CBD	Central Business District
CCC	Canadian Climate Centre
CCIRG	Climate Change Impact Review Group
CCN	Cloud Condensation Nuclei
CDC	Centers for Disease Control and Prevention
CDIAC	Carbon Dioxide Information Analysis Center
CEC	Commission of the European Communities
CECA	Consumer Energy Council of America
CEFE	Centre d'Ecologie Fonctionelle et Evolutive
CERES	Clouds and Earth's Radiant Energy System
CFL	Compact Fluorescent Lamp
CHP	Combined Heat and Power
CI	Coal-Intensive
CIAB	Coal Industry Advisory Board
CIESIN	Consortium for International Earth Science Information Network
CNG	Compressed Natural Gas
COP	Coefficient of Performance
CPA	Centrally Planned Asia, including China
CRESTA	Catastrophe Risk Evaluating and Standardized Target Accumulations
CSIRO	Commonwealth Scientific and Industrial Research Organization
CVT	Continuously Variable Transmission
CZM	Coastal Zone Management
CZMS	Coastal Zone Management Subgroup
DC	Direct Current
DDE	Direction Dèpartementale de l'Environnement
DENR-ADB	Department of Environment and Natural Resources–Asian Development Bank
DEPE	Chinese Department of Environmental Protection and Energy
DH	District Heating
DMS	Dimethyl Sulfide
DNA	Deoxyribonucleic Acid
DOC	Dissolved Organic Carbon
DOE	U.S. Department of Energy
DSM	Demand-Side Management
DTI	Danish Technological Institute
E	Evaporation
EC	European Community
ECS	Environmentally Compatible Energy Strategies
ECU	European Currency Unit
EDS	Ecologically-Driven Scenario
EEZ	Exclusive Economic Zone
EIA	Energy Information Administration

EIS	Energy Industry System
ELA	Equilibrium Line Altitude
ELA	Experimental Lakes Area
EMF	Energy Modeling Forum
EMI	Eemian Interglacial Period
EMS	Energy Management System
ENSO	El Niño-Southern Oscillation
EOR	Enhanced Oil Recovery
EPRI	Electric Power Research Institute
EPS	Energy Performance Systems, Inc.
ERB	Edmonds-Reilly-Barns Model
ERS	European Space Agency Remote Sensing Satellite
ESCAP	U.N. Economic and Social Commission for Asia and the Pacific
ETH	Swiss Federal Institute of Technology
ETSAP	Energy Technology Systems Analysis Programme
EU	European Union
EV	Electric Vehicle
FACE	Forest Absorbing Carbon Dioxide Emissions (Foundation)
FACE	Free-Air CO_2 Enrichment
FAO	Food and Agriculture Organization
FBSC	Fiji Building Standards Committee
FC	Fluorocarbon
FCCC	Framework Convention on Climate Change
FCV	Fuel Cell Vehicle
FEMA	Federal Emergency Management Agency
FFES	Fossil Free Energy System
FLORAM	Florestales Amazonia
FSI	Forest Survey of India
FSU	Former Soviet Union
FSU/EE	Former Soviet Union/Eastern Europe
FYM	Farmyard Manure
GATT	General Agreement on Tariffs and Trade
GCC	Global Climate Coalition
GCIP	GEWEX Continental-Scale International Project
GCM	General Circulation Model
GCOS	Global Climate Observing System
GCR	Gas-Cooled Reactor
GCTE	Global Change and Terrestrial Ecosystems
GDP	Gross Domestic Product
GEA	Group for Efficient Appliances
GED	Global Ecosystem Database
GEEI	Global Energy Efficiency Initiative
GEF	Global Environment Facility
GEMS	Global Environmental Monitoring System
GEWEX	Global Energy and Water Cycle Experiment
GFDL	Geophysical Fluid Dynamics Laboratory
GFDLH	High-Resolution GFDL Model
GHG	Greenhouse Gas
GISS	Goddard Institute for Space Studies
GLOBEC	Global Ocean Ecosystem Dynamics
GLOSS	Global Sea Level Observing System
GNP	Gross National Product
GOOS	Global Ocean Observation System
GPP	Gross Primary Production
GRIP	Greenland Icecore Project
GT	Gross Tons
GTOS	Global Terrestrial Observing System
GVA	Global Vulnerability Assessment
GWP	Global Warming Potential
HD	High Demand
HEC	Herbaceous Energy Crops
HHV	Higher Heating Value
HID	High-Intensity Discharge
HIV	Human Immunodeficiency Virus
HMSO	Her Majesty's Stationery Office
HVAC	Heating, Ventilating, and Air Conditioning
HWR	Heavy Water Reactor
IAEA	International Atomic Energy Agency
IAHS	International Association of Hydrological Sciences
IAM	Integrated Assessment Model
IBSNAT	International Benchmark Sites Network for Technology Transfer
ICAO	International Civil Aviation Organization
ICEV	Internal-Combustion Engine Vehicle
ICSB	Innoprise Corporation Sdn Bhd
ICZM	Integrated Coastal Zone Management
IDNDR	International Decade for Natural Disaster Reduction
IEA	International Energy Agency
IFPRI	International Food Policy Research Institute
IFREMER	Institut Francais de Recherche pour l'Exploration de la Mer
IGBP	International Geosphere-Biosphere Programme
IGCC	Integrated Gasification Combined Cycle
IIASA	International Institute for Applied Systems Analysis
IIPLR	Insurance Institute for Property Loss Reduction
IMAGE	Integrated Model to Assess the Greenhouse Effect
IMO	International Maritime Organization
INEL	Idaho National Engineering Laboratory
IOC	Intergovernmental Oceanographic Commission
IPCC	Intergovernmental Panel on Climate Change
IPCC-EIS	Intergovernmental Panel on Climate Change, Energy and Industry Subgroup
IRC	Insurance Research Council
IRP	Integrated Resource Planning
IRRI	International Rice Research Institute
ISIC	International Standard Industrial Classification
ISO	Insurance Services Office
ISOS	Impacts of Sea-Level Rise on Society
ISRIC	International Soil Reference and Information Centre
ISTIG	Inter-Cooled Steam Injected Gas Turbine
ITCZ	Inter-Tropical Convergence Zone

ITP Integrated Transportation Planning
ITQ Individual Transferable Quota
IUCN International Union for the Conservation of Nature/World Conservation Union
JGOFS Joint Global Ocean Flux Study
JV Joint Venture
KIER Korean Institute of Energy Research
LAI Leaf Area Index
LAIA Lung and Asthma Information Agency
LAM Limited Area Model
LANL Los Alamos National Laboratory
LEC Levelized Energy Cost
LERTS Laboratoire d'Etudes et de Recherches en Teledetection Spatiale
LESS Low CO_2-Emitting Energy Supply Systems
LHV Lower Heating Value
LLNL Lawrence Livermore National Laboratory
LMR Liquid Metal-Cooled Reactor
LNG Liquid Natural Gas
LOICZ Land-Ocean Interactions in the Coastal Zone
LPG Liquefied Petroleum Gas
LWR Light Water Reactor
MAE Ministère de l'Agriculture et de l'Elevage
MAI Mean Annual Increment
MAPSS Mapped Atmosphere-Plant-Soil System
MCFC Molten Carbonate Fuel Cell
MHE-DFPP Ministère de l'Hydraulique et de l'Environnement–Direction de la Faune, de la Pêche et de la Pisciculture
MINK Missouri, Iowa, Nebraska, and Kansas
MITI Ministry of International Trade and Industry
MMT Molten Metal Technologies
MOPU Ministerio de Obras y Urbanismo
MOX Mixed Oxide
MPI Max-Planck Institut für Meteorologie
MSW Municipal Solid Waste
MT Million Tons
MTBE Methyl Tertiary Butyl Ether
MTSJ Marine Technology Society Journal
MVA Megavolt-Amps
NAS National Academy of Sciences
NASA National Aeronautics and Space Administration
NBR National Biofuels Roundtable
NCAR National Center for Atmospheric Research
NCC Nature Conservancy Council
NCPI National Committee on Property Insurance
NEA Nouvelle Entreprise Agricole
NEERI National Environmental Engineering Research Institute
NEP New England Power
NEP Net Ecosystem Productivity
NFIP National Flood Insurance Program
NGI Natural Gas-Intensive
NHMRC National Health and Medical Research Council
NI Nuclear-Intensive
NIEHS National Institute of Environmental Health Services
NINR Norwegian Institute for Nature Research
NMSC Nonmelanocytic Skin Cancer
NOAA National Oceanic and Atmospheric Administration
NPP Net Primary Productivity
NPT Treaty on the Non-Proliferation of Nuclear Weapons
NRC National Research Council
NRSA National Remote Sensing Agency
NSR Not Satisfactorily Restocked
NUE Nitrogen-Use Efficiency
OECD Organisation for Economic Cooperation and Development
OIES Office for Interdisciplinary Earth Studies
OMNR Ontario Ministry of Natural Resources
ORNL Oak Ridge National Laboratory
OSTP Office of Science and Technology Policy
OSU Oregon State University
OTA Office of Technology Assessment
OTEC Ocean Thermal Energy Conversion
OTP Ozone Trends Panel
P Precipitation
PAFC Phosphoric Acid Fuel Cell
PAN Peroxyacetal Nitrate
PAO Pacific OECD
PAR Photosynthetically Active Radiation
PCS Property Claims Services
PEM Proton Exchange Membrane
PET Potential Evapotranspiration
PFBC Pressurized Fluidized-Bed Combustion
PFT Plant Functional Type
PIC Pacific Island Countries
PIC Products of Incomplete Combustion
PNGV Partnership for a New Generation of Vehicles
PNUMA Programa de Naciones Unidas para el Medio Ambiente
P:PET Precipitation to Potential Evapotranspiration Ratio
P:R Production to Respiration Ratio
PURPA Public Utilities Regulatory Policy Act
PV Photovoltaic
RAS Russian Academy of Sciences
RCW "Rapidly Changing World" Scenario
R&D Research and Development
RD&D Research, Development, and Demonstration
RIGES Renewables-Intensive Global Energy Scenario
RIVM Institute of Public Health and Environmental Protection
RMIB Royal Meteorological Institute of Belgium
RSC/CAE Royal Society of Canada/Canada Academy of Engineering
RSWG Response Strategies Working Group
RTI Research Triangle Institute
SCC Squamous Cell Carcinoma
SCOPE Scientific Committee on Problems of the Environment
SCOR Scientific Committee on Ocean Research
SCS Soil Conservation Service

SCW "Slowly Changing World" Scenario
SDSIDS Global Conference on the Sustainable Development of Small Island Developing States
SERI Solar Energy Research Institute
SFWMD South Florida Water Management District
SI International System of Units
SMR Small- and Medium-Size Reactors
SNL Sandia National Laboratories
SO Southern Oscillation
SOC Soil Organic Carbon
SOFC Solid Oxide Fuel Cell
SPEFC Solid Polymer Electrolyte Fuel Cell
SPFC Solid Polymer Fuel Cell
SRWC Short Rotation Woody Crops
SST Sea-Surface Temperature
STIG Steam Injected Gas Turbine
TERI Tata Energy Research Institute
TES Terrestrial Environment System
TRW Thompson, Ramo, and Wooldridge
TSU Technical Support Unit
TVM Terrestrial Vegetation Model
UBC University of British Columbia
UCAR Universtiy Corporation for Atmospheric Research
UKMO United Kingdom Meteorological Office
UKMOH High-Resolution UKMO Model
UN United Nations
UNCED UN Conference on Environment and Development
UNCHS UN Center for Human Settlements
UNCNRSEED UN Committee on New and Renewable Sources of Energy and on Energy for Development
UNCTAD UN Conference on Trade and Development
UNDP UN Development Programme
UNEP UN Environment Programme
UNESCO UN Educational, Scientific and Cultural Organization
UNIDO UN Industrial Development Organization
UNFPA UN Population Fund
UNSCEAR UN Scientific Committee on the Effects of Atomic Radiation
UNSEGED UN Solar Energy Group on Environment and Development
USDA U.S. Department of Agriculture
USDC U.S. Department of Commerce
USEPA U.S. Environmental Protection Agency
USGCRP U.S. Global Change Research Program
USGS U.S. Geological Survey
USLE Universal Soil Loss Equation
UVR Ultraviolet Radiation
VAV Variable Air Volume
VBD Vector-Borne Disease
VOC Volatile Organic Compound
WCC World Coast Conference
WCRP World Climate Research Programme
WEC World Energy Council
WEC Weigel Engineering Company
WEE Wind Erosion Equation
WGI Working Group I
WGMS World Glacier Monitoring Service
WHO World Health Organization
WMO World Meteorological Organization
WRI World Resources Institute
WRR Wetenschappelijke Raad voor het Regeringsbeleid (The Netherlands Scientific Council for Government Policy)
WUE Water-Use Efficiency
WWF World Wide Fund for Nature

CHEMICAL SYMBOLS USED IN THE WORKING GROUP II VOLUME

a-Si Amorphous Silicon
C Carbon
CF_4 Tetrafluoromethane, CFC-14
C_2F_6 Hexafluoroethane, CFC-116
CFC Chlorofluorocarbon
CH_4 Methane
CH_3OH Methanol
CO Carbon Monoxide
CO_2 Carbon Dioxide
CaO Calcium Oxide
$CaCO_3$ Calcium Carbonate
CdTe Cadmium Telluride
$CuInSe_2$ Copper Indium Selenide
EthOH Ethanol
H Atomic Hydrogen
H_2 Molecular Hydrogen
HCFC Hydrochlorofluorocarbon
HCO_3 Bicarbonate Ion
HFC Hydrofluorocarbon
K Potassium
MeOH Methanol
N Atomic Nitrogen
N_2 Molecular Nitrogen
NH_3 Ammonia
N_2O Nitrous Oxide
NO Nitric Oxide
NO_X Nitrogen Oxide
O Atomic Oxygen
O_2 Molecular Oxygen
O_3 Ozone
Pu Plutonium
SO_X Sulfur Oxide
SO_2 Sulfur Dioxide

Appendix D

UNITS USED THROUGHOUT THE WORKING GROUP II VOLUME

SI (Systéme Internationale) Units

Physical Quantity	Name of Unit	Symbol
length	meter	m
mass	kilogram	kg
time	second	s
thermodynamic temperature	kelvin	K
amount of substance	mole	mol

Fraction	Prefix	Symbol	Multiple	Prefix	Symbol
10^{-1}	deci	d	10	deca	da
10^{-2}	centi	c	10^{2}	hecto	h
10^{-3}	milli	m	10^{3}	kilo	k
10^{-6}	micro	μ	10^{6}	mega	M
10^{-9}	nano	n	10^{9}	giga	G
10^{-12}	pico	p	10^{12}	tera	T
10^{-15}	femto	f	10^{15}	peta	P
10^{-18}	atto	a	10^{18}	exa	E

Special Names and Symbols for Certain SI-Derived Units

Physical Quantity	Name of SI Unit	Symbol for SI Unit	Definition of Unit
force	newton	N	$kg\ m\ s^{-2}$
pressure	pascal	Pa	$kg\ m^{-1}\ s^{-2}$ (= Nm^{-2})
energy	joule	J	$kg\ m^{2}\ s^{-2}$
power	watt	W	$kg\ m^{2}\ s^{-3}$ (= Js^{-1})
frequency	hertz	Hz	s^{-1} (cycle per second)

Decimal Fractions and Multiples of SI Units Having Special Names

Physical Quantity	Name of Unit	Symbol for Unit	Definition of Unit
length	ångstrom	Å	10^{-10} m = 10^{-8}cm
length	micrometer	μm	10^{-6}m = μm
area	hectare	ha	10^{4} m^{2}
force	dyne	dyn	10^{-5} N
pressure	bar	bar	10^{5} N m^{-2}
pressure	millibar	mb	1hPa
weight	ton	t	10^{3} kg

Non-SI Units

°C	degrees Celsius (0°C = ~273K); Temperature differences are also given in °C rather than the more correct form of "Celsius degrees"
Btu	British Thermal Unit
kWh	kilowatt-hour
MW_e	megawatts of electricity
ppmv	parts per million (10^6) by volume
ppbv	parts per billion (10^9) by volume
pptv	parts per trillion (10^{12}) by volume
tce	tons of coal equivalent
toe	tons of oil equivalent
tWh	terawatt-hour

Appendix E

LIST OF MAJOR IPCC REPORTS (IN ENGLISH UNLESS OTHERWISE STATED)

Climate Change—The IPCC Scientific Assessment
The 1990 Report of the IPCC Scientific Assessment Working Group (*also in Chinese, French, Russian, and Spanish*)

Climate Change—The IPCC Impacts Assessment
The 1990 Report of the IPCC Impacts Assessment Working Group (*also in Chinese, French, Russian, and Spanish*)

Climate Change—The IPCC Response Strategies
The 1990 Report of the IPCC Response Strategies Working Group (*also in Chinese, French, Russian, and Spanish*)

Emissions Scenarios
Prepared for the IPCC Response Strategies Working Group, 1990

Assessment of the Vulnerability of Coastal Areas to Sea Level Rise–A Common Methodology
1991 (*also in Arabic and French*)

Climate Change 1992—The Supplementary Report to the IPCC Scientific Assessment
The 1992 Report of the IPCC Scientific Assessment Working Group

Climate Change 1992—The Supplementary Report to the IPCC Impacts Assessment
The 1992 Report of the IPCC Impacts Assessment Working Group

Climate Change: The IPCC 1990 and 1992 Assessments
IPCC First Assessment Report Overview and Policymaker Summaries, and 1992 IPCC Supplement

Global Climate Change and the Rising Challenge of the Sea
Coastal Zone Management Subgroup of the IPCC Response Strategies Working Group, 1992

Report of the IPCC Country Studies Workshop
1992

Preliminary Guidelines for Assessing Impacts of Climate Change
1992

IPCC Guidelines for National Greenhouse Gas Inventories
Three volumes, 1994 (*also in French, Russian, and Spanish*)

IPCC Technical Guidelines for Assessing Climate Change Impacts and Adaptations
1995 (*also in Arabic, Chinese, French, Russian, and Spanish*)

Climate Change 1994—Radiative Forcing of Climate Change and an Evaluation of the IPCC IS92 Emission Scenarios
1995

ENQUIRIES: IPCC Secretariat, c/o World Meteorological Organization, P.O. Box 2300, CH1211, Geneva 2, Switzerland.